ASTM STANDARDS RELATED TO ENVIRONMENTAL SITE CHARACTERIZATION

Sponsored by Committee D-18 on Soil and Rock

1997

ASTM Publication Code Number (PCN): 03-418297-38

ASTM
100 Barr Harbor Drive, West Conshohocken, PA 19428-2959

Editorial Staff

Director, Editorial Services:
Roberta A. Storer

Manager, Standards Publications:
Joan L. Cornillot
Paula C. Fazio-Fluehr

Senior Indexer:
H. Joel Shupak

Editors:
Jennifer Azara
Nicole C. Baldini
Lisa Bernhardt
Elizabeth L. Gutman
Joanne G. Kramer
Christine M. Leinweber

Vernice A. Mayer
Patricia A. McGee
Karen L. Riley
Todd J. Sandler
Richard F. Wilhelm

Library of Congress Cataloging-in-Publication Data

ASTM standards related to environmental site characterization /
sponsored by Committee D-18 on Soil and Rock.
    p.   cm.
  "ASTM publication code number (PCN) 03-418297-38."
  Includes bibliographical references and index.
  ISBN 0-8031-1841-4
  1. Acid precipitation (Meteorology)—Analysis—Standards.
2. Atmospheric deposition—Analysis—Standards. 3. Atmospheric
monitoring—Standards. I. ASTM Committee D-18 on Soil and Rock.
QC926.52.A88 1997
628.5'32—dc21

97-24292
CIP

Photocopy Rights

Printed in Baltimore, MD
August 1997

FOREWORD

This compilation includes more than 130 ASTM guides, practices, and test methods of potential value for environmental site characterization, drawn from the work of the following five ASTM committees:

- Committee D18 on Soil and Rock (And Fluids Contained Therein)
- Committee D19 on Water
- Committee D34 on Waste Management
- Committee E47 on Biological Effects and Environmental Fate
- Committee E50 on Environmental Assessment

Organization of the Compilation

Standards in this compilation are organized topically so that related standards are grouped together. Following the topical table of contents there is an alphanumeric table of contents which lists standards in sequence by designation number, and at the end of the compilation is a regular index.

Standards are grouped in 6 Parts:

Part I (Site Characterization) includes more that 50 standards organized into seven subtopics (general guidance, data elements, geophysical methods, geologic characterization, hydrogeologic characterization, drilling methods and surface water).

Part II (Soil, Vadose Zone and Sediment Sampling and Monitoring) includes more than 20 standards grouped into three areas (soil sampling, vadose zone sampling and monitoring, and sediment sampling).

Part III (Water Sampling and Monitoring) includes more than 20 standards organized into 4 subtopics (general, water sampling, ground water monitoring wells, and ground water sampling).

Part IV (Waste/Contaminant Characterization and Sampling) includes more than 20 standards that provide guidance for planning sampling and cover specific sampling procedures.

Part V (Atmospheric Characterization and Sampling) includes more than 10 standards that address field meteorological measurements, general sample planning and specific sampling procedures.

Part 6 (Biological Sampling) includes 5 guides and classifications that provide an overview of available methods for biological sampling.

Criteria for Inclusion in the Compilation

It was not possible to include all ASTM standards potentially relevant to environmental site characterization because there are so many. Most of the standards included here are *guides,* which provide information on a series of options without recommending a specific course of action, and *practices* which provide a set of instructions for performing one or more specific operations that do not produce a test result.

Specific test methods have generally not been included in the compilation unless needed for completeness. For example, most ASTM standards related to soil sampling are practices, but D 1586 covering the standard penetration test is a test method. Test methods for measuring water content in the field are included in Section 2.2 of the compendium, because this is such an important method for vadose zone monitoring, but laboratory test methods for measuring water content have not been included.

Identifying Other Potentially Relevant ASTM Standards

Tables X1.1 and X1.2 in Guide D 5730 (the first standard in this compendium) provide an index to more than 400 ASTM standards of potential relevance for environmental site characterization, along with a list that gives title and the volume of the annual book of standards in which they appear. Guide D 5730 is updated annually to add any newly approved ASTM standards that may be relevant to environmental site characterization. The 1997 revision will add more than 100 standards to Tables X1.1 and X1.2 of the guide.

Other ASTM standards identified in D 5730 that are not included in this compilation will be located in one of the following volumes of the annual book of standards:

Volume 4.08/Soil and Rock (I): D 420–D 4914
Volume 4.09/Soil and Rock (II): D 4943–latest; Geosynthetics
Volume 11.01/Water (I)
Volume 11.02/Water (II)
Volume 11.03/Atmospheric Analysis: Occupational Health and Safety; Protective Clothing

Volume 11.04/Environmental Assessment; Hazardous Substances and Oil Spill Response; Waste Management; Environmental Risk Management

Volume 11.05/Biological Effects and Environmental Fate; Biotechnology; Pesticides

Future Standards Development

ASTM standards development in the environmental area continues to be active, with new standards currently in the balloting process. Existing ASTM standards must be reviewed, revised if needed and reapproved every 5 years. Since many ASTM standards in the environmental field are relatively recent, they are currently going through this periodic review. ASTM is an open-membership organization, and I encourage anyone who is interested in participating in the standards development process to join by contacting ASTM at 610-832-9500 and asking the staff manager of the committee or committees that interest you (see list at the beginning of the introduction) for a membership application.

Russell Boulding

Chair, D18.01.01 (Section on Environmental Site Characterization, Subcommittee on Surface and Subsurface Characterization, Committee on Soil and Rock)

March, 1997

ASTM

ASTM, founded in 1898, is a developer and publisher of technical information designed to promote the understanding and development of technology and to ensure the quality of commodities and services and safety of products.

ASTM's primary mission is to develop voluntary full-consensus standards for materials, products, systems, and services. It provides a forum for producers, users, ultimate consumers, and those having a general interest (representatives of government and academia) to meet on common ground to write standards that best meet their goal. ASTM also publishes books containing reports on state-of-the-art testing techniques and their possible applications.

For more information, contact ASTM, 100 Barr Harbor Drive, West Conshohocken, PA 19428-2959: Phone: 610-832-9585; e-mail: service@astm.org; website: <http://www.astm.org>

ALPHANUMERIC TABLE OF CONTENTS

A complete subject index begins on p. 1385

v

COMPILATION OF STANDARDS ON ENVIRONMENTAL SITE CHARACTERIZATION

Topical Table of Contents

PART 1. SITE CHARACTERIZATION

PART 3. WATER SAMPLING AND MONITORING

PART 4. WASTE/CONTAMINANT CHARACTERIZATION AND SAMPLING

PART 5.   ATMOSPHERIC CHARACTERIZATION AND SAMPLING

PART 6.   BIOLOGICAL SAMPLING

RELATED PUBLICATIONS

ASTM has issued several publications that may be of interest to users of this compilation of standards. These are listed as follows:

Compilations:
   ASTM Standards on Environmental Sampling (1997)
   ASTM Standards on Evaluation of Sites for Septic Systems (1997)
   ASTM Standards on Environmental Site Assessments for Commercial Real Estate (1997)
   ASTM Standards on Ground Water and Vadose Zone Investigations: Drilling, Sampling, Well Installation, and Abandonment Procedures (1996)
   ASTM Standards on Analysis of Hydrologic Parameters and Ground Water Modeling (1996)
   ASTM Standards on Design and Planning for Ground Water and Vadose Zone Investigations (1996)

Data Series:
   DS 64 Cleanup Criteria for Contaminated Soil and Ground Water (1996)

Manual:
   MNL 24 User's Guide and Software Package in Support of ASTM Standard Practice for Environmental Site Assessments: Transaction Screen Process (E1528)

Special Technical Publications (STPs):
   STP 1240 Stabilization and Solidification of Hazardous, Radioactive and Mixed Wastes (1996)
   STP 1261 Volatile Organic Compounds (VOCs) in the Environment (1996)
   STP 1262 Environmental Toxicology and Risk Assessment (1996)
   STP 1282 Sampling Environmental Media (1996)
   STP 1306 Environmental Toxicology and Risk Assessment: Biomarkers and Risk Assessment (1996)

PART 1.   SITE CHARACTERIZATION

1.1 GENERAL GUIDANCE

Standard Guide for
Site Characterization for Environmental Purposes With Emphasis on Soil, Rock, the Vadose Zone and Ground Water[1]

This standard is issued under the fixed designation D 5730; the number immediately following the designation indicates the year of original adoption or, in the case of revision, the year of last revision. A number in parentheses indicates the year of last reapproval. A superscript epsilon (ε) indicates an editorial change since the last revision or reapproval.

INTRODUCTION

This guide covers the selection of the various ASTM Standards that are available for the investigation of soil, rock, the vadose zone, ground water, and other media where the investigations have an environmental purpose.[2] It is intended to improve consistency of practice and to encourage rational planning of a site characterization program by providing a checklist to assist in the design of an environmental reconnaissance/investigation plan. The subsurface conditions at a particular site are usually the result of a combination of natural geologic, topographic, hydrologic, and climatic factors, and of historical modifications both natural and manmade. An adequate and internally consistent site characterization program will allow evaluation of the results of these influences. Site characterization for engineering, design, and construction purposes are addressed in a separate guide, Guide D 420.

The understanding of environmental processes occurring in soil and rock systems depends on adequate characterization of physical, chemical, and biological properties of soil and rock. Processes of interest may include, but are not limited to, surface and subsurface hydrology, contaminant mobilization, distribution, fate and transport; chemical and biological degradation of wastes; and geomorphological/ecological processes. Although this guide focusses primarily on characterization of soil and rock, it is understood that climatic and biotic factors may also be important in understanding environmental processes in soil and rock systems.

1. Scope

1.1 This guide covers a general approach to planning field investigations that is useful for any type of environmental investigation with a primary focus on the subsurface and major factors affecting the surface and subsurface environment. Generally, such investigations should identify and locate, both horizontally and vertically, significant soil and rock masses and ground water conditions present within a given site area and establish the characteristics of the subsurface materials by sampling or in situ testing, or both. The extent of characterization and specific methods used will be determined by the environmental objectives and data quality requirements of the investigation. This guide focuses on field methods for determining site characteristics and collection of samples for further physical and chemical characterization. This guide does not address special considerations required for characterization of karst and fractured rock terrane. In such hydrogeologic settings refer to Quinlan and Guide D 5717, (1).

1.2 This guide refers to ASTM standard methods by which soil, rock, vadose zone, and ground water conditions may be determined. Laboratory testing of soil, rock, and ground-water samples is specified by other ASTM standards which are not specifically discussed in this guide. Laboratory methods for measurement of physical properties relevant to environmental investigations are included in Appendix X1.

1.3 The values stated in SI units are to be regarded as the standard.

1.4 *This standard does not purport to address all of the safety concerns, if any, associated with its use. It is the responsibility of the user of this standard to establish appropriate safety and health practices and determine the applicability of regulatory limitations prior to use.*

2. Referenced Documents

2.1 The pertinent ASTM guides for selection of field investigation methods are listed at appropriate points in the sections that follow, and a comprehensive list of guides, standards, methods, practices, and terminology is contained in Appendix X1. Tables X1.1 and X1.2 provide an index of field and laboratory standards listed in Appendix X1.

2.1.1 *ASTM Standards:*

---

[1] This guide is under the jurisdiction of ASTM Committee D-18 on Soil and Rock and is the direct responsibility of Subcommittee D18.01 on Surface and Subsurface Characterization.

Current edition approved May 10, 1996. Published November 1996. Originally published as D 5730 – 95. Last previous edition D 5730 – 95a.

[2] This guide is under the jurisdiction of Subcommittee D18.01 (Surface and Subsurface Characterization) of the Committee on Soil and Rock, and as such has a primary focus on subsurface characterization, including soil, rock, and fluids contained therein (including liquid and gaseous components), and subsurface biota. Surface hydrology, meteorology, air quality, geomorphic processes, biota, and waste materials (when present at a site) are to a greater or lesser extent linked to environmental processes in soil and rock systems. Consequently other ASTM methods of particular relevance to environmental site investigations are identified in this guide, but are addressed in less detail.

C 998 Practice for Sampling Surface Soil for Radionuclides[3]

D 420 Guide to Site Characterization for Engineering, Design, and Construction Purposes[4]

D 422 Test Method for Particle-Size Analysis of Soils[4]

D 653 Terminology Relating to Soil, Rock, and Contained Fluids[4]

D 1140 Test Method for Amount of Material in Soils Finer than the No. 200 (75-μm) Sieve[4]

D 1452 Practice for Soil Investigation and Sampling by Auger Borings[4]

D 1586 Test Method for Penetration Test and Split-Barrel Sampling of Soils[4]

D 1587 Practice for Thin-Walled Tube Geotechnical Sampling of Soils[4]

D 2113 Practice for Diamond Core Drilling for Site Investigation[4]

D 2434 Test Method for Permeability of Granular Soils (Constant Head)[4]

D 2487 Classification of Soils for Engineering Purposes (Unified Soil Classification System)[4]

D 2488 Practice for Description and Identification of Soils (Visual-Manual Procedure)[4]

D 2922 Test Methods for Density of Soil and Soil-Aggregate In Place by Nuclear Methods (Shallow Depth)[4]

D 3404 Guide for Measuring Matric Potential in the Vadose Zone Using Tensiometers[4]

D 3441 Test Method for Deep, Quasi-Static, Cone and Friction-Cone Penetration Tests of Soil[4]

D 3550 Practice for Ring-Lined Barrel Sampling of Soils[4]

D 3584 Practice for Indexing Papers and Reports on Soil and Rock for Engineering Purposes[4]

D 4043 Guide for Selection of Aquifer-Test Method in Determining of Hydraulic Properties by Well Techniques[4]

D 4044 Test Method (Field Procedure) for Instantaneous Change in Head (Slug Tests) for Determining Hydraulic Properties of Aquifers[4]

D 4050 Test Method (Field Procedure) for Withdrawal and Injection Well Tests for Determining Hydraulic Properties of Aquifer Systems[4]

D 4104 Test Method (Analytical Procedure) for Determining Transmissivity of Nonleaky Confined Aquifers by Overdamped Well Response to Instantaneous Change in Head (Slug Test)[4]

D 4105 Test Method (Analytical Procedure) for Determining Transmissivity and Storage Coefficient of Nonleaky Confined Aquifers by the Modified Theis Nonequilibrium Method[4]

D 4106 Test Method (Analytical Procedure) for Determining Transmissivity and Storage Coefficient of Nonleaky Confined Aquifers by the Theis Nonequilibrium Method[4]

D 4127 Terminology Used with Ion-Selective Electrodes[5]

D 4210 Practice for Interlaboratory Quality Control Procedures and a Discussion on Reporting Low Level Data[5]

D 4220 Practice for Preserving and Transporting Soil Samples[4]

D 4448 Guide for Sampling Groundwater Monitoring Wells[6]

D 4547 Practice for Sampling Waste and Soils for Volatile Organics[6]

D 4630 Test Method for Determining Transmissivity and Storativity of Low Permeability Rocks by In Situ Measurements Using the Constant Head Injection Test[4]

D 4631 Test Method for Determining Transmissivity and Storativity of Low Permeability Rocks by In Situ Measurements Using the Pressure Pulse Technique[4]

D 4687 Guide for General Planning of Waste Sampling[6]

D 4696 Guide for Pore-Liquid Sampling from the Vadose Zone[4]

D 4700 Guide for Soil Sampling from the Vadose Zone[4]

D 4750 Test Method for Determining Subsurface Liquid Levels in a Borehole or Monitoring Well (Observation Well)[4]

D 5079 Practices for Preserving and Transporting Rock Core Samples[7]

D 5084 Test Method of Hydraulic Conductivity of Saturated Porous Materials Using a Flexible Wall Permeameter[7]

D 5092 Practice for Design and Installation of Ground Water Monitoring Wells in Aquifers[7]

D 5093 Test Method for Field Measurement of Infiltration Rate Using a Double-Ring Infiltrometer With a Sealed Inner Ring[7]

D 5195 Test Method for Density of Soil and Rock In-Place at Depths Below the Surface by Nuclear Methods[7]

D 5254 Practice for Minimum Set of Data Elements to Identify a Ground-Water Site[7]

D 5269 Test Method for Determining Transmissivity of Nonleaky Confined Aquifers by the Theis Recovery Method[7]

D 5270 Test Method for Determining Transmissivity and Storage Coefficient of Bounded, Nonleaky, Confined Aquifers[7]

D 5314 Guide for Soil Gas Monitoring in the Vadose Zone[7]

D 5408 Guide for Set of Data Elements to Describe a Ground-Water Site; Part 1—Additional Identification Descriptors[7]

D 5409 Guide for Set of Data Elements to Describe a Ground-Water Site; Part 2—Physical Descriptors[7]

D 5410 Guide for Set of Data Elements to Describe a Ground-Water Site; Part 3—Usage Descriptors[7]

D 5447 Guide for Application of a Ground-Water Flow Model to a Site-Specific Problem[7]

D 5451 Practice for Sampling Using a Trier Sampler[6]

D 5717 Guide for the Design of Ground-Water Monitoring Systems In Karst and Fractured-Rock Aquifers[4]

E 177 Practice for Use of the Terms Precision and Bias in ASTM Test Methods[8]

[3] Annual Book of ASTM Standards, Vol 12.01.
[4] Annual Book of ASTM Standards, Vol 04.08.
[5] Annual Book of ASTM Standards, Vol 11.01.
[6] Annual Book of ASTM Standards, Vol 11.04.
[7] Annual Book of ASTM Standards, Vol 04.09.
[8] Annual Book of ASTM Standards, Vol 14.02.

E 380 Practice for Use of the International System of Units (SI) (The Modernized Metric System)[8]
E 1527 Practice for Environmental Site Assessments: Phase I Environmental Site Assessment Process[6]
G 57 Test Method for Field Measurement of Soil Resistivity Using the Wenner Four-Electrode Method[9]

2.2 *Non-ASTM References*—Appendix X2 identifies major non-ASTM references that focus on field methods for environmental site characterization. Other guidance documents covering procedures for environmental investigations with specific objectives or in particular geographic settings may be available from federal, state, and other agencies or organizations. The appropriate agency or organization should be contacted to determine the availability and most current edition of such documents.

3. Terminology

3.1 *Definitions:*[10]

3.1.1 *site, n*—a place or location designated for a specific use, function, or study.

3.1.2 *site, v*—to determine a place or location for a specific use, function, or study.

3.1.3 *characterization, n*—the delineation or representation of the essential features or qualities existing at a site.

3.1.4 *characterize, v*—the process of delineation or representation of the essential features or qualities existing at a site.

3.1.5 *conceptual site model, n*—for the purposes of this guide, a testable interpretation or working description of the relevant physical, chemical, and biological characteristics of a site.[11]

3.1.6 *environment, n*—the aggregate of conditions, influences, and circumstances that affect the existence or development of properties intrinsic to a site.

3.1.7 *environmental, adj*—having to do with the environment.

3.1.8 *environmental site characterization, n*—the delineation or representation of the essential features or qualities, including the conditions, influences, and circumstances, existing at a place or location designated for a specific use, function, or study.[12]

3.1.9 *environmental audit, n*—the investigation process to determine if the operations of an existing facility are in compliance with applicable environmental laws and regulations.

3.1.10 *environmental site assessment (ESA), n*—the process by which a person or entity seeks to determine if a particular parcel of real property (including improvements)

is subject to recognized environmental conditions.[13]

3.2 In addition to Terminology D 653, Appendix X3 identifies major references from a range of disciplines that can be used as sources for definitions of terms that are related to environmental site characterization.

4. Significance and Use

4.1 This guide provides a general approach to environmental site characterization. Environmental site characterization provides information for a wide variety of uses including:

4.1.1 Determination of ambient background or baseline conditions, including, but not limited to, geochemistry, hydrogeology, microbiology, mineralogy, and water quality.

4.1.2 Assessment of site suitability for a future use or a use which may be compromised by site characteristics, such as flooding, seismic activity, and landslides (mass wasting).

4.1.3 Protection of site quality from the detrimental effects of human activities and natural processes, or minimization of adverse environmental impacts. Specific examples of uses of environmental site characterization for these purposes include: (*1*) delineation of ground-water or wellhead protection areas, (*2*) assessing the suitability of sites for disposal of industrial and residential liquid and solid wastes, (*3*) assessing soil suitability for land treatment of wastes, and (*4*) evaluating soil suitability for agricultural practices in order to minimize soil erosion and contamination from agricultural chemicals.

4.1.4 Assessment of the type, distribution, and extent of surface and subsurface contamination to determine compliance; risk to human health and the environment; and responsibility for remediation. Such assessments include: (*1*) sites involved in real estate transactions, (*2*) controlled and uncontrolled hazardous waste sites, (*3*) controlled and uncontrolled municipal solid waste, wastewater, and other nonhazardous waste disposal sites.

4.1.5 Assessment of surface and subsurface environmental processes that affect the fate, mobilization, and rate of transport of natural and anthropogenic chemicals in the subsurface.

4.1.6 Assessment of the type, distribution, and extent of natural and anthropogenic radioactive elements in the subsurface.

4.1.7 Assessment of the degree of risk that adverse subsurface site conditions not related to 4.1.4 and 4.1.5 pose to human health and safety or the natural ecosystem.

4.1.8 Selection and design of remediation systems for

---

[9] *Annual Book of ASTM Standards*, Vol 03.02.

[10] The first seven definitions are ordered logically to illustrate construction of the definition for *environmental site characterization* rather than in alphabetical order.

[11] The meaning of conceptual site model may have more restricted or specific meanings depending on the objective or use of the model. For example, ground water flow modeling focuses on the physical characteristics as they relate to subsurface flow (see Guide D 5447), and a conceptual site model for the purpose of risk assessment will focus on contaminant sources, pathways, and receptors to exposure.

[12] Environmental audits and environmental site assessments as defined below are examples of environmental site characterization with specific objectives.

[13] This definition is taken from Practice E 1527. Other definitions of environmental site assessment may apply in other contexts. For example, EPA's Site Assessment Branch, Hazardous Site Evaluation Division in the Office of Emergency and Remedial Response defines site assessment as the decision process for identifying the most seriously contaminated uncontrolled hazardous waste sites that will receive funding for long-term remediation. Practice E 1527 defines *recognized environmental conditions* as "the presence or likely presence of any hazardous substances or petroleum products on a property under conditions that indicate an existing release, a past release, or a material threat of a release of any hazardous substances or petroleum products into structures on the property or into the ground, ground water, or surface water of the property." In other environmental site investigations, nonhazardous substances (because of their physical condition, smell, or other aesthetic properties) or substances that have hazardous characteristics but do not meet a regulatory definition of hazardous may be the focus of concern.

cleanup of subsurface contamination and of other reclamation or rehabilitation practices on disturbed land.

4.2 This guide is meant to be a flexible description of investigation requirements; methods defined by other ASTM Standards (Appendix X1) or non-ASTM techniques (Appendix X2) may be appropriate in some circumstances. The methods and amount of effort required for environmental site characterization will vary with site conditions and objectives of the investigation. This standard does not set mandatory guidelines and does not constitute a list of necessary steps or procedures for all investigations. In karst and fractured rock hydrogeologic settings, this guide should be used in conjunction with Guide D 5717.

5. Steps in Planning and Conducting Environmental Site Characterization

5.1 The following minimum elements, not necessarily in sequential order, are required for most environmental site investigations to determine project and site strategy:

5.1.1 Definition of objectives, site boundaries,[14] and other information necessary for efficient project planning.[15]

5.1.2 Collection of available existing data and information about the site, pertinent to the objectives of the investigation.[16]

5.1.3 Development of one or more conceptual site models of the site from existing information. The objectives of the investigation will affect the type and complexity of site conceptualization.

5.1.4 Performance of a reconnaissance site investigation, that may include nondestructive geophysical methods, and relatively simple field sampling and characterization methods,[17] to refine the conceptual model of the site.

5.1.5 Development of a detailed site investigation and sampling plan, that identifies methods to be used to collect and analyze required additional data, protocols for sampling and field measurements, and procedures to ensure quality assurance and quality control of site characterization data. Criteria defining the quality of data that are collected is a requirement for most environmental site investigations.[18]

5.1.6 Collection of field samples and measurements in accordance with the site investigation and sampling plan.[19]

5.1.7 Analysis of field and laboratory data to further refine the conceptual model of the site and preparation of a report that fulfills the objectives of the investigation.

5.2 Environmental site characterization is an iterative process of continually refining a conceptual site model of a site as new information becomes available. The objective of the investigation and availability of funds to carry out the investigation will affect the final number of iterations that are possible. For example, a Phase I Environmental Site Assessment (Guide E 1527) for a real estate property transaction may require only a single site visit to evaluate whether any activities may have resulted in contamination of the site. Complex sites, however, may require multiple site visits to collect data in order to develop a satisfactory conceptual model of the site, especially if a human health and environmental risk assessment is required. The number of site visits may be reduced by on-site analysis of results and real-time adjustments to the site investigation plan (see Section 7).

5.2.1 Planning for environmental site investigations should recognize that regional and local differences in climate, biota, soils, geology, and hydrology will influence the selection of appropriate methods and procedures.

5.2.2 Urban settings create special challenges for planning environmental site investigations. Difficulties include busy streets, access to public and private property, especially occupied buildings and fenced areas, and determining predisturbance background environmental conditions. Consult with local agencies and private property owners for relevant information. Select methods and schedule activities accordingly.

[14] The boundaries of a site are defined using one or more of the following considerations: (1) land ownership, (2) current and past land use, (3) natural site characteristics (topography, soils, geology, hydrology, biota). Where site boundaries are initially defined by ownership, natural site characteristics should be evaluated to determine whether the scope of at least parts of the investigation should include areas that are offsite. For example, investigations of ground water contamination should include identification of any potential sources of contamination that are upgradient from a site.

[15] This should include, but not necessarily be limited to: (1) definition of the technical and scientific approach to be used, (2) organization of a data management system, including both paper and electronic records, (3) identification of types of personnel and technical expertise, appropriate ASTM and other methods and field equipment required to meet the defined objective, (4) defining how spatial data will be recorded (see Section 7.1.3), (5) identification of applicable primary and secondary regulatory programs, and any required coordination with government agencies and other organizations, (6) development of health and safety plans, where appropriate, (7) identification of scheduling and budgetary constraints, (8) definition of data quality requirements for each stage of the investigation, (9) identification of deliverables at intermediate stages of the investigation and upon completion, (10) selection of performance measures to determine whether the objective has been achieved, and (11) definition of project decision statements.

[16] A site visit prior to extensive collection of existing data should be made unless the limited scope of a project does not allow multiple visits. The advantage of such a visit is that it may prevent preconceived ideas derived from inaccurate existing information from influencing initial conceptual site model development.

[17] When contaminated sites are being investigated, field chemical analytical methods can be valuable for identifying areas where more detailed investigations may be required, and for designing cost-effective detailed sampling and monitoring plans. Surface geophysical methods may be especially valuable for guiding placement of exploratory drillholes and placement of vadose zone and ground water monitoring installations. Any such field methods should be documented for quality assurance and quality control.

[18] The level of data accuracy and precision needed to meet the intended use for the data must be determined and this constitutes the "data quality requirements" referred to in this guide. The term data quality requirements is similar to the term data quality objectives (DQOs) used to describe the quality determination process in the U.S. EPA RCRA/Superfund program, but does not necessarily include statistically based confidence levels for assessing false negative or false positive designations. For chemical characterization, data quality requirements may vary depending on the phase of the investigation. For example, relatively inexpensive field instrumentation such as colorimetric test kits, or portable gas chromatographs can be used to analyze a relatively large number of samples initially to identify areas with different levels of contamination. Data requirements at this level would specify that instrumentation used be capable only of detecting the presence of a contaminant at or above the regulatory level of concern. These results in turn would guide sampling of a moderate number of samples for more accurate field analytical methods or a mobile laboratory. Finally, a relatively small number of samples might be selected for analysis that would meet the rigorous data quality requirements of EPA's contract laboratory program (CLP). This phased approach has the potential for significant cost and time savings as compared to the use of CLP laboratories for the analysis of all samples.

[19] Prior to commencement of any intrusive exploration the site should be checked for underground utilities or other buried materials. Should evidence of potentially hazardous or otherwise contaminated materials or conditions be encountered in the course of the investigation, work should be interrupted until the circumstances have been evaluated and revised instructions issued before resumption.

6. Collection of Existing Data and Site Reconnaissance

6.1 Collect and review available technical data from the literature and historical records, and conduct interviews before any field program is started. These include, but are not limited to, topographic maps, aerial photography, satellite imagery, geologic maps, statewide or county soil surveys, and mineral resource surveys covering the proposed project area. Reports of subsurface investigations of nearby or adjacent projects should be studied. If feasible, all existing data should be critically evaluated and independently confirmed.

6.1.1 The Natural Resource Conservation Service (formerly U.S. Soil Conservation Service) and the Agricultural Stabilization and Conservation Service (ASCS) are the primary sources of aerial photographs at the county level. The U.S. Geological Survey is the primary source for satellite imagery, and agencies other than SCS (USGS EROS Data Center).[20] Aerial photographs also may be available from the following federal agencies where lands under their jurisdiction are involved: Bureau of Land Management, Bureau of Reclamation, U.S. Army Corps of Engineers, U.S. Environmental Protection Agency Photo-Interpretation Center (EPIC), U.S. Forest Service, and U.S. Fish and Wildlife Service. State natural resource agencies and local/regional planning agencies also may be good sources of aerial photographs.[21,22] Guide D 5518 provides additional information of how to obtain air photos.

6.1.2 The United States Geological Survey and the geological surveys of the various states are the principal sources of geologic maps, reports on mineral resources, ground water, and surface water.[23] Guidebooks prepared by professional associations, technical journals, and published conference and symposium proceedings may also be important sources of information.

6.1.3 United States Department of Agriculture Soil Conservation Service soil surveys, where available and of recent date, should enable the investigator to estimate the range in soil profile characteristics to depths of 1.5 or 2 m (5 or 6 ft) for each soil mapped.[24]

6.1.4 Preliminary information on sites where commercial and industrial activity has occurred should include site history, such as type of activity, location of existing surface and subsurface building, and chemicals, manufactured, used, or stored at the site.[25]

6.2 Study the soil and rock in the vicinity of the proposed project in areas where descriptive data are limited by insufficient or inadequate geologic or soil maps. Take advantage of all exposures of the rock and soil to obtain the best understanding of them. Obtain information on both vertical and horizontal properties of the materials and their distribution. Make appropriate notes and, where appropriate, illustrate by sketches. The notes should include data outlined in 9.3.[26]

6.3 Prepare and compile preliminary maps of the project area using aerial photography and topographic maps that show ground conditions. The scale of the map should be appropriate for the area being investigated. Aerial photographs and topographic maps at scales of 1:15 000 to 1:24 000 are available for most parts of the United States and are usually adequate as a basis for a preliminary map of sites covering tens to hundreds of acres. The distribution of the predominant soil and rock deposits likely to be observed during the investigation may be shown using data obtained from geologic maps, landform analysis, and limited ground reconnaissance. Experienced photointerpreters can deduce much subsurface data from a study of black-and-white, color, and color-infrared photographs because similar soil or rock conditions, tend to have similar patterns of appearance in regions of similar climate or vegetation.[27]

6.4 Perform site reconnaissance. This provides an opportunity to check the accuracy of information compiled from existing sources, and to make site observations to assist in refining the conceptual model of the site. Site reconnaissance should be planned to identify the site characteristics needing further investigation.[28]

7. Detailed Site Investigation

7.1 Review objectives of the investigation and develop a detailed site investigation plan. The detailed site investigation plan should clearly identify the types of data that are required to meet the objectives of the investigation.[29] A complete environmental site investigation will usually en-

[20] USGS EROS Data Center, Sioux Falls, SD 57198.

[21] The following references on air-photo interpretation may be of value: Avery (2),[22] Ciciarelli (3), Denny et al (4), Drury (5), Dury (6), Johnson and Gnaedinger (7), Lattman and Ray (8), Lueder (9), Miller and Miller (10), Ray (11), SCS (12), Strandberg (13), Wright (14).

[22] The boldface numbers in parentheses refer to a list of references at the end of the text.

[23] Sun and Weeks (15) provide a comprehensive bibliography of the U.S. Geological Survey's regional aquifer systems analysis program. Data from the RASA program are being synthesized and published in the Ground Water Atlas of the United States (16).

[24] Each soil type has a distinctive soil profile due to age, parent material, relief, climatic condition, and biological activity. Consideration of these factors can assist in evaluating the potential for movement of contaminants in the vadose zone and ground water based on differing soil characteristics. Boulding (17) and Cameron (18) provide guidance on interpretation of soil properties in relation to potential for contaminant transport. Changes in soil properties in adjacent areas often indicate changes in parent material or relief.

[25] Practice E 1527 and Practice E 1528 provide guidance on the types of information that should be collected and sources from which such information can be obtained.

[26] The following major references may be useful for field geologic and hydrogeologic investigations: Bishop (19), Brassington (20), Bureau of Reclamation (21, 22), Compton (23, 24), Dutro, et al (25), Erdélyi and Gálfi (26), Fetter (27), Lahee (28), LeRoy, et al (29), Low (30), Rahn (31), Tearpocke and Bischke (32). Guides for field description of rocks include: Fry (33), Thorpe and Brown (34), and Tucker (35).

[27] This preliminary map may be expanded into a detailed site map by locating all test holes, pits, and sampling stations and by revising boundaries as determined from the detailed subsurface survey. Geographic information systems (GIS) should be considered for visualization, map preparation, data manipulation, and analysis.

[28] Exploratory sampling of soil solids, soil gas, or ground water using temporary in situ samplers may provide useful information for design of the sampling plan in the detailed site investigation plan. Surface geophysical methods may also be useful at this stage of the investigation.

[29] Considerations for identifying data requirements include: (1) data required to comply with applicable federal, state, or local regulatory programs, (2) data required as inputs to computer models expected to be used, (3) data required for selection and design of implementation measures (that is, protective measures at controlled waste disposal sites, remediation options at contaminated sites).

compass the following activities:[30]

7.1.1 Review available information, both regional and local, on the geologic history (including seismic activity and other potential geologic hazards), rock, soil, ground water, surface water, and other significant environmental and anthropogenic features (for example, buried utilities) occurring at the proposed location and in the immediate vicinity of the site.

7.1.2 Interpret aerial photography and other remote sensing data.

7.1.3 Select appropriate methods for locational data collection coincident with field observations and sampling.[31] Failure to accurately locate elevation and x-y coordinates of observation points and samples can severely compromise the accuracy of interpretations developed from the data. Where geographic information systems (GIS) are to be used for data management, analysis and visualization, locational accuracy should at a minimum satisfy the resolution of the system to be used.

7.1.4 Perform field reconnaissance for identification of surficial geologic and hydrologic conditions, mapping of stratigraphic or lithologic exposures and outcrops, mapping of vegetation and other significant ecological conditions, and examination of anthropogenic features at the site.

7.1.5 Perform on-site investigation of the surface and subsurface materials by geophysical surveys, borings, or test pits.

7.1.6 Obtain representative samples for chemical, biological, and physical analysis of soil, rock, and ground water. When required for the objectives of the investigation, these should be supplemented by samples suitable for the determination of in situ physical and chemical properties, such as hydraulic conductivity and flow-through sorption tests (see 11.2). Measurement of in situ properties requires collection of undisturbed samples (see definitions in Terminology D 653). Where present, wastes and sediments from surface water bodies should also be sampled.

7.1.7 Perform field identification of soil, sediments, and rock with particular references to physical and chemical properties, such as, but not limited to, color, odor, texture (grain-size distribution), mineralogy, zone of increased or reduced porosity and permeability, depth of occurrence, and the types and locations of structural discontinuities.

7.1.8 Identify and measure the potentiometric surface(s) of the aquifer or aquifers. Methods for determining ground-water levels are covered in Test Method D 4750. The variability of these positions in both short (minutes to days)

and long (months to years) time frames should be considered.[32]

7.1.9 Assess vertical and horizontal variations in the pedologic, geologic, and hydrologic characteristics of the subsurface, including the vadose zone, aquifers, and confining units. When the flow direction and velocity of ground water are concerns, special emphasis should be placed on evaluating the degree of aquifer anisotropy, presence of highly permeable unconsolidated materials such as sands and gravels, extent and orientation of soil joint and rock fracture development, and extent of conduit development in karst limestone. Where unconsolidated materials consist of thick loess, alluvial, lacustrine materials, or glacial till, the spacing and depth of vertical soil jointing should be evaluated as potential zones of preferential contaminant movement (42). Guide D 5717 identifies methods for characterization of karst and fractured rock aquifers.

7.1.10 Assess temporal changes in subsurface hydrologic conditions, including changes in ground-water levels using piezometers or monitoring wells, changes in soil moisture conditions using tensiometers (see Guide D 3404), neutron soil moisture probes (see Test Method D 5093) or other methods, and measurement or estimation of soil moisture flux. Changes in soil solute quality using suction lysimeters and other methods (see Guide D 4687).

7.1.11 Identify type and extent of contaminants, if present, and identify and assess the geochemical characteristics of subsurface solids and ground water that may affect the fate and transport of contaminants. Temporal changes in soil pore liquids and ground-water quality should be characterized by periodic sampling from monitoring wells installed in accordance with standard procedures (see Guide D 4696—pore liquids, and Practice D 5092—ground-water monitoring wells). Important geochemical characteristics include organic matter content, clay mineralogy, pH, Eh, specific conductance, and temperature.[33]

8. Use of Field Methods in Environmental Site Characterization

NOTE—All field procedures should be documented by identifying time, date, location, and personnel involved. Practice D 5254 identifies additional basic information required for documentation. Guide D 5408, Guide D 5409, and Guide D 5410 can serve as checklists to ensure that important additional information is not omitted. Paragraph 7.1.3 contains additional discussion of locational data considerations.

8.1 Equipment and procedures used in site characterization for environmental purposes can be classified in the following general categories:

8.1.1 Indirect observation of surface and subsurface con-

---

[30] Examples of investigations that may not require all of the activities in this section include environmental site assessments for real estate property transactions that do not identify recognized environmental conditions using Practice E 1527 Phase 1 Assessment Process or Practice E 1528 Transaction Screen Process, and Environmental Audits to assess presence or degree of lead or asbestos contamination in a building.

[31] The U.S. EPA (36) requires latitude/longitude determination for all agency-sponsored data collection and activities that define or describe environmental characteristics about a site. The U.S. EPA (37) provides guidance to selecting latitude/longitude determination methods. Global positioning systems (GPS) are used for locating field observation points when accuracy on the order of tens of meters is sufficient, which is sufficient for many GIS applications (U.S. EPA, (38)). Other coordinate systems (such as x-y coordinates referenced to a known datum, State Plane, UTM) can also be used if more convenient in the field, and converted to latitude and longitude, if required.

[32] Other methods in addition to direct water level measurements may be useful for characterizing depth to water table. For example, description of the color, depth, and patterns of mottling of soil horizons may be indicative of long-term seasonal high ground water positions. Boulding (39) and Vepraskas (40) describe methods recently adopted by the Soil Survey Staff (41) for description and interpretation of morphologic features indicative of soil wetness. At contaminated sites, color variations can also be chemically induced or the result of staining.

[33] Microorganisms in soil and ground water may play a significant role in the transport and fate of inorganic constituents (such as nitrate, sulfate, and redox sensitive species, such as iron, manganese, arsenic, chromium, and selenium) and organic chemicals. Table X1.2 identifies ASTM methods for characterization of microorganisms and biodegradation potential.

ditions using remote sensing techniques and geophysical surveys (see Section 9),

8.1.2 Direct observation of subsurface conditions and field description using visual-manual procedures (see Section 10),

8.1.3 Sampling for further physical, chemical, and biological testing and analysis (see Section 11), and

8.1.4 In situ testing of soil, rock, vadose zone, and aquifer characteristics (see Section 12).

8.2 Many test methods and procedures are potentially applicable to environmental site investigations.[34] Factors to consider when selecting equipment and procedures for a site investigation include:

8.2.1 *Objectives and Data Quality Requirements of the Investigation*—For example, a Phase I Environmental Site Assessment may require only simple, hand-held equipment. On the other hand, site investigations for suitability for long-term storage of high-level radioactive wastes involve complex and sophisticated instrumentation and testing.

8.2.2 *Characteristics of the Site*—Soils and geology of the site and properties of known or suspected contaminants will influence the type of drilling methods used and selection of sampling equipment. Different aquifer test procedures and analytical methods are required based on aquifer characteristics such as degree of confinement, presence of nearby hydrologic boundaries, and how much of an aquifer is intersected by the central test well.

8.2.3 *Characteristics of the Equipment or Method*—When several alternatives are available to achieve a given objective a number of factors should be considered in selecting equipment and methods. Equipment available from commercial sources is generally preferable to homemade equipment because specifications can be readily documented and replacement parts can usually be obtained quickly and be used without affecting the comparability of results with the broken equipment. Equipment with the greatest durability and reliability is preferable. Standard ASTM test methods or standard operating procedures and protocols established by government agencies should be used whenever applicable. The timeliness of results is another important consideration when selecting equipment or methods. Methods that provide real-time data that fulfill data quality requirements are preferable to methods that require use of offsite laboratories. Special circumstances may require use of uncommon or nonstandard equipment, methods, or procedures. The use of such equipment and procedures should be justified in the detailed site investigation plan.

8.2.4 *Cost of the Equipment or Method*—All factors affecting cost should be considered before selecting equipment or methods. For example, the higher initial cost of using dedicated samplers for ground-water monitoring wells may be more cost-effective in the long run as a result of reduced sample collection costs.

8.3 Equipment and standard operating procedures (SOPs) should be identified in the detailed site investigation plan. Equipment design and specifications should be documented and standard operating procedures defined. Any departures from equipment and procedures defined in the site investi-

gation plan should be documented and justified.

9. Field Methods: Remote Sensing and Geophysical Surveys

9.1 Remote sensing techniques may assist in mapping the geological formations and for evaluating variations in soil and rock properties. Satellite and airborne mapping methods, such as multispectral imagery obtained from the LANDSAT platform and SLAR (side-looking airborne radar imagery), may be used to find and map the areal extent of subsurface materials and geologic structure. Interpretation of aircraft photographs and satellite imagery may locate and identify significant geologic features that may be indicative of faults and fractures. Interpretation of aerial photography taken at different times can identify changes in land use and identify areas that have been disturbed by human activity. Some ground control is generally required to verify information derived from remote sensing data.

9.2 Surface geophysical investigations can be a useful guide in the placement of boring, test hole, vadose zone, and monitoring well locations. Surface geophysical methods are essential at most contaminated sites to avoid hazardous drilling locations. Surface geophysical methods, together with push technologies (see 10.1.5), are especially valuable for ensuring that permanent monitoring well locations intercept contaminant plumes and that background reference wells reflect uncontaminated conditions. Interpretation of surface geophysical surveys should be verified by borings or test excavations, or confirmed by different types of geophysical measurements. Surface and borehole geophysical measurements provide a useful supplement to borehole and outcrop data and assist in interpolation between holes. Seismic, ground penetrating radar, electrical resistivity and electromagnetic methods can be particularly valuable when distinct differences in the physical or electrochemical properties of contiguous subsurface materials are indicated.[35]

9.3 Major applications of surface geophysical methods for environmental site characterization include:[36]

9.3.1 Ground penetrating radar may be useful in defining soil and rock layers, water table, and man-made structures in the depth range from $\frac{1}{3}$ to 10 m (1 to 30 ft).

9.3.2 Electromagnetic induction, electrical resistivity, and induced polarization (or complex resistivity) techniques can be used to map variations in water content, clay horizons, stratification, depth to aquifer/bedrock, and conductive contaminant plumes.[37]

9.3.3 The shallow seismic refraction method can be used to map soil horizons, stratigraphic depth profiles, and water tables. It can be especially useful in determining depth of

---

[34] A recent EPA report, (43) provides information on more than 280 methods.

[35] Major references that address use of surface geophysical methods for environmental site characterization include: Benson et al (44), Boulding (45), Haeni (46), Ward (47,48), and Zohdy et al (49).

[36] Depth ranges indicated here and for other geophysical methods are typical ranges for instruments available at the time this guide was written and intended to give a general idea of the capabilities of different methods. Actual depth penetration at a particular site is dependent on site conditions, instrumentation used for field measurements, and methods used to analyze the signal data. Depth of penetration can also increase with improved instrumentation and signal analysis methods.

[37] Test Method G 57 is intended for use in the control of corrosion in buried structures. Use of the electrical resistivity method in environmental investigations involves field procedures and data interpretation that differ substantially from the procedures outlined in this guide.

unconsolidated material where bedrock is at a depth of 30 m or less.

9.3.4 The seismic reflection method may be useful in delineating geological units at depths below 15 to 30 m (50 to 100 ft). It is not constrained by layers of low seismic velocity and is especially useful in areas of rapid stratigraphic change.

9.3.5 Magnetic surveys may be useful for detecting the presence of subsurface ferrous materials within 1 to 3 m of the ground surface.

9.4 Borehole geophysical methods can be used to obtain information on lithology, stratigraphy and formation properties, aquifer properties, ground-water flow and direction, borehole fluid characteristics, contaminant characterization, and borehole/casing characterization. Major categories of borehole logging methods include: (1) electrical and electromagnetic, (2) nuclear, (3) acoustic and sonic, and (4) miscellaneous methods, such as caliper, temperature, and borehole flowmeters. Considerations in the selection of borehole logging methods include, but are not limited to, (1) diameter of the borehole, (2) presence or absence and type of casing, (3) presence or absence and type of borehole fluid; and (4) characteristics of the subsurface formations.[38]

9.5 All geophysical methods require site conditions that provide contrasts in the subsurface property being measured by the method, and, depending on the method, may be subject to interferences at a site, such as metal fences, powerlines, FM radio transmission, or ground vibrations. The depth of penetration and interference effects are highly site specific. Depth penetration and resolution vary with local conditions. Data collection and interpretation for geophysical surveys require skilled personnel familiar with the principles and limitations of the method being used.

10. Field Methods: Direct Observation of Subsurface Conditions[39]

10.1 The type of equipment required for an intrusive subsurface investigation depends upon various factors including, but not limited to, the type of subsurface material, the depth of exploration, the nature of the terrain, the intended use of the data, and prevention of cross-contamination of aquifers.[40] Identify the location of buried utility lines, pipes, and any other subsurface anthropogenic features using maps or geophysical methods (9.2) prior to drilling.

10.1.1 Hand augers, hole diggers, shovels, and push tube samplers are suitable for exploration of surficial soils to depths of 1 to 5 m (3 to 15 ft).

10.1.2 Earth excavation equipment, such as backhoes, draglines, and drilled pier augers (screw or bucket) can allow in situ examination of soil deposits and sampling of materials containing cobbles and boulders.

10.1.3 In soil and unconsolidated material hollow-stem augers are commonly used for collection of geoenvironmental samples and installation of monitoring wells. Hollow-stem augers commonly reach depths of up to 45 m (about 150 ft) and large auger rigs can reach depths of up to 100 m (about 330 ft) under favorable conditions.[41] Greater depths are possible in unconsolidated material if casing advancement methods are used. Sections 11.2.2 and 11.2.3 identify ASTM methods for collection of unconsolidated material samples.

10.1.4 For bedrock and for deep unconsolidated material, well drilling equipment such as rotary and cable tool systems is usually required. Normally samples are collected in the form of sand sized cuttings captured from the return flow, which generally are not adequate for environmental site characterization but other methods of determining stratigraphy, such as borehole geophysical logging methods are available (see Section 9.4). Diamond rotary coring (see Practice D 2113) is the preferred method of obtaining samples for physical property testing and geologic characterization of rocks. Diamond drills are effective to depths of up to 600 m or more, even in extremely strong igneous and metamorphic rocks. As depths in unconsolidated material increase, sampling devices attached to drill rods become increasing cumbersome due to the time required to lower and retrieve the sampler. Wireline soil sampling, in which the sampler is lowered, driven, and retrieved on a wireline through the drill pipe, allows retrieval of samples at frequent intervals and provides for prompt visual identification of soil texture and subsurface features (Section 10.3). More commonly wireline systems are used for rock coring. Various factors affect the depth at which wireline sample becomes more cost effective than drill-rod attached sampling devices, including the length of the core barrel, the height of the mast on the rig being used, the number of hoist lines available for handling sampling tools and the hardness of the rock. The breakeven point for the two methods typically ranges between 15 m (50 ft) and 45 m (150 ft). As hole depths increase, cost savings using wireline sampling methods increase. Reduced physical labor for drill crews using wireline methods help to reduce worker fatigue, improving efficiency and safety.

10.1.5 Cone penetration technologies (CPT) and other drive/push technologies (DPT) are being used increasingly for environmental field investigations to depths of 100 to 300 ft (30.5 to 100 m) depending on geology and push capacity of the system. The CPT methods can be used for lithologic characterization, measurement of hydraulic conductivity of soils, sampling of solids, soil gases and ground water, and in situ chemical detection (see 11.3). Accurate lithologic characterization using CPT requires correlation of measured resistance with one or more direct subsurface observations for each site. Advantages of CPT and DPT at contaminated sites include: no cuttings, increased worker safety, and improved siting of permanent monitoring wells.

10.2 A stratigraphic profile in complex geology is often developed by correlation of lithologic character established

---

[38] Major references that address use of borehole geophysical methods for environmental site characterization include: Boulding (45), Keys (50), Keys and MacCary (51), and Respold (52).

[39] Major references on borehole drilling methods for direct observation of the subsurface and for installation of ground-water monitoring wells include: Aller et al (53), Eggington et al (54), Clark (55), Driscoll (56), Harlan et al (57), Lehr et al (58), Roscoe Moss Company (59), Ruda and Bosscher (60), and Shuter and Teasdale (61).

[40] Plans for a program of intrusive subsurface investigation should check whether there are requirements for licensing of installers and permits for installation and proper closure of bore holes and wells at the completion of the investigation.

[41] Fluid rotary drilling can extend to greater depths in unconsolidated materials, but is generally not recommended for environmental investigations because drilling fluids can alter subsurface chemistry.

by a significant number of borings, by continuous geophysical profiling techniques, and surface mapping. This phase of the investigation may be implemented by plotting logs of soil and rock exposures in walls of excavations or cut areas and by plotting logs of the test borings. Then one may interpolate between, and extrapolate a reasonable distance beyond, these logs. The spacing of these investigations depends on the geologic complexity of the project area and on the importance of soil and rock continuity to the project design. Exploration should be deep enough to identify all strata that might be significant in assessing environmental conditions at the site.

10.3 Field description of test pits and excavations and boring logs are an essential element of environmental site investigations. Test pits allow observation of pedogenic soil features that are larger than the diameter of borings and allow the most accurate description of subsurface features in the soil zone that may influence movement of contaminants in the vadose zone. Paragraph 6.4.3 in Guide D 5409 identifies key features that should be included in field description of samples and unconsolidated material. Additional description of samples of soil and rock may be added after samples have been transported from the field. Subsurface observation records should be kept in a systematic manner for each project. Such records should include, when applicable:

10.3.1 *Description of Each Site or Area Investigated—* Each test hole, boring, test pit, or geophysical test site should be referenced to an established coordinate system, datum, or permanent monument (see 7.1.3):

10.3.2 Subsurface investigation logs of each test hole, boring, test pit, or cut surface exposure should show the field description and location of each material, and any water encountered, either by symbol or verbal description. Guide D 5434 provides guidance for field logging of subsurface explorations for soil and rock. Reference to a Munsell Soil or Rock Color Chart designation is a substantial aid to the description of soil and rock materials.[41] Various soil classification systems are available for description of soil and other unconsolidated materials in the field. The selection of the classification system depends on the purpose and data needs of the investigation and the end users of the information. The ASTM (Classification D 2487 and Practice D 2488) version of the Unified Soil Classification System (USCS) is used by most governmental agencies and geotechnical engineering firms for the description of soil for engineering purposes both in the United States and in many other countries. The USDA textural system (41,63,64),[42] developed by the Soil Conservation Service for agricultural aspects of near surface soils, is especially useful for environmental investigations because it can provide hydrologic information for soils. Using the modified Wentworth grain size scale (AGI Date Sheet 29.1; (65),[43] which is widely used by geologists, in combination with the USDA textural system will provide useful additional data on soil particle-size distribution because the Wentworth system has more subdivisions in the silt and gravel particle-size classes. Field

classification of samples and exposures should be confirmed with laboratory tests (Classification D 2487, Test Method D 422, and Test Method D 1140) if the information is critical to the investigation program.[43] Laboratory test results should clearly identify the textural system being used, because different systems may have slightly different grain-size cutoffs. Use of sieves that include all divisions of the USCS, USDA and modified Wentworth scales will provide the greatest amount of useful information for the coarse fraction, and measurement of all relevant cutoffs for the fine fraction in the three systems will ensure that environmental and engineering interpretations using all three systems will be possible.

10.3.3 Location and description of seepage and water-bearing zones, zones of low permeability, and records of potentiometric elevations found in each test hole, boring, piezometer, or test pit,

10.3.4 Grain size distribution of unconsolidated materials in the interval to be screened when installation of a ground-water monitoring well is planned,

10.3.5 The results and locations of in situ test results (see Section 12),

10.3.6 Percentage of core recovery and rock quality designation (RQD) in core drilling. Deere and Deere (69) provides guidance on RQD, and

10.3.7 Graphical presentation of field and laboratory data, such as graphic borehole logs, cross sections, potentiometric maps, and contaminant concentration isopleths, facilitates interpretation and understanding of subsurface conditions.

11. Field Methods: Sampling

11.1 The location, type of samples collected, and sampling equipment should be in accordance with the sampling plan developed in the detailed site investigation plan. A statistically valid method should be used to select sampling locations if required by the data-collection objectives. Sample equipment and procedures should be appropriate for the medium being sampled (soil, rock, soil gas, soil-pore liquids, ground water, solid/liquid waste) and be designed to minimize the introduction of error into physical and chemical test results. Barth et al (70) and Mason (71) provide guidance on development of soil sampling plans. Gilbert (72) provides a good general reference on statistical aspects of sampling and monitoring for environmental purposes. Geostatistical methods are especially useful for developing a

---

[42] Color photographs of rock cores, soil samples, and exposed strata may be of considerable value. Each photograph should include an identifying number or symbol, a date, and a reference scale. Soil Conservation Service (62) provides standardized procedures for photographic documentation of rock cores.

[43] Both USCS and USDA classification systems can be used to estimate some hydrologic properties of soils impacting the fate and transport of contaminants. The plasticity index of the USCS is directly related to colloidal content in the soil and resulting sorptive capacity. The USCS system is more applicable to very coarse-grained alluvial deposits containing appreciable gravels. Saturated hydraulic conductivity can often be estimated within two or three orders of magnitude based on both USCS and USDA classification, and estimates can sometimes be narrowed within one order of magnitude for clean sands or in fine-grained soils where descriptions of morphology (structure, macroporosity, fissures, laminations, etc.) of undisturbed cores or exposures are possible. Numerous additional hydrologic parameters have been correlated to the USDA classification system for near-surface soils and could be useful for fine-grained unconsolidated material at greater depths. These parameters include: capillary fringe (39), specific yield (66), field moisture capacity (67), and available water capacity (68). Estimation of hydrologic parameters may be useful during the initial site characterization phases, but should not replace necessary definitive field and laboratory measurements which may be required for final site evaluations.

sampling strategy and analysis of results when parameters are spatially correlated.[44]

11.2 *Solids Sampling*[45]—The type of solids sample will depend on the purpose of the sample.

11.2.1 Disturbed or grab samples using augers (see Practice D 1452) and trier samplers (see Practice D 5451) are used primarily for chemical analyses, but are also suitable for texture analysis.

11.2.2 Split barrel samplers (see Test Method D 1586) provide samples that allow description of lithology and other subsurface features, and laboratory measurements of soil parameters unaffected by sample disturbance (for example, water content, gradation or particle size, organic content, and nonsensitive and chemical constituents) but are not suitable for laboratory measurement of soil properties.

11.2.3 Thin-wall tube (see Practice D 1587) and ring-lined barrel samplers (see Practice D 3550) collect undisturbed samples of unconsolidated material that are suitable for laboratory testing of hydrologic properties, and diamond core drilling (see Practice D 2113) provides similar samples for consolidated rock. Undisturbed core samples can provide important information on structure, sedimentary features, secondary porosity, and color patterns that cannot be obtained from disturbed samples.

11.2.4 Guide D 4700 provides guidance on soil sampling in the vadose zone. Special considerations in sampling of soil for radionuclides and volatile organics are addressed in Practice C 998 and Practice D 4547, respectively. Practice D 4220 covers procedures for preservation and transportation of soil samples and Practices D 5079 does the same for rock core samples.

11.3 *Other Sampling:*

11.3.1 Characterization of the vadose zone may require sampling of soil gases and soil pore liquids. Guide D 5314 provides guidance on soil gas sampling for detection of volatile contaminants. Guide D 4696 addresses various methods for sampling soil pore liquids in the vadose zone.

11.3.2 *Ground Water Sampling*—Guide D 4448 covers ground-water sampling methods from permanent monitoring wells. Springs can also be used for sampling and monitoring ground water quality as covered by Guide D 5717. The CPT and other push technologies (see 10.1.5) allow collection of one-time or multiple samples for preliminary delineation of ground-water quality and installation of driven monitoring wells. Manually driven well points are generally restricted to shallow depths (<10 m) that can be sampled using suction-lift ground-water sampling devices. The CPT rigs can install small-diameter monitoring wells to

depths exceeding 30 m which can be sampled with small diameter (5/8-in. (15.8-mm) outside diameter typical) bailers and pumps.[46]

11.3.3 Submerged sediments can be sampled with coring devices (see Guide D 4823) or with dredges, such as the Eckman, Peterson, and Ponar dredges. Guide D 4411 provides guidance on sampling of fluvial sediments in motion.[47]

11.3.4 Waste materials often require special sampling equipment and procedures. Devices for sampling single-phase and multiphase liquids include: coliwasa, dipper, drum thief, glass tube, and peristaltic pump. Mixed liquid/solid phases are usually sampled using a coliwasa, dipper, or long-handled sludge sampler. Consolidated waste solids are usually sampled using an auger, chipper, hammer and chisel, or a rotating coring device. Unconsolidated waste solids can be sampled using a trier (see Practice D 5451), grain sampler, or conventional soil sampling equipment (see 10.1.1). Lagoons may require sludge sampling tools.

11.3.5 *Biological Sampling*—Investigation of soil and ground-water geochemistry may involve sampling for aerobic and anaerobic microorganisms and other subsurface biota.[48] Special care is required when sampling for anaerobic microorganisms to ensure that the sample is not exposed to the air (90). Table X1.1 identifies ASTM methods for microbiological analysis of water samples. Ecosystem characterization may require sampling of plants and animals at a site. Table X1.2 provides an index to ASTM methods for field sampling of phytoplankton, zooplankton, benthic macroinvertebrates, and fishes.

11.3.6 *Atmospheric Sampling*—Measurement of atmospheric parameters such as precipitation, humidity, temperature, and wind may be required for ground-water budget studies. Humidity, temperature, and wind should be routinely monitored during field investigations to document site conditions. Such monitoring is especially important for worker health and safety when it is very cold, or impermeable protective clothing is required when ambient temperatures are high. Air quality sampling may be required during environmental site investigations for: (1) identifying unsafe working conditions and monitoring worker exposure to hazardous chemicals, (2) evaluating air exposure pathways for environmental risk assessment, and (3) assessing ambient concentrations of regulated air pollutants. Table X1.1 provides an index to ASTM methods for monitoring ambient atmospheric parameters, and field sampling.

---

[44] Most soil physical and chemical properties are spatially correlated, and the location of contaminants in soil and ground water are also usually spatially correlated. Clark (73) is a good introductory text on geostatistics and Isaaks and Srivastava (74) provide advanced treatment of this subject. The ASCE Task Committee on Geostatistical Techniques in Geohydrology (75) contains a good review of basic concepts and applications. Several public domain geostatistical software packages are available from U.S. EPA (76,77,78). Geostatistical methods may not be appropriate where soil contamination exists as localized hotspots. Zirschky and Gilbert (79) provide guidance on statistical sampling designs for detecting hot spots at hazardous-waste sites.

[45] The investigation may require the collection of sufficiently large soil, rock, waste, and ground water samples of such quality as to allow adequate testing to determine the soil or rock classification or mineralogic type, or both, and chemical constituents of potential interest or concern.

[46] Driven wells are relatively easy and inexpensive to install in most unconsolidated materials where ground water is shallow. Deeper sampling and well installation methods using CPT and other push methods may be an alternative to conventional monitoring well installations. The decision to use these techniques for permanent monitoring well installations should be based on information indicating that sample representativeness and installation integrity (that is, avoidance of cross-contamination of aquifers) are comparable to conventional monitoring well installations with screens, filter pack, and grouting. To ensure comparability of ground-water quality results, all monitoring well installations for a particular phase of investigations at a site should be of the same type (that is, either all conventional or all drive/push installations).

[47] Good references for additional information on sampling of submerged bottom sediments include: Barth and Starks (80) and Palmer (81). Guy and Norman (82) provide additional information on sampling of fluvial sediment.

[48] Chappelle (83) provides a useful reference for microbial sampling of ground water. Some major references on soil microbiology and ecology, geomicrobiology and microbial biogeochemistry include: Alexander (84), Ehrlich (85), Killham (86), Kuznetsov et al (87), Paul and Clark (88), and Zajic (89).

11.4 *Other Sampling Considerations:*

11.4.1 All sampling procedures for chemical analysis require cleaning and decontamination of sampling equipment to prevent cross contamination of samples. Practice D 5088 addresses decontamination of field equipment at nonradioactive waste sites.

11.4.2 Special care is required when locating permanent vadose-zone and ground-water monitoring installations for time-series sampling. Failure to site such permanent installations in locations downgradient from potential contaminant sources or to intersect zones of preferential movement of contaminants can result in wasteful expenditure of funds for chemical analysis, or even failure to detect contaminant movement. The location of permanent monitoring installations should be thoroughly justified based on the conceptual model of the site. Surface geophysical methods, multiple piezometers for measurement of ground-water flow direction, and analysis of samples taken using in situ soil gas and ground-water samplers can provide information required for rational siting of permanent monitoring installations.

12. Field Methods: In Situ Testing and Analytical Methods

12.1 In situ testing (for example, tests that measure the in-place characteristics of subsurface materials) is useful for: measuring the hydrologic properties of a larger volume of the subsurface than is possible in laboratory tests on soil and rock cores; for rapid or closely spaced measurements, or both, of earth properties without the necessity of sampling; measuring subsurface chemical parameters to minimize chemical alterations by bringing samples to the surface for analysis; and measurement of engineering properties of soil or rock in an undisturbed condition including consideration of lateral and vertical loads associated with the surrounding mass.

12.2 Field analytical methods use conventional or adapted laboratory methods for analysis of samples that have been removed from their in-place position. Mechanical sieve analysis of soil samples is commonly used to assist in field classification of soil texture. Field analytical methods such as portable gas chromatographs, X-ray fluorescence, and enzyme immunoassay kits are being increasingly used for environmental site characterization. Field instrumentation and procedures for chemical characterization are changing rapidly and the appropriate regulatory authority should be consulted for accepted methods and procedures when investigations are performed for regulatory purposes.

12.3 *Hydrologic Properties:*

12.3.1 Aquifer tests are used to collect data for calculating hydraulic conductivity, transmissivity, and aquifer storage properties. Tests may include, but are not limited to packer tests for low-permeability rocks (see Test Method D 4630 and Test Method D 4631), slug tests (see Test Method D 4044, Test Method D 4104, and Test Method D 4050), and aquifer tests with control wells (see Test Methods D 4105, D 4106, D 5269, and D 5270). Guide D 4043 provides guidance on selection of aquifer field test methods and analysis procedures based on aquifer characteristics. In karst and fractured rock hydrogeologic settings refer to Guide D 5717 for special approaches required for characterization of hydrologic properties. In such settings conventional aquifer test methods may not be appropriate. Similar prob-

lems may also occur in heterogeneous porous media.

12.3.2 Test Method D 5084 provides a method for measuring point hydraulic conductivity of low-permeability materials using undisturbed core materials ($<1 \times 10^{-3}$ cm/s); Test Method D 2434 provides a method for measuring hydraulic conductivity of granular materials ($>1 \times 10^{-3}$ cm/s).

12.3.3 Field measurements of vadose zone hydrologic parameters include water content, matric potential, and infiltration rate and hydraulic conductivity. Guide D 5126 provides guidance on selection of methods for measuring saturated and unsaturated hydraulic conductivity in the vadose zone. Table X1.1 identifies ASTM methods for measuring infiltration rate, matric potential, and moisture content.

12.4 *Physical Properties*—In situ measurements of vertical and horizontal variations in soil density using a standard penetration test (see Test Method D 1586), cone penetrometry (see Test Method D 3441) or a gamma-gamma nuclear probe (see Test Methods D 2922 and D 5195), can provide useful information for stratigraphic interpretation within and between boreholes. Other physical properties that affect water movement can be measured with a variety of borehole geophysical logging tools, including induction, gamma, resistivity, neutron porosity, sonic, and caliper tools. In addition, downhole video and acoustic televiewer instruments may provide information on such properties as fractures, vugs, and bedding plane orientation.

12.5 Use of ion-selective electrodes to measure chemical parameters in place in the subsurface is an established technology (see Terminology D 4127). Fiber optic chemical sensors are a relatively new technology with good possibilities for making in-situ chemical measurements for environmental site characterization.[49]

12.6 *Engineering Properties*—Engineering properties are primarily of interest for environmental site characterization if design of pollution control measures, such as impoundment liners, or remediation activities are required. Numerous ASTM field methods are available for in situ measurement of soil and rock engineering properties. These are identified in Table X1.1.

13. Analysis and Interpretation of Results

13.1 Evaluate data to determine whether data quality requirements were met (see 5.1.5). Data review and analysis should include all reliable field and laboratory data from previous investigations in the same area. Review field and laboratory QA/QC procedures and measurements to assess data validity, and determine whether data quality requirements (see 5.1.5) have been satisfied. Interpretation of field- and laboratory-measured environmental parameters should include an evaluation of possible limitations of the methods used. Basic assumptions for analytical techniques and methods should be evaluated to determine if site conditions meet assumptions. For example, the analysis of aquifer test

---

[49] The term in situ here is confined to methods that measure chemical parameters in place without bringing a sample to the surface. The term may also be applied to ground water sampling devices that do not require installation of a monitoring well. In situ chemical sensors can be placed in monitoring wells or by using CPT or other push technologies.

results should identify the approximate volume of the aquifer measured by the test, and the underlying analytical or other equations used to compute aquifer parameters. If site conditions do not satisfy the assumptions of the solution method, the effect on accuracy and interpretation of results should be stated.

13.2 Develop graphical presentation of data to facilitate interpretation of spatial relationships and, when time series data are available, presence or absence of trends. Map views, cross sections, and data contouring methods are especially useful for presentation of spatial data.

13.2.1 A map of the area under investigations provides essential information about the land surface, including natural and anthropogenic features, and the locations of sampling, monitoring wells, and other observation points.

13.2.2 Cross sections should identify actual surface and subsurface observations according to elevation and location. Cross sections showing correlation of stratigraphic or lithologic units and interpretation of other conditions between direct subsurface observations should be indicated as interpretations (that is, dashed lines) and based on standard geologic procedures. When feasible, geophysical survey data, such as continuous geophysical profiles should be used to support correlations and other interpretations. The interpretive cross sections should be accompanied by notes describing anomalies or otherwise significant variations in the site conditions that might affect any interpretations.[50]

13.2.3 Contouring methods, such as structure contours of geologic strata or buried bedrock, potentiometric surfaces, and maps showing lines of equal concentration or value should be constructed using appropriate interpolation techniques.[51] The method of interpolation should be documented. When feasible, the same data should be contoured using different interpolation methods and compared.

13.3 Statistical methods used to analyze data should be appropriate for the type of data. Most conventional statistical methods assume a normal distribution around the mean. Typically, environmental data measurements do not exhibit normality because of spatial correlation, the presence of outliers, or other effects.[52] Geostatistical methods are best for analyzing spatially related data (see 11.1 and related references). When contaminants are present in low concentrations (at or below the detection limit), the report should indicate whether chemical data have been censored and if so, assess the possible effect of censoring (that is, values reported

as not detected or below the detection limit). Practice D 4210 addresses data censoring and its possible significance.

13.4 Information on topography, geomorphology, soils, climate, vegetation, surface hydrology, and anthropogenic influences should be integrated with subsurface geologic, hydrogeologic, and geochemical/hydrochemical interpretations as required by the objectives of the investigation. At contaminated sites, identification of pathways for movement of contaminants and estimation of exposure concentrations, and determining the rate of movement may be an important result of the site characterization study.[53]

14. Report

14.1 Report the following information:

14.1.1 Pertinent ASTM Standards, (Terminology D 653, Practice D 3584, Practice E 177, and Practice E 380).

14.2 The report of an environmental site characterization study should include:

14.2.1 The location of the area investigated in terms pertinent to the project. This may include sketch maps or aerial photos on which the test pits, bore holes, and sample areas are located, as well as geomorphological data relevant to the determination of the various soil and rock types. Such data include elevation contours, stream beds, sinkholes, cliffs, and all other relevant physiographic features. All significant anthropogenic features should be located on the base map or a separate map. Where feasible, include in the report a geologic map or an agronomic soils map, or both, of the area investigated,

14.2.2 Additional basic information as required by the objectives of the investigation,

14.2.3 A description of the investigation procedures including all borings and test hole logs, graphic presentation of all lithologic and well construction logs, tabulation of all test results, and graphical interpretations of geophysical measurements, and

14.2.4 A summary of the findings obtained under Sections 6, 9, 10, 11, 12, and 13, using subhead titles for the respective sections, and appropriate recommendations and disclaimers for the use of the report.

15. Keywords

15.1 conceptual site model; environmental site characterization; exploration; feasibility studies; field investigations; geological investigations; geophysical investigations; ground water; hydrologic investigations; maps; preliminary investigations; reconnaissance surveys; sampling; site characterization; site investigations; soil surveys; subsurface investigations

[50] Additional exploration should be considered if there is not sufficient information to develop interpretative cross sections, with realistic descriptions of anticipated variations in subsurface conditions, to meet project requirements.

[51] Most field geology texts identified in Footnote 16 discuss interpolation techniques. Davis (91), Hamilton and Jones (94) and Jones et al. (95) are other good references.

[52] The EPA's GRITS/STAT software for analyzing ground-water monitoring data allows testing of the appropriateness of conventional statistical analysis of time series data, and includes a number of alternative tests if conventional tests are not appropriate (92).

[53] Estimation of exposure concentrations may require use of batch sorption/ leaching tests and vadose zone contaminant mobility studies. Roy et al (93) provides U.S. EPA guidance on batch test for estimating soil sorption of chemicals, and Table X1.2 (Laboratory Fate Testing) provides an index of ASTM sorption and leachability test methods.

APPENDIXES

(Nonmandatory Information)

X1. ASTM STANDARDS PERTINENT TO ENVIRONMENTAL SITE CHARACTERIZATION

X1.1 This appendix lists more than 400 standard test methods, practices, and guides for use in the field and laboratory that could be pertinent for site characterization for environmental purposes. Two tables provide a quick reference to two categories of standards:

X1.1.1 Table X1.1 identifies more than 320 potential field and laboratory methods for sampling and characterization of soil, ground water, and waste materials. Other ASTM methods for measuring site parameters that may need to be monitored during environmental site investigations, such as humidity and wind, are also included in Table X1.1. All standards listed in this table follow in alphanumeric sequence.

X1.1.2 Table X1.2 identifies more than 80 standard test method, practices and guides that address field sampling for ecological characterization and laboratory methods, such as toxicity testing, relevant to human and ecological risk assessment.

X1.2 This guide does not specifically address laboratory methods, but Table X1.1 provides a convenient index to laboratory methods that might be useful for testing and analysis of soil, water, and waste samples collected during an environmental site investigation. This appendix does not contain a detailed listing of laboratory methods for measurement of specific chemicals that might be of concern in an environmental investigation. However, methods for measurement of chemical parameters that are routinely used in field investigations and laboratory methods that provide information relevant to the transport and fate of contaminants and other chemical constituents in the subsurface are included.

X1.3 The following ASTM compilations may be useful for Environmental Site Characterization:

X1.3.1 *Compilation of Scopes of ASTM Standards Relating to Environmental Monitoring.* 1993, 328 pp. [Contains 700 scope statements],

X1.3.2 *ASTM Standards on Ground Water and Vadose Zone Investigations, 2nd ed.* 1994. [46 Standards],

X1.3.3 *ASTM Standards on Environmental Site Assessments for Commercial Real Estate, 2nd ed.* 1994, 55 pp. [Includes E 1527 and E 1528],

X1.3.4 *ASTM Standards on Hazardous Substance and Oil Spill Response, 2nd ed.* 1994, 144 pp. [38 standards],

X1.3.5 *ASTM Standards on Lead-Based Paint Abatement in Buildings.* 1994, 174 pp. [28 standards],

X1.3.6 *ASTM Standards on Environmental Sampling,* 1995, 532 pp. [70 standards],

X1.3.7 *ASTM Standards on Analysis of Hydrologic Parameters and Ground Water Modeling,* 1996, 148 pp. [23 standards], and

X1.3.8 *ASTM Standards on Design and Planning for Ground Water and Vadose Zone Investigations,* 1996, 120 pp. [10 standards].

X1.4 *ASTM Standards*—Indexed in Table X1.1 (In Vols 04.08 or 04.09, unless otherwise specified).[54,55]

C 150 – 92	Specification for Portland Cement (Vol 04.01)
C 199 – 91	Terminology Relating to Dimension Stone
C 294 – 86	Descriptive Nomenclature for Constituents of Natural Mineral Aggregates (Vol 04.02)
C 998 – 90	Practice for Sampling Surface Soil for Radionuclides (Vol 12.01)
C 999 – 90	Practice for Soil Sample Preparation for Determination of Radionuclides (Vol 12.01)
D 140 – 88	Practice for Sampling Bituminous Materials
D 420 – 93	Guide to Site Characterization for Engineering, Design, and Construction Purposes
D 421 – 85	Practice for Dry Preparation of Soil Samples for Particle-Size Analysis and Determination of Soil Constants
D 422 – 63	Method for Particle-Size Analysis of Soils
D 425 – 88	Test Method for Centrifuge Moisture Equivalent of Soils
D 427 – 93	Test Method for Shrinkage Factors of Soils by the Mercury Method
D 653 – 90	Terminology Relating to Soil, Rock, and Contained Fluids
D 698 – 91	Test Method for Laboratory Compaction Characteristics of Soil Using Standard Effort (12 400 ft-lbf/ft³) (600 kN-m/m³)
D 854 – 92	Test Method for Specific Gravity of Soil
D 887 – 82	Practice for Sampling Water-Formed Deposits (Vol 11.02)
D 932 – 85	Test Method for Iron Bacteria in Water and Water-Formed Deposits (Vol 11.02)
D 933 – 84	Practice for Reporting Results of Examination and Analysis of Water-Formed Deposits (Vol 11.04)
D 1067 – 92	Test Methods for Acidity or Alkalinity of Water (Vol 11.01)
D 1125 – 82	Test Methods for Electrical Conductivity and Resistivity of Water (Vol 11.01)
D 1140 – 92	Test Method for Amount of Materials in Soils Finer than No. 200 (75-µm) Sieve
D 1189 – 90	Terminology Relating to Water (Vol 11.01)
D 1192 – 95	Specification for Equipment for Sampling Water and Steam in Closed Conduits (Vol 11.01)
D 1194 – 94	Test Method for Bearing Capacity of Soil for Static Load on Spread Footings
D 1292 – 86	Test Method for Odor in Water (Vol 11.01)
D 1293 – 84	Test Methods for pH in Water (Vol 11.01)
D 1356 – 95a	Terminology Relating to Atmospheric Sampling and Analysis (Vol 11.03)
D 1357 – 95	Practice for Planning the Sampling of the Ambient Atmosphere (Vol 11.03)
D 1452 – 80	Practice for Soil Investigation and Sampling by Auger Borings
D 1498 – 93	Practice for Oxidation-Reduction Potential of Water (Vol 11.01)
D 1556 – 90	Test Method for Density and Unit Weight of Soil In Place by the Sand-Cone Method
D 1557 – 91	Test Method for Laboratory Compaction Characteristics of Soil Using Modified Effort (56 000 ft-lbf/ft³) (2700 kN-m/m³)
D 1558 – 94	Test Method for Moisture Content Penetration Resistance Relationships of Fine Grained Soils
D 1586 – 84	Test Method for Penetration Test and Split-Barrel Sampling of Soils
D 1587 – 94	Method for Thin-Walled Tube Sampling of Soils
D 1704/ D 1704M – 95	Test Method for Determining the Amount of Particulate Matter in the Atmosphere by Measurement of the Absorbance of a Filter Sample (Vol 11.03)
D 1739 – 82	Test Method for Collection and Analysis of Dustfall (Settleable Particulates) (Vol 11.03)
D 1785 – 91	Specifications for Polyvinyl Chloride (PVC) Plastic Pipe Schedules 40, 80, and 120 (Vol 08.04)
D 1883 – 94	Test Method for CBR (California Bearing Ratio) of Laboratory-Compacted Soils

---

[54] Prior to 1994 Volume 04.08 contained all standards on soil and rock. Beginning in 1994 this volume was broken into two volumes: Volume 04.08; Soil and Rock (I): D 420 to D 4914, and Volume 04.09; Soil and Rock (II): D 4943 to latest; Geosynthetics.

[55] The last two digits indicate the year in which the standard was adopted or the most recent year in which substantive revisions to the standard were adopted.

TABLE X1.1 Index to ASTM Field and Laboratory Methods Possibly Pertinent to Environmental Site Characterization

Topic	ASTM Standard
General:	
Reports	Indexing papers and reports (D 3584), use of modernized metric system (E 380)
Terminology	Soil, rock and contained fluids (D 653); atmospheric sampling (D 1356); waste and waste management (D 5681); water (D 1189)
Objective-Oriented Guides	Acquisition of file aerial photography and imagery for establishing historic site use and surficial conditions (D 5518); Developing conceptual site models for contaminated sites (E 1689); real estate property transactions (E 1527, E 1528); site characterization for engineering and construction purposes (D 420); site characterization for septic systems (D 5879-surface); development and implementation of a pollution prevention program (E 1609); risk-based corrective action at petroleum release sites (E 1739); short term measures or early actions for site remediation (D 5745); assessment of buried steel tanks (ES 40); environmental condition of property area types (D 5746); environmental regulatory compliance audits (PS 11); evaluation of an organizations environmental management system (PS 12)
Sampling:	
General	Collection and preservation of information and physical items by a technical investigator (E 1188); planning water quality measurement program (D 5612); probability sampling of materials (E 105)
Air	Choosing locations and sampling methods for atmospheric deposition at nonurban locations (D 5111, D 5012); guide for laboratories (D 3614); flow rate calibration of personal sampling pumps (D 5337); planning ambient air sampling (D 1357); ambient air analyzer procedures (D 3249); sampling stationary source emissions (D 5835); *Airborne Microorganisms:* Sampling at municipal solid waste facilities (E 884); *Sampling Organic Vapors/Toxic Vapors:* Charcoal tube absorption (D 3686), canister (D 5466); detector tubes (D 4490); length-of-stain dosimeter (D 4599); *Particulate Matter Determination:* Filter absorbance method (D 1704, D 1704M); high-volume sampler (D 4096); dustfall (D 1739—settleable particulates); *Worker Protection:* Air monitoring at waste management facilities for worker protection (D 4844); air sampling strategies for worker and workplace protection (E 1370); collection of airborne particulate lead during abatement and construction activities (E 1553); activated charcoal samplers (D 4597), liquid sorbent difussional samplers (D 4598); pesticides and PCBs (D 4861)
Biological Materials	Aseptic sampling (E 1287); see also Table A.2
Soil/Rock/Sediments	Minimum set of data elements for soil sampling (D 5911); *Drilling Methods:* cable tool (D 5875); casing advancement (D 5872); diamond core drilling (D 2113); direct air-rotary (D 5782); direct fluid rotary (D 5783); direct rotary wireline (D 5876); dual-wall reverse circulation (D 5781); hollow-stem auger (D 5784); *Field Sampling and Handling Methods:* Auger sampling (D 1452); radionuclides (C 998); ring-lined barrel (D 3550); split barrel (D 1586); thin-wall tube (D 1587); trier sampler (D 5451); volatile organics (D 4547); *Sediments:* sediments (D 4411—fluvial sediment in motion, D 4823—submerged, D 3213—handling, storing and preparing soft undisturbed marine soil; E 1391—collection for toxicological testing)
Vadose Zone	*Field Methods:* **Pore liquids (D 4696); soil (D 4700);** soil gas (D 5314)
Water	Purgeable headspace sampling (D 3871); waterborne oils (D 4489); continual on-line monitoring (D 3864); on-line sampling/analysis (D 5540—flow and temperature control), water-formed deposits (D 887); *Ground Water:* **Sampling methods (D 4448);** planning a ground-water sampling event (D 5903); *Surface Water:* dipper or pond sampler (D 5358); *Closed Conduits:* equipment (D 1192); sampling (D 3370); Laboratory Practices: D 3856
Waste/Contaminants	*Field Methods:* **General planning (D 4687);** bituminous materials (D 140); COLIWASA (D 5495); drums (consolidated solids—D 5679, unconsolidated solids—D 5680); single or multilayered liquids (D 5743); pipes and other point discharges (D 5013); scoop (D 5633); unconsolidated waste from truck (D 5658); UST release detection devices (E 1430, E 1526); volatile organics (D 4547); waterborne oils (D 4489); oil/water mixtures for oil spill recovery equipment (F 1084)
Preservation/Transport	Sample chain of custody (D 4840); estimation of holding time for water samples (D 4515, D 4841); *Field Methods*—Rock core samples (D 5079); sample containers for organic constituents (D 3694); soil samples (D 4220); sediments for toxicological testing (E 1391); preservation of waterborne oil samples (D 3325); handling, storing and preparing soft undisturbed marine soil (D 3213)
Decontamination of Field Equipment	*Field Methods:* Nonradioactive waste sites (D 5088); low-level radioactive waste sites (D 5608)
Data Management/Analysis	*QA/QC:* Waste management environmental data (D 5283); waste management DQOs (D 5792); precision and bias (E 177); QC specification for organic constituents (D 5789); *Data Analysis:* Evaluation of technical data (E 678); outlying observations (E 178); reporting results of examination and analysis of water-formed deposits (D 933); *Geostatistics:* reporting geostatistical site investigations (D 5549); *Spatial Data:* digital geospatial metadata (D 5714); see also Ground Water (Data Analysis)
Soil/Rock Hydrologic Properties:	
Infiltration Rate	*Field Methods:* Double-ring infiltrometer (D 3385); sealed double-ring infiltrometer (D 5093)
Matric Potential	*Field Methods:* Tensiometers (D 3404); *Laboratory Method:* Filter paper method (D 5298)
Water Content	*Field Methods:* Calcium carbide method (D 4944); neutron probe (D 3017—shallow depth, D 5220); *Laboratory Methods:* Direct heating method (D 4959); microwave oven method (D 4643); standard oven drying method (D 2216); centrifuge moisture equivalent (D 425)
Hydraulic Conductivity	*Field Methods:* **Vadose zone (D 5126);** *Laboratory Methods:* Granular soils (D 2434—>1 × 10⁻³ cm/sec); low permeability soils (D 5084—<1 × 10⁻³ cm/sec); rigid-wall compaction-mold permeameater (D 5856); peat (D 4511)
Other Hydrologic Properties	*Laboratory Methods:* Air permeability (D 4525); Soil water retention (D 2325—medium/coarse textured, D 3152—fine-textured)
Soil/Rock Physical Properties:	
Particle Size	*Soil Laboratory Methods:* Analysis (D 422); dry preparation (D 421); <200 sieve (D 1140); wet preparation (D 2217); *Sediment:* Selection of methods for fluvial sediment (D 4822)
Soil Density	*Field Methods:* Drive cylinder (D 2937); gamma-gamma (D 2922—<12″, D 5195—>12″); (D 4531); penetration (D 1586); rubber-balloon method (D 2167), sand-cone method (D 1556); sand replacement method (D 4914); sellve method (D 4564)
Pore Volume/Specific Density	*Laboratory Methods:* pore volume (D 4404); specific gravity (D 854, D 5550—gas pycnometer)
Cone Penetration	*Field Methods:* In-situ cone penetration testing (D 3441); CPT stress wave energy measurements (D 4633)
Classification	*Field Methods:* Field logging (D 5434); noncohesive sediments (D 5387); peat (D 4544—deposit thickness, degree of humification—D 5715); sediments (D 4410); visual-manual procedure (D 2488—unified, D 4083—frozen soils); *Laboratory Methods:* Dimension stone (C 199); frozen soils (D 4083); natural mineral aggregates (C 294); peat (D 2607); unified soil classification (D 2487)
Geophysical Properties	*Field Methods:* Crosshole seismic testing (D 4428/D 4428M); seismic refraction (D 5777); soil resistivity (G 57—Wenner 4-electrode method); planning and conducting borehole geophysical logging (D 5753)

TABLE X1.1 *Continued*

Topic	ASTM Standard
Engineering Properties	*In Situ Field Methods:* Bearing capacity/ratio (D 1194, D 4429); deformability and strength of weak rock (D 4555); direct shear strength (D 4554, D 5607); extensometers (D 4403); in situ creep (D 4553); in situ modulus of deformation (D 4394—rigid plate, D 4395—flexible plate, D 4506—radial jacking test, D 4729—flatjack method, D 4791—borehole jack); in situ stress (D 4623—borehole deformation gage, D 4645—hydraulic fracturing; D 4729—flatjack method); pressure measurement (D 4719—pressuremeter, D 5720—transducer calibration); vane shear test (D 2573); *Laboratory Methods:* California bearing ratio (D 1883); classification (D 2487); compaction (D 698, D 1557, D 5080); compressive strength (D 2166, D 2938); core dimensional and shape tolerances (D 4543); dispersive characteristics (D 4221—double hydrometer; D 4647—pinhole test); elastic properties (D 2845, D 3148); liquid limit (D 4318); moisture content-penetration resistance (D 1558); one-dimensional swell (D 4546); plastic limit/plasticity index (D 4318); point load strength (D 5731); shrinkage factors (D 427; D 4943); tensile strength (D 2936; D 3967); thermal properties (D 5334, D 5335); triaxial compression (D 2850, D 2664, D 4406, D 4767, D 5311, D 5407); uniaxial compression (D 4341, D 4405); vane shear test (D 4648); *Evaluation of Laboratories:* D 3740
Miscellaneous	*Field:* Geotechnical mapping of large underground openings in rock (D 4543); *Laboratory Methods:* X-ray radiography (D 4452)
Peat/Organic Soils	*Laboratory Methods:* Bulk density (D 4531); classification (D 2607); hydraulic conductivity (D 4511); pH (D 2976); moisture/ash/organic matter (D 2974)
Frozen Soils	*Field:* Descirption (D 4083); *Laboratory:* Creep properties by uniaxial compression (D 5520)
Soil/Rock Chemistry:	
Basic Chemistry	*Field Methods:* Soil pH for corrosion testing (G 51); *Laboratory Methods:* Calcium carbonate (D 4373); pH (D 4972); soluble salt content (D 4542); diagnostic soil test for plant growth and food chain protection (D 5435); minimum requirements for laboratories engaged in chemical analysis (D 5522)
Sediments	Preparation for chemical analysis (D 3975, D 3976)
Sorption/Leachability	See fate-related procedures in Table A.1
Ground Water:	
Monitoring System Design	Karst and fractured rock aquifers (D 5717)
Data Elements	*Field Methods:* Minimum set (D 5254); additional identification descriptors (D 5408); additional physical descriptors (D 5409); additional usage description (D 5410); selection of data elements (D 5474)
Data Analysis/Presentation	*Chemical Analysis:* Diagrams for single analyses (D 5738); trilinear diagrams (D 5754); diagrams based on data analytical calculations (D 5877)
Monitoring Wells	*Field Methods:* **Design/installation** (D 5092); protection (D 5787); decommissioning (D 5299); casing (D 1785, F 480); grout (C 150—portland cement; water level measurement (D 4750); well development in granular aquifers (D 5521); well discharge (D 5716—circular orifice weir, D 5737—guide to methods)
Aquifer Hydraulic Properties	*Field Methods:* Packer tests (D 4630, D 4631); aquifer tests with control wells (D 4105, D 4106, D 5269, D 5270, D 5472, D 5473); slug tests (D 4044, D 4050, D 4104, D 5785, D 5881, D 5912); constant drawdown for flowing wells (D 5787, D 5855); partially penetrating wells (D 5850); **test selection** (D 4043)
Modeling	Site specific application (D 5447); comparing simulation to site-specific information (D 5490); documenting model application (D 5718); defining boundary conditions (D 5609); defining initial conditions (D 5610); conducting sensitivity analysis (D 5611); simulation of subsurface air flow (D 5719); subsurface flow and transport modeling (D 5880)
Chemistry	*Field Methods:* Acidity/Alkalinity (D 1067); electrical conductivity/resistivity (D 1125); ion-selective electrodes (D 4127); low-level dissolved oxygen (D 5462); odor (D 1292); pH (D 1293, D 5464); redox potential (D 1498), test kits for inorganic constituents (D 5463); turbidity (D 1889); *Extraction Methods:* purgeable organics using headspace sampling (D 3871); micro-extraction for volatiles and semivolatiles (D 5241); *Laboratory Methods:* Organic carbon (D 2579); minimum requirements for laboratories engaged in chemical analysis (D 5522); see, generally, Vols 11.01 and 11.02
Microbiology	ATP content (D 4012); iron bacteria (D 932); sulfate-reducing bacteria (D 4412); microbial respiration (D 4478); microscopy (D 4454—total respiring bacteria, D 4455—epifluorescence); plating methods (D 5465); on site screening heterotrophic bacteria (F 488)
Surface Water:	
Geometry/Flow Measurement	Depth measurement (D 5073); measurement of morphologic characteristics of surface water bodies (D 4581); operating a gaging station (D 5674); *Discharge:* Step backwater method (D 5388); *Open Channel Flow:* Selection of weirs and flumes (D 5640); acoustic methods (D 4408); acoustic velocity method (D 5389); broad-crested weirs (D 5614); culverts (D 5243); developing a stage-discharge relation (D 5541); dye tracers (D 5613); electromagnetic current meters (D 5089); Palmer-Bowles Flume (D 5390); Parshall flume (D 1941); rotating element current meters (D 4409); slope-area method (D 5130); thin-plate weirs (D 5242); velocity-area method (D 3858); width contractions (D 5129); *Open Water Bodies:* Water level measurement (D 5413)
Other Characteristics	Suspended sediment concentration (D 3977); environmental conditions relevant to spill control systems (F 625); *Chemistry:* See ground water above
Waste/Contaminants:	
Waste Properties	*Field/Screening Methods:* Compatibility (D 5059); cyanides (D 5049); flammability potential (D 4982); oxidizers (D 4981); pH (D 4980); physical description screening analysis (D 4979); sulfides (D 4878); waste specific gravity/bulk density (D 5057); *Laboratory Methods:* Waste bulk density (E 1109); biological clogging of geotextiles (D 1987); coal fly ash (D 5759); solid waste freeze-thaw resistance (D 4842); stability and miscibility (D 5232); wetting and drying (D 4843); *Extraction Methods:* Single batch extraction methods (D 5233); sequential batch extraction with water (D 4793); soxhlet extraction (D 5369); total solvent extractable content (D 5368); solvent extraction of total petroleum hydrocarbons (D 5765); shake extraction of solid waste and water (D 3987)
Contaminant Fate	See fate-related procedures in Table A.2
Radioactive Materials	*Monitoring:* Detector calibration (E 181); radiation measurement/dosimetry (E 170); radiation protection programs for decommissioning operations (E 1167); *Sampling/Preparation:* sampling surface soil for radionuclides (C 998); soil sample preparation for determination of radionuclides (C 999)
Asbestos	Screen analysis (D 2947)
Other Site Conditions:	
Field Atmospheric Conditions	Atmospheric pressure (D 3631); conversion unit and factors (D 1914); determining comparability of meteorological measurements (D 4430); *Humidity:* Dew-point hygrometer (D 4030); psychrometer (E 337); terminology (D 4023); *Wind:* Anemometers (D 4480); surface wind by acoustic means (D 5527); see Volume 11.03 generally
Solar insolation	Pyranometers (E 824, E 913, E 941); pyrheliometers (E 816)

Boldface = method selection guides

TABLE X1.2 Index to ASTM Field and Laboratory Methods for Ecological Characterization

Topic	ASTM Standard
General:	
Terminology	Biological effects and environmental fate (Terminology E 943)
Field Sampling:	
Phytoplankton	Classification for sampling (Classification D 4149); *Sampling Methods*—Depth-integrating samplers (Practice D 4135); Clarke-Bumpus sampler (Practice D 4134); conical tow nets (Practice D 4132); pumps (Practice D 4133); water sampling bottles (Practice D 4136); *Sample Preservation*—Practice D 4137; *Other Methods*—Measurement of chlorophyll content of algae in surface water (Practice D 3731); Sedgwick-Rafter method (Test Method D 4148)
Zooplankton	*Sampling Methods*—Clarke-Bumpus sampler (Practice E 1199); conical tow nets (Practice E 1201); pumps (Practice E 1198); *Sample Preservation*—Practice E 1200.
Benthic Macroinvertebrates	**Selection of sampling devices (Guide D 4387);**[A] *Grab Samplers*—Ekman (Practice D 4343); Holme scoop (Practice D 4348); Okean 50 (Practice D 4346); Orange Peel (Practice D 4407); Ponar (Practice D 4342); Peterson (Practice D 4401); Shipek scoop (Practice D 4347); Smith-McIntyre (Practice D 4344); Van Veen (Practice D 4345); *Other Samplers*: Basket (Practice E 1468); drift net (Practice D 4558); multiple plate (Practice E 1469); **selecting stream net devices (Guide D 4556);**[A] Surber and related samplers (Practice D 4557)
Fishes	Classification for sampling (Practice D 4211); sampling with rotenone (Practice D 4131)
Toxicity Testing:	
Microbial Detoxification	Chemically contaminated water and soils (D 5660)
Water	Behavioral testing in aquatic toxicology (E 1604); measurement of behavior during fish toxicity test (E 1711); estimation of median lethal concentration for fish using octonol-water partition coefficient (Practice E 1242); *Acute Toxicity Tests*—Bivalve mollusks (Guide E 724); fishes, macroinvertebrates and amphibians (Practice E 729, Guide E 1192-aqueous effluents); mosquito *Wyoemyia smithii* (Guide E 1365); polychateous annelids (Guide E 1562); rotifer *Brachionus* (Guide E 1440), west coast Mysids (Guide E 1463); echinoid embryos (E 1563); *Life Cycle/Renewal Toxicity Tests*—*Daphnia magna* (Guide E 1193); *Ceriodaphnia dubia* (Guide E 1295); polychateous annelids (Guide E 1562); saltwater Mysids (Guide E 1191); *Static Toxicity Tests*—Bivalve mollusks (Guide E 724); *Lemna gibba* G3 (Guide E 1415); microalgae (Guide E 1218); mosquito *Wyoemyia smithii* (Guide E 1365); west coast Mysids (Guide E 1463); *Other Toxicity Tests*—Algal growth potential (Practice D 3978); chronic—polychateous annelids (Guide E 1562); early life-stage fishes (Guide E 1241); toxicity-induced enzymatic inhibition in *Daphnia magna* (Provisional Test Method P 235)
Soil	Soil toxicity test with Lumbricid earthworm *EiSenia* (E 1676)
Sediment	Collection, storage, characterization and manipulation (Guide E 1391); designing biological tests (E 1525); bioaccumulation by benthic invertebrates (Guide E 1688); *Toxicity Tests*: marine and estuarine Amphipods (Guide E 1367); freshwater invertebrates (Guide E 1383, E 1706)
Other Tests	Sexual reproduction test with seaweeds (E 1498); *Avian Species*: Substrate dietary toxicity tests (E 857); reproductive studies (E 1062); use of lighting in laboratory testing (E 1733)
Fate-Related Procedures:	
Hazard Assessment	Assessing the hazard of a material to aquatic organisms (Guide D 1023); radioactive pathway methodology for release of site following decommissioning (Guide E 1278)
Modeling	Evaluating mathematical models for the environmental fate of chemicals (Practice E 978)
Microcosm Tests	Freshwater aquatic microcosm (Practice E 1366); terrestrial soil-core microcosm (Guide E 1197); chemical fate in site specific sediment/water microcosms (E 1624)
Laboratory-Fate Testing	*Bioconcentration*—Fishes and bivalve mollusks (Practice E 1022); *Biodegradation*—anaerobic (Test Method E 1196); shake-flask die-away method (Test Method E 1279); organic chemicals in semi-continuous activated sludge (E 1625); *Fate-Related Chemical Properties*—Aqueous solubility (Test Method E 1148); hydrolysis rate constants (Practice E 895); octanol/water partition coefficient (Test Method E 1147); vapor pressure (Test Method E 1194); *Sorption/Leachability*—Contaminant sorption (Test Method D 4646); 24-h batch sorption of volatile organics (Test Method D 5285); distribution ratios (Test Method D 4319); organic carbon sorption constant (Test Method E 1195); waste leaching column test (Test Method D 4874)

[A] Boldface—Method selection guide.

D 1889 – 94 Test Methods for Turbidity in Water (Vol 11.01)
D 1914 – 95 Practice for Conversion Units and Factors Relating to Atmospheric Analysis (Vol 11.03)
D 1941 – 91 Test Method for Open Channel Flow Measurements of Water with the Parshall Flume (Vol 11.01)
D 1987 – 91 Test Method for Biological Clogging of Geotextiles of Soil/Geotextile Filters
D 2113 – 83 Method for Diamond Core Drilling for Site Investigation
D 2166 – 91 Test Method for Unconfined Compressive Strength of Cohesive Soil
D 2167 – 94 Test Method for Density and Unit Weight of Soil in Place by the Rubber Balloon Method
D 2216 – 90 Test Method for Laboratory Determination of Water (Moisture) Content of Soil and Rock (Gravimetric Oven Drying)
D 2217 – 85 Practice for Wet Preparation of Soil Samples for Particle-Size Analysis and Determination Soil Constants
D 2325 – 68 Test Method for Capillary-Moisture Relationship for Coarse- and Medium-Textured Soils by Porous-Plate Apparatus (Soil Water Retention)
D 2434 – 68 Test Method for Permeability of Granular Soils (Constant Head) ($>1 \times 10^{-3}$ cm/s)
D 2487 – 93 Classification of Soils for Engineering Purposes
D 2488 – 93 Practice for Description and Identification of Soils (Visual-Manual Procedures)
D 2573 – 94 Method for Field Vane Shear Test in Cohesive Soil
D 2579 – 93 Test Methods for Total and Organic Carbon in Water (Vol 11.02)

D 2607 – 69 Classification of Peats, Mosses, Humus, and Related Products (Discontinued in 1992)
D 2664 – 88 Test Method for Triaxial Compressive Strength of Undrained Rock Core Specimen Without Pore Pressure Measurements
D 2845 – 90 Method for Laboratory Determination of Pulse Velocities and Ultrasonic Elastic Constants of Rock
D 2850 – 95 Test Method for Unconsolidated, Undrained Compressive Strength of Cohesive Soils in Triaxial Compression
D 2922 – 91 Test Methods for Density of Soil and Soil-Aggregate in Place by Nuclear Methods (Shallow Depth) (gamma-gamma, surface or <12 in. (305 mm))
D 2936 – 95 Test Method for Direct Tensile Strength of Intact Rock Core Specimens
D 2937 – 95 Test Method for Density of Soil in Place by the Drive-Cylinder Method
D 2938 – 86 Test Method for Unconfined Compressive Strength of Intact Rock Core Specimens
D 2947 – 87 Test Method for Screen Analysis of Asbestos Fibers (Vol 04.05 and Vol 08.02)
D 2974 – 87 Test Methods for Moisture, Ash, and Organic Matter of Peat and Other Organic Materials
D 2976 – 71 Test Method for pH of Peat Materials
D 3017 – 88 Test Method for Moisture Content of Soil and Rock in Place by Nuclear Methods (Shallow Depth), (neutron probe)
D 3080 – 90 Test Method for Direct Shear Test of Soils Under Consolidated Drained Conditions

D 3148 – 93	Test Methods for Elastic Moduli of Intact Rock Core Specimens in Uniaxial Compression
D 3152 – 72	Test Method for Capillary-Moisture Relationships for Fine-Textured Soils by Pressure-Membrane Apparatus
D 3213 – 91	Practice for Handling, Storing, and Preparing Soft Undisturbed Marine Soil
D 3249 – 95	Practice for General Ambient Air Analyzer Procedures (Vol 11.03)
D 3325 – 90	Practice for the Preservation of Waterborne Oil Samples (Vol 11.02)
D 3370 – 95	Practices for Sample Water from Closed Conduits (Vol 11.01)
D 3385 – 95	Test Method for Infiltration Rate of Soils in Field Using Double-Ring Infiltrometers
D 3404 – 91	Guide for Measuring Matric Potential in the Vadose Zone Using Tensiometers
D 3441 – 95	Test Method for In-Situ Cone Penetration Tests of Soil
D 3550 – 84	Practice for Ring-Lined Barrel Sampling of Soils
D 3584 – 83	Practice for Indexing Papers and Reports on Soil and Rock for Engineering Purposes
D 3614 – 90	Guide for Laboratories Engaged in Sampling and Analysis of Atmospheres and Emissions (Vol 11.03)
D 3631 – 84	Method for Measuring Surface Atmospheric Pressure (Vol 11.03)
D 3686 – 95	Practice for Sampling Atmospheres to Collect Organic Compound Vapors (Activated Charcoal Tube Adsorption Method) (Vol 11.03)
D 3694 – 95	Practice for Preparation of Sample Containers and for Preservation of Organic Constituents (Vol 11.02)
D 3740 – 95	Practice for Evaluation of Agencies Engaged in the Testing and/or Inspection of Soil and Rock Used In Engineering Design and Construction
D 3856 – 95	Guide for Good Laboratory Practices in Laboratories Engaged in Sampling and Analysis of Water (Vol 11.01)
D 3858 – 90	Practice for Open Channel Flow Measurement of Water by Velocity-Area Method (Vol 11.01)
D 3864 – 79	Guide for Continual On-Line Monitoring Systems for Water Analysis (Vol 11.01)
D 3871 – 84	Test Method for Purgeable Organic Compounds in Water Using Headspace Sampling (Vol 11.02)
D 3967 – 92	Test Method for Splitting Tensile Strength of Intact Rock Core Specimens
D 3975 – 93	Practice for Development and Use (Preparation) of Samples for Collaborative Testings of Methods for Analysis of Sediments (Vol 11.02)
D 3976 – 92	Practice for Preparation of Sediments Samples for Chemical Analysis (Vol 11.02)
D 3977 – 80	Practice for Determining Suspended Sediment Concentration in Water Sample (Vol 11.02)
D 3987 – 85	Method for Shake Extraction of Solid Waste and Water (Vol 11.04)
D 4012 – 81	Test Method for Adenosine Triphosphate (ATP) Content of Microorganisms in Water (Vol 11.02)
D 4023 – 82a	Definitions of Terms Relating to Humidity Measurements (Vol 11.03)
D 4043 – 91	Guide for Selection of Aquifer-Test Method in Determination of Hydraulic Properties by Well Techniques
D 4044 – 91	Test Method for (Field Procedures) for Instantaneous Change in Head (Slug Tests) for Determining Hydraulic Properties of Aquifers
D 4050 – 91	Test Method (Field Procedure) for Withdrawal and Injection Well Tests for Determining Hydraulic Properties of Aquifer Systems
D 4083 – 89	Practice for Description of Frozen Soils (Visual-Manual Procedure)
D 4096 – 91	Test Method for Determination of Total Suspended Particulate Matter in the Atmosphere (High-Volume Sampler Method), (Vol 11.03)
D 4104 – 91	Test Method (Analytical Procedure for) Determining Transmissivity of Confined Nonleaky Aquifers by Overdamped Well Response to Instantaneous Change in Head (Slug Test)
D 4105 – 91	Test Method (Analytical Procedure for) Determining Transmissivity and Storage Coefficient of Nonleaky Confined Aquifers by the Modified Theis Nonequilibrium Method
D 4106 – 91	Test Method for (Analytical Procedure for) Determining Transmissivity and Storativity of Nonleaky Confined Aquifers by the Theis Nonequilibrium Method
D 4127 – 92	Terminology Used with Ion-Selective Electrodes (Vol 11.01)
D 4220 – 95	Practice for Preserving and Transporting Soil Samples
D 4221 – 90	Test Method for Dispersive Characteristics of Clay Soil by Double Hydrometer
D 4230 – 83	Test Method of Measuring Humidity with Cooled-Surface Condensation (Dew Point) Hygrometer (Vol 11.03)
D 4318 – 93	Test Method for Liquid Limit, Plastic Limit, and Plasticity Index of Soils
D 4341 – 93	Test Method for Creep of Cylindrical Hard Rock Core Specimens in Uniaxial Compression
D 4373 – 84	Test Method for Calcium Carbonate Content of Soils
D 4394 – 84	Test Method for Determining the In-Situ Modulus of Deformation of Rock Mass Using the Rigid Plate Loading Method
D 4395 – 84	Test Method for Determining the In-Situ Modulus of Deformation of Rock Mass Using the Flexible Plate Loading Method
D 4403 – 84	Practice for Extensometers Used in Rock
D 4404 – 84	Test Method for Determination of the Pore Volume and Pore Volume Distribution of Soil and Rock by Mercury Intrusion Porosimetry
D 4405 – 93	Test Method for Creep of Cylindrical Soft Rock Core Specimens in Uniaxial Compression
D 4406 – 93	Test Method for Creep of Cylindrical Soft Rock Core Specimens in Triaxial Compression
D 4408 – 84	Practice for Open Channel Flow Measurements by Acoustic Means (Vol 11.01)
D 4409 – 91	Test Method for Velocity Measurement of Water in Open Channels with Rotating Element Current Meters (Vol 11.01)
D 4410 – 94	Terminology of Fluvial Sediment (Vol 11.01)
D 4411 – 93	Guide for Sampling Fluvial Sediment in Motion (Vol 11.02)
D 4412 – 84	Test Methods for Sulfate-Reducing Bacteria in Water and Water-Formed Deposits (Vol 11.02)
D 4428/ D 4428M – 91	Test Method for Crosshole Seismic Testing
D 4429 – 93	Test Method for Bearing Ratio of Soils in Place
D 4430 – 84	Practice for Determining the Operational Comparability of Meteorological Measurement (Vol 11.03)
D 4448 – 85a	Guide for Sampling Groundwater Monitoring Wells (Vol 11.04)
D 4452 – 85	Methods for X-Ray Radiography of Soil Samples
D 4454 – 85	Test Method for Simultaneous Enumeration of Total Respiring Bacteria in Aquatic Systems by Microscopy (Vol 11.02)
D 4455 – 85	Test Method for Enumeration of Aquatic Bacteria by Epifluorescence Microscopy Counting Procedure (Vol 11.02)
D 4478 – 85	Test Methods for Oxygen Uptake (Vol 11.02), (Microbial respiration)
D 4480 – 85	Test Method for Measuring Surface Wind by Means of Wind Vanes and Rotating Anemometers (Vol 11.03)
D 4489 – 95	Practices for Sampling Waterborne Oils (Vol 11.02)
D 4490 – 90	Practice for Measuring the Concentration of Toxic Gases or Vapors Using Detector Tubes (Vol 11.03)
D 4506 – 90	Test Method for Determining the In-Situ Modulus of Deformation of Rock Mass Using a Radial Jacking Test
D 4511 – 92	Test Method for Hydraulic Conductivity of Essentially Saturated Peat (Constant Head)
D 4515 – 85	Practice for Estimation of Holding Time for Water Samples Containing Organic Constituents (Vol 11.02)
D 4525 – 90	Test Method for Permeability of Rocks by Flowing Air
D 4531 – 86	Test Method for Bulk Density of Peat and Peat Products
D 4542 – 85	Test Method for Pore-Water Extraction and Determination of the Soluble Salt Content of Soils by Refractometer
D 4543 – 91	Practice for Determining Dimensional and Shape Tolerances of Rock Core Specimens
D 4544 – 86	Practice for Estimating Peat Deposit Thickness
D 4546 – 90	Test Methods for One-Dimensional Swell/Settlement Potential of Cohesive Soils
D 4547 – 91	Practice for Sampling Waste and Soils for Volatile Organics (Vol 11.04)
D 4553 – 90	Test Method for Determining the In-Situ Creep Characteristics of Rock
D 4554 – 90	Test Method for In-Situ Determination of Direct Shear Strength of Rock Discontinuities
D 4555 – 90	Test Method for Conducting an In-Situ Uniaxial Compressive Test for Determining Deformability and Strength of Weak Rock
D 4564 – 86	Test Method for Density of Soil in Place by the Sleeve Method (cohesionless, gravelly soils)
D 4581 – 86	Guide for Measurement of Morphologic Characteristics of Surface Water Bodies (Vol 11.02)
D 4597 – 92	Practice for Sampling Workplace Atmospheres to Collect Organic Gases or Vapors with Activated Charcoal Diffusional Samplers (Vol 11.03)
D 4598 – 87	Practice for Sampling Workplace Atmospheres to Collect Gases of Vapors with Liquid Sorbent Diffusional Samplers (Vol 11.03)
D 4599 – 90	Practice for Measuring the Concentration of Toxic Gases or Vapors Using Length-of-Stain Dosimeter (Vol 11.03)
D 4623 – 86	Test Method for Determination of In-Situ Stress in Rock Mass by Overcoring Method—USBM Borehole Deformation Gage

D 4630 – 86 Test Method for Determining Transmissivity and Storativity of Low Permeability Rocks by In-Situ Measurements Using the Constant Head Injection Test

D 4631 – 86 Test Method for Determining Transmissivity and Storativity of Low Permeability Rocks by In-Situ Measurements Using the Pressure Pulse Technique

D 4633 – 86 Test Method for Stress Wave Energy Measurement for Dynamic Penetrometer Testing Systems

D 4638 – 86 Guide for Preparation of Biological Samples for Inorganic Chemical Analysis (Vol 11.01)

D 4643 – 93 Method for Determination of Water (Moisture) Content of Soil by the Microwave Oven Method

D 4645 – 87 Test Method for Determination of the In-Situ Stress in Rock Using the Hydraulic Fracturing Method

D 4647 – 94 Test Method for Identification and Classification of Dispersive Clay Soils by the Pinhole Test

D 4648 – 94 Test Method for Laboratory Miniature Vane Shear Test for Saturated Fine-Grained Clayey Soil

D 4687 – 87 Guide for General Planning of Waste Sampling (Vol 11.04)

D 4696 – 92 Guide for Pore-Liquid Sampling from the Vadose Zone

D 4700 – 91 Guide for Soil Sampling from the Vadose Zone

D 4719 – 87 Test Method for Pressuremeter Testing in Soils

D 4729 – 87 Test Method for In Situ Stress and Modulus of Deformation Using the Flatjack Method

D 4750 – 87 Test Method for Determining Subsurface Liquid Levels in a Borehole or Monitoring Well (Observation Well)

D 4767 – 88 Test Method for Consolidated-Undrained Triaxial Compression Test on Cohesive Soils

D 4793 – 93 Method for Sequential Batch Extraction of Waste with Water (Vol 11.04)

D 4822 – 88 Guide for Selection of Methods of Particle Size Analysis of Fluvial Sediments (Manual Methods) (Vol 11.02)

D 4823 – 95 Guide for Core-Sampling Submerged, Unconsolidated Sediments (Vol 11.02)

D 4840 – 88 Guide for Sample Chain of Custody Procedure (Vol 11.01)

D 4841 – 88 Practice to Estimation of Holding Time for Water Samples Containing Organic and Inorganic Constituents (Vol 11.01)

D 4842 – 90 Test Method for Determining the Resistance of Solid Wastes to Freezing and Thawing (Vol 11.04)

D 4843 – 88 Test Method for Wetting and Drying Test of Solid Waste (Vol 11.04)

D 4844 – 88 Guide for Air Monitoring at Waste Management Facilities for Worker Protection (Vol 11.04)

D 4861 – 91 Practice for Sampling and Analysis of Pesticides and Polychlorinated Biphenyls in Indoor Atmospheres (Vol 11.03)

D 4878 – 89 Test Method for Screening of Sulfide in Waste (Vol 11.04)

D 4879 – 89 Guide for Geotechnical Mapping of Large Underground Openings in Rock

D 4914 – 89 Test Method for Density of Soil and Rock in Place by the Sand Replacement Method in a Test Pit, (soils with particles larger than 3 in. (76 mm))

D 4943 – 95 Test Method for Shrinkage Factors of Soils by the Wax Method

D 4944 – 89 Test Method for Field Determination of Water (Moisture) Content of Soil by the Calcium Carbide Gas Pressure Tester Method

D 4959 – 89 Test Method for Determination of Water (Moisture) Content of Soil by Direct Heating Method

D 4971 – 89 Test Method for Determining the In-Situ Modulus of Deformation of Rock Using the Diametrically Loaded 76-mm (3-in.) Borehole Jack

D 4972 – 95a Test Method for pH of Soils

D 4979 – 89 Test Method for Physical Description Screening Analysis in Waste (Vol 11.04)

D 4980 – 89 Test Method for Screening of pH in Waste (Vol 11.04)

D 4981 – 89 Test Method for Screening of Oxidizers in Waste (Vol 11.04)

D 4982 – 89 Method for Flammability Potential Screening Analysis of Waste (Vol 11.04)

D 5012 – 89 Guide for Preparation of Materials Used for the Collection and Preservation of Atmospheric Wet Deposition (Vol 11.03)

D 5013 – 89 Practices for Sampling Wastes from Pipes and Other Point Discharges (Vol 11.04)

D 5049 – 90 Test Method for Screening of Cyanides in Waste (Vol 11.04)

D 5057 – 90 Test Method for Screening of Apparent Specific Gravity and Bulk Density of Waste (Vol 11.04)

D 5059 – 90 Test Methods for Compatibility Screening Analysis of Waste (Vol 11.04)

D 5073 – 90 Practice for Depth Measurement of Surface Water (Vol 11.02)

D 5089 – 90 Test Method for Velocity Measurements of Water in Open Channels with Electromagnetic Current Meters (Vol 11.01)

D 5079 – 90 Practices for Preserving and Transporting Rock Core Samples

D 5080 – 93 Test Method for Rapid Determination of Percent Compaction

D 5084 – 90 Test Method for Hydraulic Conductivity of Saturated Porous Materials Using a Flexible Wall Permeameter (low permeability materials $<1 \times 10^{-3}$ cm/s)

D 5088 – 90 Practice for Decontamination of Field Equipment Used at NonRadioactive Waste Sites

D 5092 – 90 Recommended Practice for Design and Installation of Ground Water Monitoring Wells in Aquifers

D 5093 – 90 Test Method for Field Measurement of Infiltration Rate Using a Double-Ring Infiltrometer with a Sealed-Inner Ring

D 5111 – 95 Guide for Choosing Locations and Sampling Methods to Monitor Atmospheric Deposition at Non-Urban Locations (Vol 11.03)

D 5126 – 90 Guide for Comparison of Field Methods for Determining Hydraulic Conductivity in the Vadose Zone (saturated: single-/double-ring infiltrometer, double-tube, air-entry permeameter, borehole permeameter (constant head-borehole infiltration, Guelph permeameter), empirical; unsaturated: instantaneous profile, crust, empirical)

D 5129 – 95 Test Method for Open Channel Flow Measurement of Water Indirectly by Using Width Contractions (Vol 11.01)

D 5130 – 95 Test Method for Open Channel Flow Measurement of Water Indirectly by Slope-Area Method (Vol 11.01)

D 5195 – 91 Test Method for Determination of Density of Soil and Rock In-Place at Depths Below the Surface by Nuclear Methods (gamma-gamma, >12 in. (305 mm))

D 5220 – 92 Test Method for Water Content of Soil and Rock In-Place by the Neutron Depth Probe Method

D 5232 – 92 Test Method for Determining the Stability and Miscibility of a Solid, Semi-Solid, or Liquid Waste Material (Vol 11.04)

D 5233 – 92 Test Method for Single Batch Extraction Methods for Wastes (Vol 11.04)

D 5241 – 92 Practice for Micro-Extraction of Water for the Analysis of Volatile and Semi-Volatile Organic Compounds in Water (Vol 11.02)

D 5242 – 92 Test Method for Open Channel Flow Measurement of Water Indirectly at Culverts (Vol 11.01)

D 5243 – 92 Test Method for Open Channel Flow Measurements of Water Indirectly at Culverts (Vol 11.01)

D 5435 – 93 Test Method for Diagnostic Soil Test for Plant Growth and Food Chain Protection

D 5254 – 92 Practice for the Minimum Set of Data Elements to Identify a Ground Water Site

D 5269 – 92 Test Method for (Analytical Procedure) Determining Transmissivity of Non-Leaky Confined Aquifers by the Theis Recovery Method

D 5270 – 92 Test Method for (Analytical Procedure) Determining Transmissivity and Storage Coefficient of Bounded, Nonleaky Confined Aquifers

D 5283 – 92 Practice for Generation of Environmental Data Related to Waste Management Activities: Quality Assurance and Quality Control Planning and Implementation (Vol 11.04)

D 5298 – 94 Test Method for Measurement of Soil Potential (Suction) Using Filter Paper

D 5299 – 92 Guide for the Decommissioning of Ground Water Wells, Vadose Zone Monitoring Devices, Boreholes and Other Devices for Environmental Activities

D 5311 – 92 Test Method for Load Controlled Cyclic Triaxial Strength of Soil

D 5314 – 92 Guide for Soil Gas Monitoring in the Vadose Zone

D 5334 – 92 Test Method for Determination of Thermal Conductivity of Soil and Rock by Thermal Needle Probe Procedure

D 5335 – 92 Test Method for Linear Coefficient of Thermal Expansion of Rock Using Bonded Electric Resistance Strain Gages

D 5337 – 92 Practice for Flow Rate for Calibration of Personal Sampling Pumps (Vol 11.03)

D 5358 – 93 Practice for Sampling with a Dipper or Pond Sampler (Vol 11.04)

D 5368 – 93 Test Method for the Gravimetric Determination of Total Solvent Extractable Content (TSEC) of Solid Waste Samples (Vol 11.04)

D 5369 – 93 Test Method for the Extraction of Solid Waste Samples for Chemical Analysis Using Soxhlet Extraction (Vol 11.04)

D 5387 – 93 Guide for Elements of a Complete Data Set for Non-Cohesive Sediments (Vol 11.02)

D 5388 – 93 Method for Measurement of Discharge by Step-Backwater Method (Vol 11.01)

D 5389 – 93 Method for Open Channel Flow Measurement by Acoustic Velocity Meter Systems (Vol 11.02)

D 5390 – 93 Method for Open Channel Flow Measurement of Water with Palmer-Bowlus Flumes (Vol 11.02)

D 5407 – 93 Test Method for Elastic Moduli of Undrained Intact Rock Core Specimens in Triaxial Compression Without Pore Pressure Measurement

D 5408 – 93 Guide for the Set of Data Elements to Describe a Ground-Water Site, Part 1—Additional Identification Descriptors

D 5409 – 93 Guide for the Set of Data Elements to Describe a Ground-Water Site, Part 2—Physical Descriptors

D 5410 – 93 Guide for the Set of Data Elements to Describe a Ground-Water Site, Part 3—Usage Descriptors

D 5413 – 93 Methods for Measurement of Water Levels in Open-Water Bodies (Vol 11.02)

D 5434 – 93 Guide for Field Logging of Subsurface Explorations of Soil and Rock

D 5447 – 93 Guide for Application of a Ground-Water Flow Model to a Site Specific Problem

D 5451 – 93 Practice for Sampling Using a Trier Sampler (Vol 11.04)

D 5462 – 93 Test Method for On-Line Measurement of Low Level Dissolved Oxygen in Water (Vol 11.01)

D 5463 – 93 Guide for the Use of Test Kits to Measure Inorganic Constituents in Water (Vol 11.02)

D 5464 – 93 Test Methods for pH Measurement of Water of Low Conductivity (Vol 11.01)

D 5465 – 93 Practices for Counting, Calculating, and Reporting Microbial Colonies in Water (Vol 11.02)

D 5466 – 93 Test Methods for the Determination of Volatile Organic Chemicals in Atmospheres (Canister Sampling Methodology) (Vol 11.03)

D 5472 – 93 Test Method for Determining Specific Capacity and Estimating Transmissivity at the Control Well

D 5473 – 93 Test Method (Analytical Procedure) for Analyzing the Effects of Partial Penetration of Control Well and Determining the Horizontal and Vertical Hydraulic Conductivity in a Nonleaky Aquifer

D 5474 – 93 Guide for Selection of Data Elements for Ground-Water Investigations

D 5490 – 93 Guide for Comparing Ground-Water Flow Model Simulations to Site-Specific Information

D 5495 – 94 Practice for Sampling with a Composite Liquid Waste Sampler (COLIWASA) (Vol 11.04)

D 5518 – 94 Guide for Acquisition of File Aerial Photography and Imagery for Establishing Historic Site-Use and Surficial Conditions

D 5520 – 94 Test Method for Laboratory Determination of Creep Properties of Frozen Soil Samples by Uniaxial Compression

D 5521 – 94 Guide for Development of Ground-Water Monitoring Wells in Granular Aquifers

D 5522 – 94 Specification for Minimum Requirements for Laboratories Engaged in Chemical Analysis of Soil, Rock and Contained Fluid

D 5527 – 94 Practice for Measuring Surface Wind and/or Temperature by Acoustic Means (Vol 11.03)

D 5540 – 94a Practice for Flow Control and Temperature Control for On-Line Water Sampling and Analysis (Vol 11.01)

D 5541 – 94 Practice for Developing a Stage-Discharge Relation for Open Channel Flow (Vol 11.01)

D 5542 – 94 Test Methods for Low Level Dissolved Oxygen in Water (Vol 11.01)

D 5549 – 94 Guide for Reporting Geostatistical Site Investigations

D 5550 – 94 Test Method for Specific Gravity of Soils Solids by Gas Pycnometer

D 5607 – 94 Test Method for Performing Laboratory Direct Shear Strength Tests of Rock Specimens under Constant Normal Stress

D 5608 – 94 Standard Practice for the Decontamination of Field Equipment Used at Low Level Radioactive Waste Sites

D 5609 – 94 Guide for Defining Boundary Conditions in Ground-Water Flow Modeling

D 5610 – 94 Guide for Defining Initial Conditions in Ground-Water Flow Modeling

D 5611 – 94 Guide for Conducting a Sensitivity Analysis for a Ground-Water Flow Model Application

D 5612 – 94 Guide for the Quality Planning and Field Implementation of a Water Quality Measurement Program (Vol 11.01)

D 5613 – 94 Test Method for Open Channel Measurements of Time of Travel Using Dye Tracers (Vol 11.02)

D 5614 – 94 Test Method for Open Channel Flow Measurement of Water with Broad-Crested Weirs (Vol 11.02)

D 5633 – 94 Practice for Sampling with a Scoop (Vol 11.04)

D 5658 – 95 Practice for Sampling Unconsolidated Waste from Trucks (Vol 11.04)

D 5640 – 95 Guide for Selection of Weirs and Flumes for Open Channel Flow Measurement of Water (Vol 11.02)

D 5674 – 95 Guide for Operations of a Gaging Station (Vol 11.02)

D 5679 – 95a Practice for Sampling Consolidated Solids in Drums or Similar Containers (Vol 11.04)

D 5680 – 95a Practice for Sampling Unconsolidated Solids in Drums or Similar Containers (Vol 11.04)

D 5681 – 95 Terminology for Waste and Waste Management (Vol 11.04)

D 5714 – 95 Specification for Digital Geospatial Metadata

D 5715 – 95 Test Method for Estimating the Degree of Humification of Peat and Other Organic Soils (Visual/Manual Method)

D 5716 – 95 Test Method to Measure the Rate of Well Discharge by Circular Orifice Weir

D 5717 – 95 Guide for the Design of Ground-Water Monitoring Systems in Karst and Fractured-Rock Aquifers

D 5718 – 95 Guide for Documenting a Ground-Water Flow Model Application

D 5719 – 95 Guide to Simulation of Subsurface Air Flow Using Ground-Water Flow Modeling Codes

D 5720 – 95 Practice for Static Calibration of Electronic Transducer-Based Pressure Measurements Systems for Geotechnical Purposes

D 5731 – 95 Test Method for Determination of the Point Load Strength Index of Rock

D 5737 – 95 Guide to Methods for Measuring Well Discharge

D 5738 – 95 Guide for Displaying the Results of Chemical Analyses of Ground Water for Major Ions and Trace Elements—Diagrams for Single Analyses

D 5743 – 95 Practice for Sampling Single or Multilayered Liquids, With or Without Solids in Drums or Similar Containers (Vol 11.04)

D 5745 – 95 Guide for Developing and Implementing Short-Term Measures or Early Actions for Site Remediation (Vol 11.04)

D 5746 – 95 Classification of Environmental Condition of Property Area Types (Vol 11.04)

D 5753 – 95 Guide for Planning and Conducting Borehole Geophysical Logging

D 5754 – 95 Guide for Displaying the Results of Chemical Analyses of Ground Water for Major Ions and Trace Elements—Trilinear and Other Multi-Coordinate Diagrams

D 5759 – 95 Guide for Characterization of Coal Fly Ash and Clean Coal Combustion Fly Ash for Potential Use (Vol 11.04)

D 5765 – 95 Practice for Solvent Extraction of Total Petroleum Hydrocarbons from Soils and Sediments Using Closed Vessel Microwave Heating (Vol 11.01)

D 5777 – 95 Guide for Using the Seismic Refraction Method for Subsurface Investigations

D 5781 – 95 Guide for Use of Dual-Wall Reverse-Circulation Drilling for Geoenvironmental Exploration and Installation of Subsurface Water-Quality Monitoring Devices

D 5782 – 95 Guide for Use of Direct Air-Rotary Drilling for Geoenvironmental Exploration and Installation of Subsurface Water-Quality Monitoring Devices

D 5783 – 95 Guide for Use of Direct Rotary Drilling with Water-Based Drilling Fluid for Geoenvironmental Exploration and Installation of Subsurface Water-Quality Monitoring Devices

D 5784 – 95 Guide for Use of Hollow-Stem Augers for Geoenvironmental Exploration and Installation of Subsurface Water-Quality Monitoring Devices

D 5785 – 95 Test Method (Analytical Procedure) for Determining Transmissivity of Confined Nonleaky Aquifers by Under-Damped Well Response to Instantaneous Change in Head (Slug Test)

D 5786 – 95 Practice (Field Procedure) for Constant Drawdown Tests in Flowing Wells for Determining Hydraulic Properties of Aquifer Systems

D 5787 – 95 Practice for Monitoring Well Protection

D 5789 – 95 Practice for Writing Quality Control Specifications for Standard Test Methods for Organic Constituents (Vol 11.01)

D 5792 – 95 Practice for Generation of Environmental Data Related to Waste Management Activities: Development of Data Quality Objectives (Vol 11.04)

D 5835 – 95 Practice for Sampling Stationary Source Emissions for the Automated Determination of Gas Concentrations (Vol 11.03)

D 5850 – 95 Test Method for (Analytical Procedure) for Determining Transmissivity, Storage Coefficient, and Anisotropy Ratio from a Network of Partially Penetrating Wells

D 5855 – 95 Test Method for (Analytical Procedure) for Determining Transmissivity and Storage Coefficient of a Confined Nonleaky or Leaky Aquifer by the Constant Drawdown Method in a Flowing Well

D 5856 – 95 Test Method for Measurement of Hydraulic Conductivity of Porous Material Using a Rigid-Wall, Compaction-Mold Permeameter

D 5872 – 95 Guide for Use of Casing Advancement Drilling Methods for Geoenvironmental Exploration and Installation of Subsurface Water-Quality Monitoring Devices

D 5875 – 95 Guide for Use of Cable-Tool Drilling and Sampling Methods for Geoenvironmental Exploration and Installation of Subsurface Water-Quality Monitoring Devices

D 5876 – 95 Guide for Use of Direct Rotary Wireline Casing Advancement Drilling Methods for Geoenvironmental Exploration and Installation of Subsurface Water-Quality Monitoring Devices

D 5877 – 95 Guide for Displaying the Results of Chemical Analyses of Ground Water for Major Ions and Trace Elements—Diagrams Based on Data Analytical Calculations

D 5879 – 95 Practice for Surface Site Characterization for On-Site Septic Systems

D 5880 – 95 Guide for Subsurface Flow and Transport Modeling

D 5881 – 95 Test Method (Analytical Procedure) for Determining Transmissivity of Confined Nonleaky Aquifers by Critically Damped Well Response to Instantaneous Change in Head (Slug Test)

D 5903 – 95 Guide for Planning and Preparing for a Ground-Water Sampling Event

D 5911 – 95 Practice for a Minimum Set of Data Elements to Describe a Soil Sampling Site

D 5912 – 95 Test Method (Analytical Procedure) for Determining Hydraulic Conductivity of an Unconfined Aquifer by Overdamped Well Response to Instantaneous Change in Head (Slug Test)

E 105 – 58 Practice for Probability Sampling of Materials (Vols 07.02 and 14.02)

E 170 – 92a Terminology Relating to Radiation Measurements and Dosimetry (Vol 12.02)

E 177 – 90a Practice for Use of the Terms Precision and Bias in ASTM Test Methods (Vol 14.02)

E 178 – 80 Practice for Dealing with Outlying Observations (Vol 14.02)

E 181 – 93 General Methods for Detector Calibration and Analysis of Radionuclides (Vol 12.02)

E 337 – 84 Test Method of Measuring Humidity with a Psychrometer (the Measurement of Wet- and Dry-Bulb Temperatures) (Vol 11.03)

E 380 – 92 Practice for Use of International System of Unit (SI) (Modernized Metric System) (Vol 14.02; excerpts included in all ASTM volumes)

E 678 – 90 Practice for Evaluation of Technical Data (Vol 14.02) [Focusses on product liability matters]

E 816 – 84 Method for Calibration of Secondary Reference Pyrheliometers and Pyrheliometers for Field Use (Vol 11.03)

E 824 – 81 Test Method for Transfer of Calibration from Reference to Field Pyranometers (Vol 11.03)

E 884 – 82 Practice for Sampling Airborne Microorganisms at Municipal Solid Waste Processing Facilities (Vol 11.04)

E 913 – 82 Method for Calibration of Reference Pyranometers with Axis Vertical by the Shading Method (Vol 12.02)

E 941 – 83 Test Method for Calibration of Reference Pyranometers with Axis Tilted by the Shading Method (Vol 12.02)

E 1109 – 86 Test Method for Determining the Bulk Density of Solid Waste Fractions (Vol 11.04)

E 1167 – 87 Guide for Radiation Protection Programs for Decommissioning Operations (Vol 12.02)

E 1188 – 87 Practice for Collection and Preservation of Information and Physical Items by a Technical Investigator (Vol 14.02)

E 1287 – 89 Guide for Aseptic Sampling of Biological Materials (Vol 11.04)

E 1370 – 90 Guide to Air Sampling Strategies for Worker and Workplace Protection (Vol 11.03)

E 1391 – 90 Guide for Collection, Storage, Characterization, and Manipulation of Sediments for Toxicological Testing (Vol 11.04)

E 1430 – 91 Guide for Using Release Detection Devices with Underground Storage Tanks (Vol 11.04)

E 1526 – 93 Practice for Evaluating the Performance of Release Detection Systems for Underground Storage Tank Systems (Vol 11.04)

E 1527 – 94 Practice for Environmental Site Assessments: Phase I Assessment Process (Vol 11.04)

E 1528 – 93 Practice for Environmental Site Assessment Transaction Screen Process (Vol 11.04)

E 1553 – 93 Practice for the Collection of Airborne Particulate Lead During Abatement and Construction Activities (Vol 04.07)

E 1609 – 94 Guide for Development and Implementation of a Pollution Prevention Program (Vol 11.04)

E 1689 – 95 Guide for Developing Conceptual Site Models for Contaminated Sites (Vol 11.05)

E 1739 – 95 Guide for Risk-Based Corrective Action Applied at Petroleum Release Sites (Vol 11.04)

F 480 – 91 Specifications for Thermoplastic Water Well Casing Pipe and Couplings Made in Standard Dimension Ratios (SDR) SCH 40 and SCH 80 (Vol 08.04)

F 488 – 95 Test Method for On-Site Screening Heterotrophic Bacteria in Water (Vol 11.02)

F 625 – 79 Practice for Describing of Environmental Conditions Relevant to Spill Control Systems (Vol 11.04)

F 1084 – 90 Guide for Sampling Oil/Water Mixtures for Oil Spill Recovery Equipment (Vol 11.04)

G 51 – 92 Test Method for pH of Soil for Use in Corrosion Testing

G 57 – 78 Method for Field Measurement of Soil Resistivity Using the Wenner Four-Electrode Method

ES 40 Practice for Procedures for the Assessment of Buried Steel Tanks Prior to the Additions of Cathodic Protection (Vol 11.04)

PS 11 – 95 Practice for Environmental Regulatory Compliance Audits (Vol 11.04)

PS 12 – 95 Guide for the Study and Evaluation of an Organization's Environmental Management Systems (Vol 11.04)

X1.5 *ASTM Standards*—Indexed in Table X1.2 (In Vol 11.05, unless otherwise specified):[56]

D 3731 – 87 Practice for Measurement of Chlorophyll Content of Algae in Surface Waters

D 3978 – 80 Practice for Algal Growth Potential Testing with *Selenastrum Capricornutum*

D 4131 – 84 Practice for Sampling Fish with Rotenone

D 4132 – 82 Practice for Sampling Phytoplankton with Conical Tow Nets

D 4133 – 82 Practice for Sampling Phytoplankton with Pumps

D 4134 – 82 Practice for Sampling Phytoplankton with a Clark-Bumpus Plankton Sampler

D 4135 – 82 Practice for Sampling Phytoplankton with Depth-Integrating Samplers

D 4136 – 82 Practice for Sampling Phytoplankton with Water-Sampling Bottles

D 4137 – 82 Practice for Preserving Phytoplankton Samples

D 4148 – 82 Test Method for Analysis of Phytoplankton in Surface Water by the Sedgwick-Rafter Method

D 4149 – 82 Classification for Sampling Phytoplankton in Surface Waters

D 4211 – 82 Classification for Fish Sampling

D 4319 – 93 Test Method for Distribution Ratios by the Short-Term Batch Method (Vol 04.08)

D 4342 – 84 Practice for Collecting Benthic Macroinvertebrates with the Ponar Grab Sampler

D 4343 – 84 Practice for Collecting Benthic Macroinvertebrates with the Ekman Grab Sampler

D 4344 – 84 Practice for Collecting Benthic Macroinvertebrates with the Smith-McIntyre Grab Sampler

D 4345 – 84 Practice for Collecting Benthic Macroinvertebrates with the Van Veen Grab Sampler

D 4346 – 84 Practice for Collecting Benthic Macroinvertebrates with the Okean 50 Grab Sampler

D 4347 – 84 Practice for Collecting Benthic Macroinvertebrates with the Shipek (Scoop) Grab Sampler

D 4348 – 84 Practice for Collecting Benthic Macroinvertebrates with the Holme (Scoop) Grab Sampler

D 4387 – 87 Guide for Selecting Sampling Devices for Collection Benthic Macroinvertebrates

D 4401 – 84 Practice for Collecting Benthic Macroinvertebrates with the Peterson Grab Sampler

D 4407 – 84 Practice for Collecting Benthic Macroinvertebrates with the Orange Peel Grab Sampler

D 4556 – 85 Guide for Selecting Stream-Net Sampling Devices for Collecting Benthic Macroinvertebrates

D 4557 – 85 Practice for Collecting Benthic Macroinvertebrates with Surber and Related Type Samplers

D 4558 – 85 Practice for Collecting Benthic Macroinvertebrates with Drift Nets

D 4646 – 87 Test Method for 24-Hour Batch Type Measurement of Contaminant Sorption by Soils and Sediments

D 4874 – 89 Method for Leaching Solid Wastes in a Column Apparatus (Vol 11.04)

D 5285 – 92 Test Method for 24-Hour Batch-Type Measurements of Volatile Organic Sorption by Soils and Sediment

D 5660 – 95 Test Method for Assessing the Microbial Detoxification of Chemically Contaminated Water and Soils Using a Toxicity Test with a Luminescent Marine Bacterium (Vol 11.04)

---

[56] Prior to 1995 ASTM standards on biological effects and environmental fate were published in Volume 11.04. Beginning in 1995 that volume was divided into two separate volumes: (*1*) Volume 11.04 (Environmental Assessment; Hazardous Substances and Oil Spill Response; Waste Management) and (*2*) Volume 11.05 (Biological Effects and Environmental Fate; Biotechnology; Pesticides).

E 724 – 94 Guide for Conducting Static Acute Toxicity Tests Starting with Embryos of Four Species of Saltwater Bivalve Mollusks

E 729 – 88a Practice for Conducting Acute Toxicity Tests with Fishes, Macroinvertebrates, and Amphibians

E 857 – 87 Practice for Conducting Substrate Dietary Toxicity Tests with Avian Species

E 895 – 89 Practice for Determination of Hydrolysis Rate Constants of Organic Chemicals in Aqueous Solutions

E 896 – 92 Test Method for Conducting Aqueous Direct Photolysis Tests

E 943 – 95 Terminology Relating to Biological Effects and Environmental Fate

E 978 – 92 Practice for Evaluating Environmental Fate Models of Chemicals

E 1022 – 94 Practice for Conducting Bioconcentration Tests with Fishes and Saltwater Bivalve Mollusks

E 1023 – 84 Guide for Assessing the Hazard of a Material to Aquatic Organisms and Their Uses

E 1062 – 86 Practice for Conducting Reproductive Studies with Avian Species

E 1147 – 92 Test Method for Partition Coefficient (N-Octanol/Water) Estimation by Liquid Chromatography

E 1148 – 87 Test Method for Measurements of Aqueous Solubility

E 1191 – 90 Guide for Conducting Life-Cycle Toxicity Tests with Saltwater Mysids

E 1192 – 88 Guide for Conducting Acute Toxicity Tests on Aqueous Effluents with Fishes, Macroinvertebrates, and Amphibians

E 1193 – 94 Guide for Conducting Renewal Life-Cycle Toxicity Tests with *Daphnia magna*

E 1194 – 87 Test Method for Vapor Pressure

E 1195 – 87 Test Method for Sorption Constant (Koc) for Organic Chemicals in Soil and Sediments

E 1196 – 92 Test Method for Determining the Anaerobic Biodegradation Potential of Organic Chemicals

E 1197 – 87 Guide for Conducting a Terrestrial Soil-Core Microcosm Test

E 1198 – 87 Practice for Sampling Zooplankton with Pumps

E 1199 – 87 Practice for Sampling Zooplankton with Clarke-Pumpus Plankton Sampler

E 1200 – 87 Practice for Preserving Zooplankton Samples

E 1201 – 87 Practice for Sampling Zooplankton with Conical Tow Nets

E 1218 – 90 Guide for Conducting Static 96-h Toxicity Tests with Microalgae

E 1241 – 92 Guide to Conducting Early Life-Stage Toxicity Tests with Fishes

E 1242 – 88 Practice for Using Octanol-Water Partition Coefficient to Estimate Median Lethal Concentration for Fish Due to Narcosis

E 1278 – 88 Guide for Radioactive Pathway Methodology for Release of Site Following Decommissioning (Vol 12.02)

E 1279 – 89 Test Method for Biodegradation by a Shake-Flask Die-Away Method (Vol 11.02)

E 1295 – 89 Guide for Conducting Three-Brood Renewal Toxicity Tests with *Ceriodaphnia dubia*

E 1365 – 90 Guide for Conducting Static Acute Aquiatic Toxicity Screening Tests with the Mosquito *Wyoemyia smithii* (Coquillett) (Discontinued; 1994)

E 1366 – 91 Practice for Standardized Aquatic Microcosm: Fresh Water

E 1367 – 92 Guide for Conducting 10-Day Static Sediment Toxicity Tests with Marine and Estuarine Amphipods

E 1383 – 94a Guide for Conducting Sediment Toxicity Tests with Freshwater Invertebrates

E 1391 – 94 Guide for Collection, Storage, Characterization, and Manipulation of Sediments for Toxicological Testing

E 1415 – 91 Guide for Conducting Static Toxicity Tests with *Lemna Gibba* G3

E 1440 – 91 Guide for Acute Toxicity Test with the Rotifer *Brachionus*

E 1463 – 92 Guide for Conducting Static and Flow-Through Acute Toxicity Tests with Mysids from the West Coast of the United States

E 1468 – 92 Practice for Collecting Benthic Macroinvertebrates with the Basket Sampler

E 1469 – 92 Practice for Collecting Benthic Macroinvertebrates with Multiple-Plate Samplers

E 1498 – 92 Guide for Conducting Sexual Reproduction Test with Seaweeds

E 1525 – 94a Guide for Designing Biological Tests With Sediments (Vol 11.05)

E 1562 – 94 Guide for Conducting Acute, Chronic and Lifecycle Aquatic Toxicity Tests with Polychateous Annelids (Vol 11.04)

E 1563 – 95 Guide for Conducting Static Acute Toxicity Tests with Echinoid Embryos

E 1604 – 94 Guide for Behavioral Testing in Aquatic Toxicology

E 1624 – 94 Guide for Chemical Fate in Site-Specific Sediment/Water Microcosms

E 1625 – 94 Test Method for Determining Biodegradability of Organic Chemicals in Semi-Continuous Activated Sludge (SCAS)

E 1676 – 95 Guide for Conducting a Laboratory Soil Toxicity Test with Lumbricid Earthworm *Ei Senia Foetida*

E 1688 – 95 Guide for Determination of the Bioaccumulation of Sediment-Associated Contaminants by Benthic Invertebrates

E 1706 – 95a Test Methods for Measuring the Toxicity of Sediment-Associated Contaminates with Fresh Water Invertebrates

E 1711 – 95 Guide for the Measurement of Behavior During Fish Toxicity Tests

E 1733 – 95 Guide for the Use of Lighting in Laboratory Testing

P 235 – 93 Test Method for Fluorometric Determination of Toxicity-Induced Enzymatic Inhibition in *Daphnia magna*

X2. MAJOR NON-ASTM REFERENCES ON ENVIRONMENTAL SITE CHARACTERIZATION

X2.1 *General Applications:*

Boulding, J. R., *Subsurface Field Characterization and Monitoring Techniques: A Desk Reference Guide, Volume I: Solids and Ground Water, Volume II: The Vadose Zone, Field Screening and Analytical Methods,* EPA/625/R-93/003a and b, 1993.[57]

Boulding, J. R., *Practical Handbook of Soil, Vadose Zone, and Ground Water Contamination: Assessment, Prevention and Remediation,* Lewis Publishers, Boca Raton, FL, 1995.

Brown, E. T. (ed), *Rock Characterization Testing and Monitoring: ISRM Suggested Methods,* Pergamon Press, Oxford, 1981.

Brown, R. H., Konoplyantsev, A. A., Ineson, J., and Kovalensky, V. S., "Ground Water Studies: An International Guide for Research and Practice," *Studies and Reports in Hydrology No. 7,* UNESCO, Paris. Originally published in 1972, with supplements added in 1973, 1975, 1977, and 1983.

CCME, *Guidance Manual on Sampling, Analysis, and Data Management for Contaminated Sites, Vol. I: Main Report, Vol. II: Analytical Method Summaries,* CCME EPC-NCS62E and CCME EPC-NCS66E, Canadian Council of Ministers of the Environment, 326 Broadway, Suite 400, Winnipeg, Manitoba R3C 0S5, 1993.

CCME, *Subsurface Assessment Handbook for Contaminated Sites,* CCME EPC-NCSRP-48E, Canadian Council of Ministers of the Environment, 326 Broadway, Suite 400, Winnipeg, Manitoba R3C 0S5, 1994.

Dowding, C. H. (ed), *Site Characterization Exploration,* Proceeding of Specialty Workshop, American Society of Civil Engineers, New York, 1978.

Flanagan, F., "Description of Eight New USGS Rock Standards," *U.S. Geological Survey Professional Paper 840,* 1976.

Harrelson, C. C., Rawlins, C. L., and Putyondy, J. P., Stream Channel Reference Sites: An Illustrated Guide for Field Techniques. *General Technical Report RM-245,* Rocky Mountain Forest and Range Experiment Station, Fort Collins, CO, pp. 61, 1994.

Hanna, T. H., *Field Instrumentation in Geotechnical Engineering,* Trans Tech Publications, Clausthal, Germany, 1985.

Hathaway, A. W., *Manual on Subsurface Investigations,* American Association of State Highway and Transportation Officials, Washington, DC, 1988.

Hvorslev, M. J., *Subsurface Exploration and Sampling of Soils,* Engineering Foundation, New York, 1948.

Krajca, J. M., *Water Sampling,* Halstead Press, John Wiley & Sons, New York, 1989.

Kolm, K. E., *Conceptualization and Characterization of Hydrologic Systems,* GWMI 93-01, International Ground Water Modeling Center, Golden, CO, 1993.

---

[57] Available from ORD Publications, U.S. EPA Center for Environmental Research Information, P.O. Box 19963, Cincinnati, OH 45268.

Lambe, W. T., *Soil Testing for Engineers*, John Wiley & Sons, New York, 1951.

Mudroch, A., and Azcue, J. M., *Manual of Aquatic Sediment Sampling*, Lewis Publishers, Boca Raton, FL, 1995.

Nielsen, D. M. (ed), *Practical Handbook of Ground Water Monitoring*, Lewis Publishers, Chelsea, MI, 1991.

Rehm, B. W., Stolzenburg, T. R., and Nichols, D. G., *Field Measurement Methods for Hydrogeologic Investigations: A Critical Review of the Literature*, EPRI EA-4301, Electric Power Research Institute, Palo Alto, CA, 1985.

Stednick, J. D., *Wildland Water Quality Sampling and Analysis*, Academic Press, HBJ, San Diego, CA, 1991.

Struckmeier, W. F., and Margat, J., *Hydrogeological Maps: A Guide and a Standard Legend*, IAH International Contributions to Hydrogeology Vol. 17, Verlag Heinz Heise, Hannover, Germany, 1995.

Thompson, C. M., et al, *Techniques to Develop Data for Hydrogeochemical Models*, EPRI EN-6637, Electric Power Research Institute, Palo Alto, CA, 1989.

Wilson, L. G., Everett, L. G., and Cullen, S. J., *Handbook of Vadose Zone Characterization and Monitoring*. Lewis Publications, Boca Raton, FL, pp. 896, 1994.

Wilson, N., *Soil Water and Ground Water Sampling*. Lewis Publishers, Boca Raton, FL, pp. 192, 1995.

U.S. Army Corps of Engineers (USACE), "Engineering and Design—Geotechnical Investigation," *Engineer Manual EM 1110-1-1804*, USACE, Washington, DC, 1984.

U.S. Geological Survey (USGS), *National Handbook of Recommended Methods for Water Data Acquisition*, USGS Office of Water Data Coordination, Reston, VA (Individual chapters have come out at different dates. Pertinent chapters include: (2) Ground Water (1980); (4) Biological and Microbiological Quality of Water (1983); (5) Chemical Quality (1982); and (6) Soil Water (1982)), 1977.

U.S. Environmental Protection Agency (EPA), *Test Methods for Evaluating Solid Waste*, 3rd ed, EPA/530/SW-846 (NTIS PB88-239223); First update, 3rd edition, EPA/530/SW-846.3-1 (NTIS PB89-148076). (Second edition was published in 1982 (NTIS PB87-1200291); current edition and updates available on a subscription basis from U.S. Government Printing Office, Stock #955-001-00000-1, 1986c, Revised final draft of Chapter 11 (Ground-Water Monitoring System Design, Installation, and Operating Practices contains extensive new guidance—see U.S. EPA, 1993b) (Volumes 1A (*Metallic Analytes*), IB (*Organic Analytes*), and IC (*Miscellaneous Test Methods*) cover laboratory methods; Volume II covers field methods (Part IV defines acceptable and unacceptable designs and practice for ground-water monitoring), 1986c.

U.S. Environmental Protection Agency (EPA), *Final RCRA Comprehensive Ground-Water Monitoring Evaluation (CME) Guidance Document*, Final OSWER Directive 9950.2 (NTIS PB91-140194), (Contains detailed checklist drawing heavily from U.S. EPA (1986b)), 1986d.

U.S. Environmental Protection Agency (EPA), *A Compendium of Superfund Field Operations Methods*, EPA/540/P-87/001 (NTIS PB88-181557), 1987a.

U.S. Environmental Protection Agency (EPA), *Data Quality Objectives for Remedial Response Activities*, Vol 1: Development Process; Vol 2: RI/FS Activities at a Site with Contaminated Soils and Ground Water, Vol 1 EPA/G-87/003 (NTIS PB88-131370), Vol 2 EPA/G-87/004 (NTIS PB88-131388), both Volumes: NTIS PB90-272634, 1987b.

U.S. Environmental Protection Agency (EPA), *Guidance for Conducting Remedial Investigations and Feasibility Studies Under CERCLA*, EPA/540/G-89/004, OSWER Directive 9355.3-01 (NTIS PB89184626), 1989a.

U.S. Environmental Protection Agency (EPA), *RCRA Facility Investigation (RFI) Guidance; Interim Final*, Vol I: *Development of an RFI Work Plan and General Considerations for RCRA Facility Investigations*; Vol II: *Soil, Groundwater, and Subsurface Gas Releases*; Vol III: *Air and Surface Water Releases*; Vol IV: *Case Study Examples*, EPA/530/SW-89/001, OSWER Directive 9502.00-6D (NTIS PB89-200299), 1989b.

U.S. Environmental Protection Agency (EPA), *Site Characterization for Subsurface Remediation*, EPA/625/4-91/026, 1991a.[56]

U.S. Environmental Protection Agency (EPA), *Emergency Response Team (ERT) Standard Operating Procedures (SOPs) Compendia: Compendium of ERT Soil Sampling and Surface Geophysics Procedures* (EPA/540/P-91/006); *Compendium of ERT Groundwater Sampling Procedures* (EPA/540/P-91/007); *Compendium of ERT Waste Sampling Procedures* (EPA/540/P-91/008); *Compendium of ERT Toxicity Testing Procedures* (EPA/540/P-91/009), 1991b.

U.S. Environmental Protection Agency (EPA), *Guidance for Performing Preliminary Assessments Under CERCLA*, OSWER 9345.0-01A (NTIS PB92-963303), 1991c.

U.S. Environmental Protection Agency (EPA), *Environmental Compliance Branch Standard Operating Procedures and Quality Assurance Manual*, U.S. EPA Region IV Environmental Services Division, College Station Road, Athens, GA 30613, (Available in Wordperfect 5.1 electronic format), 1991d.

U.S. Environmental Protection Agency (EPA), *Characterization Protocol for Radioactive Contaminated Soils*, OSWER Directive 9380.1-10FS, 1992.

U.S. Environmental Protection Agency (EPA), *Field Methods Compendium (FMC) Draft*, OERR #9285.2-11, Analytical Operation Branch, Hazardous Site Evaluation Division, Office of Emergency and Remedial Response, (Available in Wordperfect 5.1 electronic format from OERR, Washington, DC), 1993.

U.S. Environmental Protection Agency (EPA), *RCRA Ground Water Monitoring: Draft Technical Guidance*, EPA/530/R-93/001 (NTIS PB93-139350), (Provides supplemental guidance to Chapter 11 of U.S. EPA (1986a) and U.S. EPA 1986b), 1993b.

U.S. Geological Survey (USGS), Various authors and dates, *Techniques of Water-Resources Investigations*. (Series of more than 50 manuals and guides describing addressing field and laboratory methods for water resources investigations).

U.S. Naval Facilities Engineering Command, *Soil Mechanics Design Manual*, Volume 7.1. NAVFAC DM-7.1, (NITS ADA123-622) Department of the Navy (Includes section on site assessment techniques), 1982.

X2.2 *Controlled and Uncontrolled Waste Sites:*

Breckenridge, R. P., Williams, J. R., and Keck, J. F., Characterizing Soils for Hazardous Waste Site Assessments. *Ground-Water Issue Paper EPA/600/8-91/008*, 1991, Available from CERI.[56]

Byrnes, M. E., *Field Sampling Methods for Remedial Investigations*, Lewis Publishers, Boca Raton, FL, 1994.

Cheremisinoff, P., *A Guide to Underground Storage Tanks: Evaluation, Site Assessment and Remediation*. Prentice-Hall, Englewood Cliffs, NJ, 1992.

Cohen, R. M., and Mercer, J. W., *DNAPL Site Evaluation*. EPA/600/R-93/002 (NTIS PB93-150217), (Also published by Lewis Publishers as C. K. Smoley edition, Boca Raton, FL), 1993.

Ford, P. J., and Turina, P. J., *Characterization of Hazardous Waste Sites—A Methods Manual: Vol I. Site Investigations*. EPA/600/4-84/075 (NTIS PB85-215960), 1985.

Ford, P. J., Turina, P. J., and Seely, D. E., *Characterization of Hazardous Waste Sites—A Methods Manual: Vol II. Available Sampling Methods*, 2nd ed. EPA 600/4-84/076 (NTIS PB85-521596), 1984.

Lipsky, D., Tusa, W., Dorrier, R., Johnson, B., and Gardner, M., *Methods for Evaluating the Attainment of Cleanup Standards, Volume 1: Soils and Solid Media*. EPA/230/2-89/042 (NTIS PB89-234959), 1989.

Oudjik, G., and Mujica, K., *Handbook for Identification, Location and Investigation of Pollution Sources Affecting Ground Water*. National Water Well Association, Dublin, OH, 1989.

Sara, M. N., *Standard Handbook of Site Assessment for Solid and Hazardous Waste Facilities*. Lewis Publishers, Boca Raton, FL, 1994.

Sisk, S. W., *NEIC Manual for Groundwater/Subsurface Investigations at Hazardous Waste Sites*, EPA/330/9-81-002 (NTIS PB82-103755), 1981.

U.S. Department of Energy (DOE), Various Dates, *The Environmental Survey Manual*. DOE/EH-0053: Vol 1 (August 1987; Chapter 8, 2nd ed. January 1989—*Sampling and Analysis Phase*); Vol 2 (August 1987—Appendixes A, B, and C); Vol 3 (2nd ed. January 1989—Appendix D, Parts 1, 2, and 3; *Organic and Inorganic Analysis Methods and Non-Target List Parameters*); Vol 4 (2nd ed. January 1989—Appendix D, Part 4; *Radiochemical Analysis Procedures*); Vol 5 (2nd ed. January 1989—Appendixes: E, *Field Sampling*; F, *Quality Assurance*; G, *Decontamination*; H, *Sample Management*; I, *Sample Handling, Transport and Documentation*; J, *Health and Safety*; and K, *Sampling and Analysis Plan*).

U.S. Environmental Protection Agency (EPA), *RCRA Facility Assessment Guidance.* EPA/530-SW-86-053, (NTIS PB87-107769), 1986a.

U.S. Environmental Protection Agency (EPA), *RCRA Ground Water Monitoring Technical Enforcement Guidance Document.* EPA/530/SW-86/055 (OSWER-9950.1) (NTIS PB87-107751), 332 pp. (Also published in NWWA/EPA Series, National Water Well Association, Dublin, OH. Final OSWER Directive 9950.2 (NTIS PB91-140194). Executive Summary: OSWER 9950.1a (NTIS PB91-140186), 17 pp. See also, U.S. EPA (1986d and 1993b), 1986b.

U.S. Environmental Protection Agency (EPA), *U.S. EPA Region VIII Standard Operating Procedures for Field Sampling Activities*, Version 2, Denver, CO, 1994.

Water Pollution Control Federation (WPCF/WEF), *Hazardous Waste Site Remediation: Assessment and Characterization*, Water Environment Federation, Alexandria, VA, 1988.

X2.3 *Environmental Audits and Site Assessments:*

Association of Ground Water Scientists and Engineers, *Guidance to Environmental Site Assessments.* National Ground Water Association, Dublin, OH, 1992.

Colangelo, R. V., *Buyer Be(a)ware: The Fundamentals of Environmental Property Assessments*, National Water Well Association, Dublin, OH, 1991.

Environmental Resource Center, *Environmental Auditing and Compliance Manual*, Van Nostrand Reinhold, New York (1992 edition published by Environmental Resource Center), 1993.

Hess, K., *Environmental Site Assessment, Phase I: A Basic Guide*, Lewis Publishers, Boca Raton, FL, 1993.

Jain, R. K., Urban, L. V., Stacey, G. S., and Balbach, H. E., *Environmental Assessment*, McGraw-Hill, New York, 1993.

Vincoli, J., *Basic Guide to Environmental Compliance*, Van Nostrand Reinhold, New York, 1993.

X2.4 *Environmental Exposure and Risk Assessment:*

Calabrese, E. J., and Kostecki, P. T., *Risk Assessment and Environmental Fate Methodologies*, Lewis Publishers, Boca Raton, FL (Description and critical review of existing software (AERIS, GEOTOX, LUFT, MYGRT, PCGEMS/SESOIL, POSSM, PPLV, PRZM, RAFT, Risk Assistant, SESOIL), and other methods developed at the state level (California, New Jersey, and Massachusetts)), 1992.

Hallenback, W. H., *Quantitative Risk Assessment for Environmental and Occupational Health*, 2nd Ed., Lewis Publishers, Chelsea, MI, 1993.

Hill, I. R., Heimbach, F., Leeuwangh, P., and Matthiessen, P. (eds.), *Freshwater Field Tests for Hazard Assessment of Chemicals*, Lewis Publishers, Boca Raton, FL, pp. 608, 1994.

Hoffman, D. J., Rattner, B. A., Burton, Jr., G. A., and Cairns, Jr., J., *Handbook of Ecotoxicology*, Lewis Publishers, Boca Raton, FL, 1995.

Newman, M. C., *Quantitative Methods in Aquatic Ecotoxicology*, Lewis Publishers, Boca Raton, FL, 1995.

Norton, S., McVey, M., Colt, J., Durda, J., and Hegner, R., *Review of Ecological Risk Assessment Methods*, EPA/230/10-88-041 (NTIS PB89-134357), (Review of sixteen methodologies), 1988.

Schaum, J., *Exposure Factors Handbook*, EPA/600/8-89/043 (NTIS PB90-106774), 1990.

Suter, II, G. W., *Ecological Risk Assessment*, Lewis Publishers, Chelsea, MI, 1993.

U.S. Environmental Protection Agency, *Ecological Risk Assessment Issue Papers*, EPA/630/R-94/009, 1994.

U.S. Environmental Protection Agency (EPA), *Superfund Exposure Assessment Manual*, EPA/540/1-88/001, OSWER Directive 9285.5-1 (NTIS PB90-135859), 1988.

U.S. Environmental Protection Agency (EPA), *Risk Assessment Guidance for Superfund, Volume 1: Human Health Evaluation Manual, Part A, Interim Final*, EPA/540/1-89/002 (NTIS PB90-155581), (1990 9-page Fact Sheet with same title: NTIS PB90-273830; 1991 *Human Health Evaluation Manual, Supplemental Guidance: Standard Default Exposure Factors*: NTIS PB91-921314), 1989a.

U.S. Environmental Protection Agency (EPA), *Risk Assessment Guidance for Superfund, Volume 2: Environmental Evaluation Manual, Interim Final*, EPA/540/1-89/001 (NTIS PB90-155599), 1989b.

U.S. Environmental Protection Agency (EPA), *Statistical Methods for Estimating Risk for Exposure Above the Reference Dose*, EPA/600/8-90/065 (NTIS PB90-261504), 1990.

U.S. Environmental Protection Agency (EPA), *Guidance for Data Useability in Risk Assessment (Parts A and B)*, Final, Part A: OSWER Directive 9285.7-09A (NTIS PB92-963356), Part B: OSWER Directive 9285.7-09B (NTIS PB92-963362), (Supercedes 1990 document by same title (EPA/540/G-90/008, OSWER Directive 9285.7-05; NTIS PB91-921208), 1992.

Wentsel, R., et al, *Procedural Guidelines for Ecological Risk Assessment at U.S. Army Sites*, Volume 1, 1994.

X2.5 *Atmospheric and Ecological Assessment:*

Britton, L. J., and Greeson, P. E. (eds), *Methods for Collection and Analysis of Aquatic Biological and Microbiological Samples*, U.S. Geological Survey Techniques of Water-Resources Investigations TWRI 5-A4, 1989.

Electric Power Research Institute (EPRI), *Sampling Design for Aquatic Ecologic Monitoring*, Five Vols, EPRI EA-4302, EPRI, Palo Alto, CA, 1985.

Euphrat, F. D., and Warkentin, B. P., *A Watershed Assessment Primer*, U.S. EPA Region 10, EPA/B-94/005, 1994.

Kuechler, A. W., *Vegetation Mapping*, The Ronald Press Company, New York, 1967.

Lodge, J. (ed), *Methods of Air Sampling and Analysis*, 3rd ed. Lewis Publishers, Chelsea, MI, 1988.

Loeb, S. L., and Spacie, A., *Biological Monitoring of Aquatic Systems*, Lewis Publishers, Boca Raton, FL, 1994.

Maslansky, C. J., and Maslansky, S. P., *Air Monitoring Instrumentation: A Manual for Emergency, Investigatory and Remedial Responders*, Van Nostrand Reinhold, New York, NY, 1992.

Ohio Environmental Protection Agency, *Biological Criteria for the Protection of Aquatic Life, Vol II, User's Manual for Biological Assessment of Ohio Surface Waters*, Ohio EPA, Columbus, OH, 1987.

Ralph, C. J., Geupel, G. R., Pyle, P., Martin, T. E., and DeSante, D. F., *Handbook of Field Methods for Monitoring Landbirds. General Technical Report PSW-GTR-144*, U.S. Forest Service, Pacific Southwest Research Station, Albany, CA, pp. 41, 1993.

Taylor, S. A., and Ashcroft, G. L., *Physical Edaphology*, W. H. Freeman and Co., San Francisco, 1972.

Warren-Hicks, W., Parkhurst, B. R., and Baker, Jr., S. S., *Ecological Assessment of Hazardous Waste Sites: A Field and Laboratory Reference.* EPA/600/3-89/013 (NTIS PB89-205967), (Covers toxicity tests, biomarkers, and ecological field assessments), 1989.

Wight, G. D., *Fundamentals of Air Sampling*, Lewis Publishers, Boca Raton, FL, pp. 272, 1994.

X3. MAJOR NON-ASTM REFERENCES ON TERMINOLOGY RELATED TO ENVIRONMENTAL SITE CHARACTERIZATION

Allaby, A., and Allaby, M., *The Concise Oxford Dictionary of Earth Sciences*. Oxford University Press, Oxford, UK, 1990.

American Society of Agricultural Engineers, *Glossary of Soil and Water Terms*. American Society of Agricultural Engineers, St. Joseph, MI, 1967.

Bates, R., and Jackson, J. (eds), *Dictionary of Geological Terms*, 3rd ed. AGI, Washington, DC, (Supersedes Weller (1960)), 1984.

Castany, G., and Margat, J., *Dictionnaire Français D'Hydrogéolgie*, BRGM, Orléans, 1977.

Interagency Advisory Committee, *Subsurface-Water Flow and Solute Transport, Glossary of Selected Terms*. Draft report prepared by Subsurface-Water Glossary Working Group, Ground Water Subcommittee, 1988.

International Society for Rock Mechanics, *Final Document on Terminology, English Versions*, Committee on Terminology, Symbols and Graphic Representation, 1972.

Laney, R. L., and Davidson, C. B., *Aquifer Nomenclature Guidelines*, U.S. Geological Survey Open File Report 86-534, 1986.

Langbein, W. B., and Iseri, K. T., *General Introduction and Hydrologic Definitions*. U.S. Geological Survey Water Supply Paper 1541-A, 1960.

Lo, S. S., *Glossary of Hydrology*, Water Resources Publications, Highland Ranch, CO, 1992.

Lohman, S. W., et al, *Definitions of Selected Ground-Water Terms—Revisions and Conceptual Refinements*, U.S. Geological Survey Water-Supply Paper 1988, 1972.

Meinzer, O. E., *Outline of Ground Water Hydrology with Definitions*, U.S. Geological Survey Water Supply Paper 494, 1923.

Michel, J.-P., and Fairbridge, R. W., *Dictionary of Earth Sciences*, John Wiley & Sons, New York, 1992.

Moore, W. G., *A Dictionary of Geography*, 4th edition, Penguin Books, Baltimore, MD, 1968.

National Geodetic Survey, *Geodetic Glossary*, U.S. Department of Commerce, 1986.

Parker, S. P. (ed), *Dictionary of Scientific and Technical Terms*, 4th ed., McGraw-Hill, New York, 1989.

Pfannkuch, H. O., *Elsevier's Dictionary of Hydrogeology*, Elsevier, NY, 1969.

Poland, J. F., et al, *Glossary of Selected Terms Useful in Studies of the Mechanics of Aquifer Systems and Land Subsidence Due to Fluid Withdrawal*, U.S. Geological Survey Water Supply Paper 2025, 1972.

Porteous, A., *Dictionary of Environmental Science and Technology*, revised edition, John Wiley & Sons, New York, 1992.

Soil Conservation Service (SCS), *Glossary of Selected Geologic and Geomorphic Terms*, U.S. Department of Agriculture, Soil Conservation Service Western Technical Service Center, Portland, OR, 1977.

Soil Science Society of America, *Glossary of Soil Science Terms*, SSSA, Madison, WI, 1987.

Stevenson, L. H., and Wynen, B., *The Facts on File Dictionary of Environmental Sciences*. Facts on File, New York, NY, 1991.

Titelbaum, O. A., *Glossary of Water Resources Terms*, Federal Water Pollution Control Administration, 1970.

UNESCO, *International Glossary of Hydrology*, WMO/OMM/BMO No. 385, 1974.

U.S. Environmental Protection Agency (EPA), *Draft Glossary of Quality Assurance Related Terms*, Office of Research and Development, Sept. 29, 1988.

U.S. Geological Survey, *Federal Glossary of Selected Terms: Subsurface-Water Flow and Solute Transport*, Office of Water Data Coordination, USGS, Reston, VA, 1989.

Whitten, D. G. A., and Brooks, J. R. V., *The Penguin Dictionary of Geology*, Penguin Books, Baltimore, MD, 1972.

Weller, J. M. (ed), *Glossary of Geology and Related Sciences with Supplement*, 2nd edition. American Geological Institute, Washington, DC, Supplement 72 pp. (Superseded by AGI (1984)), 1960.

REFERENCES

(1) Quinlan, J. F., Special Problems of Ground-Water Monitoring in Karst and Fracture Rock Terranes, In: *Ground-Water and Vadose Zone Monitoring, ASTM STP 1053*, D. M. Nielsen and A. I. Johnson (eds.), ASTM, Philadelphia, PA, pp. 275–304, 1990.

(2) Avery, T. E., *Interpretation of Aerial Photographs*, Second ed, Burgess Publishing Company, Minneapolis, MN, 1968.

(3) Ciciarelli, J., *A Practical Guide to Aerial Photography*, Van Nostrand Reinhold, New York, 1991.

(4) Denny, C. S., Warren, C. R., Dow, D. H., and Dale, W. J., "A Descriptive Catalog of Selected Aerial Photographs of Geologic Features of the United States," *U.S. Geological Survey Professional Paper 590*, 1968.

(5) Drury, S. A., *Image Interpretation in Geology*, Allen and Unwin, London, UK, 1987.

(6) Dury, G. H., *Map Interpretation*, Pitman, London, 1960.

(7) Johnson, A. I., and Gnaedinger, J. P., Bibliography, In: *Symposium on Soil Exploration, ASTM STP 351*, ASTM, Philadelphia, PA, pp. 137–155 (90 references on air photo interpretation), 1964.

(8) Lattman, L. H., and Ray, R. G., *Aerial Photographs in Field Geology*, Holt Rinehart and Winston, New York, 1965.

(9) Lueder, D. R., *Aerial Photographic Interpretation: Principles and Applications*, McGraw-Hill, New York, 1959.

(10) Miller, V. C., and Miller, C. F., *Photogeology*, McGraw-Hill, New York, 1961.

(11) Ray, R. G., "Aerial Photographs in Geologic Interpretation and Mapping," *U.S. Geological Survey Professional Paper 373*, 1960.

(12) Soil Conservation Service (SCS), "Aerial-Photo Interpretation in Classifying and Mapping Soils," *U.S. Department of Agriculture Handbook 294*, 1973.

(13) Strandberg, C. H., *Aerial Discovery Manual*, Wiley, New York, 1967.

(14) Wright, J., *Ground and Air Survey for Field Scientists*, Oxford University Press, New York, 1982.

(15) Sun, R. J., and Weeks, J. B., "Bibliography of Regional Aquifer-System Analysis Program of the U.S. Geological Survey, 1978–91" *U.S. Geological Survey Water-Resources Investigations Report 91-4122*, 1991.

(16) U.S. Geological Survey. 1991–1994. Ground-Water Atlas of the United States [14 planned chapters; three currently published: 730-G (Alabama, Florida, Georgia, South Carolina), 730-H (Idaho, Oregon, Washington), and 730-J (Iowa, Michigan, Minnesota, Wisconsin)].

(17) Boulding, J. R., *Description and Sampling of Contaminated Soils: A Field Pocket Guide*, EPA/625/12-91/002, 1991.[58]

(18) Cameron, R. E., *Guide to Site and Soil Description for Hazardous Waste Sites, Volume 1: Metals*, EPA/600/4-91/029 (NTIS PB92-146158), 1991.

(19) Bishop, M. S., *Subsurface Mapping*, Wiley, New York, 1960.

---

[58] Available from ORD Publications, U.S. EPA Center for Environmental Research Information, P.O. Box 19963, Cincinnati, OH 45268-0963.

(20) Brassington, R., *Field Hydrogeology*, Halsted Press, New York, 1988.

(21) Bureau of Reclamation, *Engineering Geology Office Manual*, U.S. Department of the Interior, Bureau of Reclamation, Denver, CO, 1988.

(22) Bureau of Reclamation, *Engineering Geology Field Manual*, U.S. Department of the Interior, Bureau of Reclamation, Denver, CO, 1988.

(23) Compton, R. R., *Manual of Field Geology*, John Wiley & Sons, New York, 1962.

(24) Compton, R. R., *Geology in the Field*, John Wiley & Sons, New York, 1985.

(25) Dietrich, R. V., Dutro, Jr., J. V., and Foose, R. M. (Compilers), *AGI Data Sheets for Geology in Field, Laboratory, and Office*, 3rd Edition, American Geological Institute, Washington, DC, 1990.

(26) Erdélyi, M., and Gálfi, J., *Surface and Subsurface Mapping in Hydrogeology*, Wiley-Interscience, New York, 1988.

(27) Fetter, C. W., *Applied Hydrogeology*, 3rd ed. Macmillan, New York, 1994.

(28) Lahee, F. H., *Field Geology (6th ed.)*, McGraw-Hill, New York, 1961.

(29) LeRoy, L. W., LeRoy, D. O., Schwochow, S. D., and Raese, J. W. (eds), *Subsurface Geology*, 5th edition. Colorado School of Mines, Golden, CO, (1st edition: LeRoy and Cran [1947], 2nd edition: LeRoy [1951], 3rd edition: Huan and LeRoy [1958], and 4th edition [1977]), 1987.

(30) Low, J. W., *Geologic Field Methods*, Harper, New York, 1957.

(31) Rahn, P., *Engineering Geology*, Elsevier, New York, 1986.

(32) Tearpock, D., and Bischke, R. E., *Applied Subsurface Geological Mapping*, Prentice Hall, Englewood Cliffs, NJ, 1991. (Focusses on construction of geological maps from various sources, including geophysical measurements).

(33) Fry, N., *The Field Description of Metamorphic Rocks*, John Wiley & Sons, New York, 1984.

(34) Thorpe, R., and Brown, G., *The Field Description of Igneous Rocks*, John Wiley & Sons, New York, 1985.

(35) Tucker, M. E., *The Field Description of Sedimentary Rocks*, John Wiley & Sons, New York, 1982.

(36) U.S. Environmental Protection Agency (EPA). 1992. *Locational Data Policy Implementation Guidance: Guide to the Policy*. EPA/220/B-92-008, Office of Administration and Resources Management (PMD-211D), Washington DC. [Note that U.S. EPA 1992a and 1992b are separate documents, but have the same document number]

(37) U.S. Environmental Protection Agency (EPA). 1992. *Locational Data Policy Implementation Guidance: Guide to Selecting Latitude/Longitude Collection Methods*. EPA/220-B-92-008, Office of Administration and Resources Management (PMD-211D), Washington DC. [Note that U.S. EPA 1992a and 1992b are separate documents, but have the same document number]

(38) U.S. Environmental Protection Agency (EPA). 1992. *Locational Data Policy Implementation Guidance—Global Positioning System Technology and Its Application In Environmental Programs—GPS Primer*. EPA/600/R-92/036. Available from CERI.*

(39) Boulding, J. R., *Description and Sampling of Contaminated Soils: A Field Guide, Revised and Expanded*, 2nd edition, Lewis Publishers, Chelsea, MI, 1994.

(40) Vepraskas, M. J., *Redoximorphic Features for Identifying Aquaic Conditions. North Carolina Agricultural Research Service Technical Bulletin 301*, Department of Agricultural Communications, Box 7603, North Carolina State University, Raleigh, NC 27695-7603, 1992.

(41) Soil Survey Staff, *Keys to Soil Taxonomy*, 5th ed, Soil Management Support Services (SMSS) Technical Monograph No. 19, Pocahontas Press, P.O. Drawer F, Blacksburg, VA, 24063-1020, 1992.

(42) Kirkaldie, L., "Potential Contaminant Movement Through Soil Joints," *Bull. Ass. Eng. Geologists IIV(4)*, 1988, pp. 520–524.

(43) Boulding, J. R. 1993. *Subsurface Field Characterization and Monitoring Techniques: A Desk Reference Guide, Volume I: Solids and Ground Water, Volume II: The Vadose Zone, Field Screening and Analytical Methods*. EPA/625/R-93/003a&b. Available from CERI.*

(44) Benson, R. C., Glaccum, R. A., and Noel, M. R., *Geophysical Techniques for Sensing Buried Wastes and Waste Migration*, EPA/600/7-84/064 (NTIS PB84-198449). Also published in NWWA/EPA series by National Water Well Association, Dublin, OH, 1984.

(45) Boulding, J. R., *Use of Airborne, Surface, and Borehole Geophysical Techniques at Contaminated Sites: A Reference Guide*, EPA/625/R-92/007, 1993.

(46) Haeni, F. P., *Application of Seismic Refraction Techniques to Hydrogeologic Studies*, U.S. Geological Survey Techniques of Water-Resources Investigations TWRI 2-D2, 1988.

(47) Ward, S. H. (ed), *Geotechnical and Environmental Geophysics, Vol I: Review and Tutorial*, Society of Exploration Geophysicists, Tulsa, OK, 1990a.

(48) Ward, S. H. (ed), *Geotechnical and Environmental Geophysics, Vol II: Environmental and Groundwater*, Society of Exploration Geophysicists, Tulsa, OK, 1990b. (34 papers, including ER, EM multiple methods, thermal, others).

(49) Zohdy, A. A., Eaton, G. P., and Mabey, D. R., *Application of Surface Geophysics to Ground-Water Investigations*, U.S. Geological Survey Techniques of Water-Resource Investigations, TWRI 2-D1, 1974, (ER, GR, MAG, SRR).

(50) Keys, W. S., *Borehole Geophysics Applied to Ground-Water Investigations*, U.S. Geological Survey Techniques of Water-Resource Investigations TWRI 2-E2, 1990, (Supersedes report originally published in 1988 under the same title as U.S. Geological Survey Open-File Report 87-539, 303 pp., which was published in 1989 with the same title by the National Water Well Association, Dublin, OH, 313 pp.), Complements Keys and MacCary (1971).

(51) Keys, W. S., and MacCary, L. M., *Application of Borehole Geophysics to Water Resource Investigations*, TWRI 2-E1, U.S. Geological Survey Techniques of Water-Resources Investigations, 1971. (Reprinted, 1990; see, also Keys, 1990).

(52) Respold, H., *Well Logging in Groundwater Development. International Contributions to Hydrogeology*, Vol 9, International Association of Hydrogeologists, Verlag Heinz Heise, Hannover, West Germany, 1989.

(53) Aller, L., et al, *Handbook of Suggested Practices for the Design and Installation of Ground-Water Monitoring Wells*, EPA/600/4-89/034, 1991.[57] (Also published in 1989 by National Water Well Association, Dublin, OH, in its NWWA/EPA series).

(54) Eggington, J. F. et al., *Australian Drilling Manual*. Australian Drilling Industry Training Committee, North Ryde, NSW, Australia, 1992.

(55) Clark, L. 1988. *The Field Guide to Water Wells and Boreholes*. Geological Society of London Professional Handbook. Halsted Press, New York, 155 pp.

(56) Driscoll, F. G., *Groundwater and Wells*, 2nd ed, Johnson Filtration Systems, Inc., St. Paul, MN, 1986.

(57) Harlan, R. L., Kolm, K. E., and Gutentag, E. D., *Water-Well Design and Construction*, Elsevier, New York, 1989.

(58) Lehr, J., Hurlburt, S., Gallagher, B., and Voyteck, J., *Design and Construction of Water Wells: A Guide for Engineers*. Van Nostrand Reinhold, New York, NY, 1988.

(59) Roscoe Moss Company, *Handbook of Ground Water Development*. John Wiley & Sons, New York, 1990.

(60) Ruda, T. C., and Bosscher, P. J. (eds), *Drillers Handbook*, National Drilling Contractors Association, Columbia, SC, 1990.

(61) Shuter, E., and Teasdale, W. E., *Application of Drilling, Coring, and Sampling Techniques to Test Holes and Wells*, U.S. Geological Survey Techniques of Water-Resource Investigations TWRI 2-F1, 1989.

(62) Soil Conservation Service (SCS), *Photographic Documentation of Rock Core Samples*, Engineering Geology Investigation Geology Note 4, 1984.

(63) Soil Survey Staff. 1975. *Soil Taxonomy: A Basic System of Soil Classification for Making and Interpreting Soil Surveys*. U.S. Department of Agriculture Agricultural Handbook No. 436, 754 pp.

(64) Soil Survey Staff. 1993. *Soil Survey Manual (new edition)*. U.S. Dept. of Agric. Agricultural Handbook No. 18. U.S. Government Printing Office Stock No. 001-000-04611-0.

(65) Dutro, Jr., J. T., Dietrich, R. M., and Foose, R. M. (Compilers), 1989. *AGI Data Sheets for Geology in Field Laboratory and Office*, 3rd Edition. American Geological Institute, Washington, DC, pp. 294.

(66) Morris, D. A., and Johnson, A. I., *Summary of Hydraulic and Physical Properties of Rock and Soil Materials as Analyzed by the Hydraulic Laboratory of the U.S. Geological Survey*, U.S. Geological Survey Water Supply Paper 1839-D, 1967, pp. D1–D42.

(67) Birkeland, P. W., *Soils and Geomorphology*, Oxford University Press, New York, NY, 1984, (Revision of Pedology, Weathering, and Geomorphological Research published in 1973).

(68) Stefferud, A., *Water, the Yearbook of Agriculture*, U.S. Department of Agriculture, 1955.

(69) Deere, D. U., and Deere, D. W., "A Rock Quality Designation (RQD) Index in Practice," In: *Rock Classification Systems for Engineering Purposes*, L. Kirkdale (ed), *ASTM STP 984*, ASTM, Philadelphia, PA, 1988, pp. 91–101.

(70) Barth, D. S., B. J. Mason, T. H. Starks, and K. W. Brown. 1989. *Soil Sampling Quality Assurance User's Guide*, 2nd ed. EPA 600/8-89/046 (NTIS PB89-189864), 225+ pp.

(71) Mason, B. J., *Preparation of Soil Sampling Protocols: Sampling Techniques and Strategies*, EPA/600/R-92/128 (NTIS PB92-220532), 1992, (Supersedes 1983 edition titled, Preparation of Soil Sampling Protocol: Techniques and Strategies, EPA-600/4-03-020 (NTIS PB83-206979).

(72) Gilbert, R. O., *Statistical Methods for Environmental Pollution Monitoring*, Van Nostrand Reinhold, New York, 1987.

(73) Clark, I., *Practical Geostatistics*, Applied Science Publishers, London, 1979.

(74) Isaaks, E. H., and Srivastava, R. M., *Applied Geostatistics*, Oxford University Press, New York, 1989.

(75) ASCE Task Committee on Geostatistical Techniques in Geohydrology, "Review of Geostatistics in Geohydrology, I. Basic Concepts, II. Applications," *ASCE Journal of Hydraulic Engineering 116(5)*, 1990, pp. 612–658.

(76) Englund, E. J., and Sparks, A. R., *Geo-EAS (Geostatistical Environmental Assessment Software) User's Guide*, EPA/600/4-88/033a, 1988, (Guide: NTIS PB89-151252, Software: PB89-151245).

(77) Englund, E., and Sparks, A., *GEO-EAS 1.2.1 User's Guide*, EPA/600/8-91/008, Available from U.S. EPA Environmental Monitoring Systems Laboratory, Las Vegas, NV, 1991.

(78) Yates, S. R., and Yates, M. V., *Geostatistics for Waste Management: A User's Manual for the GEOPACK* (Version 1.0) Geostatistical Software System, EPA/600/8-90/004 (NTIS PB90-186420/AS), 1990.

(79) Zirschky, J., and Gilbert, R. O., "Detecting Hot Spots at Hazardous-Waste Sites," *Chemical Engineering*, July 9, 1984, pp. 97–100.

(80) Barth, D. S., and Starks, T. H., *Sediment Sampling Quality Assurance User's Guide*, EPA/600/4-85/048 (NTIS PB85-233542), 1985.

(81) Palmer, M., *Methods Manual for Bottom Sediment Sample Collection*, EPA/905/4-85/004 (NTIS PB86-107414), 1985.

(82) Guy, H. P., and Norman, V. W., *Field Methods for Measurement of Fluvial Sediment*, U.S. Geological Survey Techniques of Water-Resources Investigations TWRI 3-C2, 1970, [Updated in U.S. Geological Survey Open-File Report 86-531].

(83) Chappelle, F. H., *Ground-Water Microbiology and Geochemistry*, John Wiley & Sons, New York, NY, 1993.

(84) Alexander, M., *Introduction to Soil Microbiology*, 2nd ed, John Wiley & Sons, New York, NY, 1977.

(85) Ehrlich, H. L., *Geomicrobiology*, Marcel Dekker, New York, NY, 1981, (Focuses on microbiology of natural geological systems).

(86) Killham, K., *Soil Ecology*, Cambridge University Press, New York, NY, 1994.

(87) Kuznetsov, S. I., Ivanov, M. V., and Lyalikova, N. N., *Introduction to Geological Microbiology*, McGraw-Hill, New York, NY, 1963 (Text focusing on natural microbiological activity in subsurface geologic systems).

(88) Paul, E. A., and Clark, F. E., *Soil Microbiology and Biochemistry*, Academic Press, 1989.

(89) Zajic, J. E. 1969. *Microbial Biogeochemistry*. Academic Press, New York, NY. [Text focusing on the microbial biogeochemistry of natural geological systems]

(90) Leach, L. E., Beck, F. P., Wilson, J. T., and Kampbell, D. H., "Aseptic Subsurface Sampling Techniques for Hollow-Stem Auger Drilling," In: *Proc. Third National Outdoor Action Conference on Aquifer Restoration, Ground Water Monitoring and Geophysical Methods*, National Water Well Association, Dublin, OH, 1988, pp. 31–51.

(91) Davis, J. C., *Statistics and Data Analysis in Geology*, John Wiley & Sons, New York, 1973.

(92) U.S. Environmental Protection Agency (EPA), *User Documentation: A Ground Water Information Tracking System with Statistical Capability*, GRITS/STAT Version 4.2. EPA/625/11-91/002, 1992d.[57]

(93) Roy, W. R., Krapac, I. G., Chou, S. F. J., and Griffin, R. A., *Batch-Type Procedures for Estimating Soil Adsorption of Chemicals*, EPA/530/SW-87/006F (NTIS PB92-146190), 1992.

(94) Hamilton, D. G., and T. A. Jones (eds.), *Computer Modeling of Geologic Surfaces*, American Association of Petroleum Geologists, Tulsa, OK, 1992.

(95) Jones, T. A., Hamilton, D. G., and Johnson, C. R., *Contouring Geologic Surfaces with the Computer*, Van Nostrand Reinhold, New York, NY, 1986.

Standard Guide for
Environmental Site Characterization in Cold Regions[1]

This standard is issued under the fixed designation D 5995; the number immediately following the designation indicates the year of original adoption or, in the case of revision, the year of last revision. A number in parentheses indicates the year of last reapproval. A superscript epsilon (ϵ) indicates an editorial change since the last revision or reapproval.

INTRODUCTION

Understanding environmental processes that occur in soil and rock systems in cold regions of the world depends on adequate characterization of not only the physical, chemical, and biological properties of soil and rock but also the climatic factors under which they exist. Processes of interest may include, but are not limited to, surface and subsurface hydrology, contaminant mobilization, distribution, fate and transport, chemical and biological degradation of wastes, geomorphological, and ecological processes in general.

1. Scope

1.1 Use this guide in conjunction with Guide D 5730.

1.2 This guide describes special problems to be considered when planning field investigations in cold regions. The primary focus of this guide is presenting the special problems and concerns of site characterization in the cold regions of the world.

1.3 Laboratory testing of soil, rock, and ground-water samples is specified by other ASTM standards that are not specifically discussed in this guide. Laboratory methods for measurement of physical properties relevant to environmental investigations are included in Guide D 5730.

1.4 The values stated in SI units are to be regarded as the standard.

1.5 This guide emphasizes the care that must be taken by all field personnel during operations in tundra and permafrost areas of the world.

1.6 *This standard does not purport to address all of the safety concerns, if any, associated with its use. It is the responsibility of the user of this standard to establish appropriate safety and health practices and determine the applicability of regulatory limitations prior to use.*

2. Referenced Documents

2.1 *ASTM Standards:*

D 653 Terminology Relating to Soil, Rock, and Contained Fluids[2]

D 4083 Practice for Description of Frozen Soils (Visual-Manual Procedure)[2]

D 5254 Practice for the Minimum Set of Data Elements to Identify a Ground-Water Site[3]

D 5408 Guide for Set of Data Elements to Describe a Ground-Water Site; Part One—Additional Identification Descriptors[3]

D 5409 Guide for Set of Data Elements to Describe a Ground-Water Site; Part Two—Physical Descriptors[3]

D 5410 Guide for Set of Data Elements to Describe a Ground-Water Site; Part Three—Usage Descriptors[3]

D 5730 Guide to Site Characterization for Environmental Purposes with Emphasis on Soil, Rock, the Vadose Zone and Ground Water[3]

D 5781 Guide for Use of Dual-Wall Reverse-Circulation Drilling for Geoenvironmental Exploration and Installation of Subsurface Water-Quality Monitoring Devices[3]

D 5783 Guide for Use of Direct Rotary Drilling with Water-Based Drilling Fluid for Geoenvironmental Exploration and Installation of Subsurface Water-Quality Monitoring Devices[3]

D 6001 Guide for Direct Push Water Sampling for Geoenvironmental Investigations[3]

3. Terminology

3.1 *Definitions*—Definitions of terms used in this guide are in accordance with Terminology D 653.

3.1.1 Guide D 5730 identifies major references from a range of disciplines that can be used as additional sources for definitions of terms that are related to environmental site characterization.

3.2 *Definitions of Terms Specific to This Standard:*

3.2.1 *active layer, n*—the top layer of ground above the permafrost table that thaws each summer and refreezes each fall.

3.2.2 *alpine permafrost, n*—permafrost developed in temperate climate mountainous areas of the world.

3.2.3 *continuous permafrost, n*—permafrost occurring everywhere beneath the exposed land surface throughout a geographic regional zone, with the exception of widely scattered sites, such as newly deposited unconsolidated sediments, where the climate has just begun to impose its influence on the ground thermal regime that will cause the formation of continuous permafrost.

3.2.4 *discontinuous permafrost, n*—permafrost occurring in some areas beneath the ground surface throughout a

[1] This guide is under the jurisdiction of ASTM Committee D-18 on Soil and Rock and is the direct responsibility of Subcommittee D18.01 on Surface and Subsurface Characterization.

Current edition approved July 10, 1996. Published December 1996.

[2] *Annual Book of ASTM Standards*, Vol 04.08.

[3] *Annual Book of ASTM Standards*, Vol 04.09.

geographic regional zone where other areas are free of permafrost.

3.2.5 *icing, n*—a sheet-like mass of layered ice, either on the ground surface or on the surface of river ice. Aufeis (German), Naled (Russian).

3.2.6 *permafrost, n*—the thermal condition in earth materials where temperatures below 0°C persist over at least two consecutive winters and the intervening summer; moisture in the form of water and ground ice may or may not be present. Earth materials in this thermal condition may be described as perennially frozen, irrespective of their water and ice content.

4. Significance and Use

4.1 This guide, when used in conjunction with Guide D 5730, provides direction to the selection of the various ASTM standards that are available for the investigation of soil, rock, the vadose zone, ground-water, and other media where the investigations have an environmental purpose and are conducted in cold regions of the world. It is intended to improve consistency of practice and to encourage rational planning of a site characterization program by providing information to assist in the design of an environmental reconnaissance or investigation plans. This guide is intended to provide information that will help minimize the effect of site characterization operations on areas of frozen ground or permafrost and increase the safety of environmental operations in cold regions.

4.2 This guide presents information and references for site characterization for environmental purposes in cold regions of the world.

5. Special Problems of Cold Regions

5.1 *Safety*—When working in very cold temperatures safety is of utmost importance. Weather is volatile and unpredictable. The difficulty of working under arctic conditions tends to cause frustration and increases the chance of injury. Freezing of exposed flesh and hypothermia can occur very quickly under winter conditions. Specific training in artic survival techniques in accordance with the Department of the Army or comparable training is recommended for anyone expected to work in these conditions.

5.2 *Tundra*—All operations in areas of tundra must be undertaken with special care. What causes a minor impact in a temperate region from a small environmental site characterization study will have a greater impact on tundra or areas underlain by permafrost. Special care and attention during the planning process must be given to field operations to prevent damage to the tundra surface and vegetation. Winter field operations when tundra is protected by snow and ice are less damaging than summer operations but increase difficulties created by very cold temperatures (see 5.3).

5.2.1 Give special attention to all operations using any form of vehicle in tundra areas. Because of the fragile nature of tundra only a single vehicle pass or aircraft landing may be all that is required to cause uncontrolled degradation of the vegetation and underlying permafrost.

5.2.2 Give special attention to any operation using a motorized or heat producing unit (for example, drilling equipment). These items must be insulated in order to protect permafrost or frozen surface layers against heat

transfer, which can result in irreversible degradation of the vegetation and underlying permafrost.

5.3 *Very Cold Temperatures*—Field operations during seasons of very cold temperatures require special planning and concern. Work elements that would require only an hour or so to perform in temperate climates may require several days to perform under the winter temperatures of cold regions. Site investigation planning should take into consideration and allow sufficient time to perform all steps of the investigation. Some procedures, such as tactile methods for visual-manual classification of soils, may not be feasible during cold weather.

5.4 *Permafrost*—The cold winters and short summers of the polar regions produce a layer of frozen ground or permafrost that remains frozen through the summer. Permafrost is a phenomenon of the polar and subpolar regions of the world. About 20 % of the world's land is underlain by permafrost. Permafrost and permafrost hazards uniquely affect most activities in the cold regions, and permafrost and associated hazards must be considered in the planning of all environmental site characterization operations.

5.4.1 Many permafrost areas of the world are not in equilibrium with the existing climate. Any small disturbance of the thermal regime of the permafrost, such as a tire track or drill hole, may result in a drastic change in the underlying permafrost. Therefore, extreme care must be given to prevent damage to the environment when conducting characterization operations in areas underlain by permafrost.

5.4.2 Permafrost acts as a natural barrier in some areas, containing aquifers not usually exposed to surface conditions. Penetration of the permafrost layer into underlying ground water during installation of monitoring wells or collection of deep core samples can increase and exaserbate the fate and transport of environmental contaminants. This can, in turn, change a relatively small, contained site into a much larger area of contamination with greater environmental impact in a region with fragile, highly specialized flora and fauna.

5.5 *Seepage Icings*—Ground-water that seeps or flows at ground surface often results in the formation of disruptive icings. Because many of these seepage sites are located along road cuts the icings may result in loss of use of the roadway. Seepage icings from uncontrolled artesian well flow have been known to cause disruptions. Seasonal frost moves downward more quickly along roadways than it does adjacent undisturbed areas. At times, seasonal frost will move downward to contact the underlying permafrost and form a frost dam within the soil that impedes the flow of groundwater. Hydrostatic pressure will then increase, forcing water to the surface forming an icing. Special attention must be given when undertaking environmental site investigations in cold regions to prevent the occurrence of icings, unless specifically created by design for construction of winter haul roads.

5.6 *Frost Heaving*—In areas of fine-grained sediments, such as silt and clay, frost heaving along with loss of bearing strength is a major problem that must be considered when installing recorder sites for monitoring operations in cold regions. Frost heaving may distort structures, collapse well casings, and cause changes in casing elevations of wells. If not corrected, changes in casing elevation may result in water level measurements that are not correct. During design, siting or construction of structures, frost heaving must be considered and taken into account.

5.7 *Transient Artesian Conditions*—During drilling operations, special attention must be given to possible artesian ground-water conditions below any existing permafrost layers. Drilling operations in cold regions must include plans for dealing with the artesian pressures and blow-out prevention. This may require the use of forward rotary drilling equipment and mud additives to increase the specific weight of the drilling fluid during drilling. Guide D 5783 on direct rotary drilling should be consulted for information on use of drilling fluid additives.

6. Site Investigation Plan

6.1 Review objectives of the investigation prior to final development of a detailed site investigation plan. In cold regions this requires the involvement of individuals or organizations with experience working in such regions. The detailed site investigation plan should clearly identify the types of data that are required to meet the objectives of the investigation. Considerations for identifying data requirements include:

6.1.1 Data required to comply with applicable federal, state, or local regulatory programs.

6.1.2 Data required as inputs to computer models expected to be used.

6.1.3 Data required for selection and design of any implementation measures (that is, protective measures at controlled waste disposal sites, remediation options at contaminated sites).

6.1.4 Data and information on any known geologic or hydrologic hazards at the site.

6.1.5 Data required for risk assessment or to propose alternative cleanup levels.

6.2 A site visit prior to extensive collection of existing data should be made unless the limited scope of a project does not allow multiple visits. The advantage of such a visit is that it may prevent preconceived ideas derived from inaccurate existing information from influencing initial conceptual site model development. A complete environmental site investigation will usually encompass the following activities:

6.2.1 Review available information, both regional and local, on the geologic history (including seismic activity and other potential geologic hazards), rock, soil, ground-water, surface water, and other significant environmental and anthropogenic features (for example, buried utilities) occurring at the proposed location and in the immediate vicinity of the site.

6.2.2 In cold regions, the site investigation plan should include information as to study site selection, routes of access to the site with minimum environmental damage, type and number of tests to be performed at the site, and disposal of waste produced by tests and personnel along with any special requirements needed to reduce the effects of the testing on the surrounding environment. Nonintrusive, nondestructive geophysical testing methods, such as seismic refraction, electromagnetic induction, and ground-penetrating radar may help optimize sampling programs and selection of locations for monitoring well installations.

6.2.3 A site investigation plan in cold regions usually will

require a subsurface temperature monitoring system to help assess natural seasonal changes in ground conditions and document impacts of disturbance on tundra ecosystems.

7. Field Methods

7.1 All field procedures should be documented by identifying time, date, location, meteorological conditions, and personnel involved. Practice D 5254 and Guides D 5408, D 5409, and D 5410 identify minimum and additional data elements for identifying a ground-water site, and can serve as checklists to ensure that important information is not omitted. Samples collected should be assigned a unique descriptor number to specifically identify sample location and for reference to field log data.

7.2 Practice D 4083 presents a procedure for the description of frozen soils based on visual examination and simple manual tests.

7.3 *Equipment Selection*—When several alternatives are available to achieve a given objective, all factors should be considered in selecting equipment or a method. Equipment available from commercial sources generally is preferable to homemade equipment because specifications can be documented readily and replacement parts usually can be obtained quickly and used without affecting the comparability of results secured with the defective equipment. Equipment specifications should be checked for operation under cold temperatures. Equipment with the greatest durability and reliability is preferable. ASTM test methods or standard operating procedures and protocols established by government agencies should be used whenever applicable. The timeliness of results is an important consideration when selecting equipment or methods. Methods that provide real-time data that fulfill data quality requirements are preferable to methods that require use of remote laboratories.

7.3.1 Special circumstances in cold regions may require use of uncommon or nonstandard equipment, methods, or procedures. The rationale for use of such equipment and procedures should be given in the detailed site investigation plan.

7.4 *Drilling Operations:*

7.4.1 *Drilling Operations Using Water-Based Drilling Fluids*—It is not practical in many areas of frozen ground or permafrost to use drilling methods that use water-based drilling fluids because special heat maintenance operations have to be formed to keep the drilling fluid from freezing in the pipes of the drilling rig and other areas where the drilling fluids are stored for use, such as mud pits.

7.4.2 *Direct Push and Vibratory Technology*—Truck or tractor mounted hydraulic-push or vibratory equipment may be useful for ground-water sampling in cold regions. This equipment has an overall depth capability exceeding 100 ft (30 m) under suitable geologic conditions of unconsolidated sediments. The frozen ground of the cold regions, however, may limit the use of the equipment. The technology should be useful in discontinuous permafrost areas and in areas where there is a layer of unfrozen material above the permafrost. The hydraulic hammer included with the push equipment may be used to penetrate relative thin frozen units encountered during drilling. With the use of direct push equipment and a mobile laboratory, ground-water samples can be collected and analyzed in relatively large numbers within a short time (see Guide D 6001). Direct push technology does not generate drill cuttings, and therefore, eliminates the need for disposal of any cuttings. The use of rotation with flighted augers is possible with many direct push systems currently available. Augers may provide an access hole through permafrost allowing direct push tools to be used in the penetrable soils underneath.

7.4.3 *Reverse Rotary with Multiple Well Casings*—Reverse-circulation drilling with dual-wall drill stem is a drilling method where the flow of drilling fluid (air or water) down the borehole occurs between the outer and inner casing of a dual-wall drill stem rather than between an uncased borehole and the drill stem (See Guide D 5781). Otherwise, multiple well casing drilling is similar to reverse circulation rotary

drilling. Usually the drill bit is very close to the inside diameter of the outer casing. Consequently, the outer casing partially or completely supports the borehole wall. A major advantage of using dual-wall reverse-circulation drilling over other rotary drilling methods in cold regions is that drilling is readily accomplished in frozen ground conditions with minimum drilling fluid contact with the frozen material thereby reducing the amount of melting. Also, the presence of a casing minimizes the problem of loss of circulation in porous formations or hole collapse in unconsolidated sediments.

8. Keywords

8.1 antarctic; arctic; cold regions; drilling; environmental site characterization; ground-water

---

APPENDIX

(Nonmandatory Information)

X1. SELECTED COLD REGION REFERENCES

X1.1 The references have been provided for additional information.

X1.1.1 Aldrich, H. P., 1956, "Frost Penetration Below Highway and Airfield Pavements," U.S. Highway Research Board, Bull. 135, pp. 124–149.

X1.1.2 Andersland, O. B., and Anderson, D. M. (ed.), 1978, "Geotechnical Engineering for Cold Regions," McGraw-Hill, New York, NY, 576 p.

X1.1.3 Andersland, O. B., and Ladanyi, B., 1995, "An Introduction to Frozen Ground Engineering," Chapman and Hall, New York, 352 p.

X1.1.4 Brennan, A. M., 1993, Permafrost Bibliography Update 1988–1992: *Glaciological Data Report GD-26*, Boulder, CO, World Data Center, A for Glaciology, p. 401.

X1.1.5 Brewer, M. C., 1958, "Some Results of Geothermal Investigations of Permafrost in Northern Alaska," *Trans. Amer. Geophys.* Union 39(1): 19–26.

X1.1.6 Brown, R. J. E., and Pewe, T., 1973, "Distribution of Permafrost In North America and Its Relationship To the Environment, 1968 to 1978—a review," in Permafrost, North American Contribution to the Second International Conference, Natl. Acad. Sci., Washington DC, p. 71–100.

X1.1.7 Brown, J., 1973, "Environmental Considerations for the Utilization of Permafrost Terrain," in Permafrost, North American Contribution to the Second International Conference, Natl. Acad. Sci., Washington DC, p. 587–589.

X1.1.8 Brown, R. J. E., and Kupsch, W. O., 1974, Permafrost Terminology, Canadian National Research Council, *Asso Comm. Geotechnical Research, Tech Memo* 111, 62 p.

X1.1.9 Corte, A. E., 1969, Geocryology and Engineering: *Geol. Soc. Amer. Rev. Eng. Geol.* 2:119–185.

X1.1.10 Downey, J. S., and Sinton, P. O., 1990, "Geohydrology and Ground-Water Geochemistry at a Sub-Arctic Landfill, Fairbanks, Alaska," *U.S. Geol. Survey, Water Resources Investigations Report* 90-4022, 25 p.

X1.1.11 Embleton, C., and King, C. A. M., 1968, "Glacial and Periglacial Geomorphology," Edward Arnold Publishers, London, 608 p.

X1.1.12 Ferrians, O. J., Kachadoorian, R., and Greene, G. W., 1969, "Permafrost and Related Engineering Problems in Alaska," *U.S. Geol. Survey Prof. Paper* 678, 37 p.

X1.1.13 Glen, J. W., 1974, "The Physics of Ice," *U.S. Army C.R.R.E.L.*, Hanover, NH, Monograph II-C2a, 86 p.

X1.1.14 Hamelin, L. E., and Cook, F. A., 1967, "Illustrated Glossary of Periglacial Phenomenon: Montreal," Laval Press, 237 p.

X1.1.15 Hennion, F., 1995, "Frost and Permafrost Definitions," *Highway Research Board, Bull 111*, Washington, DC.

X1.1.16 Hopkins, O. L., Karlstrom, T. D., and others, 1955, "Permafrost and Ground Water in Alaska," *U.S. Geological Survey Prof.*, Paper 264-F.

X1.1.17 Gates, W. C. B., 1989, "Protection of Ground-Water Monitoring Wells Against Frost Heave," *Bulletin AEG*, XXVI(2):241–251.

X1.1.18 Johnson, G. H. (ed.), 1981, Permafrost, Assoc. Committee Geotechnical Research, Natl. Res. Council of Canada, John Wiley & Sons, Toronto, Ont., 540 p.

X1.1.19 Koster, E. A., and Judge, A. S., 1994, "Permafrost and Climate Change: An Annotated Bibliography," in *Glaciological Data Report GD-27*, World Data Center for Glaciology (snow and ice) Univ. Colorado, Boulder, June 1994, 94 p.

X1.1.20 Lange, G. R., 1973, "An Investigation of Core Drilling in Perennially Frozen Gravels and Rock," *U.S. Army, CRREL Tech. Rpt.* 245, 31 p.

X1.1.21 Lange, G. R., 1968, "Rotary Drilling and Coring in Permafrost, Part I, Preliminary Investigation, Fort Churchill, Manitoba," *U.S. Army, CRREL Tech. Rpt.* 95, 22 p.

X1.1.22 Lange, G. R., and Smith, T. K., 1972, "Rotary Drilling and Coring in Permafrost, Part III," *U.S. Army Corp Engineers, CRREL Tech. RPT*, 95–111.

X1.1.23 Linell, K. A., 1973, "Risk of Uncontrolled Flow from Wells Through Permafrost," in North American Contribution to the Second International Conference, Natl. Acad. Sci., Washington DC, p. 462–468.

X1.1.24 MacFarlane, I. C. (ed.), 1969, *Muskeg Engineering Handbook*, Univ. of Toronto Press, Toronto, Ont., 320 p.

X1.1.25 Moore, J. P., and Ping, C. L., 1989, "Classification of Permafrost Soils: Soil Survey Horizons," Winter, p. 98–104.

X1.1.26 National Research Council Canada, 1988, "Glossary of Permafrost and Related Ground-Ice Terms," *Technical Memorandum No. 142*, 156 p.

X1.1.27 National Academy of Sciences, 1973, "Permafrost, North American Contribution to the Second International Conference" Natl. Academy of Sci., Washington DC, 782 p.

X1.1.28 Nelson, G. L., 1982, "Vertical Movement of Ground Water Under the Merrill Field Landfill, Anchorage, Alaska," *U.S. Geol. Survey Open-File Report 82-1016*, 25 p.

X1.1.29 Pewe, T. L., 1982, "Geologic Hazards of the Fairbanks Area, Alaska," *Div. Geol. and Geophys. Surveys, Spec. report 15*, 109 p.

X1.1.30 Ray, L. L., 1956, "Perennially Frozen Ground, an Environmental Factor in Alaska: 17th Inter. Geog. Cong. and 8th Gen Assembly," Washington DC, 1952, Proc., p. 260–264.

X1.1.31 Samson-Liebig, S. E., Kimble, J. M., and Ping, C. L., 1995, "Improvements in the Definition of Cryic and Pergelic Soil Temperature Regimes in Soil Taxonomy Using Daylength/Solar Radiation," Soil Survey Horizons, Spring, pp. 20–25.

X1.1.32 Scalf, M. R., Dunlap, W. J., and Kreissl, J. F., 1977, "Environmental Effects of Septic Tank Systems: U.S. EPA-600/3-77-096, 35 p.

X1.1.33 Soil Survey Staff, 1994, "Keys to Soil Taxonomy," 6th ed., U.S. Dept. of Agriculture, Soil Conservation Service, U.S. Govern. Printing Office, No. 001-000-04612-8.

X1.1.34 Stanek, W., 1977, "A List of Terms and Definitions, in Muskeg and the Northern Environment in Canada," Univ. of Toronto Press, Toronto, Ont., pp. 367–382.

X1.1.35 Straughn, R. O., 1972, "The Sanitary Landfill in the Subarctic," *Arctic*, v. 25, no. 1, p. 40–48.

X1.1.36 U.S. Army, 1966, "Arctic and Subarctic Construction: Calculation Methods for Determination of Depth of Free-Thaw in Soils," *TM 5-852-6*.

X1.1.37 U.S. Army, 1985, "Pavement Design for Seasonal Frost Conditions," *TM 5-818-2*.

X1.1.38 U.S. Army Corps of Engineers, 1960, "Core Drilling in Frozen Ground," *U.S. Army Engineers Waterways Experiment Station Tech. Rpt. No. 3534*.

X1.1.39 U.S. Army Corps of Engineers, 1966, "Description and Classification of Frozen Soils," *CRREL Technical Report 150*, Cold Regions Research Engineering Laboratory, Hanover, NH.

X1.1.40 U.S. Army Corps of Engineers, 1968, "Digital Solution of Modified Berggren Equation to Calculation Depth of Freeze or Thaw in Multi-Layered Systems," *CRREL Special Report 122*, Cold Regions Research Engineering Laboratory, Hanover, NH.

X1.1.41 Weller, G., and Holmgren, B., 1974, "The Microclimate of the Arctic Tundra," *Jour Applied Meteorology*, 13, pp. 854–862.

Standard Guide to
Site Characterization for Engineering, Design, and Construction Purposes[1]

This standard is issued under the fixed designation D 420; the number immediately following the designation indicates the year of original adoption or, in the case of revision, the year of last revision. A number in parentheses indicates the year of last reapproval. A superscript epsilon (ϵ) indicates an editorial change since the last revision or reapproval.

INTRODUCTION

Investigation and identification of subsurface materials involves both simple and complex techniques that may be accomplished by many different procedures and may be variously interpreted. These studies are frequently site specific and are influenced by geological and geographical settings, by the purpose of the investigation, by design requirements for the project proposed, and by the background, training, and experience of the investigator. This guide has been extensively rewritten and enlarged since the version approved in 1987. Material has been added for clarification and for expansion of concepts. Many new ASTM standards are referenced and a bibliography of non-ASTM references is appended.

This document is a guide to the selection of the various ASTM standards that are available for the investigation of soil, rock, and ground water for projects that involve surface or subsurface construction, or both. It is intended to improve consistency of practice and to encourage rational planning of a site characterization program. Since the subsurface conditions at a particular site are usually the result of a combination of natural, geologic, topographic, and climatic factors, and of historical modifications both natural and manmade, an adequate and internally consistent exploration program will allow evaluation of the results of these influences.

1. Scope

1.1 This guide refers to ASTM methods by which soil, rock, and ground water conditions may be determined. The objective of the investigation should be to identify and locate, both horizontally and vertically, significant soil and rock types and ground water conditions present within a given site area and to establish the characteristics of the subsurface materials by sampling or in situ testing, or both.

1.2 Laboratory testing of soil, rock, and ground water samples is specified by other ASTM standards not listed herein. Subsurface exploration for environmental purposes will be the subject of a separate ASTM document.

1.3 Prior to commencement of any intrusive exploration the site should be checked for underground utilities. Should evidence of potentially hazardous or otherwise contaminated materials or conditions be encountered in the course of the investigation, work should be interrupted until the circumstances have been evaluated and revised instructions issued before resumption.

1.4 The values stated in (SI) inch-pound units are to be regarded as the standard.

1.5 *This standard does not purport to address all of the safety problems, if any, associated with its use. It is the responsibility of the user of this standard to establish appropriate safety and health practices and determine the applica-* *bility of regulatory limitations prior to use.*

2. Referenced Documents

2.1 *ASTM Standards:*
C 119 Terminology Relating to Dimension Stone[2]
C 294 Descriptive Nomenclature for Constituents of Natural Mineral Aggregates[3]
C 851 Practice for Estimating Scratch Hardness of Coarse Aggregate Particles[3]
D 75 Practice for Sampling Aggregates[4]
D 653 Terminology Relating to Soil, Rock, and Contained Fluids[2]
D 1194 Test Method for Bearing Capacity of Soil for Static Load and Spread Footings[2]
D 1195 Test Method for Repetitive Static Plate Load Tests of Soils and Flexible Pavement Components, for Use in Evaluation and Design of Airport and Highway Pavements[2]
D 1196 Test Method for Nonrepetitive Static Plate Load Tests of Soils and Flexible Pavement Components, for Use in Evaluation and Design of Airport and Highway Pavements[2]
D 1452 Practice for Soil Investigation and Sampling by Auger Borings[2]
D 1586 Test Method for Penetration Test and Split-Barrel Sampling of Soils[2]

[1] This guide is under the jurisdiction of ASTM Committee D-18 on Soil and Rock and is the direct responsibility of Subcommittee D18.01 on Surface and Subsurface Characterization.

Current edition approved June 15, 1993. Published August 1993. Originally published as D 425 – 65 T. Last previous edition D 420 – 87.

[2] *Annual Book of ASTM Standards*, Vol 04.08.
[3] *Annual Book of ASTM Standards*, Vol 04.02.
[4] *Annual Book of ASTM Standards*, Vol 04.03.

D 1587 Practice for Thin-Walled Tube Sampling of Soils[2]

D 2113 Practice for Diamond Core Drilling for Site Investigation[2]

D 2487 Classification of Soils for Engineering Purposes (Unified Soil Classification System)[2]

D 2488 Practice for Description and Identification of Soils (Visual-Manual Procedure)[2]

D 2573 Test Method for Field Vane Shear Test in Cohesive Soil[2]

D 2607 Classification for Peats, Mosses, Humus, and Related Products[2]

D 3017 Test Method for Water Content of Soil and Rock in Place by Nuclear Methods (Shallow Depth)[2]

D 3213 Practices for Handling, Storing, and Preparing Soft Undisturbed Marine Soil[2]

D 3282 Classification of Soils and Soil-Aggregate Mixtures for Highway Construction Purposes[2]

D 3385 Test Method for Infiltration Rate of Soils in Field Using Double-Ring Infiltrometers[2]

D 3404 Guide to Measuring Matric Potential in the Vadose Zone Using Tensiometers[2]

D 3441 Test Method for Deep, Quasi-Static, Cone and Friction-Cone Penetration Tests of Soil[2]

D 3550 Practice for Ring-lined Barrel Sampling of Soils[2]

D 3584 Practice for Indexing Papers and Reports on Soil and Rock for Engineering Purposes[2]

D 4083 Practice for Description of Frozen Soils (Visual-Manual Procedure)[2]

D 4220 Practices for Preserving and Transporting Soil Samples[2]

D 4394 Test Method for Determining the In Situ Modulus of Deformation of Rock Mass Using the Rigid Plate Loading Method[2]

D 4395 Test Method for Determining the In Situ Modulus of Deformation of Rock Mass Using the Flexible Plate Loading Method[2]

D 4403 Practice for Extensometers Used in Rock[2]

D 4428 Test Methods for Crosshole Seismic Testing[2]

D 4429 Test Method for Bearing Ratio of Soils in Place[2]

D 4452 Methods for X-Ray Radiography of Soil Samples[2]

D 4506 Test Method for Determining the In Situ Modulus of Deformation of Rock Mass Using a Radial Jacking Test[2]

D 4544 Practice for Estimating Peat Deposit Thickness[2]

D 4553 Test Method for Determining the In Situ Creep Characteristics of Rock[2]

D 4554 Test Method for In Situ Determination of Direct Shear Strength of Rock Discontinuities[2]

D 4555 Test Method for Determining Deformability and Strength of Weak Rock by an In Situ Uniaxial Compressive Test[2]

D 4622 Test Method for Rock Mass Monitoring Using Inclinometers[2]

D 4623 Test Method for Determination of In Situ Stress in Rock Mass by Overcoring Method—USBM Borehole Deformation Gage[2]

D 4630 Test Method for Determining Transmissivity and Storativity of Low Permeability Rocks by In Situ Measurements Using the Constant Head Injection Test[2]

D 4631 Test Method for Determining Transmissivity and Storativity of Low Permeability Rocks by In Situ Measurements Using the Pressure Pulse Technique[2]

D 4633 Test Method for Stress Wave Energy Measurement for Dynamic Penetrometer Testing Systems[2]

D 4645 Test Method for Determination of the In Situ Stress in Rock Using the Hydraulic Fracturing Method[2]

D 4700 Guide for Soil Sampling from the Vadose Zone[2]

D 4719 Test Method for Pressuremeter Testing in Soils[2]

D 4729 Test Method for In Situ Stress and Modulus of Deformation Using the Flatjack Method[2]

D 4750 Test Method for Determining Subsurface Liquid Levels in a Borehole or Monitoring Well (Observation Well)[2]

D 4879 Guide for Geotechnical Mapping of Large Underground Openings in Rock[2]

D 4971 Test Method for Determining the In Situ Modulus of Deformation of Rock Using the Diametrically Loaded 76-mm (3-in.) Borehole Jack[3]

D 5079 Practices for Preserving and Transporting Rock Core Samples[3]

D 5088 Practice for Decontamination of Field Equipment Used at Nonradioactive Waste Sites[3]

D 5092 Practice for Design and Installation of Ground Water Monitoring Wells in Aquifers[3]

D 5093 Test Method for Field Measurement of Infiltration Rate Using a Double-Ring Infiltrometer with a Sealed Inner Ring[3]

D 5126 Guide for Comparison of Field Methods for Determining Hydraulic Conductivity in the Vadose Zone[3]

D 5195 Test Method for Density of Soil and Rock In-Place at Depths Below the Surface by Nuclear Methods[3]

E 177 Practice for the Use of the Terms Precision and Bias in ASTM Test Methods[5]

E 380 Practice for the Use of the International System of Units (SI) (the Modernized Metric System)[5]

G 51 Test Method for pH of Soil for Use in Corrosion Testing[6]

G 57 Method for Field Measurement of Soil Resistivity Using the Wenner Four-Electrode Method[6]

3. Significance and Use

3.1 An adequate soil, rock, and ground water investigation will provide pertinent information for decision making on one or more of the following subjects:

3.1.1 Optimum location of the structure, both vertically and horizontally, within the area of the proposed construction.

3.1.2 Location and preliminary evaluation of suitable borrow and other local sources of construction aggregates.

3.1.3 Need for special excavating and dewatering techniques with the corresponding need for information, even if only approximate, on the distribution of soil water content or pore pressure, or both, and on the piezometric heads and apparent permeability (hydraulic conductivity) of the various subsurface strata.

---

[5] *Annual Book of ASTM Standards*, Vol 14.02.
[6] *Annual Book of ASTM Standards*, Vol 03.02.
[7] The boldface numbers in parentheses refer to the list of references at the end of this standard.

3.1.4 Investigation of slope stability in natural slopes, cuts, and embankments.

3.1.5 Conceptual selection of embankment types and hydraulic barrier requirements.

3.1.6 Conceptual selection of alternate foundation types and elevations of the corresponding suitable bearing strata.

3.1.7 Development of additional detailed subsurface investigations for specific structures or facilities.

3.2 The investigation may require the collection of sufficiently large soil and rock samples of such quality as to allow adequate testing to determine the soil or rock classification or mineralogic type, or both, and the engineering properties pertinent to the proposed design.

3.3 This guide is not meant to be an inflexible description of investigation requirements; methods defined by other ASTM standards or non-ASTM techniques may be appropriate in some circumstances. The intent is to provide a checklist to assist in the design of an exploration/investigation plan.

4. Reconnaissance of Project Area

4.1 Available technical data from the literature or from personal communication should be reviewed before any field program is started. These include, but are not limited to, topographic maps, aerial photography, satellite imagery, geologic maps, statewide or county soil surveys and mineral resource surveys, and engineering soil maps covering the proposed project area. Reports of subsurface investigations of nearby or adjacent projects should be studied.

NOTE 1—While certain of the older maps and reports may be obsolete and of limited value in the light of current knowledge, a comparison of the old with the new will often reveal valuable information.

4.1.1 The United States Geological Survey and the geological surveys of the various states are the principal sources of geologic maps and reports on mineral resources and ground water.

4.1.2 United States Department of Agriculture Soil Conservation Service soil surveys, where available and of recent date, should enable the investigator to estimate the range in soil profile characteristics to depths of 5 or 6 ft (1.5 or 2 m) for each soil mapped.

NOTE 2—Each soil type has a distinctive soil profile due to age, parent material, relief, climatic condition, and biological activity. Consideration of these factors can assist in identifying the various soil types, each requiring special engineering considerations and treatment. Similar engineering soil properties are often found where similar soil profiles characteristics exist. Changes in soil properties in adjacent areas often indicate changes in parent material or relief.

4.2 In areas where descriptive data are limited by insufficient geologic or soil maps, the soil and rock in open cuts in the vicinity of the proposed project should be studied and various soil and rock profiles noted. Field notes of such studies should include data outlined in 11.6.

4.3 Where a preliminary map covering the area of the project is desired, it can be prepared on maps compiled from aerial photography that show the ground conditions. The distribution of the predominant soil and rock deposits likely to be encountered during the investigation may be shown using data obtained from geologic maps, landform analysis and limited ground reconnaissance. Experienced photo-interpreters can deduce much subsurface data from a study of black and white, color, and infrared photographs because similar soil or rock conditions, or both, usually have similar patterns of appearance in regions of similar climate or vegetation.

NOTE 3—This preliminary map may be expanded into a detailed engineering map by locating all test holes, pits, and sampling stations and by revising boundaries as determined from the detailed subsurface survey.

4.4 In areas where documentary information is insufficient, some knowledge of subsurface conditions may be obtained from land owners, local well drillers, and representatives of the local construction industry.

5. Exploration Plan

5.1 Available project design and performance requirements must be reviewed prior to final development of the exploration plan. Preliminary exploration should be planned to indicate the areas of conditions needing further investigation. A complete soil, rock, and ground water investigation should encompass the following activities:

5.1.1 Review of available information, both regional and local, on the geologic history, rock, soil, and ground water conditions occurring at the proposed location and in the immediate vicinity of the site.

5.1.2 Interpretation of aerial photography and other remote sensing data.

5.1.3 Field reconnaissance for identification of surficial geologic conditions, mapping of stratigraphic exposures and outcrops, and examination of the performance of existing structures.

5.1.4 On site investigation of the surface and subsurface materials by geophysical surveys, borings, or test pits.

5.1.5 Recovery of representative disturbed samples for laboratory classification tests of soil, rock, and local construction material. These should be supplemented by undisturbed specimens suitable for the determination of those engineering properties pertinent to the investigation.

5.1.6 Identification of the position of the ground water table, or water tables, if there is perched ground water, or of the piezometric surfaces if there is artesian ground water. The variability of these positions in both short and long time frames should be considered. Color mottling of the soil strata may be indicative of long-term seasonal high ground water positions.

5.1.7 Identification and assessment of the location of suitable foundation material, either bedrock or satisfactory load-bearing soils.

5.1.8 Field identification of soil sediments, and rock, with particular reference to type and degree of decomposition (for example, saprolite, karst, decomposing or slaking shales), the depths of their occurrence and the types and locations of their structural discontinuities.

5.1.9 Evaluation of the performance of existing installations, relative to their structure foundation material and environment in the immediate vicinity of the proposed site.

6. Equipment and Procedures for Use in Exploration

6.1 *Pertinent ASTM Standards*—Practices D 1452, D 2113, D 4544, D 5088, D 5092; Method D 1586; and Test Methods D 4622, D 4633, D 4750.

6.2 The type of equipment required for a subsurface investigation depends upon various factors, including the type of subsurface material, depth of exploration, the nature of the terrain, and the intended use of the data.

6.2.1 *Hand Augers, Hole Diggers, Shovels, and Push Tube Samplers* are suitable for exploration of surficial soils to depths of 3 to 15 ft (1 to 5 m).

6.2.2 *Earth Excavation Equipment*, such as backhoes, draglines, and drilled pier augers (screw or bucket) can allow in situ examination of soil deposits and sampling of materials containing very large particles. The investigator should be aware of the possiblity of permanent disturbance of potential bearing strata by unbalanced pore pressure in test excavations.

6.2.3 Soil and rock boring and drilling machines and proofing devices may be used to depths of 200 to 300 ft in soil and to a much greater depth in rock.

6.2.4 Well drilling equipment may be suitable for deep geologic exploration. Normally samples are in the form of sand-sized cuttings captured from the return flow, but coring devices are available.

7. Geophysical Exploration

7.1 *Pertinent ASTM Standards*—Test Methods D 4428, and Method G 57.

7.2 Remote sensing techniques may assist in mapping the geological formations and for evaluating variations in soil and rock properties. Satellite and aircraft spectral mapping tools, such as LANDSAT, may be used to find and map the areal extent of subsurface materials and geologic structure. Interpretation of aircraft photographs and satellite imagery can locate and identify significant geologic features that may be indicative of faults and fractures. Some ground control is generally required to verify information derived from remote sensing data.

7.3 Geophysical survey methods may be used to supplement borehole and outcrop data and to interpolate between holes. Seismic, ground penetrating radar, and electrical resistivity methods can be particularly valuable when distinct differences in the properties of contiguous subsurface materials are indicated.

7.4 Shallow seismic refraction/reflection and ground penetrating radar techniques can be used to map soil horizons and depth profiles, water tables, and depth to bedrock in many situations, but depth penetration and resolution vary with local conditions. Electromagnetic induction, electrical resistivity, and induced polarization (or complex resistivity) techniques may be used to map variations in water content, clay horizons, stratification, and depth to aquifer/bedrock. Other geophysical techniques such as gravity, magnetic, and shallow ground temperature methods may be useful under certain specific conditions. Deep seismic and electrical methods are routinely used for mapping stratigraphy and structure of rock in conjunction with logs. Crosshole shear wave velocity measurements can provide soil and rock parameters for dynamic analyses.

7.4.1 The seismic refraction method may be especially useful in determining depth to, or rippability of, rock in locations where successively denser strata are encountered.

7.4.2 The seismic reflection method may be useful in delineating geological units at depths below 10 ft (3 m). It is not constrained by layers of low seismic velocity and is especially useful in areas of rapid stratigraphic change.

7.4.3 The electrical resistivity method, Method G 57, may be similarly useful in determining depth to rock and anomalies in the stratigraphic profile, in evaluating stratified formations where a denser stratum overlies a less dense stratum, and in location of prospective sand-gravel or other sources of borrow material. Resistivity parameters also are required for the design of grounding systems and cathodic protection for buried structures.

7.4.4 The ground penetrating radar method may be useful in defining soil and rock layers and manmade structures in the depth range of 1 to 30 ft ($\frac{1}{3}$ to 10 m).

NOTE 4—Surface geophysical investigations can be a useful guide in determining boring or test hole locations. If at all possible, the interpretation of geophysical studies should be verified by borings or test excavations.

8. Sampling

8.1 *Pertinent ASTM Standards*—Practices D 75, D 1452, D 1587, D 2113, D 3213, D 3550, D 4220, D 5079; Test Method D 1586; Methods D 4452; and Guide D 4700.

8.2 Obtain samples that adequately represent each subsurface material that is significant to the project design and construction. The size and type of sample required is dependent upon the tests to be performed, the relative amount of coarse particles present, and the limitations of the test equipment to be used.

NOTE 5—The size of disturbed or bulk samples for routine tests may vary at the discretion of the geotechnical investigator, but the following quantities are suggested as suitable for most materials: (*a*) Visual classification—50 to 500 g (2 oz to 1 lb); (*b*) Soil constants and particle size analysis of non-gravelly soil—500 g to 2.5 kg (1 to 5 lb); (*c*) Soil compaction tests and sieve analysis of gravelly soils—20 to 40 kg (40 to 80 lb); (*d*) Aggregate manufacture or aggregate properties tests—50 to 200 kg (100 to 400 lb).

8.3 Accurately identify each sample with the boring, test hole, or testpit number and depth below reference ground surface from which it was taken. Place a waterproof identification tag inside the container, securely close the container, protect it to withstand rough handling, and mark it with proper identification on the outside. Keep samples for natural water content determination in sealed containers to prevent moisture loss. When drying of samples may affect classification or engineering properties test results, protect them to minimize moisture loss. Practices D 4220 and D 5079 address the transportation of samples from field to laboratory. Most of the titles of the referenced standards are self-explanatory, but some need elaboration for the benefit of the users of this guide.

8.3.1 Practice D 75 describes the sampling of coarse and fine aggregates for the preliminary investigation of a potential source of supply.

8.3.2 Practice D 1452 describes the use of augers in soil investigations and sampling where disturbed soil samples can be used. Depths of auger investigations are limited by ground water conditions, soil characteristics, and equipment used.

8.3.3 Test Method D 1586 describes a procedure to obtain representative soil samples for identification and classification laboratory tests.

8.3.4 Practice D 1587 describes a procedure to recover

relatively undisturbed soil samples suitable for laboratory testing.

8.3.5 Practice D 2113 describes a procedure to recover intact samples of rock and certain soils too hard to sample by Test Method D 1586 or Practice D 1587.

8.3.6 Practice D 3550 describes a procedure for the recovery of moderately disturbed, representative samples of soil for classification testing and, in some cases, shear or consolidation testing.

9. Classification of Earth Materials

9.1 *Pertinent ASTM Standards*—Terminology C 119; Descriptive Nomenclature C 294; Classifications D 2487, D 2607, D 3282; Practices D 2488, D 4083.

9.2 Additional description of samples of soil and rock may be added after submission to the laboratory for identification and classification tests in accordance with one or more ASTM laboratory standards or other applicable references, or both. Section 10.6.3 discusses the use, for identification and for classification purposes, of some of the standards listed in 9.1.

10. Determination of Subsurface Conditions

10.1 Subsurface conditions are positively defined only at the individual test pit, hole, boring, or open cut examined. Conditions between observation points may be significantly different from those encountered in the exploration. A stratigraphic profile can be developed by detailed investigations only where determinations of a continuous relationship of the depths and locations of various types of soil and rock can be inferred. This phase of the investigation may be implemented by plotting logs of soil and rock exposures in walls of excavations or cut areas and by plotting logs of the test borings. Then one may interpolate between, and extrapolate a reasonable distance beyond, these logs. The spacing of these investigations should depend on the geologic complexity of the project area and on the importance of soil and rock continuity to the project design. Exploration should be deep enough to identify all strata that might be significantly affected by the proposed use of the site and to develop the engineering data required to allow analysis of the items listed in Section 4 for each project.

NOTE 6—Plans for a program of intrusive subsurface investigation should consider possible requirements for permits for installation and proper closure of bore holes and wells at the completion of the investigation.

10.2 The depth of exploratory borings or test pits for roadbeds, airport paving, or vehicle parking areas should be to at least 5 ft (1.5 m) below the proposed subgrade elevation. Special circumstances may increase this depth. Borings for structures, excavations, or embankments should extend below the level of significant stress or ground water influence from the proposed load as determined by subsurface stress analysis.

10.3 When project construction or performance of the facility may be affected by either previous water-bearing materials or impervious materials that can block internal drainage, borings should extend sufficiently to determine those engineering and hydrogeologic properties that are relevant to the project design.

10.4 In all borrow areas the borings or test pits should be

sufficient in number and depth to outline the required quantities of material meeting the specified quality requirements.

10.5 Where frost penetration or seasonal desiccation may be significant in the behavior of soil and rock, borings should extend well below the depth from finished grade of the anticipated active zone.

10.6 Exploration records shall be kept in a systematic manner for each project. Such records shall include:

10.6.1 Description of each site or area investigated. Each test hole, boring, test pit, or geophysical test site shall be clearly located (horizontally and vertically) with reference to some established coordinate system, datum, or permanent monument.

10.6.2 Logs of each test hole, boring, test pit, or cut surface exposure shall show clearly the field description and location of each material and any water encountered, either by symbol or word description. Reference to a Munsell color chart designation is a substantial aid to an accurate description of soil and rock materials.

NOTE 7—Color photographs of rock cores, soil samples, and exposed strata may be of considerable value. Each photograph should include an identifying number or symbol, a date, and reference scale.

10.6.3 Identification of all soils based on Classification D 2487, Practice D 2488, Classification D 2607, or Practice D 4083. Identification of rock materials based on Terminology C 119, Descriptive Nomenclature C 294, or Practice C 851. Classification of soil and rock is discussed in Section 9.

10.6.4 Location and description of seepage and water-bearing zones and records of piezometric elevations found in each hole, boring, piezometer, or test pit.

10.6.5 The results and precise locations of in situ test results such as the penetration resistance or vane shear discussed in 8.3, plate load tests, or other in situ test-engineering properties of soils or rock.

10.6.6 Percentage of core recovery and rock quality designation in core drilling as outlined in 8.3.5.

10.6.7 Graphical presentation of field and laboratory and its interpretation facilitates comprehensive understanding subsurface conditions.

11. In Situ Testing

11.1 *Pertinent ASTM Standards*—Test Methods D 1194, D 1195, D 1196, D 1586, D 2573, D 3017, D 3441, D 3885, D 4394, D 4395, D 4429, D 4506, D 4553, D 4554, D 4555, D 4623, D 4630, D 4631, D 4645, D 4719, D 4729, D 4971, D 5093, D 5195, G 51; Guides D 3404, D 5126; and Practice D 4403.

11.2 In situ testing is useful for: (*a*) measurement of soil parameters in their undisturbed condition with all of the restraining or loading effects, or both, of the surrounding soil or rock mass active, and (*b*) for rapid or closely spaced measurements, or both, of earth properties without the necessity of sampling. Most of the titles of the various referenced standards are self-explanatory, but some need elaboration for the users of this guide.

11.2.1 Test Method D 1586 describes a penetration test that has been correlated by many authors with various strength properties of soils.

11.2.2 Test Method D 2573 describes a procedure to

measure the in situ unit shear resistance of cohesive soils by rotation of a four-bladed vane in a horizontal plane.

11.2.3 Test Method D 3441 describes the determination of the end bearing and side friction components of the resistance to penetration of a conical penetrometer into a soil mass.

11.2.4 Practice D 4403 describes the application of various types of extensometers used in the field of rock mechanics.

11.2.5 Test Method D 4429 describes the field determination of the California Bearing Ratio for soil surfaces in situ to be used in the design of pavement systems.

11.2.6 Test Method D 4719 describes an in situ stress-strain test performed on the walls of a bore hole in soil.

NOTE 8—Other standards for in situ test procedures and automated data collection are being prepared by ASTM Committee D-18 for publication at a later date.

12. Interpretation of Results

12.1 Interpret the results of an investigation in terms of actual findings and make every effort to collect and include all field and laboratory data from previous investigations in the same area. Extrapolation of data into local areas not surveyed and tested should be made only for conceptual studies. Such extrapolation can be done only where geologically uniform stratigraphic and structural relationships are known to exist on the basis of other data. Cross sections may be developed as part of the site characterization if required to demonstrate the site conditions.

12.1.1 Cross sections included with the presentation of basic data from the investigation should be limited to the ground surface profile and the factual subsurface data obtained at specific exploration locations. Stratigraphic units between the locations of intrusive explorations should only be indicated if supported by continuous geophysical profiles.

12.1.2 Cross sections showing interpretations of stratigraphic units and other conditions between intrusive explorations but without support of continuous geophysical profiles should be presented in an interpretative report appendix or in a separate interpretive report. The interpretive cross sections must be accompanied by notes describing anomalies or otherwise significant variations in the site conditions that should be anticipated for the intended design or construction activities.

NOTE 9—Additional exploration should be considered if there is not sufficient knowledge to develop interpretative cross sections, with realistic descriptions of anticipated variations in subsurface conditions, to meet project requirements.

12.2 Subject to the restrictions imposed by state licensing law, recommendations for design parameters can be made only by professional engineers and geologists specializing in the field of geotechnical engineering and familiar with purpose, conditions, and requirements of the study. Soil mechanics, rock mechanics, and geomorphological concepts must be combined with a knowledge of geotechnical engineering or hydrogeology to make a complete application of the soil, rock, and ground water investigation. Complete design recommendations may require a more detailed study than that discussed in this guide.

12.3 Delineate subsurface profiles only from actual geophysical, test-hole, test-pit, or cut-surface data. Interpolation between locations should be made on the basis of available geologic knowledge of the area and should be clearly identified. The use of geophysical techniques as discussed in 7.2 is a valuable aid in such interpolation. Geophysical survey data should be identified separately from sample data or in situ test data.

13. Report

13.1 *Pertinent ASTM Standards*—Terminology D 653; Practices D 3584, E 177, E 380; and Guide D 4879.

13.2 The report of a subsurface investigation shall include:

13.2.1 The location of the area investigated in terms pertinent to the project. This may include sketch maps or aerial photos on which the test pits, bore holes, and sample areas are located, as well as geomorphological data relevant to the determination of the various soil and rock types. Such data includes elevation contours, streambeds, sink holes, cliffs, and the like. Where feasible, include in the report a geologic map or an agronomic soils map, or both, of the area investigated.

13.2.2 A description of the investigation procedures, including all borings and testhole logs, graphic presentation of all compaction, consolidation, or load test data tabulation of all laboratory test results, and graphical interpretations of geophysical measurements.

13.2.3 A summary of the findings obtained under Sections 4, 10, and 12, using subhead titles for the respective sections and appropriate recommendations and disclaimers for the use of the report.

14. Precision and Bias

14.1 This guide provides qualitative data only; therefore, a precision and bias statement is not applicable.

15. Keywords

15.1 explorations; feasibility studies; field investigations; foundation investigations; geological investigations; geophysical investigation; ground water; hydrologic investigations; maps; preliminary investigations; reconnaissance surveys; sampling; site investigations (see Practice D 3584); soil surveys; subsurface investigations

REFERENCES

(1) *Engineering Geology Field Manual*, U.S. Bureau of Reclamation, 1989.
(2) Dietrich, R. V., Dutro, J. V., Jr., and Foose, R. M., (Compilers), "AGI Data Sheets for Geology in Field, Laboratory, and Office," Second Edition, American Geological Institute, 1982.
(3) Pelsner, A. (Ed.), "Manual on Subsurface Investigations," American Association of State Highway and Transportation Officials, Washington, DC.
(4) Shuter, E., and Teasdale, W. E., "Applications of Drilling, Coring and Sampling Techniques to Test Holes and Wells," Techniques of

Water-Resources Investigation, Book 2, U.S. Geological Survey, Washington, DC, 1989.

(5) Keys, W. S., "Borehole Geophysics Applied to Ground Water Investigations," U.S Geological Survey Open-File Report R87-539, Denver, CO, 1988.

(6) Dowding, C. H. (Ed.), "Site Characterization Exploration," American Society of Civil Engineers, Proceedings of Specialty Workshop, New York, NY, 1978.

(7) "Earth Manual," U.S. Bureau of Reclamation, Denver, CO.

(8) "Engineering and Design—Geotechnical Investigation Engineer Manual," EM 1110-1-1804, Headquarters, Department of the Army, Washington, DC, 1984.

(9) "Agricultural Handbook, No. 436, Soil Taxonomy," Soil Conservation Service, U.S. Dept. of Agriculture, U.S. Printing Office, Washington, DC, December, 1975.

Designation: D 5518 – 94

Standard Guide for
Acquisition of File Aerial Photography and Imagery for Establishing Historic Site-Use and Surficial Conditions[1]

This standard is issued under the fixed designation D 5518; the number immediately following the designation indicates the year of original adoption or, in the case of revision, the year of last revision. A number in parentheses indicates the year of last reapproval. A superscript epsilon (ϵ) indicates an editorial change since the last revision or reapproval.

1. Scope

1.1 This guide is intended to assist potential users in the search for, evaluation of, and acquisition of remotely sensed aerial photography or imagery, or both, to be used for the purpose of establishing the historic site-use and other interpretable surface or near-surface conditions regionally, locally, or at a specified project location.

1.2 The instructions given in this guide identify sources of photography and imagery, and provide information pertaining to the specifications, characteristics, and availability of these data.

1.3 The major sources considered are restricted to federal and state organizations only. The sources described do not represent all possible sources of interest for environmental and engineering applications.

1.4 The values stated in both inch-pound and SI units are to be regarded separately as the standard. The values given in parentheses are for information only.

1.5 *This standard does not purport to address all of the safety concerns, if any, associated with its use. It is the responsibility of the user of this standard to establish appropriate safety and health practices and determine the applicability of regulatory limitations prior to use.*

2. Terminology

2.1 *Definitions:*

2.1.1 *black-and-white infrared (IR) film*—film sensitive to blue-violet through reflective IR light wavelengths (0.4 to 0.9 µm), but is exposed to only green through reflective IR wavelengths (0.5 to 0.9 µm). Absence of exposure to the blue wavelengths allows for haze penetration or higher quality data collection, or both, through a greater thickness of the atmosphere or through a portion of atmosphere where light energy transmission is relatively poorer than clear, haze-free atmospheric conditions. This type of film is used for detection of different types of vegetation, diseased plants, soil/rock conditions or land/water boundaries within the constraints of the understanding of conditions of the data collection and interpretation.

2.1.2 *color film*—or conventional color or natural color film exposed to all visible wavelengths (0.4 to 0.7 µm). Uses include identifying soil types, rock outcrops, industrial stockpiles, and shorelines within the constraints of the understanding of conditions of data collection and interpretation. These data are limited due to "fogging", that is, poor haze penetration, associated with the exposure to blue wavelengths.

2.1.3 *color IR*—a form of false-color, reversal film that shows false colors for natural features and is exposed as is black-and-white IR film. Absence of exposure to the blue wavelengths allows for haze penetration or higher quality data collection, or both, through a greater thickness of the atmosphere or through a portion of atmosphere where light energy transmission is relatively poorer than clear, haze-free atmospheric conditions. Natural, healthy, deciduous foliage appears red where as painted, artificial foliage or coniferous vegetation appears purple. This film is also used for detection of diseased plants, insect infestation or other stressed vegetation, soil/rock conditions, including moisture content variations, or land/water boundaries within the constraints of the understanding of the conditions of data collection and interpretation.

2.1.4 *imagery*—usually reserved for reference to data collected by electro-mechanical methods. These methods include multi-spectral scanners (MSS), such as the instrument on U.S. LANDSAT satellites or the thematic mapper (TM) scanner, that collect reflected or emitted energy and record the magnitude of this energy.

2.1.5 *index or photo index*—usually a mosaic of photographs, uncorrected for geometric distortion, that has been collected from a flight or portion of a flight, photographically reproduced at a suitably reduced scale. A photo index will show area coverage along with adequate indexing information for ordering purposes.

2.1.6 *multi-spectral scanners*—electro-mechanical systems that simultaneously collect and record reflected or emitted energy in various wavelength ranges, spectral bands, from the same parcel of terrain. This parcel of terrain is referred to as a picture element (pixel), the size of which is a function of the optics and design of the sensor and the distance the sensor is carried above terrain.

2.1.7 *panchromatic photography*—black and white photography, the film is sensitive to all visible wavelengths (0.4 to 0.7 µm); but is often exposed only to visible red and green wavelengths (0.5 to 0.7 µm). Absence of exposure to the blue wavelengths allows for haze penetration or higher quality data collection, or both, through a greater thickness of the atmosphere or through a portion of atmosphere where light energy transmission is relatively poorer than clear, haze-free atmospheric conditions. Uses include those of color film but with much better detail available. Man-made features are easily interpreted. Such interpretations are made within the constraints of the understanding of the conditions of data collection and interpretation.

---

[1] This guide is under the jurisdiction of ASTM Committee D-18 on Soil and Rock and is the direct responsibility of Subcommittee D18.01 on Surface and Subsurface Characterization.

Current edition approved March 15, 1994. Published May 1994.

2.1.8 *photography*—reserved for reference to the type of data recorded on a film plate in proportion to the photochemical reaction to light striking the emulsion on the film plate. This term is also used in reference to the products made from processing of the exposed film plate, for example, paper prints, transparencies.

2.1.9 *resolution—for photography* the term applied to describe the smallest target which might be reliably recorded and distinguished from closely spaced objects as shown on the film plate. This is a function of a variety of factors; most importantly the chemical makeup of the film plate emulsion including grain size, the film processing, the contrast between the target and its background on the terrain and the nature of the reflectivity of the target. For scanning systems this quality is frequently called detectability and is a function of the same terrain and data collection factors in addition to the optical design, electronics/mechanics of the scanner and its data recording systems and the distance it is carried above terrain.

2.1.10 *spectral band*—used to describe the range of wavelengths over which a single datum value is collected and recorded for each picture element by a multispectral scanner. The range of wavelength for a spectral band is a function of the electronics/mechanics of the scanning system. Many bands of data may be recorded simultaneously for each pixel. Table 1 identifies typical spectral bands for scanners used on U.S. LANDSAT satellites. The information from this table may be transferred to other MSS data and used for selection of imagery components for practical applications.

3. Significance and Use

3.1 The information provided gives guidance on the procedures for acquisition of desired photographic/imagery coverage. Using these instructions as a guide, the user should be able to initiate the search process, evacuate the types of data available, and make preliminary contact with source agencies.

3.2 This guide is not meant to provide a means for obtaining file remote sensor data coverage from all available sources and for all applications. It suggests only the major sources of data from within federal and state governments, and that are a primary interest to studies of an environmental, geological, or engineering application.

3.3 No attempt has been made to describe the photography or imagery holdings of business firms, although they constitute a valuable potential source of such data. This guide recognizes uses of traditional aerial photography and aircraft or satellite collected scanner imagery, or both. Radar, videography, microwave, and other forms of remotely sensed data are not necessarily addressed by this guide.

4. Initiation of the Search Process

4.1 The area of interest must be located on a map. From this determine the longitude and latitude of the center or the corners; or geographic location, such as by the U.S. Federal Rectangular Survey, or by the Civil Land Division System, that is, township, range, and section, the universal trans mercator (UTM) system.

4.2 If a photo index is available, determine the date of photography, roll number, print number(s), project identification number or symbol, or both, or image frame number, and any other pertinent information.

4.3 Identify the scale of the coverage in order to ensure most desirable/usable data format.

4.4 Identify the medium or format of the data; for example, single- or double-weight paper, positive transparencies, digital tapes, compact disks.

4.5 Identify the size of the print desired; for example, 9-by 9-in. (229 by 229 mm) contact print or some enlargement. (Contact prints for photographs or film positive preserve maximum data and quality.)

4.6 Identify the minimum quality data and sky conditions, for example, cloud cover percentage, terrain conditions acceptable, and integrate in the most suitable time of year for these conditions to have existed in the consideration of what data to accept.

4.7 Consider stereo coverage versus pictorial coverage. The three-dimensional model of the terrain that stereo coverage provides is quite advantageous in interpretation of the photography.

4.8 Consider disclosing the purpose for which the data are to be used. The source agency may have information vital to the particular application.

5. Description of Available Remotely Sensed Data

5.1 *Federal Agency Sources of Remotely Sensed Data NB*—The Aerial Photography Summary Record System (APSRS) may be accessed by means of these agencies; particularly United States Geological Survey (USGS) offices. This library record system is inclusive of much of the whole of U.S. public record sector remotely sensed data. Use of this library may provide for efficiency in the search process. Contact should be made with the Earth Science Information Office at the nearest regional facility of USGS.

5.1.1 *Agricultural Stabilization and Conservation Service (ASCS), Salt Lake City, UT:*

5.1.1.1 Panchromatic and color IR coverage,

5.1.1.2 Range of scales from 1:10 000 to 1:120 000, and

5.1.1.3 Coverage includes approximately 80 % of the United States for panchromatic coverage and midwestern states for color IR coverage.

5.1.2 *Soil Conservation Service (SCS), Fort Worth, TX:*

5.1.2.1 Panchromatic coverage,

5.1.2.2 Range of scales from 1:3 000 to 1:75 000, and

5.1.2.3 Coverage includes parts or all of the fifty states.

5.1.3 *U.S. Forest Service (USFS), Washington, DC:*

5.1.3.1 Panchromatic, black-and-white IR, color, and color IR,

5.1.3.2 Range of scales from 1:6 000 to 1:80 000, and

5.1.3.3 Coverage includes national forest areas throughout the United States.

5.1.4 *Bureau of Land Management (BLM), Denver, CO:*

5.1.4.1 Panchromatic, color, and color IR,

5.1.4.2 Range of scales from 1:12 000 to 1:125 000, and

5.1.4.3 Coverage includes Federal lands within Arizona, California, Colorado, Idaho, Montana, Nevada, New Mexico, Oregon, Utah, and Wyoming.

5.1.5 *U.S. Bureau of Reclamation (USBR), Denver, CO:*

5.1.5.1 Panchromatic, color, and color IR,

5.1.5.2 Range of scales from 1:12 000 to 1:80 000, and

5.1.5.3 Coverage is restricted to 17 western states: Washington, Oregon, California, Idaho, Nevada, Arizona, Montana, Utah, Colorado, Wyoming, North Dakota, South Dakota, Nebraska, Kansas, Oregon, and Texas.

D 5518

TABLE 1 Spectral Bands and Characteristics of the Data as Collected by LANDSAT Multispectral Scanner and Thematic Mapper Scanner[A]

Band	Spectral Range		General Applications
	Wavelength Range, μm	Color	
LANDSAT multispectral scanner data characteristic; pixel size of 57 by 79 m, approximately 1.1 AC			
4	0.5 to 0.6	green	Greatest potential for water penetration; shows some contrast between vegetation and soil.
5	0.6 to 0.7	lower red	Best for showing topographic and overall land-use recognition, especially cultural features, such as roads and cities, bare soil, and disturbed land.
6	0.7 to 0.8	upper red to lower infrared	Tonal contrasts reflect various land-use practices; also gives good land/water contrast.
7	0.8 to 1.1	near infrared	Best for land/water discrimination, vegetation growth vigor analysis.
Thematic mapper scanner data characteristics; pixel size of 30 by 30 m (average).			
1	0.45 to 0.52	blue	Designated for water body penetration, making it useful for coastal water mapping. Also useful for differentiation of soil from vegetation, and deciduous from coniferous flora.
2	0.51 to 0.60	green	Designed to measure the visible green reflectance peak of vegetation for vigor assessment.
3	0.63 to 0.69	red	A chlorophyll absorption band important for vegetation discrimination.
4	0.76 to 0.90	reflected infrared (IR)	Useful for determining biomass content and for delineation of water bodies.
5	1.55 to 1.75	reflected IR	Indicative of vegetation moisture content and soil moisture. Also useful for differentiation of snow from clouds
6	10.4 to 12.5	thermal (emitted) IR	A thermal infrared band of use in vegetation stress analysis, soil moisture discrimination, and thermal mapping.
7	2.08 to 2.35	reflected IR	A band selected for its potential for discriminating rock types and for hydrothermal mapping.

[A] See Footnote 13.

5.1.6 *U.S. Geological Survey (USGS): Mid-continent Center, Rolla, MO:*
5.1.6.1 Panchromatic, color IR,
5.1.6.2 Range of scales from 1:11 000 to 1:80 000, and
5.2.6.3 Coverage consists of the following states: Arizona, Illinois, Iowa, Kansas, Louisiana, Michigan, Minnesota, Missouri, Oklahoma, North Dakota, South Dakota, and Wisconsin.
5.1.7 *U.S. Geological Survey, (USGS): Rocky Mountain Center, Denver, CO:*
5.1.7.1 Panchromatic, color IR,
5.1.7.2 Range of scales from 1:11 000 to 1:80 000, and
5.1.7.3 Coverage consists of the following states: Arizona, Montana, Wyoming, Utah, Colorado, New Mexico, and Texas.
5.1.8 *U.S. Geological Survey (USGS): Western Center, Menlo Park, CA:*
5.1.8.1 Panchromatic, color IR,
5.1.8.2 Range of scales from 1:6 000 to 1:80 000, and
5.1.8.3 Coverage consists of the following states: Arizona, California, Hawaii, Idaho, Nevada, Oregon, and Washington.
5.1.9 *U.S. Geological Survey (USGS): Eastern Center, Reston, VA:*
5.1.9.1 Panchromatic, color IR,
5.1.9.2 Range of scales from 1:12 000 to 1:80 000, and
5.1.9.3 Coverage consists of the following states: Alabama, Georgia, Florida, North Carolina, South Carolina, Tennessee, Kentucky, Indiana, Ohio, West Virginia, Maryland, Delaware, Pennsylvania, New York, New Jersey, Rhode Island, Connecticut, Mississippi, New Hampshire, Vermont, Maine. Coverage is also provided for Washington

DC, Puerto Rico, and the Virgin Islands.
5.1.10 *Earth Resources Observation Systems (EROS) Data Center, Sioux Falls, SD:*
5.1.10.1 Data include USGS mapping photography, National Aeronautics and Space Administration (NASA) high- and low-altitude photography, satellite imagery and photography as well as other federal agency data as cataloged. Specific types include: Panchromatic, color, and color IR, MSS and TM scanner data (digital and hard copy).
5.1.10.2 USGS photography range of scales is from 1:12 000 to 1:90 000, and NASA photography from 1:30 000 to 1:120 000, imagery at 30 and 90-m pixel size.
5.1.10.3 Coverage includes the fifty states as well as other portions of the world at a variety of time frames and ground conditions.
5.1.11 *Defense Intelligence Agency (DIA), Washington, DC:*
5.1.11.1 Data include: Panchromatic, color, color IR, black-and-white IR, thermal IR, side-looking radar (SLAR) and multi-spectral scanner data,
5.1.11.2 Range of scales from 1:1 000 to 1:100 000 and a variety of pixel sized, and
5.1.11.3 Partial to full coverage of most foreign countries, some small amount of domestic coverage, most of which is turned over to USGS. Some data might be classified.
5.1.12 *National Ocean Survey (NOS: Coastal), Rockville, MD:*
5.1.12.1 Data include: panchromatic, black-and-white IR, color, and color IR,
5.1.12.2 Range of scales from 1:5000 to 1:60 000, and
5.1.12.3 Coverage includes coastal areas and most civil airports of the United States (including HI, PR, and VI).

45

5.1.13 *National Oceanic Survey (NOS: Lake), Detroit, MI:*

5.1.13.1 Data include: panchromatic and color photography,

5.1.13.2 Range of scales from 1:10 000 to 1:30 000, and

5.1.13.3 Coverage includes shoreline areas of the Great Lakes and along connecting waterways.

5.1.14 *National Archives and Records Service, Washington, DC:*

5.1.14.1 Mostly panchromatic coverage, repository for oldest photography of all federal agencies,

5.1.14.2 Range of scales from 1:10 000 to 1:80 000,

5.1.14.3 Coverage includes approximately 85 % of the contiguous land in the United States and is increasing annually; source of the oldest of the federal agencies' photography.

5.1.15 *Tennessee Valley Authority (TVA), Chattanooga, TN:*

5.1.15.1 Data include: Panchromatic, black-and-white IR, color, color IR, and thermal IR,

5.1.15.2 Range of scales from 1:4 000 to 1:30 000, and

5.1.15.3 Coverage of areas associated with the Tennessee Valley drainage basin (includes portions of Louisiana, Tennessee, Georgia, Kentucky, Alabama, North Carolina, Virginia, and Mississippi).

5.1.16 *U.S. Army Corps of Engineers:*

5.1.16.1 Several district or division offices, or both, across the United States acquire and process data, in many cases for international locations.

5.1.16.2 Data include: Panchromatic, black-and-white IR, color, color IR, as well as thermal IR and SLAR, and

5.1.16.3 A wide range of scales is available for a variety of applications.

5.2 *State Sources of Remotely Sensed Data:*

5.2.1 Most states also have agencies what maintain photography and, in some cases, imagery libraries.

5.2.2 State highway departments generally hold most of the photography acquired within the state.

5.2.3 Other major state organizations active in the acquisition and storage of data are state and regional planning offices, environmental and natural resource departments, state geological surveys, tax commissions, universities, and water resources departments.

5.2.4 Many state agencies use remotely sensed data products that are obtained from federal agencies or cooperate in the acquisition of data that is then shared in the conduct of a given project leaving a set of data with the state agency, or both.

5.3 *Canadian Sources of Remotely Sensed Data:*

5.3.1 *Newfoundland Airphoto Center:*[2]

5.3.1.1 Panchromatic,

5.3.1.2 Range of scales from 1:12 000 to 1:80 000, and

5.3.1.3 Coverage of the Province.

5.3.2 *Ontario Airphoto Center:*[3]

5.3.2.1 Panchromatic,

5.3.2.2 Range of scales from 1:12 000 to 1:80 000, and

5.3.2.3 Coverage of the Province.

5.3.3 *Alberta Airphoto Center:*[4]

5.3.3.1 Panchromatic,

5.3.3.2 Range of scales from 1:12 000 to 1:80 000, and

5.3.3.3 Coverage of the Province.

5.3.4 *Maratimes Airphoto Center:*[5]

5.3.4.1 Panchromatic,

5.3.4.2 Range of scales from 1:12 000 to 1:80 000, and

5.3.4.3 Coverage of the Provinces.

5.3.5 *Manitoba Airphoto Center:*[6]

5.3.5.1 Panchromatic,

5.3.5.2 Range of scales from 1:12 000 to 1:80 000, and

5.3.5.3 Coverage of the Province.

5.3.6 *British Columbia Airphoto Center:*[7]

5.3.6.1 Panchromatic,

5.3.6.2 Range of scales from 1:12 000 to 1:80 000, and

5.3.6.3 Coverage of the Province.

5.3.7 *Quebec Airphoto Center:*[8]

5.3.7.1 Panchromatic,

5.3.7.2 Range of scales from 1:12 000 to 1:80 000, and

5.3.7.3 Coverage of the Province.

5.3.8 *Saskatchewan Airphoto Center:*[9]

5.3.8.1 Panchromatic,

5.3.8.2 Range of scales from 1:12 000 to 1:80 000, and

5.3.8.3 Coverage of the Province.

5.3.9 *National Airphoto Library:*[10]

5.3.9.1 Panchromatic, color, color IR, and satellite imagery,

5.3.9.2 A wide variety of scales and image picture element sizes, and

5.3.9.3 Coverage of the country of Canada, additional coverage in the radar imagery from RADARSAT.[11]

5.4 *SPOT Imagery*[12] *for the United States and Canada:*

5.4.1 Satellite collected multispectral scanner image data,

5.4.2 Nominal 30-m picture element size, some stereo capability, and

5.4.3 Coverage of much of the world, repeated on irregular basis.

6. Keywords

6.1 aerial photography; imagery; photography; remote sensing; remotely sensed data; site-use characterization; surficial conditions characterization

[4] Alberta Airphoto Center, 9945 108th Street, Edmonton, Alberta, Canada, T3K 206.

[5] Maratimes Airphoto Center, Box 310-16 Station Street, Amherst, Nova Scotia, Canada B4M 3Z5.

[6] Manitoba Airphoto Center, 1007 Century Street, Winnipeg, Manitoba. Canada.

[7] British Columbia Airphoto Center, Parliament Building, Victoria, BC, Canada VBV 1X4.

[8] Quebec Airphoto Center, 1995 Boal Charenst, Ste Foy, Quebec, Canada, G1N 4H9.

[9] Saskatchewan Airphoto Center, 2045 Broad Street, Regina, Saskatchewan, Canada, S4P 3V7.

[10] National Airphoto Library, 615 Booth Street, Ottawa, Canada, K1A 0E9.

[11] Available from RADARSAT International, 275 Slater Street, Ottawa, Ontario, Canada, K1P 5H9.

[12] Spot Image Corporation, 1897 Preston White Drive, Reston, VA 22091.

[13] Johnson, A. I., and Pettersson, C.B., eds., "Geotechnical Applications of Remote Sensing and Remote Data Transmission," *ASTM STP 976*, 1988, ASTM, Philadelphia, PA, ISBN 0-8031-0969-5. This reference contains a complete listing of file remote sensor data, a glossary of remote sensing terms and a collection of papers describing applications as related to the topic of this guide.

[2] Newfoundland Airphoto Center, Box 8700, St. Johns, Newfoundland, Canada, A1B 436.

[3] Ontario Airphoto Center, 900 Bay Street, Toronto, Ontario, Canada, M7A 2C1.

Standard Practice for
Environmental Site Assessments: Phase I Environmental Site Assessment Process[1]

This standard is issued under the fixed designation E 1527; the number immediately following the designation indicates the year of original adoption or, in the case of revision, the year of last revision. A number in parentheses indicates the year of last reapproval. A superscript epsilon (ε) indicates an editorial change since the last revision or reapproval.

1. Scope

1.1 *Purpose*—The purpose of this practice, as well as Practice E 1528, is to define good commercial and customary practice in the United States of America for conducting an *environmental site assessment*[2] of a parcel of *commercial real estate* with respect to the range of contaminants within the scope of Comprehensive Environmental Response, Compensation and Liability Act (CERCLA) and *petroleum products*. As such, this practice is intended to permit a *user* to satisfy one of the requirements to qualify for the *innocent landowner defense* to CERCLA liability: that is, the practices that constitute "all appropriate inquiry into the previous ownership and uses of the property consistent with good commercial or customary practice" as defined in 42 USC § 9601(35)(B). (See Appendix X1 for an outline of CERCLA's liability and defense provisions.)

1.1.1 *Recognized Environmental Conditions*—In defining a standard of good commercial and customary practice for conducting an *environmental site assessment* of a parcel of *property*, the goal of the processes established by this practice is to identify *recognized environmental conditions*. The term *recognized environmental conditions* means the presence or likely presence of any *hazardous substances* or *petroleum products* on a *property* under conditions that indicate an existing release, a past release, or a material threat of a release of any *hazardous substances* or *petroleum products* into structures on the *property* or into the ground, groundwater, or surface water of the *property*. The term includes *hazardous substances* or *petroleum products* even under conditions in compliance with laws. The term is not intended to include *de minimis* conditions that generally do not present a material risk of harm to public health or the environment and that generally would not be the subject of an enforcement action if brought to the attention of appropriate governmental agencies.

1.1.2 *Two Related Practices*—This practice is closely related to Practice E 1528. Both are *environmental site assessments* for *commercial real estate*. See 4.3.

1.1.3 *Petroleum Products*—*Petroleum products* are included within the scope of both practices because they are of concern with respect to many parcels of *commercial real*

estate and current custom and usage is to include an inquiry into the presence of *petroleum products* when doing an *environmental site assessment* of *commercial real estate*. Inclusion of *petroleum products* within the scope of this practice and Practice E 1528 is not based upon the applicability, if any, of CERCLA to *petroleum products*. (See Appendix X1 for discussion of *petroleum exclusion* to CERCLA liability.)

1.1.4 *CERCLA Requirements Other Than Appropriate Inquiry*—This practice does not address whether requirements in addition to *appropriate inquiry* have been met in order to qualify for CERCLA's *innocent landowner defense* (for example, the duties specified in 42 USC § 9607(b)(3)(a) and (b) and cited in Appendix X1).

1.1.5 *Other Federal, State, and Local Environmental Laws*—This practice does not address requirements of any state or local laws or of any federal laws other than the appropriate inquiry provisions of CERCLA's *innocent landowner defense*. *Users* are cautioned that federal, state, and local laws may impose environmental assessment obligations that are beyond the scope of this practice. *Users* should also be aware that there are likely to be other legal obligations with regard to *hazardous substances* or *petroleum products* discovered on *property* that are not addressed in this practice and that may pose risks of civil and/or criminal sanctions for non-compliance.

1.1.6 *Documentation*—The scope of this practice includes research and reporting requirements that support the user's ability to qualify for the *innocent landowner defense*. As such, sufficient documentation of all sources, records, and resources utilized in conducting the inquiry required by this practice must be provided in the written report (refer to 7.1.8 and 11.2).

1.2 *Objectives*—Objectives guiding the development of this practice and Practice E 1528 are (1) to synthesize and put in writing good commercial and customary practice for *environmental site assessments* for *commercial real estate*, (2) to facilitate high quality, standardized *environmental site assessments*, (3) to ensure that the standard of *appropriate inquiry* is practical and reasonable, and (4) to clarify an industry standard for *appropriate inquiry* in an effort to guide legal interpretation of CERCLA's *innocent landowner defense*.

1.3 *Considerations Beyond Scope*—The use of this practice is strictly limited to the scope set forth in this section. Section 12 of this practice, identifies, for informational purposes, certain environmental conditions (not an all-inclusive list) that may exist on a *property* that are beyond the scope of this practice but may warrant consideration by parties to a *commercial real estate* transaction.

---

[1] This practice is under the jurisdiction of ASTM Committee E-50 on Environmental Assessment and is the direct responsibility of Subcommittee E50.02 on Commercial Real Estate Transactions.

Current edition approved March 10, 1997. Published May 1997. Originally published as E 1527 – 93. Last previous edition E 1527 – 94.

[2] All definitions, descriptions of terms, and acronyms are defined in Section 3. Whenever terms defined in 3.2 or described in 3.3 are used in this practice, they are in *italics*.

1.4 *Organization of This Practice*—This practice has several parts and two appendixes. Section 1 is the Scope. Section 2 is Referenced Documents. Section 3, Terminology, has definitions of terms not unique to this practice and descriptions of terms unique to this practice and acronyms. Section 4 is Significance and Use of this practice. Section 5 describes User's Responsibilities. Sections 6 through 11 are the main body of the Phase I Environmental Site Assessment, including evaluation and report preparation. Section 12 provides additional information regarding non-scope considerations (see 1.3). The appendixes are included for information and are not part of the procedures prescribed in either this practice or Practice E 1528. Appendix X1 explains the liability and defense provisions of CERCLA that will assist the user in understanding the user's responsibilities under CERCLA; it also contains other important information regarding CERCLA and this practice. Appendix X2 provides a recommended table of contents and report format for a Phase I Environmental Site Assessment Report.

1.5 *This standard does not purport to address all of the safety concerns, if any, associated with its use. It is the responsibility of the user of this standard to establish appropriate safety and health practices and determine the applicability of regulatory limitations prior to use.*

2. Referenced Document

2.1 *ASTM Standard:*
E 1528 Practice for Environmental Site Assessments: Transaction Screen Process[3]

3. Terminology

3.1 This section provides definitions, descriptions of terms, and a list of acronyms for many of the words used in this practice and Practice E 1528. The terms are an integral part of both practices and are critical to an understanding of the practices and their use.

3.2 *Definitions:*

3.2.1 *asbestos*—six naturally occurring fibrous minerals found in certain types of rock formations. Of the six, the minerals chrysotile, amosite, and crocidolite have been most commonly used in building products. When mined and processed, asbestos is typically separated into very thin fibers. Because asbestos is strong, incombustible, and corrosion-resistant, asbestos was used in many commercial products beginning early in this century and peaking in the period from World War II into the 1970s. When inhaled in sufficient quantities, asbestos fibers can cause serious health problems.[4,5]

3.2.2 *asbestos containing material (ACM)*—any material or product that contains more than 1 % asbestos.[4]

3.2.3 *Comprehensive Environmental Response, Compensation, and Liability Information System (CERCLIS)*—the list of sites compiled by EPA that EPA has investigated or is currently investigating for potential hazardous substance contamination for possible inclusion on the National Priorities List.

3.2.4 *construction debris*—concrete, brick, asphalt, and other such building materials discarded in the construction of a building or other improvement to property.

3.2.5 *contaminated public wells*—public wells used for drinking water that have been designated by a government entity as contaminated by toxic substances (for example, chlorinated solvents), or as having water unsafe to drink without treatment.

3.2.6 *CORRACTS list*—environmental protection agencies (EPA's) list of treatment, storage, or disposal facilities subject to corrective action under RCRA.

3.2.7 *demolition debris*—concrete, brick, asphalt, and other such building materials discarded in the demolition of a building or other improvement to property.

3.2.8 *drum*—a container (typically, but not necessarily, holding 55 gal (208 L) of liquid) that may be used to store *hazardous substances* or *petroleum products*.

3.2.9 *dry wells*—underground areas where soil has been removed and replaced with pea gravel, coarse sand, or large rocks. Dry wells are used for drainage, to control storm runoff, for the collection of spilled liquids (intentional and non-intentional) and wastewater disposal (often illegal).

3.2.10 *dwelling*—structure or portion thereof used for residential habitation.

3.2.11 *environmental lien*—a charge, security, or encumbrance upon title to a *property* to secure the payment of a cost, damage, debt, obligation, or duty arising out of response actions, cleanup, or other remediation of *hazardous substances* or *petroleum products* upon a *property*, including (but not limited to) liens imposed pursuant to CERCLA 42 USC § 9607(1) and similar state or local laws.

3.2.12 *ERNS list*—EPA's emergency response notification system list of reported CERCLA hazardous substance releases or spills in quantities greater than the reportable quantity, as maintained at the National Response Center. Notification requirements for such releases or spills are codified in 40 CFR Parts 302 and 355.

3.2.13 *Federal Register, (FR)*—publication of the United States government published daily (except for federal holidays and weekends) containing all proposed and final regulations and some other activities of the federal government. When regulations become final, they are included in the Code of Federal Regulations (CFR), as well as published in the Federal Register.

3.2.14 *fire insurance maps*—maps produced for private fire insurance map companies that indicate uses of properties at specified dates and that encompass the property. These maps are often available at local libraries, historical societies, private resellers, or from the map companies who produced them. See Question 23 of the transaction screen process in Practice E 1528 and 7.3.4.2 of this practice.

3.2.15 *hazardous substance*—A substance defined as a hazardous substance pursuant to CERCLA 42 USC § 9601(14), as interpreted by EPA regulations and the courts: "(A) any substance designated pursuant to section 1321(b)(2)(A) of Title 33, (B) any element, compound, mixture, solution, or substance designated pursuant to section 9602 of this title, (C) any hazardous waste having the characteristics identified under or listed pursuant to section

---

[3] *Annual Book of ASTM Standards,* Vol 11.04.
[4] See EPA, *Managing Asbestos in Place, A Building Owner's Guide to Operations and Maintenance Programs for Asbestos-Containing Materials,* July 1990, p. 2.
[5] See also, for an additional definition of asbestos, ASTM STP 834, ASTM.

3001 of the Solid Waste Disposal Act (42 USC § 6921) (but not including any waste the regulation of which under the Solid Waste Disposal Act (42 USC § 6901 *et seq.*) has been suspended by Act of Congress), (D) any toxic pollutant listed under section 1317(a) of Title 33, (E) any hazardous air pollutant listed under section 112 of the Clean Air Act (42 USC § 7412), and (F) any imminently hazardous chemical substance or mixture with respect to which the Administrator (of EPA) has taken action pursuant to section 2606 of Title 15. The term does not include petroleum, including crude oil or any fraction thereof which is not otherwise specifically listed or designated as a hazardous substance under subparagraphs (A) through (F) of this paragraph, and the term does not include natural gas, natural gas liquids, liquefied natural gas, or synthetic gas usable for fuel (or mixtures of natural gas and such synthetic gas)." (See Appendix X1.)

3.2.16 *hazardous waste*—any hazardous waste having the characteristics identified under or listed pursuant to section 3001 of the Solid Waste Disposal Act (42 USC § 6921) (but not including any waste the regulation of which under the Solid Waste Disposal Act (42 USC § 6901 *et seq.*) has been suspended by Act of Congress). The Solid Waste Disposal Act of 1980 amended RCRA. RCRA defines a hazardous waste, in 42 USC § 6903, as: "a solid waste, or combination of solid wastes, which because of its quantity, concentration, or physical, chemical, or infectious characteristics may—(A) cause, or significantly contribute to an increase in mortality or an increase in serious irreversible, or incapacitating reversible, illness; or (B) pose a substantial present or potential hazard to human health or the environment when improperly treated, stored, transported, or disposed of, or otherwise managed."

3.2.17 *landfill*—a place, location, tract of land, area, or premises used for the disposal of solid wastes as defined by state solid waste regulations. The term is synonymous with the term *solid waste disposal site* and is also known as a garbage dump, trash dump, or similar term.

3.2.18 *local street directories*—directories published by private (or sometimes government) sources that show ownership, occupancy, and/or use of sites by reference to street addresses. Often local street directories are available at libraries of local governments, colleges or universities, or historical societies. See 7.3.4.6 of this practice.

3.2.19 *material safety data sheet (MSDS)*—written or printed material concerning a hazardous substance which is prepared by chemical manufacturers, importers, and employers for hazardous chemicals pursuant to OSHA's Hazard Communication Standard, 29 CFR 1910.1200.

3.2.20 *National Contingency Plan (NCP)*—the National Oil and Hazardous Substances Pollution Contingency Plan, found at 40 CFR § 300, that is the EPA's blueprint on how hazardous substances are to be cleaned up pursuant to CERCLA.

3.2.21 *National Priorities List (NPL)*—list compiled by EPA pursuant to CERCLA 42 USC § 9605(a)(8)(B) of properties with the highest priority for cleanup pursuant to EPA's Hazard Ranking System. See 40 CFR Part 300.

3.2.22 *occupants*—those tenants, subtenants, or other persons or entities using the *property* or a portion of the *property*.

3.2.23 *owner*—generally the fee owner of record of the *property*.

3.2.24 *petroleum exclusion*—The exclusion from CERCLA liability provided in 42 USC § 9601(14), as interpreted by the courts and EPA: "The term (hazardous substance) does not include petroleum, including crude oil or any fraction thereof which is not otherwise specifically listed or designated as a hazardous substance under subparagraphs (A) through (F) of this paragraph, and the term does not include natural gas, natural gas liquids, liquefied natural gas, or synthetic gas usable for fuel (or mixtures of natural gas and such synthetic gas)."

3.2.25 *petroleum products*—those substances included within the meaning of the *petroleum exclusion* to CERCLA, 42 USC § 9601(14), as interpreted by the courts and EPA, that is: petroleum, including crude oil or any fraction thereof which is not otherwise specifically listed or designated as a hazardous substance under Subparagraphs (A) through (F) of 42 USC § 9601(14), natural gas, natural gas liquids, liquefied natural gas, and synthetic gas usable for fuel (or mixtures of natural gas and such synthetic gas). (The word fraction refers to certain distillates of crude oil, including gasoline, kerosene, diesel oil, jet fuels, and fuel oil, pursuant to *Standard Definitions of Petroleum Statistics.*[6]

3.2.26 *Phase I Environmental Site Assessment*—the process described in this practice.

3.2.27 *pits, ponds, or lagoons*—man-made or natural depressions in a ground surface that are likely to hold liquids or sludge containing *hazardous substances* or *petroleum products*. The likelihood of such liquids or sludge being present is determined by evidence of factors associated with the pit, pond, or lagoon, including, but not limited to, discolored water, distressed vegetation, or the presence of an obvious wastewater discharge.

3.2.28 *property*—the real property that is the subject of the *environmental site assessment* described in this practice. Real property includes buildings and other fixtures and improvements located on the property and affixed to the land.

3.2.29 *property tax files*—the files kept for property tax purposes by the local jurisdiction where the property is located and includes records of past ownership, appraisals, maps, sketches, photos, or other information that is reasonably ascertainable and pertaining to the property. See 7.3.4.3.

3.2.30 *RCRA generators*—those persons or entities that generate hazardous wastes, as defined and regulated by RCRA.

3.2.31 *RCRA generators list*—list kept by EPA of those persons or entities that generate hazardous wastes as defined and regulated by RCRA.

3.2.32 *RCRA TSD facilities*—those facilities on which treatment, storage, and/or disposal of hazardous wastes takes place, as defined and regulated by RCRA.

3.2.33 *RCRA TSD facilities list*—list kept by EPA of those facilities on which treatment, storage, and/or disposal of hazardous wastes takes place, as defined and regulated by RCRA.

---

[6] *Standard Definitions of Petroleum Statistics*, American Petroleum Institute, Fourth Edition, 1988.

3.2.34 *recorded land title records*—records of fee ownership, leases, land contracts, easements, liens, and other encumbrances on or of the property recorded in the place where land title records are, by law or custom, recorded for the local jurisdiction in which the *property* is located. (Often such records are kept by a municipal or county recorder or clerk.) Such records may be obtained from title companies or directly from the local government agency. Information about the title to the property that is recorded in a U.S. district court or any place other than where land title records are, by law or custom, recorded for the local jurisdiction in which the property is located, are not considered part of recorded land title records. See 7.3.4.4.

3.2.35 *records of emergency release notifications (SARA § 304)*—Section 304 of EPCRA or Title III of SARA requires operators of facilities to notify their local emergency planning committee (as defined in EPCRA) and state emergency response commission (as defined in EPCRA) of any release beyond the facility's boundary of any reportable quantity of any extremely hazardous substance. Often the local fire department is the local emergency planning committee. Records of such notifications are "Records of Emergency Release Notifications" (SARA § 304).

3.2.36 *report*—the written record of a transaction screen process as required by Practice E 1528 or the written report prepared by the environmental professional and constituting part of a "Phase I Environmental Site Assessment," as required by this practice.

3.2.37 *solid waste disposal site*—a place, location, tract of land, area, or premises used for the disposal of solid wastes as defined by state solid waste regulations. The term is synonymous with the term *landfill* and is also known as a garbage dump, trash dump, or similar term.

3.2.38 *solvent*—a chemical compound that is capable of dissolving another substance and may itself be a *hazardous substance*, used in a number of manufacturing/industrial processes including but not limited to the manufacture of paints and coatings for industrial and household purposes, equipment clean-up, and surface degreasing in metal fabricating industries.

3.2.39 *state registered USTs*—state lists of underground storage tanks required to be registered under Subtitle I, Section 9002 of RCRA.

3.2.40 *sump*—a pit, cistern, cesspool, or similar receptacle where liquids drain, collect, or are stored.

3.2.41 *TSD facility*—treatment, storage, or disposal facility (see *RCRA TSD facilities*).

3.2.42 *underground storage tank (UST)*—any tank, including underground piping connected to the tank, that is or has been used to contain *hazardous substances* or *petroleum products* and the volume of which is 10 % or more beneath the surface of the ground.

3.2.43 *USGS 7.5 Minute Topographic Map*—the map (if any) available from or produced by the United States Geological Survey, entitled "USGS 7.5 Minute Topographic Map," and showing the property. See 7.3.4.5.

3.2.44 *wastewater*—water that (*1*) is or has been used in an industrial or manufacturing process, (*2*) conveys or has conveyed sewage, or (*3*) is directly related to manufacturing, processing, or raw materials storage areas at an industrial plant. Wastewater does not include water originating on or

passing through or adjacent to a site, such as stormwater flows, that has not been used in industrial or manufacturing processes, has not been combined with sewage, or is not directly related to manufacturing, processing, or raw materials storage areas at an industrial plant.

3.2.45 *zoning/land use records*—those records of the local government in which the *property* is located indicating the uses permitted by the local government in particular zones within its jurisdiction. The records may consist of maps and/or written records. They are often located in the planning department of a municipality or county. See 7.3.4.8.

3.3 *Definitions of Terms Specific to This Standard:*

3.3.1 *actual knowledge*—the knowledge actually possessed by an individual who is a real person, rather than an entity. Actual knowledge is to be distinguished from constructive knowledge that is knowledge imputed to an individual or entity.

3.3.2 *adjoining properties*—any real property or properties the border of which is contiguous or partially contiguous with that of the property, or that would be contiguous or partially contiguous with that of the property but for a street, road, or other public thoroughfare separating them.

3.3.3 *aerial photographs*—photographs taken from an airplane or helicopter (from a low enough altitude to allow identification of development and activities) of areas encompassing the property. Aerial photographs are often available from government agencies or private collections unique to a local area. See 7.3.4.1 of this practice.

3.3.4 *appropriate inquiry*—that inquiry constituting "all appropriate inquiry into the previous ownership and uses of the property consistent with good commercial or customary practice" as defined in CERCLA, 42 USC § 9601(35)(B), that will give a party to a *commercial real estate* transaction the *innocent landowner defense* to CERCLA liability (42 USC § 9601(A) and (B) and § 9607(b)(3)), assuming compliance with other elements of the defense. See Appendix X1.

3.3.5 *approximate minimum search distance*—the area for which records must be obtained and reviewed pursuant to Section 7 subject to the limitations provided in that section. This may include areas outside the *property* and shall be measured from the nearest *property* boundary. This term is used in lieu of radius to include irregularly shaped properties.

3.3.6 *building department records*—those records of the local government in which the property is located indicating permission of the local government to construct, alter, or demolish improvements on the property. Often building department records are located in the building department of a municipality or county. See 7.3.4.7.

3.3.7 *commercial real estate*—any real property except a dwelling or property with no more than four dwelling units exclusively for residential use (except that a dwelling or property with no more than four dwelling units exclusively for residential use is included in this term when it has a commercial function, as in the building of such dwellings for profit). This term includes but is not limited to undeveloped real property and real property used for industrial, retail, office, agricultural, other commercial, medical, or educational purposes; property used for residential purposes that has more than four residential dwelling units; and property with no more than four dwelling units for residential use when it has a commercial function, as in the building of such dwellings for profit.

3.3.8 *commercial real estate transaction*—a transfer of title to or possession of real property or receipt of a security interest in real property, except that it does not include transfer of title to or possession of real property or the receipt of a security interest in real property with respect to an individual dwelling or building containing fewer than five dwelling units, nor does it include the purchase of a lot or lots to construct a dwelling for occupancy by a purchaser, but a commercial real estate transaction does include real property purchased or leased by persons or entities in the business of building or developing dwelling units.

3.3.9 *due diligence*—the process of inquiring into the environmental characteristics of a parcel of *commercial real estate* or other conditions, usually in connection with a commercial real estate transaction. The degree and kind of due diligence vary for different properties and differing purposes. See Appendix X1.

3.3.10 *environmental audit*—the investigative process to determine if the operations of an existing facility are in compliance with applicable environmental laws and regulations. This term should not be used to describe Practice E 1528 or this practice, although an environmental audit may include an *environmental site assessment* or, if prior audits are available, may be part of an environmental site assessment. See Appendix X1.

3.3.11 *environmental professional*—a person possessing sufficient training and experience necessary to conduct a *site reconnaissance, interviews,* and other activities in accordance with this practice, and from the information generated by such activities, having the ability to develop opinions and conclusions regarding *recognized environmental conditions* in connection with the *property* in question. An individual's status as an environmental professional may be limited to the type of assessment to be performed or to specific segments of the assessment for which the professional is responsible. The person may be an independent contractor or an employee of the *user*.

3.3.12 *environmental site assessment (ESA)*—the process by which a person or entity seeks to determine if a particular parcel of real *property* (including improvements) is subject to *recognized environmental conditions*. At the option of the user, an environmental site assessment may include more inquiry than that constituting *appropriate inquiry* or, if the user is not concerned about qualifying for the *innocent landowner defense*, less inquiry than that constituting *appropriate inquiry*. See Appendix X1. An environmental site assessment is both different from and less rigorous than an *environmental audit*.

3.3.13 *fill dirt*—dirt, soil, sand, or other earth, that is obtained off-site, that is used to fill holes or depressions, create mounds, or otherwise artificially change the grade or elevation of real property. It does not include material that is used in limited quantities for normal landscaping activities.

3.3.14 *hazardous waste/contaminated sites*—sites on which a release has occurred, or is suspected to have occurred, of any *hazardous substance, hazardous waste,* or *petroleum products,* and that release or suspected release has been reported to a government entity.

3.3.15 *innocent landowner defense*—that defense to CERCLA liability provided in 42 USC § 9601(35) and § 9607(b)(3). One of the requirements to qualify for this defense is that the party make "all appropriate inquiry into the previous ownership and uses of the property consistent with good commercial or customary practice." There are additional requirements to qualify for this defense. See Appendix X1.

3.3.16 *interviews*—those portions of this practice that are contained in Section 9 and 10 thereof and address questions to be asked of *owners* and *occupants* of the *property* and questions to be asked of local government officials.

3.3.17 *key site manager*—the person identified by the *owner* of a *property* as having good knowledge of the uses and physical characteristics of the property. See 9.5.1.

3.3.18 *local government agencies*—those agencies of municipal or county government having jurisdiction over the *property*. Municipal and county government agencies include but are not limited to cities, parishes, townships, and similar entities.

3.3.19 *LUST sites*—state lists of leaking underground storage tank sites. Section 9003 (h) of Subtitle I of RCRA gives EPA and states, under cooperative agreements with EPA, authority to clean up releases from UST systems or require owners and operators to do so.

3.3.20 *major occupants*—those tenants, subtenants, or other persons or entities each of which uses at least 40 % of the leasable area of the *property* or any anchor tenant when the *property* is a shopping center.

3.3.21 *obvious*—that which is plain or evident; a condition or fact that could not be ignored or overlooked by a reasonable observer while *visually or physically observing the property*.

3.3.22 *other historical sources*—any source or sources other than those designated in 7.3.4.1 through 7.3.4.8 that are credible to a reasonable person and that identify past uses of the property. The term includes, but is not limited to: miscellaneous maps, newspaper archives, and records in the files and/or personal knowledge of the *property owner* and/or *occupants*. See 7.3.4.9.

3.3.23 *physical setting sources*—sources that provide information about the geologic, hydrogeologic, hydrologic, or topographic characteristics of a *property*. See 7.2.3.

3.3.24 *practically reviewable*—information that is practically reviewable means that the information is provided by the source in a manner and in a form that, upon examination, yields information relevant to the *property* without the need for extraordinary analysis of irrelevant data. The form of the information shall be such that the user can review the records for a limited geographic area. Records that cannot be feasibly retrieved by reference to the location of the *property* or a geographic area in which the *property* is located are not generally *practically reviewable*. Most databases of public records are *practically reviewable* if they can be obtained from the source agency by the county, city, zip code, or other geographic area of the facilities listed in the record system. Records that are sorted, filed, organized, or maintained by the source agency only chronologically are not generally practically reviewable. Listings in publicly available records which do not have adequate address information to be located geographically are not generally considered practically reviewable. For large databases with numerous facility records (such as RCRA hazardous waste generators and registered underground storage tanks), the records are not *practically reviewable* unless they can be obtained from the source agency in the smaller geographic area of zip codes. Even when information is provided by zip code for some large databases, it is common for an unmanageable number of sites to be identified within a given zip code. In these cases, it is not necessary to review the impact of all of the sites that are likely to be listed in any given zip code because that information would not be *practically reviewable*. In other words, when so much data is generated that it cannot be feasibly reviewed for its impact on the *property*, it is not *practically reviewable*.

3.3.25 *preparer*—the person preparing the *transaction screen questionnaire* pursuant to Practice E 1528, who may be either the user or the person to whom the user has delegated the preparation of the *transaction screen questionnaire*.

3.3.26 *publicly available*—information that is publicly available means that the source of the information allows access to the information by anyone upon request.

3.3.27 *reasonably ascertainable*—for purposes of both this practice and Practice E 1528, information that is (*1*) *publicly available*, (*2*) obtainable from its source within reasonable time and cost constraints, and (*3*) *practically reviewable*.

3.3.28 *recognized environmental conditions*—the presence or likely presence of any *hazardous substances* or *petroleum products* on a *property* under conditions that indicate an existing release, a past release, or a material threat of a release of any *hazardous substances* or *petroleum products* into structures on the *property* or into the ground, groundwater, or surface water of the *property*. The term includes *hazardous substances* or *petroleum products* even under conditions in compliance with laws. The term is not intended to include *de minimis* conditions that generally do not present a material risk of harm to public health or the environment and that generally would not be the subject of an enforcement action if brought to the attention of appropriate governmental agencies.

3.3.29 *records review*—that part that is contained in Section 7 of this practice addresses which records shall or may be reviewed.

3.3.30 *site reconnaissance*—that part that is contained in Section 8 of this practice and addresses what should be done in connection with the *site visit*. The site reconnaissance includes, but is not limited to, the *site visit* done in connection with such a Phase I Environmental Site Assessment.

3.3.31 *site visit*—the visit to the property during which observations are made constituting the *site reconnaissance* section of this practice and the *site visit* requirement of Practice E 1528.

3.3.32 *standard environmental record sources*—those records specified in 7.2.1.1.

3.3.33 *standard historical sources*—those sources of information about the history of uses of property specified in 7.3.4.

3.3.34 *standard physical setting source*—a current USGS 7.5 minute topographic map (if any) showing the area on which the property is located. See 7.2.3.

3.3.35 *standard practice(s)*—the activities set forth in either and both this practice and Practice E 1528.

3.3.36 *standard sources*—sources of environmental, physical setting, or historical records specified in Section 7 of this practice.

3.3.37 *transaction screen process*—the process described in Practice E 1528.

3.3.38 *transaction screen questionnaire*—the questionnaire provided in Section 6 of Practice E 1528.

3.3.39 *user*—the party seeking to use Practice E 1528 to perform an *environmental site assessment* of the *property*. A user may include, without limitation, a purchaser of *property*, a potential tenant of property, an *owner* of *property*, a lender, or a property manager.

3.3.40 *visually and/or physically observed*—during a *site visit* pursuant to Practice E 1528, or pursuant to this practice, this term means observations made by vision while walking through a *property* and the structures located on it and observations made by the sense of smell, particularly observations of noxious or foul odors. The term "walking through" is not meant to imply that disabled persons who cannot physically walk may not conduct a *site visit*; they may do so by the means at their disposal for moving through the *property* and the structures located on it.

3.4 *Acronyms:*

3.4.1 *CERCLA*—Comprehensive Environmental Response, Compensation and Liability Act of 1980 (as amended, 42 USC § 9601 *et seq.*).

3.4.2 *CERCLIS*—Comprehensive Environmental Response, Compensation and Liability Information System (maintained by EPA).

3.4.3 *CFR*—Code of Federal Regulations.

3.4.4 *CORRACTS*—TSD facilities subject to Corrective Action under RCRA.

3.4.5 *EPA*—United States Environmental Protection Agency.

3.4.6 *EPCRA*—Emergency Planning and Community Right to Know Act ((also known as SARA Title III), 42 USC § 11001 *et seq.*).

3.4.7 *ERNS*—emergency response notification system.

3.4.8 *ESA*—environmental site assessment (different than an *environmental audit*; see 3.3.12).

3.4.9 *FOIA*—U.S. Freedom of Information Act (5 USC 552 *et seq.*).

3.4.10 *FR*—Federal Register.

3.4.11 *LUST*—leaking underground storage tank.

3.4.12 *MSDS*—material safety data sheet.

3.4.13 *NCP*—National Contingency Plan.

3.4.14 *NPDES*—national pollutant discharge elimination system.

3.4.15 *NPL*—national priorities list.

3.4.16 *PCBs*—polychlorinated biphenyls.

3.4.17 *PRP*—potentially responsible party (pursuant to CERCLA 42 USC § 9607(a)).

3.4.18 *RCRA*—Resource Conservation and Recovery Act (as amended, 42 USC § 6901 *et seq.*).

3.4.19 *SARA*—Superfund Amendments and Reauthorization Act of 1986 (amendment to CERCLA).

3.4.20 *USC*—United States Code.

3.4.21 *USGS*—United States Geological Survey.

3.4.22 *UST*—underground storage tank.

4. Significance and Use

4.1 *Uses*—This practice is intended for use on a voluntary basis by parties who wish to assess the environmental condition of *commercial real estate*. While use of this practice is intended to constitute *appropriate inquiry* for purposes of CERCLA's *innocent landowner defense*, it is not intended that its use be limited to that purpose. This practice is intended primarily as an approach to conducting an inquiry designed to identify *recognized environmental conditions* in connection with a *property*, and *environmental site assessments* that are both more and less comprehensive than this practice (including, in some instances, no *environmental site assessment*) may be appropriate in some circumstances. Further, no implication is intended that a person must use this practice in order to be deemed to have conducted inquiry in a commercially prudent or reasonable manner in any particular transaction. Nevertheless, this practice is intended to reflect a commercially prudent and reasonable inquiry.

4.2 *Clarifications on Use:*

4.2.1 *Use Not Limited to CERCLA*—This practice and Practice E 1528 are designed to assist the *user* in developing information about the environmental condition of a *property* and as such has utility for a wide range of persons, including those who may have no actual or potential CERCLA liability and/or may not be seeking the *innocent landowner defense*.

4.2.2 *Residential Tenants/Purchasers and Others*—No implication is intended that it is currently customary practice for residential tenants of multifamily residential buildings, tenants of single-family homes or other residential real estate, or purchasers of dwellings for one's own residential use, to conduct an *environmental site assessment* in connection with these transactions. Thus, these transactions are not included in the term commercial real estate transactions, and it is not intended to imply that such persons are obligated to conduct an *environmental site assessment* in connection with these transactions for purposes of *appropriate inquiry* or for any other purpose. In addition, no implication is intended that it is currently customary practice for *environmental site assessments* to be conducted in other unenumerated instances (including but not limited to many commercial leasing transactions, many acquisitions of easements, and many loan transactions in which the lender has multiple remedies). On the other hand, anyone who elects to do an *environmental site assessment* of any *property* or portion of a property may, in such person's judgment, use either this practice or Practice E 1528.

4.2.3 *Site-Specific*—This practice is site-specific in that it relates to assessment of environmental conditions on a specific parcel of *commercial real estate*. Consequently, this practice does not address many additional issues raised in transactions such as purchases of business entities, or interests therein, or of their assets, that may well involve environmental liabilities pertaining to properties previously owned or operated or other off-site environmental liabilities.

4.3 *Two Related Practices*—This practice sets forth one procedure for an *environmental site assessment* known as a "Phase I Environmental Site Assessment" or a "Phase I ESA" or simply a "Phase I." This practice is a companion to Practice E 1528. These practices are each intended to meet the standard of *appropriate inquiry* necessary to qualify for the *innocent landowner defense*. It is essential to consider that these practices, taken together, provide for two alternative practices of *appropriate inquiry*.

4.3.1 *Election to Commence with Either Practice*—The *user* may commence inquiry to identify *recognized environmental conditions* in connection with a *property* by performing either the *transaction screen process*, or the Phase I Environmental Site Assessment given in this practice.

4.3.2 *Who May Conduct*—The *transaction screen process* may be conducted either by the *user* (including an agent or employee of the user) or wholly or partially by an *environmental professional*. The *transaction screen process* does not require the judgment of an *environmental professional*. Whenever a Phase I Environmental Site Assessment is conducted, it must be performed by an *environmental professional* to the extent specified in 6.5.1. Further, at the Phase I Environmental Site Assessment level, no practical standard can be designed to eliminate the role of judgment and the value and need for experience in the party performing the inquiry. The professional judgment of an *environmental professional* is, consequently, vital to the performance of appropriate inquiry at the Phase I Environmental Site Assessment level.

4.4 *Additional Services*—As set forth in 11.9, additional services may be contracted for between the *user* and the *environmental professional*.

4.5 *Principles*—The following principles are an integral part of both practices and are intended to be referred to in resolving any ambiguity or exercising such discretion as is accorded the *user* or *environmental professional* in performing an *environmental site assessment* or in judging whether a *user* or *environmental professional* has conducted *appropriate inquiry* or has otherwise conducted an adequate *environmental site assessment*.

4.5.1 *Uncertainty Not Eliminated*—No *environmental site assessment* can wholly eliminate uncertainty regarding the potential for *recognized environmental conditions* in connection with a *property*. Performance of this practice or E 1528 is intended to reduce, but not eliminate, uncertainty regarding the potential for *recognized environmental conditions* in connection with a property, and both practices recognize reasonable limits of time and cost.

4.5.2 *Not Exhaustive*—*Appropriate inquiry* does not mean an exhaustive assessment of a clean *property*. There is a point at which the cost of information obtained or the time required to gather it outweighs the usefulness of the information and, in fact, may be a material detriment to the orderly completion of transactions. One of the purposes of this practice is to identify a balance between the competing goals of limiting the costs and time demands inherent in performing an *environmental site assessment* and the reduction of uncertainty about unknown conditions resulting from additional information.

4.5.3 *Level of Inquiry Is Variable*—Not every *property* will warrant the same level of assessment. Consistent with good commercial or customary practice, the appropriate level of *environmental site assessment* will be guided by the type of *property* subject to assessment, the expertise and risk tolerance of the *user*, and the information developed in the course of the inquiry.

4.5.4 *Comparison With Subsequent Inquiry*—It should

not be concluded or assumed that an inquiry was not *appropriate inquiry* merely because the inquiry did not identify *recognized environmental conditions* in connection with a *property*. *Environmental site assessments* must be evaluated based on the reasonableness of judgments made at the time and under the circumstances in which they were made. Subsequent *environmental site assessments* should not be considered valid standards to judge the appropriateness of any prior assessment based on hindsight, new information, use of developing technology or analytical techniques, or other factors.

4.6 *Continued Viability of Environmental Site Assessment*—An *environmental site assessment* meeting or exceeding either Practice E 1528 or this practice and completed less than 180 days previously is presumed to be valid. An *environmental site assessment* meeting or exceeding either practice and completed more than 180 days previously may be used to the extent allowed by 4.7 through 4.7.5.

4.7 *Prior Assessment Usage*—Both Practice E 1528 and this practice recognize that *environmental site assessments* performed in accordance with these practices will include information that subsequent *users* may want to use to avoid undertaking duplicative assessment procedures. Therefore, the practices describe procedures to be followed to assist users in determining the appropriateness of using information in *environmental site assessments* performed previously. The system of prior assessment usage is based on the following principles that should be adhered to in addition to the specific procedures set forth elsewhere in these practices:

4.7.1 *Use of Prior Information*—Subject to 4.7.4, *users* and *environmental professionals* may use information in prior *environmental site assessments* provided such information was generated as a result of procedures that meet or exceed the requirements of this practice or Practice E 1528 and then only provided that the specific procedures set forth in the appropriate practice are met.

4.7.2 *Prior Assessment Meets or Exceeds*—Subject to 4.7.4, a prior *environmental site assessment* may be used in its entirety, without regard to the specific procedures set forth in these practices, if, in the reasonable judgment of the *user*: the prior *environmental site assessment* meets or exceeds the requirements of Practice E 1528 or this practice and the conditions at the *property* likely to affect *recognized environmental conditions* in connection with the *property* are not likely to have changed materially since the prior *environmental site assessment* was conducted. In making this judgment, the *user* should consider the type of *property* assessed and the conditions in the area surrounding the *property*.

4.7.3 *Current Investigation*—Except as provided in 4.7.2 and 4.7.2 of Practice E 1528 prior *environmental site assessments* should not be used without current investigation of conditions likely to affect *recognized environmental conditions* in connection with the *property* that may have changed materially since the prior *environmental site assessment* was conducted. At a minimum, for a Phase I Environmental Site Assessment consistent with this practice, a new *site reconnaissance*, interviews, and an update of the *records review* should be performed.

4.7.4 *Actual Knowledge Exception*—If the *user* or *environmental professional(s)* conducting an *environmental site assessment* has *actual knowledge* that the information being used from a prior *environmental site assessment* is not accurate or if it is *obvious*, based on other information obtained by means of the *environmental site assessment* or known to the person conducting the *environmental site assessment*, that the information being used is not accurate, such information from a prior *environmental site assessment* may not be used.

4.7.5 *Contractual Issues Regarding Prior Assessment Usage*—The contractual and legal obligations between prior and subsequent *users* of *environmental site assessments* or between *environmental professionals* who conducted prior *environmental site assessments* and those who would like to use such prior *environmental site assessments* are beyond the scope of this practice.

4.8 *Rules of Engagement*—The contractual and legal obligations between an *environmental professional* and a *user* (and other parties, if any) are outside the scope of this practice. No specific legal relationship between the *environmental professional* and the user is necessary for the *user* to meet the requirements of this practice.

5. User's Responsibilities

5.1 *Scope*—The purpose of this section is to describe tasks that will help identify the possibility of *recognized environmental conditions* in connection with the *property*. These tasks do not require the technical expertise of an *environmental professional* and are generally not performed by *environmental professionals* performing a Phase I Environmental Site Assessment. They may be performed by the *user*.

5.2 *Checking Title Records for Environmental Liens*—*Reasonably ascertainable recorded land title records* (see 7.3.4.4) should be checked to identify *environmental liens*, if any, that are currently recorded against the *property*. Any *environmental liens* so identified shall be reported to the *environmental professional* conducting a Phase I Environmental Site Assessment. This practice does not impose on the *environmental professional* the responsibility to check for recorded *environmental liens*. Rather the *user* should check or engage a title company or title professional to check *reasonably ascertainable recorded land title records* for *environmental liens* currently recorded against the *property*.

5.2.1 *Reasonably Ascertainable*—Environmental liens that are unrecorded or are recorded any place other than *recorded land title records* are not considered to be in *recorded land title records* that are *reasonably ascertainable*. *Recorded land title records* need not be checked if they otherwise do not meet the definition of the term *reasonably ascertainable*.

5.3 *Specialized Knowledge or Experience of the User*—If the *user* is aware of any specialized knowledge or experience that is material to *recognized environmental conditions* in connection with the *property*, it is the user's responsibility to communicate any information based on such specialized knowledge or experience to the *environmental professional*. The *user* should do so before the *environmental professional* does the *site reconnaissance*.

5.4 *Reason for Significantly Lower Purchase Price*—In a transaction involving the purchase of a parcel of *commercial real estate*, if a user has *actual knowledge* that the purchase price of the *property* is significantly less than the purchase price of comparable properties, the *user* should try to identify an explanation for the lower price and to make a written record of such explanation. Among the factors to consider will be the information that becomes known to the *user* pursuant to the Phase I Environmental Site Assessment.

6. Phase I Environmental Site Assessment

6.1 *Objective*—The purpose of this Phase I Environmental Site Assessment is to identify, to the extent feasible pursuant to the processes prescribed herein, recognized environmental conditions in connection with the property. (See 1.1.1.)

6.2 *Four Components*—A Phase I Environmental Site Assessment shall have four components, as described as follows:

6.2.1 *Records Review*—Review of records; see Section 7,

6.2.2 *Site Reconnaissance*—A visit to the property; see Section 8,

6.2.3 *Interviews:*

6.2.3.1 Interviews with current *owners* and *occupants* of the property; see Section 9, and

6.2.3.2 Interviews with local government officials; see Section 10, and

6.2.4 *Report*—Evaluation and report; see Section 11.

6.3 *Coordination of Parts:*

6.3.1 *Parts Used in Concert*—The *records review*, *site reconnaissance*, and *interviews* are intended to be used in concert with each other. If information from one source indicates the need for more information, other sources may be available to provide information. For example, if a previous use of the *property* as a gasoline station is identified through the *records review*, but the present *owner* and *occupants* interviewed report no knowledge of an underground storage tank, the person conducting the *site reconnaissance* should be alert for signs of the presence of an underground storage tank.

6.3.2 *User Obligations*—The *environmental professional* shall note in the report whether or not the *user* has reported to the *environmental professional* any *environmental liens* encumbering the *property* or any specialized knowledge or experience of the *user* that would provide important information about previous ownership or uses of the *property* that may be material to identifying *recognized environmental conditions*.

6.4 *No Sampling*—This practice does not include any testing or sampling of materials (for example, soil, water, air, building materials).

6.5 *Who May Conduct a Phase I:*

6.5.1 *Environmental Professional's Duties*—The *interviews* and *site reconnaissance*, as well as review and interpretation of information upon which the report is based and overseeing the writing of the report, are all portions of a Phase I Environmental Site Assessment that shall be performed by an *environmental professional* or *environmental professionals*. If more than one environmental professional is involved in these tasks, they shall coordinate their efforts.

6.5.2 *Environmental Professional Supervision*—Information for the *records review* needed for completion of a Phase I Environmental Site Assessment may be provided by a number of parties including government agencies, third-party vendors, the *user*, and present and past *owners* and *occupants* of the *property*, provided that the information is obtained by or under the supervision of an *environmental professional* or is obtained by a third-party vendor specializing in retrieval of the information specified in Section 7. Prior assessments may also contain information that will be appropriate for usage in a current *environmental site assessment* provided the prior usage procedures set forth in Sections 7, 8, and 9 are followed. The *environmental professional(s)* participating in the *site reconnaissance* and responsible for the report shall review all of the information provided.

6.5.2.1 *Reliance*—An environmental professional is not required to verify independently the information provided but may rely on information provided unless he or she has *actual knowledge* that certain information is incorrect or unless it is *obvious* that certain information is incorrect based on other information obtained in the Phase I Environmental Site Assessment or otherwise actually known to the *environmental professional*.

7. Records Review

7.1 *Introduction:*

7.1.1 *Objective*—The purpose of the *records review* is to obtain and review records that will help identify *recognized environmental conditions* in connection with the *property*.

7.1.2 *Approximate Minimum Search Distance*—Some records to be reviewed pertain not just to the *property* but also pertain to properties within an additional *approximate minimum search distance* in order to help assess the likelihood of problems from migrating *hazardous substances* or *petroleum products*. When the term *approximate minimum search distance* includes areas outside the *property*, it shall be measured from the nearest *property* boundary. The term *approximate minimum search distance* is used in lieu of radius in order to include irregularly shaped properties.

7.1.2.1 *Reduction of Approximate Minimum Search Distance*—When allowed by 7.2.1.1, the *approximate minimum search distance* for a particular record may be reduced in the discretion of the environmental professional. Factors to consider in reducing the *approximate minimum search distance* include: (*1*) the density (for example, urban, rural, or suburban) of the setting in which the *property* is located; (*2*) the distance that the *hazardous substances* or *petroleum products* are likely to migrate based on local geologic or hydrogeologic conditions; and (*3*) other reasonable factors. The justification for each reduction and the *approximate minimum search distance* actually used for any particular record shall be explained in the report. If the *approximate minimum search distance* is specified as "property only," then the search shall be limited to the *property* and may not be reduced unless the particular record is not *reasonably ascertainable*.

7.1.3 *Accuracy and Completeness*—Accuracy and completeness of record information varies among information sources, including governmental sources. Record information is often inaccurate or incomplete. The *user* or *environmental professional* is not obligated to identify mistakes or insufficiencies in information provided. However, the *environmental professional* reviewing records shall make a reasonable effort to compensate for mistakes or insufficiencies in the information reviewed that are *obvious* in light of other information of which the *environmental professional* has *actual knowledge.*

7.1.4 *Reasonably Ascertainable/Standard Sources*—Availability of record information varies from information source to information source, including governmental jurisdictions. The *user* or *environmental professional* is not obligated to identify, obtain, or review every possible record that might exist with respect to a *property.* Instead, this practice identifies record information that shall be reviewed from standard sources, and the *user* or *environmental professional* is required to review only record information that is reasonably ascertainable from those standard sources. Record information that is *reasonably ascertainable* means (*1*) information that is publicly available, (*2*) information that is obtainable from its source within reasonable time and cost constraints, and (*3*) information that is *practically reviewable.*

7.1.4.1 *Publicly Available*—Information that is *publicly available* means that the source of the information allows access to the information by anyone upon request.

7.1.4.2 *Reasonable Time and Cost*—Information that is obtainable within reasonable time and cost constraints means that the information will be provided by the source within 20 calendar days of receiving a written, telephone, or in-person request at no more than a nominal cost intended to cover the source's cost of retrieving and duplicating the information. Information that can only be reviewed by a visit to the source is *reasonably ascertainable* if the visit is permitted by the source within 20 days of request.

7.1.4.3 *Practically Reviewable*—Information that is *practically reviewable* means that the information is provided by the source in a manner and in a form that, upon examination, yields information relevant to the *property* without the need for extraordinary analysis of irrelevant data. The form of the information shall be such that the *user* can review the records for a limited geographic area. Records that cannot be feasibly retrieved by reference to the location of the *property* or a geographic area in which the *property* is located are not generally *practically reviewable.* Most databases of public records are practically reviewable if they can be obtained from the source agency by the county, city, zip code, or other geographic area of the facilities listed in the record system. Records that are sorted, filed, organized, or maintained by the source agency only chronologically are not generally *practically reviewable.* Listings in publicly available records which do not have adequate address information to be located geographically are not generally considered practically reviewable. For large databases with numerous facility records (such as RCRA generators and registered USTs), the records are not *practically reviewable* unless they can be obtained from the source agency in the smaller geographic area of zip codes. Even when information is provided by zip code for some large databases, it is common for an unmanageable number of sites to be identified within a given zip code. In these cases, it is not necessary to review the impact of all of the sites that are likely to be listed in any

given zip code because that information would not be *practically reviewable.* In other words, when so much data is generated that it cannot be feasibly reviewed for its impact on the *property,* it is not required to be reviewed.

7.1.5 *Alternatives to Standard Sources*—Alternative sources may be used instead of standard sources if they are of similar or better reliability and detail, or if a *standard source* is not *reasonably ascertainable.*

7.1.6 *Coordination*—If records are not *reasonably ascertainable* from *standard sources* or alternative sources, the *environmental professional* shall attempt to obtain the requested information by other means specified in this practice such as questions posed to the current *owner* or *occupant(s)* of the *property* or appropriate persons available at the source at the time of the request.

7.1.7 *Sources of Standard Source Information*—*Standard source* information or other record information from government agencies may be obtained directly from appropriate government agencies or from commercial services. Government information obtained from nongovernmental sources may be considered current if the source updates the information at least every 90 days or, for information that is updated less frequently than quarterly by the government agency, within 90 days of the date the government agency makes the information available to the public.

7.1.8 *Documentation of Sources Checked*—The report shall document each source that was used, even if a source revealed no findings. Sources shall be sufficiently documented, including name, date request for information was filled, date information provided was last updated by source, date information was last updated by original source (if provided other than by original source; see 7.1.7) so as to facilitate reconstruction of the research at a later date.

7.1.9 *Significance*—If a standard *environmental record source* (or other sources in the course of conducting the *Phase I Environmental Site Assessment*) identifies the *property* or another site within the *approximate minimum search distance,* the report shall include the *environmental professional's* judgment about the significance of the listing to the analysis of *recognized environmental conditions* in connection with the *property* (based on the data retrieved pursuant to 7.2, additional information from the government source, or other sources of information). In doing so, the *environmental professional* may make statements applicable to multiple sites (for example, a statement to the effect that none of the sites listed is likely to have a negative impact on the property except . . .).

7.2 *Environmental Information:*

7.2.1 *Standard Environmental Sources*—The following *standard environmental record sources* shall be reviewed, subject to the conditions of 7.1.1 through 7.1.7:

7.2.1.1 *Standard Environmental Record Sources: Federal and State*—The *approximate minimum search distance* may be reduced, pursuant to 7.1.2.1, for any of these *standard environmental record sources* except the Federal NPL site list and Federal RCRA TSD list.

	Approximate Minimum Search Distance, miles (kilometres)
Federal NPL site list	1.0 (1.6)
Federal CERCLIS list	0.5 (0.8)
Federal RCRA CORRACTS TSD facilities list	1.0 (1.6)
Federal RCRA non-CORRACTS TSD facilities list	0.5 (0.8)
Federal RCRA generators list	property and adjoining properties
Federal ERNS list	property only
State lists of hazardous waste sites identified for investigation or remediation:	
State-equivalent NPL	1.0 (1.6)
State-equivalent CERCLIS	0.5 (0.8)
State landfill and/or solid waste disposal site lists	0.5 (0.8)
State leaking UST lists	0.5 (0.8)
State registered UST lists	property and adjoining properties

7.2.2 *Additional Environmental Record Sources: State or Local*—One or more additional state sources or local sources of environmental records may be checked, in the discretion of the *environmental professional*, to enhance and supplement federal and state sources identified above. Factors to consider in determining which local or additional state records, if any, should be checked include (*1*) whether they are *reasonably ascertainable*, (*2*) whether they are sufficiently useful, accurate, and complete in light of the objective of the *records review* (see 7.1.1), and (*3*) whether they are generally obtained, pursuant to local good commercial or customary practice, in initial *environmental site assessments* in the type of *commercial real estate* transaction involved. To the extent additional state sources or local sources are used to supplement the same record types listed above, *approximate minimum search distances* should not be less than those specified above (adjusted as provided in 7.2.1.1 and 7.1.2.1). Some types of records and sources that may be useful include:

Types of Local Records

Lists of Landfill/Solid Waste Disposal Sites
Lists of Hazardous Waste/Contaminated Sites
Lists of Registered Underground Storage Tanks
Records of Emergency Release Reports (SARA § 304)
Records of Contaminated Public Wells

Local Sources

Department of Health/Environmental Division
Fire Department
Planning Department
Building Permit/Inspection Department
Local/Regional Pollution Control Agency
Local/Regional Water Quality Agency
Local Electric Utility Companies (for records relating to PCBs)

7.2.3 *Physical Setting Sources*—A current USGS 7.5 Minute Topographic Map (or equivalent) showing the area on which the *property* is located shall be reviewed, provided it is *reasonably ascertainable*. It is the only *standard physical setting source* and the only *physical setting source* that is required to be obtained (and only if it is *reasonably ascertainable*). One or more additional *physical setting* sources may be obtained in the discretion of the *environmental professional*. Because such sources provide information about the geologic, hydrogeologic, hydrologic, or topographic characteristics of a site, discretionary *physical setting sources* shall be sought when (*1*) conditions have been identified in which *hazardous substances* or *petroleum products* are likely to migrate to the *property* or from or within the *property* into the groundwater or soil and (*2*) more information than is provided in the current USGS 7.5 Minute Topographic Map (or equivalent) is generally obtained, pursuant to local good commercial or customary practice in initial *environmental site assessments* in the type of *commercial real estate transaction* involved, in order to assess the impact of such migration on *recognized environmental conditions* in connection with the *property*.

Mandatory Standard Physical Setting Source

USGS—Current 7.5 Minute Topographic Map (or equivalent)

Discretionary and Non-Standard Physical Setting Sources

USGS and/or State Geological Survey—Groundwater Maps
USGS and/or State Geological Survey—Bedrock Geology Maps
USGS and/or State Geological Survey—Surficial Geology Maps
Soil Conservation Service—Soil Maps
Other Physical Setting Sources that are reasonably credible (as well as reasonably ascertainable)

7.3 *Historical Use Information:*

7.3.1 *Objective*—The objective of consulting historical sources is to develop a history of the previous uses of the *property* and surrounding area, in order to help identify the likelihood of past uses having led to *recognized environmental conditions* in connection with the *property*.

7.3.2 *Uses of the Property*—All *obvious* uses of the *property* shall be identified from the present, back to the *property's obvious* first developed use, or back to 1940, whichever is <u>earlier</u>. This task requires reviewing only as many of the *standard historical sources* in 7.3.4.1 through 7.3.4.8 as are necessary and both *reasonably ascertainable* and likely to be useful (as defined under *Data Failure* in 7.3.2.3). For example, if the *property* was developed in the 1700's, it might be feasible to identify uses back to the early 1900's, using sources such as *fire insurance maps* or *USGS 7.5 minute topographic maps* (or equivalent). Although other sources such as *recorded land title records* might go back to the 1700's, it would not be required to review them unless they were both *reasonably ascertainable* and likely to be useful. As another example, if the *property* was reportedly not developed until 1960, it would still be necessary to confirm that it was undeveloped back to 1940. Such confirmation may come from one or more of the *standard historical sources* specified in 7.3.4.1 through 7.3.4.8, or it may come from *other historical sources* (such as someone with personal knowledge of the *property*; see 7.3.4.9). However, checking other historical sources (see 7.3.4.9) would not be required. For purposes of 7.3.2, the term "developed use" includes agricultural uses and placement of fill. The report shall describe all identified uses, justify the earliest date identified (for example, records showed no development of the *property* prior to the specific date), and explain the reason for any gaps in the history of use (for example, *data failure*).

7.3.2.1 *Intervals*—Review of *standard historical sources* at less than approximately five year intervals is not required by this practice (for example, if the *property* had one use in 1950 and another use in 1955, it is not required to check for a third use in the intervening period). If the specific use of the *property* appears unchanged over a period longer than five years, then it is not required by this practice to research the use during that period (for example, if *fire insurance maps* show the same apartment building in 1940 and 1960, then the period in between need not be researched).

7.3.2.2 *General Type of Use*—In identifying previous uses, more specific information about uses is more helpful than less specific information, but it is sufficient, for purposes of 7.3.2, to identify the general type of use (for example: office, retail, and residential) unless it is *obvious* from the source(s) consulted that the use may be more specifically identified. However, if the general type of use is industrial or manufacturing (for example, *zoning/land use records* show industrial zoning), then additional *standard historical sources* should be reviewed if they are likely to identify a more specific use and are *reasonably ascertainable*, subject to the constraints of *data failure* (see 7.3.2.3).

7.3.2.3 *Data Failure*—A *standard historical source* may be excluded *(1)* if the source is not *reasonably ascertainable*, or *(2)* if past experience indicates that the source is not likely to be sufficiently useful, accurate, or complete in terms of satisfying 7.3.2. *Other historical sources* specified in 7.3.4.9 may be used to satisfy 7.3.2, 7.3.2.1, and 7.3.2.2, but are not required to comply with this practice. Whatever history of previous uses is derived from checking the *standard historical sources* specified in 7.3.4.1 through 7.3.4.8 (except those excluded by *(1)* and *(2)* of 7.3.2.3) shall be deemed sufficient historical use information to comply with this practice.

7.3.3 *Uses of Properties in Surrounding Area*—Uses in the area surrounding the *property* shall be identified in the report, but this task is required only to the extent that this information is revealed in the course of researching the *property* itself (for example, an *aerial photograph* or *fire insurance map* of the *property* will usually show the surrounding area). If the *environmental professional* uses sources that include the surrounding area, surrounding uses should be identified to a distance determined at the discretion of the *environmental professional* (for example, if an aerial photo shows the area surrounding the *property*, then the *environmental professional* shall determine how far out from the *property* the photo should be analyzed). Factors to consider in making this determination include, but are not limited to: the extent to which information is *reasonably ascertainable*; the time and cost involved in reviewing surrounding uses (for example, analyzing *aerial photographs* is relatively quick, but reviewing *property tax files* for adjacent properties or reviewing *local street directories* for more than the few streets that surround the site is typically too time-consuming); the extent to which information is useful, accurate, and complete in light of the purpose of the records review (see 7.1.1); the likelihood of the information being significant to *recognized environmental conditions* in

connection with the *property*; the extent to which potential concerns are *obvious*; known hydrogeologic/geologic conditions that may indicate a high probability of *hazardous substances* or *petroleum products* migration to the property; how recently local development has taken place; information obtained from *interviews* and other sources; and local good commercial or customary practice.

7.3.4 *Standard Historical Sources:*

7.3.4.1 *Aerial Photographs*—The term "aerial photographs" means photographs taken from an airplane or helicopter (from a low enough altitude to allow identification of development and activities) of areas encompassing the *property*. Aerial photographs are often available from government agencies or private collections unique to a local area.

7.3.4.2 *Fire Insurance Maps*—The term *fire insurance maps* means maps produced by private fire insurance map companies that indicate uses of properties at specified dates and that encompass the *property*. These maps are often available at local libraries, historical societies, private resellers, or from the map companies who produced them.

7.3.4.3 *Property Tax Files*—The term *property tax files* means the files kept for property tax purposes by the local jurisdiction where the *property* is located and includes records of past ownership, appraisals, maps, sketches, photos, or other information that is *reasonably ascertainable* and pertaining to the *property*.

7.3.4.4 *Recorded Land Title Records*—The term *recorded land title records* means records of fee ownership, leases, land contracts, easements, liens, and other encumbrances on or of the *property* recorded in the place where land title records are, by law or custom, recorded for the local jurisdiction in which the *property* is located. (Often such records are kept by a municipal or county recorder or clerk.) Such records may be obtained from title companies or directly from the local government agency. Information about the title to the *property* that is recorded in a U.S. district court or any place other than where land title records are, by law or custom, recorded for the local jurisdiction in which the *property* is located, are not considered part of recorded land title records, because often this source will provide only names of previous owners, lessees, easement holders, etc. and little or no information about uses or occupancies of the *property*, but when employed in combination with another source recorded land title records may provide helpful information about uses of the *property*. *This source cannot be the sole historical source consulted. If this source is consulted, at least one additional standard historical source must also be consulted.*

7.3.4.5 *USGS 7.5 Minute Topographic Maps*—The term *USGS 7.5 Minute Topographic Maps* means the map (if any) available from or produced by the United States Geological Survey, entitled "USGS 7.5 minute topographic map," and showing the property.

7.3.4.6 *Local Street Directories*—The term *local street directories* means directories published by private (or sometimes government) sources and showing ownership and/or use of sites by reference to street addresses. Often local street directories are available at libraries of local governments, colleges or universities, or historical societies.

7.3.4.7 *Building Department Records*—The term *building*

department records means those records of the local government in which the property is located indicating permission of the local government to construct, alter, or demolish improvements on the property. Often building department records are located in the *building department* of a municipality or county.

7.3.4.8 *Zoning/Land Use Records*—The term *zoning/land use records* means those records of the local government in which the property is located indicating the uses permitted by the local government in particular zones within its jurisdiction. The records may consist of maps and/or written records. They are often located in the *planning department* of a municipality or county.

7.3.4.9 *Other Historical Sources*—The term *other historical sources* means any source or sources other than those designated in 7.3.4.1 through 7.3.4.8 that are credible to a reasonable person and that identify past uses of the *property*. This category includes, but is not limited to: miscellaneous maps, newspaper archives, and records in the files and/or personal knowledge of the property *owner* and/or *occupants*.

7.4 *Prior Assessment Usage*—Standard historical sources reviewed as part of a prior *environmental site assessment* do not need to be searched for or reviewed again, but uses of the *property* since the prior *environmental site assessment* should be identified either through *standard historical sources* (as specified in 7.3) or by alternatives to *standard historical sources*, to the extent such information is reasonably ascertainable. (See 4.7.)

8. Site Reconnaissance

8.1 *Objective*—The objective of the *site reconnaissance* is to obtain information indicating the likelihood of identifying *recognized environmental* conditions in connection with the property.

8.2 *Observation*—On a visit to the *property* (the *site visit*), the *environmental professional* shall *visually and physically* observe the *property* and any structure(s) located on the *property* to the extent not obstructed by bodies of water, adjacent buildings, or other obstacles.

8.2.1 *Exterior*—The periphery of the *property* shall be *visually and physically observed*, as well as the periphery of all structures on the property, and the *property* should be viewed from all adjacent public thoroughfares. If roads or paths with no apparent outlet are observed on the *property*, the use of the road or path should be identified to determine whether it was likely to have been used as an avenue for disposal of *hazardous substances* or *petroleum products*.

8.2.2 *Interior*—On the interior of structures on the *property*, accessible common areas expected to be used by *occupants* or the public (such as lobbies, hallways, utility rooms, recreation areas, etc.), maintenance and repair areas, including boiler rooms, and a representative sample of *occupant* spaces, should be *visually and physically observed*. It is not necessary to look under floors, above ceilings, or behind walls.

8.2.3 *Methodology*—The *environmental professional* shall document, in the report, the method used (for example, grid patterns or other systematic approaches used for large properties, which spaces for *owner* or *occupants* were observed) to observe the *property*.

8.2.4 *Limitations*—The *environmental professional* shall

document, in the report, general limitations and bases of review, including limitations imposed by physical obstructions such as adjacent buildings, bodies of water, asphalt, or other paved areas, and limiting conditions (for example, snow, rain).

8.2.5 *Frequency*—It is not expected that more than one visit to the *property* shall be made by the *environmental professional* in connection with a *Phase I Environmental Site Assessment*. The one visit constituting part of the *Phase I Environmental Site Assessment* may be referred to as the *site visit*.

8.3 *Prior Assessment Usage*—The information supplied in connection with the *site reconnaissance* portion of a prior environmental site assessment may be used for guidance but shall not be relied upon without determining through a new *site reconnaissance* whether any conditions that are material to *recognized environmental conditions* in connection with the *property* have changed since the prior *environmental site assessment*.

8.4 *Uses and Conditions*—The *environmental profesional(s)* conducting the *site reconnaissance* should note the uses and conditions specified in 8.4.1 through 8.4.4.8 to the extent *visually or physically observed* during the *site visit*. The uses and conditions specified in 8.4.1 through 8.4.4.8 should also be the subject of questions asked as part of *interviews* of *owners* and *occupants* (see Section 9). Uses and conditions to be noted shall be recorded in field notes of the *environmental professional(s)* conducting the *site reconnaissance* but are only required to be described in the report to the extent specified in 8.4.1 through 8.4.4.8. The *environmental professional(s)* performing the *Phase I Environmental Site Assessment* are obligated to identify uses and conditions only to the extent that they may be *visually and physically observed* on a *site visit*, as described in this practice, or to the extent that they are identified by the *interviews* (see Sections 9 and 10) or *record review* (see Section 7) processes described in this practice.

8.4.1 *General Site Setting:*

8.4.1.1 *Current Use(s) of the Property*—The current use(s) of the *property* shall be identified in the report. Any current uses likely to involve the use, treatment, storage, disposal, or generation of *hazardous substances* or *petroleum products* shall be identified in the report. Unoccupied occupant spaces should be noted. In identifying current uses of the *property*, more specific information is more helpful than less specific information. (For example, it is more useful to identify uses such as a hardware store, a grocery store, or a bakery rather than simply retail use.)

8.4.1.2 *Past Use(s) of the Property*—To the extent that indications of past uses of the *property* are *visually or physically observed* on the *site visit*, or are identified in the *interviews* or *record review*, they shall be identified in the report, and past uses so identified shall be described in the report if they are likely to have involved the use, treatment, storage, disposal, or generation of *hazardous substances* or *petroleum products*. (For example, there may be signs indicating a past use or a structure indicating a past use.)

8.4.1.3 *Current Uses of Adjoining Properties*—To the extent that current uses of *adjoining properties* are *visually or physically observed* on the *site visit*, or are identified in the *interviews* or *records review*, they shall be identified in the report, and current uses so identified shall be described in the report if they are likely to indicate *recognized environmental conditions* in connection with the *adjoining properties* or the *property*.

8.4.1.4 *Past Uses of Adjoining Properties*—To the extent that indications of past uses of *adjoining properties* are *visually or physically observed* on the *site visit*, or are identified in the *interviews* or *record review*, they shall be noted by the *environmental professional*, and past uses so identified shall be described in the report if they are likely to indicate *recognized environmental conditions* in connection with the *adjoining properties* or the *property*.

8.4.1.5 *Current or Past Uses in the Surrounding Area*—To the extent that the general type of current or past uses (for example, residential, commercial, industrial) of properties surrounding the *property* are *visually or physically observed* on the *site visit* or going to or from the *property* for the *site visit*, or are identified in the *interviews* or *record review*, they shall be noted by the *environmental professional*, and uses so identified shall be described in the report if they are likely to indicate *recognized environmental conditions* in connection with the *property*.

8.4.1.6 *Geologic, Hydrogeologic, Hydrologic, and Topographic Conditions*—The topographic conditions of the *property* shall be noted to the extent *visually or physically observed* or determined from *interviews*, as well as the general topography of the area surrounding the *property* that is *visually or physically observed* from the periphery of the *property*. If any information obtained shows there are likely to be *hazardous substances* or *petroleum products* on the property or on nearby properties and those *hazardous substances* or *petroleum products* are of a type that may migrate, topographic observations shall be analyzed in connection with geologic, hydrogeologic, hydrologic, and topographic information obtained pursuant to *records review* (see 7.2.3) and *interviews* to evaluate whether *hazardous substances* or *petroleum products* are likely to migrate to the *property*, or within or from the *property*, into groundwater or soil.

8.4.1.7 *General Description of Structures*—The report shall generally describe the structures or other improvements on the *property*, for example: number of buildings, number of stories each, approximate age of buildings, ancillary structures (if any), etc.

8.4.1.8 *Roads*—Public thoroughfares adjoining the *property* shall be identified in the report and any roads, streets, and parking facilities on the *property* shall be described in the report.

8.4.1.9 *Potable Water Supply*—The source of potable water for the *property* shall be identified in the report.

8.4.1.10 *Sewage Disposal System*—The sewage disposal system for the *property* shall be identified in the report. Inquiry shall be made as to the age of the system as part of the process under Sections 7, 9, or 10.

8.4.2 *Interior and Exterior Observations:*

8.4.2.1 *Current Use(s) of the Property*—The current use(s) of the *property* shall be identified in the report. Any current uses likely to involve the use, treatment, storage, disposal, or generation of *hazardous substances* or *petroleum products* shall be identified in the report. Unoccupied *occupant* spaces should be noted. In identifying current uses of the *property*,

more specific information is more helpful than less specific information. (For example, it is more useful to identify uses such as a hardware store, a grocery store, or a bakery rather than simply retail use.)

8.4.2.2 *Past Use(s) of the Property*—To the extent that indications of past uses of the *property* are *visually or physically observed* on the *site visit*, or are identified in the *interviews* or *records review*, they shall be identified in the report, and past uses so identified shall be described in the report if they are likely to have involved the use, treatment, storage, disposal, or generation of *hazardous substances* or *petroleum products*. (For example, there may be signs indicating a past use or a structure indicating a past use.)

8.4.2.3 *Hazardous Substances and Petroleum Products in Connection with Identified Uses*—To the extent that present uses are identified that use, treat, store, dispose of, or generate *hazardous substances* and *petroleum products* on the *property*: (1) the *hazardous substances* and *petroleum products* shall be identified or indicated as unidentified in the report, and (2) the approximate quantities involved, types of containers (if any) and storage conditions shall be described in the report. To the extent that past uses are identified that used, treated, stored, disposed of, or generated *hazardous substances* and *petroleum products* on the *property*, the information shall be identified to the extent it is *visually or physically observed* during the *site visit* or identified from the *interviews* or the *records review*.

8.4.2.4 *Storage Tanks*—Above ground storage tanks, or underground storage tanks or vent pipes, fill pipes or access ways indicating underground storage tanks shall be identified (for example, content, capacity, and age) to the extent *visually or physically observed* during the *site visit* or identified from the *interviews* or *records review*.

8.4.2.5 *Odors*—Strong, pungent, or noxious odors shall be described in the report and their sources shall be identified in the report to the extent *visually or physically observed* or identified from the *interviews* or *records review*.

8.4.2.6 *Pools of Liquid*—Standing surface water shall be noted. Pools or *sumps* containing liquids likely to be *hazardous substances* or *petroleum products* shall be described in the report to the extent *visually or physically observed* or identified from the *interviews* or *records review*.

8.4.2.7 *Drums*—To the extent *visually or physically observed* or identified from the *interviews* or *records review*, *drums* shall be described in the report, whether or not they are leaking, unless it is known that their contents are not *hazardous substances* or *petroleum products* (in that case the contents should be described in the report). *Drums* often hold 55 gal (208 L) of liquid, but containers as small as 5 gal (19 L) should also be described.

8.4.2.8 *Hazardous Substance and Petroleum Products Containers (Not Necessarily in Connection With Identified Uses)*—When containers identified as containing *hazardous substances* or *petroleum products* are *visually or physically observed* on the *property* and are or might be a *recognized environmental condition*: the *hazardous substances* or *petroleum products* shall be identified or indicated as unidentified in the report, and the approximate quantities involved, types of containers, and storage conditions shall be described in the report.

8.4.2.9 *Unidentified Substance Containers*—When open

or damaged containers containing unidentified substances suspected of being *hazardous substances* or *petroleum products* are *visually or physically observed* on the *property*, the approximate quantities involved, types of containers, and storage conditions shall be described in the report.

8.4.2.10 *PCBs*—Electrical or hydraulic equipment known to contain PCBs or likely to contain PCBs shall be described in the report to the extent *visually or physically observed* or identified from the *interviews* or *records review*. Fluorescent light ballast likely to contain PCBs does not need to be noted.

8.4.3 *Interior Observations:*

8.4.3.1 *Heating/Cooling*—The means of heating and cooling the buildings on the property, including the fuel source for heating and cooling, shall be identified in the report (for example, heating oil, gas, electric, radiators from steam boiler fueled by gas).

8.4.3.2 *Stains or Corrosion*—To the extent *visually or physically observed* or identified from the *interviews*, stains or corrosion on floors, walls, or ceilings shall be described in the report, except for staining from water.

8.4.3.3 *Drains and Sumps*—To the extent *visually or physically observed* or identified from the *interviews*, floor drains and *sumps* shall be described in the report.

8.4.4 *Exterior Observations:*

8.4.4.1 *Pits, Ponds, or Lagoons*—To the extent *visually or physically observed* or identified from the *interviews* or *records review*, pits, ponds, or lagoons on the *property* shall be described in the report, particularly if they have been used in connection with waste disposal or waste treatment. *Pits, ponds, or lagoons* on *properties* adjoining the *property* shall be described in the report to the extent they are *visually or physically observed* from the *property* or identified in the *interviews* or *records review*.

8.4.4.2 *Stained Soil or Pavement*—To the extent *visually or physically observed* or identified from the *interviews*, areas of stained soil or pavement shall be described in the report.

8.4.4.3 *Stressed Vegetation*—To the extent *visually or physically observed* or identified from the *interviews*, areas of stressed vegetation (from something other than insufficient water) shall be described in the report.

8.4.4.4 *Solid Waste*—To the extent *visually or physically observed* or identified from the *interviews* or *records review*, areas that are apparently filled or graded by non-natural causes (or filled by fill of unknown origin) suggesting trash or other solid waste disposal, or mounds or depressions suggesting trash or other solid waste disposal, shall be described in the report.

8.4.4.5 *Waste Water*—To the extent *visually or physically observed* or identified from the *interviews* or *records review*, waste water or other liquid (including storm water) or any discharge into a drain, ditch, or stream on or adjacent to the *property* shall be described in the report.

8.4.4.6 *Wells*—To the extent *visually or physically observed* or identified from the *interviews* or *records review*, all wells (including dry wells, irrigation wells, injection wells, abandoned wells, or other wells) shall be described in the report.

8.4.4.7 *Septic Systems*—To the extent *visually or physically observed* or identified from the *interviews* or *records review*, indications of on-site septic systems or cesspools should be described in the report.

9. Interviews With Owners and Occupants

9.1 *Objective*—The objective of *interviews* is to obtain information indicating *recognized environmental conditions* in connection with the *property*.

9.2 *Content*—Interviews with *owners* and *occupants* consist of questions to be asked in the manner and of persons as described in this section. The content of questions to be asked shall attempt to obtain information about uses and conditions as described in Section 8, as well as information described in 9.8 and 9.9.

9.3 *Medium*—Questions to be asked pursuant to this section may be asked in person, by telephone, or in writing, in the discretion of the *environmental professional*.

9.4 *Timing*—Except as specified in 9.8 and 9.9, it is in the discretion of the *environmental professional* whether to ask questions before, during, or after the *site visit* described in Section 8, or in some combination thereof.

9.5 *Who Should be Interviewed:*

9.5.1 *Key Site Manager*—Prior to the *site visit*, the *owner* should be asked to identify a person with good knowledge of the uses and physical characteristics of the *property* (the *key site manager*). Often the *key site manager* will be the property manager, the chief physical plant supervisor, or head maintenance person. (If the *user* is the current *property* owner, the user has an obligation to identify a *key site manager*, even if it is the *user* himself or herself.) If a *key site manager* is identified, the person conducting the *site visit* shall make at least one reasonable attempt (in writing or by telephone) to arrange a mutually convenient appointment for the *site visit* when the *key site manager* agrees to be there. If the attempt is successful, the *key site manager* shall be interviewed in conjunction with the *site visit*. If such an attempt is unsuccessful, when conducting the *site visit*, the *environmental professional* shall inquire whether an identified *key site manager* (if any) or if a person with good knowledge of the uses and physical characteristics of the *property* is available to be interviewed at that time; if so, that person shall be interviewed. In any case, it is within the discretion of the *environmental professional* to decide which questions to ask before, during, or after the *site visit* or in some combination thereof.

9.5.2 *Occupants*—A reasonable attempt shall be made to interview a reasonable number of *occupants* of the *property*.

9.5.2.1 *Multi-Family Properties*—For multi-family residential properties, residential *occupants* do not need to be interviewed, but if the *property* has nonresidential uses, interviews should be held with the nonresidential *occupants* based on criteria specified in 9.5.2.2.

9.5.2.2 *Major Occupants*—Except as specified in 9.5.2.1, if the *property* has five or fewer current *occupants*, a reasonable attempt shall be made to interview a representative of each one of them. If there are more than five current *occupants*, a reasonable attempt shall be made to interview the *major occupant(s)* and those other *occupants* whose operations are likely to indicate *recognized environmental conditions* in connection with the *property*.

9.5.2.3 *Reasonable Attempts to Interview*—Examples of reasonable attempts to interview those *occupants* specified in 9.5.2.2 include (but are not limited to) an attempt to interview such *occupants* when making the *site visit* or calling such *occupants* by telephone. In any case, when there are several *occupants* to interview, it is not expected that the *site visit* must be scheduled at a time when they will all be available to be interviewed.

9.5.2.4 *Occupant Identification*—The report shall identify the occupants interviewed and the duration of their occupancy.

9.5.3 *Prior Assessment Usage*—Persons interviewed as part of a prior *Phase I Environmental Site Assessment* consistent with this practice do not need to be questioned again about the content of answers they provided at that time. However, they should be questioned about any new information learned since that time, or others should be questioned about conditions since the prior *Phase I Environmental Site Assessment* consistent with this practice.

9.6 *Quality of Answers*—The person(s) interviewed should be asked to be as specific as reasonably feasible in answering questions. The person(s) interviewed should be asked to answer in good faith and to the extent of their knowledge.

9.7 *Incomplete Answers*—While the person conducting the interview(s) has an obligation to ask questions, in many instances the persons to whom the questions are addressed will have no obligation to answer them.

9.7.1 *User*—If the person to be interviewed is the user (the person on whose behalf the *Phase I Environmental Site Assessment* is being conducted), the *user* has an obligation to answer all questions posed by the person conducting the interview, in good faith, to the extent of his or her *actual knowledge* or to designate a *key site manager* to do so. If answers to questions are unknown or partially unknown to the *user* or such *key site manager*, this *interview* section of the *Phase I Environmental Site Assessment* shall not thereby be deemed incomplete.

9.7.2 *Non-user*—If the person conducting the interview(s) asks questions of a person other than a user but does not receive answers or receives partial answers, this section of the *Phase I Environmental Site Assessment* shall not thereby be deemed incomplete, provided that (*1*) the questions have been asked (or attempted to be asked) in person or by telephone and written records have been kept of the person to whom the questions were addressed and the responses, or (*2*) the questions have been asked in writing sent by first class mail or by private, commercial carrier and no answer or incomplete answers have been obtained and at least one reasonable follow up (telephone call or written request) was made again asking for responses.

9.8 *Questions About Helpful Documents*—Prior to the *site visit*, the *property owner*, *key site manager* (if any is identified), and *user* (if different from the *property owner*) shall be asked if they know whether any of the documents listed in 9.8.1 exists and, if so, whether copies can and will be provided to the *environmental professional* within reasonable time and cost constraints. Even partial information provided may be useful. If so, the *environmental professional* conducting the *site visit* shall review the available documents prior to or at the beginning of the *site visit*.

9.8.1 *Helpful Documents:*

9.8.1.1 Environment site assessment reports,

9.8.1.2 Environment audit reports,

9.8.1.3 Environmental permits (for example, solid waste disposal permits, hazardous waste disposal permits, waste-water permits, NPDES permits),

9.8.1.4 Registrations for underground and above-ground storage tanks,

9.8.1.5 Material safety data sheets,

9.8.1.6 Community right-to-know plan,

9.8.1.7 Safety plans; preparedness and prevention plans; spill prevention, countermeasure, and control plans; etc.,

9.8.1.8 Reports regarding hydrogeologic conditions on the property or surrounding area,

9.8.1.9 Notices or other correspondence from any government agency relating to past or current violations of environmental laws with respect to the property or relating to environmental liens encumbering the property,

9.8.1.10 Hazardous waste generator notices or reports, and

9.8.1.11 Geotechnical studies.

9.9 *Proceedings Involving the Property*—Prior to the *site visit*, the *property owner*, *key site manager* (if any is identified), and *user* (if different from the *property owner*) shall be asked whether they know of: (*1*) any pending, threatened, or past litigation relevant to *hazardous substances* or *petroleum products* in, on, or from the *property*; (*2*) any pending, threatened, or past administrative proceedings relevant to *hazardous substances* or *petroleum products* in, on or from the *property*; and (*3*) any notices from any governmental entity regarding any possible violation of environmental laws or possible liability relating to *hazardous substances* or *petroleum products*.

10. Interviews With Local Government Officials

10.1 *Objective*—The objective of *interviews* with *local government officials* is to obtain information indicating *recognized environmental conditions* in connection with the *property*.

10.2 *Content*—Interviews with *local government officials* consist of questions to be asked in the manner and of persons as described in this section. The content of questions to be asked shall be decided in the discretion of the *environmental professional(s)* conducting the *Phase I Environmental Site Assessment*, provided that the questions shall generally be directed towards identifying *recognized environmental conditions* in connection with the *property*.

10.3 *Medium*—Questions to be asked may be asked in person or by telephone, in the discretion of the *environmental professional*.

10.4 *Timing*—It is in the discretion of the *environmental professional* whether to ask questions before or after the *site visit* described in Section 8, or in some combination thereof.

10.5 *Who Should Be Interviewed:*

10.5.1 *Local Agency Officials*—A reasonable attempt shall be made to interview at least one staff member of any one of the following types of local government agencies:

10.5.1.1 Local fire department that serves the *property*,

10.5.1.2 Local health agency or local/regional office of state health agency serving the area in which the *property* is located, or

10.5.1.3 Local agency or local/regional office of state agency having jurisdiction over hazardous waste disposal or other environmental matters in the area in which the *property* is located.

10.6 *Prior Assessment Usage*—Persons interviewed as

part of a prior *Phase I Environmental Site Assessment* consistent with this practice do not need to be questioned again about the content of answers they provided at that time. However, they should be questioned about any new information learned since that time, or others should be questioned about conditions since the prior *Phase I Environmental Site Assessment* consistent with this practice.

10.7 *Quality of Answers*—The person(s) interviewed should be asked to be as specific as reasonably feasible in answering questions. The person(s) interviewed should be asked to answer in good faith and to the extent of their knowledge.

10.8 *Incomplete Answers*—While the person conducting the *interview(s)* has an obligation to ask questions, in many instances the persons to whom the questions are addressed will have no obligation to answer them. If the person conducting the *interview(s)* asks questions but does not receive answers or receives partial answers, this section shall not thereby be deemed incomplete, provided that questions have been asked (or attempted to be asked) in person or by telephone and written records have been kept of the person to whom the questions were addressed and their responses.

11. Evaluation and Report Preparation

11.1 *Report Format*—The report of findings for the *Phase I Environmental Site Assessment* should generally follow the recommended report format attached as Appendix X2 unless otherwise required by the user.

11.2 *Documentation*—The report should include documentation (for example, references, key exhibits) to support the analysis, opinions, and conclusions found in the report. All sources, including those that revealed no findings, should be sufficiently documented to facilitate reconstruction of the research at a later date.

11.3 *Contents of Report*—The report shall include those matters required to be included in the report pursuant to various provisions of this practice. In addition, the report shall state whether the *user* reported to the *environmental professional* any information pursuant to the user's responsibilities described in Section 5 of this practice (for example, an *environmental lien* encumbering the *property* or any relevant specialized knowledge or experience of the user).

11.4 *Credentials*—The report shall name the *environmental professional(s)* involved in conducting the *Phase I Environmental Site Assessment*. One of two options shall be available to address qualifications of those persons involved in conducting the *Phase I Environmental Site Assessment*: (*1*) the report shall include a qualifications statement of the *environmental professional(s)* responsible for the *Phase I Environmental Site Assessment* and preparation of the report, and the qualifications statement shall include relevant individual and corporate qualifications; or (*2*) a written qualifications statement of the *environmental professional(s)* responsible for the *Phase I Environmental Site Assessment* and preparation of the report, including relevant individual and corporate qualifications, shall be delivered to the *user*.

11.5 *Opinion*—All evidence of recognized environmental conditions shall be described in full. The report shall include the *environmental professional's* opinion of the impact of *recognized environmental conditions* in connection with the *property*.

11.6 *Findings and Conclusions*—The report shall have a findings and conclusions section that states one of the following:

11.6.1 "We have performed a *Phase I Environmental Site Assessment* in conformance with the scope and limitations of ASTM Practice E 1527 of [insert address or legal description], the *property*. Any exceptions to, or deletions from, this practice are described in Section [ ] of this report. This assessment has revealed no evidence of recognized environmental conditions in connection with the property," or

11.6.2 "We have performed a *Phase I Environmental Site Assessment* in conformance with the scope and limitations of ASTM Practice E 1527 of [insert address or legal description], the property. Any exceptions to, or deletions from, this practice are described in Section [ ] of this report. This assessment has revealed no evidence of recognized environmental conditions in connection with the *property* except for the following: (list)."

11.7 *Deviations*—All deletions and deviations from this practice (if any) shall be listed individually and in detail and all additions should be listed.

11.8 *Signature*—The environmental professional(s) responsible for the *Phase I Environmental Site Assessment* shall sign the report.

11.9 *Additional Services*—Any additional services contracted for between the user and the environmental professional(s), including a broader scope of assessment, more detailed conclusions, liability/risk evaluations, recommendation for Phase II testing, remediation techniques, etc., are beyond the scope of this practice, and should only be included in the report if so specified in the terms of engagement between the user and the *environmental professional*.

12. Non-Scope Considerations

12.1 *General:*

12.1.1 *Additional Issues*—There may be environmental issues or conditions at a property that parties may wish to assess in connection with *commercial real estate* that are outside the scope of this practice (the non-scope considerations). As noted by the legal analysis in Appendix X1 of this practice, some substances may be present on a property in quantities and under conditions that may lead to contamination of the property or of nearby properties but are not included in CERCLA's definition of hazardous substances (42 USC § 9601(14)) or do not otherwise present potential CERCLA liability. In any case, they are beyond the scope of this practice.

12.1.2 *Outside Standard Practices*—Whether or not a *user* elects to inquire into non-scope considerations in connection with this practice or any other environmental site assessment, no assessment of such non-scope considerations is required for appropriate inquiry as defined by this practice.

12.1.3 *Other Standards*—There may be standards or protocols for assessment of potential hazards and conditions associated with non-scope conditions developed by governmental entities, professional organizations, or other private entities.

12.1.4 *List of Additional Issues*—Following are several non-scope considerations that persons may want to assess in connection with commercial real estate. No implication is intended as to the relative importance of inquiry into such non-scope considerations, and this list of non-scope considerations is not intended to be all-inclusive:

12.1.4.1 Asbestos-Containing Materials,

12.1.4.2 Radon,

12.1.4.3 Lead-Based Paint,

12.1.4.4 Lead in Drinking Water, and

12.1.4.5 Wetlands.

APPENDIXES

(Nonmandatory Information)

X1. LEGAL BACKGROUND TO FEDERAL LAW AND THE PRACTICES ON ENVIRONMENTAL ASSESSMENTS IN COMMERCIAL REAL ESTATE TRANSACTIONS

INTRODUCTION

The legal section of Subcommittee E50.02 on Environmental Assessments In Commercial Real Estate Transactions provides the following background to the Comprehensive Environmental Response, Compensation, and Liability Act (CERCLA), as amended including amended by the Superfund Amendments and Reauthorization Act (SARA), 42 USC § 9601 *et seq.* The background to CERCLA, commonly known as the Superfund law, outlines the potential liability for the cleanup of hazardous substances, available defenses to such liability, appropriate inquiry under Superfund, statutory definition of hazardous substances, petroleum products and petroleum exclusion to CERCLA, and reasons why certain environmental-hazards are excluded from the scope of Superfund and this practice and Practice E 1528.

There are several elements of Superfund liability and the commonly termed "innocent purchaser" defense, that arises out of the statutory third-party defense, that may impact on the development and understanding of this practice and Practice E 1528.

X1.1 *Superfund Liability:*

X1.1.1 All of the following elements of liability under Superfund must be established by a plaintiff before a defendant will be held liable under superfund for a *government's* response costs:[7]

X1.1.1.1 The site is a facility, as defined at § 9601(9),

X1.1.1.2 A release or threatened release of a hazardous substance from the site occurred (release is defined at § 9601(22) as *any* amount of any hazardous substance; "hazardous substance" is defined at § 9601(14) (see statutory definition of "hazardous substance"),

X1.1.1.3 A release or threatened release caused the plaintiff to incur response costs. Response costs are defined at § 9601(25) to mean costs related to both removal actions (§ 9601(23)) and remedial actions (§ 9601(24)), and

X1.1.1.4 Defendants fall within at least one of the four classes of responsible parties.[8]

X1.1.2 In order to recover response costs, a government plaintiff must prove that the costs were not inconsistent with the National Oil and Hazardous Substances Pollution Contingency Plan (commonly referred to as the National Contingency Plan or NCP), 40 CFR § 300.[9] A private plaintiff must prove its costs were necessary costs of response and that the response action was consistent with the NCP. 42 USC § 9607 (a).[10]

X1.1.3 If there is a release or threatened release of hazardous substances on a site, private parties, even if they are not PRPs, may decide to incur response costs and seek recovery from other private parties, and PRPs may seek contribution from other PRPs.

X1.1.4 There is an important difference between government's burden to show that its response costs are "not inconsistent with the NCP" and the burden a private party bears to show that its response costs are "consistent with the NCP". See § 9607 (a)(4)(A) and (B). Courts have interpreted this statutory difference to give the government a rebuttable presumption that its response costs are consistent with the

[7] 42 USC §9607(a). (All statutory references are to Title 42 of the United States Code, unless otherwise specified.) See *United States versus Aceto Agricultural Chemicals Corp.*, 872 F.2d 1373 (8th Cir. 1989). Private plaintiffs, as well as the government, may seek response costs under Superfund from defendants. While many users of these ASTM practices or other private parties may think in terms of how to defend against Superfund liability, they should recognize that they may decide to conduct cleanup actions and seek response costs from other parties.

[8] The four classes of potentially responsible parties (PRPs) are listed as § 9607(a) as follows:

(1) Owner and operator of a facility (See § 9601 (20)(A): the term "owner or operator" does not include a person, who, without participating in the management of a facility, holds indicia of ownership primarily to protect his security interest in the facility. In *re: Bergsoe Metal Corporation*, 910 F.2d 668 (9th Cir. Aug. 9, 1990); *Guidice versus BFG Electroplating and Manufacturing Co.*, 732 F.Supp. 556 (W.D.Pa. 1989); *United States v. Mirabile*, 23 ERC 1511 (E.D.Pa. 1985). But see *United States versus Fleet Factors Corp.*, 901 F.2d 1550 (11th Cir. 1990); *United States versus Maryland Bank and Trust Co.*, 632 F. Supp. 573 (D. Md. 1986). For

clarification of the security interest exclusion, see EPA's rule on lender liability under CERCLA, 57 Federal Register 18344 (April 29, 1992),

(2) Any person who at the time of disposal of any hazardous substance owned or operated any facility at which such hazardous substances were disposed of,

(3) Any person who by contract, agreement, or otherwise arranged for disposal or treatment or transport of hazardous substances, and

(4) Any person who accepts hazardous substances for transport to a facility selected by such person.

[9] The National Contingency Plan is the federal government's blueprint on how hazardous substances are to be cleaned up pursuant to CERCLA.

[10] See *Dedham Water Co. versus Cumberland Farms Dairy, Inc.*, 889 F.2d 1146 (1st Cir. 1989); other cases cited at ABA, *Natural Resources, Energy, and Environmental Law: 1989 The Year In Review*, p. 215, Note 155.

NCP, whereas a private party who undertakes response costs and seeks recovery from responsible parties bears the burden of proving its response was consistent with the NCP.[11] The EPA takes the position that a private party who undertakes a response action must be only in "substantial compliance", rather than strict technical compliance, with the NCP, as long as a CERCLA-quality cleanup is achieved. The NCP requirements for a private party response-action are set forth at 40 CFR § 300.700.

X1.2 *Defenses to Liability:*

X1.2.1 Assuming all the elements of liability exist, a party may avoid liability only by meeting one of the defenses listed in § 9607(b). These listed defenses are exclusive of all others.[12] Section 9607(b) states (*emphasis added*):

"There shall be no liability under subsection (a) for a person otherwise liable who can establish by a preponderance of the evidence [the lowest evidentiary standard available, meaning more probable than not] that the release or threat of release of a hazardous substance and the damages resulting therefrom were *caused solely by—*
1) an act of God;
2) an act of war;
3) *an act or omission of a third party other* than an employee or agent of the defendant, or *than one whose act or omission occurs in connection with a contractual relationship* [see the definition of "contractual relationship" in X1.2.2], existing directly or indirectly, with the defendant . . ., if the defendant establishes by a preponderance of the evidence that (a) he exercised due care with respect to the hazardous substance concerned . . ., and (b) he took precautions against foreseeable acts or omissions of any such third party and the consequences that could foreseeably result from such actions or omissions."

X1.2.2 Under § 9601(35)(A), a contractual relationship "includes, but is not limited to, land contracts, deeds, or other instruments transferring title or possession . . .". These contractual relationships with third parties eliminate the defense to liability unless the defendant is an innocent purchaser. Or as stated by the statute at § 9601(35)(A)(emphasis added), a contractual relationship with the third party defeats the defense.

"unless the real property on which the facility is located was acquired by the defendant after disposal or placement of the hazardous substance . . . and one or more of the following circumstances is also established by the defendant by a preponderance of the evidence:
    (i) At the time the defendant acquired the facility the defendant *did not know and had no reason to know* that any hazardous substance which is the subject of the release or threatened release was disposed of on, in, or at the facility.
    (ii) The defendant is the government . . .
    (iii) The defendant acquired the facility by inheritance or bequest."

X1.2.3 Therefore, the so-called innocent purchaser de-

fense arises out of the third-party defense of § 9607(b)(3). Restated, this defense to Superfund liability is available only if the defendant shows the following:

X1.2.3.1 The release or threat of release was caused solely by a third party,

X1.2.3.2 The third party is not an employee or agent of the defendant,

X1.2.3.3 The acts or omissions of the third party did not occur in connection with a direct or indirect contractual relationship to the defendant, or if there was a contractual relationship, the defendant acquired the property after disposal or placement of the hazardous substance, and at the time the defendant acquired the facility the defendant *did not know and had no reason to know* that any hazardous substance that is the subject of the release or threatened release was disposed of on, in, or at the facility, and

X1.2.3.4 The defendant exercised due care with respect to the hazardous substances and took precautions against foreseeable acts or omissions of the third party.

X1.2.4 The statute then states at § 9601(35)(B) (*emphasis added*):

"To establish that the defendant had no reason to know, as provided [above], *the defendant must have undertaken, at the time of acquisition, all appropriate inquiry into the previous ownership and uses of the property consistent with good commercial or customary practice* in an effort to minimize liability. . . . [T]he court shall take into account any specialized knowledge or experience on the part of the defendant, the relationship of the purchase price to the value of the property if uncontaminated, commonly known or reasonably ascertainable information about the property, the obviousness of the presence or likely presence of contamination at the property, and the ability to detect such contamination by appropriate inspection."

X1.3 *Appropriate Inquiry in Commercial Real Estate Transactions:*

X1.3.1 One of the major questions that parties to commercial real estate transactions face when considering their potential Superfund liability is, "What level of inquiry into the previous ownership and uses of the property is appropriate to establish the innocent purchaser defense to Superfund liability?" These practices are structured to articulate the level of inquiry under Superfund that is appropriate for different situations.

X1.3.2 *The Appropriate Level of Inquiry:*

X1.3.2.1 The level of environmental inquiry that is appropriate under Superfund cannot be the same for every property or every party to a real estate transaction. The level of inquiry, in fact, will change depending on the particular property or party involved in a transaction. The statutory language, Congressional history, and common sense support this conclusion.

X1.3.2.2 First, it must be noted that little case law exists to serve as guidance about the minimum level of inquiry that will be deemed appropriate for the innocent purchaser defense. See, for example, *United States versus Serafini*, 706 F. Supp. 346 (M.D. Pa. 1988), and 1990 U.S. Dist. LEXIS 18466 (M.D. Pa. 1990) (By entertaining disputed facts as to the custom and practice of viewing land prior to purchase, the court implied that appropriate inquiry necessarily varies on a site-by-site basis); *United States versus Pacific Hide and Fur Depot, Inc.*, 716 F. Supp. 1341 (D. Idaho 1989) (No

[11] *Amland Properties Corp. versus Aluminum Co. of America*, 711 F. Supp. 784, 794 (D. N.J. 1989); *Artesian Water Co. versus New Castle County*, 659 F. Supp. 1269, 1291 (D. Del. 1987); *United States versus Northeastern Pharmaceutical and Chemical Co.*, 579 F. Supp. 823 (W.D. Mo. 1984), aff'd in part, rev'd on other grounds, 810 F.2d 726 (8th Cir. 1986).
[12] *United States versus Aceto Agricultural Chemicals Corp.*, 872 F.2d 1373 (8th Cir. 1989). But see *United States versus Marisol, Inc.*, 725 F. Supp. 833 (M.D. Pa. 1989) (equitable defenses under CERCLA may be available after the development of a factual record).

inquiry was required by those who received an ownership interest in property via corporate stock transfer and warranty deed under the facts of this case); *International Clinical Laboratories, Inc. versus Stevens*, 30 ERC 2066, 20 ELR 20,560 (E.D.N.Y. 1990). (Despite a long history of toxic wastewater disposal and presence of the site on the state's hazardous waste disposal site list, the purchaser established the innocent purchaser defense since there were no visible environmental problems at the site, the defendant had no knowledge of environmental problems at the site and the purchase price did not reflect a reduction on account of the problem.)

X1.3.2.3 While the statute does not specifically distinguish certain types of properties and uses from others, or certain types of parties from others, it does list certain factors courts should consider in determining whether one's inquiry under the circumstances is appropriate. The statute, as explained in X1.2, requires a court to consider a party's specialized knowledge or experience. The statute further mandates a court to consider what is *"reasonably ascertainable information about the property"*, what contamination is *obviously* present, and the party's *"ability to detect such contamination"*. The very use of terms such as "appropriate" and "reasonably", and the use of "specialized knowledge and experience" and "ability" in conjunction with the specific person attempting to utilize the defense signifies that Congress did not intend the appropriateness of the inquiry be judged by a bright line standard. If it so intended, Congress would have stated, but did not, that the same inquiry should be made in every case.

X1.3.2.4 What is reasonable and obvious to one party may not be so to other parties, and ability, by necessity, varies among all parties. The statute, therefore, recognizes that different properties and parties must be treated differently. That is, different parties may conduct different levels of inquiry appropriate to their circumstances.

X1.3.2.5 The statutory standard of "appropriate inquiry" suggests the level of inquiry will depend on the circumstances and the underlying facts. Since the facts are almost always different, the level of inquiry must change with them. The legislative history on this particular issue demonstrates that Congress intended that the level of inquiry changes with the type of property and party:[13]

> The duty to inquire under this provision shall be judged as of the time of acquisition. Defendants shall be held to a higher standard as public awareness of the hazards associated with hazardous releases has grown, as reflected by this Act, the 1980 Act [CERCLA] and other Federal and State statutes.
> Moreover, good commercial or customary practice with respect to inquiry in an effort to minimize liability shall mean that a reasonable inquiry must have been made in all circumstances, in light of best business and land transfer principles.
> Those engaged in commercial transactions should, however, be held to a higher standard than those who are engaged in private residential transactions.

X1.3.2.6 Because few cases address the standard of inquiry in the innocent purchaser defense and the legislative

history describing Congressional intent is sparse, common sense is a useful guide in interpreting statutory language. If CERCLA mandated that the level of inquiry be the same for every property or potential defendant, then a lay consumer (renter or buyer) of a home, a purchaser of a small environmentally benign business, and a multinational corporate buyer of an industrial complex would have to conduct the same environmental site assessment (ESA) of the different properties in question. Additionally, the statute makes no mention of a Phase I ESA or any other specific type of inquiry one is to conduct in order for the inquiry to be deemed appropriate. If all inquires had to be at the same level to be "appropriate", it would be illogical to stop at a Phase I ESA since some commercial or industrial properties routinely undergo, in the exercise of good commercial and customary practices, intrusive sampling (typically a Phase II ESA activity). Therefore, since routinely some properties undergo sampling, an inflexible standard would require sampling of all properties, no matter what its use. This could not have been the intended result of SARA.

X1.3.3 *The Minimum Inquiries to Satisfy All Appropriate Inquiry:*

X1.3.3.1 Recognizing that inquiry changes with the underlying circumstances, the next question concerns that level of inquiry, if any, that Superfund requires to utilize the innocent landowner defense.

X1.3.3.2 As noted above, in some real estate transactions a Phase II ESA is routinely conducted. A Phase I ESA is conducted in these transactions only as a necessary prerequisite to outline the scope of the Phase II ESA. A Phase II ESA typically involves taking soil, water, and air samples to determine their contaminant content or verify that no contaminants are present or likely present. Note, however, that this simplistic outline of the Phase II ESA is misleading since the party can always dig down one foot deeper, take one more sample, or conduct one more test. The problem of how much inquiry is conducted, or at what level a party should begin, involves proving a negative, that is, that no contamination is present.[14] Since, according to the statute, inquiries should be judged by the circumstances existing at the time of acquisition, then there could be some properties and parties to real estate transactions where it may be appropriate to begin the inquiry with an intrusive Phase II ESA in order to invoke the innocent purchaser defense to liability.

X1.3.3.3 At the other extreme, the minimum level of inquiry that a party would be expected to conduct is found by looking at the least environmentally obtrusive class of

---

[13] H.R. Rep. No. 962, 99th Cong., 2d Sess. 187 (1986), *reprinted at* 1986 U.S. Code Cong. and Admin. News 3276, 3280.

[14] The inability to prove a negative creates a dilemma for the potential defendant. If the party's inquiry discovers contamination, then under the statute the party will not be able to avail itself of the innocent purchaser defense. If the inquiry does not discover contamination, EPA or another private party can argue in a response action that the inquiry was not "appropriate" and, therefore, the defendant can have no defense. This dilemma is explicitly recognized by the Subcommittee E50.02 as beyond any reasonable interpretation of Congressional intent. The scope of the E50.02 Standard Practices resolves the party's dilemma in the only reasonable way by stating: "It should not be concluded or assumed that the inquiry was not appropriate inquiry merely because the inquiry did not identify existing recognized environmental conditions in connection with a property. Environmental site assessments must be evaluated based on the reasonableness of the judgments made at the time and under the circumstances in which they were made." See 4.5.4.

property and party from a CERCLA perspective. This transaction likely involves the lay buyer of a home or the renter of an apartment. Assuming these parties meet the other prerequisites for the innocent purchaser defense, what level of environmental inquiry must they conduct to avoid Superfund liability? While there are no recorded court cases on this issue, the answer is probably none, unless a particular residential purchaser or renter has some specialized knowledge about or experience with the property in question that would lead a court to conclude that some questions should have been asked. Beyond these rare situations, it is highly unlikely that Congress intended to saddle housing consumers with the burden of investigating or cleaning up contaminated sites. In fact, EPA has issued a statement of enforcement policy to the effect that it will not generally pursue owners of single family residences pursuant to CERCLA.[15] Therefore, for some properties and parties to real estate transactions, it is appropriate to conduct no environmental inquiry in order to meet one's innocent purchaser defense to liability.

X1.3.3.4 The minimum level of appropriate inquiry under Superfund, therefore, ranges from no specialized inquiry to conducting an intrusive Phase II ESA. In order to satisfy these practices, to do no specialized inquiry, such as the Transaction Screen or Phase I ESA, is not enough for commercial real estate transactions. Under current commercial and customary practice and in light of best business and land transfer principles, however, no environmental site assessments are conducted in many real estate transactions, particularly those involving smaller properties, vacant land, or transactions of low monetary value. This practice and Practice E 1528 and the minimum level of inquiry under these practices, actually raises the average level of inquiry that should be performed where the parties want to come within the protection of the innocent landowner defense.

X1.3.3.5 The burden of proof is on the defendant to sustain by a preponderance of the evidence, the innocent purchaser defense. This is the least onerous burden of proof available to a party in litigation. The defendant must show only that the evidence offered to support the level of inquiry that was taken at the time of acquisition is of greater weight or more convincing than the evidence offered in opposition to it. In other words, the evidence on the inquiry issue taken as a whole shows that the fact sought to be proved is more probable than not. There may be technical or business judgments on whether the inquiry conducted or any other fact in a particular case is sufficient to meet the needs or concerns of a party to the real estate transaction. The bottom line, however, is that the judgment on whether the specific facts of a case, in light of statutory language, are sufficient to produce liability or a viable defense to liability is a legal one and such judgments constitute the practice of law.

X1.3.3.6 Practice E 1528 is designed as the minimum level of inquiry to satisfy the practice from which a party to a commercial real estate transaction should proceed, recognizing that some parties to some commercial real estate transactions may wish to proceed by beginning with a Phase I or a Phase II ESA.

---

[15] EPA, *Policy Towards Owners of Residential Property at Superfund Sites*, OSWER Directive No. 9834.6, July 3, 1991.

X1.4 *Statutory Definition of Hazardous Substance:*

X1.4.1 The statute at 42 USC § 9601(14)(A − F) defines hazardous substance by referring to five other statutes as well as to Superfund's own § 9602. The following is a description of the relevant portions of the other statutes and § 9602 of Superfund:

42 USC § 9601(14)(A): "[A]ny substance designated pursuant to section 1321(b)(2)(A) of Title 33." Title 33 USC § 1321 lies within the Clean Water Act and refers to, among other things, hazardous substance liability. 33 USC § 1321(b)(2)(A) states that the EPA shall develop, "as may be appropriate, regulations designating as hazardous substances, other than oil as defined in this section, such elements and compounds which, when discharged in any quantity into" the navigable waters of the United States . . ., present an imminent and substantial danger to the public . . . health or welfare, including, but not limited to, fish, shellfish, wildlife, shorelines, and beaches."

42 USC § 9601(14)(B): "[A]ny element, compound, mixture, solution, or substance designated pursuant to section 9602 of this title." Section 9602 gives EPA the authority to designate as a hazardous substance "such elements, compounds, mixtures, solutions, and substances which, when released into the environment may present substantial danger to the public health or welfare or the environment. . . ."

42 USC § 9601(14)(C): "[A]ny hazardous waste having the characteristics identified under or listed pursuant to section 3001 of the Solid Waste Disposal Act [42 USCA § 6921] (but not including any waste the regulation of which under the Solid Waste Disposal Act [42 USCA § 6901 *et seq.*] has been suspended by Act of Congress)." The Solid Waste Disposal Act of 1980 amended the Resource Conservation and Recovery Act (RCRA). 42 USCA § 6921 of RCRA provides authority to the EPA to develop criteria for identifying characteristics of hazardous waste and for listing particular hazardous wastes within the meaning of 42 USCA § 6903(5) of RCRA. RCRA, § 6903(5), defines hazardous waste to mean

"a solid waste, or combination of solid wastes, which because of its quantity, concentration, or physical, chemical, or infectious characteristics may—

(A) cause, or significantly contribute to an increase in mortality or an increase in serious irreversible, or incapacitating reversible, illness; or

(B) pose a substantial present or potential hazard to human health or the environment when improperly treated, stored, transported, or disposed of, or otherwise managed."

For the identification and listing of hazardous wastes under RCRA, see 40 CFR §§ 261.1 *et seq.*

42 USC § 9601(14)(D): "[A]ny toxic pollutant listed under Section 1317(a) of Title 33." Section 1317(a) of Title 33 refers to toxic and pretreatment effluent standards under the Clean Water Act. The EPA is charged in this section with publishing and revising from time to time a list of toxic pollutants, taking "into account toxicity of the pollutant, its persistence, degradability, the usual or potential presence of the affected organisms in any waters, the importance of the affected organisms, and the nature and extent of the effect of the toxic pollutant on such organisms." Each toxic pollutant listed according to this section shall be subject to effluent limitations. For toxic pollutant effluent standards, see 40 CFR §§ 129.1 *et seq.*

42 USC § 9601(14)(E): "[A]ny hazardous air pollutant listed under Section 112 of the Clean Air Act [42 USCA § 7412]." Section 7412 of Title 42 deals with national emission standards for hazardous air pollutants. The EPA is charged here with publishing and revising from time to time "a list which includes each hazardous air pollutant for which [it] intends to establish an

emission standard under this section." The term "hazardous air pollutant" means an air pollutant that in EPA's judgment "causes, or contributes to, air pollution which may reasonably be anticipated to result in an increase in mortality or an increase in serious irreversible, or incapacitating reversible, illness." For emission standards for hazardous pollutants, see 40 CFR § 61.01 *et seq.*

42 USC § 9601(14)(F): "[A]ny imminently hazardous chemical substance or mixture with respect to which the [EPA] has taken action pursuant to Section 2606 of Title 15." Section 2606 of Title 15 deals with imminent hazards under the Toxic Substances Control Act (TSCA). The EPA is authorized under 15 USC § 2606 to seize an imminently hazardous chemical substance or mixture or seek other relief, such as requiring notice to users of the chemical substance or public notice of the risk associated with the substance or mixture. The term " 'imminently hazardous chemical substance or mixture' means a chemical substance or mixture which presents an imminent and unreasonable risk of serious or widespread injury to health or the environment."

X1.4.2 After Subsections A–F, outlined above, the Superfund definition of "hazardous substance" in § 9601(14) then goes on to state:

"The term does not include petroleum, including crude oil or any fraction thereof which is not otherwise specifically listed or designated as a hazardous substance under Subparagraphs (A) through (F) of this paragraph, and the term does not include natural gas, natural gas liquids, liquefied natural gas, or synthetic gas usable for fuel (or mixtures of natural gas and such synthetic gas)."

X1.4.3 The EPA has collected a list of "those substances in the statutes referred to in Section 101(14) of the Act [42 USC § 9601(14)]" 40 CFR § 302.1 (1989) ("List of Hazardous Substances And Reportable Quantities," 40 CFR Part 302). This list changes with notices in the Federal Register. Also, any time a new hazardous waste is listed, the waste automatically becomes a hazardous substance.

X1.5 *Petroleum Products:*

X1.5.1 Under the petroleum exclusion of CERCLA (42 USC § 9601(14)), petroleum and crude oil have been explicitly excluded from the definition of hazardous substances under CERCLA. Nevertheless, petroleum products are included within the scope of both practices because they are of concern in many commercial real estate transactions and current custom and usage is to include an inquiry into the presence of petroleum products in an environmental site assessment. Inclusion of petroleum products within the scope of the practices is not based upon the applicability, if any, of CERCLA to petroleum products.

X1.5.2 One reason to include petroleum products within the scope of the practices is because to do so reflects custom and usage: when environmental assessments are conducted in connection with commercial real estate transactions, they customarily include an assessment of the presence or likely presence of petroleum products under conditions that may lead to contamination. For example, environmental assessments ordinarily seek to assess whether there may be underground or above-ground storage tanks that may be leaking, whether those tanks contain petroleum products or some other product.

X1.5.3 In addition, although CERCLA may exclude petroleum products, other laws require cleanup of releases or

spills of petroleum products. For example, petroleum products sometimes (for example, when they cannot be reclaimed from soil) become hazardous wastes subject to RCRA Subtitle C (42 USC § 6921 *et seq.*), must be cleaned up if released from underground storage tanks pursuant to RCRA Subtitle I (42 USC § 6991 *et seq.*), must be cleaned up pursuant to the Oil Pollution Act of 1990 (33 USC § 1321 *et seq.*), and must be cleaned up if released into the navigable waters of the United States pursuant to the Clean Water Act (33 USC § 1251 *et seq.*).

X1.5.4 Moreover, case law and EPA interpretations of the petroleum exclusion require an analysis of the facts of each case to determine whether a particular petroleum product is included in CERCLA's petroleum exclusion. The exclusion has been broadly interpreted to exclude gasoline and leaded gasoline from CERCLA's definition of hazardous substances regardless of the fact that gasoline and leaded gasoline contain certain indigenous components and additives which have themselves been designated as hazardous pursuant to CERCLA. See *Wilshire Westwood Associates versus Atlantic Richfield Corporation*, 881 F.2d 801 (9th Cir. 1989). The interpretation was narrowed when a judicial distinction was made between petroleum fractions produced by distillation processes and waste products resulting from contaminated tank scale. See *United States versus Western Processing Co.*, 761 F.Supp. 713 (W.D. Wash. 1991). Another decision narrowly interpreted CERCLA's petroleum exclusion to be inapplicable to oil-related wastes containing hazardous substances because the primary purpose of the exclusion is to remove "spills or other releases strictly of oil" from the scope of CERCLA response and liability (not releases of hazardous substances mixed with oil). See *City of New York versus Exxon Corporation*, 744 F. Supp. 474 (S.D.N.Y. 1990). For additional discussion, see EPA Memorandum entitled, "The Petroleum Exclusion Under the Comprehensive Environmental Response Compensation and Liability Act," issued by EPA's General Counsel, Francis S. Blake, July 31, 1987.

X1.6 *Exclusion of Certain Hazards From Superfund:*

X1.6.1 The information that follows is provided to explain why these potential environmental hazards are not covered by Superfund's appropriate inquiry responsibilities:

X1.6.2 As a preliminary matter, it should be noted that an environmental site assessment that does not address substances excluded from CERCLA (whether those substances are excluded because they are petroleum products or by virtue of other characteristics) but that otherwise constitutes "all appropriate inquiry into the previous ownership and uses of the property consistent with good commercial or customary practice" should nevertheless entitle the user to the innocent purchaser defense, assuming that other requirements of the defense are met.

X1.6.3 *Radon:*

X1.6.3.1 A case discussing Superfund and radon is *Amoco Oil Company versus Borden, Inc.*, 889 F.2d 664 (5th Cir. 1989). This case dealt with a private cost recovery action by the buyer of a site against the seller for response costs relating to radiation from phosphogypsum wastes left on the site. Radon emanated from these radioactive wastes. The case points out that the "EPA has designated radionuclides as hazardous substances under § 9602(a) of CERCLA.... Additionally, the ... EPA under § 112 of the Clean Air Act ...

list radionuclides as a hazardous air pollutant. Radon and its daughter products are considered radionuclides, which are defined as 'any nuclide that emits radiation.'" Therefore, radon is a CERCLA hazardous substance. Also, when discussing what constitutes a release of a hazardous substance under the statute, the statute is plain that there is no quantitative requirement and that a release, broadly defined at 42 USC § 9601(22), of *any* amount constitutes a CERCLA release.

X1.6.3.2 Liability under Superfund depends on several factors, as noted in X1.1. Only one of four factors is the release or threatened release of a hazardous substance. The other three factors are (*1*) the site is a facility, (*2*) the defendant falls within at least one of four classes of potentially responsible parties (PRPs), and (*3*) the release or threatened release *caused* the plaintiff (that can be the government or another private party) to incur response costs. Further, response costs must not be inconsistent with the National Contingency Plan (NCP), *and must not be limited by § 9604(a)(3)*. And, of course, there is no need to raise the innocent purchaser defense and its appropriate inquiry requirements unless the elements of liability will be met.

X1.6.3.3 Where radon from any source occurs in a building, three of the liability elements under CERCLA are met. There is a release of a hazardous substance, the building is a facility, and we can assume the defendant is a PRP. However, under 42 USC § 9604(a)(3)(A), "[r]emedial actions taken in response to hazardous substances as they occur naturally are specifically excluded from the NCP and are therefore not recoverable." *Amoco Oil Company versus Borden, Inc.*, 889 F.2d at 570. The statute is plain.[16]

"(3) Limitations on response
   The President shall not provide for a removal or remedial action under this section in response to a release or threat of release—
   (A) *of a naturally occurring substance in its unaltered form, or altered solely through naturally occurring processes or phenomena, from a location where it is naturally found*;
   (B) from products which are part of the structure of, and result in exposure within, residential buildings or business or community structures;[17] or
   (C) into public or private drinking water supplies due to deterioration of the system through ordinary use.[18]
(4) Exception to limitations
   Notwithstanding paragraph (3) of this subsection, to the extent authorized by this section, the President may respond to any release or threat of release if in the President's discretion, it constitutes a public health or environmental emergency and no other person with the authority and capability to respond to the emergency will do so in a timely manner."

X1.6.3.4 Therefore, no liability under CERCLA attaches for naturally occurring radon. If a party to a real estate transaction wants to look for radon within a building, no amount of radon investigation will have any bearing on one's innocent purchaser defense under Superfund. Investigation of naturally occurring radon would be included, if at all, in

the portion of the practice that deals with non-scope issues.

X1.6.4 *Asbestos:*

X1.6.4.1 The analysis of asbestos is similar to that involving radon. Before considering appropriate inquiry responsibilities, the four elements of CERCLA liability must be satisfied. Once again, as with radon, they are not met.

X1.6.4.2 Section 9604(a)(3)(B) of CERCLA prohibits response actions involving a release or threat of release "from products which are part of the structure of, and result in exposure within, residential buildings or business or community structures." There are a number of cases dealing with asbestos that interpret this statutory language. One such case is *First United Methodist Church of Hyattsville versus United States Gypsum Co.* that cites to other relevant cases.

X1.6.4.3 In *First United* the church brought a private cost recovery action against the manufacturer of asbestos-containing acoustical plaster. In holding that the action was barred by a state statute of repose (a certain time allowed by statute for bringing litigation) and that CERCLA did not preempt the state statute of repose, the court stated that § 9604(a)(3)(B) of CERCLA "represents much more than a procedural limitation on the President's authority; it is a substantive limitation of the breadth of CERCLA itself."[19] Therefore, the limitations of § 9604(a)(3) apply to private parties as well.

X1.6.4.4 Citing to the legislative history, the *First United* court concluded, "[i]n view of this clear expression of Congressional intent, we wil[l] not expand CERCLA to encompass asbestos-removal actions." The court further explained:[20]

"In closing, we note that this interpretation of CERCLA fully comports with the most fundamental guide to statutory construction—common sense. To extend CERCLA's strict liability scheme to all past and present owners of buildings containing asbestos as well as to all persons who manufactured, transported, and installed asbestos products into buildings, would be to shift literally billions of dollars of removal cost liability based on nothing more than an improvident interpretation of a statute that Congress never intended to apply in this context. [FN12] . . . Certainly, if Congress had intended for CERCLA to address the monumental asbestos problem, it would have said so more directly when it passed SARA. . . .

FN12—It is for this reason, that Congress simply did not intend for CERCLA to remedy the asbestos-removal problem, that we decline to follow the reasoning of *Prudential, Knox* and *Covalt* in rejecting First United's preemption argument. Instead of recognizing the fact that CERCLA is out of context in this situation, these courts rejected similar attempts to invoke the statute by construing CERCLA's key terms in a way to exclude asbestos-removal actions. *Covalt*, 860 F.2d [1434] at 1438-39 (defining "environment" to exclude the interior of a workplace); *Knox*, 690 F. Supp at 756-57 (defining "release" in terms of "spills" or "disposal"); *Prudential*,

---

[16] 42 USC § 9604(a)(3) and (4) (*emphasis added*).

[17] This provision has implications for asbestos and lead-based paint. See X1.6.4 and X1.6.5.

[18] This provision has implications for lead from lead pipes and solder. See X1.6.5.

[19] One such case is *First United Methodist Church of Hyattsville versus United States Gypsum Co.*, 882 F.2d 862 (4th Cir. 1989), that cites to other relevant cases.

[20] The same at 869; See also *3550 Stevens Creek Associates versus Barclays Bank of California*, 915 F.2d 1355 (9th Cir. 1990).

[711 F. Supp 1244] at 1254-55 (defining "disposal" to exclude the sale of a product for consumer use). We find this analysis unsatisfactory because it runs the risk of unnecessarily restricting the scope of CERCLA merely to dispose of claims that the statute was never intended to encompass in the first place. It is far better to simply acknowledge the inapplicability of CERCLA to asbestos-removal claims than to restrict its operative terms."

X1.6.4.5 Since asbestos that is a part of the structure of, and results in exposure within, residential buildings or business or community structures is excluded from CERCLA liability, it should not be investigated pursuant to a party's innocent purchaser appropriate inquiry requirements. Like naturally occurring radon, investigation of asbestos-containing materials that are part of the structure of buildings should be included, if at all, in the portion of this practice that deals with non-scope issues. Note, however, if asbestos is disposed of on a site and, therefore, is no longer part of the structure of a building, the cleanup of the disposed asbestos is subject to Superfund response actions. Likewise, if a building is sold with the knowledge that it will be demolished, one court ruled that the sale constitutes a disposal falling under CERCLA's liability provisions.[21]

X1.6.5 *Lead in Drinking Water and Lead-Based Paint*—These hazards can be evaluated in terms of the exclusions of 42 USC § 9604(a)(3)(B) and (C), in an analysis similar to the analysis applied above to radon and asbestos. While there is no reported case law on these environmental issues as they relate to Superfund, the statutory language seems clear that these environmental hazards are not encompassed by Superfund's appropriate inquiry responsibilities. Note, however, like asbestos, where there is a disposal of these substances on the site or in a facility, CERCLA liability may arise.

---

[21] *CP Holdings, Inc. versus Goldberg-Zoino & Associates, Inc.*, 769 F. Supp. 432 (D.N.H. 1991).

X2. RECOMMENDED TABLE OF CONTENTS AND REPORT FORMAT

X2.1 Summary
X2.2 Introduction
  X2.2.1 Purpose
  X2.2.2 Special Terms and Conditions
  X2.2.3 Limitations and Exceptions of Assessment
  X2.2.4 Limiting Conditions and Methodology Used
X2.3 Site Description
  X2.3.1 Location and Legal Description
  X2.3.2 Site and Vicinity Characteristics
  X2.3.3 Descriptions of Structures, Roads, Other Improvements on the Site (including heating/cooling system, sewage disposal, source of potable water)
  X2.3.4 Information (if any) Reported by User Regarding Environmental Liens or Specialized Knowledge or Experience (pursuant to Section 5.)
  X2.3.5 Current Uses of the Property
  X2.3.6 Past Uses of the Property (to the extent identified)
  X2.3.7 Current and Past Uses of Adjoining Properties (to the extent identified)
  X2.3.8 Site Rendering, Map, or Site Plan
X2.4 Records Review
  X2.4.1 Standard Environmental Record Sources, Federal and State
  X2.4.2 Physical Setting Source(s)
  X2.4.3 Historical Use Information
  X2.4.4 Additional Record Sources (if any)
X2.5 Information from Site Reconnaissance and Interviews
  X2.5.1 Hazardous Substances in Connection with Identified Uses (including storage, handling, disposal)
  X2.5.2 Hazardous Substance Containers and Unidentified Substance Containers (including storage, handling, disposal)
  X2.5.3 Storage Tanks (including contents and assessment of leakage or potential for leakage)
  X2.5.4 Indications of PCBs (including how contained and assessment of leakage or potential for leakage)
  X2.5.5 Indications of Solid Waste Disposal
  X2.5.6 Physical Setting Analysis, if migrating Hazardous Substances are an issue
  X2.5.7 Any Other Conditions of Concern
  X2.5.8 Site Plan (if available)
X2.6 Findings and Conclusions
X2.7 Signatures of Environmental Professionals
X2.8 Qualifications of Environmental Professionals Participating in Phase I Environmental Site Assessment
X2.9 Optional Appendices (for example):
  X2.9.1 Other Maps, Figures, and Photographs
  X2.9.2 Ownership/Historical Documentation
  X2.9.3 Regulatory Documentation
  X2.9.4 Interview Documentation
  X2.9.5 Contract between User and Environmental Professional

ASTM Designation: E 1528 – 96

Standard Practice for
Environmental Site Assessments: Transaction Screen Process[1]

This standard is issued under the fixed designation E 1528; the number immediately following the designation indicates the year of original adoption or, in the case of revision, the year of last revision. A number in parentheses indicates the year of last reapproval. A superscript epsilon (ϵ) indicates an editorial change since the last revision or reapproval.

1. Scope

1.1 *Purpose*—The purpose of this practice, as well as Practice E 1527, is to define good commercial and customary practice in the United States of America for conducting an *environmental site assessment*[2] of a parcel of *commercial real estate* with respect to the range of contaminants within the scope of the Comprehensive Environmental Response Compensation and Liability Act (CERCLA) and *petroleum products*. As such, this practice is intended to permit a user to satisfy one of the requirements to qualify for the *innocent landowner defense* to CERCLA liability: that is, the practices that constitute "all appropriate inquiry into the previous ownership and uses of the property consistent with good commercial or customary practice" as defined in 42 USC § 9601(35)(B). (See Appendix X1 for an outline of CERCLA's liability and defense provisions.)

1.1.1 *Recognized Environmental Conditions*—In defining a standard of good commercial and customary practice for conducting an *environmental site assessment* of a parcel of *property*, the goal of the processes established by this practice is to identify *recognized environmental conditions*. The term *recognized environmental conditions* means the presence or likely presence of any *hazardous substances* or *petroleum products* on a *property* under conditions that indicate an existing release, a past release, or a material threat of a release of any *hazardous substances* or *petroleum products* into structures on the *property* or into the ground, groundwater, or surface water of the *property*. The term includes *hazardous substances* or *petroleum products* even under conditions in compliance with laws. The term is not intended to include *de minimis* conditions that generally do not present a material risk of harm to public health or the environment and that generally would not be the subject of an enforcement action if brought to the attention of appropriate governmental agencies.

1.1.2 *Two Related Practices*—This practice is closely related to Practice E 1527, a *Phase I Environmental Site Assessment*. Both are *environmental site assessments* for *commercial real estate*. See 4.3.

1.1.3 *Petroleum Products*—*Petroleum products* are included within the scope of both practices because they are of concern on many parcels of *commercial real estate* and current custom and usage is to include an inquiry into the presence of *petroleum products* when doing an *environmental site assessment* of *commercial real estate*. Inclusion of *petroleum products* within the scope of this practice is not based upon the applicability, if any, of CERCLA to *petroleum products*.

1.1.4 *CERCLA Requirements Other Than Appropriate Inquiry*—This practice does not address whether requirements in addition to *appropriate inquiry* have been met in order to qualify for CERCLA's *innocent landowner defense* (for example, the duties specified in 42 USC § 9607(b)(3)(a) and (b) and cited in Appendix X1).

1.1.5 *Other Federal, State, and Local Environmental Laws*—This practice does not address requirements of any state or local laws or of any federal laws other than the appropriate inquiry provisions of CERCLA's *innocent landowner defense*. Users are cautioned that federal, state, and local laws may impose environmental assessment obligations that are beyond the scope of this practice. *Users* should also be aware that there are likely to be other legal obligations with regard to *hazardous substances* or *petroleum products* discovered on *property* that are not addressed in this practice and may pose risks of civil and/or criminal sanctions for non-compliance.

1.2 *Objectives*—Objectives guiding the development of this practice and Practice E 1527 are (1) to synthesize and put in writing good commercial and customary practice for *environmental site assessments* for *commercial real estate*, (2) to facilitate high- quality, standardized *environmental site assessments*, (3) to ensure that the standard of *appropriate inquiry* is practical and reasonable, and (4) to clarify an industry standard for *appropriate inquiry* in an effort to guide legal interpretation of CERCLA's *innocent landowner defense*.

1.3 *Considerations Beyond the Scope*—The use of this practice is strictly limited to the scope set forth in this section. Section 11 of this practice identifies, for informational purposes, certain environmental conditions (not an all-inclusive list) that may exist on a *property* but may warrant consideration by parties to a *commercial real estate* transaction.

1.4 *Organization of This Practice*—This practice has several parts and two appendixes. Section 1 is the Scope. Section 2 refers to other ASTM standards in the Referenced Documents. Section 3, Terminology, has definitions of terms not unique to this practice, descriptions of terms unique to this practice, and acronyms. Section 4 is Significance and Use of this practice. Section 5 is the Introduction to the Transaction Screen Questionnaire. Section 6 sets forth the Transaction

[1] This practice is under the jurisdiction of ASTM Committee E-50 on Environmental Assessment and is the direct responsibility of Subcommittee E50.02 on Commercial Real Estate Transactions.
Current edition approved Oct. 10, 1996. Published December 1996. Originally published as E 1528 – 93. Last previous edition E 1528 – 93.
[2] All definitions, descriptions of terms, and acronyms are defined in Section 3 of Practice E 1527. Whenever terms defined in 3.2 or described in 3.3 are used in this practice, they are in *italics*.

Screen Questionnaire itself. Sections 7 through 10 contain the Guide to the Transaction Screen Questionnaire and its various parts. Section 11 provides additional information regarding non-scope considerations (see 1.3). The appendixes are included for information and are not part of the procedures prescribed in either Practice E 1527 or this practice. Appendix X1 and Practice E 1527 explain the liability and defense provisions of CERCLA that will assist the user in understanding the user's responsibilities under CERCLA; it also contains other important information regarding CERCLA and this practice. Appendix X2 provides information referred to in the guide to the transaction screen questionnaire.

1.5 *This standard does not purport to address all of the safety problems, if any, associated with its use. It is the responsibility of the user of this standard to establish appropriate safety and health practices and determine the applicability of regulatory limitations prior to use.*

2. Referenced Document

2.1 *ASTM Standard:*

E 1527 Practice for Environmental Site Assessments: Phase I Environmental Site Assessment Process[3]

3. Terminology

3.1 *Scope*—This section provides definitions, descriptions of terms, and a list of acronyms for many of the words used in this practice and Practice E 1527. The terms are an integral part of both practices and are critical to an understanding of the written practices and their use.

3.2 *Definitions:*

3.2.1 *asbestos*—six naturally occurring fibrous minerals found in certain types of rock formations. Of the six, the minerals chrysotile, amosite, and crocidolite have been most commonly used in building products. When mined and processed, asbestos is typically separated into very thin fibers. Because asbestos is strong, incombustible, and corrosion-resistant, asbestos was used in many commercial products beginning early in this century and peaking in the period from World War II into the 1970s. When inhaled in sufficient quantities, asbestos fibers can cause serious health problems.[4] (See also, for an additional definition of asbestos, ASTM STP 834.[5])

3.2.2 *asbestos-containing material (ACM)*—any material or product that contains more than one percent asbestos.[4]

3.2.3 *Comprehensive Environmental Response, Compensation and Liability Information System (CERCLIS)*—the list of sites compiled by EPA that EPA has investigated or is currently investigating for potential hazardous substance contamination for possible inclusion on the National Priorities list.

3.2.4 *construction debris*—concrete, brick, asphalt, and other such building materials discarded in the construction of a building or other improvement to property.

3.2.5 *contaminated public wells*—public wells used for drinking water that have been designated by a government entity as contaminated by toxic substances, (for example, chlorinated solvents), or as having water unsafe to drink without treatment.

3.2.6 *CORRACTS list*—EPA's list of treatment, storage, or disposal facilities subject to corrective action under RCRA.

3.2.7 *demolition debris*—concrete, brick, asphalt, and other such building materials discarded in the demolition of a building or other improvement to property.

3.2.8 *drum*—a container (typically, but not necessarily, holding 55 gal (208 L) of liquid) that may be used to store *hazardous substances* or *petroleum products*.

3.2.9 *dry wells*—underground areas where soil has been removed and replaced with pea gravel, coarse sand, or large rocks. Dry wells are used for drainage, to control storm runoff, for the collection of spilled liquids (intentional and non-intentional) and wastewater disposal (often illegal).

3.2.10 *dwelling*—structure or portion thereof used for residential habitation.

3.2.11 *environmental lien*—a charge, security, or encumbrance upon title to a *property* to secure the payment of a cost, damage, debt, obligation, or duty arising out of response actions, cleanup, or other remediation of *hazardous substances* or *petroleum products* upon a *property*, including (but not limited to) liens imposed pursuant to CERCLA 42 USC § 9607(1) and similar state or local laws.

3.2.12 *ERNS list*—EPA's Emergency Response Notification System list of reported CERCLA hazardous substance releases or spills in quantities greater than the reportable quantity, as maintained at the National Response Center. Notification requirements for such releases or spills are codified in 40 CFR Parts 302 and 355.

3.2.13 *Federal Register (FR)*—publication of the United States government published daily (except for federal holidays and weekends) containing all proposed and final regulations and some other activities of the federal government. When regulations become final, they are included in the Code of Federal Regulations (CFR), as well as published in the Federal Register.

3.2.14 *fire insurance maps*—maps produced for private fire insurance map companies that indicate uses of properties at specified dates and that encompass the property. These maps are often available at local libraries, historical societies, private resellers, or from the map companies who produced them. See Question 23 of the transaction screen process in this practice and 7.3.4.2 of Practice E 1527.

3.2.15 *hazardous substance*—a substance defined as a hazardous substance pursuant to CERCLA 42 USC § 9601(14), as interpreted by EPA regulations and the courts: "(A) any substance designated pursuant to Section 1321(b)(2)(A) of Title 33, (B) any element, compound, mixture, solution, or substance designated pursuant to Section 9602 of this title, (C) any hazardous waste having the characteristics identified under or listed pursuant to Section 3001 of the Solid Waste Disposal Act (42 USC § 6921) (but not including any waste the regulation of which under the Solid Waste Disposal Act (42 USC § 6901 *et seq.*) has been suspended by Act of Congress), (D) any toxic pollutant listed under Section 1317(a) of Title 33, (E) any hazardous air pollutant listed under Section 112 of the Clean Air Act (42

---

[3] *Annual Book of ASTM Standards*, Vol 11.04.
[4] See EPA *Managing Asbestos in Place, A Building Owner's Guide to Operations and Maintenance Programs for Asbestos-Containing Materials*, EPA, July 1990, p. 2.
[5] See also, for an additional definition of asbestos, *ASTM STP 834*, ASTM.

USC § 7412), and (F) any imminently hazardous chemical substance or mixture with respect to which the administrator (of EPA) has taken action pursuant to Section 2606 of Title 15. The term does not include petroleum, including crude oil or any fraction thereof which is not otherwise specifically listed or designated as a hazardous substance under Subparagraphs (A) through (F) of this paragraph, and the term does not include natural gas, natural gas liquids, liquefied natural gas, or synthetic gas usable for fuel (or mixtures of natural gas and such synthetic gas)," (See Appendix X1.)

3.2.16 *hazardous waste*—any hazardous waste having the characteristics identified under or listed pursuant to Section 3001 of the Solid Waste Disposal Act (42 USC § 6921) (but not including any waste the regulation of which under the Solid Waste Disposal Act (42 USC § 6901 *et seq.*) has been suspended by Act of Congress). The Solid Waste Disposal Act of 1980 amended RCRA. The RCRA defines a hazardous waste, in 42 USC § 6903, as:

"a solid waste, or combination of solid wastes, which because of its quantity, concentration, or physical, chemical, or infectious characteristics may—(A) cause, or significantly contribute to an increase in mortality or an increase in serious irreversible, or incapacitating reversible, illness; or (B) pose a substantial present or potential hazard to human health or the environment when improperly treated, stored, transported, or disposed of, or otherwise managed."

3.2.17 *landfill*—a place, location, tract of land, area, or premises used for the disposal of solid wastes as defined by state solid waste regulations. The term is synonymous with the term *solid waste disposal site* and is also known as a garbage dump, trash dump, or similar term.

3.2.18 *local street directories*—directories published by private (or sometimes government) sources that show ownership, occupancy, use of sites and/or by reference to street addresses. Often local street directories are available at libraries of local governments, colleges or universities, or historical societies. See 7.3.4.6 of the Records Review Section of Practice E 1527.

3.2.19 *material safety data sheet (MSDS)*—written or printed material concerning a *hazardous substance* which is prepared by chemical manufacturers, importers, and employers for hazardous chemicals pursuant to OSHA's Hazard Communication Standard, 29 CFR 1910.1200(g).

3.2.20 *National Contingency Plan (NCP)*—the National Oil and Hazardous Substances Pollution Contingency Plan, found at 40 CFR § 300, that is the EPA's blueprint on how hazardous substances are to be cleaned up pursuant to CERCLA.

3.2.21 *National Priorities List (NPL)*—list compiled by EPA pursuant to CERCLA 42 USC § 9605(a)(8)(B) of properties with the highest priority for cleanup pursuant to EPA's hazard ranking system. See 40 CFR Part 300.

3.2.22 *occupants*—those tenants, subtenants, or other persons or entities using the *property* or a portion of the *property*.

3.2.23 *owner*—generally the fee owner of record of the *property*.

3.2.24 *petroleum exclusion*—the exclusion from CERCLA liability provided in 42 USC § 9601(14), as interpreted by the courts and EPA: "The term (hazardous substance) does not include petroleum, including crude oil or any fraction thereof which is not otherwise specifically listed or designated as a hazardous substance under Subparagraphs (A) through (F) of this paragraph, and the term does not include natural gas, natural gas liquids, liquefied natural gas, or synthetic gas usable for fuel (or mixtures of natural gas and such synthetic gas)."

3.2.25 *petroleum products*—those substances included within the meaning of the terms within the *petroleum exclusion* to CERCLA, 42 USC § 9601(14), as interpreted by the courts and EPA, that is: petroleum, including crude oil or any fraction thereof that is not otherwise specifically listed or designated as a *hazardous substance* under Subparagraphs (A) through (F) of 42 USC § 9601(14), natural gas, natural gas liquids, liquefied natural gas, and synthetic gas usable for fuel (or mixtures of natural gas and such synthetic gas). (The word fraction refers to certain distillates of crude oil, including gasoline, kerosene, diesel oil, jet fuels, and fuel oil, pursuant to *Standard Definitions of Petroleum Statistics*.[6]

3.2.26 *Phase I Environmental Site Assessment*—the process described in Practice E 1527.

3.2.27 *pits, ponds, or lagoons*—man-made or natural depressions in a ground surface that are likely to hold liquids or sludge containing *hazardous substances* or *petroleum products*. The likelihood of such liquids or sludge being present is determined by evidence of factors associated with the pit, pond, or lagoon, including, but not limited to, discolored water, distressed vegetation, or the presence of an obvious wastewater discharge.

3.2.28 *property*—the real property that is the subject of the environmental site assessment described in this practice. Real property includes buildings and other fixtures and improvements located on the property and affixed to the land.

3.2.29 *property tax files*—the files kept for property tax purposes by the local jurisdiction where the property is located and includes records of past ownership, appraisals, maps, sketches, photos, or other information that is reasonably ascertainable and pertaining to the property. See 7.3.4.3 of the Records Review Section of Practice E 1527.

3.2.30 *RCRA generators*—those persons or entities which generate *hazardous wastes*, as defined and regulated by RCRA.

3.2.31 *RCRA generators list*—list kept by EPA of those persons or entities that generate *hazardous wastes* as defined and regulated by RCRA.

3.2.32 *RCRA TSD facilities*—those facilities on which treatment, storage, and/or disposal of *hazardous wastes* takes place, as defined and regulated by RCRA.

3.2.33 *RCRA TSD facilities list*—list kept by EPA of those facilities on which treatment, storage, and/or disposal of *hazardous wastes* takes place, as defined and regulated by RCRA.

3.2.34 *recorded land title records*—records of fee ownership, leases, land contracts, easements, liens, and other encumbrances on or of the *property* recorded in the place where land title records are, by law or custom, recorded for

[6] *Standard Definitions of Petroleum Statistics*, American Petroleum Institute, Fourth Edition, 1988.

the local jurisdiction in which the property is located. (Often such records are kept by a municipal or county recorder or clerk.) Such records may be obtained from title companies or directly from the local government agency. Information about the title to the *property* that is recorded in a U.S. district court or any place other than where land title records are, by law or custom, recorded for the local jurisdiction in which the property is located, are not considered part of recorded land title records. See 7.3.4.4 of the Records Review Section of Practice E 1527.

3.2.35 *records of emergency release notifications* (SARA §304)—Section 304 of EPCRA or Title III of SARA requires operators of facilities to notify their local emergency planning committee (as defined in EPCRA) and State Emergency Response Commission (as defined in EPCRA) of any release beyond the facility's boundary of any reportable quantity of any extremely *hazardous substance*. Often the local fire department is the local emergency planning committee. Records of such notifications are "Records of Emergency Release Notifications" (SARA §304).

3.2.36 *report*—the written record of a transaction screen process as required by Practice E 1527 or the written report prepared by the environmental professional and constituting part of a *Phase I Environmental Site Assessment*, as required by Practice E 1527.

3.2.37 *solid waste disposal site*—a place, location, tract of land, area, or premises used for the landfill disposal of solid wastes as defined by state solid waste regulations. The term is synonymous with the term *landfill* and is also known as a garbage dump, trash dump, or similar term.

3.2.38 *solvent*—a chemical compound that is capable of dissolving another substance and may itself be a *hazardous substance* used in a number of manufacturing/industrial processes including, but not limited to, the manufacture of paints and coatings for industrial and household purposes, equipment clean-up, and surface degreasing in metal fabricating industries.

3.2.39 *state registered USTs*—state lists of underground storage tanks required to be registered under Subtitle I, Section 9002 of RCRA.

3.2.40 *sump*—a pit, cistern, cesspool, or similar receptacle where liquids drain, collect, or are stored.

3.2.41 *TSD Facility*—treatment, storage, or disposal facility (see definition of *RCRA TSD facilities*).

3.2.42 *underground storage tank (UST)*—any tank, including underground piping connected to the tank, that is or has been used to contain *hazardous substances* or *petroleum products* and the volume of which is 10 % or more beneath the surface of the ground.

3.2.43 *USGS 7.5 Minute Topographic Map*—the map (if any) available from or produced by the United States Geological Survey, entitled "USGS 7.5 Minute Topographic Map," and showing the *property*. See 7.3.4.5 of Practice E 1527.

3.2.44 *wastewater*—water that is or has been used in an industrial or manufacturing process, conveys or has conveyed sewage, or is directly related to manufacturing, processing, or raw materials storage areas at an industrial plant. Wastewater does not include water originating on or passing through or adjacent to a site, such as stormwater flows, that has not been used in industrial or manufacturing processes,

has not been combined with sewage, or is not directly related to manufacturing, processing, or raw materials storage areas at an industrial plant.

3.2.45 *zoning/land use records*—those records of the local government in which the *property* is located indicating the uses permitted by the local government in particular zones within its jurisdiction. The records may consist of maps and/or written records. They are often located in the planning department of a municipality or county. See 7.3.4.8 of the Records Review Section of Practice E 1527.

3.3 *Descriptions of Terms Specific to This Standard:*

3.3.1 *actual knowledge*—the knowledge actually possessed by an individual who is a real person, rather than an entity. Actual knowledge is to be distinguished from constructive knowledge, that is, knowledge imputed to an individual or entity. (See 5.5.3.)

3.3.2 *adjoining properties*—any real property or properties the border of which is contiguous or partially contiguous with that of the property, or that would be contiguous or partially contiguous with that of the property but for a street, road, or other public thoroughfare separating them.

3.3.3 *aerial photographs*—photographs taken from an airplane or helicopter of areas encompassing the property. Aerial photographs are often available from government agencies or private collections unique to a local area. See 7.3.4.1 of Practice E 1527.

3.3.4 *appropriate inquiry*—that inquiry constituting "all appropriate inquiry into the previous ownership and uses of the property consistent with good commercial or customary practice" as defined in CERCLA, 42 USC § 9601(35)(B), that will give a party to a *commercial real estate* transaction the *innocent landowner defense* to CERCLA liability (42 USC § 9601(A) and (B) and 9607(b)(3)), assuming compliance with other elements of the defense. See Appendix X1 and Practice E 1527.

3.3.5 *approximate minimum search distance*—the area for which records must be obtained and reviewed pursuant to the Records Review Section of Practice E 1527, subject to the limitations provided in that section. The term *approximate minimum search distance* may include areas outside the *property* and shall be measured from the nearest *property* boundary. The term *approximate minimum search distance* is used instead of radius to include irregularly shaped properties.

3.3.6 *building department records*—those records of the local government in which the property is located indicating permission of the local government to construct, alter, or demolish improvements on the property. Often building department records are located in the building department of a municipality or county. See 7.3.4.7 of Practice E 1527.

3.3.7 *commercial real estate*—any real property except a dwelling or property with no more than four dwelling units exclusively for residential use (except that a dwelling or property with no more than four dwelling units exclusively for residential use is included in the term commercial real estate when it has a commercial function, as in the building of such dwellings for profit). The term *commercial real estate* includes but is not limited to undeveloped real property and real property used for industrial, retail, office, agricultural, other commercial, medical, or educational purposes; *property* used for residential purposes that has more than four residential dwelling units; and *property* with no more than four dwelling units for residential use when it has a commercial function, as in the building of such dwellings for profit.

3.3.8 *commercial real estate transaction*—a transfer of title to or possession of real property or receipt of a security interest in real property, except that it does not include transfer of title to or possession of real property or the receipt of a security interest in real property with respect to an individual dwelling or building containing fewer than five dwelling units, nor does it include the purchase of a lot or lots to construct a dwelling for occupancy by a purchaser, but a commercial real estate transaction does include real property purchased or leased by persons or entities in the business of building or developing dwelling units.

3.3.9 *due diligence*—the process of inquiring into the environmental characteristics of a parcel of *commercial real estate* or other conditions, usually in connection with a commercial real estate transaction. The degree and kind of due diligence vary for different properties and differing purposes. See Appendix X1 and Practice E 1527.

3.3.10 *environmental audit*—the investigative process to determine if the operations of an existing facility are in compliance with applicable environmental laws and regulations. The term *environmental audit* should not be used to describe this practice or Practice E 1527 although an environmental audit may include an *environmental site assessment* or, if prior audits are available, may be part of an environmental site assessment. See Appendix X1 and Practice E 1527.

3.3.11 *environmental professional*—a person possessing sufficient training and experience necessary to conduct a *site reconnaissance, interviews*, and other activities in accordance with Practice E 1527, and from the information generated by such activities, having the ability to develop opinions and conclusions regarding *recognized environmental conditions* in connection with the *property* in question. An individual's status as an environmental professional may be limited to the type of assessment to be performed or to specific segments of the assessment for which the professional is responsible. The person may be an independent contractor or an employee of the *user*.

3.3.12 *environmental site assessment (ESA)*—the process by which a person or entity seeks to determine if a particular parcel of real *property* (including improvements) is subject to *recognized environmental conditions*. At the option of the user, an environmental site assessment may include more inquiry than that constituting *appropriate inquiry* or, if the user is not concerned about qualifying for the *innocent landowner defense*, less inquiry than that constituting *appropriate inquiry*. See Appendix X1 and Practice E 1527. An environmental site assessment is both different from and less rigorous than an *environmental audit*.

3.3.13 *fill dirt*—dirt, soil, sand, or other earth, that is obtained off-site, that is used to fill holes or depressions, create mounds, or otherwise artificially change the grade or elevation of real property. It does not include material that is used in limited quantities for normal landscaping activities.

3.3.14 *hazardous waste/contaminated sites*—sites on which a release has occurred, or is suspected to have occurred, of any *hazardous substance, hazardous waste*, or *petroleum products*, and on which release or suspected

release has been reported to a government entity.

3.3.15 *innocent landowner defense*—that defense to CERCLA liability provided in 42 USC § 9601(35) and § 9607(b)(3). One of the requirements to qualify for this defense is that the party make "all appropriate inquiry into the previous ownership and uses of the property consistent with good commercial or customary practice." There are additional requirements to qualify for this defense. See discussion in Appendix X1 and Practice E 1527.

3.3.16 *interviews*—those portions of the *Phase I Environmental Site Assessment* in Practice E 1527 that are contained in Sections 9 and 10 thereof and addresses questions to be asked of *owners* and *occupants* of the *property* and questions to be asked of local government officials.

3.3.17 *key site manager*—the *key site manager* is the person identified by the *owner* of a *property* as having good knowledge of the uses and physical characteristics of the property. See 9.5.1 of Practice E 1527.

3.3.18 *local government agencies*—those agencies of municipal or county government having jurisdiction over the *property*. Municipal and county government agencies include but are not limited to cities, parishes, townships, and similar entities.

3.3.19 *LUST sites*—state lists of leaking underground storage tank sites. Section 9003 (h) of Subtitle I of RCRA gives EPA and states, under cooperative agreements with EPA, authority to clean up releases from UST systems or require owners and operators to do so.

3.3.20 *major occupants*—those tenants, subtenants, or other persons or entities each of which uses at least 40 % of the leasable area of the *property* or any anchor tenant when the *property* is a shopping center.

3.3.21 *obvious*—that which is plain or evident; a condition or fact which could not be ignored or overlooked by a reasonable observer while *visually or physically observing* the *property*.

3.3.22 *other historical sources*—any source or sources other than those designated in 7.3.4.1 through 7.3.4.8 of Practice E 1527 that are credible to a reasonable person and that identify past uses or occupancies of the *property*. The term includes records in the files, and/or personal knowledge of the *property owner* and/or *occupants*. See 7.3.4.9 of the Records Review Section of Practice E 1527.

3.3.23 *physical setting sources*—sources that provide information about the geologic, hydrogeologic, hydrologic, or topographic characteristics of a *property*. See 7.2.3 of Practice E 1527.

3.3.24 *practically reviewable*—information that is practically reviewable means that the information is provided by the source in a manner and in a form that, upon examination, yields information relevant to the *property* without the need for extraordinary analysis of irrelevant data. The form of the information shall be such that the user can review the records for a limited geographic area. Records that cannot be feasibly retrieved by reference to the location of the *property* or a geographic area in which the *property* is located are not generally *practically reviewable*. Listings in publicly available records that do not have adequate address information to be located geographically are not generally considered *practically reviewable*. Most databases of public records are *practically reviewable* if they can be obtained from the source agency by the county, city, zip code, or other geographic area of the facilities listed in the record system. Records that are sorted, filed, organized, or maintained by the source agency only chronologically are not generally practically reviewable. For large databases with numerous facility records (such as RCRA hazardous waste generators and registered underground storage tanks), the records are not *practically reviewable* unless they can be obtained from the source agency in the smaller geographic area of zip codes. Even when information is provided by zip code for some large databases, it is common for an unmanageable number of sites to be identified within a given zip code. In these cases, it is not necessary to review the impact of all of the sites that are likely to be listed in any given zip code because that information would not be practically reviewable. In other words, when so much data is generated that it cannot be feasibly reviewed for its impact on the property, it is not *practically reviewable.*

3.3.25 *preparer*—the person preparing the *transaction screen questionnaire* pursuant to this practice, who may be either the *user* or the person to whom the *user* has delegated the preparation.

3.3.26 *publicly available*—information that is publicly available means that the source of the information allows access to the information by anyone upon request.

3.3.27 *reasonably ascertainable*—for purposes of both this practice and Practice E 1527 information that is *publicly available,* obtainable from its source within reasonable time and cost constraints, and *practically reviewable.*

3.3.28 *recognized environmental conditions*—the presence or likely presence of any *hazardous substances* or *petroleum products* on a *property* under conditions that indicate an existing release, a past release, or a material threat of a release of any *hazardous substances* or *petroleum products* into structures on the *property* or into the ground, groundwater, or surface water of the *property.* The term includes *hazardous substances* or *petroleum products* even under conditions in compliance with laws. The term is not intended to include *de minimis* conditions that generally do not present a material risk of harm to public health or the environment and that generally would not be the subject of an enforcement action if brought to the attention of appropriate governmental agencies.

3.3.29 *records review*—that part of the *Phase I Environmental Site Assessment* in Practice E 1527 that is contained in Section 7 thereof and addresses which records shall or may be reviewed.

3.3.30 *site reconnaissance*—that part of the *Phase I Environmental Site Assessment* in Practice E 1527 that is contained in Section 8 thereof and addresses what should be done in connection with the *site visit.* The site reconnaissance includes, but is not limited to, the *site visit* done in connection with such as *Phase I Environmental Site Assessment.*

3.3.31 *site visit*—the visit to the property during which observations are made constituting the *site reconnaissance* section of the *Phase I Environmental Site Assessment* in Practice E 1527 and the *site visit* requirement of the transaction screen process in this practice.

3.3.32 *standard environmental record sources*—those records specified in 7.2.1.1 of the Records Review Section of the *Phase I Environmental Site Assessment* of Practice E 1527.

3.3.33 *standard historical sources*—those sources of information about the history of uses of property specified in 7.3.4 of the Records Review Section of the *Phase I Environmental Site Assessment* of Practice E 1527.

3.3.34 *standard physical setting source*—a current USGS 7.5 minute topographic map (if any) showing the area on which the property is located. See 7.2.3 of Practice E 1527.

3.3.35 *standard practice(s)*—the activities set forth in either this practice or Practice E 1527, or both, for the conduct of environmental site assessments.

3.3.36 *standard sources*—sources of environmental, physical setting, or historical records specified in the Records Review Section (Section 7) of the *Phase I Environmental Site Assessment* of Practice E 1527.

3.3.37 *transaction screen questionnaire*—the questionnaire provided in Section 6 of Practice E 1527.

3.3.38 *transaction screen process*—the process described in Practice E 1527.

3.3.39 *user*—the party seeking to use the transaction screen process of this practice or the *Phase I Environmental Site Assessment* of Practice E 1527 to perform an *environmental assessment* of the *property.* A *user* may include, without limitation, a purchaser of *property,* a potential tenant of *property,* an *owner* of *property,* a lender, or a property manager.

3.3.40 *visually and/or physically observed*—during a *site visit* pursuant to the *transaction screen process* of this practice or pursuant to a *Phase I Environmental Site Assessment* of Practice E 1527, the term *visually and physically observed* means observations made by vision upon walking through a *property* and the structures located on it and observations made by the sense of smell, particularly observations of noxious or foul odors. The term *walking through* is not meant to imply that disabled persons who cannot physically walk may not conduct a *site visit;* they may do so by the means at their disposal for moving through the *property* and the structures located on it.

3.4 *Acronyms:*

3.4.1 *CERCLA*—Comprehensive Environmental Response, Compensation and Liability of 1980 Act (as amended, 42 USC § 9601 *et seq.*).

3.4.2 *CERCLIS*—Comprehensive Environmental Response, Compensation and Liability Information System maintained by EPA.

3.4.3 *CFR*—Code of Federal Regulations.

3.4.4 *CORRACTS*—TSD Facilities subject to Corrective Action under RCRA.

3.4.5 *EPA*—United States Environmental Protection Agency.

3.4.6 *EPCRA*—Emergency Planning and Community Right to Know Act (also known as SARA Title III), (42 USC § 11001 *et seq.*).

3.4.7 *ERNS*—Emergency Response Notification System.

3.4.8 *ESA*—environmental site assessment (different than an environmental audit; see 3.3.12).

3.4.9 *FOIA*—U.S. Freedom of Information Act (5 USC § 552 *et seq.*).

3.4.10 *FR*—Federal Register.

3.4.11 *LUST*—leaking underground storage tank.

3.4.12 *MSDS*—material safety data sheet.

3.4.13 *NCP*—National Contingency Plan.

3.4.14 *NPDES*—National Pollutant Discharge Elimination System.

3.4.15 *NPL*—National Priorities List.

3.4.16 *PCBs*—polychlorinated biphenyls.

3.4.17 *PRP*—potentially responsible party (pursuant to CERCLA 42 USC § 9607(a).

3.4.18 *RCRA*—Resource Conservation and Recovery Act (as amended, 42 USC § 6901 *et seq.*).

3.4.19 *SARA*—Superfund Amendments and Reauthorization Act of 1986 (amendment to CERCLA).

3.4.20 *USC*—United States Code.

3.4.21 *USGS*—United States Geological Survey.

3.4.22 *UST*—underground storage tank.

4. Significance and Use

4.1 *Uses*—This practice is intended for use on a voluntary basis by parties who wish to assess the environmental condition of *commercial real estate*. While use of this practice is intended to constitute *appropriate inquiry* for purposes of CERCLA's *innocent landowner defense*, it is not intended that its use be limited to that purpose. This practice is intended primarily as an approach to conducting an inquiry designed to identify *recognized environmental conditions* in connection with a *property*, and *environmental site assessments* that are both more and less comprehensive than this practice (including, in some instances, no *environmental site assessment*) may be appropriate in some circumstances. Further, no implication is intended that a person must use this practice in order to be deemed to have conducted inquiry in a commercially prudent or reasonable manner in any particular transaction. Nevertheless, this practice is intended to reflect a commercially prudent and reasonable inquiry.

4.2 *Clarifications on Use:*

4.2.1 *Use Not Limited to CERCLA*—This practice and Practice E 1527 are designed to assist the *user* in developing information about the environmental condition of a *property* and as such has utility for a wide range of persons, including those who may have no actual or potential CERCLA liability and/or may not be seeking the *innocent landowner defense*.

4.2.2 *Residential Tenants/Purchasers and Others*—No implication is intended that it is currently customary practice for residential tenants of multifamily residential buildings, tenants of single-family homes or other residential real estate, or purchasers of dwellings for residential use, to conduct an *environmental site assessment* in connection with these transactions. Thus, these transactions are not included in the term commercial real estate transactions, and it is not intended to imply that such persons are obligated to conduct an *environmental site assessment* in connection with these transactions for purposes of *appropriate inquiry* or for any other purpose. In addition, no implication is intended that it is currently customary practice for *environmental site assessments* to be conducted in other unenumerated instances (including but not limited to many commercial leasing transactions, many acquisitions of easements, and many loan transactions in which the lender has multiple remedies). On the other hand, anyone who elects to do an *environmental site assessment* of any *property* or portion of a property may, in such person's judgment, use either this practice or Practice E 1527.

4.2.3 *Site-Specific*—This practice is site-specific in that it relates to assessment of environmental conditions on a specific parcel of *commercial real estate*. Consequently, this practice does not address many additional issues raised in transactions such as purchases of business entities, or interests therein, or of their assets, that may well involve environmental liabilities pertaining to properties previously owned or operated or other off-site environmental liabilities.

4.3 *Two Related Practices*—This practice sets forth one procedure for an environmental site assessment for purposes of appropriate inquiry necessary to qualify for CERCLA's *innocent landowner defense*, known as a "transaction screen process" or a transaction screen." This practice is a companion to Practice E 1527 that is called the "Phase I Environmental Site Assessment Process" or a "Phase I ESA" or simply a "Phase I." These practices are each intended to meet the standard of *appropriate inquiry* necessary to qualify for the *innocent landowner defense* of CERCLA. It is essential to consider that these two practices, taken together, provide for two alternative practices of appropriate inquiry: the *transaction screen process* described in this practice and the *Phase I Environmental Site Assessment* described in Practice E 1527.

4.3.1 *Election to Commence With Either Practice*—The *user* may commence inquiry to identify recognized *environmental conditions* in connection with a *property* by performing either the *transaction screen process* or the *Phase I Environmental Site Assessment*.

4.3.2 *Who May Conduct*—The *transaction screen process* may be conducted either by the *user* (including an agent, independent contractor or employee of the *user*) or wholly or partially by an *environmental professional*. The *transaction screen process* does not require the judgment of an *environmental professional*. Whenever a *Phase I Environmental Site Assessment* is conducted, it must be performed by an *environmental professional* to the extent specified in 6.5.1 through 6.5.2.1 of Practice E 1527. Further, at the *Phase I Environmental Site Assessment* level, no practical standard should be designed to eliminate the role of judgment and the value and need for experience in the party performing the inquiry. The professional judgment of an *environmental professional* is, consequently, vital to the performance of appropriate inquiry at the *Phase I Environmental Site Assessment* level.

4.3.3 *Completion of Transaction Screen Process*—Performance of the *transaction screen process* may allow the *user* to conclude that no further inquiry is needed to assess the potential for identifying any *recognized environmental condition* at the property and hence that performance of the *transaction screen process* constitutes *appropriate inquiry* without undertaking the *Phase I Environmental Site Assessment*. Upon completion of the *transaction screen process*, the *user* should conclude either (1) no further inquiry into recognized environmental conditions at the property is needed for purposes of *appropriate inquiry*, or (2) further inquiry is needed to assess *recognized environmental conditions* appropriately for purposes of *appropriate inquiry*. If no further inquiry is needed, the *user* has completed his or her *appropriate inquiry* of the *property*.

4.3.4 *Inquiry Beyond the Transaction Screen Process*—If further inquiry is needed after performance of the *transaction screen process* (as described in 4.3.3), the *user* must determine, in the exercise of the *user's* reasonable business judgment, whether further inquiry may be limited to those specific issues identified as of concern or should proceed to a full *Phase I Environmental Site Assessment.*

4.4 *Additional Services*—As set forth in 11.9 of Practice E 1527, additional services may be contracted for between the *user* and the *environmental professional(s).*

4.5 *Principles*—The following principles are an integral part of this practice and are intended to be referred to in resolving any ambiguity or exercising such discretion as is accorded the *user* or *environmental professional* in performing an *environmental site assessment* or in judging whether a *user* or *environmental professional* has conducted *appropriate inquiry* or has otherwise conducted an adequate *environmental site assessment.*

4.5.1 No *environmental site assessment* can wholly eliminate uncertainty regarding the potential for *recognized environmental conditions* in connection with a *property*. Performance of either this practice or Practice E 1527 is intended to reduce but not eliminate uncertainty regarding the existence of *recognized environmental conditions* in connection with a property, and both practices recognize reasonable limits of time and cost.

4.5.2 *Appropriate inquiry* does not mean an exhaustive assessment of a clean *property*. There is a point at which the cost of information obtained or the time required to gather it outweighs the usefulness of the information and, in fact, may be a material detriment to the orderly completion of transactions. One of the purposes of this practice is to identify a balance between the competing goals of limiting the costs and time demands inherent in performing an *environmental site assessment* and the reduction of uncertainty about unknown conditions resulting from additional information.

4.5.3 Not every *property* will warrant the same level of assessment. Consistent with good commercial or customary practice, the appropriate level of *environmental site assessment* will be guided by the type of property subject to assessment, the expertise and risk tolerance of the *user*, and the information developed in the course of the inquiry.

4.5.4 It should not be concluded or assumed that an inquiry was not an *appropriate inquiry* merely because the inquiry did not identify *recognized environmental conditions* in connection with a *property*. *Environmental site assessments* must be evaluated based on the reasonableness of judgments made at the time and under the circumstances in which they were made. Subsequent *environmental site assessments* should not be considered valid standards to judge the appropriateness of any prior assessment based on hindsight, new information, use of developing technology or analytical techniques, or other factors.

4.6 *Continued Viability of Environmental Site Assessment*—An *environmental site assessment* meeting or exceeding either this practice or Practice E 1527 and completed less than 180 days previously is presumed to be valid. An *environmental site assessment* meeting or exceeding either practice and completed more than 180 days previously may be used to the extent allowed by 4.7 through 4.7.5.

4.7 *Prior Assessment Usage*—Both this practice and Practice E 1527 recognize that *environmental site assessments* performed in accordance with these practices will include information which *users* or subsequent users may want to use to avoid undertaking duplicative assessment procedures. Therefore, the practices describe procedures to be followed to assist users in determining the appropriateness of using information in *environmental site assessments* performed previously. The system of prior assessment usage is based on the following principles that should be adhered to in addition to the specific procedures set forth elsewhere in these practices:

4.7.1 Subject to 4.7.4, *users* and *environmental professionals* may use information in prior *environmental site assessments* provided such information was generated as a result of procedures that meet or exceed the requirements of this practice or Practice E 1527 and then only provided that the specific procedures set forth in each practice are met.

4.7.2 Subject to 4.7.4, a prior *environmental site assessment* may be used in its entirety, without regard to the specific procedures set forth in these practices, if, in the reasonable judgment of the *user:* the prior *environmental site assessment* meets or exceeds the requirements of this practice or Practice E 1527, and the conditions at the *property* likely to affect *recognized environmental conditions* in connection with the *property* are not likely to have changed materially since the prior *environmental site assessment* was conducted. In making this judgment, the *user* should consider the type of *property* assessed and the conditions in the area surrounding the *property*.

4.7.3 Except as provided in 4.7.2 and 4.7.2 of Practice E 1527, prior *environmental site assessments* should not be used without current investigation of conditions likely to affect *recognized environmental conditions* in connection with the *property* that may have changed materially since the prior *environmental site assessment* was conducted. At a minimum, for a *transaction screen process* consistent with this practice, a new *site visit* should be performed.

4.7.4 If the *user* or *environmental professional(s)* conducting an *environmental site assessment* has *actual knowledge* that the information being used from a prior *environmental site assessment* is not accurate or if it is *obvious*, based on other information obtained by means of the *environmental site assessment* or known to the person conducting the *environmental site assessment*, that the information being used is not accurate, such information from a prior *environmental site assessment* may not be used.

4.7.5 The contractual and legal obligations between prior and subsequent *users* of *environmental site assessments* or between *environmental professionals* who conducted prior *environmental site assessments* and those who would like to use such prior *environmental site assessments* are beyond the scope of this practice.

4.8 The contractual and legal obligations between an *environmental professional* and a *user* (and other parties, if any) are beyond the scope of this practice. No specific legal relationship between the *environmental professional* and the *user* is necessary for the *user* to meet the requirements of this practice.

4.9 If the *user* is aware of any specialized knowledge or experience that is material to *recognized environmental conditions* in connection with the *property*, and the *preparer* is not the *user*, it is the *user*'s responsibility to communicate any information based on such specialized knowledge or experience to the *preparer*. The *user* should do so before the preparer makes the site visit.

4.10 In a transaction involving the purchase of a parcel of *commercial real estate*, if a *user* has *actual knowledge* that the purchase price of the property is significantly less than the purchase price of comparable properties, the *user* should try to identify an explanation for the lower price and to make a written record of such explanation. Among the factors to consider will be the information that becomes known to the *user* pursuant to the *transaction screen environmental site assessment*.

5. Introduction to Transaction Screen Questionnaire

5.1 *Process*—The *transaction screen process* consists of asking questions contained within the *transaction screen questionnaire* of *owners* and *occupants* of the *property*, observing site conditions at the *property* with direction provided by the *transaction screen questionnaire*, and, to the extent *reasonably ascertainable*, conducting limited research regarding certain *government records* and certain *standard historical sources*. The questions asked of *owners* are the same questions as those asked of *occupants*.

5.2 *Guide*—The *transaction screen questionnaire* is followed by a guide designed to assist the person completing the *transaction screen questionnaire*. The guide to the *transaction screen questionnaire* is set out in Sections 7 through 10 of this practice. The guide is divided into three sections: Guide for Owner/Occupant Inquiry, Guide to Site Visit, and Guide to Government Records/Historical Sources Inquiry.

5.2.1 To assist the *user*, its employee or agent, or the *environmental professional* in preparing a report, the guide repeats each of the questions set out in the *transaction screen questionnaire* in both the guide for *owner/occupant inquiry* and the guide to *site visit*. The questions regarding *government records/historical sources inquiry* are also repeated in the guide to that section.

5.2.2 The guide also describes the procedures to be followed to determine if reliance upon the information in a prior *environmental site assessment* is appropriate under this practice.

5.2.3 A *user*, his employee or agent, or *environmental professional* conducting the *transaction screen process* should not use the *transaction screen questionnaire* without reference to, or familiarity from prior usage with, the guide.

5.3 *User and Preparer*—The *user* conducting the *transaction screen process* is the party seeking to perform *appropriate inquiry* with respect to the *property*. The *user* may delegate the preparation of the *transaction screen questionnaire* to an employee or agent of the *user* or may contract with a third party to prepare the questionnaire on behalf of the *user*. The person preparing the questionnaire is the *preparer*, who may be either the *user* or the person to whom the *user* has delegated the preparation of the *transaction screen questionnaire*.

5.4 *Exercise of Care*—The *preparer* conducting the transaction screen process should use good faith efforts in determining answers to the questions set forth in the

transaction screen questionnaire. The *user* should take time and care to check whatever records are in the user's possession. The *preparer* should ask all persons to whom questions are directed to give answers to the best of the respondent's knowledge. As required by Section 9601(35)(B) of CERCLA, the *user* or *preparer* should discuss with a responsible person in authority in the user's organization (if any) any specialized knowledge or experience relating to *hazardous substances* on the *property* and the *preparer* should understand such information.

5.5 *Knowledge*—The owner or occupant of the *property* to which portions of the transaction screen questionnaire are directed should have sufficient knowledge and experience with respect to the property or in the owner's or occupant's particular business to understand the purpose and use of the transaction screen questionnaire. All answers should be given to the best of the owner's or occupant's actual knowledge.

5.5.1 While the person conducting the *transaction screen process* has an obligation to ask the questions set forth in the *transaction screen questionnaire*, in many instances the parties to whom the questions are addressed will have no obligation to answer them. The *user* is only required to obtain information to the extent it is *reasonably ascertainable*.

5.5.2 If the preparer asks the questions set forth in the *transaction screen questionnaire*, but does not receive any response or receives partial responses, the questions will be deemed to have been answered provided the questions have been asked, or were attempted to be asked, in person or by telephone and written records have been kept of the person to whom the questions were addressed and their responses, or the questions have been asked in writing sent by certified or registered mail, return receipt requested, postage prepaid, or by private, commercial overnight carrier and no responses have been obtained after at least two follow-up telephone calls were made or written request was sent again asking for responses.

5.5.3 The *transaction screen questionnaire* and the *transaction screen guide* sometimes include the phrase "to the best of your knowledge." Use of this phrase shall not be interpreted as imposing a constructive knowledge standard when it is not included or as imposing anything other than an *actual knowledge* standard for the person answering the questions, regardless of whether it is used. It is sometimes included as an assurance to the person being questioned that he or she is not obligated to search out information he or she does not currently have in order to answer the particular question.

5.6 *Conclusions Regarding Affirmative or Unknown Answers*—If any of the questions set forth in the transaction screen questionnaire are answered in the affirmative, the user must document the reason for the affirmative answer. If any of the questions are not answered or the answer is unknown, the user should document such nonresponse or answer of unknown and evaluate it in light of the other information obtained in the transaction screen process, including, in particular, the *site visit* and the government records/historical sources inquiry. If the *user* decides no further inquiry is warranted after receiving no response, an answer of unknown or an affirmative answer, the *user* must document the reasons for any such conclusion.

5.6.1 Upon obtaining an affirmative answer, an answer of unknown or no response, the *user* should first refer to the guide. The guide may provide sufficient explanation to allow a *user* to conclude that no further inquiry is appropriate with respect to the particular question.

5.6.2 If the guide to a particular question does not, in itself, permit a *user* to conclude that no further inquiry is appropriate, then the *user* should consider other information obtained from the *transaction screen process* relating to this question. For example, while on the site performing a *site visit*, a person may find a storage tank on the *property* and therefore answer Question 10 of the *transaction screen questionnaire* in the affirmative. However, during or subsequent to the *owner/occupant inquiry*, the *owner* may produce evidence that substances now or historically contained in the tank (for example, water) are not likely to cause contamination.

5.6.3 If either the guide to the question or other information obtained during the *transaction screen process* does not permit a *user* to conclude no further inquiry is appropriate with respect to such question, then the user must determine, in the exercise of the *user's* reasonable business judgment, based upon the totality of unresolved affirmative answers or answers of unknown received during the *transaction screen process*, whether further inquiry may be limited to those specific issues identified as of concern or should proceed with a full *Phase I Environmental Site Assessment*.

5.7 *Presumption*—A presumption exists that further inquiry is necessary if an affirmative answer is given to a question or because the answer was unknown or no response was given. In rebutting this presumption, the *user* should evaluate information obtained from each component of the transaction screen process and consider whether sufficient information has been obtained to conclude that no further inquiry is necessary. The user must determine, in the exercise of the user's reasonable business judgment, the scope

of such further inquiry: whether to proceed with a *Phase I Environmental Site Assessment* prepared in accordance with Practice E 1527 or a lesser inquiry directed at specific issues raised by the questionnaire.

5.8 *Further Inquiry Under Practice E 1527*—Upon completing the *transaction screen questionnaire*, if the *user* concludes that a *Phase I Environmental Site Assessment* is needed, the *user* should proceed with such inquiry with the advice and guidance of an *environmental professional*. Such further inquiry should be undertaken in accordance with Practice E 1527.

5.9 *Signature*—The *user* and the *preparer* of the *transaction screen questionnaire* must complete and sign the questionnaire as provided at the end of the questionnaire.

6. Transaction Screen Questionnaire

6.1 *Persons to Be Questioned*—The following questions should be asked of (*1*) the current *owner* of the *property*, (*2*) any major *occupant* of the *property* or, if the *property* does not have any major *occupants*, at least 10 % of the *occupants* of the *property*, and (*3*) in addition to the current *owner* and the *occupants* identified in (*2*), any *occupant* likely to be using, treating, generating, storing, or disposing of *hazardous substances* or *petroleum products* on or from the *property*. A major *occupant* is any *occupant* using at least 40 % of the leasable area of the property or any anchor tenant when the *property* is a shopping center. In a multifamily property containing both residential and commercial uses, the *preparer* does not need to ask questions of the residential *occupants*. The preparer should ask each person to answer all questions to the best of the respondent's *actual knowledge* and in good faith. When completing the *site visit* column, the *preparer* should be sure to observe the *property* and any buildings and other structures on the *property*. The guide provides further details on the appropriate use of this questionnaire.

Description of Site: Address:

_____

_____

_____

Question	Owner[7]			Occupants (if applicable)			Observed During Site Visit	
1a. Is the *property* used for an industrial use?	Yes	No	Unk	Yes	No	Unk	Yes	No
1b. Is any *adjoining property* used for an industrial use?	Yes	No	Unk	Yes	No	Unk	Yes	No
2a. Did you observe evidence or do you have any prior knowledge that the *property* has been used for an industrial use in the past?	Yes	No	Unk	Yes	No	Unk	Yes	No
2b. Did you observe evidence or do you have any prior knowledge that any *adjoining property* has been used for an industrial use in the past?	Yes	No	Unk	Yes	No	Unk	Yes	No
3a. Is the *property* used as a gasoline station, motor repair facility, commercial printing facility, dry cleaners, photo developing laboratory, junkyard or landfill, or as a waste treatment, storage, disposal, processing, or recycling facility (if applicable, identify which)?	Yes	No	Unk	Yes	No	Unk	Yes	No

[7] Unk = "unknown" or "no response."

Question	Owner[7]			Occupants (if applicable)			Observed During Site Visit	
3b. Is any *adjoining property* used as a gasoline station, motor repair facility, commercial printing facility, dry cleaners, photo developing laboratory, junkyard or landfill, or as a waste treatment, storage, disposal, processing, or recycling facility (if applicable, identify which)?	Yes	No	Unk	Yes	No	Unk	Yes	No
4a. Did you observe evidence or do you have any prior knowledge that the *property* has been used as a gasoline station, motor repair facility, commercial printing facility, dry cleaners, photo developing laboratory, junkyard or landfill, or as a waste treatment, storage, disposal, processing, or recycling facility (if applicable, identify which)?	Yes	No	Unk	Yes	No	Unk	Yes	No
4b. Did you observe evidence or do you have any prior knowledge that any *adjoining property* has been used as a gasoline station, motor repair facility, commercial printing facility, dry cleaners, photo developing laboratory, junkyard or landfill, or as a waste treatment, storage, disposal, processing, or recycling facility (if applicable, identify which)?	Yes	No	Unk	Yes	No	Unk	Yes	No
5a. Are there currently any damaged or discarded automotive or industrial batteries, pesticides, paints, or other chemicals in individual containers of >5 gal (19 L) in volume or 50 gal (190 L) in the aggregate, stored on or used at the *property* or at the facility?	Yes	No	Unk	Yes	No	Unk	Yes	No
5b. Did you observe evidence or do you have any prior knowledge that there have been previously any damaged or discarded automotive or industrial batteries, or pesticides, paints, or other chemicals in individual containers of >5 gal (19 L) in volume or 50 gal (190 L) in the aggregate, stored on or used at the *property* or at the facility?	Yes	No	Unk	Yes	No	Unk	Yes	No
6a. Are there currently any industrial *drums* (typically 55 gal (208 L)) or sacks of chemicals located on the property or at the facility?	Yes	No	Unk	Yes	No	Unk	Yes	No
6b. Did you observe evidence or do you have any prior knowledge that there have been previously any industrial *drums* (typically 55 gal (208 L)) or sacks of chemicals located on the property or at the facility?	Yes	No	Unk	Yes	No	Unk	Yes	No
7a. Did you observe evidence or do you have any prior knowledge that *fill dirt* has been brought onto the property that originated from a contaminated site?	Yes	No	Unk	Yes	No	Unk	Yes	No
7b. Did you observe evidence or do you have any prior knowledge that *fill dirt* has been brought onto the property that is of an unknown origin?	Yes	No	Unk	Yes	No	Unk	Yes	No
8a. Are there currently any *pits, ponds,* or *lagoons* located on the *property* in connection with waste treatment or waste disposal?	Yes	No	Unk	Yes	No	Unk	Yes	No
8b. Did you observe evidence or do you have any prior knowledge that there have been previously, any *pits, ponds,* or *lagoons* located on the *property* in connection with waste treatment or waste disposal?	Yes	No	Unk	Yes	No	Unk	Yes	No
9a. Is there currently any stained soil on the *property*?	Yes	No	Unk	Yes	No	Unk	Yes	No

Question	Owner[7]			Occupants (if applicable)			Observed During Site Visit	
9b. Did you observe evidence or do you have any prior knowledge that there has been previously, any stained soil on the *property*?	Yes	No	Unk	Yes	No	Unk	Yes	No
10a. Are there currently any registered or unregistered storage tanks (above or under-ground) located on the *property*?	Yes	No	Unk	Yes	No	Unk	Yes	No
10b. Did you observe evidence or do you have any prior knowledge that there have been previously, any registered or unregistered storage tanks (above or under-ground) located on the *property*?	Yes	No	Unk	Yes	No	Unk	Yes	No
11a. Are there currently any vent pipes, fill pipes, or access ways indicating a fill pipe protruding from the ground on the *property* or adjacent to any structure located on the *property*?	Yes	No	Unk	Yes	No	Unk	Yes	No
11b. Did you observe evidence or do you have any prior knowledge that there have been previously, any vent pipes, fill pipes, or access ways indicating a fill pipe protruding from the ground on the *property* or adjacent to any structure located on the *property*?	Yes	No	Unk	Yes	No	Unk	Yes	No
12a. Are there currently any flooring, drains, or walls located within the facility that are stained by substances other than water or are emitting foul odors?	Yes	No	Unk	Yes	No	Unk	Yes	No
12b. Did you observe evidence or do you have any prior knowledge that there have been previously any flooring, drains, or walls within the facility that were stained by substances other than water or were emitting foul odors?	Yes	No	Unk	Yes	No	Unk	Yes	No
13a. If the property is served by a private well or non-public water system, is there evidence or do you have prior knowledge that contaminants have been identified in the well or system that exceed guidelines applicable to the water system?	Yes	No	Unk	Yes	No	Unk	Yes	No
13b. If the property is served by a private well or non-public water system, is there evidence or do you have prior knowledge that the well has been designated as contaminated by any government environmental/health agency?	Yes	No	Unk	Yes	No	Unk	Yes	No
14. Does the *owner* or *occupant* of the *property* have any knowledge of *environmental liens* or governmental notification relating to past or recurrent violations of environmental laws with respect to the *property* or any facility located on the *property*?	Yes	No	Unk	Yes	No	Unk		
15a. Has the *owner* or *occupant* of the *property* been informed of the past existence of *hazardous substances* or *petroleum products* with respect to the *property* or any facility located on the *property*?	Yes	No	Unk	Yes	No	Unk		
15b. Has the *owner* or *occupant* of the *property* been informed of the current existence of *hazardous substances* or *petroleum products* with respect to the *property* or any facility located on the *property*?	Yes	No	Unk	Yes	No	Unk		
15c. Has the *owner* or *occupant* of the *property* been informed of the past existence of environmental violations with respect to the *property* or any facility located on the *property*?	Yes	No	Unk	Yes	No	Unk		
15d. Has the *owner* or *occupant* of the *property* been informed of the current existence of environmental violations with respect to the *property* or any facility located on the *property*?	Yes	No	Unk	Yes	No	Unk		

Question	Owner[7]			Occupants (if applicable)			Observed During Site Visit		
16. Does the *owner* or *occupant* of the *property* have any knowledge of any *environmental site assessment* of the *property* or facility that indicated the presence of *hazardous substances* or *petroleum products* on, or contamination of, the *property* or recommended further assessment of the *property*?	Yes	No	Unk	Yes	No	Unk			
17. Does the *owner* or *occupant* of the *property* know of any past, threatened, or pending lawsuits or administrative proceedings concerning a release or threatened release of any *hazardous substance* or *petroleum products* involving the *property* by any owner or occupant of the *property*?	Yes	No	Unk	Yes	No	Unk			
18a. Does the *property* discharge waste water, on or adjacent to the *property*, other than storm water, into a storm water sewer system?	Yes	No	Unk	Yes	No	Unk	Yes	No	
18b. Does the *property* discharge waste water, on or adjacent to the *property*, other than storm water, into a sanitary sewer system?	Yes	No	Unk	Yes	No	Unk	Yes	No	
19. Did you observe evidence or do you have any prior knowledge that any *hazardous substances* or *petroleum products*, unidentified waste materials, tires, automotive or industrial batteries, or any other waste materials have been dumped above grade, buried and/or burned on the *property*?	Yes	No	Unk	Yes	No	Unk	Yes	No	
20. Is there a transformer, capacitor, or any hydraulic equipment for which there are any records indicating the presence of PCBs?	Yes	No	Unk	Yes	No	Unk	Yes	No	

Government Records/Historical Sources Inquiry

(See guide, Section 10)

21. Do any of the following Federal government record systems list the property or any property within the circumference of the area noted below:

National Priorities List—within 1.0 mile (1.6 Km)?	Yes	No
CERCLIS List—within 0.5 mile (0.8 Km)?	Yes	No
RCRA CORRACTS Facilities—within 1.0 mile (1.6 Km)?	Yes	No
RCRA non-CORRACTS TSD Facilities—within 0.5 mile (0.8 Km)?	Yes	No

22. Do any of the following state record systems list the property or any property within the circumference of the area noted below:

List maintained by state environmental agency of hazardous waste sites identified for investigation or remediation that is the state agency equivalent to NPL—within approximately 1.0 mile (1.6 Km)?	Yes	No
List maintained by state environmental agency of sites identified for investigation or remediation that is the state equivalent to CERCLIS—within 0.5 mile (0.8 Km)?	Yes	No
Leaking Underground Storage Tank (LUST) List—within 0.5 mile (0.8 Km)?	Yes	No
Solid Waste/Landfill Facilities—within 0.5 mile (0.8 Km)?	Yes	No

23. Based upon a review of *fire insurance maps* or consultation with the local fire department serving the *property*, all as specified in the guide, are any buildings or other improvements on the *property* or on an *adjoining property* identified as having been used for an industrial use or uses likely to lead to contamination of the *property*?   Yes   No   N/A

The *preparer* of the *transaction screen questionnaire* must complete and sign the following statement. (For definition of "preparer" and "user," see 5.3 or 3.3.25.)

This questionnaire was completed by:
Name _____
Title _____
Firm _____
Address _____
_____
_____
Phone number _____
Date _____

If the preparer is different than the user, complete the following:
Name of user _____
User's address _____
_____
User's phone number _____
Preparer's relationship to site _____
Preparer's relationship to user (for example, principal, employee, agent, consultant)

Copies of the completed questionnaire have been filed at:

_____

_____

_____

Copies of the completed questionnaire have been mailed or delivered to:

_____

_____

_____

Preparer represents that to the best of the preparer's knowledge the above statements and facts are true and correct and to the best of the preparer's actual knowledge no material facts have been suppressed or misstated.

Signature	Date
Signature	Date
Signature	Date

7. Guide to Transaction Screen Questionnaire

7.1 The following sets forth the guide to the *transaction screen questionnaire*. The guide accompanies the *transaction screen questionnaire* to assist the *preparer* in completing the questionnaire. Questions found in the *transaction screen questionnaire* are repeated in the guide.

7.2 If the *preparer* completing the *transaction screen questionnaire* is familiar with the guide from prior usage, the questionnaire may be completed without reference to the guide.

7.3 The *site visit* portion of the guide considers most of the same questions set forth in the guide to *owner/occupant inquiry* because the *transaction screen process* requires both questions of *owners* and *occupants* of the *property* and observations of the *property* by the *preparer*.

7.4 Prior *environmental site assessment* usage procedures are contained in the guide to *owner/occupant inquiry* and the guide to *government records/historical sources inquiry*. The information supplied in connection with the *site visit* portion of a prior *environmental site assessment* may be used for guidance, but may not be relied upon without determining through a new *site visit* whether any conditions that are material to recognized environmental conditions in connection with the *property* have changed since the prior *environmental site assessment*. Therefore, the guide to the *site visit* does not contain any prior assessment procedures.

7.5 In performing the *site visit* portion of the *transaction screen process*, the *preparer* should *visually and physically observe* the *property* and any structure located on the *property* to the extent not obstructed by bodies of water, cliffs, adjacent buildings, or other impassable obstacles.

7.5.1 The periphery of the *property* should be *visually and physically observed*, as well as the periphery of all structures on the *property*, and the *property* should be viewed from all adjacent public thoroughfares. Any overgrown areas should be inspected, including roads or paths with no apparent outlet that should be *visually and physically* observed to their ends.

7.5.2 On the interior of structures on the *property*, accessible common areas expected to be used by building *occupants* or the public (such as lobbies, hallways, utility rooms, and recreation areas), a representative sample of *owner* and *occupant* spaces, and maintenance and repair areas, including boiler rooms, should be *visually and physically*

observed. It is not necessary to look under floors, above ceilings, or behind walls.

7.5.3 After completing the *site visit*, the *preparer* of the *transaction screen questionnaire* may obtain "yes" answers that require the *preparer* once again to ask questions of the *owner* of the *property* or *occupants* of the *property* to satisfy the *user* that no further inquiry is necessary.

7.6 In addition to asking questions of the *owner* of the *property* and *occupants* of the *property* (Section 8) and *visually and physically* observing the *property* (Section 9), the *user* completing the *transaction screen process* should determine, either from governmental agencies or through commercial services providing government environmental records, whether certain known or suspected contaminated sites or activities involving the release of *hazardous substances* or *petroleum products* occur on or near the *property*. See Section 10.

7.6.1 These records may be obtained either directly from the government agencies or from commercial services that provide the records for a fee. Because of the numerous sources that must be searched and the response time of government agencies, commercial services are available that provide a single source for federal and state records. These services may provide a quicker response than the government agencies but fees will be charged for the information.

7.6.2 If government information is obtained from a commercial service, the firm should provide assurances that its records stay current with the government agency record sources. Government information obtained from non-government sources may be considered current if the source updates the information at least every 90 days, or, for information that is updated less frequently than quarterly by the government agency, within 90 days of the date the government agency makes the updated information available to the public.

7.6.3 The identity of firms providing this type of government information may be obtained through local telephone directories or through an inquiry of environmental professionals in the area of the preparer completing the transaction screen questionnaire.

8. Guide for Owner/Occupant Inquiry

8.1 Is the property used for an industrial use?

_____Yes _____No _____Unknown

8.1.1 Is any adjoining property used for an industrial use?

_____Yes _____No _____Unknown

Land Use

Property: _____

Adjoining properties north: _____

Adjoining properties south: _____

Adjoining properties east: _____

Adjoining properties west: _____

8.1.2 *Guide:*

8.1.2.1 It is recommended that the *preparer* describe the use of the *property* and *adjoining properties*.

8.1.2.2 Certain industrial uses on the *property* may raise concerns regarding the possibility of contamination affecting the *property*. For purposes of the *transaction screen questionnaire*, an industrial use is an activity requiring the application of labor and capital for the production or distribution of a product or article, including, without limitation, manufacturing, processing, extraction, refining, warehousing, transportation, and utilities. Manufacturing is defined as a process or operation of producing by hand, machinery, or other means a finished product or article from raw material. Industrial uses may be categorized as light or heavy industrial uses, depending upon the scale of the operations and the impact upon surrounding property in terms of smoke, fumes, and noise. Regardless of such categorization, the concern for purposes of the transaction screen process is whether the use involves the processing, storage, manufacture, or transportation of *hazardous substances* or *petroleum products*. For example, further inquiry would be necessary if the industrial use concerned the manufacture of paints, oils, solvents, and other chemical products but not if the use concerned the storage of inert goods in containers.

8.1.2.3 The term *adjoining properties* means any real property or properties the border of which is contiguous or partially contiguous with that of the property, or that would be contiguous or partially contiguous with that of the property but for a street, road, or other public thoroughfare separating them. *Adjoining properties* means the *property* and include *properties* across the street or any right of way from the property.

8.1.2.4 To use the information supplied in response to this question in a prior *environmental site assessment*, the *preparer* must determine if there were changes in the use of the *property* or any *adjoining property* since the prior *environmental site assessment* that are material to *recognized environmental conditions* in connection with the *property*. If not, using information in the prior *environmental site assessment* is appropriate. If so, the information requested must be supplied for each *property* for which the use has so changed.

8.2 Did you observe evidence or do you have any prior knowledge that the *property* has been used for an industrial use in the past?

_____Yes _____No _____Unknown

8.2.1 Did you observe evidence or do you have any prior knowledge that any *adjoining property* has been used for an industrial use in the past?

_____Yes _____No _____Unknown

8.2.2 *Guide*—See guide for question 8.1.

	Owner	Use	Dates
Previous use of property			
Previous use of properties to north			
Previous use of properties to south			
Previous use of properties to east			
Previous use of properties to west			

8.3 Is the *property* used as a gasoline station, motor repair facility, commercial printing facility, dry cleaners, photo developing laboratory, junkyard, or landfill, or as a waste treatment, storage, disposal, processing, or recycling facility (if applicable, identify which)?

_____Yes _____No _____Unknown

8.3.1 Is any *adjoining property* used as a gasoline station, motor repair facility, commercial printing facility, dry cleaners, photo developing laboratory, junkyard or landfill, or as a waste treatment, storage, disposal, processing, or recycling facility (if applicable, identify which)?

_____Yes _____No _____Unknown

Land Use

Property: _____

Adjoining properties north: _____

Adjoining properties south: _____

Adjoining properties east: _____

Adjoining properties west: _____

8.3.2 *Guide:*

8.3.2.1 It is recommended that the *preparer* describe the uses of the *property* and *adjoining properties*.

8.3.2.2 Gasoline stations, motor vehicle repair facilities (with or without supplying gas for the motor vehicles), dry cleaners, photo developing laboratories, commercial printing facilities, junkyards or landfills, and waste treatment, storage, disposal, processing, or recycling facilities all involve the use of *hazardous substances* or *petroleum products* and therefore require further inquiry concerning the possible release of such substances.

8.3.2.3 The term *adjoining properties* means any real property or properties the border of which is contiguous or partially contiguous with that of the *property*, or that would be contiguous or partially contiguous with that of the *property* but for a street, road, or other public thoroughfare separating them.

8.3.2.4 To rely on the information supplied in response to this question in a prior *environmental site assessment*, the *preparer* must determine if there were changes in the use of the property or any *adjoining property* since the prior *environmental site assessment* that are material to *recognized environmental conditions* in connection with the *property*. If not, then use of information in the prior *environmental site assessment* is appropriate. If so, the information requested must be supplied for each *property* for which the use has so changed.

8.4 Did you observe evidence or do you have any prior

knowledge that the *property* has been used as a gasoline station, motor repair facility, commercial printing facility, dry cleaners, photo developing laboratory, junkyard or landfill, or as a waste treatment, storage, disposal, processing, or recycling facility (if applicable, identify which)?

_____Yes _____No _____Unknown

8.4.1 Did you observe evidence or do you have any prior knowledge that any *adjoining property* has been used as a gasoline station, motor repair facility, commercial printing facility, dry cleaners, photo developing laboratory, junkyard or landfill, or as a waste treatment, storage, disposal, processing, or recyling facility (if applicable, identify which)?

_____Yes _____No _____Unknown

8.4.2 *Guide*—See guide for question 8.3.

	Owner	Use	Dates
Previous use of property			
Previous use of properties to north			
Previous use of properties to south			
Previous use of properties to east			
Previous use of properties to west			

LAND ISSUES

8.5 Are there currently any damaged or discarded automotive or industrial batteries, pesticides, paints, or other chemicals in individual containers of >5 gal (19 L) in volume or 50 gal (190 L) in the aggregate, stored on or used at the property or at the facility?

_____Yes _____No _____Unknown

8.5.1 Did you observe evidence or do you have any prior knowledge that there have been previously any damaged or discarded automotive or industrial batteries, or pesticides, paints, or other chemicals in individual containers of >5 gal (19 L) in volume or 50 gal (190 L) in the aggregate, stored on or used at the *property* or at the facility?

_____Yes _____No _____Unknown

8.5.2 *Guide:*

8.5.2.1 Are there any containers on the site that may contain any of these items? Is there any reason to suspect that chemicals or *hazardous substances* in such quantities may be stored on the site? Sheltered areas, cartons, sacks, storage bins, large canisters, sheds, or cellars of existing improvements are examples of containers and areas where chemicals or *hazardous substances* may be stored. If the answer to this question is "yes," list the items and the location(s) where they are stored. If unfamiliar with the contents of any container located on the site, the question must be answered "yes" until the materials are identified.

8.5.2.2 *Hazardous substances* may often be unmarked. The *preparer* should never open any containers that are unmarked because they may contain explosive materials or acids.

8.5.2.3 Consumer products in undamaged containers used for routine office maintenance or business, such as copy toner, should not create a need for further inquiry unless the quantity of such products is in excess of what would be customary for such use. The Environmental Protection

Agency has published a guidance document that identifies hazardous substances that must be reported under Sections 311 and 312 of the Emergency Planning and Community Right to Know Act ("EPCRA").[8] This document lists in tabular form the CERCLA Section 103 chemicals. If a preparer has a question regarding whether the substance is a hazardous substance under CERCLA, the preparer may refer to the list of lists or 40 CFR Part 302. In addition, the Environmental Protection Agency has also published a guidance document.[9] This document sets forth the *hazardous substances* found in many common consumer products listed by trade name.

8.5.2.4 A *preparer* should not rely exclusively upon a prior *environmental site assessment* in supplying this information.

8.6 Are there currently any industrial drums (typically, 55 gal (208 L)) or sacks of chemicals located on the property or at the facility?

_____Yes _____No _____Unknown

8.6.1 Did you observe evidence or do you have any prior knowledge that there have been previously any industrial *drums* (typically 55 gal (208 L)) or sacks of chemicals located on the property or at the facility?

_____Yes _____No _____Unknown

8.6.2 *Guide:*

8.6.2.1 Chemicals are frequently stored in large 55-gal (208–L) drums and dry chemicals are often stored in 20 lb (9 kg) sacks. See Appendix X2 for examples of 55-gal (208–L) drums and for surface staining resulting from improper drum storage.

8.6.2.2 A *preparer* should not rely exclusively upon a prior *environmental site assessment* in supplying this information.

8.7 Did you observe evidence or do you have any prior knowledge that *fill dirt* has been brought onto the property that originated from a contaminated site?

_____Yes _____No _____Unknown

8.7.1 Did you observe evidence or do you have any prior knowledge that *fill dirt* has been brought onto the property that is of an unknown origin?

_____Yes _____No _____Unknown

8.7.2 *Guide:*

8.7.2.1 The origin of *fill dirt* brought onto the *property* should be investigated to determine whether such dirt originated from a contaminated site. The term *fill dirt* is defined in the definitions and the *preparer* should refer to the definitions if the *preparer* has any question concerning the meaning of the term.

8.7.2.2 If any structures have been demolished on the property, the *preparer* should investigate whether the structures were demolished in place and *fill dirt* compacted over

---

[8] "Title III List of Lists, Consolidated List of Chemicals Subject to Reporting Under Title III of the Superfund Amendments and Reauthorization Act (SARA) of 1986," U.S. EPA, Office of Toxic Substances, January 1989.

[9] "Common Synonyms for Chemicals Listed Under Section 313 of the Emergency Planning and Community Right to Know Act" Office of Toxic Substances, U.S. EPA, January 1988.

them because such demolition debris may contain asbestos or *hazardous substances*.

8.7.2.3 To use the information supplied in response to this question in a prior *environmental site assessment*, the *preparer* must determine if there has been any filling at the site since the prior *environmental site assessment*. If not, then using information in the prior *environmental site assessment* is appropriate. If so, the information requested must be supplied for any *fill dirt* brought on the property since the prior *environmental site assessment*.

8.8 Are there currently any *pits, ponds, or lagoons* located on the *property* in connection with waste treatment or waste disposal?

_____Yes _____No _____Unknown

8.8.1 Did you observe evidence or do you have any prior knowledge that there have been previously, any *pits, ponds, or lagoons* located on the *property* in connection with waste treatment or waste disposal?

_____Yes _____No _____Unknown

8.8.2 *Guide:*

8.8.2.1 The presence of *pits, ponds, or lagoons*, together with waste treatment or waste disposal may indicate contaminated property. See the definitions with respect to the definition of *pits, ponds, or lagoons* in 3.2.27.

8.8.2.2 A *preparer* should not rely exclusively upon a prior *environmental site assessment* in supplying this information.

8.9 Is there currently any stained soil on the *property*?

_____Yes _____No _____Unknown

8.9.1 Did you observe evidence or do you have any prior knowledge that there has been previously, any stained soil on the *property*?

_____Yes _____No _____Unknown

8.9.2 *Guide:*

8.9.2.1 Stained soils are frequently associated with contamination and often are an indication of either current or previous leakage associated with piping and liquid storage containers. Soils that are stained show a marked discoloration as compared to other soils in the immediate vicinity.

8.9.2.2 A *preparer* should not rely exclusively upon a prior *environmental site assessment* in supplying this information.

8.10 Are there currently any registered or unregistered storage tanks (above or underground) located on the *property*?

_____Yes _____No _____Unknown

8.10.1 Did you observe evidence or do you have any prior knowledge that there have been previously, any registered or unregistered storage tanks (above or underground) located on the *property*?

_____Yes _____No _____Unknown

8.10.2 *Guide:*

8.10.2.1 Tanks are often used to store heating fuels, chemicals, and petroleum products; while tanks may be associated with the storage of chemicals, they are most often associated with liquid fuel heating systems (for example, oil furnaces).

8.10.2.2 To use the information supplied in response to

this question in a prior *environmental site assessment*, the *user* must determine if there were storage tanks installed on the site since the prior *environmental site assessment*. If not, then using information in the prior *environmental site assessment* is appropriate. If so, the information requested must be supplied on all storage tanks installed on the site since the prior *environmental site assessment*.

8.11 Are there currently any vent pipes, fill pipes, or access ways indicating a fill pipe protruding from the ground on the *property* or adjacent to any structure located on the *property*?

_____Yes _____No _____Unknown

8.11.1 Did you observe evidence or do you have any prior knowledge that there have been previously, any vent pipes, fill pipes, or access ways indicating a fill pipe protruding from the ground on the *property* or adjacent to any structure located on the *property*?

_____Yes _____No _____Unknown

8.11.2 *Guide:*

8.11.2.1 Vent or fill pipes often signal the current or previous existence of underground storage tanks.

8.11.2.2 Additionally, in answering this question the owner and occupant should consider any asphalt or concrete patching that would indicate the possibility of previous underground storage tank removal. Examples of vent and fill pipes are illustrated in Appendix X2.

8.11.2.3 A *preparer* should not rely exclusively upon a prior *environmental site assessment* in supplying this information.

STRUCTURE ISSUES

8.12 Are there currently any flooring, drains, or walls located within the facility that are stained by substances other than water or are emitting foul odors?

_____Yes _____No _____Unknown

8.12.1 Did you observe evidence or do you have any prior knowledge that there have been previously any flooring, drains, or walls within the facility that were stained by substances other than water or were emitting foul odors?

_____Yes _____No _____Unknown

8.12.2 *Guide:*

8.12.2.1 Stains (other than water stains) or foul odors may indicate leaks of hazardous substances of contaminants. Floor drains located within a building adjacent to hazardous substance storage areas or connected to an on-site disposal system (for example, septic system) present a potential source of subsurface discharge of contaminants.

8.12.2.2 A *preparer* should not rely exclusively upon a prior *environmental site assessment* in supplying this information.

OTHER ISSUES

8.13 If the *property* is served by a private well or non-public water system, is there evidence or do you have prior knowledge that contaminants have been identified in the well or system that exceed guidelines applicable to the water system?

_____Yes _____No _____Unknown

8.13.1 If the property is served by a private well or non-public water system, is there evidence or do you have prior knowledge that the well has been designated as contaminated by any government environmental/health agency?

_____ Yes _____ No _____ Unknown

8.13.2 *Guide:*

8.13.2.1 Private wells and non-public water systems are not monitored daily for water quality as municipal systems are monitored. If the system is private, it probably has been tested for contamination or evidence that it is free from contamination, and the results of any such tests should be produced by the *owner* or *occupant* of the well. The *preparer* is not required to test the water system to conduct the *transaction screen.*

8.13.2.2 A *preparer* should not rely exclusively upon a prior *environmental site assessment* in supplying this information.

8.14 Does the *owner* or *occupant* of the property have any knowledge of *environmental liens* or governmental notification relating to past or recurrent violations of environmental laws with respect to the *property* or any facility located on the property?

_____ Yes _____ No _____ Unknown

8.14.1 *Guide:*

8.14.1.1 In most cases, the federal or state government will notify the property owner prior to filing a lien on the *property.* Sections 302, 311, 312, and 313 of The Emergency Planning and Community Right-to-Know Act and other provisions of federal and state environmental laws establish reporting requirements with respect to businesses storing or using *hazardous substances* in excess of certain quantities. These businesses should be making periodic reports to a federal, state, or local environmental department, agency, or bureau. The government may periodically inspect such facilities to ensure compliance with environmental laws. In the event of a release of a reportable quantity within a 24-h period (as defined in CERCLA and the regulations promulgated pursuant to CERCLA), the person in charge of the facility is obligated to notify the U.S. EPA of the release. Any notification or response by any governmental entity will be in writing.

8.14.1.2 The information supplied in response to this question in a prior *environmental site assessment* may be used provided it is updated to the present time.

8.15 Has the *owner* or *occupant* of the *property* been informed of the past existence of *hazardous substances* or *petroleum products* with respect to the *property* or any facility located on the *property?*

_____ Yes _____ No _____ Unknown

8.15.1 Has the *owner* or *occupant* of the *property* been informed of the current existence of *hazardous substances* or *petroleum products* with respect to the *property* or any facility located on the *property?*

_____ Yes _____ No _____ Unknown

8.15.2 Has the *owner* or *occupant* of the *property* been informed of the past existence of environmental violations with respect to the *property* or any facility located on the *property?*

_____ Yes _____ No _____ Unknown

8.15.3 Has the *owner* or *occupant* of the *property* been informed of the current existence of environmental violations with respect to the *property* or any facility located on the *property?*

_____ Yes _____ No _____ Unknown

8.15.4 *Guide:*

8.15.4.1 Consider whether any *environmental professionals* familiar with *hazardous substances* or *petroleum products* have observed or determined that contamination existed on the property. *Hazardous substances* or *petroleum products* from the *property* may have affected soils, air quality, water quality, or otherwise affected structures located on the *property.*

8.15.4.2 The information supplied in response to this question in a prior *environmental site assessment* may be used provided it is updated to the present time.

8.16 Does the *owner* or *occupant* of the *property* have any knowledge of any *environmental site assessment* of the *property* or facility that indicated the presence of *hazardous substances* or *petroleum products* on, or contamination of, the *property* or recommended further assessment of the *property?*

_____ Yes _____ No _____ Unknown

8.16.1 *Guide:*

8.16.1.1 Copies of *reasonably ascertainable* prior *environmental site assessments* of the *property* or any portion thereof should be obtained and examined to determine whether further action or inquiry is necessary in connection with any environmental problems raised by a prior *environmental site assessment.*

8.16.1.2 The information supplied in response to this question in a prior *environmental site assessment* may be used provided it is updated to the present time.

8.17 Does the *owner* or *occupant* of the *property* know of any past, threatened, or pending lawsuits or administrative proceedings concerning a release or threatened release of any *hazardous substance* or *petroleum products* involving the *property* by any owner or occupant of the *property?*

_____ Yes _____ No _____ Unknown

8.17.1 *Guide:*

8.17.1.1 The *user* is not required to make an independent investigation or search of records on file with a court or public agency in answering this question; this question is to be answered by the *owner* or *occupant* based upon their respective *actual knowledge* and review of *reasonably ascertainable* records in their possession.

8.17.1.2 The information supplied in response to this question in a prior *environmental site assessment* may be used provided it is updated to the present time.

8.18 Does the *property* discharge wastewater, on or adjacent to the property, other than storm water, into a storm water sewer system?

_____ Yes _____ No _____ Unknown

8.18.1 Does the property discharge wastewater, on or adjacent to the property, other than storm water, into a sanitary sewer system?

_____Yes _____No _____Unknown

8.18.2 *Guide:*

8.18.2.1 The *owner* and each *occupant* should be asked where drain traps lead and the purpose of drainage pipes at the facility. Domestic sewage is not a CERCLA issue and the reference to *wastewater* does not include domestic sewage.

8.18.2.2 To use the information supplied in response to this question in a prior *environmental site assessment*, the *preparer* must determine if there was any change in discharge practices at the facility since the prior *environmental site assessment*. If not, using information in the prior *environmental site assessment* is appropriate. If so, the information requested must be supplied for all new or changed discharge practices.

8.18.2.3 Some jurisdictions require facilities with large roof or paved areas and construction sites to collect and divert runoff through a treatment process prior to discharging the stormwater runoff to municipal, separate storm sewer systems, or the waters of the United States. Such units are often called stormwater treatment systems. Oil-water separators are most often found outside a building under a manhole and require routine servicing to remove oil. Oil-water separators are usually in restaurants, repair garages, and service stations. An example of an oil-water separator is shown in Appendix X2. If any such oil-water separators or treatment systems have been installed at the *property* since a prior *environmental site assessment*, the requested information must be supplied for each new installation.

8.19 Did you observe evidence or do you have any prior knowledge that any *hazardous substances* or *petroleum products*, unidentified waste materials, tires, automotive or industrial batteries, or any other waste materials have been dumped above grade, buried and/or burned on the *property*?

_____Yes _____No _____Unknown

8.19.1 *Guide:*

8.19.1.1 Past waste disposal practices should be examined because these may have resulted in *hazardous substances* or *petroleum products* being released on the *property*. Does the *property* evidence any mounds or depressions that suggest a disposal site?

8.19.1.2 To use the information supplied in response to this question in a prior *environmental site assessment*, the *preparer* must determine if there was any dumping, burying, or burning of such materials at the site since the prior *environmental site assessment*. If not, then using information in the prior *environmental site assessment* is appropriate. If so, the information requested must be supplied for all such events since the prior *environmental site assessment*.

8.20 Is there a transformer, capacitor, or any hydraulic equipment for which there are any records indicating the presence of PCBs?

_____Yes _____No _____Unknown

8.20.1 *Guide:*

8.20.1.1 The PCBs are regulated by the Toxic Substances Control Act 15 USC. Section 2601 *et seq.* and, in the absence of a release, are not regulated by CERCLA. The provisions of CERCLA do apply if there is a release of PCBs. Accordingly, if an affirmative answer is obtained to this question, the further focus should be on whether there have been any instances of insulating oil leakage and, if so, whether these are suspected of being PCB or PCB-contaminated.

8.20.1.2 Transformers containing PCBs may have many different sizes and shapes. Some of the more commonly used transformers are set forth in Appendix X2. Transformers are to be registered pursuant to 40 CFR § 761.30.

8.20.1.3 Elevators and auto lifts are often run by hydraulically controlled systems containing PCBs. If inspection or maintenance records for the elevator, capacitor, or other hydraulic equipment indicate no release has occurred or that regular, scheduled maintenance has taken place and the machinery does not appear to be damaged or leaking, no further inquiry is required.

8.20.1.4 To use the information supplied in response to this question in a prior *environmental site assessment*, the *preparer* must determine if there were any transformers installed at the site since the prior *environmental site assessment* that are not owned by a utility, cooperative, or association. If not, then using information in the prior *environmental site assessment* is appropriate, except that for any transformer identified in the prior *environmental site assessment*, the PCB status should be updated. If new transformers have been installed, their PCB status should also be verified.

9. Guide to Site Visit

9.1 Is the *property* used for an industrial use?

_____Yes _____No

9.1.1 Is any *adjoining property* used for an industrial use?

_____Yes _____No

	Land Use
Property:	_____
	_____
Adjoining properties north:	_____
	_____
Adjoining properties south:	_____
	_____
Adjoining properties east:	_____
	_____
Adjoining properties west:	_____
	_____

9.1.2 *Guide:*

9.1.2.1 It is recommended that the *preparer* describe the uses of the *property* and *adjoining properties*.

9.1.2.2 Certain industrial uses on the *property* may raise concerns regarding the possibility of contamination affecting the *property*. For purposes of the *transaction screen questionnaire*, an industrial use is an activity requiring the application of labor and capital for the production or distribution of a product or article, including, without limitation, manufacturing, processing, extraction, refining, warehousing, transportation, and utilities. Manufacturing is defined as a process or operation of producing by hand, machinery, or other means, a finished product or article from raw material. Industrial uses may be categorized as light or heavy industrial uses, depending upon the scale of the operations and the impact upon surrounding property in terms of smoke, fumes, and noise. Regardless of such categorization, the concern for purposes of the transaction screen process is whether the use involves the processing, storage, manufacture, or transportation of *hazardous substances* or *petroleum* products. For example, further inquiry would be necessary if the industrial use concerned the manufacture of paints, oils, solvents, and other chemical products but not if the use concerned the storage of inert goods in containers.

9.1.2.3 The term *adjoining properties* means any real property or properties the border of which is contiguous or partially contiguous with that of the *property*, or that would be contiguous or partially contiguous with that of the *property* but for a street, road, or other public thoroughfare separating them.

9.2 Did you observe evidence or do you have any prior knowledge that the *property* has been used for an industrial use in the past?

_____Yes _____No

9.2.1 Did you observe evidence or do you have any prior knowledge that any *adjoining property* has been used for an industrial use in the past?

_____Yes _____No

9.2.2 *Guide:*

9.2.2.1 The *user* should inspect for any indications present on the *property* that would cause the *user* to suspect an industrial facility may once have existed on the site. Old buildings, pipes, containers, or other debris are indicators of previous industrial use of the site.

9.2.2.2 See guide for 9.1.

	Owner	Use	Dates
Previous use of property			
Previous use of properties to north			
Previous use of properties to south			
Previous use of properties to east			
Previous use of properties to west			

9.3 Is the property used as a gasoline station, motor repair facility, commercial printing facility, dry cleaners, photo developing laboratory, junkyard or landfill, or as a waste treatment, storage, disposal, processing, or recycling facility (if applicable, identify which)?

_____Yes _____No

9.3.1 Is any *adjoining property* used as a gasoline station, motor repair facility, commercial printing facility, dry cleaners, photo developing laboratory, junkyard or landfill, or as a waste treatment, storage, disposal, processing, or recycling facility (if applicable, identify which)?

_____Yes _____No

	Land Use
Property:	
Adjoining properties north:	
Adjoining properties south:	
Adjoining properties east:	
Adjoining properties west:	

9.3.2 *Guide:*

9.3.2.1 It is recommended that the *preparer* describe the uses of the *property* and *adjoining properties.*

9.3.2.2 Gasoline stations, motor vehicle repair facilities (with or without supplying gas for the motor vehicles), dry cleaners, photo developing laboratories, commercial printing facilities, junkyards or landfills, and waste treatment, storage, disposal, processing, or recycling facilities all involve the use of *hazardous substances* or *petroleum products* and therefore require further inquiry concerning the possible release of such substances.

9.3.2.3 The term *adjoining properties* means any real property or properties the border of which is contiguous or partially contiguous with that of the *property*, or that would be contiguous or partially contiguous with that of the *property* but for a street, road, or other public thoroughfare separating them. Adjoining properties include those that border the property and include properties across the street or any right of way from the property.

9.4 Did you observe evidence or do you have any knowledge that the *property* has been used as a gasoline station, motor repair facility, commercial printing facility, dry cleaners, photo developing laboratory, junkyard or landfill, or as a waste treatment, storage, disposal, processing, or recycling facility (if applicable, identify which)?

_____Yes _____No

9.4.1 Did you observe evidence or do you have any prior knowledge that any *adjoining property* has been used as a gasoline station, motor repair facility, commercial printing facility, dry cleaners, photo developing laboratory, junkyard or landfill, or as a waste treatment, storage, disposal, processing, or recyling facility (if applicable, identify which)?

_____Yes _____No

9.4.2 *Guide*—See guide for 9.2 and 9.3.

	Owner	Use	Dates
Previous use of property			
Previous use of properties to north			
Previous use of properties to south			
Previous use of properties to east			
Previous use of properties to west			

LAND ISSUES

9.5 Are there currently any damaged or discarded automotive or industrial batteries, pesticides, paints, or other chemicals in individual containers of >5 gal (19 L) in volume or 50 gal (190 L) in the aggregate, stored on or used at the property or at the facility?

_____Yes _____No

9.5.1 Did you observe evidence or do you have any prior knowledge that there have been previously any damaged or discarded automotive or industrial batteries, or pesticides, paints, or other chemicals in individual containers of >5 gal (19 L) in volume or 50 gal (190 L) in the aggregate, stored on or used at the *property* or at the facility?

_____Yes _____No

9.5.2 *Guide:*

9.5.2.1 Are there any containers on the site that may contain any one of these items? Is there any reason to suspect that chemicals or *hazardous substances* or *petroleum products* in such quantities may be stored on the site? Sheltered areas, cartons, sacks, storage bins, large canisters, sheds, or cellars of existing improvements should be investigated because these are areas where chemicals or *hazardous substances* or *petroleum products* may be stored. If the answer to this question is "yes," list the items and the location(s) where they are stored. If you are unfamiliar with the contents of any container located on the site, the question must be answered "yes" until the materials are identified. The existence of any damaged or opened containers identified as containing *hazardous substances* or *petroleum products* requires further investigation.

9.5.2.2 *Hazardous substances* or *petroleum products* may often be unmarked. The *preparer* should never open any unmarked containers at the facility because they may contain explosive materials or acids.

9.5.2.3 Consumer products in undamaged containers used for routine office maintenance or business, such as copy toner, should not create a need for further inquiry unless the quantity of such products is in excess of what would be customary for such use. The Environmental Protection Agency has published a guidance document that identifies hazardous substances or petroleum products that must be reported under Section 311 and 312 of EPCRA.[8] This document lists in tabular form the CERCLA Section 103 chemicals. If a preparer has a question regarding whether the substance is a hazardous substance under CERCLA, the preparer may refer to the list of lists or 40 CFR Part 302. In addition, the Environmental Protection Agency has also published a guidance document.[9] This document sets forth the hazardous substances or petroleum products found in many common consumer products listed by trade name.

9.6 Are there currently any industrial *drums* (typically, 55 gal (208 L)) or sacks of chemicals located on the *property* or at the facility?

_____Yes _____No

9.6.1 Did you observe evidence or do you have any prior knowledge that there have been previously any industrial *drums* (typically 55 gal (208 L)) or sacks of chemicals located on the property or at the facility?

_____Yes _____No

9.6.2 *Guide*—If found, they will require further examination with respect to any *hazardous substance* associated with them.

9.7 Did you observe evidence or do you have any prior knowledge that *fill dirt* has been brought onto the *property* that originated from a contaminated site?

_____Yes _____No

9.7.1 Did you observe evidence or do you have any prior knowledge that *fill dirt* has been brought onto the property that is of an unknown origin?

_____Yes _____No

9.7.2 *Guide*—*Fill dirt* brought onto the *property* may appear as mounds or depressions that do not appear to be naturally occurring. *Fill dirt* may be added in construction of a facility. The term *fill dirt* is defined in the definitions, and the *preparer* should refer to the definitions if the *preparer* has any question concerning the meaning of the term.

9.8 Are there currently any *pits, ponds, or lagoons* located on the *property* in connection with waste treatment or waste disposal?

_____Yes _____No

9.8.1 Did you observe evidence or do you have any prior knowledge that there have been previously, any *pits, ponds, or lagoons* located on the *property* in connection with waste treatment or waste disposal?

_____Yes _____No

9.8.2 *Guide*—The presence of *pits, ponds, or lagoons*, together with waste treatment or waste disposal may indicate contaminated property. See the definitions with respect to the definition of *pits, ponds, or lagoons* in 3.2.27.

9.9 Is there currently any stained soil on the *property*?

_____Yes _____No

9.9.1 Did you observe evidence or do you have any prior knowledge that there has been previously, any stained soil on the *property*?

_____Yes _____No

9.9.2 *Guide*—Stained soils are frequently associated with contamination and often are an indication of either current or previous leakage associated with piping and liquid storage containers. Soils that are stained show a marked discoloration as compared to other soils in the immediate vicinity.

9.10 Are there currently any registered or unregistered storage tanks (above or underground) located on the *property*?

_____Yes _____No

9.10.1 Did you observe evidence or do you have any prior knowledge that there have been previously, any registered or unregistered storage tanks (above or underground) located on the *property*?

_____Yes _____No

9.10.2 *Guide*—Tanks are often used to store heating fuels, chemicals, and *petroleum products*; while tanks may be associated with storage of chemicals, they are most often associated with liquid fuel heating systems (that is, oil furnaces). Examples of tanks are illustrated in Appendix X2.

9.11 Are there currently any vent pipes, fill pipes, or access ways indicating a fill pipe protruding from the ground on the property or adjacent to any structure located on the *property*?

_____Yes _____No

9.11.1 Did you observe evidence or do you have any prior knowledge that there have been previously, any vent pipes, fill pipes, or access ways indicating a fill pipe protruding from the ground on the *property* or adjacent to any structure located on the *property*?

_____Yes _____No

9.11.2 *Guide*—Vent or fill pipes often signal the current or previous existence of underground storage tanks. Additionally, observations should be made regarding any asphalt or concrete patching that would indicate the possibility of previous underground storage tank removal. Examples of vent and fill pipes are illustrated in Appendix X2.

STRUCTURE ISSUES

9.12 Are there currently any flooring, drains, or walls located within the facility that are stained by substances other than water or are emitting foul odors?

_____Yes _____No

9.12.1 Did you observe evidence or do you have any prior knowledge that there have been previously any flooring, drains, or walls within the facility that were stained by substances other than water or were emitting foul odors?

_____Yes _____No

9.12.2 *Guide*—Stains (other than water stains) or foul odors may indicate leaks of *hazardous substances* or *petroleum products* or contaminants. Floor drains located within a building adjacent to *hazardous substance* storage areas or connected to an on-site disposal system (for example, septic system) present a potential source of subsurface discharge of contaminants.

OTHER ISSUES

9.13 If the *property* is served by a private well or non-public water system, is there evidence or do you have prior knowledge that contaminants have been identified in the well or system that exceed guidelines applicable to the water system?

_____Yes _____No

9.13.1 If the property is served by a private well or non-public water system, is there evidence or do you have prior knowledge that the well has been designated as contaminated by any government environmental/health agency?

_____Yes _____No

9.13.2 *Guide*—Evidence of well water generally consists of a 4- to 12-in. (102- to 305-mm) diameter low level pipe protruding from the ground that is capped, as illustrated in Appendix X2.

9.14 Does the *property* discharge *wastewater*, on or adjacent to the property, other than storm water, into a storm water sewer system?

_____Yes _____No

9.14.1 Does the property discharge wastewater, on or adjacent to the property, other than storm water, into a sanitary sewer system?

_____Yes _____No

9.14.2 *Guide:*

9.14.2.1 All drain traps and pipes should be examined and their end points should be determined. Any ditches or streams on or adjacent to the site should be *visually and physically observed* for *wastewater* flow.

9.14.2.2 Some jurisdictions require facilities with large roof or paved areas and construction sites to collect and divert such runoff through a treatment process prior to discharging the stormwater runoff to municipal, separate storm sewer systems, or the waters of the United States. Such units are often called stormwater treatment systems. Oil-water separators are most often found outside a building under a manhole and require routine servicing to remove oil. Oil-water separators are usually in restaurants, repair garages, and service stations. An example of an oil-water

separator is shown in Appendix X2.

9.15 Did you observe evidence or do you have any prior knowledge that any *hazardous substances* or *petroleum products*, unidentified waste materials, tires, automotive or industrial batteries, or any other waste materials have been dumped above grade, buried and/or burned, on the *property*?

_____Yes _____No

9.15.1 *Guide*—Past waste disposal practices should be examined because these may have resulted in *hazardous substances* being released on the *property*. Does the site evidence any mounds or depressions that suggest a disposal site?

9.16 Is there a transformer, capacitor, or any hydraulic equipment for which there are any records indicating the presence of PCBs?

_____Yes _____No

9.16.1 *Guide:*

9.16.1.1 The PCBs are regulated by the Toxic Substances Control Act 15 USC Section 2601 *et seq.* and, in the absence of a release, are not regulated by CERCLA. The provisions of CERCLA do apply if there is a release of PCBs. Accordingly, if an affirmative answer is obtained to this question, the further focus should be on whether there have been any instances of insulating oil leakage and, if so, whether these are suspected of being PCB or PCB-contaminated.

9.16.1.2 Elevators and auto lifts are often operated by hydraulically controlled systems containing PCBs. If inspection or maintenance records for the elevator, capacitor, or other hydraulic equipment indicate no release has occurred and the machinery does not appear to be damaged or leaking, no further inquiry is required.

9.16.1.3 Transformers containing PCBs may have many different sizes and shapes. Some of the more commonly used transformers are set forth on Appendix X2. Transformers are to be registered pursuant to 40 CFR § 761.30.

10. Guide to Government Records/Historical Sources Inquiry

10.1 Do any of the following federal government record systems list the property or any property within the circumference of the area noted below:

National Priorities List—within 1.0 mile (1.6 Km)?	_____Yes	_____No
CERCLIS List—within 0.5 mile (0.8 Km)?	_____Yes	_____No
RCRA CORRACTS Facilities—within 1.0 mile (1.6 Km)?	_____Yes	_____No
RCRA non-CORRACTS TSD Facilities—within 0.5 mile (0.8 Km)?	_____Yes	_____No

10.1.1 *Guide:*

10.1.1.1 The NPL or National Priorities List is a list compiled by EPA pursuant to CERCLA 42 USC § 9605(a)(8)(B) of properties with the highest priority for cleanup pursuant to EPA's Hazard Ranking System. See 40 CFR Part 300.

10.1.1.2 The Comprehensive Environmental Response Compensation and Liability Information System (CERCLIS) is the list of sites compiled by EPA that EPA has investigated or is currently investigating for potential hazardous substance contamination for possible inclusion on the national Priorities List.

10.1.1.3 RCRA CORRACTS Facilities are those facilities which treat, store and/or dispose of hazardous wastes on-site and at which corrective remedial action is underway, as defined and regulated by RCRA. The RCRA non-CORRACTS TSD Facilities List are those facilities on which treatment, storage, and/or disposal of hazardous wastes takes place and at which corrective remedial action has not been required by EPA, as defined and regulated by RCRA.

10.1.1.4 If the preparer elects to obtain the records directly from government agencies, those records typically must be obtained through a formal written request to the office within each agency that is responsible for maintaining the records or for responding to public requests for records. At the federal level, these requests are governed by the Freedom of Information Act (FOIA). FOIA requires a written request and the request should identify the records the preparer requires and should identify the site and geographic area for which the preparer needs the records (for example, the address of the site and the appropriate city, county, or zip code to be searched). The request should be directed to the FOIA officer for the regional EPA office responsible for the region in which the site is located. A list of the FOIA offices for each of the EPA regions may be obtained from the federal government or local library. From the federal EPA offices, the *preparer* should anticipate a response no sooner than four to eight weeks.

10.1.1.5 If government information is obtained from a commercial service, the firm should provide ensurances that its records stay current with the government agency record sources. Government information obtained from commercial sources may be considered current if the source updates the information at least every 90 days, or for information that is updated less frequently than quarterly by the government agency, within 90 days of the date the government agency makes the updated information available to the public.

10.1.1.6 The information supplied in response to this question in a prior *environmental site assessment* may be used provided it is updated to the present time.

10.2 Do any of the following state record systems list the *property* or any *property* within the circumference of the area noted below:

List maintained by state environmental agency of hazardous waste sites identified for investigation or remediation that is the state agency equivalent to NPL—within 1.0 mile (1.6 Km)?

_____Yes _____No

List maintained by state environmental agency of sites identified for investigation or remediation that is the state equivalent to CERCLIS—within 0.5 mile (0.8 Km)?

_____Yes _____No

Leaking Underground Storage Tank (LUST) List—within 0.5 mile (0.8 Km)?

_____Yes _____No

Solid Waste/Landfill Facilities—within 0.5 mile (0.8 Km)?

_____Yes _____No

10.2.1 *Guide:*

10.2.1.1 The LUST list is a list of sites containing one or more underground storage tanks that have been identified as having leaked or are potentially leaking their contents into the ground or groundwater; these sites may be involved in a state cleanup program.

10.2.1.2 The solid waste/landfill facilities list is a list of sites that currently accept, or have accepted in the past, waste of any kind for disposal on site. Solid waste/landfill facilities lists typically are obtained through a state office of solid waste management that is often a division of the primary state environmental agency.

10.2.1.3 Although many states do not have specific Freedom of Information laws, if the preparer elects to obtain the records directly from government agencies, a similar written request for state records should be made to the primary state agency responsible for environmental regulation in that state. Typically, the office responsible for maintaining the records and for responding to requests for records are the same. Once again, the written request should identify the specific records requested and identify the site and geographic area for which the preparer needs the records. The state agency response will vary from state to state and agency to agency, but the *preparer* should anticipate a minimum of four weeks for a response.

10.2.1.4 In some cases, the request should be directed to a specific state office. For example, leaking underground storage tank requests should be made through either the state agency's groundwater management division, the state Fire Marshall's office, or the state Emergency Planning and Management Agency.

10.2.1.5 The identity of the state office to which the request should be made can be obtained by contacting the primary state environmental agency. Also, there are publications listing agency sources for each state. The local public library may contain these publications.

10.3 Based upon a review of *fire insurance maps* or consultation with the local fire department serving the property all as specified in the guide, are any buildings or other improvements on the property or on an adjoining property identified as having been used for an industrial use or uses likely to lead to contamination of the property.

_____Yes _____No _____Not Applicable

10.3.1 *Guide:*

10.3.1.1 The focus of this research is to determine whether any past use of the *property* would suggest the presence of contamination associated with the *property*. If *reasonably ascertainable*, one of two sources of data should be examined in the following order of preference: *fire insurance maps* showing the *property* or the local fire department serving the property. However, if the user has first-hand knowledge of the use of the *property* from the present back to 1940 or if the *preparer* interviewed disinterested people with such knowledge, then the *preparer* may eliminate this research and answer "not applicable" to the questions above. In addition, the preparer may eliminate this research and answer "not applicable" to the question if the *preparer* is unable to find appropriate sources of *fire insurance maps* or individuals at the local fire department for the property with knowledge of the property's past use, after making a reasonable effort in good faith to locate such information or if the information is otherwise not *reasonably ascertainable.*

10.3.1.2 Subject to the previous paragraph, the *preparer* should obtain *fire insurance maps* from the period(s) not covered by the first-hand knowledge of the *user* or of those interviewed, beginning with when the maps are first available for the area or when the area was first thought to be developed. At least two maps should be ordered at points in time separated by at least ten years.

10.3.1.3 *Fire insurance maps* are defined in 3.2.14 and may be available for review from public libraries, colleges, and local historical societies, or from commercial services.

10.3.1.4 In examining a *fire insurance map*, the *user* is only required to review those areas shown in the given source. For example, if the *property* is at the edge of a map sheet, the user need not order the adjoining sheet. If a source covers a large area, the *user* need only review the area within approximately 1/8 mile (200 m) of the *property.*

10.3.1.5 *Fire insurance maps* reviewed as part of a prior *environmental site assessment* do not need to be searched for or reviewed again, but the *preparer* should make a reasonable effort to determine the uses of the *property* since the last use identified in a prior *environmental site assessment.*

11. Non-Scope Considerations

11.1 *General:*

11.1.1 There may be environmental issues or conditions at a *property* that parties may wish to assess in connection with *commercial real estate* that are outside the scope of this practice (the non-scope considerations). As noted by the legal analysis in Appendix X1 of this practice, some substances may be present on a property in quantities and under conditions that may lead to contamination of the property or of nearby properties but are not included in CERCLA's definition of *hazardous substances* (42 USC 9601(14)) or do not otherwise present potential CERCLA liability. In any case, they are beyond the scope of this practice.

11.1.2 Whether or not a *user* elects to inquire into non-scope considerations in connection with this practice or any other *environmental site assessment*, no assessment of such non-scope considerations is required for appropriate inquiry as defined by this practice.

11.1.3 There may be standards or protocols for assessment of potential hazards and conditions associated with non-scope conditions developed by governmental entities, professional organizations, or other private entities.

11.1.4 Following are several non-scope considerations that persons may want to assess in connection with commercial real estate. No implication is intended as the relative importance of inquiry into such non-scope considerations, and this list of non-scope considerations is not intended to be all-inclusive:

11.1.4.1 Asbestos-containing materials,

11.1.4.2 Radon,

11.1.4.3 Lead-based paint,

11.1.4.4 Lead in drinking water, and

11.1.4.5 Wetlands.

APPENDIXES

(Nonmandatory Information)

X1. LEGAL BACKGROUND TO FEDERAL LAW AND THE PRACTICES ON ENVIRONMENTAL ASSESSMENTS IN COMMERCIAL REAL ESTATE TRANSACTIONS

INTRODUCTION

The legal section of Subcommittee E.50.02 on Environmental Assessments in Commercial Real Estate Transactions provides the following background to the Comprehensive Environmental Response, Compensation, and Liability Act (CERCLA), as amended including amended by the Superfund Amendments and Reauthorization Act (SARA), 42 USC §§ 9601 *et seq*. The background to CERCLA, commonly known as the Superfund Law, outlines the potential liability for the cleanup of hazardous substances, available defenses to such liability, appropriate inquiry under Superfund, statutory definition of hazardous substances, petroleum products and petroleum exclusion to CERCLA, and reasons why certain environmental hazards are excluded from the scope of Superfund and this practice and Practice E 1527.

There are several elements of Superfund liability and the commonly termed "innocent purchaser" defense, that arises out of the statutory third-party defense, that may impact on the development and understanding of this practice and Practice E 1527.

X1.1 *Superfund Liability:*

X1.1.1 All of the following elements of liability under Superfund must be established by a plaintiff before a defendant will be held liable under Superfund for a *government's* response costs:[10]

X1.1.1.1 The site is a facility, as defined as § 9601(9),

X1.1.1.2 A release or threatened release of a hazardous substance from the site occurred (release is defined at § 9601(22) as *any* amount of any hazardous substance; "hazardous substance" is defined at § 9601(14) (see statutory definition of hazardous substance below),

X1.1.1.3 A release or threatened release caused the plaintiff to incur response costs. Response costs are defined at § 9601(25) to mean costs related to both removal actions (§ 9601(23)) and remedial actions (§ 9601(24)), and

X1.1.1.4 Defendants fall within at least one of the four classes of responsible parties.[11]

X1.1.2 In order to recover response costs, a government plaintiff must prove that the costs were not inconsistent with the National Oil and Hazardous Substances Pollution Contingency Plan (commonly referred to as the National Contingency Plan or NCP), 40 CFR § 300.[12] A private plaintiff must prove its costs were necessary costs of response and that the response action was consistent with the NCP 42 USC § 9607 (a).[13]

X1.1.3 If there is a release or threatened release of hazardous substances on a site, private parties, even if they are not PRPs, may decide to incur response costs and seek recovery from other private parties. The PRPs may seek contribution from other PRPs.

X1.1.4 There is an important difference between the government's burden to show that its response costs are "not inconsistent with the NCP" and the burden a private party bears to show that its response costs are "consistent with the NCP." See § 9607 (a)(4)(A) and (B). Courts have interpreted this statutory difference to give the government a rebuttable presumption that its response costs are consistent with the NCP, whereas a private party who undertakes response costs and seeks recovery from responsible parties bears the burden of proving its response was consistent with the NCP.[14] The EPA takes the position that a private party who undertakes a response action must be only in "substantial compliance,"

---

[10] USC § 9607(a). (All statutory references are to Title 42 of the United States Code, unless otherwise specified.) See *United States versus Aceto Agricultural Chemicals Corp.*, 872 F.2d 1373 (8th Cir. 1989). Private plaintiffs, as well as the government, may seek response costs under Superfund from defendants. While many users of these practices or other private parties may think in terms of how to defend against Superfund liability, they should recognize that they may decide to conduct cleanup actions and seek response costs from other parties.

[11] The four classes of potentially responsible parties (PRPs) are listed at § 9607(a) as follows:

(1) Owner and operator of a facility (see § 9601 (20)(A): the term "owner or operator" does not include a person, who, without participating in the management of a facility, holds indicia of ownership primarily to protect his security interest in the facility. In *re: Bergsoe Metal Corporation*, 910 F.2d 668 (9th Cir. Aug. 9, 1990); *Guidice versus BFG Electroplating and Manufacturing Co.*, 732 F. Supp. 556 (W.D. Pa. 1989); *United States versus Mirabile*, 23 ERC 1511 (E.D. Pa. 1985). But see *United States versus Fleet Factors Corp.*, 901 F.2d 1550 (11th Cir. 1990); *United States versus Maryland Bank and Trust Co.*, 632 F. Supp. 573 (D. Md. 1986). For clarification of the security interest exclusion, see EPA's rule on lender liability under CERCLA, 57 Federal Register 18344 (April 29, 1992);

(2) Any person who at the time of disposal of any hazardous substance owned or operated any facility at which such hazardous substances were disposed of;

(3) Any person who by contract, agreement, or otherwise arranged for disposal or

treatment or transport of hazardous substances; and

(4) Any person who accepts hazardous substances for transport to a facility selected by such person.

[12] The National Contingency Plan is the federal government's blueprint on how hazardous substances are to be cleaned up pursuant to CERCLA.

[13] See *Dedham Water Co. versus Cumberland Farms Dairy, Inc.*, 889 F.2d 1146 (1st Cir. 1989); other cases cited at ABA, *National Resources, Energy, and Environmental Law: 1989 The Year In Review*, p. 215, Note 155.

[14] *Amland Properties Corp. versus Aluminum Co. of America*, 711 F. Supp. 784, 794 (D. N.J. 1989); *Artesian Water Co. versus New Castle County*, 659 F. Supp. 1269, 1291 (D. Del. 1987); *United States versus Northeastern Pharmaceutical Chemical Co.*, 579 F. Supp. 823 (W.D. Mo. 1984), *aff'd in part, rev'd on other grounds*, 810 F.2d 726 (8th Cir. 1986).

rather than strict technical compliance, with the NCP, as long as a CERCLA-quality cleanup is achieved. The NCP requirements for a private party response-action are set forth at 40 CFR § 300.700.

X1.2 *Defenses to Liability:*

X1.2.1 Assuming all the elements of liability exist, a party may avoid liability only by meeting one of the defenses listed in § 9607(b). These listed defenses are exclusive of all others.[15] Section 9607(b) states (*emphasis added*):

"There shall be no liability under subsection (a) for a person otherwise liable who can establish by a preponderance of the evidence (the lowest evidentiary standard available, meaning more probable than not) that the release or threat of release of a hazardous substance and the damages resulting therefrom were *caused solely by*—

1) an act of God;

2) an act of war;

3) *an act or omission of a third party other* than an employee or agent of the defendant, or *than one whose act or omission occurs in connection with a contractual relationship* (see the definition of "contractual relationship" below), existing directly or indirectly, with the defendant . . ., if the defendant establishes by a preponderance of the evidence that (a) he exercised due care with respect to the hazardous substance concerned . . ., and (b) he took precautions against foreseeable acts or omissions of any such third party and the consequences that could foreseeably result from such actions or omissions."

X1.2.2 Under § 9601(35)(A), a contractual relationship "includes, but is not limited to, land contracts, deeds, or other instruments transferring title or possession . . .". These contractual relationships with third parties eliminate the defense to liability unless the defendant is an innocent purchaser. Or as stated by the status of § 9601(35)(A) (*emphasis added*), a contractual relationship with the third party defeats the defense

"unless the real property on which the facility is located was acquired by the defendant after disposal or placement of the hazardous substance . . . and one or more of the following circumstances is also established by the defendant by a preponderance of the evidence:

(i) At the time the defendant acquired the facility the defendant did not know and had no reason to know that any hazardous substance which is the subject of the release or threatened release was disposed of on, in, or at the facility.

(ii) The defendant is the government . . .

(iii) The defendant acquired the facility by inheritance or bequest."

X1.2.3 Therefore, the so-called innocent purchaser defense arises out of the third party defense of § 9607(b)(3). Restated, this defense to Superfund liability is available only if the defendant shows the following:

X1.2.3.1 The release or threat of release was caused solely by a third party,

X1.2.3.2 The third part is not an employee or agent of the defendant,

X1.2.3.3 The acts or omissions of the third party did not occur in connection with a direct or indirect contractual relationship to the defendant, or if there was a contractual relationship, the defendant acquired the property after dis-

posal or placement of the hazardous substance, and at the time the defendant acquired the facility the defendant *did not know and had no reason to know* that any hazardous substance which is the subject of the release or threatened release was disposed of on, in, or at the facility, and

X1.2.3.4 The defendant exercised due care with respect to the hazardous substances and took precautions against foreseeable acts or omissions of the third party.

X1.2.4 The statute then states at § 9601(35)(B)(*emphasis added*):

"To establish that the defendant had no reason to know, as provided (above), *the defendant must have undertaken, at the time of acquisition, all appropriate inquiry into the previous ownership and uses of the property consistent with good commercial or customary practice* in an effort to minimize liability. . . . [T]he court shall take into account any specialized knowledge or experience on the part of the defendant, the relationship of the purchase price to the value of the property if uncontaminated, commonly known or reasonably ascertainable information about the property, the obviousness of the presence or likely presence of contamination at the property, and the ability to detect such contamination by appropriate inspection."

X1.3 *Appropriate Inquiry in Commercial Real Estate Transactions:*

X1.3.1 One of the major questions that parties to commercial real estate transactions face when considering their potential Superfund liability is, "What level of inquiry into the previous ownership and uses of the property is appropriate to establish the innocent purchaser defense to Superfund liability?" These practices are structured to articulate the level of inquiry under Superfund that is appropriate for different situations.

X1.3.2 *The Appropriate Level of Inquiry:*

X1.3.2.1 The level of environmental inquiry that is appropriate under Superfund cannot be the same for every property or every party to a real estate transaction. The level of inquiry, in fact, will change depending on the particular property or party involved in a transaction. The statutory language, Congressional history, and common sense support this conclusion.

X1.3.2.2 First, it must be noted that little case law exists to serve as guidance about the minimum level of inquiry that will be deemed appropriate for the innocent purchaser defense. See, for example, *United States versus Serafini*, 706 F. Supp. 346 (M.D. Pa. 1988), and 1990 U.S. Dist. LEXIS 18466 (M.D. Pa. 1990) (By entertaining disputed facts as to the custom and practice of viewing land prior to purchase, the court implied that appropriate inquiry necessarily varies on a site-by-site basis); *United States versus Pacific Hide and Fur Depot, Inc.*, 716 F. Supp. 1341 (D. Idaho 1989) (No inquiry was required by those who received an ownership interest in property by means of corporate stock transfer and warranty deed under the facts of this case.); *International Clinical Laboratories, Inc. versus Stevens*, 30 ERC 2066, 20 ELR 20,560 (E.D.N.Y. 1990) (Despite a long history of toxic wastewater disposal and presence of the site on the state's hazardous waste disposal site list, the purchaser established the innocent purchaser defense since there were no visible environmental problems at the site, the defendant had no knowledge of environmental problems at the site, and the purchase price did not reflect a reduction on account of the problem.)

---

[15] *United States versus Aceto Agricultural Chemicals Corp.*, 872 F.2d 1373 (8th Cir. 1989). But see *United States versus Marisol, Inc.*, 725 F. Supp. 833 (M.D. Pa. 1989) (equitable defenses under CERCLA may be available after the development of a factual record).

X1.3.2.3 While the statute does not specifically distinguish certain types of properties and uses from others, or certain types of parties from others, it does list certain factors courts should consider in determining whether one's inquiry under the circumstances is appropriate. The statute, as explained above in X1.2 requires a court to consider a party's specialized knowledge or experience. The statute further mandates a court to consider what is "*reasonably* ascertainable information about the property," what contamination is *obviously* present, and the *party's "ability* to detect such contamination". The very use of terms such as "appropriate" and "reasonably", and the use of "specialized knowledge and experience" and "ability" in conjunction with the specific person attempting to utilize the defense signifies that Congress did not intend the appropriateness of the inquiry be judged by a bright line standard. If it is so intended, Congress would have stated, but did not, that the same inquiry should be made in every case.

X1.3.2.4 What is reasonable and obvious to one party may not be so to other parties, and ability, by necessity, varies among all parties. The statute, therefore, recognizes that different properties and parties must be treated differently. That is, different parties may conduct different levels of inquiry appropriate to their circumstances.

X1.3.2.5 The statutory standard of "appropriate inquiry" suggests the level of inquiry will depend on the circumstances and the underlying facts. Since the facts are almost always different, the level of inquiry must change with them. The legislative history on this particular issue demonstrates that Congress intended that the level of inquiry changes with the type of property and party.[16]

The duty to inquire under this provision shall be judged as of the time of acquisition. Defendants shall be held to a higher standard as public awareness of the hazards associated with hazardous releases has grown, as reflected by this Act, the 1980 Act [CERCLA] and other Federal and State statutes.

Moreover, good commercial or customary practice with respect to inquiry in an effort to minimize liability shall mean that a reasonable inquiry must have been made in all circumstances, in light of best business and land transfer principles.

Those engaged in commercial transactions should, however, be held to a higher standard than those who are engaged in private residential transactions.

X1.3.2.6 Because few cases address the standard of inquiry in the innocent purchaser defense and the legislative history describing Congressional intent is sparse, common sense is a useful guide in interpreting statutory language. If CERCLA mandated that the level of inquiry be the same for every property or potential defendant, then a lay consumer (renter or buyer) of a home, a purchaser of a small environmentally benign business, and a multinational corporate buyer of an industrial complex would have to conduct the same environmental site assessment (ESA) of the different properties in question. Additionally, the statute makes no mention of a Phase I ESA or any other specific type of inquiry one is to conduct in order for the inquiry to be deemed appropriate. If all inquiries had to be at the same level to be "appropriate," it would be illogical to stop at a Phase I ESA since some commercial or industrial properties routinely undergo, in the exercise of good commercial and customary practices, intrusive sampling (typically a Phase II ESA activity). Therefore, since routinely some properties undergo sampling, an inflexible standard would require sampling of all properties no matter what its use. This could not have been the intended result of SARA.

X1.3.3 *The Minimum Inquiries to Satisfy All Appropriate Inquiry:*

X1.3.3.1 Recognizing that inquiry changes with the underlying circumstances, the next question concerns that level of inquiry, if any, that Superfund requires to utilize the innocent landowner defense.

X1.3.3.2 As noted above, in some real estate transactions a Phase II ESA is routinely conducted. A Phase I ESA is conducted in these transactions only as a necessary prerequisite to outline the scope of the Phase II ESA. A Phase II ESA typically involves taking soil, water, and air samples to determine their contaminant content or verify that no contaminants are present or likely present. Note, however, that this simplistic outline of the Phase II ESA is misleading since the party can always dig down one foot deeper, take one more sample, or conduct one more test. The problem of how much inquiry is conducted, or at what level a party should begin, involves proving a negative—that is, that no contamination is present.[17] Since, according to the statute, inquiries should be judged by the circumstances existing at the time of acquisition, then there could be some properties and parties to real estate transactions where it may be appropriate to begin the inquiry with an intrusive Phase II ESA in order to invoke the innocent purchaser defense to liability.

X1.3.3.3 At the other extreme, the minimum level of inquiry that a party would be expected to conduct is found by looking at the least environmentally obtrusive class of property and party from a CERCLA perspective. This transaction likely involves the lay buyer of a home or the renter of an apartment. Assuming these parties meet the other prerequisites for the innocent purchaser defense, what level of environmental inquiry must they conduct to avoid Superfund liability? While there are no recorded court cases on this issue, the answer is probably none, unless a particular residential purchaser or renter has some specialized knowledge about or experience with the property in question that would lead a court to conclude that some questions should have been asked. Beyond these rare situations, it is highly unlikely that Congress intended to saddle housing consumers with the burden of investigating or cleaning up contaminated

[16] H.R. Rep. No. 962, 99th Cong., 2d Sess. 187 (1986), *reprint at* 1986 U.S. Code Cong. and Admin. News 3276, 3280.

[17] The inability to prove a negative creates a dilemma for the potential defendant. If the party's inquiry discovers contamination, then under the statute the party will not be able to avail itself of the innocent purchaser defense. If the inquiry does not discover contamination, EPA or another private party can argue in a response action that the inquiry was not "appropriate" and, therefore, the defendant can have no defense. The dilemma is explicitly recognized by Subcommittee E50.02 as beyond any reasonable interpretation of Congressional intent. The scope of these practices resolves the party's dilemma in the only reasonable way by stating: "It should not be concluded or assumed that the inquiry was not appropriate inquiry merely because the inquiry did not identify existing recognized environmental conditions in connection with a property. Environmental site assessments must be evaluated based on the reasonableness of the judgments made at the time and under the circumstances in which they were made". See 4.5.4.

sites. In fact, EPA has issued a statement of enforcement policy to the effect that it will not generally pursue owners of single family residences pursuant to CERCLA.[18] Therefore, for some properties and parties to real estate transactions, it is appropriate to conduct no environmental inquiry in order to meet the innocent purchaser defense to liability.

X1.3.3.4 The minimum level of appropriate inquiry under Superfund, therefore, ranges from no specialized inquiry to conducting an intrusive Phase II ESA. In order to satisfy the practices, to do no specialized inquiry, such as the transaction screen or Phase I ESA, is not enough for commercial real estate transactions. Under current commercial and customary practice and in light of best business and land transfer principles, however, no environmental site assessments are conducted in many real estate transactions, particularly those involving smaller properties, vacant land, or transactions of low monetary value. This practice and Practice E 1527 and the minimum level of inquiry under these practices, actually raises the average level of inquiry that should be performed where the parties want to come within the protection of the innocent landowner defense.

X1.3.3.5 The burden of proof is on the defendant to sustain by a preponderance of the evidence, the innocent purchaser defense. This is the least onerous burden of proof available to a party in litigation. The defendant must show only that the evidence offered to support the level of inquiry that was taken at the time of acquisition is of greater weight or more convincing than the evidence offered in opposition to it. In other words, the evidence on the inquiry issue taken as a whole shows that the fact sought to be proved is more probable than not. There may be technical or business judgments on whether the inquiry conducted or any other fact in a particular case is sufficient to meet the needs or concerns of a party to the real estate transaction. The bottom line, however, is that the judgment on whether the specific facts of a case, in light of statutory language, are sufficient to produce liability or a viable defense to liability is a legal one and such judgments constitute the practice of law.

X1.3.3.6 This practice is designed as the minimum level of inquiry to satisfy the practice from which a party to a commercial real estate transaction should proceed, recognizing that some parties to some commercial real estate transactions may wish to proceed by beginning with a Phase I or a Phase II ESA.

X1.4 *Statutory Definition of Hazardous Substance:*

X1.4.1 The statute at 42 USC § 9601(14)(A–F) defines hazardous substance by referring to five other statutes as well as to Superfund's own § 9602. The following is a description of the relevant portions of the other statutes and § 9602 of Superfund:

42 U.S.C. § 9601(14)(A): "[A]ny substance designated pursuant to section 1321(b)(2)(A) of Title 33." Title 33 USC § 1321 lies within the Clean Water Act and refers to, among other things, hazardous substance liability. 33 USC § 1321(b)(2)(A) states that the EPA shall develop, "as may be appropriate, regulations designating as hazardous substances, other than oil as defined in this section, such elements and compounds which, when discharged in any quantity into" the navigable waters of the

United States . . ., present an imminent and substantial danger to the public . . . health or welfare, including, but not limited to, fish, shellfish, wildlife, shorelines, and beaches."

42 U.S.C. § 9601(14)(B): "[A]ny element, compound, mixture, solution, or substance designated pursuant to section 9602 of this title." Section 9602 gives EPA the authority to designate as a hazardous substance "such elements, compounds, mixtures, solutions, and substances which, when released into the environment may present substantial danger to the public health or welfare or the environment . . ."

42 U.S.C. § 9601(14)(C): "[A]ny hazardous waste having the characteristics identified under or listed pursuant to section 3001 of the Solid Waste Disposal Act [42 USCA § 6921](but not including any waste the regulation of which under the Solid Waste Disposal Act [42 USCA § 6901 et seq.] has been suspended by Act of Congress)." The Solid Waste Disposal Act of 1980 amended the Resource Conservation and Recovery Act (RCRA). 42 USCA § 6921 of RCRA provides authority to the EPA to develop criteria for identifying characteristics of hazardous waste and for listing particular hazardous wastes within the meaning of 42 USCA § 6903(5) of RCRA. RCRA, § 6903(5), defines hazardous waste to mean

"a solid waste, or combination of solid wastes, which because of its quantity, concentration, or physical, chemical, or infectious characteristics may—

(A) cause, or significantly contribute to an increase in mortality or an increase in serious irreversible, or incapacitating reversible, illness; or

(B) pose a substantial present or potential hazard to human health or the environment when improperly treated, stored, transported, or disposed of, or otherwise managed."

For the identification and listing of hazardous wastes under RCRA, see 40 CFR. §§ 261.1 *et seq.*

42 USC § 9601(14)(D): "[A]ny toxic pollutant listed under section 1317(a) of Title 33." Section 1317(a) of Title 33 refers to toxic and pretreatment effluent standards under the Clean Water Act. The EPA is charged in this section with publishing and revising from time to time a list of toxic pollutants, taking "into account toxicity of the pollutant, its persistence, degradability, the usual or potential presence of the affected organisms in any waters, the importance of the affected organisms, and the nature and extent of the effect of the toxic pollutant on such organisms." Each toxic pollutant listed according to this section shall be subject to effluent limitations. For toxic pollutant effluent standards, see 40 CFR §§ 129.1 *et seq.*

42 USC § 9601(14)(E): "[A]ny hazardous air pollutant listed under section 112 of the Clean Air Act [42 USCA § 7412]." Section 7412 of the Title 42 deals with national emission standards for hazardous air pollutants. The EPA is charged here with publishing and revising from time to time "a list which includes each hazardous air pollutant for which [it] intends to establish an emission standard under this section." The term "hazardous air pollutant" means an air pollutant that in EPA's judgment "causes, or contributes to, air pollution which may reasonably be anticipated to result in an increase in mortality or an increase in serious irreversible, or incapacitating reversible, illness." For emission standards for hazardous pollutants, see 40 CFR § 61.01 *et seq.*

42 USC § 9601(14)(F): "[A]ny imminently hazardous chemical substance or mixture with respect to which the [EPA] has taken action pursuant to section 2606 of Title 15." Section 2606 of Title 15 deals with imminent hazards under the Toxic Substances Control Act (TSCA). The EPA is authorized under 15 USC § 2606 to seize an imminently hazardous chemical substance or mixture or seek other relief, such as requiring notice to users of the chemical substance or public notice of

[18] EPA, *Policy Towards Owners of Residential Property at Superfund Sites,* OSWER Directive No. 9834.6, July 3, 1991.

the risk associated with the substance or mixture. The term "'imminently hazardous chemical substance or mixture' means a chemical substance or mixture which presents an imminent and unreasonable risk of serious or widespread injury to health or the environment."

X1.4.2 After Subsections A through F, the Superfund definition of "hazardous substance" in § 9601(14) then goes on to state:

"The term does not include petroleum, including crude oil or any fraction thereof which is not otherwise specifically listed or designated as a hazardous substance under Subparagraphs (A) through (F) of this paragraph, and the term does not include natural gas, natural gas liquids, liquefied natural gas, or synthetic gas usable for fuel (or mixtures of natural gas and such synthetic gas)."

X1.4.3 The EPA has collected a list of "those substances in the statutes referred to in section 101(14) of the Act [42 USC § 9601(14)]." 40 CFR § 302.1 (1989) ("List of Hazardous Substances and Reportable Quantities," 40 CFR Part 302). This list changes with notices in the Federal Register. Also, any time a new hazardous waste is listed, the waste automatically becomes a hazardous substance.

X1.5 *Petroleum Products:*

X1.5.1 Under the petroleum exclusion of CERCLA (42 USC § 9601(14)), petroleum and crude oil have been explicitly excluded from the definition of hazardous substances under CERCLA. Nevertheless, petroleum products are included within the scope of both practices because such they are of concern in many commercial real estate transactions and current custom and usage is to include an inquiry into the presence of petroleum products in an environmental site assessment. Inclusion of petroleum products within the scope of the practices is not based upon the applicability, if any, of CERCLA to petroleum products.

X1.5.2 One reason to include petroleum products within the scope of the practice is because to do so reflects custom and usage: when environmental assessments are conducted in connection with commercial real estate transactions, they customarily include an assessment of the presence or likely presence of petroleum products under conditions that may lead to contamination. For example, environmental assessments ordinarily seek to assess whether there may be underground or above-ground storage tanks that may be leaking, whether those tanks contain petroleum products or some other product.

X1.5.3 In addition, although CERCLA may exclude petroleum products, other laws require cleanup of releases or spills of petroleum products. For example, petroleum products sometimes (for example, when they cannot be reclaimed from soil) become hazardous wastes subject to RCRA Subtitle C (42 USC § 6921 *et seq.*), must be cleaned up if released from underground storage tanks pursuant to RCRA Subtitle I (42 USC § 6991 *et seq.*), must be cleaned up pursuant to the Oil Pollution Act of 1990 (33 USC § 1321 *et seq.*), and must be cleaned up if released into the navigable waters of the United States pursuant to the Clean Water Act (33 USC § 1251 *et seq.*).

X1.5.4 Moreover, case law and EPA interpretations of the petroleum exclusion require an analysis of the facts of each case to determine whether a particular petroleum product is included in CERCLA's petroleum exclusion. The exclusion has been broadly interpreted to exclude gasoline and leaded gasoline from CERCLA's definition of hazardous substances regardless of the fact that gasoline and leaded gasoline contain certain indigenous components and additives which have themselves been designated as hazardous pursuant to CERCLA. See *Wilshire Westwood Associates versus Atlantic Richfield Corporation*, 881 F.2d 801 (9th Cir. 1989). The interpretation was narrowed when a judicial distinction was made between petroleum fractions produced by distillation processes and waste products resulting from contaminated tank scale. *United States versus Western Processing Co.*, 761 F. Supp. 713 (W.D. Wash. 1991). Another decision narrowly interpreted CERCLA's petroleum exclusion to be inapplicable to oil-related wastes containing hazardous substances because the primary purpose of the exclusion is to remove "spills or other releases strictly of oil" from the scope of CERCLA response and liability (not releases of hazardous substances mixed with oil). See *City of New York versus Exxon Corporation*, 744 F. Supp. 474 (S.D. N.Y. 1990). For additional discussion, see EPA memorandum entitled "The Petroleum Exclusion under the Comprehensive Environmental Response Compensation and Liability Act," issued by EPA's General Counsel, July 31, 1987.

X1.6 *Exclusion of Certain Hazards from Superfund:*

X1.6.1 The information that follows is provided to explain why these potential environmental hazards are not covered by Superfund's appropriate inquiry responsibilities.

X1.6.2 As a preliminary matter, it should be noted that an environmental site assessment that does not address substances excluded from CERCLA (whether those substance are excluded because they are petroleum products or by virtue of other characteristics) but that otherwise constitutes "all appropriate inquiry into the previous ownership and uses of the property consistent with good commercial or customary practice" should nevertheless entitle the user to the innocent purchaser defense, assuming that other requirements of the defense are met.

X1.6.3 *Radon:*

X1.6.3.1 A case discussing Superfund and radon is *Amoco Oil Company versus Borden, Inc.* This case dealt with a private cost recovery action by the buyer of a site against the seller for response costs relating to radiation from phosphogypsum wastes left on the site. Radon emanated from these radioactive wastes. The case points out that the "EPA has designated radionuclides as hazardous substances under § 9602(a) of CERCLA. . . . Additionally, the . . . EPA under § 112 of the Clean Air Act . . . list radionuclides as a hazardous air pollutant." "Radon and its daughter products are considered radionuclides, which are defined as 'any nuclide that emits radiation.'" Therefore, radon is a CERCLA hazardous substance. Also, when discussing what constitutes a release of a hazardous substance under the statute, the statute is plain that there is no quantitative requirement and that a release, broadly defined at 42 USC § 9601(22), of *any* amount constitutes a CERCLA release.

X1.6.3.2 Liability under Superfund depends on several factors, as noted in X1.1. Only one of four factors is the release or threatened release of a hazardous substance. The other three factors are the site is a facility, the defendant falls within at least one of four classes of potentially responsible parties (PRPs), and the release or threatened release *caused* the plaintiff (which can be the government or another private party) to incur response costs. Further, response costs must not be inconsistent with the National Contingency Plan (NCP), *and must not be limited by § 9604(a)(3)*. And, of course, there is no need to raise the innocent purchaser defense and its appropriate inquiry requirements unless the elements of liability will be met.

X1.6.3.3 Where radon from any source occurs in a building, three of the liability elements under CERCLA are met. There is a release of a hazardous substance, the building is a facility, and we can assume the defendant is a PRP. However, under 42 USC § 9604(a)(3)(A), "[r]emedial actions taken in response to hazardous substances as they occur naturally are specifically excluded from the NCP and are therefore not recoverable." A case discussing superfund and radon is *Amoco Oil Company versus Borden, Inc.*, 889 F.2d 664 (5th Cir. 1989). The statute is plain (42 USC § 9604(a)(3) and (4)(*emphasis added*):

"(3) Limitations on response

The President shall not provide for a removal or remedial action under this section in response to a release or threat of release—

(A) *of a naturally occurring substance in its unaltered form, or altered solely through naturally occurring processes or phenomena, from a location where it is naturally found;*

(B) from products which are part of the structure of, and result in exposure within, residential buildings or business or community structures;[19] or

(C) into public or private drinking water supplies due to deterioration of the system through ordinary use.[20]

(4) Exception to limitations

Notwithstanding Paragraph (3) of this subsection, to the extent authorized by this section, the President may respond to any release or threat of release if in the President's direction, it constitutes a public health or environmental emergency and no other person with the authority and capability to respond to the emergency will do so in a timely manner."

X1.6.3.4 Therefore, no liability under CERCLA attaches for naturally occurring radon. If a party to a real estate transaction wants to look for radon within a building, no amount of radon investigation will have any bearing on one's innocent purchaser defense under Superfund. Investigation of naturally occurring radon would be included, if at all, in the portion of the practice that deals with non-scope issues.

X1.6.4 *Asbestos:*

X1.6.4.1 The analysis of asbestos is similar to that involving radon. Before considering appropriate inquiry responsibilities, the four elements of CERCLA liability must be satisfied. Once again, as with radon, they are not met.

X1.6.4.2 Section 9604(a)(3)(B) of CERCLA prohibits response actions involving a release or threat of release "from products which are part of the structure of, and result in exposure within, residential buildings or business or community structures." There are a number of cases dealing with asbestos that interpret this statutory language. One such case is *First United Methodist Church of Hyattsville versus United States Gypsum Co.*,[21] that cites to other relevant cases.

X1.6.4.3 In *First United* the church brought a private cost recovery action against the manufacturer of asbestos-containing acoustical plaster. In holding that the action was barred by a state statute of repose (a certain time allowed by statute for bringing litigation) and that CERCLA did not preempt the state statute of repose, the court stated that § 9604(a)(3)(B) of CERCLA "represents much more than a procedural limitation on the President's authority; it is a substantive limitation of the breadth of CERCLA itself.[22] Therefore, the limitations of § 9604(a)(3) apply to private parties as well.

X1.6.4.4 Citing to the legislative history, the *First United* court concluded, "[i]n view of this clear expression of Congressional intent, we will not expand CERCLA to encompass asbestos-removal actions." The court further explained:[22]

"In closing, we note that this interpretation of CERCLA fully comports with the most fundamental guide to statutory construction—common sense. To extend CERCLA's strict liability scheme to all past and present owners of buildings containing asbestos as well as to all persons who manufactured, transported, and installed asbestos products into buildings, would be to shift literally billions of dollars of removal cost liability based on nothing more than an improvident interpretation of a statute that Congress never intended to apply in this context. [FN12] . . . Certainly, if Congress had intended for CERCLA to address the monumental asbestos problem, it would have said so more directly when it passed SARA. . . .

FN12—It is for this reason, that Congress simply did not intend for CERCLA to remedy the asbestos-removal problem, that we decline to follow the reasoning of *Prudential, Knox* and *Covalt* in rejecting First United's preemption argument. Instead of recognizing the fact that CERCLA is out of context in this situation, these courts rejected similar attempts to invoke the statute by construing CERCLA's key terms in a way to exclude asbestos-removal actions. *Covalt*, 860 F.2d [1434] at 1438-39 (defining "environment" to exclude the interior of a workplace); *Knox*, 690 F. Supp at 756-57 (defining "release" in terms of "spills" or "disposal"); *Prudential*, [711 F. Supp 1244] at 1254-55 (defining "disposal" to exclude the sale of a product for consumer use). We find this analysis unsatisfactory because it runs the risk of unnecessarily restricting the scope of CERCLA merely to dispose of claims that the statute was never intended to encompass in the first place. It is far better to simply acknowledge the inapplicability of CERCLA to asbestos-removal claims than to restrict its operative terms."

X1.6.4.5 Since asbestos that is a part of the structure of, and results in exposure within, residential buildings or business or community structures is excluded from CERCLA liability, it should not be investigated pursuant to a party's innocent purchaser appropriate inquiry requirements. Like naturally occurring radon, investigation of asbestos-containing materials that are part of the structure of buildings should be included, if at all, in the portion of the practice that deals with non-scope issues. Note, however, if asbestos is disposed of on a site and, therefore, is no longer part of the structure of a building, the cleanup of the disposed asbestos is subject to Superfund response actions. Likewise, if a building is sold with the knowledge that it will be demolished, one court ruled that the sale constitutes a

---

[19] This provision has implications for asbestos and lead-based paint. See X1.6.4.

[20] This provision has implications for lead from lead pipes and solder. See X1.6.5.

[21] One such case is *First United Methodist Church of Hyattsville versus United States Gypsum Co.*, 882 F.2d 862 (4th Cir. 1989), that cites to other relevant cases.

[22] See also *3550 Stevens Creek Associates versus Barclays Bank of California*, 915 F.2d 1355 (9th Cir. 1990).

disposal falling under CERCLA's liability provisions.[23]

X1.6.5 *Lead in Drinking Water and Lead-Based Paint*—These hazards can be evaluated in terms of the exclusions of 42 USC § 9604(a)(3)(B) and (C), in an analysis similar to the

analysis applied above to radon and asbestos. While there is no reported case law on these environmental issues as they relate to Superfund, the statutory language seems clear that these environmental hazards are not encompassed by Superfund's appropriate inquiry responsibilities. Note, however, like asbestos, where there is a disposal of these substances on the site or in a facility, CERCLA liability may arise.

[23] *CP Holdings, Inc. versus Goldberg-Zoino and Associates, Inc.*, 769 F. Supp. 432 (D.N.H. 1991).

X2. SUPPLEMENTAL INFORMATION FOR USE IN CONNECTION WITH THE GUIDE ON ENVIRONMENTAL SITE ASSESSMENTS OF COMMERCIAL REAL ESTATE

FIG. X2.1   Chemical Storage in 55-gal (208-L) Steel Drums

FIG. X2.2   Chemical Storage in 55-gal (208-L) Plastic Drums

FIG. X2.3   Typical Pole-Mounted Transformer

NOTE—Oil-water separators are often located under manholes outside repair garages, or at any location where it is necessary to separate oil from water prior to discharge.

FIG. X2.4   Manhole Cover Outside Repair Garage

NOTE—Floor drains come in various shapes and sizes. Shown here is one type of floor drain. It is important to know the point of discharge of any floor drain.

FIG. X2.5   Example of Floor Drain

FIG. X2.8   Single Tall Vent Pipe (Arrow) for Underground Storage Tank on Side of Building

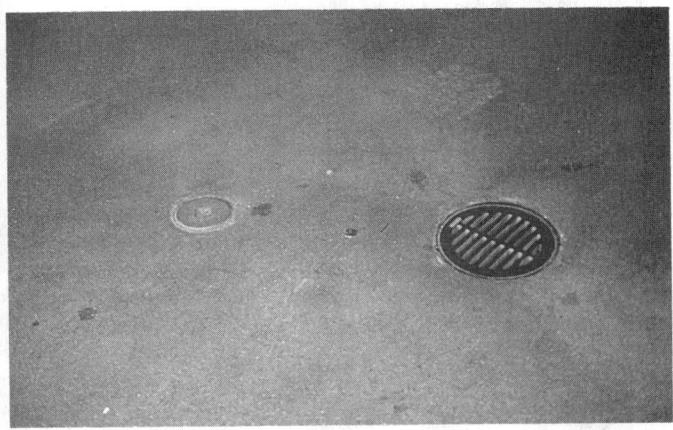

NOTE—Floor drains come in various shapes and sizes. Shown here is one type of floor drain. It is important to know the point of discharge of any floor drain.

FIG. X2.6   Example of Floor Drain

NOTE—Approximately 2½-in. (64-mm) diameter with screw cap.

FIG. X2.9   Fill Pipe for Residential Underground Fuel Oil Storage Tank

FIG. X2.7   Two Fill and Vent Pipes Leading to Two Underground Storage Tanks

NOTE—Approximately 8-in. (203-mm) diameter.

FIG. X2.10   Water Supply Well for Residential Property

NOTE—Approximately 8-in. (203-mm) diameter

FIG. X2.11   Water Supply Well for Residential Property

FIG. X2.12   Surface Staining from Improper Drum Storage

Provisional Standard Practice for
Environmental Regulatory Compliance Audits[1]

This standard is issued under the fixed designation PS 11; the number immediately following the designation indicates the year of original adoption.

INTRODUCTION

This provisional practice provides information on the terms, practices, and procedures associated with environmental regulatory compliance audits as practiced in the United States. Providing input to developing international standards on environmental management, International Organization for Standardization (ISO) TC207 Subcommittee 2 on Environmental Auditing has developed two draft standards, ISO 14010 and ISO 14012.[2] These draft standards are currently in the international balloting process with a final ISO standard publication date scheduled for October 1996. ISO TC207 currently does not intend to publish an ISO standard specifically on compliance auditing.

This provisional practice is consistent with the fundamental auditing principles in the ISO TC207 draft standards. There are some differences in specific requirements, most notably with regard to auditor qualifications. Potential users of both this provisional practice and the ISO TC207 drafts should carefully review the standards to ensure the user's goals and objectives will be met.

1. Scope

1.1 *Purpose*—This provisional practice covers a reference to which consultants; industry; local, state, and federal agencies; judicial systems; and other interested parties may consult for definition and description of accepted practices, procedures, and policies associated with Environmental Regulatory Compliance Audits (audits). As such, this provisional practice is intended to satisfy the generally accepted expectations of all parties interested in audits.

1.2 *Objectives*—The objectives of this practice are to document minimum requirements for audits, ensure that the audit is practical and reasonable, identify some of the legal issues that are associated with audits, explain how Environmental Regulatory Compliance Audits are different from other types of activities called environmental audits and assessments, and present minimum criteria for auditor qualifications.

1.3 *Focus:*

1.3.1 The use of this provisional practice is strictly limited to the scope set forth in this section. There may be valid reasons for deviation from this scope in unique situations or situations that are not normally encountered, however, deviations from this provisional practice should be documented in Audit and Protocol Reports that cite this provisional practice as the basis of the audit.

1.3.2 This provisional practice focuses on evaluating the degree of conformity with environmental laws, regulations, standards, or policies, or combination thereof, that have been established by a regulating body. Users of this provisional practice should recognize that audits generally require both legal and technical judgments. This provisional practice is directed primarily at the technical part of auditing.

1.3.3 This provisional practice does not address action plans to resolve audit findings.

1.3.4 Nonscope issues may be included in an audit as clearly identified supplements. These supplements, if included, should be clearly described in a modified audit scope and agreed upon by all relevant parties prior to conducting the audit. To avoid confusion, issues related to audit scope supplements found during an audit should not be reported as audit findings, unless agreed upon by all parties conducting the audit. The evaluation of compliance with voluntary standards, policies, and documents related to best environmental practices may be included as supplements to the audit scope.

1.4 *Organization of This Provisional Practice:*

1.4.1 This provisional practice is organized in the following manner:

	Section
Scope	1
Referenced Documents	2
Terminology	3
Significance and Use	4
Responsibilities	5
Audit Team Staffing and Auditor Qualifications	6
Main Body of the Practice (Including Evaluation and Report Writing)	7 through 11

1.4.2 An appendix, discussing legal considerations, provides information on some legal issues associated with an audit. Both the user and producer should be aware of these legal considerations. For example, to the extent that a user is concerned about maximizing the confidentiality of the audit, the user should be aware of the specific process that needs to be followed to do so. Similarly, the producer needs to be cautious not to render improper legal judgments.

---

[1] This provisional practice is under the jurisdiction of ASTM Committee E-50 on Environmental Assessment and is the direct responsibility of Subcommittee E50.04 on Performance Standards Related to Environmental Regulatory Programs. Approved Sept. 10, 1995. Published November 1995.

[2] International Organization for Standardization (ISO), 14000 Series, available from ASTM Customer Service, 100 Barr Harbor Drive, West Conshohocken, PA 19428-2959.

1.5 Provisional standards[3] achieve limited consensus through approval of the sponsoring subcommittee.

1.6 *This standard does not purport to address all of the safety concerns, if any, associated with its use. It is the responsibility of the user of this standard to establish appropriate safety and health practices and determine the applicability of regulatory limitations prior to use.*

2. Referenced Documents

2.1 *ASTM Standards:*

E 1527 Practice for Environmental Site Assessments: Phase I Environmental Site Assessment Process[4]

E 1528 Practice for Environmental Site Assessments: Transaction Screen Process[4]

3. Terminology

3.1 *Descriptions of Terms Specific to This Standard:*

3.1.1 *action plan*—a plan to address negative audit findings that contains detailed steps to be taken, assignment of responsibility to specific person(s) for completion, and specifies time frames for final resolution of all issues.

3.1.2 *audit*—an environmental regulatory compliance audit.

3.1.3 *audit criteria*—laws, regulations, standards, policies, and practices that the audited entity is evaluated against in an audit.

3.1.4 *audit data*—factual information obtained during an audit. Audit data are used by auditors to understand the audited entity's condition and may include laws, regulations, permits, consent orders, reports submitted to regulating bodies, environmental site assessments, prior audits, publicly available records of environmental performance, and all papers used and generated during the course of an audit, such as working papers.

3.1.5 *audit findings*—a factual statement of conditions observed by an auditor at the time of the audit compared to audit criteria. Audit findings must not include legal judgments or conclusions unless legal counsel is involved in the audit as part of the audit team. Audit findings may be either positive or negative with respect to audit criteria.

3.1.6 *audit protocol*—a plan that systematically documents how and when the audit will be conducted. Specific items that should be included in the audit protocol are a description of the responsibilities of the auditor, the audited entity and other users, schedules, expectations for site visit activities, auditor obligations regarding disclosure and release reporting, and whatever measures, if any, desired to maximize the confidentiality of the audit.

3.1.7 *audit questionnaire*—a set of specific questions to be answered by an auditor during an audit. The answers may come from auditor observations, interviews, or document reviews. The questions may be based upon regulatory requirements or audited entity conditions.

3.1.8 *audit report*—the written or verbal summary of audit findings. The audit report should include a summary description of the audit scope, protocol, and deviations from this provisional practice, if any.

3.1.9 *audit scope*—a statement documenting what is to be audited. This statement contains at a minimum a description of the period of review, the property to be audited, and the audit criteria.

3.1.10 *audit team*—when two or more auditors are responsible for conducting an audit, the group is called an audit team. An audit team should have an audit team leader who is qualified as a lead auditor. Outside or in-house corporate counsel must be part of the audit team if the user is seeking legal as well as factual audit findings.

3.1.11 *audit team leader*—a qualified lead auditor with management responsibility for the audit team. There may be several lead auditors on an audit team, but there should be only one audit team leader.

3.1.12 *audited entity*—a property, or part thereof, that is the subject of an audit.

3.1.13 *auditing entity*—the organization containing and responsible for auditors. The auditing entity may be the auditor's employer and may be external to the audited entity.

3.1.14 *auditor*—a person who conducts an audit according to this provisional practice.

3.1.15 *confidential information*—audit data, audit findings, or audit reports that are not intended for distribution to any entity other than a designated user.

3.1.16 *dwelling*—inhabited portions of structures and buildings used for commercial, noncommercial, or residential purposes.

3.1.17 *environmental regulatory compliance audit*—A systematic, documented, and objective review of an audited entity to evaluate the compliance status of an audited entity with regard to specified audit criteria. The audit should result in a factual statement of conditions contained in an audit report. The audit may include a review of the environmental, health, and safety aspects of solid and hazardous waste management, air emissions, wastewater discharges, storm water discharges, spill prevention and control, hazardous substances release reporting, emergency planning and community right-to-know, toxic substances control, underground tank management, registered pesticides application, portable water supply management, medical waste management, radioactive materials management, hazardous materials transportation, or other areas of environmental concern. For the purpose of this provisional practice, an audit is not an environmental site assessment, pollution prevention assessment, or other assessment, review, or inspection unrelated to environmental regulatory compliance.

3.1.18 *environmental site assessment*—the process by which a person or entity seeks to determine if a particular property is subject to recognized environmental conditions. An environmental site assessment is different from an environmental regulatory compliance audit. See Practices E 1527 and E 1528.

3.1.19 *facility*—a specific property subject to an audit.

3.1.20 *field notes*—a portion of the audit data developed

---

[3] Provisional standards exist for two years subsequent to the approval date.
[4] *Annual Book of ASTM Standards*, Vol 11.04.

by an auditor during visit activities. Typically these are considered part of working papers.

3.1.21 *generally accepted standard practices*—minimum existing requirements for an audit that are commonly and widely practiced.

3.1.22 *interviews*—a procedure whereby an auditor verbally collects audit data.

3.1.23 *lead auditor*—a lead auditor is a more highly experienced and accomplished auditor who is qualified to lead audit teams.

3.1.24 *monitoring instruments*—equipment or devices used to determine potential occupational health or safety threats to auditors by providing measurements of environmental, health, or safety conditions.

3.1.25 *objectivity*—an auditor characteristic demonstrated by the absence of bias and influences that may affect the preparation for, conduct of, and results from an audit.

3.1.26 *open lines of inquiry*—questions or issues related to property conditions that cannot be resolved without obtaining additional information.

3.1.27 *organization*—an entity united for a particular purpose.

3.1.28 *period of review*—the time interval over which property conditions are evaluated against audit criteria. Start and end dates to define the time interval should be specified.

3.1.29 *physical inspection*—first-hand viewing of property conditions during an audit.

3.1.30 *post-visit activities*—portions of the provisional practice that are conducted after the visit.

3.1.31 *pre-visit activities*—portions of this provisional practice that are conducted prior to site visit activities.

3.1.32 *producer*—an entity conducting audits, developing audit findings, and audit reports.

3.1.33 *property*—property may be land and improvements such as buildings, fixtures, and structures, including those that are offshore and airborne.

3.1.34 *quality assurance and quality control program*—measures taken to ensure the objectivity, completeness, and accuracy of audit findings.

3.1.35 *recipient of audit report*—the person or organization that receives an audit report.

3.1.36 *recognizable issues*—environmental issues that are related to audit criteria and become the basis of audit findings.

3.1.37 *regulating bodies*—federal, state, or local governmental organizations that administer programs through development, implementation, and enforcement of laws and regulations.

3.1.38 *rules of engagement*—terms and conditions of a preaudit agreement among the auditor, audited entity, and recipient of audit results. These terms and conditions may include disclosure and reporting obligations and confidentiality, if desired. Rules of engagement are especially important when an independent third party is contracted to conduct an audit.

3.1.39 *site visit activities*—portions of this provisional practice conducted at the audited entity.

3.1.40 *user*—an intentional and designated recipient of audit findings or an audit report.

3.1.41 *working papers*—documentation, developed by an auditor during an audit regarding the audit findings of

physical inspections and interviews. These may include, but are not limited to, handwritten notes, completed questionnaires, schedules, plans, maps, diagrams, and copies of audit criteria or portions of same.

4. Significance and Use

4.1 *Intended Use:*

4.1.1 This provisional practice is intended for use by parties who wish to perform an audit, direct an independent outside party to perform an audit, or rely upon audit findings or an audit report. Such use includes, but is not limited to, performance of an audit by internal or external auditors.

4.1.2 This provisional practice is intended to level expectations among all parties interested in audits. Audits may be more or less comprehensive than is specified by this provisional practice, but deviations from this provisional practice must be specified in the audit protocol and audit report if this provisional practice is referenced.

4.1.3 This provisional practice describes generally accepted standard practices that will provide an unbiased evaluation of compliance with audit criteria.

4.2 *Standard Practice Based on Laws of the United States of America*—This provisional practice is designed to assist in making factual determinations about an audited entity's compliance with applicable federal, state, and local laws and regulations for audited entities located within the United States of America and territories under the jurisdiction of the United States of America. This provisional practice may be used for audited entities located outside the boundaries of the United States of America, but the auditor is cautioned to determine whether any other practices or standards govern compliance auditing in that country.

4.3 *Related Standards*—This provisional practice provides a procedure for an audit. This provisional practice is related to companion Practices E 1527 and E 1528. A number of terms and procedures from these practices are similar to terms and procedures associated with this provisional practice but they are not the same.

4.3.1 *Standard Practices not Interchangeable*—Although related to companion Practices E 1527 and E 1528, this provisional practice is designed to achieve different results. The primary purpose of this provisional practice is to obtain a factual determination regarding the audited entity's compliance with specified audit criteria. Therefore, companion practices should not be used interchangeably with, or in place of, this provisional practice.

4.4 *Additional Services*—A number of related issues may arise as a result of an audit, but are outside the scope of this provisional practice. Such issues include, but are not limited to, development of action plans and cost estimates to remedy audit findings. These services are not part of an audit.

4.5 *Principles*—The following principles are an integral part of this provisional practice. They are intended to be referenced in resolving any ambiguity or exercising such discretion as is accorded the producer or user in performing an audit. They may be used to judge whether an adequate audit has been conducted.

4.5.1 *Not Exhaustive*—An audit generally cannot constitute an exhaustive review of the audited entity's compliance with all applicable laws, regulations, standards, and policies. Moreover, an audit conducted in accordance with this provisional practice renders factual judgments only. The user must involve outside or in-house counsel to obtain legal judgment and determinations. Auditors should evaluate information within a given framework of knowledge, and preface their conclusions by stating that all audit findings are a matter of professional opinion. The auditor should, however, possess the expertise, training, and professional judgment to perform an audit according to a previously determined audit protocol and audit scope.

4.5.2 *Level of Review is Variable*—Consistent with the user's objectives, an audit may have a variable audit scope. The audit scope may vary in emphasis or focus based on the property to be audited and the information developed during the course of the audit. For example, in some audits the audit scope may include only selected regulatory areas or only a selected period of review. Every audit, in order to claim conformance with this provisional practice, shall meet the provisions of this provisional practice. It is the responsibility of both the user and the auditor to confirm in writing, prior to commencing the audit process, the audit scope and audit protocols.

4.5.3 *Comparison With Subsequent Audits*—It should not be assumed that an audit was performed improperly because an audit report failed to list audit findings identified in a subsequent audit of the same audited entity. Differences in audit findings may be attributed to hindsight, new information, use of new technology or analytical techniques, different regulatory requirements, changed conditions at the audited entity, or other factors.

4.6 *Continued Viability of Environmental Audit*—An audit completed for a specified period of review in accordance with this provisional practice is generally valid or accurate for only that period of review. Because requirements for environmental compliance may change, or facility conditions may change, it should not be assumed that an audit report is reliable, or has continued viability as an accurate and true evaluation for other than the conditions audited during that specified period of review.

4.7 *Usage of Prior Audits*—This provisional practice recognizes that audits performed in accordance with this provisional practice will include information that may be useful to subsequent users and auditors. Therefore, this provisional practice describes procedures to be followed in determining the appropriateness of using information from previous audits. The use of such audits should be based on the following principles in addition to the specific procedures set forth elsewhere in this provisional practice.

4.7.1 *Meets the Requirements of This Provisional Practice*—Information contained within prior audit(s) should be used only if that audit was performed in accordance with procedures that meet or exceed this provisional practice.

4.7.2 *Continued Viability*—Information contained within prior audit reports should be used only if it has continued viability as noted previously and if continued use of that prior audit report is appropriate based on the specified audit scope of the prior audit.

4.8 *Rules of Engagement*—The agreed-upon contractual and legal obligations among the auditors, audited entity, and recipient of the audit results. No specific contractual or legal relationship between the auditor and the user is necessary to meet the requirements of this provisional practice.

4.9 *Confidentiality*—This provisional practice acknowledges that some users may wish to perform audits with an intent to maximize the confidentiality of the audit findings or audit report. Confidentiality requirements and measures should be outlined in the audit protocol before the audit begins, because failure to properly structure the audit process, to properly label the audit findings and report for example, may waive whatever privileges might otherwise apply. See Appendix X1 for further details on confidentiality of audit findings and reports.

4.10 *Auditor Liability*—Auditors should be careful to conduct audits in an ethical manner and to avoid making legal judgments that would constitute the unauthorized practice of law. The reliance of the user and possibly the public and business community on the information contained in an audit report imposes an obligation that auditors maintain high standards of technical competence and integrity. Failure of auditors to comply with such standards may result in possible exposure to liability. See Appendix X1 for further information about potential auditor liability.

4.11 *Use of Audit Findings by Regulating Bodies*—Regulating bodies may request, or have access to the information contained in audit reports. This information may be used by these agencies in various cases, such as when a governmental agency determines it is needed to accomplish a statutory mission, or where the government deems information to be material to a criminal investigation. The disclosure and use of audit findings and audit reports by regulating bodies are governed by various principles that are addressed in Appendix X1. This provisional practice acknowledges the possibility that regulating bodies may seek and use audit findings and reports. This provisional practice explicitly does not imply either approval or disapproval of that possibility. A growing number of states have enacted legislation restricting access to audit findings and audit reports as discussed in the appendix.

5. Responsibilities

5.1 *General Responsibilities*—All included parties have specific responsibilities for meeting the requirements of this provisional practice.

5.2 *User Responsibilities:*

5.2.1 Articulate the expected audit protocol and audit scope to the auditor and audited entity if different from the user.

5.2.2 Determine need for confidentiality and address audit report disclosure and release concerns.

5.2.3 Adhere to the specified requirements of the audit, especially any that apply to confidentiality of audit findings and audit reports and control of audit report distribution.

5.3 *Auditor Responsibilities:*

5.3.1 Clarify the audit scope and audit protocol with the user.

5.3.2 Communicate with the audited entity regarding:

5.3.2.1 The environmental history of facility to be audited,

5.3.2.2 The audit schedule and logistics,

5.3.2.3 Disclosure and confidentiality issues,

5.3.2.4 Access issues (for example, security clearances), and

5.3.2.5 Health and safety issues and equipment requirements for auditors.

5.3.3 Verify what, if any, personnel protective equipment will be needed. Specialized equipment for long-term use such as steel-toed shoes, prescription safety glasses, and fit-tested respirators should be the auditor's responsibility,

5.3.4 Ensure that audit criteria are current and understood,

5.3.5 Prepare audit, protocols, and questionnaires that reflect current audit criteria,

5.3.6 Make travel arrangements,

5.3.7 Notify user or audited entity, or both, of conditions that would prevent audit completion in accordance with this provisional practice.

5.3.8 Report audit findings to the user, and

5.3.9 Maintain working papers and audit data in a secure manner.

5.4 *Audited Entity Responsibilities:*

5.4.1 Ensure auditor's access to the property,

5.4.2 Brief auditor on significant health and safety issues, audio-visual emergency signals, and evacuation routes that are relevant to the property,

5.4.3 Provide audit team, for the duration of the audit, with a secure and clean working space equipped with lighting, telephone, climate comfort controls, electrical outlets, toilets and washrooms, and access to clean food and beverages,

5.4.4 Provide auditors with equipment and real-time monitoring of health and safety, as required,

5.4.5 Provide a facility escort(s) with knowledge about the audited entity operations, and environmental issues, to accompany auditor on the physical inspections,

5.4.6 Assist auditor to identify and schedule interviews with identified personnel, and

5.4.7 Ensure auditor(s) access to documents prepared during or otherwise relevant to the period of review.

5.5 *Auditing Entity:*

5.5.1 Ensure that auditors are competent in conformance with this provisional practice. Appropriate and documented training, tools, and monitoring programs should be in place.

5.5.2 Provide an effective quality assurance and quality control program that includes evaluations of auditor qualifications, auditor effectiveness, and audit reports.

6. Audit Team Staffing and Auditor Qualifications

6.1 *General Requirement*—Audits shall be performed by an auditor or audit team capable of meeting the requirements of the audit. An auditor should have the following attributes, skills, and qualifications.

6.2 *Attributes*—Auditors should be able to maintain independence and objectivity. Additionally they should be able to exercise tenacity, curiosity, and an ability to achieve closure by applying rational judgment. Auditors should be detail oriented and nonjudgmental with respect to cultural and business issues. Overall, auditors should instill confidence by demonstrating superior skills and knowledge related to auditing.

6.3 *Skills*—Auditors should demonstrate strong interpersonal, observational, organizational, diplomatic, listening, and communication skills. An auditor should also be able to recognize occupational hazards at audited entities.

6.4 *Qualifications*—Auditors should be knowledgeable of this provisional practice, and have specific and appropriate knowledge of the following:

6.4.1 Applicable environmental laws, regulations, and related documents,

6.4.2 Technical and environmental aspects of facilities being audited,

6.4.3 Technical, scientific, and legal terms and concepts,

6.4.4 Environmental management practices and standards, and

6.4.5 Environmental science and technology.

6.5 *Ethics:*

6.5.1 Auditors shall adhere to the professional accountability statement applicable to their profession.

6.5.2 Auditors may remove themselves from an audit at any time if they believe they cannot meet the requirements of this provisional practice.

6.6 *Staffing:*

6.6.1 Audit teams should have a designated audit team leader to provide technical and managerial direction to the audit team's efforts in conducting the audit and to coordinate the efforts of the individual auditors. The audit team leader should be a qualified lead auditor.

6.6.2 If the user is seeking legal judgment or conclusions, or if the user seeks to maximize the confidentiality of the audit findings and report, outside or in-house counsel must be part of the audit team.

6.7 *Education, Training, and Experience*—Auditors may be able to demonstrate their qualifications in auditing by meeting relevant education, training, and experience requirements and familiarity with this provisional practice.

7. The Audit

7.1 *Objective*—The audit seeks to provide the user with a factual statement of conditions observed by the auditor at the audited entity compared to audit criteria.

7.2 *Audit Scope*—The audit scope should be discussed freely between producers and users to ensure mutual expectations are met. Some items for these discussions follow.

7.2.1 *Aspects of the Audit:*

7.2.1.1 Audits make factual judgments on determinations about an audited entity's compliance with applicable laws, regulations, and requirements.

7.2.1.2 Entities that have audits completed should involve a senior manager of the organization who has authority to ensure that the audit is supported and that audit data is handled in a responsible manner.

7.2.1.3 Audits must be based on audit protocols. The audit protocol should address any concerns about report disclosure, release, and confidentiality. In addition, the audit protocol should include checklists consistent with the audit scope in order to provide a clear and consistent approach for conducting the audit.

7.2.2 *Notice*—Auditors must be notified in advance of the site visit by the audited entity regarding any possible health and safety risks that may be encountered so that those risks may be included in the audit scope. Required personal protective equipment that is for temporary use will be provided by the audited entity in a clean and functional condition. These items include disposable ear plugs, nonprescription safety glasses, gloves, and hard hats.

7.2.3 *Audit Data*—Audit data must be consistent with the defined audit scope. Although the audit report may be provided verbally, auditors should maintain audit data to verify the basis of audit findings.

7.2.4 *Period of Review*—The period of review must be clearly defined in the audit scope. Typical periods of review are twelve months coincident with calendar or fiscal years, but any clearly stated time interval is acceptable.

7.2.4.1 *Initial Audit*—It is suggested that initial audits, also known as "baseline" audits, review all past practices of which the audited entity is aware. Thus, an appropriate period of review may be several years, for an initial audit.

7.2.4.2 *Follow-Up Audits*—Follow-up audits are performed after initial audits. The audit criteria for follow-up audits may focus on previously detected audit findings and new requirements as a priority, however, all current requirements may also be used for audit criteria.

7.2.5 *Multiple Organizations at One Facility*—In defining the audit scope, the physical boundaries of the audited entry must be well defined. For example, on large properties, an audit may be completed for one facility but not the entire property. The auditor should confirm the physical boundaries prior to the audit.

7.3 *Major Components*—There are three steps to an audit as defined in this provisional practice.

7.3.1 *Pre-Visit*—Pre-visit activities are conducted prior to and in preparation for the site visit.

7.3.2 *Site Visit*—Site visit activities are conducted while auditor(s) are present at the audited entity.

7.3.3 *Post-Visit*—Post-visit activities take place after the site visit.

7.4 *Sampling*—This provisional practice does not require any physical testing or sampling of environmental media (for example, soil, water, air, waste), except that the auditor may carry monitoring instruments for the purposes of detecting potential health and safety threats to the auditor. Data created by these devices is not audit data. Information and observations that indicate a dangerous or unsafe condition must be reported immediately to the audited entity.

8. Pre-Visit

8.1 Pre-visit activities consist of the following steps:

8.1.1 *Audit Scope*—The audit protocol and scope must be developed prior to the site visit. It should clearly identify the facility to be audited, specific areas of concern, and the period of review for which a given topic or issue will be audited. It must address the issues of recordkeeping, document retention, disclosure and release reporting obligation, types of audit reports, and measures for maximizing confidentiality, if any.

8.1.2 *Management Commitment*—Audited entity management should be cooperative to the performance of an audit. The audited entity should take measures to ensure that activities and the processes evaluated by the auditor(s) accurately represent normal, and known abnormal operations. Audited entity management should be aware that civil or criminal penalties could apply if noncompliant conditions are discovered and not corrected with reasonable diligence.

8.1.3 *Schedule*—A schedule of audit activities should be developed by the auditor and shared with the audited entity during pre-visit activities. Issues such as a pre-visit site orientation meeting, identification and introduction of site contacts, the actual dates of the site visit, and resolution of logistical issues such as lodging, transportation, etc. should be scheduled and completed in advance of the site visit.

8.1.4 *Audit Team*—The audit team, formed during pre-visit activities, should have a collective knowledge of all areas needed to perform the audit. The team leader is responsible for ensuring that audit team members have adequate qualifications, training, and knowledge to perform the audit. The number of team members and their individual qualifications should reflect the complexity of the audit scope.

8.1.5 *Background Information*—Auditors should request the audited entity to provide relevant background information prior to the site visit. This should be done by administering a pre-visit questionnaire. Additional background information may be obtained from external sources as well. Additional information may consist of records, process and site descriptions, operation and maintenance manuals, emergency plans, environmental manuals, compliance inspection reports, environmental permits, environmental citations, and notices of violations and other relevant materials. They may also include accident and accident investigation reports.

8.1.6 *Records Review*—Pre-visit audit data should be reviewed by the audit team to aid in developing a revised audit protocol and audit scope and to help define appropriate site visit activities.

8.1.7 *Audit Protocol*—A written protocol should be developed prior to the site visit. It provides a plan for auditors to follow during the site visit and typically consists of questionnaires and instructions, etc. to facilitate the audit. Once an audit is underway deviations in audit protocols shall be documented along with a rationale for the changes.

8.1.8 *Systematic Updating*—A process should be included in the audit for the updating of audit criteria. Due to the changing regulations and requirements, an auditor should systematically seek to ensure that audit criteria are accurate and up-to-date.

9. Site Visit

9.1 Specific activities undertaken during a site visit follow:

9.1.1 *Opening Conference*—An opening conference should be held to bring together the auditor(s) and appropriate members of the audited entity staff or users, or both. The purpose of the opening conference is to discuss the audit protocol and objectives. The meeting should facilitate the later gathering of information by auditor(s) and encourage discussion of any questions or concerns of the audited entity staff. The audited entity should provide an orientation and overview of the operations for the audit team during the opening conference.

9.1.2 *Gather Facility Information*—Information is gathered and evaluated by auditor(s) from various sources to support the thoroughness, objectivity, and accuracy of the audit. The audit team should examine appropriate records, conduct physical inspections, interview key personnel, and observe routine and emergency operations and practices. Information gathered should be verified by independent means or corroborating information. The auditor(s) should utilize written questionnaires to ensure consistency in conducting the audit.

9.1.3 *Physical Inspections*—Inspections of the facility shall be conducted to familiarize the audit team with the physical layout of the property and to aid in identifying specific concerns and environmental, health, and safety practices. If permitted by audited entity management or their legal counsel, or both, photographs may be taken to document conditions during the audit.

9.1.4 *Interviews*—Interviews should be conducted to obtain information on facility practices and procedures that are subject to audit criteria. Key operating management and staff personnel, as well as representatives of employee groups and, where applicable, contractors, should be interviewed. As appropriate, a written questionnaire or guide should be used to ensure consistency in interviews and to aid documentation of responses. Relevant information from interviews must be included in the working papers. Summary information from multiple interviews may be used to provide anonymity, if necessary, but the details of each interview must be documented.

9.1.5 Generally, it is not advisable to use audio or video tape recording devices for conducting interviews, because they often are intimidating to interviewees. It is acceptable for an auditor to record notes to himself/herself after an interview or during the site visit. Procedures for the use of such recording devices, and their subsequent transfer and incorporation into or with the written working papers, should be established in advance of the audit and incorporated as a part of the audit protocol and audit scope.

9.1.6 *Documentation*—Documents and records of the audited entity should be reviewed to make a factual determination regarding compliance with pertinent laws and regulations. Records should also be evaluated to determine if regulatory requirements for testing and self-monitoring are being properly completed and if proper record keeping and reporting practices are being followed. Environmental permits should be evaluated to determine if operations are within allowable regulatory limits. Auditors should make copies of records and documents that are necessary to support audit findings. However, care should be exercised to avoid copying large documents when a portion (for example, table of contents rather than an entire manual) is sufficient evidence. Auditors should treat all documentation as confidential. All copies made become audit data and, as such, will be subject to the record keeping and confidentiality requirements established in the audit protocol and audit scope.

9.1.7 *Follow-Up All Open Issues*—All open lines of inquiry shall be documented and researched for closure as soon as practicable. Individuals should be assigned responsibility for following up on issues that are not resolved during the site visit. Unresolved issues must be reported as being "open."

9.1.8 *Team Meeting(s)*—A daily meeting of the audit team should be conducted to ensure the timely and consistent completion of the audit. Working papers and documentation should be reviewed to ensure a clear basis for all audit findings. The audit team may also need to collectively evaluate data and information obtained by individual team members to more accurately understand and evaluate compliance issues. Audit findings may be developed by the individual auditors, but they should be compiled and reviewed by the audit team prior to the closing conference.

9.1.9 *Audit Progress Meeting*—Prior to the closing conference, the user or audited entity, or both, may wish to have informal progress meetings with the audit team. In these meetings, preliminary audit findings are discussed as are logistical issues. These meetings may facilitate communications among all parties and eliminate surprises at the closing conference.

9.1.10 *Closing Conference*—At the conclusion of the site visit, a closing conference between the audit team and audited entity or user, or both, should be held to discuss audit findings. The meeting may include the facility manager and other facility personnel with environmental regulatory compliance responsibilities. The meeting should provide, at a minimum, a verbal description of the audit findings identified by the audit team. The team leader may present all of the audit findings, using the audit team members as backup. Alternatively, each audit team member may be given the responsibility to present his/her individual audit findings. This conference provides an opportunity for audited entity personnel to better understand and discuss any questions about the audit findings. If consistent with the audit scope, a written draft report of the audit findings may also be provided at the closing conference. Post-visit procedures should be discussed including a conflict resolution strategy for findings that are challenged for factual reasons and for closing any open lines of inquiry.

10. Post-Visit

10.1 During the post-visit, the following activities are completed:

10.1.1 *Documentation*—Audit team members should be certain of audit data relevance, validity, and completeness. The audit team leader should ensure that all audit findings are supported by appropriate audit data. Audit criteria in forms such as procedures, questionnaires, and checklists should be completed or explanations provided for all areas or topics not completed in accordance with the originally proposed audit scope. Typically, this should be done before leaving the audited entity during the site visit, but may be delayed for clearing open lines of inquiry.

10.1.2 *Resolution of Outstanding or Unresolved Issues*—Specific assignments to individuals for all unresolved issues should be made to ensure their follow-up and completion. This may result in additional audit findings or documentation supporting the lack of an audit finding. Open lines of inquiry should not be left unresolved.

10.1.3 *Evaluate Audit Findings*—Audit findings should be developed by the audit team based on information obtained during the site visit, the closing conference, and any additional information provided prior to the issuance of the audit report. Audit findings must be factual determinations, not legal conclusions or judgments. Audit findings may be reviewed by outside or in-house counsel and may be reviewed by operations personnel if specified by audit scope. Distribution of audit findings must be in accordance with the procedures set forth in the audit protocol and scope. If a written audit report has been requested, a draft audit report should be provided to the audited entity for review and comment to enhance the factual accuracy of the audit report.

10.1.4 *Report Audit Findings*—A final audit report should be prepared that accurately, clearly, and concisely represents the facts developed during the audit. Audit report formats may vary, but audit findings should clearly identify issues in a manner that provides for a clear understanding of the audit finding and the basis for the audit finding. Audit findings may be prioritized by the audit team based upon criteria agreed to in the audit scope.

10.1.5 *Record Keeping*—The requirements for distribution and maintenance of final audit reports should be established as part of pre-visit activities. Record keeping requirements should also be documented in the audit protocol and audit scope with appropriate security measures taken for audit data. Record retention policies must also be specified in the audit protocol and audit scope for all audit reports, working papers, and other supporting documents.

10.1.6 *Quality Assurance and Quality Control (QA/QC)*—An appropriate quality assurance and quality control (QA/QC) program should be developed for audits and for audit programs.

11. Evaluation and Report Preparation

11.1 *Recordkeeping*—This section discusses the retention and handling of written materials.

11.1.1 *Written Materials*—The audit protocol and scope, that is established and agreed upon as part of the pre-visit activities, define the specific procedures to be employed relating to handling and retention of all written materials. These procedures include, but are not limited to, correspondence, copies of "reference" documents, notes, and other material generated during the entire course of the audit (pre-, during, and post-visit activities). Some of these materials may be audit data.

11.1.2 *Retention Policy*—A record retention policy for working papers can vary depending on how the audit findings are to be compiled and the user's desire to preserve confidentiality. Some of the major document control issues are access authorization and policies for copying, filing, and storage for future reference, including where and how long each type of document is to be retained. The appendix provides guidance to both the audited entity and the auditor(s) as they seek to establish this portion of the audit protocol.

11.2 *Working Papers*—The following subsections identify specific issues related to recordkeeping and preparation for the audit report. The audit protocol and audit scope should clearly specify how all optional or variable procedures will be conducted.

11.2.1 *Working Papers Generation*—Multiple pages of working papers will generally be produced during the on-site activities. Working papers consist of documented notes created by the auditor(s) reflecting his/her evaluations, results of discussions, plans, and schedules. In addition, they may include copies of site maps, facility layout diagrams, process/equipment descriptions or drawings, or both, applicable permits and regulations, and other documents useful to support the audit findings and document the thoroughness of the audit. How the notes are recorded will vary depending on the personal preference of the auditor. Any type of journal (hardbound), notebook (spiral, loose-leaf), tablet, or pad may be used, however, the following information should be recorded on each page:

11.2.1.1 *Facility Identification*—The name of the audited entity.

11.2.1.2 *Date*—The specific day (and possibly time, if needed) that the working paper was created.

11.2.1.3 *Auditor Identification*—The name of the auditor who created the working paper.

11.2.1.4 *Page Number*—Any form of sequential numbering (and lettering) to aid in filing and retrieval of needed documentation.

11.2.1.5 *Margins*—It is recommended that 1.5 in. (3.81 cm) margins be used along the top, bottom, and both sides in order to provide space for additional notes or clarifying remarks, or both.

11.2.2 *Working Papers Review*—Working papers should be reviewed and organized daily. Pages subsequently corrected or revised should not be discarded. It is preferable to "line out" words or sections that are changed rather than erase or delete the item(s). The auditor should initial any such corrections. If a page is completely rewritten, care must be taken to preserve the page numbering system and avoid loss of information not intended for deletion.

11.2.3 *Audit Questionnaires*—The use of audit questionnaires promotes audit thoroughness, detail, and also ensures consistency. Completed questionnaires will become part of the auditor's working papers. They should be numbered and dated in a format consistent with the auditor's working papers. It is also acceptable to attach the completed questionnaire(s) as appendices, as long as the auditor's note or reference attachment(s) on a working paper that is dated and sequentially numbered.

11.2.4 *Disposition:*

11.2.4.1 The working papers generated on site by the audit team members will generally remain with the individual responsible for preparing the audit report. If one individual, for example, the team leader, is assigned the responsibility for preparing the entire report, then that person should retain all of the working papers. Audit team members, however, should compile their individual findings for the areas, or elements, of the audit assigned to them. These can be compiled formally, that is in final form and language, or an auditor may provide a list of the items to the team leader who will provide the appropriate language.

11.2.4.2 In many cases, the final audit report will not be completed prior to leaving the site. Depending upon the record retention and confidentiality policies developed in the audit protocol and audit scope, it may be inappropriate for individual audit team members to make copies of their working papers prior to giving them to the person responsible for writing the final audit report. In this situation, any questions that subsequently arise may be resolved verbally or through follow-up meetings.

11.2.4.3 The user may request that all of the working papers, and any copies that may have been made during the course of the audit, be submitted along with the final audit report. The audited entity, if different, may have special considerations and concerns about the disposition of documents and these issues should be addressed in the audit protocol and audit scope and as such, agreed upon by all parties prior to beginning the audit.

11.3 *Reports*—This section discusses the various types of reports that may be generated during the course of an audit.

11.3.1 *On-Site Activity Status Reports*—The audit protocol should define the specific procedures to be employed regarding audit status reports. Thus, they will be established and agreed upon prior to beginning the audit. They will be a function of how the audit program is structured and the level of confidentiality that is desired by the audited entity or user, or both.

11.3.2 *Preliminary Findings*—Communication of audit findings may occur on a daily basis during visit activities. It will generally be done verbally by the team leader to a designated member of audited entity management or user. If written progress reports are requested, these should be completed by the team leader. Each subsequent update should contain or simply build on the previous report. Upon submission of a written update, the previous written report should be returned to the team leader for retention or disposition.

11.3.3 *Initial Report of All Findings:*

11.3.3.1 An initial audit report of all findings made during the visit activities will generally be provided to the audited entity or user at the conclusion of the visit. This can be done verbally or in writing. Care should be taken to limit the number of copies of any written materials made and distributed at the concluding meeting. It is recommended that all copies made for use by the attendees during the meeting be collected for retention and disposition by the team leader.

11.3.3.2 If the user seeks to maximize the confidentiality of the audit findings and report, outside or in-house corporate counsel must determine whether it is advisable for attendees to make or retain notes, or both, during the concluding meeting when the initial report of all audit findings is presented. Legal counsel must also determine if a copy of the audit report developed for the concluding meeting may be left on-site with the audited entity.

11.3.4 *Final Audit Report of Findings:*

11.3.4.1 *Verbal Report of Findings*—The user commissioning the audit may request that no written audit report be issued. If such is the case, all working papers will be given to the user or the audited entity, or both, in accordance with the audit protocol and audit scope and the audit may be deemed completed at the conclusion of the closing meeting.

11.3.5 *Written Audit Report*—The following subsections address issues related to development and issuance of the written audit report:

11.3.5.1 *Supplemental Information*—If a written audit report is required, it is useful to allow a short period of time (14 days is suggested) for the audited entity to provide additional information to the audit team to clarify issues that may still be unresolved (open lines of inquiry) after the concluding meeting. This will generally be in the form of providing documents that were not able to be produced during the site visit. If there are questions arising from legal interpretations of regulations or verification of applicability of a regulation, outside or in-house counsel may be included to address these concerns. In no case will corrective action during this period of time be grounds for exclusion of any audit finding from the final audit report.

11.3.5.2 *Draft Final Audit Report*—If all information from the audit team members is not received by the close of the site visit activities, the individual responsible for compiling the written audit report should also receive final input within 14 days from the closing meeting. Upon receipt of all information, a draft final audit report should be compiled and submitted to the audited entity for its review and final comments. The time required for completion of this draft report will vary depending on the complexity and number of findings. However, the time should not exceed 30 days from the time supplemental input from the audited entity and the audit team members is received.

11.3.6 *Final Written Audit Report:*

11.3.6.1 The audited entity should provide comments on the draft final audit report in a timely fashion, generally within 10 to 15 days. Failure to provide comments will not prevent the issuance of the final audit report. Depending on the number of comments received, the report writer may require an additional 10 to 15 days to finalize the audit report.

11.3.6.2 Once completed, the final audit report, including all of the individual auditors' notes, audit data, and other attachments, will be presented to the user or the audited entity, or both, in accordance with the audit protocol and audit scope. Pursuant to the audit protocol and audit scope, this can be done directly to the management staff, or to legal counsel, depending upon the desire to maintain confidentiality through preservation of the attorney-client privilege and work-product doctrine. Failure to report directly to legal counsel could destroy the attorney-client privileges and work product doctrine. (See Appendix X1.) In accordance with the audit protocol and audit scope, the audited entity or user, or both, who commissioned the audit will then be responsible for retention or disposal of any and all written documents that were produced over the course of the audit and the completion of the final audit report. If the audited entity is not the user commissioning the audit, guidance from outside or in-house corporate legal counsel should be sought during the establishment of the audit protocol and audit scope to address the advisability of written audit material being retained by others.

11.4 *Report Evaluation and Follow-Up:*

11.4.1 *Corrective Action*—This provisional practice is intended to only deal with issues surrounding the actual performance of an audit. As previously noted, the user or the audited entity, or both, may request that the audit team develop recommendations for corrective actions to remedy audit findings identified in the audit report. The level of detail and scope of this work, if any, would be established as a part of the audit scope. Outside or in-house legal counsel may be involved in the development of all correction action plans. While the effort to establish the recommendations may be conducted simultaneously with the audit, it should be managed separately. That is, all corrective action documents and especially those containing recommendations, should be kept separate and distinct from the materials associated with the audit and the final audit report.

11.4.2 *Responsibilities*—The audited entity or user, or both, must assume responsibility for evaluation for the final audit report and for determination of appropriate follow-up activities.

APPENDIX

(Nonmandatory Information)

X1. LEGAL OVERVIEW

X1.1 Introduction

X1.1.1 This appendix provides an overview of certain legal issues that are associated with environmental regulatory compliance auditing. In particular, this appendix discusses the inherent benefits and risks of environmental compliance auditing, the need for confidentiality, bases for protecting audit information, the limited bases for immunity for users that report violations discovered during the performance of audits and auditor liability.

X1.1.2 The purpose of this appendix is to inform auditors and users of legal issues that frequently arise in the context of performing environmental regulatory compliance audits. It is not intended to render legal advice that can be applied to specific factual circumstances. Legal counsel should be consulted to provide case-specific guidance.

X1.1.3 This appendix is considered to be a part of, and not separate from, this provisional practice.

X1.2 Benefits of Environmental Compliance Auditing

X1.2.1 Organizations that conduct environmental compliance audits can realize a number of benefits:

X1.2.1.1 A better understanding of their compliance status, potential liabilities, and appropriate corrective actions for noncompliance;

X1.2.1.2 A quality assurance check that verifies the adequacy of environmental compliance programs;

X1.2.1.3 Long-term avoidance of fines, penalties, jail terms, and other liabilities (including personal liabilities of an organization's officers or employees) that can result from violations of environmental statutes;

X1.2.1.4 Development of better relations with government agencies through affirmative programs designed to find and correct problems;

X1.2.1.5 Independent assessments of compliance programs for the benefit of organization and plant management; and

X1.2.1.6 Education of organization and plant management and staff about environmental compliance issues.

X1.2.2 *Department of Justice Guidelines:*

X1.2.2.1 In July, 1991, the Department of Justice (DOJ) issued a statement of policy identifying factors DOJ considers in exercising its prosecutorial discretion in pursuing environmental criminal enforcement actions (the "Policy"). In an effort to "encourage critical self-auditing [and] self-policing," DOJ will consider whether a target company has a "regularized, intensive, and comprehensive environmental compliance program" that incorporates sufficient measures to identify and prevent future noncompliance; a compliance program in name only will not be considered by DOJ as a mitigating factor.

X1.2.2.2 The Policy recognizes an environmental compliance audit as an integral part of a credible compliance program and a method of regularly evaluating compliance and detecting and preventing noncompliance. As with the compliance program as a whole, however, a superficial environmental compliance audit will not be viewed favorably by DOJ. The policy provides that, in evaluating the audit portion of a compliance program, DOJ will consider the scope of the audit (that is, did it evaluate all pollution sources including cross-media transfers); whether there are safeguards to ensure the integrity of the audit; whether audit recommendations are timely implemented; and whether sufficient resources are employed to conduct the audit and implement the recommendations.

X1.2.3 *EPA Auditing Policy*—In its 1986 Environmental Auditing Policy Statement, EPA recognized the significance of auditing, encouraged regulated entities to implement environmental auditing programs, and set forth appropriate basic elements of an auditing program. Specifically, EPA stated:

> [Environmental] [a]uditing can result in improved facility environmental performance, help communicate effective solutions to common environmental problems, focus facility managers' attention on current and upcoming regulatory requirements, and generate protocols and checklists which help facilities better manage themselves. Auditing also can result in better-integrated management of environmental hazards, since auditors frequently identify environmental liabilities which go beyond regulatory compliance.[5]

In view of those, and other benefits, EPA stated:

> EPA encourages regulated entities ... to institute environmental auditing programs to help insure the adequacy of internal systems to achieve, maintain and monitor compliance. Implementation of environmental auditing programs can result in better identification, resolution, and avoidance of environmental problems, as well as improvements to management practices.[5]

X1.2.3.1 Also in 1986, EPA issued a policy statement promoting the use of audits in settlements. "EPA Policy on the Inclusion of Environmental Auditing Provisions in Enforcement Settlements" (Nov. 14, 1986). This policy states that environmental auditing is an appropriate part of an enforcement settlement where "heightened management attention could lower the potential for noncompliance to recur." In such cases an environmental audit can, in part, provide EPA with "assurances" that noncompliance will be corrected.

X1.2.3.2 It should be noted however, that EPA's documents also reflect its awareness of the need for development in this field, and the difficulty of establishing firm requirements for all industries and facilities.

X1.2.4 *EPA Investigative Discretion*—On Jan. 12, 1994, EPA's Office of Criminal Enforcement issued guidance entitled "*The Exercise of Investigative Discretion*". This

---

[5] 51 Fed. Reg. 25006 (July 9, 1986).

guidance is intended for EPA special agents and sets out specific factors that distinguish cases meriting criminal investigation from those more appropriately pursued under administrative or civil judicial authorities. The EPA guidance memorandum recognizes that "a violation that is voluntarily revealed and fully and promptly remedied as part of a corporation's systematic and comprehensive self-evaluation program generally will not be a candidate for the expenditure of scarce criminal investigative resources".

X1.3 Risks Associated With Environmental Compliance Audit

X1.3.1 Environmental compliance audits may identify situations in which an organization is not in compliance with applicable laws. Many environmental laws establish criminal offenses or allow for more severe civil penalties if an organization "knowingly" violates an environmental law or regulation. In addition, many environmental statutes provide that each day on which a violation (whether civil or criminal) continues either constitutes a separate violation or is subject to an additional penalty.

X1.3.1.1 Uncovering information through a compliance audit may trigger a legal obligation to disclose under securities laws or to report under the environmental laws. Failure to do so can subject both a corporation and its employees individually to substantial criminal and civil penalties.

X1.3.1.2 Information about noncompliance developed in an environmental compliance audit has great potential to serve as the basis for enforcement authorities to establish that a violation is "knowing" or "continuing," or both. As a result, it is vitally important that an organization that conducts an environmental audit be committed to take prompt and effective action to followup on and correct identified instances of noncompliance.

X1.3.1.3 An additional short-term risk associated with performing compliance audits is the increased cost of compliance and potential penalties that result from reporting to agencies. While the existence of an audit program and policy of disclosure may mitigate fines and penalties, they may not eliminate them entirely.

X1.3.1.4 Because of the possibility that companies may be forced to disclose audit results under certain circumstances, audits can be useful road maps for agencies in enforcement action and for private parties in citizen suits and cost recovery litigation.

X1.4 Need for Confidentiality

X1.4.1 Two main considerations often drive organizations to seek to treat business information as confidential:

X1.4.1.1 The desire to protect proprietary or other commercial information which could harm the competitive position of the organization if disclosed to others; and

X1.4.1.2 The desire to protect information that results from a privileged relationship (for example, attorney-client communications) or that is prepared for litigation.

X1.4.1.3 The concepts of legal privilege that may be relevant to environmental audits are discussed in X1.5.

X1.4.2 *Confidentiality of Business Information:*

X1.4.2.1 Auditors conducting environmental compliance audits may need to have access to proprietary or other business information that an organization desires to protect from disclosure to its competitors. Various statutes and common law principles recognize that organizations have a legitimate interest in protecting such information.

X1.4.2.2 The interests of the organization in confidentiality and of the auditor in access are potentially in conflict. This potential conflict is often resolved through a confidentiality agreement in which the organization agrees to provide the auditor with access to confidential business information, and the auditor agrees to maintain the confidentiality of such information.

X1.4.3 *Enhance Free Flow of Information:*

X1.4.3.1 The law has long recognized that the free flow of information is enhanced where there are guarantees that information communicated to another party will be held confidential, and protected from use by others in litigation. As a result, there are a number of recognized legal privileges which constitute exceptions to the general rule that relevant evidence should be admissible in court. These exceptions, such as the attorney-client and doctor-patient privileges, acknowledge that the free flow of information, however sensitive or potentially damaging, is essential to effective representation, assistance, counseling, treatment, and corrective action.

X1.4.3.2 Environmental compliance auditing also requires a free exchange of information between the organization being audited and the auditor. Without a free exchange of information, the audit may not be complete, and the audit findings may not be valid or useful. Confidential treatment of communications conducted as part of an audit helps to promote the free exchange of information.

X1.5 Basis for Protecting Environmental Compliance Audit Information

X1.5.1 *Maintaining Confidentiality:*

X1.5.1.1 Maintaining the confidentiality of the environmental compliance audit information and reports is a significant concern for the audited entity. The compliance audit will collect information relating to plant design, manufacturing processes and technology, products, and related matters. Disclosure of information collected during the audit can have adverse consequences with regard to competitive business position, labor-management relations, regulatory proceedings, future self-auditing, public perception and current or future litigation. Adherence to certain procedures, including the use of legal counsel, careful organization of the report and data generated, and restrictions on the internal and external distribution of the completed audit can reduce the risk of undesired disclosure.

X1.5.1.2 Attempts to compel disclosure will generally arise in the context of an agency investigation or litigation. Such litigation can involve citizens suits, insurance claims, and toxic tort actions. The rules governing litigation and agency investigations tend to favor disclosure of information, and there are relatively few legal means of preventing compelled disclosure under such circumstances. The primary protections that might shield an audit from disclosure are the attorney-client privilege and the work product doctrine. A developing legal doctrine providing for a qualified "self-evaluation" protection also may shield the audit from compelled disclosure.[6] In certain instances, information developed by the audit may be shared with outside parties having an alignment of interests under the "joint defense" or "common interest" doctrine.[7] However, care must be taken to avoid inadvertently waiving any protection against compelled disclosure. Whether audit information is protected from disclosure depends upon the facts and circumstances surrounding the conduct of the audit itself and the presentation and handling of any audit report.

X1.5.2 *Attorney-Client Privilege:*

X1.5.2.1 The attorney-client privilege is the oldest and best-developed protection against compelled disclosure. When properly asserted, it prevents disclosure of confidential communications between an attorney and the client that relate to legal advice, unless the privilege is waived.[8] The purpose of the privilege is to "encourage full and frank communication between attorneys and their clients and thereby promote broader public interests in the observance of law and the administration of justice".[9] The privilege exists to protect the "giving of information to the lawyer to enable him to give sound and informed advice"[10] and encourages clients to seek early legal assistance.[10] The privilege may only be asserted on behalf of the client.[11]

X1.5.2.2 *Elements of Attorney-Client Privilege*—The essential elements of the attorney-client privilege have been described as: use of an attorney (or a subordinate) to provide legal advice; communication between one who is or seeks to become a client; communication that is confidential; advice sought is not in furtherance of a crime or tort; and there is no waiver of the privilege.[12] Only communications between privileged persons, and not underlying facts discoverable from some source other than the communication, are protected.[13] Information cannot simply be transmitted through an attorney for the purpose of shielding it.[14] The burden of proving the existence of the privilege lies with the person claiming it.[15]

X1.5.3 *Use of an Attorney:*

X1.5.3.1 The privilege applies only if the attorney is acting as an attorney and not as a businessman.[16] The primary purpose of such communication must be to gain or provide legal assistance.[17]

X1.5.3.2 The privilege may extend to communications made by a client to an agent of the attorney hired to assist the attorney in providing legal advice.[18] The privilege may exist even if the client actually hires the consultant, so long as the advice sought is that of the attorney.[19] Documents prepared by the consultant to aid the attorney in providing advice may be privileged,[20] but at least one court has held to the contrary on the ground that the privilege does not cover information obtained by a witness.[21] The policy behind this extension of the privilege is that such consultants provide specific technical advice which is necessary for the attorney to render informed legal advice.

X1.5.3.3 When the purpose of conducting an audit is to seek technical environmental advice rather than legal advice, the audit will not be privileged.[22] In *In re Grand Jury Matter*,[23] a company asserted the attorney-client privilege to protect records of an environmental consultant hired to help it comply with waste disposal requirements and to develop a company-wide waste management plan. The court concluded that the client's ultimate goal was to receive environmental technical advice rather than legal advice because the company was listed as the expert's client, the expert met with the state environmental agency on several occasions when the client's attorneys were not present, and the consultant's billing breakdowns showed no charges for time spent with attorneys.[24] The court did not find persuasive that the consultant was paid from an escrow fund managed by the client's attorneys.[24] Consequently, the company was compelled to disclose the consultant's reports to a state environmental agency seeking to use the documents in a criminal investigation of the company's waste handling practices.[25]

X1.5.4 *Client Communications:*

X1.5.4.1 The attorney-client privilege does not protect client communications that relate only to business or technical data.[26] However, client communications intended to keep the attorney apprised of business matters may be privileged if they embody at the very least "an implied request for legal advice".[27]

X1.5.4.2 A corporation's claim of attorney-client privilege will be upheld where the communications are made by employees to the corporation's counsel "acting as such"; the communications are made "at the direction of the corporation's superiors"; the communications come from an investigation which is undertaken "in order to secure legal advice from counsel"; and the employees are sufficiently aware that

[6] See generally, T. H. Truitt, et al., *Environmental Audit Handbook*, Chapter 3, 1983.

[7] *John Morrell and Co. v. Local Union 304A*, 913 F. 2nd 544, (8th Cir. 1990), *cert. denied*, 111 S. Ct. 1683 (1991).

[8] 8 Wigmore, *Evidence* § 2292 (McNaughton rev. 1961).

[9] *Upjohn Co. v. United States*, 449 U.S. 383, 389 (1981).

[10] See Footnote 9 at 390.

[11] *Hunt v. Blackburn*, 128 U.S. 464, 470 (1888).

[12] *United States v. United Shoe Machinery Corp.*, 89 F. Supp. 357, 358–59 (D. Mass. 1950); *Town of Norfolk v. U.S. Army Corps of Engineers*, 968 F. 2nd 1438, 1457 (1st Cir. 1992).

[13] *Upjohn Co. v. United States*, 449 U.S. 383, 395 (1981).

[14] *Morgan v. United States, cert. denied*, 390 U.S. 962 (1968).

[15] *von Bulow Auersperg v. von Bulow*, 811 F. 2d 136, 144 (2d Cir. 1987).

[16] *United States v. United Shoe Machinery Corp.*, 89 F Supp. 357, 358–60 (D. Mass. 1950).

[17] *United States v. Chevron U.S.A., Inc.*, No. 88-6681, 1989 WL 121616, *6 (E.D. Pa. Oct. 16, 1989) (annual environmental compliance audits not privileged where in-house corporate counsel was merely present during audits and there was no indication that legal advice was being sought).

[18] *United States v. Kovel*, 296 F. 2d 918 (2d Cir. 1961) (accountant hired to aid attorney). *See generally* 81 Am. Jur. 2d *Witness* § 218 (1976).

[19] See Footnote 18 at 922.

[20] *United States v. Cote*, 456 F. 2d 142 (8th Cir. 1972) (accountant's work memoranda prepared for purposes of enabling tax attorney to provide legal advice).

[21] *United States v. McKay*, 372 F. 2d 174 (5th Cir. 1967) (property appraiser's report prepared for tax attorney).

[22] See, for example, *Bituminous Casulty Corp v. Tonka Corp.*, 140 F.R.D. 381, 387 (D. Minn. 1992) (consultant's drafts and memos circulated among client, consultant and counsel were not privileged where the client failed to establish that they contained communications made to secure legal advice, but solely were assessments of the extent and magnitude of groundwater and soil contamination).

[23] 147 F.R.D. 82 (E.D. Pa. 1992).

[24] See Footnote 23 at 85–86.

[25] See Footnote 23 at 84.

[26] *Simon v. Searle & Co.*, 816 F. 2d 397, 403 (8th Cir.) *cert. denied* 484 U.S. 917 (1987).

[27] *Jack Winter, Inc. v. Koratron Co., Inc.*, 54 F.R.D. 44, 46 (N.D. Cal. 1971).

they are being questioned "in order that the corporation could obtain legal advice."[28] Environmental audit documents prepared by an in-house environmental manager to gather information to assist company attorneys in evaluating compliance with relevant laws and regulations were protected from discovery under the attorney-client privilege.[29]

X1.5.5 *Confidential Treatment*—Failure to maintain the confidentiality of such communications and documents nullifies the privilege. Some courts have expressed concern about widespread or indiscriminate dissemination of the communication within the corporation.[30] Other courts suggest there is no privilege if confidential documents are commingled with routine documents and no effort is made to preserve the confidential documents in segregated files.[31]

X1.5.6 *Waiver and Joint Defense Privilege*—The privilege may not be asserted if it has been waived, either intentionally or involuntarily. Waiver may occur if the holder of the privilege produces documents,[32] testifies as to documents or communications for which the privilege is claimed,[33] or fails to object to the production of the material by another.[33] There is no waiver under the joint defense privilege if communications are to outside attorneys (and their clients) whose legal interests are aligned with the one claiming the privilege.[34]

X1.5.7 *Crime or Fraud Exception*—Communications made to further a crime or fraud will defeat an otherwise proper claim of privilege. A request for legal advice regarding a crime or fraud wholly past does not defeat the privilege.[35] Discovery of communications between a corporate defendant and its attorney was allowed where the evidence was such that a prudent person would suspect that they concerned a potentially fraudulent transfer of assets to new corporations after contamination was discovered at a company's operation.[36]

X1.5.8 *Work Product Doctrine:*

X1.5.8.1 The work product doctrine provides a qualified protection to materials prepared by or for counsel "in anticipation of litigation or for trial", unless the party seeking the material can show a "substantial need" for the material and that equivalent materials cannot be obtained without "undue hardship."[37] The protection applies to documents and other tangible things but does not bar discovery of the existence or location of protected material, or the facts contained in such material. The purpose of the doctrine is to allow careful and thorough preparation by an attorney without fear that a litigation opponent will discover strategy and other litigation-related thought processes.[38]

X1.5.8.2 *Scope*—Although the work product doctrine is broader in scope than the attorney-client privilege, it is conditional rather than absolute. Work product includes more than just communications between privileged persons. Memoranda, photographs, diagrams, drawings, and computer-generated data, including those created by persons working for the attorney, may be included.[39] As with the attorney-client privilege, underlying facts are not protected.[40] Factual information that is inextricably intertwined with attorney opinions may be protected.[41] Disclosure to third parties with a common interest does not waive the protection.[42] However, upon a proper showing of need, disclosure can be compelled by an adverse party, so long as the "mental impressions, conclusions, opinions, or legal theories" of the party's attorney or other representative are protected.[43]

X1.5.9 *Anticipation of Litigation:*

X1.5.9.1 The critical factor in asserting work product protection is that the material be prepared "in anticipation of litigation." Documents prepared in the regular course of business and not for litigation purposes are not protected.[44] Documents prepared in response to public requirements unrelated to litigation, such as routine exposure records necessary for OSHA to carry out its enforcement or other regulatory functions, are not prepared "in anticipation of litigation."[45]

X1.5.9.2 There is no requirement that litigation actually be underway, but the remote possibility of an adversarial proceeding is not sufficient.[46] Whether a document is prepared "in anticipation of litigation" depends upon the state of mind of the person preparing the document and whether that person's belief is objectively reasonable.[47] Courts are most comfortable applying the doctrine when litigation is "imminent," a "substantial probability," a "real

[28] *Upjohn*, 499 U.S. at 394.
[29] *Olen Properties Corp. v. Sheldahl, Inc.*, 38 Env't L. Cas. (BNA) 1887 (C.D. Cal. 1994).
[30] For example, *Natta v. Hogan*, 392 F. 2d 686, 692 (10th Cir. 1968); *Virginia Elec. & Power Co. v. Sun Shipbuilding & Dry Dock Co.*, 68 F.R.D. 397, 401 (E.D. Va. 1975); *Gorzegno v. Maguire*, 62 F.R.D. 617, 620–21 (S.D.N.Y. 1973).
[31] *United States v. Kelsey-Hayes Wheel Co.*, 15 F.R.D. 461, 465 (E.D. Mich. 1954); *Hardy v. New York News, Inc.*, 114 F.R.D. 633, 644 (S.D. N.Y. 1987).
[32] *Bituminous Casulty Corp. v. Tonka Corp.*, 140 F.R.D. 381, 387 (D. Minn. 1992) (confidential status of counsel's notes lost when circulated to a state agency or the client's environmental consulting firm).
[33] *Continental Illinois National Bank & Trust Co. v. Indemnity Insurance Co.*, No. 87-C-8439, 1989 U.S. Dist. LEXIS 13004, * 8 (N.D. Ill. Oct. 30, 1989), (waiver occured when client met with its financial analyst and attorney to discuss an internal compliance audit of one of its sites; financial analyst's notes of the meeting were not privileged because a government official was present at the meeting).
[34] See *John Morrell & Co. v. Local Union 304A*, 913 F. 2d 544, 555 (8th Cir. 1990), *cert. denied*, 111 S. Ct. 1683 (1991); *Olen Properties Corp. v. Sheldahl, Inc.*, 38 Env't L. Cas. (BNA) 1887 (C.D. Cal. 1994).
[35] In re *Grand Jury Proceedings*, 604 F. 2d 798, 803 (3rd Cir. 1979).
[36] *Kelley v. Thomas Solvent Co.*, Opinion and Order in Nos. K86-164 and K86-167 (W. D. Mich., Dec. 2, 1989). See also *Burlington Indus. v. Exxon Corp.* 65 F.R.D. 26, 41 (D. Md. 1974) (privilege cited upon *prima facie* showing of misuse of communication.

[37] Fed. R. Civ. P. 26(b)(3).
[38] See *Hickman v. Taylor*, 329 U.S. 495, 510–11 (1947).
[39] *Hickman v. Taylor*, 429 U.S. 495, 511 (1947); *Burlington Indus. v. Exxon Corp.*, 65 F.R.D. 26, 42 (D. Md. 1974).
[40] *Eoppolo v. National R.R. Passenger Corp.*, 108 F.R.D. 292, 294 (E.D. Pa. 1985).
[41] In re: *Grand Jury Subpoena dated Dec. 9, 1978*, 599 F. 2d 504, 512 (2d Cir. 1979).
[42] *Burlington Indus. v. Exxon Corp.*, 65 F.R.D. 26, 43–45 (D. Md. 1974).
[43] Fed. R. Civ. P. 26(b)(3).
[44] See *United States v. Gulf Oil Corp.*, 760 F. 2d 292 (1985) (no protection for auditors' documents underlying reports prepared pursuant to securities laws); in re *Grand Jury Matter*, 147 F.R.D. 82, 86 (E.D. Pa. 1992) (no protection where consultant prepared waste management plan for submission to federal and state authorities, not for the purpose of assisting law firm in providing legal advice).
[45] See *Martin v. Bally's Park Place Hotel & Casino*, 983 F. 2d 1252, 1260 (3rd Cir. 1993).
[46] See, for example, In re Grand Jury Proceedings, 658 F. 2d 782, 785 (10th Cir. 1981) (self-evaluation prompted by "potential" Department of Energy investigation did not trigger work product protection).
[47] See Footnote 45 and *United States v. El Paso Co.*, 682 F. 2d 530, 542–43 (5th Cir. 1982).

prospect," or an "immediate showing."[48] Such a determination is necessarily fact-specific.[49] The protection applies if the prospect of litigation is identifiable because of specific claims that have already arisen, even if litigation itself has not ensued.[50]

X1.5.9.3 Litigation was imminent, and therefore anticipated, where an environmental agency requested a property owner to assess site contamination, but was too remote before such a request even though the owner had previously notified the agency of the on-site contamination.[51] Documents prepared after an EPA request for information received work product protection, while those prepared prior to the information request did not.[52] An accident report may be prepared in anticipation of litigation after occurrence of an accident.[53]

X1.5.10 *Protection in Subsequent Litigation*—The work product protection does not necessarily end when the litigation for which the documents were prepared ends.[41] For example, documents prepared during the course of administrative proceedings before the EPA qualified for work product protection in a subsequent grand jury proceeding.[34]

X1.5.11 *Work Conducted by Non-Attorneys*—Documents and investigative reports compiled by a non-attorney for an attorney or under the attorney's general direction in anticipation of litigation, or both, are subject to work product protection.[54] Written statements of witness interviews that are adopted by the witness are not work product, though statements not so adopted will normally be protected.[55] The essential element for inclusion of materials prepared by a consultant is that the legal talent and training of the attorney have been exercised in initiating and directing the investigation. *Id.* For example, notes taken by a consultant during an environmental audit to prepare a report for counsel to assist in counsel's defense of claims were found to be prepared in anticipation of litigation and therefore work product.[29]

X1.5.12 *Common Interest and Waiver:*

X1.5.12.1 Because the purpose of the work product rule is "not to protect evidence from disclosure to the outside world but rather to protect it only from the knowledge of opposing counsel and his client, thereby preventing its use against the lawyer gathering the materials,[56] disclosure of protected documents to a non-adversarial party does not waive the protection.[57] Work product offers protection to documents that are widely disseminated in a way that the attorney-client privilege cannot.

X1.5.12.2 However, waiver occurs where protected documents are disclosed to an adversary.[58] Using an investigator as a witness waives protection of matters covered in such testimony.[59] Generally, waiver of the work product doctrine is limited to the disclosed document. This is in contrast to a waiver of attorney-client privilege, that includes all communications pertaining to the same subject matter.[60]

X1.5.13 *Substantial Need and Undue Hardship*—Disclosure of otherwise protected material may be compelled upon a showing by the party opponent that there is a substantial need for the documents sought and that deriving equivalent information would result in an undue hardship upon the opponent.[61]

X1.5.14 *Self-Evaluation Doctrine:*

X1.5.14.1 The emerging self-evaluation doctrine proposes a qualified protection for frank self-analysis. The purpose of this doctrine is to encourage a candid self-analysis ensuring compliance with statutes and policies. Without such protection, such analysis might otherwise be chilled by the threat of disclosure to adversaries. Such protection is supported by the Supreme Court's admonition that protections against discovery should be developed on a case-by-case basis.[62] In general, protection from compelled disclosure should be granted when: the communications are made with the understanding they will not be disclosed; confidentiality is essential to a full and satisfactory relationship between the parties; the relationship is one which public policy ought to foster; and the injury to the relationship from compelled disclosure outweighs the benefit of the disclosure.[63]

X1.5.14.2 The self-evaluation privilege was recognized in a medical malpractice suit between private parties to protect minutes and reports of hospital staff meetings concerning patient care.[64] The protection has been extended to employment discrimination,[65] product liability,[66] and failure to comply with stock-exchange reserve requirements.[67]

---

[48] See, for example, 8 C. Wright & A. Miller, *Federal Practice and Procedure* § 2024.

[49] *Simon v. G. D. Searle & Co.*, 816 F. 2d 397, 401 (8th Cir., *cert denied*, 484 U.S. 917 (1987)).

[50] *Stix Products, Inc. v. United Merchants and Manufacturers*, Inc., 47 F.R.D. 334, 337 (S.D.N.Y. 1969).

[51] *Bituminous Casulty Corp. v. Tonka Corp.*, 140 F.R.D. 381, 389–90 (D. Minn. 1992).

[52] *Vermont Gas Sys., Inc. v. United States Fidelity & Guaranty Co.*, No. 90–121, 1993 U.S. Dist. LEXIS 13429,* 18 (D. Vt. Sept. 14, 1993).

[53] *Waste Management Inc. v. Florida Power and Light Co.*, 571 So. 2d 507 (Fla. Dist. Ct. App. 1990).

[54] See Footnote 41 at 42 and Footnote 51 at 388.

[55] *Scourtes v. Fred W. Albrecht Grocery Co.*, 15 F.R.D. 55, 58 (N.D. Ohio 1983).

[56] Wright & Miller at § 2024.

[57] See, for example, *Burlington Indus. v. Exxon Corp.*, 65 F.R.D. 26, 43–45 (D. Md. 1974) (no waiver as to documents provided to joint licensor corporation not to a party to a patent infringement lawsuit); *West v. Marion Laboratories, Inc.* No. 90-0661-CW-W-2, 1991 U.S. Dist. LEXIS 18457 (W.D. Mo., Dec. 12, 1991), (no waiver of documents provided to agency which under regulations was prohibited from releasing documents to plaintiff); *Stix Products, Inc. v. United Merchants & Manufacturers, Inc.*, 47 F.R.D. 334, 338 (S.D.N.Y. 1969).

[58] In re: *Martin Marietta Corp. v. Pollard*, 856 F. 2d 619, 625 (4th Cir. 1988), *cert. denied*, 490 U.S. 1011 (1989).

[59] *United States v. Nobels*, 422 U.S. 225, 239–40 (1975).

[60] See Footnote 41 at 46.

[61] Fed. R. Civ. P. 26(b)(3).

[62] *Trammel v. United States*, 445 U.S. 40, 47 (1980).

[63] Wigmore, *Evidence* § 2285.

[64] *Bredice v. Doctors Hospital, Inc.*, 50 F.R.D. 249 (D.D.C. 1970), *aff'd mem.*, 479 F. 2d 920 (D.C. Cir. 1973).

[65] *Webb v. Westinghouse Electric Corp.*, 81 F.R.D. 431 (E.D. Pa. 1978).

[66] *Lloyd v. Cessna Aircraft Co.*, 74 F.R.D. 518 (E.D. Tenn. 1977).

[67] *New York Stock Exchange, Inc. v. Sloan*, 22 Fed. R. Serv. 2d (Callaghan) 500 (S.D.N.Y. 1976). See generally *Hardy v. New York News Inc.*, 114 F.R.D. 633, 640 (S.D.N.Y. 1987) (collecting cases).

X1.5.14.3 However, a number of courts have declined to recognize the doctrine,[68] or have severely limited its application in employment discrimination cases in view of the strong policy favoring private enforcement of antidiscrimination laws.[69] While the courts have not universally accepted the protection or established parameters, those that do recognize the protection tend to limit its application to "purely evaluative" material and have ordered disclosure of nonevaluative facts, statistics, or other data.[70] In the employment discrimination field, the protection seems to be limited to information or reports that are mandated by statute or regulation.[71]

X1.5.14.4 Courts have typically refused to apply the doctrine where documents are sought by government agencies because of the "strong public interest in having administrative investigations proceed expeditiously and without impediment".[72]

X1.5.14.5 In an action brought by the United States pursuant to the Clean Water Act, there was no self-evaluation protection because it would "effectively impede the Administrator's ability to enforce the Clean Water Act and would be contrary to public policy."[73] In a suit involving private parties in a CERCLA action,[74] the court upheld the self-evaluation privilege. The court found *in camera* that six of thirteen documents asserted to be protected from discovery exclusively by the self-evaluative privilege were within the scope of the privilege. In so doing, the court found a qualified privilege for retrospective analyses of past conduct, practices, and occurrences, and the resulting environmental consequences. The privilege applied only to reports that were prepared after the fact for the purpose of: candid self-evaluation and analysis of the cause and effect of past pollution, and assessment of the roles of parties in contributing to the pollution at an industrial plant site. The court also ruled that such reports would be privileged only if they were created with the expectation that they would be confidential, and if they had in fact been kept confidential. In another environmental suit involving private parties, again the protection was not rejected, but the protection was of little or no value because only irrelevant information was redacted from the environmental audit at issue.[75]

X1.5.14.6 The policy underlying the self-evaluation doctrine may entitle environmental compliance audit informational reports to some protection from disclosure if confidentiality is maintained and if the public interest is served.

However, factual material may still be discoverable.[76]

X1.5.14.7 Several states (including Oregon, Indiana, Kentucky, and Colorado) have enacted legislation that codify the self-evaluation privilege for environmental compliance audits. Many other states (including Arizona, California, Florida, Idaho, Illinois, New York, North Carolina, Ohio, Pennsylvania, Rhode Island, and Virginia) have proposed legislation for an environmental audit privilege.

X1.5.14.8 The first of the laws for the environmental audit privilege was signed into law in Oregon in July 1993. Oregon's statute provides the general framework for proposed federal legislation and all of the States' statutes and bills. The Oregon statute protects environmental audits from disclosure in civil, criminal, or administrative proceedings unless the owner or operator of the facility: waives the privilege, asserts the privilege for a fraudulent purpose, asserts the privilege for an audit not subject to the privilege, or shows evidence of noncompliance without appropriate efforts to reach compliance.

X1.5.14.9 Most states explicitly require the set of documents comprising the environmental audit report to be labeled *Environmental Audit Report: Privileged Document* and to be prepared as a result of an environmental audit. The privilege does not, however, extend to documents or information otherwise available to the government by regulation or by rule of law.

X1.5.14.10 The burden of proof for the privilege is placed on the party either asserting the privilege or seeking the disclosure. While all of the statutes or bills identify which party has the burden of proof, only a few explicitly identify the standard of review as "the preponderance of the evidence". The Colorado statute also provides a safe harbor provision against the imposition of penalties for voluntary disclosure arising from self-evaluation.

X1.6 Maximizing Compliance Audit Confidentiality

X1.6.1 Depending upon the jurisdiction and the specific facts, environmental audit information may be discoverable regardless of the precautions taken. However, meticulous regard for the requirements of the attorney-client privilege, attorney work product protection, and the self-evaluation doctrine improve the likelihood that compliance audit information and reports (or at least nonfactual portions of such information or reports) will remain confidential.

X1.6.2 Requests for a compliance audit should be directed from upper levels of management to the corporation's environmental attorney and documented in writing. The request should make clear that the audit is a request for legal advice relating to the compliance status of the audited plant. The request should authorize the use of environmental consultants, if appropriate, and request that all information and reports generated be treated confidentially. If litigation is anticipated, the documentation should set out what that action is, why it is anticipated, that the audit is sought to aid counsel in preparing for the anticipated dispute, and how the

---

[68] *Myers v. Uniroyal Chemical Co., Inc.*, No. 91 6716, 1992 U.S. Dist. LEXIS 6472 (E.D. Pa. May 5, 1992) (post-accident investigations not protected).

[69] See *Hardy v. New York News, Inc.*, 114 F.R.D. 633, 640 (S.D.N.Y. 1987) (collecting cases).

[70] See, for example, *Emerson Electric Co., v. Schlesinger*, 609 F. 2d 898 (8th Cir. 1979) (no self-evaluation protection because the information sought was not for internal use); *Banks v. Lockheed-Georgia Co.*, 53 F.R.D. 283, 285 (N.D. Ga 1971).

[71] *Steinle v. Boeing Co.*, No. 90-1377-C, 1992 U.S. Dist. LEXIS 2708,* 20–21 (D. Kan. Feb. 4, 1992) (collecting cases); *Hardy v. New York News Inc.*, 114 F.R.D. 633, 641 (S.D.N.Y.) (collecting cases).

[72] *F.T.C. v. TRW, Inc.*, 628 F. 2d 207; 210 (D.C. Cir. 1980).

[73] *United States v. Dexter Corp.*, 132 F.R.D. 8, 10 (D. Conn. 1990).

[74] *Reichold v. Textron*, No. 92-30390-RV, slip opo at 3 and 14 (N.D. Fla. Sept. 21, 1994).

[75] *Ankwright Mut. Ins. Co. v. National Union Fire Ins. Co. of Pittsburgh*, No. 90 Civ. 7811 (KC), 1993 WL 14448 (S.D.N.Y. Jan. 11, 1993).

[76] *Wylie v. Mills*, 478 A. 2d 1273, 1277 (N.J. Super. Ct. Law Div. 1989) (evaluative portion of accident report received protection under self-evaluation analysis; factual portions of report not protected). See also, *Granger v. Nat'l R. R. Passenger Corp.*, 116 F.R.D. 507, 511 (E.D. Pa. 1987).

audit relates to any claims that may be asserted in that action. Any corporate policy on environmental matters should include a discussion concerning the public benefits of conducting confidential self-evaluative environmental audits.

X1.6.3 The attorney-client privilege is available whether in-house environmental counsel or outside environmental counsel may be engaged. Where in-house counsel is employed to direct the audit, however, it is imperative that in-house counsel maintain the role of "acting as a lawyer" rather than as a businessman. Because in-house counsel frequently is a part of management, the distinction between in-house counsel as management and as the entity's attorney can become blurred. Counsel should tightly control all meetings and documents related to the audit. The attorneys should be present at and actively participate in all meetings.

X1.6.4 The audit can be conducted using in-house personnel or independent consultants. Outside consultants can be particularly useful in assisting the company in establishing a new audit program and in conducting audits in circumstances where it is not cost effective to train in-house personnel to perform the audits. Moreover, the use of outside consultants may increase the likelihood of maintaining confidentiality. On the other hand, in-house personnel generally are more familiar with company practices and goals and may have useful perspectives that outside consultants lack. If outside consultants are employed, they should be retained and directed by environmental counsel rather than nonlegal personnel of the corporation. If in-house environmental professionals are to participate in the audit process, their job description should reflect a responsibility to assist legal counsel with environmental compliance issues. Participation by such personnel in an audit should be specially tasked so as to distinguish the audit work from normal day-to-day activities.

X1.6.5 All communications, including preliminary observations, findings, and the final report should be directed to the attorney providing legal advice. The attorney should have the opportunity to personally review and revise the report in draft before it is finalized. Once finalized, all drafts should be destroyed. The final report should speak for itself and no other document should characterize the contents or recommendations of the final report.

X1.6.6 To the extent practical, the audit should be written in a form that identifies requirements but does not present factual information or judgments in a way that highlights or admits potential violations. Audit findings are best characterized as "areas of concern or improvement" rather than violations or exceedances. The report should document existing or potential environmental litigation concerning the audited plant, if applicable, as well as the fact that it was requested to aid the attorney in providing legal advice. Each page of the report and related materials should be marked "privileged and confidential".

X1.6.7 Distribution of the final report should be minimized and controlled with copies circulated only to corporate and plant management charged with direct responsibility for overseeing company efforts to ensure environmental compliance. The final report should remain the property of the attorney providing the legal advice. Client copies should be kept in segregated confidential files.

X1.7 Immunity from Criminal and Civil Prosecutions Based on Violations Identified and Reported as a Result of an Environmental Audit

X1.7.1 *Benefits of Immunity:*

X1.7.1.1 The benefits of immunity have been well-recognized by commentators. First, it encourages reporting of violations, and allows for corrective action regarding violations with proper agency oversight. Without immunity, violators still may take corrective action, but as a practical matter they are much more likely to do so without involving regulatory agencies, which may result in corrective actions of lesser quality.

X1.7.1.2 Immunity also should increase the number of environmental audits conducted. Without immunity, companies may conduct fewer, or no, audits because of fear of incurring penalties as a result of violations uncovered through the audit. By reducing potential liability for "known but uncorrected noncompliances," immunity will encourage prompt add environmentally beneficial corrective actions.

X1.7.2 *Federal Statutory Basis:*

X1.7.2.1 In several federal statutes, Congress has recognized that in some circumstances the need to protect the environment outweighs the need to prosecute someone who reports a violation of an environmental law that he or she discovers. These statutes provide that if a person promptly and properly reports certain spills of oil and hazardous substances to the authorities, he or she cannot be prosecuted based on the notification or any information derived therefrom. The Clean Water Act,[77] for example, provides that:

> Any person in charge of a vessel or of an onshore facility shall, as soon as he has knowledge of any discharge of oil or a hazardous substance from such vessel or facility in violation of [another provision of this] subsection, immediately notify the appropriate agency of the United States Government of such discharge. The Federal agency shall immediately notify the appropriate State agency of any State which is, or may reasonably be expected to be, affected by the discharge of oil or a hazardous substance ... *Notification received pursuant to this paragraph shall not be used against any such natural person in any criminal case*, except a prosecution for perjury or giving a false statement.[78]

This subsection also provides that person's failure to notify authorities of a discharge may result in the imposition of criminal sanctions.[78]

X1.7.2.2 The Comprehensive Environmental Response, Compensation and Liability Act (CERCLA), provides criminal prosecutorial immunity for persons in charge of vessels and facilities from which there is a release of hazardous substances (in reportable quantities as defined under the Environmental Protection Agency (EPA) regulations) who report such releases to the appropriate federal agency:

> Notification received pursuant to this subsection or information obtained by the exploitation of such notification shall not be used against any such person in any criminal case, except a prosecution for perjury or for giving a false statement.

X1.7.2.3 Courts have relied upon such statutory provisions to immunize companies that report violations from

[77] 33 U.S.C. §§ 1251 *et. seq.*
[78] See Footnote 77 at § 1321(b)(5) (emphasis added).

prosecution.[79] While this immunity extends only to criminal prosecutions,[80] rejecting immunity in civil penalty case, it demonstrates that Congress has used immunity to encourage self-reporting in certain situations.

X1.7.3 *Federal Policy Basis*—The United States Department of Justice and the Environmental Protection Agency have issued policy statements encouraging individuals and corporations to conduct environmental audits of the facilities and to disclose violations. While recognizing the value of audits and corporate self-policing, these statements provide no assurances against prosecution. Entities that conduct audits and promptly report violations may still be subject to criminal actions. Nonetheless these policy statements at least recognize that the environmental benefit of audits may outweigh the need for prosecution in many circumstances.

X1.7.4 *United States Environmental Protection Agency Auditing Policy Statement (EPA Audit Policy):*

X1.7.4.1 In 1986, EPA issued its final policy regarding environmental compliance audits.[81] The EPA Audit Policy encourages regulated entities to develop, implement, and upgrade environmental auditing programs.

X1.7.4.2 In devising enforcement responses to violations, the EPA will take into account, on a "case-by-case" basis, the "honest and genuine" efforts of regulated entities to remain in compliance.[82] As provided in its Audit Policy, EPA will evaluate entities as follows:

> When regulated entities take reasonable precautions to avoid noncompliance, expeditiously correct underlying environmental problems discovered through audits or other means and implement measures to prevent their recurrence, EPA may exercise its discretion to consider such actions as honest and genuine efforts to assure compliance. *Such consideration applies particularly when a regulated entity promptly reports violations* or compliance data which otherwise were not required to be recorded or reported to EPA.[83]

X1.7.4.3 The EPA clarifies in its audit policy, however, that it will not "forgo inspections, reduce enforcement responses, or offer other . . . incentives" in exchange for an entity's implementation of environmental auditing. It also provides no assurances against criminal prosecutions.

X1.7.5 *United States Department of Justice Policy Regarding Factors in Decisions on Criminal Prosecutions for Environmental Violations in the Context of Significant Voluntary Compliance or Disclosure Efforts by the Violator (DOJ Policy):*

X1.7.5.1 On July 1, 1991, the DOJ issued the guidance on the exercise of prosecutorial discretion in environmental enforcement cases where the alleged violator has made

significant compliance efforts. In determining whether and to what extent to exercise criminal enforcement discretion under this policy, prosecutors are directed to *take these efforts into account*, when determining whether and how to prosecute a case. The DOJ Policy states:

> This guidance explains the current general practice of the Department in making criminal prosecutive and other decisions after giving consideration to the criteria described above. . . .

X1.7.5.2 The criteria referenced in this statement require prosecutors to evaluate whether an alleged violator's compliance efforts were made in good faith and timely manner, and whether the measures taken were sufficient to identify and prevent future noncompliance. The DOJ Policy further directs prosecutors to consider: whether the company had a strong institutional policy to comply with all environmental requirements; whether the company had safeguards beyond those required by law; whether the company utilized regular procedures (for example, audits) to remedy instances of noncompliance; and whether the company's audits evaluated all sources of pollution for all media, and the possibility of cross-media transfers of pollutants.

X1.7.6 *State Statutory Basis*—Colorado has codified the self-evaluation privilege for environmental compliance audits, and the law requires a rebuttable presumption against the imposition of penalties for voluntary disclosure arising from self-evaluation. This rebuttable presumption only applies if:

X1.7.6.1 The disclosure is made promptly;

X1.7.6.2 The disclosure arises out of a voluntary self-evaluation;

X1.7.6.3 The entity making the disclosure initiates the appropriate effort to achieve compliance with due diligence and corrects the noncompliance within two years; and

X1.7.6.4 The entity making the disclosure cooperates with the appropriate group in the Department of Health regarding issues of disclosure.

X1.8 Consultant Liability Issues

X1.8.1 Legal liability issues surrounding the conduct of environmental compliance audits are not limited to the companies whose activities are being audited. Environmental consultants also should be aware of and consider the potential liabilities that they may be accepting when undertaking an environmental audit. These potential liabilities spring from a number of sources: the contract between the consultant and the client, environmental statutes and regulations, and the common law. In addition, while professional standards generally do not impose liability to other parties, they can create certain ethical obligations or imposed duties upon the environmental consultant, some of which can conflict with obligations to the client. Given the significant ramifications of and the potentially conflicting interests associated with many of these legal or ethical issues, environmental consultants and clients alike should be cognizant of these issues and to the extent possible undertake steps to manage them in such a way as to minimize and assign the risks.

[79] See, for example, *U.S. v. Mobil Oil Corp.*, 464 F. 2d 1124 (8th Cir. 1972) (criminal prosecution under Clean Water Act may not be based on information obtained from notice of discharge required by § 311(b)(5) of the CWA); *accord, U.S. v. General American Transportation Corporation*, 367 F. Supp. 1284 (D.N.J. 1973); *contra, U.S. v. Skil Corp.*, 351 F. Supp. 295 (N.D. Ill. 1972).
[80] See, for example, *U.S. v. Allied Towing*, 578 F. 2d 478 (4th Cir. 1978).
[81] *Environmental Auditing Policy Statement*, 51 Fed. Reg. 25004 (July 9, 1986).
[82] 51 Fed. Reg. at 25007.
[83] 51 Fed. Reg. at 25007 (emphasis added).

X1.8.2 *Potential Reporting Obligations*—Reporting of audit findings is potentially the most difficult and misunderstood liability issue facing environmental consultants and their clients. Virtually all of the environmental statues and their implementing regulations contain provisions requiring the reporting of the discovery of certain information. Generally, however, those reporting obligations apply only to the owner, operator, or person in charge of the facility or location in question. Rarely do the environmental statutes or regulations impose such a duty upon the environmental consultant. Reporting obligations also may arise from the common law. As with statutory and regulatory obligations, however, most common law obligations apply only to owners and operators or persons in charge or control of a facility or location. There may be circumstances, however, in which the common law does impose an obligation on the environmental consultant, where for example the consultant discovers a situation that presents a risk of harm to third-parties (for example, a risk of explosion). Professional obligations sometimes are viewed by consultants as imposing reporting obligations. While these professional standards do not generally present liability issues, they can create an ethical dilemma for the environmental consultant where reporting would be inconsistent with the interests or wishes of a client. The existence and extent of reporting obligations are complex, fact-specific issues that can create significant problems for both the client and the environmental consultant if the obligations are not fully understood. Thus, prior to contracting to undertake a compliance audit, the client and the consultant should fully discuss and understand potential reporting obligations and agree on clear protocol to be followed where information triggering a potential reporting requirement is discovered.

X1.8.3 *Third-Party Liability*—Liability to third parties, that is, those parties not signatories to the agreement between the environmental consultant and the client, is a common issue associated with environmental audits. Liability to third parties can arise in several ways, most commonly where an audit is being undertaken in anticipation of a sale or refinancing of a business, as a requirement of insurance or some other situation where parties other than the client will be reviewing and relying on the audit report. The contract between the environmental consultant and the client should state clearly the purpose and scope of the undertaking and any limitations thereto.

X1.8.4 *Breach of Contract/Malpractice Liability*—When a consultant enters into a contract to conduct an environmental compliance audit, the consultant not only agrees to perform as the contract specifies, but also to live up to certain professional standards of conduct in undertaking the work. Thus the environmental consultant not only faces the possibility of traditional breach of contract claims if a problem arises but also potential professional malpractice claims if the work does measure up to the standards of the profession. It is important that the environmental consultant recognize that its work will be measured against an objective industry standard and that the standard can change over time and from one locality to another. With regard to the potential for breach of contract claims, both the client and the consultant can reduce the likelihood that claims will arise by providing a clear scope and limitations provisions in their contracts.

Standard Provisional Guide for
Expedited Site Characterization of Hazardous Waste
Contaminated Sites[1]

This standard is issued under the fixed designation PS 85; the number immediately following the designation indicates the year of original adoption.

1. Scope

1.1 This provisional guide describes a process for expedited site characterization (ESC) of hazardous waste contaminated sites[2] to identify all relevant contaminant migration pathways and to determine the distribution, concentration, and fate of contaminants for the purpose of evaluating risk, determining regulatory compliance, and designing remediation systems (when required).[3] Generally, the process is applicable to larger scale projects, such as CERCLA (Superfund) remedial investigations and RCRA facility investigations.[4] It is also applicable to other contaminated sites where contaminant characteristics and heterogeneities in the geologic and hydrologic system create a risk that other site characterization approaches will fail to identify relevant contaminant migration pathways. ESC has been successfully applied at multiple sites in different states and EPA regions (Table X1.1). It typically achieves significant cost and schedule savings compared to conventional site characterization (see X1.2 and X1.3).

1.2 The ESC process operates within the framework of existing regulatory programs. It focuses on collecting only the information required to meet clearly defined objectives and ensuring that characterization ceases as soon as the objectives are met. Central to the ESC process is on-site decisionmaking by a multidisciplinary core technical team using the clearly defined framework of a dynamic work plan which provides the flexibility and responsibility to select the type and location of measurements to optimize data collection activities. Other essential features of the ESC process include: intensive compilation, quality evaluation, and independent interpretation of prior data to develop a preliminary conceptual site model, use of multiple complementary site appropriate geologic and hydrologic investigation methods, rapid site-appropriate methods for data collection and interpretation, rigorous quality control for all aspects of data collection, daily on-site reduction and archiving of field data, and daily integration, analysis, and interpretation of data to support on-site decisionmaking. (See Table 1 and Section 5.)

1.3 The process described in this provisional guide is based on good scientific practice but is not tied to any particular regulatory program, site investigation method or technique, chemical analysis method, statistical analysis method, risk evaluation method, or computer modeling code. The appropriate specific site investigation techniques in an ESC project are highly site specific, whenever feasible noninvasive and minimally invasive methods, are used in Appendix X3. Appropriate chemical analysis methods are equally site specific and may be conducted in the field or laboratory depending on data quality requirements, required turnaround time, and costs.

1.4 Generally, the ESC process is not applicable to individual petroleum release sites, real estate property transactions, and sites where contamination is limited to the near-surface and there is no basis for suspecting that contaminant movement through the vadose zone and ground water is a matter of concern, or where the cost of remedial action is likely to be less than the cost of site characterization. Guide PS 03-95 addresses accelerated site characterization (ASC) for petroleum release sites, and Guide E 1739 addresses use of the risk-based corrective action (RBCA) process used at petroleum release sites. Section X1.6.1 describes the ASC process, and X1.6.2 discusses the relationship between ESC and the RBCA process. Practices E 1527 and E 1528 address real estate property transactions, and X1.6.3 discusses the relationship between the ESC process and investigations for real estate property transactions. Classification D 5746 addresses environmental conditions of property area types, for Department of Defense Installations and Practice D 6008 provides guidance on conducting environmental baseline surveys in order to determine certain elements of the environmental condition of Federal property.

1.5 *This standard does not purport to address all of the safety concerns, if any, associated with its use. It is the responsibility of the user of this standard to establish appropriate safety and health practices and determine the applicability of regulatory limitations prior to use.*

1.6 Provisional standards[5] achieve limited consensus through approval of the sponsoring subcommittee.

2. Referenced Documents

2.1 *ASTM Standards:*

---

[1] This provisional guide is under the jurisdiction of ASTM Committee D-18 on Soil and Rock and is the direct responsibility of Subcommittee D18.01 on Surface and Subsurface Characterization.
Current edition approved Dec. 11, 1996. Published March 1997.

[2] The term hazardous waste in the title is used descriptively. The term also has specific meanings in the context of different regulatory programs. Expedited site characterization is appropriate for radiologically contaminated sites and some larger petroleum release sites, such as refineries. Section 4.2 further identifies types of contaminated sites where expedited site characterization may be appropriate.

[3] An ESC project should address all relevant contaminant migration pathways, including air, surface water, biota and submerged sediments. The text of this provisional guide emphasizes vadose zone and ground water contamination because they represent the contaminant migration pathways that are most difficult to characterize.

[4] CERCLA Preliminary Assessments/Site Investigations (PA/SIs) and RCRA Facility Assessments (RFAs) are generally required to provide information relevant to a decision whether the ESC process should be initiated. (See Appendix X2.)

[5] Provisional standards exist for two years subsequent to the approval date.

TABLE 1 Checklist of Minimum Criteria for a Project Using ASTM Expedited Site Characterization Process[A]

———— Project objectives and data quality requirements defined by some process that includes ESC client, regulatory authority, and, where appropriate, other stakeholders. See 6.3.

———— A technical team leader with a working understanding of all elements and functions of contaminated site characterization heads up the ESC project and leads the ESC core technical team. See 7.1.1.

———— An integrated multidisciplinary core technical team with expertise in geologic, hydrologic, and chemical systems works together, as areas of expertise are needed, throughout the process. See 7.1.

———— Intensive compilation, quality evaluation, and independent analysis and interpretation of prior data are used to develop a preliminary conceptual site model. See 8.1, 8.2, 8.3, and 8.5.

———— Dynamic work plan, approved by ESC client and regulatory authority, provides framework for use of multiple complementary, site appropriate, geologic, and hydrologic investigation methods, and rapid site appropriate methods for data collection and analysis. See 8.6 and 9.2.4 and Appendix X3.

———— Quality control procedures are applied to all aspects of data collection and handling, not just chemistry. See 9.2.6 and Appendix X4.

———— Field data collection initially focuses on geologic and hydrologic characterization of the system of relevant contaminant pathways (and identifying contaminants of concern, if not already known), followed by delineating the distribution, concentration and fate of contaminants, based on knowledge of the relevant contaminant pathways. This typically requires no more than two field mobilizations. See Sections 10 and 11.

———— Daily on-site reduction and archiving of field data. See 10.1.2 and Appendix X5.

———— Daily integration, analysis, and interpretation of data to support on-site decisionmaking to optimize field investigations. See 10.1.3.

[A] Other site characterization approaches may include many of the above elements, but all must be present for an investigation using the ASTM ESC process.

D 653 Terminology Relating to Soil, Rock, and Contained Fluids[6]

D 5717 Guide for the Design of Ground-Water Monitoring Systems in Karst and Fractured-Rock Aquifers[7]

D 5730 Guide to Site Characterization for Environmental Purposes With Emphasis on Soil, Rock, the Vadose Zone, and Ground Water[7]

D 5745 Guide for Developing and Implementing Short-Term Measures or Early Actions for Site Remediation[8]

D 5746 Classification of Environmental Conditions of Property Area Types[8]

D 5792 Practice for Generation of Environmental Data Related to Waste Management Activities: Development of Data Quality Objectives[8]

D 5979 Guide for Conceptualization and Characterization of Ground-Water Systems[7]

D 6008 Practice for Conducting Environmental Baseline Surveys[8]

E 1527 Practice for Environmental Site Assessments: Phase 1 Environmental Site Assessment Process[8]

E 1528 Practice for Environmental Site Assessments: Transaction Screen Process[8]

E 1689 Guide for Developing Conceptual Site Models for Contaminated Sites[9]

E 1739 Guide for Risk-Based Corrective Action Applied at Petroleum Release Sites[8]

PS 03 Provisional Guide for Accelerated Site Characterization for Confirmed or Suspected Petroleum Releases[8]

3. Terminology

3.1 *Definitions of Terms Specific to This Standard*—The following terms are specific to this provisional guide, unless otherwise indicated. Other terms are in accordance with other ASTM standards as specified.

3.1.1 *conceptual site model*—a testable interpretation or working description of the relevant physical, chemical, and biological characteristics of a site. **D 5730**

3.1.1.1 *Discussion*—This provisional guide uses the term "preliminary" conceptual site model to refer to the initial model based on regional geology and other prior data, the

———

[6] *Annual Book of ASTM Standards*, Vol 04.08.
[7] *Annual Book of ASTM Standards*, Vol 04.09.
[8] *Annual Book of ASTM Standards*, Vol 11.04.
[9] *Annual Book of ASTM Standards*, Vol 11.05.

term "evolving" conceptual site model to refer to the developing conceptual site model during an ESC project, and the term "final" conceptual site model when further refinement is no longer required to satisfy the objectives of the ESC project.

3.1.2 *contaminants of concern*—the contaminants which are known to be present at an ESC project site, and for which sampling and analysis occurs in order to delineate their spatial distribution and concentration.

3.1.2.1 *Discussion*—Identification of contaminants of concern from a larger list of suspected contaminants usually takes place as a separate effort prior to initiation of an ESC project but can also be integrated into an ESC project. As appropriate, and if approved by the ESC client and regulatory authority, deletions or additions to the list of contaminants of concern may occur during an ESC project.

3.1.3 *decision authority*—agency, organization, or individual responsible for deciding whether the ESC process should be initiated and for using information developed by the ESC process to decide what further action is required.

3.1.3.1 *Discussion*—The decision authority may be regulatory or nonregulatory, depending on the situation. The ESC client is usually the decision authority for initiating the ESC process. A regulatory authority is usually the decision authority for deciding what further action is required.

3.1.4 *dynamic work plan*—a site characterization work plan including a technical program that identifies the suite of field investigation methods and measurements that may be necessary to characterize a specific site, with the actual methods used and the location of measurements and sampling points based on on-site decisionmaking.

3.1.4.1 *Discussion*—The dynamic work plan, which must be approved by the ESC client and regulatory authority, provides a clearly defined framework (including geographic area, maximum depth, standard operating procedures for specific methods) within which the ESC technical team leader, supported by the appropriate technical core team members, has flexibility and responsibility to select the type and location of measurements to optimize data collection activities. In contrast, a conventional site characterization work plan contains prescribed numbers and locations for field measurements, samples, and monitoring wells. See Section 9 for other important elements that are included in the dynamic work plan.

3.1.5 *Environmental Site Assessment (ESA)*—the process by which a person or entity seeks to determine if a particular parcel of real property (including improvements) is subject to Recognized Environmental Conditions. **E 1527**

3.1.5.1 *Discussion*—This provisional guide refers to ESC Phase I/II investigations to differentiate them for Phase I/II ESAs. The phases are not comparable (see X1.6.3).

3.1.6 *environmental site characterization*—the delineation or representation of the essential features or qualities, including the conditions, influences, and circumstances, existing at a phase or location designated for a specific use, function, or study. **D 5730**

3.1.7 *ESC client*—the individual, agency, or organization responsible for a site or sites where expedited site characterization is being considered or has been initiated. An ESC client contracts with an ESC provider for an ESC project that characterizes a specific site.

3.1.8 *ESC core technical team*—the central multidisciplinary team that is responsible for an ESC project, consisting of a technical team leader and experienced individuals with expertise in geologic, hydrologic and chemical systems, and capable of integrating and interpreting all relevant data generated by the ESC project.

3.1.8.1 *Discussion*—The core technical team members are available, and involved as needed, for every stage of an ESC project. The technical team leader is normally present in the field at all times with other core technical team members present as needed. See 7.1 for further discussion of the responsibilities of the ESC core technical team.

NOTE 1—The core technical team should not be confused with the core team in the DOE SAFER process which consists of the key decisionmakers for the problem unit (see X1.4.5). Normally, the ESC technical team leader would be a member of the SAFER core team.

3.1.9 *ESC Phase I investigation*—phase of ESC project focusing on geologic and hydrologic characterization of the system of relevant contaminant pathways and other potential contaminant pathways, such as air, surface water, submerged sediments, as appropriate.

3.1.9.1 *Discussion*—Identification of contaminants of concern will also be carried out in Phase I, if not already known, and subsurface contaminant distribution sampling occurs to the extent that it is convenient or contributes to understanding of the geologic and hydrologic system and other relevant pathways.

3.1.10 *ESC Phase II investigation*—phase of ESC project focusing on delineating the spatial distribution, concentration, and understanding the fate of contaminants, based on knowledge of the relevant contaminant pathways identified in Phase I, with additional geologic and hydrologic characterization carried out as needed.

NOTE 2—This provisional guide describes the ESC process as involving two phases with two discrete field mobilizations because experience has shown that the amount of time required to accomplish both geologic and hydrologic characterization and delineation of contaminants based on understanding of the geologic and hydrologic system is too long for field personnel to function effectively in a single mobilization. However, when sufficient quality data are available, it may be possible to complete both activities in a single mobilization, whereas at difficult, complex sites, more than two field mobilizations might be required.

3.1.11 *ESC project*—the application of the ESC process

by an ESC provider to a specific site to give an ESC client and responsible regulators the necessary information to evaluate regulatory compliance and choose a course of action (no action, monitoring, or interim/final remedial action).

3.1.12 *ESC project team*—members of the ESC core technical team and all other individuals that provide technical and other support during an ESC investigation.

3.1.13 *ESC provider*—organization that supplies the ESC project team to an ESC client.

3.1.14 *ESC technical team leader*—an individual with training and experience in geologic and hydrologic systems (and some familiarity with chemical systems), negotiating and contracting, and a working understanding of all elements and functions of contaminated site characterization, including regulatory and health and safety requirements, who heads up an ESC project and leads the ESC core technical team. (See also 7.1.1.)

3.1.15 *Expedited site characterization (ESC)*—a process for characterizing contaminated sites led by a core technical team of experienced professional staff with expertise in geologic, hydrologic, and chemical systems as they relate to contaminant migration pathways using careful analysis of prior data, multiple complementary investigation methods, on-site data reduction, integration, and interpretation, and on-site decisionmaking to optimize field investigations for the purpose of giving an ESC client and responsible regulators the necessary information to evaluate regulatory compliance and choose a course of action (no action, monitoring, or interim/final remedial action).

3.1.16 *on-site decisionmaking*—the use of multidisciplinary interpretation and integration of field measurements and sample analyses, within a framework defined by a dynamic work plan, to select the type and location of subsequent field measurements and sampling points during the same field mobilization.

3.1.16.1 *Discussion*—On-site decisionmaking is used by the ESC core technical team for field data collection and should not be confused with decisionmaking by the decision authority, which is based on the integration, interpretation, and recommendations of the ESC core technical team.

3.1.17 *quality assurance*—measures taken to independently check and verify that quality control procedures specified in the QA/QC plan for an ESC project are being carried out.

3.1.18 *quality control*—all measures taken by the core technical team and project support team to ensure the quality of geologic, hydrologic, chemical, and other data collection and measurement.

3.1.19 *regulatory authority*—the federal, state, or local agency, or combination thereof, with primary responsibility for ensuring compliance with the environmental statutes and regulations that prompted initiation of ESC at a site.

3.1.20 *remedial action*—a course of action chosen by a decision authority that involves an engineered solution to address contamination.

3.1.21 *site, n*—a place or location designated for a specific use, function, or study. **D 5730**

3.1.22 *stakeholder*—any individual or organization other

than the ESC client and regulatory authority that may be affected by the consequences of initiating ESC at a site; generally, this will include individuals or communities which may be affected by contamination at the site.

3.1.23 *vadose zone*—the hydrogeological region extending from the soil surface to the top of the principal water table; commonly referred to as the "unsaturated zone" or "zone of aeration." The alternate names are inadequate as they do not take into account locally saturated regions above the principle water table (for example, perched water zones). **D 653**

4. Significance and Use

4.1 This provisional guide describes a process for characterizing contaminated sites within the framework of a dynamic work plan and led by a core technical team of experienced professional staff with expertise in geologic, hydrologic, and chemical systems, using careful analysis of prior data, multiple complementary geologic and hydrologic investigation methods, rapid site-appropriate methods for data collection and interpretation, rigorous quality control, on-site data reduction, integration, and interpretation and on-site decisionmaking to optimize field investigations for the purpose of ensuring regulatory compliance and providing information for correct decisionmaking concerning remedial action. See Appendix X1 for additional background on the ESC process, how it differs from conventional site characterization, and how it relates to other approaches to site characterization.

4.2 The ESC process should be initiated when a decision authority determines that: (*1*) contaminants at a site present a potential threat to human health or the environment, (*2*) decisions concerning remedial or other action must be expedited as rapidly as possible, and (*3*) contaminant characteristics and heterogeneities in the geologic and hydrologic system at the site create a risk that other site characterization approaches will fail to identify relevant contaminant migration pathways. In addition to vadose zone and ground water contamination other relevant contaminant pathways, such as air, surface water, biota, and submerged sediments may also be evaluated within the ESC process. Examples of situations where the process may be applicable include:

4.2.1 CERCLA remedial investigation/feasibility studies (RI/FS). (See Appendix X2.)

4.2.2 RCRA facility investigation/corrective measures studies (RFI/CMS). (See Appendix X2.)

NOTE 3—The ESC process can be carried through to include CERCLA feasibility studies and RCRA corrective measures studies, but this provisional guide focuses on its use for site characterization. Section X1.4.5 describes the relationship of the ESC process to the DOE SAFER and EPA SACM programs for accelerating the cleanup of contaminated sites.

4.2.3 Sites where environmental site assessments (ESAs) conducted using Practices E 1527 Practice E 1528 and a Phase II ESA identify levels of contamination that require further, more intensive characterization of the geologic and hydrologic system of contaminant migration pathways. Section X1.6.3 discusses the relationship between ESAs and ESC.

4.2.4 Large petroleum release sites, such as refineries. The user should review both this provisional guide and Provisional Guide PS 03-95 in order to evaluate whether the ESC

or ASC process is more appropriate for such sites.

4.2.5 Sites or facilities with subsurface contamination by radioactivity not regulated by RCRA or CERCLA.

4.2.6 Other sites or facilities where levels of subsurface contamination are a matter of regulatory or other concern and require thorough understanding of the geologic and hydrologic system of contaminant pathways.

4.3 The ESC process requires clearly defined objectives and data quality requirements that will satisfy the needs of the ESC client, regulatory authority, and stakeholders. Once these have been defined, the ESC process relies on the expert judgment of the core technical team, operating within the framework of an approved dynamic work plan, as the primary means for selecting the type and location of measurements and samples throughout the ESC process. An ESC project focuses on collecting only the information required to meet the predefined objectives and ceases characterization as soon as the objectives are met.

NOTE 4—This provisional guide uses the term "data quality requirements" to refer to the level of data accuracy and precision needed to meet the intended use for the data. The U.S. EPA Data Quality Objectives (DQO) process is one way to accomplish this. The ESC process applies the concepts of quality control and data quality requirement to geologic and hydrologic data as well as chemical data but within a general framework of judgment-based rather than statistical sampling methods. Section X1.4.4 discusses the DQO process in more detail and role of judgment- and statistically-based sampling methods in the ESC process. Practice D 5792 provides guidance on development of DQOs for generation of environmental data related to waste management.

4.4 The ESC process avoids a presumption that remedial action is required (that is, an engineered solution rather than no further action or ongoing monitoring). In any ESC project, remediation engineering expertise is incorporated into the process at the earliest point at which a need for remedial action is identified (see 13.3). Guide D 5745 provides guidance for developing and implementing short-term measure or early actions for site remediation.

4.5 If an ESC project is to be initiated, modification of procedures described in this provisional guide may be appropriate if required to satisfy project objectives, regulatory requirements, or other reasons. However, for an investigation to qualify as an ESC project, as formalized by ASTM, modifications should not eliminate any of the essential features of the ESC process listed in Table 1. The rest of this document provides further guidance on how these essential feature can be implemented and does not replace or relieve professional judgment in the design and implementation of an ESC project or in the selection of alternative site characterization approaches if they are appropriate for a site. ASTM expects that as the ESC process becomes more widely used, modifications, enhancements, and refinements of the process will become evident. ASTM requests that suggestions for revisions to the provisional guide based on field application of the process be addressed to: Committee D18 Staff Manager, ASTM, 100 Barr Harbor Drive, West Conshohocken, PA 19428.

NOTE 5—Users may prefer to use or develop alternative terminology for different aspects of the ESC process, depending on the regulatory context in which it is applied. However, precise or approximate equivalencies to steps or functions in the ESC process should be clearly identified. See, for example, RCRA and CERCLA equivalencies in Appendix X2.

4.6 This provisional guide can be used in conjunction with Guide D 5730 for identification of potentially applicable ASTM standards and major non-ASTM guidance. In karst and fractured rock hydrogeologic settings, this provisional guide can be used in conjunction with Guide D 5717.

5. Summary of ESC Process

5.1 ESC is a process that, when properly implemented, should provide higher quality information for decisionmaking in a shorter period of time and at a lower cost than conventional site characterization. Appendix X1 discusses the features of ESC that make this possible. Most current problems with remedial action at contaminated sites can be attributed to inadequate understanding of the geologic and hydrologic system of contaminant pathways, which results in failure to delineate the full extent of contamination, controls on contaminant migration, and suboptimal design of remedial measures. The multidisciplinary and focused nature of the ESC process results in a final conceptual model of a site where uncertainty concerning the spatial distribution and concentration of contaminants is minimized, providing a sound basis for choosing the appropriate course of action.

5.2 *Organization of an ESC Project*—A decision authority, normally the ESC client, must determine that the ESC process should be initiated based on information that potential or known chemicals of concern exceed acceptable levels and present a known or potential threat to human health or the environment, and that contaminant characteristics and heterogeneities in the geologic and hydrologic system create a risk that other site characterization approaches will fail to identify relevant contaminant migration pathways. Figure 1 illustrates key relationships in an ESC project.

5.2.1 *ESC Client, Regulatory Authority and Stakeholders*—The ESC client, regulatory authority, and, where appropriate, stakeholders, provide the overall framework for an ESC project by defining project objectives and data quality requirements. The technical team leader and other project team members, as appropriate, also participate in this process to ensure the objectives and data quality requirements are reasonable and technically feasible.

5.2.2 *Core Technical Team*—The core technical team, headed by a technical team leader and consisting of typically 2 or 3 additional individuals with expertise in geologic, hydrologic, and chemical systems appropriate to the site,

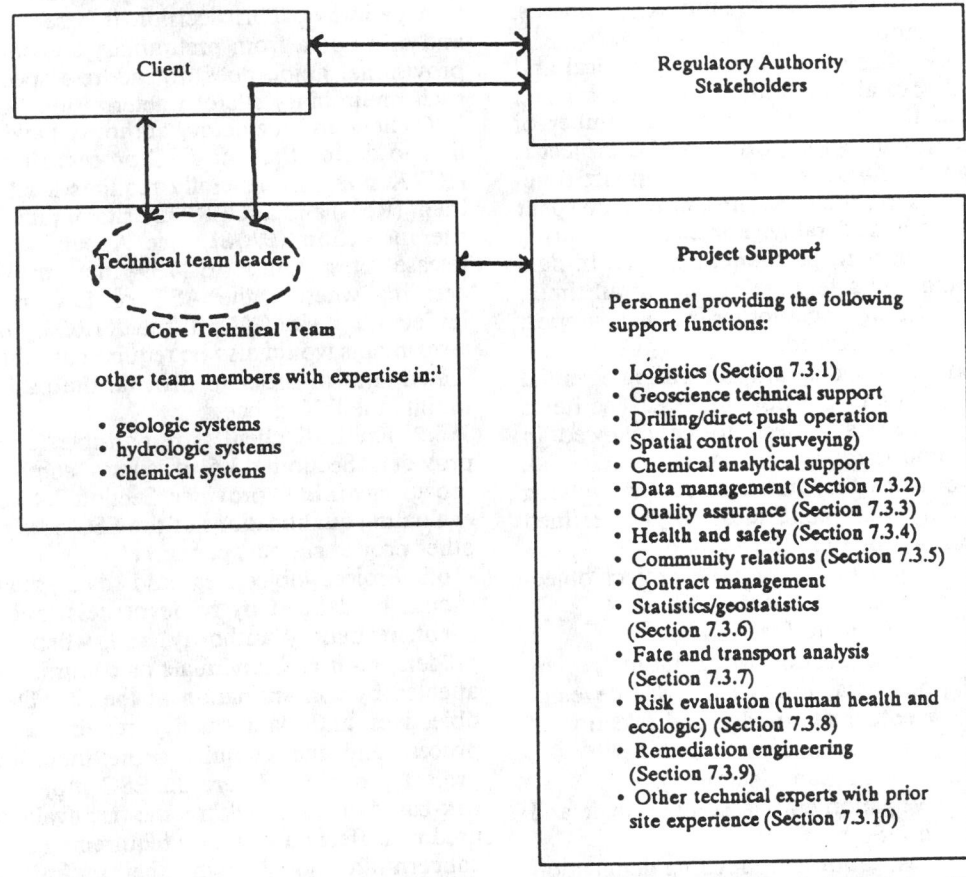

A Other areas of expertise may be included in the core technical team, as appropriate. The main criteria for determining whether a person with an area of expertise should be included on the core technical team is whether the type of expertise is required for the process from beginning to end of an ESC project, and if it is required for on-site decisionmaking to optimize field investigations.
B Involved, as needed throughout the process.

FIG. 1 ESC Project Team Relationships

provide a continuous, integrated, multidisciplinary presence throughout the process (see 7.1). The technical team leader operates in close communication with the ESC client (the individual, agency, or organization responsible for the site that is the focus of the ESC project), appropriate regulatory authorities, and stakeholders (primarily the community of affected or potentially affected people). The core technical team members are involved, as needed, in all steps of the process, and are present in the field when data collection involving their areas of expertise is taking place. The optimization of field investigation activities and the quality of the final conceptual site model depend on the interaction of the different perspectives of the core technical team members.

5.2.3 *Project Support*—The ESC core technical team operates with the support of a larger project team that includes technical personnel and equipment operators involved in data collection and sampling, and personnel providing other support functions such as logistics, data management, QA/QC, health and safety, and community relations (see 7.3). Some areas of project support expertise, such as statistics/geostatistics, fate and transport analysis (including digital modeling), risk analysis, and remediation engineering, may have a special role early in a project in defining the type of data required for the project and data quality requirements, and are involved throughout the project as needed.

5.2.4 Qualified individuals within the core technical and support team can fill several functions to lower costs and increase integration of the team. The required number of individuals to provide project support for an ESC project is site-specific. Although the number of project support functions shown in Fig. 1 is large, the total amount of time spent for each function varies considerably. For example, during field operations project support, personnel involved in data management and health and safety are present at all times, whereas personnel providing most other project support functions are present only as needed.

5.3 Figure 2 presents a flow diagram illustrating essential features and decision points in the ESC process. The items outlined in this figure generally need to be followed in sequence. However, some items are not strictly sequential. For example, item 3b is the first iteration of the evolving conceptual site model that continues to be refined throughout the process. Major steps are as follows:

5.3.1 Initiate the ESC process and define project objectives and data quality requirements (Section 6).

5.3.2 Establish ESC project team (Section 7).

5.3.3 Develop ESC project (Section 8), including review and interpretation of prior data, initial site visit, development of preliminary conceptual site model, and selection of multiple complementary investigation methods.

5.3.4 Develop dynamic work plan (Section 9).

5.3.5 ESC Phase I investigation emphasizing geologic and hydrologic characterization (Section 10).

5.3.6 ESC Phase II investigation emphasizing delineation of the distribution, concentration, and understanding the fate of contaminants (Section 11).

5.3.7 Fate and Transport Analysis and Risk Evaluation (Section 12).

5.4 Section 13 discusses considerations in the implementation of ESC as follows:

5.4.1 Relationship to regulatory process (13.1 and Appendix X2).

5.4.2 Role of risk evaluation in ESC process (13.2).

5.4.3 Relationship of remediation engineering design and implementation to ESC (13.3).

5.4.4 Role of modeling in ESC process (13.4).

5.4.5 Procurement and contracting procedures for ESC (13.5).

5.4.6 Performance indicators for evaluating the success of ESC (13.6).

5.4.7 Factors that may affect performance indicators (13.7).

6. Initiate the ESC Process and Define Objectives and Data Quality Requirements

6.1 The ESC process is initiated when a decision authority, usually an ESC client with input from a regulatory authority, determines that: (*1*) contaminants at a site present a potential threat to human health or the environment, (*2*) decisions concerning remedial or other action must be expedited as rapidly as possible, and (*3*) contaminant characteristics and heterogeneities in the geologic and hydrologic system at the site create a risk that other site characterization approaches will fail to identify relevant contaminant migration pathways. This decision is based on chemical sample and other data from preliminary site characterization. This provisional guide does not address specific procedures for such preliminary site characterization but assumes that the ESC client and regulatory authority have sufficient information to decide that the ESC process should be initiated. At RCRA sites this generally requires a RCRA facility assessment (RFA) and at CERCLA sites a preliminary assessment/site inspection (PA/SI); see Appendix X2. At petroleum release sites, Guide E 1739 may provide the basis for deciding whether the ASC or ESC processes should be initiated at a site (X1.6.1 and X1.6.2). Some form of initial assessments would also be required at other types of contaminated sites in order to provide the basis for a decision to initiate the ESC process.

6.2 The ESC client is responsible for procuring an ESC provider. Section 13.5 discusses some considerations in procuring an ESC provider. Section 7.4 describes criteria for evaluating qualifications of the ESC core technical team and other project support personnel.

6.3 Project objectives and data quality requirements should be defined by some process that includes the ESC client, regulatory authority, and, where appropriate, stakeholders, such as individuals or communities which may be affected by contamination at the site. Definition of project objectives and data quality requirements is an iterative process and may require some modification as an ESC project proceeds. Where an ESC project is to be used for risk-based decisionmaking, the risk evaluation method to be used will affect data quality requirements. If contaminants of concern are not known, their definition can occur by additional sampling and analysis as a separate activity prior to initiation of an ESC project, or this can be incorporated as an objective of the ESC project.

7. Establish ESC Project Team

7.1 *ESC Core Technical Team*—ESC will not work

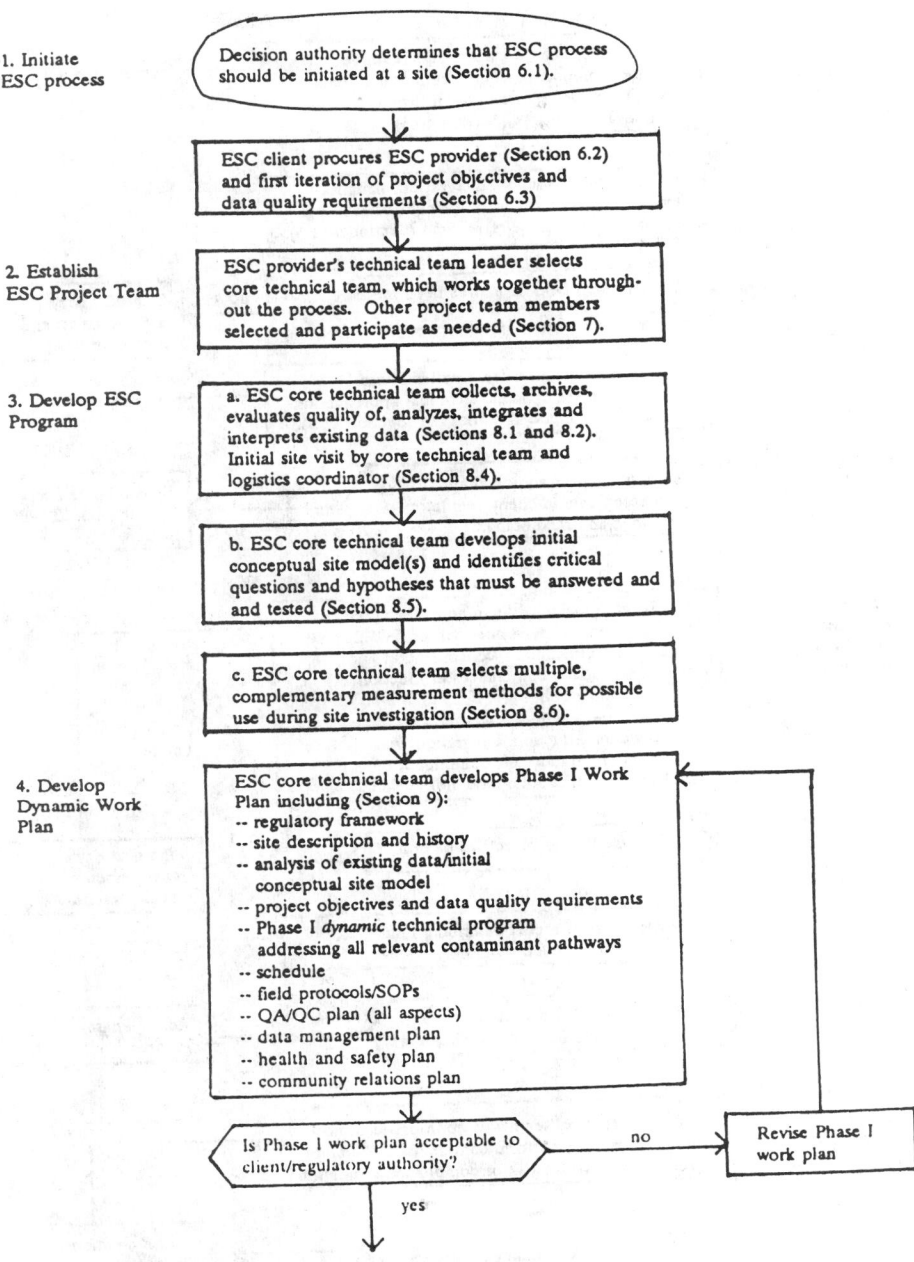

1. Initiate
ESC process

Decision authority determines that ESC process should be initiated at a site (Section 6.1).

ESC client procures ESC provider (Section 6.2) and first iteration of project objectives and data quality requirements (Section 6.3)

2. Establish
ESC Project Team

ESC provider's technical team leader selects core technical team, which works together throughout the process. Other project team members selected and participate as needed (Section 7).

3. Develop ESC
Program

a. ESC core technical team collects, archives, evaluates quality of, analyzes, integrates and interprets existing data (Sections 8.1 and 8.2). Initial site visit by core technical team and logistics coordinator (Section 8.4).

b. ESC core technical team develops initial conceptual site model(s) and identifies critical questions and hypotheses that must be answered and and tested (Section 8.5).

c. ESC core technical team selects multiple, complementary measurement methods for possible use during site investigation (Section 8.6).

4. Develop
Dynamic Work
Plan

ESC core technical team develops Phase I Work Plan including (Section 9):
-- regulatory framework
-- site description and history
-- analysis of existing data/initial
   conceptual site model
-- project objectives and data quality requirements
-- Phase I *dynamic* technical program
   addressing all relevant contaminant pathways
-- schedule
-- field protocols/SOPs
-- QA/QC plan (all aspects)
-- data management plan
-- health and safety plan
-- community relations plan

Is Phase I work plan acceptable to client/regulatory authority?

no

Revise Phase I work plan

yes

FIG. 2 Expedited Site Characterization Flow Diagram

without an effective, integrated technical team consisting of experienced individuals with expertise in geologic, hydrologic, and chemical systems. The ESC provider is responsible for establishing the ESC core technical team, which will typically consist of two or three members, in addition to the technical team leader. The core technical team members supervise all field operations in their area of expertise and are personally involved with much of the data acquisition. The technical team leader, with the support of other core technical team members is responsible for all data, ensuring proper data management, interpretation, and integration of data into a conceptual site model and reports. The core technical team is supported by appropriate personnel and on-site and off-site contracted personnel as needed (7.2 and

7.3). The members of the ESC core technical team must be qualified to perform the following functions:

7.1.1 *Technical Team Leader*—The technical team leader is ultimately responsible for all decisions related to the design and implementation of an ESC project, within the framework provided by the approved dynamic work plan. The technical team leader plays a lead role in interpreting prior data, developing the preliminary conceptual site model, selecting other core technical team members and other project support personnel, identifying multiple complementary geologic and hydrologic field investigation and chemical analytical methods, and developing the work plan. The technical team leader is the primary point of contact for the ESC client and, if authorized by the ESC client, the

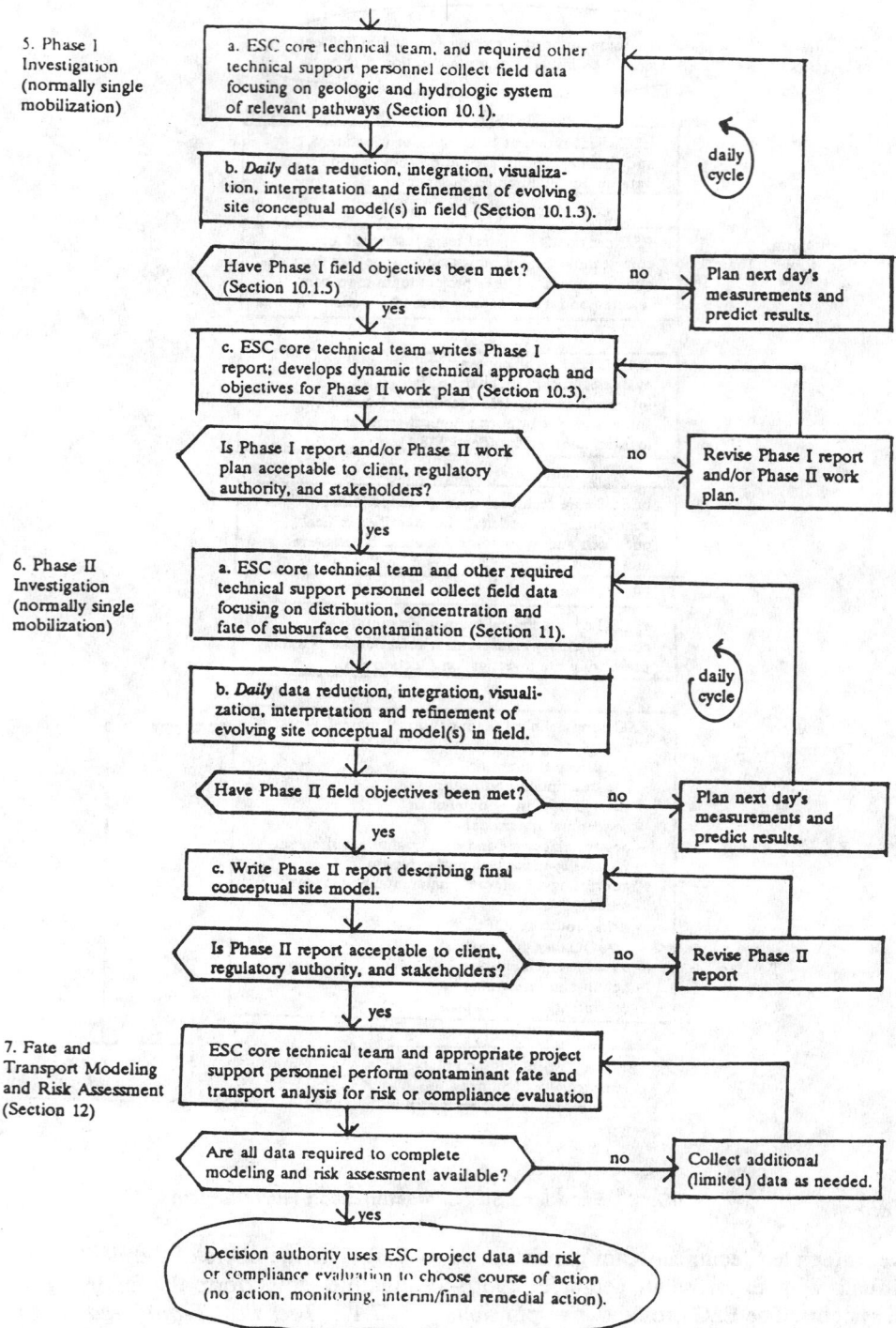

FIG. 2 Expedited Site Characterization Flow Diagram (Continued)

regulatory authority. The technical team leader must make sure that regulatory requirements specific to filing permits, securing access agreements, meeting local codes, etc., are met, although the actual tasks may be assigned to other project personnel. She or he is responsible for ensuring that the ESC client and regulatory authority are fully informed about the progress of an ESC project. The technical team leader is normally in the field during an entire field mobili-

zation, and leads interpretation of data as it is collected, modifying the evolving conceptual site model in collaboration with other technical members of the team, and making decisions concerning subsequent data collection efforts. The technical, management, and leadership responsibilities of the technical team leader require an individual with extensive training and experience in geologic and hydrologic systems (and a familiarity with chemical systems), negotiating and contracting, and a working understanding of all elements and functions of contaminated site characterization. The success or failure of an ESC project rests largely on the shoulders of the technical team leader.

7.1.2 *Core Technical Team Expertise*—The core technical team, which includes the technical team leader, requires a high level of expertise and experience in geologic, hydrologic, and chemical systems. Individuals on the core technical team need to be integrators as well as specialists, with in-depth expertise in some areas and the ability to communicate well with other team members with complementary areas of in-depth expertise. Section 7.4 discusses criteria for evaluating qualifications of core technical team members and other project personnel. The relative importance of specific areas of geologic and hydrologic expertise will vary somewhat from site to site, but most sites will require individuals on the core technical team with expertise in soil science, geology (with emphasis on stratigraphy, petrology, and structural geology), geophysics (where geophysical methods are appropriate for the site), hydrogeology, and geochemistry (broadly defined to include natural chemistry of solid, gaseous, and aqueous phases). Desirable areas of secondary expertise (or primary expertise where air and surface water are significant contaminant pathways) may include geomorphology, surface water hydrology and sedimentology, and climatology/meteorology. Areas of secondary expertise may also be provided by other project support personnel. Chemistry expertise on the core technical team needs to cover both analytical chemistry (organic and/or inorganic chemistry as appropriate) and knowledge of contaminant characteristics for evaluating their fate and transport. To this end, desirable areas of secondary expertise may include soil and water microbiology and ecotoxicology. Areas of secondary expertise may also be provided by other project support personnel. The core technical team includes such other technical or functional expertise as may be needed to address the specific site being characterized.

7.1.3 *Core Technical Team Field Operations*—The ESC process differs from conventional site characterization by the involvement of multiple senior technical personnel throughout the process, including field operations (Appendix X1). All members of the core technical team are directly involved with supervising field operations in their area(s) of expertise and are personally involved with much of the data acquisition. The technical team leader, supported by other core technical team members, is responsible for ensuring data quality, effective data management, and performs interpretation and integration of data into the evolving conceptual site model and reports. The technical team leader has the final say in decisions concerning field operations. If unavoidable circumstances require the technical team leader to be absent during field operations, another member of the core technical team should be designated as the acting technical team leader, who, in phone consultation with the technical team leader makes decisions concerning the next day's activities. Other core technical team members are in the field for data collection involving their primary area(s) of expertise and are available for phone consultation when not present in the field.

7.1.4 *Importance of Continuity*—The same individuals on the core technical team that begin an ESC project should stay with the team to completion. The success of the ESC process depends on a high level of integration and continuity on the part of the ESC core technical team. Individuals capable of performing the various functions of the ESC core technical team should be identified and regularly briefed on the progress of work as a contingency if unforeseen events require a change in the team.

7.1.5 *Importance of a Multidisciplinary Perspective*—The ESC process differs from conventional site characterization and other accelerated site characterization approaches by its emphasis on placing multiple experienced personnel (the core technical team) in the field (Appendix X1). The cost of fielding senior-level personnel is more than offset by the expert judgment that allows reduced time and total cost to obtain an accurate final conceptual model of the site (Table X1.3). This expert judgment is required because the heterogeneity of vadose zone and ground water flow systems has historically resulted to two common outcomes from site characterization efforts: (1) a conceptual model of the site that has such a high uncertainty that it does not provide a useful basis for making decisions (and when decisions are made, often resulting in bad decisions), or (2) very high costs because of the variability of geologic and hydrologic systems requires a large number of samples to reduce uncertainty using statistically-based sampling approaches. The multidisciplinary perspective of the core technical team functions as on-site peer review of the evolving conceptual model and will reveal inconsistencies that might be missed by a single experienced individual.

7.1.6 *Core Technical Team Data Management and QA/QC Responsibilities*—Although data management (7.3.2) and QA/QC (7.3.3) are specific ESC project support functions, the technical team leader, supported by the other core technical team members, is responsible for ensuring that: (1) data collection is relevant to the objectives of the project (that is, is collected to satisfy data quality requirements), (2) QA/QC procedures for data collection and processing for their respective areas of expertise are strictly followed, and (3) field data reduction and processing do not introduce errors into the evolving conceptual site model. It is especially important that core technical team members be familiar with the limitations of any software used to analyze data, such as truncation of numeric data, and the interpolation/extrapolation errors associated with gridding, contouring, and visualization programs.

7.2 *ESC Project Team*—Figure 1 identifies the major functional areas required to support an ESC core technical team, and 7.3 discusses major project support functions further. Some areas of project support expertise, such as statistics/geostatistics, fate and transport analysis (including computer modeling), risk analysis, and remediation engineering, may have a special role early in a project in defining the type of data required for the project and data quality requirements and are involved throughout the project as needed. Depending on training and experience, technical support personnel may cover more than one area of expertise. The ESC project team roster should clearly identify the expertise of the each member. As with the ESC core technical team, continuity is important. If the same discipline is required for more than one mobilization, the same individual should be involved, if possible, in all phases of the investigation, including review, planning, field, data interpretation, and report writing. The logistics coordinator plays an essential role of facilitating field operations (7.3.1). The data manager is responsible for identifying other personnel needs for data management, such as needs for computer data entry and quality control (7.3.2). The QA officer (7.3.3) and health and safety officer (7.3.4) are the only personnel who are not ultimately answerable to the technical team leader. They are independently responsible for assuring that the investigation complies with appropriate regulation and practice in their area of expertise. Two basic approaches are available for assembling an ESC field team:

7.2.1 *Inhouse Team*—Where an ESC provider has a large staff, it may be possible to establish a field team entirely, or mostly from within the organization. The close, ongoing relationship between team members allows a high degree of integration during all phases of the ESC project for a site. A single team is typically engaged in multiple projects. Inhouse expertise may be supplemented in certain areas, such as subcontracting for chemical analysis, drilling, and geophysical surveys.

7.2.2 *ESC Core Technical Team and Subcontractors*—The second approach is for the ESC provider to employ all or most of the ESC core technical team which uses individual consultants and subcontractors for the rest of the field team. The involvement of different organizations provides different perspectives with the effect of creating peer review. Integration and coordination of the field team may be more difficult than with an inhouse team, and scheduling conflicts for subcontractors with small staff may occur if the ESC project extends longer than expected. It is essential that the core technical team and subcontractors have a good working relationship when using this approach.

7.3 *Other Project Support Functions:*

7.3.1 *Logistics Coordination*—The logistics coordinator is responsible for ensuring that all aspects of a field mobilization run smoothly. Activities performed by the logistics coordinator include all mobilization and demobilization coordination, making site security arrangements, and anticipating needs/organizing supplies and equipment for technical team members. Technical training or experience are not required for a logistics coordinator, but good organizational skills are essential.

7.3.2 *Project Data Management*—The project data manager is responsible for assembling, organizing, and archiving project data (Appendix X5) and is in the field for the duration of any field work. This function is essential for allowing the technical team leader and other technical members of the field team to interpret the field data for on-site decisionmaking. Ideally the individual performing the function of field data management is fully qualified with experience in geologic and hydrologic systems and has training and experience with computer hardware and software used by the team (that is, databases and database management, geographic information systems, computer assisted software programs, contouring/cross section programs, and other visualization software). A critical capability of the data manager is the ability to convert a wide variety of data from various software programs in an expedient manner so that it can be used by the project teams members.

7.3.3 *Quality Assurance/Quality Control*—The QA manager is independently responsible for monitoring field activities and conducting audits, as appropriate, to ensure that procedures in the QA/QC plan are being followed. The personnel responsible for QA should have broad familiarity with field and analytical methodologies. As noted in 7.1.5, the technical team leader, supported by the appropriate core technical team members, has primary responsibility for data adequacy. Quality control of data begins with the individuals who are directly involved with the data acquisition, initial data processing, and interpretation. Initial data processing is commonly done by the persons making the measurements (that is, geologic boring logs, slug tests, geophysical measurements, etc.) and is often done on a separate computer with specific software. Once in the hands of the data manager, data quality control becomes his or her responsibility (Appendix X5), except that the person who originally acquired the data has a continuing responsibility to check the integrity of their data.

7.3.4 *Health and Safety*—The health and safety officer is responsible for monitoring field activities and to ensure that operations are in conformance with the health and safety plan.

7.3.5 *Stakeholder Liaison/Community Relations*—Smooth functioning and acceptance of ESC team activities require good communication with stakeholders who are concerned with the outcome of an ESC project. The liaison function is proactive, having the objective of positive and early engagement with all stakeholders in the ESC project and the future of the site. The technical team leader supports the ESC client in stakeholder liaison, and if authorized, may serve as the primary point of contact for an ESC project. Other project personnel may provide support for community relations, such as publicity and organization of public meetings.

7.3.6 *Statistics/Geostatistics*—As noted in 7.1.5, the ESC process uses sampling based on the expert judgment of the core technical team to characterize the geologic and hydrologic system and contaminant distribution. Statistical sampling approaches may be used, as appropriate.

7.3.7 *Fate and Transport Analysis*—As discussed in 13.4, full-scale fate and transport analysis occurs after the ESC Phase I and Phase II investigations are completed. The person primarily responsible for this function should provide input into the Work Plan to ensure that data necessary for fate and transport analysis are collected. Individual(s) in this functional area may also be involved in field data collection as a technical support function.

7.3.8 *Risk Evaluation*—As discussed in 13.2, an ESC project should ensure that data collected during ESC Phases I and II provides information required for risk evaluation where this provides the basis for decisionmaking (Section 12). Risk evaluation needs will usually be defined as part of the process for developing project objectives and data quality requirements (6.3) in order to ensure that the necessary data are being collected by the ESC project. This function requires expertise in toxicology, regulatory-approved risk evaluation procedures, and familiarity with EPA and state contaminant screening levels and health-based cleanup criteria.

7.3.9 *Remediation Engineering*—As discussed in 4.4 and 13.3, remediation engineering expertise is incorporated into the ESC process at the earliest point at which a need for remedial action is identified.

7.3.10 *Technical Experts with Prior Site Experience*—At many sites the ESC client or client's subcontractor may have years of experience with the site where an ESC project is being initiated. At a minimum this experience should be leveraged by using such knowledgeable individuals in an advisory capacity. Such individuals would also be potential candidates for membership in the core technical team or for project support.

7.4 *Criteria for Evaluating ESC Project Team Qualifications*—Qualifications of personnel for ESC projects should be evaluated based on four major factors: education, experience, professional registration or certification, and publications.

7.4.1 *Educational qualifications*, include academic degrees (BS, MS, PhD), course work, and continued education in the fields of geoscience and contaminant chemistry for the core technical team and relevant additional fields for other project support personnel.

7.4.2 *Experience and area of specialization*, are the most important factor in assessing an individual's capacity to do ESC work. For core technical team members such experience must be documentable, and verifiable, and include direct participation by the individual at a senior professional level. The individual's role must have been in a hands-on role, carrying the project from beginning to end. Experience should be weighed heavily when evaluating an individual's ability to carry out quality ESC work.

7.4.3 *Professional registration*, at the state level and/or certification by a recognized professional organization in the appropriate area of expertise, provide a secondary means of assessing education and experience.

7.4.4 *Publications*, directly associated with work carried out in the field of site characterization provide another useful means of assessing education and experience.

7.4.5 *Documentation of Technical Team Member Qualifications*—Documenting the qualifications of and experience of ESC core technical team and other field team members provides one means for an ESC client to compare the qualifications of several potential ESC providers. The ESC provider should maintain a file of full resumes for each core technical team member and one- to two-page vitae for each field team member, including ESC subcontractors that includes the following information: (*1*) name, (*2*) functions and disciplinary expertise (for technical field team members) performed as part of ESC team, (*3*) education, including year and type of degree received, major and minor subjects, and post-academic technical and professional training courses, (*4*) field project experience, including year(s), location, client, purpose of project, responsibilities, weeks/months in the field, and activities performed, (*5*) professional memberships, registrations, certifications, (*6*) major publications other than project reports, and (*7*) optional references (names addresses and phone numbers of professionals familiar with the individual's work). The purpose of this file is to provide a succinct summary of the most relevant information concerning an individual's competence for performing the designated functions as part of an ESC team. Any other information normally included in a vita should be eliminated. The ESC provider should maintain a file with vitae for all technical ESC project team members.

7.4.6 *Documentation of Core Technical Team Qualifications*—In addition to documenting the qualifications of individual ESC core technical team members, the overall team qualifications can be evaluated by documenting: (*1*) prior experience working as a team on ESC or other projects, (*2*) examples of integrated reports prepared by the team for ESC or other projects, and (*3*) references from ESC clients or regulators who have participated in ESC projects.

Note 6—The number of experienced ESC core technical teams is presently limited. Consequently, the collective qualifications of individual core technical team members would be the primary basis for evaluating core technical team qualifications in the absence of prior ESC experience.

8. Develop ESC Project

8.1 *Use of Prior Data*—Conventional site characterization practice, using multiple contractors and mobilizations has tended (but not always) to result in underutilization of available prior data for developing the preliminary conceptual site model and for guiding collection of new data. The ESC process breaks this cycle by emphasizing the importance of compiling original source materials and critically evaluating, integrating, and interpreting all available data for a site as part of developing the preliminary conceptual site model. When possible, integration and review of prior data is a core technical team activity.

8.2 *Archive Prior Data*—The ESC provider collects and develops an archive of all prior site data.

8.2.1 Materials in an archive should include: (*1*) available historical information about contaminant use and locations, including the results of interviews with individuals familiar with the history of the site, (*2*) available aerial photography and relevant remote sensing imagery of the site, (*3*) published and unpublished topographic, vegetation, soils, geologic, hydrologic, and other maps including the site, (*4*) originals or copies of well, other borehole, and geophysical logs, (*5*) copies of published soil survey, geologic, hydrologic, and engineering reports that provide information about the site and surrounding area, (*6*) any reports, such as CERCLA Preliminary Assessments/Site Investigations, RCRA Facility Assessments, and site investigation reports prepared by consultants or regulators, (*7*) soil and ground water sample analytical results, (*8*) any geophysical survey results, including copies of electronically stored data when available, (*9*) original or copies of data from aquifer tests, (*10*) available streamflow, fluvial sediments, and climatic data, as appropriate, and (*11*) any other information, as appropriate.

8.2.2 Each item should be given a locatable archive number and a master list with the full title of each item and a topical index (well logs, geology, soil sample data, etc.) should be prepared and updated as new information becomes available.

8.3 The ESC core technical team (7.1) reviews data in the central archive, emphasizing compilation of original source materials, such as well and borehole logs, water level measurements, and soil and water sample analyses. All such original data should be evaluated and given data quality ratings (such as low, medium, high), copied, clearly identified as to its source, and placed in notebooks for the field archive. If sample analytical data are too voluminous as a result of time series sampling, simple statistical summaries of key parameter values should be compiled (mean, coefficient of variation, etc.) and evaluated for possible trends or discontinuities. Reprocessing of data may be performed if there is a reasonable expectation that the benefits of the information obtained will exceed the cost of reprocessing. Essential interpreted maps and cross sections from existing sources, such as soils, geology, potentiometric surfaces, etc., should also be copied, clearly identified as to their source and placed in notebooks for the field archive.

8.4 *Initial Site Visit*—The ESC core technical team and other project team members, such as the logistics coordinator and health and safety officer, visit the site as a unit. The purpose of the site visit is threefold: (*1*) visually inspect the site to identify significant site features as part of developing the preliminary conceptual site model, (*2*) evaluate logistical concerns that may affect timing and efficiency of field mobilizations, including utility clearance or location and marking or all subsurface utilities for safety and planning sampling locations, and (*3*) identify site conditions that may affect suitability of field investigation methods. Viewing the site from the air may also be beneficial.

8.4.1 The technical core team members give particular attention to evaluating site conditions that may adversely affect use of specific field techniques used for their area of expertise. The initial site visit also provides the first opportunity for the core technical team to explain the ESC process and to hear the concerns of site personnel, the regulators, and other stakeholders.

8.5 *Develop Preliminary Conceptual Site Model*—The ESC technical team leader, with support from other appropriate core technical team members, develops the preliminary conceptual site model, focusing on features of the geologic and hydrologic system that exert controls on contaminant movement. Documentation of the preliminary conceptual site model should note consistency with prior data or note where prior data are contradictory or at variance. Knowledge of the direction of ground water flow and potential preferential pathways for contaminant transport allows targeted sampling for three dimensional mapping of contamination. Guide D 5979 provides a comprehensive framework for developing the preliminary conceptual site model. Guide E 1687 provides additional guidance on developing conceptual site models at contaminated sites. Section 13.4 discusses the possible role of modeling in developing the preliminary conceptual site model.

8.5.1 Develop one or more interpreted plan maps including key site features (roads, buildings, surface water features, depth to bedrock, direction of ground water flow, etc.) and cross sections that clearly identify water levels, zones of high and low permeability, other aquifer boundaries, and vertical variations in geochemical parameters (if available). The maps and cross sections should be based on original rather than interpreted sources (in this context driller's well logs are considered original sources, although they represent interpretations of the driller), and clearly identifying the data quality ratings of individual sources. Serious data gaps should be left as blank spaces on the maps and cross sections, and uncertain features should be identified with question marks or dashed lines. All spatial data should be referenced to a single site coordinate system

compatible with ESC client and regulatory authority needs.

8.5.2 The ESC core technical team identifies essential features and alternative interpretations of the interpreted plan maps and cross sections as they relate to the system pathways for contaminant movement in the subsurface. These essential features of the preliminary conceptual site model are formulated as critical questions and hypotheses that must be answered and tested. These could be formulated as specific questions, such as: (*1*) is the continuity of low permeability strata sufficient to create two separate aquifers, (*2*) are there stratigraphic controls that could cause DNAPLs to migrate in a direction different from the direction of ground water flow, (*3*) are there preferential flow paths that could cause ground water to flow in directions different from that indicated by the potentiometric surface. Alternatively, testing of the initial conceptual site model could be formulated as specific objectives, such as (*1*) define the number of aquifers and degree of connection between them, (*2*) determine the direction of ground water flow and identify potential pathways for preferential flow. In the DQO process, questions are formulated in the form of decision rules in which the answer to the question determines the next course of action (X1.4.4).

8.5.3 Refinement of the preliminary conceptual site model is primarily documented in the form of revisions to the maps and cross sections that represent the site surface and subsurface and text, where appropriate, that identifies measurements that may not fit the model and require explanation. Appendix X5.4 addresses documentation procedures for revisions to maps and cross sections.

8.6 *Select Multiple, Complementary Investigation Methods:*

8.6.1 *Identify Types of Measurements Required for ESC Project*—The core technical team is responsible for identifying the types of observations and measurements that are required to answer critical questions and test hypotheses in order to refine the conceptual site model. This may also involve collection of data to rule out alternative conceptual site models, in addition to data that support the preliminary conceptual site model and to resolve observations that are not consistent with the preliminary model. Although the emphasis on the first phase of the ESC project is on geologic and hydrologic characterization, this step also requires identification of chemical and fate-related parameters required during the contaminant characterization in ESC Phase II. Also, data required for vadose zone and ground water flow and transport computer modeling and risk analysis should be identified. For chemical analysis of samples, it is important to specify appropriate sensitivity and detection limits that will provide useful data for comparison with background, risk-based or other thresholds.

8.6.2 *Identify All Appropriate Measurement Techniques*—A key element of the ESC process is use of multiple, complementary measurement and sampling methods to characterize a site. The first step in optimizing the dynamic field characterization and sampling plan is to identify all appropriate measurement and sampling techniques for a particular site, including selection of the appropriate EPA or other analytical methods and protocols specified by the appropriate regulatory authority that will be used during the investigation. Refer to Guide D 5730 for discussion of major types of field investigation methods, an index of more than 400 ASTM standards that may be useful for environmental site characterization, and major reference sources that provide information on the characteristics of site characterization methods. Section X3.3 identifies selected major ASTM guides pertinent to selection of field investigation methods.

8.6.3 *Select Multiple Measurement Techniques*—The core technical team leader, with support from the appropriate core technical team members, selects a suite of site investigation methods that are suitable for conditions at the site and that will allow independent testing of essential features of the preliminary conceptual site model by use of multiple methods to characterize a given site feature. Appendix X3 discusses further criteria for selection of measurement techniques and their use in the ESC process.

9. Develop Dynamic Work Plan

9.1 The core technical team with appropriate support from other project team members prepares the work plan which includes all information required to conduct the ESC project. The element of the work plan that makes it dynamic is the technical approach which identifies the suite of field investigation methods and measurements that may be necessary to characterize a specific site but does not specify the number and location of observations or measurements. The technical team leader, with support from the appropriate other core technical team members, adjusts the location and type of field data collection efforts in response to previous observations and data in order to optimize the site characterization effort. The dynamic technical program operates within constraints defined in the work plan, including the geographic area within which the investigation will take place, maximum depth of penetration, and standard operating procedures for methods that are to be used. The dynamic technical program and all other aspects of the work plan create a well-defined framework from which departures are not allowed without the review and approval of the ESC client and regulatory authority.

9.2 *Work Plan Contents*—A typical dynamic work plan will include the elements described below. The organization and contents of the work plan may be modified, as appropriate, for specific regulatory programs or projects.

9.2.1 *Regulatory Framework*—The regulatory framework identifies all federal, state, and local laws, executive orders, regulations, and site-specific regulatory agreements that may be a source of environmental protection controls or performance standards for the ESC project. In this section or the section describing the dynamic technical program (9.2.4) it may be helpful to define the extent of involvement of regulatory personnel during field activities, and if a continuous presence during field activities is not feasible, to identify critical checkpoints where the regulatory authority should be consulted or present.

9.2.2 *Site Descriptions and History of Contaminant Use and Discovery*—This section includes information, such as site structures and use, topography and surface drainage, regional geology, climate, demographics and land use, history of contaminant use and discovery, environmental concerns, and previous environmental activities. Information on natural background levels and contamination from chemical discharges would be included in this section, or if

not available and required for the investigation, this would be incorporated as an objective of the Dynamic Technical Program (9.2.4).

9.2.3 *Analysis of Prior Data and Preliminary Conceptual Site Model*—This section presents an overview of regional and local geology, hydrogeology and geochemistry, including appropriate maps and cross sections. The emphasis in this section should be on the independent synthesis and interpretation of available data. The conclusion of the section should identify critical questions to be answered and hypotheses to be tested when field activities commence.

9.2.4 *ESC Phase I Dynamic Technical Program*—This section identifies the multiple investigation methods discussed in 8.6, and the area in which they may be applied. Although this is the core of the ESC process, this section of the work plan does not need to be very long. It may include a table, listing all features of the site to be characterized in one column and methods or measurements that may be used during the field mobilization. It will include clear criteria, in the form of a list of essential questions to be answered or specific objectives (see examples in 8.5.2), for determining when project objectives have been met. This section combined with the next section on field protocols and SOPs are the functional equivalent of the field sampling and analysis plan in conventional site characterization.

9.2.5 *Field Protocols and Standard Operating Procedures (SOPs)*—This section includes descriptions or copies of all field protocols and standard operating procedures, such as, but not limited to: (*1*) sample collection and analytical methods, (*2*) direct push methods, (*3*) geophysical methods, (*4*) drilling and monitoring well installation, and (*5*) aquifer test methods. This portion of the work plan may consist of field protocols and SOPs developed by the ESC team and approved by the appropriate regulatory authority, published protocols developed by regulatory agencies, and ASTM or other consensus or peer-reviewed standard methods. All field protocols and SOPs must satisfy existing regulatory requirements. If very large, this section may be attached as a separate document to the work plan.

9.2.6 *Quality Assurance/Quality Control Plan*—The QA/QC plan clearly defines responsibilities of different members of the ESC team for ensuring that protocols and standard operating procedures are followed. In addition to any regulatory program-specific sampling and analysis procedures in the QA/QC plan, the plan should define QA/QC procedures for other field activities, including: (*1*) geologic characterization, (*2*) hydrologic characterization, (*3*) geophysical surveys, (*4*) spatial control procedures (x, y, z accuracy of point measurements), and (*5*) computer records. The QA/QC plan describes procedures to be used to monitor conformance with, or documentation and justification of departures from, the field protocols and SOPs and procedures in the data management plan. Appendix X4 provides illustrative examples of QC procedures for geologic characterization.

9.2.7 *Data Management Plan*—The on-site decision-making that guides the dynamic field characterization and sampling activities requires a level of data management in the field that is not normally done during conventional field investigations. The data management plan identifies (*1*) staff and in-field computer equipment to be used, (*2*) software for field operations, and (*3*) software to be used for post field investigation analysis and interpretation. The plan also includes procedures for quality assurance of ESC project data. Measurement data are formally archived once in the field. Appendix X5 presents suggested procedures for quality control of computer records.

9.2.8 *Health and Safety Plan*—While a health and safety plan is necessary, there are no special features of the health and safety plan in an ESC project that distinguish it from any other type of environmental investigation.

9.2.9 *Community Relations Plan*—A community relations plan is desirable to enhance the public credibility of the ESC project (13.1.5). A typical community relation plan will include (*1*) site description, (*2*) community background, (*3*) community relations objectives, (*4*) timing of community relations activities, and (*5*) a contact list of key officials and other major stakeholders. The stakeholder liaison member of the ESC project team is responsible for implementation of the community relations plan. In some smaller, less complex sites, a formal community relations plan may not be required, but any stakeholders should be informed and involved in the process if possible.

9.3 *Work Plan Approval*—The ESC Phase I Work Plan is developed as a stand alone document and the ESC Phase II work plan is usually incorporated into the ESC Phase I Report (10.3). The draft work plans are reviewed by the ESC client and regulatory authority and revised by the ESC core technical team until the plan is acceptable. The ESC client or regulatory authority may also chose to have the draft plans reviewed by peer technical reviewers.

10. ESC Phase I Investigation (Focus on Geologic and Hydrologic Characterization)

10.1 *Field Mobilization*—The ESC Phase I investigation is normally completed with a single field mobilization of two to four weeks. Four weeks is about the maximum time that an ESC project team can operate effectively in the field. If site size or other conditions preclude completing a Phase I investigation in a single mobilization, two options are possible: (*1*) divide the site into smaller units and plan separate, but coordinated ESC projects for each unit or (*2*) plan a second (Phase Ib) mobilization. More than two mobilizations would be the exception rather than the rule. Acquisition of seasonally varying data, such as ground-water levels, may require additional site visits involving only a few personnel.

10.1.1 *Daily Field Data Collection Activities*—The core technical team oversees and participates in field collection activities. Each core technical team member directs field collection activities in their respective areas of expertise (see 7.1.2). The focus of the ESC Phase I investigation is on geologic and hydrologic characterization of the system of relevant contaminant pathways. Identification of contaminants of concern will also be carried out, if not already known, and subsurface contaminant distribution sampling occurs to the extent that it is convenient or contributes to understanding of the geologic and hydrologic system pathways. Other potential contaminant pathways (air, surface water, submerged sediments, biota) are also characterized, as appropriate. Characterization of near-surface contamination may incorporate the adaptive sampling and analysis approach described in X1.4.3.

10.1.2 *On-site Data Management*—The ESC data manager and supporting staff are responsible for monitoring the coordination of site activities to ensure that all data are incorporated into the computerized site database and made available to the core technical team as rapidly as possible. Integration of ESC data sets with prior site data sets should be accommodated without restricting the ESC provider from using state-of-the-practice hardware and software tools. Section X5.4.1 describes in more detail how data can be processed to allow on-site decisionmaking.

10.1.3 *On-site Decisionmaking*—The core of the ESC process is the use of multidisciplinary integration and interpretation of field measurements and sample analyses to select the type and location of subsequent field measurements and sampling points. During an ESC Phase I field mobilization the technical team leader and other core technical team members with input from other technical field team members meet on a daily basis to plan the next day's measurements. The regulatory authority and ESC client are encouraged to participate in any or all of the meetings at which the next day's activities are planned. They are always kept informed through brief notes and telephone calls, if not in attendance, and should be alerted when critical points in the investigation have been reached. The daily cycle of data collection, processing, and evaluation continues until the technical team leader determines that the objectives of the ESC Phase I investigation have been met. A useful procedure during field activities is to record in field notebooks, prior to making an observation or taking a measurement, the reason for the activity, and the expected result. This discipline helps focus data collection activities on the objectives of the investigation and provides immediate feedback concerning the accuracy of the evolving conceptual site model when the measurement result is available.

10.1.4 *Determining When ESC Phase I Field Objectives Have Been Met*—The overall objective of the ESC Phase I investigation, defined in the work plan as essential questions to be answered and hypotheses to be tested or a list of specific objectives or decision rules (8.5.2), is to characterize the geologic and hydrologic system with sufficient accuracy to allow targeted sampling to delineate the concentration and distribution of contaminants in ESC Phase II. This occurs when the field data fit into a consistent conceptual site model of the geologic and hydrologic systems that has no major unexplained anomalous observations. The technical team leader is responsible for deciding when the objectives of the ESC Phase I investigation have been met. Information used to make this decision may include an assessment of data accuracy and adequacy. Data accuracy involves defining acceptable levels, identifying, and correcting errors in data. Data adequacy may be assessed in a variety of ways which may include considerations of spatial density, temporal density, location significance, and resolution.

10.2 *Post Field Investigation Analysis and Interpretation*—Upon return from the field, additional processing and interpretation of data are performed, if required. The results of this analysis may identify aspects of the evolving conceptual site model that require further refinement during the ESC Phase II investigation. This analysis also forms the basis for developing the ESC Phase II work plan for detailed contaminant characterization.

10.3 *ESC Phase I Report and Phase II Work Plan*—The Phase I report presents the site model and supporting data. Usually the Phase II work plan is also incorporated into the Phase I report. The Phase II work plan uses the Phase I Work Plan as a reference and only includes information that is pertinent to the Phase II investigation.

11. ESC Phase II Investigation (Focus on Contaminant Distribution)

11.1 *Field Mobilization*—The ESC Phase II field mobilization focuses on delineating the spatial distribution, concentration, and understanding the fate of contaminants, based on knowledge of the relevant contaminant pathways identified in Phase I. Aquifer characterization, geochemical analyses, and geophysical surveys continue, as necessary, to guide sampling decisions. On-site decisionmaking functions similarly to that described in 10.1.3 for the Phase I investigation. Phase II field objectives have been met when the distribution and concentration of contaminants have been mapped with sufficient spatial and temporal accuracy to evaluate exposure to possible receptors and the geologic and hydrologic system have been sufficiently characterized to allow evaluation of the direction of movement and concentration of contaminants in the future using contaminant fate and transport models or other methods of analysis.

11.2 *Post Field Investigation Analysis and Interpretation*—Upon return from the field, additional processing and interpretation of data are performed, if required. Since most of the data being analyzed are chemical data, the focus is on understanding how contaminants are interacting with the physical and biological system as they move and are transformed in the subsurface.

11.3 *ESC Phase II Report*—The Phase II report presents contaminant distribution and concentrations as two-dimensional maps or maps and cross sections showing the distribution of contaminated soil and ground water and supporting data. Three-dimensional images may be used to illustrate spatial distribution of contaminants, but should be based on actual data points. It also presents any refinements to the site model that may be pertinent to fate and transport analysis for risk evaluation.

12. Fate and Transport Analysis and Risk Evaluation

12.1 *Fate and Transport Analysis*—The ESC Phase I and Phase II investigations should provide the data required to perform contaminant fate and transport analysis for exposure and risk evaluation. The ESC core technical team, with the assistance of project team members with modeling and risk evaluation expertise, perform this final stage of the ESC process. Additional data acquisition, such as water level monitoring and aquifer tests for calibration of fate and transport models may be required. This would involve a limited number of team members targeted at collecting the needed data.

12.2 *Risk Evaluation*—The ESC project team uses methods for evaluating the risk of contamination that are approved by the appropriate regulatory or decision authority, or both. Where federal or state environmental standards or criteria, or both, are the basis for remedial decisions, risk evaluation would not be performed unless justified for other reasons.

12.3 *ESC Process Closure*—The ESC process is finished when the site model/risk evaluation is sufficient for the decision authority to choose a course of action. When no action other than monitoring is required, the final site model provides a good basis for optimal location of permanent monitoring wells. If remedial action is required, the final ESC site model provides the starting point for remediation selection, design, and implementation. Further field investigations will be required, but collection of data can be targeted to the specific needs of the design engineers (13.3).

13. Considerations in Implementation of ESC

13.1 *Relationship of ESC to the Regulatory Process*—The ESC process operates within the framework defined by the agencies or organizations responsible for ensuring compliance with the environmental statutes, regulations or management practices, or both, that affect the site being investigated. The increased flexibility inherent in on-site decisionmaking requires increased accountability. It is the responsibility of the ESC provider to take seriously the elements of the ESC process that increase accountability. Elements of the ESC process that are intended to increase accountability and the quality of information provided by the process include:

13.1.1 *ESC Team Qualifications*—Multiple, experienced personnel on the ESC field team are required. Although it is difficult to quantify professional competence, recommended documentation of qualifications and field experience of ESC team members (7.4) provides some accountability.

13.1.2 *Work Plan*—The QA/QC plan, which covers all aspects of the field investigation, not just chemical sampling and analysis (9.2.6, Appendix X4), and the data management plan (9.2.7, Appendix X5) increase the accuracy of non-chemical data obtained and the practice of archiving data in the field ensures that useful information is not lost in the transition from the field to the office.

13.1.3 *Use of Multiple Complementary Geologic and Hydrologic Investigation Methods*—(See Appendix X3.)

13.1.4 *Field Investigations*—During each field investigation phase regulatory personnel have access to the same information as the ESC team and can observe and, to the extent appropriate, participate in the on-site decisionmaking that guides the investigation (10.1.3). The involvement of regulatory staff in on-site decisionmaking increases the likelihood and speed of regulatory approval of the Phase I report, Phase II Work Plan, and the final decision. It also may result in more efficient use of regulatory staff resources because familiarity with the site may reduce the amount of time required to critically review draft ESC project reports.

13.1.5 *Involvement of Stakeholders*—The community relations plan (9.2.9) improves accountability by helping the ESC project team know the principle concerns of those who may be adversely affected by contamination at a site. This may help the ESC project team collect data and present it in a way that is responsive to stakeholder concerns. Furthermore, when other stakeholders understand the ESC process and are kept informed about how the investigation is proceeding, the credibility of its results increases.

13.2 *Role of Risk Evaluation in ESC*—The primary objective of the ESC process is to provide scientifically and technically sound information for decisionmaking, for which assessment of risk is a major consideration. Risk evaluation is incorporated into the process in several ways:

13.2.1 *Risk Analysis Initiates the ESC Process*—The decision to initiate the ESC process will usually be based on a risk-based judgment that contaminants at a site present a potential threat to human health or the environment (6.1).

13.2.2 *Planning of ESC Phase I and II Field Mobilizations*—Risk evaluation methodology and any associated transport and fate computer models should be identified early in the process to ensure that field data collection provides information required for the risk evaluation that concludes the ESC process.

13.2.3 *Concluding Risk Evaluation*—The ESC process follows accepted regulatory protocols for risk evaluation (Section 12).

13.3 *Relationship of Remediation Engineering Design and Implementation to ESC*—As discussed in 4.4, the ESC process normally avoids a presumption that remedial action will be required, since it is usually not known beforehand when an ESC project is initiated whether remedial action requiring an engineered solution will be required (no action or ongoing monitoring being possible outcomes). Where the ESC process provides information for risk-based decisions, costs for data collection required only for remediation design during site characterization activities may be wasted if the ESC project results in a no action or ongoing monitoring decision. Where remedial action is required, an advantage of the ESC process is that the reduced time for site characterization, allows timely initiation of CERCLA feasibility studies or RCRA corrective measures studies. Furthermore, the thorough understanding of the geologic, hydrologic, and chemical system at a site reduces the uncertainties that remediation engineering design must address. However, remediation engineering expertise is incorporated into the ESC process at the earliest point at which a need for remedial action is identified. The time required for ESC Phase I and II investigations is normally short enough that interim corrective action are usually confined to noninvasive measures, such as access restriction, with more invasive measures optimized using information obtained by the ESC investigation. Remediation engineering expertise should be involved at the outset of an ESC project where regulatory compliance, such as RCRA closure, rather than risk-based analysis (which allows the possibility of no action or ongoing monitoring), requires remedial action. ESC projects at sites where presumptive remedies have been identified, but the decision for further action is risk-based, should involve remediation engineering expertise from the outset, but the cost of collecting data required only for remediation engineering design should be weighed against the possibility that it ends up being an unnecessary expense if the ESC investigation supports a decision for no action or ongoing monitoring. Guide D 5745 provides guidance for developing and implementing short-term measure or early actions for site remediation.

13.4 *Role of Modeling in ESC*—The ESC process relies heavily on computers in the field for compilation and management of field data and may benefit from the use of visualization software and geographic information systems for 2-dimensional and 3-dimensional presentation of spatial data to refine the evolving conceptual site model. Interpretative computer ground-water modeling or water budget analysis using relatively simple arithmetic or analytical models may be useful when developing the preliminary conceptual site model and to identify geologic and hydrologic system parameters that need better resolution by additional sampling or testing. Formal numerical computer modeling of vadose/ground-water flow and contaminant fate and transport, if used, occurs only after the ESC Phase I and Phase II investigations are completed. The final site conceptual model, based on multiple complementary investigation methods, should drive the modeling process, not the other way around. However, the person on the ESC project team responsible for the final vadose zone and ground water contaminant fate and transport analysis, should be involved in work plan development and field operations, as appropriate, to ensure that critical data necessary for definition and calibration of the models to be used are collected. The person responsible for fate and transport analysis should document all model assumptions, computer codes employed, and model input values so that an independent party can reproduce the results and, if necessary, modify the model in response to additional data.

13.5 *Procurement and Contracting Procedures for ESC*—The flexibility in the ESC process that allows time and cost reductions compared to conventional site characterization also requires use of procurement and contracting procedures that address the distinctive characteristics of the process. Specifically, the daily costs of an ESC project will tend to be high because of the use of multiple, senior, highly qualified personnel in the field, but, as discussed in X1.3, significant total cost savings can be expected. The cost of field activities contains an element of uncertainty, but upper bounds can be placed. This is because it is not possible to predict beforehand the precise combination of multiple, complementary investigation methods that will ultimately be used during an ESC project, or how extensively a particular method will be used. Ways to address these issues include but are not limited to:

13.5.1 *Criteria for Selecting ESC Provider*—Evaluation criteria for proposals by ESC providers need to be defined so that the qualifications, experience, and capabilities of an ESC provider receive as much, if not more, weight than cost.

13.5.2 *Subcontractors*—Time and materials contracts with reasonable upper and lower limits, and provisions for mobilization and demobilization costs provide the flexibility required when an ESC project team includes subcontractors during the field mobilization.

13.6 *Performance Indicators for Evaluating ESC*—Quantitative performance indicators may be useful for both ESC clients and ESC providers for the following purposes: (*1*) comparing the ESC process to conventional site characterization at a site, (*2*) comparing performance of ESC teams (taking into consideration any differences in objectives), and (*3*) evaluating effect of site characteristics on the ESC process. The key performance indicator is the ability of the final conceptual site model to predict contaminant fate and transport and realistic risk calculations for correct resolution of remedial or preventative measure, or both, including no action. Ongoing ground water monitoring provides an additional means for verifying of the final conceptual site model.

13.6.1 *Comparative Indicators*—The following indicators can be readily quantified: (*1*) length of each mobilization (days), (2) number of mobilizations, (3) cost of each phase of the investigation (dollars), (4) total time for each phase. Care should be taken when using comparative indicators between sites to consider the effect of site conditions on each performance indicator used (13.7).

13.7 *Factors that May Affect Performance Indicators*—The ESC process at a typical site requires single mobilizations of 2 to 4 weeks for Phase I and a mobilization of similar length for Phase II (see further discussion in 10.1). Various site-specific factors will affect the actual time and cost of an investigation. Planning the schedule for an ESC site investigation should take into account these factors. The following factors may contribute to either lengthening the time required or increasing the cost of an investigation:

13.7.1 *Stakeholder Relationships*—Polarization and antagonism between stakeholders creates an environment in which it is more difficult for the ESC process to function smoothly. Conflict between stakeholders will not necessarily affect the field investigation time but may affect the scheduling of field mobilizations and lengthen the total time for a project. The community relations plan and the ESC team member responsible for other stakeholder liaison have the responsibility for facilitating interactions between stakeholders in a way that reduces polarization and antagonism.

13.7.2 *Site Area and Access*—All other things being equal, time, and cost may increase as the size of the site increases. If performance indicators of sites of greatly different areas are compared, they should be in the form of unit area comparisons (that is, penetrations, days, dollars per unit area). Site access limitations may also increase characterization costs.

13.7.3 *Site Geology and Hydrogeology*—In general, as site complexity increases, performance indicators may appear less favorable as a result of increased time requirements and costs that satisfy regulatory requirements. However, for any given site, the ESC process can be expected to take less time and cost less than conventional site characterization for a given level of data and accuracy. Factors that will tend to increase time and costs include: (1) significant seasonal effects in hydrologic system (requires greater period of time to characterize seasonal variations, but does not necessarily increase time or cost required for field mobilization); (2) multiple or very deep aquifers; (3) structurally complex bedrock sedimentary aquifers (folding and faulting); and (4) fractured rock and karst aquifers.

13.7.4 *Contaminant Characteristics*—Optimal selection of chemical analytical methods should result in use of a technique or techniques that have the lowest cost and analysis time and provide the level of desired data quality for the regulatory framework of the project. For most contaminants, analytical methods are available that yield results from minutes to days and do not significantly affect the time required for field mobilization, all other things being equal. Contaminants that tend to increase times and costs include: (1) dense nonaqueous phase liquids (DNAPLs), which tend to sink to the base of an aquifer, or downward hydraulic gradients, requiring deeper sampling, (2) dioxins and furans, which have special health and safety requirements, and (3) highly radioactive contaminants which may need long time counts and slow sample collection and analysis due to health and safety considerations.

14. Keywords

14.1 environmental site characterization; exploration; feasibility studies; field investigations; geological investigations; geophysical investigations; ground water; hydrologic investigations; maps; preliminary investigations; reconnaissance surveys; sampling; site characterization; site investigations; subsurface investigations

APPENDIXES

(Nonmandatory Information)

X1. BACKGROUND ON EXPEDITED SITE CHARACTERIZATION

X1.1 *History of the ESC Process:*

X1.1.1 *Origins*—The process described in this provisional guide has its origins in the work of Dr. Jacqueline Burton's multidisciplinary team at the U.S. Department of Energy's (DOE) Argonne National Laboratory, Argonne, Illinois, starting in 1989. It was first developed for several U.S. Bureau of Land Management landfills in New Mexico. Use of multiple surface geophysical techniques (magnetic, two types of electromagnetic and seismic surveys) combined with targeted subsurface drilling and sampling at the Flora Vista landfill allowed making a determination that contaminants present at the landfill were not migrating deeper than the upper few metres and that the landfill could be closed without continuing ground water monitoring (see Appendix X6). The process was formalized by the Argonne team, and has resulted in significant cost savings and time for the following federal agencies: (1) USDA Commodity Credit Corporation at numerous sites where past practices included grain elevator storage in Nebraska and Kansas known to be contaminated by carbon tetrachloride (Burton et al., (1),[10] (2), (2) the U.S. Department of Energy at the Pantex site in Texas (2), and (3) Department of Defense at several Air Force and Navy installations. Table X1.1 provides summary information on ESC investigations at these sites.

X1.1.2 *Further Testing and Demonstration*—Dr. Al Bevolo, with a core technical team (including some consultants) at the U.S. Department of Energy's Ames Laboratory, Ames, Iowa and a project team using contract consultants has tested and demonstrated the ESC process at a manufactured gas site in Marshalltown, Iowa Bevolo et al. (3) a DOE site near the St. Louis airport with low-level radioactive contamination in 1994, an oil seepage basin at DOE's

---

[10] The boldface numbers in parentheses refer to the list of references at the end of this standard.

TABLE X1.1 Examples of the ESC Process

Client	Site Type/ Size/Depth	Location/EPA Region	Contaminants of Concern	Regulatory Setting	Problem or Objectives	Results of ESC Investigation	Selected References
USDI/BLM	3 landfills 650' by 800' to 1000' by 2000'	NM/Region VI	Metals, VOCs, SVOCs, petroleum products	CERCLA RI/FS	PA/SI results indicated likely migration to ground or surface water.	All 3 landfills closed with no remediation; cost $300 000 less than projected. See case study, Appendix X6.	Burton (30), Burton et al. (1)
USDA/CCC	20 former grain storage facilities 800' by 1000' to 23 sq mi	NE, KS/Region VII	Carbon tetrachloride (CCl₄), chloroform	CERCLA RI/FS (2 NPL sites), SDWA	CCl₄ contamination of ground water drinking water supplies potentially linked to CCCs former operations.	Streamlined RI/FS process and generated technical defensible data for rapid remedial action decisionmaking for 1/5 to 1/10 cost and 1/30 time of conventional methods.	Aggarwal et al. (31), Burton (6), Burton et al. (1), Hasting et al. (32)
DOE	Zone 12 Pantex weapons facility, 1 sq mi	Amarillo TX/Region VI	Explosive, metals, fuels, chlorinated hydrocarbons	CERCLA RI and RCRA RFI	Previous RI had not generated technically acceptable explanation for contaminant distribution, migration pathways, and potential for cross aquifer contamination.	Reevaluation, interpretation, and integration of existing (pre ESC) databases allowed for a streamlined field investigation that generated technical defensible explanations of all issues. ESC saved a minimum of $4 million and 4 years from original project estimates.	Burton et al. (2)
DOD/Navy	Closing marine air station, 3 sq mi	Western US/Region IX	Fuels, solvents, oils, paints, pesticides	CERCLA RI/FS	Previous RI had not defined aquifer systems (including flow, number of aquifers, etc.), migration pathways, sources, or contaminant distribution.	Phase I investigation delineated hydrologic systems, including flow, number of aquifers, isolation of aquifers, and migration pathways. Phase II presently being conducted by private sector firm.	Burton et al., in progress
DOD/Air Force	Air Force base, 1300' by 2600'	Southwest Region VI	TCE, 1,2 dichloroethene, carbon tetrachloride	CERCLA RI, RCRA RFI	Rapid transfer of ESC technology to agency and contractor for interim corrective action measures selection.	ESC investigation ongoing.	Burton et al., in progress
IA Department of Natural Resources	Former gas manufacturing plant, 3 acres, 50'	Marshalltown, IA Region VII	16 polyaromatic hydrocarbons (PAH)	State Regulatory Program	DOE technology demonstration: IMAs, geophysics, five PAH soil extraction methods, percussive conductivity logging (ISTTime)	Technology evaluations.	Bevolo et al. (3)
DOE	SLAPS uranium processing waste site, 52 acres, 65'	St. Louis MO, Region VII	Uranium, thorium, radium, radon isotopes	UMTRA	Delineate extent of radioactive soil contaminants (ICP/MS and ICP/AES in mobile labs); evaluate connection to karst aquifer.	Microgravity surveys and geochemical dating of karst aquifer were completed.	Ames Laboratory, in progress
DOE	Savannah River Site facility, 10 acres, 65'	SC, Region IV	TCE, PCE, vinyl chloride, a-BHC pesticide, Sb, Be, Mn, and As metals	CERCLA RI, RCRA, RFI	Determine contaminants of concern; delineate extent of ground water contamination; assess risk of soil/gw contamination.	Contaminant plume would have been missed by originally proposed monitoring well locations; discovered two shallow controlling aquitards.	Savannah River Site (4), Technos (33)
Polish government/ DOE	Oil refinery, 15 acres, 60'	Czechowice, Poland	VOCs, 5 PAHs, RCRA metals	Polish action levels	Demonstration of eastern and eastern european site characterization technologies	CPT/LIF, CPT/Hg lamp, IMA comparison; percussive conductivity logging, horizontal drilling, and sampling.	Ames Laboratory (5)

BLM = Bureau of Land Management
CCC = Commodity Credit Corporation
CERCLA = Comprehensive Environmental Response, Compensation, and Liability Act (Superfund)
CPT/LIF = cone penetrometer/laser-induced fluorescence
DOD = U.S. Department of Defense
DOE = U.S. Department of Energy
EPA = U.S. Environmental Protection Agency
ICP/AES = inductively-coupled plasma/atomic emission spectrometer
ICP/MS = inductively-coupled plasma/mass spectrometer
NPL = National Priority List (Superfund)
PAHs = polyaromatic hydrocarbons
PCE = perchlorethylene (tetrachloroethylene)
RCRA = Resource Conservation and Recovery Act
RI/FS = remedial investigation/feasibility study (Superfund)
RFI = RCRA facility investigation
SDWA = Safe Drinking Water Act
SVOCs = semivolatile organic compounds
TCE = trichloroethylene
UMTRA = Uranium Mill Tailing Remediation Act
USDI = U.S. Department of the Interior
VOCs = volatile organic compounds

Savannah River site (4) in 1995, and at an active oil refinery in the Katowice region of Poland (5). Table X1.1 summarizes information on these ESC investigations.

X1.2 *Comparison of Conventional and Expedited Site Characterization*—Conventional site characterization has benefitted from, and uses to a varying extent, the advances in technology described in Appendix X3 that allow the ESC process to compress site characterization into a limited number of field mobilizations. Perhaps the defining difference between ESC and conventional site characterization is that the ESC process is structured to produce a final conceptual site model developed by an experienced multidisciplinary team in which high confidence can be placed as a basis for deciding an appropriate course of action, whereas conventional site characterization is structured to mainly produce specified numbers of boreholes, monitoring wells and chemical sample analyses, which may or may not result in an accurate final conceptual site model. Table X1.2 summarizes and contrasts the differences between conventional and expedited site characterization for eight site characterization process components. The limited number of mobilizations results in significant time savings compared to the multiple mobilizations required by conventional site characterization. Use of multiple measurement and sampling technologies with on-site decisionmaking by a technically skilled team results in a more accurate characterization of essential features of the geologic and hydrologic system as they affect contaminant movement. Fewer invasive penetrations means increased safety for field personnel and reduces risk of inadvertently creating paths for contaminant transport.

X1.3 *Cost Savings of ESC*—The ESC process can be expected to yield significant cost savings compared to conventional site characterization, both in terms of total life cycle cost of a project, and in terms of the cost required to obtain information of comparable quality. Typically, initial costs for analysis of prior data and planning for a mobilization, and daily costs during mobilization will be higher compared to conventional site characterization. The cost savings come primarily from the shorter period of time required to complete the investigation, and the reduction in permanent monitoring well installations (which have high ongoing costs for sample analysis). Also the improved quality of information for decisionmaking may also result in reduced costs for interim corrective actions and remediation. Table X1.3 summarizes the relative costs of various cost factors in conventional and expedited site characterization. Burton et al. (1) calculated that at two ESC investigations in Nebraska where direct push ground-water sampling was used to delineate contaminant plumes, that the cost was 10 % to 20 % the amount that would have been required to obtain equivalent data using monitoring wells. Burton (6) presents data indicating the staffing costs for the ESC process are about 87 % those for the conventional approach in a typical remedial investigation. Starke et al. (7) estimated that total costs for an ESC investigation at the Pantex Plant, near Amarillo, Texas were 30 % the cost compared to the original conventional work plan.

X1.4 *Other Site Characterization Approaches:*

X1.4.1 *Multiple Working Hypotheses*—The method of multiple working hypotheses, first described by Chamberlain (8) in a paper read before the Society of Western Naturalists in 1889, and later revised to focus on applications for geological study (Chamberlain (9)), is well suited for initial stages of environmental site characterization of geologic and hydrologic systems because the relative difficulty in making direct observations means that conceptualization of potential migration pathways for contaminants in the vadose zone and ground water system is based on a relatively limited number

TABLE X1.2 Process Comparison of Conventional and Expedited Site Characterization (Adapted from 2)

Process Component	Conventional[A]	Expedited
1. Project Duration	Longer because of multiple mobilizations and intervals between where data are compiled and analyzed.	Shorter because of fewer, coordinated field mobilizations, analysis and archiving of most data in the field, and improved regulatory and community acceptance.
2. Project Leadership	Project leader typically in office; junior staff in field.	Project leader in field with experienced multidisciplinary team.
3. Use of Prior Data	Reviewed, but often not carefully evaluated or interpreted.	Carefully compiled, evaluated for quality, analyzed, and interpreted as part of developing preliminary conceptual model for the site.
4. Technical Approach	Different disciplines tend to work independently at different times for field data collection and interpretation.	All phases of field investigations—data collection, compilation, analysis and interpretation—integrated in the field.
5. Field Investigation Methods	Measurements usually not corroborated by complementary methods. Emphasis on installation of monitoring wells.	Use of multiple, complementary methods. Emphasis on noninvasive and minimally invasive investigation methods.
6. Work Plan	—Field characterization and sampling plan defined before full mobilization and generally not modified during the course of a mobilization (may be modified if DQO process used). —QA/QC plan for chemical sampling and analysis but usually not for other characterization activities. —Data management plan usually not formal part of work plan. —Community relations plan not always part of work plan (except for CERCLA program).	—Dynamic field technical program uses data as it comes in to guide type and location of field measurements and samples for analysis. —QA/QC plan for all aspects of field data collection and handling; strong emphasis on individual team member responsibility for QA/QC. —Data management plan formal part of work plan. —Community relations plan considered important part of work plan.
7. Number of Mobilizations	Multiple mobilizations, often carried out by different groups with little communication, interspersed with office analysis of data for a given investigation phase.	Normally one full mobilization for each investigation phase (two total), carried out under direct control and participation of a single core technical team.
8. Data Results and Analysis	Data results and interpretation usually interpreted in office weeks to months after field work. Computers not normally used in field for data management and analysis.	Data obtained, interpreted, and archived in the field (hours to days) as part of dynamic technical program. Computers used in field for data management as an aid in analyzing data.

[A] Certain elements included in the "expedited" column may be incorporated into a given "conventional" site investigation, but the characteristics described in this column can be considered typical of most site characterization activities during the 1980s and early 1990s.

TABLE X1.3 Cost Comparison of Conventional and Expedited Site Characterization

Cost Factor/Component	Conventional	Expedited
Initial cost	Lower because multiple, mobilizations have lower separate costs, and junior staff are used in the field.	Higher because multiple senior technical personnel and supporting junior staff are in 2-phased, coordinated multidisciplinary field investigation.
Total cost	Higher for reasons discussed below (not all factors will necessarily be higher compared to ESC but all cost factors combined can be expected to be higher).	Lower for reasons discussed below.
Field Characterization	Higher because field characterization activities tend to be proposed in terms of a fixed number of borings/wells and samples.	Lower because on-site decisionmaking allows efficient use of field methods to target collection to meet project objectives
Monitoring	Higher because monitoring wells tend to be installed before the geologic and hydrologic system is well understood resulting in a larger number of monitoring wells (higher well installation costs) and consequent high ongoing costs for sampling and analysis.	Lower because installation of permanent monitoring wells may be avoided or are limited to a few locations where they are needed the most.
Interim Corrective Action	Higher because less coordinated nature of field data collection may prevent optimal decisions for interim corrective action.	Lower because on-site decisionmaking can target data collection to minimize interim corrective action costs.
Remediation	Higher to the extent that suboptimal remedial action decisions are made compared to the ESC process (see next column).	Lower as a result of one or more effects: (1) an improved conceptual site model and risk evaluation may allow a decision of no action of monitoring instead of remedial action, (2) where remedial action is required, reduced design and implementation costs as a result of an improved conceptual site model.

of observations for which more than one explanation is possible.

X1.4.2 *Observational Method*—The observational method used by Dr. K. Terghazi for applied soil mechanics investigations from the 1920s to the 1950s and documented by Bjerrum (10) and Peck (11) is an investigation process for geotechnical characterization of soils and geotechnical engineering design, in which characterization, design, and construction proceed hand-in-hand. Observed change and response of the soil system as construction proceeds is used to modify the design, as required. A critical element of the method is an early assessment of the most probable conditions and the most unfavorable conceivable deviations from these conditions. Approaches to the U.S. EPA Superfund RI/FS process (see Appendix X2) using the observational method have been described by Mark et al. (12), Brown et al. (13), (14) and Holm (15). In this approach, the emphasis in the RI stage is to gather information to establish general site conditions and identify most probable conditions and reasonable deviations as the basis for a flexible approach to remedial design.

X1.4.3 *Adaptive Sampling and Analysis*—Adaptive sampling and analysis is an approach that has been successfully used at a number of Department of Defense facilities to characterize near surface contamination of soils (Robbat and Johnson, (16); U.S. EPA, (17)). The approach uses field chemical analytical methods and on-site decisionmaking using geostatistically-based models to guide sampling to determine the nature, extent, and level of contamination present at a site. The approach requires (1) field analytical techniques applicable to the contaminants and action levels of concern for the site, and (2) a means for rapidly making decisions in the field regarding the course of the sampling program, which is accomplished by on-site computer processing using geostatistical, visualization, and other data analysis software. The adaptive sampling approach generally focuses on characterization of near-surface soil contamination.

X1.4.4 *Data Quality Objectives (DQO) Process*—The DQO process is a quality management tool developed by the

U.S. Environmental Protection Agency to facilitate the planning of environmental data collection activities (U.S. EPA, (18)). The DQO process involves seven major steps: (1) state the problem, (2) identify the decision(s), (3) identify inputs, (4) define boundaries, (5) develop a decision rule, (6) specify acceptable decision errors, and (7) optimize data design. It brings together the right players (stakeholders and technical staff) at the right time to gain consensus and commitment about the scope of the project. This interaction results in a clear understanding of the problem, the actions needed to address that problem, and the level of uncertainty that is acceptable for making decisions. Through this process, data collection and analysis are optimized so only those data needed to address the appropriate questions are collected. The benefit of applying DQO is it improves planning efficiency, promotes defensibility of data, saves resources, is able to manage uncertainty, ensures consistency in decision analysis and reporting, and provides clear decision rules to identify the point at which further collection of data is no longer needed.

X1.4.4.1 The ESC and DQO processes share common elements, and both are flexible enough to be used together, but differences between ESC and the DQO process as it is commonly practiced should be recognized. Both ESC and DQO processes start with identifying all parties responsible for data use and decisionmaking. They develop strategies, goals, and actions designed to address the problem. Both methodologies are driven by the scientific method. The DQO process is designed to ensure each measurement is of known quality and consistent with available budgetary resources. It also usually assumes an optimal but fixed number of measurements. Information from the first measurement of the batch are usually not used to refine the next set of measurements. A completely efficient data acquisition system would incorporate all previous measurement to decide the nature and location of the next measurement, as is done in ESC. The DQO process typically uses a statistical approach, which is predicated upon a random, independent set of measurements in contrast to the ESC focus on judgmental sampling. Finally, the DQO process often focuses only on chemical measurement of the contaminants of concern in contrast to ESC where an integrated understanding of the geologic, hydrologic, and chemical system is required.

X1.4.5 *SACM and SAFER Processes*—The U.S. EPA's Superfund Accelerated Cleanup Model (SACM) and the U.S. DOE's Streamlined Approach For Environmental Restoration (SAFER) are complementary approaches for speeding up the CERCLA RI/FS process (U.S. EPA **(19)**; U.S. DOE, **(20)**). Both SACM and SAFER were developed primarily in response to the fact the site characterization efforts at CERCLA and RCRA sites were taking too long and not always providing the information required for effective remedial action. SAFER integrates the data quality objectives (DQO) process (X1.4.4) with the observational method (X1.4.2). SAFER involves regulators, stakeholders, and project managers in an integrated process that includes all activities associated with site characterization and remediation. The SAFER approach is designed to be aggressive and flexible in making decisions based on current information and also compatible and compliant with existing environmental regulations. ESC, SACM, and SAFER all emphasize the importance of including all stakeholders in the planning process. The ESC process can be readily incorporated into the site characterization component of the SAFER process. Since ESC reduces the time needed to obtain an accurate understanding of the geologic, hydrologic, and chemical system at a site, it allows less reliance on using the observational approach which focuses on identifying most probable conditions and contingency planning for reasonable deviations during remedial action. Since uncertainties concerning optimal design and operation of remedial measures are reduced by the accurate conceptual site model, contingency planning would be a less significant element when ESC is used in the framework of SACM or SAFER.

X1.6 *Relationship of ESC to Other ASTM Standards:*

X1.6.1 *Provisional Guide PS 03 – 95*—ASTM Committee E-50 has developed an accelerated site characterization (ASC) process for use at petroleum release sites. A number of the basic principles of the ASC and ESC processes are similar, namely use of highly experienced personnel in the field to lead the investigation and an on-site iterative process to guide data collection which allows a limited number of field mobilizations. Distinctive features of the ASC processes include: (*1*) ASC is oriented towards petro-

leum releases sites, (*2*) ASC is led by a single on-site field manager, and (*3*) an ASC investigation is completed in a single mobilization. The ESC process may be appropriate for large petroleum release sites, such as refineries. At such sites both this provisional guide and PS 03 – 95 should be reviewed in order to evaluate whether the ESC or ASC process is more appropriate.

X1.6.2 *Guide E 1739*—ASTM Committee E-50, with funding support from U.S. EPA, has developed a tiered approach to assessing risk for making decisions concerning remedial action at petroleum release sites. Depending on the geographic extent and amount of contamination, the expanded site assessment involved in Tier 3 of the RBCA process might justify initiation of the ESC process.

X1.6.3 ASTM Committee E-50 is developing guidance for conducting a Phase II environmental site assessment (ESA) for real estate property transactions, where a Phase I ESA has identified recognized environmental conditions that might result in liability under CERCLA or other statutes. The general focus of this guide is on chemical characterization, and Phase II ESAs can normally be expected to provide the kind of information required in Step 1a of the ESC process. If a Phase I ESA identifies the likelihood of significant soil and ground water contamination, it may be more cost effective to initiate the ESC process designed to meet the special needs of a real estate property transaction rather than conduct a Phase II ESA. In this event, care should be taken to incorporate into the ESC process the specific needs for conducting an investigation for a real estate property transaction as identified in the guide.

X1.6.4 *Development of This Guide*—In early 1996, the U.S. Department of Energy provided funding to ASTM to facilitate development of a guide that would formalize the ESC process. This provisional guide is the result of an intensive effort by a 10-member core task group. In addition to the normal ASTM balloting process for a provisional standard, pre-ballot drafts of the guide received review by selected groups of geophysicists, direct push technology practitioners, chemists, and individuals representing the perspectives of state regulatory agencies, including a 9-state review panel convened by the Interstate Technology and Regulatory Cooperation Group, the U.S. Environmental Protection Agency, the U.S. Department of Defense (Air Force, Army, and Navy), the U.S. Department of Energy, and U.S. Department of Agriculture.

X2. RELATIONSHIP OF ESC PROCESS TO RCRA AND CERCLA

X2.1 The Resource Conservation and Recovery Act (RCRA) and Comprehensive Environmental Response, Compensation, and Liability Act (CERCLA, also called Superfund) are the two main federal statutes and regulatory programs that require investigations to determine the extent and seriousness of subsurface contamination and to identify and implement appropriate remedial measures. Most ESC projects will take place in the context of either RCRA, CERCLA, or both.

X2.2 The RCRA and CERCLA programs have developed different terminology for similar phases of facility and site investigations as follows (refer to U.S. DOE, **(21)**, **(22)** for

additional information and relevant statutory and CFR citations):

X2.2.1 *RCRA Facility Assessment (RFA)*—Initial site investigation at a proposed RCRA facility or at an existing facility with a solid waste management unit with a known or potential release or discovery of a solid waste management unit which was not examined during a previous RFA. U.S. EPA **(23)** provides guidance on RFAs.

X2.2.2 *RCRA Facility Investigation (RFI)*—Detailed site investigation initiated when an RFA determines that hazardous waste or hazardous constituents have been or are likely to be released from a solid waste management unit.

U.S. EPA **(24)** provides guidance on RFIs.

X2.2.3 *Corrective Measures Study (CMS)*—Study to evaluate corrective measures and select an alternative at a RCRA facility where an RFI indicates that concentrations of one or more hazardous constituents exceed action levels or may pose a threat to human health or the environment.

X2.2.4 *Preliminary Assessment/Site Inspection (PA/SI)*—Phase of an initial site investigation for a potential Superfund (CERCLA) site generally involving review of readily available prior information (preliminary assessment) and a site visit to obtain samples for chemical analysis (site inspection). U.S. EPA **(25)**, **(26)** provide guidance on PAs, and U.S. EPA **(27)** provides guidance on SIs.

X2.2.5 *Remedial Investigation/Feasibility Study (RI/ FS)*—Detailed site investigation at a Superfund site to develop data for developing and evaluating effective remedial alternatives (RI) and selecting an alternative (FS). U.S. EPA **(28)**, **(29)** provide guidance on RI/FSs.

X2.3 The various phases of RCRA and CERCLA investigations relate to the ESC process in the following ways:

X2.3.1 Completion of a RCRA RFA or CERCLA PA/SI is necessary to make a determination that contamination is serious enough to initiate the ESC process (Step 1a in Fig. 1).

X2.3.2 ESC Phase I and II investigations are roughly equivalent to a RCRA RFI and the RI phase of a CERCLA RI/FS. The ESC process can be extended into the CERCLA FS or RCRA CMS phases, but this provisional guide does not specifically address how to do this.

X3. SELECTION AND USE OF MULTIPLE, COMPLEMENTARY INVESTIGATION METHODS

X3.1 The ESC process represents a significant shift from the approach to environmental site characterization that has evolved since the 1970s as a result of interaction between the regulatory, scientific, and applied environmental consulting communities. Individual elements of the process are currently used in conventional site characterization. ESC differs from conventional site characterization by formally combining the following relatively recent technological developments:

X3.1.1 *Advances in Noninvasive and Minimally Invasive Technologies*—Improvements in instrumentation and signal processing have reduced the cost and turnaround time for data interpretation for a variety of surface and borehole geophysical techniques. Similar advances have occurred in direct push soil and ground-water sampling equipment and by adaptation and enhancements of standard geotechnical cone penetration technology for subsurface physical and chemical characterization. The variety of noninvasive and minimally invasive techniques that are available allow use of multiple, complementary measurement and sampling methods for three-dimensional characterization of the subsurface to an extent that would be prohibitively expensive using conventional drilling and monitoring well installation.

X3.1.2 *Advances in Chemical Field and Laboratory Analytical Technologies*—Improvements and miniaturization of standard laboratory chemical analytical instruments, including automation of sample analyses and improvements in software for analyzing instrument signals and the development of a wide variety of field chemical test kits for environmental contaminants makes "real time" (minutes to days) sample data results possible during a single mobilization for three-dimensional mapping of subsurface contamination. Mobile laboratories or use of overnight delivery services to fixed laboratories mean that there are essentially no limits on the quality of chemical data that are used for on-site decisionmaking during the ESC process.

X3.1.3 *Advances in Data Analysis and Management Technologies*—Miniaturization and increased computing power, and the availability of a wide range of software for environment data management, visualization and analysis, makes field data compilation, reduction, and interpretation possible.

X3.2 The ESC process is not technology specific, but is designed to optimize selection and use of the most appropriate technologies for a particular site. The process has a bias towards noninvasive and minimally invasive methods, but use of any method that is not producing good results should be discontinued during a field mobilization. The principle of multiple, complementary measurements can be applied in two different ways: (*1*) use of multiple methods to measure the parameter or site quality of interest, and (*2*) use of the same method by more than one individual.

X3.2.1 *Multiple Methods For Geologic and Hydrologic Characterization*—Agreement between different methods or approaches used to obtain the same information increases the assurance the measurements accurately represent what is actually in the subsurface. When they disagree, it is an indication that additional investigation is required to explain the differences. Examples of complementary methods include: (*1*) use of multiple geophysical surveys (such as, ground penetrating radar, electromagnetic induction and magnetic surveys) to detect areas of buried waste, (*2*) collection of continuous soil cores adjacent to an electronic cone penetration boring to correlate ECPT log with actual lithology, (*3*) use of multiple borehole geophysical logs (resistivity, induction, gamma) in a single borehole, (*4*) use of multiple geochemical constituents and isotopes to differentiate aquifers, and (*5*) analysis of the same soil or ground water sample using different analytical methods. Multiple methods do not necessarily have to be used at all sampling points. For example, if soil cores adjacent to two or three ECPT logs show good correlation between logs and major lithologic units, complementary boring may be discontinued. Conversely, if ECPT show poor correlation, with visually logged lithologic units, more careful evaluation of the ECPT sensor outputs (tip, sleeve and pore pressure) may be required, or the lithologic log descriptions may need to be evaluated to see whether inaccurate descriptions or poor recovery are causing the poor correlation. The same would be true for any innovative technology that could potentially provide information with greater ease or at a lower cost. Complementary methods would be used along with the innovative technology enough times to determine whether it can continue as a stand-alone method at the site. Caution should be used in applying innovative technology to assure that it is based upon sound principles and carried out by reputable professionals.

X3.2.2 *Multiple Individuals Using Same Method*—The same principle applies when different individuals use the same method: agreement increases confidence in the measurement or observation and disagreement requires further investigation to resolve differences. Examples of this complementary use of the same method include: (*1*) two geophysical surveys, such as gravity, designed by two individuals or groups using the same or different instruments, (*2*) preparation of two separate lithologic logs of the same soil or rock cores by two individuals, and (*3*) analysis of the same soil or ground water sample using the same method in a field laboratory and an EPA-approved CLP laboratory.

X3.3 *ASTM Guidance on Field Investigation Methods*—Guide D 5730 includes an index of more than 400 ASTM guides, practices, and test methods that may be of use for environmental site characterization. Major ASTM guides that may be of value in selecting field characterization methods and conducting field investigations include (in Annual Book of Standards Vol 04.09, unless otherwise indicated):

D 4043 Guide for Selection of Aquifer—Test Field and Analytical Procedures in Determination of Hydraulic Properties by Well Techniques (Vol 04.08)

D 4448 Guide for Sampling Groundwater Monitoring Wells (Vol 11.04)

D 4687 Guide for General Planning of Waste Sampling (Vol 11.04)

D 4696 Guide for Pore-Liquid Sampling From the Vadose Zone (Vol 04.08)

D 4700 Guide for Soil Sampling from the Vadose Zone (Vol 04.08)

D 5088 Practice for Decontamination of Field Equipment Used at Nonradioactive Waste Sites

D 5092 Recommended Practice for Design and Installation of Ground Water Monitoring Wells in Aquifers

D 5126 Guide for Comparison of Field Methods for Determining Hydraulic Conductivity in the Vadose Zone

D 5254 Practice for the Minimum Set of Data Elements to Identify a Ground Water Site

D 5314 Guide for Soil Gas Monitoring in the Vadose Zone

D 5408 Guide for the Set of Data Elements to Describe a Ground-Water Site, Part 1—Additional Identification Descriptors

D 5409 Guide for the Set of Data Elements to Describe a Ground-Water Site, Part 2—Physical Descriptors

D 5410 Guide for the Set of Data Elements to Describe a Ground-Water Site, Part 3—Usage Descriptors

D 5434 Guide for Field Logging of Subsurface Explorations of Soil and Rock

D 5474 Guide for Selection of Data Elements for Ground-Water Investigations

D 5518 Guide for Acquisition of File Aerial Photography and Imagery for Establishing Historic Site—Use and Surficial Conditions

D 5521 Guide for Development of Ground-Water Monitoring Wells in Granular Aquifers

D 5753 Guide for Planning and Conducting Borehole Geophysical Logging

D 5777 Guide for Using the Seismic Refraction Method for Subsurface Investigations

D 5792 Practice for Generation of Environmental Data Related to Waste Management Activities: Development of Data Quality Objectives (Vol 11.04)

D 5903 Guide for Planning and Preparing for a Ground-Water Sampling Event

D 5911 Practice for a Minimum Set of Data Elements to Describe a Soil Sampling Site

D 5980 Guide for Selection and Documentation of Existing Wells for Use in Environmental Site Characterization and Monitoring

D 6001 Guide for Direct Push Ground Water Sampling for Geoenvironmental Investigations

D 6067 Guide for Using the Electronic Cone Penetrometer for Environmental Site Characterization

X3.4 *Field Sampling and Analysis*—The importance of rapid turnaround of chemical sample analysis for on-site decisionmaking in the ESC process requires special attention to chemical QA/QC issues. The optimization of sampling locations allowed by the ESC process is generally accompanied by selection of analytical methods that provide high levels of data quality. Appendix X2 in Provisional Guide PS 3 discusses data quality levels for field chemical characterization in more detail. It is the responsibility of the ESC provider to address and resolve issues of field data quality with the appropriate regulatory authority in the Phase I and Phase II Work Plans. Examples of such issues include:

X3.4.1 *Comparability of Direct Push and Conventional Monitoring Well Ground-Water Samples*—Direct push ground water samples may be more turbid than samples from properly developed monitoring wells so issues of filtration, including size of filters need to be addressed in the ESC Work Plan protocols.

X3.4.2 *Comparability/Quality of Field, Mobile Laboratory, and CLP Laboratory Analytical Results*—Where volatile contaminants are involved, field analytical methods may well provide more accurate measurements than fixed laboratory. For other constituents this may or may not be the case. Also, some state regulatory authorities require use of certified laboratories for chemical analyses. The field sampling and analysis protocols in the ESC work plan may need to specifically address these issues.

X4. QUALITY CONTROL FOR GEOLOGIC AND HYDROLOGIC CHARACTERIZATION

X4.1 An important feature of an ESC dynamic work plan is incorporation of quality control (QC) procedures for geologic and hydrologic characterization activities into the QA/QC plan, in addition to the more established QA/QC procedures for collection, handling and analysis of samples for chemical analysis. The variability of geologic and hydrologic systems means that it is difficult to assign statistically-based confidence levels to measurements. Consequently, quality control focuses on the use of multiple complementary methods or multiple individuals using the same method, as described in X3.2. In addition to the usual QA/QC procedures for chemical sampling and analysis, an ESC QA/QC plan should include sections on quality control and quality assurance for other field activities including data management (Appendix X5), geologic, hydrologic, geophysical, and spatial control procedures (that is, accuracy of horizontal and vertical location of measurement and sampling points). These procedures should be formulated in such a way as to all allow the person on the project team responsible for quality assurance to determine whether procedures were followed. In the illustrative examples for geologic characterization that follow, the QC procedure is described first, followed by the QA check in the form of one or more questions that can be answered yes or no. Quality control and QA checks can be developed in a similar manner for hydrologic and geophysical characterization and spatial control. Once an ESC provider has developed QC procedures in these areas, the appropriate ones can generally be incorporated into the QA/QC plan for a particular site with little or no modification.

X4.2 *Illustrative Examples of Geologic Characterization Quality Control*[11]

X4.2.1 Identification of marker stratigraphic units and drawing of geologic cross sections based on original interpretation by core technical team members with expertise in geology and hydrology of all aerial photographs, geologic surface outcrops, wells, borings, and test holes within the immediate area of the site (within at least two- or three-mile radius). QA check: have original cross sections been prepared, and how many people were involved in preparing them?

X4.2.2 Initial use of continuous coring (including appro-

priate measures to prevent possible cross contamination, with selected cores logged separately in detail by at least two qualified project team members, and differences reconciled to confirm or modify hypotheses for the stratigraphic sequence and to identify water-bearing units. QA check: are there two separate core logs and a reconciled one if they differ significantly?

X4.2.3 Use of standard borehole log forms with identification of the minimum information that will be recorded. QA check: does a random check of one or more log forms indicate that all minimum information has been recorded?

X4.2.4 Initial field or laboratory particle-size analysis of selected logged units to check accuracy of field soil textural descriptions (USDA, unified soil classification, Wentworth) in field logs, until there is a good match between manual textural classification and particle size analyses. Borehole log descriptions revised when field textural classification is in error. QA check: have initial field logs been corroborated and corrected by particle-size analysis?

X4.2.5 Use of the continuous geologic logs as initial controls for multiple geophysical borehole logs and electronic cone penetrometer (ECPT) logs. QA check: are geophysical logs available for at least one continuous geologic log, and is there an adjacent ECPT log?

X4.2.6 Use of surface geophysical measurements, if technically appropriate, to identify continuities and discontinuities of major stratigraphic boundaries or units. QA check: have surface geophysical measurements been used identify continuities and discontinuities of major stratigraphic boundaries or units? If no, have surface geophysical measurements been determined to be technically inappropriate?

X4.2.7 Use of continuous coring or direct push geophysical measurements to confirm possible stratigraphic discontinuities identified by surface geophysical measurements. QA check: have any surface geophysical surveys identified possible stratigraphic discontinuities, and if so, have they been checked by coring or borehole geophysical observations?

---

[11] Adapted from: *Master Work Plan: Expedited Site Characterizations at CCC/USDA Sites in Nebraska*, Applied Geosciences and Environmental Management Division, Environmental Research Division, Argonne National Laboratory, January, 1994.

X5. QUALITY CONTROL FOR FIELD DATA AND COMPUTER RECORDS[12]

X5.1 Computer records are files that contain either prior data, data acquired in the course of ESC field work, or the results of processing and interpreting these data. The primary quality objective with regard to computer records is to maintain the security and integrity of characterization data, from initial acquisition through final archival, while providing characterization team members with ready access to information contained in computer records.

X5.2 *Quality Assurance Responsibility*—The data manager will be responsible for verifying that computer records of field data are assembled and maintained according to the QA standards defined in this Appendix. Each team member who is involved in acquiring computer-recorded field data will be responsible for recording the original computer records according to procedures defined in X5.3 and for making these records available to the data manager in the

prescribed formats. After field data have been verified by the data manager assuring the continued quality of computer records is the responsibility of the computer group.

X5.3 *Field Data*—The following are requirements for field records that are generated, stored, or transferred in the form of computer files. This includes data recorded directly by means of computer systems, such as geophysical well logs, seismic recordings, and electromagnetic soundings, and data that are transferred to computer systems for analysis or storage, or both, such as survey information, geologic well logs, processed geophysical data, and transcribed chemical analysis data.

X5.3.1 *Original Records*—The primary QA concern with regard to original field data is that accurate and secure copies of all data are preserved and that no data are lost or altered because of transcribing data, mishandling, transfer of computer files, or computer system failure,. Any original records that are entered directly onto a computer hard disk or memory will be transcribed in a timely fashion after collection onto a removable storage medium, such as floppy discs or tapes, and these copies will be handled as original or

---

[12] Taken, with minor modifications from: *Master Work Plan: Expedited Site Characterizations at CCC/USDA Sites in Nebraska*, Applied Geosciences and Environmental Management Division, Environmental Research Division, Argonne National Laboratory, January 1994.

master data files. If the original recording is made on a removable medium, these originals will be considered to be original data master files, and all further processing of the data will be done with working copies made from these master files.

X5.3.2 *Transcribed Data*—Field data that are transcribed from written form into computer files such as survey locations, analytical results, or geologic logs will be checked and verified for accuracy at the time of transcription by the team's data manager. Following verification, a copy of the computer file will be saved on a removable medium such as a master file together with copies of paper records from which the master files were transcribed. These master files will then be handled in the same manner as original master data files.

X5.3.3 *Master Data Files*—Master data files, defined as either original data recorded on transportable media or transcriptions of original data onto transportable media, will be stored in a secure work or storage location that is separate from working copies of these files. Copies of original supporting paperwork such as field notes, laboratory reports, observer reports, and paper logs will be maintained with the master data files, either in the form of duplicate paper copies or as digital images. Master data files will be labeled in a manner that includes the location name of the data source. If corrections are made to original paperwork that has been used to generate a master file, the team member initiating the change will be responsible for informing the data manager who will document the correction of the original master file, initial the original paperwork to ensure that all changes are made, and retain a copy. The corrected master file will be saved with the other project master files and labeled as a corrected version. The original master file will be saved in a separate folder of corrected master files.

X5.4 *Data Processing*—To ensure that master data files are not altered or damaged all processing of field data will be performed by using duplicate copies of master files for input, not master data files. Quality assurance of processed data consists of ensuring that processing is conducted with approved methods that preserve data integrity throughout the processing sequence and that processed data are quickly made available to technical team leaders for review and verification. The QA procedures are discussed below.

X5.4.1 *Data Processing Standards*—Data will be processed by technical or computer group personnel using benchmark-tested software and systems. Documentation of software used in processing will be maintained and available for reference. The computer group will be responsible for keeping track of software and upgrades that meet quality standards and for providing documentation of software. Field data will initially be preprocessed into a common format that can be imported to processing, display, and database programs. This normally is a spreadsheet format, except in the case of certain geophysical data, for which industry-standard formats are specified. Data from these files will be imported to analysis and display programs to produce graphic displays or results in the field. Results will be submitted to technical team leaders for validation. During all stages of data processing, records will be backed up daily on a storage medium that is independent of the computer on which the processing takes place. This medium may be an

external hard drive, a removable disk, a floppy disk, or a tape.

X5.4.2 *Review and Correction of Errors*—To minimize errors in processed field data, processing results will be distributed daily for review by team members. Where possible, results will be presented in graphic format to facilitate review and detection of errors. Maps will be distributed showing the locations of field activities such as soil borings, cone penetrometer sites, and soil or water samples. Analytical results will normally be in the form of listings, but values will also be shown on maps and cross sections where possible. Subsurface geologic and hydrologic data will be displayed in the form of well logs and cross sections. Surface geophysical data will be displayed on maps or cross sections, or both, as appropriate. These displays may be made from interim or temporary processing results, provided that all displays indicate the version or date, or both, of the processing file from which the display was generated. Team leaders will be responsible for notifying computer support personnel immediately of any erroneous or inconsistent results seen in these displays. Corrections to master data files will be documented and approved by the data manager by following the procedures in X5.3.

X5.5 *Final Data Processing Results*—Whenever final results of data processing are reported in a form other than listings of raw field data, the computer files that generated these results will be saved with appropriate labeling to indicate a final result. This includes files that generate graphic displays, database files, spreadsheet listings, and AutoCAD maps, for example.

X5.6 *Quality Control Tracking*—At various stages in the acquisition, processing, and reporting sequence data files will be processed into database formats for compilation and display. To ensure that only accurate data are included, each data entry or each group of entries from a single source into the database will be accompanied by an index indicating the level of QC checks through which the data have been passed. Database entries include the names of the data manager or other individual who verified the entries. The following indices will be used:

X5.6.1 *QC Level 1*—Data entries are tentative and subject to change (for example, unsurveyed boring locations, unverified analytical results, extrapolated stratigraphic boundaries). This category is used in the field to allow immediate display of daily results.

X5.6.2 *QC Level 2*—The data entry has been verified by the data manager.

X5.6.3 *QC Level 3*—The data entry has been verified by the data manager, and the ground location of the data source has been confirmed by an official survey.

X5.6.4 *QC Level 4*—All checks have been completed. Processed results have been reviewed and approved by technical team leaders for inclusion in reports.

X5.7 *Data Archival*—All master files, files of final results, files of significant intermediate or auxiliary results, and copies of databases will be archived in permanent storage together with supporting paperwork and documents. Archival is on a removable medium which is stored in addition to data saved on standard backup systems.

X6. EXAMPLE OF THE ESC PROCESS

X6.1 *Introduction:*

X6.1.1 The following example illustrates the use of the ESC process by Argonne National Laboratory at a landfill in New Mexico where unauthorized dumping of hazardous waste was suspected (Burton (**30**), Burton et al., (**1**)). With minimal prior information, on-site decisionmaking and use of multiple, complementary characterization methods allowed identification of contaminant sources, potential migration pathways, and contaminant distribution in two major field mobilizations without the installation of a single monitoring well. The ESC investigation at this particular site demonstrated no significant movement of contaminants from the landfill trenches and pits or risk of ground water contamination. The appropriate regulatory agencies accepted a recommendation of no remedial action or ongoing monitoring.

X6.2 *Landfill Setting and History:*

X6.2.1 The Flora Vista landfill, located on federal land in northwestern New Mexico and owned by the Bureau of Land Management, covers 13 acres. It was leased and used as a modified sanitary landfill (not covered daily) by San Juan County from July 1978 through 1989. In 1986 the New Mexico Environmental Improvement Division alleged that large quantities of petroleum, industrial, or other hazardous waste were deposited in septage waste pits at the site without BLM authorization between the 1970s and August 1985.

X6.2.2 A preliminary assessment (PA) in 1986 and a site investigation (SI) by separate private contractors generated conflicting maps of previously covered trenches and pits. Surface and subsurface soil samples collected during the PA and SI programs indicated the presence of hazardous compounds (including tetrachloroethylene, toluene, benzene, acetone, 1,1,1-trichloroethane) in one septage pit area of the landfill. The SI results were interpreted by the contractor as indicating possible lateral and vertical migration from this pit and presented a potential hazard to human health and the environment. The contractor recommended installation of five monitoring wells at the site. On the basis of the results of the PA and SI, operation of the landfill was closed. At this stage Argonne was asked by BLM to initiate an ESC project to determine if remediation of the landfill was required.

X6.3 *Phase I Investigation:*

X6.3.1 *Prior Data and Initial Site Visit*—Prior data on the disposal history of the landfill were extremely sparse at the beginning of the ESC project. Maps of previously covered trenches and pits generated during the PA and SI were conflicting. Information on regional geology indicated that thick alluvium (>100 ft) with clays at 30 to 50 ft, covered the Tertiary Nacimiento Formation, a hard, lithified, fairly impermeable sandstone. The nearest subdivision and the Animals River were about 1.8 miles southeast of the site and the nearest residence 1.2 miles south. At the initial site visit the core technical team found that the area had been fenced and covered with approximately 2 ft of sandy soil. The only surface indication of the location of pits and trenches at the site was an interior fence in the southwest corner where a septage pit was supposedly located.

X6.3.2 *Phase I Objectives*—Two major objectives based on the preliminary conceptual site model were defined for the Phase I investigation: (*1*) map the now-buried source areas (trenches and pits), and (*2*) define migration pathways for contaminant movement from these sources areas (bedrock surfaces, clay lenses, surface drainage patterns, depth to water table). When the source and migration routes were delineated, an optimal sampling project in Phase II could be designed to fully test migration from the sources.

X6.3.3 *Source Area Characterization*—Surface geophysical surveys used to map trenches and pits included magnetometry to detect possible buried metallic waste and trench boundaries (Fig. X6.1), and two frequency domain electromagnetic (EM 31 and EM 34) surveys to detect possible contaminant plumes associated with oil field wastes that might have been placed in the landfill. The chief emphasis was on the EM 31, with a penetration of about 15 ft, for mapping landfill trenches and pits (Fig. X6.2). The results of the EM 34 survey (Fig. X6.3), with a depth of exploration of about 50 ft, could be interpreted as either conductive contaminant plumes or clay layers, based on the preliminary conceptual site model. The three surveys were combined to establish probable locations of shallow and deep trenches and pits (Fig. X6.4). Significant interpretations by the ESC core technical team of the surface geophysical surveys included: (*1*) the possibility of deep contaminant source areas in the north central and southeast areas of the landfill, and (*2*) evidence of shallow contamination that extended beyond the interior fenced area in the southwest corner that was supposed to delineate the area of the septage disposal pit.

X6.3.4 *Migration Pathway Characterization*—Seismic refraction, electrical resistivity, and time-domain electromagnetic (TDEM) surveys were used to evaluate potential subsurface migration pathways. After two days in the field, it was evident that the seismic refraction survey was mapping a bedrock surface at fairly shallow levels (15 to 45 ft) beneath the landfill surface. This relatively shallow bedrock had not been predicted by the regional geology data, but inspection of several nearby bedrock outcrops indicated that the Nacimiento Formation could be present as an erosional bedrock surface at the landfill site. Because this surface might play an important role in unsaturated flow from the trenches and pits the field project was altered to include a more comprehensive seismic refraction survey within and immediately around the landfill. Seismic survey profiles provided 76 depths to bedrock, which were contoured to reveal an erosional surface with potential subsurface migration pathways at the interface between the less permeable bedrock and overlying unconsolidated materials having a generally southwesterly direction (Fig. X6.5). The seismic surveys combined with electrical resistivity, and time-domain electromagnetic (TDEM) studies indicated that the water table was about 150 ft beneath the surface. These survey were confirmed when two existing wells were located near the site and the depth to water was measured at 150 ft. The TDEM and EM-31 surveys identified clay layers (aquitards) in the subsurface,

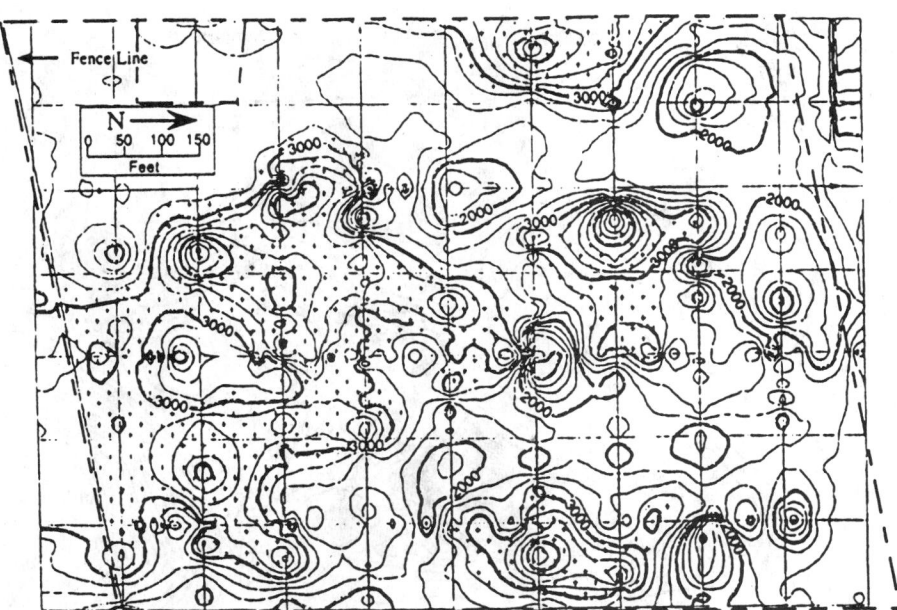

FIG. X6.1 Total Field Magnetic Contour Map for Flora Vista Landfill (Values are in gammas. Areas 200 gammas above background are marked with + symbol and represent solid waste trench locations.)

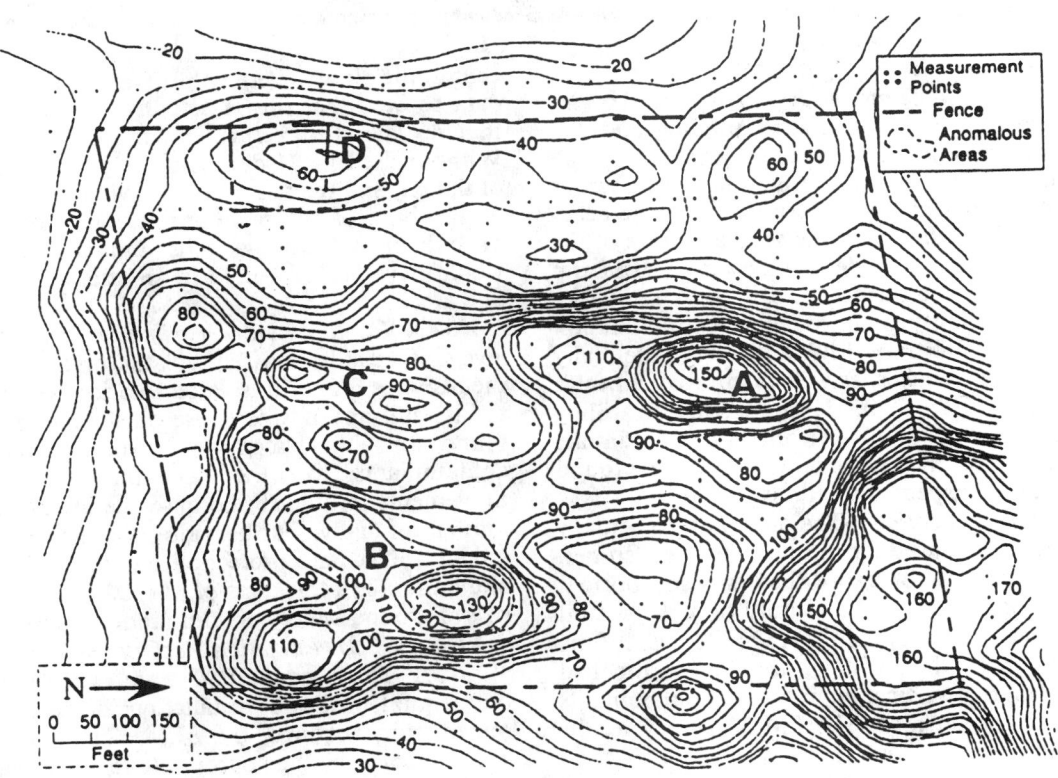

FIG. X6.2 EM 31 Contour Map for Flora Vista Landfill (Values are in millimhos per meter. Anomalous areas A–D are possible pit or trench areas.)

dipping in a direction generally similar to that of the buried bedrock surface. A surface topographic survey of the landfill indicated a general southerly direction of surface runoff with runoff from the eastern edge to the east (Fig. X6.6).

X6.4 *Phase II Investigation:*

X6.4.1 *Phase II Objectives*—The conceptual site model, represented as a map showing potential trench and pit sources, potential surface, and subsurface migration pathways, was used to identify soil and soil boring sampling points to determine the distribution and concentration of

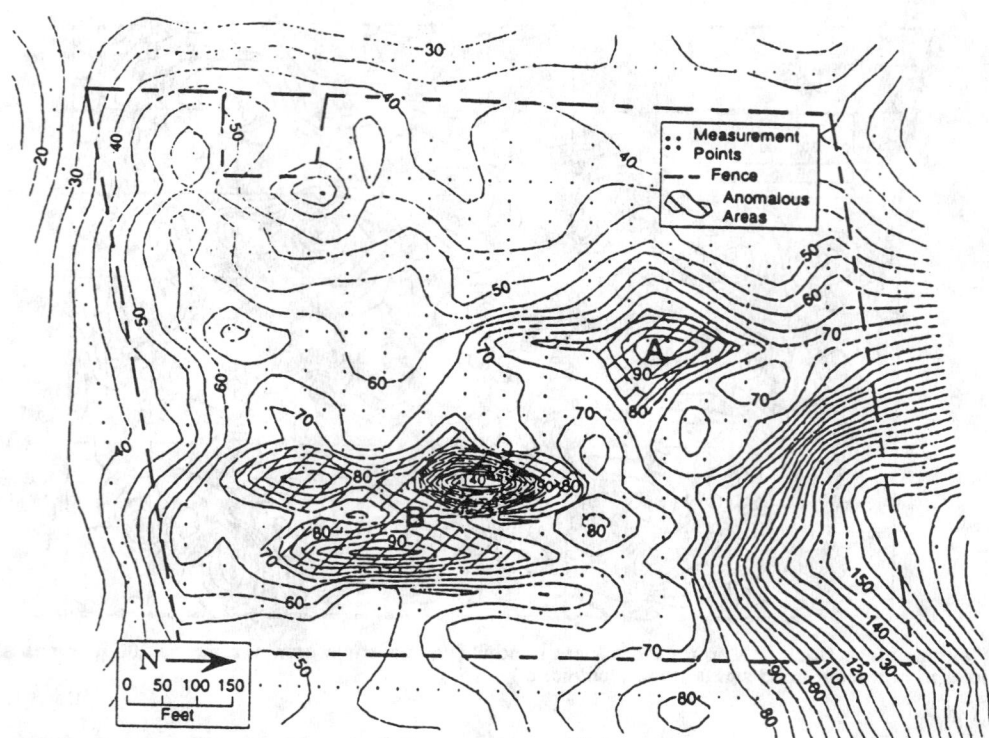

FIG. X6.3   EM 34 Contour Map for Flora Vista Landfill (Values are in millimhos per meter. Anomalous areas A and B are trench locations.)

contaminants in the subsurface (Fig. X6.7).

X6.4.2 *Soil Boring and Sampling Results*—At the beginning of the Phase II field investigation sampling and soil boring locations were chosen to observe and sample all major migration pathways without drilling into a trench or pit. Bedrock depths determined by deep drilling were accurate within 10 ft and substantiated the general nature of the bedrock surface. The clay layers identified by the TDEM and EM 31 surveys were also confirmed by drilling. This restriction avoided the possibility of introducing contamination into the subsurface by arbitrary drilling through potentially contaminated pits and trenches, as well as improving health and safety protection of the staff. Hand augering or drilling carefully into the near surface was performed to substantiate the presence of a pit or trench boundaries which were found to be within 5 ft of the expected locations.

X6.4.3 *Waste Pit Delineation*—The only pit area actually drilled to depth was the fenced area in the southwest corner. During the SI, the contractor had drilled two soil borings (WB1 and WB2, Fig. X6.8) in the vicinity of a septage pit that was open to the surface at the time. On the basis of the analytical results from this drilling, the contractor had proposed that lateral and vertical migration of contaminants had occurred from the pit area. The EM 31 survey results combined with information for 8 borings within and adja-

cent to the pit, coupled with reexamination of the contractor data showed that this is actually one large septage pit and that no lateral or vertical migration has occurred from it. Without the EM 31 data, outlining the possible boundaries of the pit, and the seismic data giving an idea of the location of the base of the pit, considerable time and money for drilling of monitoring wells and additional soil borings would have been required in an attempt to understand this one area.

X6.4.4 *Contaminant Distribution and Risk*—Only low levels of volatile and semivolatile organic compounds and metals were found in isolated locations of the septage pit of the landfill and nowhere else on the site. No evidence was found for either lateral or vertical movement of these contaminants from the waste pit or landfill, and future movements is unlikely because the area receives <8 in. of rainfall annually. Furthermore, the depth to ground water (about 150 ft) and protection by 10 to 20 ft of clay in the alluvium plus a thickness of 85–100 ft of bedrock meant the risk of ground water contamination was negligible.

X6.5 *ESC Investigation Recommendations:*

X6.5.1 The Argonne ESC team recommended closure of the landfill with no installation of monitoring wells or remediation required by BLM. These recommendations were accepted by the appropriate regulatory agencies.

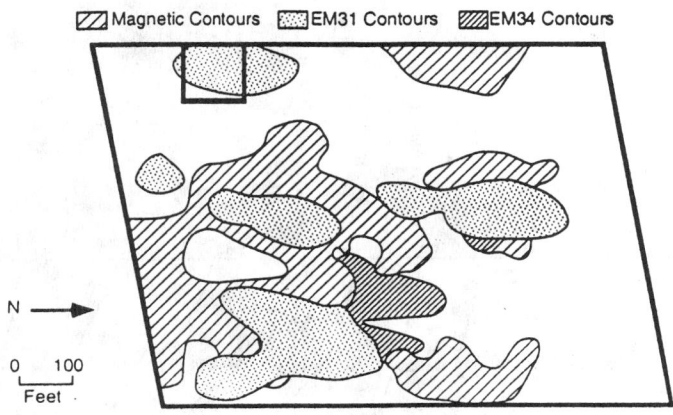

□ Magnetic Contours   ▦ EM31 Contours   ▨ EM34 Contours

FIG. X6.4 Magnetic and Electromagnetic Data Were Combined to Establish Probable Locations of Shallow and Deep Waste Materials

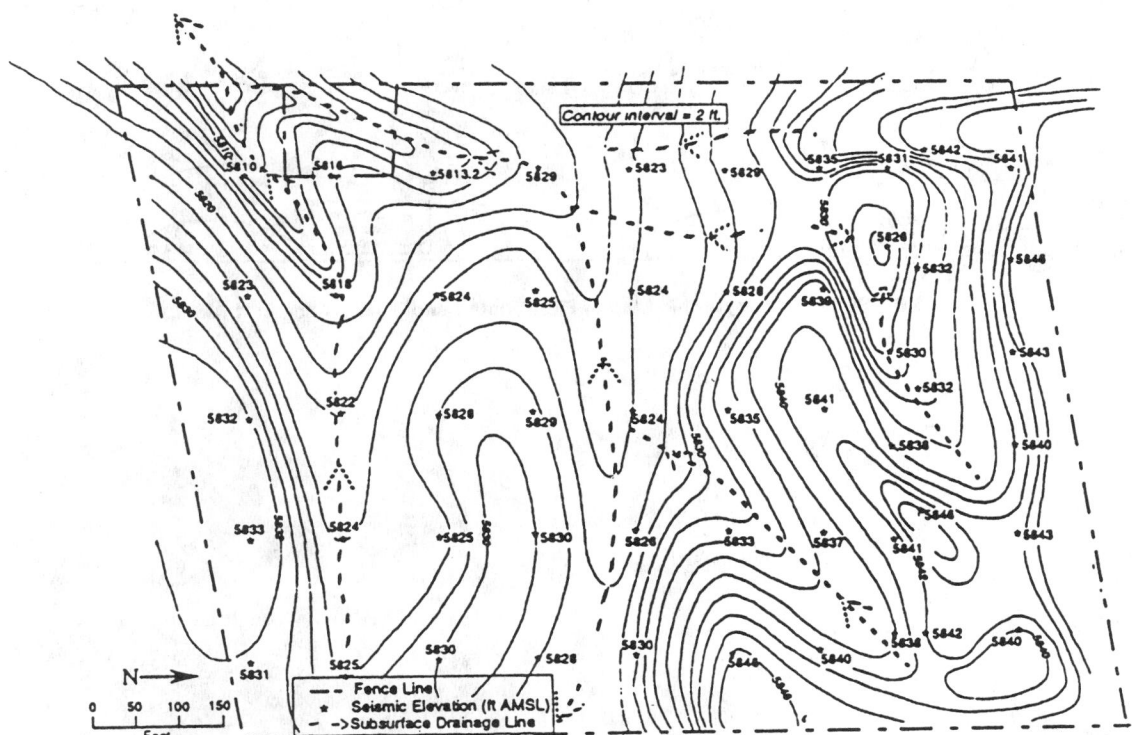

FIG. X6.5 Bedrock Topography Map of Flora Vista Landfill Showing Potential Subsurface Migration Pathways (Elevation in ft AMSL)

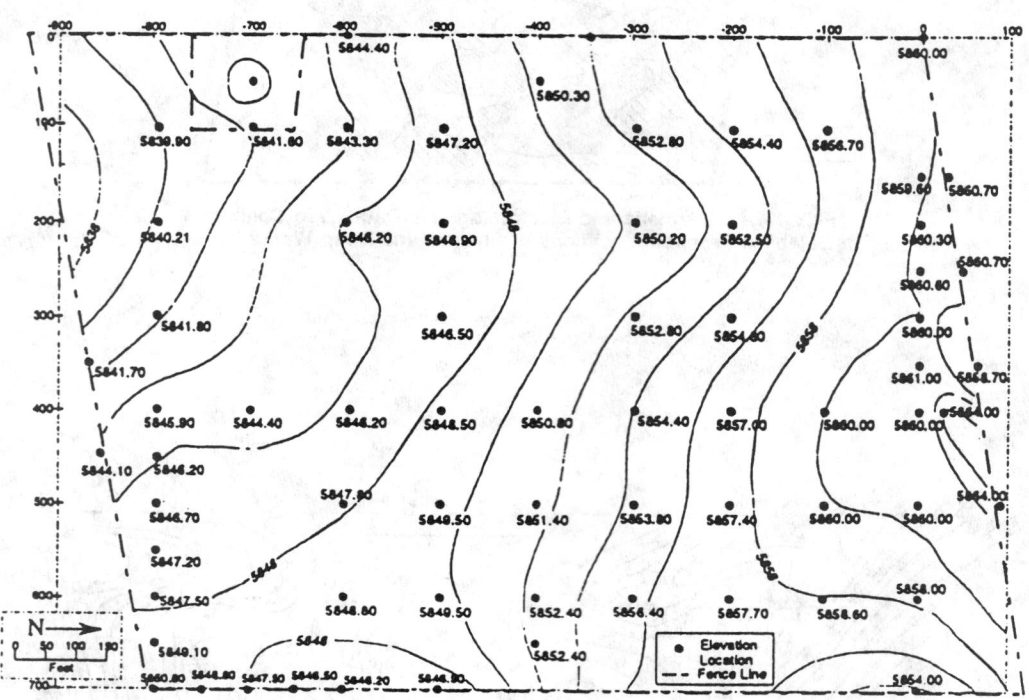

FIG. X6.6   Surface Topographic Map of Flora Vista Landfill (Elevation in ft AMSL)

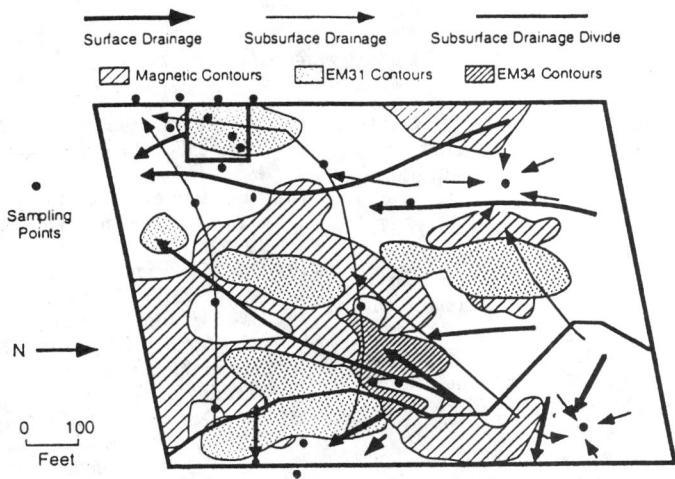

FIG. X6.7 Source Areas and Potential Surface and Subsurface Migration Pathways at Flora Vista Landfill as Defined by Geologic and Geophysical Surveys, with Soil and Soil Boring Sampling Points for Testing Contaminant Migration

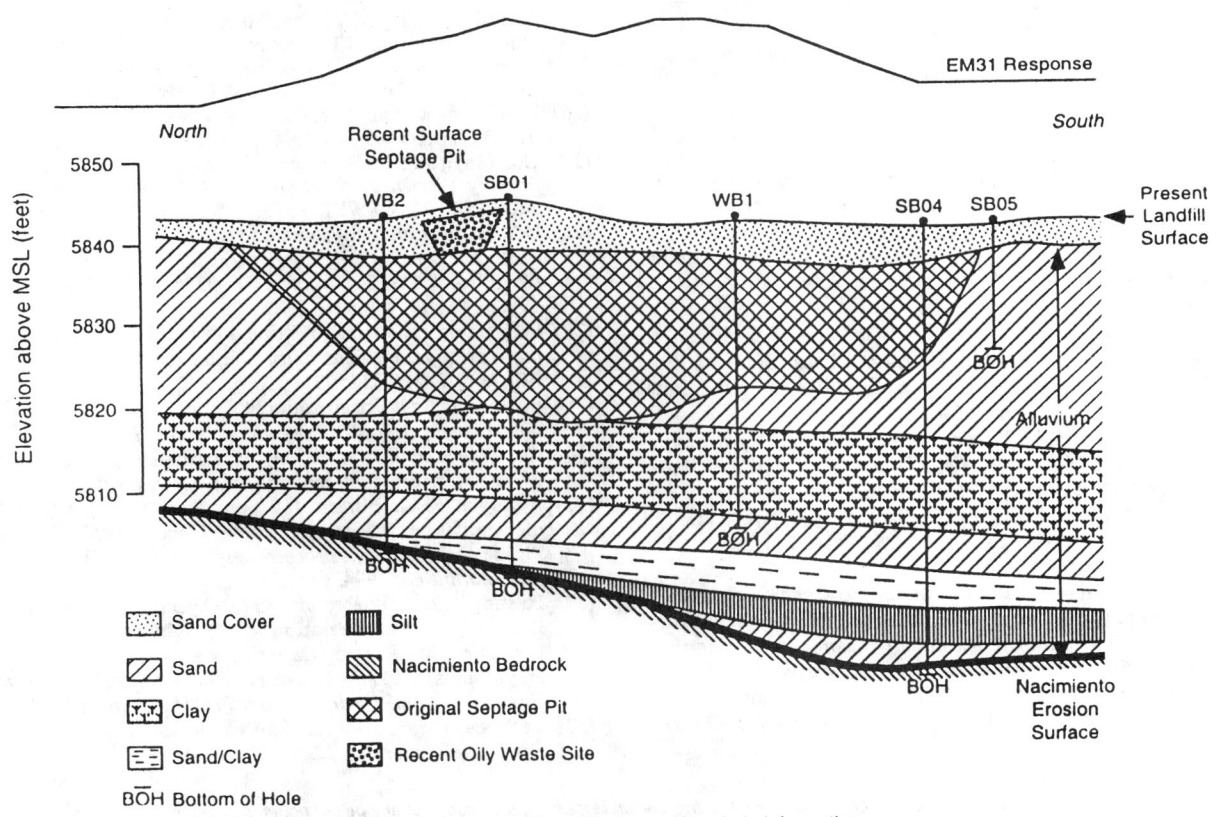

FIG. X6.8 Drilling Confirmation of the One Waste Pit Predicted by the EM 31 Survey at Flora Vista Landfill

REFERENCES

(1) Burton, J. C., et al., "Expedited Site Characterization: A Rapid, Cost-Effective Process for Preremedial Site Characterization," *Superfund XIV*, Vol II, Hazardous Materials Research and Control Institute, Greenbelt, MD, 1993, pp. 809–826.

(2) Burton, J. C., Walker, J. L., Aggarwal, P. K., and Meyer, W. T., "Argonne's Expedited Site Characterization: An Integrated Approach to Cost- and Time-Effective Remedial Investigation," paper presented at 88th Annual Meeting, Air and Waste Management Association, 1995, 27 pp.

(3) Bevolo, A. J., Kjartanson, B. H., and Wonder, D., Ames Expedited Site Characterization Demonstration at the Former Manufactured Gas Site, Marshalltown, Iowa. Ames National Laboratory Document IS-5121, 1996.

(4) Savannah River Site, RFI/RI and BRA Report for the D-Area Oil Seepage Basin (631-G), Rev 0, (NRSC-RP-96-00154, First) Savannah River Site, Aiken, S. C., 1996.

(5) Ames Laboratory, Report on the Results of the Expedited Site Characterization at the Czechowice Refinery, Ames Laboratory, Ames, IA, 1996.

(6) Burton, J. C., "Expedited Site Characterization for Remedial Investigations at Federal Facilities," *Proceedings FER III & WM II Conference and Exhibition*, 1994, pp. 1407–1415.

(7) Starke, T. P., Purdy, C., Belencan, H., Ferguson, D., and Burton, J. C., "Expedited Site Characterization at the Pantex Plant," paper presented at ER-95 Conference, Denver, CO, 1995.

(8) Chamberlain, T. C., "The Method of Multiple Working Hypotheses," *Science*, 15(92); 1890 reprinted in *Science* May 7, 1965, Vol 148, pp. 754–759.

(9) Chamberlain, T. C., "The Method of Multiple Working Hypotheses," *J. Geology* Vol 5, 1897 pp. 837–848.

(10) Bjerrum, L., "Some Notes of Terzaghi's Method of Working," *From Theory to Practice in Soil Mechanics: Selections from the Writings of Karl Terzaghi*, Wiley, New York, 1960, pp. 22–25.

(11) Peck, R. B., "Advantages and Limitations of the Observational Method in Applied Soil Mechanics," *Géotechnique*, Vol 19(2), 1969, pp. 171–187.

(12) Mark, D. L., et al., "Application of the Observational Method to an Operable Unit Feasibility Study—A Case Study," *Proc. Superfund '89*, Hazardous Material Control Research Institute, Silver Spring, MD, 1989, pp. 436–442.

(13) Brown, S. M., Lincoln, D. R., and Wallace, W. A., *Application of the Observational Method to Remediation of Hazardous Waste Sites*, CH2M Hill, Bellevue, WA, 1989, 16 pp.

(14) Brown, S. M., Lincoln, D. R., and Wallace, W. A., "Application of the Observational Method to Hazardous Waste Engineering," *J. Management Eng.* (ASCE), Vol 6, 1990, pp. 479–500.

(15) Holm, I. A., "Strategies for Remediation," *Geotechnical Practice in Waste Disposal*, Chapman & Hall, London, 1993, pp. 289–310.

(16) Robbat, Jr., A., and Johnson, R., *Adaptive Sampling and Analysis Programs for Soils Contaminated with Explosives, Case Study: Joliet Army Ammunition Plant*, Environmental Technology Division, Army Environmental Center, Aberdeen Proving Ground, MD, 1996, 16 pp.

(17) U.S. Environmental Protection Agency (EPA), Record of Decision: Marine Corps Air Station, Yuma, Arizona, U.S. EPA, Region IX, 1995.

(18) U.S. Environmental Protection Agency (EPA), Guidance for the Data Quality Objectives Process, EPA/QA/G-4, Final, September 1994.

(19) U.S. Environmental Protection Agency (EPA), RI/FS Streamlining, OSWER Directive No. 9344-3-06, 1989.

(20) U.S. Department of Energy (DOE), *Remedial Investigation/Feasibility Studies (RI/FS): Process, Elements and Techniques*, Office of Environmental Guidance, RCRA/CERCLA Division (EH-231), Washington, DC, 1993.

(21) U.S. Department of Energy (DOE), *RCRA Corrective Action and CERCLA Remedial Action Reference Guide*, DOE/EH-0001, Office of Environmental Guidance, RCRA/CERCLA Division, EH-231, Washington, DC, 1994.

(22) U.S. Department of Energy (DOE), *A Comparison of the RCRA Corrective Action and CERCLA Remedial Action Processes*, DOE/EH-0365, Office of Environmental Guidance, RCRA/CERCLA Division, EH-231, Washington, DC, 1994.

(23) U.S. Environmental Protection Agency (EPA), *RCRA Facility Assessment Guidance*, EPA/530-SW-86-053, (NTIS PB87-107769), 1986, 174 pp.

(24) U.S. Environmental Protection Agency (EPA), *RCRA Facility Investigation Guidance*, 4 Vol, EPA/530-SW-89-031, OSWER Directive 9502.6C (NTIS PB90-200299), 1989, 1221 pp.

(25) U.S. Environmental Protection Agency (EPA), *Guidance for Performing Preliminary Assessment Under CERCLA*, EPA/540/G-91/103, OSWER 9345.0-01A (NTIS PB92-963303), 1991, 286 pp.

(26) U.S. Environmental Protection Agency (EPA), *PA Review Checklist*, OSWER 9345.0-08 (NTIS PB93-963342), 1993, 15 pp.

(27) U.S. Environmental Protection Agency (EPA), *Guidance for Performing Site Inspections Under CERCLA* (Interim Final), EPA/540-R-92-021, OSWER 9345.1-05 (NTIS PB92-963375), 1992.

(28) U.S. Environmental Protection Agency (EPA), *Guidance for Conducting Remedial Investigation/Feasibility Study (RI/FS) Under CERCLA*, OSWER Directive No. 9355.3-01 (NTIS PB89-184625), 1988, 145 pp.

(29) U.S. Environmental Protection Agency (EPA), *Conducting Remedial Investigation/Feasibility Studies for CERCLA Landfill Sites*, EPA/540/P-91/001, 1991, 300 pp.

(30) Burton, J. C., "Prioritization to Limit Sampling and Drilling in Site Investigations," *Proceedings 1992 Federal Environmental Conference and Exhibition*, Hazardous Material Control and Research Institute, Greenbelt, MD, 1992, pp. 242–251.

(31) Aggarwal, P. K., Burton, J. C., and Rose, C. R., "Characterization of Aquifer Relationships Using Geochemical Techniques for Plume Delineation," *Proceedings FER III & WM II Conference and Exhibition*, 1994, pp. 1311–1317.

(32) Hastings, B., Hildebrandt, G., Meyers, T., Saunders, W., and Burton, J. C., "Integration of Geophysics With the Argonne Expedited Site Characterization Program at a Site in the Southern High Plains," *1995 Symposium on the Application of Geophysics to Engineering and Environmental Problems* (SAGEEP '95), 1995.

(33) Technos, Inc., *Expedited Site Characterization at the Savannah River Site*, Technos, Miami, FL, 1995, 7 pp.

Designation: E 1689 – 95

Standard Guide for
Developing Conceptual Site Models for Contaminated Sites[1]

This standard is issued under the fixed designation E 1689; the number immediately following the designation indicates the year of original adoption or, in the case of revision, the year of last revision. A number in parentheses indicates the year of last reapproval. A superscript epsilon (ε) indicates an editorial change since the last revision or reapproval.

1. Scope

1.1 This guide is intended to assist in the development of conceptual site models to be used for the following: (*1*) integration of technical information from various sources, (*2*) support the selection of sample locations for establishing background concentrations of substances, (*3*) identify data needs and guide data collection activities, and (*4*) evaluate the risk to human health and the environment posed by a contaminated site. This guide generally describes the major components of conceptual site models, provides an outline for developing models, and presents an example of the parts of a model. This guide does not provide a detailed description of a site-specific conceptual site model because conditions at contaminated sites can vary greatly from one site to another.

1.2 The values stated in either inch-pound or SI units are to be regarded as the standard. The values given in parentheses are for information only.

1.3 This guide is intended to apply to any contaminated site.

1.4 *This standard does not purport to address all of the safety concerns, if any, associated with its use. It is the responsibility of the user of this standard to establish appropriate safety and health practices and determine the applicability of regulatory limitations prior to use.*

2. Referenced Documents

2.1 *ASTM Standard:*
D 2216 Test Method for Laboratory Determination of Water (Moisture) Content of Soil and Rock[2]

2.2 *EPA Documents:*[3]
Guidance for Data Useability in Risk Assessment (Part A) Final, Publication 9285.7-09A, PB 92-963356, April 1992
Guidance for Data Useability in Risk Assessment (Part B), OSWER Directive 9285.7-09B, May 1992
Guidance for Conducting Remedial Investigations and Feasibility Studies Under CERCLA, OSWER Directive 9355.3-01, October 1988

3. Terminology

3.1 *Definitions:*

3.1.1 *background concentration, n*—the concentration of a substance in ground water, surface water, air, sediment, or soil at a source(s) or nearby reference location, and not attributable to the source(s) under consideration. Background samples may be contaminated, either by naturally occurring or manmade sources, but not by the source(s) in question.

3.1.2 *conceptual size model, n*—for the purpose of this guide, a written or pictorial representation of an environmental system and the biological, physical, and chemical processes that determine the transport of contaminants from sources through environmental media to environmental receptors within the system.

3.1.3 *contaminant, n*—any substance, including any radiological material, that is potentially hazardous to human health or the environment and is present in the environment at concentrations above its background concentration.

3.1.4 *contaminant release, n*—movement of a substance from a source into an environmental medium, for example, a leak, spill, volatilization, runoff, fugitive dust emission, or leaching.

3.1.5 *environmental receptor, n*—humans and other living organisms potentially exposed to and adversely affected by contaminants because they are present at the source(s) or along contaminant migration pathways.

3.1.6 *environmental transport, n*—movement of a chemical or physical agent in the environment after it has been released from a source to an environmental medium, for example, movement through the air, surface water, ground water, soil, sediment, or food chain.

3.1.7 *exposure route, n*—the process by which a contaminant or physical agent in the environment comes into direct contact with the body, tissues, or exchange boundaries of an environmental receptor organism, for example, ingestion, inhalation, dermal absorption, root uptake, and gill uptake.

3.1.8 *migration pathway, n*—the course through which contaminants in the environment may move away from the source(s) to potential environmental receptors.

3.1.9 *source, n*—the location from which a contaminant(s) has entered or may enter a physical system. A primary source, such as a location at which drums have leaked onto surface soils, may produce a secondary source, such as contaminated soils; sources may hence be primary or secondary.

4. Summary of Guide

4.1 The six basic activities associated with developing a conceptual site model (not necessarily listed in the order in which they should be addressed) are as follows: (*1*) identification of potential contaminants; (*2*) identification and characterization of the source(s) of contaminants; (*3*) delin-

---

[1] This guide is under the jurisdiction of ASTM Committee E-47 on Biological Effects and Environmental Fate and is the direct responsibility of Subcommittee E47.13 on Assessment of Risk to Human Health and the Environment from Hazardous Waste Sites.

Current edition approved March 15, 1995. Published May 1995.

[2] *Annual Book of ASTM Standards*, Vol 04.08.

[3] Available from Standardization Documents Order Desk, Bldg 4 Section D, 700 Robbins Ave., Philadelphia, PA 19111-5094, Attn: NPODS.

eation of potential migration pathways through environmental media, such as ground water, surface water, soils, sediment, biota, and air; (4) establishment of background areas of contaminants for each contaminated medium; (5) identification and characterization of potential environmental receptors (human and ecological); and (6) determination of the limits of the study area or system boundaries.

4.2 The complexity of a conceptual site model should be consistent with the complexity of the site and available data. The development of a conceptual site model will usually be iterative. Model development should start as early in the site investigation process as possible. The model should be refined and revised throughout the site investigation process to incorporate additional site data. The final model should contain sufficient information to support the development of current and future exposure scenarios.

4.3 The concerns of ecological risk assessment are different from those of human-health risk assessment, for example, important migration pathways, exposure routes, and environmental receptors. These differences are usually sufficient to warrant separate descriptions and representations of the conceptual site model in the human health and ecological risk assessment reports. There will be elements of the conceptual site model that are common to both representations, however, and the risk assessors should develop these together to ensure consistency.

5. Significance and Use

5.1 The information gained through the site investigation is used to characterize the physical, biological, and chemical systems existing at a site. The processes that determine contaminant releases, contaminant migration, and environmental receptor exposure to contaminants are described and integrated in a conceptual site model.

5.2 Development of this model is critical for determining potential exposure routes (for example, ingestion and inhalation) and for suggesting possible effects of the contaminants on human health and the environment. Uncertainties associated with the conceptual site model need to be identified clearly so that efforts can be taken to reduce these uncertainties to acceptable levels. Early versions of the model, which are usually based on limited or incomplete information, will identify and emphasize the uncertainties that should be addressed.

5.3 The conceptual site model is used to integrate all site information and to determine whether information including data are missing (data gaps) and whether additional information needs to be collected at the site. The model is used furthermore to facilitate the selection of remedial alternatives and to evaluate the effectiveness of remedial actions in reducing the exposure of environmental receptors to contaminants.

5.4 This guide is not meant to replace regulatory requirements for conducting environmental site characterizations at contaminated (including radiologically contaminated) sites. It should supplement existing guidance and promote a uniform approach to developing conceptual site models.

5.5 This guide is meant to be used by all those involved in developing conceptual site models. This should ideally include representatives from all phases of the investigative and remedial process, for example, preliminary assessment,

remedial investigation, baseline human health and ecological risk assessments, and feasibility study. The conceptual site model should be used to enable experts from all disciplines to communicate effectively with one another, resolve issues concerning the site, and facilitate the decision-making process.

5.6 The steps in the procedure for developing conceptual site models include elements sometimes referred to collectively as site characterization. Although not within the scope of this guide, the conceptual site model can be used during site remediation.

6. Procedure

6.1 *Assembling Information*—Assemble historical and current site-related information from maps, aerial images, cross sections, environmental data, records, reports, studies, and other information sources. A visit(s) to the site by those preparing the conceptual site model is recommended highly. The quality of the information being assembled should be evaluated, preferably including quantitative methods, and the decision to use the information should be based on the data's meeting objective qualitative and quantitative criteria. For more information on assessing the quality and accuracy of data, see *Guidance for Data Useability in Risk Assessment (Part A)* and *Guidance for Data Useability in Risk Assessment (Part B)*. Methods used for obtaining analytical data should be described, and sources of information should be referenced. A conceptual site model should be developed for every site unless there are multiple sites in proximity to one another such that it is not possible to determine the individual source or sources of contamination. Sites may be aggregated in that case. A conceptual model should then be developed for the aggregate.

6.2 *Identifying Contaminants*—Identify contaminants in the ground water, surface water, soils, sediments, biota, and air. If no contaminants are found, the conceptual site model should be used to help document this finding.

6.3 *Establishing Background Concentrations of Contaminants*—Background samples serve three major functions: (1) to establish the range of concentrations of an analyte attributable to natural occurrence at the site; (2) to establish the range of concentrations of an analyte attributable to source(s) other than the source(s) under consideration; and (3) to help establish the extent to which contamination exceeds background levels.

6.3.1 The conceptual site model should include the naturally occurring concentrations of all contaminants found at the site. The number and location of samples needed to establish background concentrations in each medium will vary with specific site conditions and requirements. The model should include sufficient background samples to distinguish contamination attributable to the source(s) under consideration from naturally occurring or nearby anthropogenic contamination. The procedures mentioned in 6.2 and 6.3 are sometimes grouped under the general heading of contaminant assessment and may be performed as a separate activity prior to the development of a conceptual site model.

6.4 *Characterizing Sources*—At a minimum, the following source characteristics should be measured or estimated for a site:

6.4.1 Source location(s), boundaries, and volume(s).

Sources should be located accurately on site maps. Maps should include a scale and direction indicator (for example, north arrow). They should furthermore show where the source(s) is located in relationship to the property boundaries.

6.4.2 The potentially hazardous constituents and their concentrations in media at the source.

6.4.3 The time of initiation, duration, and rate of contaminant release from the source.

6.5 *Identifying Migration Pathways*—Potential migration pathways through ground water, surface water, air, soils, sediments, and biota should be identified for each source. Complete exposure pathways should be identified and distinguished from incomplete pathways. An exposure pathway is incomplete if any of the following elements are missing: (1) a mechanism of contaminant release from primary or secondary sources, (2) a transport medium if potential environmental receptors are not located at the source, and (3) a point of potential contact of environmental receptors with the contaminated medium. The potential for both current and future releases and migration of the contaminants along the complete pathways to the environmental receptors should be determined. A diagram (similar to that in Fig. X1.4) of exposure pathways for all source types at a site should be constructed. This information should be consistent with the narrative portion and tables in the exposure assessment section of an exposure or risk assessment. Tracking contaminant migration from sources to environmental receptors is one of the most important uses of the conceptual site model.

6.5.1 *Ground Water Pathway*—This pathway should be considered when hazardous solids or liquids have or may have come into contact with the surface or subsurface soil or rock. The following should be considered further in that case: vertical distance to the saturated zone; subsurface flow rates; presence and proximity of downgradient seeps, springs, or caves; fractures or other preferred flow paths; artesian conditions; presence of wells, especially those for irrigation or drinking water; and, in general, the underlying geology and hydrology of the site. Other fate and transport phenomena that should be considered include hydrodynamic dispersion, interphase transfers of contaminants, and retardation. Movement through the vadose zone should be considered.

6.5.2 *Surface Water and Sediment Pathway*—This pathway should always be investigated in the following situations: (1) a perennial body of water (river, lake, continuous stream, drainage ditch, etc.) is in direct contact with, or is potentially contaminated by a source or contaminated area, (2) an uninterrupted pathway exists from a source or contaminated area to the surface water, (3) sampling and analysis of the surface water body or sediments indicate contaminant concentrations substantially above background, (4) contaminated ground water or surface water runoff is known or suspected to discharge to a surface water body, and (5) under arid conditions in which ephemeral drainage may convey contaminants to downstream points of exposure.

6.5.3 *Air Pathway*—Contaminant transport through the air pathway should be evaluated for contaminants in the surface soil, subsurface soil, surface water, or other media capable of releasing gasses or particulate matter to the air.

The migration of contaminants from air to other environmental compartments should be considered, for example, deposition of particulates resulting from incineration onto surface waters and soil.

6.5.4 *Soil Contact Pathway*—Contaminated soils that may come into direct contact with human or ecological receptors should be investigated. This includes direct contact with chemicals through dermal absorption and direct exposure to gamma radiation from radioactively contaminated soil. There is a potential for human and ecological receptors to be exposed to contaminants at different soil depths (for example, humans may be exposed to only surface and subsurface soils, whereas plants and animals may encounter contaminants that are buried more deeply). This should be considered when contaminated soils are being evaluated.

6.5.5 *Biotic Pathway*—Bioconcentration and bioaccumulation in organisms and the resulting potential for transfer and biomagnification along food chains and environmental transport by animal movements should be considered. For example, many organic, lipophilic contaminants found in soils or sediments can bioaccumulate and bioconcentrate in organisms such as plankton, worms, or herbivores and biomagnify in organisms such as carnivorous fish and mammals or birds. The movement of contaminated biota can transport contaminants.

6.6 *Identifying Environmental Receptors*—Identify environmental receptors currently or potentially exposed to site contaminants. This includes humans and other organisms that are in direct contact with the source of contamination, potentially present along the migration pathways, or located in the vicinity of the site. It is advisable to compile a list of taxa representative of the major groups of species present at the site. It will rarely be possible or desirable to identify all species present at a site. It is recommended that the conceptual site model include species or guilds representative of major trophic levels. The complexity and iterative nature of the conceptual site model has already been mentioned in 4.2.

6.6.1 *Human Receptors*—The conceptual site model should include a map or maps indicating the physical boundaries of areas within which environmental receptors are potentially or currently exposed to the source(s) or migration pathways; separate maps may be prepared to illustrate specific contaminants or groups of contaminants. In addition, the human receptors should be represented in a figure similar to Fig. X1.4, which is based on *Guidance for Conducting Remedial Investigations and Feasibility Studies Under CERCLA*. Fig. X1.4 shows the potentially exposed populations, sources, and exposure routes. It represents a clear and concise method of displaying exposure information.

6.6.2 *Ecological Receptors*—The conceptual site model should include a map or maps identifying and locating terrestrial and aquatic habitats for plants and animals within and around the study area or associated with the source(s) or migration pathways. Consult local and state officials, U.S. Environmental Protection Agency regional specialists, and Natural Resource Trustees to determine whether any of the areas identified are critical habitats for federal- or state-listed threatened or endangered species or sensitive environments. Identify all dominant, important, declining, threatened, endangered, or rare species that either inhabit (permanently, seasonally, or temporarily) or migrate through the study area.

7. Keywords

7.1 conceptual site model; ecological; hazardous waste site; human health; risk assessment; site characterization

APPENDIX

(Nonmandatory Information)

X1. OUTLINE FOR A CONCEPTUAL SITE MODEL FOR CONTAMINATED SITES

X1.1 The conceptual site model should include a narrative and set of maps, figures, and tables to support the narrative. An outline of the narrative sections, along with an example for each section, is given below. The example is based on an hypothetical landfill site at which only preliminary sampling data are available. *The landfill site example is intentionally simplified and is for illustrative purposes only. Conceptual site models may contain considerably more detail than provided in this example.*

X1.1.1 *Brief Site Summary*—Summarize the information available for the site as this information relates to the site contaminants, source(s) of the contaminants, migration pathways, and potential environmental receptors. A brief description of the current conditions at the site (photographs optional) should be included. The inclusion of a standard 7.5-min United States Geological Survey topographic quadrangle map or geologic quadrangle map, or both, that shows the location of the site is recommended. All maps should contain directional information (for example, north arrow) and a scale.

Example—Geophysical surveys, aerial photographs, and subsurface exploration at Landfill No. 1 (LF-1) reveal the presence of at least one northeast-southwest trending waste trench. The trench is 300-ft (91-m) long and 100-ft (30-m) wide. Maximum depth of the trench indicated by the soil borings is 22 ft (7 m). As determined from the soil boring program, the waste material samples indicated that metal concentrations were at or below background concentrations, with the exception of cadmium and manganese in one sample. However, solvents (methylene chloride and trichloroethene (TCE) and pesticides (DDE, DDT, and DDD) were found at concentrations above background in soil boring samples. Soil samples taken from beneath the fill indicate that downward migration of contaminants has occurred. The surficial aquifer (ABC Formation) contains naturally high dissolved solids (>2000 mg/L) with yields of less than 4 gpm. Ground water flow in the surficial aquifer is toward the southeast at a rate of approximately 15 ft (5 m) per year. The terrain is flat with seeded and natural grasses and small (15-ft (5-m)), widely spaced loblolly pine tress covering the site. The site is fenced and unused currently.

X1.1.2 *Historical Information Concerning the Site:*

X1.1.2.1 *Site Description*—Describe the history of the site, paying particular attention to information affecting the present environmental condition of the site.

Example—LF-1, operated from 1960 to 1968. This trench-type landfill was reportedly used for the disposal of construction rubble and debris, packing material, paper, paints, thinners, unrinsed pesticide containers, oils, solvents, and contaminated fuels. Most of the trenches for waste disposal were reportedly oriented east-west and were 75-ft (23-m) wide, 350-ft (107-m) long, and an estimated 20-ft (6-m) deep. A few empty containers presumably buried in the landfill have worked their way to the surface and are partially exposed at the site. The site was partly covered by an unpaved industrial haulage road. The site was fenced in 1985 and has been unused since.

X1.1.2.2 *Source Characterization*—Present site-specific information to identify and define the location, size, and condition of the source(s) of contamination at the site.

Example—Four soil borings were used to characterize the waste disposal units at LF-1. Fig. X1.1 illustrates the soil boring locations. The depth of the soil borings were SB05 = 28 (9 m), SB06 = 30 ft (9 m), SB07 = 30 ft (9 m) and SB08 = 30 ft (9 m) below ground surface. Two of the borings, SB07 and SB08, encountered refuse/waste material. In SB08, the refuse was encountered from approximately 8 to 22 ft (2 to 7 m) below ground surface. The material was noted to be burnt debris, glass, and organic matter. A much dryer and thinner waste zone was encountered at SB07. The base of the excavation at this location was approximately 10 ft (3 m). Material that appeared to be burnt trash was noted in the backfill. The remaining two borings, SB05 and SB06, did not encounter waste. One sample was collected from each of these borings (SB05 and -06). These samples were used as background samples. Additional samples were collected from SB07 and SB08, within the landfill, to characterize the source. Analytical results are summarized in Table X1.1.

Petroleum hydrocarbons, which were suspected of being contaminants based on the site history, were not detected in any of the samples.

Volatile organic compounds found in the samples included methylene chloride and TCE. Methylene chloride was found in all soil samples in trace amounts (0.005 to 0.008 mg/kg).

The field quality control information suggests that methylene chloride may be a field artifact. The chlorinated solvent, TCE, was found significantly above background only at SB08 at a concentration of 0.05 mg/kg.

Organochlorine pesticides (DDE, DDD, and DDT), which were suspected of being present based on the site history, were not present above the detection limit in any of the samples.

Comparing metal concentrations of soil samples from SB05 and SB06 (background samples) with the remaining soil samples (SB07 and SB08) reveals that SB08 metals data exceeded the background soils data substantially for one analyte. That analyte was manganese (4320 mg/kg).

X1.1.2.3 *Migration Pathway Descriptions*—Describe the route(s) potentially taken by contaminants from the site as they migrate away from the source through the environmental media (ground water, surface water, air, sediment, soils, and food chain).

Example: *Ground Water Migration*—Three monitor wells (MWs) were installed at LF-1. The bedrock formation is typically nonwater-bearing and consists of thick clay and clay-stone (Fig. X1.2). The unconsolidated materials above the bedrock include a layer of fluvial terrace deposits. The sand and gravels that lie above the bedrock contain water with flow velocities of approximately 13 to 18 ft/year (4 to 5 m/year). Flow velocities were estimated from permeability tests conducted at MW06. Recharge at the site is from runoff associated with the nearby area that pools and stagnates at and near the site. Table X1.2 contains the water quality analyses from samples of MW05, MW06 (upgradient), and MW07 (downgradient). The upgradient samples contained no contaminants at concentrations above the detection limits, while the downgradient sample contained organic contaminants (pesticides). A comparison of metals from the downgradient and upgradient

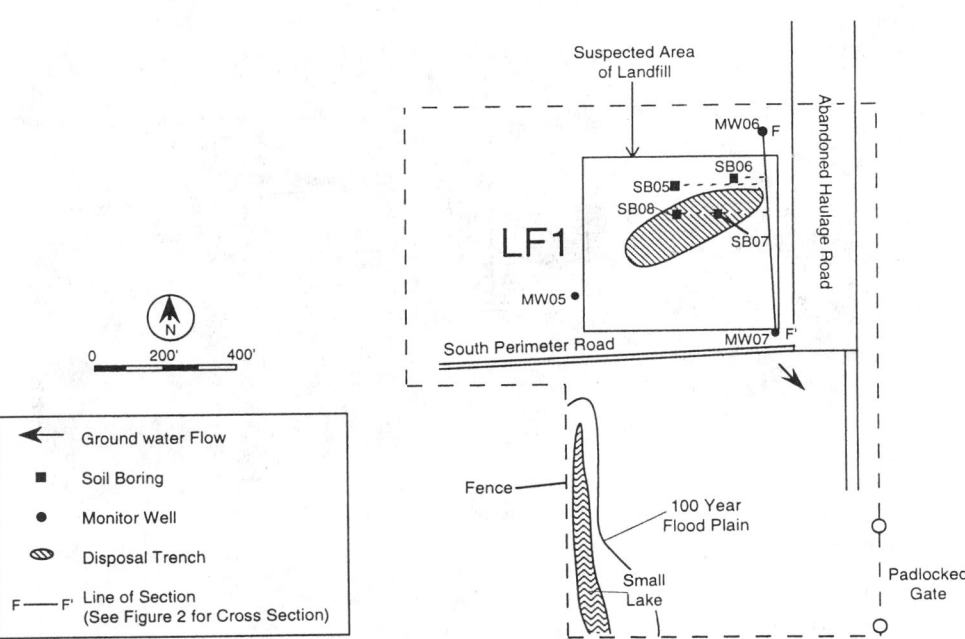

FIG. X1.1 Location Map for Landfill Number 1; Contours Showing the Potentiometric Surface from which Ground Water Flow Direction was Determined Could be Included in a Separate Figure to Avoid Clutter

TABLE X1.1 Summary of Analytical Results at LF-1[A]

Parameter (Method)	DL[B]	Units	SB05[C]	SB06	SB07	SB08
				Field Identification Number		
Moisture (Test Method D 2216)	N/A[D]	%	20.6	19.1	12.7	21.1
Petroleum hydrocarbons (SW3550/E418.1)	25	mg/kg	ND_{25}[E]	ND_{25}	ND_{25}	ND_{25}
Volatile organics (SW8240)						
Methylene chloride[F]	0.005	mg/kg	0.008	$ND_{0.0050}$	$ND_{0.0050}$	$ND_{0.0050}$
Trichloroethene	0.005	mg/kg	0.006	$ND_{0.0050}$	$ND_{0.0050}$	0.05
Organochlorine pesticides (SW3550/8080)		mg/kg				
4,4-DDE	0.0033	mg/kg	$ND_{0.0033}$	$ND_{0.0033}$	$ND_{0.0033}$	$ND_{0.0033}$
4,4-DDD	0.0033	mg/kg	$ND_{0.0033}$	$ND_{0.0033}$	$ND_{0.0033}$	$ND_{0.0033}$
4,4-DDT	0.0033	mg/kg	$ND_{0.0033}$	$ND_{0.0033}$	$ND_{0.0033}$	$ND_{0.0033}$
Metals (SW3050/6010)						
Cadmium	0.5	$ND_{0.5}$	$ND_{0.5}$	$ND_{0.5}$	$ND_{0.5}$	$ND_{0.5}$
Manganese	2	mg/kg	284	178	228	4320

[A] All results are expressed on a dry weight basis.
[B] DL = detection limit.
[C] SB = soil boring.
[D] N/A = not applicable.
[E] ND_x = not detected at concentration x.
[F] Suspected laboratory contaminant.

samples indicates that the concentration of metals in the downgradient ground water does not exceed background (upgradient) concentrations.

Example: *Surface Water and Sediment Migration*—The site surface water drainage map is shown in Fig. X1.3. Three surface water runoff samples and three sediment samples were collected at locations shown on the map. Samples SW-02 and SD-02 were collected to determine background, while SW-03, SW-04, SD-03, and SD-04 were placed downstream of the site. The analytical results given in Table X1.2 indicate that no contaminants are present above background in any of the samples. There appears to be no contamination entering the surface water pathway from the site.

Example: *Air Migration*—No air samples were taken since there was no indication that vapor or dust can enter the air pathway. The contamination is buried and effectively prevented from reaching the air pathway, and the site is covered by a thick layer of vegetation, which effectively acts as a natural cap and prevents dust from becoming airborne. Qualitative air monitoring showed no evidence of any organic vapors being present at the site during the initial stages of the site investigation.

Example: *Soils*—This pathway is not complete for humans because

the site is surrounded by a 6-ft (2-m) fence with a padlocked gate and posted with no trespassing signs. Soil and sediment samples taken for the surface water pathway did not indicate the presence of contamination above background concentrations. Also, there was no loose soil at the site since the site was covered by a thick layer of vegetation. Exposed, empty containers have been tested for the presence of contaminant residues, and none have been found. The site was inspected for evidence of burrowing mammals and other small mammals, reptiles, amphibians, or birds that might not be deterred by the fence. There was no evidence of any threat to ecological receptors from the soils or direct contact.

Example: *Food Chain Transfer*—Samples collected from surface water, sediment, and soils indicate that there are no contaminants present at concentrations above background. There is therefore no concern for food chain transfer (biomagnification) in and around the landfill.

X1.1.2.4 *Environmental Receptor Identification and Discussion*—Current and future human and ecological receptor groups should be identified and located on site maps. The migration pathways and source(s) that place or potentially

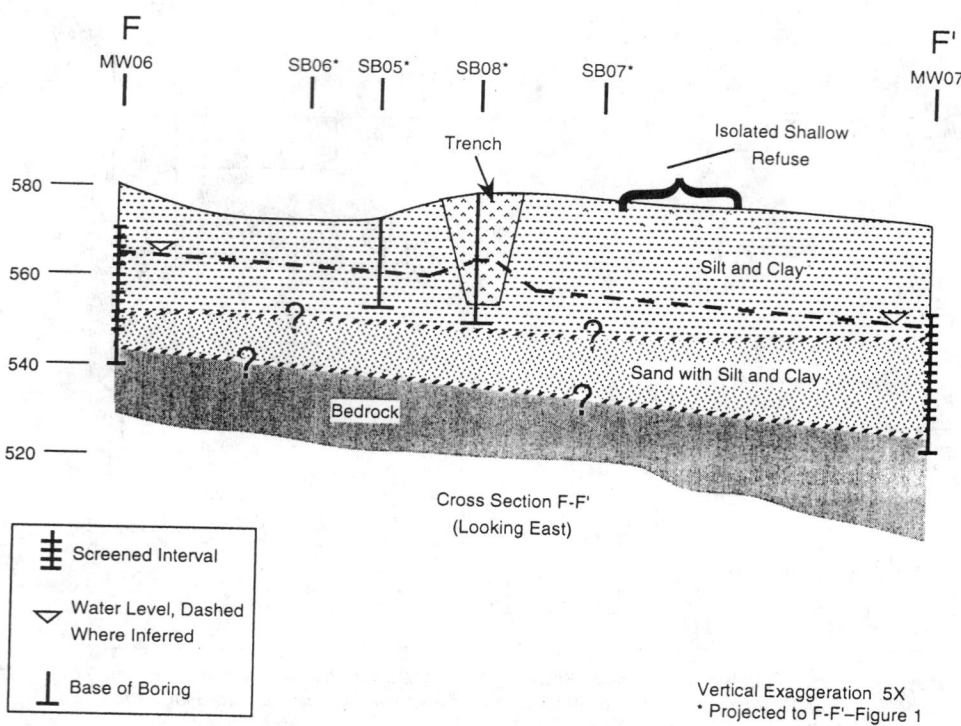

FIG. X1.2   Cross Section of Landfill Number 1

TABLE X1.2   Ground and Surface Water Quality Analysis at LF-1

Parameter	DL[A]	MW-05 µg/L	MW-06 µg/L	MW-07 µg/L		
Volatile organics						
Trichloroethene	5	ND$_5$[B]	ND$_5$	ND$_5$		
Methylene chloride	5	ND$_5$	ND$_5$	ND$_5$		
Organochlorine pesticides						
4,4-DDE	0.1	ND$_{0.1}$	ND$_{0.1}$	1		
4,4-DDD	0.1	ND$_{0.1}$	ND$_{0.1}$	3		
4,4-DDT	0.1	ND$_{0.1}$	ND$_{0.1}$	4		
Metals						
Cadmium	5	ND$_5$	ND$_5$	ND$_5$		
Manganese	15	ND$_{15}$	ND$_{15}$	ND$_{15}$		
	DL Water	µg/L SW-02	µg/L SW-03	µg/L SW-04	mg/kg SD-02	mg/kg SD-03	mg/kg SD-04
Petroleum hydrocarbons	1000	ND$_{1000}$	ND$_{1000}$	ND$_{1000}$	ND$_{1000}$	ND$_{1000}$	ND$_{1000}$
Volatile organics							
Trichloroethene	1	ND$_1$	ND$_1$	ND$_1$	ND$_1$	ND$_1$	ND$_1$
Methylene chloride	2	ND$_2$	ND$_2$	ND$_2$	ND$_2$	ND$_2$	ND$_2$
Organochlorine pesticides							
4,4-DDE	0.04	ND$_{0.04}$	ND$_{0.04}$	ND$_{0.04}$	ND$_{0.04}$	ND$_{0.04}$	ND$_{0.04}$
4,4-DDD	0.1	ND$_{0.1}$	ND$_{0.1}$	ND$_{0.1}$	ND$_{0.1}$	ND$_{0.1}$	ND$_{0.1}$
4,4-DDT	0.1	ND$_{0.1}$	ND$_{0.1}$	ND$_{0.1}$	ND$_{0.1}$	ND$_{0.1}$	ND$_{0.1}$
Metals							
Cadmium	5	ND$_5$	ND$_5$	ND$_5$	ND$_{0.5}$	ND$_{0.5}$	ND$_{0.5}$
Manganese	20	ND$_{20}$	ND$_{20}$	ND$_{20}$	ND$_2$	ND$_2$	ND$_2$

[A] DL = detection limit.
[B] ND$_x$ = not detected at concentration x.

place the environmental receptors at risk should be discussed.

Example: The only residential housing in the vicinity of the site is approximately 2100 ft northwest of the landfill. The surficial aquifer is not used as a source of drinking water by the residents, and the ground water flow is toward the southeast and away from the residential housing. There is an active golf course just to the west of the residential housing. Golf Course Lake is recharged from north of the lake and is not influenced by LF-1. The golf course does not use the surficial aquifer for a drinking water source or for irrigating the golf course. There are no other human receptors in the vicinity of the site. There are no local, state, or federally designated declining, endangered, or rare species that inhabit or migrate through the vicinity of the study area. Other wildlife species that were observed on-site show no evidence of harm from the site. Plants on-site include seeded, cool-season grasses, and volunteer native grasses; herbian vegetation; upland shrubs; and coniferous trees. None of the vegetation shows signs of stress. The most likely potentially threatened aquatic habitats are Small Lake and Big River, south of the landfill. However, environmental sampling of surface water and sediments (Table X1.2) has not shown any evidence of contaminant migration from the landfill to the lake or river. Fig. X1.4 illustrates the

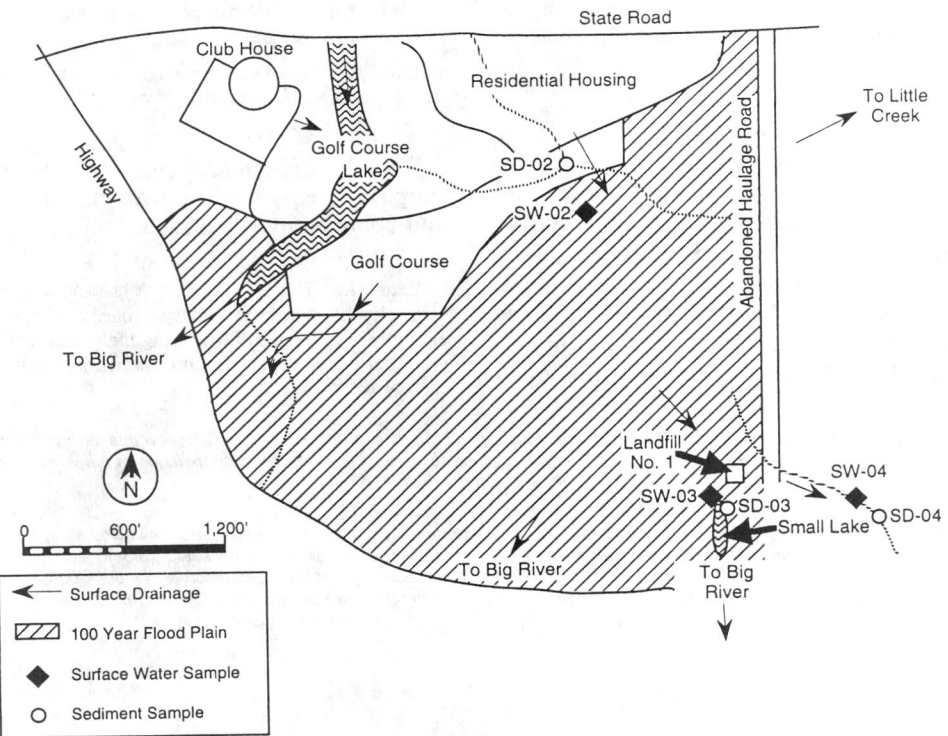

FIG. X1.3 Surface Drainage Pattern around Landfill Number 1

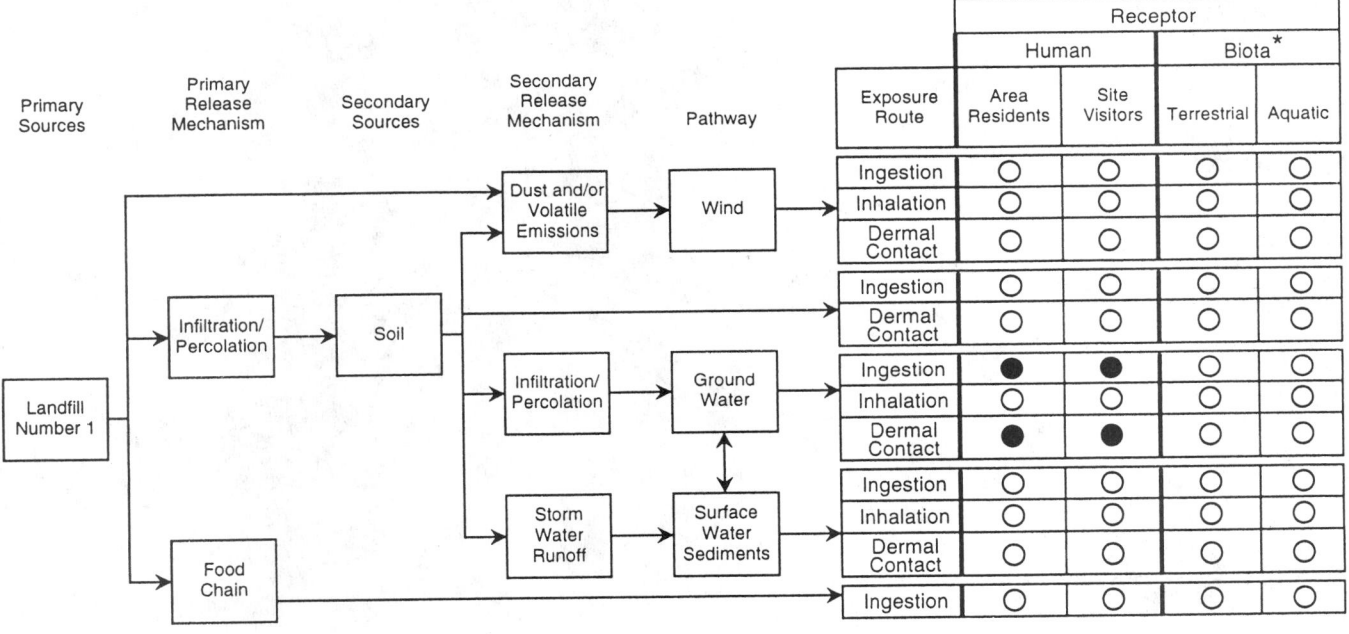

● = Pathway complete, further evaluation recommended

○ = Pathway evaluated and found incomplete, no further evaluation recommended

✳ = The terrestrial and aquatic columns can be subdivided as appropriate.

    Examples of terrestrial receptors are: plants, insects, worms, mammals, and birds.

    Examples of aquatic receptors are: periphyton, benthic invertebrates, insects, and fish

NOTE—This example is based on Figure 2-2 of *Guidance for Conducting Remedial Investigations and Feasibility Studies Under CERCLA.*

FIG. X1.4 Example Diagram for a Conceptual Model at Landfill Number 1

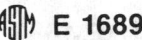

relationships among the elements of the conceptual site model, including the sources, release mechanisms, pathways, and environmental receptors.

X1.2 *Examples of Maps, Tables, and Figures:*

X1.2.1 *Maps*—The use of maps in a conceptual site model is important. The maps may include United States Geological Survey topographic and geologic maps, site sketch maps, and maps drawn to scale. The maps should identify and locate key elements of the conceptual site model including source(s); ground water, surface water, sediment, soil and air pathway routes (direction of flow); and areas covered by environmental receptor populations and migration pathways. Morphological and geological features rele-

vant to the environmental assessment of the site should be included on a map.

Example: Figs. X1.1 through X1.3 are examples of sketch maps that contain a scale, a north arrow, and a legend.

X1.2.2 *Tables and Figures*—Tables and figures should be simple and easy to read, with explanations of qualified data and abbreviations. All tables and figures should be referred to in the narrative.

Examples: Tables X1.1 and X1.2 and Figs. X1.1 through X1.3 are examples of simple summary tables and site maps. Fig. X1.4 is an example of a diagram illustrating the relationships between primary and secondary sources, release mechanisms, exposure routes, and environmental receptors.

ASTM Designation: D 5745 – 95

Standard Guide for
Developing and Implementing Short-Term Measures or Early Actions for Site Remediation[1]

This standard is issued under the fixed designation D 5745; the number immediately following the designation indicates the year of original adoption or, in the case of revision, the year of last revision. A number in parentheses indicates the year of last reapproval. A superscript epsilon (ε) indicates an editorial change since the last revision or reapproval.

1. Scope

1.1 The purpose of this guide is to provide guidance for assisting in the development, selection, design, and implementation of partial, short-term, or early action remedies undertaken at sites of waste contamination for the purpose of managing, controlling, or reducing risk posed by environmental site contamination. Early action remedies and strategies are applicable to the management of other regulatory processes (for example, state UST programs are equally applicable) in addition to the Comprehensive Environmental Response, Compensation and Liability Act (CERCLA)/NCP process. This guide identifies and describes a standard process, technical requirements, information needs, benefits, and strategy for early actions.

1.2 This guide is applicable to both nonhazardous and hazardous sites of contamination as defined by CERCLA as amended by the Superfund Amendments and Reauthorization Act of 1986 (SARA) and the Resource Conservation and Recovery Act (RCRA) as amended by the Hazardous and Solid Waste Amendments (HSWA) of 1986.

1.3 To the extent that this guide may be used for hazardous materials actions operations, it does not address the applicability of regulatory limitations and local requirements.

1.4 *This standard does not purport to address all of the safety concerns, if any, associated with its use. It is the responsibility of the user of this standard to establish appropriate safety and health practices and determine the applicability of regulatory limitations prior to use.*

2. Referenced Documents

2.1 *ASTM Standard:*
E 1528 Practice for Environmental Site Assessment: Transaction Screen Process[2]

2.2 *Code of Federal Regulations:*[3]
Corrective Action or Solid Waste Management Units at Hazardous Waste Management Facilities, Proposed Rule, 27 July 1990, 40 CFR Parts 264, 265, 270, and 271 (55 FR 30797)
Corrective Action Management Units and Temporary Units; Corrective Action Provisions; Final Rule, 16 February 1993, 58 FR 8658

National Oil and Hazardous Substances Pollution Contingency Plan, Final Rule, 8 March 1990, 40 CFR Part 300

2.3 *EPA Documents:*[3]
CERCLA, Compliance with Other Laws Manual, Part I (Interim Final), August 1988, EPA/9234.1-01
CERCLA, Compliance with Other Laws Manual, Part II: Clean Air Act and Other Environmental Statutes and State Requirements, August 1989, EPA/9234.1-02
Guidance for Performing Preliminary Assessments under CERCLA, September 1991, EPA/9345.0-01A
Guidance for Performing Site Inspections under CERCLA, September 1992, EPA/9345.1-05
Data Quality Objectives for Remedial Response Activities: Development Process, EPA/540/G-87/003
Guidance for Conducting Remedial Investigations and Feasibility Studies (RI/FS) under CERCLA, October 1988, EPA/9355.3-01
RCRA Corrective Action Interim Measures Guidance, Interim Final, June 1988, EPA/9902.4

3. Terminology

3.1 *Definitions:*

3.1.1 *applicable or relevant and appropriate requirements (ARAR)*—those requirements, cleanup standards, standards of control, and other substantive environmental protection requirements, criteria, or limitations promulgated under federal or state law that show either a direct correspondence or address problems or situations sufficiently similar at a site to show that they are well suited for application.

3.1.2 *conceptual site model, n*—a mental or physical representation of the physical system and the iterative characterization of the physical and chemical processes and conditions that affect the transport of contaminants from sources through environmental media to receptors or potential receptors.

3.1.3 *contaminant, n*—any substance potentially hazardous to human health or the environment and present in the environment above background concentration.

3.1.4 *early action, n*—any remedial plan initiated in advance of a complete or final characterization of a contaminated site.

3.1.5 *final remedy, n*—site restoration.

3.1.6 *interim remedial measure, n*—a remedial action that implements a partial solution prior to the selection of a final complete remedy. Interim remedial measures may be early actions, but they are often not.

3.1.7 *migration, n*—the movement of contaminant(s) away from a source through permeable subsurface media (such as the movement of a groundwater plume of contamination) or the movement of contaminant(s) by a combina-

[1] This guide is under the jurisdiction of ASTM Committee D-34 on Waste Management and is the direct responsibility of Subcommittee D34.11 on Site Remediation.
Current edition approved June 15, 1995. Published August 1995.
[2] *Annual Book of ASTM Standards*, Vol 11.04.
[3] Available from Superintendent of Documents, U.S. Government Printing Office, Washington, DC 20402.

tion of surficial and subsurface processes.

3.1.8 *partial remedy, n*—an interim or incomplete solution intended to be consistent with the expected permanent remedy for treatment, control, elimination, or management of risk associated with the release of a contaminant to the environment.

3.1.9 *potential migration pathway, n*—the route that may be taken by contaminants in the environment as they move or are transported from the source(s), usually in a downgradient direction.

3.1.10 *preliminary assessment (PA), n*—a review of existing information and an off-site reconnaissance, if appropriate, to determine whether a release may require additional investigation or action. A preliminary assessment may include an on-site reconnaissance, if appropriate. See ASTM Guidance for Transaction Screen Questionnaire (Practice E 1528).

3.1.11 *receptor, n*—humans or other species potentially at risk from exposure to contaminant(s) at the point(s) of exposure.

3.1.12 *release, n*—any spilling, leaking, pumping, emitting, emptying, discharging, injecting, escaping, leaching, dumping, or disposing into the environment (including the abandonment or discarding of barrels, containers, and other closed receptacles) of any hazardous chemical, extremely hazardous substance, or CERCLA hazardous substance.

3.1.13 *removal, n*—the cleanup or removal of released hazardous substances from the environment; such actions as may be necessary to take in the event of the threat of release of hazardous substances into the environment; such actions as may be necessary to monitor, assess, and evaluate the release or threat of release of hazardous substances; the disposal of removed material; or the taking of such other actions as may be necessary to prevent, minimize, or mitigate damage to the public health or welfare or to the environment, which may otherwise result from a release or threat of release.

3.1.14 *short-term measure, n*—an early action designed to have an authorized duration of less than one year for the effective control or management of a contaminant released to the environment.

3.1.15 *site characterization, n*—the process by which information relating to the nature, extent, potential migration pathways, and receptors of environmental contaminants is gathered, interpreted, and documented. Site characterization efforts provide a basis for the following: (1) the development of a conceptual site model (CSM), (2) the selection and design of a site remediation plan, or (3) the measuring point against which the effectiveness of a remedy can be evaluated, or some combination thereof.

3.1.16 *site inspection (SI), n*—an on-site investigation to determine whether a release or potential release exists and the nature of the associated threats. The purpose is to augment the data collected in the preliminary assessment and to generate, if necessary, sampling and other field data to determine whether further action or investigation is appropriate.

3.1.17 *site remediation, n*—those actions taken in the event of a release or threatened release of a hazardous substance into the environment, to prevent or minimize the impact of the release, or to mitigate a substantial hazard to

present or future environmental conditions. This early action may or may not lead to ultimate restoration of the site.

3.1.18 *source, n*—the location at which contamination has entered the natural environment.

3.2 *Description of Term Specific to This Standard:*

3.2.1 *significantly above background, adj.*—the mean concentration of a site contaminant can be shown (by statistical analysis or other methods) to be greater than nearby background samples from the same pathway.

4. Summary of Guide

4.1 The basic activities associated with implementing an early action are as follows: (1) construction of a CSM and estimation of risk(s); (2) identification of exposure control pathways amenable to engineered control; (3) development of interim or partial solutions, estimation of engineered risk, and identification and negotiation of required action levels; (4) selection of the desired solution(s); (5) attainment of legal authority for implementation of the planned solution(s); (6) design and execution of the selected solution(s); and (7) post-implementation monitoring of the conceptual site model.

4.2 Five common objectives for an early action are to achieve the following: (1) minimize the human or environmental risk exposure, or both; (2) minimize the time required to implement a final remedy; (3) protect resources (for example, financial, mineral, and ecological); (4) minimize the complexity of a final remedy; or (5) provide a solution-oriented project focus, or combination thereof.

4.3 There are three basic types of partial, short-term, or early action remedies: (1) source control remedies, (2) pathway control remedies, and (3) receptor control remedies. It is more common for early actions to be of the source or receptor control type since pathway controls usually require a sophisticated understanding of the dynamics of a conceptual site model.

4.4 The development of a final remedy is often an iterative process that evolves frequently with the compilation of new data in the CSM. The development and implementation of early actions that support the final remediation objectives of a project provides maximum benefit when performed as promptly as practical.

4.5 Early actions, short-term remedies, or interim remedial measures are effective risk management tools when designed and executed properly. Some common benefits derived from early actions are as follows: (1) human, ecological, and financial resources are protected; (2) the actual time required to remediate an unacceptable environmental condition is minimized or reduced; (3) the geometric magnitude or scale of an unacceptable environmental condition is reduced; (4) the complexity and scope of a final remedial solution is reduced; and (5) environmental projects become "solution" oriented.

4.6 A successful strategy for the application of early actions has been developed. The strategy consists of phases or steps that are as follows:

4.6.1 Development of a list of potential proactive early action remedies.

4.6.2 Identification of candidate sites for early action.

4.6.3 Identification of site-specific and easily definable CSM component(s).

4.6.3.1 Establishing and prioritizing early action objecives for each CSM component.

4.6.3.2 Identifying early action alternatives to address each objective and identifying anticipated or expected results and their impact on final regulations and remedy.

4.6.3.3 Seeking regulatory and public comment, as appropriate.

4.6.4 Establishment of funding availability for early actions.

4.6.5 Prioritization of early action solutions consistent with the objectives, public response, expected results, and funding availability.

4.6.6 Selection and integration of early action solutions.

4.6.6.1 Selecting criteria for management and measurement of the results and progress of early action.

4.6.6.2 Establishing documentation and recording procedures and requirements for early action implementation and effective final remedy implementation.

4.6.6.3 Analyzing the validation approach prior to the implementation of early action.

4.6.7 Preparation and finalization of the early action remedial plan.

4.6.8 Implementation and documentation of early action activities.

4.6.8.1 Validating early action results in comparison to the early action plan and the final remedial action frequently and periodically.

4.6.8.2 Reviewing the documentation of all early action activities frequently and periodically.

5. Significance and Use

5.1 This guide is intended to provide a systematic approach for the application and execution of early actions for purposes of remediating sites of both hazardous and non-hazardous contamination. Fundamental to the use of this guide is the iterative development of a CSM.

5.2 Anticipated users of this guide are owners or operators at sites of environmental contamination; technical professionals involved in the field of environmental site characterization and remediation; environmental regulators, property owners, employees, and residents adjacent to sites of environmental contamination; and lenders, sureties, and persons of general interest within an affected community.

5.3 This guide is not intended to replace legal requirements for remediating sites of environmental contamination. This guide should be used to supplement existing regulatory guidance and to focus remedial efforts toward final remedy solutions.

6. Procedure

6.1 *Assembling Required Information*—Assemble all available information, including the following: historical records, interviews, previous studies, environmental analytical data, permits, regulatory guidance and requirements, maps, geologic cross sections, engineering infrastructure as-built plans, and drawings. At least one site visit by technical personnel tasked with the responsibility of designing and implementing an early action is required prior to the development of a remedial plan.

6.2 *Development of the Conceptual Site Model*—An initial concept of the site(s) conceptual site model should be developed using all assembled information. The quality and accuracy of all information should be assessed both quantitatively and qualitatively, and the use of the information should be focused on the following:

6.2.1 *Identification of Contaminants*—Identify the environmental contaminants for all pathways of a conceptual site model. Particular emphasis should be placed on identifying the contaminants for any suspected exposure pathways of concern.

6.2.2 *Characterization of Background Conditions*—The natural and secondary (modified) background concentration of contaminants in all conceptual site model pathways must be characterized or estimated in order to design a useful early action. This information is necessary in order to develop appropriate action levels, identify possible synergism, estimate environmental risk, and identify and design remedial solutions.

6.2.3 *Contaminant Source Characterization*—An understanding of contaminant source characteristics is essential in developing a successful early action remedy. At a minimum, the following source characteristics should be measured or estimated for a site:

6.2.3.1 Source location, boundaries, volume, and mass;

6.2.3.2 Hazardous constituents and their concentration at a source;

6.2.3.3 Time, duration, rate of volume, and mass contaminant release from a source; and

6.2.3.4 Suspected areas (three dimensional) of contaminant migration within a pathway from a point or source release.

6.2.4 *Migration Pathway Characterization*—Potential contaminant migration pathways through the soil, surface water, air, and ground water must be identified and characterized primarily for each source of contamination at a site. The minimum information or characterization requirements for developing an early action for each migration pathway type is as follows: (*1*) an evaluation and estimate of the contaminant mass released and its release mechanism to a pathway, (*2*) identification of the transport mechanism and an estimate of contaminant transport rate or dispersion within a pathway, or both; and (*3*) identification of the human and ecological receptors at potential points of exposure above levels of acceptable risk on a contaminant migration pathway.

6.2.5 *Contaminant Mass Estimate*—An estimate of contaminant mass and contaminant distribution is required for developing successfully focused early action remedies.

6.2.6 *Receptor Exposure Characterization*—Estimates of the concentration and duration of both human and ecological contaminant exposure should be developed for each exposure point within a migration pathway.

6.2.7 *Estimation of Human and Ecological and Other Risk*—Early actions are engineered risk management solutions. An estimate or perception of unacceptable risk should exist before an early action is considered and developed. There are many categories of environmental risks; some examples are human and ecological risk, financial risk, community relations, etc.

6.3 *Identification of Early Action Strategy*—Most successful early actions or interim remedial measures incorporate a strategy that emphasize a technical approach that expediously balances and expedites the technical requirements and needs of a project risk and available resources. The elements of a proven strategy for developing and implementing early actions, as summarized in 4.6, are discussed as follows.

6.3.1 *Proactive Development of Early Action Remedies*— It is important for all affected parties to provide input within the framework of a "positive" forum to identify their concerns, risks, resources, and objectives for an early action. The development and implementation of an optimum early action will be delayed unless a proactive and technically focused environment of cooperation is developed among the parties affected by environmental contamination concerns. It is especially important for time and resource critical projects to foster proactive interaction on technical issues. ASTM advocates the early solicitation and consideration of community concerns. Some examples of early action remedies are listed in Appendix X1.

6.3.2 *Identification of Early Action Candidate Sites*—Not all sites of environmental contamination are appropriate candidates for early action. Sites that are dynamic and contain complex migration pathways commonly require sophisticated and detailed site characterization before sufficient technical information is available to design an appropriate partial remedy. Usually, the more simplistic an environmental problem, the more likely the site is to be a candidate for an early action remedy.

6.3.3 *Identification of Manageable CSM Components and Early Action Solution Alternatives:*

6.3.3.1 Each site of environmental contamination has a CSM component appropriate to manage for the control of human or ecological risk. Three examples are as follows: (*1*) as a pathway control, surface water diversion and runoff control from a contaminant release area may be a useful CSM component to manage for risk control; (*2*) source control or removal of a contaminant release to the environment may prevent migration of contaminant mass through a pathway to a receptor; and (*3*) fencing or warning signs of hazardous contaminants. Identification of the CSM components appropriate for engineered risk management is often the most critical element for developing a successful early action. Regulatory agency involvement is recommended to communicate the evaluation of the CSM components. Early agreement to the strategy by the regulatory agencies is essential.

6.3.3.2 Each CSM component identified should have well-defined risk management and mitigation objectives, each with associated desired and anticipated results from the potential early action solutions. These CSM components and objectives should be prioritized as the primary basis for evaluating alternatives and desired results. To the extent practical at this stage in the strategy, the possible impact on projected final remedies should be considered while the CSM components, objectives, and expected results are being identified and prioritized.

6.3.3.3 Public participation should be solicited and evaluated whether or not legally required. Early public/citizen participation may reveal objectives and concerns in addition to technical and site issues that could jeopardize the future success of the early action unless considered in all phases of the strategy.

6.3.3.4 At many sites where early actions have been

implemented, often only one potential technical remedy was considered. The identification of several potential multiple technical solutions *targeted at the most appropriate CSM components* is essential if the most flexible, timely, and technically responsive remedy(ies) is to be developed for that site.

6.3.4 *Funding of Early Actions*—Few sites have been remediated successfully using early actions alone and seldom are all contaminant migration pathways and risks understood at the early stages of a remedial project, the time when many early actions are performed. For these reasons, it is advisable to identify and allocate (budget) only a reasonable portion of the available funding for early action, which is balanced between cost and risk management benefits. The available funding level should be used to guide and focus the following steps toward a realistic early action solution. If the human or ecological risks identified in the CSM component(s) cannot be addressed adequately by available funding, other or additional funding alternatives should be considered.

6.3.5 *Prioritization of Early Action Solutions*—The alternative elements, including desired results and technical components, of a proposed early action should be prioritized by the affected parties. It is important that the prioritization be performed in a proactive fashion to ensure that most critical and beneficial elements of an early action are implemented. The resulting priority should be consistent with the technical and risk management objectives, public response, expected results, and available funding.

6.3.6 *Selection and Integration of Early Actions*—Based on the priority of alternative solutions, selection of the most beneficial solution should be conducted before formulating a remedial implementation plan. Performance criteria should be selected to document and measure progress toward the expected results in order to integrate the selected early action with follow-on remedial activities and a final remedy. These criteria should be incorporated into the remedial plan and include, as a minimum, recording and reporting procedures by the responsible party, interim technical objectives and schedule, budgetary objectives and constraints, reporting format for public participation, and documentation of early action activities useful for final remedy preparation and implementation. The criteria resulting from the selection process should also include an analysis to validate that the selected early action approach does, in fact, satisfy the risk management objectives and the CSM components.

6.3.7 *Preparation and Finalization of Remedial Plan*— Regulatory agencies often have format and content requirements for remedial plans; however, the regulatory agency requirements may be minimal for many of the example early actions listed in Appendix X1. The preparation of early action remedial plans must meet these regulatory requirements to receive approval. The remedial plan should be sufficiently detailed to provide guidance for implementation but simple enough to allow flexibility to respond to changing technical and site conditions. Specifically, it should be noted that site characterizations activities may be ongoing during early action activities on complicated contaminated sites with complex CSMs. This ongoing site characterization will contribute developments and refinements to the CSM that may require changes to the early action remedial plan.

6.3.8 *Implementation and Documentation of Early Action Activities*—During implementation of the plan, the results must be documented faithfully and compared to the original objectives frequently. Actual results and progress during the early action must be validated as achieving the targeted objectives. Consistency with a projected final remedy must be validated frequently during implementation of the early action.

6.3.9 At some sites where early actions have been implemented, valuable technical information has been lost or not properly documented, recorded, and reported by the responsible party. For example, early and undocumented removal and disposal of contaminated soil resulted in lack of contaminant characterization chemical data and knowledge concerning the volume of the removed soil. This lack of information made it difficult and more costly to plan and implement a final remedy. Extreme care and extra expense may be needed to ensure proper documentation. There must be proper documentation and record-keeping in order for the early action strategy to benefit the final remedy.

6.4 *Identification of Requirements for Early Actions*—Some requirements for developing an early action are site specific. The following sections discuss those requirements that must be considered for any interim remedial measure or early action. Often, although not always required, written documents describing the following topics are developed and submitted to legal entities for approval.

6.4.1 *Legal Authority*—Early actions must meet those cleanup standards, standards of control, and other substantive requirements, criteria, or limitations promulgated under federal environmental or state environmental or facility citing laws that specifically address a hazardous substance, pollutant, contaminant, remedial action, location, or other circumstances found at a hazardous waste, RCRA, or CERCLA site. Only those state standards that are identified by a state in a timely manner and that are more stringent than federal requirements may be applicable. Many different legal requirements may impact the design and implementation of an early action. Early actions are commonly authorized by the following: (*1*) letters of agreement, (*2*) interim records of decision, (*3*) engineering estimates and cost analysis (EECAs), and (*4*) permit amendments. There are many other types of legal mechanisms that may also be used to authorize or approve early actions.

6.4.2 *Health and Safety Plan*—The operational health and safety aspects of implementing an early action must be considered. Typically, emergency response plans, site evacuation plans, worker safety, and alternate pathway contaminant transport (for example, soils contamination transported in the air pathway during waste excavation) control are topics that are considered and discussed by a health and safety plan.

6.4.3 *Sampling and Analysis Plan*—Most early actions incorporate some sampling and analytical testing; however, it is not always required. Samples are frequently collected to monitor treatment efficiency, characterize wastes for disposal, and characterize a site further as components of an early action. All sampling and analysis plans (SAPs) should identify the following: sampling procedures; sampling frequency; preservation, transport, and handling techniques; decontamination procedures; analytical methods; and

quality assurance/quality control (QA/QC) systems that are associated with the environmental data generation process. Additional guidance for these efforts is available by reviewing the following:

6.4.4 *Early Action Plan*—All early actions must have a remedial plan. The remedial plan describes how an early action will be implemented. Operational items that should be addressed by the remedial plan are discussed as follows:

(*1*) Security,
(*2*) Mobilization/demobilization,
(*3*) Unit/system operation,
(*4*) Unit/system test/performance monitoring,
(*5*) Community relations,
(*6*) Site analysis,
(*7*) Contaminant mass balance,
(*8*) Waste characterization/management plan, and
(*9*) Permits.

Early action plans address complex technical issues affecting operation and execution of the remedy, but they are often relatively short and simple documents.

6.4.5 *Execution and Implementation of Early Action Plan*—It is usually not possible to deviate from an approved early action plan. Operations should conform to the plan unless circumstances require change and written authorization for plan modification is obtained. Documentation of all operations and activities must be maintained to verify that the early action plan was implemented correctly and fully and to demonstrate what was accomplished at a future date. Documentation such as the following is necessary so that the final remedy can be selected and implemented without delay and question:

6.4.5.1 *Records of Public Participation*—Notes of all public meetings; records of responses to public comments and meetings.

6.4.5.2 *Field Logs*—Daily records of activities, site manager; health and safety meeting notes; transportation, disposal, treatment records; field sampling records and sample identification; site entry logs; all other records to document field activities.

6.4.5.3 *Analytical Data Records*—Chain of custody records to document field sampling records; purchase orders for traceability of laboratory data and expenses to project; analytical data results and QC data/records.

6.4.5.4 *Early Action Results*—Chronological comparison of remedial plan with actual activities; soil excavation history and verification sampling results to verify that post excavation remedial goals were met; volume and disposition (location) of contaminated material removed from site; analytical data associated with disposal and waste management activities; all waste management and disposal manifests and bill of latent (solid, sludge, liquids, etc.); documentation of all post remedial site restoration activities; copies/records of all written and verbal correspondence with property owners, public media, and regulatory agencies; recommendations/lessons learned for final remedy.

6.4.6 *Documentation Retention*—Responsibility for record keeping rests with the property owner. Documentation for early actions taken under CERCLA should be maintained in the administrative record. This documentation should be maintained for other sites as part of the legal records for the site.

6.4.7 *Post Remedy Monitoring Plan*—Many early actions will require that the success of the remedy be monitored during its operation. It is advisable to develop an operation and monitoring plan prior to, or in conjunction with, development of the remedial plan.

6.4.8 *Early Action Performance Assessment*—The success of an early action should be assessed by comparing its actual result to the predicted goal or desired objective.

6.5 *Other Considerations*—In addition to the previously discussed requirements, other factors must often be consid-ered when developing a remedial plan for an early action. Some of the more common factors of this type are as follows: (*1*) funding limitations, (*2*) time constraints, (*3*) community acceptance, and (*4*) technology availability.

7. Keywords

7.1 conceptual site model; early action; environmental risk management; hazardous waste; interim-remedial measure; nonhazardous waste; short-term remedy; site characterization; site remediation; waste management

APPENDIX

(Nonmandatory Information)

X1. EXAMPLES OF EARLY ACTION REMEDIES

X1.1 Some examples of early action remedies are as follows: fences; site access controls; warning signs; physical security; covers; barriers; underground barrier walls; drainage controls; runoff diversion barriers; berms; dikes; impound-ment areas; capping; neutralizing chemicals; removal of debris; removal of drums, tanks, containers; removal of soil or solid materials; removal of liquids; in-situ treatments; bioremediation; alternate water treatment process; provision of alternate potable water sources or supplies; and provision of alternate habitat.

Provisional Standard Guide for
Accelerated Site Characterization for Confirmed or Suspected Petroleum Releases[1]

This standard is issued under the fixed designation PS 3; the number immediately following the designation indicates the year of original adoption.

1. Scope

1.1 This provisional guide covers a process to rapidly and accurately characterize a confirmed or suspected petroleum release site. This provisional guide is intended to provide a framework for responsible parties, contractors, consultants, and regulators to streamline and accelerate the site characterization process. The accelerated site characterization (ASC) approach may be incorporated in state and local regulations as a cost-effective method of making informed corrective-action decisions sooner.

1.2 This provisional guide describes a process for collecting site characterization information in one mobilization, using rapid sampling techniques; on-site analytical methods; and on-site interpretation and iteration of field data to refine the conceptual model for understanding site conditions as the characterization proceeds. This information can be used to determine the need for interim remedial actions (IRA); site classification or prioritization, or both; further corrective actions; and active remediation. The process outlined in this provisional guide can be incorporated into existing corrective action programs, and is organized to be used in conjunction with Guides E 1599 and E 1739.

1.3 For guidance concerning contractor health and safety issues, appropriate federal, state, and local regulations (for example, Occupational Safety and Health Administration) and industry standards should be consulted. For sampling quality assurance/quality control (QA/QC) practices, see references in Section 2. Considerations for field analytical method quality assurance/quality control are discussed in Section 5.

1.4 This provisional guide is organized as follows:

1.4.1 Section 1 describes the scope and purpose,

1.4.2 Section 2 lists reference documents,

1.4.3 Section 3 defines terminology,

1.4.4 Section 4 identifies the significance and use,

1.4.5 Section 5 describes the accelerated site characterization process,

1.4.6 Appendix X1 identifies additional reference documents,

1.4.7 Appendix X2 provides an example of a data quality classification system,

1.4.8 Appendix X3 contains a list of physical and chemical properties and geologic/hydrogeologic characteristics applicable to site characterizations, and

1.4.9 Appendix X4 contains a case study example of an application of the ASC process.

1.5 Provisional standards[2] achieve limited consensus through approval of the sponsoring subcommittee.

2. Referenced Documents

2.1 *ASTM Standards:*

D 5730 Guide to Site Characterization for Environmental Purposes With Emphasis on Soil, Rock, the Vadose Zone, and Ground Water[3]

E 1599 Guide for Corrective Action for Petroleum Releases[4]

E 1739 Guide for Risk-Based Corrective Action Applied at Petroleum Release Sites[4]

2.2 *EPA Standards:*[5]

USEPA SW 846, Recommended Analytical Procedures, Test Methods for Evaluating Solid Waste—Physical/Chemical Methods

USEPA, Draft Field Methods Compendium, OER 9285.2-11.

USEPA Publication No. USGPO 055-000-00368-8, Field Measurement Techniques: Dependable Data When You Need It

USEPA, Subsurface Characterization and Monitoring Techniques: A Desk Reference Guide—Vols I and II, EPA 625/R-93/003a and b.

USEPA, Description and Sampling of Contaminated Soils: A Field Pocket Guide, EPA 625/12-91/002.

New Jersey Department of Environmental Protection and Energy, Field Analysis Manual.

New Jersey Department of Environmental Protection and Energy, Alternative Groundwater Sampling Techniques Guide.

3. Terminology

3.1 *Descriptions of Terms Specific to This Standard:*

3.1.1 *accelerated site characterization (ASC)*—a process for collecting and evaluating information pertaining to site geology/hydrogeology, nature, and distribution of the chemical(s) of concern, potential exposure pathways, and receptors in one mobilization. The ASC employs rapid sampling techniques, on-site chemical analysis and geological/hydrogeological evaluation, and field decision making to provide a comprehensive "snap-shot" of subsurface conditions.

3.1.2 *active remediation*—actions taken to reduce the concentrations of chemical(s) of concern. Active remediation

---

[1] This provisional guide is under the jurisdiction of ASTM Committee E-50 on Environmental Assessment and is the direct responsibility of Subcommittee E50.01 on Storage Tanks.

Current edition approved Nov. 10, 1995. Published January 1996.

[2] Provisional standards exist for two years subsequent to the approval date.

[3] *Annual Book of ASTM Standards*, Vol 04.08.

[4] *Annual Book of ASTM Standards*, Vol 11.04.

[5] Available from Superintendent of Documents, U.S. Government Printing Office, Washington, DC 20402.

could be implemented when the no further action and passive remediation courses of action are not appropriate.

3.1.3 *chemical(s) of concern*—specific constituents that are identified for evaluation in the site characterization process.

3.1.4 *conceptual model*—a summary of information that is known about a site. Available site information is compiled onto simple graphics to develop an understanding of the site conditions. The conceptual model is not an analytical or numerical computer model.

3.1.5 *corrective action*—activities performed in response to a suspected or confirmed release, which include one or more of the following: site characterization, monitoring of natural attenuation, interim remedial action, remedial action, operation and maintenance of equipment, monitoring of progress, and termination of remedial action. The sequence of actions that include site assessment (characterization), interim remedial action, remedial action, operation and maintenance of equipment, monitoring of progress, and termination of remedial action.

3.1.6 *exposure pathway*—the course a chemical(s) of concern takes from the source area(s) to an exposed organism. An exposure pathway describes a unique mechanism by which an individual or population is exposed to a chemical(s) of concern originating from a site. Each exposure pathway includes a source or release from a source, a point of exposure, and an exposure route. If the exposure point differs from the source, a transport/exposure medium (for example, air) or media also is included.

3.1.7 *facility*—the property containing the source of the chemical(s) of concern where a release has occurred.

3.1.8 *field-generated analytical data*—information generated on site immediately after sample acquisition that is used to direct the site characterization process. Included are contaminant concentrations in air; soil; soil vapor or ground water, or both; and geologic/hydrogeologic conditions.

3.1.9 *indicator compounds*—compounds in ground water, soil, or air, specific to the petroleum product released, used to confirm the existence of the petroleum product, define the extent of the chemical(s) of concern, define the target levels, monitor progress of the remedial action, and identify the termination point of the remedial action.

3.1.10 *interim remedial action*—the course of action to mitigate fire and safety hazards and to prevent further migration of hydrocarbons in their vapor, dissolved, or liquid phase.

3.1.11 *mobilization*—the movement of equipment and personnel to the site, conducted during a continuous time frame to prepare for, collect, and evaluate site characterization data. These activities, when conducted as one continuous event (from one day to several weeks), are referred to as a single mobilization. Activities that are not conducted continuously are referred to as multiple-site mobilizations.

3.1.12 *on-site analytical methods*—methods or techniques that measure physical properties or chemical presence in soil, soil vapor, and ground water immediately or within a relatively short period of time to be used during a site characterization. Measurement capabilities range from qualitative (positive/negative) response to below parts per billion (sub-ppb) quantitation. Accuracy and precision of data from

these methods depends on the method detection limits and QA/QC procedures.

3.1.13 *on-site field manager*—an individual who is on site during field activities, and is responsible for directing field activities and decision-making during the site characterization. The on-site field manager should be familiar with the purpose of the site characterization, pertinent existing data, and the data collection and analysis program. The on-site field manager is the principal investigator, developing and refining the conceptual understanding/model of site conditions. This individual should have the necessary experience and background to perform the required site characterization activities and to accurately interpret the results and direct the investigation. For the purposes of this provisional guide, sufficient qualification criteria includes knowledge and experience in the following areas:

3.1.13.1 Soil and ground water sampling and analytical methods to be used at the site;

3.1.13.2 Fate and transport of petroleum hydrocarbons in the subsurface;

3.1.13.3 Local geology/hydrogeology;

3.1.13.4 Local regulations and ordinances, including knowledge of state-specific certification requirements;

3.1.13.5 Personal health and safety requirements; and

3.1.13.6 Evaluation and interpretation of site characterization results.

3.1.14 *petroleum*—including crude oil or any fraction thereof that is liquid at standard conditions of temperature and pressure (60°F at 14.7 psia). The term includes petroleum-based substances comprised of a complex blend of hydrocarbons derived from crude oil though processes of separation, conversion, upgrading, and finishing, such as motor fuels, jet oils, lubricants, petroleum solvents, and used oils.

3.1.15 *points of exposure*—the point(s) at which an individual or population may come in contact with a chemical(s) of concern originating from a site.

3.1.16 *quality assurance/quality control (QA/QC)*—the use of standards and procedures to ensure that samples collected and data generated are reliable, reproducible, and verifiable.

3.1.17 *rapid sampling tools*—equipment and techniques that allow personnel to collect samples from different media, in a relatively short period of time, for on-site chemical analysis and geologic/hydrogeologic evaluation within the same mobilization. Provided in Tables 1 and 2 are examples of rapid sampling tools. These tables will expand as additional rapid sampling equipment is developed.

3.1.18 *receptors*—persons, structures, utilities, surface waters, and water supply wells that are or may be adversely affected by a release.

3.1.19 *regulatory agency*—any state or local program responsible for overseeing or implementing underground storage tank (or other petroleum/hazardous material source) site characterization and corrective action activities.

3.1.20 *release*—any spilling, leaking, emitting, discharging, escaping, leaching, or disposing of petroleum products into ground water, surface water, soils, or air.

3.1.21 *remediation/remedial action*—activities conducted to protect human health, safety, and the environment. These activities include evaluating risk, making no-further-action

TABLE 1 Example Sample Collection Tools[A]

Method	Access[B]	Suitable Media			Sample Depth,[C] m	Comments
		Soil	Soil Vapor	Ground Water		
Grab samplers (trowels, scoops, shovel, post-hole digger)	M, B	X	...	...	<1	Low cost, loss of volatiles, ease of use.
Hand augers Slam bar and tubing	M	X	...	...	<3	Slow, labor intensive, shallow depth, can be used near located utility/product lines.
Split spoon	DP, DR	X	...	...	<100	Minimal sample disturbance, difficult to use below water table without auger.
Sample sleeve	DP	X	...	...	<100	Difficult in cobbles or hardpan, visual observation of sample, can be used below water table, minimal sample dist.
Other core samplers[D]	M	X	...	...	<2	Equipment-specific capabilities and limitations.
	DP	X	...	...	<100	
	DR	X	...	...	<100	
Active gas samplers (vacuum pumps and tubing)	OH, DP, DR	...	X	...	<100	Larger sample volume, loss of volatiles, low cost.
Passive gas samplers	M	...	X	...	<1	Time intensive.
Pneumatic depth-specific samplers	OH	...	X	X	<100	...
Check valve and tubing	OH	...	...	X	<100	Limited sample volume, low cost.
Exposed-screen sampler	DP	...	...	X	<100	...
Bailer	OH	...	...	X	<100	Labor-intensive.
Sheathed wellpoint	DP, DR	...	...	X	<100	...
Peristaltic pump	OH	...	...	X	<10	...
Gas-drive/displacement pump	OH	...	...	X	<100	...
Gas-drive/piston pump	OH	...	...	X	<100	...
Bladder pump	OH	...	...	X	<100	...
Helical rotor pump	OH	...	...	X	<100	...

[A] Some commonly used tools for shallow and intermediate depth investigations (generally <50 m) are listed. Many other tools are available. Refer to "Subsurface Characterization Monitoring Techniques: A Desk Reference Guide, Vols I and II," (EPA/625/R-93/003a and b), USEPA, May 1993, for additional information about these and other methods.

[B] Access to the sample for collection or installation of sample tool by means of the listed approaches.

M = manual (hand-operated equipment).

B = backhoe (mechanical excavating equipment).

OH = open hole (unobstructed access to the sample medium by means of a pit or cavity, a cased well, or narrow-diameter sampling point).

DR = drill rig (mechanical boring equipment, such as hollow-stem auger, mud/air rotary).

DP = direct-push (mechanical, hydraulic, pneumatic, or vibratory devices which push or drive narrow-diameter sampling points into the subsurface).

[C] Sample depth refers to practical depth limitation range, depending upon the sampling device used and the lithologic conditions.

[D] Numerous types and sizes available for different soil conditions. Drill rig is the only sample access equipment listed in this table that can be used readily to sample consolidated material.

determinations, monitoring, institutional controls, engineering controls, and designing and operating cleanup systems.

3.1.22 *site characterization*—an evaluation of subsurface geology/hydrogeology, and surface characteristics to determine if a release has occurred, the levels of the chemical(s) of concern, and the extent of the migration of the chemical(s) of concern. The data collected on soil, soil vapor and ground water quality, and potential exposure pathways and receptors may be used to generate information to support remedial action decisions.

3.1.23 *source area(s)*—the location(s) of liquid hydrocarbons or the zone(s) of highest soil or ground water concentrations, or both, of the chemical(s) of concern.

3.1.24 *user*—an individual or group involved in the ASC process including owners, operators, regulators, petroleum fund managers, attorneys, consultants, legislators, and so forth.

4. Significance and Use

4.1 The ASC process described in this provisional guide is intended for use in situations where the potential exists that petroleum has been released. The same principles may be applicable to other indicator compounds or chemicals of concern, and sources (for example, chlorinated solvent releases).

4.2 A conventional site characterization approach most often involves several mobilizations. Each mobilization typically includes a predefined sampling and analysis plan, where analysis and interpretation of results are performed off-site after demobilization. A conventional site characterization can provide high-quality data; however, multiple mobilizations often prolong the process required to adequately characterize subsurface conditions.

4.3 The unique goal of an ASC is to complete a site characterization in one mobilization. This can be accomplished by utilizing rapid sampling tools and techniques, field-generated analytical data, and on-site interpretation of results. Evaluation of data concurrent with the investigation allows the on-site field manager to select subsequent sampling points based on actual subsurface conditions, resulting in a more comprehensive and cost-effective "snapshot" of subsurface conditions. The ASC process has the following advantages:

4.3.1 Immediate identification of potential risks to human or environmental receptors or potential liabilities, or both;

4.3.2 Rapid determination of the need for interim remedial actions, site classification, and prioritization;

4.3.3 Rapid sample collection and analysis, near contem-

TABLE 2 Example Sample Analytical Techniques[A]

Method	Analyte	Media			Detection Range			Limitations	Result Time
		Soil Vapor	Soil	Ground Water	Soil Vapor	Soil	Ground Water		
PID- or FID-headspace	TOV[B]	X	X	X	ppmv	ppmv	ppmv	Temperature, humidity,	immediate
Indicator tube	Specified compound	X	X	...	ppmv	ppmv	...	instrument flowrate,	
O₂	Oxygen	X	...	...	%	...	...	cross-sensitivity	
CO₂	Carbon dioxide	X	...	...	ppmv	...	...	issues.	...
pH meter	pH	...	...	X	...	...	1–14	None.	
DO meter[C]	Dissolved oxygen	...	...	X	...	...	mg/L	Temperature, active	
REDOX meter	REDOX potential	...	...	X	...	...	...	fouling by materials	
Conductivity meter	Electrical conductivity	...	...	X	...	...	...	that react, coat, or	
Ion-specific meter	Indicator compounds	...	...	X	...	...	mg/L	clog.	
Infrared (IR) spectrometer	Indicator compounds	...	X	X	...	mg/kg	mg/L	Low bias for aromatics.	minutes
Colorimetric methods	Indicator compounds	...	X	X	...	mg/kg	mg/L	...	
Immunoassay kits[D]	Indicator and specific compounds	...	X	X	...	mg/kg	µg/L	Cross-reactivity.	
Portable GC	Specific compounds	X	X	X	ppbv	µg/kg	µg/L	Moderate peak resolution.	
Laboratory grade GC (on-site)	Specific compounds	X	X	X	ppbv	µg/kg	µg/L	Negligible.	minutes to hours
Laboratory grade mass spectrometer (on-site)	Specific compounds	X	X	X	ppbv	µg/kg	µg/L	Negligible.	
Laboratory grade GC (off-site)	Specific compounds	X	X	X	ppbv	µg/kg	µg/L	Negligible.	days to weeks
Laboratory grade mass spectrometer (off-site)	Specific compounds	X	X	X	ppbv	µg/kg	µg/L	Negligible.	

[A] Some commonly used techniques for analyzing environmental media are listed. Many other techniques are available. This list was generated using "Field Analysis Manual," New Jersey Department of Environmental Protection and Energy, May 1994, and "Subsurface Characterization and Monitoring Techniques: A Desk Reference Guide, Vols I and II," (EPA/625/R-93/003a and b), USEPA, May 1993.

[B] TOV refers to total organic vapors.

[C] Most "down-hole" dissolved oxygen (DO) probes deplete oxygen during measurement. Considerations should be given to sampling and analysis procedures that provide more accurate readings (for example, using flow-through cells).

[D] Analytical techniques that utilize a field extraction may provide less accuracy and precision for silty and clayey soils.

poraneous analytical results, and maximum data comparability;

4.3.4 Optimization of sample point locations and analytical methods;

4.3.5 Greater number of data points for resources expended;

4.3.6 Near immediate data availability for accelerating corrective action decisions; and

4.3.7 Collection of vertical and horizontal data, allowing for three-dimensional delineation of chemical(s) of concern in soil, soil vapor, or ground water.

4.4 The ASC process requires the use of an on-site field manager to make decisions to guide the characterization. Without an individual on site who is able to interpret data as it is generated, and is authorized to adjust sample locations or scope of the investigation, or both, an ASC has little chance of meeting its stated objective of full characterization in one mobilization.

5. Accelerated Site Characterization Process

5.1 The unique feature of the ASC process is the collection, analysis, and evaluation of geologic/hydrogeologic and chemical data while on-site. A flowchart of the ASC process is presented in Fig. 1, and a discussion of each activity begins in 5.2. While many of the steps in an ASC are similar to those in a conventional assessment, the following activities, as illustrated in the shaded area in Fig. 1, are performed on-site during an ASC:

5.1.1 Interpretation and evaluation of field-generated data as it is collected;

5.1.2 Continuous refinement of the conceptual model, or the understanding of site conditions;

5.1.3 Modification of the sampling and analysis program to address any necessary adjustments in the scope of work; and

5.1.4 Collection of additional data necessary to complete the characterization.

5.2 *Identify the Site Characterization:*

5.2.1 *Purpose*—The objectives of any environmental site characterization, as noted previously, are to understand the site geology/hydrogeology, the nature and extent of the chemicals of concern, the migration pathways and location of point(s) of exposure. The scope of work; however, will vary depending upon the purpose of the specific characterization. Typical purposes include one or more of the following: hazard determination, initial response action, release confirmation, risk determination, corrective action evaluation, regulatory compliance, or real estate transaction. The purpose will dictate the priority of the type of specific information to be collected. For example, a corrective action evaluation will require that a higher priority be placed on understanding subsurface geologic/hydrogeologic conditions, whereas a risk determination will focus first on receptors, exposure pathways, and points of exposure, in addition to levels of contaminants of concern.

5.2.2 The scope of the ASC is determined prior to mobilization, but will often be revised based upon interpretation of the field-generated data.

5.3 *Review Existing Site Information:*

5.3.1 There is a variety of regional and site-specific

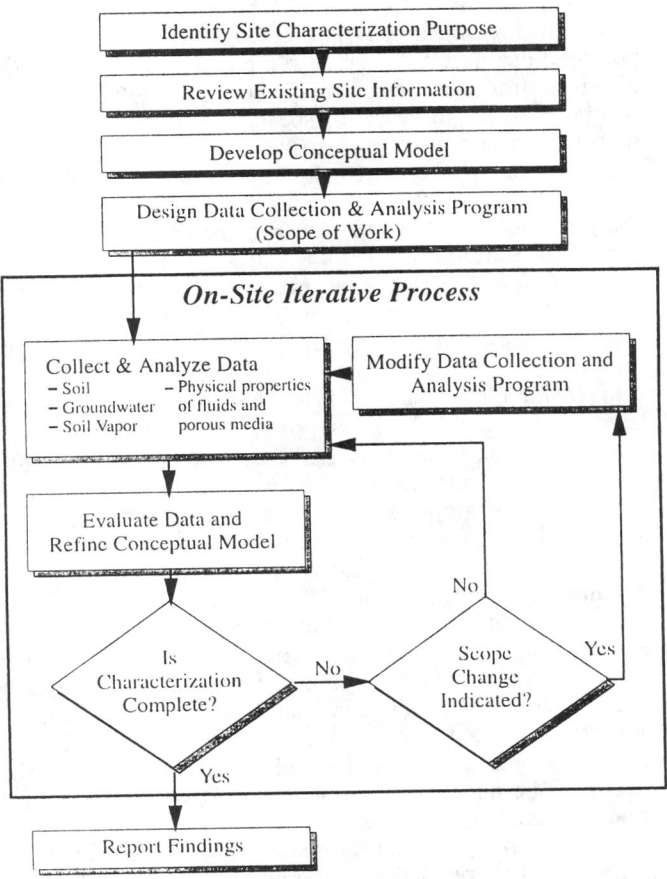

FIG. 1 **Accelerated Site Characterization Process Flowchart**

information that should be obtained prior to mobilization. A review of existing information, and a site visit, are important in the design of a data collection and analysis program, and in the development of the conceptual model. Information obtained through the site visit, interviews, and records search include the following:

5.3.1.1 Local and regional geologic/hydrogeologic maps to identify general soil types/regional depth to bedrock, rock type, depth to ground water, aquifer properties, and so forth;

5.3.1.2 Past and current land use history of the site and adjacent properties (including future land use if known);

5.3.1.3 Location of potential sources (for example, current and former storage tank systems);

5.3.1.4 Releases, spills, and overfill incidents on the site and adjacent properties;

5.3.1.5 Previous or on-going corrective action activities, or both, on-site and on nearby properties (that is, existing monitoring wells);

5.3.1.6 Potential human health and environmental receptors, such as basements, private and public water supply wells, surface waters, and streams within a given proximity of the site;

5.3.1.7 Potential transport to exposure pathways or specific points of exposure, or both, (ground water flow, vapor migration through soils and utilities, and so forth);

5.3.1.8 Other potential off-site sources of chemical(s) of conern; and

5.3.1.9 Site conditions that may affect the health and safety plan.

5.4 *Develop Conceptual Model:*

5.4.1 The initial conceptual model is the starting point of the investigation, and is used as a basis for planning field activities. This model is the result of the compilation and interpretation of all information obtained from the existing site information review, and may include the following:

5.4.1.1 Anticipated locations and depths of subsurface geologic units;

5.4.1.2 Anticipated ground water depth and flow direction(s) and possible interaction with surface water bodies;

5.4.1.3 Layout of the site, including areas and depths of artificial fill (tank and trench backfill), subsurface utility lines, and subsurface piping;

5.4.1.4 Existing soil and ground water analytical data and information regarding the location and volume of the release;

5.4.1.5 Potential releases in the vicinity of the site (especially upgradient from the site); and

5.4.1.6 Location of potential receptors.

5.4.2 The on-site field manager should summarize this information onto simple graphics such as a large-scale base map, structure contour maps, ground water elevation contour maps, isoconcentration contour maps, and geologic/hydrogeologic cross sections. These graphics can easily be hand drawn or can be generated using computerized graphics programs before actual field work begins. These documents should be used on-site and updated as the characterization progresses.

5.5 *Design Data Collection and Analysis Program:*

5.5.1 The data collection and analysis program is developed based on the initial conceptual model, prior to mobilization. This program does not need to be a formalized document, but should be agreed upon between the on-site field manager and the responsible party prior to initiation of field activities (in some cases, the regulatory agency is involved as well). The exact number and location of data collection points are left somewhat flexible, and are determined in the field based upon the actual site conditions. Levels of communication and authority between the on-site manager and the responsible party should be established to keep all parties informed as the ASC progresses.

5.5.2 Proper implementation of the data collection program requires that the on-site field manager be familiar with the capabilities and limitations of the sampling tools and on-site analytical methods, and that he or she interpret the field-generated data as it becomes available.

5.5.3 The data collection and analysis program should incorporate the following:

5.5.3.1 Purpose of the ASC;

5.5.3.2 Initial conceptual model, including site historical information, geologic/hydrogeologic characteristics of the site, and physical properties of fluids and porous media;

5.5.3.3 Methods to collect and analyze data;

5.5.3.4 General location and number of initial samples and the decision process for locating additional samples;

5.5.3.5 Media to be analyzed;

5.5.3.6 Sample collection and analysis criteria (depth, interval, sampling protocol, chemical(s) of concern, data quality levels, analytical methods, and data validation);

5.5.3.7 Specific qualifications of the on-site field manager(s);

5.5.3.8 Site constraints (for example, USTs, structures, canopy, limited space, utilities, property boundaries, depth to bedrock, and access constraints);

5.5.3.9 Need for data collection for fate and transport modeling, risk evaluations, or corrective action design;

5.5.3.10 Level of communication between the on-site field manager and the responsible party (for example agreement on changes to the scope of work or the data collection and analysis program);

5.5.3.11 Contingencies based upon reasonably anticipated deviations from expected site conditions, such as shallow bedrock, depth to ground water, disposal of investigatory wastes, change in equipment requirements, and the appearance or detection of unanticipated chemical(s) of concern; and

5.5.3.12 Determination of the possible need for off-site access.

5.5.4 *Data Collection Methods:*

5.5.4.1 The selection of sampling tools should be based upon the following:

(*1*) Purpose and anticipated scope of the ASC

(*2*) Capabilities, limitations, and cost of each tool;

(*3*) Speed by which samples can be obtained;

(*4*) Advantages of using a combination of tools;

(*5*) Site features and layout;

(*6*) Anticipated geologic site conditions;

(*7*) Anticipated chemical(s) of concern and concentrations; and

(*8*) Disturbance to site operations and neighboring properties.

5.5.4.2 Table 1 presents several common methods and devices that can be used to obtain samples. This provisional guide recognizes that additional techniques exist and continue to be developed, and sample collection during an ASC is not limited to the techniques listed in Table 1. There is a role in an accelerated site characterization for these types of tools, both in terms of defining subsurface structures (potential receptor pathways) or barriers at a site, and in selection of areas for further investigation. The ASC concept can utilize these and other methods, as previously mentioned. It is important to note the special contribution nonintrusive techniques such as geophysical methods can make in pre-survey planning for site characterizations.

5.5.5 *Analysis:*

5.5.5.1 *Geologic/Hydrogeologic Conditions and Physical Properties*—Understanding the geology/hydrogeology and physical characteristics of the subsurface is essential to properly evaluate migration potential, and to develop an appropriate corrective action plan. Physical properties that may be evaluated are listed in Appendix X3.

5.5.5.2 *Chemical Analysis*—On-site analytical methods are used in an ASC to analyze soil, soil vapor, ground water or air, or a combination thereof. On-site analysis allows the on-site field manager to determine the location of, or need for additional samples. On-site analytical methods can typically provide more data at lower cost than sending samples to an off-site laboratory. Key considerations in selecting field analytical methods are as follows:

(*1*) *Analyte*—The analytical method(s) selected will depend on the chemical(s) of concern or indicator compound(s) of interest. When gasoline is the suspected release the indicator compound may be total volatile organic compounds. In many cases, specific chemicals of concern such as benzene, may need to be measured. Depending on the chemical(s) of concern, it may be necessary to use an on-site analytical method capable of providing chemical-specific results. Table 2 is a summary of commonly used analytical techniques.

(*2*) *Media*—Consideration must be given to the targeted sample media (soil, soil vapor, ground water, air) and the method's capability of measuring concentrations in that medium.

(*3*) *Data Quality Level*—The reliability of results is related to the data quality level of the method used. An example of a data quality classification system for commonly used analytical methods is presented in Appendix X2. As shown in the example, several of the field analytical methods are capable of measuring chemical(s) of concern or indicator compounds, or both, at differing data quality levels. Depending on the chemical(s) of concern or indicator compound(s) of interest and the intended use of the data (for example, well placement or delineation of indicator compounds), field analytical methods with a lower data quality level may provide results in a qualitative or semiquantitative manner. If a specific chemical of concern such as benzene is desired, a field analytical method of a higher data quality level may be needed. When determining what level of data quality is most appropriate, the following is considered:

(*a*) The quality level selected should be consistent with the purpose and scope of the ASC, and the intended use of the data.

(*b*) Often, many points containing lower quality level data can provide a better understanding of site conditions than fewer data points at a higher data quality level. A combination of data quality levels along with an appropriate number of data points may be needed to provide a more complete understanding or meet regulatory requirements.

(*c*) A lower data quality level is often sufficient to locate source areas or to determine the placement of borings or monitoring wells, or both. Higher data quality levels may be used to determine low concentrations of specific chemicals of concern in soil or ground water or to locate delineation borings or monitoring wells. Regulatory requirements should be considered with respect to the detection limit of the selected on-site analytical method.

(*4*) *Limitations*—All analytical methods and instruments have limitations that may affect results. These include affects of temperature or humidity, cross-sensitivity issues, masking of certain constituents, and operational expertise of persons handling the equipment. These limitations should be considered when selecting analytical methods or instruments.

(*5*) *Regulatory Acceptance*—On-site analytical methods are changing rapidly and the appropriate regulatory authority should be consulted for accepted methods and procedures when an ASC is performed for regulatory purposes.

5.5.5.3 *Method Protocol and QA/QC Considerations*—Each analytical method has a standard protocol established either by the United States Environmental Protection Agency (USEPA), a state regulatory agency, an industry consensus group or manufacturer, or has a protocol specifically developed for use on-site. A quality assurance/quality control (QA/QC) plan should be developed for the methods used (for example, instrument calibration, review of instrument maintenance log and field logs, blank results, reproducibility, review of deviations, and field standards). The equipment for analyzing soil vapor, soil, and ground water in Table 2 is listed in the order of increasing capabilities and time required for analysis.

5.6 *Collect and Analyze Data*—The established data collection and analysis program is implemented to perform an intensive, short-term field investigation. Flexibility is a key component for a successful ASC, therefore, the data collection and analysis program should be used to guide the site characterization to completion. As data is collected and analyzed, it may be necessary to adjust the data collection and analysis program to refine the conceptual model and satisfy the purpose of the site characterization.

5.7 *Evaluate Data and Refine Conceptual Model:*

5.7.1 Geologic/hydrogeologic, and analytical data collected during the field investigation are continually interpreted on-site by the field manager. Compilation of the data onto simple graphics is essential for on-site data interpretation. This is best done by updating the maps and cross sections prepared to develop the initial conceptual model. As the investigation proceeds, the maps and cross sections are continually revised (geologic contacts are erased and moved, borehole lithologic data are plotted on cross sections, new isoconcentration contour lines are drawn, and so forth), by incorporating the new data. Using the field-generated graphics, the on-site field manager directs the investigation to fill in data gaps or resolve anomalies, or both. New data are collected, and the investigation proceeds in an iterative, scientific manner, until the site geology/hydrogeology, and nature and extent of soil and ground water contamination are accurately defined.

5.7.2 The degree of detail and accuracy of the graphical representation of site conditions varies according to the purpose of the characterization, complexity of the site geology/hydrogeology, and the type and volume of the released contaminant.

5.7.3 *Data Validation*—To ensure that it is useful, field-generated data must be validated. Considerations for data validation include the following:

5.7.3.1 Quality assurance/quality control (QA/QC) results (for example, duplicates, multi-point calibration curves, calibration checks, blanks, and so forth);

5.7.3.2 Comparison of higher quality level data to check lower quality level data;

5.7.3.3 Consistency of results among analytical methods and sampling techniques;

5.7.3.4 Comparison with results from other media;

5.7.3.5 Comparison with other chemical(s) of concern or indicator compounds;

5.7.3.6 Comparison against previous data, if available; and

5.7.3.7 The data should make sense in the context of the site conditions and previously generated data.

5.7.4 Once the validity of the data has been assessed, it can be used to determine whether data quality requirements have been satisfied.

5.8 *Termination of Data Collection:*

5.8.1 The data collection and evaluation should continue until the on-site field manager has determined that the site is fully characterized or that constraints prevent complete characterization. Typically, the ASC is complete and no further data collection is required when the following have been satisfied:

5.8.1.1 The conceptual model of the site geology/hydrogeology, the nature and extent of chemicals of concern, and indicator compounds fit the regional geologic/hydrogeologic setting; and

5.8.1.2 The conceptual model of the site generally incorporates/fits all of the site data; and

5.8.1.3 The conceptual model can be used to make accurate predictions, and

5.8.1.4 Sufficient detail and delineation of the chemicals of concern have been achieved to fulfill the requirements of the responsible party; or

5.8.1.5 Constraints prevent collection of any additional data.

5.9 *Report Findings:*

5.9.1 Upon completion of the field work, a report of findings is provided to the user. The report should contain at a minimum: the purpose of the characterization, a statement of objectives, the background data, a description of the data collection and analysis program, a presentation or summary of the data, and quality assurance/quality control measures. The report may be used to identify the appropriate course of action, which may include the following:

5.9.1.1 No further action;

5.9.1.2 Compliance monitoring;

5.9.1.3 Further evaluation under the risk-based corrective action (RBCA) process (Tier 2 or Tier 3 analysis: the ASC should be able to meet the requirements of a Tier 1 analysis); or

5.9.1.4 Evaluation of remedial action alternatives, and subsequent selection of technologies, or combination thereof.

5.9.2 For further information on these courses of action, please refer to Guides E 1599 and E 1739.

6. Keywords

6.1 accelerated; analytical methods; borings; characterization; chemicals of concern; corrective action; data quality; exposure pathways; field methods; ground water; LUST; mobilization; parameters; petroleum; risk based approach; sampling tools

APPENDIXES

(Nonmandatory Information)

X1. OTHER REFERENCES

X1.1 *ASTM Standards:*
D 1452 Practice for Soil Investigation and Sampling by Auger Borings[3]
D 1586 Test Method for Penetration Test and Split-Barrel Sampling of Soils[3]
D 1587 Practice for Thin-Walled Tube Geotechnical Sampling of Soils[3]
D 2488 Practice for Description and Identification of Soils (Visual—Manual Procedure)[3]
D 3550 Practice for Ring-Lines Barrel Sampling of Soils[3]
D 4447 Guide for Disposal of Laboratory Chemicals and Samples[4]
D 4448 Guide for Sampling Groundwater Monitoring Wells[4]
D 4700 Guide for Soil Sampling from the Vadose Zone[3]
D 4750 Test Method for Determining Subsurface Liquid Levels in a Borehole or Monitoring Well (Observation Well)[3]
D 4823 Guide for Core-Sampling Submerged, Unconsolidated Sediments[6]
D 5092 Practice for Design and Installation of Ground Water Monitoring Wells in Aquifers[7]
D 5299 Guide for the Decommissioning of Ground Water Wells, Vadose Zone Monitoring Devices, Boreholes and Other Devices for Environmental Activities[7]
D 5314 Guide for Soil Gas Monitoring in the Vadose Zone[7]
D 5730 Guide to Site Characterization for Environmental Purposes With Emphasis on Soil, Rock, the Vadose Zone, and Ground Water[3]

[6] *Annual Book of ASTM Standards*, Vol 11.02.
[7] *Annual Book of ASTM Standards*, Vol 04.09.

X2. AN EXAMPLE OF A DATA QUALITY CLASSIFICATION SYSTEM

X2.1 *Introduction:*

X2.1.1 This appendix describes an example of a four-tiered data quality hierarchy modified from New Jersey Department of Environmental Protection Field Analysis Manual, July, 1994. Two significant modifications to the New Jersey Department of Environmental Protection and Energy (NJDEPE) Manual have been incorporated into the example data quality level hierarchy. First, the applications are for petroleum products only. The second modification designates Level 1 as screening levels, either qualitative or semiquantitative, that may require confirmatory analyses with higher data quality methods. Levels 2, 3, and 4 are considered to be essentially quantitative, with Level 2 being less quantitative than Levels 3 or 4. These levels can produce data of sufficient quality that does not necessarily need laboratory confirmation on a routine basis. An overview of these data quality levels are presented in Table X2.1.

X2.1.2 The USEPA utilizes a two-tiered approach to data quality. The first category "Screening Data With Definitive Confirmation" would include data quality Levels 1 and 2. The second category "Definitive Data" would include data quality Levels 3 and 4.

X2.1.3 State regulatory programs may develop their own definitions for data quality for the methods listed in this appendix, and may have specific reporting requirements when using these methods. Details on data quality levels, use of field analytical methods, and specific reporting requirements can be obtained by contacting the appropriate state environmental regulatory agency or fire marshal, or other local jurisdictions.

X2.2 *Data Quality Level 1:*

X2.2.1 Level 1A methods are intended to be used for health and safety evaluations, initial contaminant screening of soil and ground water. The measurements made with these methods (1A) are qualitative and only provide an indication of the presence of contamination above a specified value (for example, pass or fail, positive or negative, and so forth). Because measurements made with these methods may not always be consistent, the data shall only be used as an initial screening for sample locations for analysis using higher level methods. Clean samples cannot be determined from these methods at this level.

X2.2.1.1 Instruments used for data quality Level 1 include: photoionization detector (PID) survey instruments, flameionization detector (FID) survey instruments, colorimetric analysis, and headspace analysis.

TABLE X2.1 Data Quality Classifications

Data Quality Level	Field Applications	Example Methods or Instruments
1 Screening: 1A—Qualitative, 1B—Semiquantitative	health and safety, qualitative contaminant screening (DQL 1A); contaminant mass location (DQL 1B)	portable PID, portable FID, colorimetric analysis, PID/FID headspace analysis
2 Delineate: Quantitative	contaminant plume delineation, well placement, remediation process monitoring	portable GC, portable IR, immunoassay, USEPA SW-846 field methods, mobile laboratories
3 Clean Zone: Quantitative	clean zone confirmation, regulatory monitoring	standard laboratory analyses with SW-846 QA/QC, mobile laboratories with certified methods
4 Nonstandard: Quantitative	constituent surveys of unknown contamination; specialty analysis	survey instrumentation (for example, GC/MS), modified laboratory methods, with full QA/QC

X2.2.1.2 Quality control procedures are limited primarily to instrument calibration, consistency in method procedure, and background level checks. Since relatively few quality control procedures are employed compared to higher-level field methods, data quality is very much a function of sample handling techniques and analyst skill.

X2.2.2 Level 1B methods can be used for qualitative and semiquantitative screening and defining the location of known types of contamination (that is, orders of magnitude or ranges). Level 1B data can be generated when PIDs and FIDs are used with controlled sample preparation and analysis procedures that include additional QA/QC such as polyethylene bag headspace.

X2.2.2.1 Quality Assurance (QA) procedures include multipoint calibration curves using matrix-spiked field standards, a calibration check using matrix spike duplicates, and a field blank/background sample.

X2.2.2.2 Depending on regulatory requirements, laboratory confirmation may be needed for establishing laboratory-field correlation over the concentration ranges measured for confirming the achievable lower detection limit.

X2.3 *Data Quality Level 2:*

X2.3.1 Level 2 methods are intended to be used for contaminant delineation. These methods can achieve a high degree of reproducibility when required QA/QC procedures are employed.

X2.3.2 Level 2 methods are typically laboratory methods that have been adapted for field use (that is, field gas chromatograph (GC), portable infrared (IR)) or are EPA-derived methods from SW-846 (for example, immunoassay). These methods may not be as rigorous because field extractions are not directly comparable to laboratory extraction methods.

X2.3.3 Quality assurance (QA) requirements include initial multi-point calibration curves, continuing calibration checks, matrix spike duplicates, background/blank samples, laboratory confirmation of clean samples, and possibly contaminated samples depending on the objective. A matrix spike recovery should be performed on a site-specific basis.

X2.3.4 Level 2 methods are semiquantitative in that they provide a direct numerical value for the contaminant indicator measured but do not definitively identify the contaminants present (for example, immunoassay, portable IR). Level 2 methods that measure specific constituents (for example, transportable GCs) are considered quantitative.

X2.3.5 Depending on regulatory requirements, laboratory confirmation may be needed for establishing laboratory-field correlation over the concentration ranges measured for confirming the achievable lower detection limit.

X2.3.6 Level 2 methods also include EPA field screening and laboratory methods from SW-846. The laboratory methods considered to be Level 2 have limited QA information documented. The quality of the data generated using Level 2 laboratory methods depends on the sample handling, storage, and preservation procedures, and analytical procedure and QC used.

X2.4 *Data Quality Level 3*—Level 3 methods are approved laboratory methods with complete QA/QC (for example, EPA SW-846 Laboratory Methods, third or more recent edition). Level 3 analyses can be performed at off-site laboratories or at on-site mobile laboratories that perform EPA methods. Certain regulatory agencies may require these laboratories to be certified.

X2.5 *Data Quality Level 4:*

X2.5.1 Level 4 methods are generally "State of the Art" methods developed specifically for a particular site or contaminant. Level 4 methods are used when standard laboratory methods are either unavailable or impractical.

X2.5.2 Generation of Level 4 data may necessitate the use of a laboratory that specializes in methods development.

X3. LIST OF PHYSICAL AND CHEMICAL PROPERTIES AND GEOLOGIC/HYDROGEOLOGIC CHARACTERISTICS

X3.1 This list is intended to provide an example of a broad range of information that may be collected during a site characterization. It is not comprehensive nor does it imply that all of this information should be collected for every site characterization. A user applying the ASC approach would consider this list, when determining the benefits of collecting information while the mobilization is in place.

X3.2 Listed in Guide D 5730 on Site Characterization for Environmental Purposes With Emphasis on Soil, Rock, Vadose Zone and Ground Water. There are additional ASTM standards and references for methods that may apply but have not been listed in this provisional guide.

X3.3 *Fluid Properties (for Example, Liquid, Dissolved, and Vapor-Phase Contaminants):*

X3.3.1 Density.

X3.3.2 Viscosity.

X3.3.3 Interfacial tension.

X3.3.4 Solubility.

X3.3.5 Sorptive properties.

X3.3.6 Vapor transport properties.

X3.3.7 Chemical composition.

X3.4 *Fluid-Media Properties:*

X3.4.1 Wettability.

X3.4.2 Capillary pressure-saturation relations.

X3.4.3 Moisture content.

X3.4.4 Relative permeabilities (air permeability).

X3.5 *Porous Media Properties:*

X3.5.1 Intrinsic permeabilities.

X3.5.2 Porosities (total and effective).

X3.5.3 Bulk density.

X3.5.4 Pore volume.

X3.5.5 Hydraulic conductivity.

X3.5.6 Grain size distribution.

X3.5.7 Organic carbon content.

X3.5.8 Clay content (soil classification).

X3.5.9 Infiltration rate.

X3.5.10 Oxygen and carbon dioxide content.

X3.5.11 Soil pH.

X3.5.12 Storativity.

X3.6 *Local Geology/Hydrogeology:*

X3.6.1 Heterogeneities.

X3.6.2 Stratigraphy/lithology/soil type.

X3.6.3 Presence, type, and relative abundance of consolidated media.

X3.6.4 Preferential contaminant migration pathways (for example, utilities, fractures, and so forth).

X3.6.5 Depth to ground water.

X3.6.6 Depth to bedrock.

X3.6.7 Aquifer thickness.

X3.6.8 Hydraulic gradient.

X3.6.9 Ground water flow direction.

X3.6.10 Dissolved oxygen.

X3.6.11 REDOX potential.

X3.6.12 Dissolved metals.

X3.6.13 Ground water pH.

X3.6.14 Hydraulic conductivity.

X3.7 *Contaminant Distribution:*

X3.7.1 Presence of nonaqueous phase liquid.

X3.7.2 Depth to impacted soil in the unsaturated or saturated zones.

X3.7.3 Zone of contamination (that is, depth, base, and areal extent of impacted soils or water-bearing stratum).

X3.7.4 Areal extent of ground water plume.

X4. EXAMPLE OF THE ACCELERATED SITE CHARACTERIZATION PROCESS

X4.1 Introduction

X4.1.1 The following example illustrates the ASC process at a petroleum release site. A hypothetical site with relatively complex geologic conditions is presented in this example to show that the ASC process works for complex sites as well as simple ones. For complex sites, the ASC process can be used to quickly identify gaps in the subsurface data and then fill those gaps while the subsurface sampling tools are still on site.

X4.1.2 In this example, the site is being characterized to provide data necessary to make corrective action decisions following a method similar to Guide E 1739. An ASC can be used at any UST (or other petroleum source) site for accurate and rapid site characterization.

X4.2 Background

X4.2.1 *Release Scenario:*

X4.2.1.1 A release of petroleum hydrocarbons at a closed service station was suspected after a contractor noticed a strong gasoline odor in the sanitary sewer adjacent to the station (see Fig. X4.1). The contractor contacted the local fire department who determined that the gas vapors did not constitute an immediate explosion hazard. The fire department filed an inspection report with the State Environmental Department (SED). The SED subsequently sent a letter to the current property owners and owners of nearby gas stations, calling for an investigation of the source of the gasoline release.

X4.2.1.2 The former station began dispensing gasoline in the mid-1950's. Gasoline was stored in two 10 000-gal tanks in the southern portion of the property. Two pump islands were located adjacent to the UST area. The station was closed in the early 1980s, and, according to a note in the fire department's file, the tanks and associated piping were removed and the excavation was filled with clean soil. The dispensing pumps were removed, and the site is now operated as a tune up shop and auto repair garage. No records are available regarding the construction of the tanks, the tank removal, or the location of underground piping. No inventory records were found that might have helped define the source and magnitude of the subsurface petroleum hydrocarbon release.

X4.2.2 *Previous Investigations:*

X4.2.2.1 An initial environmental investigation was performed at the site in the mid-1980s by the contractor who removed the tanks. The investigation consisted of installing two ground water monitoring wells along the northern boundary of the site, that was presumed to be downgradient of the UST area and the pump islands. Analyses of ground water samples collected soon after the wells were installed detected 80 µg/L (ppb) of benzene in the well downgradient from the UST area. No benzene was detected in the well located downgradient of the pump islands. No chemical analyses of soil samples from the borings was performed.

X4.2.2.2 Additional ground water samples were collected infrequently from the wells after the initial sampling event. Benzene concentrations in subsequent samples collected from the well downgradient from the UST area were erratic, ranging from non-detect (ND) to 150 ppb. The erratic range of analytical results was attributed to several factors including laboratory error, sampling bias, or seasonal ground water level fluctuations. Samples from the well downgradient from the pump islands consistently yielded ND results.

X4.2.2.3 The site is located in an urban area. Several other gas stations and industrial facilities are located within ½ mile of the site. Consequently, the possibility of upgradient sources of petroleum hydrocarbons (and other contaminants) certainly exists. Moreover, several environmental investigations have been performed at nearby sites, providing valuable information about the geologic and ground water conditions beneath the subject site.

X4.2.3 *Site Characterization Purpose:*

X4.2.3.1 The site is being investigated because the property owner is interested in selling the property. The site owner has retained a consultant who has proposed using a risk-based approach to determine the appropriate corrective action, an approach that is supported by the SED. The guidelines for risk-based corrective action are outlined in a new state UST corrective action manual, which closely follows the three-tiered Guide E 1739.

X4.2.3.2 A Tier 1 initial assessment has just been completed. Information regarding probable receptors was obtained from the SED, County Health Department, and from the reports of other consultants who have investigated nearby contaminated properties. Information about the nature and extent of contamination at the site and pathways of migration was limited to data from the earlier environmental investigation and anecdotal information from a previous employee at the site. Results of the Tier 1 assessment indicated the need for a Tier 2 assessment.

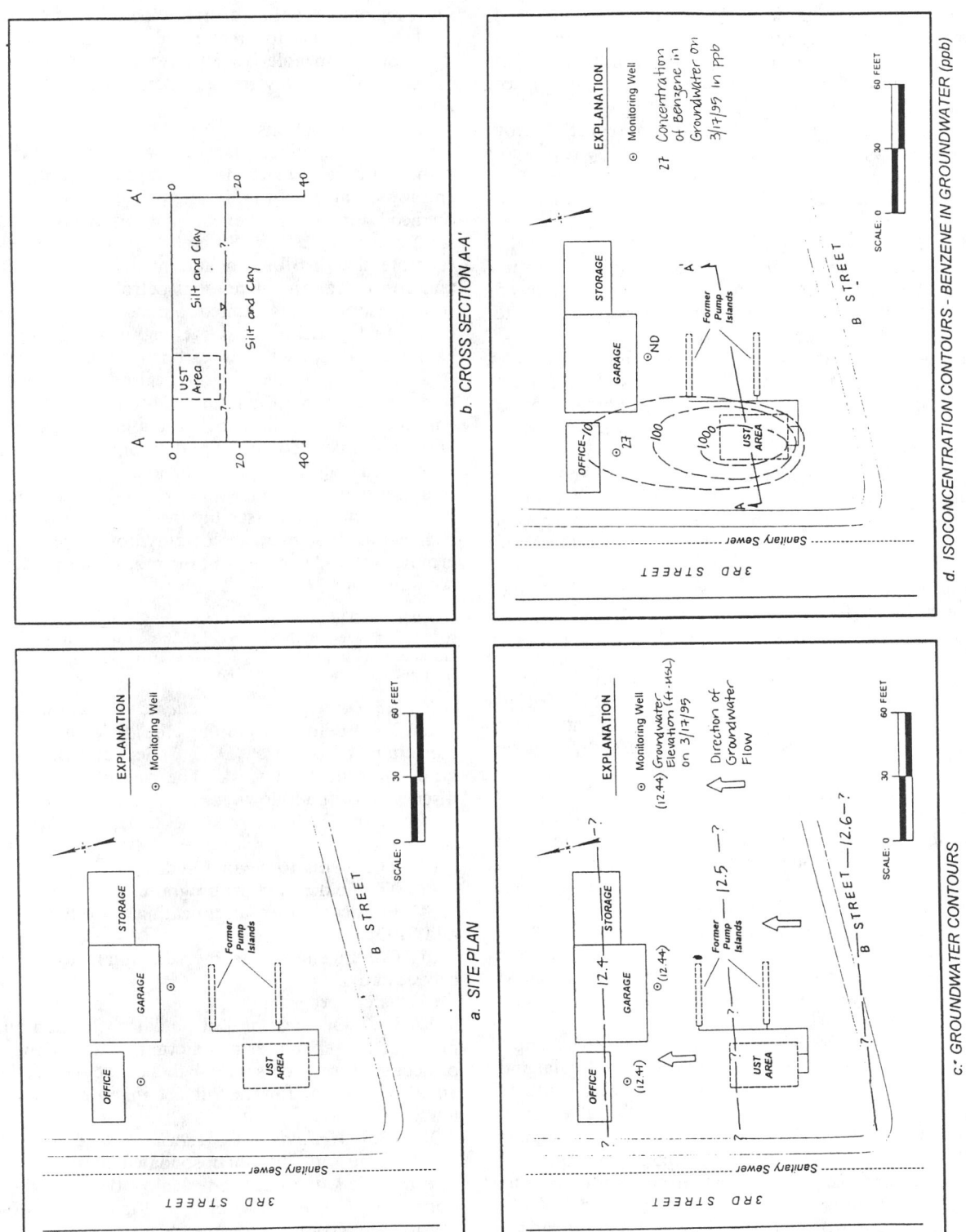

FIG. X4.1 Initial Conceptual Model (Prior to Beginning Field Work)

X4.3 Pre-Field Tasks

X4.3.1 *Review Existing Information:*

X4.3.1.1 Before the field activities were defined, available information about the site geology, ground water conditions,

and nature and extent of contamination was collected and reviewed. Reports published by the United States Geological Survey (USGS), State Geologic Survey, Soil Conservation Service, and the local Water Service provided valuable information regarding the subsurface geology and ground water conditions (for example, water table elevation, hydraulic properties, general flow direction) near the site. Regional information was also obtained from subsurface investigation reports of nearby sites.

X4.3.1.2 The 1984 site well installation report, boring logs, and previous analytical results were carefully reviewed. Geologic data shown on the boring logs was compared to published data contained in the USGS and Soil Conservation Service maps.

X4.3.1.3 Published reports and project files at the USEPA, SED, and the County provided valuable information on likely downgradient receptors. The files also showed several petroleum release sites and one superfund site upgradient from the subject property.

X4.3.1.4 As-built plans of the utility lines adjacent to the site were reviewed at the County Public Works Department in order to identify potential pathways of vapor or ground water migration from the site to the sanitary sewer line where the hydrocarbon vapors were initially detected.

X4.3.1.5 Finally, a former employee who worked at the site during its operation was interviewed by the field manager. The former employee provided valuable anecdotal information regarding the location of the USTs and piping systems, unreported inventory losses, tank and piping upgrades, and the removal of the underground tanks. Information provided by the former employee was summarized on a scaled base map of the site (see Fig. X4.1).

X4.3.2 *Develop Conceptual Model:*

X4.3.2.1 Based on the review of existing regional and site data, the field manager began to formulate an initial conceptual model of the site geology/hydrogeology, and nature and extent of contamination. Regional geologic/hydrogeologic data were compiled on the site base map. The site boring logs and water level data were then reviewed to see if they were consistent with the regional information. Available data regarding the nature and extent of contamination were compared with the anecdotal information obtained from the former employee and the as-built utility map obtained from the County Public Works Department. The field manager synthesized all of the available data and developed working hypotheses about the subsurface distribution of geologic materials, ground water flow direction, source areas, release volumes, and distribution of hydrocarbons in soil and ground water.

X4.3.2.2 Regional geologic data indicate that the shallow subsurface materials near the site are a mixture of fluvial and estuarine sediments. Granitic bedrock occurs at a depth of approximately 500 ft below ground surface (bgs). The fluvial deposits in the area are typically 2 to 15-ft thick, elongated sand beds (that is, buried stream channels) encapsulated within finer-grained silt and clay estuarine sediments. The buried stream channels regionally are oriented N40W. Boring logs from the two site monitoring wells do not show a sand unit beneath the site. However, a boring drilled across the street to the west during a previous investigation encountered a 12-ft thick sand bed.

X4.3.2.3 Unconfined ground water occurs regionally within the unconsolidated sediments at depths ranging from 25 to 30 ft bgs. This is consistent with water levels measured in the site ground water monitoring wells. Ground water

flows regionally to the north, but localized ground water flow patterns exist due to preferential ground water flow within the more permeable buried stream channels. The inferred direction of ground water flow beneath the site is toward the north.

X4.3.2.4 Locations of the former tanks, subsurface piping, on-site utility lines, and areas of artificial fill were compiled onto the site map by the field manager. A former employee indicated that strong petroleum odors and discolored soil were evident beneath the supply lines leading to the southern pump island. Several scenarios were developed to estimate the distribution and relative magnitude (volume and concentration) of residual petroleum hydrocarbons resulting from the presumed piping release. Based on the conceptual model of the site geology/hydrogeology, the likely extent of a dissolved hydrocarbon plume was estimated.

X4.3.2.5 Thus, even before beginning the field investigation, the field manager has learned much about the site conditions. A large amount of data has been compiled from various sources. To keep the data organized and accessible, the field manager summarized and compiled the information (representing the conceptual model) onto some simple, hand-drawn graphics (see Fig. X4.1). These graphics include a large-scale base map, geologic/hydrogeologic cross sections, ground water elevation contour maps, and isoconcentration contour maps.

NOTE 1—The graphics depicted in the figures are simplified and reduced in size for the purpose of this provisional guide. The actual graphics were larger working drawings that could easily be revised in the field as new data were collected.

X4.3.3 *Design Data Collection and Analysis Program:*

X4.3.3.1 Before beginning the field investigation, the consultant prepared a "Data Collection and Analysis Program" for the field work. The program included a short discussion of the following:

(1) Methods that would be used to collect subsurface samples,

(2) The media to be analyzed,

(3) The field analytical program,

(4) Protocol for communicating project status to client and SED,

(5) Contingency plans, including plans to procure off-site access, and

(6) Safety program.

X4.3.3.2 On most projects, a data collection and analysis program is prepared as an internal guide for the use of the project team members. Highlights of the data collection and analysis program for the subject investigation were as follows:

X4.3.3.3 *Methods to Collect Subsurface Samples:*

(1) Because of the unconsolidated nature of the subsurface materials, the relatively shallow depth of the investigation, and the need to sample multiple media (that is, soil, vapor, and ground water) a direct-push (DP) method of sample collection was selected. The DP sampling tools are small-diameter steel probes that are pushed or pushed and pounded, or both, into the ground. These sampling tools can be used to collect samples of soil, ground water, and soil vapor. The DP sampling tools collect a greater number of depth-discrete samples per day than conventional drilling methods. In addition, small-diameter monitoring wells

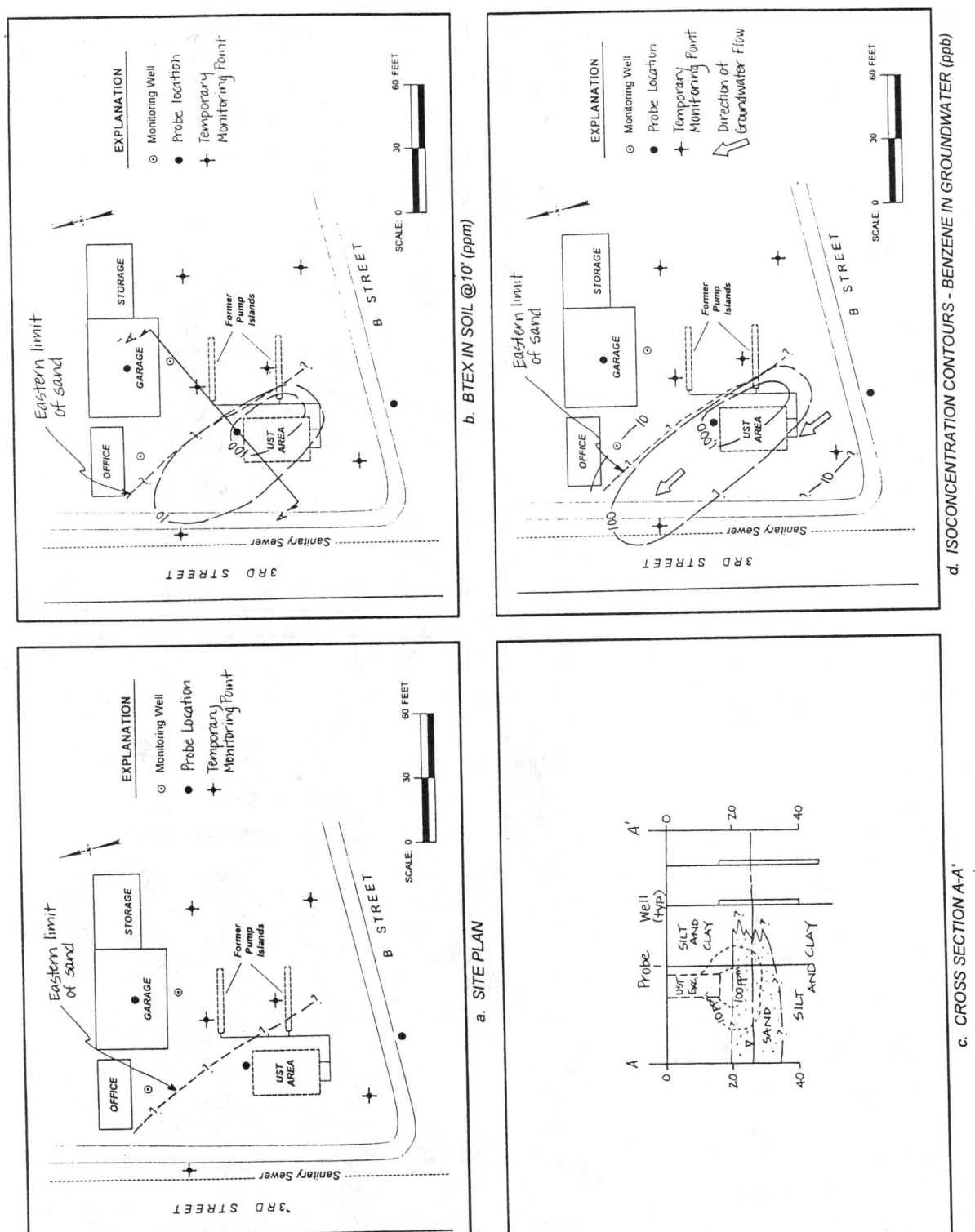

FIG. X4.2 Conceptual Model After Day 1

(monitoring points or microwells) can be installed with most DP rigs, which do not generate drill cuttings, eliminating the cost of waste soil disposal.

(2) If the site was underlain by consolidated sediments or if the sampling depths were much greater, another method of collecting samples, such as conventional hollow stem auger or rotary drilling, would have been necessary. Accelerated site characterization (ASC) is an approach (not a set of

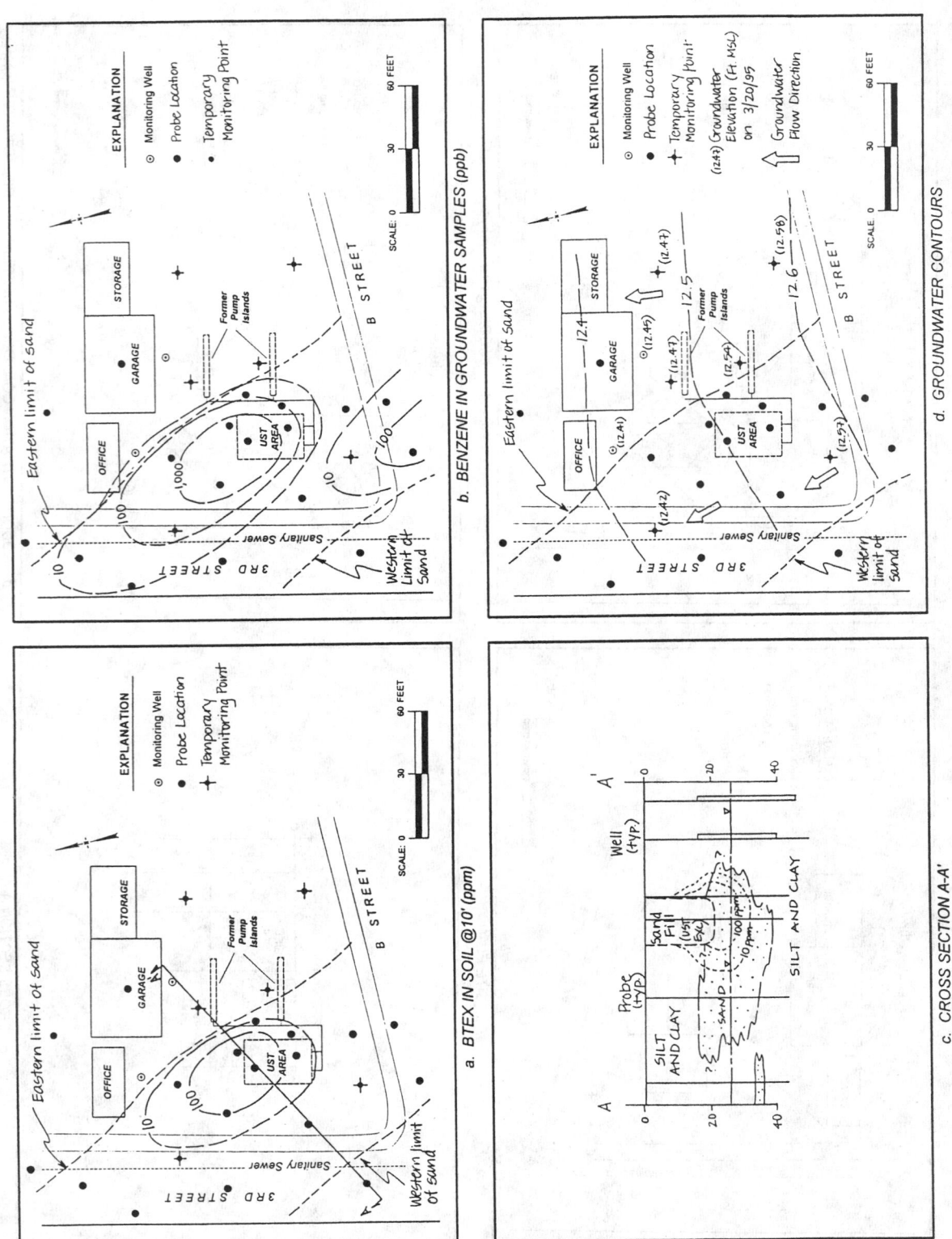

FIG. X4.3   Final Conceptual Model (After Day 3)

sampling tools) that is fully compatible with conventional sampling methods.

X4.3.3.4 *The Field Analytical Program:*

(*1*) After discussions with the SED, a mobile laboratory was contracted to perform the analytical testing. The mobile laboratory performs the analyses of soil and ground water on site, providing real-time analytical data to the field manager. The mobile laboratory selected was able to process up to 35 samples/day, and is certified by the state to perform analyses for petroleum hydrocarbons.

(2) All soil samples were to be screened in the field with a portable flame ionization detector (FID). Soil samples between depths of 10 and 30 ft were to be analyzed by the mobile laboratory every 5 ft, and at intervals where significant hydrocarbon concentrations were indicated by the portable FID. Soil samples were to be analyzed for benzene, toluene, ethylbenzene, and xylenes (BTEX) by EPA Method 8020 and total petroleum hydrocarbons as gasoline (TPH-G) by Modified EPA Method 8015 in accordance with SED requirements. Ground water samples were to be analyzed for BTEX by EPA Method 8020, and for pH, Eh, and dissolved oxygen, using portable field instruments.

X4.3.3.5 *Protocol for Communicating Project Status*— The consultant agreed to update the client and SED with the status of the ASC at the end of each field day. The field manager had a pager and portable telephone to communicate with all project participants whenever necessary.

X4.3.3.6 *Other Pre-Field Tasks:*

(1) *Permits*—Before beginning the field program, the field manager obtained permits for drilling borings (DP probes are considered to be borings by the SED) and installing monitoring wells at the site. One permit was sufficient for any number of borings and wells. Based on the data review, it seemed likely that off-site exploration beneath Third Street would be necessary. Therefore, the field manager obtained an encroachment permit from the City and filed a traffic plan with the county public works department. As part of this permit, the field manager also included *B* Street in the permit and plan.

(2) *Utility Clearance*—A private underground utility locating company was hired to locate subsurface utility lines beneath the property and Third and *B* Streets. This was done in order to avoid penetrating the utility lines with the DP sampling equipment. The utility locating was directly supervised by the field manager, because information regarding the location of subsurface utility lines gave the field manager valuable information about the location of potential pathways of contaminant migration.

X4.4 Field Investigation

X4.4.1 *Day 1:*

X4.4.1.1 On the first day of the field investigation, samples were collected at the locations shown in Fig. X4.2(a). Soil samples were collected at a minimum of every 5 ft. The assistant geologist logged the soil samples in detail, and screened the samples with a portable FID. Soil and ground water samples were submitted to the on-site mobile laboratory for chemical analysis. Several soil samples were collected and preserved for later analysis of total organic carbon (TOC), bulk density, and moisture content. Site-specific values of those parameters would be necessary for making more accurate estimates of contaminant fate and transport during the Tier 2 risk evaluation. Geologic information, depth to ground water, and soil and ground water analytical results were compiled throughout the day onto the field drawings shown in Fig. X4.2.

X4.4.1.2 To investigate the suspected source area, soil and ground water samples were collected from locations immediately north of the former UST area and former tank islands.

Soil and ground water immediately north of the UST area was strongly impacted with petroleum hydrocarbons, however the samples adjacent to the pump islands were relatively unimpacted. The UST area, therefore, seemed to be the likely source of the release.

X4.4.1.3 The boring drilled near the former UST area penetrated a native sand bed between the depths of 20 and 33 ft bgs, which was thought to extend beneath the neighboring property to the west (a northwest orientation of this sand bed was consistent with the regional geologic setting).

X4.4.1.4 Soil and ground water samples were then collected from other portions of the site to identify significant geologic units or obvious zones of contamination. Because the subsurface geology was more heterogeneous than initially thought, continuous soil cores were collected below a depth of 15 ft bgs. (Continuous soil cores are necessary to accurately identify geologic contacts and thin beds.) During the course of the day, the sand bed was penetrated in three additional locations, confirming its northwest-southeast orientation. Unconfined ground water was encountered within the sand bed at a depth of approximately 26 ft bgs.

X4.4.1.5 Once the presence of the buried stream channel was identified, the field manager suspected that it may control the movement of ground water, and hence contaminant migration, beneath the site. Indeed, isoconcentration contours of benzene in the ground water samples clearly indicated a northwest alignment of dissolved contaminants within the unit (see Fig. X4.2(d)). Benzene concentrations in ground water in the southwest portion of the site upgradient from the former UST area, however, were anomalously high. Also, chromatograms of water samples collected in the southwestern area had a different "fingerprint" than chromatograms of water samples in the UST area, leading the field manager to speculate about the likelihood of an upgradient source(s) of dissolved petroleum hydrocarbons.

X4.4.1.6 Finally, six small-diameter temporary monitoring points, consisting of ¾-in. diameter slotted PVC, were installed in the DP probe holes shown in Fig. X4.2(a). The temporary monitoring points were installed to provide a way to measure the ground water elevation at many locations beneath the site. Installing numerous temporary monitoring points allowed complete definition of ground water flow direction and hydraulic gradient beneath the site. Holes that were not converted to temporary monitoring points were filled with bentonite grout.

X4.4.2 *Day 2:*

X4.4.2.1 Characterization of the site continued on Day 2, with the field manager collecting additional subsurface data to refine the conceptual model of the site geology/hydrogeology, and nature and extent of contamination. The elevation of the tops of the temporary monitoring points were surveyed relative to the site datum (mean sea level) in order to convert depth-to-water measurements to ground water elevations.

X4.4.2.2 The eastern and western limits of the buried stream channel were refined by collecting additional soil samples from locations in between DP probe holes advanced during Day 1. The western limit of the buried stream channel was found to coincide with the southwestern corner of the site (see Fig. X4.3(a) and (b)). Additional DP probes were advanced to confirm that the eastern and northern portion of the site was underlain entirely by silt and clay, and that ground water there was not impacted by the petroleum hydrocarbon release.

X4.4.2.3 The areal and vertical extent of soil contamination began to be clearly defined. Contours of BTEX in soil showed that the highest levels of contamination were directly beneath the former UST excavation (see Fig. X4.3(a) and (b)). Also, analyses of additional ground water samples showed that the dissolved plume of benzene extended off site, beneath Third Street.

X4.4.3 *Day 3:*

X4.4.3.1 On the third and last day of the investigation, the remaining gaps in the site characterization were filled. In particular, soil samples were collected in and around the former UST excavation to further define the source of the release. The material used to fill the excavation was found to be permeable, medium-grained sand. In addition, original tank backfill material (also medium-grained sand) underlay the excavation fill and extended several feet deeper than was originally thought, into the buried stream channel (see Fig. X4.3(c)), thus providing a direct pathway from the UST backfill to the permeable sand bed.

X4.4.3.2 Several soil samples were collected adjacent to the location of the former supply lines where the former employee recalled seeing discolored soil. Analyses of soil samples from these locations indicated that the contamination was limited to shallow depths. This confirmed that the former supply lines were not a significant source of the petroleum release. The true location of the hydrocarbon release was identified using geologic data from two DP probes that penetrated the tank excavation. The floor of the original tank excavation slopes towards the north, therefore, petroleum hydrocarbons that leaked from the USTs accumulated and seeped from the northern end of the original tank excavation. This hypothesis is supported by the distribution of residual petroleum hydrocarbons in soil (see Fig. X4.3(b)).

X4.4.3.3 Several soil and ground water samples were collected in the southwestern portion of the site and beneath *B* Street to investigate the anomalous analytical data collected during Day 1. The presence of relatively high benzene concentrations in ground water beneath *B* Street, different chromatographic fingerprint of samples collected in that area, and pattern of decreasing benzene concentrations in ground water samples collected closer to the former UST area indicates that a plume of dissolved petroleum hydrocarbons has migrated onto the subject property from an upgradient source.

X4.4.3.4 Water elevations measured in the temporary monitoring points and the existing monitoring wells indicate that ground water within the silt and clay flows toward the north, consistent with the regional ground water flow direction. Not surprisingly, within the buried stream channel, ground water flow toward the northwest is indicated.

X4.4.3.5 Finally, by the afternoon of the third day, the conceptual model had been developed in sufficient detail to meet the objectives of the project (that is, to perform a Tier 2 assessment). No anomalies remained, and new DP probes yielded expected geologic information and analytical results. Moreover, the site data, including the geologic units, ground water depth and flow direction, and upgradient impacts, were consistent with the regional setting. The final conceptual model of the site is depicted in Fig. X4.3.

X4.4.3.6 Before demobilizing from the site, two of the temporary monitoring points were removed and the resulting holes were filled with bentonite grout. The remaining four temporary monitoring points were left in place for one month (to provide additional ground water elevation and analytical data) before they were removed.

X4.5 Conclusions and Discussion

X4.5.1 Through the use of the ASC process, site characterization was completed in a fraction of the time needed for a conventional investigation. The use of DP sampling technology and on-site analysis allowed the field manager to direct the investigation intelligently, filling in gaps in the subsurface data, until the characterization was complete.

X4.5.2 A key to the success of the investigation was the up-front, pre-field review of available data and development of the initial conceptual model. Because of this work, the presence of the buried stream channel and upgradient petroleum release was not unexpected, and the investigation was not delayed when these anomalies were encountered. The significance of these discoveries should not be minimized. The presence of an upgradient source of dissolved hydrocarbons would certainly complicate efforts to remediate the subject property (if required). The results of this investigation would be more than sufficient for the SED to request an environmental investigation by the owner of the upgradient property.

X4.5.3 The primary pathway for contaminant migration beneath the site is certainly different than what was initially thought. The ground water velocity and adsorptive properties of the buried stream channel are significantly different than the values used in the Tier 1 assessment. Moreover, because of the localized northwest ground water flow direction, the probable receptors of the contaminated ground water are altogether different than those used in the Tier 1 assessment.

X4.5.4 In this example, ASC provided accurate data to perform a Tier 2 risk evaluation. If active remediation is deemed necessary, the thorough understanding of the site conditions ensures that effective remedial measures will be undertaken. Ongoing ground water monitoring will likely be required at the example site. The location of permanent monitoring wells can be selected based on the clear understanding of the site ground water flow patterns.

Standard Guide for
Risk-Based Corrective Action Applied at Petroleum Release Sites[1]

This standard is issued under the fixed designation E 1739; the number immediately following the designation indicates the year of original adoption or, in the case of revision, the year of last revision. A number in parentheses indicates the year of last reapproval. A superscript epsilon (ε) indicates an editorial change since the last revision or reapproval.

ε1 NOTE—Editorial changes were made throughout in December 1996.

1. Scope

1.1 This is a guide to risk-based corrective action (RBCA), which is a consistent decision-making process for the assessment and response to a petroleum release, based on the protection of human health and the environment. Sites with petroleum release vary greatly in terms of complexity, physical and chemical characteristics, and in the risk that they may pose to human health and the environment. The RBCA process recognizes this diversity, and uses a tiered approach where corrective action activities are tailored to site-specific conditions and risks. While the RBCA process is not limited to a particular class of compounds, this guide emphasizes the application of RBCA to petroleum product releases through the use of the examples. Ecological risk assessment, as discussed in this guide, is a qualitative evaluation of the actual or potential impacts to environmental (nonhuman) receptors. *There may be circumstances under which a more detailed ecological risk assessment is necessary* (see Ref (1).[2]

1.2 The decision process described in this guide integrates risk and exposure assessment practices, as suggested by the United States Environmental Protection Agency (USEPA), with site assessment activities and remedial measure selection to ensure that the chosen action is protective of human health and the environment. The following general sequence of events is prescribed in RBCA, once the process is triggered by the suspicion or confirmation of petroleum release:

1.2.1 Performance of a site assessment;

1.2.2 Classification of the site by the urgency of initial response;

1.2.3 Implementation of an initial response action appropriate for the selected site classification;

1.2.4 Comparison of concentrations of chemical(s) of concern at the site with Tier 1 Risk Based Screening Levels (RBSLs) given in a look-up table;

1.2.5 Deciding whether further tier evaluation is warranted, if implementation of interim remedial action is warranted or if RBSLs may be applied as remediation target levels;

1.2.6 Collection of additional site-specific information as

necessary, if further tier evaluation is warranted;

1.2.7 Development of site-specific target levels (SSTLs) and point(s) of compliance (Tier 2 evaluation);

1.2.8 Comparison of the concentrations of chemical(s) of concern at the site with the Tier 2 evaluation SSTL at the determined point(s) of compliance or source area(s);

1.2.9 Deciding whether further tier evaluation is warranted, if implementation of interim remedial action is warranted, or if Tier 2 SSTLs may be applied as remediation target levels;

1.2.10 Collection of additional site-specific information as necessary, if further tier evaluation is warranted;

1.2.11 Development of SSTL and point(s) of compliance (Tier 3 evaluation);

1.2.12 Comparison of the concentrations of chemical(s) of concern at the site at the determined point(s) of compliance or source area(s) with the Tier 3 evaluation SSTL; and

1.2.13 Development of a remedial action plan to achieve the SSTL, as applicable.

1.3 The guide is organized as follows:

1.3.1 Section 2 lists referenced documents,

1.3.2 Section 3 defines terminology used in this guide,

1.3.3 Section 4 describes the significance and use of this guide,

1.3.4 Section 5 is a summary of the tiered approach,

1.3.5 Section 6 presents the RBCA procedures in a step-by-step process,

1.3.6 Appendix X1 details physical/chemical and toxicological characteristics of petroleum products,

1.3.7 Appendix X2 discusses the derivation of a Tier 1 RBSL Look-Up Table and provides an example,

1.3.8 Appendix X3 describes the uses of predictive modeling relative to the RBCA process,

1.3.9 Appendix X4 discusses considerations for institutional controls, and

1.3.10 Appendix X5 provides examples of RBCA applications.

1.4 This guide describes an approach for RBCA. It is intended to compliment but not supersede federal, state, and local regulations. Federal, state, or local agency approval may be required to implement the processes outlined in this guide.

1.5 The values stated in either inch-pound or SI units are to be regarded as the standard. The values given in parentheses are for information only.

1.6 *This standard does not purport to address all of the safety concerns, if any, associated with its use. It is the responsibility of the user of this standard to establish appro-*

---

[1] This guide is under the jurisdiction of ASTM Committee E-50 on Environmental Assessment and is the direct responsibility of Subcommittee E50.01 on Storage Tanks.

Current edition approved Sept. 10, 1995. Published November 1995. Originally published as ES 38 – 94. Last previous edition ES 38 – 94.

[2] The boldface numbers in parentheses refer to the list of references at the end of this guide.

priate safety and health practices and determine the applicability of regulatory limitations prior to use.

2. Referenced Documents

2.1 *ASTM Standard:*
E 1599 Guide for Corrective Action for Petroleum Releases[3]
2.2 *NFPA Standard:*
NFPA 329 Handling Underground Releases of Flammable and Combustible Liquids[4]

3. Terminology

3.1 *Descriptions of Terms Specific to This Standard:*

3.1.1 *active remediation*—actions taken to reduce the concentrations of chemical(s) of concern. Active remediation could be implemented when the no-further-action and passive remediation courses of action are not appropriate.

3.1.2 *attenuation*—the reduction in concentrations of chemical(s) of concern in the environment with distance and time due to processes such as diffusion, dispersion, absorption, chemical degradation, biodegradation, and so forth.

3.1.3 *chemical(s) of concern*—specific constituents that are identified for evaluation in the risk assessment process.

3.1.4 *corrective action*—the sequence of actions that include site assessment, interim remedial action, remedial action, operation and maintenance of equipment, monitoring of progress, and termination of the remedial action.

3.1.5 *direct exposure pathways*—an exposure pathway where the point of exposure is at the source, without a release to any other medium.

3.1.6 *ecological assessment*—a qualitative appraisal of the actual or potential effects of chemical(s) of concern on plants and animals other than people and domestic species.

3.1.7 *engineering controls*—modifications to a site or facility (for example, slurry walls, capping, and point of use water treatment) to reduce or eliminate the potential for exposure to a chemical(s) of concern.

3.1.8 *exposure*—contact of an organism with chemical(s) of concern at the exchange boundaries (for example, skin, lungs, and liver) and available for absorption.

3.1.9 *exposure assessment*—the determination or estimation (qualitative or quantitative) of the magnitude, frequency, duration, and route of exposure.

3.1.10 *exposure pathway*—the course a chemical(s) of concern takes from the source area(s) to an exposed organism. An exposure pathway describes a unique mechanism by which an individual or population is exposed to a chemical(s) of concern originating from a site. Each exposure pathway includes a source or release from a source, a point of exposure, and an exposure route. If the exposure point differs from the source, a transport/exposure medium (for example, air) or media also is included.

3.1.11 *exposure route*—the manner in which a chemical(s) of concern comes in contact with an organism (for example, ingestion, inhalation, and dermal contact).

3.1.12 *facility*—the property containing the source of the

chemical(s) of concern where a release has occurred.

3.1.13 *hazard index*—the sum of two or more hazard quotients for multiple chemical(s) of concern or multiple exposure pathways, or both.

3.1.14 *hazard quotients*—the ratio of the level of exposure of a chemical(s) of concern over a specified time period to a reference dose for that chemical(s) of concern derived for a similar exposure period.

3.1.15 *incremental carcinogenic risk levels*—the potential for incremental carcinogenic human health effects due to exposure to the chemical(s) of concern.

3.1.16 *indirect exposure pathways*—an exposure pathway with at least one intermediate release to any media between the source and the point(s) of exposure (for example, chemicals of concern from soil through ground water to the point(s) of exposure).

3.1.17 *institutional controls*—the restriction on use or access (for example, fences, deed restrictions, restrictive zoning) to a site or facility to eliminate or minimize potential exposure to a chemical(s) of concern.

3.1.18 *interim remedial action*—the course of action to mitigate fire and safety hazards and to prevent further migration of hydrocarbons in their vapor, dissolved, or liquid phase.

3.1.19 *maximum contaminant level (MCL)*—a standard for drinking water established by USEPA under the Safe Drinking Water Act, which is the maximum permissible level of chemical(s) of concern in water that is delivered to any user of a public water supply.

3.1.20 *Monte Carlo simulation*—a procedure to estimate the value and uncertainty of the result of a calculation when the result depends on a number of factors, each of which is also uncertain.

3.1.21 *natural biodegradation*—the reduction in concentration of chemical(s) of concern through naturally occurring microbial activity.

3.1.22 *petroleum*—including crude oil or any fraction thereof that is liquid at standard conditions of temperature and pressure (60°F and 14.7 lb/in.² absolute; (15.5°C and 10 335.6 kg/m²)). The term includes petroleum-based substances comprised of a complex blend of hydrocarbons derived from crude oil through processes of separation, conversion, upgrading, and finishing, such as motor fuels, jet oils, lubricants, petroleum solvents, and used oils.

3.1.23 *point(s) of compliance*—a location(s) selected between the source area(s) and the potential point(s) of exposure where concentrations of chemical(s) of concern must be at or below the determined target levels in media (for example, ground water, soil, or air).

3.1.24 *point(s) of exposure*—the point(s) at which an individual or population may come in contact with a chemical(s) of concern originating from a site.

3.1.25 *qualitative risk analysis*—a nonnumeric evaluation of a site to determine potential exposure pathways and receptors based on known or readily available information.

3.1.26 *reasonable maximum exposure (RME)*—the highest exposure that is reasonably expected to occur at a site. RMEs are estimated for individual pathways or a combination of exposure pathways.

3.1.27 *reasonable potential exposure scenario*—a situation with a credible chance of occurence where a receptor

---

[3] *Annual Book of ASTM Standards,* Vol 11.04.
[4] Available from National Fire Protection Association, 1 Batterymarch Park, P.O. Box 9101, Quincy, MA 02269.

may become directly or indirectly exposed to the chemical(s) of concern without considering extreme or essentially impossible circumstances.

3.1.28 *reasonably anticipated future use*—future use of a site or facility that can be predicted with a high degree of certainty given current use, local government planning, and zoning.

3.1.29 *receptors*—persons, structures, utilities, surface waters, and water supply wells that are or may be adversely affected by a release.

3.1.30 *reference dose*—a preferred toxicity value for evaluating potential noncarcinogenic effects in humans resulting from exposure to a chemical(s) of concern.

3.1.31 *remediation/remedial action*—activities conducted to protect human health, safety, and the environment. These activities include evaluating risk, making no-further-action determinations, monitoring institutional controls, engineering controls, and designing and operating cleanup equipment.

3.1.32 *risk assessment*—an analysis of the potential for adverse health effects caused by a chemical(s) of concern from a site to determine the need for remedial action or the development of target levels where remedial action is required.

3.1.33 *risk reduction*—the lowering or elimination of the level of risk posed to human health or the environment through interim remedial action, remedial action, or institutional or engineering controls.

3.1.34 *risk-based screening level/screening levels (RBSLs)*—risk-based site-specific corrective action target levels for chemical(s) of concern developed under the Tier 1 evaluation.

3.1.35 *site*—the area(s) defined by the extent of migration of the chemical(s) of concern.

3.1.36 *site assessment*—an evaluation of subsurface geology, hydrology, and surface characteristics to determine if a release has occurred, the levels of the chemical(s) of concern, and the extent of the migration of the chemical(s) of concern. The site assessment collects data on ground water quality and potential receptors and generates information to support remedial action decisions.

3.1.37 *site classification*—a qualitative evaluation of a site based on known or readily available information to identify the need for interim remedial actions and further information gathering. Site classification is intended to specifically prioritize sites.

3.1.38 *site-specific target level (SSTL)*—risk-based remedial action target level for chemical(s) of concern developed for a particular site under the Tier 2 and Tier 3 evaluations.

3.1.39 *site-specific*—activities, information, and data unique to a particular site.

3.1.40 *source area(s)*—either the location of liquid hydrocarbons or the location of highest soil and ground water concentrations of the chemical(s) of concern.

3.1.41 *target levels*—numeric values or other performance criteria that are protective of human health, safety, and the environment.

3.1.42 *Tier 1 evaluation*—a risk-based analysis to develop non-site-specific values for direct and indirect exposure pathways utilizing conservative exposure factors and fate and transport for potential pathways and various property use categories (for example, residential, commercial, and industrial uses). Values established under Tier 1 will apply to all sites that fall into a particular category.

3.1.43 *Tier 2 evaluation*—a risk-based analysis applying the direct exposure values established under a Tier 1 evaluation at the point(s) of exposure developed for a specific site and development of values for potential indirect exposure pathways at the point(s) of exposure based on site-specific conditions.

3.1.44 *Tier 3 evaluation*—a risk-based analysis to develop values for potential direct and indirect exposure pathways at the point(s) of exposure based on site-specific conditions.

3.1.45 *user*—an individual or group involved in the RBCA process including owners, operators, regulators, underground storage tank (UST) fund managers, attorneys, consultants, legislators, and so forth.

4. Significance and Use

4.1 The allocation of limited resources (for example, time, money, regulatory oversight, qualified professionals) to any one petroleum release site necessarily influences corrective action decisions at other sites. This has spurred the search for innovative approaches to corrective action decision making, which still ensures that human health and the environment are protected.

4.2 The RBCA process presented in this guide is a consistent, streamlined decision process for selecting corrective actions at petroleum release sites. Advantages of the RBCA approach are as follows:

4.2.1 Decisions are based on reducing the risk of adverse human or environmental impacts,

4.2.2 Site assessment activities are focussed on collecting only that information that is necessary to making risk-based corrective action decisions,

4.2.3 Limited resources are focussed on those sites that pose the greatest risk to human health and the environment at any time,

4.2.4 The remedial action achieves an acceptable degree of exposure and risk reduction,

4.2.5 Compliance can be evaluated relative to site-specific standards applied at site-specific point(s) of compliance,

4.2.6 Higher quality, and in some cases faster, cleanups than are currently realized, and

4.2.7 A documentation and demonstration that the remedial action is protective of human health, safety, and the environment.

4.3 Risk assessment is a developing science. The scientific approach used to develop the RBSL and SSTL may vary by state and user due to regulatory requirements and the use of alternative scientifically based methods.

4.4 Activities described in this guide should be conducted by a person familiar with current risk and exposure assessment methodologies.

4.5 In order to properly apply the RBCA process, the user should avoid the following:

4.5.1 Use of Tier 1 RBSLs as mandated remediation standards rather than screening levels,

4.5.2 Restriction of the RBCA process to Tier 1 evaluation only and not allowing Tier 2 or Tier 3 analyses,

4.5.3 Placing arbitrary time constraints on the corrective action process; for example, requiring that Tiers 1, 2, and 3 be completed within 30-day time periods that do not reflect the actual urgency of and risks posed by the site,

4.5.4 Use of the RBCA process only when active remediation is not technically feasible, rather than a process that is applicable during all phases of corrective action,

4.5.5 Requiring the user to achieve technology-based remedial limits (for example, asymptotic levels) prior to requesting the approval for the RBSL or SSTL,

4.5.6 The use of predictive modelling that is not supported by available data or knowledge of site conditions,

4.5.7 Dictating that corrective action goals can only be achieved through source removal and treatment actions, thereby restricting the use of exposure reduction options, such as engineering and institutional controls,

4.5.8 The use of unjustified or inappropriate exposure factors,

4.5.9 The use of unjustified or inappropriate toxicity parameters,

4.5.10 Neglecting aesthetic and other criteria when determining RBSLs or SSTLs,

4.5.11 Not considering the effects of additivity when screening multiple chemicals,

4.5.12 Not evaluating options for engineering or institutional controls, exposure point(s), compliance point(s), and carcinogenic risk levels before submitting remedial action plans,

4.5.13 Not maintaining engineering or institutional controls, and

4.5.14 Requiring continuing monitoring or remedial action at sites that have achieved the RBSL or SSTL.

5. Tiered Approach to Risk-Based Corrective Action (RBCA) at Petroleum Release Sites

5.1 RBCA is the integration of site assessment, remedial action selection, and monitoring with USEPA-recommended risk and exposure assessment practices. This creates a process by which corrective action decisions are made in a consistent manner that is protective of human health and the environment.

5.2 The RBCA process is implemented in a tiered approach, involving increasingly sophisticated levels of data collection and analysis. The assumptions of earlier tiers are replaced with site-specific data and information. Upon evaluation of each tier, the user reviews the results and recommendations and decides whether more site-specific analysis is warranted.

5.3 *Site Assessment*—The user is required to identify the sources of the chemical(s) of concern, obvious environmental impacts (if any), any potentially impacted humans and environmental receptors (for example, workers, residents, water bodies, and so forth), and potentially significant transport pathways (for example, ground water flow, utilities, atmospheric dispersion, and so forth). The site assessment will also include information collected from historical records and a visual inspection of the site.

5.4 *Site Classification*—Sites are classified by the urgency of need for initial response action, based on information collected during the site assessment. Associated with site classifications are initial response actions that are to be implemented simultaneously with the RBCA process. Sites should be reclassified as actions are taken to resolve concerns or as better information becomes available.

5.5 *Tier 1 Evaluation*—A look-up table containing screening level concentrations is used to determine whether site conditions satisfy the criteria for a quick regulatory closure or warrant a more site-specific evaluation. Ground water, soil, and vapor concentrations may be presented in this table for a range of site descriptions and types of petroleum products ((for example, gasoline, crude oil, and so forth). The look-up table of RBSL is developed in Tier 1 or, if a look-up table has been previously developed and determined to be applicable to the site by the user, then the existing RBSLs are used in the Tier 1 process. Tier 1 RBSLs are typically derived for standard exposure scenarios using current RME and toxicological parameters as recommended by the USEPA. These values may change as new methodologies and parameters are developed. Tier 1 RBSLs may be presented as a range of values, corresponding to a range of risks or property uses.

5.6 *Tier 2 Evaluation*—Tier 2 provides the user with an option to determine SSTLs and point(s) of compliance. It is important to note that both Tier 1 RBSL and Tier 2 SSTLs are based on achieving similar levels of protection of human health and the environment (for example, 10^{-4} to 10^{-6} risk levels). However, in Tier 2 the non-site-specific assumptions and point(s) of exposure used in Tier 1 are replaced with site-specific data and information. Additional site-assessment data may be needed. For example, the Tier 2 SSTL can be derived from the same equations used to calculate the Tier 1 RBSL, except that site-specific parameters are used in the calculations. The additional site-specific data may support alternate fate and transport analysis. At other sites, the Tier 2 analysis may involve applying Tier 1 RBSLs at more probable point(s) of exposure. Tier 2 SSTLs are consistent with USEPA-recommended practices.

5.7 *Tier 3 Evaluation*—Tier 3 provides the user with an option to determine SSTLs for both direct and indirect pathways using site-specific parameters and point(s) of exposure and compliance when it is judged that Tier 2 SSTLs should not be used as target levels. Tier 3, in general, can be a substantial incremental effort relative to Tiers 1 and 2, as the evaluation is much more complex and may include additional site assessment, probabilistic evaluations, and sophisticated chemical fate/transport models.

5.8 *Remedial Action*—If the concentrations of chemical(s) of concern at a site are above the RBSL or SSTL at the point(s) of compliance or source area, or both, and the user determines that the RBSL or SSTL should be used as remedial action target levels, the user develops a remedial action plan in order to reduce the potential for adverse impacts. The user may use remediation processes to reduce concentrations of the chemical(s) of concern to levels below or equal to the target levels or to achieve exposure reduction (or elimination) through institutional controls discussed in Appendix X4, or through the use of engineering controls, such as capping and hydraulic control.

6. Risk-Based Corrective Action (RBCA) Procedures

6.1 The sequence of principal tasks and decisions associated with the RBCA process are outlined on the flowchart shown in Fig. 1. Each of these actions and decisions is discussed as follows.

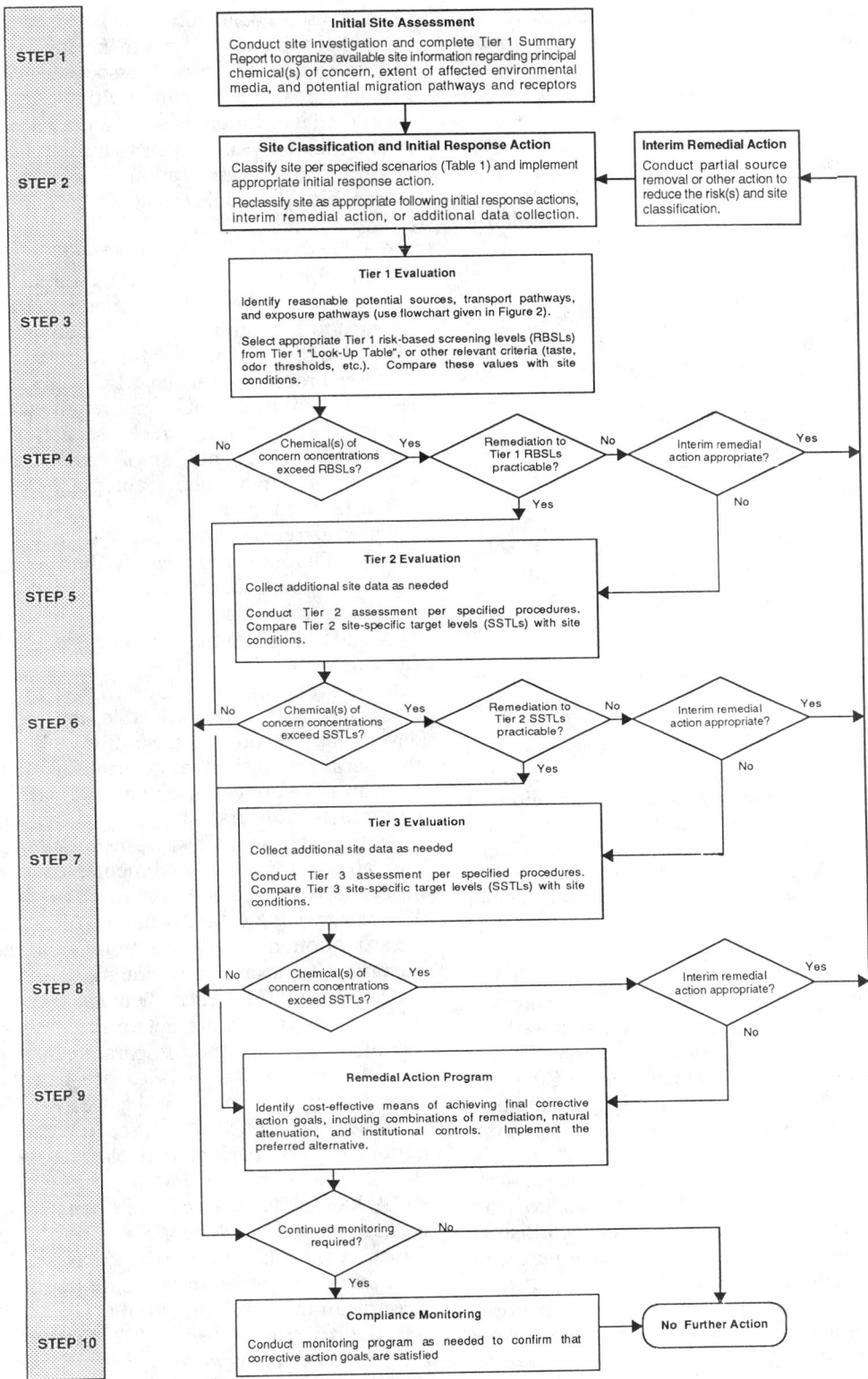

FIG. 1 Risk-Based Corrective Action Process Flowchart

6.2 *Site Assessment*—Gather the information necessary for site classification, initial response action, comparison to the RBSL, and determining the SSTL. Site assessment may be conducted in accordance with Guide E 1599. Each successive tier will require additional site-specific data and information that must be collected as the RBCA process proceeds. The user may generate site-specific data and information or estimate reasonable values for key physical

189

characteristics using soil survey data and other readily available information. The site characterization data should be summarized in a clear and concise format.

6.2.1 The site assessment information for Tier 1 evaluation may include the following:

6.2.1.1 A review of historical records of site activities and past releases;

6.2.1.2 Identification of chemical(s) of concern;

6.2.1.3 Location of major sources of the chemical(s) of concern;

6.2.1.4 Location of maximum concentrations of chemical(s) of concern in soil and ground water;

6.2.1.5 Location of humans and the environmental receptors that could be impacted (point(s) of exposure);

6.2.1.6 Identification of potential significant transport and exposure pathways (ground water transport, vapor migration through soils and utilities, and so forth);

6.2.1.7 Determination of current or potential future use of the site and surrounding land, ground water, surface water, and sensitive habitats;

6.2.1.8 Determination of regional hydrogeologic and geologic characteristics (for example, depth to ground water, aquifer thickness, flow direction, gradient, description of confining units, and ground water quality); and

6.2.1.9 A qualitative evaluation of impacts to environmental receptors.

6.2.2 In addition to the information gathered in 6.2.1, the site assessment information for Tier 2 evaluation may include the following:

6.2.2.1 Determination of site-specific hydrogeologic and geologic characteristics (for example, depth to ground water, aquifer thickness, flow direction, gradient, description of confining units, and ground water quality);

6.2.2.2 Determination of extent of chemical(s) of concern relative to the RBSL or SSTL, as appropriate;

6.2.2.3 Determination of changes in concentrations of chemical(s) of concern over time (for example, stable, increasing, and decreasing); and

6.2.2.4 Determination of concentrations of chemical(s) of concern measured at point(s) of exposure (for example, dissolved concentrations in nearby drinking water wells or vapor concentrations in nearby conduits or sewers).

6.2.3 In addition to the information gathered in 6.2.1 and 6.2.2, the site assessment information for Tier 3 evaluation includes additional information that is required for site-specific modeling efforts.

6.3 *Site Classification and Initial Response Action*—As the user gathers data, site conditions should be evaluated and an initial response action should be implemented, consistent with site conditions. This process is repeated when new data indicate a significant change in site conditions. Site urgency classifications are presented in Table 1, along with example classification scenarios and potential initial responses. *Note that the initial response actions given in Table 1 may not be applicable for all sites. The user should select an option that best addresses the short-term health and safety concerns of the site while implementing the RBCA process.*

6.3.1 The classification and initial response action scheme given in Table 1 is an example. It is based on the current and projected degree of hazard to human health and the environment. This is a feature of the process that can be customized

by the user. "Classification 1" sites are associated with immediate threats to human health and the environment; "Classification 2" sites are associated with short-term (0 to 2-year) threats to human health, safety, and the environment; "Classification 3" sites are associated with long-term (greater than 2-year) threats to human health, safety, and the environment; "Classification 4" sites are associated with no reasonable potential threat to human health or to the environment.

6.3.2 Associated with each classification scenario in Table 1 is an initial response action; the initial response actions are implemented in order to eliminate any potential immediate impacts to human health and the environment as well as to minimize the potential for future impacts that may occur as the user proceeds with the RBCA process. Note that initial response actions do not always require active remediation; in many cases the initial response action is to monitor or further assess site conditions to ensure that risks posed by the site do not increase above acceptable levels over time. The initial response actions given in Table 1 are examples, and the user is free to implement other alternatives.

6.3.3 The need to reclassify the site should be evaluated when additional site information is collected that indicates a significant change in site conditions or when implementation of an interim response action causes a significant change in site conditions.

6.4 *Development of a Tier 1 Look-Up Table of RBSL*—If a look-up table is not available, the user is responsible for developing the look-up table. If a look-up table is available, the user is responsible for determining that the RBSLs in the look-up table are based on currently acceptable methodologies and parameters. The look-up table is a tabulation for potential exposure pathways, media (for example, soil, water, and air), a range of incremental carcinogenic risk levels (10E-4 to 10E-6 are often evaluated as discussed in Appendix X1 paragraph X1.7, Discussion of Acceptable Risk) and hazard quotients equal to unity, and potential exposure scenarios (for example, residential, commercial, industrial, and agricultural) for each chemical(s) of concern.

6.4.1 The RBSLs are determined using typical, non-site-specific values for exposure parameters and physical parameters for media. The RBSLs are calculated according to methodology suggested by the USEPA. For each exposure scenario, the RBSLs are based on current USEPA RME parameters and current toxicological information given in Refs (2, 3) or peer-reviewed source(s). Consequently, the RBSL look-up table is updated when new methodologies and parameters are developed. For indirect pathways, fate and transport models can be used to predict RBSLs at a source area that corresponds to exposure point concentrations. An example of the development of a Tier 1 Look-Up Table and RBSL is given in Appendix X2. *Figure 2 and Appendix X2 are presented solely for the purpose of providing an example development of the RBSL, and the values should not be viewed as proposed RBSLs.*

6.4.2 Appendix X2 is an example of an abbreviated Tier 1 RBSL Look-Up Table for compounds of concern associated with petroleum releases. The exposure scenarios selected in the example case are for residential and industrial/commercial scenarios characterized by USEPA RME parameters for

TABLE 1 Example Site Classification and Initial Response Actions[A]

Criteria and Prescribed Scenarios	Example Initial Response Actions[B]
1. Immediate threat to human health, safety, or sensitive environmental receptors	Notify appropriate authorities, property owners, and potentially affected parties, and only evaluate the need to
• Explosive levels, or concentrations of vapors that could cause acute health effects, are present in a residence or other building.	• Evacuate occupants and begin abatement measures such as subsurface ventilation or building pressurization.
• Explosive levels of vapors are present in subsurface utility system(s), but no building or residences are impacted.	• Evacuate immediate vicinity and begin abatement measures such as ventilation.
• Free-product is present in significant quantities at ground surface, on surface water bodies, in utilities other than water supply lines, or in surface water runoff.	• Prevent further free-product migration by appropriate containment measures, institute free-product recovery, and restrict area access.
• An active public water supply well, public water supply line, or public surface water intake is impacted or immediately threatened.	• Notify user(s), provide alternate water supply, hydraulically control contaminated water, and treat water at point-of-use.
• Ambient vapor/particulate concentrations exceed concentrations of concern from an acute exposure or safety viewpoint.	• Install vapor barrier (capping, foams, and so forth), remove source, or restrict access to affected area.
• A sensitive habitat or sensitive resources (sport fish, economically important species, threatened and endangered species, and so forth) are impacted and affected.	• Minimize extent of impact by containment measures and implement habitat management to minimize exposure.
2. Short-term (0 to 2 years) threat to human health, safety, or sensitive environmental receptors	Notify appropriate authorities, property owners, and potentially affected parties, and only evaluate the need to
• There is potential for explosive levels, or concentrations of vapors that could cause acute effects, to accumulate in a residence or other building.	• Assess the potential for vapor migration (through monitoring/modeling) and remove source (if necessary), or install vapor migration barrier.
• Shallow contaminated surface soils are open to public access, and dwellings, parks, playgrounds, day-care centers, schools, or similar use facilities are within 500 ft (152 m) of those soils.	• Remove soils, cover soils, or restrict access.
• A non-potable water supply well is impacted or immediately threatened.	• Notify owner/user and evaluate the need to install point-of-use water treatment, hydraulic control, or alternate water supply.
• Ground water is impacted, and a public or domestic water supply well producing from the impacted aquifer is located within two-years projected ground water travel distance down gradient of the known extent of chemical(s) concern.	• Institute monitoring and then evaluate if natural attenuation is sufficient, or if hydraulic control is required.
• Ground water is impacted, and a public or domestic water supply well producing from a different interval is located within the known extent of chemicals of concern.	• Monitor ground water well quality and evaluate if control is necessary to prevent vertical migration to the supply well.
• Impacted surface water, storm water, or ground water discharges within 500 ft (152 m) of a sensitive habitat or surface water body used for human drinking water or contact recreation.	• Institute containment measures, restrict access to areas near discharge, and evaluate the magnitude and impact of the discharge.
3. Long-term (>2 years) threat to human health, safety, or sensitive environmental receptors	Notify appropriate authorities, property owners, and potentially affected parties, and only evaluate the need to
• Subsurface soils (>3 ft (0.9 m) BGS) are significantly impacted, and the depth between impacted soils and the first potable aquifer is less than 50 ft (15 m).	• Monitor ground water and determine the potential for future migration of the chemical(s) concerns to the aquifer.
• Ground water is impacted, and potable water supply wells producing from the impacted interval are located >2 years ground water travel time from the dissolved plume.	• Monitor the dissolved plume and evaluate the potential for natural attenuation and the need for hydraulic control.
• Ground water is impacted, and non-potable water supply wells producing from the impacted interval are located >2 years ground water travel time from the dissolved plume.	• Identify water usage of well, assess the effect of potential impact, monitor the dissolved plume, and evaluate whether natural attenuation or hydraulic control are appropriate control measures.
• Ground water is impacted, and non-potable water supply wells that do not produce from the impacted interval are located within the known extent of chemical(s) of concern.	• Monitor the dissolved plume, determine the potential for vertical migration, notify the user, and determine if any impact is likely.
• Impacted surface water, storm water, or ground water discharges within 1500 ft (457 m) of a sensitive habitat or surface water body used for human drinking water or contact recreation.	• Investigate current impact on sensitive habitat or surface water body, restrict access to area of discharge (if necessary), and evaluate the need for containment/control measures.
• Shallow contaminated surface soils are open to public access, and dwellings, parks, playgrounds, day-care centers, schools, or similar use facilities are more than 500 ft (152 m) of those soils.	• Restrict access to impact soils.
4. No demonstrable long-term threat to human health or safety or sensitive environmental receptors	Notify appropriate authorities, property owners, and potentially affected parties, and only evaluate the need to
Priority 4 scenarios encompass all other conditions not described in Priorities 1, 2, and 3 and that are consistent with the priority description given above. Some examples are as follows:	
• Non-potable aquifer with no existing local use impacted.	• Monitor ground water and evaluate effect of natural attenuation on dissolved plume migration.
• Impacted soils located more than 3 ft (0.9 m) BGS and greater than 50 ft (15 m) above nearest aquifer.	• Monitor ground water and evaluate effect of natural attenuation on leachate migration.
• Ground water is impacted, and non-potable wells are located down gradient outside the known extent of the chemical(s) of concern, and they produce from a nonimpacted zone.	• Monitor ground water and evaluate effect of natural attenuation on dissolved plume migration.

[A] Johnson, P. C., DeVaull, G. E., Ettinger, R. A., MacDonald, R. L. M., Stanley, C. C., Westby, T. S., and Conner, J., "Risk-Based Corrective Action: Tier 1 Guidance Manual," Shell Oil Co., July 1993.

[B] Note that these are potential initial response actions that may not be appropriate for all sites. The user is encouraged to select options that best address the short-term health and safety concerns of the site, while the RBCA process progresses.

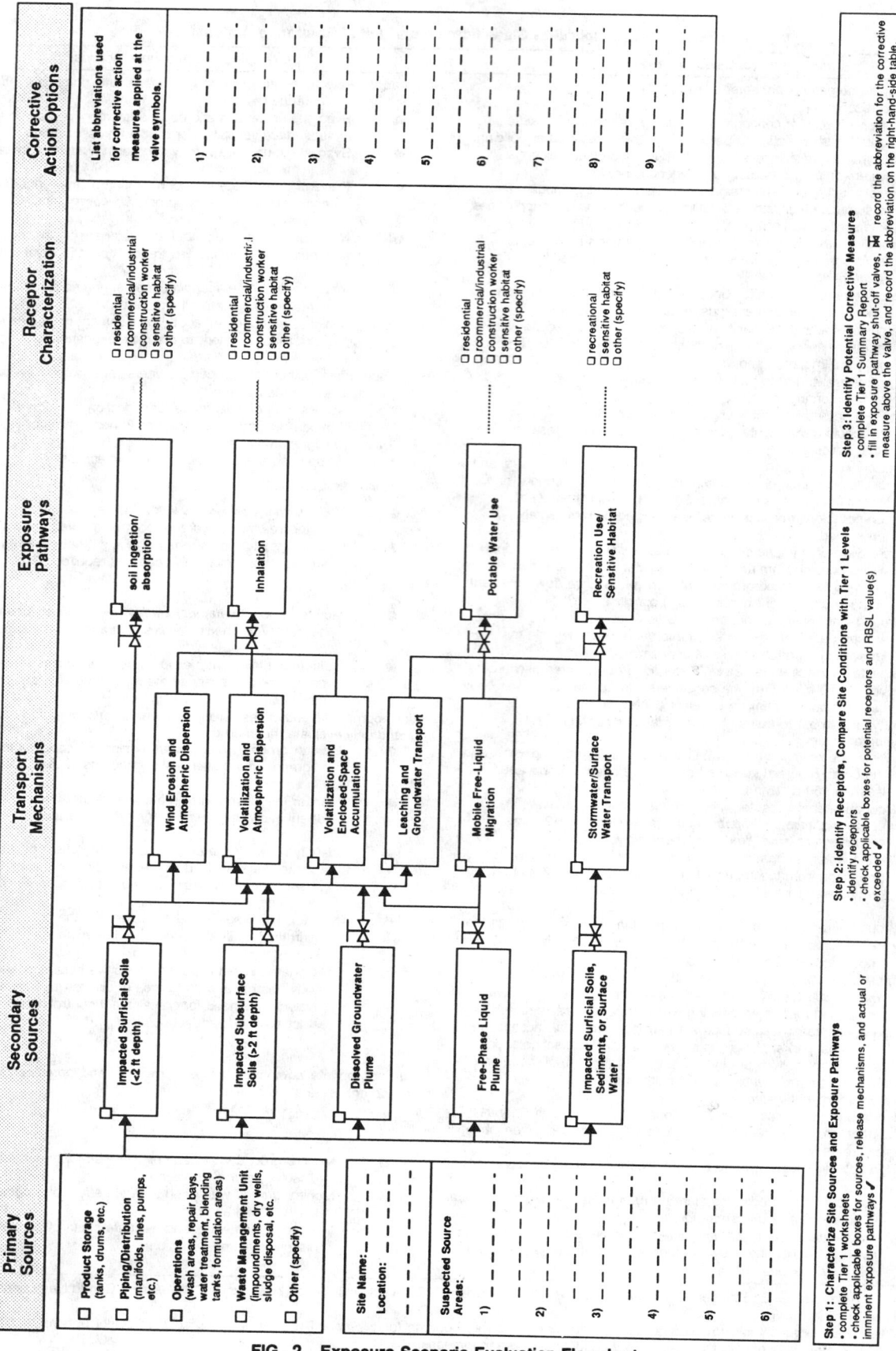

FIG. 2 Exposure Scenario Evaluation Flowchart

adult males. The assumptions and methodology used in deriving the example are discussed in Appendix X2. Note that not all possible exposure pathways are considered in the derivation of the example. *The user should always review the assumptions and methodology used to derive values in a look-up table to make sure that they are consistent with reasonable exposure scenarios for the site being considered as well as currently accepted methodologies.* The value of creating a look-up table is that users do not have to repeat the exposure calculations for each site encountered. The look-up table is only altered when RME parameters, toxicological information, or recommended methodologies are updated. Some states have compiled such tables for direct exposure pathways that, for the most part, contain identical values (as they are based on the same assumptions). Values for the cross-media pathways (for example, volatilization and leaching), when available, often differ because these involve coupling exposure calculations with predictive equations for the fate and transport of chemicals in the environment. As yet, there is little agreement in the technical community concerning non-site-specific values for the transport and fate model parameters, or the choice of the models themselves. *Again, the reader should note that the example is presented here only as an abbreviated example of a Tier 1 RBSL Look-Up Table for typical compounds of concern associated with petroleum products.*

6.4.3 *Use of Total Petroleum Hydrocarbon Measurements*—Various chemical analysis methods commonly referred to as total petroleum hydrocarbons (TPHs) are often used in site assessments. These methods usually determine the total amount of hydrocarbons present as a single number and give no information on the types of hydrocarbon present. The TPHs should not be used for risk assessment because the general measure of TPH provides insufficient information about the amounts of individual chemical(s) of concern present.

6.5 *Comparison of Site Conditions with Tier 1 Risk-Based Screening Levels (RBSL)*—In Tier 1, the point(s) of exposure and point(s) of compliance are assumed to be located within close proximity to the source area(s) or the area where the highest concentrations of the chemical(s) of concern have been identified. Concentrations of the chemical(s) of concern measured at the source area(s) identified at the site should be compared to the look-up table RBSL. If there is sufficient site assessment data, the user may opt to compare RBSLs with statistical limits (for example, upper confidence levels) rather than maximum values detected. Background concentrations should be considered when comparing the RBSLs, to the site concentrations as the RBSLs may sometimes be less than background concentrations. Note that additivity of risks is not explicitly considered in the Tier 1 evaluation, as it is expected that the RBSLs are typically for a limited number of chemical(s) of concern considered at most sites. Additivity may be addressed in Tier 2 and Tier 3 analyses. To accomplish the Tier 1 comparison:

6.5.1 Select the potential exposure scenario(s) (if any) for the site. Exposure scenarios are determined based on the site assessment information described in 6.2;

6.5.2 Based on the impacted media identified, determine the primary sources, secondary sources, transport mechanisms, and exposure pathways;

6.5.3 Select the receptors (if any) based on current and anticipated future use. Consider land use restrictions and surrounding land use when making this selection.

6.5.4 Identify the exposure scenarios where the measured concentrations of the chemical(s) of concern are above the RBSL.

6.6 *Exposure Evaluation Flowchart*—During a Tier 1 evaluation, the risk evaluation flowchart presented in Fig. 2 may be used as a tool to guide the user in selecting appropriate exposure scenarios based on site assessment information. This worksheet may also be used in the evaluation of remedial action alternatives. To complete this flowchart:

6.6.1 Characterize site sources and exposure pathways, using the data summarized from Tier 1 to customize the risk evaluation flowchart for the site by checking the small checkbox for every relevant source, transport mechanism, and exposure pathway.

6.6.2 Identify receptors, and compare site conditions with Tier 1 levels: For each exposure pathway selected, check the receptor characterization (residential, commercial, and so forth) where the concentrations of the chemical(s) of concern are above the RBSL. Consider land use restrictions and surrounding land use when making this selection. Do not check any boxes if there are no receptors present, or likely to be present, or if institutional controls prevent exposure from occurring and are likely to stay in place.

6.6.3 Identify potential remedial action measures. Select remedial action options to reduce or eliminate exposure to the chemical(s) of concern.

6.6.4 The exposure evaluation flowchart (Fig. 2) can be used to graphically portray the effect of the Tier 1 remedial action. Select the Tier 1 remedial action measure or measures (shown as valve symbols) that will break the lines linking sources, transport mechanisms, and pathways leading to the chemical(s) of concern above the RBSL. Adjust the mix of remedial action measures until no potential receptors have concentrations of chemical(s) of concerns above the RBSL with the remedial action measures in place. Show the most likely Tier 1 remedial action measure(s) selected for this site by marking the appropriate valve symbols on the flowchart and recording a remedial action measure on the right-hand-side of this figure.

6.7 *Evaluation of Tier Results*—At the conclusion of each tier evaluation, the user compares the target levels (RBSLs or SSTLs) to the concentrations of the chemical(s) of concern at the point(s) of compliance.

6.7.1 If the concentrations of the chemical(s) of concern exceed the target levels at the point(s) of compliance, then either remedial action, interim remedial action, or further tier evaluation should be conducted.

6.7.1.1 *Remedial Action*—A remedial action program is designed and implemented. This program may include some combination of source removal, treatment, and containment technologies, as well as engineering and institutional controls. Examples of these include: soil venting, bioventing, air sparging, pump and treat, and natural attenuation/passive remediation. When concentrations of chemical(s) of concern no longer exceed the target levels at the point of compliance, then the user may elect to move to 6.7.3.

6.7.1.2 *Interim Remedial Action*—If achieving the desired risk reduction is impracticable due to technology or resource limitations, an interim remedial action, such as removal or treatment of "hot spots," may be conducted to address the most significant concerns, change the site classification, and facilitate reassessment of the tier evaluation.

6.7.1.3 *Further Tier Evaluation*—If further tier evaluation is warranted, additional site assessment information may be collected to develop SSTLs under a Tier 2 or Tier 3 evaluation. Further tier evaluation is warranted when:

(*1*) The basis for the RBSL values (for example, geology, exposure parameters, point(s) of exposure, and so forth) are not representative of the site-specific conditions; or

(*2*) The SSTL developed under further tier evaluation will be significantly different from the Tier 1 RBSL or will significantly modify the remedial action activities; or

(*3*) Cost of remedial action to RBSLs will likely be greater than further tier evaluation and subsequent remedial action.

6.7.2 If the concentrations of chemicals of concern at the point of compliance are less than the target levels, but the user is not confident that data supports the conclusion that concentrations will not exceed target levels in the future, then the user institutes a monitoring plan to collect data sufficient to confidently conclude that concentrations will not exceed target levels in the future. When this data is collected, the user moves to 6.7.3.

6.7.3 If the concentrations of chemicals of concern at the point of compliance are less than target levels, and the user is confident that data supports the conclusion that concentrations will not exceed target levels in the future, then no additional corrective action activities are necessary, and the user has completed the RBCA process. In practice, this is often accompanied by the issuing of a no-further-action letter by the oversight regulatory agency.

6.8 *Tier 2*—Tier 2 provides the user with an option to determine the site-specific point(s) of compliance and corresponding SSTL for the chemical(s) of concern applicable at the point(s) of compliance and source area(s). Additional site assessment data may be required; however, the incremental effort is typically minimal relative to Tier 1. If the user completes a Tier 1 evaluation, in most cases, only a limited number of pathways, exposure scenarios, and chemical(s) of concern are considered in the Tier 2 evaluation since many are eliminated from consideration during the Tier 1 evaluation.

6.8.1 In Tier 2, the user:

6.8.1.1 Identifies the indirect exposure scenarios to be addressed and the appropriate site-specific point(s) of compliance. A combination of assessment data and predictive modeling results are used to determine the SSTL at the source area(s) or the point(s) of compliance, or both; or

6.8.1.2 Applies Tier 1 RBSL Look-Up Table values for the direct exposure scenarios at reasonable point(s) of exposure (as opposed to the source area(s) as is done in Tier 1). The SSTLs for source area(s) and point(s) of compliance can be determined based on the demonstrated and predicted attenuation (reduction in concentration with distance) of compounds that migrate away from the source area(s).

6.8.1.3 An example of a Tier 2 application is illustrated in Appendix X5.

6.8.2 Tier 2 of the RBCA process involves the development of SSTL based on the measured and predicted attenuation of the chemical(s) of concern away from the source area(s) using relatively simplistic mathematical models. The SSTLs for the source area(s) are generally not equal to the SSTL for the point(s) of compliance. The predictive equations are characterized by the following:

6.8.2.1 The models are relatively simplistic and are often algebraic or semianalytical expressions;

6.8.2.2 Model input is limited to practicably attainable site-specific data or easily estimated quantities (for example total porosity, soil bulk density); and

6.8.2.3 The models are based on descriptions of relevant physical/chemical phenomena. Most mechanisms that are neglected result in predicted concentrations that are greater than those likely to occur (for example, assuming constant concentrations in source area(s)). Appendix X3 discusses the use of predictive models and presents models that might be considered for Tier 2 evaluation.

6.8.3 *Tier 2 Evaluation*—Identify the exposure scenarios where the measured concentrations of the chemical(s) of concern are above the SSTL at the point(s) of compliance, and evaluate the tier results in accordance with 6.7.

6.9 *Tier 3*—In a Tier 3 evaluation, SSTLs for the source area(s) and the point(s) of compliance are developed on the basis of more sophisticated statistical and contaminant fate and transport analyses, using site-specific input parameters for both direct and indirect exposure scenarios. Source area(s) and the point(s) of compliance SSTLs are developed to correspond to concentrations of chemical(s) of concern at the point(s) of exposure that are protective of human health and the environment. Tier 3 evaluations commonly involve collection of significant additional site information and completion of more extensive modeling efforts than is required for either a Tier 1 or Tier 2 evaluation.

6.9.1 Examples of Tier 3 analyses include the following:

6.9.1.1 The use of numerical ground water modeling codes that predict time-dependent dissolved contaminant transport under conditions of spatially varying permeability fields to predict exposure point(s) of concentrations;

6.9.1.2 The use of site-specific data, mathematical models, and Monte Carlo analyses to predict a statistical distribution of exposures and risks for a given site; and

6.9.1.3 The gathering of sufficient data to refine site-specific parameter estimates (for example, biodegradation rates) and improve model accuracy in order to minimize future monitoring requirements.

6.9.2 *Tier 3 Evaluation*—Identify the exposure scenarios where the measured concentrations of the chemical(s) of concern are above the SSTL at the point(s) of compliance, and evaluate the tier results in accordance with 6.7 except that a tier upgrade (6.7.5) is not available.

6.10 *Implementing the Selected Remedial Action Program*—When it is judged by the user that no further assessment is necessary, or practicable, a remedial alternatives evaluation should be conducted to confirm the most cost-effective option for achieving the final remedial action target levels (RBSLs or SSTLs, as appropriate). Detailed design specifications may then be developed for installation and operation of the selected measure. The remedial action must continue until such time as monitoring indicates that concentrations of the chemical(s) of concern are not above the RBSL or SSTL, as appropriate, at the points of compliance or source area(s), or both.

6.11 *RBCA Report*—After completion of the RBCA activities, a RBCA report should be prepared and submitted to the regulatory agency. The RBCA report should, at a minimum, include the following:

6.11.1 An executive summary;

6.11.2 A site description;

6.11.3 A summary of the site ownership and use;

6.11.4 A summary of past releases or potential source areas;

6.11.5 A summary of the current and completed site activities;

6.11.6 A description of regional hydrogeologic conditions;

6.11.7 A description of site-specific hydrogeologic conditions;

6.11.8 A summary of beneficial use;

6.11.9 A summary and discussion of the risk assessment (hazard identification, dose response assessment, exposure assessment, and risk characterization), including the methods and assumptions used to calculate the RBSL or SSTL, or both;

6.11.10 A summary of the tier evaluation;

6.11.11 A summary of the analytical data and the appropriate RBSL or SSTL used;

6.11.12 A summary of the ecological assessment;

6.11.13 A site map of the location;

6.11.14 An extended site map to include local land use and ground water supply wells;

6.11.15 Site plan view showing location of structures, aboveground storage tanks, underground storage tanks, buried utilities and conduits, suspected/confirmed sources, and so forth;

6.11.16 Site photos, if available;

6.11.17 A ground water elevation map;

6.11.18 Geologic cross section(s); and

6.11.19 Dissolved plume map(s) of the chemical(s) of concern.

6.12 *Monitoring and Site Maintenance*—In many cases, monitoring is necessary to demonstrate the effectiveness of implemented remedial action measures or to confirm that current conditions persist or improve with time. Upon completion of this monitoring effort (if required), no further action is required. In addition, some measures (for example, physical barriers such as capping, hydraulic control, and so forth) require maintenance to ensure integrity and continued performance.

6.13 *No Further Action and Remedial Action Closure*—When RBCA RBSLs or SSTLs have been demonstrated to be achieved at the point(s) of compliance or source area(s), or both, as appropriate, and monitoring and site maintenance are no longer required to ensure that conditions persist, then no further action is necessary, except to ensure that institutional controls (if any) remain in place.

APPENDIXES

(Nonmandatory Information)

X1. PETROLEUM PRODUCTS CHARACTERISTICS: COMPOSITION, PHYSICAL AND CHEMICAL PROPERTIES, AND TOXICOLOGICAL ASSESSMENT SUMMARY

X1.1 *Introduction:*

X1.1.1 Petroleum products originating from crude oil are complex mixtures of hundreds to thousands of chemicals; however, practical limitations allow us to focus only on a limited subset of key components when assessing the impact of petroleum fuel releases to the environment. Thus, it is important to have a basic understanding of petroleum properties, compositions, and the physical, chemical, and toxicological properties of some compounds most often identified as the key chemicals or chemicals of concern.

X1.1.2 This appendix provides a basic introduction to the physical, chemical, and toxicological characteristics of petroleum products (gasoline, diesel fuel, jet fuel, and so forth)[5] and other products focussed primarily towards that information which is most relevant to assessing potential impacts due to releases of these products into the subsurface. Much of the information presented is summarized from the references listed at the end of this guide. For specific topics, the reader is referred to the following sections of this appendix:

X1.1.2.1 *Composition of Petroleum Fuels*—See X1.2.

X1.1.2.2 *Physical, Chemical, and Toxicological Properties of Petroleum Fuels*—See X1.3.

X1.1.2.3 *Chemical of Concern*—See X1.4.

X1.1.2.4 *Toxicity of Petroleum Hydrocarbons*—See X1.5.

X1.1.2.5 *Profiles of Select Compounds*—See X1.6.

X1.2 *Composition of Petroleum Products:*

X1.2.1 Most petroleum products are derived from crude oil by distillation, which is a process that separates compounds by volatility. Crude oils are variable mixtures of thousands of chemical compounds, primarily hydrocarbons; consequently, the petroleum products themselves are also variable mixtures of large numbers of components. The biggest variations in composition are from one type of product to another (for example, gasoline to motor oil); however, there are even significant variations within different samples of the same product type. For example, samples of gasoline taken from the same fuel dispenser on different days, or samples taken from different service stations, will have different compositions. These variations are the natural result of differing crude oil sources, refining processes and conditions, and kinds and amount of additives used.

---

[5] "Alternative products," or those products not based on petroleum hydrocarbons (or containing them in small amounts), such as methanol or M85, are beyond the scope of the discussion in this appendix.

X1.2.2 *Components of Petroleum Products*—The components of petroleum products can be generally classified as either hydrocarbons (organic compounds composed of hydrogen and carbon only) or as non-hydrocarbons (compounds containing other elements, such as oxygen, sulfur, or nitrogen). Hydrocarbons make up the vast majority of the composition of petroleum products. The non-hydrocarbon compounds in petroleum products are mostly hydrocarbon-like compounds containing minor amounts of oxygen, sulfur, or nitrogen. Most of the trace levels of metals found in crude oil are removed by refining processes for the lighter petroleum products.

X1.2.3 *Descriptions and Physical Properties of Petroleum Products*—In order to simplify the description of various petroleum products, boiling point ranges and carbon number (number of carbon atoms per molecule) ranges are commonly used to describe and compare the compositions of various petroleum products. Table X1.1 summarizes these characteristics for a range of petroleum products. Moving down the list from gasoline, increases in carbon number range and boiling range and decreases in volatility (denoted by increasing flash point) indicate the transition to "heavier products." Additional descriptions of each of these petroleum products are provided as follows.

X1.2.4 *Gasoline*—Gasoline is composed of hydrocarbons and "additives" that are blended with the fuel to improve fuel performance and engine longevity. The hydrocarbons fall primarily in the C4 to C12 range. The lightest of these are highly volatile and rapidly evaporate from spilled gasoline. The C4 and C5 aliphatic hydrocarbons rapidly evaporate from spilled gasoline (hours to months, depending primarily on the temperature and degree of contact with air). Substantial portions of the C6 and heavier hydrocarbons also evaporate, but at lower rates than for the lighter hydrocarbons.

X1.2.4.1 Figure X1.1 shows gas chromatograms of a fresh gasoline and the same gasoline after simulated weathering; air was bubbled through the gasoline until 60 % of its initial volume was evaporated. In gas chromatography, the mixture is separated into its components, with each peak representing different compounds. Higher molecular weight components appear further to the right along the *x*-axis. For reference, positions of the *n*-aliphatic hydrocarbons are indicated in Fig. X1.1. The height of, and area under, each peak are measures of how much of that component is present in the mixture. As would be expected by their higher volatilities, the lighter hydrocarbons (up to about C7) evaporate first and are greatly reduced in the weathered gasoline. The gas chromatogram of a fuel oil is also shown for comparison.

X1.2.4.2 The aromatic hydrocarbons in gasoline are primarily benzene (C_6H_6), toluene (C_7H_8), ethylbenzene (C_8H_{10}), and xylenes (C_8H_{10}); these are collectively referred to as "BTEX." Some heavier aromatics are present also, including low amounts of polyaromatic hydrocarbons (PAHs). Aromatics typically comprise about 10 to 40 % of gasoline.

X1.2.4.3 Oxygenated compounds ("oxygenates") such as alcohols (for example, methanol or ethanol) and ethers (for example, methyl tertiarybutyl ether—MTBE) are sometimes added to gasoline as octane boosters and to reduce carbon monoxide exhaust emissions. Methyl tertiarbutyl ether has been a common additive only since about 1980.

X1.2.4.4 Leaded gasoline, which was more common in the past, contained lead compounds added as octane boosters. Tetraethyl lead (TEL) is one lead compound that was commonly used as a gasoline additive. Other similar compounds were also used. Sometimes mixtures of several such compounds were added. Because of concerns over atmospheric emissions of lead from vehicle exhaust, the EPA has reduced the use of leaded gasolines. Leaded gasolines were phased out of most markets by 1989.

X1.2.4.5 In order to reduce atmospheric emissions of lead, lead "scavengers" were sometimes added to leaded gasolines. Ethylene dibromide (EDB) and ethylene dichloride (EDC) were commonly used for this purpose.

X1.2.5 *Kerosene and Jet Fuel*—The hydrocarbons in kerosene commonly fall into the C11 to C13 range, and distill at approximately 150 to 250°C. Special wide-cut (that is, having broader boiling range) kerosenes and low-flash kerosenes are also marketed. Both aliphatic and aromatic hydrocarbons are present, including more multi-ring compounds and kerosene.

X1.2.5.1 Commercial jet fuels JP-8 and Jet A have similar compositions to kerosene. Jet fuels JP-4 and JP-5 are wider cuts used by the military. They contain lighter distillates and have some characteristics of both gasoline and kerosene.

X1.2.5.2 Aromatic hydrocarbons comprise about 10 to 20 % of kerosene and jet fuels.

X1.2.6 *Diesel Fuel and Light Fuel Oils*—Light fuel oils include No. 1 and No. 2 fuel oils, and boil in the range from 160 to 400°C. Hydrocarbons in light fuel oils and diesel fuel typically fall in the C10 to C20 range. Because of their higher molecular weights, constituents in these products are less volatile, less water soluble, and less mobile than gasoline- or kerosene-range hydrocarbons.

X1.2.6.1 About 25 to 35 % of No. 2 fuel oil is composed of aromatic hydrocarbons, primarily alkylated benzenes and naphthalenes. The BTEX concentrations are generally low.

X1.2.6.2 No. 1 fuel oil is typically a straight run distillate.

X1.2.6.3 No. 2 fuel oil can be either a straight run distillate, or else is produced by catalytic cracking (a process in which larger molecules are broken down into smaller ones). Straight run distillate No. 2 is commonly used for home heating fuel, while the cracked product is often used for industrial furnaces and boilers. Both No. 1 and No. 2 fuel oils are sometimes used as blending components for jet fuel or diesel fuel formulations.

X1.2.7 *Heavy Fuel Oils*—The heavy fuel oils include Nos. 4, 5, and 6 fuel oils. They are sometimes referred to as "gas oils" or "residual fuel oils." These are composed of hydrocarbons ranging from about C19 to C25 and have a boiling range from about 315 to 540°C. They are dark in color and considerably more viscous than water. They typically contain 15 to 40 % aromatic hydrocarbons, dominated by alkylated phenanthrenes and naphthalenes. Polar compounds containing nitrogen, sulfur, or oxygen may comprise 15 to 30 % of the oil.

X1.2.7.1 No. 6 fuel oil, also called "Bunker Fuel" or "Bunker C," is a gummy black product used in heavy industrial applications where high temperatures are available to fluidize the oil. Its density is greater than that of water.

X1.2.7.2 Nos. 4 and 5 fuel oils are commonly produced by blending No. 6 fuel oil with lighter distillates.

ASTM E 1739

X1.2.8 Motor Oils and Other Lubricating Oils—Lubricating oils and motor oils are predominately comprised of compounds in the C20 to C45 range and boil at approximately 425 to 540°C. They are enriched in the most complex molecular fractions found in crude oil, such as cycloparaffins and PNAs having up to three rings or more. Aromatics may make up to 10 to 30 % of the oil. Molecules containing nitrogen, sulfur, or oxygen are also common. In addition, used automative crankcase oils become enriched with PNAs and certain metals.

X1.2.8.1 These oils are relatively viscous and insoluble in ground water and relatively immobile in the subsurface.

X1.2.8.2 Waste oil compositions are even more difficult to predict. Depending on how they are managed, waste oils may contain some portion of the lighter products in addition to heavy oils. Used crankcase oil may contain wear metals from engines. Degreasing solvents (gasoline, naphtha, or light chlorinated solvents, or a combination thereof) may be present in some wastes.

X1.3 Physical, Chemical, and Toxicological Characteristics of Petroleum Products:

X1.3.1 Trends in Physical/Chemical Properties of Hydrocarbons—In order to better understand the subsurface behavior of hydrocarbons it is helpful to be able to recognize trends in important physical properties with increasing number of carbon atoms. These trends are most closely followed by compounds with similar molecular structures, such as the straight-chained, single-bonded aliphatic hydrocarbons. In general, as the carbon number (or molecule size) increases, the following trends are observed:

X1.3.1.1 Higher boiling points (and melting points),

X1.3.1.2 Lower vapor pressure (volatility),

X1.3.1.3 Greater density,

X1.3.1.4 Lower water solubility, and

X1.3.1.5 Stronger adhesion to soils and less mobility in the subsurface.

X1.3.2 Table X1.2 lists physical, chemical, and toxicological properties for a number of hydrocarbons found in petroleum products. In general:

X1.3.2.1 Aliphatic petroleum hydrocarbons with more than ten carbon atoms are expected to be immobile in the subsurface, except when dissolved in nonaqueous phase liquids (NAPLs), due to their low water solubilities, low vapor pressures, and strong tendency to adsorb to soil surfaces.

X1.3.2.2 Aromatic hydrocarbons are more water soluble and mobile in water than aliphatic hydrocarbons of similar molecular weight.

X1.3.2.3 Oxygenates generally have much greater water solubilities than hydrocarbons of similar molecular weight, and hence are likely to be the most mobile of petroleum fuel constituents in leachate and ground water. The light alcohols, including methanol and ethanol, are completely miscible with water in all proportions.

X1.3.3 Properties of Mixtures—It is important to note that the partitioning behavior of individual compounds is affected by the presence of other hydrocarbons in the subsurface. The maximum dissolved and vapor concentrations achieved in the subsurface are always less than that of any pure compound, when it is present as one of many constituents of a petroleum fuel. For example, dissolved benzene concentrations in ground water contacting gasoline-impacted soils rarely exceed 1 to 3 % of the ~1800-mg/L pure component solubility of benzene.

X1.3.4 Trends in Toxicological Properties of Hydrocarbons—A more detailed discussion of toxicological assessment is given in X1.5 (see also Appendix X3), followed by profiles for select chemicals found in petroleum products given in X1.6. Of the large number of compounds present in petroleum products, aromatic hydrocarbons (BTEX, PAHs, and so forth) are the constituents that human and aquatic organisms tend to be most sensitive to (relative to producing adverse health impacts).

X1.4 Chemicals of Concern for Risk Assessments:

X1.4.1 It is not practicable to evaluate every compound present in a petroleum product to assess the human health or environmental risk from a spill of that product. For this reason, risk management decisions are generally based on assessing the potential impacts from a select group of "indicator" compounds. It is inherently assumed in this approach that a significant fraction of the total potential impact from all chemicals is due to the chemicals of concern. The selection of chemicals of concern is based on the consideration of exposure routes, concentrations, mobilities, toxicological properties, and aesthetic characteristics (taste, odor, and so forth). Historically, the relatively low toxicities and dissolved-phase mobilities of aliphatic hydrocarbons have made these chemicals of concern of less concern relative to aromatic hydrocarbons. When additives are present in significant quantities, consideration should also be given to including these as chemicals of concern.

X1.4.2 Table X1.3 identifies chemicals of concern most often considered when assessing impacts of petroleum products, based on knowledge of their concentration in the specific fuel, as well as their toxicity, water solubility, subsurface mobility, aesthetic characteristics, and the availability of sufficient information to conduct risk assessments. The chemicals of concern are identified by an "X" in the appropriate column.

X1.5 Toxicity of Petroleum Hydrocarbons:

X1.5.1 The following discussion gives a brief overview of origin of the toxicity parameters (reference doses (RfDs)), and slope factors (SFs), a justification for common choices of chemicals of concern and then, in X1.6, a brief summary of the toxicological, physical, and chemical parameters associated with these chemicals of concern.

X1.5.2 How Toxicity Is Assessed: Individual Chemicals Versus Mixtures—The toxicity of an individual chemical is typically established based on dose-response studies that estimate the relationship between different dose levels and the magnitude of their adverse effects (that is, toxicity). The dose-response data is used to identify a "safe dose" or a toxic level for a particular adverse effect. For a complex mixture of chemicals, the same approach can be used. For example, to evaluate the toxicity of gasoline, a "pure" reference gasoline would be evaluated instead of the individual chemical. This "whole-product" approach to toxicity assessment is strictly applicable only to mixtures identical to the evaluated mixture; gasolines with compositions different from the reference gasoline might have toxicities similar to the reference, but some differences would be expected. In addition, as the composition of gasoline released to the environment changes

197

through natural processes (volatilization, leaching, biodegradation), the toxicity of the remaining portion may change also.

X1.5.3 An alternative to the "whole-product" approach for assessing the toxicity of mixtures is the "individual-constituent" approach. In this approach, the toxicity of each individual constituent (or a selected subset of the few most toxic constituents, so-called chemicals of concern) is separately assessed and the toxicity of the mixture is assumed to be the sum of the individual toxicities using a hazard index approach. This approach is often used by the USEPA; however, it is inappropriate to sum hazard indices unless the toxicological endpoints and mechanisms of action are the same for the individual compounds. In addition, the compounds to be assessed must be carefully selected based on their concentrations in the mixture, their toxicities, how well their toxicities are known, and how mobile they are in the subsurface. Lack of sufficient toxicological information is often an impediment to this procedure.

X1.5.4 *Use of TPH Measurements in Risk Assessments*—Various chemical analysis methods commonly referred to as TPH are often used in site assessments. These methods usually determine the total amount of hydrocarbons present as a single number, and give no information on the types of hydrocarbon present. Such TPH methods may be useful for risk assessments where the whole product toxicity approach is appropriate. However in general, *TPH should not be used for "individual constituent" risk assessments because the general measure of TPH provides insufficient information about the amounts of individual compounds present.*

X1.5.5 *Toxicity Assessment Process*—Dose-response data are used to identify a "safe dose" or toxic level for a particular observed adverse effect. Observed adverse effects can include whole body effects (for example, weight loss, neurological observations), effects on specific body organs, including the central nervous system, teratogenic effects (defined by the ability to produce birth defects), mutagenic effects (defined by the ability to alter the genes of a cell), and carcinogenic effects (defined by the ability to produce malignant tumors in living tissues). Because of the great concern over risk agents which may produce incremental carcinogenic effects, the USEPA has developed weight-of-evidence criteria for determining whether a risk agent should be considered carcinogenic (see Table X1.4).

X1.5.6 Most estimates of a "safe dose" or toxic level are based on animal studies. In rare instances, human epidemiological information is available on a chemical. Toxicity studies can generally be broken into three categories based on the number of exposures to the risk agent and the length of time the study group was exposed to the risk agent. These studies can be described as follows:

X1.5.6.1 *Acute Studies*—Acute studies typically use one dose or multiple doses over a short time frame (24 h). Symptoms are usually observed within a short time frame and can vary from weight loss to death.

X1.5.6.2 *Chronic Studies*—Chronic studies use multiple exposures over an extended period of time, or a significant fraction of the animal's (typically two years) or the individual's lifetime. The chronic effects of major concern are carcinogenic, mutagenic, and teratogenic effects. Other

chronic health effects such as liver and kidney damage are also important.

X1.5.6.3 *Subchronic Studies*—Subchronic studies use multiple or continuous exposures over an extended period (three months is the usual time frame in animal studies). Observed effects include those given for acute and chronic studies.

X1.5.6.4 Ideally, safe or acceptable doses are calculated from chronic studies, although, due to the frequent paucity of chronic data, subchronic studies are used.

X1.5.6.5 For noncarcinogens, safe doses are based on no observed adverse effect levels (NOAELs) or lowest observed adverse effect levels (LOAELs) from the studies.

X1.5.6.6 Acceptable doses for carcinogens are determined from mathematical models used to generate dose-response curves in the low-dose region from experimentally determined dose-response curves in the high-dose region.

X1.5.7 Data from the preceding studies are used to generate reference doses (RfDs), reference concentrations (RfCs), and slope factors (SFs) and are also used in generating drinking water maximum concentration levels (MCLs) and goals (MCLGs), health advisories (HAs), and water quality criteria. These terms are defined in Table X1.5 and further discussed in X3.8.

X1.5.8 *Selection of Chemicals of Concern*—The impact on human health and the environment in cases of gasoline and middle distillate contamination of soils and ground water can be assessed based on potential receptor (that is, aquatic organisms, human) exposure to three groups of materials: light aromatic hydrocarbons, PAHs, and in older spills, lead. Although not one of the primary contaminants previously described, EDB and EDC were used as lead scavengers in some leaded gasolines and may be considered chemicals of concern, when present.

X1.5.9 The light aromatics, benzene, toluene, xylenes, and ethylbenzene have relatively high water solubility and sorb poorly to soils. Thus, they have high mobility in the environment, moving readily through the subsurface. When released into surface bodies of water, these materials exhibit moderate to high acute toxicity to aquatic organisms. Although environmental media are rarely contaminated to the extent that acute human toxicity is an issue, benzene is listed by the USEPA as a Group A Carcinogen (known human carcinogen) and, thus, exposure to even trace levels of this material is considered significant.

TABLE X1.1 Generalized Chemical and Physical Characterization of Petroleum Products

	Predominant Carbon No. Range	Boiling Range, °C	Flash Point,[A] °C
Gasoline	C4 to C12	25 to 215	−40
Kerosene and Jet Fuels	C11 to C13	150 to 250	<21,[B] 21 to 55,[C] >55[D]
Diesel Fuel and Light Fuel Oils	C10 to C20	160 to 400	>35
Heavy Fuel Oils	C19 to C25	315 to 540	>50
Motor Oils and Other Lubricating Oils	C20 to C45	425 to 540	>175

[A] Typical values.
[B] Jet-B, AVTAG and JP-4.
[C] Kerosene, Jet A, Jet A-1, JP-8 and AVTUR.
[D] AVCAT and JP-5.

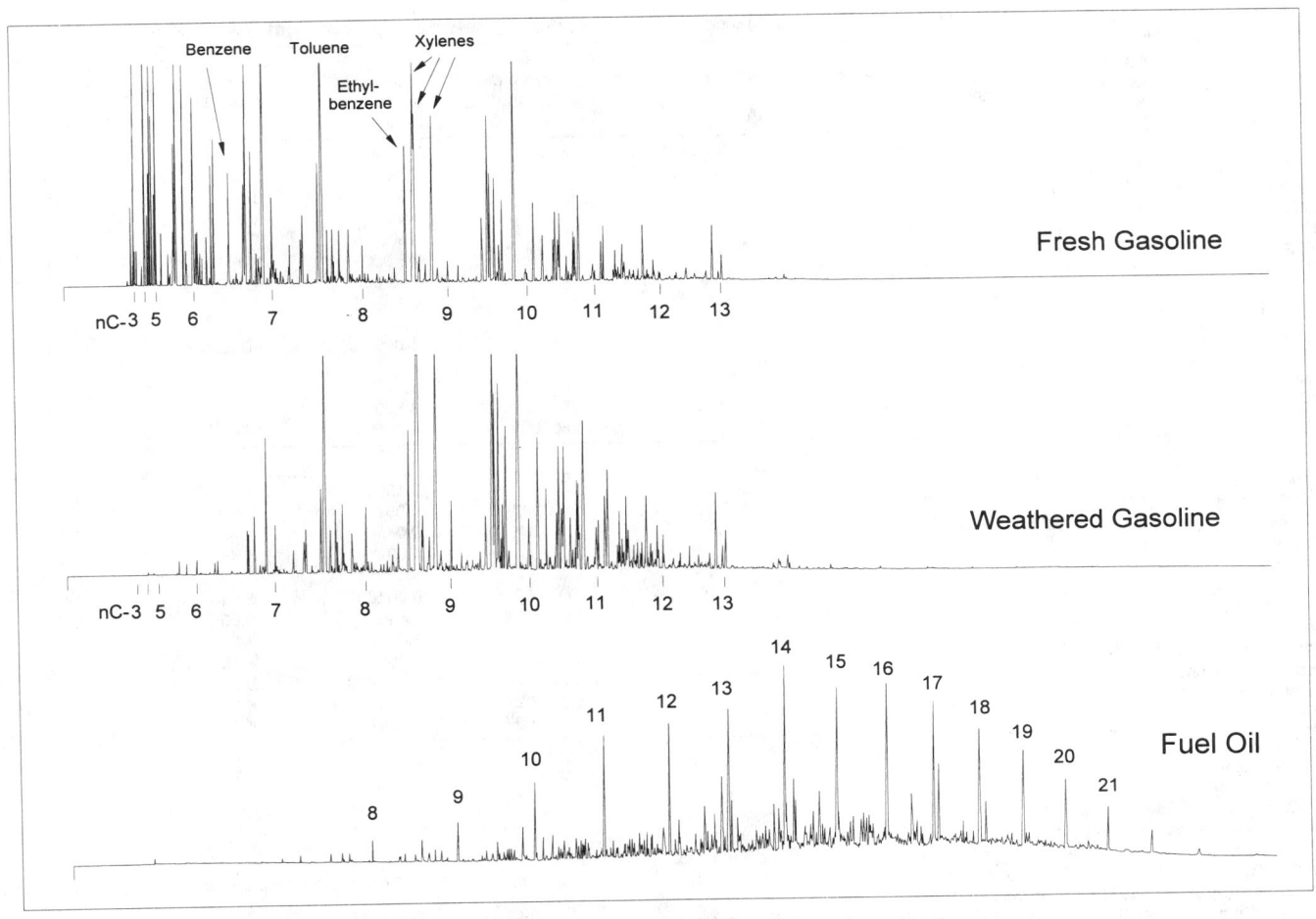

FIG. X1.1 Gas Chromatograms of Some Petroleum Fuels

X1.5.10 Polycyclic aromatics can be broken into two categories: naphthalenes and methylnaphthalenes (diaromatics) have moderate water solubility and soil sorption potential and, thus, their movement through the subsurface tends to be less than monoaromatics, but substantial movement can still occur. When released into surface bodies of water, these materials have moderate to high toxicity to aquatic organisms. The PAHs with three or more condensed rings have very low solubility (typically less than 1 mg/L) and sorb strongly to soils. Thus, their movement in the subsurface is minimal. Several members in the group of three to six-ring PAHs are known or suspected carcinogens and, thus, exposure to low concentrations in drinking water or through the consumption of contaminated soil by children is significant. In addition, materials containing four to six-ring PAHs are poorly biodegradable and, coupled with the potential to bioaccumulate in tissues of aquatic organisms, these materials have the potential to bioconcentrate (be found at levels in living tissue far higher than present in the general surroundings) in the environment.

X1.5.11 Although almost totally eliminated from use in gasolines in the United States, lead is found associated with older spills. Lead was typically added to gasoline either as tetraethyl or tetramethyl lead and may still be found in its original form in areas containing free product. Typically outside the free product zones, these materials have decom-

posed into inorganic forms of lead. Lead is a neurotoxin and lead in the blood of children has been associated with reduced intellectual development. The ingestion by children of lead-contaminated soils is an exposure route of great concern, as is the consumption of lead-contaminated drinking water. Ethylene dibromide and ethylene dichloride, used as lead scavengers in gasolines, are of concern because of their high toxicity (potential carcinogens) and their high mobility in the environment.

X1.5.12 In summary, benzene and benzo(a)pyrene (and in some cases EDB and EDC) are chemicals of concern because of their carcinogenicity. Other PAHs may also be grouped with B(a)P because of uncertainties in their carcinogenicity and because they may accumulate (bioconcentrate) in living tissue.

X1.5.13 *Toxicity and Physical/Chemical Properties for Chemicals of Concern*—A summary of health effects and physical/chemical properties for a number of chemicals of concern is provided in Table X1.2. This table provides toxicological data from a variety of sources, regardless of data quality. A refined discussion for selected chemicals of concern is given as follows. The reader is cautioned that this information is only current as of the dates quoted, and the sources quoted may have been updated, or more recent information may be available in the peer-reviewed literature.

TABLE X1.2 Chemical and Toxicological Properties of Selected Hydrocarbons

Compounds	Weight of Evidence Class[A]	Oral RfD, mg/kg-day	Inhalation RfC, mg/m³	Oral Slope Factor,[A] mg/kg-day⁻¹	Drinking Water MCL,[A] mg/L	Solubility,[B] mg/L	Octanol/Water Partition Coefficient,[B] log K_{ow}	Organic Carbon Adsorption Coefficient,[B] log K_{oc}
Benzene	A	c	c,F	0.029[F]	0.005	1750	2.13	1.58
Toluene	D	0.2[A]	0.4[A]	...	1	535	2.65	2.13
Ethylbenzene	D	0.1[A]	1[A]	...	0.7	152	3.13	1.98
Xylenes	D	2[A]	0.3[C,D]	...	10.0	198	3.26	2.38
n-Hexane	c	0.06[D], 0.6[E]	0.2[E]	...	...	13[K]	...	...
MTBE	...	c	3[A]	...	...	48 000[L]	1.06–1.30[M]	1.08[N]
MEK	D	0.6[A]	1[A]	...	H	268 000	0.26	0.65
MIBK	...	0.05[D], 0.5[E]	0.08[C,D], 0.8[E]	...	...	...	...	...
Methanol	...	0.5[A]	c	...	...	...	...	...
Ethanol	...	...	...	...	...	1 000 000	−0.032	0.34
TBA	...	...	...	...	...	...	...	...
Lead	B2	...	...	...	0.015[I]	...	...	...
EDC	B2	...	...	0.091	0.006	8 520	1.48	1.15
EDB	B2	...	c	85	0.00006	4 300	1.76	1.64
PNAs:								
Pyrene	D	0.03[A]	...	...	...	0.132	4.88	4.58
Benzo(a)pyrene	B2	...	...	7.3	0.0002[J]	0.00120	5.98	5.59
Anthracene	D	0.3[A]	...	...	...	0.0450	4.45	4.15
Phenanthrene	D	...	...	...	...	1.00	4.46	4.15
Naphthalene	D[C]	0.004[D], 0.04[E]	...	...	...	31.0[K]	3.28[K]	3.11[O]
Chrysene	B2	...	...	1.15[G]	0.0002	0.00180	5.61	5.30
Benzo(k)fluoranthene	B2	...	...	...	0.0002[J]	0.430	6.06	5.74
Fluorene	D	0.04[A]	...	...	...	1.69	4.20	3.86
Fluoranthene	D	0.04[A]	...	...	...	0.206	4.90	4.58
Benzo(g,h,i)perylene	D	...	...	...	...	0.000700	6.51	6.20
Benzo(b)fluoranthene	B2	...	...	...	0.0002[J]	0.0140	6.06	5.74
Benz(a)anthracene	B2	...	...	...	0.0002[J]	0.00670	5.60	6.14

[A] See Ref (2).

[B] See Ref (4).

[C] The data is pending in the EPA-IRIS database.

[D] Chronic effect. See Ref (5).

[E] Subchronic effect. See Ref (5).

[F] The inhalation unit risk for benzene is 8.3×10^{-3} (mg/m³)⁻¹. The drinking water unit is 8.3×10^{-4} (mg/L).

[G] See Ref (6). Health-based criteria for carcinogenic polycyclic aromatic compounds (PAHs) with the exception of dibenzo(a,h)anthracene are set at one tenth of the level of benzo(a)pyrene due to their recognized lesser potency.

[H] Listed in the January 1991 Drinking Water Priority List and may be subject to future regulation (56 FR 1470, 01/14/91).

[I] USEPA. May 1993. Office of Drinking Water. 15 µg/L is an action level; standard for tap water.

[J] Proposed standard.

[K] See Ref (7).

[L] See Ref (8).

[M] See Ref (9).

[N] Estimation Equation (from (10)):
  (1) log K_{oc} = −0.55 log S + 3.64, where S = water solubility (mg/L)
  (2) log K_{oc} = 0.544 log P + 1.377

[O] See Ref (11).

X1.5.13.1 The RfD or SF values are generally obtained from a standard set of reference tables (for example, Integrated Risk Information System, IRIS (2), or the Health Effects Assessment Summary Tables, HEAST (3)). Except as noted, the toxicity evaluations that follow were taken from IRIS (2) because these are EPA-sanctioned evaluations. The information in IRIS (2), however, has typically only been peer-reviewed within the EPA and may not always have support from the external scientific community. The information in IRIS may also be subject to error (as exampled by recent revisions in the slope factor for B(a)P and RfC for MTBE).

X1.5.13.2 HEAST (3) is a larger database than IRIS (2) and is often used as a source of health effects information. Whereas the information in IRIS (2) has been subject to data quality review, however, the information in the HEAST (3) tables has not. The user is expected to consult the original assessment documents to appreciate the strengths and limi-

TABLE X1.3 Commonly Selected Chemicals of Concern for Petroleum Products

	Unleaded Gasoline	Leaded Gasoline	Kerosene/ Jet Fuels	Diesel/ Light Fuel Oils	Heavy Fuel Oils
Benzene	X	X	X	...	...
Toluene	X	X	X	...	...
Ethylbenzene	X	X	X	...	...
Xylene	X	X	X	...	...
MTBE, TBA, MEK, MIBK, methanol, ethanol	when suspected[A]	when suspected[A]	...	...	...
Lead, EDC, EDB	...	X	...	...	...
PNAs[B]	...	...	X	X	X

[A] For example, when these compounds may have been present in the spilled gasoline. These additives are not present in all gasolines.

[B] A list of selected PNAs for consideration is presented in Table X1.2.

tations of the data in HEAST (3). Thus, care should be exercised in using the values in HEAST (3).

X1.5.13.3 References for the physical/chemical properties

are provided in Table X1.2. All Henry's law constants quoted in text are from Ref **(11)** except MTBE which is from estimation: $H = (V_p)(MW)/760(S)$, where MW is the molecular weight, $V_p = 414$ mmHg at 100°F, and $S = 48\,000$ mg/L.

X1.6 *Profiles of Select Compounds:*

X1.6.1 *Benzene:*

X1.6.1.1 *Toxicity Summary*—Based on human epidemiological studies, benzene has been found to be a human carcinogen (classified as a Group A carcinogen, known human carcinogen by the USEPA). An oral slope factor of 2.9×10^{-2} (mg/kg/day)$^{-1}$ has been derived for benzene based on the observance of leukemia from occupational exposure by inhalation. The USEPA has set a drinking water maximum contaminant level (MCL) at 5 µg/L. The maximum contaminant level goal (MCLG) for benzene is set at zero.

X1.6.1.2 Although the EPA does not usually set long-term drinking water advisories for carcinogenic materials (no exposure to carcinogens is considered acceptable), a ten-day drinking water health advisory for a child has been set at 0.235 mg/L based on hematological impairment in animals. The EPA is in the process of evaluating noncancer effects and an oral RfD for benzene is pending.

X1.6.1.3 In situations in which both aquatic life and water are consumed from a particular body of water, a recommended EPA water quality criterion is set at 0.66 µg/L. When only aquatic organisms are consumed, the criterion is 40 µg/L. These criteria were established at the one-in-one-million risk level (that is, the criteria represent a one-in-one-million estimated incremental increase in cancer risk over a lifetime).

X1.6.1.4 *Physical/Chemical Parameter Summary*—Benzene is subject to rapid volatilization (Henry's law constant $= 5.5 \times 10^{-3}$ m³-atm/mol) under common above-ground environmental conditions. Benzene will be mobile in soils due to its high water solubility (2.75×10^6 µg/L) and relatively low sorption to soil particles (log $K_{oc} = 1.92$) and, thus, has the potential to leach into ground water. Benzene has a relatively low log K_{ow} value (2.12) and is biodegradable. Therefore, it is not expected to bioaccumulate. In laboratory tests, when a free gasoline phase was in equilibrium with water, typical benzene concentrations in water ranged from 2.42×10^4 to 1.11×10^5 µg/L.

X1.6.2 *Toluene:*

X1.6.2.1 *Toxicity Summary*—Using data from animal studies, the USEPA has set an oral RfD for toluene at 0.2 mg/kg/day. In converting a NOAEL from an animal study, in which the critical effect observed was changes in liver and kidney weights, an uncertainty factor of 1000 and a modifying factor of 1 were used. The EPA has assigned an overall medium level of confidence in the RfD because, although the principal study was well performed, the length of the study corresponded to only subchronic rather than a chronic evaluation, and reproductive aspects were lacking. Based on the RfD and assuming 20 % exposure from drinking water, the EPA has set both drinking water MCL and MCLG of 1000 µg/L. Drinking water health advisories range from 1 mg/L (lifetime equivalent to the RfD) to 20 mg/L (one-day advisory for a child).

X1.6.2.2 In situations in which both aquatic life and water are consumed from a particular body of water, the recommended water quality criterion is set at 1.43×10^4 µg/L. When only aquatic organisms are consumed, the criterion is 4.24×10^5 µg/L.

X1.6.2.3 An inhalation RfC of 0.4 mg/m³ was derived based on neurological effects observed in a small worker population. An uncertainty factor of 300 and a modifying factor of 1 were used to convert the lowest observed adverse effect level (LOAEL) to the RfC. The overall confidence in the RfC was established as medium because of the use of a LOAEL and because of the paucity of exposure information.

X1.6.2.4 *Physical/Chemical Parameter Summary*—Toluene is expected to volatilize rapidly, under common above-ground environmental conditions, due to its relatively high Henry's law constant (6.6×10^{-3} m³-atm/mol). It will be mobile in soils based on an aqueous solubility of 5.35×10^5 µg/L and relatively poor sorption to soils (estimated log $K_{oc} = 2.48$) and, hence, has a potential to leach into ground water. Toluene has a relatively low log K_{ow} (2.73) and is biodegradable. Bioaccumulation of toluene is, therefore, expected to be negligible. In laboratory tests, when a free gasoline phase was in equilibrium with water, typical toluene concentrations in water ranged from 3.48×10^4 to 8.30×10^4 µg/L.

X1.6.3 *Xylenes:*

X1.6.3.1 *Toxicity Summary*—Using data from animal studies, the USEPA has set an oral RfD for xylenes at 2.0 mg/kg/day. In converting a NOAEL from the animal study, in which the critical effects observed were hyperactivity, decreased body weight, and increased mortality (among male rats), an uncertainty factor of 100 and a modifying factor of 1 were used. The EPA has assigned an overall medium level of confidence in the RfD because, although the principal study was well designed and performed, supporting chemistry was not performed. A medium level of confidence was also assigned to the database. Based on the RfD and assuming 20 % exposure from drinking water, the EPA has set both drinking water MCL and MCLG of 10 mg/L. Drinking water health advisories of 10 mg/L (lifetime, adult) and 40 mg/L (one-day, ten-day, and long-term child) are quoted by the EPA's Office of Drinking Water. No USEPA ambient water criteria are available for xylenes at this time. Evaluation of an inhalation RfC is pending.

X1.6.3.2 *Physical/Chemical Parameter Summary*—Xylenes are expected to rapidly volatilize under common above-ground environmental conditions based on their Henry's law constants (for *o*-xylene, $H = 5.1 \times 10^{-3}$ m³-atm/mol). Xylenes have a moderate water solubility (1.46 to 1.98×10^5 µg/L) (pure compound) as well as moderate capacities to sorb to soils (estimated log K_{oc} 2.38 to 2.79) and, therefore, they will be mobile in soils and may leach into ground water. Xylenes are biodegradable, and with log K_{ow} values in the range from 2.8 to 3.3, they are not expected to bioaccumulate.

X1.6.4 *Ethylbenzene:*

X1.6.4.1 *Toxicity Summary*—Using data from animal studies, the USEPA has set an oral RfD for ethylbenzene at 0.1 mg/kg/day. In converting a NOAEL from the animal study, in which the critical effects observed were liver and kidney toxicity, an uncertainty factor of 1000 and a modifying factor of 1 were used. The EPA has assigned an overall

TABLE X1.4 Weight of Evidence Criteria for Carcinogens

Category	Criterion
A	Human carcinogen, with sufficient evidence from epidemiological studies
B1	Probable human carcinogen, with limited evidence from epidemiological studies
B2	Probable human carcinogen, with sufficient evidence from animal studies and inadequate evidence or no data from epidemiological studies
C	Possible human carcinogen, with limited evidence from animal studies in the absence of human data
D	Not classifiable as to human carcinogenicity, owing to inadequate human and animal evidence
E	Evidence of noncarcinogenicity for humans, with no evidence of carcinogenicity in at least two adequate animal tests in different species, or in both adequate animal and epidemiological studies

low level of confidence in the RfD because the study was poorly designed and confidence in the supporting database is also low. Based on the RfD and assuming 20 % exposure from drinking water, the EPA has set both drinking water MCL and MCLG of 700 µg/L. Drinking water health advisories range from 700 µg/L (lifetime equivalent to the RfD) to 32 mg/L (one-day advisory for a child). In situations in which both aquatic life and water are consumed from a particular body of water, a recommended ambient water criterion is set at 1400 µg/L. When only aquatic organisms are consumed, the criterion is 3280 µg/L. An inhalation RfC of 1 mg/m^3 was derived based on developmental toxicity effects observed in rats and rabbits. An uncertainty factor of 300 and a modifying factor of 1 were used to convert the NOAEL to the RfC. Both the study design and database were rated low and, thus, the overall confidence in the RfC was established as low.

X1.6.4.2 *Physical/Chemical Parameter Summary*— Ethylbenzene has a relatively high Henry's law constant (8.7 × 10^{-3} m^3-atm/mol) and, therefore, can rapidly volatilize under common above-ground environmental conditions. Based on its moderate water solubility (1.52 × 10^5 µg/L) and moderate capacity to sorb to soils (estimated log K_{oc} = 3.04), it will have moderate mobility in soil and may leach into ground water. In laboratory tests, when a free gasoline phase was in equilibrium with water, typical combined ethylbenzene and xylenes concentrations in water ranged from 1.08 × 10^4 to 2.39 × 10^4 µg/L, due to partitioning effects. Ethylbenzene has a moderate low K_{ow} value (3.15) and is biodegradable. Therefore, it is not expected to bioaccumulate. In laboratory tests, when a free gasoline phase was in equilibrium with water, typical combined ethylbenzene and xylenes

concentrations in water ranged from 1.08 × 10^4 to 2.39 × 10^4 µg/L.

X1.6.5 *Naphthalenes:*

X1.6.5.1 *Toxicity Summary*—In general, poisoning may occur by ingestion of large doses, inhalation, or skin adsorption of naphthalene. It can cause nausea, headache, diaphoresis, hematuria, fever, anemia, liver damage, vomiting, convulsions, and coma. Methylnaphthalenes are presumably less acutely toxic than naphthalene. Skin irritation and skin photosensitization are the only effects reported in man. Inhalation of the vapor may cause headache, confusion, nausea, and sometimes vomiting. The environmental concerns with naphthalenes are primarily attributed to effects on aquatic organisms. As a consequence, the EPA has not set any human health criteria for these materials (that is, there is no RfD or RfC, no drinking water MCL or MCLG or ambient water quality criteria). A risk assessment to define a RfD for these materials is presently under review by the EPA. Drinking water health advisories range from 20 µg/L (lifetime, adult) to 500 µg/L (one-day advisory for a child).[6]

X1.6.5.2 *Physical/Chemical Parameter Summary: Naphthalene*—Naphthalene has a relatively high Henry's law constant (1.15 × 10^{-3} m^3-atm/mol) and, thus, has the capacity to volatilize rapidly under common above-ground environmental conditions. It has a moderate water solubility (3.10 × 10^4 µg/L) and log K_{oc} (3.11) and has the potential to leach to ground water. A moderate log K_{ow} value of 3.01 has been reported, but because naphthalene is very biodegradable, it is unlikely to bioconcentrate to a significant degree.

X1.6.5.3 *Methylnaphthalenes*—Henry's law constants (2.60 × 10^{-4} m^3-atm/mol and 5.18 × 10^{-4} m^3-atm/mol for 1- and 2-methylnaphthalene, respectively) suggest that these materials have the potential to volatilize under common above-ground environmental conditions. 1-Methylnaphthalene exhibits a water solubility similar to naphthalene (2.60 × 10^4 µg/L to 2.8 × 10^4 µg/L). However, solubility decreases with increasing alkylation (dimethylnaphthalenes: 2.0 × 10^3 µg/L to 1.1 × 10^4 µg/L, 1,4,5-trimethylnaphthalene: 2.0 × 10^3 µg/L). These materials are, therefore, expected to be slightly mobile to relatively immobile in soil (for example, log K_{oc} is in the range from 2.86 to 3.93 for 1- and 2-methylnaphthalenes). In aquatic systems, methylnaphthalenes may partition from the water column to

---

[6] Office of Water, USEPA, Washington, DC.

TABLE X1.5 Definitions of Important Toxicological Characteristics

Reference Dose—A reference dose is an estimate (with an uncertainty typically spanning an order of magnitude) of a daily exposure (mg/kg/day) to the general human population (including sensitive subgroups) that is likely to be without an appreciable risk of deleterious effects during a lifetime of exposure.

Reference Concentration—A reference concentration is an estimate (with an uncertainty spanning perhaps an order of magnitude) of a continuous exposure to the human population (including sensitive subgroups) that is likely to be without appreciable deleterious effects during a lifetime.

Slope Factor—The slope of the dose-response curve in the low-dose region. When low-dose linearity cannot be assumed, the slope factor is the slope of the straight line from zero dose to the dose at 1 % excess risk. An upper bound on this slope is usually used instead of the slope itself. The units of the slope factor are usually expressed as (mg/kg/day).$^{-1}$

Drinking Water MCLs and MCLGs—Maximum contaminant levels (MCLs) are drinking water standards established by the EPA that are protective of human health. However, these standards take into account the technological capability of attaining these standards. The EPA has, therefore, also established MCL goals (MCLGs) which are based only on the protection of human health. The MCL standards are often used as clean-up criteria.

Drinking Water Health Advisories—The Office of Drinking Water provides health advisories (HAs) as technical guidance for the protection of human health. They are not enforceable federal standards. The HA's are the concentration of a substance in drinking water estimated to have negligible deleterious effects in humans, when ingested for specified time periods.

Water Quality Criteria—These criteria are not rules and they do not have regulatory impact. Rather, these criteria present scientific data and guidance of the environmental effects of pollutants which can be useful to derive regulatory requirements based on considerations of water quality impacts.

organic matter contained in sediments and suspended solids. Methylnaphthalenes have high log K_{ow} values (greater than 3.5) and have the potential to bioaccumulate. They do, however, exhibit a moderate degree of biodegradation, which typically decreases with increased alkylation.

X1.6.6 *Three to Six-Ringed PAHs*—The most significant health effect for this class of compounds is their carcinogenicity, which is structure-dependent. Anthracene and phenanthrene have not been shown to cause cancer in laboratory animals. The available data does not prove pyrene to be carcinogenic to experimental animals. On the other hand, benz[a]-anthracene, benzo[a]pyrene, dibenz[a,h]anthracene, and 7,12-dimethylbenz[a]-anthracene have been shown to be carcinogenic in laboratory animals. B(a)P and pyrene are discussed in X1.6.7 and X1.6.8 as representatives of carcinogenic and noncarcinogenic effects of this class.

X1.6.7 *Benzo(a)pyrene (BaP):*

X1.6.7.1 *Toxicity Summary*—Based on animal data, B(a)P has been classified as a probable human carcinogen (B2 carcinogen) by the USEPA. A range of oral slope factors from 4.5 to 11.7 $(mg/kg/day)^{-1}$ with a geometric mean of 7.3 $(mg/kg/day)^{-1}$ has been derived for B(a)P based on the observance of tumors of the forestomach and squamous cell carcinomas in mice. The data was considered less than optimal but acceptable (note that the carcinogenicity assessment for B(a)P may change in the near future pending the outcome of an on-going EPA review). The EPA has proposed a drinking water MCL at 0.2 µg/L (based on the analytical detection limits). The MCLG for B(a)P is set at zero. In situations in which both aquatic life and water are consumed from a particular body of water, a recommended EPA water quality criterion is set at 2.8×10^{-3} µg/L. When only aquatic organisms are consumed, the criterion is 3.11×10^{-2} µg/L.

X1.6.7.2 *Physical/Chemical Parameter Summary*— When released to water, PAHs are not subject to rapid volatilization (Henry's law constants are on the order of 1.0 $\times 10^{-4}$ m³-atm/mol or less) under common environmental conditions. They have low aqueous solubility values and tend to sorb to soils and sediments and remain fixed in the environment. Three ring members of this group such as anthracene and phenanthrene have water solubilities on the order of 1000 µg/L. The water solubilities decrease substantially for larger molecules in the group, for example, benzo[a]pyrene has a water solubility of 1.2 µg/L. The log K_{oc} values for PAHs are on the order of 4.3 and greater, which suggests that PAHs will be expected to adsorb very strongly to soil. The PAHs with more than three rings generally have high log K_{ow} values (6.06 for benzo[a]pyrene), have poor biodegradability characteristics and may bioaccumulate.

X1.6.8 *Pyrene:*

X1.6.8.1 *Toxicity Summary*—Using data from animal studies, the USEPA has set an oral RfD for pyrene at 3 $\times 10^{-2}$ mg/kg/day. In converting a NOAEL from the animal study, in which the critical effects observed were kidney toxicity, an uncertainty factor of 3000 and a modifying factor of 1 were used. The EPA has assigned an overall low level of confidence in the RfD because although the study was well-designed, confidence in the supporting database is low. No drinking water MCLs or health advisories have been set. In situations in which both aquatic life and water are consumed from a particular body of water, a recommended EPA water quality criterion is set at 2.8×10^{-3} µg/L. When only aquatic organisms are consumed, the criterion is 3.11×10^{-2} µg/L.

X1.6.8.2 *Physical/Chemical Parameter Summary*—Refer to X1.6.7.2 for BaP. Also see Table X1.2.

X1.6.9 *MTBE:*

X1.6.9.1 *Toxicity Summary*—Using data from animal studies, the USEPA has set an inhalation RfC for MTBE at 3 mg/m³. In converting a NOAEL from the animal study, in which the critical effects observed included increased liver and kidney weight and increased severity of spontaneous renal lesions (females), increased prostration (females) and swollen pericolar tissue, an uncertainty factor of 100 and a modifying factor of 1 were used. The EPA has assigned an overall medium level of confidence in the RfC because although the study was well-designed, some information on the chemistry was lacking. The confidence in the supporting database is medium to high. No drinking water MCLs or ambient water quality criteria have been set. However, a risk assessment, which may define a RfD for this material, is presently under review by EPA. Drinking water health advisories range from 40 µg/L (lifetime, adult) to 3000 µg/L (one-day advisory for a child).[6]

X1.6.9.2 *Physical/Chemical Parameter Summary*—The Henry's law constant for MTBE is estimated to be approximately 1.0 $\times 10^{-3}$ m³-atm/mol. It is, therefore, expected to have the potential to rapidly volatilize under common above-ground environmental conditions. It is very water soluble (water solubility is 4.8 $\times 10^7$ µg/L), and with a relatively low capacity to sorb to soils (estimated log K_{oc} = 1.08), MTBE will migrate at the same velocity as the water in which it is dissolved in the subsurface. The log K_{ow} value has been estimated to be between 1.06 and 1.30, indicating MTBE's low bioaccumulative potential. It is expected to have a low potential to biodegrade, but no definitive studies are available.

X1.6.10 *Lead:*

X1.6.10.1 *Toxicity Summary—(The following discussion is for inorganic lead—not the organic forms of lead (tetraethyllead, tetramethyllead) that were present in petroleum products.)* A significant amount of toxicological information is available on the health effects of lead. Lead produces neurotoxic and behavioral effects particularly in children. However, the EPA believes that it is inappropriate to set an RfD for lead and its inorganic compounds because the agency believes that some of the effects may occur at such low concentrations as to suggest no threshold. The EPA has also determined that lead is a probable human carcinogen (classified as B2). The agency has chosen not to set a numeric slope factor at this time, however, because it is believed that standard procedures for doing so may not be appropriate for lead. At present, the EPA has set an MCLG of zero but has set no drinking water (MCL) or health advisories because of the observance of low-level effects, the overall Agency goal of reducing total lead exposure and because of its classification as a B2 carcinogen. An action level of 15 µg/L has been set for water distribution systems (standard at the tap). The recommended EPA water quality criterion for consumption of both aquatic life and water is set at 50 µg/L.

X1.6.10.2 *Physical/Chemical Parameter Summary*—Organic lead additive compounds are volatile (estimated Henry's law constant for tetraethyl lead = 7.98×10^{-2} m³-atm/mol) and may also sorb to particulate matter in the air. Tetraethyl lead has an aqueous solubility of 800 µg/L and an estimated log K_{oc} of 3.69 and, therefore, should not be very mobile in the soil. It decomposes to inorganic lead in dilute aqueous solutions and in contact with other environmental media. In free product (gasoline) plumes, however, it may remain unchanged. Inorganic lead compounds tightly bind to most soils with minimal leaching under natural conditions. Aqueous solubility varies depending on the species involved. The soil's capacity to sorb lead is correlated with soil pH, cation exchange capacity, and organic matter. Lead does not appear to bioconcentrate significantly in fish but does in some shellfish, such as mussels. Lead is not biodegradable.

X1.7 *Discussion of Acceptable Risk* (**12**)—Beginning in the late 1970s and early 1980s, regulatory agencies in the United States and abroad frequently adopted a cancer risk criteria of one-in-one-million as a negligible (that is, of no concern) risk when fairly large populations might be exposed to a suspect carcinogen. Unfortunately, theoretical increased cancer risks of one-in-one-million are often incorrectly portrayed as serious public health risks. As recently discussed by Dr. Frank Young (**13**), the current commissioner of the Food and Drug Administration (FDA), this was not the intent of such estimates:

X1.7.1 In applying the de minimis concept and in setting other safety standards, the FDA has been guided by the figure of "one-in-one-million." Other Federal agencies have also used a one-in-one-million increased risk over a lifetime as a reasonable criterion for separating high-risk problems warranting agency attention from negligible risk problems that do not.

X1.7.2 The risk level of one-in-one-million is often misunderstood by the public and the media. It is not an actual risk, that is, we do not expect one out of every million people to get cancer if they drink decaffeinated coffee. Rather, it is a mathematical risk based on scientific assumptions used in risk assessment. The FDA uses a conservative estimate to ensure that the risk is not understated. We interpret animal test results conservatively, and we are extremely careful when we extrapolate risks to humans. When the FDA uses the risk level of one-in-one-million, it is confident that the risk to humans is virtually nonexistent.

X1.7.3 In short, a "one-in-one-million" cancer risk estimate, which is often tacitly assumed by some policy-makers to represent a trigger level for regulatory action, actually represents a level of risk that is so small as to be of negligible concern.

X1.7.4 Another misperception within the risk assessment arena is that all occupational and environmental regulations have as their goal a theoretical maximum cancer risk of 1 in 1 000 000. Travis, et al (**14**) recently conducted a retrospective examination of the level of risk that triggered regulatory action in 132 decisions. Three variables were considered: (*1*) individual risk (an upper-bound estimate of the probability at the highest exposure), (*2*) population risk (an upper-limit estimate of the number of additional incidences of cancer in the exposed population), and (*3*) population size. The

findings of Travis, et al (**14**) can be summarized as follows:

X1.7.4.1 Every chemical with an individual lifetime risk above 4×10^{-3} received regulation. Those with values below 1×10^{-6} remained unregulated.

X1.7.4.2 For small populations, regulatory action never resulted for individual risks below 1×10^{-4}.

X1.7.4.3 For potential effects resulting from exposures to the entire United States population, a risk level below 1×10^{-6} never triggered action; above 3×10^{-4} always triggered action.

X1.7.5 Rodricks, et al (**15**) also evaluated regulatory decisions and reached similar conclusions. In decisions relating to promulgation of National Emission Standards for Hazardous Air Pollutants (NESHAPS), the USEPA has found the maximum individual risks and total population risks from a number of radionuclide and benzene sources too low to be judged significant. Maximum individual risks were in the range from 3.6×10^{-5} to 1.0×10^{-3}. In view of the risks deemed insignificant by USEPA, Rodricks, et al (**15**) noted that 1×10^{-5} (1 in 100 000) appears to be in the range of what USEPA might consider an insignificant average lifetime risk, at least where aggregate population risk is no greater than a fraction of a cancer yearly.

X1.7.6 Recently, final revisions to the National Contingency Plan (**16**) have set the acceptable risk range between 10^{-4} and 10^{-6} at hazardous waste sites regulated under CERCLA. In the recently promulgated *Hazardous Waste Management System Toxicity Characteristics Revisions* (**17**), the USEPA has stated that:

"For drinking water contaminants, EPA sets a reference risk range for carcinogens at 10^{-6} excess individual cancer risk from lifetime exposure. Most regulatory actions in a variety of EPA programs have generally targeted this range using conservative models which are not likely to underestimate the risk."

X1.7.7 Interestingly, the USEPA has selected and promulgated a single risk level of 1 in 100 000 (1×10^{-5}) in the *Hazardous Waste Management System Toxicity Characteristics Revisions* (**17**). In their justification, the USEPA cited the following rationale:

The chosen risk level of 10^{-5} is at the midpoint of the reference risk range for carcinogens (10^{-4} to 10^{-6}) generally used to evaluate CERCLA actions. Furthermore, by setting the risk level at 10^{-5} for TC carcinogens, EPA believes that this is the highest risk level that is likely to be experienced, and most if not all risks will be below this level due to the generally conservative nature of the exposure scenario and the underlying health criteria. For these reasons, the Agency regards a 10^{-5} risk level for Group A, B, and C carcinogens as adequate to delineate, under the Toxicity Characteristics, wastes that clearly pose a hazard when mismanaged."

X1.7.8 When considering these limits it is interesting to note that many common human activities entail annual risks greatly in excess of one-in-one-million. These have been discussed by Grover Wrenn, former director of Federal Compliance and State Programs at OSHA, as follows:

X1.7.9 State regulatory agencies have not uniformly adopted a one-in-one-million (1×10^{-6}) risk criterion in making environmental and occupational decisions. The states of Virginia, Maryland, Minnesota, Ohio, and Wisconsin have employed or proposed to use the one-in-one-hundred-thousand (1×10^{-5}) level of risk in their risk management decisions (**18**). The State of Maine Department of Human Services (DHS) uses a lifetime risk of one in one hundred thousand as a reference for non-threshold (carcinogenic) effects in its risk management decisions regarding exposures to environmental contaminants (19). Similarly, a lifetime incremental cancer risk of one in one hundred thousand is used by the Commonwealth of Massachusetts as a cancer risk limit for exposures to substances in more than one medium at hazardous waste disposal sites (20). This risk limit represents the total cancer risk at the site associated with exposure to multiple chemicals in all contaminated media. The State of California has also established a level of risk of one in one hundred thousand for use in determining levels of chemicals and exposures that pose no significant risks of cancer under the Safe Drinking Water and Toxic Enforcement Act of 1986 (Proposition 65) (21). Workplace air standards developed by the Occupational Safety and Health Administration (OSHA) typically reflect theoretical risks of one in one thousand (1×10^{-3}) or greater (15).

X1.7.10 Ultimately, the selection of an acceptable and de minimis risk level is a policy decision in which both costs and benefits of anticipated courses of action should be thoroughly evaluated. However, actuarial data and risk estimates of common human activities, regulatory precedents, and the relationship between the magnitude and variance of background and incremental risk estimates all provide compelling support for the adoption of the de minimis risk level of 1×10^{-5} for regulatory purposes.

X1.7.11 In summary, U.S. Federal and state regulatory agencies have adopted a one-in-one-million cancer risk as being of negligible concern in situations where large populations (for example, 200 million people) are involuntarily exposed to suspect carcinogens (for example, food additives). When smaller populations are exposed (for example, in occupational settings), theoretical cancer risks of up to 10^{-4} (1 in 10 000) have been considered acceptable.

X2. DEVELOPMENT OF RISK-BASED SCREENING LEVELS (RBSLs) APPEARING IN SAMPLE LOOK-UP TABLE X2.1

X2.1 *Introduction:*

X2.1.1 This appendix contains the equations and parameters used to construct the example "Look-Up" (Table X2.1). This table was prepared solely for the purpose of presenting an example Tier 1 matrix of RBSLs, and these values should not be viewed, or misused, as proposed remediation "standards." The reader should note that not all possible pathways have been considered and a number of assumptions concerning exposure scenarios and parameter values have been made. These should be reviewed for appropriateness before using the listed RBSLs as Tier 1 screening values.

X2.1.2 The approaches used to calculate RBSLs appearing in Table X2.1 are briefly discussed as follows for exposure to vapors, ground water, surficial soils, and subsurface soils by means of the following pathways:

X2.1.2.1 Inhalation of vapors,

X2.1.2.2 Ingestion of ground water,

X2.1.2.3 Inhalation of outdoor vapors originating from dissolved hydrocarbons in ground water,

X2.1.2.4 Inhalation of indoor vapors originating from dissolved hydrocarbons in ground water,

X2.1.2.5 Ingestion of surficial soil, inhalation of outdoor vapors and particulates emanating from surficial soils, and dermal absorption resulting from surficial soil contact with skin,

X2.1.2.6 Inhalation of outdoor vapors originating from hydrocarbons in subsurface soils,

X2.1.2.7 Inhalation of indoor vapors originating from subsurface hydrocarbons, and

X2.1.2.8 Ingestion of ground water impacted by leaching of dissolved hydrocarbons from subsurface soils.

X2.1.3 For the pathways considered, approaches used in this appendix are consistent with guidelines contained in Ref (26).

X2.1.4 The development presented as follows focuses only on human-health RBSLs for chronic (long-term) exposures.

X2.1.4.1 In the case of compounds that have been classi-

fied as carcinogens, the RBSLs are based on the general equation:

risk = average lifetime intake [mg/kg-day]
$\times$ *potency factor* [mg/kg-day]$^{-1}$

where the intake depends on exposure parameters (ingestion rate, exposure duration, and so forth), the source concentration, and transport rates between the source and receptor. The potency factor is selected after reviewing a number of sources, including the USEPA Integrated Risk Information System (IRIS) (2) database, USEPA Health Effects Assessment Summary Tables (HEAST) (3), and peer-reviewed sources. The RBSL values appearing in Table X2.1 correspond to probabilities of adverse health effects ("risks") in the range from 10^{-6} to 10^{-4} resulting from the specified exposure. Note that this risk value does not reflect the probability for the specified exposure scenario to occur. Therefore, the actual potential risk to a population for these RBSLs is lower than the 10^{-6} to 10^{-4} range.

X2.1.4.2 In the case of compounds that have not been classified as carcinogens, the RBSLs are based on the general equation:

hazard quotient = average intake [mg/kg-day]/
reference dose [mg/kg-day]

where the intake depends on exposure parameters (ingestion rate, exposure duration, and so forth), the source concentration, and transport rates between the source and receptor. The reference dose is selected after reviewing a number of sources, including the USEPA Integrated Risk Information System (IRIS) (2) database, USEPA Health Effects Assessment Summary Tables (HEAST) (3), and peer-reviewed sources. The RBSL values appearing in Table X2.1 correspond to hazard quotients of unity resulting from the specified exposure. Note that this hazard quotient value does not reflect the probability for the specified exposure scenario to occur. Therefore, the actual potential impact to a population for these RBSLs is lower than a hazard quotient of unity.

X2.1.5 Tables X2.2 through X2.7 summarize the equa-

TABLE X2.1 Example Tier 1 Risk-Based Screening Level (RBSL) Look-up Table[A]

NOTE—This table is presented here only as an example set of Tier 1 RBSLs. It is not a list of proposed standards. The user should review all assumptions prior to using any values. Appendix X2 describes the basis of these values.

Exposure Pathway	Receptor Scenario	Target Level	Benzene	Ethylbenzene	Toluene	Xylenes (Mixed)	Napthalenes	Benzo (a)pyrene
				Air				
Indoor air screening levels for inhalation exposure, μ/m³	residential	cancer risk = 1E-06	3.92E-01					1.86E-03
		cancer risk = 1E-04	3.92E+01					1.86E-01
		chronic HQ = 1		1.39E+03	5.56E+02	9.73E+03	1.95E+01	
	commercial/ industrial	cancer risk = 1E-06	4.93E-01					2.35E-03
		cancer risk = 1E-04	4.93E+01					2.35E-01
		chronic HQ = 1		1.46E+03	5.84E+02	1.02E+04	2.04E+01	
Outdoor air screening levels for inhalation exposure, μg/m³	residential	cancer risk = 1E-06	2.94E-01					1.40E-03
		cancer risk = 1E-04	2.94E+01					1.40E-01
		chronic HQ = 1		1.04E+03	4.17E+02	7.30E+03	1.46E+01	
	commercial/ industrial	cancer risk = 1E-06	4.93E-01					2.35E-03
		cancer risk = 1E-04	4.93E+01					2.35E-01
		chronic HQ = 1		1.46E+03	5.84E+02	1.02E+04	2.04E+01	
OSHA TWA PEL, μg/m³			3.20E+03	4.35E+05	7.53E+05	4.35E+06	5.00E+04	2.00E+02[A]
Mean odor detection threshold, μg/m³ [B]			1.95E+05		6.00E+03	8.70E+04	2.00E+02	
National indoor background concentration range, μg/m³ [C]			3.25E+00 to 2.15E+01	2.20E+00 to 9.70E+00	9.60E-01 to 2.91E+01	4.85E+00 to 4.76E+01		
				Soil				
Soil volatilization to outdoor air, mg/kg	residential	cancer risk = 1E-06	2.72E-01					RES[D]
		cancer risk = 1E-04	2.73E+01					RES
		chronic HQ = 1		RES	RES	RES	RES	
	commercial industrial	cancer risk = 1E-06	4.57E-01					RES
		cancer risk = 1E-04	4.57E+01					RES
		chronic HQ = 1		RES	RES	RES	RES	
Soil-vapor intrusion from soil to buildings, mg/kg	residential	cancer risk = 1E-06	5.37E-03					RES
		cancer risk = 1E-04	5.37E-01					RES
		chronic HQ = 1		4.27E+02	2.06E+01	RES	4.07E+01	
	commercial/ industrial	cancer risk = 1E-06	1.69E-02					RES
		cancer risk = 1E-04	1.69E+00					RES
		chronic HQ = 1		1.10E+03	5.45E+01	RES	1.07E+02	
Surficial soil (0 to 3 ft) (0 to 0.9 m) ingestion/ dermal/ inhalation, mg/kg	residential	cancer risk = 1E-06	5.82E+00					1.30E-01
		cancer risk = 1E-04	5.82E+02					1.30E+01
		chronic HQ = 1		7.83E+03	1.33E+04	1.45E+06	9.77E+02	
	commercial/ industrial	cancer risk = 1E-06	1.00E+01					3.04E-01
		cancer risk = 1E-04	1.00E+03					3.04E+01
		chronic HQ = 1		1.15E+04	1.87E+04	2.08E+05	1.50E+03	
Soil-leachate to protect ground water ingestion target level, mg/kg		MCLs	2.93E-02	1.10E+02	1.77E+01	3.05E+02	N/A	9.42E+00
	residential	cancer risk = 1E-06	1.72E-02					5.50E-01
		cancer risk = 1E-04	1.72E+00					RES
		chronic HQ = 1		5.75E+02	1.29E+02	RES	2.29E+01	
	commercial/ industrial	cancer risk = 1E-06	5.78E-02					1.85E+00
		cancer risk = 1E-04	5.78E+00					RES
		chronic HQ = 1		1.61E+03	3.61E+02	RES	6.42E+01	
				Ground Water				
Ground water volatilization to outdoor air, mg/L	residential	cancer risk = 1E-06	1.10E+01					>S[E]
		cancer risk = 1E-04	1.10E+03					>S
		chronic HQ = 1		>S	>S	>S	>S	
	commercial/ industrial	cancer risk = 1E-06	1.84E+01					>S
		cancer risk = 1E-04	>S					>S
		chronic HQ = 1		>S	>S	>S	>S	
Ground water ingestion, mg/L		MCLs	5.00E-03	7.00E-01	1.00E+00	1.00E+01	N/A	2.00E-04
	residential	cancer risk = 1E-06	2.94E-03					1.17E-05
		cancer risk = 1E-04	2.94E-01					1.17E-03
		chronic HQ = 1		3.65E+00	7.30E+00	7.30E+01	1.46E-01	
	commercial/ industrial	cancer risk = 1E-06	9.87E-03					3.92E-05
		cancer risk = 1E-04	9.87E-01					>S
		chronic HQ = 1		1.02E+01	2.04E+01	>S	4.09E-01	
Ground water—vapor intrusion from ground water to buildings, mg/L	residential	cancer risk = 1E-06	2.38E-02					>S
		cancer risk = 1E-04	2.38E+00					>S
		chronic HQ = 1		7.75E+01	3.28E+01	>S	4.74E+00	
	commercial/ industrial	cancer risk = 1E-06	7.39E-02					>S
		cancer risk = 1E-04	7.39E+00					>S
		chronic HQ = 1		8.50E+01	>S	>S	1.23E+01	

[A] As benzene soluble coal tar pitch volatiles.

[B] See Ref (22).

[C] See Refs (23–25).

[D] RES—Selected risk level is not exceeded for pure compound present at any concentration.

[E] >S—Selected risk level is not exceeded for all possible dissolved levels (≦ pure component solubility).

tions and parameters used to prepare the example look-up Table X2.1. The basis for each of these equations is discussed in X2.2 through X2.10.

X2.2 *Air—Inhalation of Vapors (Outdoors/Indoors)*—In this case chemical intake results from the inhalation of vapors. It is assumed that vapor concentrations remain constant over the duration of exposure, and all inhaled chemicals are absorbed. Equations appearing in Tables X2.2 and X2.3 for estimating RBSLs for vapor concentrations in the breathing zone follow guidance given in Ref **(26)**. Should the calculated RBSL exceed the saturated vapor concentration for any individual component, "$>P_{vap}$" is entered in the table to indicate that the selected risk level or hazard quotient cannot be reached or exceeded for that compound and the specified exposure scenario.

X2.3 *Ground Water—Ingestion of Ground Water*—In this case chemical intake results from ingestion of ground water. It is assumed that the dissolved hydrocarbon concentrations remain constant over the duration of exposure. Equations appearing in Tables X2.2 and X2.3 for estimating RBSLs for drinking water concentrations follow guidance given in Ref **(26)** for ingestion of chemicals in drinking water. Should the calculated RBSL exceed the pure component solubility for any individual component, "$>S$" is entered in the table to indicate that the selected risk level or hazard quotient cannot be reached or exceeded for that compound and the specified exposure scenario (unless free-phase product is mixed with the ingested water).

TABLE X2.2 Equations Used to Develop Example Tier 1 Risk-Based Screening Level (RBSLs) Appearing in "Look-Up" Table X2.1—Carcinogenic Effects[A]

NOTE—See Tables X2.4 through X2.7 for definition of parameters.

Medium	Exposure Route	Risk-Based Screening Level (RBSL)
Air	inhalation[B]	$RBSL_{air}\left[\dfrac{\mu g}{m^3\text{-}air}\right] = \dfrac{TR \times BW \times AT_c \times 365\frac{days}{years} \times 10^3\frac{\mu g}{mg}}{SF_i \times IR_{air} \times EF \times ED}$
Ground water	ingestion (potable ground water supply only)[B]	$RBSL_w\left[\dfrac{mg}{L\text{-}H_2O}\right] = \dfrac{TR \times BW \times AT_c \times 365\frac{days}{years}}{SF_o \times IR_w \times EF \times ED}$
Ground water[C]	enclosed-space (indoor) vapor inhalation[D]	$RBSL_w\left[\dfrac{mg}{L\text{-}H_2O}\right] = \dfrac{RBSL_{air}\left[\frac{\mu g}{m^3\text{-}air}\right]}{VF_{wesp}} \times 10^{-3}\frac{mg}{\mu g}$
Ground water[C]	ambient (outdoor) vapor inhalation[D]	$RBSL_w\left[\dfrac{mg}{L\text{-}H_2O}\right] = \dfrac{RBSL_{air}\left[\frac{\mu g}{m^3\text{-}air}\right]}{VF_{wamb}} \times 10^{-3}\frac{mg}{\mu g}$
Surficial soil	ingestion of soil, inhalation of vapors and particulates, and dermal contact[B]	$RBSL_s\left[\dfrac{mg}{kg\text{-}soil}\right] =$ $\dfrac{TR \times BW \times AT_c \times 365\frac{days}{years}}{EF \times ED\left[\left(SF_o \times 10^{-6}\frac{kg}{mg} \times (IR_{soil} \times RAF_o + SA \times M \times RAF_d)\right) + (SF_i \times IR_{air} \times (VF_{ss} + VF_p))\right]}$ For surficial and excavated soils (0 to 1 m)
Subsurface soil[C]	ambient (outdoor) vapor inhalation[D]	$RBSL_s\left[\dfrac{mg}{kg\text{-}soil}\right] = \dfrac{RBSL_{air}\left[\frac{\mu g}{m^3\text{-}air}\right]}{VF_{samb}} \times 10^{-3}\frac{mg}{\mu g}$
Subsurface soil[C]	enclosed space (indoor) vapor inhalation[D]	$RBSL_s\left[\dfrac{mg}{kg\text{-}soil}\right] = \dfrac{RBSL_{air}\left[\frac{\mu g}{m^3\text{-}air}\right]}{VF_{sesp}} \times 10^{-3}\frac{mg}{\mu g}$
Subsurface soil[C]	leaching to ground water[D]	$RBSL_s\left[\dfrac{mg}{kg\text{-}soil}\right] = \dfrac{RBSL_w\left[\frac{mg}{L\text{-}H_2O}\right]}{LF_{sw}}$

[A] Note that all RBSL values should be compared with thermodynamic partitioning limits, such as solubility levels, maximum vapor concentrations, and so forth. If a RBSL exceeds the relevant partitioning limit, this is an indication that the selected risk or hazard level will never be reached or exceeded for that chemical and the selected exposure scenario.

[B] Screening levels for these media based on other considerations (for example, aesthetic, background levels, environmental resource protection, and so forth) can be derived with these equations by substituting the selected target level for $RBSL_{air}$ or $RBSL_w$ appearing in these equations.

[C] These equations are based on Ref **(26)**.

[D] These equations simply define the "cross-media partitioning factors," VF_{ij} and LF_{sw}.

TABLE X2.3 Equations Used to Develop Example Tier 1 Risk-Based Screening Level (RBSLs) Appearing in "Look-Up" Table X2.1— Noncarcinogenic Effects[A]

NOTE—See Tables X2.4 through X2.7 for definition of parameters.

Medium	Exposure Route	Risk-Based Screening Level (RBSL)
Air	inhalation[B]	$RBSL_{air}\left[\dfrac{\mu g}{m^3\text{-}air}\right] = \dfrac{THQ \times RfD_i \times BW \times AT_n \times 365\,\frac{days}{years} \times 10^3\,\frac{\mu g}{mg}}{IR_{air} \times EF \times ED}$
Ground water	ingestion (potable ground water supply only)[B]	$RBSL_{w}\left[\dfrac{mg}{L\text{-}H_2O}\right] = \dfrac{THQ \times RfD_o \times BW \times AT_n \times 365\,\frac{days}{years}}{IR_{w} \times EF \times ED}$
Ground water[C]	enclosed-space (indoor) vapor inhalation[D]	$RBSL_{w}\left[\dfrac{mg}{L\text{-}H_2O}\right] = \dfrac{RBSL_{air}\left[\frac{\mu g}{m^3\text{-}air}\right]}{VF_{wesp}} \times 10^{-3}\,\frac{mg}{\mu g}$
Ground water[C]	ambient (outdoor) vapor inhalation[D]	$RBSL_{w}\left[\dfrac{mg}{L\text{-}H_2O}\right] = \dfrac{RBSL_{air}\left[\frac{\mu g}{m^3\text{-}air}\right]}{VF_{wamb}} \times 10^{-3}\,\frac{mg}{\mu g}$
Surficial soil	ingestion of soil, inhalation of vapors and particulates, and dermal contact[B]	$RBSL_{s}\left[\dfrac{mg}{kg\text{-}soil}\right] =$ $\dfrac{THQ \times BW \times AT_n \times 365\,\frac{days}{years}}{EF \times ED\left[10^{-6}\,\frac{kg}{mg} \times \frac{IR_{soil} \times RAF_o + SA \times M \times RAF_d}{RfD_o} + \frac{IR_{air} \times (VF_{ss} + VF_p)}{RfD_i}\right]}$ For surficial and excavated soils (0 to 1 m)
Subsurface soil[C]	ambient (outdoor) vapor inhalation[D]	$RBSL_{s}\left[\dfrac{mg}{kg\text{-}soil}\right] = \dfrac{RBSL_{air}\left[\frac{\mu g}{m^3\text{-}air}\right]}{VF_{samb}} \times 10^{-3}\,\frac{mg}{\mu g}$
Subsurface soil[C]	enclosed space (indoor) vapor inhalation[D]	$RBSL_{s}\left[\dfrac{mg}{kg\text{-}soil}\right] = \dfrac{RBSL_{air}\left[\frac{\mu g}{m^3\text{-}air}\right]}{VF_{sesp}} \times 10^{-3}\,\frac{mg}{\mu g}$
Subsurface soil[C]	leaching to ground water[D]	$RBSL_{s}\left[\dfrac{mg}{kg\text{-}soil}\right] = \dfrac{RBSL_{w}\left[\frac{mg}{L\text{-}H_2O}\right]}{LF_{sw}}$

[A] Note that all RBSL values should be compared with thermodynamic partitioning limits, such as solubility levels, maximum vapor concentrations, and so forth. If a RBSL exceeds the relevant partitioning limit, this is an indication that the selected risk or hazard level will never be reached or exceeded for that chemical and the selected exposure scenario.

[B] Screening levels for these media based on other considerations (for example, aesthetic, background levels, environmental resource protection, and so forth) can be derived with these equations by substituting the selected target level for RBSL_air or RBSL_w appearing in these equations.

[C] These equations are based on Ref **(26)**.

[D] These equations simply define the "cross-media partitioning factors," VF_{ij} and LF_{sw}.

X2.4 Ground Water—Inhalation of Outdoor Vapors:

X2.4.1 In this case chemical intake is a result of inhalation of outdoor vapors which originate from dissolved hydrocarbons in ground water located some distance below ground surface. Here the goal is to determine the dissolved hydrocarbon RBSL that corresponds to the target RBSL for outdoor vapors in the breathing zone, as given in Tables X2.2 and X2.3. If the selected target vapor concentration is some value other than the RBSL for inhalation (that is, odor threshold or ecological criterion), this value can be substituted for the RBSL_air parameter appearing in the equations given in Tables X2.2 and X2.3.

X2.4.2 A conceptual model for the transport of chemicals from ground water to ambient air is depicted in Fig. X2.1. For simplicity, the relationship between outdoor air and dissolved ground water concentrations is represented in Tables X2.2 and X2.3 by the "volatilization factor," VF_{wamb} [(mg/m³-air)/(mg/L-H₂O)], defined in Table X2.5. It is based on the following assumptions:

X2.4.2.1 A constant dissolved chemical concentration in ground water,

X2.4.2.2 Linear equilibrium partitioning between dissolved chemicals in ground water and chemical vapors at the ground water table,

X2.4.2.3 Steady-state vapor- and liquid-phase diffusion through the capillary fringe and vadose zones to ground

TABLE X2.4 Exposure Parameters Appearing in Tables X2.2 and X2.3

Parameters	Definitions, Units	Residential	Commercial/Industrial
AT_c	averaging time for carcinogens, years	70 years	70 years[A]
AT_n	averaging time for noncarcinogens, years	30 years	25 years[A]
BW	adult body weight, kg	70 kg	70 kg[A]
ED	exposure duration, years	30 years	25 years[A]
EF	exposure frequency, days/years	350 days/year	250 days/year[A]
IR_{soil}	soil ingestion rate, mg/day	100 mg/day	50 mg/day[A]
IR_{air}-indoor	daily indoor inhalation rate, m³/day	15 m³/day	20 m³/day[A]
IR_{air}-outdoor	daily outdoor inhalation rate, m³/day	20 m³/day	20 m³/day[A]
IR_w	daily water ingestion rate, L/day	2 L/day	1 L/day[A]
LF_{sw}	leaching factor, (mg/L-H$_2$O)/(mg/kg-soil)—see Table X2.5	chemical-specific	chemical-specific
M	soil to skin adherence factor, mg/cm²	0.5	0.5[B]
RAF_d	dermal relative absorption factor, volatiles/PAHs	0.5/0.05	0.5/0.05[B]
RAF_o	oral relative absorption factor	1.0	1.0
$RBSL_i$	risk-based screening level for media i, mg/kg-soil, mg/L-H$_2$O, or µg/m³-air	chemical-, media-, and exposure route-specific	chemical-, media-, and exposure route-specific
RfD_i	inhalation chronic reference dose, mg/kg-day	chemical-specific	chemical-specific
RfD_o	oral chronic reference dose, mg/kg-day	chemical-specific	chemical-specific
SA	skin surface area, cm²/day	3160	3160[A]
SF_i	inhalation cancer slope factor, (mg/kg-day)⁻¹	chemical-specific	chemical-specific
SF_o	oral cancer slope factor, (mg/kg-day)⁻¹	chemical-specific	chemical-specific
THQ	target hazard quotient for individual constituents, unitless	1.0	1.0
TR	target excess individual lifetime cancer risk, unitless	for example, 10^{-6} or 10^{-4}	for example, 10^{-6} or 10^{-4}
VF_i	volatilization factor, (mg/m³-air)/(mg/kg-soil) or (mg/m³-air)/(mg/L-H$_2$O)—see Table X2.5	chemical- and media-specific	chemical- and media-specific

[A] See Ref **(27)**.
[B] See Ref **(28)**.

surface,

X2.4.2.4 No loss of chemical as it diffuses towards ground surface (that is, no biodegradation), and

X2.4.2.5 Steady well-mixed atmospheric dispersion of the emanating vapors within the breathing zone as modeled by a "box model" for air dispersion.

X2.4.3 Should the calculated $RBSL_w$ exceed the pure component solubility for any individual component, ">S" is entered in the table to indicate that the selected risk level or hazard quotient cannot be reached or exceeded for that compound and the specified exposure scenario.

X2.5 *Ground Water—Inhalation of Enclosed-Space (Indoor) Vapors:*

X2.5.1 In this case chemical intake results from the inhalation of vapors in enclosed spaces. The chemical vapors originate from dissolved hydrocarbons in ground water located some distance below ground surface. Here the goal is to determine the dissolved hydrocarbon RBSL that corresponds to the target RBSL for vapors in the breathing zone, as given in Tables X2.2 and X2.3. If the selected target vapor concentration is some value other than the RBSL for inhalation (that is, odor threshold or ecological criterion), this value can be substituted for the $RBSL_{air}$ parameter appearing in the equations given in Tables X2.2 and X2.3.

X2.5.2 A conceptual model for the transport of chemicals from ground water to indoor air is depicted in Fig. X2.2. For simplicity, the relationship between enclosed-space air and dissolved ground water concentrations is represented in Tables X2.2 and X2.3 by the "volatilization factor" VF_{wesp} [(mg/m³-air)/(mg/L-H$_2$O)] defined in Table X2.5. It is based on the following assumptions:

X2.5.2.1 A constant dissolved chemical concentration in ground water,

X2.5.2.2 Equilibrium partitioning between dissolved chemicals in ground water and chemical vapors at the ground water table,

X2.5.2.3 Steady-state vapor- and liquid-phase diffusion

through the capillary fringe, vadose zone, and foundation cracks,

X2.5.2.4 No loss of chemical as it diffuses towards ground surface (that is, no biodegradation), and

X2.5.2.5 Steady, well-mixed atmospheric dispersion of the emanating vapors within the enclosed space, where the convective transport into the building through foundation cracks or openings is negligible in comparison with diffusive transport.

X2.5.3 Should the calculated $RBSL_w$ exceed the pure component solubility for any individual component, ">S" is entered in the table to indicate that the selected risk level or hazard quotient cannot be reached or exceeded for that compound and the specified exposure scenario.

X2.6 *Surficial Soils—Ingestion, Dermal Contact, and Vapor and Particulate Inhalation:*

X2.6.1 In this case it is assumed that chemical intake results from a combination of intake routes, including: ingestion, dermal absorption, and inhalation of both particulates and vapors emanating from surficial soil.

X2.6.2 Equations used to estimate intake resulting from ingestion follow guidance given in Ref **(26)** for ingestion of chemicals in soil. For this route, it has been assumed that surficial soil chemical concentrations and intake rates remain constant over the exposure duration.

X2.6.3 Equations used to estimate intake resulting from dermal absorption follow guidance given in Ref **(26)** for dermal contact with chemicals in soil. For this route, it has been assumed that surficial soil chemical concentrations and absorption rates remain constant over the exposure duration.

X2.6.4 Equations used to estimate intake resulting from the inhalation of particulates follow guidance given in Ref **(26)** for inhalation of airborne chemicals. For this route, it has been assumed that surficial soil chemical concentrations, intake rates, and atmospheric particulate concentrations remain constant over the exposure duration.

X2.6.5 Equations used to estimate intake resulting from

E 1739

the inhalation of airborne chemicals resulting from the volatilization of chemicals from surficial soils follow guidance given in Ref (26) for inhalation of airborne chemicals.

X2.6.6 A conceptual model for the volatilization of chemicals from surficial soils to outdoor air is depicted in Fig. X2.3. For simplicity, the relationship between outdoor air and surficial soil concentrations is represented in Tables X2.2 and X2.3 by the "volatilization factor" VF_{ss} [(mg/m³-air)/(mg/kg-soil)] defined in Table X2.5. It is based on the following assumptions:

X2.6.6.1 Uniformly distributed chemical throughout the depth 0—d (cm) below ground surface,

X2.6.6.2 Linear equilibrium partitioning within the soil matrix between sorbed, dissolved, and vapor phases, where

TABLE X2.5 Volatilization Factors (VF_i), Leaching Factor (LF_{sw}), and Effective Diffusion Coefficients (D_i^{eff})

Symbol	Cross-Media Route (or Definition)	Equation
VF_{wesp}	Ground water → enclosed-space vapors	$VF_{wesp}\left[\dfrac{(mg/m^3\text{-}air)}{(mg/L\text{-}H_2O)}\right] = \dfrac{H\left[\dfrac{D_{ws}^{eff}/L_{GW}}{ER\,L_B}\right]}{1+\left[\dfrac{D_{ws}^{eff}/L_{GW}}{ER\,L_B}\right]+\left[\dfrac{D_{ws}^{eff}/L_{GW}}{(D_{crack}^{eff}/L_{crack})\eta}\right]}\times 10^3\,\dfrac{L}{m^3}$ A
VF_{wamb}	Ground water → ambient (outdoor) vapors	$VF_{wamb}\left[\dfrac{(mg/m^3\text{-}air)}{(mg/L\text{-}H_2O)}\right] = \dfrac{H}{1+\left[\dfrac{U_{air}\delta_{air}L_{GW}}{WD_{ws}^{eff}}\right]}\times 10^3\,\dfrac{L}{m^3}$ B
VF_{ss}	Surficial soils → ambient air (vapors)	$VF_{ss}\left[\dfrac{(mg/m^3\text{-}air)}{(mg/kg\text{-}soil)}\right] = \dfrac{2W\rho_s}{U_{air}\delta_{air}}\sqrt{\dfrac{D_s^{eff}H}{\pi[\theta_{ws}+k_s\rho_s+H\theta_{as}]\tau}}\times 10^3\,\dfrac{cm^3\text{-}kg}{m^3\text{-}g}$ C or: $VF_{ss}\left[\dfrac{(mg/m^3\text{-}air)}{(mg/kg\text{-}soil)}\right] = \dfrac{W\rho_s d}{U_{air}\delta_{air}\tau}\times 10^3\,\dfrac{cm^3\text{-}kg}{m^3\text{-}g}$; whichever is less D
VF_p	Surficial soils → ambient air (particulates)	$VF_p\left[\dfrac{(mg/m^3\text{-}air)}{(mg/kg\text{-}soil)}\right] = \dfrac{P_e W}{U_{air}\delta_{air}}\times 10^3\,\dfrac{cm^3\text{-}kg}{m^3\text{-}g}$ E
VF_{samb}	Subsurface soils → ambient air	$VF_{samb}\left[\dfrac{(mg/m^3\text{-}air)}{(mg/kg\text{-}soil)}\right] = \dfrac{H\rho_s}{[\theta_{ws}+k_s\rho_s+H\theta_{as}]\left(1+\dfrac{U_{air}\delta_{air}L_S}{D_s^{eff}W}\right)}\times 10^3\,\dfrac{cm^3\text{-}kg}{m^3\text{-}g}$ F
VF_{sesp}	Subsurface soil → enclosed-space vapors	$VF_{sesp}\left[\dfrac{(mg/m^3\text{-}air)}{(mg/kg\text{-}soil)}\right] = \dfrac{\dfrac{H\rho_s}{[\theta_{ws}+k_s\rho_s+H\theta_{as}]}\left[\dfrac{D_s^{eff}/L_S}{ER\,L_B}\right]}{1+\left[\dfrac{D_s^{eff}/L_S}{ER\,L_B}\right]+\left[\dfrac{D_s^{eff}/L_s}{(D_{crack}^{eff}/L_{crack})\eta}\right]}\times 10^3\,\dfrac{cm^3\text{-}kg}{m^3\text{-}g}$ A
LF_{sw}	Subsurface soils → ground water	$LF_{sw}\left[\dfrac{(mg/L\text{-}H_2O)}{(mg/kg\text{-}soil)}\right] = \dfrac{\rho_s}{[\theta_{ws}+k_s\rho_s+H\theta_{as}]\left(1+\dfrac{U_{gw}\delta_{gw}}{IW}\right)}\times 10^0\,\dfrac{cm^3\text{-}kg}{L\text{-}g}$ B
D_s^{eff}	Effective diffusion coefficient in soil based on vapor-phase concentration	$D_s^{eff}\left[\dfrac{cm^2}{s}\right] = D^{air}\dfrac{\theta_{as}^{3.33}}{\theta_T^2}+D^{wat}\dfrac{1}{H}\dfrac{\theta_{ws}^{3.33}}{\theta_T^2}$ A
D_{crack}^{eff}	Effective diffusion coefficient through foundation cracks	$D_{crack}^{eff}\left[\dfrac{cm^2}{s}\right] = D^{air}\dfrac{\theta_{acrack}^{3.33}}{\theta_T^2}+D^{wat}\dfrac{1}{H}\dfrac{\theta_{wcrack}^{3.33}}{\theta_T^2}$ A
D_{cap}^{eff}	Effective diffusion coefficient through capillary fringe	$D_{cap}^{eff}\left[\dfrac{cm^2}{s}\right] = D^{air}\dfrac{\theta_{acap}^{3.33}}{\theta_T^2}+D^{wat}\dfrac{1}{H}\dfrac{\theta_{wcap}^{3.33}}{\theta_T^2}$ A
D_{ws}^{eff}	Effective diffusion coefficient between ground water and soil surface	$D_{ws}^{eff}\left[\dfrac{cm^2}{s}\right] = (h_{cap}+h_v)\left[\dfrac{h_{cap}}{D_{cap}^{eff}}+\dfrac{h_v}{D_s^{eff}}\right]^{-1}$ A
C_s^{sat}	Soil concentration at which dissolved pore-water and vapor phases become saturated	$C_s^{sat}\left[\dfrac{mg}{kg\text{-}soil}\right] = \dfrac{S}{\rho_s}\times[H\theta_{as}+\theta_{ws}+k_s\rho_s]\times 10^0\,\dfrac{L\text{-}g}{cm^3\text{-}kg}$ F

A See Ref (29).
B See Ref (30).
C See Ref (31).
D Based on mass balance.
E See Ref (32).
F See Ref (33).

210

TABLE X2.6 Soil, Building, Surface, and Subsurface Parameters Used in Generating Example Tier 1 RBSLs

NOTE—See X2.10 for justification of parameter selection.

Parameters	Definitions, Units	Residential	Commercial/Industrial
d	lower depth of surficial soil zone, cm	100 cm	100 cm
D^{air}	diffusion coefficient in air, cm²/s	chemical-specific	chemical-specific
D^{wat}	diffusion coefficient in water, cm²/s	chemical-specific	chemical-specific
ER	enclosed-space air exchange rate, 1/sec	0.00014 s⁻¹	0.00023 s⁻¹
f_{oc}	fraction of organic carbon in soil, g-C/g-soil	0.01	0.01
H	henry's law constant, (cm³-H₂O)/(cm³-air)	chemical-specific	chemical-specific
h_{cap}	thickness of capillary fringe, cm	5 cm	5 cm
h_v	thickness of vadose zone, cm	295 cm	295 cm
I	infiltration rate of water through soil, cm/years	30 cm/year	30 cm/year
k_{oc}	carbon-water sorption coefficient, cm³-H₂O/g-C	chemical-specific	chemical-specific
k_s	soil-water sorption coefficient, cm³-H₂O/g-soil	$f_{oc} \times k_{oc}$	$f_{oc} \times k_{oc}$
L_B	enclosed-space volume/infiltration area ratio, cm	200 cm	300 cm
L_{crack}	enclosed-space foundation or wall thickness, cm	15 cm	15 cm
L_{GW}	depth to ground water $= h_{cap} + h_v$, cm	300 cm	300 cm
L_s	depth to subsurface soil sources, cm	100 cm	100 cm
P_e	particulate emission rate, g/cm²-s	6.9×10^{-14}	6.9×10^{-14}
S	pure component solubility in water, mg/L-H₂O	chemical-specific	chemical-specific
U_{air}	wind speed above ground surface in ambient mixing zone, cm/s	225 cm/s	225 cm/s
U_{gw}	ground water Darcy velocity, cm/year	2500 cm/year	2500 cm/year
W	width of source area parallel to wind, or ground water flow direction, cm	1500 cm	1500 cm
δ_{air}	ambient air mixing zone height, cm	200 cm	200 cm
δ_{gw}	ground water mixing zone thickness, cm	200 cm	200 cm
η	areal fraction of cracks in foundations/walls, cm²-cracks/cm²-total area	0.01 cm²-cracks/cm²-total area	0.01 cm²-cracks/cm²-total area
θ_{acap}	volumetric air content in capillary fringe soils, cm³-air/cm³-soil	0.038 cm³-air/cm³-soil	0.038 cm³-air/cm³-soil
θ_{acrack}	volumetric air content in foundation/wall cracks, cm³-air/cm³ total volume	0.26 cm³-air/cm³ total volume	0.26 cm³-air/cm³ total volume
θ_{as}	volumetric air content in vadose zone soils, cm³-air/cm³-soil	0.26 cm³-air/cm³-soil	0.26 cm³-air/cm³-soil
θ_T	total soil porosity, cm³/cm³-soil	0.38 cm³/cm³-soil	0.38 cm³/cm³-soil
θ_{wcap}	volumetric water content in capillary fringe soils, cm³-H₂O/cm³-soil	0.342 cm³-H₂O/cm³-soil	0.342 cm³-H₂O/cm³-soil
θ_{wcrack}	volumetric water content in foundation/wall cracks, cm³-H₂O/cm³ total volume	0.12 cm³-H₂O/cm³ total volume	0.12 cm³-H₂O/cm³ total volume
θ_{ws}	volumetric water content in vadose zone soils, cm³-H₂O/cm³-soil	0.12 cm³-H₂O/cm³-soil	0.12 cm³-H₂O/cm³-soil
ρ_s	soil bulk density, g-soil/cm³-soil	1.7 g/cm³	1.7 g/cm³
τ	averaging time for vapor flux, s	9.46×10^8 s	7.88×10^8 s

the partitioning is a function of constant chemical- and soil-specific parameters,

X2.6.6.3 Diffusion through the vadose zone,

X2.6.6.4 No loss of chemical as it diffuses towards ground surface (that is, no biodegradation), and

X2.6.6.5 Steady well-mixed atmospheric dispersion of the emanating vapors within the breathing zone as modeled by a "box model" for air dispersion.

X2.6.7 In the event that the time-averaged flux exceeds that which would occur if all chemical initially present in the surficial soil zone volatilized during the exposure period,

then the volatilization factor is determined from a mass balance assuming that all chemical initially present in the surficial soil zone volatilizes during the exposure period.

X2.7 *Subsurface Soils—Inhalation of Outdoor Vapors:*

X2.7.1 In this case chemical intake is a result of inhalation of outdoor vapors which originate from hydrocarbons contained in subsurface soils located some distance below ground surface. Here the goal is to determine the RBSL for subsurface soils that corresponds to the target RBSL for outdoor vapors in the breathing zone, as given in Table X2.1. If the selected target vapor concentration is some value

TABLE X2.7 Chemical-Specific Properties Used in the Derivation Example Tier 1 RBSLs

Chemical	CAS Number	M_w, g/mol	H, L-H₂O/L-air	D^{air}, cm²/s	D^w, cm²/s	$\log(K_{oc})$, L/kg	$\log(K_{ow})$, L/kg
Benzene	71-43-2	78[A]	0.22[A]	0.093[A]	1.1×10^{-5A}	1.58[A]	2.13[A]
Toluene	108-88-3	92[A]	0.26[A]	0.085[A]	9.4×10^{-6D}	2.13[A]	2.65[A]
Ethyl benzene	100-41-4	106[A]	0.32[A]	0.076[A]	8.5×10^{-6D}	3.11[A]	3.13[A]
Mixed xylenes	1330-20-7	106[A]	0.29[A]	0.072[D]	8.5×10^{-6D}	2.38[A]	3.26[A]
Naphthalene	91-20-3	128[A]	0.049[A]	0.072[D]	9.4×10^{-6A}	3.11[A]	3.28[A]
Benzo(a)pyrene	50-32-8	252[C]	5.8×10^{-8B}	0.050[D]	5.8×10^{-6D}	5.59[E]	5.98[B]

Chemical	CAS Number	SF_o, kg-day/mg	SF_i, kg-day/mg	RfD_o, mg/kg-day	RfD_i, mg/kg-day
Benzene	71-43-2	0.029[F]	0.029[F]	. . .	. . .
Toluene	108-88-3	. . .	. . .	0.2[F]	0.11[F]
Ethyl benzene	100-41-4	. . .	. . .	0.1[F]	0.29[F]
Mixed xylenes	1330-20-7	. . .	. . .	2.0[F]	2.0[F]
Naphthalene	91-20-3	. . .	. . .	0.004[G]	0.004[G]
Benzo(a)pyrene	50-32-8	7.3[F]	6.1[F]	. . .	. . .

[A] See Ref (34).
[B] See Ref (35).
[C] See Ref (7).
[D] Diffusion coefficient calculated using the method of Fuller, Schettler, and Giddings, from Ref (11).
[E] Calculated from K_{ow}/K_{oc} correlation: $\log(K_{oc}) = 0.937 \log(K_{ow}) - 0.006$, from Ref (11).
[F] See Ref (2).
[G] See Ref (3).

other than the RBSL for inhalation (that is, odor threshold or ecological criterion), this value can be substituted for the $RBSL_{air}$ parameter appearing in the equations given in Tables X2.2 and X2.3.

X2.7.2 A conceptual model for the transport of chemicals from subsurface soils to ambient air is depicted in Fig. X2.4. For simplicity, the relationship between outdoor air and soil concentrations is represented in Tables X2.2 and X2.3 by the "volatilization factor," VF_{samb} [(mg/m³-air)/(mg/kg-soil)], defined in Table X2.5. It is based on the following assumptions:

X2.7.2.1 A constant chemical concentration in subsurface soils,

X2.7.2.2 Linear equilibrium partitioning within the soil matrix between sorbed, dissolved, and vapor phases, where the partitioning is a function of constant chemical- and soil-specific parameters,

X2.7.2.3 Steady-state vapor- and liquid-phase diffusion through the vadose zone to ground surface,

X2.7.2.4 No loss of chemical as it diffuses towards ground surface (that is, no biodegradation), and

X2.7.2.5 Steady well-mixed atmospheric dispersion of the emanating vapors within the breathing zone as modeled by a "box model" for air dispersion.

X2.7.3 Should the calculated $RBSL_s$ exceed the value for which the equilibrated vapor and dissolved pore-water phases become saturated, C_s^{sat} [mg/kg-soil] (see Table X2.5 for calculation of this value), "RES" is entered in the table to indicate that the selected risk level or hazard quotient cannot be reached or exceeded for that compound and the specified exposure scenario (even if free-phase product or precipitate is present in the soil).

X2.8 *Subsurface Soils—Inhalation of Enclosed-Space (Indoor) Vapors:*

X2.8.1 In this case chemical intake is a result of inhalation of enclosed-space vapors which originate from hydrocarbons contained in subsurface soils located some distance below ground surface. Here the goal is to determine the RBSL for subsurface soils that corresponds to the target RBSL for indoor vapors, as given in Tables X2.2 and X2.3. If the selected target vapor concentration is some value other than the RBSL for inhalation (that is, odor threshold or

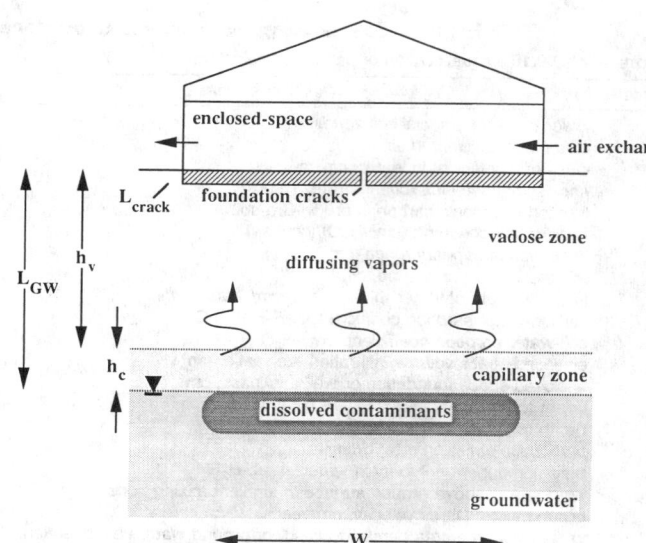

FIG. X2.2 Volatilization from Ground Water to Enclosed-Space Air

ecological criterion), this value can be substituted for the $RBSL_{air}$ parameter appearing in the equations given in Tables X2.2 and X2.3.

X2.8.2 A conceptual model for the transport of chemicals from subsurface soils to enclosed spaces is depicted in Fig. X2.5. For simplicity, the relationship between indoor air and soil concentrations is represented in Tables X2.2 and X2.3 by the "volatilization factor," VF_{sesp} [(mg/m³-air)/(kg-soil)], defined in Table X2.5. It is based on the following assumptions:

X2.8.2.1 A constant chemical concentration in subsurface soils,

X2.8.2.2 Linear equilibrium partitioning within the soil matrix between sorbed, dissolved, and vapor phases, where the partitioning is a function of constant chemical- and soil-specific parameters,

X2.8.2.3 Steady-state vapor- and liquid-phase diffusion through the vadose zone and foundation cracks,

X2.8.2.4 No loss of chemical as it diffuses towards ground surface (that is, no biodegradation), and

X2.8.2.5 Well-mixed atmospheric dispersion of the emanating vapors within the enclosed space.

X2.8.3 Should the calculated $RBSL_s$ exceed the value C_s^{sat} [mg/kg-soil] for which the equilibrated vapor and dissolved pore-water phases become saturated (see Table X2.5 for calculation of this value), "RES" is entered in the table to indicate that the selected risk level or hazard

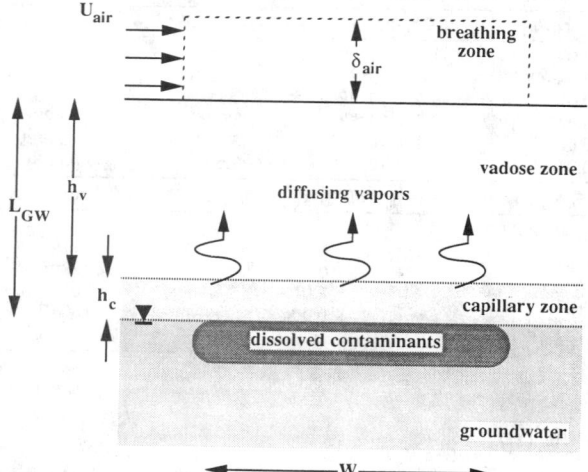

FIG. X2.1 Volatilization from Ground Water to Ambient Air

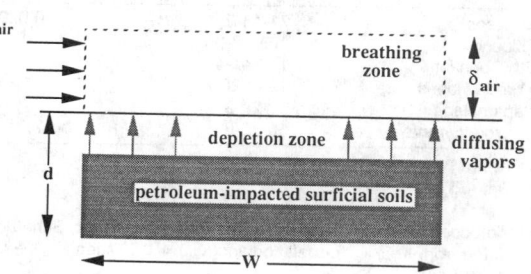

FIG. X2.3 Volatilization from Surficial Soils

quotient cannot be reached or exceeded for that compound and the specified exposure scenario (even if free-phase product or precipitate is present in the soil).

X2.9 *Subsurface Soils—Leaching to Ground Water:*

X2.9.1 In this case chemical intake is a result of chemicals leaching from subsurface soils, followed by inhalation of enclosed-space vapors, inhalation of outdoor vapors, or ingestion of ground water as discussed in X2.1 through X2.3. Here the goal is to determine the RBSL for subsurface soils that corresponds to the target RBSLs for the inhalation or ingestion routes. If the selected target ground water concentration is some value other than an RBSL for ground water (that is, odor threshold or ecological criterion), this value can be substituted for the $RBSL_w$ parameter appearing in the equations given in Tables X2.2 and X2.3.

X2.9.2 A conceptual model for the leaching of chemicals from subsurface soils to ground water is depicted in Fig. X2.6. For simplicity, the relationship between ground water and soil concentrations is represented in Tables X2.2 and X2.3 by the "leaching factor," LF_{sw} [(mg/L-H$_2$O)/ (mg/kg-soil)], defined in Table X2.5. It is based on the following assumptions:

X2.9.2.1 A constant chemical concentration in subsurface soils,

X2.9.2.2 Linear equilibrium partitioning within the soil matrix between sorbed, dissolved, and vapor phases, where the partitioning is a function of constant chemical- and soil-specific parameters,

X2.9.2.3 Steady-state leaching from the vadose zone to ground water resulting from the constant leaching rate I [cm/s],

X2.9.2.4 No loss of chemical as it leaches towards ground water (that is, no biodegradation), and

X2.9.2.5 Steady well-mixed dispersion of the leachate within a ground water "mixing zone."

X2.9.3 Should the calculated $RBSL_s$ exceed the value C_s^{sat}, for which the equilibrated vapor and dissolved pore-water phases become saturated (see Table X2.5 for calculation of this value), "RES" is entered in the table to indicate that the selected risk level or hazard quotient cannot be reached or exceeded for that compound and the specified exposure scenario (even if free-phase product or precipitate is present in the soil).

X2.9.4 In some regulatory programs, "dilution attenuation factors" (DAFs) are currently being proposed based on

fate and transport modeling results. A DAF is typically defined as the ratio of a target ground water concentration divided by the source leachate concentration, and is inherently very similar to the leachate factor, LF_{sw}, discussed here. The difference between these two terms is that LF_{sw} represents the ratio of the target ground water concentration divided by the source area soil concentration. Should a regulatory program already have a technically defensible DAF value, it can be equated to a leachate factor by the following expression:

$$LF_{sw} = \frac{DAF \times \rho_s}{[\theta_{ws} + k_s\rho_s + H\theta_{as}]} \times 10^0$$

where the parameters are defined in Table X2.6.

X2.10 *Parameter Values:*

X2.10.1 Table X2.4 lists exposure parameters used to calculate the RBSLs appearing in sample Look-Up Table X2.1. All values given are based on adult exposures only. With the exception of the dermal exposure parameters (SA, M, and RAF_d), the values given are reasonable maximum exposure (RME) values presented in Ref (27) and are regarded as upper bound estimates for each individual exposure parameter.

X2.10.2 The skin surface area, $SA = 3160$ cm^2/day, is based on the average surface area of the head, hands, and

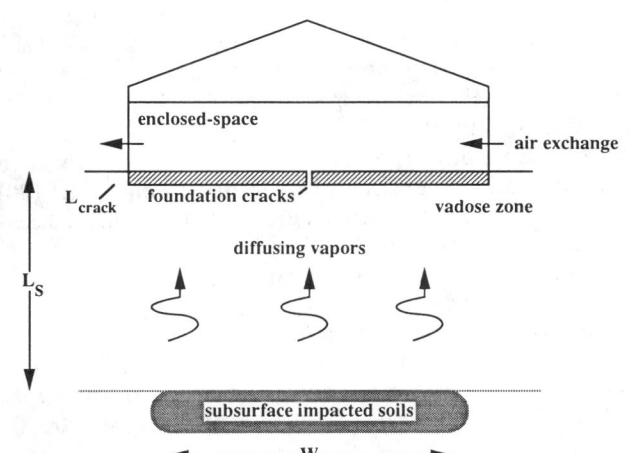

FIG. X2.5 Volatilization from Subsurface Soils to Enclosed-Space Air

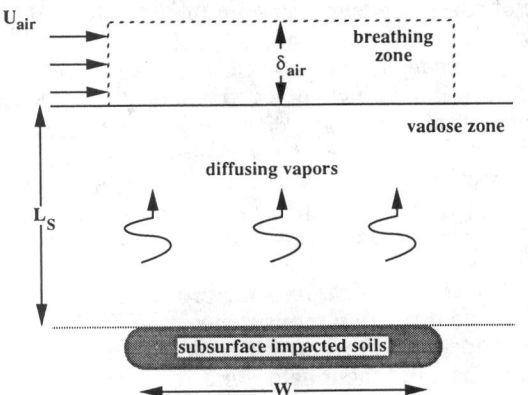

FIG. X2.4 Volatilization from Subsurface Soils to Ambient Air

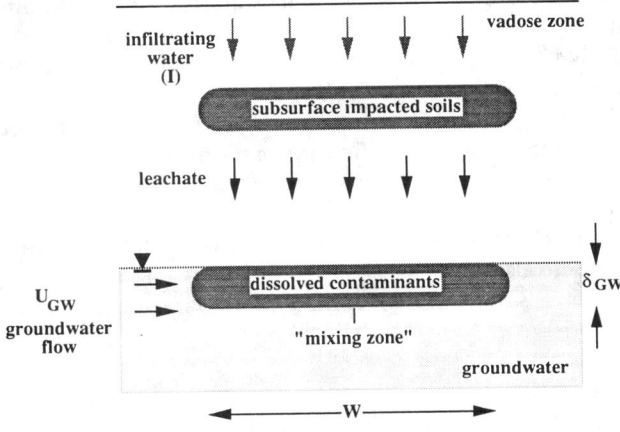

FIG. X2.6 Leaching from Subsurface Soils to Ground Water

forearms for adult males given in Ref (27). The soil-to-skin adherence factor, M [mg/cm^2], and dermal relative absorption factor, RAF_d [mg-absorbed/mg-applied], are based on guidance issued by Ref (28).

X2.10.3 Soil properties are based on typical values for sandy soils and are consistent with values given in Ref (30).

X2.10.4 Physical dimensions are consistent with the scale of typical underground fuel tank releases.

X2.10.5 Particulate emission rates were estimated by the approach presented by Cowherd, et al (32). It was assumed that the mode of the surficial soil size distribution was 2 mm, the erosion potential was unlimited, there was no vegetative cover, and the mean average annual wind speed was 4 m/s.

X2.10.6 The chemical-specific parameters used are defined in Table X2.7.

X2.10.7 In this development, surficial soils are defined as those soils present within 1 m of ground surface. Subsurface soil RBSLs are based on assumed source depths of 1 m. Ground water is assumed to be located 3 m below ground surface.

X2.10.8 Once again, the reader is reminded that the parameter (and corresponding RBSL) values are presented here as examples only, and are not intended to be used as standards. At best, the parameters presented are reasonable values based on current information and professional judgment. The reader should review and verify all assumptions prior to using any of the example RBSLs as screening level values.

X3. USE OF PREDICTIVE MODELING IN THE RISK-BASED CORRECTIVE ACTION PROCESS

X3.1 *Scope:*

X3.1.1 Predictive modeling is a valuable tool that can provide information to the risk management process. In a RBCA, modeling is used to predict the location and concentration contaminants and to interpret, or extrapolate, site characterization data, historical monitoring data, and toxicological information. In addition, predictive modeling may be used in evaluation of remedial alternatives and in evaluating compliance targets in monitoring plans. This appendix discusses the following:

X3.1.1.1 Significance and use of predictive modeling in the RBCA process;

X3.1.1.2 Interpretation of predictive modeling results;

X3.1.1.3 Procedures for predictive migration models; and

X3.1.1.4 Procedures for exposure, risk, and dose-response assessment.

X3.1.2 This appendix is not intended to be all inclusive. Each predictive model is unique and may require additional procedures in its development and application. All such additional analyses should be documented in the RBCA process.

X3.2 *Referenced Documents:*

X3.2.1 *ASTM Standards:*

D 653 Terminology Relating to Soil, Rock, and Contained Fluids[7]

D 5447 Guide for Application of a Ground-Water Flow Model to a Site-Specific Problem[8]

D 5490 Guide for Comparing Ground-Water Flow Model Simulations to Site-Specific Information[8]

E 943 Terminology Relating to Biological Effects and Environmental Fate[9]

E 978 Practice for Evaluating Environmental Fate Models of Chemicals[9]

D 5609 Guide for Defining Boundary Conditions in Ground-Water Flow Modeling[8]

D 5610 Guide for Defining Initial Conditions in Ground-Water Flow Modeling[8]

D 5611 Guide for Conducting a Sensitivity Analysis for a Ground-Water Flow Model Application[8]

X3.3 *Terminology:*

X3.3.1 *Definitions*—For definitions of terms used in this appendix, see Terminologies D 653 and E 943.

X3.3.2 *Descriptions of Terms Specific to This Appendix:*

X3.3.2.1 *analytical model*—a model that uses mathematical solutions to governing equations that are continuous in space and time and applicable to the flow and transport process.

X3.3.2.2 *application verification*—using the set of parameter values and boundary conditions from a calibrated model to approximate acceptably a second set of field data measured under similar conditions.

DISCUSSION—Application verification is to be distinguished from code verification, which refers to software testing, comparison with analytical solutions, and comparison with other similar codes to demonstrate that the code represents its mathematical foundation.

X3.3.2.3 *boundary condition*—a mathematical expression of a state of the physical system that constrains the equations of the mathematical model.

X3.3.2.4 *calibration (model application)*—the process of refining the model representation of the fluid and media properties and boundary conditions to achieve a desired degree of correspondence between the model simulation and observations of the real system.

X3.3.2.5 *code validation*—the process of determining how well a modeling code's theoretical foundation and computer implementation describe actual system behavior in terms of the "degree of correlation" between calculated and independently observed cause-and-effect responses of the prototype fluid flow system (for example, research site or laboratory experiment) for which the code has been developed.

X3.3.2.6 *code verification*—the procedure aimed at establishing the completeness, consistency, correctness, and accuracy of modeling software with respect to its design criteria by evaluating the functionality and operational characteristics of the code and testing embedded algorithms and data transfers through execution of problems for which indepen-

[7] *Annual Book of ASTM Standards*, Vol 04.08.
[8] *Annual Book of ASTM Standards*, Vol 04.09.
[9] *Annual Book of ASTM Standards*, Vol 11.04.

dent benchmarks are available.

X3.3.2.7 *computer code (computer program)*—the assembly of numerical techniques, bookkeeping, and control language that represents the model from acceptance of input

data and instructions to delivery of output.

X3.3.2.8 *conceptual model*—an interpretation or working description of the characteristics and dynamics of the physical system.

TABLE X3.1 Example Screening Level Transport Models

Description	Mathematical Approximation	Parameters
Dissolved Phase Transport: Maximum transport rate $u_{d,max}$ [cm/day] of dissolved plume	$u_{d,max} = \dfrac{K_s i}{\theta_s R_c}$	$C(x)$ = dissolved hydrocarbon concentration along centerline (x, $y = 0$, $z = 0$) of dissolved plume [g/cm³-H₂O]
Minimum time $\tau_{d,min}$ [d] for leading edge of dissolved plume to travel distance, L [cm]	$\tau_{d,min} = \dfrac{L}{u_{d,max}}$	C_{source} = dissolved hydrocarbon concentration in dissolved plume source area [g/cm³-H₂O]
		i = ground water gradient [cm/cm]
		K_s = saturated hydraulic conductivity [cm/day]
		k_s = sorption coefficient [(g/g-soil)/(g/cm³-H₂O)]
		L = distance downgradient [cm]
		R_c = retardation factor = $[1 + k_s \rho_s / \theta_s]$
		S_w = source width (perpendicular to flow in the horizontal plane) [cm]
Steady-state attenuation [(g/cm³-H₂O)/(g/cm³-H₂O)] along the centerline (x, $y = 0$, $z = 0$) of a dissolved plume	$\dfrac{C(x)}{C_{source}} = \exp\left\{\dfrac{x}{2\alpha_x}\left[1 - \sqrt{\left(1 + \dfrac{4\lambda\alpha_x}{u}\right)}\right]\right\}$ $\cdot \left(erf\left[\dfrac{S_w}{4\sqrt{\alpha_y x}}\right]\right)\left(erf\left[\dfrac{S_d}{4\sqrt{\alpha_z x}}\right]\right)$ where: $u = K_s i / \theta_s$	S_d = source width (perpendicular to flow in the vertical plane) [cm]
		u = specific discharge [cm/day]
		$u_{d,max}$ = maximum transport rate of dissolved plume [cm/day][A]
		x = distance along centerline from downgradient edge of dissolved plume source zone [cm]
		y = depth below water table [cm]
		z = lateral distance away from dissolved plume centerline [cm]
		α_x = longitudinal dispersivity [cm] ≈ $0.10\ x$
		α_y = transverse dispersivity [cm] ≈ $\alpha_x/3$
		α_z = vertical dispersivity [cm] ≈ $\alpha_x/20$
		λ = first-order degradation constant [d^{-1}]
		θ_s = volumetric water content of saturated zone [cm³-H₂O/cm³-soil]
		ρ_s = soil bulk density [g-soil/cm³-soil]
		$\tau_{d,min}$ = minimum convective travel time of dissolved hydrocarbons to distance L [d][A]
		$erf(\eta)$ = error function evaluated for value η
Immiscible Phase Transport: Maximum depth D_{max} [cm] of immiscible phase penetration	$D_{max} = \dfrac{V_{spill}}{\theta_R \pi R_{spill}^2}$	C_{soil} = total soil hydrocarbon concentration [g/g-soil]
Equilibrium Partitioning: **Vapor Concentration:** $C_{v,eq}$ [g/cm³-vapor]		$C_{v,eq}$ = equilibrium vapor concentration [g/cm³-vapor][A]
		$C_{w,eq}$ = equilibrium dissolved concentration [g/cm³-H₂O][A]
Maximum vapor concentration above dissolved hydrocarbons	$C_{v,eq} = H C_{w,eq}$	D_{max} = maximum depth of immiscible phase penetration [cm][A]
		H = Henry's Law Constant [(g/cm³-vapor)/(g/cm³-H₂O)]
Maximum vapor concentration when immiscible hydrocarbon is present	$C_{v,eq} = \dfrac{x_i P_v^i M_w}{RT}$	k_s = sorption coefficient [(g/g-soil)/(g/cm³-H₂O)]
		M_w = molecular weight [g/mol]
		P_v^i = vapor pressure of compound i [atm]
		R = gas constant = 82 cm³-atm/mol-K
Maximum vapor concentrations in soil pores (no immiscible phase present)	$C_{v,eq} = \dfrac{H C_{soil}\ \rho_s}{[\theta_w + k_s\rho_s + H\theta_v]}$	R_{spill} = radial extent of hydrocarbon impact [cm]
		S_i = pure component solubility [g/cm³-H₂O]
		T = absolute temperature [K]
		V_{spill} = volume of hydrocarbon released [cm³]
Dissolved Concentration: $C_{w,eq}$ [g/cm³-H₂O]		x_i = mol fraction of component i
		θ_R = volumetric residual content of hydrocarbon under drainage conditions [cm³-hydrocarbon/cm³-soil]
Maximum dissolved concentration when immiscible hydrocarbon is present	$C_{w,eq} = x_i S_i$	θ_w = volumetric content of soil pore water [cm³-H₂O/cm³-soil]
Maximum dissolved concentration in soil pores (no immiscible phase present)	$C_{w,eq} = \dfrac{C_{soil}\rho_s}{[\theta_w + k_s\rho_s + H\theta_v]}$	θ_v = volumetric content of soil vapor [cm³-vapor/cm³-soil]
		π = 3.1416
		ρ_s = soil bulk density [g-soil/cm³-soil]
Equilibrium Partioning: **Soil Concentrations [g/g-soil]:** Soil concentration [C_{soil}] [g/g-soil] at which immiscible hydrocarbon phase forms in soil matrix	$(C_{soil}) = \dfrac{S_i}{\rho_s}[\theta_w + k_s\rho_s + H\theta_v]$	(C_{soil}) = concentration at which immiscible phase forms in soil [g/g-soil][A]
		D^{air} = pure component diffusion coefficient in air [cm²/day]
		D^{eff} = effective diffusion coefficient for combined vapor and solute transport, expressed as a vapor phase diffusion coefficient (no immiscible hydrocarbon present outside of source area) [cm²/day][A]
Vapor Phase Transport: Effective porous media diffusion coefficient D^{eff} [cm²/day] for combined vapor and solute transport, expressed as a vapor phase diffusion coefficient (no immiscible hydrocarbon present outside of source area)	$D^{eff} = \dfrac{\theta_v^{3.33}}{\theta_T^2}\cdot D^{air} + \dfrac{1}{H}\dfrac{\theta_w^{3.33}}{\theta_T^2}\cdot D^{wat}$	D^w = pure component diffusion coefficient in water [cm²/day]
		H = Henry's Law Constant [(g/cm³-vapor)/(g/cm³-H₂O)]
		k_s = sorption coefficient [(g/g-soil)/(g/cm³-H₂O)]
		k_v = permeability to vapor flow [cm²]
		L = distance [cm]
		R_v = porous media "retardation" factor (no immiscible hydrocarbon present outside of source area)
		S_i = pure component solubility [g/cm³-H₂O]
		$u_{v,max}$ = maximum convective transport rate of vapors [cm/day][A]
Porous media "retardation" factor R_v (no immiscible hydrocarbon present outside of source area)	$R_v = \left[\dfrac{\theta_w}{H} + \dfrac{k_s\rho_s}{H} + \theta_v\right]$	∇P = vapor phase pressure gradient [g/cm²-s²]
		θ_w = volumetric content of soil pore water [cm³-H₂O/cm³-soil]

TABLE X3.1 *Continued*

Description	Mathematical Approximation	Parameters
Maximum convective transport rate $u_{v,max}$ [cm/day] of vapors	$u_{v,max} = \dfrac{1}{R_v}\dfrac{k_v}{\mu_v}\nabla P$	θ_v = volumetric content of soil vapor [cm^3-vapor/cm^3-soil] θ_T = total volumetric content of pore space in soil matrix [cm^3/cm^3-soil]
Minimum time $\tau_{v,min}$ [d] for vapors to travel a distance L [cm] from source area by convection[A]	$\tau_{c,min} = \dfrac{L}{u_{v,max}}$	μ_v = vapor viscosity [g/cm-s] ρ_s = soil bulk density [g-soil/cm^3-soil] $\tau_{c,min}$ = minimum time for vapors to travel a distance L [cm] by convection [day][A]
Minimum time $\tau_{v,min}$ [d] for vapors to travel a distance L [cm] from source area by diffusion	$\tau_{d,min} = \dfrac{L^2}{(D^{eff}/R_v)}$	$\tau_{d,min}$ = minimum time for vapors to travel a distance L [cm] by diffusion [day][A] C_{soil} = total soil hydrocarbon concentration [g/g-soil] $C_{v,eq}$ = equilibrium vapor concentration [g/cm^3-vapor][A]
Vapor Emissions from Subsurface Vapor Sources to Open Surfaces:		d = distance below ground surface to top of hydrocarbon vapor source [cm]
Maximum diffusive vapor flux F_{max} [g/cm^2-day] from subsurface vapor source located a distance d [cm] below ground surface (steady-state, constant source)	$F_{max} = D^{eff}\dfrac{C_{v,eq}}{d}$	D^{eff} = effective diffusion coefficient for combined vapor and solute transport, expressed as a vapor phase diffusion coefficient (no immiscible hydrocarbon present outside of source area) [cm^2/day][A]
Maximum time-averaged diffusive vapor flux $<F_{max}>$ [g/cm^2-day] from subsurface soils over period from time = 0 to time = τ, single-component immiscible phase present	$<F_{max}> = \dfrac{\rho_s C_{soil}}{\tau}\left\{\sqrt{\left[d^2 + \dfrac{2C_{v,eq}D^{eff}\tau}{\rho_s C_{soil}}\right]} - d\right\}$	R_v = porous media "retardation" factor (no immiscible hydrocarbon present outside of source area)[A] $u_{v,max}$ = maximum convective transport rate $u_{v,max}$ of vapors [cm/day][A] ρ_s = soil bulk density [g-soil/cm^3-soil] τ = averaging time [s] A_B = total area of enclosed space exposed to vapor intrusion (area of foundation) [cm^2] A_{crack} = area of foundation through which vapors are transported (area of cracks, open seams, and so forth) [cm^2]
Maximum combined convective and diffusive vapor flux F_{max} [g/cm^2-day] from subsurface vapor source located a distance d [cm] below ground surface	$F_{max} = R_v u_{v,max} C_{v,eq} - \dfrac{R_v u_{v,max} C_{v,eq}}{\left[1 - \exp\left(\dfrac{R_v u_{v,max} d}{D^{eff}}\right)\right]}$	C_{soil} = total soil hydrocarbon concentration [g/g-soil] $C_{v,eq}$ = equilibrium vapor concentration [g/cm^3-vapor][A] d = distance between foundation/walls and hydrocarbon vapor source [cm] D^{eff} = effective diffusion coefficient through soil for combined vapor and solute transport, expressed as a vapor phase diffusion coefficient (no immiscible hydrocarbon present outside of source area) [cm^2/day][A]
Vapor Emissions from Surface Soils to Open Spaces:		D^{crack} = effective diffusion coefficient through foundation cracks [cm^2/day][A] L_{crack} = thickness of foundation/wall [cm]
Maximum time-averaged diffusive vapor flux $<F_{max}>$ [g/cm^2-day] from surface soils over period from time = 0 to time = τ, single-component immiscible phase present	$<F_{max}> = \rho_s C_{soil}\sqrt{\dfrac{2C_{v,eq}D^{eff}}{\rho_s C_{soil}\tau}}$	$M_{w,I}$ = molecular weight of i [g/mol] $M_{w,T}$ = average molecular weight of the hydrocarbon mixture [g/mol] P_i^v = vapor pressure of pure component i [atm] Q_B = volumetric flow rate of air within enclosed space [cm^3/s] Q_{soil} = volumetric infiltration flow rate of soil gas into enclosed space [cm^3/s]
Maximum time-averaged diffusive vapor flux $<F_{max}>$ [g/cm^2-day] from surface soils over period from time = 0 to time = τ, no immiscible phase present	$<F_{max}> = 2\rho_s C_{soil}\sqrt{\dfrac{D^{eff}}{\pi R_v \tau}}$	R = gas constant = 82 atm-cm^3/mol-K R_v = porous media "retardation" factor[A] T = absolute temperature [K] x_I = mol fraction of component i θ_v = volumetric content of soil vapor [cm^3-vapor/cm^3-soil] ρ_s = soil bulk density [g-soil/cm^3-soil]
Maximum time-averaged diffusive vapor flux $<F_{max}>$ [g/cm^2-day] from surface soils over period from time = 0 to time = τ, volatile components from relatively nonvolatile immiscible phase (for example, benzene from gasoline)	$<F_{max}> = \dfrac{2D^{eff}\left(\dfrac{x_I P_i^v M_{w,I}}{RT}\right)}{\sqrt{\pi\alpha\tau}}$ where: $\alpha = \dfrac{D^{eff}}{\theta_v + \dfrac{\rho_s RT(C_{soil}/M_{w,T})}{P_i^v}}$	π = 3.1416 τ = averaging time [s] $C_{w,eq}$ = equilibrium dissolved concentration in leachate source area [g/cm^3-H$_2$O][A] E_B = enclosed space air exchange rate [l/d] E_{max} = vapor emission rate into enclosed space [g/day][A] F = vapor flux [g/cm^2-day][A] i = ground water gradient [cm/cm] K_s = saturated hydraulic conductivity [cm/day] L = downwind length of vapor emissions source area [cm] M = ground water mixing zone thickness [cm] q_I = water infiltration rate [cm/day]

X3.3.2.9 *ground water flow model*—application of a mathematical model to represent a site-specific ground water flow system.

X3.3.2.10 *mathematical model*—mathematical equations expressing the physical system and including simplifying assumptions. The representation of a physical system by mathematical expressions from which the behavior of the system can be deduced with known accuracy.

X3.3.2.11 *migration model*—application of a mathematical model to represent a site-specific fluid flow system.

X3.3.2.12 *model*—an assembly of concepts in the form of mathematical equations that portray understanding of a natural phenomenon.

X3.3.2.13 *sensitivity (model application)*—the degree to

TABLE X3.1 *Continued*

Description	Mathematical Approximation	Parameters
Vapor Emissions to Enclosed Spaces: Maximum vapor emission rate E_{max} [g/cm^2-d] to enclosed spaces from subsurface vapor sources located a distance d [cm] away from the enclosed spaces	$E_{max} = Q_B C_{v,eq} \left(\dfrac{D^{eff} A_B}{Q_B d} \right) \exp\left(\dfrac{Q_{soil} L_{crack}}{D^{crack} A_{crack}} \right)$ $/ \left[\exp\left(\dfrac{Q_{soil} L_{crack}}{D^{crack} A_{crack}} \right) \right.$ $\left. + \left(\dfrac{D^{eff} A_B}{Q_{soil} d} \right) \left(\exp\left(\dfrac{Q_{soil} L_{crack}}{D^{crack} A_{crack}} \right) - 1 \right) \right]$	u_w = wind speed [cm/day] V_B = volume of enclosed space [cm^3] W = width of impacted soil zone [cm] δ = height of breathing zone [cm]
Hydrocarbon Vapor Dispersion: Ambient hydrocarbon vapor concentration resulting from area vapor source $C_{outdoor}$ [g/cm^3] Enclosed space vapor concentration C_{indoor} [g/cm^3]	$C_{outdoor} = \dfrac{FL}{u_w \delta}$ $C_{indoor} = \dfrac{E_{max}}{V_B E_B}$	
Leachate Transport: *Leaching Impact on Ground Water:* Ground water source area concentration C_{source} [g/cm^3-H$_2$O] resulting from leaching through vadose zone hydrocarbon-impacted soils Ground water source area concentraiton C_{source} [g/cm^3-H$_2$O] resulting from hydrocarbon-impacted soils in direct contact with ground water	$C_{source} = C_{w,eq} \dfrac{q_l W}{(K_s j M + q_l W)}$ $C_{source} = C_{w,eq}$	

[A] Equation for this parameter given in this table.

TABLE X3.2   Reported Degradation Rates for Petroleum Hydrocarbons

Reference	Source of Data	Chemical Decay Rates (day^{-1}, [half-life days])							
		Benzene	Toluene	Ethyl-Benzene	Xylenes	O-Xylene	MTBE	Naphthalene	Benzo (a)Pyrene
Barker, et al[A]	Borden Aquifer, Canada	0.007 [99]	0.011 [63]	...	...	0.014 [50]	...	...	...
Kemblowski[B]	Eastern Florida Aquifer	0.0085 [82]	...	...	...	...	...	...	...
Chiang, et al[C]	Northern Michigan Aquifer	0.095 [7]	...	...	...	...	...	...	...
Wilson, et al[D]	Traverse City, MI Aquifer	0.007 to 0.024 [99] to [29]	0.067 [10]	...	0.004 to 0.014 [173] to [50]	...	...	...	...
Howard, et al[E]	Literature	0.0009 [730] to 0.069 [10]	0.025 [28] to 0.099 [7]	0.003 [228] to 0.116 [6]	0.0019 [365] to 0.0495 [14]	...	0.0019 [365] to 0.0866 [8]	0.0027 [258]	0.0007 [1058] to 0.0061 [114]

[A] See Ref (**36**).
[B] See Ref (**37**).
[C] See Ref (**38**).
[D] See Ref (**39**).
[E] See Ref (**40**).

which the model result is affected by changes in a selected model input representing fluid and media properties and boundary conditions.

X3.3.2.14 *simulation*—in migration modeling, one complete execution of a fluid flow modeling computer program, including input and output.

DISCUSSION—for the purposes of this appendix, a simulation refers to an individual modeling run. However, simulation is sometimes also used broadly to refer to the process of modeling in general.

X3.4 *Significance and Use:*

X3.4.1 Predictive modeling is significant in many phases of RBCA, including the following:

X3.4.1.1 Determining the potential urgency of response based on estimated migration and attenuation rates of compounds of concern,

X3.4.1.2 Determining the extent of corrective action

based on estimated migration and attenuation rates of compounds of concern,

TABLE X3.3   Results of Exponential Regression for Concentration Versus Time[A]

Site	Compound	k, % per day
Campbell, CA	benzene	1.20
	ethylbenzene	0.67
	xylene	1.12
	benzene	0.42
Palo Alto, CA	benzene	0.30
Virginia Beach, VA	PCE	0.46
	TCE	0.30
Montrose County, CO	benzene	0.42
Provo, UT	benzene	0.23
San Jose, CA	benzene	0.16
	benzene	0.10
Chemical facility	toluene	0.39
	PCE	0.34
	TCE	0.26

[A] Source: Ref (**41**).

X3.4.1.3 Establishing relationships between administered doses and adverse impacts to humans and sensitive environmental receptors, and

X3.4.1.4 Determining RBSLs concentrations at points of exposure.

X3.4.2 Examples of predictive modeling uses in the RBCA process include the following:

X3.4.2.1 The prediction of contaminant concentration distributions for future times based on historical trend data, as in the case of ground water transport modeling,

X3.4.2.2 The recommendation of sampling locations and sampling frequency based on current interpretation and future expectations of contaminant distributions, as in the design of ground water monitoring networks,

X3.4.2.3 The design of corrective action measures, as in the case of hydraulic control systems, and

X3.4.2.4 The calculation of site-specific exposure point concentrations based on assumed exposure scenarios, as in the case of direct exposure to surficial soils.

X3.4.3 Predictive modeling is not used in the RBCA process as a substitute for validation of site-specific data.

X3.5 *Interpretation of Predictive Modeling Results:*

X3.5.1 Predictive models are mathematical approximations of real processes, such as the movement of chemicals in the subsurface, the ingestion of chemicals contained in drinking water, and adverse impacts to human health and environmental resources resulting from significant exposures. One key step towards evaluating model results is to assess the accuracy and uncertainty, and to verify the model used.

X3.5.2 The accuracy of modeling-based predictions is evaluated using a post audit and is dependent upon a number of factors, including the following:

X3.5.2.1 The approximations used when describing the real system by mathematical expressions,

X3.5.2.2 The model setup, that is, the input parameters (for example, boundary conditions) used to generate the results, and

X3.5.2.3 The mathematical methods used to solve the governing equations (for example, user selection of numerical solution methods, expansion approximations, numerical parameters, and so forth).

X3.5.3 Predictive modeling results are always subject to some degree of uncertainty. It is important to quantify this uncertainty to properly interpret the results. Many times this is done with a sensitivity analysis in which the user identifies those parameters that most significantly influence the results. If most of all of the parameters do not produce "sensitivity," then the model may need to be reevaluated because it is possible that the key parameters are missing from the model.

X3.5.4 A postaudit may be performed to determine the accuracy of the predictions. While model calibration and verification demonstrate that the model accurately simulates past behavior of the system, the postaudit tests whether the model can predict future system behavior. Postaudits are normally performed several years after the initial assessment and corrective action.

X3.5.5 In the RBCA process, "conservative" is an important criterion of predictive modeling. In the initial evaluation, Tier 1, the most conservative approach, is used, which provides a worst case scenario for potential exposure and

risk. Models that, because of their simplicity, neglect factors that yield conservative results are used. Input may include conservative values such as the USEPA RME values. When a more rigorous approach is warranted, such as in Tier 2 of the RBCA process, conservative values are often used, but in conjunction with a more reasonable case scenario. This level requires more specific information about the site and may involve the use of either simple or moderately complex mathematical models. It may involve the use of most likely exposure scenario (that is, USEPA MLE values). This information is used to set conservative corrective action objectives that are still regarded as overly protective. At some sites a comprehensive assessment is required (Tier 3) where SSTLs are determined using a site-specific transport and exposure model and, in some cases, parameter distributions. Tier 3 provides the most realistic evaluation of potential exposure and risk.

X3.6 *Types of Predictive Migration and Risk Assessment Models:*

X3.6.1 Predictive models typically used in the RBCA process can be grouped into two broad categories:

X3.6.1.1 Migration models, and

X3.6.1.2 Exposure, risk, and dose-response assessment models.

X3.6.2 The determination of Tier 1 RBSLs or Tiers 2 and 3 SSTLs generally involves the use of combinations of both types of models. A more detailed description of each type of model is given in X3.7 and X3.8.

X3.7 *Procedures for Predictive Migration Models:*

X3.7.1 Migration (fate and transport) models predict the movement of a petroleum release through soil, ground water, or air, or combination thereof, over time. Most models focus on specific phenomena (for example, ground water transport) and vary in complexity, depending on assumptions made during model development. In RBCA, simplistic screening-level migration models are utilized in Tiers 1 and 2, while more complex models are utilized in Tier 3.

X3.7.2 References to many simplistic models suitable for screening-level evaluations for a number of pathways relevant to hydrocarbon contaminant releases are listed in Table X3.1. Most of the screening-level migration models have a simple mathematical form and are based on multiple limiting assumptions rather than on actual phenomena. For example, a simple model is the use of estimated ground water flow velocity to assess the travel time between the leading edge of a dissolved hydrocarbon plume and a ground water well. The travel time is approximated by the following:

[distance to well (ft)/flow velocity (ft/years)] = travel time (years)

X3.7.2.1 In the case of a relatively light compound such as benzene dissolved in ground water, the flow velocity may best be equated with the ground water flow velocity. Heavier compounds such as naphthalene may be retarded so that a flow velocity lower than the ground water velocity may be used. If miscible liquids are present on the ground water surface, such as gasoline, the liquid flow velocity may actually exceed the ground water velocity.

X3.7.3 The use of more complex models is not precluded in the RBCA process; however, given limited data and assumptions that must be made, many complex numerical models reduce to the analytical expressions given in Table X3.1.

X3.7.4 *Migration Model Data Requirements*—Predictive migration models require input of site-specific characteristics. Those most commonly required for various simplistic models include the following:

X3.7.4.1 Soil bulk density (for a typical soil: ≈ 1.7 g/cm^3),

X3.7.4.2 Total soil porosity (for a typical soil: ≈ 0.38 cm^3/cm^3),

X3.7.4.3 Soil moisture content can be conservatively estimated in many cases. It is approximately equal to the total soil porosity beneath the water table, and typically >0.05 cm^3-H$_2$O/cm^3-soil in the vadose zone; this can be a critical input parameter in the case of diffusion models and may require site-specific determination unless conservative values are used,

X3.7.4.4 Fraction organic matter in soil particles (=0.00d − 0.01: sandy soil is often conservatively assumed); this can also be a critical parameter requiring site-specific determination unless conservative values are used),

X3.7.4.5 Hydraulic conductivity (generally site-specific determination required),

X3.7.4.6 Ground water gradient and flow direction (requires site-specific determination), and

X3.7.4.7 First-order decay-rate (generally requires site-specific calibration as models are very sensitive to this parameter); see Tables X3.2 and X3.3 and Ref **(41)** for a summary of measured values currently available from the literature. The data in Table X3.3 include retardation and dispersion as well as natural biodegradation in attenuation rates measured. However, sensitivity studies indicate that natural biodegradation is the dominant factor. The sensitivity studies use Ref **(42)**. According to these sensitivity studies, an order of magnitude increase in natural biodegradation rate is 3.5 times as effective as an order of magnitude increase in retardation and 12 times as effective as an order of magnitude increase in dispersion in attenuating concentration over distance. Therefore, approximately 80 % of the attenuation shown in the Ref **(41)** data can be attributed to natural biodegradation.

X3.7.4.8 A similar analysis of the sensitivity of attenuation parameters for the vapor transport pathway also indicates that natural biodegradation is the predominant attenuation mechanism **(43)**. Soil geology is not considered an attenuation mechanism directly, but is a stronger determinant of how far contamination travels than even natural biodegradation. Gasoline contamination does not travel very far in clay (less than 30 ft (9 m)) according to the vapor transport model **(43)**.

X3.7.5 Depending on the models selected, other information may be required, such as meteorological information (wind speed, precipitation, temperature), soil particle size distributions, and nearby building characteristics.

X3.7.6 In most cases, measurements of the attenuation (decrease in concentrations) of compounds with distance away from the contaminant source area will be required to calibrate and verify the predictive capabilities of the selected models. The amount of data required varies depending on the following:

X3.7.6.1 The model code used,

X3.7.6.2 The model's sensitivity to changes in input parameters, and

X3.7.6.3 The contribution of the pathway of concern to the total incremental exposure and risk.

X3.7.7 Generally, site-specific physical and chemical properties for the most sensitive parameters are required for migration models to obtain accurate results. However, instead of site-specific data, conservative values selected from the literature may be used with appropriate caution.

3.7.8 *Migration Modeling Procedure*—The procedure for applying a migration model includes the following steps: definition of study objectives, development of a conceptual model, selection of a computer code or algorithm, construction of the model, calibration of the model and performance of sensitivity analysis, making predictive simulations, documentation of the modeling process, and performing a postaudit. These steps are generally followed in order; however, there is substantial overlap between steps, and previous steps are often revisited as new concepts are explored or as new data are obtained. The iterative modeling approach may also require the reconceptualization of the problem. The basic modeling steps are discussed as follows.

X3.7.8.1 *Modeling Objectives*—Modeling objectives must first be identified (that is, the questions to be answered by the model). The objectives aid in determining the level of detail and accuracy required in the model simulation. Complete and detailed objectives would ideally be specified prior to any modeling activities. Objectives may include interpreting site characterization and monitoring data, predicting future migration, determining corrective action requirements, or predicting the effect of proposed corrective action measures.

X3.7.8.2 *Conceptual Model*—A conceptual model of a subsurface contaminant release, such as a hydrocarbon release from an underground tank, is an interpretation or working description of the characteristics and dynamics of the physical system. The purpose of the conceptual model is to consolidate site and regional data into a set of assumptions and concepts that can be evaluated quantitatively. Development of the conceptual model requires the collection and analysis of physical data pertinent to the system under investigation.

(*1*) The conceptual model identifies and describes important aspects of the physical system, including the following: geologic and hydrologic framework; media type (for example, fractured or porous); physical and chemical processes; and hydraulic, climatic, and vapor properties. The conceptual model is described in more detail for ground water flow systems in Guide D 5447.

(*2*) Provide an analysis of data deficiencies and potential sources of error with the conceptual model. The conceptual model usually contains areas of uncertainty due to the lack of field data. Identify these areas and their significance to the conceptual model evaluated with respect to project objectives.

X3.7.8.3 *Computer Code Selection*—Computer code selection is the process of choosing the appropriate software algorithm, or other analysis technique, capable of simulating the characteristics of the physical system, as identified in the conceptual model. The types of codes generally used in the RBCA process are analytical and numerical models. The selected code should be appropriate to fit the available data and meet the modeling objectives. The computer code must also be tested for the intended use and be well documented.

(1) Analytical models are generally based on assumptions of uniform properties and regular geometries. Advantages include quick setup and execution. Disadvantages include, in many cases, that analytical models are so simplistic that important aspects of a given system are neglected.

(2) Numerical models allow for more complex heterogeneous systems with distributed properties and irregular geometries. Advantages include the flexibility to simulate more complex physical systems and natural parameter variability. Disadvantages include that the approach is often very time-intensive and may require much more data and information to be collected.

(3) Other factors may also be considered in the decision-making process, such as the model analyst's experience and those described as follows for model construction process; factors such as dimensionality will determine the capabilities of the computer code required for the model.

X3.7.8.4 *Model Construction*—Model construction is the process of transforming the conceptual model into a mathematical form. The model typically consists of two parts, the data set and the computer code. The model construction process includes building the data set used by the computer code. Fundamental components of a migration model are dimensionality, discretization, boundary and initial conditions, contaminant, and media properties.

X3.7.8.5 *Model Calibration*—Calibration of a model is the process of adjusting input for which data are not available within reasonable ranges to obtain a match between observed and simulated values. The range over which model parameters and boundary conditions may be varied is determined by data presented in the conceptual model. In the case where parameters are well characterized by field measurements, the range over which that parameter is varied in the model should be consistent with the range observed in the field. The degree of fit between model simulations and field measurements can be quantified using statistical techniques.

(1) In practice, model calibration is frequently accomplished through trial-and-error adjustment of the model's input data to match field observations. The calibration process continues until the degree of correspondence between the simulation and the physical system is consistent with the objectives of the project.

(2) Calibration of a model is evaluated through analysis of residuals. A residual is the difference between the observed and simulated variable. Statistical tests and illustrations showing the distribution of residuals are described for ground water flow models in Guide D 5490.

(3) Calibration of a model to a single set of field measurements does not guarantee a unique solution. To minimize the likelihood of nonuniqueness, the model should be tested to a different set of boundary conditions or stresses. This process is referred to as application verification. If there is poor correspondence to a second set of field data, then additional calibration or data collection are required. Successful verification of an application results in a higher degree of confidence in model predictions. A calibrated but unverified model may still be used to perform predictive simulations when coupled with a sensitivity analysis.

X3.7.8.6 *Sensitivity Analysis*—Sensitivity analysis is a quantitative method of determining the effect of parameter variation on model results. Two purposes of a sensitivity analysis are (1) to quantify the uncertainty in the calibrated model caused by uncertainty in the estimates of parameters, stresses, and boundary conditions, and (2) to identify the model inputs that have the most influence on model calibration and predictions.

(1) Sensitivity of a model parameter is often expressed as the relative rate of change of a selected model calculation during calibration with respect to that parameter. If a small change in the input parameter or boundary condition causes a significant change in the output, the model is sensitive to that parameter or boundary condition.

(2) Whether a given change in the model calibration is considered significant or insignificant is a matter of judgment. However, changes in the model's conclusions are usually able to be characterized objectively. For example, if a model is used to determine whether a contaminant is captured by a potable supply well, then the computed concentration is either detectable or not at the location. If, for some value of the input that is being varied, the model's conclusions are changed but the change in model calibration is insignificant, then the model results may be invalid because, over the range of that parameter in which the model can be considered calibrated, the conclusions of the model change. More information regarding conducting a sensitivity analysis for a ground water flow model application is presented in Guide D 5611.

X3.7.8.7 *Model Predictions*—Once these steps have been conducted, the model is used to satisfy the modeling objectives. Predictive simulations should be documented with appropriate illustrations, as necessary, in the model report.

X3.8 *Procedures for Risk, Exposure, and Dose-Response Assessment Models:*

X3.8.1 "Exposure models" are used to estimate the chemical uptake, or dose, while "risk assessment models" are used to relate human health or ecological impacts to the uptake of a chemical. Risk and exposure assessment models are often combined to calculate a target exposure point concentration of a compound in air, water, or soil.

X3.8.1.1 In the case of compounds that have been classified as carcinogens, exposure and risk assessment models are generally linked by the expression:

risk = average lifetime intake [mg/kg-day]
$$\times \ slope \ factor \ [\text{mg/kg-day}]^{-1}$$

where the intake depends on exposure parameters (ingestion rate, exposure duration, and so forth) and the concentration at point-of-exposure. The slope factor (sometimes called the "potency factor") is itself based on a model and set of underlying assumptions, which are discussed as follows.

X3.8.1.2 In the case of compounds that have not been classified as carcinogens, exposure and risk assessment models are generally linked by the expression:

hazard quotient =
$$average \ intake \ [\text{mg/kg-day}]/reference \ dose \ [\text{mg/kg-day}]$$

where the intake depends on exposure parameters (ingestion rate, exposure duration, and so forth) and the concentration at point-of-exposure. The reference dose is itself based on a model and set of underlying assumptions, which are discussed as follows.

X3.8.2 *Toxicity Assessment: Dose-Response Models*—Toxicity assessments use dose-estimates of a "safe dose" or toxic level based on animal studies. In some instances, human epidemiological information is available on a chemical. Toxicologists generally make two assumptions about the effects of risk agents at the low concentrations typical of environmental exposures:

X3.8.2.1 Thresholds exist for most biological effects; in other words, for noncarcinogenic, nongenetic toxic effects, there are doses below which no adverse effects are observed in a population of exposed individuals, and

X3.8.2.2 No thresholds exist for genetic damage or incremental carcinogenic effects. Any level of exposure to the genotoxic or carcinogenic risk agent corresponds to some non-zero increase in the likelihood of inducing genotoxic or incremental carcinogenic effects.

X3.8.3 The first assumption is widely accepted in the scientific community and is supported by empirical evidence. The threshold value for a chemical is often called the NOAEL. Scientists usually estimate NOAELs from animal studies. An important value that typically results from a NOAEL or LOAEL value is the RfD. A reference dose is an estimate (with an uncertainty typically spanning an order of magnitude) of a daily exposure (mg/kg/day) to the general human population (including sensitive subgroups) that is likely to be without an appreciable risk of deleterious effects during a lifetime of exposure. The RfD value is derived from the NOAEL or LOAEL by application of uncertainty factors (UF) that reflect various types of data used to estimate RfDs and an additional modifying factor (MF), which is based on a professional judgment of the quality of the entire database of the chemical. The oral RfD, for example, is calculated from the following equation:

$$RfD = \frac{NOAEL}{(UF \times MF)}$$

X3.8.4 The second assumption regarding no threshold effects for genotoxic or carcinogenic agents is more controversial but has been adopted by the USEPA. For genotoxic and carcinogenic agents, extrapolations from high experimental doses to low doses of environmental significance require the use of mathematical models to general low dose-response curves. It should be noted that although the EPA uses the linear multi-state model to describe incremental carcinogenic effect, there is no general agreement in the scientific community that this is the appropriate model to use.

X3.8.5 The critical factor determined from the dose-response curve is the slope factor (SF), which is the slope of the dose-response curve in the low-dose region. The units of the slope factor are expressed as $(mg/kg-day)^{-1}$ and relate a given environmental intake to the risk of additional incidence of cancer above background.

X3.8.6 The RfD or SF values are generally obtained from a standard set of reference tables (for example, Ref (**2**) or Ref (**3**)). It is important to note that the information in IRIS has typically only been peer-reviewed within the EPA and may not always have support from the external scientific community. Whereas the information in IRIS has been subject to

agency-wide data quality review, the information in the HEAST tables has not. The user is expected to consult the original assessment documents to appreciate the strengths and limitations of the data in HEAST. Thus, care should be exercised in using the values in HEAST. Some state and local agencies have toxicity factors they have derived themselves or preferences for factors to use if neither IRIS nor HEAST lists a value. Values for a range of hydrocarbons typically of interest are presented in Table X3.1.

X3.8.7 It is important to note that in extrapolating the information obtained in animal studies to humans, a number of conservative assumptions are made.

X3.8.7.1 For noncarcinogens, an arbitrary system of default safety and uncertainty factors, as discussed (in multiples of ten), is used to convert observations, in animals to estimates in humans.

X3.8.7.2 For carcinogens, some of the most important assumptions include: (*1*) the results of the most sensitive animal study are used to extrapolate to humans, (*2*) in general, chemicals with any incremental carcinogenic activity in animals are assumed to be potential human carcinogens, and (*3*) no threshold exists for carcinogens.

X3.8.8 The uncertainty in the RfD and SF values are often neglected in deference to single point values which are then typically summarized in databases such as IRIS and HEAST and assumptions described are risk management policy decisions made by the USEPA. These assumptions are not explicitly defined and further obscure the conservatism in the safe dose estimate. Thus, care must be exercised in interpreting results which have as a basis these conservative toxicity evaluations.

X3.8.9 *Exposure Assessment Modeling*—The goal of exposure assessment modeling is to estimate the chemical uptake that occurs when a receptor is exposed to compounds present in their environment. In principal, the process for developing and using migration models presented in X3.7 is directly applicable to exposure assessment modeling. In this case the user:

X3.8.9.1 Develops a conceptual model by identifying significant exposure pathways and receptors,

X3.8.9.2 Selects a model to describe the contact rate and subsequent uptake of the chemical(s),

X3.8.9.3 Performs a sensitivity analysis to identify critical parameters,

X3.8.9.4 Selects appropriate exposure parameters (breathing rates, and so forth),

X3.8.9.5 Generates estimates of exposure and uptake, and

X3.8.9.6 Assesses the uncertainty in the estimates.

X3.8.10 There are differences between the process outlined in X3.7 and that which can be practically applied to exposure assessment modeling. For example, with the exception of exposures and impacts to environmental resources, it is difficult to calibrate exposure assessment models unless very expensive epidemiological studies are conducted.

X3.8.11 Typically, the models used to estimate uptake are simplistic algebraic expressions, such as those contained in Ref (**27**). Application of these equations is illustrated in Appendix X2.

X3.8.12 In many cases, exposure parameter values are available in Ref (**27**), but other more recent information is also available in peer-reviewed publications, and all sources should be carefully reviewed. While point values are often selected for simplicity, statistical distributions for many of the exposure parameters are readily available for Tier 3 analyses.

X3.8.13 It is common for USEPA RME values to be used in exposure assessment calculation, as is done for the example Tier 1 Look-Up Table discussed in Appendix X2. The RME value is generally defined as a statistical upper limit of available data (generally 85 to 90 % of all values are less than the RME value). Therefore, by consistently selecting and multiplying conservative RME values the user models a scenario that is very improbable and always more

conservative than the "true" RME exposure scenario. Thus, great care must be exercised, when using combinations of these default values in risk assessments, to avoid a gross overestimation of exposure for a specific site.

X3.9 *Report*—The purpose of the model report is to communicate findings, to document the procedures and assumptions inherent in the study, and to provide detailed information for peer review. The report should be a complete document allowing reviewers and decision makers to formulate their own opinion as to the credibility of the model. The report should describe all aspects of the modeling study outlined in this appendix.

X4. INSTITUTIONAL CONTROLS

X4.1 *Introduction:*

X4.1.1 The purpose of this appendix is to provide a review of generally used institutional controls. For purposes of this appendix, "institutional controls" are those controls that can be used by responsible parties and regulatory agencies in remedial programs where, as a part of the program, certain concentrations of the chemical(s) of concern will remain on site in soil or ground water, or both. Referenced in this appendix are examples of programs from California, Connecticut, Illinois, Indiana, Iowa, Massachusetts, Michigan, Missouri, and New Jersey. In addition, federal programs, such as Superfund settlements and RCRA closure plans have used the following techniques described for some years as a mechanism to ensure that exposure to remaining concentrations of chemical(s) of concern is reduced to the degree necessary.

X4.1.2 The types of institutional controls discussed in this appendix are as follows:

X4.1.2.1 Deed restrictions, or restrictive covenants,

X4.1.2.2 Use restrictions (including well restriction areas),

X4.1.2.3 Access controls,

X4.1.2.4 Notice, including record notice, actual notice, and notice to government authorities,

X4.1.2.5 Registry act requirements,

X4.1.2.6 Transfer act requirements, and

X4.1.2.7 Contractual obligations.

X4.1.3 Institutional controls for environmental remedial programs vary in both form and content. Agencies and landowners can invoke various authorities and enforcement mechanisms, both public and private, to implement any one or a combination of the controls. For example, a state could adopt a statutory mandate (see X4.2) requiring the use of deed restrictions (see X4.3) as a way of enforcing use restrictions (see X4.4) and posting signage (a type of access control, see X4.5). Thus, the institutional controls listed as follows are often used as overlapping strategies, and this blurs the distinctions between them.

X4.2 *Statutory Mandates*—Some states' emergency response programs mandate post-remediation institutional controls and impose civil penalties for noncompliance. The schemes vary from state to state, but all impose obligations on landowners to use one or more institutional controls listed in this appendix.

X4.3 *Deed Restrictions:*

X4.3.1 Deed restrictions place limits and conditions on the use and conveyance of land. They serve two purposes: (*1*) informing prospective owners and tenants of the environmental status of the property and (*2*) ensuring long-term compliance with the institutional controls that are necessary to maintain the integrity of the remedial action over time. Restraining the way someone can use their land runs counter to the basic assumptions of real estate law, so certain legal rules must be satisfied in order to make a deed restriction binding and enforceable.

X4.3.2 There are four requirements for a promise in a deed restriction (also called a "restrictive covenant") to be held against current and subsequent landowners: (*1*) a writing, (*2*) intention by both original parties that particular restrictions be placed on the land in perpetuity, (*3*) "privity of estate," and (*4*) that the restrictions "touch and concern the land."

X4.3.2.1 The first requirement is that of a writing. It is a rule of law that conveyances of land must be documented in a writing. The same rule holds for deed restrictions affecting land. Ideally, a deed restriction used as an institutional control would be written down with particularity and then recorded in the local land records office, in much the same fashion as the documentation and recordation of a sale of land. Parties may also encounter the requirement that the deed restriction be executed "under seal," a legal formality that has been abandoned in most states.

X4.3.2.2 The second requirement is that the deed restriction should precisely reflect what the parties' intentions are in regard to the scope and the duration of the restrictions. Explicitly stating in the deed restriction that the parties intend the restriction to "run with the land" (that is, last forever and bind subsequent owners) is strongly recommended.

X4.3.2.3 The third requirement, privity of estate, arises from a concern that only persons with a certain relationship to the land should be able to enforce a deed restriction. Normally, deed restrictions are promises between the buyer and the seller or between neighbors; therefore, the state or a third party may not enforce a deed restriction. However, even in states that require privity of estate, this concern is addressed if the landowner took the land with knowledge that the restrictions existed and might be enforced by these third parties. Thus, it is also strongly recommended that the deed restriction explicitly state that the state environmental authority may enforce the restriction. Recording of the deed restriction serves as notice to anyone who later purchases or acquires an interest in the land. Therefore, privity of estate should not be a barrier to state enforcement of the deed restriction if the proper steps are taken.

X4.3.2.4 Finally, a deed restriction is only enforceable if the promise "touches and concerns the land." A rough rule of thumb to decide this point is whether the landowner's legal interest in the land is decreased in value due to the deed restriction. If the land is devalued in this way, then the restriction could be said to "touch and concern the land." Note that the focus of the inquiry is on the land itself; promises that are personal in nature and merely concern human activities that happen to take place on the land are least likely to be enforceable. Thus, any deed restriction used as an institutional control should be written so that it centers on the land and the use of the land.

X4.3.3 Due to the potential enforcement hurdles encountered by a governmental agency in enforcing a deed restriction, it may be appropriate for an individual state to seek statutory and regulatory amendments to ensure that such authority exits in regard to all deed restrictions for environmental purposes.

X4.3.4 Remedies for noncompliance with deed restrictions comes in two forms: (1) persons or agencies may sue to obtain a court order (injunction) requiring compliance or (2) if the state statute allows for it, the state's attorney general can seek enforcement of civil penalties, such as fines, for noncompliance.

X4.3.5 A state program can require a landowner to continue monitoring activities and to allow state environmental officials access to the site to monitor compliance with institutional controls. These arrangements may have to be put in a deed restriction in order to run with the land from owner to owner, but responsible parties can also be required to sign a contract making these promises. Of course, almost every state has authority to issue administrative orders to accomplish some or all of these arrangements.

X4.3.6 The preceding arrangements can also set out procedures that will be followed if some emergency requires that the remediation site be disturbed. If, for example, underground utility lines must be repaired, the landowner would follow this protocol for handling the soil and alerting the state authority.

X4.4 *Use Restrictions:*

X4.4.1 Use restrictions are usually the heart of what is in a deed restriction. Use restrictions describe appropriate and inappropriate uses of the property in an effort to perpetuate the benefits of the remedial action and ensure property use that is consistent with the applicable cleanup standard. Such techniques also prohibit any person from making any use of the site in a manner that creates an unacceptable risk of human or environmental exposure to the residual concentrations of chemical(s) of concern.

X4.4.2 Use restrictions address uses that may disturb a containment cap or any unremediated soils under the surface or below a building. A prohibition on drinking on-site (or off-site by means of well restriction areas discussed as follows) ground water may also be appropriate.

X4.4.3 As an example, a program may allow a restriction

of record to include one or more of the following:

X4.4.3.1 Restriction on property use;

X4.4.3.2 Conditioning the change of use from nonresidential on compliance with all applicable cleanup standards for a residential property;

X4.4.3.3 Restricting access; or

X4.4.3.4 Restricting disturbance of department-approved remedial effects.

X4.4.4 Well restriction areas can be a form of institutional control by providing notice of the existence of chemical(s) of concern in ground water, and by prohibiting or conditioning the construction of wells in that area.

X4.4.4.1 This technique preserves the integrity of any ground water remedial action by prohibiting or conditioning the placement and use of any or all types of wells within the area.

X4.4.4.2 Well restrictions of this nature would be subject to agency approval and public notice, and may include the restriction on constructing or locating any wells within a particular designated area. Notice of the well restriction is recorded on the land records and with various health officials and municipal officials. The restrictions can only be released upon a showing that the concentrations of the chemical(s) of concern in the well restriction area is remediated in accordance with state standards.

X4.5 *Access Controls:*

X4.5.1 Another subset of institutional controls is the control of access to any particular site. The state uses the following criteria to determine the appropriate level and means of access control:

X4.5.1.1 Whether the site is located in a residential or mixed use neighborhood;

X4.5.1.2 Proximity to sensitive land-use areas including day-care centers, playgrounds, and schools; and

X4.5.1.3 Whether the site is frequently traversed by neighbors.

X4.5.2 Access can be controlled by any of the following: fencing and gates, security, or postings or warnings.

X4.6 *Notice*—Regulations of this type generally provide notice of specific location of chemical(s) of concern on the site, and disclose any restrictions on access, use, and development of part or all of the contaminated site to preserve the integrity of the remedial action.

X4.6.1 *Record Notice:*

X4.6.1.1 Some states require that sites having releases of hazardous waste file a notice on the land records providing to any subsequent purchaser of the property information regarding the past or current activities on the site.

X4.6.1.2 The record notice requirement may be broad; the program may require any property subject to a response action to obtain a professional opinion and then prepare and record a Grant of Environmental Restriction that is supported by that opinion.

X4.6.1.3 The record notice requirement can be ancillary to a transfer act (see X4.8), in which case recording of an environmental statement is only required in conjunction with a land transaction.

X4.6.2 *Actual Notice:*

X4.6.2.1 States may require direct notice of environmental information to other parties to a land transaction.

These laws protect potential buyers and tenants, and they also help ensure that use restrictions and other institutional controls are perpetuated.

X4.6.2.2 Actual notice of an environmental defect or failure to provide notice may give a party the right to cancel the transaction and result in civil penalties. For example, landlords and sellers who do not give notice as required by the state may be liable for actual damages plus fines. Nonresidential tenants who fail to notify landowners of suspected or actual hazardous substance releases can have their leases canceled and are subject to fines.

X4.6.3 *Notice to Government Authorities*—Parties to a land transaction may also be required to file the environmental statement with various environmental authorities. Notice to the government may be required before the transaction takes place.

X4.7 *Registry Act Requirements:*

X4.7.1 Some states have registry act programs that provide for the maintenance of a registry of hazardous waste disposal sites and the restriction of the use and transfer of listed sites.

X4.7.2 A typical registry act provides that the state environmental agency establish and maintain a registry of all real property which has been used for hazardous substance disposal either illegally or before regulation of hazardous waste disposal began in that state.

X4.7.3 The state agency is responsible for investigating potential sites for inclusion on the registry. The registry includes the location of the site and a listing of the hazardous wastes on the property, and may also include a classification of the level of health or environmental danger presented by the conditions on the property. The state agency may be required to perform detailed inspections of the site to determine its priority relative to other registered sites.

X4.7.4 Owners of sites proposed for inclusion on the registry have rights of hearing and appeal, and owners of sites on the registry have rights to modify or terminate their listing. In some cases, the owner of a site proposed for inclusion on the registry may obtain the withdrawal of the proposed registration by entering into a consent agreement with the state. Such a consent agreement establishes a timetable and responsibility for remedial action.

X4.7.5 When a site appears on the state registry, the owner must comply with regulatory requirements in regard to use and transfer of the site. The use of a site listed on the registry may not be changed without permission of the state agency. In negotiations for a conveyance of a registered site, the owner may be obligated to disclose the registration early in the process, and permission of the state agency may be required to convey a registered property. Under other schemes, permission to convey is not required, but the seller must notify the state agency of the transaction.

X4.7.6 Finally, registry acts require that the listing of a property on a hazardous materials site registry be recorded in the records of the appropriate locality so that the registration will appear in the chain of title.

X4.8 *Transfer Act Requirements:*

X4.8.1 Some states have transfer act programs that require full evaluation of all environmental issues before or after the transfer occurs. It may be that within such program, institutional controls can be established by way of consent order, administrative order, or some other technique that establishes implementation and continued responsibility for institutional controls.

X4.8.2 A typical transfer act imposes obligations and confers rights on parties to a land transaction arising out of the environmental status of the property to be conveyed. Transfer acts impose information obligations on the seller or lessor of a property (see X4.6.3). That party must disclose general information about strict liability for cleanup costs as well as property-specific information, such as presence of hazardous substances, permitting requirements and status, releases, and enforcement actions and variances.

X4.8.3 Compliance with transfer act obligations in the manner prescribed is crucial for ensuring a successful conveyance. Sometimes the transfer act operates to render a transaction voidable before the transfer occurs. Failure to give notice in the required form and within the time period required or the revelation of an environmental violation or unremediated condition will relieve the transferee and the lender of any obligation to close the transaction, even if a contract has already been executed. Moreover, violation of the transfer act can be the basis for a lawsuit to recover consequential damages.

X4.9 *Contractual Obligations:*

X4.9.1 One system for ensuring the future restriction on use of a site, or the obligation to remediate a site, is to require private parties to restrict use by contract. While this method is often negotiated among private parties, it will be difficult, if not impossible, to institutionalize some control over that process without interfering with the abilities and rights of private parties to freely negotiate these liabilities.

X4.9.2 Another avenue is for the landowner or responsible party to obligate itself to the state by contract. The state may require a contractual commitment from the party to provide long-term monitoring of the site, use restrictions, and means of continued funding for remediation.

X4.10 *Continued Financial Responsibility*—Another aspect of institutional controls is the establishment of financial mechanisms by which a responsible party ensures continued funding of remediation measures and assurance to the satisfaction of the state.

X4.11 *References:*

X4.11.1 The following references serve as examples and are current as of the fourth quarter of 1993:

X4.11.1.1 *References for Deed Restrictions:*

24 New Jersey Regulations 400 (1992) (New Jersey Administration Code § 7.26D-8.2 (e) (2))
24 New Jersey Regulations 400-02 (1992) (New Jersey Administration Code §§ 7.26D-8.1–8.4)
24 New Jersey Regulations 401 (1992) (New Jersey Administration Code § 7.26D Appendix A, Model Document, Declaration of Environmental Restrictions and Grant of Ease ment, Item 8)
Illinois Responsible Property Transfer Act § 7(c) (1985)
Massachusetts Regulations Code Title _____ § 40.1071 (2) (1) & (k)
Massachusetts Regulations Code, Title _____ § 40.1071(4)
Michigan Administration Code 299.5719 (3) (e) (1990)
Michigan Rules 299.5719 (2), (3) (d)

X4.11.1.2 *References for Use Restrictions:*

24 New Jersey Regulations 400 (New Jersey Administration Code § 7.26D-8.2 (d))
Michigan Administration Code 299.5719 (3) (a), (b), (g)
New Jersey Regulation 7.26D-8.4

X4.11.1.3 *References for Access Controls:*

Iowa Administration Code r. 133.4 (2) (b)
Michigan Rule 299.4719 (3) (f)
New Jersey Regulations § 7.26D-8.2

X4.11.1.4 *References for Notice:*

California Health and Safety Code § 25359.7 (1981)
Illinois Responsible Property Transfer Act (1985)
Indiana Code §§13-7-22.5-1–22 (1989) ("Indiana Environmental Hazardous Disclosure and Responsible Party Transfer Law")
Massachusetts Regulations Code Title _____ §§ 40.1071-1090 (1993)
Michigan Rule 299.5719 (3) (c)

X4.11.1.5 *References for Registry Act Requirements:*

Iowa Code Ann. §§ 455B.426–455B.432, 455B.411 (1) (1990)
Missouri Code Regulations Title 10, §§ 25-10.010, 25-3.260 (1993)

X4.11.1.6 *References for Transfer Act Requirements:*

Connecticut General Stat. §22a-134 *et seg*
Illinois Responsible Property Transfer Act (1985)
Indiana Code §§ 13-7-22.5-1–22 (1989) ("Indiana Environmental Hazardous Disclosure and Responsible Party Transfer Law")
New Jersey Senate Bill No. 1070, the Industrial Site Recovery Act, amending the environmental cleanup Responsibility Act, N.J.S.A. 13:1K-6 *et seg*
New Jersey Spill Compensation and Control Act, N.J.S.A. 58:10-23.11 *et seg*

X4.11.1.7 *Reference for Contractual Obligations:*

Michigan Rule 299.5719 (2)

X4.11.1.8 *Reference for Continued Financial Responsibility:*

Michigan Rule 299.5719 (2)

X5. EXAMPLE APPLICATIONS OF RISK-BASED CORRECTIVE ACTION

X5.1 *Introduction*—The following examples illustrate the use of RBCA at petroleum release sites. The examples are hypothetical and have been simplified in order to illustrate that RBCA leads to reasonable and protective decisions; nevertheless, they do reflect conditions commonly encountered in practice.

X5.2 *Example 1—Corrective Action Based on Tier 1 Risk-Based Screening Levels:*

X5.2.1 *Scenario*—A release from the underground storage tank (UST), piping, and dispenser system at a service station is discovered during a real estate divestment assessment. It is known that there are petroleum-impacted surficial soils in the area of the tank fill ports; however, the extent to which the soils are impacted is unknown. In the past, both gasoline and diesel have been sold at the facility. The new owner plans to continue operating the service station facility.

X5.2.2 *Site Assessment*—The responsible party completes an initial site assessment focussed on potential source areas (for example, tanks, lines, dispensers) and receptors. Based on historical knowledge that gasoline and diesel have been dispensed at this facility, chemical analyses of soil and ground water are limited to benzene, toluene, ethylbenzene, xylenes, and naphthalene. Site assessment results are summarized as follows:

X5.2.2.1 Field screening instruments and laboratory analyses indicate that the extent of petroleum-impacted soils is confined to the vicinity of the fill ports for the tanks. A tank and line test reveals no leaks; therefore, evidence suggests that soils are impacted due to spills and overfills associated with filling the storage tank,

X5.2.2.2 The current tanks and piping were installed five years ago,

X5.2.2.3 The concrete driveway is highly fractured,

X5.2.2.4 No other sources are present,

X5.2.2.5 The site is underlain by layers of fine to silty sands,

X5.2.2.6 Ground water, which is first encountered at 32 ft (9.7 m) below ground surface, is not impacted,

X5.2.2.7 Maximum depth at which hydrocarbons are detected is 13 ft (3.9 m). Maximum detected soil concentrations are as follows:

Compound	Depth Below Ground Surface, ft (m)	Concentration, mg/kg
Benzene	8 (2.4)	10
Ethylbenzene	4 (1.2)	4
Toluene	6.5 (1.9)	55
Xylenes	3.5 (1.01)	38
Naphthalene	2 (0.6)	17

X5.2.2.8 A receptor survey indicates that two domestic water wells are located within 900 ft (273.6 m) of the source area. One well is located 500 ft (152.4 m) hydraulically down-gradient from the impacted soil zone, the other well is hydraulically up-gradient. Both wells produce water from the first encountered ground water zone.

X5.2.3 *Site Classification and Initial Response Action*—Based on classification scenarios given in Table 1, this site is classified as a Class 3 site because conditions are such that, at worst, it is a long-term threat to human health and environmental resources. The appropriate initial response is to evaluate the need for a ground water monitoring program (see Table X5.1). At most, this would consist of a single well located immediately down-gradient of the impacted petroleum soils. The responsible party recommends deferring the decision to install a ground water monitoring system until the Tier 1 evaluation is complete, and justifies this recommendation based on no detected ground water impact, the limited extent of impacted soils, and the separation between impacted soils and first-encountered ground water. The regulatory agency concurs with this decision.

X5.2.4 *Development of Tier 1 Look-Up Table of Risk-Based Screening Level (RBSL)*—Assumptions used to derive example Tier 1 RBSL Look-Up Table X2.1 in Appendix X2 are reviewed and presumed valid for this site. A comparison of RBSLs for both pathways of concern indicates that RBSLs associated with the leaching pathway are the most restrictive of the two. As this aquifer is currently being used as a drinking water supply, RBSL values based on meeting drinking water MCLs are selected. In the case of naphthalene, for which there is no MCL, the RBSL value corresponding to a residential scenario and a hazard quotient of unity is used.

X5.2.5 *Exposure Pathway Evaluation*—Based on current and projected future use, the only two potential complete exposure pathways at this site are: (1) the inhalation of ambient vapors by on-site workers, or (2) the leaching to ground water, ground water transport to the down-gradient drinking-water well, and ingestion of ground water (see Fig. X5.1).

X5.2.6 *Comparison of Site Conditions With Tier 1 RBSLs*—Based on the data given in X5.2.2.7 and the RBSLs given in Look-Up Table X2.1 in Appendix X2, exceedances of Tier 1 RBSLs are noted only for benzene and toluene.

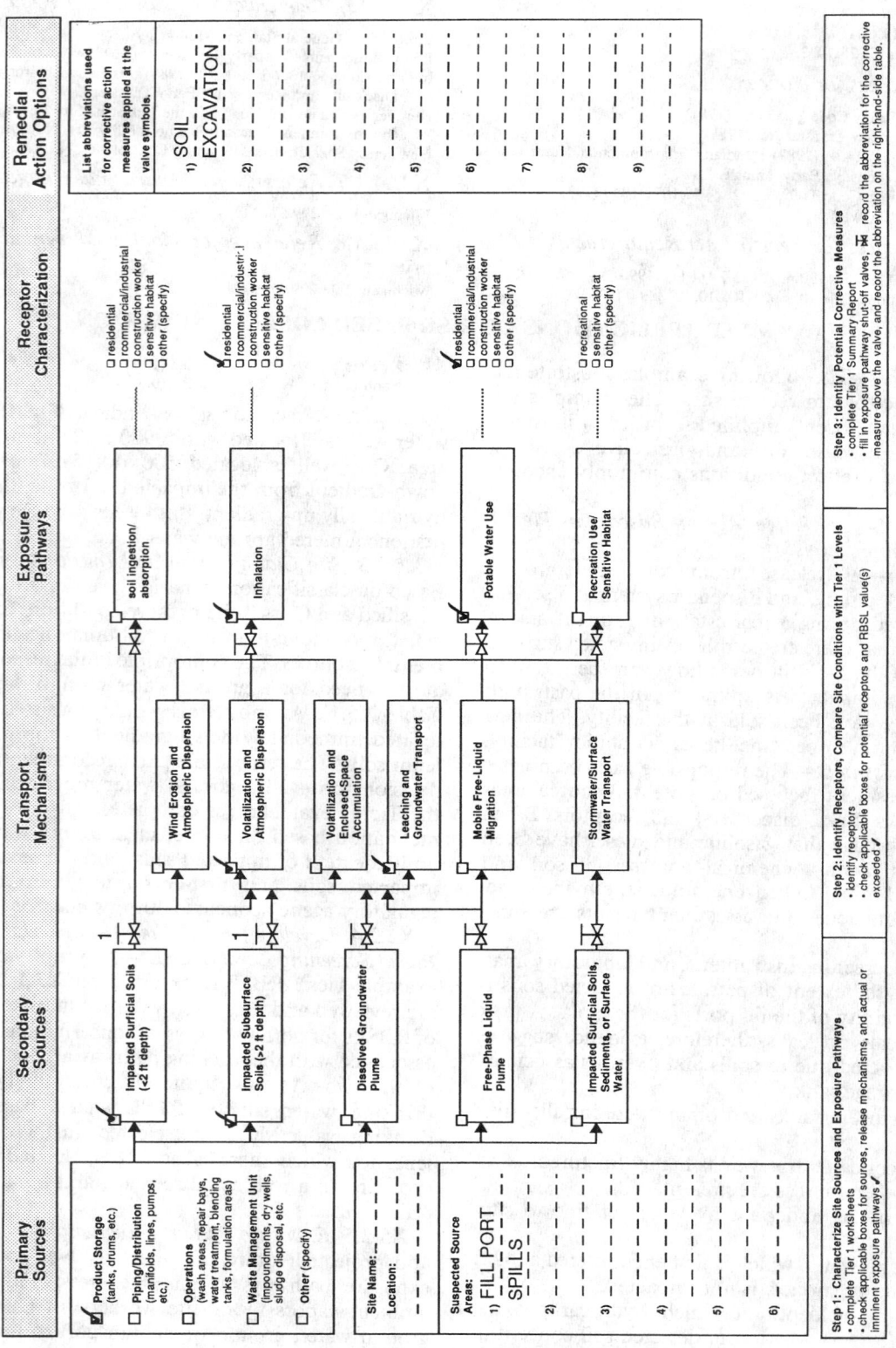

FIG. X5.1 Example 1—Exposure Evaluation Flowchart

X5.2.7 *Evaluation of Tier 1 Results*—The responsible party decides to devise a corrective action plan to meet Tier 1 standards after considering the following factors:

X5.2.7.1 The shallow aquifer is not yet affected,

X5.2.7.2 Quick (relative to rate of chemical migration) removal of the source will eliminate the need for ground water monitoring,

X5.2.7.3 The new owner plans to install new tanks within six months,

X5.2.7.4 Limited excavation of soils to meet Tier 1 criteria could be performed quickly and inexpensively when the tanks are removed, relative to the cost of proceeding to a Tier 2 analysis, and

X5.2.7.5 An excavation proposal will facilitate the real estate deal.

X5.2.8 *Tier 1 Remedial Action Evaluation*—Excavate all impacted soils with concentrations above the Tier 1 RBSLs when the current tanks are replaced. Subsequently resurface the area with new concrete pavement to reduce future infiltration and leaching potential through any remaining impacted soils. It is agreed that ground water monitoring is not necessary and the governing regulatory agency agrees to issue a No Further Action and Closure letter following implementation of the corrective action plan.

X5.3 *Example 2—RBCA Based on Tier 2 Evaluation:*

X5.3.1 *Scenario*—During the installation of new double-contained product transfer lines, petroleum-impacted soils are discovered in the vicinity of a gasoline dispenser at a service station located close to downtown Metropolis. In the past, both gasoline and diesel have been sold at this facility, which has been operating as a service station for more than twenty years.

X5.3.2 *Site Assessment*—The owner completes an initial site assessment focussed on potential source areas (for example, tanks, lines, dispensers) and receptors. Based on historical knowledge that gasoline and diesel have been dispensed at this facility, chemical analyses of soil and ground water are limited to benzene, toluene, ethylbenzene, xylenes, and naphthalene. Results of the site investigation are as follows:

X5.3.2.1 The extent of petroleum-impacted soils is confined to the vicinity of the tanks and dispensers. A recent tank and line test revealed no leaks; therefore, evidence suggests that the releases occurred sometime in the past,

X5.3.2.2 The current tanks, lines, and dispensers were installed three years ago,

X5.3.2.3 The asphalt driveway is competent and not cracked,

X5.3.2.4 Another service station is located hydraulically down gradient, diagonally across the intersection,

X5.3.2.5 The site is underlain by silty sands with a few thin discontinuous clay layers,

X5.3.2.6 Ground water, which is first encountered at 32 ft (9.7 m) below ground surface, is impacted, with highest dissolved concentrations observed beneath the suspected source areas. Dissolved concentrations decrease in all directions away from the source areas, and ground water samples taken hydraulically down gradient from a well located in the center divider of the street (about 100 ft (30.4 m) from the source area) do not contain any detectable levels of dissolved hydrocarbons,

X5.3.2.7 Ground water flow gradient is very shallow, and ground water flow velocities are at most tens of feet per year,

X5.3.2.8 Ground water yield from this aquifer is estimated to be in excess of 5 gal/min (18.9 L/min), and total dissolved solids levels are less than 700 mg/L. Based on this information, this aquifer is considered to be a potential drinking water supply,

X5.3.2.9 A shallow soil gas survey indicates that no detectable levels of hydrocarbon vapors are found in the utility easement running along the southern border of the property, or in soils surrounding the service station kiosk,

X5.3.2.10 Impacted soils extend down to the first encountered ground water. Maximum concentrations detected in soil and ground water are as follows:

Compound	Soil, mg/kg	Ground water, mg/L
Benzene	20	2
Ethylbenzene	4	0.5
Toluene	120	5
Xylenes	100	5.0
Napthalene	2	0.05

X5.3.2.11 A receptor survey indicates that no domestic water wells are located within one-half mile of the site; however, there is an older residential neighborhood located 1200 ft (365.7 m) hydraulically down gradient of the site. Land use in the immediate vicinity is light commercial (for example, strip malls). The site is bordered by two streets and a strip mall parking lot.

X5.3.3 *Site Classification and Initial Response Action*—Based on classification scenarios given in Table 1, this site is classified as a Class 3 site because conditions are such that, at worst, it is a long-term threat to human health and environmental resources (see Table X5.2). The appropriate initial response is to evaluate the need for a ground water monitoring program. The owner proposes that the ground water monitoring well located hydraulically down gradient in the street divider be used as a sentinel well, and be sampled yearly. The regulatory agency concurs, provided that the well be sampled every six months.

X5.3.4 *Development of Tier 1 Look-Up Table of Risk-Based Screening Level (RBSL) Selection*—Assumptions used to derive example Tier 1 RBSL Look-Up Table X2.1 in Appendix X2 are reviewed and presumed valid for this site. Due to the very low probability of the exposure pathway actually being completed in the future, MCLs are not used and the site owner is able to negotiate Tier 1 RBSLs based on a 10^{-5} risk to human health for carcinogens and hazard quotients equal to unity for the noncarcinogens (based on ground water ingestion).

X5.3.5 *Exposure Pathway Evaluation*—Based on current and projected future use, and the soil gas survey results, there are no potential complete exposure pathways at this site. The down gradient residential neighborhood is connected to a public water supply system, and there is no local use of the impacted aquifer. However, being concerned about future uncontrolled use of the aquifer, the regulatory agency requests that the owner evaluate the ground water transport to residential drinking water ingestion pathway, recognizing that there is a low potential for this to occur (see Fig. X5.2).

X5.3.6 *Comparison of Site Conditions With Tier 1 RBSLs*—Based on the data given in X5.3.2.10 and the RBSLs given in example Look-Up Table X2.1 in Appendix X2,

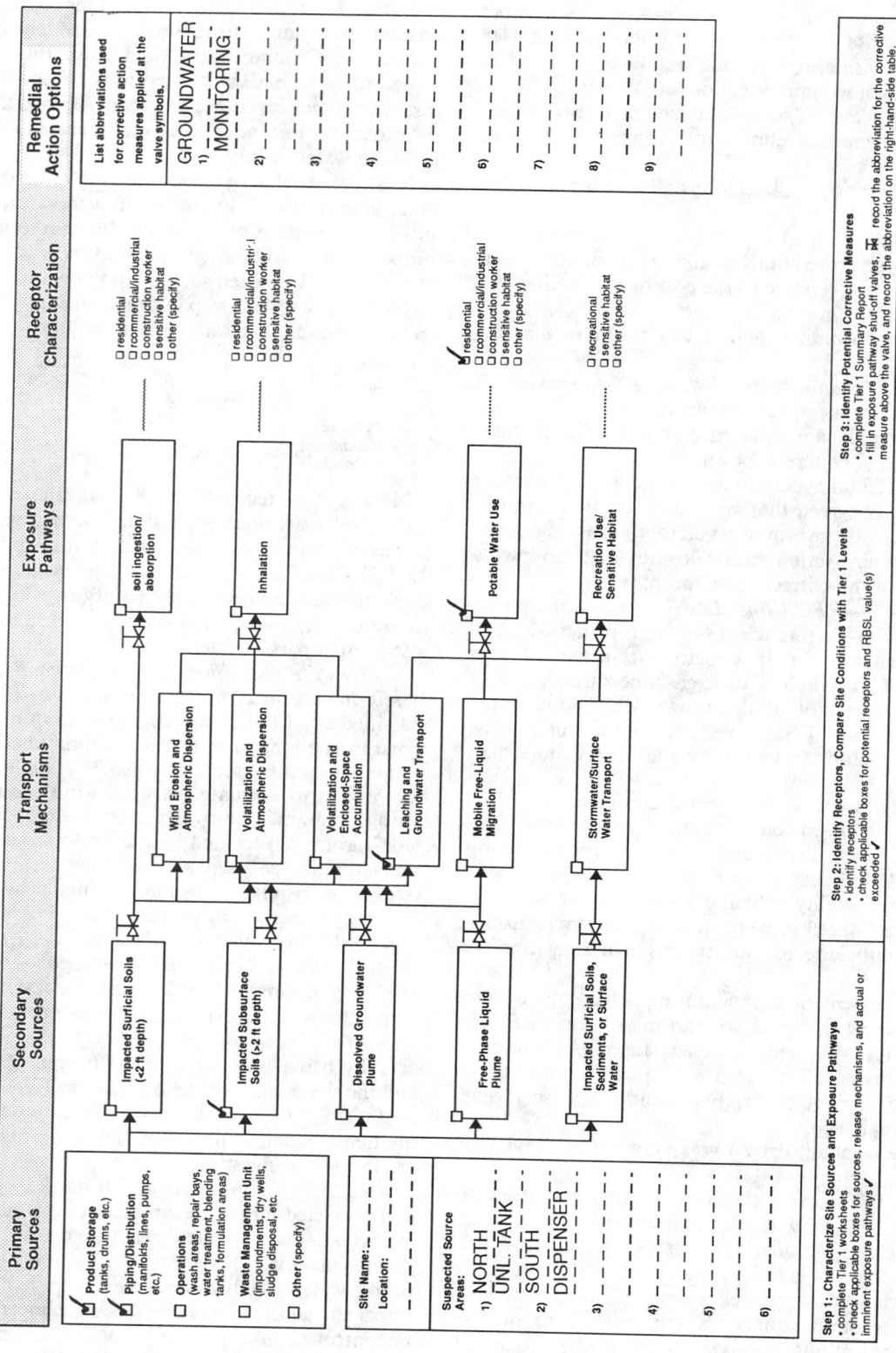

FIG. X5.2   Example 2—Exposure Evaluation Flowchart

exceedences of Tier 1 soil and ground water RBSLs are noted only for benzene.

X5.3.7 *Evaluation of Tier 1 Results*—The responsible party decides to proceed to a Tier 2 evaluation for benzene and the pathway of concern, rather than devise a corrective action plan to meet Tier 1 standards after considering the following factors:

X5.3.7.1 The shallow aquifer is impacted, but the dissolved plume appears to be stable and ground water movement is very slow,

X5.3.7.2 Excavation of soils to meet Tier 1 criteria would be expensive, due to the depth of impacted soils. Excavation would shut down the facility, and require all tanks and new lines to be removed and reinstalled,

X5.3.7.3 Costs for application of other conventional treatment methods, such as vapor extraction and pump and treat, are estimated to exceed $300 000 over the life of the remediation, and

X5.3.7.4 A tier 2 analysis for this site is estimated to require minimal additional data, and is anticipated to result in equally protective, but less costly corrective action.

X5.3.8 *Tier 2 Evaluation*—The owner collects additional ground water monitoring data and verifies that:

X5.3.8.1 No mobile free-phase product is present,

X5.3.8.2 The dissolved plume is stable and ground water concentrations appear to be decreasing with time,

X5.3.8.3 Extent of the dissolved plume is limited to within 50 ft (15.2 m) of the property boundaries,

X5.3.8.4 Dissolved oxygen concentrations are higher outside of the dissolved plume, indicating some level of aerobic biodegradation,

X5.3.8.5 Ground water movement is less than 50 ft/year (15.2 m), and

X5.3.8.6 Simple ground water transport modeling indicates that observations are consistent with expectations for the site conditions.

X5.3.9 *Remedial Action Evaluation*—Based on the demonstration of dissolved plume attenuation with distance, the owner negotiates a corrective action plan based on the following: (*1*) compliance with the Tier 1 RBSLs at the monitoring well located in the street center divider, provided that deed restrictions are enacted to prevent the use of ground water within that zone until dissolved levels decrease below drinking water MCLs, (*2*) deed restrictions are enacted to ensure that site land use will not change significantly, (*3*) continued sampling of the sentinel/compliance ground water monitoring well on a yearly basis, (*4*) should levels exceed Tier 1 RBSLs at that point for any time in the future, the corrective action plan will have to be revised, and (*5*) closure will be granted if dissolved conditions remain stable or decrease for the next two years.

X5.4 *Example 3—RBCA With Emergency Response and In Situ Remediation:*

X5.4.1 *Scenario*—A 5 000-gal (18 925-L) release of super unleaded gasoline occurs from a single-walled tank after repeated manual gaging with a gage stick. Soils are sandy at this site, ground water is shallow, and free-product is observed in a nearby monitoring well within 24 h. The site is located next to an apartment building that has a basement where coin-operated washers and dryers are located for use by the tenants.

X5.4.2 *Site Assessment*—In this case the initial site assessment is conducted rapidly and is focussed towards identifying if immediately hazardous conditions exist. It is known from local geological assessments that the first encountered ground water is not potable, as it is only about 2 ft (0.6 m) thick and is perched on a clay aquitard. Ground water monitoring wells in the area (from previous assessment work) are periodically inspected for the appearance of floating product, and vapor concentrations in the on-site utility corridors are analyzed with an explosimeter. While this flurry of activity begins, a tenant of the apartment building next door informs the station operator that her laundry room/basement has a strong gasoline odor. Explosimeter readings indicate vapor concentrations are still lower than explosive levels, but the investigation team notes that "strong gasoline odors" are present.

X5.4.3 *Site Classification and Initial Response Action*—This limited information is sufficient to classify this site as a Class 2 site (strong potential for conditions to escalate to immediately hazardous conditions in the short term), based on the observed vapor concentrations, size of the release, and geological conditions (see Table X5.3). The initial response implemented is as follows:

X5.4.3.1 Periodic monitoring of the apartment basement begins to ensure that levels do not increase to the point where evacuation is necessary (either due to explosion or acute health effects). In addition, the fire marshall is notified and building tenants are informed of the activities at the site, potential hazards, and abatement measures being implemented,

X5.4.3.2 A free-product recovery/hydraulic control system is installed to prevent further migration of the mobile liquid gasoline, and

X5.4.3.3 A subsurface vapor extraction system is installed to prevent vapor intrusion to the building.

X5.4.4 *Development of Tier 1 Look-Up Table of Risk-Based Screening Level (RBSL) Selection*—Assumptions used to derive example Tier 1 RBSL Look-Up Table X2.1 in Appendix X2 are reviewed and presumed valid for this site. Target soil and ground water concentrations are determined based on the vapor intrusion scenario. After considering health-based, OSHA PEL, national ambient background, and aesthetic vapor concentrations, target soil levels are based on achieving a 10^{-4} chronic inhalation risk for benzene, and hazard quotients of unity for all other compounds. The agency agrees to base compliance on the volatile monoaromatic compounds in gasoline (benzene, toluene, xylenes, and ethylbenzene), but reserves the right to alter the target levels if aesthetic effects persist in the building basement at the negotiated levels.

X5.4.5 *Exposure Pathway Evaluation*—Given that: (*1*) there is a very low potential for ground water usage, (*2*) a 20-ft (6.1-m) thick aquitard separates the upper perched water from any potential drinking water supplies, and (*3*) the close proximity of the apartment building, the owner proposes focusing on the vapor intrusion—residential inhalation scenario (see Fig. X5.3). The agency concurs, but in order to eliminate potential ground water users as receptors of concern, requests that a down-gradient piezometer be installed in the lower aquifer. The owner concurs.

X5.4.6 *Comparison of Site Conditions With Tier 1*

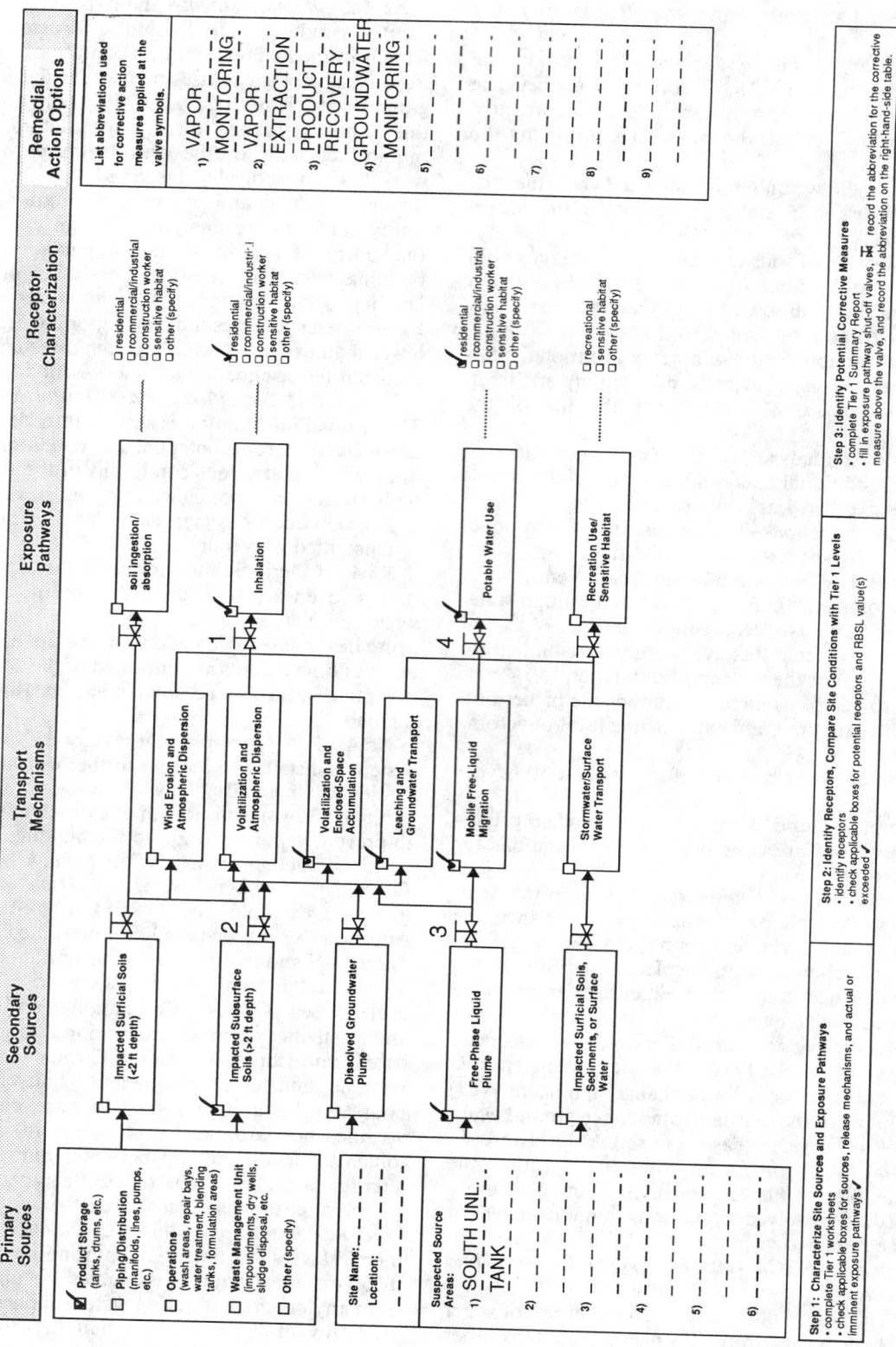

FIG. X5.3   Example 3—Exposure Evaluation Flowchart

RBSLs—While a complete initial site investigation has yet to be conducted, all parties agree that currently the RBSLs are likely to be exceeded.

X5.4.7 *Evaluation of Tier 1 Results*—The owner decides to implement an interim corrective action plan based on Tier 1 RBSLs, but reserves the right to propose a Tier 2 evaluation in the future.

X5.4.8 *Tier 1 Remedial Action Evaluation*—The owner proposes expanding the vapor extraction system to remediate source area soils. In addition he proposes continuing to operate the free-product recovery/hydraulic control system until product recovery ceases. Monitoring of the piezometer placed in the lower aquifer will continue, as well as periodic monitoring of the apartment building basement. Additional assessments will be conducted to ensure that building vapors are not the result of other sources. After some period of operation, when hydrocarbon removal rates decline, a soil and ground water assessment plan will be instituted to collect data to support a Tier 2 evaluation.

X5.5 *Example 4—RBCA Based on Use of a Tier 2 Table Evaluation*—In circumstances where site-specific data are similar among several sites, a table of Tier 2 SSTL values can be created. The following example uses such a table.

X5.5.1 *Scenario*—Petroleum-impacted ground water is discovered in monitoring wells at a former service station. The underground tanks and piping were removed, and the site is now occupied by an auto repair shop.

X5.5.2 *Site Assessment*—The responsible party completes an initial site assessment to determine the extent of hydrocarbon-impacted soil and ground water. Because gasoline was the only fuel dispensed at the site, the assessment focussed on benzene, toluene, ethylene benzene, and xylenes (BTEX) as the chemicals of concern. Site assessment results are summarized as follows:

X5.5.2.1 The area of hydrocarbon-impacted soil is approximately 18 000 ft² (1672 m²) and the depth of soil impaction is less than 5 ft (1.5 m); The plume is off site,

X5.5.2.2 The site is covered by asphalt or concrete,

X5.5.2.3 The site is underlain by clay,

X5.5.2.4 Hydrocarbon-impacted perched ground water is encountered at 1 to 3 ft (0.3 to 0.9 m) below grade. This water is non-potable. The first potable aquifer is located over 100 ft (30 m) below grade and is not impacted. There is no free product,

X5.5.2.5 Maximum detected concentrations are as follows:

Compound	Soil, mg/kg	Ground water, mg/L
Benzene	39	1.8
Toluene	15	4.0
Ethylbenzene	12	0.5
Xylenes	140	9.0

X5.5.2.6 Ground water velocity is 0.008 ft/day (0.0024 m/day) based on slug tests and ground water elevation survey and assumed soil porosity of 50 %,

X5.5.2.7 A receptor survey indicates that the nearest down gradient water well is greater than 1.0 mile (1.6 km) away and the nearest surface water body is 0.5 miles (0.8 km). The distance to the nearest sensitive habitat is greater than 1.0 mile; however, there is a forest preserve frequented by day hikers and picnickers next to the site. The nearest home is 1000 ft (305 m) away. The commercial building on site is 25 ft (7.6 m) from the area of hydrocarbon-impacted soil.

X5.5.3 *Site Classification and Initial Response Action*—Based on the classification scenarios given in Table 1, this site is classified as a Class 4 site, with no demonstrable long-term threat to human health, safety, or sensitive environmental receptors, because the hydrocarbon-impacted soils are covered by asphalt or concrete and cannot be contacted, only non-potable perched water with no existing local use is impacted, and there is no potential for explosive levels or concentrations that could cause acute effects in nearby buildings. The appropriate initial response is to evaluate the need for a ground water monitoring program.

X5.5.4 *Development of Tier 1 Look-Up Table of Risked-Based Screening Level (RBSL)*—The assumptions used to derive the example Tier 1 RBSL Look-Up Table are presumed valid for this site.

X5.5.5 *Exposure Pathway Evaluation*—The complete pathways are ground water and soil volatilization to enclosed spaces and to ambient air, and direct exposure to impacted soil or ground water by construction workers. A comparison of RBSLs for these pathways of concern indicates that

TABLE X5.1 Example 1—Site Classification and Initial Response Actions

Criteria and Prescribed Scenarios	Example Initial Response Actions
3. Long-term (>2 years) threat to human health, safety, or sensitive environmental receptors	Notify appropriate authorities, property owners, and potentially affected parties, and evaluate the need to
• Subsurface soils (>3 ft (0.9 m) BGS) are significantly impacted, and the depth between impacted soils and the first potable aquifer is less than 50 ft (15 m).	• Monitor ground water and determine the potential for future migration of the chemical(s) of concern to the aquifer.
• Ground water is impacted, and potable water supply wells producing from the impacted interval are located >2 years ground water travel time from the dissolved plume.	• Monitor the dissolved plume and evaluate the potential for natural attenuation and the need for hydraulic control.
• Ground water is impacted, and non-potable water supply wells producing from the impacted interval are located >2 years ground water travel time from the dissolved plume.	• Identify water usage of well, assess the effect of potential impact, monitor the dissolved plume, and evaluate whether natural attenuation or hydraulic control are appropriate control measures.
• Ground water is impacted, and non-potable water supply wells that do not produce from the impacted interval are located within the known extent of chemical(s) of concern.	• Monitor the dissolved plume, determine the potential for vertical migration, notify the user, and determine if any impact is likely.
• Impacted surface water, storm water, or ground water discharges within 1500 ft (457 m) of a sensitive habitat or surface water body used for human drinking water or contact recreation.	• Investigate current impact on sensitive habitat or surface water body, restrict access to area of discharge (if necessary), and evaluate the need for containment/control measures.
• Shallow contaminated surface soils are open to public access, and dwellings, parks, playgrounds, day-care centers, schools, or similar-use facilities are more than 500 ft (152 m) of those soils.	• Restrict access to impact soils.

TABLE X5.2 Example 2—Site Classification and Initial Response Actions

Criteria and Prescribed Scenarios	Example Initial Response Actions
3. Long-term (>2 years) threat to human health, safety, or sensitive environmental receptors	Notify appropriate authorities, property owners, and potentially affected parties, and evaluate the need to
• Subsurface soils (>3 ft (0.9 m) BGS) are significantly impacted, and the depth between impacted soils and the first potable aquifer is less than 50 ft (15 m).	• Monitor ground water and determine the potential for future contaminant migration to the aquifer.
• Ground water is impacted, and potable water supply wells producing from the impacted interval are located >2 years ground water travel time from the dissolved plume.	• Monitor the dissolved plume and evaluate the potential for natural attenuation and the need for hydraulic control.
• Ground water is impacted, and non-potable water supply wells producing from the impacted interval are located >2 years ground water travel time from the dissolved plume.	• Identify water usage of well, assess the effect of potential impact, monitor the dissolved plume, and evaluate whether natural attenuation or hydraulic control are appropriate control measures.
• Ground water is impacted, and non-potable water supply wells that do not produce from the impacted interval are located within the known extent of chemical(s) of concern.	• Monitor the dissolved plume, determine the potential for vertical migration, notify the user, and determine if any impact is likely.
• Impacted surface water, storm water, or ground water discharges within 1500 ft (457 m) of a sensitive habitat or surface water body used for human drinking water or contact recreation.	• Investigate current impact on sensitive habitat or surface water body, restrict access to area of discharge (if necessary), and evaluate the need for containment/control measures.
• Shallow contaminated surface soils are open to public access, and dwellings, parks, playgrounds, day-care centers, schools, or similar-use facilities are more than 500 ft (152 m) of those soils.	• Restrict access to impact soils.

TABLE X5.3 Example 3—Site Classification and Initial Response Actions

Criteria and Prescribed Scenarios	Example Initial Response Actions
2. Short-term (0 to 2 years) threat to human health, safety, or sensitive environmental receptors	Notify appropriate authorities, property owners, and potentially affected parties, and evaluate the need to
• There is potential for explosive levels, or concentrations of vapors that could cause acute effects, to accumulate in a residence or other building.	• Assess the potential for vapor migration (through monitoring/modeling) and remove source (if necessary), or install vapor migration barrier.
• Shallow contaminated surface soils are open to public access, and dwellings, parks, playgrounds, day-care centers, schools, or similar use facilities are within 500 ft (152 m) of those soils.	• Remove soils, cover soils, or restrict access.
• A non-potable water supply well is impacted or immediately threatened.	• Notify owner/user and evaluate the need to install point-of-use water treatment, hydraulic control, or alternate water supply.
• Ground water is impacted, and a public or domestic water supply well producing from the impacted aquifer is located within two-years projected ground water travel distance down gradient of the known extent of chemical(s) of concern.	• Institute monitoring and then evaluate if natural attenuation is sufficient, or if hydraulic control is required.
• Ground water is impacted, and a public or domestic water supply well producing from a different interval is located within the known extent of chemicals of concern.	• Monitor ground water well quality and evaluate if control is necessary to prevent vertical migration to the supply well.
• Impacted surface water, storm water, or ground water discharges within 500 ft (152 m) of a sensitive habitat or surface water body used for human drinking water or contact recreation.	• Institute containment measures, restrict access to areas near discharge, and evaluate the magnitude and impact of the discharge.

RBSLs associated with soil volatilization to an enclosed space are the most restrictive RBSLs.

X5.5.6 *Comparison of Site Conditions with Tier 1 RBSLs*—Based on the data given in X5.5.2 and the RBSLs given in Table X2.1, exceedances of Tier 1 RBSLs are noted for benzene in soil and ground water and toluene for ground water.

X5.5.7 *Evaluation of Tier 1 Results*—The responsible party decided to proceed to a Tier 2 evaluation for the pathways of concern rather than develop a corrective action plan for the following reasons:

X5.5.7.1 Only shallow perched water is impacted, and the dissolved plume is moving very slowly in tight clay,

X5.5.7.2 Excavation of soils to meet Tier 1 criteria would be expensive and would disrupt activities of the on-site business. Off-site excavation would be impractical and may not be able to clean up ground water to Tier 1 criteria,

X5.5.7.3 Other conventional treatment methods, such as pump and treat and vapor extraction, would be relatively ineffective in the heavy clay, and

X5.5.7.4 A Tier 2 evaluation for this site requires no additional data and is expected to be an equally protective but less costly corrective action.

X5.5.8 *Development of a Tier 2 Table of Site-Specific Target Levels (SSTLs)*—The Tier 2 table is similar to the Tier 1 Look-Up Table with the exception that SSTLs for the pathways of concern are presented as functions of both the distance from the source to the receptor and the soil type.

X5.5.8.1 For the pathways considered, approaches for the Tier 2 table are consistent with guidelines contained in Ref (26).

X5.5.8.2 The equations, assumptions, and parameters used to construct the Tier 1 Look-Up Table and Tier 2 table are similar, except as noted as follows:

(1) *Ground Water: Ingestion of Ground Water*—A one-dimensional analytical mass balance equation with attenuation mechanisms of retardation, dispersivity, and first-order biological decay (in sandy soil only) was applied in conjunction with the equations in Tables X2.2 and X2.3 to calculate SSTLs. The analytical model is limited to steady-state conditions and longitudinal dispersion. The analytical solution to the mass balance equation is presented in Ref (44).

TABLE X5.4 Example Tier 2 Site-Specific Target Level (SSTL) Table—Soil and Ground Water

Exposure Pathway	Receptor Scenario	Distance to Source, ft (m)	SSTLs at Source Sandy Soil, Natural Biodegradation Carcinogenic Risk = 1 × 10⁻⁵, HQ = 1				SSTLs at Source Clay Soil, No Natural Biodegradation Carcinogenic Risk = 1 × 10⁻⁵, HQ = 1			
			Benzene	Ethylbenzene	Toluene	Xylene	Benzene	Ethylbenzene	Toluene	Xylene
Soil Soil vapor intrusion from soil to buildings, mg/kg	residential	10 (3)	0.052	18	11	450	1.7	570	300	9500
		25 (7.6)	0.47	160	160	1.7[A]	65	11[A]	10[A]	RES[B]
		100 (30)	3.1[A]	RES	RES	RES	RES	RES	RES	RES
	commercial/industrial	10 (3)	0.13	39	24	980	4.3	1200	650	2.0[A]
		25 (7.6)	1.2	340	340	3.6[A]	950	24[A]	22.5[A]	RES
		100 (30)	8.0[A]	RES	RES	RES	RES	RES	RES	RES
Surficial soil ingestion and dermal, mg/kg	residential		22	5100	5400	280	22	5100	5400	280
	commercial/industrial		120	9600	1.7[A]	1500	117	9600	1.7[A]	1500
Soil lechate to protect ground water ingestion target level, mg/kg	residential	0 (0)	0.17	47	130	2200	0.17	47	130	2200
		100 (30)	0.32	88	250	4200	0.20	130	760	RES
		500 (152)	4.0	1200	6300	RES	RES	RES	RES	RES
	commercial/industrial	0 (0)	0.58	130	350	6200	0.58	130	350	6200
		100 (30)	1.1	250	670	1.2[A]	0.70	380	2100	RES
		500 (152)	13	3300	1.75[A]	RES	RES	RES	RES	RES
Ground Water Ground water ingestion, mg/L	residential	0	0.029	3.6	7.3	73	0.029	3.6	7.3	>S[C]
		100	0.054	6.8	14	140	0.035	10	43	>S
		500	0.68	90	350	>S	>S	>S	>S	>S
	commercial/industrial	0	0.099	10	20	200	0.099	10	20	200
		100	0.185	19	38	>S	0.12	29	120	>S
		500	2.3	250	>S	>S	>S	>S	>S	>S
Ground water vapor intrusion from ground water to buildings, mg/L	residential	10	0.11	32	17	510	5.0	>S	>S	>S
		25	0.72	210	160	>S	1200	>S	>S	>S
		100	>S	>S	>S	>S	>S	>S	>S	>S
	commercial/industrial	10	0.28	70	36	>S	13	>S	>S	>S
		25	1.9	>S	350	>S	>S	>S	>S	>S
		100	>S	>S	>S	>S	>S	>S	>S	>S

[A] Weight percent.
[B] RES—Selected risk level is not exceeded for pure compound present at any concentration.
[C] >S—Selected risk level is not exceeded for all possible dissolved levels.

(2) Ground Water: Inhalation of Outdoor Vapors—This pathway was not considered because exposure concentrations were very low.

(3) Ground Water: Inhalation of Enclosed-Space (Indoor) Vapors—A one-dimensional mass balance equation following Jury, et al (**31**) has been used to model vapor transport (**43**). This model was used in conjunction with the equations in Table X2.2 and X2.3 to calculate SSTLs. The model includes concentration attenuation between the source and the building by partitioning into immobile pore water, adsorption onto soil, and biological degradation (in sandy soil only).

(4) Subsurface Soils: Inhalation of Outdoor Vapors—This pathway was not considered because exposure concentrations were very low.

(5) Subsurface Soils: Inhalation of Enclosed-Space (Indoor) Vapors—The SSTLs were calculated using the Jury model (**31**) as discussed in Paragraph (*3*) of X5.5.8.2.

(6) Subsurface Soils: Leaching to Ground Water—The SSTLs were calculated using the one-dimensional mass-balance equation described in Paragraph (*1*) of X5.5.8.2, in conjunction with the lechate factor, LF_{SW}, as discussed in X2.9.4.1.

(7) All exposure parameter values listed in Table X2.4, soil, building surface, and subsurface parameter values listed in Table X2.6, and chemical-specific properties listed in Table X2.7 have not been changed.

(8) First-order decay rates in sandy soil were assumed to be 0.2 % per day for all BTEX compounds. These rates are considered conservative. Chiang, et al (**38**) determined that a

DO of 2.0 mg/L is required for rapid and complete biodegradation of benzene. Chiang, et al (**38**) measured a biodegradation rate of 0.95 % per day, and Barker, et al (**36**) measured a biodegradation rate of 0.6 % per day for benzene. In general, published biodegradation rates range from 0.6 to 1.25 % per day. Chiang, et al (**38**) also determined that biodegradation rates may be slower and incomplete at DO concentrations below 2.0 mg/L. This is a conservative value since aerobic biodegradation continues at DO concentrations as low as 0.7 mg/L (**44**).

(9) Clay properties are as follows:

Total soil porosity, cm³/cm³	0.05
Volumetric water content, cm³/cm³	0.40
Ground water Darcy velocity, cm/s	25

X5.5.8.3 Assumptions used to derive the example Tier 2 SSTL table are reviewed and presumed valid for this site. Due to the very conservative assumptions used to calculate exposure and the small number of people potentially exposed, the Tier 2 SSTLs are based on a 10⁻⁵ risk to human health for carcinogens and hazard quotients equal to unity for noncarcinogens.

X5.5.9 *Comparison of Site Conditions with Tier 2 Table SSTLs*—Based on the data given in X5.5.2 and the SSTLs given in the example of Table X5.4, no exceedances of Tier 2 soil or ground water SSTLs are noted.

X5.5.10 *Tier 2 Remedial Action Evaluation*—Based on the fact that Tier 2 soil or ground water SSTLs are not exceeded, the responsible party negotiates a corrective action plan based on the following:

X5.5.10.1 Annual compliance monitoring of ground

water at down gradient monitoring wells will be performed to demonstrate decreasing concentrations,

X5.5.10.2 Should levels exceed Tier 2 SSTLs at any of these monitoring points at any future time, the corrective

action plan will be reevaluated, and

X5.5.10.3 Closure will be granted if dissolved concentrations remain stable or decrease for the next two years.

REFERENCES

(1) *Ecological Assessment of Hazardous Waste Sites: A Field and Laboratory Reference Document*, EPA/600/3-89/013, NTIS No. PB-89205967, Environmental Protection Agency, Washington, DC, March 1989.

(2) *Integrated Risk Information System (IRIS)*, Environmental Protection Agency, Washington, DC, October 1993.

(3) *Health Effects Assessment Summary Tables (HEAST)*, OSWER OS-230, Environmental Protection Agency, Washington, DC, March 1992.

(4) *Superfund Public Health Evaluation Manuals*, NTIS No. PB87-183125, Environmental Protection Agency, Washington, DC, October 1986.

(5) *Health Effects Assessment Summary Tables (HEAST)*, USEPA/OERR 9200.6-303(91.1), NTIS No. PB91-921199, Environmental Protection Agency, Washington, DC, January 1991.

(6) *Technical Basis and Background for Cleanup Standards for Contaminated Sites*, New Jersey, 1993.

(7) Verschueren, K., *Handbook of Environmental Data on Organic Chemicals*, 2nd Edition, Van Nostrand Reinhold Co., Inc., New York, NY, 1983.

(8) *CHEM-BANK, Hazardous Chemical Databank on Compact Disk—HSDB*, U.S. National Library of Medicine.

(9) *Information Review Tert-Butyl Methyl Ether*, EPA Contract No. 68-01-6650, CRCS., Inc., Environmental Protection Agency, Washington, DC, March 1986.

(10) Dragun, J., *The Soil Chemistry of Hazardous Materials*, Hazardous Materials Control Research Institute, Silver Springs, MD, 1988.

(11) Lyman, W. J., Reehl, W. F., Rosenblatt, D. H., *Handbook of Chemical Property Estimation Methods*, McGraw Hill, New York, NY, 1982.

(12) Paustenbach, D. J., Jernigan, J. D., Bass, R. Kalmes, R., and Scott, P., "A Proposed Approach to Regulating Contaminated Soil: Identify Safe Concentrations for Seven of the Most Frequently Encountered Exposure Scenarios," *Regulatory Toxicology and Pharmacology*, Vol 16, 1992, pp. 21–56.

(13) Young, F. A., "Risk Assessment: The Convergence of Science and Law," *Regulatory Toxicology and Pharmacology*, Vol 7, 1987, pp. 179–184.

(14) Travis, C. C., Richter, S. A., Crouch, E. A., Wilson, R., and Wilson, E., "Cancer Risk Management: A Review of 132 Federal Regulatory Decisions," *Environmental Science and Technology*, Vol 21, No. 5, 1987, pp 415–420.

(15) Rodricks, J. V., Brett, S. M., and Wrenn, G. C., "Significant Risk Decisions in Federal Regulatory Agencies," *Toxicology Pharmacology*, Vol 7, 1987, pp. 307–320.

(16) *National Oil and Hazardous Substances Pollution Contingency Plan*, 40 CRF Part 300, Environmental Protection Agency, Washington, DC, 1990.

(17) *Hazardous Waste Management System Toxicity Characteristics Revisions* 55 FR 11798-11863, Environmental Protection Agency, Washington, DC.

(18) Personal communications, State Agencies, 1990.

(19) *Policy for Identifying and Assessing the Health Risks of Toxic Substances*, Environmental Toxicology Program, Division of Disease Control, Bureau of Health, Maine Department of Human Services (DHS), February 1988.

(20) *Draft Interim Guidance for Disposal Site Risk Characterization—In Support of the Massachusetts Contingency Plan*, Office of Research and Standards, Massachusetts Department of Environmental Quality Engineering (DEQE), October 1988.

(21) *Safe Drinking Water and Toxic Enforcement Act of 1986 (Proposition 65)*, Health and Welfare Agency, Office of the Secretary, Sacramento, CA, 1986.

(22) *Odor Thresholds for Chemicals with Established Occupational Health Standards*, American Industrial Hygiene Association, 1989.

(23) Shah and Singh, *Environmental Science and Technology*, Vol 22, No. 12, 1988.

(24) *Toxicological Profiles*, ATSDR, U.S. Public Health Services, 1988.

(25) Wallace, L. A., *Journal of Occupational Medicine*, Vol 28, No. 5, 1986.

(26) *Risk Assessment Guidance for Superfund*, Vol 1, *Human Health Evaluation Manual, Part A*, EPA/540/1-89/002, Environmental Protection Agency, Washington, DC, December 1989.

(27) *Exposure Factors Handbook*, EPA 600/8-89/043, Environmental Protection Agency, Washington, DC, July 1989.

(28) *Supplemental Risk Assessment Guidance for the Superfund Program*, EPA/901/5-89/001, Environmental Protection Agency Region I, Washington, DC, 1989.

(29) Johnson, P. C., and Ettinger, R. A., "Heuristic Model for Predicting the Intrusion Rate of Contaminant Vapors into Buildings," *Environmental Science and Technology*, Vol 25, No. 8, 1991, pp. 1445–1452.

(30) *Superfund Exposure Assessment Manual*, EPA/540/1-88/001, Environmental Protection Agency, Washington, DC, 1988.

(31) Jury, W. A., Spencer, W. F., and Farmer, W. J., "Behavior Assessment Model for Trace Organics in Soil: I, Model Description," *Journal of Environmental Quality*, Vol 12, 1983, pp. 558–564.

(32) Cowherd, C., Muleski, G. E., Englehart, P. J., and Gillett, D. A., *Rapid Assessment of Exposure to Particulate Emissions from Surface Contamination Sites*, Midwest Research Institute, PB85-192219, 1985.

(33) Johnson, P., Hertz, M. B., and Byers, D. I., "Estimates for Hydrocarbon Vapor Emissions Resulting from Service Stations Remediations and Buried Gasoline-Contaminated Soils," *Petroleum Contaminated Soils*, Vol III, Kostecki, P. T., and Calabrese, E. J., eds., Lewis Publishers, Chelsea, MI, 1990, pp. 295–326.

(34) Mullens, M., and Rogers, T., *AIECHE/DIPPR Environmental, Safety, and Health Data*, Design Institute for Physical Property Research—Research Project 911, American Institute for Chemical Engineers, June 1, 1993.

(35) *Hazardous Waste Treatment, Storage, and Disposal Facilities (TSDF)*, OAQPS, Air Emissions Models, EPA/450/3-87/026, Environmental Protection Agency, Washington, DC, 1989.

(36) Barker, J. F., Patrick, G. C., and Major, D., "Natural Attenuation of Aromatic Hydrocarbons in a Shallow Sand Aquifer," *Ground Water Monitoring Review*, Vol 7, 1987, pp. 64–71.

(37) Kemblowski, M. W., Salanitro, J. P., Deeley, G. M., and Stanley, C. C., "Fate and Transport of Residual Hydrocarbons in Ground Water: A Case Study," *Proceedings of the Petroleum Hydrocarbons and Organic Chemicals in Ground Water Conference*, National Well Water Association and American Petroleum Institute, Houston, TX, 1987, pp. 207–231.

(38) Chiang, C. Y., Salanitro, J. P., Chai, E. Y., Colthart, J. D., and Klein, C. L., "Aerobic Biodegradation of Benzene, Toluene, and Xylene in a Sandy Aquifer—Data Analysis and Computer Modeling," *Ground Water*, Vol 27, No. 6, 1989, pp. 823–834.

(39) Wilson, B. H., Wilson, J. T., Kampbell, D. H., Bledsoe, B. E., and Armstrong, J. M., "Biotransformation of Monoaromatic and Chlorinated Hydrocarbons at an Aviation Gasoline Spill Site," *Geomicrobiology Journal*, Vol 8, 1991, pp. 225–240.

(40) Howard, P., et al, *Handbook of Environmental Degradation Rates*, Lewis Publishers Inc., Chelsea, MI, 1991.

(41) Chevron Research and Technology Company, "Evaluation of Intrinsic Bioremediation at Field Sites," *Proceedings of the 1993 Petroleum Hydrocarbons and Organic Chemicals in Ground Water: Prevention, Retention, and Restoration*, Westing Galleria, Houston, TX, Nov. 10–12, 1993.

(42) Van Genuchten and Alves, *Analytical Solutions of the One-Dimensional Convective-Dispersive Solute Transport Equation*, Technical Bulletin No. 1661, U.S. Department of Agriculture, 1982.

(43) Jeng, C. Y., Kremesec, V. J., Primack, H. S., and Olson, C. B. (Amoco Oil Company), "Predicting the Risk in Buildings Posed by Vapor Transport of Hydrocarbon Contaminants," *Hydrocarbon Contaminated Soil and Ground Water, Proceedings for the 5th West Coast Conference: Contaminated Soils and Ground Water*, Vol 5, Association for Environmental Health of Soils, 1994.

(44) Wilson, J. T., "Natural Bioattenuation of Hazardous Organic Compounds in the Subsurface," R. S. Kerr Laboratory, Environmental Protection Agency, Draft Paper, 1993.

Standard Classification of
Environmental Condition of Property Area Types[1]

This standard is issued under the fixed designation D 5746; the number immediately following the designation indicates the year of original adoption or, in the case of revision, the year of last revision. A number in parentheses indicates the year of last reapproval. A superscript epsilon (ε) indicates an editorial change since the last revision or reapproval.

1. Scope

1.1 *Purpose*—The purpose of this classification is to define seven standard environmental condition of property area types for Department of Defense (DoD) real property at a closing military installation with respect to the requirements of the Comprehensive Environmental Response, Compensation and Liability Act (CERCLA) of 1980 Section 120(h), as amended by the Community Environmental Response Facilitation Act (CERFA) of 1992. As such, this classification is intended to permit a DoD component to classify property into seven area types, in order to facilitate and support findings of suitability to transfer (FOSTs), findings of suitability to lease (FOSLs), and uncontaminated parcel determinations pursuant to the requirements of CERFA. *Users of this classification should note that it does not address (except where noted explicitly) requirements for appropriate and timely regulatory consultation or concurrence, or both, during the identification and use of these environmental condition of property area types.*

1.1.1 *Seven Recognized Standard Environmental Condition of Property Area Types*—The goal of this classification is to permit DoD components to classify properties on closing DoD installations in order to support determinations of which properties are suitable and unsuitable for transfer by lease or by deed. The term "standard environmental condition of property area type" refers to one of the seven area types defined in this classification. An identification of an area type on an environmental condition of property map means that a DoD component has conducted sufficient studies to make a determination of the recognized environmental conditions of installation real property or has complied with the identification requirements of uncontaminated property under CERFA, or both, and has categorized the property into one of the following seven area types:

1.1.1.1 *Standard Environmental Condition of Property Area Type 1*—An area or parcel of real property where no storage, release, or disposal of hazardous substances or petroleum products or their derivatives has occurred (including no migration of these substances from adjacent properties).

1.1.1.2 *Standard Environmental Condition of Property Area Type 2*—An area or parcel of real property where only the storage of hazardous substances or petroleum products or their derivatives has occurred (but no release, disposal, or migration from adjacent properties has occurred).

1.1.1.3 *Standard Environmental Condition of Property*

Area Type 3—An area or parcel of real property where storage and release, release, disposal, or migration, or some combination thereof, of hazardous substances or petroleum products or their derivatives has occurred, but at concentrations that do not require a removal or remedial action.

1.1.1.4 *Standard Environmental Condition of Property Area Type 4*—An area or parcel of real property where storage and release, release, disposal, or migration, or some combination thereof, of hazardous substances or petroleum products or their derivatives has occurred, and all remedial actions necessary to protect human health and the environment have been taken.

1.1.1.5 *Standard Environmental Condition of Property Area Type 5*—An area or parcel of real property where storage and release, release, disposal, or migration, or some combination thereof, of hazardous substances or petroleum products or their derivatives has occurred and removal or remedial actions, or both, are under way, but all required actions have not yet been taken.

1.1.1.6 *Standard Environmental Condition of Property Area Type 6*—An area or parcel of real property where storage and release, release, disposal, or migration, or some combination thereof, of hazardous substances or petroleum products or their derivatives has occurred, but required response actions have not yet been initiated.

1.1.1.7 *Standard Environmental Condition of Property Area Type 7*—An area or parcel of real property that is unevaluated or requires additional evaluation.

1.1.2 *CERCLA Section 120(h) Requirements*—This classification of environmental condition of property area types is consistent with CERCLA § 120(h) requirements relating to the transfer of contaminated federal real property (42 USC 9601 and following). Areas classified as Area Types 1 through 4, as defined in this classification, are suitable, with respect to CERCLA § 120(h) requirements, for deed transfer to a non-federal recipient.

1.1.3 *CERFA Requirements*—This classification of environmental condition of property area types can be used in conjunction with the reporting requirements of CERFA, which amended CERCLA (Public Law 102-426, 106 Statute 2174). As defined in this classification, areas classified as Type 1 areas are eligible for reporting as "uncontaminated property" under the provisions of CERFA. Additionally, certain areas classified as Type 2 areas, where evidence indicates that storage occurred for less than one year, may also be reported as uncontaminated property. At installations listed on the national priorities list, Environmental Protection Agency (EPA) concurrence must be obtained for

[1] This classification is under the jurisdiction of ASTM Committee D-34 on Waste Management and is the direct responsibility of Subcommittee D34.11 on Site Remediation.

Current edition approved June 15, 1995. Published August 1995.

a parcel to be considered uncontaminated and therefore transferable under CERCLA § 120(h)(4). EPA has stated as a matter of policy that there may be instances in which it would be appropriate to concur with the military service that certain parcels can be identified as uncontaminated under CERCLA § 120(h)(4), although some limited quantity of hazardous substances or petroleum products have been stored, released, or disposed of on the parcel. If the information available indicates that the storage, release, or disposal was associated with activities that would not be expected to pose a threat to human health or the environment (for example, housing areas, petroleum-stained pavement areas, and areas having undergone routine application of pesticides), such parcels should be eligible for expeditious reuse.

1.1.4 *Petroleum Products*—Petroleum products and their derivatives are included within the scope of this classification. Areas on which "petroleum products or their derivatives were stored for one year or more, known to have been released or disposed of" (CERCLA § 120(h)(4)) are not eligible to be reported as uncontaminated property under CERFA. Additionally, under DoD policy, such areas may not be suitable for deed transfer until a response action has been completed.

1.2 *Objectives*—The objectives guiding the development of this classification are as follows: (*1*) to synthesize and put in writing a standard classification of environmental condition of property area types; (*2*) to facilitate the development of high-quality, standardized environmental condition of property maps that can be used to support FOSTs and FOSLs; (*3*) to facilitate the development of a standard practice for conducting environmental baseline surveys; and (*4*) to facilitate the development of a standard guide for preparing environmental baseline survey reports.

2. Referenced Documents

2.1 *ASTM Standards:*
E 1527 Practice for Environmental Site Assessments: Phase I Environmental Site Assessment Process[2]
E 1528 Practice for Environmental Site Assessments: Transaction Screen Process[2]
2.2 *Department of Defense Policies:*[3]
DoD Policy on the Environmental Review Process to Reach a Finding of Suitability to Lease (FOSL), September 1993
DoD Policy on the Environmental Review Process to Reach a Finding of Suitability to Transfer (FOST) for Property Where No Release or Disposal Has Occurred, June 1994
DoD Policy on the Environmental Review Process to Reach a Finding of Suitability to Transfer (FOST) for Property Where Release or Disposal Has Occurred, June 1994
DoD Policy on the Implementation of the Community Environmental Response Facilitation Act (CERFA), September 1993

---

[2] *Annual Book of ASTM Standards*, Vol 11.04.
[3] Available from National Technical Information Sevices, 5285 Port Royal Road, Springfield, VA 22161.

3. Terminology

3.1 This section provides definitions, descriptions of terms, and a list of acronyms for many of the words used in this classification. The terms are an integral part of this classification and are critical to an understanding of this classification and its use.

3.2 *Definitions:*

3.2.1 *environmental baseline survey (EBS)*—a survey of DoD real property based on all existing environmental information related to the storage, release, treatment, or disposal of hazardous substances or petroleum products or derivatives on the property to determine or discover the obviousness of the presence or likely presence of a release or threatened release of any hazardous substance or petroleum product. Additional data, including sampling and analysis, may be needed in certain cases in the EBS to support classification of the property into one of the standard environmental condition of property area types. Additionally, an EBS may also satisfy the uncontaminated property identification requirements of CERFA. An EBS will consider all sources of available information concerning environmentally significant current and past uses of the real property and shall, at a minimum, consist of the following: (*1*) a detailed search and review of available information and records in the possession of the DoD components or records made available by the regulatory agencies or other involved Federal agencies. DoD components are responsible for requesting and making reasonable inquiry into the existence and availability of relevant information and records to include any additional study information (for example, surveys for radioactive materials, asbestos, radon, lead-based paint, transformers containing PCB, Resource Conservation and Recovery Act Facility Assessments and Investigations (RFA and RFI), and underground storage tank cleanup program) to determine the environmental condition of the property; (*2*) a review of all reasonably obtainable Federal, state, and local government records for each adjacent facility where there has been a release or likely release of any hazardous substance or any petroleum product, and that is likely to cause or contribute to a release or threatened release of any hazardous substance or any petroleum product on the DoD real property; (*3*) an analysis of aerial photographs that may reflect prior uses of the property, which are in the possession of the Federal government or are reasonably obtainable through state or local government agencies; (*4*) interviews with current or former employees, or both, involved in operations on the real property; (*5*) visual inspections of the real property; any buildings, structures, equipment, pipe, pipeline, or other improvements on the real property; and of properties immediately adjacent to the real property, noting sewer lines, runoff patterns, evidence of environmental impacts (for example, stained soil, stressed vegetation, and dead or ill wildlife), and other observations that indicate the actual or potential release of hazardous substances or petroleum products; (*6*) the identification of sources of contamination on the installation and on adjacent properties that could migrate to the parcel during Federal government ownership; (*7*) ongoing response actions or actions that have been taken at or adjacent to the parcel; and (*8*) physical inspection of the property adjacent to the real property, to the extent permitted by owners or operators of such property.

3.2.2 *environmental baseline survey (EBS) report*—the written record of an EBS that includes the following: (*1*) an executive summary briefly stating the areas of real property (or parcels) evaluated and the conclusions of the EBS; (*2*) the property identification (for example, the address, assessor parcel number, or legal description); (*3*) any relevant information obtained from a detailed search of Federal government records pertaining to the property, including available maps; (*4*) any relevant information obtained from a review of the recorded chain of title documents regarding the real property. The review should address those prior ownerships and uses that could reasonably have contributed to an environmental concern, and, at a minimum, cover the preceding 60 years; (*5*) a description of past and current activities, including all past DoD uses to the extent such information is reasonably available, on the property and on adjacent properties; (*6*) a description of hazardous substances or petroleum products management practices (to include storage, release, treatment, or disposal) at the property and adjacent properties; (*7*) any relevant information obtained from records reviews and visual and physical inspections of adjacent properties; (*8*) a description of ongoing response actions or actions that have been taken at or adjacent to the property; (*9*) an evaluation of the environmental suitability of the property for an intended lease or deed transaction, if known, including the basis for determination of such suitability; and (*10*) references to key documents examined (for example, aerial photographs, spill incident reports, and investigation results).

3.2.3 *environmental condition of property map*—a map, prepared on the basis of all environmental investigation information conducted to date, that shows the environmental condition of a DoD installation's real property in terms of the seven standard environmental condition of property area types defined in this classification.

3.2.4 *hazardous substance*—a substance defined as a hazardous substance pursuant to CERCLA 42 USC § 9601(14), as interpreted by EPA regulations and the courts: "(*A*) any substance designated pursuant to section 1321(b)(2)(A) of Title 33, (*B*) any element, compound, mixture, solution, or substance designated pursuant to section 9602 of this title, (*C*) any hazardous waste having the characteristics identified under or listed pursuant to section 3001 of the Solid Waste Disposal Act (42 USC § 6921) (but not including any waste the regulation of which under the Solid Waste Disposal Act (42 USC § 6921 *et seq.*) has been suspended by Act of Congress), (*D*) any toxic pollutant listed under section 1317(a) of Title 33, (*E*) any hazardous air pollutant listed under section 112 of the Clean Air Act (42 USC § 7412), and (*F*) any imminently hazardous chemical substance or mixture with respect to which the Administrator (of EPA) has taken action pursuant to section 2606 of Title 15. The term does not include petroleum, including crude oil or any fraction thereof which is not otherwise specifically listed or designated as a hazardous substance under subparagraphs (*A*) through (*F*) of this paragraph, and the term does not include natural gas, natural gas liquids, liquefied natural gas, or synthetic gas usable for fuel (or mixtures of natural gas and such synthetic gas)."

3.2.5 *petroleum products*—those substances included within the meaning of the petroleum exclusion to CERCLA 42 USC § 9601(14) as interpreted by the courts and EPA, that is: "petroleum, including crude oil or any fraction thereof which is not otherwise specifically listed or designated as a hazardous substance under subparagraphs (*A*) through (*F*) of this paragraph, and the term does not include natural gas, natural gas liquids, liquefied natural gas, synthetic gas usable for fuel (or mixtures of natural gas and such synthetic gas)."

3.2.6 *property*—the real DoD property subject to classification under the classification of environmental condition of property area types.

3.2.7 *recorded land title records*—records to be searched during a chain of title search, including records of fee ownership, leases, land contracts, easements, liens, and other encumbrances on or of the property recorded in the place where land title records are recorded, by law or custom, for the local jurisdiction in which the property is located. (Such records are commonly kept by a municipal or county recorder or clerk.) Such records may be obtained from title companies or from the local government agency directly.

3.2.8 *release*—any spilling, leaking, pumping, pouring, emitting, emptying, discharging, injecting, escaping, leaching, dumping, or disposing into the environment (including the abandonment or discarding of barrels, containers, and other closed receptacles) of any hazardous chemical, extremely hazardous substance, or CERCLA hazardous substance.

3.2.9 *relevant and appropriate requirements*—those cleanup standards, standards of control, and other substantive requirements, criteria, or limitations promulgated under federal environmental or state environmental or facility siting laws that, while not "applicable" to a hazardous substance, pollutant, contaminant, remedial action, location, or other circumstance at a CERCLA site, address problems or situations sufficiently similar to those encountered at the CERCLA site that their use is well suited to the particular site. Only those state standards that are identified in a timely manner and are more stringent than federal requirements may be relevant and appropriate.

3.2.10 *remedial actions*—those actions consistent with a permanent remedy taken instead of, or in addition to, removal action in the event of a release or threatened release of a hazardous substance into the environment, to prevent or minimize the release of hazardous substances so that they do not migrate to cause substantial danger to the present or future public health or welfare or the environment.

3.2.11 *removal*—the cleanup or removal of released hazardous substances from the environment; such actions as may be necessary to take in the event of the threat of release of hazardous substances into the environment; such actions as may be necessary to monitor, assess, and evaluate the release or threat of release of hazardous substances; the disposal of removed material; or the taking of such other actions as may be necessary to prevent, minimize, or mitigate damage to the public health or welfare or to the environment, which may otherwise result from a release or threat of release.

3.2.12 *required remedial actions*—remedial actions determined necessary to comply with the requirements of CERCLA § 120(h)(3)(B)(i).

3.2.13 *required response actions*—removal or remedial actions, or both, determined necessary to comply with the requirements of CERCLA § 120(h)(3)(B)(i).

3.2.14 *risk-based criteria*—cleanup levels intended to meet a predetermined level of acceptable risk to human health or the environment.

3.2.15 *site inspection (SI)*—an on-site investigation to determine whether a release or potential release exists and the nature of the associated threats. The purpose is to augment the data collected in the preliminary assessment and to generate, if necessary, sampling and other field data to determine whether further action or investigation is appropriate.

3.3 *Descriptions of Terms Specific to This Standard:*

3.3.1 *adjacent properties*—those properties contiguous or partially contiguous to the boundaries of the property being surveyed during an EBS or other activity intended to classify the property into a standard environmental condition of property area type, or other properties relatively near the installation that could pose significant environmental concern or have a significant impact on the results of an EBS or on the classification of installation property into standard environmental condition of property area types, or both.

3.3.2 *aerial photographs*—photographs taken from an airplane or helicopter (from a low enough altitude to allow the identification of development and activities) of areas encompassing the property. Aerial photographs are commonly available from government agencies or private collections unique to a local area.

3.3.3 *all remedial action taken*—for the purposes of this classification, all remedial action, as described in CERCLA § 120(h)(3)(B)(i), has been taken if "the construction and installation of an approved remedial design has been completed, and the remedy has been demonstrated to the Administrator [of EPA] to be operating properly and successfully. The carrying out of long-term pumping and treating, or operation and maintenance, after the remedy has been demonstrated to the Administrator to be operating properly and successfully does not preclude the transfer of the property" (42 USC § 9620(h)(3)).

3.3.4 *applicable requirements*—those cleanup standards, standards of control, and other substantive requirements, criteria, or limitations promulgated under federal environmental or state environmental or facility siting laws that specifically address a hazardous substance, pollutant, contaminant, remedial action, location, or other circumstances found at a CERCLA site. Only those state standards that are identified by a state in a timely manner and that are more stringent than federal requirements may be applicable.

3.3.5 *biased field sampling*—by any technique, field sampling of environmental media, which aids in the delineation of standard environmental condition of property area types.

3.3.6 *BRAC statutes*—Title II of the Defense Authorization Amendments and Base Closure and Realignment Act of 1988 (Public Law 100-526, 10 USC 2687) and the Defense Base Closure and Realignment Act of 1990 (Part A of Title XXIX of Public Law 101-510, 10 USC 2687), collectively.

3.3.7 *carcinogenic*—a cancer-causing substance.

3.3.8 *chain of title review*—a review of recorded land title

records, conducted as part of an EBS.

3.3.9 *chemical-specific*—associated with the definition of applicable, or relevant and appropriate, requirements (ARARs), chemical-specific ARARs are those that may define acceptable exposure levels and can therefore be used in establishing primary remediation goals.

3.3.10 *closing military installation*—installations identified for closure pursuant to BRAC statutes.

3.3.11 *disposal*—the discharge, deposit, injection, dumping, spilling, leaking, or placing of any hazardous substances, or petroleum products or their derivatives, into or on any land or water so that such hazardous substances, or petroleum products or their derivatives or any constituent thereof, may enter the environment or be emitted into the air or discharged into any waters including groundwater.

3.3.12 *environmental investigation*—any investigation intended to determine the nature and extent of environmental contamination or to determine the environmental condition of property at a BRAC installation. Environmental investigations may include, but are not limited to, environmental site assessments, preliminary assessments, site inspections, remedial investigations, EBSs, Resource Conservation and Recovery Act (RCRA) facility assessments, and RCRA facility investigations.

3.3.13 *environmental site assessment*—the process by which a person or entity seeks to determine whether a particular parcel of real property (including improvements) is subject to recognized environmental conditions. This is the same meaning as provided in Practice E 1527.

3.3.14 *exposure pathway*—the route from a contaminant source to a human or any other environmental receptor.

3.3.15 *interviews*—sessions with current or former employees involved in operations on the real property, conducted to ascertain whether the storage, release, treatment, or disposal of hazardous substances, petroleum products, or their derivatives has occurred or is occurring on the real property.

3.3.16 *migration*—the movement of contaminant(s) away from a source through permeable subsurface media (such as the movement of a groundwater plume of contamination), or the movement of contaminant(s) by a combination of surficial and subsurface processes.

3.3.17 *obviousness*—the condition of being plain or evident. A condition or fact that could not be ignored or overlooked by a reasonable observer while conducting a records search or while physically or visually observing the property in conjunction with an EBS.

3.3.18 *practically reviewable*—information that is practically reviewable, that is, provided by the source in a manner and form that, upon examination, yields information relevant to the property without the need for extraordinary analysis of irrelevant data. The form of the information shall be such that the user can review the records for a limited geographic area. Records that cannot be retrieved feasibly by reference to the location of the property or a geographic area in which the property is located are not generally practically reviewable. Most data bases of public records are practically reviewable if they can be obtained from the source agency by the county, city, zip code, or other geographic area of the facilities listed in the record system. Records that are sorted, filed, organized, or maintained by the source agency only chronologically are not generally practically reviewable. This term has the same meaning as provided in Practice E 1527.

3.3.19 *preliminary assessment (PA)*—the review of existing information and an off-site reconnaissance, if appropriate, to determine whether a release or potential release may require additional investigation or action. A preliminary assessment may include an on-site reconnaissance, if appropriate.

3.3.20 *reasonably obtainable*—information that is (*1*) publicly available, (*2*) obtainable from its source within reasonable time and cost constraints, and (*3*) practically reviewable. This term has the same meaning as the term "reasonably ascertainable," as provided in Practice E 1527.

3.3.21 *recognized environmental conditions*—the presence or likely presence of any hazardous substances or petroleum products on any DoD real property under conditions that indicate an existing release, a past release, or the material threat of a release of any hazardous substances or petroleum products into structures on the property or into the ground, groundwater, or surface water of the property. The term includes hazardous substances or petroleum products even under conditions in compliance with laws. The term is not intended to include de minimus conditions that generally do not present a material risk of harm to the public health or environment and that generally would not be the subject of an enforcement action if brought to the attention of appropriate governmental agencies.

3.3.22 *records search*—a detailed search and review of available information and records in the possession of the DoD components and records made available by the regulatory agencies or other involved Federal agencies, including, but not limited to, installation restoration program studies and analyses, surveys for radioactive materials, asbestos, radon, lead-based paint, electrical devices (that is, transformers) containing polychlorinated biphenyl (PCB), RCRA facility assessments and investigations to determine which, if any, hazardous substances or petroleum products may be present on the property. For the purposes of adjacent facilities, a records search includes the review of all reasonably obtainable Federal, state, and local government records for each adjacent facility where there has been a release or likely release of any hazardous substance or any petroleum product, and which is likely to cause or contribute to a release or threatened release of any hazardous substance or any petroleum product on the DoD real property.

3.3.23 *site property*—property within the boundaries of the DoD installation.

3.3.24 *standard environmental condition of property area type*—one of the seven environmental condition of property area types defined in this classification.

3.3.25 *standards-based criteria*—cleanup criteria intended to meet the performance standards for the selected remedial technology.

3.3.26 *storage*—the containment of hazardous substances, petroleum products, or their derivatives, either on a temporary basis or for a period of years, in such a manner as not to constitute the disposal of such hazardous substances, petroleum products, or their derivatives.

3.3.27 *unevaluated*—not previously evaluated during any type of environmental investigation. This may also be used to designate areas that are unevaluated regarding CERFA reporting requirements.

3.3.28 *visual or physical inspection, or both*—actions taken during an EBS to include observations made by vision while walking through or otherwise traversing a property and structures located on it and observations made by the sense of smell, particularly observations of noxious or foul odors.

3.4 *Acronyms:*

3.4.1 *ARARs*—applicable, or relevant and appropriate, requirements.

3.4.2 *ASTM*—American Society for Testing and Materials.

3.4.3 *CERCLA*—Comprehensive Environmental Response, Compensation and Liability Act of 1980, as amended (42 USC 9620 and following).

3.4.4 *CERFA*—Community Environmental Response Facilitation Act of 1992 (102 Public Law 426, 106 Statute 2174).

3.4.5 *DoD*—Department of Defense.

3.4.6 *EPA*—United States Environmental Protection Agency.

3.4.7 *FOSL*—finding of suitability to lease, as described in the applicable DoD policy.

3.4.8 *FOST*—finding of suitability to transfer, as described in the applicable DoD policy.

3.4.9 *PCBs*—polychlorinated biphenyls.

3.4.10 *RCRA*—Resource Conservation and Recovery Act, as amended, 42 USC 6901 and following.

3.4.11 *USC*—United States Code.

4. Significance and Use

4.1 *Uses*—This classification is intended for use by DoD components in order to direct EBS efforts. It is also intended for use by preparers and reviewers of environmental condition of property maps and EBS reports used to support CERFA uncontaminated parcel identifications and parcels suitable for transfer by lease or by deed. This classification should be used to facilitate standardized determinations of the environmental condition of a DoD installation's real property. Such environmental condition of property determinations are necessary to assess the progress of ongoing environmental restoration, identify areas where further response may be required, identify areas where further evaluation is necessary, and to support FOSTs and FOSLs. An environmental condition of property map, which should be prepared using this classification, provides a consolidated view of a DoD installation's environmental investigation data, including sampling information.

5. Basis of Classification

5.1 *Classification*—Classification of the seven standard environmental condition of property area types is according to statutory requirements for (*1*) the identification of uncontaminated property within the provisions of CERFA and (*2*) for designating parcels of DoD installations as being suitable or unsuitable for transfer by deed within the provisions of CERCLA § 120(h)(3)(B)(i). Standard environmental condition of property area types are ranked in order of their suitability for transfer, with Area Types 1 through 4 being suitable for transfer by deed and Area Types 5 through 6 being unsuitable for transfer by deed until all remedial actions have been taken. Areas classified as Area Type 7 are either unevaluated or require further evaluation in order to classify them as one of the other environmental condition of property area types.

6. Standard Classification of Environmental Condition of Property Area Types

6.1 *Standard Environmental Condition of Property Area Type 1*—Areas where no storage, release, or disposal of hazardous substances or petroleum products or their derivatives has occurred, including no migration of these substances from adjacent areas. This is a geographically contiguous area or parcel of real property where the results of investigations reveal that no hazardous substances or petroleum products or their derivatives were stored, released, or disposed of on site property. A determination of this area type cannot be made, however, unless a minimum level of information gathering and assessment has been completed. In accordance with CERCLA § 120(h)(4), "the identification [of Type 1 areas] shall consist, at a minimum, of a review of each of the following sources of information concerning the current and previous uses of the real property: (i) A detailed search of Federal Government records pertaining to the property. (ii) Recorded chain of title documents regarding the real property. (iii) Aerial photographs that may reflect prior uses of the real property and that are reasonably obtainable through State or local government agencies. (iv) A visual inspection of the real property and any buildings, structures, equipment, pipe, pipeline, or other improvements on the real property, and a visual inspection of properties immediately adjacent to the real property. (v) A physical inspection of property adjacent to the real property, to the extent permitted by owners or operators of such property. (vi) Reasonably obtainable Federal, state, and local government records of each adjacent facility where there has been a release of any hazardous substance or any petroleum product or its derivatives, including aviation fuel and motor oil, and which is likely to cause or contribute to a release or threatened release of any hazardous substance or any petroleum product or its derivatives, including aviation fuel and motor oil, on the real property. (vii) Interviews with current or former employees involved in operations on the real property. Such identification shall also be based on sampling, if appropriate under the circumstances." These efforts (i through vii) can be functionally accomplished during the conduct of an EBS or a properly scoped preliminary assessment of the property being classified. However, if information gathered from these efforts indicates that hazardous substances, petroleum products, or their derivatives have been released, disposed of, or stored on the property, the property should be classified as an Area Type 2, 3, 4, 5, 6, or 7.

6.2 *Standard Environmental Condition of Property Area Type 2*—Areas where only the storage of hazardous substances or petroleum products or their derivatives has occurred, but no release, disposal, or migration from adjacent properties has occurred. This is a geographically contiguous area or parcel of real property where the results of investigations reveal that only the storage of hazardous substances, petroleum products, or their derivatives has occurred. In accordance with CERCLA § 120(h)(4), "the identification [of Type 2 areas] shall consist, at a minimum, of a review of each of the following sources of information concerning the current and previous uses of the real property: (i) A detailed search of Federal Government records pertaining to the property. (ii) Recorded chain of title documents regarding the real property. (iii) Aerial photographs that may reflect prior uses of the real property and that are reasonably obtainable through State or local government agencies. (iv) A visual inspection of the real property and any buildings, structures, equipment, pipe, pipeline, or other improvements on the real property, and a visual inspection of properties immediately adjacent to the real property. (v) A physical inspection of property adjacent to the real property, to the extent permitted by owners or operators of such property. (vi) Reasonably obtainable Federal, State, and local government records of each adjacent facility where there has been a release of any hazardous substance or any petroleum product or its derivatives, including aviation fuel and motor oil, and which is likely to cause or contribute to a release or threatened release of any hazardous substance or any petroleum product or its derivatives, including aviation fuel and motor oil, on the real property. (vii) Interviews with current or former employees involved in operations on the real property. Such identification shall also be based on sampling, if appropriate under the circumstances." These efforts (i through vii) can be functionally accomplished during the conduct of an EBS or a properly scoped preliminary assessment of the property being classified. However, if information gathered from these efforts indicates that the storage of hazardous substances or petroleum products or their derivatives has resulted in releases to the property, the property should be classified as an Area Type 3, 4, 5, or 6.

6.3 *Standard Environmental Condition of Property Area Type 3*—Areas where the storage and release, release, disposal, or migration, or some combination thereof, of hazardous substances or petroleum products or their derivatives has occurred, but at concentrations that do not require a removal or remedial action. This is a geographically contiguous area or parcel of real property where environmental evidence demonstrates that hazardous substances or petroleum products or their derivatives have been stored, released, or disposed of, but are present in quantities that require no response action to protect human health and the environment. Such quantities of hazardous substances or petroleum products can be below defensible detection limits or can be above detection limits but below action levels. In the absence of installation-specific risk-based or standards-based criteria that have received regulatory concurrence, below action levels means that the concentration of any hazardous substance or petroleum constituent in any medium does not exceed chemical-specific ARARs. A designation of this area type also means that risk estimates completed for contamination do not do the following: (*1*) exceed 10^{-6} for any carcinogenic hazardous substance or petroleum constituent detected in any medium; (*2*) result in a hazard index above 1 for any noncarcinogenic hazardous substance or petroleum constituent detected in any medium; (*3*) exceed 10^{-6} for all carcinogenic hazardous substances and petroleum constituents, taken together, in any exposure pathway; (*4*) result in a hazard index above 1 for all noncarcinogenic hazardous substances and petroleum constituents, taken together, in any exposure pathway; (5) exceed 10^{-4} for all carcinogenic hazardous substances and petroleum constituents accumulated across all pathways; or (6) result in a hazard index above 1 for all noncarcinogenic hazardous substances and petroleum constituents cumulated across all pathways. An Area Type 3 classification cannot be made with confidence unless a minimum level of information gathering and assessment has been completed. As such, all such determinations should be made on the basis of a site inspection or equivalent level of effort, which includes biased field sampling and laboratory analysis to support a conceptual understanding of the area. However, if information gathered from these efforts indicates that hazardous substances or petroleum products or their derivatives are on the property above action levels, the property should be classified as an Area Type 5 or 6.

6.4 *Standard Environmental Condition of Property Area Type 4*—Areas where storage and release, release, disposal, or migration, or some combination thereof, of hazardous substances or petroleum products or their derivatives has occurred, and all remedial actions necessary to protect human health and the environment have been taken. This is a geographically contiguous area or parcel of real property where all remedial actions necessary to protect human health and the environment have been taken. Type 4 areas include those areas in which an EBS report or other environmental investigation documents evidence that hazardous substances are known to have been released or disposed of on the property, but all remedial actions necessary to protect human health and the environment regarding any hazardous substances remaining on the property have already been taken to meet the covenant requirements of CERCLA § 120(h)(3)(B)(i).

6.5 *Standard Environmental Condition of Property Area Type 5*—Areas where storage and release, release, disposal, or migration, or some combination thereof, of hazardous substances or petroleum products or their derivatives has occurred and removal or remedial actions, or both, are under way, but all required remedial actions have not yet been taken. This is a geographically contiguous area or parcel of real property where the presence of sources or releases of hazardous substances or petroleum products or their derivatives has been confirmed by the results of sampling and analysis efforts. The results of such sampling and analysis efforts may be contained in electronic data bases or other

environmental investigation or environmental compliance reports, or both. This area type contains contaminant concentrations above action levels. Such concentrations do not meet the criteria of a Type 3 area classification. Removal actions are under way but are not yet demonstrated to have met the criteria of an Area Type 4. Remedial systems for Type 5 areas may be partially or entirely in place, but they have not been demonstrated to EPA to be "operating properly and successfully" within the meaning of CERCLA § 120(h)(3)(B)(i).

6.6 *Standard Environmental Condition of Property Area Type 6*—Areas where storage and release, release, disposal, or migration, or some combination thereof, of hazardous substances or petroleum products or their derivatives has occurred, but required response actions have not yet been initiated. This is a geographically contiguous area or parcel of real property where the presence of sources or releases of hazardous substances or petroleum products or their derivatives has been confirmed by the results of sampling and analysis efforts. The results of such sampling and analysis efforts may be contained in electronic data bases or environmental investigation or environmental compliance reports, or both. This area type contains concentrations of contaminants above action levels. Such concentrations do not meet the criteria of a Type 3 area classification. Required remedial systems or other response actions have not been initiated.

6.7 *Standard Environmental Condition of Property Area Type 7*—Areas that are unevaluated or that require additional evaluation. This is a geographically contiguous area or parcel of real property that is unevaluated, or a geographically contiguous area or parcel of real property where the presence of sources or releases of hazardous substances or petroleum products or their derivatives is suspected, but not well characterized, based on the results of a properly scoped records search, chain of title review, aerial photography review, visual inspection, set of employee interviews, and possibly sampling and analysis. They do not fit any of the previous area types with certainty because evaluation efforts have not occurred, are ongoing, or are inconclusive.

7. Keywords

7.1 Community Environmental Response Facilitation Act (CERFA); Comprehensive Environmental Response and Liability Act; environmental baseline survey (EBS); environmental condition of property; FOSL; FOST; hazardous substance; property area type; real estate; recognized environmental condition; remediation

Standard Practice for
Conducting Environmental Baseline Surveys[1]

This standard is issued under the fixed designation D 6008; the number immediately following the designation indicates the year of original adoption or, in the case of revision, the year of last revision. A number in parentheses indicates the year of last reapproval. A superscript epsilon (ε) indicates an editorial change since the last revision or reapproval.

. Scope

1.1 *Purpose*—The purpose of this practice is to define ood commercial and customary practice in the United states for conducting an environmental baseline survey EBS) in order to determine certain elements of the environmental condition of federal real property, including excess nd surplus property at closing and realigning military installations. This effort is conducted to fulfill certain requirements of the Comprehensive Environmental Response, Compensation and Liability Act of 1980 (CERCLA) section 20(h), as amended by the Community Environmental Response Facilitation Act of 1992 (CERFA). As such, this practice is intended to help a user to gather and analyze data nd information in order to classify property into seven nvironmental condition of property area types (in accordance with the Standard Classification of Environmental Condition of Property Area Types). Once documented, the EBS is used to support Findings of Suitability to Transfer FOSTs), Findings of Suitability to Lease (FOSLs), or ncontaminated property determinations, or a combination hereof, pursuant to the requirements of CERFA. Users of his practice should note that it does not address (except vhere explicitly noted) requirements for appropriate and imely regulatory consultation or concurrence, or both, uring the conduct of the EBS or during the identification nd use of the standard environmental condition of property rea types.

1.1.1 *Environmental Baseline Survey*—In accordance vith the Department of Defense (DoD) policy, an EBS will e prepared or evaluated for its usefulness (and updated if ecessary) for any property to be transferred by deed or eased. The EBS will be based on existing environmental nformation related to storage, release, treatment, or disposal f hazardous substances or petroleum products on the roperty to determine or discover the obviousness of the resence or likely presence of a release or threatened release f any hazardous substance or petroleum product. In certain ases, additional data, including sampling, if appropriate nder the circumstances, may be needed in the EBS to upport the FOST or FOSL. A previously conducted EBS nay be updated as necessary and used for making a FOST or FOSL. An EBS also may help to satisfy other environmental equirements (for example, to satisfy the requirements of CERFA or to facilitate the preparation of environmental ondition reports). In addition, the EBS provides a useful reference document and assists in compliance with hazard abatement policies related to asbestos and lead-based paint. The EBS process consists of discrete steps. This practice principally addresses EBS-related information gathering and analysis.

1.1.2 *CERCLA Section 120(h) Requirements*—This practice is intended to assist with the identification of installation areas subject to the notification and covenant requirements of CERCLA § 120(h) relating to the deed transfer of contaminated Federal real property (42 USC 9601 *et seq.*).

1.1.3 *CERFA Requirements*—This practice can be used to provide information that can be used to partially fulfill the identification requirements of CERFA [Pub. L. 102-426, 106 Stat. 2174], which amended CERCLA. Property classified as area Type *1*, in accordance with Classification D 5746 is eligible for reporting as "uncontaminated" under the provisions of CERFA. Additionally, certain property classified as area Type *2*, where evidence indicates that storage occurred for less than one year, may also be identified as uncontaminated. At installations listed on the National Priorities List, Environmental Protection Agency (EPA) concurrence must be obtained for the property to be considered "uncontaminated" and therefore transferable under CERCLA § 120(h)(4). The EPA has stated that there may be instances in which it would be appropriate to concur with the DoD Component that certain property can be identified as uncontaminated under CERCLA § 120(h)(4) although some limited quantity of hazardous substances or petroleum products have been stored, released, or disposed of on the property. If the information available indicates that the storage, release, or disposal was associated with activities that would not be expected to pose a threat to human health or the environment (for example, housing areas, petroleum-stained pavement areas, and areas having undergone routine application of pesticides), such property should be eligible for expeditious reuse.

1.1.4 *Petroleum Products*—Petroleum products and their derivatives are included within the scope of this practice. Areas on which petroleum products or their derivatives were stored for one year or more, known to have been released or disposed of [CERCLA § 120(h)(4)] are not eligible to be reported as "uncontaminated property" under CERFA.

1.1.5 *Other Federal, State, and Local Environmental Laws*—This practice does not address requirements of any federal, state, or local laws other than the applicable provisions of CERCLA identified in 1.1.2 and 1.1.3. Users are cautioned that federal, state, and local laws may impose additional EBS or other environmental assessment obligations that are beyond the scope of this practice. Users should also be aware that there are likely to be other legal obligations with regard to hazardous substances or petroleum products

[1] This practice is under the jurisdiction of ASTM Committee D-34 on Waste Management and is the direct responsibility of Subcommittee D34.11 on Site Remediation.

Current edition approved Oct. 10, 1996. Published December 1996. Originally published as PS 37. Last previous edition PS 37 – 95.

discovered on property that are not addressed in this practice and that may pose risks of civil or criminal sanctions, or both, for noncompliance.

1.1.6 *Other Federal, State, and Local Real Property and Natural and Cultural Resources Laws*—This practice does not address requirements of any federal, state or local real property or natural and cultural resources laws. Users are cautioned that numerous federal, state, and local laws may impose additional environmental and other legal requirements that must be satisfied prior to deed transfer of property that are beyond the scope of this practice.

1.2 *Objectives*—Objectives guiding the development of this practice are (*1*) to synthesize and put in writing a standard practice for conducting a high quality EBS, (*2*) to facilitate the development of high quality, standardized environmental condition of property maps to be included in an EBS that can be used to support FOSTs, FOSLs, and other applicable environmental condition reports, (*3*) to facilitate the use of the standard classification of environmental condition of property area types, and (*4*) to facilitate the development of a standard guide for preparing and updating EBS reports.

1.3 *Limitations*—Users of this practice should note that, while many of the elements of an EBS are performed in a manner consistent with other "due diligence" functions, an EBS is not prepared to satisfy a purchaser of real property's duty to conduct an "appropriate inquiry" in order to establish an "innocent landowner defense" to CERCLA § 107 liability. Any such use of any EBS by any party is outside the control of the United States Department of Defense and its components and beyond the scope of any EBS. No warranties or representations are made by the United States Department of Defense, its components, its officers, employees, or contractors that any EBS Report satisfies any such requirement for any party.

1.4 *Organization of This Practice*—This practice has 15 sections. Section 1 is the scope. Section 2 identifies referenced documents. Section 3, Terminology, includes definitions of terms not unique to this practice, descriptions of terms unique to this practice, and acronyms and abbreviations. Section 4 is the significance and use of this practice. Section 5 describes user's responsibilities. Sections 6 through 13 are the main body of the data gathering analysis steps of the EBS process. Section 14 briefly describes the EBS Step 3 classification of environmental condition of property area types. Section 15 contains a list of keywords.

1.5 *This standard does not purport to address all of the safety concerns, if any, associated with its use. It is the responsibility of the user of this standard to establish appropriate safety and health practices and determine the applicability of regulatory limitations prior to use.*

2. Referenced Documents

2.1 *ASTM Standards:*
E 1527 Practice for Environmental Site Assessments: Phase I Environmental Site Assessment Process[2]
E 1528 Practice for Environmental Site Assessments: Transaction Screen Process[2]

D 5746 Classification of Environmental Condition of Property Area Types[2]
2.2 *Department of Defense Policies:*
DoD Policy on the Environmental Review Process to Reach a Finding of Suitability to Lease (FOSL), September 1993[3]
DoD Policy on the Environmental Review Process to Reach a Finding of Suitability to Transfer (FOST) for Property Where No Release or Disposal Has Occurred, June 1994[3]
DoD Policy on the Environmental Review Process to Reach a Finding of Suitability to Transfer (FOST) for Property Where Release or Disposal Has Occurred, June 1994[3]
DoD Policy on the Implementation of the Community Environmental Response Facilitation Act (CERFA), September 1993[3]
2.3 *Department of Defense Guidance Document:*
BRAC Cleanup Plan Guidebook, Fall 1993[3]
2.4 *Federal Standards:*[4]
Title 40, Code of Federal Regulations (CFR), Part 300, National Oil and Hazardous Substances Pollution Contingency Plan
Title 40, Code of Federal Regulations (CFR), Part 302, Designation Reportable Quantities and Notification
Title 40, Code of Federal Regulations (CFR), Part 355, Emergency Planning and Notification

3. Terminology

3.1 This section provides definitions (of terms not unique to this practice), descriptions of terms specific to this practice, and a list of acronyms and abbreviations used herein. The terms are an integral part of this practice and are critical to its understanding and use. Many of these terms are also found in Practice E 1527.

3.2 *Definitions:*
3.2.1 *asbestos*—six naturally occurring fibrous minerals found in certain types of rock formations. Of the six, the minerals chrysotile, amosite, and crocidolite have been most commonly used in building products. When mined and processed, asbestos is typically separated into very thin fibers. Because asbestos is strong, incombustible, and corrosion-resistant, asbestos was used in many commercial products beginning early in this century and peaking in the period from World War II into the 1970s. When inhaled in sufficient quantities, asbestos fibers can cause serious health problems.

3.2.2 *asbestos-containing material (ACM)*—any material or product that contains more than 1 % asbestos.

3.2.3 *Comprehensive Environmental Response, Compensation, and Liability Information System (CERCLIS)*—the list of sites compiled by EPA that EPA has investigated or is currently investigating for potential hazardous substance contamination for possible inclusion on the National Priorities List.

3.2.4 *contaminated public wells*—public wells used for

---

[2] *Annual Book of ASTM Standards*, Vol 11.04.

[3] Available from Department of Defense, Office of Environmental Security, 3400 Defense Pentagon, Washington, DC 20301-3400.
[4] Available from the Superintendent of Documents, U.S. Government Printing Office, Washington, DC 20402.

drinking water that have been designated by a government entity as contaminated by toxic substances (for example, chlorinated solvents), or as having water unsafe to drink without treatment.

3.2.5 *drum*—a container (typically, but not necessarily, holding 55 gal [208 L] of liquid) that may have been used to store hazardous substances or petroleum products.

3.2.6 *dwelling*—structure or portion thereof used for residential habitation.

3.2.7 *environmental lien*—a charge, security, or encumbrance upon title to a property to secure the payment of a cost, damage, debt, obligation, or duty arising out of response actions, cleanup, or other remediation of hazardous substances or petroleum products upon a property, including (but not limited to) liens imposed pursuant to CERCLA 42 USC § 9607(1) and similar state or local laws.

3.2.8 *ERNS list*—EPA's Emergency Response Notification System list of reported CERCLA hazardous substance releases or spills in quantities equal to or greater than the reportable quantity, as maintained by the National Response Center. Notification requirements for such releases or spills are codified in 40 CFR Parts 302 and 355.

3.2.9 *Federal Register (FR)*—publication of the United States government published daily (except for Federal holidays and weekends) containing all proposed and final regulations and some other activities of the Federal government. When regulations become final, they are included in the Code of Federal Regulations (CFR) as well as published in the Federal Register.

3.2.10 *hazardous substance*—a substance defined as a hazardous substance pursuant to CERCLA 42 USC § 9601(14), as interpreted by EPA regulations and the courts: "(A) any substance designated pursuant to section 1321(b)(2)(A) of Title 33, (B) any element, compound, mixture, solution, or substance designated pursuant to Section 9602 of this title, (C) any hazardous waste having the characteristics identified under or listed pursuant to Section 3001 of the Solid Waste Disposal Act (42 USC § 6921) (but not including any waste the regulation of which under the Solid Waste Disposal Act (42 USC § 6921 *et seq.*) has been suspended by Act of Congress), (D) any toxic pollutant listed under Section 1317(a) of Title 33, (E) any hazardous air pollutant listed under Section 112 of the Clean Air Act (42 USC § 7412), and (F) any imminently hazardous chemical substance or mixture with respect to which the Administrator (of EPA) has taken action pursuant to Section 2606 of Title 15. The term does not include petroleum, including crude oil or any fraction thereof which is not otherwise specifically listed or designated as a hazardous substance under subparagraphs (A) through (F) of this paragraph, and the term does not include natural gas, natural gas liquids, liquefied natural gas, or synthetic gas usable for fuel (or mixtures of natural gas and such synthetic gas)." Users of this practice should note that certain states may expand this definition to include other substances not meeting the above definition. The user or environmental professional should consider whether the state in which the installation is located has identified such identified substances.

3.2.11 *hazardous waste*—any hazardous waste having the characteristics identified under or listed pursuant to section 3001 of the Solid Waste Disposal Act (42 USC § 6901 *et seq.*) (but not including any waste the regulation of which under the Solid Waste Disposal Act has been suspended by Act of Congress) and so forth.

3.2.12 *landfill*—a place, location, tract of land, area, or premises used for the disposal of solid wastes as defined by state solid waste regulations. The term is synonymous with the term solid waste disposal site and is also known as a garbage dump, trash dump, or similar term.

3.2.13 *local street directories*—directories published by private (or sometimes government) sources that show ownership, occupancy, or use of sites, or combination thereof, by reference to street addresses. Often local street directories are available at libraries of local governments, colleges or universities, or historical societies.

3.2.14 *material safety data sheet (MSDS)*—written or printed material concerning a hazardous substance which is prepared by chemical manufacturers, importers, and employers for hazardous chemicals pursuant to OSHA's Hazard Communication Standard, 29 CFR 1910.1200.

3.2.15 *National Contingency Plan (NCP)*—the National Oil and Hazardous substances Pollution Contingency Plan found at 40 CFR § 300, which is the EPA's regulations for how hazardous substances are to be cleaned up pursuant to CERCLA.

3.2.16 *National Priorities List*—list compiled by EPA pursuant to CERCLA 42 USC § 9605(a)(8)(B) of properties with the highest priority for cleanup pursuant to EPA's Hazard Ranking System. See 40 CFR Part 300.

3.2.17 *occupants*—those tenants, subtenants, or other persons or entities using the property or a portion of the property.

3.2.18 *owner*—generally the fee owner of record of the property.

3.2.19 *petroleum exclusion*—the exclusion from CERCLA liability provided in 42 USC § 9601(14), as interpreted by the courts and EPA: "The term (hazardous substance) does not include petroleum, including crude oil or any fraction thereof which is not otherwise specifically listed or designated as a hazardous substance under subparagraphs (A) through (F) of this paragraph, and the term does not include natural gas, natural gas liquids, liquefied natural gas, or synthetic gas usable for fuel (or mixtures of natural gas and such synthetic gas)."

3.2.20 *petroleum products*—those substances included within the meaning of the petroleum exclusion to CERCLA, 42 USC § 9601(14) as interpreted by the courts and EPA, that is: "petroleum, including crude oil or any fraction thereof which is not otherwise specifically listed or designated as a hazardous substance under subparagraphs (A) through (F) of this paragraph, and the term does not include natural gas, natural gas liquids, liquefied natural gas, or synthetic gas usable for fuel (or mixtures of natural gas and such synthetic gas)."

3.2.21 *Phase I Environmental Site Assessment*—the process described in Practice E 1527.

3.2.22 *pits, ponds, or lagoons*—man-made or natural depressions in a ground surface that are likely to hold liquids or sludge containing hazardous substances or petroleum products. The likelihood of such liquids or sludge being present is determined by evidence of factors associated with the pit, pond, or lagoon, including, but not limited to, discolored water, distressed vegetation, or the presence of an obvious wastewater discharge.

3.2.23 *property*—the real property that is the subject of the EBS described in this practice as well as the real property adjacent to the subject property (which may be privately owned). Real property includes buildings and other fixtures and improvements located on the property and affixed to the land.

3.2.24 *property tax files*—the files kept for property tax purposes by the local jurisdiction where the property is located and includes records of past ownership, appraisals, maps, sketches, photos, or other information that is reasonably ascertainable and pertaining to the property.

3.2.25 *RCRA generators*—those persons or entities that generate hazardous wastes, as defined and regulated by RCRA.

3.2.26 *RCRA generators list*—list kept by EPA of those persons or entities that generate hazardous wastes, as defined and regulated by RCRA.

3.2.27 *RCRA TSD facilities*—those facilities on which treatment, storage, or disposal, or a combination thereof, of hazardous wastes takes place, as defined and regulated by RCRA.

3.2.28 *RCRA TSD facilities list*—list kept by EPA of those facilities on which treatment, storage, or disposal, or a combination thereof, of hazardous wastes takes place, as defined and regulated by RCRA.

3.2.29 *recorded land title records*—records of fee ownership, leases, land contracts, easements, liens, and other encumbrances on or of the property recorded in the place where land title records are, by law or custom, recorded for the local jurisdiction in which the property is located. (Commonly, such records are kept by a municipal or county recorder or clerk.) Such records may be obtained from title companies or directly from the local government agency. Information about the title to the property that is recorded in a U.S. district court or any place other than where land title records are, by law or custom, recorded for the local jurisdiction in which the property is located, are not considered part of recorded land title records. See 3.3.33 and 7.2.4.

3.2.30 *records of emergency release notifications* (SARA § 304)—Section 304 of EPCRA or Title III of SARA requires operators of facilities to notify their local emergency planning committee (as defined in EPCRA) and State emergency response commission (as defined in EPCRA) of any release beyond the facility's boundary of any reportable quantity of any extremely hazardous substance. Often the local fire department is the local emergency planning committee. Records of such notifications are "records of emergency release notifications" (SARA § 304).

3.2.31 *solid waste disposal site*—a place, location, tract of land, area, or premises used for the disposal of solid wastes as defined by state solid waste regulations. The term is synonymous with the term landfill and is also known as a garbage dump, trash dump, or similar term.

3.2.32 *solvent*—a chemical compound that is capable of dissolving another substance and a hazardous substance, used in a number of manufacturing/industrial processes including but not limited to the manufacture of paints and coatings for industrial and household purposes, equipment

clean-up, and surface degreasing in metal fabricating industries.

3.2.33 *State registered USTs*—State lists of underground storage tanks required to be registered under Subtitle I, Section 9002 of RCRA.

3.2.34 *sump*—a pit, cistern, cesspool, or similar receptacle where liquids drain, collect, or are stored.

3.2.35 *underground storage tank (UST)*—any tank, including underground piping connected to the tank that is or has been used to contain hazardous substances or petroleum products and the volume of which is 10 % or more beneath the surface of the ground.

3.2.36 *USGS 7.5 Minute Topographic Map*—the map (if any) available from or produced by the United States Geological Survey, entitled "USGS 7.5 Minute Topographic Map" and showing the property.

3.2.37 *wastewater*—water that (*1*) is or has been used in an industrial or manufacturing process, (*2*) conveys or has conveyed sewage, or (*3*) is directly related to manufacturing, processing, or raw materials storage areas at an industrial plant. Wastewater does not include water originating on or passing through or adjacent to a site, such as stormwater flows, that has not been used in industrial or manufacturing processes, has not been combined with sewage, or is not directly related to manufacturing, processing, or raw materials storage areas at an industrial plant.

3.3 *Definitions of Terms Specific to This Standard:*

3.3.1 *adjacent properties*—those properties contiguous or partially contiguous to the boundaries of the property being surveyed during an EBS or other activity intended to classify the property into a standard environmental condition of property area type, or other properties relatively near the installation that could pose significant environmental concern and/or have a significant impact on the results of an EBS or on the classification of installation property into standard environmental condition of property area types.

3.3.2 *aerial photographs*—photographs, taken from an aerial platform, having sufficient resolution to allow identification of development and activities of areas encompassing the property. Aerial photographs are commonly available from government agencies or private collections unique to a local area.

3.3.3 *all remedial action taken*—for the purposes of this practice, all remedial action, as described in CERCLA § 120(h)(3)(B)(i), has been taken if "the construction and installation of an approved remedial design has been completed, and the remedy has been demonstrated to the administrator [of EPA] to be operating properly and successfully. The carrying out of long-term pumping and treating, or operation and maintenance, after the remedy has been demonstrated to the administrator to be operating properly and successfully does not preclude the transfer of the property." [42 USC § 9620(h)(3)]. Alternatively, in circumstances where a remedy has been constructed, but no ongoing treatment or operation and maintenance is required, for example, "clean closure" or excavation of soil with off-site treatment, all remedial action means that all action required to meet applicable state or federal regulatory standards, including, as required, state or federal regulatory approval, has been taken.

3.3.4 *applicable requirements*—those cleanup standards,

standards of control, and other substantive requirements, criteria, or limitations promulgated under federal environmental or State environmental or facility siting laws that specifically address a hazardous substance, pollutant, contaminant, remedial action, location, or other circumstances found at a CERCLA site. Only those state standards that are identified by a state in a timely manner and that are more stringent than federal requirements may be applicable.

3.3.5 *approximate minimum search distance*—the area for which records must be obtained and reviewed pursuant to Section 7 subject to the limitations provided in that section. This may include areas outside the property and shall be measured from the nearest property boundary. This term is used instead of radius to include irregularly shaped properties.

3.3.6 *BRAC statutes*—Title II of the Defense Authorization Amendments and Base Closure and Realignment Act of 1988 (Pub. L. 100-526, 10 USC 2687, note.) and the Defense Base Closure and Realignment Act of 1990 (Part A of Title XXIX of Pub. L. 101-510, 10 USC 2687, note.), collectively.

3.3.7 *closing military installation*—installations identified for closure pursuant to BRAC statutes, or installations previously closed under the authority of 10 USC 2687.

3.3.8 *DoD Component*—collectively, the Office of the Secretary of Defense, the Military Departments, the Chairman of the Joint Chiefs of Staff, the Inspector General of the Department of Defense, the Defense Agencies and the DoD Field Activities.

3.3.9 *disposal*—the discharge, deposit, injection, dumping, spilling, leaking, or placing of any hazardous substances, or petroleum products or their derivatives into or on any land or water so that such hazardous substances, or petroleum products or their derivatives or any constituent thereof may enter the environment or be emitted into the air or discharged into any waters including ground water.

3.3.10 *due diligence*—the process of inquiring into the environmental characteristics of a parcel of commercial real estate or other conditions, usually in connection with a commercial real estate transaction. The degree and kind of due diligence vary for different properties and differing purposes.

3.3.11 *environmental audit*—the investigative process to determine if the operations of an existing facility are in compliance with applicable environmental laws and regulations. This term should not be used to describe Practices E 1527, E 1528, or this practice, although an environmental audit may include an EBS or, if prior audits or EBSs are available, may be part of an EBS.

3.3.12 *environmental baseline survey* (EBS)—a survey of federal real property based on all existing environmental information related to storage, release, treatment, or disposal of hazardous substances or petroleum products or derivatives on the property to determine or discover the obviousness of the presence or likely presence of a release or threatened release of any hazardous substance or petroleum product. In certain cases, additional data, including sampling and analysis, may be needed in the EBS to support the classification of the property into one of the standard environmental condition of property area types. Additionally, an EBS may also satisfy the uncontaminated property identification requirements of CERFA. An EBS will consider all sources of available information concerning environmentally significant current and past uses of the real property, and shall, at a minimum, consist of the following:

3.3.12.1 Detailed search and review of available information and records in the possession of the DoD Components or records made available by the regulatory agencies or other involved federal agencies. The DoD Components are responsible for requesting and making reasonable inquiry into the existence and availability of relevant information and records to include any additional study information (for example, surveys for radioactive materials, asbestos, radon, lead-based paint, transformers containing PCB, RCRA Facility Assessments and Investigations, Underground Storage Tank Cleanup Program) to determine the environmental condition of the property;

3.3.12.2 Review of all reasonably obtainable federal, state, and local government records for each adjacent facility where there has been a release or likely release of any hazardous substance or any petroleum product, and which is likely to cause or contribute to a release or threatened release of any hazardous substance or any petroleum product on the federal real property;

3.3.12.3 Analysis of aerial photographs that may reflect prior uses of the property, which are in the possession of the federal government or are reasonably obtainable through state or local government agencies;

3.3.12.4 Interviews with current or former employees, or both, involved in operations on the real property;

3.3.12.5 Visual inspections of the real property; any buildings, structures, equipment, pipe, pipeline, or other improvements on the real property; and of properties immediately adjacent to the real property, noting sewer lines, runoff patterns, evidence of environmental impacts (for example, stained soil, stressed vegetation, dead or ill wildlife) and other observations which indicate actual or potential release of hazardous substances or petroleum products;

3.3.12.6 Identification of sources of contamination on the installation and on adjacent properties which could migrate to the parcel during federal government ownership;

3.3.12.7 Ongoing response actions or actions that have been taken at or adjacent to the parcel; and

3.3.12.8 A physical inspection of property adjacent to the real property, to the extent permitted by owners or operators of such property.

3.3.13 *environmental baseline survey (EBS) report*—the written record of an EBS that includes the following:

3.3.13.1 An executive summary briefly stating the areas of real property (or parcels) evaluated and the conclusions of the EBS;

3.3.13.2 The property identification (for example, address, assessor parcel number, legal description);

3.3.13.3 Any relevant information obtained from a detailed search of federal government records pertaining to the property, including available maps;

3.3.13.4 Any relevant information obtained from a review of the recorded chain of title documents regarding the real property. The review should address those prior ownerships/uses that could reasonably have contributed to an environmental concern, and, at a minimum, cover the preceding 60 years;

3.3.13.5 A description of past and current activities,

including all past DoD uses to the extent such information is reasonably available, on the property and on adjacent properties,

3.3.13.6 A description of hazardous substances or petroleum products management practices (to include storage, release, treatment, or disposal) at the property and at adjacent properties;

3.3.13.7 Any relevant information obtained from records reviews and visual and physical inspections of adjacent properties;

3.3.13.8 Description of ongoing response actions or actions that have been taken at or adjacent to the property;

3.3.13.9 An evaluation of the environmental suitability of the property for an intended lease or deed transaction, if known, including the basis for the determination of such suitability; and

3.3.13.10 Reference to key documents examined (for example, aerial photographs, spill incident reports, investigation results).

3.3.14 *environmental condition of property area type*—any of the seven standard environmental condition of property area types defined in the Standard Classification of Environmental Condition of Property Area Types.

3.3.15 *environmental condition of property map*—a map, prepared on the basis of all environmental investigation information conducted to date, that shows the environmental condition of a DoD installation's real property in terms of the seven standard environmental condition of property area types as defined in the standard classification.

3.3.16 *environmental investigation*—any investigation intended to determine the nature and extent of environmental contamination or to determine the environmental condition of property at a BRAC installation. Environmental investigations may include, but are not limited to, environmental site assessments, preliminary assessments, site inspections, remedial investigations, EBSs, RCRA facility assessments, and RCRA facility investigations.

3.3.17 *environmental professional*—a person possessing sufficient training and experience necessary to conduct an EBS including all activities related to this practice, and from the information and data gathered by such activities, having the ability to develop conclusions regarding environmental condition of property and recognized environmental conditions in connection with the property being evaluated. An individual's status as an environmental professional may be limited to the type of EBS to be performed or to specific steps of the EBS for which the professional is responsible. The person may be an independent contractor of an employee of the Department of Defense or its components.

3.3.18 *fill dirt*—dirt, soil, sand, or other earth, that is obtained off-site, that is used to fill holes or depressions, create mounds, or otherwise artificially change the grade or elevation of real property. It does not include material that is used in limited quantities for normal landscaping activities.

3.3.19 *innocent landowner defense*—that defense to CERCLA liability provided in 42 USC § 9601(35) and 42 USC § 9607(b)(3). One of the requirements to qualify for this defense is that the party make "all appropriate inquiry into the previous ownership and uses of the property consistent with good commercial or customary practice." There are additional requirements to qualify for this defense.

3.3.20 *installation restoration program (IRP)*—the DoD program, mandated by 10 USC § 2407 to assess and respond to releases of hazardous substances on military property under the control of the military services. Additionally, based upon policy decisions, the IRP serves as an umbrella program for environmental response in all media, including RCRA corrective action, LUST corrective action, as well as CERCLA removals and remedial actions. Generally, where field sampling or intrusive environmental testing is required, the IRP will serve as a vehicle for such testing. The IRP is also known as the Defense Environmental Restoration Program.

3.3.21 *interviews*—sessions with current or former employees involved in operations on the real property, conducted to ascertain if storage, release, treatment, or disposal of hazardous substances, petroleum products or their derivatives occurred or is occurring on the real property.

3.3.22 *local government agencies*—those agencies of municipal or county government having jurisdiction over the property. Municipal and county government agencies include, but are not limited to, cities, parishes, townships, and similar entities. Local government agencies may also include, where appropriate, state agencies with local jurisdiction which perform functions commonly performed in other locations by local government agencies.

3.3.23 *migration*—the movement of contaminant(s) away from a source through permeable subsurface media (such as the movement of a ground water plume of contamination), or movement of contaminant(s) by a combination of surficial and subsurface processes.

3.3.24 *obviousness*—the condition of being plain or evident. A condition or fact which could not be ignored or overlooked by a reasonable observer while conducting a records search or while physically or visually observing the property in conjunction with an EBS.

3.3.25 *other historical sources*—any source or sources other than those designated in 7.2.1 through 7.2.4 that are credible to a reasonable person and that identify past uses of the property. The term includes, but is not limited to: miscellaneous maps, newspaper archives, and records in the files and/or personal knowledge of the property owner and/or occupants.

3.3.26 *physical setting sources*—sources that provide information about the geologic, hydrogeologic, hydrologic, or topographic characteristics of a property.

3.3.27 *practically reviewable*—information that is practically reviewable is information provided by the source in a manner and in a form that, upon examination, yields information relevant to the property without the need for extraordinary analysis of irrelevant data. The form of the information shall be such that the user can review the records for a limited geographic area. Records that cannot be feasibly retrieved by reference to the location of the property or a geographic area in which the property is located are not generally practically reviewable. Most data bases of public records are practically reviewable if they can be obtained from the source agency by the county, city, zip code, or other geographic area of the facilities listed in the record system. Records that are sorted, filed, organized, or maintained by the source agency only chronologically are not generally practically reviewable. This term has the same meaning as provided in Practice E 1527.

3.3.28 *preliminary assessment (PA)*—review of existing information and an off-site reconnaissance, if appropriate to determine if a release or potential release may require additional investigation or action. A PA may include an on-site reconnaissance, if appropriate.

3.3.29 *publicly available*—information that is publicly available means that the source of the information allows access to the information by anyone upon request.

3.3.30 *reasonably available*—information that is (*1*) publicly available, (*2*) obtainable from its source within reasonable time and cost constraints, and (*3*) practically reviewable. This term has the same meaning as the term "reasonably ascertainable" as provided in Practice E 1527.

3.3.31 *reasonably obtainable*—information that is (*1*) publicly available, (*2*) obtainable from its source within reasonable time and cost constraints, and (*3*) practically reviewable. This term has the same meaning as the term "reasonably ascertainable" as provided in Practice E 1527. Reasonably available and reasonably obtainable are synonyms.

3.3.32 *recognized environmental conditions*—the presence or likely presence of any hazardous substances or petroleum products on any federal real property under conditions that indicate an existing release, a past release, or a material threat of a release of any hazardous substances or petroleum products into the environment. The term includes hazardous substances or petroleum products even under conditions in compliance with laws. The term is not intended to include *de minimis* conditions that generally do not present a material risk of harm to public health or the environment and that generally would not be the subject of an enforcement action if these conditions were brought to the attention of appropriate governmental agencies. This term is introduced in Practice E 1527, and is used herein only in conjunction with EBS Steps *1* and *2* (see 6.2), as an intermediate outcome prior to the Step *3* classification of environmental condition of property area types. A Phase I Site Assessment results in recognized environmental conditions, but not environmental condition of property area types.

3.3.33 *recorded chain of title documents*—this term has the same meaning as recorded land title records.

3.3.34 *records search and/or review*—detailed search and review of available information and records in the possession of the DoD components and records made available by the regulatory agencies or other involved federal agencies, including, but not limited to IRP studies and analyses, surveys for radioactive materials, asbestos, radon, lead-based paint, electrical devices (that is, transformers) containing PCB, RCRA facility assessments and Investigations to determine what, if any, hazardous substances or petroleum products may be present on the property. For the purposes of adjacent facilities, a records search includes the review of all reasonably obtainable federal, state, and local government records for each adjacent facility where there has been a release or likely release of any hazardous substance or any petroleum product, and which is likely to cause or contribute to a release or threatened release of any hazardous substance or any petroleum product on the federal real property.

3.3.35 *release*—any spilling, leaking, pumping, pouring, emitting, emptying, discharging, injecting, escaping, leaching, dumping, or disposing into the environment (including the abandonment or discarding of barrels, containers, and other closed receptacles) of any hazardous chemical, extremely hazardous substance, or CERCLA hazardous substance.

3.3.36 *relevant and appropriate requirements*—those cleanup standards, standards of control, and other substantive requirements, criteria, or limitations promulgated under federal environmental or State environmental or facility siting laws that, while not "applicable" to a hazardous substance, pollutant, contaminant, remedial action, location, or other circumstance at a CERCLA site, address problems or situations sufficiently similar to those encountered at the CERCLA site that their use is well suited to the particular site. Only those state standards that are identified in a timely manner and are more stringent than federal requirements may be relevant and appropriate.

3.3.37 *remedial actions*—those actions consistent with a permanent remedy taken instead of, or in addition to, removal action in the event of a release or threatened release of a hazardous substance into the environment, to prevent or minimize the release of hazardous substances so that they do not migrate to cause substantial danger to present or future public health or welfare or the environment.

3.3.38 *removal*—the cleanup or removal of released hazardous substances from the environment; such actions as may be necessary to take in the event of the threat of release of hazardous substances into the environment; such actions as may be necessary to monitor, assess, and evaluate the release or the threat of release of hazardous substances; the disposal of removed material; or the taking of such other actions as may be necessary to prevent, minimize, or mitigate damage to the public health or welfare or to the environment, which may otherwise result from a release or threat of release.

3.3.39 *required remedial actions*—remedial actions determined necessary to comply with the requirements of CERCLA § 120(h)(3)(B)(i).

3.3.40 *required response actions*—removal and/or remedial actions determined necessary to comply with the requirements of CERCLA § 120(h)(3)(B)(i).

3.3.41 *significant and significance*—in this practice, significant and significance connote the opposite of trivial or *de minimis*. An event or condition is considered significant if it has the potential to present a nontrivial risk to human health and the environment, using the risk range established by the NCP. A probability is considered significant when an environmental professional estimates the probability as nontrivial. For example, in the hypothetical case of an underground tank that was installed and removed prior to the existence of regulatory requirements for tank closure, the environmental professional must evaluate the possibility of release from the tank in the absence of soil testing results. If such an evaluation, based upon observed site conditions and documented soil corrosivity characteristics were to conclude that the probability of release is trivial or very close to zero, then no soil testing would be undertaken in the absence of a specific regulatory requirement for such testing. On the other hand, there is a significant probability of release if such an evaluation were to determine that the probability of release were greater than the extremely low probability encompassed by the concept of trivial. The evaluation of significance, in this sense, is a matter of professional judgment on the part of the environmental professional and should be so documented.

3.3.42 *site inspection (SI)*—an on-site investigation to determine whether there is a release or potential release and the nature of the associated threats. The purpose is to augment the data collected in the PA and to generate, if necessary, sampling and other field data to determine if further action or investigation is appropriate.

3.3.43 *standard classification*—the Standard Classification of Environmental Condition of Property Area Types.

3.3.44 *standard environmental condition of property area type*—one of the seven environmental condition of property area types defined in the Standard Classification.

3.3.45 *standard practice*—the activities set forth in this practice and Practices E 1527 and E 1528, where referenced.

3.3.46 *storage*—the containment of hazardous substances, petroleum products or their derivatives, either on a temporary basis or for a period of years, in such a manner as not to constitute disposal of such hazardous substances, petroleum products, or their derivatives.

3.3.47 *transaction screen process*—the process described in Practice E 1528.

3.3.48 *transaction screen questionnaire*—the questionnaire provided in Practice E 1528.

3.3.49 *user*—the party seeking to use this practice to perform an EBS of the property. A user may include, without limitation, a DoD component (acting as owner of the property).

3.3.50 *visual and/or physical inspection*—actions taken during an EBS to include observations made by vision while walking through or otherwise traversing a property and structures located on it and observations made by the sense of smell, particularly observations of noxious or foul odors.

3.4 *Acronyms and Abbreviations:*

3.4.1 *ARARs*—applicable or relevant and appropriate requirements.

3.4.2 *ASTM*—American Society for Testing and Materials.

3.4.3 *BRAC*—Base Realignment and Closure.

3.4.4 *CERCLA*—Comprehensive Environmental Response, Compensation and Liability Act of 1980, as amended (42 USC 9620 *et seq.*).

3.4.5 *CERCLIS*—Comprehensive Environmental Response, Compensation and Liability Information System.

3.4.6 *CERFA*—Community Environmental Response Facilitation Act of 1992 (102 Pub. L. 426, 106 Stat. 2174).

3.4.7 *CFR*—Code of Federal Regulations.

3.4.8 *DoD*—Department of Defense.

3.4.9 *EBS*—environmental baseline survey.

3.4.10 *ECP*—environmental condition of property.

3.4.11 *USEPA*—United States Environmental Protection Agency.

3.4.12 *EPCRA*—Emergency Planning and Community Right to Know Act, 42 USC.

3.4.13 *ERNS*—Emergency Response Notification System.

3.4.14 *ESA*—environmental site assessment.

3.4.15 *FOIA*—U.S. Freedom of Information Act (5 USC 552 *et seq.*).

3.4.16 *FOSL*—Finding of Suitability to Lease as described in applicable DoD Policy.

3.4.17 *FOST*—Finding of Suitability to Transfer as described in applicable DoD Policy.

3.4.18 *FR*—Federal Register.

3.4.19 *IRP*—Installation Restoration Program.

3.4.20 *LUST*—leaking underground storage tank.

3.4.21 *MSDS*—material safety data sheet.

3.4.22 *NCP*—National Contingency Plan.

3.4.23 *PA*—preliminary assessment.

3.4.24 *PCBs*—polychlorinated biphenyls.

3.4.25 *RCRA*—The Resource Conservation and Recovery Act, as amended, 42 USC 6901 *et seq.*

3.4.26 *SARA*—Superfund Amendments and Reauthorization Act of 1986.

3.4.27 *SI*—site inspection.

3.4.28 *TSD*—treatment, storage, and disposal.

3.4.29 *USC*—United States Code.

3.4.30 *USGS*—United States Geological Survey.

3.4.31 *UST*—underground storage tank.

4. Significance and Use

4.1 *Uses*—This practice is intended for use by DoD components and environmental professionals in order to facilitate EBS efforts. It is also intended for use by preparers and reviewers of environmental condition of property maps and EBS Reports used to support CERFA uncontaminated property identifications and property suitable for transfer by lease or by deed.

4.2 *Clarifications on Use:*

4.2.1 *Use Not Limited to CERCLA*—This practice is designed to assist the user in developing information about the environmental condition of a property and as such has utility for a wide range of persons, including those who may have no actual or potential CERCLA liability.

4.2.2 *Residential Tenants/Purchasers and Others*—No implication is intended that it is currently customary practice for residential tenants of multifamily residential buildings, tenants of single-family homes or other residential real estate, or purchasers of dwellings for one's own residential use, to conduct an EBS in connection with these transactions. Thus, these transactions are not included in the term commercial real estate transactions. Thus, although such property may be included within the scope of an EBS, their occupants shall not be treated as key site personnel with regard to the housing occupied for the purpose of conducting an EBS.

4.2.3 *Site-Specific*—This practice is site-specific in that it relates to assessment of environmental conditions of federal real property. Consequently, this practice does not address many additional issues raised in transactions such as purchases of business entities; or interests therein, or of their assets, that may well involve environmental liabilities pertaining to properties previously owned or operated or other off-site environmental liabilities.

4.3 *Related Practices*—See Practices E 1527 and E 1528.

4.4 *Principles*—The following principles are an integral part of this practice and all related practices and are intended to be referred to in resolving any ambiguity or exercising such discretion as is accorded the user or environmental professional in performing an EBS or in judging whether a user or environmental professional has conducted appropriate inquiry or has otherwise conducted an adequate EBS.

4.4.1 *Uncertainty Not Eliminated*—No EBS can wholly eliminate uncertainty regarding the potential for recognized environmental conditions in connection with a property. Performance of this practice is intended to reduce uncertainty regarding the potential for recognized environmental conditions in connection with a property to the minimum practicable level, but not eliminate such uncertainty altogether, as well as to recognize reasonable limits of time and cost for property information (see 7.1.3.2).

4.4.2 *Not Exhaustive*—Appropriate inquiry does not mean an exhaustive assessment of an uncontaminated property. There is a point at which the cost of information obtained or the time required to gather it outweighs the usefulness of the information and, in fact, may be a material detriment to the orderly completion of transactions. One of the purposes of this practice is to identify a balance between the competing goals of limiting the costs and time demands inherent in performing an EBS and the reduction of uncertainty about unknown conditions resulting from additional information.

4.4.3 *Level of Inquiry Is Variable*—Not every property will warrant the same level of EBS effort. Consistent with good practice, the appropriate level of EBS will be guided by the type of property subject to EBS and the information developed in its conduct.

4.4.4 *Comparison With Subsequent Inquiry*—It should not be concluded or assumed that an inquiry was not an appropriate inquiry merely because the inquiry did not identify recognized environmental conditions in connection with a property. The EBSs must be evaluated based on the reasonableness of judgments made at the time and under the circumstances in which they were made. Subsequent EBSs should not be considered valid standards to judge the appropriateness of any prior EBS based on hindsight, new information, use of developing technology or analytical techniques, or other factors.

4.5 *Continued Viability of Environmental Baseline Survey*—An EBS meeting or exceeding this practice and completed less than 180 days prior to the date of a subsequent use is presumed to be valid for that use. An EBS not meeting or exceeding this practice or completed more than 180 days previously may be used to the extent allowed by 4.6 through 4.6.5.

4.6 *Prior EBS Usage*—This practice recognizes that EBSs performed in accordance with this practice or otherwise containing information which was reasonably accurate at the time prepared will include information that subsequent users may want to use to avoid undertaking duplicative EBS procedures. Therefore, this practice describes procedures to be followed to assist users in determining the appropriateness of using information in EBSs performed previously. The system of prior EBS usage is based on the following principles that should be adhered to in addition to the specific procedures set forth elsewhere in this practice:

4.6.1 *Use of Prior Information*—Subject to 4.6.4, users and environmental professionals may use information in prior EBSs provided such information was generated as a result of procedures that meet or exceed the requirements of

this practice or accurately state the limitations of the information presented. When using information from an EBS which, as a whole, fails to meet or exceed the requirements of this practice, the use shall be limited to those portions of the EBS which, based upon the limitations and methodology of the EBS Report, the environmental professional finds to be reasonably accurate.

4.6.2 *Prior EBS Meets or Exceeds*—Subject to 4.6.4, a prior EBS may be used in its entirety, without regard to the specific procedures set forth in these practices if, in the reasonable judgment of the user, the prior EBS meets or exceeds the requirements of this practice and the conditions at the property likely to affect environmental condition of property area types in connection with the property are not likely to have changed materially since the prior EBS was conducted. In making this judgment, the user should consider the type of property subject to the EBS and the conditions in the area surrounding the property.

4.6.3 *Current Investigation*—Except as specifically provided in 4.6.2, prior EBSs should not be used without current investigation of conditions likely to affect the environmental condition of property in connection with the property that may have changed materially since the prior EBS was conducted. For an EBS to be consistent with this practice, a new visual inspection, interviews, an update of the records review, and other appropriate activities may have to be performed.

4.6.4 *Actual Knowledge Exception*—If the user or environmental professional(s) conducting an EBS has actual knowledge that the information being used from a prior EBS is not accurate or if it is obvious, based on other information obtained by means of the EBS or known to the person conducting the EBS, that the information being used is not accurate, such information from a prior EBS may not be used.

4.6.5 *Contractual Issues Regarding Prior EBS Usage*—The contractual and legal obligations between prior and subsequent users of EBSs or between environmental professionals who conducted prior EBSs and those who would like to use such prior EBSs are beyond the scope of this practice.

5. **User's Responsibilities**

5.1 *Scope*—This section is limited to the responsibilities of users of this practice. Users may be either DoD component staff or environmental professionals contractually engaged to perform EBSs. Users of this practice should be familiar with its entire contents before conducting or documenting an EBS, and to use best professional judgment regarding its applicability to a particular situation.

5.1.1 *DoD Component Staff*—DoD component staff who have both the requisite specialized knowledge and experience and appropriate training can use this practice as a starting point for conducting or updating EBSs. Although this practice has been designed to help DoD components meet certain legal and policy requirements, it should not be used as a substitute for meeting environmental, BRAC statute, or health and safety legal requirements that exist under various laws, regulations, and DoD and DoD component policies and guidance.

5.1.2 *Environmental Professionals*—Environmental professionals who have both the requisite specialized knowledge and experience, and appropriate training can use this practice as a starting point for conducting or updating EBSs. Although this practice has been designed to help environmental professionals contractually engaged by DoD components to conduct EBSs, in accordance with applicable legal and policy requirements, it should not be used as a substitute for meeting environmental, BRAC statute, or health and safety legal requirements that exist under various laws, regulations, and DoD and DoD component policies and guidance. Contractually engaged environmental professionals should not use this practice to perform tasks that are inherently governmental functions.

5.2 *Specialized Knowledge or Experience of the User*— Users of this practice are expected to have the requisite environmental and health and safety training necessary to conduct the tasks identified in this practice. The DoD components are responsible for identifying appropriate staff for conducting these functions, and are also responsible for contractually ensuring that environmental professionals engaged to perform EBSs have appropriate qualifications. These qualifications should be identified in the contract or scope of work.

6. Environmental Baseline Survey Process

6.1 *Objective*—In accordance with DoD policy, the purpose of the EBS is to determine or discover and to document the obviousness of the presence or likely presence of a release or threatened release of any hazardous substance or petroleum product. In certain cases, additional data, including sampling and analysis, may be needed in the EBS or EBS supplement to support the classification of the property into one of the standard environmental condition of property area types. Additionally, an EBS may also satisfy the uncontaminated property identification requirements of CERFA. Users are cautioned that elements of this practice pertain to an initial EBS conducted by a DoD component as well as to EBS updates, supplemental EBSs, or site-specific EBSs, or a combination thereof, (however termed by the DoD component). As such, it is anticipated that it will only be necessary to complete all steps and tasks identified in this practice for the initial EBS for the property. The user or environmental professional should obtain the input of the DoD component end user regarding the level of effort to be used during any supplemental EBS efforts.

6.2 *Five Steps*—Within the limitations described in 6.1, it is anticipated that the EBS process will commonly consist of at least four and possibly five discrete steps. These are summarized as follows:

6.2.1 *EBS Step 1*—Gathering of data and information in accordance with the process described in the applicable DoD policy referenced in 2.2 and as further elaborated in Sections 7 through 13 of this practice.

6.2.2 *EBS Step 2*—Analysis of data and information in accordance with the process described in Sections 7 through 13 of this practice.

6.2.3 *EBS Step 3*—Determination of the environmental condition of property area type for the real property being evaluated by the EBS, in accordance with the process described in this practice and DoD policy.

6.2.4 *EBS Step 4*—Preparation of an EBS Report in accordance with the format described in the applicable DoD policy.

6.2.5 *EBS Step 5*—Updating and enhancing, as necessary, an EBS Report to support property transfer transactions (for example, FOSLs, FOSTs, or environmental condition reports). This process may require repeating Steps 1 through 3 to incorporate additional information or data, or both, generated between the time an initial EBS report is issued and the time an updated version is used to support a property transfer transaction.

6.3 *Additional Explanation of EBS Steps 1 and 2*—Steps 1 and 2 of the EBS process will consider all sources of available information concerning environmentally significant current and past uses of the real property, and shall, at a minimum, consist of the following eight components, as described in overview as follows (more detailed descriptions of each component are found in Sections 7 through 13):

6.3.1 *Records Search and Review Scope*—Detailed search and review of available information and records in the possession of the DoD components or records made available by the regulatory agencies or other involved federal agencies. Department of Defense (DoD) components are responsible for requesting and making reasonable inquiry into the existence and availability of relevant information and records to include any additional study information (for example, surveys for radioactive materials, asbestos, radon, lead-based paint, drinking water quality, indoor air quality, transformers containing PCBs, RCRA Facility Assessments and Investigations, and Underground Storage Tank Cleanup Program) to help support the determination of the environmental condition of property area type.

6.3.2 *Adjacent Facility Records Search and Review Scope*—Review of all reasonably obtainable federal, state, and local government records for each adjacent facility where there has been a release or likely release of any hazardous substance or any petroleum product, and which is likely to cause or contribute to a release or threatened release of any hazardous substance or any petroleum product on the real property.

6.3.3 *Aerial Photography Analysis*—Analysis of aerial photographs that are in the possession of the federal government or are reasonably obtainable through state or local government agencies that may reflect prior uses of the property.

6.3.4 *Interviews*—Interviews with key current or former employees, or both, involved in operations on the real property.

6.3.5 *Visual Inspections*—Nonintrusive visual inspections of the real property; any buildings, structures, equipment, pipe, pipeline, or other improvements on the real property; and of properties immediately adjacent to the real property, noting sewer lines, runoff patterns, evidence of environmental impacts (for example, stained soil, stressed vegetation, dead or ill wildlife), and other observations which indicate actual or potential release of hazardous substances or petroleum products.

6.3.6 *Contamination Source Identification*—Identification of sources of contamination on the installation and on adjacent properties which could migrate to the real property.

6.3.7 *Ongoing Response Actions*—Ongoing response actions or actions that have been taken at or adjacent to the property will be identified and documented.

6.3.8 *Physical and Visual Inspection of Adjacent Property*—A physical inspection of property adjacent to the real property, to the extent permitted by owners or operators of such property. A visual inspection will be accomplished from areas of public access if a physical inspection is not authorized by the owners or operators of such property.

7. Records Search and Review

7.1 *Introduction*—Reasonable prudence, CERFA requirements (in the case of an EBS performed to support the identification of uncontaminated property), and DoD guidance mandate that the federal real property be evaluated in order to support real property transactions. One component of this evaluation is the review of all reasonably obtainable federal, state, and local government records to determine where, on the installation, there has been storage, release or likely release of any hazardous substance or any petroleum product, and which is likely to cause or contribute to a release or threatened release of any hazardous substance or any petroleum product on the real property.

7.1.1 *Objective*—The objective of the records review is to perform those parts of Steps 1 and 2 of the EBS process pertaining to obtaining and reviewing adequate and complete records that will help the user or environmental professional make an environmental condition of property area type determination regarding the federal real property.

7.1.2 *Accuracy and Completeness*—Accuracy and completeness of record information varies among information sources, including governmental sources. Record information is often inaccurate or incomplete. The user or environmental professional is not obligated to identify mistakes or insufficiencies in information provided. However, the environmental professional reviewing the records shall make a reasonable effort to compensate for mistakes or insufficiencies in the information reviewed that are obvious in light of other information of which the environmental professional has actual knowledge.

7.1.3 *Reasonably Obtainable/Standard Sources*—Availability of record information varies from information source to information source, including governmental jurisdictions. The user or environmental professional is not obligated to identify, obtain, or review every possible record that might exist with respect to a property. Instead, this practice identifies record information that shall be reviewed from standard sources, and the user or environmental professional is required to review only record information that is reasonably ascertainable from those standard sources. Record information that is reasonably obtainable means: (*1*) information that is publicly available, (*2*) information that is obtainable from its source within reasonable time and cost constraints, and (*3*) information that is practically reviewable.

7.1.3.1 *Publicly Available*—Information that is publicly available means that the source of the information allows access to the information by anyone upon request.

7.1.3.2 *Reasonable Time and Cost*—Information that is obtainable within reasonable time and cost constraints means that the information will be provided by the source within a reasonable amount of time of receiving a written, telephone, or in-person request at no more than a nominal cost intended to cover the source's cost of retrieving and duplicating the information. Information that can only be reviewed by a visit to the source is reasonably ascertainable if the visit is permitted by the source within a reasonable amount of time of the request.

7.1.3.3 *Practically Reviewable*—Information that is practically reviewable means that the information is provided by the source in a manner and in a form that, upon examination, yields information relevant to the property without the need for extraordinary analysis of irrelevant data. The form of the information shall be such that the user can review the records for a limited geographic area. Records that cannot be feasibly retrieved by reference to the location of the property or a geographic area in which the property is located are not generally practically reviewable. Most databases of public records are practically reviewable if they can be obtained from the source agency by the county, city, zip code, or other geographic area of the facilities listed in the record system. Records that are sorted, filed, organized, or maintained by the source agency only chronologically are not generally practically reviewable. For large databases with numerous facility records (such as RCRA generators and registered USTs), the records are not practically reviewable unless they can be obtained from the source agency in the smaller geographic area of zip codes. Even when information is provided by zip code for some large databases, it is common for an unmanageable number of sites to be identified within a given zip code. In these cases, it is not necessary to review the impact of all of the sites that are likely to be listed in any given zip code because that information would not be practically reviewable. In other words, when so much data is generated that it cannot be feasibly reviewed for its impact on the property, it is not required to be reviewed.

7.1.4 *Alternatives to Standard Sources*—Alternative sources may be used instead of standard sources if they are of similar or better reliability and detail, or if a standard source is not reasonably ascertainable.

7.1.5 *Coordination*—If records are not reasonably ascertainable from standard sources or alternative sources, the environmental professional shall attempt to obtain the requested information by other means specified in this practice such as questions posed to the current owner or occupant(s) of the property or appropriate persons available at the source at the time of the request.

7.1.6 *Sources of Standard Source Information*—Standard source information or other record information from government agencies may be obtained directly from appropriate government agencies or from commercial services. Government information obtained from nongovernmental sources may be considered current if the source updates the information at least every 90 days or, for information that is updated less frequently than quarterly by the government agency, within 90 days of the date the government agency makes the information available to the public.

7.1.7 *Documentation of Sources Checked*—The EBS report shall document each source that was used, even if a source revealed no findings. Sources shall be sufficiently described, including name, date request for information was filled, date information provided was last updated by source, date information was last updated by original source (if provided other than by original source; see 7.1.4) so as to facilitate reconstruction of the research at a later date.

7.1.8 *Significance*—If a standard environmental record source (or other sources in the course of conducting the EBS) identifies the property or another site within the approximate minimum search distance, the EBS report shall include the environmental professional's judgment about the significance of the listing to the analysis of recognized environmental conditions in connection with the property (based on the data retrieved pursuant to this section, additional information from the government source, or other sources of information). In doing so, the environmental professional may make statements applicable to multiple sites (for example, a statement to the effect that none of the sites listed is likely to have a negative impact on the property except . . .).

7.2 *EBS Step 1: Records Gathering*—In accordance with 6.2.1, this section specifies the general level of effort required to complete EBS Step 1 tasks associated with records gathering. At a minimum, records to be gathered and reviewed when an EBS is initially conducted include the following:

7.2.1 *Background and Physical Setting Records.*

7.2.1.1 *Physical Setting Sources*—A current USGS 7.5 Minute Topographic Map showing the area on which the property is located shall be reviewed, provided it is reasonably obtainable. If a current USGS 7.5 Minute Topographic Map is not readily obtainable, a current 15 Minute Topographic Map showing the area on which the property is located shall be reviewed, provided it is reasonably obtainable. It is the only standard physical setting source and the only physical setting source that is required to be obtained (and only if it is reasonably obtainable). One or more additional physical setting sources may be obtained in the discretion of the environmental professional. Because such sources provide information about the geologic, hydrogeologic, hydrologic, or topographic characteristics of a site, discretionary physical setting sources shall be sought when: (1) conditions have been identified in which hazardous substances or petroleum products are likely to migrate to the property or from or within the property into the ground water or soil and (2) more information than is provided in the current USGS 7.5 Minute Topographic Map is generally obtained, pursuant to local good commercial or customary practice in initial environmental site assessments in the type of commercial real estate transaction involved, in order to assess the impact of such migration on recognized environmental conditions in connection with the property.

Standard Physical Setting Source: Current USGS
7.5 Minute Topographic Map

If this map is unavailable, obtain the USGS 15 Minute Map, if available. If neither the USGS 7.5 Minute Topographic Map nor the 15 Minute Map are available, a larger scale (for example, 1:250 000) USGS topographic map should be considered. Where appropriate, a comparable topographic map, prepared by the Defense Mapping Agency, may be used instead of the USGS 7.5 Minute Topographic Map.

Other Physical Setting Sources:

The following sources may also be utilized if requested:

• USGS and/or State Geological Survey—Groundwater Maps
• USGS and/or State Geological Survey—Bedrock Geology Maps
• USGS and/or State Geological Survey—Surficial Geology Maps
• Soil Conservation Service—Soil Maps
• Other physical setting sources that are reasonably credible (as well as reasonably ascertainable)

7.2.2 Department of Defense (DoD) Component records maintained on the property or elsewhere, but reasonably obtainable, which are relevant to classification of environmental condition of property area types, including, but not limited to records of:

7.2.2.1 Ongoing and completed site remediation and environmental response activities, including IRP activities, corrective action programs, LUST responses, and similar activities. This includes all records in the Administrative Record maintained under CERCLA.

7.2.2.2 Records of reported spills of hazardous substances and responses.

7.2.2.3 Records of hazardous waste accumulation, storage, treatment, or disposal, including satellite accumulation records, manifests, and records maintained in connection with permitted hazardous waste activities.

7.2.2.4 Records of hazardous substance and petroleum usage and/or storage.

7.2.2.5 Records of potential hazard surveys, including, but not limited to asbestos surveys, lead-based paint surveys, radioactive materials surveys, mercury surveys, PCB surveys, and radon surveys.

7.2.2.6 Environmental compliance records not specifically included in other required records. This includes, but is not limited to Safe Drinking Water Act reports, Clean Water Act permits and discharge reports, Clean Air Act permits and discharge and emission reports, EPCRA reports, hazardous waste minimization plans and reports, and pollution prevention plans and reports.

7.2.2.7 Additional records to include planning maps, base historian records, the base comprehensive plan or base master plan, military construction records, real property records, fire department records, historical photographs, and facility and utility records.

7.2.3 Federal, state, and local agency records.

7.2.4 Recorded chain of title documents.

7.3 *EBS Step 2: Records Analysis*—Upon review of the required records gathered to complete EBS Step 1, the user or environmental professional shall indicate in the EBS Report whether the search revealed any of the following on the property:

7.3.1 Spills of hazardous substances or petroleum products, or both,

7.3.2 Leaks of hazardous substances or petroleum products, or both,

7.3.3 Discharges of hazardous substances or petroleum products, or both,

7.3.4 Leaching of hazardous substances or petroleum products, or both,

7.3.5 Injection of hazardous substances or petroleum products, or both,

7.3.6 Dumping of hazardous substances or petroleum products, or both,

7.3.7 Abandoned or discarded barrels, containers, or other,

7.3.8 Receptacles containing hazardous substances or petroleum products,

7.3.9 Automotive batteries,

7.3.10 Industrial batteries,

7.3.11 Pesticides in containers, cartons, sacks, storage bins, canisters,

7.3.12 Paints in containers, cartons, sacks, storage bins, canisters,

7.3.13 Drums containing hazardous substances or petroleum products, or both,

7.3.14 Tanks containing hazardous substances or petroleum products, or both,

7.3.15 Fill dirt from a contaminated site,

7.3.16 Fill pipes,

7.3.17 PCBs in transformers or capacitors,

7.3.18 Heavy industrial equipment, including hydraulic equipment in storage or use,

7.3.19 Ditches subject to contaminated runoff or discharges,

7.3.20 Railroad loading/unloading areas,

7.3.21 Ordinance,

7.3.22 Medical/biohazardous waste,

7.3.23 Radioactive materials and mixed wastes, or

7.3.24 Mercury, for example, seals.

7.4 If the records review revealed any of the above, the EBS Report shall document it/them.

8. Adjacent Facility Records Search and Review

8.1 *Introduction*—Reasonable prudence, CERFA requirements (in the case of an EBS performed to support the identification of uncontaminated property), and DoD guidance mandate that the federal real property be evaluated in order to categorize real property into applicable environmental condition of property area types. One component of this evaluation is the review of all reasonably obtainable federal, state, and local government records for each adjacent facility where there has been a release or likely release of any hazardous substance or any petroleum product, and which is likely to cause or contribute to a release or threatened release of any hazardous substance or any petroleum product on the real property or which might migrate to the federal real property. In this connection, adjacent has the meaning provided in 3.3.1 and includes those properties near enough to the federal real property to present a reasonable probability of affecting the environmental condition of property on the federal real property.

8.1.1 *Objective*—The objective of the adjacent facility records search and review is to perform those parts of Steps 1 and 2 of the EBS process pertaining to identifying, obtaining, and reviewing those reasonably available Federal, State, and local agency records that might disclose information which would affect the environmental condition of property area type determination regarding the federal real property.

8.1.2 *Approximate Minimum Search Distance*—Adjacent facility records pertain not only to facilities adjacent to the federal real property, but also pertain to properties within an additional approximate minimum search distance in order to help assess the likelihood of problems from migrating hazardous substances or petroleum products. When the term approximate minimum search distance includes areas outside the property, it shall be measured from the nearest

property boundary. The term approximate minimum search distance is used instead of radius in order to include irregularly shaped properties.

8.1.2.1 *Reduction of Approximate Minimum Search Distance*—When allowed by 8.2.1.1, the approximate minimum search distance for a particular record may be reduced at the discretion of the environmental professional. Factors to consider in reducing the approximate minimum search distance include: (*1*) the density (for example, urban, rural, or suburban) of the setting in which the property is located; (*2*) the distance that the hazardous substances or petroleum products are likely to migrate based on local geologic or hydrogeologic conditions; and (*3*) other reasonable factors. The justification for each reduction and the approximate minimum search distance actually used for any particular record shall be explained in the EBS report.

8.1.3 *Accuracy and Completeness*—See 7.1.2.

8.1.4 *Reasonably Obtainable/Standard Sources*—See 7.1.3.

8.1.4.1 *Publicly Available*—See 7.1.3.1.

8.1.4.2 *Reasonable Time and Cost*—See 7.1.3.2.

8.1.4.3 *Practically Reviewable*—See 7.1.3.3.

8.1.5 *Alternatives to Standard Sources*—See 7.1.4.

8.1.6 *Coordination*—See 7.1.5.

8.1.7 *Sources of Standard Source Information*—See 7.1.6.

8.1.8 *Documentation of Sources Checked*—See 7.1.7.

8.1.9 *Significance*—See 7.1.8.

8.2 *EBS Step 1: Adjacent Facility Records Gathering*—In accordance with 6.2.1, this section specifies the level of effort required to complete EBS Step 1 tasks associated with adjacent facility records gathering. At a minimum, the following records are to be searched:

8.2.1 *Standard Environmental Sources*—The following standard environmental record sources shall be reviewed, subject to the conditions of 7.1.1 through 7.1.8:

8.2.1.1 *Standard Environmental Record Sources: Federal and State*—The approximate minimum search distance should be established for each installation or portion of an installation, based on the physical setting and surrounding land use. Table 1 includes recommended approximate minimum search distances. An approximate minimum search distance for a particular record may be reduced at the discretion of the environmental professional. Factors to consider in reducing the approximate minimum search distance include: (*1*) the density (for example, urban, rural,

TABLE 1 Recommended Approximate Minimum Search Distances

Record(s) Source	Approximate Minimum Search Distance, miles (kilometres)
Federal NPL site list	1.0 (1.6)
Federal CERCLIS list	0.5 (0.8)
Federal RCRA TSD facilities list	1.0 (1.6)
Federal RCRA generators list	property and adjoining properties
Federal ERNS list	property only
State lists of hazardous waste sites Identified for investigation or remediation (NPL and CERCLIS equivalents)	1.0 (1.6)
State landfill and/or solid waste disposal site lists	0.5 (0.8)
State leaking UST lists	0.5 (0.8)
State registered UST lists	property and adjoining properties

or suburban) of the setting in which the property is located; (2) the distance that the hazardous substances or petroleum products are likely to migrate based on local geologic or hydrogeologic conditions; and (3) other reasonable factors. The justification for each reduction and the approximate minimum search distance actually used for any particular record should be explained in the EBS report.

8.2.2 *Additional Environmental Record Sources: State or Local*—One or more additional state sources or local sources of environmental records may be checked, at the discretion of the environmental professional, to enhance and supplement federal and state sources identified in Table 1. Factors to consider in determining which local or additional state records, if any, should be checked include: (1) whether they are reasonably ascertainable, (2) whether they are sufficiently useful, accurate, and complete in light of the objective of the records review (see 7.1.1), and (3) whether they are generally obtained, pursuant to local good commercial or customary practice, in initial environmental site assessments in the type of commercial real estate transaction involved. To the extent additional state sources or local sources are used to supplement the same record types listed in Table 1, approximate minimum search distances should not be less than those specified (adjusted as provided in 8.1.2.1 and 8.2.1.1). Some types of records and sources that may be useful include:

Types of Local Records
Lists of Landfill/Solid Waste Disposal Sites
Lists of Hazardous Waste/Contaminated Sites
Lists of Registered Underground Storage Tanks
Records of Emergency Release Reports (SARA Part 304)
Records of Contaminated Public Wells

Local Sources
Department of Health/Environmental Division
Fire Department
Planning Department
Building Permit/Inspection Department
Local/Regional Pollution Control Agency
Local/Regional Water Quality Agency
Local Electric Utility Companies (for records relating to PCBs)

8.2.3 *Physical Setting Sources*—See 7.2.1.

8.3 *EBS Step 2: Adjacent Facility Records Analysis*—Upon review of adjacent facility records listed in 8.2, the environmental professional shall indicate in the EBS report whether or not the search revealed any of the following on the property:

8.3.1 Potential or actual migration of hazardous substances or petroleum products, or both, into the area in question from sources of these substances.

8.3.2 The presence of actual sources of hazardous substances or petroleum products, or both, on adjacent property and facilities with suspected migration that has not been evaluated or characterized.

8.3.3 Uncontrolled migration of hazardous substances or petroleum products, or both, in the immediate vicinity of a boundary of the area in question.

8.4 If the records review revealed any of the preceding, the EBS report shall document it/them.

9. Aerial Photography Analysis

9.1 *Introduction*—Analysis of aerial photography can provide an extremely useful source of supplemental information

regarding both land use and the environmental condition of property area type. This analysis may encompass both the federal real property and the adjacent facilities. Analysis of aerial photography should be focused on patterns of land use and human activities as well as direct and indirect evidence of the potential existence of a recognized environmental condition.

9.1.1 *Objective*—The objective of the aerial photography analysis is to perform those parts of Steps 1 and 2 of the EBS process pertaining to assembling, if reasonably available, an adequately complete set of adjacent properties encompassing both the federal real property and the adjacent facilities, analyzing those photographs for patterns of land use and human activities as well as direct and indirect evidence of the potential existence of a recognized environmental condition, and thereafter, incorporating that information into the overall environmental condition of property area type. Aerial photographs will be sought for the period encompassing the past 60 years of facility use.

9.2 *EBS Step 1: Aerial Photography Gathering*—In accordance with 6.2.1, this section specifies the level of effort required to complete EBS Step 1 tasks associated with identifying and gathering adjacent properties. At a minimum, the following sources are to be searched:

9.2.1 *Standard Sources: Aerial Photographs*—Aerial photographs are commonly available from government agencies or private collections unique to a local area, and may also be obtained from universities, colleges, and history museums.

9.2.2 *Optional Sources*—Other sources of similar information, including, but not limited to satellite reconnaissance may be sought where the environmental professional feels that these records would be reasonably available and would materially enhance the analysis required in this section.

9.3 *EBS Step 2: Aerial Photography Analysis*—The environmental professional shall perform the analysis of photographs using an accepted photo interpretation land use and land cover classification system. Analyzed photographs should preferably be at a 1:24 000 scale or smaller scale (that is, more magnified) to reveal adequate surface detail and necessary spatial coverage. Upon review of adjacent properties in accordance with the previously stated guidelines, the environmental professional shall specify the analysis method used in the EBS Report and also indicate whether any of the following were identified on the property:

9.3.1 Evidence of excavation activities of unknown type, or of industrial operations.

9.3.2 Evidence of dumping or disposing of waste materials.

9.3.3 Evidence of significant storage activities involving drums, tanks, or pipelines containing hazardous substances or petroleum products.

9.3.4 Evidence of staining associated with industrial activities or activities of unknown origin or type.

9.4 If the aerial photography analysis revealed any of the preceding, the EBS report shall document it/them.

10. Interviews

10.1 *Introduction*—Reasonable prudence, CERFA requirements (in the case of an EBS performed to support the identification of uncontaminated property), and DoD guidance mandate that the federal real property be evaluated in order to categorize real property into applicable environmental condition of property area types. One component of this evaluation are interviews of current or former occupants, or both, involved in operations on the property that are conducted to aid in identifying a recognized environmental condition on the federal real property and other information necessary to determine standard environmental condition of property area types. Interviews of current or former personnel, or both, on the federal real property and interviews of appropriate local government officials supplement documented information and may also provide keys to effective interpretation of such information. Where available information indicates the presence of recognized environmental conditions, it may be necessary to conduct interviews relating to those portions of the installation that have been, or are currently being, actively investigated under the IRP or similar investigation effort.

10.1.1 *Objective*—The objective of conducting interviews is to perform those parts of Steps 1 and 2 of the EBS process pertaining to obtaining information indicating recognized environmental conditions in connection with the property, so that environmental condition of property area type determinations can be made.

10.2 *EBS Step 1: Interviews with Site Personnel*—In accordance with 6.2.1, this section specifies the level of effort required to complete EBS Step 1 tasks associated with interviewing current or former personnel, or both, and appropriate local governmental officials. At a minimum, the following should be incorporated into this process:

10.2.1 *Content*—Interviews with site personnel consist of questions to be asked in the manner and of persons as described in this section. The content of questions to be asked shall attempt to obtain information about uses and conditions as described in Section 9, as well as the information described in Sections 11 through 13.

10.2.2 *Medium*—Questions to be asked pursuant to this section may be asked in person, by telephone, or in writing, at the discretion of the environmental professional.

10.2.3 *Timing*—Except as specified in 9.8 and 9.9, it is at the discretion of the environmental professional whether to ask questions before, during, or after the site visit described in Section 11, or in some combination thereof.

10.2.4 *Who Should be Interviewed:*

10.2.4.1 *Key Site Manager*—Prior to the site visit, the owner can be asked to identify a person with good knowledge of the uses and physical characteristics of the property (the key site manager). Often the key site manager will be the installation or base commander, base civil engineer, public works commander or other property manager, chief physical plant supervisor or head maintenance person. (If the user is the current property owner, the user has an obligation to identify a key site manager, even if it is the user himself or herself.) If a key site manager is identified, the person conducting the site visit shall make at least one reasonable attempt (in writing or by telephone) to arrange a mutually convenient appointment for the site visit when the key site manager agrees to be there. If the attempt is successful, the key site manager shall be interviewed in conjunction with the site visit. If such an attempt is unsuccessful, when conducting the site visit, the environmental professional shall

inquire whether an identified key site manager (if any) or if a person with good knowledge of the uses and physical characteristics of the property is available to be interviewed at that time; if so, that person shall be interviewed. In any case, it is within the discretion of the environmental professional to decide which questions to ask before, during, or after the site visit or in some combination thereof.

10.2.4.2 *Occupants, Including Current and Former Employees*—A reasonable attempt shall be made to interview a reasonable number of occupants of, if any, and current employees involved in operations on the property.

(1) *Residential Properties*—For residential properties, residential occupants do not need to be interviewed, but if the property has nonresidential uses, interviews can be held with the nonresidential occupants based on criteria specified in 10.2.4.2.

(2) *Major Occupants*—Except as specified in *residential properties*, if the property has five or fewer current occupants, a reasonable attempt shall be made to interview a representative of each one of them. If there are more than five current occupants, a reasonable attempt shall be made to interview the major occupant(s) and those other occupants whose operations are likely to indicate recognized environmental conditions in connection with the property.

(3) *Reasonable Attempts to Interview*—Examples of reasonable attempts to interview those occupants specified in *major occupants* include (but are not limited to) an attempt to interview such occupants when making the site visit or calling such occupants by telephone. In any case, when there are several occupants to interview, it is not expected that the site visit must be scheduled at a time when they will all be available to be interviewed.

(4) *Occupant Identification*—The EBS report shall identify the occupants interviewed and the duration of their occupancy.

10.2.4.3 *Local Agency Officials*—A reasonable attempt shall be made to interview at least one staff member of any one of the following types of local government agencies:

(1) Local fire department that serves the property,

(2) Local health agency or local/regional office of state health agency serving the area in which the property is located, or

(3) Local agency or local/regional office of state agency having jurisdiction over hazardous waste disposal or other environmental matters in the area in which the property is located.

10.2.5 *Prior Assessment Usage*—Persons interviewed as part of a prior EBS consistent with this practice do not need to be questioned again about the content of answers they provided at that time. However, they can be questioned about any new information learned since that time, or others can be questioned about conditions since the prior EBS consistent with this practice.

10.2.6 *Quality Of Answers*—The person(s) interviewed should be asked to be as specific as reasonably feasible in answering questions. The person(s) interviewed should be asked to answer in good faith and to the extent of their knowledge.

10.2.7 *Incomplete Answers*—In accordance with 10.2.7.1 and 10.2.7.2, the person conducting the interview(s) has an obligation to ask questions, in certain instances the persons to whom the questions are addressed may not have an obligation to answer them.

10.2.7.1 *User/DoD Component Personnel*—If the person to be interviewed is the user (an employee of a DoD component on whose behalf the EBS is being conducted), the user has an obligation to answer all questions posed by the person conducting the interview, in good faith, to the extent of his or her actual knowledge, or to designate a key site manager to do so. If answers to questions are unknown or partially unknown to the user or such key site manager, this interview section of the EBS shall not thereby be deemed incomplete.

10.2.7.2 *Non-User*—If the person conducting the interview(s) asks questions of a person other than a user but does not receive answers or receives partial answers, this section of the EBS shall not thereby be deemed incomplete, provided that (*1*) the questions have been asked (or attempted to be asked) in person or by telephone and written records have been kept of the person to whom the questions were addressed and the responses, or (*2*) the questions have been asked in writing sent by first class mail or by private, commercial carrier and no answer or incomplete answers have been obtained and at least one reasonable follow-up (telephone call or written request) was made again asking for responses.

10.2.8 *Questions About Helpful Documents*—Prior to the site visit, the property owner, key site manager (if any is identified), and user (if different from the property owner) shall be asked if they know whether any of the documents listed in 10.2.8.1 exist and, if so, whether copies can and will be provided to the environmental professional within reasonable time and cost constraints. Even partial information provided may be useful. If so, the environmental professional conducting the site visit shall review the available documents prior to or at the beginning of the site visit.

10.2.8.1 *Helpful Documents:*

(*1*) Environment site assessment reports, including PA, SI, or other similar reports.

(*2*) Environment audit reports.

(*3*) Environmental permits (for example, solid waste disposal permits, hazardous waste disposal permits, waste-water permits, NPDES permits).

(*4*) Registrations for underground and above-ground storage tanks.

(*5*) Material safety data sheets.

(*6*) Community right-to-know plan.

(*7*) Safety plans; preparedness and prevention plans; spill prevention, countermeasure, control plans, and so forth.

(*8*) Reports regarding hydrogeologic conditions of the property or surrounding area.

(*9*) Notices or other correspondence from any government agency relating to past or current violations of environmental laws with respect to the property or relating to environmental liens encumbering the property.

(*10*) Hazardous waste—generator notices or reports.

(*11*) Geotechnical studies.

10.2.9 *Proceedings Involving the Property*—Prior to the site visit, the property owner, key site manager (if any is identified), and user (if different from the property owner) shall be asked whether they know of: (*1*) any pending, threatened, or past litigation relevant to hazardous sub-

stances or petroleum products in, on, or from the property, (*2*) any pending, threatened, or past administrative proceedings relevant to hazardous substances or petroleum products in, on, or from the property, and (*3*) any notices from any governmental entity regarding any possible violation of environmental laws or possible liability relating to hazardous substances or petroleum products.

10.2.10 *Interview Questions*—The person conducting each interview shall, at a minimum, ask and document responses to the following questions:

10.2.10.1 Was or is the area in question used as a gasoline station, motor repair facility, dry cleaners, photo developing laboratory, plating shop, medical or dental facility, junkyard or landfill, training area, or as a waste treatment, storage, disposal, processing, or recycling facility?

10.2.10.2 Has there been any damaged or discarded automotive or industrial batteries, or pesticides, paints, or other chemical or individual containers stored or used in the area in question?

10.2.10.3 Are there drums, sacks, cartons, or other containers of chemicals located on the property in question?

10.2.10.4 Was or is the area in question used for any waste generation or disposal activities?

10.2.10.5 Was or is the area in question used as a firing or bombing range, or both?

10.2.10.6 Have there been or are there storage tanks containing hazardous substances or petroleum products located on the property in question?

10.2.10.7 Have spills, leaks, or other releases of hazardous substance or petroleum products occurred to the best of your knowledge?

10.2.10.8 Have unidentified waste materials, tires, automotive or industrial batteries, ordnance or any other waste materials been dumped, buried, or burned, or a combination thereof, in the area in question?

10.3 *EBS Step 2: Interview Analysis*—Upon completion of Step 1 of the EBS process related to interviews, the environmental professional shall refer to 7.3 and indicate in the EBS report whether or not the responses revealed any of the items in 7.3 on the property. If the responses revealed any of these items, the EBS report shall document it/them.

11. Visual and Physical Inspections

11.1 *Introduction*—Reasonable prudence, CERFA requirements (in the case of an EBS performed to support the identification of uncontaminated property), and DoD guidance mandate that the federal real property be evaluated in order to categorize real property into applicable environmental condition of property area types. One element of this evaluation is the requirement that visual inspections be conducted of the installation property and adjacent property. The visual inspection supplements the documentary record, including interviews, records developed during records search, aerial photography analysis, and the other components of the EBS. The terms "visual inspection" and "visual site inspection" as used in this practice are synonymous. For the purpose of this practice, visual inspection includes two similar, but distinct inspections, the visual inspection of the federal real property and the visual inspection of the adjacent property. The visual inspection of the federal real property will be conducted as described in 11.3. The visual and physical inspection of the adjacent property will be conducted as described in 11.4. The environmental professional and the user should exercise reasonable prudence in scoping and executing the tasks described in this section, recognizing that it is not the purpose of the EBS visual inspection to take the place of other regulatory program requirements, for example, asbestos surveys.

11.2 *Objective*—The objective of the visual inspection is to perform those parts of Steps 1 and 2 of the EBS process pertaining to visually obtaining information indicating the likelihood of recognized environmental conditions in connection with the property, so that environmental condition of property area type determinations can be made.

11.3 *EBS Step 1: Visual Inspection*—In accordance with 5.2.1, this section specifies the level of effort required to complete EBS Step 1 tasks associated with making a visual inspection of the installation property. At a minimum, the following should be incorporated into this process:

11.3.1 *Observation*—On a visit to the property (the site visit), the environmental professional shall visually and physically observe the property and any structure(s) located on the property to the extent not obstructed by bodies of water, adjacent buildings, or other obstacles. The environmental professional shall also prepare a visual inspection record of the visit. The visual inspection record, which can be field notes and the like, will be used to prepare the EBS report preserved to supplement the EBS report.

11.3.1.1 *Exterior*—The periphery of the property shall be visually and physically observed, as well as the periphery of all structures on the property, and the property should be viewed from all adjacent public thoroughfares. If roads or paths with no apparent outlet are observed on the property, the use of the road or path should be identified to determine whether it was likely to have been used as an avenue for disposal of hazardous substances or petroleum products. For each road without an apparent outlet, the environmental professional shall travel or overfly the length of the road and observe whether areas adjacent to the road appear to have been used for waste disposal.

11.3.1.2 *Interior*—On the interior of structures on the property, accessible common areas expected to be used by occupants or the public (such as lobbies, hallways, utility rooms, recreation areas, and so forth), maintenance and repair areas, including boiler rooms, and a representative sample of occupant spaces, should be visually and physically observed. It is generally not necessary to look under floors, above ceilings, or behind walls. The environmental professional shall exercise best professional judgment and appropriately coordinate with the DoD component when observing the interior of structures on the property.

11.3.1.3 *Methodology*—The environmental professional shall document, in the EBS report, the method used (for example, grid patterns or other systematic approaches used for large properties, which spaces for owner or occupants were observed) to observe the property. For example, representative inspections may be appropriate for structures: (*1*) that were built using like construction methods, during a fixed time period, (*2*) are geographically co-located, (*3*) were subjected to similar categorical use throughout their history, and (*4*) have no significant differences in their other investigations criteria (for example, records search, interviews, and

aerial photos). Common examples would include: military housing units, barracks, officer and enlisted quarters. In consultation with the user, the environmental professional may determine the percentage of structures/properties to be inspected and establish an inspection pattern that provides sufficient representation of the area in question.

11.3.1.4 *Limitations*—The environmental professional shall document, in the EBS report, general limitations and bases of review, including limitations imposed by physical obstructions such as adjacent buildings, bodies of water, asphalt, or other paved areas, and limiting conditions (for example, snow and rain).

11.3.1.5 *Frequency*—It is not expected that more than one visit to the property shall be made by the environmental professional in connection with an EBS. The one visit constituting part of the EBS may be referred to as the site visit.

11.3.2 *Prior EBS Usage*—The information supplied in connection with the visual inspection portion of a prior EBS may be used for guidance but shall not be relied upon without determining through a new visual inspection whether any conditions that are material to recognized environmental conditions in connection with the property have changed since the prior EBS.

11.3.3 *Uses and Conditions*—The environmental professional(s) conducting the visual inspection should note the uses and conditions specified in 11.3.3.1 through 11.3.3.4 to the extent visually or physically observed during the site visit. The uses and conditions specified in 11.3.3.1 through 11.3.3.4 should also be the subject of questions asked as part of interviews of owners and occupants (see Section 10). Uses and conditions to be noted shall be recorded in field notes of the environmental professional(s) conducting the visual inspection but are only required to be described in the EBS report to the extent specified in 11.3.3.1 through 11.3.3.4. The environmental professional(s) performing the EBS are obligated to identify uses and conditions only to the extent that they may be visually and physically observed on a site visit, as described in this practice, or to the extent that they are identified by the interviews (see Sections 9 and 10) or records search (see Section 7) processes described in this practice.

11.3.3.1 *General Site Setting:*

(*1*) *Current Use(s) of the Property*—The current use(s) of the property shall be identified in the EBS report. Any current uses likely to involve the use, treatment, storage, disposal, or generation of hazardous substances or petroleum products shall be identified in the EBS report. Unoccupied occupant spaces should be noted. In identifying current uses of the property, more specific information is more helpful than less specific information. (For example, it is more useful to identify uses such as a commissary or base exchange rather than simply retail use.)

(*2*) *Past Use(s) of the Property*—To the extent that indications of past uses of the property are visually or physically observed on the site visit, or are identified in the interviews or record review, they shall be identified in the EBS report, and past uses so identified shall be described in the EBS report if they are likely to have involved the use, treatment, storage, disposal, or generation of hazardous substances or petroleum products. (For example, there may be signs indicating a past use or a structure indicating a past use.)

(3) *Geologic, Hydrogeologic, Hydrologic, and Topographic Conditions*—The topographic conditions of the property shall be noted to the extent visually or physically observed or determined from interviews, as well as the general topography of the area surrounding the property that is visually or physically observed from the periphery of the property. If any information obtained shows there are likely to be hazardous substances or petroleum products on the property or on nearby properties and those hazardous substances or petroleum products are of a type that may migrate, topographic observations shall be analyzed in connection with geologic, hydrogeologic, hydrologic, and topographic information obtained pursuant to records review (see 7.2.3) and interviews to evaluate whether hazardous substances or petroleum products are likely to migrate to the property, or within or from the property, into ground water or soil.

(4) *General Description of Structures*—Generally, the EBS report shall describe the structures or other improvements on the property for example: number of buildings, number of stories each, approximate age of buildings, ancillary structures (if any), and so forth.

(5) *Thoroughfares*—Public thoroughfares adjoining the property shall be identified in the EBS Report, and any roads, streets, railroads, and parking facilities on the property shall be described in the EBS Report.

(6) *Potable Water Supply*—The source of potable water for the property shall be identified in the EBS report.

(7) *Sewage Disposal System*—The sewage disposal system for the property shall be identified in the EBS report. Inquiry shall be made as to the age of the system as part of the process under Sections 7, 9, or 10.

(8) *Storm Drains*—The storm drains for the property shall be identified in the EBS report.

(9) *Access Vaults*—Access vaults occurring on the property shall be identified in the EBS report.

11.3.3.2 *Interior and Exterior Observations:*

(1) *Current Use(s) of the Property*—The current use(s) of the property shall be identified in the EBS report. Any current uses likely to involve the use, treatment, storage, disposal, or generation of hazardous substances or petroleum products shall be identified in the EBS Report. Unoccupied occupant spaces should be noted. In identifying current uses of the property, more specific information is more helpful than less specific information. (For example, it is more useful to identify uses such as a commissary or base exchange rather than simply retail use.)

(2) *Past Use(s) of the Property*—To the extent that indications of past uses of the property are visually or physically observed on the site visit, or are identified in the interviews or records review, they shall be identified in the EBS report, and past uses so identified shall be described in the EBS report if they are likely to have involved the use, treatment, storage, disposal, or generation of hazardous substances or petroleum products. (For example, there may be signs indicating a past use or a structure indicating a past use.)

(3) *Hazardous Substances and Petroleum Products in Connection with Identified Uses*—To the extent that present uses are identified that use, treat, store, dispose of, or generate hazardous substances and petroleum products on the property. The hazardous substances and petroleum products shall be identified or indicated as unidentified in the EBS report, and the approximate quantities involved, types of containers (if any) and storage conditions shall be described in the EBS report. To the extent that past uses are identified that used, treated, stored, disposed of, or generated hazardous substances and petroleum products on the property, the information shall be identified to the extent it is visually or physically observed during the site visit or identified from the interviews or the records review.

(4) *Storage Tanks*—Above-ground storage tanks, or solvents or vent pipes, fill pipes or access ways indicating solvents shall be identified (for example, content, capacity, and age) to the extent visually or physically observed during the site visit or identified from the interviews or records review.

(5) *Odors*—Strong, pungent, or noxious odors shall be described in the EBS report and their sources shall be identified in the EBS report to the extent visually or physically observed or identified from the interviews or records review.

(6) *Pools of Liquid*—Standing surface water (other than common rain puddles) shall be noted. Pools or sumps containing liquids likely to be hazardous substances or petroleum products shall be described in the EBS report to the extent visually or physically observed or identified from the interviews or records review.

(7) *Drums*—To the extent visually or physically observed or identified from the interviews or records review, drums shall be described in the EBS report, whether or not they are leaking, unless it is known that their contents are not hazardous substances or petroleum products (in that case the contents should be described in the EBS report). Drums commonly hold 55 g (208 L) of liquid, but containers as small as 5 g (19 L) should also be described.

(8) *Hazardous Substance and Petroleum Products Containers (Not Necessarily in Connection With Identified Uses)*—When containers identified as containing hazardous substances or petroleum products are visually or physically observed on the property and are or might be a recognized environmental condition: the hazardous substances or petroleum products shall be identified or indicated as "unidentified" in the EBS report, and the approximate quantities involved, types of containers (for example, cartons, tanks, cans), and storage conditions shall be described in the EBS report.

(9) *Unidentified Substance Containers*—When open or damaged containers containing unidentified substances suspected of being hazardous substances or petroleum products are visually or physically observed on the property, the approximate quantities involved, types of containers, and storage conditions shall be described in the EBS report.

(10) *PCBs*—Electrical or hydraulic equipment known to contain PCBs or likely to contain PCBs shall be described in the EBS report to the extent visually or physically observed or identified from the interviews or records review. Fluorescent light ballast likely to contain PCBs does not need to be noted.

11.3.3.3 *Interior Observations:*

(1) *Heating/Cooling*—The means of heating and cooling

the buildings on the property, including the fuel source for heating and cooling, shall be identified in the EBS report (for example, heating oil, gas, electric, radiators from steam boiler fueled by gas).

(2) *Stains or Corrosion*—To the extent visually or physically observed or identified from the interviews, stains or corrosion on floors, walls, or ceilings shall be described in the EBS report, except for staining from water.

(3) *Drains and Sumps*—To the extent visually or physically observed or identified from the interviews, floor drains and sumps shall be identified in the EBS report. Floor drains and sumps where a release or suspected release occurred should be further described in the EBS report.

11.3.3.4 *Exterior Observations:*

(1) *Pits, Ponds, or Lagoons*—To the extent visually or physically observed or identified from the interviews or records review, pits, ponds, or lagoons on the property shall be identified in the EBS report, and further described if they have been used in connection with waste disposal or waste treatment. Pits, ponds, or lagoons on properties adjoining the property shall be described in the EBS report to the extent they are visually or physically observed from the property or identified in the interviews or records review.

(2) *Stained Soil or Pavement*—To the extent visually or physically observed or identified from the interviews, areas of stained soil or pavement shall be described in the EBS report.

(3) *Stressed Vegetation*—To the extent visually or physically observed or identified from the interviews, areas of stressed vegetation (from something other than insufficient water) shall be described in the EBS report.

(4) *Solid Waste*—To the extent visually or physically observed or identified from the interviews or records review, areas that are apparently filled or graded by nonnatural causes (or filled by fill of unknown origin) suggesting trash or other solid waste disposal, or mounds or depressions suggesting trash or other solid waste disposal, shall be described in the EBS report.

(5) *Waste Water*—To the extent visually or physically observed or identified from the interviews or records review, waste water or other liquid (including storm water) or any discharge into a drain, ditch, or stream on or adjacent to the property shall be described in the EBS report.

(6) *Wells*—To the extent visually or physically observed or identified from the interviews or records review, all wells (including dry wells, irrigation wells, injection wells, abandoned wells, or other wells) shall be described in the EBS report.

(7) *Septic Systems*—To the extent visually or physically observed or identified from the interviews or records review, indications of on-site septic systems or cesspools should be described in the EBS report.

11.4 *Visual Inspection of Adjacent Property*—The visual inspection of adjacent property will be accomplished when consent has been obtained to enter the adjacent facility. The user or environmental professional shall visually and physically observe the adjacent property to evaluate and identify, if possible, conditions which could give rise to recognized environmental conditions on the property, through migration or other transport of petroleum products or hazardous substances.

11.4.1 *Consent of Owner/Operator of Adjacent Property*—

Access to the adjacent property is required to undertake a visual and physical inspection of the adjacent property. Written consent will be sought by either the DoD Component or the environmental professional. Where voluntary consent is obtained, a visual and a physical inspection will be conducted, as described in this section. If consent is not given, a visual inspection will be conducted from outside the facility boundaries, for example, from public rights of way, or adjacent facilities where consent has been given and from other suitable sites. If voluntary consent is not obtained, no direct access to the adjacent property will be sought through involuntary means, for example, condemnation of an easement or right of entry or obtaining an administrative warrant.

11.4.2 *General Site Setting:*

11.4.2.1 *Current Uses of Adjacent Properties*—To the extent that current uses of adjacent properties are visually or physically observed on the site visit, or are identified in the interviews or records review, they shall be identified in the EBS report, and current uses so identified shall be described in the EBS report if they are likely to indicate recognized environmental conditions in connection with the adjoining properties or the property.

11.4.2.2 *Past Uses of Adjacent Properties*—To the extent that indications of past uses of adjacent properties are visually or physically observed on the site visit, or are identified in the interviews or record review, they shall be noted by the environmental professional, and past uses so identified shall be described in the EBS report if they are likely to indicate recognized environmental conditions in connection with the adjacent properties or the property.

11.4.3 *Current or Past Uses in the Surrounding Area*—To the extent that the general type of current or past uses (for example, residential, commercial, industrial) of properties surrounding the property are visually or physically observed on the site visit or going to or from the property for the site visit, or are identified in the interviews or record review, they shall be noted by the environmental professional, and uses so identified shall be described in the EBS report if they are likely to indicate recognized environmental conditions in connection with the property.

11.5 *EBS Step 2: Visual Inspection Analysis*—If the visual inspection revealed any conditions indicating the storage, release, or disposal of hazardous substances or petroleum products, the EBS report shall document them.

12. Contamination Source Identification

12.1 *Introduction*—The identification of sources of actual or potential contamination is required by published DoD policy statements relating to each of the actions that require an EBS, particularly the DoD FOST guidance. It is anticipated that, in the ordinary course of events, all relevant records to accomplish this task will be obtained through the records search and review tasks, interview tasks, site visit tasks, and other data collection efforts prescribed in this practice. The principal additional goal of this section is to fulfill Step 1 and Step 2 EBS requirements, by analyzing information and data, where necessary to specifically identify sources of actual or potential contamination (recognized environmental conditions), leading to the identification of an environmental condition of property area type.

12.2 *Objective*—Identify sources of actual or potential contamination so as to evaluate whether they affect or may affect the categorization of the environmental condition of the property, including whether all required remedial action has been taken.

12.3 *EBS Step 1: Contamination Source Identification*—In accordance with 6.2.1, this section specifies the level of effort required to complete EBS Step 1 tasks associated with contamination source identification. At a minimum, this step should be completed by following the procedures described in Sections 7 through 11. If current records are incomplete regarding the nature and extent of contaminant sources, the user or environmental professional will note the status of records search efforts and which records remain incomplete.

12.4 *EBS Step 2: Contamination Source Identification Analysis*—If records are complete for a contaminant source, its nature and extent will be used to determine the environmental condition of the real property. If current records are incomplete regarding the nature and extent of contaminant sources, the user or environmental professional will note the strategy for completing relevant records so that a determination other than environmental condition of property area Type 7 can be made (for example, "contaminant source currently being characterized through remedial investigation; anticipated completion in month/year").

13. Ongoing Response Actions

13.1 *Introduction*—The identification of ongoing response actions is required by published DoD policy statements relating to each of the actions that require an EBS, particularly the DoD FOST guidance. It is anticipated that, in the ordinary course of events, all relevant records to accomplish this task will be obtained through the records search and review tasks, interview tasks, site visit tasks, and other data collection efforts prescribed in this practice. The principal additional goal of this section is to fulfill Step 1 and Step 2 EBS requirements, by analyzing information and data, where necessary to specifically identify ongoing response actions and determine relevant information regarding effectiveness and completeness of any ongoing response actions, so that an accurate environmental condition of property area type determination can be made.

13.2 *Objective*—Identify ongoing response actions and determine what portion of the property is affected and whether the environmental condition of the property satisfies the requirements of CERCLA § 120(h)(3), including whether all required remedial action has been taken.

13.3 *EBS Step 1: Ongoing Response Actions*—In accordance with 6.2.1, this section specifies the level of effort required to complete EBS Step 1 tasks associated with identification of ongoing response actions at the installation or on properties that would affect environmental condition of property area type determinations for all or parts of the installation. At a minimum, this step should be completed by following the procedures described in Sections 7 through 11. If current records are incomplete regarding ongoing response actions, the user or environmental professional will note the status of records search efforts and which records remain incomplete.

13.4 *EBS Step 2: IRP and Other Ongoing Response Action Analysis*—The purpose of the Step 2 analysis is to support the objective of 13.2. This analysis can be critical to the differentiation between or among certain environmental condition of property area types (for example, between area Type 5 areas, which are not transferrable by deed, and area Type 4 areas, which are), as performed as EBS Step 3.

14. Determining Environmental Condition of Property Area Type

14.1 *Introduction*—The user or environmental professional will generally take information obtained from the activities described in Sections 7 through 13 and classify federal real property into the seven standard environmental condition of property area types identified in the Standard Classification. Steps 1 and 2 of the EBS process are intended to collect and analyze information that will provide the basis for assigning, in EBS Step 3, an environmental condition of property (ECP) area type for each portion of the installation. The actual classification of any portion of installation property is based on both the information and analyses at hand and on a variety of site-specific conditions (for example, geologic conditions, hydrologic conditions, nature, and extent of contamination, and so forth) and, with the exception of the general guidelines presented in this section, is beyond the scope of this practice. Ideally, the classification of any property into an ECP area type should reflect a consensus among the Military Department and the federal and state regulators. In general, Steps 1 and 2 of the EBS process will have accomplished their purpose if they provide all information and analyses required to make accurate Step 3 ECP area type determinations.

14.2 *Objective*—The objective of this section is to identify a general process to guide the EBS Step 3 classification of areas of federal real property into one of seven standard ECP area types. The ECP area type classification is a tool that is intended to provide an EBS user with a complete and accurate "snapshot" of relevant aspects of the environmental condition of installation property in support of property transfer and reuse decisions (see also Section 1).

14.3 *General Process for EBS Step 3—Determining ECP Area Types:*

14.3.1 *Process for Determining ECP Area Type 1 Property*—Complete Steps 1 and 2 of the EBS process (see 6.2.1 and 6.2.2). Identify ECP area Type 1 property in accordance with CERFA criteria (see 1.1.3). The ECP area Type 1 property may be identified as either part of the CERFA process or subsequent to the CERFA statutory deadline.

14.3.2 *Process for Determining ECP Area Type 2 Property*—Complete Steps 1 and 2 of the EBS process (see 6.2.1 and 6.2.2). Identify ECP area Type 2 property in accordance with CERFA criteria (see 1.1.3). Identify real property where only storage for less than one year occurred (CERFA uncontaminated) and where only storage occurred for more than one year. Real property may be proposed as ECP area Type 2 through the CERFA process or subsequent to the CERFA statutory deadline.

14.3.3 *Process for Determining ECP Area Type 3 Property*—Complete Steps 1 and 2 of the EBS process (see 6.2.1 and 6.2.2). If a determination can be made that concentrations of hazardous substances or petroleum products are below action levels, in accordance with the criteria contained in the Standard Classification, the real property may be classified as ECP area Type 3. The ECP area Type 3 real property may be identified at any time after completion of Steps 1 and 2 of the EBS process.

14.3.4 *Process for Determining ECP Area Type 4 Property*—Complete Steps 1 and 2 of the EBS process (see 6.2.1 and 6.2.2). If the determination can be made that all required remedial actions have been taken, in accordance with CERCLA and the criteria contained in the Standard Classification, the real property may be classified as ECP area Type 4. The ECP area Type 4 real property may be identified at any time after completion of Steps 1 and 2 of the EBS process.

14.3.5 *Process for Determining ECP Area Type 5 Property*—Complete Steps 1 and 2 of the EBS process (see 6.2.1 and 6.2.2). If the determination can be made that a remedy has been selected but that all required remedial actions have not yet been taken, in accordance with CERCLA and the criteria contained in the Standard Classification, the real property may be classified as ECP area Type 5. The ECP area Type 5 real property may be identified at any time after completion of Steps 1 and 2 of the EBS process.

14.3.6 *Process for Determining ECP Area Type 6 Property*—Complete Steps 1 and 2 of the EBS process (see 6.2.1 and 6.2.2). If the determination can be made that concentrations of hazardous substances or petroleum products are above action levels, in accordance with the criteria contained in federal, state, and local statutes and the Standard Classification, the real property should be designated as ECP area Type 6. The ECP area Type 6 real property may be identified at any time after completion of Steps 1 and 2 of the EBS process.

14.3.7 *Process for Determining ECP Area Type 7 Property*—Complete Steps 1 and 2 of the EBS process (see 6.2.1 and 6.2.2). If the real property cannot be conclusively categorized into area Types 1 through 6, as defined in the Standard Classification, the real property should be designated as ECP area Type 7. The ECP area Type 7 real property may be identified at any time after completion of Steps 1 and 2 of the EBS process.

15. Keywords

15.1 environment; environmental assessment; environmental baseline; environmental condition; environmental condition of property; restoration; site assessment; site characterization; site remediation

Standard Practice for
Surface Site Characterization for On-Site Septic Systems[1]

This standard is issued under the fixed designation D 5879; the number immediately following the designation indicates the year of original adoption or, in the case of revision, the year of last revision. A number in parentheses indicates the year of last reapproval. A superscript epsilon (ϵ) indicates an editorial change since the last revision or reapproval.

1. Scope

1.1 This practice covers procedures for the characterization of surface conditions at a site for evaluating suitability for an on-site septic system for disposal and treatment of wastewater. This practice provides a method for identifying potentially suitable areas for soil absorption of septic tank wastewater.

1.2 This practice can be used at any site where on-site treatment of residential and nonhazardous commercial wastewaters using septic tanks and natural soils or constructed filter beds is required or an option under consideration. This practice may also be useful when constructed wetlands are used as an alternative wastewater treatment method.

1.3 This practice should be used in conjunction with Practices D 5921 and D 5925.

2. Referenced Documents

2.1 *ASTM Standards:*
D 5921 Practice for Subsurface Characterization of Test Pits for On-Site Septic Systems[2]
D 5925 Practice for Preliminary Sizing and Delineation of Soil Absorption Field Areas for On-Site Septic System[2]

3. Terminology

3.1 *clinometer, n*—an instrument for measuring inclination, as in topographic slope.

3.2 *constructed filter bed, n*—a material, usually of a sandy texture, placed above or in an excavated portion of the natural soil for filtration and purification of wastewater from an on-site septic system.

3.3 *on-site septic system, n*—any wastewater treatment and disposal system that uses a septic tank or functionally equivalent device for collecting waste solids and treats wastewater using natural soils, or constructed filter beds with disposal of the treated wastewater into the natural soil.

3.4 *potentially suitable field area, n*—the portions of a site that remain after observable limiting surface features, such as excessive slope, unsuitable landscape position, proximity to water supplies, and applicable setbacks, have been excluded.

3.5 *recommended field area, n*—the portion of the potentially suitable field area at a site that has been determined to be most suitable for an on-site septic system soil absorption field or filter bed based on surface and subsurface observations.

3.6 *soil absorption area, n*—an area of natural soil use for filtration and purification of wastewater from an on-sit septic system.

3.7 *soil absorption field area, n*—an area that includes so absorption trenches and any soil barriers between th trenches. Also called a *leachfield*.

3.8 *soil absorption trench, n*—an excavated trench, usu ally 1.5 to 3 ft wide that receives wastewater for treatmen Also called a *lateral* or *leachline*.

4. Summary of Practice

4.1 This practice describes a procedure using existin information about a site, simple field equipment, and visu observation for identifying and evaluating all significar conditions at the surface of a site, including climate, vegeta tion, topography, surface drainage, water sources, an human influences (structures, property lines), that may affe(the suitability for design and construction of an on-site septi system. The procedure involves exclusion of areas that a unsuitable for natural soil absorption or constructed filte beds as a result of topography, landscape position, an proximity to surface drainage, water sources, and othe limiting surface characteristics (structures, utilities, propert lines). If no areas at a site comply with applicable regulator requirements, no additional field investigations are requirec This procedure also provides guidance on selection of th specific area or areas at a site for subsurface investigation a covered in Practice D 5921.

5. Significance and Use

5.1 This practice should be used as the initial step f(evaluating a site for its potential to support an on-site septi system and to determine the best location for subsurfac observations as covered in Practice D 5921.

5.2 This practice should be used by individuals involve with the evaluation of properties for the use of on-site septi systems. Such individuals may be required to be licensec certified, or meet minimum educational requirements by th local or state regulatory authority. Generally, such indivic uals should be familiar with the appropriate regulator requirements governing the design and placement of on-sit septic systems for the area of the site being investigated, an at least some experience or training in geomorphology, soil: geology, and hydrology.[3]

5.3 This practice is one step in the design of an on-sit

[1] This practice is under the jurisdiction of ASTM Committee D-18 on Soil and Rock and is the direct responsibility of Subcommittee D18.01 on Surface and Subsurface Characterization.
Current edition approved Dec. 10, 1995. Published February 1996.
[2] *Annual Book of ASTM Standards,* Vol 04.08.

[3] National Small Flows Clearinghouse (NFSC), 1995. *Site Evaluation from t(State Regulations.* NFSC, Morgantown, WV.

septic system that also includes subsurface characterization, see Practice D 5921, staking and protection of the soil absorption or constructed filter bed area, see Practice D 5925, selection of system type, and design of the system size and configuration. Typically, the same individual will perform the surface and subsurface characterization of a site. Local regulation and practice will determine whether the same individual is responsible for all steps in the process of locating and designing an on-site septic system. Effective surface and subsurface characterization of a site for on-site septic systems, however, requires some knowledge of the following for the county or state in which the site is located: (1) on-site septic system types typically used for different soil conditions, and (2) typical soil absorption/filter bed areas required for different wastewater flow rates and areal soil wastewater loading rates.

6. Field Equipment

6.1 In addition to equipment identified in Practice D 5925, additional equipment useful for site surface investigations include the following:

6.1.1 *Clinometer* or *Hand Level*, and a *Surveyor's* or other rod for slope measurements;

6.1.2 *Hammer, Stakes* and *Flagging*, for marking probe or auger holes and the recommended field area. If an extendable surveyor's rod is used, a tripod for stabilizing the rod may also be useful. Accurate measurement of distances requires a tape measure (30 m or 100 ft), although for many investigations pacing may be adequate for measuring approximate distances.

6.2 At some sites, surveying equipment may be required to determine more definitively suitability for an on-site septic system or to provide additional information at the design stages. Examples of such situations include marginal sites where accurate measurements of a recommended field area are required to determine if the suitable area is large enough and sites where accurate topographic contours are required for engineering design of constructed filter beds. This practice does not address the use of surveying equipment for such purposes.

7. Procedure

7.1 *Preliminary Documentation*—All readily available information about the site should be obtained and reviewed prior to visiting the site.

7.1.1 A survey showing the boundaries of the site is the preferred method for locating the site because it can also serve as a base map for field observations. A legal description of the property can also be used to plot the site on other available maps or for drawing a sketch map of the site. A topographic survey with contour intervals of 1 to 5 ft will facilitate preliminary identification of potentially suitable field areas and final map preparation. Usually, such maps will not be available unless the site is part of a larger planned subdivision.

7.1.2 The following information concerning local or state regulatory on-site septic system siting requirements should be available for field reference, if required:

7.1.2.1 Minimum separation distance between soil absorption or constructed filter fields and water supply, prop-

erty lines and other surface and subsurface features,[4]

7.1.2.2 Wastewater hydraulic loading rates for different soil texture, structure and other field observable soil properties,[5]

7.1.2.3 Selection criteria for alternative on-site septic system designs (that is, depth to seasonal high water table, depth to limiting soil layer, slope, and so forth), and

7.1.2.4 Other site-specific features that may affect design of on-site septic systems, such as perimeter drain clearances, and wastewater loading rates.

7.1.3 If the site is undeveloped, the following information should be obtained, prior to visiting the site:

7.1.3.1 Planned location and size of the house or commercial structure,

7.1.3.2 Planned location of water well, if applicable, water lines, and other buried utilities, and

7.1.3.3 Information required for determining wastewater load rates and strength for septic system design (that is, number of bedrooms, number of full-time employee equivalents and shifts per day, biological/chemical oxygen demand). Practice D 5925 addresses in more detail wastewater hydraulic loading and strength considerations in sizing on-site septic systems.

7.1.4 A published soil survey prepared by the U.S. Natural Resource Conservation Service (formerly Soil Conservation Service) is the best single background reference on subsurface conditions for an on-site septic system field investigation. Plotting the site boundaries on the soil map and reviewing information in the soil survey report provide a preliminary indication of climate, topography, geology, hydrology, and types of limiting soil conditions that may be encountered, such as shallow bedrock or ground water.

7.1.5 Potentially useful supplemental materials include: (1) USGS 7.5-ft topographic maps, (2) aerial photographs, (3) well logs, (4) wetland inventories, (5) state and USGS geologic and hydrologic reports, and (6) adjacent or previous septic system evaluations, designs, or permits.

7.2 *Scheduling*—The investigation should be scheduled for a time and date that allows all parties interested or required for the investigation to be present. People who may need to be present for part or all of the investigation include the property owner, the construction contractor, a backhoe operator, and a representative of the on-site septic system permitting authority.

7.3 *Identification of Unsuitable Areas*—At a site, the characterization process begins with identification of all areas of the site that a clearly unsuitable for a wastewater soil absorption field or constructed filter bed. Specific exclusionary features and criteria for defining them will depend upon regulatory requirements and guidance identified in 7.1.2. Such exclusionary features typically fall into three categories: (1) water supply separation distances, (2) other buffer zones, and (3) limiting physiographic features. When most of the area at a site is potentially suitable, it may be possible to go directly to the subsurface investigation phase described in 7.4.

---

[4] National Small Flows Clearinghouse (NFSC), 1995. *Location and Separation Guidelines from the State Regulations.* NFSC, Morgantown, WV.

[5] National Small Flows Clearinghouse (NFSC), 1995. *Application Rates and Sizing of Fields from the State Regulations.* NFSC, Morgantown, WV.

7.3.1 *Water Supply*—Identify and mark on the investigation map water supply sources (drinking water and irrigation wells, reservoirs) and water supply lines. Include both existing and planned locations for new sources. Note minimum required separation distance from on-site septic systems for all identified features.

7.3.2 *Other Buffer Zones*—Identify and mark on the investigation map all other features requiring separation distances, such as building foundations, property lines, buried utility lines, cuts or embankments, large trees, irrigation ditches, streams, lakes, and wetlands. Include both existing and planned locations for new sources. Note minimum required separation distance from on-site septic systems for all identified features.

7.3.3 *Limiting Physiographic Features*—Identify and delineate on the investigation map all areas that are physiographically unsuitable, such as severely eroded or gullied soils, disturbed soils (cut and fill), excessively steep slopes, unsuitable landscape position (toe slopes, concave slopes, depressional areas), and flood plains. Actual criteria for identifying limiting physiographic features will be based on regulatory requirements identified in 7.1.2.

7.4 *Subsurface Investigations*—The area that remains after all minimum separation distances, buffer zones, and unsuitable physiographic features have been excluded represents the potentially suitable field area for an on-site septic system. Subsurface observations, as covered in Practice D 5921 may identify unsuitable or limiting subsurface conditions that will limit further the potentially suitable field area.

7.5 *Recommended Field Area*—The portion of the potentially suitable field area at a site that is most suitable for an on-site septic system soil absorption field or filter bed based on surface and subsurface observations should be delineated on the investigation map as the recommended field area. This area should be staked and protected from disturbance during construction activities as covered in Practice D 5925. Practice D 5925 also provides guidance on the size of area that should be included in the recommended field area.

7.5.1 The recommended field area should include the area that, taking into account limiting surface and subsurface conditions at the site, provides the greatest flexibility in selection and design of an on-site septic system. Placing the field at a lower topographic position than the septic tank outfall allows the option of either gravity or pumped distribution of wastewater where soils are suitable for drainfields.

7.5.2 The recommended field area usually will represent a smaller area than the potentially suitable field area and the area to which subsurface observations in accordance with Practice D 5921 can be extrapolated reasonably. Moving the actual field area to a different location generally will require additional subsurface observations to confirm suitability.

8. Report

8.1 Reporting of results of the surface investigations should be integrated with the results of the subsurface investigation. The local or state regulatory authority may have developed forms or formats for investigation reports, in which case, these should be used.

8.2 Basic elements of an on-site septic system site investigation report include:

8.2.1 A vicinity map and directions to the site,

8.2.2 General site information,

8.2.3 A sketch map,

8.2.4 Identification of surface and subsurface features that limit suitability for an on-site septic system, and

8.2.5 Detailed information about the surface and subsurface characteristics of the recommended field area that are pertinent to the design of the on-site septic system.

8.3 Generally, unless desired by the appropriate septic system permitting agency, the report should not contain recommendations for possible options to overcome limiting features in the recommended field area or recommend the type or types of septic system that might be suitable for the site.

9. Keywords

9.1 field investigations; preliminary investigations; septic systems; site characterization; site investigations

APPENDIX

(Nonmandatory Information)

X1. Related Publications

X1.1 American Society of Agricultural Engineers. 1975–1994. *On-Site Waste Water Treatment Proceedings Series.* Proc. of the 1st Nat. Home Sewage Treatment Symposium (1975), 2nd (1977, 292 pp.); Proc. 3rd Nat. Symp. on Individual and Small Community Sewage Treatment (1981, 352 pp.); 4th (ASAE Pub. 07-85, 1984, 381 pp.); 5th (ASAE Pub. 10-87, 1987, 411 pp.); 6th (ASAE Pub. 10-91, 1991, 375 pp.); Proc. 7th Int. Symp. on Individual and Small Community Sewage Systems (E. Collins, ed., 1994, 578 pp.)

X1.2 Burks, B. D., and Minnis, M. M., *Onsite Waste-water Treatment Systems.* Hogarth House, Madison, WI 1994, 248 pp.

X1.3 Kaplan, O. B. 1991. *Septic Systems Handbook,* Second Edition. Lewis Publishers, Chelsea, MI, 434 pp.

X1.4 Canter, L. W. and R. C. Knox. 1985. *Septic Tank Systems Effects on Ground Water Quality.* Lewis Publishers, Chelsea, MI.

X1.5 Perkins, R. J. 1989. *Onsite Wastewater Disposal.* Lewis Publishers, Chelsea, MI 251 pp. [Chapter 3 covers selection of site and system]

X1.6 National Small Flows Clearinghouse (NFSC). 1995.

State Regulation Compilations (updated annually): Site Evaluation from the State Regulations (Pub. No. WWPCRG27); Location and Separation Guidelines from the State Regulations (Pub. No. WWPCRG20); Application Rates and Sizing of Fields from the State Regulations (Pub. No. WWPCRG19); Percolation Tests from the State Regulations (WWPCRG22). NFSC, West Virginia University, P.O. Box 6064, Morgantown, WV 26506-8301, 800/624-8301.

X1.7 U.S. Environmental Protection Agency. 1980. Design Manual: Onsite Wastewater Treatment and Disposal Systems. EPS/625/1-80-012. [Chapter 3 covers site evaluation procedures]

X1.8 U.S. Environmental Protection Agency (EPA). 1986. Septic Systems and Groundwater Protection: A Program Manager's Guide and Reference Book. EPA/440/6-86/005 (NTIS PB88-112123), 134 pp.

X1.9 University of Washington College of Engineering. 1976–1992. *Proceedings of the Northwest On-Site Wastewater Disposal Short Course*: 1st (1976); 2nd (1978, R. W. Seabloom, ed., 287 pp., 16 papers); 3rd (1980, R. W. Seabloom, ed., 374 pp., 21 papers); 4th (1982, R. W. Seabloom, ed., 382 pp., 19 papers); 5th (1985, R. W. Seabloom and D. Lenning, and D. Stenset, eds., 299 pp., 18 papers); 6th (1989, R. W. Seabloom and D. Lenning, eds., 431 pp., 24 papers); 7th (1992, R. W. Seabloom, ed., 380 pp., 26 papers). Office of Engineering Continuing Education, University of Washington, 4725 30th Ave., NE, Seattle, WA 98105.

X1.10 Winneberger, J. T. 1984. Septic Tank Systems. Butterworth Publishers, Stoneham, MA.

Designation: D 5921 – 96

Standard Practice for
Subsurface Site Characterization of Test Pits for On-Site Septic Systems[1]

This standard is issued under the fixed designation D 5921; the number immediately following the designation indicates the year of original adoption or, in the case of revision, the year of last revision. A number in parentheses indicates the year of last reapproval. A superscript epsilon (ε) indicates an editorial change since the last revision or reapproval.

INTRODUCTION

Many State and local jurisdictions have requirements for evaluating sites for approval of on-site septic systems. This practice provides a method to describe and interpret subsurface characteristics to evaluate sites for septic systems. All characteristics used in this practice influence the ability of a site to provide treatment and disposal of septic tank effluent. However, this practice is not meant to be an inflexible description of investigation requirements. State and local jurisdictions may require fewer or greater numbers of subsurface features to evaluate a site.

This practice primarily follows the U.S. Department of Agriculture, Soil Conservation Service (SCS) soil classification system, which encompasses a systematic framework for soil morphological characterization. The SCS classification the most prevalent system in use for on-site septic systems. This practice can be complemented by application of other soil description techniques as appropriate, such as the Unified Soil Classification System (D 2485).

1. Scope

1.1 This practice covers procedures for the characterization of subsurface soil conditions at a site as part of the process for evaluating suitability for an on-site septic system. This practice provides a method for determining the usable unsaturated soil depth for septic tank effluent to infiltrate for treatment and disposal.

1.2 This practice describes a procedure for classifying soil by field observable characteristics within the United States Department of Agriculture, Soil Conservation Service (SCS) classification system.[2] The SCS classification system is defined in Refs (1–4),[3] not in this practice. This practice is based on visual examination and manual tests that can be performed in the field. This practice is intended to provide information about soil characteristics in terms that are in common use by soil scientists, public health sanitarians, geologists, and engineers currently involved in the evaluation of soil conditions for septic systems.

1.3 This procedure can be augmented by Test Method D 422, when verification or comparison of field techniques is required. Other standard test methods that may be used to augment this practice include: Test Methods D 2325, D 3152, D 5093, D 3385, and D 2434.

1.4 This practice is not intended to replace Practice D 2488 which can be used in conjunction with this practice if construction engineering interpretations of soil properties are required.

1.5 This practice should be used in conjunction with D 5879 to determine a recommended field area for an on-site septic system. Where applicable regulations define loading rates-based soil characteristics, this practice, in conjunction with D 5925, can be used to determine septic tank effluent application rates to the soil.

1.6 This practice should be used to complement standard practices developed at state and local levels to characterize soil for on-site septic systems.

1.7 The values stated in SI units are to be regarded as the standard.

1.8 *This standard does not purport to address all of the safety concerns, if any, associated with its use. It is the responsibility of the user of this standard to establish appropriate safety and health practices and determine the applicability of regulatory limitations prior to use.*

2. Referenced Documents

2.1 *ASTM Standards:*
D 422 Standard Test Method for Particle-Size Analysis of Soils[4]
D 653 Terminology Relating to Soil, Rock, and Contained Fluids[4]
D 2325 Test Method for Capillary-Moisture Relationships for Coarse- and Medium-Textured Soils by Porous-Plate Apparatus[4]
D 2434 Test Method for Permeability of Granular Soils (Constant Head)[4]
D 2488 Practice for Description and Identification of Soils (Visual-Manual Procedure)[4]

---

[1] This practice is under the jurisdiction of ASTM Committee D-18 on Soil and Rock and is the direct responsibility of Subcommittee D18.01 on Surface and Subsurface Characterization.
Current edition approved Feb. 10, 1996. Published November 1996.
[2] In 1995, the name of the SCS was changed to Natural Resource Conservation Service. This guide uses SCS rather than NRCS because referenced documents were published before the name change.
[3] The boldface numbers given in parentheses refer to a list of references at the end of the text.

---

[4] *Annual Book of ASTM Standards*, Vol 04.08.

D 3152 Test Method for Capillary-Moisture Relationships for Fine-Textured Soils by Pressure-Membrane Apparatus[4]

D 3385 Test Method for Infiltration Rate of Soils in Field Using Double-Ring Infiltrometer[4]

D 5093 Test Method for Field Measurement of Infiltration Rate Using a Double-Ring Infiltrometer with a Sealed-Inner Ring[5]

D 5879 Practice for Surface Site Characterization for On-Site Septic Systems[5]

D 5925 Practice for Preliminary Sizing and Delineation of Soil Absorption Field Areas for On-Site Septic Systems[5]

3. Terminology

3.1 *Definitions:*

3.1.1 *limiting depth*—for the purpose of determining suitability for on-site septic systems, the depth at which the flow of water, air, or the downward growth of plant roots is restricted.

3.1.2 *mottle*—spots or blotches of different colors or shades of color interspersed with the dominant color (5). In SCS (3) practice mottles associated with wetness in the soil are called redox concentrations or redox depletions.

3.1.3 *pocket penetrometer*—a hand operated calibrated spring instrument used to measure resistance of the soil to compressive force.

3.1.4 *potentially suitable field area*—the portions of a site that remain after observing limiting surface features such as excessive slope, unsuitable landscape position, proximity to water supplies, and applicable setbacks have been excluded.

3.1.5 *recommended field area*—the portion of the potentially suitable field area at a site that has been determined to be most suitable as a septic tank soil absorption field or filter bed based on surface and subsurface observations.

3.1.6 *unsaturated*—soil water condition at which the void spaces that are able to be filled are less than full.

3.1.7 *vertical separation*—the depth of unsaturated, native, undisturbed soil between the bottom of the disposal component of the septic system and the limiting depth.

4. Summary of Practice

4.1 This practice describes a field technique using visual examination and simple manual tests for characterizing and evaluating soils and identifying any limiting depth.

5. Significance and Use

5.1 This practice should be used as part of the evaluation of a site for its potential to support an on-site septic system in conjunction with Practice D 5879 and Practice D 5925.

5.2 This practice should be used after applicable steps in Practice D 5879 have been performed to document and identify potentially suitable field areas.

5.3 This practice should be used by those who are involved with the evaluation of properties for the use of on-site septic systems. They may be required to be licensed, certified, meet minimum educational requirements by the area governing agencies, or all of these.

5.4 This practice requires exposing the soil to an appropriate depth (typically 1.5 to 1.8 m, or greater as site conditions or project objectives require) for examining the

soil morphologic characteristics related to the performance of on-site septic systems.

6. Limitations

6.1 The water content of the soil will affect its properties. The soil should be evaluated in the moist condition because the normal operating state of the septic system is a moist condition. If the soil is dry, moisten it.

6.2 This practice is not applicable to frozen soil.

6.3 Optimum lighting conditions for determining soil color are full sunlight from mid-morning to mid-afternoon. Less favorable lighting conditions exist when sun is low or skies are cloudy or smoky. If artificial light is used, it should be as near the light of mid-day as possible.

7. Apparatus

7.1 Tools typically used are a soil knife or a flat blade screw driver, tape measure, pencil and paper, Munsell soil color charts (6), water bottle, wash rag, and a sack to carry samples if required. A pocket penetrometer may also be useful. When the presence of carbonate may be significant in soils, dilute hydrochloric acid (10 % HCl) should be used.

7.2 A backhoe will facilitate excavation of the test pits for examination. However, if the site is inaccessible or funds are limited, one may excavate by hand with a shovel. Depending on site conditions, power driven or hand held soil augers may also be suitable. Tube samplers allow description of soil morphologic features providing the size of the feature does not exceed the diameter of the core. Augers generally destroy such morphologic features as soil structure and porosity. The advantage of augers and tube samplers is that they are generally faster and less expensive than excavated pits. Their disadvantage is that they sample a smaller area of soil, preventing characterization of lateral changes in horizon boundaries and description of larger-scale morphologic features. Use of probes or augers as an alternative to excavated pits requires a higher degree of experience and knowledge about soils in an area.

7.3 For preliminary examination of a site, one may probe vertically into the soil to get a feel for the presence and depth to a compacted layer, or a water table. Tools that might be used include a digging bar, tile probe, post hole digger, or hand soil auger.

8. Location of Sampling Points

8.1 Test pits or other subsurface sampling points should be located in the potentially suitable field area as determined using Practice D 5879, taking into consideration proximity of source of waste water and down slope of source, if possible. Locating down slope gives most flexibility in system design by allowing either gravity flow or pressure distribution. A preliminary sizing of the field should be performed in accordance with Practice D 5925 to determine placement of the sample points. Generally, sample points should be located on diagonal corners of the preliminary drainfield area so as to avoid disturbing the soil within the recommended field area. Depending on site conditions, additional sample points may be required to identify a recommended field area.

9. Procedure

9.1 Orient the excavation to expose the vertical face to the best light.

[5] *Annual Book of ASTM Standards*, Vol 04.09.

TABLE 1 Definitions and Designations for Soil Horizons (1), (3)

Master Horizons and Layers:

O — Horizons—Layers dominated by organic material, except limnic layers that are organic.

A — Horizons—Mineral horizons that form at the surface or below an O horizon and (1) are characterized by an accumulation of humified organic matter intimately mixed with the mineral fraction and not dominated by properties characteristic of E or B horizons; or (2) have properties resulting from cultivation, pasturing, or similar kinds of disturbance.

E — Horizons—Mineral horizons in which the main feature is loss of silicate clay, iron, aluminum, or some combination of these, leaving a concentration of sand and silt particles of quartz or other resistant materials.

B — Horizons—Horizons that formed below an A, E, or O horizon and are dominated by (1) carbonates, gypsum, or silica, alone or in combination; (2) evidence of removal of carbonates; (3) concentrations of sesquioxides; (4) alterations that form silicate clay; (5) formation of granular, blocky, or prismatic structure; or (6) a combination of these.

C — Horizons—Horizons or layers, excluding hard bedrock, that are little affected by pedogenic processes and lack properties of O, A, E, or B horizons. Most are mineral layers, but limnic layers, whether organic or inorganic are included.

R — Layers—Hard bedrock including granite, basalt, quartzite, and indurated limestone or sandstone that is sufficiently coherent to make hand digging impractical.

Transitional Horizons:

Two kinds of transitional horizons occur. In one, the properties of an overlying or underlying horizon are superimposed on properties of the other horizon throughout the transition zone (that is, AB, BC, etc.). In the other, distinct parts that are characteristic of one master horizon are recognizable and enclose parts characteristic of a second recognizable master horizon (that is, E/B, B/E, and B/C).

Alphabetical Designation of Horizons:

Capital letters designate master horizons (see definitions above).

Lowercase letters are used as suffixes to indicate specific characteristics of the master horizons (see definitions below). The lowercase letter immediately follows the capital letter designation.

Numeric Designation of Horizons:

Arabic numerals are used as (1) suffixes to indicate vertical subdivisions within a horizon and (2) prefixes to indicate discontinuities.

Prime Symbol:

The prime symbol (′) is used to identify the lower of two horizons having identical letter designations that are separated by a horizon of a different kind. If three horizons have identical designations, a double prime (″) is used to indicate the lowest.

Subordinate Distinctions within Horizons and Layers:

a— Highly decomposed organic material where rubbed fiber content averages <1/6 of the volume.

b— Identifiable buried genetic horizons in a mineral soil.

c— Concretions or hard nonconcretionary nodules of iron, aluminum, manganese, or titanium cement.

d— Physical root restriction, such as dense basal till, plow pans, and other mechanically compacted zones.

e— Organic material of intermediate decomposition in which rubbed fiber content is 1/6 to 2/5 of the volume.

f— Frozen soil in which the horizon or layer contains permanent ice.

g— Strong gleying in which iron has been reduced and removed during soil formation or in which iron has been preserved in a reduced state because of saturation with stagnant water.

h— Illuvial accumulation of organic matter in the form of amorphous, dispersible organic matter-sesquioxide complexes, where sesquioxides are in very small quantities and the value and chroma of the horizons are < 3.

i— Slightly decomposed organic material in which rubbed fiber content is more than about 2/5 of the volume.

k— Accumulation of pedogenic carbonates, commonly calcium carbonate.

m— Continuous or nearly continuous cementation or induration of the soil matrix by carbonates (km), silica (qm), iron (sm), gypsum (ym), carbonates and silica (kqm), or salts more soluble than gypsum (zm).

n— Accumulation of sodium on the exchange complex sufficient to yield a morphological appearance of a natric horizon.

o— Residual accumulation of sesquioxides.

p— Plowing or other disturbance of the surface layers for cultivation, pasturing, or similar uses.

q— Accumulation of secondary silica.

r— Weathered or soft bedrock including saprolite; partly consolidated soft sandstone, siltstone, or shale; or dense till that roots penetrate only along joint planes and which is sufficiently incoherent to permit hand digging with a spade.

s— Illuvial accumulation of sesquioxides and organic matter in the form of illuvial, amorphous dispersible organic matter-sesquioxide complexes, if both organic matter and sesquioxide components are significant and the value and chroma of the horizon are > 3.

ss— Presence of slickensides.

t— Accumulation of silicate clay that either has formed in the horizon and is subsequently translocated or has been moved into it by illuviation.

v— Plinthite which is composed of iron-rich, humus-poor, reddish material that is firm or very firm when moist and that hardens irreversibly when exposed to the atmosphere under repeated wetting and drying.

w— Development of color or structure in a horizon with little or no apparent illuvial accumulation of materials.

x— Fragic or fragipan characteristics that result in genetically developed firmness, brittleness, or high bulk density.

y— Accumulation of gypsum.

z— Accumulation of salts more soluble than gypsum.

9.2 Excavate the test pit to a depth sufficient to satisfy the vertical separation required by the governing agency. If the limiting depth is too shallow to meet the vertical separation requirement, it may be desirable to excavate deeper to determine if the layer is underlain by permeable material.

9.3 Enter the test pit using all applicable safety requirements and examine the soil layers, or horizons. Select a representative area to examine in detail.[6]

9.4 Using a soil knife or other tool, expose the natural soil structure in an area approximately 0.5 m in width the full height of the test pit.

9.5 Describe master soil horizons following the criteria in Table 1. Horizons are separated by boundaries. Locate these boundaries by changes in color, texture, or structure.

9.6 For each layer describe and test as follows:

9.6.1 Measure the depth of the layer from the soil-air interface. Positive numerical values indicate increasing depth.

9.6.2 Describe color of soil with soil in the moist state. Use Munsell color chart (6) designation for hue, value, and chroma. Include the color name. Indicate lighting conditions, if other than direct sunlight.

9.6.3 Estimate the volumetric percentage of rock fragments (see Fig. 1).

---

[6] Test pits should comply with applicable Federal, State and Local safety regulations. Generally, test pits 1.5 meters or less in depth do not require special protection if the soil is cohesive.

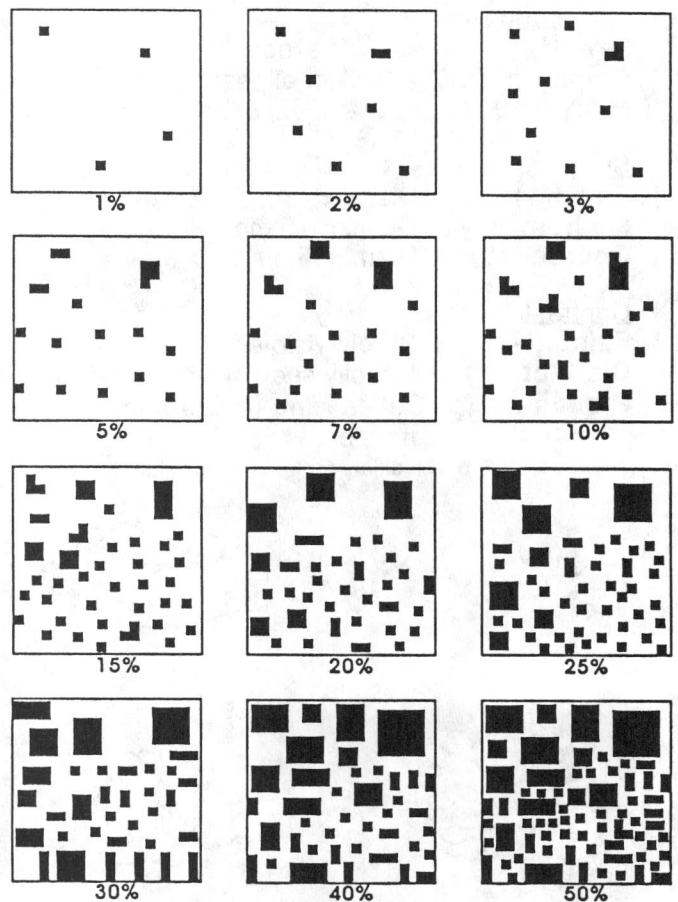

FIG. 1 Chart for Estimating Proportions of Mottles or Rock Fragments (6), (7), (8), (9)

TABLE 2 Abbreviations and Designations for Rock Fragment Classes (1), (3), and (7)

Modifier (Volume %)[A]		Adjective/Noun	Shape/Size Rounded, Subrounded, Angular, or Irregular (diameter, mm)
<15 %	none	GR—gravelly/pebbles	2 to 75
>15 to 35 %	dominant rock		
35 to 60 %	dominant rock + very (v)		
> 60 %	(>10 % fines) dominant rock + extremely (x)	CB—cobbly/cobbles	75 to 250
> 60 %	(<10 % fines) dominant rock noun	ST—stony/stones	250 to 600
		B—bouldery/boulders	>600 flat (long, mm)
		CN—channery/channers	2 to 150
		FL—flaggy/flagstones	150 to 380
		ST—stony/stones	380 to 600
		B—bouldery/boulders	> 600

[A] Classes for application of rock fragment modifiers (that is, gravelly loam would have >15 to 35 % pebbles by volume).

9.6.4 Describe size, shape, and percentage of rock fragments (see Table 2).

9.6.5 Describe the texture of the < 2 mm fraction of the layer using the flow chart in Fig. 2 as a guide. See Table 3 for

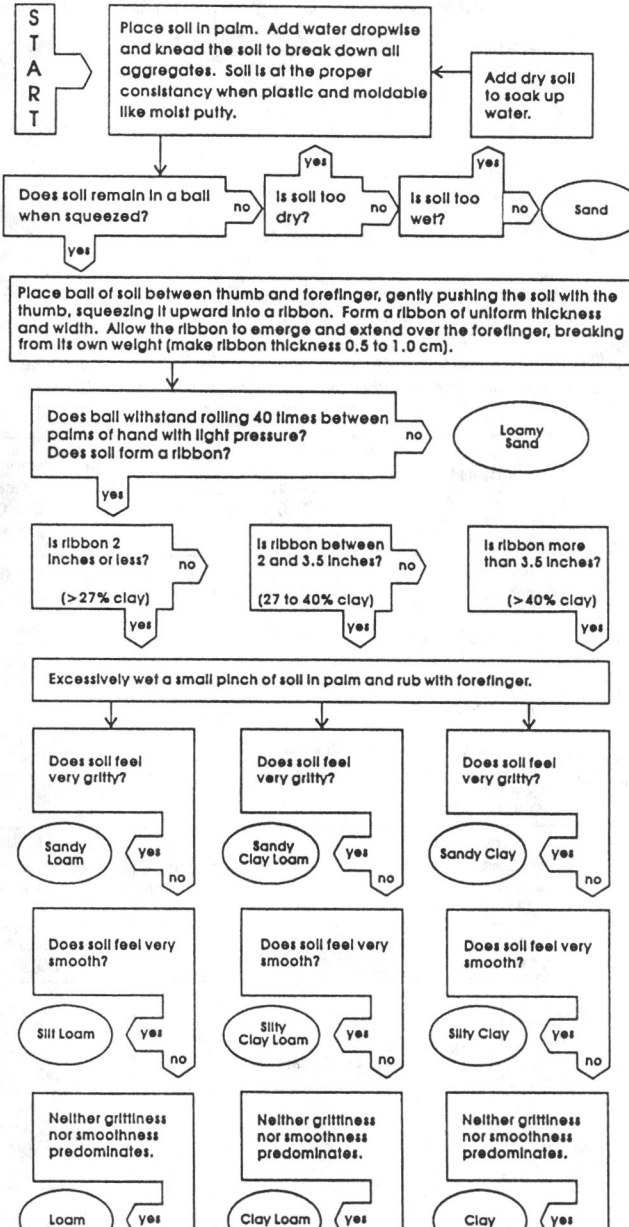

NOTE—Local clay mineralogy may require modifications in the above procedure. Field texture determinations should be periodically corroborated by laboratory analyses (weight %).

FIG. 2 Flow Chart for Estimating Soil Texture (6), (7), (11)

abbreviations. For sandy soils, (that is, less than 20 % clay and greater than 50 % sand by weight), a field sieve analysis allows more precise texture classification using Table 4.

9.6.6 Note the presence or absence of mottles. Describe color (6); proportion (see Fig. 1); and abundance, size, and contrast of mottles (see Table 5).

9.6.7 Describe soil structure by grade using Table 6 and shape and size using Figs. 3 and 4.

9.6.8 Describe soil-rupture resistance using criteria in Table 5.

9.6.9 If cementation is suspected, bring an intact soil clod

271

TABLE 3 Abbreviations and Designations for USDA Soil Texture Classes (1), (3), and (7)

s—sand
ls—loamy sand
sl—sandy loam
l—loam
si—silt
sil—silt loam
cl—clay loam
sicl—silty clay loam
sc—sandy clay
sic—silty clay
c—clay

TABLE 4 Percentage of Sand Sizes in Subclasses of Sand, Loamy Sand, and Sandy Loam Basic Classes (12). (Weight %)

Basic soil class	Subclass (abbreviation)	Very coarse sand, 2.0–1.0 mm	Coarse sand, 1.0–0.5 mm	Medium sand, 0.5–0.25 mm	Fine sand, 0.25–0.1 mm	Very fine sand, 0.1–0.05 mm
		Soil separates				
Sands	Coarse sand (COS)	25 % or more		Less than 50 %	Less than 50 %	Less than 50 %
	Sand (S)	25 % or more			Less than 50 %	Less than 50 %
	Fine sand (FS)	—or—			50 % or more	
		Less than 25 %				Less than 50 %
	Very fine sand (VFS)					50 % or more
Loamy Sands	Loamy coarse sand (LCOS)	25 % or more		Less than 50 %	Less than 50 %	Less than 50 %
	Loamy sand (LS)	25 % or more			Less than 50 %	Less than 50 %
	Loamy fine sand (LFS)	—or—			50 % or more	
		Less than 25 %				Less than 50 %
	Loamy very fine sand (LVFS)					50 % or more
Sandy Loams	Coarse sandy loam (COSL)	25 % or more		Less than 50 %	Less than 50 %	Less than 50 %
	Sandy loam (SL)	30 % or more				
		—and—				
		Less than 25 %			Less than 30 %	Less than 30 %
	Fine sandy loam (FSL)	—or—			30 % or more	Less than 30 %
		Between 15 and 30 %				
	Very fine sandy loam (VFSL)					30 % or more
		—or—				
		Less than 15 %				More than 40 %

* Half of fine sand and very fine sand must be very fine sand.

from the site for further testing. Air dry the clod. Submerge the clod in water for at least 1 h. Perform the same tests for

<u>Abundance</u>
Few (f) <2% of exposed surface.
Common (c) 2-20% of exposed surface.
Many (m) >20% of exposed surface.

<u>Size</u>
Fine (1) Diam. <5 mm.
Medium (2) Diam. 5-15 mm.
Coarse (3) Diam >15 mm.

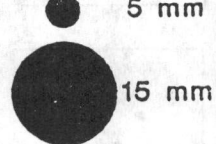
5 mm
15 mm

<u>Contrast</u>
Faint (f) Barely visible.
Distinct (d) Readily seen but not striking.
Prominent (p) Outstanding visible feature of horizon.

TABLE 5 Modifiers for Mottles (3, 6, and 7)

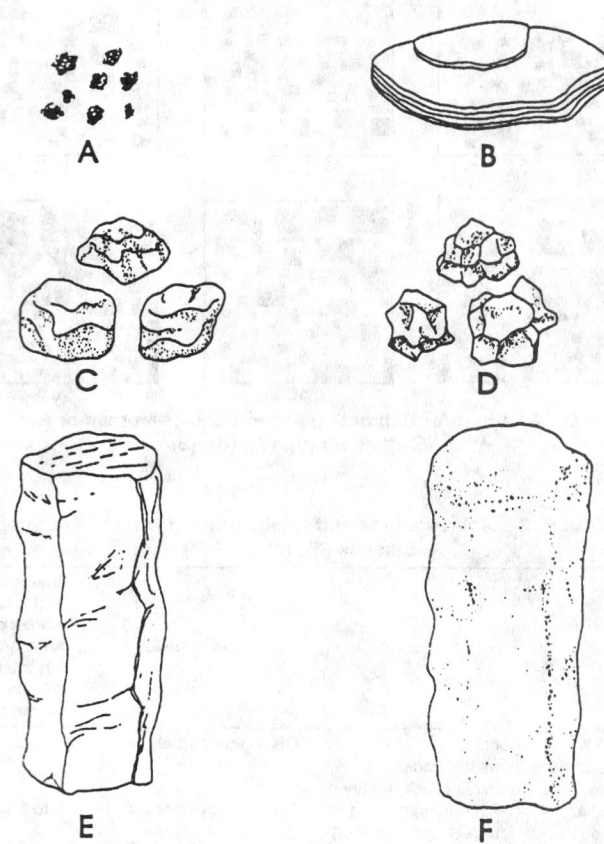

NOTE—Not shown, massive (MA), single grain (SGR).

FIG. 3 Drawings Illustrating Some of the Types of Soil Structure: A, Prismatic; B, Columnar; C, Angular Blocky; D, Subangular Blocky; E, Platy; and F, Granular (3)

rupture resistance as shown in Table 7. The sample is cemented if it meets the very hard classification test. Describe the degree of cementation using classes given in Table 7.

9.6.10 Measure soil penetration resistance with a pocket penetrometer and describe the condition of the soil following the criteria in Table 8.

9.6.11 Describe abundance, size, and distribution of roots

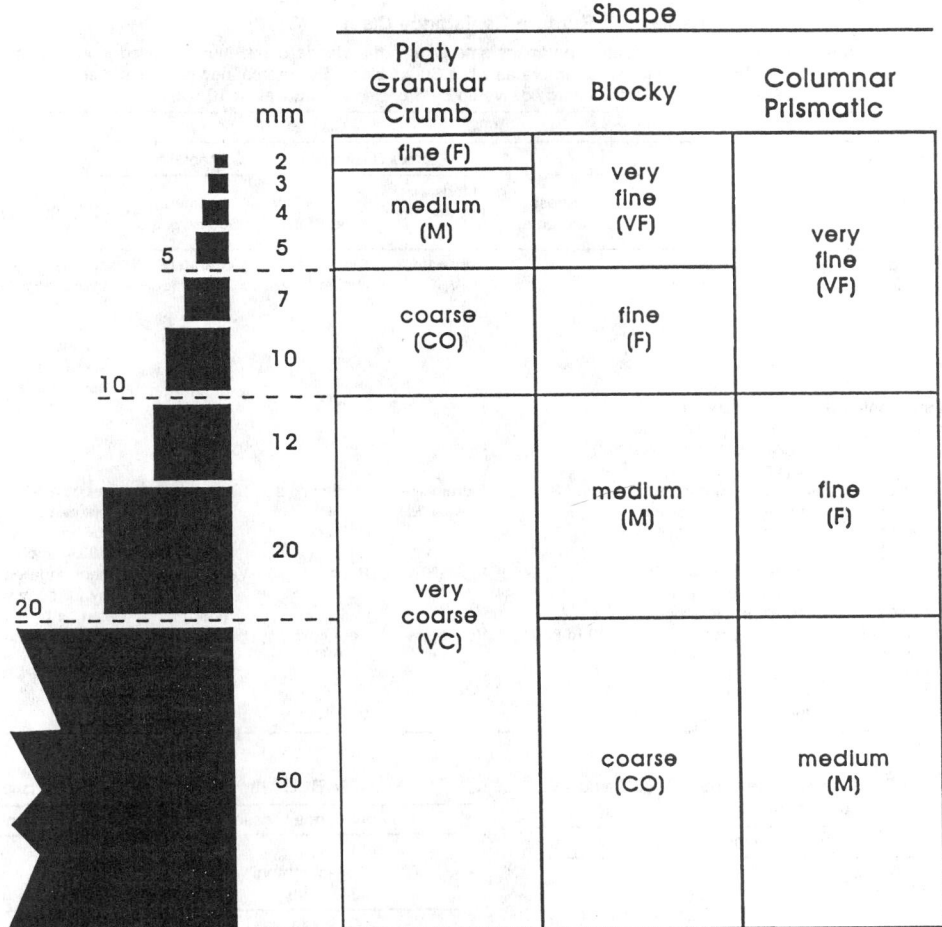

NOTE—Based on classes defined in Ref **(3)**.

FIG. 4 Charts for Estimating Size Class of Different Structural Units (8)

using modifier criteria given in Table 9 and Fig. 5.

9.6.12 Describe abundance, size, distribution and type of soil pores using criteria in Table 10 and Fig. 5.

9.6.13 If presence or absence of carbonates is a diagnostic soil property, use hydrochloric acid to determine depth to free carbonate. Describe effervescence as follows: (0) very slightly effervescent (few bubbles), (1) slightly effervescent (bubbles readily), (2) strongly effervescent (bubbles form low foam), (3) violently effervescent (thick foam forms quickly), and (4) noneffervescent.

9.6.14 Describe layer boundaries according to its distinctness and topography as shown in Table 11.

9.6.15 Estimate moisture conditions of the soil as dry, moist, or wet using the guidelines in Table 12. Measure the depth to zone of saturation, if encountered, immediately and remeasure periodically during evaluation of the site.

9.7 Evaluate changes in soil profile laterally within each pit and between the test pits, augmented by hand auger

TABLE 6 Grades of Soil Structure (3)

Grade
1—Weak (poorly defined individual peds)
2—Moderate (well formed individual peds)
3—Strong (durable peds, quite evident in place; will stand displacement)

borings, as necessary, to determine if more test pits are needed to fully characterize the site.

10. Interpretation of Results

10.1 Identify limiting depth at each sampling point based on applicable regulatory criteria or definitions. Major types of limiting depths include depth to saturation, depth to a very slowly permeable layer that restricts downward movement of water, depth to an excessively permeable layer, and depth to a layer of strongly contrasting texture that impedes downward movement of water. Interpretation of limiting depth is a matter of judgement involving consideration of various observable soil features.

10.2 Depth to saturation. Soil morphologic indicators of depth to saturation include gleyed horizons, redox related mottles (redox concentrations and depletions, that is, zones indicative of oxidizing and reducing conditions), and iron and manganese concentrations (coatings, concretions and nodules).

10.2.1 Gleyed horizons (hues of 5GY, 5G, 5BG, 5B, and N **(6)**) and depleted matrices (generally two chroma or less **(6)**) indicate permanent saturation.

10.2.2 Mottled horizons characterized by areas of redox concentrations and redox depletions generally indicate sea-

TABLE 7 Rupture Resistance Classes (3)

NOTE—Specimens should be block-like and 25 to 30 mm on edge. If specimens smaller than the standard size must be used, corrections should be made for class estimates (that is, a 10-cm block will require about one-third the force to rupture as will a 30-cm block. Both force, newton (N) and energy, joule (J), are employed. The number of newtons is ten times the kilograms of force. One joule is the energy delivered by dropping a 1 kg weight 10 cm.

Classes			Test Description		Classes			Test Description	
Rupture Resistance		Cementation			Rupture Resistance		Cementation		
Moderately Dry and Very Dry	Slightly Dry and Wetter	Air Dried, Submerged	Operation	Stress Applied a/	Moderately Dry and Very Dry	Slightly Dry and Wetter	Air Dried, Submerged	Operation	Stress Applied a/
Loose (L)	Loose (L)	Not applicable	Specimen not obtainable		Very hard (VH)	Extremely firm (EFI)	Moderately cemented (MC)	Cannot be failed between thumb and forefinger but can be between both hands or by placing on a nonresilent surface and applying gentle force underfoot.	80 to 160N
Soft (S)	Very friable (VFR)	Noncemented (CO)	Fails under very slight force applied slowly between thumb and fore-finger	< 8N					
Slightly hard (SH)	Friable (FR)	Extremely weakly cemented (XWC)	Fails under slight force applied slowly between thumb and forefinger	8 to 20 N	Extremely hard (EH)	Slightly rigid (SR)	Strongly cemented (SC)	Cannot be failed in hands but can be underfoot by full body weight (ca 800N) applied slowly.	160 to 800N
Moderately hard (MH)	Firm (FI)	Very weakly cemented (VWC)	Fails under moderate force applied slowly between thumb and forefinger	20 to 40 N	Rigid (R)	Rigid (R)	Very strongly cemented (VSC)	Cannot be failed underfoot by full body weight but can be by < 3J blow.	800N to 3J
Hard (H)	Very firm (VFI)	Weakly cemented (WC)	Fails under strong force applied slowly between thumb and forefinger (80N about maximum force can be applied)	40 to 80 N	Very rigid (VR)	Very rigid (VR)	Indurated (I)	Cannot be failed by blow of < 3J.	≥3J

TABLE 8 Soil Penetration Resistance Classes (3), (8), and (9), (MPa, Megapascal)

Classes		Penetration Resistance (MPa)
Small		< 0.1
	Extremely low (EL)	< 0.01
	Very low (VL)	0.01 to 0.1
Intermediate		0.1 to 2
	Low (L)	0.1 to 1
	Moderate (M)	1 to 2
Large		>2
	High (H)	2 to 4
	Very high (VH)	4 to 8
	Extremely high (EH)	> 8

TABLE 9 Modifiers for Roots (3), (8)

Abundance Classes		Number per Unit Area
v1—very few		<0.2
1—few		<1
2—moderately few		0.2 to 1
3—common		1 to 5
4—many		≥5
Size Classes	Diameter	Unit Area
v1—very fine	<1 mm	1 cm²
1—fine	1 to 2 mm	1 cm²
2—medium	2 to 5 mm	100 cm²
3—coarse	5 to 10 mm	100 cm²
4—very coarse	≥10 mm	1 m²

Distribution Within Horizons:
P—Between peds
C—In cracks
M—In mat at top of horizon
S—Matted around stones
T—Throughout

TABLE 10 Modifiers for Soil Pores (3) (8)

Abundance Classes		Number/Unit Area
1—few		<1
2—common		1 to 5
3—many		> 5
Size Classes	Diameter	Unit Area
V1—very fine	<1 mm	1 cm²
1—fine	1 to 2 mm	1 cm²
2—medium	2 to 5 mm	100 cm²
3—coarse	5 to 10 mm	100 cm²
4—very coarse	>10 mm	1 m²

Distribution Within Horizons

in—inped (most pores are within peds)
ex—exped (most pores follow interfaces between peds)

Types of Pores

v—vesicular (approximately spherical or elliptical)
t—tubular (approximately cylindrical and elongated)
i—irregular

TABLE 11 Classes of Soil Water (3), (8), (9)

Dry (D)—Very little visual or tactile change between field observation and after air-dried samples.
Moist (M)—Visual or tactile change between field observation and after air drying.
Wet (W)—Water films evident, or free water.

sonal saturation. A common rule of thumb is the depth to two chroma mottles (redox depletions) represents the seasonal high water table. In some geographic areas and soil types, three chroma mottles may also indicate seasonal saturation. Generally, the percentage of the soil that is gray serves as an indicator of length of saturation, with more gray indicating longer periods of saturation. Soil morphologic features do not always correlate well with seasonal fluctuations in saturation, and the confidence in interpretations can be increased by studies that demonstrate a correlation for soils in an area. When evaluating soil mottling, consideration

TABLE 12 Guide for Estimation of Capillary Fringe (10)

USDA Texture Class	Est. Capillary Fringe, cm
Coarse sand	1 to 7
Sand	1 to 9
Fine sand	3 to 10
Very fine sand	4 to 12
Loamy coarse sand	5 to 14
Loamy sand	6 to 14
Loamy fine sand	8 to 18
Coarse sandy loam	8 to 18
Loamy very fine sand	10 to 20
Sandy loam	10 to 20
Fine sandy loam	14 to 24
Very fine sandy loam	16 to 26
Loam	20 to 30
Silt loam	25 to 40
Silt	35 to 50
Sandy clay loam	20 to 30
Clay loam	25 to 35
Silty clay loam	35 to 55
Sandy clay	20 to 30
Silty clay	40 to 60
Clay	25 to 40

Very Fine (VF)
(less than 1 mm diameter)

Fine (1)
(1 to <2 mm diameter)

Medium (2)
(2 to <5 mm diameter)

Coarse (3)
(5 to <10 mm diameter)

Very Coarse (4)
(more than 10 mm diameter)

0 5 10
mm

NOTE—Modified from Ref (9).

FIG. 5 Charts for Estimating Pore and Root Size

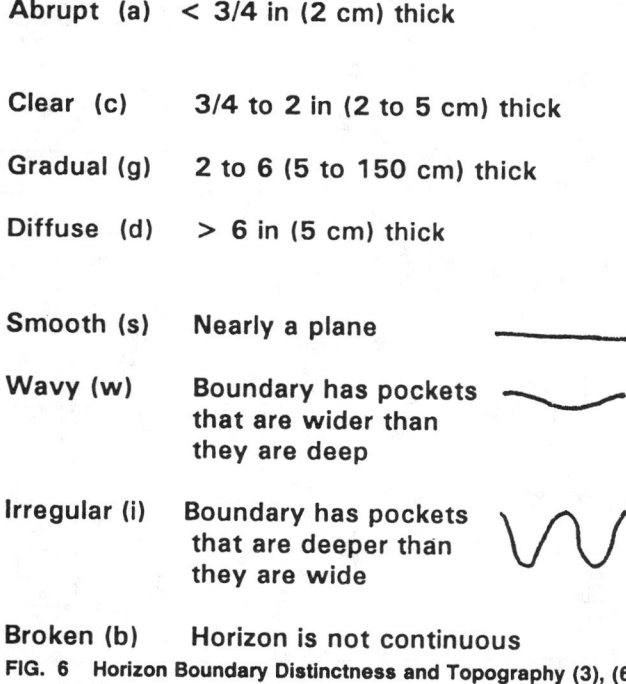

Abrupt (a)   < 3/4 in (2 cm) thick

Clear (c)   3/4 to 2 in (2 to 5 cm) thick

Gradual (g)   2 to 6 (5 to 150 cm) thick

Diffuse (d)   > 6 in (5 cm) thick

Smooth (s)   Nearly a plane

Wavy (w)   Boundary has pockets that are wider than they are deep

Irregular (i)   Boundary has pockets that are deeper than they are wide

Broken (b)   Horizon is not continuous

FIG. 6 Horizon Boundary Distinctness and Topography (3), (6), (7)

practice in the area. Also, the absence of redox depletions does not necessarily prove lack of saturation. Redox depletions may not be evident where ground water is well oxygenated, soils are very low in dissolved organic carbon, and low in iron oxides. Also, redoximorphic features do not develop where soils or ground water is less than 5°C and in soils with high pH (generally >8).

10.2.3 Horizons with iron and manganese concretions may indicate seasonal saturation or capillary fringe. Depth to iron and manganese concentrations will generally provide the most conservative estimate to depth to seasonal high water table.

10.2.4 Where the capillary fringe is also considered as part of the saturated zone for defining the limiting depth, soil texture can be used to estimate the thickness of the capillary fringe as shown in Table 13.

10.3 *Depth to Impermeable Layers*—Observable soil features that indicate layers that limit downward movement of water include slowly permeable soil genetic horizons, such as fragipans, duripans, and caliche, soil horizons with very weak, platy or massive structure, very firm or very hard rupture resistance, layers that are moderately cemented, strongly cemented or indurated, and high penetration resistance.

10.4 *Depth to Excessively Permeable Layers*—Coarse sand, very gravelly, extremely gravelly or soils with greater than 15 % rock fragments larger than gravel generally do not provide adequate treatment of wastewater effluent. Such layers are identified based on the size class and amount of sand in the < 2 mm fraction, and the percentage of rock fragments in the >2 mm fraction.

10.5 Strong textural contrasts between soil layers (fine-grained over coarse grained, or coarse-grained over fine-grained) impede both unsaturated and saturated flow. Where

should be given to the possibility that they are relict features, especially when agricultural tile drainage is a common

SOILS EVALUATION FORM PER ASTM D 5921

PROPERTY OWNER: _____

LOCATION: _____

DATE: _____

INVESTIGATORS: _____

EXCAVATION METHOD: _____

WEATHER, LIGHTING: _____

Test Pit #	Horizon	Depth	Color	Rock Fragments		Texture		Rock	Mottles				Structure			RUPTURE RESISTANCE		Cementation	Penetration	Resistance	Roots			Pores				Carbonates	Boundary		Water	Limiting Depth
					SAND SIZE				ABUNDANCE	SIZE	CONTRAST		GRADE	SIZE	SHAPE	DRY	MOIST				ABUNDANCE	SIZE	LOCATION	ABUNDANCE	SIZE	LOCATION	SHAPE		DISTINCTNESS	TOPOGRAPHY		
O A E B C R	a b c d e f g h i k l m n		H,V/C C≤2 H=N,5	GR CB CN FL ST B	V X	CO F VF	C SIC SC SICL CL SIL SI L SL LS S	UWB WB	f c m	1 2 3	F D P	Color H V/C	1 2 3	VF F M CO VC	PR COL ABK SBK GR PL MA SGR	L SH MH H VH EH R VR	L VFR FR FI VFI EFI SR R VR	CO XWC VWC WC MC SC VSC I	EL VL L M H VH EH		VI 1 2 3 4	VI 1 2 3 4	P C M S T	1 2 3	VI 1 2 3 4	IN EX	VS TU	4 0 1 2 3	A C G D	S W I B	D M W	

*Bold print indicates soil features that may limit depth.

FIG. 7 Soil Evaluation for On-Site Septic System

opq r s ss t v w x y z

excess soil water percolates through the soil, such contrasts will also be indicated by mottling, whereas mottling may not be evident in areas where evapotranspiration exceeds precipitation.

11. Report

11.1 Reporting of results of the subsurface investigation should be integrated with the results of the surface investigation. The local or state regulatory authority may have developed forms or formulas for investigation reports, in which case, these should be used.

11.2 The report on the results of the subsurface soils examination should include the following:

11.2.1 Site map prepared for the surface site characterization investigation (see D 5879) with locations of the test pits or soil borings located and identified.

11.2.2 Completed field data from each test pit on a standard form. A sample form and its headings is shown in

Fig. 6. An example of a completed form for a site is shown in Fig. 7. A summary of abbreviations is shown in Fig. 8.

11.2.3 A narrative of each soil profile describing the major features and interpreting the limiting depths.

12. Precision and Bias

12.1 This practice provides qualitative information only, therefore, a precision and bias statement is not applicable.

12.2 Because the analysis is based on visual and manual tests, the observer should maintain proficiency of visual and manual testing ability by periodic review of standards and standard materials and by collecting random samples for laboratory analysis for comparison with visual and manual analysis.

13. Keywords

13.1 septic system; site characterization; soil classification; soil description; visual classification

SOILS EVALUATION FORM PER ASTM D 5921

PROPERTY OWNER: _Sarah & Rocky Shale, 37 Stone Road, Mountain View, AR 72000 501-234-5678_
LOCATION: _Green Meadows Road, Port Angeles, Clallam County, WA, Parcel # 03-30-29-41'0030_
DATE: _3/25/95   10:00 AM - 11:30 A.M._
INVESTIGATORS: _Sam S. Spade CPSS PE, Soils-R-Us Inc. 111 East 3rd St. Seattle WA 98000 206 555 1234_
EXCAVATION METHOD: _Case 580C Backhoe J & C Excavating, Inc._
WEATHER, LIGHTING: _Clear, 60°F, Full Sunlight_

Test Pit #	Horizon			Depth (CM)	Color	Rock Fragments		Texture		Rock	Mottles				Structure			Rupture Resistance		Cementation	Penetration Resistance	Roots			Pores				Carbonates	Boundary		Water	Limiting Depth
								SAND SIZE			ABUNDANCE	SIZE	CONTRAST		GRADE	SIZE	SHAPE	DRY	MOIST			ABUNDANCE	SIZE	LOCATION	ABUNDANCE	SIZE	LOCATION	SHAPE		DISTINCTNESS	TOPOGRAPHY		
1	A	P		0-20	GR BRN 10YR 5/2	—	—	F	SL	—	—	—	—	—	1	W	GR	—	VFR	CO	VL	4	12	T	2	VI	IN	VS	—	A	S	M	—
	B	—		20-75	DK YEL BRN 10YR 4/4	—	GR	—	SL	—	—	—	—	—	2	M	SBK	—	FR	CO	L	3	12	P	2	1	IN	VS	—	C	W	M	—
	C	—		75-120	GRAY 10YR 5/1	V	GR	—	LS	—	C	1	P	DK BRN 7.5YR 4/4	—	—	SGR	—	EFI	WC	H	3	12	M	—	—	—	—	—	—	—	W	YES
																																	—
2	A	P		0-18	GR BRN 10YR 5/2	—	—	F	SL	—	—	—	—	—	1	W	GR	—	VFR	CO	VL	3	12	T	2	VI	IN	VS	—	A	S	M	—
	B	—		18-80	DK YEL BRN 10YR 4/4	—	GR	—	SL	—	—	—	—	—	2	M	SBK	—	FR	CO	L	3	12	P	2	1	IN	—	—	C	W	M	—
	C	—		80-95	GRAY 10YR 5/1	Y	GR	—	LS	—	C	2	P	RED DISH 5YR 4/3	—	—	SGR	—	EFI	WC	H	4	12	M	—	—	—	—	—	—	—	W	YES
3	A	—		0-10	DK BRN 10YR 4/2	—	—	—	SIL	—	—	—	—	—	1	M	GR	—	VFR	CO	VL	3	1	T	2	VI	IN	VS	—	C	S	M	—
	B	—		10-70	DK GR BRN 10YR 4/2	—	—	—	SIL	—	f	2	F	YEL BRN 2.5Y 4/4	2	M	PR	—	FR	CO	M	1	1	P	2	2	EX	TU	—	G	W	M	YES
	C	—		70-150	GR BRN 2.5YR 3/2	—	GR	—	SIL	—	f	2	D	STR BRN 7.5YR 5/6	—	—	MA	—	FI	CO	H	2	1	M	—	—	—	—	—	—	—	W	YES

*Bold print indicates soil features that may limit depth.

FIG. 8   Example Soil Evaluation Form for Typical Site

Horizon
a - organic <1/6
b - buried
c - concretions
d - root restriction
e - organic 1/6 -2/5
f - frozen
g - gleyed
h - fluvial organic, v, c<3
i - organic > 2/5
k - carbonates
m - cemented
n - sodium
o - sesquioxides
p - plowed
q - silica
r - rock
s - fluvial organic, v, c>3
ss - slickensides
t - clay
v - plinthite
w - color and structure
x - fragipan
y - gypsum
z - salts

Rock Fragments
V - very
X - extremely
GR - gravelly
CB - cobbly
ST - stony
B - bouldery
CN - channery
FL - flaggy

Texture
CO - Coarse
F - Fine
VF - Very Fine
C - Clay
SIC - Silty Clay
SC - Sandy Clay
SICL - Silty Clay Loam
SI - Silt
L - Loam
SL - Sandy Loam
LS - Loamy Sand
S - Sand

Rock
UWB - Unweathered Bedrock
WB - Weathered Bedrock

Mottles
f - Few
C - Common
M - Many
1 - Fine
2 - Medium
3 - Coarse
F - Faint
D - Distinct
P - Prominent

Structure
1 - Weak
2 - Moderate
3 - Strong
VF - Very Fine
F - Fine
M - Medium
CO - Coarse
VC - Very Coarse
PR - Prismatic
COL - Columnar
ABK - Angular Blocky
SBK - Subangular Blocky
GR - Granular
PL - Platy
MA - Massive
SGR - Single Grain

Rupture Resistance
DRY:
L - Loose
S - Soft
SH - Slightly Hard
MH - Moderately Hard
VH - Very Hard
H - Hard
EH - Extremely Hard
R - Rigid
VR - Very Rigid

MOIST:
L - Loose
VFR - Very Friable
FR - Friable
FI - Firm
VFI - Very Firm
EFI - Extremely Firm
SR - Slightly Rigid
R - Rigid
VR - Very Rigid

Cementation
CO - Non Cemented
XWC - Extremely Weakly Cemented
VWX - Very Weakly Cemented
WC - Weakly Cemented
MC - Moderately Cemented
SC - Strongly Cemented
VSC - Very Strongly Cemented
I - Indurated

Penetration Resistance
EL - Extremely Low
VL - Very Low
L - Low
M - Moderate
H - High
VH - Very High
EH - Extremely High

Roots
V1 - Very Few
1 - Few
2 - Moderately Few
3 - Common
4 - Many
V1 - Very Fine
1 - Fine
2 - Medium
3 - Coarse
4 - Very Coarse
P - Between Peds
C - In Cracks
M - Matted On Top
S - Matted On Stones
T - Throughout

Pores
1 - Few
2 - Common
3 - Many
V1 - Very Fine
1 - Fine
2 - Medium
3 - Coarse
4 - Very Coarse
IN - In Ped
EX - Ex Ped
VS - Vesicular
TU - Tubular

Carbonates
4 - Non Effervescent
0 - Very Slightly Effervescent
1 - Slightly Effervescent
2 - Strongly Effervescent
3 - Violently Effervescent

Boundary
a - Abrupt
c - Clear
g - Gradual
d - Diffuse
s - Smooth
w - Wavy
I - Irregular
b - Broken

Water
D - Dry
M - Moist
W - Wet

FIG. 9 Definitions for Abbreviations

REFERENCES

(1) "Keys to Soil Taxonomy," 6th Edition, *Soil Survey Staff*, U.S. Government Printing Office, 1994.

(2) Buol, S. W., Hole, F. D., and McCracken, R. J., *Soil Genesis and Classification*, 2nd Edition, The Iowa State University Press, Ames, 1980.

(3) "Soil Survey Manual," *U.S.D.A. Agricultural Handbook No. 18*, 1993.

(4) "Soil Taxonomy: A Basic System of Soil Classification for Making and Interpreting Soil Surveys," *U.S.D.A. Agricultural Handbook No. 436*, 1975.

(5) *Glossary of Soil Science Terms*, Soil Science Society of America, July 1987.

(6) *Munsell Soil Color Charts*, Munsell Color Company, 2441 N. Calvert St., Baltimore, MD 21218.

(7) Cogger, C. G., *Detailed Soils Descriptions for Onsite Sewage (Soils II)*, Washington State University Extension, Puyallup, 1992.

(8) Boulding, J. R., "Description and Sampling of Contaminated Soils," *U.S. E.P.A. Document 625/12-9/002*, November, 1991.

(9) *Description and Sampling of Contaminated Soils, A Field Guide*, 2nd Ed., J. R. Boulding, Lewis Publishers, Boca Raton, FL, 1994.

(10) Mausbach, M. J., "Soil Survey Interpretations for Wet Soils," in *Proc. 8th Int. Soil Correlation Meeting (VIII ISCOM): Characterization, Classification, and Utilization of Wet Soils*, J. M. Kimble (ed.), USDA Soil Conservation Service, National Soil Survey Center, Lincoln, NE, pp. 172–178, 1992.

(11) McRae, S. G., *Practical Pedology: Studying Soils in the Field*, Halsted Press, Chichester, NY, 1988.

Standard Practice for
Preliminary Sizing and Delineation of Soil Absorption Field Areas for On-Site Septic Systems[1]

This standard is issued under the fixed designation D 5925; the number immediately following the designation indicates the year of original adoption or, in the case of revision, the year of last revision. A number in parentheses indicates the year of last reapproval. A superscript epsilon (ε) indicates an editorial change since the last revision or reapproval.

1. Scope

1.1 This practice covers procedures for estimating the dimensions and marking the boundaries of a soil absorption area for an on-site septic system involving residential-strength wastewater. It can also be used to estimate the dimensions of commercial on-site septic systems where wastewater strengths are similar to residential wastewater.

1.2 This practice can also be used for marking the boundaries of the area for a septic system constructed filter bed.

1.3 This practice can be used at any site where a potentially suitable or recommended field area has been identified in accordance with Practices D 5879 and D 5921.

1.4 Non-metric units remain the common practice in design and installation of on-site waste disposal systems, and are used in this practice. Use of SI units given in parentheses is encouraged, if acceptable to the appropriate permitting agency.

1.5 *This standard does not purport to address all of the safety concerns, if any, associated with its use. It is the responsibility of the user of this standard to establish appropriate safety and health practices and determine the applicability of regulatory limitations prior to use.*

2. Referenced Documents

2.1 *ASTM Standards:*
D 5879 Practice for Surface Site Characterization for On-Site Septic Systems[2]
D 5921 Practice for Subsurface Characterization of Test Pits for On-Site Septic Systems[2]

3. Terminology

3.1 *Descriptions of Terms Specific to This Standard:*

3.1.1 *clinometer*—an instrument for measuring inclination, as in topographic slope.

3.1.2 *constructed filter bed (CFB)*—for the purposes of this practice, material, usually of a sandy texture, placed above or in an excavated portion of the natural soil for filtration and purification of wastewater from an on-site septic system.

3.1.3 *on-site septic system*—for the purposes of this practice, any wastewater treatment and disposal system that uses a septic tank or functionally equivalent device for collecting waste solids and treats wastewater using natural soils, or constructed filter beds with disposal of the treated wastewater into the natural soil.

3.1.4 *potentially suitable field area*—the portions of a site that remain after observable limiting surface features, such as excessive slope, unsuitable landscape position, proximity to water supplies, and applicable setbacks, have been excluded.

3.1.5 *recommended field area*—the portion of the potentially suitable field area at a site that has been determined to be most suitable for an on-site septic system soil absorption field or filter bed based on surface and subsurface observations.

3.1.6 *soil absorption (SA) area*—an area of natural soil used for filtration and purification of wastewater from an on-site septic system.

3.1.7 *soil absorption field area (SAF)*—an area that includes soil absorption trenches and any soil barriers between the trenches. Also called a leachfield.

3.1.8 *soil absorption trench*—an excavated trench, usually 1.5 to 3 ft wide that receives wastewater for treatment. Also called a lateral or leachline.

4. Significance and Use

4.1 This practice should be used in conjunction with a surface and subsurface site investigation to delineate a recommended field area that is adequate for any septic system that can reasonably be anticipated for the site. If actual design results in a smaller field area, the boundaries can be modified accordingly.

4.2 Staking and flagging procedures in the practice help prevent accidental disturbance of a recommended septic system field area by equipment traffic and other construction activities prior to installation of the system. Soil disturbance resulting in compaction from heavy equipment traffic or removal by excavation equipment usually invalidates the results of the surface and subsurface investigation that led to recommendation of a field area.

4.3 In the event of suspected disturbance or removal of natural soil in the recommended field area, soil elevation benchmarks established by this practice allow assessment of the actual extent of disturbance or soil removal.

4.4 This practice should also be used where topographic limitations create uncertainty as to whether a potentially suitable field area for a septic system will provide a large enough absorption area to treat anticipated wastewater flows. In such situations clear demarcation of the suitable areas will also provide greater assurance of proper system installation.

5. Field Equipment

5.1 A clinometer or hand level and rod that is marked in

---

[1] This practice is under the jurisdiction of ASTM Committee D-18 on Soil and Rock and is the direct responsibility of Subcommittee D18.01 on Surface and Subsurface Characterization.

Current edition approved April 10, 1996. Published November 1996.

[2] *Annual Book of ASTM Standards*, Vol 04.09.

TABLE 1 Soil Absorption/Filter Bed Area Requirements for Different Wastewater Flow and Soil Loading Rates

| Soil Loading, R | | Wastewater Flow | | | | | | |
| | | 150 gal | | 300 gal | | 450 gal | | 600 gal/day |
gal/day ft²	sq gal/day ft²	ft²	Square Root	ft²	Square Root	ft²	Square Root	ft²	Square Root
0.2	5.0	750	...	1500	...	2250	...	3000	...
0.25	4.0	600	...	1200	...	1800	...	2400	...
0.3	3.3	500	...	1000	...	1500	...	2000	...
0.35	2.9	429	...	857	...	1286	...	1714	...
0.4	2.5	375	...	750	...	1125	...	1500	...
0.45	2.2	333	...	667	...	1000	...	1333	...
0.5	2.0	300	...	600	...	900	...	1200	...
0.6	1.7	250	16	500	22	750	27	1000	32
0.7	1.4	214	15	429	21	643	25	857	29
0.8	1.3	188	14	375	19	563	24	750	27
0.9	1.1	167	13	333	18	500	22	667	26
1.0	1.0	150	12	300	17	450	21	600	24
1.1	0.9	136	12	273	17	409	20	545	23
1.2	0.8	125	11	250	16	375	19	500	22

feet/metric increments and at the eye level of the investigator are used for measuring slope and delineating topographic contours. A compass may be useful for defining position of the field area. A single person can take measurements if the rod has a point that can be driven into the ground so that it stands vertically, as described in 7.1. An extendible surveyor's rod with a tripod can also be used by a single person and may facilitate the elevation benchmarking procedure described in 7.2.

5.2 A 100 ft tape or longer can be used to measure the length and width of the field area. A screwdriver or spike is also useful for anchoring one end of the tape when making measurements. Where there are no concerns about the adequacy of the available suitable area, pacing can be used as an alternative to a tape. In this case, the investigator should periodically check the accuracy of his or her pace against a known distance.

5.3 Stakes and flagging are used to mark the corners and other boundaries of the field area. Stakes can be of any material (wood, fiberglass, metal) that is durable enough to remain standing during the period from staking until installation of the system. If the area is to be mowed, the stakes should be tall enough and sturdy enough to prevent accidental damage to the stake or the mower. If there is any possibility that the stakes might be confused with other markers at the site, colored flagging coded for different purposes can be used. Generally, actual fencing is not required unless heavy equipment traffic is expected to run regularly by the area.

6. Procedure for Estimating Field Dimensions

6.1 Use this procedure for preliminary sizing of the area required for a soil absorption field or constructed filter bed for the purpose of staking a field areas as described in Section 7. This procedure should also be used whenever marginal site surface and subsurface conditions indicate doubt as to whether there is a large enough area that is suitable for an on-site septic system.

6.2 Factors that affect the soil absorption field (SAF) area requirements include wastewater quantity (typically expressed as gallons per day (gpd)), loading rate (typically expressed as gpd/ft²) that is derived from soil characteristics or percolation test results (see 6.3.3), and trench spacing, that determines the area of soil between absorption trenches.[3] Dividing Factor 1 by Factor 2 gives the total required soil absorption (SA) area.[4] The SA area plus the area represented by soil barriers between trenches yields the SAF area. In some jurisdictions the SA area may be determined by the number of bedrooms based on assumptions concerning wastewater flow and loading rates. Factors affecting the area required for constructed filter beds are the same as for soil absorption fields.

6.3 *Method for Estimating Soil Absorption Field Dimensions*—The method described here assumes residential-strength wastewater and includes tables that should be generally applicable to most parts of the United States. Alternative tables using other wastewater flow and soil loading rates can easily be developed. The method for estimating a SAF area involves the following steps: determining wastewater flow, determining soil loading rate, determining required SA area, determining the number of trenches and their length to provide the required SA area, and determining the width of the field based on the number of trenches.[5]

6.3.1 *Wastewater Flow*—Typically wastewater flow is determined by the number of bedrooms in a residence. 150 gpd per bedroom, recommended by U.S. EPA (1),[6] is widely used. Table 1 includes rates of 150, 300, 450, and 600 gpd that correspond to a 1-, 2-, 3-, and 4-bedroom house, respectively. Some jurisdictions may use different loading rates (120 to 200 gpd per bedroom, 60 to 150 gpd per person). Reference (2) compiles design flow rates specified in state regulations in the United States.

6.3.2 *Soil Loading Rate*—Increasingly, septic system design is being based on soil loading rates based on soil texture and structure as determined by subsurface site characterization as covered in Practice D 5879. Table 1 includes loading rates from 0.2 to 1.2 gpd/ft². Loading rates may also be determined by percolation test results (3). Figure 1 can be used to convert percolation rates measured as minutes per

---

[3] Other factors that may need to be considered include: wastewater strength (suspended solids, biological/chemical oxygen demand, nitrogen, phosphorus, etc.), potential for ground-water mounding under absorption trenches or constructed filter beds, and evapotranspiration. Standard loading rates based on soil characteristics or percolation test results usually assume residential-strength wastewater. Wastewaters with parameters that differ significantly from residential wastewater require special design procedures that are not addressed in this practice. Section 6.3.5 discusses situations where ground-water mounding analysis may need to be considered. In temperate climates evapotranspiration is usually not considered when determining the required SA area because it is zero during winter months. In areas where evapotranspiration is significant throughout the year, it may be possible to reduce SA area requirements. This requires a water budget analysis for the time of year when evapotranspiration is at a minimum and adjusting the field size accordingly. For example, if 20 % of the wastewater entering the soil could be expected to be transpired, the field size could be reduced by one-fifth.

[4] This actually gives only the absorption area of the bottom of the trench. Depending on the depth of effluent in a trench additional absorption area is provided by the sidewalls. Normal practice is to ignore this area when calculating required soil absorption area. However, some jurisdictions allow credit for sidewall area.

[5] This method can also be used to estimate field dimensions for grade soil absorption fields, and trench systems where the lower portions are filled with filter bed material.

[6] The boldface numbers given in parentheses refer to a list of references at the end of the text.

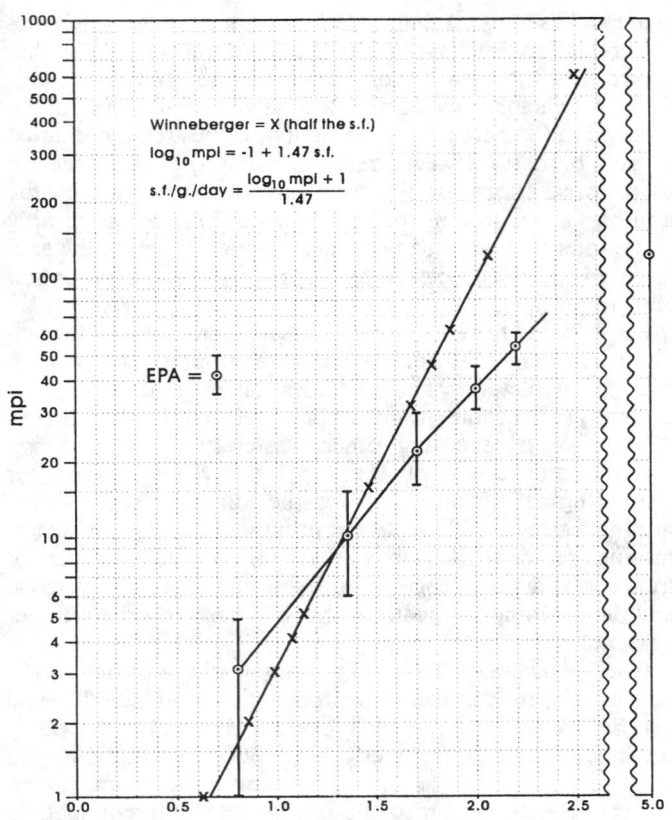

FIG. 1 EPA's and Winneberger's Recommendations for Absorptions Area (Square Feet per Gallon of Sewage per Day) Versus Measured Percolation (Minutes per Inch) (5)

Winneberger = X (half the s.f.)

$$\log_{10} mpi = -1 + 1.47 \, s.f.$$

$$s.f./g./day = \frac{\log_{10} mpi + 1}{1.47}$$

EPA = ○

TABLE 2 Trench Number and Length Chart

Trench Length	Number of Trenches					
	2	3	4	5	6	7
			3-Ft Wide Trench			
ft²	ft²	ft²	ft²	ft²	ft²	ft²
45	270	405	540	675	810	945
50	300	450	600	750	900	1050
55	330	495	660	825	990	1155
60	360	540	720	900	1080	1260
65	390	585	780	975	1170	1365
70	420	630	840	1050	1260	1470
75	450	675	900	1125	1350	1575
80	480	720	960	1200	1440	1680
85	510	765	1020	1275	1530	1785
90	540	810	1080	1350	1620	1890
95	570	855	1140	1425	1710	1995
100	600	900	1200	1500	1800	2100
			1.5-Ft Wide Trench			
ft²	ft²	ft²	ft²	ft²	ft²	ft²
45	135	203	270	338	405	473
50	150	225	300	375	450	525
55	165	248	330	413	495	578
60	180	270	360	450	540	630
65	195	293	390	488	585	683
70	210	315	420	525	630	735
75	225	338	450	563	675	788
80	240	360	480	600	720	840
85	255	383	510	638	765	893
90	270	405	540	675	810	945
95	285	428	570	713	855	998
100	300	450	600	750	900	1050

inch (mpi) to recommended SA area loading rates as suggested by U.S. EPA (7) and Winneberger (4). If percolation test results are reported in inches/hour, convert to minutes per inch (mpi = 60/in./h). Soil loading rates are lower than the saturated hydraulic conductivity of soil in order to take into account reduced infiltration resulting from development of a biological clogging mat on absorption trench surfaces and to allow for unsaturated flow. Some jurisdictions may require consideration of climatic factors such as precipitation and evapotranspiration when determining soil loading rates for soil. Reference (2) compiles application rates defined in state regulations in the United States.

6.3.3 *Soil Absorption Area*—Table 1 shows soil absorption areas required for commonly used wastewater flows and soil loading rates. If the applicable wastewater flow and soil loading rate is not included in Table 1, multiply the actual values determined in 6.3.1 and 6.3.2.

6.3.4 *Trench Width*—The width of a soil absorption trench determines how many square feet of soil absorption are available on the trench bottom per lineal foot of trench. Typical trench widths range from 1.5 to 3.0 ft (5). Use the trench width that represents common installation practice in the area of the site.

6.3.5 *Number and Length of Trenches*—Use Table 2 to determine the possible combinations of number and length of trenches that will provide the soil absorption area deter-

mined in 6.3.3. For example, if the required soil absorption area is 600 ft², and the trenches are 3 ft wide, there are three possible configurations: (*1*) two trenches 100 ft long, (*2*) three trenches 70 ft long (that provide a little more than the required area), and (*3*) four trenches 50 ft long. Select the configuration that best fits the site, giving preferences for the configuration that minimizes the number of trenches.[7] This determines the length of the field for staking, as described in Section 7. Where the vertical separation between the bottom of the disposal component of the on-site septic system and a limiting layer is at or near the minimum allowed, groundwater mounding calculations may be required, especially if more than two or three trenches are required.[8]

6.3.6 *Width of Field*—Whenever two or more trenches are required, the width of a soil absorption field will be larger than for a single trench to account for the soil areas between trenches. Typically soil barriers range from 4.5 to 7.0 ft (7.5 to 10.0 ft on center). Table 3 shows field widths for different combinations of trench spacing, width, and number of trenches. Wider spacings may be required if potential for ground-water mounding is a concern. Use a trench width and spacing that represents common installation practices in

---

[7] Minimizing the number of trenches both simplifies installation and improves wastewater treatment by increasing soil oxygen availability and reducing ground water mounding.

[8] References that may be useful for ground-water mounding calculations in homogeneous horizontal aquifers include: ((6)—Chapter 4), (7,8,9,10) Chapter 13), ((11)—Section 5.7.2), ((12)—Section 4.3.2). The U.S. EPA ground-water mounding analysis procedures (11,12) are based on mound height analysis developed by Glover (13) and summarized by Bianchi and Muckel (14). Finnemore (15) describes procedures for ground-water mounding calculations in layered horizontal aquifers, and for four types of sloping aquifers (uniform flow/homogeneous, uniform flow/layered, nonuniform flow/homogeneous, and nonuniform flow/layered).

TABLE 3 Soil Absorption Field Width Chart

NOTE—All units in feet except number of trenches.

Soil Barrier Width	Trench Distance Width on Center		Number of Trenches					
			2	3	4	5	6	7
			Field Width					
ft	ft	ft	ft	ft	ft	ft	ft	ft
4.5	7.5	3.0	10.5	18.0	25.5	33.0	40.5	48.0
		1.5	7.5	13.5	19.5	25.5	31.5	37.5
5.0	8.0	3.0	11.0	19.0	27.0	35.0	43.0	51.0
		1.5	8.0	14.5	21.0	27.5	34.0	40.5
5.5	8.5	3.0	11.5	20.0	28.5	37.0	45.5	54.0
		1.5	8.5	15.5	22.5	29.5	36.5	43.5
6.0	9.0	3.0	12.0	21.0	30.0	39.0	48.0	57.0
		1.5	9.0	16.5	24.0	31.5	39.0	46.5
7.0	10.0	3.0	13.0	23.0	33.0	43.0	53.0	63.0
		1.5	10.0	18.5	27.0	35.5	44.0	52.5

the area of the site, and the number of trenches as determined in 6.3.5.[9] The width determined in this step represents the minimum width of the field area. Depending on site topography, requirements that trenches follow the contour may result in staked areas larger than the area indicated by multiplying length determined in 6.3.5 and the width determined in this step.

6.4 *Method for Estimating Above-Grade Constructed Filter Bed (CFB) Dimensions*—Above-grade constructed filter beds typically require less area than soil absorption fields because soil loading rates are reduced by use of optimum filter material[10] and no soil barriers are required between trenches. To estimate dimensions of above-grade constructed filter beds, follow steps described in 6.3.1 through 6.3.3. In Table 1 the second column under each wastewater flow amount give the square root of the soil absorption area, which is the length in feet of a square providing the required filter area. If topography dictates a different shape, dividing the soil absorption area by the desired width or length provides the dimensions for staking.[11] Depending on the design of the constructed filter bed, additional area should be added to account for any berming around the margins of the filter bed.

7. General Procedure for Staking Field Area

7.1 The general procedure for marking the boundaries of a field area described here applies to a recommended field area where contours are parallel and do not curve significantly. Additional procedures for other topographic situations are described in Section 8. The procedure assumes that field width and length have been determined using the

procedures described in Section 6. The procedure is written for a single person using a clinometer and rod (see 5.1). The procedure is somewhat simpler if two people are involved, one with the clinometer and the other with the rod because less walking back and forth is required for setting the rod and taking sitings. An alternative approach is to set the rod at benchmark point that is visible from the whole area, as described in 7.2, measuring relative elevations at each corner of the field area, and calculating slope gradients.

7.1.1 *Rectangular Field*—Figure 2(*a*) shows the sequence for marking the corners of a rectangular field (any corner can be used as a starting point, and the sequence of steps can be modified as a matter of convenience): (*1*) Place stake and the rod at corner 1 and measure or pace along the contour a distance equal to the trench length (Corner 2); (*2*) sight back to the rod with the clinometer, and adjust position upslope or downslope if required to get a reading of zero (level); (*3*) place stake at Corner 2 and walk back to Corner 1; (*4*) measure or pace a distance equal to the field width upslope (or downslope) being sure to cross the slope at right angles to the contour (Corner 3); (*5*) place a stake at Corner 3 and site back to the rod with the clinometer to measure and record the slope; (*6*) go back to Corner 1, retrieve the rod and return to Corner 3; (*7*) place the rod at Corner 3 and measure or pace a distance equal to the trench length along the contour (Corner 4); (*8*) sight back to the rod with the clinometer, and adjust position upslope or downslope if required to get a reading of zero (level); (*9*) place stake at Corner 4, and measure or pace the distance between Corner 4 and Corner 2; (*10*) if the distance between Corner 4 and Corner 2 is equal to or greater than the distance between Corners 1 and 3, the procedure is completed; (*11*) if the distance between Corners 4 and 2 is less than the distance between Corners 1 and 3, move either stake to obtain the required field width and move Stake 1 or 3 accordingly so that lengths 1 and 2 and 3 and 4 are on the contour.

7.1.2 *Nonrectangular Field*—Figure 2(*b*) shows the sequence for marking a field where the field boundaries do not meet at right angles. This situation may arise when limiting surface or subsurface features do not allow a square or rectangular field area. The steps are basically the same as for a rectangular field (see 7.1.1), except that in Step 4 it is necessary to measure or pace a distance equal to the field width from a point between Corners 1 and 2 such that the line connecting Corner 3 and Trench 1 and 2 is perpendicular (see Fig. 2(*b*)).

7.2 *Establishing Benchmark Elevations*—The elevation and location of corner stakes should be defined with reference to property lines or a permanent feature of the site. This allows relocating the field in the event that the markers delineating the field are removed, and determining whether the original land surface has been altered by excavation or fill if there is concern that the recommended field area has been subsequently disturbed.

7.2.1 A benchmark point, such as a property corner or other permanent landmark on the site should be selected for establishing the relative elevations of the recommended field area's corners. A benchmark point that allows direct horizontal siting on the rod with a clinometer or hand level from the highest and lowest points of the recommended field area is best. An extensible surveyor's rod with a tripod may

---

[9] Where more than three trenches are required, the field can also be divided into two areas on either side of the wastewater distribution box in order to reduce ground water mounding. For example, an SAF area requiring six 3-ft wide trenches 75 ft long with a 6 ft soil barrier width could either be configured as an area 75 by 48 ft (six parallel trenches) or 150 by 21 ft (two areas on either side of the distribution box with three trenches). In the latter case, when staking the field area, the length should be increased to provide an additional soil barrier between the two field areas where the distribution lines will be laid.

[10] Some jurisdiction require sizing of CFBs based on the loading rate of the soil upon which the filter bed is constructed if the CFB is in direct contact with the natural soil.

[11] Constructed filter beds wider than ten feet tend to have inadequate oxygen supply in the soil below the center of the bed. Wider beds conserve space; narrower beds increase wastewater renovation performance and decrease ground-water mounding.

ASTM D 5925

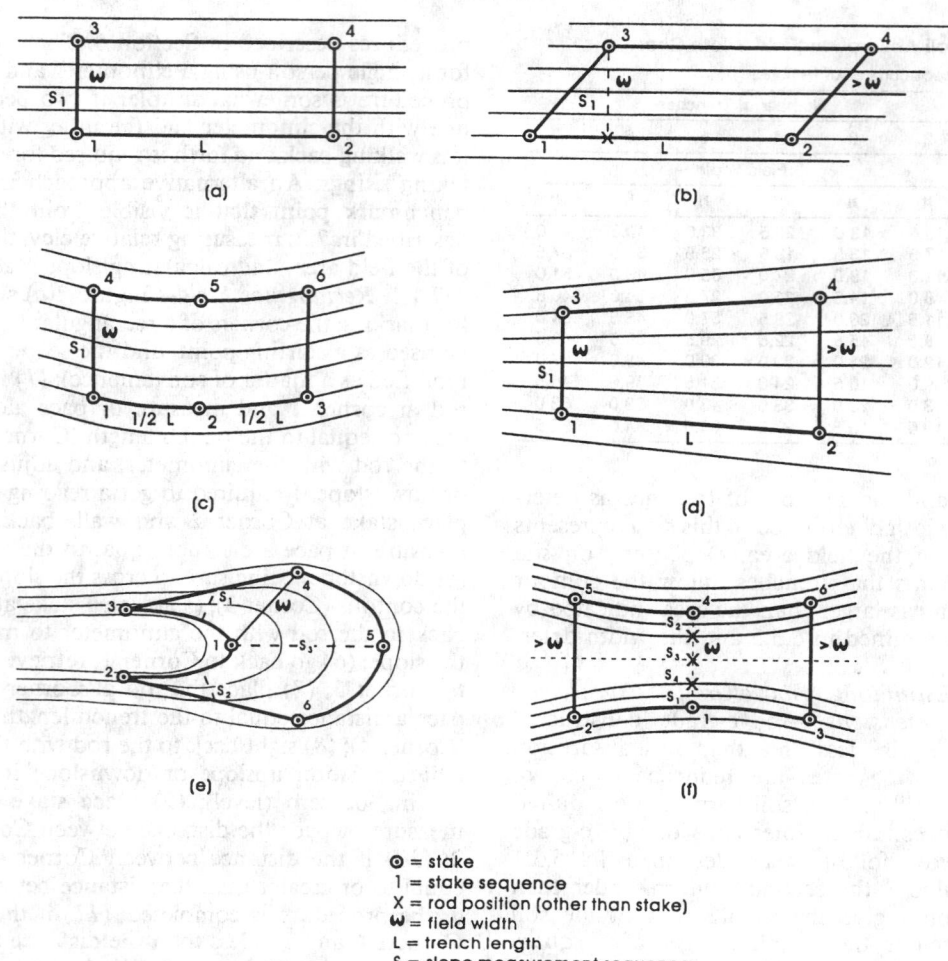

⊙ = stake
1 = stake sequence
X = rod position (other than stake)
W = field width
L = trench length
S_1 = slope measurement sequence

FIG. 2 Staking and Slope Measurement Sequences for Different Absorption Field Shapes and Topographic Settings

facilitate establishing elevations on sloping sites. Setting the ground surface of the benchmark point as zero and subtracting the eye-level height of the observer from the height on the rod intersected by the horizontal siting (that is, zero gradient) gives the elevation relative to the benchmark ground surface. Since the upper and lower lengths of the field area are laid out on the contour, elevations of only one upper and one lower corner need to be established. For example, in Fig. 2(a) measuring the elevation relative to the benchmark at Points 1 and 3 will establish the lowest and highest elevations of the field.

7.2.2 At a minimum, one corner of the field area should be located accurately with respect to the benchmark used to determine elevations and compass readings taken to define the position of the other corners.

8. Special Topographic Situations

8.1 Figures 2(c) through 2(f) illustrate shapes of fields that result when contours are either not parallel or curve sig-

nificantly.[12] Special considerations in marking field areas in these situations are described below. Special considerations in measuring slope are covered in 8.2.

8.1.1 *Parallel Curving Contours (see Fig. 2(c))*—Where contours are parallel, staking procedures are basically the same as described in 7.1.1, except that it is helpful to place additional stakes along the contour to delineate the field boundaries more clearly, as shown in Fig. 2(c).

8.1.2 *Diverging Contours (see Fig. 2(d))*—Where contours are not parallel, the procedures are the same as described in Section 7.1.1, with the proviso that it is essential that the starting point for staking be at the location where contours are closest together. Failure to do this will result in the field width being less than the required amount at the other end of the field.

---

[12] If the limiting depth identified in Practice D 5921 is deep, some departure of trenches from the contour may be acceptable. For example, if the allowed variation in trend depth were 18 to 36 in., there could be as much as a 1.5 ft difference in elevation from one end of the trench to the other at the ground surface. Any departures from staking along the contour should be documented and justified.

284

8.1.3 *Strongly Curving Contours (see Fig. 2(e)*—The noseslope of narrow ridgetops generally have strongly curving contours which often quickly converge away from the noseslope to form slopes that a too steep to be suitable for soil absorption fields. This requires use of additional stakes as described in 8.2, but the staking sequence is different as shown in Fig. 2(*e*). Slope measurements in this situation should be measured at the point they are steepest, which will usually be at Stakes 4 and 6 in Fig. 2(*e*).

8.1.4 *Saddle (see Fig. 2(f)*—Normally convex contours are considered unfavorable topographic positions for soil absorption fields because subsurface flow concentrates in such areas. On narrow ridgetops this is not as much of a concern as in other landscape positions, but special care is required when marking such areas. As shown in Fig. 2(*f*) the staking sequence begins where the saddle is narrowest, and proceeds accordingly.

8.2 *Fields With Unequal Lateral Lengths*—Where site surface and subsurface conditions do not allow delineating a soil absorption field with trenches of equal length, alternative designs, such as serial trenches or a pressure distribution system, may be feasible. The procedures described in the previous section can be adapted by first calculating the total linear feet of trench that is required. This can be obtained by dividing soil absorption area (see 6.3.3) by the trench width. Trenches of varying lengths can then be staked out on the contour until the required total length of trenches is reached.

8.3 *Measuring Slope*—Where the contours of the slope on which a soil absorption field is staked are evenly spaced (see Figs. 2(*a*), 2(*b*), and 2(*c*)) slope measurements can be taken over the full width of the field. In these situations, a single slope measurement (S_1) along the distance w is adequate for documenting compliance with any applicable slope restrictions. A single slope measurement where contours diverge (Fig. 2(*d*)) is also adequate, provided the measurement is taken where contours are closest together.

8.3.1 *Single Slope Break*—The rod should be positioned at the point a slope changes and the slope measured above and below the rod. If the field area is on a slope close to the maximum allowed, and spacing of contours changes over the width of the field, measure the maximum slope rather than the average to demonstrate compliance with any applicable slope limitations. For example, Fig. 3 illustrates how taking an average reading over a distance of 50 ft can give the mistaken impression that the slope is suitable. In this

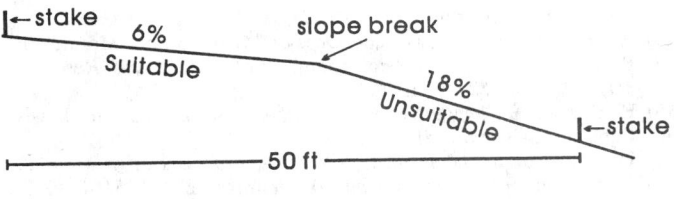

Average Slope = 12%
Maximum Allowed Slope = 15%

FIG. 3 Slope Measurements for Soil Absorption Field Areas Should Include Maximum Rather Than Average Slopes

example, if a field area had been staked out and only the average slope measured, half of the field area would have been out of compliance.

8.3.2 *Complex Slope Breaks*—Complex slopes may require moving the rod a number of times to adequately characterize slope. With strongly curving contours (see Fig. 2(*e*)) several slope measurements may be required to determine maximum slope (S_1 and S_2) and another to measure minimum slope (S_3). Measuring maximum slope in a topographic saddle (see Fig. 2(*f*)) requires measuring slope on both sides (S_1 and S_2). In this case the rod can be left at Stake 1 and Stake 4 and slope measurements taken from the top of the ridge or any break before the top of the ridge. If there is a slope break before the top of the ride, the rod should also be placed at the top of the ridge and slopes measured from the break on either side (S_3 and S_4).

9. Report

9.1 The locations of all stakes used to delineate the recommended field area and all slope measurements should be recorded on the sketch map developed for the site investigation report (see Section 8 in Practice D 5879). If the sketch map is not drawn accurately to scale, the following information should be provided on the sketch map: the distance and compass direction from the benchmark point (see 7.2) to one corner of the field, length of each side of the field, and compass directions of each side. As an alternative to compass directions, distances from each corner to property boundaries or other permanent landmarks on the site can be noted.

10. Keywords

10.1 field investigations; preliminary investigations; septic systems; site characterization; site investigations

REFERENCES

(1) U.S. Environmental Protection Agency, *Design Manual: Onsite Wastewater Treatment and Disposal Systems*, EPA/625/1-80-012, 1980.

(2) National Small Flows Clearinghouse (NFSC), *Location and Separation Guidelines from the State Regulations*, NFSC, West Virginia University, P.O. Box 6064, Morgantown, WV 26506-8301, 800/624-8301, 1995.

(3) National Small Flows Clearinghouse (NFSC), *Application Rates and Sizing of Fields from the State Regulations*, NFSC, Morgantown, WV, 1995.

(4) Winneberger, J. T., *Septic Tank Systems*, Butterworth Publishers, Stoneham, MA, 1984.

(5) Perkins, R. J., *Onsite Wastewater Disposal*, Lewis Publishers, Chelsea, MI, 1989.

(6) Canter, L. W. and Knox, R. C., *Septic Tank Systems Effects on Ground Water Quality*, Lewis Publishers, Chelsea, MI, 1985.

(7) Finnemore, E. J., "Estimation of Ground-Water Mounding Beneath Septic Drainfields," *Ground Water 31(6)*, 1993, pp. 884–889.

(8) Finnemore, E. J. and Hantzsche, N. N., "Ground-Water Mounding Due to On-Site Sewage Disposal," *J. Irrigation and*

Drainage Engineering, ASCE 109(2), 1983, pp. 199–210.

(9) Hantush, M. S., "Growth and Decay of Groundwater-Mounds in Response to Uniform Percolation," *Water Resources Research* 3(1), 1967, pp. 227–234.

(10) Kaplan, O. B., *Septic Systems Handbook*, Second Edition, Lewis Publishers, Chelsea, MI, 1991.

(11) U.S. Environmental Protection Agency, *Process Design Manual: Land Treatment of Municipal Wastewater*, EPA/625/1-81-013, 1981.

(12) U.S. Environmental Protection Agency, *Process Design Manual: Land Treatment of Municipal Wastewater, Supplement on Rapid*

Infiltration and Overland Flow, EPA/625/1-81-013a, 1984.

(13) Glover, R. E., *Mathematical Derivations as Pertaining to Ground Water Recharge*, USDA Agricultural Research Service, 1961.

(14) Bianchi, W. C. and Muckel, C., *Ground Water Recharge Hydrology*, ARS 41-161, USDA, Agricultural Research Service.

(15) Finnemore, E. J., "Water Table Rise in Layered Soils Due to Onsite Sewage Disposal," In *On-Site Wastewater Treatment (Proc. 7th Int. Symp. on Individual and Small Community Sewage Systems), ASAE Publ. No. 18-94*, American Society of Agricultural Engineers, St. Joseph, MI, 1994, pp. 200–208.

1.2 DATA ELEMENTS

Standard Specification for
Content of Digital Geospatial Metadata[1]

This standard is issued under the fixed designation D 5714; the number immediately following the designation indicates the year of original adoption or, in the case of revision, the year of last revision. A number in parentheses indicates the year of last reapproval. A superscript epsilon (ε) indicates an editorial change since the last revision or reapproval.

1. Scope

1.1 This specification covers the information content of metadata for a set of digital geospatial data. This specification provides a common set of terminology and definitions for concepts related to these metadata.

1.2 The use of the term "geographic information system" and its definition in this specification is not intended to introduce a standard definition.

1.3 This specification covers minimum content and processing requirements for geospatial metadata.

1.4 There are at least three categories of use for geospatial metadata: (1) to accompany data transfers as documentation, (2) internal, on-line documentation of processing steps and data lineage, and (3) as stand-alone data set synopses for use by spatial data catalogs, indexes, and referral services.

2. Referenced Documents

2.1 *ANSI Standards:*
ANSI X3.51 Representations of Universal Time, Local Time Differentials, and United States Time Zone Reference for Information Interchange[2]
ANSI X3.30 Representation for Calendar Date and Ordinal Date for Information Interchange[2]
ANSI Z39.50 Information Retrieval Service Protocol for Open Systems Interconnection[2]
2.2 *SDTS Standard:*
Federal Information Processing Standard 173 in SDTS 70-1[3]
2.3 *Military Standards:*
MIL-STD-600006 Vector Product Format[4]
MIL-A-89007 Military Specification ARC Digitized Raster Graphics (ADRG)[4]

3. Terminology

3.1 *abscissa*—the coordinate of a point in a plane cartesian coordinate system obtained by measuring parallel to the *x*-axis ("the '*x*' value").

3.2 *accuracy*—the degree of conformity of a measured or calculated value to some recognized standard or specified value. This concept involves the systematic and random error of an operation.

3.3 *altitude*—elevation above or below a reference datum, as defined in Federal Information Processing Standard 70-1. See also *elevation*.

3.4 *area*—a generic term for a bounded, continuous, two-dimensional object that may or may not include its boundary.

3.5 *area chain*—a chain that explicitly references left and right polygons and not start and end nodes. It is a component of a two-dimensional manifold.

3.6 *area point*—a representative point within an area usually carrying attribute information about that area.

3.7 *attribute*—a defined characteristic of an entity type (for example, composition).

3.8 *attribute value*—a specific quality or quantity assigned to an attribute (for example, steel), for a specific entity instance.

3.9 *chain*—a directed non-branching sequence of nonintersecting line segments or arcs bounded by nodes, or both, not necessarily distinct, at each end. Area chain, complete chain, and network chain are special cases of chain, and share all characteristics of the general case as defined above.

3.10 *complete chain*—a chain that explicitly references left and right polygons and start and end nodes. It is a component of a two-dimensional manifold.

3.11 *compound element*—a group of data elements and other compound elements. Compound elements represent higher-level concepts that cannot be represented by individual data elements.

3.12 *coordinates*—pairs of numbers expressing horizontal distances along orthogonal axes; alternatively, triplets of numbers measuring horizontal and vertical distances.

3.13 *data element*—a logically primitive item of data.

3.14 *data set*—a file or files that contain related geometric and attribute information; a collection of related data.

3.15 *depth*—perpendicular distance of an interior point from the surface of an object.

3.16 *developable surface*—a surface that can be flattened to form a plane without compressing or stretching any part of it. Examples include cones and cylinders.

3.17 *digital image*—a two-dimensional array of regularly spaced picture elements (pixels) constituting a picture.

3.18 *digital volume*—a three-dimensional array of regularly spaced volume elements (voxels) constituting a volume.

3.19 *domain*—in the definition of the elements in this specification, the domain identifies valid values for a data element.

3.20 *elevation*—conforming to Federal Information Processing Standard 70-1, the term "altitude" is used in this specification, rather than the common term elevation.

3.21 *entity instance*—a spatial phenomenon of a defined

---

[1] This specification is under the jurisdiction of ASTM Committee D-18 on Soil and Rock and is the direct responsibility of Subcommittee D18.01 on Surface and Subsurface Characterization.

Current edition approved April 15, 1995. Published January 1996.

[2] Available from American National Standards Institute, 11 W. 42nd St., 13th Floor, New York, NY 10036.

[3] Available from Spatial Data Transfer Standard, Washington Department of Commerce NIST, 11 W. 42nd St., 13th Floor, New York, NY 10036. (Supportive Terminology)

[4] Available from Standardization Documents Order Desk, Bldg. 4 Section D, 700 Robbins Ave., Philadelphia, PA 19111-5094, Attn: NPODS.

type that is embedded in one or more phenomena of different type, or that has at least one key attribute value different from the corresponding attribute values of surrounding phenomena (for example, the 10th Street Bridge).

3.22 *entity point*—a point used for identifying the location of point features (or areal features collapsed to a point), such as towers, buoys, buildings, places, etc.

3.23 *entity type*—the definition and description of a set into which similar entity instances are classified (for example, bridge).

3.24 *explicit position*—method of identifying positions directly by pairs (for horizontal positions) or triplets (for horizontal and vertical positions) of numbers.

3.25 *G-polygon*—an area consisting of an interior area, one outer G-ring and zero or more nonintersecting, nonnested inner G-rings. No ring, inner or outer, shall be collinear with or intersect any other ring of the same G-polygon.

3.26 *G-ring*—a ring created from strings or arcs, or both.

3.27 *geoid*—a mathematical representation of the surface of the earth accounting for local geodetic and gravity measurements.

3.28 *geospatial data*—information that identifies the geographic location and characteristics of natural or constructed features and boundaries on the earth. This information may be derived from, among other things, remote sensing, mapping, and surveying technologies.

3.29 *graph*—a set of topologically interrelated zero-dimensional (node), one-dimensional (link or chain), and sometimes two-dimensional (GT-polygon) objects that conform to a set of defined constraint rules. Numerous rule sets can be used to distinguish different types of graphs. Three such types, planar graph, network, and two-dimensional manifold, are used in this specification. All three share the following rules: each link or chain is bounded by an ordered pair of nodes, not necessarily distinct; a node may bound one or more links or chains; and links or chains may only intersect at nodes. Planar graphs and networks are two specialized types of graphs, and a two-dimensional manifold is an even more specific type of planar graph.

3.30 *grid*—(1) a set of grid cells forming a regular, or nearly regular, tessellation of a surface; (2) a set of points arrayed in a pattern that forms a regular, or nearly regular, tesselation of a surface. The tessellation is regular if formed by repeating the pattern of a regular polygon, such as a square, equilateral triangle, or regular hexagon. The tessellation is nearly regular if formed by repeating the pattern of an "almost" regular polygon such as a rectangle, non-square parallelogram, or non-equilateral triangle.

3.31 *grid cell*—a two-dimensional object that represents the smallest non-divisible element of a grid.

3.32 *GT-polygon*—an area that is an atomic two-dimensional component of one and only one two-dimensional manifold. The boundary of a GT-polygon may be defined by GT-rings created from its bounding chains. A GT-polygon may also be associated with its chains (either the bounding set, or the complete set) by direct reference to these chains. The complete set of chains associated with a GT-polygon may also be found by examining the polygon references on the chains.

3.33 *GT-ring*—a ring created from complete or area chains, or both.

3.34 *horizontal*—tangent to the geoid or parallel to a plane that is tangent to the geoid.

3.35 *implicit position*—method of identifying positions by a place in an array of values.

3.36 *interior area*—an area not including its boundary.

3.37 *label point*—a reference point used for displaying map and chart text (for example, feature names) to assist in feature identification.

3.38 *layer*—an integrated, areally distributed, set of spatial data usually representing entity instances within one theme, or having one common attribute or attribute value in an association of spatial objects. In the context of raster data, a layer is specifically a two-dimensional array of scaler values associated with all or part of a grid or image.

3.39 *line segment*—a direct line between two points.

3.40 *link*—a topological connection between two nodes. A link may be directed by ordering its nodes.

3.41 *media*—the physical devices used to record, store, or transmit data, or combination thereof.

3.42 *metadata*—data about the content, quality, condition, and other characteristics of data.

3.43 *network*—a graph without two-dimensional objects. If projected onto a two-dimensional surface, a network can have either more than one node at a point or intersecting links or chains, or both, without corresponding nodes.

3.44 *network chain*—a chain that explicitly references start and end nodes and not left and right polygons. It is a component of a network.

3.45 *node*—a zero-dimensional object that is a topological junction of two or more links or chains, or an end point of a link or chain.

3.46 *object*—a digital representation of all or part of an entity instance.

3.47 *ordinate*—the coordinate of a point in a plane cartesian coordinate system obtained by measuring parallel to the y-axis ("the 'y' value").

3.48 *phenomenon*—a fact, occurrence, or circumstance. Route 10, George Washington National Forest, and Chesterfield County are all phenomena.

3.49 *pixel*—two-dimensional picture element that is the smallest non-divisible element of a digital image.

3.50 *planar graph*—the node and link or chain objects of the graph occur or can be represented as though they occur upon a planar surface. Not more than one node may exist at any given point on the surface. Links or chains may only intersect at nodes.

3.51 *point*—a zero-dimensional object that specifies geometric location. One coordinate pair or triplet specifies the location. Area point, entity point, and label point are special implementations of the general case.

3.52 *primitive*—the quality of not being subdivided; atomic.

3.53 *processing step*—a discrete unit of processing that affects either the data or metadata in a data set.

3.53.1 *Discussion*—Different GISs may discretize processes differently, and so the definition of processing step depends somewhat on the particular GIS. Processing steps shall include all steps followed to automate the data set, such as digitizing or scanning. Processing steps shall also include data-set reviews. A data set review typically will not alter the basic data, but the review with its results should be documented in the metadata.

3.54 *quality*—an essential or distinguishing characteristic necessary for cartographic data to be fit for use.

3.55 *raster*—one or more overlapping layers for the same grid or digital image.

3.56 *raster object*—one or more images or grids, or both, each grid or image representing a layer, such that corresponding grid cells or pixels, or both, between layers are congruent and registered.

3.57 *resolution*—the minimum difference between two independently measured or computed values which can be distinguished by the measurement or analytical method being considered or used.

3.58 *ring*—sequence of nonintersecting chains or strings or arcs, or both, with closure. A ring represents a closed boundary but not the interior area inside the closed boundary.

3.59 *schema*—the definition of table columns, relations, data, Domain, and other elements of a data base, often illustrated using an entity-relationship diagram.

3.60 *SDTS*—the Spatial Data Transfer Standard (see 2.2).

3.61 *spatial data*—see geospatial data.

3.62 *stratum*—one of a series of layers, levels, or gradations in an ordered system. For this specification, the term is used in the sense of (1) a region of sea, atmosphere, or geology that is distinguished by natural or arbitrary limits; (2) a socioeconomic level of society comprised of persons of the same or similar status, especially with regard to education or culture; or (3) a layer of vegetation, usually of the same or similar height.

3.63 *string*—a connected non-branching sequence of line segments specified as the ordered sequence of points between those line segments.

3.63.1 *Discussion*—A string may intersect itself or other strings.

3.64 *two-dimensional manifold*—a planar graph and its associated two-dimensional objects. Each chain bounds two and only two, not necessarily distinct, GT-polygons. The GT-polygons are mutually exclusive and completely exhaust the surface.

3.65 *type*—in the definition of the elements in the metadata standard, a compound element has the type "compound" to provide a unique way to identify compound elements. For a data element, the type identifies the kind of value that can be assigned to the data element. The choices are "integer" for integer numbers, "real" for real numbers, "text" for ASCII characters, "date" for day of the year, and "time" for time of the day.

3.66 *universe polygon*—defines the part of the universe that is outside the perimeter of the area covered by other GT-polygons ("covered area") and completes the two-dimensional manifold. This polygon completes the adjacency relationships of the perimeter links. The boundary of the universe polygon is represented by one or more inner rings and no outer ring. Attribution of the universe polygon may not exist, or may be substantially different from the attribution of the covered area.

3.67 *vector*—composed of directed lines.

3.68 *vertical*—at right angles to the horizontal; includes altitude and depth.

3.69 *VPF*—the vector product format (see 2.3).

3.70 *void polygon*—defines a part of the two-dimensional manifold that is bounded by other GT-polygons, but otherwise has the same characteristics as the universe polygon. The geometry and topology of a void polygon are those of a GT-polygon. Attribution of a void polygon may not exist, or may be substantially different from the attribution of the covered area.

3.71 *voxel*—a three-dimensional element that is the smallest non-divisible element of a digital volume.

3.72 *Definitions*—These definitions are provided to clarify terms used in this specification. Definitions are from SDTS, FIPS 173.

4. Data Element Description

4.1 A data element is a logically primitive item of data. The entry for a data element includes the name of the data element, the definition of the data element, a description of the values that can be assigned to the data element. The form for the definition of the data elements is:

> Data element name—definition.
> Type:
> Domain:
> Tag Name:
> Tag Value:

4.2 The information about the values for the data elements include a description of the type of the value and a description of the domain of the valid values. The type of the data element describes the kind of value to be provided. The choices are "integer" for integer numbers, "real" for real numbers, "text" for ASCII characters, "date" for day of the year, and "time" for time of the day.

4.3 The domain describes valid values that can be assigned to the data element. The domain may specify a list of valid values, references to lists of valid values, or restrictions on the range of values that can be assigned to a data element.

4.4 The domain also may note that the domain is free from restrictions, and any values that can be represented by the "type" of the data element can be assigned. These unrestricted, domains are represented by the use of the word "free" followed by the type of the data element (that is, free text, free date, free real, free time, free integer). Some domains can be partly, but not completely, specified. For example, there are several widely used data transfer formats, but there may be many more that are less well known. To allow a producer to describe its data in these circumstances, the convention of providing a list of values followed by the designation of a "free" domain is used. In these cases, assignments of values shall be made from the provided domain when possible. When not possible, providers may create and assign their own value. A created value shall not redefine a value provided by this specification.

4.5 The descriptor "Tag Name" contains a character string tag for the specified data element whose maximum length is ten characters. This descriptor may be used by implementors to internally name data elements within a database or software system where long text of the data element name would not be feasible. For reporting and display purposes, the full data element name is the preferred form of presentation. Tag name will also be used by metadata management software to declare the format tags in both Standard Graphics Markup Language (SGML) and Hyper Text Markup Language (HTML). Metadata may be exchanged between software systems using ASCII text files in which all metadata elements are encoded using the markup tags using the following example:

<TAGNAME> metadata entry value text </TAGNAME>

4.6 The descriptor "Tag Value" contains a unique integer tag value to be used to describe and manipulate these data elements within the Information Retrieval Service Protocol (see ANSI/ISO Z39.50).

5. Data Format

5.1 *Introduction*—This specification does not require specific internal formats for data elements but does specify the data types required for data exchange. Internal data formats are a design issue for GIS developers. A compliant GIS must provide access to the required metadata for interactive query and update, where appropriate, for use in data processing procedures, and for transfer by means of the Spatial Data Transfer Standard (SDTS) and by means of a spatial metadata transfer file, formatted using Standard Graphics Markup Language (SGML) using the provided tag names as text markers.

5.2 This specification specifies only that data elements are one of numeric, date, code, or text for ease in data transfer between software systems. In addition, data elements may be described as coordinates or coordinate pairs, in which case, a pair of numeric elements is indicated.

5.2.1 Numeric elements shall be made available for data exchange as a character representation, conforming to ISO 6093 with the provision that FULL STOP (that is, period) shall be used for the decimal mark. A compliant GIS would store numeric elements internally using numeric or character format, but the data must be made available for standard numeric processing, preserving the precision of the data values. Numeric elements may be further characterized as real or integer.

5.2.2 Date elements shall be expressed for access and transfer in the format YYYYMMDD, where Y is year, M is month, and D is day (see ANSI X3.30) ((adopted as Federal Information Processing Standard 4-1). Time elements are to be represented using hours, minutes, seconds, and decimal fractions of a second (to the precision desired) without separators convention, with the general form of HHMMSSSS (see ANSI X3.43). Information with a differential factor is to be represented using the general form HHMMSSSShhmm, where HHMMSSSS is the local time using 24-h timekeeping (expressed to the precision desired), "s" is the plus or minus sign for the time differential factor, and hhmm is the time differential factor (see ANSI X3.51).

5.2.3 Text elements are used for nonnumeric elements such as names, descriptions, keywords, and commentary. This specification does not require specific lengths for text elements. Ideally, unlimited-length strings should be supported by a compliant GIS. In practice, however, a compliant GIS could implement these elements with fixed sizes providing enough space for reasonable use of the elements.

5.2.4 Code elements are used for referencing a limited set of valid attribute values. Codes may be expanded into full text for display. Code values (domain) are described in Section 11 of this specification. Additions to these attributes, domains may be made as required by the user community and this specification will be amended by the ASTM mapping and GIS section to include them.

5.2.5 Values for latitude and longitude shall be expressed as decimal fractions of degrees. Whole degrees of latitude shall be represented by a two-digit decimal number ranging from 0 through 90. Whole degrees of longitude shall be represented by a three-digit decimal number ranging from 0 through 180. When a decimal fraction of a degree is specified, it shall be separated from the whole number of degrees by a decimal point. Decimal fractions of a degree may be expressed to the precision desired. Latitudes north of the equator shall be specified by a plus sign (+) or by the absence of a minus sign (−) preceding the two digits designating degrees. Latitudes south of the Equator shall be designated by a minus sign (−) preceding the two digits designating degrees. Coordinate elements are used to store positional information with respect to the earth's surface.

5.2.6 For purposes of data transfer, this specification requires that a compliant GIS be able to automatically transfer full or subsets of metadata in conformance with the SDTS and using SGML formatting.

6. Integration of Metadata and Data Set

6.1 The metadata for a data set shall be treated by a compliant GIS as an integral part of the data. Operations performed on a data set by GIS software shall not render the data set's metadata invalid and shall update the metadata where possible. Metadata shall be retained when a data set is copied, imported, or exported. Software should also support the extraction and transmission of metadata from a spatial data set, independent from the data set, for purposes of data indexing and information exchange.

7. Metadata Contents

7.1 This section describes the individual metadata elements to be included in this specification. Major headings (7.1, 7.2, etc.) denote major groupings of elements that are conceptually associated at a high level. Dependencies and optionality of data elements are described in Section 8.

7.1 *Identification Information*—Basic information about the data set. Type: compound, Tag Name: IDINFO, Tag Value: 3100.

7.1.1 *Citation*—Information to be used to reference the data set. Type: compound, Tag Name: CITATION, Tag Value: 3101.

7.1.2 *Description*—A characterization of the data set, including its intended use and limitations. Type: compound, Tag Name: DESCR, Tag Value: 3102.

7.1.2.1 *Abstract*—A brief narrative summary of the data set. Type: text, Domain: free text, Tag Name: ABSTRACT, Tag Value: 62 (bib1).

7.1.2.2 *Purpose*—A summary of the intentions with which the data set was developed. Type: text, Domain: free text, Tag Name: PURPOSE, Tag Value: 3104.

7.1.2.3 *Supplemental Information*—Other descriptive information about the data set. Type: text, Domain: free text, Tag Name: SUPPLINF, Tag Value: 3105.

7.1.3 *Time Period of Content*—Time period(s) for which the data set corresponds to the ground. Type: compound, Tag Name: TIMEPDCTNT, Tag Value: 3103.

7.1.3.1 *Currentness Reference*—The basis on which the time period of content information is determined. Type: text, Domain: "ground condition" "publication date" free text, Tag Name: CURRENT, Tag Value: 3106.

7.1.4 *Status*—The state of and maintenance information for the data set. Type: compound, Tag Name: STATUS, Tag Value: 3107.

7.1.4.1 *Progress*—The state of the data set. Type: text, Domain: "Complete" "In work" "Planned," Tag Name: PROGRESS, Tag Value: 3108.

7.1.4.2 *Maintenance and Update Frequency*—The frequency with which changes and additions are made to the data set after the initial data set. Type: text, Domain: "Unknown" "As needed" "Irregular" "None planned" "Continually" "Daily" "Weekly" "Monthly" "Annually," Tag Name: UPDATE, Tag Value: 3109.

7.1.5 *Spatial, Domain*—The geographic areal, domain of the data set. Type: compound, Tag Name: SPDOM, Tag Value: 3110.

7.1.5.1 *Bounding Coordinates*—The limits of coverage of a data set expressed by latitude and longitude values in the order western-most, eastern-most, northern-most, and southern-most. For data sets that include a complete band of latitude around the earth, the West Bounding Coordinate shall be assigned the value −180.0, and the East Bounding Coordinate shall be assigned the value 180.0 Type: compound, Tag Name: BOUNDING, Tag Value: 3111.

7.1.5.2 *West Bounding Coordinate*—Western-most coordinate of the limit of coverage expressed in longitude. Type: real, Domain: −180.0 ≤ West Bounding Coordinate < 180.0, Tag Name: WBNDGCOORD, Tag Value: 3112.

7.1.5.3 *East Bounding Coordinate*—Eastern-most coordinate of the limit of coverage expressed in longitude. Type: real, Domain: −180.0 ≤ East Bounding Coordinate ≤ 180.0, Tag Name: EBNDGCOORD, Tag Value: 3113.

7.1.5.4 *North Bounding Coordinate*—Northern-most coordinate of the limit of coverage expressed in latitude. Type: real, Domain: −90.0 ≤ North Bounding Coordinate ≤ 90.0; North Bounding Coordinate ≧ South Bounding Coordinate, Tag Name: NBNDGCOORD, Tag Value: 3114.

7.1.5.5 *South Bounding Coordinate*—Southern-most coordinate of the limit of coverage expressed in latitude. Type: real, Domain: −90.0 ≤ South Bounding Coordinate ≤ 90.0; South Bounding Coordinate ≤ North Bounding Coordinate, Tag Name: SBNDGCOORD, Tag Value: 3115.

7.1.5.6 *Data Set G-Polygon*—Coordinates defining the outline of an area covered by a data set. Type: compound, Tag Name: DSGPOLY, Tag Value: 3116.

7.1.5.7 *Data Set G-Polygon Outer G-Ring*—The closed nonintersecting boundary of an interior area. Type: compound, Tag Name: DSGPOLYO, Tag Value: 3117.

7.1.5.8 *G-Ring Latitude*—The latitude of a point of the G-ring. Type: real, Domain: −90.0 ≤ G-Ring Latitude ≤ 90.0, Tag Name: GRINGLAT, Tag Value: 3118.

7.1.5.9 *G-Ring Longitude*—The longitude of a point of the G-ring. Type: real, Domain: −180.0 ≤ G-Ring Latitude < 180.0, Tag Name: GRINGLON, Tag Value: 3119.

7.1.5.10 *Data Set G-Polygon Exclusion G-Ring*—The closed nonintersecting boundary of a void area (or "hole") in an interior area. Type: compound, Tag Name: DSGPOLYX, Tag Value: 3120.

7.1.6 *Keywords*—Words or phrases summarizing an aspect of the data set. Type: compound, Tag Name: KEYWORDS, Tag Value: 3121.

7.1.6.1 *Theme*—Subjects covered by the data set (for a list of some commonly used thesauri, see Part IV: Subject/index term sources in Network Development and MARC Standards Office, 1988, USMARC code list for relators, sources, and description conventions: Washington, Library of Congress). Type: compound, Tag Name: THEME, Tag Value: 3122.

7.1.6.2 *Theme Keyword Thesaurus*—Reference to a formally registered thesaurus or a similar authoritative source of theme keywords. Type: text, Domain: "None" free text, Tag Name: THMKWTHSRS, Tag Value: 3123.

7.1.6.3 *Theme Keyword*—Common-use word or phrase used to describe the subject of the data set. Type: text, Domain: free text, Tag Name: THEMEKEY, Tag Value: 3124.

7.1.6.4 *Place*—Geographic locations characterized by the data set. Type: compound, Tag Name: GEOGPLACE, Tag Value: 58 (bibl).

7.1.6.5 *Place Keyword Thesaurus*—Reference to a formally registered thesaurus or a similar authoritative source of place keywords. Type: text, Domain: "None" "Geographic Names Information System" free text, Tag Name: PLCKWTHSRS, Tag Value: 3126.

7.1.6.6 *Place Keyword*—The geographic name of a location covered by a data set. Type: text, Domain: free text, Tag Name: PLCKEYWORD, Tag Value: 3127.

7.1.6.7 *Stratum*—Layered, vertical locations characterized by the data set. Type: compound, Tag Name: STRATUM, Tag Value: 3128.

7.1.6.8 *Stratum Keyword Thesaurus*—Reference to a formally registered thesaurus or a similar authoritative source of stratum keywords. Type: text, Domain: "None" free text, Tag Name: STRATKT, Tag Value: 3129.

7.1.6.9 *Stratum Keyword*—The name of a vertical location used to describe the locations covered by a data set. Type: text, Domain: free text, Tag Name: STRATKEY, Tag Value: 3130.

7.1.6.10 *Temporal*—Time period(s) characterized by the data set. Type: compound, Tag Name: TEMPORAL, Tag Value: 3131.

7.1.6.11 *Temporal Keyword Thesaurus*—Reference to a formally registered thesaurus or a similar authoritative source of temporal keywords. Type: text, Domain: "None" free text, Tag Name: TEMPKEYT, Tag Value: 3132.

7.1.6.12 *Temporal Keyword*—The name of a time period covered by a data set. Type: text, Domain: free text, Tag Name: TMPKEYWORD, Tag Value: 3133.

7.1.7 *Access Constraints*—Restrictions and legal prerequisites for accessing the data set. These include any access constraints applied to ensure the protection of privacy or intellectual property, and any special restrictions or limitations on obtaining the data set. Type: text, Domain: "None" free text, Tag Name: ACCESSCONS, Tag Value: 3134.

7.1.8 *Use Constraints*—Restrictions and legal prerequisites for using the data set after access is granted. These include any access constraints applied to ensure the protection of privacy or intellectual property and any special restrictions or limitations on obtaining the data set. Type: text, Domain: "None" free text, Tag Name: USECONSTR, Tag Value: 3135.

7.1.9 *Point of Contact*—Contact information for an individual or organization that is knowledgeable about the data set. Type: compound, Tag Name: PTCONTAC, Tag Value: 3136.

7.1.10 *Browse Graphic*—A graphic that provides an illustration of the data set. The graphic should include a legend for interpreting the graphic. Type: compound, Tag Name: BROWSE, Tag Value: 3137.

7.1.10.1 *Browse Graphic File Name*—Name of a related graphic file that provides an illustration of the data set. Type: text, Domain: free text, Tag Name: BROWSEN, Tag Value: 3138.

7.1.10.2 *Browse Graphic File Description*—A text description of the illustration. Type: text, Domain: free text, Tag Name: BROWSED, Tag Value: 3139.

7.1.10.3 *Browse Graphic File Type*—Graphic file type of a related graphic file. Type: text, Domain:, Domain values in the following table; free text, Tag Name: BROWSET, Tag Value: 3140, Domain: "CGM" Computer Graphics Metafile "EPS" Encapsulated Postscript format "GIF" Graphic Interchange Format "JPEG" Joint Photographic Experts Group format "PBM" Portable Bit Map format "PS" Postscript format "TIFF," Tagged Image File Format "XWD" X-Windows Dump.

7.1.11 *Data Set Credit*—Recognition of those who contributed to the data set. Type: text, Domain: free text, Tag Name: DATACRED, Tag Value: 3141.

7.1.12 *Security Information*—Handling restrictions imposed on the data set because of national security, privacy, or other concerns. Type: compound, Tag Name: SEINFO, Tag Value: 3142.

7.1.12.1 *Security Classification System*—Name of the classification system Type: text, Domain: free text, Tag Name: SECSYS, Tag Value: 3143.

7.1.12.2 *Security Classification*—Name of the handling restrictions on the data set. Type: text, Domain: "Top secret" "Secret" "Confidential" "Restricted" "Unclassified" "Sensitive" free text, Tag Name: SECCLAS, Tag Value: 3144.

7.1.12.3 *Security Handling Description*—Additional information about the restrictions on handling the data set. Type: text, Domain: free text, Tag Name: SECHANDL, Tag Value: 3145.

7.1.13 *Native Data Set Environment*—A description of the data set in the producer's processing environment, including items such as the name of the software (including version), the computer operating system, file name (including host-, path-, and file names), and the data set size. Type: text, Domain: free text, Tag Name: NATIVE, Tag Value: 3146.

7.1.14 *Cross Reference*—Information about other, related data sets that are likely to be of interest. Type: compound, Tag Name: CROSSREF, Tag Value: 3147.

7.2 *Data Quality Information*—A general assessment of the quality of the data set. (Recommendations on information to be reported and tests to be performed are found in "Spatial Data Quality," that is Chapter 3 of Part 1 in

Department of Commerce, 1992, Spatial Data Transfer Standard (SDTS) (Federal Information Processing Standard 173). Type: compound, Tag Name: DATAQUAL, Tag Value: 3200.

7.2.1 *Attribute Accuracy*—An assessment of the accuracy of the identification of entities and assignment of attribute values in the data set. Type: compound, Tag Name: ATTRACC, Tag Value: 3201.

7.2.1.1 *Attribute Accuracy Report*—An explanation of the accuracy of the identification of the entities and assignments of values in the data set and a description of the tests used. Type: text, Domain: free text, Tag Name: ATTRACCR, Tag Value: 3202.

7.2.1.2 *Quantitative Attribute Accuracy Assessment*—A value assigned to summarize the accuracy of the identification of the entities and assignments of values in the data set and the identification of the test that yielded the value. Type: compound, Tag Name: QATTRACC, Tag Value: 3203.

7.2.1.3 *Attribute Accuracy Value*—An estimate of the accuracy of the identification of the entities and assignments of attribute values in the data set. Type: text, Domain: "Unknown" free text, Tag Name: ATTRACCV, Tag Value: 3204.

7.2.1.4 *Attribute Accuracy Explanation*—The identification of the test that yielded the Attribute Accuracy Value. Type: text, Domain: free text, Tag Name: ATTRACCE, Tag Value: 3205.

7.2.2 *Logical Consistency Report*—An explanation of the fidelity of the relationships in the data set and the tests used. Type: text, Domain: free text, Tag Name: LOGIC, Tag Value: 3206.

7.2.3 *Completeness Report*—Information about omissions, selection criteria, generalization, definitions used, and other rules used to derive the data set. Type: text, Domain: free text, Tag Name: COMPLETE, Tag Value: 3207.

7.2.4 *Positional Accuracy*—An assessment of the accuracy of the positions of spatial objects. Type: compound, Tag Name: POACCC, Tag Value: 3208.

7.2.4.1 *Horizontal Positional Accuracy*—An estimate of accuracy of the horizontal positions of the spatial objects. Type: compound, Tag Name: HORIZPA, Tag Value: 3209.

7.2.4.2 *Horizontal Positional Accuracy Report*—An explanation of the accuracy of the horizontal coordinate measurements and a description of the tests used. Type: text, Domain: free text, Tag Name: HORIZPAR, Tag Value: 3210.

7.2.4.3 *Quantitative Horizontal Positional Accuracy Assessment*—Numeric value assigned to summarize the accuracy of the horizontal coordinate measurements and the identification of the test that yielded the value. Type: compound, Tag Name: QHORIZPA, Tag Value: 3211.

7.2.4.4 *Horizontal Positional Accuracy Value*—An estimate of the accuracy of the horizontal coordinate measurements in the data set expressed in (ground) meters. Type: real, Domain: free real, Tag Name: HORIZPAV, Tag Value: 3212.

7.2.4.5 *Horizontal Positional Accuracy Explanation*—The identification of the test that yielded the Horizontal Positional Accuracy Value. Type: text, Domain: free text, Tag Name: HORIZPAE, Tag Value: 3213.

7.2.4.6 *Vertical Positional Accuracy*—An estimate of accuracy of the vertical positions in the data set. Type: compound, Tag Name: VERTACC, Tag Value: 3214.

7.2.4.7 *Vertical Positional Accuracy Report*—An explanation of the accuracy of the vertical coordinate measurements and a description of the tests used. Type: text, Domain: free text, Tag Name: VERTACCR, Tag Value: 3215.

7.2.4.8 *Quantitative Vertical Positional Accuracy Assessment*—Numeric value assigned to summarize the accuracy of vertical coordinate measurements and the identification of the test that yielded the value. Type: compound, Tag Name: QVERTPA, Tag Value: 3216.

7.2.4.9 *Vertical Positional Accuracy Value*—An estimate of the accuracy of the vertical coordinate measurement in the data set expressed in (ground) meters. Type: real, Domain: free real, Tag Name: VERTACCV, Tag Value: 3217.

7.2.4.10 *Vertical Positional Accuracy Explanation*—The identification of the test that yielded the Vertical Positional Accuracy Value. Type: text, Domain: free text, Tag Name: VERTACCE, Tag Value: 3218.

7.2.5 *Lineage*—Information about the events, parameters, and source data which constructed the data set, and information about the responsible parties. Type: compound, Tag Name: LINEAGE, Tag Value: 3219.

7.2.5.1 *Source Information*—List of sources and a short discussion of the information contributed by each. Type: compound, Tag Name: SRCINFO, Tag Value: 3220.

7.2.5.2 *Source Citation*—Reference for a source data set. Type: compound, Tag Name: SRCCIT, Tag Value: 3221.

7.2.5.3 *Source Scale Denominator*—The denominator of the representative fraction on a map (for example, on a 1:24,000-scale map, the Source Scale Denominator is 24000). Type: integer, Domain: Source Scale Denominator > 1, Tag Name: SRCSCALE, Tag Value: 1024 (bibl).

7.2.5.4 *Type of Source Media*—The medium of the source data set. Type: text, Domain: "paper" "stable-base material" "microfiche" "microfilm" "audiocassette" "chart" "filmstrip" "transparency" "videocassette" "videodisc" "videotape" "physical model" "computer program" "disk" "cartridge tape" "magnetic tape" "online" "CD-ROM" "electronic bulletin board" "electronic mail system" free text, Tag Name: TYPESRC, Tag Value: 1031 (bibl).

7.2.5.5 *Source Time Period of Content*—Time period(s) for which the source data set corresponds to the ground. Type: compound, Tag Name: SRCTIME, Tag Value: 3223.

7.2.5.6 *Source Currentness Reference*—The basis on which the source time period of content information of the source data set is determined. Type: text, Domain: "ground condition" "publication date" free text, Tag Name: SRCCURR, Tag Value: 3224.

7.2.5.7 *Source Citation Abbreviation*—Short-form alias for the source citation. Type: text, Domain: free text, Tag Name: SRCCITCA, Tag Value: 3225.

7.2.5.8 *Source Contribution*—Brief statement identifying the information contributed by the source to the data set. Type: text, Domain: free text, Tag Name: SRCCONTR, Tag Value: 3226.

7.2.5.9 *Process Step*—Information about a single event. Type: compound, Tag Name: PROCSTEP, Tag Value: 3227.

7.2.5.10 *Process Description*—An explanation of the event and related parameters or tolerances. Type: text,

Domain: free text, Tag Name: PROCDESC, Tag Value: 3228.

7.2.5.11 *Source Used Citation Abbreviation*—The source citation abbreviation of a data set used in the processing step. Type: text, Domain: Source Citation Abbreviations from the Source Information entries for the data set., Tag Name: SRCUSED, Tag Value: 3229.

7.2.5.12 *Process Date*—The date when the event was completed. Type: date, Domain: "Unknown" "Not complete" free date, Tag Name: PROCDATE, Tag Value: 3230.

7.2.5.13 *Process Time*—The time when the event was completed. Type: time, Domain: free time, Tag Name: PROCTIME, Tag Value: 3231.

7.2.5.14 *Source Produced Citation Abbreviation*—The source citation abbreviation of an intermediate data set that (1) is significant in the opinion of the data producer, (2) is generated in the processing step, and (3) is used in later processing steps. Type: text, Domain: Source Citation Abbreviations from the Source Information entries for the data set, Tag Name: SRCPROD, Tag Value: 3232.

7.2.5.15 *Process Contact*—The party responsible for the processing step information. Type: compound, Tag Name: PROCCONT, Tag Value: 3233.

7.2.6 *Cloud Cover*—Area of a data set obstructed by clouds, expressed as a percentage of the spatial extent. Type: integer, Domain: $0 \leq$ Cloud Cover ≤ 100 "Unknown," Tag Name: CLOUD, Tag Value: 3234.

7.3 *Spatial Data Organization Information*—The mechanism used to represent spatial information in the data set. Type: compound, Tag Name: SPDOINFO, Tag Value: 3300.

7.3.1 *Indirect Spatial Reference*—Name of types of geographic features, addressing schemes, or other means through which locations are referenced in the data set. Type: text, Domain: free text, Tag Name: INDSPREF, Tag Value: 3301.

7.3.2 *Direct Spatial Reference Method*—The system of objects used to represent space in the data set. Type: text, Domain: "Point" "Vector" "Raster," Tag Name: DIRECT, Tag Value: 3302.

7.3.3 *Point and Vector Object Information*—The types and numbers of vector or non-gridded point spatial objects in the data set. Type: compound, Tag Name: PTVCTCNT, Tag Value: 3314.

7.3.3.1 *SDTS Terms Description*—Point and vector object information using the terminology and concepts from "Spatial Data Concepts," that is Chapter 2 of Part 1 in Department of Commerce, 1992, Spatial Data Transfer Standard (SDTS) (Federal Information Processing Standard 173). (Note that this reference to the SDTS is used ONLY to provide a set of terminology for the point and vector objects.) Type: compound, Tag Name: SDTSTERM, Tag Value: 3303.

7.3.3.2 *SDTS Point and Vector Object Type*—Name of point and vector spatial objects used to locate zero-, one-, and two-dimensional spatial locations in the data set. Type: text, Domain: (The domain is from "spatial data concepts," that is Chapter 2 of Part 1 in Department of Commerce, 1992, Spatial Data Transfer Standard (SDTS) (Federal Information Processing Standard 173): "Point" "Entity point" "Label point" "Area point" "Node, planar graph" "Node, network" "String" "Link" "Complete chain" "Area chain"

"Network chain, planar graph" "Network chain, nonplanar graph" "Circular arc, three point center" "Elliptical arc" "Uniform B-spline" "Piecewise Bezier" "Ring with mixed composition" "Ring composed of strings" "Ring composed of chains" "Ring composed of arcs" "G-polygon" "GT-polygon composed of rings" "GT-polygon composed of chains" "Universe polygon composed of rings" "Universe polygon composed of chains" "Void polygon composed of rings" "Void polygon composed of chains," Tag Name: SDTSTYPE, Tag Value: 3304.

7.3.3.3 *Point and Vector Object Count*—The total number of the point or vector object type occurring in the data set. Type: integer, Domain: Point and Vector Object Count > 0, Tag Name: PTVCTCNT, Tag Value: 3305.

7.3.3.4 *VPF Terms Description*—Point and vector object information using the terminology and concepts from MIL-STD-600006. (Note that this reference to the VPF is used ONLY to provide a set of terminology for the point and vector objects.) Type: compound, Tag Name: VPFTERM, Tag Value: 3306.

7.3.3.5 *VPF Topology Level*—The completeness of the topology carried by the data set. The levels of completeness are defined in MIL-STD-600006. Type: integer, Domain: 0 $\leq$ VPF Topology Level $\leq$ 3, Tag Name: VPFLEVEL, Tag Value: 3307.

7.3.3.6 *VPF Point and Vector Object Type*—Name of point and vector spatial objects used to locate zero-, one-, and two-dimensional spatial locations in the data set. Type: text, Domain: (The Domain is from MIL-STD-600006). "Node" "Edge" "Face" "Text," Tag Name: VPFTYPE, Tag Value: 3308.

7.3.4 *Raster Object Information*—The types and numbers of raster spatial objects in the data set. Type: compound, Tag Name: RASTINFO, Tag Value: 3309.

7.3.4.1 *Raster Object Type*—Raster spatial objects used to locate zero-, two-, or three-dimensional locations in the data set. Type: text, Domain: (With the exception of "voxel", the domain is from "spatial data concepts," that is Chapter 2 of Part 1 in Department of Commerce, 1992 Spatial Data Transfer Standard (SDTS) (Federal Information Processing Standard 173). "Point" "Pixel" "Grid Cell" "Voxel," Tag Name: RASTINFO, Tag Value: 3310.

7.3.4.2 *Row Count*—The maximum number of raster objects along the ordinate (y) axis. For use with rectangular raster objects. Type: Integer, Domain: Row Count > 0, Tag Name: ROWCOUNT, Tag Value: 3311.

7.3.4.3 *Column Count*—The maximum number of raster objects along the abscissa (x) axis. For use with rectangular raster objects. Type: Integer, Domain: Column Count > 0, Tag Name: COLUMNS, Tag Value: 3312.

7.3.4.4 *Vertical Count*—The maximum number of raster objects along the vertical (z) axis. For use with rectangular volumetric raster objects (voxels). Type: Integer, Domain: Depth Count > 0, Tag Name: VERTCNT, Tag Value: 3313.

7.4 *Spatial Reference Information*—The description of the reference frame for, and the means to encode, coordinates in the data set. Type: compound, Tag Name: SPREF, Tag Value: 3400.

7.4.1 *Horizontal Coordinate System Definition*—The reference frame or system from which linear or angular quantities are measured and assigned to the position that a point occupies. Type: compound, Tag Name: HORIZSYS, Tag Value: 3401.

7.4.1.1 *Geographic*—The quantities of latitude and longitude which define the position of a point on the Earth's surface with respect to a reference spheroid. Type: compound, Tag Name: GEOGRAPH, Tag Value: 3402.

7.4.1.2 *Latitude Resolution*—The minimum difference between two adjacent latitude values expressed in Geographic Coordinate Units of measure. Type: real, Domain: Latitude Resolution > 0.0, Tag Name: LATRES, Tag Value: 3403.

7.4.1.3 *Longitude Resolution*—The minimum difference between two adjacent longitude values expressed in Geographic Coordinate Units of measure. Type: real, Domain: Longitude Resolution > 0.0, Tag Name: LONGRES, Tag Value: 3404.

7.4.1.4 *Geographic Coordinate Units*—Units of measure used for the latitude and longitude values. Type: text, Domain: "Decimal degrees" "Decimal minutes" "Decimal seconds" "Degrees and decimal minutes" "Degrees, minutes, and decimal seconds" "Radians" "Grads," Tag Name: GEOGUNIT, Tag Value: 3405.

7.4.1.5 *Planar*—The quantities of distances, or distances and angles, which define the position of a point on a reference plane to which the surface of the earth has been projected. Type: compound, Tag Name: PLANAR, Tag Value: 3406.

7.4.1.6 *Map Projection*—The systematic representation of all or part of the surface of the earth on a plane or developable surface. Type: compound, Tag Name: MAPPROJ, Tag Value: 3407.

7.4.1.7 *Map Projection Name*—Name of the map projection. Type: text, Tag Name: MAPPRO, Tag Value: 3408, Domain: "Albers Conical Equal Area" "Azimuthal Equidistant" "Equidistant Conic" "Equirectangular" "General Vertical Near-sided Projection" "Gnomomic" "Lambert Azimuthal Equal Area" "Lambert Conformal Conic" "Mercator" "Modified Stereographic for Alaska" "Miller Cylindrical" "Oblique Mercator" "Orthographic" "Polar Stereographic" "Polyconic" "Robinson" "Sinusoidal" "Space Oblique Mercator" "Stereographic" "Transverse Mercator" "van der Grinten" "other projection."

7.4.1.8 *Map Projection Parameters*—Parameters required for a specific map projection, each having a unique mathematical relationship between the earth and the plane or developable surface. Type: compound, Tag Name: MAPPRJPARM, Tag Value: 3409.

7.4.1.9 *Standard Parallel*—Line of constant latitude at which the surface of the Earth and the plane or developable surface intersect. Type: real, Domain: −90.0 $\leq$ Standard Parallel $\leq$ 90.0, Tag Name: STDPARLL, Tag Value: 3410.

7.4.1.10 *Longitude of Central Meridian*—The line of longitude at the center of a map projection generally used as the basis for constructing the projection. Type: real, Domain: −180.0 $\leq$ Longitude of Central Meridian < 180.0, Tag Name: LONGCM, Tag Value: 3411.

7.4.1.11 *Latitude of Projection Origin*—Latitude chosen as the origin of rectangular coordinates for a map projection. Type: real, Domain: −90.0 $\leq$ Latitude of Projection Origin $\leq$ 90.0, Tag Name: LATPRJO, Tag Value: 3412.

7.4.1.12 *False Easting*—The value added to all "x" values in the rectangular coordinates for a map projection. This value frequently is assigned to eliminate negative numbers. Expressed in the unit of measure identified in Planar Coordinate Units. Type: real, Domain: free real, Tag Name: FEAST, Tag Value: 3413.

7.4.1.13 *False Northing*—The value added to all "y" values in the rectangular coordinates for a map projection. This value frequently is assigned to eliminate negative numbers. Expressed in the unit of measure identified in Planar Coordinate Units. Type: real, Domain: free real, Tag Name: FNORTH, Tag Value: 3414.

7.4.1.14 *Scale Factor at Equator*—A multiplier for reducing a distance obtained from a map by computation or scaling to the actual distance along the equator. Type: real, Domain: Scale Factor at Equator > 0.0, Tag Name: SFEQUAT, Tag Value: 3415.

7.4.1.15 *Height of Perspective Point Above Surface*—Height of viewpoint above the Earth, expressed in metres. Type: real, Domain: Height of Perspective Point Above Surface > 0.0, Tag Name: HEIGHTPT, Tag Value: 3416.

7.4.1.16 *Longitude of Projection Center*—Longitude of the point of projection for azimuthal projections. Type: real, Domain: −180.0 ≤ Longitude of Projection Center < 180.0, Tag Name: LONGPC, Tag Value: 3417.

7.4.1.17 *Latitude of Projection Center*—Latitude of the point of projection for azimuthal projections. Type: real, Domain: −90.0 ≤ Latitude of Projection Center ≤ 90.0, Tag Name: LATPRJC, Tag Value: 3418.

7.4.1.18 *Scale Factor at Center Line*—A multiplier for reducing a distance obtained from a map by computation or scaling to the actual distance along the center line. Type: real, Domain: Scale Factor at Center Line > 0.0, Tag Name: SFCTRLIN, Tag Value: 3419.

7.4.1.19 *Oblique Line Azimuth*—Method used to describe the line along which an oblique mercator map projection is centered using the map projection origin and an azimuth. Type: compound, Tag Name: OBQLAZIM, Tag Value: 3420.

7.4.1.20 *Azimuthal Angle*—Angle measured clockwise from north, and expressed in degrees. Type: real, Domain: 0.0 ≤ Azimuthal Angle < 360.0, Tag Name: AZIMANGL, Tag Value: 3421.

7.4.1.21 *Azimuth Measure Point Longitude*—Longitude of the map projection origin. Type: real, Domain: −180.0 ≤ Azimuth Measure Point Longitude < 180.0, Tag Name: AZIMPTL, Tag Value: 3422.

7.4.1.22 *Oblique Line Point*—Method used to describe the line along which an oblique mercator map projection is centered using two points near the limits of the mapped region that define the center line. Type: compound, Tag Name: OBQLPT, Tag Value: 3423.

7.4.1.23 *Oblique Line Latitude*—Latitude of a point defining the oblique line. Type: real, Domain: −90.0 ≤ Oblique Line Latitude ≤ 90.0, Tag Name: OBQLLAT, Tag Value: 3424.

7.4.1.24 *Oblique Line Longitude*—Longitude of a point defining the oblique line. Type: real, Domain: −180.0 ≤ Oblique Line Longitude < 180.0, Tag Name: OBQLLONG, Tag Value: 3425.

7.4.1.25 *Straight Vertical Longitude from Pole*—Longitude to be oriented straight up from the North or South Pole. Type: real, Domain: −180.0 ≤ Straight Vertical Longitude from Pole < 180.0, Tag Name: SVLONG, Tag Value: 3426.

7.4.1.26 *Scale Factor at Projection Origin*—A multiplier for reducing a distance obtained from a map by computation or scaling to the actual distance at the projection origin. Type: real, Domain: Scale Factor at Projection Origin > 0.0, Tag Name: SFPRJORG, Tag Value: 3427.

7.4.1.27 *Landsat Number*—Number of the Landsat satellite. (Note: This data element exists solely to provide a parameter needed to define the space oblique mercator projection. It is not used to identify data originating from a remote sensing vehicle.) Type: Integer, Domain: 0 < Landsat Number < 5, Tag Name: LANDSAT, Tag Value: 3428.

7.4.1.28 *Path Number*—Number of the orbit of the Landsat satellite. (Note: This data element exists solely to provide a parameter needed to define the space oblique mercator projection. It is not used to identify data originating from a remote sensing vehicle.) Type: integer, Domain: 0 < Path Number < 251 for Landsats 1, 2, or 3 0 < Path Number < 233 for Landsats 4 or 5, Tag Name: PATHNUM, Tag Value: 3429.

7.4.1.29 *Scale Factor at Central Meridian*—A multiplier for reducing a distance obtained from a map by computation or scaling to the actual distance along the central meridian. Type: real, Domain: Scale Factor at Central Meridian > 0.0, Tag Name: SFCTRMER, Tag Value: 3430.

7.4.1.30 *Other Projection's Definition*—A complete description of a projection, not defined elsewhere in this specification, that was used for the data set. The information provided shall include the name of the projection, the names of the parameters and values used for the data set, and the citation of the specification for the algorithms that describe the mathematical relationship between the Earth and the plane or developable surface for the projection. Type: text, Domain: free text, Tag Name: OTHERPRJ, Tag Value: 3431.

7.4.1.31 *Grid Coordinate System*—A plane-rectangular coordinate system usually based on, and mathematically adjusted to, a map projection so that geographic positions can be readily transformed to and from plane coordinates. Type: compound, Tag Name: GRIDSYS, Tag Value: 3432.

7.4.1.32 *Grid Coordinate System Name*—Name of the grid coordinate system. Type: text, Domain: "Universal Transverse Mercator" "Universal Polar Stereographic" "State Plane Coordinate System 1927" "State Plane Coordinate System 1983" "ARC Coordinate System" "other grid system," Tag Name: GRIDSYSN, Tag Value: 3433.

7.4.1.33 *Universal Transverse Mercator (UTM)*—A grid system based on the transverse mercator projection, applied between latitudes 84° north and 80° south on the earth's surface. Type: compound, Tag Name: UTM, Tag Value: 3434.

7.4.1.34 *UTM Zone Number*—Identifier for the UTM zone. Type: integer, Domain: 1 ≤ UTM Zone Number ≤ 60 for the northern hemisphere; −60 ≤ UTM Zone Number ≤ −1 for the southern hemisphere, Tag Name: UTMZONE, Tag Value: 3435.

7.4.1.35 *Universal Polar Stereographic (UPS)*—A grid system based on the polar stereographic projection, applied to the Earth's polar regions north of 84 degrees north and south of 80 degrees south. Type: compound, Tag Name: UPS, Tag Value: 3436.

7.4.1.36 *UPS Zone Identifier*—Identifier for the UPS zone. Type: text, Domain: "A" "B" "Y" "Z," Tag Name: UPSZONE, Tag Value: 3437.

7.4.1.37 *State Plane Coordinate System (SPSC)*—A plane-rectangular coordinate system established for each state in the United States by the National Geodetic Survey. Type: compound, Tag Name: SPCS, Tag Value: 3438.

7.4.1.38 *SPCS Zone Identifier*—Identifier for the SPCS zone. Type: text, Domain: Four-digit numeric codes for the State Plane Coordinate Systems based on the North American Datum of 1927 are found in Department of Commerce, 1986, Representation of geographic point locations for information interchange (Federal Information Processing Standard 70-1): Washington: Department of Commerce, National Institute of Standards and Technology. Codes for the State Plane Coordinate Systems based on the North American Datum of 1983 are found in Department of Commerce, 1989 (January), State Plane Coordinate System of 1983 (National Oceanic and Atmospheric Administration Manual NOS NGS 5): Silver Spring, Maryland, National Oceanic and Atmospheric Administration, National Ocean Service, Coast and Geodetic Survey, Tag Name: SPCSZONE, Tag Value: 3439.

7.4.1.39 *ARC Coordinate System*—The Equal Arc-second Coordinate System, a plane-rectangular coordinate system established in MIL-A-89007. Type: compound, Tag Name: ARCSYS, Tag Value: 3440.

7.4.1.40 *ARC System Zone Identifier*—Identifier for the ARC Coordinate System Zone. Type: integer, Domain: 1 ≦ ARC System Zone Identifier ≦ 18, Tag Name: ARCZONE, Tag Value: 3441.

7.4.1.41 *Other Grid System's Definition*—A complete description of a grid system, not defined elsewhere in this specification that was used for the data set. The information provided shall include the name of the grid system, the names of the parameters and values used for the data set, and the citation of the specification for the algorithms that describe the mathematical relationship between the earth and the coordinates of the grid system. Type: text, Domain: free text, Tag Name: OTHERGRD, Tag Value: 3442.

7.4.1.42 *Local Planar*—Any right-handed planar coordinate system of which the z-axis coincides with a plumb line through the origin that locally is aligned with the surface of the Earth. Type: compound, Tag Name: LOCALP, Tag Value: 3443.

7.4.1.43 *Local Planar Description*—A description of the local planar system. Type: text, Domain: free text, Tag Name: LOCALPD, Tag Value: 3444.

7.4.1.44 *Local Planar Georeference Information*—A description of the information provided to register the local planar system to the earth (for example, control points, satellite ephemeral data, inertial navigation data). Type: text, Domain: free text, Tag Name: LOCALPGI, Tag Value: 3445.

7.4.1.45 *Planar Coordinate Information*—Information about the coordinate system developed on the planar surface. Type: compound, Tag Name: PLANCI, Tag Value: 3446.

7.4.1.46 *Planar Coordinate Encoding Method*—The means used to represent horizontal positions. Type: text,

Domain: "coordinate pair" "distance and bearing" "row and column," Tag Name: PLANCE, Tag Value: 3447.

7.4.1.47 *Coordinate Representation*—The method of encoding the position of a point by measuring its distance from perpendicular reference axes (the "coordinate pair" and "row and column" methods). Type: compound, Tag Name: COORDREP, Tag Value: 3448.

7.4.1.48 *Abscissa Resolution*—The (nominal) minimum distance between the "x" or column values of two adjacent points, expressed in Planar Distance Units of measure. Type: real, Domain: Abscissa Resolution > 0.0, Tag Name: ABSRES, Tag Value: 3449.

7.4.1.49 *Ordinate Resolution*—The (nominal) minimum distance between the "y" or row values of two adjacent points, expressed in planar distance units of measure. Type: real, Domain: Ordinate Resolution > 0.0, Tag Name: ORDRES, Tag Value: 3450.

7.4.1.50 *Distance and Bearing Representation*—A method of encoding the position of a point by measuring its distance and direction (azimuth angle) from another point. Type: compound, Tag Name: DISTBREP, Tag Value: 3451.

7.4.1.51 *Distance Resolution*—The minimum distance measurable between two points, expressed Planar Distance Units of measure. Type: real, Domain: Distance Resolution > 0.0, Tag Name: DISTRES, Tag Value: 3452.

7.4.1.52 *Bearing Resolution*—The minimum angle measurable between two points, expressed in Bearing Units of measure. Type: real, Domain: Bearing Resolution > 0.0, Tag Name: BEARRES, Tag Value: BRNGRESOL, Tag Value: 3453.

7.4.1.53 *Bearing Units*—Units of measure used for angles. Type: text, Domain: "Decimal degrees" "Decimal minutes" "Decimal seconds" "Degrees and decimal minutes" "Degrees, minutes, and decimal seconds" "Radians" "Grads," Tag Name: BEARUNIT, Tag Value: 3454.

7.4.1.54 *Bearing Reference Direction*—Direction from which the bearing is measured. Type: text, Domain: "North" "South," Tag Name: BEARREFD, Tag Value: 3455.

7.4.1.55 *Bearing Reference Meridian*—Axis from which the bearing is measured. Type: text, Domain: "Assumed" "Grid" "Magnetic" "Astronomic" "Geodetic," Tag Name: BEARREFM, Tag Value: 3456.

7.4.1.56 *Planar Distance Units*—Units of measure used for distances. Type: text, Domain: "meters" "international feet" "survey feet" free text, Tag Name: PLANDU, Tag Value: 3457.

7.4.1.57 *Local*—A description of any coordinate system that is not aligned with the surface of the earth. Type: compound, Tag Name: LOCAL, Tag Value: 3458.

7.4.1.58 *Local Description*—A description of the coordinate system and its orientation to the surface of the earth. Type: text, Domain: free text, Tag Name: LOCALDES, Tag Value: 3459.

7.4.1.59 *Local Georeference Information*—A description of the information provided to register the local system to the earth (for example, control points, satellite ephemeral data, inertial navigation data). Type: text, Domain: free text, Tag Name: LOCALGEO, Tag Value: 3460.

7.4.1.60 *Geodetic Model*—Parameters for the shape of the earth. Type: compound, Tag Name: GEODETIC, Tag Value: 3461.

7.4.1.61 *Horizontal Datum Name*—The identification given to the reference system used for defining the coordinates of points. Type: text, Domain: "North American Datum of 1927" "North American Datum of 1983" free text, Tag Name: HORIZDN, Tag Value: 3462.

7.4.1.62 *Ellipsoid Name*—Identification given to established representations of the earth's shape. Type: text, Domain: "Clarke 1866" "Geodetic Reference System 80" free text, Tag Name: ELLIPS, Tag Name: 3463.

7.4.1.63 *Semi-Major Axis*—Radius of the equatorial axis of the ellipsoid. Type: real, Domain: Semi-major Axis > 0.0, Tag Name: SEMIAXIS, Tag Value: 3464.

7.4.1.64 *Denominator of Flattening Ratio*—The denominator of the ratio of the difference between the equatorial and polar radii of the ellipsoid when the numerator is set to 1. Type: real, Domain: Denominator of Flattening > 0.0, Tag Name: DENFLAT, Tag Value: 3465.

7.4.2 *Vertical Coordinate System Definition*—The reference frame or system from which vertical distances (altitudes or depths) are measured. Type: compound, Tag Name: VERTDEF, Tag Value: 3466.

7.4.2.1 *Altitude System Definition*—The reference frame or system from which altitudes (elevations) are measured. The term "altitude" is used instead of the common term "elevation" to conform to the terminology in Federal Information Processing Standards 70-1 and 173. Type: compound, Tag Name: ALTSYS, Tag Value: 3467.

7.4.2.2 *Altitude Datum Name*—The identification given to the level surface taken as the surface of reference from which altitudes are measured. Type: text, Domain: "National Geodetic Vertical Datum of 1929" "North American Vertical Datum of 1988" free text, Tag Name: ALTDATUM, Tag Value: 3468.

7.4.2.3 *Altitude Resolution*—The minimum distance possible between two adjacent altitude values, expressed in Altitude Distance Units of measure. Type: real, Domain: Altitude Resolution > 0.0, Tag Name: ALTRES, Tag Value: 3469.

7.4.2.4 *Altitude Distance Units*—Units in which altitudes are recorded. Type: text, Domain: "meters" "feet" free text, Tag Name: ALTUNITS, Tag Value: 3470.

7.4.2.5 *Altitude Encoding Method*—The means used to encode the altitudes. Type: text, Domain: "Explicit elevation coordinate included with horizontal coordinates" "Implicit coordinate" "Attribute values," Tag Name: ALTENC, Tag Value: 3471.

7.4.2.6 *Depth System Definition*—The reference frame or system from which depths are measured. Type: compound, Tag Name: DEPTHSYS, Tag Value: 3472.

7.4.2.7 *Depth Datum Name*—The identification given to surface of reference from which depths are measured. Type: text, Domain: "Local surface" "Chart datum; datum for sounding reduction" "Lowest astronomical tide" "Highest astronomical tide" "Mean low water" "Mean high water" "Mean sea level" "Land survey datum" "Mean low water springs" "Mean high water springs" "Mean low water neap" "Mean high water neap" "Mean lower low water" "Mean lower low water springs" "Mean higher high water" "Mean higher low water" "Mean lower high water" "Spring tide" "Tropic lower low water" "Neap tide" "High water" "Higher high water" "Low water" "Low-water datum" "Lowest low water" "Lower low water" "Lowest normal low water" "Mean tide level" "Indian spring low water" "High-water full and charge" "Low-water full and charge" "Columbia River datum" "Gulf Coast low water datum" "Equatorial springs low water" "Approximate lowest astronomical tide" "No correction" free text, Tag Name: DEPTHDN, Tag Value: 3473.

7.4.2.8 *Depth Resolution*—The minimum distance possible between two adjacent depth values, expressed in Depth Distance Units of measure. Type: real, Domain: Depth Resolution > 0.0, Tag Name: DEPTHRES, Tag Value: 3474.

7.4.2.9 *Depth Distance Units*—Units in which depths are recorded. Type: text, Domain: "meters" "feet" free text, Tag Name: DEPTHDU, Tag Value: 3475.

7.4.2.10 *Depth Encoding Method*—The means used to encode depths. Type: text, Domain: "Explicit depth coordinate included with horizontal coordinates" "Implicit coordinate" "Attribute values," Tag Name: DEPTHEM, Tag Value: 3476.

7.5 *Entity and Attribute Information*—Information about the information content of the data set, including the entities types, their attributes, and the domains from which attribute values may be assigned. Type: compound, Tag Name: EAINFO, Tag Value: 3500.

7.5.1 *Detailed Description*—Description of the entities, attributes, attribute values, and related characteristics encoded in the data set. Type: compound, Tag Name: DETAILED, Tag Value: 3501.

7.5.1.1 *Entity Type*—The definition and description of a set into which similar entity instances are classified. Type: compound, Tag Name: ENTTYP, Tag Value: 3502.

7.5.1.2 *Entity Type Label*—The name of the entity type. Type: text, Domain: free text, Tag Name: ENTTYPL, Tag Value: 3503.

7.5.1.3 *Entity Type Definition*—The description of the entity type. Type: text, Domain: free text, Tag Name: ENTTYPD, Tag Value: 3504.

7.5.1.4 *Entity Type Definition Source*—The authority of the definition. Type: text, Domain: free text, Tag Name: ENTTYPDS, Tag Value: 3505.

7.5.1.5 *Attribute*—A defined characteristic of an entity. Type: compound, Tag Name: ATTR, Tag Value: 3506.

7.5.1.6 *Attribute Label*—The name of the attribute. Type: text, Domain: free text, Tag Name: ATTRLABL, Tag Value: 3507.

7.5.1.7 *Attribute Definition*—The description of the attribute. Type: text, Domain: free text, Tag Name: ATTRDEF, Tag Value: 3508.

7.5.1.8 *Attribute Definition Source*—The authority of the definition. Type: text, Domain: free text, Tag Name: ATTRDEF, Tag Value: 3509.

7.5.1.9 *Attribute Domain Values*—The valid values that can be assigned for an attribute. Type: compound, Tag Name: ATTRDOMV, Tag Value: 3510.

7.5.1.10 *Enumerated Domain*—The members of an established set of valid values. Type: compound, Tag Name: EDOM, Tag Value: 3511.

7.5.1.11 *Enumerated Domain Value*—The name or label of a member of the set. Type: text, Domain: free text, Tag Name: EDOMV, Tag Value: 3512.

7.5.1.12 *Enumerated Domain Value Definition*—The description of the value. Type: text, Domain: free text, Tag Name: EDOMVD, Tag Value: 3513.

7.5.1.13 *Enumerated Domain Value Definition Source*—The authority of the definition. Type: text, Domain: free text, Tag Name: EDOMVDC, Tag Value: 3514.

7.5.1.14 *Range Domain*—The minimum and maximum values of a continuum of valid values. Type: compound, Tag Name: RDOM, Tag Value: 3515.

7.5.1.15 *Range Domain Minimum*—The least value that the attribute can be assigned. Type: text, Domain: free text, Tag Name: RDOMMIN, Tag Value: 3516.

7.5.1.16 *Range Domain Maximum*—The greatest value that the attribute can be assigned. Type: text, Domain: free text, Tag Name: RDOMMAX, Tag Value: 3517.

7.5.1.17 *Codeset Domain*—Reference to a standard or list that contains the members of an established set of valid values. Type: compound, Tag Name: CODESETD, Tag Value: 3518.

7.5.1.18 *Codeset Name*—The title of the codeset. Type: text, Domain: free text, Tag Name: CODESETN, Tag Value: 3519.

7.5.1.19 *Codeset Source*—The authority for the codeset. Type: text, Domain: free text, Tag Name: CODESETS, Tag Value: 3520.

7.5.1.20 *Unrepresentable Domain*—Description of the values and reasons why they cannot be represented. Type: text, Domain: free text, Tag Name: UDOM, Tag Value: 3521.

7.5.1.21 *Attribute Units of Measurement*—The standard of measurement for an attribute value. Type: text, Domain: free text, Tag Name: ATTRUNIT, Tag Value: 3522.

7.5.1.22 *Attribute Measurement Resolution*—The smallest unit increment to which an attribute value is measured. Type: real, Domain: Attribute Measurement Resolution > 0.0, Tag Name: ATTRMRES, Tag Value: 3523.

7.5.1.23 *Beginning Date of Attribute Values*—Earliest or only date for which the attribute values are current. In cases when a range of dates are provided, this is the earliest date for which the information are valid. Type: date, Domain: free date, Tag Name: BEGDATEA, Tag Value: 3524.

7.5.1.24 *Ending Date of Attribute Values*—Latest date for which the information are current. Used in cases when a range of dates are provided. Type: date, Domain: free date, Tag Name: ENDDATEA, Tag Value: 3525.

7.5.1.25 *Attribute Value Accuracy Information*—An assessment of the accuracy of the assignment of attribute values. Type: compound, Tag Name: ATTRVAI, Tag Value: 3526.

7.5.1.26 *Attribute Value Accuracy*—An estimate of the accuracy of the assignment of attribute values. Type: real, Domain: free real, Tag Name: ATTRVA, Tag Value: 3527.

7.5.1.27 *Attribute Value Accuracy Explanation*—The definition of the Attribute Value Accuracy measure and units, and a description of how the estimate was derived. Type: text, Domain: free text, Tag Name: ATTRVAE, Tag Value: 3528.

7.5.1.28 *Attribute Measurement Frequency*—The frequency with which attribute values are added. Type: real, Domain: "Unknown" "As needed" "Irregular" "None planned" free text, Tag Name: ATTRMFRQ, Tag Value: 3529.

7.5.2 *Overview Description*—Summary of, and citation to detailed description of, the information content of the data set. Type: compound, Tag Name: OVERVIEW, Tag Value: 3530.

7.5.2.1 *Entity and Attribute Overview*—Detailed summary of the information contained in a data set. Type: text, Domain: free text, Tag Name: EAOVER, Tag Value: 3531.

7.5.2.2 *Entity and Attribute Detail Citation*—Reference to the complete description of the entity types, attributes, and attribute values for the data set. Type: text, Domain: free text, Tag Name: EADETCIT, Tag Value: 3532.

7.6 *Distribution Information*—Information about the distributor of and options for obtaining the data set. Type: compound, Tag Name: DISTINFO, Tag Value: 3600.

7.6.1 *Distributor*—The party from whom the data set may be obtained. Type: compound, Tag Name: DISTRIB, Tag Value: 3601.

7.6.2 *Resource Description*—The identifier by which the distributor knows the data set. Type: text, Domain: free text, Tag Name: RESRCDESCR, Tag Value: 3602.

7.6.3 *Distribution Liability*—Statement of the liability assumed by the distributor. Type: text, Domain: free text, Tag Name: DISTRLIAB, Tag Value: 3603.

7.6.4 *Standard Order Process*—The common ways in which the data set may be obtained or received, and related instructions and fee information. Type: compound, Tag Name: STDORDER, Tag Value: 3604.

7.6.4.1 *Non-Digital Form*—The description of options for obtaining the data set on non-computer-compatible media. Type: text, Domain: free text, Tag Name: NONDIG, Tag Value: 3605.

7.6.4.2 *Digital Form*—The description of options for obtaining the data set on computer-compatible media. Type: compound, Tag Name: DIGFORM, Tag Value: 3606.

7.6.4.3 *Digital Transfer Information*—Description of the form of the data to be distributed. Type: compound, Tag Name: DIGTINFO, Tag Value: 3607.

7.6.4.4 *Format Name*—The name of the data transfer format. Type: text, Domain: domain values from the following table; free text, Tag Name: FORMNAME, Tag Value: 3608 "ARCE" ARC/INFO Export format "ARCG" ARC/INFO Generate format "ASCII" ASCII file, formatted for text attributes, declared format "BIL" Imagery, band interleaved by line "BIP" Imagery, band interleaved by pixel "BSQ" Imagery, band interleaved sequential "CDF" Common Data Format "CFF" Cartographic Feature File (U.S. Forest Service) "COORD" User-created coordinate file, declared format "DEM" Digital Elevation Model format (U.S. Geological Survey) "DFAD" Digital Feature Analysis Data (Defense Mapping Agency) "DGN" Microstation format (Intergraph Corporation) "DIGEST" Digital Geographic Information Exchange Standard "DLG" Digital Line Graph (U.S. Geological Survey) "DTED" Digital Terrain Elevation Data (MIL-D-89020 "DWG" AutoCAD Drawing format "DX90" Data Exchange '90 "DXF" AutoCAD Drawing Exchange Format "ERDAS" ERDAS image files (ERDAS Corporation) "GRASS" Geographic Resources Analysis Support System "HDF" Hierarchical Data Format "IGDS" Interactive Graphic Design System format (Intergraph Corporation) "IGES" Initial Graphics Exchange Standard "MOSS" Multiple Overlay Statistical System export file "netCDF" network Common Data Format "NITF" National Imagery Transfer Format "RPF" Raster Product Format (Defense Mapping Agency) "RVC" Raster Vector Converted Format (MicroImages) "RVF" Raster Vector Format (MicroImages) "SDTS" Spatial Data Transfer Standard (Federal Information Processing Standard 173) "SIF" Standard Interchange Format (DOD Project 2851) "SLF" Standard Linear Format (Defense Mapping Agency) "TIFF," Tagged Image File Format "TGRLN" Topologically Integrated Geographic Encoding and Referencing (TIGER) Line Format (Bureau of the Census) "VPF" Vector Product Format (Defense Mapping Agency).

7.6.4.5 *Format Version Number*—Version number of the format. Type: text, Domain: free text, Tag Name: FORMVERN, Tag Value: 3609.

7.6.4.6 *Format Version Date*—Date of the version of the format. Type: date, Domain: free date, Tag Name: FORMVERD, Tag Value: 3610.

7.6.4.7 *Format Specification*—Name of a subset, profile, or product specification of the format. Type: text, Domain: free text, Tag Name: FORMSPEC, Tag Value: 3611.

7.6.4.8 *Format Information Content*—Description of the content of the data encoded in a format. Type: text, Domain: free text, Tag Name: FORMCONT, Tag Value: 3612.

7.6.4.9 *File Decompression Technique*—Recommendations of algorithms or processes (including means of obtaining these algorithms or processes) that can be applied to read or expand data sets to which data compression techniques have been applied. Type: text, Domain: "No compression applied" free text, Tag Name: FILEDEC, Tag Value: 3613.

7.6.4.10 *Transfer Size*—The size, or estimated size, of the transferred data set in megabytes. Type: real, Domain: Transfer Size > 0.0, Tag Name: TRANSIZE, Tag Value: 3614.

7.6.4.11 *Digital Transfer Option*—The means and media by which a data set is obtained from the distributor. Type: compound, Tag Name: DIGTOPT, Tag Value: 3615.

7.6.4.12 *Online Option*—Information required to directly obtain the data set electronically. Type: compound, Tag Name: ONLINOPT, Tag Value: 3616.

7.6.4.13 *Computer Contact Information*—Instructions for establishing communications with the distribution computer. Type: compound, Tag Name: COMPUTER, Tag Value: 3617.

7.6.4.14 *Network Address*—The electronic address from which the data set can be obtained from the distribution computer. Type: compound, Tag Name: NETWORKA, Tag Value: 3618.

7.6.4.15 *Network Resource Name*—The name of the file or service from which the data set can be obtained. Type: text, Domain: free text, Tag Name: NETWORKR, Tag Value: 3619.

7.6.4.16 *Dialup Instructions*—Information required to access the distribution computer remotely through telephone lines. Type: compound, Tag Name: DIALINST, Tag Value: 3620.

7.6.4.17 *Lowest BPS*—Lowest or only speed for the connection's communication, expressed in bits per second. Type: integer, Domain: Lowest BPS $\geq$ 110, Tag Name: LOWBPS, Tag Value: 3621.

7.6.4.18 *Highest BPS*—Highest speed for the connection's communication, expressed in bits per second. Used in cases when a range of rates are provided. Type: integer, Domain: Highest BPS > Lowest BPS, Tag Name: HIGHBPS, Tag Value: 3622.

7.6.4.19 *Number DataBits*—Number of databits in each character exchanged in the communication. Type: integer, Domain: 7 $\leq$ Number DataBits $\leq$ 8, Tag Name: NUMDATA, Tag Value: 3623.

7.6.4.20 *Number StopBits*—Number of stopbits in each character exchanged in the communication. Type: integer, Domain: 1 $\leq$ Number StopBits $\leq$ 2, Tag Name: NUMSTOP, Tag Value: 3624.

7.6.4.21 *Parity*—Parity error checking used in each character exchanged in the communication. Type: text, Domain: "None" "Odd" "Even" "Mark" "Space" Tag Name: PARITY, Tag Value: 3625.

7.6.4.22 *Compression Support*—Data compression available through the modem service to speed data transfer. Type: text, Domain: "V.32" "V.32bis" "V.42" "V.42bis" free text, Tag Name: COMPRESS, Tag Value: 3626.

7.6.4.23 *Dialup Telephone*—The telephone number of the distribution computer. Type: text, Domain: free text, Tag Name: DIALTEL, Tag Value: 3627.

7.6.4.24 *Dialup File Name*—The name of a file containing the data set on the distribution computer. Type: text, Domain: free text, Tag Name: DIALFILE, Tag Value: 3628.

7.6.4.25 *Access Instructions*—Instructions on the steps required to access the data set. Type: text, Domain: free text, Tag Name: ACCSINST, Tag Value: 3629.

7.6.4.26 *Online Computer and Operating System*—The brand of distribution computer and its operating system. Type: text, Domain: free text, Tag Name: ONCOMP, Tag Value: 3630.

7.6.4.27 *Offline Option*—Information about media-specific options for receiving the data set. Type: compound, Tag Name: OFFOPTN, Tag Value: 3631.

7.6.4.28 *Offline Media*—Name of the media on which the data set can be received. Type: text, Domain: "CD-ROM" "3-1/2 inch floppy disk" "5-1/4 inch floppy disk" "9-track tape" "4 mm cartridge tape" "8 mm cartridge tape" "1/4-inch cartridge tape" free text, Tag Name: OFFMEDIA, Tag Value: 3632.

7.6.4.29 *Recording Capacity*—The density of information to which data are written. Used in cases where different recording capacities are possible. Type: compound, Tag Name: RECCAP, Tag Value: 3633.

7.6.4.30 *Recording Density*—The density in which the data set can be recorded. Type: real, Domain: Recording Density > 0.0, Tag Name: RECDEN, Tag Value: 3634.

7.6.4.31 *Recording Density Units*—The units of measure for the recording density. Type: text, Domain: free text, Tag Name: RECDENU, Tag Value: 3635.

7.6.4.32 *Recording Format*—The options available or method used to write the data set to the medium. Type: text, Domain: "cpio" "tar" "High Sierra" "ISO 9660" "ISO 9660 with Rock Ridge extensions" "ISO 9660 with Apple HFS extensions" free text, Tag Name: RECFMT, Tag Value: 3636.

7.6.4.33 *Compatibility Information*—Description of other limitations or requirements for using the medium. Type:

text, Domain: free text, Tag Name: COMPAT, Tag Value: 3637.

7.6.4.34 *Fees*—The fees and terms for retrieving the data set. Type: text, Domain: free text, Tag Name: FEES, Tag Value: 3638.

7.6.4.35 *Ordering Instructions*—General instructions and advice about, and special terms and services provided for, the data set by the distributor. Type: text, Domain: free text, Tag Name: ORDERINSTR, Tag Value: 3639.

7.6.4.36 *Turnaround*—Typical turnaround time for the filling of an order. Type: text, Domain: free text, Tag Name: TURNAROUND, Tag Value: 3640.

7.6.5 *Custom Order Process*—Description of custom distribution services available, and the terms and conditions for obtaining these services. Type: text, Domain: free text, Tag Name: CUSTOM, Tag Value: 3641.

7.6.6 *Technical Prerequisites*—Description of any technical capabilities that the consumer must have to use the data set in the form(s) provided by the distributor. Type: text, Domain: free text, Tag Name: TECHPREREQ, Tag Value: 3642.

7.6.7 *Available Time Period*—The time period when the data set will be available from the distributor. Type: compound, Tag Name: AVAILABL, Tag Value: 3643.

7.7 *Metadata Reference Information*—Information on the currentness of the metadata information, and the responsible party. Type: compound, Tag Name: METAINFO, Tag Value: 3700.

7.7.1 *Metadata Date*—The date that the metadata were created or last updated. Type: date, Domain: free date, Tag Name: METD, Tag Value: 3701.

7.7.2 *Metadata Review Date*—The date of the latest review of the metadata entry. Type: date, Domain: free date; Metadata Review Date later than Metadata Date, Tag Name: METRD, Tag Value: 3702.

7.7.3 *Metadata Future Review Date*—The date by which the metadata entry should be reviewed. Type: date, Domain: free date; Metadata Future Review Date later than Metadata Review Date, Tag Name: METFRD, Tag Value: 3703.

7.7.4 *Metadata Contact*—The party responsible for the metadata information. Type: compound, Tag Name: METC, Tag Value: 3704.

7.7.5 *Metadata Standard Name*—The name of the metadata standard used to document the data set. Type: text, Domain: "FGDC Content Standards for Digital Geospatial Metadata" free text, Tag Name: METSTDN, Tag Value: 3705.

7.7.6 *Metadata Standard Version*—Identification of the version of the metadata standard used to document the data set. Type: text, Domain: free text, Tag Name: METSTDV, Tag Value: 3706.

7.7.7 *Metadata Time Convention*—Form used to convey time of day information in the metadata entry. Used if time of day information is included in the metadata for a data set. Type: text, Domain: "local time" "local time with time differential factor" "universal time," Tag Name: METTC, Tag Value: 3707.

7.7.8 *Metadata Access Constraints*—Restrictions and legal prerequisites for accessing the metadata. These include any access constraints applied to ensure the protection of privacy or intellectual property, and any special restrictions

or limitations on obtaining the metadata. Type: text, Domain: free text, Tag Name: METAC, Tag Value: 3708.

7.7.9 *Metadata Use Constraints*—Restrictions and legal prerequisites for using the metadata after access is granted. These include any access constraints applied to ensure the protection of privacy or intellectual property, and any special restrictions or limitations on obtaining the metadata. Type: text, Domain: free text, Tag Name: METUC, Tag Value: 3709.

7.7.10 *Metadata Security Information*—Handling restrictions imposed on the metadata because of national security, privacy, or other concerns. Type: compound, Tag Name: METSI, Tag Value: 3710.

7.7.10.1 *Metadata Security Classification System*—Name of the classification system for the metadata. Type: text, Domain: free text, Tag Name: METSCS, Tag Value: 3711.

7.7.10.2 *Metadata Security Classification*—Name of the handling restrictions on the metadata. Type: text, Domain: "Top secret" "Secret" "Confidential" "Restricted" "Unclassified" "Sensitive" free text, Tag Name: METSC, Tag Value: 3712.

7.7.10.3 *Metadata Security Handling Description*—Additional information about the restrictions on handling the metadata. Type: text, Domain: free text, Tag Name: METSHD, Tag Value: 3713.

7.8 *Citation Information*—The recommended reference to be used for the data set. Type: compound, Tag Name: CITEINFO, Tag Value: 3800.

NOTE 1—This section provides a means of stating the citation of a data set, and is used by other sections of the metadata standard. This section is never used alone.

7.8.1 *Originator*—The name of the organization or individual that developed the data set. If the names of editors or compilers are provided, the names must be followed by "(ed.)" ("(eds.)" is the plural form) or "(comp.)" ("(comps.)" is the plural form) respectively. Type: text, Domain: "Unknown" free text, Tag Name: ORIGIN, Tag Value: 1003 (bibl).

7.8.2 *Publication Date*—The date when the data set is published or otherwise made available for release. Type: date, Domain: "Unknown" "Unpublished material" free date, Tag Name: PUBLDATE, Tag Value: 31 (bibl).

7.8.3 *Publication Time*—The time of day when the data set is published or otherwise made available for release. Type: time, Domain: "Unknown" free time, Tag Name: PUBTIME, Tag Value: 3803.

7.8.4 *Title*—The name by which the data set is known. Type: text, Domain: free text, Tag Name: THETITLE, Tag Value: 4 (bibl).

7.8.5 *Edition*—The version of the title. Type: text., Domain: free text., Tag Name: EDITION, Tag Value: 3815.

7.8.6 *Geospatial Data Presentation Form*—The mode in which the geospatial data are represented. Type: text, Domain: (the Domain is from Footnote 5) "atlas" "diagram" "globe" "map" "model" "profile" "remote-sensing image" "section" "view," Tag Name: GEOFORM, Tag Value: 3805.

---

[5] Anglo-American Committee on Cataloging of Cartographic Materials, *Cartographic Materials: A Manual of Interpretation for AACR2*, American Library Association, Chicago, IL 1982.

7.8.7 *Series Information*—The identification of the serial publication of which the data set is a part. Type: compound, Tag Name: SERINFO, Tag Value: 3806.

7.8.7.1 *Series Name*—The name of the serial publication of which the data set is a part. Type: text, Domain: free text, Tag Name: SERIESNAME, Tag Value: 5 (bibl).

7.8.7.2 *Issue Identification*—Information identifying the issue of the serial publication of which the data set is a part. Type: text, Domain: free text, Tag Name: ISSUE, Tag Value: 3808.

7.8.8 *Publication Information*—Publication details for published data sets. Type: compound, Tag Name: PUBINFO, Tag Value: 3809.

7.8.8.1 *Publication Place*—The name of the city (and state or province, and country, if needed to identify the city) where the data set was published or released. Type: text, Domain: free text, Tag Name: PUBLPLACE, Tag Value: 59 (bibl).

7.8.8.2 *Publisher*—The name of the individual or organization that published the data set. Type: text, Domain: free text, Tag Name: PUBLISHER, Tag Value: 1018 (bibl).

7.8.9 *Other Citation Details*—Other information required to complete the citation. Type: text, Domain: free text, Tag Name: OTHERCIT, Tag Value: 3812.

7.8.10 *Online Linkage*—The name of an online computer resource that contains the data set. Entries should follow the Uniform Resource Locator convention of the Internet. Type: text, Domain: free text, Tag Name: ONLINK, Tag Value: 3813.

7.8.11 *Larger Work Citation*—The information identifying a larger work in which the data set is included. Type: compound, Tag Name: LWORKCIT, Tag Value: 3814.

7.9 *Time Period Information*—Information about the date and time of an event. Type: compound, Tag Name: TIMEINFO, Tag Value: 3900.

NOTE 2—This section provides a means of stating temporal information, and is used by other sections of the metadata standard. This section is never used alone.

7.9.1 *Single Date/Time*—Means of encoding a single date and time. Type: compound, Tag Name: SNGD, Tag Value: 3902.

7.9.1.1 *Calendar Date*—The year (and optionally month, or month and day). Type: date, Domain: "Unknown" free date, Tag Name: CALDATE, Tag Value: 3903.

7.9.1.2 *Time of Day*—The hour (and optionally minute, or minute and second) of the day. Type: time, Domain: "Unknown" free time, Tag Name: TIME, Tag Value: 3904.

7.9.2 *Multiple Dates/Times*—Means of encoding multiple individual dates and times. Type: compound, Tag Name: MDATTIM, Tag Value: 3905.

7.9.3 *Range of Dates/Times*—Means of encoding a range of dates and times. Type: compound, Tag Name: RNGDATES, Tag Value: 3906.

7.9.3.1 *Beginning Date*—The first year (and optionally month, or month and day) of the event. Type: date, Domain: "Unknown" free date, Tag Name: BEGDATE, Tag Value: 3907.

7.9.3.2 *Beginning Time*—The first hour (and optionally minute, or minute and second) of the day for the event. Type: time, Domain: "Unknown" free time, Tag Name: BEGTIME, Tag Value: 3908.

7.9.3.3 *Ending Date*—The last year (and optionally month, or month and day) for the event. Type: date, Domain: "Unknown" "Present" free date, Tag Name: ENDDATE, Tag Value: 3909.

7.9.3.4 *Ending Time*—The last hour (and optionally minute, or minute and second) of the day for the event. Type: time, Domain: "Unknown" free time, Tag Name: ENDTIME, Tag Value: 3910.

7.10 *Contact Information*—Identity of, and means to communicate with, person(s) and organization(s) associated with the data set. Type: compound, Tag Name: CNTINFO, Tag Value: 3000.

NOTE 3—This section provides a means of identifying individuals and organizations, and is used by other sections of the metadata standard. This section is never used alone.

7.10.1 *Contact Person Primary*—The person, and the affiliation of the person, associated with the data set. Used in cases where the association of the person to the data set is more significant than the association of the organization to the data set. Type: compound, Tag Name: CNTINFO, Tag Value: 3001.

7.10.1.1 *Contact Person*—The name of the individual to which the contact type applies. Type: text, Domain: free text, Tag Name: CNTCTPERSN, Tag Value: 3002.

7.10.1.2 *Contact Organization*—The name of the organization to which the contact type applies. Type: text, Domain: free text, Tag Name: CONTACTORG, Tag Value: 3003.

7.10.2 *Contact Organization Primary*—The organization, and the member of the organization, associated with the data set. Used in cases where the association of the organization to the data set is more significant than the association of the person to the data set. Type: compound, Tag Name: CNTPERP, Tag Value: 3004.

7.10.3 *Contact Position*—The title of individual. Type: text, Domain: free text, Tag Name: CNTPOS, Tag Value: 3005.

7.10.4 *Contact Address*—The address for the organization or individual. Type: compound, Tag Name: CNTADDR, Tag Value: 3006.

7.10.4.1 *Address Type*—The information provided by the address. Type: text, Domain: "mailing address" "physical address" "mailing and physical address," Tag Name: ADDRTYPE, Tag Value: 3007.

7.10.4.2 *Address*—An address line for the address. Type: text, Domain: free text, Tag Name: ADDRESS, Tag Value: 3008.

7.10.4.3 *City*—The city of the address. Type: text, Domain: free text, Tag Name: CITY, Tag Value: 3009.

7.10.4.4 *State or Province*—The state or province of the address. Type: text, Domain: free text, Tag Name: STATEPRVNC, Tag Value: 3010.

7.10.4.5 *Postal Code*—The ZIP or other postal code of the address. Type: text, Domain: free text, Tag Name: POSTALCODE, Tag Value: 3011.

7.10.4.6 *Country*—The country of the address. Type: text, Domain: free text, Tag Name: COUNTRY, Tag Value: 3012.

7.10.5 *Contact Voice Telephone*—The telephone number by which individuals can speak to the organization or individual. Type: text, Domain: free text, Tag Name: CONTACTVPH, Tag Value: 3013.

7.10.6 *Contact TDD/TTY Telephone*—The telephone number by which hearing-impaired individuals can contact the organization or individual. Type: text, Domain: free text, Tag Name: CNTTDD, Tag Value: 3014.

7.10.7 *Contact Facsimile Telephone*—The telephone number of a facsimile machine of the organization or individual. Type: text, Domain: free text, Tag Name: CONTACTFPH, Tag Value: 3015.

7.10.8 *Contact Electronic Mail Address*—The address of the electronic mailbox of the organization or individual. Type: text, Domain: free text, Tag Name: CNTEMAIL, Tag Value: 3016.

7.10.9 *Hours of Service*—Time period when individuals can speak to the organization or individual. Type: text, Domain: free text, Tag Name: HOURSOFSVC, Tag Value: 3017.

7.10.10 *Contact Instructions*—Supplemental instructions on how or when to contact the individual or organization. Type: text, Domain: free text, Tag Name: CNTINST, Tag Value: 3018.

8. Metadata Content Syntax

8.1 The dependencies and optionality of the geospatial metadata elements from Section 7 are described in this section using the following production rules using Yourdan syntax.

8.2 *Production Rule Definitions:*

8.2.1 A production rule specifies the relationship between a compound element, and data elements and other (lower-level) compound elements. Each production rule has a left side (identifier) and a right side (expression) connected by the symbol "=," meaning that the term on the left side is replaced by or produces the term on the right side. Terms on the right side are either other compound elements or individual data elements. By making substitutions using matching terms in the production rules, one can explain higher-level concepts using data elements.

8.2.2 The symbols used in the production rules have the following meaning:

Symbol	Meaning
=	is replaced by, produces, consists of
+	and
[|]	selection—select one term from the list of enclosed terms (exclusive or). Terms are separated by " | ".
m{}n	iteration—the term(s) enclosed is(are) repeated from "m" to "n" times
()	optional—the term(s) enclosed is(are) optional

8.2.2.1 *Examples:*

a = b + c "a consists of b and c"
a = [b | c] "a consists of one of b or c"
a = 4{b}6 "a consists of four to six occurrences of b"
a = b + (c) "a consists of b and optionally c"

8.3 *Interpreting the Production Rules:*

8.3.1 The terms bounded by parentheses, "("and")", are optional and are provided at the discretion of the data producer. If a producer chooses to provide information enclosed by parentheses, the producer shall follow the production rules for the enclosed information. For example, if the producer decides to provide the optional information described in the term:

(a + b + c)

the producer shall provide a and b and c.

8.3.2 Only for terms bounded by parentheses does the producer have the discretion of deciding whether or not to provide the information.

8.3.3 The variation among the ways in which geospatial data are produced and distributed, the fact that all geospatial data do not have the same characteristics, and the issue that all details of data sets that are in work or are planned may not be decided, caused the need to express the concept of "mandatory if applicable". This concept means that if the data set exhibits (or, for data sets that are in work or planned, it is known that the data set will exhibit) a defined characteristic, then the producer shall provide the information needed to describe that characteristic. This concept is described by the production rule:

0{term}1

8.4 *Metadata Production Rules:*

```
Metadata =
    Identification_Information +
    0{Data_Quality_Information}1 +
    0{Spatial_Data_Organization_Information}1 +
    0{Spatial_Reference_Information}1 +
    0{Entity_and_Attribute_Information}1 +
    0{Distribution_Information}1 +
    Metadata_Reference_Information
Identification_Information =
    Citation +
    Description +
    Time_Period_of_Content +
    Status +
    Spatial_Domain +
    Keywords +
    Access_Constraints +
    Use_Constraints +
    (Point_of_Contact) +
    (1{Browse_Graphic}n) +
    (Data_Set_Credit) +
    (Security_Information) +
    (Native_Data_Set_Environment) +
    (1{(Cross_Reference)}n)
Citation =
    Citation_Information (see section for production rules)
Description =
    Abstract +
    Purpose +
    (Supplemental_Information)
Time_Period_of_Content =
    Time_Period_Information + (see section for production rules)
    Currentness_Reference
Status =
    Progress +
    Maintenance_and_Update_Frequency
Spatial_Domain =
    Bounding_Coordinates +
    (1{Data_Set_G-Polygon}n)
Bounding_Coordinates =
    West_Bounding_Coordinate +
    East_Bounding_Coordinate +
    North_Bounding_Coordinate +
    South_Bounding_Coordinate
Data_Set_G-Polygon =
    Data_Set_G-Polygon_Outer_G-Ring +
    0{Data_Set_G-Polygon_Exclusion_G-Ring}n
Data_Set_G-Polygon_Outer_G-Ring =
    4{G-Ring_Latitude +
    G-Ring_Longitude}n
Data_Set_G-Polygon_Exclusion_G-Ring =
    4{G-Ring_Latitude +
    G-Ring_Longitude}n
Keywords =
    Theme +
```

(Place) +
(Stratum) +
(Temporal)
Theme =
1{Theme_Keyword_Thesaurus +
1{Theme_Keyword}n }n
Place =
1{Place_Keyword_Thesaurus +
1{Place_Keyword}n }n
Stratum =
1{Stratum_Keyword_Thesaurus +
1{Stratum_Keyword}n }n
Temporal =
1{Temporal_Keyword_Thesaurus +
1{Temporal_Keyword}n }n
Point_of_Contact =
Contact_Information (see section for production rules)
Browse_Graphic =
Browse_Graphic_File_Name +
Browse_Graphic_File_Description +
Browse_Graphic_File_Type
Security_Information =
Security_Classification_System +
Security_Classification +
Security_Handling_Description
Cross_Reference =
Citation_Information (see section for production rules)
Data_Quality_Information =
0{Attribute_Accuracy}1 +
Logical_Consistency_Report +
Completeness_Report +
0{Positional_Accuracy}1 +
Lineage +
(Cloud_Cover)
Attribute_Accuracy =
Attribute_Accuracy_Report +
(1{Quantitative_Attribute_Accuracy_Assessment}n)
Quantitative_Attribute_Accuracy_Assessment =
Attribute_Accuracy_Value +
Attribute_Accuracy_Expanation
Positional_Accuracy =
0{Horizontal_Positional_Accuracy}1 +
0{Vertical_Positional_Accuracy}1
Horizontal_Positional_Accuracy =
Horizontal_Positional_Accuracy_Report +
(1{Quanttative_Horizontal_Positional_Accuracy_Assessment}n)
Quantitative_Horizontal_Poitional_Accuracy_Assessment =
Horizontal_Positional_Accuracy_Value +
Horizontal_Positional_Accuracy_Explanation
Vertical_Positional_Accuracy =
Vertical_Positional_Accuracy_Report +
(1{Quantitative_Vertical_Positional_Accuracy_Assessment}n)
Quantitative_Vertical_Positional_Accuracy_Assessment =
Vertical_Positional_Accuracy_Value +
Vertical_Positional_Accuracy_Explanation
Lineage =
0{Source_Information}n +
1{Process_Step}n
Source_Information =
Source_Citation +
0{Source_Scale_Denominator}1 +
Type_of_Source_Media +
Source_Time_Period_of_Content +
Source_Citation_Abbreviation +
Source_Contribution
Source_Citation =
Citation_Information (see section for production rules)
Source_Time_Period_of_Content =
Time_Period_Information + (see section for production rules)
Source_Currentness_Reference
Process_Step =
Process_Description +
0{Source_Used_Citation_Abbreviation}n +
Process_Date +
(Process_Time) +
0{Source_Produced_Citation_Abbreviation}n +
(Process_Contact)
Process_Contact =
Contact_Information (see section for production rules)

Spatial_Data_Organization_Information =
0{Indirect_Spatial_Reference}1 +
0{Direct_Spatial_Reference_Method +
([Point_and_Vector_Object_Information |
Raster_Object_Information])}1
Point_and_Vector_Object_Information =
[SDTS_Terms_Description |
VPF_Terms_Description]
SDTS_Terms_Description =
1{SDTS_Point_and_Vector_Object_Type +
(Point_and_Vector_Object_Count) }n
VPF_Terms_Description =
VPF_Topology_Level +
1{VPF_Point_and_Vector_Object_Type +
(Point_and_Vector_Object_Count) }n
Raster_Object_Information =
Raster_Object_Type +
(Row_Count +
Column_Count +
0{Vertical_Count}1)
Spatial_Reference_Information =
0{Horizontal_Coordinate_System_Definition}1 +
0{Vertical_Coordinate_System_Definition}1
Horizontal_Coordinate_System_Definition =
[Geographic |
1{Planar}n |
Local] +
0{Geodetic_Model}1
Geographic =
Latitude_Resolution +
Longitude_Resolution +
Geographic_Coordinate_Units
Planar =
[Map_Projection |
Grid_Coordinate_System |
Local_Planar] +
Planar_Coordinate_Information
Map_Projection =
Map_Projection_Name +
[Albers_Conical_Equal_Area |
Azimuthal_Equidistant |
Equidistant_Conic |
Equirectangular |
General_Vertical_Near-sided_Perspective |
Gnomonic |
Lambert_Azimuthal_Equal_Area |
Lambert_Conformal_Conic |
Mercator |
Modified_Stereographic_for_Alaska |
Miller_Cylindrical |
Oblique_Mercator |
Orthographic |
Polar_Stereographic |
Polyconic |
Robinson |
Sinusoidal |
Space_Oblique_Mercator_(Landsat)|
Stereographic |
Transverse Mercator |
van_der_Grinten |
Other_Projection's_Definition]
Albers_Conical_Equal_Area =
1{Standard_Parallel}2 +
Longitude_of_Central_Meridian +
Latitude_of_Projection_Origin +
False_Easting +
False_Northing
Azimuthal_Equidistant =
Longitude_of_Central_Meridian +
Latitude_of_Projection_Origin +
False_Easting +
False_Northing
Equidistant_Conic =
1{Standard_Parallel}2 +
Longitude_of_Central_Meridian +
Latitude_of_Projection_Origin +
False_Easting +
False_Northing
Equirectangular =

```
        Standard_Parallel +
        Longitude_of_Central_Meridian +
        False_Easting +
        False_Northing
General_Vertical_Near-sided_Perspective =
        Height_of_Perspective_Point_Above_Surface +
        Longitude_of_Projection_Center +
        Latitude_of_Projection_Center +
        False_Easting +
        False_Northing
Gnomonic =
        Longitude_of_Projection_Center +
        Latitude_of_Projection_Center +
        False_Easting +
        False_Northing
Lambert_Azimuthal_Equal_Area =
        Longitude_of_Projection_Center +
        Latitude_of_Projection_Center +
        False_Easting +
        False_Northing
Lambert_Conformal_Conic =
        1{Standard_Parallel}2 +
        Longitude_of_Central_Meridian +
        Latitude_of_Projection_Origin +
        False_Easting +
        False_Northing
Mercator =
        [Standard_Parallel |
        Scale_Factor_at_Equator] +
        Longitude_of_Central_Meridian +
        False_Easting +
        False_Northing
Modified_Stereographic_for_Alaska =
        False_Easting +
        False_Northing
Miller_Cylindrical =
        Longitude_of_Central_Meridian +
        False_Easting +
        False_Northing
Oblique_Mercator =
        Scale_Factor_at_Center_Line +
        [Oblique_Line_Azimuth |
        Oblique_Line_Point] +
        Latitude_of_Projection_Origin +
        False_Easting +
        False_Northing
Oblique_Line_Azimuth =
        Azimuthal_Angle +
        Azimuth_Measure_Point_Longitude
Oblique_Line_Point =
        2{Oblique_Line_Latitude +
        Oblique_Line_Longitude}2
Orthographic =
        Longitude_of_Projection_Center +
        Latitude_of_Projection_Center +
        False_Easting +
        False_Northing
Polar_Stereographic =
        Straight-Vertical_Longitude_from_Pole +
        [Standard_Parallel |
        Scale_Factor_at_Projection_Origin] +
        False_Easting +
        False_Northing
Polyconic =
        Longitude_of_Central_Meridian +
        Latitude_of_Projection_Origin +
        False_Easting +
        False_Northing
Robinson =
        Longitude_of_Projection_Center +
        False_Easting +
        False_Northing
Sinusoidal =
        Longitude_of_Central_Meridian +
        False_Easting +
        False_Northing
Space_Oblique_Mercator_(Landsat) =
        Landsat_Number +
        Path_Number +

        False_Easting +
        False_Northing
Stereographic =
        Longitude_of_Projection_Center +
        Latitude_of_Projection_Center +
        False_Easting +
        False_Northing
Transverse_Mercator =
        Scale_Factor_at_Central_Meridian +
        Longitude_of_Central_Meridian +
        Latitude_of_Projection_Origin +
        False_Easting +
        False_Northing
van_der_Grinten =
        Longitude_of_Central_Meridian +
        False_Easting +
        False_Northing
Grid_Coordinate_System =
        Grid_Coordinate_System_Name +
        [Universal_Transverse_Mercator |
        Universal_Polar_Stereographic |
        State_Plane_Coordinate_System |
        ARC_Coordinate_System |
        Other_Grid_System's_Definition]
Universal_Transverse_Mercator =
        UTM_Zone_Number +
        Transverse_Mercator
Universal_Polar_Stereographic =
        UPS_Zone_Identifier +
        Polar_Stereographic
State_Plane_Coordinate_System =
        SPCS_Zone_Identifier +
        [Lambert_Conformal_Conic |
        Transverse_Mercator |
        Oblique_Mercator |
        Polyconic]
ARC_Coordinate_System =
        ARC_System_Zone_Identifier +
        [Equirectangular |
        Azimuthal_Equidistant]
Local_Planar =
        Local_Planar_Description +
        Local_Planar_Georeference_Information
Planar_Coordinate_Information =
        Planar_Coordinate_Encoding_Method +
        [Coordinate_Representation |
        Distance_and_Bearing_Representation] +
        Planar_Distance_Units
Coordinate_Representation =
        Abscissa_Resolution +
        Ordinate_Resolution
Distance_and_Bearing_Representation =
        Distance_Resolution +
        Bearing_Resolution +
        Bearing_Units +
        Bearing_Reference_Direction +
        Bearing_Reference_Meridian
Local =
        Local_Description +
        Local_Georeference_Information
Geodetic_Model =
        0{Horizontal_Datum_Name}1 +
        Ellipsoid_Name +
        Semi-major_Axis +
        Denominator_of_Flattening_Ratio
Vertical_Coordinate_System_Definition =
        0{Altitude_System_Definition}1 +
        0{Depth_System_Definition}1
Altitude_System_Definition =
        Altitude_Datum_Name +
        1{Altitude_Resolution}n +
        Altitude_Distance_Units +
        Altitude_Encoding_Method
Depth_System_Definition =
        Depth_Datum_Name +
        1{Depth_Resolution}n +
        Depth_Distance_Units +
        Depth_Encoding_Method
Entity_and_Attribute_Information =
```

[Detailed_Description |
  Overview_Description |
  Detailed_Description +
  Overview_Description]
Detailed_Description =
  1{Entity_Type +
  0{Attribute}n }n
Entity_Type =
  Entity_Type_Label +
  Entity_Type_Definition +
  Entity_Type_Definition_Source
Attribute =
  Attribute_Label +
  Attribute_Definition +
  Attribute_Definition_Source
  1{Attribute_Domain_Values}n +
  0{Attribute_Units_of_Measure}1 +
  (Attribute_Measurement_Resolution) +
  (1{Beginning_Date_of_Attribute_Values +
  0{Ending_Date_of_Attribute_Values}1 }n) +
  (Attribute_Value_Accuracy_Information) +
  (Attribute_Measurement_Frequency)
Attribute_Domain_Values =
  [Enumerated_Domain |
  Range_Domain |
  Codeset_Domain |
  Unrepresentable_Domain]
Enumerated_Domain =
  1{Enumerated_Domain_Value +
  Enumerated_Domain_Value_Definition +
  Enumerated_Domain_Value_Definition_Source +
  0{Attribute}n }n
Range_Domain =
  Range_Domain_Minimum +
  Range_Domain_Maximum +
  0{Attribute}n
Codeset_Domain =
  Codeset_Name +
  Codeset_Source
Attribute_Value_Accuracy_Information =
  Attribute_Value_Accuracy +
  Attribute_Value_Accuracy_Explanation
Overview_Description =
  1{Entity_and_Attribute_Overview +
  1{Entity_and_Attribute_Detail_Citation}n }n
Distribution_Information =
  1{Distributor +
  0{Resource_Description}1 +
  Distribution_Liability +
  0{Standard_Order_Process}n +
  0{Custom_Order_Process}1 +
  (Technical_Prerequisites) +
  (Available_Time_Period) }n
Distributor =
  Contact_Information (see section for production rules)
Standard_Order_Process =
  [Non-digital_Form |
  1{Digital_Form} n] +
  Fees +
  (Ordering_Instructions) +
  (Turnaround)
Digital_Form =
  Digital_Transfer_Information +
  Digital_Transfer_Option
Digital_Transfer_Information =
  Format_Name +
  ([Format_Version_Number |
  Format_Version_Date] +
  (Format_Specification)) +
  (Format_Information_Content) +
  0{File_Decompression_Technique}1 +
  (Transfer_Size)
Digital_Transfer_Option =
  1{ [Online_Option |
  Offline_Option] }n
Online_Option =
  1{Computer_Contact_Information}n +
  (Access_Instructions) +
  (Online_Computer_and_Operating_System)

Computer_Contact_Information =
  [Network_Address |
  Dialup_Instructions]
Network_Address =
  1{Network_Resource_Name}n
Dialup_Instructions =
  Lowest_BPS +
  0{Highest_BPS}1 +
  Number_DataBits +
  Number_StopBits +
  Parity +
  0{Compression_Support}1 +
  1{Dialup_Telephone}n +
  1{Dialup_File_Name}n
Offline_Option =
  Offline_Media +
  0{Recording_Capacity}1 +
  0{Recording_Format}n +
  0{Compatibility_Information}1
Recording_Capacity =
  1{Recording Density}n +
  Recording_Density_Units
Available_Time_Period =
  Time_Period_Information (see section for production rules)
Metadata_Reference_Information =
  Metadata_Date +
  (Metadata_Review_Date +
  (Metadata_Future_Review_Date)) +
  Metadata_Contact +
  Metadata_Standard_Name +
  Metadata_Standard_Version +
  0{Metadata_Time_Convention}1 +
  (Metadata_Access_Constraints) +
  (Metadata_Use_Constraints) +
  (Metadata_Security_Information)
Metadata_Contact =
  Contact_Information (see section for production rules)
Metadata_Security_Information =
  Metadata_Security_Classification_System +
  Metadata_Security_Classification +
  Metadata_Security_Handling_Description
Citation_Information =
  1{Originator }n+
  Publication_Date +
  (Publication_Time) +
  Title +
  0{Edition}1 +
  0{Geospatial_Data_Presentation_Form}1 +
  0{Series_Information}1 +
  0{Publication_Information}1 +
  0{Other_Citation_Details}1 +
  (1{Online_Linkage}n) +
  0{Larger_Work_Citation}1
Series_Information =
  Series_Name +
  Issue_Identification
Publication_Information =
  Publication_Place +
  Publisher
Larger_Work_Citation =
  Citation_Information
Time_Period_Information =
  [Single_Date/Time |
  Multiple_Dates/Times |
  Range_of_Dates/Times]
Single_Date/Time =
  Calendar_Date +
  (Time_of_Day)
Multiple_Dates/Times =
  2{Calendar_Date +
  (Time_of_Day) }n
Range_of_Dates/Times =
  Beginning_Date +
  (Beginning_Time) +
  Ending_Date +
  (Ending_Time)
Contact_Information =
  [Contact_Person_Primary |
  Contact_Organization_Primary] +

```
        (Contact_Position) +
        1{Contact_Address}n +
        1{Contact_Voice_Telephone}n +
        (1{Contact_TDD/TTY_Telephone}n) +
        (1{Contact_Facsimile_Telephone}n) +
        (1{Contact_Electronic_Mail_Address}n) +
        (Hours_of_Service) +
        (Contact_Instructions)
Contact_Person_Primary =
        Contact_Person +
        (Contact_Organization)
Contact_Organization_Primary =
        Contact_Organization +
```

```
        (Contact_Person)
Contact_Address =
        Address_Type +
        0(Address}n +
        City +
        State_or_Province +
        Postal_Code +
        (Country)
```

9. Keywords

9.1 data base; geographic information systems; geospatial; GIS; information management; library; network; on-line; raster; vector

Standard Practice for
Minimum Set of Data Elements to Identify a Soil Sampling Site[1]

This standard is issued under the fixed designation D 5911; the number immediately following the designation indicates the year of original adoption or, in the case of revision, the year of last revision. A number in parentheses indicates the year of last reapproval. A superscript epsilon (ε) indicates an editorial change since the last revision or reapproval.

1. Scope

1.1 This practice covers what information should be obtained to uniquely identify any soil sampling or examination site where an absolute and recoverable location is necessary for quality control of the study, such as a waste disposal project. The minimum set of data elements was developed considering the needs for informational data bases, such as geographic information systems (GIS). Other distinguishing details, such as individual site characteristics help in singularly cataloging the site. For studies that are not environmentally regulated, such as for an agricultural or preconstruction survey, the data specifications established by a client and the project manager may be different from that of the minimum set.

1.2 As used in this practice, a soil sampling site is meant to be a single point, not a geographic area or property, located by an X, Y, and Z coordinate position at land surface or a fixed datum. All soil data collected for the site are directly related to the coordinate position, for example, sample from x feet (or metres) or sample from interval x^1 to x^2 ft (or metres) below the X, Y, and Z coordinate position. A soil sampling site can include a test well, augered or bored hole, excavation, grab sample, test pit, sidewall sample, stream bed, or any other site where samples of the soil can be collected or examined for the purpose intended.

1.3 The collection of soil samples is a disruptive procedure as the material is usually extracted from its natural environment and then transported from the site to a laboratory for analysis. Normally, in this highly variable type of material, the adjacent soil profile will not be precisely the same as the sampled soil. For these reasons, when soil samples are removed the same material cannot be collected from the site later. Therefore, it is essential that the minimum set of data elements be documented with a high degree of accuracy.

1.4 Samples of soil (sediment) filtered from the water of streams, rivers, or lakes are not in the scope of this practice.

NOTE 1—There are many additional data elements that may be necessary to identify and to describe a soil sampling site, but are not included in the minimum set of data elements. An agency or company may require additional data elements as a part of their minimum set for a specific project or program.

1.5 This practice includes those data elements that will distinguish a site's geographical location on the Earth, political regimes, source identifiers, and individual site characteristics. These elements apply to all soil and geotechnical sampling sites involved in environmental assessment studies. Each category of site, such as a bore hole or excavation, may require additional data elements to be complete.

1.6 Some suggested components and representative codes for coded data elements, for example, "setting", are those established by Ref (1),[2] by ASCE in Ref (2), by the Water Resources Division of the U.S. Geological Survey in Ref (3), and by Boulding in Ref (4) and (5).

NOTE 2—The data elements presented in this practice do not uniquely imply a computer data base, but the minimum set of soil data elements that should be collected for entry into any type of permanent file.

2. Referenced Documents

2.1 *ASTM Standards:*
D 420 Guide for Investigating and Sampling Soil and Rock[3]
D 653 Terminology Relating to Soil, Rock, and Contained Fluids[3]
D 2487 Classification of Soils for Engineering Purposes (Unified Soil Classification System)[3]
D 2488 Practice for Description and Identification of Soils (Visual-Manual Procedure)[3]
D 5254 Practice for Minimum Set of Data Elements to Identify a Ground-Water Site[4]

3. Terminology

3.1 *Definitions of Terms Specific to This Standard:*
3.1.1 Soils are sediments or other unconsolidated solid particles of rock produced by the physical and chemical disintegration of rock, and which may or may not contain organic matter (see Terminology D 653).
3.1.1.1 *Discussion*—Soil consists of any individual or combination of gravel (passes a 3-in. or 75-mm screen), sand, clay, silt, organic clay, organic silt, and peat as categorized in the Unified Soil Classification System (1, 2, 4, 5) (see Classification D 2487). Materials larger than gravel, including cobbles (between 3 and 12 in. or 75 and 300 mm) and boulders (more than 12 in. or 300 mm), are not included in the definition of soil. Soil is found above the consolidated rocks and can be unsaturated (vadose zone) or saturated

---

[1] This practice is under the jurisdiction of ASTM Committee D-18 on Soil and Rock and is the direct responsibility of Subcommittee D18.21 on Ground Water and Vadose Zone Investigations.
Current edition approved Feb. 10, 1996. Published May 1996.

[2] The boldface numbers given in parentheses refer to a list of references at the end of the text.
[3] *Annual Book of ASTM Standards*, Vol 04.08.
[4] *Annual Book of ASTM Standards*, Vol 04.09.

(capillary fringe and water table) with water or other liquids.

NOTE 3—Soil, as defined by soils engineers, is all unconsolidated material above bedrock (6); or the natural medium for growth of land plants (7). The pedologic definition is, the unconsolidated mineral or organic matter on the surface of the earth subjected to and influenced by genic and environmental factors of: parent material, climate (including water and effects), macro- and micro-organisms, and topography, all acting over a period of time and producing a product-soil-that differs from material from which it is derived in many physical, chemical, biological, and morphological properties and characteristics (8).

3.2 Sediment (for geology) is a mass of organic or inorganic solid fragmented material, or the solid fragment itself, which comes from weathering of rock and is carried by, suspended in, or dropped by air, water, or ice; or a mass accumulated by any other natural agent and that forms in layers on the Earth's surface such as sand, gravel, silt, mud, till, or loess (6,9). These materials are "soils" for the purpose of this practice.

4. Summary of Practice

4.1 This practice includes the following data elements to identify a subsurface soil site:
4.1.1 *Geographic Location:*
4.1.1.1 Latitude,
4.1.1.2 Longitude,
4.1.1.3 Coordinate precision,
4.1.1.4 Altitude, and
4.1.1.5 Altitude precision.
4.1.2 *Political Regimes:*
4.1.2.1 State or country identification, and
4.1.2.2 County or county equivalent.
4.1.3 *Source Identifiers:*
4.1.3.1 Project identification,
4.1.3.2 Owner's name,
4.1.3.3 Source agency or company and address,
4.1.3.4 Unique identification, and
4.1.3.5 Date of first record for the soil sampling site.
4.1.4 *Individual Site Characteristics:*
4.1.4.1 Setting,
4.1.4.2 Type of soil sampling site,
4.1.4.3 Use of site, and
4.1.4.4 Reason for data collection or examination.

5. Significance and Use

5.1 Normally, the basic soil data are gathered by trained personnel during the field investigation phase of a study. Each agency or company has its own methods of obtaining, recording, and storing the information. Usually, these data are recorded onto forms that serve both in organizing the information in the field and the office, and often as entry forms for a computer data base. For soil data to be of maximum value to the current project and any future studies, especially those involved in the assessment of the environment, it is essential that a minimum set of key data elements be recorded for each site.

5.2 When obtaining basic data concerning a subsurface soil site, it is necessary to thoroughly identify that site so that it may be readily located again with minimal uncertainty and may be accurately plotted and interpreted for data parameters in relationship to other sites. For example, information can be presented on maps and in summary tables.

6. Documentation

6.1 *Geographic Location:*
6.1.1 *Introduction*—The universally accepted coordinate defining the absolute two-dimensional location of a site on the Earth's surface are latitude and longitude. The coordinates are determined by careful measurement from an accurate map, by survey, for example, Geographical Positioning System (GPS) or by conversion from another coordinate system, for example, Universal Transverse Mercator (UTM) System or State Plane Coordinate System (SPCS). The third-dimension of the location is established by determining the altitude at the site, usually from topographic maps or by surveying techniques. The U.S. Environmental Protection Agency (EPA) has guidance documents concerning their policy for locating data points or sites (10–13). In addition, the publication (14) can be obtained by the address given in Footnote 5.

NOTE 4—If sites are located by property, local, State, or Federal boundaries or by soil sampling grid lines, other grid coordinates, plane coordinates, plant location grids, referenced to recoverable benchmarks, their locations should be readily convertible to absolute latitude longitude coordinates by an acceptable method.

6.1.2 *Latitude*—Latitude is a coordinate representation that indicates locations on the surface of the Earth using the Earth's equator as the respective latitudinal origin. Record the best available value for the latitude of the site in degrees, minutes, seconds and fractions of a second (DDMMSSss). If latitude of the site is south of the Equator, precede the numbers with a minus sign (−). The use of N or S is also appropriate (3,13–15).

6.1.3 *Longitude*—Longitude is a coordinate representation that indicates locations on the surface of the Earth using the prime meridian (Greenwich, England) as the longitudinal origin. Record the best available value for the longitude of the site, in degrees, minutes, seconds, and fractions of a second (DDDMMSSss). If longitude of the site is measured east of the Greenwich Meridian, precede the numbers with a minus sign (−). The use of E or W is also appropriate (3, 13–15).

6.1.4 *Coordinate Precision*—Record the precision or accuracy of the coordinate values. The precision values may be measured in linear distance (feet or metres) or in coordinate degree values (stated as decimal values or as minutes and seconds). The method specified by EPA is the coordinate degree values (13).

NOTE 5—For most soil surveys the precision of the coordinate values is dependent upon the size of the sample. In most subsurface drilling operations, the highest level of attainable accuracy is about ±0.05 ft (1.5 cm), therefore surveys of greater precision should not be required.

6.1.5 *Altitude*—Record the altitude of land surface or measuring point. Altitude of the land surface is the vertical distance in feet (or metres) either above or below a reference datum surface. The reference datum surface must be noted (3,13,15).

NOTE 6—In the United States, this reference surface should be the North American Vertical Datum (NAVD) of 1988 or National Geodetic Vertical Datum (NGVD) of 1929. If another vertical reference datum is

5 Available from National Technical Information Service, U.S. Department of Commerce, 5285 Port Royal Road, Springfield, VA 22161.

used to determine the altitude, describe the system.

NOTE 7—The measuring point is usually a carefully surveyed and permanently fixed object near a soil sampling site used for determining the altitude of the collected or examined material at the site.

6.1.6 *Altitude Precision*—Record the precision or accuracy of the altitude. As an example, Record 1, for an accuracy of ±1 ft (or m) or 0.1 for ±0.1 th ft (or m) to denote the judged error of the measurement (**3**).

6.2 *Political Regimes:*

6.2.1 *Introduction*—The description of the soil sampling site in some political jurisdictions helps in the proper identification of the site.

6.2.2 *State or Country Identification*—Record the state or country in which the site is physically located. The common systems for identifying States and countries are the Federal Information Processing Standard code (FIPS), a two-digit numeric code or the American National Standard Abbreviation two-letter code. The country codes are a two-character and a set of three-character alphabetic codes (**3,13,16–18**).

NOTE 8—The publications (FIPS PUB 5-2, FIPS PUB 6-4 and FIPS PUB 104-1) containing the codes for countries, states, and counties are available from the address in Footnote 5.

6.2.3 *County and County Equivalent*—Record the county or county equivalent in which the site is physically located. The common code system for identifying counties is the FIPS code, a three-digit numeric code. The documentation of political subdivisions will depend on the system used in each individual country (**3,13,15,18**).

NOTE 9—In many cases it is necessary to record a subdivision of the local government to further identify the area where the soil sampling site is located. Some local subdivisions are a city, town, village, municipality, township, or borough. Identify the local subdivision, for example "City of Rockville", to clearly denote the unit.

6.3 *Source Identifiers:*

6.3.1 *Introduction*—The soil sampling site must be identified as to the project, owner, the agency or company that recorded data, and its distinctive identification.

6.3.2 *Project Identification*—Record the name of the project that includes the soil sampling site, for example, Coralville Dam, Johnson County Soil Survey, or Cedar Low-level Waste Disposal (**3–5**).

6.3.3 *Owner's Name*—Record the name of the property owner of the soil sampling site. The recommended format for an individual's name is: last name, first name, middle initial. If a company's name is lengthy, use meaningful abbreviations. The owner's address can be included for further identification (**3,15**).

6.3.4 *Source Agency or Company and Address*—Record the name and address of the agency or company that collected the data for the soil sampling site. This data element is necessary to determine the original source of the data for the site (**19**).

6.3.5 *Unique Identification*—Record the unique naming that the agency or company uses to identify the soil sampling site. This identification is called by several terms such as "local site number", "site identification", and "well number" (if the site was finished as a well), etc. The description is commonly a combination of letters and numbers that could represent a land-net location or a sequential assignment for a site in a county, city, company, or project. This identification is important to precisely differentiate a site in the

records of an agency or company (**2–5,15**).

6.3.6 *Date of First Record for the Soil Sampling Site*—Record the date that the first valid transaction occurred for any element of the specified site. This could be the date of a permit application or start of construction. This element is important to facilitate the proper identification of the record (**2–5,15**).

6.4 *Individual Characteristics of the Site:*

6.4.1 *Introduction*—Each soil sampling site has specific features that, in combination, uniquely identify that site. These characteristics should be recorded for further defining the site.

6.4.2 *Setting*—Record the information that best describes the setting in which the site is located. Setting refers to the topographic, landform, or geomorphic features near the site. Suggested setting components and representative codes are (**2–5, 15**):

6.4.2.1 A—Alluvial fan,
6.4.2.2 B—Playa,
6.4.2.3 C—Stream channel,
6.4.2.4 D—Local depression,
6.4.2.5 E—Dunes,
6.4.2.6 F—Flat surface,
6.4.2.7 G—Flood plain,
6.4.2.8 H—Hilltop,
6.4.2.9 I—Inland wetlands,
6.4.2.10 J—River delta,
6.4.2.11 K—Sinkhole,
6.4.2.12 L—Lake,
6.4.2.13 M—Mangrove swamp or coastal wetlands,
6.4.2.14 N—Estuary,
6.4.2.15 P—Pediment,
6.4.2.16 S—Hillside (slope),
6.4.2.17 T—Alluvial or marine terrace,
6.4.2.18 U—Undulating,
6.4.2.19 V—Valley flat (valleys of all sizes),
6.4.2.20 W—Upland draw,
6.4.2.21 X—Unknown,
6.4.2.22 Y—Wetlands, and
6.4.2.23 Z—Other—describe.

NOTE 10—Components and codes given for "setting", "type of soil sampling site", "use of site", and "reason for data collection or examination" are only suggestions and are not considered absolute or complete lists. The agency or company that uses the Standard may want to alter these lists by deleting, adding, or fully explaining each individual component. The use of codes for the components may not be desirable for the purposes intended by the agency or company, as shown in Fig. 2. The important factor is that the information is included as a part of the data set.

6.4.3 *Type of Soil Sampling Site*—This data element helps to identify the physical type of soil sampling site. Record the type of site to which these data apply. Suggested site type components and representative codes are (**3–5**) (see Note 10):

6.4.3.1 A—Augered hole, hand, specify type,
6.4.3.2 B—Bored hole, mechanical, specify type,
6.4.3.3 C—Cone penetration,
6.4.3.4 D—Trench,
6.4.3.5 E—Excavated hole, for example, construction location,
6.4.3.6 F—Test pit,
6.4.3.7 G—Geophysical test hole,

6.4.3.8 O—Outcrop, natural slopes and embankments,

6.4.3.9 P—Push tube, hand, specify type,

6.4.3.10 Q—Push tube, mechanical, specify type,

6.4.3.11 R—Road cut,

6.4.3.12 S—Surface, sampled with shovel, scoop, spoon, pick, etc.,

6.4.3.13 T—Tunnel, shaft, or mine,

6.4.3.14 W—Test hole, drilled, completed as well,

6.4.3.15 X—Test hole, drilled, not completed as a well, and

6.4.3.16 Z—Other—describe.

6.4.4 *Use of Site*—Record the use of the site or the purpose for which the site was constructed (the former always holds precedence over the latter). If site is used for more than one purpose, also record the subordinate uses. Suggested site use components and representative codes are (**1,2,4,5**) (see Note 10):

6.4.4.1 C—Cut for road construction,

6.4.4.2 F—Dam construction,

6.4.4.3 M—Mine or road tunnel or shaft,

6.4.4.4 Q—Quarry or mine embankment,

6.4.4.5 B—Soil sampling—boring,

6.4.4.6 E—Soil sampling—excavation,

6.4.4.7 S—Soil sampling—surface extraction,

6.4.4.8 T—Test hole for water,

6.4.4.9 G—Test hole for oil and gas,

6.4.4.10 H—Test hole for exploration of minerals,

6.4.4.11 L—Test hole for liquid contaminate extraction,

6.4.4.12 D—Test boring for contaminate detection,

6.4.4.13 A—Test boring for construction,

6.4.4.14 W—Hazardous and non-hazardous release site excavation,

6.4.4.15 U—Unknown, and

6.4.4.16 Z—Other—describe.

6.4.5 *Reason for Data Collection or Examination*—Record the reason for which data were collected from or examined at the site. If the data were collected or examined for more than one purpose, record the subordinate reasons. Suggested data components and representative codes are (see Note 10):

6.4.5.1 A—Agricultural survey,

6.4.5.2 G—Construction design,

6.4.5.3 B—Research,

6.4.5.4 C—Comprehensive Environmental Response, Compensation, and Liability Act (CERCLA), amended by Superfund Amendments Reauthorization Act (SARA),

6.4.5.5 R—Resource Conservation and Recovery Act (RCRA),

6.4.5.6 D—Drinking water regulations,

6.4.5.7 E—Exploration (water),

6.4.5.8 L—Local ordinance,

6.4.5.9 S—State regulations, other than CERCLA/SARA or RCRA,

6.4.5.10 F—Federal regulations, other than CERCLA, SARA or RCRA,

6.4.5.11 I—Environmental issues,

6.4.5.12 J—Judicial/litigation,

6.4.5.13 M—Mining regulations,

6.4.5.14 N—Natural resources exploration,

6.4.5.15 P—Property transfer,

6.4.5.16 V—Reconnaissance,

6.4.5.17 U—Unknown, and

6.4.5.18 Z—Other—describe.

7. Sample Form

7.1 An example of a generalized form for recording a minimum set of data elements for a soil sampling site is shown in Fig. 1. An example of a filled-out form is shown in Fig. 2.

8. Keywords

8.1 key data elements; sediment; site coordinates; site identification; site location; soils; soil sample collection

ASTM STANDARD PRACTICE

MINIMUM SET OF DATA ELEMENTS TO IDENTIFY A SOIL SAMPLING SITE

Site Identification _____ Date Prepared _____
Sampled by _____
Prepared by _____
Map or Topo Quad and Series _____

Geographic Location:
Latitude _____ Latitude Precision _____
Longitude _____ Longitude Precision _____
Altitude of Measuring Point--
    Land Surface _____ metres/feet _____
    Other _____ metres/feet _____
Altitude Precision _____ Datum Reference _____

Political Regimes:
State or Country Identification _____
County or County Equivalent Identification _____
Additional Identification _____

Source Identifiers:
Project Identification _____
Owner's Name _____
Address of Owner _____
Source Agency or Company _____
Address of Source Agency _____
_____
Unique Identification of Site _____
_____
Date of First Record for Site _____

Individual Characteristics of the Site:
Setting _____
Type of Soil Sampling Site _____
Use of Site _____
Reason for Data Collection _____

FIG. 1 Example of Minimum Set of Data Elements Form

ASTM STANDARD PRACTICE

MINIMUM SET OF DATA ELEMENTS TO IDENTIFY A SOIL SAMPLING SITE

Site Identification _391826074370901_   Date Prepared _Sept.4,1991_

Sampled by _Roger J.Henning_

Prepared by _Roger J.Henning_

Map or Topo Quad and Series _Jobs Point Quad, 7.5-minute_

Geographic Location:

   Latitude _39 18 26_   Latitude Precision _+/- 1 sec._

   Longitude _74 37 09_   Longitude Precision _+/- 1 sec._

   Altitude of Measuring Point--

     Land Surface _10.00_   metres/feet _feet_

     Other _____   metres/feet _____

   Altitude Precision _0.01 feet_   Datum Reference _NAVD - 1929_

Political Regimes:

   State or Country Identification _New Jersey_

   County or County Equivalent Identification _Atlantic County_

   Additional Identification _Garden Township_

Source Identifiers:

   Project Identification _Jobs Point Dam_

   Owner's Name _State of New Jersey_

   Address of Owner _Trenton, NJ_

   Source Agency or Company _U.S. Geological Survey_

   Address of Source Agency _810 Bear Tavern Road, Suite 206_

     West Trenton, NJ 08628

   Unique Identification of Site _Jobs Point Construction Project_

     Test Hole #3

   Date of First Record for Site _October, 1959_

Individual Characteristics of the Site:

   Setting _Unknown; Near Exit 29 - Garden State Prkwy._

   Type of Soil Sampling Site _Bored Hole_

   Use of Site _Dam Construction_

   Reason for Data Collection _Construction Design_

FIG. 2   Example of Filled-Out Minimum Set of Data Elements Form

REFERENCES

(1) U.S. Department of the Interior, "Earth Manual, Water Resources Technical Publication, Second Edition," *Water and Power Resources Service*, 1980.

(2) Casagrande, A., "Classification and Identification of Soils," *Transactions, ASCE*, 1948.

(3) Mathey, S. B., ed., *National Water Information System User's Manual*, Vol 2, Chapter 4. "Ground-Water Site Inventory System," U.S. Geological Survey, Open-File Report 89-587, 1990.

(4) Boulding, J. R., "Description and Sampling of Contaminated Soils, A Field Pocket Guide," *Center for Environmental Research Information*, U.S. EPA, EPA/625/12-91/002, Cincinnati, OH, 1991.

(5) Boulding, J. R., *Description and Sampling of Contaminated Soils, A Field Guide, Second Edition*, Lewis Publishers, Boca Ratan, FL, 1994.

(6) McGraw-Hill, *Dictionary of Scientific and Technical Terms*, Fourth Edition, McGraw-Hill, 1989.

(7) Bates, R. L., and Jackson, J. A., *Glossary of Geology*, Third Edition, American Geological Institute, Alexandria, VA, 1987.

(8) Soil Science Society of America, *Glossary of Soil Science Terms*, SSSA, Madison, WI, 1987.

(9) U.S. Geological Survey, *National Handbook of Recommended Methods for Water-Data Acquisition*," Chapter 3—"Sediment" Office of Data Coordination, Reston, VA, 1978.

(10) U.S. Environmental Protection Agency (EPA), *Locational Data Policy Implementation Guidance: Guide to the Policy*, EPA/220/B-92-008, U.S. EPA Office of Administrative and Resources Management (PMD-211D), Washington, DC, 1992.

(11) U.S. Environmental Protection Agency (EPA), *Locational Data Policy Implementation Guidance: Guide to Selecting Latitude/Longitude Collection Methods*, EPA/220/B-92-008, U.S. EPA Office of Administrative and Resources Management (PMD-211D), Washington, DC, 1992.

(12) U.S. Environmental Protection Agency (EPA), *Locational Data Policy Implementation Guidance: Guide—Global Positioning System Technology and Its Application In Environmental Programs—GPS Primer*, EPA/600/R-92/036, U.S. EPA Center for Environmental Research Information, Cincinnati, OH, 1992.

(13) U.S. Environmental Protection Agency (EPA), *Definitions for the Minimum Set of Data Elements for Ground Water Quality*, EPA 813/B-92-002, U.S. EPA Office of Ground Water and Drinking Water, Washington, DC, 1992.

(14) U.S. Department of Commerce, "Representation of Geographic Point Locations for Information Interchange," *Federal Information Standards (FIPS) Publication 70-1*, National Institute for Standards and Technology, Washington, DC, June 23, 1986.

(15) Texas Natural Resources Information System, *Ground-Water Data INTERFACE, Users Reference Manual*, Texas Natural Resources Information System, November 20, 1986.

(16) U.S. Department of Commerce, "American National Standard Codes for the Representation of Names of Countries, Dependencies, and Areas of Special Sovereignty for Information Interchange," *Federal Information Standards (FIPS) Publication 104-1*, National Institute for Standards and Technology, Washington, DC, May 12, 1986.

(17) U.S. Department of Commerce, "Codes for the Identification of the States, the District of Columbia and Outlying Areas of the United States, and Associated Areas," *Federal Information Standards (FIPS) Publication 5-2*, National Institute for Standards and Technology, Washington, DC, May 28, 1987.

(18) U.S. Department of Commerce, "Counties and Equivalent Entities the United States, Its Possessions, and Associated Areas," *Federal Information Standards (FIPS) Publication 6-4*, National Institute for Standards and Technology, Washington, DC, August 31, 1990.

(19) Edwards, M. D., and Josefson, B. M., *Identification Codes for Organizations Listed in Computerized Data Systems of the U.S. Geological Survey*, U.S. Geological Survey, Open-File Report 82-921, 1982.

Standard Guide for
Elements of a Complete Data Set for Non-Cohesive Sediments[1]

This standard is issued under the fixed designation D 5387; the number immediately following the designation indicates the year of original adoption or, in the case of revision, the year of last revision. A number in parentheses indicates the year of last reapproval. A superscript epsilon (ε) indicates an editorial change since the last revision or reapproval.

1. Scope

1.1 This guide covers criteria for a complete sediment data set.

1.2 This guide provides guidelines for the collection of non-cohesive sediment alluvial data.

1.3 This guide describes what parameters should be measured and stored to obtain a complete sediment and hydraulic data set that could be used to compute sediment transport using any prominently known sediment-transport equations.

1.4 *This standard does not purport to address all of the safety problems, if any, associated with its use. It is the responsibility of the user of this standard to establish appropriate safety and health practices and determine the applicability of regulatory limitations prior to use.*

2. Referenced Documents

2.1 *ASTM Standards:*
D 1129 Terminology Relating to Water[2]
D 4410 Terminology for Fluvial Sediment[2]
D 4411 Guide for Sampling Fluvial Sediment in Motion[3]
D 4822 Guide for Selection of Methods of Particle Size Analysis of Fluvial Sediments (Manual Methods)[3]
D 4823 Guide for Core-Sampling Submerged, Unconsolidated Sediments[3]

3. Terminology

3.1 *Definitions*—For definitions of terms used in this guide, refer to Terminology D 1129 and D 4410.

3.2 *Descriptions of Terms Specific to This Standard:*

3.2.1 *diameter, intermediate axis*—the diameter of a sediment particle determined by direct measurement of the axis normal to a plane containing the longest and shortest axes.

3.2.2 *diameter, nominal*—the diameter of a sphere of the same volume as the given particle (1).[4]

3.2.3 *diameter, sieve*—the size of sieve opening through which a given particle of sediment will just pass.

3.2.4 D_x—the diameter of the sediment particle that has x percent of the sample less than this size (diameter is determined by method of analysis; that is, sedimentation, size, nominal, etc.).

3.2.4.1 *Discussion*—Example: D_{45} is the diameter that has 45 % of the particles that have diameters finer than the specified diameter. The percent may be by mass, volume, or numbers and is determined from a particle size distribution analysis.

4. Summary of Guide

4.1 This guide establishes criteria for a complete sediment data set and provides guidelines for the collection of data about non-cohesive sediments.

5. Significance and Use

5.1 This guide describes what parameters should be measured and stored to obtain a complete sediment and hydraulic data set that could be used to compute sediment transport using any prominently known sediment-transport equations.

5.2 The criteria will address only the collection of data on noncohesive sediment. A noncohesive sediment is one that consists of discrete particles and whose movement depends on the particular properties of the particles themselves (1). These properties can include particle size, shape, density, and position on the streambed with respect to other particles. Generally, sand, gravel, cobbles, and boulders are considered to be noncohesive sediments.

6. Procedure

6.1 Parameters discussed here are divided into three major categories: sediment, hydraulic, and others. Within each of these categories there is a listing of the minimum parameters that should be collected or analyzed for and some additional parameters that, although are not critical, would add significant information to the data set if recorded.

6.2 *Sediment Parameters (Minimal):*

6.2.1 There are give basic sediment parameters that must be collected in order to have a complete data set. They are: concentration, bedload, bed material, particle-size distribution, and specific gravity.

6.2.1.1 *Concentration*—Report concentration of suspended-sediment or total-sediment samples in milligrams per litre (mg/L) or in parts per million (ppm). Collect these samples in such a way that they represent either the point, vertical, or cross section sampled. Follow sampling guides set forth in Guide D 4411 or in Ref (2) when collecting suspended-sediment or total-load samples.

6.2.1.2 *Bedload*—Report discharge of bedload in megagrams per day (Mg/d) or some other form of mass per time unit. The procedures for the collection of bedload samples, both in a flume and in the field, have not been standardized as well as those for suspended sediment. This is in part

[1] This guide is under the jurisdiction of ASTM Committee D-19 on Water and is the direct responsibility of Subcommittee D19.07 on Sediments.
Current edition approved April 15, 1993. Published August 1993.
[2] *Annual Book of ASTM Standards*, Vol 11.01.
[3] *Annual Book of ASTM Standards*, Vol 11.02.
[4] The boldface numbers in parentheses refer to the list of references at the end of this guide.

because the sampler development has not achieved the state of uniformity that the suspended-sediment samplers have and because not enough is currently known about bedload transport in open channels to accurately define a protocol for data collection. However, the procedure outlined in Ref (2) appears to be a reasonable approach to the problem and gives the state of knowledge and equipment at the present time.

6.2.1.3 *Bed Material*—Because the bed material is the primary source of noncohesive sediments, collect detailed samples. Most field bed-material sampling programs have been restricted to sampling sand-bed streams because of the overall lack of knowledge and the practical problems associated with sampling gravel-bed streams (3). References (2) and (3), as well as Guide D 4823, present several methods for collection of bed-material samples from gravel-bed streams. Also, some of the equipment and procedures given in Ref (2) and Guide D 4823 can be used to collect samples from sand bed streams.

6.2.1.4 *Particle-Size Distribution*—Record the particle-size distribution in percent finer than a given diameter size. The most generally used size grading system for sediment work in the United States is the grade scale proposed by the Subcommittee on Sediment Terminology of the American Geophysical Union (AGU), which is an extension of the Wentworth scale (1). Determine as an absolute minimum the percent finer than and greater than 0.062 mm. Ideally, determine all applicable breaks given on the AGU scale (1). Determine particle size either as a physical size (sieve) or as a sedimentation (fall) diameter. Whichever method is used, record the method of determination. Guide D 4822 presents a way to help choose which method might work best given the particle sizes to be sampled and the units of the distribution desired. Several of the more common particle-size analysis methods are given in Ref (4).

6.2.1.5 Preform particle-size distribution analysis on suspended-sediment, total-load, bedload, and bed-material samples. Results should indicate whether the diameters determined are sieve, fall, intermediate axis, or nominal diameters, and whether they are percent finer than by mass, volume, or number of particles.

6.2.1.6 Record the method or specific piece of equipment, or both, used to determine particle-size distribution.

6.2.1.7 *Specific Gravity*—The specific gravity of a particle effects to how the particle reacts in the flow. Most of the time the specific gravity is assumed to be 2.65. Although this is true most of the time, Brownlie (5) points out that about half of J. J. Franco's data has a specific gravity of 1.30 and that the following data sets have these ranges in specific gravity: Pang-Yung Ho, 2.45 to 2.70; C. R. Neill, 1.36 to 2.59; and U.S. Waterways Experiment Station, 1936c, 1.03 to 1.85.

6.3 *Sediment Parameters (Additional):*

6.3.1 The following parameters are considered to be ones that are not absolutely necessary for a complete data set but would give significant additional information and clarification to the data.

6.3.1.1 *Specific Diameters*—Calculated diameters such as D_{16}, D_{35}, D_{50}, D_{65}, D_{84}, and D_{90} are quite often used in sediment transport equations. Having these computed diameter sizes stored in the data bases will allow everyone using the data in the future to use the same values for these

percentiles, thus avoiding some additional sources of errors when comparing their results to the original developer's results. Store diameters in millimetres and give the type, that is, fall, sieve, etc.

6.3.1.2 *Method of Collection*—Document how the samples were collected. It is often very important to know if the samples were collected from single vertical or multiverticals, surface dipped, or point samples. This not only is important for suspended-sediment and total-load samples, but also is important for bedload and bed-material samples. If multiple verticals are used to collect the sample, note the number of verticals used and some general description of their placement in the cross section. If the sample is collected from a single point or vertical, identify the collection point.

6.3.1.3 *Sampler*—Record the type of sampler and nozzle size. The US-D, US-DH, and US-P series samplers (1) are depth integrating and point integrating samplers that collect samples of the water sediment mixture isokinetically. This ensures the proper concentration of sand is sampled from the stream. When collecting bedload samples, in addition to the sampler type and nozzle size, record the bag mesh opening size and nozzle flare if appropriate for the sampler being used.

6.4 *Hydraulic Parameters (Minimal):*

6.4.1 There are four major hydraulic parameters that should be collected to provide a complete sediment-transport data set. They are water discharge, width, depth, and slope.

6.4.1.1 *Discharge, Water*—The amount or rate of water flowing past the sampling point or cross section at the time of sampling is extremely critical to understanding the interpretation of the sediment data collected. Chapter 1 of Ref (6) gives a good summary of how surface-water discharge data can be collected. Record water discharge in cubic metres per second (m^3/s).

6.4.1.2 *Width*—Channel or flume width is important in computing other hydraulic parameters, such as area and mean velocity, and for determining depth to width ratios that are used, among other things, to assess the bank or boundary effects. In addition, repeated measurements of channel width at the same location over a period of time can be useful, when used with other data, in determining bank and channel stabilization.

6.4.1.3 *Depth*—Record the average depth of flow. This depth is normally calculated by dividing the area of flow by the channel or flume width.

6.4.1.4 *Slope*—There are three common types of slope that are used: bed, water surface, and energy. For whichever slope is measured, or computed, record the value and type.

6.5 *Hydraulic Parameters (Additional):*

6.5.1 *Area*—Cross sectional area of flow is normally one of the parameters computed when making discharge measurements, especially in the field. It is used in computing average stream depth in natural channels.

6.5.2 *Gage Height*—Record gage height, or stage, when repetitious measurements are made at a site over a long period of time or when flow conditions might be changing during the time taken to collect the sediment data, or both. Reference gage height to some fixed point at the site. By periodically recording gage height, water discharge, and cross sectional area, overall change of scour or fill in a channel. Also, assessment can be made of any changes in flow that occur during and between collection of sediment data, for example between the time the suspended-sediment samples were collected and the bedload discharge was measured, can be assessed.

6.5.3 *Hydraulic Radius*—Compute hydraulic radius from the area and wetted perimeter. Sometimes it is computed as, and assumed to be equivalent to, the average stream depth. It is always good to record what was used as the hydraulic radius and to describe how it was computed.

6.5.4 *Roughness Coefficient*—Record a roughness coefficient, usually either Manning's "*n*" or Chezy's "*C*". Estimate either in the field (7, 8) or compute using other hydraulic information.

6.6 *Other Parameters:*

6.6.1 In addition to the parameters listed above, record the following.

6.6.1.1 *Temperature*—Record temperature for each sediment data set collected. The concentration and distribution of sand particles with depth is affected by water temperature (1). Lane and others (9) found that sediment transport for the same water discharge was approximately 2.5 times greater in the winter than in the summer on the lower Colorado River.

6.6.1.2 *Sample Information*—Record information about the sample. As a minimum, record the date, time, and sampling location (that is, stream name and location of sampling point on the stream). Record any information pertinent to the sample, such as any angle between the cross section and the perpendicular of the flow. If the samples were collected from a flume, note this as well as the location of the flume.

6.6.1.3 *Bed Forms*—If possible, record a description of the bed forms present at the time of data collection. If the bed forms cannot be observed, record a description of the water surface, that is, standing waves, boils, smooth, etc. Bed form can be a major contribution to the overall bed roughness of a stream. They also can cause alternating increases and decreases in stream depth and thus can cause locally strong eddies, which can bring about larger, short-term variations in sediment concentration.

6.6.1.4 *Conductivity/Dissolved Solids*—Like temperature, changes in dissolved solids can affect sediment-transport rates. Increases in dissolved solid can cause increases in sediment-transport rates for the same flow conditions.

6.6.1.5 *Site Description*—Whether the samples are collected in a flume or in the field, give a general description of the sampling site. Special note should be made of flow conditions, weather, sampling apparatus used, anything upstream or downstream that might have affected the sample collection process, and any tributary inflow that might have affected flow or mixing at the sampling cross section.

6.6.1.6 *Particle Shape*—Size alone may not be sufficient to adequately describe sediment particles (1), also, use shape and roughness (p. 21 of Ref (1)). Shape describes the form of a particle. Roughness is a measure of the sharpness of radius of curvature of the edges.

6.6.1.7 *Collector*—Record the name of the individual(s) that collected the sample. This will allow others analyzing the data to evaluate the experience of the collector and therefore be better able to evaluate the data.

7. Precision and Bias

7.1 The precision is a function of the conditions encountered and the measurement techniques used for each measurement.

8. Keywords

8.1 data elements; sampling; sediment; surface-water

REFERENCES

(1) Vanoni, V. A., "Sedimentation Engineering," *Manuals and Reports on Engineering Practice*, No. 54, ASCE, 1975.
(2) Edwards, T. K., and Glysson, G. D., "Field Methods for Measurement of Fluvial Sediment," *U.S. Geological Survey Open-File*, Report 86-531, 1988.
(3) Yuzyk, T. R., "Bed Material Sampling in Gravel-Bed Streams, Environment Canada, Water Survey of Canada," *Sediment Survey Section*, Report No. IWD-HQ-WRB-SS-86-8, 1986.
(4) Guy, H. P., *Laboratory Theory and Methods for Sediment Analysis, Techniques of Water-Resources Investigations of the U.S. Geological Survey*, Book 5, Chapter C1, 1969.
(5) Brownlie, W. R., *Compilation of Alluvial Channel Data: Laboratory and Field*, California Institute of Technology, Pasadena, California, Report No. KH-R-43B, 1981.
(6) OWDC, *National Handbook of Recommended Methods for Water Data Acquisition*, Chapter 1, Surface Water, U.S. Geological Survey, 1980.
(7) Barnes, H. H., "Roughness Characteristics of Natural Channels," *U.S. Geological Survey Water-Supply Paper 1849*, 1967.
(8) Limerinos, J. T., "Determination of the Manning Coefficient from Measured Bed Roughness in Natural Channels," *U.S. Geological Survey Water-Supply Paper 1898-B*, 1970.
(9) Lane, E. W., Carlson, E. J., and Hanson, O. S., "Low Temperature Increases Sediment Transport in Colorado River," *Civil Engineering, ASCE*, Vol 19, No. 9, 1949, pp. 45–46.

Standard Guide for
Selection of Data Elements for Ground-Water Investigations[1]

This standard is issued under the fixed designation D 5474; the number immediately following the designation indicates the year of original adoption or, in the case of revision, the year of last revision. A number in parentheses indicates the year of last reapproval. A superscript epsilon (ε) indicates an editorial change since the last revision or reapproval.

1. Scope

1.1 This guide covers the selection of data elements for the documentation of ground-water sites. The data elements are described in four ASTM standards outlining information that may be collected at ground-water sites. Examples of specific investigations are given with the logic of why to select individual and combinations of data elements to meet the requirements of the studies.

NOTE 1—A ground-water site is any source, location, or sampling station capable of producing water or hydrologic data from a natural stratum from below the surface of the earth. A source or facility can include a well, spring or seep, and drain or tunnel (nearly horizontal in orientation). Other sources, such as excavations, driven devices, bore holes, ponds, lakes, and sinkholes, that can be shown to be hydraulically connected to the ground water, are appropriate for the use intended.

NOTE 2—The four ASTM standards that describe the data elements for ground water are Practice D 5254 and Guides D 5408, D 5409, and D 5410.

1.2 Systematic and consistent data collection are necessary for the investigation of the availability and the protection or restoration of ground-water resources. The level of detail, precision and bias, and the type of data that need to be collected depend on the objective of the study, the expected complexity of the system, and the resources available for the investigation. This guide presents ideas on what information should be collected for specific studies, why certain data elements are mandatory, and the importance to current and future investigations of maintaining quality control on the collection and retention of these data. This guide focuses on those data elements that are gathered at the field-site location and are used to assist in interpreting the hydrology of the ground-water source and to meet regulatory requirements. Other analytical and quality assurance/quality control (QA/QC) considerations are addressed in other standards and beyond the scope of this guide.

1.3 *This standard does not purport to address all of the safety problems, if any, associated with its use. It is the responsibility of the user of this standard to establish appropriate safety and health practices and determine the applicability of regulatory limitations prior to use.*

2. Referenced Documents

2.1 *ASTM Standards:*
D 653 Terminology Relating to Soil, Rock, and Contained Fluids[2]

D 5254 Practice for the Minimum Set of Data Elements to Identify a Ground-Water Site[2]
D 5408 Guide for the Set of Data Elements to Describe a Ground-Water Site; Part 1—Additional Identification Descriptors[2]
D 5409 Guide for the Set of Data Elements to Describe a Ground-Water Site; Part 2—Physical Descriptors[2]
D 5410 Guide for the Set of Data Elements to Describe a Ground-Water Site; Part 3—Usage Descriptors[2]

3. Terminology

3.1 *Definitions:*

3.1.1 Except as listed as follows, all definitions are in accordance with Terminology D 653.

3.1.1 *code*—a suggested abbreviation for a component, for example, "G" is the code suggested for the galvanized iron component of data element casing material. The data element is in the "casing record" record.

3.1.2 *component*—a subdivision of a data element, for example, galvanized iron is one of 30 components suggested for data element casing material. The data element is in the casing record record.

3.1.3 *data element*—an individual segment of information about a ground-water site, for example, casing material. The data element is in the casing record record.

3.1.4 *record*—denotes a set of related data elements that may need to be repeated to fully describe a ground-water site. For example, a well that consists of several diameters of casing from the top end to the bottom will need more than one casing record record (the record includes data elements depth to top, depth to bottom, diameter, casing material, and casing thickness) to fully describe the construction of the well. However, if only a single size of casing is used in the well, the record is utilized once.

3.1.5 *record group*—a set of related records. For example, the lift record group includes the lift record, power record, and standby record. Some record groups consist of only one record, for example, the spring record group includes only the spring record.

4. Summary of Guide

4.1 This guide describes four representative categories of investigations to demonstrate the logic of selecting data elements for the documentation of ground-water data. Included in this guide is a series of four tables that list the records (groups of data elements) used for the examples. The tables cross-reference the sections in this guide where specific explanations for data elements are found. A complete list of the individual data elements for each record is included in the text of this guide. The minimum set of data elements is standard and mandatory with all types of ground-water

[1] This guide is under the jurisdiction of ASTM Committee D-18 on Soil and Rock and is the direct responsibility of Subcommittee D18.21 on Ground Water and Vadose Zone Investigations.

Current edition approved Nov. 15, 1993. Published January 1994.

[2] *Annual Book of ASTM Standards*, Vol 04.08.

investigations and is presented in 6.1.3.

5. Significance and Use

5.1 Data are gathered at ground-water sites for many purposes. Each purpose requires a different combination of data elements. However, it is mandatory that every ground-water site include a minimum set of data elements to uniquely identify that site by precisely locating with coordinates and political regimes, absolutely identifying the owner and data source, and clearly defining the basic site characteristics. This information is described in Practice D 5254.

5.2 As a part of a ground-water project, each site requires additional data elements, beyond the minimum set, to assist in the interpretation of the local and areal hydrology. As an example, for a hydrologic reconnaissance study of a ground-water basin, each well or spring site requires basic information concerning construction, water level, yield, geology, and water chemistry. Additional information is needed if the project is a waste facility investigation, usually to satisfy local, state, and federal environmental regulations.

6. Documentation

6.1 Introduction:

6.1.1 Four representative hydrologic projects with very different objectives are provided as examples to demonstrate what data elements may be selected for a comprehensive ground-water data file. When designing a ground-water data file, data elements from all four ASTM guides should be considered (see Note 2). Agencies or companies that engage in widely diverse projects involving ground-water resources may require nearly all of the data elements described in the four standards. Those organizations should design a permanent file system to their specifications that includes these data.

NOTE 3—A ground-water data file can be stored as various media such as flat files in cabinets or as digital records on a computer. No matter which system is used, the data elements retained are the same information. An advantage of using a computerized file is that the data base containing the ground-water information can be easily displayed, duplicated, and transferred to another computer. Advantages of paper flat files include low cost, easy access without equipment, and transportability to field locations and meetings.

NOTE 4—For the explanation of ground-water investigations in this guide, the term "well" is used to mean any test or finished hole (that is, casing, screen, pump, etc.) that penetrates the surface of the earth to access the ground-water source. These include drilled, bored, driven, and dug holes.

6.1.2 Some agencies or companies may be very specialized in the objective of their projects and require only a finite number of data elements beyond the minimum data set. However, a limited data file may be expanded at a later date by adding additional data elements to satisfy the requirements of more extensive projects.

6.1.3 The minimum set of data elements (see Practice D 5254) is mandatory to uniquely locate, identify, and describe each individual ground-water site. In addition, photographs, sketches, and maps of the site and associated facility, including the measuring point, are valuable pictorial material to enhance the site description.

6.1.3.1 *Geographic Location*—Including latitude, longitude, latitude-longitude coordinate accuracy, altitude, and altitude accuracy.

6.1.3.2 *Political Regimes*—State or country identification, and county or county equivalent.

6.1.3.3 *Source Identifiers*—Owner's name, source agency

TABLE 1 General Resource Appraisal Investigation of an Area [A]

Minimum Set of Data Elements (see 6.1.3):
  Geographic Location (see 6.1.3.1)
  Political Regimes (see 6.1.3.2)
  Source Identifiers (see 6.1.3.3)
  Individual Site Characteristics (see 6.1.3.4)
Additional Data Elements (see X1.6):
  Geographic Location Record (see X1.6.1)
  Owner Record (see X1.6.2)
  Site Visits Record (see X1.6.3)
  Other Identification Record (see X1.6.4)
  Remarks Record (see X1.6.5)
  Individual Site Characteristics Record (see X1.6.6)
  Construction Record (see X1.6.7)
  Casing Record (see X1.6.8)
  Opening/Screen Record (see X1.6.9)
  Lift Record (see X1.6.10)
  Power Record (see X1.6.11)
  Geophysical Log Record (see X1.6.12)
  Geohydrologic Unit Record (see X1.6.13)
  Hydraulics Record (see X1.6.14)
  Aquifer Parameters Record (see X1.6.15)
  Well Clusters Record (see X1.6.16)
  Collector Well/Laterals Record (see X1.6.17)
  Ponds Record (see X1.6.18)
  Tunnel or Drain Record (see X1.6.19)
  Spring Record (see X1.6.20)
  Measuring-Point Record (see X1.6.21)
  Water-level Record (see X1.6.22)
  Discharge Record (see X1.6.23)
  Water-Quality Record (see X1.6.24)
  Field Water-Quality Record (see X1.6.25)

[A] See Appendix X1.

TABLE 2 Monitoring Project for a Waste-Disposal Facility [A]

Minimum Set of Data Elements (see 6.1.3):
  Geographic Location (see 6.1.3.1)
  Political Regimes (see 6.1.3.2)
  Source Identifiers (see 6.1.3.3)
  Individual Site Characteristics (see 6.1.3.4)
Additional Data Elements (see X2.5):
  Geographic Location Record (see X2.5.1)
  Political Regime Record (see X2.5.2)
  Source Identifiers Record (see X2.5.3)
  Owner Record (see X2.5.4)
  Site Visits Record (see X2.5.5)
  Other Identification Record (see X2.5.6)
  Remarks Record (see X2.5.7)
  Individual Site Characteristics Record (see X2.5.8)
  Construction Record (see X2.5.9)
  Casing Record (see X2.5.10)
  Opening/Screen Record (see X2.5.11)
  Lift Record (see X2.5.12)
  Geophysical Log Record (see X2.5.13)
  Geohydrologic Unit Record (see X2.5.14)
  Sample/Unconsolidated Material Record (see X2.5.15)
  Sample/Consolidated Material Record (see X2.5.16)
  Hydraulics Record (see X2.5.17)
  Aquifer Parameters Record (see X2.5.18)
  Measuring-Point Record (see X2.5.19)
  Network Record (see X2.5.20)
  Water-level Record (see X2.5.21)
  Discharge Record (see X2.5.22)
  Water-Quality Record (see X2.5.23)
  Field Water-Quality Record (see X2.5.24)
  Monitoring Site/Waste-Facility Record (see X2.5.25)
  Decommissioning Record (see X2.5.26)

[A] See Appendix X2.

TABLE 3 Contamination Assessment and Remediation[A]

Minimum Set of Data Elements (see 6.1.3):
  Geographic Location (see 6.1.3.1)
  Political Regimes (see 6.1.3.2)
  Source Identifiers (see 6.1.3.3)
  Individual Site Characteristics (see 6.1.3.4)
Additional Data Elements (see X3.5):
  Geographic Location Record (see X3.5.1)
  Political Regime Record (see X3.5.2)
  Remarks Record (see X3.5.3)
  Individual Site Characteristics Record (see X3.5.4)
  Construction Record (see X3.5.5)
  Casing Record (see X3.5.6)
  Opening/Screen Record (see X3.5.7)
  Lift Record (see X3.5.8)
  Geophysical Log Record (see X3.5.9)
  Geohydrologic Unit Record (see X3.5.10)
  Sample/Unconsolidated Material Record (see X3.5.11)
  Sample/Consolidated Material Record (see X3.5.12)
  Hydraulics Record (see X3.5.13)
  Aquifer Parameters Record (see X3.5.14)
  Ponds Record (see X3.5.15)
  Measuring-Point Record (see X3.5.16)
  Water-level Record (see X3.5.17)
  Discharge Record (see X3.5.18)
  Water-Quality Record (see X3.5.19)
  Field Water-Quality Record (see X3.5.20)
  Decommissioning Record (see X3.5.21)

[A] See Appendix X3.

TABLE 4 Underground Storage Tank Ground-Water Assessment[A]

Minimum Set of Data Elements (see 6.1.3):
  Geographic Location (see 6.1.3.1)
  Political Regimes (see 6.1.3.2)
  Source Identifiers (see 6.1.3.3)
  Individual Site Characteristics (see 6.1.3.4)
Additional Data Elements (see X4.6):
  Geographic Location Record (see X4.6.1)
  Owner Record (see X4.6.2)
  Site Visits Record (see X4.6.3)
  Remarks Record (see X4.6.4)
  Individual Site Characteristics Record (see X4.6.5)
  Construction Record (see X4.6.6)
  Casing Record (see X4.6.7)
  Opening/Screen Record (see X4.6.8)
  Lift Record (see X4.6.9)
  Repairs Record (see X4.6.10)
  Ponds Record (see X4.6.11)
  Geophysical Log Record (see X4.6.12)
  Sample/Unconsolidated Material Record (see X4.6.13)
  Sample/Consolidated Material Record (see X4.6.14)
  Measuring-Point Record (see X4.6.15)
  Water-level Record (see X4.6.16)
  Field Water-Quality Record (see X4.6.17)

[A] See Appendix X4.

or company and address, unique identification, and date of first record for the site.

6.1.3.4 *Individual Site Characteristics*—Hydrologic unit, setting, type of ground-water site, use of site, use of water from site, and reason for data collection.

7. Keywords

7.1 contamination assessment; data base; data elements; documentation; ground water; monitoring network; record; resource appraisal; site characterization; underground storage tank; waste disposal facility

APPENDIXES

(Nonmandatory Information)

X1. EXAMPLE 1—GENERAL GROUND-WATER INVESTIGATION OF AN AREA

X1.1 General ground-water investigations are commonly funded by local, state, and federal agencies in order to encourage economic development of an area. These investigations are typically conducted where insufficient data are known or documented about the quantity and quality of the ground water. The areas of concern are usually distinct political jurisdictions (such as municipal utility authority service districts, cities, or counties), ground-water basins, or aquifer units.

X1.2 Ground-water studies commonly define the water container (aquifer), the quantity of water obtainable for withdrawal, the quality of water available for specific uses, and the renewability or recharge rate for water that may be removed.

X1.3 A general ground-water study may require from a few months to several years to complete, usually depending upon the size of the area, extent of development, urgency of need (that is, severe drought conditions causing a local water supply emergency), complexity of the aquifer system, amount of information available from record-keeping agencies or water-development companies, and total objective of the project. Much of the time is utilized by the project personnel for collecting available information, conducting field surveys, gathering time-related information (that is, water levels), interpreting basic data, and drafting of the report.

X1.4 Additional data elements compiled for ground-water sites, beyond the minimum set of data elements, are a matter of need by the collecting agency or company to meet the objective of the total project. Therefore, the data file should be designed to include all possible information that are collected from all sources and to be retained for current and future ground-water projects. Comparison of the following records and data elements with the three guides will show that some records were not included and several data elements were deleted from the selected records because of the scope of the general ground-water investigation.

NOTE X1.1—The accuracy or confidence level should be documented for each measured data element by the agency or company that gathered and recorded the information.

NOTE X1.2—Specific types of ground-water sites have mandatory information, for example, a depth value is required with a well or bore hole. Some data elements are mandatory when a related data element is used, for example, a water level or water yield requires that the data of measurement be included. In addition, a number of data elements may be designated as mandatory by the agency or company to meet their specific research or management needs.

Note X1.3—The list of possible data elements that can be collected for a ground-water study is lengthy, however, an individual site normally will use only part of the total number of elements that can be included in the entire data file, for example, a spring site will not have the geophysical log record.

Note X1.4—Several of the data elements are repeated in more than one record (that is, primary aquifer and static water level). This duplication is important so that the key data elements for the record can be directly associated with critical information. For example, the geohydrologic unit record has a description of the lithology of an aquifer, while the hydraulics record defines the basic water-bearing characteristics of the same aquifer.

X1.5 *Minimum Set of Data Elements*—See 6.1.3.

X1.6 *Additional Data Elements*—From Guide D 5408:

X1.6.1 *Geographic Location Record*—Including land-net location, location map, map scale, and method altitude determined (see 6.1.3.1).

X1.6.1.1 *Explanation*—Additional information is desirable to further define the location of the site. The land-net location is commonly used, in many areas of the country, as a primary ground-water site identifier and locator. The location map is critical as a spatial locator for ground-water site data and is an absolute necessity for interpretation of the ground-water hydrology (see Note X1.5). A map scale is required with the location map.

Note X1.5—Geographical information systems (GIS) are becoming very common for agencies and companies using computer systems for data storage, retrieval, and manipulation. Computerized U.S. Geological Survey topographic maps are available for most of the country. The capability to retrieve data from a computer data base and locate these data precisely on a GIS map is the state-of-the practice.

X1.6.2 *Owner Record*—Including date of ownership and owner's name.

X1.6.2.1 *Explanation*—The documentation of the history of ownership of sites that have had several owners is important in searching for past information and determining management responsibility. For these sites, the date of proprietorship is required with each owner's name. The name of the current owner of the site is one of the components in the minimum set of data elements.

X1.6.3 *Site Visits Record*—Including date of visit and person who made visit.

X1.6.3.1 *Explanation*—Documentation of the identification of the person that gathered information at the site is important for quality control. The date of visitation is required with each site visit record.

X1.6.4 *Other Identification Record*—Including other name, number or identification, and assigner.

X1.6.4.1 *Explanation*—Documentation of site identifications assigned by other agencies or companies that collect data at the site is important in tracing information residing in their files. Data can be gathered from the same site for various purposes (that is, resource evaluation, health control, and pollution detection) by several organizations. Each organization (that is, the owner and state health department) commonly have their own system of site identification for filing purposes.

X1.6.5 *Remarks Record*—Including remark date and remark.

X1.6.5.1 *Explanation*—Many sites have important and unique text information that does not fit as a data element into the file. A space should be reserved in the file for this information.

X1.6.6 *Individual Site Characteristics Record*—Including drainage basin/watershed, hole depth, well depth, source of depth data, and primary aquifer (see Guide D 5409).

X1.6.6.1 *Explanation*—The hydrologic unit codes and primary aquifer data elements apply to all ground-water sites and are important components in understanding the hydrology of the project area. Whereas, the hole and well construction data, which indicates the open intervals, are point data that relate to the site where the ground-water aquifer was accessed by penetrating the surface of the earth (see Note 4). For monitoring projects and contamination investigations, the relationship of the site to surface-water bodies is important in understanding nearby water sources that could allow water to flow into or away from the studied area. In addition, the use of land in the vicinity of the contaminated area is critical information needed for determining the method to use for delineating the extent of contamination and the technique to employ for removal of the contaminate (that is, highly developed urban or sparsely populated agricultural area).

X1.6.7 *Construction Record*—Including date construction began, date construction ended, name of contractor, source of construction data, method of construction, type of finish, type of seal, and depth to bottom of seal.

X1.6.7.1 *Explanation*—The construction record applies primarily to wells, however, several of the data elements (construction dates, contractor, source of data, construction method, and type of finish) can be used to describe improvements at ponds, springs, and tunnels. The record, along with the casing and opening records, gives a complete description of construction at the ground-water site prior to and during the project (including resource investigations and contamination studies). This information is important for understanding the origin of the water, method of ground-water development in the project area, and zone of monitoring.

X1.6.8 *Casing Record*—Including depth to top of the cased interval, depth to bottom of the cased interval, diameter of the cased interval, casing material, and casing thickness.

X1.6.8.1 *Explanation*—The casing record applies primarily to wells, however a few of the data elements (casing diameter, material, and thickness) can be used to document the characteristic of the pipes used to extract water from ponds, springs and tunnels. In combination with the construction and opening records, the casing record gives a complete description of construction at the site prior to and during the project (including resource investigations and contamination studies).

X1.6.9 *Opening/Screen Record*—Including depth to top of the open interval, depth to bottom of the open interval, diameter of the open interval, type of material in the open interval, type of openings in the open interval, length of openings, and width of openings.

X1.6.9.1 *Explanation*—The opening or screen data delineates the zone that yields water to the well (see Note 4). However, screens are used in horizontally wide holes (ponds, sinks) that are open to below the water-table surface and can be described by this record. In combination with the construction and casing records, the opening/screen record gives a complete description of construction at the site prior to and during the project (including resource investigations and contamination studies).

X1.6.10 *Lift Record*—Including type of lift, date permanent lift was installed, depth of intake, manufacturer of lift device, serial number, and pump rating.

X1.6.10.1 *Explanation*—The lift record is used to describe the equipment employed to remove water (or water and contaminants) from the aquifer. This record is important in understanding the capacity of the aquifer to yield water, also, as a means of further identifying the ground-water site. In addition, this record can be used to describe the technique used for removal of contaminated water.

X1.6.11 *Power Record*—Including type of power, horsepower rating, name of power company, power company account number, and power meter number.

X1.6.11.1 *Explanation*—The power record data are needed when gathering power usage information (that is, kilowatts of electricity or gallons of fuel) from the power companies. Power utilization data, in many areas, are used to compute an estimate of the amount of water withdrawn from the site and, in combination with the other sites in the project, for determining the total amount of water withdrawn from the area.

X1.6.12 *Geophysical Log Record*—Including type of log, depth to top of logged interval, depth to bottom of logged interval, and source of log data.

X1.6.12.1 *Explanation*—Geophysical log data are important in interpreting the geology and hydrology at the site and of the project area. Some logs are written records by drillers or hydrologists of materials encountered during the drilling or digging of the hole. Other logs are run by lowering a cable-suspended probe into the well or test hole. The variations in sediment, rock, and water characteristics are detected by the probe and logged by recording equipment at the surface. Usually, the logs are in a graphical format (either on paper or in the computer) and stored in a separate file. This record is a summary of the available log data.

X1.6.13 *Geohydrologic Unit Record*—Including aquifer unit(s), contributing unit, depth to top of interval, depth to bottom of interval, lithology, and description of material.

X1.6.13.1 *Explanation*—The geohydrologic unit record identifies the water-yielding zone by a formal geologic name, locates the vertical position of the unit in reference to the altitude datum, and describes the rock material of the zone. Not only can this record be used to describe the lithology of the aquifer units, but also can be used as a descriptive log of the entire site (that is, well or test hole). This information is critical in correlating and interpreting the geology and hydrology of the project area.

X1.6.14 *Hydraulics Record*—Including hydraulic/aquifer unit, hydraulic/aquifer unit type, depth to top of unit, depth to bottom of unit, static water level, date of measurement, and unit contribution.

X1.6.14.1 *Explanation*—The hydraulics record includes data elements that describe some of the basic water-bearing parameters of the individual aquifers. Although several data elements are the same as in the geohydrologic record, it is important that the water-bearing parameters be associated directly with the individual aquifer unit. The agency or company that maintains the ground-water data file may want to combine the hydrogeologic and hydraulics records

for their purpose. These data are critical in correlating and interpreting the hydrology of the investigated area.

X1.6.15 *Aquifer Parameters Record*—Including transmissivity, horizontal hydraulic conductivity, vertical hydraulic conductivity, coefficient of storage, leakance, diffusivity, specific storage, specific yield, barometric efficiency, porosity, specific capacity, method used to determine aquifer characteristics, and availability of file of detailed result.

X1.6.15.1 *Explanation*—The aquifer parameters record includes a summary of field-determined or estimated values. These values are commonly determined by controlled tests conducted at field locations, customarily identified as aquifer tests. However, a number of these data elements can be estimated from previously gathered information concerning the aquifer properties within the studied area and by ground-water studies in adjacent or geologically similar areas. These data are critical for completely understanding the total water-producing potential (also, the path of movement of contaminants) of the investigated area. However, in most basic areal studies, these data only can be obtained at a limited number of sites. Detailed analysis results used for determining these aquifer parameters are usually filed separately in a graphical or tabular format.

X1.6.16 *Well Clusters Record*—Including number of wells in cluster, depth of deepest well in cluster, depth of shallowest well in cluster, and diameter of well cluster.

X1.6.16.1 *Explanation*—This record is for expanding the documentation of a ground-water site that consists of well clusters. Well clusters are specialized arrangements of more than one well with a single centralized pumping system. Normally, this type of multiple-well system is used where the aquifer is thin, near the land's surface or low yielding, or both.

X1.6.17 *Collector Well/Laterals Record*—Including number of laterals in collector well, depth of laterals in collector well, length of laterals in collector well, diameter of laterals in collector well, and mesh of screen in laterals.

X1.6.17.1 *Explanation*—This record is for expanding the documentation of a ground-water site that consists of a collector well. Collector wells (well with a series of lateral collector pipes) are specialized arrangements that have been developed for areas where the aquifer is thin, near the land's surface, or low yielding, or both. The main usage of this type of well has been along river valleys where the aquifer is composed of a thin alluvium that is bounded beneath and at the valley walls by a low-yielding bedrock material.

X1.6.18 *Ponds Record*—Including length of pond, width of pond, depth of pond, and volume of pond.

X1.6.18.1 *Explanation*—The pond record is for expanding the documentation of a ground-water site where the level and quantity of water in the pond is controlled by the ground-water table in the vicinity of the pond. Therefore, water withdrawn from the pond is replaced by water from the adjacent aquifer. For contaminated areas, ponds need to be documented and dealt with the same as any other ground-water site.

X1.6.19 *Tunnel or Drain Record*—Including length of tunnel or drain, width of tunnel or drain, depth of tunnel or drain, bearing (azimuth) tunnel or drain, and dip of tunnel or drain.

X1.6.19.1 *Explanation*—The tunnel or drain record is for expanding the documentation of a ground-water site that consists of a nearly horizontal tunnel or drain. The ground-water table (also a permeable fractured or faulted zone) is intentionally or unintentionally intersected and the downward slope of the tunnel away from the water source allows the water to flow to the land's surface. However, a pump may be used to move the water from the source to point of usage.

X1.6.20 *Spring Record*—Including name of spring, type of spring, permanence of spring, sphere of discharge, discharge, date of discharge, improvements, number of spring openings, flow variability, accuracy of flow variability, and magnitude of spring.

X1.6.20.1 *Explanation*—The spring record is for expanding the documentation of a ground-water site that consists of naturally occurring springs and seeps. These sites exist where the water table intersects the surface or where the combination of a positive hydraulic head in the aquifer and openings in the confining rock allows the water to flow to the land's surface.

X1.6.21 *Measuring-Point Record (from Guide D 5410)*—Including date interval of measuring-point utilization, weight in reference to datum, and description.

X1.6.21.1 *Explanation*—The measuring point is required for a ground-water site where the measured data need to be converted to an altitude value, such as water levels, for direct correlation with other sites in the studied area. Also, an established measuring point allows for consistency in comparing multiple water levels at a single site that are measured over a period of time. The location of the measuring point is usually a convenient and fixed position for conducting the measurements. The height value can be directly related to an altitude datum and would be defined as lower than, equal to, or higher than the datum. The date interval is required because the location of the measuring point can be changed by repairs and general maintenance at the site, therefore requiring establishment of another measuring point having a different height relationship to the altitude datum.

X1.6.22 *Water-Level Record*—Including measurement date, water level, water-level accuracy, status, method of measurement and instrumentation, instrumentation, source of data, and statistics method.

X1.6.22.1 *Explanation*—Water levels are basic data routinely (usually mandatory) gathered at ground-water sites. Many hydrologic interpretations are based on a project-wide combination of point water-level data, for example, the direction of water movement through the aquifer at a waste disposal site or the effects of withdrawals by wells in the aquifer. The effect of climatic fluctuations on the level of water in the aquifer can be evaluated by a series of water-level measurements over time at a single site. Many of the hydraulic characteristics of the aquifer are determined by

the measurement of water levels (and yields) during controlled test procedures.

X1.6.23 *Discharge Record*—Including measurement date, discharge, type of discharge, source of data, method of discharge measurement, instrumentation, production or pumping level, static level, method of water-level measurement, pumping period, specific capacity, drawdown, and source of water-level data.

X1.6.23.1 *Explanation*—The discharge is a value (volume per time unit) for water yield that is routinely (usually mandatory) measured at ground-water sites. The discharge of water can be by artificial means, such as an electrical-powered pump or a hand-powered bailer. Also, some sites, such as springs and artesian wells where the hydraulic head is above the discharge point, yield water by natural flow. The discharge data are important in determining the capacity of the aquifer to yield water for the use intended, and also as a component needed for calculating the hydraulic characteristics of the aquifer.

X1.6.24 *Water-Quality Record*—Including sample date, agency that analyzes samples, type of analyses, sample depth/interval, water-quality file containing analysis, collecting agency or company, sampling purpose, site condition, sample appearance, sample odor, presence of immiscible stratum, thickness of immiscible stratum, sample preservation method, sample filtration material, pumped period, casing volume, amount water purged, sampling method or sampler type, sampler material, and aquifer sampled.

X1.6.24.1 *Explanation*—The water-quality record includes basic data gathered at the ground-water site during the time the water sample was collected. Also, relevant information about the ground-water site and the agencies involved in the collection and sample analysis are retained for additional quality control. The laboratory-determined chemical analysis values are usually stored in a separate file, for example EPA's STORET or the company's own file. The possible water-quality chemical elements and compounds that can be determined number in the hundreds. This is the most important category of information for determining the geochemistry of the water in the aquifer and the evaluation of contamination of the ground water by numerous anthropogenic constituents created by man's development. Care must be exercised to gather comprehensive information to satisfy regulatory guidelines.

X1.6.25 *Field Water-Quality Record*—Including field sample date, parameter code, value of parameter, unit for parameter, and instrumentation or method of determination.

X1.6.25.1 *Explanation*—The field water-quality record includes chemical and related data that are determined at the field location. Many of these data are parameters that change rapidly after removal from the natural conditions, for example the water temperature, the pH, and volatile compounds. Usually these data are determined at the same time as a water sample is collected for laboratory analysis.

X2. EXAMPLE 2—MONITORING PROJECT FOR A WASTE-DISPOSAL FACILITY

X2.1 Waste-disposal facilities are land areas developed for the purpose of disposing and storage of various forms of

unwanted materials, both hazardous and nonhazardous. The amount of land utilized by these facilities usually is small relative to resource evaluation studies. Before final construction of the facility, it is common for a considerable amount of information to be gathered concerning the proposed disposal area. These data are obtained from land surface surveys, geophysical techniques, and drilling of test holes. Many of the modern facilities that are constructed for the storage of potentially hazardous wastes are underlain with impermeable material, such as a dense clay or plastic liner. These facilities commonly have drains above the impermeable zone to carry off any liquid wastes and have monitoring wells below the impermeable material to identify potential leaks of wastes or leachates into the underlying sediments and eventually, the local ground-water system.

X2.2 Some disposal facilities have no prepared underlying barrier and rely on natural sediments to retard movement of any waste products. Arid-zone facilities have a limited amount of water to serve as a potential carrier for wastes.

X2.3 The amount of ground-water data that are gathered for the range of disposal facilities are quite variable. Thus, a great deal depends upon the materials that are disposed of at the facility as to the amount and refinement of ground-water related data that are gathered at the waste-disposal facility. The following example of ground-water data elements assumes that the facility is for disposal and storage of hazardous wastes and requires enforcement of the appropriate regulations for its long-term containment and monitoring. The accuracy or confidence level should be documented for each measured data element by the agency or company that gathered and recorded the information.

X2.4 *Minimum Set of Data Elements*—See 6.1.3.

X2.5 *Additional Data Elements*—See Guide D 5408:

X2.5.1 *Geographic Location Record*—Including land-net location, location map, map scale, and method altitude determined (see 6.1.3.1 and X1.6.1 for explanation).

X2.5.2 *Political Regime Record*—Including congressional district.

X2.5.2.1 *Explanation*—The documentation of the political area of responsibility for the ground-water monitoring sites at waste-disposal facilities and in contaminated areas is important because of legislated regulatory control and intense public interest in these areas.

X2.5.3 *Source Identifiers Record*—Including site data used in report, site information in a computer data base, and photograph/sketch available of site.

X2.5.3.1 *Explanation*—Supporting components add to the extent of knowledge concerning the availability of data for the ground-water site. Each of these components are valuable additions to the general data base.

X2.5.4 *Owner Record*—Including date of ownership and owner's name (see X1.6.2 for explanation).

X2.5.5 *Site Visits Record*—Including date of visit, and person who made visit (see X1.6.3 for explanation).

X2.5.6 *Other Identification Record*—Including other name, number or identification, and assigner (see X1.6.4 for explanation).

X2.5.7 *Remarks Record*—Including remark date and remark (see X1.6.5 for explanation) (from Guide D 5409).

X2.5.8 *Individual Site Characteristics Record*—Including land use (in vicinity of site), drainage basin/watershed, relationship to surface stream/lake, hole depth, well depth,

source of depth data, and primary aquifer (see X1.6.6 for explanation).

X2.5.9 *Construction Record*—Including date construction began, date construction ended, name of contractor, source of construction data, method of construction, type of finish, type of seal, and depth to bottom of seal (see X1.6.7 for explanation).

X2.5.10 *Casing Record*—Including depth to top of the cased interval, depth to bottom of the cased interval, diameter of the cased interval, casing material, and casing thickness (see X1.6.8 for explanation).

X2.5.11 *Opening/Screen Record*—Including depth to top of the open interval, depth to bottom of the open interval, diameter of the open interval, type of material in the open interval, type of openings in the open interval, length of openings, and width of openings (see X1.6.9 for explanation).

X2.5.12 *Lift Record*—Including type of lift, date permanent lift was installed, depth of intake, manufacturer of lift device, serial number, and pump rating (see X1.6.10 for explanation).

X2.5.13 *Geophysical Log Record*—Including type of log, depth to top of logged interval, depth to bottom of logged interval, and source of log data (see X1.6.12 for explanation).

X2.5.14 *Geohydrologic Unit Record*—Including aquifer unit(s), contributing unit, depth to top of interval, depth to bottom of interval, lithology, and description of material (see X1.6.13 for explanation).

X2.5.15 *Sample/Unconsolidated Material Record*—Including sample weight, sample interval, particle size, percent of total sample, particle shape, and mineralogy.

X2.5.15.1 *Explanation*—The unconsolidated material encountered when drilling the monitor well or test hole is important in understanding the circulation of water in the vadose and saturated zone at the waste facility or spill location. This information is important in proper placement of the waste facility and location of the monitoring sites for detecting liquid and vapor leaks. In addition, there could be a chemical interaction between the sediment and hazardous material and this knowledge could dictate the disposal or remedical technique.

X2.5.16 *Sample/Consolidated Material Record*—Including drill cuttings or core, sample size (weight), sample interval, mineralogy, core length, core diameter, core recovery-percent, bedding, structure, and porosity.

X2.5.16.1 *Explanation*—As with the unconsolidated material, the consolidated material encountered when drilling the monitor well, test hole, or remedial well is important in understanding the circulation of water in the vadose and saturated zone in the project area. Again, this information is important in proper placement of the waste facility and location of the monitoring sites. In addition, there could be a chemical interaction between the rock substance and hazardous material and this knowledge could dictate the disposal or remedial technique.

X2.5.17 *Hydraulics Record*—Including hydraulic/aquifer unit, hydraulic/aquifer unit type, depth to top of unit, depth to bottom of unit, static water level, date of measurement, and unit contribution (see X1.6.14 for explanation).

X2.5.18 *Aquifer Parameters Record*—Including transmissivity, horizontal hydraulic conductivity, vertical hydraulic conductivity, coefficient of storage, leakance, diffusivity, specific storage, specific yield, barometric efficiency, porosity, specific capacity, method used to determine aquifer characteristics, and availability of file of detailed result (see X1.6.15 for explanation) (from Guide D 5410).

X2.5.19 *Measuring-Point Record*—Including date interval of measuring-point utilization, height in reference to datum, and description (see X1.6.21 for explanation).

X2.5.20 *Network Record*—Including data type, date interval of network utilization, source agency for network data, frequency of data collection, method of data acquisition, power type of instruments, and network.

X2.5.20.1 *Explanation*—Monitoring networks are normally established at hazardous-waste facilities to systematically gather required data for the evaluation of ground-water movement and for the detection of any unforeseen leaks of wastes into the sediments and ground-water system. Proper placement of each sampling site and selection of the time-frequency interval for data collection is important in management of the monitoring network for regulatory purposes.

X2.5.21 *Water-Level Record*—Including measurement date, water level, water-level accuracy, status, method of measurement and instrumentation, instrumentation, source of data, and statistics method (see X1.6.22 for explanation).

X2.5.22 *Discharge Record*—Including measurement date, discharge, type of discharge, source of data, method of discharge measurement, instrumentation, production or pumping level, static level, method of water-level measurement, pumping period, specific capacity, drawdown, and source of water-level data (see X1.6.23 for explanation).

X2.5.23 *Water-Quality Record*—Including sample date, agency that analyzes samples, type of analyses, sample depth/interval, water-quality file containing analysis, collecting agency or company, sampling purpose, site condition, sample appearance, sample odor, presence of immiscible

stratum, thickness of immiscible stratum, sample preservation method, sample filtration material, pumped period, casing volume, amount of water purged, sampling method or sampler type, sampler material, and aquifer sampled (see X1.6.24 for explanation).

X2.5.24 *Field Water-Quality Record*—Including field sample date, parameter code, value of parameter, unit for parameter, and instrumentation or method of determination (see X1.6.25 for explanation).

X2.5.25 *Monitoring Site at Waste-Facility Record*—Including date interval in service, state regulatory agency, state registration identification, EPA registration identification, responsible company, company's site identification, site location in relationship to waste facility, status of site, and sampled interval.

X2.5.25.1 *Explanation*—The record documents data elements that distinctively apply to waste-disposal facilities. This record has some information that is duplicated in other records, whereas, other data are included that are unique to a waste-disposal facility. However, this duplication is important so that the key data elements for the record can be directly associated with critical information.

X2.5.26 *Decommissioning Record*—Including date decommissioned, method used for decommissioning, reason for decommissioning, plugging material, name and address of decommissioner, step-by-step procedures, availability of decommissioning report, and regulations followed, federal, state, local.

X2.5.26.1 *Explanation*—The decommissioning record can be used to document the destruction of wells and test holes at the waste-disposal facility or the contaminated area. This information is important for tracing the history of the locality in order to certify that proper procedures were fulfilled in managing the facility or for remedial action undertaken in the project area.

X3. EXAMPLE 3—CONTAMINATION ASSESSMENT AND REMEDIATION

X3.1 The contaminated area is highly variable in physical size and commonly has not been intensively evaluated for ground-water resources. Consequently, the area probably lacks information concerning the geologic sediments and hydrologic characteristics, that are needed to estimate the extent of the contamination.

X3.2 The approximate range of contamination can be estimated from nearby hydrologic, geologic, and soil studies (perhaps previous projects in the area). All possible local, state, and federal agencies, as well as private companies with interests in the area, should be contacted in an effort to compile relevant information concerning the contaminated area. As soon as these data are gathered, the contaminated area can be evaluated as to what additional physical information should be gathered to assist in determining the extent of pollution and the proper remedial action to be undertaken.

X3.3 The following arbitrary selection of ground-water data elements assumes that the polluted area requires the removal of the contaminated liquid or material for the enforcement of the appropriate remedial regulations. The accuracy or confidence level should be documented for each measured data element by the agency or company that

gathered and recorded the information.

NOTE X3.1—The following list of information may include individual records and data elements that are not essential for the evaluation of some contamination assessment and remediation projects. For example, there may be no ground-water-influenced ponds or streams in the vicinity. The selection of the records and data elements within each record are at the discretion of the project manager and requirements of the regulating agency.

X3.4 *Minimum Set of Data Elements*—See 6.1.3.

X3.5 *Additional Data Elements*—See Guide D 5408.

X3.5.1 *Geographic Location Record*—Including location map, map scale, and method altitude determined (see 6.1.3.1) (see X1.6.1 for explanation).

X3.5.2 *Political Regime Record*—Including congressional district (see X1.6.2 for explanation).

X3.5.3 *Remarks Record*—Including remark date and remark (see X1.6.5 for explanation) (from Guide D 5409).

X3.5.4 *Individual Site Characteristics Record*—Including land use (in vicinity of site), drainage basin/watershed, relationship to surface stream/lake, hole depth, well depth, source of depth data, and primary aquifer (see X1.6.6 for explanation).

X3.5.5 *Construction Record*—Including date construction began, date construction ended, name of contractor, source of construction data, method of construction, type of finish, type of seal, and depth to bottom of seal (see X1.6.7 for explanation).

NOTE X3.2—The construction record, casing record, and opening/screen record can be combined into one record for most shallow wells and test holes. The purpose intended for separating these records is that in order to describe multiple sizes of casings and opening/screens in a single well more than one record is required.

X3.5.6 *Casing Record*—Including depth to top of the cased interval, depth to bottom of the cased interval, diameter of the cased interval, casing material, and casing thickness (see X1.6.8 for explanation).

X3.5.7 *Opening/Screen Record*—Including depth to top of the open interval, depth to bottom of the open interval, diameter of the open interval, type of material in the open interval, type of openings in the open interval, length of openings, and width of openings (see X1.6.9 for explanation).

X3.5.8 *Lift Record*—Including type of lift, date permanent lift was installed, depth of intake, manufacturer of lift device, serial number, and pump rating (see X1.6.10 for explanation).

X3.5.9 *Geophysical Log Record*—Including type of log, depth to top of logged interval, depth to bottom of logged interval, and source of log data.

X3.5.9.1 *Explanation*—Geophysical log data of groundwater sites constructed before the occurrence of the contamination event are of great value for planning additional data collection and the remediation of the polluted area. All new test holes and wells need detailed log data in order to determine the extent and remedial techniques to use in the removal of the pollution. See X1.6.12 for additional explanation.

X3.5.10 *Geohydrologic Unit Record*—Including aquifer unit(s), contributing unit, depth to top of interval, depth to bottom of interval, lithology, and description of material.

X3.5.10.1 *Explanation*—The geohydrologic unit record is critical for identifying the lithologic material that will allow the contamination to move away from the initial spill or disposal location. Normally, the lithologic units can be traced between sites and the more permeable units are the paths through which the ground-water and contaminants move. See X1.6.13 for additional explanation.

X3.5.11 *Sample/Unconsolidated Material Record*—Including sample weight, sample interval, particle size, percent of total sample, particle shape, and mineralogy (see X2.5.15 for explanation).

X3.5.12 *Sample/Consolidated Material Record*—Including drill cuttings or core, sample size (weight), sample interval, mineralogy, core length, core diameter, core recovery-percent, bedding, structure, and porosity (see X2.5.16 for explanation).

X3.5.13 *Hydraulics Record*—Including hydraulic/aquifer unit, hydraulic/aquifer unit type, depth to top of unit, depth to bottom of unit, static water level, date of measurement, and unit contribution (see X1.6.14 for explanation).

X3.5.14 *Aquifer Parameters Record*—Including transmissivity, horizontal hydraulic conductivity, vertical hydraulic conductivity, coefficient of storage, leakance, diffusivity, specific storage, specific yield, barometric efficiency, porosity, specific capacity, method used to determine aquifer characteristics, and availability of file of detailed result (see X1.6.15 for explanation).

X3.5.15 *Ponds Record*—Including length of pond, width of pond, depth of pond, and volume of pond (see X1.6.18 for explanation) (from Guide D 5410).

X3.5.16 *Measuring-Point Record*—Including date interval of measuring-point utilization, height in reference to datum, and description.

X3.5.16.1 *Explanation*—Carefully selected, clearly identified, and datum (altitude or local reference level) correlated measuring points of individual wells and test holes in the area are important for the precision needed in determining the hydraulic gradient and, thus, the direction of pollutant movement. As many of the contaminated zones are small in physical area, the gradient is likely to be slight, therefore accurate water levels converted to the datum are critical in interpreting the direction of water movement. See X1.6.21 for additional explanation.

X3.5.17 *Water-Level Record*—Including measurement date, water level, water-level accuracy, status, method of measurement and instrumentation, instrumentation, source of data, and statistics method.

X3.5.17.1 *Explanation*—The direction of water and pollutant movement through the contaminated aquifer is based on the evaluation of water-level data from individual wells and test holes. These water levels are converted to a common plane (such as altitude) for direct correlation. These data are critical in the evaluation of the extent and determining the method for removal of the contaminate. See X1.6.22 for additional explanation.

X3.5.18 *Discharge Record*—Including measurement date, discharge, type of discharge, source of data, method of discharge measurement, instrumentation, production or pumping level, static level, method of water-level measurement, pumping period, specific capacity, drawdown, and source of water-level data.

X3.5.18.1 *Explanation*—The discharge record is critical for determining the capacity of the aquifer to yield water in order to calculate the time required and amount of water to withdrawal for the complete removal of the contaminate. See X1.6.23 for additional explanation.

X3.5.19 *Water-Quality Record*—Including sample date, agency that analyzes samples, type of analyses, sample depth/interval, water-quality file containing analysis, collecting agency or company, sampling purpose, site condition, sample appearance, sample odor, presence of immiscible stratum, thickness of immiscible stratum, sample preservation method, sample filtration material, pumped period, casing volume, amount water purged, sampling method or sampler type, sampler material, and aquifer sampled.

X3.5.19.1 *Explanation*—This is the most important category of information for determining the background geochemistry of the water in the aquifer and the evaluation of contamination of the ground water. Care must be exercised to gather comprehensive information to satisfy regulatory guidelines. See X1.6.24 for additional explanation.

X3.5.20 *Field Water-Quality Record*—Including field sample date, parameter code, value of parameter, unit for parameter, and instrumentation or method of determination.

X3.5.20.1 *Explanation*—A large amount of the water-quality data gathered for the evaluation of a contaminated area will probably be analyzed at the field location. Many of these data are parameters that change rapidly after removal from the natural conditions, for example, volatile compounds. Complete documentation of these analyses is a

necessity for interpretation of the extent and developing the method of removal of the contaminates. See X1.6.25 for additional explanation.

X3.5.21 *Decommissioning Record*—Including date decommissioned, method used for decommissioning, reason for decommissioning, plugging material, name and address of decommissioner, step-by-step procedures, availability of decommissioning report, and regulations followed, federal, state, local (see X2.5.26 for explanation).

X4. EXAMPLE 4—UNDERGROUND STORAGE TANK GROUND-WATER ASSESSMENT

X4.1 Ground-water contamination by leakage from and spills adjacent or along associated delivery pipes to underground storage tanks (UST) (usually for petroleum fuels or organic solvents) can cause a considerable amount of damage to nearby water supplies, both ground and surface. Although the contaminated area is usually small in physical size, contaminants from an undetected long-term leak can migrate an extended distance from the source. In arid areas, where development (such as a rural service station) has occurred over limited water source, leaks from UST can easily pollute all of the local potable ground-water supply. On occasion, leaks from UST that are not monitored are first detected by seepage into a residential basement, ground-water sustained pond (such as a gravel pit), or surface stream, thereby creating a hazardous situation.

NOTE X4.1—For an UST assessment, the type of wells and holes considered include: (*1*) production wells in the vicinity of the UST, especially if they produce water from the ground-water aquifer that is most likely to be damaged by a leak or spill; (*2*) test wells constructed to investigate and monitor the ground-water aquifer (saturated zone) contiguous to the UST; and (*3*) test holes to investigate and monitor the vadose zone (unsaturated sediment) adjacent to the UST. In addition, springs and ground-water-influenced surface water bodies (ponds and streams) in the area of the UST must be appraised.

X4.2 UST normally are present at commercial facilities that distribute petroleum products. A number of these facilities have ceased operations or gone out of business leaving behind UST, some with petroleum still remaining in the tanks. Other UST are located at private operations for vehicle and heating fuel (such as large farms) and commercial or public facilities that maintain a fleet of motorized vehicles. A number of UST were installed when fuel was in short supply, quite often as backup rather than a primary source for vehicles and equipment. These emergency supply tanks are easily overlooked if they are not a routine part of a business or operation. The number of UST in existence, both in use and abandoned, is in the hundreds of thousands.

X4.3 New commercial facilities that have UST require monitoring of the surrounding sediments (that may include both the vadose and saturated zones) to detect vapors and liquid products resulting from accidental spills and leaks. Regulations require that UST at existing facilities be monitored; many of the older tanks are replaced to meet the current regulations. Detailed documentation of the monitoring network is mandatory for satisfying regulatory conditions.

X4.4 For new facilities and UST that have not leaked, the area of concern is very small, perhaps an acre in size. The assumption used to select records and data elements for the

following list is that the UST assessment will examine the area in the vicinity of the facility, with no significant leaks or spills expected. If a major leak is detected at an existing facility, then a more extensive investigation should be undertaken and an expanded data element list used, similar to that of the contamination assessment and remediation section (see Note X3.1). The accuracy or confidence level should be documented for each measured data element by the agency or company that gathered and recorded the information.

X4.5 *Minimum Set of Data Elements*—See 6.1.3.

X4.6 *Additional Data Elements*—From Guide D 5408.

X4.6.1 *Geographic Location Record*—Including land-net location, location map, map scale, and method altitude determined (see 6.1.3.1).

X4.6.1.1 *Explanation*—The location map is critical as a spatial locator for ground-water site and UST data and is an absolute necessity for interpretation of the ground-water hydrology. For UST, the map could be a local plat map. See X1.6.1 for additional explanation.

X4.6.2 *Owner Record*—Including date of ownership and owner's name (see X1.6.2 for explanation).

X4.6.3 *Site Visits Record*—Including date of visit and person who made the visit (see X1.6.3 for explanation).

X4.6.4 *Remarks Record*—Including remark date and remark (see X1.6.5 for explanation) (from Guide D 5409).

X4.6.5 *Individual Site Characteristics Record*—Including land use (in vicinity of site), drainage basin/watershed, relationship to surface stream/lake, hole depth, well depth, source of depth data, and primary aquifer.

X4.6.5.1 *Explanation*—The use of land in the vicinity of the UST is critical information needed for determining how problems can be handled assuming a leak is detected. Whereas, the hole (vapor monitoring point) and well depth data indicates the lowest extent of testing and monitoring at the UST. See X1.6.6 for additional explanation.

X4.6.6 *Construction Record*—Including date construction began, date construction ended, name of contractor, source of construction data, method of construction, type of finish, type of seal, and depth to bottom of seal (see X1.6.7 for explanation).

X4.6.7 *Casing Record*—Including depth to top of the cased interval, depth to bottom of the cased interval, diameter of the cased interval, casing material, and casing thickness (see X1.6.8 for explanation).

X4.6.8 *Opening/Screen Record*—Including depth to top of the open interval, depth to bottom of the open interval, diameter of the open interval, type of material in the open interval, type of openings in the open interval, length of openings, and width of openings (see X1.6.9 for explanation).

X4.6.9 *Lift Record*—Including type of lift, date permanent lift was installed, depth of intake, manufacturer of lift device, serial number, and pump rating.

X4.6.9.1 *Explanation*—This record can be used to describe vapor extraction equipment used at monitoring points. See X1.6.10 for additional explanation.

X4.6.10 *Repairs Record*—Including date of repairs, nature of repairs, and name of contractor who made repairs.

X4.6.10.1 *Explanation*—Periodic maintenance and repairs are required of UST monitoring systems. A complete documentation of these activities is mandatory for quality control and regulatory management.

X4.6.11 *Ponds Record*—Including length of pond, width of pond, depth of pond, and volume of pond (see X1.6.18 for explanation).

X4.6.12 *Geophysical Log Record*—Including type of log, depth to top of logged interval, depth to bottom of logged interval, and source of log data.

X4.6.12.1 *Explanation*—Geophysical log data of groundwater sites and monitoring points are important for understanding the type of material, both unconsolidated and consolidated, that are in the area encompassing the UST. These data are needed in order to assess the UST and to properly design a monitoring system. See X1.6.12 for additional explanation.

X4.6.13 *Sample/Unconsolidated Material Record*—Including sample weight, sample interval, particle size, percent of total sample, particle shape, and mineralogy (see X2.5.15 for explanation).

X4.6.14 *Sample/Consolidated Material Record*—Including drill cuttings or core, sample size (weight), sample interval, mineralogy, core length, core diameter, core recovery-percent, bedding, structure, and porosity (see X2.5.16 for explanation) (from Guide D 5410).

X4.6.15 *Measuring-Point Record*—Including date interval of measuring-point utilization, height in reference to datum, and description.

X4.6.15.1 *Explanation*—The measurements of concern for UST's include liquid levels and vapor pressures. See X1.6.21 for additional explanation.

X4.6.16 *Water-Level Record*—Including measurement date, water level, water-level accuracy, status, method of measurement and instrumentation, instrumentation, source of data, and statistics method (see X1.6.22 for explanation).

X4.6.17 *Field Water-Quality Record*—Including field sample date, parameter code, value of parameter, unit for parameter, and instrumentation or method of determination.

X4.6.17.1 *Explanation*—A major portion of the water-quality data gathered for the assessment of leaks from and spills in the vicinity of the UST probably will be analyzed at the field location. Complete documentation of these analyses is necessary for the assessment of conditions in the area of the UST. See X1.6.25 for additional explanation.

Designation: D 5254 – 92

Standard Practice for
Minimum Set of Data Elements to Identify a Ground-Water Site[1]

This standard is issued under the fixed designation D 5254; the number immediately following the designation indicates the year of original adoption or, in the case of revision, the year of last revision. A number in parentheses indicates the year of last reapproval. A superscript epsilon (ε) indicates an editorial change since the last revision or reapproval.

1. Scope

1.1 This practice specifies what information should be obtained for any individual ground-water site, also known as monitoring location or sampling station. As used in this practice, a site is meant to be a single point, not a geographic area or property. A ground-water site is defined as any source, location, or sampling station capable of producing water or hydrologic data from a natural stratum from below the surface of the earth. A source or facility can include a well, spring or seep, and drain or tunnel (nearly horizontal in orientation). Other sources, such as excavations, driven devices, bore holes, ponds, lakes, and sinkholes, that can be shown to be hydraulically connected to the ground water, are appropriate for the use intended (see 6.4.2.3).

NOTE 1—There are many additional data elements that may be necessary to identify a site, but are not included in the minimum set of data elements. An agency or company may require additional data elements as a part of their minimum set.

1.2 This practice includes those data elements that will distinguish a site as to its geographical location on the surface of the earth, political regimes, source identifiers, and individual site characteristics. These elements apply to all ground-water sites. Each category of site, such as a well or spring, may individually require additional data elements to be complete. Many of the suggested components and representative codes for coded data elements are those established by the Water Resources Division of the U.S. Geological Survey and used in the National Water Information Systems computerized data base (1).[2]

NOTE 2—The data elements presented in this practice do not uniquely imply a computer data base, but rather the minimum set of ground-water data elements that should be collected for entry into any type of permanent file.

1.3 The values stated in SI units are to be regarded as the standard. The inch-pound units given in parentheses are for information only.

1.4 *This standard does not purport to address all of the safety problems, if any, associated with its use. It is the responsibility of the user of this standard to establish appropriate safety and health practices and determine the applicability of regulatory limitations prior to use.*

2. Referenced Documents

2.1 *ASTM Standards:*
D 653 Terminology Relating to Soil, Rock, and Contained Fluids[3]

3. Terminology

3.1 *Definitions:*
3.1.1 For definitions of terms applicable to this practice refer to Terminology D 653.

4. Summary of Practice

4.1 This practice includes the following data elements to identify a ground-water site:
4.1.1 *Geographic Location*—Including latitude, longitude, latitude-longitude coordinate accuracy, altitude, and altitude accuracy.
4.1.2 *Political Regimes*—Including state or country identification, and county or county equivalent.
4.1.3 *Source Identifiers*—Including owner's name, source agency or company and address, unique identification, and date of first record for the ground-water site.
4.1.4 *Individual Site Characteristics*—Including hydrologic unit, setting, type of ground-water site, use of site, use of water from site, reason for data collection.

5. Significance and Use

5.1 Normally, the basic ground-water data are gathered by trained personnel during the field investigation phase of a study. Each agency or company has its own methods of obtaining, recording, and storing the information. Usually, these data are recorded onto forms that serve both in organizing the information in the field and the office, and many times as entry forms for a computer data base. For ground-water data to be of maximum value to the current project and any future studies, it is essential that a minimum set of key data elements be recorded for each site. The data elements presented in this practice do not uniquely imply a computer data base, but rather the minimum set of ground-water data elements that should be collected for entry into any type of permanent file.

5.2 When obtaining basic data concerning a ground-water site, it is necessary to identify thoroughly that site so that it may be readily field located again with minimal uncertainty and that it may be accurately plotted and interpreted for data parameters in relationship to other sites. For example,

---

[1] This practice is under the jurisdiction of ASTM Committee D-18 on Soil and Rock and is the direct responsibility of Subcommittee D18.21 on Ground Water and Vadose Zone Investigations.
Current edition approved July 15, 1992. Published November 1992.
[2] The boldface numbers given in parentheses refer to a list of references at the end of the text.

[3] *Annual Book of ASTM Standards*, Vol 04.08.

information can be presented on scientific maps and in summary tables.

6. Documentation

6.1 *Geographic Location:*

6.1.1 *Introduction*—The universally accepted coordinates defining the absolute two-dimensional location of a site on the Earth's surface are latitude and longitude. The coordinates are determined by careful measurement from an accurate map or by survey. The third-dimension of the location is established by determining the altitude at the site, usually from topographic maps or by surveying techniques (2).[4]

NOTE 3—If sites are located by plane coordinates, plant location grids, or referenced to recoverable benchmarks, they may be recorded if the position is converted to absolute location coordinates by an acceptable method.

6.1.2 *Documentation Procedures:*

6.1.2.1 *Latitude*—Latitude is a coordinate representation that indicates locations on the surface of the earth using the earth's equator as the respective latitudinal origin. Record the best available value for the latitude of the site in degrees, minutes, seconds and fractions of a second (DDMMSSss). If latitude of the site is south of the Equator, precede the numbers with a minus sign (−). The use of *N* or *S* is also appropriate (1, 2, 3, 4, 5, 6, 7, 8).

6.1.2.2 *Longitude*—Longitude is a coordinate representation that indicates locations on the surface of the Earth using the prime meridian (Greenwich, England) as the longitudinal origin. Record the best available value for the longitude of the site, in degrees, minutes, seconds, and fractions of a second (DDDMMSSss). If longitude of the site is measured east of the Greenwich Meridian, precede the numbers with a minus sign (−). The use of *E* or *W* is also appropriate (1, 2, 3, 4, 5, 6, 7, 8).

6.1.2.3 *Latitude-Longitude Coordinate Accuracy*—Record the accuracy of the latitude and longitude values. Suggested coordinate accuracy components and representative codes are as follows (1, 6, 7, 8):

H — The measurement is accurate to ±0.01 s.
U — The measurement is accurate to ±0.1 s.
S — The measurement is accurate to ±1 s.
F — The measurement is accurate to ±5 s.
T — The measurement is accurate to ±10 s.
M — The measurement is accurate to ±1 min.

NOTE 4—Components and corresponding codes listed under data elements, such as latitude-longitude coordinate accuracy and setting, are only suggestions. An agency or company may require additional components to fully describe their ground-water sites. Also, having the data element components written out, for example, "accurate to within 1 s" for the latitude-longitude accuracy, may be preferred to the use of codes. The important factor is that each data element in the "minimum set of data elements" be included with every ground-water site.

6.1.2.4 *Altitude*—Record the altitude of land surface or measuring point. Altitude of the land surface is the vertical distance in feet (or metres) either above or below a reference datum surface. The reference datum surface must be noted.

NOTE 5—In the United States, this reference surface should be the North American Vertical Datum (NAVD) of 1988 or National Geodetic Vertical Datum (NGVD) of 1929. If another vertical reference datum is used to determine the altitude, describe the system. Altitudes below the reference datum must be preceded by a minus sign (−) (1, 2, 4, 7, 8).

NOTE 6—The measuring point is usually a clearly defined mark or permanently fixed object at a ground-water site that is used for conducting repeated evaluations, such as water levels in a monitoring well.

6.1.2.5 *Altitude Accuracy*—Record the accuracy of the altitude. As an example, record 1.0 for an accuracy of ± 1 m or 0.1 for ± 0.1"th" m to denote the judged error of the measurement (1, 3).

6.2 *Political Regimes:*

6.2.1 *Introduction*—The placement of the ground-water site into a political jurisdiction assists in the proper identification of the site.

6.2.2 *Documentation Procedures:*

6.2.2.1 *State or Country Identification*—Record the state or country in which the site is physically located. The common systems for identifying states and countries are the Federal Information Processing Standard code (FIPS), a two-digit numeric code or the American National Standard abbreviation two-letter code. The country codes are a two-character and a set of three-character alphabetic codes (1, 3, 9, 10, 11).

NOTE 7—The publications (9, 10, 12) containing the codes for countries, states, and counties are available from the address given in Footnote 4.

6.2.2.2 *County and County Equivalent*—Record the county or county equivalent in which the site is physically located. The common code system for identifying counties is the FIPS code, a three-digit numeric code. The documentation of political subdivisions will depend on the system used in each individual country (1, 3, 7, 11).

NOTE 8—In many cases it is necessary to record a subdivision of the local government to further identify the area where the ground-water site is located. Some of the local subdivisions are a city, town, village, municipality, township, or borough. Identify the local subdivision, for example "City of Rockville", to clearly denote the unit.

6.3 *Source Identifiers:*

6.3.1 *Introduction*—The ground-water site must be identified as to the owner, the agency or company that recorded data, and its distinctive classification.

6.3.2 *Documentation Procedures:*

6.3.2.1 *Owner's Name*—Record the name of the property owner of the ground-water site. The recommended format for an individual's name is: last name, first name, middle initial. If a company's name is lengthy, use meaningful abbreviations (1, 8).

6.3.2.2 *Source Agency or Company and Address*—Record the name and address of the agency or company that collected the data for the ground-water site. This data element is necessary to determine the original source of the data for the site. A coded list of agency and company names is available through National Water Data Exchange (NAWDEX);[5] the list has over 1200 organizations that actively collect and store water data throughout the United States (1, 3, 4, 5, 6, 12).

6.3.2.3 *Unique Identification*—Record the unique naming that the agency or company uses to identify the ground-water

---

[4] Available from National Technical Information Service, U.S. Department of Commerce, 5285 Port Royal Road, Springfield, VA 22161.

[5] Available from National Water Data Exchange, U.S. Geological Survey, 421 National Center, Reston, VA 22092.

site. This identification is called by several terms such as "local site number", "site identification", "well number", etc. The description is commonly a combination of letters and numbers that could represent a land-net location or a sequential assignment for a site in a county, city, or company. This identification is very important to precisely differentiate a site in the records of an agency or company (1, 5, 6, 7, 8).

6.3.2.4 *Date of First Record for the Ground-Water Site*— Record the date that the first valid transaction occurred for any element of the specified site. This could be the date of permit application, start of construction, or first used as a monitoring site. This element is important to facilitate in the proper identification of the record (1, 3, 12).

6.4 *Individual Characteristics of the Site:*

6.4.1 *Introduction*—Each ground-water site has very specific features that, in combination, uniquely identify that site, that is, water from a ground-water sustained pond used for aquaculture. These characteristics should be recorded as a means of further defining the site.

6.4.2 *Documentation Procedures:*

6.4.2.1 *Hydrologic Unit*—Record the hydrologic unit code for the Office of Water Data Coordination (OWDC) cataloging unit in which the site is located. This eight-digit code consists of four 2-digit parts (1, 4, 5, 6, 13, 14): hydrographic region code, subregion code designated by the Water Resources Council, accounting unit within the National Water Data Network, and cataloging unit of the USGS's "Catalog of Information on Water Data".

NOTE 9—An explanation of a hydrologic unit code, for example Code 07080107, is the following; Region Code "07" is the Upper Mississippi River Basin above the confluence with the Ohio River, Subregion Code "08" is the Mississippi River Basin below Lock and Dam 13 to the confluence with the Des Moines River Basin, excluding the Rock River Basins, Accounting Unit Code "01" is the Mississippi River Basin below Lock and Dam 13 to the confluence with the Des Moines River Basin, excluding the Iowa and Rock River Basins, and Catalog Unit Code "07" is the Skunk River Basin of Iowa.

NOTE 10—State hydrologic unit maps delineating the hydrographic boundaries of these units are available[6] (see Ref (13)).[4]

6.4.2.2 *Setting*—Record the information that best describes the setting in which the site is located. Setting refers to the topographic or geomorphic features in the vicinity of the site. Suggested setting components and representative codes are as follows (1, 8):

A — Alluvial fan
B — Playa
C — Stream channel
D — Local depression
E — Dunes
F — Flat surface
G — Flood plain
H — Hilltop
I — Inland wetlands
J — River delta
K — Sinkhole
L — Lake
M — Mangrove swamp or coastal wetlands
O — Offshore (estuary)
P — Pediment
S — Hillside (slope)
T — Alluvial or marine terrace

---

[6] Available from USGS Books and Reports Sales Federal Center, P.O. Box 25425, Denver, CO, 80225.

U — Undulating
V — Valley flat (valleys of all sizes)
W — Upland draw
X — Unknown
Y — Wetlands
Z — Other (describe)

6.4.2.3 *Type of Ground-Water Site*—This data element helps to identify the physical type of ground-water site. Record the type of site to which these data apply. Suggested site type components and representative codes are as follows (1, 8):

C — Collector (radial-collector) well
D — Drain dug to intercept the water table or potentiometric surface to either lower the ground-water level or serve as a water supply
E — Excavation
H — Sinkhole
I — Interconnected wells, also called connector or drainage wells; that is, a well interconnected via an underground lateral
M — Multiple wells—Use only for well field consisting of a group of wells that are pumped through a single header and for which little or no data about the individual wells are available
O — Outcrop
P — Pond that intercepts the water table or potentiometric surface
S — Spring
T — Tunnel, shaft, or mine from which ground water is obtained
W — Well, for single wells other than wells of the collector (radial collector) type
X — Test hole, not completed as a well
Z — Other (describe)

6.4.2.4 *Use of Site*—Record the use of the site or the purpose for which the site was constructed (the former always holds precedence over the latter). If site is used for more than one purpose, also record the subordinate uses. Suggested site use components and representative codes are as follows (1, 4, 7, 8):

A — Electrical anode
C — Standby emergency supply
D — Drain or dry well
E — Geothermal—for geothermal extraction or injection
G — Seismic exploration
H — Heat reservoir—fluid circulated in closed system
M — Mine—primary use for extraction of minerals
O — Observation/monitoring
P — Oil or gas well
R — Recharge
S — Repressurize—to increase pressure in aquifer
T — Test—for hydrologic testing
U — Unused
W — Withdrawal of water
X — Waste disposal
Y — Other (describe)
Z — Destroyed

6.4.2.5 *Use of Water from Site*—Record the use of the water from the site. If water from the site is used for more than one purpose, also record the subordinate uses. Suggested water use components and representative codes are as follows (1):

A — Air conditioning
B — Bottling
C — Commercial
D — Dewater
E — Power
F — Fire
G — Hydrogeologic interpretation
H — Domestic
I — Irrigation
J — Industrial (cooling)
K — Mining
L — Chemical screening for contaminants
M — Medicinal
N — Industrial (manufacturing)

Ground–Water Site Minimum Set of Data Elements	Date prepared:
GEOGRAPHIC LOCATION	
Latitude:	Accuracy:
Longitude:	Accuracy:
ALTITUDE	
Land Surface: (meters/feet)	Accuracy:
Other (Specify): (meters/feet)	Accuracy:
Altitude Reference Datum:	
POLITICAL REGIMES	
State (or Country) Identification:	
County (or County Equivalent) Identification:	
SOURCE IDENTIFIERS	
Owner's Name:	
Source Agency (or Company):	
Address of Source Agency:	
Unique Identification of Site:	
Date of First Record for Site:	
INDIVIDUAL CHARACTERISTICS OF THE SITE	
Hydrologic Unit:	
Setting:	
Type of Ground–Water Site:	
Use of Site:	
Use of Water from Site:	
Reason for Data Collection:	

FIG. 1   Example of Minimum Set of Data Elements Form

Ground–Water Site Minimum Set of Data Elements	Date prepared: SEPT 4, 1991
GEOGRAPHIC LOCATION (5.1)	
Latitude: 39° 15' 26"	Accuracy: ± 1 SEC
Longitude: 74° 37' 09"	Accuracy: ± 1 SEC
ALTITUDE (5.1.2.4)	
Land Surface: 10.00 FT (meters/feet)	Accuracy: 0.01 FT
Other (Specify): 9.34 FT RECORDER SHELF (meters/feet)	Accuracy: 0.01 FT
Altitude Reference Datum: NGVD (1929)	
POLITICAL REGIMES (5.2)	
State (or Country) Identification: NEW JERSEY	
County (or County Equivalent) Identification: ATLANTIC COUNTY	
SOURCE IDENTIFIERS (5.3)	
Owner's Name: US GEOLOGICAL SURVEY	
Source Agency (or Company): US GEOLOGICAL SURVEY	
Address of Source Agency: 810 BEAR TAVERN ROAD, SUITE 206 WEST TRENTON, NJ 08628	
Unique Identification of Site: JOBS POINT CP5, NJ - WRD WELL #01-0578	
Date of First Record for Site: OCTOBER, 1959	
INDIVIDUAL CHARACTERISTICS OF THE SITE (5.4)	
Hydrologic Unit: 02040302	
Setting: UNKNOWN; NEAR EXIT 29 - GARDEN STATE PKWY	
Type of Ground–Water Site: WELL	
Use of Site: OBSERVATION/MONITORING	
Use of Water from Site: HYDROGEOLOGIC INTERPRETATION	
Reason for Data Collection: HYDROLOGIC BENCHMARK/RESEARCH	

FIG. 2   Example of Filled-Out Minimum Set of Data Elements Form

P — Public supply
Q — Aquaculture
R — Recreation
S — Stock
T — Institutional
U — Unused
Y — Desalination
Z — Other (describe)

6.4.2.6 *Reason for Data Collection*—Record the reason for which data were collected from the site. If the data were collected for more than one purpose, record the subordinate reasons. Suggested data collection components and representative codes are as follows:

A — Construction/dewatering
B — Research
C — CERCLA
R — RCRA
D — Drinking water regulations
E — Exploration (water)
L — Local ordinance
S — State regulations, other than CERCLA or RCRA

F — Federal regulations, other than CERCLA or RCRA
G — Geothermal
H — Hydrologic benchmark
I — Environmental issues
J — Judicial/litigation
M — Mining regulations
N — Natural resources exploration
U — Unknown
Z — Other (describe)

7. Sample Form

7.1 An example of a generalized form for recording a minimum set of data elements for a ground-water site is shown in Fig. 1. An example of a filled-out form is shown in Fig. 2.

8. Keywords

8.1 ground water; ground-water sampling site hydrologic unit; key data elements; site identification; site location; setting monitoring location; site coordinates

REFERENCES

(1) Mathey, S. B., ed., *National Water Information System User's Manual*, Vol 2, Chapter 4, "Ground-Water Site Inventory System," U.S. Geological Survey, Open-File Report 89-587, 1990.

(2) U.S. Department of Commerce, Representation of Geographic Point Locations for Information Interchange; *Federal Information Standards (FIPS) Publication 70-1*, National Institute for Standards and Technology, Washington, DC, June 23, 1986.

(3) Perry, R. A., and Williams, O. O., *Data Index Maintained by the National Water Data Exchange*, U.S. Geological Survey, Open-File Report 82-327.

(4) Texas Natural Resources Information System, *Ground-Water Data INTERFACE, Users Reference Manual*, Texas Natural Resources Information System, November 20, 1986.

(5) U.S. Environmental Protection Agency, *STORET Users Handbook*, Vols 1 and 2; U.S. EPA, Washington, DC, February 1982.

(6) U.S. Environmental Protection Agency, *Ground-Water Data Management With STORET*, Office of Ground-Water Protection, U.S. EPA, Washington, DC, March 1986.

(7) U.S. Environmental Protection Agency, *Definitions for the Minimum Set of Data Elements for Ground Water Quality*, U.S. Environmental Protection Agency (Draft), July 22, 1991.

(8) U.S. Geological Survey, *National Handbook of Recommended Methods for Water-Data Acquisition*, Chapter 2, "Ground Water," Office of Data Coordination, Reston, VA, 1980.

(9) U.S. Department of Commerce, "American National Standard Codes for the Representation of Names of Countries, Dependencies, and Areas of Special Sovereignty for Information Interchange," *Federal Information Standards (FIPS) Publication 104-1*, National Institute for Standards and Technology, Washington, DC, May 12, 1986.

(10) U.S. Department of Commerce, "Codes for the Identification of the States, the District of Columbia and Outlying Areas of the United States, and Associated Areas," *Federal Information Standards (FIPS) Publication 5-2*, National Institute for Standards and Technology, Washington, DC, May 28, 1987.

(11) U.S. Department of Commerce, "Counties and Equivalent Entities the United States, Its Possessions, and Associated Areas," *Federal Information Standards (FIPS) Publication 6-4*, National Institute for Standards and Technology, Washington, DC, August 31, 1990.

(12) Edwards, M. D., and Josefson, B. M., "Identification Codes for Organizations listed in Computerized Data Systems of the U.S. Geological Survey," U.S. Geological Survey, Open-File Report 82-921, 1982.

(13) U.S. Geological Survey, "Codes for the Identification of Hydrologic Units in the United States and the Caribbean Outlying Areas," U.S. Geological Survey, Circular 878-A, Reston, VA, (also *FIPS PUB 103*), 1982.

(14) Seaber, P. R., Kapinos, F. P., and Knapp, G. L., State Hydrologic Unit Maps, U.S. Geological Survey, Open-File Report 84-708, Reston, VA, 1984.

Designation: D 5408 – 93

Standard Guide for
Set of Data Elements to Describe a Ground-Water Site;
Part One—Additional Identification Descriptors[1]

This standard is issued under the fixed designation D 5408; the number immediately following the designation indicates the year of original adoption or, in the case of revision, the year of last revision. A number in parentheses indicates the year of last reapproval. A superscript epsilon (ε) indicates an editorial change since the last revision or reapproval.

1. Scope

1.1 This guide is Part One of three guides to be used in conjunction with Practice D 5254 that delineates the data desirable to describe a ground-water data collection or sampling site. This guide describes additional information beyond the minimum set of data elements that may be needed to identify a ground-water site. Part Two identifies physical descriptors, such as construction, for a site, while Part Three identifies usage descriptors, such as monitoring, for an individual ground-water site.

NOTE 1—A ground-water site is defined as any source, location, or sampling station capable of producing water or hydrologic data from a natural stratum from below the surface of the earth. A source or facility can include a well, spring or seep, and drain or tunnel (nearly horizontal in orientation). Other sources, such as excavations, driven devices, bore holes, ponds, lakes, and sinkholes, that can be shown to be hydraulically connected to the ground water, are appropriate for the use intended.

NOTE 2—Part Two (Guide D 5409) includes individual site characteristic descriptors (7 data elements), construction descriptors (56 data elements), lift descriptors (16 data elements), geologic descriptors (26 data elements), hydraulic descriptors (20 data elements), and spring descriptors (11 data elements). Part Three (Guide D 5410) includes monitoring descriptors (77 data elements), irrigation descriptors (4 data elements), waste site descriptors (9 data elements), and decommissioning descriptors (8 data elements). For a list of descriptors in this guide, see Section 4.

1.2 These data elements are described in terms used by ground-water hydrologists. Standard references, such as the Glossary of Geology and various hydrogeologic professional publications, are used to determine these definitions. Many of the suggested elements and their representative codes are those established by the Water Resources Division of the U.S. Geological Survey and used in the National Water Information Systems computerized data base (1–9).[2]

NOTE 3—The purpose of this guide is to suggest data elements that can be collected for ground-water sites. This does not uniquely imply a computer data base, but rather data elements for entry into any type of permanent file.

NOTE 4—Component and code lists given with some of the data elements, for example "Format of Other Data," are only suggestions. These lists can be modified, expanded, or reduced for the purpose intended by the company or agency maintaining the ground-water data file.

NOTE 5—Use of trade names in this guide is for identification purposes only and does not constitute endorsement by ASTM.

1.3 This guide includes the data elements desirable to identify a ground-water site beyond those given in the "Minimum Set of Data Elements." Some examples of the data elements are map identification, permitting facts, and supporting information. No single site will need every data element, for example, many ground-water sites do not need the data elements described in the legal record group. Each record (group of related data elements) for a site has mandatory data elements, such as the date for the ownership record. However, these elements are considered necessary only when that specific record is gathered for the site.

1.4 The values stated in inch-pound units are to be regarded as the standard. The SI units given in parentheses are for information only.

1.5 *This standard does not purport to address all of the safety problems, if any, associated with its use. It is the responsibility of the user of this standard to establish appropriate safety and health practices and determine the applicability of regulatory limitations prior to use.*

2. Referenced Documents

2.1 *ASTM Standards:*
D 653 Terminology Relating to Soil, Rock, and Contained Fluids[3]
D 5254 Practice for the Minimum Set of Data Elements to Identify a Ground-Water Site[4]
D 5409 Guide for Set of Data Elements to Describe a Ground-Water Site; Part Two—Physical Descriptors[4]
D 5410 Guide for Set of Data Elements to Describe a Ground-Water Site; Part Three—Usage Descriptors[4]

3. Terminology

3.1 *Definitions:*
3.1.1 For definitions of terms applicable to this guide, see Terminology D 653.

3.2 *Descriptions of Terms Specific to This Standard:*
3.2.1 *code*—a suggested abbreviation for a component, for example, "F" is the code suggested for the "Files (Raw Data)" component of data element "Format of Other Data."

3.2.2 *component*—a subdivision of a data element, for example, "Files (Raw Data)" is one of four components suggested for data element "Format of Other Data."

3.2.3 *data element*—an individual segment of information about a ground-water site, for example, "Format of Other Data." The data element is in the "Other Data Record" record.

---

[1] This guide is under the jurisdiction of ASTM Committee D-18 on Soil and Rock and is the direct responsibility of Subcommittee D18.21 on Ground Water and Vadose Zone Investigations.
Current edition approved May 15, 1993. Published November 1993.
[2] The boldface numbers in parentheses refer to a list of references at the end of the text.

[3] *Annual Book of ASTM Standards*, Vol 04.08.
[4] *Annual Book of ASTM Standards*, Vol 04.09.

3.2.4 *record*—a set of related data elements that may need to be repeated to fully describe a ground-water site. For example, a ground-water site that has a series of separate data files will need more than one "Other Data Record" record (the record includes data elements, other data type, other data location, and format of other data) to fully document the history of the site. However, if only a single separate data file exists for the well, the record is utilized once.

3.2.5 *record group*—a set of related records. For example, the "Supporting Information Record Group" includes the owner record, site visits record, other identification record, other data record, and remarks record. Some record groups consist of only one record, for example, the "Legal Record Group" includes only the legal record.

4. Summary of Guide

4.1 This guide includes the following additional identification descriptor data elements to describe a ground-water site. The universal element accompanies any data element requiring a confidence classification. Single elements usually need one entry for a site, while repeated elements commonly require several records to fully describe the conditions and history of the site:

Universal Element
  Data Confidence Classification
Single Elements
  Geographic Location:
    Land-Net Location
    Location Map
    Map Scale
    Method Altitude Determined
  Political Regimes
    Congressional District
  Source Identifiers:
    Mean Greenwich Time Offset
    Site Reference in Report
    Site in a Computer Data Base
    Photography/Sketch Available of Site
Repeated Elements
  Legal Record Group:
    Legal Record:
      Permitting Agency
      Priority Date
      Application Number
      Application Date
      Certification Number
      Certification Date
      Permit Number
      Permit Date
      Water Allocation
Supporting Information Record Group
  Owner Record:
    Date of Ownership
    Owner's Name
  Site Visits Record:
    Date of Visit
    Person Who Made Visit
    Purpose of Visit
  Other Identification Record:
    Other Name, Number, or Identification
    Assigner
  Other Data Record:
    Other Data Type
    Other Data Location
    Format of Other Data
  Remarks Record:
    Remark Date
    Remark
    Remark Source

5. Significance and Use

5.1 Data at ground-water sites are gathered for many purposes, each of which generally requires a specific set of data elements. For example, when ground-water quality is a concern, not only are the minimum set of data elements required for the site, but information concerning the sample collection depth interval, method of collection, and date and time of collection are needed to fully qualify the data. Another group of elements are recommended for each use of the data, such as aquifer characteristics or water-level records. Normally the more information that is gathered about a site by field personnel, the easier it is to understand the ground-water conditions and to reach valid conclusions and interpretations regarding the site.

5.2 The data elements listed in this guide and Guides D 5409 and D 5410 should assist in planning what information can be gathered for a ground-water site and how to document these data.

NOTE 6—Some important data elements may change during the existence of a site. For example, the elevation of the measuring point used for the measurement of water levels may be modified because of repair or replacement of equipment. This frequently occurs when the measuring point is an opening in the pump and the pump is modified or replaced. Because changes cannot always be anticipated, it is preferable to reference the height of the measuring point to a nearby, permanent altitude datum. The measuring point is referenced by being the same altitude (zero correction) or above (negative correction) or below (plus correction) the altitude datum. All appropriate measurements should be corrected in reference to the altitude datum before entry into the permanent record. Care must be exercised to keep the relationship of these data elements consistent throughout the duration of the site.

5.3 Some data elements have an extensive list of components. For example, the aquifer identification list described in Guide D 5409, has over 5000 components. Lengthy lists of possible components are not included in this guide, however, information on where to obtain these components is included with the specific data element.

NOTE 7—This guide identifies many sources, lists, etc. of information required to completely document information about any ground-water site.

6. Documentation of Universal Element

6.1 For any element that requires a Confidence Classification, document the data confidence classification for that specified critical data element for the ground-water site. Field-measured or laboratory-determined values have varying degrees of accuracy depending upon the methods used to obtain the information. This subjective or judged confidence should be documented for each measured data element by the agency or company that gathered or recorded the information, or both. Suggested components for the data confidence classification and representative codes are as follows:

A—Value is accurate to within the tolerance of the measurement instrument.

I—Value is judged to be inaccurate due to improper instrumentation or bias instrumentation or laboratory methods.

N—Not verified, value was obtained from another source and due to the nature of the data, cannot be verified.

NOTE 8—At a minimum, it is important, and often sufficient, that data be classified subjectively by experienced professionals. It is not always possible, or necessary to objectively quantify the confidence that a data user might have in a data value, but a professional classification can be useful. For the purposes of the three guides, the word confidence refers to a subjective professional judgment on data accuracy as represented by the three data confidence classification components, and does not imply the more rigorous confidence limits or interval as used by statisticians.

NOTE 9—A critical data element is one that the value can be field measured or laboratory determined with an instrument that has a statistically resolved degree of precision. Many data elements gathered for ground-water sites require no accompanying confidence classification, for example, owner's name, location map, type of lift, etc. Each data element that generally requires an accompanying confidence classification will be so noted in these guides.

7. Documentation of Miscellaneous Singular Data Elements

7.1 *Introduction*—A vast number of data elements can be documented about a ground-water site to thoroughly describe its location, physical features, relationship to other features on the earth's surface, and to designate what information is gathered at the site. These data elements typically are transcribed once for a site, in contrast to data elements that may be repetitive, such as water levels. Many of these data are extremely valuable in the characterization of sites that fall into certain categories, for example wells, for which the location map is an essential element to assist in properly positioning the well.

7.2 *Geographic Location:*

7.2.1 *Land-Net Location*—In addition to the locational data required by the minimum set of data elements, land-net location may be a general land office description of the site's position on the surface of the earth. This description is used in many parts of the United states to subdivide the land into sections, townships, and ranges for the purpose of governmental administration and originally was used (beginning in 1786) as a systematic method for the disposal of unoccupied land (10). An abbreviated form of this description is used by many water agencies, in the many parts of the country, as the primary method of systematically cataloging ground-water sites. The method allows for the location of sites to a minimum of a 2½-acre (one hectare) tract (1/256th of a section) within a specified section, township, range, and meridian. The meridian designation must be included to denote where the township and range are located in the National grid system. An example of a 2½-acre (one hectare) location is "Northeast ¼ of the Southeast ¼ of the Northwest ¼ of the Southwest ¼, Section 22, Township 45 South, Range 87 West, Boise meridian." This location is usually abbreviated to a form similar to "NESENWSW Sec. 22, T45S, R87W B." A number of formats comparable to this abbreviation have been established by the various agencies that use the system, however, they basically communicate the same results (5, 6, 11, 12).

NOTE 10—The accuracy of this location method for the minimum 2½-acre (one hectare) area is about 230 ft (70.104 m), that corresponds to between 2 and 3 s of latitude or longitude. Surveying errors are common in the original measurements. See FIPS PUB 70-1.[5]

NOTE 11—To supplement the description of the location of a ground-water site, a common method used is to draw a sketch showing the relationship of the site to other features in the immediate area, such as roads, buildings, etc. In addition, a sketch of the measuring point can assist in defining its exact location at the site. Photographs of the site and measuring point commonly are used as a part of the description.

7.2.2 *Location Map*—The location map name that is documented is that or the best available map of the area where the site is located. Much of the United States is covered by U.S. Geological Survey (USGS) topographic quadrangles. However, for those areas without USGS maps, the name of the map that shows the site's location should be documented. In addition, record the map's source, such as county highway or Army Map Service. The availability and identification of the USGS maps are given on individual State topographic map indexes. These indexes and the individual topographic maps can be obtained from USGS Public Inquires Office (5, 9, 13).[6]

NOTE 12—Many mapped areas are available on a computer-stored Geographical Information System (GIS). Document information required to identify and obtain the GIS map of the area where the site is located.

7.2.3 *Map Scale*—Document the scale of the map that is used to locate the site. This value helps to define the accuracy of the site location data (5).

NOTE 13—The map scale is the ratio between the linear distance on a map and the corresponding distance on the surface being mapped. For example, 1 in. = 1 mile (1 mm = 1 m) or the equivalent 1:63 360, are ways of expressing the same ratio.

7.2.4 *Method Altitude Determined*—Document the method used to determine the altitude of the reference datum at the ground-water site. Suggested method altitude determined components and representative codes are as follows (5):

A—Altimeter,
L—Level or other surveying method,
M—Interpolated from topographic map, and
Z—Other, explain (for example, historical local datum).

7.3 *Political Regimes*—Document the political regime (for example, Congressional district) where the site is physically located. The date of documentation should be included because of changes that are commonly made in the boundaries of the districts. This allows for determining the legislative responsibility of the site. A guide to these districts is defined in FIPS PUB 9-1 (14).[7]

NOTE 14—Congressional district boundaries can be modified over time because of population changes. Care must be exercised in using this data element to ensure that the ground-water site is still in the originally assigned District.

7.4 *Source Identifiers:*

7.4.1 *Mean Greenwich Time Offset*—Much of the data collected at a ground-water site is time related, such as water-level measurements or water-quality samples. Document, where applicable, the mean Greenwich time offset or United States time zone of the site, so that the time-dimension can be reduced to a common denominator.

7.4.2 *Site Referenced in Report*—If this site has been used or is referenced in a report, document the data concerning the published or unpublished report(s) and, if available, the identification of the report and the address of where to obtain a copy.

[5] FIPS PUB 70-1, *Representation of Geographic Point Locations for Information Interchange*, is available from National Technical Information Service, U.S. Department of Commerce, 5285 Port Royal Road, Springfield, VA 22161.

[6] Public Inquires Office, U.S. Geological Survey, 503 National Center, Room 1-C-402, 12201 Sunrise Valley Drive, Reston, VA 22092.

[7] FIPS PUB 9-1, *Congressional Districts of the United States* is available from National Technical Information Service, U.S. Department of Commerce, 5285 Port Royal Road, Springfield, VA 22161.

7.4.3 *Site in a Computer Data Base*—Document whether or not the information concerning the site has been entered into a computer data base and, if in a data base, the location. Show the data base management system (DBMS) used to organize the data base, for example, "INGRES," a relational DBMS. Give the name assigned to the data base containing the site, for example, "WATSTORE," the U.S. Geological Survey water data base.

7.4.4 *Photography/Sketch Available of Site*—Document the existence of a photograph or sketch of the site, or both. Photographs and sketches of the site and associated facility, including the measuring point, are valuable pictorial material to enhance the site description.

NOTE 15—An example of a form (see Fig. 1) for documenting the data elements as described under "Miscellaneous Singular Data Elements" is illustrated here to show a method of design for this tool. These forms are commonly known as field forms or as coding forms (for computer entry). This type of form is routinely used for transcribing field data while at the ground-water site and entering non-field information at the agency's or company's office. It should be noted that each form has the site identification (primary identification as used by the agency or company), date of field visit, and person that recorded the data as the first entries. These three data items are mandatory to ensure correct filing of the information, either in cabinets or in a computer data base, and for quality control.

8. Documentation of Miscellaneous Repetitive Data Elements

8.1 *Introduction:*

8.1.1 Many of the ground-water elements require multiple records to completely describe a site. Time-related elements, such as water levels, discharge measurements, and water chemistry, may need hundreds or thousands of records for a period of many years to document measurements at a single site. These time-related data help to determine historical trends and serve to establish bench-mark standards for the site.

8.1.2 Other data elements that are not time related, such as casing, lengths, spring openings, and an array of geophysical logs, require a sequence of records to thoroughly describe the site. These data are extremely valuable in site characterization, for example, wells for which the construction components are essential to understand the source of the water.

8.2 *Legal Record Group:*

8.2.1 *Legal Record*—The legal record includes information about any regulatory agencies or authorities, such as for establishment of the right-to-use water and the amount of water allocated for use at a ground-water site. This legal record is normally administered by a government agency or government authorized agency (for example, ground-water management district or health department) within the specific state. Some states use the method of permitting to assemble site records.

8.2.1.1 *Permitting Agency*—If applicable, document the name and address of the agency that is responsible for issuing the permit for the legal development of water at the site.

8.2.1.2 *Priority Date*—If applicable, document the date in year, month, and day (YYYYMMDD), that establishes the legal priority for use of water at the ground-water site. If necessary, show the time of day that the priority was authorized.

8.2.1.3 *Application Number*—If applicable, document the number or identification assigned by the agency to the application for the permit.

8.2.1.4 *Application Date*—If applicable, document the application date, in year, month, and day (YYYYMMDD), for the ground-water site. If necessary, document the time of day.

8.2.1.5 *Certification Number*—If applicable, document the number or identification assigned by the agency to the certification credential.

8.2.1.6 *Certification Date*—If applicable, document the certification date, in year, month, and day (YYYYMMDD), for the ground-water site. If necessary, document the time of day.

8.2.1.7 *Permit Number*—If applicable, document the number or identification assigned by the agency to the permit for the use of water at the site.

8.2.1.8 *Permit Date*—If applicable, document the date, in year, month, and day (YYYYMMDD), the permit was issued for the ground-water site by the responsible agency. If necessary, document the time of day.

8.2.1.9 *Water Allocation*—If applicable, document the amount of water allocated by the permitting agency to the permit holder for the subject permit. Include the measurement unit utilized for the water allocation.

8.3 *Supporting Information Record Group:*

8.3.1 *Owner Record*—The owner's record is used to document a history of ownership of the ground-water site. This record is important to aid in the proper identification of the site and to assign the responsibility for the facility. The following data elements are required to document the history of ownership (5).

8.3.1.1 *Date of Ownership*—If applicable, document the date, in year, month, and day (YYYYMMDD), that the owner acquired possession of the ground-water site.

8.3.1.2 *Owner's Name*—Document the name of the owner and owner's address that corresponds with the date of

FIG. 1 Example Form

ownership for the event record.

8.3.2 *Site Visits Record*—The sites visits record is used to document data collection, verification, and quality-control visits to the ground-water site. The following data elements are required to document the history of these site visits (5).

8.3.2.1 *Date of Visit*—If applicable, document the date, in year, month, and day (YYYYMMDD), that the site was visited. If necessary, document the time of day.

8.3.2.2 *Person Who Made Visit*—If applicable, document the name, title, and address of person who made the visit to the ground-water site for the record event.

8.3.2.3 *Purpose of Visit*—If applicable, document a description of the purpose of the visit to the ground-water site.

8.3.3 *Other Identification Record*—Many ground-water sites have more than one identification. These identifiers can be assigned by a company, state agency or federal agency to conform with an internal file system. To aid in the tracking of data for a site the following data elements may be required (5).

8.3.3.1 *Other Name, Number, or Identification*—If applicable, document the ground-water site identification that was assigned by the other company or agency.

8.3.3.2 *Assigner*—Document the name and address of the person, company, or agency that assigned the other identification for this event record.

8.3.4 *Other Data Record*—The other data available record is used to indicate the availability of additional data pertinent to the ground-water site. Many sites have detailed information, such as continuous water-level recorder charts, geophysical logs, detailed geological logs, and extensive water-quality analyses, that may not be filed at a central location. These data are valuable in understanding conditions at the site (5).

8.3.4.1 *Other Data Type*—If applicable, describe the type of other data available for the ground-water site.

8.3.4.2 *Other Data Location*—If applicable, document the location of the other data for the ground-water site. The complete name and address of the holder of the data should be documented.

8.3.4.3 *Format of Other Data*—If applicable, document the format of the other data available. Suggested other data available components and representative codes are as follows:

> F—Files (raw data),
> M—Machine readable (computer),
> P—Published (report or basic-data release), and
> Z—Other (describe).

8.3.5 *Remarks Record*—The remarks record is used for documenting meaningful information about the site for which no specific data elements are defined (5).

8.3.5.1 *Remark Date*—If applicable, document the date, in year, month, and day (YYYYMMDD), of the origin of the remark.

8.3.5.2 *Remark*—If applicable, document information concerning the site that does not conform to any of the data elements that are listed in these guides.

8.3.5.3 *Remark Source*—Document the name and address of the person, company, or agency that wrote the remark for this event record.

9. Keywords

9.1 data confidence classification; data element; ground water; monitoring location; sampling site; site identification; site location; water allocation; water quality

REFERENCES

(1) Bates, Robert L., and Jackson, Julia A., *Glossary of Geology*, Third Edition; American Geological Institute, Alexandria, Virginia, 1987.

(2) Bureau of Reclamation, *Ground-Water Manual, A Water Resources Technical Publication*, Revised Reprint; U.S. Department of Interior, Bureau of Reclamation, Washington, DC, 1981.

(3) Campbell, Michael D., and Lehr, Jay H., *Water Well Technology*, McGraw-Hill, New York, NY, 1973.

(4) Heath, Ralph C., *Basic Ground-Water Hydrology; U.S. Geological Survey Water-Supply Paper 2220*, 1983.

(5) Mathey, Sharon B., Editor, *National Water Information System User's Manual*, Vol 2, Chapter 4, Ground-Water Site Inventory System; U.S. Geological Survey, Open-File Report 89-587, 1990.

(6) Texas Natural Resources Information System, *Ground-Water Data INTERFACE, Users Reference Manual*; Texas Natural Resources Information System, Nov. 20, 1986.

(7) U.S. Environmental Protection Agency, *Handbook of Suggested Practices for the Design and Installation of Ground-Water Monitoring Wells*; Office of Research and Development, U.S. EPA, March 1991, Washington, DC, 1991.

(8) U.S. Geological Survey, *National Handbook of Recommended Methods for Water-Data Acquisition*, Chapter 2—Ground Water; Office of Data Coordination, Reston, Virginia, 1980, pp. 2-1 to 2-149.

(9) van der Leedan, Frits, Troise, Fred L., and Todd, David Keith, 1990, *The Water Encyclopedia*, Geraghty and Miller Ground-Water Series, 2nd Edition, Third Printing, Lewis Publishers, Inc., Chelsea, Michigan, 1991.

(10) Stewart, Lowell O., Public Land Survey; Iowa State University Press, Ames, Iowa, 1936.

(11) Morgan, Charles O., and McNellis, Jesse M., *FORTRAN IV Program KANS, for the Conversion of General Land Office Locations to Latitude and Longitude Coordinates*; Kansas State Geological Survey Special Distribution Publication 42, 1969.

(12) U.S. Department of Commerce, *Representation of Geographic Point Locations for Information Interchange*, Federal Information Standards (FIPS) Publication 70-1, National Institute for Standards and Technology, Washington, DC, June 23, 1986.

(13) U.S. Geological Survey, *Guide to Obtaining USGS Information*, U.S. Geological Survey Circular 900, 1989.

(14) U.S. Department of Commerce, *Congressional Districts of the United States*, Federal Information Standards (FIPS) Publication 9-1, National Institute for Standards and Technology, Washington, DC, Nov. 30, 1990.

$\text{\small{Ⓐ}}$ **D 5408**

Standard Guide for
Set of Data Elements to Describe a Ground-Water Site; Part Two—Physical Descriptors[1]

This standard is issued under the fixed designation D 5409; the number immediately following the designation indicates the year of original adoption or, in the case of revision, the year of last revision. A number in parentheses indicates the year of last reapproval. A superscript epsilon (ε) indicates an editorial change since the last revision or reapproval.

1. Scope

1.1 This guide is Part Two of three guides to be used in conjunction with Practice D 5254 that delineates the data desirable to describe a ground-water data collection or sampling site. This guide identifies physical descriptors, such as construction and geologic elements, for a site. Part One (Guide D 5408) describes additional information beyond the minimum set of data elements that may be specified to identify any individual ground-water site, while Part Three identifies usage descriptors, such as monitoring, for an individual ground-water site.

NOTE 1—A ground-water site is defined as any source, location, or sampling station capable of producing water or hydrologic data from a natural stratum from below the surface of the earth. A source or facility can include a well, spring or seep, and drain or tunnel (nearly horizontal in orientation). Other sources, such as excavations, driven devices, bore holes, ponds, lakes, and sinkholes, that can be shown to be hydraulically connected to the ground water, are appropriate for the use intended.

NOTE 2—Part One (Guide D 5408) includes data confidence classification descriptor (one element), geographic location descriptors (four elements), political regime descriptor (one element), source identifier descriptors (four elements), legal descriptors (nine elements), owner descriptors (two elements), site visit descriptors (three elements), other identification descriptors (two elements), other data descriptors (three elements), and remarks descriptors (three elements). Part Three (Guide D 5410) includes monitoring descriptors (77 data elements), irrigation descriptors (four data elements), waste site descriptors (nine data elements), and decommissioning descriptors (eight data elements). For a list of descriptors in this guide, see Section 3.

1.2 These data elements are described in terms used by ground-water hydrologists. Standard references, such as the Glossary of Geology (1)[2] and various hydrogeologic professional publications, are used to determine these definitions. Many of the suggested elements and their representative codes are those established by the Water Resources Division of the U.S. Geological Survey and used in the National Water Information Systems computerized data base (1–19).[2]

NOTE 3—The purpose of this guide is to suggest data elements that can be collected for ground-water sites. This does not uniquely imply a computer data base, but rather data elements for entry into any type of permanent file.

NOTE 4—Component and code lists given with some of the data elements, for example "Type of Spring," are only suggestions. These lists can be modified, expanded, or reduced for the purpose intended by the company or agency maintaining the ground-water data file.

NOTE 5—Use of trade names in this guide is for identification purposes only and does not constitute endorsement by ASTM.

1.3 This guide includes the data elements desirable to document a ground-water site beyond those given in the "Minimum Set of Data Elements." Some examples of the data elements are well depth, contributing aquifer, and permanence of spring. No single site will need every data element, for example, springs do not need well depth and well casing data. Each record (group of related data elements) for a site has mandatory data elements, such as the type of lift for the lift record. However, these elements are considered necessary only when that specific record is gathered for the site.

1.4 The values given in either inch-pound units or SI units are to be regarded separately as the standard. The values given in parentheses are for information only.

1.5 *This standard does not purport to address all of the safety problems, if any, associated with its use. It is the responsibility of the user of this standard to establish appropriate safety and health practices and determine the applicability of regulatory limitations prior to use.*

2. Referenced Documents

2.1 *ASTM Standards:*
D 653 Terminology Relating to Soil, Rock, and Contained Fluids[3]
D 2488 Practice for Description and Identification of Soils (Visual—Manual Procedure)[3]
D 5254 Practice for the Minimum Set of Data Elements to Identify a Ground-Water Site[4]
D 5408 Guide for Set of Data Elements to Describe a Ground-Water Site; Part One—Additional Identification Descriptors[4]
D 5410 Guide for Set of Data Elements to Describe a Ground-Water Site; Part Three—Usage Descriptors[4]

3. Terminology

3.1 *Definitions:*
3.1.1 For definitions of terms applicable to this guide, see Terminology D 653.
3.2 *Descriptions of Terms Specific to This Standard:*
3.2.1 *code*—a suggested abbreviation for a component, for example, "G" is the code suggested for the galvanized iron component of data element casing material.
3.2.2 *component*—a subdivision of a data element, for example, galvanized iron is one of 30 components suggested

---

[1] This guide is under the jurisdiction of ASTM Committee D-18 on Soil and Rock and is the direct responsibility of Subcommittee D18.21 on Ground Water and Vadose Zone Investigations.
Current edition approved May 15, 1993. Published November 1993.
[2] The **boldface** numbers in parentheses refer to a list of references at the end of the text.

[3] *Annual Book of ASTM Standards*, Vol 04.08.
[4] *Annual Book of ASTM Standards*, Vol 04.09.

for data element casing material.

3.2.3 *data element*—an individual segment of information about a ground-water site, for example, casing material. The data element is in the casing record record.

3.2.4 *record*—a set of related data elements that may need to be repeated to fully describe a ground-water site. For example, a well that consists of several diameters of casing from the top end to the bottom will need more than one Casing Record record (the record includes data elements depth to top, depth to bottom, diameter, casing material, and casing thickness) to fully describe the construction of the well. However, if only a single size of casing is used in the well, the record is utilized once.

3.2.5 *record group*—a set of related records. For example, the lift record group includes the lift record, power record, and standby record. Some record groups consist of only one record, for example, the spring record group includes only the spring record.

4. Summary of Guide

4.1 This guide includes the following physical descriptor data elements to describe a ground-water site. Single elements usually need one entry for a site, while repeated elements commonly require several records to fully describe the conditions and history of the site.

Single Elements:
  Individual Site Characteristics:
    Land Use (in vicinity of site)
    Drainage Basin/Watershed
    Relationship to Surface Stream/Lake, etc.
    Hole Depth
    Well Depth
    Source of Depth Data
    Primary Aquifer
Repeated Elements:
  Construction Record Group:
    Construction Record:
      Date Construction Began
      Date Construction Ended
      Name of Contractor
      Source of Construction Data
      Method of Construction
      Type of Drilling Fluid
      Volume of Drilling Fluid
      Type of Finish
      Type of Seal
      Depth to Bottom of Seal
      Method of Development
      Length of Time of Development
      Volume of Liquid Removed During Development
      Special Treatment
    Hole Record:
      Depth to Top of the Hole Interval
      Depth to Bottom of the Hole Interval
      Diameter of the Hole Interval
    Casing Record:
      Depth to Top of the Cased Interval
      Depth to Bottom of the Cased Interval
      Diameter of the Cased Interval
      Casing Material
      Casing Thickness
    Opening or Screen Record:
      Depth to Top of the Open Interval
      Depth to Bottom of the Open Interval
      Diameter of the Open Interval
      Type of Material in the Open/Screened Interval
      Type of Openings in the Open Interval
      Length of Openings
      Width of Openings
      Mesh of Screen
      Packing Material

      Size of Packing Material
      Thickness of Packing Material
      Depth to Top and Bottom of Packing Material
    Repairs Record:
      Date of Repairs
      Nature of Repairs
      Name of Contractor Who Made Repairs
      Percent Change in Performance After Repairs
  Special Cases Record:
    Well Clusters:
      Number of Wells in Cluster
      Depth of Deepest Well in Cluster
      Depth of Shallowest Well in Cluster
      Diameter of Well Cluster
    Collector Well/Laterals:
      Number of Laterals in Collector Well
      Depth of Laterals in Collector Well
      Length of Laterals in Collector Well
      Diameter of Laterals in Collector Well
      Mesh of Screen in Laterals
    Ponds:
      Length of Pond
      Width of Pond
      Depth of Pond
      Volume of Pond
    Tunnel or Drain:
      Length of Tunnel or Drain
      Width of Tunnel or Drain
      Depth of Tunnel or Drain
      Bearing (Azimuth) Tunnel or Drain
      Dip of Tunnel or Drain
  Lift Record Group:
    Lift Record:
      Type of Lift
      Date Permanent Lift was Installed
      Depth of Intake
      Manufacturer of Lift Device
      Serial Number
      Pump Rating
    Power Record:
      Type of Power
      Horsepower Rating
      Name of Power Company
      Power-Company Account Number
      Power-Meter Number
    Standby Lift Record:
      Additional Lift
      Name of Company that Maintains Lift
      Rated Pump Capacity
      Type of Standby Power
      Horsepower of Standby Power Source
  Geologic Record Group:
    Geophysical Log Record:
      Date of Log
      Type of Log
      Depth to Top of Logged Interval
      Depth to Bottom of Logged Interval
      Source of Log Data
    Geohydrologic Unit Record:
      Aquifer Unit(s)
      Contributing Unit
      Depth to Top of Interval
      Depth to Bottom of Interval
      Lithology
      Description of Material
    Sample/Unconsolidated Material Record:
      Sample Weight
      Sample Interval
      Particle Size
      Percent of Total Sample
      Particle Shape
      Mineralogy
    Sample/Consolidated Material Record:
      Drill Cuttings or Core
      Sample Size (Weight)
      Sample Interval
      Mineralogy
      Core Length
      Core Diameter

Core Recovery-Percent
Bedding
Structure
Porosity
Hydraulic Record Group:
  Hydraulics Record:
    Hydraulic/Aquifer Unit
    Hydraulic/Aquifer Unit Type
    Depth to Top of Unit
    Depth to Bottom of Unit
    Static Water Level
    Measurement Date and Time
    Unit Contribution
  Aquifer Parameters Record:
    Transmissivity
    Horizontal Hydraulic Conductivity
    Vertical Hydraulic Conductivity
    Coefficient of Storage
    Leakance
    Diffusivity
    Specific Storage
    Specific Yield
    Barometric or Tidal Efficiency
    Porosity
    Specific Capacity
    Method Used to Determine Aquifer Characteristics
    Availability of File of Detailed Results
Spring Record Group:
  Spring Record:
    Name of Spring
    Type of Spring
    Permanence of Spring
    Sphere of Discharge
    Discharge
    Date of Discharge
    Improvements
    Number of Spring Openings
    Flow Variability
    Accuracy of Flow Variability
    Magnitude of Spring

5. Significance and Use

5.1 Data at ground-water sites are gathered for many purposes. Each of these purposes generally requires a specific set of data elements. For example, when the ground-water quality is of concern not only are the 'minimum set of data elements' required for the site, but information concerning the sample collection depth interval, method of collection, and date and time of collection are needed to fully qualify the data. Another group of elements are recommended for each use of the data, such as aquifer characteristics or water-level records. Normally the more information that is gathered about a site by field personnel, the easier it is to understand the ground-water conditions and to reach valid conclusions and interpretations regarding the site.

5.2 The data elements listed in this guide and Guides D 5408 and D 5410 should assist in planning what information can be gathered for a ground-water site and how to document these data.

NOTE 6—Some important data elements may change during the existence of a site. For example, the elevation of the measuring point used for the measurement of water levels may be modified because of repair or replacement of equipment. This frequently occurs when the measuring point is an opening in the pump and the pump is modified or replaced. Because changes cannot always be anticipated, it is preferable to reference the height of the measuring point to a permanent nearby altitude datum. The measuring point is referenced by being the same altitude (zero correction) or above (negative correction) or below (plus correction) the altitude datum. All appropriate measurements should be corrected in reference to the altitude datum before entry into the permanent record. Care must be exercised to keep the relationship of

these data elements consistent throughout the duration of the site.

5.3 Some data elements have an extensive list of components or possible entries. For example, the aquifer identification list described in 6.1.8 has over 5000 entries. Lengthy lists of possible entries are not included in this guide, however, information on where to obtain these components is included with the specific data element.

NOTE 7—This guide identifies other sources, lists, etc. of information required to completely document information about any ground-water site.

6. Documentation of Individual Site Characteristics

6.1 *Introduction:*

6.1.1 A vast number of data elements can be documented about a ground-water site to thoroughly describe its location, physical features, relationship to other features on the earth's surface, and to designate what information is gathered at the site. These data elements typically are transcribed once for a site, in contrast to data elements that may be repetitive, such as water levels. Many of these data are extremely valuable in the characterization of sites that fall into certain categories, for example wells, for which the primary aquifer is an essential element to assist in the identification of the source of water at the site (2–5, 7, 8, 10–17, 19).

6.1.2 *Land Use (in Vicinity of Site)*—Document the use of the land in the area surrounding the ground-water site. This data element is important if there is a possibility of the use affecting the availability or quality of the water. If more than one significant land use is nearby, such as industrial and farming, document each purpose (5, 16).

6.1.3 *Drainage Basin/Watershed*—Document the name or other identification of the watershed and drainage basin where the site is located. Maps with watersheds delineated are available from the State Conservationists, U.S. Department of Agriculture, Soil Conservation Service located in each of the states, possessions, and associated areas. Information about river basins is available on maps in "Atlas of River Basins of the United States" published by the U.S. Department of Agriculture, Soil Conservation Service (20).[5]

6.1.4 *Relationship to Surface Stream/Lake, etc.*—Document information concerning the influence of any nearby surface-water source upon the ground-water site. For example, the ground-water source for the site could be directly connected to a surface-water body (recharging the aquifer or discharging to the surface-water body) or have no connection and be influenced by a seasonal variation in loading of the surface water body upon the aquifer (4, 7, 8, 16).

NOTE 8—This information is more useful if a quantitative estimate of the amount of connection is given rather than a yes, there is a connection, or no, there is no connection notation. For example, a ground-water body that is only influenced by seasonal loading of a

---

[5] Regional contacts for obtaining information are as follows: Northeastern United States contact Director, Northeastern National Technical Center, USDA, SCS, 160 East 7th Street, Chester, PA 19013. Southern United States contact Director, Southern National Technical Center, USDA, SCS, Fort Worth Federal Center, Bldg. 23, Room 60, Felix and Hemphill Streets, PO Box 6567, Fort Worth, TX 76115. Midwestern United States contact Director, Midwest National Technical Center, USDA, SCS, Federal Bldg., Room 345, 100 Centennial Mall North, Lincoln, NB 68508-3866. Western United States contact Director, Western National Technical Center, USDA, SCS, Federal Bldg., Room 248, 511 N.W. Broadway, Portland, OR 97209-3489.

surface-water body would have 0 % connection. While a stream or lake that is partially or completely linked to the ground-water body could have from 1 to 100 % connection, however, a quantitative value seldom can be determined. Usually, the range of thickness of the aquifer penetrated by the surface water body or thickness and lithology of the material between the aquifer and surface water body is all that is known about the connection.

6.1.5 *Hole Depth*—If applicable, document the total depth that the hole was drilled, in feet or metres below a datum at or near land surface. Many times the hole is drilled deeper in order to explore stratum below the completed depth of the final well. This number is always equal to or greater than the well depth. The hole depth is important because the information concerning the stratum below the final well can be critical in understanding ground-water conditions at the site. Document the accuracy or confidence classification for this data element (4, 5, 7, 8, 13, 14, 16). If applicable, note orientation and angle of hole if not vertical.

6.1.6 *Well Depth*—If applicable, document the depth of the finished well, in feet or metres below a datum at or near land surface. This depth is important as a means to delineate the maximum depth at which water is entering the well bore. Document the accuracy or confidence classification for this data element (4, 5, 7, 8, 13, 14, 16).

6.1.7 *Source of Depth Data*—If applicable, document the source of the hole and well depth information. Suggested source of depth data components and representative codes are as follows(13):

A—Reported by a government agency
D—From driller's log or report
G—Private geologist-consultant or university associate
L—Depth interpreted from geophysical logs by personnel of source agency
M—Memory (owner, operator, driller)
O—Reported from records by owner of well
R—Reported by person other than owner, driller, or another government agency
S—Measured by personnel of reporting agency
Z—Other source (describe)

6.1.8 *Primary Aquifer*—Document the identification of the primary aquifer unit from which the water is withdrawn or monitoring data are collected. A convenient and systematic method of coding geologic units was described by Cohee (6) in the American Association of Petroleum Geologists Bulletin. This method is used by the U.S. Geological Survey to code aquifer and geologic unit names in a National file (Catalog of Aquifer Names and Geologic Unit Codes used by the Water Resources Division) (for example, Edwards Limestone of Texas is coded 218EDRD). Information needed to obtain an ordered list of aquifers and related codes is available from the following (6, 13):[6]

NOTE 9—An example of a form (see Fig. 1) for documenting the data elements as described under "Individual Site Characteristics" is illustrated here to show a method of design for this tool. The forms are commonly known as field forms or as coding forms (for computer entry). This type of form is routinely used for transcribing field data while at the ground-water site and entering non-field information at the agency's or company's office. It should be noted that each form has the site identification (primary identification as used by the agency or company), date of field visit, and person that recorded the data as the first entries. These three data items are mandatory to ensure correct filing of the information, either in cabinets or in a computer data base, and for quality control.

[6] Geologic Names Unit, U.S. Geological Survey, 439 National Center, Reston, VA 22092.

FIG. 1 Example Form

7. Documenting of Miscellaneous Repetitive Data Elements

7.1 *Introduction:*

7.1.1 Many of the ground-water data elements require multiple records or entries to completely describe a site. Time-related elements, such as water levels, discharge measurements, and water chemistry, may present hundreds or thousands of records over a period of many years that answer a specific question about a single site. These time-related data help to determine historical trends and serve to establish bench-mark conditions for the site (4, 5, 13, 14).

7.1.2 Other data elements that are not time related, such as casing lengths, spring openings, and some geophysical logs, require a sequence of records to thoroughly describe the site. These data are extremely valuable in site characterization, for example, wells for which the construction components are required to understand the source of the water (4, 5, 13, 14).

7.2 *Construction Record Group*—The construction record group includes records for documenting data elements relating to any type of structure built for withdrawal of water or monitoring at a ground-water site, including construction, hole, casing, openings or screen, repairs, and special cases, such as well clusters, collector wells, ponds, tunnels, and drains (2, 3, 5, 7, 8, 13–17). If applicable, any construction that may have modified the ambient ground water conditions should be documented. Examples include grouting, blasting, hydrofracturing, and local disruption such as tunnels, underground chambers, or excavations.

7.2.1. *Construction Record*—The construction record includes data elements relating to the date of construction, contractor, construction method, drilling fluids, finish, and development. Data elements that are included in the construction record are the following:

7.2.1.1 *Date Construction Began*—If applicable, document the date (year, month, day in YYYYMMDD format) on which the construction work was initiated at the ground-water site.

NOTE 10—Although this guide is written to be used with any type of data file, date information should be arranged in year, month, day, and time (24-h clock) format (for example, 19910822094158 for 1991, August 22nd, 9 h, 41 min, and 58 s AM), especially for ease of interchanging information among data systems (computerized files). This is the format recommended by the American National Standard for Information Systems (ANSI) and adopted as a Federal Information Processing Standards (FIPS) system (21, 22).

7.2.1.2 *Date Constructed Ended*—If applicable, document

344

he date (year, month, day in YYYYMMDD format) on
which the construction work was completed at the ground-
water site.

7.2.1.3 *Name of Contractor*—If applicable, document the
name and address of the principal individual or company
that did the construction work at the ground-water site (for
example, drilled the well).

7.2.1.4 *Source of Construction Data*—If applicable, document the source of the information concerning the construction at the ground-water site (for example, driller's log or
geologist's log). Suggested source of construction data components and representative code are as follows:

A—Reported by a government agency
D—From driller's log or report
G—Private geologist-consultant or university associate
L—Depth interpreted from geophysical logs by personnel of source agency
M—Memory (owner, operator, driller)
O—Reported from records by owner of well
R—Reported by person other than owner, driller, or another government agency
S—Measured by personnel of reporting agency
Z—Other source (describe)

7.2.1.5 *Method of Construction*—If applicable, document
the method by which the ground-water site was constructed.
Suggested method of construction components and representative codes are as follows:

B—Bored or augered, generalized
L—Wash boring
M—Hollow stem auger
N—Solid stem auger
E—Bucket auger
A—Direct air-rotary method, with bit
K—Direct air-rotary method, with downhole hammer
H—Direct mud rotary
R—Reverse circulation rotary (no casing)
F—Dual-wall reverse circulation, generalized
G—Dual-wall reverse rotary
I—Dual-wall reverse percussion
C—Cable-tool
P—Air-percussion drill
Q—Hydraulic percussion
S—Jet percussion
J—Jetted by water
D—Dug or excavated
T—Trenching, dammed pond, or drain
V—Driven pipe
U—Cone penetration
W—Combined driven and jetting
Z—Other (describe)

NOTE 11—Several of the method of construction components are the
same or similar methods (jetted by water and wash boring), but with
different name identifications. In addition, several of the components
that have generalized names, for example, bored or augered also have
the specific methods (hollow stem auger, solid stem auger, etc.) included
in the list.

7.2.1.6 *Type of Drilling Fluid*—If applicable, document
the type and amount of additives (in pounds or kilograms)
used in the drilling fluid (water) for the construction of the
ground-water site. Suggested additive components and representative codes are as follows:

A—Acrylic polymers
D—Attapulgite
E—Baking soda
B—Barites
F—Biodegradable material
G—Caustic soda
C—Cellulosic polymers
H—Chromelignosulfonates
I—Chrysotile asbestos
J—Complex phosphates

K—Lignitic materials
L—Lime
Q—Lubricants
M—Modified guar gum products
R—Modified polysaccharide
N—Native clay
O—Organic polymers
P—Peptized bentonite
U—Pregelatinized starch
V—Soda ash
W—Sodium carboxymethylcellulose
S—Standard bentonite
X—Surfactants
T—Tannins
Z—Other (describe)

7.2.1.7 *Volume of Drilling Fluid*—If applicable, document the volume (in gallons or litres) of drilling fluid lost in
the drilled hole. Specify the unit of measurement. Document
the accuracy or confidence classification for this data element. It may be difficult to quantify losses in air drilling.
Estimates may be made by comparing output versus compressor capacity.

7.2.1.8 *Type of Finish*—If applicable, document the
method of finish or the nature of the openings that allow
water to enter the well. Suggested type of finish components
and representative codes are as follows:

C—Porous concrete
G—Gravel-packed screen
H—Horizontal gallery or collector
O—Open-ended casing
P—Perforated or slotted casing
S—Screen, commercial
T—Sand point, driven screen
W—Walled or shored
X—Open-hole in aquifer
Z—Other (describe)

7.2.1.9 *Type of Seal*—If applicable, document the type
and amount (in pounds or kilograms) of material used to seal
the well against the entry of surface water and the leakage of
water between aquifers having different hydraulic pressures.
Suggested type of seal components and representative codes
are as follows:

B—Bentonite
C—Clay or cuttings
G—Cement grout
N—None
Z—Other (describe)

7.2.1.10 *Depth to Bottom of Seal*—If applicable, document the depth to the bottom of the seal in feet or metres
below a datum at or near land surface. Document the
accuracy or confidence classification for this data element.

7.2.1.11 *Method of Development*—If applicable, document the primary method used to develop the well. Suggested method of development components and representative codes are as follows:

A—Pumped with air lift
B—Bailed
D—Chemical, for example, dry ice
C—Surged, compressed air
J—Jetted, air or water
N—None
P—Overpumped
S—Surge block
Z—Other (describe)

7.2.1.12 *Length of Time of Development*—If applicable,
document the number of hours and minutes that the well
was bailed, pumped, or surged for development. Document

the accuracy or confidence classification for this data element.

7.2.1.13 *Volume of Liquid Removed During Development*—If applicable, document the volume of liquid (in gallons or litres) removed from well during development. Specify the unit of measurement. Document the accuracy or confidence classification for this data element.

7.2.1.14 *Special Treatment*—If applicable, document any special treatment that was applied during development of the well. Suggested special treatment components and representative codes are as follows:

C—Chemical (acid, and so forth)
D—Dry ice
E—Explosives
F—Deflocculent
H—Hydrofracturing
M—Mechanical abrasion
Z—Other (describe)

7.2.2 *Hole Record*—The hole record includes data elements that relate to the description of the opening constructed for emplacement of hardware into the ground for the development of a monitoring or production well at a ground-water site. For many sites, several distinct hole length and size intervals are required for the completion of the well. Data elements that are included in the hole record are the following:

7.2.2.1 *Depth to Top of the Hole Interval*—If applicable, document the depth to the top of the hole interval, in feet or metres below a datum at or near land surface. The first or uppermost section of the hole starts at or near the datum. Document the accuracy or confidence classification for this data element.

7.2.2.2 *Depth to Bottom of the Hole Interval*—If applicable, document the depth to the bottom of the hole interval, in feet or metres below the datum. Document the accuracy or confidence classification for this data element.

7.2.2.3 *Diameter of the Hole Interval*—If applicable, document the nominal diameter of that interval of the hole, in inches or millimetres. Document the accuracy or confidence classification for this data element. Caliper logs may be very useful as documentation.

7.2.3 *Casing Record*—The casing record includes all information that relates to the description of the casing material placed into the ground for the construction of a monitoring or production well at a ground-water site. For many sites, several distinct length and size intervals are required for the completion of the well. Data elements that are included in the casing record are the following:

7.2.3.1 *Depth to Top of the Cased Interval*—If applicable, document the depth to the top of the cased interval, in feet or metres below a datum at or near land surface. The first or uppermost section of the casing starts at or near the datum. Document the accuracy or confidence classification for this data element.

7.2.3.2 *Depth to Bottom of the Cased Interval*—If applicable, document the depth to the bottom of the cased interval, in feet or metres below the datum. Document the accuracy or confidence classification for this data element.

7.2.3.3 *Diameter of the Cased Interval*—If applicable, document the inside diameter of that interval of the casing, in inches or centimetres. Document the accuracy or confidence classification for this data element.

7.2.3.4 *Casing Material*—If applicable, document the type of casing material used for the construction of the well. Note if casing joint or other components are different than casing material. Suggested casing material components and representative codes are as follows:

E—Acrylonitrile butadiene styrene (ABS)
A—Aluminum
H—Asbestos cement
B—Brick
J—Carbon structural steel
L—Chlorotrifluoroethylene (CTFE)
N—Coal tar epoxy coated steel
U—Coated steel
C—Concrete
D—Copper
O—Cupro-nickel
F—Fiberglass-reinforced epoxy
Q—Fluorinated ethylene propylene (FEP)
G—Galvanized iron
K—Kai-well
V—Perfluoroalkoxy (PFA)
X—Polytetrafluoroethylene (PTFE)
Y—Polyvinyl chloride (PVC)
1—Polyvinylidene fluoride (PVDF)
P—PVC, fiberglass, other plastic (general term)
R—Rock or stone
2—Rubber-modified polystyrene
3—Silicon bronze
4—Stainless steel
S—Steel
T—Tile
5—Transite
W—Wood
I—Wrought iron
M—Other metal (describe)
X—Other material, not metal (describe)

7.2.3.5 *Casing Thickness*—If applicable, document the thickness of the casing wall, in inches or centimetres. Document the accuracy or confidence classification for this data element.

7.2.4 *Opening or Screen Record*—The opening or screen record includes all information that relates to the description of the open or screened area that allows for the passage of water into a well at a ground-water site. For some sites, several distinct length and size intervals of open or screened areas are required for the completion of the well. Data elements that are included in the opening or screen record are the following:

7.2.4.1 *Depth to Top of the Open Interval*—If applicable, document the depth to the top of the open interval, in feet or metres below a datum at or near land surface. Document the accuracy or confidence classification for this data element.

7.2.4.2 *Depth to Bottom of the Open Interval*—If applicable, document the depth to the bottom of the open interval, in feet or metres below the datum. Document the accuracy or confidence classification for this data element.

7.2.4.3 *Diameter of the Open Interval*—If applicable, document the diameter of the open interval, in inches or centimetres. The diameter documented normally would be the inside diameter for a screen and the hole diameter for an open hole. Document the accuracy or confidence classification for this data element.

7.2.4.4 *Type of Material in the Open/Screened Interval*—If applicable, document the type of material used for the construction of the open/screened interval. Suggested type of material in the open interval components and representative codes are as follows:

E—Acrylonitrile butadiene styrene (ABS)
A—Aluminum
H—Asbestos cement
B—Brick
J—Carbon structural steel
L—Chlorotrifluoroethylene (CTFE)
N—Coal tar epoxy coated steel
U—Coated steel
C—Concrete
D—Copper
O—Cupro-nickel
F—Fiberglass-reinforced epoxy
Q—Fluorinated ethylene propylene (FEP)
G—Galvanized iron
K—Kai-well
V—Perfluoroalkoxy (PFA)
X—Polytetrafluoroethylene (PTFE)
Y—Polyvinyl chloride (PVC)
1—Polyvinylidene fluoride (PVDF)
P—PVC, fiberglass, other plastic (general term)
R—Rock or stone
2—Rubber-modified polystyrene
3—Silicon bronze
4—Stainless steel
S—Steel
T—Tile
5—Transite
W—Wood
I—Wrought iron
M—Other metal (describe)
X—Other material, not metal (describe)

7.2.4.5 *Type of Openings in the Open Interval*—If applicable, document the type of openings in this interval. Suggested type of openings in the open interval components and representative codes are as follows:

B—Bridge slot
C—Continuous slot wire-wound
F—Open hole in fractured rock
L—Louvered or shutter-type screen
M—Mesh screen
P—Perforated, porous, or slotted casing
R—Wire-wound screen
S—Screen, type not known
T—Sand point
W—Walled or shored
X—Open hole, undefined rock condition
Z—Other (describe)

7.2.4.6 *Length of Openings*—If applicable, document the length or long dimension of the perforations, slots, or mesh of the screen, in inches or centimetres. Document the accuracy or confidence classification for this data element.

7.2.4.7 *Width of Openings*—If applicable, document the short dimension of the perforations, slots, or mesh of the screen, in inches or centimetres. Document the accuracy or confidence classification for this data element.

7.2.4.8 *Mesh of Screen*—If applicable, document the slot or mesh size of the screen, in inches or centimetres. Using the mesh of screen data element may be preferable to using the length and width of openings elements. Document the accuracy or confidence classification for this data element.

7.2.4.9 *Packing Material*—If applicable, document the type of material and supplier (or analysis) used to pack the space or void on the outside of the screened interval. Suggested packing material components and representative codes are as follows:

B—Beads, glass
C—Crushed stone—describe type
G—Gravel, sorted
H—Gravel, unsorted
I—Gravel, graded pack
M—Mixture of sand and gravel

N—Natural formation material
S—Sand, sorted
R—Sand, unsorted
T—Sand, graded pack
Z—Other (describe)

7.2.4.10 *Size of Packing Material*—If applicable, document the grain size of the sorted material or the range of grain size of the unsorted or graded packing material, in inches or millimetres. Document the accuracy or confidence classification for this data element.

7.2.4.11 *Thickness of Packing Material*—If applicable, document the thickness of the material packed between the screen and the natural formation, in inches or centimetres (thickness = (hole size-screen size)/2). Document the accuracy or confidence classification for this data element.

7.2.4.12 *Depth to Top and Bottom of Packing Material*—If applicable, document the depth to the top and to the bottom of the packing material, in feet or metres below the datum. Document the accuracy or confidence classification for this data element.

7.2.5 *Repairs Record*—The repairs record includes all information that relates to repair work done on previously constructed facilities at the ground-water site. For many sites, several occurrences of repairs are normal. Data elements that are included in the repairs record are the following:

7.2.5.1 *Date of Repairs*—If applicable, document the date (year, month, day in YYYYMMDD format) that the repairs were completed.

7.2.5.2 *Nature of Repairs*—If applicable, document the type of repairs that occurred at the ground-water site. Suggested nature of repair components and representative codes are as follows:

B—Blocked off
C—Cleaned
D—Deepened
I—Pump intake lowered
L—Liner installed
O—Slotted or perforated
P—Plugged back
S—Screen replaced
Z—Other (describe)

7.2.5.3 *Name of Contractor Who Made Repairs*—If applicable, document the name and address of the contractor that performed the repairs.

7.2.5.4 *Percent Change in Performance After Repairs*—If applicable, document the percent change in the performance (plus or minus) of the ground-water site. For example, percent change = ((new yield − old yield)/old yield) (100 %). Document the accuracy or confidence classification for this data element.

7.2.6 *Well Cluster*—A well cluster is multiple wells or a gallery of wells that are connected to one pumping source. This type of withdrawal system is used in areas of thin and shallow aquifers where each well in the cluster produces a small amount of water. However, in combination, the wells in the cluster yield the amount of water needed for the use intended. Data elements that are not given below and are needed to document the construction details for the well cluster are found under the construction record, hole record, casing record, and opening or screen record.

NOTE 12—Cluster wells are intended for the production of water (for example, a small public water supply or commercial facility) where potable water sources are limited. This type of withdrawal system is not initially intended for hydrologic investigations or for use in the determination of the water-quality characteristics of an aquifer.

NOTE 13—Well cluster, collector well/laterals, ponds, and tunnel or drain fall in the special cases record for documenting information that is unique to the other sources of ground water (other than wells and springs).

7.2.6.1 *Number of Wells in Cluster*—If applicable, document the number of wells in the cluster that are connected into one pumping system.

7.2.6.2 *Depth of Deepest Well in Cluster*—If applicable, document the depth of the deepest well in the cluster. Document the accuracy or confidence classification for this data element.

7.2.6.3 *Depth of Shallowest Well in Cluster*—If applicable, document the depth of the shallowest well in the cluster. Document the accuracy or confidence classification for this data element.

7.2.6.4 *Diameter of Well Cluster*—If applicable, document the largest diameter or dimension, in feet or metres, of the well cluster field. Document the accuracy or confidence classification for this data element.

7.2.7 *Collector Well/Laterals*—A collector well or radial collector well consists of a large central caisson (for example, 13-ft (3.96-m) inside diameter) with laterals and horizontal screens projecting away (for example, 240 ft (13.15 m)) from the bottom of the central caisson. These radials can be in a radial or linear pattern, depending upon the configuration of the aquifer. This type of water-withdrawal system allows for the optimum development of some low-yielding aquifers and more economical development of large supplies from thin, high-production aquifers (such as under rivers). Data elements that are not given below and are needed to document the construction details for the collector well are found under the construction record, hole record, casing record, and opening or screen record. See Note 12 for an explanation of the purpose of a similar type of withdrawal system.

7.2.7.1 *Number of Laterals in Collector Well*—If applicable, document the number of laterals that are connected to the central caisson.

7.2.7.2 *Depth of Laterals in Collector Well*—If applicable, document the depth or average depth of the laterals connected to the collector well. Document the accuracy or confidence classification for this data element.

7.2.7.3 *Length of Laterals in Collector Well*—If applicable, document the length of the laterals that extend away from the collector well. If there is a large difference in the lengths, document the range in lengths or the length of each individual lateral. Document the accuracy or confidence classification for this data element.

7.2.7.4 *Diameter of Laterals in Collector Well*—If applicable, document the diameter of the laterals, in inches or centimetres. Document the accuracy or confidence classification for this data element.

7.2.7.5 *Mesh of Screen in Laterals*—If applicable, document the slot or mesh size of screens, in inches or centimetres. If there is a large difference in the mesh size of the various screens, document the range in size or the size of each individual screen. Document the accuracy or confidence classification for this data element.

7.2.8 *Ponds*—This category of the ground-water withdrawal system includes natural or constructed ponds that intercept the water table. In areas of shallow water tables, natural ponds occur or ponds can be dug into the water-bearing aquifer and the water pumped from the pond to the area of use.

7.2.8.1 *Length of Pond*—If applicable, document the longest dimension of the pond, in feet or metres. Document the accuracy or confidence classification for this data element.

7.2.8.2 *Width of Pond*—If applicable, document the width of the pond (usually the dimension at right angle to the length), in feet or metres. Document the accuracy or confidence classification for this data element.

7.2.8.3 *Depth of Pond*—If applicable, document the maximum or average depth of the pond (include whether the depth given is the maximum or average), in feet or metres. Document the accuracy or confidence classification for this data element.

7.2.8.4 *Volume of Pond*—If applicable, document the average volume of water contained in the pond, in gallons, litres, or other volume unit. Document the volume unit used. Document the accuracy or confidence classification for this data element.

7.2.9 *Tunnel or Drain*—This category of the ground-water withdrawal system includes tunnels constructed principally to intercept the water table and drains constructed primarily to lower the water table in the vicinity of mines or man-made structures (23).

NOTE 14—Tunnels (called falaj, qanat, karez, and foggara in the Middle East) are used as a water collection and distribution system in many parts of the world, especially in arid regions. This nearly horizontal tunnel system is a very conservative method of skimming the upper surface of the water table. Water from drains, that are used to lower the water table at man-made structures (for example, mines), is commonly used for other purposes (for example, processing of ore). Water from drains that are used for the purpose of lowering the near-surface water table of poorly drained agricultural lands normally are discharged to nearby surface water bodies.

7.2.9.1 *Length of Tunnel or Drain*—If applicable, document the length of the tunnel or drain, in feet or metres. Document the accuracy or confidence classification for this data element.

7.2.9.2 *Width of Tunnel or Drain*—If applicable, document the width of the channel where the water flows, in feet or metres. Document the accuracy or confidence classification for this data element.

7.2.9.3 *Depth of Tunnel or Drain*—If applicable, document the average depth of the tunnel or drain, in feet or metres. Document the accuracy or confidence classification for this data element.

7.2.9.4 *Bearing (Azimuth) Tunnel or Drain*—If applicable, document the orientation in degrees bearing from due north of the tunnel or drain, beginning at the origin and going in direction of the terminus. Document the accuracy or confidence classification for this data element.

7.2.9.5 *Dip of Tunnel or Drain*—If applicable, document the dip in degrees from the horizontal of the tunnel or drain, beginning at the origin and ending at the terminus. Document the accuracy or confidence classification for this data element.

7.3 *Lift Record Group*—The lift record group includes records for documenting data elements relating to any type of equipment or method used for withdrawal of water at a ground-water site, including lift technique, power method, and backup or standby lift and power system (2, 4, 5, 7, 8, 13, 14, 16, 19).

7.3.1 *Lift Record*—The lift record includes all information that relates to the method and equipment that is used to remove the ground water from the aquifer. Commonly, several arrangements and types of lift systems are used over the history of a ground-water site because of maintenance and replacement of worn equipment. In rare cases, several lift systems are used at the site at the same time. Information concerning the pump rating or yield and power consumption may be used to estimate the water usage. Data elements that are included in the lift record are the following:

7.3.1.1 *Type of Lift*—The type of lift is the specific method used to remove the water from the aquifer, either by mechanical or natural means. Suggested type of lift components and representative codes are as follows:

A—Air lift
B—Bucket or bailer
C—Centrifugal pump
G—Natural flow or gravity
J—Jet pump
P—Piston pump
R—Rotary pump
S—Submergible pump
T—Turbine pump
N—None
U—Unknown
Z—Other (describe)

7.3.1.2 *Date Permanent Lift Was Installed*—If applicable, document the date (year, month, day in YYYYMMDD format) that the lift unit was installed. This information is used to identify the age of the lift unit and to further identify the site.

7.3.1.3 *Depth of Intake*—If applicable, document the depth below a datum, in feet or metres, to the bottom of the pump intake. Document the accuracy or confidence classification for this data element.

7.3.1.4 *Manufacturer of Lift Device*—If applicable, document the name and address of the company that manufactured the pump.

7.3.1.5 *Serial Number*—If applicable, document the serial number of the pump. This data element allows for additional identification of the pump and ground-water site.

7.3.1.6 *Pump Rating*—If applicable, document the rating of the pump as the volume of the water lifted per unit of power consumed. Tables are normally available for determining the efficiency of each type of pump according to the amount of lift involved. This pump efficiency table must be used for determining the pump rating. The value should be expressed as million of gallons or litres of water per kilowatt-hour of electricity, cubic foot or metres of natural gas, gallon or litre of liquid fuel or engine hour, depending upon type of power. Document the accuracy or confidence classification for this data element.

7.3.2 *Power Record*—The power record includes all information that relates to the type of power used to drive a lift unit or to remove water from the aquifer. Commonly, several arrangements and types of power are used over the history of a ground-water site because of maintenance and replacement of worn equipment. In rare cases, several types of power are used at the site at the same time. Data elements that are included in the power record are the following:

7.3.2.1 *Type of Power*—Document the type of energy used to power the pump or to remove the water from the aquifer. Suggested type of power components and representative codes are as follows:

A—Animal
C—Compressed air
D—Diesel engine
E—Electric motor
F—Natural flow or gravity
G—Gasoline engine
H—Hand or human
L—LP gas (propane or butane engine)
N—Natural-gas engine
W—Windmill
Z—Other (describe)

7.3.2.2 *Horsepower Rating*—If applicable, document the horsepower rating of the power component given under "Type of Power." For example, 10 hp for the rating of an electric motor used to drive a turbine pump. Document the accuracy or confidence classification for this data element.

7.3.2.3 *Name of Power Company*—If applicable, document the name and address of the company that furnishes the electricity, natural gas, or other fuel for the power source.

7.3.2.4 *Power-Company Account Number*—If applicable, document the account number under which the power company stores information on power consumption at the site.

7.3.2.5 *Power-Meter Number*—If applicable, document the meter number of the electric or gas meter which logs the power consumption of the power source.

7.3.3 *Standby Lift Record*—The standby lift record includes information that relates to the type of lift and power used as a backup to the primary lift and power system. Data elements that are included in the standby lift record are the following:

7.3.3.1 *Additional Lift*—If applicable, document the additional head (above land-surface datum) against which the pump work, in feet or metres of water.

7.3.3.2 *Name of Company that Maintains Lift*—If applicable, document the name and address of the company that is responsible for the maintenance of the pump.

7.3.3.3 *Rated Pump Capacity*—If applicable, document the manufacturer's pump capacity rating. Document the accuracy or confidence classification for this data element.

7.3.3.4 *Type of Standby Power*—If applicable, document the type of standby power available. Suggested type of power components and representative codes are as follows:

A—Animal
C—Compressed air
D—Diesel engine
E—Electric motor
F—Natural flow or gravity
G—Gasoline engine
H—Hand or human
L—LP gas (propane or butane engine)
N—Natural-gas engine
W—Windmill
Z—Other (describe)

7.3.3.5 *Horsepower of Standby Power Source*—If applicable, document the horsepower rating of the standby power source. Document the accuracy or confidence classification for this data element.

7.4 *Geologic Record Group*—The geologic record group includes records for documenting data elements relating to geophysical logs, geologic units, and geologic samples of both unconsolidated and consolidated materials (**4, 5, 7, 8, 12–14, 16**).

7.4.1 *Geophysical Log Record*—The log record is used to enter information about types of geophysical or other logs available for the site. Data elements that are included in the geophysical log record are the following:

7.4.1.1 *Date of Log*—If applicable, document the date (year, month, day in YYYYMMDD format) that the geophysical log was completed at the ground-water site.

7.4.1.2 *Type of Log*—If applicable, document the type of log available for the hole. If more than one type of log was run on the well, document those with the corresponding depth intervals. Suggested type of log components and representative codes are as follows (**12**):

A—Drilling time
B—Casing collar
C—Caliper
D—Drillers
E—Electric
R—Single-point resistance
W—Spontaneous potential
Y—Multi-electrode
1—Acoustic velocity
2—Acoustic televiewer
F—Conductivity, fluid
G—Geologists or sample
H—Magnetic
I—Induction
J—Gamma ray
K—Dipmeter survey
L—Lateral log
M—Microlog
N—Neutron
O—Microlateral log
P—Photographic
Q—Radioactive-tracer
S—Sonic
T—Temperature
U—Gamma-gamma
V—Fluid velocity (flow)
X—Core
Z—Other (describe)

7.4.1.3 *Depth to Top of Logged Interval*—Enter the depth to the top of the logged interval, in feet or metres below a datum at or near land surface. Document the accuracy or confidence classification for this data element.

7.4.1.4 *Depth to Bottom of Logged Interval*—Enter the depth to the bottom of the logged interval, in feet or metres below a datum at or near land surface. Document the accuracy or confidence classification for this data element.

7.4.1.5 *Source of Log Data*—If applicable, document the source of the log information. Suggested source of depth data components and representative codes are as follows:

A—Reported by a government agency
D—From driller's log or report
G—Private geologist-consultant or university associate
L—Depth interpreted from geophysical logs by personnel of source agency
M—Memory (owner, operator, driller)
O—Reported from records by owner of well
R—Reported by person other than owner, driller, or another government agency
S—Measured by personnel of reporting agency
Z—Other source (describe)

7.4.2 *Geohydrologic Units Record*—The geohydrologic units record is used to document information about the rock material that yields water or is monitored at the ground-water site. Normally, information is gathered for all rock material encountered in drilling the well, both above and below the water-bearing aquifer unit. Geophysical log data (see 7.4.1) are commonly used to assist in the interpretation of rock material, especially in accurately defining depth intervals and fluid characteristics. The more common data elements that are included in the geohydrologic units record are the following:

NOTE 15—Usually, data describing rock material are documented sequentially (top to bottom) on strip charts. These strip charts can be simple field-compiled logs generated from visual examination of the rock material by a hydrogeologist or drilling engineer. Detailed strip charts are laboratory compiled from the examination of all the properties of the field-collected rock (drilling) samples. These properties include such characteristics as color, mineralogy, luster, structure, induration, inclusions, cementation, sorting, grain size, grain shape, porosity, hardness of material, solubility, etc.

7.4.2.1 *Aquifer Unit(s)*—Document the identification of the water-bearing aquifer unit or units from which the water is withdrawn or monitoring data are collected and the non-water-bearing units above and below the aquifer. See 6.1.8 for additional information on aquifer unit identification.

7.4.2.2 *Contributing Unit*—In combination with the aquifer identification, indicate how the unit is categorized as an aquifer. Suggested contributing unit components and representative codes are as follows:

P—Principal contributing aquifer
S—Secondary contributing aquifer
N—Contributes no water
U—Unknown contribution

7.4.2.3 *Depth to Top of Interval*—If applicable, document the depth, in feet or metres below a datum at or near land surface, to the top of this aquifer or non-water-bearing unit. Document the accuracy or confidence classification for this data element.

7.4.2.4 *Depth to Bottom of Interval*—If applicable, document the depth, in feet or metres below a datum or non-water-bearing unit. Document the accuracy or confidence classification for this data element.

7.4.2.5 *Lithology*—If applicable, document the lithology of the aquifer or non-water-bearing unit. Suggested lithology components and representative codes are as follows:

Rock Term	Abbreviation
Alluvium	ALVM
Anhydrite	ANDR
Anorthosite	ANRS
Arkose	ARKS
Basalt	BSLT
Bentonite	BNTN
Boulders	BLDR
Boulders and sand	BLSD
Boulders, silt, and clay	BLSC
Breccia	BRCC
Calcite	CLCT
Caliche (hard pan)	CLCH
Chalk	CHLK
Chert	CHRT
Clay	CLAY
Clay, some sand	CLSD
Claystone	CLSN
Coal	COAL
Cobbles	COBB
Cobbles and sand	COSD
Cobbles, silt, and clay	COSC
Colluvium	CLVM
Conglomerate	CGLM
Coquina	CQUN
Diabase	DIBS
Diorite	DORT

Rock Term	Abbreviation
Dolomite	DLMT
Drift	DRFT
Evaporite	EVPR
Gabbro	GBBR
Glacial (undifferentiated)	GLCL
Gneiss	GNSS
Granite	GRNT
Granite, gneiss	GRGN
Gravel	GRVL
Gravel and clay	GRCL
Gravel, cemented	GRCM
Gravel, sand, and silt	GRDS
Gravel, silt, and clay	GRSC
Graywacke	GRCK
Greenstone	GNST
Gypsum	GPSM
Hard pan	HRDP
Igneous (undifferentiated)	IGNS
Lignite	LGNT
Limestone	LMSN
Limestone and Dolomite	LMDM
Loam	LOAM
Loess	LOSS
Marble	MRBL
Marl	MARL
Marlstone	MRLS
Metamorphic (undifferentiated)	MMPC
Muck	MUCK
Mud	MUD
Mudstone	MDSN
Other	OTHR
Outwash	OTSH
Overburden	OBDN
Peat	PEAT
Quartzite	QRTZ
Residium	RSDM
Rhyolite	RYLT
Rock	ROCK
Rubble	RBBL
Sand	SAND
Sand and clay	SDCL
Sand and gravel	SDGL
Sand and silt	SDST
Sand, gravel, and clay	SGVC
Sand, some clay	SNCL
Sandstone	SNDS
Sandstone and shale	SDSL
Saprolite	SPRL
Schist	SCST
Sedimentary (undifferentiated)	SDMN
Serpentine	SRPN
Shale	SHLE
Silt	SILT
Silt and clay	STCL
Siltstone	SLSN
Slate	SLTE
Soil	SOIL
Syenite	SYNT
Till	TILL
Travertine	TRVR
Tuff	TUFF
Volcanic (undifferentiated)	VLCC

7.4.2.6 *Description of Material*—If applicable, in combination with the lithology described in 7.4.2.5, document the adjective modifiers needed to describe the rock type of the aquifer or non-water-bearing unit. Use of meaningful abbreviations assists in condensing the description. Standard color guides are valuable in dictating consistency in describing the rock and soil material **(9, 24)**. Various guides are available as aid to standardizing the descriptions of rock and soil materials **(3)**. The following are examples of lithologic descriptions:

Example 1—For soft, chalky grey limestone, suggested description: LMSN, GREY, SOFT, CHALKY.

Example 2—For hard red sandstone, iron stained, suggested description: SNDS, HARD, RED, FE STND.

7.4.3 *Sample/Unconsolidated Material Record*—Samples of geologic materials commonly are collected when drilling holes are to be completed as monitoring or water wells. These samples are used to assist in the determination of the aquifer, vadose zone, and underlying material characteristics for evaluating the movement of water through these materials. Undisturbed unconsolidated samples are collected from driven or cored holes, while disturbed unconsolidated samples are collected from bored, rotary, and cable tool holes. This record is used to describe the geologic material in combination with data given in the sections on construction (see 7.2) and geohydrologic units (see 7.4.2).

NOTE 16—If samples are collected of soil materials that are located in the weathered zone (1.5 to 2.0 metres or 6 ft in depth) and below the weathered zone (an additional 1.5 to 2.0 metres or about 12 ft in depth) a description that is more comprehensive than the following may be required for the features that are found in these zones. Those additional features, when present, are: (*1*) texture (USDA and Unified estimated textures, coarse fragments), (*2*) sorting and roundness, (*3*) moisture condition (moist, wet, dry, presence of water table), (*4*) color and mottling, (*5*) consistency (rupture resistance, cementation), (*6*) secondary porosity features, (*7*) sedimentary structure, (*8*) presence of organic matter, and (*9*) effervescence in dilute HCl. The field pocket guide by Boulding **(3)** and Practice D 2488 presents an excellent summary of these features.

7.4.3.1 *Sample Weight*—If applicable, document the weight, in ounces or grams, of the sample of geologic material. Indicate whether this is wet or dry weight. The volume of the sample may be included so that bulk density can be determined. Document the accuracy or confidence classification for this data element.

7.4.3.2 *Sample Interval*—If applicable, document the depth interval, in feet or metres below a datum at or near land surface, of the sample of geologic material. Document the accuracy or confidence classification for this data element.

7.4.3.3 *Particle Size*—If applicable, document the particle or grain sizes of the unconsolidated geologic material. Indicate whether this is a visual determination using a hand lens or microscope, or a mechanical analysis using calibrated sieves. The grain sizes of sand or larger-sized materials are normally recorded in millimetres. While silt and clay-sized materials are recorded in micrometres (μm). Document the accuracy or confidence classification for this data element.

7.4.3.4 *Percent of Total Sample*—If applicable, document the percentage, by weight, of each particle size (see 7.4.3.3) contained in the total sample (see 7.4.3.1). Document the accuracy or confidence classification for this data element.

7.4.3.5 *Particle Shape*—If applicable, document the particle shape or roundness of the sampled particles. Common shape descriptions are rounded, sub-rounded, subangular, and angular. The shape is usually determined visually by use of a hand lens or microscope.

7.4.3.6 *Mineralogy*—If applicable, document the mineralogy of the particles in the sample. The mineralogy is usually determined visually by use of a hand lens or microscope, however, the exact composition must be determined in the laboratory by powder X-ray diffraction. The most common visually recognizable mineral is silica (SiO_2).

7.4.4 *Sample/Consolidated Material Record*—Samples of

geologic materials commonly are collected when drilling holes to be completed as monitoring or water wells. These samples are used to assist in the determination of the aquifer or vadose zone characteristics for evaluating the movement of water through these materials. Undisturbed consolidated samples are collected from cored holes, while disturbed consolidated samples are collected from bored, rotary, and cable tool holes. This record is used to describe the geologic material in combination with data given in the sections on construction (see 7.2) and geohydrologic units (see 7.4.2).

7.4.4.1 *Drill Cuttings or Core*—If applicable, document whether the samples of consolidated geologic material is undisturbed (cored) or disturbed (for example, rotary tool).

7.4.4.2 *Sample Size (Weight)*—If applicable, document the weight, in ounces or grams, of the sample of geologic material. Indicate whether this is wet or dry weight. The volume of the sample may be included so that bulk density can be determined. Document the accuracy or confidence classification for this data element.

7.4.4.3 *Sample Interval*—If applicable, document the depth interval, in feet or metres below a datum at or near land surface, of the sample of geologic material. Document the accuracy or confidence classification for this data element.

7.4.4.4 *Mineralogy*—If applicable, document the mineralogy of the sampled interval. Indicate whether the mineralogy was field (visually with hand lens or microscope) or laboratory determined by powder X-ray diffraction. This information supplements data given in 7.4.2.

7.4.4.5 *Core Length*—If applicable, document the length of core, if feet and inches or metres and centimetres, recovered from the sample interval (see 7.4.4.3). Document the accuracy or confidence classification for this data element.

7.4.4.6 *Core Diameter*—If applicable, document the core diameter, in inches or centimetres, recovered in this interval. Document the accuracy or confidence classification for this data element.

7.4.4.7 *Core Recovery-Percent*—If applicable, document the percentage of core recovery (core length/distance of sample interval). Document the accuracy or confidence classification for this data element.

7.4.4.8 *Bedding*—If applicable, document the type of bedding or plane of stratification of the geologic materials of the recovered core. This characteristic of the rock material can be critical in the interpretation of the movement of water.

7.4.4.9 *Structure*—If applicable, document any structure or microstructure apparent in the recovered core. This characteristic of the rock material can be critical in the interpretation of the movement of water.

7.4.4.10 *Porosity*—If applicable, document the apparent porosity of the recovered geologic material. Indicate whether the porosity was field (estimated visually with a hand lens or microscope) or laboratory determined. Document the accuracy or confidence classification for this data element.

7.5 *Hydraulic Record Group*—The hydraulic record group includes records for documenting data elements relating to hydraulic characteristics of the aquifer, including both the basic description and hydraulic parameters of the aquifer unit (**4, 5, 7, 8, 10–13, 16, 25**).

NOTE 17—Information given in 7.4 is commonly used to assist in the interpretation of hydraulic characteristics, especially in accuracy defining depth intervals and aquifer properties.

7.5.1 *Hydraulics Record*—The hydraulics record is used to document information about the rock unit that yields water or is monitored at the ground-water site. The more common data elements that are included in the hydraulics record are the following:

7.5.1.1 *Hydraulic/Aquifer Unit*—If applicable, document the identification of the aquifer unit or units from which the hydraulic data were determined. See 6.1.7 for additional information on aquifer unit identification.

7.5.1.2 *Hydraulic/Aquifer Unit Type*—If applicable, document the type of aquifer(s) tested for hydraulic characteristics at the ground-water site. These types normally are confined, unconfined, or a combination of confined and unconfined.

7.5.1.3 *Depth to Top of Unit*—If applicable, enter the depth to the top of the tested interval, in feet or metres below a datum at or near land surface. Document the accuracy or confidence classification for this data element.

7.5.1.4 *Depth to Bottom of Unit*—If applicable, enter the depth to the bottom of the tested interval, in feet or metres below a datum at or near land surface. Document the accuracy or confidence classification for this data element.

7.5.1.5 *Static Water Level*—If applicable, document the water level, in feet or metres below a datum at or near land surface, prior to the hydraulic test. For those water levels that are above the measuring-point datum (normally artesian wells), precede the value with a minus (−) sign to distinguish those water levels from ones at or below the measuring point. Document the accuracy or confidence classification for this data element.

7.5.1.6 *Measurement Date and Time*—If applicable, document the date (year, month, day, and time of day in YYYYMMDDHHMM format) of the measurement of the static water level.

7.5.1.7 *Unit Contribution*—If applicable, enter the percentage of the total yield of the site that is contributed by this hydraulic/aquifer unit, if known. If part of the water that the site would otherwise produce is lost to this unit, enter the percentage of the water lost preceded by a minus sign (−). Document the accuracy or confidence classification for this data element.

7.5.2 *Aquifer Parameters Record*—The aquifers parameters record is for the documentation of hydraulic characteristics as determined by testing of the aquifer or estimation by using data facts about the aquifer.

NOTE 18—Data gathered during aquifer testing and generated as a result of the interpretation process are sometimes extensive. The information discussed here are the final results.

7.5.2.1 *Transmissivity*—If applicable, document the transmissivity of the aquifer. Indicate whether this data element is estimated or is determined by an aquifer test. Document the aquifer test method and the accuracy or confidence classification for this data element.

NOTE 19—Transmissivity is the capability of an aquifer to transmit water of the prevailing kinematic viscosity in a unit time through a unit width of the aquifer under a unit hydraulic gradient, expressed in metres squared per day.

7.5.2.2 *Horizontal Hydraulic Conductivity*—If applicable,

document the hydraulic conductivity parallel to bedding of the aquifer. Indicate whether this data element is estimated or is determined by an aquifer test. Document the aquifer test method and the accuracy or confidence classification for this data element.

NOTE 20—Hydraulic conductivity is the capacity of the rock to transmit water, expressed as the volume of water at the existing kinematic viscosity that will move in unit time under a unit hydraulic gradient through a unit area measured at right angles to the direction of flow.

7.5.2.3 *Vertical Hydraulic Conductivity*—If applicable, document the vertical hydraulic conductivity of the aquifer. Indicate whether this data element is estimated or is determined by an aquifer test. Document the aquifer test method and the accuracy or confidence classification for this data element.

7.5.2.4 *Coefficient of Storage*—If applicable, document the coefficient of storage of the aquifer. Indicate whether this data element is estimated or is determined by an aquifer test. Document the aquifer test method and the accuracy or confidence classification for this data element.

NOTE 21—Coefficient of storage is the volume of water an aquifer releases from or takes into storage per unit surface area of the aquifer per unit change in head. For a confined aquifer, it is equal to the product of specific storage and aquifer thickness. For an unconfined aquifer, the storage coefficient is equal to the specific yield.

7.5.2.5 *Leakance*—If applicable, document the leakance of the confining unit in 1/day. Indicate whether this data element is from estimated hydraulic characteristics or from those determined by an aquifer test. Document the aquifer test method and the accuracy or confidence classification for this data element.

NOTE 22—Leakance (K′/b′) is the vertical hydraulic conductivity of the confining unit (K′) divided by the thickness of the confining unit (b′).

7.5.2.6 *Diffusivity*—If applicable, document the diffusivity of the aquifer. Indicate whether this data element is from estimated hydraulic characteristics or from those determined by an aquifer test. Document the aquifer test method and the accuracy or confidence classification for this data element.

NOTE 23—Diffusivity is the transmissivity divided by the storage coefficient (T/S in feet (metres) squared per day).

7.5.2.7 *Specific Storage*—If applicable, document the specific storage of the aquifer. Indicate whether this data element is from estimated hydraulic characteristics or from those determined by an aquifer test. Document the aquifer test method and the accuracy or confidence classification for this data element.

NOTE 24—Specific storage is the storage coefficient divided by the thickness of the aquifer. It is the volume of water the aquifer released from or taken into storage per unit volume of porous medium per unit change in head.

7.5.2.8 *Specific Yield*—If applicable, document the specific yield of the aquifer. Indicate whether this data element is from estimated hydraulic characteristics or from those determined by an aquifer test. Document the aquifer test method and the accuracy or confidence classification for this data element.

NOTE 25—Specific yield is the ratio of the volume of water that the saturated rock or soil will yield by gravity to the volume of the rock or soil. In the field, specific yield is generally determined by tests of unconfined aquifers and represents the change that occurs in the volume of water in storage per unit area of unconfined aquifer as the result of a unit change in head. Such a change in storage is produced by the draining or filling of pore space and is, therefore, mainly dependent on particle size, rate of change of the water table, and time of drainage.

7.5.2.9 *Barometric or Tidal Efficiency*—If applicable, document the barometric or tidal efficiency of the aquifer. Document the accuracy or confidence classification for this data element.

NOTE 26—Barometric efficiency is the ratio of the change in depth to water in a well to the inverse of water-level change in a water barometer.

7.5.2.10 *Porosity*—If applicable, document the porosity of the aquifer. Indicate whether this data element is estimated or determined by an aquifer test. Document the aquifer test method and the accuracy or confidence classification for this data element.

NOTE 27—Porosity of the aquifer is its property of containing interstices or voids and may be expressed quantitatively as the ratio of the volume of the interstices to the total volume.

7.5.2.11 *Specific Capacity*—If applicable, document the specific capacity of the ground-water site. Indicate whether this data element is estimated or determined by a test. Document the aquifer test method and the accuracy or confidence classification for this data element.

NOTE 28—Specific capacity is the rate of discharge from a well divided by the drawdown of the water level within the well at a specific time since pumping started.

7.5.2.12 *Method Used to Determine Aquifer Characteristics*—If applicable, document the method used to determine the hydraulic characteristics of the aquifer. The following are some generalized descriptions for the method used to determine aquifer characteristics components and their corresponding representative codes. If the specific aquifer test method is known, that information should be documented.

B—Bail test
C—Controlled single-well test methods
D—Controlled multiple-well test methods
N—Natural ground-water fluctuation
P—Cyclic pumping—single well
W—Cyclic pumping—multiple wells
S—Slug test
E—Estimated (describe)

7.5.2.13 *Availability of File of Detailed Results*—If applicable, document the availability and format of a file of detailed aquifer test results. Suggested availability and format of a file of aquifer test components and representative codes are as follows:

F—Files (raw data)
M—Machine readable
P—Published (report or basic-data release)
Z—Other (describe)

7.6 *Spring Record Group*—The spring record group is a single record for documenting data elements relating to properties of a spring, including both the basic description and flow characteristics of the ground-water source (**4, 5, 7, 8, 13, 19, 26**).

7.6.1 *Spring Record*—The spring record includes all information that relates to the description of a ground-water site that is determined to be a spring. A spring is defined as a place where ground water flows naturally from a rock or the soil onto the land surface or into a body of surface water. Data elements that are included in the spring record are the following:

NOTE 29—The spring record includes the data elements that relate directly to the properties of a spring, however, to completely document a spring site additional data elements need to be selected from those described in the three "Standard Guides." For example, the aquifer identification and lithology of the unit(s), along with the location map and photographs or sketches of the site, contribute additional information in interpreting the hydrology of the area. Most of the data elements narrated in Guide D 5408 can apply to springs, as well as applicable portions of this guide and Guide D 5410.

7.6.1.1 *Name of Spring*—If applicable, document the name by which the spring is known locally or, preferably, displayed on a published map.

7.6.1.2 *Type of Spring*—If applicable, document the type of spring at the ground-water site. Suggested type of spring components and representative codes are as follows (13):

A—Artesian
J—Artesian and depression
K—Artesian and seepage or filtration
C—Contact
D—Depression
F—Fracture or fault
L—Fracture fault and depression
G—Geyser
B—Perched and contact
E—Perched and depression
H—Perched and tubular
O—Perched and fracture
P—Perched
R—Perched and seepage or filtration
S—Seepage or filtration
T—Conduit or tubular (cave)
Z—Other (describe)

7.6.1.3 *Permanence of Spring*—If applicable, document the permanence of the spring at the ground-water site. Suggested permanence of spring components and representative codes are as follows:

E—Periodic—Ebb and flow, normally have periods of relatively greater discharge at regular and frequent intervals.
G—Geyser—Discharge at more or less regular intervals. Nature of discharge is caused by expansive force of highly heated steam.
P—Perennial—Springs that discharge continuously.
I—Intermittent—Springs that discharge only during certain periods but at other times are dry.
R—Response to precipitation—Exist only after periods of rainfall.
S—Seasonal—Exist only during periods of high-water levels.
T—Estavelle—A cave that is a spring during some periods and a sinking stream (swallet) during other periods.
Z—Other (describe).

7.6.1.4 *Sphere of Discharge*—If applicable, document the sphere of discharge of the spring. Suggested sphere of discharge components and representative codes are as follows:

A—Subaerial—Discharges at lands surface
W—Subaqueous—Discharges under water

7.6.1.5 *Discharge*—If applicable, document the discharge value for the spring in gallons per minute, cubic feet (or metres) per second, litres per second or any other standard volume/time unit. Document the volume/time unit used. Note whether the discharge is clear or sometimes turbid. Document the accuracy or confidence classification for this data element.

7.6.1.6 *Date of Discharge*—If applicable, document the date (year/month/day/time of day in YYYYMMDD-HHMM format) that the discharge was measured. If the

discharge value is averaged over a period of time, explain.

7.6.1.7 *Improvements*—If applicable, document the type of improvements at the spring of the ground-water site. Suggested improvement components and representative codes are as follows:

B—Boxed or small covered basin
C—Concrete basin
G—Gallery
H—Spring house
L—Lined
N—None
P—Pond
R—Pipe (not for conduction of water from spring)
T—Trough
Z—Other (describe)

7.6.1.8 *Number of Spring Openings*—If applicable, document the number of openings through which water discharges from the spring. Document the accuracy or confidence classification for this data element.

7.6.1.9 *Flow Variability*—If known or applicable, document the discharge variability of the spring, in percent, as expressed by the formula:

$$V = 100 \times [(a - b)/c]$$

where:
V = variability, %,
a = maximum discharge,
b = minimum discharge, and
c = average discharge.
Document the accuracy or confidence classification for this data element.

7.6.1.10 *Accuracy of Flow Variability*—If applicable, document the basis on which the variability of the spring was determined. Suggested accuracy of flow variability components and representative codes are as follows:

A—Calculated from less than one year of continuous discharge record.
B—Calculated from one to five years of continuous discharge record.
C—Calculated from more than five years of continuous discharge record.
D—Calculated from intermittent measurements made over a period of more than one year.
E—Calculated from less than one year of record, or estimated.
Z—Determined by other method (describe).

7.6.1.11 *Magnitude of Spring*—If applicable, document the magnitude of the spring. Document the accuracy or confidence classification for this data element. The following is a magnitude classification that has been used in the United States and was published by Meinzer (19, 26).

Magnitude	Inch-Pound Units	SI (Metric) Units
First	Greater than 100 ft³/s	>10 m³/s
Second	10 to 100 ft³/s	1 to 10 m³/s
Third	1 to 10 ft³/s	0.1 to 1 m³/s
Fourth	100 gal/min to 1 ft³/s	10 to 100 l/s
Fifth	10 to 100 gal/min	1 to 10 l/s
Sixth	1 to 10 gal/min	0.1 to 1 l/s
Seventh	1 pint to 1 gal/min	10 to 100 ml/s
Eighth	Less than 1 pint/min	<10 ml/s

NOTE 30—Spring magnitudes were commonly used in the older hydrogeologic literature to classify the approximate discharge from a spring. This terminology is rarely used in the current hydrology reports. The common unit for measurement in recent literature is a cubic-foot per second, that translates to 646 317 gal (2 446 310 L) per day. Meinzer used the second-foot measurement that is equal to 646 000 gal (2 445 110 L) of water per day.

8. Keywords

8.1 aquifer test; confidence classification; drainage basin;

geohydrologic unit; geological sample; geophysical log; ground water; ground-water discharge; ground-water site; ground-water site construction; monitoring site; source agency; spring site; surface-water site; well site

REFERENCES

(1) Bates, Robert L., and Jackson, Julia A., *Glossary of Geology*, Third Edition; American Geological Institute, Alexandria, Virginia, 1987.

(2) Anderson, Keith E., *Water Well Handbook*; Fourth Edition: Missouri Water Well and Pump Contractors Association, Inc., Rolla, Missouri, 1971.

(3) Boulding, J. R., *Description and Sampling of Contaminated Soils, A Field Pocket Guide:* Center for Environmental Research Information, U.S. EPA, EPA/625/12-91/002, Cincinnati, Ohio, November 1991.

(4) Bureau of Reclamation, *Ground-Water Manual, A Water Resources Technical Publication*, Revised Reprint: U.S. Department of Interior, Bureau of Reclamation, Washington, DC, 1981.

(5) Campbell, Michael D., and Lehr, Jay H., *Water Well Technology*: McGraw-Hill, New York, NY, 1973.

(6) Cohee, George V., *Standard Stratigraphic Code Adopted by AAPG, Committee on Standard Stratigraphic Coding: American Association of Petroleum Geologists Bulletin*, Vol 51, No. 10, October 1967, p. 2146–2151.

(7) Fetter, C. W., *Applied Hydrogeology*, Second Edition: Macmillan Publishing Company, 866 Third Avenue, New York, NY, 1988.

(8) Freeze, R. A., and Cherry, J. A., *Groundwater*, Prentice Hall, Englewood Cliffs, NJ, 1975.

(9) Geological Society of America, *Rock Color Chart*, 7th Printing: Geological Society of America, 3300 Penrose Place, PO Box 9140, Boulder, Colorado 80301-9140, 1991.

(10) *Handbook of Ground-Water Development:* Roscoe Moss Co., Los Angeles, CA, Published by John Wiley and Sons, Inc., New York, NY, 1990.

(11) Heath, Ralph C., *Basic Ground-Water Hydrology: U.S. Geological Survey Water-Supply Paper 2220*, 1983.

(12) Keys, W. S., *Borehole Geophysics Applied to Ground-Water Investigations: U.S. Geological Survey Techniques of Water-Resource Investigations*, TWRI 2-E2, 1990.

(13) Mathey, Sharon B. (Editor), *National Water Information System User's Manual*, Vol 2, Chapt. 4. Ground-Water Site Inventory System: U.S. Geological Survey, Open-File Report, 1990, pp. 89–587.

(14) Texas Natural Resources Information System, *Ground-Water Data INTERFACE, Users Reference Manual*, Texas Natural Resources Information System, Nov. 20, 1986.

(15) U.S. Environmental Protection Agency, *Ground Water Monitoring in SW-846, Field Manual Physical/Chemical Methods, Test Methods for Evaluating Soil Wastes*, Vol II, Chapt. 11, Third Edition; Office of Solid Wastes and Emergency Responses, U.S. EPA, Washington, DC, 1991.

(16) U.S. Environmental Protection Agency, *Handbook of Suggested Practices for the Design and Installation of Ground-Water Monitoring Wells*: Office of Research and Development, U.S. EPA, EPA/600/4-89/034 Washington, DC, March 1991.

(17) U.S. Geological Survey, *National Handbook of Recommended Methods for Water-Data Acquisition*, Chapter 2—Ground Water: Office of Data Coordination, Reston, Virginia, 1980.

(18) U.S. Geological Survey, *Guide to Obtaining USGS Information: U.S. Geological Survey Circular 900*, 1989.

(19) van der Leedan, Frits, Troise, Fred L., and Todd, David Keith, *The Water Encyclopedia*, Geraghty and Miller Ground-Water Series, 2nd Edition, Third Printing, Lewis Publishers, Inc., Chelsea, Michigan, 1991.

(20) U.S. Department of Agriculture, *Atlas of River Basins of the United States: Soil Conservation Service*, 82 Maps, 1970.

(21) American National Standards Institute, Inc., American National Standard for Information Systems—*Representations of Local Time of Day for Information Interchange*; American National Standards Institute, Inc. Publication ANSI X3.43, 1430 Broadway, New York, NY 10018, 1986.

(22) U.S. Department of Commerce, *Representation of Local Time of Day for Information Interchange*, Federal Information Standards (FIPS) Publication 58-1, National Bureau of Standards, Washington, DC, Jan. 27, 1988.

(23) Morgan, Charles O., *Transition from the Ancient Underground Falaj to the Modern Pumped Well in Oman: In Minimizing Risk to the Hydrologic Environment*, Proceedings of the American Institute of Hydrology Conference held in Las Vegas, NV, March 13–15, 1990, pp. 155–160.

(24) MacBeth Division of Collmorgen Instrument Corp., *Munsell Soil Color Charts*, revised edition: MacBeth Division of Collmorgen Instrument Corp., PO Box 230, Newburgh, NY 12551-0230, 1990.

(25) Lohman, S. W., Ground-Water Hydraulics: U.S. Geological Survey Professional Paper 708, 1972.

(26) Meinzer, O. E., *Large Springs in the United States: U.S. Geological Survey Water-Supply Paper 557*, 1927.

Standard Guide for
Set of Data Elements to Describe a Ground-Water Site; Part Three—Usage Descriptors[1]

This standard is issued under the fixed designation D 5410 ; the number immediately following the designation indicates the year of original adoption or, in the case of revision, the year of last revision. A number in parentheses indicates the year of last reapproval. A superscript epsilon (ε) indicates an editorial change since the last revision or reapproval.

1. Scope

1.1 This guide is Part Three of three guides to be used in conjunction with Practice D 5254 that delineates the data desirable to describe a ground-water data collection or sampling site. This guide identifies usage descriptors, such as monitoring, for an individual ground-water site. Guide D 5408 describes additional information beyond the minimum set of data elements that may be specified to identify a ground-water site, while Guide D 5409 identifies physical descriptors, such as construction, for a site.

NOTE 1—A ground-water site is defined as any source, location, or sampling station capable of producing water or hydrologic data from a natural stratum from below the surface of the earth. A source or facility can include a well, spring or seep, and drain or tunnel (nearly horizontal in orientation). Other sources, such as excavations, driven devices, boreholes, ponds, lakes, and sinkholes, that can be shown to be hydraulically connected to the ground water, are appropriate for the use intended.

NOTE 2—Guide D 5408 includes data confidence classification descriptor (1 element), geographic location descriptors (4 elements), political regime descriptor (1 element), source identifier descriptors (4 elements), legal descriptors (9 elements), owner descriptors (2 elements), site visit descriptors (3 elements), other identification descriptors (2 elements), other data descriptors (3 elements), and remarks descriptors (3 elements). Guide D 5409 includes individual site characteristics (7 data elements), construction descriptors (56 data elements), lift descriptors (16 data elements), geologic descriptors (26 data elements), hydraulic descriptors (20 data elements), and spring descriptors (11 data elements). For a list of descriptors in this guide, see Section 3.

1.2 These data elements are described in terms used by ground-water hydrologists. Standard references, such as Ref (1)[2] and various hydrogeologic professional publications, are used to determine these definitions. Many of the suggested elements and their representative codes are those established by the Water Resources Division of the U.S. Geological Survey and used in the National Water Information Systems computerized data base[3] (1–21).

NOTE 3—The purpose of this guide is to suggest data elements that can be collected for ground-water sites. This does not uniquely imply a computer data base, but rather data elements for entry into any type of permanent file.

NOTE 4—Component and code lists given with some of the data elements, for example "method of discharge measurement," are only suggestions. These lists can be modified, expanded, or reduced for the purpose intended by the company or agency maintaining the ground-water data file.

NOTE 5—Use of trade names in this guide is for identification purposes only and does not constitute endorsement by ASTM.

1.3 This guide includes the data elements desirable to document a ground-water site beyond those given in the minimum set of data elements. Some examples of the data elements are water level, discharge, and water-quality sample collection date. No single site will need every data element, for example, a monitoring site may not need a long-term water use record. Each record (group of data elements) for a site has mandatory data elements, such as the date for the water level record. However, these elements are considered necessary only when that specific record is gathered for the site.

1.4 The values stated in both inch-pound and SI units are to be regarded separately as the standard. The values given in parentheses are for information only.

1.5 *This standard does not purport to address all of the safety problems, if any, associated with its use. It is the responsibility of the user of this standard to establish appropriate safety and health practices and determine the applicability of regulatory limitations prior to use.*

2. Referenced Documents

2.1 *ASTM Standards:*
D 653 Terminology Relating to Soil, Rock, and Contained Fluids[4]
D 5254 Practice for the Minimum Set of Data Elements to Identify a Ground-Water Site[5]
D 5408 Guide for Set of Data Elements to Describe a Ground-Water Site, Part One—Additional Identification Descriptors[5]
D 5409 Guide for Set of Data Elements to Describe a Ground-Water Site, Part Two—Physical Descriptors[5]

3. Terminology

3.1 *Definitions:*
3.1.1 For definitions of terms applicable to this guide, see Terminology D 653.

3.2 *Descriptions of Terms Specific to This Standard:*
3.2.1 *code*—a suggested abbreviation for a component, for example, "T" is the code suggested for the "electric tape" component of data element method of measurement.

3.2.2 *component*—a subdivision of a data element, for example, "electric tape" is one of 14 components suggested for data element method of measurement.

3.2.3 *data element*—an individual segment of informa-

---

[1] This guide is under the jurisdiction of ASTM Committee D-18 on Soil and Rock and is the direct responsibility of Subcommittee D18.21 on Ground Water and Vadose Zone Investigations.
Current edition approved May 15, 1993. Published November 1993.
[2] The boldface numbers given in parentheses refer to the list of references at the end of the text.
[3] Guide for the Decommissioning of Ground-Water Wells, Vadose Zone Monitoring Devices, Boreholes, and Other Devices for Environmental Activities (draft); ASTM Subcommittee D18.21.06 on Well Maintenance, Rehabilitation, and Abandonment Section, August 1991.

[4] *Annual Book of ASTM Standards*, Vol 04.08.
[5] *Annual Book of ASTM Standards*, Vol 04.09.

tion about a ground-water site, for example, "method of measurement." The data element is in the water-level record record.

3.2.4 *record*—denotes a set of related data elements that may need to be repeated to fully describe a ground-water site. For example, a ground-water monitoring site where water levels are measured periodically will need more than one water-level record record (the record includes data elements date of measurement, date accuracy, water level, water-level accuracy, status, method of measurement, instrumentation, and statistics method) to fully document the water-level history of the site. However, if only a single water level was measured for the site, the record is utilized once.

3.2.5 *record group*—a set of related records. For example the "monitoring record group" includes the measuring point record, network record, water level record, discharge record, water use record, water quality record, and field water quality record. Some record groups consist of only one record, for example, the "irrigation record group" includes only the irrigation record.

4. Summary of Guide

4.1 This guide includes the following usage descriptor data elements to describe a ground-water site. This guide includes only repeated elements that commonly require several records to fully describe the conditions and history of the site:

Monitoring Record Group:
  Network Record:
    Data Type
    Date Interval of Network Utilization
    Source Agency for Network Data
    Frequency of Data Collection
    Method of Data Acquisition
    Power Type of Instruments
    Network
  Measuring-Point Record:
    Date Interval of Measuring-Point Utilization
    Height in Reference to Datum
    Description
  Water-Level Record:
    Measurement Date and Time
    Water Level
    Water-Level Accuracy
    Status
    Method of Measurement and Instrumentation
    Instrumentation
    Source of Data
    Statistics Method
  Discharge Record:
    Measurement Date and Time
    Discharge
    Type of Discharge
    Source of Data
    Method of Discharge Measurement
    Instrumentation
    Production or Pumping Level
    Static Level
    Method of Water-Level Measurement
    Pumping Period
    Specific Capacity
    Drawdown
    Source of Water-Level Data
  Water-Use Record:
    Date Range of Water-Use Record
    Data Collection Interval
    Long-Term Water Use
    Method Used to Determine Long-Term Water Use
  Water-Quality Record:
    Sample Date and Time
    Agency That Analyzes Samples

    Type of Analyses
    Parameters Requested for Analysis
    Sample Depth/Interval
    Water-Quality File Containing Analysis
    Laboratory Number
    Laboratory Name
    Replicate Sequence Number
    Collecting Agency or Company
    Agency or Company Code
    Chain of Custody
    Sampling Purpose
    Site Condition
    Sample Appearance
    Sample Odor
    Presence of Immiscible Stratum
    Thickness of Immiscible Stratum
    Sensors
    Sample Preservation Method
    Sample Filtration Material
    Pumped Period
    Casing Volume
    Amount of Water Purged
    Sampling Method or Sampler Type
    Sampler Material
    Aquifer Sampled
    Regulating Agency
  Field Water-Quality Record:
    Field Sample Date and Time
    Parameter Code
    Value of Parameter
    Unit for Parameter
    Instrumentation or Method of Determination
  Monitoring Site at Waste-Facility Record:
    Date Interval in Service
    State Regulatory Agency
    State Registration Identification
    EPA Registration Identification
    Responsible Company
    Company's Site Identification
    Site Location in Relationship to Waste Facility
    Status of Site
    Sampled Interval
Irrigation Record Group:
  Irrigation Record:
    Irrigated Land Area
    Allowance for Irrigating
    Date Legal Irrigation Begins
    Date Legal Irrigation Ends
Decommissioning Record Group:
  Decommissioning Record:
    Date Decommissioned
    Method Used for Decommissioning
    Reason for Decommissioning
    Plugging Material
    Name and Address of Decommissioner
    Step-by-Step Procedures
    Availability of Decommissioning Report
    Regulations Followed, Federal, State, Local

5. Significance and Use

5.1 Data at ground-water sites are gathered for many purposes. Each of these purposes generally requires a specific set of data elements. For example, when the ground-water quality is of concern not only are the minimum set of data elements required for the site, but information concerning the sample collection depth interval, method of collection, and date and time of collection are needed to fully qualify the data. Another group of elements are recommended for each use of the data, such as aquifer characteristics or water-level records. Normally the more information that is gathered about a site by field personnel, the easier it is to understand the ground-water conditions and to reach valid conclusions and interpretations regarding the site.

5.2 The data elements listed in this guide and Guides

D 5408 and D 5409 should assist in planning what information can be gathered for a ground-water site and how to document these data.

NOTE 6—Some important data elements may change during the existence of a site. For example, the elevation of the measuring point used for the measurement of water levels may be modified because of repair or replacement of equipment. This frequently occurs when the measuring point is an opening in the pump and the pump is modified or replaced. Because changes cannot always be anticipated. It is preferable to reference the height of the measuring point to a nearby, permanent altitude datum. The measuring point is referenced by being the same altitude (zero correction) or above (negative correction) or below (plus correction) the altitude datum. All appropriate measurements should be corrected in reference to the altitude datum before entry into the permanent file. Care must be exercised to keep the relationship of these data elements consistent throughout the duration of the site.

5.3 Some data elements have an extensive list of components or possible entries. For example the aquifer identification list described in Guide D 5409 has over 5000 entries. Lengthy lists of possible entries are not included in this guide, however, information on where to obtain these components is included with the specific data element.

NOTE 7—This guide identifies sources, lists, etc. of information required to completely document information about any ground-water site.

6. Documentation

6.1 Introduction:

6.1.1 Many of the ground-water data elements require multiple records or entries to completely describe a site. Time-related elements, such as water levels, discharge measurements, and water chemistry, may present hundreds or thousands of records for a period of many years that document measurements at a single site. These time-related data help to determine historical trends and serve to establish bench-mark facts for the site.

6.1.2 Other data elements that are not time related, such as casing lengths, spring openings, and an array of geophysical logs, require a sequence of records to thoroughly describe the site. These data are extremely valuable in site characterization, for example, wells for which the construc-

tion components are required to understand the source of the water.

6.2 Monitoring Record Group:

6.2.1 Introduction—The monitoring record group includes records for documenting data elements relating to any type of information gathered at a ground-water site for the purpose of monitoring hydrologic, usage, and water-quality trends.

6.2.2 Network Record—The network record includes the data elements that describes a ground-water site as it relates to a node in a hydrologic, usage, or water-quality network (10).

NOTE 8—An example of a form (see Fig. 1) for documenting the data elements as described for two records of the monitoring record group is illustrated to show a method of design for this tool. These forms are commonly known as "field forms" or as "coding forms" (for computer entry). This type of form is routinely used for transcribing field data while at the ground-water site and entering non-field information at the agency's or company's office. It should be noted that each form has the site identification (primary identification as used by the agency or company), date of field visit, and person that recorded the data as the first entries. These three data items are mandatory to ensure correct filing of the information, either in cabinets or in a computer data base, and for quality control.

6.2.2.1 Data Type—If applicable, document the type of monitoring conducted or information collected at the ground-water network site. The information identified as continuous are from automatic recording devices, such as water-level recorders. Suggested network data type components and representative codes are as follows:

A—Water quality, analyzed in the laboratory
B—Water quality, analyzed in the field
C—Water levels, continuous
D—Water levels, intermittent
E—Water discharge, continuous
F—Water discharge, intermittent
G—Prediction and detection of earthquakes
H—Land compaction for subsidence
I—Vadose-zone water pressure
Z—Other (describe)

NOTE 9—A ground-water site can be used as a node in a network for monitoring phenomenon other than the common water-quality, water-level, and discharge parameters. For example, in some areas of the

FIG. 1 Sample Form

country, wells are used indirectly as part of programs to monitor the subsidence of the land surface as a result of collapse or compaction resulting from extraction of liquid from the underground material. Wells and springs have been used in past research as monitoring sites to predict and detect earthquakes by chemical changes in the water and by water-level fluctuations.

6.2.2.2 *Date Interval of Network Utilization*—If applicable, document the date interval, in year, month, and day (YYYYMMDD), that the ground-water site was used for network monitoring. If still in use, document the date that the site was put into service.

6.2.2.3 *Source Agency for Network Data*—If applicable, document the name and address of the agency responsible for the collection of the ground-water data at the network monitoring site.

6.2.2.4 *Frequency of Data Collection*—If applicable, document the frequency of data collection at the network monitoring site. Suggested frequency of data collection components and representative codes are as follows:

C—Continuously (analog recorder)
J—Fixed interval (digital recorder) give interval
K—Variable interval (digital recorder) give interval
H—Hourly
D—Daily
W—Weekly
F—Semimonthly (twice a month)
M—Monthly
B—Bimonthly (every two months)
Q—Quarterly
S—Semiannually
A—Annually
2—Biennially (every two years)
3—Every three years
4—Every four years
5—Every five years
X—Every ten years
O—One time only
I—Intermittently or variable time scale
Z—Other (describe)

6.2.2.5 *Method of Data Acquisition*—If applicable, document the method of data collection at the network monitoring site. Suggested method of data collection components and representative codes are as follows:

A—Automated instruments accessed by field personnel
B—Automated instruments accessed by direct line
D—Automated instruments accessed by radio
G—Automated instruments accessed by remote data transmission
F—Periodic field visits by agency or company personnel
C—Calculated from records of owner
E—Estimated from other records
U—Unknown
Z—Other (describe)

6.2.2.6 *Power Type of Instruments*—If applicable, document the power type of permanently mounted data collection instruments at the ground-water monitoring site in the network. A detailed description of the instruments should be included with each specific type of monitoring (for example, digital recorder with water levels). Suggested power type of instrument components and representative codes are as follows:

M—Mechanical
B—Battery operated
S—Spring driven
E—Electrical

R—Solar batteries
Z—Other (describe)

6.2.2.7 *Network*—If applicable, document the areal extent or management level of the network that includes the ground-water monitoring site. Document any additional networks that may include the site. Suggested network data components and representative codes are as follows:

N—National
S—State or Province
R—Regional, multiple state or county
C—County
D—Drainage basin
P—Project
Z—Other (describe)

6.2.3 *Measuring-Point Record*—The measuring point represents a convenient position at a ground-water facility to reference repeated measurements, such as water levels. For some ground-water sites the measuring point is at the same location as the altitude datum (10).

NOTE 10—The altitude of the datum is described as one of the geographic locational data elements in the "Standard Practice for the Minimum Set of Data Elements to Identify a Ground-Water Site." This record describes the relationship between the measuring point and the datum, thereby linking measurements made at the site to the third dimension, the altitude. The datum normally remains the same for the life of the site; the measuring point can change as the water-withdrawal facilities are modified.

6.2.3.1 *Date Interval of Measuring-Point Utilization*—If applicable, document the date interval, in year, month, and day (YYYYMMDD), that the measuring point was used for conducting measurements at the ground-water site. If still in use, document the date that the measuring point was first used.

NOTE 11—The measuring point can be modified because of changing conditions at the ground-water site, therefore, several different measuring points may be used over the life of the site. These changes must be dated in order to relate to time-significant information, such as ground-water levels, that are collected over the history of the site.

6.2.3.2 *Height in Reference to Datum*—Document the height of the measuring point above or below the datum, in feet or metres. If the position of the measuring point is the same as the datum, the value is 0.0. Document the accuracy or confidence classification for this data element.

6.2.3.3 *Description*—Document a detailed description of the measuring point in relationship to the ground-water withdrawal facilities. A sketch or photograph of the facility and measuring point is valuable as an aid to future identification of the location.

6.2.4 *Water-Level Record*—This record is used to document water-level measurements of ground-water sites. Each water-level record requires most of the following data elements to thoroughly document the event. A single site may have thousands of records (3–6, 9, 10, 12, 18, 19).

6.2.4.1 *Measurement Date and Time*—Document the date and time of day (standard time and 2400 clock) of the water-level measurement. Many historical measurements do not have the time of day information and are accurate only to the nearest day. Some measurements are only accurate to the nearest year, however are extremely valuable in documenting long-term water-level trends. Unless obvious, document the exactness of the date of measurement. An example is "193800000000" for the year 1938 without the month, day, and time (2, 13).

6.2.4.2 *Water Level*—If applicable, document the water level, in feet or metres, in reference to the measuring point, for the ground-water site. For water levels that are above the measuring point (normally artisan wells), precede the value with a minus (−) sign to distinguish those water levels from ones at or below the measuring point. For those events where the condition at the site is dry, plugged, discontinued, certain flowing situations, or the site is destroyed, the water level is not transcribed (see 6.2.4.4). Document the accuracy or confidence classification for this data element (see 6.2.4.3 for suggestions).

6.2.4.3 *Water-Level Accuracy*—Document the accuracy of the water level as an aid for interpretation. Suggested water-level accuracy components and representative codes are as follows:

    0—Accurate to nearest foot or metre
    1—Accurate to nearest 1/10th ft or 1 cm
    2—Accurate to nearest 1/100th ft or 1 mm
    9—Unknown
    Z—Other (describe)

6.2.4.4 *Status*—If applicable, document the status of the ground-water site as it relates to the water level. Suggested status components and representative codes are as follows:

A—Static water level (site is in equilibrium)
D—Site dry (no water level recorded)
E—Flowed recently
F—Site flowing (water level could not be measured) (no water level recorded)
G—Nearby site tapping same aquifer was flowing
H—Nearby site tapping same aquifer flowed recently
I—Injector site (recharge water being injected)
J—Injector site monitor (nearby site tapping same aquifer injecting recharge water)
N—Measurements discontinued
O—Obstruction encountered in well above water surface (no water level recorded)
P—Site being pumped
R—Site pumped recently
S—Nearby site tapping same aquifer being pumped
T—Nearby site tapping same aquifer pumped recently
V—Foreign substance present on surface of water
W—Site destroyed
X—Affected by stage in nearby surface-water body
Z—Other conditions affect water level (describe)

6.2.4.5 *Method of Measurement and Instrumentation*—If applicable, document the instruments and method designating the means by which the water level was measured. Suggested method of measurement components and representative codes are as follows:

    A—Air line
    B—Analog or graphic recorder
    C—Calibrated air line
    E—Estimated
    F—Fiberglass tape
    G—Pressure gage, mechanical
    H—Calibrated pressure gage
    I—Interface probe
    L—Interpreted from geophysical logs
    M—Manometer
    N—Nonrecording gage (for example, staff gage)
    P—Pressure transducer with data logger
    R—Reported, method not known
    Q—Sonar sounder

    S—Steel tape
    T—Electric tape
    V—Calibrated electric tape
    Z—Other (describe)

6.2.4.6 *Instrumentation*—If applicable, document the type of all permanently mounted data collection instruments at the ground-water site used for water-level monitoring. Suggested instrument components and representative codes are as follows:

    B—Bubble gage
    C—Crest-stage gage
    D—Digital recorder (mechanical or electronic)
    E—Continuous-record type recorder
    G—Graphic or analog recorder
    I—In situ, without readout or data logger
    M—Data logger
    P—Pressure transducer
    R—Radio relay
    S—Satellite relay
    T—Telemetry
    Z—Other (describe)

6.2.4.7 *Source of Data*—If applicable, document the source of the discharge data. Suggested source of data components and representative codes are as follows:

A—Government agency
C—Consultant
D—Driller's log or report
L—Personnel of source agency or company
M—Memory (owner, operator, driller)
O—Records by owner
R—Person other than owner, driller, or another government agency
S—Personnel of reporting agency or company
U—University associate
Z—Other source (describe)

6.2.4.8 *Statistics Method*—If applicable, document the method which describes how the measurement was selected from a continuous recorder (for example, analog, digital or micrologger) available for that day. Suggested statistics code components and representative codes are as follows:

M—Water level shown is a daily maximum (for example, deepest water level for the day)
N—Water level shown is a daily minimum (for example, shallowest water level for the day)
A—Water level is 12:00 noon reading
P—Water level is 12:00 midnight reading
N—Mean, daily, monthly, etc. (specify)
O—Other (describe)

6.2.5 *Discharge Record*—This record is used to document instantaneous discharge measurements for ground-water sites. Each discharge record requires most of the following data elements to thoroughly document the measurement. A single site may have many records (3–6, 10, 12, 18, 19).

6.2.5.1 *Measurement Date and Time*—Document the date and time of day (standard time and 2400 clock) of the instantaneous discharge measurement. Many historical measurements do not have the time of day information and are accurate only to the nearest day. Some measurements, accurate to only the nearest year, are extremely valuable in documenting long-term water usage. Unless obvious, document the accuracy of the date of measurement.

6.2.5.2 *Discharge*—Document the discharge value for the ground-water site in gallons per minute, cubic feet (or metres) per second, litres per second or any other standard volume/time unit. This value must correspond to the remainder of the discharge event record for the site. Document the volume/time unit used. Document the accuracy or confidence classification for this data element.

6.2.5.3 *Type of Discharge*—If applicable, document the method of discharge. Suggested type of discharge components and representative codes are as follows:

P—Pumped
F—Flow
Z—Other (describe)

6.2.5.4 *Source of Data*—If applicable, document the source of the discharge and related (water level) data. Suggested source of the data components and representative codes are as follows:

A—Government agency
C—Consultant
D—Driller's log or report
L—Personnel of source agency or company
M—Memory (owner, operator, driller)
O—Records by owner
R—Person other than owner, driller, or another government agency
S—Personnel of reporting agency or company
U—University associate
Z—Other source (describe)

6.2.5.5 *Method of Discharge Measurement*—If applicable, document the method used to measure the discharge. Suggested method of discharge measurement components and representative codes are as follows:

A—Acoustic or sonic meter (transient-time meter)
B—Bailer
C—Current meter; either propeller-type meter in the discharge pipe, or propeller- or cup-type meter in the discharge channel
D—Doppler meter
E—Estimated
F—Flume
M—Totaling meter
O—Orifice
P—Pitot-tube meter, includes Cox meter, Collins meter, and the like
R—Reported, method not known
T—Trajectory method (free-fall method)
U—Venturi meter
V—Volumetric; bucket or barrel and stopwatch
W—Weir
Z- -Other (describe)

6.2.5.6 *Instrumentation*—Ground-water sites used for discharge monitoring can have permanently mounted data collection instruments. Document the type and use of all instruments at the site as follows:

D—Digital recorder (mechanical and electronic)
E—Continuous-record type recorder
G—Graphic or analog recorder
M—Data logger
R—Radio relay
S—Satellite relay
T—Telemetry
Z—Other (describe)

6.2.5.7 *Production or Pumping Level*—If applicable, document the water level, in feet or metres in reference to the measuring point, measured while the ground-water site was discharging at the amount transcribed for this discharge record. If the water level is above the measuring point, for example a flowing artesian well, precede the value with a minus (−) sign. Document the accuracy or confidence

classification for this data element.

6.2.5.8 *Static Level*—Document the water level, in feet or metres in reference to the measuring point, that relates to the remainder of the even record at the ground-water site. The measurement should be made before production begins or after the production has been stopped and the water level has reached equilibrium. For those water levels that are above the measuring point (normally flowing artesian wells), precede the value with a minus (−) sign to distinguish those water levels from one at or below the measuring point. Document the accuracy or confidence classification for this data element.

6.2.5.9 *Method of Water-Level Measurement*—If applicable, document the method indicating how the water level was measured. Suggested method of water-level measurement components and representative codes are as follows:

A—Airline
B—Analog or graphic recorder
C—Calibrated airline
E—Estimated
G—Pressure gage
H—Calibrated pressure gage
L—Interpreted from geophysical logs
M—Manometer
N—Nonrecording gage
R—Reported, method not known
S—Steel tape
T—Electric tape
V—Calibrated electric tape
Z—Other (describe)

6.2.5.10 *Pumping Period*—If applicable, document length of time, in hours and minutes, that the ground-water site was pumped or allowed to flow prior to the measurement of the production water level. Document the accuracy or confidence classification for this data element.

6.2.5.11 *Specific Capacity*—If applicable, document the specific capacity of the ground-water site. The value is computed by dividing the yield of the well in gallons per minute by the drawdown in feet ((yield)/(production water level−static water level)). Document the time of test if different than pumping period and the accuracy or confidence classification for this data element.

6.2.5.12 *Drawdown*—If applicable, document the drawdown, in feet or metres, of the water level of the pumping or flowing ground-water site. The drawdown is equal to the production level minus the static level. Document the accuracy or confidence classification for this data element.

6.2.5.13 *Source of Water-Level Data*—If applicable, document the source of the water-level data. Suggested source of data components and representative codes are as follows:

A—Government agency
C—Consultant
D—Driller's log or report
L—Personnel of source agency or company
M—Memory (owner, operator, driller)
O—Records by owner
R—Person other than owner, driller, or another government agency
S—Personnel of reporting agency or company
U—University associate
Z—Other source (describe)

6.2.6 *Water-Use Record*—The water-use record is used to document a history of long-term water withdrawals from a ground-water site. These withdrawals are usually reported daily, monthly, quarterly, or yearly (10, 20).

6.2.6.1 *Date Range of Water-Use Record*—If applicable, document the date interval, in year, month, and day (YYYYMMDD), that the water-use data were gathered for the ground-water site.

6.2.6.2 *Data Collection Interval*—If applicable, document the data collection interval for the cumulative long-term water use for the ground-water site. Suggested time-increment components and representative codes are as follows:

D—Daily
M—Monthly
Q—Quarterly
R—Seasonal
S—Semiannually
Y—Yearly
Z—Other (describe)

6.2.6.3 *Long-Term Water Use*—If applicable, document the cumulative long-term water use, in gallons or litres (for large amounts, use units of thousands, millions, or acre-feet), for the ground-water site. This value can be stated in volume per day, month, quarter, year, etc. If other volume units are used, identify the unit. Document the accuracy or confidence classification for this data element.

6.2.6.4 *Method Used to Determine Long-Term Water Use*—If applicable, document the method used to determine the long-term water use. Suggested method used to determine long-term water use components and representative codes are as follows:

E—Estimated from Periodic Measurements
M—Totaling Meter
O—Owners Meter Records
P—Estimated from Power Records
R—Reported, Method Not Known
Z—Other (describe)

6.2.7 *Water-Quality Record*—The process of collecting and analyzing water-quality samples requires that complete information be gathered to confirm that the sample meets quality assurance procedures that may be required by law. The following data elements may be used to document the water samples (3, 4, 10, 14, 15, 18).

6.2.7.1 *Sample Date and Time*—If applicable, document the date (year, month, and day) and time of day (standard time and 2400 clock) that the sample was collected at the ground-water site. Many historical samples do not have the time of day information and are accurate only to the nearest day. Some measurements, accurate to only the nearest year, are extremely valuable in documenting long-term water-quality trends. Unless obvious, document the accuracy of the date of measurement (2, 13).

6.2.7.2 *Agency That Analyzes Samples*—If applicable, document the name and address of the agency that performed the quality analyses on the water sample collected for the monitoring site.

6.2.7.3 *Type of Analyses*—If applicable, document the type of quality analyses conducted on the water collected at the monitoring site. Suggested type of analyses components and representative codes are as follows:

A—Physical properties
B—Common ions (major cations and anions)
C—Trace elements
D—Pesticides
E—Nutrients

F—Sanitary analysis (organisms)
H—Herbicides
R—Radioactive
T—Biological taxa
V—Volatile organic compounds
Z—Other (describe)

NOTE 12—The Environmental Protection Agency (EPA) has compiled lists of components required to fulfill the needs of various federal regulations. These lists are available through EPA publications (16).

6.2.7.4 *Parameters Requested for Analysis*—If applicable, document the EPA's five-digit STORET or chemical name or symbol for those parameters requested for analysis. Normally, the results of the analyses will be stored in a related and separate file, for example, EPA's STORET data base.[6]

6.2.7.5 *Sample Depth/Interval*—If applicable, document the maximum depth or preferably, interval, in feet or metres below the datum, to specify the zone of origin of the water sample at the ground-water site. Document the accuracy or confidence classification for this data element.

6.2.7.6 *Water-Quality File Containing Analysis*—If applicable, document the location of the file that contains the final water-quality analysis. If stored in a computer data base, identify the file (for example, STORET). If stored in file cabinets or published, identify the file type or publication.

6.2.7.7 *Laboratory Number*—If applicable, document the number assigned by the analyzing laboratory to the water sample.

6.2.7.8 *Laboratory Name*—If applicable, document the name and address of the laboratory that analyzed the water sample.

6.2.7.9 *Replicate Sequence Number*—If applicable, document the sequence number for the replicate sample.

6.2.7.10 *Collecting Agency or Company*—If applicable, document the name and address of the agency or company that collected the sample.

6.2.7.11 *Agency or Company Code*—If applicable, document the code of the agency or company that collected the sample. This code is assigned by EPA's STORET User Assistance staff and is required before entry of the ground-water site data into EPA's computerized Water Quality File (WQF).[6]

6.2.7.12 *Chain of Custody*—If applicable, document the names of the people and agency or company that signed the chain of custody form and the dates signed.

6.2.7.13 *Sampling Purpose*—If applicable, document the purpose or reason for collecting the sample. Suggested sampling purpose components and representative codes are as follows:

B—Research
C—CERCLA
R—RCRA
D—Drinking water regulations
E—Exploration (water)
L—Local ordinance
S—State regulations, other than CERCLA or RCRA
F—Federal regulations, other than CERCLA or RCRA
H—Hydrologic benchmark
I—Environmental issues

---

[6] STORET Users Assistance, USEPA, Mail Code PM-2180, 401 M Street, S.W., Washington, DC 20406.

J—Judicial/litigation
M—Mining regulations
N—Natural resources exploration
U—Unknown
Z—Other (describe)

6.2.7.14 *Site Condition*—If applicable, document the condition of the ground-water site at the time the sample was collected. Suggested sampling condition components and representative codes are as follows:

P—Pond—fenced, but open to atmosphere
U—Pond—unprotected
S—Spring or tunnel—protected
V—Spring or tunnel—unprotected
W—Well—sealed
Y—Well—open casing
Z—Other (describe)

6.2.7.15 *Sample Appearance*—If applicable, document the appearance of the sample at the time of collection as to color and turbidity. Suggested sample appearance components and representative codes are as follows:

A—Clear or colorless
C—Colored, not turbid (give color)
S—Turbid, suspended matter, particles visible (give color)
T—Turbid, suspended matter, particles not visible (give color)
U—Unknown
Z—Other (describe)

6.2.7.16 *Sample Odor*—If applicable, document the odor given off by the ground water during the collection of the sample. Suggested sample odor components and representative codes are as follows:

C—Chemical, unknown
D—Chlorine
H—Hydrogen-sulfide
M—Methane
N—None
P—Petroleum
Z—Other (describe)

6.2.7.17 *Presence of Immiscible Stratum*—If applicable, document the presence of an immiscible stratum that may be at the top or bottom of the water column at the time of sample collection. Suggested presence of immiscible stratum components and representative codes are as follows:

T—Top presence of immiscible stratum
B—Bottom presence of immiscible stratum
I—Indeterminable
M—Mixed as globules in water
N—None
S—Sheen
U—Unknown
Z—Other (describe)

6.2.7.18 *Thickness of Immiscible Stratum*—If applicable, document the thickness in inches or millimetres of an immiscible stratum that may be at the top or bottom of the water column at the time of sample collection.

6.2.7.19 *Sensors*—If applicable, document the type of sensors, manufacturer, and model numbers of sensors at the ground-water site used for sensing of water-quality parameters. Suggested sensor components and representative codes are as follows:

E—Electrical
I—Ion specific
O—Sensors, using fiber optics

U—Unknown
Z—Other (describe)

6.2.7.20 *Sample Preservation Method*—If applicable, document the method used to preserve the sample. Several methods of preservation may be used that depend upon the parameter analyzed. Each method should be documented. Suggested sample preservation method components and representative codes are as follows:

C—Cooled or iced
H—Hydrochloric acid
P—Hydrogen peroxide
N—Nitric acid
R—Phosphoric acid
S—Sulfuric acid
X—None
Z—Other (describe)

6.2.7.21 *Sample Filtration Material*—If applicable, document the filter size and sample filtration material used in-line or in the field following the sample collection. If more than one filter is used, document each material and filter size. Suggested sample filtration material components and representative codes are as follows:

D—Cellulose acetate
C—Cellulose nitrate
G—Glass
E—Polycarbonate
T—Teflon
X—None
Z—Other (describe)

6.2.7.22 *Pumped Period*—If applicable, document length of time, in hours and minutes, that the ground-water site was pumped or allowed to flow prior to the collection of the water sample. Document the accuracy or confidence classification for this data element.

6.2.7.23 *Casing Volume*—If applicable, document the amount of water that needs to be pumped to purge the casing one time, in gallons, cubic feet (or metres), litres, or any other standard volume/time unit. Document the volume unit used. For example, the amount of water needed to purge the system is in the well casing above the screen or open area. Document the accuracy or confidence classification for this data element.

6.2.7.24 *Amount Water Purged*—If applicable, document the amount of water that was pumped or allowed to flow before the water sample was collected at the ground-water site, in gallons per minute, cubic feet (or metres) per second, litres per second, or any other standard volume/time unit. Document the volume/time unit used. Document the accuracy or confidence classification for this data element.

6.2.7.25 *Sampling Method or Sampler Type*—If applicable, document the sampling method or sampler type used to collect the water sample at the ground-water site. Numerous sampling devices have been and are being developed. The following list was patterned after a draft USEPA reference guide (**4**). Also see Note 5. Suggested type of sampling method or sampler type components and representative codes are as follows:

Portable Grab/Depth Specific Samplers
O—Open bailer
P—Point-source bailer
B—Bucket
S—Syringe sampler

W—Westbay sampler
K—Kemmerer/Van Dorn sampler
C—Collwasa sampler
T—Stratified thief sampler
A—Swabbing
D—Packer pumps

Portable Positive Displacement (Submersible) Samplers
E—Bladder pump
G—Grundfos centrifugal pump
F—Other centrifual pump
H—Helical rotor pump
I—Gas-drive piston pump
J—Gear-drive pump
L—Submersible rod pump

Other Portable Samplers
M—Peristaltic suction lift
N—Centrifugal suction
Q—Gas-drive/displacement
R—Inertial pump
V—Gas-lift pump
Y—Jet pump

Portable In Situ Samplers
1—Hydropunch
2—BAT sampler
3—Other cone penetrometer samplers
4—Other in situ samplers

Other or Unknown Methods
X—Natural flowing, spring or well
U—Unknown
Z—Other (describe)

6.2.7.26 *Sampler Material*—If applicable, document the material used in construction of the sampler. Suggested sampler material components and representative codes are as follows:

A—Aluminum
B—ABS (Plastic)
C—Copper
G—Galvanized iron
L—Steel
N—Nylon
O—Polypropylene
P—Polyvinyl chloride (PVC)
Q—Polyalkene
R—Rubber
S—Stainless steel
T—Teflon
W—Wood
U—Unknown
Z—Other (describe)

6.2.7.27 *Aquifer Sampled*—If applicable, document the identification of the aquifer or aquifers from where the water was obtained at the ground-water site. A convenient and systematic method of coding aquifer and geologic units is used by the U.S. Geological Survey in a national file (Catalog of Aquifer Names and Geologic Unit Codes used by the Water Resources Division) (for example, Edwards Limestone of Texas is coded 218EDRD). Information needed to obtain an ordered list of aquifers and related codes is available (10):[7]

6.2.7.28 *Regulating Agency*—If applicable, document the name and address of the agency that regulates water quality

[7] Geologic Names Unit, U.S. Geological Survey, 439 National Center, Reston, VA 22092.

at the ground-water site, for example, State Health Department.

6.2.8 *Field Water-Quality Record*—The field water-quality data denote those constituents or characteristics where the final values are determined in other than the laboratory. Many of the values for these parameters, such as temperature and pH, only can be measured at the time of sample collection because of rapidly changing conditions in the water upon removal from the natural environment. Properly measured, these parameters are extremely valuable in determining the true hydrologic conditions in the aquifer at the ground-water site (3, 7, 10, 14, 15).

NOTE 13—Many of the supporting data elements listed under the water-quality record also apply and can be used for documenting the field water-quality record. Normally, field water-quality characteristics (for example, temperature, pH) are determined during the same site visit that water-quality samples are collected for laboratory analysis. Therefore, when samples are collected for laboratory analysis, those supporting data elements (for example, sample depth/interval, site condition, aquifer sampled) need not be repeated for the field water-quality record.

6.2.8.1 *Field Sample Date and Time*—Document the date (year, month, and day) and time of day (standard time and 2400 clock) that the water-quality characteristic was determined at the ground-water site. Many historical determinations do not have the time of day information and are accurate only to the nearest day. Some measurements, accurate to only the nearest year, are extremely valuable in documenting long-term water-quality trends. Unless obvious, document the accuracy of the date of measurement (2, 13).

6.2.8.2 *Parameter Code*—Document the EPA's five-digit STORET number, CAS number, or chemical name or symbol for those water-quality parameters analyzed or measured in the field. The following are some of the common field-determined water-quality characteristics and the corresponding STORET number code:

00010—Temperature, water (°C)
00095—Specific conductance (microsiemens/cm at 25°C)
00300—Oxygen, dissolved (DO), milligrams per litre (mg/L)
00400—pH (standard units)
00405—Carbon dioxide, dissolved (mg/L as CO_2)
00410—Alkalinity, water, whole, total, field, as $CaCO_3$, mg/L
00430—Alkalinity, carbonate (mg/L as $CaCO_3$)
00440—Bicarbonate, water, whole, total, field, as HCO_3, mg/L
00445—Carbonate, water, whole, total, field, as CO_3, mg/L
00900—Hardness, total (mg/L as $CaCO_3$)
00940—Chloride, dissolved (mg/L as Cl)
00945—Sulfate, dissolved (mg/L as SO_4
00950—Fluoride, dissolved (mg/L as F)
01045—Iron, total (micrograms per litre (μg/L) as Fe)
31501—Coliform, membrane filter, immediate M-endo medium (colonies/100 mL)
31625—Coliform, fecal, 0.7 μm membrane filter (UM-MF) (colonies/100 mL)
31673—Streptococci, fecal, membrane filter, KF agar (colonies/100 mL)
71820—Density—grams per millilitre (g/mL) at 20°C
71830—Hydroxide, water, whole, total, field, as OH, mg/L

6.2.8.3 *Value of Parameter*—Document the value for the field determined water-quality constituent or characteristic. Document the accuracy or confidence classification for this water-quality parameter.

6.2.8.4 *Unit for Parameter*—Include with each water-quality constituent or characteristic the unit used for recording the value, for example, degrees celsius for temperature, micromhos (or microsiemens) per centimetre at 25° (25° is the laboratory standard) celsius for specific conductance or milligrams per litre (mg/L) for chloride.

6.2.8.5 *Instrumentation or Method of Determination*—Document the method or function, manufacturer, model, and accuracy of the instruments used to obtain field water-quality characteristics at the ground-water site. A number of instruments are currently available to measure these data. Some generic examples of the more common field instruments are as follows (17, 19):

    pH meter
    Thermometer
    Specific-conductance meter
    Dissolved-oxygen metre
    Kits for field measurement of constituent values
    Kits for field determination of biological values

NOTE 14—New apparatus become available daily and many instruments of laboratory quality have become portable and are suitable for use in the field. Whether the field-determined chemical analyses can be classed as laboratory equivalent depends upon the quality control maintained on the instruments used for the measurement.

6.2.9 *Monitoring Site at Waste-Facility Record*—The monitoring site at waste-facility record includes the documentation of data elements of ground-water sites that were constructed for the primary purpose of monitoring the pollutants and hydraulics at hazardous and solid-waste facilities (11, 19).

NOTE 15—Many of the other records listed under the monitoring record group and other standards describing ground-water data components also apply and can be used for documenting the ground-water sites at waste facilities.

6.2.9.1 *Date Interval in Service*—If applicable, document the date interval, in year, month, and day (YYYYMMDD), that the ground-water site was in service at the waste facility. If still in use, document the date that the site was put into service.

6.2.9.2 *State Regulatory Agency*—If applicable, document the name and address of the state regulatory agency that has jurisdiction over the site.

6.2.9.3 *State Registration Identification*—If applicable, document the number or identification assigned to the site by the state regulatory agency.

6.2.9.4 *EPA Registration Identification*—If applicable, document the number or identification assigned to the site by the U.S. Environmental Protection Agency.

6.2.9.5 *Responsible Company*—If applicable, document the name and address of the company or agency that owns or has control of the ground-water site.

6.2.9.6 *Company's Site Identification*—If applicable, document the company or agency number or identification of the ground-water site.

6.2.9.6 *Site Location in Relationship to Waste Facility*—If applicable, document the relative location of the ground-water site at the waste facility. Common site locations in relationship to the ground-water hydraulic gradient and waste facility and representative codes are as follows:

    A—Upgradient
    D—Downgradient
    S—Side

    W—Within boundaries
    U—Unknown
    Z—Other (describe)

6.2.9.8 *Status of Site*—If applicable, document the status of the ground-water site at the waste facility. Suggested status components and representative codes are as follows:

    A—Active
    D—Destroyed
    E—Damaged
    I—Inactive
    U—Unknown
    Z—Other (describe)

6.2.9.9 *Sampled Interval*—If applicable, document the monitored interval, in feet or metres. Document the accuracy or confidence classification for this data element.

NOTE 16—For example, the interval might be 15 to 18 ft (4.57 to 5.49 m) below the measuring point. The measuring point is defined under the measuring-point record and is related to the altitude, therefore, if the supporting records are utilized, the sampled interval can be converted to altitude values.

6.3 *Irrigation Record Group:*
6.3.1 *Introduction*—The irrigation record group includes the record for documenting the data elements pertaining to irrigation at a ground-water site.

6.3.2 *Irrigation Record*—The irrigation record includes data elements that relate to the withdrawal of water to be used for irrigation at the ground-water site (10).

6.3.2.1 *Irrigated Land Area*—If applicable, document the land area, in acres or hectares, that is irrigated by water from the ground-water site.

6.3.2.2 *Allowance for Irrigating*—If applicable, document the maximum amount of water, in acre-feet per year or another unit, that is allocated for withdrawal from the ground-water site. Document the measurement unit and accuracy or confidence classification for this data element.

6.3.2.3 *Date Legal Irrigation Begins*—If applicable, document the date, in year, month, and day (YYYYMMDD), that irrigation can begin in the spring or at the start of the irrigation season (for example, 19870301).

6.3.2.4 *Date Legal Irrigation Ends*—If applicable, document the date, in year, month, and day (YYYYMMDD), that irrigation ends in the fall or at the end of the irrigation season (for example, 19871001).

6.4 *Decommissioning Record Group:*
6.4.1 *Introduction*—The decommissioning record group includes the record for documenting data elements that pertain to the permanent decommissioning or closure of a ground-water site.

6.4.2 *Decommissioning Record*—The decommissioning record includes data elements that pertain to permanent closure of a ground-water site. Although the decommissioning normally applies to sites at solid or hazardous waste facilities, the record can be used for any ground-water site, such as a public supply well[3] (18).

6.4.2.1 *Date Decommissioned*—If applicable, document the date, in year, month, and day (YYYYMMDD), that the ground-water site was decommissioned or closed.

6.4.2.2 *Method Used for Decommissioning*—If applicable, document the method used to decommission or close the ground-water site. Suggested method used for decommissioning components and representative codes are as follows:

L—Casing left in place and plugged
R—Casing removed and hole plugged
U—Unknown
Z—Other (describe)

6.4.2.3 *Reason for Decommissioning*—If applicable, document the reason for decommissioning or closing the ground-water site. Suggested reason for decommissioning components and representative codes are as follows:

C—Remove chance of unauthorized use
P—Prevent migration of contaminants
M—Reduce chance of vertical or horizontal migration
N—No longer required
F—Failed or damaged
R—Regulatory requirement
U—Unknown
Z—Other (describe)

6.4.2.4 *Plugging Material*—If applicable, document the material used to fill the opening, for example the borehole opening. Suggested generic plugging material components and representative codes are as follows:

P—Portland cement (describe)
E—Expansive cement (describe)
A—API cement (describe)
B—Other cement (describe)
D—Gypsum cement (describe)
F—Epoxy resin cement (describe)
C—Clay (describe)
G—Coarse-grained material (describe)

U—Unknown
Z—Other (describe)

NOTE 17—Plugging materials, such as Portland cement, have a number of different types. In addition, each type can have a number of possible additives that are used for extenders, accelerators, retarders, density improver, fluid loss controllers, and friction reducers. The various materials and additives should be described.

6.4.2.5 *Name and Address of Decommissioner*—If applicable, document the company or agency name and address of the decommissioner of the ground-water site.

6.4.2.6 *Step-by-Step Procedures*—If applicable, document the step-by-step procedure used to decommission the ground-water site.

6.4.2.7 *Availability of Decommissioning Report*—If applicable, document the availability of the decommissioning report. Include the name and address of the source of the report.

6.4.2.8 *Regulations Followed, Federal, State, Local*—If applicable, document the regulations followed in decommissioning the ground-water site. Include the name and address of the regulator.

7. Keywords

7.1 confidence classification; decommissioning of ground-water site; field water-quality sample; ground water; ground-water site; irrigation; monitoring network; regulatory agency; site identification; waste facility; water-quality sample; water use

D 5410

REFERENCES

(1) Bates, R. L., and Jackson, J. A., *Glossary of Geology*, Third Edition; American Geological Institute, Alexandria, Virginia, 1987.

(2) *American National Standard for Information Systems—Representations of Local Time of Day for Information Interchange*, American National Standards Institute, Inc. Publication ANSI X3.43-1986, 1430 Broadway, New York, NY 10018.

(3) Barcelona, M. J., Gibb, J. P., Helfrich, J. A., and Garske, E. E., *Practical Guide for Ground-Water Sampling*, Office of Ground-Water Protection, U.S. EPA, EPA/600/2-85/104, Robert S. Kerr Environmental Research Laboratory, Ada, OK, February 1985.

(4) Boulding, J. R., (in preparation), *Subsurface Field Characterization and Monitoring Techniques: A Desk Reference Guide, Vol I: Solids and Ground Water, Vol II: The Vadose Zone Chemical Field Screening and Analysis*, Prepared for U.S. EPA Center for Environmental Research Information, Cincinnati, OH.

(5) Bureau of Reclamation, *Ground-Water Manual*, A Water Resources Technical Publication, Revised Reprint, U.S. Department of Interior, Bureau of Reclamation, Washington, DC, 1981.

(6) Campbell, M. D., and Lehr, J. H., *Water Well Technology*, New York, NY, McGraw-Hill, 1973.

(7) Fishman, M. J., and Bradford, W. L., *Methods of Determination of Inorganic Substances in Water and Fluvial Substances*, 3rd Edition, Techniques of Water-Resources Investigation of the United States Geological Survey, Chapter A1, Book 5, 1989.

(8) Edwards, M. D., and Josefson, B. M., *Identification Codes for Organizations Listed in Computerized Data Systems of the U.S. Geological Survey*, U.S. Geological Survey, Open-File Report 82-921, 1982.

(9) Heath, Ralph C., *Basic Ground-Water Hydrology*, U.S. Geological Survey Water-Supply Paper 2220, 1983.

(10) Mathey, Sharon B. (editor), *National Water Information System User's Manual, Vol 2, Chapter 4, Ground-Water Site Inventory System*, U.S. Geological Survey, Open-File Report, 1990, pp. 89–587.

(11) Perry, R. A., and Williams, O. O., *Data Index Maintained by the National Water Data Exchange*, U.S. Geological Survey, Open-File Report 82-327, 1982.

(12) *Ground-Water Data INTERFACE, Users Reference Manual*, Texas Natural Resources Information System, Nov. 20, 1986.

(13) U.S. Department of Commerce, *Representation of Local Time of Day for Information Interchange, Federal Information Standards (FIPS) Publication 58-1*, National Institute for Standards and Technology, Washington, DC, Jan. 27, 1988.

(14) U.S. Environmental Protection Agency, *STORET Users Handbook*, Vols 1 and 2, U.S. EPA, Washington, D.C., February 1982.

(15) U.S. Environmental Protection Agency, *Ground-Water Data Management with STORET*, Office of Ground-Water Protection, U.S. EPA, Washington, DC, March 1986.

(16) U.S. Environmental Protection Agency, *Ground Water Monitoring in SW-846, Field Manual Physical/Chemical Methods, Test Methods for Evaluating Soil Wastes*, Vol II, Chapter Eleven, Third Edition; Office of Solid Wastes and Emergency Responses, U.S. EPA, Washington, DC, November 1986.

(17) U.S. Environmental Protection Agency, *Field Screening Methods for Hazardous Waste Site Investigations, Proceedings of the First International Symposium*, presented by U.S. Environmental Protection Agency, Environmental Monitoring Systems Laboratory-Las Vegas and U.S. Army Toxic and Hazardous Materials Agency at Las Vegas, NV, Oct. 11–13, 1988.

(18) U.S. Environmental Protection Agency, *Handbook of Suggested Practices for the Design and Installation of Ground-Water Monitoring Wells*, Office of Research and Development, U.S. EPA, Washington, DC, March 1991.

(19) U.S. Geological Survey, *National Handbook of Recommended Methods for Water-Data Acquisition, Chapter 2—Ground Water*, Office of Data Coordination, Reston, VA, 1980, pp. 2-1 to 2-149.

(20) U.S. Geological Survey, *Guide to Obtaining USGS Information*, U.S. Geological Survey Circular 900, 1989.

(21) van der Leedan, F., Troise, F. L., and Todd, D. K., *The Water Encyclopedia, Geraghty and Miller Ground-Water Series*, 2nd Edition, Third Printing, Lewis Publishers, Inc., Chelsea, MI, 1991.

1.3 GEOPHYSICAL METHODS

Standard Guide for
Planning and Conducting Borehole Geophysical Logging[1]

This standard is issued under the fixed designation D 5753; the number immediately following the designation indicates the year of original adoption or, in the case of revision, the year of last revision. A number in parentheses indicates the year of last reapproval. A superscript epsilon (ε) indicates an editorial change since the last revision or reapproval.

1. Scope

1.1 This guide covers the documentation and general procedures necessary to plan and conduct a geophysical log program as commonly applied to geologic, engineering, ground-water, and environmental (hereafter referred to as geotechnical) investigations. It is not intended to describe the specific or standard procedures for running each type of geophysical log and is limited to measurements in a single borehole. It is anticipated that standard guides will be developed for specific methods subsequent to this guide.

1.2 Surface or shallow-depth nuclear gages for measuring water content or soil density (that is, those typically thought of as construction quality assurance devices), measurements while drilling (MWD), cone penetrometer tests, and logging for petroleum or minerals are excluded.

1.3 Borehole geophysical techniques yield direct and indirect measurements with depth of the (1) physical and chemical properties of the rock matrix and fluid around the borehole, (2) fluid contained in the borehole, and (3) construction of the borehole.

1.4 To obtain detailed information on operating methods, publications (for example, 2, 5, 7, 18, 24, 29, 34, 35, and 36)[2] should be consulted. A limited amount of tutorial information is provided, but other publications listed herein, including a glossary of terms and general texts on the subject, should be consulted for more complete background information.

1.5 This guide provides an overview of the following: (1) the uses of single borehole geophysical methods, (2) general logging procedures, (3) documentation, (4) calibration, and (5) factors that can affect the quality of borehole geophysical logs and their subsequent interpretation. Log interpretation is very important, but specific methods are too diverse to be described in this guide.

1.6 Logging procedures must be adapted to meet the needs of a wide range of applications and stated in general terms so that flexibility or innovation are not suppressed.

1.7 *This standard does not purport to address all of the safety and liability concerns, if any, (for example, lost or lodged probes and radioactive sources[3]) associated with its use. It is the responsibility of the user of this standard to establish appropriate safety and health practices and deter-* mine the applicability of regulatory limitations prior to use.

2. Referenced Documents

2.1 *ASTM Standards:*
D 653 Terminology Relating to Soil, Rock, and Contained Fluids[4]
D 5088 Practice for the Decontamination of Field Equipment Used at Non-Radioactive Waste Sites[5]
D 5608 Practice for the Decontamination of Field Equipment Used at Low Level Radioactive Waste Sites

3. Terminology

3.1 *Definitions*—Definitions shall be in accordance with Terminology D 653.

3.2 *Descriptions of Terms Specific to This Standard*—Terms shall be in accordance with Ref (1).

4. Summary of Guide

4.1 This guide applies to borehole geophysical techniques that are commonly used in geotechnical investigations. This guide briefly describes the significance and use, apparatus, calibration and standardization, procedures and reports for planning and conducting borehole geophysical logging. These techniques are described briefly in Table 1 and their applications in Table 2.[6]

4.2 Many other logging techniques and applications are described in the textbooks in the reference list. There are a number of logging techniques with potential geotechnical applications that are either still in the developmental stage or have limited commercial availability. Some of these techniques and a reference on each are as follows: buried electrode direct current resistivity (37), deeply penetrating electromagnetic techniques (38), gravimeter (39), magnetic susceptibility (40), magnetometer, nuclear activation (41), dielectric constant (42), radar (50), deeply penetrating seismic (39), electrical polarizability (45), sequential fluid conductivity (46), and diameter (48). Many of the guidelines described in this guide also apply to the use of these newer techniques that are still in the research phase. Accepted practices should be followed at the present time for these techniques.

5. Significance and Use

5.1 An appropriately developed, documented, and executed guide is essential for the proper collection and application of borehole geophysical logs.

---

[1] This guide is under the jurisdiction of ASTM Committee D-18 on Soil and Rock and is the direct responsibility of Subcommittee D18.01 on Surface and Subsurface Characteristics.
Current edition approved July 15, 1995. Published October 1995.
[2] The boldface numbers in parentheses refer to the list of references at the end of this standard.
[3] The use of radioactive materials required for some log measurements is regulated by federal, state, and local agencies. Specific requirements and restrictions must be addressed prior to their use.

[4] *Annual Book of ASTM Standards*, Vol 04.08.
[5] *Annual Book of ASTM Standards*, Vol 04.09.
[6] The references indicated in these tables should be consulted for detailed information on each of these techniques and applications.

TABLE 1 Common Geophysical Logs

Type of Log (References)	Varieties and Related Techniques	Properties Measured	Required Hole Conditions	Other Limitations	Typical Measuring Units and Calibration or Standardization	Brief Probe Description
Spontaneous potential (7, 8, 12)	differential	electric potential caused by salinity differences in borehole and interstitial fluids, streaming potentials	uncased hole filled with conductive fluid	salinity difference needed between borehole fluid and interstitial fluids; needs correction for other than NaCl fluids	mV; calibrated power supply	records natural voltages between electrode in well and another at surface
Single-point resistance (7)	conventional, differential	resistance of rock, saturating fluid, and borehole fluid	uncased hole filled with conductive fluid	not quantitative; hole diameter effects are significant	Ω; V-Ω meter	constant current applied across lead electrode in well and another at surface of well
Multi-electrode resistivity (7, 8, 13)	various normal focused, guard, lateral arrays	resistivity and saturating fluids	uncased hole filled with conductive fluid	reverses or provides incorrect values and thickness in thin beds	Ω-m; resistors across electrodes	current and potential electrodes in probe and remote current and potential electrodes
Induction (10, 11)	various coil spacings	conductivity or resistivity of rock and saturating fluids	uncased hole or nonconductive casing; air or fluid filled	not suitable for high resistivities	mS or Ω-m; standard dry air zero check or conductive ring	transmitting coil(s) induce eddy currents in formation; receiving coil(s) measures induced voltage from secondary magnetic field
Gamma (5, 7, 22)	gamma spectral (44)	gamma radiation from natural or artificial radioisotopes	any hole conditions	may be problem with very large hole, or several strings of casing and cement	pulses per second or API units; gamma source	scintillation crystal and photomultiplier tube measure gamma radiation
Gamma-gamma (23, 24)	compensated (dual detector)	electron density	optimum results in uncased hole; can be calibrated for casing	severe hole-diameter effects; difficulty measuring formation density through casing or drill stem	gs/cm^3; Al, Mg, or Lucite blocks	scintillation crystal(s) shielded from radioactive source measure Compton scattered gamma
Neutron (7, 14, 25)	epithermal, thermal, compensated sidewall, activation, pulsed	hydrogen content	optimum results in uncased hole; can be calibrated for casing	hole diameter and chemical effects	pulses/s or API units; calibration pit or plastic sleeve	crystal(s) or gas-filled tube(s) shielded from radioactive neutron source
Acoustic velocity (5, 26, 27)	compensated, waveform, cement bond	compressional wave velocity or transit time, or compressional wave amplitude	fluid filled, uncased, except cement bond	does not detect secondary porosity; cement bond and wave form require expert analysis	velocity units, for example, ft/s or m/s or μs/ft; steel pipe	1 or more transmitters and 2 or more receivers
Acoustic televiewer (28, 7)	acoustic caliper	acoustic reflectivity of borehole wall	fluid filled, 3 to 16-in. diameter; problems in deviated holes	heavy mud or mud cake attenuate signal; very slow logging speed	orientated image-magnetometer must be checked	rotating transducer sends and receives high-frequency pulses
Borehole video	axial or side view (radial)	visual image on tape	air or clean water; clean borehole wall	may need special cable	NA[A]	video camera and light source
Caliper (29, 7)	oriented, 4-arm high-resolution, x-y or max-min bow spring	borehole or casing diameter	any conditions	deviated holes limit some types; significant resolution difference between tools	distance units, for example, in.; jig with holes or rings	1 to 4 retractable arms contact borehole wall
Temperature (30, 31, 32)	differential	temperature of fluid near sensor	fluid filled	large variation in accuracy and resolution of tools	°C or °F; ice bath or constant temperature bath	thermistor or solid-state sensor
Fluid conductivity (7)	fluid resistivity	most measure resistivity of fluid in hole	fluid filled	accuracy varies, requires temperature correction	μS/cm or Ω-m; conductivity cell	ring electrodes in a tube
Flow (12, 33, 7)	impellers, heat pulse	vertical velocity of fluid column	fluid filled	impellers require higher velocities. Needs to be centralized.	velocity units, for example, ft/min; lab flow column or log in casing	rotating impellers; thermistors detect heated water; other sensors measure tagged fluid
Deviation (4, 7, 47)	magnetic, gyroscopic, or mechanical	horizontal and vertical displacement of borehole	any conditions (see limitations)	magnetic methods orientation not valid in steel casing	degrees and depth units; orientation and inclination must be checked	various techniques to measure inclination and bearing of borehole

[A] NA = not applicable.

5.1.1 The benefits of its use include improving the following:

5.1.1.1 Selection of logging methods and equipment,

5.1.1.2 Log quality and reliability, and

5.1.1.3 Usefulness of the log data for subsequent display and interpretation.

5.1.2 This guide applies to commonly used logging methods (see Tables 1 and 2) for geotechnical investigations.

5.1.3 It is essential that personnel (see 7.3.3) consult up-to-date textbooks and reports on each of the logging techniques, applications, and interpretation methods. A partial list of selected publications is given at the end of this guide.

5.1.4 This guide is not meant to describe the specific or standard procedures for running each type of geophysical log and is limited to measurements in a single borehole.

6. Apparatus

6.1 *Geophysical Logging System*, including probes, cable, draw works, depth measurement system, interfaces and surface controls, and digital and analog recording equipment.

TABLE 2 Log Selection Chart for Geotechnical Applications Using Common Geophysical Logs[A]

Information Desired	Acoustic		Electric and Induction					Fluid Logs				Radioactive or Nuclear				Other Methods			
	Acoustic Televiewer	Acoustic Velocity, Δt, CBL, VDL, FWS	Induced Polarization	Multi-electrode Resistivity, Normal, Lateral, Micro Guard Resistivity	Single-Point Resistance	Spontaneous Potential	Induction (Conductivity)	Flow Meter	Fluid Resistivity	Fluid Sampler	Temperature, Differential Temperature	Gamma-Gamma Density	Gamma	Neutron	Spectral Gamma	Borehole Video	Caliper	Casing Collar Locator	Deviation
Lithology and Correlation																			
Bed/aquifer thickness; correlation, structure	●	●		●	●		★					Δ	✓	Δ	✓	◊	✓		
Lithology—depositional environment	?	●		●	●		★					Δ	✓	Δ	✓	◊	✓		
Shale or clay content			●	●		●	★					Δ	✓	Δ	✓				
Bulk density												Δ							
Formation resistivity				●			★												
Injection/production profiles			?				?	□	□		□	Δ		Δ					
Permeability estimates		●						□	□		□		✓						
Porosity (amount and type)	●	●		●			★					Δ		Δ					
Mineral identification			●												✓				
Potassium-uranium thorium content (KUT)															✓				
Rock Structure																			
Strike and dip of bedding	●															◊			✓
Fracture detection (number of fractures), RQD	●	●		●	●											◊	✓		
Fracture orientation and character	●															◊			✓
Thin bed resolution	●			?	●											◊	✓		
Fluid Parameters																			
Borehole fluid characteristics								?	□	□	□								
Fluid flow						●		□	□		□					◊			
Formation water quality				●		●	★			□									
Moisture content—water saturation				?			?					Δ		Δ					
Temperature		?									□								
Water level and water table	●	●		●	●	●	?	□			□	Δ		Δ		◊			
Borehole Parameters																			
Casing evaluation integrity, leaks, damage, screen location	■	■			■		?				■					◆	✓	†	
Deviation of borehole																			✓
Diameter of borehole	●																✓		
Examination behind casing		●					★					Δ		Δ					
Location of debris in wells	●															◆	✓	✓	
Well completion evaluation, for example, cement bond, seal location, grout location	?	■					★					Δ	✓	Δ					

[A] Required hole conditions: ■ = cased fluid-filled hole, ◆ = clear fluid or dry cased hole, □ = screened or open fluid-filled hole, ◊ = clear fluid or dry open hole, † = steel casing only, Δ = active nuclear log to be run in stable holes, ★ = open or nonconductive cased holes, dry or fluid filled, ✓ = no restrictions, ● = open fluid-filled hole only, and ? = possible applications.

6.1.1 Logging probes, also called sondes or tools, enclose the sensors, sources, electronics for transmitting and receiving signals, and power supplies.

6.1.2 Logging cable routinely carries signals to and from the logging probe and supports the weight of the probe.

6.1.3 The draw works move the logging cable and probe up and down the borehole and provide the connection with the interfaces and surface controls.

6.1.4 The depth measurement system provides probe depth information for the interfaces and surface controls and recording systems.

6.1.5 The surface interfaces and controls provide some or all of the following: electrical connection, signal conditioning, power, and data transmission between the recording system and probe.

6.1.6 The recording system includes the digital recorder and an analog display or hard copy device.

7. Calibration and Standardization of Geophysical Logs

7.1 *General:*

7.1.1 National Institute of Standards and Technology (NIST) calibration and operating procedures do not exist for the borehole geophysical logging industry. However, calibration or standardization physical models are available (see Appendix X1).

7.1.2 Geophysical logs can be used in a qualitative (for example, comparative) or quantitative manner, depending on the project objectives. (For example, a gamma-gamma log can be used to indicate that one rock is more or less dense than another, or it can be expressed in density units.)

7.1.3 The calibration and standardization scope and frequency shall be sufficient for project objectives.

7.1.3.1 Calibration or standardization should be performed each time a logging probe is modified or repaired or at periodic intervals.

7.2 *Calibration:*

7.2.1 Calibration is the process of establishing values for log response. It can be accomplished with a representative physical model or laboratory analysis of representative samples. Calibration data values related to the physical properties (for example, porosity) may be recorded in units (for example, pulses/s or μm/ft) that can be converted to apparent porosity units.

7.2.1.1 At least three, and preferably more, values are needed to establish a calibration curve, and the interface or contact between different values in the model should be recorded. Because of the variability in subsurface conditions, many more values are needed if sample analyses are used for calibration.

7.2.1.2 The statistical scatter in regression of core analysis against geophysical log values may be caused by the difference between the sample size and geophysical volume of investigation and may not represent measurement error.

7.2.2 *Physical Models*—A representative model simulates the chemical and physical composition of the rock and fluids to be measured.

7.2.2.1 Physical models include calibration pits, coils, resistors, rings, temperature baths, etc.

7.2.2.2 The calibration of nuclear probes should be performed in a physical model that is nearly infinite with respect to probe response.

7.2.2.3 Some probes have internal devices such as resistors, but this does not substitute for checking the probe response in an environment that simulates borehole conditions, and the use of such devices is considered standardization.

7.2.2.4 *Calibration Facilities*—Commonly used calibration pits or models for use by anyone at the present time are listed in Appendix X1 (14–18). The user should inquire concerning the present validity of any facility.

7.2.3 *Sample Analyses:*

7.2.3.1 Representative samples from boreholes in the project area that have been collected carefully and analyzed quantitatively also may be used to calibrate log response.

7.2.3.2 To reduce depth errors, the sample recovery of rock cores in calibration holes needs to approach 100 % for the intervals used for calibration. Log response should be used to select sample depths to span the range of desired log calibration values and to be within thick units to minimize the effects of potential depth errors. Samples need to be analyzed immediately or steps taken to preserve them for later analysis.

7.2.3.3 Samples to be used for log calibration should be analyzed only from depth intervals at which the log response is relatively uniform for a depth interval considerably greater than the vertical dimension of the volume of investigation of the logging probe. Samples near lithologic contacts or fluid interfaces should not be used because of possible boundary effects or depth errors.

7.3 *Standardization:*

7.3.1 Standardization is the process of checking the log response to reveal evidence of repeatability and consistency.

7.3.2 Standardization is needed to establish comparability between logs made with different equipment or at different times and to ensure the accuracy of measurements.

7.3.2.1 Standardization checks should include at least two different measurement values approximating the range of interest (For example, aluminum and magnesium or plastic blocks are used commonly to check the response of gamma-gamma density logging systems in the field.)

7.3.3 Standardization uses some type of a standard that may be used in the field or laboratory and repeat logs.

7.3.3.1 Log response needs to be checked using field standards often enough to satisfy the project objectives. Standardization of the log response provides the basis for correcting for changes (for example, changes in output with time due to system drift or changes of equipment).

7.3.3.2 Selected log intervals should be repeated (that is, re-logged). Repeat logs provide information on the stability of logging equipment.

7.3.3.3 A representative borehole may be used to check log response periodically. This borehole environment and the rocks and fluids penetrated may change with time.

8. Procedure

8.1 *Planning the Logging Program:*

8.1.1 A work plan should be developed prior to implementing the logging program.

8.1.2 The key steps in developing a logging work plan should include the following:

8.1.2.1 *Log Selection*—See Tables 1 and 2.

8.1.2.2 *Personnel Selection*—See 8.3.2.

8.1.2.3 *Quality Control and Documentation*—See 8.4.

8.1.2.4 *Calibration and Standardization Procedures*—See Section 7.

8.1.2.5 *Equipment Liability*—See 1.7.

8.1.2.6 *Equipment Decontamination*—In environmental investigations, equipment decontamination may be required before, after, and between individual wells. Equipment decontamination may involve a number of standardized procedures, depending on the nature of the project (see Practices D 5088 and D 5608). A decontamination program should be agreed upon by all parties before logging commences, and procedures specified by the work plan should be followed.[7]

8.1.2.7 *Log Interpretation*—See 8.5.

8.2 *Field Assessment of Borehole Conditions:*

---

[7] Equipment decontamination procedures may have specific safety and equipment limitations that must be addressed prior to their use.

8.2.1 Borehole conditions can have a profound influence on the quality of log data and subsequent interpretation. Important parameters to consider include the following:

8.2.1.1 Drilling method, casing, drill hole history, and well completion materials.

8.2.1.2 *Borehole Fluid Properties*—Resistivity, temperature, density, viscosity, and chemistry at the time of logging.

8.2.1.3 Borehole diameter, rugosity, and stability.

8.2.1.4 Deviation of borehole.

8.2.1.5 Wellhead pressure.

8.2.2 *Logging Operations:*

8.2.2.1 Determine the sequence and direction of logging. The sequence in which a suite of logs is run is important from both a data quality and operational viewpoint. Because logging operations mix the borehole fluid, logs of fluid properties (for example, temperature, fluid resistivity, and fluid sampling should be run prior to other logs). Consideration should also be given to when borehole video surveys are performed because some logging tools may degrade borehole clarity. Tools that have arms or bowsprings that contact the borehole wall should be run late in the logging sequence because of the greater possibility of material from the borehole wall falling into the borehole. Because of the consequences of losing a tool with a radioactive source, these tools should be run last, and after a caliper log. Unstable boreholes should not be logged with radioactive source probes. All logs except fluid properties and video should be run with the probe moving up the borehole to reduce depth errors.

8.2.2.2 Select the depth reference. The selected depth reference needs to be stable and accessible.

8.2.2.3 Select horizontal and vertical scales.

8.2.2.4 Select the digitizing interval. See 8.3.1.2.

8.3 *Other Considerations:*

8.3.1 *Data Formats*—There are two methods of recording log data, digital and analog. Digital recording of logs should be used because of the numerous benefits of data manipulation. Digital recording is not yet practical for some logs such as video or acoustic televiewer.

8.3.1.1 An analog display should be available to be viewed in the field to verify the correct tool operation. Depth scales and units of measurement for the horizontal scale must be indicated clearly on each log.

8.3.1.2 The digital data are recorded at an operator-selected depth interval that should be as small as possible, at most, half the thickness of the smallest rock unit that can be resolved. The time interval for digital samples can also be selected by the operator. ASCII is the recommended format except for such logs as spectral gamma, full waveform sonic, borehole video, and acoustic televiewer. The digital file header should include all of the necessary information to reconstruct the logging procedures accurately and should duplicate the information included in the written header of the log.

8.3.1.3 Unprocessed data should be available. Nonproprietary processing algorithms shall be furnished if processed data is provided.

8.3.2 *Personnel:*

8.3.2.1 Personnel not having specialized training or experience should be cautious about using borehole geophysics and should solicit assistance from qualified practitioners or

attend courses on borehole geophysics.

8.3.2.2 Personnel operating logging equipment should have an understanding of the theory, field procedures, and methods of log interpretation.

8.3.2.3 A geoscientist, with experience in borehole geophysics, who understands the project objectives and local geohydrology may need to be available to examine logging results during logging operations when consistent with objectives of the program. This geoscientist is responsible for determining whether the instructions selected in the pre-logging conference are being followed and whether changes should be made.

8.3.2.4 Log interpretation should be performed by a geoscientist with experience in borehole geophysics and knowledge of the site geology and hydrology.

8.4 *Field Documentation*—A documentation plan for both the analog plot and digital data file should be established and become part of the work plan. Documentation of the following procedures is needed: calibration of logging probes, field operation of geophysical logging equipment, applicable decontamination, and format for presenting geophysical well log data. Repair, standardization, and calibration information should also be documented. Probes should be numbered to simplify the identification of associated documentation. Document all field problems including equipment malfunctions. This should include the steps taken to solve the problem and how the logs might have been affected. Repeat runs and field standardization should be more frequent when equipment problems occur. The use of one borehole on the project to check the probe response may aid in the identification of equipment or other problems. Probes should be recalibrated in a physical model after major repairs have been made.

8.4.1 *Log Headings (Headers)*—The log heading should contain all of the information that is necessary to analyze the log trace. Because auxiliary documents are frequently unavailable to other users of the log, all of the critical information concerning the log should be included on the final log heading. The header information should also be included in the same computer file as the log data. The following items listed are necessary and should be included on the log headings and computer files when appropriate. If information is not available or applicable, it should be noted on the heading. The following information should be included:

8.4.1.1 *Background Well Information:* owner of well and address, location of well (UTM coordinates, ¼ section, etc.); date; logging contractor and address; logging operator; drilling contractor and address; client and address; observer and address; elevation of top casing and distance above ground; and drilling history, methods etc.

8.4.1.2 *Borehole Conditions:* casing description; description of log depth datum; elevation of log depth datum; type of drilling fluid; resistivity and temperature of borehole fluid; depth of origin of borehole fluid samples; fluid level; time since last mud circulation; bottom hole temperature; and problems and unusual conditions.

8.4.1.3 *Equipment Data and Logging Parameters:* description of probe reference point; model and manufacturer of logging tools; logging company tool number; date and type of last calibration; date, type, and response of field standardization; top and bottom of logged interval; logging speed and direction; vertical depth error after logging; time constant or the time interval of digital samples; identification of disk containing digitized logs; and equipment problems.

8.4.1.4 *Specific Information for Nuclear Logging Probes:* Source description, initial source strength, and date determined; source to detector or receiver spacing; detector description; and data filtering or enhancement parameters.

8.4.1.5 *Specific Information for Acoustic and Electric Logging Probes:* source or transmitter description and signal output; source or transmitter to detector or receiver spacing; detector or receiver description; and data filtering or enhancement parameters.

8.4.2 *Quality Control During Logging Operations:* request changes in logging speed and time constant; repeat logs or log intervals based on field log analysis; check depth readout against log; note errors or changes on the log; and verify documentation listed above.

8.5 *Log Interpretation*—The full potential of a logging program cannot be realized until the logging measurements are interpreted. Log interpretation should start at the time of data acquisition and should continue as an iterative process throughout the project.

8.5.1 Logs should be analyzed and described as a suite and combined with information on lithology and fluid quality because of the synergistic nature of log data. The nonunique response of logs dictates the use of data from other sources to check the log interpretation, and this background data must be included in the report. A computer will be used in most cases to aid analysis of the logs, and information on the software and algorithms used should be included in the report.

8.5.2 Important interpretation steps include the following:

8.5.2.1 Establishing database (for example, format conversion, depth corrections, editing, and filtering).

8.5.2.2 Applying borehole corrections (for example, correct electric logs for borehole diameter and fluid resistivity).

8.5.2.3 Performing initial data inversion-conversion log units to values appropriate for investigation (for example, density units to porosity).

8.5.2.4 Performing large-scale data inversion (for example, cross sections, regional correlation, and model parameters).

9. Report

9.1 Depending on the project objective, report only data or data and interpretations.

9.1.1 Both types of reports should include the following:

9.1.1.1 Objectives and scope.

9.1.1.2 Field Documentation (for example, site conditions, borehole conditions, data collection procedures, calibration and standardization of logging probes, field operation of geophysical logging equipment, and format for recording geophysical log data, including any filtering or processing of the data, problems, and unusual conditions; see 8.4).

9.1.1.3 Both the digital log data and log plots.

9.1.1.4 Abstract, executive summary, or conclusions.

9.1.2 Interpretation reports should include the following:

9.1.2.1 Log composites (for example, summary plots showing logs, lithology, well construction, and water quality zones). These composites are commonly annotated to indicate the features of interest and correlated with lithologic descriptions.

9.1.2.2 Brief description of the geologic and hydrologic setting.

9.1.2.3 Specific information on log analysis, that is, depth corrections and recalibration of logs, physical models or sample analyses that were used for calibration, methods of log interpretation, software used, and copies of cross-plots or other plots of data resulting from log analysis.

9.1.2.4 Well-to-well correlation sections and comparison to surface geophysical and other testing data, when available.

10. Keywords

10.1 acoustic logging; acoustic televiewer; borehole geophysics; borehole video; caliper logging; chemical properties and physical properties; deviation; electric logging; environmental; fluid conductivity/resistivity logging; fluid logging; gamma logging; gamma-gamma logging; geology; geophysics; geotechnical; ground water; hydrology; induction logging; log calibration and standardization; log headings; neutron logging; nuclear logging; resistivity logging; single-point resistance logging; spontaneous potential logging; temperature logging; well logging

APPENDIX

(Nonmandatory Information)

X1. CALIBRATION FACILITIES AVAILABLE FOR PUBLIC USE (1989)

X1.1 *Name and Location*—American Petroleum Institute Calibration Facility, University of Houston, Houston, TX: four pits **(14, 19, 20)**.

X1.2 *Who to Contact*: University of Houston, Cullen College of Engineering, (713) 749-3423.

X1.3 *Probes That Can Be Calibrated*—Pit 1: neutron and gamma-gamma; Pit 2: gamma (simulated shale); Pits 3 and 4: spectral gamma.

X1.3.1 *Name and Location*—U.S. Department of Energy, Grand Junction, CO: 20 models or pits **(18)**.

X1.3.2 *Who to Contact*—U.S. Department of Energy, Grand Junction Operations Office, or the prime contractor at the U.S. Department of Energy office, (303) 248-7768 or 6702.

X1.4 *Probes That Can Be Calibrated*—Gamma, gamma spectral, neutron, gamma-gamma, and magnetic susceptibility. Also, wet and dry borehole size factors and a 300-ft borehole with radium foil at known depths for check of depth measurements.

X1.4.1 *Name and Location*—U.S. Bureau of Mines density pits Pit 1: six holes and magnetic susceptibility (Pits 2). Denver Federal Center, Lakewood, CO: Pit six holes; Pit 2: three holes **(17)**.

X1.4.2 *Who to Contact*—U.S. Geological Survey, Water Resources Division, Borehole Geophysics Project, Building 25, Denver Federal Center, (303) 236-5913.

X1.5 *Probes That Can Be Calibrated*—Pit 1: gamma-gamma, acoustic, resistivity; and Pit 2: magnetic susceptibility.

X1.5.1 *Name and Location*—U.S. Department of Energy, Fractured igneous rock calibration models, Denver Federal Center, Lakewood, CO: Three models or pits **(16)**.

X1.5.2 *Who to Contact*—U.S. Geological Survey, Water Resources Division, Borehole Geophysics Project, Building 25, Denver Federal Center, (303) 236-5913.

X1.6 *Probes That Can Be Calibrated*—Fracture detection probes, neutron, gamma-gamma, short-spaced resistivity, and acoustic velocity.

X1.7 *Other Facilities*—The Geological Survey of Canada is developing a system of deep test holes and calibration facilities that are presently available at several locations in Canada. Gamma, gamma spectral, and coal property models are completed, and other physical property models are under construction **(15)**. Calibration facilities at universities, private logging companies, and government agencies may also be available at other locations for use by outside logging groups.

REFERENCES

The following is a partial list of references intended to provide basic information on the various logging methods. There are many more pertinent references, but they are too numerous for listing in this guide **(34, 36, 51)**.

(1) *Glossary of Terms and Expressions Used in Well Logging*, 2nd Ed., Society of Professional Well Log Analysts, Houston, TX, 1984, p. 74.

(2) Bateman, R. M., *Log Quality Control*, IHRDC, Boston, MA, 1985, p. 398.

(3) Doveton, J. H., *Log Analysis of Subsurface Geology—Concepts and Computer Methods*, John Wiley and Sons, Inc., New York, NY, 1986, p. 273.

(4) Hallenberg, J. K., *Geophysical Logging for Mineral and Engineering Applications*, Penn Well Books, p. 264.

(5) Hearst, J. R., and Nelson, P. H., *Well Logging for Physical Properties*, McGraw-Hill Book Co., 1985, p. 576.

(6) Hilchie, D. W., *Applied Open Hole Log Interpretation for Geologists and Engineers*, Douglas W. Hilchie Inc., 1978.

(7) Keys, W. S., *Borehole Geophysics Applied To Ground-Water Investigations*, National Water Well Association, 1989, p. 313.

(8) Lynch, E. J., *Formation Evaluation*, Harper and Row, New York, NY, 1962, p. 422.

(9) Guyod, H., "Interpretation of Electric and Gamma Ray Logs in Water Wells," *The Log Analyst*, Vol 6, No. 5, 1966, pp. 29–44.

(10) Taylor, K. C., Hess, J. W., and Mazzela, A., "Field Evaluation of a Slim-Hole Borehole Induction Tool," *Ground Water Monitoring Review*, Vol 9, No. 1, 1989.

(11) Darr, P. S., Gilkeson, R. H., and Yearsley, E. N., "Intercomparison of Borehole Geophysical Techniques in a Complex Depositional Environment," *Proceedings of the Fourth Outdoor Action Conference on Aquifer Restoration, Ground Water Monitoring and Geophysical Methods*, Las Vegas, NV, May 14–17, 1990.

(12) Patten, E. P., and Bennett, G. D., "Methods of Flow Measurement in Well Bores," *U.S. Geological Survey Water-Supply Paper 1544-C*, 1962, p. 28.

(13) Society of Professional Well Log Analysts, *The Art of Ancient Log Analysis*, Houston, TX, 1979, p. 131.

(14) Belknap, W. B., Dewan, J. F., Kirkpatrick, C. V., Mott, W. E., Pearson, A. J., and Robson, W. R., "API Calibration Facility for Nuclear Logs," *Drilling, and Production Practice*: American Petroleum Institute, 1959, pp. 289–316.

(15) Killeen, P. G., "A System of Deep Test Holes and Calibration Facilities for Developing and Testing New Borehole Geophysical Techniques," *Borehole Geophysics for Mining and Geotechnical Applications*, Paper 85-27, Geological Survey of Canada, 1986, pp. 29–46.

(16) Mathews, M. A., Scott, J. H., and LaDelfe, C. M., *Test Pits for Calibrating Well Logging Equipment in Fractured Hard-Rock Environment*, Los Alamos National Laboratory Report LA-UR-85-859, 1985, p. 84.

(17) Snodgrass, J. J., *Calibration Models for Geophysical Borehole Logging*, U.S. Bureau of Mines Report of Investigations 8148, p. 21.

(18) Stromswold, D. C., and Wilson, R. D., "Calibration and Data Correction Techniques for Spectral Gamma-Ray Logging," *Society of Professional Well Log Analysts 22nd Annual Logging Symposium Transactions*, 1981, pp. M 1–18.

(19) Bryant, T. M., and Gage, T. D., "API Calibration of MWD

Gamma Ray Tools," *Society of Professional Well Log Analysts 29th Annual Logging Symposium Transactions*, 1988, pp. B 1–14.

(20) Scott, H. D., "Analysis of Samples from the API K-U-TH Logging Calibration Facility," *Society of Professional Well Log Analysts 30th Annual Logging Symposium Transactions*, 1989, pp. MM 1–25.

(21) Wahl, J. S., "Gamma-Ray Logging," *Geophysics*, Vol 48, No. 11, 1983, pp. 1536–1550.

(22) Killeen, P. G., "Gamma-Ray Logging and Interpretation," *Developments in geophysical exploration methods: Barking, Essex, England*, A. A. Fitch, ed., Applied Science Publishers, Book 3, Chapter 7, 1982, pp. 95–150.

(23) Tittman, J., and Wahl, J. S., "The Physical Foundations of Formation Density Logging (Gamma-Gamma)," *Geophysics*, Vol 30, No. 2, 1965, pp. 284–294.

(24) Scott, J. H., "Borehole Compensation Algorithms for a Small-Diameter, Dual-Detector Density Well-Logging Probe," *Society of Professional Well Log Analysts Annual Logging Symposium 18th Symposium Transactions*, 1977, pp. S1–S17.

(25) Arnold, D. M., and Smith, H. D., Jr., "Experimental Determination of Environmental Corrections for a Dual-Spaced Neutron Porosity Log," *Society of Professional Well Log Analysts Annual Logging Symposium Transactions*, Mexico City, Vol 2, 1981, pp. VV1–VV24.

(26) Guyod, H., and Shane, L. E., "Introduction to Geophysical Well Logging—Acoustical Logging," *Geophysical Well Logging: Houston, Texas*, Vol 1, Hubert Guyod, 1969, p. 256.

(27) Pirson, S. J., *Handbook of Well Log Analysis*, Prentice Hall, Englewood Cliffs, NJ, 1963, p. 326.

(28) Zemanek, J., Caldwell, R. L., Glenn, E. E., Jr., Holcomb, S. V., Norton, L. J., and Straus, A. J. D., "The Borehole Televiewer—A New Logging Concept for Fracture Location and Other Types of Borehole Inspection," *Journal of Petroleum Technology*, Vol 21, No. 6, 1969, pp. 762–774.

(29) Hilchie, D. W., "Caliper Logging—Theory and practice," *The Log Analyst*, Vol 9, No. 1, 1968, pp. 3–12.

(30) Stevens, H. H., Jr., Ficke, J. F., and Smoot, G. F., "Water Temperature-Influential Factors, Field Measurement, and Data Presentation," *U.S. Geological Survey Techniques of Water-Resources Investigations*, Book 1, Chapter D1, 1975.

(31) Sammel, E. A., "Convective Flow and Its Effect on Temperature Logging in Small-Diameter Wells," *Geophysics*, Vol 33, No. 6, 1968, pp. 1004–1012.

(32) Conaway, J. G., "Deconvolution of Temperature Gradient Logs," *Geophysics*, Vol 42, No. 4, 1977, pp. 823–837.

(33) Hess, A. E., *A Heat-Pulse Flowmeter for Measuring Low Velocities in Boreholes*, U.S. Geological Survey Open-File Report 82-699, 1982, p. 44.

(34) Prensky, S. E., "Geological Applications of Well Logs—An Introductory Bibliography and Survey of Well Logging Literature Through September 1986, Arranged by Subject and First Author," *The Log Analyst*, Parts A and B, Vol 28, No. 1, 1987, pp. 71–107; Part C, Vol 28, No. 2, 1987, pp. 219–248.

(35) Prensky, S. E., "Geological Applications of Well Logs—An Introductory Bibliography and Survey of Well Logging Literature; Annual Update, October 1986 through September 1987," *The Log Analyst*, Vol 28, No. 6, 1987, pp. 558–575. Bibliographic update

for October 1987 through September 1988, *The Log Analyst*, Vol 29, No. 6, 1988, pp. 426–443.

(36) Prensky, S. E., "Bibliography of Well Log Applications," October 1988–September 1989, annual update; *The Log Analyst*, Vol 30, No. 6, 1989, pp. 448–470. October 1989–September 1990, annual update: *The Log Analyst*, Vol 31, No. 6, 1990, pp. 395–424.

(37) Daniels, J. J., "Extending the Range of Investigation of Borehole Electrical Measurements," *Transactions of the SPWLA 18th Annual Logging Symposium*, 1977, 17 pp.

(38) Dyck, A. V., *A Method for Quantitative Interpretation of Wideband Drill-Hole EM Surveys in Mineral Exploration*, University of Toronto PhD Thesis, 1981.

(39) Labo, J., "A Practical Introduction to Borehole Geophysics," *Geophysical References, Soc. Explor. Geophysicists*, Vol 2, Chapter 9, 1987, pp. 179–195.

(40) Scott, J. H., Seeley, R. L., and Barth, J. J., "A Magnetic Susceptibility Well Logging System for Mineral Exploration," *Transactions of the SPWLA 22nd Annual Logging Symposium*, 1981.

(41) Senftle, F. E., "Application of Gamma Ray Spectral Analysis to Subsurface Mineral Exploration," *A Short Course Handbook for Neutron Activation Analysis in the Geosciences*, Mineralogical Association of Canada, Halifax, N.S., 1980.

(42) Freedman, R., and Vogiatzis, J. P., "Theory of Microwave Dielectric Constant Logging Using the Electromagnetic Wave Propagation Method," *Geophysics*, Vol 44, No. 5, 1979, pp. 969–986.

(43) Wright, D. L., Watts, R. D., and Bramsoe, E., "A Short-Pulse Electromagnetic Transponder for Hole-to-hole Use," *IEEE Transactions on Geoscience and Remote Sensing*, Vol GE-22, No. 6, 1984, pp. 720–725.

(44) Quirein, J. A., Gardner, J. S., and Watson, J. T., "Combined Natural Gamma Ray Spectral/Lith-Density Measurements Applied to Complex Lithologies," *Society of Petroleum Engineering of AIME Paper SPE 11143*, 1982, 14 pp.

(45) Olhoeft, G. R., and Scott, J. H., "Nonlinear Complex Resistivity Logging," *Transactions of SPWLA 21st Annual Logging Symposium*, 1980.

(46) Tsang, C., Hufschmied, P., and Hale, F. V., "Determination of Fracture Inflow Parameters with a Borehole Fluid Conductivity Logging Method," *Water Resources Research*, Vol 26, No. 4, 1990, pp. 561–578.

(47) Craig, J. T., Jr., and Randall, B. V., "Directional Survey Calculation," *Pet. International*, 1976, pp. 38–54.

(48) Bigelow, E. L., "Making More Intelligent Use of Log Derived Dip Information, Parts I–V," *Log Analyst*, Vol 26, 1985, No. 1, pp. 41–51; No. 2, pp. 25–41; No. 3, pp. 18–31; No. 4, pp. 21–43; and No. 5, pp. 25–64.

(49) Hodges, R. E., and Teasdale, W. E., *Considerations Related To Drilling Methods in Planning and Performing Borehole-Geophysical Logging for Ground-Water Studies*, U.S. Geological Survey Water-Resources Investigations Report 91-4090, Denver, CO, 1991.

(50) Sandberg, E. V., Olsson, O. L., and Falk, L. R., "Combined Interpretation of Fracture Zones in Crystalline Rock Using Single-Hole and Crosshole Tomography and Directional Borehole-Radar Data," *The Log Analyst*, Vol 32, No. 2, 1991, pp. 108–119.

(51) Boulding, J. R., *Use of Airborne, Surface, and Borehole Geophysical Techniques at Contaminated Sites: A Reference Guide*, U.S. EPA/625/R-92/007, 295 pp.

Standard Guide for
Using the Seismic Refraction Method for Subsurface Investigation[1]

This standard is issued under the fixed designation D 5777; the number immediately following the designation indicates the year of original adoption or, in the case of revision, the year of last revision. A number in parentheses indicates the year of last reapproval. A superscript epsilon (ϵ) indicates an editorial change since the last revision or reapproval.

1. Scope

1.1 *Purpose and Application*—This guide summarizes the equipment, field procedures, and interpretation methods for the assessment of subsurface materials using the seismic refraction method. Seismic refraction measurements as described in this guide are applicable in mapping subsurface conditions for various uses including geologic, geotechnical, hydrologic, environmental, mineral exploration, petroleum exploration, and archaeological investigations. The seismic refraction method can sometimes be used to map geologic conditions including depth to bedrock, or to water table, lithology, structure, and fractures or all of these. The calculated seismic wave velocity is related to mechanical material properties. Therefore, characterization of the material (type of rock, degree of weathering, and rippability) can sometimes be made on the basis of seismic velocity and other geologic information.

1.2 *Limitations:*

1.2.1 This guide provides an overview of the seismic refraction method using compressional (P) waves. It does not address the details of the seismic refraction theory, field procedures, or interpretation of the data. Numerous references are included for that purpose and are considered an essential part of this guide. It is recommended that the user of the seismic refraction method be familiar with the relevant material within this guideline and the references provided.

1.2.2 This guide is limited to the commonly used approach to seismic refraction measurements made on land. The seismic refraction method can be adapted for a number of special uses, on land, within a borehole and on water. However, a discussion of these other adaptations of seismic refraction measurements is not included in this guide.

1.2.3 There are certain cases in which shear waves need to be measured to satisfy project requirements. The measurement of seismic shear waves is a subset of seismic refraction. This guide is not intended to include this topic and focuses only on P wave measurements.

1.2.4 The approaches suggested in this guide for the seismic refraction method are most commonly used, widely accepted, and proven; however, other approaches or modifications to the seismic refraction method that are technically sound may be substituted.

1.2.5 Technical limitations and interferences of the seismic refraction method are discussed in 5.4.

1.3 *Precautions:*

1.3.1 It is the responsibility of the user of this guide to follow any precautions within the equipment manufacturer's recommendations, establish appropriate health and safety practices, and consider the safety and regulatory implications when explosives are used.

1.3.2 If the method is applied at sites with hazardous materials, operations, or equipment, it is the responsibility of the user of this guide to establish appropriate safety and health practices and determine the applicability of any regulations prior to use.

1.4 *This standard does not purport to address all of the safety concerns, if any, associated with its use. It is the responsibility of the user of this standard to establish appropriate safety and health practices and determine the applicability of regulatory limitations prior to use.*

2. Referenced Documents

2.1 *ASTM Standards:*
D 853 Terminology Relating to Soil, Rock, and Contained Fluids[2]
D 2845 Test Method for Laboratory Determination of Pulse Velocities and Ultrasonic Elastic Constants of Rock[2]
D 4428/D 4428M Test Method for Crosshole Seismic Testing[2]

3. Terminology

3.1 *Definitions:*

3.1.1 The majority of the technical terms used in this guide are defined in Refs (1) and (2).[3] Also see Terminology D 853.

4. Summary of Practice

4.1 *Summary of the Method*—Measurements of the travel time of a compressional (P) wave from a seismic source to a geophone(s) are made from the land surface and are used to interpret subsurface conditions and materials. This travel time, along with distance between the source and geophone(s), can also be interpreted to yield the depth to refracting layer(s). The calculated seismic velocities of the layers can often be used to characterize some of the properties of natural or man-made man subsurface materials.

4.2 *Complementary Data*—Geologic and water table data

[1] This guide is under the jurisdiction of ASTM Committee D-18 on Soil and Rock and is the direct responsibility of Subcommittee D18.01 on Surface and Subsurface Characterization.
Current edition approved Sept. 10, 1995. Published February 1996.

[2] *Annual Book of ASTM Standards*, Vol 04.08.
[3] The boldface numbers given in parentheses refer to a list of references at the end of the text.

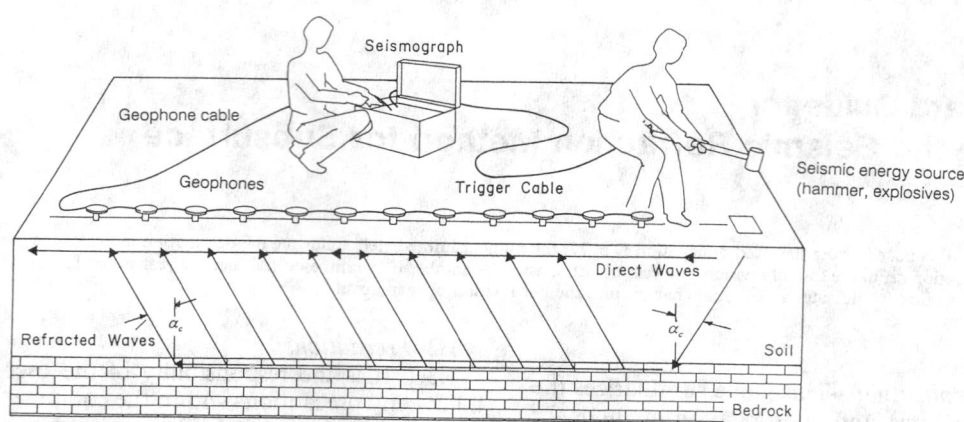

FIG. 1 Field Layout of a Twelve-Channel Seismograph Showing the Path of Direct and Refracted Seismic Waves in a Two-Layer Soil/Rock System (α_c = Critical Angle)

obtained from borehole logs, geologic maps, data from outcrops or other complementary surface and borehole geophysical methods may be necessary to properly interpret subsurface conditions from seismic refraction data.

5. Significance and Use

5.1 *Concepts:*

5.1.1 This guide summarizes the equipment, field procedures, and interpretation methods used for the determination of the depth, thickness and the seismic velocity of subsurface soil and rock or engineered materials, using the seismic refraction method.

5.1.2 Measurement of subsurface conditions by the seismic refraction method requires a seismic energy source, trigger cable (or radio link), geophones, geophone cable, and a seismograph (see Fig. 1).

5.1.3 The geophone(s) and the seismic source must be placed in firm contact with the soil or rock. The geophones are usually located in a line, sometimes referred to as a geophone spread. The seismic source may be a sledge hammer, a mechanical device that strikes the ground, or some other type of impulse source. Explosives are used for deeper refractors or special conditions that require greater energy. Geophones convert the ground vibrations into an electrical signal. This electrical signal is recorded and processed by the seismograph. The travel time of the seismic wave (from the source to the geophone) is determined from

the seismic wave form. Fig. 2 shows a seismograph record using a single geophone. Fig. 3 shows a seismograph record using twelve geophones.

5.1.4 The seismic energy source generates elastic waves which travel through the soil or rock from the source, or both. When the seismic wave reaches the interface between two materials of different seismic velocities, the waves are refracted according to Snell's Law (3, 7). When the angle of incidence equals the critical angle at the interface, the refracted wave moves along the interface between two materials, transmitting energy back to the surface (Fig. 1). This interface is referred to as a refractor.

5.1.5 A number of elastic waves are produced by a seismic energy source. Because the compressional P-wave has the highest seismic velocity, it is the first wave to arrive at each geophone (see Figs. 2 and 3).

5.1.6 The P-wave velocity V_p is dependent upon the bulk

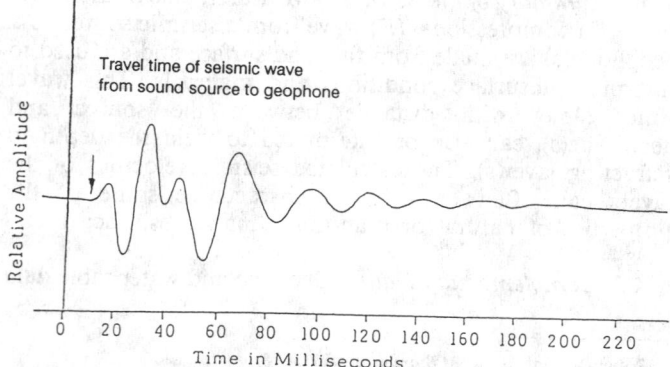

NOTE—Arrow marks arrival of first compressional wave.

FIG. 2 A Typical Seismic Waveform from a Single Geophone

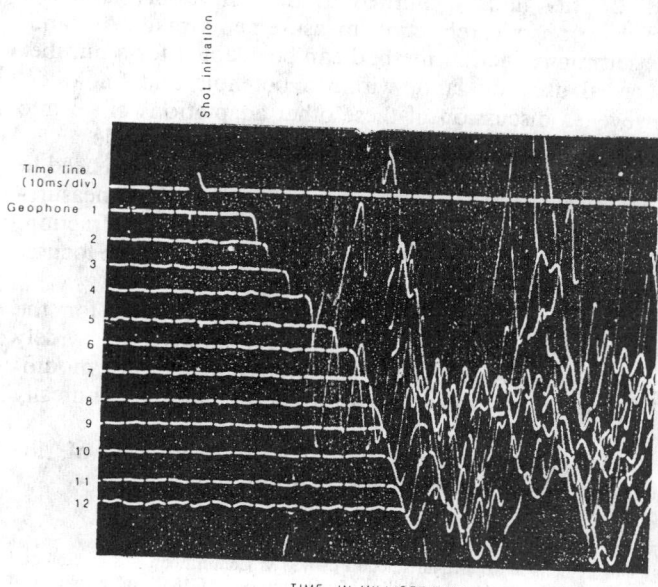

FIG. 3 Twelve-Channel Analog Seismograph Record Showing Good First Breaks Produced by an Explosive Sound Source (26)

modulus, the shear modulus and the density in the following manner (4):

$$V_p = \sqrt{[(K + 4/3\mu)/\rho]}$$

where:
V_p = compressional wave velocity,
K = bulk modulus,
μ = shear modulus, and
ρ = density.

5.1.7 The arrival of energy from the seismic source at each geophone is recorded by the seismograph (Fig. 3). The travel time (the time it takes for the seismic *P*-wave to travel from the seismic energy source to the geophone(s)) can be determined from each waveform. The unit of time is usually milliseconds (1 ms = 0.001 s).

5.1.8 The travel times are plotted against the distance between the source and the geophone to make a time distance plot. Fig. 4 shows the source and geophone layout and the resulting idealized time distance plot for a horizontal two-layered earth.

5.1.9 The travel time of the seismic wave between the seismic energy source and a geophone(s) is a function of the distance between them, the depth to the refractor(s) and the seismic velocities of the materials through which the wave passes.

5.1.10 The depth to a refractor can be calculated by knowing the source to geophone geometry (spacing and

V_1 = seismic velocity in layer 1

V_2 = seismic velocity in layer 2

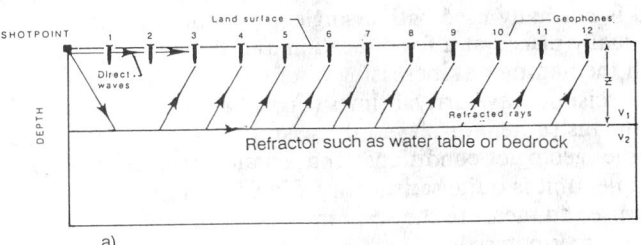

a).

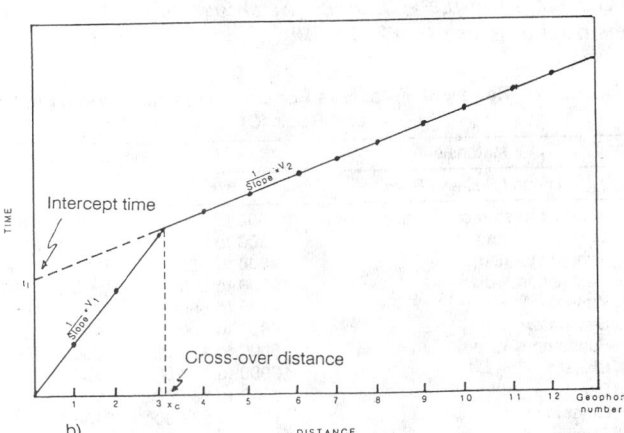

b).

FIG. 4 (*a*) Seismic Raypaths and (*b*) Time-Distance Plot for a Two-Layer Earth With Parallel Boundaries (26)

elevation), determining the apparent seismic velocities (which are the reciprocals of the slopes of the plotted lines in the time distance plot), and the intercept time or crossover distances on the time distance plot (see Fig. 4). Intercept time and crossover distance-depth formulas have been derived in the literature (5–8). These derivations are straightforward inasmuch as the total travel time of the seismic wave is measured, the velocity in each layer is calculated from the time-distance plot, and the raypath geometry is known. The only unknown is the depth of the high-velocity refractor. These interpretation formulas are based on the following assumptions: (*1*) the boundaries between layers are planes that are either horizontal or dipping at a constant angle, (*2*) there is no land-surface relief, (*3*) each layer is homogeneous and isotropic, (*4*) the seismic velocity of the layers increases with depth, and (*5*) intermediate layers must be of sufficient velocity contrast, thickness and lateral extent to be detected. Reference (9) provides an excellent summary of these equations for two and three layer cases. The formulas for a two-layered case (see Fig. 4) are given below.

5.1.10.1 Intercept-time formula:

$$z = \frac{t_i}{2} \frac{V_2 V_1}{\sqrt{(V_2)^2 - (V_1)^2}}$$

where:
z = depth to Layer two at point,
t_i = intercept time,
V_2 = seismic velocity in Layer two, and
V_1 = seismic velocity in Layer one.

5.1.10.2 Crossover distance formula:

$$z = \frac{x_c}{2} \sqrt{\frac{V_2 - V_1}{V_2 + V_1}}$$

where:
z, V_2 and V_1 are as defined above and x_c = crossover distance.

5.1.11 Three to four layers are usually the most that can be resolved by seismic refraction measurements. Fig. 5 shows the source and geophone layout along with the resulting time distance plot for an idealized three-layer case.

5.1.12 The refraction method is used to define the depth or profile of the top of one or more refractors, or both, for example, depth to water table or bedrock.

5.1.13 The source of energy is usually located at or near each end of the geophone spread; a refraction measurement is made in each direction. These are referred to as forward and reverse measurements, sometimes incorrectly called reciprocal measurements, from which separate time distance plots are made. Fig. 6 shows the source and geophone layout and the resulting time distance plot for a dipping refractor. The velocity obtained for the refractor from either of these two measurements alone is the apparent velocity of the refractor. Both measurements are necessary to resolve the true seismic velocity and the dip of layers (9) unless other data are available that indicate a horizontal layered earth. These two apparent velocity measurements and the intercept time or crossover distance can be used to calculate the true velocity, depth and dip of the refractor. Note that only two depths of the planar refractor are obtained using this approach (see Fig. 7). Depth of the refraction surface can be obtained under each geophone by using a more sophisticated data collection and interpretation approach.

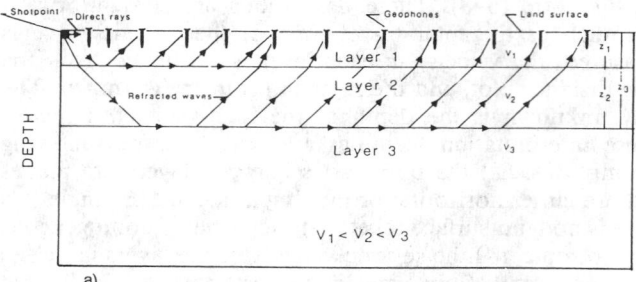

V_1 = seismic velocity in layer 1

V_2 = seismic velocity in layer 2

V_3 = seismic velocity in layer 3

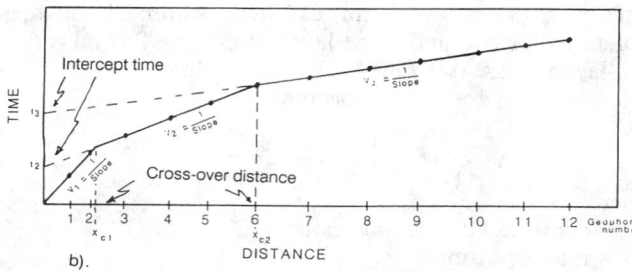

FIG. 5 **(a) Seismic Raypaths and (b) Time-Distance Plot for a Three-Layer Model With Parallel Boundaries (26)**

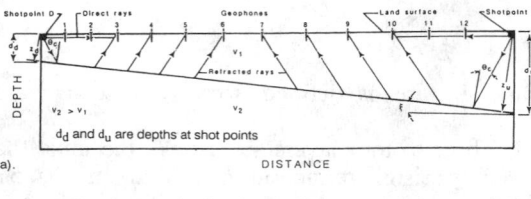

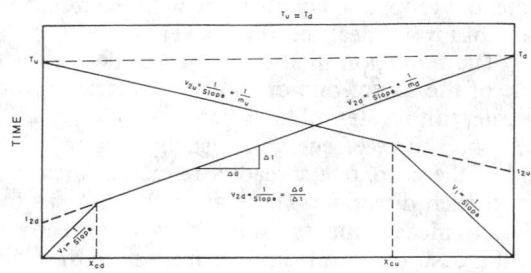

FIG. 6 **(a) Seismic Raypaths and (b) Time-Distance Plot for a Two-Layer Model With A Dipping Boundary (26)**

5.1.14 Most refraction surveys for geologic, engineering, hydrologic and environmental applications are carried out to determine depths of refractors that are less than a 100 m (about 300 ft). However, with sufficient energy, refraction measurements can be made to depths of 1000 ft (300 m) and more **(5)**.

5.2 *Parameter Measured and Representative Values:*

5.2.1 The seismic refraction method provides the velocity of compressional *P*-waves in subsurface materials. Although the *P*-wave velocity can be a good indicator of the type of soil or rock, it is not a unique indicator. Table 1 shows that each type of sediment or rock has a wide range of seismic velocities, and many of these ranges overlap. While the seismic refraction technique measures the seismic velocity of seismic waves in earth materials, it is the interpreter who, based on knowledge of the local conditions or other data, or both, must interpret the seismic refraction data and arrive at a geologically reasonable solution.

5.2.2 According to Mooney **(8)**, *P*-wave velocities are generally greater for:

5.2.2.1 Denser rocks than lighter rocks;

5.2.2.2 Older rocks than younger rocks;

5.2.2.3 Igneous rocks than sedimentary rocks;

5.2.2.4 Solid rocks than rocks with cracks or fractures;

5.2.2.5 Unweathered rocks than weathered rocks;

5.2.2.6 Consolidated sediments than unconsolidated sediments;

5.2.2.7 Water-saturated unconsolidated sediments than dry unconsolidated sediments; and

5.2.2.8 Wet soils than dry soils.

5.3 *Equipment*—Geophysical equipment used for surface seismic refraction measurement includes a seismograph, geophones, geophone cable, an energy source and a trigger cable or radio link. A wide variety of seismic geophysical equipment is available and the choice of equipment for a seismic refraction survey should be made in order to meet or exceed the objectives of the survey.

5.3.1 *Seismographs*—A wide variety of seismographs are available from different manufacturers. They range from relatively simple, single-channel units to very sophisticated multichannel units. Most engineering seismographs sample, record and display the seismic wave digitally.

5.3.1.1 *Single Channel Seismograph*—A single channel seismograph is the simplest seismic refraction instrument and is normally used with a single geophone. The geophone is usually placed at a fixed location and the ground is struck with the hammer at increasing distances from the geophone. First seismic wave arrival times (Figs. 2 and 3) are identified on the instrument display of the seismic waveform. For some simple geologic conditions and small projects a single-channel unit is quite satisfactory. Single channel systems are also used to measure the seismic velocity of rock samples or engineered materials.

5.3.1.2 *Multiple Channel Seismograph*—Multi-channel seismographs use 6, 12, 24, 48 or more geophones. With a

TABLE 1 Range of Velocities For Compressional Waves in Soil and Rock (3)

Materials	Velocity	
Natural Soil and Rock	ft/s	m/s
Weathered surface material	800 to 2000	240 to 610
Gravel or dry sand	1500 to 3000	460 to 915
Sand (saturated)	4000 to 6000	1220 to 1830
Clay (saturated)	3000 to 9000	915 to 2750
Water[A]	4700 to 5500	1430 to 1665
Sea water[A]	4800 to 5000	1460 to 1525
Sandstone	6000 to 13 000	1830 to 3960
Shale	9000 to 14 000	2750 to 4270
Chalk	6000 to 13 000	1830 to 3960
Limestone	7000 to 20 000	2134 to 6100
Granite	15 000 to 19 000	4575 to 5800
Metamorphic rock	10 000 to 23 000	3050 to 7000

[A] Depending on temperature and salt content.

multi-channel seismograph, the seismic wave forms are recorded simultaneously for all geophones (see Fig. 3).

5.3.1.3 The simultaneous display of waveforms enables the operator to observe trends in the data and helps in making reliable picks of first arrival times. This is especially useful in areas that are seismically noisy and in areas with complex geologic conditions. Computer programs are available that help the interpreter pick the first arrival time.

5.3.1.4 *Signal Enhancement*—Signal enhancement or energy stacking that improves the signal to noise ratio is available in most seismographs. It is a significant aid when working in noisy areas or with small energy sources. Signal enhancement is accomplished by adding the refracted seismic signals for a number of impacts. This process increases the signal to noise ratio by summing the amplitude of the coherent seismic signals while reducing the amplitude of the random noise by averaging.

5.3.2 *Geophone and Cable:*

5.3.2.1 A geophone transforms the *P*-wave energy into a voltage that can be recorded by the seismograph. For refraction work, the frequency of the geophones varies from 8 to 14 Hz. The geophones are connected to a geophone cable that is connected to the seismograph (see Fig. 1). The geophone cable has electrical connection points (take outs) for each geophone, usually located at uniform intervals along the cable. Geophone placements are spaced from about 1 m to hundreds of meters (2 or 3 ft to hundreds of feet) apart depending upon the level of detail needed to describe the surface of the refractor and the depth of the refractor(s). In some cases, the geophone intervals may be adjusted at the shot end of a cable to provide additional seismic velocity information in the shallow subsurface.

5.3.2.2 If connections between geophones and cables are not waterproof, care must be taken to assure they will not be shorted out by wet grass, rain, etc. Special waterproof geophones (mersh geophones), geophone cables and connectors are required for areas covered with shallow water.

5.3.3 *Energy Sources:*

5.3.3.1 The selection of seismic refraction energy sources is dependent upon the depth of investigation and geologic conditions. Four types of energy sources are commonly used in seismic refraction surveys: sledge hammers, mechanical weight drop or impact devices, projectile (gun) sources, and explosives.

5.3.3.2 For shallow depths of investigation, 5 to 10 m (15 to 30 ft), a 4 to 7 kg (10 to 15 lb) sledge hammer may be used. Usually three to five hammer blows using signal enhancement capabilities of the seismograph will usually be sufficient. A strike plate on the ground can be used to improve the coupling of energy from the hammer to the soil.

5.3.3.3 For deeper investigations in dry and loose materials, more seismic energy is required, and a mechanized or a projectile (gun) source may be selected. Projectile sources may be discharged at or below the ground surface. Mechanical seismic sources use a large weight (of about 100 to 500 lb or 45 to 225 kg) that is dropped or driven downward under power. Mechanical weight drops are usually trailer mounted because of their size.

5.3.3.4 A small amount of explosives can provide a substantial increase in energy levels. Explosive charges are usually buried to improve reduce energy losses as well as for safety reasons. Burial of small amounts of explosives (less than 1 lb or 0.5 kg) as shallow as 3 to 6 ft (1 to 2 m) can be effective for shallow depths (less than 300 ft or 100 m) if properly backfilled and tamped. For greater depth of investigation (below 300 ft or 100 m), larger explosives charges (greater than 1 lb or 0.5 kg) may be required and are usually buried 6 ft (2 m) deep or more. Use of explosives requires specialized personnel and procedures.

5.3.4 *Timing*—A timing signal at the time of impact ($t = 0$) must be sent to the seismograph (see Fig. 1). The time of impact ($t = 0$) can be detected with mechanical switches, piezoelectric devices or a geophone (or accelerometer), or with a signal from a blasting unit. If electric blasting caps are used, they should be of the seismic type for accurate timing.

5.4 *Limitations and Interferences:*

5.4.1 *General Limitations Inherent to Geophysical Methods:*

5.4.1.1 A fundamental limitation of all geophysical methods is that a given set of data cannot be associated with a unique set of subsurface conditions. In most situations, surface geophysical measurements alone cannot resolve all ambiguities, and some additional information, such as borehole data, is required. Because of this inherent limitation in the geophysical methods, a seismic refraction survey alone can never be considered a complete assessment of subsurface conditions. Properly integrated with other geologic information, seismic refraction surveying is a highly effective, accurate, and cost-effective method of obtaining subsurface information.

5.4.1.2 In addition, all surface geophysical methods are inherently limited by decreasing resolution with depth.

5.4.2 *Limitations Specific to the Seismic Refraction Method:*

5.4.2.1 When refraction measurements are made over a layered earth, the seismic velocity of the layers are generally assumed to be uniform and isotropic. If actual conditions in the subsurface layers deviate significantly from this idealized model, then any interpretation will also deviate from the ideal. An increasing error is introduced in the depth calculations as the angle of dip of the layer increases. The error is a function of dip angle and the velocity contrast between dipping layers **(10, 11)**.

5.4.2.2 Another set of limitations inherent to seismic refraction surveys are referred to as blind-zone problems **(3, 9, 12)**. There must be a sufficient contrast between the seismic velocity of the overlying material and that of the refractor for the refractor to be detected. Some significant geologic or hydrogeologic boundaries may have no field-measurable seismic velocity contrast across them, and consequently cannot be detected with this technique.

5.4.2.3 A layer must also have a sufficient thickness in order to be detected **(12)**.

5.4.2.4 If a layer has a seismic velocity lower than that of the layer above it (a velocity reversal), the low seismic velocity layer cannot be detected. As a result, the computed depths of deeper layers will be greater than the actual depths (although the most common geologic condition is that of increasing seismic velocity with depth, there are situations in which seismic velocity reversals can occur). Interpretation methods are available to address this problem in some instances **(13)**.

5.4.3 *Interferences Caused by Natural and by Cultural Conditions:*

5.4.3.1 The seismic refraction method is sensitive to ground vibrations (time-variable noise) from a variety of sources. Spatial variables caused by geologic factors and cultural factors may also produce unwanted noise superimposed upon the data.

5.4.3.2 *Ambient Sources*—Ambient sources of noise include any vibration of the ground due to wind, water movement (for example, waves breaking on a nearby beach), natural seismic activity, or by rainfall on the geophones.

5.4.3.3 *Geologic Sources*—Geologic sources of noise may include unsuspected variations in travel time due to lateral and vertical variations in seismic velocity of subsurface layers (for example, the presence of large boulders within a soil matrix).

5.4.3.4 *Cultural Sources*—Cultural sources of noise include vibration due to movement of the field crew, nearby vehicles, and construction equipment, aircraft, or blasting. Cultural factors such as buried structures under or near the survey line may also cause spatial variable noise leading to unsuspected variations in travel time. In some cases, electrical noise from nearby powerlines may induce noise in long geophone cables.

5.4.3.5 During the course of designing and carrying out a refraction survey, sources of ambient, geologic, and cultural noise should be considered and its time of occurrence or location noted, or both. The exact form of the interference is not always predictable as it not only depends upon the magnitude of the noises but also upon the geometry and spacing of the geophones and source.

5.5 *Alternative Methods*—The limitations discussed above may prevent the effective use of the seismic refraction method, and other geophysical or non-geophysical methods may be required to investigate subsurface conditions.

6. Procedure

6.1 This section includes a discussion of personnel qualification (see 6.1.1), considerations for planning and implementing the seismic refraction survey (see 6.2 and 6.3) and interpretation of seismic refraction data (see 6.4).

6.1.1 *Qualification of Personnel*—The success of a seismic refraction survey, as with most geophysical techniques, is dependent upon many factors. One of the most important factors is the competency of the person(s) responsible for planning, carrying out the survey, and interpreting the data. An understanding of the theory, field procedures, and methods for interpretation of seismic refraction data along with an understanding of the site geology is necessary to successfully complete a seismic refraction survey. Personnel not having specialized training or experience, or both, should be cautious about using this technique and solicit assistance from qualified practitioners.

6.2 *Planning the Survey*—Successful use of the surface seismic refraction method depends to a great extent on careful and detailed planning as discussed in this section.

6.2.1 *Objective(s) of the Seismic Refraction Survey:*

6.2.1.1 Planning and design of a seismic refraction survey should be done with due consideration of the objectives of the survey and the characteristics of the site. These factors will determine the survey design, the equipment used, the

level of effort, the interpretation method selected, and budget necessary to achieve the desired results. Important considerations include site geology, depth of investigation, topography, and access. The presence of noise-generating activities (for example, on-site utilities, man-made structures), and operational constraints (for example, restrictions on the use of explosives), must also be considered. It is good practice to obtain as much relevant information (for example, data from any previous seismic refraction work, boring, geologic and geophysical logs in the study area, topographic maps or aerial photos, or both) as possible about the site prior to designing a survey and mobilization to the field.

6.2.1.2 A simple geologic/hydrologic model of the subsurface conditions at the site should be developed early in the design phase and should include the thickness and type of soil cover, depth and type of rock, depth to water table and a stratigraphic section along with horizons to be mapped with the seismic refraction method.

6.2.1.3 The objective of the survey may simply be a reconnaissance of subsurface conditions or it may be to provide the most detailed subsurface information possible. In reconnaissance surveys, such as regional geologic or ground water studies and preliminary engineering studies, the spacing between the geophone spreads, or geophone spacing, or both, may be large, only a few shot-points are used, and topographic maps or hand-level elevations may be usually sufficient. Under these conditions, the cost of obtaining seismic refraction data is relatively low, but the resulting subsurface data are not very detailed. In a detailed survey, the spacing between the geophone spreads, or geophone spacing, or both, is usually small, multiple shot-points are used, and elevations and locations of geophones and shot-points are more accurately determined. Under these conditions, the cost of obtaining seismic refraction data is higher, but can still be cost-effective because the resulting subsurface data is more detailed.

6.2.2 *Assess Whether or Not There is a Seismic Velocity Contrast:*

6.2.2.1 One of the most critical elements in planning a seismic refraction survey is the determination of whether there is an adequate seismic velocity contrast between the two geologic or hydrologic units of interest.

6.2.2.2 Assuming that no previous seismic refraction surveys have been made in the area, one is forced to rely upon knowledge of the geology, published references containing the seismic velocities of earth materials, and published reports of seismic refraction studies performed under similar conditions.

6.2.2.3 When there is doubt that sufficient seismic velocity contrast exists, a pre-survey test is desirable at a control point, such as a borehole or well, where the stratigraphy is known and the seismic velocities can be determined. Two types of tests may be considered: a vertical seismic profile (VSP) or another type of borehole log (such as a density log or sonic log) that provides an indication of subsurface velocity layering. From this information, the feasibility of using the seismic refraction method at the site can be assessed.

6.2.2.4 Forward modeling using mathematical equations (**6, 7, 9**) can be used to develop theoretical time distance plots. Given the thickness and the seismic velocity of the subsurface layers, these plots can be used to assess the feasibility of conducting a seismic refraction survey and to determine the geometry of the field-survey. However, all too often, sufficient information about layer thickness and seismic velocities will not be available to accurately model a site before field work is carried out. In that case, initial field measurements should be taken to assess whether an adequate seismic velocity contrast exists between the subsurface layers of interest.

6.2.3 *Selection of the Approach:*

6.2.3.1 The desired level of detail and ability to cope with unusual geologic conditions will determine the interpretation method to be used for a refraction survey, that in turn will determine the field procedures to be followed. General field considerations are given by Refs (3, 7, 9, 13–15).

6.2.3.2 Numerous approaches can be used to quantitatively interpret seismic refraction data; however, the most commonly used interpretation methods can be classified into two general groups: methods that are used to define planar refractors and methods that are used to define nonplanar refractors.

6.2.4 *Methods That Are Used To Define Planar Refractors:*

6.2.4.1 The intercept time method (ITM) and crossover distance method are the simplest and probably the best known of all the methods for the interpretation of seismic refraction data (8, 11). They can be described as the rigorous application of Snell's law to a subsurface model consisting of homogeneous layers and planar interfaces. These planar interfaces can be either horizontal or dipping. The intercept time method requires that a constant seismic velocity exists in the overburden and in the refractor within a single geophone spread (between the shot points). The intercept time method uses simple field and interpretation procedures. Measurements are usually made from each end of the seismic refraction line (a minimum of one off-end shot-point on each end of the geophone spread). The results obtained using this method include the thickness of the overburden and the dip of the refractor at two points (see Fig. 6). It is also common to make one shot in the middle of the geophone spread. Shots off of each end of the spread may also be made to provide additional data. Additional shot-points can increase the number of points along the refractor where depth can be determined.

6.2.4.2 The intercept time method or crossover distance method can be applied where a limited number of refractor depth determinations are required within a single geophone spread; the surface of the refractor can be satisfactorily approximated by a plane (horizontal or dipping); lateral variations in seismic velocity of the subsurface layers (over the length of the geophone spread) can be neglected; and thin intermediate seismic velocity layers and seismic velocity inversions can be neglected.

6.2.4.3 Additional discussion of survey design and field considerations for this method are given by Refs (3, 8, and 9).

6.2.5 *Methods That Are Used To Define Nonplanar Refractors*—A number of methods can be viewed as an extension of the intercept time method, whereby the depth to the refractor is calculated at the shot-points and at each geophone location. These methods require a greater under-standing of the seismic refraction theory, as well as a greater level of effort in data acquisition, processing, and interpretation.

6.2.6 *Common Reciprocal Methods:*

6.2.6.1 A group of methods (referred to as the common reciprocal methods (CRM) by Palmer (12)). These methods can provide a more detailed interpretation of nonplanar refractors. Depths are obtained under each geophone, thereby accounting for irregular refracting surfaces (nonplanar refractors). The CRM has many variations including the plus-minus method, the ABC Method and Hagiwaras Method (12). Most, but not all, of the methods are based on the assumption that within a single geophone spread, a constant seismic velocity exists both in the overlying units and in the refractor. Fig. 7 shows an interpreted seismic refraction section of an irregular rock surface using this approach. All these methods usually require that travel times be measured in both forward and reverse directions from at least three to seven shot-points per single geophone

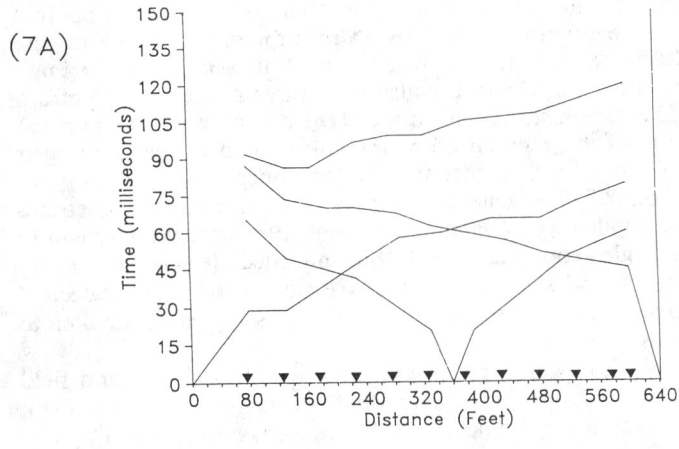

(7A)

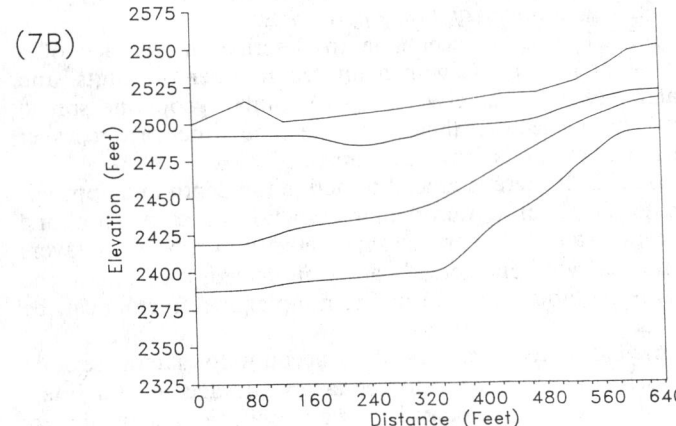

(7B)

**FIG. 7 Time Distance Plot (a) and Interpreted Seismic Section (b)
(30)**

spread. The resolution of the refractor topography obtained by the survey is dependent on the spacing between the geophones and the number of shot-points. Additional discussion of survey design and field considerations for these methods are given in Refs (3) and (9).

6.2.6.2 These methods can be applied where depths to the refractor are required at each geophone; the surface of the refractor has some relief; lateral variations in seismic velocity of the subsurface layers (over the length of the spread) can be neglected; and thin intermediate seismic velocity layers and seismic velocity inversions can be neglected.

6.2.7 *Generalized Reciprocal Method:*

6.2.7.1 The generalized reciprocal method (GRM), as described by Palmer (11, 16–18) and Lankston (13, 19), can sometimes aid in resolving complex conditions including undetected layers, lateral changes in seismic velocity and anisotropy. The GRM includes as special cases the delay time method and Hales method (11). The GRM method requires well sampled data (in time and space) to achieve the necessary resolution; therefore, a relatively small geophone spacing is required. This method usually requires that travel times be measured in both forward and reverse directions from five to seven shot-points per geophone spread. The generalized reciprocal method survey incorporates the strengths of most other seismic refraction methods and can provide the most detailed profile of a refractor, but requires considerably more effort in field data collection and interpretation. The full use of the generalized reciprocal method, that has been demonstrated by Palmer for model data and case histories, has still to achieve routine acceptance in engineering geophysics because it requires a greater field effort. The case histories in Palmer (18) demonstrate the application of the generalized reciprocal method to shallow targets of considerable geotechnical significance.

6.2.7.2 The generalized reciprocal method can sometimes be applied where lateral variations in seismic velocity within a single geophone spread, thin intermediate seismic velocity layers, and seismic velocity inversions cannot be neglected. Geophone spacing for this method is generally smaller to provide sufficient spatial data.

6.2.7.3 Additional discussions of survey design and field considerations for this method are given by Palmer (16); Lankston and Lankston (19); and Lankston (13, 15).

6.2.8 *Summary of Two Approaches:*

6.2.8.1 If it is acceptable to describe the surface of a refractor as a plane with a limited number of points, and lateral seismic velocity changes within a geophone spread can be neglected, then the intercept time or crossover distance methods may be sufficient.

6.2.8.2 If there is a need to define the depth and approximate shape of a non-planar refractor at each geophone location, and the lateral seismic velocity in subsurface layers within a geophone spread can be neglected, then one of the many methods that can define nonplanar refractors can be used.

6.2.8.3 If there is a need to account for lateral seismic velocity changes in subsurface layers and account for intermediate seismic velocity layers and seismic velocity inversions, then the generalized reciprocal method can be used.

6.2.8.4 Table 2 summarizes the features and limitations of each of these methods. It is modified from Palmer (11).

TABLE 2 Features and Limitations of Methods

Methods Used For Defining Planar Refractors
Include the Time Intercept and Crossover Distance Methods
These methods require the least field and interpretation effort and are, therefore, the lowest cost.

They can be applied where:
- Depth computations are provided near shot-points;
- The refractor is approximated by a plane (horizontal or dipping);
- Lateral variations in seismic velocity within a single geophone spread are neglected; and
- Thin intermediate velocity layers and velocity inversions are neglected.

Methods Used for Defining Non-Planar Refractors
The Common Reciprocal Method (CRM) Including Plus-Minus Method, the ABC Method, and the Hagiwaras Method
These CRM methods require additional field and interpretation effort and are intermediate in cost.

They can be applied where:
- Depth computations are provided at geophones;
- The refractor has some relief;
- Lateral variations in seismic velocity within a single geophone spread are neglected; and
- Thin intermediate velocity layers and velocity inversions are neglected.

The Generalized Reciprocal Method (GRM)
The Delay Time Method and Hales Method are special cases of the GRM

In addition to all the features of the CRM methods, the Generalized Reciprocal Method (GRM) may account for:
- Lateral variation in seismic velocity within a single geophone spread;
- Thin intermediate velocity layers and velocity inversions.

The GRM requires the greatest level of field and interpretation effort and is the most costly.

6.2.8.5 The choice of interpretation method may vary from site to site and will depend upon the detail required from the seismic refraction survey and the complexity of the geology at the site. The interpretation method will in turn determine the approach and level of effort required in the field.

6.2.8.6 When selecting the approach for data acquisition the specific processing and interpretation method that will be used must be considered since most processing and interpretation methods have specific requirements for data acquisition.

6.2.8.7 There are many field and interpretation methods that fall under the broad categories listed above. No attempt has been made to list all of the individual field and interpretation methods. Each one has strengths and weaknesses and must be selected to meet the project needs. The use of other field and interpretation methods not specifically mentioned are not precluded by this guide.

6.2.9 *Survey Design:*

6.2.9.1 *Location of Survey Lines*—Preliminary location of survey lines is usually done with the aid of topographic maps and aerial photos if an on-site visit is not possible. Consideration should be given to: the need for data at a given location; the accessibility of the area; the proximity of wells or test holes for control data; the extent and location of any asphalt or concrete surface, buried structures and utilities and other sources of cultural noise that will prevent measurements from being made, or introduce noise into the data (see 5.7.3); and adequate space for the refraction line.

6.2.9.2 The geophone stations should lie along as straight

a line as possible. Large deviations from a straight path will result in inaccuracies unless the line is carefully surveyed and appropriate geometric corrections are applied to the data. Often the location of the line will be determined by topography. Line locations should be selected so that the ground surface along each geophone spread (cable) is as flat as possible or an interpretation method should be selected that accounts for topography.

6.2.9.3 *Coverage*—Survey coverage and orientation of survey lines should be designed to meet survey objectives. The area of survey should usually be larger than the area of interest so that measurements are taken in both "background" conditions and over any anomalous conditions. Consideration should be given to the orientation of lines with respect to geologic features of interest, for example, buried channel, fault, fractures, etc. For example, in mapping a buried channel, the refraction survey line should cross over the channel so that its boundaries can be determined. The number and locations of shot-points will depend upon the method chosen to collect and interpret the seismic refraction data. Generally, geophone spacing is determined by two factors: the expected depth of the refractor(s) and desired degree of definition (lateral resolution) of the surface of the refractor. Generally, the geophone to shot-point separation will be larger for deeper refractors and smaller for shallow refractors. For reconnaissance measurements that do not require extensive detailed mapping of the top of the refractor, widely spaced geophones may be used. For detailed mapping of the top of a refractor, more closely-spaced geophones are required. To define the surface of a refractor in detail, the geophone spacing must be smaller than the size of the spatial changes in the refractor. Geophone spacing can be varied from less than 1 m (3 ft) to more than 100 m (300 ft) depending upon the depth to the refractor and lateral resolution needed to define the top of a refractor. Examples of geophone spacing and shot distance needed to define various geologic conditions are given by Haeni (9). A refraction survey line may require a source-to-geophone distance of up to three to five times the required depth of investigation (Haeni (9)). Therefore, adequate space for the refraction line is a consideration. If the length of the geophone spread and the source to geophone offset are not sufficient to reach the maximum depth of investigation, then the source to geophone offset distance must be increased until a sufficient depth is obtained. If the length of the line to be surveyed is longer than a single geophone spread, data can be obtained by using multiple geophone spreads.

6.2.9.4 Refraction data along a line with a series of geophone spreads may be reconnaissance or detailed. For reconnaissance work, a gap may be left between the ends of successive spreads. As more detailed data is required, the gap will decrease until the geophone spreads are overlapping and provide a continuous profile of the refractor being mapped. The geophone spacing and the amount of overlap of the geophones from each cable spread will depend upon the detail and continuity required to map the desired refractor. Since the common reciprocal method and generalized reciprocal method are used to obtain depth to a refractor under individual geophones, the geophone spreads must be overlapped if continuous coverage of the refractor is desired. The overlap will commonly range from one to two geophones for common reciprocal method and from two to five geophones for generalized reciprocal method. Greater overlaps may be necessary for deeper refractors. The time-distance plots for the seismic refraction measurements can be constructed by combining and plotting together the data from each geophone spread by a process called phantoming. Phantoming is discussed by Lankston and Lankston (13).

6.2.10 *Data Acquisition Format*—A recommended standard written under the guidance of the Society of Exploration Geophysicists (SEG)—Engineering and Ground Water Geophysics Committee for Seismic—data files used in the personal computer (PC) environment is given by Pullan (20).

6.3 *Implementation of Survey:*

6.3.1 *On Site Check of Survey Plan:*

6.3.1.1 A systematic visual inspection of the site should be made upon arrival to determine if the initial survey plan is reasonable. At this point, modifications to the survey plan may be required.

6.3.1.2 If a feasibility test has not been previously done, the results of initial measurements can be used to confirm the existence of an adequate seismic velocity contrast and can also be used to assess signal to noise ratio at the site. Results of these initial measurements may require that changes be made to the original survey plan.

6.3.2 *Layout the Survey Lines*—Locate the best position for the refraction lines based on the survey design described in 6.2.4 and the on-site visit in 8.3.1.

6.3.3 *Conducting the Survey:*

6.3.3.1 Check for adequate space to lay out as straight a line as possible.

6.3.3.2 Locate the position of the first geophone.

6.3.3.3 Lay out the geophone cable.

6.3.3.4 Place geophones firmly in the ground and connect them to the cable. The geophone must be vertical and in contact with the soil or rock. Improper placement of geophones is a common problem resulting in poor detection of the seismic *P*-wave. Each geophone spike should be pushed firmly into the ground to make the contact between the soil and the geophone as tight as possible. Often the top few inches (10 cm) of soil is very loose and should be scraped off so that the geophone can be implanted into firm soil. Where rock is exposed at the surface the geophone spike may be replaced by a tripod base on the geophone. In both soil and rock, one must assume that there is a good coupling between the ground and the geophones.

6.3.3.5 Test the geophones and geophone cable for short and open circuits if possible (see seismograph instruction manual).

6.3.3.6 Set up the source at the first shot-point or a test point.

6.3.3.7 Test the seismic source and trigger cable.

6.3.3.8 Test for noise level and set gains and filters (see seismograph instruction manual).

6.3.3.9 The required degree of accuracy of the position and elevation of shot-points and geophones varies with the objectives of the project. If the ground is relatively flat or the accuracy of the refraction survey is not critical, the distance between source and geophone measured with a tape measure will be sufficient. If there are considerable changes in surface elevation, shot-point and geophone elevations and their horizontal locations must be surveyed and referenced to the project datum. Measurements (made by tape) to within 15- to 20-cm (about 0.5 ft) are adequate for most purposes.

6.3.3.10 Proceed with the refraction measurements, making sure that an adequate signal-to-noise ratio exists so that the first arrivals can be determined.

6.3.4 *Quality Control (QC)*—Quality control can be appropriately applied to seismic refraction measurements in the field. Good quality-control procedures require that reasonable standard procedures be followed and appropriate documentation be made. The following items are recommended to provide QC of field operations and data acquisition:

6.3.4.1 Documentation of the field procedures and interpretation method that are planned to be used in the study. The method of interpretation will often dictate the field procedures, and the field procedures as well as site conditions used may limit the level and method of interpretation.

6.3.4.2 A field log in which field operational procedures used for the project are recorded.

6.3.4.3 Any changes to the planned field procedures should be documented.

6.3.4.4 Any conditions that could reduce the quality of the data (weather conditions, sources of natural and cultural noise, etc.) should be documented.

6.3.4.5 If data are being recorded (by a computer or digital-acquisition system) with no visible means of observing the data, it is recommended that the data be reviewed as soon as possible to check its quality.

6.3.4.6 Care should be taken to maintain accurate and repeatable timing.

6.3.4.7 Ensure that a uniform method of picking first arrival time is employed.

6.3.4.8 During or after data acquisition, time-distance plots should be made to assure that the data is of adequate quality and quantity (for example, a sufficient number of data points) to support the method of interpretation and define the refractor of interest.

6.3.4.9 Both forward and reverse measurements are necessary to properly resolve dipping layers.

6.3.4.10 In addition to the time distance curves, three additional tools can be used as a means of quality control of seismic refraction data: the irregularity test, the reciprocal time test, and the parallelism test.

6.3.4.11 The irregularity test checks for travel time consistency along the refraction profile. If time differences (deviations from the straight line slope) are great, then the time picks may be in error, time distance curves may have an error in data entry or plotting, data may be noisy, or geologic conditions may be highly variable.

6.3.4.12 The reciprocal time test is used to check reciprocal time differences between forward and reverse profile curves. If the differences between reciprocal times are excessive, then the time picks may be in error or the time distance curves may have an error in data entry or plotting.

6.3.4.13 The parallelism test is used to check the relative parallelism between selected forward or reverse time distance curves and another curve from the same refractor. If the slopes of the two curves are sufficiently different, then time picks for one of the sets of data may be in error or the time distance curves may have an error in data entry or plotting.

6.3.4.14 Finally, a check should be made to determine if the depths and seismic velocities obtained using the seismic refraction method make geologic sense.

6.3.5 *Calibration and Standardization*—In general, the manufacturer's recommendation should be followed for calibration and standardization. If no such recommendations are provided, a periodic check of equipment should be made. A check should also be made after each equipment problem and repair. An operational check of equipment should be carried out before each project and before starting field work each day.

6.4 *Interpretation of Seismic Refraction Data:*

6.4.1 *Method of Interpretation:*

6.4.1.1 In some limited cases, quantitative interpretation of the data may not be required and a simple qualitative interpretation may be sufficient. Examples of qualitative and semi-quantitative interpretation may include the lateral location of a buried channel without a concern for its depth or minimum depth to rock calculations (8). In most cases, however, a quantitative interpretation will be necessary.

6.4.1.2 The level of effort involved in the interpretation will depend upon the objectives of the survey and the detail desired that in turn will determine the method of interpretation. A number of manual methods and computer programs are available for interpretation. While the solutions for these methods can be carried out manually, the process can be labor intensive for the more sophisticated methods.

6.4.1.3 A problem inherent in all geophysical studies is the non-unique correlation between possible geologic models and a single set of field data. This ambiguity can be resolved only through the use of sufficient geologic data and by an experienced interpreter.

6.4.1.4 The first step in the interpretation process is to determine the time interval from the impact of the seismic source to the first arrival of energy at each geophone. When the first arrivals are sharp and there is no ambient noise, this procedure is straightforward (see Figs. 2 and 3). In many cases, noise in the data will make picking the first arrival times difficult. To minimize errors, a consistent approach to the picking of the arrival times must be used. Care should be taken to ensure that each trace is picked at the same point, that is, at the first point of movement or the point of maximum curvature. This procedure will make the interpretation a more uniform process, as the data will be consistent from one trace to the next. In some cases, a first arrival pick from one or more geophones may be uncertain; then, one must rely upon the experience of the interpreter. If this occurs, these picks should be noted. If a computer program is used to make first arrival picks, these picks must be checked (and re-adjusted as needed) by the individual(s) doing the processing and interpretation.

6.4.1.5 Corrections to travel time for elevation or other geometric factors are then made. The two main types of corrections are elevation corrections and weathering corrections. Both are used to adjust field-derived travel times to some selected datum, so that straight-line segments on the time distance plot can be associated with subsurface refractors. These corrections can be applied manually (6) or by computer (21).

6.4.1.6 With the corrected travel-time data, a time-distance plot of arrival times versus shotpoint-to-geophone distance can be constructed. Lines are then fitted to these points to complete a time-distance plot. These time-distance plots are the foundation of seismic refraction interpretation. Examples of time-distance plots and their relationships to geologic models are shown by Mooney (8), Zohdy (5), and by Crice (22). Anyone undertaking seismic refraction measurements should be familiar with time-distance plots over a variety of geologic conditions and recognize the lack of uniqueness of these plots.

6.4.2 *Preliminary Interpretation*—Preliminary interpretation of field data should be labeled as draft or preliminary, and treated with caution since it is easy to make errors in an initial field interpretation and a preliminary analysis is never a complete and thorough interpretation. Preliminary analysis done in the field is done mostly as a means of QC.

6.4.3 *Programs for Interpreting Planar Refractors:*

6.4.3.1 A wide variety of formulas, nomograms, and computer programs are available for solving seismic refraction problems using the intercept time method (or the crossover distance method).

6.4.3.2 For manual interpretation techniques see Mooney (8), U.S. Army Corps of Engineers (24); Palmer (11); and Haeni (9). Hand-held programmable calculator programs are available for solving the various seismic refraction equations (24). A number of computer programs are commercially available that are based on intercept time method.

6.4.4 *Programs for Interpreting Non-Planar Refractors:*

6.4.4.1 Manual interpretation techniques are given by Pakhiser and Black (25); Redpath (3); and Dobrin and Savit (6). Computer-assisted interpretation techniques are presented by Haeni, et al (26) and are discussed in Scott, et al (21, 27) and Haeni (9). A number of computer programs are commercially available that are based on the common reciprocal method.

6.4.4.2 Manual-interpretation techniques for the generalized reciprocal method are described by Palmer (16). However, due to the volume of data required for the method, interpretation is usually carried out on a computer. Computer programs are commercially available that are based on the generalized reciprocal method.

6.4.5 *Verification of Seismic Refraction Interpretation*—Seismic refraction interpretation can be verified by comparison with drilling data or other subsurface information. If such data is not available, this fact should be mentioned within the report.

6.4.6 *Presentation of Data:*

6.4.6.1 In some cases, there may be little need for a formal presentation of data or interpreted results. A statement of findings may be sufficient.

6.4.6.2 The final seismic refraction interpretation generally leads to a geologic or hydrologic model of site conditions. Such a model is a simplified characterization of a site that incorporates all the essential features of the physical system under study. This model is usually represented as a cross-section, a contour map, or other drawings that illustrate the general geologic and hydrogeologic conditions and any anomalous conditions at a site.

6.4.6.3 If the original data is to be provided to the client, the data and related survey grid maps must be labeled so that another competent practitioner can review the data.

7. Report

7.1 *Components of the Report*—The following is a list of the key items that should be contained within most reports. In some cases, there may be no need for an extensive formal report:

7.1.1 The report should include a discussion of:

7.1.1.1 The purpose and scope of the seismic refraction survey;

7.1.1.2 The geologic setting;

7.1.1.3 Any limitations of the seismic refraction survey;

7.1.1.4 Any assumptions that were made;

7.1.1.5 The field approach used, including a description of the equipment and the data acquisition parameters used;

7.1.1.6 The location of the seismic refraction line(s) along with a site map;

7.1.1.7 The shot-point/geophone layout;

7.1.1.8 The approach used to pick first arrivals;

7.1.1.9 Any corrections applied to field data, along with justification for their use;

7.1.1.10 The results of field measurements, copies of typical raw records (optional), and time-distance plots (optional);

7.1.1.11 The method of interpretation used (intercept time method, common reciprocal method or generalized reciprocal method), and specifically what analytical method(s), or software program(s), were used;

7.1.1.12 The interpreted results along with any qualifications and alternate interpretations;

7.1.1.13 The format of recording data (for example, notebook, hardcopy analog recorder, digital format, SEG, other);

7.1.1.14 If conditions occurred where a variance from this ASTM guide is necessary, the reason for the variance should be given;

7.1.1.15 Provide appropriate references for any supporting data used in the interpretation; and

7.1.1.16 Identify the person(s) responsible for the refraction survey and data interpretation.

7.2 *Quality Assurance of the Seismic Refraction Work and Report*—To provide quality assurance of the seismic refraction work, it is generally good practice to have the entire seismic refraction work, including the report, reviewed by a person knowledgeable with the seismic refraction method and the site geology but not directly involved with the project.

8. Precision and Bias

8.1 *Bias*—Bias is defined as a measure of the closeness to the truth.

8.1.1 The bias with which the depth and the shape of a refractor can be determined by seismic refraction methods depends on many factors. Some of these factors are:

8.1.1.1 Human errors in field procedures, record-keeping, picking of first arrivals, corrections to data, processing and interpretation;

8.1.1.2 Instrument errors in measuring, recording;

8.1.1.3 Geometry limitations, relating to geophone spacing, line location, topography, and noise;

8.1.1.4 Variation of the earth from simplifying assumptions used in the field and interpretation procedure;

8.1.1.5 Site-specific geologic limitations, such as dip,

joints, fractures and highly weathered rock with gradual changes in seismic velocities with depth; and

8.1.1.6 Ability and experience of the field crew and interpreter.

8.1.2 Published references (**4, 5, 9, 28, 29**), indicate that the depth to a refractor can reasonably be determined to within ±10 % of the true depth. Large errors are usually due to difficult field situations or improper interpretation due to blind zone problems.

8.1.3 Arrival times must be picked with an accuracy of a millisecond. This is done using what appears to be the onset of the pulse (see Figs. 2 and 3). A 1 ms error could translate to a depth error of 1 to 10 ft (0.3 to 3 m) depending upon geometry and seismic velocities of the subsurface layers.

8.2 *Differences Between Depths Determined Using Seismic Refraction and Those Determined by Drilling:*

8.2.1 The bias of a seismic refraction survey is commonly thought of as how well the refraction results agree with borehole data. In many cases, the depth obtained by refraction agrees with the borehole data. In other cases, there will be considerable disagreement between the refraction results and boring data. While a refraction measurement may be quite accurate, the interpreted results may disagree with a depth obtained from drilling for the reasons discussed in 8.2.2 through 8.2.4. It is important that the user of seismic refraction results be aware of these concepts and understand that the results of a seismic refraction survey will not always agree with drilling data.

8.2.2 *The Fundamental Differences Between Refraction and Drilling Measurements:*

8.2.2.1 The seismic refraction method is based upon a measure of travel time of the *P*-wave. In order to measure depth to a refractor, such as a soil-to-rock interface, a significant change in seismic velocity must exist between the two layers.

8.2.2.2 When the top of rock is defined by drilling it is often based upon refusal of the drill bit to continue to penetrate, the number of blow counts with split-spoon, or the first evidence of rock fragments. None of these necessarily agree with each other or the top of the rock surface measured by the seismic refraction method. The differences between seismic refraction and drilling interpretation can yield considerable differences in depth even when the top of rock is relatively flat. In general, the top of rock interpreted from refraction measurements will usually be deeper than that determined by drilling.

8.2.3 *Lateral Geologic Variability*—Agreement between refraction and boring measurements may vary considerably along the seismic refraction line depending upon lateral geologic changes, such as dip as well as the degree of weathering and fracturing in the rock. The refraction measurements may not account for small lateral geologic changes and may only provide an average depth over them. In addition, the presence of a water table near the bedrock surface can in some cases lead to an error in interpretation. Therefore, it is not always possible to have exact agreement between refraction and boring data along a survey line.

8.2.4 *Positioning Differences*—The drilling location and the refraction measurement may not be made at exactly the same point. It is common to find that the boreholes are located on the basis of drill-rig access and may not be located along the seismic refraction line. Differences in position can easily account for anywhere from a few feet to tens of feet (1 to 10 m) of difference in depth where top of rock is highly variable, for example, karst.

8.3 *Precision*—Precision is the repeatability between measurements, that is, the degree to which the travel times from two identical measurements in the same location with the same equipment match one another. Precision of a seismic refraction measurement will be affected by the sources used, the repeatability of the trigger signal timing, placement of geophones, soil conditions, the care involved in picking arrival times, and the level and variations of the noise impacting the measurements. If a refraction survey is repeated under identical conditions, the measurements would be expected to have a high level of precision.

8.4 *Resolution:*

8.4.1 *Lateral Resolution*—Lateral resolution of a seismic refraction survey is determined by geophone spacing and shot-point spacing. Close spacing of geophones will provide higher lateral resolution, for example, greater definition of the shape of the top of the refractor.

8.4.2 *Vertical Resolution:*

8.4.2.1 Vertical resolution can be thought of in three ways: how small a change in depth can be determined by the refraction method; how thin a layer can be detected by the seismic refraction method; and how much relief or dip can be accurately mapped without smoothing or errors in depth determination.

8.4.2.2 The answers to all three of these questions is a complex function of the geophone spacing, the depth to the refractors and the seismic velocity contrasts and near surface conditions such as freezing, changes in materials on which sources and receivers are placed and fluctuating of water tables.

9. Keywords

9.1 geophysics; refraction; seismic refraction; surface geophysics

REFERENCES

(1) Sheriff, Robert E., *Encyclopedic Dictionary of Exploration Geophysics*, 3rd edition, Tulsa, Soc. Explor. Geophysics, 1991.

(2) Bates, R. L., and Jackson, J. A., *Glossary of Geology*, 1980.

(3) Redpath, B. B., *Seismic Refraction Exploration for Engineering Site Investigations: Technical Report E-73-4*, U.S. Army Engineering Waterways Experiment Station, Explosive Excavation Research Lab., Livamore, California, 1973.

(4) Griffiths, D. H., and King, R. F., *Applied Geophysics for Engineers and Geologists*, Second Edition, Pergamon Press, 1981.

(5) Zohdy, A. A., Eaton, G. P., and Mabey, D. R., "Application of Surface Geophysics to Ground Water Investigations," *U.S. Geological Survey, Techniques of Water Resources Investigation*, Book 2, Chapter D1, 1974.

(6) Dobrin, M. B., and Sawl, C. H., *Introduction to Geophysical Prospecting*, Fourth Edition, McGraw-Hill, New York, 1988.

(7) Telford, W. M., Geldart, L. P., Sheriff, R. E., and Keys, D. A., *Applied Geophysics*, Cambridge University Press, New York, New York, 1990.

(8) Mooney, H. M., *Handbook of Engineering Geophysics, Volume 1: Engineering Seismology*, Second Edition, Bison Instruments, Minneapolis, Minnesota, 1984.

(9) Haeni, F. P., "Application of Seismic-Refraction Techniques to Hydrologic Studies," *U.S. Geological Survey Techniques of Water Resources Investigations, Book 2*, Chapter D2, 1988.

(10) Sjogren, Bengl, *Shallow Refraction Seismics*, Chapman and Hall, New York, 1984.

(11) Palmer, D., "Refraction Seismics, the Lateral Resolution of Structure and Seismic Velocity," *Handbook of Geophysical Exploration Vol 13 Section 1 Seismic Refraction*, K. Helbig and S. Tredel (eds.), Geophysical Press, London, 1988.

(12) Saska, J. L., *The Blind Zone Problem in Engineering Geophysics, Geophysics*, Vol 24, No. 2, 1959, pp. 359–385.

(13) Lankston, R., "High Resolution Refraction Seismic Data Acquisition and Interpretation. In: Geotechnical and Environmental Geophysics, Vol 1," Reviewer and Tutorial, S. Ward (ed.), *Investigations in Geophysics No. 5*, Society of Exploration Geophysicists, Tulsa, OK, 1990, pp. 45–73.

(14) Ackermann, H. D., Pankratz, L. W., and Dansereau, D. A., "A Comprehensive System for Interpreting Seismic Refraction Arrival-Time Data Using Interactive Computer Methods," *U.S. Geological Survey Open File Report 82-1065*, 1983, p. 265.

(15) Lankston, Robert, "The Seismic Refraction Method: A Viable Tool for Mapping Shallow Targets in the 1990's," *Geophysics*, Vol 54, No. 2, 1989, pp. 1–6.

(16) Palmer, D., "The GRM an Integrated Approach to Shallow Refraction Seismology," *Expl. Geophys. Bulletin of the Australian Society of Exploration Geophysics*, Vol 21, 1990.

(17) Palmer, D., *The Generalized Reciprocal Method of Seismic Refraction Interpretation*, Soc. of Expl. Geophysicists, Tulsa, Oklahoma, 1980.

(18) Palmer, D., "The Resolution of Narrow Low Velocity Zones with the Generalized Reciprocal Method," *Geophysical Prospecting*, Vol 39, 1991, pp. 1031–1060.

(19) Lankston, Robert and Lankston, M., "Obtaining Multi-Layer Reciprocal Times Through Phantoming," *Geophysics*, Vol 51, No. 1, 1986, pp. 45–49.

(20) Pullan, S. E., "Recommended Standard for Seismic (/radar) Data Files in the Personal Computer Environment," *Geophysics*, Vol 55, No. 9, 1990, pp. 1260–1271.

(21) Scott, J. H., Tibbetts, B. L., and Burdick, R. G., "Computer Analysis of Seismic Refraction Data," *U.S. Bureau of Mines Report of Investigation 7595*, 1972.

(22) Crice, Douglas B., "Applications for Shallow Exploration Seismographs," In *Practical Geophysics for the Exploration Geologist*, Northwest Mining Association, Spokane, Washington, 1980.

(23) U.S. Army Corps of Engineers (USACOE), *Geophysical Exploration Engineering Manual EM 111D-1-1802*, Washington, DC, 1979.

(24) Balfantyne, Edwing J., Jr., Campbell, D. L., Mantameaier, S. H., and Wiggins, R., *Manual of Geophysical Hand-Calculator Programs*, Soc. of Explt. Geophysicists, Tulsa, Oklahoma, 1981.

(25) Pekhiser, L. C., and Black, R. A., "Exploration for Ancient Channels with the Exploration Seismograph," *Geophysics*, Vol 22, 1957, pp. 32–47.

(26) Haeni, F. P., Grantham, D. G., and Eliefsen, K., "Microcomputer-Based Version of SiPT—A Program for the Interpretation of Seismic-Refraction Data (Text)," *U.S. Geological Survey, Open File Report 87-183-A*, 1987.

(27) Scott, J. H., and Marldewicz, "Dips and Chips—PC Programs for Analyzing Seismic Refraction Data," *Proceedings of the Symposium on the Application of Geophysics to Environmental and Engineering Problems*, Society of Engineering and Mineral Exploration Geophysicists, Goldan, Colorado, 1990, pp. 175–200.

(28) Eaton, G. P., and Wallukins, J. S., "The Use of Seismic Refraction and Gravity Methods in Hydrogeological Investigations," In: *Mining and Groundwater Geophysics 1967*, L. W. Morely (ed.), Geological Survey of Canada Economic Geology Report 26, 1970, pp. 544–568.

(29) Wallace, D. E., "Some Limitations of Seismic Refraction Methods in Geohydrological Surveys of Deep Alluvial Basins," *Ground Water*, Vol 8, No. 6, 1970, pp. 8–13.

(30) Scott, J. H., "Seismic-Refraction Modeling By Computer," *Geophysics*, Vol 38, No. 2, 1973, pp. 274–284.

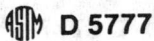

1.4 GEOLOGIC CHARACTERIZATION

Designation: D 5434 – 93

Standard Guide for
Field Logging of Subsurface Explorations of Soil and Rock[1]

This standard is issued under the fixed designation D 5434; the number immediately following the designation indicates the year of original adoption or, in the case of revision, the year of last revision. A number in parentheses indicates the year of last reapproval. A superscript epsilon (ε) indicates an editorial change since the last revision or reapproval.

1. Scope

1.1 This guide describes the type of information that should be recorded during field subsurface explorations in soil and rock.

1.2 This guide is not intended to specify all of the information required for preparing field logs. Such requirements will vary depending on the purpose of the investigation, the intended use of the field log, and particular needs of the client or user.

1.3 This guide is applicable to boreholes, auger holes, excavated pits, or other subsurface exposures such as road side cuts or stream banks. This guide may serve as a supplement to Guide D 420.

1.4 This guide may not be suited to all types of subsurface exploration such as mining, agricultural, geologic hazardous waste, or other special types of exploration.

1.5 *This standard does not purport to address all of the safety problems, if any, associated with its use. It is the responsibility of the user of this standard to establish appropriate safety and health practices and determine the applicability of regulatory limitations prior to use.*

2. Referenced Documents

2.1 *ASTM Standards:*

D 420 Guide for Investigating and Sampling Soil and Rock[2]

D 653 Terminology Relating to Soil, Rock, and Contained Fluids[2]

D 1452 Practice for Soil Investigation and Sampling by Auger Borings[2]

D 1586 Method for Penetration Test and Split-Barrel Sampling of Soils[2]

D 1587 Practice for Thin-Walled Tube Sampling of Soils[2]

D 2113 Practice for Diamond Core Drilling for Site Investigation[2]

D 2488 Practice for Description and Identification of Soils (Visual-Manual Procedure)[2]

D 2573 Test Method for Field Vane Shear Test in Cohesive Soil[2]

D 3441 Test Method for Deep, Quasi-Static, Cone and Friction Cone Penetration Tests of Soil[2]

D 3550 Practice for Ring-Lined Barrel Sampling of Soils[2]

D 4083 Practice for Description of Frozen Soils (Visual-Manual Procedure)[2]

D 4220 Practices for Preserving and Transporting Soil Samples[2]

D 4403 Practice for Extensometers Used in Rock[2]

D 4544 Practice for Estimating Peat Deposit Thickness[2]

D 4622 Test Method for Rock Mass Monitoring for Using Inclinometers[2]

D 4623 Test Method for Determination of In Situ Stress in Rock Mass by Overcoring Method—USBM Borehole Deformation Gage[2]

D 4633 Test Method for Stress Wave Energy Measurement for Dynamic Penetrometer Testing Systems[2]

D 4645 Test Method for Determination of the In-Situ Stress in Rock Using the Hydraulic-Fracturing Method[2]

D 4719 Test Method for Pressuremeter Testing in Soils[2]

D 4750 Test Method for Determining Subsurface Liquid Levels in a Borehole or Monitoring Well (Observation Well)[2]

D 4879 Guide for Geotechnical Mapping of Large Underground Openings in Rock[2]

D 5079 Practices for Preserving and Transporting Rock Core Samples[3]

3. Terminology

3.1 *Definitions:*

3.1.1 Except as listed below, all definitions are in accordance with Terminology D 653.

3.2 *Description of Term Specific to This Standard:*

3.2.1 *field log*—a record prepared during subsurface explorations of soil and rock to document procedures used, test data, descriptions of materials and depths where encountered, ground water conditions, and other information.

4. Summary of Guide

4.1 This guide describes the type of information that should be recorded during the execution of field subsurface explorations in soil and rock. The information described relates to the project, personnel, methods of investigation and equipment used, visual description of subsurface materials and ground water conditions, in-situ testing, installation of monitoring equipment, and other data that may be appropriate.

5. Significance and Use

5.1 The preparation of field logs provides documentation of field exploration procedures and findings for geotechnical, geologic, hydrogeologic, and other investigations of subsurface site conditions. This guide may be used for a broad range of investigations.

5.2 The recorded information in a field log will depend on the specific purpose of the site investigation. All of the information given in this guide need not appear in all field logs.

[1] This guide is under the jurisdiction of ASTM Committee D-18 on Soil and Rock and is the direct responsibility of Subcommittee D18.07 on Structural Properties of Soils.

Current edition approved July 15, 1993. Published September 1993.

[2] *Annual Book of ASTM Standards*, Vol 04.08.

[3] *Annual Book of ASTM Standards*, Vol 04.09.

6. Summary of Work

6.1 Soil and rock field logs should include the following written information:

6.1.1 Project information should include:

6.1.1.1 Name and location of the project or project number, or both,

6.1.1.2 Name of personnel onsite during the exploration, such as drilling crew, supervisor, geologist, engineer, and technicians,

6.1.1.3 Names and addresses of organizations involved,

6.1.1.4 Name of person(s) preparing log,

6.1.1.5 Reference datum for project if available and description of datum, and

6.1.1.6 General remarks as appropriate.

6.1.2 Exploration information should include:

6.1.2.1 Exploration number and location (station and coordinates if available and applicable, position relative to a local permanent reference which is identified, or markings of exploration location),

6.1.2.2 Type of exploration, such as drill hole, auger hole, test pit, or road cut,

6.1.2.3 Date and time of start and finish,

6.1.2.4 Weather conditions including recent rain or other events that could affect subsurface conditions,

6.1.2.5 Depth and size of completed exploration,

6.1.2.6 The condition of exploration prior to and after backfilling or sealing, or both, and

6.1.2.7 Method of backfilling or sealing exploration, or both.

6.1.3 Explorations by drill hole or auger hole should include the following drilling information:

6.1.3.1 Type and make (manufacture and model if known) of drilling machine or description and name of contractor,

6.1.3.2 Method of drilling or advancing and cleaning the borehole. State if air, water, or drilling fluid is used. Describe type, source of water and additives, concentration, and tests performed on fluid,

6.1.3.3 Size, type, and section length of drilling rods (rod designations should conform with Table 3 of Method D 2113) and drilling bits used.

6.1.3.4 Dates and times of each stage of operation and time to complete intervals,

6.1.3.5 Size of hole (diameter and depth),

6.1.3.6 Ground elevation at top of borehole,

6.1.3.7 Orientation of drill hole, if not vertical (azimuth or bearing and angle),

6.1.3.8 Size and description of casing, if appropriate, method of casing installation (driven, drilled, or pushed) and depth of cased portion of boring (casing size designations should conform with Table 2 of Method D 2113), hollow-stem augers,

6.1.3.9 Methods used for cleaning equipment or drilling tools, or both, when required, and

6.1.3.10 Describe and state depth of any drilling problems such as borehole instability (cave in, squeezing hole, flowing sands), cobbles, lost drilling fluid, lost ground, obstruction, fluid return color changes, and equipment problems.

6.1.4 Exploration by test pit, road cut, stream cut, etc., should include:

6.1.4.1 Method of exploration,

6.1.4.2 Equipment used for excavation,

6.1.4.3 Type of shoring used, and

6.1.4.4 Excavation problems: instability of cut (sloughing, caving, etc.), depth of refusal, difficulty of excavating, etc.

6.1.5 Subsurface information should include:

6.1.5.1 Depth of changes and discontinuities in geologic material and method used to establish change (such as Practice D 4544).

6.1.5.2 Description of material encountered with origin or formation name, if possible, and type of samples used for description. The system or method of soil (such as Practices D 2488 and D 4083) or rock description should be referenced.

6.1.5.3 Description of nature of boundary between strata (gradual or abrupt, as appropriate) and other relevant structural features such as breccia, slickensides, solution zones, discolorations by weathering or hydrothermal fluids, and other stratigraphic information.

6.1.6 Soil or rock sampling and testing information should include:

6.1.6.1 Depth of each sample and number (if used),

6.1.6.2 Method of sampling (reference to appropriate ASTM standard, for example, Practice D 1452, Test Method D 1586, Method D 1587, Practice D 3550, or other method).

6.1.6.3 Description of sampler: inside and outside dimensions, length, type of metal, type of coating, and type of liner,

6.1.6.4 Method of sampler insertion: pushed, cored, or driven,

6.1.6.5 Sampler penetration and recovery lengths of samples (rock quality designation (RQD) and rate of coring in the case of rock),

6.1.6.6 Method of sample extrusion. Mark direction of extrusion,

6.1.6.7 Method of preserving samples and preparing for transport (refer to Practices D 4220 or D 5079),

6.1.6.8 Mark top and bottom of samples and orientation, if possible,

6.1.6.9 Depth and description of any in-situ test performed (reference to applicable ASTM standard, for example, Test Methods D 1586, D 2573, D 3441, D 4623, D 4633, D 4645, D 4719, or other tests if applicable),

6.1.6.10 Description of any other field tests conducted on soil and rock during the exploration such as pH, hydraulic conductivity, pressuremeter geophysical, pocket penetrometer, soil gas/vapor analysis, or other tests, and

6.1.6.11 Destination or recipient of samples and method of transportation.

6.1.7 Ground water information should include:

6.1.7.1 Depths and times at which ground water is encountered, including seepage zones, if appropriate,

6.1.7.2 In the case of drilling using drilling fluid, depth of fluid surface in boring and drilling depth at the time of a noted loss or gain in drilling fluid,

6.1.7.3 Depth to ground water level at the completion of drilling and removal of drill steel and description of datum (note condition of borehole, for example, cased or uncased). Date and time measured,

6.1.7.4 Depth to ground water level at some reported time period following completion of drilling and description of datum, when possible.

6.1.7.5 Method or equipment used to determine depth of

ground water level, such as Test Method D 4750,

6.1.7.6 Method and depth of ground water samples obtained, including size of samples taken and description of sampler, and

6.1.7.7 Description of any field tests conducted on ground water samples such as pH, temperature, conductivity, turbidity, or odor.

6.1.8 Information regarding installation of instrumentation or monitoring equipment should include:

6.1.8.1 Type of equipment installed, for example, piezometers, monitoring well screens, inclinometer, including sizes and types of materials,

6.1.8.2 Depth and description of equipment installed (reference to applicable ASTM standard, for example, Test Method D 4622, Practice D 4403, or other standards or procedures),

6.1.8.3 Methods used for installation of equipment and method used for sealing annular space, and

6.1.8.4 Methods used to protect equipment (casing cap or locks).

6.2 Soil and rock field logs should include the following pictorial information:

6.2.1 Maps, drawings, or sketches of area of exploration and subsurface surfaces observed. Include pertinent surface information such as neighboring outcrops, as appropriate. Describe system of mapping, such as Guide D 4879 for rock, or legend for symbols of materials. Include dimensions, directions, and slopes, and

6.2.2 Photographs of activities, surfaces, or core. Describe sequence, dates and time, direction, objects used for scale, and subject.

7. Keywords

7.1 drilling; explorations; geologic investigations; ground water; logging; preliminary investigations; sampling; soil investigations; subsurface investigations

Designation: D 6067 – 96

Standard Test Method for
Using the Electronic Cone Penetrometer for Environmental
Site Characterization[1]

This standard is issued under the fixed designation D 6067; the number immediately following the designation indicates the year of original adoption or, in the case of revision, the year of last revision. A number in parentheses indicates the year of last reapproval. A superscript epsilon (ε) indicates an editorial change since the last revision or reapproval.

1. Scope

1.1 The electronic cone penetrometer test often is used to determine subsurface stratigraphy for geotechnical and environmental site characterization purposes (1).[2] The geotechnical application of the electronic cone penetrometer test is discussed in detail in Test Method D 5778, however, the use of the electronic cone penetrometer test in environmental site characterization applications involves further considerations that are not discussed.

1.2 The purpose of this test method is to discuss aspects of the electronic cone penetrometer test that need to be considered when performing tests for environmental site characterization purposes.

1.3 The electronic cone penetrometer test for environmental site characterization projects often requires steam cleaning the push rods and grouting the hole. There are numerous ways of cleaning and grouting depending on the scope of the project, local regulations, and corporate preferences. It is beyond the scope of this test method to discuss all of these methods in detail. A detailed explanation of grouting procedures is discussed in Guide D 6001.

1.4 *This standard does not purport to address all of the safety concerns, if any, associated with its use. It is the responsibility of the user of this standard to establish appropriate safety and health practices and determine the applicability of regulatory limitations prior to use.*

1.5 This test method is applicable only at sites where chemical (organic and inorganic) wastes are a concern and is not intended for use at radioactive or mixed (chemical and radioactive) waste sites.

1.6 The values stated in either SI units or inch-pound units are to be regarded as standard. Within the text, the inch-pound units are shown in brackets. The values stated in each system are not equivalents, therefore, each system must be used independently of the other.

2. Referenced Documents

2.1 *ASTM Standards:*
C 150 Specification for Portland Cement[3]
D 653 Terminology Relating to Soil, Rock, and Contained Fluids[4]

D 2488 Practice for Description and Identification of Soils (Visual-Manual Procedure)[4]
D 3441 Test Method for Deep, Quasi-Static, Cone and Friction-Cone Penetration Tests of Soil[4]
D 5088 Practice for Decontamination of Field Equipment Used at Nonradioactive Waste Sites[5]
D 5092 Practice for Design and Installation of Ground Water Monitoring Wells in Aquifers[5]
D 5730 Guide to Site Characterization for Environmental Purposes[5]
D 5778 Test Method for Performing Electronic Friction Cone and Piezocone Penetration Testing of Soils[5]
D 6001 Guide for Direct Push Water Sampling for Geoenvironmental Investigations[5]

3. Terminology

3.1 *Definitions*—The definitions of terms in this test method are in accordance with Terminology D 653. Terms that are not included in Terminology D 653 are described as follows.

3.2 *Definitions of Terms Specific to This Standard:*

3.2.1 *baseline, n*—a set of zero load readings, expressed in terms of apparent resistance, that are used as reference values during performance of testing and calibration.

3.2.2 *bentonite, n*—the common name for drilling fluid additives and well construction products consisting mostly of naturally occurring sodium montmorillonite. Some bentonite products have chemical additives that may affect water quality analyses.

3.2.3 *cone, n*—the conical point of a cone penetrometer on which the end bearing component of penetration resistance is developed.

3.2.4 *cone resistance, q_c, n*—the end bearing component of penetration resistance.

3.2.5 *cone sounding, n*—a series of penetration readings performed at one location over the entire depth when using a cone penetrometer.

3.2.6 *electronic cone penetrometer, n*—a friction cone penetrometer that uses force transducers, such as strain gage load cells, built into a nontelescoping penetrometer tip for measuring within the penetrometer tip, the components of penetration resistance.

3.2.7 *electronic piezocone penetrometer, n*—an electronic cone penetrometer equipped with a low-volume fluid chamber, porous element, and pressure transducer for determination of pore pressure at the porous element soil interface.

---

[1] This test method is under the jurisdiction of ASTM Committee D-18 on Soil and Rock and is the direct responsibility of Subcommittee D18.21 on Ground Water and Vadose Zone Investigations.
Current edition approved Dec. 10, 1996. Published June 1997.
[2] The boldface numbers in parentheses refer to the list of references at the end of this test method.
[3] *Annual Book of ASTM Standards*, Vol 04.01.
[4] *Annual Book of ASTM Standards*, Vol 04.08.

[5] *Annual Book of ASTM Standards*, Vol 04.09.

3.2.8 *end bearing resistance, n*—same as cone resistance or tip resistance, q_c.

3.2.9 *equilibrium pore water pressure, u_o, n*—at rest water pressure at depth of interest. Same as hydrostatic pressure.

3.2.10 *excess pore water pressure, $u-u_o$, n*—the difference between pore pressure measured as the penetratoin occurs, u, and estimated equilibrium pore water pressure, u_o. Excess pore pressure can be either positive or negative.

3.2.11 *friction ratio, R_f, n*—the ratio of friction sleeve resistance, f, to cone resistance, q_c, measured with the middle of the friction sleeve at the same depth as the cone point. It is usually expressed as a percentage.

3.2.12 *friction reducer, n*—a narrow local protuberance on the outside of the push rod surface, placed at a certain distance above the penetrometer tip, which is provided to reduce the total side friction on the push rods and allow for greater penetration depths for a given push capacity.

3.2.13 *friction sleeve resistance, f_s, n*—the friction component of penetration resistance developed on a friction sleeve, equal to the shear force applied to the friction sleeve divided by its surface area.

3.2.14 *friction sleeve, n*—an isolated cylindrical sleeve section on a penetrometer tip upon which the friction component of penetration resistance develops.

3.2.15 *local friction, n*—same as friction sleeve resistance.

3.2.16 *penetrometer, n*—an apparatus consisting of a series of cylindrical push rods with a terminal body (end section) called the penetrometer tip and measuring devices for determination of the components of penetration resistance.

3.2.17 *penetrometer tip, n*—the terminal body (end section) of the penetrometer which contains the active elements that sense the components of penetration resistance.

3.2.18 *piezocone, n*—same as electronic piezocone penetrometer.

3.2.19 *piezocone pore pressure, u, n*—fluid pressure measured using the piezocone penetration test.

3.2.20 *push rods, n*—the thick walled tubes or rods used to advance the penetrometer tip.

3.2.21 *sleeve friction or resistance, n*—same as friction sleeve resistance, f.

3.2.22 *stratigraphy, n*—a classification of soil behavior type that categorizes soils of lateral continuity (4).

3.3 *Acronyms:*

3.3.1 *CPT*—Cone Penetration Test.

3.3.2 PPT_u—Piezocone Penetration Test.

3.3.3 *ECP*—Electronic Cone Penetrometer (used when referring to the cone penetrometer).

4. Significance and Use

4.1 Environmental site characterization projects almost always require information regarding subsurface soil stratigraphy. Soil stratigraphy often is determined by various drilling procedures and bore logs. Although drilling is very accurate and useful, the electronic cone penetrometer test may be faster, less expensive, and provide greater resolution, and does not generate contaminated cuttings that may present other disposal problems (2,3,4,5). Investigators may obtain soil samples from adjacent borings for correlation purposes, but prior information or experience in the same area may preclude the need for borings (1).

4.2 The electronic cone penetration test is an in situ investigation method involving:

4.2.1 Pushing an electronically instrumented probe into the ground (see Fig. 1 for a diagram of a typical cone penetrometer). The position of the pore pressure element may vary.

4.2.2 Recording force resistances, such as tip resistance, local friction, and sometimes pore pressure.

4.2.3 Data interpretation.

4.2.1 The most common use of the interpreted data is stratigraphy. Several charts are available. A typical CPT stratigraphic chart is shown in Fig. 2 (1). The first step in determining the extent and motion of contaminants is to determine the subsurface stratigraphy. Since the contaminants will migrate with ground water flowing through the more permeable strata, it is impossible to characterize an environmental site without valid stratigraphy. Cone penetrometer data has been used as a stratigraphic tool for many years. The pore pressure channel of the cone can be used to determine the depth to the water table or to locate perched water zones.

4.2.2 When attempting to retrieve a soil gas or water sample, it is advantageous to know where the bearing zones (permeable zones) are located. Although soil gas and water can be retrieved from on-bearing zones such as clays, the length of time required usually makes it impractical. Soil gas and water samples can be retrieved much faster from bearing zones, such as sands. The cone penetrometer tip and friction data generally can identify and locate the bearing zones and nonbearing zones less than a foot thick. Since the test is run at a constant rate, the pore pressure data can often identify layers less than 20 mm thick.

4.2.3 The electronic cone penetrometer test is used in a variety of soil types. Lightweight equipment with reaction weights of less than 10 tons generally are limited to soils with relatively small grain sizes. Typical depths obtained are 20 to 40 m, but depths to over 70 m with heavier equipment weighing 20 tons or more are not uncommon. Since penetration is a direct result of vertical forces and does not include rotation or drilling, it cannot be utilized in rock or heavily cemented soils. Depth capabilities are a function of many factors including:

4.2.3.1 The force resistance on the tip,

4.2.3.2 The friction along the push rods,

4.2.3.3 The force and reaction weight available,

4.2.3.4 Rod support provided by the soil, and

4.2.3.5 Large grained materials causing nonvertical deflection or unacceptable tool wear.

4.2.4 Depth is always site dependent. Local experience is desirable.

4.3 *Pore Pressure Data:*

4.3.1 The pore pressure data often is used in environmental site characterization projects to identify thin soil layers that will either be aquifers or aquitards. The pore pressure channel often can detect these thin layers even if they are less than 20 mm thick.

4.3.2 Pore pressure data also is used to provide an indication of relative hydraulic conductivity. Excess pore pressure is generated during an electronic cone penetrometer test. Generally, high excess pore pressure indicates the presence of aquitards, and low excess pore pressure indicates

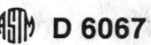

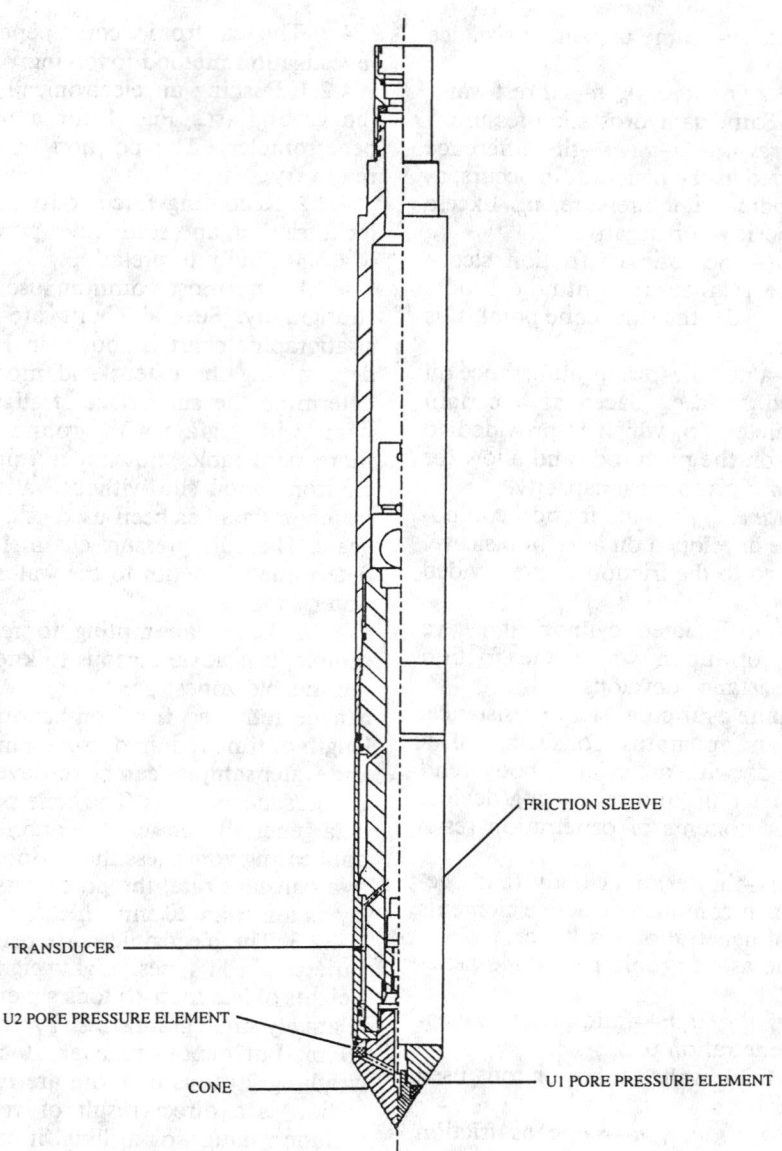

FIG. 1 Electronic Cone Penetrometer

the presence of aquifers. This is not always the case, however. For example, some silty sands and over-consolidated soils generate negative pore pressures if monitored above the shoulder of the cone tip. See Fig. 2. The balance of the data, therefore, also must be evaluated.

4.3.3 In general, since the ground water flows primarily through sands and not clays, modeling the flow through the sands is most critical. The pore pressure data also can be monitored with the sounding halted. This is called a pore pressure dissipation test. A rapidly dissipating pore pressure indicates the presence of an aquifer while a very slow dissipation indicates the presence of an aquitard.

4.3.4 A pore pressure decay in a sand is almost instantaneous. The permeability (hydraulic conductivity), therefore, is very difficult to measure in a sand with a cone penetrometer. As a result, the cone penetrometer is not used very often for measuring the permeability of sands in environmental applications.

4.3.5 A thorough study of ground water flow also includes determining where the water cannot flow. Cone penetrometer pore pressure dissipation tests can be used very effectively to study the permeability of aquitards.

4.3.6 The pore pressure data also can be used to estimate the depth to the water table or identify perched water zones. This is accomplished by allowing the pressure to equilibrate and then subtract the appropriate head pressure. Due to excess pore pressures being generated, typical pore pressure transducers are configured to measure pressures up to 3.5 MPa [500 psi]. Since transducer accuracy is a function of maximum range, this provides a relative depth to water level accuracy of about ±150 mm. Better accuracy can be achieved if the operator allows sufficient time for the transducer to dissipate the heat generated while penetrating dry soil above the water table. Lower pressure transducers are sometimes used just for the purpose of determining the depth to the water table more accurately. For example, a

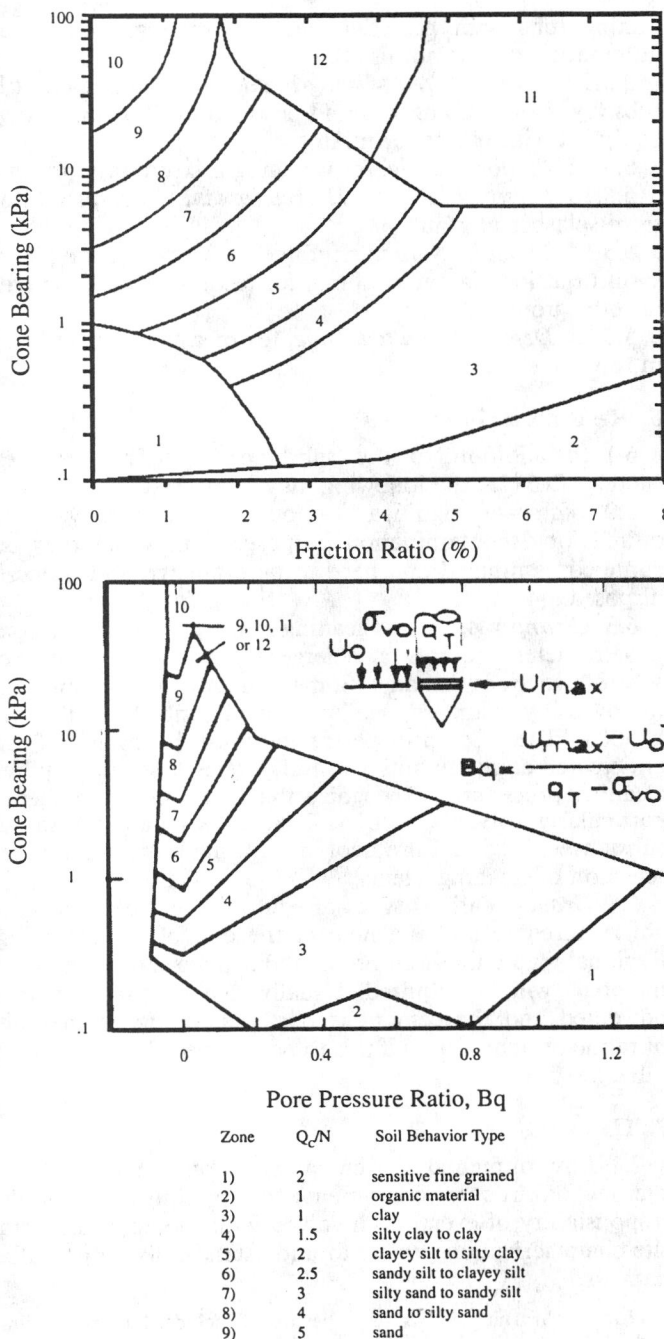

Zone	Q_c/N	Soil Behavior Type
1)	2	sensitive fine grained
2)	1	organic material
3)	1	clay
4)	1.5	silty clay to clay
5)	2	clayey silt to silty clay
6)	2.5	sandy silt to clayey silt
7)	3	silty sand to sandy silt
8)	4	sand to silty sand
9)	5	sand
10)	6	gravelly sand to sand
11)	1	very stiff fine grained (•)
12)	2	sand to clayey sand (•)

(•) over consolidated or cemented

FIG. 2 Simplified Soil Classification Chart for Standard Electric Friction Cone (Robertson and Campanella 1985)

175-KPa [25-psi] transducer would provide accuracy that is better than 10 mm. Caution must be used, however, to prevent these transducers from being damaged due to a quick rise in excess pressure.

4.4 For a complete description of a typical geotechnical electronic cone penetrometer test, see Test Method D 5778.

4.5 This test method tests the soil in situ. Soil samples are not obtained. The interpretation of the results from this test method provides estimates of the types of soil penetrated. Investigators may obtain soil samples from adjacent borings for correlation purposes, but prior information or experience in the same area may preclude the need for borings.

4.6 Certain subsurface conditions may prevent cone penetration. Penetration is not possible in hard rock and usually not possible in softer rocks, such as claystones and shales. Coarse particles, such as gravels, cobbles, and boulders may be difficult to penetrate or cause damage to the cone or push rods. Cemented soil zones may be difficult to penetrate depending on the strength and thickness of the layers. If layers are present which prevent direct push from the surface, rotary or percussion drilling methods can be employed to advance a boring through impeding layers to reach testing zones.

5. Apparatus

5.1 Most apparatus required is discussed in Test Method D 5778. When using the electronic cone penetrometer test for environmental site characterization purposes, however, other items often are necessary.

5.2 *Safety Equipment*—Environmental site characterization often involves exposure to potentially hazardous substances. Detection equipment to determine oxygen content and the presence of combustible or toxic materials may be required. Numerous air monitors are available to detect harmful situations, such as the lack of oxygen, excess carbon monoxide or carbon dioxide, the presence of methane, or other combustible gasses. Other devices, such as flame-ionization or photoionization detectors and LELs can be used to monitor vapors form the rods or the hole, or both, to forewarn the operators of potential contamination. Operator protective equipment, such as breathing apparatus and bodily protection, also may be required.

5.3 *Laboratory Equipment*—The electronic cone penetrometer often is used in conjunction with sampling devices and field laboratory equipment (see Guide D 6001). Since many cone penetrometer systems are deployed from enclosed, air conditioned, and heated trucks, these vehicles can also be used as a mobile laboratory. This unique capability provides rapid on-site analysis. First, the cone penetrometer data eliminates most guess work in determining where to retrieve samples. Second, the on-site laboratory analysis can provide important information, such as where to retrieve subsequent samples and avoids many unnecessary samples. On-site laboratory instruments range from simple portable devices, such as photoionization devices, to sophisticated gas chromatographs and mass spectrometers (GC-MS). This approach often is called Expedited Site Characterization.

5.4 *Steam Cleaning Equipment*—When the push rods are withdrawn from the ground, they may be contaminated by toxic, combustible, or corrosive compounds. If this is the case, the push rods will need to be steam cleaned. Many dedicated purpose systems have built in chambers that automatically steam clean the rods while they are being withdrawn from the ground and before they enter the vehicle. A typical diagram of an automatic decon assembly is shown in Fig. 3.

5.4.1 *Steam Cleaner/High-Pressure Washer*—Portable or trailer-mounted for cleaning the rods after grouting, with

 ASTM D 6067

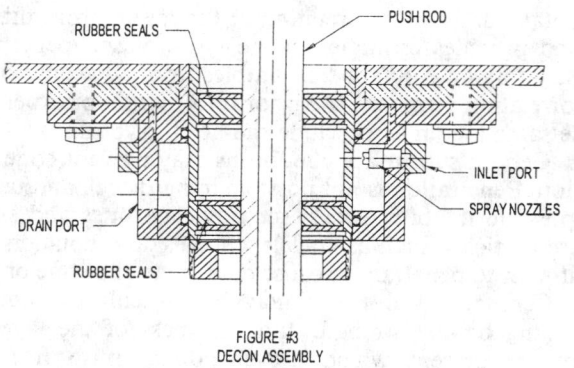

FIG. 3 Decon Assembly

appropriate hoses for connection to the steam cleaning unit.

5.4.2 *Personal Protective Equipments*, such as boots, gloves, glasses, and so forth.

5.4.3 *Water Trough Cleaning Tub*, for cleaning grout rods and containing grey water.

5.4.4 *Shotgun Bristle Brush*, for cleaning inside of cone or grout rods.

5.5 *Grouting Equipment*—When multiple ground water aquifers have been penetrated, grouting the hole closed after the test is completed may also be required to prevent cross contamination of one aquifer by another. A detailed explanation of grouting procedures is discussed in guide D 6001. The equipment required includes, but may not be limited to the following:

5.5.1 *Expendable Grout Tips for the Grout Rods*—These tips should be conical in shape and have an outside diameter larger than the grout rods. Tip size will be varied to enlarge the hole and to reduce friction on the push rods.

5.5.2 *Suitable Small-Diameter Grout Rods*—These rods may be steel or PVC. The type and size depends on the capability of being pushed back down the same CPT hole.

5.5.3 *Grout Line Connector Assembly*™this assembly is screwed into the top of the grout rods. A pressure fitting may be required if gravity placement is unacceptable and pressure is required to force grout down the rods. Since this fitting must be attached and unattached many times it may be preferable to have a quick-connect coupling.

5.5.4 *Foot Clamps/Retraction Jack*, for holding and manually extracting rods.

5.5.5 *Hoisting Plug*, for holding rods by overhead rope or cable.

5.5.6 *Grout Mixing Equipment*—Grouting quantities for cone holes are small with only 20 to 60 L [5 to 15 gal] of grout required for filling. In many cases grout mixing in small tubs with mechanical agitation devices are acceptable.

5.5.7 *Cement*, see Specification C 150. Either Type I or Type II cements are acceptable. Cement should be supplied in sacks.

5.5.8 *Bentonite*, powdered high-yield sodium montmorillonite or prehydrated bentonite. The bentonite should not contain any particular additives.

5.5.9 *Potable Water*, for mixing grout. As long as an acceptable supply of drinking water is found, the chemical analysis may not be required.

5.5.10 *Mixing Tubs*, 40 to 60-L [10 to 15-gal] plastic mixing tubs with rope handles (for mixing grout if an automatic mixer is not used).

5.5.11 *Drill-Powered Mixing Paddles*, for mixing grout in tubs by hand. Using a hand-powered drill with a stem equipped with blades for mixing.

5.5.12 *Platform Scale*, for weighing mixture proportions.

5.5.13 *Personal Protective Equipment*, eye protection from splashes of grout.

5.5.14 *Flexible Nylon*, reinforced 13 or 10-mm [½ or ⅜-in.] outside diameter tubing, for feeding grout by gravity into the grout rods.

5.5.15 *Depth-Graduated Tape*, for measuring grout levels in rods.

6. Reagents and Materials[6]

6.1 In addition to the substances described in Test Method D 5778, the following may be necessary:

6.2 *Water*—A significant amount of water may be required for decontamination purposes. This water may become contaminated and need to be evaluated and properly disposed.

6.3 *Cleaning Agents*—Cleaning of the push rods and cone penetrometer requires a detergent, such as alconox, or solvent, such as hexane. Some contaminants cannot be removed by standard methods as described in Practice D 5088. The operating personnel must be aware of the anticipated contaminants and fully understand the required cleaning procedures. Recognize that some cleaning agents, particularly solvents such as hexane, also are hazardous substances. Regional protocol and regulations influence the selection of cleaning agents.

6.4 *Grout*—Various types of bentonite and cement often are required to seal the hole at the end of the sounding. Regional regulations and protocol dictate exactly what grout materials will be required. Usually, the bentonite used is powdered and the cement is portland, though sometimes ultrafine cement is used if it is to be pumped through a small tube. See 5.5.

7. Hazards

7.1 Environmental site characterization can present numerous hazards to equipment and personnel. It is the responsibility of everyone involved with the environmental site characterization project to understand fully all potential hazards.

NOTE 1: **Warning**—Hazards to personnel include, but are not limited to, fire; toxicity; heat exhaustion; local vegetation, such as poison ivy; local animals, such as snakes, or simply accidents due to the cumbersome aspects of safety equipment. A complete understanding of the OSHA 40-h safety course and the Health and Safety Plan[7] is required.

---

[6] *Reagent Chemicals, American Chemical Society Specifications*, American Chemical Society, Washington, DC. For suggestions on the testing of reagents not listed by the American Chemical Society, see *Analar Standards for Laboratory Chemicals*, BDH Ltd., Poole, Dorset, U.K., and the *United States Pharmacopeia and National Formulary*, U.S. Pharmacopeial Convention, Inc. (USPC), Rockville, MD.

[7] Follow NIOSH/OSHA Pocket Guide to Chemical Hazards, NIOSH/OSHA Occupational Health Guidelines for Chemical Hazards, and NIOSH/OSHA Occupational Safety and Health Guidance Manual for Hazardous Waste Site Activities available from U.S. Dept. of Health and Human Services, Centers for Disease Control, U.S. Government Printing Office.

402

NOTE 2: **Caution**—Hazards to equipment include, but are not limited to, fire or chemical attack. Seals for the cone penetrometer must be compatible with the local contaminants.

Procedure

8.1 The first step of any environmental site characterization project is to understand fully safety issues, such as the Health and Safety Plan, and having the area cleared and marked for utilities. A proper Health and Safety Plan addresses the aspects that apply specifically to the cone penetration operation and not just to drilling.

8.2 Upon arrival at the site, review the definition of the project to determine if any safety issues have been overlooked. If any unanticipated hazardous situation exists, notify the proper authorities immediately. An exclusion zone around the vehicle must be established to prevent unauthorized entry in the area. Appropriate flagmen, warning signs, cones, and street markings is required if the work is near a street or parking lot.

8.3 Regulations and safety specifications often are generic in nature and are intended to cover a wide variety of environmental site characterization projects. It is possible that one or more of these procedures could be counterproductive or even present an alternative hazard. If this is the case, notify the appropriate authorities immediately.

8.4 All cleaning, grouting, and safety equipment must be in good working order and fully prepared before starting each cone penetration test.

8.5 Perform the electronic cone penetrometer test in accordance with Test Method D 5778. Calibrate the cone penetrometer in accordance with Test Method D 5778, as well. Note any variances to the test due to environmental conditions.

8.6 Monitor the pore pressure dissipation. If monitoring is done, the cone should be saturated fully in accordance with the manufacturer's recommendations. The data acquisition system should begin timing automatically the dissipation the instant the rod motion is stopped.

8.7 During the extraction process, monitor for volatile organic compounds with a PID or FID at the top of rods and in the breathing zone and note readings in the scientific notebook. Take wipe samples as required in the health and safety plans. If any chemical constituent exceeds safe limits, as determined by the health and safety plan, respirators or other appropriate action will be required in the breathing zone. Use double gloves at all times while handling rods.

8.8 During disassembly of rods, if there is any free water within the rod column, these rods must be treated carefully. Check free water with an FID or PID. If this water registers PID readings or appears discolored, remove the end rods in the string from the cone truck or cleaned appropriately.

8.9 Clean the equipment according to predetermined appropriate methods. Inspect the equipment regularly for chemical attack and seal deterioration. First, externally clean and dry the cone. This will help prevent contaminants from intruding during disassembly for a more thorough cleaning. If only limited contamination exists and no cleaning is required, store the penetrometer in plastic or foil, and do not handle the penetrometer without protective gloves. The O-rings in the cone may need to be inspected or changed, or both, after every sounding. Change the O-rings if they appear

to be swollen, stuck to the metal surfaces, or spongy. If they appear to be deteriorating rapidly, use a more impervious compound. O-ring deterioration may cause erroneous friction data. A different compound, however, also may alter the data.

8.10 Normally, the dirt seals in the joints around the sleeve jacket contain only a minor amount of soil (less than 1 g) such that there is usually no concern for cross contamination between sounding sites. In cases where cleaning is required, the soil and fluids that the cone was exposed to may be considered contaminated; therefore, take the following measures to clean and decontaminate the cone.

8.11 Contamination will only be present on the cone body and the seals around the piezo element and friction sleeve. Place a protective cap over the electrical connector. Wash the cone with a brush and warm water and nonphosphate detergent, such as Alconox, and rinse with deionized water. Repeat as necessary to remove any visible soil from o-ring and quad-ring areas. O-rings, quad-rings, and piezo elements will be discarded during disassembly.

8.12 After pore pressure soundings, the pore pressure element may require special attention. Whereas, in geotechnical cone penetrometer tests, the elements can be used more than one time. In environmental tests, the elements may need to be replaced and discarded after each test if the material has been chemically degraded.

8.13 Grout the holes closed according to the appropriate predetermined method. A complete discussion of grouting holes resulting from direct push tools is discussed in Guide D 6001. Reentry grouting is the most common method of grouting (6). The grout rod could be a PVC pipe, a plastic tube, or another steel pipe, depending on how well the hole stays open. A grout rod will follow the path of least resistance and often can be pushed to the complete depth of the original CPT hole. This method is simple and usually very effective.

8.14 There are several methods of grouting during rod retraction without reentry. The following discussion is intended to discuss advantages and disadvantages of each method. Not all methods are discussed, but most methods include the following principles.

8.14.1 Grouting through the CPT push rod is possible. The grout can be pumped down the rod and out special ports near or at the cone tip. Pumping the grout inside the rod smears grout on the inside of the rod and on the signal cable. The extra cleaning time often makes this method impractical. The grout also can be pumped down an inner tube inside the rod. This usually requires a thinner grout mix to flow through a thin tube or it requires a larger tube with a larger diameter push rod requiring additional force to push.

8.14.2 It is possible to grout during rod retraction by pushing an expendable ring under the friction reducer. The cone penetrometer would be pushed through a reservoir of grout, dragging the grout down the annular area of the expendable ring. The expendable ring drops off the end when the rods are retracted allowing the grout to fill the hole from the bottom up. The ring increases the hole size, however, requiring additional push force. The outside of the push rods will need to be cleaned of the grout, but this may take less time than cleaning the inside of reentry rods.

8.14.3 It is possible to grout during rod retraction by

pushing a casing over the push rod, withdrawing the push rod, and then routing through the open casing. This, too, requires a larger hole, but is often used in soft soils where the outer casing also can provide lateral rod support for the CPT rod.

8.14.4 In some cases, it is possible to simply grout the hole by pouring grout into the open hole. This is normally only permitted if the hole does not extend into the water table.

9. Report

9.1 Where possible, the data and calibration reports should conform to the test methods and information described in Test Method D 5778.

9.2 Include information that may alter the cone penetrometer data.

9.2.1 Chemical attack of seals may cause failure and leaks or high friction if the seals become sticky. If alternate seals are used to prevent this from happening, document this information since different seal types have different friction characteristics that may affect the data.

9.2.2 The data obtained by the electronic cone penetrometer test is assumed to pertain to normal soils. This may not be the case, necessarily, in environmental site characterization projects. Report the presence of known tar, waste, debris, landfill deposits, and so forth, that are not normally deposited soils. For example, oily and greasy soils have less local friction, and landfills may have voids and numerous items that are not soils. Voids will be indicated by zero tip and friction values and should be identified as such so the engineer does not think the data indicates depth counter problems. Some landfill items can be identified by sound. The breaking of metal objects or timbers, or both, produces distinct noises that can be identified and noted in the report.

9.3 The cone penetrometer process is a valuable method for deploying alternative sensors. An in-depth discussion of alternative sensors is beyond the scope of this test method. Report the type of sensor and data from the sensor. Report the location and physical shape of the sensor since this may affect the cone penetrometer data. Include a complete description of the sensor technology, equipment, and procedures.

9.4 As indicated in 8.8, include the grouting procedure and anomalies.

10. Keywords

10.1 cone penetrometer; cone penetrometer test; direct push; explorations; ground water; penetration tests; piezocone; soil investigations; soundings; water sampling; well point

REFERENCES

(1) Manchon, R. G., "Introduction to Cone Penetrometer Testing and Ground Water Samplers," Workshop presented at the Fifth National Outdoor Action Conference on Aquifer Restoration, Ground Water Monitoring and Geophysical Methods, IT Corp., Martinez, CA, May 1991.

(2) Auxt, J. A., and Wright, D. E., "Environmental Site Characterization in the United States Using the Cone Penetrometer," *Proceedings of CPT '95—International Symposium on Cone Penetration Testing*, Vol 2, 1995, pp. 387–392.

(3) Lutennegger, A. J., and DeGroot, D. J., "Techniques for Sealing Cone Penetrometer Holes," *Canadian Geotechnical Journal*, Vol 32, No. 5, pp. 880–891.

(4) Smolley, M., and Kappemyer, J., "Cone Penetrometer Tests and Hydropunch Sampling—An Alternative to Monitoring Wells for Plume Detection," *Proceedings of the Hazmacon 1989 Conference*, April 18–20, 1989, Santa Clara, CA, pp. 71–80.

(5) Strutynsky, A., and Bergen, C., "Use of Piezometric Cone Penetration Testing and Penetrometer Ground Water Sampling for Volatile Organic Contaminant Plume Detection," *Proceedings of the National Water Well Association Petroleum Hydrocarbons Conference*, Oct. 31–Nov. 2, 1990, Houston, TX, pp. 71–84.

(6) Berzins, N. A., "Use of the Cone Penetration Test and BAT Ground Water Monitoring System to Assess Deficiencies in Monitoring Well Data," *Proceedings of the National Ground Water Association Outdoor Action Conference*, 1992, pp. 327–341.

Standard
Classification of Soils for Engineering Purposes
(Unified Soil Classification System)[1]

This standard is issued under the fixed designation D 2487; the number immediately following the designation indicates the year of original adoption or, in the case of revision, the year of last revision. A number in parentheses indicates the year of last reapproval. A superscript epsilon (ε) indicates an editorial change since the last revision or reapproval.

This standard has been approved for use by agencies of the Department of Defense. Consult the DOD Index of Specifications and Standards for the specific year of issue which has been adopted by the Department of Defense.

1. Scope

1.1 This standard describes a system for classifying mineral and organo-mineral soils for engineering purposes based on laboratory determination of particle-size characteristics, liquid limit, and plasticity index and shall be used when precise classification is required.

NOTE 1—Use of this standard will result in a single classification group symbol and group name except when a soil contains 5 to 12 % fines or when the plot of the liquid limit and plasticity index values falls into the crosshatched area of the plasticity chart. In these two cases, a dual symbol is used, for example, GP-GM, CL-ML. When the laboratory test results indicate that the soil is close to another soil classification group, the borderline condition can be indicated with two symbols separated by a slash. The first symbol should be the one based on this standard, for example, CL/CH, GM/SM, SC/CL. Borderline symbols are particularly useful when the liquid limit value of clayey soils is close to 50. These soils can have expansive characteristics and the use of a borderline symbol (CL/CH, CH/CL) will alert the user of the assigned classifications of expansive potential.

1.2 The group symbol portion of this sytem is based on laboratory tests performed on the portion of a soil sample passing the 3-in. (75-mm) sieve (see Specification E 11).

1.3 As a classification system, this standard is limited to naturally occurring soils.

NOTE 2—The group names and symbols used in this test method may be used as a descriptive system applied to such materials as shale, claystone, shells, crushed rock, etc. See Appendix X2.

1.4 This standard is for qualitative application only.

NOTE 3—When quantitative information is required for detailed designs of important structures, this test method must be supplemented by laboratory tests or other quantitative data to determine performance characteristics under expected field conditions.

1.5 This standard is the ASTM version of the Unified Soil Classification System. The basis for the classification scheme is the Airfield Classification System developed by A. Casagrande in the early 1940's.[2] It became known as the Unified Soil Classification System when several U.S. Government Agencies adopted a modified version of the Airfield System in 1952.

1.6 *This standard does not purport to address all of the* safety problems, if any, associated with its use. It is the responsibility of the user of this standard to establish appropriate safety and health practices and determine the applicability of regulatory limitations prior to use.

2. Referenced Documents

2.1 *ASTM Standards:*

C 117 Test Method for Materials Finer Than 75-μm (No. 200) Sieve in Mineral Aggregates by Washing[3]

C 136 Test Method for Sieve Analysis of Fine and Coarse Aggregates[3]

C 702 Practice for Reducing Field Samples of Aggregate to Testing Size[3]

D 420 Guide for Investigating and Sampling Soil and Rock[4]

D 421 Practice for Dry Preparation of Soil Samples for Particle-Size Analysis and Determination of Soil Constants[4]

D 422 Test Method for Particle-Size Analysis of Soils[4]

D 653 Terminology Relating to Soil, Rock, and Contained Fluids[4]

D 1140 Test Method for Amount of Material in Soils Finer than the No. 200 (75-μm) Sieve[4]

D 2216 Test Method for Laboratory Determination of Water (Moisture) Content of Soil and Rock[4]

D 2217 Practice for Wet Preparation of Soil Samples for Particle-Size Analysis and Determination of Soil Constants[4]

D 2488 Practice for Description and Identification of Soils (Visual-Manual Procedure)[4]

D 4083 Practice for Description of Frozen Soils (Visual-Manual Procedure)[4]

D 4318 Test Method for Liquid Limit, Plastic Limit, and Plasticity Index of Soils[4]

D 4427 Classification of Peat Samples by Laboratory Testing[4]

E 11 Specification for Wire-Cloth Sieves for Testing Purposes[3]

3. Terminology

3.1 *Definitions*—Except as listed below, all definitions are in accordance with Terminology D 653.

---

[1] This standard is under the jurisdiction of ASTM Committee D-18 on Soil and Rock and is the direct responsibility of Subcommittee D18.07 on Identification and Classification of Soils.
Current edition approved Sept. 15, 1993. Published November 1993. Originally published as D 2487 – 66 T. Last previous edition D 2487 – 92.
[2] Casagrande, A., "Classification and Identification of Soils," *Transactions*, ASCE, 1948, p. 901.

[3] *Annual Book of ASTM Standards*, Vol 04.02.
[4] *Annual Book of ASTM Standards*, Vol 04.08.

NOTE 4—For particles retained on a 3-in. (75-mm) U.S. standard sieve, the following definitions are suggested:

Cobbles—particles of rock that will pass a 12-in. (300-mm) square opening and be retained on a 3-in. (75-mm) U.S. standard sieve, and

Boulders—particles of rock that will not pass a 12in. (300-mm) square opening

3.1.1 *gravel*—particles of rock that will pass a 3-in. (75-mm) sieve and be retained on a No. 4 (4.75-mm) U.S. standard sieve with the following subdivisions:

Coarse—passes 3-in. (75-mm) sieve and retained on ¾-in. (19-mm) sieve, and

Fine—passes ¾-in. (19-mm) sieve and retained on No. 4 (4.75-mm) sieve.

3.1.2 *sand*—particles of rock that will pass a No. 4 (4.75-mm) sieve and be retained on a No. 200 (75-μm) U.S. standard sieve with the following subdivisions:

Coarse—passes No. 4 (4.75-mm) sieve and retained on No. 10 (2.00-mm) sieve,

Medium—passes No. 10 (2.00-mm) sieve and retained on No. 40 (425-μm) sieve, and

Fine—passes No. 40 (425-μm) sieve and retained on No. 200 (75-μm) sieve.

3.1.3 *clay*—soil passing a No. 200 (75-μm) U.S. standard sieve that can be made to exhibit plasticity (putty-like properties) within a range of water contents and that exhibits considerable strength when air dry. For classification, a clay is a fine-grained soil, or the fine-grained portion of a soil, with a plasticity index equal to or greater than 4, and the plot of plasticity index versus liquid limit falls on or above the "A" line.

3.1.4 *silt*—soil passing a No. 200 (75-μm) U.S. standard sieve that is nonplastic or very slightly plastic and that exhibits little or no strength when air dry. For classification, a silt is a fine-grained soil, or the fine-grained portion of a soil, with a plasticity index less than 4 or if the plot of plasticity index versus liquid limit falls below the "A" line.

3.1.5 *organic clay*—a clay with sufficient organic content to influence the soil properties. For classification, an organic clay is a soil that would be classified as a clay except that its liquid limit value after oven drying is less than 75 % of its liquid limit value before oven drying.

3.1.6 *organic silt*—a silt with sufficient organic content to influence the soil properties. For classification, an organic silt is a soil that would be classified as a silt except that its liquid limit value after oven drying is less than 75 % of its liquid limit value before oven drying.

3.1.7 *peat*—a soil composed of vegetable tissue in various stages of decomposition usually with an organic odor, a dark-brown to black color, a spongy consistency, and a texture ranging from fibrous to amorphous.

3.2 *Descriptions of Terms Specific to This Standard:*

3.2.1 *coefficient of curvature*, Cc—the ratio $(D_{30})^2/(D_{10} \times D_{60})$, where D_{60}, D_{30}, and D_{10} are the particle diameters corresponding to 60, 30, and 10 % finer on the cumulative particle-size distribution curve, respectively.

3.2.2 *coefficient of uniformity*, Cu—the ratio D_{60}/D_{10}, where D_{60} and D_{10} are the particle diameters corresponding to 60 and 10 % finer on the cumulative particle-size distribution curve, respectively.

4. Summary

4.1 As illustrated in Table 1, this classification system identifies three major soil divisions: coarse-grained soils, fine-grained soils, and highly organic soils. These three divisions are further subdivided into a total of 15 basic soil groups.

4.2 Based on the results of visual observations and prescribed laboratory tests, a soil is catalogued according to the basic soil groups, assigned a group symbol(s) and name, and thereby classified. The flow charts, Fig. 1 for fine-grained soils, and Fig. 2 for coarse-grained soils, can be used to assign the appropriate group symbol(s) and name.

5. Significance and Use

5.1 This standard classifies soils from any geographic location into categories representing the results of prescribed laboratory tests to determine the particle-size characteristics, the liquid limit, and the plasticity index.

5.2 The assigning of a group name and symbol(s) along with the descriptive information required in Practice D 2488 can be used to describe a soil to aid in the evaluation of its significant properties for engineering use.

5.3 The various groupings of this classification system have been devised to correlate in a general way with the engineering behavior of soils. This standard provides a useful first step in any field or laboratory investigation for geotechnical engineering purposes.

5.4 This standard may also be used as an aid in training personnel in the use of Practice D 2488.

5.5 This standard may be used in combination with Practice D 4083 when working with frozen soils.

6. Apparatus

6.1 In addition to the apparatus that may be required for obtaining and preparing the samples and conducting the prescribed laboratory tests, a plasticity chart, similar to Fig. 3, and a cumulative particle-size distribution curve, similar to Fig. 4, are required.

NOTE 5—The "U" line shown on Fig. 3 has been empirically determined to be the approximate "upper limit" for natural soils. It is a good check against erroneous data, and any test results that plot above or to the left of it should be verified.

7. Sampling

7.1 Samples shall be obtained and identified in accordance with a method or methods, recommended in Recommended Guide D 420 or by other accepted procedures.

7.2 For accurate identification, the minimum amount of test sample required for this test method will depend on which of the laboratory tests need to be performed. Where only the particle-size analysis of the sample is required, specimens having the following minimum dry weights are required:

Maximum Particle Size, Sieve Opening	Minimum Specimen Size, Dry Weight
4.75 mm (No. 4)	100 g (0.25 lb)
9.5 mm (⅜ in.)	200 g (0.5 lb)
19.0 mm (¾ in.)	1.0 kg (2.2 lb)
38.1 mm (1½ in.)	8.0 kg (18 lb)
75.0 mm (3 in.)	60.0 kg (132 lb)

Whenever possible, the field samples should have weights two to four times larger than shown.

TABLE 1 Soil Classification Chart

Criteria for Assigning Group Symbols and Group Names Using Laboratory Tests[A]				Soil Classification	
				Group Symbol	Group Name [B]
COARSE-GRAINED SOILS More than 50 % retained on No. 200 sieve	Gravels More than 50 % of coarse fraction retained on No. 4 sieve	Clean Gravels Less than 5 % fines[C]	$Cu \geq 4$ and $1 \leq Cc \leq 3$[E]	GW	Well-graded gravel[F]
			$Cu < 4$ and/or $1 > Cc > 3$[E]	GP	Poorly graded gravel[F]
		Gravels with Fines More than 12 % fines[C]	Fines classify as ML or MH	GM	Silty gravel[F,G,H]
			Fines classify as CL or CH	GC	Clayey gravel[F,G,H]
	Sands 50 % or more of coarse fraction passes No. 4 sieve	Clean Sands Less than 5 % fines[D]	$Cu \geq 6$ and $1 \leq Cc \leq 3$[E]	SW	Well-graded sand[I]
			$Cu < 6$ and/or $1 > Cc > 3$[E]	SP	Poorly graded sand[I]
		Sands with Fines More than 12 % fines[D]	Fines classify as ML or MH	SM	Silty sand[G,H,I]
			Fines classify as CL or CH	SC	Clayey sand[G,H,I]
FINE-GRAINED SOILS 50 % or more passes the No. 200 sieve	Silts and Clays Liquid limit less than 50	inorganic	$PI > 7$ and plots on or above "A" line[J]	CL	Lean clay[K,L,M]
			$PI < 4$ or plots below "A" line[J]	ML	Silt[K,L,M]
		organic	$\dfrac{\text{Liquid limit} - \text{oven dried}}{\text{Liquid limit} - \text{not dried}} < 0.75$	OL	Organic clay[K,L,M,N]
					Organic silt[K,L,M,O]
	Silts and Clays Liquid limit 50 or more	inorganic	PI plots on or above "A" line	CH	Fat clay[K,L,M]
			PI plots below "A" line	MH	Elastic silt[K,L,M]
		organic	$\dfrac{\text{Liquid limit} - \text{oven dried}}{\text{Liquid limit} - \text{not dried}} < 0.75$	OH	Organic clay[K,L,M,P]
					Organic silt[K,L,M,Q]
HIGHLY ORGANIC SOILS		Primarily organic matter, dark in color, and organic odor		PT	Peat

[A] Based on the material passing the 3-in. (75-mm) sieve.

[B] If field sample contained cobbles or boulders, or both, add "with cobbles or boulders, or both" to group name.

[C] Gravels with 5 to 12 % fines require dual symbols:
GW-GM well-graded gravel with silt
GW-GC well-graded gravel with clay
GP-GM poorly graded gravel with silt
GP-GC poorly graded gravel with clay

[D] Sands with 5 to 12 % fines require dual symbols:
SW-SM well-graded sand with silt
SW-SC well-graded sand with clay
SP-SM poorly graded sand with silt
SP-SC poorly graded sand with clay

[E] $Cu = D_{60}/D_{10}$ $Cc = \dfrac{(D_{30})^2}{D_{10} \times D_{60}}$

[F] If soil contains ≥ 15 % sand, add "with sand" to group name.

[G] If fines classify as CL-ML, use dual symbol GC-GM, or SC-SM.

[H] If fines are organic, add "with organic fines" to group name.

[I] If soil contains ≥ 15 % gravel, add "with gravel" to group name.

[J] If Atterberg limits plot in hatched area, soil is a CL-ML, silty clay.

[K] If soil contains 15 to 29 % plus No. 200, add "with sand" or "with gravel," whichever is predominant.

[L] If soil contains ≥ 30 % plus No. 200, predominantly sand, add "sandy" to group name.

[M] If soil contains ≥ 30 % plus No. 200, predominantly gravel, add "gravelly" to group name.

[N] $PI \geq 4$ and plots on or above "A" line.

[O] $PI < 4$ or plots below "A" line.

[P] PI plots on or above "A" line.

[Q] PI plots below "A" line.

7.3 When the liquid and plastic limit tests must also be performed, additional material will be required sufficient to provide 150 g to 200 g of soil finer than the No. 40 (425-µm) sieve.

7.4 If the field sample or test specimen is smaller than the minimum recommended amount, the report shall include an appropriate remark.

8. Classification of Peat

8.1 A sample composed primarily of vegetable tissue in various stages of decomposition and has a fibrous to amorphous texture, a dark-brown to black color, and an organic odor should be designated as a highly organic soil and shall be classified as peat, PT, and not subjected to the classification procedures described hereafter.

8.2 If desired, classification of type of peat can be performed in accordance with Classification D 4427.

9. Preparation for Classification

9.1 Before a soil can be classified according to this standard, generally the particle-size distribution of the minus 3-in. (75-mm) material and the plasticity characteristics of the minus No. 40 (425-µm) sieve material must be deter-

mined. See 9.8 for the specific required tests.

9.2 The preparation of the soil specimen(s) and the testing for particle-size distribution and liquid limit and plasticity index shall be in accordance with accepted standard procedures. Two procedures for preparation of the soil specimens for testing for soil classification purposes are given in Appendixes X3 and X4. Appendix X3 describes the wet preparation method and is the preferred method for cohesive soils that have never dried out and for organic soils.

9.3 When reporting soil classifications determined by this standard, the preparation and test procedures used shall be reported or referenced.

9.4 Although the test procedure used in determining the particle-size distribution or other considerations may require a hydrometer analysis of the material, a hydrometer analysis is not necessary for soil classification.

9.5 The percentage (by dry weight) of any plus 3-in. (75-mm) material must be determined and reported as auxiliary information.

9.6 The maximum particle size shall be determined (measured or estimated) and reported as auxiliary information.

9.7 When the cumulative particle-size distribution is required, a set of sieves shall be used which include the

ASTM D 2487

GROUP
SYMBOL

GROUP NAME

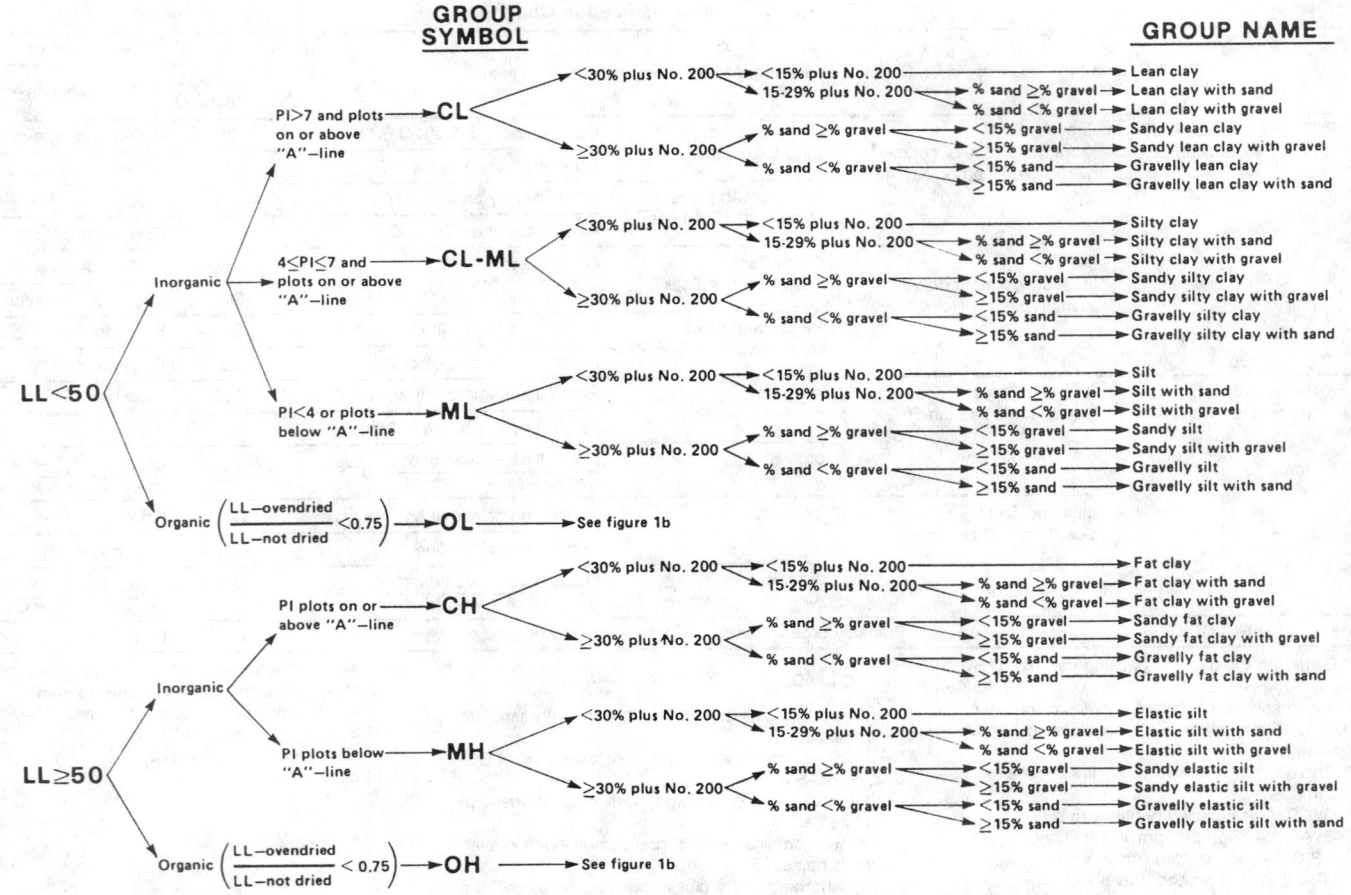

FIG. 1a Flow Chart for Classifying Fine-Grained Soil (50 % or More Passes No. 200 Sieve)

following sizes (with the largest size commensurate with the maximum particle size) with other sieve sizes as needed or required to define the particle-size distribution:

3-in. (75-mm)
¾-in.(19.0-mm)
No. 4 (4.75-mm)
No. 10 (2.00-mm)
No. 40 (425-μm)
No. 200 (75-μm)

9.8 The tests required to be performed in preparation for classification are as follows:

9.8.1 For soils estimated to contain less than 5 % fines, a plot of the cumulative particle-size distribution curve of the fraction coarser than the No. 200 (75-μm) sieve is required. The cumulative particle-size distribution curve may be plotted on a graph similar to that shown in Fig. 4.

9.8.2 For soils estimated to contain 5 to 15 % fines, a cumulative particle-size distribution curve, as described in 9.8.1, is required, and the liquid limit and plasticity index are required.

9.8.2.1 If sufficient material is not available to determine the liquid limit and plasticity index, the fines should be estimated to be either silty or clayey using the procedures described in Practice D 2488 and so noted in the report.

9.8.3 For soils estimated to contain 15 % or more fines, a determination of the percent fines, percent sand, and percent gravel is required, and the liquid limit and plasticity index

are required. For soils estimated to contain 90 % fines or more, the percent fines, percent sand, and percent gravel may be estimated using the procedures described in Practice D 2488 and so noted in the report.

10. Preliminary Classification Procedure

10.1 Class the soil as fine-grained if 50 % or more by dry weight of the test specimen passes the No. 200 (75-μm) sieve and follow Section 11.

10.2 Class the soil as coarse-grained if more than 50 % by dry weight of the test specimen is retained on the No. 200 (75-μm) sieve and follow Section 12.

11. Procedure for Classification of Fine-Grained Soils (50 % or more by dry weight passing the No. 200 (75-μm) sieve)

11.1 The soil is an inorganic clay if the position of the plasticity index versus liquid limit plot, Fig. 3, falls on or above the "A" line, the plasticity index is greater than 4, and the presence of organic matter does not influence the liquid limit as determined in 11.3.2.

NOTE 6—The plasticity index and liquid limit are determined on the minus No. 40 (425 μm) sieve material.

11.1.1 Classify the soil as a *lean clay*, CL, if the liquid limit is less than 50. See area identified as CL on Fig. 3.

11.1.2 Classify the soil as a *fat clay*, CH, if the liquid limit

408

GROUP SYMBOL

GROUP NAME

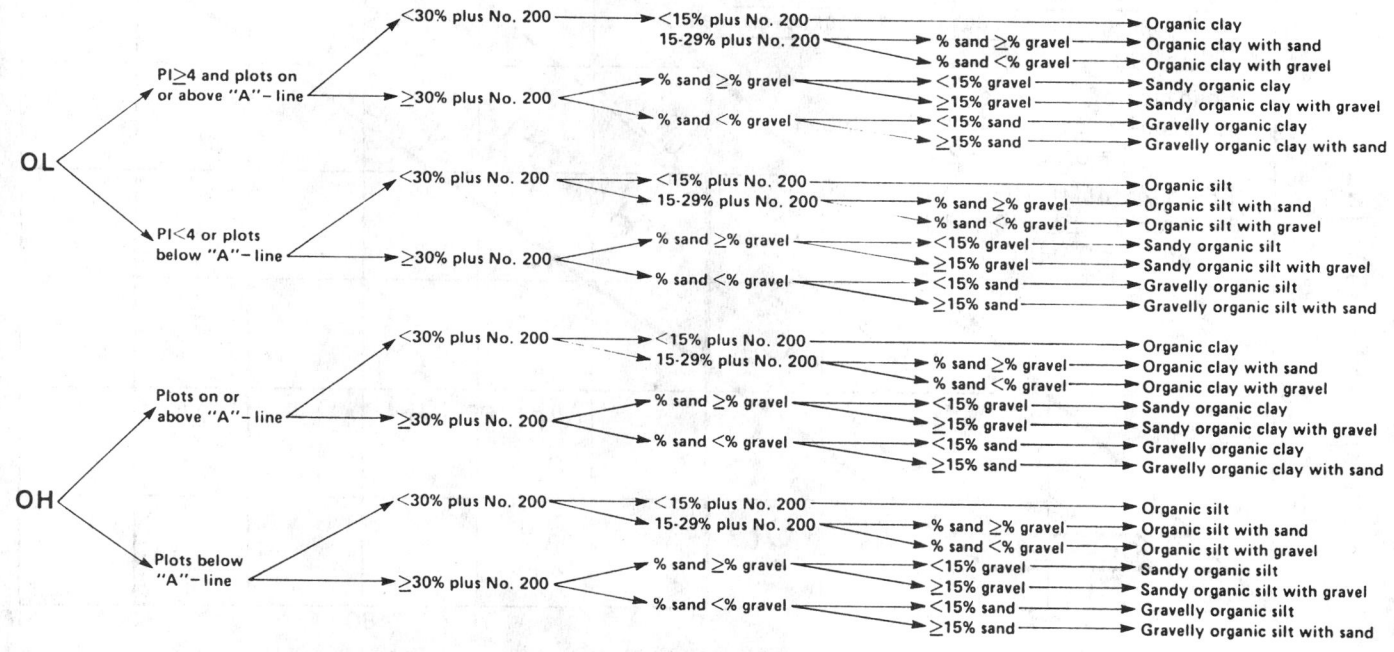

FIG. 1b   Flow Chart for Classifying Organic Fine-Grained Soil (50 % or More Passes No. 200 Sieve)

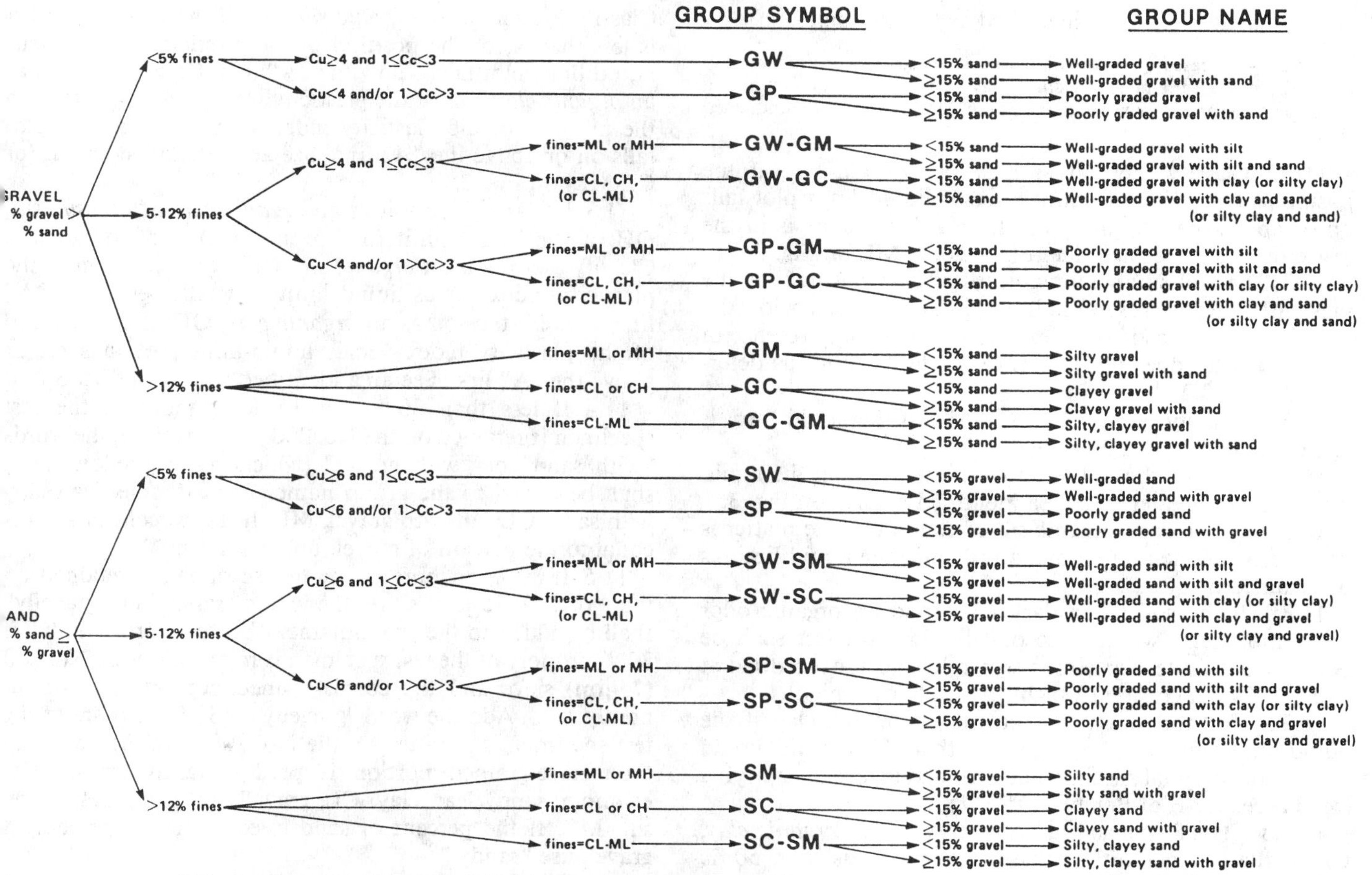

GROUP SYMBOL

GROUP NAME

<5% fines — $Cu \geq 4$ and $1 \leq Cc \leq 3$ — **GW** — <15% sand → Well-graded gravel
≥15% sand → Well-graded gravel with sand
$Cu < 4$ and/or $1 > Cc > 3$ — **GP** — <15% sand → Poorly graded gravel
≥15% sand → Poorly graded gravel with sand

GRAVEL
% gravel >
% sand

5-12% fines — $Cu \geq 4$ and $1 \leq Cc \leq 3$
fines=ML or MH — **GW-GM** — <15% sand → Well-graded gravel with silt
≥15% sand → Well-graded gravel with silt and sand
fines=CL, CH, (or CL-ML) — **GW-GC** — <15% sand → Well-graded gravel with clay (or silty clay)
≥15% sand → Well-graded gravel with clay and sand (or silty clay and sand)

$Cu < 4$ and/or $1 > Cc > 3$
fines=ML or MH — **GP-GM** — <15% sand → Poorly graded gravel with silt
≥15% sand → Poorly graded gravel with silt and sand
fines=CL, CH, (or CL-ML) — **GP-GC** — <15% sand → Poorly graded gravel with clay (or silty clay)
≥15% sand → Poorly graded gravel with clay and sand (or silty clay and sand)

>12% fines
fines=ML or MH — **GM** — <15% sand → Silty gravel
≥15% sand → Silty gravel with sand
fines=CL or CH — **GC** — <15% sand → Clayey gravel
≥15% sand → Clayey gravel with sand
fines=CL-ML — **GC-GM** — <15% sand → Silty, clayey gravel
≥15% sand → Silty, clayey gravel with sand

<5% fines — $Cu \geq 6$ and $1 \leq Cc \leq 3$ — **SW** — <15% gravel → Well-graded sand
≥15% gravel → Well-graded sand with gravel
$Cu < 6$ and/or $1 > Cc > 3$ — **SP** — <15% gravel → Poorly graded sand
≥15% gravel → Poorly graded sand with gravel

SAND
% sand ≥
% gravel

5-12% fines — $Cu \geq 6$ and $1 \leq Cc \leq 3$
fines=ML or MH — **SW-SM** — <15% gravel → Well-graded sand with silt
≥15% gravel → Well-graded sand with silt and gravel
fines=CL, CH, (or CL-ML) — **SW-SC** — <15% gravel → Well-graded sand with clay (or silty clay)
≥15% gravel → Well-graded sand with clay and gravel (or silty clay and gravel)

$Cu < 6$ and/or $1 > Cc > 3$
fines=ML or MH — **SP-SM** — <15% gravel → Poorly graded sand with silt
≥15% gravel → Poorly graded sand with silt and gravel
fines=CL, CH, (or CL-ML) — **SP-SC** — <15% gravel → Poorly graded sand with clay (or silty clay)
≥15% gravel → Poorly graded sand with clay and gravel (or silty clay and gravel)

>12% fines
fines=ML or MH — **SM** — <15% gravel → Silty sand
≥15% gravel → Silty sand with gravel
fines=CL or CH — **SC** — <15% gravel → Clayey sand
≥15% gravel → Clayey sand with gravel
fines=CL-ML — **SC-SM** — <15% gravel → Silty, clayey sand
≥15% gravel → Silty, clayey sand with gravel

FIG. 2   Flow Chart for Classifying Coarse-Grained Soils (More Than 50 % Retained on No. 200 Sieve)

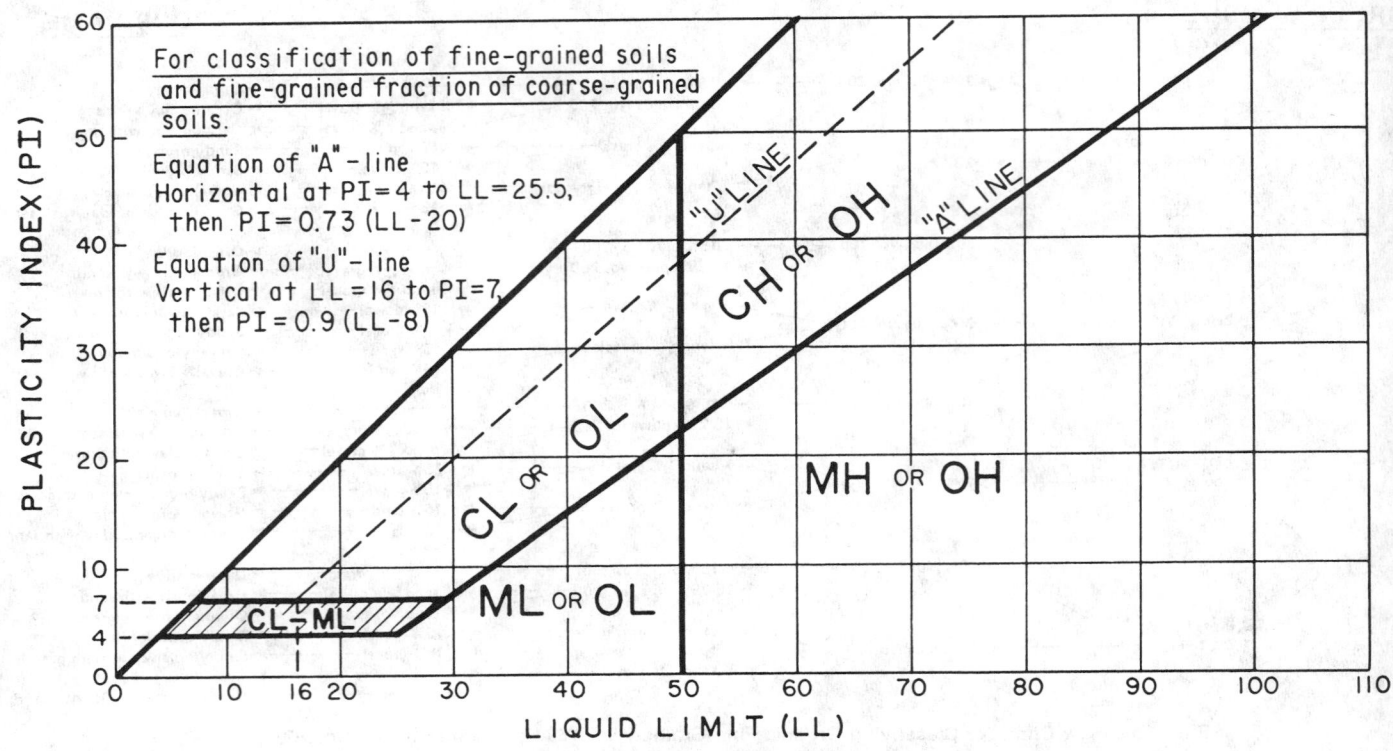

FIG. 3 Plasticity Chart

is 50 or greater. See area identified as CH on Fig. 3.

NOTE 7—In cases where the liquid limit exceeds 110 or the plasticity index exceeds 60, the plasticity chart may be expanded by maintaining the same scale on both axes and extending the "A" line at the indicated slope.

11.1.3 Classify the soil as a *silty clay*, CL-ML, if the position of the plasticity index versus liquid limit plot falls on or above the "A" line and the plasticity index is in the range of 4 to 7. See area identified as CL-ML on Fig. 3.

11.2 The soil is an inorganic silt if the position of the plasticity index versus liquid limit plot, Fig. 3, falls below the "A" line or the plasticity index is less than 4, and presence of organic matter does not influence the liquid limit as determined in 11.3.2.

11.2.1 Classify the soil as a *silt*, ML, if the liquid limit is less than 50. See area identified as ML on Fig. 3.

11.2.2 Classify the soil as an *elastic silt*, MH, if the liquid limit is 50 or greater. See area identified as MH on Fig. 3.

11.3 The soil is an organic silt or clay if organic matter is present in sufficient amounts to influence the liquid limit as determined in 11.3.2.

11.3.1 If the soil has a dark color and an organic odor when moist and warm, a second liquid limit test shall be performed on a test specimen which has been oven dried at 110 ± 5°C to a constant weight, typically over night.

11.3.2 The soil is an organic silt or organic clay if the liquid limit after oven drying is less than 75 % of the liquid limit of the original specimen determined before oven drying (see Procedure B of Practice D 2217).

11.3.3 Classify the soil as an *organic silt* or *organic clay*, OL, if the liquid limit (not oven dried) is less than 50 %.

Classify the soil as an *organic silt*, OL, if the plasticity index is less than 4, or the position of the plasticity index versus liquid limit plot falls below the "A" line. Classify the soil as an *organic clay*, OL, if the plasticity index is 4 or greater and the position of the plasticity index versus liquid limit plot falls on or above the "A" line. See area identified as OL (or CL-ML) on Fig. 3.

11.3.4 Classify the soil as an *organic clay* or *organic silt*, OH, if the liquid limit (not oven dried) is 50 or greater. Classify the soil as an *organic silt*, OH, if the position of the plasticity index versus liquid limit plot falls below the "A" line. Classify the soil as an *organic clay*, OH, if the position of the plasticity index versus liquid-limit plot falls on or above the "A" line. See area identified as OH on Fig. 3.

11.4 If less than 30 % but 15 % or more of the test specimen is retained on the No. 200 (75-μm) sieve, the words "with sand" or "with gravel" (whichever is predominant) shall be added to the group name. For example, lean clay with sand, CL; silt with gravel, ML. If the percent of sand is equal to the percent of gravel, use "with sand."

11.5 If 30 % or more of the test specimen is retained on the No. 200 (75-μm) sieve, the words "sandy" or "gravelly" shall be added to the group name. Add the word "sandy" if 30 % or more of the test specimen is retained on the No. 200 (75-μm) sieve and the coarse-grained portion is predominantly sand. Add the word "gravelly" if 30 % or more of the test specimen is retained on the No. 200 (75-μm) sieve and the coarse-grained portion is predominantly gravel. For example, sandy lean clay, CL; gravelly fat clay, CH; sandy silt, ML. If the percent of sand is equal to the percent of gravel, use "sandy."

SIEVE ANALYSIS

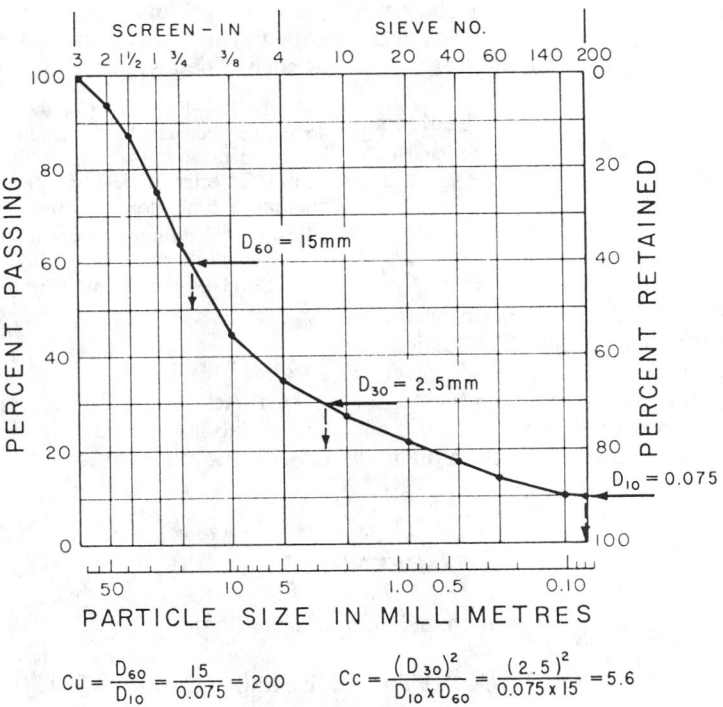

$$Cu = \frac{D_{60}}{D_{10}} = \frac{15}{0.075} = 200 \qquad Cc = \frac{(D_{30})^2}{D_{10} \times D_{60}} = \frac{(2.5)^2}{0.075 \times 15} = 5.6$$

FIG. 4 Cumulative Particle-Size Plot

12. Procedure for Classification of Coarse-Grained Soils (more than 50 % retained on the No. 200 (75-μm) sieve)

12.1 Class the soil as gravel if more than 50 % of the coarse fraction [plus No. 200 (75-μm) sieve] is retained on the No. 4 (4.75-mm) sieve.

12.2 Class the soil as sand if 50 % or more of the coarse fraction [plus No. 200 (75-μm) sieve] passes the No. 4 (4.75-mm) sieve.

12.3 If 12 % or less of the test specimen passes the No. 200 (75-μm) sieve, plot the cumulative particle-size distribution, Fig. 4, and compute the coefficient of uniformity, Cu, and coefficient of curvature, Cc, as given in Eqs 1 and 2.

$$Cu = D_{60}/D_{10} \qquad (1)$$
$$Cc = (D_{30})^2/(D_{10} \times D_{60}) \qquad (2)$$

where:
D_{10}, D_{30}, and D_{60} = the particle-size diameters corresponding to 10, 30, and 60 %, respectively, passing on the cumulative particle-size distribution curve, Fig. 4.

NOTE 8—It may be necessary to extrapolate the curve to obtain the D_{10} diameter.

12.3.1 If less than 5 % of the test specimen passes the No. 200 (75-μm) sieve, classify the soil as a *well-graded gravel*, GW, or *well-graded sand*, SW, if Cu is greater than 4.0 for gravel or greater than 6.0 for sand, and Cc is at least 1.0 but not more than 3.0.

12.3.2 If less than 5 % of the test specimen passes the No. 200 (75-μm) sieve, classify the soil as *poorly graded gravel*, GP, or *poorly graded sand*, SP, if either the Cu or the Cc criteria for well-graded soils are not satisfied.

12.4 If more than 12 % of the test specimen passes the No. 200 (75-μm) sieve, the soil shall be considered a coarse-grained soil with fines. The fines are determined to be either clayey or silty based on the plasticity index versus liquid limit plot on Fig. 3. (See 9.8.2.1 if insufficient material available for testing). (See NOTE 6)

12.4.1 Classify the soil as a *clayey gravel*, GC, or *clayey sand*, SC, if the fines are clayey, that is, the position of the plasticity index versus liquid limit plot, Fig. 3, falls on or above the "A" line and the plasticity index is greater than 7.

12.4.2 Classify the soil as a *silty gravel*, GM, or *silty sand*, SM, if the fines are silty, that is, the position of the plasticity index versus liquid limit plot, Fig. 3, falls below the "A" line or the plasticity index is less than 4.

12.4.3 If the fines plot as a silty clay, CL-ML, classify the soil as a *silty, clayey gravel*, GC-GM, if it is a gravel or a *silty, clayey sand*, SC-SM, if it is a sand.

12.5 If 5 to 12 % of the test specimen passes the No. 200 (75-μm) sieve, give the soil a dual classification using two group symbols.

12.5.1 The first group symbol shall correspond to that for a gravel or sand having less than 5 % fines (GW, GP, SW, SP), and the second symbol shall correspond to a gravel or sand having more than 12 % fines (GC, GM, SC, SM).

12.5.2 The group name shall correspond to the first group symbol plus "with clay" or "with silt" to indicate the plasticity characteristics of the fines. For example, well-graded gravel with clay, GW-GC; poorly graded sand with silt, SP-SM (See 9.8.2.1 if insufficient material available for testing).

NOTE 9—If the fines plot as a *silty clay*, CL-ML, the second group symbol should be either GC or SC. For example, a poorly graded sand with 10 % fines, a liquid limit of 20, and a plasticity index of 6 would be classified as a poorly graded sand with silty clay, SP-SC.

12.6 If the specimen is predominantly sand or gravel but contains 15 % or more of the other coarse-grained constituent, the words "with gravel" or "with sand" shall be added to the group name. For example, poorly graded gravel with sand, clayey sand with gravel.

12.7 If the field sample contained any cobbles or boulders or both, the words "with cobbles," or "with cobbles and boulders" shall be added to the group name. For example, silty gravel with cobbles, GM.

13. Report

13.1 The report should include the group name, group symbol, and the results of the laboratory tests. The particle-size distribution shall be given in terms of percent of gravel, sand, and fines. The plot of the cumulative particle-size distribution curve shall be reported if used in classifying the soil. Report appropriate descriptive information according to the procedures in Practice D 2488. A local or commercial name or geologic interpretation for the material may be added at the end of the descriptive information if identified as such. The test procedures used shall be referenced.

NOTE 10—*Example: Clayey Gravel with Sand and Cobbles (GC)*— 46 % fine to coarse, hard, subrounded gravel; 30 % fine to coarse, hard, subrounded sand; 24 % clayey fines, LL = 38, PI = 19; weak reaction with HCl; original field sample had 4 % hard, subrounded cobbles; maximum dimension 150 mm.

In-Place Conditions—firm, homogeneous, dry, brown,
Geologic Interpretation—alluvial fan.

NOTE 11—Other examples of soil descriptions are given in Appendix X1.

14. Keywords

14.1 Atterberg limits; classification; clay; gradation; gravel; laboratory classification; organic soils; sand; silt; soil classification; soil tests

APPENDIXES

(Nonmandatory Information)

X1. EXAMPLES OF DESCRIPTIONS USING SOIL CLASSIFICATION

X1.1 The following examples show how the information required in 13.1 can be reported. The appropriate descriptive information from Practice D 2488 is included for illustrative purposes. The additional descriptive terms that would accompany the soil classification should be based on the intended use of the classification and the individual circumstances.

X1.1.1 *Well-Graded Gravel with Sand (GW)*—73 % fine to coarse, hard, subangular gravel; 23 % fine to coarse, hard, subangular sand; 4 % fines; Cc = 2.7, Cu = 12.4.

X1.1.2 *Silty Sand with Gravel (SM)*—61 % predominantly fine sand; 23 % silty fines, LL = 33, PI = 6; 16 % fine, hard, subrounded gravel; no reaction with HCl; (field sample smaller than recommended). *In-Place Conditions*—Firm, stratified and contains lenses of silt 1 to 2 in. thick, moist, brown to gray; in-place density = 106 lb/ft^3 and in-place moisture = 9 %.

X1.1.3 *Organic Clay (OL)*—100 % fines, LL (not dried) = 32, LL (oven dried) = 21, PI (not dried) = 10; wet, dark brown, organic odor, weak reaction with HCl.

X1.1.4 *Silty Sand with Organic Fines (SM)*—74 % fine to coarse, hard, subangular reddish sand; 26 % organic and silty dark-brown fines, LL (not dried) = 37, LL (oven dried) = 26, PI (not dried) = 6, wet, weak reaction with HCl.

X1.1.5 *Poorly Graded Gravel with Silt, Sand, Cobbles and Boulders (GP-GM)*—78 % fine to coarse, hard, subrounded to subangular gravel; 16 % fine to coarse, hard, subrounded to subangular sand; 6 % silty (estimated) fines; moist, brown; no reaction with HCl; original field sample had 7 % hard, subrounded cobbles and 2 % hard, subrounded boulders with a maximum dimension of 18 in.

X2. USING SOIL CLASSIFICATION AS A DESCRIPTIVE SYSTEM FOR SHALE, CLAYSTONE, SHELLS, SLAG, CRUSHED ROCK, ETC.

X2.1 The group names and symbols used in this standard may be used as a descriptive system applied to materials that exist in situ as shale, claystone, sandstone, siltstone, mudstone, etc., but convert to soils after field or laboratory processing (crushing, slaking, etc.).

X2.2 Materials such as shells, crushed rock, slag, etc., should be identified as such. However, the procedures used in this standard for describing the particle size and plasticity characteristics may be used in the description of the material. If desired, a classification in accordance with this standard may be assigned to aid in describing the material.

X2.3 If a classification is used, the group symbol(s) and group names should be placed in quotation marks or noted with some type of distinguishing symbol. See examples.

X2.4 Examples of how soil classifications could be incorporated into a description system for materials that are not naturally occurring soils are as follows:

X2.4.1 *Shale Chunks*—Retrieved as 2 to 4-in. pieces of shale from power auger hole, dry, brown, no reaction with HCl. After laboratory processing by slaking in water for 24 h, material classified as "Sandy Lean Clay (CL)"—61 % clayey fines, LL = 37, PI = 16; 33 % fine to medium sand; 6 % gravel-size pieces of shale.

X2.4.2 *Crushed Sandstone*—Product of commercial crushing operation; "Poorly Graded Sand with Silt (SP-SM)"—91 % fine to medium sand; 9 % silty (estimated) fines; dry, reddish-brown, strong reaction with HCl.

X2.4.3 *Broken Shells*—62 % gravel-size broken shells;

31 % sand and sand-size shell pieces; 7 % fines; would be classified as "Poorly Graded Gravel with Sand (GP)".

X2.4.4 *Crushed Rock*—Processed gravel and cobbles from Pit No. 7; "Poorly Graded Gravel (GP)"—89 % fine, hard, angular gravel-size particles; 11 % coarse, hard, angular sand-size particles, dry, tan; no reaction with HCl; Cc = 2.4, Cu = 0.9.

X3. PREPARATION AND TESTING FOR CLASSIFICATION PURPOSES BY THE WET METHOD

X3.1 This appendix describes the steps in preparing a soil sample for testing for purposes of soil classification using a wet-preparation procedure.

X3.2 Samples prepared in accordance with this procedure should contain as much of their natural water content as possible and every effort should be made during obtaining, preparing, and transporting the samples to maintain the natural moisture.

X3.3 The procedures to be followed in this standard assume that the field sample contains fines, sand, gravel, and plus 3-in. (75-mm) particles and the cumulative particle-size distribution plus the liquid limit and plasticity index values are required (see 9.8). Some of the following steps may be omitted when they are not applicable to the soil being tested.

X3.4 If the soil contains plus No. 200 (75-μm) particles that would degrade during dry sieving, use a test procedure for determining the particle-size characteristics that prevents this degradation.

X3.5 Since this classification system is limited to the portion of a sample passing the 3-in. (75-mm) sieve, the plus 3-in. (75-mm) material shall be removed prior to the determination of the particle-size characteristics and the liquid limit and plasticity index.

X3.6 The portion of the field sample finer than the 3-in. (75-mm) sieve shall be obtained as follows:

X3.6.1 Separate the field sample into two fractions on a 3-in. (75-mm) sieve, being careful to maintain the natural water content in the minus 3-in. (75-mm) fraction. Any particles adhering to the plus 3-in. (75-mm) particles shall be brushed or wiped off and placed in the fraction passing the 3-in. (75-mm) sieve.

X3.6.2 Determine the air-dry or oven-dry weight of the fraction retained on the 3-in. (75-mm) sieve. Determine the total (wet) weight of the fraction passing the 3-in. (75-mm) sieve.

X3.6.3 Thoroughly mix the fraction passing the 3-in. (75-mm) sieve. Determine the water content, in accordance with Test Method D 2216, of a representative specimen with a minimum dry weight as required in 7.2. Save the water-content specimen for determination of the particle-size analysis in accordance with X3.8.

X3.6.4 Compute the dry weight of the fraction passing the 3-in. (75-mm) sieve based on the water content and total (wet) weight. Compute the total dry weight of the sample and calculate the percentage of material retained on the 3-in. (75-mm) sieve.

X3.7 Determine the liquid limit and plasticity index as follows:

X3.7.1 If the soil disaggregates readily, mix on a clean, hard surface and select a representative sample by quartering in accordance with Practice C 702.

X3.7.1.1 If the soil contains coarse-grained particles coated with and bound together by tough clayey material, take extreme care in obtaining a representative portion of the No. 40 (425-μm) fraction. Typically, a larger portion than normal has to be selected, such as the minimum weights required in 7.2.

X3.7.1.2 To obtain a representative specimen of a basically cohesive soil, it may be advantageous to pass the soil through a ¾-in. (19-mm) sieve or other convenient size so the material can be more easily mixed and then quartered or split to obtain the representative specimen.

X3.7.2 Process the representative specimen in accordance with Procedure B of Practice D 2217.

X3.7.3 Perform the liquid-limit test in accordance with Test Method D 4318, except the soil shall not be air dried prior to the test.

X3.7.4 Perform the plastic-limit test in accordance with Test Method D 4318, except the soil shall not be air dried prior to the test, and calculate the plasticity index.

X3.8 Determine the particle-size distribution as follows:

X3.8.1 If the water content of the fraction passing the 3-in. (75-mm) sieve was required (X3.6.3), use the water-content specimen for determining the particle-size distribution. Otherwise, select a representative specimen in accordance with Practice C 702 with a minimum dry weight as required in 7.2.

X3.8.2 If the cumulative particle-size distribution including a hydrometer analysis is required, determine the particle-size distribution in accordance with Test Method D 422. See 9.7 for the set of required sieves.

X3.8.3 If the cumulative particle-size distribution without a hydrometer analysis is required, determine the particle-size distribution in accordance with Method C 136. See 9.7 for the set of required sieves. The specimen should be soaked until all clayey aggregations have softened and then washed in accordance with Test Method C 117 prior to performing the particle-size distribution.

X3.8.4 If the cumulative particle-size distribution is not required, determine the percent fines, percent sand, and percent gravel in the specimen in accordance with Test Method C 117, being sure to soak the specimen long enough to soften all clayey aggregations, followed by Method C 136 using a nest of sieves which shall include a No. 4 (4.75-mm) sieve and a No. 200 (75-μm) sieve.

X3.8.5 Calculate the percent fines, percent sand, and percent gravel in the minus 3-in. (75-mm) fraction for classification purposes.

X4. AIR-DRIED METHOD OF PREPARATION OF SOILS FOR TESTING FOR CLASSIFICATION PURPOSES

X4.1 This appendix describes the steps in preparing a soil sample for testing for purposes of soil classification when air-drying the soil before testing is specified or desired or when the natural moisture content is near that of an air-dried state.

X4.2 If the soil contains organic matter or mineral colloids that are irreversibly affected by air drying, the wet-preparation method as described in Appendix X3 should be used.

X4.3 Since this classification system is limited to the portion of a sample passing the 3-in. (75-mm) sieve, the plus 3-in. (75-mm) material shall be removed prior to the determination of the particle-size characteristics and the liquid limit and plasticity index.

X4.4 The portion of the field sample finer than the 3-in. (75-mm) sieve shall be obtained as follows:

X4.4.1 Air dry and weigh the field sample.

X4.4.2 Separate the field sample into two fractions on a 3-in. (75-mm) sieve.

X4.4.3 Weigh the two fractions and compute the percentage of the plus 3-in. (75-mm) material in the field sample.

X4.5 Determine the particle-size distribution and liquid limit and plasticity index as follows (see 9.8 for when these tests are required):

X4.5.1 Thoroughly mix the fraction passing the 3-in. (75-mm) sieve.

X4.5.2 If the cumulative particle-size distribution including a hydrometer analysis is required, determine the particle-size distribution in accordance with Test Method D 422. See 9.7 for the set of sieves that is required.

X4.5.3 If the cumulative particle-size distribution without a hydrometer analysis is required, determine the particle-size distribution in accordance with Test Method D 1140 followed by Method C 136. See 9.7 for the set of sieves that is required.

X4.5.4 If the cumulative particle-size distribution is not required, determine the percent fines, percent sand, and percent gravel in the specimen in accordance with Test Method D 1140 followed by Method C 136 using a nest of sieves which shall include a No. 4 (4.75-mm) sieve and a No. 200 (75-μm) sieve.

X4.5.5 If required, determine the liquid limit and the plasticity index of the test specimen in accordance with Test Method D 4318.

X5. ABBREVIATED SOIL CLASSIFICATION SYMBOLS

X5.1 In some cases, because of lack of space, an abbreviated system may be useful to indicate the soil classification symbol and name. Examples of such cases would be graphical logs, databases, tables, etc.

X5.2 This abbreviated system is not a substitute for the full name and descriptive information but can be used in supplementary presentations when the complete description is referenced.

X5.3 The abbreviated system should consist of the soil classification symbol based on this standard with appropriate lower case letter prefixes and suffixes as:

Prefix	Suffix
s = sandy	s = with sand
g = gravelly	g = with gravel
	c = cobbles
	b = boulders

X5.4 The soil classification symbol is to be enclosed in parenthesis. Some examples would be:

Group Symbol and Full Name	Abbreviated
CL, Sandy lean clay	s(CL)
SP-Sm, Poorly graded sand with silt and gravel	(SP-SM)g
GP, poorly graded gravel with sand, cobbles, and boulders	(GP)scb
ML, gravelly silt with sand and cobbles	g(ML)sc

X6. RATIONALE

X6.1 Changes in this version from the previous D 2488 – 92 include the addition of X5 on Abbreviated Soil Classification Symbols.

 **D 2487**

Standard Practice for
Description and Identification of Soils (Visual-Manual Procedure)[1]

This standard is issued under the fixed designation D 2488; the number immediately following the designation indicates the year of original adoption or, in the case of revision, the year of last revision. A number in parentheses indicates the year of last reapproval. A superscript epsilon (ε) indicates an editorial change since the last revision or reapproval.

This standard has been approved for use by agencies of the Department of Defense. Consult the DoD Index of Specifications and Standards for the specific year of issue which has been adopted by the Department of Defense.

1. Scope

1.1 This practice covers procedures for the description of soils for engineering purposes.

1.2 This practice also describes a procedure for identifying soils, at the option of the user, based on the classification system described in Test Method D 2487. The identification is based on visual examination and manual tests. It must be clearly stated in reporting an identification that it is based on visual-manual procedures.

1.2.1 When precise classification of soils for engineering purposes is required, the procedures prescribed in Test Method D 2487 shall be used.

1.2.2 In this practice, the identification portion assigning a group symbol and name is limited to soil particles smaller than 3 in. (75 mm).

1.2.3 The identification portion of this practice is limited to naturally occurring soils (disturbed and undisturbed).

NOTE 1—This practice may be used as a descriptive system applied to such materials as shale, claystone, shells, crushed rock, etc. (See Appendix X2).

1.3 The descriptive information in this practice may be used with other soil classification systems or for materials other than naturally occurring soils.

1.4 *This standard does not purport to address all of the safety problems, if any, associated with its use. It is the responsibility of the user of this standard to establish appropriate safety and health practices and determine the applicability of regulatory limitations prior to use. For specific precautionary statements see Section 8.*

1.5 The values stated in inch-pound units are to be regarded as the standard.

2. Referenced Documents

2.1 *ASTM Standards:*
D 653 Terminology Relating to Soil, Rock, and Contained Fluids[2]
D 1452 Practice for Soil Investigation and Sampling by Auger Borings[2]
D 1586 Method for Penetration Test and Split-Barrel Sampling of Soils[2]
D 1587 Practice for Thin-Walled Tube Sampling of Soils[2]
D 2113 Practice for Diamond Core Drilling for Site Investigation[2]
D 2487 Classification of Soils for Engineering Purposes (Unified Soil Classification System)[2]
D 4083 Practice for Description of Frozen Soils (Visual-Manual Procedure)[2]

3. Terminology

3.1 *Definitions:*

3.1.1 Except as listed below, all definitions are in accordance with Terminology D 653.

NOTE 2—For particles retained on a 3-in. (75-mm) US standard sieve, the following definitions are suggested:
Cobbles—particles of rock that will pass a 12-in. (300-mm) square opening and be retained on a 3-in. (75-mm) sieve, and
Boulders—particles of rock that will not pass a 12-in. (300-mm) square opening.

3.1.1.2 *clay*—soil passing a No. 200 (75-μm) sieve that can be made to exhibit plasticity (putty-like properties) within a range of water contents, and that exhibits considerable strength when air-dry. For classification, a clay is a fine-grained soil, or the fine-grained portion of a soil, with a plasticity index equal to or greater than 4, and the plot of plasticity index versus liquid limit falls on or above the "A" line (see Fig. 3 of Test Method D 2487).

3.1.1.3 *gravel*—particles of rock that will pass a 3-in. (75-mm) sieve and be retained on a No. 4 (4.75-mm) sieve with the following subdivisions:
coarse—passes a 3-in. (75-mm) sieve and is retained on a ¾-in. (19-mm) sieve.
fine—passes a ¾-in. (19-mm) sieve and is retained on a No. 4 (4.75-mm) sieve.

3.1.1.4 *organic clay*—a clay with sufficient organic content to influence the soil properties. For classification, an organic clay is a soil that would be classified as a clay, except that its liquid limit value after oven drying is less than 75 % of its liquid limit value before oven drying.

3.1.1.5 *organic silt*—a silt with sufficient organic content to influence the soil properties. For classification, an organic silt is a soil that would be classified as a silt except that its liquid limit value after oven drying is less than 75 % of its liquid limit value before oven drying.

3.1.1.6 *peat*—a soil composed primarily of vegetable tissue in various stages of decomposition usually with an organic odor, a dark brown to black color, a spongy consistency, and a texture ranging from fibrous to amorphous.

3.1.1.7 *sand*—particles of rock that will pass a No. 4

[1] This practice is under the jurisdiction of ASTM Committee D-18 on Soil and Rock and is the direct responsibility of Subcommittee D18.07 on Identification and Classification of Soils.
Current edition approved Sept. 15, 1993. Published November 1993. Originally published as D 2488 – 66 T. Last previous edition D 2488 – 90.
[2] *Annual Book of ASTM Standards*, Vol 04.08.

GROUP SYMBOL

GROUP NAME

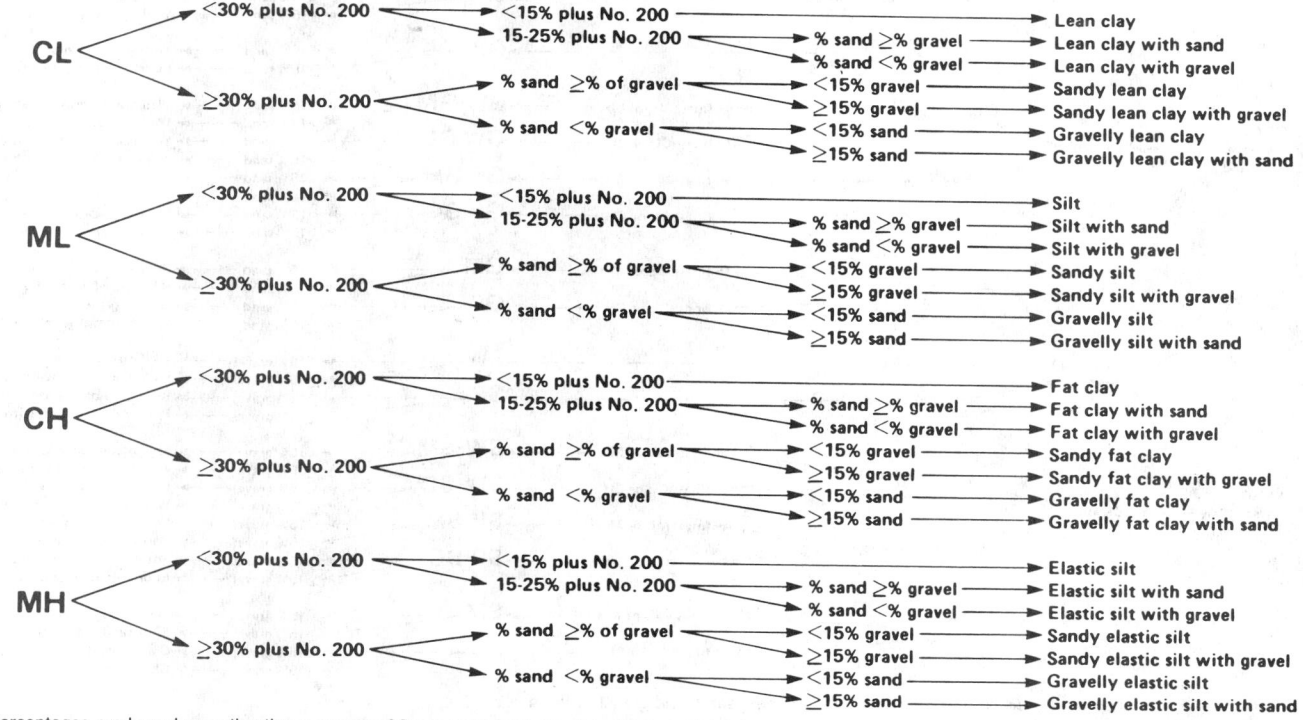

NOTE—Percentages are based on estimating amounts of fines, sand, and gravel to the nearest 5 %.

FIG. 1a Flow Chart for Identifying Inorganic Fine-Grained Soil (50 % or more fines)

(4.75-mm) sieve and be retained on a No. 200 (75-μm) sieve with the following subdivisions:

coarse—passes a No. 4 (4.75-mm) sieve and is retained on a No. 10 (2.00-mm) sieve.

medium—passes a No. 10 (2.00-mm) sieve and is retained on a No. 40 (425-μm) sieve.

fine—passes a No. 40 (425-μm) sieve and is retained on a No. 200 (75-μm) sieve.

3.1.1.8 *silt*—soil passing a No. 200 (75-μm) sieve that is nonplastic or very slightly plastic and that exhibits little or no strength when air dry. For classification, a silt is a fine-grained soil, or the fine-grained portion of a soil, with a plasticity index less than 4, or the plot of plasticity index versus liquid limit falls below the "A" line (see Fig. 3 of Test Method D 2487).

4. Summary of Practice

4.1 Using visual examination and simple manual tests, this practice gives standardized criteria and procedures for describing and identifying soils.

4.2 The soil can be given an identification by assigning a group symbol(s) and name. The flow charts, Figs. 1a and 1b for fine-grained soils, and Fig. 2, for coarse-grained soils, can be used to assign the appropriate group symbol(s) and name. If the soil has properties which do not distinctly place it into a specific group, borderline symbols may be used, see Appendix X3.

NOTE 3—It is suggested that a distinction be made between *dual symbols* and *borderline symbols*.

Dual Symbol—A dual symbol is two symbols separated by a hyphen, for example, GP-GM, SW-SC, CL-ML used to indicate that the soil has been identified as having the properties of a classification in accordance with Test Method D 2487 where two symbols are required. Two symbols are required when the soil has between 5 and 12 % fines or

GROUP SYMBOL

GROUP NAME

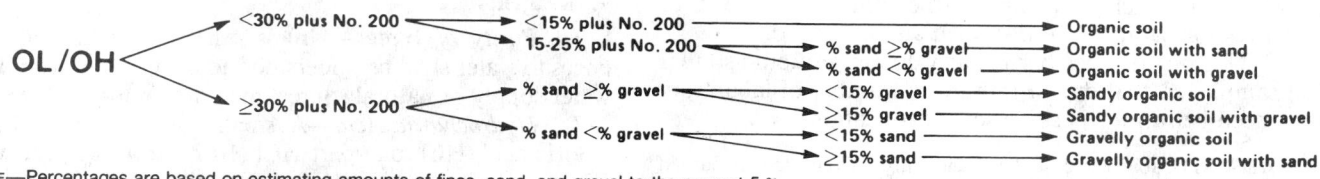

NOTE—Percentages are based on estimating amounts of fines, sand, and gravel to the nearest 5 %.

FIG. 1b Flow Chart for Identifying Organic Fine-Grained Soil (50 % or more fines)

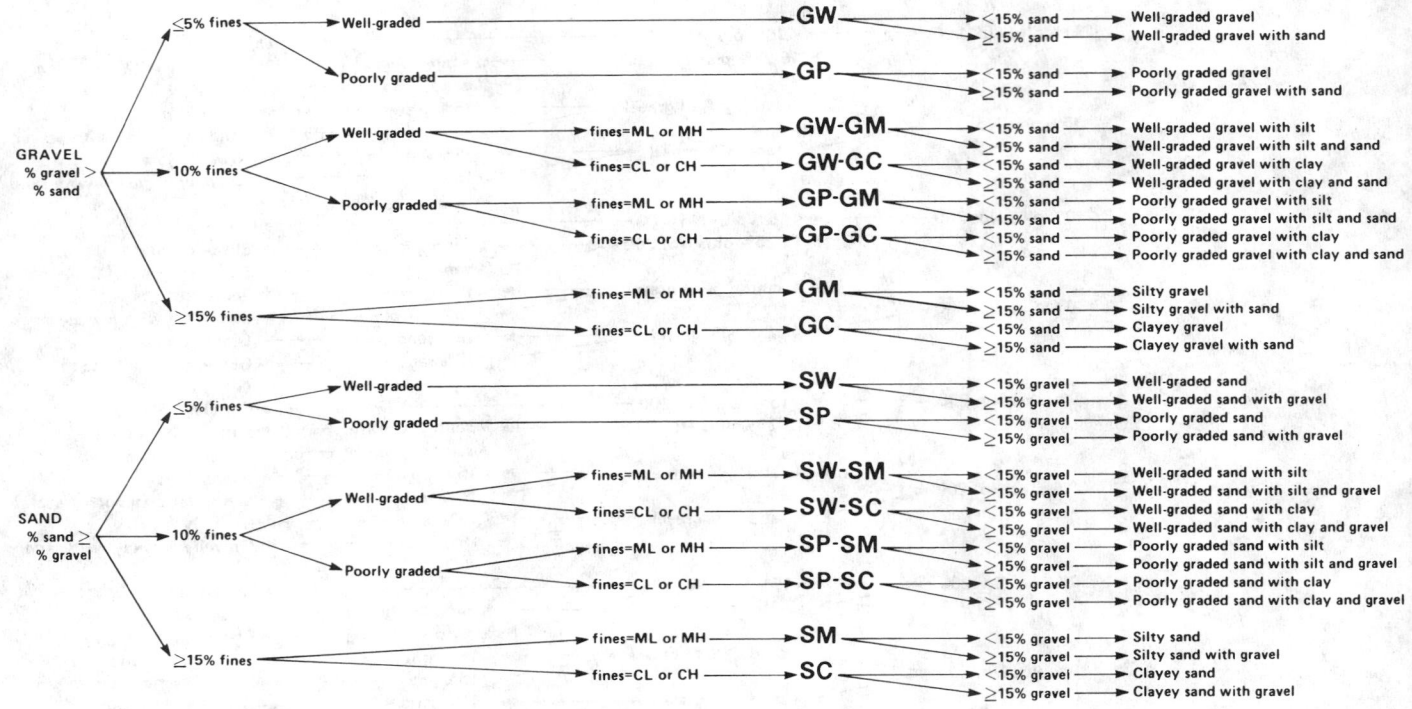

NOTE—Percentages are based on estimating amounts of fines, sand, and gravel to the nearest 5 %.

FIG. 2 Flow Chart for Identifying Coarse-Grained Soils (less than 50 % fines)

when the liquid limit and plasticity index values plot in the CL-ML area of the plasticity chart.

Borderline Symbol—A borderline symbol is two symbols separated by a slash, for example, CL/CH, GM/SM, CL/ML. A borderline symbol should be used to indicate that the soil has been identified as having properties that do not distinctly place the soil into a specific group (see Appendix X3).

5. Significance and Use

5.1 The descriptive information required in this practice can be used to describe a soil to aid in the evaluation of its significant properties for engineering use.

5.2 The descriptive information required in this practice should be used to supplement the classification of a soil as determined by Test Method D 2487.

5.3 This practice may be used in identifying soils using the classification group symbols and names as prescribed in Test Method D 2487. Since the names and symbols used in this practice to identify the soils are the same as those used in Test Method D 2487, it shall be clearly stated in reports and all other appropriate documents, that the classification symbol and name are based on visual-manual procedures.

5.4 This practice is to be used not only for identification of soils in the field, but also in the office, laboratory, or wherever soil samples are inspected and described.

5.5 This practice has particular value in grouping similar soil samples so that only a minimum number of laboratory tests need be run for positive soil classification.

NOTE 4—The ability to describe and identify soils correctly is learned more readily under the guidance of experienced personnel, but it may also be acquired systematically by comparing numerical laboratory test

results for typical soils of each type with their visual and manual characteristics.

5.6 When describing and identifying soil samples from a given boring, test pit, or group of borings or pits, it is not necessary to follow all of the procedures in this practice for every sample. Soils which appear to be similar can be grouped together; one sample completely described and identified with the others referred to as similar based on performing only a few of the descriptive and identification procedures described in this practice.

5.7 This practice may be used in combination with Practice D 4083 when working with frozen soils.

6. Apparatus

6.1 *Required Apparatus:*

6.1.1 *Pocket Knife or Small Spatula.*

6.2 *Useful Auxiliary Apparatus:*

6.2.1 *Small Test Tube and Stopper* (or jar with a lid).

6.2.2 *Small Hand Lens.*

7. Reagents

7.1 *Purity of Water*—Unless otherwise indicated, references to water shall be understood to mean water from a city water supply or natural source, including non-potable water.

7.2 *Hydrochloric Acid*—A small bottle of dilute hydrochloric acid, HCl, one part HCl (10 N) to three parts water (This reagent is optional for use with this practice). See Section 8.

ASTM D 2488

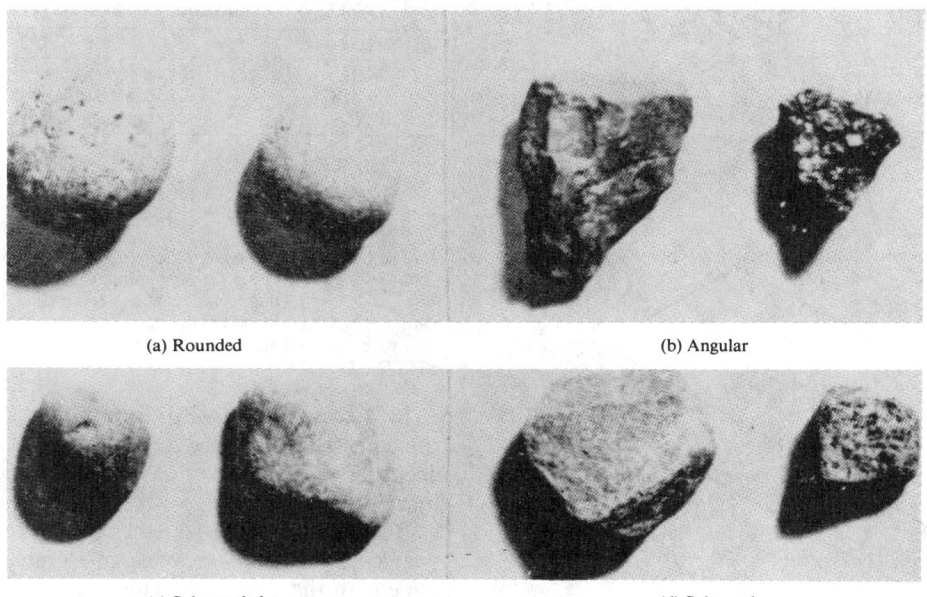

(a) Rounded (b) Angular
(c) Subrounded (d) Subangular

FIG. 3 Typical Angularity of Bulky Grains

8. Safety Precautions

8.1 When preparing the dilute HCl solution of one part concentrated hydrochloric acid (10 N) to three parts of distilled water, slowly add acid into water following necessary safety precautions. Handle with caution and store safely. If solution comes into contact with the skin, rinse thoroughly with water.

8.2 **Caution**—Do not add water to acid.

9. Sampling

9.1 The sample shall be considered to be representative of the stratum from which it was obtained by an appropriate, accepted, or standard procedure.

NOTE 5—Preferably, the sampling procedure should be identified as having been conducted in accordance with Practices D 1452, D 1587, or D 2113, or Method D 1586.

9.2 The sample shall be carefully identified as to origin.

NOTE 6—Remarks as to the origin may take the form of a boring number and sample number in conjunction with a job number, a geologic stratum, a pedologic horizon or a location description with respect to a permanent monument, a grid system or a station number and offset with respect to a stated centerline and a depth or elevation.

9.3 For accurate description and identification, the minimum amount of the specimen to be examined shall be in accordance with the following schedule:

Maximum Particle Size, Sieve Opening	Minimum Specimen Size, Dry Weight
4.75 mm (No. 4)	100 g (0.25 lb)
9.5 mm (⅜ in.)	200 g (0.5 lb)
19.0 mm (¾ in.)	1.0 kg (2.2 lb)
38.1 mm (1½ in.)	8.0 kg (18 lb)
75.0 mm (3 in.)	60.0 kg (132 lb)

NOTE 7—If random isolated particles are encountered that are significantly larger than the particles in the soil matrix, the soil matrix can be accurately described and identified in accordance with the preceeding schedule.

9.4 If the field sample or specimen being examined is smaller than the minimum recommended amount, the report shall include an appropriate remark.

10. Descriptive Information for Soils

10.1 *Angularity*—Describe the angularity of the sand (coarse sizes only), gravel, cobbles, and boulders, as angular, subangular, subrounded, or rounded in accordance with the criteria in Table 1 and Fig. 3. A range of angularity may be stated, such as: subrounded to rounded.

10.2 *Shape*—Describe the shape of the gravel, cobbles, and boulders as flat, elongated, or flat and elongated if they meet the criteria in Table 2 and Fig. 4. Otherwise, do not mention the shape. Indicate the fraction of the particles that have the shape, such as: one-third of the gravel particles are flat.

10.3 *Color*—Describe the color. Color is an important property in identifying organic soils, and within a given

TABLE 1 Criteria for Describing Angularity of Coarse-Grained Particles (see Fig. 3)

Description	Criteria
Angular	Particles have sharp edges and relatively plane sides with unpolished surfaces
Subangular	Particles are similar to angular description but have rounded edges
Subrounded	Particles have nearly plane sides but have well-rounded corners and edges
Rounded	Particles have smoothly curved sides and no edges

TABLE 2 Criteria for Describing Particle Shape (see Fig. 4)

The particle shape shall be described as follows where length, width, and thickness refer to the greatest, intermediate, and least dimensions of a particle, respectively.

Flat	Particles with width/thickness > 3
Elongated	Particles with length/width > 3
Flat and elongated	Particles meet criteria for both flat and elongated

PARTICLE SHAPE

W = WIDTH
T = THICKNESS
L = LENGTH

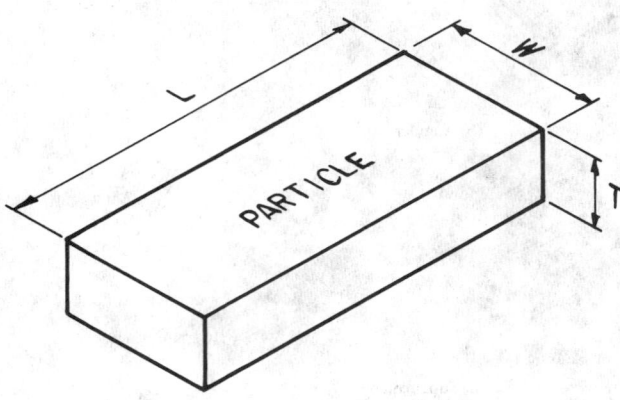

FLAT: W/T > 3
ELONGATED: L/W > 3
FLAT AND ELONGATED:
— meets both criteria

FIG. 4   Criteria for Particle Shape

TABLE 3   Criteria for Describing Moisture Condition

Description	Criteria
Dry	Absence of moisture, dusty, dry to the touch
Moist	Damp but no visible water
Wet	Visible free water, usually soil is below water table

TABLE 4   Criteria for Describing the Reaction With HCl

Description	Criteria
None	No visible reaction
Weak	Some reaction, with bubbles forming slowly
Strong	Violent reaction, with bubbles forming immediately

TABLE 5   Criteria for Describing Consistency

Description	Criteria
Very soft	Thumb will penetrate soil more than 1 in. (25 mm)
Soft	Thumb will penetrate soil about 1 in. (25 mm)
Firm	Thumb will indent soil about ¼ in. (6 mm)
Hard	Thumb will not indent soil but readily indented with thumbnail
Very hard	Thumbnail will not indent soil

10.7 *Consistency*—For intact fine-grained soil, describe the consistency as very soft, soft, firm, hard, or very hard, in accordance with the criteria in Table 5. This observation is inappropriate for soils with significant amounts of gravel.

10.8 *Cementation*—Describe the cementation of intact coarse-grained soils as weak, moderate, or strong, in accordance with the criteria in Table 6.

10.9 *Structure*—Describe the structure of intact soils in accordance with the criteria in Table 7.

10.10 *Range of Particle Sizes*—For gravel and sand components, describe the range of particle sizes within each component as defined in 3.1.2 and 3.1.6. For example, about 20 % fine to coarse gravel, about 40 % fine to coarse sand.

10.11 *Maximum Particle Size*—Describe the maximum particle size found in the sample in accordance with the following information:

10.11.1 *Sand Size*—If the maximum particle size is a sand size, describe as fine, medium, or coarse as defined in 3.1.6. For example: maximum particle size, medium sand.

10.11.2 *Gravel Size*—If the maximum particle size is a gravel size, describe the maximum particle size as the smallest sieve opening that the particle will pass. For example, maximum particle size, 1½ in. (will pass a 1½-in. square opening but not a ¾-in. square opening).

10.11.3 *Cobble or Boulder Size*—If the maximum particle size is a cobble or boulder size, describe the maximum dimension of the largest particle. For example: maximum dimension, 18 in. (450 mm).

10.12 *Hardness*—Describe the hardness of coarse sand and larger particles as hard, or state what happens when the particles are hit by a hammer, for example, gravel-size particles fracture with considerable hammer blow, some gravel-size particles crumble with hammer blow. "Hard" means particles do not crack, fracture, or crumble under a hammer blow.

10.13 Additional comments shall be noted, such as the presence of roots or root holes, difficulty in drilling or augering hole, caving of trench or hole, or the presence of mica.

10.14 A local or commercial name or a geologic interpre-

locality it may also be useful in identifying materials of similar geologic origin. If the sample contains layers or patches of varying colors, this shall be noted and all representative colors shall be described. The color shall be described for moist samples. If the color represents a dry condition, this shall be stated in the report.

10.4 *Odor*—Describe the odor if organic or unusual. Soils containing a significant amount of organic material usually have a distinctive odor of decaying vegetation. This is especially apparent in fresh samples, but if the samples are dried, the odor may often be revived by heating a moistened sample. If the odor is unusual (petroleum product, chemical, and the like), it shall be described.

10.5 *Moisture Condition*—Describe the moisture condition as dry, moist, or wet, in accordance with the criteria in Table 3.

10.6 *HCl Reaction*—Describe the reaction with HCl as none, weak, or strong, in accordance with the critera in Table 4. Since calcium carbonate is a common cementing agent, a report of its presence on the basis of the reaction with dilute hydrochloric acid is important.

TABLE 6   Criteria for Describing Cementation

Description	Criteria
Weak	Crumbles or breaks with handling or little finger pressure
Moderate	Crumbles or breaks with considerable finger pressure
Strong	Will not crumble or break with finger pressure

TABLE 7 Criteria for Describing Structure

Description	Criteria
Stratified	Alternating layers of varying material or color with layers at least 6 mm thick; note thickness
Laminated	Alternating layers of varying material or color with the layers less than 6 mm thick; note thickness
Fissured	Breaks along definite planes of fracture with little resistance to fracturing
Slickensided	Fracture planes appear polished or glossy, sometimes striated
Blocky	Cohesive soil that can be broken down into small angular lumps which resist further breakdown
Lensed	Inclusion of small pockets of different soils, such as small lenses of sand scattered through a mass of clay; note thickness
Homogeneous	Same color and appearance throughout

tation of the soil, or both, may be added if identified as such.

10.15 A classification or identification of the soil in accordance with other classification systems may be added if identified as such.

11. Identification of Peat

11.1 A sample composed primarily of vegetable tissue in various stages of decomposition that has a fibrous to amorphous texture, usually a dark brown to black color, and an organic odor, shall be designated as a highly organic soil and shall be identified as peat, PT, and not subjected to the identification procedures described hereafter.

12. Preparation for Identification

12.1 The soil identification portion of this practice is based on the portion of the soil sample that will pass a 3-in. (75-mm) sieve. The larger than 3-in. (75-mm) particles must be removed, manually, for a loose sample, or mentally, for an intact sample before classifying the soil.

12.2 Estimate and note the percentage of cobbles and the percentage of boulders. Performed visually, these estimates will be on the basis of volume percentage.

NOTE 8—Since the percentages of the particle-size distribution in Test Method D 2487 are by dry weight, and the estimates of percentages for gravel, sand, and fines in this practice are by dry weight, it is recommended that the report state that the percentages of cobbles and boulders are by volume.

12.3 Of the fraction of the soil smaller than 3 in. (75 mm), estimate and note the percentage, by dry weight, of the gravel, sand, and fines (see Appendix X4 for suggested procedures).

NOTE 9—Since the particle-size components appear visually on the basis of volume, considerable experience is required to estimate the percentages on the basis of dry weight. Frequent comparisons with laboratory particle-size analyses should be made.

12.3.1 The percentages shall be estimated to the closest 5 %. The percentages of gravel, sand, and fines must add up to 100 %.

12.3.2 If one of the components is present but not in sufficient quantity to be considered 5 % of the smaller than 3-in. (75-mm) portion, indicate its presence by the term *trace*, for example, trace of fines. A trace is not to be considered in the total of 100 % for the components.

13. Preliminary Identification

13.1 The soil is *fine grained* if it contains 50 % or more fines. Follow the procedures for identifying fine-grained soils of Section 14.

13.2 The soil is *coarse grained* if it contains less than 50 % fines. Follow the procedures for identifying coarse-grained soils of Section 15.

14. Procedure for Identifying Fine-Grained Soils

14.1 Select a representative sample of the material for examination. Remove particles larger than the No. 40 sieve (medium sand and larger) until a specimen equivalent to about a handful of material is available. Use this specimen for performing the dry strength, dilatancy, and toughness tests.

14.2 *Dry Strength:*

14.2.1 From the specimen, select enough material to mold into a ball about 1 in. (25 mm) in diameter. Mold the material until it has the consistency of putty, adding water if necessary.

14.2.2 From the molded material, make at least three test specimens. A test specimen shall be a ball of material about ½ in. (12 mm) in diameter. Allow the test specimens to dry in air, or sun, or by artificial means, as long as the temperature does not exceed 60°C.

14.2.3 If the test specimen contains natural dry lumps, those that are about ½ in. (12 mm) in diameter may be used in place of the molded balls.

NOTE 10—The process of molding and drying usually produces higher strengths than are found in natural dry lumps of soil.

14.2.4 Test the strength of the dry balls or lumps by crushing between the fingers. Note the strength as none, low, medium, high, or very high in accorance with the criteria in Table 8. If natural dry lumps are used, do not use the results of any of the lumps that are found to contain particles of coarse sand.

14.2.5 The presence of high-strength water-soluble cementing materials, such as calcium carbonate, may cause exceptionally high dry strengths. The presence of calcium carbonate can usually be detected from the intensity of the reaction with dilute hydrochloric acid (see 10.6).

14.3 *Dilatancy:*

14.3.1 From the specimen, select enough material to mold into a ball about ½ in. (12 mm) in diameter. Mold the material, adding water if necessary, until it has a soft, but not sticky, consistency.

14.3.2 Smooth the soil ball in the palm of one hand with the blade of a knife or small spatula. Shake horizontally, striking the side of the hand vigorously against the other hand several times. Note the reaction of water appearing on

TABLE 8 Criteria for Describing Dry Strength

Description	Criteria
None	The dry specimen crumbles into powder with mere pressure of handling
Low	The dry specimen crumbles into powder with some finger pressure
Medium	The dry specimen breaks into pieces or crumbles with considerable finger pressure
High	The dry specimen cannot be broken with finger pressure. Specimen will break into pieces between thumb and a hard surface
Very high	The dry specimen cannot be broken between the thumb and a hard surface

ASTM D 2488

TABLE 9 Criteria for Describing Dilatancy

Description	Criteria
None	No visible change in the specimen
Slow	Water appears slowly on the surface of the specimen during shaking and does not disappear or disappears slowly upon squeezing
Rapid	Water appears quickly on the surface of the specimen during shaking and disappears quickly upon squeezing

TABLE 10 Criteria for Describing Toughness

Description	Criteria
Low	Only slight pressure is required to roll the thread near the plastic limit. The thread and the lump are weak and soft
Medium	Medium pressure is required to roll the thread to near the plastic limit. The thread and the lump have medium stiffness
High	Considerable pressure is required to roll the thread to near the plastic limit. The thread and the lump have very high stiffness

the surface of the soil. Squeeze the sample by closing the hand or pinching the soil between the fingers, and note the reaction as none, slow, or rapid in accordance with the criteria in Table 9. The reaction is the speed with which water appears while shaking, and disappears while squeezing.

14.4 *Toughness:*

14.4.1 Following the completion of the dilatancy test, the test specimen is shaped into an elongated pat and rolled by hand on a smooth surface or between the palms into a thread about ⅛ in. (3 mm) in diameter. (If the sample is too wet to roll easily, it should be spread into a thin layer and allowed to lose some water by evaporation.) Fold the sample threads and reroll repeatedly until the thread crumbles at a diameter of about ⅛ in. The thread will crumble at a diameter of ⅛ in. when the soil is near the plastic limit. Note the pressure required to roll the thread near the plastic limit. Also, note the strength of the thread. After the thread crumbles, the pieces should be lumped together and kneaded until the lump crumbles. Note the toughness of the material during kneading.

14.4.2 Describe the toughness of the thread and lump as low, medium, or high in accordance with the criteria in Table 10.

14.5 *Plasticity*—On the basis of observations made during the toughness test, describe the plasticity of the material in accordance with the criteria given in Table 11.

14.6 Decide whether the soil is an *inorganic* or an *organic* fine-grained soil (see 14.8). If inorganic, follow the steps given in 14.7.

14.7 *Identification of Inorganic Fine-Grained Soils:*

TABLE 11 Criteria for Describing Plasticity

Description	Criteria
Nonplastic	A ⅛-in. (3-mm) thread cannot be rolled at any water content
Low	The thread can barely be rolled and the lump cannot be formed when drier than the plastic limit
Medium	The thread is easy to roll and not much time is required to reach the plastic limit. The thread cannot be rerolled after reaching the plastic limit. The lump crumbles when drier than the plastic limit
High	It takes considerable time rolling and kneading to reach the plastic limit. The thread can be rerolled several times after reaching the plastic limit. The lump can be formed without crumbling when drier than the plastic limit

14.7.1 Identify the soil as a *lean clay*, CL, if the soil has medium to high dry strength, no or slow dilatancy, and medium toughness and plasticity (see Table 12).

14.7.2 Identify the soil as a *fat clay*, CH, if the soil has high to very high dry strength, no dilatancy, and high toughness and plasticity (see Table 12).

14.7.3 Identify the soil as a *silt*, ML, if the soil has no to low dry strength, slow to rapid dilatancy, and low toughness and plasticity, or is nonplastic (see Table 12).

14.7.4 Identify the soil as an *elastic silt*, MH, if the soil has low to medium dry strength, no to slow dilatancy, and low to medium toughness and plasticity (see Table 12).

NOTE 11—These properties are similar to those for a lean clay. However, the silt will dry quickly on the hand and have a smooth, silky feel when dry. Some soils that would classify as MH in accordance with the criteria in Test Method D 2487 are visually difficult to distinguish from lean clays, CL. It may be necessary to perform laboratory testing for proper identification.

14.8 *Identification of Organic Fine-Grained Soils:*

14.8.1 Identify the soil as an *organic soil*, OL/OH, if the soil contains enough organic particles to influence the soil properties. Organic soils usually have a dark brown to black color and may have an organic odor. Often, organic soils will change color, for example, black to brown, when exposed to the air. Some organic soils will lighten in color significantly when air dried. Organic soils normally will not have a high toughness or plasticity. The thread for the toughness test will be spongy.

NOTE 12—In some cases, through practice and experience, it may be possible to further identify the organic soils as organic silts or organic clays, OL or OH. Correlations between the dilatancy, dry strength, toughness tests, and laboratory tests can be made to identify organic soils in certain deposits of similar materials of known geologic origin.

14.9 If the soil is estimated to have 15 to 25 % sand or gravel, or both, the words "with sand" or "with gravel" (whichever is more predominant) shall be added to the group name. For example: "lean clay with sand, CL" or "silt with gravel, ML" (see Figs. 1a and 1b). If the percentage of sand is equal to the percentage of gravel, use "with sand."

14.10 If the soil is estimated to have 30 % or more sand or gravel, or both, the words "sandy" or "gravelly" shall be added to the group name. Add the word "sandy" if there appears to be more sand than gravel. Add the word "gravelly" if there appears to be more gravel than sand. For example: "sandy lean clay, CL", "gravelly fat clay, CH", or "sandy silt, ML" (see Figs. 1a and 1b). If the percentage of sand is equal to the percent of gravel, use "sandy."

15. Procedure for Identifying Coarse-Grained Soils (Contains less than 50 % fines)

15.1 The soil is a *gravel* if the percentage of gravel is estimated to be more than the percentage of sand.

TABLE 12 Identification of Inorganic Fine-Grained Soils from Manual Tests

Soil Symbol	Dry Strength	Dilatancy	Toughness
ML	None to low	Slow to rapid	Low or thread cannot be formed
CL	Medium to high	None to slow	Medium
MH	Low to medium	None to slow	Low to medium
CH	High to very high	None	High

15.2 The soil is a *sand* if the percentage of gravel is estimated to be equal to or less than the percentage of sand.

15.3 The soil is a *clean gravel* or *clean sand* if the percentage of fines is estimated to be 5 % or less.

15.3.1 Identify the soil as a *well-graded gravel*, GW, or as a *well-graded sand*, SW, if it has a wide range of particle sizes and substantial amounts of the intermediate particle sizes.

15.3.2 Identify the soil as a *poorly graded gravel*, GP, or as a *poorly graded sand*, SP, if it consists predominantly of one size (uniformly graded), or it has a wide range of sizes with some intermediate sizes obviously missing (gap or skip graded).

15.4 The soil is either a *gravel with fines* or a *sand with fines* if the percentage of fines is estimated to be 15 % or more.

15.4.1 Identify the soil as a *clayey gravel*, GC, or a *clayey sand*, SC, if the fines are clayey as determined by the procedures in Section 14.

15.4.2 Identify the soil as a *silty gravel*, GM, or a *silty sand*, SM, if the fines are silty as determined by the procedures in Section 14.

15.5 If the soil is estimated to contain 10 % fines, give the soil a dual identification using two group symbols.

15.5.1 The first group symbol shall correspond to a clean gravel or sand (GW, GP, SW, SP) and the second symbol shall correspond to a gravel or sand with fines (GC, GM, SC, SM).

15.5.2 The group name shall correspond to the first group symbol plus the words "with clay" or "with silt" to indicate the plasticity characteristics of the fines. For example: "well-graded gravel with clay, GW-GC" or "poorly graded sand with silt, SP-SM" (see Fig. 2).

15.6 If the specimen is predominantly sand or gravel but contains an estimated 15 % or more of the other coarse-grained constituent, the words "with gravel" or "with sand" shall be added to the group name. For example: "poorly graded gravel with sand, GP" or "clayey sand with gravel, SC" (see Fig. 2).

15.7 If the field sample contains any cobbles or boulders, or both, the words "with cobbles" or "with cobbles and boulders" shall be added to the group name. For example: "silty gravel with cobbles, GM."

16. Report

16.1 The report shall include the information as to origin, and the items indicated in Table 13.

NOTE 13—*Example: Clayey Gravel with Sand and Cobbles, GC—* About 50 % fine to coarse, subrounded to subangular gravel; about 30 % fine to coarse, subrounded sand; about 20 % fines with medium plasticity, high dry strength, no dilatancy, medium toughness; weak

TABLE 13 Checklist for Description of Soils

1. Group name
2. Group symbol
3. Percent of cobbles or boulders, or both (by volume)
4. Percent of gravel, sand, or fines, or all three (by dry weight)
5. Particle-size range:
   Gravel—fine, coarse
   Sand—fine, medium, coarse
6. Particle angularity: angular, subangular, subrounded, rounded
7. Particle shape: (if appropriate) flat, elongated, flat and elongated
8. Maximum particle size or dimension
9. Hardness of coarse sand and larger particles
10. Plasticity of fines: nonplastic, low, medium, high
11. Dry strength: none, low, medium, high, very high
12. Dilatancy: none, slow, rapid
13. Toughness: low, medium, high
14. Color (in moist condition)
15. Odor (mention only if organic or unusual)
16. Moisture: dry, moist, wet
17. Reaction with HCl: none, weak, strong
For intact samples:
18. Consistency (fine-grained soils only): very soft, soft, firm, hard, very hard
19. Structure: stratified, laminated, fissured, slickensided, lensed, homogeneous
20. Cementation: weak, moderate, strong
21. Local name
22. Geologic interpretation
23. Additional comments: presence of roots or root holes, presence of mica, gypsum, etc., surface coatings on coarse-grained particles, caving or sloughing of auger hole or trench sides, difficulty in augering or excavating, etc.

reaction with HCl; original field sample had about 5 % (by volume) subrounded cobbles, maximum dimension, 150 mm.

In-Place Conditions—Firm, homogeneous, dry, brown

Geologic Interpretation—Alluvial fan

NOTE 14—Other examples of soil descriptions and identification are given in Appendixes X1 and X2.

NOTE 15—If desired, the percentages of gravel, sand, and fines may be stated in terms indicating a range of percentages, as follows:

Trace—Particles are present but estimated to be less than 5 %

Few—5 to 10 %

Little—15 to 25 %

Some—30 to 45 %

Mostly—50 to 100 %

16.2 If, in the soil description, the soil is identified using a classification group symbol and name as described in Test Method D 2487, it must be distinctly and clearly stated in log forms, summary tables, reports, and the like, that the symbol and name are based on visual-manual procedures.

17. Precision and Bias

17.1 This practice provides qualitative information only, therefore, a precision and bias statement is not applicable.

18. Keywords

18.1 classification; clay; gravel; organic soils; sand; silt; soil classification; soil description; visual classification

APPENDIXES

(Nonmandatory Information)

X1. EXAMPLES OF VISUAL SOIL DESCRIPTIONS

X1.1 The following examples show how the information required in 16.1 can be reported. The information that is included in descriptions should be based on individual circumstances and need.

X1.1.1 *Well-Graded Gravel with Sand (GW)*—About 75 % fine to coarse, hard, subangular gravel; about 25 % fine to coarse, hard, subangular sand; trace of fines; maximum size, 75 mm, brown, dry; no reaction with HCl.

X1.1.2 *Silty Sand with Gravel (SM)*—About 60 % predominantly fine sand; about 25 % silty fines with low plasticity, low dry strength, rapid dilatancy, and low toughness; about 15 % fine, hard, subrounded gravel, a few gravel-size particles fractured with hammer blow; maximum size, 25 mm; no reaction with HCl (Note—Field sample size smaller than recommended).

In-Place Conditions—Firm, stratified and contains lenses of silt 1 to 2 in. (25 to 50 mm) thick, moist, brown to gray; in-place density 106 lb/ft³; in-place moisture 9 %.

X1.1.3 *Organic Soil (OL/OH)*—About 100 % fines with low plasticity, slow dilatancy, low dry strength, and low toughness; wet, dark brown, organic odor; weak reaction with HCl.

X1.1.4 *Silty Sand with Organic Fines (SM)*—About 75 % fine to coarse, hard, subangular reddish sand; about 25 % organic and silty dark brown nonplastic fines with no dry strength and slow dilatancy; wet; maximum size, coarse sand; weak reaction with HCl.

X1.1.5 *Poorly Graded Gravel with Silt, Sand, Cobbles and Boulders (GP-GM)*—About 75 % fine to coarse, hard, subrounded to subangular gravel; about 15 % fine, hard, subrounded to subangular sand; about 10 % silty nonplastic fines; moist, brown; no reaction with HCl; original field sample had about 5 % (by volume) hard, subrounded cobbles and a trace of hard, subrounded boulders, with a maximum dimension of 18 in. (450 mm).

X2. USING THE IDENTIFICATION PROCEDURE AS A DESCRIPTIVE SYSTEM FOR SHALE, CLAYSTONE, SHELLS, SLAG, CRUSHED ROCK, AND THE LIKE

X2.1 The identification procedure may be used as a descriptive system applied to materials that exist in-situ as shale, claystone, sandstone, siltstone, mudstone, etc., but convert to soils after field or laboratory processing (crushing, slaking, and the like).

X2.2 Materials such as shells, crushed rock, slag, and the like, should be identified as such. However, the procedures used in this practice for describing the particle size and plasticity characteristics may be used in the description of the material. If desired, an identification using a group name and symbol according to this practice may be assigned to aid in describing the material.

X2.3 The group symbol(s) and group names should be placed in quotation marks or noted with some type of distinguishing symbol. See examples.

X2.4 Examples of how group names and symbols can be incororated into a descriptive system for materials that are not naturally occurring soils are as follows:

X2.4.1 *Shale Chunks*—Retrieved as 2 to 4-in. (50 to 100-mm) pieces of shale from power auger hole, dry, brown, no reaction with HCl. After slaking in water for 24 h, material identified as "Sandy Lean Clay (CL)"; about 60 % fines with medium plasticity, high dry strength, no dilatancy, and medium toughness; about 35 % fine to medium, hard sand; about 5 % gravel-size pieces of shale.

X2.4.2 *Crushed Sandstone*—Product of commercial crushing operation; "Poorly Graded Sand with Silt (SP-SM)"; about 90 % fine to medium sand; about 10 % nonplastic fines; dry, reddish-brown, strong reaction with HCl.

X2.4.3 *Broken Shells*—About 60 % gravel-size broken shells; about 30 % sand and sand-size shell pieces; about 10 % fines; "Poorly Graded Gravel with Sand (GP)."

X2.4.4 *Crushed Rock*—Processed from gravel and cobbles in Pit No. 7; "Poorly Graded Gravel (GP)"; about 90 % fine, hard, angular gravel-size particles; about 10 % coarse, hard, angular sand-size particles; dry, tan; no reaction with HCl.

X3. SUGGESTED PROCEDURE FOR USING A BORDERLINE SYMBOL FOR SOILS WITH TWO POSSIBLE IDENTIFICATIONS.

X3.1 Since this practice is based on estimates of particle size distribution and plasticity characteristics, it may be difficult to clearly identify the soil as belonging to one category. To indicate that the soil may fall into one of two possible basic groups, a borderline symbol may be used with the two symbols separated by a slash. For example: SC/CL or CL/CH.

X3.1.1 A borderline symbol may be used when the

percentage of fines is estimated to be between 45 and 55 %. One symbol should be for a coarse-grained soil with fines and the other for a fine-grained soil. For example: GM/ML or CL/SC.

X3.1.2 A borderline symbol may be used when the percentage of sand and the percentage of gravel are estimated to be about the same. For example: GP/SP, SC/GC, GM/SM. It is practically impossible to have a soil that would have a borderline symbol of GW/SW.

X3.1.3 A borderline symbol may be used when the soil could be either well graded or poorly graded. For example: GW/GP, SW/SP.

X3.1.4 A borderline symbol may be used when the soil could either be a silt or a clay. For example: CL/ML, CH/MH, SC/SM.

X3.1.5 A borderline symbol may be used when a fine-grained soil has properties that indicate that it is at the boundary between a soil of low compressibility and a soil of high compressibility. For example: CL/CH, MH/ML.

X3.2 The order of the borderline symbols should reflect similarity to surrounding or adjacent soils. For example: soils in a borrow area have been identified as CH. One sample is considered to have a borderline symbol of CL and CH. To show similarity, the borderline symbol should be CH/CL.

X3.3 The group name for a soil with a borderline symbol should be the group name for the first symbol, except for:

CL/CH lean to fat clay
ML/CL clayey silt
CL/ML silty clay

X3.4 The use of a borderline symbol should not be used indiscriminately. Every effort shall be made to first place the soil into a single group.

X4. SUGGESTED PROCEDURES FOR ESTIMATING THE PERCENTAGES OF GRAVEL, SAND, AND FINES IN A SOIL SAMPLE

X4.1 *Jar Method*—The relative percentage of coarse- and fine-grained material may be estimated by thoroughly shaking a mixture of soil and water in a test tube or jar, and then allowing the mixture to settle. The coarse particles will fall to the bottom and successively finer particles will be deposited with increasing time; the sand sizes will fall out of suspension in 20 to 30 s. The relative proportions can be estimated from the relative volume of each size separate. This method should be correlated to particle-size laboratory determinations.

X4.2 *Visual Method*—Mentally visualize the gravel size particles placed in a sack (or other container) or sacks. Then, do the same with the sand size particles and the fines. Then, mentally compare the number of sacks to estimate the percentage of plus No. 4 sieve size and minus No. 4 sieve size present. The percentages of sand and fines in the minus sieve size No. 4 material can then be estimated from the wash test (X4.3).

X4.3 *Wash Test (for relative percentages of sand and fines)*—Select and moisten enough minus No. 4 sieve size material to form a 1-in (25-mm) cube of soil. Cut the cube in half, set one-half to the side, and place the other half in a small dish. Wash and decant the fines out of the material in the dish until the wash water is clear and then compare the two samples and estimate the percentage of sand and fines. Remember that the percentage is based on weight, not volume. However, the volume comparison will provide a reasonable indication of grain size percentages.

X4.3.1 While washing, it may be necessary to break down lumps of fines with the finger to get the correct percentages.

X5. ABBREVIATED SOIL CLASSIFICATION SYMBOLS

X5.1 In some cases, because of lack of space, an abbreviated system may be useful to indicate the soil classification symbol and name. Examples of such cases would be graphical logs, databases, tables, etc.

X5.2 This abbreviated system is not a substitute for the full name and descriptive information but can be used in supplementary presentations when the complete description is referenced.

X5.3 The abbreviated system should consist of the soil classification symbol based on this standard with appropriate lower case letter prefixes and suffixes as:

Prefix:	Suffix:
s = sandy	s = with sand
g = gravelly	g = with gravel
	c = with cobbles
	b = with boulders

X5.4 The soil classification symbol is to be enclosed in parenthesis. Some examples would be:

Group Symbol and Full Name	Abbreviated
CL, Sandy lean clay	s(CL)
SP-SM, Poorly graded sand with silt and gravel	(SP-SM)g
GP, poorly graded gravel with sand, cobbles, and boulders	(GP)scb
ML, gravelly silt with sand and cobbles	g(ML)sc

 **D 2488**

X6. RATIONALE

Changes in this version from the previous version, D 2488 – 90, include the addition of X5 on Abbreviated Soil Classification Symbols.

Designation: D 4083 – 89 (Reapproved 1994)[ε1]

Standard Practice for
Description of Frozen Soils (Visual-Manual Procedure)[1]

This standard is issued under the fixed designation D 4083; the number immediately following the designation indicates the year of original adoption or, in the case of revision, the year of last revision. A number in parentheses indicates the year of last reapproval. A superscript epsilon (ε) indicates an editorial change since the last revision or reapproval.

This standard has been approved for use by agencies of the Department of Defense. Consult the DoD Index of Specifications and Standards for the specific year of issue which has been adopted by the Department of Defense.

ε1 NOTE—Keywords were added in January 1994.

1. Scope

1.1 This practice presents a procedure for the description of frozen soils based on visual examination and simple manual tests.

1.2 It is intended to be used in conjunction with Test Method D 2487 and Practice D 2488, which describe and classify soils, but do not cover their frozen state.

1.3 This procedure is based on "Guide to Field Description of Permafrost for Engineering Purposes," National Research Council of Canada, 1963, and MIL-STD-619.

2. Referenced Documents

2.1 *ASTM Standards:*
D 420 Guide for Investigating and Sampling Soil and Rock[2]
D 653 Terminology Relating to Soil, Rock, and Contained Fluids[2]
D 1452 Practice for Soil Investigation and Sampling by Auger Borings[2]
D 2487 Classification of Soils for Engineering Purposes (Unified Soil Classification System)[2]
D 2488 Practice for Description and Identification of Soils (Visual-Manual Procedure)[2]
2.2 *Military Standard:*
MIL-STD-619 Unified Soil Classification System for Roads, Airfields, Embankments and Foundations[3]

3. Terminology

3.1 *Definitions:*

3.1.1 Definitions of the soil components of a frozen soil mass, that is, boulders, cobbles, gravel, sand, fines (silt and clay), and organic soils and peat shall be in accordance with Terminology D 653.

3.1.2 The following terms are used in conjunction with the description of frozen ground areas (Fig. 1):[4]

3.1.2.1 *annual frost zone (active layer)*—the top layer of ground subject to annual freezing and thawing.

3.1.2.2 *frost table*—the frozen surface, usually irregular, that represents the level, to which thawing of seasonally frozen ground has penetrated. See Fig. 1.

3.1.2.3 *frozen zone*—a range of depth within which the soil is frozen. The frozen zone may be bounded both top and bottom by unfrozen soil, or at the top by the ground surface.

3.1.2.4 *ground ice*—a body of more or less clear ice within frozen ground.

3.1.2.5 *ice wedge*—a wedge-shaped mass in permafrost, usually associated with fissures in polygons.

3.1.2.6 *icing*—a surface ice mass formed by freezing of successive sheets of water.

3.1.2.7 *permafrost*—the thermal condition in soil or rock, wherein the materials have existed at a temperature below 0°C (32°F) continuously for a number of years. Pore fluids or ice may or may not be present.

3.1.2.8 *permafrost table*—the surface that represents the upper limit of permafrost.

3.1.2.9 *polygons (polygonal ground)*—more or less regular-sized surface patterns created by thermal contraction of the ground. Two types are common: (*a*) those with depressed centers and (*b*) those with raised centers.

3.1.2.10 *residual thaw zone*—a layer of unfrozen ground between the permafrost and the annual frost zone. This layer does not exist where annual frost extends to permafrost.

3.1.3 The following terms are used to describe the characteristics of the frozen earth:

3.1.3.1 *candled ice*—ice that has rotted or otherwise formed into long columnar crystals, very loosely bonded together.

3.1.3.2 *clear ice*—ice that is transparent and contains only a moderate number of air bubbles.

3.1.3.3 *cloudy ice*—ice that is translucent or relatively opaque due to the content of air or for other reasons, but which is essentially sound and nonpervious.

3.1.3.4 *excess ice*—ice in excess of the fraction that would be retained as water in the soil voids after thawing.

3.1.3.5 *friable*—a condition under which the material is easily broken up under light to moderate pressure.

3.1.3.6 *granular ice*—ice that is composed of coarse, more or less equidimensional, crystals weakly bonded together.

3.1.3.7 *ice coatings on particles*—discernible layers of ice found on or below the larger soil particles in a frozen soil mass. They are sometimes associated with hoarfrost crystals,

---

[1] This practice is under the jurisdiction of ASTM Committee D-18 on Soil and Rock and is the direct responsibility of Subcommittee D18.19 on Frozen Soils and Rock.
Current edition approved Feb. 24, 1989. Published October 1989. Originally published as D 4083 – 82. Last previous edition D 4083 – 83.
[2] *Annual Book of ASTM Standards*, Vol 04.08.
[3] Available from Naval Publications and Forms Center, 5801 Tabor Ave., Philadelphia, PA 19120.
[4] For more complete lists of generally accepted terms used in the description of frozen ground see: Hennion, F., "Frost and Permafrost Definitions," *Bulletin 111*, Highway Research Board, Washington, DC 1955; and Brown, R. J. E., and Kupsch, W. D., "Permafrost Terminology," *Technical Memorandum No. 111*, National Research Council of Canada, 1974.

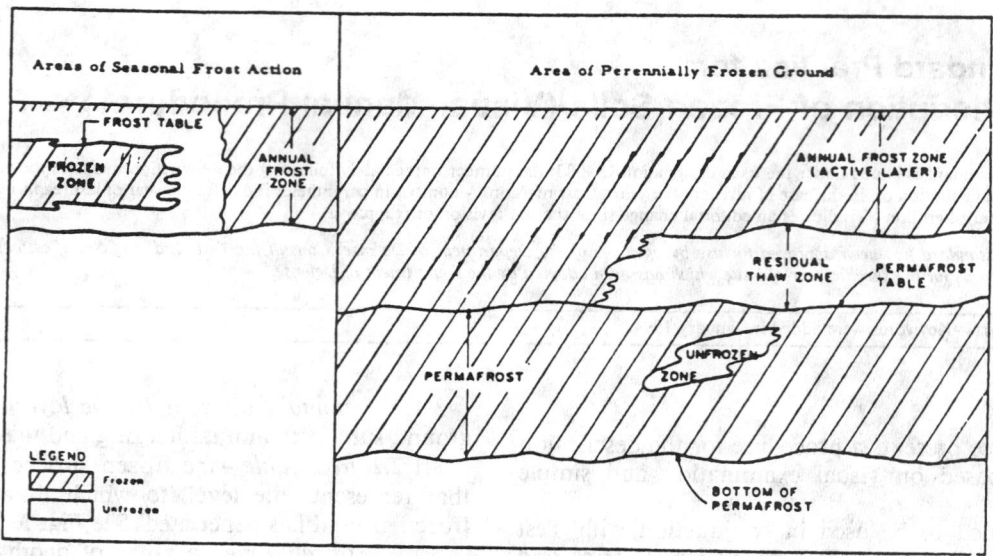

FIG. 1 Illustration of Frozen Soil Terminology

which have grown into voids produced by the freezing action.

3.1.3.8 *ice crystal*—a very small individual ice particle visible in the face of a soil mass. Crystals may be present alone or in combination with other ice formations.

3.1.3.9 *ice lenses*—lenticular ice formations in soil occurring essentially parallel to each other, generally normal to the direction of heat loss, and commonly in repeated layers.

3.1.3.10 *ice segregation*—the growth of ice within soil in excess of the amount that may be produced by the in-place conversion of the original void moisture to ice. Ice segregation occurs most often as distinct lenses, layers, veins, and masses, commonly, but not always, oriented normal to the direction of heat flow.

3.1.3.11 *poorly bonded*—a condition in which the soil particles are weakly held together by the ice so that the frozen soil has poor resistance to chipping and breaking.

3.1.3.12 *porous ice*—ice that contains numerous voids, usually interconnected and usually resulting from melting at air bubbles or along crystal interfaces from presence of salt or other materials in the water, or from the freezing of saturated snow. Though porous, the mass retains its structural unity.

3.1.3.13 *thaw stable*—the characteristic of frozen soils that, upon thawing, do not show loss of strength in comparison to normal, long-time thawed values nor produce detrimental settlement.

3.1.3.14 *thaw unstable*—the characteristic of frozen soils that, upon thawing, show significant loss of strength in comparison to normal, long-time thawed values or produce significant settlement, or both, as a direct result of the melting of excess ice in the soil.

3.1.3.15 *well bonded*—a condition in which the soil particles are strongly held together by the ice so that the frozen soil possesses relatively high resistance to chipping or breaking.

4. Significance and Use

4.1 This practice is intended primarily for use by geotechnical engineers and technicians and geologists in the field, where the soil profile or samples from it may be observed in a relatively undisturbed (frozen) state.

4.2 It may also be used in the laboratory to describe the condition of relatively undisturbed soil samples that have been maintained in a frozen condition following their acquisition in the field.

4.3 The practice is not intended to be used in describing unfrozen soils or disturbed samples of frozen soil.

5. Apparatus

5.1 *Required Apparatus:*
5.1.1 Pocket knife or small spatula.
5.1.2 Low-power magnifying hand lens.
5.1.3 Pint-size graduated jars.
5.2 *Useful Auxiliary Apparatus:*
5.2.1 Camera.
5.2.2 Small bottle of dilute hydrochloric acid.
5.2.3 Small test tube and stopper.
5.2.4 Munsell Soil Color Chart or Rock Color Chart, or both.
5.2.5 Thermometer.

6. General Procedure for Identification

6.1 The system for describing and classifying frozen soil is based on an identification procedure which involves three steps designated as Parts I, II, and III. Part I consists of a description of the soil phase, Part II consists of the addition of soil characteristics resulting from the frozen state, and Part III consists of a description of the important ice strata associated with the soil.

NOTE 1—In addition to the description of the soil profile at a given site, it is normally advantageous to describe the local terrain features. Particularly useful are descriptions of the type of vegetation cover, depth and type of snow cover, local relief and drainage conditions, and depth of thaw. One or more photos of the area also can be very helpful. The terminology given in 3.1.2 should be used to describe any special conditions which can be recognized. To these should be added any

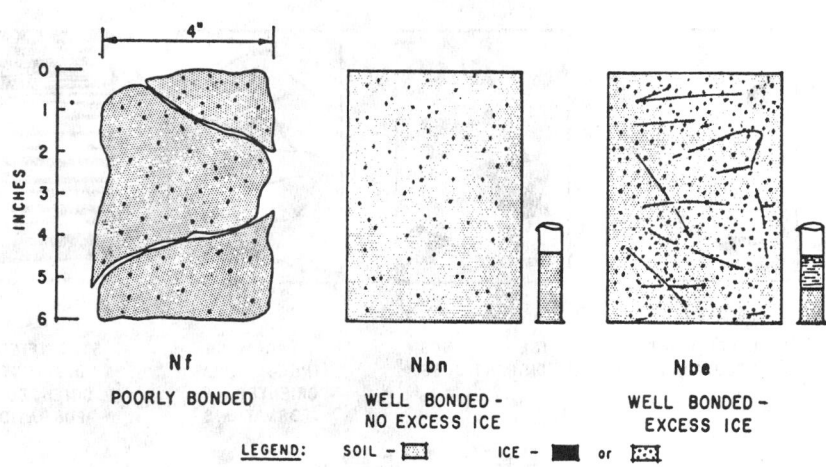

N f
POORLY BONDED

N bn
WELL BONDED -
NO EXCESS ICE

N be
WELL BONDED -
EXCESS ICE

LEGEND: SOIL – ☐ ICE – ■ or ▨

NOTE 1–Frozen soils in the *N* group may, on close examination, indicate presence of ice within the voids of the material by crystalline reflections or by a sheen on fractured or trimmed surfaces. The impression received by the unaided eye, however, is that none of the frozen water occupies space in excess of the original voids in the soil. The opposite is true of frozen soils in the *V* group.

NOTE 2–When visual methods may be inadequate, a simple field test to aid evaluation of volume of excess ice can be made by placing some frozen soil in a small jar, allowing it to melt, and observing the quantity of supernatant water as a percentage of total volume.

Group Symbol	Subgroup		Field Identification
	Description	Symbol	
N	Poorly bonded or friable	N_f	Identify by visual examination. To determine presence of excess ice, use procedure under Note 2 and hand magnifying lens as necessary. For soils not fully saturated, estimate degree of ice saturation; medium, low. Note presence of crystals or of ice coatings around larger particles.
	No excess ice Well-bonded Excess ice	N_b ⋮ N_{bn} ⋮ N_{be}	

FIG. 2   Description of Frozen Soils—Ice Not Visible

available information on the depth of thaw as estimated from borings and test pits at the site.

7. Part I, Description of the Soil Phase[5]

7.1 The soil phase, whether thawed or frozen, is first described in accordance with Practice D 2488.

8. Part II, Description of the Frozen Soil

8.1 Frozen soils in which ice is *not visible to the unaided eye* are designated by the symbol *N* and are divided into two main subgroups as shown in Fig. 2.

8.1.1 *Poorly bonded or friable* material in which segregated ice is not visible to the unaided eye is designated by the symbol N_f. This condition exists when the degree of saturation is low.

8.1.2 *Well-bonded* frozen soil in which the ice cements the material into a hard solid mass, but in which segregated ice is not visible to the unaided eye is designated by the symbol N_b. It may further be described on the basis of detailed examination and assigned to one of two subtypes. See Fig. 2.

8.1.2.1 If no excess ice is present as indicated by the absence of segregation even under magnified viewing, the material is designated by the symbol N_{bn}.

8.1.2.2 If *excess ice* is present, but is so *uniformly distributed* that it is not easily apparent to the unaided eye, the material is designated by the symbol N_{be}. This condition may

occur in very fine silty sands or coarse silts and can be verified by placing some frozen soil in a graduated jar, allowing it to melt, and observing the quantity of supernatant water as a percentage of the total volume. See Fig. 2.

8.2 Frozen soils in which *significant segregated ice is visible* to the unaided eye, but individual ice masses or layers are *less than 1 in.* (25 mm) in thickness are designated by the symbol *V*. These are divided into five subgroups as shown in Fig. 3.

8.2.1 The symbol V_x designates those frozen soils which contain *individual ice crystals* or inclusions. See Fig. 3.

8.2.2 The symbol V_c designates those frozen soils in which the *ice occurs as coatings* on particles.

8.2.3 The symbol V_r designates frozen soil masses with *random or irregularly oriented ice* formations.

8.2.4 The symbol V_s designates that the frozen soil is interspersed with *stratified or distinctly oriented ice* formations.

8.2.5 The symbol V_u designates *visible* ice, *uniformly* distributed throughout the soil mass.

NOTE 2—When more than one subgroup characteristic is present in the same material, multiple subgroup designations such as $V_{s,r}$ may be used.

9. Part III, Description of Substantial Ice Strata

9.1 Ice strata that are *greater than 1 in.* (25 mm) in thickness are designated by the symbol ICE and divided into two subgroups as shown in Fig. 4.

9.1.1 If the ice stratum contains *soil inclusions*, it is designated as ICE + Soil Type.

---

[5] When the surface soils are mostly organic (peat) a more complete description can be achieved through use of the "Guide to a Field Description of Muskeg," I. C. McFarlane, in *Special Procedures for Testing Soil and Rock for Engineering Purposes*, 5th Ed., *ASTM STP 479*, 1970.

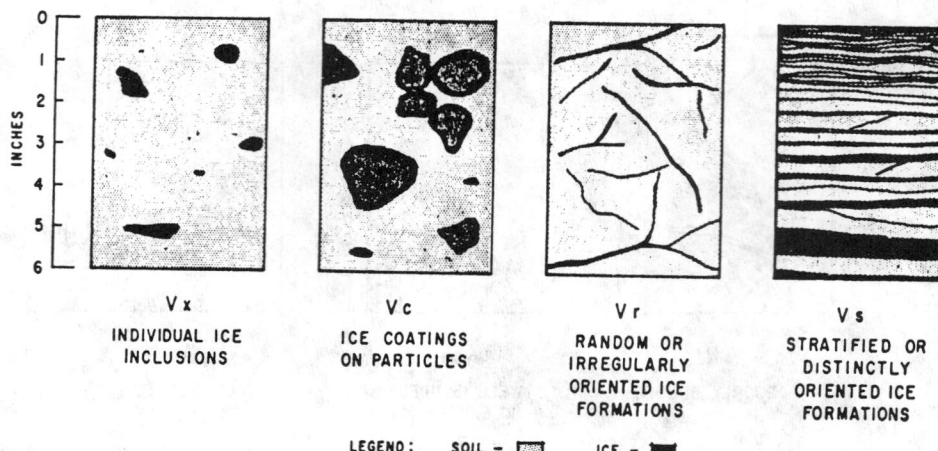

LEGEND: SOIL - ▨   ICE - ▰

NOTE—Frozen soils in the N group may, on close examination, indicate presence of ice within the voids of the material by crystalline reflections or by a sheen on fractured or trimmed surfaces. The impression received by the unaided eye, however, is that none of the frozen water occupies space in excess of the original voids in the soil. The opposite is true of frozen soils in the V group.

Group Symbol	Subgroup		Field Identification
	Description	Symbol	
	Individual ice crystal or inclusions	V_x	For ice phase, record the following when applicable:
	Ice coatings on particles	V_c	Location Size
V	Random or irregularly oriented ice formations	V_r	Orientation Shape Thickness Pattern Length
	Stratified or distinctly oriented ice formations	V_s	Spacing Hardness } Structure }
	Uniformly distributed ice	V_u	Color
			Estimate volume of visible segregated ice present as percentage of total sample volume.

FIG. 3 Description of Frozen Soils—Visible Ice Less Than 1 in. (25 mm) Thick

9.1.2 If the ice stratum contains *no soil inclusions*, it is designated simply as ICE.

10. Identification of Frozen Soils

10.1 Figures 2 to 4 also contain information that is helpful in determining the proper identification of a frozen soil mass. The various items listed which pertain to the ice phase should be recorded whenever applicable.

10.2 When greater detail and more specific information are desired than is obtainable from visual inspection, additional physical tests and measurements may be performed on the frozen or thawed soil, or both. These may include in-place temperature, density, water content, stress-strain characteristics, thermal properties, and ice crystal structure.

11. Keywords

11.1 frozen; ice; permafrost; soil; undisturbed

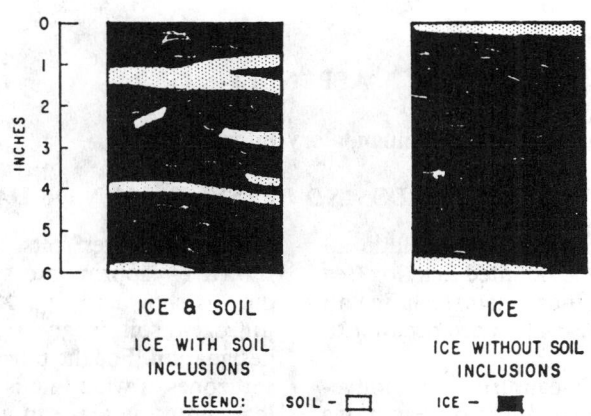

ICE & SOIL

ICE WITH SOIL
INCLUSIONS

ICE

ICE WITHOUT SOIL
INCLUSIONS

LEGEND: SOIL - ☐   ICE - ■

NOTE 1–Where special forms of ice such as hoarfrost can be distinguished, more explicit description should be given.
NOTE 2–Observer should be careful to avoid being misled by surface scratches or frost coating on the ice.

Group Symbol	Subgroup		Field Identification
	Description	Symbol	
ICE	Ice with soil inclusions	ICE + soil type	Designate material as ICE (Note 1) and use descriptive terms as follows, usually one item from each group, when applicable:
	Ice without soil inclusions	ICE	

Hardness   Structure (Note 2)

HARD   CLEAR
SOFT   CLOUDY
(of mass, not individual   POROUS
crystals)   CANDLED
                                 GRANULAR
                                 STRATIFIED

Color   Admixtures

(Examples):   (Examples):
COLORLESS   CONTAINS
GRAY   FEW THIN
BLUE   SILT INCLUSIONS

FIG. 4   Description of Visible Ice Strata Greater Than 1 in. (25 mm) Thick

APPENDIX

(Nonmandatory Information)

X1. FIELD RECORDS AND PRESENTATION OF DATA

X1.1 The record of site exploration should include all items normally contained in a well-documented field log (see Practices D 420 and D 1452) plus those items relating to terrain, permafrost, and thaw conditions that are peculiar to frozen soil areas.

X1.2 The results of the exploration can usually be conveniently presented on drawings as schematic representations of borings or test pits, with the various soils encountered shown by appropriate symbols. The recommended procedure is shown in Fig. X1.1. Note that the symbol for the unfrozen soil is given first, followed by the frozen soil designation. For the purpose of readily identifying the frozen soil zones, a wide line is drawn down the left of the graphic log within the range in which the frozen material occurs.

Depth, m	Depth, ft	Symbol	Soil Description	Ice Features
0.0	0.0*			
0.15	0.5	OL	Organic, sandy SILT, not frozen	None
0.6	1.8	GW	Brown, well-graded, sandy GRAVEL, medium compact, moist, not frozen	None
1.1	3.7	GW N_f	Brown well-graded, sandy GRAVEL, frozen, poorly bonded	No visible segregation, negligible thin ice film on gravel sizes and within larger voids
1.6	5.4	GW N_{bn}	Brown, well-graded, sandy GRAVEL, frozen, well bonded	No visible segregation
2.4	7.7	ML V_s	Black, micaceous, sandy SILT, frozen	Stratified horizontal ice lenses averaging 4 in. (10 cm) in horizontal extent, hairline to ¼ in. (0.6 cm) in thickness, ½ to ¾-in. (1.2 to 1.9-cm) spacing. Visible excess ice ~20 ± % of total volume. Ice lenses hard, clear, colorless.
2.8	9.1	ICE		Hard, slightly cloudy, colorless, few scattered inclusions of silty SAND
3.2	10.5		Dark brown PEAT, frozen, well bonded, high degree of saturation	~5 % visible ice
4.4	14.3	MH V_r	Light brown SILT, frozen	Irregularly oriented ice lenses and layers ¼ to ¾ in. (0.6 to 1.9 cm) thick on random pattern grid approx. 3 to 4-in. (7.6 to 10-cm) spacing. Visible ice ~10 ± % of total volume. Ice moderately soft, porous, gray-white.
4.9	16.0			
6.3	20.6		Bedrock. Laminated SHALE. Top few feet weathered	1/16-in. (0.2-cm) thick ice lenses in fissures to 16.0 ft (4.9 m). None below.
			Bottom of exploration	

* Surface elevation 963.2 ft

FIG. X1.1 Graphic Log of Field Exploration Illustrating Use of the Frozen Soil Descriptive Nomenclature and Symbols

Standard Guide for
Using Rock-Mass Classification Systems for Engineering Purposes[1]

This standard is issued under the fixed designation D 5878; the number immediately following the designation indicates the year of original adoption or, in the case of revision, the year of last revision. A number in parentheses indicates the year of last reapproval. A superscript epsilon (ε) indicates an editorial change since the last revision or reapproval.

1. Scope

1.1 This guide covers the selection of a suitable system of classification of rock mass for specific engineering purposes, such as tunneling and shaft-sinking, excavation of rock chambers, ground support, modification and stabilization of rock slopes, and preparation of foundations and abutments. These classification systems may also be of use in work on rippability of rock, quality of construction materials, and erosion resistance. Although widely used classification systems are treated in this guide, systems not included here may be more appropriate in some situations, and may be added to subsequent editions of this standard.

1.2 The valid, effective use of this guide is contingent upon the prior complete definition of the engineering purposes to be served and on the complete and competent definition of the geology and hydrology of the engineering site. Further, the person or persons using this guide must have had field experience in studying rock-mass behavior. An appropriate reference for geological mapping in the underground is provided by Guide D 4879.

1.3 This guide identifies the essential characteristics of each of the five included classification systems. It does not include detailed guidance for application to all engineering purposes for which a particular system might be validly used. Detailed descriptions of the five systems are presented in STP 984 (1),[2] with abundant references to source literature.

1.4 The range of applications of each of the systems has grown since its inception. This guide summarizes the major fields of application up to this time of each of the five classification systems.

1.5 *This standard does not purport to address all of the safety concerns, if any, associated with its use. It is the responsibility of the user of this standard to establish appropriate safety and health practices and determine the applicability of regulatory limitations prior to use.*

2. Referenced Documents

2.1 *ASTM Standards:*
D 653 Terminology Relating to Soil, Rock, and Contained Fluids[3]
D 2938 Test Method for Unconfined Compressive Strength of Intact Rock Core Specimens[3]

D 4879 Guide for Geotechnical Mapping of Large Underground Openings in Rock[3]

3. Terminology

3.1 *Definitions:*

3.1.1 *classification, n*—a systematic arrangement or division of materials, products, systems, or services into groups based on similar characteristics such as origin, composition, properties, or use *Regulations Governing ASTM Technical Committees.*[4]

3.1.2 *rock mass (in situ rock), n*—rock as it occurs in situ, including both the rock material and its structural discontinuities (Modified after Terminology D 653 [ISRM]).

3.1.3 *rock material (intact rock, rock substance, rock element), n*—rock without structural discontinuities; rock on which standardized laboratory property tests are run.

3.1.4 *structural discontinuity (discontinuity), n*—an interruption or abrupt change in a rock's structural properties, such as strength, elasticity, or density, usually occurring across internal surfaces or zones, such as bedding, parting, cracks, joints, faults, or cleavage.

NOTE 1—To some extent, 3.1.1, 3.1.2, and 3.1.4 are scale-related. A rock's microfractures might be structural discontinuities to a petrologist, but to a field geologist the same rock could be considered intact. Similarly, the localized occurrence of jointed rock (rock mass) could be inconsequential in regional analysis.

3.1.5 For the definition of other terms that appear in this guide, refer to STP 984, Guide 4879, and Terminology D 653.

3.2 *Definitions of Terms Specific to This Standard:*

3.2.1 *classification system, n*—a group or hierarchy of classifications used in combination for a designated purpose, such as evaluating or rating a property or other characteristic of a rock mass.

4. Significance and Use

4.1 The classification systems included in this guide and their respective applications are as follows:

4.1.1 *Rock Mass Rating System (RMR) or Geomechanics Classification*—This system has been applied to tunneling, hard-rock mining, coal mining, stability of rock slopes, rock foundations, borability, rippability, dredgability, weatherability, and rock bolting.

4.1.2 *Rock Structure Rating System (RSR)*—This system has been used in tunnel support and excavation and in other ground support work in mining and construction.

[1] This test method is under the jurisdiction of ASTM Committee D-18 on Soil and Rock and is the direct responsibility of Subcommittee D18.12 on Rock Mechanics.
Current edition approved Dec. 10, 1995. Published February 1996.
[2] The boldface numbers given in parentheses refer to a list of references at the end of the text.
[3] *Annual Book of ASTM Standards*, Vol 04.08.

[4] Available from ASTM Headquarters, 100 Barr Harbor Drive, West Conshohocken, PA 19428.

4.1.3 *The Q System or Norwegian Geotechnical Institute System (NGI)*—This system has been applied to work on tunnels and chambers, rippability, excavatability, hydraulic erodibility, and seismic stability of roof-rock.

4.1.4 *The Unified Rock Classification System (URCS)*—This system has been applied to work on foundations, methods of excavation, slope stability, uses of earth materials, blasting characteristics of earth materials, and transmission of ground water.

4.1.5 *The Rock Material Field Classification Procedure (RMFC)*—This system has been used mainly for applications involving shallow excavation, particularly with regard to resistance to erosion, excavatability, construction quality of rock, fluid transmission, and rock-mass stability.

4.2 Other classification systems are described in detail in the general references listed in the appendix.

4.3 Using this guide, the classifier should be able to decide which system appears to be most appropriate for the specified engineering purpose at hand. The next step should be the study of the source literature on the selected classification system and on case histories documenting the application of that system to real-world situations and the degree of success of each such application. Appropriate but by no means exhaustive references for this purpose are provided in the appendix and in STP 984 (**1**). *The classifier should realize that taking the step of consulting the source literature might lead to abandonment of the initially selected classification system and selection of another system, to be followed again by study of the appropriate source literature.*

5. Bases for Classification

5.1 The parameters used in each classification system follow. In general, the terminology used by the respective author or authors of each system is listed, to facilitate reference to STP 984 (**1**) or source documents.

5.1.1 *Rock Mass Rating System (RMR) or Geomechanics Classification*

    Uniaxial compressive strength (see Test Method D 2938)
    Rock quality designation (RQD)
Spacing of discontinuities
    Condition of discontinuities
    Ground water conditions
    Orientation of discontinuities

5.1.2 *Rock Structure Rating System (RSR)*

    Rock type plus rock strength
    Geologic structure
    Spacing of joints
    Orientation of joints
    Weathering of joints
    Ground water inflow

5.1.3 *Q-System or Norwegian Geotechnical Institute System (NGI)*

    Rock quality designation (RQD)
    Number of joint sets
    Joint roughness
    Joint alteration
    Joint water-reduction factor
    Stress-reduction factor

5.1.4 *Unified Rock Classification System (URCS)*

    Degree of weathering

    Uniaxial compressive strength (see Test Method D 2938)
    Discontinuities
    Unit weight

5.1.5 *Rock Material Field Classification Procedure (RMFC)*

    Discrete rock-particle size
    Uniaxial compressive strength (see Test Method D 2938)
    Joint orientation
    Joint-aperture width
    Geologic structure
    Rock-unit thickness
    Seismic velocity
    URCS rating
    Rock quality designation (RQD)
    Mineralogy
    Porosity and voids
    Hydraulic conductivity and transmissivity

5.2 Comparison of parameters among these systems indicates some strong similarities. It is not surprising, therefore, that paired correlations have been established between RMR, RSR, and Q (**2**). Some of the references in the appendix also present procedures for estimating some in situ engineering properties from one or more of these indexes (**2, 3, 4,** and **5**).

NOTE 2—Reference (**2**) presents step-by-step procedures for calculating and applying RSR, RMR, and Q values. Applications of all five systems are discussed in STP 984 (**1**), as is a detailed treatment of RQD.

6. Procedures for Determining Parameters

6.1 The annex of this guide contains tabled and other material for determining the parameters needed to apply each of the classification systems. These materials should be used in conjunction with detailed, instructive references such as STP 984 (**1**) and Ref (**2**). The annexed materials are as follows:

6.1.1 *RMR System (RMR = adjusted sum)*

    Classification parameters (five) and their ratings (Sum ratings)
    Rating adjustment for joint orientations (Parameter No. 6)

    Effect of discontinuity strike and dip in tunneling
    Adjustments for mining applications
    Input data

6.1.2 *RSR System*

    Schematic of the six parameters
    Rock type plus strength, geologic structure ("A")
    Joint spacing and orientation ("B")
    Weathering of joints and ground water inflow ("C")

$$(RSR = A + B + C)$$

6.1.3 *Q-System:*

    RQD
    Joint set number, J_n
    Joint roughness number, J_r
    Joint alteration number, J_a
    Joint water reduction factor, J_w
    Stress reduction factor SRF

$$(Q = (RQD/J_n) \times (J_r/J_a) \times (J_w/SRF)$$

6.1.4 *URCS*

Degree of weathering (*A–E*)
Estimated strength (*A–E*)
Discontinuities (*A–E*)
Unit weight (*A–E*)
Schematic of notation (*results = AAAA* through *EEEE*)

6.1.5 *RMFCP*

Schematic of procedure through performance assessment

Classification (description and definitions),
Rock unit
Classification Elements—Including rock material properties, rock mass properties, and hydrogeologic properties.
Performance Assessment—Performance objectives
Erosion resistance

Excavation Characteristics
Construction Quality
Fluid Transmission
Rock Mass Stability

7. Precision

7.1 Precision statements will be available for some components of some of the classification systems, such as uniaxial compressive strength and rock quality designation.

8. Keywords

8.1 classification; classification system; Q-system (NGI); rock mass, rock mass rating system (RMR), rock material field classification procedure (RMFCP); rock quality designation (RQD); rock structure rating system (RSR); unified rock classification system (URCS)

ANNEX

(Mandatory Information)

A1.1 The materials presented in this Annex for RMR, RSR, URCS, and RMFCP have been extracted from STP 984 (**1**). The materials for Q (NGI) are from Ref (**4**).

RMR

TABLE 1—*Geomechanics Classification of jointed rock masses.*

A. CLASSIFICATION PARAMETERS AND THEIR RATINGS

PARAMETER			RANGES OF VALUES				
1	Strength of intact roack material	Point-load strength index	> 10 MPa	4 - 10 MPa	2 - 4 MPa	1 - 2 MPa	For this low range – uniaxial compressive test is preferred
		Uniaxial compressive strength	>250 MPa	100 - 250 MPa	50 - 100 MPa	25 - 50 MPa	5-25 MPa / 1-5 MPa / <1 MPa
		Rating	15	12	7	4	2 / 1 / 0
2	Drill core quality RQD		90% - 100%	75% - 90%	50% - 75%	25% - 50%	< 25%
	Rating		20	17	13	8	3
3	Spacing of discontinuities		>2 m	0,6 - 2 m	200 - 600 mm	60 -200 mm	<60 mm
	Rating		20	15	10	8	5
4	Condition of discontinuities		Very rough surfaces. Not continuous No seperation Unweathered wall rock.	Slightly rough surfaces. Separation < 1 mm Slightly weathered walls	Slightly rough surfaces. Separation < 1 mm Highly weathered walls	Slickensided surfaces OR Gouge < 5 mm thick OR Separation 1-5 mm. Continuous	Soft gouge > 5 mm thick OR Separation > 5 mm. Continous
	Rating		30	25	20	10	0
5	Ground water	Inflow per 10 m tunnel length	None	<10 litres/min	10-25 litres/min	25 - 125 litres/min	> 125
		Ratio (joint water pressure / major principal stress)	0	0,0-0,1	0,1-0,2	0,2-0,5	> 0,5
		General conditions	Completely dry	Damp	Wet	Dripping	Flowing
	Rating		15	10	7	4	0

B. RATING ADJUSTMENT FOR JOINT ORIENTATIONS

Strike and dip orientations of joints		Very favourable	Favourable	Fair	Unfavourable	Very unfavourable
Ratings	Tunnels	0	-2	-5	-10	-12
	Foundations	0	-2	-7	-15	-25
	Slopes	0	-5	-25	-50	-60

C. ROCK MASS CLASSES DETERMINED FROM TOTAL RATINGS

Rating	100 — 81	80 — 61	60 — 41	40 — 21	< 20
Class No.	I	II	III	IV	V
Description	Very good rock	Good rock	Fair rock	Poor rock	Very poor rock

D. MEANING OF ROCK MASS CLASSES

Class No	I	II	III	IV	V
Average stand-up time	10 years for 15 m span	6 months for 8 m span	1 week for 5 m span	10 hours for 2,5 m span	30 minutes for 1 m span
Cohesion of the rock mass	> 400 kPa	300 - 400 kPa	200 - 300 kPa	100 - 200 kPa	< 100 kPa
Friction angle of the rock mass	> 45°	35° - 45°	25° - 35°	15° - 25°	< 15°

RMR

TABLE 2—*Effect of discontinuity strike and dip orientations in tunneling.*

Strike Perpendicular to Tunnel Axis			
Drive with Dip		Drive against Dip	
Dip 45–90°	Dip 20–45°	Dip 45–90°	Dip 20–45°
Very favorable	Favorable	Fair	Unfavorable

Strike Parallel to Tunnel Axis		Irrespective of Strike
Dip 20–45°	Dip 45–90°	Dip 0–20°
Fair	Very unfavorable	Fair

TABLE 3—*Adjustments to the Geomechanics Classification for mining applications.*

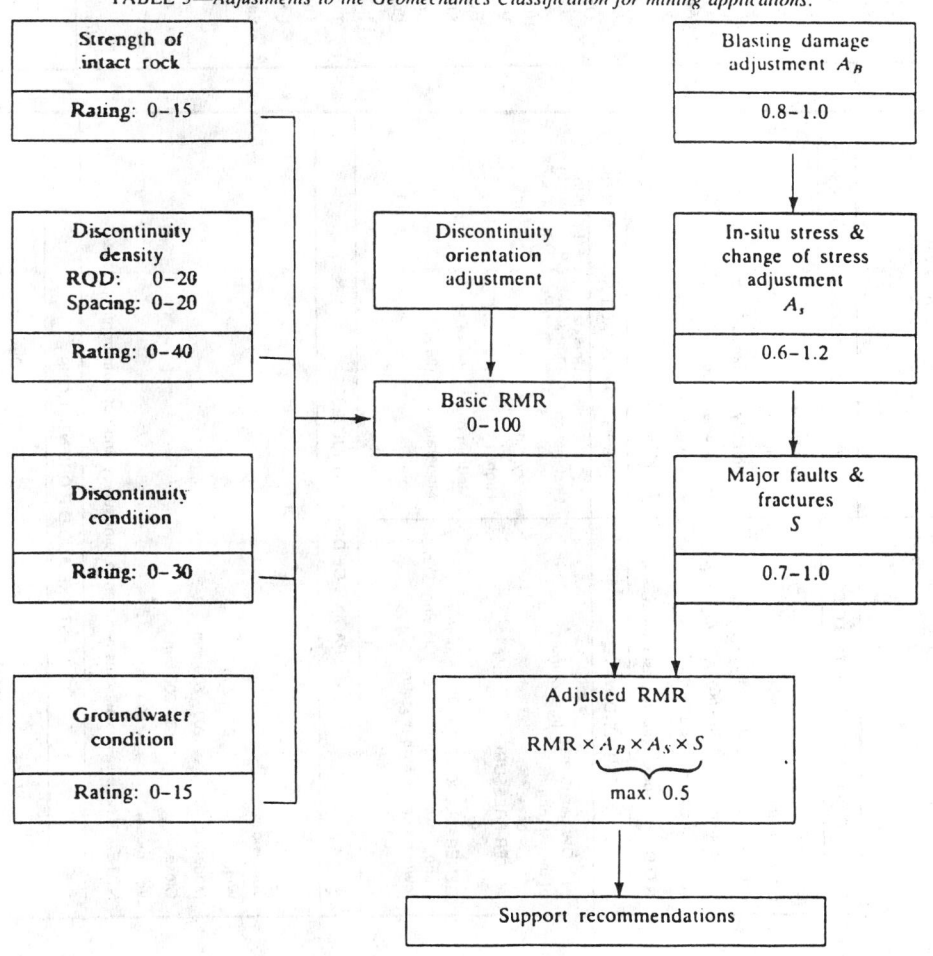

RMR

Input data form for the Geomechanics Classification (RMR System)

Name of project:
Site of survey:
Conducted by:
Date:

STRUCTURAL REGION	ROCK TYPE AND ORIGIN

DRILL CORE QUALITY R.Q.D.*

Excellent quality:	90 - 100%
Good quality:	75 - 90%
Fair quality:	50 - 75%
Poor quality:	25 - 50%
Very poor quality:	<25%

*R.Q.D. = Rock Quality Designation

GROUND WATER

INFLOW per 10 m   litres/minute
of tunnel length

or

WATER PRESSURE   kPa

or

GENERAL CONDITIONS (completely dry, damp, wet, dripping or flowing under low/medium or high pressure:

SPACING OF DISCONTINUITIES

	Set 1	Set 2	Set 3	Set 4
Very wide:	Over 2 m			
Wide:	0.6 - 2 m			
Moderate:	200 - 600 mm			
Close:	60 - 200 mm			
Very close:	<60 mm			

NOTE: These values are obtained from a joint survey and not from borehole logs.

STRIKE AND DIP ORIENTATIONS

Set 1	Strike:	(average)	(from	to	) Dip:	(angle)	(direction)
Set 2	Strike:		(from	to	) Dip:		
Set 3	Strike:		(from	to	) Dip:		
Set 4	Strike:		(from	to	) Dip:		

NOTE: Refer all directions to magnetic north.

WALL ROCK OF DISCONTINUITIES

Unweathered
Slightly weathered
Moderately weathered
Highly weathered
Completely weathered
Residual soil

STRENGTH OF INTACT ROCK MATERIAL

Designation	Uniaxial compressive strength, MPa	OR	Point-load strength index, MPa
Very high:	Over 250		>10
High:	100 - 250		4-10
Medium high:	50 - 100		2-4
Moderate:	25 - 50		1-2
Low:	5 - 25		<1
Very low:	1 - 5		

CONDITION OF DISCONTINUITIES

	Set 1	Set 2	Set 3	Set 4

PERSISTENCE (CONTINUITY)

Very low:	<1 m
Low:	1 - 3 m
Medium:	3 - 10 m
High:	10 - 20 m
Very high:	> 20 m

SEPARATION (APERTURE)

Very tight joints:	<0.1 mm
Tight joints:	0.1 - 0.5 mm
Moderately open joints:	0.5 - 2.5 mm
Open joints:	2.5 - 10 mm
Very wide aperture	> 10 mm

ROUGHNESS (state also if surfaces are stepped, undulating or planar)

Very rough surfaces:
Rough surfaces:
Slightly rough surfaces:
Smooth surfaces:
Slickensided surfaces:

FILLING (GOUGE)

Type:
Thickness:
Uniaxial compressive strength, MPa
Seepage:

MAJOR FAULTS OR FOLDS

Describe major faults and folds specifying their locality, nature and orientations.

GENERAL REMARKS AND ADDITIONAL DATA

NOTE:
(1) For definitions and methods consult ISRM document: 'Quantitative description of discontinuities in rock masses.'
(2) The data on this form constitute the minimum required for engineering design. The geologist should, however, supply any further information which he considers relevant.

RSR

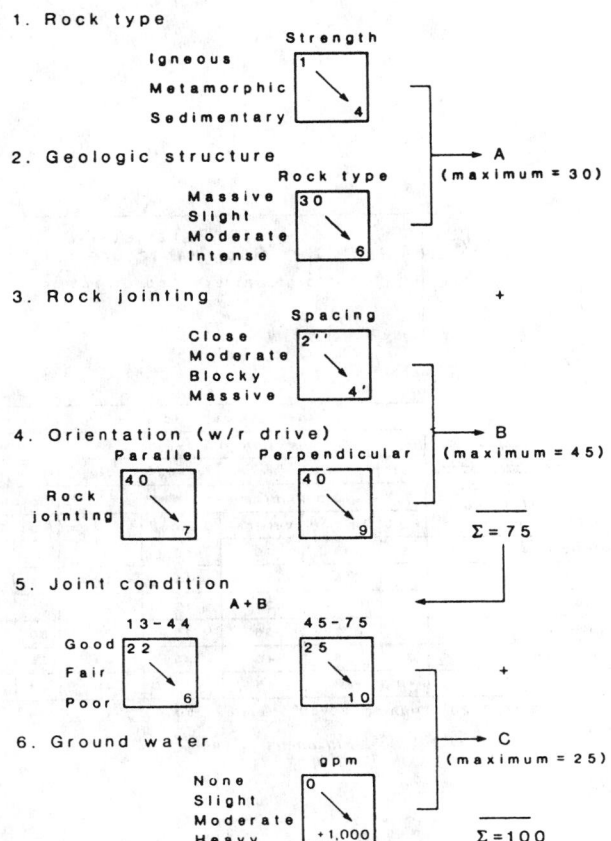

1. Rock type

    Igneous   Strength

    Metamorphic

    Sedimentary

2. Geologic structure

              Rock type

    Massive

    Slight

    Moderate

    Intense

A (maximum = 30)

3. Rock jointing

             Spacing

    Close

    Moderate

    Blocky

    Massive

4. Orientation (w/r drive)

    Parallel   Perpendicular

Rock jointing

B (maximum = 45)

$\Sigma = 75$

5. Joint condition

    A + B

    13 – 44   45 – 75

    Good

    Fair

    Poor

6. Ground water

             gpm

    None

    Slight

    Moderate

    Heavy

C (maximum = 25)

$\Sigma = 100$

FIG. 1—*Schematic of Rock Structure Rating.*

Parameter A
Rock structure rating
Rock type, strength index and geologic structure
Maximum value 30

Basic rock type					Geological structure			
	Hard	Medium	Soft	Decomp				
Igneous	1	2	3	4				
Metamorphic	1	2	3	4	Massive	Slightly faulted or folded	Moderately faulted or folded	Intensely faulted or folded
Sedimentary	2	3	4	4				
Type 1					30	22	15	9
Type 2					27	20	13	8
Type 3					24	18	12	7
Type 4					19	15	10	6

FIG. 2—*Parameter A.*

RSR

Parameter B
Rock structure rating
Joint pattern and direction of drive

Maximum value 45

	Strike perpendicular to axis					Strike parallel to axis		
	Direction of drive					Direction of drive		
	Both	With dip		Against dip		Both		
	Dip of prominent joints					Dip of prominent joints		
	Flat	Dipping	Vertical	Dipping	Vertical	Flat	Dipping	Vertical
① Very closely jointed	9	11	13	10	12	9	9	7
② Closely jointed	13	16	19	15	17	14	14	11
③ Moderately jointed	23	24	28	19	22	23	23	19
④ Moderate to blocky	30	32	36	25	28	30	28	24
⑤ Blocky to massive	36	38	40	33	35	36	34	28
⑥ Massive	40	43	45	37	40	40	38	34

Flat 0-20°, Dipping 20-50°, Vertical 50 90°

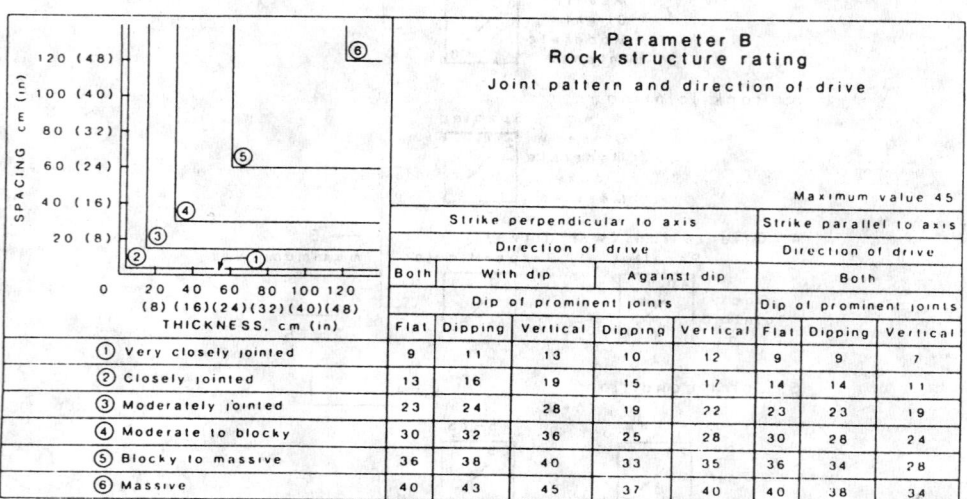

FIG. 3—*Parameter B.*

Parameter C
Rock structure rating
Ground water and joint condition

Maximum value 25

Anticipated water inflow m3/min/300m (gpm/1,000 ft)	Sum of parameters A + B					
	13-44			45-75		
	Joint condition					
	Good	Fair	Poor	Good	Fair	Poor
None	22	18	12	25	22	18
Slight <0.75 m3/min (<200 gpm)	19	15	9	23	19	14
Moderate 0.75-3.8 m3/min (200-1,000 gpm)	15	11	7	21	16	12
Heavy >3.8 m3/min (>1,000 gpm)	10	8	6	18	14	10

Joint condition: Good = Tight or cemented; Fair = Slightly weathered or altered; Poor = Severely weathered, altered or open

FIG. 4—*Parameter C.*

Q (NGI)

Ratings for the six Q-system parameters

1. Rock Quality Designation — RQD

		RQD
A	Very poor	0 - 25
B	Poor	25 - 50
C	Fair	50 - 75
D	Good	75 - 90
E	Excellent	90 - 100

Note: i) Where RQD is reported or measured as ≤ 10 (including 0), a nominal value of 10 is used to evaluate Q.
  ii) RQD intervals of 5, i.e., 100, 95, 90, etc., are sufficiently accurate.

2. Joint Set Number — J_n

		J_n
A	Massive, no or few joints	0.5 - 1.0
B	One joint set	2
C	One joint set plus random joints	3
D	Two joint sets	4
E	Two joint sets plus random joints	6
F	Three joint sets	9
G	Three joint sets plus random joints	12
H	Four or more joint sets, random, heavily jointed, "sugar cube", etc.	15
J	Crushed rock, earthlike	20

Note: i) For intersections, use ($3.0 \times J_n$)
  ii) For portals, use ($2.0 \times J_n$)

3. Joint Roughness Number — J_r

		J_r
a) Rock-wall contact, and b) rock-wall contact before 10 cm shear		
A	Discontinuous joints	4
B	Rough or irregular, undulating	3
C	Smooth, undulating	2
D	Slickensided, undulating	1.5
E	Rough or irregular, planar	1.5
F	Smooth, planar	1.0
G	Slickensided, planar	0.5

Note: i) Descriptions refer to small scale features and intermediate scale features, in that order.

	c) No rock-wall contact when sheared	
H	Zone containing clay minerals thick enough to prevent rock-wall contact	1.0
J	Sandy, gravelly or crushed zone thick enough to prevent rock-wall contact	1.0

Note: i) Add 1.0 if the mean spacing of the relevant joint set is greater than 3m.
  ii) J_r = 0.5 can be used for planar slickensided joints having lineations, provided the lineations are oriented for minimum strength.

4. Joint Alteration Number

		ϕ_r approx.	J_a
a) Rock-wall contact (no mineral fillings, only coatings)			
A	Tightly healed, hard, non-softening, impermeable filling, i.e., quartz or epidote		0.75
B	Unaltered joint walls, surface staining only	25-35°	1.0
C	Slightly altered joint walls. Non-softening mineral coatings, sandy particles, clay-free disintegrated rock, etc.	25-30°	2.0
D	Silty- or sandy-clay coatings, small clay fraction (non-softening)	20-25°	3.0
E	Softening or low friction clay mineral coatings, i.e., kaolinite or mica. Also chlorite, talc, gypsum, graphite, etc., and small quantities of swelling clays.	8-16°	4.0
b) Rock-wall contact before 10 cm shear (thin mineral fillings)			
F	Sandy particles, clay-free disintegrated rock, etc.	25-30°	4.0
G	Strongly over-consolidated non-softening clay mineral fillings (continuous, but < 5mm thickness)	16-24°	6.0
H	Medium or low over-consolidation, softening, clay mineral fillings (continuous, but < 5mm thickness)	12-16°	8.0
J	Swelling-clay fillings, i.e., montmorillonite (continuous, but < 5mm thickness). Value of J_a depends on percent of swelling clay-size particles, and access to water, etc.	6-12°	8-12
c) No rock-wall contact when sheared (thick mineral fillings)			
KLM	Zones or bands of disintegrated or crushed rock and clay (see G, H, J for description of clay condition)	6-24°	6, 8, or 8-12
N	Zones or bands of silty- or sandy-clay, small clay fraction (non-softening)		5.0
OPR	Thick, continuous zones or bands of clay (see G, H, J for description of clay condition)	6-24°	10, 13, or 13-20

5. Joint Water Reduction Factor

		approx. water pres. (kg/cm²)	J_w
A	Dry excavations or minor inflow, i.e., < 5 l/min locally	< 1	1.0
B	Medium inflow or pressure, occasional outwash of joint fillings	1-2.5	0.66
C	Large inflow or high pressure in competent rock with unfilled joints	2.5-10	0.5
D	Large inflow or high pressure, considerable outwash of joint fillings	2.5-10	0.33
E	Exceptionally high inflow or water pressure at blasting, decaying with time	> 10	0.2-0.1
F	Exceptionally high inflow or water pressure continuing without noticeable decay	> 10	0.1-0.05

Note: i) Factors C to F are crude estimates. Increase J_w if drainage measures are installed.
  ii) Special problems caused by ice formation are not considered.

6. Stress Reduction Factor — SRF

a) Weakness zones intersecting excavation, which may cause loosening of rock mass when tunnel is excavated

		SRF
A	Multiple occurrences of weakness zones containing clay or chemically disintegrated rock, very loose surrounding rock (any depth)	10
B	Single weakness zones containing clay or chemically disintegrated rock (depth of excavation ≤ 50m)	5
C	Single weakness zones containing clay or chemically disintegrated rock (depth of excavation > 50m)	2.5
D	Multiple shear zones in competent rock (clay-free), loose surrounding rock (any depth)	7.5
E	Single shear zones in competent rock (clay-free) (depth of excavation ≤ 50m)	5.0
F	Single shear zones in competent rock (clay-free) (depth of excavation > 50m)	2.5
G	Loose, open joints, heavily jointed or "sugar cube", etc. (any depth)	5.0

Note: ii) Reduce these values of SRF by 25-50% if the relevant shear zones only influence but do not intersect the excavation.

b) Competent rock, rock stress problems

		σ_c/σ_1	σ_θ/σ_c	SRF
H	Low stress, near surface, open joints	> 200	< 0.01	2.5
J	Medium stress, favourable stress condition	200-10	0.01-0.3	1
K	High stress, very tight structure. Usually favourable to stability, may be unfavourable for wall stability.	10-5	0.3-0.4	0.5-2
L	Moderate slabbing after > 1 hour in massive rock	5-3	0.5-0.65	5-50
M	Slabbing and rock burst after a few minutes in massive rock	3-2	0.65-1	50-200
N	Heavy rock burst (strain-burst) and immediate dynamic deformations in massive rock	< 2	> 1	200-400

Note: iii) For strongly anisotropic virgin stress field (if measured): when $5 \leq \sigma_1/\sigma_3 \leq 10$, reduce σ_c to $0.75\sigma_c$. When $\sigma_1/\sigma_3 > 10$, reduce σ_c to $0.5\sigma_c$, where σ_c = unconfined compression strength, σ_1 and σ_3 are the major and minor principal stresses, and σ_θ = maximum tangential stress (estimated from elastic theory).
  iii) Few case records available where depth of crown below surface is less than span width. Suggest SRF increase from 2.5 to 5 for such cases (see H).

c) Squeezing rock: plastic flow of incompetent rock under the influence of high rock pressure

		σ_θ/σ_c	SRF
O	Mild squeezing rock pressure	1-5	5-10
P	Heavy squeezing rock pressure	> 5	10-20

Note: iv) Cases of squeezing rock may occur for depth H > 350 $Q^{1/3}$ (Singh et al., 1992). Rock mass compression strength can be estimated from $q = 0.7 \gamma Q^{1/3}$ (MPa) where γ = rock density in kN/m³ (Singh, 1993).

d) Swelling rock: chemical swelling activity depending on presence of water

		SRF
R	Mild swelling rock pressure	5-10
S	Heavy swelling rock pressure	10-15

Note: J_r and J_a classification is applied to the joint set or discontinuity that is least favourable for stability both from the point of view of orientation and shear resistance, τ (where $\tau = \sigma_n \tan^{-1} (J_r/J_a)$). Choose the most likely feature to allow failure to initiate.

$$Q = \frac{RQD}{J_n} \times \frac{J_r}{J_a} \times \frac{J_w}{SRF}$$

Q (NGI)

Logging chart for assembling Q-parameter statistics

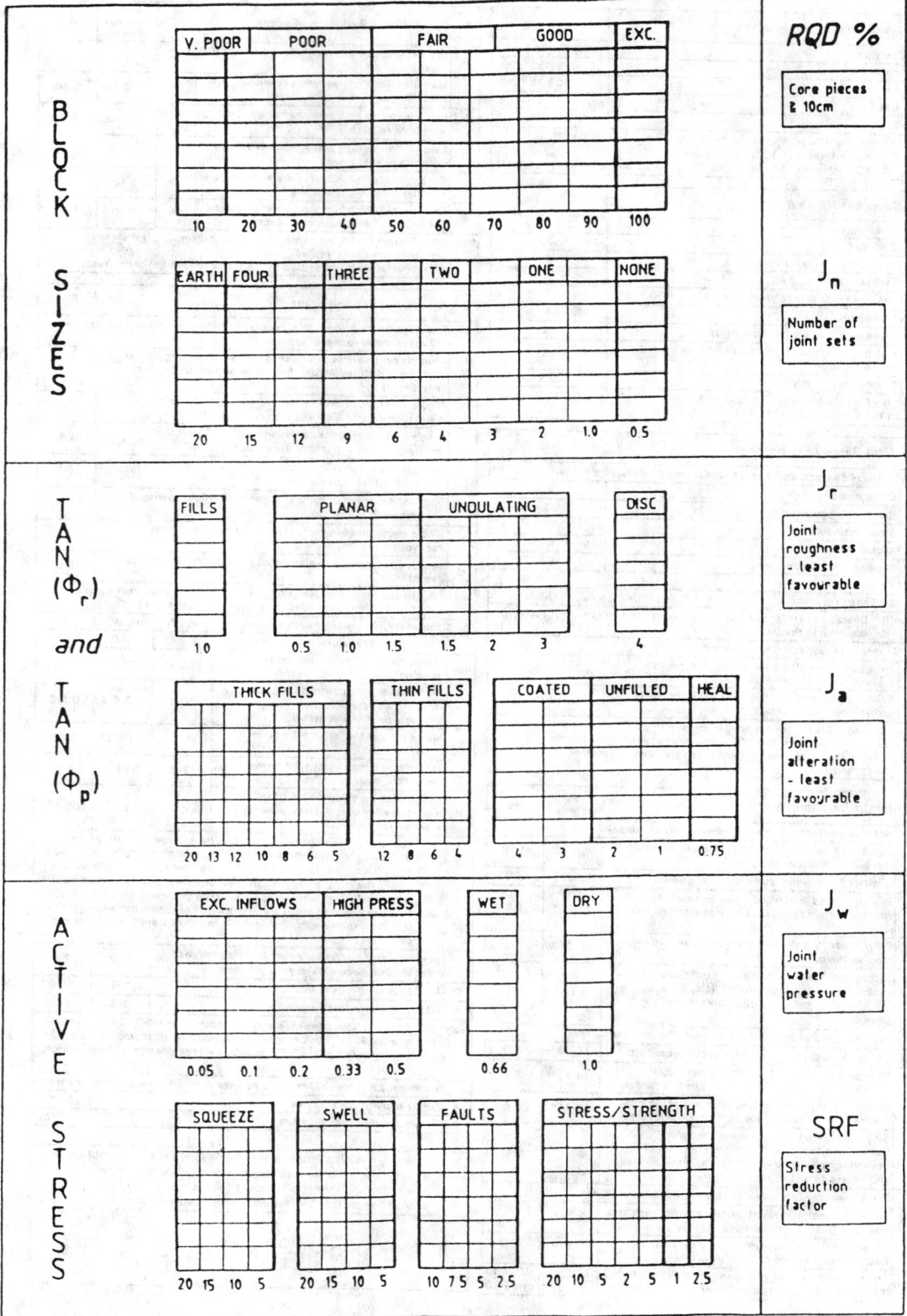

URCS

DEGREE OF WEATHERING

REPRESENTATIVE		ALTERED	WEATHERED	
			>GRAVEL SIZE	<SAND SIZE
Micro Fresh State (MFS)	Visually Fresh State (VFS)	Stained State (STS)	Partly Decomposed State (PDS)	Completely Decomposed State (CDS)
A	B	C	D	E
UNIT WEIGHT RELATIVE ABSORPTION		COMPARE TO FRESH STATE	NON-PLASTIC / PLASTIC	NON-PLASTIC / PLASTIC

ESTIMATED STRENGTH

REACTION TO IMPACT OF 1 LB. BALLPEEN HAMMER				REMOLDING[1]
"Rebounds" (Elastic) (RQ)	"Pits" (Tensional) (PQ)	"Dents" (Compression) (DQ)	"Craters" (Shears) (CQ)	Moldable (Friable) (MQ)
A	B	C	D	E
>15000 psi[2] >103 MPa	8000-15000 psi[2] 55-103 MPa	3000-8000 psi[2] 21-55 MPa	1000-3000 psi[2] 7-21 MPa	<1000 psi[2] <7 MPa

(1) Strength Estimated by Soil Mechanics Techniques
(2) Approximate Unconfined Compressive Strength

DISCONTINUITIES

VERY LOW PERMEABILITY			MAY TRANSMIT WATER	
Solid (Random Breakage) (SRB)	Solid (Preferred Breakage) (SPB)	Solid (Latent Planes Of Separation) (LPS)	Nonintersecting Open Planes (2-D)	Intersecting Open Planes (3-D)
A	B	C	D	E
			ATTITUDE	INTERLOCK

UNIT WEIGHT

Greater Than 160 pcf 2.55 g/cc	150-160 pcf 2.40-2.55 g/cc	140-150 pcf 2.25-240 g/cc	130-140 pcf 2.10-2.25 g/cc	Less Than 130 pcf 2.10 g/cc
A	B	C	D	E

DESIGN NOTATION

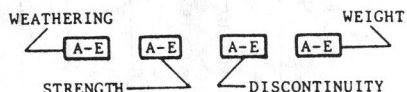

WEATHERING — A-E A-E A-E A-E — WEIGHT

STRENGTH — DISCONTINUITY

Figure 1. Basic elements of the unified rock classification system.

RMFCP

CLASSIFICATION

ROCK UNIT. A rock unit is an identifiable rock that is consistent in mineral, structural, and hydraulic characteristics. A rock unit can be considered essentially homogeneous for project analysis and for descriptive and mapping purposes. The degree of homogeneity of the rock units at the site of investigation is indicated by assignment of an "outcrop confidence level".

Rock units are delineated by observable and measurable physical features. When a rock unit has been established it can be defined by classification elements and analyzed for performance in relation to selected performance objectives.

CLASSIFICATION ELEMENTS. Classification elements are objective physical properties of the rock unit that define the characteristics of the material. Engineering classification of a rock unit reflects not only the material properties of the rock itself but the structural characteristics of the rock mass in the field, and the interactions between the rock and its system of discontinuities.

(1) Rock Material Properties: The lithologic properties of the rock that can be evaluated in hand specimen (and in many instances, in outcrop) and thus can be subject to meaningful inquiry in the laboratory. They include characteristics such as mineralogic composition, grain size, rock hardness, degree of weathering, unconfined compressive strength, porosity, unit weight, and other index properties.

(2) Rock Mass Properties: The lithologic properties of the rock that must be evaluated on a macroscopic scale in the field. They include description of tectonic features that are too large to be observed directly in their entirety, such as regional structure, karst features, and lineaments. Rock mass properties include features that cannot be sampled for laboratory analysis, such as fractures, joints, and faults, bedding, schistosity, lineations, as well as the lateral and vertical extent of the rock unit.

(3) Geohydrologic Properties: The lithologic properties of the rock that affect the mode of occurrence, location, distribution and flow characteristics of subsurface waters; these properties may include primary and secondary porosity, hydraulic conductivity, transmissivity, and other fluid transmission characteristics.

The following diagram illustrates the procedure:

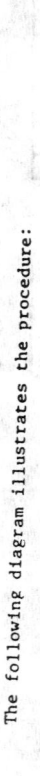

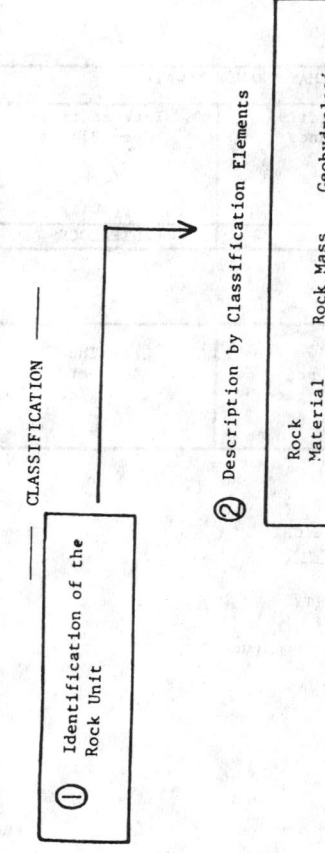

——— CLASSIFICATION ———

① Identification of the Rock Unit

② Description by Classification Elements

| Rock Material Properties | Rock Mass Properties | Geohydrologic Properties |

——— PERFORMANCE ASSESSMENT ———

③ Selection of Performance Objectives

- Erosion Resistance
- Excavation Characteristics
- Construction Quality
- Fluid Transmission
- Rock Mass Stability

④ Performance Assessment

RMFCP

ROCK CLASSIFICATION PROCESS

The rock classification process involves identifying the rock units at the site of investigation and describing appropriate classification elements. The following outline can be used as a guide in the process. Primary and secondary levels of description are indicated. Additional levels or factors may be added as required for further clarification. Appropriate appendixes are referred to. See Appendix IV for an example of a completed outline.

ROCK UNIT CLASSIFICATION

Project: _____
Date: _____
Geologist: _____

1. ROCK UNIT IDENTIFICATION.

The description of each rock unit should include the location and extent of the unit in outcrop or in stratigraphic section which, in turn, should provide an indication of outcrop confidence level. The rock unit can be identified either by name or alpha-numeric designation.

(a) Designation: (Vishnu schist, Rock Unit L-6, etc.)
(b) Location: (geographic, station, depth, etc.)
(c) Outcrop Confidence Level:

2. CLASSIFICATION ELEMENTS.

(a) Rock Material Properties: To be determined by examination and classification of hand specimens, core sections, drill cuttings, outcroppings, and disturbed samples, using standard geological terminology. Typical elements may include:

Rock formation name: (primary, secondary). See Appendix III
Mineralogy: (principal and accessory minerals, estimate percent; type of cement; note presence of alterable minerals)
Texture and fabric:
Primary porosity: (free-draining or not)
Discrete rock particle size: (See: Definitions)
Rock hardness: (See NEH-8, p. 1-13)
Micro structures: (bedding, foliation, etc.)
Degree of weathering (URCS): See Appendix I
Estimated strength (URCS): See Appendix I
Unit weight (URCS): See Appendix I

(b) Rock Mass Properties: To be determined by geologic mapping, geophysical survey, remote imagery interpretation, core sample analysis, and geomorphic evaluation. Typical elements may include:

Discontinuities (URCS): See Appendix I
Strike and dip of formation: (show where measured)
Joint analysis: (spacing, orientation, separation, description of wall rock: wavy, rough, smooth, or slickensided)
Joint tightness: (open, cemented, filled, cavernous)
Other structures: (folds, faults, unconformities, rock unit contacts, random fractures, etc.)
Geomorphic features: (karst topography, lava flows, lineaments, etc.)
Voids: (caverns, vugs, sinkholes, lava tubes, etc.: include shape, orientation, type of filling)
Rock quality designation (RQD):
Seismic velocity:
Unified Rock Class: See Appendix I

(c) Geohydrologic Properties: To be determined by pressure testing; water wells, observation wells, drill holes, and/or piezometer data; review of published maps and reports; interpretation of rock material and rock mass properties; dye tests. Typical elements may include:

Primary porosity: (see: Rock Material Properties)
Secondary porosity: (see: Rock Material Properties)
Hydraulic conductivity: See Appendix II
Transmissivity: See Appendix II
Storativity/specific yield:
Soluble rock: (occurrence of limestone, gypsum, or dolomite; also see: Rock Material Properties)
Water table/potentiometric surface: (contour map, dated)
Aquifer type: (confined or unconfined)

RMFCP

PERFORMANCE ASSESSMENT

PERFORMANCE OBJECTIVES. Performance objectives are selected operational elements or conditions that require an assessment of rock material performance. Five performance objectives are considered.

1. **Erosion Resistance:** Evaluation of the rock to resist erosion in spillways, channels, or other areas where rock material must withstand the stress of flowing water.

2. **Excavation Characteristics:** Evaluation of rock excavation characteristics, including the type of procedure required (rock, common, etc.) and the fragmentation characteristics and blasting response anticipated.

3. **Construction Quality:** Analysis of rock quality for riprap, aggregate, embankment fill, foundation, and other construction requirements.

4. **Fluid Transmission:** Evaluation of rock unit potential for fluid transmission through primary and secondary pores; for investigations concerning reservoir, canal, and dam foundation seepage losses, excavation dewatering, engineering subdrainage for slope stability, point and non-point source pollution, ground water yield for development (water wells, springs, aquifers, and basins), ground water recharge or disposal, and other ground water conditions of concern.

5. **Rock Mass Stability:** Evaluation of rock mass stability in relation to natural and constructed slopes, adequacy as a foundation material, seismic effects, and other construction requirements.

The performance assessment of rock material is developed through the following process:
1. Classification of the rock unit in terms of the CLASSIFICATION ELEMENTS.

2. Selection of appropriate PERFORMANCE OBJECTIVES based upon project requirements or structure conditions.

3. Identification of the levels of rock capability and limitations using the Performance Assessment Tables 1-5.

4. Further description or amplification of the rock capabilities and limitations as required to provide specific performance assessments in support of planning, design, and construction of project elements.

APPENDIX

(Nonmandatory Information)

X1. ADDITIONAL REFERENCES

Afrouz, A. A., *Practical Handbook of Rock Mass Classification Systems and Modes of Ground Failure*, CRC Press, Boca Raton, 1992.

Bell, F. G., *Engineering Properties of Soils and Rocks*, Butterworth-Heinemann, Oxford, 1992.

Bieniawski, Z. T., "Engineering Classification of Jointed Rock Masses", *Transactions of the South African Institution of Civil Engineers*, Vol 15, 1973, pp. 335–344.

Deere, D. U., Hendron, A. J., Jr., Patton, F. D., and Cording, E. J., "Design of Surface and Near-Surface Construction in Rock", in *Failure and Breakage of Rock*, Fairhurst, C., Ed., Society of Mining Engineers of AIME, New York, 1967, pp. 237–302.

Wickham, G. E., Tiedemann, H. R., and Skinner, E. H., "Ground Support Prediction Model, RSR Concept," in *Proceedings*, Second Rapid Excavation and Tunneling Conference, San Francisco, June 1974, Vol I, pp. 691–707.

Williamson, D. A., "Uniform Rock Classification for Geotechnical Engineering Purposes," *Transportation Research Record 783*, National Academy of Sciences, Washington, DC, 1980, pp. 9–14.

REFERENCES

(1) *Rock Classification Systems for Engineering Purposes*, ASTM STP 984, ASTM, 1988.

(2) Bieniawski, Z. T., *Rock Mechanics Design in Mining and Tunneling*, Balkema, A. A., Rotterdam, 1984.

(3) Barton, N., Lien, R., and Lunde, J., "Engineering Classification of Rock Masses for the Design of Tunnel Support," *Rock Mechanics*, Vol 6, No. 4, 1974, pp. 189–236.

(4) Barton, N., and Grimstad, E., "The Q-System Following Twenty Years of Application in NMT Support Selection," *Felsbau*, Vol 12, No. 6, 1994, pp. 428–436.

(5) Bieniawski, Z. T., *Engineering Rock Mass Classifications*, Wiley-Interscience, New York, 1989.

1.5 HYDROGEOLOGIC CHARACTERIZATION

Standard Guide for
Conceptualization and Characterization of Ground-Water Systems[1]

This standard is issued under the fixed designation D 5979; the number immediately following the designation indicates the year of original adoption or, in the case of revision, the year of last revision. A number in parentheses indicates the year of last reapproval. A superscript epsilon (ε) indicates an editorial change since the last revision or reapproval.

1. Scope

1.1 This guide covers an integrated, stepwise method for the qualitative conceptualization and quantitative characterization of ground-water flow systems, including the unsaturated zone, for natural or human-induced behavior or changes.

1.2 This guide may be used at any scale of investigation, including site-specific, subregional, and regional applications.

1.3 This guide describes an iterative process for developing multiple working hypotheses for characterizing ground-water flow systems. This process aims at reducing uncertainty with respect to conceptual models, observation, interpretation, and analysis in terms of hypothesis and refinement of the most likely conceptual model of the ground-water flow system. The process is also aimed at reducing the range of realistic values for parameters identified during the characterization process. This guide does not address the quantitative uncertainty associated with specific methods of hydrogeologic and ground-water system characterization and quantification, for example, the effects of well construction on water-level measurement.

1.4 This guide addresses the general procedure, types of data needed, and references that enable the investigator to complete the process of analysis and interpretation of each data type with respect to geohydrologic processes and hydrogeologic framework. This guide recommends the groups of data and analysis to be used during each step of the conceptualization process.

1.5 This guide does not address the specific methods for characterizing hydrogeologic and ground-water system properties.

1.6 This guide does not address model selection, design, or attribution for use in the process of ground-water flow system characterization and quantification. This guide does not address the process of model schematization, including the simplification of hydrologic systems and the representation of hydrogeologic parameters in models.

1.7 This guide does not address special considerations required for characterization of karst and fractured rock terrain. In such hydrogeologic settings, refer to Quinlan (1)[2] and Guide D 5717 for additional guidance.

1.8 This guide does not address special considerations regarding the source, fate, and movement of chemicals in the subsurface.

1.9 *This standard does not purport to address all of the safety concerns, if any, associated with its use. It is the responsibility of the user of this standard to establish appropriate safety and health practices and determine the applicability of regulatory limitations prior to use.*

2. Referenced Documents

2.1 *ASTM Standards:*

2.1.1 This procedure is used in conjunction with the following ASTM Standards:

D 653 Terminology Relating to Soil, Rock, and Contained Fluids[3]

D 5254 Practice for the Minimum Set of Data Elements to Identify a Ground Water Site[5]

D 5408 Guide for the Set of Data Elements to Describe a Ground Water Site; Part 1—Additional Identification Descriptors[5]

D 5409 Guide for the Set of Data Elements to Describe a Ground Water Site; Part 2—Physical Descriptors[5]

D 5410 Guide for the Set of Data Elements to Describe a Ground Water Site; Part 3—Usage Descriptors[5]

D 5447 Guide for Application of a Ground-Water Flow Model to a Site-Specific Problem[5]

D 5474 Guide for Selection of Data Elements for Ground-Water Investigations[5]

D 5609 Guide for Defining Boundary Conditions in Ground Water Flow Modeling[5]

D 5717 Guide to Design of Ground-Water Monitoring Systems in Karst and Fractured Rock Aquifers[5]

D 5730 Guide to Site Characterization for Environmental Purposes With Emphasis on Soil, Rock, the Vadose Zone, and Ground Water[5]

3. Terminology

3.1 *Definitions:*

3.1.1 *conceptual model*—an interpretation or working description of the characteristics and dynamics of the physical system.

3.1.2 *ground-water flow model*—application of a mathematical model to represent a regional or site-specific ground-water flow system.

3.1.3 *hydrologic system*—the general concepts of the hy-

---

[1] This guide is under the jurisdiction of ASTM Committee D-18 on Soil and Rock and is the direct responsibility of Subcommittee D18.21 on Ground Water and Vadose Zone Investigations.

Current edition approved July 10, 1996. Published November 1996.

[2] The boldface numbers given in parentheses refer to a list of references at the end of the text.

[3] *Annual Book of ASTM Standards*, Vol 04.08.

[4] *Annual Book of ASTM Standards*, Vol 11.01.

[5] *Annual Book of ASTM Standards*, Vol 04.09.

drologic elements, active hydrologic processes, and the interlinkages and hierarchy of elements and processes.

3.1.4 For definitions of other terms used in this guide, see Terminology D 653 and Guide D 5447.

4. Summary of Guide

4.1 This guide presents an integrated approach for conceptualizing and characterizing ground-water systems. The conceptualization and characterization process includes: Problem Definition and Data Base Development (Section 6); Preliminary Conceptualization (Section 7); Surface Characterization (Section 8); Subsurface Characterization (Section 9); Hydrogeologic Characterization (Section 10); Ground-Water System Characterization (Section 11); and Ground-Water System Quantification (Section 12) (see Fig. 1). Conceptualization and characterization is an iterative process beginning with a theoretical understanding of the ground-water system followed by data collection and refinement of the understanding. Additional data collection and analysis, and the refinement of the ground-water system conceptual model occurs during the entire process of conceptualization and characterization, and during ground-water model development and use (see Fig. 1).

4.2 This guide presents an approach that can be used at any scale. The nature of the problem to be solved will determine the type and scale of data collected.

5. Significance and Use

5.1 Conceptualization and characterization of a ground-water system is fundamental to any qualitative or quantitative analysis. This conceptualization begins with simple abstractions in the investigator's mind, emphasizing the major components of the studied system, that can be rendered in qualitative terms or simple illustrations. The extent of further development of the representation of the system depends on the character of the ground-water problem and the project objective. The abstract concept may suffice, or it may be further defined and quantified through use of analytical models of increasing complexity, and, in some cases, numerical models may be employed. If numerical models are used, the level of detail and sophistication of features represented in the model is likely to increase as the project develops. Evolution of conceptualization of a ground-water flow system should be terminated when the results of the related analyses are sufficient for the problem being addressed.

5.2 This guide may be used in the following:

5.2.1 Evaluating natural variations in ground-water flow systems.

5.2.2 Evaluating anthropogenic stresses on ground-water flow systems, such as pumping for water supply, irrigation, induced infiltration, or well injection.

5.2.3 Evaluating presence and velocity of ground-water contaminants.

5.2.4 Designing and selecting mathematical models to simulate ground-water systems; and completing model schematization and attribution based on the problem defined, characterized ground-water flow system, and model(s) selected.

5.2.5 Designing ground-water remediation systems.

5.3 This guide is a flexible description of specific tech-

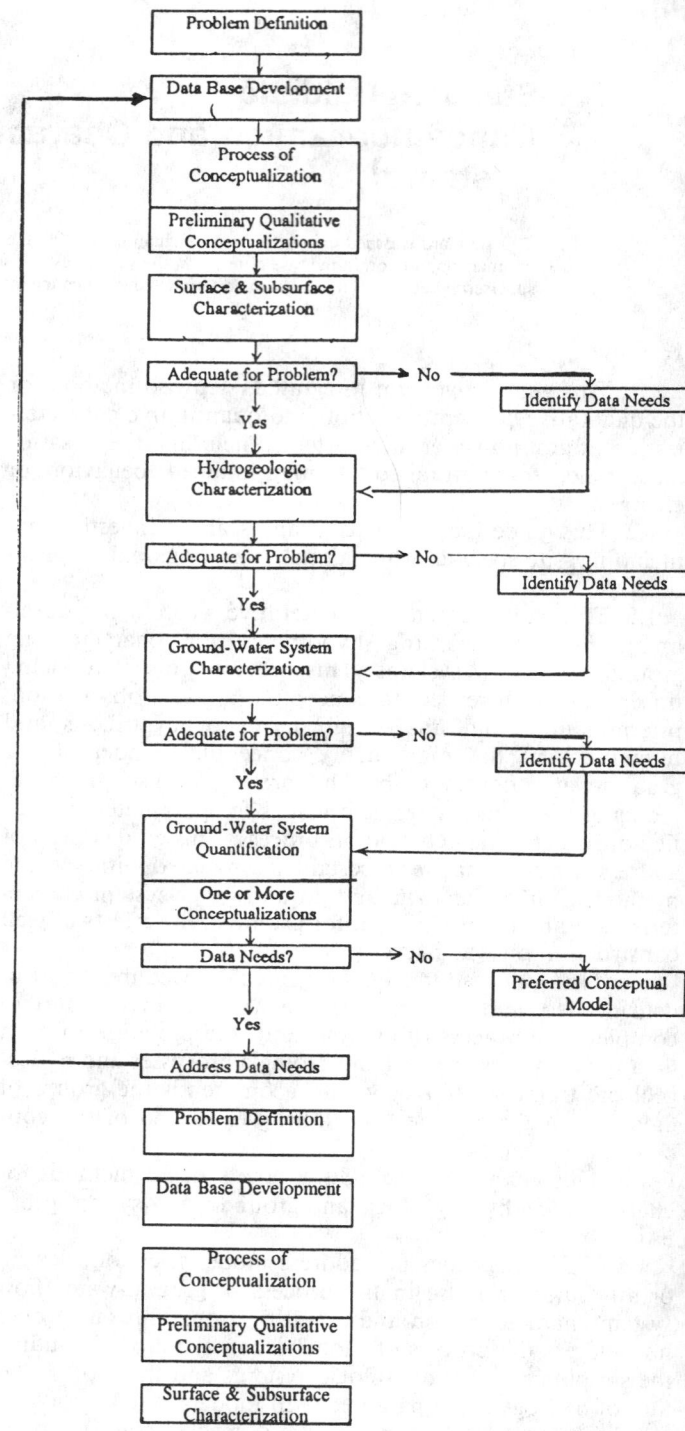

NOTE—Conceptualization and characterization is an iterative process beginning with a theoretical understanding of the ground-water system followed by data collection and refinement of the understanding. Additional data collection and analysis, and the refinement of the ground-water system conceptual model occurs during the process of conceptualization and characterization, and during ground-water model development and use.

FIG. 1 Procedure for Conceptualization and Characterization of Ground-Water Flow Systems (32)

niques and investigation requirements; methods defined by other ASTM Standards or non-ASTM techniques may be appropriate in some circumstances and, after due consideration, some of the techniques herein may be omitted, altered, or enhanced.

5.3.1 A comprehensive list of items to be considered conceptualization and characterization are included in the main headings (Sections 6 through 13) and first subheadings (for example, 7.1 and 8.1).

5.3.2 In karst and fractured rock hydrogeologic settings, this guide should be used in conjunction with Guide D 5717.

5.4 The methods and amount of effort required for conceptualization, characterization, and quantification of ground-water systems for modeling or other applications will vary with site conditions, objectives of investigation, and investigator experience. This guide does not replace proper academic training and experience in hydrogeologic principles, or in ground-water system analysis and quantification. This guide does not set mandatory guidelines and does not constitute a list of necessary steps or procedures for all investigations.

5.5 This guide may be used for project planning and data collection, but does not provide specific aspects for field characterization techniques. Refer to Table X1.1 in Guide D 5730, Practice D 5254, and Refs (2, 3, 4, and 5) for further guidance regarding field characterization techniques.

5.6 This guide may be used to generate the necessary information as part of the process for model selection, design, and as input to model schematization, including the simplification of hydrologic systems and the representation of hydrogeologic parameters in models. Refer to Ref (6) for further guidance.

6. Problem Definition and Data Base Development

6.1 *Define the Objectives of the Project*—Once the objectives are defined, identify the appropriate facets and scale of the ground-water system for characterization.

6.2 *Define the Site*—The boundaries of a site are defined using one or more of the following considerations: natural site characteristics (topography, soils, geology, hydrology, biota), current and past land use and ownership, or known or suspected extent of current or anticipated project-related stresses, which may include cones of depression or contaminant migration. If site boundaries are initially defined by ownership, natural site characteristics of a broader scale should be evaluated to determine whether the scope of at least parts of the investigation should include areas that are off-site. For example, investigations of ground-water contamination should include areas of potential sources upgradient and potential migration paths down-gradient from a site.

6.3 *Gather Data from Existing Sources*—This step involves locating, collecting, and organizing the data needed (see Table 1) to solve the problem into a manageable data base. See Guides D 5254, D 5408, D 5409, D 5410, D 5474, and D 5730 for data elements to identify a ground-water site.

6.3.1 Collect data, such as maps, tables, and reports, from available published and unpublished sources, and field and laboratory studies. Note the methods used to collect and analyze the data. Note levels of quality assurance and quality control as required by the project.

6.3.2 Collect data from interviews of local and regionally knowledgeable people. This may include, but is not limited to, worker histories, former practices, and engineering activities that either changed the site or provide historical data

TABLE 1 Data Topics and Types

Topography and Remote Sensing:
(a) Topography
(b) Aerial photography
(c) Satellite imagery
(d) Multispectral data
(e) Thermal imagery
(f) Radar, side-looking airborne radar, microwave imagery
Geomorphology:
(a) Surficial geology or geomorphology maps
(b) Engineering geology maps
(c) Surface water inventory maps
(d) Hydrography digital line graphs
Geology:
(a) Geologic maps and cross sections
(b) Lithologic or drillers logs, or both
Geophysics:
(a) Gravity, electromagnetic magnetics, resistivity, and seismic survey data or interpretations, or both
(b) Natural seismic activity data
(c) Borehole geophysical data
Climate:
(a) Precipitation data
(b) Temperature, humidity, and wind data
(c) Evaporation data
(d) Effects of climate change on hydrologic system information
Vegetation:
(a) Communities or species maps, or both
(b) Density map
(c) Agricultural species, crop calendars, consumptive use data
(d) Land use—Land cover maps
Soils:
(a) Soil surveys
(b) Soil properties determined from laboratory analysis
Hydrology:
(a) Potentiometric head data
(b) Subsurface test information
(c) Subsurface properties determined from laboratory analyses
(d) Previous work regarding modeling studies, hydrogeologic and ground-water system maps
(e) Spring and seep data
(f) Surface water data
(g) Well design, construction, and development information
Hydrochemistry/Geochemistry (Related to Ground-Water Flow System):
(a) Subsurface chemistry derived from well samples
(b) Surface water chemistry
(c) Rock and soil chemistry
(d) Water quality surveys
Anthropogenic Aspects:
(a) Planimetric maps
(b) Land use—Land cover maps
(c) Roads, transportation, political boundary DLGs
(d) Land ownership maps include historical information, if available
(e) Resource management maps

(location of old wells, contaminant history, and so forth).

6.4 *Organize and Prepare Data Bases Based on Project Objectives*—This step involves organizing the data into appropriate data bases that could include, but are not limited to: geomorphology, geology, geophysics, climate, vegetation, soils, hydrology, hydrochemistry/geochemistry, and anthropogenic aspects (see Table 2).[6]

7. Preliminary Conceptualization

7.1 Conduct field conceptualization using data bases developed under Section 6. In areas where field data are sparse,

[6] Quality assurance/quality control should be maintained throughout the project. Data may be organized into three types: 1) raw, original data collected in the field or laboratory, or both; 2) extracted data produced from the original, raw data base to solve the study purposes, goals, and objectives; and 3) interpretations and analyses of both raw or extracted data as applied to solving the problem.

TABLE 2 Data Bases

Geomorphology:
   (a)   Topographic map or digital elevation model, or both
   (b)   Drainage trace map
Geology:
   (a)   Geologic map and stratigraphic column
   (b)   Surficial geology map and stratigraphic column
   (c)   Geologic cross sections
   (d)   Lithologic or driller's logs, or both
Geophysics:
   (a)   Gravity maps and data
   (b)   Magnetic maps and data
   (c)   Resistivity maps and data
   (d)   Seismic and earthquake activity maps and data
   (e)   electromagnetic induction data
Meteorology and Climate:
   (a)   Precipitation data
   (b)   Temperature data
   (c)   Evaporation data
   (d)   Solar radiation data
Vegetation:
   (a)   Vegetation type and distribution maps
   (b)   Consumptive water use data
Soils:
   (a)   Soil type and characteristics maps
   (b)   Soil properties data
Hydrology:
   (a)   Water well data
   (b)   Potentiometric surface maps
   (c)   Springs and seeps data
   (d)   Surface water data
   (e)   Aquifer properties data
Hydrochemistry/Geochemistry (as Related to Ground-Water Flow Systems):
   (a)   Isotope hydrochemistry
   (b)   Organic hydrochemistry
   (c)   Inorganic hydrochemistry
   (d)   Soil, chemical precipitates, and rock geochemistry
Anthropogenic Aspects:
   (a)   Political boundaries maps
   (b)   Land ownership maps
   (c)   Land use—Land cover maps including historical information, if available
Hydrogeologic Characterization:
   (a)   Hydrogeologic table of attributes
   (b)   Hydrogeologic map
   (c)   Hydrogeologic cross-sections and stratigraphic columns
Ground-Water System Characterization:
   (a)   Ground-water system tables for recharge and discharge types and amounts
   (b)   Ground-water system maps showing recharge, discharge, and flow system
   (c)   Ground-water system cross sections showing recharge, discharge, and flow system
   (d)   Potentiometric surface maps for each hydrologic layer

basic photointerpretation and terrain analysis techniques may be applied to remote sensing data, aerial photography, and topographic maps to acquire information, and may be used to quantify and distribute hydrogeologic and ground-water system parameters.

7.1.1 Analyze existing data. This includes both the natural and anthropogenic features of the site. This preliminary analysis may include land cover patterns (vegetation, soils, surface water type and distribution, topography, geology), landforms (surficial geology and geography), and drainage analysis.[7]

7.1.2 Conduct field reconnaissance to relate the preliminary analysis of the information collected to study site conditions.[8]

7.2 Conduct qualitative ground-water system conceptualization. This results in the development of one or more initial conceptual models that will be used for characterization and quantification. This qualitative analysis uses the same logic presented in Sections 8 through 12 for quantitative analysis.

7.2.1 Qualitatively characterize the study area surface using procedures stated in Section 8.

7.2.2 Qualitatively characterize the study area subsurface geologic framework using procedures stated in Section 9.

7.2.3 Qualitatively characterize the study area hydrogeologic framework using procedures stated in Section 10.

7.2.4 Qualitatively characterize the study area ground-water system using procedures stated in Section 11. The resulting ground-water system conceptual model, to be used for quantitative characterization, includes a qualitative assessment of how water enters, moves through or is stored in, and leaves the ground-water system. The potentiometric surfaces and boundary conditions of each aquifer in the ground-water system are conceptualized at this time.

7.2.5 Describe and visualize the ground-water system conceptual model using cross sections and plan view illustrations. This ground-water system conceptual model may be modified at any stage of quantitative characterization (see Sections 8 through 12).

8. Surface Characterization

8.1 Conduct surface characterization of anthropogenic and natural features and processes at or near ground surface.

8.1.1 Conduct anthropogenic effects analysis to show hydrologic land use. Anthropogenic effects analysis includes, but is not limited to, irrigation or agricultural consumptive use of water; and industrial, municipal, and domestic water use.

8.1.2 Conduct vegetation analysis including vegetation type and distribution, consumptive water use data, and the hydrologic land use. See Ref (9) for guidance and references.

8.1.3 Conduct topography analysis including terrain, slope characteristics, hydrologic system continuity, and boundary locations. See Ref (10) for guidance and references.

8.1.4 Conduct surface water classification and distribution analysis, including classification and distribution of surface water flow (gaining and losing streams, constant or ephemeral stream flow; baseflow analysis), springs, lakes, and oceans. See Ref (11) for guidance and references.

8.1.5 Conduct climate analysis, including types and distribution of precipitation and temperatures, wind effects, and evapotranspiration potential. See Ref (3) for guidance and references.

8.1.6 Conduct pedogenic process and deposits analysis, including soil framework (horizons) and thickness, and soil permeability analysis, using standard pedogenic methods (12,13,14,15,16,17). It may be possible to use existing soil information for this analysis.

8.1.7 Conduct a geomorphologic process and deposits

---

[7] See Ref (7) and Ref (8) for interpretations related to drainage density, drainage network patterns, valley morphological patterns, and channel patterns and longitudinal profiles.

[8] The importance of this step will vary depending on site conditions and investigator experience. This step is especially important when site conditions are complex or the investigator's experience is limited regarding site conditions.

analysis, including maps depicting the type, properties, and distribution of geomorphic materials; geologic outcrops; landforms and slope; or other geomorphic characteristics needed to understand and solve the problem.[9]

9. Subsurface Characterization

9.1 Determine stratigraphic and lithologic units (soil and rock) using the soils, geology, and geophysics data bases and analysis, and surface characterization results. The stratigraphy or lithology of the subsurface framework, or both, is determined for the study area using standard geologic methods (19,20), and geophysical methods (2,3,4,5, and 21).

9.1.1 Geologic maps and cross sections, subsurface investigation logs, and stratigraphic columns are used, in conjunction with surface characterization and geophysical data and analysis, to develop a part of the geologic framework that represents the distribution of lithologic units.

9.1.2 Stratigraphic continuity of the geologic units may be evaluated using cross sections derived from geologic maps, well logs, and geophysical data.

9.2 Determine structural and geomorphologic discontinuities and stress history of the framework (for example, faults, fracture zones, karst) in the study area using the geology (geologic maps and cross sections) and geophysics analysis, surface characterization, geologic stratigraphic columns, and standard geologic and hydrogeologic methods (see Refs (2,3,5, and 22)).

9.3 Develop subsurface geologic framework geometries and cross sections using all of the soils, geology, geomorphology, and geophysics databases constructed during the preliminary conceptualization and surface characterization process, and the on-going subsurface characterization process. See Refs (19) and (20) for guidance.

10. Hydrogeologic Characterization

10.1 Characterize, quantify, and evaluate the uncertainty of the hydrostratigraphic units in terms of thickness, porosity, permeability, hydraulic conductivity (or soil moisture characteristic functions), transmissivity, and storativity. Primary, or matrix, porosity and permeability values, or hydraulic conductivity (or soil moisture characteristic functions), transmissivity, and storativity values may be quantified based on aquifer tests, laboratory analysis, or parameter estimation. Refer to 2.1 for ASTM Standards and the Reference Section for major non-ASTM references for information on characterization and quantification procedures. For specific vadose zone references and procedures, see Ref (23) and Guide D 5730.

10.1.1 Determine the continuity, geometry and spatial distribution, and thickness (total and saturated) of the hydrostratigraphic units.

10.1.2 Determine the isotropy/anisotropy of the hydrostratigraphic units.

10.1.3 Determine the homogeneity/heterogeneity of the hydrostratigraphic units.

10.1.4 Determine the hydrologic response of the hydrostratigraphic units (aquifer or confining unit). These units may be aquifers (conduits) or confining units (barriers) on the basis of hydraulic conductivity, saturated thickness, and continuity.

10.2 Characterize, quantify, and evaluate the uncertainty of the hydrostructural units, such as faults, fracture zones, fractured materials and karst conduits, in terms of thickness, porosity, permeability, hydraulic conductivity, transmissivity, and storativity. Fracture and fracture/karst porosity and permeability values, or hydraulic conductivity, transmissivity, and storativity values may be quantified based on aquifer tests, laboratory analysis, or parameter estimation.[10] Refer to 2.1 for ASTM Guides and Standards. For specific vadose zone references and procedures, see Ref (23) and Guide D 5730. For fractured rock characterization, see Ref (16) and Ref (27).

10.2.1 Determine the continuity, geometry and spatial distribution, and thickness (total and saturated) of the hydrostructural units.

10.2.2 Determine the isotropy/anisotropy of the hydrostructural units.

10.2.3 Determine the homogeneity/heterogeneity of the hydrostructural units.

10.2.4 Determine the hydrologic response of the hydrostructural units (aquifer or confining unit). These units may be aquifers (conduits) or confining units (barriers) on the basis of hydraulic conductivity, saturated thickness, and continuity.

10.3 Characterize and quantify the hydrogeologic framework.

10.3.1 Each hydrostratigraphic and hydrostructural unit, defined as a discrete volume element of the subsurface geologic framework, is evaluated based on the scale (site, local, or regional evaluation), temporal aspects (steady-state or transient analysis; daily, seasonal, annual analysis), and scope (saturated or unsaturated zone evaluation, two-dimensional or three-dimensional analysis) of the problem.

10.3.2 The hydrostratigraphic and hydrostructural units have been assigned numerical attributes, and can now be combined as units of relatively uniform character such as aquifers (conduits) or confining layers (barriers), isotropic or anisotropic, homogeneous or heterogeneous, and confined or unconfined parts of the hydrogeologic framework. As a result, these combined hydrogeologic units may be classified in any combination of these characteristics, and may be combined together, or further differentiated, based on the constraints of the problem.

10.3.3 In karst and fractured rock hydrogeologic settings, refer to Guide D 5717 for special approaches required for characterization and quantification of hydrogeologic proper-

---

[9] The geomorphologic processes, such as weathering, mass wasting, fluvial, eolian, glacial, oceanic, and ground water; and responses, such as: landforms and deposits, are interpreted using the landform, drainage, and land cover analyses derived from both on-site observations and databases created from remote sensing data, aerial photographs, and topographic maps. The general geomorphic process and response systems are described in more detail in geomorphology texts, such as Ref (18).

[10] For additional information on the effects of subsurface geochemical processes on the hydrologic system, see Ref (10). The methods used to evaluate these effects may include fault and fracture zone analysis (24), the hydrochemistry and geochemistry of the aquifer material and related flow system (25), and the general surface and subsurface evaluation of karst terrains (including regolith) (26).

ties. Refer to 2.1 for ASTM Standards for information on characterization and quantification procedures.

11. Ground-Water System Characterization

11.1 Characterize and evaluate uncertainty of the recharge areas and determine the type, amount, and distribution of recharge using surface, subsurface, and hydrogeologic analysis. Recharge is evaluated based on the scale, temporal aspects, and scope of the problem. Estimate recharge derived from: infiltration of precipitation; infiltration of surface water; return-flow from irrigated lands; inter-aquifer leakage; flux through natural or study area boundaries; and anthropogenic "sources." See Refs (28) and (29) for further guidance.

11.2 Characterize and evaluate uncertainty of the discharge areas and determine the type, amount, and distribution of discharge using surface, subsurface, and hydrogeologic analysis. Discharge is evaluated based on the scale, temporal aspects, and scope of the problem. Estimate discharge derived from: springs and seeps; surface water bodies; evapotranspiration from vegetation; well discharge; inter-aquifer leakage; flux through natural and study area boundaries; and anthropogenic "sources." See Ref (5) for further guidance.

11.3 Characterize and quantify the chemical constituents, both natural and anthropogenic, that pertain to characterizing the ground-water flow system. Chemical constituents are evaluated based on the scale, temporal aspects, and scope of the problem. See Refs (5) and (30) for further guidance.

11.3.1 Analyze the natural and anthropogenic chemical inputs to the subsurface hydrologic system from atmospheric, vegetation, and surface water sources.

11.3.2 Analyze the natural and anthropogenic chemical inputs to the subsurface hydrologic system from soil and rock materials.

11.3.3 Use the natural and anthropogenic chemical information and knowledge of the chemical processes, including, but not limited to, the biochemical, geochemical, and hydrochemical processes, to determine subsurface flow paths and estimate flow velocities, and to aid in characterizing the ground-water flow system. Useful information may include, but is not limited to, isotopes, natural or anthropogenic tracers, and ground-water chemical species evolution. For further guidance, see Boulding (5).

11.4 Characterize and quantify the ground-water system using Problem Definition and Data Base Development (see Section 6); Preliminary Conceptualization (see Section 7); Surface Characterization (see Section 8); Subsurface Characterization (see Section 9); Hydrogeologic Characterization (see Section 10); and Ground-Water System Characterization (see Section 11).[11] Ground-water system characteriza-

tion and quantification is based on the scale (site, local, or regional evaluation), temporal aspects (steady-state or transient analysis; daily, seasonal, annual analysis), and scope (saturated or unsaturated zone evaluation, two-dimensional or three-dimensional analysis) of the problem. See Refs (7) and (31) for guidance.

11.4.1 Characterize and quantify the initial conditions and boundary conditions of the ground-water system. See Guide D 5609; Guide D 5610; and Ref (6) for guidance.

11.4.2 Characterize the flow paths and construct the potentiometric surfaces for each hydrogeologic unit or layer of the ground-water system. See Ref (5) for further guidance.

11.4.3 Characterize, quantify, and balance the ground-water system water budget. Steady-state water budgets may be quantified by balancing the estimates of recharge (see 11.1) with the estimates of discharge (see 11.2). See Ref (5) for further guidance.

12. Ground-Water System Quantification

12.1 Exploratory ground-water modeling of one or more conceptual models, particularly the matching of the results of numerical models, with observations of heads and fluxes, may be used for the quantification of the hydrodynamics of the characterized ground-water system, for checking the ground-water flow system characterization for deficiencies (conceptual model or attributes), and for determining subsequent field sampling programs.[12]

12.2 Collect additional field data as required to address the identified data gaps.

12.3 As additional hydrogeologic and ground-water system data are collected or become available, the ground-water flow system conceptualization and characterization is refined (see Fig. 1).

12.4 If appropriate, select a modeling code, construct and calibrate a model and perform sensitivity analysis (see Guide D 5447 for guidance).

13. Report

13.1 If the characterized and quantified ground-water flow system is a product unto itself, then a report summarizing the data collected and the analyses performed should be prepared. If the characterized and quantified ground-water flow system is a part of another study or site characterization report, then the report for this activity can be included as a section.

13.2 The report should include a review of the assumptions used to characterize and quantify the ground-water flow system. Each of the sections mentioned previously that are relevant to the project should be mentioned in the report.

14. Keywords

14.1 characterization; conceptualization; ground water; ground-water modeling; ground-water systems

---

[11] The data bases needed for this analysis include recharge maps, discharge maps, topographic maps (or DEMs), water level data (heads), and hydrogeologic characterization. Hydrochemistry data, including the distribution of isotopes and hydrochemical species, and flow path chemistry, may help to confirm flow path vector distribution.

[12] Ground-water modeling may be used for checking the hydrogeologic attributes, such as hydraulic conductivity, transmissivity, and storativity, and the ground-water system attributes, such as water budgets, potentiometric surfaces, recharge, and discharge amounts, of the characterized ground-water system.

REFERENCES

(1) Quinlan, J. F., "Special Problems of Ground-Water Monitoring in Karst Terranes," *Ground Water and Vadose Zone Monitoring, ASTM STP 1053*, D. M. Nielsen and A. I. Johnson, eds., ASTM, Philadelphia, PA, 1990, pp. 275–304.

(2) Boulding, J. R., *Subsurface Characterization and Monitoring Techniques: A Desk Reference Guide. Volume I: Solids and Ground Water*, Appendices A and B, EPA/625/R-93/003a, 1993a.

(3) Boulding, J. R., *Subsurface Characterization and Monitoring Techniques: A Desk Reference Guide. Volume II: The Vadose Zone, Field Screening and Analytical Methods*, Appendices C and D, EPA/625/R-93/003b, 1993b.

(4) Boulding, J. R., *Use of Airborne, Surface, and Borehole Geophysical Techniques at Contaminated Sites*, EPA/625/R-92, 007, 1993c.

(5) Boulding, J. R., *Practical Handbook of Soil, Vadose Zone, and Ground-Water Contamination: Assessment, Prevention, and Remediation*, Lewis Publishers, Chelsea, MI, 1995.

(6) Anderson, M. P., and Woessner, W. W., *Applied Groundwater Modeling, Simulation of Flow and Advective Transport*, Academic Press, San Diego, CA, 1992.

(7) Kolm, K. E., *Conceptualization and Characterization of Hydrologic Systems: International Ground-Water Modeling Center Technical Report 93-01*, Colorado School of Mines, Golden, CO, 1993.

(8) Way, D. S., *Terrain Analysis: A Guide to Site Selection Using Aerial Photographic Interpretation*, Dowden, Hutchinson, and Ross, Inc., Stroudsburg, PA, 1973.

(9) Kuechler, A. W., *Vegetation Mapping*, The Ronald Press Company, NY, 1967.

(10) Fetter, C. W., *Applied Hydrogeology*, 3rd ed., Macmillan, NY, 1994.

(11) V. T. Chow, D. R. Maidment, and L. W. Mays, eds., *Applied Hydrology*, McGraw-Hill, New York, NY 1988.

(12) Birkeland, P. W., *Soils and Geomorphological Research*, Oxford University Press, New York, 1984.

(13) Soil Survey Staff, *Soil Survey Manual*, Revised Edition, U.S. Department of Agriculture, *Agricultural Handbook 18*, U.S. Government Printing Office. Washington, DC, 1993.

(14) Soil Survey Staff, *Keys to Soil Taxonomy*, 6th ed., 1994.

(15) Boulding, J. R., *Description and Sampling of Contaminated Soils: A Field Pocket Guide*, EPA/625/12-91/002, 1991.

(16) Boulding, J. R., *Description and Sampling of Contaminated Soils: A Field Guide*, Revised and Expanded 2nd ed., Lewis Publishers, Chelsea, MI, 1994.

(17) Boulding, J. R., *Ground Water and Wellhead Protection*, EPA/625/R-94/001, 1994b.

(18) Ritter, D. F., *Process Geomorphology*, Wm. C. Brown Publishers, Dubuque, Iowa, 1994.

(19) Compton, R. R., *Manual of Field Geology*, John Wiley & Sons, Inc., New York, 1962.

(20) Compton, R. R., *Geology in the Field*, John Wiley & Sons, New York, 1985.

(21) Keys, W. S., "Borehole Geophysics Applied to Ground-Water Investigations," *U.S. Geological Survey Techniques of Water-Resources Investigations*, TWRI 2-E2, 1990.

(22) Brassington, R., *Field Hydrogeology*, Halsted Press, New York, 1988.

(23) Wilson, L. G., Everett, L. G., and Cullen, S. J., *Handbook of Vadose Zone Characterization and Monitoring*, Lewis Publishers, Boca Raton, FL, 1994.

(24) Parizek, R. R., "On the Nature and Significance of Fracture Traces and Lineaments in Carbonate and Other Terrains," *Karst Hydrology and Water Resources*, V. Yevjevich, ed., Water Resources Publications, Ft. Collins, CO, 1976, pp. 47–108.

(25) Drever, J. I., *The Geochemistry of Natural Waters*, Prentice-Hall, Inc., Englewood Cliffs, NJ, 1982.

(26) Ford, D. C. and Williams, P. W., *Karst Geomorphology and Hydrology*, Unwin Hyman, Boston, MA, 1989.

(27) Sara, M. N., *Standard Handbook of Site Characterization for Solid and Hazardous Waste Facilities*, Lewis Publishers, Boca Raton, FL, 1994.

(28) Lerner, D. N., Issar, A. S., and Simmers, I., *Groundwater Recharge: A Guide to Understanding and Estimating Natural Recharge*, IAH International Contributions to Hydrogeology, Vol 8, Verlag Heinz Heise, Hannover, Germany, 1990.

(29) Simmers, I., ed., *Estimation of Natural Groundwater Recharge*, D. Reidel Publishing Co., Boston, MA, 1987.

(30) CCME, *Subsurface Assessment Handbook for Contaminated Sites*, Report CCME EPC-NCSRP-48E, Canadian Council of Ministers of the Environment, Winnipeg, Manitoba, 1994.

(31) G. B. Engelen, and G. P. Jones, eds., *Developments in the Analysis of Groundwater Flow Systems*, IAHS Publication No. 163, 1986.

(32) Kolm, K. E., van der Heijde, P. K. M., Downey, J. S., and Gutentag, E. D., "Conceptualization and Characterization of Ground-Water Systems," *Subsurface Fluid-Flow (Ground Water and Vadose Zone) Modeling, ASTM STP 1288*, J. D. Ritchey and J. O. Rumbaugh, eds., ASTM, 1996.

Standard Guide for
Selection of Methods for Assessing Ground Water or Aquifer Sensitivity and Vulnerability[1]

This standard is issued under the fixed designation D 6030; the number immediately following the designation indicates the year of original adoption or, in the case of revision, the year of last revision. A number in parentheses indicates the year of last reapproval. A superscript epsilon (ε) indicates an editorial change since the last revision or reapproval.

1. Scope

1.1 This guide covers information needed to select one or more methods for assessing the sensitivity of ground water or aquifers and the vulnerability of ground water or aquifers to water-quality degradation by specific contaminants.

1.2 This guide may not be all-inclusive; it offers a series of options and does not specify a course of action. It should not be used as the sole criterion or basis of comparison, and does not replace professional judgment.

1.3 This guide is to be used for evaluating sensitivity and vulnerability methods for purposes of land-use management, water-use management, ground-water protection, government regulation, and education. This guide incorporates descriptions of general classes of methods and selected examples within these classes but does not advocate any particular method.

1.4 *Limitations*—The utility and reliability of the methods described in this guide depend on the availability, nature, and quality of the data used for the assessment; the skill, knowledge, and judgment of the individuals selecting the method; the size of the site or region under investigation; and the intended scale of resulting map products. Because these methods are being continually developed and modified, the results should be used with caution. These techniques, whether or not they provide a specific numeric value, provide a relative ranking and assessment of sensitivity or vulnerability. However, a relatively low sensitivity or vulnerability for an area does not preclude the possibility of contamination, nor does a high sensitivity or vulnerability necessarily mean that ground water or an aquifer is contaminated.

1.5 The values stated in SI units are to be regarded as standard.

1.6 *This standard does not purport to address all of the safety concerns, if any, associated with its use. It is the responsibility of the user of this standard to establish appropriate safety and health practices and determine the applicability of regulatory limitations prior to use.*

2. Referenced Documents

2.1 *ASTM Standards:*
D 653 Terminology Relating to Soil, Rock, and Contained Fluids[2]

D 5447 Guide for Application of a Ground-Water Flow Model to a Site-Specific Problem[3]
D 5490 Guide for Comparing Ground-Water Flow Model Simulations to Site-Specific Information[3]
D 5549 Guide for Reporting Geostatistical Site Investigations[3]
D 5717 Guide for the Design of Ground-Water Monitoring Systems in Karst and Fractured-Rock Aquifers[3]
D 5880 Guide for Subsurface Flow and Transport Modeling[3]

3. Terminology

3.1 *Definitions*—Many of the terms discussed in this guide are contained in Terminology D 653. The reader should refer to this guide for definitions of selected terms.

3.1.1 *ground-water region, n*—an extensive area where relatively uniform geology and hydrology controls ground water movement.

3.1.2 *hydrogeologic setting, n*—a composite description of all the major geologic and hydrologic features which affect and control ground-water movement into, through, and out of an area (1).[4]

3.1.3 *sensitivity, n—in ground water*, the potential for ground water or an aquifer to become contaminated based on intrinsic hydrogeologic characteristics. Sensitivity is not dependent on land-use practices or contaminant characteristics. Sensitivity is equivalent to the term "*intrinsic ground-water vulnerability*" (2).

3.1.3.1 *Discussion*—Hydrogeologic characteristics include the natural properties of the soil zone, unsaturated zone, and saturated zone.

3.1.4 *vulnerability, n—in ground water*, the relative ease with which a contaminant can migrate to ground water or an aquifer of interest under a given set of land-use practices, contaminant characteristics, and sensitivity conditions. Vulnerability is equivalent to "specific ground-water vulnerability."

4. Significance and Use

4.1 Sensitivity and vulnerability methods can be applied to a variety of hydrogeologic settings, whether or not they contain specifically identified aquifers. However, some methods are best suited to assess ground water within aquifers, while others assess ground water above aquifers or

---

[1] This guide is under the jurisdiction of ASTM Committee D-18 on Soil and Rock and is the direct responsibility of Subcommittee D18.21 on Ground Water and Vadose Zone Investigations.
Current edition approved Oct. 10, 1996. Published May 1997.
[2] *Annual Book of ASTM Standards*, Vol 04.08.

[3] *Annual Book of ASTM Standards*, Vol 04.09.
[4] The boldface numbers in parentheses refer to a list of references at the end of this guide.

ground water in areas where aquifers have not been identified.

4.1.1 Intergranular media systems, including alluvium and terrace deposits, valley fill aquifers, glacial outwash, sandstones, and unconsolidated coastal plain sediments are characterized by intergranular flow, and thus generally exhibit slower and more predictable ground-water velocities and directions than in fractured media. Such settings are amenable to assessment by the methods described in this guide. Hydrologic settings dominated by fracture flow or flow in solution openings are generally not amenable to such assessments, and application of these techniques to such settings may provide misleading or totally erroneous results.

4.2 The methods discussed in this guide provide users with information for making land- and water-use management decisions based on the relative sensitivity or vulnerability of underlying ground water or aquifers to contamination. Most sensitivity and vulnerability assessment methods are designed to evaluate broad regional areas for purposes of assisting federal, state, and local officials to identify and prioritize areas where more detailed assessments are warranted, to design and locate monitoring systems, and to help develop optimum ground-water management, use and protection policies. However, some of these methods are independent of the size of the area evaluated and, therefore, can be used to evaluate the aquifer sensitivity and vulnerability of any specific area.

4.3 Many methods for assessing ground-water sensitivity and vulnerability require information on soils, and for some types of potential ground-water contaminants, soil is the most important factor affecting contaminant movement and attenuation from the land surface to ground water. The relatively large surface area of the clay-size particles in most soils and the soils' content of organic matter provide sites for the retardation and degradation of contaminants. Unfortunately, there are significant differences in the definition of soil between the sciences of hydrogeology, engineering, and agronomy. For the purposes of this guide, soils are considered to be those unconsolidated organic materials and solid mineral particles that have been derived from weathering and are characterized by significant biological activity. In the United States, these typically include unconsolidated materials that occur to a depth of 2 to 3 m or more.

4.3.1 In many areas, significant thicknesses of unconsolidated materials may occur below the soil. Retardation, degradation, and other chemical attenuation processes are typically less than in the upper soil horizons. These underlying materials may be the result of depositional processes or may have formed in place by long-term weathering processes with only limited biological activity. Therefore, when compiling the data required for assessing ground-water sensitivity and vulnerability, it is important to distinguish between the soil zone and the underlying sediments and to recognize that the two zones have significantly different hydraulic and attenuation properties.

5. Description of Methods

5.1 *Hydrogeologic Settings and Scoring Methods*—This group of methods includes those that involve geologic mapping, evaluation, and scoring of hydrogeologic characteristics to produce a composite sensitivity map or composite vulnerability map, or both. The methods range from purely descriptive of hydrogeologic settings to methods incorporating numerical scoring. They can include descriptive information or quantitative information, or both, and the maps can be applied as a "filter" to exclude specific hydrogeologic units from further consideration or select sensitive areas for further study.

5.1.1 The concept of assessing ground-water sensitivity and vulnerability is relatively recent and still developing. Thus, the methods presented differ because they have been developed for different purposes by different researchers using various types of data bases in several hydrogeologic settings. These methods have been divided into three groups: assessments using hydrogeologic settings without scoring or rankings, assessments in which hydrogeologic setting information is combined with ranking or scoring of hydrologic factors, and assessments using scoring methods applied without reference to the hydrogeologic setting. The groups are not exclusive but overlap. Each of these methods produces relative, not absolute, results whether or not it produces a numerical score. Sensitivity analyses can be used as the basis for a vulnerability assessment by adding the information on potential point and non-point contaminant sources.

5.1.2 *Hydrogeologic Settings, No Scoring or Ranking*—Hydrogeologic mapping has been widely used to provide aquifer sensitivity information. This subgroup of methods includes those that generally present information as composite hydrogeologic maps that can be used for multiple purposes. The maps can be used individually to make a variety of land-use decisions or used as a basis for ground-water and aquifer sensitivity evaluations. Although derivative ground-water and aquifer sensitivity maps can be prepared, any geologic or hydrogeologic map could potentially be used to assess sensitivity. In settings where quantitative data are lacking, hydrogeologic maps can allow the same conclusions, with the same level of confidence, as scoring methods. Hydrogeologic settings were mapped in detail without scoring or ranking in the Denver Colorado, United States area by Hearne and others (3).

5.1.2.1 Sensitivity assessments based on hydrogeologic settings with no scoring or ranking can be used to assess ground-water or aquifer vulnerability by overlaying information on potential point or non-point contamination sources. For example, the sensitivity map included in Ref (3) has been used in combination with a series of maps entitled "Land Uses Which Affect Ground-Water Management" (4) to conduct vulnerability assessments at specific sites within the greater Denver area.

5.1.3 *Hydrogeologic Settings with Ranking or Scoring, or Both*—This group of methods includes those which assess ground-water or aquifer sensitivity within or among various hydrogeologic settings using specific criteria to rank or score areas beneath which the ground water or aquifers have different potentials for becoming contaminated. The assessment is usually based on two or more hydrogeologic criteria. For example, material texture and depth to aquifer are parameters that are commonly used to establish criteria (5–10). Criteria, once defined, can then be ranked or scored, or both.

5.1.3.1 Assessing vulnerability from point and non-point

sources of potential contamination (for example, leaking tanks, waste generators, landfills, and abandoned hazardous waste sites) is accomplished by mapping their location on a sensitivity map (for example, numerous waste-generation sites in an area of low sensitivity would result in a relatively low vulnerability rank, all other factors being equal). This mapping method is particularly useful for evaluating the vulnerability of a large region. However, it can also be used to target smaller areas of particular concern where more detailed investigations may be needed. For example, Shafer (11) mapped regional aquifer vulnerability based on sensitivity analysis. Bhagwat and Berg (12) defined aquifer sensitivity according to depth to aquifers and the characteristics of the geologic materials. The sensitivity map was combined with information showing the distribution of waste-source sites per zip code per square mile. Highly vulnerable areas have aquifers at or near the surface and contain numerous point sources of potential contamination with mobile contaminants. Areas of low vulnerability have deep ground water or no aquifers and contain few potential contaminant sources or relatively immobile contaminants. This vulnerability information was then used to establish ground-water protection planning regions.

5.1.4 *Scoring, Without Hydrogeologic Settings*—This category includes those methods that use qualitative ranking or quantitative scoring with hydrogeologic information, but without subdividing the area on the basis of hydrogeologic settings. Methods were developed to have universal application and were intended to be used consistently to provide uniform results regardless of location. The methods are useful for applications that require a consistent approach over large areas, however, these methods can be complex and may require much unnecessary data preparation. Furthermore, because criteria selection and ranking are subjective, the final scores may be misleading.

5.1.4.1 These methods classify a site or region based on a ranking or a numerical score derived from hydrogeological information irrespective of the different hydrogeologic settings that may be present within the mapped area. Scores are calculated from equations based on criteria assumed to apply to different geographic areas and different hydrogeologic conditions (1,13–14). For example, in South Dakota (15), drilling logs and soil survey maps were used to prepare maps based on hydraulic conductivity which was inferred from the percent and thickness of surface organic matter. Attenuation potentials of soil in selected Wisconsin counties (16) were mapped based on soil depth, permeability, drainage class, organic matter content, pH, and texture.

5.2 *Process-Based Simulation Models*—These methods for assessment of ground-water sensitivity and vulnerability use a variety of models, each of which simulates some combination of the physical, chemical, and biological processes that control the movement of water and chemicals from land surface through the unsaturated zone to and through the saturated zone. These processes are formulated in terms of equations that are derived theoretically or empirically. Analytical or numerical techniques are used, usually within a computer program, to solve the equations. The solutions take the form of predicted rates of water and chemical movement as a function of location and time. Models differ greatly in the degree of complexity used to

incorporate actual processes, the amount of data required, the intended scale of the application, and the domain simulated. The latter criterion is arbitrarily selected here to categorize different simulation models. The three categories are: Root Zone Models, which simulate water and chemical movement through the portion of the unsaturated zone that is affected by vegetation; Unsaturated Zone Models, which simulate transport through the entire thickness of the unsaturated zone; and Saturated Zone Models which deal with processes occurring beneath the water table. Within each category there can be a wide range of model complexity with some models overlapping between different categories. Unsaturated-zone and root-zone models have been cataloged by van der Heijde (17,18) and van der Heijde and Elnawawy (19).

5.2.1 Model complexity, data requirements, and scale of application are closely related and should be considered in conjunction with each other. As models increase in complexity, it is expected that the accuracy of their predicted results would be improved. However, there would also be a commensurate increase in the amount of data required by the models. The lack of requisite data often limits the scale at which complex models may be applied, and many model codes are restricted to field-scale applications.

NOTE 1—The term "field-scale" as used here refers to the typical size of an agricultural field. In general, this is an area of 65 hectares (160 acres) or less that is planted to a single crop. "Local scale" refers to an area the size of a 1:24 000-scale quadrangle or the area of a typical county, while "regional scale" refers to an area of from several counties to one or more states.

5.2.2 *Root-Zone Models*—Models in this category were developed primarily for the agricultural industry to assess and compare the effects of agronomic best management practices (BMPs) on the management, protection, and enhancement of the chemical quality of ground- and surface-water resources. These simulation models provide a relative prediction of the fate and transport of sediments, salts, pesticides, fertilizers, and organic wastes applied to crop production systems. Because of the specificity of these models, they are generally applied at the scale of a single farm field although they can be used for areal management in combination with regional sensitivity maps.

5.2.2.1 Model components include the hydrology of the site (weather, surface runoff, return flow, percolation, evapotranspiration, lateral subsurface flow, and snow melt), erosion (water and wind), nitrogen and phosphorus cycling (loss in runoff, leaching, transport on sediment, mineralization, immobilization, and crop uptake as well as denitrification and nitrogen fixation), pesticide fate and transport, crop management factors (growth, yield, rotation, tillage, drainage, irrigation, fertilization, furrow diking, liming, and waste management), and economic accounting. Some models contain default values that allow them to be used for general planning, however, the user may supply site-specific values to improve the applicability of the result to the site of interest. These root-zone models usually calculate the amount of each pollutant of concern delivered out of the bottom of the root zone or unsaturated zone, but do not account for reactions in the saturated zone.

5.2.2.2 Examples of models in this category are the Pesticide Root Zone Model, PRZM (20), the Groundwater

Loading Effects of Agricultural Management Systems Model, GLEAMS (21), the Chemical Movement in Layered Soils Model, CMLS (22), and EPIC (Erosion Production/Impact Calculator) (23). An application of the EPIC model is given in Williams (24).

5.2.3 *Unsaturated-Zone Models*—Models in this category are capable of simulating processes throughout the entire unsaturated zone. Some models were developed specifically for agricultural applications, others were developed for more general problems of water and contaminant transport. In general, these models offer more sophistication in the treatment of the physical process of water movement than the root-zone models. Water movement through the unsaturated zone is usually described by Richard's Equation and the advection-dispersion equation is employed to describe solute transport. The equations are solved in one or two dimensions with primary consideration given to vertical water movement. Some models are capable of solving three-dimensional problems and others can account for both unsaturated and saturated movement of water and chemicals. Additional data are required for solving a more complex equation. For example, information on the relations between water and soil (that is, moisture-retention and relative permeability data) may be required.

5.2.3.1 Two problems limit the scale at which these models may be applied: the aforementioned lack of requisite data, and the fact that Richard's Equation is difficult to accurately solve for large regions. Application of these models is usually limited to areas less than or equal to the size of a single field. These models also require a certain amount of expertise to operate and to interpret results. Examples of these models include: LEACHM (25), VS2DT (26), RZWQM (27,28), and SWMS_3D (29). These models are used primarily for vulnerability assessment, although they can also be used for sensitivity analysis. A summary of commonly used unsaturated zone models, and their data requirements, is presented by Kramer and Cullen (30).

5.2.4 *Saturated-Zone Models*—This category of models is limited to processes in the saturated zone. Effects of unsaturated zone processes such as recharge and evapotranspiration are often incorporated in an ad hoc fashion. For ground-water sensitivity studies, a ground-water flow model such as MODFLOW (31), is often applied. Flow rates, position in the flow system, ground- and surface-water interaction, and recharge rates can be identified through model analysis. For example, regions with high simulated recharge rates may be considered to be highly sensitive to ground-water contamination. Data requirements are generally less stringent than for the previous category because Richard's Equation is not involved and chemical transport is often not addressed.

5.2.4.1 Ground-water modeling studies to evaluate sensitivity of a particular site should be developed in accordance with the procedures described in Guides D 5447 and D 5490. Ordinarily, these models are used to simulate primarily horizontal ground-water flow in two or three dimensions. These models have the advantage of also being applicable at large scales (regional analysis). A vulnerability analysis may be performed using a solute-transport model such as MOC (32,33) or MT3D (34) in conjunction with the guidance of Guide D 5880.

5.2.5 *Limitations*—Process-based simulation models are

powerful and useful tools, but their application can be problematic. Uncertainty in simulation results can arise from two major causes: model-related errors and data-related errors. Modeling errors can arise from improper conceptualization of the problem or inappropriate application of a model on the part of the modeler. Also of concern is failure of the selected model to accurately and completely represent system processes. This matter is often a question of scale; while some very detailed processes can be addressed at the scale of a laboratory column experiment, it would not be practical to incorporate that detail into a regional-scale model. An example of such a process is preferential water flow through soils, such as flow through root or worm holes, desiccation cracks, and joints. The importance of this process is widely recognized, but because of the large amount of detailed data required to understand it, it is not practical to deterministically account for it in large-scale models.

NOTE 2—In karst or fractured-rock aquifers, velocity, turbulence, boundary conditions, directions of flow, and contaminant transport cannot be adequately simulated using currently available code (50).

5.2.5.1 Data are needed in order to determine parameter values and to evaluate the accuracy of model results. A large constraint on model application is the availability of representative data. Representativeness refers to both the quality (all methods of data collection have some degree of error) and the quantity of data required to adequately represent the modeled region. Various approaches have been taken to study the effects of uncertainty in parameter values upon simulation results (35). One approach is to use Monte Carlo techniques (36) and a large number of model simulations to assess parameters. Carsel and others (36) used this approach to assess leaching potential by applying PRZM in conjunction with probability distributions of soil properties in a simple screening procedure.

5.3 *Statistical Methods*—Statistical methods provide estimates of the likelihood of contamination based on the relationship of soil, hydrogeologic, or cultural factors to known or calculated contaminant distributions. Statistical methods include discriminant analysis, regression analysis, and spatial estimation. These techniques are specific for hydrogeologic settings for which they were developed. Successful application of these methods to other sites has not been demonstrated.

5.3.1 *Discriminant Analysis*—Ground-water contamination by pesticides has been predicted using the Soil Conservation Service's Cooperative Soil Survey and a regional inventory of water-quality analyses from wells. The method has been applied to areas as large as a county and as small as 0.01 km² (37).

5.3.2 *Regression Analysis*—If adequate data are available, the frequency of occurrence of an individual contaminant in excess of a specified detection limit can be estimated using multiple-regression techniques. An example is a study of triazine-herbicide and nitrate concentrations in Nebraska (38,39). Independent variables describing soil, hydraulic, and well properties were used to predict the concentration of triazine herbicides and nitrates in wells. Similarly, nitrate concentrations were predicted by Steichen and others (40) who related pesticide concentrations to the age of the well, land use, and the distance to the nearest possible source of pesticides.

5.3.3 *Geostatistics*—Contaminants with erratic spatial variability can be analyzed through the application of spatial estimation, using least-squares estimators such as kriging **(41,42)**. If information about the variability of values at a sampled point is to be presented, geostatistical simulation methods may be used **(43)**. The results for a specific site should be presented in accordance with Guide D 5549. For example, public domain software to assist in spatial analysis is presented in Englund and Sparks **(44)**.

6. Procedure

6.1 The procedure for the selection of methods for determining sensitivity and vulnerability is based on determining the appropriate type of method for the intended use. This requires an understanding of the scale of the problem and intended map products, the type of geologic setting, soil characteristics and distribution, and aquifer geometry and hydrology. For vulnerability methods, mappable data on the contaminants of concern as well as land use is necessary. Individual methods vary widely in their specific data requirements.

6.2 *Determine the Purpose of the Assessment*—Determine whether the assessment will be used to (*1*) assist policy analysis, planning, development, and program management; (*2*) make informed land-use decisions; or (*3*) improve general education **(35)** as stated in 1.3.

6.3 *Determine the Area to Be Assessed*—Maps should always be prepared at a scale that is appropriate for the density of the data. Three general classifications of scales are appropriate: regional, local, and field (see Note 2). Regional studies should be those presented at scales at or smaller than 1:100 000 (such as on a state-base map). Local studies are those presented at larger scales, typically about 1:24 000, such as for county studies or those based on a USGS quadrangle. Field-scale studies are those presented at an appropriate scale for the field in question, such as 1:6000 or less. The scales of maps or other graphic products determine the potential uses of the maps.

6.3.1 The validity of regional sensitivity and vulnerability assessments are particularly influenced by the density of the data and provide limited information for evaluating potential contamination at a specific field. Therefore, data from regional or local studies should only be used at the field level to give an indication of what to expect. Similarly, a field-scale sensitivity or vulnerability assessment should not be extrapolated to a larger area unless the hydrogeologic setting is the same and the regional variability of the physical setting is similar to that measured at the field. If field-scale conditions are to be assessed, then field-scale data are required.

6.3.2 County soil survey reports, for example, are useful sources of information for both regional and local assessments. Hydraulic properties and organic matter classifications are given for each soil series and for specific soil horizons within each series. For smaller-scale regional assessments requiring soils information, the procedure outlined by Keefer **(45)** could be followed. In that study the State Soil Geographic Data Base (STATSGO) was used to evaluate water movement through surface soils **(46)**. In addition, the presence of soil joints can affect the ease with which contaminants can move through the soil zone **(47)**. Jointing is best evaluated on a local scale, however, once established,

the effects may be generalized to larger areas of similar hydgeologic setting.

6.4 *Determine the Availability and Quality of the Data Required to Assess the Area*—The methods that can be used to prepare sensitivity or vulnerability maps depend on the availability and quality of the resource data. For example, small-scale assessments using map overlays require less detailed information than simulation methods that require detailed information on hydraulic properties, geology, and soils. Table 1 shows the data required for the methods discussed in this guide. This table can be used to narrow the choice of methods; however, the documentation and examples of the methods should be reviewed in detail before a final selection is made.

6.4.1 Information from geologic reports and maps; field observations; water-well logs and samples; driller's records; engineering records, logs, and core samples; and test drilling data can be used to determine the stratigraphy, construct cross-sections, and identify the continuity of subsurface units, particularly aquifers and confining layers. A stack-unit map can be made based on the succession of geologic materials in their order of occurrence over specified areas and to a specified depth. It is important to show how earth materials are distributed both horizontally and vertically.

6.4.2 Delineate where aquifer materials (for example, unconsolidated sands and gravels; permeable sandstones and carbonates; and jointed or fractured rocks) and non-aquifer materials (diamictons, silts, shale, and other low-permeability rocks) lie in the vertical succession. Successions subsequently can be rated according to the proximity of aquifer materials to the surface and the thickness of confining layers. The closer the aquifer to the surface, and the thinner the confining layers, the greater the likelihood of it becoming contaminated.

6.4.3 Glacial terranes composed of porous rocks and similar hydrogeologic settings can be classified using the techniques of Berg and others **(5,6)**, Soller and Berg **(10)**, Berg **(48)**, and modified by Keefer **(45)** for specific land uses. For regional assessments, soils in glacial terrains may be classified according to the parent materials from which they formed and a soil-geologic or surficial geologic map can be constructed. This surficial geologic map would show the succession of materials to a depth of about 2 to 3 m.

TABLE 1 Summary of Data Requirements for Methods that Can Be Used for Assessing Sensitivity and Vulnerability

Methods	Data Required			
	Geology	Soils	Hydrology	Chemistry[D]
Hydrogeologic settings, no scoring or ranking	A[A]	M[B]	S[C]	M
Hydrogeologic settings, with scoring or ranking	A	M	M	M
Scoring, without Hydrogeologic settings	M	M	A	M
Root-zone models[E]	S	A	M	M
Unsaturated-zone models	M	A	A	A
Saturated-zone models	M	S	A	A

[A] A—Abundant, detailed data are required.
[B] M—Moderate amounts of less-detailed data are required.
[C] S—The assessment can be performed with sparse data.
[D] Information about the chemistry of contaminants is not required for sensitivity assessments, but is needed for vulnerability assessments.
[E] Root-zone models are not used for sensitivity assessments.

6.5 *Determine Whether to Do a Sensitivity or Vulnerability Assessment, or Both*—The decision of whether to do a sensitivity or vulnerability analysis depends on the purpose of the project and the availability of information. A sensitivity assessment will provide a general framework for considering any contaminants. A vulnerability assessment provides information relative to a specific contaminant or group of contaminants. Ground-water and aquifer sensitivity assessments, done regionally or on a site-specific basis, evaluate the contamination potential or potential for ground-water and aquifer degradation within various hydrogeologic settings and can aid in identifying areas where more detailed assessments are warranted. Regions or specific sites that have been determined to be sensitive and that have been or may be subjected to adverse land-use practices can be assessed additionally using a vulnerability model. Therefore, ground-water and aquifer vulnerability assessments usually require sensitivity assessments.

6.5.1 In order to conduct a regional vulnerability assessment, the nature and distribution of actual or potential contaminant sources, contaminant characteristics, loading information, and land-use practices, together with a measure of aquifer sensitivity, need to be considered (2). Detailed recharge information, piezometric surfaces, the ground-water flow regime, rates of contaminant loading, as well as chemical and biological reactions that may degrade a contaminant need to be considered for field-scale assessments.

6.6 *Select Appropriate Method*—Choose a method based on whether or not the contaminant is introduced at the land surface (such as agricultural chemicals, sewage sludge, septage, or accidental spills) or beneath the land surface (such as for pipeline breaks, landfills, leaking underground storage tanks, and septic tanks). If the purpose of the sensitivity assessment is to evaluate potential contamination from surface point or non-point sources, information such as the organic matter content and hydraulic conductivity of the soil, and information on other soil and vadose zone factors must be available at an appropriate scale and level of detail (45). Sensitivity and vulnerability assessments focusing on potential contamination from subsurface sources usually do not require information on soils. Likewise, if a contaminant is introduced below the water table, it is not generally necessary to select a method that incorporates information on the vadose zone.

6.6.1 Tables 2 and 3 summarize the appropriateness of the methods that can be used for sensitivity and vulnerability analyses at various scales. Statistical methods were consid-

TABLE 2 Methods of Conducting Sensitivity Assessments Depending on the Scale of the Assessment

Sensitivity Methods	Regional	Local	Field
Hydrogeologic settings, no scoring or ranking	4[A]	3	2
Hydrogeologic settings, with scoring or ranking	5	5	4
Scoring, without Hydrogeologic settings	1	3	4
Root-zone models	N[B]	N	N
Unsaturated-zone models	N	N	4
Saturated-zone models	3	4	4

[A] 1–5 indicates relative ranking of appropriateness, where 5 is most appropriate for the scale indicated.
[B] N—not appropriate.

TABLE 3 Methods of Conducting Vulnerability Assessments Depending on the Scale of the Assessment

Vulnerability Methods	Regional	Local	Field
Hydrogeologic settings, no scoring or ranking	3[A]	2	1
Hydrogeologic settings, with scoring or ranking	4	5	4
Scoring, without Hydrogeologic settings	1	3	4
Root-zone models	N[B]	N	3
Unsaturated-zone models	N	N	4
Saturated-zone models	1	3	3

[A] 1–5 indicates relative ranking of appropriateness, where 5 is most appropriate for the scale indicated.
[B] N—not appropriate.

ered too diverse and specialized to tabulate. These tables should only be considered as a general guide: other considerations such as purpose or data availability should also influence the selection. References and examples of the various methods should be reviewed in detail before a final selection is made.

6.7 *Determine Whether Special Conditions Exist*—Special conditions may exist which preclude the use of some or all aquifer sensitivity or vulnerability methods, or require their modification. Such conditions include settings where water may move from the surface to the aquifer with little interaction with the soil, sediments, or rocks, such as where karst or fractured rocks are present; and settings where ground-water flow is modified by interaction with streams or lakes. Ground water may be affected by leaky abandoned or improperly constructed wells. These and other special conditions require that the hydrology and potential flow paths of contaminants be understood in much greater detail for an assessment to be done.

6.7.1 Karst, volcanic, and fractured rock settings are generally very susceptible to potential contamination. Basalt and other extrusive volcanic rocks are characterized by fractures, interflow breccias, and lava tubes. Karst systems developed in soluble rocks contain large channels that provide little opportunity for interaction between the contaminated water and soil or the surrounding rock. Ground-water flow is often rapid in these systems and attenuation of contaminants insignificant. Also, in many cases, contaminants do not follow the apparent regional flow paths and discharge at unexpected locations and times. Whether a technique can be applied to a karst or fractured rock setting depends on whether, at the scale of interest, that setting can be approximated as an equivalent porous medium. Guidance in making this determination is provided in Guide D 5717, in Quinlan and Ewers (49), and in Quinlan and others (50). Modifications to scoring methods applied to a karst setting are discussed in Davis and others (51).

6.7.2 Aquifers exchange water with surface-water sources in many hydrogeologic settings. Streams flowing through alluvial or glacio-fluvial valley-fill deposits often have a significant hydraulic connection with those deposits. Such streams may receive poor-quality water from surface sources or from discharge from contaminated aquifers. This stream water may then recharge the aquifer elsewhere along the stream, either under natural ground-water gradients or gradients caused by pumping. In these hydrogeologic settings, the contact between the stream and the aquifer is often a chemically active zone. Because of this, the zone can affect the quality of water moving into the aquifer, either increasing or decreasing the level of contamination. The actual sensitivity of the aquifer may therefore be more or less than that determined using the hydrogeologic methods in this guide. The user of this guide thus should consider this interaction when evaluating the results of any method applied to hydrogeologic settings containing streams.

6.7.3 Abandoned wells that have not been properly plugged, and improperly constructed withdrawal or injection wells can provide pathways for the rapid movement of contaminants between the surface and an aquifer or between aquifers. Some wells have been constructed for the drainage of surface runoff, or for disposal of water from field tile drains. When these and similar wells are present, their role in the movement of contaminants must be understood to properly perform an assessment.

7. Keywords

7.1 aquifers; contamination; ground water flow; pollution

REFERENCES

(1) Aller, L. T., Bennet, T., Lehr, J. H., and Petty, R. J., *DRASTIC: A Standardized System for Evaluating Ground Water Pollution Potential Using Hydrogeologic Settings*, U.S. EPA Robert S. Kerr Environmental Research Laboratory, EPA/600/287/035, Ada, OK, 1987.

(2) U.S. Environmental Protection Agency (USEPA), *Ground Water Resource Assessment*, U.S. EPA Office of Ground Water and Drinking Water, EPA 813-R-93-003, 1993.

(3) Hearne, G. A., Wireman, M., Campbell, A., Turner, S., and Ingersoll, G. P., "Vulnerability of the Uppermost Ground Water to Contamination in the Greater Denver Area, Colorado," *U.S. Geological Survey Water Resources Investigations Report* 92-4143, 1995.

(4) Wireman, M., Campbell, A., and Marr, P., *Land Uses Which Affect Ground-Water Management*, USEPA and Colorado Department of Health, 1994.

(5) Berg, R. C., Kempton, J. P., and Cartwright, K., "Potential for Contamination of Shallow Aquifers in Illinois," *Illinois State Geological Survey Circular 532*, 1984a.

(6) Berg, R. C., Kempton, J. P., and Stecyk, A. N., "Geology for Planning in Boone and Winnebago Counties," *Illinois State Geological Survey Circular 531*, 1984b.

(7) Keefer, D. A., and Berg, R. C., with contributions by Day, W. S., *Potential for Aquifer Recharge in Illinois (Appropriate Recharge Areas)*, Illinois State Geol. Survey, 1 map sheet, 1990.

(8) Pettyjohn, W. A., Savoca, M., and Self, D., *Regional Assessment of Aquifer Vulnerability and Sensitivity in the Conterminous United States*, U.S. EPA Robert S. Kerr Environmental Research Laboratory, EPA/600/2-91/043, 1991.

(9) Lusch, P. P., Rader, C. P., Barrett, L. R., and Barrett, K., *Aquifer Vulnerability to Surface Contamination in Michigan*, Michigan State University, Center for Remote Sensing and Department of Geography, East Lansing, MI, 1:1,500 000-scale map, 1992.

(10) Soller, D. R., and Berg, R. C., "A Model for the Assessment of Aquifer Contamination Potential Based on Regional Geologic Framework," *Environmental Geology and the Water Sciences*, Vol 19, 1992, pp. 205–213.

(11) Shafer, J. M., "An Assessment of Ground Water Quality and Hazardous Substances for a Statewide Monitoring Strategy," *Illinois State Water Survey Contract Report 367*, 1985.

(12) Bhagwat, S. B., and Berg, R. C., "Benefits and Costs of Geologic Mapping Programs in Illinois," *Case study of Boone and Winnebago Counties and its statewide applicability: Illinois State Geological Survey Circular 549*, 1991.

(13) Moore, J. S., "SEEPPAGE: A System for Early Evaluation of Pollution Potential of Agricultural Ground Water Environments," *Geology Technical Note 5 (Revision 1)*, U.S. Department of Agriculture, Soil Conservation Service, 1988.

(14) Soller, D. R., "Applying the DRASTIC Model—A Review of County-Scale Maps," *U.S. Geological Survey Open-File Report*, 92–297, 1992.

(15) Lemme, G., Carlson, C. G., Dean, R., and Khakural, B., "Contamination Vulnerability Indexes: A Water-Quality Planning Tool," *Journal of Soil and Water Conservation 2*, 1990, pp. 349–351.

(16) Cates, K. J., and Madison, F. W., *Soil-Attenuation Potential Map of Pepin County, Wisconsin*, University of Wisconsin, Extension Soil Map 10, 1990.

(17) van der Heijde, P. K. M., *Identification and Compilation of Unsaturated/Vadose Zone Models*, Environmental Protection Agency EPA/600/R-94/028, 1994.

(18) van der Heijde, P. K. M., *Compilation of Saturated and Unsaturated Zone Modeling Software*, Update of EPA/600-/R-93/118, EPA/600/R-96/009, 1996.

(19) van der Heijde, P. K. M., and Elnawawy, O. A., *Compilation of Ground-Water Models*, EPA/600/R-93/118, 1993.

(20) Carsel, R. F., Smith, C. N., Mulkey, L. A., Dean, J. D., and Jowise, P., *User's Manual for the Pesticide Root-Zone Model (PRZM)*, Athens, Georgia: U.S. Environmental Protection Agency, Environmental Research Laboratory, 1984.

(21) Leonard, R. A., Knisel, W. G., and Still, D. A., "GLEAMS: Groundwater Loading Effects of Agricultural Management Systems," *Transactions of the ASCE 30*, 1987, pp. 1403–1418.

(22) Nofziger, D. L., and Hornsby, A. G., *Chemical Movement in Soil, User's Guide*, University of Florida, Gainsville, 1985.

(23) A. N., Sharpley, and J. R., Williams, eds., "Epic-Erosion/Productivity Impact Calculator: 1. Model Documentation," *USDA Technical Bulletin 1768*, 1990.

(24) Williams, J. R., "The Erosion-Productivity Impact Calculator (EPIC) Model: A Case History," *Phil. Trans. Royal Soc. London B. 329*, 1990, pp. 421–428.

(25) Wagenet, R. J., and Hutson, J. L, *LEACHM: A Finite-Difference Model for Simulating Water, Salt, and Pesticide Movement in the Plant Root Zone, Continuum 2*, New York State Resources Institute, Cornell University, 1987.

(26) Healy, R. W., "Simulation of Solute Transport in Variably Saturated Porous Media with Supplemental Information on Modifications to the U.S. Geological Survey's Computer Program VS2D," *U.S. Geological Survey Water Resources Investigation Report 90-4025*, 1990.

(27) RZWQM Team, "Root Zone Water Quality Model, Version 1.0, User's Manual," *GPSR Technical Report No. 3*, USDA-ARS-GPSR, Fort Collins, CO, 1992.

(28) Ma, Q. L., Ahjua, L. R., Rojas, K. W., Ferreira, V. F., and DeCoursey, D. F., "Measures and RZWQM Predicted Atrazine Dissipation and Movement in a Field Soil," *Transactions of the American Society of Agricultural Engineers 38*, 1995, pp. 471–479.

(29) Simunek, J., Huang, K., and van Genuchten, M. Th., "The SWMS-3D Code for Simulating Water Flow and Solute Transport in Three-Dimensional Variably Saturated Media," *U.S. Salinity Laboratory Research Report 139*, Agricultural Research Service, Riverside, CA, 1995.

(30) Kramer, J. H., and Cullen, S. J., "Review of Vadose Zone Flow and Transport Models," *Handbook of Vadose Zone Characterization and Monitoring*, L. G. Wilson, L. G. Everett, and S. J. Cullen, eds., Lewis Publishers, 1995, pp. 267–289.

(31) McDonald, J. M., and Harbaugh, A. W., "A Modular Three-Dimensional Finite-Difference Ground-Water Flow Model," *Tech-*

niques of *Water Resources Investigations of the U.S. Geological Survey, Book 6*, 1988.

(32) Konikow, L. F., Granato, G. E., and Hornberger, G. Z., "User's Guide to Revised Method-of-Characteristics Solute-Transport Model," *U.S. Geological Survey Water Resources Investigations Report, 94-4115*, 1994.

(33) Konikow, L. F., and Bredehoeft, J. D., "Computer Model of Two-Dimensional Solute Transport and Dispersion in Ground Water," *Techniques of Water Resources Investigations of the U.S. Geological Survey, Book 7, C2*, 1978.

(34) Zheng, C., *MT3D, A Modular Three-Dimensional Transport Model*, S. S. Papadopulos and Assoc., Bethesda, MD, 1992.

(35) National Research Council, *Ground Water Vulnerability Assessment, Predicting Relative Contamination Potential under Conditions of Uncertainty*, National Academy Press, Washington DC, 1993.

(36) Carsel, R. F., Parrish, R. S., Jones, R. L., Hansen, J. L., and Lamb, R. L., "Characterizing the Uncertainty of Pesticide Leaching in Agricultural Soils," *Journal of Contaminant Hydrology 2*, 1988, pp. 111–124.

(37) Teso, R. R., Younglove, T., Peterson, M. R., Sheeks, D. L., III, and Gallavan, R. E., "Soil Taxonomy and Surveys: Classification of Areal Sensitivity to Pesticide Contamination of Groundwater," *Journal of Soil and Water Conservation 43*, 1988, pp. 348–352.

(38) Chen, H., and Druliner, A. D., "Agricultural Chemical Contamination of Ground Water in Six Areas of the High Plains Aquifer, Nebraska," *National Water Summary 1986—Hydrologic Events and Ground-Water Quality: U.S. Geological Survey Water-Supply Paper 2325*, 1988.

(39) Druliner, A. D., "Overview of the Relations of Nonpoint Source Agricultural Chemical Contamination to Local Hydrologic, Soil, Land-Use, and Hydrochemical Characteristics of the High Plains Aquifer of Nebraska," *U.S. Geological Survey Open-File Report 88-4220*, 1989.

(40) Steichen, J., Koelliker, J., Grosh, D., Heiman, A., Yearout, R., and Robbins, V., "Contamination of Farmstead Wells by Pesticides, Volatile Organics, and Inorganic Chemicals in Kansas," *Ground-Water Monitoring Review 8*, 1988, pp. 153–160.

(41) Delhomme, J. P., "Kriging in the Hydrosciences," *Advances in Water Resources 1*, 1978, pp. 475–499.

(42) Hoeksema, R. J., and Kitanidis, P. K., "Analysis of the Spatial Structure of Properties of Selected Aquifers," *Water Resources Research 21*, 1985, pp. 536–572.

(43) McBratney, A. B., Webster, R., and Burgess, T. M., "The Design of Optimal Sampling Schemes for Local Estimation and Mapping of Regionlized Variables—I & II," *Computer and Geosciences 7*, 1981, pp. 331–365.

(44) Englund, E. J., and Sparks, A. R., *Geo-EAS (Geostatistical Environmental Assessment Software) User's Guide*, Environmental Monitoring Systems Lab, Las Vegas, NV EPA/600/4-88/033, 1988.

(45) Keefer, D. A., *Potential for Agricultural Chemical Contamination of Aquifers in Illinois: 1995 Revision*, Illinois State Geological Survey, Environmental Geology 148, 1995.

(46) U.S. Department of Agriculture (USDA), *State Soil Geographic Data Base (STATSGO)*, U.S. Department of Agriculture, Soil Conservation Service, Miscellaneous Publication 1492, 1991.

(47) Kirkaldie, L., "Potential Contaminant Movement Through Soil Joints," *Bull. Assoc. Eng. Geol. 25*, 1988, pp. 520–524.

(48) Berg, R. C., "Geologic Aspects of a Groundwater Protection Needs Assessment for Woodstock, Illinois: A Case Study," *Illinois State Geological Survey Environmental Geology 146*, 1994.

(49) Quinlan, J. F., *Special Problems of Ground-Water Monitoring in Karst Terranes, Ground Water and Vadose Zone Monitoring, ASTM STP 1053*, D. M. Nielsen and A. I. Johnson, eds., ASTM, 1990, pp. 275–304.

(50) Quinlan, J. F., Davies, G. J., Jones, S. W., and Huntoon, P. W., *The Applicability of Numerical Models to Characterize Ground-Water Flow in Karstic and Other Triple-Porosity Aquifers, Substance Fluid-Flow (Ground-Water) Modeling, ASTM STP 1288*, J. D. Ritchey and J. O. Rumbough eds., ASTM, 1996.

(51) Davis, A. D., Long, A. J., Nazir, M., and Tan, X., "Ground Water Vulnerability in the Rapid Creek Basin above Rapid City, South Dakota," *South Dakota School of Mines and Technology Final Technical Report: U.S. EPA Contract X008788-01-0*, 1994.

Designation: D 5980 – 96

Standard Guide for
Selection and Documentation of Existing Wells for Use in
Environmental Site Characterization and Monitoring[1]

This standard is issued under the fixed designation D 5980; the number immediately following the designation indicates the year of original adoption or, in the case of revision, the year of last revision. A number in parentheses indicates the year of last reapproval. A superscript epsilon (ϵ) indicates an editorial change since the last revision or reapproval.

1. Scope

1.1 This guide covers the use of existing wells for environmental site characterization and monitoring. It covers the following major topics: criteria for determining the suitability of existing wells for hydrogeologic characterization and ground-water quality monitoring, types of data required to document the suitability of an existing well, and the relative advantages and disadvantages of existing large- and small-capacity wells.

1.2 This guide should be used in conjunction with Guide D 5730, that provides a general approach for environmental site investigations.

1.3 This guide does not specifically address design and construction of new monitoring or supply wells. Refer to Practices D 5092 and D 5787.

1.4 This guide does not specifically address ground-water sampling procedures. Refer to Guide D 5903.

1.5 The values stated in SI units are to be regarded as the standard. However, dimensions of materials used in the water well industry are given in inch-pound (English) units by convention, therefore, inch-pound units are used where necessary in this guide.

1.6 *This standard does not purport to address all of the safety concerns, if any, associated with its use. It is the responsibility of the user of this standard to establish appropriate safety and health practices and determine the applicability of regulatory limitations prior to use.*

2. Referenced Documents

2.1 Pertinent ASTM guides addressing specific information necessary to utilize existing wells for hydrologic and water-quality data for environmental site characterization. A comprehensive list of guides, standards, methods, practices, and terminology is contained in Guide D 5730. Other guidance documents covering procedures for environmental site investigations with specific objectives or in particular geographic settings may be available from federal, state, and other agencies or organizations. The appropriate agency or organization should be contacted to determine the availability and most current edition of such documents.

2.1.1 *ASTM Standards:*
D 653 Terminology Relating to Soil, Rock, and Contained Fluids[2]

D 4750 Test Method for Determining Subsurface Liquid Levels in a Borehole or Monitoring Well (Observation Well)[2]
D 5092 Practice for Design and Installation of Ground Water Monitoring Wells in Aquifers[3]
D 5254 Practice for a Minimum Set of Data Elements to Identify a Ground-Water Site[3]
D 5408 Guide for Set of Data Elements to Describe a Ground-Water Site; Part One—Additional Identification Descriptors[3]
D 5409 Guide for Set of Data Elements to Describe a Ground-Water Site; Part Two—Physical Descriptors[3]
D 5410 Guide for Set of Data Elements to Describe a Ground Water Site; Part Three—Usage Descriptors[3]
D 5474 Guide for Selection of Data Elements for Ground-Water Investigations[3]
D 5521 Guide for Development of Ground-Water Monitoring Wells in Granular Aquifers[3]
D 5730 Guide to Site Characterization for Environmental Purposes with Emphasis on Soil, Rock, the Vadose Zone, and Ground Water[2]
D 5753 Guide for Planning and Conducting Borehole Geophysical Logging[2]
D 5787 Practice for Monitoring Well Protection[2]
D 5903 Guide for Planning and Preparing for a Ground Water Sampling Event[2]
D 5978 Guide for Maintenance and Rehabilitation of Ground Water Monitoring Wells[3]
D 5979 Guide for Conceptualization and Characterization of Ground-Water Systems[3]

3. Terminology

3.1 *Definitions*—Except as noted below, all definitions are in accordance with Terminology D 653:

3.1.1 *aquifer, n*—a geologic formation, group of formations, or part of a formation that is saturated and is capable of providing a significant quantity of water (see Practice D 5092).

3.1.2 *monitoring well (observation well), n*—a special well drilled in a selected location for observing parameters such as liquid level or pressure changes or for collecting liquid samples. The well may be cased or uncased, but if cased the casing should have openings to allow flow of borehole liquid into or out of the casing.

3.1.3 *observation well, n*—for the purposes of this guide, an existing well constructed for other purposes that is also used to measure water levels and to collect ground-water

[1] This guide is under the jurisdiction of ASTM Committee D-18 on Soil and Rock and is the direct responsibility of Subcommittee D18.21 on Ground Water and Vadose Zone Investigations.
Current edition approved July 10, 1996. Published November 1996.
[2] *Annual Book of ASTM Standards*, Vol 04.08.
[3] *Annual Book of ASTM Standards*, Vol 04.09.

quality samples. Observation well may be referred to as "well" in this guide.

3.1.4 *supply (production) well, n*—well primarily installed for public supply, irrigation, and industrial use. Supply wells may be used as an observation well.

4. Significance and Use

4.1 This guide describes a general approach for the use of existing wells in environmental investigations with a primary focus on the subsurface and major factors affecting the surface and subsurface environment.

4.2 Existing wells represent a valuable source of information for subsurface environmental investigations. Specific uses of existing wells include:

4.2.1 Well driller logs provide information on subsurface lithology and major water-bearing units in an area. Existing wells can also offer access for downhole geophysical logging for stratigraphic and aquifer interpretations. Examples include natural gamma logs in cased wells and an entire suite of methods in uncased bedrock wells (see Guide D 5753). This information can assist in developing the preliminary conceptual model of the site.

4.2.2 Well tests using existing wells may provide information on the hydrologic characteristics of an aquifer.

4.2.3 Monitoring of water levels in existing wells, provided that they are cased in the aquifer of interest, allow development of potentiometric maps and interpretations of ground-water flow directions.

4.2.4 Existing wells are the primary means by which regional drinking water quality is evaluated and monitored.

4.2.5 Existing wells may assist in the mapping of contaminant plumes, and in ongoing monitoring of ground-water quality changes at the site-specific level.

4.3 Data from existing wells should only be used when characteristics of the well have been sufficiently documented to determine that they satisfy criteria for the purpose for which the data are to be used.

5. General Considerations in Selection and Use of Existing Wells

5.1 Selection and use of existing wells should take place in the context of a conceptual framework consisting of a description of the system, including, as necessary, physical and cultural characteristics such as climate, hydrology, ecology, physiography, population, water use and land use, and hypotheses about processes of interest that occur within that system. A step-wise approach for conceptualization and characterization is a direct approach to develop the framework for Hydrologic Systems as described in Kolm (1), (see Guide D 5979). Conceptualization of hydrologic and regional ground-water quality systems can be formulated using the methods outlined in Alley (2). The framework is reviewed and refined by an iterative process of data collection and analysis, testing hypotheses with data collected, and identifying data needs to further revise the framework. Refinement must be made within the limits established by the accuracy, precision, and completeness of the data. Methods for data collection are selected that will provide data appropriate for testing hypotheses which evaluate the conceptual framework.

5.2 Well design and installation can critically affect the quality of water level measurements and ground-water

samples. Such effects apply both to existing wells and to wells specifically installed for a purpose. The effects of well design and installation, therefore, need to be considered regardless of whether existing wells are selected or if wells are specifically installed for a specific purpose. The most common feature of an existing well that may render it unsuitable for water level measurement or water-quality monitoring is that multiple hydrogeologic units are connected causing water levels and water-quality parameters to reflect a mixing of multiple hydrogeologic units. Such data cannot be reliably compared with data from wells completed in the individual hydrogeologic units.

5.3 Major steps in the selection of existing wells for environmental investigations include: developing specific criteria for evaluating the suitability of existing wells in relation to the objectives of the investigations (see Section 6), conducting an inventory of existing wells in the area of interest (see 8.1), documenting the characteristics of the wells identified in the inventory that are relevant to the selection criteria (see 8.2), and identification of wells that satisfy the selection criteria (see Section 9).

6. Well-Selection Criteria

6.1 Assessing the suitability of existing wells for hydrological and ground-water quality studies requires development of specific well-selection criteria. The criteria are based on considerations of project objectives by defining the problem to be solved, the conceptual framework, and data-collection requirements.

6.2 *Specific Well-Selection Criteria*—Specific criteria will depend on the objectives of the investigation. The following general criteria will apply to most situations:

6.2.1 The well is suitably located for use in relation to the conceptual framework.

6.2.2 The well must be completed in the targeted hydrogeologic unit or units.

6.2.3 Well design and construction must not bias water level measurements or water-quality sampling results (see Note 1). Section 7 provides information on the general characteristics of major types of existing wells.

NOTE 1—Gillham et al. (3), provides information on the suitability of materials coming in contact with water samples and that table provides information on the compatibility of well casing materials with different organic contaminants.

6.2.4 The well is accessible for measurements and sampling.

6.2.5 The well's maintenance condition may not compromise it as a sampling point; however, there are examples that may compromise it as a sampling point, that is, a cracked casing allowing non-screened water into the well.

6.3 *Examples of Well Selection Criteria*—The following are illustrative examples of criteria for specific investigation objectives (see Note 2).

NOTE 2—These are illustrative examples and should not be construed as recommended criteria.

6.3.1 A project to determine the quality of potable ground water might require the following selection criteria: wells selected must be used for public water supply, must be geographically distributed over the entire aquifer of interest, and must be able to be sampled prior to any water treatment.

6.3.2 All wells or a subset of wells down-gradient from a hazardous-waste site would be unsuitable to include in a network designed for a study to determine non-point source ground-water quality.

6.3.3 Choosing a well located down-gradient of a complex mix of land uses would be inappropriate in a study designed to assess the effects of specific land uses on ground-water quality.

6.3.4 A supply well screened over a long interval would not be appropriate for investigating small-scale vertical variations in water quality down-gradient of a landfill, or for potentiometric mapping.

6.3.5 A well constructed of PVC (polyvinylchloride) with glued joints would not be suitable for sampling if the volatile-organic compounds of interest in the ground water also are found in the glue used to join the sections of well casing. Similarly, a well constructed of steel may not be suitable for the sampling of metals.

6.3.6 Selecting an observation well in an area undergoing rapid development would be avoided in constructing a network of wells for evaluating long-term trends in ground-water quality because of the possibility of the well being destroyed by later development.

7. General Characteristics of Major Types of Existing Wells

7.1 There are two general categories of existing wells available for hydrologic and ground-water quality studies: large- and small-capacity supply or production wells installed for drinking, irrigation, and industrial use (see 7.2 and 7.3); and wells specially designed and installed to monitor hydrologic or water-quality studies, or both, (see 7.4). Each type of well has its own general advantages and disadvantages.

7.2 *Large-Capacity Supply Wells*—Large capacity supply wells are usually developed for drinking water systems that supply multiple households, and for irrigation and industrial purposes.

7.2.1 *Advantages:*

7.2.1.1 Documentation of well construction commonly is good.

7.2.1.2 Large-capacity wells generally are well developed and fully purged.

7.2.1.3 Long-term access may be possible, particularly for municipal wells.

7.2.1.4 Large-capacity wells generally provide a larger vertical mix of water in an aquifer or aquifer system than small-capacity wells, and thus can provide a more integrated measure of regional ground-water quality than small-capacity wells.

7.2.1.5 Much of the water produced for irrigation and municipal water is from large-capacity wells equipped with taps which allow a direct sample of the pumped water.

7.2.1.6 Long-term water-quality and quantity data may be available.

7.2.2 *Disadvantages:*

7.2.2.1 Large-capacity wells may not have flow-rate controls and a sampling point near the well head.

7.2.2.2 High pumping rates may entrain artifacts, such as colloids or suspended material, into the sample stream.

7.2.2.3 Pumping schedules could be irregular: for example, irrigation wells generally are pumped seasonally, and could lead to seasonal variations in water quality that actually are an artifact of the pumping regime.

7.2.2.4 Large capacity wells may have a long vertical gravel pack, screened or open intervals might span more than one aquifer or aquifer system, making them unsuitable for potentiometric mapping or water quality monitoring. For example, dilution of contaminant concentrations wells with long screen intervals may result in large errors if concentrations are used for detailed delineation of the geometry and concentrations of contaminant plumes or for detection of contaminants in low concentrations (Pohlmann and Alduino (4)).

7.2.2.5 Wells with high pumping rates may draw water from water-bearing units other than those screened even if the well is screened solely within one unit, thus, the vertical integration of water from water-bearing units might be unknown.

7.2.2.6 Local hydraulics may be atypical of regional ground-water movement as a result of compaction or enhanced downward flow.

7.2.2.7 Municipal wells that produce water not meeting water-quality standards are usually abandoned, implying that the remaining population of municipal wells is biased toward acceptable water quality.

7.2.2.8 Down-hole chlorination or other chemical treatment might affect water chemistry, so that samples do not reflect ambient ground-water composition.

7.2.2.9 Depth-dependent differences in water quality could be lost, as water sampled could reflect a mixture of water obtained at different depths.

7.3 *Small-Capacity Supply Wells*—Small-capacity supply wells are usually developed for domestic water use involving a single household.

7.3.1 *Advantages:*

7.3.1.1 Domestic wells are a major source of drinking-water supply for rural population, so wells reflect this resource use.

7.3.1.2 Good to excellent areal and depth coverage in some areas, particularly for water-table aquifers.

7.3.1.3 Small-capacity pumping rates limit withdrawal of water from water-bearing formations other than those screened.

7.3.1.4 The low pumping rates of small-capacity wells are less likely to entrain artifacts, such as colloids or suspended material, into the sample stream than the high pumping rates of large-capacity wells.

7.3.2 *Disadvantages:*

7.3.2.1 Domestic wells may not be available in urban and suburban areas.

7.3.2.2 Documentation of well-construction characteristics may not be available.

7.3.2.3 Well construction, pressure tanks and treatment, and/or pumps may preclude being able to collect a sample at the well head.

7.3.2.4 The relation between well locations, septic systems, and other potential processes that could affect ground-water chemistry must be established in order to correctly assess what conditions water-quality data truly reflect.

7.3.2.5 The open interval may provide connections for more than one water-bearing unit, making a well unsuitable for potentiometric mapping or water quality monitoring.

7.4 *Existing Monitoring Wells*—Existing monitoring wells may be available that have been installed for purposes other than the current investigations.

7.4.1 *Advantages:*

7.4.1.1 Well construction details are usually well documented.

7.4.1.2 Well construction usually avoids interconnection of different water-bearing units.

7.4.1.3 Well construction is usually designed to optimize quality of ground-water samples.

7.4.2 *Disadvantages*—Well location, screen interval, or well construction details may not be suitable for the purpose of the current environmental investigation.

8. Well Inventory and Documentation

8.1 *Well Inventory*—Selection of wells begins with an inventory of existing production wells or previously installed monitoring wells in the locale of interest (see Guide D 5521). In order to collect information related to well selection criteria (see Section 6 and 8.2).

8.1.1 Well records of municipal, irrigation, and industrial wells can be obtained from the appropriate state agency, local ground-water management district, natural resource district, and local offices of the U.S. Geological Survey, Water Resources District. Ganley (5) identifies the primary sources of domestic well records and identifies other locations where records may be available as of 1989.

8.1.2 If the locale of interest is not too large, a property-owner survey may identify additional wells that have not been recorded elsewhere. Interviews with well owners may also provide additional information that is missing from state or local agency records.

8.2 *Well Documentation*—Well documentation involves collection of all available data that are relevant to the well selection criteria. Practice D 5254 identifies the minimum set of data elements to identify a ground water site and Guides D 5408, D 5409 and D 5410 identify additional data elements. Table 1 and Fig. 1 identifies types of information available from state agency well record forms as of 1989 (5). Figure 3 provides a checklist for documenting well information. Well documentation should include, but not necessarily be limited to:

8.2.1 Well identification number, type, location, elevation, and depth.

8.2.2 Lithologic log describing character and depths of different materials encountered during well drilling.

8.2.3 Hydrogeologic unit or units that supply water to the well. If accurate well logs are not available, additional investigations, such as borehole geophysical logging may be required. The major hydrogeologic units in an area must be well established in order to correlate water-bearing units in an existing well to hydrogeologic units in the area.

8.2.4 Well construction details, such as casing type, depth, screened interval, filter pack, grouts, and seals. Refer to Practice D 5092 for additional information on important elements of well design and construction. Other major sources of information on this topic include: Aller et al. (6), Driscoll (7), Harlan et al. (8), Lehr et al. (9), Nielsen and Schalla (10), and McCaulou et al. (11).

8.2.5 Well water levels. Measurements should reflect water levels without the influence of pumping if unstressed potentiometric mapping is desired.

8.2.6 Well yields, and any other well test results.

8.2.7 Land use in the vicinity of the well (see Fig. 2) (13).

8.3 If existing information sources are not sufficient to provide the information required to adequately document a well, it should be removed from further consideration or subject to further investigations. Borehole geophysical methods, as described in Guide D 5753, may be useful for obtaining additional data on lithology and well construction details.

9. Well Selection

9.1 Application of the well selection criteria to wells identified in the well inventory will usually result in elimination of wells that do not meet the criteria. The set or subset of wells selected may be modified as a result of site visits to evaluate and obtain permission to use wells. If existing wells fail the selection process then installing wells to meet the well selection criteria is necessary to solve the problem (12).

9.2 Selection of a supply well or existing monitoring well for water-level measurements and/or sampling should be based on the ability to document that the following are true:

9.2.1 The hydrogeologic unit(s) have been identified with reasonable certainty for those intervals: in which the water level is being measured or from which samples are collected.

9.2.2 Possible biases caused by use and location of the well that may compromise meeting project objectives have been considered during selection.

9.2.3 The design of the well, the materials from which the well is constructed, and methods of well installation are not likely to affect the water-level measurement or the water-quality constituents of interest.

9.2.4 Well-construction integrity has been verified.

9.2.5 Possible biases caused by the pumping rate have been considered. For example, wells might be selected that have a pumping rate sufficiently low to avoid entrainment into the sample stream of sampling artifacts, such as colloids or suspended materials, or to ensure that water will not be drawn from units other than those of interest.

9.2.6 Materials used in the construction of pumps may effect the water-quality constituents of interest and should be considered when planning the collection of water samples.

9.2.7 The type of pump may also effect the validity of collecting water samples. Air-lift pumps most likely may effect the collection of many water constituents by introducing air to the samples. Wells with oil lubricated pumps should not be used to collect constituents effected by hydrocarbons.

9.2.8 Possible biases caused by the sampling-point location that may compromise meeting project objectives have been considered.

10. Report

10.1 An environmental investigation report that includes data from existing wells should include a section that identifies the well selection criteria and includes the documentation necessary to show that the existing wells from which data were used satisfied the criteria.

11. Keywords

11.1 environmental site characterization; field investigations; ground water; hydrologic investigations; monitoring wells; sampling; site characterization; site investigations; subsurface investigations

TABLE 1 Technical Information Items and Tabulations for Agency Well Record Forms

	STATES								Florida[b]																		
	AL	AK	AZ	AR	CA	CO	CT	DE	A	B	C	D	E	HI	ID	IL	IN	KS	KY	LA[c]	ME	MD	MA	MI	MN	MS	MO
GENERAL																											
1 Owner's Name	•	•	•	•	•	•	•	•	•	•	•	•	•	•	•	•	•	•	•	•	•	•	•	•	•	•	•
2 Driller's Name	•	•	•	•	•	•	•	•	•	•	•	•	•	•	•	•	•	•	•	•	•	•	•	•	•	•	•
3 Registration/License Number	•	•	•		•	•	•	•	•	•	•	•	•	•			•	•	•			•	•	•	•	•	•
4 Permit Number					•	•	•	•	•	•	•	•	•													•	•
5 Construction Date	•	•	•	•	•		•	•	•	•		•	•	•		•		•	•	•	•	•	•	•	•		•
6 Well Log Confidentiality					•																						
7 Well Use	•	•	•		•	•	•	•	•	•	•	•		•		•		•	•	•				•	•		•
8 Nearest Septic, Sewer or Possible Contamination Source								•											•				•		•	•	
9 Replacement or Abandoned Wells						•		•	•		•	•		•				•		•	•	•		•	•		•
10 Comments/Remarks		•	•	•				•	•			•		•		•			•	•	•	•		•	•	•	
CONSTRUCTION DETAILS																											
11 Well Service					•		•		•	•				•			•		•	•					•		
12 Drilling Method	•	•	•			•		•	•	•			•	•	•		•		•	•		•		•	•	•	
13 Drilling Fluid								•		•										•							
14 Open Hole	•										•	•	•														
15 Total Well Depth	•		•	•		•	•	•	•	•		•		•			•		•	•		•		•			
16 Borehole Diameter	•				•	•	•	•						•					•	•					•		•
17 Grout	•		•	•	•		•	•	•	•		•		•		•		•	•	•		•		•	•		•
18 Grout Material		•			•				•			•		•		•		•	•	•		•		•	•		•
19 Grout Interval	•				•	•	•	•		•		•		•		•		•	•	•		•		•	•		•
20 Grout Method					•	•	•							•								•			•		
21 Packers	•														•												
22 Gravel Pack	•				•	•	•	•						•		•		•	•	•		•	•	•		•	•
23 Gravel Pack Material								•											•			•					
24 Gravel Pack Size								•																			
25 Gravel Pack Interval					•	•	•	•																			•
26 Drive Shoe								•						•	•									•	•		•
27 Casing Gauge		•			•	•		•								•		•		•	•			•	•		
28 Casing Diameter	•		•	•	•	•		•	•	•	•	•	•			•		•	•	•		•	•	•	•	•	•
29 Casing Material	•		•	•	•		•	•	•	•		•				•		•	•	•		•	•	•	•		•
30 Casing Interval or Length	•	•	•	•	•	•	•	•	•	•		•				•		•	•	•		•	•	•	•		•
31 Casing Stick Up		•														•			•			•		•	•		
32 Couplings	•		•					•								•			•			•		•			
33 Perforated or Screen Material	•	•	•	•	•		•	•								•		•	•	•		•		•	•		•
34 Slot or Perforation Size	•	•	•	•	•		•	•						•		•		•	•	•		•		•	•		•
35 Perforation or Screen Interval or Length	•	•	•	•	•	•	•	•		•		•				•		•	•	•		•		•	•		•
36 Well Sanitation			•													•											
37 Pump Information	•		•			•	•	•	•	•	•		•			•		•	•	•	•			•	•		•
38 Surface Completion								•							•		•								•		
HYDROGEOLOGY																											
39 Static Water Level	•	•	•	•	•	•	•	•	•	•		•	•	•		•		•	•	•		•	•	•	•	•	•
40 Pumping Water Level	•	•	•	•		•	•	•	•	•		•		•		•		•	•	•		•	•	•	•	•	•
41 Aquifer Media					•		•		•	•		•				•		•				•	•	•			
42 Geologic Formation																					•	•		•	•		
43 Driller's Log	•	•	•	•	•	•	•	•		•		•		•		•		•	•	•		•	•	•	•	•	•
44 Well Test	•		•	•	•	•	•	•	•	•		•				•		•	•	•		•	•	•			•
45 Estimated Yield	•	•	•	•	•	•	•	•	•	•		•		•		•		•	•	•		•	•	•	•	•	•
46 Water Quality	•				•																						
LOCATION																											
47 Owner Mailing Address	•	•	•	•	•		•		•													•		•	•	•	•
48 Well Location Address	•	•	•	•	•	•			•	•	•	•		•		•		•	•	•		•		•	•	•	•
49 Well Location TRS	•	•	•		•	•	•		•	•	•		•	•				•	•	•			•	•		•	•
50 Well Location Lat. and Long.		•							•	•		•	•										•				
51 Other Location	•		•	•	•		•		•	•				•		•		•	•	•				•	•		•
52 Well Elevation			•		•	•								•		•		•	•	•				•			•

NOTE: Various factors may contribute to what are apparent inaccuracies in the data presented in this table. Misinterpretations or errors noticed by readers would be of great interest to the author.

[a] Data are not available for Georgia, Iowa, and Nebraska.

[b] Florida's water management districts are:

A St. John's River
B South Florida
C Northwest Florida
D Southwest Florida
E Suwannee

[c] Louisiana has a separate form for plugging and abandonment.

TABLE 1 Technical Information Items and Tabulations for Agency Well Record Forms (continued)

	MT^d	NV	NH	NJ	NM^e	NY^f	NC	ND^g	OH	OK	OR	PA	RI	SC	SD	TN	TX	UT	VT	VA	WA	WV	WI	WY
GENERAL																								
1. Owner's Name	•	•	•	•	•	•	•	•	•	•	•	•	•	•	•	•	•	•	•	•	•	•	•	•
2. Driller's Name	•	•	•	•	•	•	•	•	•	•	•	•	•	•	•	•	•	•	•	•	•	•	•	•
3. Registration/License Number	•	•	•	•	•	•	•	•	•	•	•	•		•		•	•	•	•	•	•	•		
4. Permit Number		•		•	•		•			h	•			i			•	•		•	•	•	•	•
5. Construction Date	•	•	•	•	•	•	•	•		•		•		•		•	•	•	•	•	•	•	•	•
6. Well Log Confidentiality																	•							
7. Well Use	•	•	•	•		•	•			•	•		•	•	•	•	•	•	•	•	•	•	•	•
8. Nearest Septic, Sewer or Possible Contamination Source					•									•						•			•	
9. Replacement or Abandoned Wells	•	•	•		•		•	•		•	•					•	•			•	•			•
10. Comments/Remarks			•		•			•						•	•	•				•		•	•	•
CONSTRUCTION DETAILS																								
11. Well Service		•	•			•				•	•			•	•	•	•	•	•	•	•	•	•	•
12. Drilling Method	•			•	•	•		•		•	•			•	•	•	•	•		•	•		•	•
13. Drilling Fluid								•			•	•	•	•		•				•			•	
14. Open Hole	•							•			•	•	•	•		•	•			•				
15. Total Well Depth		•	•	•	•	•	•	•		•	•		•	•		•	•		•	•		•	•	•
16. Borehole Diameter	•	•		•	•			•		•	•		•			•	•			•		•	•	•
17. Grout	•	•		•	•		•	•		•	•	•		•		•	•	•		•	•	•	•	•
18. Grout Material	•	•		•	•		•	•		•	•	•		•		•	•	•	•		•	•	•	
19. Grout Interval	•	•		•	•		•	•		•	•		•		•	•				•			•	
20. Grout Method				•	•		•				•				•		•							
21. Packers	•							•							•		•							
22. Gravel Pack	•	•	•	•			•	•		•	•			•			•		•	•				•
23. Gravel Pack Material			•	•			•				•			•			•	•	•	•				
24. Gravel Pack Size		•	•	•			•				•			•			•	•	•	•		•		
25. Gravel Pack Interval		•		•			•			•	•			•			•	•	•	•				
26. Drive Shoe			•	•		•					•									•				
27. Casing Gauge	•	•	•		•		•			•	•		•	•		•	•	•	•	•			•	•
28. Casing Diameter	•	•	•	•	•	•	•		•	•	•		•	•		•	•	•	•	•	•	•	•	•
29. Casing Material		•	•	•	•		•	•		•	•	•		•		•	•	•	•	•	•	•	•	•
30. Casing Interval or Length	•	•	•	•	•	•	•	•	•	•	•	•	•	•		•	•	•	•	•	•	•	•	•
31. Casing Stick Up	•				•			•			•					•	•	•		•				
32. Couplings				•		•					•									•				
33. Perforated or Screen Material	•	•	•	•	•	•	•	•	•	•	•		•	•	•	•	•	•	•	•	•	•	•	•
34. Slot or Perforation Size	•	•	•	•	•	•	•	•		•	•		•	•	•	•	•	•	•	•	•	•	•	•
35. Perforation or Screen Interval or Length	•	•	•	•	•	•	•	•	•	•	•		•	•	•	•	•	•	•	•	•	•	•	•
36. Well Sanitation								•						•	•	•				•			•	
37. Pump Information				•			•		•	•				•	•	•		•		•				•
38. Surface Completion	•							•						•		•		•		•	•		•	•
HYDROGEOLOGY																								
39. Static Water Level	•	•	•	•	•	•	•	•	•	•	•	•		•	•	•	•	•		•	•	•	•	•
40. Pumping Water Level	•	•		•	•	•	•		•		•	•		•		•	•	•		•	•	•	•	•
41. Aquifer Media			•	•	•					•	•		•				•			•				
42. Geologic Formation					•								•							•				
43. Driller's Log	•	•	•	•	•	•	•		•		•	•		•	•	•	•	•	•	•	•	•	•	•
44. Well Test	•	•	•	•	•		•		•		•			•	•	•	•	•		•	•	•	•	•
45. Estimated Yield	•	•	•	•	•	•	•		•	•	•			•	•	•	•	•	•	•	•	•	•	•
46. Water Quality		•	•	•			•		•	•				•	•	•		•		•	•	•	•	•
LOCATION																								
47. Owner Mailing Address	•	•	•	•	•						•	•	•	•			•			•			•	•
48. Well Location Address		•	•	•	•	•					•	•	•			•	•	•		•		•	•	•
49. Well Location TRS	•	•	•	•	•					•	•			•										•
50. Well Location Lat. and Long.				•	•			•			•	•			•	•			•	•				
51. Other Location	•	•		•	•	•	•		•	•	•	•		•	•			•	•	•				•
52. Well Elevation			•	•	•	•	•		•			•			•						•	•		

^d Montana has two forms; the form used here is the one for wells with yields greater than 100 GPM.
^e Includes "Application to Appropriate Underground Waters" form.
^f Applies to Long Island only.
^g Separate form for "Pitless Unit and Pump Record" included.
^h Does not apply to domestic wells.
^i Request for public supply only.

NOTE—Data not available for Georgia, Iowa, and Nebraska.

WELL-INFORMATION CHECK LIST
PROJECT:_____

Station ID:_____ Local well ID:_____

Well type (public-supply, domestic-supply, observation, etc.):_____

Item	Date item filed
Ground-Water Site Inventory (GWSI) form	_____
Paper copt of GWSI form in file:	_____
Well-owner information	_____
Copies of permission to drill, sample, etc. (list permits):_____	_____
_____	_____
Copies of field notes and logs:	
Drillers log	_____
Lithologic log	_____
Cuttings	_____
Cores	_____
Well-construction record	_____
Well-development record	_____
Geophysical logs: List logs:_____	_____
Checks on well-construction integrity	_____
Other_____	_____
Well-location information:	
Location map(s)	_____
Site-sketch map	_____
Written description of location	_____
Well-casing elevation (elevation, and method and date of determination)	_____
Photographs of well and vicinity, if taken, with measuring point identified:	_____
Land use—land cover form (Figure 2) (enter dates of updates):	_____
Water-quality records for each sampling event (for example, purging, field measurements, field forms)	_____
Water-level measurements:	_____
Other information (for example, historical water-level records, type pf pump in well, location of sampling point, sampling history):	_____

Remarks:

NOTE—Source: U.S. Geological Survey.

FIG. 1 Example of a Well-Information Check List Modified from (12) GWSI USGS Ground-Water Site Inventory

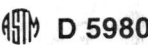

LAND-USE/LAND-COVER FIELD SHEET Page 1 (04/93)

1. Project name_____

   Field-check date____/____/____   Person conducting fied inspection:_____

   Well-station ID:_____ Latitude:_____ Loongitude:_____

   Entered by_____ Date____/____/____ Checked by_____ Date____/____/____

2. LAND USE AND LAND COVER CLASSIFICATION—(modified from Anderson and others. 1976, p.8). Check all land uses that occur within each approximate distance range from the sampled well. Identify the predominant land use within each distance range and estimate its percentage of the total area within a 1/4-mile radius of the well.

Land use and land cover	within 100 ft	100ft– 1/4 mi	Comments
I. URBAN LAND			
—Residential			
—Commercial			
—Industrial			
—Other (Specify)_____			
II. AGRICULTURAL LAND			
—Nonirrigated cropland			
—Irrigated cropland			
—Pasture			
—Orchard, grove, vineyard, or nursery			
—Confined feeding			
—Other (Specify)_____			
III. RANGELAND			
IV. FOREST LAND			
V. WATER			
VI. WETLAND			
VII. BARREN LAND			
Predominant land use			
Approximate percentage of area covered by predominant land use			

3. AGRICULTURAL PRACTICES within 1/4 mile of the sampled well.

   a. Extent of irrigation—indicate those that apply.
      Nonirrigated____ Supplimental irrigation in dry years only_____ Irrigated_____

   b. Method of irrigation—indicate those that apply.
      Spray_____ Flood_____ Furrow_____ Drip_____ Chemigation_____ Other_____(Specify)_____

   C. Source of irrigation water—indicate those that apply.
      Ground water_____ Surface water_____ Spring_____
      Sewage effluent_____(treatment): Primary_____ Secondary_____ Tertiary_____

   d. Pesticide and fertilizer application—Provide information about present and past pesticides and fertilizers used, application rates, and application methods.
      _____
      _____

   e. Crop and animal types—Provide information about present and past crop and animal types, and crop rotation practices.
      _____
      _____

NOTE—Source: U.S. Geological Survey

FIG. 2   Example of a Land-Use and Land-Cover Field Sheet Modified from (12)

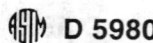
LAND-USE/LAND-COVER FIELD SHEET Page 2

Well-station ID:_____   Field-check date_____/_____/_____

4. LOCAL FEATURES—Indicate all local features that may affect ground-water quality whih occur within each approximate distance range from the sampled well.

Feature	within 100 ft	100ft– 1/4 mi	Comments
Gas station			
Dry cleaner			
Chemical plant or storage facility			
Airport			
Military base			
Road			
Pipeline or fuel storage facility			
Septic field			
Waste disposal pond			
Landfill			
Golf course			
Stream, river, or creak			
Perennial			
Ephemeral			
Irrigation canal			
Lined_____ Unlined_____			
Drainage ditch			
Lined_____ Unlined_____			
Lake			
Natural_____ Manmade_____			
Reservoir			
Lined_____ Unlined_____			
Bay or estuary			
Spring			
Geothermal (>25C)_____			
Nongeothermal_____			
Salt flat or playa			
Dry_____ Wet_____			
Mine, quarry, or pit			
Active_____ Abandoned_____			
Oil Well			
Major withdrawn well			
Waste injection well			
Recharge injection well			
Other_____			

5. LAND-USE CHANGES—Have there been major changes in the last 10 years in land use within 1/4 mile of the sampled well? Yes_____ Probably_____ Probably not_____ No_____ If yes, describe major changes.

_____

_____

_____

_____

6. ADDITIONAL COMMENTS— Emphasize factors that might influence local ground-water quality.

_____

_____

_____

Remarks:

FIG. 2   Land-Use and Land-Cover Field Sheet (continued)

REFERENCES

(1) Kolm, K. E., *Conceptualization and Characterization of Hydrologic Systems: International Ground-Water Modeling Center Technical Report GMWI 93-01*, Colorado School of Mines, Golden, CO, 1993.

(2) Alley, W. M., ed., *Regional Ground-Water Quality*, Van Nostrand Reinhold, New York, NY, 1993.

(3) Gillham, R. W., Robin, M. J. L., Barker, J. F., and Cherry, J. A., "Groundwater Monitoring and Sample Bias," *American Petroleum Institute Publication 4367*, Washington, DC, 1983.

(4) Pohlmann, K. F., and Alduino, A. J., *Potential Sources of Error in Ground-Water Sampling at Hazardous Waste Sites*, EPA/540/S-92/019, 1992.

(5) Ganley, M. C., "Availability of Domestic Well Records in the United States," *Ground Water Monitoring Review 9(4)*, 1989.

(6) Aller, L., et al., *Handbook of Suggested Practices for the Design and Installation of Ground-Water Monitoring Wells*, EPA/600/4-89/034, 1991. [Also published in 1989 by National Water Well Association, Dublin, OH, in its NWWA/EPA series.] See Nielsen and Schalla (10) for an updated version of the material in this handbook that is related to design and installation of ground-water monitoring wells.

(7) Driscoll, F. G., *Groundwater and Wells*, 2nd ed., Johnson Filtration Systems, Inc., St. Paul, MN, 1986.

(8) Harlan, R. L., Kolm, K. E., Gutentag, E. D., *Water-Well Design and Construction: Developments in Geotechnical Engineering, 60*, Elsevier, Amsterdam, 1989.

(9) Lehr, J., Hurlburt, S., Gallagher, B., and Voyteck, J., *Design and Construction of Water Wells: A Guide for Engineers*, Van Nostrand Reinhold, New York, NY, 1988.

(10) Nielsen, D. M., and Schalla, R., "Design and Installation of Ground-Water Monitoring Wells," In: *Practical Handbook of Ground-Water Monitoring*, D. M. Nielsen (ed.), Lewis Publishers, Chelsea, MI, 1991, pp. 239–331.

(11) McCaulou, D. R., Jewett, D. G., and Huling, S. G., *Nonaqueous Phase Liquids Compatibility with Materials Used in Well Construction, Sampling and Remediation*, EPA/540/5-95/503, 1995.

(12) Lapham, W. W., F. W. Wilde and M. T. Koterba, 1995. *Groundwater Data Collection Protocols and Procedures for the National Water-Quality Assessment Program-Selection, Installation, and Documentation of Wells and Collection of Related Data*; USGS Open File Report-45-348, 69 pp.

(13) Anderson, J. R., E. E. Hardy, J. T. Roach, and R. E. Witmer, 1976. *A Land Use and Land Classification System for Use With Remote Data*. U. S. Geological Survey Professional Paper 964, 28 pp

ASTM Designation: D 5126 – 90

Standard Guide for
Comparison of Field Methods for Determining Hydraulic Conductivity in the Vadose Zone[1]

This standard is issued under the fixed designation D 5126; the number immediately following the designation indicates the year of original adoption or, in the case of revision, the year of last revision. A number in parentheses indicates the year of last reapproval. A superscript epsilon (ϵ) indicates an editorial change since the last revision or reapproval.

1. Scope

1.1 This guide provides a review of the test methods for determining hydraulic conductivity in unsaturated soils and sediments. Test methods for determining both field-saturated and unsaturated hydraulic conductivity are described.

1.2 Measurement of hydraulic conductivity in the field is used for estimating the rate of water movement through clay liners to determine if they are a barrier to water flux, for characterizing water movement below waste disposal sites to predict contaminant movement, and to measure infiltration and drainage in soils and sediment for a variety of applications. Test methods are needed for measuring hydraulic conductivity ranging from 1×10^{-2} to 1×10^{-8} cm/s, for both surface and subsurface layers, and for both field-saturated and unsaturated flow.

1.3 For these field test methods a distinction must be made between "saturated" (K_s) and "field-saturated" (K_{fs}) hydraulic conductivity. True saturated conditions seldom occur in the vadose zone except where impermeable layers result in the presence of perched water tables. During infiltration events or in the event of a leak from a lined pond, a "field-saturated" condition develops. True saturation does not occur due to entrapped air (1).[2] The entrapped air prevents water from moving in air-filled pores that, in turn, may reduce the hydraulic conductivity measured in the field by as much as a factor of two compared to conditions when trapped air is not present (2). Field test methods should simulate the "field-saturated" condition.

1.4 Field test methods commonly used to determine field-saturated hydraulic conductivity include various double-ring infiltrometer test methods, air-entry permeameter test methods, and borehole permeameter tests. Many empirical test methods are used for calculating hydraulic conductivity from data obtained with each test method. A general description of each test method, and special characteristics affecting applicability is provided.

1.5 Field test methods used to determine unsaturated hydraulic conductivity in the field include direct measurement techniques and various estimation methods. Direct measurement techniques for determining unsaturated hydraulic conductivity include the instantaneous profile (IP) test method, and the gypsum crust method. Estimation techniques have been developed using borehole permeameter data, and using data obtained from desorption curves (a curve relating water content to matric potential).

1.6 The values stated in SI units are to be regarded as standard.

1.7 *This standard does not purport to address the safety problems associated with its use. It is the responsibility of the user of this standard to establish appropriate safety and health practices and determine the applicability of regulatory limitations prior to use.*

2. Referenced Documents

2.1 *ASTM Standards:*
D 653 Terms and Symbols Relating to Soil and Rock[3]
D 2434 Test Method for Permeability of Granular Soils (Constant Head)[3]
D 3385 Test Method for Infiltration Rate of Soils in the Field Using Double-Ring Infiltrometers[3]
D 4643 Test Method for Determination of Water (Moisture) Content of Soil by the Microwave Oven Method[3]

3. Terminology

3.1 *Definitions:*
3.1.1 Definitions shall be in accordance with Terms and Symbols D 653.
3.2 *Descriptions of Terms Specific to This Standard:*
3.2.1 Descriptions of terms shall be in accordance with Ref (2).

4. Summary of Guide

4.1 *Test Methods for Measuring Saturated Hydraulic Conductivity Above the Water Table*—There are several test methods available for determining the field saturated hydraulic conductivity of unsaturated materials above the water table. Most of these methods involve measurement of the infiltration rate of water into the soil from an infiltrometer or permeameter device. Infiltrometers typically measure conductivity at the soil surface, whereas permeameters may be used to determine conductivity at different depths within the soil profile. A representative list of the most commonly used equipment includes the following: infiltrometers, (single and double ring infiltrometers); double tube method; air-entry permeameter; borehole permeameter methods, (constant and multiple head methods).

4.1.1 *Infiltrometer Test Method:*
4.1.1.1 Infiltrometer test methods measure the rate of infiltration at the soil surface, (see Test Method D 2434), that is influenced both by saturated hydraulic conductivity as well

[1] This guide is under the jurisdiction of ASTM Committee D-18 on Soil and Rock and is the direct responsibility of Subcommittee D18.21.02 on Vadose Zone Monitoring.
Current edition approved Oct. 26, 1990. Published December 1990.
[2] The boldface numbers in parentheses refer to a list of references at the end of the text.
[3] *Annual Book of ASTM Standards*, Vol 04.08.

as capillary effects of soil (4). Capillary effect refers to the ability of dry soil to pull or wick water away from a zone of saturation faster than would occur if soil were uniformly saturated. The magnitude of the capillary effect is determined by initial moisture content at the time of testing, the pore size, soil physical characteristics (texture, structure), and a number of other factors. By waiting until steady-state infiltration is reached the capillary effects are minimized.

4.1.1.2 Most infiltrometers generally employ the use of a metal cylinder placed at shallow depths into the soil, and include the single ring infiltrometer, the double ring infiltrometer, and the infiltration gradient method. Various adaptations to the design and implementation of these methods have been employed to determine the field-saturated hydraulic conductivity of material within the unsaturated zone (5). The principles of operation of these methods are similar in that the steady volumetric flux of water infiltrating into the soil enclosed within the infiltrometer ring is measured. Saturated hydraulic conductivity is derived directly from solution of Darcy's Equation for saturated flow. Primary assumptions are that the volume of soil being tested is field-saturated and that the saturated hydraulic conductivity is a function of the flow rate and the applied hydraulic gradient across the soil volume.

4.1.1.3 Additional assumptions common to infiltrometer tests are as follows:

(a) The movement of water into the soil profile is one-dimensional downward.

(b) Equipment compliance effects are minimal and may be disregarded or easily accounted for.

(c) The pressure of soil gas does not offer any impedance to the downward movement of the wetting front.

(d) The wetting front is distinct and easily determined.

(e) Dispersion of clays in the surface layer of finer soils is insignificant.

(f) The soil is non-swelling, or the effects of swelling can easily be accounted for.

4.1.2 *Single Ring Infiltrometer:*

4.1.2.1 The single ring infiltrometer typically consists of a cylindrical ring 30 cm or larger in diameter that is driven several centimetres into the soil. Water is ponded within the ring above the soil surface. The upper surface of the ring is often covered to prevent evaporation. The volumetric rate of water added to the ring sufficient to maintain a constant head within the ring is measured. Alternatively, if the head of water within the ring is relatively large, a falling head type test may be used wherein the flow rate, as measured by the rate of decline of the water level within the ring, and the head for the later portion of the test are used in the calculations. Infiltration is terminated after the flow rate has approximately stabilized. The infiltrometer is removed immediately after termination of infiltration, and the depth to the wetting front is determined either visually, with a penetrometer-type probe, or by moisture content determination for soil samples (see Test Method D 4643).

4.1.2.2 A special type of single ring infiltrometer is the ponded infiltration basin. This type of test is conducted by ponding water within a generally rectangular basin that may be as large as several metres on a side. The flow rate required to maintain a constant head of water within the pond is measured. If the depth of ponding is negligible compared to the depth of the wetting front, the steady state flux of water across the soil surface within the basin is presumed to be equal to the saturated hydraulic conductivity of the soil.

4.1.2.3 Another variant of the single ring infiltrometer is the air-entry permeameter. The air-entry permeameter is discussed in 4.1.4.

4.1.3 *Double Ring Infiltrometer:*

4.1.3.1 The underlying principles and method of operation of the double ring infiltrometer are similar to the single ring infiltrometer, with the exception that an outer ring is included to ensure that one-dimensional downward flow exists within the tested horizon of the inner ring. Water that infiltrated through the outer ring acts as a barrier to lateral movement of water from the inner ring. Double ring infiltrometers may be either open to the atmosphere, or most commonly, the inner ring may be covered to prevent evaporation. For open double ring infiltrometers the flow rate is measured directly from the rate of decline of the water level within the inner ring for falling head tests, or from the rate of water input necessary to maintain a stable head within the inner ring for the constant head case; for sealed double ring infiltrometers, the flow rate is measured by weighing a sealed flexible bag that is used as the supple reservoir for the inner ring (6).

4.1.3.2 Refer to Test Method D 3385 for measuring infiltration rates in the range of 10^{-2} to 10^{-5} cm/s. A modified double-ring infiltrometer test method for infiltration rates from 10^{-5} to 10^{-8} cm/s is also being developed.

4.1.4 *Double Tube Test Method:*

4.1.4.1 The double tube test method proposed by Bouwer (6, 7, 8) has been described by Boersma (9) as a means of measuring the horizontal, as well as the vertical, field-saturated hydraulic conductivity of material in the vadose zone.

4.1.4.2 This test method as proposed by Bouwer (6, 7, 8) utilizes two coaxial cylinders positioned in an auger hole. The difference between the rate of flow in the inner cylinder and the simultaneous rate of combined flow from in the inner and outer cylinders is used to calculate K_{fs}.

4.1.4.3 A borehole is augered to the desired depth and a hole conditioning device is used to square the bottom of the hole. The hole is then cleaned and a 1 to 2 cm layer of coarse protective sand is placed in the bottom of the hole. An outer tube is then placed in the hole and sunken about 5 cm into the soil. The outer tube is then filled with water and a smaller inner tube is placed at the center of the outer tube. It is then driven into the soil. A top plate assembly (see Fig. 2) consisting of water supply valves and standpipes for the inner and outer cylinders is installed. Water is then supplied to both cylinders. The standpipe for the outer cylinder is allowed to overflow and the standpipe gage for the inner cylinder is set at 0 by adjusting the appropriate water supply values. After an equilibrium period of approximately 1 h, the hole is saturated.

4.1.4.4 After saturation is achieved, the level of fall of water in the inner standpipe, H, is recorded at given time intervals, t. H is recorded at least every 5 cm, for a total of at least 30 cm (Test 2). During this test, water in the outer standpipe remains at a constant head.

4.1.4.5 After the data is recorded, the inner reservoir is again filled and the inner standpipe water level is set to 0.

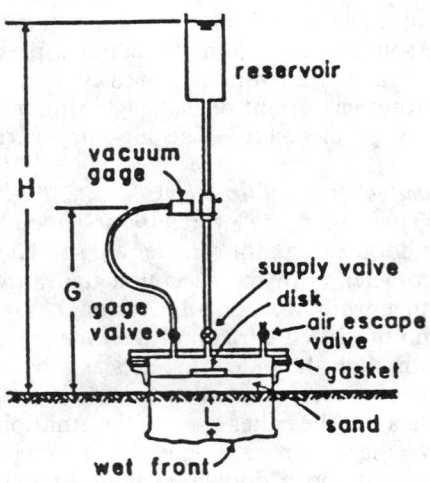

FIG. 1 Diagram of the Equipment for the Air-Entry Permeameter Technique (from Klute, 1986)

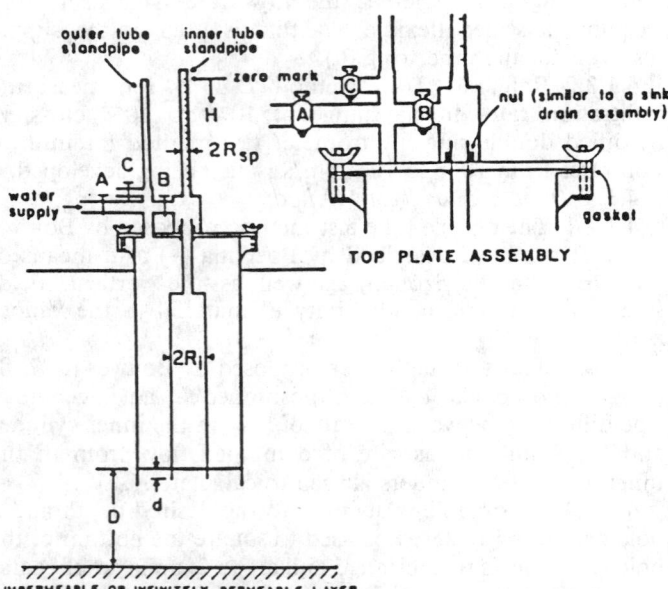

FIG. 2 Diagram of the Equipment Used for Double-Tube Test Method (from Klute, 1986)

The system is allowed to re-equilibrate for a period of time at least ten times as long as the time required to collect the first data set.

4.1.4.6 After waiting, Test 2 is performed. The levels in the outer standpipe and inner standpipe are both brought to 0. Once again the drop in the inner standpipe in cm, H, is recorded as a function of time, t. During the second test, however, water levels in both tubes drop simultaneously. Both tests are then performed a second time or until the results of two consecutive runs are consistent.

4.1.5 *Air-Entry Permeameter:* (Fig 1.)

4.1.5.1 The air-entry permeameter is similar to a single ring infiltrometer in design and operation in that the volumetric flux of water into the soil within a single permeameter ring is used to calculate field-saturated hydraulic conductivity. The primary differences between the two test methods are that the air-entry permeameter typically penetrates deeper into the soil profile and measures the air-entry pressure of the soil. Air-entry pressure is used as an approximation of the wetting front pressure head for determination of the hydraulic gradient, and consequently field-saturated hydraulic conductivity.

4.1.5.2 The air-entry permeameter consists of a single ring, typically 30 cm in diameter, sealed at the top, that is driven into the soil approximately 15 to 25 cm. Water is introduced into the permeameter through a standpipe, to the top of which is attached a water supply reservoir. Water is allowed to infiltrate into the soil within the permeameter ring, and the flow rate is measured by observing the decline of the water level within the reservoir. After a predetermined amount of water has infiltrated (based upon the estimated available storage of the soil interval contained within the ring), and the flow rate is relatively stable, infiltration is terminated and the wetted profile is allowed to drain. The air-entry value is the minimum pressure measured over the standing water inside of the permeameter ring attained during drainage. Once the minimum pressure is achieved, the permeameter is removed, and the depth to the wetting front is determined (10).

4.1.6 *Borehole Permeameter:*

4.1.6.1 Borehole permeameter test methods encompass a wide range of test designs, methods of operation, and methods of solution. The common feature among the different types of borehole tests is that the rate of water infiltration into a cylindrical borehole is used to determine field-saturated hydraulic conductivity. One of the most popular borehole infiltration tests is the constant-head borehole infiltration test, wherein the flow rate necessary to maintain a constant water level within a borehole is measured. The steady state flow rate, borehole geometry, borehole radius (r), and depth of ponding within the borehole (h), and along with certain capillary parameters are typically used in the solution. Hence, by accounting for capillary effects, borehole test methods attempt to measure field-saturated hydraulic conductivity rather than infiltration rate. Another variation of this test consists of conducting multiple constant head borehole infiltration tests within with the same borehole. Different water levels are established within the borehole for each individual test. Results from one or more tests at different ponded heights are solved simultaneously to independently find hydraulic conductivity and capillarity.

4.1.6.2 Borehole infiltration tests are the only currently available tests which can measure field-saturated hydraulic conductivity at depth within the unsaturated zone. Borehole tests may be conducted at great depth within the unsaturated zone, and are frequently used to measure the variability of conductivity with depth by conducting tests at selected horizons within an advancing borehole.

4.1.6.3 During constant head borehole tests water is introduced into a cylindrical borehole and maintained at a predetermined level. This may be accomplished by use of a float valve connected to an external water supply reservoir, or with a Mariotte-siphon device (2, 10). The flow rate into the borehole necessary to maintain the water at the prescribed level is measured at various times. The flow rate at steady state is used in the solution of field-saturated hydraulic conductivity. The dimensions and geometry of the borehole and the depth to the water table are also required for the solution.

4.1.7 *Empirical Methods—Saturated Hydraulic Conductivity:*

4.1.7.1 A number of empirical methods have been developed for estimation of hydraulic conductivity from grain size data (Shepard (11)). Shepard suggested that hydraulic conductivity could be predicted from the following:

$$K = cd^a$$

where:
c = a dimensionless constant found through regression analysis,
d = the mean pore throat or particle diameter, and
a = an exponent generally ranging from 1.65 to 1.85.

4.1.7.2 Values for c and a were found to vary substantially depending on the degree of sorting of particles and the amount of induration. Both c and a decreased as the degree of sorting became poorer and as the induration increased. The amount of secondary porosity ("structure" in soils, or "fractures" in rock and sediment) is also expected to affect the values for c and a. Estimates of K for a particular value of d varied by nearly three orders of magnitude depending on the choice of values for c and a (11).

4.2 *Test Methods for Measuring Unsaturated Hydraulic Conductivity:*

4.2.1 *Instantaneous Profile Test Method (IP):*

4.2.1.1 Several references describe the IP test method including Watson (12). The relationship between water potential and hydraulic conductivity can be determined by measuring the rate of drainage and water potential and then solving a form of the Richards equation. The Richards equation solves for the change in water content through time for non-steady, uniform unsaturated flow by relating water potential and unsaturated hydraulic conductivity.

4.2.1.2 To conduct an IP test a small basin is constructed in which water is ponded. Neutron access tubing and a nest of tensiometers at varying depths are installed in the center of the basin. Water is ponded in the basin until the wetting front passes the bottom of the horizon being investigated. Movement of the wetting front is detected with a neutron probe. The soil basin is then covered to reduce evaporation and water content and water potential are measured periodically as water drains downward under the influence of gravity.

4.2.2 *Gypsum Crust Test Method:*

4.2.2.1 The gypsum crust test method is similar to infiltrometer methods in that the rate of water flux across an infiltrative surface is measured. A crust composed of varying mixtures of gypsum and coarse sand is poured over the surface of an exposed excavated cylinder of soil. After the crust cures water is ponded on the crust. The presence of the crust causes unsaturated conditions to form in the soil beneath the crust.

4.2.2.2 The cylinder of soil is instrumented with a nest of tensiometers to measure water potential below the gypsum crust. The rate of flux of water necessary to maintain a constant head over the gypsum crust and the diameter of the cylinder is also recorded (13, 14).

4.2.3 *Empirical Test Methods—Unsaturated Hydraulic Conductivity:*

4.2.3.1 A number of empirical test methods have been developed to estimate unsaturated hydraulic conductivity from other hydraulic parameters. Van Genuchten (15) and Mualem (16) developed methods for predicting unsaturated hydraulic conductivity from the desorption curve (that relates water content to water potential) and from K_s measurements. Reynolds and Elrick (2) developed a borehole permeameter method for measuring a fitting parameter used for estimating unsaturated hydraulic conductivity according to a model proposed by Gardner. The fitting parameter is found by solving simultaneous equations developed from borehole water flux data for two ponded heights. The two ponded height test method is discussed further in 6.4. Infiltration data can be used to estimate hydraulic conductivities by solving the Green-Ampt or Philips Eq. (4).

5. Significance and Use

5.1 Saturated hydraulic conductivity measurements are made for a variety of purposes varying from design of landfills, construction of clay liners, to assessment of irrigation systems. Infiltrometers are commonly used where infiltration or percolation rates through a surface or subsurface layer are desired. Evaluation of the rate of water movement through a pond liner is one example of this kind of measurement. Penetration of the liner by a borehole would invalidate the measurement of liner permeability. It has been noted that small-ring infiltrometers are subject to error due to lateral divergence of flow. Therefore, techniques using very large (1 to 2 m diameter) infiltration basins have been recommended for measuring the very slow percolation rates typically required for clay liners. The air-entry permeameter can be used instead of infiltrometer tests to avoid lateral divergence of flow. However, because a cylinder must be driven into the media tested, the actual soil column tested may be disrupted by introduction of the cylinder, especially in structured soils.

5.2 Borehole tests for determining saturated hydraulic conductivity are applicable for evaluating the rate of water movement through subsurface layers. For slowly permeable layers, an accurate method of measuring the rate of water movement into the borehole must be developed. Use of a flexible bag as a reservoir that can be periodically weighed is advisable for these conditions. A number of mathematical solutions for borehole outflow data are available (Stephens et al. (17), Reynolds et al. (18), and Philip (19)).

5.3 Information on unsaturated flow rates is needed to design hazardous waste landfills and impoundments where prevention of flow of contaminants into ground water is required. Of the test methods available, the primary differences are cost and resultant bias and precision. The instantaneous profile test method appears to provide very reliable data because it uses a large volume of soil (several cubic metres) and is performed on undisturbed soils in the field. However, a single test can cost several thousand dollars. The gypsum crust test method, although more rapid than the instantaneous profile test method, sacrifices precision of results due to the smaller spatial extent of the tested area. Methods for estimating unsaturated hydraulic conductivity from fundamental soil hydraulic functions like the desorption curve may readily deviate from true values by an order of magnitude, but may be of use where relative differences in permeability between materials or across water content ranges is of interest.

6. Report

6.1 The reporting requirements for each test vary substantially. However, the variability of hydraulic conductivity in soils, and the sensitivity of some test methods to factors such as textural stratifications, anisotropic conditions, changes in temperature or barometric pressure, initial and final water contents, and depth to groundwater, suggest that a detailed description of each test site be recorded. Record the following:

6.1.1 Soil series (for comparison to existing data).

6.1.2 Soil horizon characteristics above and below layer tested (to help interpret deviations from theoretical response).

6.1.3 Initial and final water content (measure or describe subjectively depending upon method and to identify which numerical solution is most applicable).

6.1.4 General climatic conditions (for example, barometric pressure, temperature, precipitation, cloud cover to estimate possible evaporation, pressure responses, accumulation of prescription that might bias results).

6.1.5 Diameter of borehole, or infiltration ring (parameter used in solution).

6.1.6 Rate of outflow, infiltration, or drainage (parameter used in solution).

6.1.7 Water potential (tensiometer) readings as required (parameter used in solution).

6.1.8 Temperature of water used.

6.1.9 Chemical composition of water used.

6.2 *Infiltrometer Tests:*

6.2.1 Infiltrometer tests are useful for measuring the rate of infiltration but do not provide a direct measure of field-saturated hydraulic conductivity. Since entrapped air exists within the wetting front, true saturated conditions do not form during infiltration tests. Experience indicates that field saturated K_{fs} is approximately 50 to 75 % less than K_s (1, 2).

6.2.2 Infiltration data can be fitted to empirical models such as those developed by Green and Ampt and Philip (described by Bouwer (4)).

$$I = S_i t^{1/2} + At$$

where:
I = cumulative infiltration (cm of H_2O),
S_i = sorbtivity of soil (determined from plot of cumulative infiltration against $t^{1/2}$),
t = time increment in seconds, and
A = approximates $1/2 K_{fs}$.

6.3 *Air-Entry Permeameter:*

6.3.1 As soon as minimum pressure is reached, air begins to bubble up through the wetting front. Field-saturated K_{fs} can be calculated from the critical "air-entry value" or minimum pressure. Field-saturated K_{fs} is approximately equal to $1/2$ of K_s in most soils or $1/4$ of K_s in fine-textured (clayey) soils.

6.3.2 Field saturated K_{fs} is calculated (from Amoozegar and Warrick (12)) as follows:

$$K_{fs} = L(dH/dt)(R/Rc)^2 / (H + L - (P/2\ pg))$$

where:
K_{fs} = field-saturated hydraulic conductivity (cm/s),
L = depth of wetting front (cm),
H = ponded height of water above the soil (cm),
dH/dt = rate of fall just before water supply was shut off (cm/s),
R/Rc = Radius of the reservoir divided by the cylinder radius, and
$P/2\ pg$ = air entry value (minimum pressure divided by the unit weight of liquid (cm)).

6.4 *Double-Tube Test Method:*

6.4.1 Data from both tests are plotted on a graph of H versus t (H is on the y axis). Due to the decrease in head in the inner tube and the greater head in the outer tube, in Test 1, H decreases more rapidly through time than in Test 2. A curve of H verses t data for Test 2 will lie above the curve for Test 1 because in Test 2 the head is the same in both the inner and outer tubes.

6.4.2 Saturated hydraulic conductivity (K) is calculated using the H versus t graphs and the following equation (Amoozegar and Warrick, (11)):

$$K = R^2_{sp}\ dHt_1/(FR_i \int_{t0}^{t1} Hdt)$$

where:
R_{sp} = radius inner tube standpipe,
R_i = radius inner tube,
dHt_1 = vertical distance between the two curves (Fig. 2) at $t = t_1$,
$\int_{t0}^{t1} Hdt$ = areas under the lower curve between $t = 0$ and $t = t_1$, and
F = a dimensionless quantity dependent on the geometry of the flow system. See Fig. 3 (nomograph).

6.5 *Borehole Permeameter Test Methods:*

6.5.1 Unlike the previous described infiltrometer and permeameter test methods, borehole permeameters account for three-dimensional flow as a result of lateral, as well as downward, flow components. The actual configuration of the flow field around the borehole is highly dependent on the geometry of the borehole, the hydraulic properties of the soil and the capillary suction of the soil. Many of the earlier solutions for falling-head and constant-head type borehole tests ignore the effects of unsaturated flow away from the borehole. Several authors (Glover (20)); U.S. Bureau of

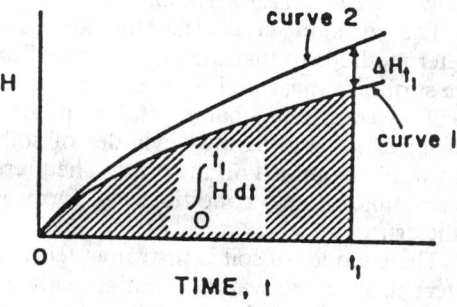

FIG. 3 Graph of H versus t for Double-Tube Procedure (from Klute, 1986)

ASTM D 5126

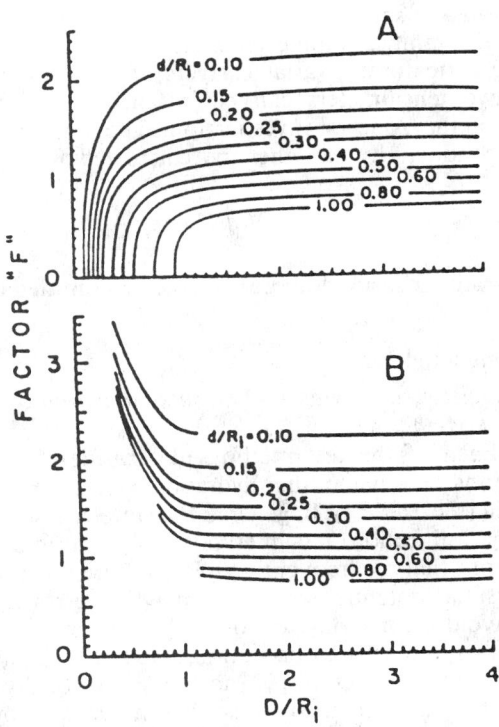

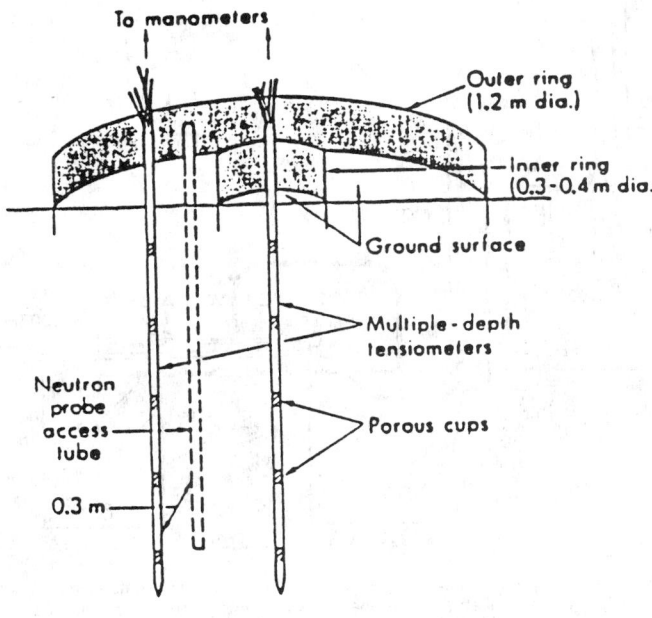

FIG. 4 Values of *F* for the Double-Tube Test Method, (*A*) An Impermeable Layer Below the Hole; and (*B*) An Infinitely Permeable Layer Below the Hole (from Klute, 1986)

Reclamation (21); have proposed borehole test methods that are entirely dependent on "free surface" solutions that ignore capillarity. More recently, Stephens et al. (17); Philip (19); and Reynolds and Elrick (18), have shown that unsaturated flow can greatly affect the infiltration rate from a borehole—especially in fine-textured soils, and must be considered in the solution for hydraulic conductivity. Each of these workers has proposed testing methods and/or solutions which account for unsaturated flow away from a wetted bulb around the borehole.

6.5.2 The solution methods of Stephens et al. (17) and Philip (19) require that certain capillary parameters be either determined separately or be estimated based on soil texture.

6.5.3 The methods of solution proposed by Stephens et al. (17) account for capillary effects and are based on multivariate regression equations developed from numerical simulations. Capillary parameters are determined from a catalog of soil hydraulic properties based on soil texture (for example, Mualem (16)), or by a fit to moisture retention curves using a model developed by Van Genuchten (16).

6.5.4 The Philip (19) method is an approximate quasi-analytical solution that accounts for unsaturated flow from a borehole. The solution is based on an approximation of the borehole geometry as an elongate half-spheroid. The capillary parameter must be either known *a priori*; or estimated from a catalog of soil hydraulic properties based on soil texture.

6.5.6 Reynolds and Elrick (18) described an analytical solution for borehole permeameter data that involves a simultaneous solution for data collected at two different ponded heights. This approach was found to be sensitive to

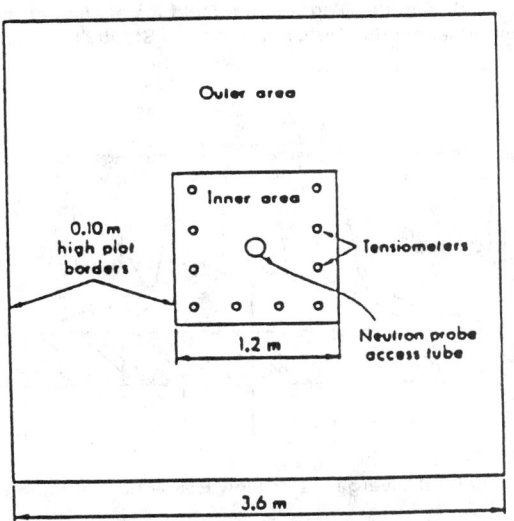

FIG. 5 Sectional and Plan View of Double-Ring Infiltrometer With Instruments Installed for Unsteady Drainage-Flux Method (from Klute, 1986)

slight field measurement error and to texturally stratified systems with the result that negative values for K_{fs} are frequently obtained. Reynolds and Elrick (18) suggested an alternative analytical solution where capillary effects are estimated based on soil texture and structure.

6.6 *Instantaneous Profile (IP) Test Method:*

6.6.1 A detailed description of calculating unsaturated hydraulic conductivity (or diffusivity) for different depth increments is provided in Green and others (22). Graphical plots of tensiometric data, and soil water content data through time are used to estimate instantaneous water flux at known levels of water content and water potential. An alternative analytical solution was described by Hillel (23). Unsaturated hydraulic conductivity data are subject to

481

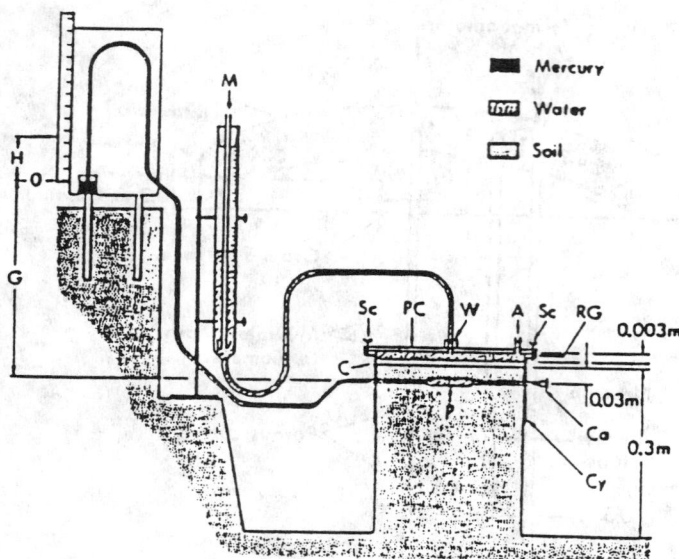

NOTE—M = constant-head device; Sc = wing nut; PC = plastic cover; W = water inlet; A = air outlet; RG = rubber gasket; C = gypsum-sand crust; Ca = tensiometer cap; Cy = metal cylinder with sharpened edge; H = height of mercury column above mercury pool; and H = height of mercury pool above tensiometer porous cup, P.

FIG. 6 Schematic Diagram of a Field Installation of the Measurement Apparatus for Crust-Imposed Steady Flux Method

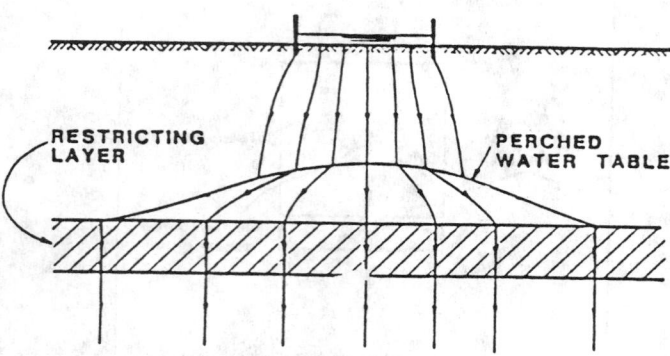

FIG. 7 Lateral Divergence of Flow Below an Infiltrometer

hysteresis. The IP test method provides data from the desorption loop.

6.7 Gypsum Crust Test Method:

6.7.1 The gypsum crust test method yields a single measurement of unsaturated hydraulic conductivity as a function of measured water potential for each crust constructed. The unsaturated hydraulic conductivity values are associated with the absorption loop rather than the desorption loop obtained with drainage methods.

6.7.2 In the crust test method a steady unsaturated flux of water is attained with a unit hydraulic gradient (influenced only by gravity). Under these conditions the measured water flux is equal to the hydraulic conductivity:

$$K(h) = Q/A$$

where:
$K(h)$ = unsaturated hydraulic conductivity at head = h,
Q = outflow of water (cm³/s), and

A = area of soil cylinder (cm²).

6.7.3 If a unit hydraulic gradient does not form due to presence of texturally variable layers or due to compacted zones, two tensiometers can be installed at the top and bottom of the zone of investigation. The unsaturated hydraulic conductivity can then be calculated from the following:

$$K(h) = Q/(dH/dz)A$$

where:
dH/dz = the measured hydraulic gradient (unitless).

7. Precision and Bias

7.1 *Precision and Bias of Saturated Hydraulic Conductivity Test Methods:*

7.1.1 Each of the test methods described make certain assumptions to enable the hydraulic conductivity to be calculated (see Table 1). In general, the simpler test methods, especially infiltration test methods, rely on many more assumptions that are frequently violated. Care must be taken to understand potential source of analytical error and bias, and to avoid errors while conducting field tests. Most test methods assume soils to be homogeneous (the same in all directions) or at least isotropic (no changes vertically) that is seldom the case in field soils. Errors caused by non-homogeneous or anisotropic conditions vary from test to test. Appropriate test methods for a particular situation should be chosen based on the cost of testing, the bias and precision required, the depth at which a layer is to be tested, the characteristics of the soil profile (uniform or layered), and the approximate hydraulic conductivity range expected.

7.1.1.1 The accuracy of hydraulic conductivity tests is highly dependent on the spatial variability of soils or sediments to be tested. Studies indicate that hydraulic properties of field soils are highly variable (Neilsen et al. (**24**)) and that numerous readings (at closely spaced locations) would typically be needed to characterize a "site" or field-sized area. Field-saturated hydraulic conductivity values tend to be log-normally distributed rather than normally distributed meaning that a majority of the net flux of water may occur in a few permeable spots (**24**).

7.1.1.2 Infiltrometer test methods should be used cautiously if the saturated hydraulic conductivity is to be determined. Infiltration is affected by both hydraulic conductivity as well as by capillary effects. Infiltration measurements are sensitive to: disruption of the infiltration surface (for example, compaction, sealing by rain splash); presence of textural stratification; chemistry of the water used; and water temperature. Water that is low in salts or high in sodium is dispersive and may result in lower calculated values of K_{fs}.

7.1.2 *Single-Ring Infiltrometer:*

7.1.2.1 The single ring infiltrometer is subject to divergent flow due to the effects of unsaturated flow heterogeneities, and anisotropy. Because with most applications the wetting front is allowed to propagate below the bottom of the ring. These effects may lead to inaccuracies in the determination of saturated hydraulic conductivity.

7.1.3 *Double-Ring Infiltrometers:*

7.1.3.1 As with single ring infiltrometers the wetting front is allowed to advance below the bottom of the ring, but it's

TABLE 1 Review and Comparison of Test Methods for Measuring Hydraulic Conductivity in the Vadose Zone

Note: Columns "Free Surface (20)", "Capillarity Fixed (2)", "Capillarity Predicted (16)", "2-Head Simultaneous Solution (2)", and "IP (21)" fall under the spanning heading **Borehole Permeameter Methods**.

Characteristics	Single Ring Infiltrometer (4)	Double Ring Infiltrometer (4)	Double-Tube Test Method (8)	Air-Entry Permeameter (10)	Free Surface (20)	Capillarity Fixed (2)	Capillarity Predicted (16)	2-Head Simultaneous Solution (2)	IP (21)	Crust (21)	Empirical (14, 15)
Relative accuracy	Low	Fair	Fair	Good	Poor	Good	Good	Variable	Good	Good	Low
Relative cost	Low	Low-Moderate	Moderate	Moderate	Low to Moderate				High	High	Low
Time required (at K_{fs} = 10^{-5} cm/s)	<4 h	<4 h	4 h to 1 day	<4 h	<4 h			1 day	1 day	4 h to 1 day	4 h
Depth of Testing Possible	Surface	Surface	0 to 1 ft	0 to 1 ft	Any				0 to 5 ft	0 to 1 ft	Any
Range of K_{fs} (cm/s) for which test is suited	10^{-2} to 10^{-6}	10^{-2} to 10^{-6}; 10^{-6} to 10^{-8} (with flexible bag for inner reservoir)	$<10^{-6}$	$<10^{-8}$	$<10^{-8}$ (with precautions for temperature effects on reservoir volume)		$<10^{-6}$	—	—	—	—
Advantages	Simple apparatus, rapid, can estimate K_{fs} from infiltration data, can increase diameter to reduce scale effects and edge effect	Similar to single ring	....	Measures vertical K_{fs} only, accounts for capillary effects	Simple numerical solution, good approximation for sands	Accounts for capillary effects	—	Simple solution, accounts for capillarity	Excellent method for deriving K (unset) curve	Good for values of Y near zero	Simple, rapid
Drawbacks	Lateral flow affects accuracy, measures infiltration not K_{fs}, surface crust reduces infiltration, measured on surface of soil only	Similar to single ring	Cumbersome apparatus, time-consuming numerical solution	Sometimes difficult to drive tube, difficult to identify wetting front in wet soil	Does not account for capillary effects, high error for medium to fine unstructured soil	Must assume ratio of capillary to flux effects, difficult to predict	Requires description data	Occasionally gives negative values	Time-consuming, affected by barometric changes	Time-economical, only one value of K(unset) and water potential per test	Low accuracy

assumed that infiltration through the outer ring functions as an effective barrier to lateral flow beneath the ring. However, the accuracy of this assumption may be limited.

7.1.3.2 Bouwer (4) discussed the ratio of the inner ring to the outer ring necessary to maintain vertical flow in the inner ring. He suggested that an error of several hundred percent can occur unless cylinders of very large diameter are used because of edge effects. For large diameter (1 m or more) cylinders the "edge" effects become small enough that a "double-ring" system is not necessary. Edge effects are not corrected through use of a double-ring infiltrometer though they are somewhat reduced. Bouwer (4) mentioned that true vertical infiltration below a ring infiltrometer only occurs after a surface soil crust forms that limits the rate of water intake. Hence, the rate of infiltration thus measured does not represent fully saturated flow.

7.1.4 *Air-Entry Permeameter:*

7.1.4.1 The same restrictions and assumptions apply for the air-entry permeameter as for the infiltrometer test methods. However, since the wetting front is not allowed to advance below the bottom of the permeameter ring, one-dimensional vertically downward flow is ensured. In addition, since the hydraulic gradient is measured during the test, the infiltration rate need not necessarily reach steady state during the first portion of the test. One potential problem with the air-entry permeameter test method is determining the depth of the wetting front after completion of the test. Visual determination is especially difficult in soils with higher initial moisture content.

7.1.5 *Double-Tube Test Method:*

7.1.5.1 Depending on the permeability of the soil, the double-tube method requires over 200 L of water and 2 to 6 h for completion.

7.1.5.2 The test method is not suitable for rocky soils because of the difficulty in driving tubes into the ground. Due to soil disturbance around the inner tube, the diameter of the outer tube should be at least two times that of the inner tube (Bouwer and Riel, 1967). An inner tube with a diameter <10 cm is not recommended. The K value obtained utilizing this method is affected by both the horizontal and vertical conductivities of soil.

7.1.6 *Borehole Permeameter Test Methods:*

7.1.6.1 Permeameter test methods rely on an accurate measure of steady-state flow. The length of time required to establish steady flow can range from minutes, for small-diameter borehole tests in coarse-textured soils, to months, for large-diameter borehole tests in clays. The analytical

solution used will also affect the accuracy of the results. Stephens and others (25) compared large and small-diameter borehole test methods to the air-entry permeameter method for several geologic materials. Various analytical solutions were used to find K_{fs} from the borehole data. It was found that accuracy of permeameter test methods is sensitive to the spatial variability of soils both vertically and laterally. Test results are sensitive to the condition of the sidewall of the test borehole. Care should be taken to avoid smearing of the borehole that creates a hydraulic barrier.

7.2 *Precision and Bias of Unsaturated Hydraulic Conductivity Measurements:*

7.2.1 Test methods for unsaturated hydraulic conductivity are subject to the same limitations as methods for saturated hydraulic conductivity. Little comparative information is available concerning precision and bias of hydraulic conductivity tests, hence no one method can be clearly judged to be superior to another.

7.2.2 *Instantaneous Profile Test Method (IP):*

7.2.2.1 The IP test method is thought to be an accurate test of unsaturated hydraulic conductivity. However, it is costly and time-consuming to perform. Errors in measurement of water content or water potential will affect the accuracy of calculated unsaturated hydraulic conductivity. Mercury manometer-type tensiometers are suggested for measuring water potential. Rapid changes in barometric pressure may affect soil water potential readings. As with all methods, stratification within the soil profile being measured will also affect accuracy. Presence of a water table within about three to four feet of the base of the zone of measurement should be avoided (12).

7.2.3 *Crust Test Method:*

7.2.3.1 The crust test method provides a single measurement of unsaturated hydraulic conductivity at a specific water potential which is read off of a tensiometer installed below the crust. The water potentials that evolve below the crust are a function of the crust material used (for example, specific gypsum/sand mixture). A steady-state flow rate must be measured. This may take hours. Tensiometer readings must be made accurately. The geometry of the excavated block of soil is critical to the solution, hence soil cylinders with a consistent diameter must be accurately excavated. This is difficult in highly-structured or rocky soils (13).

8. Keywords

8.1 Air-entry permeameter, air-entry value, borehole permeameter, hydraulic conductivity, infiltrometer, vadose zone monitoring

REFERENCES

(1) Bouwer, H. "Rapid Field Measurement of Air Entry Value and Hydraulic Conductivity of Soil as Significant Parameters in Flow System Analysis" Water Resources Research, Vol 2, No. 4, 1966, pp. 729–738.

(2) Reynolds, D. and Elrick, D. E. "A Method for Simultaneous *In-Situ* Measurement in the Vadose Zone of Field-Sa Hydraulic Conductivity, Sorptivity and the Conductivity-Pressure Head Relationship" Groundwater Monitoring Review, Vol 6, No. 4, 1986, pp. 84.

(3) *Glossary of Science Terms*, Soil Science Society of America, Madison, WI, 1987.

(4) Bouwer, H. "Intake Rate: Cylinder Infiltrometer." Methods of Soil Analysis, Part 1: Physical and Mineralogical Methods, Agronomy Monograph No. 9, American Society of Agronomy, Madison, WI, 1986, pp. 825–844.

(5) Philip, J. R. "The Theory of Infiltration: The Infiltration Equation and its Solution" Soil Science, 1957, 83:345–357.

(6) Bouwer, H. "A Double Tube Method for Measuring Hydraulic Conductivity of Soil in Sites Above a Water Table" Soil Science Soc. Amer. Proc. 25:334–342, 1961.

(7) Bouwer, H. "Field Determination of Hydraulic Conductivity Above a Water Table with a Double-Tube Method" Soil Science

Soc. Amer. Proc. 26:330–335, 1962.

(8) Bouwer, H. "Measuring Horizontal and Vertical Hydraulic Conductivity of Soil with the Double-Tube Method" Soil Science Soc. Amer. Proc. 28:19–23, 1964.

(9) Boersma, L. "Field Measurement of Hydraulic Conductivity Below a Water Table" Methods of Soil Analysis Part 1: Physical and Mineralogical Methods. Agronomy Monograph No. 9. American Society of Agronomy, Madison, WI, 1965.

(10) Amoozegar, A. and Warrick A. W. "Hydraulic Conductivity of Saturated Soils—Field Methods" Methods of Soil Analysis Part 1: Physical and Mineraological Methods, Agronomy Monograph 9, American Society of Agronomy, Madison, WI, 1986.

(11) Shepard, R. G. Correlations of Permeability and Grain Size, Ground Water. 27(5):633–638, 1989.

(12) Watson, K. K. "An Instantaneous Profile Method for Determining the Hydraulic Conductivity of Unsaturated Porous Materials" Water Resources Res. 2:709–715, 1966.

(13) Bouma, J., Hillel, D. I., Hole, F. D., and Amerman, C. R. "Field Measurement of Hydraulic Conductivity by Infiltration through Artificial Crusts" Soil Science Soc. Amer. Proc. 33:362–344, 1971.

(14) Bouma, J. and Denning, J. C. "Field Measurement of Unsaturated Hydraulic Conductivity by Infiltration through Gypsum Crusts" Soil Science Soc. Amer. Proc. 36:846–847, 1972.

(15) Van Genuchten, M. T. "A Closed Form Equation for Predicting the Hydraulic Conductivity of Unsaturated Soil" Soil Sci. Soc. Amer. J. 44:892–898, 1980.

(16) Mualem, Y. "A New Model for Predicting the Hydraulic Conductivity of Unsaturated Porous Media" Water Resources Res. 12:513–522, 1976.

(17) Stephens, D. B., Lambet, K., and Watson, D. "Regression Models for Hydraulic Conductivity and Field Test of the Borehole Permeameter" Water Resources Res. 23:2207–2214, 1987.

(18) Reynolds, W. D. and Elrick, D. E. "A Laboratory and Numerical Assessment of the Guelph Permeameter Method" Soil Science 144:282–299, 1987.

(19) Philip, J. R. "Approximate Analysis of the Borehole Permeameter in Unsaturated Soil" Water Resour. Res. 21(7):1025–1033, 1985.

(20) Glover, R. E. "Flow from a Test-Hole Located Above Groundwater Level" U.S. Bur. Rec. Eng. Meng. 8:69–71, 1953.

(21) Bureau of Reclamation. Drainage Manual U.S. Govt. Print Offc. Wash. DC, 1978, pp. 74–97.

(22) Green, R. E., Ahuja, L. R., and Chong, S. K. "Hydraulic Conductivity, Diffusivity, and Sorptivity of Unsaturated Soils: Field Methods." Methods of Soil Analysis Part 1: Physical and Mineralogical Methods. Agron. Mono. No. 9. American Soc. of Agron. Madison, WI, 1986, pp. 771–798.

(23) Hillel, D. Fundamentals of Soil Physics. Academic Press. New York, 1980.

(24) Neilsen, D. R., Biggar, J. W., and Erh, K. T. "Spatial Variability of Field-Measured Soil Water Properties: Hilgardia 42:215–260, 1973.

(25) Stephens, D. B., Unruh, M., Havlena, J., Knowlton, R. G., Mattson, E., and Cox, W. "Vadose Zone Characterization of Low-Permeability Sediments Using Field Permeameters" Ground Water Monitoring Res. Vol 8, No. 2, 1988, pp. 59–66.

Standard Guide for
Selection of Aquifer-Test Method in Determining of Hydraulic Properties by Well Techniques[1]

This standard is issued under the fixed designation D 4043; the number immediately following the designation indicates the year of original adoption or, in the case of revision, the year of last revision. A number in parentheses indicates the year of last reapproval. A superscript epsilon (ε) indicates an editorial change since the last revision or reapproval.

1. Scope

1.1 This guide is an integral part of a series of standards that are being prepared on the in situ determination of hydraulic properties of aquifer systems by single- or multiple-well tests. This guide provides guidance for development of a conceptual model of a field site and selection of an analytical test method for determination of hydraulic properties. This guide does not establish a fixed procedure for determination of hydrologic properties.

1.2 The values stated in SI units are to be regarded as standard.

1.3 *Limitations*—Well techniques have limitations in the determination of hydraulic properties of ground-water flow systems. These limitations are related primarily to the simplifying assumptions that are implicit in each test method. The response of an aquifer system to stress is not unique; therefore, the system must be known sufficiently to select the proper analytical method.

1.4 *This standard does not purport to address all of the safety problems, if any, associated with its use. It is the responsibility of the user of this standard to establish appropriate safety and health practices and determine the applicability of regulatory limitations prior to use.*

2. Referenced Documents

2.1 *ASTM Standards:*

D 653 Terminology Relating to Soil, Rock, and Contained Fluids[2]

D 4044 Test Method (Field Procedure) for Instantaneous Change in Head (Slug Tests) for Determining Hydraulic Properties of Aquifers[2]

D 4104 Test Method (Analytical Procedure) for Determining Transmissivity of Nonleaky Confined Aquifers by Overdamped Well Response to Instantaneous Change in Head (Slug Test)[2]

D 4105 Test Method (Analytical Procedure) for Determining Transmissivity and Storage Coefficient of Nonleaky Confined Aquifers by the Modified Theis Nonequilibrium Method[1]

D 4106 Test Method (Analytical Procedure) for Determining Transmissivity and Storage Coefficient of Nonleaky Confined Aquifers by the Theis Nonequilibrium Method[2]

3. Terminology

3.1 *Definitions:*

3.1.1 *aquifer, confined*—an aquifer bounded above and below by confining beds and in which the static head is above the top of the aquifer.

3.1.2 *unconfined aquifer*—an aquifer that has a water table.

3.1.3 *barometric efficiency*—the ratio of the change in depth to water in a well to the inverse of water-level change in barometric pressure, expressed in length of water.

3.1.4 *conceptual model*—a simplified representation of the hydrogeologic setting and the response of the flow system to stress.

3.1.5 *confining bed*—a hydrogeologic unit of less permeable material bounding one or more aquifers.

3.1.6 *control well*—well by which the aquifer is stressed, for example, by pumping, injection, or change of head.

3.1.7 *hydraulic conductivity (field aquifer tests)*—the volume of water at the existing kinematic viscosity that will move in a unit time under unit hydraulic gradient through a unit area measured at right angles to the direction of flow.

3.1.8 *observation well*—a well open to all or part of an aquifer.

3.1.9 *piezometer*—a device so constructed and sealed as to measure hydraulic head at a point in the subsurface.

3.1.10 *specific capacity*—the rate of discharge from a well divided by the drawdown of the water level within the well at a specific time since pumping started.

3.1.11 *specific storage*—the volume of water released from or taken into storage per unit volume of the porous medium per unit change in head.

3.1.12 *specific yield*—the ratio of the volume of water that the saturated rock or soil will yield by gravity to the volume of the rock or soil. In the field, specific yield is generally determined by tests of unconfined aquifers and represents the change that occurs in the volume of water in storage per unit area of unconfined aquifer as the result of a unit change in head. Such a change in storage is produced by the draining or filling of pore space and is, therefore, mainly dependent on particle size, rate of change of the water table, and time of drainage.

3.1.13 *storage coefficient*—the volume of water an aquifer releases from or takes into storage per unit surface area of the aquifer per unit change in head. For a confined aquifer the storage coefficient is equal to the product of specific storage and aquifer thickness. For an unconfined aquifer, the storage coefficient is approximately equal to the specific yield.

3.1.14 *transmissivity*—the volume of water at the existing kinematic viscosity that will move in a unit time under a unit

[1] This guide is under the jurisdiction of ASTM Committee D-18 on Soil and Rock and is the direct responsibility of Subcommittee D18.21 on Ground Water and Vadose Zone Investigations.

Current edition approved Aug. 15, 1991. Published December 1991.

[2] *Annual Book of ASTM Standards*, Vol 04.08.

hydraulic gradient through a unit width of the aquifer.

3.1.15 For definitions of other terms used in this guide, see Terminology D 653.

4. Significance and Use

4.1 An aquifer test is a controlled field experiment made to determine the approximate hydraulic properties of water-bearing material. The hydraulic properties determined are specific to the test method. The hydraulic properties that can be determined are also dependent upon the instrumentation of the field test, the knowledge of the aquifer system at the field site, and conformance of the hydrogeologic conditions at the field site to the assumptions of the test method. Hydraulic conductivity and storage coefficient of the aquifer are the basic properties determined by most test methods. Some tests are designed to determine vertical and horizontal anisotropy, aquifer discontinuities, and vertical hydraulic conductivity of confining beds.

5. Procedure

5.1 The procedure for selection of an aquifer test method or methods is primarily based on selection of a test method that is compatible with the hydrogeology of the proposed test site. Secondarily, the test method is selected on the basis of the testing conditions specified by the test method, such as the method of stressing or causing water-level changes in the aquifer and the requirements of a test method for observations of water level response in the aquifer. The decision tree in Fig. 1 is designed to assist, first, in selecting test methods applicable to specific hydrogeologic site characteristics. Secondly, the decision tree will assist in selecting a test method on the basis of the nature of the stress on the aquifer imposed by the control well. The decision tree references the sections in this guide where the test methods are cited.

5.2 *Pre-test-selection procedures*—Aquifer test methods are highly specific to the assumptions of the analytical solution of the test method. Reliability of determination of hydraulic properties depends upon conformance of the hydrologic site characteristics to the assumptions of the test method. A prerequisite for selecting an aquifer test method is knowledge of the hydrogeology of the test site. A conceptual understanding of the hydrogeology of the aquifer system at the prospective test site should be gained in as much detail as possible from existing literature and data, and a site reconnaissance. In developing a site characterization, incorporate geologic mapping, driller's logs, geophysical logs, records of existing wells, water-level and water-quality data, and results of geophysical surveys. Include information on the thickness, lithology, stratification, depth, attitude, continuity, and extent of the aquifer and confining beds.

5.3 *Select Applicable Aquifer Test Methods*—Select a test method based on conformation of the site hydrogeology to assumptions of the test model and the parameters to be determined. A summary of principal aquifer-test methods and their applicability to hydrogeologic site conditions is given in the following paragraphs. The decision tree for aquifer test selection, Table 1, provides a graphic display of the hydrogeologic site conditions for each test method and references to the section where each test method is cited.

5.3.1 *Extensive, Isotropic, Homogeneous, Confined, Nonleaky Aquifer:*

5.3.1.1 *Constant Discharge*—Methods for tests in which the discharge or injection rate in the control well is constant are given by the nonequilibrium method of Theis (1)[3] for the drawdown and recovery phases. The Theis method is the most widely referenced and applied aquifer-test method and is the basis for the solution to other more complicated boundary condition problems. The Theis test method for the pumping or injection phase is given in Test Method D 4106. Cooper and Jacob (2) and Jacob (3) recognized that for large values of time and small values of distance from the control well the Theis solution yields a straight line on semilogarithmic plots of various combinations of drawdown and distance from the control well. The solution of the Theis equation can therefore be simplified by the use of semilogarithmic plots. The modified Theis nonequilibrium method is given in Test Method D 4105.

5.3.1.2 *Variable Discharge*—Test methods for a variably discharging control well have been presented by Stallman (4) and Moench (5). These test methods simulate pumpage as a sequence of constant-rate stepped changes in discharge. The test methods utilize the principle of superposition in constructing type curves by summing the effects of successive changes in discharge. The type curves may be derived for control wells discharging from extensive, leaky, and nonleaky confined aquifers or any situation where the response to a unit stress is known. Hantush (6) developed drawdown functions for three types of decreases in control-well discharge. Abu-Zied and Scott (7) presented a general solution for drawdown in an extensive confined aquifer in which the discharge of the control well decreases at an exponential rate. Aron and Scott (8) proposed an approximate test method of determining transmissivity and storage from an aquifer test in which discharge decreases with time during the early part of the test. Lai et al (9) presented test methods for determining the drawdown in an aquifer taking into account storage in the control well and having an exponentially and linearly decreasing discharge.

5.3.1.3 *Constant Drawdown*—Test methods have been presented to determine hydraulic-head distribution around a discharging well in a confined aquifer with near constant drawdown. Such conditions are most commonly achieved by shutting in a flowing well long enough for the head to fully recover, then opening the well. The solutions of Jacob and Lohman (10) and Hantush (6) apply to aerially extensive, nonleaky aquifers. Rushton and Rathod (11) used a numerical model to analyze aquifer-test data.

5.3.1.4 *Slug Test Methods*—Test methods for estimating transmissivity by injecting a given quantity or *slug* of water into a well were introduced by Hvorslev (12) and Ferris and Knowles (13). Solutions to overdamped well response to slug tests have also been presented by Cooper et al (14). The solution presented by Cooper et al (14) is given in Test Method D 4104. Solutions for slug tests in wells that exhibit oscillatory water-level fluctuations caused by a sudden injection or removal of a volume of water have been presented by Krauss (15), van der Kamp (16) and Shinohara and Ramey (17). Kipp (18) analyzed the complete range of response of

---

[3] The boldface numbers in parentheses refer to the list of references at the end of this guide.

D 4043

TABLE 1   Decision Tree for Selection of Aquifer Test Method

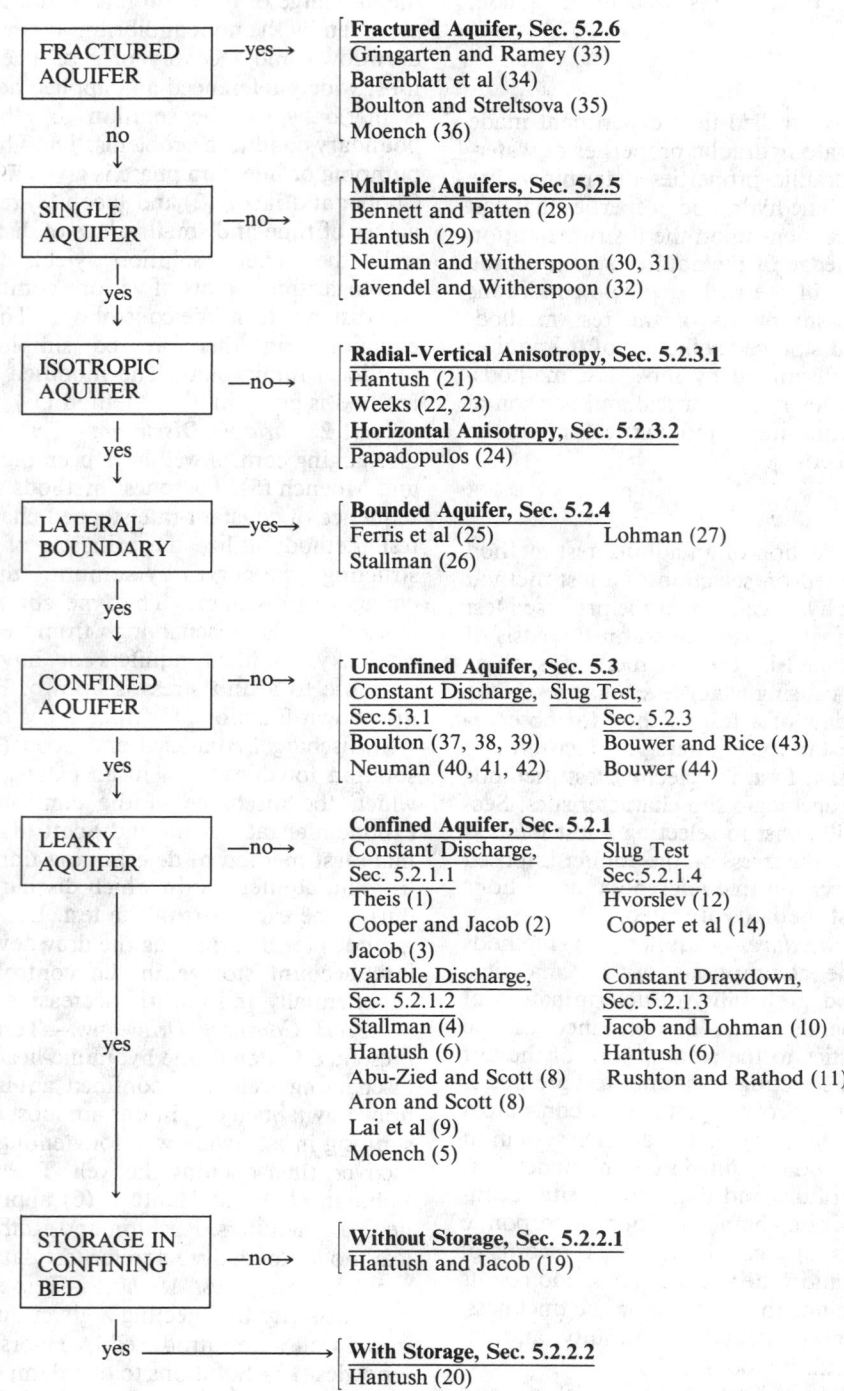

wells ranging from those having negligible inertial effects through full oscillatory behavior and developed type curves for the analysis of slug-test data. The field procedure for slug test methods is given in Test Method D 4044.

5.3.2 *Extensive, Isotropic, Homogeneous, Confined, Leaky Aquifers*—Confining beds above or below the aquifer commonly allow transmission of water to the aquifer by leakage. Test methods that account for this source of water

have been presented for several aquifer-confining bed situations.

5.3.2.1 *Leaky Confining Bed, Without Storage*—Hantush and Jacob (19) presented a solution for the situation in which a confined aquifer is overlain, or underlain, by a leaky confining layer having uniform properties. Radial flow is assumed in a uniform aquifer. The hydraulic properties of the aquifer and confining bed are determined by matching

488

logarithmic plots of aquifer-test data to a family of type curves.

5.3.2.2 *Leaky Confining Bed, With Storage*—Solutions for determining the response of a leaky confined aquifer where the release of water in the confining bed is taken into account were presented by Hantush (20). Flow in the uniform confined aquifer is assumed to be radial and flow in the leaky confining beds is assumed to be vertical.

5.3.3 *Extensive, Confined, Anisotropic Aquifer:*

5.3.3.1 *Radial-Vertical Anisotropy*—Solutions to the head distribution in a homogeneous confined aquifer with radial-vertical anisotropy in response to constant discharge of a partially penetrating well are presented by Hantush (21). Weeks (22, 23) presented test methods to determine the ratio of horizontal to vertical hydraulic conductivity.

5.3.3.2 *Horizontal Anisotropy*—Papadopulos (24) presented a test method for determination of horizontal plane anisotropy in an aerially extensive homogeneous confined aquifer.

5.3.4 *Aerially Bounded Aquifers*—Aquifer test methods discussed previously are based on the assumption that the aquifer is aerially extensive. Effects of limitations in aerial extent of aquifers by impermeable boundaries or by source boundaries, such as hydraulically connected streams, may preclude the direct application of an aquifer-test method. The method of images, described by Ferris et al (25), Stallman (26) and Lohman (27), provide solutions to head distribution in aerially finite aquifers.

5.3.5 *Multiple Aquifers*—Test methods for multiple aquifers, that is, two or more aquifers separated by a leaky confining bed and penetrated by a control well, require special methods for analysis. Bennett and Patten (28) presented a method for testing a multiaquifer system using downhole metering and constant drawdown. Hantush (29) presented solutions for two aquifers separated by a leaky confining bed. Neuman and Witherspoon (30) provided solutions for drawdown in leaky confining beds above and below an aquifer being pumped. Neuman and Witherspoon (31) developed an analytical solution for the flow in a leaky confined system of two aquifers separated by a leaky

confining bed with storage. Javendel and Witherspoon (32) presented a finite-element method of analyzing anisotropic multiaquifer systems.

5.3.6 *Fractured Media*—Solutions for the flow in a single finite fracture are presented by Gringarten and Ramey (33). Barenblatt et al (34) presented a test method for solving a double-porosity model. Boulton and Streltsova (35) presented a solution for a system of porous layers separated by fractures. Moench (36) developed type curves for a double-porosity model with a fracture skin that may be present at the fracture-block interface as a result of mineral deposition or alteration.

5.4 *Extensive, Isotropic, Homogeneous, Unconfined Aquifer*—Conditions governing drawdown due to discharge from an unconfined aquifer differ markedly from those due to discharge from a nonleaky confined aquifer. Difficulties in deriving analytical solutions to the hydraulic-head distribution in an unconfined aquifer result from the following characteristics: (*1*) transmissivity varies in space and time as the water table is drawn down and the aquifer is dewatered, (*2*) water is derived from storage in an unconfined aquifer mainly at the free water surface and, to a lesser degree, from each discrete point within the aquifer, and (*3*) vertical components of flow exist in the aquifer in response to withdrawal of water from a well in an unconfined aquifer.

5.4.1 Boulton (37, 38, 39) introduced a mathematical solution to the head distribution in response to discharge at a constant rate from an unconfined aquifer. Boulton's solution invokes the use of a semi-empirical delay index that was not defined on a physical basis. Neuman (40, 41, 42) presented solutions for unconfined aquifer tests utilizing fully penetrating and partially penetrating control and observation wells hypothesized on well-defined physical properties of the aquifer.

5.4.2 Bouwer and Rice (43) and Bouwer (44) present a slug test method for unconfined aquifer conditions.

6. Keywords

6.1 aquifers; aquifer tests; confining beds; control wells; discharging wells; hydraulic conductivity; observation wells; piezometers; storage coefficient; storativity; transmissivity

REFERENCES

(1) Theis, C. V., "The Relation Between The Lowering of the Piezometric Surface and the Rate and Duration of Discharge of a Well Using Ground-Water Storage," *American Geophysical Union Transactions*, Vol 16, Pt. 2, 1935, pp. 519–524.

(2) Cooper, H. H., Jr., and Jacob, C. E., "Generalized Graphical Method for Evaluating Formation Constants and Summarizing Well-Field History," *American Geophysical Union Transactions*, Vol 27, No. 4, 1946, pp. 526–534.

(3) Jacob, C. E., "Flow of Ground Water," *Engineering Hydraulics*, Proceedings of the Fourth Hydraulics Conference, June 12–15, 1949, John Wiley & Sons, Inc., New York, NY, 1950, pp. 321–386.

(4) Stallman, R. W., "Variable Discharge without Vertical Leakage (Continuously Varying Discharge)," *Theory of Aquifer Tests*, U. S. Geological Survey Water-Supply Paper 1536-E, 1962, pp. 118–122.

(5) Moench, A. E., "Ground-Water Fluctuations in Response to Arbitrary Pumpage," Ground Water Vol 9, No. 2, 1971, pp. 4–8.

(6) Hantush, M. S., "Hydraulics of Wells," *Advances in Hydroscience*, Vol 1, Academic Press, Inc., New York, 1964, pp. 281–442.

(7) Abu-Zied, M., and Scott, V. H., "Nonsteady Flow for Wells with Decreasing Discharge," *Proceedings*, American Society of Civil Engineers, Vol 89, No. HY3, 1963, pp. 119–132.

(8) Aron, G., and Scott, V. H., "Simplified Solutions for Decreasing Flow in Wells," *Proceedings*, American Society of Civil Engineers Vol 91, No. HY5, 1965, pp. 1–12.

(9) Lai, R. Y., Karadi, G. M., and Williams, R. A., "Drawdown at Time-Dependent Flowrate," *Water Resources Bulletin*, Vol 9, No. 5, 1973, pp. 854–859.

(10) Jacob, C. E., and Lohman, S. W., "Nonsteady Flow to a Well of Constant Drawdown in an Extensive Aquifer," *American Geophysical Union Transactions*, Vol 33, No. 4, 1952, pp. 552–569.

(11) Rushton, K. R., and Rathod, K. S., "Overflow Tests Analyzed by Theoretical and Numerical Methods," *Ground Water*, Vol 18, No. 1, 1980, pp. 61–69.

(12) Hvorslev, M. J., "Time Lag and Soil Permeability in Ground-Water Observations," U. S. Army Corps of Engineers, 1951, p. 49.

(13) Ferris, J. G., and Knowles, D. B., "The Slug Test for Estimating Transmissibility," *Ground Water Note 26*, U. S. Geological Survey, 1954, p. 26.

(14) Cooper, H. H., Jr., Bredehoeft, J. D., and Papdopulos, I. S., "Response of a Finite-Diameter Well to an Instantaneous Charge of Water," *Water Resources Research*, Vol 3, 1967, pp. 263–269.

(15) Krauss, I., "Die Bestimmung der Transmissivitat von Grundwasserleitern aus dem Einschwingverhalten des Brunnen-Grundwasserleitersystems," *Journal of Geophysics*, Vol 40, 1974, pp. 381–400.

(16) van der Kamp, Garth, "Determining Aquifer Transmissivity by Means of Well Response Tests—The Underdamped Case," *Water Resources Research*, Vol 12, No. 1, 1976, pp. 71–77.

(17) Shinohara, K., and Ramey, H. J., "Slug Test Data Analysis, including the Inertial Effect of the Fluid in the Well Bore," *54th Annual Fall Conference and Exhibition of the Society of Petroleum Engineers*, 1979.

(18) Kipp, K. L., "Type Curve Analysis of Inertial Effects in the Response of a Well to a Slug Test," *Water Resources Research*, Vol 21, No. 9, 1985, pp. 1399–1408.

(19) Hantush, M. S., and Jacob, C. E., "Non-Steady Radial Flow in an Infinite Leaky Aquifer," *American Geophysical Union Transactions*, Vol 36, No. 1, 1955, pp. 95–100.

(20) Hantush, M. S., "Modification of the Theory of Leaky Aquifers," *Journal of Geophysical Research*, Vol 65, No. 11, 1960, pp. 3713–3725.

(21) Hantush, M. S., "Drawdown Around a Partially Penetrating Well," *Proceedings*, American Society of Civil Engineers, Vol 87, No. HY4, 1961, pp. 83–98.

(22) Weeks, E. P., "Field Methods for Determining Vertical Permeability and Aquifer Anisotropy," *U. S. Geological Survey Professional Paper 501-D*, 1964, pp. D193–D198.

(23) Weeks, E. P., "Determining the Ratio of Horizontal to Vertical Permeability by Aquifer-Test Analysis" *Water Resources Research*, Vol 5, No. 1, 1969, pp. 196–214.

(24) Papadopulos, I. S., "Nonsteady Flow to a Well in an Infinite Anisotropic Aquifer," *Symposium of Dubrovnik*, International Association of Scientific Hydrology, 1965, pp. 21–31.

(25) Ferris, J. G., Knowles, D. B., Brown, R. H., and Stallman, R. W., "Theory of Aquifer Tests," *U. S. Geological Survey Water-Supply Paper 1536-E*, 1962, pp. 69–174.

(26) Stallman, R. W., "Type Curves for the Solution of Single-Bound Problems," in Bentall, Ray, Compiler, *Short Cuts and Special Problems in Aquifer Tests, U. S. Geological Survey Water-Supply Paper 1545-C*, 1963, pp. 45–47.

(27) Lohman, S. W., "Ground-Water Hydraulics," *Professional Paper 708*, U. S. Geological Survey, 1972, p. 70.

(28) Bennett, G. D., and Patten, E. P., Jr., "Constant-Head Pumping Test of a Multiaquifer Well to Determine Characteristics of Individual Aquifers," *U. S. Geological Survey Water-Supply Paper 1536-G*, 1962, p. 203.

(29) Hantush, M. S., "Flow to Wells Separated by Semipervious Layer," *Journal of Geophysical Research*, Vol 72, No. 6, 1967, pp. 1709–1720.

(30) Neuman, S. P., and Witherspoon, P. A., "Field Determination of the Hydraulic Properties of Leaky Multiple Aquifer Systems," *Water Resources Research*, Vol 8, No. 5, 1972, pp. 1284–1298.

(31) Neuman, S. P., and Witherspoon, P. A., "Theory of Flow in a Confined Two Aquifer System," *Water Resources Research*, Vol 5, No. 4, 1969, pp. 803–816.

(32) Javandel, I., and Witherspoon, P. A., "Method of Analyzing Transient Fluid Flow in Multilayered Aquifers," *Water Resources Research*, Vol 5, No. 4, 1969, pp. 856–869.

(33) Gringarten, A. C., and Ramey, H. J., Jr., "Unsteady-State Pressure Distributions Created by a Well with a Single Horizontal Fracture, Partial Penetration, or Restricted Entry," *Society of Petroleum Engineers Journal*, Vol 14, No. 4, 1974, pp. 413–426.

(34) Barenblatt, G. I., Zheltov, Iu P., and Kochina, I. N., "Basic Concepts in the Theory of Seepage of Homogeneous Liquids in Fissured Rocks [Strata]," *Journal of Applied Mathematics and Mechanics*, Vol 24, 1960, pp. 1286–1301.

(35) Boulton, N. S., and Streltsova, T. D., "Unsteady Flow to a Pumped Well in a Fissured Water-Bearing Formation," *Journal of Hydrology*, Vol 35, 1977, pp. 257–269.

(36) Moench, A. F., "Double Porosity Model for a Fissured Groundwater Reservoir with Fracture Skin," *Water Resources Research*, Vol 20, No. 7, 1984, pp. 831–846.

(37) Boulton, N. S., "Drawdown of the Water Table Under Non-Steady Conditions Near a Pumped Well in an Unconfined Formation," *Proceedings*, Institution of Civil Engineers, Vol 3, Pt. 3, 1954, pp. 564–579.

(38) Boulton, N. S., "Unsteady Flow to a Pumped Well Allowing for Delayed Yield from Storage," *International Association of Scientific Hydrology*, Publication 37, 1954, pp. 472–477.

(39) Boulton, N. S., "Analysis of Data from Non-Equilibrium Pumping Tests Allowing for Delayed Yield from Storage," *Proceedings*, Institution of Civil Engineers, Vol 26, 1963, pp. 469–482.

(40) Neuman, S. P., "Theory of Flow in Unconfined Aquifers Considering Delayed Response of the Water Table," *Water Resources Research*, Vol 8, No. 4, 1972, pp. 1031–1045.

(41) Neuman, S. P., "Supplementary Comments on 'Theory of Flow in Unconfined Aquifers Considering Delayed Response of the Water Table'," *Water Resources Research*, Vol 9, No. 4, 1973, pp. 1102.

(42) Neuman, S. P., "Analysis of Pumping Test Data from Anisotropic Unconfined Aquifers Considering Delayed Gravity Response," *Water Resources Research*, Vol 11, No. 2, 1975, pp. 329–342.

(43) Bouwer, H., and Rice, R. C., "A Slug Test for Determining Hydraulic Conductivity of Unconfined Aquifers with Completely or Partially Penetrating Wells," *Water Resources Research*, Vol 12, No. 3, pp. 423–428.

(44) Bouwer, H., "The Bouwer-Rice Slug Test—An Update," *Ground Water*, Vol 27, No. 3, 1989, pp. 304–309.

Designation: D 5737 – 95

Standard Guide for
Methods for Measuring Well Discharge[1]

This standard is issued under the fixed designation D 5737; the number immediately following the designation indicates the year of original adoption or, in the case of revision, the year of last revision. A number in parentheses indicates the year of last reapproval. A superscript epsilon (ε) indicates an editorial change since the last revision or reapproval.

1. Scope

1.1 This guide covers an overview of methods to measure well discharge. This guide is an integral part of a series of standards prepared on the in-situ determination of hydraulic properties of aquifer systems by single- or multiple-well tests. Measurement of well discharge is a common requirement to the determination of aquifer and well hydraulic properties.

1.2 This guide does not establish a fixed procedure for any method described. Rather, it describes different methods for measuring discharge from a pumping or flowing well. A pumping well is one type of control well. A control well can also be an injection well or a well in which slug tests are conducted.

1.3 This guide does not address borehole flow meters that are designed for measuring vertical or horizontal flow within a borehole.

1.4 The values stated in SI units are to be regarded as standard.

1.5 *This standard does not purport to address all of the safety concerns, if any, associated with its use. It is the responsibility of the user of this standard to establish appropriate safety and health practices and determine the applicability of regulatory limitations prior to use.* Furthermore, it is the user's responsibility to properly dispose of water discharged.

2. Referenced Documents

2.1 *ASTM Standards:*
D 653 Terminology Relating to Soil, Rock, and Contained Fluids[2]
D 1941 Test Method for Open Channel Flow Measurement of Water with the Parshall Flume[3]
D 4043 Guide for Selection of Aquifer-Test Method in Determining Hydraulic Properties by Well Techniques[4]
D 5242 Test Method for Open-Channel Flow Measurement of Water Indirectly at Culverts[3]
D 5390 Test Method for Open Channel Flow Measurement with Palmer-Bowlus Flumes[5]
D 5716 Test Method to Measure the Rate of Well Discharge by Circular Orifice Weir
2.2 *ISO Standard:*

Recommendation R541 Measurement of Fluid Flow by Means of Orifice Plates and Nozzles[6]
2.3 *ANSI Standard:*
Standard 1042 Part 1 Methods for the Measurement of Fluid Flow in Pipes, 1, Orifice Plates, Nozzles and Venturi Tubes[6]
2.4 *ASME Standard:*
Standard MFC-3M-1989 Measurement of Fluid Flow in Pipes Using Orifice, Nozzle, and Venturi[7]

3. Terminology

3.1 *Definitions:*
3.1.1 *conceptual model*—an interpretation or description of the characteristics, interactions, and dynamics of a physical system.
3.1.2 *control well*—a well by which the head and flow in the aquifer is changed, by pumping, injection, or imposing a change of head.
3.1.3 *discharge*—or rate of flow, is the volume of water that passes a particular reference section in a unit of time.
3.1.4 *totalizing flow meter*—a flow meter that indicates the cumulative flow displayed as a volume. The flow rate is calculated based on the time between two readings.
3.2 For definitions of other terms used in this guide, see Terminology D 653.

4. Significance and Use

4.1 This guide is limited to the description of test methods typical for measurement of ground-water discharge from a control well.

4.1.1 Controlled field tests are the primary means of determining aquifer properties. Most mathematical equations developed for analyzing field tests require measurement of control well discharge.

4.1.2 Discharge may be needed for evaluation of well design and efficiency.

4.1.3 For aquifer tests, a conceptual model should be prepared to evaluate the proper test method and physical test requirements, such as well placement and design (see Guide D 4043). Review the site data for consistency with the conceptual model. Revise the conceptual model as appropriate and consider the implications on the planned activities.

4.1.4 For aquifer tests, the discharge rate should be sufficient to cause significant stress of the aquifer without violating test assumptions. Conditions that may violate test

---

[1] This guide is under the jurisdiction of ASTM Committee D-18 on Soil and Rock and is the direct responsibility of Subcommittee D18.21 on Ground Water and Vadose Zone Investigations.
Current edition approved June 15, 1995. Published August 1995.
[2] *Annual Book of ASTM Standards,* Vol 04.03.
[3] *Annual Book of ASTM Standards,* Vol 11.01.
[4] *Annual Book of ASTM Standards,* Vol 04.08.
[5] *Annual Book of ASTM Standards,* Vol 11.02.

[6] Available from American National Standards Institute, 11 W. 42nd St., 13th Floor, New York, NY 10036.
[7] Available from American Society of Mechanical Engineers, 345 E. 47th Street, New York, NY 10017.

assumptions include conversion of the aquifer from confined to unconfined conditions, lowering the water level in the control well to below the top of the well screen, causing a well screen entrance velocity that promotes well development during the test, or decreasing the filter pack permeability characteristics.

4.1.5 Some test methods described here are not applicable to injection well tests.

4.2 This guide does not apply to test methods used in measurement of flow of other fluids used in industrial operations, such as waste water, sludge, oil, and chemicals.

5. Test Methods

5.1 *Selection of a Well Discharge Rate Measurement Method*—Select a well discharge measurement method based on the desired discharge rate or rates, the desired pumping method, the required accuracy and frequency of measurement, the type of pump discharge and the water conveyance method.

5.2 *Principal Well Discharge Rate Measurement Methods*—A summary of principal methods is given below for typical hydrogeologic testing. Additional information may be found in a publication of the National Institute of Standards and Technology (NIST) (1)[8], the American Society of Mechanical Engineers (ASME) (2) or in a comprehensive book on the subject of flow meter engineering (3). Discharge methods can be classified as open channel flow and closed conduit flow. Open channel flow is limited to calibrated control structures, such as weirs and flumes. Closed conduit flow includes methods such as turbine meters and magnetic meters. Also included are methods that measure the discharge of water from the closed conduit to the air, such as the orifice tube.

5.3 *Open Channel Flow Methods:*

5.3.1 *Weirs*—A weir is a vertical obstruction that restricts the total flow of water in channel. Weirs fall into three general classifications, sharp crested, broad crested, and suppressed. Sharp crested weirs use a flat plate that is configured in a triangular "V" or rectangular shape; they are described in 5.3.1.1. See Test Method D 5242. Broad crested weirs are wide rectangular restrictions that are usually only used as spillways in dams. They are not described here. More information on broad crested weirs may be found in Ref (4). A third classification of weirs, called suppressed weirs, are more commonly known as flumes. Flumes are discussed in 5.3.2.

5.3.1.1 *Sharp Crested Weirs*—The weir is placed flush against the flowing stream, and the notch is made as sharp as possible using a flat piece of metal with sharp edges forming the weir notch. The relation between the head and the discharge of a weir varies according to the shape of the weir notch. A weir is inexpensive to construct, easy to install and highly accurate when installed and used properly.

5.3.2 *Flume*—A flume is a device that restricts flow in the channel which causes the water to accelerate, producing a corresponding change in the water level. The head can then be related to discharge. Several types of flumes have been developed; the most common flume for measuring well discharge is the Parshall flume, originally designed by R. L. Parshall of the U.S. Soil Conservation Service (5). See Test Methods D 1941 and D 5390.

5.3.2.1 Flumes have several advantages over weirs. The most important of these is the self-cleaning capacity of flumes compared with sharp-edged weirs. Head losses through a flume are also much less than for a weir, so when the available head is limited, flumes are more desirable. Flumes can function over a wide range of discharges and still require only a single upstream head measurement. However, flumes require more time to set up than weirs.

5.4 *Closed Conduit Methods:*

5.4.1 *Invasive Methods:*

5.4.1.1 *Turbine-Type (Propeller) Flow Meters*—A totalizing flow meter is a device used in measuring water in most domestic and commercial potable water uses. This flow meter consists of a flow tube in which a rotor blade is mounted together with either a means of generating an electrical signal proportional to the angular velocity of the rotor or a mechanical system of gears that rotates proportional to the flow volume. The meter is installed as a section of the water line between the pump and the point of discharge. Turbine-type flow meters have a limited operating range. The meter must be calibrated and the pipe must be full. Mechanical turbine meters typically only totalize flow.

5.4.2 *Non-Invasive Methods:*

5.4.2.1 *Magnetic Flow Meters*—The magnetic flow meter operates on the same general principle as an electric generator (2). The pipe is placed such that the fluid path is normal to the magnetic field. The motion of the fluid through the magnetic field induces an electromotive force across the fluid. By placing insulated electrodes in the pipe in a plane normal to the magnetic field the strength of the field can be measured using a special voltmeter. An electromotive force is induced in the flowing water (that is, the conductor) across a pair of electrodes. Advantages are that there is no added head loss. Disadvantages include their relatively high cost and potential errors due to scaling.

5.4.2.2 *Venturi Meters*—The venturi meter uses the relationship between pressure and flow velocity across a throat. Advantages include less head loss relative to orifice meters and less required maintenance. More information may be gained from British Standard 1042.

5.4.2.3 *Acoustic Meters*—The acoustic meter, also called the sonic meter or doppler flow meter, uses sound waves in conjunction with knowledge of pipe wall thickness to allow estimation of flow rate. Many acoustic meters require suspended material or entrapped air to obtain a quality reading. An advantage of acoustic meters over other types is their limited mechanical parts and thus longer equipment life. A disadvantage is sensitivity to pipe encrustation.

5.4.3 *Discharge to Air Methods:*

5.4.3.1 *Bucket and Stop Watch*—The bucket and stop watch method is simply the collection of discharged water in a container of known volume for a measured period of time. The volume collected divided by the time the water is collected is the discharge rate over that time period. Alternately, the volume can be determined by measuring the weight and using a density conversion. The rate is measured periodically over the course of the test. Advantages include

---

[8] The boldface numbers given in parentheses refer to a list of references at the end of the text.

its ease to set up and the simplicity of taking readings. Disadvantages include its manual operation and inability to obtain continuous measurements.

5.4.3.2 *Orifice Bucket*—The orifice bucket is a container with precisely cut holes in the bottom and a calibrated piezometer tube on the side. The well discharge is directed into the top of the bucket where water then accumulates as it is delayed in flowing out the holes located in the base of the container. The accumulated head can be read on the piezometer tube. The discharge is read from a chart relating discharge to head. The device is especially useful in measuring rates of production of reciprocating and airlift pumps where the flow is not at a uniform rate. An orifice bucket 30.5 cm (12 in.) in diameter can be constructed to measure discharge rates from 8 to 151 L/min (2 to 40 gal/min). An orifice bucket 61.0 cm (24 in.) in diameter can be constructed to measure discharge rates from 38 to 680 L/min (10 to 180 gal/min). Advantages include its ease in setup and use and its ability to be configured with a float water level recorder. The orifice bucket was described by the Illinois State Water Survey (5).

5.4.3.3 *Circular Orifice Weir*—Also called the orifice tube and orifice meter, the circular orifice weir is the device often used to measure the discharge rate from a high-capacity pump. The orifice meter is a circular restriction in a pipe that causes back pressure that can be measured in a piezometer tube. The water level in the piezometer tube is the pressure head in the approach pipe when water is being pumped through the orifice. For any size of orifice diameter and approach pipe diameter, the rate of flow through the orifice varies with the pressure head in the piezometer tube. For example, a discharge of 208 L/min (55 gal per minute) will cause 12.7 cm (5 in.) of head due to a 6.35-cm (2½-in.) orifice and a 10.16-cm (4-in.) approach pipe. Similarly, a discharge of 20 100 L/min (5 310 gal/min) will cause 177.8 cm (70 in.) of head due to a 30.48-cm (12-in.) orifice and a

40.64 cm (16-in.) approach pipe. Advantages are the low cost and relatively high accuracy of measurement. Disadvantages include the relatively low range of measurement, high head loss, and physical constraints, for example, the orifice meter must be level and must always run full. For more information see I.S.O. Recommendation R541 and Layne Western Company, Inc. (6) and Test Method D 5716.

5.4.3.4 *Trajectory Method*—The flow rate can be determined from measurement of the trajectory of a horizontal pipe discharge to the air. Curves were prepared from Purdue University experiments in 1948 on pipes from 5.08 to 15.24 cm (2 to 6 in.) in diameter (1). Tabularized data also exist for discharge from vertical pipes.

5.5 *Procedures Specific to Measurement of Well Discharge for Aquifer Test Analysis:*

5.5.1 Certain aquifer tests require variation of the discharge rate over the course of the test. Therefore, select a test method that is capable of accurate measurement over the required discharge rate range.

5.5.2 Methods that allow continuous measurement of discharge are desirable to establish any change or trend. Most test methods for analysis of aquifers properties include an assumption that the discharge is constant. Some variation is acceptable. The specific aquifer test analysis method should be considered in determining acceptable variation in discharge.

6. Report

6.1 Prepare a report presenting the purposes of discharge measurement and the criteria used in selecting a particular method. Discuss the frequency of discharge measurements required and the anticipated accuracy of the measurement method.

7. Keywords

7.1 aquifer; aquifer test methods; discharge rate; flume; ground water; weir

REFERENCES

(1) National Institute of Standards and Technology Special Publication 421, *A Guide to Methods and Standards for the Measurement of Water Flow*, Kulin, G. and Compton, P. R., Institute for Basic Standards, NIST, Washington, D.C. 20234, May 1975.

(2) *Fluid Meters, Their Theory and Application, A Report of ASME Research Committee on Fluid Meters*, Sixth Edition, edited by H. S. Bean, American Society of Mechanical Engineers, 345 East 47th Street, New York, NY 10017, 1971.

(3) Spink, L. K., *Principles and Practice of Flow Meter Engineering*, Ninth Edition, The Foxboro Co. Foxboro, MA, 1967.

(4) *Water Measurement Manual*, Bureau of Reclamation, U.S. Dept. of the Interior, U.S. Government Printing Office, 1984.

(5) Parshall, R. L., *Measuring Water in Irrigation Channels with Parshall Flumes and Small Weirs*, Soil Conservation Circular No. 843, U.S. Department of Agriculture, May, 1950.

(6) *Measurement of Water Flow Through Pipe Orifice With Free Discharge*, Bulletin 501, Published by Layne & Bowler, Inc., Memphis, TN, available from Layne Western Company, Inc. 5800 Foxridge Dr., Mission, KS 66202, 1958.

Standard Guide for
Presentation of Water-Level Information From Ground-Water Sites[1]

This standard is issued under the fixed designation D 6000; the number immediately following the designation indicates the year of original adoption or, in the case of revision, the year of last revision. A number in parentheses indicates the year of last reapproval. A superscript epsilon (ε) indicates an editorial change since the last revision or reapproval.

1. Scope

1.1 This guide covers a series of options, but does not specify a course of action. It should not be used as the sole criterion or basis of comparison, and does not replace or relieve professional judgment.

1.2 This guide summarizes methods for the presentation of water-level data from ground-water sites.

NOTE 1—As used in this guide, a site is meant to be a single point, not a geographic area or property, located by an X, Y, and Z coordinate position with respect to land surface or a fixed datum. A ground-water site is defined as any source, location, or sampling station capable of producing water or hydrologic data from a natural stratum from below the surface of the earth. A source or facility can include a well, spring or seep, and drain or tunnel (nearly horizontal in orientation). Other sources, such as excavations, driven devices, bore holes, ponds, lakes, and sinkholes, which can be shown to be hydraulically connected to the ground water, are appropriate for the use intended.

1.3 The study of the water table in aquifers helps in the interpretation of the amount of water available for withdrawal, aquifer tests, movement of water through the aquifers, and the effects of natural and human-induced forces on the aquifers.

1.4 A single water level measured at a ground-water site gives the height of water at one vertical position in a well or borehole at a finite instant in time. This is information that can be used for preliminary planning in the construction of a well or other facilities, such as disposal pits.

NOTE 2—Hydraulic head measured within a short time from a series of sites at a common (single) horizontal location, for example, a specially constructed multi-level test well, indicate whether the vertical hydraulic gradient may be upward or downward within or between the aquifer (see 7.2.1).

NOTE 3—The phrases "short time period" and "finite instant in time" are used throughout this guide to describe the interval for measuring several project-related ground-water levels. Often the water levels of ground-water sites in an area of study do not change significantly in a short time, for example, a day or even a week. Unless continuous recorders are used to document water levels at every ground-water site of the project, the measurement at each site, for example, use of a steel tape, will be at a slightly different time (unless a large staff is available for a coordinated measurement). The judgment of what is a critical time period must be made by a project investigator who is familiar with the hydrology of the area.

1.5 Where hydraulic heads are measured in a short period of time, for example, a day, from each of several horizontal locations within a specified depth range, or hydrogeologic unit, or identified aquifer, a potentiometric surface can be drawn for that depth range, or unit, or aquifer. Water levels from different vertical sites at a single horizontal location may be averaged to a single value for the potentiometric surface when the vertical gradients are small compared to the horizontal gradients.

NOTE 4—The potentiometric surface assists in interpreting the gradient and horizontal direction of movement of water through the aquifer. Phenomena such as depressions or sinks caused by withdrawal of water from production areas and mounds caused by natural or artificial recharge are illustrated by these potentiometric maps.

1.6 Essentially all water levels, whether in confined or unconfined aquifers, fluctuate over time in response to natural- and human-induced forces.

NOTE 5—The fluctuation of the water table at a ground-water site is caused by several phenomena. An example is recharge to the aquifer from precipitation. Changes in barometric pressure cause the water table to fluctuate because of the variation of air pressure on the ground-water surface, open bore hole, or confining sediment. Withdrawal of water from or artificial recharge to the aquifer should cause the water table to fluctuate in response. Events such as rising or falling levels of surface water bodies (nearby streams and lakes), evapotranspiration induced by phreatophytic consumption, ocean tides, moon tides, earthquakes, and explosions cause fluctuation. Heavy physical objects that compress the surrounding sediments, for example, a passing train or car or even the sudden load effect of the starting of a nearby pump, can cause a fluctuation of the water table (1).[2]

1.7 This guide covers several techniques developed to assist in interpreting the water table within aquifers. Tables and graphs are included.

1.8 This guide includes methods to represent the water table at a single ground-water site for a finite or short period of time, a single site over an extended period, multiple sites for a finite or short period in time, and multiple sites over an extended period.

NOTE 6—This guide does not include methods of calculating or estimating water levels by using mathematical models or determining the aquifer characteristics from data collected during controlled aquifer tests. These methods are discussed in Guides D 4043, D 5447, and D 5490, Test Methods D 4044, D 4050, D 4104, D 4105, D 4106, D 4630, D 4631, D 5269, D 5270, D 5472, and D 5473.

1.9 Many of the diagrams illustrated in this guide include notations to help the reader in understanding how these diagrams were constructed. These notations would not be required on a diagram designed for inclusion in a project document.

---

[1] This guide is under the jurisdiction of ASTM Committee D-18 on Soil and Rock and is the direct responsibility of Subcommittee D18.21 on Ground-Water and Vadose-Zone Investigations.

Current edition approved August 10, 1996. Published December 1996.

[2] The boldface numbers in parentheses refer to a list of references at the end of this guide.

NOTE 7—Use of trade names in this guide is for identification purposes only and does not constitute endorsement by ASTM.

2. Referenced Documents

2.1 *ASTM Standards:*

D 653 Terminology Relating to Soil, Rock, and Contained Fluids[3]

D 4043 Guide for Selection of Aquifer-Test Method in Determining of Hydraulic Properties by Well Techniques[3]

D 4044 Test Method (Field Procedure) for Instantaneous Change in Head (Slug) Tests for Determining Hydraulic Properties of Aquifers Systems[3]

D 4050 Test Method (Field Procedure) for Withdrawal and Injection Well Tests for Determining Hydraulic Properties of Aquifer Systems[3]

D 4104 Test Method (Analytical Procedure) for Determining Transmissivity of Nonleaky Confined Aquifers by Overdamped Well Response to Instantaneous Change in Head (Slug Tests)[3]

D 4105 Test Method (Analytical Procedure) for Determining Transmissivity and Storage Coefficient of Nonleaky Confined Aquifers by the Modified Theis Nonequilibrium Method[3]

D 4106 Test Method (Analytical Procedure) for Determining Transmissivity and Storage Coefficient of Nonleaky Confined Aquifers by the Modified Theis Nonequilibrium Method[3]

D 4630 Test Method for Determining Transmissivity and Storage Coefficient of Low Permeability Rocks by in Situ Measurements Using the Constant Head Injection Test[3]

D 4631 Test Method for Determining Transmissivity and Storativity of Low Permeability Rocks by in Situ Measurements Using the Pressure Pulse Technique[3]

D 4750 Test Method for Determining Subsurface Liquid Levels in a Borehole or Monitoring Well (Observation Well)[3]

D 5092 Practice for Design and Installation of Ground Water Monitoring Wells in Aquifers[4]

D 5254 Practice for the Minimum Set of Data Elements to Identify a Ground-Water Site[4]

D 5269 Test Method for Determining Transmissivity of Nonleaky Confined Aquifers by the Theis Recovery Method[4]

D 5270 Test Method for Determining Transmissivity and Storage Coefficient of Bounded, Nonleaky, Confined Aquifers[4]

D 5408 Guide for the Set of Data Elements to Describe a Ground-Water Site; Part 1—Additional Identification Descriptors[4]

D 5409 Guide for the Set of Data Elements to Describe a Ground-Water Site; Part 2—Physical Descriptors[4]

D 5410 Guide for the Set of Data Elements to Describe a Ground-Water Site; Part 3—Usage Descriptors[4]

D 5447 Guide for Application of a Ground-Water Flow Model to a Site–Specific Problem[4]

D 5472 Test Method for Determining Specific Capacity and Estimating Transmissivity at the Control Well[4]

D 5473 Test Method for (Analytical Procedure for) Analyzing the Effects of Particle Penetration of Control Well and Determining the Horizontal and Vertical Hydraulic Conductivity in a Nonleaky Confined Aquifer[4]

D 5474 Guide for Selection of Data Elements for Ground-Water Investigations[4]

D 5490 Guide for Comparing Ground-Water Flow Model Simulations to Site-Specific Information[4]

D 5609 Guide for Defining Boundary Conditions in Ground-Water Flow Modeling[4]

3. Terminology

3.1 All definitions appear in Terminology D 653.

3.1.1 *aquifer, n*—a geologic formation, group of formations, or part of a formation that is saturated and is capable of providing a significant quantity of water. **D 653, D 5092**

3.1.2 *aquitard, n*—a confining bed that retards but does not prevent the flow of water to or from an adjacent aquifer; a leaky confining bed. **D 653**

3.1.3 *confined or artesian aquifer, n*—an aquifer bounded above and below by confining beds and in which the static head is above the top of the aquifer. **D 4050, D 4104, D 4105, D 4106, D 5269, D 5609**

3.1.4 *hydrograph, n*—for ground water, a graph showing the water level or head with respect to time (2).

3.1.5 *unconfined or water-table aquifer, n*—an aquifer that has a water table (3). **D 4050, D 4105, D 4106, D 5609**

3.1.6 *water level, n*—for ground water, the level of the water table surrounding a borehole or well. The ground-water level can be represented as an elevation or as a depth below the ground surface. **D 4750**

3.1.7 *water table (ground-water table), n*—the surface of a ground-water body at which the water pressure equals atmospheric pressure. Earth material below the ground-water table is saturated with water. **D 653, D 4750**

4. Summary of Guide

4.1 The Significance and Use section presents the relevance of the tables and diagrams of the water table and related parameters.

4.2 A description is given of the selection process for data presentation along with a discussion on water level data preparation.

4.3 Tabular methods of presenting water-levels:

4.3.1 Tables with single water levels, and

4.3.2 Tables with multiple water levels (4).

4.4 Graphical methods for presenting water levels:

4.4.1 Vertical gradient at a single site,

4.4.2 Hydrographs,

4.4.3 Temporal trends in hydraulic head,

4.4.4 Potentiometric maps,

4.4.5 Change maps,

4.4.6 Water-table cross sections, and

4.4.7 Statistical comparisons of water levels.

4.5 Sources for automated procedures (computer-aided graphics) for basic calculations and the construction of the water-level tables and diagrams are identified.

4.6 Keywords.

---

[3] *Annual Book of ASTM Standards*, Vol 04.08.

[4] *Annual Book of ASTM Standards*, Vol 04.09.

4.7 A list of references is given for additional information.

5. Significance and Use

5.1 Determining the potentiometric surface of an area is essential for the preliminary planning of any type of construction, land use, environmental investigations, or remediation projects that may influence an aquifer.

5.1.1 The potentiometric surface in the proposed impacted aquifer must be known to properly plan for the construction of a water withdrawal or recharge facility, for example, a well. The method of construction of structures, such as buildings, can be controlled by the depth of the ground water near the project. Other projects built below land surface, such as mines and tunnels, are influenced by the hydraulic head.

5.2 Monitoring the trend of the ground-water table in an aquifer over a period of time, whether for days or decades, is essential for any permanently constructed facility that directly influences the aquifer, for example, a waste disposal site or a production well.

5.2.1 Long-term monitoring helps interpret the direction and rate of movement of water and other fluids from recharge wells and pits or waste disposal sites. Monitoring also assists in determining the effects of withdrawals on the stored quantity of water in the aquifer, the trend of the water table throughout the aquifer, and the amount of natural recharge to the aquifer.

5.3 This guide describes the basic tabular and graphic methods of presenting ground-water levels for a single ground-water site and several sites over the area of a project. These methods were developed by hydrologists to assist in the interpretation of hydraulic-head data.

5.3.1 The tabular methods help in the comparison of raw data and modified numbers.

5.3.2 The graphical methods visually display seasonal trends controlled by precipitation, trends related to artificial withdrawals from or recharge to the aquifer, interrelationship of withdrawal and recharge sites, rate and direction of water movement in the aquifer, and other events influencing the aquifer.

5.4 Presentation techniques resulting from extensive computational methods, specifically the mathematical models and the determination of aquifer characteristics, are contained in the ASTM standards listed in Section 2.

6. Selection and Preparation of Water-Level Data

6.1 Water levels should be subject to rigorous quality-control standards. Correct procedures must be followed and properly recorded in the field and the office in order for the water table to represent that in the aquifer.

6.1.1 Field-quality controls include the use of an accurate and calibrated measuring device, a clearly marked and unchanging measuring point, an accurate determination of the altitude of the measuring point for relating this site to other sites or facilities in the project area, notation of climatic conditions at the time of measurement, a system of validating the water-level measurement, and a straightforward record keeping form or digital device.

6.1.2 Digital recording devices must be checked regularly to ensure that a malfunction has not occurred. A properly operating device that transfers the data directly to a digital computer should alleviate any problems with the transposing of numbers.

NOTE 8—Many permanently installed digital devices record water levels at fixed intervals, for example every 15 min. Unless the device is designed to be activated when sudden changes occur, events that cause an instantaneous and short term fluctuation in the water table may not be recorded, for example, earthquakes and explosions. Continuous recording analog devices are used to detect these types of events.

6.1.3 Much of the problem in preparation of water-level measurements occurs in the office as the result of transposing numbers. This transposition can result when the numbers are manually transferred from a field form to an office data file, perhaps another form or a digital computer data bank. The accuracy of this transfer, and any succeeding transfers or computations, must be verified, preferably by a co-worker, or an independent QA/QC (quality assurance/quality control) officer.

6.2 To interpret the significance of the raw water-level data, usually the information is prepared by adjusting to other values by using simple mathematics. For example, the water-level values in relationship to the measuring point are reduced to the altitude of the water table by subtracting the water level (+ or −) from the altitude of the measuring point. This procedure applied to all water levels from sites in the project area reduces these water levels to a common plane for comparison.

6.2.1 Preparation of water-level data for interpreting upward or downward trends over a period of time may require the use of simple regression or moving average/mean computations. A common analysis of the water-level data involves the selection of yearly highs and lows for use in computing high and low trends.

6.2.2 A technique of presenting water levels is to give the value as below or above land surface. This method requires that the numerical relationship of the measuring point and land surface be determined and the value of the measuring point be subtracted (+ or −) from the water-level measurement. This information gives the relationship of a single water level to the land surface at a finite instant in time. At a long-termed monitoring site the fluctuations and trends are shown. These water levels cannot be completely related to other sites in the area without additional computation (determining altitude of water level).

6.2.3 On occasion, the interpretations of human-induced water-table fluctuations at a site are masked by natural events, such as oscillations caused by barometric pressure or ocean tide. The magnitude and frequency of these fluctuations can be determined by monitoring the barometric pressure, ocean tide, and water levels in wells outside the radius of influence of the principal monitored site.

7. Presentation of Water-Level Information

7.1 *Tabular Methods of Presenting Water Levels*—Tables of ground-water levels in project reports vary from single measurements included in lists of related information, for example, well inventory data (Practice D 5254, Guides D 5408, D 5409, D 5410, and D 5474), to tables that represent a long-term comprehensive record of the water levels at a site. The water levels can be presented as values in feet or metres as related to land surface or the altitude as related to mean sea level or other common level. These values can be

TABLE 1 Example Table—Sites With A Single Water Level[A]

				Ground-Water Site Inventory		
Site ID	Owner	Geologic Unit	Altitude (in feet above msl)		Date	Water Level (in feet below lsd)
404240116025001	CARLIN TOWN GOVT	110VLFL	5950.		03/31/81	11.37
402100116352001	BEOWAWE FARMS	110VLFL	5650.		03/23/81	77.89
412421117303301	SHELTON SCHOOL	110VLFL	4582.		03/18/81	6.11
404940117475001	J BALLARD	110VLFL	4317.		12/11/80	22.30
374638087054101	OWENSBORO, CITY	1120TSH	405.		10/12/82	53.23

[A] Table adapted from Ref (**5**).

for a time-interval, for example, daily or weekly, giving the high, low, mean, or median water level for each period. Other methods include presenting water levels for a specific time, for example, noon or midnight (**4**).

7.1.1 *Tables with Single Water Levels*—A single water level is normally included as one of the data items in a table entitled the "description of selected wells" or "ground-water site-inventory data" in many project reports. This table contains pertinent information from selected ground-water sites of the studied area. Table 1 is an abbreviated example of a "ground-water site-inventory data."

NOTE 9—The data included with the water level varies depending upon the priorities of the project, however, the site identification is standard information in most tables. Computerized tabular procedures are normally designed to print any data item in any order from the ground-water site files.

7.1.2 *Tables of Multiple Water Levels from Single Sites*—The following are common types of tables used to present ground-water levels from single sites. The format usually depends upon the method and frequency of data collection.

NOTE 10—Each individual table commonly includes a heading of information that describes the ground-water site. This heading normally contains the site location, owner, aquifer, site or well characteristics, instrumentation, datum and measuring point, relevant remarks, period of record, and extremes for the period of record.

7.1.2.1 *Tables of High and Low Water Levels for a Selected Period*—The water levels are retrieved from the continuous analog or digital recorders. The period for selecting the water levels can be of any length, for example, daily, weekly, monthly, seasonally, semiannually, yearly, and

for the total period of record. For aquifer testing, for example, it can be for a background period and stress period separately. The table of water levels can be the high, low, or both values for the selected period of record (see Table 2).

7.1.2.2 *Mean Water Levels for a Selected Period*—The water levels are retrieved from digital recording media and the mean water levels determined for a specific period by computer procedures. The mean water level can be determined from the analog recorders by use of electronic scanners or, with more difficulty, manually. The period for determining each water level may be daily, five-day, monthly, etc., and should be determined based on the objective of the project (see Table 3).

7.1.2.3 *Periodic Fixed-time Reading*—Periodic water levels can be selected from the records of analog or digital recorders. The interval between each selected water level may be daily, every fifth day and end of month, weekly, or monthly, with the selected time-of-day constant, for example, the noon reading (see Table 4).

7.1.2.4 *Intermittent Water-level Measurements*—Water levels are considered intermittent when determined manually by instruments such as a steel tape or an electronic water-detection device. These measurements are usually collected by field personnel on a periodic time schedule at ground-water sites where there is no continuous recorder (see Table 5).

7.1.3 *Tables of Water Levels from Multiple Sites*—Tables that include water levels from more than one ground-water site allow for comparison of data from related locations (see Table 6).

TABLE 2 Example Table—Lowest Water Levels For A Site[A]

382150078424001. Local number, 41Q1.
LOCATION.—Lat 38°21′50″, long 78°42′40″, Hydrologic Unit 02070005, at Virginia Department of Highways and Transportation garage near McGaheysville. Owner: U.S. Geological Survey.
AQUIFER.—Conococheague limestone of Late Cambrian age.
WELL CHARACTERISTICS.—Drilled observation water well, diameter 6¼ in., depth 310 ft, cased to 131 ft, open hole 131 to 310 ft.
INSTRUMENTATION.—Water-level recorder.
DATUM.—Elevation of land-surface datum is 1105 ft above National Geodetic Vertical Datum of 1929, from topographic map. Measuring point: Top edge of recorder shelf, 3.50 ft above land-surface datum.
PERIOD OF RECORD.—August 1970 to current year.
EXTREMES FOR PERIOD OF RECORD.—Highest water level recorded, 60.38 ft below land-surface datum, Dec. 26, 1972; lowest recorded, 87.18 ft below land-surface datum, Oct. 26, 1977.

	Water Level, in Feet Below Land-Surface Datum, Water Year October 1982 to September 1983 Lowest Values											
DAY	OCT	NOV	DEC	JAN	FEB	MAR	APR	MAY	JUN	JUL	AUG	SEP
5	73.32	76.01	76.07	71.52	72.79	68.43	65.68	64.46	64.70	66.09	68.04	71.10
10	73.87	76.11	75.60	71.48	71.81	68.14	65.54	64.81	65.09	66.35	68.42	71.72
15	74.39	76.33	75.27	71.69	71.07	68.03	64.41	65.04	65.41	66.62	68.86	72.28
20	74.90	76.60	75.11	72.14	70.34	65.85	64.39	64.53	65.55	66.93	69.32	72.86
25	75.36	76.94	72.94	72.55	69.14	65.88	64.07	64.18	65.60	67.25	69.86	73.48
EOM	75.75	76.98	71.94	73.00	68.76	66.10	64.08	64.54	65.88	67.67	70.52	74.04
WTR YR 1983			HIGHEST 63.81 APR 27, 1983						LOWEST 76.98 NOV 28, 1982			

[A] Table adapted from Ref (**5**).

TABLE 3 Example Table—Mean Water Levels For A Site[A]

402208074145201. Local I.D., Marlboro 1 Obs. NJ-WRD Well Number, 25-0272.
LOCATION.—Lat 40°22′08″, long 74°14′52″, Hydrologic Unit 02030104, on the west side of New Jersey Route 79, 0.9 ml south of Morganville, Monmouth County, New Jersey. Owner: Marlboro Township Municipal Utilities Authority.
AQUIFER.—Farrington aquifer, Potomac-Raritan-Magothy aquifer system of Cretaceous age.
WELL CHARACTERISTICS.—Drilled artesian observation well, diameter 6 in., depth 680 ft, screened 670 to 680 ft.
INSTRUMENTATION.—Digital water-level recorder—60-minute punch.
DATUM.—Land-surface datum is 116.73 ft above National Geodetic Vertical Datum of 1929. Measuring point: Top edge of recorder shelf, 2.50 ft above land-surface datum.
REMARKS.—Water level affected by nearby pumping. Missing record from May 19 to July 4 was due to recorder malfunction.
PERIOD OF RECORD.—March 1977 to current year. Records for 1973 to 1977 are unpublished and are available in files of New Jersey District Office.
EXTREMES FOR PERIOD OF RECORD.—Highest water level, 144.06 ft below land-surface datum, Apr. 4, 1973; lowest, 190.49 ft below land-surface datum, July 29, 1983.

| | Water Level, in Feet Below Land Surface Datum, Water Year October 1983 to September 1984 Mean Values | | | | | | | | | | | |
DAY	OCT	NOV	DEC	JAN	FEB	MAR	APR	MAY	JUN	JUL	AUG	SEP
5	178.44	168.09	161.50	159.63	158.03	158.25	157.72	156.94	...	170.00	169.37	172.95
10	177.44	166.41	161.52	159.12	158.47	158.16	158.17	156.95	...	169.11	168.93	172.67
15	173.78	166.48	160.28	158.45	158.27	157.79	158.00	157.42	...	171.58	168.45	171.39
20	172.68	165.34	160.07	158.25	158.09	157.50	157.99	...	...	170.39	169.50	171.09
25	171.04	164.31	159.81	157.83	158.05	157.69	157.39	...	...	169.74	171.15	172.76
EOM	170.22	163.51	160.20	157.95	157.94	156.78	157.81	...	...	167.63	174.11	171.45
MEAN	174.70	166.15	160.77	158.63	158.27	157.75	157.88	...	...	169.50	169.99	172.60

WTR YR 1984   MEAN 164.15   HIGH 155.71 MAY 5   LOW 182.94 OCT 1

[A] Table adapted from Ref (5).

TABLE 4 Abbreviated Table—Noon Water Levels For A Site[A]

374638087054101. Map number 1.
LOCATION.—Lat 37°46′38″, long 87°05′41″, Hydrologic Unit 05140201, County Code 059, Owensboro East quadrangle, at Owensboro Municipal Utilities water treatment plant, 100 ft (30 m) south of south bank of Ohio River, 0.1 ml (0.2 km) northeast of Davies County High School. 0.3 ml (0.5 km) north of U.S. Highway 60, in Owensboro, Daviess County, Kentucky. Owner: Owensboro Municipal Utilities.
AQUIFER.—Glacial sand and gravel of Quaternary age. Aquifer code: 112OTSH.
WELL CHARACTERISTICS.—Drilled unused water-table well, diameter 12 in. (0.30 m), depth 104 ft (32 m), screened 74–104 ft (22.6–31.7 m).
DATUM.—Altitude of land-surface datum (from topographic map) is about 405 ft (123 m). Measuring point: Floor of recorder shelter 4.33 ft (1.32 m) above land-surface datum.
REMARKS.—Water level affected by pumping from nearby wells.
PERIOD OF RECORD.—February 1951 to current year.
EXTREMES FOR PERIOD OF RECORD.—Highest water level, 18.16 ft (5.54 m) below land-surface datum, May 5, 1983; lowest, 63.21 ft (19.27 m) below land-surface datum, Sept. 17, 1970.

| | Depth Below Land Surface (Water Level), (ft), Water Year October 1982 to September 1983 Instantaneous Observations at 1200 | | | | | | | | | | | |
DAY	OCT	NOV	DEC	JAN	FEB	MAR	APR	MAY	JUN	JUL	AUG	SEP
1	54.51	48.09	44.14	45.05	55.92	46.52	49.32	30.32	39.39	49.56	50.40	52.97
2	49.52	48.78	44.89	42.32	55.71	47.08	46.04	37.11	43.03	48.96	49.74	52.09
3	49.65	49.20	42.17	48.59	50.84	50.39	46.03	30.69	43.46	43.70	47.87	50.16
4	50.29	47.12	41.20	...	54.38	48.90	50.79	23.20	40.92	43.12	50.86	49.67
5	51.37	47.45	40.22	51.32	49.47	49.12	49.06	18.16	39.86	43.78	49.27	49.56
6	51.73	45.38	45.11	51.86	47.42	44.92	49.22	28.90	44.66	46.53	46.02	51.96
7	50.62	46.26	46.60	54.53	49.47	50.32	48.96	28.47	45.58	46.70	45.89	52.22
Water Levels for Days 8th through 28th Deleted for This Illustration												
29	49.24	45.13	45.73	54.57	...	46.92	41.06	31.82	46.42	51.62	52.73	52.46
30	47.34	48.89	45.69	54.85	...	47.53	36.55	34.78	47.30	49.14	51.46	52.77
31	47.37	...	44.73	55.99	...	50.07	...	36.29	...	48.82	52.22	...
MAX	54.51	49.71	53.19	58.00	55.92	51.26	56.44	38.75	50.57	54.70	54.38	53.72
MIN	46.74	43.70	40.22	42.32	44.76	44.76	36.55	18.16	39.39	43.12	45.89	45.21

WTR YR 1983   HIGH 16.16 MAY 5   LOW 58.00 JAN 20

[A] Table adapted from Ref (5).

7.2 *Graphical Methods of Presenting Water Levels*— Methods to represent water levels include those at a single ground-water site for a finite or short period of time, a single site over an extended period of time, multiple sites for a finite or short period in time, and multiple sites over an extended period of time.

NOTE 11—The simplest category of the presentation of a water level is from a single ground-water site for a finite instant or short period in time. Water levels measured at a single ground-water site over a period of time give climatic trends and the effects of human and natural stresses on water in the aquifer. Water levels can be measured continuously by analog recorders or digital recorders and intermittently by a steel tape or electronic devices.

NOTE 12—To interpret hydraulic-head data over the area of a project or political entity, multiple ground-water sites need to be included in the analysis. These sites should be in the same aquifer, widely distributed, and the water levels measured during a short period.

NOTE 13—Multiple sites where ground-water levels are measured by a continuous recorder or periodically by other methods are valuable for interpreting changes in aquifers caused by discharge and recharge events. These changes can be illustrated by maps and cross sections, and by the comparison of hydrographs.

7.2.1 *Vertical Gradient at a Single Site*—Multiple water levels can be measured within a short period of time from a

TABLE 5 Example Table—Intermittent Water Levels For A Site[A]

424202087542301. Local Number, RA-03/22E/21-0005.
LOCATION.—Lat 42°42′02″, long 87°54′23″, Hydrologic Unit 04040002. Owner: Chicago, Milwaukee, St. Paul, and Pacific Railroad Co., Racine County, Wisconsin.
AQUIFER.—Sandstone.
WELL CHARACTERISTICS.—Drilled unused artesian well, diameter 12 in. (0.30 m), depth 1,176 ft (358 m), cased to 586 ft (179 m), 10 in. (0.25 m) liner 976-1083 ft (297–330 m).
DATUM.—Altitude of land-surface is 730 ft (225 m) National Geodetic Vertical Datum of 1929. Measuring point: top of casing, 1.00 ft (0.30 m) above land-surface datum.
REMARKS.—Water level affected by regional pumping of wells.
PERIOD OF RECORD.—July 1946 to current year.
EXTREMES FOR PERIOD OF RECORD.—Highest water level measured, 109.00 ft (33.25 m) below land-surface datum, July 29, 1946; lowest water level measured, 264.70 ft (80.68 m) below land-surface datum, Mar. 3, 1981.

				Water Level, in Feet Below Land-Surface Datum, Water Year October 1980 to September 1981							
DATE	WATER LEVEL	DATE	WATER LEVEL	DATE	WATER LEVEL	DATE	WATER LEVEL	DATE	WATER LEVEL	DATE	WATER LEVEL
FEB 12	257.00	MAR 17	256.63	MAY 1	262.50	JUN 1	263.30	JUN 29	262.70	SEP 15	263.30
MAR 3	264.70	APR 6	257.40								

[A] Table adapted from Ref (5).

TABLE 6 Abbreviated Table—Water Levels From Multiple Sites[A]

LOCATION.—State of Nevada.
WELL DEPTH.—Depths are referenced to Land-surface Datum (LSD).
PERIOD OF RECORD.—Interval shown spans period from earliest measurement to latest measurement, and may include intervals with no record.
WATER LEVELS.—Levels above LSD are listed as negative values.

Site ID	Well Depth (Ft)	Period of Record	Water Levels (Feet Below Land Surface)					
			Highest	Date	Lowest	Date	Current	Date
415800118370001	200.	1968-	45.58	03/20/68	56.80	05/01/69	51.55	03/17/81
413630119520001	70.	1968-	10.22	03/13/72	14.66	04/10/79	12.34	04/07/81
403200119490001	111.	1966-	37.91	09/15/66	54.97	04/17/79	54.41	03/24/81
402700119250001	109.	1966-	45.20	04/09/69	50.11	03/26/81	50.11	03/23/81
405211119202901	134.	1979-	29.53	04/17/79	31.25	03/23/81	31.25	03/23/81
405208119161501	15.	1967-	3.77	04/16/73	14.21	03/23/81	14.21	03/23/81
405208119161502	66.	1967-	−2.25	06/14/67	9.37	03/23/81	9.37	03/23/81
412954117495001	250.	1971-	50.96	04/30/73	78.11	04/29/71	58.24	03/17/81
413310117482002	95.	1948-	36.54	04/21/48	116.58	03/23/77	72.17	03/17/81
413320117482001	160.	1949-	16.55	01/20/50	123.19	03/23/77	91.85	03/17/81

[A] Table adapted from Ref (5).

series of vertical positions in different aquifers at a specially constructed ground-water site. The data gathered indicates the hydraulic gradient of the water (5,6). Examples of the three gradient possibilities from tightly spaced piezometers in a single unit (7) are given in Fig. 1. An example of a downward gradient in eight aquifers (8) is given in Fig. 2.

NOTE 14—In Fig. 2, water levels at 143 ft (43.58 m), 305 ft (92.96 m), and 460 ft (140.21 m) were measured in 1961, others in 1959. These data are from an area where little development had taken place at the time of the water-level measurements.

NOTE 15—An example of a specially constructed well is a test hole where the water level is measured at progressively deeper positions in the aquifer or a series of aquifers. The well is open to the aquifer at progressively deeper depths and each opening is uniquely accessible for measurement of the water level by a pipe to the surface, or several piezometers or wells that are tightly spaced and each open at a different depth in the aquifer.

7.2.2 *Hydrographs*—The hydrograph is used to illustrate the fluctuation of the hydraulic head over a period of time at a ground-water site. Interpolated lines (areas of missing or indeterminate record) on hydrographs should be clearly identified. The hydrograph is accompanied commonly with time-related phenomena to help in the interpretation of fluctuations, for example, precipitation. Recession curves of surface-water hydrographs are used to determine ground-water baseflow in the streams. Some examples of the

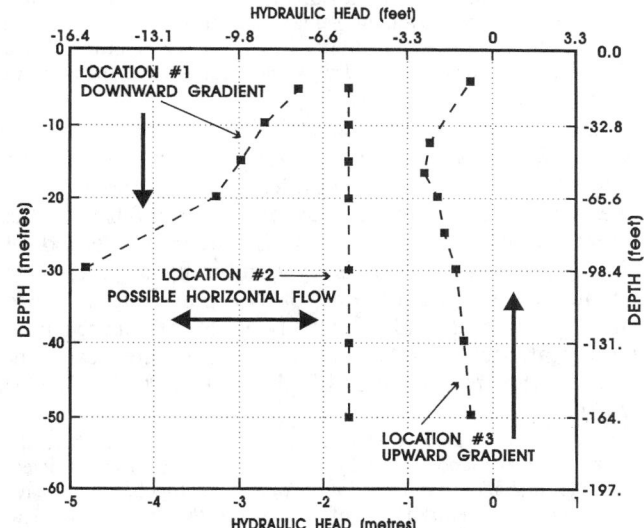

NOTE—Location No. 2 is fabricated to simulate horizontal flow.

FIG. 1 Hydraulic Gradient at Three Ground-Water Locations (adapted from Ref (8))

hydrographs and combined phenomena for a ground-water site follow.

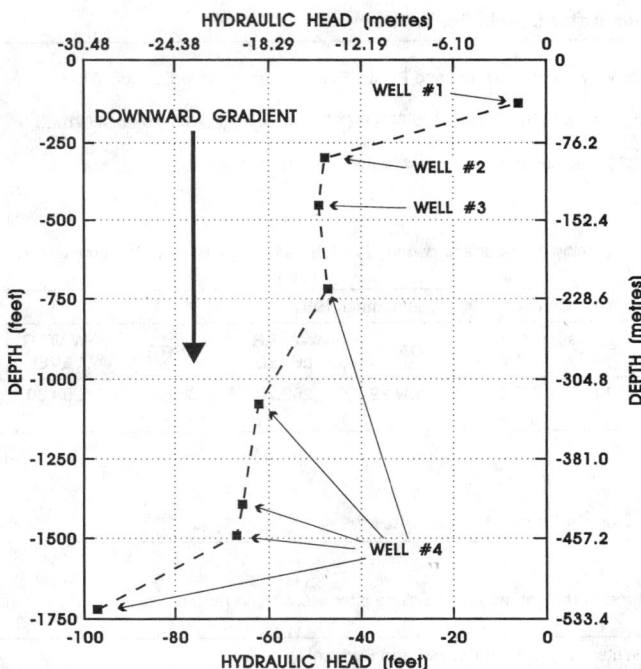

FIG. 2 Hydraulic Gradient at a Ground-Water Location (data from four wells) (adapted from Ref (9))

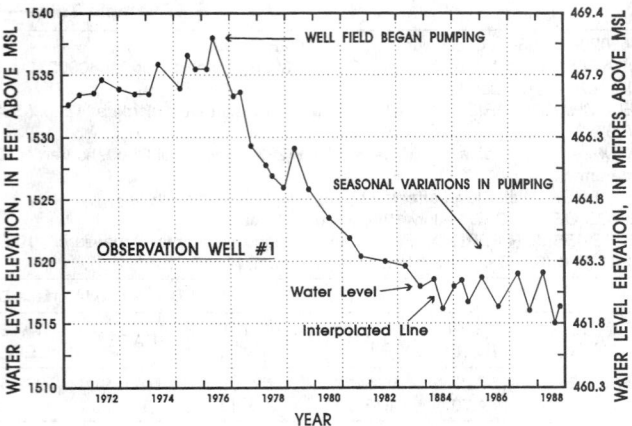

FIG. 3 Example of Simple Hydrograph (adapted from Ref (10))

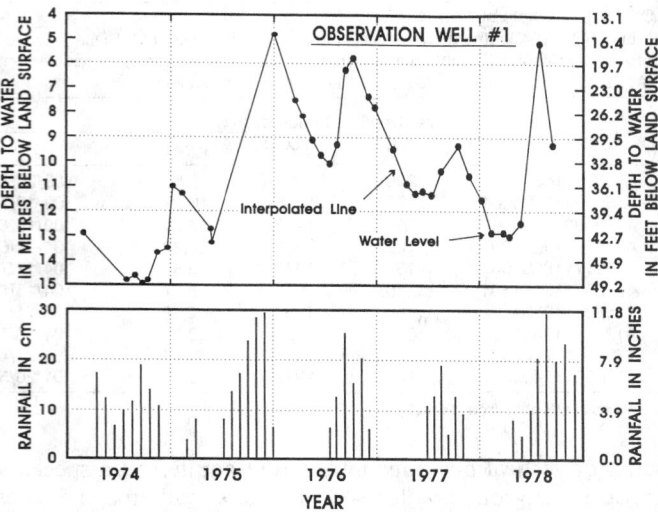

FIG. 4 Hydrograph and Precipitation Plot (adapted from Ref (13))

7.2.2.1 *Simple Hydrograph*—The basic hydrograph of the water table at a ground-water site displays the natural and human-induced fluctuations over a period of time. The example hydrograph shows fluctuations controlled by natural conditions from 1971 to 1976, those resulting from pumping withdrawals that began in 1976, and those caused by seasonal variations in pumping that are apparent from 1984 to 1988 (see Fig. 3) (9–11).

NOTE 16—The water level measurements in Fig. 3 average two values per year. These intermittent values are connected by interpolated lines to simulate a continuous hydrograph. Water levels determined by a nearly continuous digital recorder would result in a continuous hydrograph.

7.2.2.2 *Hydrograph Compared with Precipitation that Results in Natural Recharge*—Precipitation that results in recharge to an unconfined aquifer can be analyzed by comparison of the timing and amount of rainfall with the hydrographs of shallow wells in the area. A method of displaying this relationship is by combining a water-table graph and a precipitation line or bar plot onto a single illustration. The time scales for the two sets of data are equal, and the water-table and precipitation data are scaled to emphasize the relationship of the values (see Fig. 4) (1,12–18).

NOTE 17—Rapid response to recharge events is evident where the travel path from the land surface to the aquifer is short or unrestricted, for example, a shallow sand formation or a karst topography. Heavy rainstorms can cause entrapment of air between the recharge water at the surface and a shallow water table. This recharge surge can increase the pressure of the trapped air creating a rapid decline in the water table and a resultant rise of water in open observation wells. The water table will rise when the entrapped air escapes by breaching the recharged water and continue to rise as the recharge water reaches the water table. In aquifers where restrictions occur, for example, intermediate clay layers or aquitards, the response can be dampened or delayed because of a much longer travel time.

7.2.2.3 *Hydrograph Compared with Artificial Recharge to*

the Aquifer—Artificial recharge to aquifers can occur from methods that spread water on the land's surface, for example, irrigation, or from techniques that direct the water below the land's surface, for example, recharge wells and pits. This type of recharge can be monitored by wells in the area and illustrated by hydrographs (see Fig. 5) (19–21).

7.2.2.4 *Hydrograph Compared with Barometric Pressure*—A change in barometric pressure causes water levels to fluctuate in open wells. The effects of barometric pressure often mask other influences that cause fluctuations of the water table. By plotting the hydrograph and barometric pressure on an equal time scale, the correlation of oscillations can be demonstrated (see Fig. 6) (22–28).

NOTE 18—The effect of barometric pressure can be removed from the water-table fluctuations by subtracting the value determined from multiplying the "barometric efficiency" (BE) times the amount of water-table fluctuation. The BE is a decimal number determined by dividing the change in water level (ΔW) by the change in barometric pressure (ΔB) over an interval of time ($BE = \Delta W/\Delta B$). These two values must be in the same units to calculate the BE, for example, if the water levels are in metres, then convert the barometric pressure to metres of water at 4°C (1000 millibars pressure = 10.197 m of water at 4°C).

7.2.2.5 *Hydrograph Compared with Withdrawals from the Aquifer*—Water withdrawals from an aquifer can result in

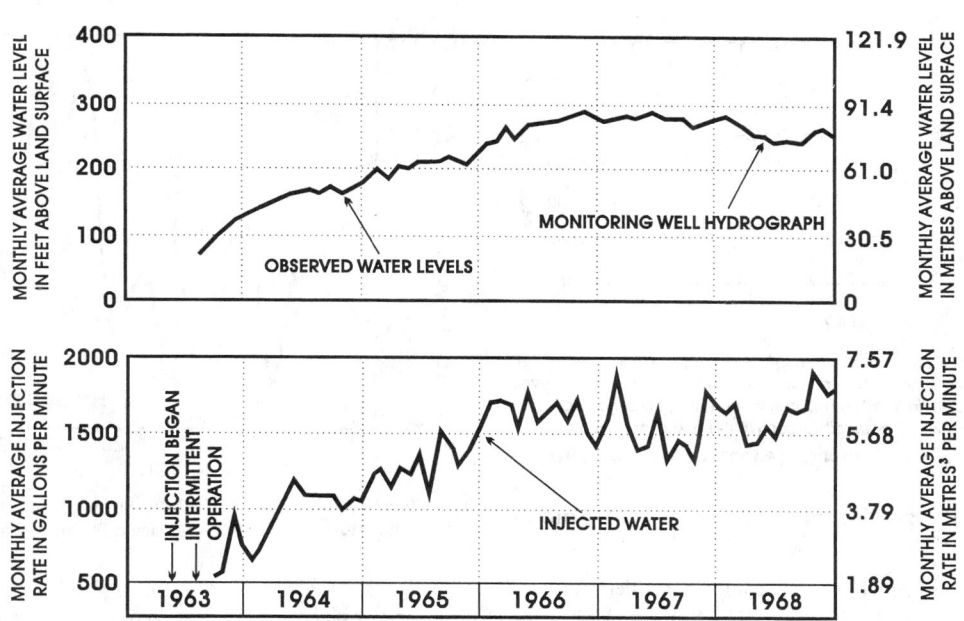

FIG. 5 Hydrograph Showing Effects of Artificial Recharge by Injection Well (adapted from Ref (20))

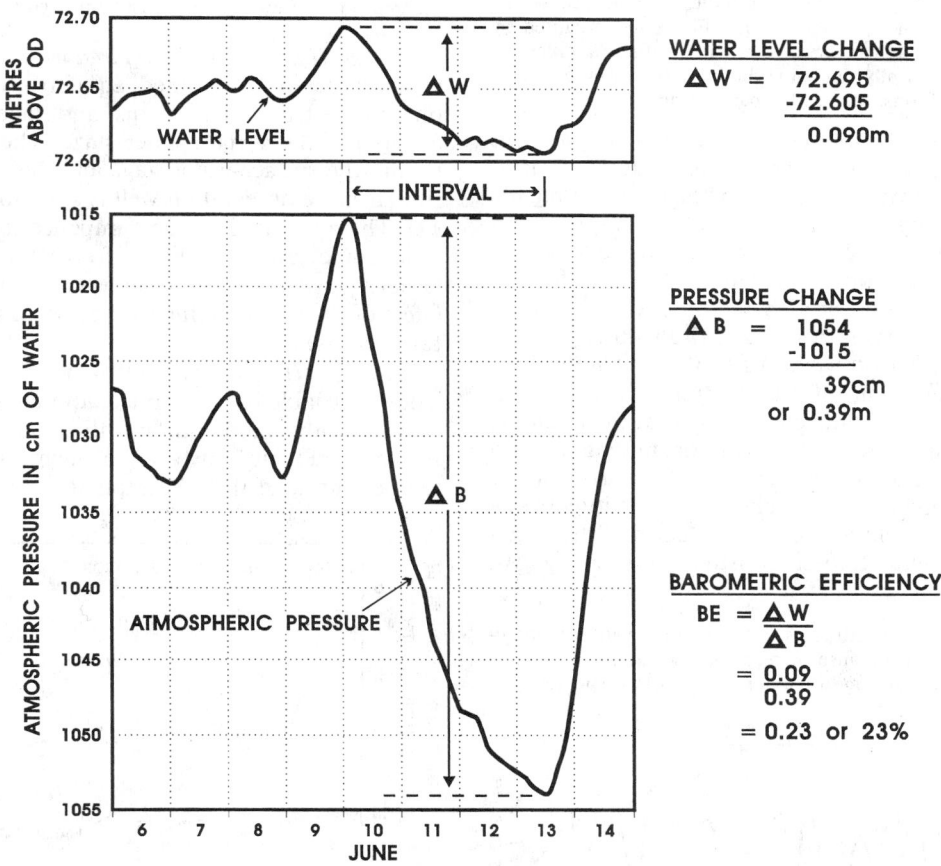

FIG. 6 Hydrograph with Barometric Efficiency (adapted from Refs (23,24))

the fluctuation and decline of the hydraulic head. The hydraulic head fluctuates depending upon the periodic oscillation in the amount of water withdrawn and decline when the water removed is more than water recharged to the aquifer. A hydrograph from a ground-water site compared with the withdrawal amounts displays the effect on the hydraulic head in the aquifer (see Fig. 7) (22,29–34).

7.2.2.6 *Hydrograph Compared with Tidal Effects*—The hydraulic head fluctuates semidiurnally in response to tides in the solid earth and in large bodies of surface water. The

501

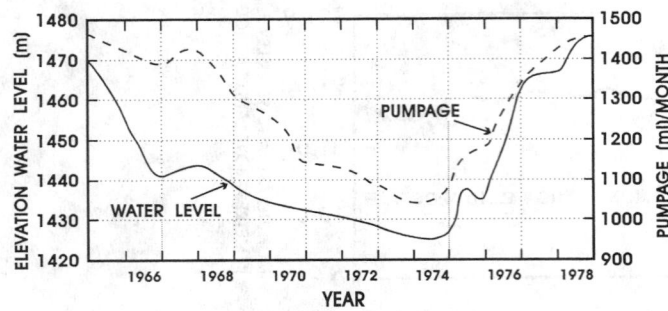

NOTE—This is a mined area where pumpage is for dewartering the mine. Pumpage exceeded recharge before 1975 resulting in a decline of the water level. Abnormally high rainfall beginning in 1975 resulted in increase recharge and a rise of the water level. Pumpage was increased to control the rise of the water level.

FIG. 7 Hydrograph with Pumpage (adapted from Ref (30))

tides are caused by the gravitational attraction of the moon and sun upon the earth (see Fig. 8) (22,35–38).

NOTE 19—Fluctuations are obvious in confined aquifers that are next to an ocean where a rising tide compresses the underlying sediments (rising hydraulic head) and a falling tide allows the underlying sediments to expand (falling hydraulic head). The water table in unconfined aquifers near large surface water bodies fluctuates caused by the actual movement of water in the aquifer. Fluctuations caused by earth tides are obscure, but can be detected in confined aquifers of inland areas by mathematically removing the influence of other causes of hydraulic-head oscillations, such as the barometric pressure.

7.2.2.7 *Hydrograph Compared Earthquakes, Explosions, and Loading Effects*—Shock waves radiating out from earthquakes and explosions travel through the earth and along the earth's surface causing the elastic crust to compress and expand, resulting in a fluctuation of the hydraulic head (see Fig. 9). Loading effects on underlying sediments, for example, a train that moves through the area, can cause the hydraulic head to oscillate in response (37,39–45).

7.2.2.8 *Hydrograph Compared with Water Quality Parameters*—The fluctuation of the hydraulic head in an aquifer can indicate the movement of water containing natural- and human-induced chemical constituents toward an area of lower hydraulic pressure. A comparison of the hydrograph and a time-plot of the chemical constituents at a ground-water site can help in the interpretation of the origin and rate of movement of these constituents (see Fig. 10) (20,31,46–51).

NOTE 20—Some of the constituents in the ground water can originate from natural leaching because of recharge oscillations caused by climatic cycles. Artificial recharge of water from surface spreading or

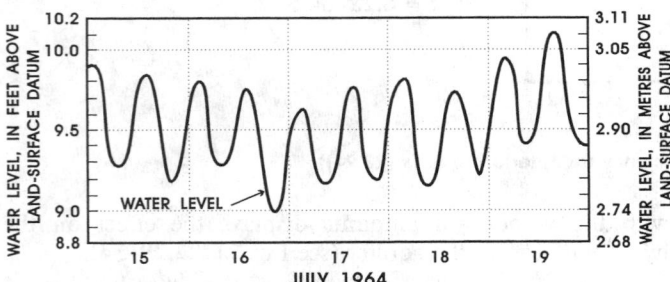

NOTE—This is an artesian aquifer.

FIG. 8 Hydrograph Showing Tidal Effects (adapted from Ref (36))

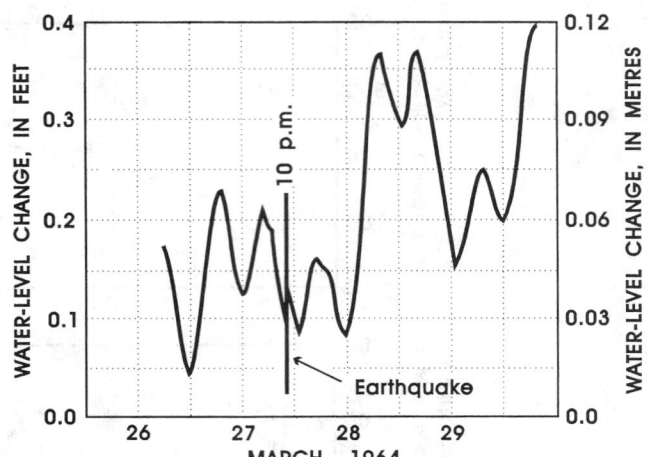

NOTE—March 27, 1964 Alaskan earthquake, well at Vincent Dome, Iowa.

FIG. 9 Hydrograph With Seismic Fluctuation (adapted from Ref (40))

injection by pits or wells can leach or induce ions into the ground water. Water that has a high concentration of dissolved solids, for example, seawater, is denser than fresh water and, therefore, will have a slight difference in the water table when compared to bordering fresh water.

7.2.2.9 *Hydrograph Compared with Surface Stream*—The water table in unconfined aquifers that are next to and interconnected with streams and lakes, react rapidly to changes in the surface-water stage. The amount of fluctuation in the surface-water stage and the ground-water table is similar if the observation well is close to the stream (see Fig. 11). These fluctuations are dampened if the observation well is at some greater distance from the surface-water body. Oscillations in confined aquifers are caused by the loading effect of rising and falling surface-water stages (see 7.2.2.6 on tidal effects) (33,52–54).

7.2.2.10 *Hydrograph Compared with Air Temperature*—The water table in unconfined aquifers that are a few feet or metres below lands surface fluctuate in response to the thermal gradient between the mean air and ground-water temperatures, in that the capillary moisture and soil vapor

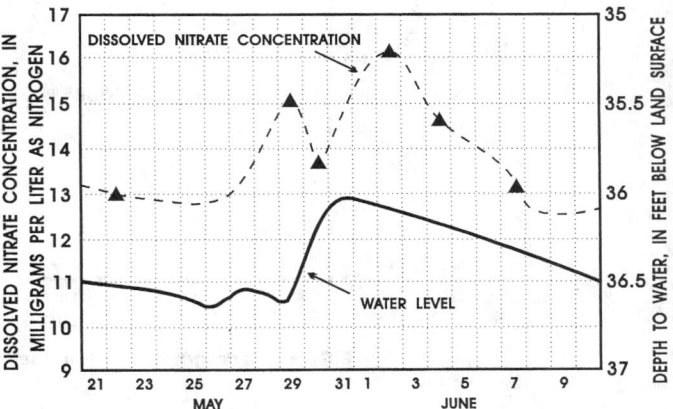

NOTE—The rise in water level and nitrate concentration is the result of a storm. Graph lines are interpolated. To convert to metres, multiply feet value times 0.3048. ▲ = Analysed dissolved nitrate concentration.

FIG. 10 Hydrograph and Graph of Dissolved Nitrate Concentration (adapted from Ref (47))

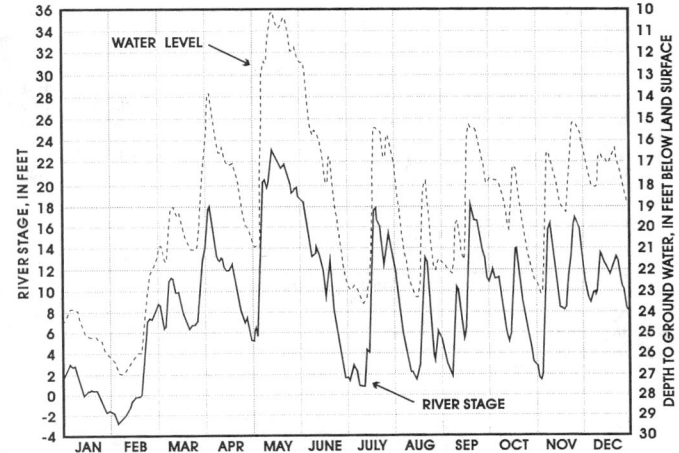

NOTE—Well, screened in alluvium, is 1700 ft from the river. To convert to metres, multiply feet value times 0.3048.

FIG. 11 Hydrographs of River Stage and Water Levels in a Well (adapted from Ref (53))

move toward the medium having the lowest temperature (see Fig. 12) (1,55,56).

NOTE 21—When the mean daily air temperature remains below freezing over time, the upward moving water freezes in the near surface soil material, forming a frost layer. Because of this water transfer, the ground-water table declines. Soon after the mean daily temperature rises above freezing, melted water from the frost layer moves downward as recharge causing a rise in the ground-water table. During the spring and summer months, evapotranspiration causes diurnal fluctuations of the shallow water table. If no recharge occurs during this period, the general trend of the water table will be downward.

7.2.2.11 *Hydrograph with Fluctuations Caused by Unusual Phenomenon*—The sudden rise of a hydraulic head may be a clue to a problem that has affected the aquifer, for

example, a defective casing of a gas well that has allowed natural gas to escape into the aquifer (see Fig. 13). An undefined change of the hydraulic head may indicate a movement of water from one aquifer to another having a lower water table, perhaps from a failed casing or improperly constructed well (57).

7.2.2.12 *Hydrograph with Boxplots of Water Levels, Precipitation, Surface Water, and Evaporation*—An association of ground water, surface water, and precipitation time-series graphs with statistical boxplots offers a useful combination for data interpretation. The boxplots concisely illustrate the median, 25th percentile, 75th percentile, skewness, and the outside and far-outside values for each of those data sets (see Fig. 14) (58).

7.2.2.13 *Multiple Hydrographs*—Hydrographs from multiple ground-water sites of an area can be compared to interpret the rate of water movement in an aquifer and between several aquifers (see Fig. 15) (18,59–65).

NOTE 22—Hydrographs from precisely positioned ground-water sites in an aquifer of a project area can be compared to determine the effect of distance from an impacted locality on the water table, for example, the water levels of monitoring wells for a recharge pit. The elapse-time effects of natural or artificial recharge can be evaluated by comparing hydrographs from a shallow and the underlying aquifers. The effects of distance from fluctuating surface-water bodies on adjacent aquifers can be shown by comparing the hydrographs.

7.2.3 *Temporal Trends in Hydraulic Head*—The temporal trend of hydraulic head is dictated by many factors that contribute to the stress of an aquifer, for example, recharge of water to and discharge of water from the aquifer. All longer-term hydrographs exhibit a trend, either downward, level, upward, or cyclical.

7.2.3.1 *Trend Hydrograph*—At ground-water sites where the water level is measured by a continuous recorder, the trend can be determined by selecting the high, computing the mean, or selecting the low water level from a fixed period, for example, a day, week, month, or year, and plotting these values as a hydrograph. At ground-water sites where water levels are measured intermittently, the trend can be determined by selecting water levels from the same yearly period, for example, January or June, and plotting these values as a hydrograph (see Fig. 16) (4,61,66–73).

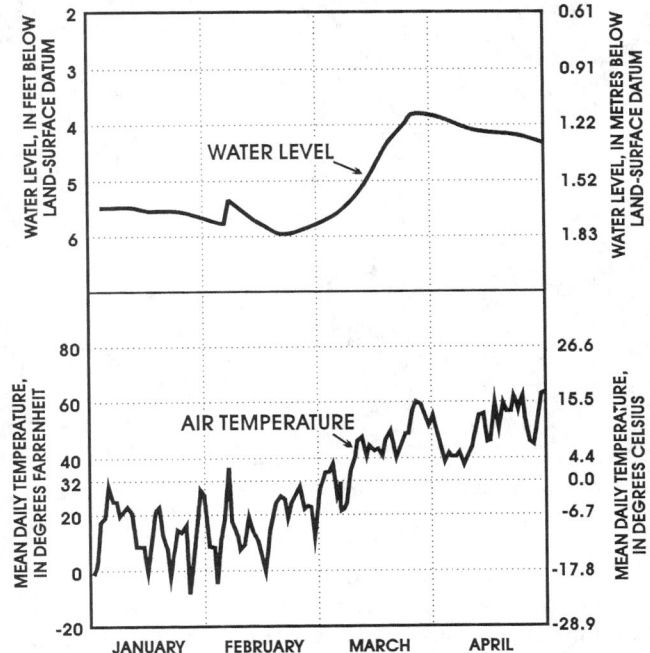

FIG. 12 Water Levels and Air Temperatures (adapted from Ref (56))

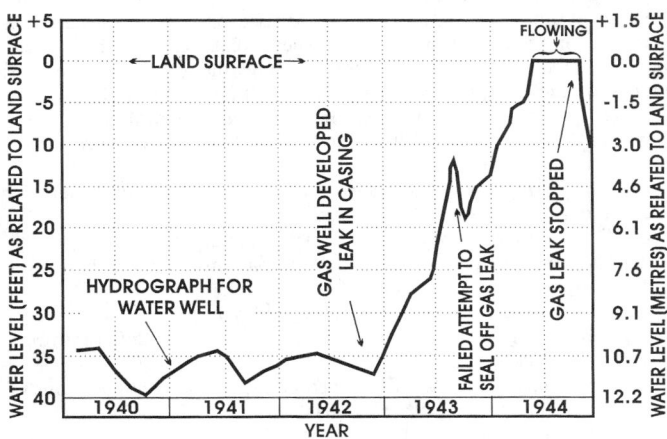

NOTE—Gas well was located five miles from water well.

FIG. 13 Hydrograph With Fluctuations Caused by Unusual Phenomenon (adapted from Ref (58))

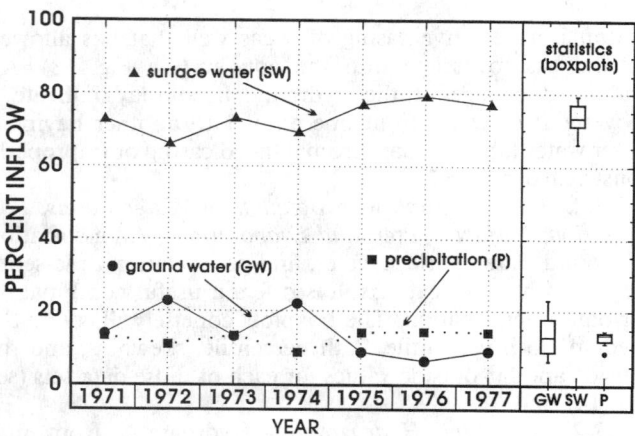

FIG. 14 Inflow to a Lake from Ground Water, Surface Water, and Precipitation Sources with Statistics Given by Boxplots (adapted from Ref (59))

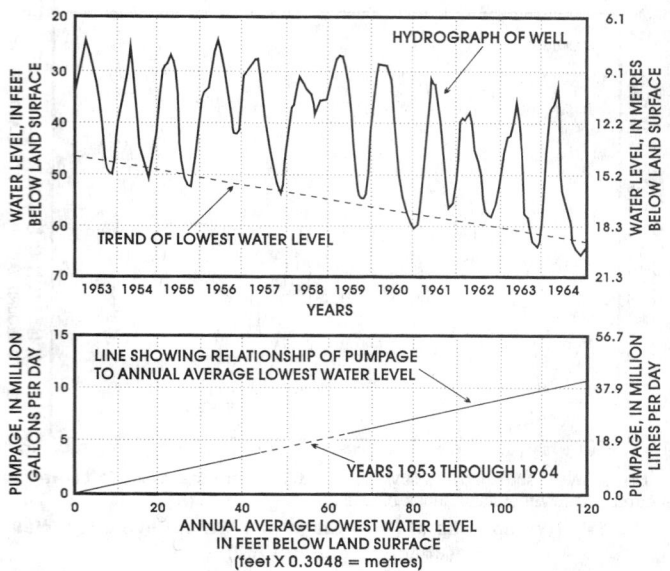

FIG. 16 Hydrograph Showing Water Level Trend and Graph Showing Relationship of Trend to Pumpage (adapted from Ref (67))

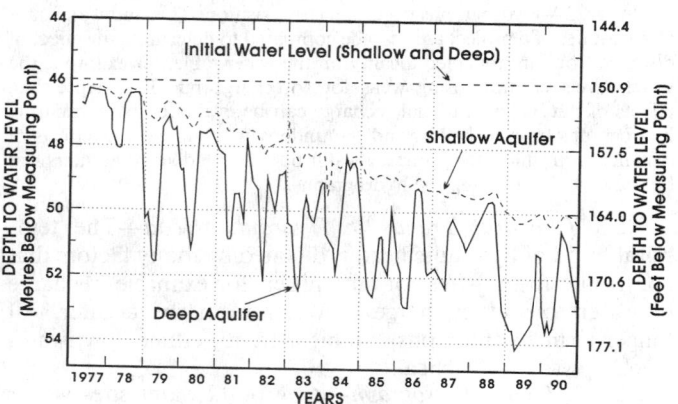

FIG. 15 Multiple Hydrographs Comparing Shallow and Deep Aquifers (adapted from Ref (60))

7.2.4 *Potentiometric Maps*—Maps that illustrate the potentiometric surface commonly show the altitude of the hydraulic head as related to mean sea level (msl) or a fixed level in the vicinity of the project (see Fig. 17) **(9,74–80)**.

NOTE 23—Potentiometric maps help in the interpretation of the hydraulic gradient, direction of water movement, and losing and gaining of surface-water bodies. The water levels used on the map need to be measured in a short-time period. These plots can be drawn on topographic maps or aerial photos to show the relationship of the hydraulic head to surface topography and cultural features **(81)**.

NOTE 24—In addition to the consideration of QA/QC items discussed in Section 6, some factors that must be avoided in constructing potentiometric maps include:

(1) Contouring of water levels from wells screened at different depths in aquifers with vertical hydraulic gradients,

(2) Over-simplified contours, for example, straight-line,

(3) Over-interpreted contours, for example, more curves than justified by number of data points,

(4) Extrapolation of contours well beyond data points,

(5) Contouring of data values from substantially different time periods,

(6) Contours adjusted to "fit" the contaminant plume as a means of justifying a contaminant pathway, and

(7) Contouring of water levels impacted by liquid phase contaminants without proper adjustment of the contours.

7.2.5 *Depth to Water Maps*—Maps that illustrate the

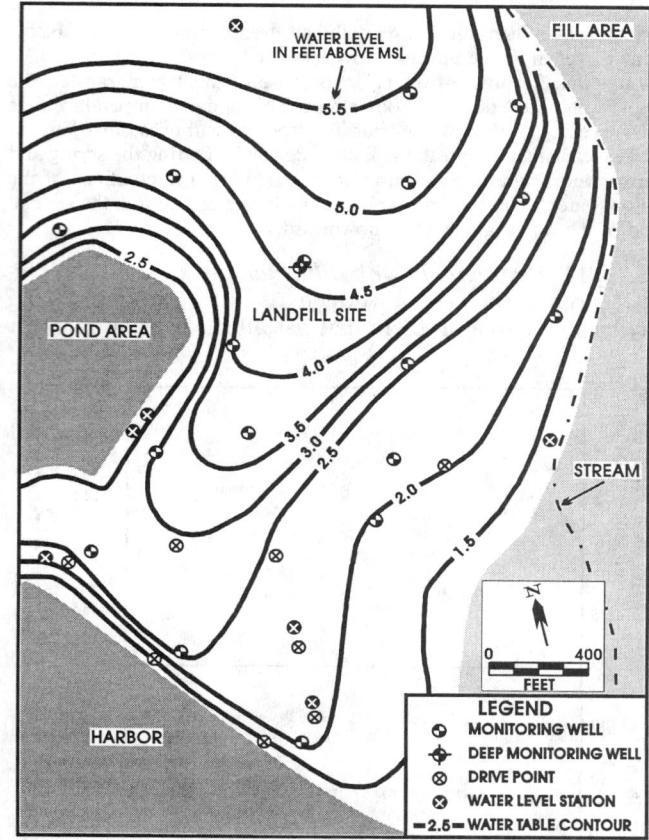

FIG. 17 Potentiometric Map at Landfill Facility (adapted from Ref (75))

depth of water below the land surface are useful for construction projects where the concern is intersecting the unconfined water surface, for example, by basements, disposal pits, or mines. These maps can also provide information about natural features, including the relationship of

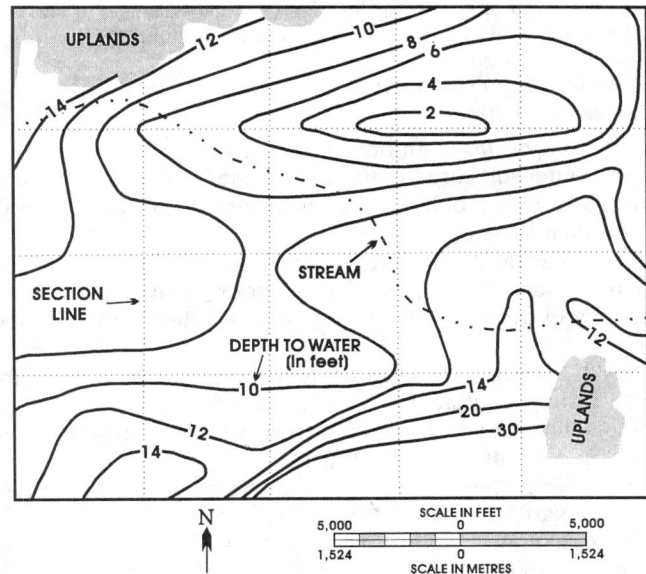

FIG. 18 Map Showing Depth to Water Below Land Surface (adapted from Ref (83))

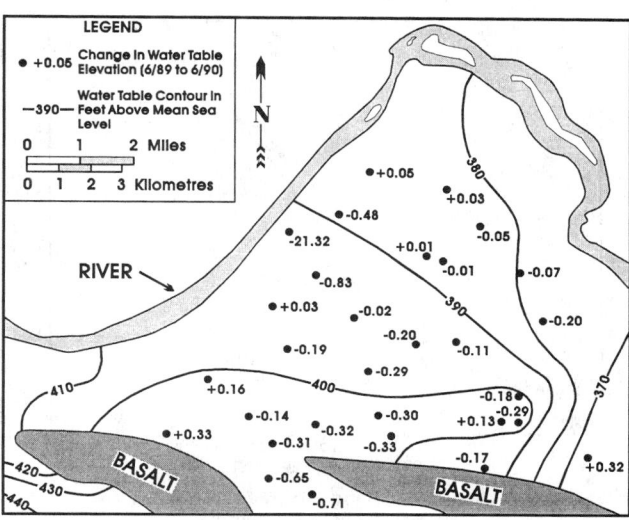

FIG. 19 Water Level Change Map and Potentiometric Surface (adapted from Ref (29))

surface-water bodies and wetland areas to the ground-water table (see Fig. 18) (82).

7.2.6 *Change Maps*—The change of the potentiometric surface over time, for example, one or ten years, helps in the interpretation of the effects of natural- and human-induced stresses on the aquifer (see Fig. 19) (28,82–85).

NOTE 25—The change map (Fig. 19) is a plot of the difference in the hydraulic head of an area over a period of one year. The map is constructed by subtracting the water levels from a potentiometric surface of the later time (June 1990) from those of the earlier time (June 1989). Positive plotted values on the change map show a rising hydraulic head (indicating recharge) and negative values show a falling hydraulic head (indicating discharge).

7.2.7 *Water-table Cross Sections*—A vertically oriented cross-section through several sites shows an exaggerated shape of the aquifers, the ground-water table, and the

hydraulic gradient as they relate to land surface features (see Fig. 20) (8,9,22,76,82,86–89).

NOTE 26—Cross-sections of unconfined aquifers commonly show the relationship of surface features, for example, pits, lakes, streams, and cultural structures, with the sub-surface materials, for example, aquifer configuration, depth and gradient of water surface, ground-water flow net, location and construction features of wells, and chemical characteristics of the water (22). Cross-sections of confined aquifers tend to place less emphasis on the surface features that have little effect on conditions in the aquifer.

7.2.8 *Statistical Comparisons of Water Levels*—Ground-water table data can be analyzed by many common statistical methods to determine trends and to correlate these data with related natural and human-caused factors (see Fig. 21) (4,28,90–94).

NOTE 27—Basic statistics, for example, mean, median, high, and low values are commonly used to determine the long-term trends of the hydraulic head. A long-term average hydrograph, for example, from 20 wells, can be determined for a project area or these same water levels can be shown on a hydrograph for a single year (93). Probability plots for

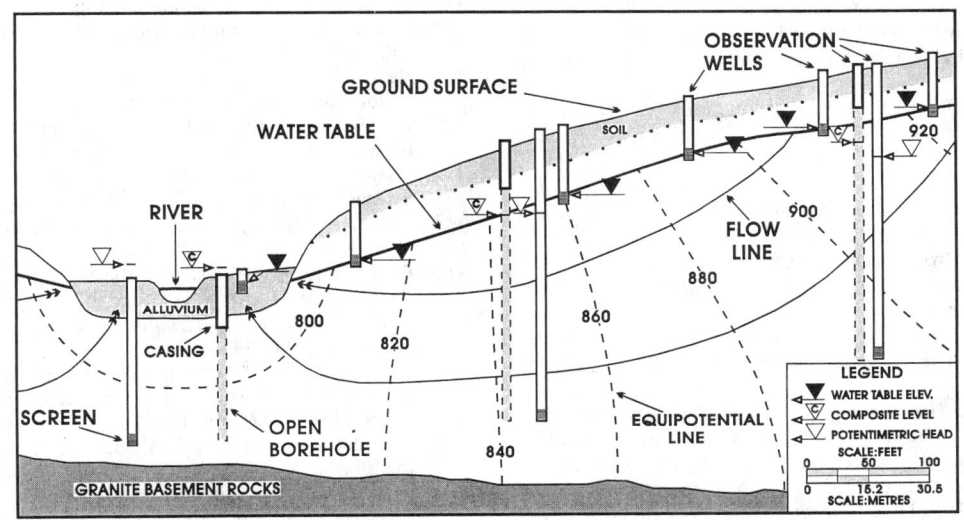

FIG. 20 Water-level Cross Section (adapted from Ref (77))

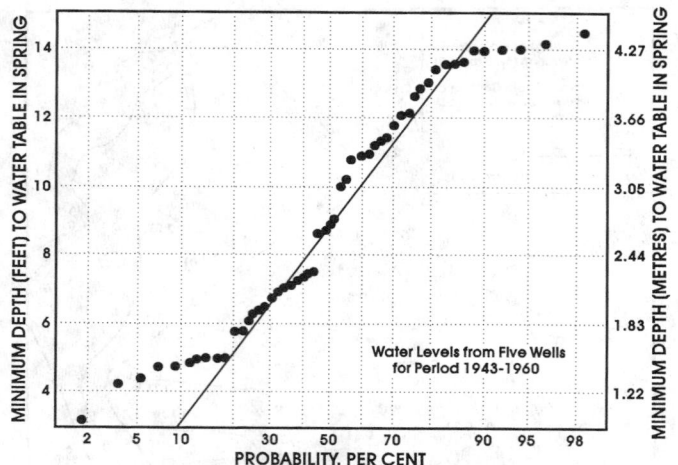

**FIG. 21 Probability of Spring Minimum Depths to Water Table
(adapted from Ref (91))**

minimum spring time or differences between springtime minimum and fall-time maximum water table can be determined from long-term records (90). Cumulative departures in pumpage and precipitation rates versus average water table can be plotted to compare interdependence of the data (91). Correlation analyses between water table fluctuations and related data, for example, river stages, precipitation, or barometric pressure, can be valuable in detecting the cause of the fluctuations (28). Maps showing the seasonal deviation of the water table from the long-term mean of selected shallow wells can indicate areas of drought and above normal precipitation conditions, for example, for a state (4).

8. Automated Procedures for Water-Level and Hydraulic Head Graphics

8.1 *Introduction*—Information concerning the availability of computer software for displaying water level and hydraulic head data in a tabular and graphic format can be obtained from scientific software clearing houses.

8.1.1 Packages of software marketed by Rockware contain routines for plotting graphs, contour maps, and cross-sections of ground-water level data on a desktop computer (95).

8.1.2 Packages of software marketed by Scientific Software Group contain routines for plotting graphs, contour maps, and cross-sections of ground-water level data on a desktop computer (see Ref (96)).

8.1.3 The International Ground Water Modeling Center supplies various types of ground-water software.

8.1.4 Donley Technology, a software information company, documents numerous environmental and hydrologic packages.

9. Keywords

9.1 aquifer; confined aquifer; ground water; hydraulic head; hydrograph; potentiometric surface; unconfined aquifer; water level; water table

REFERENCES

(1) Freeze, R. A., and Cherry, J. A., *Groundwater*, Prentice-Hall, Inc., Englewood Cliffs, New Jersey, 1979, p. 230.

(2) Bates, R. L., and Jackson, J. A., *Glossary of Geology*, Third Edition, American Geological Institute, Alexandria, Virginia, 1987, p. 787.

(3) Subsurface-Water Glossary Working Group, Ground-Water Subcommittee, Interagency Advisory Committee on Water Data, *Subsurface-water Flow and Solute Transport Federal Glossary of Selected Terms*, Office of Water Data Coordination, U.S. Geological Survey, 1989, p. 38.

(4) Novak, C. E., "WRD Data Reports Preparation Guide," 1985 edition, *U.S. Geological Survey*, 1993, p. 199.

(5) Ruland, W. W., Cherry, J. A., and Feenstra, S., "The Depth of Fractures and Active Ground-Water Flow in a Clayey Till Plain in Southwestern Ohio," *Ground Water*, Vol 29, No. 3, 1991, pp. 405–417.

(6) Siegel, D. I., "The Recharge-discharge Function of Wetlands Near Juneau, Alaska: Part I, Hydrogeological Investigations," *Ground Water*, Vol 26, No. 4, 1988, pp. 427–434.

(7) Ortega-Guerrero, A., Cherry, J. A., and Rudolph, D. L., "Large-scale Aquitard Consolidation Near Mexico City," *Ground Water*, Vol 31, No. 5, 1993, pp. 708–718.

(8) Morgan, C. O., "Ground-water Resources of East Feliciana and West Feliciana Parishes, Louisiana," *Louisiana Department of Public Works*, 1963, p. 58.

(9) Shaver, R. B., and Pusc, S. W., "Hydraulic Barriers in Pleistocene Buried-valley Aquifers," *Ground Water*, Vol 30, No. 1, 1992, pp. 21–28.

(10) Hardt, W. F., and Hutchinson, C. B., "Model Aids Planners in Predicting Rising Ground-water Levels in San Bernardino, California," *Ground Water*, Vol 16, No. 6, 1978, pp. 424–431.

(11) Rosenberry, D. O., "Effect of Sensor Error on Interpretation of Long-term Water-level Data," *Ground Water*, Vol 28, No. 6, 1990, pp. 927–936.

(12) Briz-Kishore, B. H., and Bhimasankaram, V. L. S., "Analysis of Ground-water Hydrographs for Defining a Crystalline Hydrogeological Environment," *Ground Water*, Vol 19, No. 5, 1981, pp. 476–481.

(13) Gburek, W. J., and Urban, J. B., "The Shallow Weathered Fracture Layer in the Near-stream Zone," *Ground Water*, Vol 28, No. 6, 1990, pp. 875–883.

(14) Bredenkamp, D. B., "Quantitative Estimation of Groundwater Recharge by Means of a Simple Rainfall-recharge Relationship in Groundwater Recharge, A Guide to Understanding and Estimating Natural Recharge," *International Association of Hydrogeologists*, Vol 8, 1990, pp. 247–256.

(15) Hoyle, B. L., "Ground-water Quality Variations in a Silty Alluvial Soil Aquifer, Oklahoma," *Ground Water*, Vol 27, No. 4, 1989, pp. 540–549.

(16) Vegter, J. R., and Foster, M. B. J., "The Hydrogeology of Dolomitic Formations in the Southern and Western Transvaal in Hydrogeology of Selected Karst Regions," *International Contributions to Hydrogeology*, Vol 13, 1992, pp. 355–376.

(17) Llamas, M. R., "Wetlands: An Important Issue in Hydrogeology in Selected Papers on Aquifer Overexploitation," *International Association of Hydrogeologists*, 23rd International Congress, Vol 3, 1992, pp. 69–86.

(18) Rushton, K. R., "Recharge in the Mehsana Alluvial Plain, India in Groundwater Recharge, A Guide to Understanding and Estimating Natural Recharge," *International Association of Hydrogeologists*, Vol 8, 1990, pp. 297–312.

(19) Goolsby, D. A., "Hydrogeochemical Effects of Injecting Wastes into a Limestone Aquifer near Pensacola, Florida," *Ground Water*, Vol 9, No. 1, 1971, pp. 13–19.

(20) Van der Goot, H. A., Zielbauer, E. J., Bruington, A. E., and Ingram, A. A., "Sea Water Intrusion in California, Appendix B, Part II, Report by Los Angeles County Flood Control District on Investigational Work for Prevention and Control of Sea Water Intrusion, West Coast Basin Experimental Project, Los Angeles

County," *State of California, Department of Water Resources,* Bulletin No. 63, Appendix B, Report by Los Angeles County Flood Control District, December 20, 1954, 1957, pp. 21–87.

(21) Kelly, T. E., "Artificial Recharge at Valley City, North Dakota, 1932 to 1965," *Ground Water,* Vol 5, No. 2, 1967, pp. 20–25.

(22) Sara, M. N., *Standard Handbook for Solid and Hazardous Waste Facility Assessments,* Lewis Publishers, Boca Raton, 1993, pp. 4-64 to 4-71.

(23) Brassington, R., *Field Hydrology,* John Wiley & Sons, Inc., Somerset, 1988, p. 175.

(24) Vacher, H. L., "Hydrology of Small Oceanic Islands—Influence of Atmospheric Pressure on the Water Table," *Ground Water,* Vol 16, No. 6, 1978, pp. 417–423.

(25) Van Hylckama, T. E. A., "Water Level Fluctuation in Evapotranspirometers," *Water Resources Research,* Vol 4, No. 4, 1968, pp. 761–768.

(26) Keller, C. K., and Van der Kamp, G., "Slug Tests with Storage Due to Entrapped Air," *Ground Water,* Vol 30, No. 1, 1992, pp. 2–7.

(27) Robson, S. G., and Banta, E. R., "Determination of Specific Storage by Measurement of Aquifer Compression Near a Pumping Well," *Ground Water,* Vol 28, No. 6, 1990, pp. 868–874.

(28) Gilmore, T. J., Borghese, J. V., and Newcomer, D. R., "Effects of River Stage and Waste Water Discharges on the Unconfined Aquifer, Hanford, Washington," *Ground Water Monitoring Review,* Vol XII, No. 1, 1993, pp. 130–138.

(29) Beukes, J. H. T., and du Plessis, A., "Surface Subsidences and Sinkhole Formation Due to Partial Recharge of a Dewatered Area on the Far West Rand, Republic of South Africa, in Selected Papers on Hydrogeology, 28th International Geological Congress," *International Association of Hydrogeologists,* Vol 1, 1990, pp. 43–52.

(30) Piper, L. M., "Analysis of Unexpected Results of Water Level Study to Determine Aquifer Interconnection in Proceedings of the Fifth National Outdoor Action Conference on Aquifer Restoration, Ground Water Monitoring, and Geophysical Methods," *Ground Water Management,* Book 5, 1991, pp. 205–219.

(31) Blaszyk, T., and Gorski, J., "Ground-water Quality Changes During Exploitation," *Ground Water,* Vol 19, No. 1, 1981, pp. 28–33.

(32) Morgan, C. O., "Ground-water Conditions in the Baton Rouge Area, 1954–59, with Special Reference to Increased Pumpage," *Louisiana Department of Conservation, Geological Survey, and Department of Public Works, Water Resources Bulletin,* No. 2, 1961, p. 78.

(33) Thomas, H. E., "A Water Budget for the Artesian Aquifer in Ogden Valley, Weber County, Utah," *U.S. Geological Survey, Water-Supply Paper 1544-H,* 1963, pp. H63–H97.

(34) Jellali, M., Geanah, M., and Bichara, S., "Groundwater Mining and Development in the Souss Valley, (Morocco) in Selected Papers on Aquifer Overexploitation," *International Association of Hydrogeologists,* 23rd International Congress, Vol 3, 1992, pp. 337–348.

(35) Gregg, D. O., "An Analysis of Ground-water Fluctuations Caused by Ocean Tides in Glynn County, Georgia," *Ground Water,* Vol 4, No. 3, 1966, pp. 24–32.

(36) Marine, I. W., "Water Level Fluctuations Due to Earth Tides in a Well Pumping from a Slightly Fractured Crystalline Rock," *Water Resources Research,* Vol 11, No. 1, 1975, pp. 165–173.

(37) Ferris, J. G., Knowles, D. B., Brown, R. H., and Stallman, R. W., "Theory of Aquifer Tests," *U.S. Geological Survey, Water-Supply Paper 1536-E,* 1962, pp. 69–174.

(38) Hsieh, P. A., Bredehoeft, J. D., and Rojstaczer, S. A., "Response of Well Aquifer Systems to Earth Tides; Problem Revisited," *Water Resources Research,* Vol 24, No. 3, 1988, pp. 468–472.

(39) Vorhis, R. C., "Hydrologic Effects of the Earthquake of March 27, 1964, Outside Alaska," *U.S. Geological Survey, Professional Paper 544-C,* 1967, p. 54.

(40) DaCosta, J. A., "Effect of Hebgen Lake Earthquake on Water Levels in Wells in the United States," *U.S. Geological Survey, Professional Paper 435-0,* 1964, pp. 167–178.

(41) Grantz, A., and others, "Alaska's Good Friday Earthquake, March 27, 1964, a Preliminary Geologic Evaluation," *U.S. Geological Survey, Circular 491,* 1964, p. 35.

(42) Jacob, C. E., "Fluctuations in Artesian Pressure Produced by Passing Railroad-Trains as Shown in a Well on Long Island, New York," *U.S. Geological Survey, Ground Water Notes,* No. 16, 1953, p. 13.

(43) Garber, M. S., and Koopman, F. C., "Methods of Measuring Water Levels in Deep Wells in Techniques of Water-Resources Investigations, Chapter A1," *U.S. Geological Survey,* Book 8, 1968, pp. 23.

(44) Garber, M. S., and Wollitz, L. E., "Measuring Underground-explosion Effects on Water Levels in Surrounding Aquifers," *Ground Water,* Vol 7, No. 4, 1969, pp. 3–7.

(45) Andreasen, G. E., and Brookhart, J. W., "Reverse Water-level Fluctuations," *U.S. Geological Survey, Water-Supply Paper 1544-H,* 1963, pp. H30–H35.

(46) Gerhart, J. M., "Ground-water Recharge and Its Effects on Nitrate Concentration Beneath a Manured Field Site in Pennsylvania," *Ground Water,* Vol 24, No. 4, 1986, pp. 483–489.

(47) Grassi, S., Celati, R., Bolognesi, L., Calore, C., D'Amore, F., Squarci, P., and Taffi, L., "Long-term Observations of a Low-temperature Hydrothermal System (Campiglia, Central Italy)," in *Selected Papers on Hydrogeology, 28th International Geological Congress,* International Association of Hydrogeologists, Vol 1, 1990, pp. 181–198.

(48) Gregg, D. O., "Protective Pumping to Reduce Aquifer Pollution, Glynn County, Georgia," *Ground Water,* Vol 9, No. 5, 1971, pp. 21–29.

(49) Davidson, C. C., and Vonhof, J. A., "Spatial and Temporal Hydrochemical Variations in a Semiconfined Buried Channel Aquifer: Esterhazy, Saskatchewan, Canada," *Ground Water,* Vol 16, No. 5, 1978, pp. 341–351.

(50) Hall, D. W., "Effects of Nutrient Management on Nitrate Levels in Ground Water near Ephrata, Pennsylvania," *Ground Water,* Vol 30, No. 5, 1992, pp. 720–730.

(51) Salama, R. B., Otto, C. J., Bartle, G. A., and Watson, G. D., "Management of Saline Groundwater Discharge by Long-term Windmill Pumping in the Wheatbelt, Western Australia," *Applied Hydrology, Journal of the International Association of Hydrologists,* Vol 2, No. 1, 1994, pp. 19–33.

(52) Bedinger, M. S., and Reed, J. E., "Computing Stream-induced Ground-water Fluctuation in Geological Survey Research 1964, Chapter B," *U.S. Geological Survey, Professional Paper 501,* 1964, pp. B177–B180.

(53) Rorabaugh, M. I., "Streambed Percolation in Development of Water Supplies," *U.S. Geological Survey, Water-Supply Paper 1544-H,* 1963, pp. H47–H62.

(54) O'Brien, A. L., "The Role of Ground Water in Stream Discharge from Two Small Wetland Controlled Basins in Eastern Massachusetts," *Ground Water,* Vol 18, No. 4, 1980, pp. 359–365.

(55) Schneider, R., "Correlation of Ground-Water Levels and Air Temperatures in the Winter and Spring in Minnesota," *U.S. Geological Survey, Water-Supply Paper 1539-D,* 1961, p. 14.

(56) Atwood, D. F., and Lamb, B., "Resolution Problems with Obtaining Accurate Ground Water Elevation Measurements in a Hydrogeologic Site Investigation" in *Proceedings of the First National Outdoor Action Conference on Aquifer Restoration, Ground Water Monitoring, and Geophysical Methods,* National Water Well Association, 1987, pp. 185–192.

(57) Rose, N. A., and Alexander, W. H., Jr., "Relation of Phenomenal Rise of Water Levels to a Defective Gas Well, Harris County, Texas," *Am. Assoc. Petroleum Geologists,* Vol 29, 1945, pp. 253–279.

(58) Crowe, A. S., "Numerical Modelling of the Groundwater Contribution to the Hydrological Budget of Lakes," in *Selected Papers on Hydrogeology, 28th International Geological Congress,* International Association of Hydrogeologists, Vol 1, 1990, pp. 283–300.

(59) El Baruni, S. S., "Earth Fissures Caused by Groundwater Withdrawals in Sarir South Agricultural Project Area, Libya," *Applied Hydrology, Journal of the International Association of Hydrolo-*

gists, Vol 2, No. 1, 1994, pp. 45–52.

(60) Booth, C. J., "Strata-movement Concepts and the Hydrogeological Impact of Underground Coal Mining," *Ground Water*, Vol 24, No. 4, 1986, pp. 507–515.

(61) Sulam, D. J., "Analysis of Changes in Ground-water Levels in a Sewered and an Unsewered Area of Nassau County, Long Island, New York," *Ground Water*, Vol 17, No. 5, 1979, pp. 446–455.

(62) Faulkner, G. L., and Pascale, C. A., "Monitoring Regional Effects of High Pressure Injection of Industrial Waste Water in a Limestone Aquifer," *Ground Water*, Vol 13, No. 2, 1975, pp. 197–208.

(63) Pankow, J. F., Johnson, R. L., Houck, J. E., Brillante, S. M., and Bryan, W. J., "Migration of Chloropnenolic Compounds at the Chemical Waste Disposal Site at Alkali Lake, Oregon—1. Site Description and Ground-water Flow," *Ground Water*, Vol 22, No. 5, 1984, pp. 593–601.

(64) Korte, N. E., and Kearl, P. M., "The Utility of Multiple-completion Monitoring Wells for Describing a Solvent Plume," *Ground Water Monitoring Review*, Vol XI, No. 2, 1991, pp. 153–156.

(65) MacFarlane, P. A., Townsend, M. A., and Evans, D. L., "Construction and Use of Multiple Completion Monitoring Wells in Hydrogeologic Studies of Deep Aquifer Systems: a Case Study from the Dakota Aquifer in Central Kansas" in *Proceedings of the Second National Outdoor Action Conference on Aquifer Restoration, Ground Water Monitoring, and Geophysical Methods*, National Water Well Association, Vol I, 1988, pp. 391–413.

(66) Heath, R. C., "Design of Ground-water Level Observation-well Program," *Ground Water*, Vol 14, No. 2, 1976, pp. 71–77.

(67) Singh, K. P., "Groundwater Overdraft in North West Parts of Indo-Gangetic Alluvial Plains, India," *Feasibility of Artificial Recharge in Selected Papers on Aquifer Overexploitation*, International Association of Hydrogeologists, 23rd International Congress, Vol 3, 1992, pp. 183–189.

(68) Ball, D. F., and Herbert, R., "The Use and Performance of Collector Wells Within the Regolith Aquifer of Sri Lanka," *Ground Water*, Vol 30, No. 5, 1992, pp. 683–689.

(69) Helsel, D. R., and Hirsch, R. M., "Statistical Methods in Water Resources," *Studies in Environmental Science 49*, Elsevier, Amsterdam, 1992, p. 522.

(70) Parizek, R. R., and Siddiqui, S. H., "Determining the Sustained Yields of Wells in Carbonate and Fractured Aquifers," *Ground Water*, Vol 8, No. 5, 1970, pp. 12–20.

(71) Visocky, A. P., "Estimating the Ground-water Contribution to Storm Runoff by the Electrical Conductance Method," *Ground Water*, Vol 8, No. 2, 1970, pp. 5–10.

(72) Dial, D. C., "Water-level Trends in Southeastern Louisiana," *Louisiana Department of Conservation, Geological Survey, and Department of Public Works, Water Resources Pamphlet*, No. 22, 1968, p. 11.

(73) Marbury, R. E., and Brazie, M. E., "Ground Water Monitoring in Tight Formations" in *Proceedings of the Second National Outdoor Action Conference on Aquifer Restoration, Ground Water Monitoring, and Geophysical Methods*," National Water Well Association, Vol I, 1988, pp. 483–492.

(74) Johansen, E., Meadows, J. K., and Jones, J., "Joint Interpretation of Geophysical, Soil Gas, and Hydrogeologic Data to Characterize Hydrochemical Facies Within a Partially Submerged Landfill" in *Proceedings of the Seventh National Outdoor Action Conference and Exposition, Ground Water Management*, Book 15, 1993, pp. 287–301.

(75) Osiensky, J. L., Winter, G. V., and Williams, R. E., "Monitoring and Mathematical Modeling of Contaminated Ground-water Plumes in Fluvial Environments," *Ground Water*, Vol 22, No. 3, 1984, pp. 298–306.

(76) Saines, M., "Errors in Interpretation of Ground-Water Level Data," *Ground Water Monitoring Review*, Vol 1, No. 1, 1981, pp. 56–61.

(77) Fetter, C. W., *Applied Hydrogeology*, second edition, Merrill Publishing Company, Columbus, 1988, pp. 325–366.

(78) Meyer, K. A., Jr., "Ground Water Monitoring Strategies at the Weldon Spring Site, Weldon Spring, Missouri" in *Proceedings of the Second National Outdoor Action Conference on Aquifer Restoration, Ground Water Monitoring, and Geophysical Methods*, National Water Well Association, Vol III, 1988, pp. 1257–1268.

(79) Kehew, A. E., Schwindt, F. J., and Brown, D. J., "Hydrogeochemical Interaction Between a Municipal Waste Stabilization Lagoon and a Shallow Aquifer," *Ground Water*, Vol 22, No. 6, 1984, pp. 746–754.

(80) Houston, J. F. T., and Lewis, R. T., "The Victoria Province Drought Relief Project, II. Borehole yield relationships," *Ground Water*, Vol 26, No. 4, 1988, pp. 418–426.

(81) Boulding, J. R., *Practical Handbook of Soil, Vadose Zone, and Ground Water Contaminational Assessment, Prevention, and Remediation*, Lewis Publishers, Boca Raton, FL, 1995.

(82) Bureau of Reclamation, *Ground-Water Manual, A Water Resources Technical Publication, Revised Reprint*, U.S. Department of Interior, Bureau of Reclamation, Washington, DC, 1981, p. 480.

(83) Zhaoxin, W., "Environmental Effects Related to Aquifer Overexploitation in Arid and Semi-arid Areas of China" in Selected Papers on Aquifer Overexploitation, *International Association of Hydrogeologists*, 23rd International Congress, Vol 3, 1992, pp. 155–164.

(84) Mukhopadhyay, A., Al-Sulaimi, J., and Barrat, J. M., "Numerical Modeling of Ground-water Resource Management Options in Kuwait," *Ground Water*, Vol 32, No. 6, 1994, pp. 917–928.

(85) Meyer, W. R., Gutentag, E. D., and Lobmeyer, D. H., "Geohydrology of Finney County, Southwestern Kansas," *U.S. Geological Survey, Water-Supply Paper*, 1891, 1970.

(86) Jaquet, N. G., "Ground-water and Surface-water Relationships in the Glacial Province of Northern Wisconsin—Snake Lake," *Ground Water*, Vol 14, No. 4, 1976, pp. 194–199.

(87) Kroitoru, L., Mazor, E., and Issar, A., "Flow Regimes in Karstic Systems: the Judean Anticlinorium, Central Israel," in *Hydrogeology of Selected Karst Regions, International Contributions to Hydrogeology*, Vol 13, 1992, pp. 339–354.

(88) Motz, L. H., "Well-field Drawdowns Using Coupled Aquifer Model," *Ground Water*, Vol 19, No. 2, 1981, pp. 172–179.

(89) Stone, R., Raber, E., and Winslow, A. M., "Effects of Aquifer Interconnection Resulting from Underground Coal Gasification," *Ground Water*, Vol 21, No. 5, 1983, pp. 606–618.

(90) Orsborn, J. F., "The Prediction of Piezometric Groundwater Levels in Observation Wells Based on Prior Occurrences," *Water Resources Research*, Vol 2, No. 1, 1966, pp. 139–144.

(91) Hagerty, D. J., and Lippert, K., "Rising Ground Water—Problem or Resource?," *Ground Water*, Vol 20, No. 2, 1982, pp. 217–223.

(92) Hamlin, S. N., "Hydraulic/Chemical Changes During Ground-water Recharge by Injection," *Ground Water*, Vol 25, No. 3, 1987, pp. 267–274.

(93) Norris, S. E., and Eagon, H. B., Jr., "Recharge Characteristics of a Watercourse Aquifer System at Springfield, Ohio," *Ground Water*, Vol 9, No. 1, 1971, pp. 30–41.

(94) Hodgson, F. D. I., "The Use of Multiple Linear Regression in Simulating Ground-water Level Responses," *Ground Water*, Vol 16, No. 4, 1978, pp. 249–253.

(95) Rockware, *The Scientific Software Catalog*, RockWare Scientific Software, Wheat Ridge, CO, 1993.

(96) Scientific Software, *Environmental, Engineering, and Water Resources Software*, Scientific Software Group, Washington, DC, 1994.

 **D 6000**

1.6 DRILLING METHODS

Standard Guide for
Use of Dual-Wall Reverse-Circulation Drilling for Geoenvironmental Exploration and the Installation of Subsurface Water-Quality Monitoring Devices[1]

This standard is issued under the fixed designation D 5781; the number immediately following the designation indicates the year of original adoption or, in the case of revision, the year of last revision. A number in parentheses indicates the year of last reapproval. A superscript epsilon (ε) indicates an editorial change since the last revision or reapproval.

1. Scope

1.1 This guide covers how dual-wall reverse-circulation drilling may be used for geoenvironmental exploration and installation of subsurface water-quality monitoring devices.

NOTE 1—The term *reverse circulation* with respect to dual-wall drilling in this guide indicates that the circulating fluid is forced down the annular space between the double-wall drill pipe and transports soil and rock particles to the surface through the inner pipe.

NOTE 2—This guide does not include considerations for geotechnical site characterizations that are addressed in a separate guide.

1.2 Dual-wall reverse-circulation for geoenvironmental exploration and monitoring-device installations will often involve safety planning, administration, and documentation. This guide does not purport to specifically address exploration and site safety.

1.3 The values stated in SI units are to be regarded as the standard. The inch-pound units given in parentheses are for information only.

1.4 *This standard does not purport to address all of the safety concerns, if any, associated with its use. It is the responsibility of the user of this standard to establish appropriate safety and health practices and determine the applicability of regulatory limitations prior to use.*

2. Referenced Documents

2.1 *ASTM Standards:*

D 653 Terminology Relating to Soil, Rock, and Contained Fluids[2]

D 1452 Practice for Soil Investigation and Sampling by Auger Borings[2]

D 1586 Test Method for Penetration Test and Split-Barrel Sampling of Soils[2]

D 1587 Test Method for Thin-Walled Tube Sampling of Soils[2]

D 2487 Classification of Soils for Engineering Purposes[2]

D 3550 Practice for Ring-Lined Barrel Sampling of Soils[2]

D 4428/D 4428M Test Method for Crosshole Seismic Testing[2]

D 5088 Practice for Decontamination of Field Equipment Used at Non-Radioactive Waste Sites[3]

D 5092 Practice for Design and Installation of Ground Water Monitoring Wells in Aquifers[3]

D 5099 Test Method for Rubber—Measurement of Processing Properties Using Capillary Rheometry[4]

D 5254 Practice for Minimum Set of Data Elements to Identify a Ground-Water Site[3]

D 5434 Guide for Field Logging of Subsurface Explorations of Soil and Rock[3]

3. Terminology

3.1 *Definitions:*

3.1.1 Terminology used within this guide is in accordance with Terminology D 653. Definitions of additional terms may be found in Terminology D 653.

3.2 *Descriptions of Terms Specific to This Standard:*

3.2.1 *bentonite*—common name for drilling-fluid additives and well-construction products consisting mostly of naturally-occurring montmorillonite. Some bentonite products have chemical additives that may affect water-quality analyses.

3.2.2 *bentonite granules and chips*—irregularly-shaped particles of bentonite (free from additives) that have been dried and separated into a specific size range.

3.2.3 *bentonite pellets*—roughly spherical- or disc-shaped units of compressed bentonite powder (some pellet manufacturers coat the bentonite with chemicals that may affect the water quality analysis).

3.2.4 *coefficient of uniformity*—$C_u(D)$, the ratio D_{60}/D_{10}, where D_{60} is the particle diameter corresponding to 60 % finer on the cumulative particle-size distribution curve, and D_{10} is the particle diameter corresponding to 10 % finer on the cumulative particle-size distribution curve.

3.2.5 *drawworks*—a power-driven winch, or several winches, usually equipped with a clutch and brake system(s) for hoisting or lowering a drilling string.

3.2.6 *drill hole*—a cylindrical hole advanced into the subsurface by mechanical means. Also known as a borehole or boring.

3.2.7 *filter pack*—also known as a gravel pack or a primary filter pack in the practice of monitoring-well installations. The gravel pack is usually granular material, having selected grain size characteristics, that is placed between a monitoring device and the borehole wall. The basic purpose of the filter pack or gravel envelope is to act as: (*1*) a non-clogging filter when the aquifer is not suited to natural

[1] This guide is under the jurisdiction of ASTM Committee D-18 on Soil and Rock and is the direct responsibility of Subcommittee D18.21 on Ground Water and Vadose Zone Investigations.

Current edition approved Oct. 10, 1995. Published December 1995.

[2] *Annual Book of ASTM Standards*, Vol 04.08.

[3] *Annual Book of ASTM Standards*, Vol 04.09.

[4] *Annual Book of ASTM Standards*, Vol 09.01.

development or, (2) act as a formation stabilizer when the aquifer is suitable for natural development.

3.2.7.1 *Discussion*—Under most circumstances a clean, quartz sand or gravel should be used. In some cases a pre-packed screen may be used.

3.2.8 *hoisting line*—or drilling line, is wire rope used on the drawworks to hoist and lower the drill string.

3.2.9 *in-situ testing devices*—sensors or probes, used for obtaining mechanical or chemical-test data, that are typically pushed, rotated or driven below the bottom of a borehole following completion of an increment of drilling. However, some in-situ testing devices (such as electronic pressure transducers, gas-lift samplers, tensiometers, and etc.) may require lowering and setting of the device(s) in a pre-existing borehole by means of a suspension line or a string of lowering rods or pipe. Centralizers may be required to correctly position the device(s) in the borehole.

3.2.10 *intermittent-sampling devices*—usually barrel-type samplers that are driven or pushed below the bottom of a borehole following completion of an increment of drilling. The user is referred to the following ASTM Standards relating to suggested sampling methods and procedures: Practice D 1452, Test Method D 1586, Practice D 3550, and Practice D 1587.

3.2.11 *mast*—or derrick, on a drilling rig is used for supporting the crown block, top drive, pulldown chains, hoisting lines, etc. It must be constructed to safely carry the expected loads encountered in drilling and completion of wells of the diameter and depth for which the rig manufacturer specifies the equipment.

3.2.11.1 *Discussion*—To allow for contingencies, it is recommended that the rated capacity of the mast should be at least twice the anticipated weight load or normal pulling load.

3.2.12 *piezometer*—an instrument for measuring pressure head.

3.2.13 *subsurface water-quality monitoring device*—an instrument placed below ground surface to obtain a sample for analysis of the chemical, biological or radiological characteristics of subsurface-pore water or to make in-situ measurements.

4. Significance and Use

4.1 Dual-wall reverse-circulation drilling can be used in support of geoenvironmental exploration and for installation of subsurface water-quality monitoring devices in unconsolidated and consolidated materials. Dual-wall reverse-circulation drilling methods permit the collection of water-quality samples at any depth(s), allows the setting of temporary casing during drilling, cuttings samples can be taken continuously as circulation is maintained at all times during drilling. Other advantages of the dual-wall reverse-circulation drilling method include: (1) the capability of drilling without the introduction of any drilling fluid(s) to the subsurface; (2) maintenance of hole stability for sampling purposes and monitor-well installation/construction in poorly-indurated to unconsolidated materials.

NOTE 3—The user of dual-wall reverse-circulation drilling for geoenvironmental exploration and monitoring-device installations should be cognizant of both the physical (temperature and airborne particles) and chemical (compressor lubricants and possible fluid addi-

tives) qualities of compressed air that may be used as the circulating medium.

4.2 The application of dual-wall reverse-circulation drilling to geoenvironmental exploration may involve soil or rock sampling, or in-situ soil, rock, or pore-fluid testing.

NOTE 4—The user may install a monitoring device within the same borehole wherein sampling, in-situ or pore-fluid testing, or coring was performed.

4.3 The subsurface water-quality monitoring devices that are addressed in this guide consist generally of a screened- or porous-intake device and riser pipe(s) that are usually installed with a filter pack to enhance the longevity of the intake unit, and with isolation seals and low-permeability backfill to deter the movement of fluids or infiltration of surface water between hydrologic units penetrated by the borehole (see Practice D 5092). Inasmuch as a piezometer is primarily a device used for measuring subsurface hydraulic heads, the conversion of a piezometer to a water-quality monitoring device should be made only after consideration of the overall quality and integrity of the installation to include the quality of materials that will contact sampled water or gas.

NOTE 5—Both water-quality monitoring devices and piezometers should have adequate casing seals, annular isolation seals and backfills to deter communication of contaminants between hydrologic units.

5. Apparatus

5.1 The basic mechanical components of dual-wall reverse-circulation drilling systems include dual-wall pipe, drill compressor and filter(s), water pump, discharge hose, cleaning device (cyclone separator). The dual-wall drill advanced by the percussive action of an above-ground pile hammer or by rotation from a rotary-drive unit.

NOTE 6—Other methods, such as vibratory equipment sonic resonators, may be used to apply the energy required to advance the dual-wall drill pipe.

5.1.1 *dual-wall drill pipe*, consists of an inner pipe secured concentrically within an outer pipe. Inner-pipe connections utilize pin and box components with seals. Outer-pipe connections are flush threaded.

NOTE 7—Drill pipes usually require lubricants on the threads to allow easy unthreading (breaking) of the connecting joints. Some lubricants have organic or metallic constituents, or both, that could be interpreted as contaminants if detected in a sample. Various lubricants are available that have components of known chemistry. The effect of pipe-thread lubricants on chemical analyses of samples should be considered and documented when using dual-wall reverse-circulation drilling. The same consideration and documentation should be given to lubricants used with water swivels, hoisting swivels, or other devices used near the drilling axis.

5.1.2 The drill bit is attached to the bottom of the dual-wall drill pipe and provides the soil- or rock-cutting capability. Drill bit types include tricone roller, down-the-hole (DTH) hammer or, open faced. Drill bit selection should be based upon the character of the soils or rocks penetrated. DTH lubricants should be documented.

NOTE 8—In North America, the sizes of casings bits, drill rods and core barrels are standardized by American Petroleum Institute (API) and the Diamond Core Drill Manufacturers Association (DCDMA). Refer to the DCDMA technical manual and to published materials of API for available sizes and capacities of drilling tools equipment.

5.1.3 The air compressor and filter(s) should provide an adequate volume of air for removal of cuttings without significant contamination generated at the bit. Air requirements will vary depending upon the size and configuration of the drill pipe used, and the character of the soil and rock penetrated. The air-flow rates are usually based on maintaining an upflow air velocity of about 1,400 m/min (4200 ft/min).

NOTE 9—The quality of compressed air entering the borehole and the quality of air discharged from the borehole and air-cleaning devices must be considered. If not adequately filtered, the air produced by most oil-lubricated air compressors inherently introduces a significant quantity of oil into the circulation system. High-efficiency, in-line, air filters are usually required to prevent significant contamination of the borehole. Air-quality monitoring may be required and, if performed, results should be documented.

5.1.4 A water pump may be used to inject water into the circulating air stream or may be used to inject water without air as the circulating fluid. If water is injected, the approximate volumes and locations should be reported.

5.1.5 A discharge hose conducts discharged drill cuttings and circulation-return air away from the borehole.

5.1.6 *Air-Cleaning Device System*, generally called a cyclone separator, separates cuttings from the air returning from the borehole.

NOTE 10—A properly-sized cyclone separator can remove practically all of the cuttings from the return air. A small quantity of fine particles, however, are usually discharged to the atmosphere with the "cleaned" air. Some air-cleaning devices consist of a cyclone separator alone. In special cases, the cyclone separator can be combined with a HEPA (high-efficiency particulate air) filter for removing dust particles that might be radioactive. In other special situations, the cyclone separator may be used in conjunction with a charcoal-filtering arrangement for removal of organic volatiles. Samples of drill cuttings can be collected for analyses of materials penetrated. If samples are obtained, the depth(s) and interval(s) of sample collection should be documented.

5.1.7 *Pile Hammer*, is commonly used to advance dual-wall drill pipe. The percussive force of the pile hammer is applied only to the outer pipe.

5.1.8 *Rotary-Drive Unit*, may be used to advance dual-wall drill pipe by rotation. Torque generated from a rotary-drive unit is applied only to the outer pipe.

6. Drilling Procedures

6.1 *Dual-Wall Percussion-Hammer Method (see Fig. 1):*

6.1.1 As a prelude to and throughout the drilling process stabilize the drill rig, and raise the drill-rig mast and position the cyclone separator. If air-monitoring operations are performed the prevalent wind direction relative to the exhaust from the drill rig should be considered. Also, the location of the cyclone relative to the rig exhaust should be considered since air-quality monitoring will be performed at the cyclone separator discharge point.

6.1.2 Thread an open-faced bit to the drill pipe.

6.1.3 Force compressed air down the annular space formed between the inner pipes and outer pipes as the percussive action of the pile hammer advances the dual-wall drill pipe. Conduct drill cuttings to the surface through the inner pipe.

6.1.4 Continue air circulation and the percussive action until drilling progresses to a depth where sampling or in-situ testing is to be performed or until the length of the drill-pipe

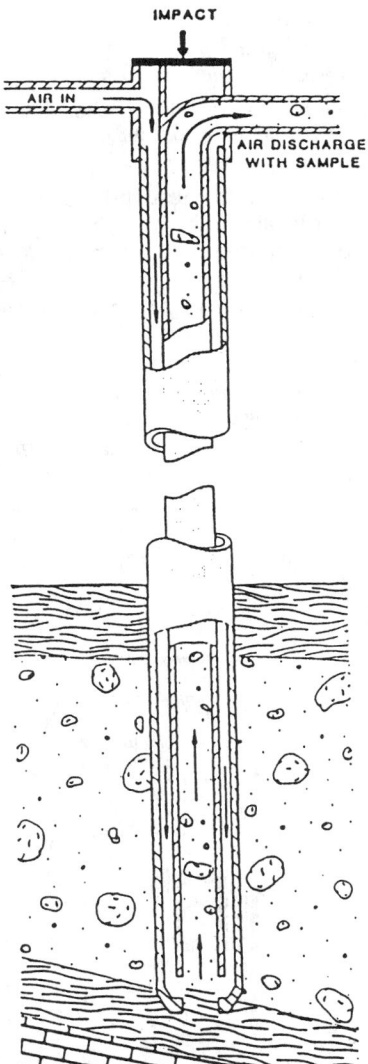

FIG. 1 Drilling with the Dual Wall Percussion Hammer Method

section limits further penetration.

NOTE 11—At a minimum, the following information should be documented: number of impacts or driving conditions (i.e. hard, soft, rapid/slow penetration rate), air pressures, water added, volume of cuttings or cuttings return, air quality data, samples taken, water losses, heaving, and any observed unusual occurrences. Drilling rates depend on many factors such as the density or stiffness of unconsolidated material and the existence of cobbles or boulders, the hardness and/or durability of the rock, the swelling activity of clays or shales encountered in the borehole and the erosiveness of the borehole wall. Drilling rates can vary from a few mm (less than an in./min) to about 1 m (3 ft)/min, depending on subsurface conditions. Other factors influencing drilling rates include the weight of the drill string. These data as well as any other drilling-rate information should be recorded.

6.1.5 The percussive action is then stopped. Maintain air circulation, however, for a short time until the drill cuttings are removed from the inner pipe.

6.1.6 Increase drilling depth by attaching an additional section of dual-wall drill pipe to the top of the previously-advanced section of dual-wall drill pipe.

6.1.7 Sampling or in-situ testing can be performed at any

depth. Insert the sampling or in-situ testing device through the open inner pipe and open-faced bit and lower to the material at the bottom of the borehole.

NOTE 12—Sampling and testing devices should be decontaminated according to Practice D 5088 prior to testing.

6.2 *"Triple-Wall" Percussion Method (see Fig. 2):*

6.2.1 As a prelude to and throughout the drilling process, stabilize the drill rig, and raise the drill rig mast with the cyclone separator positioned. If air-monitoring operations are performed, the prevalent wind direction relative to the exhaust from the drill rig should be considered. Also, the location of the cyclone relative to the rig exhaust should be considered since air-quality monitoring will be performed at the cyclone separator discharge point.

6.2.2 Place a single-wall, flush-threaded pipe over the outside of the dual-wall drill pipe, thus making a triple-wall drilling assembly.

6.2.3 Advance the triple-wall drilling assembly as a single unit by the percussive action of the pile hammer as described in 6.1. Drill cuttings are removed only through the dual-wall

part of this drill-pipe assembly.

6.2.4 Perform sampling or in-situ testing at any depth. Insert the sampling or in-situ testing device through the open inner pipe and open-faced bit and thence into the material at the bottom of the borehole.

NOTE 13—Sampling and testing devices should be decontaminated according to Practice D 5088 prior to testing.

6.3 *Dual-Wall Rotary Method (see Fig. 3):*

6.3.1 As a prelude to and throughout the drilling process first stabilize the drill rig, raise rig mast, and position the cyclone separator. If air-monitoring operations are performed, consider the prevalent wind direction relative to the exhaust from the drill rig. Also, consider the location of the cyclone separator relative to the rig exhaust since air-quality monitoring may be performed at the cyclone separator discharge point.

6.3.2 Thread an open-faced multicone roller bit or DTH-hammer bit (using appropriate crossover sub) to the drill pipe.

6.3.3 Force compressed air down the annular space

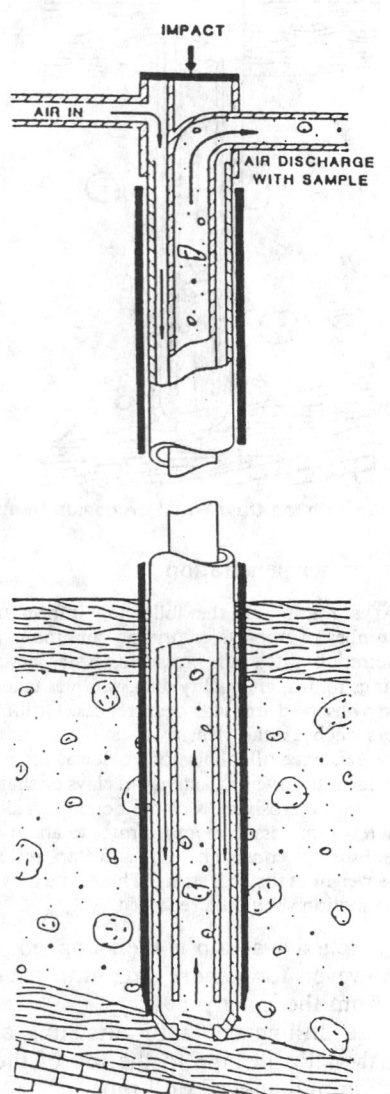

FIG. 2 Drilling with the "Triple Wall" Percussion Hammer Method

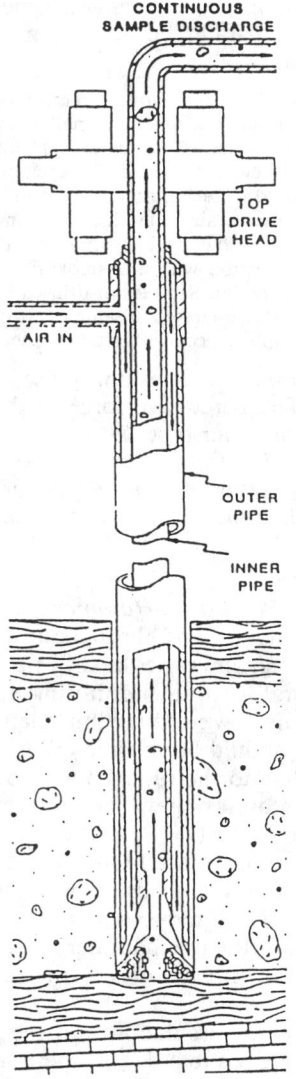

FIG. 3 Drilling with the Dual Wall Rotary Method

formed between the inner pipes and outer pipes as the rotation from the top-head drive unit advances the dual-wall drill pipe. Conduct drill cuttings to the surface through the inner pipe. Drill the borehole and temporarily case in one pass.

6.3.4 Continue air circulation and rotation until drilling progresses to a depth where sampling or in-situ testing is to be performed or until the length of the drill-pipe section limits further penetration.

6.3.5 Then stop the rotation. Maintain air circulation, however, for a short time until the drill cuttings are removed from the inner pipe.

6.3.6 Drilling depth can be increased by attaching an additional section of dual-wall drill pipe to the top of the previously-advanced section of dual-wall drill pipe.

6.4 *"Triple-Wall" for Dual-Wall Rotary Method:*

6.4.1 As a prelude to and throughout the drilling process, stabilize the drill rig and raise the drill rig mast. Position the cyclone separator and "seal" to the ground surface. If air-monitoring operations are performed, consider the prevalent wind direction relative to the exhaust from the drill rig. Also, consider the location of the cyclone relative to the rig exhaust since air-quality monitoring will be performed at the cyclone-separator discharge point.

6.4.2 Thread an open-faced, tricone roller bit or down-the-hole (DTH) hammer bit to the dual-wall drill pipe.

6.4.3 Force compressed air down the annular space between the inner pipe and the outer pipe as the rotation from the top-head-drive unit advances the dual-wall drill pipe assembly. Conduct drill cuttings to the surface through the inner pipe.

6.4.4 Continue air circulation and rotation until drilling progresses to a depth where sampling or in-situ testing is to be conducted or until the length of drill-pipe section limits further penetration.

6.4.5 Stop the rotation. Maintain air circulation, however, for a short time until the drill cuttings are removed from the inner pipe.

6.4.6 Place a single-wall, flush-threaded drill pipe over the outside of the dual-wall drill pipe, thus making a triple-wall drilling assembly.

6.4.7 Then advance this triple-wall drill pipe to the same depth as the bit on the dual-wall pipe by rotating and washing it over the dual-wall string.

6.4.8 To facilitate downhole testing, remove the dual-wall drill-pipe assembly, leaving the triple-wall pipe temporarily in place to support the borehole wall. Then insert the sampling or in-situ testing device into the formation at the bottom of the borehole.

6.4.9 Increase drilling depth by placing the dual-wall drill string into the triple-wall pipe and attaching an additional section of dual-wall pipe to the top of the previously-advanced section of dual-wall drill pipe.

6.4.10 Repeat the sampling procedure as outlined in the above section describing use of the triple-wall procedure.

NOTE 14—In all of the drilling methods discussed above, compressed air alone can often transport drilled cuttings to the surface. For some geologic conditions, injection of water into the air stream will help control dust or aid in the recovery of some types of materials. Water may also be circulated without air to remove drilled cuttings or control flowing sand conditions. The chemical makeup and quantity of water added to the air stream during the drilling process should be docu-mented because it may affect the mechanical and chemical characteristics of the soil and water samples collected. Containment and disposal of contaminated and potentially-contaminated drilling fluids and associated cuttings should be in accordance with applicable regulations.

7. Installation of Monitoring Devices

7.1 Subsurface water-quality monitoring devices are generally installed in boreholes drilled by dual-wall percussion-hammer method using the four-step procedure. The four steps consist of: (1) drilling, with or without sampling, (2) the dual-wall drill pipe is temporarily left in place to support the borehole wall after total depth of the borehole is reached, (3) insertion of the monitoring device through the inside of the inner pipe, and (4) addition of well-completion materials such as filter packs, annular seals and grouts as the dual-wall drill pipe is extracted from the borehole.

NOTE 15—Practical tooling dimensions commonly employed when utilizing the dual-wall percussion-hammer drilling method to install nominal 10.16 cm (4 in.) diameter instrumentation devices typically include: dual-wall hammer pipe at 22.86 cm OD by 15.24 cm ID, or 16.83 cm OD by 10.79 cm ID (9 in. OD by 6 in. ID, or 6⅝ in. OD by 4¼ in. ID).

7.2 *Triple-Wall Percussion-Hammer Drilling Method*—Subsurface water-quality monitoring devices are generally installed in boreholes drilled by the "triple-wall" percussion-hammer drilling method using a four-step procedure. The four steps consist of: (1) drilling, with or without sampling, (2) removal of the inner, dual-wall drill pipe after total depth of the borehole is reached, and temporarily, leaving the outer pipe in place to support the borehole wall, (3) insertion of the monitoring device inside the cased borehole, and (4) addition of well-completion materials such as filter packs, annular seals and grouts as the outer pipe is hydraulically extracted from the borehole.

NOTE 16—Practical tooling dimensions commonly employed when utilizing the dual-wall percussion-hammer drilling method with the triple-wall casing to install nominal 15.24 cm (6 in.) diameter and larger instrumentation devices typically include: dual-wall hammer pipe at 22.86 cm OD by 15.24 cm ID (9 in. OD by 6 in. ID); and triple-wall casings at 27.30 cm OD by 24.76 cm ID (10¾ in. OD by 9¾ in. ID). In certain applications, triple-wall casings to 45.72 cm (18 in.) in diameter can be practically employed. In most cases, centralizers are only used to center monitoring devices in boreholes drilled by the triple-wall percussion hammer drilling method when larger triple-wall casings are used.

7.3 *Dual-Wall Rotary-Drilling Method*—Subsurface water-quality monitoring devices are generally installed in boreholes drilled by dual-wall rotary-drilling method using the four-step procedure. The four steps consist of: (1) drilling, (2) removal of the dual-wall drill pipe, (3) insertion of the monitoring device, and (4) addition of well-completion materials such as filter packs, annular seals and grouts.

7.4 *Triple-Wall Reverse-Rotary Drilling Method*—Instrumentation devices are generally installed in boreholes drilled by the dual-wall reverse-rotary drilling method and utilizing a triple-wall casing. The installation procedure involved uses the following steps: (1) drilling with or without sampling, (2) removal of the dual-wall drill pipe, (3) insertion of the monitoring device, and (4) addition of well-completion materials such as filter packs, annular seals, and grouts as the outer triple-wall casing is removed.

NOTE 17—Practical tooling dimensions commonly employed when utilizing the dual-wall reverse-rotary drilling method with the triple-wall casing, to install nominal 5.08 cm (2 in.) diameter instrumentation devices, typically include: dual-wall drill pipe at 11.43 cm OD by 5.40 cm ID (4½ in. OD by 2⅛ in. ID); drill bits at 12.38 cm (4⅞ in.) to 12.70 cm (5 in.); and triple-wall casing at 12.70 cm ID by 13.97 cm OD (5 in. ID by 5½ in. OD). The drilling shoe, or bit, on the triple-wall casing is then a nominal 15.24 cm (6 in.) OD.

NOTE 18—In most cases, a centralizer should be used to center a monitoring device in the borehole drilled by the triple-wall percussion-hammer drilling method or the dual-wall reverse-rotary drilling method. If caving overburden conditions occur, temporary surface casing may be needed to prevent hole collapse. The user is referred to Practice D 5092 for monitoring-well installation methods and Practice D 5088 for suggested methods of field-equipment decontamination.

7.5 Assemble water-quality monitoring devices with attached fluid conductors (risers) and insert into the borehole with the least possible addition of contaminants.

7.5.1 Some materials, such as screens and risers, may require cleaning or decontamination, or both, at the job site (see Practice D 5088).

7.5.2 Prior to installation, store all monitoring-device materials undercover and place upwind and well away from the drill rig and any other sources of contamination such as electrical generators, air compressors, or industrial machinery.

7.5.3 Clean hoisting tools, particularly wire rope and hoisting swivels and decontaminate according to Practice D 5088 before using.

7.6 Select filter materials, bentonite pellets, granules and chips, and grouts and install according to subsurface monitoring or instrumentation requirements.

NOTE 19—Filter packs, for monitoring devices are usually installed in borings drilled with dual-wall reverse-circulation methods by placing the materials through the casing-riser annulus. This annular area then serves the same function as a separate tremie pipe for placing the annular materials. In some cases, it may be appropriate to use a tremie pipe inserted in the annulus between the inner pipe and the monitoring-device riser provided it is sufficiently large. Monitoring devices installed in a saturated zone ordinarily have sand size filter packs that are selected primarily on the basis of the grain size characteristics of the hydrologic unit adjacent to the screened intake. The coefficient of uniformity of the filter-pack sand is usually less than 2.5. Filter packs for monitoring devices installed in a vadose zone may be predominantly silt sized. These filter materials are often mixed with water of known quality, inserted through a tremie pipe, and tamped into place around the device. Care should be taken when adding backfill or filter material(s), or both, so that the materials do not bridge. However, if bridging does occur during the installation procedure, tamping rods or other tamping devices may be used to dislodge the "bridge".

7.7 Sealing materials, consisting of either bentonite pellets, chips, or granules, are usually placed directly above the filter pack of a monitoring device.

NOTE 20—It may be effective, when granular filter packs are used, to install a thin, fine sand, secondary filter either below the annular seal or both above and below the seal. These secondary filters protect both the monitoring-device filter and the seal from intrusion of grout installed above the seal.

7.8 The backfill that is placed above the annular seal is usually a bentonite or cement-base grout.

NOTE 21—Grouts should be designed and installed in consideration of the ambient hydrogeologic conditions. The constituents should be selected according to specific performance requirements and these data documented. Typical grout mixtures are given in Practice D 5092 and Practice D 4428.

NOTE 22—Grouting equipment should be cleaned and decontaminated prior to use according to Practice D 5088. Also, the equipment used for grouting should be constructed from materials that do not "leach" significant amounts of contaminants to the grout.

7.8.1 The initial position of the tremie pipe and grouting pressures should be controlled to prevent materials from being jetted into underlying seal(s) and filter(s) (use of a tremie pipe having a plugged bottom and side-discharge ports should be considered to minimize bottom-jetting problems).

7.8.2 When it is appropriate to use a grout line the grout should be discharged at a depth of approximately 1.5 to 3 m (5 to 10 ft) below the grout surface within the annulus (after the initial 1.5 to 3 m (5 to 10 ft) of grout has been deposited above the uppermost filter or seal).

NOTE 23—The need for chemical analysis of samples of each grout component and the final mixture should be documented. Also, it should be noted that if cements are used for grouting, they generate hydroxides and heat thereby, causing a localized increase in the alkalinity and temperature of the surrounding ground water.

7.8.3 The grout should be installed from the bottom of the borehole to the top of the borehole so as to displace fluids in the borehole.

8. Development

8.1 Most monitoring-device installations should be developed to remove any air that may have been introduced into the formation by the drilling method, suspended solids from drilling fluids, and disturbance of geologic materials during installation and to improve the hydraulic characteristics of the filter pack and the geohydrologic unit adjacent to the intake. For suggested well-development methods and techniques the user is referred to Test Method D 5099. The method(s) selected and time expended to develop the installation and the changes in water quality discharged at the surface should be carefully observed and documented.

NOTE 24—Under most circumstances, development should be initiated as soon as possible following completion however, time should be allowed for initial setting of grout.

9. Field Report and Project Control

9.1 The field report should include information recommended under Guide D 5434, and identified as necessary and pertinent to the needs of the exploration program.

9.2 Other information in addition to Guide D 5434 should be considered if deemed appropriate and necessary to the needs of the exploration program. Additional information should be considered as follows:

9.2.1 *Drilling Methods:*

9.2.1.1 Description of the dual-wall drilling method system.

9.2.1.2 Type, quantities, and locations in the borehole of use of additives added to the circulation media.

9.2.1.3 Description of circulation rates, cuttings return, including quantities, over intervals used.

9.2.1.4 Descriptions of drilling conditions related to drilling pressures, rotation rates, and general ease of drilling as related to subsurface materials encountered.

9.2.2 *Sampling*—Document conditions of the bottom of the borehole prior to sampling and report any slough or cuttings present in the recovered sample.

9.2.3 *In-situ Testing:*

9.2.3.1 For devices inserted below the bottom of the borehole document the depths below the bottom of the hole and any unusual conditions during testing.

9.2.3.2 For devices testing or seating at the borehole wall, report any unusual conditions of the borehole wall such as inability to seat borehole packers.

9.2.4 *Installations*—A description of well-completion materials and placement methods, approximate volumes placed, depth intervals of placement, methods of confirming placement, and areas of difficulty of material placement or unusual occurrences.

10. Keywords

10.1 down-the-hole hammer (DTH) drilling; drilling; dual-wall reverse-circulation drilling method(s); geoenvironmental exploration; ground water; percussion-hammer drilling method; triple-wall percussion-hammer drilling method; vadose zone

APPENDIX

(Nonmandatory Information)

X1. REFERENCES

Aller, L., et al., *Handbook of Suggested Practices for the Design and Installation of Ground-Water Monitoring Wells*, EPA/600/4-89/034, NWWA/EPA Series, National Water Well Association, Dublin, OH, 1989.

American Petroleum Institute, *API Specifications for Casing, Tubing, and Drill Pipe*, API Spec 5A, American Petroleum Institute, Dallas, TX, 1978.

Australian Drilling Manual, Australian Drilling Industry Training Committee Limited, P.O. Box 1545, Macquarie Centre, NSW 2113, Australia, 1992.

Baroid, *Baroid Drilling Fluid Products for Minerals Exploration, NL Baroid/NL Industries*, Houston, TX, 1980.

Bowen, R., *Grouting in Engineering Practice*, 2nd Edition, Applied Science Publishers, Halstad Press, New York, NY, 1981.

Campbell, M. D., and Lehr, J. H., *Water Well Technology*, McGraw-Hill Book Company, New York, NY, 1973.

DCDMA Technical Manual, Drilling Equipment Manufacturers Association, 3008 Millwood Avenue, Columbia, South Carolina, 29205, 1991.

Diamond Drilling Handbook, Heinz, W. F., First Edition, South African Drilling Association, Johannesburg, Republic of South Africa, 1985.

Drillers Handbook, Ruda, T. C., and Bosscher, P. J., editors, National Drilling Contractors Association, 3008 Millwood Avenue, Columbia, South Carolina, 29205, June 1990.

Driscoll, F. G., *Groundwater and Wells*, Johnson Filtration Systems, Second Edition, St. Paul, MN, 1989.

Heinz, W. F., *First Edition*, South African Drilling Association, Johannesburg, Republic of South Africa, 1985.

Hix, G. L., "Casing Advancement Methods for Drilling Monitoring Wells," *Water Well Journal 45(5):60–64*, 1991.

Morrison, Robert D., *Ground Water Monitoring Technology, Procedures, Equipment and Applications*, Timco Mfg., Inc., Prairie Du Sac, WI, 1983.

Roscoe Moss Company, *Handbook of Ground Water Development*, Roscoe Moss Company, Los Angeles, CA, John Wiley and Sons, Inc., New York, NY, 1990.

Shuter, E., and Teasdale, W. E., "Application of Drilling, Coring, and Sampling Techniques to Test Holes and Wells," *U.S. Geological Survey Techniques of Water-Resource Investigations*, TWRI 2-F1, 1989.

Strauss, M. F., Story, S. L., and Mehlhorn, N. E., "Application of Dual Wall Reverse Circulation Drilling in Ground Water Exploration and Monitoring," *Ground Water Monitoring Review 9(2):63–71*, 1989.

Standard Guide for
Use of Direct Air-Rotary Drilling for Geoenvironmental Exploration and the Installation of Subsurface Water-Quality Monitoring Devices[1]

This standard is issued under the fixed designation D 5782; the number immediately following the designation indicates the year of original adoption or, in the case of revision, the year of last revision. A number in parentheses indicates the year of last reapproval. A superscript epsilon (ϵ) indicates an editorial change since the last revision or reapproval.

1. Scope

1.1 This guide covers how direct (straight) air-rotary drilling procedures may be used for geoenvironmental exploration and installation of subsurface water-quality monitoring devices.

NOTE 1—The term direct with respect to the air-rotary drilling method of this guide indicates that compressed air is injected through a drill-rod column to a rotating bit. The air cools the bit and transports cuttings to the surface in the annulus between the drill-rod column and the borehole wall.

NOTE 2—This guide does not include considerations for geotechnical site characterizations that are addressed in a separate guide.

1.2 Direct air-rotary drilling for geoenvironmental exploration will often involve safety planning, administration, and documentation. This guide does not purport to specifically address exploration and site safety.

1.3 The values stated in SI units are to be regarded as the standard. The inch-pound units given in parentheses are for information only.

1.4 *This standard does not purport to address all of the safety concerns, if any, associated with its use. It is the responsibility of the user of this standard to establish appropriate safety and health practices and determine the applicability of regulatory limitations prior to use.*

2. Referenced Documents

2.1 *ASTM Standards:*

D 420 Guide for Site Characterization for Engineering Design and Construction Purposes[2]

D 653 Terminology Relating to Soil, Rock, and Contained Fluids[2]

D 1586 Test Method for Penetration Test and Split-Barrel Sampling of Soils[2]

D 1587 Test Method for Thin-Walled Tube Sampling of Soils[2]

D 2113 Test Method for Diamond Core Drilling for Site Investigation[2]

D 3550 Practice for Ring-Lined Barrel Sampling of Soils[2]

D 4428/D 4428M Test Methods for Crosshole Seismic Testing[2]

D 5088 Practice for Decontamination of Field Equipment Used at Non-Radioactive Waste Sites[3]

D 5092 Practice for Design and Installation of Ground Water Monitoring Wells in Aquifers[3]

D 5099 Test Method for Rubber—Measurement of Processing Properties Using Capillary Rheometry[4]

D 5434 Guide for Field Logging of Subsurface Explorations of Soil and Rock[3]

3. Terminology

3.1 *Definitions*—Terminology used within this guide is in accordance with Terminology D 653. Definitions of additional terms may be found in Terminology D 653.

3.2 *Descriptions of Terms Specific to This Standard:*

3.2.1 *bentonite*—the common name for drilling fluid additives and well-construction products consisting mostly of naturally occurring montmorillonite. Some bentonite products have chemical additives which may affect water-quality analyses.

3.2.2 *bentonite granules and chips*—irregularly shaped particles of bentonite (free from additives) that have been dried and separated into a specific size range.

3.2.3 *bentonite pellets*—roughly spherical- or disk-shaped units of compressed bentonite powder (some pellet manufacturers coat the bentonite with chemicals that may affect the water-quality analysis).

3.2.4 *cleanout depth*—the depth to which the end of the drill string (bit or core barrel cutting end) has reached after an interval of cutting. The cleanout depth (or drilled depth as it is referred to after cleaning out of any sloughed material in the bottom of the borehole) is usually recorded to the nearest 0.1 ft (0.03 m).

3.2.5 *coefficient of uniformity*—$C_u(D)$, the ratio D_{60}/D_{10}, where D_{60} is the particle diameter corresponding to 60 % finer on the cumulative particle-size distribution curve, and D_{10} is the particle diameter corresponding to 10 % finer on the cumulative particle-size distribution curve.

3.2.6 *drawworks*—a power-driven winch, or several winches, usually equipped with a clutch and brake system(s) for hoisting or lowering a drilling string.

3.2.7 *drill hole*—a cylindrical hole advanced into the subsurface by mechanical means. Also known as a borehole or boring.

3.2.8 *drill string*—the complete rotary-drilling assembly under rotation including bit, sampler/core barrel, drill rods,

---

[1] This guide is under the jurisdiction of ASTM Committee D-18 on Soil and Rock and is the direct responsibility of Subcommittee D18.21 on Ground Water and Vadose Zone Investigations.

Current edition approved Oct. 10, 1995. Published December 1995.

[2] *Annual Book of ASTM Standards*, Vol 04.08.

[3] *Annual Book of ASTM Standards*, Vol 04.09.

[4] *Annual Book of ASTM Standards*, Vol 09.01.

and connector assemblies (subs). The total length of this assembly is used to determine drilling depth by referencing the position of the top of the string to a datum near the ground surface.

3.2.9 *drill string*—the complete direct air-rotary drilling assembly under rotation including bit, sampler/core barrel, drill rods, and connector assemblies (subs). The total length of this assembly is used to determine drilling depth by referencing the position of the top of the string to a datum near the ground surface.

3.2.10 *filter pack*—also known as a gravel pack or a primary filter pack in the practice of monitoring-well installations. The gravel pack is usually granular material, having specified grain size characteristics, that is placed between a monitoring device and the borehole wall. The basic purpose of the filter pack or gravel envelope is to act as: (*1*) a nonclogging filter when the aquifer is not suited to natural development or, (*2*) act as a formation stabilizer when the aquifer is suitable for natural development.

3.2.10.1 *Discussion*—Under most circumstances a clean, quartz sand or gravel should be used. In some cases a pre-packed screen may be used.

3.2.11 *grout packer*—an inflatable or expandable annular plug attached to a tremie pipe, usually just above the discharge end of the pipe.

3.2.12 *grout shoe*—a drillable plug containing a check valve positioned within the lowermost section of a casing column. Grout is injected through the check valve to fill the annular space between the casing and the borehole wall or another casing.

3.2.12.1 *Discussion*—The composition of the drillable plug should be known and documented.

3.2.13 *hoisting line*—or drilling line, is wire rope used on the drawworks to hoist and lower the drill string.

3.2.14 *in-situ testing devices*—sensors or probes, used for obtaining mechanical or chemical test data, that are typically pushed, rotated, or driven below the bottom of a borehole following completion of an increment of drilling. However, some in situ testing devices (such as electronic pressure transducers, gas-lift samplers, tensiometers, and so forth) may require lowering and setting of the device(s) in a preexisting borehole by means of a suspension line or a string of lowering rods or pipe. Centralizers may be required to correctly position the device(s) in the borehole.

3.2.15 *intermittent-sampling devices*—usually barrel-type samplers that are driven or pushed below the bottom of a borehole following completion of an increment of drilling. The user is referred to the following ASTM standards relating to suggested sampling methods and procedures: Practice D 1452, Test Method D 1586, Practice D 3550, and Practice D 1587.

3.2.16 *mast*—or derrick, on a drilling rig is used for supporting the crown block, top drive, pulldown chains, hoisting lines, and so forth. It must be constructed to safely carry the expected loads encountered in drilling and completion of wells of the diameter and depth for which the rig manufacturer specifies the equipment.

3.2.16.1 *Discussion*—To allow for contingencies, it is recommended that the rated capacity of the mast should be at least twice the anticipated weight load or normal pulling load.

3.2.17 *piezometer*—an instrument for measuring pressure head.

3.2.18 *subsurface water-quality monitoring device*—an instrument placed below ground surface to obtain a sample for analysis of the chemical, biological, or radiological characteristics of subsurface pore water or to make in situ measurements.

4. Significance and Use

4.1 The application of direct air-rotary drilling to geoenvironmental exploration may involve sampling, coring, in situ or pore-fluid testing, installation of casing for subsequent drilling activities in unconsolidated or consolidated materials, and for installation of subsurface water-quality monitoring devices in unconsolidated and consolidated materials. Several advantages of using the direct air-rotary drilling method over other methods may include the ability to drill rather rapidly through consolidated materials and, in many instances, not require the introduction of drilling fluids to the borehole. Air-rotary drilling techniques are usually employed to advance drill hole when water-sensitive materials (that is, friable sandstones or collapsible soils) may preclude use of water-based rotary-drilling methods. Some disadvantages to air-rotary drilling may include poor borehole integrity in unconsolidated materials without using casing, and the possible volitization of contaminants and air-borne dust.

NOTE 3—Direct-air rotary drilling uses pressured air for circulation of drill cuttings. In some instances, water or foam additives, or both, may be injected into the air stream to improve cuttings-lifting capacity and cuttings return. The use of air under high pressures may cause fracturing of the formation materials or extreme erosion of the borehole if drilling pressures and techniques are not carefully maintained and monitored. If borehole damage becomes apparent, consideration to other drilling method(s) should be given.

NOTE 4—The user may install a monitoring device within the same borehole in which sampling, in situ or pore-fluid testing, or coring was performed.

4.2 The subsurface water-quality monitoring devices that are addressed in this guide consist generally of a screened or porous intake and riser pipe(s) that are usually installed with a filter pack to enhance the longevity of the intake unit, and with isolation seals and a low-permeability backfill to deter the movement of fluids or infiltration of surface water between hydrologic units penetrated by the borehole (see Practice D 5092). Inasmuch as a piezometer is primarily a device used for measuring subsurface hydraulic heads, the conversion of a piezometer to a water-quality monitoring device should be made only after consideration of the overall quality of the installation to include the quality of materials that will contact sampled water or gas.

NOTE 5—Both water-quality monitoring devices and piezometers should have adequate casing seals, annular isolation seals, and backfills to deter movement of contaminants between hydrologic units.

5. Apparatus

5.1 Direct air-rotary drilling systems consist of mechanical components and the drilling fluid.

5.1.1 The basic mechanical components of a direct air-rotary drilling system include the drill rig with rotary table and kelly or top-head drive unit, drawworks drill rods, bit or core barrel, casing (when required to support the hole and prevent wall collapse when drilling unconsolidated deposits), air compressor and filter(s), discharge hose, swivel, dust collector, and air-cleaning device (cyclone separator).

Note 6—In general, in North America, the sizes of casings, casing bits, drill rods, and core barrels are usually standardized by manufacturers according to size designations set forth by the American Petroleum Institute (API) and the Diamond Drill Core Manufacturers Association (DCDMA). Refer to the DCDMA technical manual and to published materials of API for available sizes and capacities of drilling tools equipment.

5.1.1.1 *Drill Rig*, with rotary table and kelly or top-head drive unit should have the capability to rotate a drill-rod column and apply a controllable axial force on the drill bit appropriate to the drilling and sampling requirements and the geologic conditions.

5.1.1.2 *Kelly*, a formed or machined section of hollow drill steel that is joined to the swivel at the top and the drill rods below. Flat surfaces or splines of the kelly engage the rotary table so that its rotation is transmitted to the drill rods.

5.1.1.3 *Drill Rods*, (that is, drill stems, drill string, drill pipe) transfer force and rotation from the drill rig to the bit or core barrel. Drill rods conduct drilling fluid to the bit or core barrel. Individual drill rods should be straight so they do not contribute to excessive vibrations or "whipping" of the drill-rod column. All threaded connections should be in good repair and not leak significantly at the internal air pressure required for drilling. Drill rods should be made up securely by wrench tightening at the threaded joint(s) at all times to prevent rod damage.

Note 7—Drill rods used for air drilling jointed to ensure that the cutting's-laden return air will not be deflected to the borehole wall as it passes the return air were deflected against the borehole blasting and erosion of the borehole wall would occur.

Note 8—Drill rods usually require lubricants on the thread to allow easy unthreading (breaking) of the drill-rod tool joints. Some lubricants have organic or metallic constituents, or both, that could be interpreted as contaminants if detected in a sample. Various lubricants are available that have components of known chemistry. The effect of drill-rod lubricants on chemical analyses of samples should be considered and documented when using direct air-rotary drilling. The same consideration and documentation should be given to lubricants used with water swivels, hoisting swivels, or other devices used near the drilling axis.

5.1.1.4 *Rotary Bit or Core Bit*, provides material cutting capability for advancing the hole. Therefore, a core barrel can also be used to advance the hole.

Note 9—The bit is usually selected to provide a borehole of sufficient diameter for insertion of monitoring-device components such as the screened intake and filter pack and installation devices such as a tremie pipe. It should be noted that if bottom-discharge bits are used in loose cohesionless materials, jetting or erosion of test intervals could occur. The borehole opening should permit easy insertion and retraction of a sampler, or easy insertion of a pipe with an inside diameter large enough for placing completion materials adjacent to the screened intake and riser of a monitoring device. Core barrels may also be used to advance the hole. Coring bits are selected to provide the hole diameter or core diameter required. Coring of rock should be performed in accordance with Practice D 2113. The user is referred to Test Method D 1586, Practice D 1587, and Practice D 3550 for techniques and soil-sampling equipment to be used in sampling unconsolidated materials. Consult the DCDMA technical manual and published materials of API for matching sets of nested casings and rods if nested casing must be used for drilling in incompetent formation materials.

5.1.1.5 *Air Compressor*, should provide an adequate volume of air, without significant contamination, for re-

moval of cuttings. Air requirements will depend upon the drill rod and bit configuration, the character of the material penetrated, the depth of drilling below ground water level, and the total depth of drilling. The airflow rate requirements are usually based on an annulus upflow air velocity of about 1000 to 1300 m/min (about 3000 to 4000 ft/min) even though air-upflow rates of less than 1000 m/min are often adequate for cuttings transport. For some geologic conditions, air-blast erosion may increase the borehole diameter in easily eroded materials such that 1000 m/min may not be appropriate for cuttings transport. Should air-blast erosion occur, the depth(s) of the occurrence(s) should be noted and documented so that subsequent monitoring-equipment installation quality may be evaluated accordingly.

Note 10—The quality of compressed air entering the borehole and the quality of air discharged from the borehole and the cyclone separator must be considered. If not adequately filtered, the air produced by most oil-lubricated air compressors inherently introduces a significant quantity of oil into the circulation system. High-efficiency, in-line air filters are usually required to prevent significant contamination of the borehole.

5.1.1.6 *Pressure Hose*, conducts the air from the air compressor to the swivel.

5.1.1.7 *Swivel*, directs the air to the rotating kelly or drill-rod column.

5.1.1.8 *Dust Collector*, conducts air and cuttings from the borehole annulus past the drill rod column to an air-cleaning device (cyclone separator).

5.1.1.9 *Air-Cleaning Device*, (cyclone separator) separates cuttings from the air returning from the borehole by means of the dust collector.

Note 11—A properly sized cyclone separator can remove practically all of the cuttings from the return air. A small quantity of fine particles, however, are usually discharged to the atmosphere with the "cleaned" air. Some air-cleaning devices consist of a cyclone separator alone; whereas, some utilize a cyclone separator combined with a power blower and sample-collection filters. It is virtually impossible to direct the return "dry" air past the drill rods without some leakage of air and return cuttings. Samples of drill cuttings can be collected for analysis of materials penetrated. If samples are obtained, the depth(s) and interval(s) should be documented.

Note 12—Zones of low air return and also zones of no air return should be documented. Likewise, the depth(s) of sampled interval(s) and quality of samples obtained should be documented.

Note 13—Compressed air alone can often transport cuttings from the borehole and cool the bit. For some geologic conditions, injection of water into the air stream will help control dust or break down "mud rings" that tend to form on the drill rods. If water is injected the depth(s) of water injection should be documented. Under other circumstances, for example, if the borehole starts to produce water, the injection of a foaming agent may be required. The depth when a foaming agent is added should also be recorded. When foaming agents are used, a cyclone-type cuttings separator is not used and foam discharge accumulates near the top of the borehole. When contaminants are encountered during drilling and returning from the borehole at geoenvironmental-exploration sites, special measures should be taken to contain the foam and protect personnel and the environment. Therefore, added water and some available foaming agents could affect water-quality analyses. The need for chemical analysis of added water or foaming agents should be considered and documented.

6. Drilling Procedures

6.1 As a prelude to and throughout the drilling process, stabilize the drill rig and raise the drill-rig mast. Position the cyclone separator and seal it to the ground surface. If air-monitoring operations are performed consider the prevalent wind direction relative to the exhaust from the drill rig. Also, consider the location of the cyclone relative to the rig exhaust since air-quality monitoring will be performed at the cyclone separator discharge point.

NOTE 14—Under some circumstances surface casing may be required to prevent hole collapse. Deeper casing(s) (nested casings) may also be required to facilitate adequate downhole air circulation and hole control. All casing used should be decontaminated according to Practice D 5088 prior to use.

6.2 Drilling usually progresses as follows:

6.2.1 Attach an initial assembly of a bit or core barrel, often with a single section of drill rod, below the rotary table or top-head drive unit with the bit placed below the top of the dust collector.

NOTE 15—The drill rig, drilling, hoisting and sampling tools, drilling rod and bits, the rotary gear or chain case, the spindle, and all components of the rotary drive above the drilling axis should be cleaned and decontaminated according to Practice D 5088 prior to commencing drilling and sampling operations.

6.2.2 Activate the air compressor, causing compressed air to circulate through the system.

6.2.3 Initiate rotation of the bit.

6.2.4 Continue air circulation and rotation of the drill-rod column until drilling progresses to a depth where sampling or in-situ testing will be performed or until the length of the drill-rod section limits further penetration. Air pressures at the bit should be low to prevent fracturing of the surrounding material. Monitor all air pressures during drilling. Note and document any abrupt changes or anomalies in the air pressure including the depth(s) of occurrence(s). Air-quality monitoring may be required. If air-quality monitoring is performed document the sampled intervals and air-quality data.

6.2.5 Stop rotation and lift the bit slightly off the bottom of the hole to facilitate drill-cuttings removal, and continue air circulation for a short time until the drill cuttings are removed from the borehole annulus. If sampling is to be done, stop air circulation and rest the bit on the hole bottom to determine hole depth. Document the hole depth and amount of any caving that occurred. If caving is apparent, set decontaminated casing to protect the boring.

6.2.6 Increase drilling depth by attaching an additional drill-rod section to the top of the previously advanced drill-rod column and resuming drilling operations according to 6.2.2 through 6.2.5. Record drilling behavior as drilling progresses. This recorded information should include (as a minimum): air-circulation pressures, depth(s) of low or lost circulation, depth(s) of water-/foam-additive injection(s), air-quality data, drill-cuttings description, depths of and type of sample(s)/core(s) taken from the hole, and any other data identified as necessary and pertinent to the needs of the exploration program.

NOTE 16—Drilling rates depend on many factors such as the density or stiffness of unconsolidated material and the existence of cobbles or boulders, the hardness or durability of the rock, or both, the swelling activity of clays or shales encountered in the borehole, and the erosiveness of the borehole wall. Drilling rates can vary from a few millimetres (less than an inch/minute) to about 1 m (3 ft)/min, depending on subsurface conditions. Other factors influencing drilling rates include the weight of the drill string, collar(s) weight and size of

drill pipe, and the rig pulldown or holdback pressure. These data as well as any other drilling rate information should be recorded.

6.2.7 Sampling or in-situ testing can be performed at any depth in the hole by interrupting the advance of the bit, cleaning the hole of cuttings according to 6.2.5, stopping air circulation, and removing the drill-rod column from the borehole. Drill-rod removal is not necessary when a sample may be obtained or an in-situ test can be performed through the hollow axis of the drill rods and bit. Compare the sampling depth to the cleanout depth. Verify the depth comparison data by first resting the sampler on the bottom of the hole and comparing that measurement with the cleanout-depth measurement. If bottom-hole contamination is apparent (determined by comparing the hole-cleanout depth with the sampling depth) it is recommended that a minimum depth below the sampler/bit be at least 18-in. for testing. This should be done before every sampling or in-situ testing is performed in the hole. Record the depth of in-situ testing or sampling as well as the depth below the sampler/bit for evaluation of data quality. Decontaminate sampling and testing devices according to Practice D 5088 prior to testing.

6.3 When drilling must progress through material suspected of being contaminated, installation of single or multiple (nested) casings may be required to isolate zones of suspected contamination. Isolation casings are usually installed in a predrilled borehole or by using a casing advancement method. A grout seal is then installed, usually by applying the grout at the bottom of the annulus with the aid of a grout shoe or a grout packer and a tremie pipe. The grout should be allowed to set before drilling activities are continued. Document complete casing and grouting records, including location(s) of nested casings for the hole.

7. Installation of Monitoring Devices

7.1 Subsurface water-quality monitoring devices are generally installed in boreholes drilled by direct air-rotary methods using the three-step procedure shown in Fig. 1. The three steps are: (1) drilling, with or without sampling, (2) removal of the drill-rod column assembly and insertion of the instrumentation or monitoring device, and (3) addition of completion materials such as filter packs, seals, and grouts. If protective casings are present in the borehole they are usually removed in incremental fashion as completion materials are added.

7.2 Assemble water-quality monitoring devices, with attached fluid conductors (risers), and insert into the borehole with the least possible addition of contaminants. The user is referred to Practice D 5092 for monitoring-well installation methods and Practice D 5088 for suggested methods of field-equipment decontamination.

7.2.1 Some materials, such as screens and risers, may require cleaning or decontamination, or both, at the job site. The user is referred to Practice D 5088 for equipment decontamination procedures.

7.2.2 Prior to installation, store all monitoring-device materials undercover and place upwind and well away from the drill rig and any other sources of potential contamination, such as electrical generators, air compressors, or industrial machinery.

7.2.3 Clean hoisting tools, particularly wire rope and

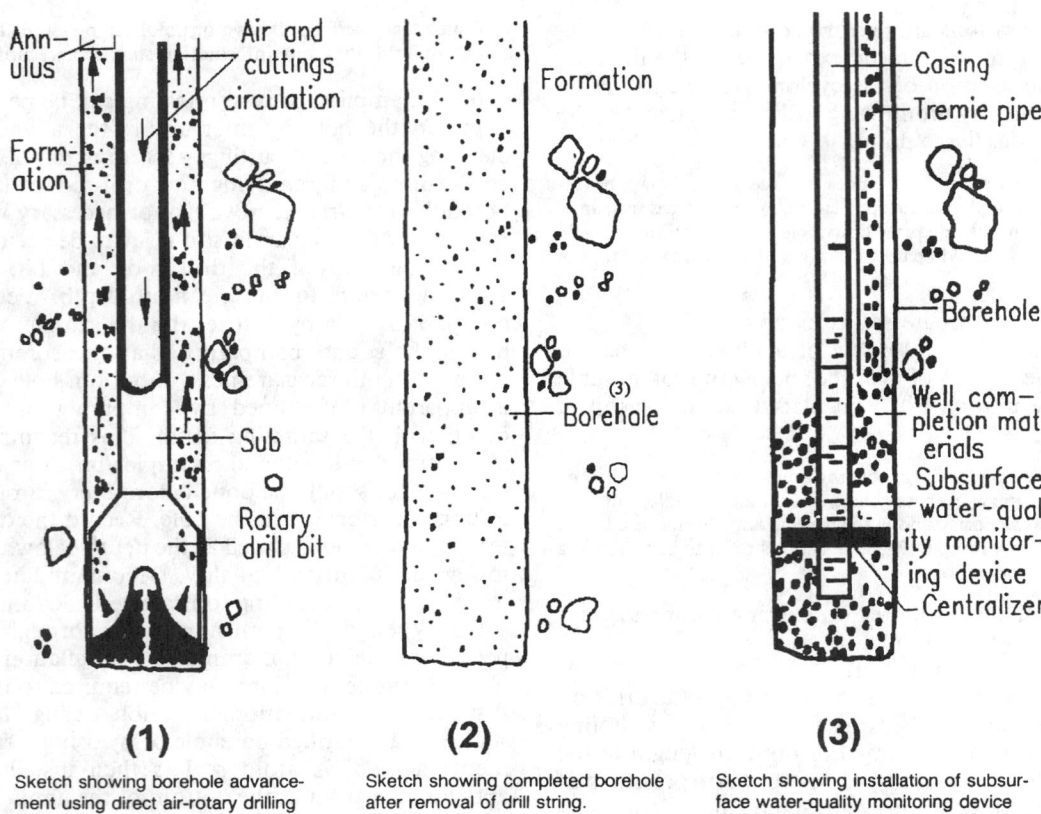

(1)

Sketch showing borehole advancement using direct air-rotary drilling method.

(2)

Sketch showing completed borehole after removal of drill string.

(3)

Sketch showing installation of subsurface water-quality monitoring device with centralizer, tremie-pipe installation of completion materials.

FIG. 1 Sketch Showing Basic Three-Step Procedure for Installation of Subsurface Water-Quality Monitoring Device Using Direct Air-Rotary Drilling Method

hoisting swivels, and decontaminate according to Practice D 5088, before using.

7.3 Select filter materials, bentonite pellets, granules and chips, and grouts and install according to specific subsurface-monitoring or instrumentation requirements.

NOTE 17—Filter packs for monitoring devices, are usually installed in air-rotary drilled holes using a tremie pipe inserted in the annulus between the borehole wall and the monitoring device (minimum annulus between riser pipe and hole wall should be about 1 in. (25 mm) completely around the riser pipe). However, unless needed for silt control or seal separation between water-bearing zones, filter packing monitoring wells in competent rock adds an unnecessary source of sample contamination due to the fouling of the sand interstices by the invasion of the filter-pack material. Monitoring devices installed in a saturated zone typically have sand-sized filter packs that are selected mainly on the basis of the grain-size characteristics of the hydrologic unit adjacent to the screened intake. The coefficient of uniformity of the filter pack sand is usually less than 2.5. In most cases, a centralizer should be used to center a monitoring device requiring a filter pack in an uncased borehole. Filter packs for vadose-zone monitoring devices may be predominantly silt sized however, soil-gas monitoring devices should not use silt-sized filter packs but typically use coarse sand or gravel packs. These filter materials are often mixed with water of known quality and then inserted through a tremie pipe and tamped into place around the device. The type(s) and volumes of filter materials used and the quality and quantities of mixing water should be documented. In most cases, a centralizer should be used to position the monitoring device in the borehole. The intake device and riser(s) should be suspended above the bottom of the borehole during installation of the filter pack(s), seal(s), and backfill to keep the riser(s) as straight as possible. Care should be taken when adding backfill or filter material(s),

or both, so that the materials do not bridge. However, if bridging does occur during the installation procedure, tamping rods or other tamping devices may be used to dislodge the bridge.

7.4 Sealing materials consisting of either bentonite pellets, chips, or granules are usually placed directly above the filter pack.

NOTE 18—It may be effective, when granular filter packs are used, to install a thin, fine sand, secondary filter either below the annular seal or both, above and below the seal. These secondary filters protect the principal filter and the seal from intrusion of grout installed above the seal.

7.5 The backfill that is placed above the annular seal of a monitoring device is usually a bentonite or cement-base grout.

NOTE 19—Grouts should be designed and installed in consideration of the ambient hydrogeologic conditions. The constituents should be selected according to specific performance requirements and these data documented. Typical grout mixtures are given in Practice D 5092 and Test Methods D 4428.

7.5.1 In most cases, the grout should be pumped into the annulus between the borehole wall and the monitoring device(s) riser(s) using a tremie pipe.

NOTE 20—Grouting equipment should be cleaned and decontaminated prior to use according to Practice D 5088. Also, the equipment used for grouting should be constructed from materials that do not leach significant amounts of contaminants to the grout.

7.5.2 Control the initial position of the tremie pipe and

grouting pressures to prevent materials from being jetted into underlying seal(s) and filter(s) (use of a tremie pipe having a plugged bottom and side-discharge ports should be considered to minimize bottom-jetting problems).

7.5.3 In most cases, the grout should be discharged at a depth of approximately 1.5 to 3 m (5 to 10 ft) below the grout surface within the annulus (after the placement of the initial 1.5 to 3 m of grout has been deposited above the underlying filter or seal). Additional grout should be discharged at a depth of approximately 1.5 to 3 m below the grout surface within the annulus. The tremie pipe should be periodically raised as grout is discharged to maintain the appropriate depth below the grout surface.

NOTE 21—The need for chemical analysis of samples of each grout component and chemical analysis of the final mixture should be documented. Also, it should be noted that if cements are used for grouting, they generate hydroxides and thereby, can cause a localized increase in the alkalinity and pH of the surrounding ground water.

7.5.4 Install the grout from the bottom of the borehole to the top of the borehole so as to displace fluids in the borehole.

8. Development

8.1 Most monitoring device installations should be developed to remove any air that may have been introduced into the formation by the drilling method, suspended solids from drilling fluids, and disturbance of geologic materials during installation and to improve the hydraulic characteristics of the filter pack and the hydrologic unit adjacent to the monitoring device intake. The method(s) selected and time expended to develop the installation and the changes in quality of water discharged at the surface should be carefully observed and documented. For suggested well-development methods and techniques the user is referred to Test Method D 5099.

NOTE 22—Under most circumstances, development should be initiated as soon as possible following completion, however, time should be allowed for setting of grout.

9. Field Report and Project Control

9.1 The field report should include information recommended under Guide D 5434, and identified as necessary and pertinent to the needs of the exploration program.

9.2 Other information in addition to Guide D 5434 should be considered if deemed appropriate and necessary to the needs of the exploration program. Additional information should be considered as follows:

9.2.1 *Drilling Methods:*

9.2.1.1 Description of the air-rotary system including the air compressor, air-circulation, and discharge system.

9.2.1.2 Type, quantities, and locations in the borehole of use of additives such as water or foaming agent(s) added to the circulation media.

9.2.1.3 Description of circulation rates and cuttings return, including quantities, over intervals used. Locations and probable cause of loss of circulation in the borehole.

9.2.1.4 Descriptions of drilling conditions related to drilling pressures, rotation rates, and general ease of drilling as related to subsurface materials encountered.

9.2.2 *Sampling*—Document conditions of the bottom of the borehole prior to sampling and report any slough or cuttings present in the recovered sample.

9.2.3 *In Situ Testing:*

9.2.3.1 For devices inserted below the bottom of the borehole, document the depths below the bottom of the hole and any unusual conditions during testing.

9.2.3.2 For devices testing or seating at the borehole wall, report any unusual conditions of the borehole wall such as inability to seat borehole packers.

9.2.4 *Installations*—A description of well-completion materials and placement methods, approximate volumes placed, depth intervals of placement, methods of confirming placement, and areas of difficulty of material placement or unusual occurrences.

10. Keywords

10.1 air-rotary drilling method; drilling; geoenvironmental exploration; ground water; vadose zone

APPENDIX

(Nonmandatory Information)

X1. ADDITIONAL REFERENCES

Aller, L., et al, *Handbook of Suggested Practices for the Design and Installation of Ground-Water Monitoring Wells*, EPA/600/4-89/034, NWWA/EPA Series, National Water Well Assn., Dublin, OH, 1989.

American Petroleum Institute, *API Specifications for Casing, Tubing, and Drill Pipe, API Spec 5A*, American Petroleum Institute, Dallas, TX, 1978.

Australian Drilling Manual, Australian Drilling Industry Training Committee Limited, P.O. Box 1545, Macquarie Centre, NSW 2113, Australia, 1992.

Baroid, *Baroid Drilling Fluid Products for Minerals Exploration*, NL Baroid/NL Industries, Houston, TX, 1980.

Bowen, R., *Grouting in Engineering Practice*, 2nd Edition, Applied Science Publishers, Halstad Press, New York, NY, 1981.

Campbell, M. D., and Lehr, J. H., *Water Well Technology*, McGraw-Hill Book Co., New York, NY, 1973.

DCDMA Technical Manual, Drilling Equipment Manufacturers Assn., 3008 Millwood Ave., Columbia, SC, 29205, 1991.

Drillers Handbook, T. C., Ruda and P. J., Bosscher, eds., National Drilling Contractors Assn., 3008 Millwood Ave., Columbia, SC, 29205, June 1990.

Driscoll, F. G., *Groundwater and Wells*, Johnson Filtration Systems, 2nd Edition, St. Paul, MN, 1989.

Heinz, W. F., *First Edition*, South African Drilling Association, Johannesburg, Republic of South Africa, 1985.

Morrison, Robert D., *Ground Water Monitoring Technology, Procedures, Equipment and Applications*, Timco Manufacturing, Inc., Prairie Du Sac, WI, 1983.

Handbook of Ground Water Development, Roscoe Moss Co., Los Angeles, CA, John Wiley and Sons, Inc., New York, NY, 1990.

Russell, B. F., Hubbell, J. M., and Minkin, S. C., "Drilling and Sampling Procedures to Minimize Borehole Cross-Contamination." *Proc. Third Nat. Outdoor Action Conf. on Aquifer Restoration, Ground Water Monitoring and Geophysical Methods,*

National Water Well Assn., Dublin, OH, 1989, pp. 81–93. (Air and mud rotary.)

Shuter, E., and Teasdale, W. E., "Application of Drilling, Coring, and Sampling Techniques to Test Holes and Wells," *U.S. Geological Survey Techniques of Water-Resource Investigations*, TWRI 2-F1, 1989.

Standard Guide for
Use of Direct Rotary Drilling with Water-Based Drilling Fluid for Geoenvironmental Exploration and the Installation of Subsurface Water-Quality Monitoring Devices[1]

This standard is issued under the fixed designation D 5783; the number immediately following the designation indicates the year of original adoption or, in the case of revision, the year of last revision. A number in parentheses indicates the year of last reapproval. A superscript epsilon (ε) indicates an editorial change since the last revision or reapproval.

1. Scope

1.1 This guide covers how direct (straight) rotary-drilling procedures with water-based drilling fluids may be used for geoenvironmental exploration and installation of subsurface water-quality monitoring devices.

NOTE 1—The term direct with respect to the rotary-drilling method of this guide indicates that a water-based drilling fluid is pumped through a drill-rod column to a rotating bit. The drilling fluid transports cuttings to the surface through the annulus between the drill-rod column and the borehole wall.

NOTE 2—This guide does not include considerations for geotechnical site characterization that are addressed in a separate guide.

1.2 Direct-rotary drilling for geoenvironmental exploration and monitoring-device installations will often involve safety planning, administration and documentation. This standard does not purport to specifically address exploration and site safety.

1.3 The values stated in SI units are to be regarded as the standard. The inch-pound units given in parentheses are for information only.

1.4 *This standard does not purport to address all of the safety concerns, if any, associated with its use. It is the responsibility of the user of this standard to establish appropriate safety and health practices and determine the applicability of regulatory limitations prior to use.*

2. Referenced Documents

2.1 *ASTM Standards:*

D 653 Terminology Relating to Soil, Rock, and Contained Fluids[2]

D 1452 Practice for Soil Investigation and Sampling by Auger Borings[2]

D 1586 Test Method for Penetration Test and Split-Barrel Sampling of Soils[2]

D 1587 Test Method for Thin-Walled Tube Sampling of Soils[2]

D 2113 Test Method for Diamond Core Drilling for Site Investigation[2]

D 2487 Test Method for Classification of Soils for Engineering Purposes[2]

D 2488 Practice for Description and Identification of Soils (Visual-Manual Procedure)[2]

D 3550 Practice for Ring-Lined Barrel Sampling of Soils[2]

D 5088 Practice for Decontamination of Field Equipment Used at Non-Radioactive Waste Sites[3]

D 5092 Practice for Design and Installation of Ground Water Monitoring Wells in Aquifers[3]

D 5099 Test Method for Rubber—Measurement of Processing Properties Using Capillary Rheometry[4]

3. Terminology

3.1 *Definitions:*

3.1.1 Terminology used within this guide is in accordance with Terminology D 653. Definitions of additional terms may be found in Terminology D 653.

3.2 *Descriptions of Terms Specific to This Standard:*

3.2.1 *bentonite*—the common name for drilling-fluid additives and well-construction products consisting mostly of naturally-occurring montmorillonite. Some bentonite products have chemical additives that may affect water-quality analyses.

3.2.2 *bentonite granules and chips*—irregularly-shaped particles of bentonite (free from additives) that have been dried and separated into a specific size range.

3.2.3 *bentonite pellets*—roughly spherical- or disc-shaped units of compressed bentonite powder (some pellet manufacturers coat the bentonite with chemicals that may affect the water quality analysis).

3.2.4 *cleanout depth*—the depth to which the end of the drill string (bit or core barrel cutting end) has reached after an interval of cutting. The cleanout depth (or drilled depth as it is referred to after cleaning out of any sloughed material in the bottom of the borehole) is usually recorded to the nearest 0.1 ft (0.03 m).

3.2.5 *coefficient of uniformity*—$C_u(D)$, the ratio D_{60}/D_{10}, where D_{60} is the particle diameter corresponding to 60 % finer on the cumulative particle-size distribution curve, and D_{10} is the particle diameter corresponding to 10 % finer on the cumulative particle-size distribution curve.

3.2.6 *drawworks*—a power-driven winch, or several winches, usually equipped with a clutch and brake system(s) for hoisting or lowering a drilling string.

3.2.7 *drill hole*—a cylindrical hole advanced into the subsurface by mechanical means. Also known as a borehole or boring.

3.2.8 *drill string*—the complete direct rotary-drilling assembly under rotation including bit, sampler/core barrel,

---

[1] This guide is under the jurisdiction of ASTM Committee D-18 on Soil and Rock and is the direct responsibility of Subcommittee D18.21 on Ground Water and Vadose Zone Investigations.

Current edition approved Oct. 10, 1995. Published December 1995.

[2] *Annual Book of ASTM Standards*, Vol 04.08.

[3] *Annual Book of ASTM Standards*, Vol 04.09.

[4] *Annual Book of ASTM Standards*, Vol 09.01.

drill rods and connector assemblies (subs). The total length of this assembly is used to determine drilling depth by referencing the position of the top of the string to a datum near the ground surface.

3.2.9 *filter pack*—also known as a gravel pack or a primary filter pack in the practice of monitoring-well installations. The gravel pack is usually granular material, having selected grain size characteristics, that is placed between a monitoring device and the borehole wall. The basic purpose of the filter pack or gravel envelope is to act as: (*1*) a non-clogging filter when the aquifer is not suited to natural development or, (*2*) act as a formation stabilizer when the aquifer is suitable for natural development.

3.2.9.1 *Discussion*—Under most circumstances a clean, quartz sand or gravel should be used. In some cases a pre-packed screen may be used.

3.2.10 *grout packer*—an inflatable or expandable annular plug attached to a tremie pipe, usually just above the discharge end of the pipe.

3.2.11 *grout shoe*—a drillable "plug" containing a check valve positioned within the lowermost section of a casing column. Grout is injected through the check valve to fill the annular space between the casing and the borehole wall or another casing.

3.2.11.1 *Discussion*—The composition of the drillable "plug" should be known and documented.

3.2.12 *hoisting line*—or drilling line, is wire rope used on the drawworks to hoist and lower the drill string.

3.2.13 *in-situ testing devices*—sensors or probes, used for obtaining mechanical or chemical-test data, that are typically pushed, rotated or driven below the bottom of a borehole following completion of an increment of drilling. However, some in-situ testing devices (such as electronic pressure transducers, gas-lift samplers, tensiometers, and etc.) may require lowering and setting of the device(s) in a pre-existing borehole by means of a suspension line or a string of lowering rods or pipe. Centralizers may be required to correctly position the device(s) in the borehole.

3.2.14 *intermittent-sampling devices*—usually barrel-type samplers that are driven or pushed below the bottom of a borehole following completion of an increment of drilling. The user is referred to the following ASTM standards relating to suggested sampling methods and procedures: Practice D 1452, Test Method D 1586, Practice D 3550, and Practice D 1587.

3.2.15 *mast*—or derrick, on a drilling rig is used for supporting the crown block, top drive, pulldown chains, hoisting lines, etc. It must be constructed to safely carry the expected loads encountered in drilling and completion of wells of the diameter and depth for which the rig manufacturer specifies the equipment.

3.2.15.1 *Discussion*—To allow for contingencies, it is recommended that the rated capacity of the mast should be at least twice the anticipated weight load or normal pulling load.

3.2.16 *piezometer*—an instrument for measuring pressure head.

3.2.17 *subsurface water-quality monitoring device*—an instrument placed below ground surface to obtain a sample for analysis of the chemical, biological or radiological character-istics of subsurface-pore water or to make in-situ measurements.

4. Significance and Use

4.1 Direct-rotary drilling may be used in support of geoenvironmental exploration and for installation of subsurface water-quality monitoring devices in unconsolidated and consolidated materials. Direct-rotary drilling may be selected over other methods based on advantages over other methods. In drilling unconsolidated sediments and hard rock, other than cavernous limestones and basalts where circulation cannot be maintained, the direct-rotary method is a faster drilling method than the cable-tool method. The cutting samples from direct-rotary drilled holes are usually as representative as those obtained from cable-tool drilled holes however, direct-rotary drilled holes usually require more well-development effort. If however, drilling of water-sensitive materials (that is, friable sandstones or collapsible soils) is anticipated, it may preclude use of water-based rotary-drilling methods and other drilling methods should be considered.

4.1.1 The application of direct-rotary drilling to geoenvironmental exploration may involve sampling, coring, in-situ or pore-fluid testing, or installation of casing for subsequent drilling activities in unconsolidated or consolidated materials. Several advantages of using the direct-rotary drilling method are stability of the borehole wall in drilling unconsolidated formations due to the buildup of a filter cake on the wall. The method can also be used in drilling consolidated formations. Disadvantages to using the direct-rotary drilling method include the introduction of fluids to the subsurface, and creation of the filter cake on the wall of the borehole that may alter the natural hydraulic characteristics of the borehole.

NOTE 3—The user may install a monitoring device within the same borehole wherein sampling, in-situ or pore-fluid testing, or coring was performed.

4.2 The subsurface water-quality monitoring devices that are addressed in this guide consist generally of a screened or porous intake and riser pipe(s) that are usually installed with a filter pack to enhance the longevity of the intake unit, and with isolation seals and low-permeability backfill to deter the movement of fluids or infiltration of surface water between hydrologic units penetrated by the borehole (see Practice D 5092). Inasmuch as a piezometer is primarily a device used for measuring subsurface hydraulic heads, the conversion of a piezometer to a water-quality monitoring device should be made only after consideration of the overall quality of the installation, including the quality of materials that will contact sampled water or gas.

NOTE 4—Both water-quality monitoring devices and piezometers should have adequate casing seals, annular isolation seals and backfills to deter movement of contaminants between hydrologic units.

5. Apparatus

5.1 Direct-rotary drilling systems consist of mechanical components and the drilling fluid.

5.1.1 The basic mechanical components of a direct-rotary drilling system include the drill rig with derrick, rotary table and kelly or top-head drive unit, drill rods, bit or core barrel, casing (when required to protect the hole and prevent wall collapse when drilling unconsolidated deposits), mud pit, suction hose, cyclone desander(s), drilling-fluid circulation pump, pressure hose, and swivel.

NOTE 5—In general, in North America, the sizes of casings, casing bits, drill rods, and core barrels are usually standardized by manufacturers according to size designations set forth by the American Petroleum Institute (API) and the Diamond Drill Core Manufacturers Association (DCDMA). Refer to the DCDMA technical manual and to published materials of API for available sizes and capacities of drilling tools equipment.

5.1.1.1 *Drill Rig*, with rotary table and kelly or top-head drive unit should have the ability to rotate a drill-rod column and apply a controllable axial force on the drill bit appropriate to the drilling and sampling requirements and the geologic conditions.

5.1.1.2 *Kelly*, a formed or machined section of hollow drill steel, used with some rotary-drilling systems, that is joined to the swivel at the top and the drill rods below. Flat surfaces or splines of the kelly engage the rotary table so that rotation is transmitted to the drill rods.

5.1.1.3 *Drill Rods*, (that is, drill stems, drill string, drill pipe) transfer force and rotation from the drill rig to the bit or core barrel. Drill rods conduct drilling fluid to the bit or core barrel. Individual drill rods should be straight so they do not contribute to excessive vibrations or "whipping" of the drill-rod column. All threaded connections should be in good repair and not leak significantly at the internal fluid pressure required for drilling. Drill rods should be made up securely by wrench tightening at the threaded joint(s) at all times to prevent rod damage.

NOTE 6—Drill rods usually require lubricants on the threads to allow easy unthreading of the drill-rod tool joints. Some lubricants have organic or metallic constituents, or both, that could be interpreted as contaminants if detected in a sample. Various lubricants are available that have components of known chemistry. The effect of drill-rod lubricants on chemical analyses of samples should be considered and documented when using direct-rotary drilling. The same consideration and documentation should be given to lubricants used with water swivels, hoisting swivels, or other devices used near the drilling axis.

5.1.1.4 *Rotary Bit or Core Bit*, provides the material cutting capability. Therefore, a core barrel can also be used to advance the hole.

NOTE 7—The bit is usually selected to provide a borehole of sufficient diameter for insertion of monitoring-device components such as the screened intake and filter pack and installation devices such as a tremie pipe. It should be noted that if bottom-discharge bits are used in loose cohesionless materials, jetting or erosion of test intervals could occur. The borehole opening should permit easy insertion and retraction of a sampler, or easy insertion of a pipe with an inside diameter large enough for placing completion materials adjacent to the screened intake and riser of a monitoring device. Core barrels may also be used to advance the hole. Coring bits are selected to provide the hole diameter or core diameter required. Coring of rock should be performed in accordance with Practice D 2113. The user is referred to Test Method D 1586, Practice D 1587, and Practice D 3550 for techniques and soil-sampling equipment to be used in sampling unconsolidated materials. Consult the DCDMA technical manual and published materials of API for matching sets of nested casings and rods if nested casings must be used for drilling in incompetent formation materials.

5.1.1.5 *Mud Pit*, is a reservoir for the drilling fluid and, if properly designed and utilized, provides sufficient flow-velocity reduction to allow separation of drill cuttings from the fluid before recirculation. The mud pit is usually a shallow, open metal tank with baffles; however, for some circumstances, an excavated pit with some type of liner, designed to prevent loss of drilling fluid and to contain potential contaminants that may be present in the cuttings and recirculated fluids may be used. The mud pit can be used as a mixing reservoir for the initial quantity of drilling fluid and, in some circumstances, for adding water and additives to the drilling fluid as drilling progresses.

NOTE 8—Some drilling-fluid components must be added to the composite mixture before other components; consequently, an auxiliary mixing reservoir may be required to premix these components with water before adding to the mud pit. All quantities, chemical composition and types of drilling-fluid components and additives used in the composite drilling-fluid mixture should be documented.

5.1.1.6 *Suction Hose*, sometimes equipped with a foot valve or strainer, or both, conducts the drilling fluid from the mud pit to the drilling-fluid circulation pump.

5.1.1.7 *Drilling-Fluid Circulation Pump*, must have the capability to lift the drilling fluid from the mud pit and move it through the system against variable pumping heads and provide an annular velocity adequate to transport drill cuttings out of the borehole.

NOTE 9—Drilling-fluid pressures at the bit should be low to prevent fracturing of the surrounding material. All drilling-fluid pressures should be monitored during drilling. Any abrupt changes or anomalies in the drilling-fluid pressure should be duly noted and documented including the depth(s) of occurrence(s).

5.1.1.8 *Pressure Hose*, conducts the drilling fluid from the circulation pump to the swivel.

5.1.1.9 *Swivel*, directs the drilling fluid to the rotating kelly or drill-rod column.

5.1.2 *Drilling Fluid*, usually consists of a water base and one or more additives that increase viscosity or provide other desirable physical or chemical properties. Principal functions of drilling fluid include: (*1*) sealing the borehole wall to minimize loss of drilling fluid, (*2*) providing a hydraulic pressure against the borehole wall to support the open borehole, (*3*) removing cuttings generated at the bit and (*4*) lubricating and cooling of the bit.

NOTE 10—Particular attention should be given to the drilling-fluid makeup-water source and the means used to transport the makeup water to the drilling site as potential sources of contamination to the drilling fluid. If the chemical makeup of the water is determined the test results should be documented.

5.1.3 Some commonly used additives for water base drilling fluids are listed below:

5.1.3.1 Beneficiated bentonite, a primary viscosifier and borehole sealer, consists of montmorillonite with other naturally-occurring minerals and various additives such as sodium carbonate or polyacrylates, or both.

5.1.3.2 Unbeneficiated bentonite, a primary viscosifier and borehole sealer, consists of montmorillonite with other naturally-occurring minerals but without additives such as sodium carbonate or polyacrylates.

5.1.3.3 Sodium carbonate powder (soda ash) is used to precipitate calcium carbonate hardness from the drilling fluid water base before adding other components. An increase in pH will occur with the addition of sodium carbonate. Sodium hydroxide (caustic soda) generally should not be used in this application.

5.1.3.4 Carboxylmethylcellulose powder (CMC) is sometimes used in a water based fluid as a viscosifier and as an inhibitor to clay hydration.

NOTE 11—Some additives to water-based drilling fluid systems retard clay hydration, inhibiting swelling of clays on the borehole wall and inhibiting "balling" or "smearing" of the bit.

5.1.3.5 Potassium chloride (muriated potash) or diammonium phosphate can be used as an inhibitor to clay hydration.

5.1.3.6 Polyacrylamide, a primary viscosifier and clay-hydration inhibitor, is a polymer that is mixed with water to create a drilling fluid.

5.1.3.7 Barium sulfate increases the density of water-based drilling fluids. It is a naturally occurring high specific gravity mineral processed to a powder for rotary drilling-fluid applications.

5.1.3.8 Lost-circulation materials are used to seal the borehole wall when fluids are being lost through large pores, cracks or joints. These additives usually consist of various coarse textured materials such as shredded paper or plastic, bentonite chips, wood fibers, or mica.

5.1.3.9 Attapulgite, a primary viscosifier for rotary drilling in high-salinity environments, is a clay mineral drilling-fluid additive.

NOTE 12—The listing and discussion of the above drilling-fluid additives does not imply general acceptance for geoenvironmental exploration. Some of the additives listed above may impact water-quality analyses. Some readily available, but not as common, drilling-fluid additives, not listed above, could cause significant contamination in a borehole or hydrologic unit. Each additive should be evaluated for each specific application. The types, amounts, and chemical compositions of all additives used should be documented. In addition, a hole log should document the depths where any new additives were introduced. Methods to break revertible fluids should be documented.

6. Drilling Procedures

6.1 As a prelude to and throughout the drilling process stabilize the drill rig and raise the drill-rig mast. Position the mud pit and install surface casing and "seal" at the ground surface.

NOTE 13—Under some circumstances, surface casing may be required to prevent hole collapse. Deeper casing(s) (nested casings) may also be required to facilitate adequate downhole fluid circulation and hole control. All casing used should first be decontaminated according to Practice D 5088 prior to use and the casing information documented.

6.2 Mix an initial quantity of drilling fluid, usually using the mud pit as the primary mixing reservoir.

NOTE 14—The need for chemical analysis of samples of each drilling-fluid component and the final mixture should be documented.

6.3 Drilling usually progresses as follows:

6.3.1 Attach an initial assembly of a bit or core barrel, often with a single section of drill rod, below the rotary table or top-head drive unit with the bit or drill head placed within the top of the surface casing.

NOTE 15—The drill rig, drilling, hoisting and sampling tools, the rotary gear or chain case, the spindle and all components of the rotary drive above the drilling axis should be cleaned and decontaminated according to Practice D 5088 prior to commencing drilling and sampling operations.

6.3.2 Activate the drilling-fluid circulation pump, causing drilling fluid to circulate through the system.

6.3.3 Initiate rotation of the bit and apply axial force to the bit.

6.3.4 Continue drilling-fluid circulation as rotation and axial force are applied to the bit until drilling progresses to a depth where: (1) sampling or in-situ testing will be performed, (2) the length of the drill-rod column limits further penetration, or (3) (when core drilling) the core specimen has entered the core barrel.

6.3.5 Stop rotation. Lift the bit slightly off hole bottom while drilling-fluid circulation is continued to facilitate removal of the drill cuttings from the borehole annulus. If sampling is to be done, stop drilling-fluid circulation and rest the bit on the hole bottom to ascertain hole depth. If, after making a depth measurement, it is apparent that caving has caused hole-depth loss, document the hole depth and amount of caving that had occurred. If caving has occurred, set decontaminated casing to support the boring.

NOTE 16—The time required to remove the cuttings from the borehole will depend mainly upon the pumping rate, the cross-sectional area of the borehole annulus, the borehole depth, the viscosity of the drilling fluid, and the size of the cuttings. If determined that caving occurred and casing had to be set, this information should be documented.

6.3.6 Increase drilling depth by attaching an additional drill-rod section to the top of the previously advanced drill-rod column and resuming drilling operations according to 6.3.2 through 6.3.5. Drilling behavior should be documented as drilling progresses. This recorded information should include (as a minimum): drilling-fluid circulation pressures, depth(s) of occurrence of low or lost drilling-fluid circulation, drill-cuttings description, depths of and type of sample(s)/core(s) taken from the hole, and any other data identified as necessary and pertinent to the needs of the exploration program.

NOTE 17—Drilling rates depend on many factors such as the density or stiffness of unconsolidated material and the existence of cobbles or boulders, the hardness or durability of the rock, or both, the swelling activity of clays or shales encountered in the borehole and the erosiveness of the borehole wall. Drilling rates can vary from a few mm (less than an in./min) to about 1 m (3 ft)/min, depending on subsurface conditions. Other factors influencing drilling rates include the weight of the drill string, collar(s) weight and size of drill pipe, and the rig pulldown or holdback pressure. These data as well as any other drilling-rate information should be recorded.

6.3.7 Perform sampling or in-situ testing at any depth by interrupting the advance of the bit, cleaning the hole of cuttings according to 6.3.5, stopping the fluid circulation, and removing the drill-rod column from the borehole. Drill-rod removal is not necessary when a sample can be obtained or an in-situ test can be performed through the hollow axis of the drill rods and bit. Sampling depth should be compared to the cleanout depth. Verify the depth-comparison data by first resting the sampler on the bottom of the hole and comparing that measurement with the cleanout-depth measurement. This should be done before every sampling or in-situ testing is performed in the hole. If bottom-hole sloughing is apparent from a depth measurement made prior to sampling (determined by comparing the hole-cleanout depth with the sampling depth) it is recommended that the hole be cleaned in order that a minimum depth of undisturbed material extend at least 18-in. below the sampler/bit for testing. Record the depth of in-situ testing or sampling as well as the depth below the sampler/bit for evaluation of data quality for later evaluation of sample quality or in-situ testing data validity. Decontaminate sampling and testing devices according to Practice D 5088 prior to testing.

6.4 When drilling must progress through material suspected of being contaminated, installation of single or multiple (nested) casings may be required to isolate zones of suspected contamination. Isolation casings are usually installed in a predrilled borehole or by using a casing-advancement method. A grout seal is then installed, usually by applying the grout at the bottom of the annulus with the aid of a grout shoe or a grout packer and a tremie pipe. Allow the grout to set before drilling activities are continued. Document complete casing and grouting records, including location(s) of nested casings for the hole.

7. Installation of Monitoring Devices

7.1 Subsurface water-quality monitoring devices are generally installed in boreholes drilled by direct-rotary methods using the three-step procedure shown on Fig. 1. The three steps are: (*1*) drilling, with or without sampling, (*2*) removal of the drill-rod column assembly and placement of the monitoring device, and (*3*) addition of other materials such as filter packs, seals and grouts. If protective casings are present in the borehole they are usually removed in incremental fashion as completion materials are added.

NOTE 18—The volumes of sand packs and seals should be documented and compared to calculated values based on hole diameter for evaluation of hole quality.

7.2 Assemble water-quality monitoring devices, with attached fluid conductors (risers), and insert into the borehole with the least possible addition of contaminants. The user is referred to Practice D 5092 for monitoring-well installation methods and Practice D 5088 for suggested methods of field-equipment decontamination.

NOTE 19—If the integrity of the borehole wall will not be compromised by removing the wall cake from the borehole in the vicinity of the screened intake, the drilling mud should be removed using a well-development procedure as suggested in Guide D XXX1 prior to inserting a monitoring well or a water-quality monitoring device in the borehole.

7.2.1 Some materials, such as screens and risers, require cleaning or decontamination, or both, at the job site (see Practice D 5088).

7.2.2 Prior to installation, store all monitoring-device materials undercover and place upwind and well away from the drill rig and any other sources of contamination such as electrical generators, air compressors, or industrial machinery.

7.2.3 Clean hoisting tools, particularly wire rope and hoisting swivels, and decontaminate according to Practice D 5088 before using.

7.3 Select filter materials, bentonite pellets, granules and chips and grouts and install according to specific subsurface-monitoring requirements. Document this information.

NOTE 20—Filter packs for monitoring devices are usually installed in rotary-drilled holes using a tremie pipe inserted in the annulus between the borehole wall and the monitoring device (minimum annulus between riser pipe and hole wall should be about 1 in. (25 mm) completely around the riser pipe). Monitoring devices installed in a saturated zone typically have sand-sized filter packs selected on the basis of the grain-size characteristics of the hydrologic unit adjacent to the screened intake. Filter-pack sands are usually selected with a coefficient of uniformity of less than 2.5. Filter packs for vadose-zone monitoring devices may be predominantly silt sized (however, soil-gas monitoring

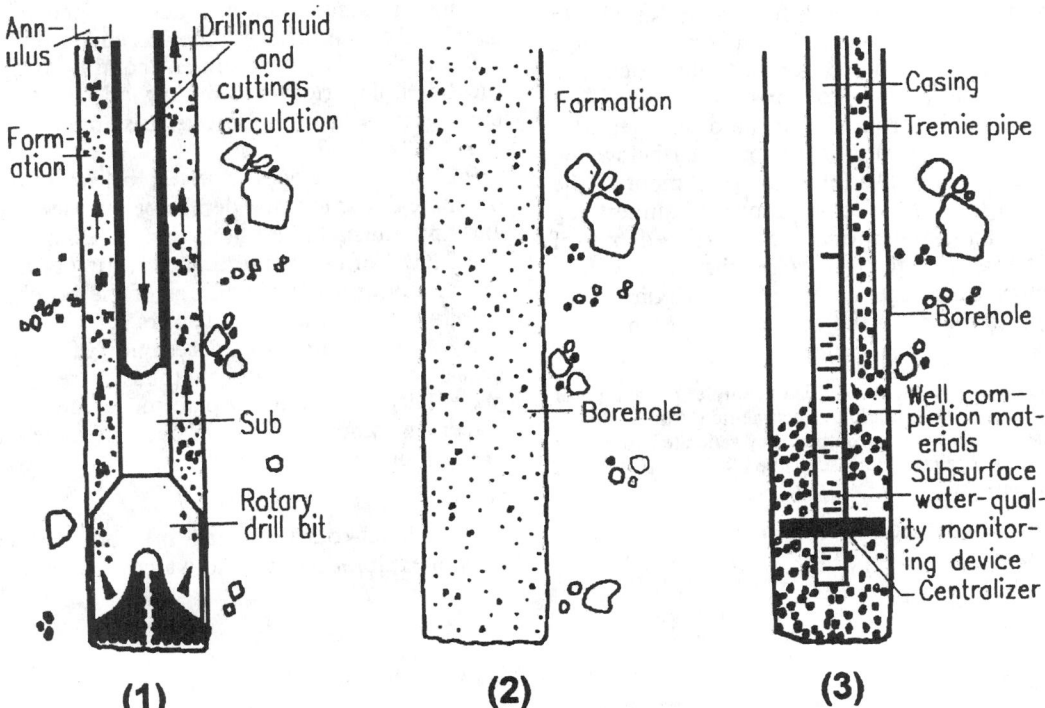

FIG. 1 Sketch Showing Basic Three-Step Procedure for Installation of Subsurface Water-Quality Monitoring Device Using Direct-Rotary Drilling With Water-Based Drilling Fluid

devices should not use silt-sized filter packs but typically use coarse sand or gravel filter packs). These filter materials are often mixed with water of known quality and then inserted through a tremie pipe and tamped into place around the device. The type(s) and volumes of filter materials used and the quality and quantities of mixing water should be documented. In most cases, a centralizer should be used to position the monitoring device in the borehole. The intake device and riser(s) should be suspended above the bottom of the borehole during installation of the filter pack(s), seal(s) and backfill to keep the riser(s) as straight as possible. Care should be taken when adding backfill or filter material(s), or both, so that the materials do not bridge. However, if bridging does occur during the installation procedure, tamping rods or other tamping devices may be used to dislodge the "bridge".

7.4 Sealing materials, consisting of either bentonite pellets, chips, or granules, are usually placed directly above the filter pack.

NOTE 21—It may be effective, when granular filter packs are used, to install a thin, fine sand, secondary filter either below the annular seal or both above and below the seal. These secondary filters protect both the monitoring device filter and the seal from intrusion of grout installed above the seal.

7.5 The backfill that is placed above the annular seal is usually a bentonite or cement-base grout.

NOTE 22—Grouts should be designed and installed in consideration of the ambient hydrogeologic conditions. The constituents should be selected according to specific performance requirements and these data documented.

7.5.1 In most cases, the grout should be pumped into the annulus using a tremie pipe.

NOTE 23—Grouting equipment should be cleaned and decontaminated prior to use according to Practice D 5088. Also, the equipment used for grouting should be constructed from materials that do not "leach" significant amounts of contaminants to the grout.

7.5.2 Control the initial position of the tremie pipe and grouting pressures to prevent materials from being jetted into underlying seal(s) and filter(s) (use of a tremie pipe having a plugged bottom and side-discharge ports should be considered to minimize bottom-jetting problems).

7.5.3 In most cases, the grout should be discharged at a depth of approximately 1.5 to 3 m (5 to 10 ft) below the grout surface within the annulus (after the placement of the initial 1.5 to 3 m (5 to 10 ft) of grout above the underlying filter or seal). Discharge additional grout at a depth of approximately 1.5 to 3 m (5 to 10 ft) below the grout surface within the annulus. Raise the tremie pipe periodically as grout is discharged to maintain the appropriate depth below the grout surface.

NOTE 24—The need for chemical analysis of samples of each grout component and the final mixture should be documented. Also, it should be noted that if cements are used for grouting, they generate hydroxides and thereby, can cause a localized increase in the alkalinity and pH of the surrounding groundwater.

7.5.4 Install the grout in such a manner as to displace fluids in the borehole.

8. Development

8.1 Most monitoring-device installations should be developed to remove suspended solids from drilling fluids and disturbance of geologic materials during installation and to improve the hydraulic characteristics of the filter pack and the geologic unit adjacent to the intake. For suggested well-development methods and techniques the user is referred to Test Method D 5099. The method(s) selected and time expended to develop the installation and the changes in water quality discharged at the surface should be carefully observed and documented. For suggested well-development methods and techniques the user is referred to Test Method D 5099.

NOTE 25—Under most circumstances, development should be initiated as soon as possible following completion, however, time should be allowed for setting of grout.

9. Field Report and Project Control

9.1 The field report should include information recommended under Guide D 5434, and identified as necessary and pertinent to the needs of the exploration program.

9.2 Other information in addition to Guide D 5434 should be considered if deemed appropriate and necessary to the needs of the exploration program. Additional information should be considered as follows:

9.2.1 *Drilling Methods:*

9.2.1.1 Description of the direct-rotary system,

9.2.1.2 Type, quantities, and locations in the borehole of use of additives added to the circulation media,

9.2.1.3 Description of circulation rates, cuttings return, including quantities, over intervals used. Locations and probable cause of loss of circulation in the borehole. Drilling-fluid loss quantities should be documented, and

9.2.1.4 Descriptions of drilling conditions related to drilling pressures, rotation rates, and general ease of drilling as related to subsurface materials encountered.

9.2.2 *Sampling*—Document conditions of the bottom of the borehole prior to sampling and report any slough or cuttings present in the recovered sample.

9.2.3 *In-situ Testing:*

9.2.3.1 For devices inserted below the bottom of the borehole document the depths below the bottom of the hole and any unusual conditions during testing.

9.2.3.2 For devices testing or seating at the borehole wall, report any unusual conditions of the borehole wall such as inability to seat borehole packers.

9.2.4 *Installations*—A description of well-completion materials and placement methods, approximate volumes placed, depth intervals of placement, methods of confirming placement, and areas of difficulty of material placement or unusual occurrences.

10. Keywords

10.1 direct-rotary drilling method; drilling; geoenvironmental exploration; ground water; vadose zone

APPENDIX

(Nonmandatory Information)

X1. ADDITIONAL REFERENCES

Aller, L., et al., *Handbook of Suggested Practices for the Design and Installation of Ground-Water Monitoring Wells*, EPA/600/4-89/034, NWWA/EPA Series, National Water Well Association, Dublin, OH, 1989.

American Petroleum Institute, *API Specifications for Casing, Tubing, and Drill Pipe*, API Spec 5A, American Petroleum Institute, Dallas, TX, 1978.

Australian Drilling Manual, Australian Drilling Industry Training Committee Limited, P.O. Box 1545, Macquarie Centre, NSW 2113, Australia, 1992.

Baroid, *Baroid Drilling Fluid Products for Minerals Exploration*, NL Baroid/NL Industries, Houston, TX, 1980.

Bowen, R., *Grouting in Engineering Practice*, 2nd Edition, Applied Science Publishers, Halstad Press, New York, NY, 1981.

Campbell, M. D., and Lehr, J. H., *Water Well Technology*, McGraw-Hill Book Company, New York, NY, 1973.

DCDMA Technical Manual, Drilling Equipment Manufacturers Association, 3008 Millwood Avenue, Columbia, South Carolina, 29205, 1991.

Drillers Handbook, Ruda, T. C., and Bosscher, P. J., editors, National Drilling Contractors Association, 3008 Millwood Avenue, Columbia, South Carolina, 29205, June 1990.

Driscoll, F. G., *Groundwater and Wells*, Johnson Filtration Systems, Second Edition, St. Paul, MN, 1989.

Heinz, W. F., *First Edition*, South African Drilling Association, Johannesburg, Republic of South Africa, 1985.

Morrison, Robert D., *Ground Water Monitoring Technology, Procedures, Equipment and Applications*, Timco Mfg., Inc., Prairie Du Sac, WI, 1983.

Roscoe Moss Company, *Handbook of Ground Water Development*, Roscoe Moss Company, Los Angeles, CA, John Wiley and Sons, Inc., New York, NY, 1990.

Shuter, E., and Teasdale, W. E., *Application of Drilling, Coring, and Sampling Techniques to Test Holes and Wells*, U.S. Geological Survey Techniques of Water-Resource Investigations, TWRI 2-F1, 1989.

Standard Guide for
Use of Hollow-Stem Augers for Geoenvironmental Exploration and the Installation of Subsurface Water-Quality Monitoring Devices[1]

This standard is issued under the fixed designation D 5784; the number immediately following the designation indicates the year of original adoption or, in the case of revision, the year of last revision. A number in parentheses indicates the year of last reapproval. A superscript epsilon (ϵ) indicates an editorial change since the last revision or reapproval.

1. Scope

1.1 This guide covers how hollow-stem auger-drilling systems may be used for geoenvironmental exploration and installation of subsurface water-quality monitoring devices.

1.2 Hollow-stem auger drilling for geoenvironmental exploration and monitoring device installations often involves safety planning, administration, and documentation. This guide does not purport to specifically address exploration and site safety.

NOTE 1—This guide does not include considerations for geotechnical site that are addressed in a separate Guide.

1.3 The values stated in SI units are to be regarded as the standard. The inch-pound units given in parentheses are for information only.

1.4 *This standard does not purport to address all of the safety concerns, if any, associated with its use. It is the responsibility of the user of this standard to establish appropriate safety and health practices and determine the applicability of regulatory limitations prior to use.*

2. Referenced Documents

2.1 *ASTM Standards:*

D 653 Terminology Relating to Soil, Rock, and Contained Fluids[2]
D 1452 Practice for Soil Investigation and Sampling by Auger Borings[2]
D 1586 Test Method for Penetration Test and Split-Barrel Sampling of Soils[2]
D 1587 Test Method for Thin-Walled Tube Sampling of Soils[2]
D 2113 Test Method for Diamond Core Drilling for Site Investigation[2]
D 2487 Test Method for Classification of Soils for Engineering Purposes[2]
D 2488 Practice for Description and Identification of Soils (Visual-Manual Procedure)[2]
D 3550 Practice for Ring-Lined Barrel Sampling of Soils[2]
D 4220 Practices for Preserving and Transporting Soil Samples[2]
D 4428/D 4428M Test Methods for Crosshole Seismic Testing[2]
D 4700 Guide for Soil Sampling from the Vadose Zone[2]

D 4750 Test Method for Determining Subsurface Liquid Levels in a Borehole or Monitoring Well (Observation Well)[2]
D 5079 Practices for Preserving and Transporting of Rock Core Samples[3]
D 5088 Practice for Decontamination of Field Equipment Used at Non-Radioactive Waste Sites[3]
D 5092 Practice for Design and Installation of Ground Water Monitoring Wells in Aquifers[3]
D 5099 Test Method for Rubber—Measurement of Processing Properties Using Capillary Rheometry[4]
D 5254 Practice for Minimum Set of Data Elements to Identify a Ground-Water Site[3]

3. Terminology

3.1 *Definitions:*

3.1.1 Terminology used within this guide is in accordance with Terminology D 653. Definitions of additional terms may be found in Terminology D 653.

3.2 *Descriptions of Terms Specific to This Standard:*

3.2.1 *bentonite*—the common name for drilling fluid additives and well-construction products consisting mostly of naturally occurring montmorillonite. Some bentonite products have chemical additives that may affect water-quality analyses.

3.2.2 *bentonite granules and chips*—irregularly shaped particles of bentonite (free from additives) that have been dried and separated into a specific size range.

3.2.3 *bentonite pellets*—roughly spherical- or disk-shaped units of compressed bentonite powder (some pellet manufacturers coat the bentonite with chemicals that may affect the water-quality analysis).

3.2.4 *coefficient of uniformity*—$C_u(D)$, the ratio D_{60}/D_{10}, where D_{60} is the particle diameter corresponding to 60 % finer on the cumulative particle-size distribution curve, and D_{10} is the particle diameter corresponding to 10 % finer on the cumulative particle-size distribution curve.

3.2.5 *continuous-sampling devices*—barrel-type samplers that fit within the lead auger of the hollow-auger column. The sampler barrel fills with material as the augers advance.

3.2.6 *drill hole*—a cylindrical hole advanced into the subsurface by mechanical means. Also known as borehole or boring.

3.2.7 *drawworks*—a power-driven winch, or several winches, usually equipped with a clutch and brake system(s)

---

[1] This guide is under the jurisdiction of ASTM Committee D-18 on Soil and Rock and is the direct responsibility of Subcommittee D18.21 on Ground Water and Vadose Zone Investigations.
Current edition approved Oct. 10, 1995. Published December 1995.
[2] *Annual Book of ASTM Standards*, Vol 04.08.

[3] *Annual Book of ASTM Standards*, Vol 04.09.
[4] *Annual Book of ASTM Standards*, Vol 09.01.

for hoisting or lowering a drilling string.

3.2.8 *filter pack*—also known as a gravel pack or a primary filter pack in the practice of monitoring-well installations. The gravel pack is usually granular material, having specified grain-size characteristics, that is placed between a monitoring device and the borehole wall. The basic purpose of the filter pack or gravel envelope is to act as: (*1*) a nonclogging filter when the aquifer is not suited to natural development or, (*2*) act as a formation stabilizer when the aquifer is suitable for natural development.

3.2.8.1 *Discussion*—Under most circumstances a clean, quartz sand or gravel should be used. In some cases a pre-packed screen may be used.

3.2.9 *fluid-injection devices*—usually consist of various auger components or drill-rig attachments that may be used to inject a fluid within a hollow-auger column during drilling.

3.2.10 *grout packer*—an inflatable or expandable annular plug that is attached to a tremie pipe, usually positioned immediately above the discharge end of the pipe.

3.2.11 *grout shoe*—a drillable plug containing a check valve that is positioned within the lowermost section of a casing column. Grout is injected through the check valve to fill the annular space between the casing and the borehole wall or another casing.

3.2.11.1 *Discussion*—The composition of the drillable plug should be known and documented.

3.2.12 *hoisting line*—or drilling line, is wire rope used on the drawworks to hoist and lower the drill string.

3.2.13 *in situ testing devices*—sensors or probes, used to obtain mechanical or chemical-test data, that are typically pushed, rotated, or driven below the bottom of a borehole following completion of an increment of drilling. However, some in situ testing devices (such as electronic pressure transducers, gas-lift samplers, tensiometers, and and so forth) may require lowering and setting of the device(s) in a preexisting borehole by means of a suspension line or a string of lowering rods or pipe. Centralizers may be required to correctly position the device(s) in the borehole.

3.2.14 *intermittent-sampling devices*—usually barrel-type samplers that may be rotated, driven, or pushed below the bottom of a borehole with drill rods or with a wireline system to lower, drive, and retrieve the sampler following completion of an increment of drilling. The user is referred to the following ASTM standards relating to suggested sampling methods and procedures: Practice D 1452, Test Method D 1586, Practice D 3550, and Practice D 1587.

3.2.15 *mast*—or derrick, on a drilling rig is used for supporting the crown block, top drive, pulldown chains, hoisting lines, and so forth. It must be constructed to safely carry the expected loads encountered in drilling and completion of wells of the diameter and depth for which the rig manufacturer specifies the equipment.

3.2.15.1 *Discussion*—To allow for contingencies, it is recommended that the rated capacity of the mast should be at least twice the anticipated weight load or normal pulling load.

3.2.16 *piezometer*—an instrument for measuring pressure head.

3.2.17 *subsurface water-quality monitoring device*—an instrument placed below ground surface to obtain a sample for analyses of the chemical, biological, or radiological characteristics of subsurface pore water or to make in-situ measurements.

4. Significance and Use

4.1 Hollow-stem auger drilling may be used in support of geoenvironmental exploration (Practice D 3550, Test Method D 4428) and for installation of subsurface water-quality monitoring devices in unconsolidated materials. Hollow-stem auger drilling may be selected over other methods based on the advantages over other methods. These advantages include: the ability to drill without the addition of drilling fluid(s) to the subsurface, and hole stability for sampling purposes (see Test Methods D 1586, D 1587, D 2487, and D 2488) and monitor-well construction in unconsolidated to poorly indurated materials. This drilling method is generally restricted to the drilling of shallow, unconsolidated materials or softer rocks. The hollow-stem drilling method is a favorable method to be used for obtaining cores and samples and for the installation of monitoring devices in many, but not all geologic environments.

NOTE 2—In many geologic environments the hollow-stem auger drilling method can be used for drilling, sampling, and monitoring-device installations without the addition of fluids to the borehole. However, in cases where heaving water-bearing sands or silts are encountered, the addition of water or drilling mud to the hollow-auger column may become necessary to inhibit the piping of these fluid-like materials into the augers. These drilling conditions, if encountered, should be documented.

4.1.1 The application of hollow-stem augers to geoenvironmental exploration may involve ground water and soil sampling, in-situ or pore-fluid testing, or utilization of the hollow-auger column as a casing for subsequent drilling activities in unconsolidated or consolidated materials (Test Method D 2113).

NOTE 3—The user may install a monitoring device within the same auger borehole wherein sampling or in-situ or pore-fluid testing was performed.

4.1.2 The hollow-stem auger column may be used as a temporary casing for installation of a subsurface water-quality monitoring device. The monitoring device is usually installed as the hollow-auger column is removed from the borehole.

4.2 The subsurface water-quality monitoring devices that are addressed in this guide consist generally of a screened or porous intake device and riser pipe(s) that are usually installed with a filter pack to enhance the longevity of the intake unit, and with isolation seals and low-permeability backfill to deter the movement of fluids or infiltration of surface water between hydrologic units penetrated by the borehole (see Practice D 5092). Inasmuch as a piezometer is primarily a device used for measuring subsurface hydraulic heads, the conversion of a piezometer to a water-quality monitoring device should be made only after consideration of the overall quality and integrity of the installation, to include the quality of materials that will contact sampled water or gas.

NOTE 4—Both water-quality monitoring devices and piezometers should have adequate casing seals, annular isolation seals, and backfills to deter the movement of fluids between hydrologic units.

5. Apparatus

5.1 Each auger section of the hollow-stem auger-column assembly consists of a cylindrical tube with continuous helical flighting rigidly attached to the outer surface of the tube (see Fig. 1). The hollow-auger section has a coupling at each end for attachment of a hollow-auger head to the bottom end of the lead auger section and for attachment of additional auger sections at the top end to make up the articulated hollow-stem auger column.

NOTE 5—The inside diameter of the hollow-stem auger column is usually selected to provide an opening large enough for insertion of monitoring-device components such as the screened intake and filter pack and installation devices such as a tremie pipe. When media sampling is required, the optimum opening should permit easy insertion and retraction of a sampler or core barrel. When a monitoring device is installed, the annular opening should provide easy insertion of a pipe with an inside diameter large enough for placing completion materials adjacent to the riser.

5.1.1 *Hollow-Auger Head*, attached to the lead auger of the hollow-auger column and usually contains replaceable, abrasion-resistant cutters or teeth (see Fig. 1). As the hollow-auger head is rotated, it cuts and directs the cuttings to the auger flights which convey the cuttings to the surface.

5.1.2 *Auger-Drive Assembly*, attaches to the uppermost hollow-auger section and transfers rotary power and axial force from the drill rig to the auger-column assembly.

5.1.3 *Pilot Assembly*, may consist of: (*1*) an auger head aperture-plugging device with or without a center cutting head, or (*2*) a sampling device that is used to sample simultaneously with advancement of the auger column.

5.1.4 *Auxiliary Components of a Hollow-Auger Drilling System*, consist of various devices such as auger-connector wrenches, auger forks, hoisting hooks, and fluid-injection swivels or adapters.

5.2 *Drill Rig*, used to rotate and advance the auger column. The drill rig should be capable of applying the rated power at a rotary velocity of 50 to 100 r/min. The drill rig should have a feed stroke of at least the effective length of the

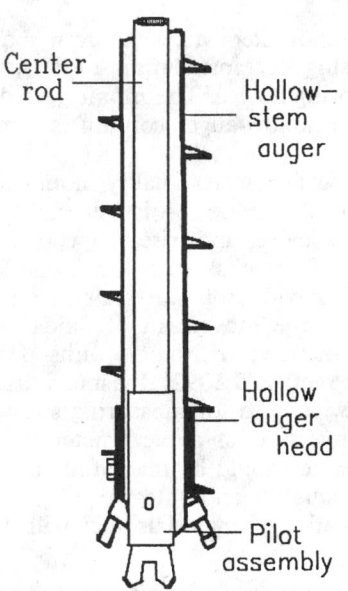

NOTE—Various pilot assemblies not shown here may vary.

FIG. 1 Sketch Showing Basic Hollow-Stem Auger Components

auger sections plus the effective length of the auger couplings plus about 100 mm (4 in.).

6. Drilling Procedures

6.1 As a prelude to and throughout the drilling process stabilize the drill rig and raise the drill-rig mast. Attach an initial assembly of hollow-auger components (see Fig. 1) to the rotary drive of the drill rig.

NOTE 6—The drill rig, drilling and sampling tools, the rotary gear or chain case, the spindle, and all components of the rotary drive above the auger column should be cleaned and decontaminated prior to drilling according to Practice D 5088. All lubricated rotary gear or chain cases should be monitored for leaks during drilling. Any lubricants used should be documented. Lubricants with organic or metallic constituents that could be interpreted as contaminants if detected in a soil or water sample should not be used on auger couplings. Any instances of possible contamination should be documented.

6.2 Push the auger-column assembly below the ground surface and initiate rotation at a low velocity.

NOTE 7—If surface contamination is suspected, special drilling procedures may be required to deter transport of contaminated materials downhole. For example, the augers and auger head may be removed and cleaned according to Practice D 5088 following drilling of the initial increments. Complete removal of the augers from a boring may allow caving and cross contamination of materials (especially below the water table). When augers are reinserted, attempts should be made to note if caving or sloughing, or both, has occurred in the borehole and the information documented.

6.3 Then continue drilling, usually at a rotary velocity of about 50 to 100 r/min, and to a depth where intermittent sampling or in situ testing is required, or until the drive assembly is advanced to within about 0.15 to 0.45 m (6 to 18 in.) of the ground surface. Soil sampling is usually accomplished by either of two methods: (*1*) removing the pilot assembly, if being used, and inserting and driving a sampler through the hollow stem of the auger column, or (*2*) using a continuous sampling device within the lead auger section. In the latter case the sampler barrel fills with material as the hollow-auger column is advanced. It should be noted that the pilot assembly and any sampling devices should be cleaned and decontaminated according to Practice D 5088 after each use and prior to reinsertion in the hollow-auger column. Water sampling can also be done through the hollow-stem augers when using augers with watertight connections to prevent fluid leakage from occurring at the connections: (*1*) by allowing the auger column to fill with water through the use of a screened lead auger section; (*2*) by allowing the auger column to fill from the bottom; (*3*) by using a soil-penetrating water sampling device that can be lowered into the hollow-auger column and either driven, rotated, or pushed out through the bottom or lead auger into the undisturbed material below the auger head.

NOTE 8—Under some circumstances it may be effective to drill without using a pilot assembly. If a pilot assembly is not used, however, and water is not injected into the auger column simultaneously with advancement, material will often enter the hollow stem of the auger column. The addition of water to the auger column during drilling may deter material entrance but, on the other hand, may also affect both the mechanical and chemical characteristics of soil samples and the quality of water samples. Therefore, if water is added and the chemistry determined, the approximate volume(s) added over specific intervals and the water chemistry should be documented.

6.4 Accomplish drilling at greater depths by attaching

additional hollow-auger sections to the top of the previously advanced hollow-auger column assembly.

NOTE 9—Cuttings are removed periodically from around the top of the auger column. Soil cuttings above the ground water may be representative of deposits being penetrated if proper cuttings-return rates are maintained. Cuttings from below the ground water surface are likely to be mixed from varying formations in the hole and are usually not representative of deposits at the end of the auger if cuttings are sampled for classification (see Practice D 2488) and relation to lithology report and document the intervals sampled. If drilling is performed in contaminated soil and cuttings control is required, drilling through a hole in a sheet of plywood or similar material held securely above the borehole by the stabilizing jacks of the drill rig will usually facilitate cuttings control. Containment and disposal of contaminated and potentially contaminated drilling fluids and associated cuttings should be in accordance with applicable regulations.

6.5 When drilling must progress through material suspected of being contaminated, installation of single or multiple (nested) casings may be required to isolate zones of suspected contamination. Install isolation casings in a predrilled borehole or by using a casing advancement method. However, when attempting to auger inside the casing, the column of cuttings return may cause the augers to bind in the casing. Then install a grout seal usually by applying the grout at the bottom of the annulus with the aid of a tremie pipe, and a grout shoe or a grout packer. Allow the grout to set before drilling activities are continued.

7. Installation of Monitoring Devices

7.1 Subsurface water-quality monitoring devices are generally installed using hollow-stem augers following the three-step procedure shown in Fig. 2. The three steps are: (1) drilling, with or without sampling, (2) removal of the pilot

assembly, if being used, and insertion of the monitoring device, and (3) incremental removal of the hollow-auger column as completion materials such as filter pack, annular seals, and backfill are installed as required.

NOTE 10—Removal of the pilot assembly following an increment of drilling or prior to installation of a monitoring device should be performed so that the entrance of material into the bottom of the hollow-auger stem is minimized. The efficacy of pilot assembly removal will depend upon several principal factors: (1) the character of the soil at the auger head, (2) the water levels inside and outside the auger prior to removal of the pilot assembly, (3) the type of pilot assembly used (special designs of pilot assemblies can be used to reduce the suction effect of removing the pilot bit), and (4) the speed of removal. As drilling progresses in saturated, granular materials, it usually becomes progressively more difficult to maintain the stability of the material below the auger column because of unbalanced hydraulic heads. The stability of the material below the auger head may be enhanced by using special pilot assemblies or injecting water of known chemistry into the hollow auger during drilling. The injection of water into a borehole usually requires consideration and documentation of the effects of injected water on (1) quality of subsequent chemical analyses of sampled water, and (2) the possible addition of moisture or contaminants to sampled materials.

7.1.1 If materials enter the bottom of the auger hollow stem during removal of the pilot assembly, remove it with a bailer, drive sampler, or other device.

NOTE 11—If heaving occurs, the amount of material entering the hollow-stem auger column should be documented. The effective use of a bailer may require the addition of a fluid to the auger stem.

7.1.2 If sampling or in situ testing is not required during drilling for installation of a monitoring device, advance the boring (for some geologic conditions) by using an expendable knockout plate or plug of known chemistry instead of a pilot assembly.

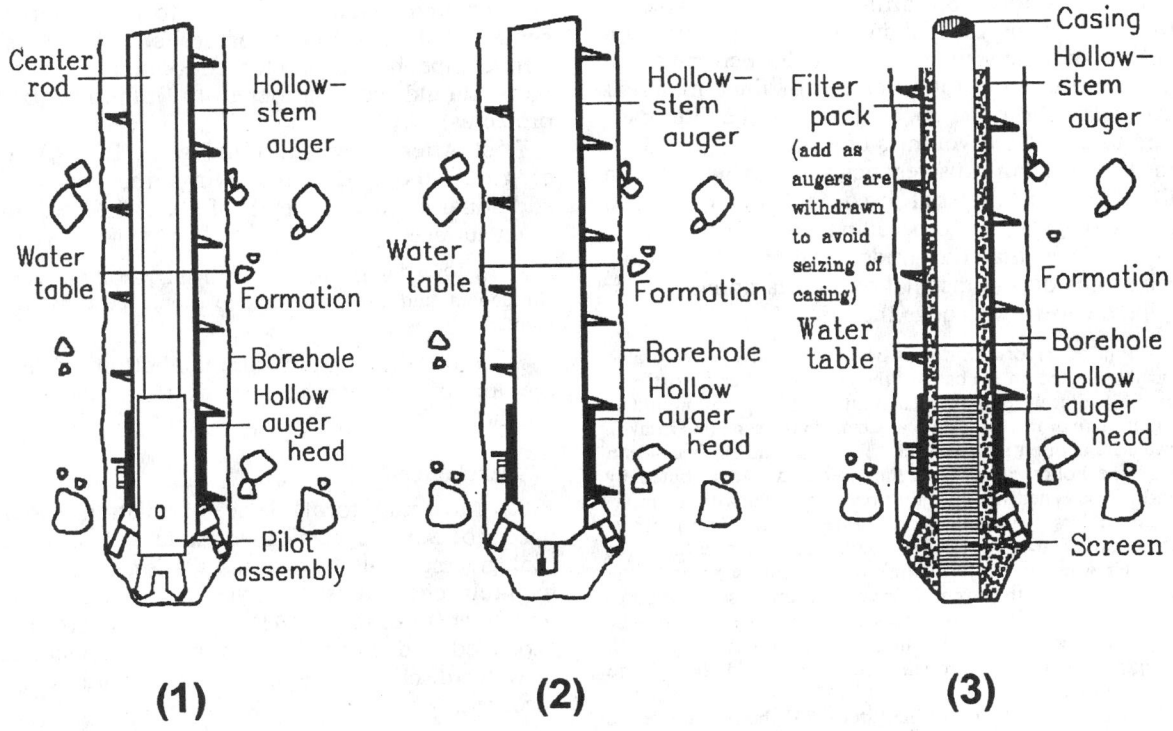

(1)   (2)   (3)

FIG. 2 Sketch Showing Basic Three-Step Procedure for Installation of Subsurface Water-Quality Monitoring Device Using the Hollow-Stem Auger Drilling Method

NOTE 12—Knockout plates or plugs usually remain in the ground close to the monitoring device. Therefore, the material components for knockout plates or plugs should be selected based on their possible effects on subsequent measurements or analyses and the information documented. It may be necessary to fill or partially fill the auger stem with water of known chemistry to prevent blow-in, piping, or sanding in at the time of the plate or plug removal. Refer to Note 7 for considerations regarding adding water to the hollow-auger stem.

7.1.3 Use an auger head with an integral, hinged aperture cover to deter entrance of materials into the auger stem.

7.2 Assemble water-quality monitoring devices, with attached fluid conductors (risers), and suspend in tension prior to placement of filter pack and during placement of filter pack in the borehole (with the least possible addition of contaminants).

7.2.1 Some materials, such as screens and risers, may require cleaning or decontamination, or both, at the job site (see Practice D 5088).

7.2.2 Prior to installation, store all monitoring device materials under cover and place upwind and well away from the drill rig and other sources of potential contamination such as electrical generators, air compressors, or industrial machinery.

7.2.3 Clean hoisting tools, particularly wire rope and hoisting swivels, and decontaminate according to Practice D 5088 before using.

7.3 Select filter materials, bentonite pellets, granules and chips, and grouts and install to specific subsurface monitoring requirements. The thickness of the emplaced materials and extension of the materials above the top of the screen should be sufficient to adequately seal the well and monitoring device(s) against contamination effects of fluid movement between hydrologic units and infiltration of surface contaminants.

7.3.1 Filter packs for monitoring devices are typically installed by withdrawing the hollow augers in small increments, while simultaneously adding small increments of filter material. Record the total volume of filter materials installed and the depth to the upper surface of the filter pack and compare to calculated volumes of material required for completion. Consider any discrepancies occurring between the actual volume of material used and the calculated volume required prior to proceeding to ensure proper completion. If filter material bridges within the hollow auger-riser annulus during installation, use tamping rods or other tamping devices to dislodge the bridge.

NOTE 13—Filter packs for monitoring devices installed in a saturated zone are typically selected on the basis of the grain size characteristics of the hydrologic unit adjacent to the screened intake (screen size should be less than the grain size of the formation adjacent to the screened intake). Filter-pack material is often inserted from above ground surface within the annulus of the hollow auger and the riser and is distributed by gravity around the screened intake. Filter-pack material with a uniformity coefficient of less than 2.5 is ordinarily selected to minimize in-place segregation of grain sizes. For some circumstances, such as installations under water in uniform, fine to very fine sand soils, the filter should be installed with a tremie pipe to minimize segregation of particle sizes. Filter packs for vadose-zone monitoring devices may be predominantly silt sized. These filter materials are often mixed with water of known quality, inserted through a tremie pipe, and tamped into place around the device.

NOTE 14—Effective installation of the filter pack, the seal above the filter pack, and the grout above the seal may be difficult to achieve. Consideration should be given to allow for sufficient annular space

between the monitoring device and the hollow-stem auger to accommodate the tremie pipe. Under some circumstances, the filter pack may be more successfully installed by injecting or inserting water of known chemistry into the hollow-auger annulus either before or during incremental pull-back of the auger column. Enough water should be injected to both fill the space previously occupied by the auger flights and to maintain or slightly increase the head within the auger-hollow stem. This additional head within the auger-hollow stem provides an outward seepage force on the wall of the borehole as the augers are retracted. The additional head deters caving prior to installation of filter or seal materials. Approximate volumes of water used and water losses should be documented.

7.4 Usually place sealing materials consisting of either bentonite pellets, chips, or granules directly above the filter pack.

NOTE 15—It may be effective, when granular filters are used, to install a thin, fine sand, secondary filter either below the annular seal or both above and below the seal. These secondary filters are installed to protect the monitoring device, primary filter pack, and seal from intrusion of grout installed above the seal.

NOTE 16—A measured volume of water of known chemistry is often placed in the annulus on top of a dry bentonite seal to initiate hydration however, hydration of a seal may require from 6 to 24 h.

7.5 The backfill that is placed above the annular seal is usually a bentonite- or cement-base grout.

NOTE 17—Grouts should be designed and installed in consideration of the ambient hydrogeologic conditions. The constituents should be selected and documented according to specific performance requirements. Typical grout mixtures are given in Practice D 5092 and Test Method D 4428.

NOTE 18—Grouting equipment and pipes should be cleaned and decontaminated according to Practice D 5088 prior to use and should be constructed of materials that do not "leach" significant amounts of contaminants to the grout.

7.5.1 When a tremie pipe is used, control its initial position and grouting pressures to prevent materials from being jetted into underlying seal(s) and filter(s) (use of a tremie pipe having a plugged bottom and side-discharge ports should be considered to minimize bottom-jetting problems).

7.5.2 After placement of the initial 1.5 to 3 m (5 to 10 ft) of grout above the underlying filter or seal, discharge additional grout at a depth of about 1.5 to 3 m below the grout surface.

NOTE 19—The need for chemical analysis of samples of each grout component and the final mixture should be considered and documented.

7.5.3 Install the grout from the bottom of the borehole to the top of the borehole so as to displace fluids in the borehole.

8. Development

8.1 Most monitoring device installations should be developed to remove suspended solids from disturbance of geologic materials during installation and to improve the hydraulic characteristics of the filter pack and the hydrologic unit adjacent to the intake. The method(s) selected and time expended to develop the installation and changes in quality of water discharged at the surface should be observed and recorded.

NOTE 20—Under most circumstances, development should be initiated as soon as possible following grouting and well completion operations. For suggested well-development methods and techniques, the user is referred to Test Method D 5099. However, time should be allowed for setting of grout.

9. Field Report and Project Control

9.1 The field report should include information recommended under Guide D 5434, and identified as necessary and pertinent to the needs of the exploration program.

9.2 Other information in addition to Guide D 5434 should be considered if deemed appropriate and necessary to the requirements of the exploration program. Additional information should be considered as follows:

9.2.1 *Drilling Methods:*

9.2.1.1 Description of the hollow-stem auger system,

9.2.1.2 Type, quantities, and locations in the borehole of use of water or additives added,

9.2.1.3 Description of cuttings return, including quantities, and

9.2.1.4 Descriptions of drilling conditions related to rotation rates, and general ease of drilling as related to subsurface materials encountered.

9.2.2 *Sampling*—Document conditions of the bottom of the borehole prior to sampling and report any slough or cuttings present in the recovered sample.

9.2.3 *In Situ Testing:*

9.2.3.1 For devices inserted below the bottom of the borehole, document the depths below the bottom of the hole and any unusual conditions during testing, and

9.2.3.2 For devices testing or seating at the borehole wall, report any unusual conditions of the borehole wall.

9.2.4 *Installations*—A description of well-completion materials and placement methods, approximate volumes placed, depth intervals of placement, methods of confirming placement, and areas of difficulty of material placement or unusual occurrences.

10. Keywords

10.1 drilling; geoenvironmental exploration; ground water; vadose zone

APPENDIX

(Nonmandatory Information)

X1. ADDITIONAL REFERENCES

Acker, W. L., III, *Basic Procedures for Soil Sampling and Core Drilling*, Acker Drill Co., P.O. Box 830, Scranton, PA 18501, 1974.

Aller, L., et al, *Handbook of Suggested Practices for the Design and Installation of Ground-Water Monitoring Wells*, EPA/600/4-89/034, NWWA/EPA Series, National Water Well Assn., Dublin, OH, 1989.

Australian Drilling Manual, Australian Drilling Industry Training Committee Limited, P.O. Box 1545, Macquarie Centre, NSW 2113, Australia, 1992.

Bowen, R., *Grouting in Engineering Practice*, 2nd Edition, Applied Science Publishers, Halstad Press, New York, NY, 1981.

Campbell, M. D., and Lehr, J. H., *Water Well Technology*, McGraw-Hill Book Co., New York, NY, 1973.

DCDMA Technical Manual, Drilling Equipment Manufacturers Assn., 3008 Millwood Ave., Columbia, SC, 29205, 1991.

Drillers Handbook, T. C. Ruda and P. J. Bosscher, eds., National Drilling Contractors Assn., 3008 Millwood Ave., Columbia, SC, 29205, June 1990.

Driscoll, F. G., *Groundwater and Wells*, Johnson Filtration Systems, Second Edition, St. Paul, MN, 1989.

Hackett, G., "Drilling and Constructing Monitoring Wells with Hollow Stem Auger: Part 1," *Ground Water Monitoring Review* 7(4): 1987, pp. 51–62.

Hackett, G., "Drilling and Constructing Monitoring Wells with Hollow Stem Augers: Part 2, Monitoring Well Installation," *Ground Water Monitoring Review* 8(1): 1988, pp. 60–68.

Leach, L. W., Beck, F. P., Wilson, J. T., and Campbell, D. H., "Aseptic Subsurface Sampling Techniques for Hollow-Stem Auger Drilling," Second National Outdoor Action Conference on Aquifer Restoration, Ground Water Monitoring and Geophysical Methods, National Water Well Assn., Dublin, OH, 1988, pp. 31–51.

Morrison, Robert D., *Ground Water Monitoring Technology, Procedures, Equipment and Applications*, Timco Manufacturing, Inc., Prairie Du Sac, WI, 1983.

Shuter, E., and Teasdale, W. E., *Application of Drilling, Coring, and Sampling Techniques to Test Holes and Wells*, U.S. Geological Survey Techniques of Water-Resource Investigations, TWRI 2-F1, 1989.

Standard Guide for
Use of Casing Advancement Drilling Methods for Geoenvironmental Exploration and Installation of Subsurface Water-Quality Monitoring Devices[1]

This standard is issued under the fixed designation D 5872; the number immediately following the designation indicates the year of original adoption or, in the case of revision, the year of last revision. A number in parentheses indicates the year of last reapproval. A superscript epsilon (ε) indicates an editorial change since the last revision or reapproval.

1. Scope

1.1 This guide covers how casing-advancement drilling and sampling procedures may be used for geoenvironmental exploration and installation of subsurface water-quality monitoring devices.

1.2 Different methods exist to advance casing for geoenvironmental exploration. Selection of a particular method should be made on the basis of geologic conditions at the site. This guide does not include procedures for wireline rotary casing advancer systems which are addressed in Guide D 5786.

1.3 The values stated in inch-pound or SI units are to be regarded separately as the standard. The values given in parentheses are for information only.

1.4 Casing-advancement drilling methods for geoenvironmental exploration and monitoring-device installations will often involve safety planning, administration and documentation. This guide does not purport to specifically address exploration and site safety.

1.5 *This standard does not purport to address all of the safety concerns, if any, associated with its use. It is the responsibility of the user of this standard to establish appropriate safety and health practices and determine the applicability of regulatory limitations prior to use.*

2. Referenced Documents

2.1 *ASTM Standards:*

D 653 Terminology Relating to Soil, Rock, and Contained Fluids[2]

D 1452 Practice for Soil Investigation and Sampling by Auger Borings[2]

D 1586 Test Method for Penetration Test and Split-Barrel Sampling of Soils[2]

D 1587 Practice for Thin-Walled Tube Geotechnical Sampling of Soils[2]

D 2113 Practice for Diamond Core Drilling for Site Investigation[2]

D 2487 Classification of Soils for Engineering Purposes (Unified Soil Classification System)[2]

D 2488 Practice for Description and Identification of Soils (Visual-Manual Procedure)[2]

D 3550 Practice for Ring-Lined Barrel Sampling of Soils[2]

D 4220 Practice for Preserving and Transporting Soil Samples[2]

D 4428/D 4428M Test Methods for Crosshole Seismic Testing[2]

D 4700 Guide for Soil Sampling from the Vadose Zone[2]

D 4750 Test Method for Determining Subsurface Liquid Levels in a Borehole or Monitoring Well (Observation Well)[2]

D 5079 Practices for Preserving and Transporting of Rock Core Samples[3]

D 5088 Practice for Decontamination of Field Equipment Used at Non-Radioactive Waste Sites[3]

D 5092 Practice for Design and Installation of Ground Water Monitoring Wells in Aquifers[3]

D 5254 Practice for Minimum Set of Data Elements to Identify a Ground-Water Site[3]

D 5299 Guide for Decommissioning of Ground Water Wells, Vadose Zone Monitoring Devices, Boreholes, and Other Devices for Environmental Activities[3]

D 5408 Guide for the Set of Data Elements to Describe a Ground-Water Site; in manuscript: Part 1—Additional Identification Descriptors[3]

D 5409 Guide for the Set of Data Elements to Describe a Ground-Water Site; in manuscript: Part 2—Physical Descriptors[3]

D 5410 Guide for the Set of Data Elements to Describe a Ground-Water Site; in manuscript: Part 3—Usage Descriptors[3]

D 5434 Guide for Field Logging of Subsurface Explorations of Soil and Rock

D 5474 Guide for Selection of Data Elements for Ground-Water Investigations[3]

D 5521 Guide Development of Ground Water Monitoring Wells in Granular Aquifers[3]

D 5730 Guide for Site Characterization for Environmental Purposes with Emphasis on Soil, Rock, the Vadose Zone and Ground Water

D 5781 Guide for the Use of Dual-Wall Reverse-Circulation Drilling for Geoenvironmental Exploration and the Installation of Subsurface Water-Quality Monitoring Devices[3]

D 5782 Guide for the Use of Direct Air-Rotary Drilling for Geoenvironmental Exploration and the Installation of Subsurface Water-Quality Monitoring Devices[3]

---

[1] This guide is under the jurisdiction of ASTM Committee D-18 on Soil and Rock and is the direct responsibility of Subcommittee D18.21 on Ground Water and Vadose Zone Investigations.

Current edition approved Dec. 10, 1995. Published February 1996.

[2] *Annual Book of ASTM Standards*, Vol 04.08.

[3] *Annual Book of ASTM Standards*, Vol 04.09.

D 5783 Guide for the Use of Direct-Rotary Drilling with Water-Based Drilling Fluid Geoenvironmental Exploration and the Installation of Subsurface Water-Quality Monitoring Devices[3]

D 5784 Guide for the Use of Hollow-Stem Augers for Geoenvironmental Exploration and the Installation of Subsurface Water-Quality Monitoring Devices[3]

D 5876 Guide for the Use of Direct Rotary Wireline Casing Advancement Drilling Methods for Geoenvironmental Exploration and the Installation of Subsurface Water-Quality Monitoring Devices[3]

3. Terminology

3.1 Terminology used within this guide is in accordance with Terminology D 653 with the addition of the following:

3.2 *Descriptions of Terms Specific to This Standard:*

3.2.1 *bentonite*—the common name for drilling fluid additives and well-construction products consisting mostly of naturally occurring montmorillonite. Some bentonite products have chemical additives that may affect water-quality analyses.

3.2.2 *bentonite granules and chips*—irregularly-shaped particles of bentonite (free from additives) that have been dried and separated into a specific size range.

3.2.3 *bentonite pellets*—roughly spherical- or disc-shaped units of compressed bentonite powder (some pellet manufacturers coat the bentonite with chemicals that may affect the water-quality analysis).

3.2.4 *cleanout depth*—the depth to which the end of the drill string (bit or core barrel cutting end) has reached after an interval of cutting. The cleanout depth (or drilled depth as it is referred to after cleaning out of any sloughed material in the bottom of the borehole) is usually recorded to the nearest 0.1 ft (0.03 m).

3.2.5 *coefficient of uniformity*—C_u (D), the ratio D_{60}/D_{10}, where D_{60} is the particle diameter corresponding to 60 % finer on the cumulative particle-size distribution curve, and D_{10} is the particle diameter corresponding to 10 % finer on the cumulative particle-size distribution curve.

3.2.6 *drawworks*—a power-driven winch, or several winches, usually equipped with a clutch and brake system(s) for hoisting or lowering a drilling string.

3.2.7 *drill hole*—a cylindrical hole advanced into the subsurface by mechanical means. Also known as a borehole or boring.

3.2.8 *drill string*—the complete rotary drilling assembly under rotation including bit, sampler/core barrel, drill rods and connector assemblies (subs). The total length of this assembly is used to determine drilling depth by referencing the position of the top of the string to a datum near the ground surface.

3.2.9 *filter pack*—also known as a gravel pack or primary filter pack in the practice of monitoring-well installations. The gravel pack is usually granular material, having selected grain-size characteristics, that is placed between a monitoring device and the borehole wall. The basic purpose of the filter pack or gravel envelope is to act as: a non-clogging filter when the aquifer is not suited to natural development or, act as a formation stabilizer when the aquifer is suitable for natural development.

3.2.9.1 *Discussion*—Under most circumstances a clean,

quartz sand or gravel should be used. In some cases a pre-packed screen may be used.

3.2.10 *hoisting line—or drilling line*, is wire rope used on the drawworks to hoist and lower the drill string.

3.2.11 *in-situ testing devices*—sensors or probes, used for obtaining mechanical- or chemical-test data, that are typically pushed, rotated or driven below the bottom of a borehole following completion of an increment of drilling. However, some *in-situ testing devices* (such as electronic pressure transducers, gas-lift samplers, tensiometers, and so forth) may require lowering and setting of the device(s) in pre-existing boreholes by means of a suspension line or a string of lowering rods or pipes. Centralizers may be required to correctly position the device(s) in the borehole.

3.2.12 *mast*—or derrick, on a drilling rig is used for supporting the crown block, top drive, pulldown chains, hoisting lines, etc. It must be constructed to safely carry the expected loads encountered in drilling and completion of wells of the diameter and depth for which the rig manufacturer specifies the equipment.

3.2.12.1 *Discussion*—To allow for contingencies, it is recommended that the rated capacity of the mast should be at least twice the anticipated weight load or normal pulling load.

3.2.13 *piezometer*—an instrument placed below ground surface to measure hydraulic head at a point.

3.2.14 *subsurface water-quality monitoring device*—an instrument placed below ground surface to obtain a sample for analyses of the chemical, biological, or radiological characteristics of subsurface pore water or to make in-situ measurements.

4. Significance and Use

4.1 Casing advancement may be used in support of geoenvironmental exploration and for installation of subsurface water-quality monitoring devices in both unconsolidated and consolidated materials. Casing-advancement systems and procedures used for geoenvironmental exploration and instrumentation installations consist of direct air-rotary drilling utilizing conventional rotary bits or a down-the-hole hammer drill with underreaming capability, in combination with a drill-through casing driver.

NOTE 1—Direct air-rotary drilling uses pressured air for circulation of drill cuttings. In some instances, water or foam additives, or both, may be injected into the air stream to improve cuttings-lifting capacity and cuttings return. The use of air under high pressures may cause fracturing of the formation materials or extreme erosion of the borehole if drilling pressures and techniques are not carefully maintained and monitored. If borehole damage becomes apparent, consideration to other drilling method(s) should be given.

4.1.1 Casing-advancement methods allow for installation of subsurface water-quality monitoring devices and collection of water-quality samples at any depth(s) during drilling.

4.1.2 Other advantages of casing-advancement drilling methods include: the capability of drilling without the introduction of any drilling fluid(s) to the subsurface; maintenance of hole stability for sampling purposes and monitor-well installation/construction in poorly-indurated to unconsolidated materials.

4.1.3 The user of casing-advancement drilling for geoenvironmental exploration and monitoring-device installations should be cognizant of both the physical (temperature and airborne particles) and chemical (compressor lubricants and possible fluid additives) qualities of compressed air that may be used as the circulating medium.

4.2 The application of casing-advancement drilling to geoenvironmental exploration may involve soil or rock sampling, or in-situ soil, rock, or pore-fluid testing. The user may install a monitoring device within the same borehole wherein sampling, in-situ or pore-fluid testing, or coring was performed.

4.3 The subsurface water-quality monitoring devices that are addressed in this guide consist generally of a screened- or porous-intake device and riser pipe(s) that are usually installed with a filter pack to enhance the longevity of the intake unit, and with isolation seals and low-permeability backfill to deter the movement of fluids or infiltration of surface water between hydrologic units penetrated by the borehole (see Practice D 5092). Inasmuch as a piezometer is primarily a device used for measuring subsurface hydraulic heads, the conversion of a piezometer to a water-quality monitoring device should be made only after consideration of the overall quality and integrity of the installation to include the quality of materials that will contact sampled water or gas. Both water-quality monitoring devices and piezometers should have adequate casing seals, annular isolation seals and backfills to deter communication of contaminants between hydrologic units.

5. Apparatus

5.1 Casing-advancement systems and procedures used for geoenvironmental exploration and instrumentation installations include: direct air rotary in combination with a drill-through casing driver, and conventional rotary bits or down-the-hole hammer drill with or without underreaming capability. Each of these methods requires a specific type of drill rig and tools.

NOTE 2—In North America, the sizes of casings bits, drill rods and core barrels are standardized by American Petroleum Institute (API)[4] and the Diamond Core Drill Manufacturers Association (DCDMA). Refer to the DCDMA Technical Manual[5] and to published materials of API for available sizes and capacities of drilling tools equipment.

5.1.2 Direct air-rotary drill rigs equipped with drill-through casing drivers have a mast-mounted, percussion driver that is used to set casing while simultaneously utilizing a top-head rotary-drive unit. The drill string is generally advanced with bit being slightly ahead of the casing. Figure 1 shows the various components of the drill-through casing driver system. Other mechanical components include casings, drill rods, drill bits, air compressors, pressure lines, swivels, dust collectors, and air-cleaning device (cyclone separator).

5.1.2.1 *Mast-Mounted Casing Driver*, using a piston activated by air pressure to create driving force. Casing drivers are devised to principally drive casing down while drilling but they can also be used to drive the casing upward for casing removal.

---

[4] American Petroleum Institute, "API Specifications for Casing, Tubing, and Drill Pipe," *API Spec 5A*, API, Dallas, TX 1978.

[5] *DCDMA Technical Manual*, Drilling Equipment Manufacturers Association, 3008 Millwood Avenue, Columbia, SC 29205, 1991.

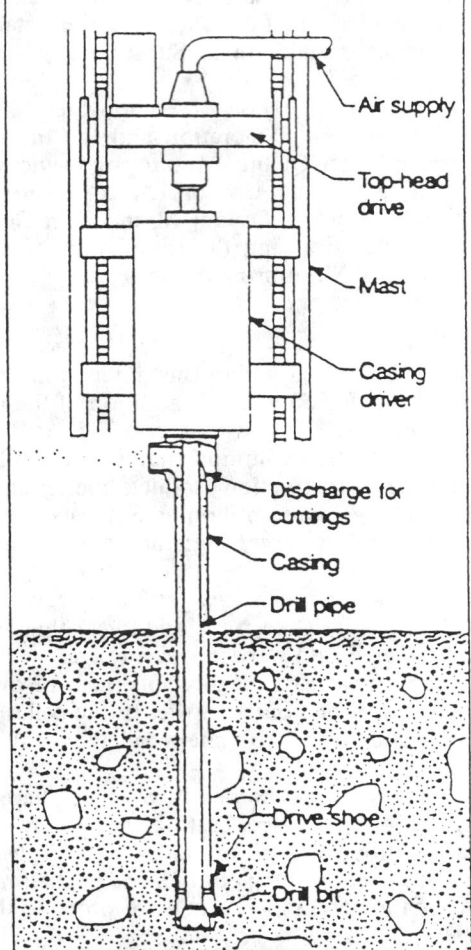

FIG. 1 Casing Drivers can be Fitted to Top-Head Drive Rotary Rigs to Simultaneously Drill and Drive Casing

5.1.2.2 *Standard Casings*, driven with the casing driver. The bottom of the casing is equipped with a forged or cast alloy drive shoe. The top of the casing fits into the casing driver by means of an anvil. In hard formations casings may be welded at connections for added stability. The casing size is usually selected to provide a drill hole of sufficient diameter for the required sampling or testing or for insertion of instrumentation device components such as the screened intake and filter pack and installation devices such as a tremie pipe.

5.1.2.3 Other considerations for selection of casing size are borehole depth and formation type. The casing size should allow for adequate annulus between the casing and the drill rod for upward discharge of cuttings. Also, consideration should be made when difficult formations are expected to require telescoping from larger to smaller casing diameters.

5.1.2.4 *Drill Rods*, used inside the casing for rotary air drilling. The rods extend through the casing driver and are connected to a top-head drive motor for rotation and transfer of rotational force from the drill rig to the bit or core barrel. Drill rod and casing are usually assembled as a unit and raised into position on the mast. Individual drill rods should be straight so they do not contribute to excessive vibrations or "whipping" of the drill-rod column. All threaded connections should be in good repair and not leak significantly at the internal air pressure required for drilling. Drill rods should be made up securely by wrench tightening at the threaded joint(s) at all times to prevent rod damage. Drill pipes usually require lubricants on the threads to allow easy unthreading (breaking) of the connecting joints. Some lubricants have organic or metallic constituents, or both, that could be interpreted as contaminants if detected in a sample. Various lubricants are available that have components of known chemistry. The effect of pipe-thread lubricants on chemical analyses of samples should be considered and documented when using casing-advancement drilling. The same consideration and documentation should be given to lubricants used with water swivels, hoisting swivels, or other devices used near the drilling axis.

5.1.2.5 *Rotary Bit*, attached to the bottom of the drill rod and provides material-cutting capability. Core barrels may be used to obtain sample cores and during this operation the casing can be advanced up to the length of the core barrel. Numerous bit types can be selected depending on the formation properties. Some types successfully used include roller-cone rock bits and drag bits. In hard formations down-the-hole hammers can be substituted for rotary drill bits. Bit selecton can be aided by review of referenced literature or consultation with manufacturers, or both.

5.1.2.6 Perform coring of rock in accordance with Practice D 2113. Soil sampling or coring methods, some of which are listed in 2.2 can also be used to obtain samples and advance the hole. Simultaneously coring and advancing the casing with the casing driver would normally be considered incompatible.

5.1.2.7 Direct-rotary bits have discharge ports that are in close proximity with the bottom of the hole. When these are used in loose cohesionless materials, jetting or excessive erosion of the test intervals could occur.

5.1.3 Casing-advancement drill rigs may be equipped with either standard or underreaming down-the-hole hammers. Standard down-the-hole hammers can be used in unconsolidated deposits to break up highly abrasive particles such as cobbles and boulders. Underreaming down-the-hole hammers operate by drilling and underreaming the drill hole using an air-activated down-the-hole percussion hammer so that the casing falls or can be pushed downward directly behind the hammer bit. Cuttings are removed from the drill hole by air exiting the down-the-hole hammer. In stable rock formations casings may not be required. Down-the-hole hammers may also be used with direct air-rotary drilling procedures discussed in Guide D 5782.

5.1.4 *Down-the-Hole Hammer*, is a pneumatic drill operated on the end of the drill rods. The bit at the end of the hammer is constructed of alloy steel and tungsten-carbide inserts to provide cutting or chipping surfaces. The pneumatic hammer impacts the rock surface while the drill pipe is slowly rotated. Rotation of the bit helps ensure even penetration and straight holes in rock. Proper rotational speed is 10 to 30 rpm with lower speeds used in harder rock. Down-the-hole hammers require air pressures ranging from 100 to 200 lb/in.^{2} and volumes of 100 to 300 cfm.

5.2 *Air Compressors*, required to operate the casing driver

and the down-the-hole hammer and to provide air to circulate the drill cuttings out of the borehole.

5.2.1 *Air Compressor and Filter(s)*, providing adequate air without significant contamination, for removal of cuttings generated at the bit. Air requirements for casing drivers can be evaluated from manufacturers' literature. Air requirements for rotary drilling bits or down-the-hole hammers will depend upon the drill rod and bit configuration, the character of the material penetrated, the depth of drilling below ground-water level, and the total depth of drilling. The flow-rate requirements are usually based on an annulus upflow velocity of about 1000 to 1300 m/min (about 3000 to 4000 ft/min) even though upflow rates of less than 1000 m/min (about 3000 to 4000 ft/min) are often adequate for cuttings transport. Guidance for design of air-pressure circulation systems can be found in referenced literature.

5.2.1.1 The quality of compressed air entering the borehole and the quality of air discharged from the borehole and air-cleaning devices must be considered. If not adequately filtered, the air produced by most oil-lubricated air compressors can introduce some oil into the circulation system. High-efficiency, inline, air filters are usually required to minimize contamination of the borehole. Air-quality monitoring may be required and, if performed, results should be documented.

5.2.2 *Pressure Hose*, conducting the air from the air compressor to the swivel.

5.2.3 *Swivel*, directing the air to the drill-rod column.

5.2.4 *Discharge Hose*, conducting air and cuttings from the drill-hole annulus to an air-cleaning device.

5.2.5 *Air-Cleaning Device*, generally called a cyclone separator—separates cuttings from the air returning from the drill hole via the discharge hose. A properly-sized cyclone separator can remove practically all of the cuttings from the return air. A small quantity of fine particles, however, are usually discharged to the atmosphere with the "cleaned" air. Some air-cleaning devices consist of a cyclone separator alone. In special cases, the cyclone separator can be combined with a HEPA (high-efficiency particulate air) filter for removing dust particles that might be radioactive. In other special situations, the cyclone separator may be used in conjunction with a charcoal-filtering arrangement for removal of organic volatiles. Samples of drill cuttings can be collected for analyses of materials penetrated. If samples are obtained, the depth(s) and interval(s) of sample collection should be documented.

5.2.6 Compressed air alone can often transport cuttings from the drill hole and cool the bit. For some geologic conditions, injection of water into the air stream will help control dust or break down "mud rings" that tend to form on the drill rods. Under other circumstances, for example if the drill hole starts to produce water, the injection of a foaming agent may be required. If changes to the circulating medium are made, such as addition of water, the depth(s) or interval(s) of these changes should be documented. It is important to observe the quantity and quality of return air and cuttings. Zones of low air return as well as zones of no air return should be documented. If circulation is lost and input air pressure is allowed to increase, the possibility of fracturing the geologic materials exists. In order not to raise pressures excessively the compressor can be equipped with pressure relief valves.

6. Drilling Procedures

6.1 As a prelude to and throughout the drilling process stabilize the drill rig and raise the drill-rig mast and position the cyclone separator. If air-monitoring operations are performed, consider the prevalent wind direction relative to the exhaust from the drill rig. Also, consider the location of the cyclone separator relative to the rig exhaust since air-quality monitoring will be performed at the cyclone-separator discharge point. Establish and document a datum for measuring hole depth. This datum is normally the top of the surface casing or the drilling deck. If the hole is to be later surveyed for elevation, record and report the height of the datum to the ground surface.

6.1.1 Clean and decontaminate the drill rig, drill rods and bits, and hoisting and sampling tools according to Practice D 5088 prior to commencing drilling and sampling operations and at periods during the drilling operation when deemed appropriate such as when the drill string is removed from the hole to permit intermittent sampling.

Note 3—It is extremely important to check above the drilling rig for overhead obstructions or hazards, such as power lines, prior to lifting the mast. In most cases it is required to perform a survey of underground and all other utilities prior to drilling to evaluate possible hazards.

6.2 Drilling is usually done as follows:

6.2.1 Attach an initial assembly of a bit, often with a single section of drill rod and casing, to the top-head drive unit.

6.2.2 Activate the air compressor, causing compressed air to circulate through the system.

6.2.3 Drill through casing driver.

6.2.3.1 Several drilling methods can be used with direct air-rotary drill rigs equipped with drill-through casing drivers: the drill bit and casing are advanced as a unit, in unconsolidated materials the casing is driven first and then the plug in the casing is drilled out, and the drill bit advances beyond the casing, and then is withdrawn into the casing and then the casing is driven. Air exiting the bit removes the cuttings uphole. Separate cuttings from the return air with an air-cleaning device such as a cyclone separator. Air pressures at the bit should be as low as necessary to maintain circulation in order to minimize hydraulic fracturing or excessive erosion of the surrounding materials. Monitor air pressures during drilling. Changes in cuttings return and circulation pressures may indicate occurrence of excessive erosion. Should excessive erosion occur, note and document the depth(s) of the occurrence(s). Duly note and document any abrupt changes or anomalies in the air pressures including the depth(s) of occurrence(s).

6.2.3.2 In most operations the casing and bit are advanced as a unit. The drill bit or down-the-hole hammer generally protrudes out through the bottom of the casing usually not more than 50 cm (12 in.). Bit-lead distance is adjusted according to drilling conditions. In loose unstable formations it may be necessary to keep the bit near the end of the casing to prevent excessive erosion of the formation. Occasionally, casing advance is stopped, the bit is retracted inside the casing and circulation is maintained to clear cuttings.

6.2.3.3 Compressed air from the drilling system, when allowed to enter geologic zones to be monitored or sampled, may not be easily developed out and may, in turn, affect the chemistry of water samples obtained. Take care in the drilling process to minimize introduction of air into the formation by not allowing the drill bit to advance beyond the casing shoe in unconsolidated formations where it may be possible to do so.

6.2.3.4 In drilling extremely unstable deposits it may be necessary to use the second method of driving casings without rotary drilling. Intermittently remove the plug in the casing by rotary drilling performed inside the casing.

6.2.3.5 In drilling stable deposits, such as when drilling with down-the-hole hammers, it may be possible to use the method of advancement mentioned above that entails drilling ahead of casing advancement. However, if circulation is lost and input air pressure is allowed to increase, the possibility of fracturing the geologic materials exists when the bit is below the casing.

Note 4—Consideration should be given to the speed of drilling so that the borehole can be maintained as close to vertical as practicable.

6.2.4 Underreaming down-the-hole air hammers:

6.2.4.1 Use direct air-rotary drill rigs equipped with underreaming down-the-hole air hammers to advance a casing without applying percussion forces to the casing.

6.2.4.2 Special down-the-hole air hammers open a hole slightly larger than the outside diameter of the casing so that the casing will fall, or can be pushed downward, immediately behind the bit. After advancing the casing, retract the radial dimension of the drill bit to facilitate removal of the down-the-hole air hammer and drill tools from inside the casing.

6.2.4.3 Cuttings are removed from the drill hole with the air that operates the hammer and can be separated from the discharged air with a cyclone separator. While the drilling proceeds, downfeed pressure, cuttings returned, and ease of drilling as it relates to the geologic strata being penetrated should be monitored.

Note 5—Discharge of the cuttings from the cyclone separator affords an excellent point for collection and logging of the cuttings during the drilling process.

6.2.5 Continue air circulation and rotation of the drill-rod column until drilling progresses to a depth where sampling or in-situ testing will be performed or until the length of the drill-rod section limits further penetration. Then, stop rotation and lift the bit slightly off the bottom of the hole to facilitate cuttings removal while air circulation is continued for a short time until the drill cuttings are removed from the drill-hole annulus. If a check of the hole quality is required, stop circulation and rest the bit on the bottom of the hole. If hole cave or other instability of the boring is suspected, check the depth at which the bit rests against the previous maximum cleanout depth.

6.2.6 Increase drilling depth by attaching an additional drill rod and casing section to the top of the previously advanced drill rod column and resuming drilling operations according to 6.2.2 through 6.2.5. In some cases, it may be necessary to remove the rotary drill string to change bits. If the string is removed and replaced, check the depth to the base of the boring where the end of the string rests and compare to the clean out depth to evaluate hole quality. As the drilling progresses, note and document drilling procedures such as circulation rates or losses, intervals where equipment is changed, intervals where casing is installed or drilling method is changed.

6.2.7 Sampling or in-situ testing can be performed at any depth by interrupting the advance of the bit and casing, stopping air circulation, and removing the drill rod column from the drill hole. Compare the sampling depth to the cleanout depth if the sampler is attached to rods. This comparison is accomplished by resting the sampler at the bottom of the hole and comparing the apparent depth with the cleanout depth. If the samples taken appear to have been contaminated and contain sloughed materials from the bottom of the hole, the apparent amount of contamination in the sample(s) and the depth of occurrence should be documented. In-situ testing can be performed at the base of the drill hole or at shallower depths if the hole is uncased. If in-situ testing is performed at the base of the drill hole performed, perform similar depth checks to accurately document the location of the test.

7. Installation of Monitoring Devices

7.1 Subsurface water-quality monitoring devices are generally installed in boreholes drilled by casing advancement methods using the three-step procedure shown in Fig. 2. The three steps are as follows: (1) casing advancement in increments, with or without sampling; (2) removal of the drill rods and the attached drill bit while the casing is temporarily left in place to support the borehole wall; and (3) insertion of the monitoring device through the inside of the casing, and followed by addition of well-completion materials (that is, filter packs, annular seals and grouts) as the casing is withdrawn from the borehole. (Under some circumstances, part of the casing may be left in the drilled hole.)

7.2 Assemble water-quality monitoring devices with attached fluid conductors (risers) and insert them into the borehole with the least possible addition of contaminants.

7.2.1 Some materials, such as screens and risers, may require cleaning or decontamination at the job site, or both (see Practice D 5088).

7.2.2 Prior to installation, store all monitoring-device materials undercover and place upwind and well away from the drill rig and any other sources of contamination such as electrical generators, air compressors, or industrial machinery.

7.2.3 Clean and decontaminate hoisting tools, particularly wire rope and hoisting swivels according to Practice D 5088 before using.

7.3 Select and install filter materials, bentonite pellets, bentonite granules and chips and grouts, according to subsurface monitoring or instrumentation requirements.

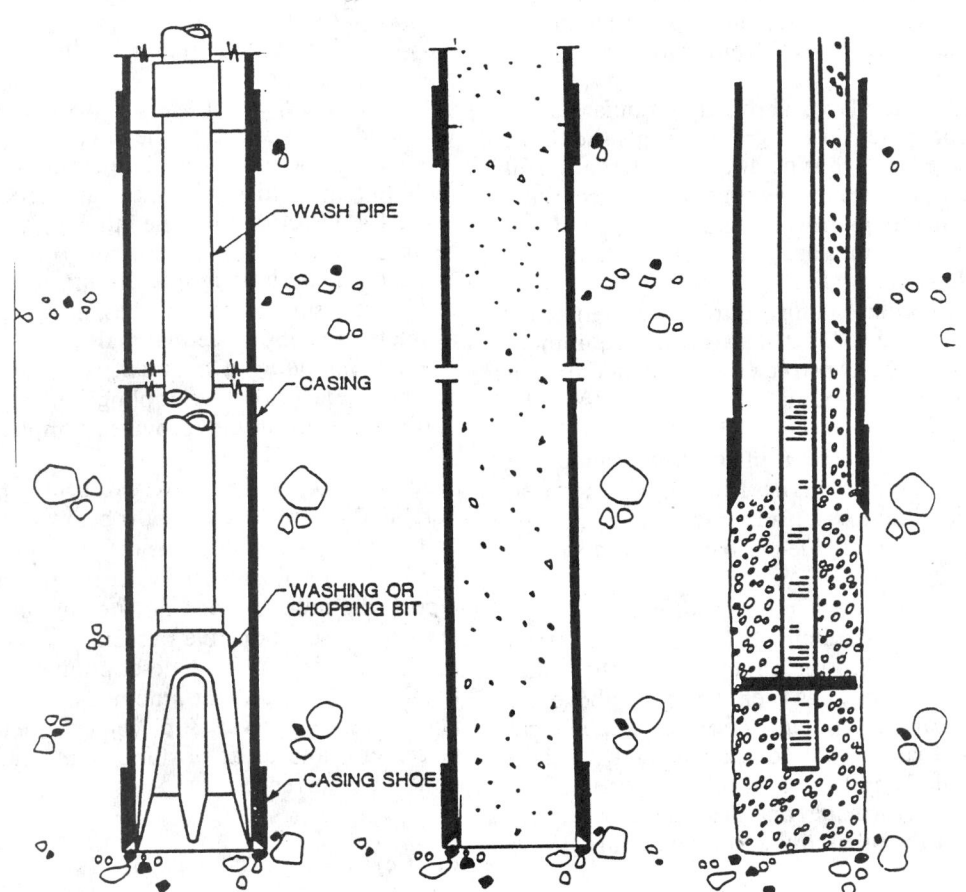

NOTE—The method shown is with using the wash boring method. Drilling procedures and tools other than those shown may be used.

FIG. 2   Basic Three-Step Procedure for Installation of Instrumentation Devices Using Casing Advancement Methods

7.3.1 Filter packs, for monitoring devices are usually installed in borings drilled with casing advancement methods by placing the materials through the casing-riser annulus. This annular area then serves the same function as a separate tremie pipe for placing the annular materials. In some cases, it may be appropriate to use a tremie pipe inserted in the annulus between the inner pipe and the monitoring-device riser, provided it is sufficiently large. Monitoring devices installed in a saturated zone ordinarily have sand size filter packs that are selected primarily on the basis of the grain size characteristics of the hydrologic unit adjacent to the screened intake. The coefficient of uniformity of the filter-pack sand is usually less than 2.5. Filter packs for monitoring devices installed in a vadose zone may be predominantly silt sized. These filter materials are often mixed with water of known quality, inserted through a tremie pipe, and tamped into place around the device. Take care when adding backfill or filter material(s), or both, so that the materials do not bridge. However, if bridging does occur during the installation procedure, tamping rods or other tamping devices may be used to dislodge the "bridge".

7.4 Sealing materials, consisting of either bentonite pellets, chips, or granules, are usually placed directly above the filter pack of a monitoring device. It may be effective, when granular filter packs are used, to install secondary filter of thin, fine sand either below the annular seal or both above and below the seal. These secondary filters protect both the monitoring-device filter and the seal from intrusion of grout installed above the seal.

7.5 The backfill that is placed above the annular seal is usually a bentonite or cement-based grout. Grouts should be designed and installed in consideration of the ambient hydrogeologic conditions. Select the constituents according to specific performance requirements and these data documented. Typical grout mixtures are given in Practice D 5092 and Test Methods D 4428.

7.5.1 Clean and decontaminate grouting equipment prior to use according to Practice D 5088. Also, the equipment used for grouting should be constructed from materials that do not "leach" significant amounts of contaminants to the grout.

7.5.2 Control the initial position of the tremie pipe and grouting pressures to prevent materials from being jetted into underlying seal(s) and filter(s). Consider use of a tremie pipe having a plugged bottom and side-discharge ports to minimize bottom-jetting problems.

7.5.3 When it is appropriate to use a grout line, discharge the grout at a depth of approximately 1.5 to 3 m (5 to 10 ft) below the grout surface within the annulus (after the initial 1.5 to 3 m (5 to 10 ft) of grout has been deposited above the uppermost filter or seal). Document the need for chemical analysis of samples of each grout component and the final mixture. Also, it should be noted that if cements are used for grouting, they generate hydroxides and heat, thereby causing a localized increase in the alkalinity and temperature of the surrounding ground water.

7.5.4 Using a tremie install the grout from the bottom of the borehole to the top of the borehole so as to displace fluids in the borehole.

8. Development

8.1 Most monitoring-device installations should be developed to remove any air that may have been introduced into the formation by the drilling method, suspended solids from drilling fluids, and disturbance of geologic materials during installation and to improve the hydraulic characteristics of the filter pack and the geohydrologic unit adjacent to the intake. For suggested well-development methods and techniques the user is referred to Guide D 5521. The methods(s) selected and time expended to develop the installation and the changes in water quality discharged at the surface should be carefully observed and documented. Under most circumstances, development should be initiated as soon as possible following completion however, time should be allowed for initial setting of grout.

9. Field Report and Project Control

9.1 The field report should include all information and identified as necessary and pertinent to the needs of the exploration program.

9.2 Other information in addition to Guide D 5434 should be considered if deemed appropriate and necessary to the needs of the exploration program. Additional information should be considered as follows:

9.2.1 *Drilling Method:*

9.2.1.1 Description of the casing advancement method system,

9.2.1.2 Type, quantities, and locations in the borehole of use of additives added to the circulation media,

9.2.1.3 Description of circulation rates, cuttings return, including quantities, over intervals used,

9.2.1.4 Description of the lithology of the cuttings that are returned, and

9.2.1.5 Descriptions of drilling conditions related to drilling pressures, rotation rates, and general ease of drilling as related to subsurface materials encountered.

9.2.2 *Sampling*—Document conditions of the bottom of the borehole prior to sampling and report any slough or cuttings present in the recovered sample.

9.2.3 *In-situ Testing:*

9.2.3.1 For devices inserted below the bottom of the borehole document the depths below the bottom of the hole and any unusual conditions during testing.

9.2.3.2 For devices testing or seating at the borehole wall, report any unusual conditions of the borehole wall such as inability to seat borehole packers.

9.2.4 *Installations*—A description of well-completion materials and placement methods, approximate volumes placed, depth intervals of placement, methods of confirming placement, and areas of difficult of material placement or unusual occurrences.

10. Keywords

10.1 casing-advancement drilling method(s); direct air rotary drilling; down-the-hole hammer (DTH); drill through casing hammer; drilling; ground water

REFERENCES

Aller, L., et al., *Handbook of Suggested Practices for the Design and Installation of Ground-Water Monitoring Wells, EPA/600/4-89/034 NWWA/EPA Series*. National Water Well Association, Dublin, OH, 1989.

American Petroleum Institute, *API Specifications for Casing Tubing and Drill Pipe API Spec 5A*, American Petroleum Institute, Dallas, TX, 1978.

Australian Drilling Manual, Australian Drilling Industry Training Committee Limited, P.O. Box 1545, Macquarie Centre, NSW 2113, Australia, 1992.

Boroid, *Baroid Drilling Fluid Products for Minerals Exploration*, NL Baroid/NL Industries, Houston, TX, 1980.

Bowden, R., *Grouting in Engineering Practice*, 2nd Edition Applied Science Publishers, Halstad Press, New York, NY, 1981.

Cambell, M. D., and Lehr, J. H., *Water Well Technology*, McGraw-Hill Book Company, New York, NY, 1973.

DCDMA Technical Manual, Drilling Equipment Manufacturers Association, 3008 Millwood Avenue, Columbia, South Carolina, 29205, 1991.

Heinz, W. F., *Diamond Drilling Handbook*, First Edition, South African Drilling Association, Johannesburg, Republic of South Africa, 1985.

Drillers Handbook, T. C. Ruda, and P. J. Bosscher, editors, National Drilling Contractors Association, 3008 Millwood Avenue, Columbia, South Carolina, 29205, June 1990.

Driscoll, F. G., *Groundwater and Wells*, Johnson Filtration Systems, Second Edition, St. Paul, MN, 1989.

Heinz, W. F., *First Edition*, South African Drilling Association, Johannesburg, Republic of South Africa, 1985.

Hix, G. L., "Casing Advancement Methods for Drilling Monitoring Wells," *Water Well Journal 45(5)*: 1991, pp 60-64.

Morrison, Robert D., *Ground Water Monitoring Technology, Procedures, Equipment and Applications*, Timco Mfg., Inc., Prairie Du Sac, WI, 1983.

Roscoe Moss Company, *Handbook of Ground Water Development*, Roscoe Moss Company, Los Angeles, CA, John Wiley and Sons, Inc., New York, NY, 1990.

Shuter, E., and Teasdale, W. E., "Application of Drilling, Coring, and Sampling Techniques to Test Holes and Wells," *U.S. Geological Survey Techniques of Water-Resource Investigations, TWRI 2-F1*, 1989.

Designation: D 5875 – 95

Standard Guide for
Use of Cable-Tool Drilling and Sampling Methods for Geoenvironmental Exploration and Installation of Subsurface Water-Quality Monitoring Devices[1]

This standard is issued under the fixed designation D 5875; the number immediately following the designation indicates the year of original adoption or, in the case of revision, the year of last revision. A number in parentheses indicates the year of last reapproval. A superscript epsilon (ε) indicates an editorial change since the last revision or reapproval.

1. Scope

1.1 This guide covers cable-tool drilling and sampling procedures used for geoenvironmental exploration and installation of subsurface water-quality monitoring devices.

1.2 Several sampling methods exist for obtaining samples from drill holes for geoenvironmental purposes and subsequent laboratory testing. Selection of a particular drilling procedure should be made on the basis of sample types needed and geohydrologic conditions observed at the study site.

1.3 Drilling procedures for geoenvironmental exploration often will involve safety planning, administration and documentation. This guide does not purport to specifically address exploration and site safety.

NOTE 1—This guide does not include considerations for geotechnical site characterizations that are addressed in a separate guide.

1.4 The values stated in inch-pound units are to be regarded as the standard. The SI units given in parentheses are for information only.

1.5 *This standard does not purport to address all of the safety concerns, if any, associated with its use. It is the responsibility of the user of this standard to establish appropriate safety and health practices and determine the applicability of regulatory limitations prior to use.*

2. Referenced Documents

2.1 *ASTM Standards:*

D 420 Guide for Site Characterization for Engineering Design and Construction Purposes[2]

D 653 Terminology Relating to Soil, Rock, and Contained Fluids[2]

D 1452 Practice for Soil Investigation and Sampling by Auger Borings[2]

D 1586 Test Method for Penetration Test and Split-Barrel Sampling of Soils[2]

D 1587 Practice for Thin-Walled Tube Geotechnical Sampling of Soils[2]

D 2113 Practice for Diamond Core Drilling for Site Investigation[2]

D 2487 Test Method for Classification of Soils for Engineering Purposes (Unified Soil Classification System)[2]

D 2488 Practice for Description and Identification of Soils (Visual-Manual Procedure)[2]

D 3550 Practice for Ring-Lined Barrel Sampling of Soils[2]

D 4220 Practices for Preserving and Transporting Soil Samples[2]

D 4428/D 4428M Test Methods for Crosshole Seismic Testing[2]

D 4700 Guide for Soil Sampling from the Vadose Zone[2]

D 4750 Test Method for Determining Subsurface Liquid Levels in a Borehole or Monitoring Well (Observation Well)[2]

D 5079 Practices for Preserving and Transporting of Rock Core Samples[3]

D 5088 Practice for Decontamination of Field Equipment Used at Non-Radioactive Waste Sites[3]

D 5092 Practice for Design and Installation of Groundwater Monitoring Wells in Aquifers[3]

D 5299 Guide for Decommissioning of Ground Water Wells, Vadose Zone Monitoring Devices, Boreholes, and Other Devices for Environmental Activities[3]

D 5434 Guide for Field Logging of Subsurface Explorations of Soil and Rock

D 5730 Guide for Site Characterization for Environmental Purposes

D 5521 Development of Ground Water Monitoring Wells in Granular Aquifers

D 5782 Guide for the Use of Casing Advancement Drilling Methods for Geoenvironmental Exploration and the Installation of Subsurface Water-Quality Monitoring Devices

D 5783 Guide for the Use of Direct-Rotary Drilling with Water-Based Drilling Fluid for Geoenvironmental Exploration and the Installation of Subsurface Water-Quality Monitoring Devices

D 5784 Guide for the Use of Hollow-Stem Augers for Geoenvironmental Exploration and the Installation of Subsurface Water-Quality Monitoring Devices

3. Terminology

3.1 Terminology used within this guide is in accordance with Terminology D 653 with the addition of the following:

3.2 *Descriptions of Terms Specific to This Standard:*

3.2.1 *bailer*—a long, narrow bucket, made from a piece of large-diameter pipe with a dart valve in the bottom, used to remove cuttings from the borehole.

[1] This guide is under the jurisdiction of ASTM Committee D-18 on Soil and Rock and is the direct responsibility of Subcommittee D18.21 on Ground Water and Vadose Zone Investigations.

Current edition approved Dec. 10, 1995. Published February 1996.

[2] *Annual Book of ASTM Standards*, Vol 04.08.

[3] *Annual Book of ASTM Standards*, Vol 04.09.

3.2.2 *bentonite*—the common name for drilling-fluid additives and well-construction products consisting mostly of naturally occurring montmorillonite. Some bentonite products have chemical additives which may affect water-quality analyses.

3.2.3 *bentonite granules and chips*—irregularly-shaped particles of bentonite (free from additives) that have been dried and separated into a specific size range.

3.2.4 *bentonite pellets*—roughly spherical- or disc-shaped units of compressed bentonite powder (some pellet manufacturers coat the bentonite with chemicals that may affect the water-quality analysis).

3.2.5 *coefficient of uniformity*—C_u (D), the ratio D_{60}/D_{10}, where D_{60} is the particle diameter corresponding to 60 % finer on the cumulative particle-size distribution curve, and D_{10} is the particle diameter corresponding to 10 % finer on the cumulative particle-size distribution curve.

3.2.6 *collar*—the section of a drill tool between the wrench square and the pin or box joint.

3.2.7 *dart valve*—a type of valve used on a bailer, that opens when the bailer drops through the cuttings at the bottom of the borehole.

3.2.8 *drill bit*—the steel tool on the lower end of the string of tools which does the actual drilling; shaped to perform the operations of penetration, reaming, crushing, and mixing.

3.2.9 *drill hole*—a cylindrical hole advanced into the subsurface by mechanical means. Also known as a borehole or boring.

3.2.10 *drill stem*—a steel tool composed of a round bar of steel with a pin joint at its upper end and a box joint at its lower end that is placed below the jars in a string of drilling tools to furnish the necessary weight to the tool string.

3.2.11 *drill string*—the complete cable-tool drilling assembly including bit, drill rods and connector assemblies (subs). The total length of this assembly is used to determine drilling depth by referencing the position of the top of the string to a datum near the ground surface.

3.2.12 *drive shoe*—a forged- or machined-steel collar either a threaded- or drop-type attached to the upper joint of casing to protect the casing threads during driving operations.

3.2.13 *filter pack*—also known as a gravel pack or primary filter pack in the practice of monitoring-well installations. The gravel pack is usually granular material, having specified grain-size characteristics, that is placed between a monitoring device and the borehole wall. The basic purpose of the filter pack or gravel envelope is to act as: a non-clogging filter when the aquifer is not suited to natural development or, a formation stabilizer when the aquifer is suitable for natural development.

3.2.13.1 *Discussion*—Under most circumstances a clean, quartz sand or gravel should be used. In some cases a pre-packed screen may be used.

3.2.14 *grout shoe*—a drillable "plug" containing a check valve that is positioned within the lowermost section of a casing column. Grout is injected through the check valve to fill the annular space between the casing and the borehole wall or another casing.

3.2.14.1 *Discussion*—The composition (or mix) of the drillable "plug" should be known and documented. A grout shoe would probably only be installed in a cable-tool drilled hole if the hole was to be continued on by a rotary-type drilling rig.

3.2.15 *grout packer*—a reusable inflatable or expandable annular plug that is attached to a tremie pipe, usually positioned immediately above the discharge end of the pipe.

3.2.16 *intermittent sampling devices*—usually barrel-type samplers that are driven below the bottom of a borehole with drill rods or with a wireline system to lower, drive, and retrieve the sampler following completion of an increment of drilling. The user is referred to the following standards relating to suggested sampling methods and procedures: Practice D 1452, Test Method D 1586, Practice D 3550, and Practice D 1587.

3.2.17 *in-situ testing devices*—sensors or probes, used to obtain mechanical- or chemical-test data, that are typically pushed, rotated or driven below the bottom of a borehole following completion of an increment of drilling. However, some *in-situ testing devices* (such as electronic pressure transducers, gas-lift samplers, tensiometers, and etc.) may require lowering and setting of the device(s) in pre-existing boreholes by means of a suspension line or a string of lowering rods or pipes. Centralizers may be required to correctly position the device(s) in the borehole.

3.2.18 *jars*—a tool composed of two connected links or reins with vertical play between them (see Fig. 3 (4)).[4] Drilling jars have a stroke of 9- to 18-in. whereas, fishing jars have a stroke of 18- to 36-in. (7 mm). Jars permit a sudden upward load or shock to loosen a string of tools stuck in the borehole.

3.2.19 *sand pump*—bailer made of tubing with a hinge-flap valve and a plunger that works inside the barrel. It is used in sand and gravel where the dart-valve bailer will not pick up the materials adequately.

3.2.20 *spear*—a fishing tool used when the drilling line or sand line breaks leaving the drilling tools or bailer in the hole with the line on top of the lost tools.

3.2.21 *swivel socket*—a socket that permits the tool string to spin or turn during the drilling action (sometimes referred to as a "rope socket").

3.2.22 *subsurface water-quality monitoring device*—an instrument placed below ground surface to obtain a sample for analysis of the chemical, biological or radiological characteristics of subsurface pore water or to make in-situ measurements.

3.2.23 *wrench square*—a square section on any drilling tool by which the joints are set up or broken.

4. Significance and Use

4.1 Cable-tool rigs (also referred to as churn rigs, water-well drilling rigs, spudders, or percussion rigs) are used in the oil fields and in the water-well industry. The Chinese developed the percussion method some 4000 years ago.

4.2 Cable-tool drilling and sampling methods may be used in support of geoenvironmental exploration and for installation of subsurface water-quality monitoring devices in both unconsolidated and consolidated materials. Cable-tool drilling and sampling may be selected over other methods

---

[4] The boldface numbers given in parentheses refer to a list of references at the end of the text.

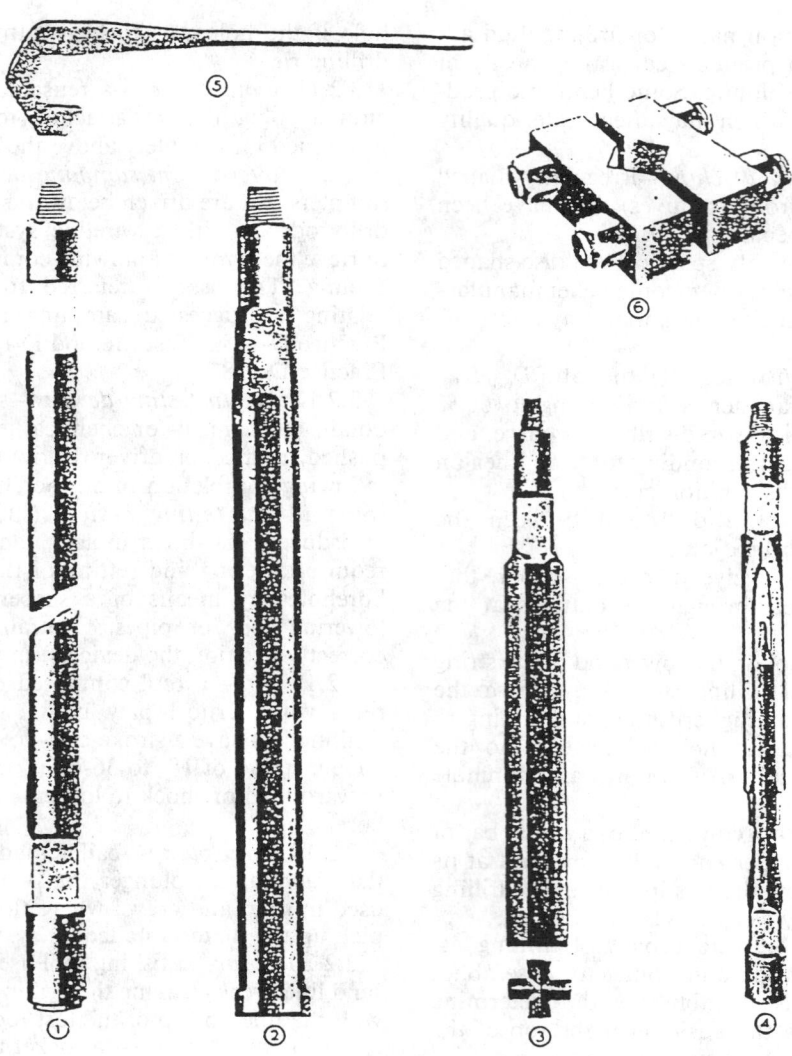

1. Drill Stem.
2. Regular pattern bit.
3. Star or four-wing bit.
4. Jars.
5. Wrench for tightening drive clamps.
6. Drive clamps.

FIG. 1 Drilling Tools

based on its advantages, some of which are its high mobility, low water use, low operating cost, and low maintenance. Cable-tool drilling is the most widely available casing-advancement method that is restricted to the drilling of unconsolidated materials and softer rocks.

4.2.1 The application of cable-tool drilling and sampling to geoenvironmental exploration may involve sampling unconsolidated materials. Depth of drill holes may exceed 3000 ft (914 m) and may be limited by the length of cable attached to the bull reel. However, most drill holes for geoenvironmental exploration rarely are required to go that deep. Rates for cable-tool drilling and sampling can vary from a general average of as much as 25 to 30 ft/h (7.6 to 9 m/h) including setting 8 in. (2.4 m) diameter casing to considerably less than that depending on the type(s) of material drilled, and the type and condition of the equip-ment and rig used.

NOTE 2—As a general rule, cable-tool rigs are used to sample the surficial materials, and to set surface casing in order that rotary-core rigs subsequently may be set up on the drill hole to core drill hard rock if coring is required.

4.2.2 The cable-tool rig may be used to facilitate the installation of a subsurface water-quality monitoring device(s) including in-situ testing devices. The monitoring device(s) may be installed through the casing as the casing is removed from the borehole. The sand line can be used to raise, lower, or set in-situ testing device(s), or all of these. If necessary, the casing may also be left in the borehole as part of the device.

NOTE 3—The user may install a monitoring device within the same borehole wherein sampling, in-situ, or pore-fluid testing, or coring was performed.

5. Apparatus

5.1 Cable-tool rigs have a string of drill tools with a drive clamp (see Fig. 3 (6)) on the drill string connected by wire rope that periodically can be hoisted and allowed to "fall" for percussion drilling in unconsolidated and consolidated materials and for driving/retrieving casing. The full string of drilling equipment consists of drill bit (see Fig. 3 (2)—bit used for all-around general drilling and, (3)—bit used for chopping and breaking hard materials and rock), drilling jars (optional), and a drill stem (see Fig. 3 (1)), with a swivel socket (see Fig. 2) connected by a wire rope fastened to a drum called a bull reel that raises and lowers the drilling tools and permits percussion drilling either by crushing the material or by drive sampling. The spudding beam, commonly referred to as the walking beam, that is driven by the pitman and crank, imparts a reciprocating motion to the drilling line (see Fig. 1 (6)).

NOTE 4—All cable-tool rigs have the capacity to lift and drop heavy drive clamps for installing large-diameter casing in unconsolidated materials.

5.2 Water-well drilling rigs have been converted (for the purpose of geoenvironmental-engineering explorations) by replacing the jars and stem, and replacing the chopping bit (see Fig. 3 (3)) with a drive barrel, (see Fig. 4) that is used for sampling purposes. If the bit becomes stuck in the borehole it can normally be freed by upward blows of the drilling jars (jars can also be used in the same mode to extract casing). The primary function of the drilling jars is to transmit the energy from the bull wheel to the drill stem and the sample barrel. The stroke of the drilling jars is 9 to 18 in. (0.23 to 0.46 m) and distinguishes them from fishing jars that have a stroke 18 to 36 in. (0.46 to 0.91 m). Jars are often not used when hard-rock drilling (6, 7).

5.3 The swivel socket connects the drill string to the cable and, in addition, the weight of the socket supplies part of the weight of the drill tools. The socket also imparts part of the upward energy to the jars when their use becomes necessary. The socket transmits the rotation of the cable to the tool string and bit (drive barrel) so that the drive is completed on the downstroke, thereby assuring that a round, straight hole will result. The elements of the tool string are coupled together with right-hand threaded tool joints of standard API

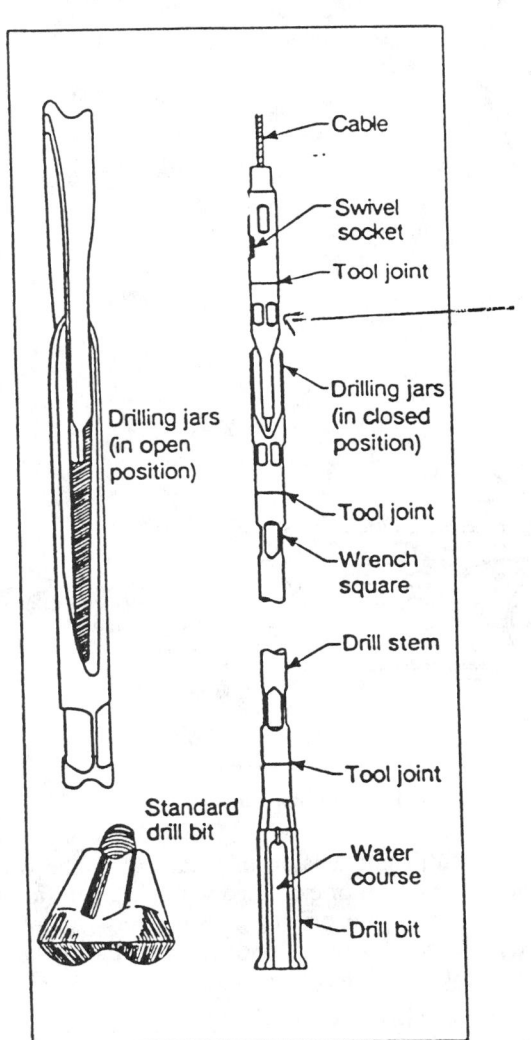

Cable

Swivel socket

Tool joint

#6 from Figure 1 goes here during casing drive-released for drilling

Drilling jars (in closed position)

Tool joint

Wrench square

Drill stem

Tool joint

Water course

Drill bit

Drilling jars (in open position)

Standard drill bit

FIG. 2 A Full String of Cable Tools Consists of Five Components That are Necessary for Drilling

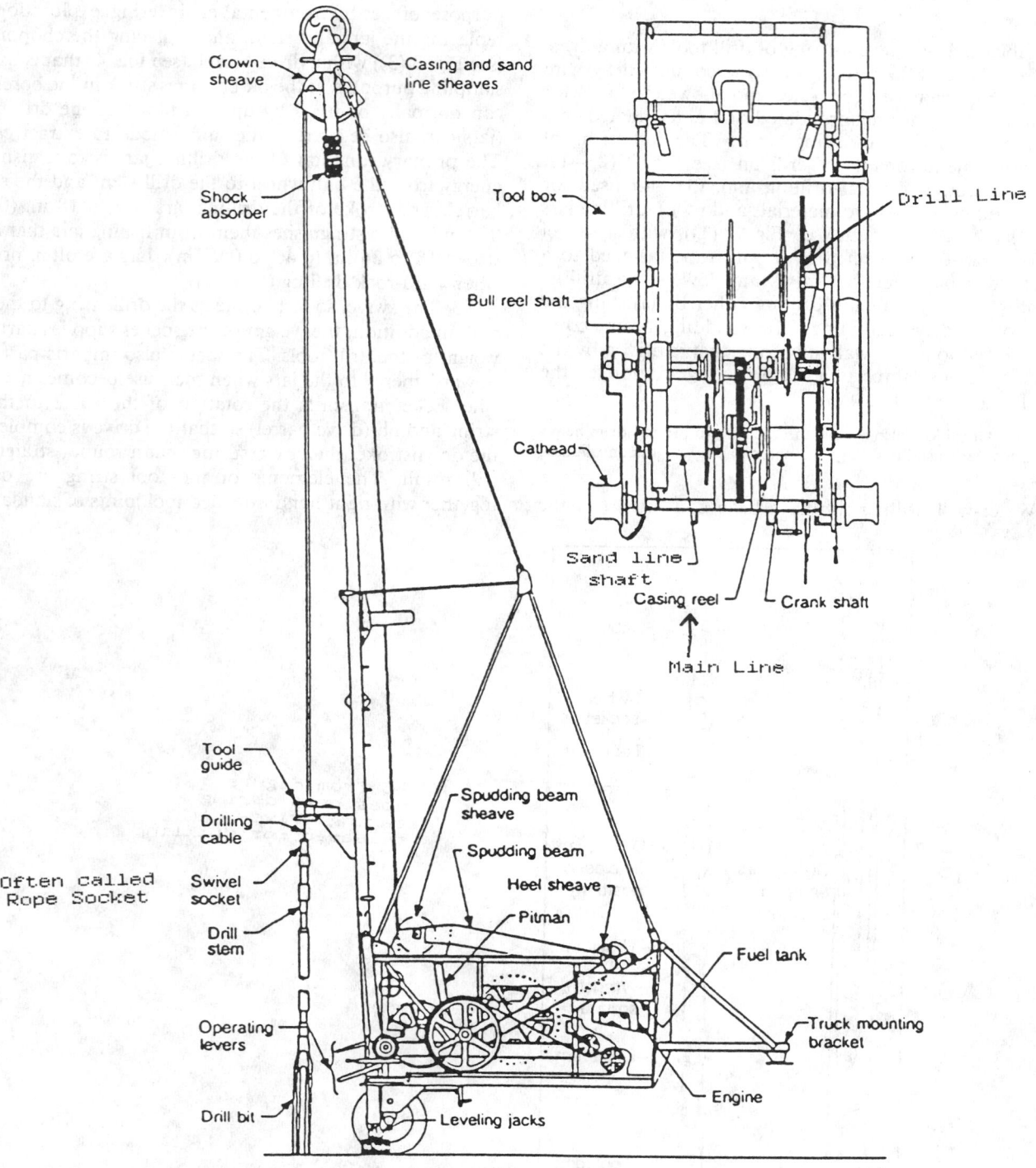

Crown
sheave

Casing and sand
line sheaves

Shock
absorber

Tool box

Drill Line

Bull reel shaft

Cathead

Sand line
shaft

Casing reel

Crank shaft

Main Line

Tool
guide

Spudding beam
sheave

Often called
Rope Socket

Drilling
cable

Spudding beam

Swivel
socket

Heel sheave

Pitman

Drill
stem

Fuel tank

Operating
levers

Truck mounting
bracket

Drill bit

Leveling jacks

Engine

FIG. 3   Diagram of a Cable Tool Drilling System

(American Petroleum Institute) design and dimension (7).

5.4 The wire cable that carries and rotates the drilling tool is called the drill line. It is a ⅝-in. (16-mm) to 1-in. (25-mm) left-hand lay cable that twists the tool joint on each upward stroke to prevent it from unscrewing. The drill line is reeved over a crown sheave at the top of the mast, down to the spudding sheave on the walking beam, to the heel sheave, and then to the working-line side of the bull-reel (see Fig. 1). The stroke of the cable-tool rig should be controlled and

sufficient tension maintained on the wire cable to keep the jars open or extended when in operation (often referred to as "tight-line drilling"). Bull reels generally are set-up with a separator on the drum to provide a working-line and a storage-line side (7).

NOTE 5—The mast must be constructed safely to carry the required loads for drilling, sampling, and completion of boreholes of the diameter and depth for which the rig manufacturer specifies the equipment. To allow for contingencies it is recommended that the rated capacity of the

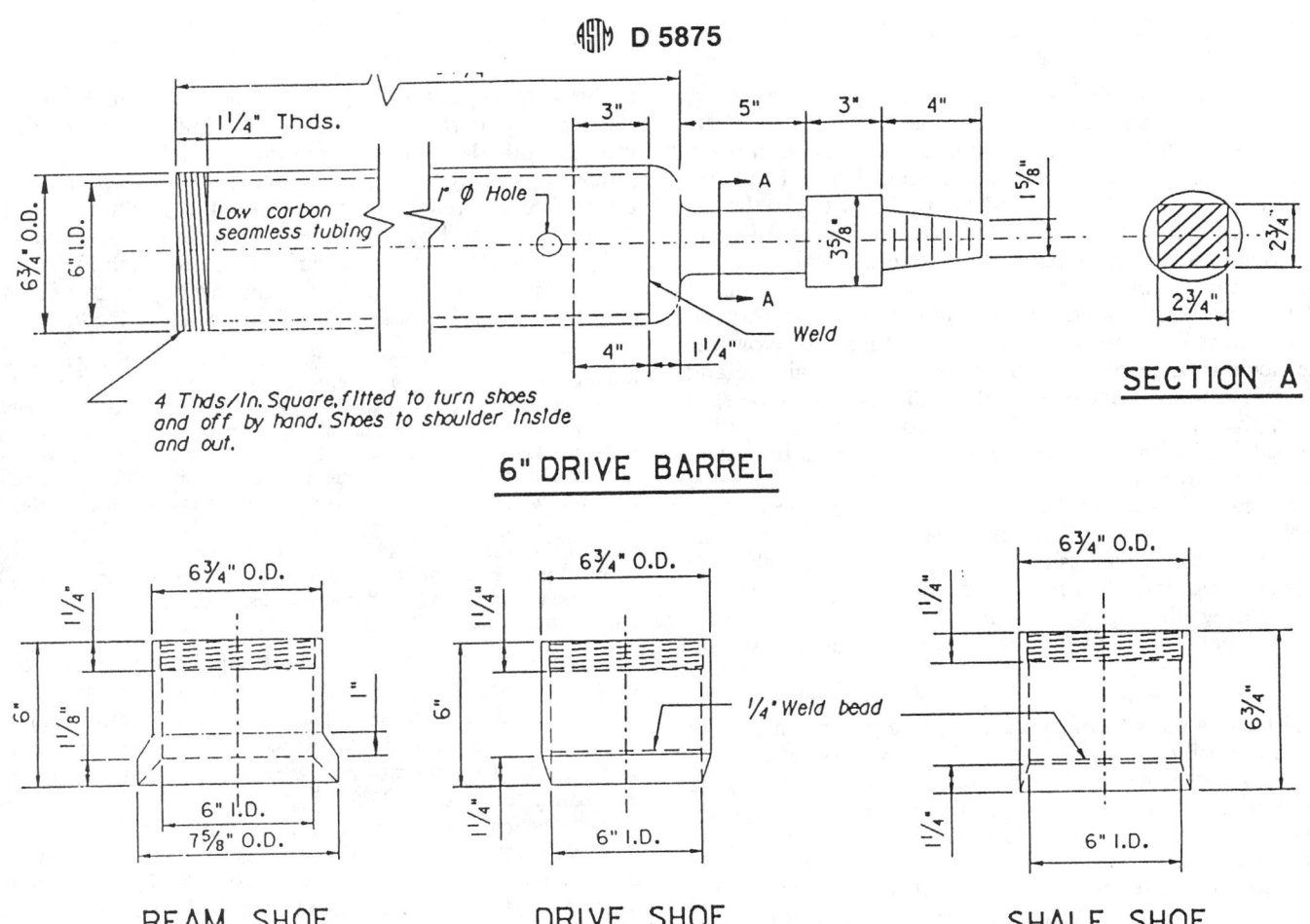

FIG. 4 Schematic Showing a 6-in. Drive Barrell and Three Shoe Configurations

mast should be at least twice the anticipated weight load or normal pulling load.

5.5 The characteristic up and down or spudding action of a cable-tool rig is imparted to the drill line and drilling tools by the walking beam. The walking beam pivots at one end while its out end, which carries the sheave for the drill line, is moved up and down by a single or double pitman connected to a crankshaft. The vertical stroke of the walking beam, and thus the drill tools, can be varied by adjusting the position of the pitman on the bull gear and the connection to the walking beam. The number of strokes per minute can be varied by changing the speed of the driveshaft. The bull gears are driven by a pinion mounted on a clutch. This clutch, the friction drive for sand line (on smaller cable tool rigs only), and the drive pinion for the drill-line reel are all mounted on the same shaft assembly.

5.6 Another drum, called a casing reel, frequently is added to the basic machine assembly. The casing reel is capable of exerting a powerful pull on a third cable, the casing- or main-line. This cable is used for handling heavy casing, tools, and pumps, or other heavy hoisting. It may be used to pull a string of casing when the cable is reeved with blocks to make two-, three-, or four-part lines (7).

5.7 Another commonly used hoisting device on a cable-tool rig is called a cathead. Use of this drum requires that a heavy line of manila rope be carried on a separate sheave at the top of the derrick. This line may be used for handling light loads at shallow depths (usually 10 ft (3 m) or less) and alternately lifting and dropping tools such as a drive block or bumper, spears, heads for driving casing, and individual lengths of casing so they may be stood on end and joined to the last piece in the ground (7). The cathead and line is often used to shake the sample from large-diameter drive barrels. Should standard-penetration tests be required for specific geoenvironmental studies, standard rotary drill rods and drop hammer can be manipulated using the cathead and line.

5.8 Depending on the length of the drive barrels, the drive of the sampling is usually 2 ft (6 m). A schematic showing a 6-in. drive barrel and three shoe configurations is shown in Fig. 4. Prior to any drilling all tools must be measured and the measurements recorded. Drive samples are usually disturbed. Therefore, laboratory testing is normally limited to obtaining only Atterberg Limits, mechanical analysis or chemical analysis of the disturbed samples. Poly(methyl methacrylate) or plastic liners can be inserted within the drive barrel, and the complete sample can be reexamined in the laboratory. In addition, sampling with the drive barrel provides the user with a complete geological sequence and field classification of the sampled materials at the time of drilling. When cohesionless materials are reached, and if sampling is required, the recovery is best while sampling is conducted from inside the casing. Attempts to sample below the bottom of the casing, especially when these materials are below the water table or the materials totally saturated, are usually futile. (When such conditions are observed the drive barrel acts analogous to a piston and creates a suction when the bottom of the casing has been knocked out. The drive barrel will "suck" in the sand and can pull it nearly to the top of the casing; subsequent cleanout is very slow and expensive.) When using a drive barrel in cohesionless soil, casing is required. Cohesionless materials may also be sampled using a dart-valve bailer, sand-pump bailer, or a flat-bottomed bailer attached to the sand line when sampling below the water level in the drill hole. Other bailers are available and this guide is not restrictive as to the trade name or type(s) of bailers that may be used. The unconsolidated samples obtained by bailing are usually dumped into a bucket, a small barrel (drum), or onto a piece of plywood and the water decanted in order that the sample can be placed in a sample container or bag. The use of drilling fluids (bentonite, etc.) as a substitute for casing may be used, but usually with limited success. If drilling fluids or other additives are used in lieu of casing, the type(s), compositions, quantities, depths, and other pertinent information to the occurrence(s) should be fully documented.

5.9 Standard body diameters and jaws showing nominal stroke lengths are shown in Table 1. The jars are normally run on the drilling string between the swivel socket and the drill stem except where used for fishing out lost tools. Jars should never be included in the drill string when starting the borehole (8). The function of the jars is to create a jarring action when needed to loosen tools from a formation in which they have a tendency to stick while drilling. The stroke of the rig should be controlled and sufficient tension maintained on the wire to keep the jars extended when in operation. A loose drilling line permits the jars to open and close on each stroke and results in faster wear and metal fatigue on the jars.

6. Drilling Procedures

6.1 As a prelude to and throughout the drilling process stabilize the cable-tool drill rig and raise the mast. If air-monitoring operations are performed the prevalent wind direction relative to the exhaust from the drill rig should be considered. Connect the drive barrel to the drill stem and jars. Reference the top of the drill hole to the survey stake that reflects its elevation and location. Put the drill in motion and begin the drive sampling. Drill the upper 2 to 8 ft (0.6 to 2.4 m) rather slowly until the top of the drive barrel is below the top of the ground surface. Perform drive sampling on a continuous basis and at 2-ft (0.6-m) intervals. Select samples

TABLE 1 Weldless Jars for Cable-Tool Drilling

Body Diameter, in.	API Joint Size	Weight 4.5 in. Stroke	Weight 8 in. Stroke	Weight 12 in. Stroke	Weight 18 in. Stroke	Weight 24 in. Stroke	Weight 36 in. Stroke
3.25	2¼	100	103	107	114	121	135
3.625	2⅝	140	145	149	158	167	185
4.375	3	237	248	256	275	294	332
4.75	3¼	285	296	312	339	366	420
5.5	3.5 to 3.75	400	421	442	474	506	570
6.5	4	605	630	650	685	720	790
6.5	4.25 to 4.5	670	694	718	754	790	862
7	5	770	808	840	890	960	1080
7	6	960	998	...	...	...	...

at the direction of the design team, and perform the appropriate testing. Perform drilling and sampling to a predetermined depth. The reference datum is established (usually by a licensed surveyor) then the elevation (or location data) is transferred to a reference at the top of the drill hole or to the "deck" of the drill rig. Document these data. Most cable-tool drill rigs do not have a drill platform from which the drilling operations are performed, and the drilling and reference elevations are referenced to land-surface elevation.

NOTE 6—The drill rig, drive barrel, drill stem, cable(s), jars, and all drilling and sampling components above the drilling axis should be decontaminated according to Practice D 5088 prior to commencing drilling and sampling operations. The user of the cable-tool drilling method for geoenvironmental work should be aware that the decontamination of this type of equipment could be difficult to accomplish perhaps, making it a less desirable method to employ for a particular study.

NOTE 7—The user must locate and advise the drill crew about the location of overhead utilities as well as underground utility lines and a safety plan formulated in order to establish a safe drilling and working distance from them.

6.2 In conjunction with drive sampling it is often necessary to first set casing. To better facilitate obtaining "uncontaminated" drive samples, casing can first be driven to seal off "contaminated zones" prior to hole advancement and further sampling operations. Decontaminate all casing used according to Practice D 5088 prior to use. Set the appropriate size casing (nominal 5-ft lengths (1.5-m)) with the drive shoe attached to the bottom of the casing. (Remove the drive barrel prior to driving any casing so that it does not unscrew and fall to the bottom of the drill hole.) Set a heavy, machined drive head over or threaded on the top of the casing, and the stem goes inside the casing. Attach the drive clamp to the jars, and the percussion of the drill stem drives the casing to the required depth. When the casing has reached the required depth, remove the drive, reattach the drive barrel, and resume drilling and sampling. Repeat this procedure until the predetermined depth is reached. The driller marks the drive barrel and drill tools, and the cable in 2-ft (0.6-m) increments, so that the drilling depths may be recorded. At the discretion of the responsible person on site, measure the bottom of the hole with a tape to verify the drillers' depth. For depth-measurement reliability, the degree of precision required and measuring tool should be considered and compatible. Document this information.

6.2.1 Drilling with cable-tool methods for geoenvironmental exploration and the installation of subsurface water-quality monitoring devices require that accurate measurements be made and documented using a steel surveyors' chain, marked in 0.01-ft (or 0.003 mm) increments. Tenths of a foot (or millimetres) may be measured using a typical engineering rule.

6.3 When drilling by the cable-tool method, the rock is broken up and pulverized by the percussive drill action of the bit. However, when drilling unconsolidated materials the bit primarily loosens the materials. In both instances, the reciprocating action of the tools mixes the crushed or loosened particles with water to form a slurry or sludge at the bottom of the borehole. If water is not present in the borehole, add water to form the slurry. As slurry accumulation increases as the drilling progresses it eventually reduces the impact of the drill tools. It is then removed from the borehole either by a sand pump or bailer (7). If water is added to form the slurry, document the source of the water, quantity of water added, chemical makeup of the water, and depth added. If necessary, cuttings are decanted, in order to be examined by the responsible person on site. Set and pull casing according to 6.2. The sand line or main line is often used to expedite pulling the casing. Decontaminate all tools used in this drilling procedure according to Practice D 5088.

7. Borehole Completion and the Installation of Subsurface Monitoring Devices

7.1 Monitor ground-water levels periodically and document the water levels both during drilling and after the predetermined depth has been reached. If ground water is not present or if the measurement of the ground-water level is of doubtful reliability, document that also.

7.2 Subsurface water-quality monitoring devices are generally installed in boreholes drilled by cable-tool rigs using a four-step procedure. The four steps are: (1) drilling without sampling, (2) temporarily leaving the casing in the drill hole to support the sidewalls after the predetermined sampling depth has been reached, (3) inserting the monitoring device to the desired depth, and (4) the simultaneous extraction of the casing (using the cathead and a drive block or drive clamp to bump the casing back) with addition of completion materials (that is, filter pack, bentonite pellets, bentonite granules and chips, annular seals and grout added either above or below the bottom of the casing).

7.3 *Other Completion Methods*—Depending on the purpose of the investigation(s) it may be necessary to perform special completion(s) with protective casing(s) or other method(s) of backfilling. An example of other completions is the seismic crosshole test (see Test Method D 4428) that requires grouted PVC casings. These completions are also performed using those methods in 7.2. Document the completion method(s) used.

8. Field Report and Project Control

8.1 The field report should include all information and identified as necessary and pertinent to the needs of the exploration program. Information normally required for the project include: exploration type and execution, drilling equipment and methods used, subsurface conditions encountered, groundwater conditions, sampling events, and type of installations made in the borehole(s) including

well-completion techniques used.

8.2 Additional information should be considered as follows:

8.2.1 *Drilling Methods:*

8.2.1.1 Description and documentation of the rig/system including weight of drill stem and jars, sizes of drive barrels used and size of casing set, if any. Record when drive barrels are changed to smaller diameters as well as the depth obtained for each drive barrel and the casing set. Document reasons why changes were made.

8.2.1.2 Note and document type, quantities, and locations in the drill hole of use of additives such as water or drilling fluids added.

8.2.1.3 Description of cuttings.

8.2.1.4 Descriptions of drilling conditions and general ease of drilling as related to subsurface materials encountered should be noted and documented.

8.2.2 *Sampling*—Document conditions of the bottom of the drill hole prior to sampling and report any slough or cuttings present in the recovered sample.

8.2.3 *In-situ Testing:*

8.2.3.1 For devices inserted below the bottom of the drill hole document the depths below the bottom of the hole and any unusual conditions during testing.

8.2.3.2 For devices testing or seating at the borehole wall, document any unusual conditions of the borehole wall such as inability to seat borehole packers.

8.2.4 *Installations*—Document a description of well-completion materials and placement methods, approximate volumes placed, depth intervals of placement, methods of confirming placement, and areas of difficulty of material placement or unusual occurrences.

8.2.5 *Site Conditions:*

8.2.5.1 *Site Description*—Description of the site and any unusual circumstances,

8.2.5.2 *Personnel*—Documentation of all personnel at the site during the drilling process; driller, helpers, geologist or logger, engineer, and other monitors or visitors,

8.2.5.3 Weather conditions during drilling, and

8.2.5.4 Working hours, operating times, breakdown times, and sampling times. Report any unusual occurrences that may have happened during the investigation.

9. Keywords

9.1 cable-tool drilling; drilling method; geoenvironmental exploration; ground water; vadose zone

REFERENCES

(1) *Drillers Handbook*, Ruda, T. C., and Bosscher, P. J., editors, National Drilling Contractors Association, 3008 Millwood Avenue, Columbia, South Carolina 29205, June 1990.

(2) Lehr, Jay, Hurlburt, Scott, Gallahger, Betsy, and Voytek, John, *Design and Construction of Water Wells*, National Water Well Association; published by Van Nostrand Reinhold, New York.

(3) American Petroleum Institute, *API Specifications for Casing, Tubing, and Drill Pipe, API Spec 5A*, American Petroleum Institute, Dallas, TX, 1978.

(4) *DCDMA Technical Manual*, Drilling Equipment Manufacturers Association, 3008 Millwood Avenue, Columbia, South Carolina 29205, 1991.

(5) Aller, L., et al., *Handbook of Suggested Practices for the Design and Installation of Ground-Water Monitoring Wells.* EPA/600/4-89/034, NWWA/EPA Series, National Water Well Association, Dublin, OH, 1989.

(6) Roscoe Moss Company, *Handbook of Ground Water Development*, Roscoe Moss Company, Los Angeles, CA, John Wiley and Sons, Inc., New York, NY, 1990.

(7) Driscoll, F. G., *Groundwater and Wells*, Johnson Filtration Systems, Second Edition, St. Paul, MN, 1989.

(8) Gordon, Raymond W., *Water Well Drilling With Cable Tools*, Bucyrus-Erie Co., S. Milwaukee, WI, 1958.

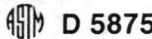

Standard Guide for
Use of Direct Rotary Wireline Casing Advancement Drilling Methods for Geoenvironmental Exploration and Installation of Subsurface Water-Quality Monitoring Devices[1]

This standard is issued under the fixed designation D 5876; the number immediately following the designation indicates the year of original adoption or, in the case of revision, the year of last revision. A number in parentheses indicates the year of last reapproval. A superscript epsilon (ε) indicates an editorial change since the last revision or reapproval.

1. Scope

1.1 This guide covers how direct (straight) wireline rotary casing advancement drilling and sampling procedures may be used for geoenvironmental exploration and installation of subsurface water-quality monitoring devices.

Note 1—The term "direct" with respect to the rotary drilling method of this guide indicates that a water-based drilling fluid or air is injected through a drill-rod column to rotating bit(s) or coring bit. The fluid or air cools the bit(s) and transports cuttings to the surface in the annulus between the drill rod column and the borehole wall.

Note 2—This guide does not include all of the procedures for fluid rotary systems which are addressed in a separate guide, Guide D 5783.

1.2 The term "casing advancement" is sometimes used to describe rotary wireline drilling because at any time, the center pilot bit or core barrel assemblies may be removed and the large inside diameter drill rods can act as a temporary casing for testing or installation of monitoring devices. This guide addresses casing-advancement equipment in which the drill rod (casing) is advanced by rotary force applied to the bit with application of static downforce to aid in the cutting process.

1.3 This guide includes several forms of rotary wireline drilling configurations. General borehole advancement may be performed without sampling by using a pilot roller cone or drag bit until the desired depth is reached. Alternately, the material may be continuously or incrementally sampled by replacing the pilot bit with a core-barrel assembly designed for coring either rock or soil. Rock coring should be performed in accordance with Practice D 2113.

1.4 The values stated in both inch-pound and SI units are to be regarded separately as the standard. The values given in parentheses are for information only.

1.5 Direct rotary wireline drilling methods for geoenvironmental exploration will often involve safety planning, administration, and documentation. This guide does not purport to specifically address exploration and site safety.

1.6 *This standard does not purport to address all of the safety concerns, if any, associated with its use. It is the responsibility of the user of this standard to establish appropriate safety and health practices and determine the applicability of regulatory limitations prior to use.*

2. Referenced Documents

2.1 *ASTM Standards:*

D 420 Guide to Site Characterization for Engineering, Design, and Construction Purposes[2]

D 653 Terminology Relating to Soil, Rock, and Contained Fluids[2]

D 1452 Practice for Soil Investigation and Sampling by Auger Borings[2]

D 1586 Test Method for Penetration Test and Split-Barrel Sampling of Soils[2]

D 1587 Practice for Thin-Walled Tube Geotechnical Sampling of Soils[2]

D 2113 Practice for Diamond Core Drilling for Site Investigation[2]

D 2488 Practice for Description and Identification of Soils (Visual-Manual Procedure)[2]

D 3550 Practice for Ring-Lined Barrel Sampling of Soils[2]

D 4220 Practice for Preserving and Transporting Soil Samples[2]

D 4428/D 4428M Test Methods for Crosshole Seismic Testing[2]

D 4630 Test Method for Determining Transmissivity and Storage Coefficient of Low-Permeability Rocks by In Situ Measurements Using the Constant Head Injection Test[2]

D 4631 Test Method for Determining Transmissivity and Storativity of Low-Permeability Rocks by In Situ Measurements Using the Pressure Pulse Technique[2]

D 4700 Guide for Soil Sampling from the Vadose Zone[2]

D 4750 Test Method for Determining Subsurface Liquid Levels in a Borehole or Monitoring Well (Observation Well)[2]

D 5079 Practices for Preserving and Transporting Rock Core Samples[3]

D 5088 Practice for Decontamination of Field Equipment Used at NonRadioactive Waste Sites[3]

D 5092 Practice for Design and Installation of Ground Water Monitoring Wells in Aquifers[3]

D 5099 Practice for Development of Ground Water Monitoring Wells in Aquifers[3]

D 5254 Practice for Minimum Set of Data Elements to Identify a Ground-Water Site[3]

D 5434 Guide for Field Logging of Subsurface Explorations of Soil and Rock[3]

---

[1] This guide is under the jurisdiction of ASTM Committee D-18 on Soil and Rock and is the direct responsibility of Subcommittee D18.21 on Ground Water and Vadose Zone Investigations.
Current edition approved Dec. 10, 1995. Published February 1996.

[2] *Annual Book of ASTM Standards*, Vol 04.08.
[3] *Annual Book of ASTM Standards*, Vol 04.09.

SgtOCR

D 5730 Guide to Site Characterization for Environmental Purposes with Emphasis on Soil, Rock, the Vadose Zone, and Ground Water[2]

D 5781 Guide for Use of Dual-Wall Reverse-Circulation Drilling for Geoenvironmental Exploration and Installation of Subsurface Water-Quality Monitoring Devices[2]

D 5782 Guide for Use of Direct Air-Rotary Drilling for Geoenvironmental Exploration and Installation of Subsurface Water-Quality Monitoring Devices[2]

D 5783 Guide for Use of Direct Rotary Drilling with Water-Based Drilling Fluid for Geoenvironmental Exploration and Installation of Subsurface Water-Quality Monitoring Devices[2]

D 5784 Guide for Use of Hollow-Stem Augers for Geoenvironmental Exploration and Installation of Subsurface Water-Quality Monitoring Devices[2]

D 5876 Guide for the Use of Direct Rotary Wireline Casing Advancement Drilling Methods for Geoenvironmental Exploration and the Installation of Subsurface Water-Quality Monitoring Devices[3]

3. Terminology

3.1 *Definitions*—Terminology used within this guide is in accordance with Terminology D 653 with the addition of the following:

3.2 *Definitions of Terms Specific to This Standard:*

3.2.1 *bentonite*—the common name for drilling fluid additives and well-construction products consisting mostly of naturally occurring montmorillonite. Some bentonite products have chemical additives that may affect water-quality analyses.

3.2.2 *bentonite pellets*—roughly spherical- or disk-shaped units of compressed bentonite powder (some pellet manufacturers coat the bentonite with chemicals that may affect the water-quality analysis).

3.2.3 *cleanout depth*—the depth to which the end of the drill string (bit or core barrel cutting end) has reached after an interval of cutting. The cleanout depth (or drilled depth as it is referred to after cleaning out of any sloughed material in the bottom of the borehole) is usually recorded to the nearest 0.1 ft (0.03 m).

3.2.4 *coefficient of uniformity*—C_u (D), the ratio D_{60}/D_{10}, where D_{60} is the particle diameter corresponding to 60 % finer on the cumulative particle-size distribution curve, and D_{10} is the particle diameter corresponding to 10 % finer on the cumulative particle-size distribution curve.

3.2.5 *drill hole*—a cylindrical hole advanced into the subsurface by mechanical means. Also known as a borehole or boring.

3.2.6 *drill string*—the complete rotary drilling assembly under rotation including bit, sampler/core barrel, drill rods, and connector assemblies (subs). The total length of this assembly is used to determine drilling depth by referencing the position of the top of the string to a datum near the ground surface.

3.2.7 *filter pack*—also known as a gravel pack or primary filter pack in the practice of monitoring-well installations. The gravel pack is usually granular material, having selected grain-size characteristics, that is placed between a monitoring device and the borehole wall. The basic purpose of the filter pack or gravel envelope is to act as: a nonclogging filter when

the aquifer is not suited to natural development or, as a formation stabilizer when the aquifer is suitable for natural development.

3.2.7.1 *Discussion*—Under most circumstances, a clean quartz sand or gravel should be used. In some cases, a prepacked screen may be used.

3.2.8 *head space*—on a double- or triple-tube wireline core barrel it is the spacing adjustment made between the pilot-shoe leading edge and the inner kerf of the outer-tube cutting bit. Spacing should be about 1/16 in. or roughly, the thickness of a matchbook. (The head-space adjustment is made by removing the inner-barrel assembly, loosening the lock nut on the hanger-bearing shaft and either tightening or loosening the threaded shaft until the inner barrel is moved the necessary distance, up or down, to obtain the correct setting. Reassemble the inner- and outer-barrel assemblies, attach the barrel to the drill rod or a wireline and suspend vertically allowing the inner-barrel assembly to hang freely inside the outer barrel on the inner hanger-bearing assembly. Check the head space. It is imperative that the adjustment is correct to ensure that the inner barrel is free to rotate without contacting the outer barrel. If incorrectly adjusted, the inner barrel will "hang up" and rotate with the outer barrel as the core is being cut. This will cause the core to break and block entry of core into the inner barrel.)

3.2.9 *grout shoe*—a drillable "plug" containing a check valve that is positioned within the lowermost section of a casing column. Grout is injected through the check valve to fill the annular space between the casing and the borehole wall or another casing.

3.2.9.1 *Discussion*—The composition of the drillable "plug" should be known and documented.

3.2.10 *grout packer*—an inflatable or expandable annular plug that is attached to a tremie pipe, usually positioned immediately above the discharge end of the pipe.

3.2.11 *intermittent sampling devices*—usually barrel-type samplers that are driven or pushed below the bottom of a borehole with drill rods or with a wireline system to lower, drive, and retrieve the sampler following completion of an increment of drilling. The user is referred to the following standards relating to suggested sampling methods and procedures: Practice D 1452, Test Method D 1586, Practice D 3550, and Practice D 1587.

3.2.12 *in-situ testing devices*—sensors or probes, used for obtaining mechanical- or chemical-test data, that are typically pushed, rotated, or driven below the bottom of a borehole following completion of an increment of drilling. However, some in-situ testing devices (such as electronic pressure transducers, gas-lift samplers, tensiometers, and so forth) may require lowering and setting of the device(s) in preexisting boreholes by means of a suspension line or a string of lowering rods or pipes. Centralizers may be required to correctly position the device(s) in the borehole.

3.2.13 *lead distance*—the mechanically adjusted length or distance that the inner-barrel cutting shoe is set to extend beyond the outer core-barrel cutting bit in order to minimize possible core-erosion damage that can be caused by the circulating drilling-fluid media. Lead distance is checked by vertically suspending the entire core-barrel assembly from a wireline or from a section of drill rod so that the inner-barrel can hang freely from the upper inner-barrel swivel assembly

The cutting shoe extension below the outer core-barrel cutting bit can then be checked. The "stiffer" or more competent the formation to be cored, the less the extension of the inner-barrel cutting shoe is necessary to avoid core erosion.

3.2.14 *overshot*—a latching mechanism located at the end of the hoisting line. It is specially designed to latch onto or release pilot bit or core-barrel assemblies.

3.2.15 *pilot bit assembly*—design to lock into the end section of drill rod for drilling without sampling. The pilot bit can be either drag, roller cone, or diamond plug types. The bit can be set to protrude from the rod coring bit depending on formation conditions.

3.2.16 *sub*—a substitute or adaptor used to connect from one size or type of threaded drill rod or tool connection to another.

3.2.17 *subsurface water-quality monitoring device*—an instrument placed below ground surface to obtain a sample for analyses of the chemical, biological, or radiological characteristics of subsurface pore water or to make in-situ measurements.

3.2.18 *wireline drilling*—a rotary drilling process which uses special enlarged inside diameter drilling rods with special latching pilot bits or core barrels which are raised or lowered inside the rods with a wireline and overshot latching mechanism.

4. Summary of Practice

4.1 Wireline drilling is a rotary drilling process that uses special enlarged inside diameter drilling rods with special latching pilot bits or core barrels which are raised or lowered inside the rods with a wireline and overshot latching mechanism. The bottom section of rod has either a diamond or carbide coring bit at the end and is specially machined to accommodate latching of either pilot bits or core barrels. The overshot mechanism is designed to latch and unlatch bit or barrel assemblies. Bit cutting is accomplished by application of the combination rotary and static down forces to the bit. General drill-hole advancement may be performed without sampling by using either a pilot roller cone or drag bit until the desired depth is reached. Alternately, the material may be continuously or incrementally sampled by replacing the pilot bit with a core-barrel assembly designed for coring either rock or soil.

4.2 The pilot bit or core barrel can be inserted or removed at any time during the drilling process and the large inside diameter rods can act as a temporary casing for testing or installation of monitoring devices.

5. Significance and Use

5.1 Wireline casing advancement may be used in support of geoenvironmental exploration and for installation of subsurface monitoring devices in both unconsolidated and consolidated materials. Use of direct-rotary wireline casing-advancement drilling methods with fluids are applicable to a wide variety of consolidated or unconsolidated materials as long as fluid circulation can be maintained. Wireline casing-advancement drilling offers the advantages of high drilling-penetration rates in a wide variety of materials with the added benefit of the large-diameter drilling rod serving as protective casing. Wireline coring does not require tripping

in and out of the hole each time a core is obtained. The drill rods need only be removed when the coring bit is worn or damaged or if the inner core barrel becomes stuck in the outer barrel.

5.1.1 Wireline casing advancers may be adapted for use with circulating air under pressure for sampling water-sensitive materials where fluid exposure may alter the core or in cavernous materials or lost circulation occurs (1, 2).[4] Several advantages of using the air-rotary drilling method over other methods may include the ability to drill rather rapidly through consolidated materials and, in many instances, not require the introduction of drilling fluids to the borehole. Air-rotary drilling techniques are usually employed to advance the borehole when water-sensitive materials (that is, friable sandstones or collapsible soils) may preclude use of water-based rotary-drilling methods. Some disadvantages to air-rotary drilling may include poor borehole integrity in unconsolidated materials when casing is not used and the possible volatilization of contaminants and air-borne dust. Air drilling may not be satisfactory in unconsolidated or cohesionless soils, or both, when drilling below the groundwater table. In some instances, water or foam additives, or both, may be injected into the air stream to improve cuttings-lifting capacity and cuttings return. Use of water or other additives, or both, should be documented. The use of air under high pressures may cause fracturing of the formation materials or extreme erosion of the borehole if drilling pressures and techniques are not carefully maintained and monitored. If borehole damage becomes apparent, other drilling method(s) should be considered.

5.1.2 When air is used as the circulating fluid, the user should consult Refs (1, 2) and Guide D 5782.

5.2 The application of wireline casing advancement to geoenvironmental exploration may involve sampling of ground water, soil, or rock; or in-situ or pore-fluid testing; or installation of other casings for subsequent drilling activities in unconsolidated or consolidated materials.

5.3 The wireline drill rod can act as a temporary casing and may be used to facilitate the installation of a monitoring device. The monitoring device may be installed as the drill rod is removed from the drill hole.

NOTE 3—The user may install a monitoring device within the same drill hole wherein sampling or in-situ testing was performed.

5.4 Wireline casing-advancement rotary-drilling methods use fluid or air circulation to lubricate cutting bits and for removal of drill cuttings. In many cases, additives are added to improve circulation, cuttings return, borehole wall stabilization, and sealing of the borehole wall from fluid loss. The use of fluid or air under high pressures may allow for damage to formation materials by fracturing or excessive erosion if drilling conditions are not carefully maintained and monitored. If undesirable formation damage is occurring or evident, other drilling method(s) should be considered.

6. Apparatus

6.1 *General*—Direct rotary wireline casing advancement systems and procedures used for geoenvironmental explora-

---

[4] The boldface numbers given in parentheses refer to a list of references at the end of the text.

tion and subsurface water-quality monitoring device installations include direct air or mud-rotary drilling using wireline drill rods. The wireline drill rod has a large inside diameter and is equipped with either a wireline-retrievable center pilot bit for general hole advancement, or a rock- or soil-core barrel for sampling the borehole as it is advanced. Figure 1(a through d) shows basic schematics of the components of a wireline-drilling assembly using a rock core-barrel assembly to sample formations as drilling progresses.

6.2 The basic mechanical components of wireline casing-advancement drilling systems include the drill rig with either hollow spindle or top-head drive, drill rods, coring or casing bits, overshot assembly, pilot bit, or core barrel. Water-based fluid circulation systems require drill fluid, mud pit, suction hose, drill fluid circulation pump, pressure hose, and swivel. Air circulation systems require an air compressor, dust collector, air cleaning device, pressure hose, and swivel.

6.2.1 *Drill Rig*—Most top-head drive or hollow-spindle drills are suitable for performing rotary wireline drilling. Rock-coring drills with smooth hydraulic operation and high RPM capability are desirable for rock-coring operations.

Rotary table and kelly drills generally are not acceptable for wireline-drilling use due to difficulty or inability to raise and lower wireline assemblies. The drill unit should have the ability to rotate a drill-rod column and apply a controllable axial force on the drill bit appropriate to the drilling and sampling requirements and the geologic conditions.

6.2.2 *Drill Rods*, transfer force and rotation from the drill rig to the bit or core barrel. When rotary drilling is stopped, the large inside diameter wireline drill rod acts can as casing, that is, by preventing against hole collapse—to allow for testing or installation of monitoring devices for hole protection. Drill rods conduct drilling fluid to the bit or core barrel. Individual drill rods (that is, drill stem, drill string, drill pipe) should be straight so they do not contribute to excessive vibrations or "whipping" of the drill-rod column. All threaded connections should be in good repair and not leak significantly at the internal fluid pressure required for drilling. Drill rods should be made up securely by wrench tightening at the threaded joint(s) at all times to prevent rod damage.

6.2.2.1 Wireline drill rod dimensions are not fully stan-

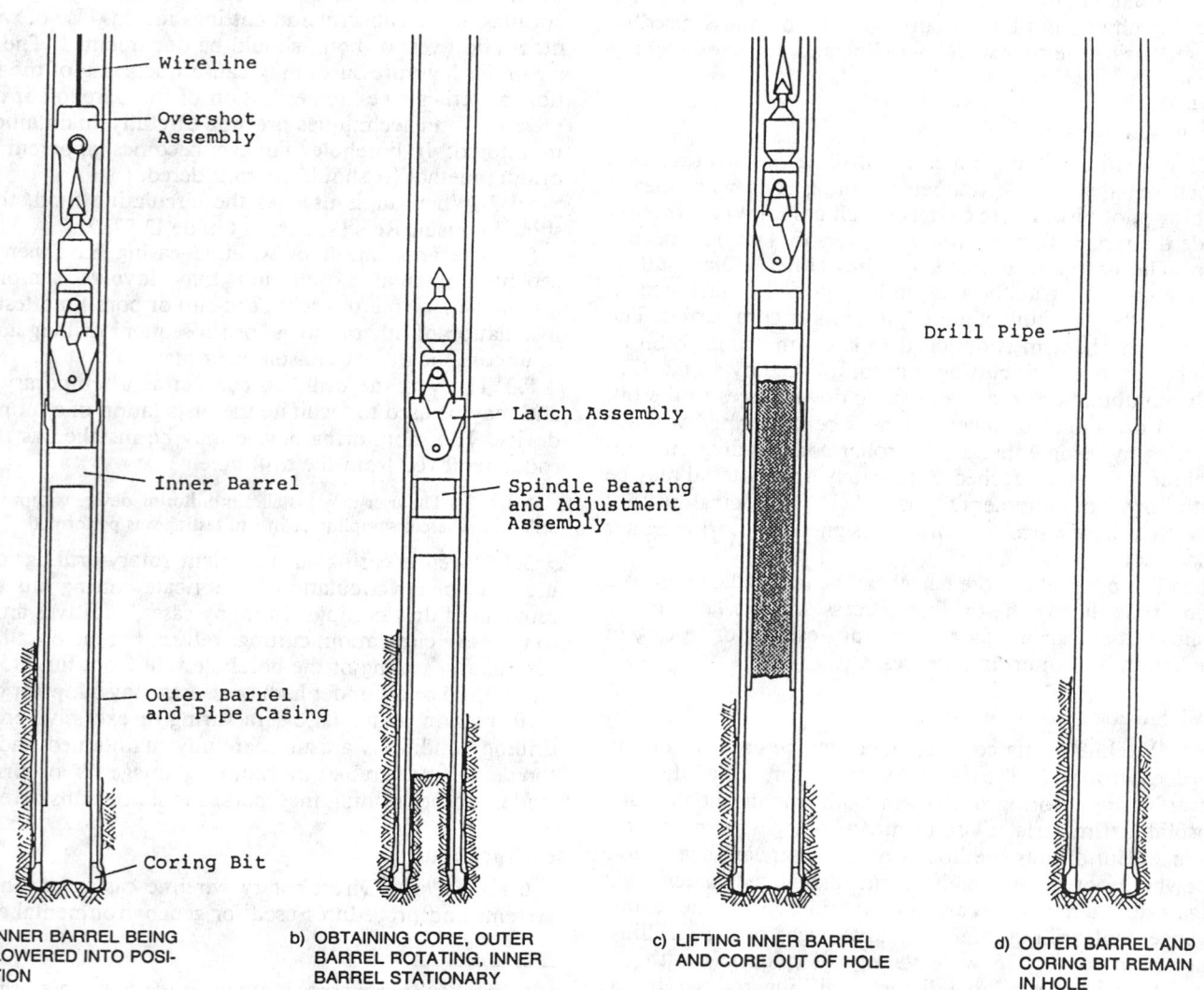

a) INNER BARREL BEING LOWERED INTO POSITION

b) OBTAINING CORE, OUTER BARREL ROTATING, INNER BARREL STATIONARY

c) LIFTING INNER BARREL AND CORE OUT OF HOLE

d) OUTER BARREL AND CORING BIT REMAIN IN HOLE

FIG. 1 Schematics of Wire-Line Drilling Assembly

ardized. The available sizes depend on the manufacturing sources (3). General hole diameter available from most manufacturers follows the Diamond Core Drill Manufacturers Association (DCDMA) size conventions of A (48-mm), B (59.9-mm), N (75.7-mm), H (96.3-mm), and P (122.6-mm) sizes (4). Inside diameter varies depending on the manufacturer.

NOTE 4—Sizes of casings, casing bits, drill rods, and core barrels are standardized by the DCDMA and the American Petroleum Institute (API). Refer to Ref (4) for available sizes and capacities of common use drilling equipments for soil and rock exploration.

6.2.2.2 The wireline lead rod contains shoulders for latching of pilot bits or core barrels. All wireline lead-rod sections are equipped with coring or casing bits. There are many configurations of wireline drilling equipment possible depending on the manufacturer. With rock-coring systems, the wireline lead rod is equipped with a reaming shell to insure circulation and act as a stabilizer. Some multipurpose systems allow for latching of pilot bit, rock-core barrel, or soil-core barrel to the lead wireline rod section. In most coring operations the lead wireline rod is considered to be the outer barrel in a double- or triple-tube core barrel design (see Practice D 2113). The bit is referred to as a core bit.

6.2.2.3 The wireline rod size is usually selected to provide a drill hole of sufficient diameter for the required sampling, testing, or insertion of instrumentation-device components, such as, the screened intake and filter pack and installation devices such as a tremie pipe. The inside diameter of the wireline rod should permit easy insertion and retraction of a sampler or a pipe with a sufficiently large inside diameter to accommodate the placement of completion materials adjacent to the screened intake and riser of an instrumentation device. Coring bits are selected to provide adequate required hole or core diameter, or both. Selection of protective casings, bits, and core barrels is made by considering size requirements listed above combined with: annulus circulation capabilities of drill rod used, and the need for tapering from larger to smaller diameters casings ("nesting" of casing) if difficult drilling conditions occur (lost circulation, zones of contamination, and so forth) requiring these problem zones in the borehole to be cased off.

NOTE 5—Drill rods usually require lubricants on the threads to allow easy threading and unthreading of the drill-rod tool joints. Some lubricants have organic or metallic constituents, or both, that could be interpreted as contaminants if detected in a sample. Various lubricants are available that have components of known chemistry. The effect of drill-rod lubricants on chemical analyses of samples should be considered and documented when using direct-rotary drilling. The same consideration and documentation should be given to lubricants used with water swivels, hoisting swivels, or other devices used near the drilling axis.

6.2.3 The casing bit or core bit provides the material cutting capability. In coring operations, the bit is referred to as a core bit. Rock coring should be performed in accordance with Practice D 2113. Soil sampling or coring methods, some of which are listed in 2.2, can also be used to obtain samples. When drilling in a casing-advancement mode using a pilot bit, without sampling, the bit is referred to as a casing bit.

6.2.3.1 Numerous coring or casing bits can be selected depending on the properties of the formation to be drilled or cored. Since it is undesirable to remove the outer bit, design

of this bit is extremely important. When coring, particularly in unconsolidated materials, it is important that the bit cuts the material and not merely tears it and pushes it aside. Some bit types successfully used include either carbide inserts coring bits or diamond core bits. In rock-coring operations the kerf design inner gage of bit, matrix cutting capacity, and location of drilling-fluid circulation ports are important. The inner gage of the bit can be selected so that the core is slightly undercut, thereby allowing it to move freely up the inner tube and not cause core blockage. If the core is over cut and air is present in the barrel, consideration should be given to possible alteration of the core that may occur during subsequent sealing and storage of the core obtained. It is beyond the scope of this guide to recommend bit styles. Bit selection can be aided by review of literature (1, 2, 3, 5, 6) and consultation with manufacturers.

6.2.3.2 The dimensions of the coring or casing bits often control the maximum diameter of testing or sampling device that can be inserted through the wireline drill rods. As mentioned previously, the bit size is usually selected to provide a drill hole of sufficient diameter for required borehole sampling or testing to be accomplished or for insertion of instrumentation device components such as the screened intake, riser pipe, filter pack, and well-completion installation devices such as a tremie pipe in the borehole.

6.2.4 *Wireline Retrievable Pilot-Bit Assembly*, used when no borehole coring/sampling is desired. The assembly is equipped with a receiver for pickup by the overshot latching assembly. Several pilot-bit styles are available including roller cone and drag bit configurations. Bit selection can be aided by review of literature (1, 2, 3, 5, 6) and consultation with manufacturers.

NOTE 6—Bottom-discharge bits are those having drill-fluid circulation discharge ports in direct contact with the base of the hole. If these bits are used to drill loose cohesionless materials, jetting or excessive erosion of the test intervals could occur.

6.2.5 *Overshot*, a latching retrieval assembly that is lowered into the hole with a wireline hoist cable to either retrieve or lower core-barrel inner assemblies or bits equipped with an upper retrieval spear and downhole latching assemblies.

NOTE 7—When lowering a latching bit assembly or retrievable inner-core barrel assembly into a dry hole, a retrievable dry-hole lowering tool should be employed to prevent damaging the outer bit kerf, matrix, or latching assemblies, or combination thereof. Inner tools should not be allowed to free-fall down the drill rod in a dry hole.

6.2.6 *Wireline Core Barrels*, available for obtaining continuous samples of soil or rock. The barrels for use in coring rock vary in design and manufacture from those barrels used for coring unconsolidated materials. The best core recovery in rock usually requires a double- or triple-tube, swivel-type design. The inner tube of the rock core barrel consists of a core-lifter case and core lifter that threads onto the lower end of the inner tube. On the upper end of the inner tube is a removable threaded inner-tube head swivel-bearing assembly with an inner-tube latching device and release mechanism. The inner-tube latching device locks into a complementary recess in the wall of the lead outer drill rod such that the drill rod may be rotated while the inner tube remains stationary. The use of split inner tubes or split inner-tube liners inside a solid inner-tube barrel facilitates easier handling, inspection, and removal of the core from the core barrel.

6.2.6.1 Several types of soil core barrels are also available. Most barrels have a cutting shoe that is either flush with the outer tube cutting bit or, it is made to extend past the outer tube core bit. Sample barrels may be of either the solid- or split-tube configuration. Some barrels may also be equipped with either a split tube or solid tube inner liner to minimize exposure of the core to fluids or other materials. Important considerations for optimum sampling results and maximum core recovery are: use of the "correct" lead distance of the inner barrel cutting shoe, using the "optimum" clearance ratio or head space of the cutting shoe, and prevention of inner-barrel rotation. (For optimum core recovery and minimum core damage the user is referred to 3.2.7 and 3.2.13 for making proper lead-distance or head-space adjustments of the inner-barrel cutting shoe.) The lead distance of the cutting shoe ahead of the cutting bit depends on the stiffness of the formations to be sampled. Stiffer materials require less lead distance.

6.2.6.2 The clearance ratio or head space of the cutting shoe with respect to inner barrel should be selected to result in core that fills the barrel without excessive compression of the core due to friction. If the clearance ratio is too great and the core is over cut, core-erosion damage may occur. If core is over cut and air is present in the barrel, alteration of the soil may occur during subsequent packaging, sealing, and storage of the core(s). Use of single-tube split barrels below the water table may expose soil cores to fluids present in the drill rod.

6.2.7 *Pressure Hose*, conducting the drilling fluid from the circulation pump to the swivel.

6.2.8 *Swivel*, directing the drilling fluid to the rotating kelly or drill-rod column.

6.3 *Rotary Wireline Drilling*, with water-based drilling fluids.

6.3.1 *Mud Pit*, a reservoir for the drilling fluid and, if properly designed and utilized, provides sufficient flow velocity reduction to allow separation of drill cuttings from the fluid before recirculation. The mud pit is usually a shallow, open metal tank with baffles; however, for some circumstances, an excavated pit with some type of liner, designed to prevent loss of drilling fluid and to contain potential contaminants that may be present in the cuttings, may be used. The mud pit can be used as a mixing reservoir for the initial quantity of drilling fluid and, in some circumstances, for adding water and additives to the drilling fluid as drilling progresses.

NOTE 8—Some drilling-fluid components must be added to the composite mixture before other components; consequently, an auxiliary mixing reservoir may be required to premix these components with water before adding to the mud pit. All quantities and types of drilling-fluid components and additives used in the composite drilling-fluid mixture should be documented.

6.3.2 *Suction Hose*, sometimes equipped with a foot valve or strainer, or both, conducts the drilling fluid from the mud pit to the drilling-fluid circulation pump.

6.3.3 *Drilling-Fluid Circulation Pump*, having the capability to lift the drilling fluid from the mud pit and move it through the system against variable pumping heads at a flow rate to provide an annular velocity that is adequate to transport drill cuttings out of the drill hole.

6.3.3.1 Fluid pressures at the bit should be as low as necessary to maintain circulation in order to minimize hydraulic fracturing or excessive erosion of the surrounding materials. Fluid pressures should be monitored during drilling. Normally, injection fluid pressures are readily monitored. Changes in fluid return and circulation pressures may indicate occurrence of excessive erosion, formation fluid loss, or formation fracturing. Any abrupt changes or anomalies in the fluid pressures should be duly noted and documented including the depth(s) of occurrence(s).

6.3.4 *Drilling Fluid*, usually consisting of a water-based circulation media and one or more additives that increase viscosity or provide other desirable physical or chemical properties. Principal functions of drilling fluid include sealing the drill hole wall to minimize loss of drilling fluid, providing a hydraulic pressure against the drill-hole wall to support the open drill hole, removing cuttings generated at the bit, and lubricating and cooling of the bit. Drilling-fluid management requires considerable experience for successful use. Drilling-fluid program design can be aided by review of literature (1, 3, 4, 5, 6) and consultation with manufacturers. If changes to the circulating medium are made, the depth(s) or interval(s) of these changes should be documented. Samples of cuttings can be collected for analysis of material being penetrated. If samples are taken, the depth(s) and interval(s) should be documented.

NOTE 9—Particular attention should be given to the drilling-fluid makeup-water source and the means used to transport the makeup water to the drilling site as potential sources of contamination in the drilling fluid. If the chemical makeup of the water is determined the test result should be documented.

6.3.4.1 Some commonly used additives for water-based drilling fluids are listed in 6.3.4.2 through 6.3.4.10.

6.3.4.2 Beneficiated bentonite, a primary viscosifier and borehole sealer, consists of montmorillonite with other naturally occurring minerals and various additives such as sodium carbonate or polyacrylates, or both.

6.3.4.3 Unbeneficiated bentonite, a primary viscosifier and drill hole sealer, consists of montmorillonite with other naturally occurring minerals but without additives such as sodium carbonate or polyacrylates.

6.3.4.4 Sodium carbonate powder (soda ash) is used to precipitate calcium carbonate hardness from the drilling fluid water-base before adding other components. An increase in pH will occur with the addition of sodium carbonate. Sodium hydroxide (caustic soda) generally should not be used in this application.

6.3.4.5 Carboxylmethylcellulose powder (CMC) is sometimes used in a water-based fluid as a viscosifier and as an inhibitor to clay hydration.

NOTE 10—Some additives to water-based drilling-fluid systems retard clay hydration, thus inhibiting swelling of clays on the drill-hole wall and inhibiting "balling" or "smearing" of the bit.

6.3.4.6 Potassium chloride (muriated potash) or diammonium phosphate can be used as an inhibitor to clay hydration.

6.3.4.7 Polyacrylamide, a primary viscosifier and clay hydration inhibitor, is a polymer that is mixed with water to create a drilling fluid.

6.3.4.8 Barium sulfate increases the density of water-based drilling fluids. It is a naturally occurring high specific gravity mineral processed to a powder for rotary drilling-fluid applications.

6.3.4.9 Lost-circulation materials are used to seal the borehole wall when fluids are being lost through large pores, cracks, or joints. These additives usually consist of various coarse-textured materials such as shredded paper or plastic, bentonite chips, wood fibers, or mica.

6.3.4.10 Attapulgite, a primary viscosifier for rotary drilling in high-salinity environments, is a clay mineral drilling-fluid additive.

NOTE 11—The listing and discussion of the above drilling-fluid additives does not imply general acceptance for geoenvironmental exploration. Some of the additives listed above may impact water-quality analyses. Some readily available, but not as common, drilling-fluid additives, not listed above, could cause significant contamination in a borehole or hydrologic unit. Each additive should be evaluated for each specific application. The types, amounts, and chemical compositions of all additives used should be documented. In addition, a hole log should document the depths where any new additives were introduced. Methods to break revertible fluids should be documented.

If drilling-fluid additives, such as any of those previously mentioned, cannot be tolerated it is recommended that other drilling method(s) be considered by the user.

6.4 *Rotary Wireline Drilling*, using air as the circulation medium:

6.4.1 *Air Compressor*, providing an adequate volume of air, without significant contamination, for removal of cuttings. Air requirements will depend upon the drill rod and bit configuration, the character of the material penetrated, the depth of drilling below ground water level, and the total depth of drilling. The airflow rate requirements are usually based on an annulus upflow air velocity of about 1000 to 1300 m/min (about 3000 to 4000 ft/min) even though air-upflow rates of less than 1000 m/min are often adequate for cuttings transport. In order to maintain air circulation between the annulus of the hole wall and large-diameter drill rods, special reaming shells may be required (2). For some geologic conditions, air-blast erosion may increase the borehole diameter in easily eroded materials such that 1000 m/min may not be appropriate for cuttings transport. Should air-blast erosion occur, the depth(s) of the occurrence(s) should be noted and documented so that subsequent monitoring equipment installation quality may be evaluated accordingly.

NOTE 12—The quality of compressed air entering the borehole and the quality of air discharged from the borehole and the cyclone separator must be considered. If not adequately filtered, the air produced by most oil-lubricated air compressors inherently introduces a significant quantity of oil into the circulation system. High-efficiency, in-line, air filters are usually required to prevent significant contamination of the borehole.

6.4.2 *Pressure Hose*, conducting the air from the air compressor to the swivel.

6.4.3 *Swivel*, directing the air to the rotating kelly or drill-rod column.

6.4.4 *Dust Collector*, conducting air and cuttings from the borehole annulus past the drill-rod column to an air cleaning device (cyclone separator).

6.4.5 *Air-Cleaning Device*, (cyclone separator) separates cuttings from the air returning from the borehole by means of the dust collector.

NOTE 13—A properly sized cyclone separator can remove practically all of the cuttings from the return air. A small quantity of fine particles, however, are usually discharged to the atmosphere with the "cleaned" air. Some air-cleaning devices consist of a cyclone separator alone; whereas, some utilize a cyclone separator combined with a power blower and sample-collection filters. It is virtually impossible to direct the return "dry" air past the drill rods without some leakage of air and return cuttings. Samples of drill cuttings can be collected for analysis of materials penetrated. If samples are obtained, the depth(s) and interval(s) should be documented.

NOTE 14—Zones of low air return also zones of no air return should be documented. Likewise, the depth(s) of sampled interval(s) and quality of samples obtained should be documented.

NOTE 15—Compressed air alone can often transport cuttings from the borehole and cool the bit. For some geologic conditions, injection of water into the air stream will help control dust or break down "mud rings" that tend to form on the drill rods. If water is injected the depth(s) of water injection should be documented. Under other circumstances, for example, if the borehole starts to produce water, the injection of a foaming agent may be required. The depth when a foaming agent is added should also be recorded. When foaming agents are used, a cyclone-type cuttings separator is not used and foam discharge accumulates near the top of the borehole. When contaminants are encountered during drilling and returning from the borehole at geoenvironmental-exploration sites, special measures should be taken to contain the foam and protect personnel and the environment. Therefore, added water and some available foaming agents could affect water-quality analyses. The need for chemical analysis of added water or foaming agents should be considered and documented.

7. Drilling Procedures

7.1 As a prelude to and throughout the drilling process, stabilize the drill rig and raise the drill-rig mast. Surface casings can be installed using a variety of drilling methods. The surface casing is normally backfilled, pressed, or sealed in place with bentonite or cement, or both. Establish and document a datum for measuring hole depth. This datum normally consists of a stake driven into the stable ground surface, the top of the surface casing, or the drilling deck. If there is a possibility for movement of the surface casing it should not be used as a datum. If the hole is to be later surveyed for elevation, record and report the height of the datum to the ground surface.

7.1.1 For water-based fluid drilling operation, position a mud pit to collect and filter fluid return flow. Mix an initial quantity of drilling fluid, usually using the mud pit as the primary mixing reservoir.

NOTE 16—The need for chemical analysis of samples of each drilling-fluid component and the final mixture should be documented.

7.1.2 For air-based circulation systems, position the dust collector or cyclone separator and "seal" to the ground surface. If air-monitoring operations are performed, consider the prevalent wind direction relative to the exhaust from the drill rig. Also, consider the location of the cyclone relative to the rig exhaust since air-quality monitoring will be performed at the cyclone separator discharge point.

NOTE 17—Deeper casing(s) or nested casing(s) may be required to facilitate adequate downhole fluid circulation and hole control. Records of casing(s) lengths and depth intervals installed should be maintained and documented if required. All casing used should be decontaminated according to Practice D 5088 prior to use.

NOTE 18—Check above the drilling rig for overhead obstructions or hazards, such as power lines prior to raising the mast. In most cases, it is required to perform a survey of underground and all other utilities prior to drilling to evaluate possible hazards.

7.2 Drilling usually progresses as follows:

7.2.1 Attach an initial assembly of lead drill rod and a bit or core barrel below the top-head drive unit with the bit or drill head placed within the top of the surface casing. Record hole depth by knowing the length of the rod-bit assemblies and comparing its position relative to the established surface datum.

NOTE 19—The drill rig, drilling, hoisting and sampling tools, the rotary gear or chain case, the spindle and all components of the rotary drive above the drilling axis should be cleaned and decontaminated according to Practice D 5088 prior to commencing drilling and sampling operations.

7.2.2 Activate the drilling-fluid circulation pump, causing drilling fluid to circulate through the system including the mud pit.

7.2.3 Initiate rotation of the bit and apply axial force to the bit.

7.2.4 Initiate drilling fluid or air circulation and apply rotation and axial force to the bit until drilling progresses to a depth where sampling or in-situ testing will be performed, the length of the drill-rod column limits further penetration, or (when core drilling) the core specimen has fully entered the core barrel or blockage is apparent. Monitor downfeed pressures, fluid/gas pressure, and circulations return during drilling. Observe the ease of drilling during drilling, and drill cuttings as they relate to the geologic strata being penetrated. Document occurrences of any significant abrupt changes and anomalies that occur during drilling.

7.2.4.1 During air drilling operations air-quality monitoring may be required. If air-quality monitoring is performed, document the sampled intervals and air-quality data.

NOTE 20—If circulation is lost and input fluid pressure is allowed to increase, the possibility of fracturing materials and excessive drill hole erosion exists.

NOTE 21—The time required to remove the cuttings from the drill hole will depend mainly upon the pumping rate, the cross-sectional area of the borehole annulus, the borehole depth, the viscosity of the drilling fluid, and the size of the cuttings.

7.2.5 Stop rotation, lift the bit slightly off the bottom of the hole to facilitate drill-cuttings removal, and continue circulation for a short time until the drill cuttings are removed from the borehole annulus. If sampling is to be done, stop circulation and rest the bit on the hole bottom to determine hole depth. Document the hole depth and amount of any caving that occurred.

7.2.6 Increase drilling depth by attaching an additional drill-rod section to the top of the previously advanced drill-rod column and resuming drilling operations according to 7.2.2 through 7.2.4.

NOTE 22—Drilling rates depend on many factors such as the density or stiffness of unconsolidated material and the existence of cobbles or boulders, the hardness or durability of the rock, or both, the swelling activity of clays or shales encountered in the borehole, and the erosiveness of the wall. Drilling rates can vary from a few mm/min (less than an in./min) to about a m/min (3 ft/min), depending on subsurface conditions. Other factors influencing drilling rates include the weight of the drill string, collar(s) weight and size of drill pipe, and the rig pulldown or holdback pressure. These data as well as any other drilling-rate information should be recorded.

7.2.7 In some cases it may be necessary to remove the pilot bit or core barrel during the drilling process. Remove core barrels when full or there is evidence of core blockage and loss of circulation. Remove pilot bits for sampling events, when worn, or when there is evidence of circulation blockage. When replacing the pilot bit or core barrel note the condition of the base of the boring and any difficulty latching the assembly. If heave is present it may be necessary to lift the drill rod to engage the locking mechanism. When drilling in unconsolidated materials it is recommended to use a vented hoisting plug when removing the inner bit or continued circulation to avoid "heaving," "piping," or "sanding in" problems.

7.2.8 In some cases it may be necessary to remove the drill-rod string to change the lead-rod bit. If the string is removed and replaced, check the depth to the base of the boring where the end of the string rests and compare to the cleanout depth to evaluate hole quality. If excessive hole cave or erosion is suspected, casing may be required to protect the borehole wall. As the drilling progresses, note and document drilling procedures such as circulation rates or losses, intervals where equipment is changed, intervals where casing is installed, or drilling method is changed.

7.2.9 *Continuous Sampling with Rock- or Soil-Core Barrels*—Continuous coring can be performed by using core barrels designed for rock or soil coring. Rock coring is performed in accordance with Practice D 2113. Soil coring follows similar procedures. When replacing the core barrel, note the elevation of the base of the boring. Record the sampling length and measure core recovery. If blockage of core and circulation loss is noted during the sampling process, the sample should be recovered.

7.2.10 Sampling or in-situ testing can be performed at any depth by interrupting the advance of the bit, stopping the fluid circulation (after cleaning the annulus of cuttings) and removing pilot bit from the drill-rod column. If there are apparent cuttings or evidence of caving, these should be noted as part of the sample taken. Drill-rod removal is not necessary when a sample can be obtained or an in-situ test can be performed through the hollow axis of the drill rods and bit.

7.2.10.1 Some sampling method may be performed through the drill-rod column (see Test Method D 1586, Practice D 1587, Practice D 2113, and Practice D 3550). Compare sampling depth to the cleanout depth. Verify the depth-comparison data by first resting the sampler on the bottom of the hole and comparing that measurement with the cleanout-depth measurement. This should be done before every sampling or in-situ testing is performed in the hole. If bottom-hole sloughing is apparent from a depth measurement made prior to sampling (determined by comparing the hole-cleanout depth with the sampling depth) it may be necessary to reclean the hole by rotary recirculation. Record the depth of in-situ testing or sampling as well as the depth below the sampler/bit for evaluation of data quality. Decontaminate sampling and testing devices according to Practice D 5088 prior to testing.

7.2.10.2 Perform in-situ testing at the base of the drill hole or at shallower depths if the hole is uncased. If in-situ testing is performed at the base of the borehole similar depth, perform checks to accurately document the location of the test.

7.2.10.3 Water testing is frequently performed in consolidated deposits by pulling back on the drill rods and passing inflatable packer(s) with pressure fittings for water injection through the drill rod to test the open borehole wall (see Test Methods D 4630 and D 4631). Record the depths or intervals of water tests performed.

7.2.10.4 Air drilling may not be satisfactory in unconsolidated or cohesionless soils, or both, under the ground water table. One problem is "heaving" or "sanding in" of sands during pilot-bit removal. Another problem with wet clays is bit plugging and formation of mud rings which may cause loss of circulation. If the purpose of the drilling is to perform in-situ tests or undisturbed sampling, and soil instability is evident from unbalanced hydrostatic pressure or pilot-bit disturbance, give consideration to using fluids or other casing-advancement methods where disturbance may be more readily controlled.

7.3 When drilling must progress through material suspected of being contaminated, installation of single or multiple (nested) casings may be required to isolate zones of suspected contamination. Isolation casings are usually installed in a predrilled borehole or by using a casing-advancement method. A grout seal is then installed, usually by applying the grout at the bottom of the annulus with the aid of a grout shoe or a grout packer and a tremie pipe. Allow the grout to set before drilling activities are continued. Document complete casing and grouting records, including location(s) of nested casings for the hole.

8. Installation of Monitoring Devices

8.1 Subsurface water-quality monitoring devices are generally installed in boreholes drilled by wireline casing-advancement methods using a three-step procedure. The three steps are: (1) drilling, with or without sampling, (2) removal of the pilot bit or core barrel and placement of the monitoring device, and (3) addition of other materials such as filter packs, seals, and grouts as the drill rods are removed. During placement of filter packs and seals the drill rods may be lifted in increments to be backfilled. Place tremie rods between riser and drill-rod wall to extend past the end of the drill rods for placement increments. The retraction increments depend on the anticipated borehole wall stability. Alternately, if borehole conditions are stable the complete drill-rod column may be removed prior to placement of the monitoring device.

NOTE 23—Document the volumes of emplaced filter packs and seals and compare the results to actual volumes of cuttings return based on hole diameter for evaluation of hole quality.

8.2 Assemble water-quality monitoring devices, with attached fluid conductors (risers), and insert into the borehole with the least possible addition of contaminants. The user is referred to Practice D 5092 for monitoring-well installation methods and Practice D 5088 for suggested methods of field equipment decontamination.

NOTE 24—If the integrity of the borehole wall will not be compromised by removing the filter cake from the borehole in the vicinity of the screened-intake device the drilling mud should be removed using a well-development procedure as suggested in Practice D 5099 prior to inserting a monitoring well or a water-quality monitoring device in the borehole.

8.2.1 Some materials, such as screens and risers, require cleaning or decontamination, or both, at the job site (see Practice D 5088).

8.2.2 Prior to installation, store all monitoring-device materials undercover and place upwind and well away from the drill rig and any other sources of contamination such as electrical generators, air compressors, or industrial machinery.

8.2.3 Clean and decontaminate hoisting tools, particularly wire rope and hoisting swivels, according to Practice D 5088 before using.

8.3 Select and install filter materials, bentonite pellets, granules and chips and grouts according to specific subsurface-monitoring requirements. Document this information.

NOTE 25—Filter packs for monitoring devices are usually installed in wireline casing-advancement drill holes using a tremie pipe inserted in the annulus formed between the drill rod or borehole wall (when drill rods are removed) and the monitoring device. The minimum annulus between riser pipe and drill rod or borehole wall should be about 1-in. (25.4 mm) completely around the riser pipe. The sizes of wireline drill rod are limited and the riser size may have to be reduced to accommodate placement clearances (see 6.2.2).

8.3.1 Monitoring devices installed in a saturated zone typically have sand-sized filter packs selected on the basis of the grain-size characteristics of the hydrologic unit adjacent to the screened intake. Filter-pack materials are usually selected with a coefficient of uniformity of less than 2.5. Filter packs for vadose-zone monitoring devices may be predominantly slit sized (however, soil-gas monitoring devices should not use silt-sized filter packs but typically use coarse sand or gravel filter packs). These filter materials are often mixed with water of known quality and then inserted through a tremie pipe and tamped into place around the device. The type(s) and volumes of filter materials used and the quality and quantities of mixing water should be documented. In most cases, a centralizer should be used to position the monitoring device in the borehole. The intake device and riser(s) should be suspended above the bottom of the borehole during installation of the filter pack(s), seal(s), and backfill to keep the riser(s) as straight as possible. Care should be taken when adding backfill or filter material(s), or both, so that the materials do not bridge. However, if bridging does occur during the installation procedure, tamping rods or other tamping devices may be used to dislodge the "bridge".

8.4 Place sealing materials, consisting of either bentonite pellets, chips, or granules, directly above the filter pack.

NOTE 26—It may be effective, when granular filter packs are used, to install a thin, fine sand, secondary filter either below the annular seal or both above and below the seal. These secondary filters protect both the monitoring-device filter and the seal from intrusion of grout installed above the seal.

8.5 The backfill that is placed above the annular seal is usually a bentonite or cement-base grout.

NOTE 27—Grouts should be designed and installed in consideration of the ambient hydrogeologic conditions. The constituents should be selected according to specific performance requirements and these data documented. Typical grout mixtures are given in Practice D 5092 and Test Method D 4428/D 4428M.

8.5.1 In most cases, the grout should be pumped into the annulus formed between the borehole wall and the monitoring device(s) riser(s) using a tremie pipe.

NOTE 28—Grouting equipment should be cleaned and decontaminated prior to use according to Practice D 5088. Also, the equipment used for grouting should be constructed from materials that do not "leach" significant amounts of contaminants to the grout.

8.5.2 The initial position of the tremie pipe and grouting pressures should be controlled to prevent materials from being jetted into underlying seal(s) and filter(s) (use of a tremie pipe having a plugged bottom and side-discharge ports should be considered to minimize bottom-jetting problems).

8.5.3 In most cases, the grout should be discharged at a depth of approximately 1.5 to 3 m (5 to 10 ft) below the grout surface within the annulus (after the placement of the initial 1.5 to 3 m (5 to 10 ft) of grout above the underlying filter or seal). Additional grout should be discharged at a depth of approximately 1.5 to 3 m (5 to 10 ft) below the grout surface within the annulus. The tremie pipe should be periodically raised as grout is discharged to maintain the appropriate depth below the grout surface.

NOTE 29—The need for chemical analysis of samples of each grout component and chemical analysis of the final mixture should be documented. Also, it should be noted that if cements are used for grouting, they generate hydroxides and thereby, can cause a localized increase in the alkalinity and pH of the surrounding ground water.

8.5.4 Install the grout from the bottom of the borehole to the top of the borehole so as to displace fluids in the borehole.

9. Development

9.1 Most monitoring-device installations should be developed to remove suspended solids from drilling fluids and disturbance of geologic materials during installation and to improve the hydraulic characteristics of the filter pack and the geologic unit adjacent to the monitoring-device intake. The method(s) selected and time expended to develop the installation and the changes in water quality discharged at the surface should be carefully observed and documented. For suggested well-development methods and techniques the user is referred to Practice D 5099.

NOTE 30—Under most circumstances, development should be initiated as soon as possible following completion, however, time should be allowed for setting of grout.

10. Field Report and Project Control

10.1 The field report should include all information identified as necessary and pertinent to the needs of the exploration program. Information normally required for the project include exploration type and execution, drilling equipment and methods used, and subsurface conditions encountered including: groundwater conditions, sampling events, and installations.

10.2 Additional information should be considered as follows:

10.2.1 *Drilling Methods:*

10.2.1.1 Description of the wireline casing-advancer system including type, sizes, pilot bits, core barrels, fluid pump, fluid circulation, and discharge systems. Note and document intervals of equipment change or drilling method changes and reasons for change.

10.2.1.2 Type, quantities, and locations in the borehole of use of additives such as water or foaming agent(s) added to the circulation media.

10.2.1.3 Descriptions of circulation rates and cuttings return, including quantities, over intervals used. Locations and probable cause of loss of circulation in the borehole.

10.2.1.4 Descriptions of drilling conditions related to drilling pressures, rotation rates, and general ease of drilling related to subsurface materials encountered.

10.2.2 *Sampling*—Document conditions of the bottom of the borehole at the base prior to sampling, and report any slough or cuttings present in the recovered sample.

10.2.2.1 Samples of fluid circulation cuttings can be collected for analysis of materials being penetrated. If samples are taken, the depth(s) and interval(s) should be documented.

10.2.3 *In-Situ Testing:*

10.2.3.1 For devices inserted below the bottom of the borehole, document the depths below the bottom of the hole and any unusual conditions during testing.

10.2.3.2 For devices testing or seating at the borehole wall, report any unusual conditions of the borehole wall such as inability to seat borehole packers.

10.2.4 *Installations*—A description of well-completion materials and placement methods, approximate volumes placed, depth-intervals of placement, methods of confirming placement, and areas of difficulty of material placement or unusual occurrences.

11. Keywords

11.1 drilling; geoenvironmental exploration; ground water; vadose zone; wireline drilling method

REFERENCES

(1) Shuter, E., and Teasdale, W. E., "Application of Drilling, Coring, and Sampling Techniques to Test Holes and Wells," *U.S. Geological Survey Techniques of Water-Resource Investigations, TWRI 2-F1*, 1989.

(2) Teasdale, W. E., and Pemberton, R. R., "Wireline-Rotary Air Coring of Bandeiler Tuff, Los Alamos, New Mexico," *U.S. Geological Survey, Water-Resources Investigations Report 84-4176*, 1984.

(3) *Diamond Drilling Handbook*, Heinz, W. F., First Edition, South African Drilling Association, Johannesburg, Republic of South Africa, 1985.

(4) *DCDMA Technical Manual*, Drilling Equipment Manufacturers Association, 3008 Millwood Ave., Columbia, SC 29205, 1991.

(5) O'Rourke, J. E., Gibbs, H. J., and O'Connor, K. O., "Core Recovery Techniques for Soft or Poorly Consolidated Materials," *Final Report, Contract No. J0275003*, U.S. Bureau of Mines, Department of Interior, Washington, DC, April 28, 1978.

(6) *Drillers Handbook*, Ruda, T. C., and Bosscher, P. J., editors, National Drilling Contractors Association, 3008 Millwood Ave., Columbia, SC 29205, June 1990.

Standard Practice for
Diamond Core Drilling for Site Investigation[1]

This standard is issued under the fixed designation D 2113; the number immediately following the designation indicates the year of original adoption or, in the case of revision, the year of last revision. A number in parentheses indicates the year of last reapproval. A superscript epsilon (ε) indicates an editorial change since the last revision or reapproval.

This standard has been approved for use by agencies of the Department of Defense. Consult the DoD Index of Specifications and Standards for the specific year of issue which has been adopted by the Department of Defense

[ε1] NOTE—Editorial changes were made and Section 9 added editorially in September 1993.

1. Scope

1.1 This practice describes equipment and procedures for diamond core drilling to secure core samples of rock and some soils that are too hard to sample by soil-sampling methods. This method is described in the context of obtaining data for foundation design and geotechnical engineering purposes rather than for mineral and mining exploration.

1.2 *This standard does not purport to address all of the safety problems, if any, associated with its use. It is the responsibility of the user of this standard to establish appropriate safety and health practices and determine the applicability of regulatory limitations prior to use.*

2. Referenced Documents

2.1 *ASTM Standards:*
D 1586 Method for Penetration Test and Split-Barrel Sampling of Soils[2]
D 1587 Practice for Thin-Walled Tube Sampling of Soils[2]
D 3550 Practice for Ring-Lined Barrel Sampling of Soils[2]

3. Significance and Use

3.1 This practice is used to obtain core specimens of superior quality that reflect the in-situ conditions of the material and structure and which are suitable for standard physical-properties tests and structural-integrity determination.

4. Apparatus

4.1 *Drilling Machine,* capable of providing rotation, feed, and retraction by hydraulic or mechanical means to the drill rods.

4.2 *Fluid Pump or Air Compressor,* capable of delivering sufficient volume and pressure for the diameter and depth of hole to be drilled.

4.3 *Core barrels,* as required:

4.3.1 *Single Tube Type, WG Design,* consisting of a hollow steel tube, with a head at one end threaded for drill rod, and a threaded connection for a reaming shell and core bit at the other end. A core lifter, or retainer located within the core bit is normal, but may be omitted at the discretion of the geologist or engineer.

4.3.2 *Double Tube, Swivel-Type, WG Design*—An assembly of two concentric steel tubes joined and supported at the upper end by means of a ball or roller-bearing swivel arranged to permit rotation of the outer tube without causing rotation of the inner tube. The upper end of the outer tube, or removable head, is threaded for drill rod. A threaded connection is provided on the lower end of the outer tube for a reaming shell and core bit. A core lifter located within the core bit is normal but may be omitted at the discretion of the geologist or engineer.

4.3.3 *Double-Tube, Swivel-Type, WT Design,* is essentially the same as the double tube, swivel-type, WG design, except that the WT design has thinner tube walls, a reduced annular area between the tubes, and takes a larger core from the same diameter bore hole. The core lifter is located within the core bit.

4.3.4 *Double Tube, Swivel Type, WM Design,* is similar to the double tube, swivel-type, WG design, except that the inner tube is threaded at its lower end to receive a core lifter case that effectively extends the inner tube well into the core bit, thus minimizing exposure of the core to the drilling fluid. A core lifter is contained within the core lifter case on the inner tube.

4.3.5 *Double Tube Swivel-Type, Large-Diameter Design,* is similar to the double tube, swivel-type, WM design, with the addition of a ball valve, to control fluid flow, in all three available sizes and the addition of a sludge barrel, to catch heavy cuttings, on the two larger sizes. The large-diameter design double tube, swivel-type, core barrels are available in three core per hole sizes as follows: 2¾ in. (69.85 mm) by 3⅞ in. (98.43 mm), 4 in. (101.6 mm) by 5½ in. (139.7 mm), and 6 in. (152.4 mm) by 7¾ in. (196.85 mm). Their use is generally reserved for very detailed investigative work or where other methods do not yield adequate recovery.

4.3.6 *Double Tube, Swivel-Type, Retrievable Inner-Tube Method,* in which the core-laden inner-tube assembly is retrieved to the surface and an empty inner-tube assembly returned to the face of the borehole through the matching, large-bore drill rods without need for withdrawal and replacement of the drill rods in the borehole. The inner-tube assembly consists of an inner tube with removable core lifter case and core lifter at one end and a removable inner-tube head, swivel bearing, suspension adjustment, and latching

[1] This practice is under the jurisdiction of ASTM Committee D-18 on Soil and Rock and is the direct responsibility of Subcommittee D18.02 on Sampling and Related Field Testing for Soil Investigations.
Current edition approved June 24, 1983. Published August 1983. Originally published as D 2113 – 62 T. Last previous edition D 2113 – 70 (1976).
[2] *Annual Book of ASTM Standards,* Vol 04.08.

device with release mechanism on the opposite end. The inner-tube latching device locks into a complementary recess in the wall of the outer tube such that the outer tube may be rotated without causing rotation of the inner tube and such that the latch may be actuated and the inner-tube assembly transported by appropriate surface control. The outer tube is threaded for the matching, large-bore drill rod and internally configured to receive the inner-tube latching device at one end and threaded for a reaming shell and bit, or bit only, at the other end.

4.4 *Longitudinally Split Inner Tubes*—As opposed to conventional cylindrical inner tubes, allow inspection of, and access to, the core by simply removing one of the two halves. They are not standardized but are available for most core barrels including many of the retrievable inner-tube types.

4.5 *Core Bits*—Core bits shall be surface set with diamonds, impregnated with small diamond particles, inserted with tungsten carbide slugs, or strips, hard-faced with various hard surfacing materials or furnished in saw-tooth form, all as appropriate to the formation being cored and with concurrence of the geologist or engineer. Bit matrix material, crown shape, water-way type, location and number of water ways, diamond size and carat weight, and bit facing materials shall be for general purpose use unless otherwise approved by the geologist or engineer. Nominal size of some bits is shown in Table 1.

Note 1—Size designation (letter symbols) used throughout the text and in Tables 1, 2, and 3 are those standardized by the Diamond Core Drill Manufacturers' Assoc. (DCDMA). Inch dimensions in the tables have been rounded to the nearest hundredth of an inch.

4.6 *Reaming Shells*, shall be surface set with diamonds,

impregnated with small diamond particles, inserted with tungsten carbide strips or slugs, hard faced with various types of hard surfacing materials, or furnished blank, all as appropriate to the formation being cored.

4.7 *Core Lifters*—Core lifters of the split-ring type, either plain or hard-faced, shall be furnished and maintained, along with core-lifter cases or inner-tube extensions or inner-tube shoes, in good condition. Basket or finger-type lifters, together with any necessary adapters, shall be on the job and available for use with each core barrel if so directed by the geologist or engineer.

4.8 *Casings:*

4.8.1 *Drive Pipe or Drive Casing,* shall be standard weight (schedule 40), extra-heavy (schedule 80), double extra-heavy (schedule 160) pipe or W-design flush-joint casing as required by the nature of the overburden or the placement method. Drive pipe or W-design casing shall be of sufficient diameter to pass the largest core barrel to be used, and it shall be driven to bed rock or to firm seating at an elevation below water-sensitive formation. A hardened drive shoe is to be used as a cutting edge and thread protection device on the bottom of the drive pipe or casing. The drive shoe inside diameter shall be large enough to pass the tools intended for use, and the shoe and pipe or casing shall be free from burrs or obstructions.

4.8.2 *Casing*—When necessary to case through formations already penetrated by the borehole or when no drive casing has been set, auxiliary casing shall be provided to fit inside the borehole to allow use of the next smaller core barrel. Standard sizes of telescoping casing are shown in Table 2. Casing bits have an obstruction in their interior and will not pass the next smaller casing size. Use a casing shoe if additional telescoping is anticipated.

4.8.3 *Casing Liner*—Plastic pipe or sheet-metal pipe may be used to line an existing large-diameter casing. Liners, so used, should not be driven, and care should be taken to maintain true alignment throughout the length of the liner.

4.8.4 *Hollow Stem Auger*—Hollow stem auger may be used as casing for coring.

4.9 *Drill Rods:*

4.9.1 *Drill Rods of Tubular Steel Construction* are normally used to transmit feed, rotation, and retraction forces from the drilling machine to the core barrel. Drill-rod sizes that are presently standardized are shown in Table 3.

4.9.2 Large bore drill rods used with retrievable inner-tube core barrels are not standardized. Drill rods used with retrievable inner-tube core barrels should be those manufac-

TABLE 1 Core Bit Sizes

Size Designation	Outside Diameter		Inside Diameter	
	in.	mm	in.	mm
RWT	1.16	29.5	0.375	18.7
EWT	1.47	37.3	0.905	22.9
EWG, EWM	1.47	37.3	0.845	21.4
AWT	1.88	47.6	1.281	32.5
AWG, AWM	1.88	47.6	1.185	30.0
BWT	2.35	59.5	1.750	44.5
BWG, BWM	2.35	59.5	1.655	42.0
NWT	2.97	75.3	2.313	58.7
NWG, NWM	2.97	75.3	2.155	54.7
2¾ × 3⅞	3.84	97.5	2.69	68.3
HWT	3.89	98.8	3.187	80.9
HWG, . . .	3.89	98.8	3.000	76.2
4 × 5½	5.44	138.0	3.97	100.8
6 × 7¾	7.66	194.4	5.97	151.6

TABLE 2 Casing Sizes

Size Designation	Outside Diameter		Inside Diameter		Threads per in.	Will Fit Hole Drilled with Core Bit Size
	in.	mm	in.	mm		
RW	1.144	36.5	1.19	30.1	5	EWT, EWG, EWM
EW	1.81	46.0	1.50	38.1	4	AWT, AWG, AWM
AW	2.25	57.1	1.91	48.4	4	BWT, BWG, BWM
BW	2.88	73.0	2.38	60.3	4	NWT, NWG, NWM
NW	3.50	88.9	3.00	76.2	4	HWT, HWG
HW	4.50	114.3	4.00	101.6	4	4 × 5½
PW	5.50	139.7	5.00	127.0	3	6 × 7¾
SW	6.63	168.2	6.00	152.4	3	6 × 7¾
UW	7.63	193.6	7.00	177.8	2	. . .
ZW	8.63	219.0	8.00	203.2	2	. . .

TABLE 3 Drill Rods

Size Designation	Rod and Coupling Outside Diameter		Rod Inside Diameter		Coupling Bore, Threads		
	in.	mm	in.	mm	in.	mm	per in.
RW	1.09	27.7	0.72	18.2	0.41	10.3	4
EW	1.38	34.9	1.00	25.4	0.44	11.1	3
AW	1.72	43.6	1.34	34.1	0.63	15.8	3
BW	2.13	53.9	1.75	44.4	0.75	19.0	3
NW	2.63	66.6	2.25	57.1	1.38	34.9	3
HW	3.50	88.9	3.06	77.7	2.38	60.3	3

tured by the core-barrel manufacturer specifically for the core barrel.

4.9.3 *Composite Drill Rods* are specifically constructed from two or more materials intended to provide specific properties such as light weight or electrical nonconductivity.

4.9.4 *Nonmagnetic Drill Rods* are manufactured of nonferrous materials such as aluminum or brass and are used primarily for hole survey work. Some nonmagnetic rods have left-hand threads in order to further their value in survey work. No standard exists for nonmagnetic rods.

4.10 *Auxiliary Equipment,* shall be furnished as required by the work and shall include: roller rock bits, drag bits, chopping bits, boulder busters, fishtail bits, pipe wrenches, core barrel wrenches, lubrication equipment, core boxes, and marking devices. Other recommended equipment includes: core splitter, rod wicking, pump-out tools or extruders, and hand sieve or strainer.

5. Transportation and Storage of Core Containers

5.1 *Core Boxes,* shall be constructed of wood or other durable material for the protection and storage of cores while enroute from the drill site to the laboratory or other processing point. All core boxes shall be provided with longitudinal separators and recovered cores shall be laid out as a book would read, from left to right and top to bottom, within the longitudinal separators. Spacer blocks or plugs shall be marked and inserted into the core column within the separators to indicate the beginning of each coring run. The beginning point of storage in each core box is the upper left-hand corner. The upper left-hand corner of a hinged core box is the left corner when the hinge is on the far side of the box and the box is right-side up. All hinged core boxes must be permanently marked on the outside to indicate the top and the bottom. All other core boxes must be permanently marked on the outside to indicate the top and the bottom and additionally, must be permanently marked internally to indicate the upper-left corner of the bottom with the letters UL or a splotch of red paint not less than 1 in.[2] Lid or cover fitting(s) for core boxes must be of such quality as to ensure against mix up of the core in the event of impact or upsetting of the core box during transportation.

5.2 Transportation of cores from the drill site to the laboratory or other processing point shall be in durable core boxes so padded or suspended as to be isolated from shock or impact transmitted to the transporter by rough terrain or careless operation.

5.3 Storage of cores, after initial testing or inspection at the laboratory or other processing point, may be in cardboard or similar less costly boxes provided all layout and marking requirements as specified in 5.1 are followed. Additional spacer blocks or plugs shall be added if necessary

at time of storage to explain missing core. Cores shall be stored for a period of time specified by the engineer but should not normally be discarded prior to completion of the project for which they were taken.

6. Procedure

6.1 Use core-drilling procedures when formations are encountered that are too hard to be sampled by soil-sampling methods. A 1-in. (25.4-mm) or less penetration for 50 blows in accordance with Method D 1586 or other criteria established by the geologist or engineer, shall indicate that soil-sampling methods are not applicable.

6.1.1 Seat the casing on bedrock or in a firm formation to prevent raveling of the borehole and to prevent loss of drilling fluid. Level the surface of the rock or hard formation at the bottom of the casing when necessary, using the appropriate bits. Casing may be omitted if the borehole will stand open without the casing.

6.1.2 Begin the core drilling using an N-size double-tube swivel-type core barrel or other size or type approved by the engineer. Continue core drilling until core blockage occurs or until the net length of the core barrel has been drilled in. Remove the core barrel from the hole and disassemble it as necessary to remove the core. Reassemble the core barrel and return it to the hole. Resume coring.

6.1.3 Place the recovered core in the core box with the upper (surface) end of the core at the upper-left corner of the core box as described in 5.1. Continue boxing core with appropriate markings, spacers, and blocks as described in 5.1. Wrap soft or friable cores or those which change materially upon drying in plastic film or seal in wax, or both, when such treatment is required by the engineer. Use spacer blocks or slugs properly marked to indicate any noticeable gap in recovered core which might indicate a change or void in the formation. Fit fracture, bedded, or jointed pieces of core together as they naturally occurred.

6.1.4 Stop the core drilling when soft materials are encountered that produce less than 50 % recovery. If necessary, secure samples of soft materials in accordance with the procedures described in Method D 1586, Practice D 1587, or Practice D 3550, or by any other method acceptable to the geologist or engineer. Resume diamond core drilling when refusal materials as described in 6.1 are again encountered.

6.2 Subsurface structure, including the dip of strata, the occurrence of seams, fissures, cavities, and broken areas are among the most important items to be detected and described. Take special care to obtain and record information about these features. If conditions prevent the continued advance of the core drilling, the hole should be cemented and redrilled, or reamed and cased, or cased and advanced with the next smaller-size core barrel, as required by the geologist or engineer.

6.3 Drilling mud or grouting techniques must be approved by the geologist or engineer prior to their use in the borehole.

6.4 *Compatibility of Equipment:*

6.4.1 Whenever possible, core barrels and drill rods should be selected from the same letter-size designation to ensure maximum efficiency. See Tables 1 and 3.

6.4.2 Never use a combination of pump, drill rod, and core barrel that yields a clear-water up-hole velocity of less than 120 ft/min.

6.4.3 Never use a combination of air compressor, drill rod, and core barrel that yields a clear-air up-hole velocity of less than 3000 ft/min.

7. Boring Log

7.1 The boring log shall include the following:

7.1.1 Project identification, boring number, location, date boring began, date boring completed, and driller's name.

7.1.2 Elevation of the ground surface.

7.1.3 Elevation of or depth to ground water and raising or lowering of level including the dates and the times measured.

7.1.4 Elevations or depths at which drilling fluid return was lost.

7.1.5 Size, type, and design of core barrel used. Size, type, and set of core bit and reaming shell used. Size, type, and length of all casing used. Description of any movements of the casing.

7.1.6 Length of each core run and the length or percentage, or both, of the core recovered.

7.1.7 Geologist's or engineer's description of the formation recovered in each run.

7.1.8 Driller's description, if no engineer or geologist is present, of the formation recovered in each run.

7.1.9 Subsurface structure description, including dip of strata and jointing, cavities, fissures, and any other observations made by the geologist or engineer that could yield information regarding the formation.

7.1.10 Depth, thickness, and apparent nature of the filling of each cavity or soft seam encountered, including opinions gained from the feel or appearance of the inside of the inner tube when core is lost. Record opinions as such.

7.1.11 Any change in the character of the drilling fluid or drilling fluid return.

7.1.12 Tidal and current information when the borehole is sufficiently close to a body of water to be affected.

7.1.13 Drilling time in minutes per foot and bit pressure in pound-force per square inch gage when applicable.

7.1.14 Notations of character of drilling, that is, soft, slow, easy, smooth, etc.

8. Precision and Bias

8.1 This practice does not produce numerical data; therefore, a precision and bias statement is not applicable.

NOTE 2—Inclusion of the following tables and use of letter symbols in the foregoing text is not intended to limit the practice to use of DCDMA tools. The table and text references are included as a convenience to the user since the vast majority of tools in use do meet DCDMA dimensional standards. Similar equipment of approximately equal size on the metric standard system is acceptable unless otherwise stipulated by the engineer or geologist.

9. Keywords

9.1 borehole; coring; rock; subsurface investigation

1.7 SURFACE WATER

Standard Guide for
Measurement of Morphologic Characteristics of Surface Water Bodies[1]

This standard is issued under the fixed designation D 4581; the number immediately following the designation indicates the year of original adoption or, in the case of revision, the year of last revision. A number in parentheses indicates the year of last reapproval. A superscript epsilon (ϵ) indicates an editorial change since the last revision or reapproval.

$^{\epsilon 1}$ NOTE—Keywords were added editorially in May 1996.

1. Scope

1.1 This guide describes the methods used for defining the morphologic characteristics of surface water bodies. This guide references manuals that provide various rationale and procedures necessary to conduct a morphologic survey.

1.2 The references were written for specific agency use and may not be applicable in all cases (1–6).[2]

1.3 The values stated in inch-pound units are to be regarded as the standard. The SI units in parentheses are provided for information only.

1.4 *This standard does not purport to address all of the safety concerns, if any, associated with its use. It is the responsibility of the user of this standard to establish appropriate safety and health practices and determine the applicability of regulatory limitations prior to use.*

2. Referenced Document

2.1 *ASTM Standard:*
D 1129 Terminology Relating to Water[3]

3. Terminology

3.1 *Definitions*—For definitions of terms used in this guide, refer to Terminology D 1129.

3.2 *Definitions of Terms Specific to This Standard:*

3.2.1 *large water bodies*—water areas large enough to require use of electronic horizontal positioning devices.

3.2.2 *morphologic surveys*—surveys made to determine shape, depth, and volume of water bodies; also density, distribution, and volume of sediment and characteristics of watersheds contributing to the water body.

3.2.3 *small water bodies*—water areas that can be surveyed using stretched cables or visual triangulation for horizontal positioning.

4. Summary of Guide

4.1 This standard provides guidance for conducting measurements and assembly of data into a standard format that facilitates comparative analysis of water body morphology on a national basis.

5. Significance and Use

5.1 No other standards presently exist for the survey of water body morphologic characteristics. The techniques described in the references represent the present state-of-art and contain sufficient information to inform geologists and engineers of the kinds of information to be gathered and the techniques to be used.

5.2 The major categories of methodologies described in the references are: sounding, positioning, land surveys, sediment properties, sediment sampling techniques, photogrammetric methods, calculating volume and area, morphologic base data, weighted sediment dry weight, reservoir operations, equipment, and reporting results.

5.3 The references are intended as operational manuals and do not describe experimental design.

6. Procedure

6.1 The references provide detailed information and procedures as follows:

6.1.1 *Field Investigations*—Section 3, Chapter 7 of the *SCS National Engineering Handbook* describes field investigations and survey techniques (1). Pages 1 to 31 specifically describe equipment, methods, notekeeping, computations, and reports for small water bodies.

6.1.2 *Sedimentation Surveys*—Specifications were prepared by the Soil Conservation Service to allow contracting for services to perform reservoir sedimentation surveys on small water bodies (2). These specifications are intended to meet SCS needs and should be used by others only as a guide in preparing their own material.

6.1.3 *Methods for Water-Data Acquisition*—Descriptions of various techniques for measuring sediment are contained in Ref (3). The section on reservoir surveys provides guidance about the kinds of work to be done and features to be considered. However, it is not intended as a detailed operational manual. The scope of small and large reservoirs is covered.

6.1.4 *Monitoring Reservoir Sedimentation*—Detailed descriptions of most aspects of performing reservoir sedimentation surveys that are applicable to small and large reservoirs are found in Ref (4). Main topics include base reservoir data, selection of surveying method, and hydrographic surveys.

6.1.5 *Hydrographic Parameters in Large Sand-Bed Streams*—Many techniques are described in varying detail (5). The techniques are suitable for reservoirs, lakes and streams. The main topics are hydrographic investigation

[1] This guide is under the jurisdiction of ASTM Committee D-19 on Water and is the direct responsibility of Subcommittee D19.07 on Sediments, Geomorphology, and Open-Channel Flow.
Current edition approved April 25, 1986. Published November 1986.
[2] The boldface numbers in parentheses refer to the list of references at the end of this guide.
[3] *Annual Book of ASTM Standards*, Vol 11.01.

programs, horizontal positioning equipment and techniques, soundings, velocity measurements and suspended-sediment, bedload and bed-material sampling.

6.1.6 *Hydrographic Manual*—Detailed description of procedures applicable to large water bodies is provided **(6)**. Many described techniques, such as echo sounding, are applicable to small water bodies. The manual is divided into three major elements: hydrographic field operations, final data processing, and appendices. Detailed information is provided on equipment, instruments, and special survey techniques.

7. Report

7.1 All morphologic surveys should report their results in a format sufficient to include the necessary elements to describe where and when the survey was performed, what methods were used and how precisely they were applied, and what were the measured results. Appendix X1 contains a data summary form currently used by several U.S. government and state agencies. Appendix X2 is a list of additional factors to be considered when reporting results.

8. Precision

8.1 The precision is a function of the conditions encountered and the measurement techniques used for each individual survey.

9. Keywords

9.1 bathymetric survey; reservoir sediment; reservoir survey; sedimentation

APPENDIX

(Nonmandatory Information)

X1. SUBCOMMITTEE ON SEDIMENTATION (ICWR) INSTRUCTIONS FOR COMPILING THE RESERVOIR SEDIMENT DATA SUMMARY FORM[4]

X1.1 The following instructions were prepared by members of the Subcommittee as a guide for use in the completion of Reservoir Sediment Data Summary forms. The purpose of the summary form is to provide for the uniform compilation and dissemination of pertinent basic data obtained from reservoir sedimentation surveys. A summary is desired for each reservoir on which one or more sedimentation surveys have been made. New summaries should be prepared when additional sedimentation surveys are made and should carry forward the results of previous surveys, as indicated in the instructions. A typed copy of each new summary in condition suitable for offset printing should be furnished for publication. After a summary is prepared it will be reproduced by the Subcommittee in sufficient numbers to meet the needs of each agency represented on the Subcommittee. This will permit each agency to maintain a file of basic data prepared in a uniform manner suitable for analysis and interpretation. The Subcommittee recognizes that all items of data provided for on the summary will not be readily available for every reservoir. The early compilation and dissemination of available data is preferable to postponement until all items can be completed. However, it is important that every item be filled out for which data are obtainable. The following instructions are based on the instructions issued by the Subcommittee on Sedimentation in 1961 but are revised to apply to the new summary form.

X1.1.1 Figure X1.1 provides a reservoir data summary form. Figure X1.2 is a reservoir data summary form in SI units. Figure X1.3 is an example of a completed reservoir data summary form. A complete description of each item on the form is given in X1.3.

X1.2 *General Notes:*

X1.2.1 In all cases where data are estimated or assumed, insert an asterisk, and show an asterisk with the word "assumed" at the bottom of the front page of the form.

X1.2.2 Where other information is presented that needs clarification, footnotes should be used and shown by numbers, as [1], [2], etc. All footnotes are to be explained in the space provided under Item 47.

X1.2.3 All data should be shown to at least three significant figures, if available, and if accuracy of the survey warrants. However, it is common practice and permissible to show all items of data to the nearest whole number, even though the accuracy of the survey may not give significance to the last one or two whole numbers. For example, for Item 14: 167 624, 16 762, 1676, 168, 16.8, 1.68.

X1.2.4 *Items 31, 32, 33, 37, 38, 40, 41*—Where the sedimentation survey of a multiple-purpose reservoir has covered only the pool level of levels used for storage most of the year (as irrigation, power, inactive) and has not covered the flood-control pool above such levels, the data should be shown for the pool levels surveyed. However, any data obtained concerning sedimentation in the controllable flood-control pool (not including surcharge storage) should be shown under the above items with a footnote reference of explanation under Item 47.

X1.2.5 Use continuation sheets when all data cannot be placed on one sheet.

X1.3 *Specific Items*—Descriptions of the numbered items as they appear in Figs. X1.1, X1.2, and X1.3 are given as follows:

X1.3.1 *Name of Reservoir:* Give the official or most commonly used name. If the dam has another name, give it in parentheses, for example, Lake Mead (Hoover Dam).

[4] Prepared by the following agencies represented on the Subcommittee on Sedimentation Inter-Agency Committee on Water Resources: Department of Agriculture: Agricultural Research Service, Forest Service, Soil Conservation Service; Department of Commerce: Bureau of Public Roads, Environmental Science Services Administration; Department of Defense: Corps of Engineers, Naval Oceanographic Office; Department of Health, Education and Welfare: Water Pollution Control Administration; Department of the Interior: Bureau of Mines, Bureau of Reclamation, Geological Survey; Federal Power Commission; and Tennessee Valley Authority.

RESERVOIR SEDIMENT
DATA SUMMARY

NAME OF RESERVOIR

DATA SHEET NO.

DAM	1. OWNER		2. STREAM		3. STATE	
	4. SEC. TWP. RANGE		5. NEAREST P.O.		6. COUNTY	
	7. LAT. ° ′ ″ LONG. ° ′ ″		8. TOP OF DAM ELEVATION		9. SPILLWAY CREST ELEV	
RESERVOIR	10. STORAGE ALLOCATION	11. ELEVATION TOP OF POOL	12. ORIGINAL SURFACE AREA, ACRES	13. ORIGINAL CAPACITY, ACRE-FEET	14. GROSS STORAGE, ACRE-FEET	15. DATE STORAGE BEGAN
	a. FLOOD CONTROL					
	b. MULTIPLE USE					
	c. POWER					16 DATE NORMAL OPER BEGAN
	d. WATER SUPPLY					
	e. IRRIGATION					
	f. CONSERVATION					
	g. INACTIVE					
	17. LENGTH OF RESERVOIR		MILES	AV. WIDTH OF RESERVOIR		MILES
WATERSHED	18. TOTAL DRAINAGE AREA		SQ. MI.	22. MEAN ANNUAL PRECIPITATION		INCHES
	19. NET SEDIMENT CONTRIBUTING AREA		SQ. MI.	23. MEAN ANNUAL RUNOFF		INCHES
	20. LENGTH MILES	AV. WIDTH	MILES	24. MEAN ANNUAL RUNOFF		AC.-FT.
	21. MAX. ELEV.	MIN. ELEV.		25. ANNUAL TEMP: MEAN	RANGE	

SURVEY DATA

26. DATE OF SURVEY	27. PERIOD YEARS	28. ACCL. YEARS	29. TYPE OF SURVEY	30. NO. OF RANGES OR CONTOUR INT.	31. SURFACE AREA, ACRES	32. CAPACITY, ACRE-FEET	33. C/I. RATIO, AC.-FT. PER AC.-FT.

26. DATE OF SURVEY	34. PERIOD ANNUAL PRECIPITATION	35. PERIOD WATER INFLOW, ACRE-FEET			36 WATER INFL. TO DATE, AC.-FT	
		a. MEAN ANNUAL	b. MAX. ANNUAL	c. PERIOD TOTAL	a. MEAN ANNUAL	b. TOTAL TO DATE

26. DATE OF SURVEY	37. PERIOD CAPACITY LOSS, ACRE-FEET			38. TOTAL SED. DEPOSITS TO DATE, ACRE-FEET		
	a. PERIOD TOTAL	b. AV. ANNUAL	c. PER SQ. MI.-YEAR	a. TOTAL TO DATE	b. AV. ANNUAL	c. PER SQ. MI.-YEAR

26. DATE OF SURVEY	39. AV. DRY WGT., LBS. PER CU. FT.	40. SED. DEP., TONS PER SQ. MI.-YR.		41. STORAGE LOSS, PCT.		42. SED. INFLOW, PPM	
		a. PERIOD	b. TOTAL TO DATE	a. AV. ANN.	b. TOT. TO DATE	a. PERIOD	b. TOT. TO DATE

FIG. X1.1 Reservoir Sediment Data Summary Form, Inch-Pound Units

X1.3.2 *Data Sheet No.*—The data sheet number is composed of two parts. The first is the river basin map number as shown in the hydrologic atlas compiled under the auspices of the Subcommittee on Hydrology (ICWR), and the second is the sheet reference number periodically supplied by the Subcommittee on Sedimentation when data are compiled for publication. If the map number for the river basin in which the reservoir is located is available, it should be shown here. The data sheet reference number will be supplied later by the Subcommittee on Sedimentation.

X1.3.3 *Item 1*—The name of the person or the organization that owns or operates the structure. If a federal or state government, give both the department and agency having supervision or control over the operation of the dam. (Abbreviate as necessary.)

X1.3.4 *Item 2*—If the reservoir is located on a small stream, the name of which is not known, list the stream as a tributary of the next largest stream, for example, "Trib. of Rock R."

X1.3.5 *Item 3*—If the dam lies in two states, both states should be listed. List first the state that is the location for dam operation headquarters.

X1.3.6 *Item 4*—Give the location of the dam by section, township, and range.

X1.3.7 *Item 5*—Give the name of the nearest post office. If space permits, help pinpoint the location of the dam by adding the distance in miles and the direction of the dam from the nearest post office, such as Tulsa 2 SE.

X1.3.8 *Item 6*—Give the county in which the dam is located. If the dam is in two counties, list first the county that is the location for dam operation headquarters, followed by a

26. DATE OF SURVEY	43. DEPTH DESIGNATION RANGE IN FEET BELOW, AND ABOVE, CREST ELEVATION												
	PERCENT OF TOTAL SEDIMENT LOCATED WITHIN DEPTH DESIGNATION												

26. DATE OF SURVEY	44. REACH DESIGNATION PERCENT OF TOTAL ORIGINAL LENGTH OF RESERVOIR														
	0-10	10-20	20-30	30-40	40-50	50-60	60-70	70-80	80-90	90-100	-105	-110	-115	-120	-125
	PERCENT OF TOTAL SEDIMENT LOCATED WITHIN REACH DESIGNATION														

45. RANGE IN RESERVOIR OPERATION

WATER YEAR	MAX. ELEV.	MIN. ELEV.	INFLOW, AC.-FT.	WATER YEAR	MAX. ELEV.	MIN. ELEV.	INFLOW, AC.-FT.

46. ELEVATION-AREA-CAPACITY DATA

ELEVATION	AREA	CAPACITY	ELEVATION	AREA	CAPACITY	ELEVATION	AREA	CAPACITY

47. REMARKS AND REFERENCES

48. AGENCY MAKING SURVEY
49. AGENCY SUPPLYING DATA
50. DATE _____

FIG. X1.1 *Continued*

hyphen and the name of the second county.

X1.3.9 *Item 7*—Give the latitude and longitude of the dam in degrees and minutes (seconds, if known).

X1.3.10 *Item 8*—The elevation of the top of the dam that is equal to the highest spillway elevation (Item 9) plus freeboard.

NOTE—In items 8, 9 and 21, if no actual sea level datum elevation is available, an assumed elevation or local datum plane should be given for these items wherever possible so that the height of the dam and the spillway above stream bed can be determined. (Observe X1.2.1 under General Notes.)

X1.3.11 *Item 9*—This is the elevation of the highest spillway. If the spillway is topped by movable gates, give the elevation of the top of the gates in closed position, with an explanatory footnote in Item 47 "Remarks and References."

(See X1.2.2 under General Notes.)

X1.3.12 *Items 10 to 14*—All data corresponding to storage allocations 10a to g refer to original storages in the reservoir, if these data are available, or otherwise, to the first accurate capacities determined after the beginning of storage. Show revisions of the initial storages if recent surveys yield more accurate data than the early surveys.

X1.3.13 *Item 10a and b*—These items designate the purpose of storage space allocation. Multiple-use storage space (Item 10b) is purposely varied, seasonally or alternately, as required to serve two or more purposes. Use a footnote to explain the specific uses in Item 47.

X1.3.14 *Item 10c*—This item ordinarily refers to storage for hydroelectric or direct power development. However, storage developed or allocated specifically for cooling pur-

RESERVOIR SEDIMENT
DATA SUMMARY

NAME OF RESERVOIR

DATA SHEET NO.

DAM					
1. OWNER		2. STREAM		3. STATE	
4. SEC. TWP. RANGE		5. NEAREST P.O.		6. COUNTY	
7. LAT. ° ' " LONG. ° ' "		8. TOP OF DAM ELEVATION m		9. SPILLWAY CREST ELEV m	

RESERVOIR

10. STORAGE ALLOCATION	11. ELEVATION TOP OF POOL, m	12. ORIGINAL SURFACE AREA, Km^2	13. ORIGINAL CAPACITY, m^3	14. GROSS STORAGE, m^3	15. DATE STORAGE BEGAN
a. FLOOD CONTROL					
b. MULTIPLE USE					
c. POWER					16. DATE NORMAL OPER. BEGAN
d. WATER SUPPLY					
e. IRRIGATION					
f. CONSERVATION					
g. INACTIVE					
17. LENGTH OF RESERVOIR Km			AV. WIDTH OF RESERVOIR		Km

WATERSHED

18. TOTAL DRAINAGE AREA	Km^2	22. MEAN ANNUAL PRECIPITATION	mm
19. NET SEDIMENT CONTRIBUTING AREA	Km^2	23. MEAN ANNUAL RUNOFF	mm
20. LENGTH Km AV. WIDTH Km		24. MEAN ANNUAL RUNOFF	m^3
21. MAX. ELEV. m MIN. ELEV. m		25. ANNUAL TEMP: MEAN RANGE	°C

SURVEY DATA

26. DATE OF SURVEY	27. PERIOD YEARS	28. ACCL. YEARS	29. TYPE OF SURVEY	30. NO. OF RANGES OR CONTOUR INT.	31. SURFACE AREA, Km^2	32. CAPACITY, m^3	33. C/I. RATIO, m^3 PER m^3

26. DATE OF SURVEY	34. PERIOD ANNUAL PRECIPITATION	35. PERIOD WATER INFLOW, m^3			36. WATER INFL. TO DATE, m^3	
		a. MEAN ANNUAL	b. MAX. ANNUAL	c. PERIOD TOTAL	a. MEAN ANNUAL	b. TOTAL TO DATE

26. DATE OF SURVEY	37. PERIOD CAPACITY LOSS, m^3			38. TOTAL SED. DEPOSITS TO DATE, m^3		
	a. PERIOD TOTAL	b. AV. ANNUAL	c. PER Km^2/YEAR	a. TOTAL TO DATE	b. AV. ANNUAL	c. PER Km^2/YEAR

26. DATE OF SURVEY	39. AV. DRY WGT., Kg/m^3	40. SED. DEP., $Mg/Km^2/yr$		41. STORAGE LOSS, PCT.		42. SED. INFLOW, mg/L	
		a. PERIOD	b. TOTAL TO DATE	a. AV. ANN.	b. TOT. TO DATE	a. PERIOD	b. TOT. TO DATE

FIG. X1.2 Reservoir Sediment Data Summary Form, SI Units

poses in steam power plant operation should be listed under this item with a footnote explanation in Item 47.

X1.3.15 *Item 10d*—This item refers to water supply for municipal, industrial, domestic or livestock use, and fire protection.

X1.3.16 *Item 10e*—This item refers to storage space allocated specifically for water used to irrigate agricultural land.

X1.3.17 *Item 10f*—This item refers to storage allocated for regulation of low-water flow of streams, navigation pools, recharge of ground water, recreation, fish and wildlife, etc. Specify with a footnote.

X1.3.18 *Item 10g*—This refers to storage below the lowest outlet in the dam that cannot be withdrawn for any consumptive or beneficial use and is not generally considered to be of significant value for any purposes listed under Item

10f, "Conservation". This pool elevation in small reservoirs generally is considered by the Department of Agriculture to be the sediment pool elevation. It is the level below which sediment is generally continually submerged and above which the sediment deposits tend to be more compacted due to periodic exposure to the air.

X1.3.19 *Items 11a to g*—These elevations should correspond to the top of pools listed under Item 10, in terms of mean sea level, if known. Otherwise, an assumed elevation or local datum should be given, as relative elevation to the streambed level, the top of the dam or the spill-way crest. If regulation schedules provide for variation (seasonal or otherwise) in the top-of-pool levels, the maximum elevation should be shown with a reference to the footnote explanation of the other pertinent pool levels.

X1.3.20 *Items 12a to g*—Give the original surface area in

26. DATE OF SURVEY	43. DEPTH DESIGNATION RANGE IN m BELOW, AND ABOVE, CREST ELEVATION												
	PERCENT OF TOTAL SEDIMENT LOCATED WITHIN DEPTH DESIGNATION												

26. DATE OF SURVEY	44. REACH DESIGNATION PERCENT OF TOTAL ORIGINAL LENGTH OF RESERVOIR														
	0-10	10-20	20-30	30-40	40-50	50-60	60-70	70-80	80-90	90-100	-105	-110	-115	-120	-125
	PERCENT OF TOTAL SEDIMENT LOCATED WITHIN REACH DESIGNATION														

45. RANGE IN RESERVOIR OPERATION

WATER YEAR	MAX. ELEV.	MIN. ELEV.	INFLOW, m3	WATER YEAR	MAX. ELEV.	MIN. ELEV.	INFLOW, m3

46. ELEVATION–AREA–CAPACITY DATA

ELEVATION	AREA	CAPACITY	ELEVATION	AREA	CAPACITY	ELEVATION	AREA	CAPACITY

47. REMARKS AND REFERENCES

48. AGENCY MAKING SURVEY

49. AGENCY SUPPLYING DATA

50. DATE _____

FIG. X1.2 *Continued*

acres (square kilometres) at the elevation at the top of each pool shown in Item 11.

X1.3.21 *Items 13a to g*—Give the original storage capacity in acre-feet (cubic metres) for each allocation.

X1.3.22 *Items 14a to g*—Give the total original accumulated storage in acre-feet (cubic metres) from the bottom of the reservoir to the top of each pool elevation indicated. Thus, the uppermost item recorded should be the original capacity of the reservoir below the spillway crest elevation shown in Item 9.

X1.3.23 *Item 15*—Give the date when water was first impounded (month, day, and year, if possible).

X1.3.24 *Item 16*—Give the date (month, day, and year, if possible) that the initial operation for any function started.

X1.3.25 *Item 17*—Give the length of the reservoir, from the dam to the head of the backwater of the contributing stream. If the reservoir is composed of two or more principal arms, give the sum of the lengths and specify the length of each main arm in a footnote in Item 47. Give the average width by dividing the surface area by the summation of the lengths.

X1.3.26 *Item 18*—Give the entire flow-contributing drainage area above the dam.

X1.3.27 *Item 19*—Give the drainage area exclusive of the surface area of the reservoir at the spillway crest elevation (Item 9) and exclusive of the upstream non-contributing

RESERVOIR SEDIMENT
DATA SUMMARY
SCS-34 Rev. 6-66

Six Mile Creek, Site No. 3
NAME OF RESERVOIR

U. S. DEPARTMENT OF AGRICULTURE
SOIL CONSERVATION SERVICE

23-
DATA SHEET NO.

DAM						
1. OWNER Enlo Conserv. District		2. STREAM Six Mile Creek		3. STATE New State		
4. SEC 25 TWP 2N RANGE 4W		5. NEAREST P O 2 mi. E of Nebo		6. COUNTY Carroll		
7. LAT 37° 17'24"N LONG 87° 34' 15"W		8. TOP OF DAM ELEVATION 131.0		9. SPILLWAY CREST ELEV. 123.0		

RESERVOIR	10. STORAGE ALLOCATION	11. ELEVATION TOP OF POOL	12. ORIGINAL SURFACE AREA, ACRES	13. ORIGINAL CAPACITY. ACRE-FEET	14. GROSS STORAGE, ACRE-FEET	15. DATE STORAGE BEGAN
	a. FLOOD CONTROL	123.0	198.0	2091.9	3584.9	April 18, 1948
	b. MULTIPLE USE					
	c. POWER					16. DATE NOR MAL OPER. BEGAN
	d. WATER SUPPLY	111.0	124.8	1002.0	1493.0	
	e. IRRIGATION					April 28, 1948
	f. CONSERVATION					
	g. INACTIVE 1/	97.0	60.2	491.0	491.0	

WATERSHED				
17. LENGTH OF RESERVOIR 1.34 MILES	AV. WIDTH OF RESERVOIR 0.23 MILES			
18. TOTAL DRAINAGE AREA 10.14 SQ. MI.	22. MEAN ANNUAL PRECIPITATION 25.13 (25 yr) INCHES			
19. NET SEDIMENT CONTRIBUTING AREA 9.83 SQ. MI.	23. MEAN ANNUAL RUNOFF 1.6 (12 yr) INCHES			
20. LENGTH 5.17 MILES AV. WIDTH 1.96 MILES	24. MEAN ANNUAL RUNOFF 855 (12 yr) AC.-FT			
21. MAX. ELEV. 398.0 MIN. ELEV. 76.0	25. ANNUAL TEMP: MEAN 58°F RANGE -3° to 100°F			

SURVEY DATA							
26. DATE OF SURVEY	27. PERIOD YEARS	28. ACCL. YEARS	29. TYPE OF SURVEY	30. NO. OF RANGES OR CONTOUR INT.	31. SURFACE AREA, ACRES	32. CAPACITY, ACRE-FEET	33. C/I RATIO, AC.-FT. PER AC.-FT.
4-18-48	-	-		-	60.2 1/ 198.0 2/	491.0 1/ 3584.9 2/	4.14
6-23-64	16.18	16.18	Range-Contour(D)	21 R 2 CI	50.3 1/ 198.0 2/	293.2 1/ 3322.4 2/	3.84

26. DATE OF SURVEY	34. PERIOD ANNUAL PRECIPITATION	35. PERIOD WATER INFLOW. ACRE-FEET			36. WATER INFL. TO DATE, AC.-FT.	
		a. MEAN ANNUAL	b. MAX. ANNUAL	c. PERIOD TOTAL	a. MEAN ANNUAL	b. TOTAL TO DATE
6-23-64	24.81	860	1033	13,930	860	13,930

26. DATE OF SURVEY	37. PERIOD CAPACITY LOSS, ACRE-FEET			38. TOTAL SED. DEPOSITS TO DATE, ACRE-FEET		
	a. PERIOD TOTAL	b. AV. ANNUAL	c. PER SQ MI.-YEAR	a. TOTAL TO DATE	b. AV. ANNUAL	c. PER SQ MI.-YEAR
6-23-64	197.80 1/ 262.44 2/	12.22 1/ 16.22 2/	1.24 1/ 1.65 2/	197.80 1/ 262.44 2/	12.22 1/ 16.22 2/	1.24 1/ 1.65 2/

26. DATE OF SURVEY	39. AV. DRY WGT., LBS. PER CU. FT.	40. SED. DEP., TONS PER SQ. MI.-YR.		41. STORAGE LOSS, PCT.		42. SED. INFLOW. PPM	
		a. PERIOD	b. TOTAL TO DATE	a. AV. ANN.	b. TOT. TO DATE	a. PERIOD	c. TOT TO DATE
6-23-64	67.4 (8)	1820 1/ 2422 2/	1820 1/ 2422 2/	2.48 1/ 0.45 2/	40.28 1/ 7.32 2/	20,350	20,350

FIG. X1.3   An Example of a Completed Reservoir Data Summary Form

basins or the watersheds above the larger reservoirs that are effective sediment traps.

X1.3.28 *Item 20*—Give the length of the total drainage area along the center line of the main stream valley. The average width is the area in Item 18 divided by the length in Item 20.

X1.3.29 *Item 21*—The maximum elevation would be the highest point of the watershed boundary. The minimum elevation of the watershed should be the lowest original stream-bed elevation at the axis of the dam. This elevation is used to determine the height of the dam.

X1.3.30 *Items 22 to 24*—Give the longest available re-corded mean value. If known, include in parentheses the number of years of record.

X1.3.31 *Item 22*—Give the average annual precipitation value for the total drainage area. If the mean annual precipitation varies widely for different parts of the watershed, record the range of values, for example, "18 – 35".

X1.3.32 *Item 23*—Mean annual runoff in inches (millimetres) may be obtained: from direct measurement; from published reports such as USGS Water Supply Papers; by transposing known data from similar adjacent watersheds; or from average annual runoff maps such as USGS Circular 52. The source of data may be shown by footnote with

26. DATE OF SURVEY	43	DEPTH DESIGNATION RANGE IN FEET BELOW, AND ABOVE, CREST ELEVATION											
		123-120	120-11	116-112	112-108	108-10	104-100	100-97	97-96	96-92	92-88	88-8	8-76
		PERCENT OF TOTAL SEDIMENT LOCATED WITHIN DEPTH DESIGNATION											
6-23-64					1	6	19	19	4	10	12	25	4

26. DATE OF SURVEY	44.	REACH DESIGNATION PERCENT OF TOTAL ORIGINAL LENGTH OF RESERVOIR												
		0-10	10-20	20-30	30-40	40-50	50-60	60-70	70-80	80-90	90-100	-105	-110	-115	-120	-125
		PERCENT OF TOTAL SEDIMENT LOCATED WITHIN REACH DESIGNATION												
		2	17	19	14	17	10	9	7	10	5					

45. RANGE IN RESERVOIR OPERATION

WATER YEAR	MAX ELEV	MIN ELEV	INFLOW, AC·FT.	WATER YEAR	MAX ELEV	MIN ELEV	INFLOW, AC·FT
1949	123	111	1011	1957	115	83	694
50	120	113	863	58	117	92	912
51	118	112	996	59	119	96	892
52	123	111	1024	60	123	112	1033
53	123	108	989	61	123	111	943
54	119	106	1002	62	119	109	862
55	114	97	868	63	123	109	834
56	117	84	623				

46. ELEVATION-AREA-CAPACITY DATA

ELEVATION	AREA	CAPACITY	ELEVATION	AREA	CAPACITY	ELEVATION	AREA	CAPACITY
Original	Capacity	1948	96	58.0	442.3	112	127.4	1587.0
123	198.0	3554.9	92	45.7	330.0	108	104.5	1125.0
120	178.4	2832.0	88	32.1	265.7	104	83.4	750.9
116	151.8	2394.2	84	21.3	170.0	100	62.1	461.6
112	128.9	1679.0	80	11.7	73.0	97	50.3	293.2
108	109.0	1228.3		1964 Capacity		96	43.1	247.0
104	94.2	931.9	123	198.0	3322.4	92	26.4	109.6
100	75.3	658.0	120	167.5	2774.8	88	17.2	23.2
97	60.2	491.0	116	150.5	2140.8	84	1.27	0.0

47. REMARKS AND REFERENCES

1/ Sediment pool only

2/ Total reservoir below crest elevation (123.0')

Land Use in Watershed: 21 percent Woodland; 47 percent Pasture; 18 percent Cropland; 6 percent Idle; 8 percent Residential.

Geology: 25 percent Chaco shale; 18 percent Thomas ls.; 57 percent Orville ss.

48. AGENCY MAKING SURVEY	New State Watershed Planning Party, Soil Conservation Service
49. AGENCY SUPPLYING DATA	Soil Conservation Service

50. DATE Sept. 3, 1966

USDA SCS HYATTSVILLE MD 1966

Apr 1966

FIG. X1.3 *Continued*

explanation under Item 47.

X1.3.33 *Item 24*—The mean annual runoff in acre-feet (cubic metres) may be obtained by multiplying Item 23, mean annual runoff in inches (millimetres), by Item 18, total drainage area in square miles (square kilometres), times the conversion factor 53.33 (1000).

X1.3.34 *Item 25*—The mean annual temperature and the average annual range in temperature should be given in degrees Fahrenheit (degrees Celsius).

X1.3.35 *Item 26*—Give the date of the beginning of storage, if used to compute sedimentation, or the average date (month, day, and year) of the first reservoir survey, and of all succeeding surveys used in computing sedimentation. The original data from which the sedimentation record begins and the subsequent data should be given under Items 26, 29, 30, 31, 32, and 33, but the original data should not be repeated under Item 26 below or in parallel boxes from Item 34 through Item 42, inclusive.

X1.3.36 *Item 27*—Give the elapsed period between the beginning of storage or the first survey used to compute sedimentation (whichever is the more recent date) and between the average dates of each succeeding sedimentation survey. Compute to the nearest 0.1 year. If computations are calculated to the nearest 0.01 year, two decimal places may be shown.

X1.3.37 *Item 28*—Give the accumulative period from the beginning of storage or the first survey used to compute the sedimentation (whichever is the more recent date) to each succeeding sedimentation survey. Compute to the nearest 0.01 year, and show two decimal places.

X1.3.38 *Item 29*—Indicate "Range" or "Contour" and "Detailed" or "Reconnaissance" as applicable. Detailed may be shown by the symbol (*D*), and reconnaissance by (*R*). A detailed range survey is defined as one in which instrumental control of all sounding and spudding positions in the lake was maintained. If this was not done, the survey should be labeled as (*R*). In a few cases where instrumental control was not maintained, but the number of ranges and observations per range were substantially the same as those made on a detailed survey, the designation "Semi-Detailed" may be used. The symbol for this should be (*S*). A contour survey to be labeled (*D*) should conform with at least standards of third order accuracy for topographic mapping (1 in 5000). If the contouring was of a sketchy or very generalized nature, designation should be (*R*). All contouring done with Kelsh Plotters and similar equipment shall be considered (*D*), but sketching contours with portable stereoscope shall be considered (*R*).

X1.3.39 *Item 30*—Give the number of ranges or the contour interval. If a reconnaissance survey, give the number of individual measurements. The letter (*M*) should follow to indicate that they are measurements and not ranges. If a combination range and contour survey is made, the symbol (*R*) should follow the number of ranges and (*CI*) should follow the contour interval.

X1.3.40 *Item 31*—The surface area at the spillway crest elevation (use the elevation of Item 9 to obtain the first entry). If the areas of different allocated storages have been determined, each should be referenced with a footnote to be shown in Item 47.

X1.3.41 *Item 32*—The first figure entered should be the original capacity (below the spillway crest elevation, Item 9). If the capacities for different allocated storages have been determined these should be shown and each referenced with a footnote in Item 47. If the original capacity was not determined, give the first accurate capacity determined after the beginning of storage and note the date.

X1.3.42 *Item 33*—Capacity-Inflow ratio (C/I) which equals Item 32 divided by Item 24. Use the maximum capacity for the date (Item 32) for which the C/I ratio is being calculated and divide by the mean annual runoff in acre-feet (cubic metres) (Item 24). This ratio should be adjusted if there are one or more upstream reservoirs that have a significant trap efficiency and control a substantial part of the drainage area (usually more than 25 %).

X1.3.43 *Item 34*—Give the mean annual precipitation over the drainage area for each period of years given in Item 27. If there is a substantial variation in precipitation for different parts of the drainage area, give the range as "10 to 23".

X1.3.44 *Item 35*—In 35a give the average annual water inflow to the reservoir, in acre-feet (cubic metres), for each period of years given in Item 27. The highest annual for each period in acre-feet (cubic metres), is to be given in Item 35b, and the total for each period is given in Item 35c.

X1.3.45 *Item 36*—Give the water inflow in acre-feet (cubic metres), to the reservoir for the accumulated periods of years given in Item 28.

X1.3.46 *Item 37*—In Item 37a, give the volume of capacity loss below crest (Item 9) for the periods of years given in Item 27. Item 37b is obtained by dividing the volume given in Item 37a by the corresponding period of years shown in Item 27. Item 37c is obtained by dividing the value in 37b by the net sediment contributing area shown in Item 19.

X1.3.47 *Item 38*—In Item 38a give the accumulative total sediment deposits below crest for the period or periods of years given in Item 28. Item 38b is obtained by dividing the value of Item 38a by the corresponding accumulative years shown in Item 28. Item 38c is determined by dividing Item 38b by the net sediment contributing area shown in Item 19. If the above-crest deposits exist and are measured, add their volume to the below-crest deposits in Items 38a, b, and c, and also give these total values just under the other values. Where above-crest deposits are included, they should be referenced with a footnote and explained in Item 47, Remarks and References. (See Notes X1.2.3 and X1.2.4.)

X1.3.48 *Item 39*—Weighted average dry weight in pounds per cubic foot (kg/m^3) of sediment in place in the reservoir. Since the dry weight of deposits *tends to increase with time due to compaction*, an average dry weight for the total deposit should be measured or estimated at the time of each survey. If assumed values are used, indicate by an asterisk. (See X1.2.1.) Subsequent dry weights must be equal or greater than preceding weights.

X1.3.49 *Item 40*—Compute the values as follows:

Item 40a = for first survey, item 38c × item 39 × 21.78
Item 40a = subsequent surveys:

$$[(\text{Item 38a for latest survey} \times \text{item 39 for latest survey})$$
$$- (\text{item 38a for preceding survey} \times \text{item 39 for preceding survey})]$$
$$\times 21.78/(\text{item 27 for latest period} \times \text{item 19})$$
(Item 40a = for first survey, item 38c × item 39 × 0.001)
(Item 40a = subsequent surveys:
$$[(\text{Item 38a for latest survey} \times \text{item 39 for latest survey})$$
$$- (\text{item 38a for preceding survey} \times \text{item 39 for preceding survey})]$$
$$\times 0.001/(\text{item 27 for latest survey} \times \text{item 19}))$$

It is imperative that samples of the sediment representative of the entire period of sediment accumulation be obtained at the time of each survey.

$$\text{Item 40b} = \text{item 38c} \times \text{item 39} \times 21.78$$
$$(\text{Item 40b} = \text{item 38c} \times \text{item 39} \times 0.001)$$

X1.3.50 *Item 41*—Compute the values as follows:

Item 41a = item 38b × 100/item 14 (maximum value in item)
Item 41b = item 38a × 100/item 14 (maximum value in item)

X1.3.51 *Item 42*—Compute as follows:

Item 42a = [(item 40a × item 27 × item 19
$$\times 10^6)/(\text{item 35c} \times 1359)] = \text{ppm by weight}$$

(Item 42a = [(item 40a × item 27 × item 19

$$\times\ 10^6)/\text{item 35c}] = \text{mg/L})$$

Item 42b = [(item 38a × item 39 × 10^6)/(item 36b

$$\times\ 62.4)] = \text{ppm by weight}$$

(Item 42b = [(item 38a × item 39 × 10^3)/item 36b] = mg/L)

X1.3.52 *Item 43*—If elevation-capacity curves are developed, select the appropriate intervals in feet (metres) below and above the crest. Give the percentage of the total sediment deposits located within each depth designation (elevation zone). For example:

$$\frac{(\text{depth range})}{(\text{sediment, \%})} = \frac{122\text{–}100}{4}\ \frac{100\text{–}85}{5}\ \frac{85\text{–}70}{6}$$

$$\frac{70\text{–}60}{7}\ \frac{60\text{–}50}{7}\ \frac{50\text{–}40}{9}\ \frac{40\text{–}30}{10}\ \frac{30\text{–}20}{12}\ \frac{20\text{–}10}{15}$$

$$\frac{10\text{–crest}}{18}\ \frac{\text{crest–}+15}{5}\ \frac{+15\text{–}+25}{2}$$

X1.3.53 *Item 44*—The sediment distribution in percent according to distance from the dam. The reach designation is the percent of the distance from the dam to the maximum upstream extent of the spillway-crest contour at the elevation given in Item 9 at the date of the beginning of storage. Thus, 20 % would be $\frac{1}{5}$ of the distance from the dam to the head of backwater at the original crest stage.

X1.3.54 *Item 45*—List the maximum and minimum water elevations and the total inflow in acre-feet (cubic metres) for each water year of record.

X1.3.55 *Item 46*—Give data from the elevation-capacity curve for the latest survey shown on Item 26. Be sure to label each survey date on the form. If space permits, give data from the elevation-capacity curve for the original survey.

X1.3.56 *Item 47*—List here all published and unpublished reports on sedimentation surveys of this reservoir. All footnote explanations are to be shown in this space. Also note and give any pertinent data, including dates of abnormal operational occurrences, such as reservoir evacuation; sluicing out sediment; releasing density currents; extreme floods and droughts; changes in spillway-crest elevation; use of flash boards; and the installation of upstream control structures. Briefly describe the sediment and any available textural analyses. If needed, use continuation sheets.

X1.3.57 *Item 48*—Give the department, agency, and division, branch, or field office responsible for each survey.

X1.3.58 *Item 49*—Give the agency and department reporting the data.

X1.3.59 *Item 50*—Give the date this form was prepared by the office listed in Item 49.

X2. COMMENTS ON RESERVOIR MORPHOLOGIC SURVEYS

X2.1 When tonnage of accumulated sediment is reported, the unit weight of the sediment should be indicated as measured (number of samples) or estimated. Details concerning measurement technique should be provided; also, if a weighted unit weight is reported, indicate how this was determined.

X2.2 The method of measuring water depth, the number of stations and the number of ranges should be reported. Dry land surveys should indicate precision and the number of points surveyed per unit area.

X2.3 Map contouring for volume computations based on the contour-area method should be performed by computer methods.

X2.4 Reservoir sediment design life volume and significant runoff elements should be reported.

X2.5 The investigation report should include the storage location of the survey data and documentation of watershed characteristics (if appropriate to the purpose of the survey).

REFERENCES

(1) *National Engineering Handbook*, Section 3, Chapter 7, U.S. Department of Agriculture, Soil Conservation Service, pp. 1–31.
(2) *Contracting Specifications for Reservoir Sedimentation Surveys*, U.S. Department of Agriculture, Soil Conservation Service.
(3) *National Handbook of Recommended Methods for Water-Data Acquisition:* Office of Water Data Coordination, USGS, 1978, pp. 3–57 to 3–69.
(4) Blanton, James, III, "Procedures for Monitoring Reservoir Sedimentation," *U.S. Bureau of Reclamation Technical Guideline*, October 1982.
(5) *Measurement of Hydrographic Parameters in Large Sand-Bed Streams from Boats*, Task Committee on Hydrographic Investigations of the Committee on Waterways of the Waterway, Port, Coastal, and Ocean Division, 1983, American Society of Civil Engineers, New York, New York 10017.
(6) *Hydrographic Manual*, Fourth Edition, National Oceanic and Atmospheric Administration, 1976.

ASTM Designation: D 5906 – 96

Standard Guide for
Measuring Horizontal Positioning During Measurements of Surface Water Depths[1]

This standard is issued under the fixed designation D 5906; the number immediately following the designation indicates the year of original adoption or, in the case of revision, the year of last revision. A number in parentheses indicates the year of last reapproval. A superscript epsilon (ε) indicates an editorial change since the last revision or reapproval.

1. Scope

1.1 This guide covers the selection of procedures commonly used to establish a measurement of horizontal position during investigations of surface water bodies that are as follows:

	Sections
Procedure A—Manual Measurement	7 to 12
Procedure B—Optical Measurement	13 to 17
Procedure C—Electronic Measurement	18 to 27

1.1.1 The narrative specifies horizontal positioning terminology and describes manual, optical, and electronic measuring equipment and techniques.

1.2 The references cited contain information that may help in the design of a high quality measurement program.

1.3 The information provided on horizontal positioning is descriptive in nature and not intended to endorse any particular item of manufactured equipment or procedure.

1.4 This guide pertains to determining horizontal position of a depth measurement in quiescent or low velocity flow.

1.5 The values stated in inch-pound units are to be regarded as the standard. The SI units in parentheses are for information only.

1.6 *This standard does not purport to address all of the safety concerns, if any, associated with its use. It is the responsibility of the user of this standard to establish appropriate safety and health practices and determine the applicability of regulatory limitations prior to use.*

2. Referenced Documents

2.1 *ASTM Standards:*
D 1129 Terminology Relating to Water[2]
D 3858 Test Method for Open-Channel Flow Measurement of Water by Velocity Area Method[2]
D 4410 Terminology for Fluvial Sediment[2]
D 4581 Guide for Measurement of Morphologic Characteristics of Surface Water Bodies[2]
D 5073 Practice for Depth Measurement of Surface Water[3]

3. Terminology

3.1 *Definitions*—For definitions of terms used in this guide, refer to Terminology D 1129.

3.2 *Descriptions of Terms Specific to This Standard:*

3.2.1 *accuracy*—refers to how close a measurement is to the true or actual value. (See Terminology D 1129.)

3.2.2 *baseline*—the primary reference line for use in measuring azimuth angles and positioning distances.

3.2.3 *continuous wave system*—an electronic positioning system in which the signal transmitted between the transmitter and responder stations travels as a wave having constant frequency and amplitude.

3.2.4 *electronic distance measurement (EDM)*—measurement of distance using pulsing or phase comparison systems.

3.2.5 *electronic positioning system (EPS)*—a system that receives two or more EDM to obtain a position.

3.2.6 *global positioning system (GPS)*—a global positioning system (GPS) is a satellite-based EDM system used in determining Cartesian coordinates (x, y, z) of a position by means of radio signals from NAVSTAR satellites.

3.2.7 *horizontal control*—a series of connected lines whose azimuths and lengths have been determined by triangulation, trilateration, and traversing.

3.2.8 *line of position (LOP)*—locus of points established along a rangeline.

3.2.9 *precision*—refers to how close a set of measurements can be repeated. (See Terminology D 1129.)

3.2.10 *pulsed wave system*—an electronic positioning system in which the signal from the transmitting station to the reflecting station travels in an electromagnetic wave pulse.

3.2.11 *range*—distance to a point measured by physical, optical, or electronic means.

3.2.12 *range line*—an imaginary, straight line extending across a body of water between fixed shore markings.

3.2.13 *range line markers*—site poles or other identifiable objects used for positioning alignment on a range line.

3.2.14 *shore markings*—any object, natural or artificial, that can be used as a reference for maintaining boat alignment or establishing the boats position as it moves along it course. Examples include range line markers, sight poles, trees, power poles, land surface features, structures, and etc.

3.2.15 *site poles*—metal or wood poles used as a sighting rod.

3.2.16 *stadia*—telescopic instrument equipment with horizontal hairs and used for measuring the vertical intercept on a graduated vertical rod held vertically and at some distance to and in front of the instrument.

3.2.17 *total station*—an electronic surveying instrument which digitally measures and displays horizontal distances and vertical angles to a distant object.

[1] This guide is under the jurisdiction of ASTM Committee D-19 on Water and is the direct responsibility of Subcommittee D19.07 on Sediments, Geomorphology, and Open-Channel Flow.
Current edition approved Feb. 10, 1996. Published May 1996.
[2] *Annual Book of ASTM Standards*, Vol 11.01.
[3] *Annual Book of ASTM Standards*, Vol 11.02.

4. Summary of Guide

4.1 This guide includes three general procedures for determining the location or horizontal position in surveying of surface water bodies. The first determines position by a manual procedure. The equipment to perform this procedure may be most readily available and most practical under certain conditions.

4.2 The second determines position by a optical procedure.

4.3 The third determines position by a electronic procedure.

4.4 Horizontal control stations shall be in accordance with Third Order, Class I, Federal Geodetic Control Committee Classification (FGCC) Standards,[4] with traverses for such controls beginning and ending at existing first- or second-order stations (1).[5]

5. Significance and Use

5.1 This guide is intended to provide instructions for the selection of horizontal positioning equipment under a wide range of conditions encountered in measurement of water depth in surface water bodies. These conditions, that include physical conditions at the measuring site, the quality of data required, the availability of appropriate measuring equipment, and the distances over which the measurements are to be made (including cost considerations), that govern the selection process. A step-by-step procedure for obtaining horizontal position is not discussed. This guide is to be used in conjunction with standard guide on measurement of surface water depth (such as standard Practice D 5173.)

6. Horizontal Positioning Criteria

6.1 The level of accuracy required in horizontal positioning can be defined in three general classes:

6.1.1 Class One pertains to precise positioning demanding a high degree of repeatability.

6.1.2 Class Two is for medium accuracy requirements typical of project condition studies or offshore/river hydraulic investigations, or both.

6.1.3 Class Three is for general reconnaissance investigations requiring only approximate measurements of positions.

6.1.4 Table 1 provides an estimate of the suitability by Class for the different horizontal positioning discussed within this guide (2).

PROCEDURE A—MANUAL MEASUREMENT

7. Scope

7.1 This procedure explains the measurement of horizontal position using manual techniques and equipment. These include use of tagline positioning techniques and application of shore marks.

7.2 Description of techniques and equipment are general in nature and may need to be modified for use in specific field conditions.

---

[4] Available from NOAA, National Geodetic Survey, 1315 East West Highway, Room 8657, Silver Spring, MD 20910-3282.
[5] The boldface numbers given in parentheses refer to a list of references at the end of this standard.

TABLE 1 Allowable Horizontal Positioning System Error (7)

System	Estimated Positional Accuracy		Suitable for Survey Class		
	±1 ft RMS	(±1 m) (RMS)	1	2	3
Visual range intersection	10 to 66	(3 to 20)	No	No	Yes
Sextant angle resection	7 to 30	(2 to 10)	No	Yes	Yes
Transit/theodolite angle intersection	3 to 16	(1 to 5)	Yes	Yes	Yes
Range-azimuth intersection	1.6 to 10	(0.5 to 3)	Yes	Yes	Yes
Tagline high frequency EPS	3 to 13	(1 to 4)	Yes	Yes	Yes

8. Significance and Use

8.1 Prior to the development of optical and electronic positioning equipment, manual equipment and techniques were the only means of measuring horizontal position. These techniques and equipment are still widely used where precise controlled measurements may be required (for example, taut cable method), or where limitations in equipment availability, site conditions and cost considerations prohibit use of more modern equipment.

9. Tagline Positioning Techniques

9.1 Tagline positioning techniques makes use of a measuring line having markings at fixed intervals along its length to indicate distance. These can be either a taut cable in which the line is anchored firmly at opposite banks and stretched taut, or a boat mounted cable in which one end of the line is firmly anchored at the bank and the other is attached to a boat with the line fed out as the boat proceeds along its course. Both methods are frequently used low cost positioning techniques. The taut cable is most commonly used for obtaining streamflow measurements and sediment sampling data at non-bridge locations on rivers and streams, but is equally applicable for controlled boat positioning when obtaining river or lake bed profiles for other purposes. In this regard it has proven especially useful for positioning on small lakes or reservoirs, usually where distances involved are less than 1000 ft (305 m), and where sheer walls exist at both ends of the range, or where the presence of dense vegetation along the shoreline precludes use of optical or electronic positioning methods. The boat mounted tagline, in contrast, is much easier to set up and use since only one end of the line is anchored at the shore, but this method can be considerably less accurate due to the increased possibility of misalignment errors.

9.1.1 *Taut Cable Method (Manual Procedure):*

9.1.1.1 For the taut cable method (see Fig. 1), firmly anchor the ends of the cable on both banks (see 9.1.1.2 for installation) and the line then pulled as taut as possible without pulling the anchors out of the bank. This method of positioning is recognized as accurate for use on streams where the flow velocity does not exceed more than a few feet per second so that the drag induced by the flow, on any boat or other attachment, does not substantially deflect the line. The taut cable method is time consuming when compared to other more modern optical and electronic positioning equipment and techniques; take this into consideration when deciding on which equipment and techniques best apply (3).

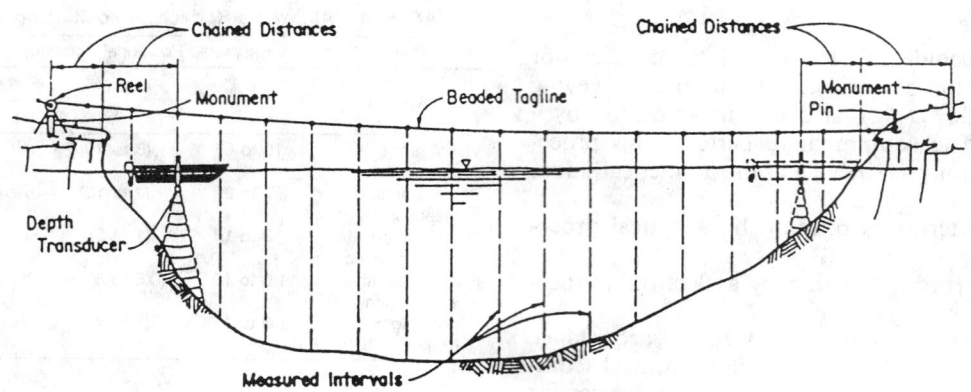

FIG. 1 Taut Cable Method

9.1.1.2 Installation of the taut cable should be done either in one of two ways: either securely anchor the cable to one bank and the line fed from a boat mounted reel as the boat proceeds across the body of water; or securely anchor the reel to one bank near the water's edge with the loose end towed across. Shore markings can be used for visual alignment, but the normal procedure is to place a transit or theodolite on line for this purpose. The transmit person, equipped with a two-way radio, relays alignment directions to the boat operator (also equipped with a two-way radio), as the line is transported to the opposite bank. A power or hand winch or hand cranked reel, skid mounted on locally fabricated support assemblies, can be attached to a tree or other firm support on shore and used to take slack out of the line and to minimize sag associated errors in distance. For safety, the reel should come equipped with a spring-loaded pin lock brake assembly. Buoys may be placed at optimum locations along the line to help reduce sag as well as provide an indicator of boat alinement.

9.1.1.3 Taglines for the taut cable method are commonly stainless steel or galvanized 7 by 7 cable, although a fiber line is increasingly being used. The stainless steel lines generally come pre-beaded at 2 ft intervals for the first 50 ft, at 5 ft for the next 100 ft (30 m), and at 10-ft intervals for to the end of the line. Sizes vary in diameter with the length of the cable used. For a length less than 400 ft (122 m), a 1/32 in. (0.79 mm) diameter line is recommended; for lengths up to 800 ft (244 m), a 1/16 in. (1.59 mm) diameter is recommended; for greater lengths, the diameter should be at least 1/8 in. (3.18 mm). The fiber line is normally 3/16 in. (4.76 mm) diameter, is normally yellow with black markings and generally comes available in any length up to 1000 ft (305 m). It is usually pre-marked with one mark every 10 ft (3 m) and two marks every 100 ft (30 m). To prevent damage when attaching the tagline to a tree, connect the free end of the tagline (the end not connected to a reel), to a 30 ft length of 3/32-in. (2.37 mm) diameter cable. One end of this cable should have a harness snap and the other should have a pelican hook. The free end of the tagline should be equipped with a sleeve and thimble, of size matching the tagline diameter (4).

9.1.1.4 Attachments for holding the boat in position at a fixed location along the tag line will vary depending on the specific needs of the data collection effort. Normally the attachment is some form of clamp arrangement. If velocity measurements or sediment sampling is being done along with the water depth measurements, the standard procedure is to equip the boat with a crosspiece (I-beam), normally a little longer than the width of the boat, and set perpendicular to the boat's centerline. The crosspiece is either clamped or bolted in place and has guide sheaves at each end and a clamp arrangement somewhere along the length of the crosspiece. With the tagline fed through the sheaves, the boat can be held in place or moved along the tagline from station to station. The mid-point clamp permits the boat to be fixed to a one location and not move laterally along the line as measurements are being taken. For safety, fasten a small rope to the clamp to permit quick release in the event of an emergency (4).

9.1.1.5 Position a standby person near the quick-release end of the cable, to release the cable, if there is a possibility of a boat, barge, or other large obstruction colliding with the cable. In addition, a chase boat should also be present in traffic locations, to warn boaters of the cable's presence. For locations where flow velocities are high and boats and obstructions are present upstream, it is recommended that all work boats be kept downstream of the cable.

9.1.1.6 Positions along the line are determined either through the use of a calibrated measuring wheel, or by keeping track of markings attached to the line. Weight, sag, and strength of the line limit its range to less than 1000 ft (305 m). Markings must be clearly visible and easy to understand. The markings are normally crimped brass beads set at 20 ft (6 m) intervals, but this can vary significantly depending on the field application and specific measurement requirements. Attach fluorescent flagging in 4 to 5 ft (1.2 to 1.5 m) lengths at 50 ft (15.2 m) intervals along the length of the cable to assist in ease of distance measurements. Fluorescent flagging provides the best visibility for all types of lighting and is recommended for use in visually marking the cable for safety purposes.

9.1.1.7 Begin measurements by positioning the boat near the bank at a fixed mark on the cable. This initial or starting mark should be determined by chain measurements from a surveyed control point near the water's edge. The boat proceeds from this initial mark along the cable to the opposite shore with a person on board calling out "fix" marks as each marking on the line is reached (5).

9.1.1.8 Maintain alignment of a taut cable within at least 1 to 2 ft (0.3 to 0.6 m) of accuracy.

9.1.2 *Boat Mounted Tagline (Manual Procedure):*

9.1.2.1 This method (see Fig. 3), also referred to as tethered piano-wire method, is similar in principle to the taut cable with the distinction that only one end of the line is anchored at the bank. The opposite end is attached to a reel mounted on the boat and the line is fed out as the boat proceeds along its path. Boat mounted taglines can be used over substantial distances but are not suitable for Class 1 survey use unless the length of the cable is less than 1500 ft (457 m) when stationary measurements are made or less than 1000 ft (305 m) if the boat is moving. Table 2 provides an estimate of positional accuracies for different cable lengths **(2)**. The cable should be constructed of 0.059 in. (0.15 cm) diameter, steel piano wire or steel cable, and have markings attached that permit distance to be read by the length of line released from the reel. The standard guide is to use a reel with a calibrated measuring wheel and mechanical counter that activates as the wheel rotates. Locate the reel in the rear of the boat for safety and position as near as possible to the desired point of measurement (that is, sounding device). If the reel and the point of measurement do not coincide, record the distances between the two and add to the tagline measurements. Boat draft should be less than 1 ft (0.3 m) for accurate measurements in shallow water areas.

9.1.2.2 Establish positions along the boat's path by setting the boat on line and as near to the shore as practical. Then set the counter to zero and pull the loose end of the wire out and fasten it to a pin or other anchoring device driven on-line near the shore. Record distance from the pin to the water's edge on a chart or notebook kept in the boat. With this done, the boat begins to proceed along its designated path. Attach styrofoam floats to the line and dropped overboard at 100 to 300 ft (30 to 91 m) intervals that reduces sag by holding the wire near the surface. Release the line from its starting point as the boat reaches the opposite shore and retrieve it using the motor driven winch aboard the boat **(6)**.

9.1.2.3 Special care should be made to certify that the lateral alignment of the boat is held as the boat proceeds along its path and that the line is held taut. The tagline method maintains alignment through use of visual shore marks. But accuracy can be improved appreciably with a transit or theolodite person relaying alignment directions by a two-way radio.

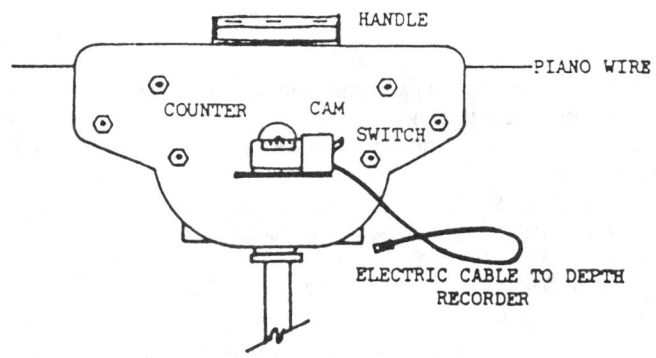

FIG. 2(a)   Calibrated Wheel—Front Cover

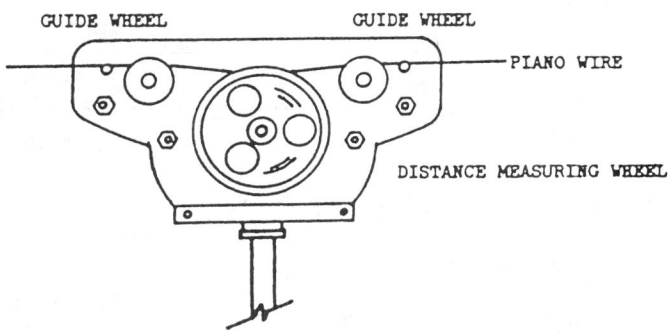

FIG. 2(b)   Calibrated Wheel (Cover Removed)

9.1.2.4 If tagline markings are used in lieu of a measuring wheel, they should be easy to see and understand to avoid errors in determining the readings.

9.2 *Calibrated Wheel (Manual Procedure):*

9.2.1 A 2 ft (0.6 m) diameter calibrated measuring wheel (see Fig. 2) can be used with the taut cable as a replacement for line markings. A counter attached to the wheel registers the revolutions of the wheel as the boat moves along its prescribed path. Anchor the piano wire on both banks and hold taut with enough sag to permit the piano wire to encircle the wheel, but provide enough friction to prevent slippage. Repair breaks that might occur in the line through the use of a compressed-sleeve type wire splicer.

9.2.2 The procedure for the calibrated wheel method is the same as that described for the taut cable, except that the calibrated wheel is used in lieu of markings permanently attached to the line. The positioning begins with the boat stationed at a fixed measured point on the line (established by distance measurement from a control station on shore), the counter on the wheel set to zero, and the boat's position relative to this starting position measured as the boat moves along its path.

10. Cable Reels (Manual Procedure)

10.1 Cable reels must be constructed of sturdy material and be equipped with a manual, electrical, or gasoline powered winch, including a clutching and braking assembly. The brake is used both for controlling rotation of the reel as the cable is let out and to serve as a safety feature. For hand operated reels, the crank should be hinged to allow the crank to be disengaged from the shaft while the wire is let out and engaged for reeling in. Various devices are employed to drive a counter to register the amount of cable released from the reel **(5)**.

10.2 The cable for the boat mounted tagline should be made of galvanized, commercially available aircraft cable with $3/32$ in. (2.38 mm) outside diameter (O.D.) and seven by seven construction. Some cables may also come with a nylon coating to reduce fraying typical of uncoated steel cables. The coating, however, increases the outside diameter to $1/8$ in. (3.18 mm) and reduces the length of cable that can be wound on a reel by about 50 %. Plastic water-ski tow cable of $1/4$ in. (6.35 mm) diameter, 1 lb/100 ft weight (0.45 kg/30.5 m), is available from sporting goods stores and may also be used occasionally with good success but is not recommended for use during windy conditions because of the tendency to

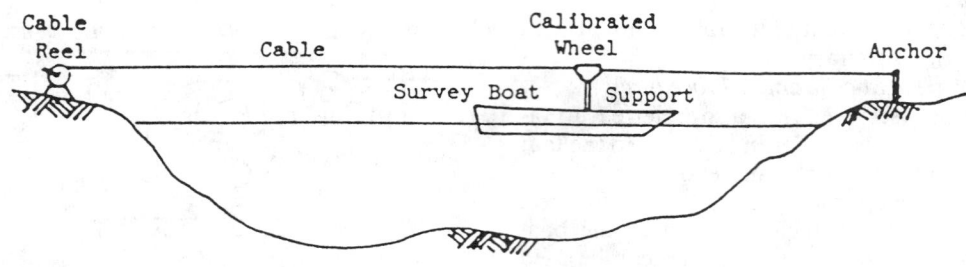

FIG. 2(c)   Calibrated Wheel Method

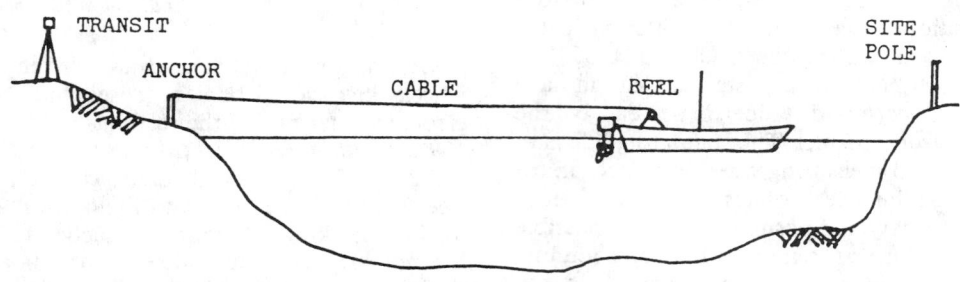

FIG. 3   Tag Line

TABLE 2   Allowable Tagline Positioning Procedures/Systems Criteria (7)

Note—Tagline distance range limits shown for Class 1 and Class 2 surveys are contingent on the tagline being pulled clear of the water and held taut during measurement. Distances for Class 1 surveys shall be adjusted downward depending on the capabilities of the boat and equipment used.

System	Estimated Positional Accuracy		Suitable for Survey Class		
	±1 ft RMS	±1 m RMS	1	2	3
Tagline (Boat Stationary)					
<1500 ft from baseline	1 to 3	(0.3 to 1)	Yes	Yes	Yes
>1500 to <3000 ft	3 to 16	(1 to 5)	No	Yes	Yes
>3000 ft from baseline	16 to 164	(5 to 50+)	No	No	Yes
Tagline (Boat Moving)					
<1000 ft from baseline	3 to 9	(1 to 3)	Yes	Yes	Yes
>1000 to <2000 ft	8 to 20	(3 to 6)	No	Yes	Yes
>2000 from baseline	20 to 164+	(6 to 50+)	No	No	Yes
Tagline Calibration Frequency (in months)			1	6	12
Accuracy of Independent Calibration (feet)			0.1	1	5
Accuracy of Independent Calibration (meters)			0.03	0.01	1.52

be deflected. The plastic cable has the advantage of being easily repaired in the field by telescoping one end into the other.

11. Current Meters (Manual Procedure)

11.1 A standard "Price current meter" suspended from a boat, in which the propeller blade rotates because of movement of the boat, can be used to measure the distance the boat has traveled by keeping track of the accumulative revolutions of the meter times a calibration constant for the mounted meter. When coupled with the sonic-sound "fix" switch, the current meter can be used in a semi-automatic operation for recording and documenting positions along the range line.

12. Shore Marks (Manual Procedure)

12.1 Shore marks may be used with taglines, alone in pairs, or with optical and electronic procedures. Their function is to help maintain boat alignment as the boat moves along its prescribed course. The boat operator can visually align the boat's movement on target with the two or more shore marks, or an instrument man on shore can convey instructions by two-way radio as the boat proceeds between shore marks on the opposite bank. A transit works well for stadia positioning since the transit can be used both for alignment and stadia measurements. Shore marks, placed in a perpendicular alignment to the boats path, can serve as an indicator of position and, as such, provide a good, rough measurement of position for reconnaissance surveys.

12.2 Accuracy in use of shore marks depends on ease of visibility and how sharp the delineation is between the two or more objects being used for line of sight. Place the shore marks far enough apart to enable alignment to be clearly distinguishable.

PROCEDURE B—OPTICAL MEASUREMENT AND ALIGNMENT

13. Scope

13.1 This procedure explains the use of optical equipment in horizontal positioning.

13.2 Equipment includes transits, theodolites, and alidades, along with sextants, or range poles. The techniques applied include stadia positioning, transit intersection, transit-stadia positioning, and sextant positioning.

14. Stadia Measurements (Optical Procedure)

14.1 The stadia method (see Fig. 4) uses transit or alidade standard guides similar to that applied in standard land surveying applications. For boat positioning, however, the

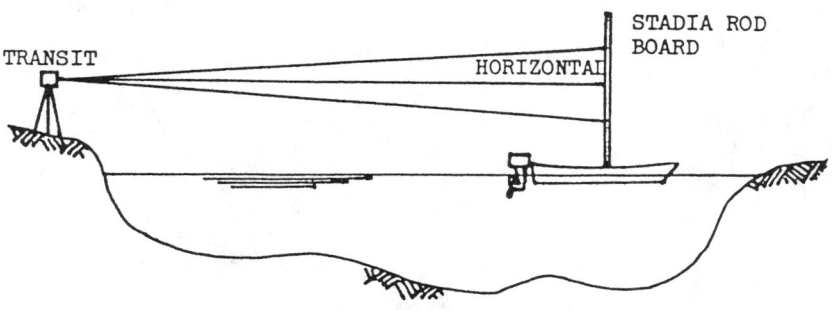

FIG. 4 Stadia Method

stadia board is much larger in size and often has coded markings in lieu of numbers. Figure 5 illustrates a section of a typical board used by some field offices of the U.S. Army Corps of Engineers. The board must be securely mounted in the sounding boat and positioned as near to the required point of measurement as possible. Distances are normally read at 100 ft (30.5 m) intervals, unless field needs warrant otherwise. The transit operator is stationed on shore at a surveyed control point and conveys alignment and distance instructions to the boat operator through use of two-way radios. Each distance measurement is usually preceded by a "stand-by" message (including an indication of the distance),

followed in a few seconds by the signal "fix" as the actual distance is read. The spacing of fix marks can be made either by the transit or sounder operator. If made by the transit operator, the spacing is more uniform and the chance for erroneous readings is reduced. If made by the sounding operator, the operator of the water depth measurement instrument, the spacing can be determined by the change in boat speed and the need for distance reading is indicated when abrupt changes in depth are detected.

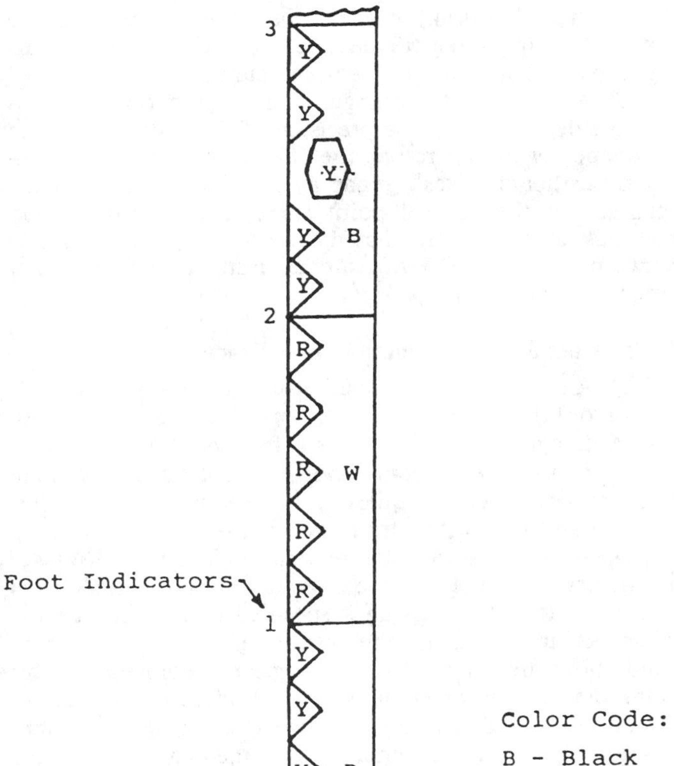

14.2 The limit for distance measurements using stadia depends on the length of the stadia board and the telescopic power of the transit. This limit should not normally exceed 1000 to 1500 ft (305 to 457 m). Longer distances are possible by stationing an instrument person at each end of the range and overlapping and averaging several stadia readings.

14.3 A stadia board should be at least 15 ft (4.6 m) in length, hinged into two 7½ ft (2.3 m) sections for easy transport, have a triangular cross section with 6 in. (1.5 mm) side widths, and be constructed of aluminum. The markings on the stadia board should be in an alternating pattern of white markings with red points and black markings with yellow points. This enables it to be distinguishable against a variety of backgrounds and more visible at greater distances. In addition the board should be equipped with special marking to signify 250, 500 and 750 ft distances.

14.4 The standard procedure in obtaining stadia measurements is to set the transit on the rangeline alignment, on land, and near the water's edge. The transit should be plumbed over a surveyed control point and be at normal eye level but as near to the water level as possible to avoid the need for excessively long stadia rods. Care must be taken to assure that the stadia rod is firmly mounted in the boat. Stadia readings are best obtained when waves are not present because of the difficulty in obtaining readings in a timely and accurate manner under conditions where the boat is subject to motions of pitch, roll, yaw, and heave. A tape, chain, stadia reading, or other suitable form of measurement will have to be made between the vertical centerline of the transit's position and the stadia rod before the boat can begin to proceed along its course. If possible the boat should proceed at a constant speed. A "fix" signal should be given when the boat has to slow down on reaching the shore with a final reading when the boat reaches the opposite bank.

15. Transit Intersection (Optical Procedure)

15.1 This is a two-transit method (see Fig. 6): one transit

Foot Indicators

Color Code:

B – Black
R – Red
W – White
Y – Yellow

FIG. 5 Stadia Board

ASTM D 5906

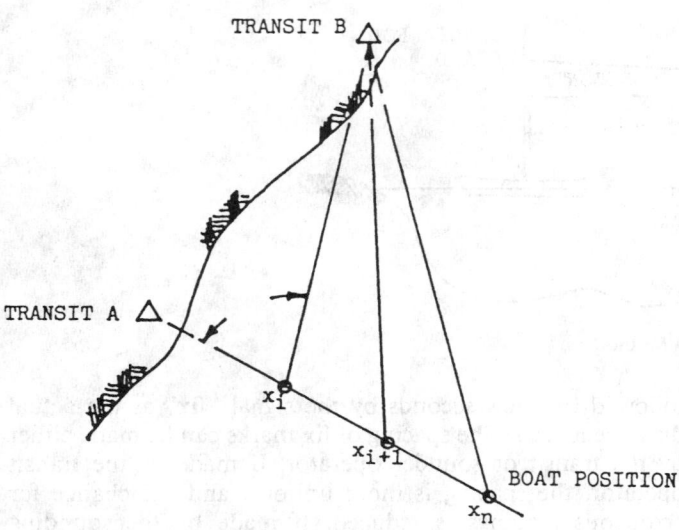

FIG. 6 Transit Intersection

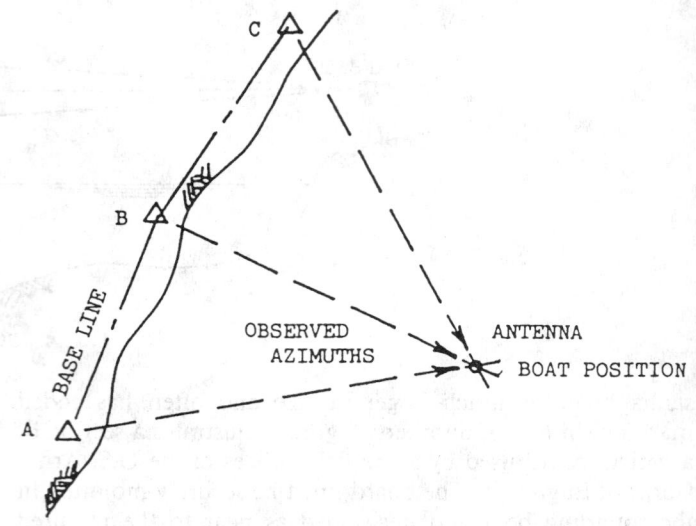

FIG. 7 Triangulation/Intersection

is placed on line with the boat's prescribed path and used for alignment instructions and the second is located at some known relative position upstream or downstream to permit the determination of an angular position on the boat as it proceeds on its course. The boat's positions along the path is determined by measuring the straight or baseline distance between the two transit stations, A and C, observing the interior angle formed between this baseline and the boat at Station x_i, then solving trigonometrically for the leg of the triangle represented by the distance between the boat and the transit at Station A. Make position readings at 100-ft (30-m) intervals or as essential to permit an accurate profile of the lake bed surface. The angles may be predetermined to intersect fixed positions along the course. Examples would be 100-ft (30 m) or 200-ft (61 m) spacings. "Fix" marks are transmitted to the boat by two-way radios as the boat reaches each mark.

15.2 The angle of intersection at the boat should be such that a directional error of 1 min in arc from a transit station will not cause the position of the boat to be in error by more than 1.0 min at the scale of the maps being used. Angles should be maintained between 30 and 150°. Table 3 provides an estimate of positional error due to azimuth misalignments (7).

16. Triangulation/Intersection Positioning (Optical Procedure)

16.1 The intersecting lines of sight (see Fig. 7) from two or more shore based transit or theodolite stations provide an

TABLE 3 Positional Error Due to Azimuth Misalignment (7)

Distance, ft (m)		Distance Error per Alignment Procedure			
		Sextant (±20 min), ft (m)		Transit (±2 min), ft (m)	
100	(30)	0.6	(0.2)	0.1	(0.0)
500	(152)	3	(0.9)	0.3	(0.1)
1000	(305)	6	(1.8)	0.6	(0.2)
2000	(610)	12	(3.7)	1.2	(0.4)
5000	(1524)	29	(8.8)	2.9	(0.9)
10 000	(3048)	58	(17.7)	6	(1.8)

accurate measurement of boat position, suitable for all three classes of horizontal positioning. Each observation point can be determined graphically, by plotting the angular data, or mathematically, by using the known distance between any two shore stations and the inclosed angles formed by the intersecting lines. Read angles to the nearest 0.01° and make at least two backsight checks when the instruments are initially set up. Make frequent rechecks if the instruments are to be used for several sets of measurements.

16.2 Accuracy of the triangulation/intersection method is strongly dependent on the precision of the instruments, the experience of the operators, the effectiveness of the coordinated positional "fixes" (made by two way radio), and the accuracy of the control point survey. As in intersection methods, the angle formed by the intersecting lines should be between 30 to 150° (3). Measure azimuths to within 30 s of arc (1).

17. Sextant Measurements (Optical Procedure)

17.1 A sextant is a hand-held instrument used to measure vertical or horizontal angles up to a maximum spread of 140° (1). Although once widely used for boat positioning, a sextant is no longer recommended for positioning of small boats, except under very quiescent stream or lake conditions. This is due to the difficulty in keeping the boat stable long enough to measure the required angles. The sextant angles are observed by holding the sextant in a horizontal plane with the left hand, viewing a specified object through the telescope, and adjusting the mirror reflection of the right hand object until the two objects appear to coincide. At this point, the angle between the two objects is read at the vernier position of a graduated arc attached to the sextant. Sextant positioning should not be made where the angle between the two objects is less than 15° or for short distances of less than 1000 ft (305 m) (6).

17.2 Sextant Resectioning (Optical Procedure):

17.2.1 Sextant resectioning (see Fig. 8) is similar in concept to transit-intersection, except that the observation station is on the boat in lieu of the shore, and baseline distances between shore markings are not normally required.

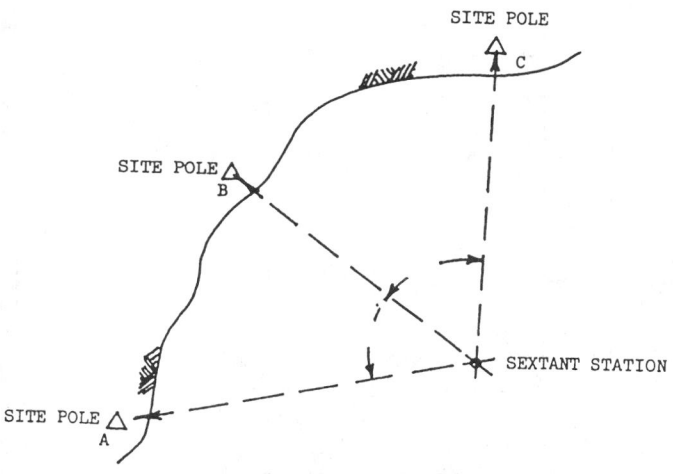

SITE POLE

C

SITE POLE
B

SEXTANT STATION

SITE POLE
A

FIG. 8 Sextant Resectioning

Positions along the boat's path are determined by manually plotting sextant angles between two or more shore marks. Read observations to the nearest 0.1 min of arc. The major drawback to sextant resectioning stems from the fact that it is labor intensive; requiring a boat operator, possibly two sextant observers, a depth recorder operator (if water depths are being measured), and a data logger/plotter (7).

17.2.2 Error sources in sextant resectioning include instability of the observation platform (boat motion), imprecision in the sextant angles, lack of synchronization in observations being made simultaneously, plotting errors, observer fatigue, and target delineation. The potential for error increases noticeably with decreases in boat size because of difficulty in maintaining a stable observation platform with smaller size boats. Precision in the angle measurement depends on the resolution of the instruments, the skills of the observers, how quickly the angles are changing, and the distinction of the shore based targets. The weight of the hand held sextants and the difficulty in holding a stable platform (boat) long enough to obtain an observation adds an additional factor that must be taken into consideration. Such errors make it essential that sextants be calibrated periodically between observations. An estimate of positional error due to sextant azimuth misalignments is given in Table 3 (7).

17.2.3 Angles of observation should ideally be between 20 and 110° of arc. If more than one sextant observation is being made simultaneously, position the observers on the boat as close together as possible to minimize parallax into the adjacent angles of observation. Also, the offset between the position point and the position where the water depth measurement is being made must be taken into account (1).

PROCEDURE C—ELECTRONIC POSITIONING

18. Scope

18.1 This procedure explains the measurement of horizontal position using electronic positioning systems (EPS). It is electronic distance measurement (EDM) equipment using microwave, visible light, laser light, and infrared light bandwidths.

18.2 Description of techniques and equipment are general in nature. Techniques and equipment may need to be modified for use in specific field conditions.

19. Significance and Use

19.1 Electronic positioning systems offer a broader variety of positioning techniques than available with mechanical and optical methods. This includes automation of angle and distance measurements, position measurement over much longer distances, and potential for on-site microcomputer data storage and processing. In addition to using positioning procedures that are adaptions of those used in optical positioning, the electronic positioning systems permit range-range boat tracking and fully integrated positioning systems.

20. Signal Frequency Selection

20.1 All EPS equipment discussed within this guide is limited to that operating in the super high frequency range (3 to 30 GHz), in contrast to lower frequency systems commonly used for ocean navigation. A classification of various electronic positioning systems is given in Table 4. The higher frequency range provides the needed level of accuracy required for measuring boat positions over distances up to 3 to 5 miles (5 to 8 KM). Most of the techniques applied are adaptations of those used with optically based systems. The techniques vary from shore based range-azimuth systems (in which the stadia rod is replaced with a reflector station) to the more sophisticated boat-based electronic transponder based systems in which distance is measured electronically from the boat to two or more transponders stationed on shore. The position of the boat is calculated on the basis of a trigonometric solution using the measurements between the EDM equipment and the shore based transponders (8). The output from electronic positioning system can be in digital or analog form, but nearly always in a form compatible with automated processing systems. Some EDM's are also equipped with a microprocessor capable of preprocessing the raw data. This includes removing skew in the data stemming from the boats travel during the measurement, as well as smoothing of the data to eliminate poor signals due to low signal-to-noise ratio, and errors associated with instrument instability (1).

21. Measurement Principals

21.1 EDM instruments operate by transmitting electromagnetic wave energy (the carrier signal) to a receiver station or responder where it is retransmitted as a return signal back

TABLE 4 Electronic Positioning Systems (7)

Bandwidth	Symbol	Frequency	System
Very low frequency	VLF	10 to 30 KHz	Omega
Low frequency	LF	30 to 300 KHz	Loran-C
Medium frequency	MF	300 to 3000 KHz	Raydist, Decca
High frequency	HF	30 to 30 MHz	
Very high frequency	VHF	30 to 300 MHz	
Ultra high frequency	UHF	300 to 3000 MHz	Del Norte
L-Band			NAVSTAR GPS
Super high frequency	SHF	3 to 30 GHz	(Microwave EPS)
C-Band			Motorola
S-Band			Cubic
X-Band			Del Norte
Visible light	. . .	. . .	(EDM)[A]
Laser light	. . .	. . .	(EDM)[A]
Infrared light	. . .	. . .	(EDM),[A] Polarfix

[A] EDM—Electronic distance measuring instrument.

to the transmitting station. The delay time or phase shift in the signals (depending on the type of equipment used) provides a measure of distance between the two stations. Special care must be made to avoid the influence of intervening objects between the transmitter and receiving stations, such as trees, hills, buildings and other structures, due to high absorption of the wave energy. EDM's operate either as pulsed wave systems or as carrier phase systems.

21.2 In pulse systems (see Fig. 9) the transmitting station sends out a coded electromagnetic pulse traveling at the velocity of sound to a repeater station where the signal is then amplified and retransmitted to the transmitting station. Distance between the two stations should be computed by the time comparison techniques and is equal to the velocity of sound, approximately 1050 ft/s (350 m/s), times one half the sum of the round trip travel time plus equipment delays. The master device is set up at one end of the distance to be measured. This device creates a beam of electromagnetic radiation, either incandescent light, laser, infrared, or microwave, that serves as a carrier for the waves used for measurement. The beam is directed toward a reflector (in the case of light beams) or a repeater (in the case of microwave) at the other end of the distance where it is reflected to the master station. Signals from the master station are coded such that each specific transponder responds only to the specific transmission intended for it. This coding occurs in the number or spacing of single pulses in a series, or by the interval in the pulses (1).

21.3 Carrier phase systems (see Fig. 10) operate by transmitting a series of waves of constant amplitude and frequency in the form of infra-red or laser light. The receiving stations retransmit the signal back to the originating stations but with a slight phase shift. Distances between the two stations is determined by the phase comparison technique in which the distance is determined as a function of the difference between the outgoing and incoming signals, and the wave length of the signal (1).

21.4 EDM equipment of this nature is complex and requires not only an experienced operator but adherence to equipment limitations regarding accuracy, repeatability, and resolution. On-board equipment should be placed directly over or as near as possible to the measurement point. The shore-based transponders should be set so that the two range lines will intersect the boat with an angle as close to 90° as possible; moreover, the transponders must be accurately tied into an appropriate grid system (5). If a transit or theodolite is mounted on the tripod with the EDM, the two need to be in alignment. Sighting objects used in the alignment must be at least 1000 ft (305 m) from the instruments when the

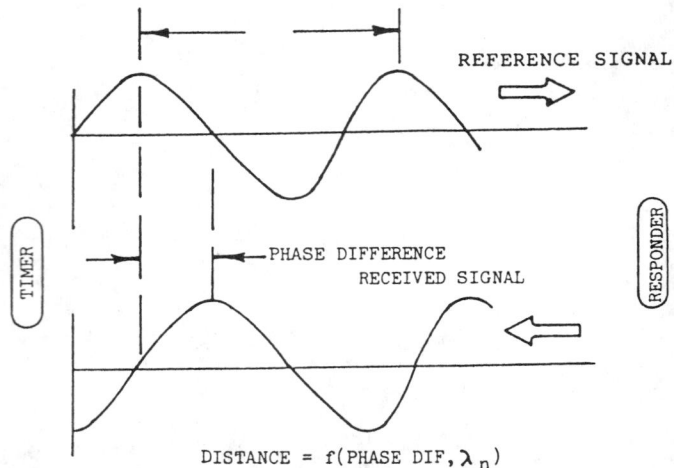

$$\text{DISTANCE} = f(\text{PHASE DIF}, \lambda_n)$$

FIG. 10 Phase Comparison Measurement

adjustments are made. The length of time to make the distance measurement needs to be kept to a minimum and as close to the other types of data being collected as possible. Care needs to be taken to compensate the distance readings for temperature and density effects, although this isn't normally a major error source (9).

21.5 Shore repeater stations are referred to as transponders, trisponders or responders, depending on the equipment manufacturer. Some EDM's also use a passive radar reflector—a specially shaped reflector designed to send back a high percentage of the incident energy (9).

21.6 The advent of the microprocessor has made it possible for electronic instruments (referred to as total-stations) to measure both angles and distance, with the added capability in some instruments to calculate coordinates from azimuth and distance. Some also offer the capability of internally storing data in built-in memory units, externally in data recorders, or microcomputers. This includes the capability to key-in atmospheric temperature and pressure for automatic calculation of atmospheric correction factor and automatic correction of distance readings. Others can also correct distance and angle readings for errors stemming from unlevel instruments. Transmitting signals may be microwave, those that employ phase-comparison techniques, or laser which use the pulse mode of operation (9).

21.7 Total-station survey systems come in a wide range of instruments, but three general types are available: manual, semi-automatic, and automatic. Differences depend on the whether the vertical angle (or zenith angle) is read manually or automatically and how the data is recorded. For manual total-stations, distance and angle measurements are read using a conventional theodolite. Slope reduction is accomplished by keying the vertical angle into an onboard or hand held calculator. In the semi-automatic total station, distances are read optically and the vertical angle is measured by a sensor. Slope reduction is accomplished automatically within the unit. The automatic total station electronically reads both the horizontal and vertical angle, makes the necessary computations automatically, and stores the results in either internal or external computer (10). The choice of which instrument to use depends on the accuracy requirements of

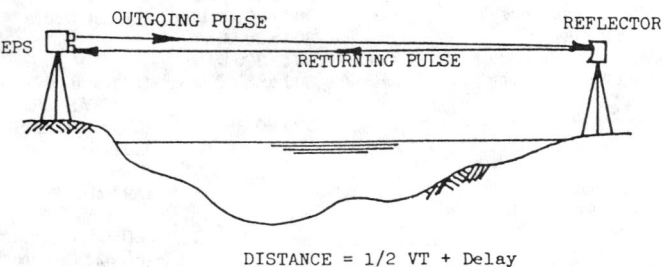

$$\text{DISTANCE} = 1/2 \text{ VT} + \text{Delay}$$

FIG. 9 Pulsing Distance Measurement

the project, the availability of equipment, cost, and the extent of automation affordable.

21.8 Components of a total station generally consist of (*a*) a distancer for measuring slope distance between two points; (*b*) a conventional or electronic theodolite for measuring horizontal and vertical angles; (*c*) a microprocessor for performing limited computational tasks such as determining horizontal measurements from slope measurements and calculating coordinate measurements from a bearing and a distance; (*d*) a digital control panel using a dot-matrix digital display and a 3 to 15-key keyboard to facilitate input of a variety of measurement functions; and (*e*) a data recorder for automatically collecting field data. The data recorder may be a 32 kilobyte or better replaceable memory card mounted to the theodolite for automatic data storage; an external electronic hand held notebook (similar in appearance to a small calculator) for storing field observations; or an external microcomputer used both for data storage and specialized field processing of the data (**10**). All total-stations include a battery pack capable of operating the equipment for several hours.

21.9 Some total-stations come equipped with a continuous tangent drive that simplifies the process of tracking a moving boat for hydrographic survey work. However, a conventional theodolite with mounted EDM equipment, coupled with an electronic field notebook, seems to be the preferred system for this type of work. The latter is not considered to be a total-station unless equipped with an electronic (digitized) theodolite (**11**). Automatic total stations, while having greater accuracy, involve substantial computational time in updating coordinates to a point and are best suited for measurements at a stationary point.

22. Range-Range Tracking (Electronic Positioning Systems)

22.1 Shore stations used for range-range tracking (see Fig. 11) should be placed at locations on high points of land and within clear view of the boat as the boat proceeds along the full length of its course. Each station should be established with at least third order survey accuracy, and each should have a common grid coordinate system and a common reference datum plane. The angle between the recorder on

the boat and the imaginary lines extending from the receiver to any two shore stations should not be less than 30° or greater than 150°. For maximum accuracy, the angle of intersection should be the largest angle possible. The boat must not be permitted to approach or cross the baseline while the system is in operation. Care also needs to be taken to ensure that the boat remains within an 80° horizontal signal beam broadcast from the shore stations. The vertical angle observed between a horizontal line and an on-board receiver should not exceed 7.5°. Distances between the shore stations and the boat should not be less than 300 ft (91 m) during the time the system is being operated.

23. Instrument Operator Errors

23.1 Calibrate EDM equipment frequently to assure reliable and accurate results. Systematic errors associated with EDM instrument maladjustments include scale, zero, and cyclic errors. These are normally considered to be relatively small, however, when compared to operator and other non-instrument errors.

23.2 Scale errors result when the modulation and design frequencies of the instrument do not correspond correctly. The magnitude of this error is expressed in parts per million (ppm) and should be negligible (<1 ppm). It varies in proportion to measurement distance and, under extreme cases, can be as much as 20 to 30 ppm (20 to 30 mm/km) (**6**).

23.3 Zero error is equivalent to the difference between the true distance and the measured distance between two points minus errors associated with scale, cycle, and atmospheric sources. It represents a constant error and does not increase with distance. The error source stems from the fact that the instruments or reflectors internal measurement center, or both, do not coincide with the instrument's physical center (**6**).

23.4 Cyclic errors are cyclic over the modulation wavelength and thus derive their name. Generally speaking they are small compared to other sources of error and originate from internal electronic interference between the transmitter and receiver circuitry (**6**).

23.5 Instrument calibration is best left to a laboratory specialist but can be done in the field using known baseline length and unknown baseline length techniques. Of the two, only the latter provides a value for scale error (**6**).

24. Fully Integrated Systems

24.1 Fully integrated positioning systems consist of a boat equipped with electronic positioning equipment, an onboard computer processor (either to serve as a control center for tracking the boats position at any instant of time or as a data logger for storing range data), track plotting equipment, and an electronic data storage unit, either magnetic tape, microcomputer disk, or cassette recorder. All electronic positioners utilize a propagated wave transmitted between the survey boat and a permanent point such as an electromagnetic wave transponder located near the water's edge, or an acoustic wave transponder actually in the water (**5**).

24.2 System components include a console, data processor, a system operator's terminal, an on-board receiver-transmitter for sending and receiving microwave signals, a

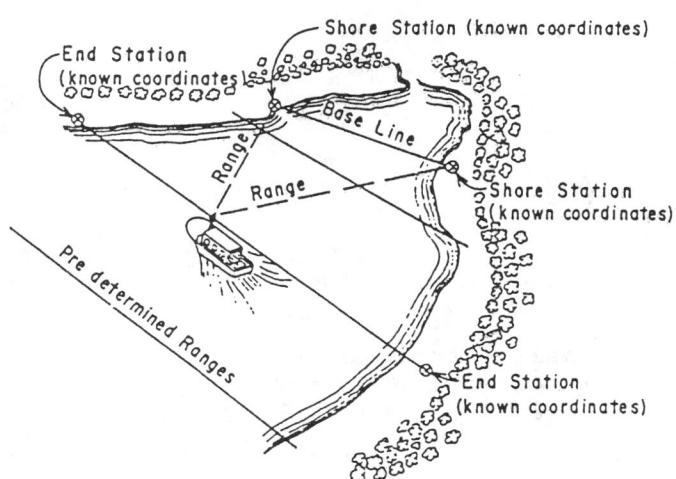

FIG. 11 Range-Range Tracking

tape recorder for storing range profile data, and a track indicator for positioning the boat on-line. Electronic measurements of water depth may also be included as part of the system. Two or more shore stations serve as the transmitting units, each powered with a portable, 24-volt ac power source or rechargeable battery pack. The batteries should be capable of operating for 8 to 10 h between charges.

25. Calibration of Electronic Positioning Equipment

25.1 All commercial electronic positioning equipment is calibrated in the laboratory at standardized air conditions for temperature, pressure and humidity. Changes in these parameters in the field result in very minor changes in calibration (9).

25.2 Regularly check positioning equipment in the field to ensure equipment accuracy by comparing readings from the equipment with known locations of boat position. Make the calibration procedure at the beginning and end of the workday by simultaneously establishing a positional "fix" using EPS and optical equipment, and then comparing the results.

26. EPS Accuracy and Repeatability

26.1 Positioning errors associated with electronic positioning equipment stem from (a) errors in the accuracy of the positioning equipment, (b) errors associated with the location of the antenna, transmitter, or stadia rod in relationship to the point of measurement being considered, that is, depth of water, sediment sample, velocity measurement, and etc; (c) errors in the relative position of the boat with respect to shore transponders; (d) errors relating to timing effects; (e) errors associated with location of shore station transformers; and (f) errors stemming from temperature and density effects on the velocity of electromagnetic wave propagation. Additional sources of error stem from the pitch, roll, yaw and heave motions of the boat.

26.2 Care should be taken to correct for slant range errors that result when the boat is too near an elevated shoreline. This should be done trigonometrically by correcting the horizontal distance, if the magnitude of the correction exceeds 2 m.

27. Global Positioning Systems

27.1 A Global Positioning System (GPS) is a satellite-based EDM system used in determining Cartesian coordinates (x, y, z) of a position by means of radio signals from a NAVSTAR satellite. NAVSTAR is an acronym for Navigation Satellite Timing and Ranging Satellite (12) with the first satellite of this type placed in orbit in 1978. Although relatively new and still undergoing technological advancements, GPS-based systems are generally considered to be more accurate and less costly to use than conventual survey methods for obtaining positions of either static monuments or moving platforms. They have the added advantage of being uneffected by weather conditions other than for safety considerations during severe electrical storms. GPS-based systems are rapidly becoming the "referred positioning systems for establishing control monument locations for use in hydrographic surveying and topographic mapping, boundary demarcation, and construction alignment work may be performed using conventional surveying instruments and procedures" (13).

27.2 Positioning of a ground receiver station relies on the processing of signals using two different methods: point positioning (see Fig. 12) and relative positioning (see Fig. 13). These are also referred to, respectively, as absolute and differential positioning. In both methods, range measurements from the receiver station to the satellite station is performed by tracking either the phase of the satellite's carrier signal or the phase code modulated on the carrier signal.

27.3 For point or absolute positioning, a single ground-based GPS receiver is set up at a location of interest and measures the travel time for signals to pass from a minimum

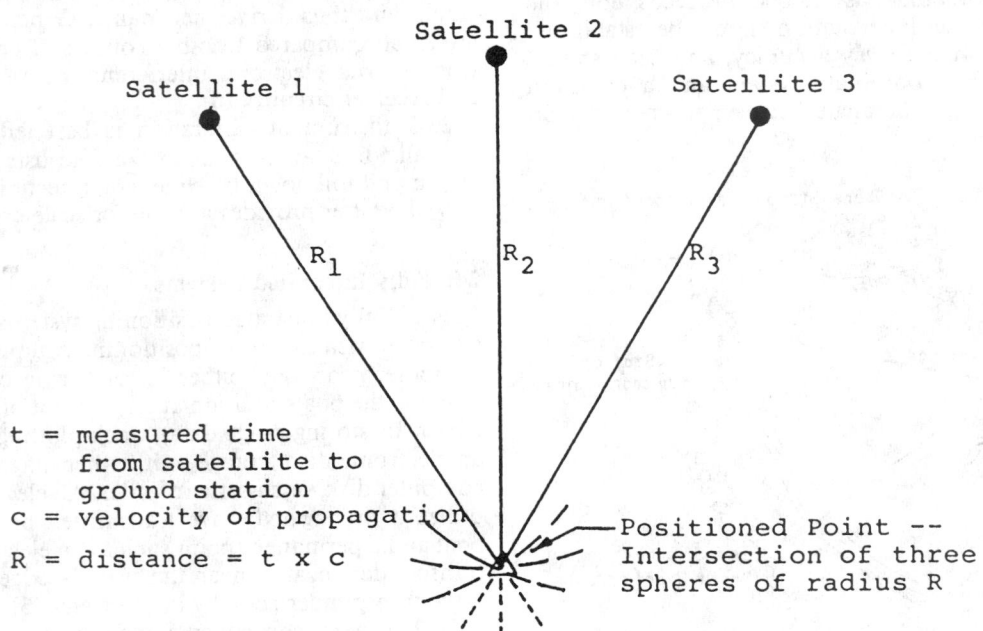

FIG. 12   General GPS Positioning Concept

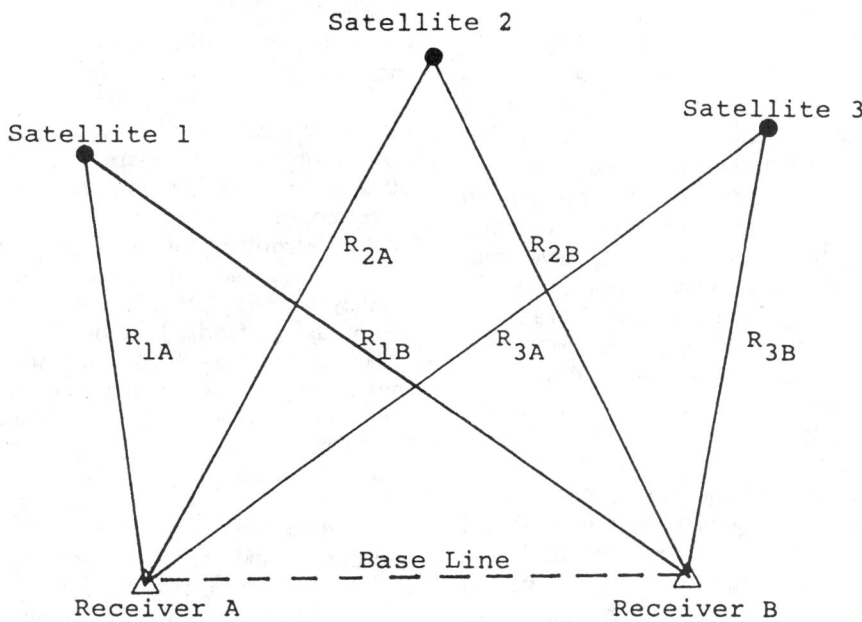

FIG. 13 Relative GPS Positioning

of three satellites to the receiver. This process is similar in concept to phase comparison EDM land or hydrographic surveying techniques. At least three satellites are required to provide the x, y, z coordinates of the receiver stations and a fourth is needed to resolve timing variations. These measurements permit approximate ranges (that is, distances from the satellites to the receiver) to be computed and reprocessed mathematically through intersection geometry to arrive at the Cartesian (x, y, z) coordinates of the GPS station relative to the center of the earth. These coordinates, in turn, can be further processed with computer software to provide the geodetic coordinates (that is, latitude, longitude, and height) of the position. Corrections for clock errors between the satellites and the receiver and signal propagation delays due to atmospheric conditions are included as part of the computations.

27.4 In relative or differential positioning, two or more receiver stations obtain data from a satellite simultaneously. One station (the reference unit) is placed over a known point (often an NGS coordinate station) and programmed with coordinates of the known point. The second (or mobile unit) is placed at some point of interest. This differential process involves measuring the differences in coordinates between the two receiver stations to arrive at distances and an azimuth of the line between the two stations. It is the relative distance (coordinate difference) between the stations that is of primary interest and not the absolute position of a station. Both stations are simultaneously observing the same satellite, so errors in the satellite position, clock differences, and atmospheric delay estimates are essentially canceled. Although a station might have poor absolute coordinates this is of little concern since one station is set up over a known point and the coordinate difference is applied to that point to obtain the coordinates of the second point.

27.5 Information transmitted from the NAVSTAR satellites includes two carrier signals referred to as L1 and L2, plus coded messages to enable a receiver station to process the satellite's celestial position and velocity, and other data such as the system time, clock correction parameters, ionospheric delay model parameters, and the satellite's position and health. Both the L1 and the L2 carrier signals correspond to the L-band of the electromagnetic spectrum and have frequencies of 1574.42 and 1227.60 MHz, respectively. The L1 signal is modulated with a precise code (P-Code), having precise timing marks every 30 m, and a coarse acquisition code (C/A-Code), having marks every 300 m. The L2 signal, in contrast, is modulated only with the P-Code. Both the C/A and the P-Codes are referred to as pseudo-random noise (PRN). The satellite's celestial position at any point in time is computed or predicted using past tracking data continuously monitored by a master control station on earth and uploaded to the satellite normally at 8-h intervals. Precise information on the satellite's position can be obtained but only at a later date from the National Geodetic Survey or from private sources that maintain their own tracking networks and provide information for a fee. For most hydrographic survey needs the predicted positions are adequate to obtain the needed accuracies (13).

27.6 For precise geodetic and engineering surveying, the distance from a satellite to a receiver station is determined using the L1 and twelve carrier signals transmitted form the satellite, with corrections applied to adjust for error sources and differences in clock timing. The process involves the satellite and the receiver station transmitting an identical (replica) signals at the same time. The two signals are cross-correlated until they are exactly in phase, and the amount by which the receiver signal is moved to achieve this phase shift is the travel time it took for the satellite signal to reach the receiver. When multiplied by the speed of light this value yields the distance between the satellite and the receiver (12).

27.7 Accuracies achieved with a GPS-based system generally range from 100 m (328.1 ft) down to the millimeter (less than a half-inch) level depending on (a) the operating mode,

(b) the process used in tracking the satellite, and (c) the static or dynamic (kinematic) environment under which the receiver is operating. Satellite tracking is performed by monitoring either the phase of the satellite's carrier signal or the pulse codes modulated on the carrier signal. Those positioning systems operating in the absolute mode and using code measurements are in the 100 m (328.1 ft) or upper end of the accuracy range. Those using phase measurements and operating kinematically in the differential mode have accuracies generally in the 1 to 5 cm range, with even greater accuracies possible if operating in the static mode (13).

27.8 GPS error sources include ephemerides error (error associated with the celestial position of the satellite), orbit perturbations, clock stability, ionospheric delays, tropospheric delays, multipath errors, and receiver noise. Table 5 lists the more significant error sources and correlates them to the segment source. The total error budget is summarized as the User Equivalent Range Error (USER) and is estimated using a root mean square (RMS) radial error statistic (13).

27.9 Unlike static GPS-based positioning, the receivers in kinematic positioning can not be turned off between positioning points nor can they be allowed to lose contact with the satellites. The latter is referred to as "holding lock." If lock is lost at any time during the positioning then a complete initialization procedure must be undertaken. This includes either (a) returning the remote receiver to its previously occupied position; (b) swapping antennas between the master and the remote receivers, after they have been in operation and permitted to exchange data for 45 to 90 s; (c) positioning both receivers over known stations whose coordinate difference has been determined to within 5 cm and permitting the two receivers to exchange data for 45 to 90 s; or (d) using the somewhat time consuming (approximately 45 min) procedure of deriving an initial baseline vector using standard, static GPS positioning techniques. Other cautions to consider in kinematic positioning include making certain that the measurement of the antenna height above the point of interest is measured precisely, that this height does not change between observation stations, and that the antenna is properly plummed over the point. In addition, care must be taken to make certain that the observation time is of sufficient length, that approximated coordinated stations (if these are used) are not too far from the correct values, that correct satellite combinations have been selected, and that the satellite's ephemerial (celestial position) data stored in the receiving units are updated prior to the positioning (13).

TABLE 5 GPS Range Measurement Accuracy

Segment Source/ Error Source	Absolute Positioning				Differential Positioning (P-CODE), ft (m)	
	C/A-Code Pseudo-Range, ft (m)		P-Code Pseudo-Range, ft (m)			
Space						
Clock stability	9.8	(3.0)	9.8	(3.0)	Negative	(Negative)
Orbit perturbations	3.3	(1.0)	3.3	(1.0)	Negative	(Negative)
Other	0.2	(0.5)	0.2	(0.5)	Negative	(Negative)
Control						
Ephemeris predictions	13.8	(4.2)	13.8	(4.2)	Negative	(Negative)
Other	3.0	(0.9)	3.0	(0.9)	Negative	(Negative)
User						
Ionosphere	24.6	(7.5)	7.5	(2.3)	Negative	(Negative)
Troposphere	6.6	(2.0)	6.6	(2.0)	Negative	(Negative)
Receiver noise	24.6	(7.5)	4.9	(1.5)	4.9	(1.5)
Multipath	3.9	(1.2)	3.9	(1.2)	3.9	(1.2)
Other	1.6	(0.5)	1.6	(0.5)	1.6	(0.5)
1-δ UERE	±41.3	(±12.1)	±21.3	(±6.5)	±6.6	(±2.0)

REFERENCES

(1) Ingham, A. E., *Hydrography for the Surveyor and Engineer*, Second Edition, John Wiley & Sons, New York, NY 1984.

(2) *Hydrographic Manual*, Fourth Edition, U.S. Department of Commerce, National Oceanic and Atmospheric Administration, 1976.

(3) Blanton, James III, "Procedures for Monitoring Reservoir Sedimentation," *U.S. Bureau of Reclamation Technical Standard Guides*, October 1982.

(4) U.S. Geological Survey Parts Catalog, 1301-07, *Taglines and Reels*.

(5) U.S. Geological Survey, *National Handbook of Recommended Procedures for Water-Data Acquisition*, Chapter 3—Sediment, 1978.

(6) Davis, Foote, and Kelly, *Surveying-Theory and Standard Guide*, Fifth Ed., Chapter 30, McGraw-Hill, New York, NY, 1966.

(7) *Hydrographic Engineering Manual*, Engineering Manual (EM) 1110-2-1003, U.S. Army Corps of Engineers, 1990.

(8) *Measurement of Hydrographic Parameters in Large Sand-Bed Streams from Boats*, Task Committee on Hydrographic Investigations of the Committee on Waterways of the Waterway, Port, Coastal, and Ocean Division, American Society of Civil Engineers, New York, NY, 10017, 1983.

(9) *Hydrographic Manual*, Publication 20-2, Third Edition, U.S. Department of Commerce and Coast and Geodetic Survey, Washington, DC, 1963.

(10) Total Survey Station, *Point of Beginning*, Volume 17, Number 4, April–May 1992.

(11) Total Survey Station, *Point of Beginning*, Volume 10, Number 6, August–September 1985.

(12) *Engineering Surveying Technology*, Blackie and Son Ltd., John Wiley & Sons, New York, NY, 1990.

(13) NAVSTAR Global Positioning System Surveying, *Engineering Manual, EM1110-1-1003*, U.S. Army Corps of Engineers, 1991.

Designation: D 5073 – 90 (Reapproved 1996)[ε1]

Standard Practice for
Depth Measurement of Surface Water[1]

This standard is issued under the fixed designation D 5073; the number immediately following the designation indicates the year of original adoption or, in the case of revision, the year of last revision. A number in parentheses indicates the year of last reapproval. A superscript epsilon (ε) indicates an editorial change since the last revision or reapproval.

[ε1] NOTE—Keywords were added editorially in July 1996.

1. Scope

1.1 This practice guides the user in selection of procedures commonly used to measure depth in water bodies that are as follows:

	Sections
Procedure A—Manual Measurement	6 through 11
Procedure B—Electronic Sonic-Echo Sounding	12 through 13
Procedure C—Electronic Nonacoustic Measurement	14 through 15

The text specifies depth measuring terminology, describes measurement of depth by manual and electronic equipment, outlines specific uses of electronic sounders, and describes an electronic procedure for depth measurement other than using sonar.

1.2 The references cited and listed at the end of this practice contain information that may help in the design of a high quality measurement program.

1.3 The information provided on depth measurement is descriptive in nature and not intended to endorse any particular item of manufactured equipment or procedure.

1.4 This practice pertains to depth measurement in quiescent or low-velocity flow. For depth measurement related to stream gaging see Test Method D 3858. For depth measurements related to reservoir surveys see Guide D 4581.

1.5 *This standard does not purport to address all of the safety concerns, if any, associated with its use. It is the responsibility of the user of this standard to establish appropriate safety and health practices and determine the applicability of regulatory limitations prior to use.*

2. Referenced Documents

2.1 *ASTM Standards:*
D 1129 Terminology Relating to Water[2]
D 3858 Test Method for Open-Channel Flow Measurement of Water by Velocity-Area Method[2]
D 4410 Terminology for Fluvial Sediment[2]
D 4581 Guide for Measurement of Morphologic Characteristics of Surface Water Bodies[3]

3. Terminology

3.1 *Definitions*—For definition of terms used in this practice refer to Terminologies D 1129 and D 4410.

3.2 *Definitions of Terms Specific to This Standard:*

3.2.1 *bar-check, n*—a method for determining depth below a survey vessel by means of a long, narrow metal bar or beam suspended on a marked line beneath a sounding transducer.

3.2.2 *bar sweep, n*—a bar or pipes, suspended by wire or cable beneath a floating vessel, used to search for submerged snags or obstructions hazardous to navigation.

3.2.3 *beam width, n*—the angle in degrees made by the main lobe of acoustical energy emitted from the radiating face of a transducer.

3.2.4 *bottom profile, n*—a line trace of the bottom surface beneath a water body.

3.2.5 *sonar, n*—a method for detecting and locating objects submerged in water by means of the sound waves they reflect or produce.

3.2.6 *sound, vt*—to determine the depth of water (1).[4]

3.2.7 *sounding line, n*—a rope or cable used for supporting a weight while the weight is lowered below the water surface to determine depth.

3.2.8 *sounding weight, n*—a heavy object usually of lead, that may be bell-shaped, for use in still water and soft bottom materials or torpedo shaped with stabilizing fins, for use in flowing water.

3.2.9 *stray, n*—spurious marks on the graphic depth records caused by surfaces other than the bottom surface of a water body below the sounding vessel.

3.2.10 *subbottom profile, n*—a trace of a subsurface horizon due to a change in the acoustic properties of the medium through which the sound energy has traveled.

3.2.11 *towfish, n*—a streamlined container, containing acoustical equipment for sounding depth, and designed to be pulled behind or beneath a survey vessel.

3.2.12 *transducer, n*—a device for translating electrical energy to acoustical energy and acoustical energy back to electrical energy.

3.2.13 *transducer draft, n*—the distance from the water surface to the radiating face of a transducer.

3.2.14 *vertical control, n*—a horizontal plane of reference used to convert measured depth to bottom elevation.

4. Summary of Practices

4.1 These practices include the following three general techniques for acquiring depth measurements in surface water:

---

[1] This practice is under the jurisdiction of ASTM Committee D-19 on Water and is the direct responsibility of Subcommittee D19.07 on Sediments, Geomorphology, and Open-Channel Flow.
Current edition approved May 25 and Oct. 26, 1990. Published December 1990.
[2] *Annual Book of ASTM Standards*, Vol 11.01.
[3] *Annual Book of ASTM Standards*, Vol 11.02.

[4] The boldface numbers in parentheses refer to a list of references at the end of this practice.

4.1.1 The first general technique is to determine depth by manual procedures. The equipment to perform these procedures may be most readily available and most practical under certain conditions.

4.1.2 The second general technique is to determine depth by electronic sonic-echo sounding procedures. These procedures are most commonly used because of their reliability and the variety of instruments available that meet specific measuring requirements.

4.1.3 The third general technique is to determine depth by an electronic procedure other than acoustic sounding. A procedure using ground penetrating radar is currently being used for measuring water depth for specific applications.

5. Significance and Use

5.1 This is a general practice intended to give direction in the selection of depth measuring procedures and equipment for use under a wide range of conditions encountered in surface water bodies. Physical conditions at the measuring site, the quality of data required, and the availability of appropriate measuring equipment govern the selection process. A step-by-step procedure for actually obtaining a depth measurement is not discussed. This practice is to be used in conjunction with a practice on positioning techniques and another practice on bathymetric survey procedures to obtain horizontal location and bottom elevations of points on a water body.

PROCEDURE A—MANUAL MEASUREMENT

6. Scope

6.1 This procedure explains the measurement of water depth using manual techniques and equipment. These include the use of sounding rods, sounding lines, sounding reels, or a bar sweep.

6.2 Description of techniques and equipment are general in nature. Techniques and equipment may need to be modified for use in specific field conditions.

7. Significance and Use

7.1 Prior to the development of acoustic sounding equipment, manual techniques provided the only means of depth measurement. Some circumstances may still require sounding by manual techniques such as shallow areas where depth is not sufficient for acoustic sounding. Manual procedures continue to serve several useful purposes such as the following:

7.1.1 To search for and confirm the minimum depths over shallow area of sunken obstacles.

7.1.2 To confirm bottom soundings in areas with submerged vegetation, or other soft bottom materials.

7.1.3 To assist in obtaining bottom samples.

7.1.4 To calibrate electronic sounding equipment.

7.1.5 To suspend other measuring instruments to known depths for making various physical or chemical water quality measurements (2).

8. Sounding Rod (Manual Procedure)

8.1 The sounding rod (or sounding pole) can be used to measure depth over extensive flat, shallow areas more easily and more accurately than by other means. Use of the sounding rod should be restricted to still water or where the velocity is relatively low, and to depths less than 12 ft (3.7 m). Sounding rods are usually not used in depths over 6 ft (1.8 m) except to provide supplemental soundings to aid in interpreting analog depth records. A weighted, flat shoe (see Fig. 1) should be attached to the bottom of the rod to prevent it from penetration of the bottom sediments. The rod may be graduated in feet and tenths of a foot; zero being at the bottom of the shoe (3).

8.2 Modern sounding rods may be made of light-weight metals for strength, neutral buoyancy, and sound transmitting capability. An experienced operator can measure the water depth and can distinguish the relative firmness of the bottom material by the feel of the rod and the tone produced by the metal pole as it contacts the bottom (4).

8.3 When sounding in still water the operator should lower the rod into the water until the bottom plate makes contact with the bottom surface. After determining that a firm bottom material has been encountered, the water surface level is visually read on the rod. When sounding in flowing water, to achieve vertical sounding, a long wire or cable anchored upstream and attached to the lower end of the rod may be necessary.

9. Sounding Line (Manual Procedure)

9.1 The sounding line (see Fig. 2) can be used to measure depths of large magnitude but is seldom used for depths greater than 15 ft (4.57 m). The sounding line should be of a material that does not shrink or stretch, or lengthen from wear or corrosion of the material as will occur in chain links over several years of use. Though manila rope and cotton, or other materials that require prestretching before use, have

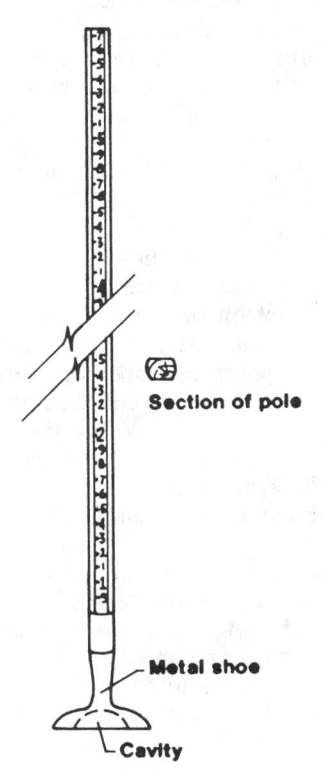

FIG. 1 Graduated Sounding Rod with Shoe Attached

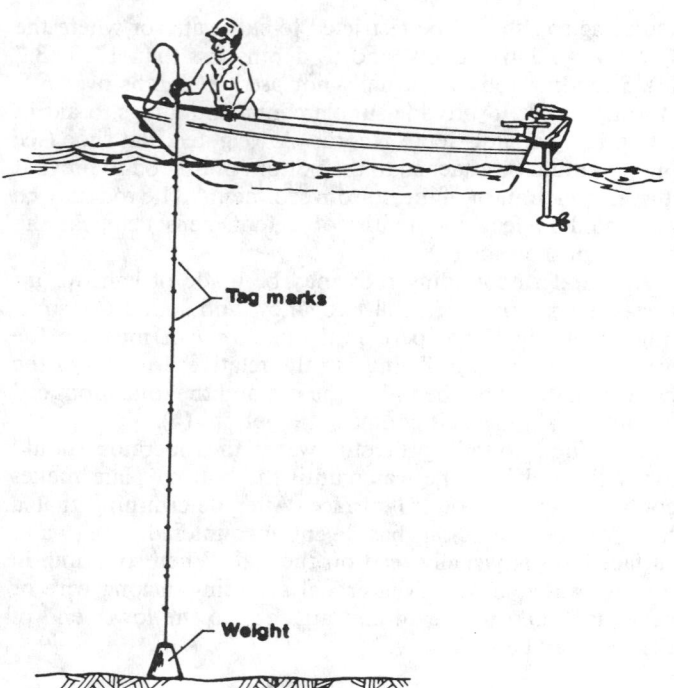

FIG. 2 Sounding Line Used from Small Boat

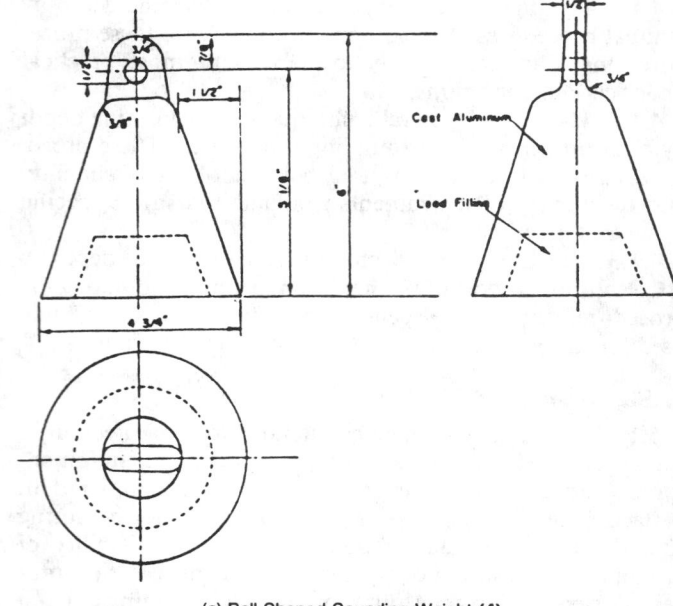

(a) Bell Shaped Sounding Weight (4)

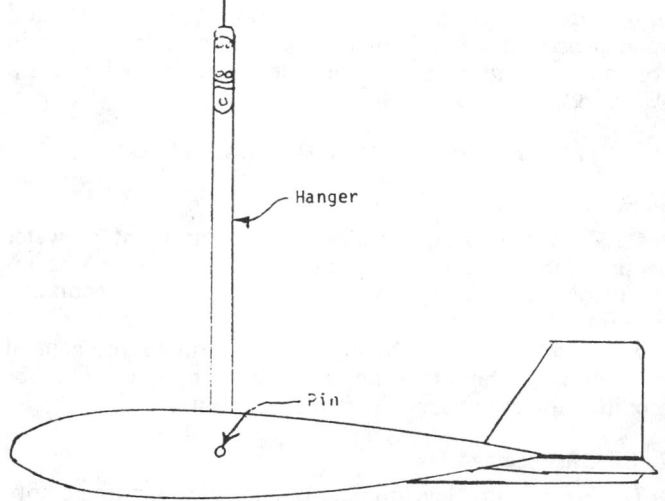

(b) Torpedo Columbus-Type Sounding Weight

FIG. 3 Typical Weights Used with Sounding Line

been employed for large depths, small-diameter high-strength steel cable wound and released from a reel with a gear driven depth indicator are readily available and greatly simplify the work (1). The stretch of the high-strength cable is very small for its intended use, and therefore, a considerable length of cable may be used without introducing significant error. Depth indicators, calibrated in either inch-pound or metric units, or both, are available (5).

9.2 Markings on the sounding line should be easy to see and understand to avoid making errors in determining the readings. For sounding relatively shallow depths, marking at 0.5-ft intervals with different colors to identify the 1, 2, and 10-ft intervals is recommended. Care must be exercised so that the first marker is the correct distance from the bottom of the sounding weight when the weight is attached. When sounding, depths are obtained from the difference in readings at an index point on the bridge or boat rail, when the base of the sounding weight is at the water surface, and when it is at the bottom. A short steel tape or folding rule is usually employed to measure the fractional distance from the line markers to the reference point. Within the minimum 0.5-ft markings depths are estimated and recorded to the nearest 0.1 ft. For sounding in deep water, a sounding reel with depth indicator and an unmarked high-strength steel cable is recommended (4).

9.2.1 When the metric system of units is used, the sounding line for use in shallow depths is usually marked at 0.5-m intervals with different colors to identify the 1 and 2-m intervals. Depths are recorded to the nearest 0.01 m.

9.3 Weights used in sounding are usually of lead, aluminum, or brass. For application in still water, the weights are bell-shaped (see Fig. 3a) and made of cast aluminum or lead. The amount of weight should be from 5 to 10 lb (2.3 to 4.5 kg).

9.3.1 For application in flowing water, the weight should be of circular cross section and steamlined with fins (see Fig. 3b) to turn the weight nose first into the current to offer a minimum of resistance to the flow. The amount of weight should be varied, depending on the water depth and flow velocity at a cross section. A rule of thumb is that the weight in pounds should be greater than the maximum product of velocity and depth in the cross section. If debris or ice is flowing or the stream is shallow or swift, use a heavier weight than the rule designates. A variety of sizes of sounding weights from 15 to 300 lb (7 to 136 kg) should be available with appropriate means of attaching to the sounding line (1). Sounding weights should always be attached to the sounding line using a hanger bar, clevis, snap hook, or thimble of brass or stainless steel to protect the line from wear or damage.

9.4 The procedure for making soundings will vary depending on depth, current velocity, and means of locating where the soundings are taken. Once at the location where a depth measurement is needed, the basic procedure is to lower the weight until the bottom of the weight is at the water surface. When using a marked sounding line, the distance is read from the sounding line at a reference point on the bridge or boat after which the weight is lowered to the bottom, and a new distance is read from the line and recorded. When using a sounding reel the indicator is set to zero after which the weight is lowered to the bottom and the depth is read and recorded. It is usually of some importance, especially when sounding an uneven bottom, to have the locations of the soundings accurately known relative to the surroundings. When sounding from a boat using weighted line, the boat should be stationary and should remain at that position until the sounding has been completed and the location is determined.

9.5 Sounding through the ice cover of a lake or river may be taken after boring holes in the ice with an ice auger. In this case, a marked sounding line with an appropriate sounding weight attached at the end, is lowered through the hole and the determined depth is recorded.

10. Sounding Reels (Manual Procedure)

10.1 Sounding reels (see Fig. 4) are used with high strength cable where heavy weights are required or where depths are great. These reels are usually very sturdily constructed having a braking system for controlling rotation of the reel as the cable is let out. For hand operated reels, the hand cranks are hinged to allow the crank to be disengaged from the shaft while the wire is let out and engaged for reeling in. Various devices are employed to drive a counter registering the amount of cable let out from which the depth below water surface is determined. These sounding reels may also be electrically driven, in that case, they may have a depth capacity of more than 5000 ft (1524 m) (1).

11. Bar Sweep (Manual Procedure)

11.1 The bar sweep is commonly used to search for and

locate any shoal or obstruction within or above navigation depth that may present a hazard to navigation. It augments the hydrographic survey in navigable waters by locating shallow submerged areas that may go undetected by the usual hydrographic procedures. The bar sweep (see Fig. 5) consists of a bar (steel pipe) suspended beneath the survey vessel by graduated wire or cable from hand operated drums. The drums may be mounted either off the stern or at the port and starboard gunwale. Each end of the bar should be packed with lead to add weight and to reduce lift when underway. Pipe weight is the major factor in allowable vessel speed. Trial and error variations are usually necessary to determine the best combination. In a normal operation, the bar is lowered to navigation depth and the vessel moves forward to sweep an area. Whenever a shoal is encountered, the operator raises the bar until it clears the obstruction. The shoal depth and position is then recorded. The bar is then returned to navigation depth and the survey continues (2).

PROCEDURE B—ELECTRONIC SONIC-ECHO SOUNDING

12. Scope

12.1 This procedure is applicable to the measurement of water depth using electronic sonic-echo sounding techniques and equipment. Because of the large variety of instrumentation currently available, this discussion is limited to types of equipment in most common use.

12.2 Discussions of the techniques used include methods of measurement, criteria for selection of sounding frequency and recording equipment, means for achieving quality assurance, and factors to consider in interpreting depth records.

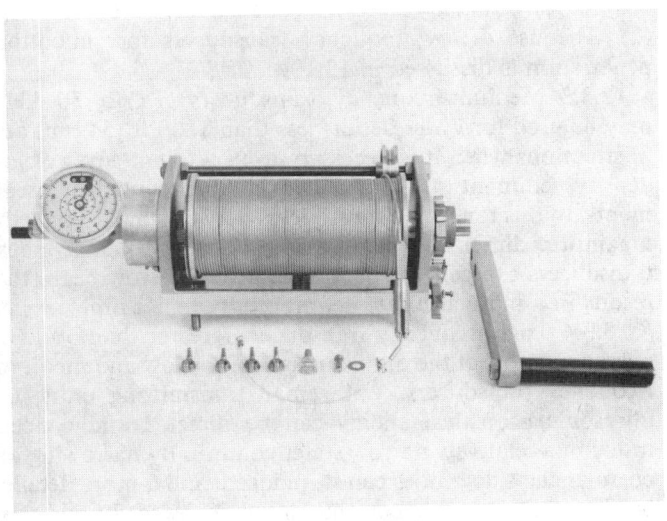

FIG. 4 Hand-operated Sounding Reel (1)

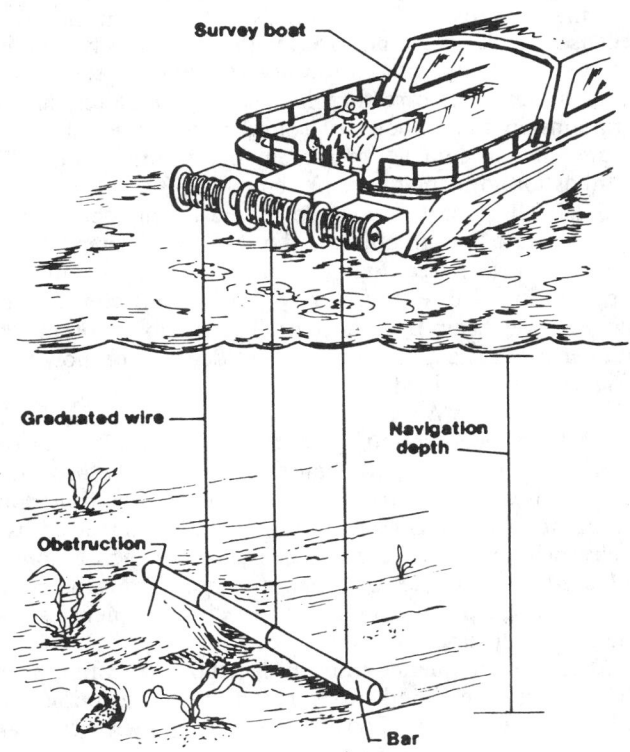

FIG. 5 Bar Sweep for Locating Shoals

13. Sonic-Echo Sounding (Electronic Procedure)

13.1 Water depths are most commonly obtained by echo sounders that record a continuous profile of the bottom surface of the water body under the vessel. Echo sounders measure the time required for a sound wave to travel from its point of origin to the bottom and the reflected wave to return. The sounder then converts this time interval to distance or depth below the face of the transducer. The transmission of sound is dependent on certain properties of the water and the reflecting surface. For a sound wave to travel at a constant velocity from the surface to bottom and be completely reflected off the bottom, the water must have the same physical characteristics throughout its entire depth and the bottom must be a perfect reflector. Because such conditions do not exist in nature, echo sounders are usually designed to permit adjustments for variations in the velocity of sound in water and wave attenuation (2).

13.2 *Measuring Principles:*

13.2.1 Echo sounding equipment is designed to generate the sound wave, receive and amplify the returning echo, measure the intervening time interval, convert the time interval into units of depth, and record the results graphically, digitally, or both. The echo sounder only measures time (that is, the time it takes for a sound wave to travel from the transmitter to the bottom or other reflecting surface and back again). The time interval is converted mechanically or electronically to depth beneath the transmitter by the following equation:

$$depth = \tfrac{1}{2}\,vt$$

where:

v = the velocity of sound in water, ft/s (m/s), and
t = the time for the pulse to travel from the transmitter to the reflective surface and back to the transmitter, s.

Because velocity of sound varies with water density, that is a function of temperature, salinity, suspended solids, and depth, a means of correcting the resulting measurements for variations in the velocity of sound must be employed to ensure an acceptable measurement accuracy (2). The methods for adjustment are presented in 13.6.

13.2.2 The sound waves transmitted by an echo sounder may be varied in frequency, duration, and shape of the acoustical beam (see Fig. 6). The sound wave may be dispersed in all directions, or contained and concentrated into a narrow beam by a reflector. The suitability of an echo sounder to meet a given requirement depends on how these variables are combined (2).

13.3 *Frequency Selection:*

13.3.1 An echo sounding transducer is used to convert electrical energy pulses to acoustical energy. The acoustical energy pulses are then transmitted through a liquid medium and the returning echoes are detected and reconverted back to electrical energy. These energy pulses are then amplified and used to compute and record depth. Transducers are usually designed to operate on specific frequencies, depending on the application and depth range (2).

13.3.2 Low-frequency transducers, those operating below 15 kHz, produce sound waves having low absorption rates and high penetrating power. These characteristics make them useful for deep soundings and penetration of the fine deposited material on the bottom of a river or lake. These

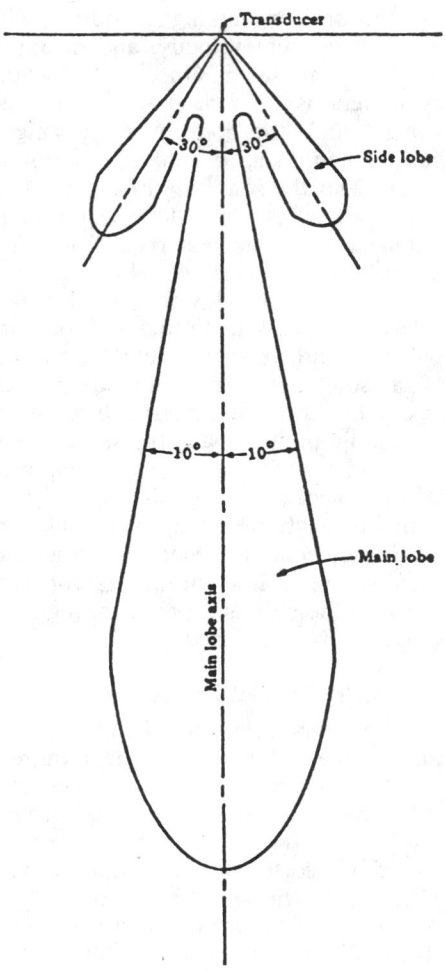

FIG. 6 Shape of a General-purpose Echo Sounders' Acoustical Beam (2)

transducers cannot be used to accurately measure very shallow depths, and they are very susceptible to noise interference in the more audible frequency range. Because of their long wavelengths the lower frequency pulses cannot be beamed directionally unless the transducers are very large (2). The use of low-frequency transducers for subbottom penetration is discussed in 13.10.

13.3.3 Medium-frequency transducers (15 to 50 kHz) may be used for water depths less than 1800 ft (549 m) and in situations when it is necessary to penetrate a layer of low density sediment suspended above more compacted sediments. In this range, the transducers may be small in size, the maximum dimension being 8 in. (20.3 cm) or less. These transducers can generate a comparatively narrow beam that results in a more accurate definition of the bottom.

13.3.4 High-frequency transducers (greater than 50 kHz) overcome most of the disadvantages of the low and medium-frequency transducers. With small transmitting units, the ultrasonic acoustical energy can be directed and concentrated in a relatively narrow water column. By narrowing the beam angle, side echoes can be reduced, and a more detailed profile of an irregular bottom can be achieved. In addition, shallow depths can be measured more accurately. Due to greater attenuation of the sound wave, the high frequencies are ineffective in very deep water (2).

13.4 *Recording Soundings:*

13.4.1 Analog recorders usually employ one of two methods for registering depth on a chart.

13.4.1.1 In the first method, the depth is recorded by a stylus mounted on a rotating arm that makes a mark on dry, electrosensitive, calibrated paper. The stylus passes over the chart paper at a constant speed marking the chart at the zero (initial) point, at a point designating the draft of the transducer, and at a point representing bottom depth. As continuing echoes are received from the bottom, a bottom profile is recorded (see Fig. 7). The horizontal scale of the plot is determined by the chart speed set by the operator.

13.4.1.2 In the second method, the depth is recorded by a fixed-head thermal recording device (6). The printing mechanism consists of a nonmoving print head containing hundreds of thermal dots heated precisely at the proper time to print the chart. The only moving parts on thermal print recorders are the motor and roller assembly that moves the paper across the printhead. Unlike moving-stylus type recorders, the chart and motor timing on the thermal print recorders have no effect on depth measurement accuracy. Thermal print recorders begin with blank thermal paper. The scale grids and other chart features are preprogrammed to be generated by these units, allowing for a variety of chart formats (see Fig. 8).

13.5 *Errors in Measurement:*

13.5.1 Factors that lead to error in depth measurement are numerous and should be recognized when conducting a bathymetric survey or analyzing graphic depth recordings. For a detailed description of these errors see Ref (2). The most significant factors are described in 13.5.2 to 13.5.9.

13.5.2 *Velocity of Sound Wave Propagation*—The velocity of a sound wave traveling through water varies with temperature and density. It is, therefore, necessary to check the effective velocity of sound in a given body of water to achieve the depth accuracy required. In a deep reservoir, temperatures may vary as much as 45°F (25°C) between the surface and the bottom. In an estuary, salinity may also vary in both the vertical and horizontal direction, thus causing

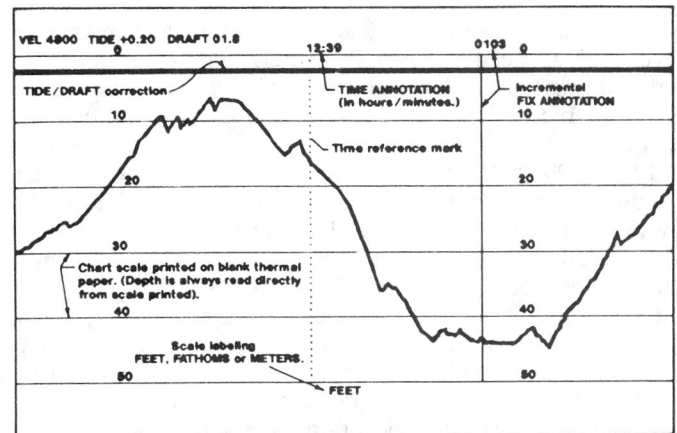

FIG. 8   Analog Bottom Profile Charted by Thermal Printer

density to vary. Calibration of the sounding instrument should be made by the survey crew, at appropriate times to adjust the depth readings for changes in water temperature and density (see 13.6).

13.5.3 *Signal Transit Time*—Water depth is determined by the time required for a signal to travel from the transducer, strike a reflective surface, and return to the transducer. With the high quality instruments currently available, errors in time measurement are insignificant (7).

13.5.4 *Transducer Location*—The draft or vertical location of the transducer with respect to water surface can be set into most high precision echo sounders. The vertical location, when set for static conditions, will change with the motion of the boat. The effect of boat motion on draft, may be corrected during calibration of the instrument.

13.5.5 *Wave Action*—The vertical and rotational motion of a boat due to wave action can result in severe fluctuations in the bottom trace. Some smoothing of the trace may be necessary during data processing to eliminate the fluctuations.

13.5.6 *Bottom Conditions*—The condition of the reflective surface of a reservoir or river bottom may vary widely, resulting in a sounding chart that gives an erroneous impression of the actual bottom profile. Vegetation attached to or suspended above the bottom, isolated boulders, or submerged man-made objects, may produce a nonrepresentative bottom profile. Depending on the purpose of the survey, the cause of these bottom reflections may have to be determined by other means before choosing to eliminate them from the trace.

13.5.7 *Nature of Bottom Sediments*—Very low density sediment, suspended as a nepheloid layer or zone above more compacted sediments, can result in an erroneous depth reading when a transducer with a frequency higher than 50 kHz is used. A waterway bottom may be described in nautical terms as any water/solid interface level that blocks or impedes the passage of ships, boats, or barges. A low or medium-frequency transducer may be used to determine depth to the more consolidated sediment layer.

13.5.8 *Tidal Effects*—When surveying in tidal zones of rivers and estuaries, a continuous record must be kept of tidal fluctuations within the area during the surveys in order to adjust the depth readings for the changing water surface.

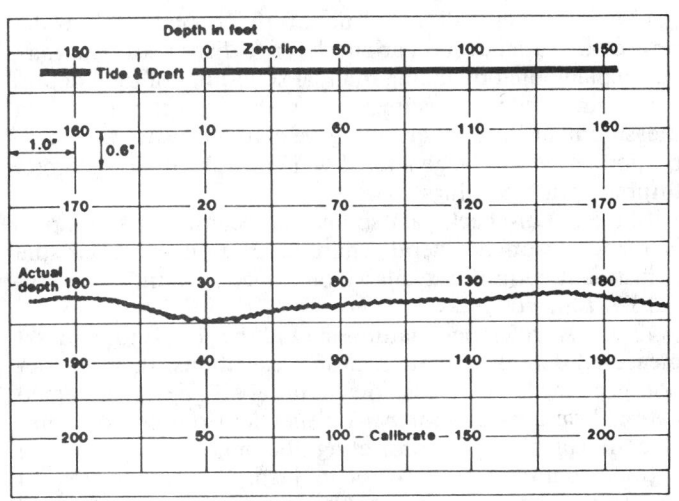

FIG. 7   Analog Bottom Profile Charted with Stylus

The measured depths are generally referred to a reference level, such as mean sea level. By exercising good technique in determining tidal changes and making tidal corrections, the errors in measuring bottom elevations can be significantly reduced.

13.5.9 *Other Causes*—Errors may occur due to special conditions during a survey that may be either unknown or overlooked by the survey crew. Examples of these conditions are as follows: a reservoir water surface elevation may fluctuate appreciably due to inflow or outflow, thus changing the conditions for vertical control during the survey; when downstream flow occurs in narrow canyon areas or in river portions of a reservoir, a water surface slope may extend in the upstream direction and produce error when a constant reservoir water surface elevation is assumed for vertical control; a constant wind blowing from one side of a water body to another may raise the water surface on the down-wind side of the water body and introduce error should a specific water surface elevation be assumed for vertical control (8).

13.6 *Calibration:*

13.6.1 Depth measurements by an echo sounder require a number of corrections. The largest correction results from the variability of sound velocity in water. The velocity varies with the temperature, salinity, and depth of water. In fresh water at 60°F, echo sounders are generally calibrated for a sound velocity of 4800 ft/s (1463 m/s). The indicated depth given by the echo sounder needs to be corrected for the difference between the calibrated velocity and the actual velocity determined by the water temperature and salinity. This can be accomplished by several methods. One method is to measure the temperature and salinity of the water at various depths, and using predeveloped tables and graphs, correct the depth readings on that basis. A more direct method is to construct calibration curves from bar-check data for a particular instrument and using these curves to make corrections. A third method is available on many echo sounders where adjustment control is offered to adjust velocity of sound to local conditions.

13.6.2 A bar-check (see Fig. 9) is the preferred method used to verify the accuracy of an echo sounder and to determine corrections for instrument and velocity error.

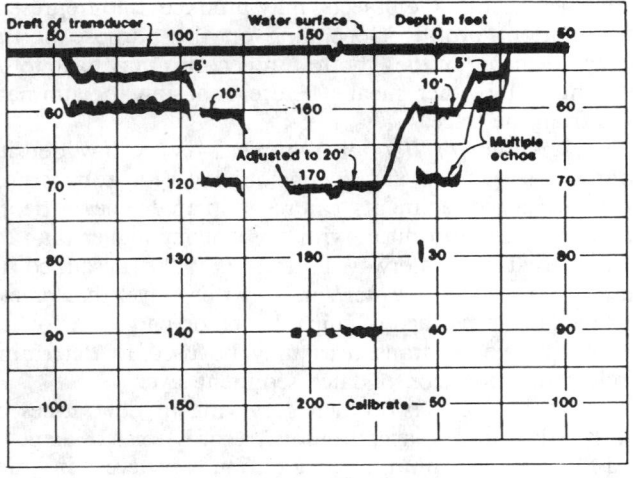

FIG. 9 Results of a Bar-check

However, reliable and accurate bar-checks can be made only under favorable conditions. When the water surface is calm and there is little differential current or wind effect near the vessel, bar-checks can be obtained in depths as great as 200 ft (61 m). Under less favorable conditions, accurate bar-check depths may be reduced to 10 ft (3.05 m). In moderate depths where bar-checks can be obtained over the full depth range of a survey, corrections to soundings can compensate for the difference between the calibration velocity of the instrument, the actual velocity of sound in the water, and for instrumental errors (2).

13.6.3 The bar-check apparatus must be a sound-reflecting surface that can be lowered to a known depth below the transducer. A variety of calibration targets have been used such as a section of standard pipe sealed at both ends, a rectangular section of sheet steel, a spherical metal ball, or a section of I-beam or T-beam. The overall dimensions of the target should depend on the type of survey vessel, location of the transducer, and the depth range to be covered in the bar-check. For transducers mounted in the hull of a vessel, the overall length of the target should be about the same as the beam width of the vessel, allowing the bar to be passed over the stern and lowered beneath the transducer. A metal ball used as the target may be lowered through a well in the hull. Safety precautions should be employed to keep the target and cables away from the boat propeller (2).

13.6.4 Flexible wire or line with a wire core is used to suspend the target below the sounding transducer. The lines should be marked in a clear readable manner at desired measuring intervals.

13.6.5 The survey crew using echo sounders for hydrographic surveying should make bar-checks and record the results.

13.6.5.1 In protected waters where conditions are favorable and survey depths lie within the bar-check range, bar-checks should be made at least twice daily, prior to beginning depth measurements and at the end of the day. Comparisons are recorded during both descent and ascent of the bar, at predetermined intervals below the surface, and throughout the depth range of the survey. An additional observation is made at a depth of 5 to 6 ft (1.5 to 1.8 m), if the soundings can be recorded.

13.6.5.2 Where all or some of the depths within the project area are beyond bar-check range, bar-check data may be supplemented by taking manual soundings of total depth, comparing these measurements with the echo sounding measurement, and determining velocity correction. A calibration curve can be generated which can be used to correct future depth recordings.

13.6.5.3 Bar-check data should be recorded in a comparison log, or on the depth chart in those cases where the adjustment control is available on the echo sounder.

13.7 *Datum:*

13.7.1 A reference datum should be used to convert measured depths to bottom profile elevations. In most, but not necessarily all cases, the datum will be an accepted national vertical datum. An independent datum is acceptable in cases where connecting the measurement to the national datum is too costly to justify, or where control structures related to the water body are already constructed to a different datum. Converting depths to elevations referenced to the national datum will ensure that measurements can be properly repeated even if control points are destroyed (9).

13.7.2 The vertical control for converting depth measurements is generally determined by measuring the water level in the survey area over the survey time period. The important factor in using a vertical datum is to ensure that all depths are referenced to the same datum elevation (9).

13.8 *Interpreting Depth Records:*

13.8.1 A correct interpretation of bottom profiles remains a major problem, although echo sounders have been used for many years. A general caution should always be observed: when recorded traces on the graphic or digital record cannot be attributed with reasonable certainty as reflections from the bottom, the traces should not be recorded as soundings. In hydrographic surveys performed for producing navigational charts, it is important that all "stray" or spurious soundings be examined with care and identified as to source, since any unidentified source could present a threat to navigation.

13.8.2 One basic factor to consider when interpreting bottom traces is that a hard bottom will reflect an echo more strongly than a soft bottom. On instruments with sensitivity control, the sensitivity should be set to the minimum position which produces a good, consistent bottom trace.

13.8.3 A relatively flat bottom composed of rock, sand, or consolidated sediments usually will produce a thin, dark trace on an analog chart. Such a bottom will often create multiple echoes in shallow water caused by the signal bouncing back and forth between the bottom and the water surface. The echoes appear as multiples of the actual depth, that is always the most shallow reading on the trace.

13.8.4 A relatively soft bottom composed of unconsolidated silt, clay, or organic materials, or all three, will produce a broad echo trace of light intensity. The broad trace is caused by the reflection of the transmitted signal from both the top of the material and any firmer surfaces of consolidated material beneath the top. The thickness of the fluff or soft layer can sometimes be determined by a split in this type of echo trace on the graphic record.

13.8.5 Another check on the type of bottom is the relative setting of the sensitivity control required to obtain recordings at various depths. The air-water interface generally produces the strongest echoes. Rock, sand, metal, wood, fish, and plankton produce echoes in a diminishing order of intensity.

13.8.6 The width of the bottom trace may also be related to water depth. In deep water soundings, the conical-shaped beam produced by the transducer is reflected from a large area of the bottom resulting in a wide trace. In shallow water, the conical beam is reflected from a smaller area resulting in a narrow trace. For very narrow beam transducers, this difference may be negligible. For wide beam transducers, the difference is large.

13.8.7 Actual bottom profiles cannot always be determined with certainty because echo traces do not always represent actual physical conditions. Correct interpretation of analog or digital depth data is sometimes rendered very difficult by the presence of heavy marine growth, floating objects, bottom projections, or depressions representing sudden bottom changes, or steep bottom slopes. For a more detailed discussion on data interpretation see Ref (2).

13.9 *Specific Use of Sonic-Echo Sounders:*

13.9.1 *Single Transducer Channel Sweep System*—The most common use of an echo sounder is for charting or mapping the boundary of water bodies. A single transducer, mounted either over the side or within a well in the boat hull, can take a continuous series of depth readings below the survey boat as the boat moves along a given line (see Fig. 10). One pass of the boat provides one line measurement of depth, usually either parallel or perpendicular to the flow direction (3). The depth recorder produces a graphic plot of depth versus time on chart paper designed for the recorder's printing capabilities (see Fig. 7). With interfacing digital equipment, the depths may either be shown numerically in a digital display, or transferred to a tape or disk for future use.

13.9.2 *Multiple Transducer Channel Sweep Systems (Sonic-Echo Sounding Technique):*

13.9.2.1 In situations where detection of navigational obstacles is of major concern, multiple transducer channel sweep systems may be employed. These systems utilize arrays of transducers that transmit overlapping acoustic beams, providing complete coverage of the bottom (see Fig. 11). The transducers can be mounted on booms projecting from the side of the survey boat, or they can be mounted on a floating boom pushed by the survey boat. The optimum width of boom depends on operating conditions in a given area. One pass with a channel sweep system equals many passes with a single transducer system, thereby saving many hours of operating time where full coverage is required (see Fig. 10).

13.9.2.2 Data can be displayed either in analog or digital form. Use of digital techniques permits the display of many more transducer signals on one recorder. In addition, the existence of shallow depths in certain areas can be emphasized by automatically outlining or shading these critical regions (3).

13.9.3 *Fan-Beam Acoustic Sounding System (Sonic Echo*

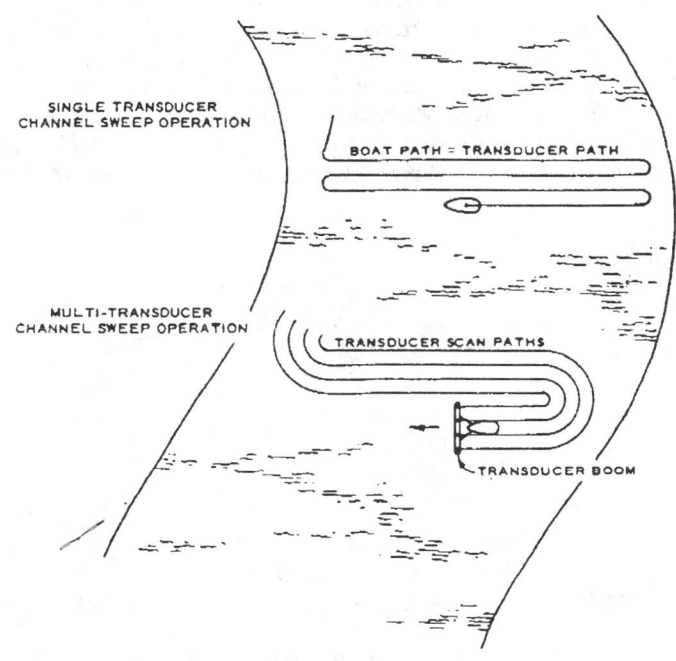

FIG. 10 Single Beam vs. Channel Sweep Acoustical Sounding (3)

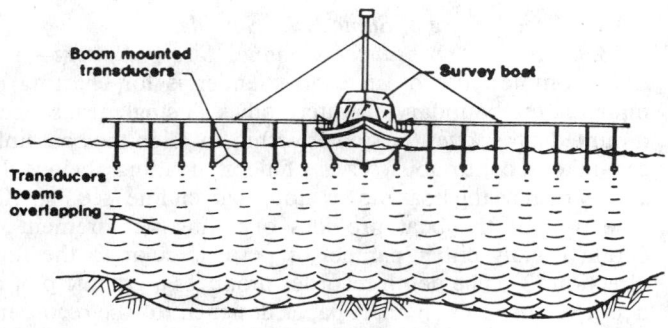

FIG. 11 Channel Sweep System

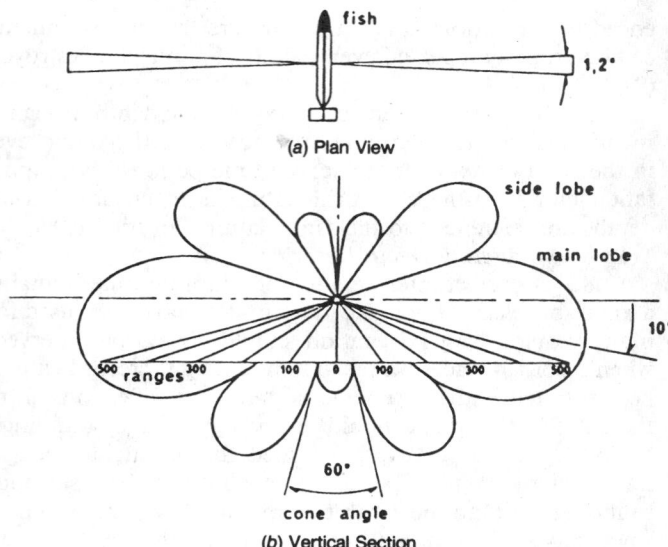

FIG. 13 Towfish Tethered to Survey Vessel Producing Lobe-figure of Side Scan Sonar Beam (14)

Sounding Technique)—The path covered by a single transducer depth recorder may also be enlarged by using a fan-beam sounding system. This system contains a number of narrow beam transducers mounted in close proximity, and focused at equally spaced angles under the survey boat (see Fig. 12). Each transducer acts as a separate acoustic-distance measuring unit beamed at a given angle with respect to the bolt-hole vertical. By electronic computation, the depth component for each beam can be derived from the slant-distance signal and a gyro vertical reference signal. Signals are displayed on a cathode-ray tube aboard the survey boat in a manner that gives a cross section of the channel perpendicular to the direction of boat travel. Signals are also digitized and recorded on magnetic tape for later automatic data processing (3).

13.9.4 *Side-Scan Sonar Systems (Sonic Echo Sounding Technique):*

13.9.4.1 Side-scan sonar systems are designed to emit fan shaped patterns of acoustic pulses from a towfish tethered to a survey vessel (see Fig. 13). A set of transducers mounted in a compact towfish generate high-powered, short-duration, acoustic pulses required for very high resolution bottom profiles. The transducers are set in the towfish, declined slightly from the horizontal, causing the pulses to project downward and outward to either side of the fish in a plane perpendicular to its path. The qualitative picture of the bottom provides a means of detecting natural and manmade objects and bottom sediment characteristics. Good acoustic reflectors such as rocks, ledges, metal objects, and sand ripples are represented as darkened areas on the record.

Depressions and other features scanned from the acoustic beam are indicated by light areas. An experienced observer can interpret most records at a glance, recognizing not only significant features and objects, but often more difficult data such as the composition and relative hardness of the bottom, and the shape and condition of submerged objects (3).

13.9.4.2 A side-scan sonar system generally consists of a recorder, transceiver, and a towfish containing transducers and pre-amplifier circuitry. The towfish consists of a streamlined, hydrodynamically balanced body about 3 ft (1 m) long containing two sets of transducers that scan the waterway bottom on either side. The high frequency beam is slightly depressed from the horizontal with the vertical axis of the main lobe pointing about 10° downward (see Fig. 13 (*b*)). The width of the beam is quite narrow in the horizontal plane (2° or less) (see Fig. 13 (*a*)). The towfish is usually operated at a distance above the bottom equal to 10 to 20 % of the range scale. The nose of the towfish contains the transmitting and receiving circuitry. The transmitter energizes the transducers when a trigger pulse is received from the shipboard recorder. The receiving circuitry amplifies the received echos and relays them through the tow cable to the recorder (3).

13.10 *Seismic (Subbottom) Reflection (Sonic Echo Sounding Technique):*

13.10.1 Continuous seismic-reflection profiling systems have made it possible to study the structures of rock and sediments beneath the bottom surface of water bodies including oceans, estuaries, lakes, and rivers. Because of its ability to penetrate surface materials on the bottom, the technique is sometimes referred to as subbottom profiling (10).

13.10.1.1 The technique requires a vessel with either a hull-mounted or towed device that emits a sonic pulse at regular intervals as the vessel travels along a selected course or survey track (see Fig. 14). The sonic pulse strikes the bottom and a portion of the energy is reflected back (return echo) to the surface where it is received by a hydrophone

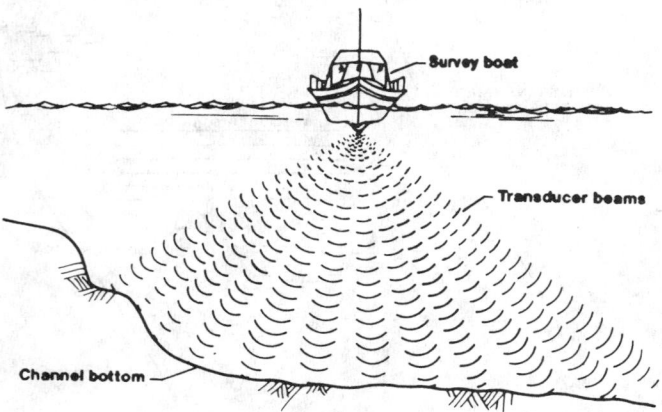

FIG. 12 Fan-beam Acoustic System

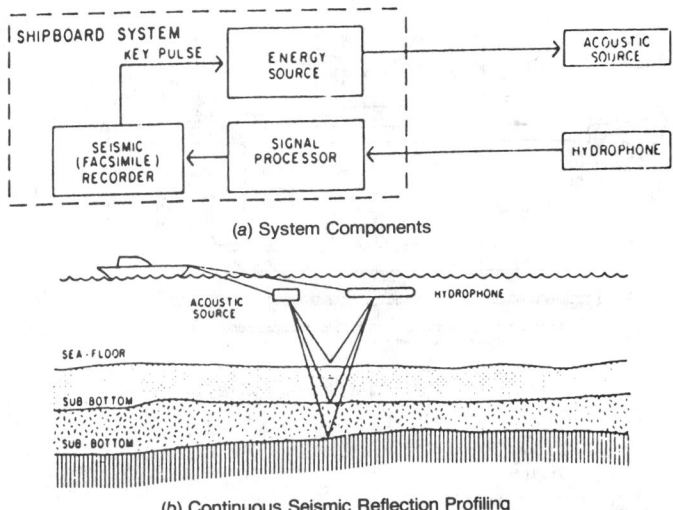

(a) System Components

(b) Continuous Seismic Reflection Profiling

FIG. 14 Seismic Reflection of Bottom (10)

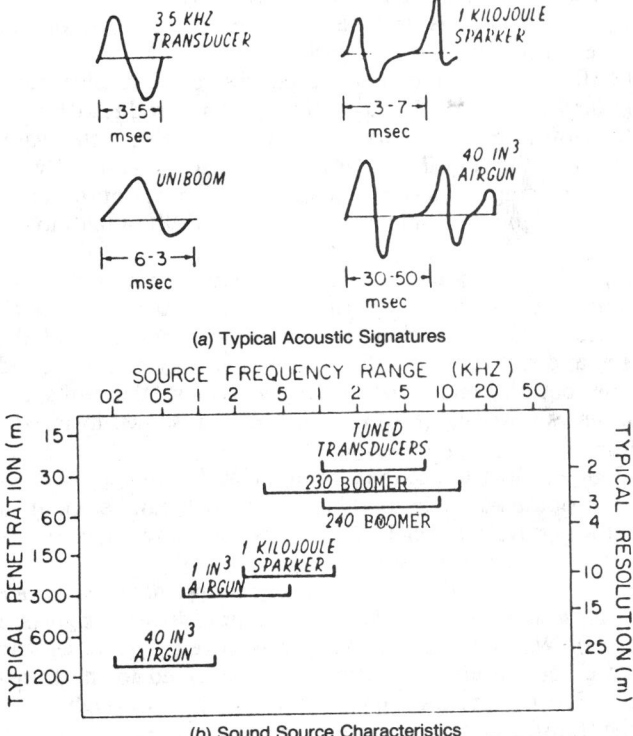

(a) Typical Acoustic Signatures

(b) Sound Source Characteristics

FIG. 15 Acoustic Signature and Relative Effect of Sound
Sources (10)

streamer. The hydrophone converts the pressure pulse to an electronic signal that is sent back to a shipboard preamplifier/filter unit where it is processed. The signal then goes to a chart recorder where it is displayed as a dark mark on the chart. The distance from the beginning edge of the chart paper (time 0) to the mark on the chart represent the water depth. As the sonic pulse continues into the bottom sediment, a portion of its energy is reflected back whenever a difference in sediment composition or density is encountered. A greater contrast in material will reflect greater energy and consequently a stronger return echo. These subbottom echoes are received and processed in the same manner as the leading (bottom) echo. When the writing stylus on the recorder completes its travel, and returns to the beginning edge of the chart, the cycle is repeated. This display is an acoustical image of the bottom surface and subbottom structure along the survey track. The depth of penetration into the subbottom is directly related to the characteristics of the profiling system and to the physical and environmental parameters of the water and sediment column (10).

13.10.1.2 A variety of seismic profiling systems are available. The four most commonly used are the tuned transducer, the boomer, the sparker, and the air gun. The names indicate the acoustic source employed by the systems. The typical pulse length, general range of penetration, and resolution that may be achieved by these systems are shown in Fig. 15. As shown, the greatest penetration is achieved by low frequency sound sources. The disadvantage of low frequency sources is that they tend to be very large, with an inherent inefficiency that requires large and expensive power sources for operation (10).

13.10.2 A tuned transducer is made up of piezoelectric crystalline materials formed into various shapes and placed into a mechanically rugged housing. When the material is subjected to an electrical signal it is deformed, thus creating a pressure pulse. Conversely, when the same material is deformed by a pressure wave, it generates an electrical signal. The transducer is therefore capable of emitting and receiving acoustic signals at the frequency for which it was designed and constructed. Other important characteristics that depend

on design are the conversion efficiency, the power handling capability, and the beam width of the transducer (10).

13.10.2.1 A variety of subbottom systems are available using the tuned transducer. They have frequencies that vary from 0.40 to 14 kHz. Resolution varies directly with frequency from about 0.5 to 3 ft (0.15 to 0.91 m). Depth of penetration is inversely proportional to frequency. The penetration depth also decreases as sediment particle size and density increase. Subbottom penetration has been recorded for depths well over 200 ft (61 m). Figure 16 shows an example of a tuned transducer system (10).

13.10.2.3 Transducer mounting configurations may be either within a well that is constructed in the hull of the

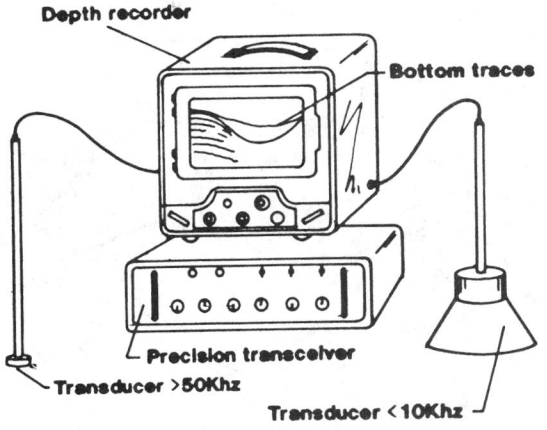

FIG. 16 Tuned Transducer System

vessel, in an over-the-side transducer array, in a towed fish array, or on catamaran-type float normally towed on the surface behind the survey vessel.

13.10.3 The boomer can be considered a displacement type device because the acoustic signal is produced by the sudden movement of a flat, circular plate against the water surface (see Fig. 17). The acoustic signatures produced by the plate displacing the water beneath it is of short duration and relatively high amplitude, having a band width of 0.40 to 4.0 kHz.

13.10.3.1 A boomer device is usually mounted on a catamaran that is towed astern or alongside the survey vessel. Because of the limitations this places on the speed of the vessel, and the wave-like distortions that may be transferred to the bottom and subbottom reflections, the units are sometimes towed beneath the surface, or installed inside the hull.

13.10.3.2 Because of the amplitude and frequency characteristics of the boomer, good bottom records can be obtained in sand, gravel, or glacial deposits that are acoustically opaque to a tuned transducer (10).

13.10.4 The sparker is a relatively simple acoustic device with regard to its power supply and transducer design (see Fig. 18). When the power supply is keyed by the seismic recorder, stored electrical energy is discharged to the transducer. Because the discharge is forced to pass through water to the transducer ground return, the heat causes the water to suddenly vaporize into steam and ionized particles. This rapid vaporization produces the initial acoustic pulse. A second acoustic pulse is produced as the steam bubble cools and eventually collapses.

13.10.4.1 A sparker system is capable of fair resolution, 15 to 30 ft (4.6 to 9.1 m), and good penetration, 650 to 1000 ft (198 to 305 m). It is particularly applicable in continental shelf regions where good penetration in hard sands and semi-consolidated material is required (10).

13.10.5 The air gun provides a pneumatic acoustic sound source. The sound production system (see Fig. 19) consists of an onboard air compressor and storage bottle, a shipboard electrical firing circuit controlled by the seismic recorder, and one or more air guns towed behind the survey vessel. When triggered from the seismic recorder, an air gun releases

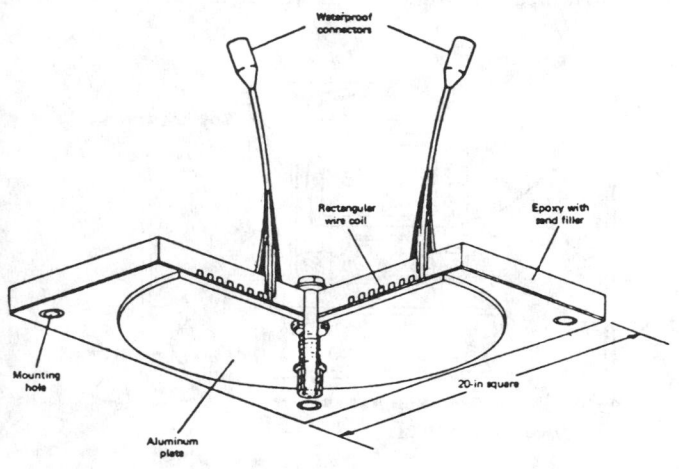

FIG. 17 Boomer Electrical-mechanical Transducer (10)

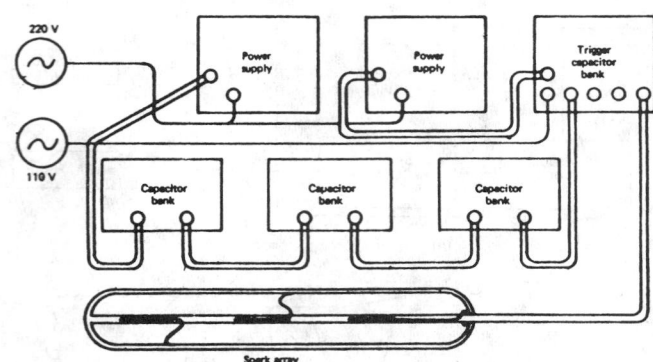

FIG. 18 The Sparker Array

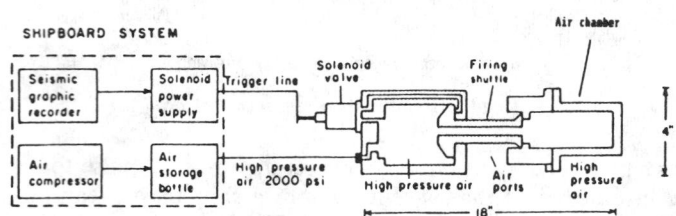

FIG. 19 Components of the Air Gun System (10)

a specified volume of compressed air into the water. This sudden release of air into the water produces a steep-front shock wave followed by several oscillations caused by the repeated contraction and expansion of the air bubble.

13.10.5.1 Air gun systems have poor resolution, 50 to 100 ft (15.2 to 30.4 m) relative to previously discussed acoustic sources. However, large penetration depths, 650 to 6500 ft (18 to 1981 m) can be achieved by the use of large air guns or air gun arrays (10).

PROCEDURE C—ELECTRONIC NONACOUSTIC MEASUREMENT

14. Scope

14.1 This practice is applicable to the measurement of water depth by electronic nonacoustic techniques including ground-penetrating radar and air-borne laser equipment.

14.2 These techniques are still in the process of development for surface water applications. Certain techniques and equipment are not available for public use.

15. Ground-Penetrating Radar (GPR) (Sonic Nonacoustic Technique)

15.1 Ground-penetrating radar (GPR) is an impulse radar system primarily designed as a reconnaissance tool for shallow subsurface site investigations. A special application of the system has been the profiling of the thickness of freshwater and sea ice as well as water depth and ground surface beneath ice cover. The system functions by radiating short electromagnetic pulses into the ice or ground from a transmitting antenna. The transmitted pulse consists of a spectrum of frequencies that are distributed around the central frequency of the antenna. As the pulse strikes an interface separating layers of different electrical properties, a portion of the pulse energy is reflected back to a receiving

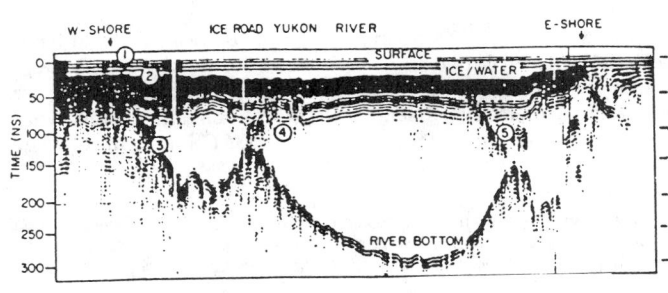

(a) Radar Section of the Ice Bridge Across the Yukon River

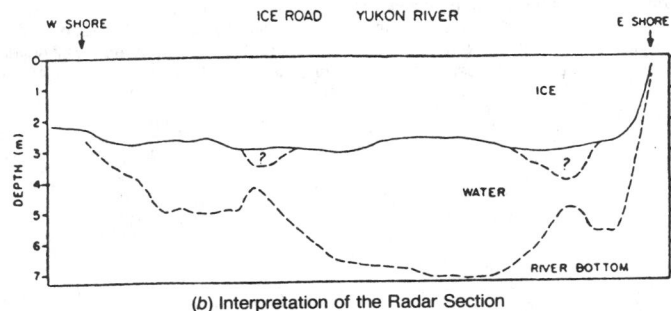

(b) Interpretation of the Radar Section

NOTE—Reproduced with the permission of the Minister of Supply and Services Canada, 1990.

FIG. 20   Interpretation of Ground-penetrating Radar Sections (12)

antenna. The receiving unit amplifies the reflected energy and converts it into a similarly shaped wave form in the audio frequency range. These processed waveforms are then displayed on a graphic recorder, or recorded on tape for future playback or recording. When displayed on a graphic recorder, a variable gray scale is employed, the strong reflections displayed as black images and the intermediate, weaker reflections in various shades of gray (11). Figure 20 shows a radar section and interpretation of an ice bridge across the Yukon River obtained with a 10 ns impulse antenna (12).

15.2 The components of a GPR system may consist of a control unit with a microprocessor, a power distribution unit, a graphic recorder, a tape recorder, and a variety of antennas for transmitting and receiving. The lower fre-

quency antennas (80 to 120 MHz) have greater powers of radiation, and longer pulse width, and therefore, emit signals that are less rapidly attenuated by earthen materials than the signals emitted from higher frequency antennas. Therefore, lower frequency antennas can penetrate to greater depths. However, when depth of penetration is not a critical factor, the higher frequency antennas (300 to 500 MHz) produce better resolution of subsurface features (13).

15.3 Radar profiling of a water body through ice cover is possible, given the appropriate physical conditions of the medium and a proven method of calibrating the instruments. The present state of the art requires an experienced operator and observer to produce accurate bottom profiles.

16. Keywords

16.1 depth measurement; echo sounder; profile; sounding; transducer

REFERENCES

(1) Rantz, S. E., et al., "Measurement and Computation of Streamflow: Volume 1, Measurement of Stage and Discharge," U.S. Geological Survey Water Supply Paper 2175, 1982.
(2) Hydrographic Manual, Fourth Edition, National Oceanic and Atmospheric Administration, Washington, DC, 1976.
(3) Measurement of Hydrographic Parameters in Large Sand-Bed Streams from Boats, Task Committee on Hydrographic Investigations of the Committee on Waterways of the Waterway, Port, Coastal, and Ocean Division, 1983, American Society of Civil Engineers, New York, NY 10017.
(4) National Engineering Handbook, Section 3, Chapter 7, U.S. Department of Agriculture, Soil Conservation Service, pp. 1–31.
(5) Davis, Foote, and Kelley, Surveying-Theory and Practice, Fifth Edition, 1966, Chapter 30, McGraw Hill, New York, NY.
(6) Keen, B. L., "A New Generation Portable Echo Sounder," Proceedings, U.S. Army Corps of Engineers Surveying Require-

ments Meeting, Feb. 2–5, 1982, U.S. Army Waterways Experiment Station, CE, Vicksburg, MS, April 1982.
(7) Hart, E. D., and Downing, G. C., "Positioning Techniques and Equipment for U.S. Army Corps of Engineers Hydrographic Surveys," TR H-77-10, U.S. Army Engineer Waterways Experiment Station, CE, Vicksburg, MS, May 1977.
(8) Blanton, James, III, "Procedures for Monitoring Reservoir Sedimentation," U.S. Bureau of Reclamation Technical Guideline, October 1982.
(9) Davis, R. E., Foote, F. S., Anderson, J. M., and Mikhail, E. M., Surveying-Theory and Practice, Sixth Edition, 1981, Chapter 21, McGraw Hill, New York, NY.
(10) Sylvester, R. E., "Single Channel, High-Resolution Seismic-Reflection Profiling: A Review of the Fundamentals and Instrumentation," Handbook of Geophysical Exploration at Sea, Geyer, R. A. ed., CRC Press, Boca Raton, FL, 1983.

 **D 5073**

(11) *Soil Science Society of America Journal*, Soil Science of America, Vol 49, No. 6, November–December 1985, pp. 1490–1498.

(12) Annon, A. P., and Davis, J. L., "Impulse Radar Applied to Ice Thickness Measurements and Freshwater Bathymetry," Energy, Mines, and Resources Canada, Report of Activities, Part B; Geological Survey of Canada, Paper 77-1B, 1977.

(13) "Soil Survey Techniques," *SSSA Special Publication No. 20*, Soil Science Society of America, Madison, WI, 1987.

(14) Fleming, B. W., *Side-Scan Sonar: A Comprehensive Presentation*, EG&G Environmental Equipment Division, Waltham, MA, 1980.

Standard Test Methods for
Measurement of Water Levels in Open-Water Bodies[1]

This standard is issued under the fixed designation D 5413; the number immediately following the designation indicates the year of original adoption or, in the case of revision, the year of last revision. A number in parentheses indicates the year of last reapproval. A superscript epsilon (ε) indicates an editorial change since the last revision or reapproval.

1. Scope

1.1 These test methods cover equipment and procedures used in obtaining water levels of rivers, lakes, and reservoirs or other water bodies. Three types of equipment are available as follows:

Test Method A—Nonrecording water-level measurement devices
Test Method B—Recording water-level measurement devices
Test Method C—Remote-interrogation water-level measurement devices

1.2 The procedures detailed in these test methods are widely used by those responsible for investigations of streams, lakes, reservoirs, and estuaries, for example, the U.S. Agricultural Research Service, the U.S. Army Corp of Engineers, and the U.S. Geological Survey.[2] The referenced ISO standard also furnishes useful information.

1.3 It is the responsibility of the user of these test methods to determine the acceptability of a specific device or procedure to meet operational requirements. Compatibility between sensors, recorders, retrieval equipment, and operational systems is necessary, and data requirements and environmental operating conditions must be considered in equipment selection.

1.4 The values stated in inch-pound units are to be regarded as the standard. The values given in parentheses are for information only.

1.5 *This standard does not purport to address all of the safety problems, if any, associated with its use. It is the responsibility of the user of this standard to establish appropriate safety and health practices and determine the applicability of regulatory limitations prior to use.*

2. Referenced Documents

2.1 *ASTM Standards:*
D 1129 Terminology Relating to Water[3]
D 1941 Test Method for Open Channel Flow Measurement of Water with the Parshall Flume[3]
D 2777 Practice for Determination of Precision and Bias of Applicable Methods of Committee D-19 on Water[3]
D 5242 Test Method for Open Channel Flow Measurement of Water with Thin-Plate Weirs[3]
2.2 *ISO Standard:*
ISO 4373 Measurement of Liquid Flow in Open Channels—Water Level Measuring Devices[4]

3. Terminology

3.1 *Definitions*—For definitions of terms used in this test method, refer to Terminology D 1129.

3.2 *Description of Terms Specific to This Standard:*

3.2.1 *elevation*—the vertical distance from a datum to a point.

3.2.2 *datum*—a level plane that represents a zero or some defined elevation.

3.2.3 *gage*—a generic term that includes water level measuring devices.

3.2.4 *gage datum*—a datum whose surface is at the zero elevation of all the gages at a gaging station; this datum is often at a known elevation referenced to National Geodetic Vertical Datum of 1929 (NGVD).

3.2.5 *gage height*—the height of a water surface above an established or arbitrary datum at a particular gaging station; also termed *stage*.

3.2.6 *gaging station*—a particular site on a stream, canal, lake, or reservoir where systematic observations of hydrologic data are obtained.

3.2.7 *National Geodetic Vertical Datum of 1929 (NGVD)* —prior to 1973 known as mean sea level datum; a spheroidal datum in the conterminous United States and Canada that approximates mean sea level but does not necessarily agree with sea level at a specific location.

4. Significance and Use

4.1 These test methods are used to determine the gage height or elevation of a river or other body of water above a given datum.

4.2 Water level data can serve as an easily recorded parameter, and through use of a stage-discharge relation provide an indirect value of stream discharge, often at a gaging station.

4.3 These test methods can be used in conjunction with other determinations of biological, physical, or chemical properties of waters.

TEST METHOD A—NONRECORDING WATER-LEVEL MEASUREMENT DEVICES

5. Summary of Test Method

5.1 This test method is usually applicable to conditions where continuous records of water level or discharge are not required. However, in some situations, daily or twice daily observations from a nonrecording water-level device can provide a satisfactory record of daily water levels or discharge. Water levels obtained by the nonrecording devices described in this test method can be used to calibrate recording water-level devices described in Test Methods B and C.

5.2 Devices included in this test method are of two

[1] These test methods are under the jurisdiction of ASTM Committee D-19 on Water and are the direct responsibility of Subcommittee D19.07 on Sediments.
Current edition approved May 15, 1993. Published November 1993.
[2] Buchanan, T. J., and Somers, W. P., "Stage Measurement at Gaging Stations," *Techniques of Water Resources Investigations*, Book 3, Chapter A-7, U.S. Geological Survey, 1968.
[3] *Annual Book of ASTM Standards*, Vol 11.01.
[4] Available from American National Standards Institute, 11 W. 42nd St., 13th Floor, New York, NY, 10036.

general types: those that are read directly, such as a staff gage; and those that are read by measurement to the water surface from a fixed point, such as wire-weight, float-tape, electric-tape, point and hook gages.

5.2.1 Staff, wire-weight, and chain gages are commonly used as both outside auxiliary and reference gages. Vertical- and inclined-staff, float-tape, electric-tape, hook and point gages are commonly used as inside auxiliary and reference gages.

5.3 Documentation of observations must be manually recorded.

6. Apparatus

6.1 *Staff Gages:*

6.1.1 *Vertical Staff Gages*—Staff gages are usually graduated porcelain-enameled plates attached to wooden piers or pilings, bridge piers, or other hydraulic structures. They may also be installed on the inside of gaging station stilling wells as inside reference gages. They are precisely graduated, usually to 0.02 ft or 2 mm, although other markings may be used for specific applications (see Fig. 1).

6.1.2 *Inclined Staff Gages*—Inclined staff gages usually consist of markings on heavy timbers, steel beams, or occasionally concrete beams built partially embedded into the natural streambed slope. Since they are essentially flush with the adjoining streambed, floating debris and ice are less likely to cause damage than for a vertical staff gage. Individual graduation and marking of the installed gages by engineering levels are required due to the variability of bank slope.

6.2 *Wire-Weight Gage*—An instrument that is mounted on a bridge or other structure above a water body. Water levels are obtained by direct measurement of the distances between the device and the water surface. A wire-weight gage consists of a drum wound with a single layer of cable, a bronze weight attached to the end of the cable, a graduated disk, a counter, and a check bar, all contained within a protective housing (see Fig. 2). The disk is graduated and is permanently connected to the counter and the shaft of the drum. The cable is guided to its position on the drum by a threading sheave. The reel is equipped with a pawl and ratchet for holding the weight at any desired elevation. A horizontally mounted check bar is mounted at the lower edge of the instrument. Differential levels are run to the check bar. When the weight is lowered to touch the check bar, readings of the counter are compared to its known elevation as a calibration procedure. The gage is set so that when the bottom of the weight is at the water surface, the gage height is indicated by the combined readings of the counter and the graduated disk.

6.3 *Needle Gages*—Frequently referred to as point or hook gages. A needle gage consists of a vertically-mounted pointed metallic, small-diameter rod, which can be lowered until an exact contact is made with the water surface. A vernier or graduated scale is read to indicate a gage height. A needle-type gage offers high measurement accuracy, but requires some skill and good visibility (light conditions) in lowering and raising the device to a position where the point just pierces the water surface. These gages are most commonly used in applications where the water surface is calm.

6.3.1 *Point Gage*—A form of needle gage where the tip or point approaches the water surface from above.

6.3.2 *Hook Gage*—A form of needle gage made in the shape of a hook, where the tip or point approaches the water surface from below (see Fig. 3). The hook gage is easier to use in a stilling well application. As the point contacts the water surface, overhead light will reflect from a dimple on the water surface.

6.4 *Float-Tape Gage*—Consists of a float attached to a stainless steel graduated tape that passes over a suitable pulley with a counterweight to maintain tension. A pointer or other index is frequently fabricated as an integral part of the pulley assembly (see Fig. 4). Float-tape gages frequently are combined with water-level recorders in a manner whereby the pulley is the stage drive wheel for the recorder.

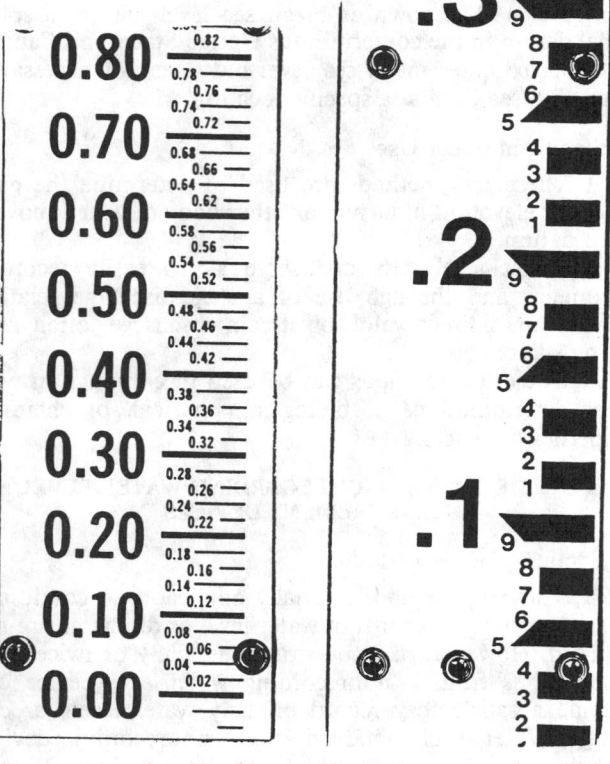

FIG. 1   Staff Gages

FIG. 2   Type A Wire-Weight Gage

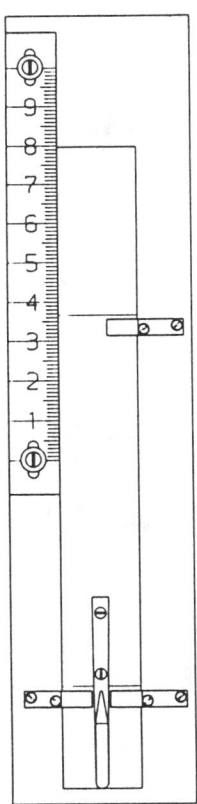

FIG. 3 Hook Gage

6.5 *Electric-Tape Gage*—Consists of a graduated steel tape and weight attached to a combined tape reel, voltmeter,

datum index and electrical circuit, powered by a 4½ to 6 volt battery (see Fig. 5). The gage frame is mounted on a shelf or bracket over the water surface, usually in a stilling well. The weight is lowered until the weight touches the water surface closing the electrical circuit that is indicated by the voltmeter. The gage height is read on the tape at the index.

6.6 A reference point is frequently selected on a stable member of a bridge, stilling well, or other structure from which distance vertical measurements to the water surface are made by steel tape and weight. The reference point is a clearly defined location, frequently a file mark or paint mark to ensure that all readings are from the same location.

7. Calibration

7.1 Establish a datum. The datum may be a recognized datum such as National Geodetic Vertical Datum of 1929 (NGVD), a datum referenced to a recognized datum such as 580.00 ft NGVD 1929, a local datum, or an arbitrary datum. A datum is usually selected that will give readings of small positive numbers.

7.2 Establish at least three reference marks (RMs). Reference marks must be located on independent permanent structures that have a good probability of surviving a major flood or other event that may destroy the gage. Reference marks should be close enough to the water-level measuring device that the leveling circuit not require more than two or three instrument setups to complete elevation verification. If the NGVD datum is used, determine the elevation of the reference marks by differential leveling from the nearest NGVD benchmark.

7.3 Set the gages to correct datum by differential leveling from the reference marks. Use leveling procedures described in a surveying text or "Levels at Streamflow Gaging

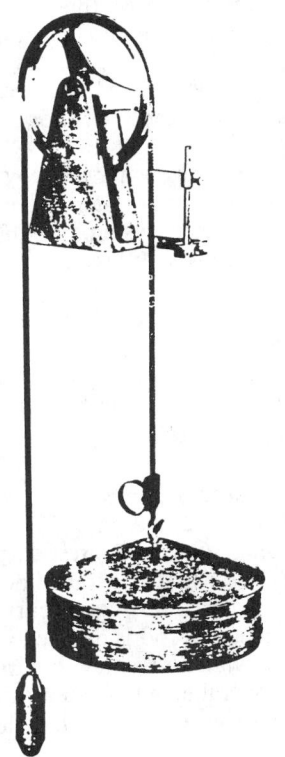

FIG. 4 Float-Type Gage

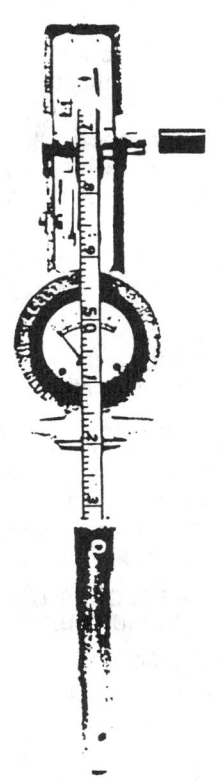

FIG. 5 Electric-Type Gage

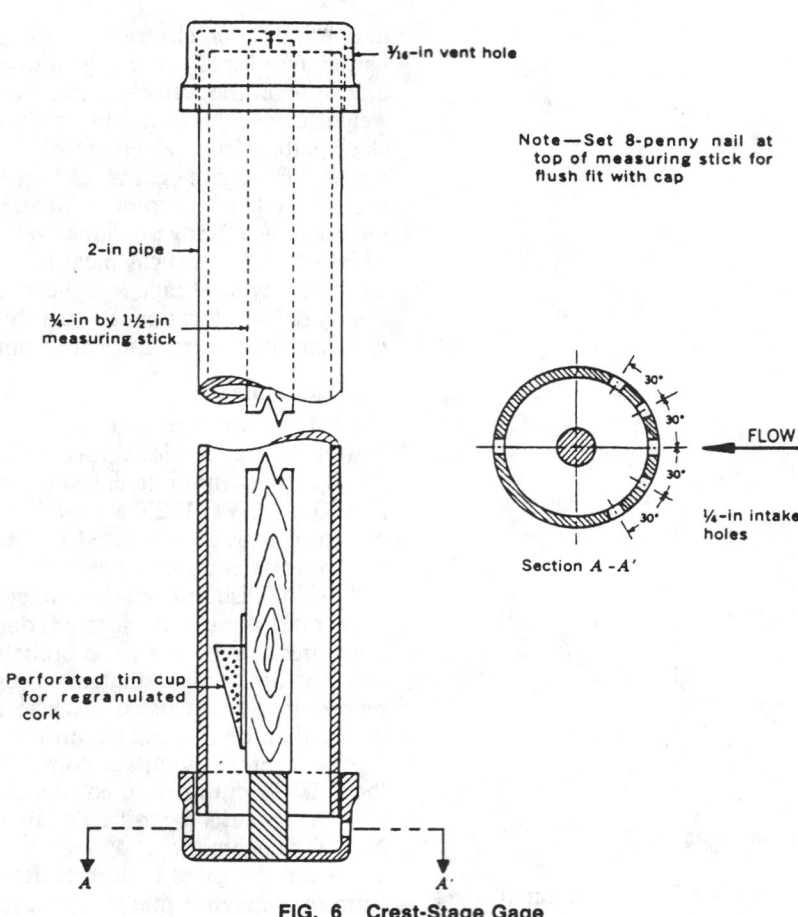

FIG. 6 Crest-Stage Gage

Stations."[5]

7.4 Run levels to gages from RMs annually for the first 3 to 5 years, then if stability is evident, a level frequency of 3 to 5 years is acceptable. Rerun levels at any time that a gage has been disturbed or has unresolved gage reading inconsistencies. Run levels to all RMs, reference points, index points, and to each staff gage, and to the water surface. Read the water surface at each gage at the time levels are run. Document differences found and changes made in a permanent record.

8. Procedure

8.1 Read direct reading gages by observing the water surface on the gage scale. Manually record this value on an appropriate form.

8.2 Gages that require measurement from a fixed point to the water surface must follow procedures provided by manufacturers of the specific instrument.

8.3 Make a visual inspection of gages at each reading to detect apparent damage, which could affect accuracy.

TEST METHOD B—RECORDING WATER-LEVEL MEASUREMENT DEVICES

9. Summary of Test Method

9.1 This test method is applicable where continuous unattended records of water level or discharge are required.

Procedures described in Test Method A are usually used to set these recording devices to the correct datum.

9.2 Devices, generically referred to as water-level recorders, or recorders, included in this test method must be capable of recording stage and the time and date at which the stage occurred.

9.3 Recorders may sense water level by direct mechanical connection, usually by float-counterweight and tape or cable, by gas purge manometer systems (bubble gages), or by electronic water level sensors (pressure transducer or acoustic devices).

9.4 Recorders may retain data in graphical, analog, digital, or other format.

9.5 Recorders are available that can remain unattended for periods from one week to longer than six months.

10. Apparatus

10.1 *Types of Sensing Systems:*

10.1.1 *Direct Reading Systems:*

10.1.1.1 *Crest Stage Gage*—A crest stage gage is a simple sensing-recording device that is installed near a water body to record the highest water level that occurs between visits of field personnel. A wooden rod is encased in a steel or plastic pipe with holes for water to enter and rise to the outside water level. A recoverable high-water mark is left on the device by particles of ground cork that float to the highest water level (Fig. 6).

10.1.1.2 *Tape Gage Maximum-Minimum Indicators—*

[5] Kennedy, E. J., "Levels at Streamflow Gaging Stations," Techniques of Water Resources Investigations, Book 3, Chapter A-19, U.S. Geological Survey, 1990.

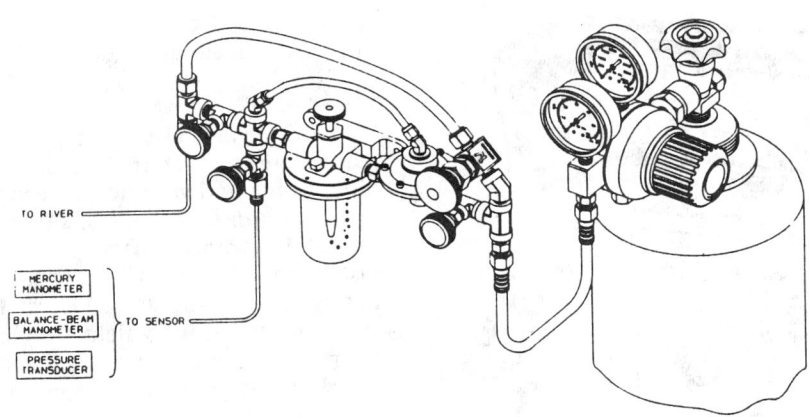

FIG. 7   Gas-Purge System

These indicators include magnetic or mechanical accessories that record maximum or minimum travel of float-drive tape gages or recorder-drive tapes.

10.1.2 *Mechanical Sensing Systems:*

10.1.2.1 *Float Tape*—Consists of a float that floats on the water surface, usually in a stilling well, and a steel tape or cable which passes over a recorder drive pulley. A weight on the opposite end of the tape maintains tension in the tape or cable. The rise and fall of the water surface is thus directly transmitted to the recorder.

10.1.2.2 *Shaft Encoders*—These devices consist of a float-tape driven shaft and pulley assembly that converts the angular shaft position to an electronic signal compatible with electronic recorders. Analog output potentiometers and several digital format output encoding systems are available.

10.1.3 *Gas-Purge System*—This system is commonly known as a bubble gage. A gas, usually nitrogen, is fed from a supply tank and pressure regulator through a tube and bubbled freely into the water body through an orifice at a fixed location on or near the bottom of the water body. The gas pressure in the tube is equal to the piezometric head on the bubble orifice corresponding to the water level over the orifice. Several methods of sensing this line pressure and converting it to a recordable format are used (Fig. 7).

10.1.3.1 *Mercury Manometer*—The manometer assembly converts the gas purge line pressure to a shaft rotation for driving a recorder. Mercury is used because its specific gravity is 13.6 times that of water, and thus shortens the length of the manometer. The theory and application of these devices are given in "Installation and Service Manual for U.S. Geological Survey Manometers."[6] These devices are being phased out of service because of potential damage to the environment should mercury spills occur.

10.1.3.2 *Balance Beam Manometer*—This form of manometer employs a bellows system coupled with a balance beam and traveling weight. Pressure changes are transmitted to the bellows, which moves the balance beam. This movement causes the traveling weight on the balance beam to adjust to a new position to put the system back in balance. The change in position of the traveling weight corresponds to

the change in water level over the orifice and is converted to a shaft rotation for recording.

10.1.3.3 *Nonsubmersible Pressure Transducer*—A pressure transducer converts gas-purge line pressure to gage heights, and transmits this data in analog or serial digital format to a compatible electronic recorder.

10.1.4 *Electronic Sensing Systems:*

10.1.4.1 *Submersible Pressure Transducer*—A pressure transducer is attached below the surface at a known datum elevation of the water body, and measures the distance above the transducer and transmits this data in analog or serial digital format by electrical cable to a compatible electronic recorder.

10.1.4.2 *Downward-Looking Acoustic Transducers*—These devices are mounted above the water surface at a fixed datum and measure the distance to the water surface by measuring the elapsed time of the reflected acoustic signal off the water surface. The data is transmitted in analog or serial digital format by electrical cable to a compatible electronic recorder.

10.1.4.3 *Upward-Looking Acoustic Transducers*—These devices are mounted on or near the bed of a water body, and measure the distance to the water surface with an acoustic signal. This data is transmitted in analog or serial digital format by electrical cable to a compatible electronic recorder.

10.2 *Recorders:*

10.2.1 *Graphic Recorders*—These recorders are also known as analog and strip-chart recorders (Fig. 8). Many designs of these devices have been used for many years. Most have a weight- or spring-driven clock or an ac or 12-volt battery-powered electric clock to move a paper chart at known speed. A mechanical float-tape or manometer unit moves a pen or pencil up and down over the chart in response to the water body level fluctuations. These units provide a constant recording of time and stage, are easy to manually observe changes in stage, but are not easily amenable to fully automated data processing, however, semi-automated processing is possible through digitization of the pen trace.

10.2.2 *Digital Recorders*—This system is a battery- or ac-powered slow-speed paper-tape punch that records a 4-digit number on a 16-channel paper tape at preset time intervals of 1, 5, 6, 15, 30, or 60 min. The stage input is a shaft rotation from the float-tape or manometer drive. The

[6] Craig, J. D., "Installation and Service Manual for U.S. Geological Survey Manometers," *Techniques of Water Resources Investigations*, Book 8, Chapter A-2, U.S. Geological Survey, 1983.

can be carried easily back to an office retrieval-computer interface to facilitate data transfer from field locations.

11. Calibration

11.1 Mechanical sensors and recorders generally do not require pre-installation calibration. Field calibration requires setting time and gage height recording units to observations made on nonrecording gages (see Section 8).

11.2 Electronic sensors and recorders may require pre-installation laboratory calibrations such as deadweight testing of pressure transducers. Preassembly of electronic components prior to field installation is recommended to verify operational compatibility. Follow manufacturers' or users' quality-assurance procedures.

12. Procedure

12.1 Install unit at field site following installation instructions.

12.2 Set gage height to nonrecording gages (see Section 8) and time and date with a suitable time reference.

12.3 Retrieval of record and required maintenance at end of unattended period as follows:

12.3.1 Observe and record gage height from nonrecording gage and correct time and date for record verifications and possible adjustment.

12.3.2 Remove data record following manufacturer's or user's instructions, or both.

12.3.3 Observe condition of sensors, recorder, and associated equipment, wind clock or replace battery, as applicable.

12.3.4 Reset gage-height datum if necessary, set time and date, and restart recorder (see Section 12.3.1).

TEST METHOD C—REMOTE INTERROGATION WATER-LEVEL MEASUREMENT DEVICES

13. Summary of Test Method

13.1 This test method is applicable where remote interrogation of water level or discharge is required. Procedures described in Test Method A are generally used to set these devices to the correct datum. Devices described in this section may be accessories to devices described in Test Method B.

13.2 Remote interrogation devices may be nonrecording, or may be able to store and then transmit data collected over a period of time.

13.3 Remote interrogation may be by telephone, direct wire, satellite, radio, or meteorburst transmission, as described in Section 14.

13.4 Information on transmission or receiving equipment and procedures it not included in the test method.

14. Apparatus

14.1 *Telephone Interrogation Systems*—Systems that transmit water levels over standard telephone lines in either signals audible to the human ear or in coded digital or pulse format are commercially available.

14.1.1 *Audible Transmission Systems*—A telephone call from any regular telephone to a telephone at the instrument triggers the devices to transmit a series of beeps to indicate the instantaneous water level in audible hearing range or voice synthesizers that can report in words and numbers, "the stage is 10.50 feet." The operable stage range is usually

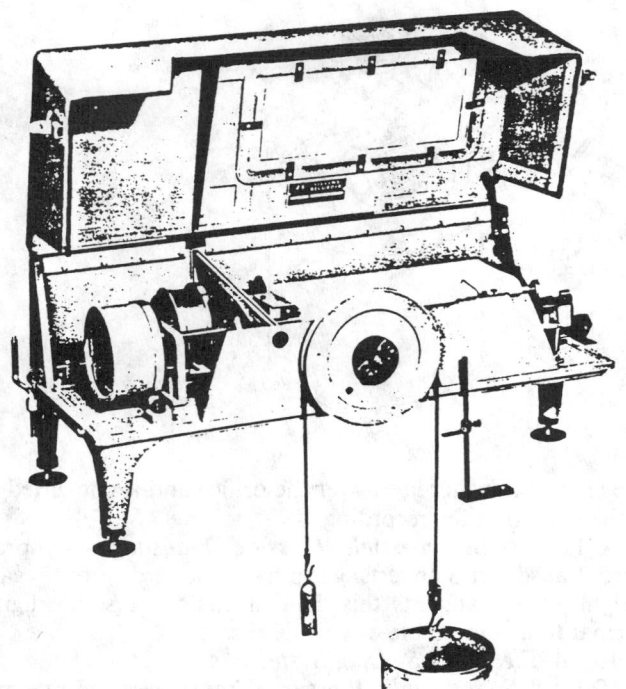

FIG. 8 Graphic Recorder

unit was developed in the early 1960's as a means of automating data processing (Fig. 9).

10.2.3 *Electronic Recorders*—A number of battery-powered electronic recorders are presently commercially available that are suitable for the field collection of hydrologic data. They range from inexpensive single-channel to complex sensor- and sampler-controlling multichannel configurations. Many types require an electronic retriever or laptop computer to retrieve stored data and transport it to an office computer for data processing, or connection to a remote interrogation device (Test Method C). Several manufacturers offer units with removable memory devices that

FIG. 9 Digital Recorder

from 00.00 to 99.99 ft. The advantage of this system is that any phone can be used to interrogate the system, making this device particularly useful during flood events, when many users can obtain data from multiple locations (Fig. 10).

14.1.2 *Coded Digital Devices*—These devices transmit instantaneous water levels in high-speed serial ASCII format over regular phone lines to a remote computer or other decoding device. These devices may either be stand-alone sensing units or as attachments to other recording devices, frequently digital-punch tape or electronic recorders.

14.2 *Direct Wire Systems*—This type of equipment consists of a sensing-transmitting unit at the water-level sensing location which is connected by electrical cable to a receiving unit that typically drives a recorder or other readout device. Common applications are short distances such as headwater and tailwater elevations at a dam to an operating house. Maximum distance for practical application would be about 15 miles.

14.2.1 *Position Motor Systems*—This system consists of a pair of self-synchronizing motors, one driven by a mechanical sensing system such as a float-tape system that serves as a transmitting unit, and a second whose rotor follows the rotary motions of the transmitter that drives a recorder or other indicating device.

14.2.2 *Impulse System*—An impulse-sending device sends incremental electrical signals over the connecting lines to a receiving device that provides a rotational output to operate a mechanical recorder or directly to an electronic recording device. This system can operate over dedicated phone lines over longer distances than a position motor system.

14.3 *Satellite Telemetry*—Satellite telemetry systems transfer data from water and sensing devices to remote field office locations via a geostationary operational environmental satellite (GOES). A water level sensor (see Section 10)

is connected to a GOES transmitter, usually known as a data collection platform (DCP). The GOES transmitter, actually a radio transmitting at 401 to 402 MH_z, transfers the water level in binary form to the GOES satellite, which retransmits the data to a direct readout ground station (DRGS). The DRGS receives, reformats the data in engineering units, and stores it in a computer for distribution or additional processing. Authorization and allocation of satellite channels must be obtained from the National Oceanic and Atmospheric Administration, National Environmental Satellite Data and Information Services. Most DCP's require a plug-compatible program unit, often a laptop computer to program and initialize the unit.

14.4 *Radio Transmission Systems*—In this type of system, data from a water-level sensor is transmitted via a line-of-sight radio transmitter to a receiving and processing device at a field office location. Repeater stations are required for long distances or mountainous terrain. Power requirements for the on-site transmission require commercial electric service or solar battery-charger systems.

14.5 *Meteor-Burst Transmission Systems*—Meteor-burst telemetry uses ionized meteor trails as reflectors for VHF radio signals to overcome the line-of-sight limitations of standard VHF radio and microwave communications. Billions of meteors enter the earth's atmosphere daily, burning up and leaving an ionized trail of gasses that remain for periods of a few microseconds to a few seconds. The altitude of useful trails is 80 to 120 km above the earth's surface, which limits the range of communications to about 2,000 km. There are also diurnal and seasonal variations in meteor trail density that can affect transmission reliability.

14.5.1 A meteor-burst system is composed of one or more master stations and a number of remote sites. Remote sites are microprocessor controlled data collection and transmission stations that collect data at preselected time intervals, for example, every 15 min, and process and store these data for transmission to the master station one or more times per day. To retrieve data from a remote site, the master station continuously transmits until a meteor trail occurs at the correct location to reflect the signal to the remote site. Upon receiving the master station signal, the remote site immediately transmits its data using the same meteor trail. Remote sites can also be programmed to initiate communications to the master station when selected sensor output exceeds a specified threshold.

14.5.2 Advantages of meteor-burst technology include access to data in near real time, relatively low system costs, operation on a common radio frequency, and communication security due to the random nature of the meteor trail.

14.5.3 Limitations are related to the short duration of meteor trails that limit the message length, and the diurnal and seasonal variation in the density of meteor trails.

15. Calibration

15.1 Remote interrogation water-level transmission devices do not usually require precalibration prior to field installation; however, connect all system components tested as a system prior to installation in remote field sites.

15.2 For field calibration, set stage and time outputs to observations of the base nonrecording gage (see Section 8).

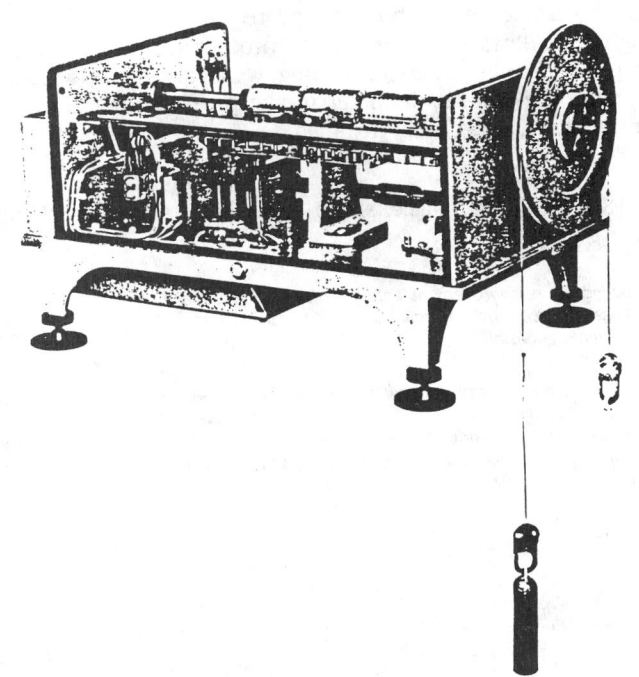

FIG. 10 Telemark Gage

16. Procedure

16.1 Install system at field site, following manufacturer's or other applicable instructions. This may require specialized programming devices for certain electronically based units.

16.2 Prior to leaving field site, verify that data transmissions are being received at remote-interrogation location.

16.3 At subsequent station visits, verify that proper gage height, time, and possible additional parameters are correctly reported. Use procedures described in Section 15.

17. Precision and Bias

17.1 Determination of the precision and bias for these test methods is not possible, at either the multiple or single operator level, due to the high degree of instability of water surfaces of open-channel flow. Temporal and spatial variability of the boundary and flow conditions do not allow for a consent standard to be used for representative sampling. Any estimate of errors would be misleading to users.

17.2 In accordance with paragraph 1.6 of Practice D 2777, an exception to the precision and bias statement required by D 2777 was recommended by the results advisor and concurred with by the Technical Operations Section of the Committee D-19 Executive Subcommittee on June 24, 1992.

17.3 The accuracy of a water-level measurement is directly related to the following.

17.3.1 Errors are caused by the instability of structures supporting the gage-sensing device. Staff gages mounted on bridge structures or piers are subject to being damaged or moved by floating debris, ice, being bumped by boats, as well as deterioration of supporting fixtures. Gages mounted in stilling wells may be subject to instability of these structures, siltation of intake pipes connecting to the water body, and manufacturing, installation, or calibration errors.

17.3.2 Errors caused by inadequate leveling procedures in establishing and checking datum.

17.3.3 In direct reading devices, observational errors include ripple or wave effect of the water surface, distance from observer to gage, cleanliness and color contrast of the gage, angle of observation, light conditions, and observer's eyesight, including the use or non-use of binoculars or other visual aids. In contact gages, this includes the ripple or wave effect of the water surface, wind effect on measuring devices, distance to water surface from gage platforms, light conditions, and observer's eyesight.

17.3.4 Recording gages are subject to mechanical errors such as gear lash, chain sprocket lash, temperature or mechanical dimensional changes, leaking floats, mechanical tolerance caused by lack of lubrication, or infiltration of dust or dirt, and numerous other mechanical problems. Additional errors in stilling well data are caused by the changing position of the float and counterweight, the amount of submerged float in the water, especially after the counterweight submerges at high stages, and errors or stage lags in intake pipes connecting the stilling well to the water body.

17.3.5 Electronic devices are subject to recording errors caused by power surges, electromagnetic interference, electronic component drift, and other problems that can affect data display, transmission, and recording.

17.3.6 When water-level measuring devices are installed in gaging stations or other hydraulic structures using stilling wells and intakes, intake errors are possible. The intake pipe should be large enough for the water in the well to follow the rise and fall of stage without significant delay, but still dampen wave fluctuations. The following relationship may be used to calculate the lag for an intake pipe for a given rate of change of stage.

$$\Delta h = \frac{0.01}{g}\frac{L}{D}\left(\frac{A_w}{A_p}\right)^2\left(\frac{dh}{dt}\right)^2$$

where:
Δh = lag, ft (m),
g = acceleration of gravity, ft/s/s (m/s/s),
L = intake length, ft (m),
D = intake diameter, ft (m),
A_w = area of stilling well, ft² (m²),
A_p = area of intake pipe, ft² (m²), and
$\frac{dh}{dt}$ = rate of change of stage, ft/s (m/s).

17.3.7 Water levels in open channels are read and recorded to the nearest 0.01 ft (2 mm). Visual observations in rough-surfaced water bodies may require estimation of the highs and lows of water-surface profiles to estimate mean stage. Readings taken in stilling wells are more accurate because the smaller-diameter intake pipes provide a dampening effect, if properly installed.

18. Keywords

18.1 elevation; gages; stage; water-level recorders; water-level sensors

Standard Guide for
Operation of a Gaging Station[1]

This standard is issued under the fixed designation D 5674; the number immediately following the designation indicates the year of original adoption or, in the case of revision, the year of last revision. A number in parentheses indicates the year of last reapproval. A superscript epsilon (ε) indicates an editorial change since the last revision or reapproval.

1. Scope

1.1 The guide covers procedures used commonly for the systematic collection of streamflow information. Continuous streamflow information is necessary for understanding the amount and variability of water for many uses, including water supply, waste dilution, irrigation, hydropower, and reservoir design.

1.2 The procedures described in this guide are used widely by those responsible for the collection of streamflow data, for example, the U.S. Geological Survey, Bureau of Reclamation, U.S. Army Corps of Engineers, U.S. Department of Agriculture, Water Survey Canada, and many state and provincial agencies. The procedures are generally from internal documents of the preceding agencies, which have become the defacto standards used in North America.

1.3 It is the responsibility of the user of the guide to determine the acceptability of a specific device or procedure to meet operational requirements. Compatibility between sensors, recorders, retrieval equipment, and operational systems is necessary, and data requirements and environmental operating conditions must be considered in equipment selection.

1.4 The values stated in inch-pound units are to be regarded as the standard. The values given in parentheses are for information only.

1.5 *This standard does not purport to address all of the safety concerns, if any, associated with its use. It is the responsibility of the user of this standard to establish appropriate safety and health practices and determine the applicability of regulatory limitations prior to use.*

2. Referenced Documents

2.1 *ASTM Standards:*
D 1129 Terminology Relating to Water[2]
D 1941 Test Method for Open Channel Flow Measurement of Water with the Parshall Flume[2]
D 3858 Practice for Open-Channel Flow Measurement of Water by Velocity-Area method[2]
D 5129 Test Method for Open Channel Flow Measurement of Water Indirectly by Using Width Contractions[2]
D 5130 Test Method for Open-Channel Flow Measurement of Water Indirectly by Slope-Area Method[2]
D 5242 Test Method for Open-Channel Flow Measurement of Water with Thin-Plate Weirs[2]
D 5243 Test Method for Open-Channel Flow Measurement of Water Indirectly at Culverts[2]

D 5388 Test Method for Indirect Measurements of Discharge by Step-Backwater Method[2]
D 5389 Test Method for Open Channel Flow Measurement by Acoustic Velocity Meter Systems[2]
D 5390 Test Method for Open Channel Flow Measurement of Water with Palmer-Bowlus Flumes[2]
D 5413 Test Method for Measurement of Water Levels in Open-Water Bodies[2]
D 5541 Practice for Developing Stage-Discharge Relation for Open-Channel Flow[2]
2.2 *ISO Standards:*[3]
ISO 1100 Liquid Flow Measurement in Open Channels—Part I: Establishment and Operation of a Gauging Station
ISO 6416 Measurement of Discharge by Ultrasonic (Acoustic) Method

3. Terminology

3.1 *Definitions*—For definitions of terms used in this guide, refer to Terminology D 1129.

3.2 *Descriptions of Terms Specific to This Standard:*

3.2.1 *control*—the physical properties of a channel, which determine the relationship between the stage and discharge of a location in the channel.

3.2.2 *datum*—a level plane that represents zero elevation.

3.2.3 *elevation*—the vertical distance from a datum to a point; also termed *stage* or *gage height*.

3.2.4 *gage*—a generic term that includes water level measuring devices.

3.2.5 *gage datum*—a datum whose surface is at the zero elevation of all of the gages at a gaging station. This datum is often at a known elevation referenced to the national geodetic vertical datum (NGVD) of 1929.

3.2.6 *gage height*—the height of a water surface above an established or arbitrary datum at a particular gaging station; also termed *stage*.

3.2.7 *gaging station*—a particular site on a stream, canal, lake, or reservoir at which systematic observations of hydrologic data are obtained.

3.2.8 *discharge*—the volume of water flowing through a cross-section in a unit of time, including sediment or other solids that may be dissolved in or mixed with the water; usually cubic feet per second (f^3/s) or metres per second (m/s).

3.2.9 *national geodetic vertical datum (NGVD) of 1929*—prior to 1973 known as *mean sea level datum*, a spheroidal datum in the conterminous United States and Canada that

---

[1] This guide is under the jurisdiction of ASTM Committee D-19 on Water and is the direct responsibility of Subcommittee D19.07 on Sediments, Geomorphology, and Open-Channel Flow.
Current edition approved February 15, 1995. Published June 1995.
[2] *Annual Book of ASTM Standards*, Vol 11.01.

[3] *Measurement of Liquid Flow in Open Channels*, ISO Standards Handbook 16, 1983. Available from American National Standards Institute, 11 W. 42nd St., 13th Floor, New York, NY 10036.

approximates mean sea level but does not necessarily agree with sea level at a specific location.

3.2.10 *stilling well*—a well connected to the stream with intake pipes in such a manner that it permits the measurement of stage in relatively still water.

4. Summary of Guide

4.1 A gaging station is usually installed where a continuous record of stage or discharge is required. A unique relationship exists between water surface elevation and discharge (flow rate) in most freely flowing streams. Water-level recording instruments continuously record the water surface elevation, usually termed stage or gage height. Discharge measurements are taken of the stream discharge to develop a stage-discharge curve. The discharge data are computed from recorded stage data by a stage-discharge rating curve.

5. Significance and Use

5.1 This guide is useful when a systematic record of water surface elevation or discharge is required at a specific location. Some gaging stations may be operated for only a few months; however, many have been operated for a century.

5.2 Gaging station records are used for many purposes:

5.2.1 Resource appraisal of long-term records to determine the maximum, minimum, and variability of flows of a particular stream. These data can be used for the planning and design of a variety of surface water-related projects such as water supply, flood control, hydroelectric developments, irrigation, recreation, and waste assimilation.

5.2.2 Management, where flow data are required for the operation of a surface-water structure or other management decision.

6. Site Location

6.1 The general location of the station will be dependent on the purpose for which the station is established. Location constraints for a resource appraisal-type station may be quite broad, for example, between major tributaries. Constraints for a management-type station may require a location just below a dam, contaminant discharge point, or other point at which discharge information is required specifically.

6.2 *Site Requirements*—Certain hydraulic characteristics of the stream channel are desirable for collecting high-accuracy data of minimal cost. Hydraulically difficult sites can still be gaged; however, accuracy and cost are affected adversely. Desirable conditions include the following:

6.2.1 The general course of the river should be straight for approximately 300 ft (100 m) above and below the gage.

6.2.2 The flow is confined to one channel at all stages.

6.2.3 The stream bed is stable, not subject to frequent scour and fill, and is free of aquatic growth.

6.2.4 The banks are sufficiently high to contain flow at all stages.

6.2.5 A natural feature such as ledge rock outcrop or stable gravel riffle, known as a "control," is present in the stream. It is necessary and practical in some cases to install a low-head dam or artificial control to provide this feature. Additional information on man-made structures is given in Test Methods D 1941, D 5242, and D 5390.

6.2.6 A pool is present behind the control where water-level instruments or stilling well intakes can be installed at a location below the lowest stream stage. The velocity of water passing sensors in a deep pool also eliminates or minimizes draw-down effects on stage sensors during high flow conditions.

6.2.7 The site is not affected by the hydraulic effects of a bridge, tributary stream entering the gaged channel, downstream impoundment, or tidal conditions.

6.2.8 A suitable site for making discharge measurements at all stages is available near the gage site.

6.2.9 There is accessibility for construction and operation of the gage.

6.3 *Site Selection*—An ideal site is rarely available, and judgement must be exercised when choosing between possible sites to determine that meeting the best combination of features.

6.3.1 *Office Reconnaissance*—The search for a gaging station begins with defining the limits along the stream at which the gage must be located on topographic maps of the area. The topographic information will indicate approximate bank heights or overflow areas, general channel width, constrictions, slope, roads, land use, locations of buildings, and other useful information so that promising locations can be checked out in the field.

6.3.2 *Field Reconnaissance*—If the range of possible gage locations is large, flying over the stream at a low altitude in a small aircraft is an efficient way of checking for promising sites. The view from the air on a clear day is much more helpful than peering off of a few highway bridges. Traversing the channel in a canoe or small boat is an alternative method. Field reconnaissance is best performed during low flow conditions; however, additional reconnaissance at high flow conditions and under ice-covered conditions for northern streams adds data that result in improved site selection.

6.3.3 *Logistical Reconnaissance*—Once a site has been selected that meets hydraulic considerations, and before design or construction begins, the following should occur:

6.3.3.1 Property ownership must be ascertained and legal permission secured to install and maintain the gage. This may include multiple landowners, especially if a cableway is required from which to make discharge measurements.

6.3.3.2 Necessary permits must be obtained from applicable governing agencies for, but not limited to, building and excavation, stream bank permits, and FAA notification for cableways or other local requirements.

6.3.3.3 Where electrical or phone service is required for operation, the availability of this service should be verified.

6.3.3.4 Most gaging stations are intended to record over the range of stream stages. It is therefore important to obtain any local information available on historical flood levels and to make estimates of stage for a 100-year event using locally used flood-frequency equations. A cross-section survey of the channel should be obtained during field reconnaissance to aid in estimating high flow stage.

6.4 More detailed information is available in Refs (**1–3**)[4] and ISO-1100.

[4] The boldface numbers in parentheses refer to the list of references at the end of this standard.

7. Types of Gaging Stations

7.1 Non-recording stations can be as simple as a permanent staff gage attached to a bridge, pier, or other structure, which is read and recorded manually in an appropriate notebook once or more each day. For details on non-recording gages, see Test Methods D 5413, ISO 1100, and Refs (1–4).

7.2 Recording gages are usually nonattended installations that require a sensor in direct contact with the water that is connected mechanically or electrically to a recording device.

7.2.1 Stilling well-type gages use a vertical well installed in the stream bank with small-diameter intake pipes connecting the river to the well. In this type of installation, a float on the water surface in the well drives a recorder housed in a shelter over the well by mechanical means (Fig. 1). Stilling well gages tend to provide more reliable data because water-level sensing as well as recording components of the system are protected from direct installation in the stream. Disadvantages are locations with unstable stream channels that may move away from the intakes and higher initial cost. For details on stilling well gages, see Test Methods D 5413, ISO 1100, and Refs (1–3, 5).

7.2.2 Bubbler-type gages consist of a gas supply, usually nitrogen, which is fed through a controller and tube to an orifice attached near the bed of a stream. The gas pressure is equal to the liquid head in the stream. A pressure transducer, mercury, or balance-beam manometer senses this pressure and passes this information either mechanically or electronically to a compatible recorder (Fig. 2). The advantage to this system is less expensive construction costs, which is especially desirable for short-term gages or in locations in which stilling well installations are difficult. Disadvantages are maintaining the orifice in a stable mounting on the river bed.

Keeping the orifice from being buried in silty streams is also a problem. For details on bubble-gages, see Test Methods D 5413, ISO 1100, and Refs (1–3, 5, 6).

7.2.3 Acoustic Velocity Meter (AVM) stations directly sense and record the velocity observed between two transducers at fixed elevations in the channel cross section. The AVM gages are used in locations in which stage-discharge relations are unreliable, usually in deep, slow-moving channels or where tidal or bidirectional flow occurs. Additional information is given in Test Method D 5389.

8. Gaging Station Structures

8.1 *Stilling Well Functional Requirements*—A stilling well must provide a water surface at the same elevation as that of the stream at any point in time, dampen out the effect of surface waves, and provide a sensor, usually a float and recording system.

8.1.1 The stilling well must be sufficiently long to cover the entire range of stages that might occur reasonably.

8.1.2 The stilling well can be any shape in plan view; however, most are either round or square. Permanent long-term gages should have a large enough area to allow personnel to work inside them for servicing; the most common size is approximately 4 by 4 ft (1.2 by 1.2 m). Some semipermanent stilling wells may be as small as 1 ft (0.3 m).

8.1.3 Stilling wells may be fabricated from poured concrete, concrete blocks, galvanized steel, concrete culvert pipe, or other suitable material. The well must have a sealed bottom to preclude the interchange of water from the stream and ground water.

8.1.4 Stilling wells are usually installed in a stream bank for protection and to minimize freezing in northern climates. They may be attached to bridge piers or wing walls in some

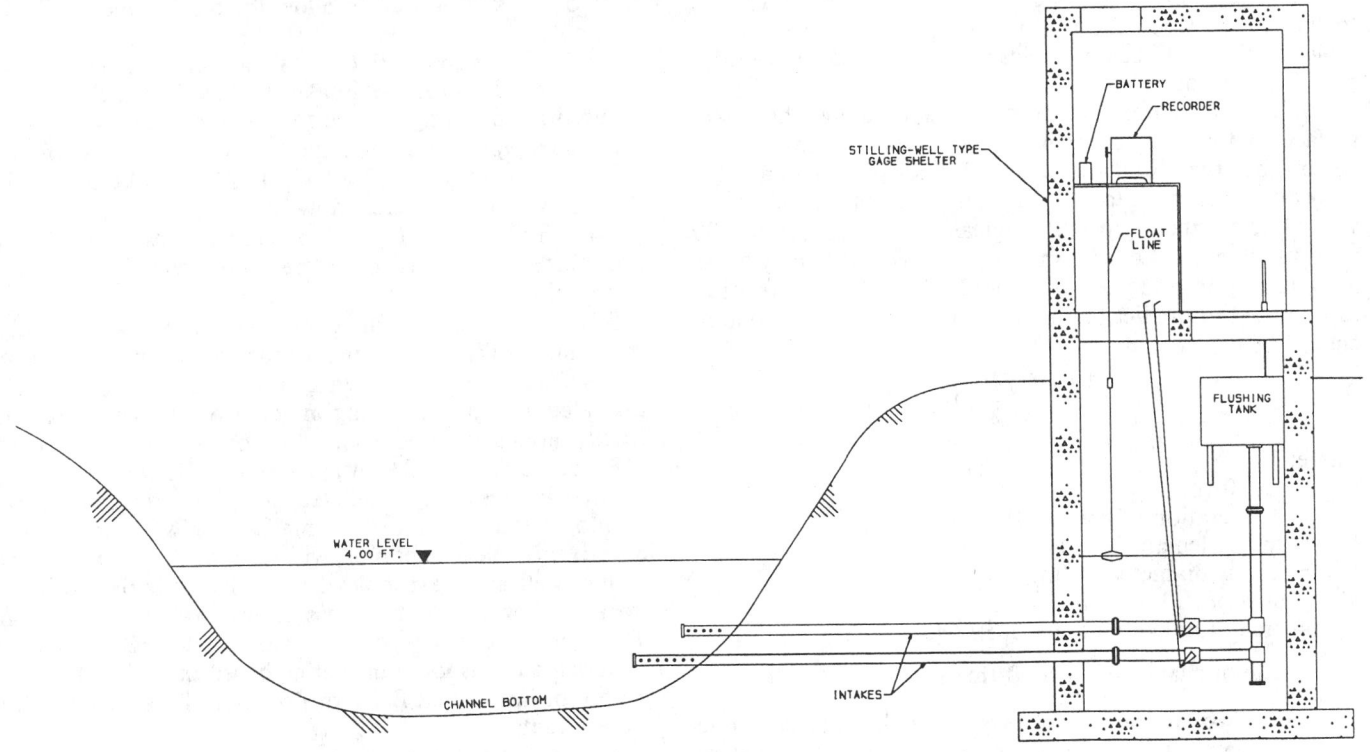

FIG. 1 Stilling Well Gage

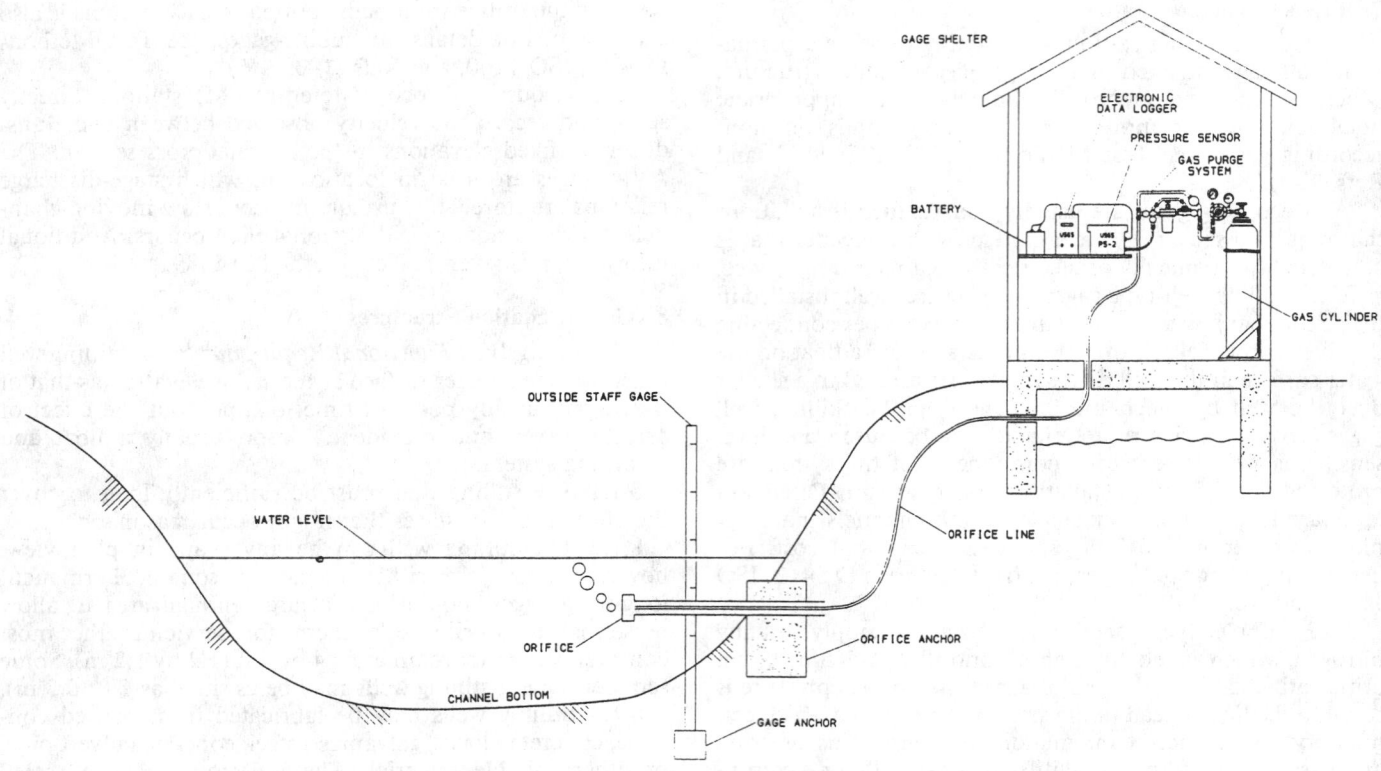

FIG. 2 Bubble Gage

applications but must be protected from damage by floating debris and must not interfere with flow patterns in the channel.

8.1.5 Intake pipes are required to connect the stilling well to the stream when the well is buried in the stream bank. Holes in the well usually suffice when installed on a bridge pier or wing wall.

8.1.5.1 Intake pipes must be sized to allow the water surface in the well to be at the same level as that in the stream, but they limit the effect of wind- or boat-generated waves or other transitory or artificial fluctuations of stream water levels. Intake pipes are typically 2 to 4 in. (50 to 100 mm) in diameter. Long or small-diameter intakes may cause a lag in response in the stilling well. The following relation can be used to predict the intake pipe lag for a given rate of change of stage (1).

$$\Delta h = \frac{0.01}{g} \frac{L}{D} \left(\frac{A_w}{A_p}\right)^2 \left(\frac{dh}{dt}\right)^2$$

where:
Δh = lag, ft (m),
g = acceleration of gravity, ft (m)/s/s,
L = intake length, ft (m),
D = intake diameter, ft (m),
A_w = area of stilling well, ft² (m²),
A_p = area of intake pipe, ft² (m²), and
$\frac{dh}{dt}$ = rate of change of stage, ft (m)/s.

8.1.5.2 Two or more intakes are usually installed, one vertically above the other, in case an intake is damaged or silted shut.

8.1.5.3 The invert elevation of the lowest intake should be at least 6 in. (150 mm) below the lowest expected stream level. The intake should be at least 1 ft (300 mm) above the floor of the stilling wall to allow for the storage of silt that may enter the structure.

8.1.5.4 Drawdown in the stilling well can occur where stream velocity past the intake is high. Drawdown can be reduced by installing a static tube to the streamward end of the intake pipe. A typical static tube is a piece of perforated pipe with the capped end attached to the intake pipe with a 90° elbow so that it points downstream.

8.1.6 Stilling wells located in cold climates require special procedures to prevent the freeze-up of water in the well or intake pipes, or both.

8.1.6.1 Stilling wells in cold climates are usually installed in stream banks where much of the well is ground covered. Wells should be constructed of nonconductive materials or insulated with an insulating material on the well's exterior. Intake pipes should be installed lower to prevent freezing.

8.1.6.2 Stilling wells with good ground cover can be kept ice-free by installing insulated subfloors at ground level. Subfloors must be above normal winter water levels to be effective. Typical subfloors will be attached rigidly to the stilling well and have holes slightly larger than instrument floats to allow the floats to pass through at high water events. These holes can be covered with light-weight insulating materials such as foam insulating board that will either float on top of instrument floats or float out of place during high water events.

8.1.6.3 Electric or propane heaters can be used to prevent freezing. Electric heat bulbs hanging in the center of the well under an instrument shelf can be quite effective for heating the air above the water surface. Submergible heaters can be placed in the well to heat the water. Heat tape can be installed in intake pipes, if necessary.

8.1.6.4 Bubbler systems, allowing a gas, usually nitrogen, to be bubbled from an open-ended tube placed on the well floor under recorder floats, will circulate warmer water from the bottom and prevent surface ice formation.

8.1.7 Instrument shelters can vary from large walk-in shelters installed on large stilling wells to small weatherproof boxes attached on small-diameter pipe wells. The shelter's functional requirements depend on the type and quantity of instrumentation, climate, and environmental and security conditions. Walk-in shelters with a 4 by 4-ft (1.2 by 1.2-m) minimum are desirable for installations with complex equipment, which require lengthy servicing during inclement weather. Some shelters are equipped with electricity, phones, telemetry, and other operational support systems.

8.2 Bubbler-type station-functional requirements basically require an instrument shelter to house pressure-sensing and recorder systems, a source of compressed gas, gas pressure regulators, and associated tubing. More information is available in Refs (1, 6) and ISO 1100.

8.2.1 Instrument shelter characteristics are similar to those described in 8.1.7.

8.2.2 The orifice from which the compressed gas exits into the stream must be mounted at least 6 in. (150 mm) below the lowest expected water levels. In locations at which ice cover is present, placing the orifice lower in the water column will minimize the damage caused by ice breakup or icing over of the orifice.

8.2.3 The orifice must be mounted in a stable structure that will not move in the channel. Suitable mountings include poured blocks of concrete, attachments to bridge structures, and pipes or pilings driven into the streambed.

8.2.3.1 Orifice mountings must have a reference point that can be checked periodically by differential leveling to discern whether movement has occurred.

8.2.4 Orifice positioning in moving sand-channel streams requires special techniques for obtaining satisfactory water-level data. A number of techniques have been devised to overcome these problems, such as using water well drive points and multiple orifice installations (1, 5, 6).

8.2.5 A constant supply of gas is required. This is typically supplied by commercially available compressed gas cylinders.

8.2.6 A regulator mechanism is required to control and reduce the pressure between the gas source and orifice and regulate the bubble discharge rate.

8.2.7 Suitable tubing is required to connect the gas source, regulators, and orifice. Neoprene tubing with an inside diameter of 1/8 in. (3 mm) is typically used. The tubing must be protected from physical damage between the instrument shelter and orifice. It is often installed in steel pipe or conduit. It should have a downward slope, with no low spots where water can collect and freeze.

8.3 Structural supports are required for outside reference gages, such as vertical staff gages and wire-weight gages. It is impossible to describe specific requirements since each installation is different. Primary considerations include stability, protection from floating debris or ice, boat traffic, or

other forms of damage or areas of hydraulic disturbance. The gage placement must be sufficiently close to the intake pipes or pressure sensor locations to represent comparable water levels.

8.4 Cableways are used frequently as a platform for obtaining high-flow discharge measurements. See Ref (4) for more detailed information.

9. Instrumentation

9.1 Gaging station instrumentation generally consists of water-level sensor and recorder systems. The remote transmission of data by landline, satellite, or other forms of radio transmission may also be used. It is not the purpose of this guide to describe this equipment in detail. This information is available in Test Methods D 5413 and Refs (1–3, 5, 6).

9.2 A limited number of gaging stations may sense and record velocity data directly. Information on this equipment is given in Test Method D 5389, ISO 6416, and Ref (4).

10. Gaging Station Datum

10.1 Each gaging station must have a datum plane as a known and constant reference for all gages and recording devices. This datum should remain unchanged throughout the life of a gaging station, even though the types of gage recorder and reference gages may change over time. The gage datum should be selected so that all readings are small, positive numbers.

10.1.1 The datum may be referred to a national datum system, usually NGVD of 1929, which is used for all national mapping activities in the United States and Canada.

10.1.2 In some cases, the datum may be tied to an independent, "local" datum maintained by a state, province, or municipal datum for specific reasons.

10.1.3 An arbitrary datum may be established for a single gaging station in some remote locations, where levels would have to be run many miles to an established datum. This may be referenced to an approximate NGVD datum by interpretation from a topographic map.

10.2 Gaging station reference marks (RMs) are permanent markers installed in the vicinity of a gaging station in order to set and maintain datum and check the various gages and recorders. The RMs are typically brass markers or bolts set in concrete posts installed in stable soil, permanent structures such as bridge abutments, cableway anchors, or large lag-bolts set in mature and stable trees. The RM locations should be selected so that they will not be destroyed or moved by activity in the area or washed away during floods. A minimum of three marks is recommended, and they should not all be in the same area or structure.

10.3 The elevations of RMs and gages are established and checked by differential leveling techniques using standard surveying equipment. Detailed leveling procedures are given in standard surveying texts and in Ref (8). Levels will typically be run to all RMs and gages once per year for the first few years and then at 2 to 5-year frequencies thereafter.

11. Operation of a Gaging Station

11.1 The objective of gaging station operation is to obtain a complete and accurate record of stream stage or discharge, or both. As with most scientific endeavors, the more time, attention, and experience exercised in the selection of instruments and the installation, calibration, and servicing of this equipment, the better the stream flow record will be.

11.2 Periodic visits are required at all gaging stations for the following: to verify that the system is operating properly; to make repairs if it is not; to remove the recording data; to check and reset the recording or transmitting devices, or both, if used; and to make discharge measurements for the development of a stage discharge rating.

11.2.1 Gaging station visits are usually made every 4 to 6 weeks; however, more frequent visits may be required with new or complex stations, by inexperienced personnel, or when the gaging station is known to have problems or a discharge measurement is necessary.

11.2.2 The technician should verify at each visit that the sensor and recording system is operating properly and make and record notations regarding the station status on forms developed for that purpose.

11.2.2.1 Read and record the date, watch time, and record time.

11.2.2.2 Read and record all gages, including outside gages, stage sensor, and recorder stages.

11.2.2.3 Read and record the values from other equipment, bubble rate, water quality, and temperature information.

11.2.2.4 Note conditions of the gaging station that could affect the data quality and channel and streamflow conditions, specifically the control conditions.

11.2.3 Remove the recorded information since the last visit. This may require the removal of a paper chart or electronic transfer of data by means of a personal computer or other electronic device.

11.2.4 Reset or recalibrate sensing and recording systems if not in agreement with the gage readings, if required. Make suitable notations to document for future data analysis.

11.2.5 Repeat the information noted in 11.2.2.1 and 11.2.2.2.

11.2.6 Make a discharge measurement, if required, in accordance with Practice D 3858, ISO 1100, and Refs (1–3, 9–11).

11.2.7 Repeat the information noted in 11.2.2.1 and 11.2.2.2.

11.3 The AVM-type gaging stations require the performance of additional procedures, as noted in Test Method D 5389, ISO 6416, and Refs (1, 7).

11.4 General maintenance is required at least once per year, usually during summer low-flow periods, to check the condition and repair or clean, as needed, the following: the stilling well and intakes, orifice attachment and lines, AVM transducers, gaging station structures, instruments, gages, cut grass, and check gages by differential levels, if applicable. Batteries and nitrogen tanks must be changed throughout the year, as required.

12. Calibration

12.1 Water-level sensing gaging stations are calibrated by making discharge measurements over the entire range of stage occurring at a particular station. A semipermanent relation will exist between stage and discharge if the station has been located carefully behind a stable control (see 6.2.5). A curve is drawn through plots of stage and discharge obtained from discharge measurements. Indirect measurements of discharge at various stages, usually from flood peak surveys, can also be used to define rating curves. See Test Methods D 5129, D 5130, D 5243, D 5388, and Refs (1, 10–16). Detailed information on rating development is given in Practice D 5541, ISO 1100, and Refs (1, 17).

12.2 The AVM-type gaging stations are calibrated by making discharge measurements over the entire range of flow conditions occurring at a particular station. The AVM-type stations are typically calibrated by developing a relation between the average velocity from discharge measurements and the line velocity between AVM transducers. A stage-area relation is also developed for computational purposes.

13. Computation

13.1 Present-day stream flow computations are usually performed by the input of data from paper or electronic means into a computer system that performs the basic calculations. Operators of a small number of gaging stations may find manual computations cost effective. See ISO 1100 and Refs (1, 18).

13.2 Typical data requirements include mean daily gage-height or stage, mean daily discharge, instantaneous maximum and minimum stage and discharge, and stage or discharge at a particular point in time.

13.3 Datum corrections are frequently necessary to correct recorded values for slippage or damage to gages, drift, or other recorder errors. Datum corrections are based on differences between a gaging station's base gage and the recorder values observed by servicing personnel or are determined from levels to permanent RMs, indicating that a gage has moved. Datum corrections should be listed chronologically on a suitable form for a permanent record. Analysis of this listing will often indicate equipment problems that can be corrected. Datum corrections are applied to stage recordings before other calculations are performed.

13.4 Shift adjustments may be used to correct for temporary changes from the stage-discharge rating curve. Those adjustments are based on the technician's visual observations and discharge measurements plotting the rating curve. Some common causes of shifting include weed growth in the channel, debris catching on a control, backwater from a downstream stream or tributary, moss buildup on a structural control or scour, or fill of a streambed, or some combination thereof. In the case of sand or other unstable channels, shift adjustments may be necessary on a constant basis to a theoretical stage-discharge rating. Shift adjustments should be listed on a suitable form and analyzed based on changes in stream stage, experience at the site, and weather records. Shift adjustments may be applied to individual stage recordings or to mean daily (stage-computed) values, depending on the magnitude and variability of shift adjustments. Shift adjustments are always applied after datum corrections.

13.5 The calculation of daily mean discharge is accomplished as indicated by the following steps, either manually or by computer program:

13.5.1 For each gage-height recording interval (usually 5, 15, 30, and 60 min) within a day, algebraically add any applicable datum correction.

13.5.2 For each datum corrected value algebraically, add any applicable shift adjustment.

13.5.3 For each value in 13.5.2, look up discharge from the applicable stage-discharge rating curve, usually converted to a table for simplicity.

13.5.4 Add all of the incremental discharge values for the day, and divide them by the number of recorded units to obtain the mean daily discharge.

13.5.5 After completing 13.5.1 through 13.5.4, the data can be tabulated to meet data needs for a period of time, usually a week, month, or year. Data are typically presented by monthly columns for a yearly reporting period.

13.5.6 Daily, or instantaneous, values of stage or discharge can be extracted from the calculated values (13.5.3) to meet user requirements.

13.6 Ice buildup on the bed, edges, or surface of flowing streams, as well as ice jams, disrupts the stage-discharge relation. The computation of daily discharge during ice-affected periods is an inexact and subjective art. The most common method is based on discharge measurements, weather records, the pattern of recorded gage-heights, comparisons with other nearby gaging stations, and the experience and judgment of the analyst. More information is available in ISO 1100 and Refs (1, 18).

13.7 The computation of discharge from AVM gaging stations requires a curve, constant, or equation, often referred to as "K," to relate the recorded line velocity to the mean section velocity and a stage area table. Computer computations are common; however, little standardization between programs exists presently. The basic computational requirements are given in the following:

13.7.1 For each incremental unit of recorded line velocity, look up the appropriate K, and compute the equivalent channel velocity.

13.7.2 For the corresponding recorded value of gage height, algebraically add any applicable datum correction.

13.7.3 For each value determined in 13.7.2, look up the applicable area value from the stage-area table.

13.7.4 For each increment, multiply the equivalent mean channel velocity (13.7.1) by the corresponding area (13.7.3) to obtain the incremental discharge.

13.7.5 Add all of the incremental discharge values for the day, and divide them by the number of recorded units to obtain the mean daily discharge.

13.7.6 Data summaries are the same as those described in 13.5.5 and 13.5.6.

13.8 Stage values are generally recorded to the nearest 0.01 ft (2 mm).

13.9 Discharge values are generally computed to three significant figures, except for extremely low flows, in which case two significant figures may be used.

13.10 The quality assurance (QA) of computations is usually performed by having one individual input the original data and perform the analysis and computation and having a second, more experienced person check the work independently.

13.10.1 The comparison of daily mean discharges provides some quality check where several gaging stations are operated in a region or river basin since nearby streams usually reflect similar runoff events and general trends. This is easily accomplished through the use of an on-screen or paper printout of daily discharge hydrographs.

13.11 Documentation of all aspects of the data collection, datum corrections, shift adjustments, analytical and computational methods, and the reasoning behind decisions should be provided in some written form, usually on an annual basis. The documentation should be kept indefinitely.

14. Precision and Bias

14.1 The accuracy of discharge data depends primarily on the following: (1) the stability of the stage-discharge relation or, if the control is unstable, the frequency of discharge measurements; and (2) the accuracy of observations of stage, measurements of discharge, and interpretation of records.

14.2 The precision and bias of gaging station data are difficult to evaluate in absolute terms since so many variables are involved. The evaluation of this many factors requires a large amount of judgment based largely on the experience and training of the operator. Agencies that operate large networks of gaging stations typically give subjective accuracy statements for each station for each year's data. Generally, "Excellent" means that approximately 95 % of the daily discharges is within 5 %, "good" within 10 %, and "fair" within 15 %. "Poor" means that the daily discharges have a less than "fair" accuracy.

15. Keywords

15.1 gaging station; open-channel flow; water discharge; water level

REFERENCES

(1) Rantz, S. E., "Measurement and Computation of Streamflow," U.S. Geological Survey Water Supply Paper 2175, Vol 2, 1982.

(2) *Water Measurement Manual*, 2nd ed., Revised Reprint 1984, U.S. Bureau of Reclamation, U.S. Government Printing Office, 1974.

(3) *Field Manual for Research in Agriculture Hydrology*, Agriculture Handbook No. 224, U.S. Department of Agriculture, U.S. Government Printing Office, 1979.

(4) Wagner, C. R., "Streamgaging Cableways," Open-file Report 91-84, U.S. Geological Survey, 1991.

(5) Buchanan, T. J., and Somers, W. P., "Stage Measurement at Gaging Stations," *Techniques of Water Resources Investigations of the U.S. Geological Survey*, Book 3, Chapter A-7, 1968.

(6) Craig, J. D., "Installation and Service Manual for U.S. Geological Survey Manometers" *Techniques of Water Resources Investigations of the U.S. Geological Survey*, Book 8, Chapter A-2, 1983.

(7) Laenen, A., "Acoustic Velocity Meter Systems," *Techniques of Water Resources Investigations of the U.S. Geological Survey*, Book 3, chapter A-17, 1985.

(8) Kennedy, E. J., "Levels at Streamflow Gaging Stations," *Techniques of Water Resources Investigations of the U.S. Geological Survey*, Book 3, Chapter A-19, 1990.

(9) Carter, R. W., and Davidian, Jr., "General Procedures for Gaging Streams," *Techniques of Water Resources Investigations of the U.S. Geological Survey*, Book 3, Chapter A-6, 1968.

(10) Buchanan, T. J., and Somers, W. P., "Discharge Measurements at Gaging Stations," *Techniques of Water Resources Investigations of the U.S. Geological Survey*, Book 3, Chapter A-8, 1969.

(11) Smoot, G. F., and Novak, C. E., "Measurement of Discharge by Moving Boat Method," *Techniques of Water Resources Investigations of the U.S. Geological Survey*, Book 3, Chapter A-11, 1969.

(12) Benson, M. A., and Dalrymple, T., "General Field and Office Procedures for Indirect Discharge Measurements," *Techniques of Water Resources Investigations of the U.S. Geological Survey*, Book 3, Chapter A-1, 1967.

(13) Dalrymple, T., and Benson, M. A., "Measurement of Peak Discharge by the Slope-Area Method," *Techniques of Water Resources Investigations of the U.S. Geological Survey*, Book 3, Chapter A-2, 1967.

(14) Bodhaine, G. L., "Measurement of Peak Discharge at Culverts by Indirect Methods," *Techniques of Water Resources Investigations of the U.S. Geological Survey*, Book 3, Chapter A-3, 1968.

(15) Matthai, H. F., "Measurement of Peak Discharge at Width Contractions by Indirect Methods," *Techniques of Water Resources Investigations of the U.S. Geological Survey*, Book 3, Chapter A-4, 1967.

(16) Hulsing, H., "Measurement of Peak Discharge at Dams by Indirect Means," *Techniques of Water Resources Investigations of the U.S. Geological Survey*, Book 3, Chapter A-5, 1967.

(17) Kennedy, E. J., "Discharge Ratings at Gaging Stations," *Techniques of Water Resources Investigations of the U.S. Geological Survey*, Book 3, Chapter A-10, 1984.

(18) Kennedy, E. J., "Computations of Continuous Records of Streamflow," *Techniques of Water Resources Investigations of the U.S. Geological Survey*, Book 3, Chapter A-13, 1983.

Standard Guide for
Selection of Weirs and Flumes for Open-Channel Flow Measurement of Water[1]

This standard is issued under the fixed designation D 5640; the number immediately following the designation indicates the year of original adoption or, in the case of revision, the year of last revision. A number in parentheses indicates the year of last reapproval. A superscript epsilon (ϵ) indicates an editorial change since the last revision or reapproval.

1. Scope

1.1 This guide covers recommendations for selection of weirs and flumes for the measurement of the volumetric flow rate of water and wastewater in open channels under a variety of field conditions.

1.2 This guide emphasizes the weirs and flumes for which ASTM standards are available, namely, thin-plate weirs, broad-crested weirs, Parshall flumes, and Palmer-Bowlus (and other long-throated) flumes. However, reference is also made to other measurement devices and methods that may be useful in specific situations.

1.3 *This standard does not purport to address all of the safety concerns, if any, associated with its use. It is the responsibility of the user of this standard to establish appropriate safety and health practices and determine the applicability of regulatory limitations prior to use.*

2. Referenced Documents

2.1 *ASTM Standards:*
D 1129 Terminology Relating to Water[2]
D 1941 Test Method for Open Channel Flow Measurement of Water with the Parshall Flume[2]
D 3858 Test Method for Open-Channel Flow Measurement of Water by Velocity-Area Method[2]
D 5242 Test Method for Open-Channel Flow Measurement of Water with Thin-Plate Weirs[2]
D 5389 Test Method for Open-Channel Flow Measurement of Water by Acoustic Velocity Meter Systems[2]
D 5390 Test Method for Open-Channel Flow Measurement of Water with Palmer-Bowlus Flume[2]
D 5614 Test Method for Open-Channel Flow Measurement of Water with Broad-Crested Weirs[2]

2.2 *ISO Standard:*
ISO 555-1973: Liquid Flow Measurement in Open Channels—Dilution Methods for Measurement of Steady Flow—Constant-Rate Injection Method[3]

3. Terminology

3.1 *Definitions*—For definitions of terms used in this guide, refer to Terminology D 1129.

3.2 *Descriptions of Terms Specific to This Standard:*

3.2.1 *blackwater*—an increase in the depth of flow upstream of a channel obstruction, in this case a weir or flume.

3.2.2 *contracted weirs*—contractions of thin-plate weirs refer to the widths of weir plate between the notch and the sidewalls of the approach channel. In fully contracted weirs, the ratio of the notch area to the cross-sectional area of the approach channel is small enough for the shape of the channel to have little effect. In suppressed (full-width) rectangular weirs, the contractions are suppressed, and the weir crest extends the full width of the channel.

3.2.3 *crest*—in rectangular thin-plate weirs, the horizontal bottom of the overflow section; in broad-crested weirs and flumes, the plane, level floor of the flow section.

3.2.4 *critical flow*—open-channel flow in which the energy, expressed in terms of depth plus velocity head, is a minimum for a given flow rate and channel.

DISCUSSION—The Froude number is unity at critical flow.

3.2.5 *Froude number*—a dimensionless number expressing the ratio of inertial to gravity forces in free-surface flow. It is equal to the average velocity divided by the square root of the product of the average depth and the acceleration due to gravity.

3.2.6 *head*—in this context, the depth of flow referenced to the crest of the weir or flume and measured at a specified location; this depth plus the velocity head are often termed the total head or total energy head.

3.2.7 *hydraulic jump*—an abrupt transition from supercritical to subcritical or tranquil flow, accompanied by considerable turbulence or gravity waves, or both.

3.2.8 *long-throated flume*—a flume in which the prismatic throat is long enough relative to the head for a region of essentially critical flow to develop on the crest.

3.2.9 *nappe*—the curved sheet or jet of water overfalling a weir.

3.2.10 *notch*—the overflow section of a triangular weir or of a rectangular weir with side contractions.

3.2.11 *primary instrument*—the device (in this case, a weir or flume) that creates a hydrodynamic condition that can be sensed by the secondary instrument.

3.2.12 *rangeability*—the spread between the maximum, Q_{max}, and minimum, Q_{min}, flow rates that a measuring instrument can usefully and reliably accommodate; this may be described as the ratio Q_{max}/Q_{min}.

3.2.13 *secondary instrument*—in this case, a device that measures the head on the weir or flume; it may also convert this measured head to an indicated flowrate or could totalize the flow.

3.2.14 *subcritical flow*—open-channel flow that is deeper and at lower velocity than critical flow for the same flow rate; sometimes called tranquil flow.

[1] This guide is under the jurisdiction of ASTM Committee D-19 on Water and is the direct responsibility of Subcommittee D19.07 on Sediments, Geomorphology, and Open-Channel Flow.
Current edition approved April 15, 1995. Published June 1995.
[2] *Annual Book of ASTM Standards*, Vol 11.01.
[3] Available from American National Standards Institute, 11 W. 42nd St., 13th Floor, New York, NY 10036.

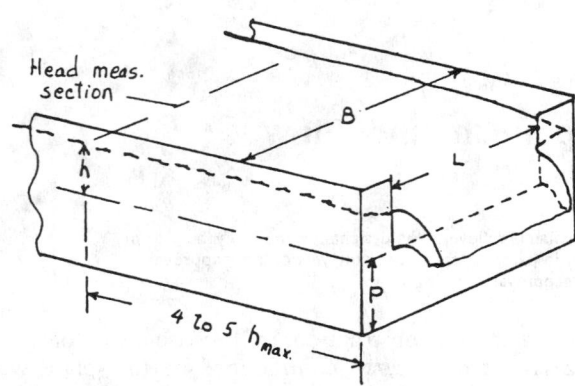

Limitations on geometry:[A] $h/P \leq 2$
$h \geq 0.1$ ft (0.03 m)
$L \geq 0.5$ ft (0.15 m)
$P \geq 0.3$ ft (0.1 m)

Approximate discharge equation, for selection purposes only:[B]

$$Q = (2/3)(2g)^{1/2} CLh^{3/2}$$

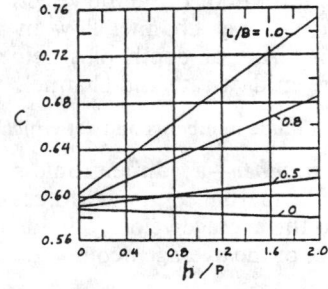

[A] These limitations are for partially contracted weirs. See Test Method D 5242 for conditions on fully contracted weirs.
[B] See Test Method D 5242 for more accurate standard equations. In compatible units, Q is flow rate, ft³/s (m³/s); g is acceleration due to gravity, ft/s² (m/s²); h is head in ft. (m); C is coefficient shown in Fig. 1.

FIG. 1 Rectangular Thin-Plate Weirs

DISCUSSION—The Froude number is less than unity for this flow.

3.2.15 *submergence*—the ratio of downstream head to upstream head on a weir or flume. Submergence greater than a critical value affects the discharge for a given upstream head.

3.2.16 *supercritical flow*—open-channel flow that is shallower and at higher velocity than critical flow for the same flow rate.

DISCUSSION—The Froude number is greater than unity for this flow.

3.2.17 *throat*—the constricted portion of a flume.

3.2.18 *velocity head*—the square of the average velocity divided by twice the acceleration due to gravity.

4. Significance and Use

4.1 Each type of weir and flume possesses advantages and disadvantages relative to the other types when it is considered for a specific application; consequently, the selection process often involves reaching a compromise among several features. This guide is intended to assist the user in making a

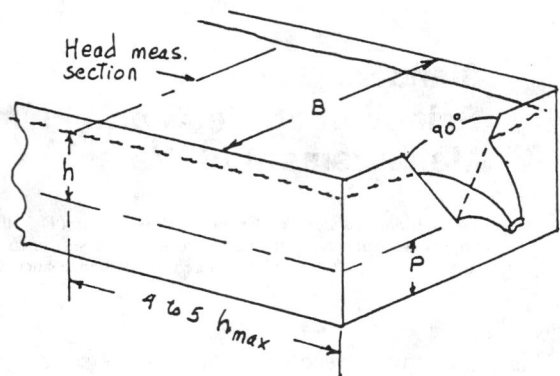

Limitations on geometry:[B] $h/P \leq 1.2$
$h/B \leq 0.4$
$P \geq 0.3$ ft (0.1 m)
$B \geq 2$ ft (0.6 m)
0.15 ft (0.05 m) $\leq h \leq 2$ ft (0.6 m)

Approximate discharge equation, for selection purposes only:[C]

$$C = (8/15)C(2g)^{1/2} (\tan \theta/2)h^{5/2}$$

[A] A 90° notch is shown; information is also available on notch angles from 20 to 100; see Test Method D 5242.
[B] These limits pertain to partially contracted 90° notches only. See Test Method D 5242 for other notch angles and full contractions.
[C] C is approximately 0.58. See Test Method D 5242 for more accurate values of C and for the complete standard equation. Here θ is the notch angle, and the other terms are as described in footnote B of Fig. 1.

FIG. 2 Triangular Thin-Plate Weir[A]

selection that is hydraulically, structurally, and economically appropriate for the purpose.

4.2 It is recognized that not all open-channel situations are amenable to flow measurement by weirs and flumes and that in some cases, particularly in large streams, discharges may best be determined by other means. (See 6.2.2.)

5. Weirs and Flumes

5.1 *Weirs:*

5.1.1 Weirs are overflow structures of specified geometries for which the volumetric flow rate is a unique function of a single measured upstream head, the other elements in the head-discharge relation having been experimentally or analytically determined. Details of the individual weirs may be found in the ASTM standards cited as follows:

5.1.2 *Standard Weirs*—The following weirs, for which ASTM standards are available, are considered in this guide:

5.1.2.1 Thin-plate weirs (see Test Method D 5242).

(1) Rectangular weirs (see Fig. 1).

(2) Triangular (V-notch) weirs (see Fig. 2).

5.1.2.2 Broad-crested weirs (see Test Method D 5614).

(1) Square-edge (rectangular) weirs (see Fig. 3).

(2) Rounded-edge weirs (see Fig. 4).

5.1.3 The quantitative information on weirs presented in Figs. 1–4 is intended to give the user only an overview and assist in the preliminary assessments for selection. To that end, some approximations and omissions were necessary for the sake of brevity and convenience, and the published standards must be consulted for exact and complete information on requirements, conditions, and equations.

5.2 *Flumes:*

5.2.1 Flumes use sidewall constrictions or bottom shapes

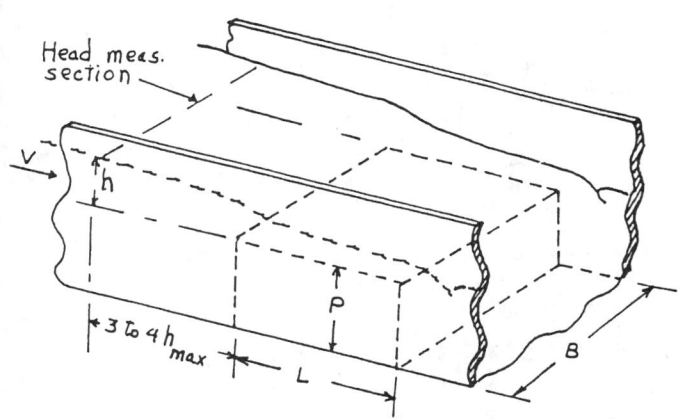

Limitations on geometry:[A]
$h \geq 0.2$ ft (0.06 m)
$b \geq 1.0$ ft (0.3 m)
$P \geq 0.5$ ft (0.15 m)
$0.1 < L/P < 4$
$0.1 < h/L < 1.6$
$h/P < 1.6$

Discharge equation:[B] $Q = (2/3)^{3/2}(g)^{1/2} CBh^{3/2}$

Approximate C, for selection purposes only[C]

h/P	h/L			
	0.2	0.6	1.0	1.6
0.1	0.85	0.88	0.97	(1.11)
0.5	0.88	0.91	1.01	1.15
1.0	(0.93)	0.96	1.06	1.19
1.5	(0.96)	1.02	1.11	1.24

[A] See Test Method D 5390 for upstream transitions, allowable downstream-to-upstream head ratios, and other requirements.
[B] Q, g, and h as described in Fig. 1, Footnote B.
[C] See Test Method D 5390 for complete table; figures in parentheses are outside of the recommended geometric limits.

FIG. 3 Rectangular (Square-Edge) Broad-Crested Weirs

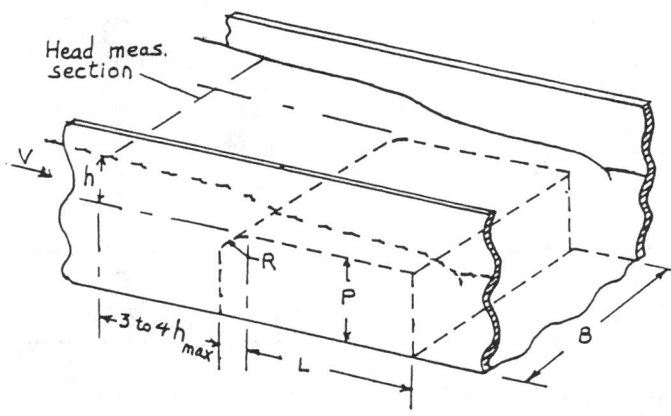

Limitations on geometry:[A]
$h \geq 0.2$ ft. (0.06 m)
$R \geq 0.2\, h_{max}$
$L \geq 1.75\, h_{max}$; $L + R \geq 2.25\, h_{max}$
$0.05 \leq H/L \leq 0.57$ ($H = h + V^2/2g$)
$P \geq 0.5$ ft (0.15 m)
$H/P < 1.5$
$B \geq 1$ ft (0.3 m); $\geq H_{max}$; $\geq L/5$

Approximate discharge equation, for selection purposes only:[B]

$$Q = 0.54\,(g)^{1/2}\, C_d B H^{3/2} \quad (H = h + V^2/2g)$$

C_d varies from about 0.9 at lowest h/L and B/L to about 0.99 at highest h/L and B/L.

[A] See Test Method D 5390 for upstream transitions, allowable downstream-to-upstream head ratios, and other requirements.
[B] See Test Method D 5390 for standard equations, and determination of C_d, Q, g, and h are as described in Fig. 1, Footnote B.

FIG. 4 Rounded Broad-Crested Weirs

found in the reference section.

5.2.4.1 *H-Series Flumes* (1, 2)—This flume, which was developed for use on agricultural watersheds, is actually a combination of flume and triangular weir and consequently exhibits very high rangeability along with good sediment transport capability.

5.2.4.2 *Portable Parshall Flume* (1)—This 3-in. (7.6-cm) flume closely resembles the 3-in. standard Parshall flume with the downstream divergent section removed. Its small size makes it convenient to transport and install in some low-flow field applications.

5.2.4.3 *Supercritical-Flow Flumes* (1)—These flumes were developed for use in streams with heavy loads of coarse sediment. The depth measurement is made in the supercritical-flow portion of the flume rather than upstream.

6. Selection Criteria

6.1 *Accuracy:*

6.1.1 The error of a flow-rate measurement results from a combination of individual errors, including errors in the coefficients of the head-discharge relations; errors in the measurement of the head; and errors due to nonstandard shape or installation or other departures from the practices recommended in the various weir or flume standards, or both. This guide considers the accuracy of the primary devices only, based on their accuracy potential under optimum or standard conditions; from information included in the individual standards, users can estimate secondary-system errors and other errors to obtain an estimate of the total measurement error.

6.1.2 The errors inherent in the basic head-discharge

or slopes of specified geometries, or both, to cause the flow to pass through the critical condition; this permits determination of the flow rate from a measured head and a head-discharge relation that has been experimentally or analytically obtained. Details of the individual flumes may be found in the ASTM standards cited as follows:

5.2.2 *Standard Flumes*—The following flumes, for which ASTM standards are available, are emphasized in this guide. Other flumes, which may be useful in specific situations, are cited in 5.2.4.

5.2.2.1 Parshall flumes (see Test Method D 1941 and Fig. 5).

5.2.2.2 Palmer-Bowlus (and other long-throated) flumes (see Test Method D 5390 and Fig. 6).

5.2.3 The quantitative information on flumes presented in Figs. 5 and 6 is intended to give the user only an overview and assist in the preliminary assessments for selection. To that end, some approximations and omissions were necessary for the sake of brevity and convenience, and the published standards must be consulted for exact and complete information on requirements, conditions, and equations.

5.2.4 *Other Flumes*—The following flumes are not covered by ASTM standards but are listed here because they were developed for specific situations that may be of interest to users of this guide. Detailed information on them can be

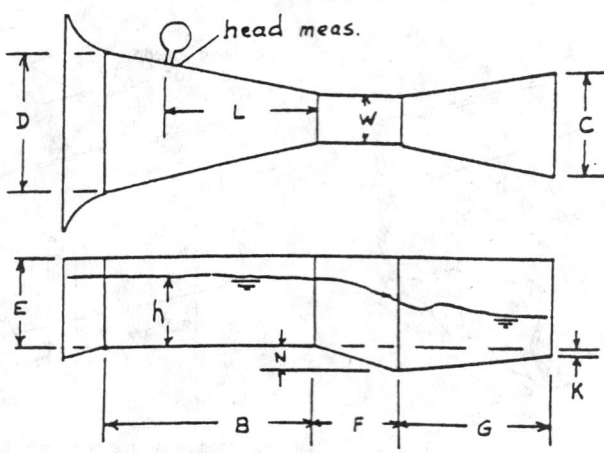

Geometry: Parshall flumes are designated by throat width, W; Standard sizes and capacities are listed as follows:

Discharge equations:[a] $Q = Ch^n$

Flume Size, W		C[b] (Inch-Pound)	C[b] (SI)	n	Maximum Q		Minimum Q	
ft-in.	cm				ft³/s	m³/s	ft³/s	m³/s
0-1	2.54	0.338	0.0479	1.55	0.19	0.0054	0.005	0.00014
0-2	5.08	0.676	0.0959	1.55	0.45	0.013	0.019	0.00054
0-3	7.62	0.972	0.141	1.55	1.60	0.045	0.028	0.00079
0-6	15.24	2.06	0.264	1.58	3.90	0.11	0.054	0.0015
0-9	22.86	3.07	0.393	1.53	8.90	0.25	0.09	0.0026
1-0	30.48	4.00	0.624	1.522	16.1	0.46	0.17	0.0034
1-6	45.72	6.00	0.887	1.538	24.6	0.88	0.17	0.0049
2-0	60.96	8.00	1.135	1.550	33.1	0.94	0.42	0.012
3-0	91.44	12.00	1.612	1.566	50.4	1.43	0.61	0.017
4-0	121.9	16.00	2.062	1.578	67.9	1.92	1.26	0.036
5-0	152.4	20.00	2.500	1.587	85.6	2.42	1.6	0.045
6-0	182.9	24.00	2.919	1.595	103.5	2.93	2.6	0.074
7-0	213.4	28.00	3.337	1.601	121.4	3.44	3.0	0.085
8-0	243.8	12.00	3.736	1.607	139.5	3.95	3.5	0.099
10-0	304.8	19.38	4.709	1.6	300	8.50	6	0.17
12-0	365.8	46.75	5.590	1.6	500	14.2	7	0.20
15-0	457.2	57.81	6.912	1.6	800	22.6	8	0.23
20-0	609.6	76.26	9.117	1.6	1340	37.9	11	0.31
25-0	762.0	94.69	11.32	1.6	1660	47.0	14	0.40
30-0	914.4	113.1	13.53	1.6	1890	56.4	16	0.45
40-0	1219.2	150.0	17.94	1.6	2640	74.8	22	0.62
50-0	1524.0	186.9	22.35	1.6	3280	92.9	27	0.76

[a] Equation and table for free (unsubmerged) flow only. See Test Method D 1941 for conditions for free flow and submerged-flow discharge curves.

[b] Use C (inch-pound) and h in feet for flowrate in cubic feet per second; use C (SI) and h in centimetres for flow rate in litres per second..

Major (Approximate) Dimensions for Parshall Flumes[c]

Flume Size, W		n		n		C		F		G		N		K		E		L	
ft-in.	cm	ft	cm	ft	cm	ft	cm	ft	cm	ft	cm	ft	cm	ft	cm	ft	cm	ft	cm
0-1	2.54	1.17	36	0.55	16.7	0.30	9.3	0.25	7.6	0.67	20.3	0.094	2.9	0.062	1.9	0.5-0.75	15-23	0.78	23.8
0-2	5.08	1.33	41	0.70	21.3	0.44	13.5	0.38	11.4	0.83	25.4	0.14	4.3	0.073	2.2	0.5-0.83	15-25	0.89	27.1
0-3	7.62	1.50	46	0.85	25.9	0.58	17.8	0.50	15.2	1.00	30.5	0.19	5.7	0.83	2.5	1.0-1.5	30-46	1.00	30.5
0-6	15.24	2.00	61	1.30	39.7	1.29	39.4	1.00	30.5	2.00	61.0	0.38	11.4	0.25	7.6	2.0	61	1.33	40.6
0-9	22.86	2.83	86	1.88	57.5	1.25	38.1	1.00	30.5	1.50	45.7	0.38	11.4	0.25	7.6	2.5	76	1.89	57.6
1-0	30.48	4.41	134	2.77	84	2.00	61	2.00	61.0	3.0	91	0.75	22.9	0.25	7.6	3.0	91	2.94	89.7
1-6	45.72	4.66	142	3.36	103	2.50	79	2.00	61.0	3.0	91	0.75	22.9	0.25	7.6	3.0	91	3.11	94.7
2-0	60.96	4.91	150	3.96	121	3.00	91	2.00	61.0	3.0	91	0.75	22.9	0.25	7.6	3.0	91	3.27	99.7
3-0	91.44	5.40	164	5.16	157	4.00	122	2.00	61.0	3.0	91	0.75	22.9	0.25	7.6	3.0	91	3.60	110
4-0	121.9	5.88	179	6.35	194	5.00	152	2.00	61.0	3.0	91	0.75	22.9	0.25	7.6	3.0	91	3.92	120
5-0	152.4	6.38	194	7.55	230	6.00	183	2.00	61.0	3.0	91	0.75	22.9	0.25	7.6	3.0	91	4.25	130
6-0	182.9	6.86	209	8.75	267	7.00	213	2.00	61.0	3.0	91	0.75	22.9	0.25	7.6	3.0	91	4.58	140
7-0	213.4	7.35	224	9.95	303	8.00	244	2.00	61.0	3.0	91	0.75	22.9	0.25	7.6	3.0	91	4.91	150
8-0	243.8	7.84	239	11.15	340	9.00	274	2.00	61.0	3.0	91	0.75	22.9	0.25	7.6	3.0	91	5.23	159
10-0	304.8	14.0	427	15.6	476	12.0	366	3.00	91.4	6.0	183	1.12	34.3	0.50	15.2	4.0	122	5.89	179
10-0	365.8	16.0	488	18.4	561	14.7	447	3.00	91.4	8.0	244	1.12	34.3	0.50	15.2	5.0	152	6.54	199
15-0	457.2	25.0	762	25.0	762	18.3	559	4.00	122	10.0	305	1.50	46.7	0.75	22.9	6.0	183	7.52	229
20-0	609.6	25.0	762	30.0	914	24.0	732	6.00	183	12.0	366	2.25	68.6	1.00	30.5	7.0	213	9.16	279
25-0	762.0	25.0	762	35.0	1067	29.3	894	6.00	183	13.0	396	2.25	68.6	1.00	30.5	7.0	213	10.79	329
30-0	914.4	26.0	792	40.4	1231	34.7	1057	6.00	183	14.0	427	2.25	68.6	1.00	30.5	7.0	213	12.43	379
40-0	1219.2	27.0	823	50.8	1548	45.3	1382	6.00	183	16.0	488	2.25	68.6	1.00	30.5	7.0	213	15.70	478
50-0	1524.0	27.0	823	60.8	1853	56.7	1727	6.00	183	20.0	610	2.25	68.6	1.00	30.5	7.0	213	18.97	578

[c] For selection purposes only. See Test Method D 1941 for complete standard dimensions and for information on upstream wingwalls and on locations for additional head measurement for submerged flow.

FIG. 5 Parshall Flumes

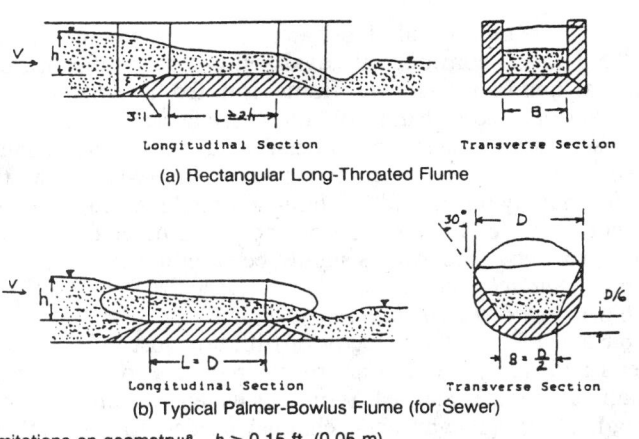

(a) Rectangular Long-Throated Flume

(b) Typical Palmer-Bowlus Flume (for Sewer)

Limitations on geometry:[a] $h \geq 0.15$ ft. (0.05 m)
$0.1 \leq h/L \leq 0.5$
$B \geq 0.33$ ft. (0.1 m)
$h < 6$ ft. (2 m)

Approximate discharge equation, for selection purposes only:[b]

$Q = 0.54 \, (g)^{1/2} \, C_d C_s B H^{3/2}, \quad (H = h + V^2/2g)$
C_d typically about 0.95; varies with viscous effects.
C_s (rectangular throat) = 1.0

[a] See D 5390 for details on location of heat measurement, allowable downstream-to-upstream head ratios, and other requirements.
[b] See D 5390 for complete standard equations and accurate determination of C_d and for variation of C_s with shape and head. Q, g, and h are as described in Fig. 1, footnote b.

FIG. 6   Palmer-Bowlus and Other Long-Throated Flumes

relations of the primary devices are as follows:

6.1.2.1 *Thin-Plate weirs:*
(*1*) Triangular, fully contracted, ±1 to 2 %.
(*2*) 90° notch, partially contracted, ±2 to 3 %.
(*3*) Rectangular, fully contracted, ±1 to 2 %.
(*4*) Rectangular, partially contracted, ±2 to 3 %.

6.1.2.2 *Broad-crested weirs:*
(*1*) Square-edge, ±3 to 5 % (depending on head-to-weir height ratio).
(*2*) Rounded, ±3 % (in the optimum range of head-to-length ratio).

6.1.2.3 *Flumes:*
(*1*) Parshall flumes, ±5 %.
(*2*) Palmer Bowlus and long-throated flumes, ±3 to 5 % (depending on head-to-length ratio).

6.1.2.4 This listing indicates that, with no consideration of other selection criteria, thin-plate weirs are potentially the most accurate of the devices.

6.1.3 *Sensitivity*—The discharge of weirs and flumes depends upon the measured head to the three-halves power for rectangular control sections (this is an approximation in the case of Parshall flumes), to the five-halves power for triangular sections, and to intermediate powers for intermediate trapezoidal sections. Consequently, the accuracy of a flow-rate measurement is sensitive to errors in head measurement and particularly so in the case of triangular control sections. It follows that in all weirs and flumes operating at or near minimum head, even a modest error or change in head can have a significant effect on the measured flow rate. Therefore, it is important to select sizes or combinations of devices that avoid prolonged operation near minimum head.

6.2 *Flow rate:*

6.2.1 This criterion includes the maximum anticipated

flow rate and the range of flow rate from minimum to maximum. The latter consideration includes not only daily or seasonal variations but also a flow chronology in which, for example, an area under development generates an initially low waste-water discharge followed in subsequent years by increasing flow rates.

6.2.2 *Flow capacities:*

6.2.2.1 *Small and Moderate Flows*—Apart from considerations of head loss (6.3) and sediment or debris transport (6.4), thin-plate weirs are most suitable for lower flow rates, with the triangular notches most appropriate for the smallest flows. Small Parshall and Palmer-Bowlus flumes are also available for low flows; these improve on the thin-plate weirs in sediment passage and head loss but at some sacrifice of potential accuracy (6.1).

6.2.2.2 *Large Flows*—Large discharges are best measured with flumes and broad-crested weirs, which can accommodate large heads and flows and, given proper construction, are inherently sturdy enough to withstand them. For example, the 50-ft (15.24-m) Parshall flume can be used for flow rates up to about 3200 ft³/s (90 m³/s). However, flumes and broad-crested weirs that are adequate for very large flows require major construction, and users may wish to consider establishing a measuring station (**3, 4**) with other methods of discharge measurement, for example, velocity-area method (Test Method D 3858), acoustic velocity meters (Test Method D 5389), or tracer dilution (ISO 555).

6.2.3 *Range of Flow Rate:*

6.2.3.1 Triangular thin-plate weirs have the largest rangeability of the standard devices because of their 2.5-power dependence on head. This rangeability can vary from slightly under 200 for fully contracted weirs to about 600 for partially contracted 90° notches that can utilize the allowable range of head.

6.2.3.2 For rectangular thin-plate weirs, the rangeability varies somewhat with the crest length-to-channel width ratio and is typically about 90, increasing to about 110 for full-width weirs. These results are based on a minimum head of 0.1 ft (0.03 m) and a suggested (although not absolute) maximum head of 2 ft (0.6 m). However, the rangeability of smaller rectangular weirs can be significantly less.

6.2.3.3 The rangeability of the rounded broad-crested weir is close to 40. However, large square-edge weirs, if used to the geometric limits of the standard, exhibit a rangeability of about 90.

6.2.3.4 The rangeability of Parshall flumes varies widely with size. (See Fig. 5.)

6.2.3.5 For Palmer-Bowlus and other long-throated flumes, the rangeability depends on the shape of the throat cross section, increasing as that shape varies from rectangular toward triangular. For the typical commercial Palmer-Bowlus flume of trapezoidal section, at least one manufacturer cites maximum-to-minimum flow-rate ratios up to, and in some cases exceeding, 100; (**5**) however, the head range is often beyond the recommendations of Test Method D 5390.

6.2.3.6 In cases in which there is a need for extreme rangeability along with sediment-transport capability, users may wish to consider the H-series flumes (5.2.4).

6.2.3.7 In cases in which low flows are expected to prevail for an extended period but will ultimately be superseded by much larger flow rates, users may wish to consider the interim use of removable small flumes nested inside larger ones.

6.3 *Head Loss:*

6.3.1 The upstream-to-downstream head difference that is required for the weir or flume to operate properly may be a selection criterion in many cases: for example, when sufficient elevation difference is not available to maintain the required flow, when the upstream channel cannot contain the backwater, or the reduced velocity in the backwater region causes excessive deposition of solids.

6.3.2 Some devices, notably the Parshall flume, can operate under partially submerged conditions with consequently reduced upstream-to-downstream head difference; however, this is done at the cost of reduced accuracy and an additional (downstream) head measurement.

6.3.3 For the same flow conditions, thin-plate weirs usually require the largest head difference, while long-throated flumes and rounded broad-crested weirs require the least. Parshall flumes are usually intermediate between these extremes, unless they are operated in the submerged regime.

6.4 *Sediment and Debris:*

6.4.1 Flumes are superior to weirs for use in flows with bed loads and coarse sediments. Finer sediments, even though they would not necessarily settle out upstream of a weir, may over time abrade the sharp edges of a thin-plate weir or the sharp corner of a rectangular broad-crested weir and thus affect the discharge coefficients.

6.4.2 When floating debris is present, the use of thin-plate weirs, particularly those with triangular notches, should be avoided.

6.4.3 In cases of exceptionally heavy loads of coarse sediment, users may wish to consider supercritical-flow flumes (5.2.4); H-series flumes (5.2.4.1) also exhibit good sediment passing behavior.

6.5 *Construction Requirements:*

6.5.1 This criterion takes into account the anticipated difficulty and expense of constructing and installing a weir or flume that meets standard specifications.

6.5.1.1 The Parshall flume is probably the most difficult to construct, owing primarily to its relatively complex shape. Also, the sharp downward slope of the throat may require some excavation of the channel floor. Because the Parshall flume is an empirical device, it is important to adhere closely to the prescribed dimensions (see also 6.5.2).

6.5.1.2 The difficulty in constructing thin-plate weirs arises from the strict requirements for fabrication of the notch edges; this difficulty can be expected to increase with weir size.

6.5.1.3 The shapes of the Palmer-Bowlus and other long-throated flumes are often moderately complex, although less so than those of the Parshall flume. However, their discharge coefficients can be obtained theoretically and consequently some departures from planned or prescribed dimensions can be accommodated (see also 6.5.2).

6.5.1.4 Broad-crested weirs have relatively simple geometry and are in principle perhaps the easiest to construct, particularly when the existing channel is rectangular in cross section. The rounded broad-crested weir, like the long-throated flumes, can be analyzed theoretically; the square-edge weir, on the other hand, is empirical, and the square

corner must be carefully fabricated and maintained.

6.5.2 The commercial availability of prefabricated Parshall and Palmer-Bowlus flumes is noted here. Sizes up to several feet (metres) can be obtained, often in a form suitable for use in sewer lines; the Palmer-Bowlus flumes are usually identified by the diameter of the sewer into which they fit rather than by throat width. Manufacturers' literature should be consulted for information on these and other flume and weir products. Dimensions should be carefully checked.

6.6 *Channel Conditions:*

6.6.1 *Velocity Distribution:*

6.6.1.1 Ideally, the velocity distribution just upstream of weirs and flumes should approach that in a long, straight, relatively smooth channel. Standards usually recommend the length of straight approach channel needed to accomplish this result; however, users should confirm the approach conditions on a case-by-case basis. Where the water is relatively clear, upstream baffles can be used to improve the velocity distribution, provided they do not affect the head measurement.

6.6.1.2 In cases in which the ideal approach flow cannot be attained, it is noted that flumes generally are affected less than weirs by moderately skewed velocity distributions. Also, the velocity distribution of the approach flow is less important when this flow is of very low velocity.

6.6.2 *Supercritical Flow*—Weirs and flumes can be used in supercritical-flow channels, provided that a hydraulic jump is formed upstream so that the approach flow is subcritical. The individual standards specify the upstream distance of the jump.

6.6.3 *Channel Shape:*

6.6.3.1 Rectangular approach channels are preferred for thin-plate weirs, with exceptions for fully contracted weirs as described in Test Method D 5242; they are required for full-width (or suppressed) rectangular weirs.

NOTE—Full-width rectangular thin-plate weirs require special provision for aeration of the nappe.

6.6.3.2 Rectangular approach channels are required for broad-crested weirs. However, this condition can be satisfied in a non-rectangular channel by construction of vertical sidewalls extending a prescribed distance upstream of the head measurement location and with appropriate transitions as described in Test Method D 5614.

6.6.3.3 Circular channels (sewers) (see 6.5.2).

7. Secondary System

7.1 The requirements for head measurement and associated instrumentation are basically similar for all weirs and flumes and therefore are not considered among the selection criteria in this guide. (An exception is the Parshall flume, which requires an additional head measurement if used in the submerged mode; other flumes and broad-crested weirs in some instances may require downstream head monitoring to ensure unsubmerged operation.) Refer to the individual standards for secondary-system requirements.

7.2 The secondary measurement is a significant contributor to the total error of a flow measurement (see 6.1.1 and 6.1.3).

8. Keywords

8.1 flumes; open-channel flow; water discharge; weirs

REFERENCES

(1) Kilpatrick, F. A., and Schneider, V. R., "Use of Flumes in Measuring Discharge," Book 3, Chapter A14 of *Techniques of Water-Resources Investigations of the U.S. Geological Survey*, 1983, pp. 1–46.

(2) "U.S. Agricultural Research Service, Field Manual for Research in Agricultural Hydrology," *USDA Handbook 724*, 1962, pp. 43–80.

(3) Carter, R. W., and Davidian, J., "General Procedure for Gaging Streams," *Techniques of Water Resources Investigations of the U.S. Geological Survey*, Book 3, Chapter A6, 1969, p. 7.

(4) Buchanan, T. J., and Somers, W. P., "Discharge Measurements at Gaging Stations," *Techniques of Water Resources Investigations of the U.S. Geological Survey*, Book 3, Chapter A8, 1969, pp. 57–63.

(5) Grant, D. M., *ISCO Open-Channel Flow Measurement Handbook*, 3rd Ed., 1989, pp. 40–46.

Designation: D 5541 – 94

Standard Practice for
Developing a Stage-Discharge Relation for Open Channel Flow[1]

This standard is issued under the fixed designation D 5541; the number immediately following the designation indicates the year of original adoption or, in the case of revision, the year of last revision. A number in parentheses indicates the year of last reapproval. A superscript epsilon (ϵ) indicates an editorial change since the last revision or reapproval.

1. Scope

1.1 This practice covers the development of a curve relating stage (elevation) to discharge. Standard test methods have been documented for measuring discharge and for measuring stage (see Practice D 3858, and Test Methods D 5129, D 5130, D 5243, D 5388, and D 5413). This practice takes the discharge and stage determined by each respective test method and shows a relation between them using a curved line. This curved line is called a stage-discharge relation or rating curve.

1.2 The procedures described in this practice are used commonly by those responsible for investigations of streamflow, for example, the U.S. Geological Survey, Army Corps of Engineers, Bureau of Reclamation, and U.S Agriculture Research Service. For the most part, these procedures are adapted from reports of the U.S. Geological Survey.[2,3]

1.3 The procedures described in this practice apply only to simple freely flowing open-channel flow. Ratings for complex hydraulic conditions of extremely low slope channels using multiple-stage inputs, channels affected by man-induced regulation, or tidal conditions are not described. These types of ratings are described in detail in Refs (2) and (3).

1.4 This practice uses the results of current-meter discharge measurements or indirect discharge measurements and the corresponding measured stage to define as much of the stage-discharge relation curve as possible. A theoretical curve is developed for the full range of stage and discharge to shape the curve.

1.5 The values stated in inch-pound units are to be regarded as the standard. The values given in parentheses are for information only.

1.6 *This standard does not purport to address all of the safety concerns, if any, associated with its use. It is the responsibility of the user of this standard to establish appropriate safety and health practices and determine the applicability of regulatory limitations prior to use.*

2. Referenced Documents

2.1 *ASTM Standards:*

D 1129 Terminology Relating to Water[4]
D 3858 Practice for Open-Channel Flow Measurement of Water by Velocity-Area Method[4]
D 5129 Test Method for Open Channel Flow Measurement of Water Indirectly by Using Width Contractions[4]
D 5130 Test Method for Open-Channel Flow Measurement of Water Indirectly by Slope-Area Method[4]
D 5243 Test Method for Open-Channel Flow Measurement of Water Indirectly at Culverts[4]
D 5388 Test Method for Measurement of Discharge by Step-Backwater Method[4]
D 5413 Test Method for Measurement of Water Levels in Open Water Bodies[4]

2.2 *ISO Standard:*

ISO 1100/2 Liquid Flow Measurement in Open Channels—Part 2, Determination of Stage-Discharge Relation[5]

3. Terminology

3.1 *Definitions*—For definitions of terms used in this practice, refer to Terminology D 1129.

3.2 *Symbols:*

GH = gage height or stage, ft (m).
Q = discharge, ft³/s (m³/s).

4. Summary of Practice

4.1 The stage-discharge relation is developed by plotting stage versus discharge from discharge measurements or other determinations of flow, either manually or through the use of computer programs and fitting a curve to these points. The stage should be determined at a single gage datum for the entire range in stage. Stages determined in stilling wells, at outside gages, and at bridge abutments can be significantly different and should not be interchanged. Discharge measurements may not be available for the entire range in stage of the stage-discharge relation. A theoretical rating curve should be developed for the entire range in stage using Test Method D 5388. This theoretical curve is used as a guide to shape the stage-discharge relation at places where discharge measurements are not available.

5. Significance and Use

5.1 This practice is particularly useful for determining the discharge at a gaging station or a location where discharge information is repeatedly needed.

---

[1] This practice is under the jurisdiction of ASTM Committee D-19 on Water and is the direct responsibility of Subcommittee D19.07 on Sediments, Geomorphology, and Open-Channel Flow.
Current edition approved April 15, 1994. Published July 1994.
[2] Kennedy, E. J., "Discharge Ratings at Gaging Stations: U.S. Geological Survey," *Techniques of Water-Resource Investigations*, Book 3, Chapt. A10, 1984, p. 59.
[3] Rantz, S. E., et al., *Measurement and Computation of Streamflow: Vol 2, Computation of Discharge*, U.S. Geological Survey, Water-Supply Paper No. 2175, 1982, p. 631.

[4] *Annual Book of ASTM Standards*, Vol 11.01.
[5] Available from American National Standards Institute, 11 West 42nd St., 13th Floor, New York, NY 10036.

5.2 This practice is applicable only for open-channel flow conditions where channel hydraulics permit a stable relation between stage and discharge.

6. Channel Hydraulics

6.1 The stage-discharge relation for open-channel flow at a gaging station or other stage reference point is governed by channel conditions downstream from that point, referred to as a control. Knowledge of the channel features that control the stage-discharge relation is important. The development of stage-discharge curves where more than one control is effective, control features change, and the number of measurements is limited usually requires judgment in interpolating between measurements and in extrapolating beyond the highest or lowest measurements.

6.1.1 *Section Controls*—A section control is a specific cross section of the stream channel that controls the relation between stage and discharge at that point in the channel. A section control can be a natural feature such as a rock ledge, sand bar, or severe constriction in the channel. A section control can likewise be a manmade feature such as a small dam, weir, flume, or overflow spillway. Section controls can frequently be identified visually in the field by observing a riffle, or pronounced drop in the water surface, as the flow passes over the control. As stage increases because of higher flows, the section control will frequently become submerged to the extent that it no longer controls the relation between stage and discharge. At this point, the riffle is no longer observable, and flow is then controlled by either another section control further downstream or by channel control.

6.1.2 *Channel Controls*—A channel control consists of a combination of features throughout a reach downstream from a gage. These features include channel size, shape, curvature, slope, and roughness. The length of channel reach that controls a stage-discharge relation varies. The stage-discharge relation for relatively steep channels may be controlled by a relatively short channel reach, whereas the relation for a relatively flat channel may be controlled by a much longer channel reach. In addition, the length of a channel control will vary depending on the magnitude of flow. Precise definition of the length of a channel-control reach is usually not possible or necessary.

6.1.3 *Combination Controls*—The stage-discharge relation may be governed by a combination of section and channel controls. This usually occurs for a short range in stage between a section-controlled segment of the rating and a channel-controlled segment of the rating. This part of the rating is commonly referred to as a transition zone of the rating and represents the change from section control to channel control. In other instances, a combination control may consist of two section controls, where each has partial controlling effect. Combination controls or transition zones, or both, occur for very limited parts of a stage-discharge relation and can usually be defined by plotting procedures. In particular, transition zones represent changes in the slope or shape of a stage-discharge relation.

6.2 Low flows are usually controlled by a section control, whereas high flows are usually controlled by a channel control. Medium flows may be controlled by either type of control. A combination of section and channel control may occur at some stages. These are general rules, and exceptions can and do occur.

7. Interferences

7.1 The stage-discharge relation may be affected by the deposition or removal of stream bed or bank material by flowing water, usually at high flow conditions or manmade changes. Large changes may require a redefinition of the rating curve. Small, transitory changes may be facilitated by adjustments to the stage observations. An example of a temporary shift would be a beaver dam on a section control or debris deposited on a dam or bridge piling that would be expected to be removed or eventually wash away.

7.2 Aquatic growth may develop in a stream during the growing season. This growth would result in a temporary backwater situation. Adjustments to stage observations would normally be made during these periods.

7.3 Ice cover changes river hydraulics and alters the stage-discharge relation.

7.4 Hysteresis may affect the high flow stage-discharge relation when the water surface slope changes due to either rapidly rising or rapidly falling water levels in a channel control reach. Hysteresis is sometimes referred to as loop ratings and is most pronounced in relatively flat sloped streams. The water surface slope on rising stages is significantly steeper than that for steady flow conditions, resulting in greater discharge than indicated by the steady flow rating. The reverse is true for falling stages. If discharge measurements are made at both rising and falling stages, a single curve splitting these measurements will generally result in satisfactory accuracy. It may be necessary to use separate curves for rising and falling conditions in extreme cases.

8. Sampling

8.1 Sampling as defined in Terminology D 1129 is not applicable in this practice.

9. Calibration

9.1 Verify the stage-discharge relation periodically with current-meter or indirect discharge measurements to ascertain that the relation has not changed. Large floods are most likely to cause erosion or filling of the channel and cause the relation to shift. The frequency of current meter measurements depends on the stability of a stream and is based in part on past experience. As a rule of thumb, monthly measurements should be made at a new site, at least until the range of stage is covered.

10. Procedure

10.1 If sufficient current-meter discharge measurements are available for the entire range in stage and discharge that is necessary, develop the entire rating curve by plotting stage versus discharge on logarithmic or rectangular coordinate plotting paper. Logarithmic plotting paper is preferred because, in the usual situation of compound controls, changes in the slope of the logarithmically plotted rating identify the range in stage for which the effective controls exist. Select a convenient stage scale on the logarithmic paper so that all of the discharge measurements below bankfull stage plot in a relatively straight line. There are three segments for a rating curve as a general rule, and they are identified by the changes

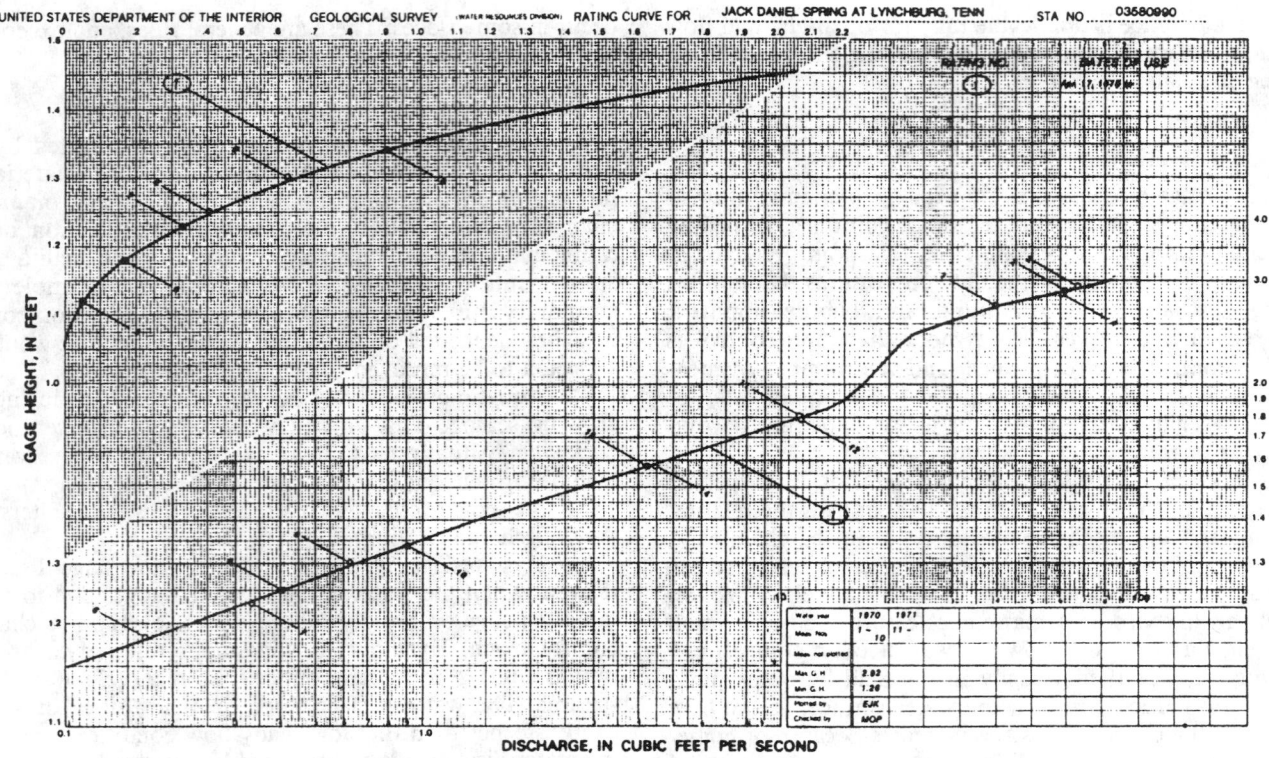

UNITED STATES DEPARTMENT OF THE INTERIOR   GEOLOGICAL SURVEY (WATER RESOURCES DIVISION)   RATING CURVE FOR ___ JACK DANIEL SPRING AT LYNCHBURG, TENN ___ STA NO ___ 03580990

GAGE HEIGHT, IN FEET

DISCHARGE, IN CUBIC FEET PER SECOND

FIG. 1 Typical rating-curve sheet.

in slope of the curve. A typical rating curve is shown in Fig. 1. At low stages, the curve is straight and relatively flat until the channel width is full (1.8 ft (0.55 m)). From this point until bankfull (2.34 ft (0.71 m)), the curve is much steeper. Above bankfull, the water will spread out and the curve will be flat and straight.

10.1.1 It is often desirable to plot the low-flow component of the rating on rectangular coordinate plotting paper. This presents an opportunity to plot at an expanded scale. For small streams that go dry or nearly so, the point of zero flow can be plotted to help shape the extreme low-flow portion of the curve. A rectangular plot is shown in Fig. 1.

10.2 If sufficient discharge measurements are not available for the entire range in stage and discharge that is necessary, develop a theoretical rating curve using the step-backwater test method. This theoretical curve is used as a guide to shape the rating curve. Plot the theoretical rating curve on logarithmic plotting paper. All of the current-meter discharge measurements are plotted on the same paper. Adjust the theoretical curve to go through the current-meter measurements. The adjustments to the theoretical curve may not be the same at the upper, middle, and lower sections of the curve.

10.3 Discharge measurements are sometimes made under undesirable conditions. The hydrographer making the measurement may rate the measurement excellent, good, fair, or poor. A measurement that is rated excellent, good, fair, or poor is believed to be within 2, 5, 8, and over 8 % of the correct value, respectively. When adjusting the theoretical rating to go through the measurements, give consideration to how accurate the measurements are believed to be.

11. Precision and Bias

11.1 Determination of the precision and bias for this practice is not possible due to the high degree of instability of open-channel flow. A minimum bias, measured under ideal conditions, is related directly to the bias of the equipment used to obtain stage and discharge values. A maximum precision and bias cannot be estimated due to the variability of the sources of potential errors and the temporal and spatial variability of open-channel flow. Any estimate of these errors could be very misleading to the user.

11.2 Stage-discharge relations represent hydraulic functions that are subject to frequent changes, as described in Section 7. Each discharge measurement represents a variable range of precision as well as defining a unique hydraulic condition. Various statistical tests have been used to test for bias. Users should always consider what is happening to controlling hydraulic characteristics and make decisions on this basis rather than arbitrarily using statistical techniques.

11.3 A comprehensive discussion of tests for bias is presented in ISO 1100/2.

12. Keywords

12.1 discharge; rating curve; stage; stage-discharge relation

₲ D 5541

PART 2. SOIL, VADOSE ZONE AND SEDIMENT SAMPLING AND MONITORING

2.1 SOIL SAMPLING
(See also D 4547, Section 4.2)

Standard Guide for
Soil Sampling from the Vadose Zone[1]

This standard is issued under the fixed designation D 4700; the number immediately following the designation indicates the year of original adoption or, in the case of revision, the year of last revision. A number in parentheses indicates the year of last reapproval. A superscript epsilon (ϵ) indicates an editorial change since the last revision or reapproval.

1. Scope

1.1 This guide addresses procedures that may be used for obtaining soil samples from the vadose zone (unsaturated zone). Samples can be collected for a variety of reasons including the following:

1.1.1 Stratigraphic description,

1.1.2 Hydraulic conductivity testing,

1.1.3 Moisture content measurement,

1.1.4 Moisture release curve construction,

1.1.5 Geotechnical testing,

1.1.6 Soil gas analyses,

1.1.7 Microorganism extraction, or

1.1.8 Pore liquid and soils chemical analyses.

1.2 This guide focuses on methods that provide soil samples for chemical analyses of the soil or contained liquids or contaminants. However, comments on how methods may be modified for other objectives are included.

1.3 This guide does not describe sampling methods for lithified deposits and rocks (for example, sandstone, shale, tuff, granite).

1.4 In general, it is prudent to perform all field work with at least two people present. This increases safety and facilitates efficient data collection.

1.5 The values stated in inch-pound units are to be regarded as the standard. The SI units given in parentheses are for information only.

1.6 *This standard does not purport to address all of the safety problems, if any, associated with its use. It is the responsibility of the user of this standard to establish appropriate safety and health practices and determine the applicability of regulatory limitations prior to use.*

2. Referenced Documents

2.1 *ASTM Standards:*

D 420 Practice for Investigating and Sampling Soil and Rock for Engineering Purposes[2]

D 653 Terminology Relating to Soil, Rock, and Contained Fluids[2]

D 1452 Practice for Soil Investigation and Sampling by Auger Borings[2]

D 1586 Method for Penetration Test and Split-Barrel Sampling of Soils[2]

D 1587 Method for Thin-Walled Tube Sampling of Soils[2]

D 2488 Practice for Description and Identification of Soils (Visual-Manual Procedure)[2]

D 2607 Classification of Peats, Mosses, Humus, and Related Products[2]

D 3550 Method for Ring-Lined Barrel Sampling of Soils[2]

D 4083 Practice for Description of Frozen Soils (Visual-Manual Procedure)[2]

D 4220 Practice for Preserving and Transporting Soil Samples[2]

3. Terminology

3.1 *Definitions:*

3.1.1 Except where noted, all terms and symbols in this guide are in accordance with the following publications. In order of consideration they are:

3.1.1.1 Terminology D 653.

3.1.1.2 *Compilation of ASTM Standard Terminology,*[3] and

3.1.1.3 *Webster's New Collegiate Dictionary.*[4]

3.1.2 For definitions and classifications of soil related terms used, refer to Practice D 2488 and Terminology D 653. Additional terms that require clarification are defined in 3.2.

3.2 *Descriptions of Terms Specific to This Standard:*

3.2.1 *cascading water*—perched ground water that enters a well casing via cracks or uncovered perforations, trickling, or pouring down the inside of the casing.

3.2.2 *sludge*—a water charged sedimentary deposit.

3.2.2.1 *Discussion*—The water-formed sedimentary deposit may include all suspended solids carried by the water and trace elements that were in solution in the water. Sludge usually does not cohere sufficiently to retain its physical shape when mechanical means are used to remove it from the surface on which it deposits, but it may be baked in place and be adherent.

4. Summary of Guide

4.1 Sampling vadose zone soil involves inserting into the ground a device that retains and recovers a sample. Devices and systems for vadose zone sampling are divided into two general groups, namely the following: samplers used in conjunction with hand operated devices; and samplers used

---

[1] This guide is under the jurisdiction of ASTM Committee D-18 on Soil and Rock and is the direct responsibility of Subcommittee D18.21 on Ground Water and Vadose Zone Investigations.

Current edition approved July 15, 1991. Published September 1991.

[2] *Annual Book of ASTM Standards,* Vol 04.08.

[3] *Compilation of ASTM Standard Terminology,* Sixth edition, ASTM, 1916 Race St., Phila., PA 19103, 1986.

[4] *Webster's New Collegiate Dictionary,* Fifth edition, _____ 1977.

Type of Sampler	Obtains Core Sample		Most Suitable Core Types		Operation in Stoney Soils		Most Suitable Soil Moisture Conditions			Access to Sample Sites During Poor Soil Conditions		Relative Sample Size		Labor Req'mts	
	Yes	No	Coh.	Cohless	Fav.	Unfav.	Wet	Dry	Inter.	Yes	No	Sm.	Lg.	Sngl.	2/More
A. Drill Rig Samplers															
1. Multipurpose Drill Rig	♦		♦	♦	♦			♦	♦	♦			♦		♦
2. Split-barrel Drive Sampler	♦		♦	♦	♦			♦	♦	♦		♦			♦
3. Thin-Walled Tube Sampler	♦		♦				♦					♦			♦
4. Piston Sampler	♦		♦				♦	♦	♦			♦			♦
5. Continuous Sample Tube system	♦		♦					♦					♦		♦
6. Hand-Held Power Auger		♦						♦				♦		♦	
B. Hand Operated Samplers															
1. Screw-Type Auger		♦						♦				♦		♦	
2. Barrel Auger															
a. Post-Hole Auger		♦	♦					♦		♦		♦		♦	
b. Dutch Auger		♦	♦				♦			♦		♦		♦	
c. Regular Barrel Auger		♦	♦					♦		♦		♦		♦	
d. Sand Auger		♦		♦			♦			♦		♦		♦	
e. Mud Auger		♦	♦				♦			♦		♦		♦	
3. Tube-Type Sampler															
a. Soil Sampling Tube															
(1) Wet Tip	♦						♦					♦		♦	
(2) Dry Tip	♦							♦	♦			♦		♦	
b. Veihmeyer Tube	♦							♦				♦		♦	

FIG. 1 Criteria for Selecting Soil Sampling Equipment

in conjunction with multipurpose or auger drill rigs. This guide discusses these groups and their associated practices.

4.2 The discussion of each device is organized into three sections, describing the device, describing sampling methods, and limitations and advantages of its use.

4.3 This guide identifies and describes a number of sampling methods and samplers. It is advisable to consult available site-specific geological and hydrological data to assist in determining the sampling method and sampler best suited for a specific project. It is also advisable to contact a local firm providing the services required as not all sampling and drilling methods described in this guide are available nationwide.

5. Significance and Use

5.1 Chemical analyses of liquids, solids, and gases from the vadose zone can provide information on the presence, possible source, migration route, and physical-chemical behavior of contaminants. Remedial or mitigating measures can be formulated based on this information. This guide describes devices and procedures that can be used to obtain vadose zone soil samples.

5.2 Soil sampling is useful for the reasons presented in Section 1. However, it should be recognized that the general method is destructive, and that resampling at an exact location is not possible. Therefore, if a long term monitoring program is being designed, other methods for obtaining samples should be considered.

6. Criteria for Selecting Soil Samplers

6.1 Important criteria to consider when selecting devices for vadose zone soil sampling include the following:

6.1.1 Type of sample: An encased core sample, an uncased core sample, a depth-specific representative sample, or a sample according to requirements of the analyses,

6.1.2 Sample size requirements,

6.1.3 Suitability for sampling various soil types,

6.1.4 Maximum sampling depth,

6.1.5 Suitability for sampling soils under various moisture conditions,

6.1.6 Ability to minimize cross contamination,

6.1.7 Accessibility to the sampling site, and

6.1.8 Personnel requirements.

6.2 The sampling devices described in this guide have been evaluated for these criteria. The results are summarized in Fig. 1.

7. Sampling with Hand Operated Devices

7.1 These devices, that have mostly been developed for agricultural purposes, include:

7.1.1 Screw-type augers,

7.1.2 Barrel augers,

7.1.3 Tube-type samplers,

7.1.4 Hand held power augers, and

7.1.5 Trench sampling with shovels in conjunction with machine excavations.

7.2 The advantages of using hand operated devices over drill rigs are the ease of equipment transport to locations with poor vehicle access, and the lower costs of setup and decontamination. However, a major disadvantage is that these devices are limited to shallower depths than drill rigs.

7.3 *Screw-Type Augers:*

7.3.1 *Description*—The screw or ship auger is essentially a small diameter (for example, 1.5 in. (3.81 cm)) wood auger from which the cutting side flanges and tip have been removed (1)[5] (see Fig. 2(a)). According to the Soil Survey Staff (1), the spiral part of the auger should be about 7 in. (18 cm) long, with the distances between flights about the same as the diameter (for example, 1.5 in.) of the auger. This facilitates measuring the depth of penetration of the tool. Variations on this design include the closed spiral auger and the Jamaica open spiral auger (2) (see Fig. 2(b) and 1(c)). The

---

[5] The boldface numbers in parentheses refer to the list of references at the end of the text.

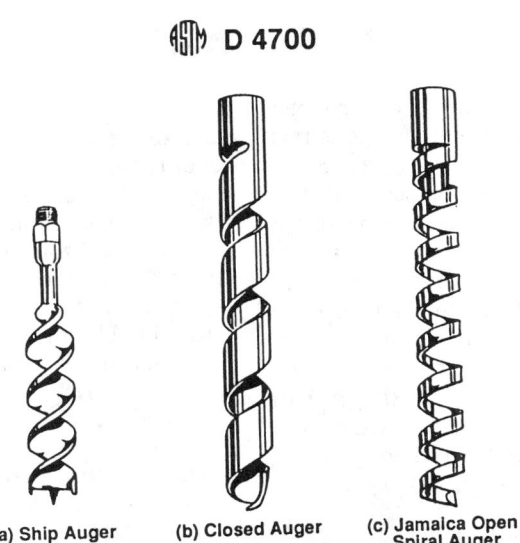

(a) Ship Auger (b) Closed Auger (c) Jamaica Open Spiral Auger

FIG. 2 Screw Type Augers

auger is welded onto a length of solid or tubular rod. The upper end of this rod is threaded, to accept a handle or extension rods. As many extensions are used as are required to reach the target sampling depth. The rod and the extensions are marked in even increments (for example, in 6-in. (15.24-cm) increments) above the base of the auger to aid in determining drilling depth. A wooden or metal handle fits into a tee-type coupling, screwed into the uppermost extension rod.

7.3.2 *Sampling Method*—For drilling, the auger is rotated manually. The operator may have to apply downward pressure to start and embed the auger; afterwards, the auger screws itself into the soil. The auger is advanced to its full length, and then pulled up and removed. Soil from the deepest interval penetrated by the auger is retained on the auger flights. A sample can be collected from the flights using a spatula. A foot pump operated hydraulic system has been developed to advance augers up to 4.5 in. (11.43 cm) in diameter. This larger diameter allows insertion of other sampling devices into the drill hole, once the auger is removed, if desired (**3**).[6]

7.3.3 *Comments*—Samples obtained with screw-type samplers are disturbed and are not truly core samples. Therefore, the samples are not suitable for tests requiring undisturbed samples, such as hydraulic conductivity tests. In addition, soil structures are disrupted and small scale lithologic features cannot be examined. Nevertheless, screw-type samplers are still suitable for use in collecting samples for the purpose of detecting contaminants. However, it is difficult to avoid transporting shallow soils downward when reentering a drill hole. When representative samples are desired from a discrete interval, the borehole must be made large enough to insert a sampler and extend it to the bottom of the borehole without touching the sides of the borehole. It is suggested that a larger diameter auger be used to advance and clear the borehole, then a smaller diameter auger sampler be used to obtain the sample. Screw-type augers work better in wet, cohesive soils than in dry, loose soils. Sampling in very dry (for example, powdery) soils may not be possible with these augers as soils will not be retained on the auger flights. Also, if the soil contains gravel or rock fragments larger than about one tenth of the hole diameter, drilling may not be possible (**4**).

7.4 *Barrel Augers:*

7.4.1 *Description*—The barrel auger consists of a bit with cutting edges welded to a short tube or barrel within which the soil sample is retained, welded in turn to shanks. The shanks are welded to a threaded rod at the other end. Extension rods are attached as required to reach the target sampling depth. Extensions are marked in increments above the base of the tool. The uppermost extension rod contains a tee-type coupling for a handle. The auger is available in carbon steel and stainless steel with hardened steel cutting edges (**5, 6**).

7.4.2 *Sampling Method*—The auger is rotated to advance the barrel into the ground. The operator may have to apply downward pressure to keep the auger advancing. When the barrel is filled, the unit is withdrawn from the soil cavity and a sample may be collected from the barrel.

---

[6] This reference is manufacturer's literature, and it has not been subjected to technical review.

7.4.3 *Comments*—Barrel augers generally provide larger samples than screw-type augers. The augers can penetrate shallow clays, silts, and fine grained sands (7).[6] The augers do not work well in gravelly soils, caliche, or semi-lithified deposits. Samples obtained with barrel augers are disturbed and are not core samples. Therefore, the samples are not suitable for tests requiring undisturbed samples, such as hydraulic conductivity tests. Nevertheless, the samplers are still suitable for use in collecting samples for the purpose of detecting contaminants. Because the sample is retained inside the barrel, there is less of a chance of mixing it with soil from a shallower interval during insertion or withdrawal of the sampler. The following are five common barrel augers:

7.4.3.1 Post-hole augers (also called Iwan-type augers),

7.4.3.2 Dutch-type augers,

7.4.3.3 Regular or general purpose barrel augers,

7.4.3.4 Sand augers, and

7.4.3.5 Mud augers.

7.4.4 *Post-Hole Augers*—The most readily available barrel auger is the post-hole auger (also called the Iwan-type auger) (8). As shown in Fig. 3, the barrel consists of two-part cylindrical leaves rather than a complete cylinder and is slightly tapered toward the cutting bit. The taper and the cupped bit help to retain soils within the barrel. The barrel is available with a 3 to 12-in. (7.62 to 30.48-cm) diameter. There are two types of drilling systems, one has a single rod and handle, and the other has two handles. In stable, cohesive soils, the auger can be advanced up to 25 ft (7.62 m) (8).

7.4.5 *Dutch-Type Augers*—The Dutch-type auger (commercially developed by Eijkelkamp) is a smaller variation of the post-hole auger design. As shown in Fig. 4, the pointed bit is continuous with two, narrow part-cylindrical barrel segments, welded onto the shanks. The barrel generally has a 3 in. (7.62 cm) outside diameter. This tool is best suited for sampling wet, clayey soils.

7.4.6 *Regular or General Purpose Barrel Augers*—A version of the barrel auger commonly used by soil scientists and county agricultural agents is depicted in Fig. 5(*a*) and (*b*). As shown, the barrel is a complete cylinder. As with the post-hole auger, the cutting blades are cupped so that soil is loosened and forced into the barrel as the unit is rotated and pushed into the ground. Each filling of the barrel corresponds to a depth of penetration of 3 to 5 in. (7.62 to 12.70 cm) (1). The most popular barrel diameter is 3.5 in. (8.89 cm), but sizes ranging from 1.5 to 7 in. (3.81 to 17.78 cm) are available (6).[6] Plastic, stainless steel, PTFE (polytetrafluoroethylene) or aluminum liners can also be used (6).[6] Extension rods are available in 4 ft (1.22 m) lengths. The rods can be made from standard black pipe, from lightweight conduit or from seamless steel tubing. The extensions have evenly spaced marks to facilitate determining sample depth. The regular barrel auger is suitable for use in loam type soils.

7.4.7 *Sand Augers*—For dry, sandy soils it may be necessary to use a variation of the regular barrel auger that includes a specially-formed bit to retain the sample in the barrel (see Fig. 5(*c*)). Sand augers with 2, 3, or 4-in. (5.08, 7.62, or 10.16-cm) diameters are available (5).[6]

7.4.8 *Mud Augers*—Another variation on the regular barrel auger design is available for sampling wet, clayey soils. As shown in Fig. 5(*d*), the barrel is designed with open sides to facilitate extraction of samples. The bits are the same as those used on the regular barrel auger (6).[6] Mud augers with 2, 3, or 4-in. (5.08, 7.62, or 10.16-cm) diameters are available (5).[6]

7.5 *Tube-Type Samplers:*

7.5.1 Tube-type samplers generally have proportionally smaller diameters and greater body lengths than those of barrel augers.

7.5.2 For sampling, these units are perched into the soil causing the tube to fill with material from the interval penetrated. The assembly is then pulled to the surface and a sample can be collected from the tube. Since the device is not

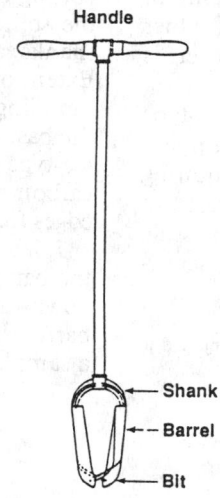

FIG. 3 Post-Hole Type Barrel Auger

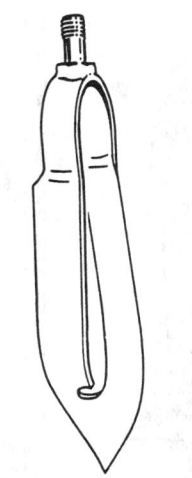

FIG. 4 Dutch Type Auger

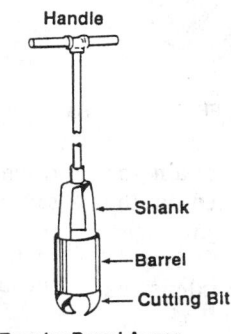

Handle

Shank

Barrel

Cutting Bit

(a) Regular Barrel Auger

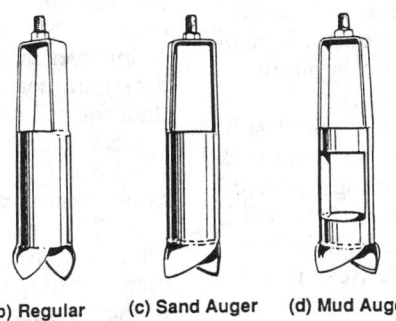

(b) Regular
Barrel Auger

(c) Sand Auger

(d) Mud Auger

FIG. 5 Barrel Auger Variations and Soil Moisture

rotated, a nearly undisturbed sample can be obtained. Commercial units are available with foot lever attachments, a hydraulic apparatus, or drop-hammers to aid in driving the sampler into the ground (5).[6] Vibratory heads have also been developed to advance tube-type samplers (9).[6]

7.5.3 These units are not as suitable for sampling in compacted, gravelly soils as are the barrel augers. They are preferred if an undisturbed sample is required. Commonly used varieties of the tube type samplers include:

7.5.3.1 Soil sampling tubes (also called Lord samplers),

7.5.3.2 Veihmeyer tubes (also called King tubes),

7.5.3.3 Thin-walled tube samplers (also called Shelby tubes),

7.5.3.4 Ring-lined barrel samplers, and

7.5.3.5 Piston samplers.

7.5.4 *Soil Sampling Tubes:*

7.5.4.1 *Description*—As depicted in Fig. 6, the soil sampling tube consists of a hardened cutting tip, a cut-away barrel, and an uppermost threaded segment. The cut-away barrel allows textural examination and easy removal of soil samples. Generally, the tube is constructed from high strength alloy steel (10).[6] The samplers are available with 6, 12, 15, 18, and 24-in. (15.24, 30.48, 38.10, 45.72, 60.96-cm) lengths (5, 6). The tubes are available with 1.13 or 0.88-in. (2.87 or 2.22-cm) outside diameter. Two modified versions of the tip are available, for sampling in wet or dry soils. The

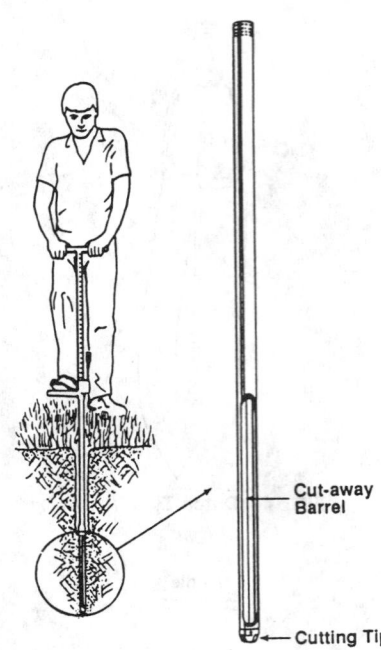

Cut-away
Barrel

Cutting Tip

FIG. 6 Soil Sampling Tube

sampling tube is attached to extension rods to attain the target sampling depth. A cross-handle is attached to the uppermost rod. Extension rods are made of lightweight, durable metal. They are available in a variety of lengths depending on the manufacturer. Markings on the extensions and the sampler facilitate determining sample depths.

7.5.4.2 *Sampling Method*—The sampler is pushed into the ground by leaning on the unit's handle. Once the sampler has reached the bottom of the sampling interval, it is twisted to break soil continuity at the tip. Depending on the type of cutting edge, the tube sampler may obtain samples varying in diameter from 0.69 to 0.75 in. (1.75 to 1.91 cm).

7.5.4.3 *Comments*—The soil sampling tube works best in soft, clayey, cohesive soils. If the soil contains cobbles or rock fragments larger than about one-half the cutting tip diameter, satisfactory sampling may not be possible. If the soil is cohesionless, it will not be retained in the tube. With time, the cutting tip will be damaged and worn dull. Most units are designed so that this part can be replaced.

7.5.5 *Veihmeyer Tubes:*

7.5.5.1 *Description*—The Veihmeyer tube is a long, complete cylinder. As shown in Fig. 6, this unit consists of a bevelled tip, that is threaded into the lower end of the tube, and a drive head threaded onto the upper end of the tube. The sampler is constructed of hardened steel. The tube is generally marked in even increments (for example, 1 ft or 0.30 m). These samplers are available in 4 to 16-ft (1.22 to 4.88-m) lengths with a 0.75-in. (1.91-cm) inside diameter.

7.5.5.2 *Sampling Method*—The lower guide rod of the drop hammer is slipped into the upper tube, through the drive head (see Fig. 7). The hammer is used to pound the sampler into the ground. The sampler is then retrieved by pulling or jerking up on the hammer to force the sampler out of the soil cavity. Samples are extruded by forcing a rod through the tube.

7.5.5.3 *Comments*—Prior to sampling, the inside of the tube is sometimes coated with a lubricant to facilitate extrusion. However, the types of analyses to be performed on the samples should be considered to determine if the presence of lubricant will cause interference. Because the Veihmeyer sampler is a solid-walled tube and is fitted with a drop hammer, it can generally be used in more resistant soils than the soil sampling tube.

7.5.6 *Thin-Walled Tube Samplers:*

7.5.6.1 *Description*—Thin-walled tube (Shelby Tube) samplers are readily available with 2, 3, and 5-in. (5.08, 7.62, and 12.70-cm) outside diameters and are commonly 30 in. (76.20 cm) long. The 3 by 30-in. (7.62 by 76.20-cm) outside diameter long sampler is most common. The advancing end of the sampler is rolled inwardly and has a cutting edge with a smaller diameter than the tube inside diameter. The cutting edge inside diameter reduction, defined as a "clearance ratio," is usually in the range of 0.0050 to 0.0150 or 0.50 to 1.50 % (Refer to Practice D 1587). The sampler tube is usually connected with set screws to a sampler head that in turn is threaded to connect with extension rods. Plastic and PTFE sealing caps for use after sampling are readily available for the 2, 3, and 5-in. (5.08, 7.62, and 12.70-cm) diameter tubes (refer to Practice D 4220). Shelby tubes are commonly available in carbon steel but can be manufactured from other metal (see Fig. 8).

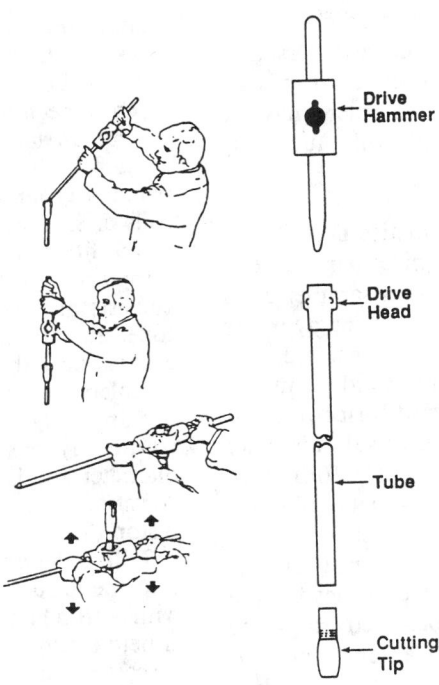

FIG. 7   Veihmeyer Tube

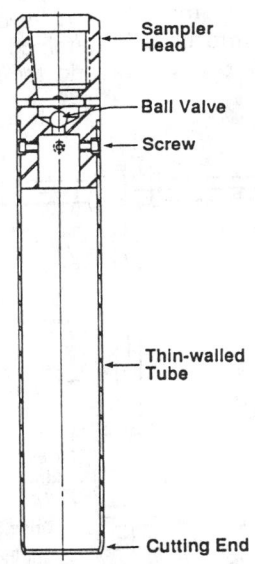

FIG. 8   Thin-Walled Tube Sampler

7.5.6.2 *Sampling Method*—The Shelby tube is pushed into soil by hand, with a jack-like system or with a hydraulic piston. The sample recovered is often less than the distance pushed, that is, the recovery ratio is less than 1.0. The recovery ratio is less than 1.0 because of soil compaction during sampling, and because friction between soil and the inner tube walls becomes greater than the shear strength of the soil in front of the tube. Consequently, soil in front of the advancing end of the tube is displaced laterally rather than entering the tube (11). In general, shorter tubes provide less-disturbed samples than longer tubes. Samples are extruded from the Shelby tube with a hydraulic ram. As with all sampling devices, the most disturbed portion of the sample in contact with the tube is considered unrepresentative. Wilson et al. (12) developed a paring device to remove this outer layer of the core during extrusion.

7.5.6.3 *Comments*—Shelby tubes are best used in clays, silts, and fine-grained sands. If the soils are cohesionless, they may not be retained in the tube. If firm to very hard soils are encountered, driving (hammering) the sampler may be required. However, this should be avoided as the tube may buckle under the drive stress.

7.5.7 *Ring-Lined Barrel Samplers:*

7.5.7.1 *Description*—As described in Practice D 3550, the ring-lined barrel sampler consists of a one piece barrel or two split barrel halves, a drive shoe, rings, and a sampler head (see Fig. 9). The rings, that are usually brass, fit snugly inside the barrel and are designed to be directly inserted into geotechnical testing apparatuses when removed from the barrel. Most samplers are designed to hold at least two rings. The barrel is commonly 3.5 in. (8.89 cm) inside diameter and 3.94 to 5.91 in. (10 to 15 cm) long (5).[6] With these lengths, the barrel can be fitted with a variety of liners ranging in length from 1 to 2.36 in. (2.54 to 6 cm).

7.5.7.2 *Sampling Method*—The ring-lined barrel sampler can be driven or pushed into soil. Once retrieved, the sampler is disassembled, and the sample-filled rings are removed. The rings are usually removed as one unit and placed into a capped container. Alternately, the individual soil-filled rings can be capped with plastic or PTFE and then sealed with wax or adhesive tape (refer to Practice D 4220).

7.5.7.3 *Comments*—Because barrel samplers are more rigid than thin-walled tubes, they can be driven into hard soils and soils containing sands and gravels that might damage thin-walled tubes. The sampler provides samples in rings which can be handled without further disturbance of the soil. Because of this, these devices are most often used when geotechnical or chemical analyses are to be performed.

7.5.8 *Piston Samplers:*

7.5.8.1 *Description*—Locally saturated (for example, by perched ground water), or cohesionless soils, and very soft soils or sludges may not be retained in most samplers, even when fitted with retainer baskets or flap valves. Piston samplers can be used in these situations. The sampler consists of a sampling tube, extension pipe attached to the tube, an internal piston, and rods connected to the piston and running through the extension pipe (see Fig. 10). These samplers are often built, as needed, out of common PVC (for use in sludge) or steel pipe fittings. The sampling tube commonly has a 0.75 to 3-in. (1.91 to 7.62-cm) inside diameter and is 8 in. to 9 ft (20.32 cm to 2.74 m) long (13). A variation designed for sampling peat has a cone shaped piston (8).

7.5.8.2 *Sampling Method*—The sampler can be pushed into the ground with the handle or driven into the ground with a drop hammer (13). As the tube is advanced, the piston is held stationary or pulled upward with the attached rods. Once the tube has been advanced through the sampling interval, it is rotated to break suction that might have developed between the soil and the outside wall of the tube. The sampler is then pulled to the surface keeping the piston rod fixed with respect to the extension pipe. The sample is retained because of suction that develops between the piston and the sample. Upon retrieval, the sample is extruded by

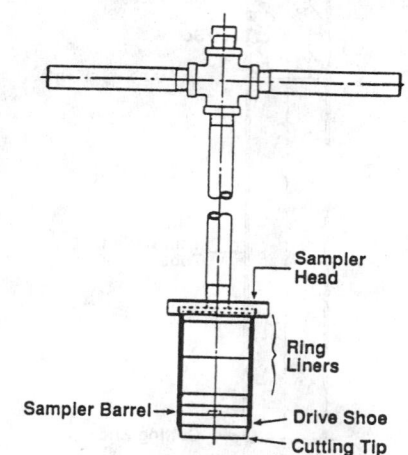

FIG. 9 Hand Operated Ring-Lined Barrel Sampler

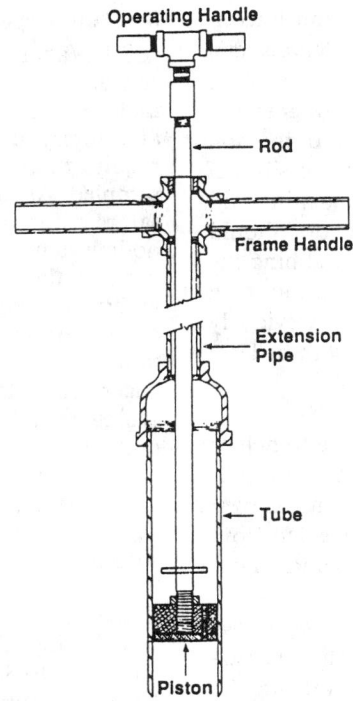

FIG. 10 Hand Operated Piston Sampler

using the piston to force the sample out of the tube. Sharma and De Dalta (14) described a cylindrical sampler for use in puddled soils that would flow back out of most samplers. The design includes a basal shutter that retains the sample while the sampler is withdrawn from the soil.

7.5.8.3 *Comments*—Because the sampler depends on development of suction between the sample and the piston, it may not work in unsaturated, coarse-grained sands and gravels. This is due to the high air permeability of such material that prevents the creation of high suction.

7.6 *Hand Held Power Augers:*

7.6.1 *Description*—A very simple, commercially available auger consists of a solid flight auger attached to and driven by a small air-cooled engine (see Fig. 11). Two handles on the head assembly allow two operators to guide the auger

into the soil. Throttle and clutch controls are integrated into grips on the handles. Augers are available with diameters ranging from 2 to 16 in. (5.08 to 40.64 cm). The auger sections are commonly 3 ft (0.91 m) long.

7.6.2 *Sampling Method*—As the auger rotates into soil, cuttings advance up the flights and are discharged at the surface. Soil samples can be collected from the surface discharge, or from the auger flights after pulling the auger out of the ground. Alternatively, samples can be collected with other samplers (for example, a thin-walled tube) after auger removal.

7.6.3 *Comments*—As discussed in 7.3, if samples are collected from surface discharge or from the flights, they are disturbed and are not suitable for some uses. In addition, if samples are collected from surface discharge, it is difficult to

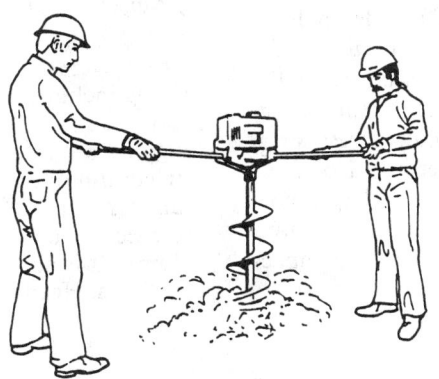

FIG. 11 Hand Held Power Auger

determine the depth from which the soil came and uncontrolled mixing of soil from different depth intervals can occur. The auger operates well in most soils. However, if the soil is cohesionless, it may not be retained on the flights and sampling in that fashion may not be possible. If the soil contains cobbles or boulders, drilling may not be possible. If the auger "hangs up" on an obstruction, the machine will start to rotate at the surface. Otherwise, the operator should not attempt to stop rotation of the machine by grabbing the handles. An alternate design that transfers the torque to a separate engine prevents this problem (15).[6] As previously stated, it is prudent to perform the field work with at least two people present.

7.7 *Trench Sampling:*

7.7.1 *Description*—Soils may be sampled from a trench or pit excavated for that purpose. Excavation is usually performed by a backhoe, and samples are collected with knives, trowels, or shovels. Occasionally, samples are collected from the sides or the bottom of the trench or pit with hand augers or tube-type samplers.

7.7.2 *Sampling Method*—Excavation is performed under the guidance of the sampling technician. Sampling is performed only after the backhoe has moved away from the trench or pit. When the trench or pit is in unstable material or is more than a few feet deep, the sampling technician should only enter the trench or pit after it has been shored up or the sidewalls have been cut back to within the angle of repose (see Occupational Safety and Health Administration regulations). Otherwise, samples are more commonly collected at the surface from the bucket of the backhoe as excavation occurs.

7.7.3 *Comments*—The maximum sampling depth for the trench or pit method is dictated by the reach of the backhoe, the soil type and the moisture content of the soil. Maximum depths of up to 20 ft (6.10 m) can be obtained in moist clays. Maximum depths of less than 10 ft (3.05 m) are common in dry sands. Samples collected from the backhoe bucket should be taken from the center of the material to prevent collecting soil contaminated by the bucket surface, and to prevent inclusion of materials that may have fallen from above the desired sampling interval. However, when this is done, it is difficult to accurately estimate the depth from which the sample was obtained. Trenches are useful for obtaining lithologic information since cross sections of the vadose zone can be studied and photographed. Trench or pit sampling is often used in areas with difficult access since backhoes are designed to travel on rough terrain. However, because the process involves excavating a much larger hole than drilling methods, chances of encountering underground utilities are increased, and proper backfilling and compaction of the trench is often very difficult.

8. Multipurpose and Auger Drill Rigs

8.1 Vadose zone samplers used in conjunction with drill rigs are identical to those used to sample below the water table. However, commonly used drill rigs such as cable tool and rotary units are not recommended as they generally require the introduction of drilling fluids to the soils to be sampled. Air rotary drilling is also undesirable for obtaining samples for pore liquid or gas extraction. In most cases, hollow-stem augers with some type of cylindrical sampler provide the greatest level of assurance that soil sampled within the vadose zone was not carried downward by the drilling or sampling process. For some situations, such as sampling firm to very hard ground, using multipurpose auger-core-rotary drill rigs will be necessary. For some geologic circumstances the use of solid stem augers will provide an adequate drilling method.

8.2 *Multipurpose Auger-Core-Rotary Drill Rigs:*

8.2.1 Multipurpose auger-core-rotary drill rigs are generally equipped with rotary power and vertical feed control to advance both hollow-stem augers and continuous flight (solid stem) augers to depths greater than 100 ft (30.48 m). These same drills have secondary capability for rotary and core drilling. The larger of these drills are typically mounted on 20 000 to 30 000-lb (9070 to 13605-kg) GVW trucks. The same multipurpose drill rigs are available on both rubber-tired and track-driven all-terrain carriers. The smaller of the multipurpose drills are typically mounted on trailers or one-ton, 4 by 4 trucks.

8.2.2 When equipped with augers, the sampling process is identical to that for auger drill rigs. When multipurpose auger-core-rotary drill rigs or auger drill rigs are used, the speed of drilling and sampling is much greater than with hand operated equipment. Therefore it is useful to have a larger crew to efficiently handle, log, identify, and preserve the samples.

8.3 *Auger Drill Rigs*—Auger drill rigs are similar to multipurpose auger-core-rotary drill rigs. They are manufactured specifically for efficient auger drilling but do not have the pumps and hoists that are required for efficient core or rotary drilling. The rigs can be equipped with either solid stem or hollow stem augers. There are relatively few auger drills available in comparison to multipurpose auger-core-rotary drills.

9. Auger Drilling and Sampling

9.1 *Solid Stem Auger Drilling and Sampling:*

9.1.1 *Description*—The tools used for solid-stem auger drilling include: auger sections, the drive cap, and the cutter head (see Fig. 12). Auger sections are typically 5 ft (1.52 m) long and are interchangeable for assembly in an articulated but continuously flighted column. Augers are available in diameters up to 24 in. (60.96 cm). The cutter head is attached to the lowermost or leading flight of the auger column. It is about 0.5 in. (1.25 cm) larger in diameter than the flights. Head types include fish tail or drag bits for use in

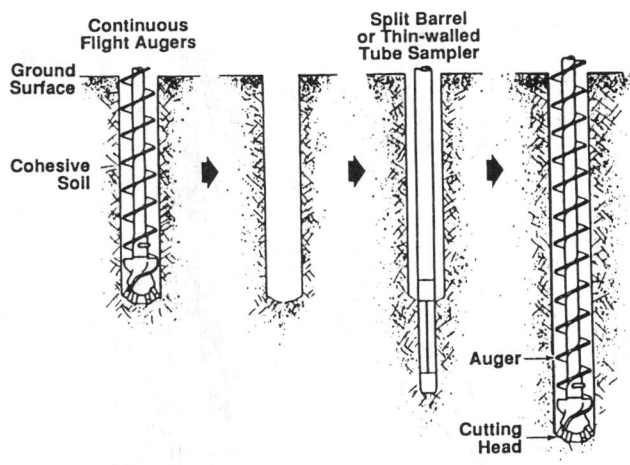

FIG. 12   Solid Stem Auger Sampling

cohesionless materials, and clay or stinger bits for use in more consolidated material **(16)**.

9.1.2 *Sampling Method*—As the auger column is rotated into soil, cuttings are retained on the flights. The augers are then removed from the hole and samples are taken from the retained soil. Samples obtained with solid stem augers are disturbed and are not core samples. Therefore, the samples are not suitable for analyses requiring undisturbed samples, such as hydraulic conductivity tests. This sampling method can provide an adequately clean borehole in some clayey and silty soils. However, when using the method in caving or squeezing ground, the quality and the origin of the recovered samples are questionable because soils from different intervals may have mixed. Therefore, when representative samples from discrete depths are desired, the borehole should be made large enough to insert a smaller diameter auger or another sampler (for example, a thin-walled tube) to the bottom of the borehole, without touching the sides of the borehole (see Fig. 11), to collect a discrete sample from the interval ahead.

9.1.3 *Comments*—Typical drilling depths with solid stem augers range from 50 to 120 ft (15.24 to 36.58 m). The greater drilling depths are attained in firm, silty and clayey soils. However, the depth to which the hole will remain open for sampling once the auger column has been removed is usually less than the maximum drilling depth. If cascading water or cohesionless soils are encountered, it can be expected that the hole will cave at that depth. The sample depth measurement, as taken from its location on an auger, is not precise. This is because soil may move up the flights in an uneven fashion as the auger column is advanced. As with hollow-stem augers, solid stem augers are often painted by the driller or manufacturer. It is prudent to remove this paint before drilling. The majority of the paint can be removed by drilling through sandy soils or by sand blasting. As with all

sampling devices, decontamination (for example, steam cleaning) should be performed between holes when chemical analyses are to be performed on the samples. This is especially important with the solid stem auger as it doubles as the drilling and sampling tool.

9.2 *Bucket Auger Drilling and Sampling:*

9.2.1 *Description*—The bucket auger is a large diameter cylindrical bucket with auger-type cutting blades on the bottom. The bucket can have a diameter ranging from 12 in. (30.48 cm) up to 6 ft (1.83 m) with lengths varying from 24 to 48 in. (60.96 to 121.92 cm) **(17)**. The bottom is hinged to allow cuttings to be emptied out (see Fig. 13).

9.2.2 *Sampling Method*—The bucket is rotated to depth in the vadose zone until the bucket is full. Therefore, depending on the bucket length, sampling intervals can range from 24 to 48 in. (60.96 to 121.92 cm). Sampling consists of extracting small diameter core samples from the interior of the bucket after lowering the full bucket to the ground (see Section 7). This approach minimizes problems with undiscrete mixing of discrete portions to be sampled.

9.2.3 *Comments*—The bucket auger is best suited for sampling from relatively stable clays as the caving problems discussed in 9.1.3 are amplified by the larger hole diameter. Boulders can impede drilling and may have to be individually removed from the hole before sampling can continue **(15)**[6]. Generally, boulders up to one-third or one-fourth the bucket diameter can be picked up by the bucket. Common sampling depths are less than 50 ft (15.24 m) but holes up to 250 ft (76.20 m) deep have been drilled **(16, 17)**.

9.3 *Hollow Stem Auger Drilling and Sampling:*

9.3.1 *Description*—Outer components of the hollow stem auger system include: hollow auger sections, the hollow auger head, and the drive cap. Inner components include: the pilot assembly, the center rod column, and the rod-to-cap adaptor (see Fig. 14). The auger head contains replaceable carbide teeth that pulverize the formation during flight

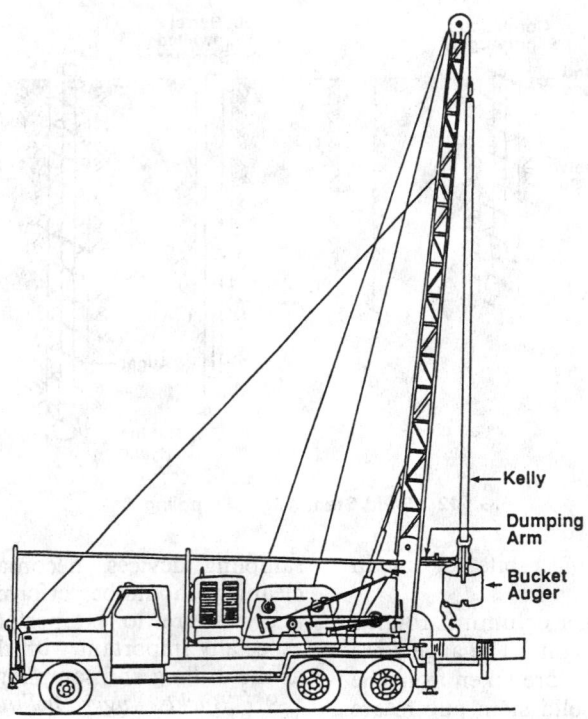

FIG. 13 Bucket Auger and Drilling Rig

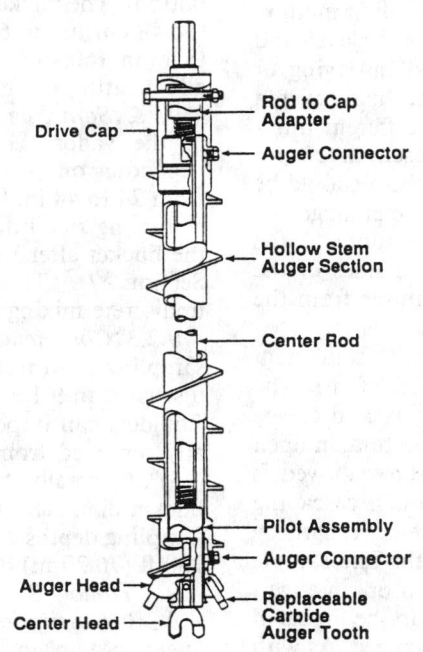

FIG. 14 Hollow-Stem Auger Components

column rotation. The cutting diameter is somewhat greater than the flighting diameter because of the protruding teeth. Auger sections are typically 5 ft (1.52 m) long and are interchangeable for assembly in an articulated but continuously flighted column. Drilling progresses in 5 ft (1.52 m) or shorter increments and sampling can be accomplished at any depth within that increment. Upon advancement of a 5 ft (1.52 m) increment, another 5 ft (1.52 m) section of hollow-stem auger and center rod is added. Hollow-stem augers are readily available with 2.25, 2.75, 3.25, 3.75, 4.25, 6.25, and 8.25-in. (5.72, 6.99, 8.26, 9.53, 10.80, 15.88, and 20.96-cm) inside diameters.

9.3.2 *Sampling Method*—The auger column and pilot assembly are advanced to the top of the desired sampling interval. Sampling is accomplished by removing the pilot assembly and center rod, if they are used, and inserting the sampler through the hollow stem of the auger column (see Fig. 15). The sampler may be lowered to the sampling depth by attaching it to center rods or by using a wireline assembly (**12**). When the sampler is attached to center rods, a sample is collected by pushing or driving the sampler into undisturbed soil with the rig hydraulic system or with a drop hammer. When a wireline is used, the sampler is locked into place ahead of the lower-most auger and advanced into the sampling interval by rotating the auger column (**18**).[6] Hollow stem augers with a 6.25-in. (15.88-cm) inside diameter allow the use of 5-in. (12.70-cm) outside diameter Shelby tubes and 4.5-in. (11.43-cm) outside diameter split barrel samplers (see 9.4).

9.3.3 *Comments*—The purpose of the center head (pilot) assembly is to prevent soils from entering the auger column as it is advanced (**19**). Driscoll (**17**) suggests that the assembly may be omitted when drilling through hard, silty and clayey soils as these materials will usually form a 2 to 4 in. (5.08 to 10.16 cm) long plug at the auger opening. However, Hackett (**19**) recommends that the pilot assembly be used when detailed samples are required. When perched water is encountered, "heaving sands" that move up into the auger column upon pilot assembly removal during sampling, may be a concern. Various one-way plugs that allow sampling, but that prevent sand from moving into the auger column, are described in Hackett (**19**). The important capability of being able to obtain samples that do not contain mixed material from shallow sources in the hole is enhanced by using the hollow-stem auger method. However, because the sections are hollow, decontamination of the auger interiors between holes to prevent cross contamination is difficult. High pressure steam cleaners are usually necessary to remove caked-on soils and contaminants. Hollow stem augers may advance rapidly through unconsolidated materials.

9.4 *Sampling Devices:*

9.4.1 Sampling devices used in conjunction with hollow stem augers and occasionally in holes advanced by solid stem augers include:

9.4.1.1 Thin-walled tube samplers (also called Shelby tubes),

9.4.1.2 Split-barrel drive samplers (also called Split spoons),

9.4.1.3 Ring-lined barrel samplers,

9.4.1.4 Continuous sample tube systems, and

9.4.1.5 Piston samplers.

9.4.2 These samplers are either pushed or driven in sequence with an increment of drilling or advanced simultaneously with the advance of a hollow stem auger column.

9.4.3 *Thin-Walled Tube Samplers:*

9.4.3.1 *Description*—The thin-walled tube sampler consists of a tube connected to a head with screws. The head is threaded to connect with standard drill rods. The head contains a ball check valve. Thin-walled tube (Shelby tube) samplers are readily available with 2, 3, and 5-in. (5.08, 7.62, and 12.70-cm) outside diameter and are commonly 30 in. (76.20 cm) long. The 3 by 30 in. (7.62 by 76.20 cm) outside diameter long sampler is most common. The advancing end of the sampler is constructed with an inward lip, machined to a cutting edge, that has a smaller diameter than the tube inside diameter. The cutting edge inside diameter reduction, defined as a "clearance ratio," is usually in the range of 0.0050 to 0.0150 or 0.50 to 1.50 % (refer to Practice D 1587). PTFE or plastic sealing caps and other sealing devices for use after sampling are readily available for the 2, 3, and 5-in. (5.08, 7.62 and 12.70-cm) diameter tubes (refer to Practice D 4220). Shelby tubes are commonly available in carbon steel but can be manufactured from other metal (see Fig. 8).

9.4.3.2 *Sampling Methods*—When a Shelby tube is pushed into soil, the length of the sample recovered is often less than the distance pushed, that is, the recovery ratio is less than 1.0 (see 7.5.6.2). In addition, a portion of the sample frequently remains in the borehole after retrieval of the sampler. This is due to suction that develops at the sampler-soil interface. This suction may be broken by twisting the

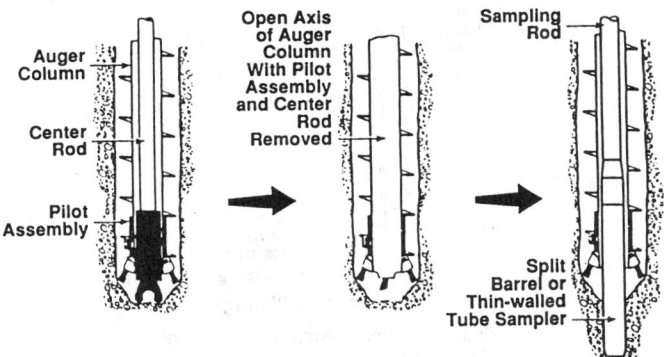

FIG. 15 Hollow-Stem Auger Sampling

sampler prior to retrieval or by advancing the auger column below the base of the sampler before retrieval (20). Samples are extruded from the Shelby tube with a hydraulic ram. As with all sampling devices, the portion of the sample in contact with the tube is considered disturbed and unrepresentative. Wilson et al. (12) developed a paring device to remove this outer layer of the core during extrusion.

9.4.3.3 *Comments*—The ball check valve was originally intended to provide a vent for drilling fluids when pushing the tube into soil, and also to prevent the column of fluid within the drill stem from forcing the sample out of the tube during retrieval. Since drilling fluids are not used when sampling in the vadose zone, these considerations are not important. However, the valve does provide a vent for air displaced as the sampler is pushed into soil. Shelby tubes are best used in clays, silts, and fine grained sands. They can be pushed with the hydraulic system of most drill rigs in fine grained sands that are loose to moderately consolidated or in clays and silts that are soft to firm. If the soils are cohesionless, they may not be retained in the tube. If consolidated or hard soils are encountered, driving the sampler may be required. However, some tubes may buckle under the drive stress. A spring-loaded barrel has been developed to protect the Shelby tube from buckling when sampling these soils (21).[6]

9.4.4 *Split-Barrel Drive Samplers:*

9.4.4.1 *Description*—The split-barrel drive sampler consists of two split-barrel halves, a drive shoe, and a sampler head containing a ball check valve, all of which are threaded together (see Fig. 15). The most common size has a 2-in. (5.08-cm) outside diameter and a 1.5-in. (3.81-cm) inside diameter split barrel with a 1.375-in. (3.49-cm) inside diameter drive shoe. This sampler is used extensively in geotechnical exploration (Refer to Method D 1586). When fitted with a 16 gage liner for encased cores, the sampler has a 1.375-in. (3.49-cm) inside diameter throughout. A 3-in. (7.62-cm) outside diameter by 2.5-in. (6.35-cm) inside diameter split-barrel sampler with a 2.375-in. (6.03-cm) inside diameter drive shoe is also available (22).[6] Other split-barrel samplers in the size range of 2.5-in. (6.35-cm) to 4.5-in. (11.43-cm) outside diameter are manufactured but are less common. A plastic or metal retainer basket, or a flap valve is often fitted into the drive shoe to prevent samples from falling out during retrieval.

9.4.4.2 *Sampling Method*—As described in Method D 1586 the sampler is threaded onto drilling rods and is lowered to the bottom of the boring. The sampler is then driven into the soil with blows from a drop hammer attached to the drill rig. The hammer usually weighs 140 lb and is operated by the driller. The sampler is extracted from the soil in a manner that will ensure maximum sample recovery. A

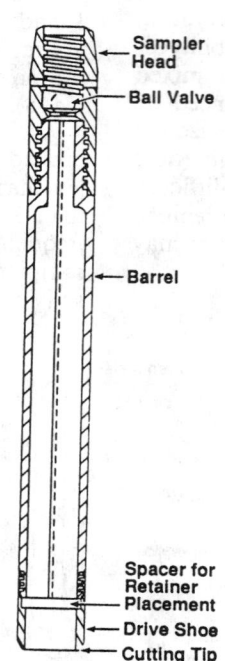

Sampler Head

Ball Valve

Barrel

Spacer for Retainer Placement

Drive Shoe

Cutting Tip

FIG. 16 Split-Barrel Drive Sampler

sample is obtained by disassembling the drive shoe and head, and splitting the barrel to expose the core of soil. Material disturbed by contact with the barrel can be scraped away, or a less disturbed interior portion collected with a spatula.

9.4.4.3 *Comments*—Split barrel drive samplers can be used in all soil types if the larger grain sizes can enter through the opening of the drive shoe. Because the sampler can be fitted with a retainer basket, it is typically used in place of thin-walled tubes when cohesionless soils are to be sampled.

9.4.5 *Ring-Lined Barrel Samplers:*

9.4.5.1 *Description*—As described in Practice D 3550, the ring-lined barrel sampler consists of a one piece barrel or two split-barrel halves, a drive shoe, rings, a waste barrel and a sampler head containing a ball check valve (see Fig. 17). The rings fit snugly inside the barrel and are designed to be directly inserted into geotechnical testing apparatus when removed from the barrel. Most samplers are designed to hold at least six rings. The waste barrel provides a space above the rings into which disturbed soil, originally at the bottom of the hole, can move. The samplers are commonly available with 2, 3, and 4-in. (5.08, 7.62, and 10.16-cm) outside diameter.

9.4.5.2 *Sampling Method*—The ring-lined barrel sampler can be driven or pushed into soil. It is important to insert the sampler deep enough to allow all disturbed soil to move through the rings and into the waste barrel. Once retrieved, the sampler is disassembled, and the sample filled rings are carefully removed. The rings are usually removed as one unit and placed into a capped container. Alternately, the individual soil filled rings can be capped with plastic or PTFE and even sealed with wax or adhesive tape (refer to Practice D 4220).

9.4.5.3 *Comments*—Because ring-lined barrel samplers are more rigid than thin-walled tubes, they can be driven into soils containing sands and gravels that might damage thin-walled tubes. The sampler provides samples in rings that can be handled without further disturbance of the soil. Because of this, these devices are most often used when geotechnical or chemical analyses are to be performed.

9.4.6 *Continuous Sample Tube System:*

9.4.6.1 *Description*—Continuous sample tube systems that fit within a hollow-stem auger column are readily available in North America. The barrel is typically 5 ft (1.52 m) long, and fits within the lead auger of the hollow auger column. The sampler is prevented from rotating as the auger column is turned (20). For many conditions the sampler provides continuous, 5-ft (1.52-m) samples (see Fig. 18). The assembly can be split- or solid-barrel and can be used with or without liners of various metallic and nonmetallic materials (20). Two clear, plastic, 30 in. (76.20 cm) long liners are often used. The sampler may also be fitted with a plastic or metal retainer basket, or a falp valve to prevent cohesionless soils from falling out of the sampler during retrieval (20).

9.4.6.2 *Sampling Method*—The sampler is locked in place inside the auger column with its open end protruding a short distance beyond the end of the column. While advancing the column, soil enters the non-rotating sampling barrel. After a 5-ft (1.52-m) advance, the sampler is withdrawn, and the liner (if used) is removed and capped.

9.4.6.3 *Comments*—The continuous sample tube system replaces the pilot head assembly in the hollow-stem auger column. Because of this, sampling speed is greatly increased since the pilot assembly does not have to be removed before taking a sample. The continuous sample tube system is best used in clays, silts, and in fine grained sands. It can be used to sample soils that are much more consolidated or harder than can be sampled with Shelby tubes.

9.4.7 *Piston Samplers:*

9.4.7.1 *Description*—Locally saturated (for example, perched ground water), or cohesionless soils, and very soft soils or sludges may not be retained in most samplers, even when they have been fitted with retainer baskets or flap valves. Piston samplers are often used under these conditions. The sampler consists of a sampling tube, an internal piston, and a drive head. The piston fits snugly inside the tube. The piston is attached to a rod assembly or a cable that leads to the surface. Tubes made of steel are available in 5.5 and 30-in. (13.97 and 76.20-cm) and 5-ft (1.5-m) lengths with 0.75, 2, 3, 4, and 5-in. (1.91, 5.08, 7.62, 10.16, and 12.70-cm) inside diameter (22, 23). When equipped with a hardened steel drive shoe, the tube can be fitted with a liner made of aluminum clear PVC, or another material (see Fig. 19) (24). A version of the sampler designed for peat sampling has a cone shaped piston (8).

9.4.7.2 *Sampling Method*—Prior to sampling, the piston is placed at the base (advancing end) of the tube. The sampler is then attached to drill rods and lowered down the borehole or hollow-stem auger column to the bottom of the hole (top of the sampling interval). The sampler is then pushed or driven into the sampling interval. As the tube

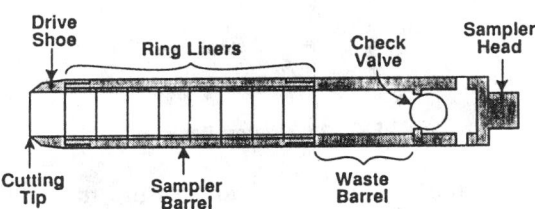

FIG. 17 Ring-Lined Barrel Sampler

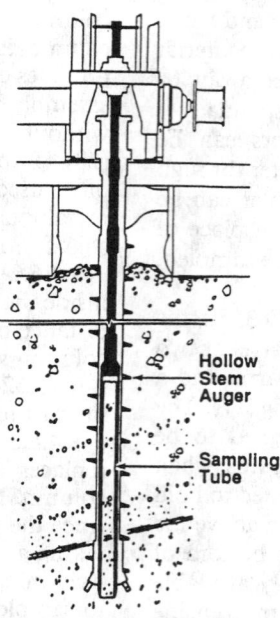

FIG. 18 Continuous Sample Tube System

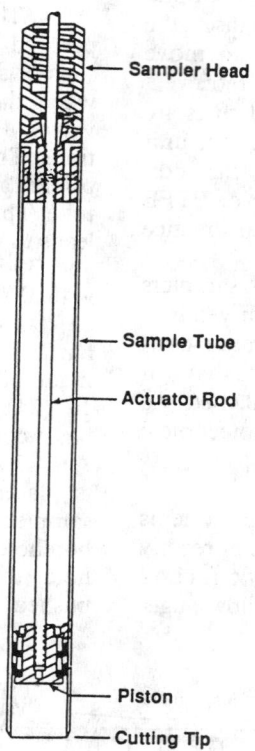

FIG. 19 Piston Sampler

moves downward, the piston remains stationary and in contact with the top of the soil sample. When the sampler is withdrawn, soil is retained because of suction that develops between the piston and the soil core within the sampler. This suction is stronger than the suction at the bottom of the sampler that would tend to extract soil from the sampler. Even so, it is often useful to twist the sampler with the drill rods prior to retrieval, to break suction at the bottom end and ensure that the sample will not be pulled out of the sampler.

9.4.7.3 *Comments*—Average recovery ratios greater than 0.9 can be attained with this sampling tool (24, 25). However, because the sampler depends on development of suction between the sample and the piston, it may not work in unsaturated, coarse grained sands and gravels. This is due to the high air permeability of such material that prevents the creation of suction with the sampler. Samples collected with piston samplers are relatively undisturbed. Zapico et al. **(24)**

described techniques for extracting fluid samples directly from liners, and for converting liners into permeameters.

REFERENCES

(1) Soil Survey Staff, *Soil Survey Manual*, U.S. Dept. of Agriculture, Superintendent of Documents, Washington, D.C., 1951.

(2) Acker, W. L., *Basic Procedures for Soil Sampling and Core Drilling*, Acker Drill Co. Inc., Scranton, Pa, 1974.

(3) Materials Testing Division, *Catalog of Products*, Soiltest Inc., Evanston, Illinois, 1983.

(4) Bureau of Reclamation, *Earth Manual*, U.S. Dept. of the Interior, United States Government Printing Office, Washington, D.C., 1974.

(5) Sales Division, *Catalog of Products*, Soilmoisture Equipment Corp., Santa Barbara, California, 1988.

(6) Sales Division, *Catalog of Products*, Art's Manufacturing and Supply, American Falls, Idaho, 1988.

(7) Sales Division, *Catalog of Products*, Brainard Kilman, Stone Mountain, Georgia, 1988.

(8) Everett, L. G., and Wilson, L. G., *Permit Guidance Manual on Unsaturated Zone Monitoring For Hazardous Waste Land Treatment Units*, EPA/530-SW-86-040, 1986.

(9) Sales Division, *Catalog of Products*, VI-COR Technologies Inc., Bellevue, Washington, 1988.

(10) Sales Division, *J.M.C. Soil Investigation Equipment, Catalog No. 6*, Clements Associates Inc.

(11) Hvorslev, M. J., *Subsurface Exploration and Sampling of Soils for Civil Engineering Purposes*, U.S. Army Corp of Engineers, Waterways Experiment Station, Vicksburg, Mississippi, 1949.

(12) Wilson, J. T., and McNabb, J. F., "Biological Transformation of Organic Pollutants in Ground Water," *EOS-Transactions*, AGU, Vol 64, pp. 505–507.

(13) Brakensiek, D. L., Osborn, H. B., and Rawls, W. L., *Field Manual for Research in Agricultural Hydrology, Agriculture Handbook No. 224*, Science and Education Administration, United States Dept. of Agriculture, Washington, D.C., 1979 (revised), pp. 258–275.

(14) Sharma, P. K., and DeDalta, S. K., "A Core Sampler for Puddled Soils," *Soil Science Society of America Journal*, Vol 49, 1985, pp. 1069–1070.

(15) Sales Division, *Catalog of Products*, Little Beaver Inc., Livingston, Texas, 1988.

(16) Scalf, M. R., McNabb, J. F., Dunlap, W. J., Cosby, R. L., and Fryberger, J., *Manual of Groundwater Sampling Procedures*, National Water Well Association, Dublin, Ohio, 1981.

(17) Driscoll, F. G., *Groundwater and Wells*, Johnson Division, St. Paul, Minnesota, (2nd ed.), 1986.

(18) Sales Division, *Catalog of Products*, Mobile Drilling Co. Inc., Indianapolis, Indiana, 1988.

(19) Hackett, G., "Drilling and Constructing Monitoring Wells with Hollow-Stem Augers Part 1: Drilling Considerations," *Ground Water Monitoring Review*, NWWA, Fall 1987, pp. 51–62.

(20) Riggs, C. O., "Soil Sampling in the Vadose Zone," *Proceedings of the NWWA/U.S. EPA Conference on Characterization and Monitoring of the Vadose Zone*, NWWA, Las Vegas, Nevada, 1983, pp. 611–622.

(21) Sales Division, *Product Literature*, Pitcher Drilling Co., Palo Alto, California, 1986.

(22) Sales Division, *Catalog of Products*, Diedrich Drilling Equipment, LaPorte, Indiana, 1988.

(23) Sales Division, *Instrumentation for Soil and Rocks, Catalog of Products*, Solinst, Burlington, Ontario, Canada, 1988.

(24) Zapico, M. M., Vales, S., and Cherry, J. A., "A Wireline Core Barrel for Sampling Cohesionless Sand and Gravel Below the Water Table," *Ground Water Monitoring Review*, NWWA, Spring 1987, pp. 75–82.

(25) Shuter, E., Teasdale, W. E., "Application of Drilling, Coring, and Sampling Techniques to Test Holes and Wells," *U.S. Geol. Survey Techniques of Water Resource Investigations*, Book 2, Chapter F-1, 1989.

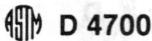 **D 4700**

ASTM Designation: D 1452 – 80 (Reapproved 1995)

Standard Practice for
Soil Investigation and Sampling by Auger Borings[1]

This standard is issued under the fixed designation D 1452; the number immediately following the designation indicates the year of original adoption or, in the case of revision, the year of last revision. A number in parentheses indicates the year of last reapproval. A superscript epsilon (ε) indicates an editorial change since the last revision or reapproval.

This standard has been approved for use by agencies of the Department of Defense. Consult the DoD Index of Specifications and Standards for the specific year of issue which has been adopted by the Department of Defense.

1. Scope

1.1 This practice covers equipment and procedures for the use of earth augers in shallow geotechnical exploration. This practice does not apply to sectional continuous flight augers.

1.2 *This standard does not purport to address all of the safety concerns, if any, associated with its use. It is the responsibility of the user of this standard to establish appropriate safety and health practices and determine the applicability of regulatory limitations prior to use.*

2. Significance and Use

2.1 Auger borings often provide the simplest method of soil investigation and sampling. They may be used for any purpose where disturbed samples can be used and are valuable in connection with ground water level determination and indication of changes in strata and advancement of hole for spoon and tube sampling. Equipment required is simple and readily available. Depths of auger investigations are, however, limited by ground water conditions, soil characteristics, and the equipment used.

3. Apparatus

3.1 *Hand-Operated Augers*:

3.1.1 *Helical Augers*—Small lightweight augers generally available in sizes from 1 through 3 in. (25.4 through 76.2 mm).

3.1.1.1 *Spiral-Type Auger*, consisting of a flat thin metal strip, machine twisted to a spiral configuration of uniform pitch; having at one end, a sharpened or hardened point, with a means of attaching a shaft or extension at the opposite end.

3.1.1.2 *Ship-Type Auger*—Similar to a carpenter's wood bit. It is generally forged from steel and machined to the desired size and configuration. It is normally provided with sharpened and hardened nibs at the point end and with an integral shaft extending through its length for attachment of a handle or extension at the opposite end.

3.1.2 *Open Tubular Augers*, ranging in size from 1.5 through 8 in. (38.1 through 203.2 mm) and having the common characteristic of appearing essentially tubular when viewed from the digging end.

3.1.2.1 *Orchard-Barrel Type*, consisting essentially of a tube having cutting lips or nibs hardened and sharpened to penetrate the formation on one end and an adaptor fitting for an extension or handle on the opposite end.

3.1.2.2 *Open-Spiral Type*, consisting of a flat thin metal strip that has been helically wound around a circular mandrel to form a spiral in which the flat faces of the strip are parallel to the axis of the augered hole. The lower helix edges are hard-faced to improve wear characteristics. The opposite end is fitted with an adaptor for extension.

3.1.2.3 *Closed-Spiral Type*—Nearly identical to the open-spiral type except the pitch of the helically wound spiral is much less than that of the open-spiral type.

3.1.3 *Post-Hole Augers*, generally 2 through 8 in. (50.8 through 203.2 mm), and having in common a means of blocking the escape of soil from the auger.

3.1.3.1 *Clam-Shell Type*, consisting of two halves, hinged to allow opening and closing for alternately digging and retrieving. It is not usable deeper than about 3.5 ft (1.07 m).

3.1.3.2 *Iwan Type*, consisting of two tubular steel segments, connected at the top to a common member to form a nearly complete tube, but with diametrically opposed openings. It is connected at the bottom by two radial blades pitched to serve as cutters which also block the escape of contained soil. Attachment of handle or extension is at the top connector.

3.2 *Machine-Operated Augers*:

3.2.1 *Helical Augers*, generally 8 through 48 in. (203.2 through 1219 mm), consisting essentially of a center shaft fitted with a shank or socket for application of power, and having one to three complete 360° (6.28-rad) spirals for conveyance and storage of cut soil. Cutter bits and pilot bits are available in moderate and hard formation types and normally replaceable in the field. They are normally operated by heavy-duty, high-torque machines, designed for heavy construction work.

3.2.2 *Stinger Augers*, generally 6 through 30 in. (152.4 through 762 mm), are similar to the helical auger in 3.2.1, but lighter and generally smaller. They are commonly operated by light-duty machines for post and power pole holes.

3.2.3 *Disk Augers*, generally 10 through 30 in. (254 through 762 mm), consisting essentially of a flat, steel disk with diametrically opposed segments removed and having a shank or socket located centrally for application of power. Replaceable cutter bits, located downward from the leading edges of the remaining disk, dig and load soil that is held on the disk by valves or shutters hinged at the disk in order to close the removed segments. The disk auger is specifically designed to be operated by machines having limited vertical clearance between spindle and ground surface.

---

[1] This practice is under the jurisdiction of ASTM Committee D-18 on Soil and Rock and is the direct responsibility of Subcommittee D18.02 on Sampling and Related Field Testing for Soil Investigations.

Current edition approved June 12, 1980. Published August 1980. Originally published as D 1452 – 57 T. Last previous edition D 1452 – 65 (1972).

3.2.4 *Bucket Auger*, generally 12 through 48 in. (304.8 through 1219 mm), consisting essentially of a disk auger, without shank or socket, but hinge-mounted to the bottom of a steel tube or bucket of approximately the same diameter as the disk auger. A socket or shank for power application is located in the top center of the bucket diametral cross piece provided for the purpose.

3.3 *Casing* (when needed), consisting of pipe of slightly larger diameter than the auger used.

3.4 *Accessory Equipment*—Labels, field log sheets, sample jars, sealing wax, sample bags, and other necessary tools and supplies.

4. Procedure

4.1 Make the auger boring by rotating and advancing the desired distance into the soil. Withdraw the auger from the hole and remove the soil for examination and test. Return the empty auger to the hole and repeat the procedure. Continue the sequence until the required depth is reached.

4.2 Casing is required in unstable soil in which the bore hole fails to stay open and especially when the boring is extended below the ground-water level. The inside diameter of the casing must be slightly larger than the diameter of the auger used. The casing shall be driven to a depth not greater than the top of the next sample and shall be cleaned out by means of the auger. The auger can then be inserted into the bore hole and turned below the bottom of the casing to obtain a sample.

4.3 The soil auger can be used both for boring the hole and for bringing up disturbed samples of the soil encountered. The structure of a cohesive soil is completely destroyed and the moisture may be changed by the auger. Seal all samples in a jar or other airtight container and label appropriately. If more than one type of soil is picked up in the sample, prepare a separate container for each type of soil.

4.4 *Field Observations*—Record complete ground water information in the field logs. Where casing is used, measure ground water levels both before and after the casing is pulled. In sands, determine the water level at least 30 min after the boring is completed; in silts, at least 24 h. In clays, no accurate water level determination is possible unless pervious seams are present. As a precaution, however, water levels in clays shall be taken after at least 24 h.

5. Report

5.1 The data obtained in boring shall be recorded in the field logs and shall include the following:

5.1.1 Date of start and completion of boring,

5.1.2 Identifying number of boring,

5.1.3 Reference datum including direction and distance of boring relative to reference line of project or other suitable reference points,

5.1.4 Type and size of auger used in boring,

5.1.5 Depth of changes in strata,

5.1.6 Description of soil in each major stratum,

5.1.7 Ground water elevation and location of seepage zones, when found, and

5.1.8 Condition of augered hole upon removal of auger, that is, whether the hole remains open or the sides cave, when such can be observed.

6. Keywords

6.1 auger borings; sampling; soil investigations

Standard Test Method for
Penetration Test and Split-Barrel Sampling of Soils[1]

This standard is issued under the fixed designation D 1586; the number immediately following the designation indicates the year of original adoption or, in the case of revision, the year of last revision. A number in parentheses indicates the year of last reapproval. A superscript epsilon (ε) indicates an editorial change since the last revision or reapproval.

This standard has been approved for use by agencies of the Department of Defense. Consult the DOD Index of Specifications and Standards for the specific year of issue which has been adopted by the Department of Defense.

[ε1]NOTE—Editorial changes were made throughout October 1992.

1. Scope

1.1 This test method describes the procedure, generally known as the Standard Penetration Test (SPT), for driving a split-barrel sampler to obtain a representative soil sample and a measure of the resistance of the soil to penetration of the sampler.

1.2 *This standard does not purport to address all of the safety problems, if any, associated with its use. It is the responsibility of the user of this standard to establish appropriate safety and health practices and determine the applicability of regulatory limitations prior to use.* For a specific precautionary statement, see 5.4.1.

1.3 The values stated in inch-pound units are to be regarded as the standard.

2. Referenced Documents

2.1 *ASTM Standards:*
D 2487 Test Method for Classification of Soils for Engineering Purposes[2]
D 2488 Practice for Description and Identification of Soils (Visual-Manual Procedure)[2]
D 4220 Practices for Preserving and Transporting Soil Samples[2]
D 4633 Test Method for Stress Wave Energy Measurement for Dynamic Penetrometer Testing Systems[2]

3. Terminology

3.1 *Descriptions of Terms Specific to This Standard*

3.1.1 *anvil*—that portion of the drive-weight assembly which the hammer strikes and through which the hammer energy passes into the drill rods.

3.1.2 *cathead*—the rotating drum or windlass in the rope-cathead lift system around which the operator wraps a rope to lift and drop the hammer by successively tightening and loosening the rope turns around the drum.

3.1.3 *drill rods*—rods used to transmit downward force and torque to the drill bit while drilling a borehole.

3.1.4 *drive-weight assembly*—a device consisting of the hammer, hammer fall guide, the anvil, and any hammer drop system.

3.1.5 *hammer*—that portion of the drive-weight assembly consisting of the 140 ± 2 lb (63.5 ± 1 kg) impact weight which is successively lifted and dropped to provide the energy that accomplishes the sampling and penetration.

3.1.6 *hammer drop system*—that portion of the drive-weight assembly by which the operator accomplishes the lifting and dropping of the hammer to produce the blow.

3.1.7 *hammer fall guide*—that part of the drive-weight assembly used to guide the fall of the hammer.

3.1.8 *N-value*—the blowcount representation of the penetration resistance of the soil. The *N*-value, reported in blows per foot, equals the sum of the number of blows required to drive the sampler over the depth interval of 6 to 18 in. (150 to 450 mm) (see 7.3).

3.1.9 *ΔN*—the number of blows obtained from each of the 6-in. (150-mm) intervals of sampler penetration (see 7.3).

3.1.10 *number of rope turns*—the total contact angle between the rope and the cathead at the beginning of the operator's rope slackening to drop the hammer, divided by 360° (see Fig. 1).

3.1.11 *sampling rods*—rods that connect the drive-weight assembly to the sampler. Drill rods are often used for this purpose.

3.1.12 *SPT*—abbreviation for Standard Penetration Test, a term by which engineers commonly refer to this method.

4. Significance and Use

4.1 This test method provides a soil sample for identification purposes and for laboratory tests appropriate for soil obtained from a sampler that may produce large shear strain disturbance in the sample.

4.2 This test method is used extensively in a great variety of geotechnical exploration projects. Many local correlations and widely published correlations which relate SPT blowcount, or *N*-value, and the engineering behavior of earthworks and foundations are available.

5. Apparatus

5.1 *Drilling Equipment*—Any drilling equipment that provides at the time of sampling a suitably clean open hole before insertion of the sampler and ensures that the penetration test is performed on undisturbed soil shall be acceptable. The following pieces of equipment have proven to be

---

[1] This method is under the jurisdiction of ASTM Committee D-18 on Soil and Rock and is the direct responsibility of Subcommittee D18.02 on Sampling and Related Field Testing for Soil Investigations.
Current edition approved Sept. 11, 1984. Published November 1984. Originally published as D 1586 – 58 T. Last previous edition D 1586 – 67 (1974).
[2] *Annual Book of ASTM Standards*, Vol 04.08.

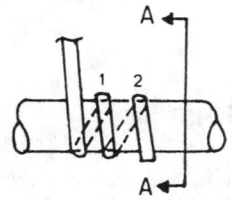

(a) counterclockwise rotation approximately 1¾ turns

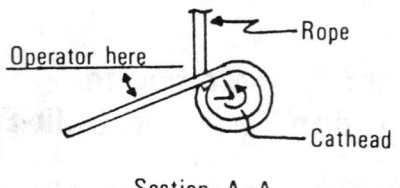

Section A–A

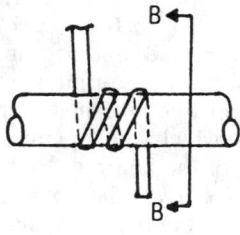

(b) clockwise rotation approximately 2¼ turns

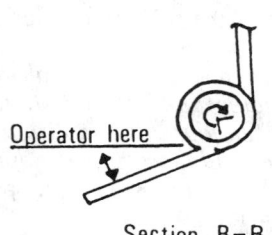

Section B–B

FIG. 1 Definitions of the Number of Rope Turns and the Angle for (a) Counterclockwise Rotation and (b) Clockwise Rotation of the Cathead

suitable for advancing a borehole in some subsurface conditions.

5.1.1 *Drag, Chopping, and Fishtail Bits*, less than 6.5 in. (162 mm) and greater than 2.2 in. (56 mm) in diameter may be used in conjuction with open-hole rotary drilling or casing-advancement drilling methods. To avoid disturbance of the underlying soil, bottom discharge bits are not permitted; only side discharge bits are permitted.

5.1.2 *Roller-Cone Bits*, less than 6.5 in. (162 mm) and greater than 2.2 in. (56 mm) in diameter may be used in conjunction with open-hole rotary drilling or casing-advancement drilling methods if the drilling fluid discharge is deflected.

5.1.3 *Hollow-Stem Continuous Flight Augers*, with or without a center bit assembly, may be used to drill the boring. The inside diameter of the hollow-stem augers shall be less than 6.5 in. (162 mm) and greater than 2.2 in. (56 mm).

5.1.4 *Solid, Continuous Flight, Bucket and Hand Augers*, less than 6.5 in. (162 mm) and greater than 2.2 in. (56 mm) in diameter may be used if the soil on the side of the boring does not cave onto the sampler or sampling rods during sampling.

5.2 *Sampling Rods*—Flush-joint steel drill rods shall be used to connect the split-barrel sampler to the drive-weight assembly. The sampling rod shall have a stiffness (moment of inertia) equal to or greater than that of parallel wall "A" rod (a steel rod which has an outside diameter of 1⅝ in. (41.2 mm) and an inside diameter of 1⅛ in. (28.5 mm).

NOTE 1—Recent research and comparative testing indicates the type rod used, with stiffness ranging from "A" size rod to "N" size rod, will usually have a negligible effect on the N-values to depths of at least 100 ft (30 m).

5.3 *Split-Barrel Sampler*—The sampler shall be constructed with the dimensions indicated in Fig. 2. The driving shoe shall be of hardened steel and shall be replaced or repaired when it becomes dented or distorted. The use of liners to produce a constant inside diameter of 1⅜ in. (35 mm) is permitted, but shall be noted on the penetration record if used. The use of a sample retainer basket is permitted, and should also be noted on the penetration record if used.

NOTE 2—Both theory and available test data suggest that N-values may increase between 10 to 30 % when liners are used.

5.4 *Drive-Weight Assembly:*

5.4.1 *Hammer and Anvil*—The hammer shall weigh 140 ± 2 lb (63.5 ± 1 kg) and shall be a solid rigid metallic mass. The hammer shall strike the anvil and make steel on steel contact when it is dropped. A hammer fall guide permitting a free fall shall be used. Hammers used with the cathead and rope method shall have an unimpeded overlift capacity of at least 4 in. (100 mm). For safety reasons, the use of a hammer assembly with an internal anvil is encouraged.

NOTE 3—It is suggested that the hammer fall guide be permanently marked to enable the operator or inspector to judge the hammer drop height.

5.4.2 *Hammer Drop System*—Rope-cathead, trip, semi-automatic, or automatic hammer drop systems may be used, providing the lifting apparatus will not cause penetration of

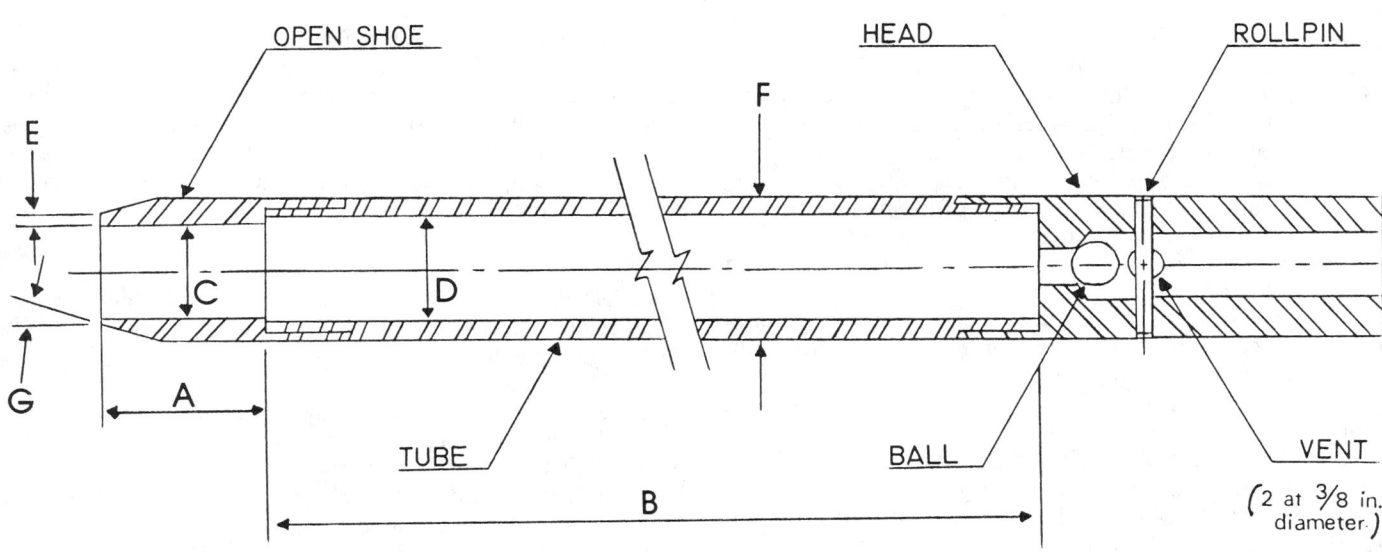

ASTM D 1586

OPEN SHOE   HEAD   ROLLPIN

E   F

C   D

G   A

TUBE   BALL   VENT

B

(2 at ⅜ in. diameter)

A = 1.0 to 2.0 in. (25 to 50 mm)
B = 18.0 to 30.0 in. (0.457 to 0.762 m)
C = 1.375 ± 0.005 in. (34.93 ± 0.13 mm)
D = 1.50 ± 0.05 − 0.00 in. (38.1 ± 1.3 − 0.0 mm)
E = 0.10 ± 0.02 in. (2.54 ± 0.25 mm)
F = 2.00 ± 0.05 − 0.00 in. (50.8 ± 1.3 − 0.0 mm)
G = 16.0° to 23.0°
   The 1½ in. (38 mm) inside diameter split barrel may be used with a 16-gage wall thickness split liner. The penetrating end of the drive shoe may be slightly rounded. Metal or plastic retainers may be used to retain soil samples.

FIG. 2 Split-Barrel Sampler

the sampler while re-engaging and lifting the hammer.

5.5 *Accessory Equipment*—Accessories such as labels, sample containers, data sheets, and groundwater level measuring devices shall be provided in accordance with the requirements of the project and other ASTM standards.

6. Drilling Procedure

6.1 The boring shall be advanced incrementally to permit intermittent or continuous sampling. Test intervals and locations are normally stipulated by the project engineer or geologist. Typically, the intervals selected are 5 ft (1.5 mm) or less in homogeneous strata with test and sampling locations at every change of strata.

6.2 Any drilling procedure that provides a suitably clean and stable hole before insertion of the sampler and assures that the penetration test is performed on essentially undisturbed soil shall be acceptable. Each of the following procedures have proven to be acceptable for some subsurface conditions. The subsurface conditions anticipated should be considered when selecting the drilling method to be used.

6.2.1 Open-hole rotary drilling method.

6.2.2 Continuous flight hollow-stem auger method.

6.2.3 Wash boring method.

6.2.4 Continuous flight solid auger method.

6.3 Several drilling methods produce unacceptable borings. The process of jetting through an open tube sampler and then sampling when the desired depth is reached shall not be permitted. The continuous flight solid auger method shall not be used for advancing the boring below a water table or below the upper confining bed of a confined non-cohesive stratum that is under artesian pressure. Casing

may not be advanced below the sampling elevation prior to sampling. Advancing a boring with bottom discharge bits is not permissible. It is not permissible to advance the boring for subsequent insertion of the sampler solely by means of previous sampling with the SPT sampler.

6.4 The drilling fluid level within the boring or hollow-stem augers shall be maintained at or above the in situ groundwater level at all times during drilling, removal of drill rods, and sampling.

7. Sampling and Testing Procedure

7.1 After the boring has been advanced to the desired sampling elevation and excessive cuttings have been removed, prepare for the test with the following sequence of operations.

7.1.1 Attach the split-barrel sampler to the sampling rods and lower into the borehole. Do not allow the sampler to drop onto the soil to be sampled.

7.1.2 Position the hammer above and attach the anvil to the top of the sampling rods. This may be done before the sampling rods and sampler are lowered into the borehole.

7.1.3 Rest the dead weight of the sampler, rods, anvil, and drive weight on the bottom of the boring and apply a seating blow. If excessive cuttings are encountered at the bottom of the boring, remove the sampler and sampling rods from the boring and remove the cuttings.

7.1.4 Mark the drill rods in three successive 6-in. (0.15-m) increments so that the advance of the sampler under the impact of the hammer can be easily observed for each 6-in. (0.15-m) increment.

7.2 Drive the sampler with blows from the 140-lb (63.5-

667

kg) hammer and count the number of blows applied in each 6-in. (0.15-m) increment until one of the following occurs:

7.2.1 A total of 50 blows have been applied during any one of the three 6-in. (0.15-m) increments described in 7.1.4.

7.2.2 A total of 100 blows have been applied.

7.2.3 There is no observed advance of the sampler during the application of 10 successive blows of the hammer.

7.2.4 The sampler is advanced the complete 18 in. (0.45 m) without the limiting blow counts occurring as described in 7.2.1, 7.2.2, or 7.2.3.

7.3 Record the number of blows required to effect each 6 in. (0.15 m) of penetration or fraction thereof. The first 6 in. is considered to be a seating drive. The sum of the number of blows required for the second and third 6 in. of penetration is termed the "standard penetration resistance," or the "N-value." If the sampler is driven less than 18 in. (0.45 m), as permitted in 7.2.1, 7.2.2, or 7.2.3, the number of blows per each complete 6-in. (0.15-m) increment and per each partial increment shall be recorded on the boring log. For partial increments, the depth of penetration shall be reported to the nearest 1 in. (25 mm), in addition to the number of blows. If the sampler advances below the bottom of the boring under the static weight of the drill rods or the weight of the drill rods plus the static weight of the hammer, this information should be noted on the boring log.

7.4 The raising and dropping of the 140-lb (63.5-kg) hammer shall be accomplished using either of the following two methods:

7.4.1 By using a trip, automatic, or semi-automatic hammer drop system which lifts the 140-lb (63.5-kg) hammer and allows it to drop 30 ± 1.0 in. (0.76 m ± 25 mm) unimpeded.

7.4.2 By using a cathead to pull a rope attached to the hammer. When the cathead and rope method is used the system and operation shall conform to the following:

7.4.2.1 The cathead shall be essentially free of rust, oil, or grease and have a diameter in the range of 6 to 10 in. (150 to 250 mm).

7.4.2.2 The cathead should be operated at a minimum speed of rotation of 100 RPM, or the approximate speed of rotation shall be reported on the boring log.

7.4.2.3 No more than 2¼ rope turns on the cathead may be used during the performance of the penetration test, as shown in Fig. 1.

NOTE 4—The operator should generally use either 1¾ or 2¼ rope turns, depending upon whether or not the rope comes off the top (1¾ turns) or the bottom (2¼ turns) of the cathead. It is generally known and accepted that 2¾ or more rope turns considerably impedes the fall of the hammer and should not be used to perform the test. The cathead rope should be maintained in a relatively dry, clean, and unfrayed condition.

7.4.2.4 For each hammer blow, a 30-in. (0.76-m) lift and drop shall be employed by the operator. The operation of pulling and throwing the rope shall be performed rhythmically without holding the rope at the top of the stroke.

7.5 Bring the sampler to the surface and open. Record the percent recovery or the length of sample recovered. Describe the soil samples recovered as to composition, color, stratification, and condition, then place one or more representative portions of the sample into sealable moisture-proof containers (jars) without ramming or distorting any apparent stratification. Seal each container to prevent evaporation of soil moisture. Affix labels to the containers bearing job designation, boring number, sample depth, and the blow count per 6-in. (0.15-m) increment. Protect the samples against extreme temperature changes. If there is a soil change within the sampler, make a jar for each stratum and note its location in the sampler barrel.

8. Report

8.1 Drilling information shall be recorded in the field and shall include the following:

8.1.1 Name and location of job,

8.1.2 Names of crew,

8.1.3 Type and make of drilling machine,

8.1.4 Weather conditions,

8.1.5 Date and time of start and finish of boring,

8.1.6 Boring number and location (station and coordinates, if available and applicable),

8.1.7 Surface elevation, if available,

8.1.8 Method of advancing and cleaning the boring,

8.1.9 Method of keeping boring open,

8.1.10 Depth of water surface and drilling depth at the time of a noted loss of drilling fluid, and time and date when reading or notation was made,

8.1.11 Location of strata changes,

8.1.12 Size of casing, depth of cased portion of boring,

8.1.13 Equipment and method of driving sampler,

8.1.14 Type sampler and length and inside diameter of barrel (note use of liners),

8.1.15 Size, type, and section length of the sampling rods, and

8.1.16 Remarks.

8.2 Data obtained for each sample shall be recorded in the field and shall include the following:

8.2.1 Sample depth and, if utilized, the sample number,

8.2.2 Description of soil,

8.2.3 Strata changes within sample,

8.2.4 Sampler penetration and recovery lengths, and

8.2.5 Number of blows per 6-in. (0.15-m) or partial increment.

9. Precision and Bias

9.1 *Precision*—A valid estimate of test precision has not been determined because it is too costly to conduct the necessary inter-laboratory (field) tests. Subcommittee D18.02 welcomes proposals to allow development of a valid precision statement.

9.2 *Bias*—Because there is no reference material for this test method, there can be no bias statement.

9.3 Variations in N-values of 100 % or more have been observed when using different standard penetration test apparatus and drillers for adjacent borings in the same soil formation. Current opinion, based on field experience, indicates that when using the same apparatus and driller, N-values in the same soil can be reproduced with a coefficient of variation of about 10 %.

9.4 The use of faulty equipment, such as an extremely massive or damaged anvil, a rusty cathead, a low speed cathead, an old, oily rope, or massive or poorly lubricated rope sheaves can significantly contribute to differences in N-values obtained between operator-drill rig systems.

 D 1586

9.5 The variability in *N*-values produced by different drill rigs and operators may be reduced by measuring that part of the hammer energy delivered into the drill rods from the sampler and adjusting *N* on the basis of comparative energies. A method for energy measurement and *N*-value adjustment is given in Test Method D 4633.

10. Keywords

10.1 blow count; in-situ test; penetration resistance; split-barrel sampling; standard penetration test

Standard Practice for
Thin-Walled Tube Geotechnical Sampling of Soils[1]

This standard is issued under the fixed designation D 1587; the number immediately following the designation indicates the year of original adoption or, in the case of revision, the year of last revision. A number in parentheses indicates the year of last reapproval. A superscript epsilon (ε) indicates an editorial change since the last revision or reapproval.

This practice has been approved for use by agencies of the Department of Defense. Consult the DoD Index of Specifications and Standards for the specific year of issue which has been adopted by the Department of Defense.

1. Scope

1.1 This practice covers a procedure for using a thin-walled metal tube to recover relatively undisturbed soil samples suitable for laboratory tests of structural properties. Thin-walled tubes used in piston, plug, or rotary-type samplers, such as the Denison or Pitcher, must comply with the portions of this practice which describe the thin-walled tubes (5.3).

NOTE 1—This practice does not apply to liners used within the above samplers.

1.2 The values stated in both inch-pound and SI units are to be regarded as the standard.

1.3 *This standard does not purport to address all of the safety concerns, if any, associated with its use. It is the responsibility of the user of this standard to establish appropriate safety and health practices and determine the applicability of regulatory limitations prior to use.*

2. Referenced Documents

2.1 *ASTM Standards:*
D 2488 Practice for Description and Identification of Soils (Visual-Manual Procedure)[2]
D 3550 Practice for Ring-Lined Barrel Sampling of Soils[2]
D 4220 Practices for Preserving and Transporting Soil Samples[2]

3. Summary of Practice

3.1 A relatively undisturbed sample is obtained by pressing a thin-walled metal tube into the in-situ soil, removing the soil-filled tube, and sealing the ends to prevent the soil from being disturbed or losing moisture.

4. Significance and Use

4.1 This practice, or Practice D 3550, is used when it is necessary to obtain a relatively undisturbed specimen suitable for laboratory tests of structural properties or other tests that might be influenced by soil disturbance.

5. Apparatus

5.1 *Drilling Equipment*—Any drilling equipment may be used that provides a reasonably clean hole; that does not disturb the soil to be sampled; and that does not hinder the penetration of the thin-walled sampler. Open borehole diameter and the inside diameter of driven casing or hollow stem auger shall not exceed 3.5 times the outside diameter of the thin-walled tube.

5.2 *Sampler Insertion Equipment,* shall be adequate to provide a relatively rapid continuous penetration force. For hard formations it may be necessary, although not recommended, to drive the thin-walled tube sampler.

5.3 *Thin-Walled Tubes,* should be manufactured as shown in Fig. 1. They should have an outside diameter of 2 to 5 in. and be made of metal having adequate strength for use in the soil and formation intended. Tubes shall be clean and free of all surface irregularities including projecting weld seams.

5.3.1 *Length of Tubes*—See Table 1 and 6.4.

5.3.2 *Tolerances,* shall be within the limits shown in Table 2.

5.3.3 *Inside Clearance Ratio,* should be 1 % or as specified by the engineer or geologist for the soil and formation to be sampled. Generally, the inside clearance ratio used should increase with the increase in plasticity of the soil being sampled. See Fig. 1 for definition of inside clearance ratio.

5.3.4 *Corrosion Protection*—Corrosion, whether from galvanic or chemical reaction, can damage or destroy both the thin-walled tube and the sample. Severity of damage is a function of time as well as interaction between the sample and the tube. Thin-walled tubes should have some form of protective coating. Tubes which will contain samples for more than 72 h shall be coated. The type of coating to be used may vary depending upon the material to be sampled. Coatings may include a light coat of lubricating oil, lacquer, epoxy, Teflon, and others. Type of coating must be specified by the engineer or geologist if storage will exceed 72 h. Plating of the tubes or alternate base metals may be specified by the engineer or geologist.

5.4 *Sampler Head,* serves to couple the thin-walled tube to the insertion equipment and, together with the thin-walled tube, comprises the thin-walled tube sampler. The sampler head shall contain a suitable check valve and a venting area to the outside equal to or greater than the area through the check valve. Attachment of the head to the tube shall be concentric and coaxial to assure uniform application of force to the tube by the sampler insertion equipment.

6. Procedure

6.1 Clean out the borehole to sampling elevation using whatever method is preferred that will ensure the material to

---

[1] This practice is under the jurisdiction of ASTM Committee D-18 on Soil and Rock and is the direct responsibility of Subcommittee D18.02 on Sampling and Related Field Testing for Soil Investigations.
Current edition approved Sept. 15, 1994. Published November 1994. Originally published as D 1587 – 58 T. Last previous edition D 1587 – 83.
[2] *Annual Book of ASTM Standards,* Vol 04.08.

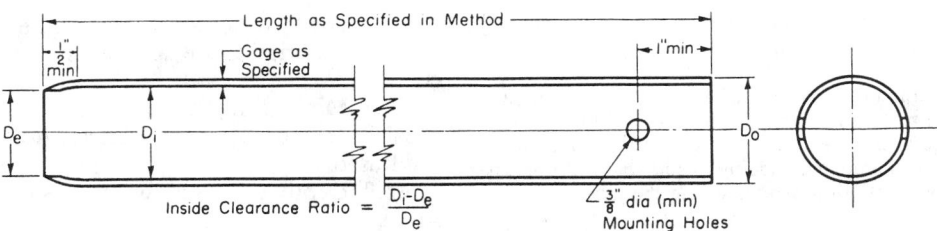

NOTE 1—Minimum of two mounting holes on opposite sides for 2 to 3½ in. sampler.
NOTE 2—Minimum of four mounting holes spaced at 90° for samplers 4 in. and larger.
NOTE 3—Tube held with hardened screws.
NOTE 4—Two-inch outside-diameter tubes are specified with an 18-gage wall thickness to comply with area ratio criteria accepted for "undisturbed samples." Users are advised that such tubing is difficult to locate and can be extremely expensive in small quantities. Sixteen-gage tubes are generally readily available.

Metric Equivalents

in.	mm
⅜	9.53
½	12.7
1	25.4
2	50.8
3½	88.9
4	101.6

FIG. 1 Thin-Walled Tube for Sampling

TABLE 1 Suitable Thin-Walled Steel Sample Tubes[A]

Outside diameter:			
in.	2	3	5
mm	50.8	76.2	127
Wall thickness:			
Bwg	18	16	11
in.	0.049	0.065	0.120
mm	1.24	1.65	3.05
Tube length:			
in.	36	36	54
m	0.91	0.91	1.45
Clearance ratio, %	1	1	1

[A] The three diameters recommended in Table 1 are indicated for purposes of standardization, and are not intended to indicate that sampling tubes of intermediate or larger diameters are not acceptable. Lengths of tubes shown are illustrative. Proper lengths to be determined as suited to field conditions.

TABLE 2 Dimensional Tolerances for Thin-Walled Tubes

	Nominal Tube Diameters from Table 1[A] Tolerances, in.		
Size Outside Diameter	2	3	5
Outside diameter	+0.007 −0.000	+0.010 −0.000	+0.015 −0.000
Inside diameter	+0.000 −0.007	+0.000 −0.010	+0.000 −0.015
Wall thickness	±0.007	±0.010	±0.015
Ovality	0.015	0.020	0.030
Straightness	0.030/ft	0.030/ft	0.030/ft

[A] Intermediate or larger diameters should be proportional. Tolerances shown are essentially standard commercial manufacturing tolerances for seamless steel mechanical tubing. Specify only two of the first three tolerances; that is, O.D. and I.D., or O.D. and Wall, or I.D. and Wall.

be sampled is not disturbed. If groundwater is encountered, maintain the liquid level in the borehole at or above ground water level during the sampling operation.

6.2 Bottom discharge bits are not permitted. Side discharge bits may be used, with caution. Jetting through an open-tube sampler to clean out the borehole to sampling elevation is not permitted. Remove loose material from the center of a casing or hollow stem auger as carefully as possible to avoid disturbance of the material to be sampled.

NOTE 2—Roller bits are available in downward-jetting and diffused-jet configurations. Downward-jetting configuration rock bits are not acceptable. Diffuse-jet configurations are generally acceptable.

6.3 Place the sample tube so that its bottom rests on the bottom of the hole. Advance the sampler without rotation by a continuous relatively rapid motion.

6.4 Determine the length of advance by the resistance and condition of the formation, but the length shall never exceed 5 to 10 diameters of the tube in sands and 10 to 15 diameters of the tube in clays.

NOTE 3—Weight of sample, laboratory handling capabilities, transportation problems, and commercial availability of tubes will generally limit maximum practical lengths to those shown in Table 1.

6.5 When the formation is too hard for push-type insertion, the tube may be driven or Practice D 3550 may be used. Other methods, as directed by the engineer or geologist, may be used. If driving methods are used, the data regarding weight and fall of the hammer and penetration achieved must be shown in the report. Additionally, that tube must be prominently labeled a "driven sample."

6.6 In no case shall a length of advance be greater than the sample-tube length minus an allowance for the sampler head and a minimum of 3 in. for sludge-end cuttings.

NOTE 4—The tube may be rotated to shear bottom of the sample after pressing is complete.

6.7 Withdraw the sampler from the formation as carefully as possible in order to minimize disturbance of the sample.

7. Preparation for Shipment

7.1 Upon removal of the tube, measure the length of sample in the tube. Remove the disturbed material in the upper end of the tube and measure the length again. Seal the upper end of the tube. Remove at least 1 in. of material from the lower end of the tube. Use this material for soil description in accordance with Practice D 2488. Measure the overall sample length. Seal the lower end of the tube. Alternatively, after measurement, the tube may be sealed without removal of soil from the ends of the tube if so directed by the engineer or geologist.

NOTE 5—Field extrusion and packaging of extruded samples under the specific direction of a geotechnical engineer or geologist is permitted.

NOTE 6—Tubes sealed over the ends as opposed to those sealed with expanding packers should contain end padding in end voids in order to prevent drainage or movement of the sample within the tube.

7.2 Prepare and immediately affix labels or apply markings as necessary to identify the sample. Assure that the markings or labels are adequate to survive transportation and storage.

8. Report

8.1 The appropriate information is required as follows:
8.1.1 Name and location of the project,
8.1.2 Boring number and precise location on project,
8.1.3 Surface elevation or reference to a datum,
8.1.4 Date and time of boring—start and finish,
8.1.5 Depth to top of sample and number of sample,
8.1.6 Description of sampler: size, type of metal, type of coating,
8.1.7 Method of sampler insertion: push or drive,
8.1.8 Method of drilling, size of hole, casing, and drilling fluid used,
8.1.9 Depth to groundwater level: date and time measured,
8.1.10 Any possible current or tidal effect on water level,
8.1.11 Soil description in accordance with Practice D 2488,
8.1.12 Length of sampler advance, and
8.1.13 Recovery: length of sample obtained.

9. Precision and Bias

9.1 This practice does not produce numerical data; therefore, a precision and bias statement is not applicable.

10. Keywords

10.1 sampling; soil exploration; undisturbed

Designation: D 3550 – 84 (Reapproved 1995)[ε1]

Standard Practice for
Ring-Lined Barrel Sampling of Soils[1]

This standard is issued under the fixed designation D 3550; the number immediately following the designation indicates the year of original adoption or, in the case of revision, the year of last revision. A number in parentheses indicates the year of last reapproval. A superscript epsilon (ε) indicates an editorial change since the last revision or reapproval.

This standard has been approved for use by agencies of the Department of Defense. Consult the DoD Index of Specifications and Standards for the specific year of issue which has been adopted by the Department of Defense.

ε1 NOTE—Editorial changes were made in March 1995.

1. Scope

1.1 This practice covers a procedure for using a ring-lined barrel sampler to obtain representative samples of soil for identification purposes and other laboratory tests. In cases where it has been established that the quality of the sample is adequate, this practice provides shear and consolidation specimens that can be used directly in the test apparatus without prior trimming. Some types of soils may gain or lose significant shear strength or compressibility, or both, as a result of sampling. In cases like these, suitable comparison tests should be made to evaluate the effect of sample disturbance on shear strength and compressibility.

1.2 This practice is not intended to be used as a penetration test; however, the force required to achieve penetration or a blow count, when driving is necessary, is recommended as supplemental information.

1.3 *This standard does not purport to address all of the safety concerns, if any, associated with its use. It is the responsibility of the user of this standard to establish appropriate safety and health practices and determine the applicability of regulatory limitations prior to use.*

2. Referenced Documents

2.1 *ASTM Standards:*
D 1586 Test Method for Penetration Test and Split-Barrel Sampling of Soils[2]
D 1587 Practice for Thin-Walled Tube Sampling of Soils[2]
D 2113 Practice for Diamond Core Drilling for Site Investigation[2]
D 2488 Practice for Description and Identification of Soils (Visual-Manual Procedure)[2]

3. Significance and Use

3.1 This practice is used where soil condition and resistance to advance of the sampler do not permit the use of a thin-wall tube (Practice D 1587) and where the formation does not require diamond coring (Practice D 2113).

4. Apparatus

4.1 *Drilling Equipment*—Any drilling equipment may be used that provides a reasonably clean hole before insertion of the sampler and that does not disturb the soil to be sampled. However, in no case shall a bottom-discharge bit be permitted. Side-discharge bits are permissible.

4.2 *Drive Weight Assembly*—Any drive weight assembly that will provide penetration in the range from 1 to 20 blows per foot (0.30-m) may be used. Whenever possible, soils are to be sampled by pushing instead of driving (see Section 5).

4.3 *Ring-Lined Barrel Sampling Assembly*—This shall consist of a shoe, sampler, and waste barrel, as shown in Fig. 1.

4.4 *Ring-Lined Sampler*—Test specimens shall be obtained using a suitable one piece or split sampling barrel lined on the inside with removable rings. These rings shall be thin-walled and shall conform to the size requirements of the particular laboratory test determinations employed. They shall fit snugly inside the sampler with no discernible free play in any direction. The sampler may be sectionalized to allow end-to-end make-up of sections as necessary. Each section shall be designed so that addition or removal of sections will not loosen, permit movement, or otherwise adversely affect retention of the rings within the sampler. The sampler and rings shall be free of bumps, dents, scratches, rust, dirt, and corrosion.

NOTE 1—It is recommended that the sampler contain at least six rings in order to provide samples for a variety of tests.

4.5 *Waste Barrel*—A waste barrel that can be removed from the sampler in the field shall be provided to contain space for disturbed soil originally at the bottom of the hole. The length of the waste barrel shall be at least three times its interior diameter, and the inside diameter shall be the same, or slightly larger than, the inside diameter of the rings.

4.5.1 An attachment, check valve, and one or more vents is required. The design of these items is optional.

4.6 *Shoe*—The shoe shall be machined as shown in Fig. 1. The inside of the assembled shoe and ring-lined sampler shall be smooth, straight, and uniform. The thin-walled extension of the shoe shall be 2 to 4 in. (51 to 102 mm) in outside diameter and made of any materials of adequate strength and resistance to corrosion. The length of the thin-walled extension shall be equal to three times the diameter of its opening, but shall not exceed 8 in. (203 mm). The inside clearance ratio shall be between 0.5 and 3.0 %. (See Fig. 1 for

---

[1] This practice is under the jurisdiction of ASTM Committee D-18 on Soil and Rock and is the direct responsibility of Subcommittee D18.02 on Sampling and Related Field Testing for Soil Investigations.

Current edition approved Jan. 27, 1984. Published April 1984. Originally published as D 3550 – 77. Last previous edition D 3550 – 77ε1.

[2] *Annual Book of ASTM Standards,* Vol 04.08.

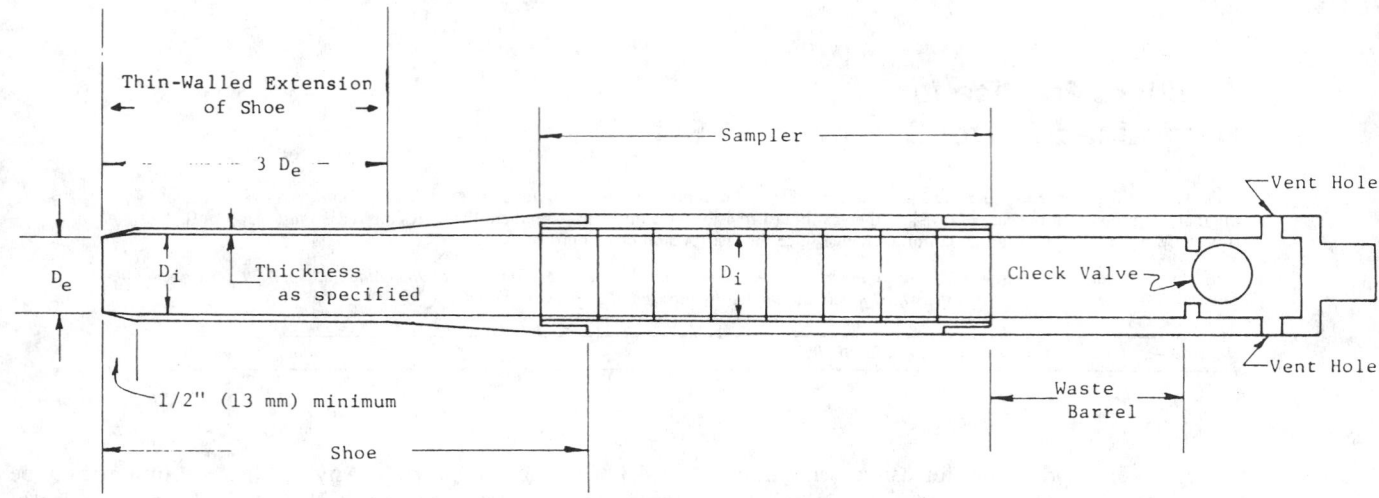

NOTE 1—Inside clearance ratio = $(D_i - D_e)/D_e$
NOTE 2—Dimensional tolerance of D_i = ±0.003 in. (±0.08 mm)

FIG. 1 Ring-Lined Barrel Sampling Assembly

inside clearance ratio formula.) The wall thickness of the thin-walled extension shall conform to Table 1.

4.6.1 The thin-walled extension of the shoe shall be perfectly round. Shoes that have become out-of-round for any reason shall not be used. If the thin-walled extension of the shoe deforms during sampling, the sample obtained shall not be used for tests, such as shear strength, where soil disturbance is a factor.

NOTE 2—The thin-walled extension of the shoe is not suitable for stiff or gravelly soils. In cases such as these, a shoe similar to the type specified in Method D 1586 is required for penetration. The use of this type of shoe, however, may result in excessive disturbance of the soil so that it is no longer suitable for shear or consolidation determinations, or both.

4.7 *Sample Extractor*—Specimen-filled rings shall be removed from the sampler by pressing them out or alternatively by the use of a split barrel. The extractor disk shall be at least 0.5 in. (13 mm) thick and shall bear solidly against the sample rings at all points. It shall slide easily inside the sampler barrel without jamming and without free play.

4.8 *Containers for Specimen-Filled Rings*—These shall be snug fitting, tightly sealed (watertight), rigid containers that will not permit movement of the specimen-filled rings inside. They shall be noncorrosive.

4.9 *Miscellaneous Equipment*—This includes a pipe vise, pipe wrenches, spatulas, cleaning brushes, buckets, rags, data sheets, transporting boxes, etc. Water must be available for cleaning the equipment.

5. Procedure

5.1 Clean the hole to sampling elevation using whatever method is preferred that will ensure that the material to be sampled is not disturbed. In saturated sands and silts, withdraw the drill bit slowly to prevent loosening of the soil around the hole. When casing is used, it shall not be driven below sampling elevation. Water or drilling liquid within the boring must be maintained at all times at or above the natural ground water level; it is preferable to keep the hole filled.

5.2 Keep a careful record of drill penetration and sampler depth to ensure that the soil being sampled is the original soil at the bottom of the hole and is not contaminated by soil falling down from the sides of the hole. If there is any significant tendency for soil to fall from the sides of the hole to the bottom, use water, drilling mud, or casing, as necessary, in order to prevent this from happening. The process of jetting through an open-tube sampler and then sampling when the desired depth is reached shall not be permitted. The use of bottom-discharge bits shall not be allowed.

5.3 Assemble the sampling assembly and lower it carefully into the hole. With the cutting edge of the shoe resting on the bottom of the hole and the water level in the boring at the ground water level or above, push the sampling assembly into the soil by a continuous and rapid motion without impact or twisting. Push the assembly in far enough so that all cuttings, sludge, and soil disturbed by drilling are in the waste barrel; however, in no case push the assembly farther than the total length of the shoe, sampler, and waste barrel. Take care that none of the sample is lost due to improper operation of the check valve.

5.4 When the soils are so hard that they cannot be penetrated by pushing, using generally acceptable field procedures, and where recovery by pushing in sands is poor, use a driving hammer to drive the sampling assembly. In such a case, record the hammer weight, height of drop, and number of blows.

5.5 Carefully disassemble the sampling assembly in such a manner as to minimize soil disturbance as much as possible. Trim the soil flush with the ends of the sampling barrel, and remove the specimen (consisting of soil plus rings). Slip the container over the specimen-filled rings and cap both ends. Be certain that there is no movement of the specimen-filled rings inside the container and that the specimen was not disturbed while being removed from the barrel and placed in the container. Label the container in a suitable manner. If the soil in the bottom end ring does not protrude from the ring after removing the shoe, do not use the soil in the bottom ring for tests other than soil classification and moisture content. If the top ring or rings contain voids, depressions, or any material other than the soil which is being sampled, do not use the soil in this ring (or rings) for any purpose whatsoever. The filling of depressions in the end rings with additional soil shall not be permitted. Discard samples that appear to be disturbed or questionable.

5.6 Examine the soil remaining in the shoe for structure, consistency, color, and condition. Record these observations and include them in the report (see 6.1.8).

NOTE 3—The soil remaining in the shoe is relatively undisturbed and therefore may be suitable for a variety of laboratory tests.

6. Report

6.1 Data obtained in each boring shall be recorded in the field and shall contain the following:
6.1.1 Name and location of job,
6.1.2 Date of boring and times of start and finish,
6.1.3 Boring number and location,
6.1.4 Surface elevation, if available,
6.1.5 Sample number and depth,
6.1.6 Method of advancing sampler, penetration, and recovery lengths,
6.1.7 Description and size of sampler,
6.1.8 Description of soil (see Practice D 2488),
6.1.9 Thickness of layer,
6.1.10 Depth to water table or depth of overlying water and time of reading,
6.1.11 Size of casing, depth of cased hole,
6.1.12 Type of drilling equipment—description,
6.1.13 Names of personnel: crewman, field engineer, technician, etc.,
6.1.14 Weather conditions, and
6.1.15 General remarks.

7. Precision and Bias

7.1 This practice does not produce numerical or repeatable data and therefore a precision and bias statement is not applicable.

8. Keywords

8.1 consolidation; direct shear; identification; liner; representative; ring; sampling

Standard Practices for
Preserving and Transporting Soil Samples[1]

This standard is issued under the fixed designation D 4220; the number immediately following the designation indicates the year of original adoption or, in the case of revision, the year of last revision. A number in parentheses indicates the year of last reapproval. A superscript epsilon (ε) indicates an editorial change since the last revision or reapproval.

1. Scope*

1.1 These practices cover procedures for preserving soil samples immediately after they are obtained in the field and accompanying procedures for transporting and handling the samples.

1.2 *Limitations*—These practices are not intended to address requirements applicable to transporting of soil samples known or suspected to contain hazardous materials.

1.3 *This standard does not purport to address all of the safety concerns, if any, associated with its use. It is the responsibility of the user of this standard to establish appropriate safety and health practices and determine the applicability of regulatory limitations prior to use.* See Section 7.

2. Referenced Documents

2.1 *ASTM Standards:*

D 420 Guide to Site Characterization for Engineering, Design, and Construction Purposes[2]

D 653 Terminology Relating to Soil, Rock, and Contained Fluids[2]

D 1452 Practice for Soil Investigation and Sampling by Auger Borings[2]

D 1586 Test Method for Penetration Test and Split-Barrel Sampling of Soils[2]

D 1587 Practice for Thin-Walled Tube Sampling of Soils[2]

D 2488 Practice for Description and Identification of Soils (Visual-Manual Procedure)[2]

D 3550 Practice for Ring-Lined Barrel Sampling of Soils[2]

D 4564 Test Method for Density of Soil in Place by the Sleeve Method[2]

D 4700 Guide for Soil Sampling from the Vadose Zone[2]

3. Terminology

3.1 Terminology in these practices is in accordance with Terminology D 653.

4. Summary of Practices

4.1 The various procedures are given under four groupings as follows:

4.1.1 *Group A*—Samples for which only general visual identification is necessary.

4.1.2 *Group B*—Samples for which only water content and classification tests, proctor and relative density, or profile logging is required, and bulk samples that will be remolded or compacted into specimens for swell pressure, percent swell, consolidation, permeability, shear testing, CBR, stabilimeter, etc.

4.1.3 *Group C*—Intact, naturally formed or field fabricated, samples for density determinations; or for swell pressure, percent swell, consolidation, permeability testing and shear testing with or without stress-strain and volume change measurements, to include dynamic and cyclic testing.

4.1.4 *Group D*—Samples that are fragile or highly sensitive for which tests in Group C are required.

4.2 The procedure(s) to be used should be included in the project specifications or defined by the designated responsible person.

5. Significance and Use

5.1 Use of the various procedures recommended in these practices is dependent on the type of samples obtained (Practice D 420), the type of testing and engineering properties required, the fragility and sensitivity of the soil, and the climatic conditions. In all cases, the primary purpose is to preserve the desired inherent conditions.

5.2 The procedures presented in these practices were primarily developed for soil samples that are to be tested for engineering properties, however, they may be applicable for samples of soil and other materials obtained for other purposes.

6. Apparatus

6.1 The type of materials and containers needed depend upon the conditions and requirements listed under the four groupings A to D in Section 4, and also on the climate and transporting mode and distance.

6.1.1 *Sealing Wax*, includes microcrystalline wax, paraffin, beeswax, ceresine, carnaubawax, or combinations thereof.

6.1.2 *Metal Disks*, about 1/16 in. (about 2 mm) thick and having a diameter slightly less than the inside diameter of the tube, liner, or ring and to be used in union with wax or caps and tape, or both.

6.1.3 *Wood Disks*, prewaxed, 1 in. (25 mm) thick and having a diameter slightly less than the inside diameter of the liner or tube.

6.1.4 *Tape*, either waterproof plastic, adhesive friction, or duct tape.

6.1.5 *Cheesecloth*, to be used in union with wax in alternative layers.

6.1.6 *Caps*, either plastic, rubber or metal, to be placed

---

[1] These practices are under the jurisdiction of ASTM Committee D-18 on Soil and Rock and are the direct responsibility of Subcommittee D18.02 on Sampling and Related Field Testing for Soil Investigations.

Current edition approved April 15, 1995. Published June 1995. Originally published as D 4220 – 83. Last previous edition D 4220 – 89.

[2] *Annual Book of ASTM Standards*, Vol 04.08.

* A Summary of Changes section appears at the end of these practices.

over the end of thin-walled tubes (Practice D 1587), liners and rings (Practice D 3550), in union with tape or wax.

6.1.7 *O'ring (Sealing End Caps)*, used to seal the ends of samples within thin-walled tubes, by mechanically expanding an O'ring against the tube wall.

NOTE 1—Plastic expandable end caps are preferred. Metal expandable end caps seal equally well; however, long-term storage may cause corrosion problems.

6.1.8 *Jars*, wide mouthed, with rubber-ringed lids or lids lined with a coated paper seal and of a size to comfortably receive the sample, commonly ½ pt (250 mL), 1 pt (500 mL) and quart-sized (1000 mL).

6.1.9 *Bag*, either plastic, burlap with liner, burlap or cloth type (Practice D 1452).

6.1.10 *Packing Material*, to protect against vibration and shock.

6.1.11 *Insulation*, either granule (bead), sheet or foam type, to resist temperature change of soil or to prevent freezing.

6.1.12 *Sample Cube Boxes*, for transporting cube (block) samples. Constructed with ½ to ¾ in. (13 to 19 mm) thick plywood (marine type).

6.1.13 *Cylindrical Sample Containers*, somewhat larger in dimension than the thin-walled tube or liner samples, such as cylindrical frozen food cartons.

6.1.14 *Shipping Containers*, either box or cylindrical type and of proper construction to protect against vibration, shock, and the elements, to the degree required.

NOTE 2—The length, girth and weight restrictions for commercial transportation must be considered.

6.1.15 *Identification Material*—This includes the necessary writing pens, tags, and labels to properly identify the sample(s).

7. Precautions

7.1 Special instructions, descriptions, and marking of containers must accompany any sample that may include radioactive, chemical, toxic, or other contaminant material.

7.2 Interstate transportation containment, storage, and disposal of soil samples obtained from certain areas within the United States and the transportation of foreign soils into or through the United States are subject to regulations established by the U.S. Department of Agriculture, Animal, and Plant Health Service, Plant Protection and Quarantine Programs, and possibly to regulations of other federal, state, or local agencies.

7.2.1 Samples shipped by way of common carrier or U.S. Postal Service must comply with the Department of Transportation Hazardous Materials Regulation, 49CRF Part 172.

7.3 Sample traceability records (see Fig. 1) are encouraged and should be required for suspected contaminated samples.

7.3.1 The possession of all samples must be traceable, from collection to shipment to laboratory to disposition, and should be handled by as few persons as possible.

7.3.2 The sample collector(s) should be responsible for initiating the sample traceability record; recording the project, sample identification and location, sample type, date, and the number and types of containers.

7.3.3 A separate traceability record shall accompany each shipment.

7.3.4 When transferring the possession of samples the person(s) relinquishing and receiving the samples shall sign, date, record the time, and check for completeness of the traceability record.

8. Procedure

8.1 *All Samples*—Properly identify samples with tags, labels, and markings prior to transporting them as follows:
8.1.1 Job name or number, or both,
8.1.2 Sampling date,
8.1.3 Sample/boring number and location,
8.1.4 Depth or elevation, or both,
8.1.5 Sample orientation,
8.1.6 Special shipping or laboratory handling instructions, or both, including sampling orientation, and
8.1.7 Penetration test data, if applicable (Test Method D 1586).
8.1.8 Subdivided samples must be identified while maintaining association to the original sample.
8.1.9 If required, sample traceability record.

8.2 *Group A*—Transport samples in any type of container by way of available transportation. If transported commercially, the container need only meet the minimum requirements of the transporting agency and any other requirements necessary to assure against sample loss.

8.3 *Group B:*

8.3.1 Preserve and transport these samples in sealed, moistureproof containers. All containers shall be of sufficient thickness and strength to ensure against breakage and moisture loss. The container types include: plastic bags or pails, glass or plastic (provided they are waterproof) jars, thin walled tubes, liners, and rings. Wrap cylindrical and cube samples in suitable plastic film or aluminum foil, or both, (Note 3) and coat with several layers of wax, or seal in several layers of cheesecloth and wax.

8.3.2 Transport these samples by any available transportation. Ship these samples as prepared or placed in larger shipping containers, including bags, cardboard, or wooden boxes or barrels.

NOTE 3—Some soils may cause holes to develop in aluminum foil, due to corrosion. Avoid direct contact where adverse affects to sample composition are a concern.

8.3.3 *Plastic Bags*—Place the plastic bags as tightly as possible around the sample, squeezing out as much air as possible. They shall be 3 mil or thicker to prevent leakage.

8.3.4 *Glass-Plastic Jars*—If the jar lids are not rubber ringed or lined with new waxed paper seals, seal the lids with wax.

8.3.5 *Plastic Pails*—If the plastic pail lids are not air tight, seal them with wax or tape.

8.3.6 *Thin-Walled Tubes:*

8.3.6.1 *Expandable Packers*—The preferred method of sealing sample ends within tubes is with plastic, expandable packers.

8.3.6.2 *Wax With Disks*—For short-term sealing, paraffin wax is acceptable. For long term sealing (in excess of 3 days) use microcrystalline waxes or combine with up to 15 % beeswax or resin, for better adherence to the wall of the tube and to reduce shrinkage. Several thin layers of wax are preferred over one thick layer. The minimum final thickness shall be 0.4 in. (10 mm).

8.3.6.3 *End Caps*—Seal metal, rubber, or plastic end caps with tape. For long term storage (longer than 3 days), also dip them in wax, applying two or more layers of wax.

8.3.6.4 *Cheesecloth and Wax*—Use alternating layers (a minimum of two each) of cheesecloth and wax to seal each end of the tube and stabilize the sample.

NOTE 4—Where necessary, spacers or appropriate packing materials, or both, must be placed prior to sealing the tube ends to provide proper confinement. Packing material must be nonabsorbent and must maintain its properties to provide the same degree of continued sample support.

8.3.7 *Liners and Rings*—Refer to 8.3.6.3 or 8.3.6.4.

8.3.8 *Exposed Samples:*

8.3.8.1 *Cylindrical, Cubical or Other Samples Wrapped in Plastic*, such as polyethylene and polypropylene, or foil should be further protected with a minimum of three coats of wax.

8.3.8.2 *Cylindrical and Cube Samples Wrapped in Cheesecloth and Wax*, shall be sealed with a minimum of three layers of each, placed alternatively.

8.3.8.3 *Carton Samples (Frozen Food Cartons)*—Samples placed in these containers must be situated so that wax can be poured completely around the sample. The wax should fill the void between the sample and container wall. The wax should be sufficiently warm to flow, but not so hot that it penetrates the pores of the soil. Generally, the samples should be wrapped in plastic or foil before being surrounded with wax.

8.4 *Group C:*

8.4.1 Preserve and seal these samples in containers as covered in 8.3. In addition, they must be protected against vibration and shock, and protected from extreme heat or cold.

8.4.2 Samples transported by the sampling or testing agency personnel on seats of automobiles and trucks need only be placed in cardboard boxes, or similar containers into which the sealed samples fit snugly, preventing bumping, rolling, dropping, etc.

8.4.3 For all other methods of transporting samples, including automobile trunk, bus, parcel services, truck, boat, air, etc., place the sealed samples in wood, metal, or other type of suitable shipping containers that provide cushioning or insulation, or both, for each sample and container. Avoid transporting by any agency whose handling of containers is suspect.

8.4.4 The cushioning material (sawdust, rubber, polystyrene, urethane foam, or material with similar resiliency) should completely encase each sample. The cushioning between the samples and walls of the shipping containers should have a minimum thickness of 1 in. (25 mm). A minimum thickness of 2 in. (50 mm) shall be provided on the container floor.

8.4.5 When required, the samples should be shipped in the same orientation in which they were sampled. Otherwise, special conditions shall be provided such as freezing, controlled drainage, or sufficient confinement, or a combination thereof, to maintain sample integrity.

8.5 *Group D:*

8.5.1 The requirements of 8.4 must be met, in addition to the following:

8.5.1.1 Samples should be handled in the same orientation in which they were sampled, including during transportation or shipping, with appropriate markings on the shipping container.

8.5.1.2 For all modes of private or commercial transportation, the loading, transporting and unloading of the shipment containers should be supervised as much as possible by a qualified person.

NOTE 5—A qualified person may be an engineer, geologist, soil scientist, soils technician or responsible person designated by the project manager.

8.6 *Shipping Containers* (see Figs. 2 to 7 for typical containers):

8.6.1 The following features should be included in the design of the shipping container for Groups C and D.

8.6.1.1 It should be reuseable,

8.6.1.2 It should be constructed so that the samples can be maintained, at all times, in the same position as when sampled or packed, or both,

8.6.1.3 It should include sufficient packing material to cushion or isolate, or both, the tubes from the adverse effect of vibration and shock, and

8.6.1.4 It should include sufficient insulating material to prevent freezing, sublimation and thawing, or undesirable temperature changes.

8.6.2 *Wood Shipping Containers:*

8.6.2.1 Wood is preferred over metal. Outdoor (marine) plywood having a thickness of ½ and ¾ in. (13 to 19 mm) may be used. The top (cover) should be hinged and latched, or fastened with screws.

8.6.2.2 The cushioning requirements are given in 8.4.4.

8.6.2.3 For protection against freezing or extreme temperature variation, the entire shipping container should be lined with a minimum insulation thickness of 2 in. (50 mm).

8.6.3 *Metal Shipping Containers*—The metal shipping containers must incorporate cushioning and insulation material to minimum thicknesses in accordance with 8.6.2, although slightly greater thicknesses would be appropriate. Alternatively, the cushion effect could be achieved with a spring suspension system, or any other means that would provide similar protection.

8.6.4 *Styrene Shipping Containers*—Bulk styrene with slots cut to the dimensions of the sample tube or liner. A protective outer box of plywood or reinforced cardboard is recommended.

8.6.5 *Other Containers*—Containers constructed with laminated fiberboard, plastic or reinforced cardboard outer walls, and properly lined, may also be used.

9. Reporting

9.1 The data obtained in the field shall be recorded and should include the following:

9.1.1 Job name or number, or both,

9.1.2 Sampling date(s),

9.1.3 Sample/boring number(s) and location(s),

9.1.4 Depth(s) or elevation(s), or both,

9.1.5 Sample orientation,

Sample Identification/Traceability Record
(Controlled Document)

Project: _____ W.O. ⌀ _____

Shipped by: _____

Shipped to: _____ Attention of:

Comments: _____ Hazardous materials suspected?
_____ (yes/no)

Sampling Point	Location	Field ID ⌀	Date	Sample Type	No. of Containers	Analysis/Test Required	(optional) Lab ID

Sampler(s) (signature) _____

Field ID	Relinquished by: (signature)	Date/Time	Received by: (signature)	Date/Time	Comments

Shipment prepared by: (signature) _____ Date/Time _____ Shipment method: _____

Received for Lab by: (signature) _____ Date/Time _____ Comments _____

Receiving Laboratory: Please return original form after signing for receipt of samples.

FIG. 7 Example Layout of Sample Traceability Form

FIG. 1   Example Layout of Record Form

9.1.6 Groundwater observation, if any,

9.1.7 Method of sampling, and penetration test data, if applicable,

9.1.8 Sample dimensions,

9.1.9 Soil description (Practice D 2488),

9.1.10 Names of technician/crewman, engineer, project chief, etc.,

9.1.11 Comments regarding contaminated or possible contaminated samples,

9.1.12 If used, a copy of traceability records,

9.1.13 Weather conditions, and

9.1.14 General remarks.

10. Precision and Bias

10.1 This practice provides qualitative and general information only. Therefore, a precision and bias statement is not applicable.

11. Keywords

11.1 preservation; soil samples; transportation

SUMMARY OF CHANGES

This section identifies the location of changes to these practices that have been incorporated since the last issue. Committee D-18 has highlighted those changes that affect the technical interpretation or use of these practices.

(1) Section 11 was added since the last revision.

(2) Section 2 was expanded since the last revision.

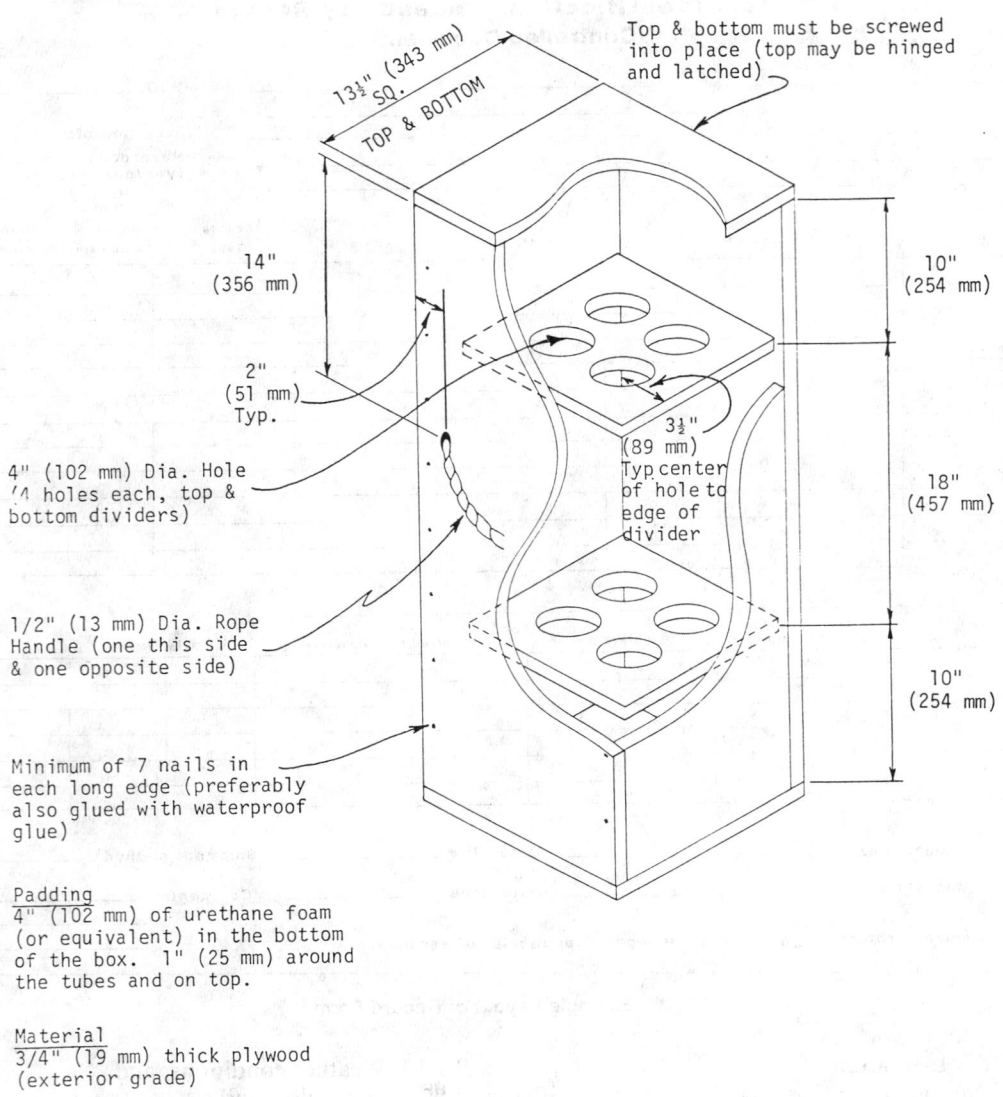

Top & bottom must be screwed into place (top may be hinged and latched)

13½" (343 mm) SQ.
TOP & BOTTOM

14" (356 mm)

2" (51 mm) Typ.

10" (254 mm)

3½" (89 mm) Typ center of hole to edge of divider

18" (457 mm)

10" (254 mm)

4" (102 mm) Dia. Hole (4 holes each, top & bottom dividers)

1/2" (13 mm) Dia. Rope Handle (one this side & one opposite side)

Minimum of 7 nails in each long edge (preferably also glued with waterproof glue)

Padding
4" (102 mm) of urethane foam (or equivalent) in the bottom of the box. 1" (25 mm) around the tubes and on top.

Material
3/4" (19 mm) thick plywood (exterior grade)

FIG. 2   Shipping Box for 3-in. (76-mm) Thin-Walled Tubes

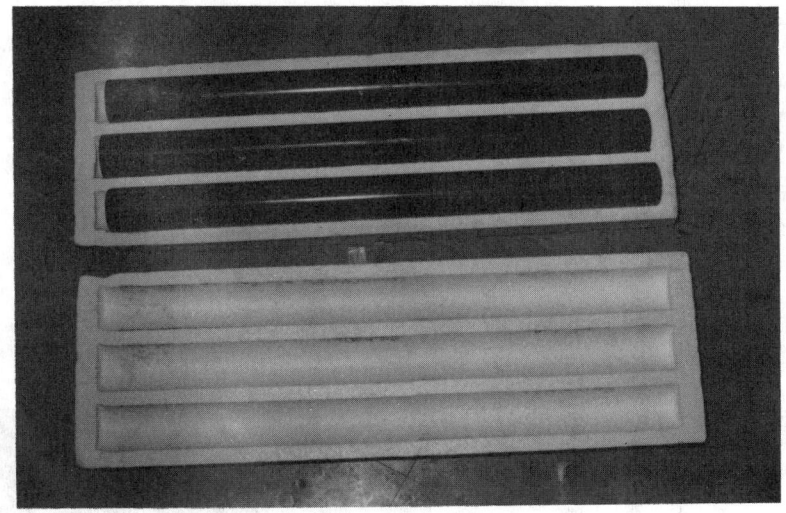

(a) Photo of Open Box For 5'' (127 mm) Tubes

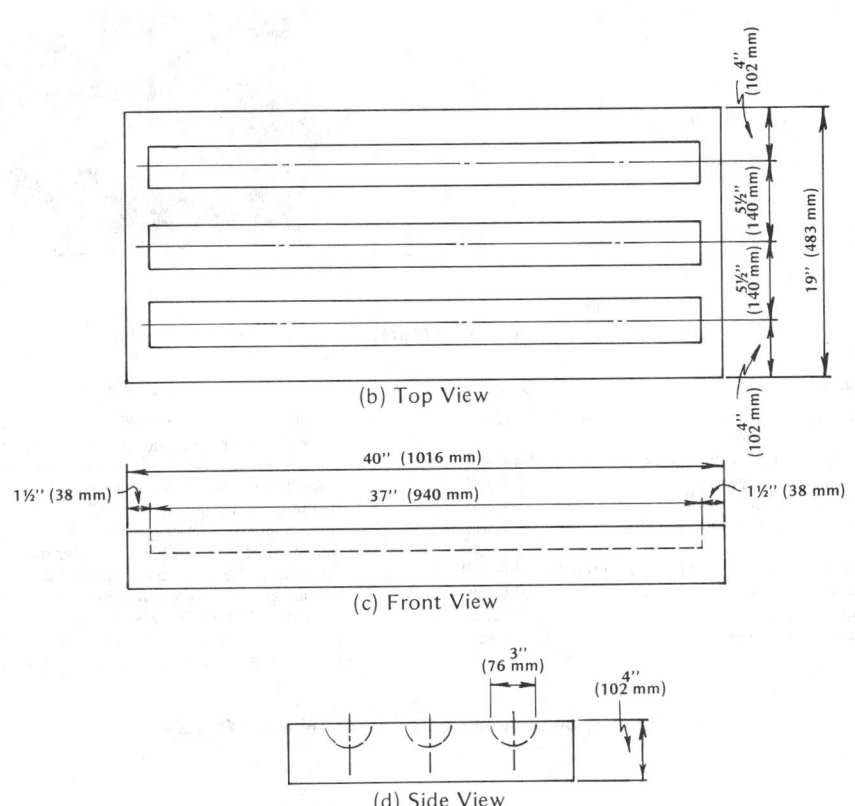

(b) Top View

(c) Front View

(d) Side View

NOTE—Top and bottom halves are identical.

FIG. 3 Styrene Shipping Container for 3-in. (76-mm) Thin-Walled Tubes

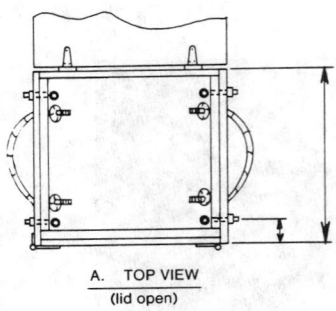

A. TOP VIEW
(lid open)

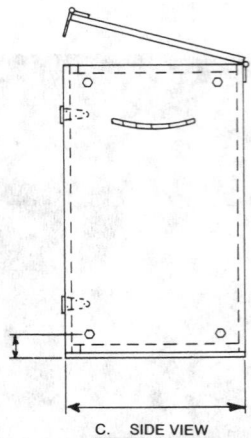

C. SIDE VIEW

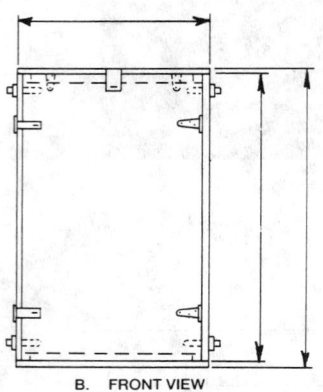

B. FRONT VIEW

D. PHOTOGRAPH OF OPEN BOX

BILL OF MATERIALS

Item No.	Description of Item	Quantity
1	*Plywood*, 4 ft by 8 ft by ¾ in. (1220 mm by 2440 mm by 19.1 mm) exterior, Grade AC	1 Sheet
2	*Hinge*, strap, 4 in. (102 mm), heavy duty with screws	4 Each
3	*Hasp*, hinged, 4½ in. (114 mm), with screws	3 Each
4	*Screw, Wood, Steel, Flathead*, No. 10 by 1¾ in. (44.5 mm)	72 Each
5	*Bolt, Machine*, ⅜ in. (9.5 mm), with nut to secure hasps	3 Each
6	*Washer*, flat, ⅜ in. (9.5 mm)	3 Each
7	*Eye Bolt*, ½ by 2 in. (6.4 mm by 51 mm), zinc-plated, with nut	8 Each
8	*Washer*, flat, ¼ in. (6.4 mm), for hasp bolt	8 Each
9	*S Hooks*, 2 in. (51 mm), open, zinc-plated	8 Each
10	*Clamp*, adjustable, hose, steel, worm screw adjustment	2 Each
11	*Spring*, expansion	8 Each
12	*Adhesive*, woodworking	1 lb (454 g)

Item No.	Description of Item	Quantity
13	*Rope*, nylon, ½-in. (12.7-mm) diameter, solid braided	5 ft (1524 mm)
14	*Cushioning Material*, expanded polystyrene foam	10 ft³ (0.28 m³)

NOTES–(a) All wooden components can be sawed from one sheet of plywood.

(b) This shipping box will accommodate approximately three 3-in. (76-mm) diameter tubes or two 5-in (127-mm) diameter tubes up to 30 in. (762 mm) in length. For longer tubes the inside height of the box must be a minimum of 6-in. (152 mm) greater than the length of the tube.

(c) All joints to be glued and fastened with screws.

(d) Stencil all sides as follows (See Views B and C).
TO PROTECT FROM FREEZING

(e) After suspending samples as indicated above, all void space must be filled with a suitable resilient packing material.

FIG. 4 Suspension System Container for Thin-Walled Tubes

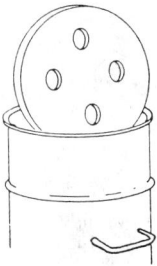 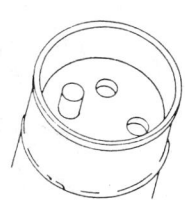

(a) 55-gallon (0.21 m^3) oil barrels
with sections of styrofoam insulation; welded handles on each side.

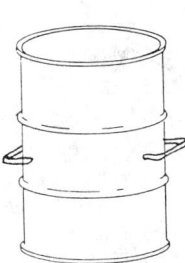 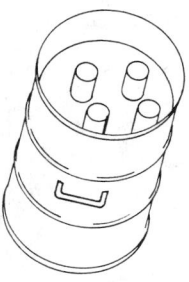

(b) Same as (a) showing barrel ready
for shipment. Steel lids bolted
on to provide tight seal.

NOTE—Two in. (51 mm) of foam rubber covers 2 in. of styrofoam at the base.
One in. (25 mm) of foam rubber overlays the top of the tubes, and the remaining
space to the lid is filled with styrofoam.

FIG. 5   Shipping Barrel for Thin-Walled Tubes

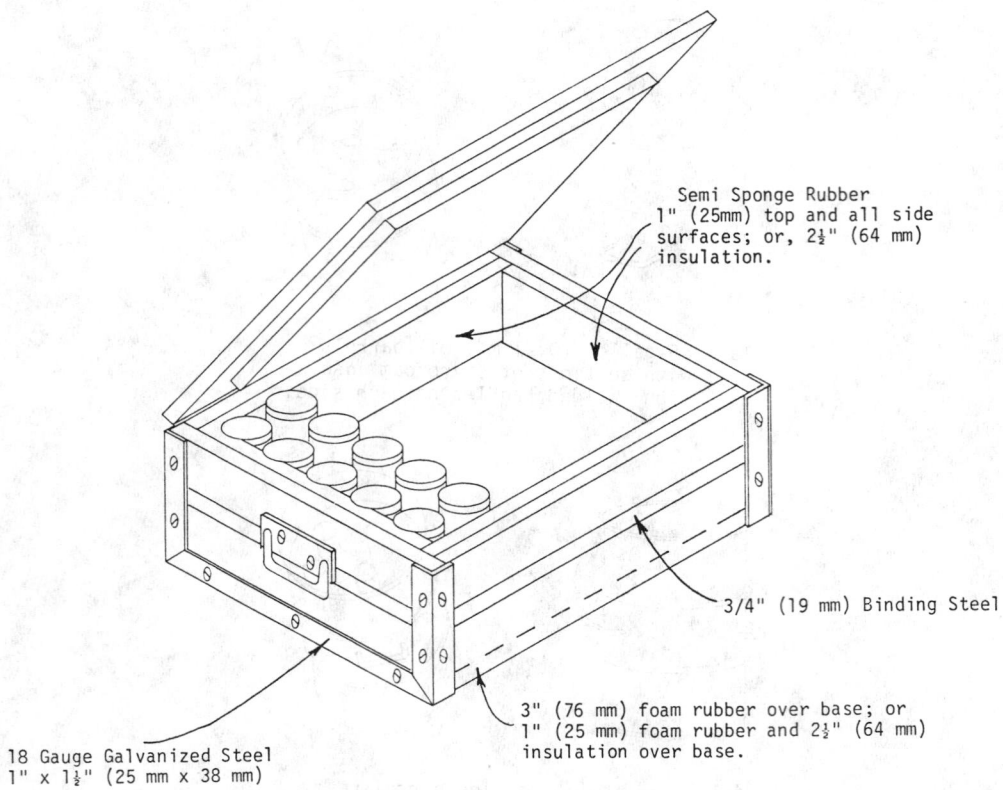

Semi Sponge Rubber 1" (25mm) top and all side surfaces; or, 2½" (64 mm) insulation.

3/4" (19 mm) Binding Steel

3" (76 mm) foam rubber over base; or 1" (25 mm) foam rubber and 2½" (64 mm) insulation over base.

18 Gauge Galvanized Steel 1" x 1½" (25 mm x 38 mm)

FIG. 6   Shipping Box for Liner (Short Tube) or Ring Samples

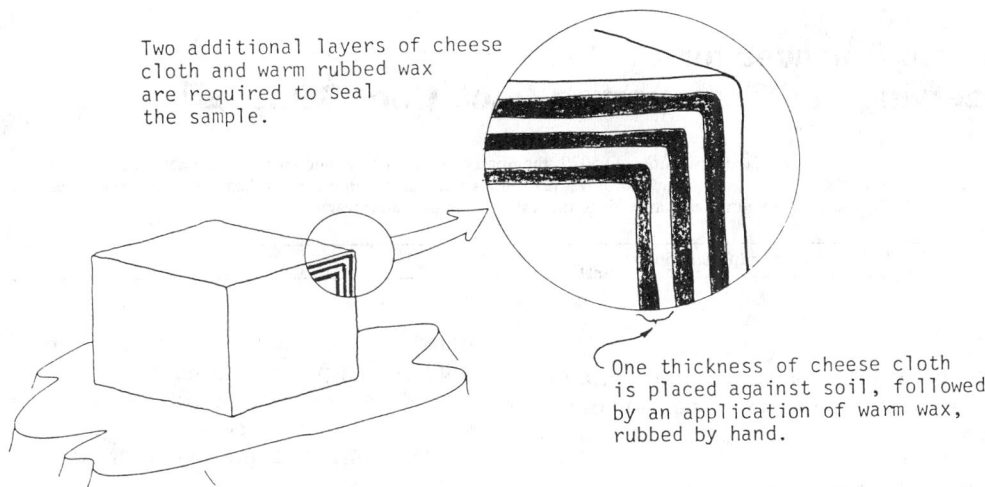

Two additional layers of cheese cloth and warm rubbed wax are required to seal the sample.

One thickness of cheese cloth is placed against soil, followed by an application of warm wax, rubbed by hand.

A. METHOD FOR SEALING HAND-CUT UNDISTURBED SAMPLES

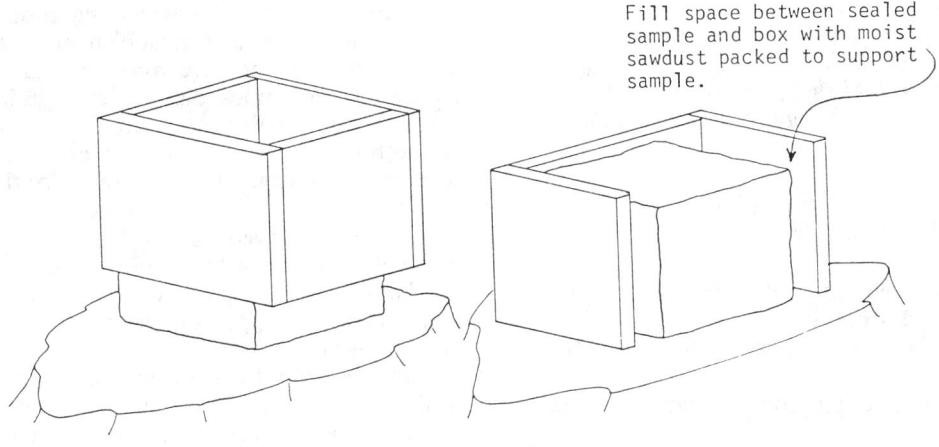

Fill space between sealed sample and box with moist sawdust packed to support sample.

B. ENCASE EASILY DISTURBED SAMPLES IN BOX PRIOR TO CUTTING

Box constructed with 1/2"-3/4" (13 - 19 mm) exterior plywood.

FIG. 7 Preparing and Packaging a Block Sample

Designation: D 5079 – 90$^{\epsilon 1}$

Standard Practices for
Preserving and Transporting Rock Core Samples[1]

This standard is issued under the fixed designation D 5079; the number immediately following the designation indicates the year of original adoption or, in the case of revision, the year of last revision. A number in parentheses indicates the year of last reapproval. A superscript epsilon (ϵ) indicates an editorial change since the last revision or reapproval.

$^{\epsilon 1}$ NOTE—Section 13 was changed editorially in December 1991.

1. Scope

1.1 These practices cover the preservation, transportation, storage, cataloging, retrieval, and post-test disposition of rock core samples obtained for testing purposes and geologic study.

1.2 These practices apply to both hard and soft rock, but exclude ice and permafrost.

1.3 These practices do not apply to those situations in which changes in volatile gas components, contamination of the pore fluids, or mechanical stress relaxation affect the intended use for the core.

1.4 *This standard does not purport to address the safety problems associated with its use. It is the responsibility of the user of this standard to establish appropriate safety and health practices and determine the applicability of regulatory limitations prior to use.*

2. Referenced Documents

2.1 *ASTM Standards:*

D 420 Practice for Investigating and Sampling Soil and Rock for Engineering Purposes[2]

D 2113 Practice for Diamond Core Drilling for Site Investigation[2]

D 4220 Practice for Preserving and Transporting Soil Samples[2]

2.2 *API Standard:*

API RP-40 Recommended Practice for Core Analysis Procedure[3]

3. Terminology

3.1 *Descriptions of Terms Specific to This Standard:*

3.1.1 *critical care*—samples which are fragile or fluid or temperature sensitive. This protection level includes the requirements prescribed for routine and special care.

3.1.2 *routine care*—non-sensitive, non-fragile samples for which only general visual identification is necessary, and samples which will not change or deteriorate before laboratory testing.

3.1.3 *soil-like care*—materials which are so poorly consol-

idated that soil sampling procedures must be employed to obtain intact pieces of core.

3.1.4 *special care*—fluid sensitive samples and those which must later be subjected to testing. Requirements for this level of protection include those prescribed for routine care.

4. Significance and Use

4.1 The geologic characteristics and the intended use of the rock core samples determine the extent and type of preservation required. If engineering properties are to be determined for the core, it must be handled and preserved in such a way that the measured properties are not significantly influenced by mechanical damage, changes in chemistry, and environmental conditions of moisture and temperature, from the time that the core is recovered from the core drill until testing is performed. Drill core is also the sample record for the subsurface geology at the borehole location, and as such must be preserved for some period of timed, in some cases indefinitely, for future geologic study.

4.2 These practices present a selection of curatorial requirements which apply to the majority of projects. The requirements are given for a variety of rock types and project types ranging from small to large and from noncritical to critical. Noncritical projects are those in which failure of an element or the structure would result in negligible risk of injury and property loss, while there is great risk to property and life after failure of critical structures and projects. Guidance is given for the selection of those specific requirements which should be followed for a given project.

5. Guide for Implementation

5.1 A qualified person shall be assigned to have curatorial management responsibility for a given project. This person shall be technically competent in the management of rock core samples and shall have a knowledge of the various end uses for the cores and their associated preservation requirements. This responsible person shall have the authority to implement the requirements selected from these practices. In some cases, he or she may also have to decide between competing uses for the same core.

5.2 The responsible person shall select from Sections 6 through 11 those requirements and procedures that should be applied for the core from a particular project. The curatorial manager shall then see that these procedures are implemented, and also see that the records specified in Section 12 are kept.

5.3 The following factors should be considered when

[1] These practices are under the jurisdiction of ASTM Committee D-18 on Soil and Rock and are the direct responsibility of Subcommittee D18.12 on Rock Mechanics.

Current edition approved June 29, 1990. Published August 1990.

[2] *Annual Book of ASTM Standards*, Vol 04.08.

[3] Available from American Petroleum Institute, 1220 L Street, Washington, DC 20005.

selecting the curatorial requirements from Sections 6 through 11:

5.3.1 Project requirements for use of the core range from simple ones, in which the only need is to identify and locate the various lithologic units, to complex and critical ones in which detailed property testing of the core is required for engineering design. Priorities for multiple uses or different types of tests must sometimes be established when available core lengths are limited and when one use or test precludes another. For example, splitting a core for detailed geologic study prevents later strength testing, which requires an intact core.

5.3.2 Mechanical property tests for structural design purposes should be performed on a core in its natural moisture state, particularly if the rocks are argillaceous. Irreversible changes occur when such rocks are allowed to dry out, often resulting in invalid design data. The initial moisture content of such a core should therefore be preserved.

5.3.3 Freezing of pore water in the core may reduce the strength of the rock. The high temperature associated with unventilated storage sheds in summer, and temperatures alternating between hot and cold, may cause moisture migration from the core and weakening of the rock due to differential thermal expansion and contraction between grains. Such temperature extremes should therefore be avoided, particularly for weak sedimentary rock types.

5.3.4 A weak rock core may be broken or further weakened by careless handling, such as dropping a core box, or by mechanical vibration and shock during transportation. Breaking of the core reduces sample lengths available for testing. Weakening caused by such mechanical stressing may lower measured strength parameters and may affect other properties.

5.3.5 The required preservation time may vary from as short as three months to several years, and sometimes core may need to be stored indefinitely. A core taken simply to identify the bedrock lithology beneath a small structure may be needed for a few months only. For large and critical structures, it may be necessary to retain the core for many years as re-examination and testing may be required at some later time for additional geologic study or re-evaluation of property data. Some states have regulations governing the disposition and storage of core obtained within the state.

5.4 Figure 1 is a flow chart that shows the various core handling, use, and storage activities and the corresponding section numbers in these practices. Note that four care or protection levels are defined in Section 3 to account for the great variety of rock sensitivities and core uses encountered in practice.

5.5 The person assigned curatorial management responsibility should study the flow chart in Fig. 1 as it relates to the designated Sections 6 through 11 in these practices. Note in particular, that a selection of the required protection must be made in 7.5, where four levels of protection are specified, namely routine care, special care, critical care, and soil-like care.

5.6 Special attention is also directed to records requirements in Section 12, that document the history of the core handling, preservation, and storage.

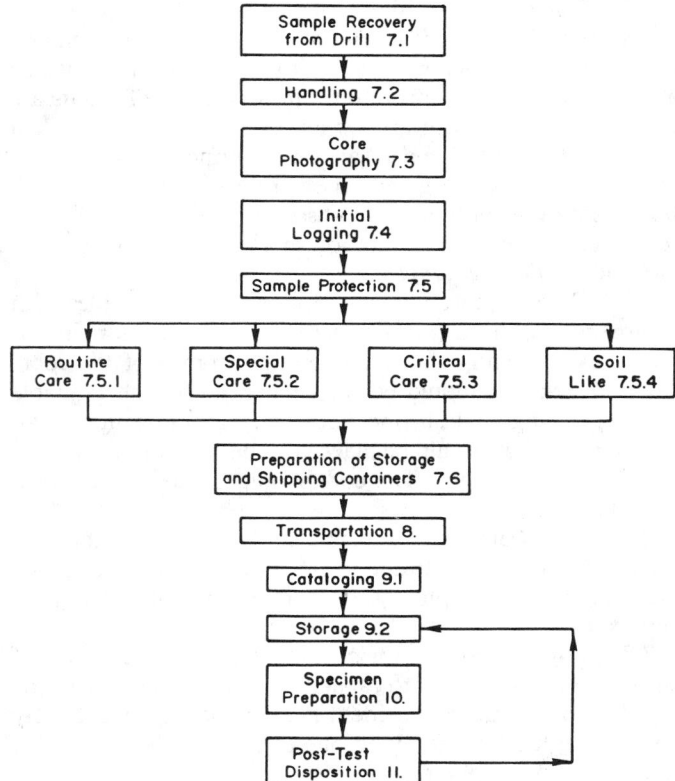

NOTE—Numbers refer to corresponding sections of this practice.

FIG. 1 Flow Chart for Core Handling, Use, and Storage Activities

6. Apparatus

6.1 *Camera,* for taking photographs of cores for logging.

6.2 *Controlled Humidity Room.*

6.3 *Core Boxes*—See 7.6.1.

6.4 *Vinylidene Chloride Plastic Film, Aluminum Foil, Plastic Microcrystalline Wax,* for sealing in moisture content of cores.

6.5 *Polyethylene Layflat Plastic Tubing.*

6.6 *Poly(vinyl chloride) Tubing.*

6.7 *Sawdust, Rubber, Polystyrene,* or material of similar resiliency to cushion the core.

6.8 *Miscellaneous Equipment,* such as adhesive tape and waterproof felt-tip markers.

7. Requirements and Procedures at the Drilling Site

7.1 *Sample Recovery:*

7.1.1 Accomplish sample recovery in accordance with Practice D 2113 or API RP-40.

7.1.2 Whichever approved drilling method is used, remove the samples from the core barrel with a minimum of disturbance.

7.2 *Handling:*

7.2.1 Each borehole shall be given full-time attention by a qualified inspector constantly available for observing, directing, photographing, and field logging. The inspector shall not perform simultaneously the same duties for more than one boring unless the borings are close enough to each other so that the entire inspection process can be done for each boring.

7.2.2 For relatively solid pieces of core that will not be adversely affected, the inspector shall use a marker, such as a felt-tip, to orient each piece so that later users will always be able to distinguish top from bottom. Acceptable formats are a continuous line with arrows or parallel solid and dashed lines with the dashed line always on the same side of the solid line. The direction convention shall be recorded in the log book. Locations of known depths should be marked directly on the core when the orientation marks are drawn.

7.3 *Core Photography:*

7.3.1 Perform core photography on all core samples with a camera of 35 mm (minimum) format using color film to record permanently the unaltered appearance of the rock. The film selected should be color balanced for the available lighting (daylight, flash, incandescent, or florescent), or an appropriate filter should be placed on the camera to compensate for the difference. The core should be cleaned prior to any photography.

7.3.2 A commercially available color strip chart should be included in the photo frame to serve as a reference to check the accuracy of the photographic reproduction of the rock core colors.

7.3.3 For rock placed in core boxes, take one photo of each box once it is filled to capacity. Include the inside of the box lid with appropriate identification data and a clearly visible length scale laid along one edge of the box so that it also shows in the photo.

7.3.4 Where very long, intact cores are being preserved in single plastic tubes, make detail-revealing close-ups of each core interval in addition to a single photo showing the complete core.

7.3.5 Take photographs before the core is obscured by protective sealants and wraps, and before any deterioration begins in particularly fragile or sensitive rock types.

7.3.6 For a boxed core that is not particularly sensitive and for which maintenance of in-situ moisture content is not important, two photos should be made: one with the core in a surface dry condition and one with the core in a wet condition to bring out optical properties that would not otherwise be apparent.

7.3.7 This procedure may require photography both in the field and then later in the storage facility, but it must be completed before any test core removal and before damage from mishandling has a chance to occur.

7.3.8 Where it is impossible for a photo to show identification data marked directly on the sample or its container, then mount appropriately marked placards so as to be included in the frame.

7.3.9 Organize the photographs and mount in a folder for easy access and preservation.

7.4 *Initial Logging:*

7.4.1 The boring inspector must complete at least a preliminary field log of the core before it is packed away to be transported. Suggested procedures for logging are given in the literature.[4-7] The preliminary log must include all identification data for the borehole and personnel and equipment involved, notations of coring run depths, recovery percentages, lithologic contact depths, types and locations of protection applied to samples, and any facts that would otherwise be unknown to whomever may complete a more detailed log at a later time. It is desirable that detailed logs be completed by the same inspector who does the field logging. It is advisable for the inspector immediately to make notations on the depths at which, in his judgment, any core losses occurred. Sometimes it is possible later to fill in gaps in the initial log by interpretations from wireline logs.

7.4.2 The inspector is to complete a detailed log on the drill site (see the literature[4-7]) in cases where the core is likely to deteriorate or otherwise change before being examined again.

7.4.3 For fragile core that must be immediately protected by wrapping and sealing, preliminary logging should take place in the field, but application of protective measures are to take precedence over time-consuming detailed logging.

NOTE 1—It is permissible later to make changes in detailed logs when laboratory analysis indicates original misidentification of rock type or other geologic features.

7.5 *Sample Protection*—Four levels of sample protection are covered (see Section 3): routine care, special care, critical care, and soil-like care. The level of protection chosen will depend on the geologic character of the rock and the intended use for the core.

7.5.1 *Routine care (see Fig. 2):*

7.5.1.1 For rock cored in 5 to 10-ft. runs, samples are sufficiently protected if placed in structurally sound core boxes. Enclosing the core in a loose-fitting polyethylene sleeve (layflat tubing) prior to placing the core in the core box is recommended.

7.5.1.2 Where very long solid cores have been recovered and need to be preserved intact, place each core in a reasonably stiff tube (poly(vinyl chloride) (PVC) tubing is recommended) of equal or slightly greater length and secure both ends to prevent slippage. The inside diameter of the tube should be slightly greater than the core diameter; the wall thickness of the tube must be sufficient to provide the rigidity to prevent core breakage due to bending.

7.5.2 *Special Care:*

7.5.2.1 The moisture state of some rocks, and even the moisture-state history of rocks such as shales, affects their properties. If tests are to be performed on the core, and if it is possible that a change in the moisture state may influence the test results, then the core must be sealed to prevent changes in the moisture state until the time of testing. This same procedure also applies to other samples where it is important to maintain fluids other than water (for example, hydrocarbons).

7.5.2.2 Seal samples requiring special care. Such sealing

[4] Association of Engineering Geologist, Core Logging Committee, South Africa Section, "A Guide to Core Logging for Rock Engineering," *Bulletin of the Association of Engineering Geologist*, Vol 15, No. 3, 1978, pp. 295–328.

[5] Deere, D. U., Dunn, J. R., Fickies, R. H., and Proctor, R. J., "Geologic Logging and Sampling of Rock Core for Engineering Purposes (Tentative)," Association of Professional Geological Scientists, 1977.

[6] The Geological Society, Engineering Group Working Party, "The Logging of Rock Core for Engineering Purposes," *Quarterly Journal of Engineering Geology*, Vol 3, 1970, pp. 1–24.

[7] International Society for Rock Mechanics, "Basic Geotechnical Description of Rock Masses, "*International Journal of Rock Mechanics and Mining Sciences and Geomechanics Abstracts*, Vol 18, 1980, pp. 85–110.

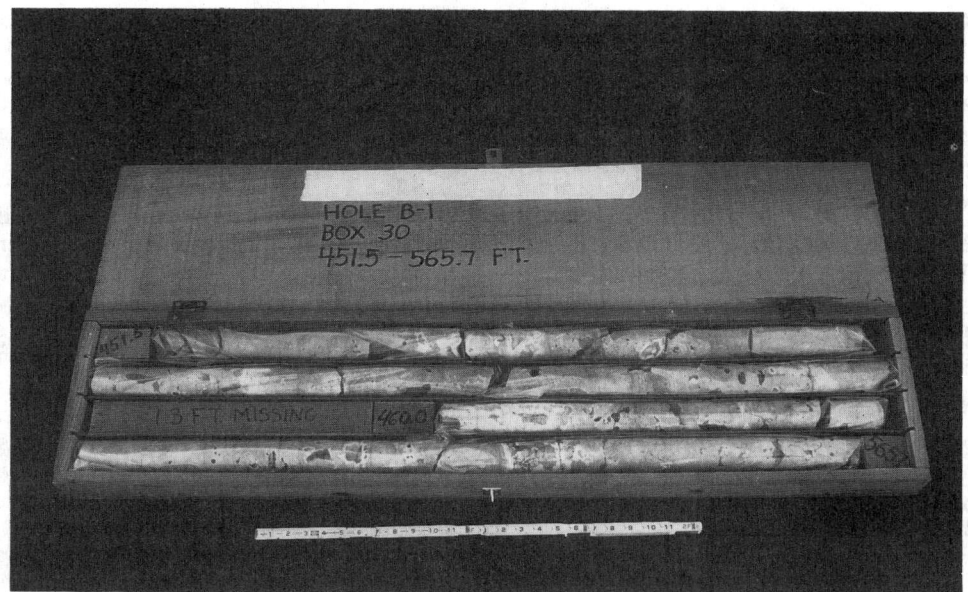

FIG. 2   Typical Wooden Core Box Showing Core Placement, Labeling, and Polyethylene Layflat Tubing Placed Over Core as Suggested for Routine Care (as in 7.5.1)

shall consist of a tightly fitting wrapping of a plastic film, such as vinylidene chloride (for example, saran or similar material). Over this, place another tight wrapping of aluminum foil (Note 2). Apply both of these wrappings so that as little air as possible is trapped beneath the wrappings. Overlap the ends of the wrappings over the ends of the sample and fold over so as to seal the ends of the sample. Finally, apply a few coats of a plastic microcrystalline wax (Note 3), preferably with a paint brush, although rapid dipping in molten wax is also acceptable. Apply a minimum of ⅛ in. (3 mm) of this plastic microcrystalline wax over the entire surface of the sample. This thickness of wax shall consist of at least two coatings of wax and preferably more. For long periods of storage, apply a minimum of ¼ in. (6 mm) of wax.

NOTE 2—In some instances where the aluminum foil wrapping chemically reacts with either the rock or its fluids, it should be replaced with some other metal foil which is nonreactive. A less preferable method is to delete the metal foil and increase the thickness of the sealing wax. If the metal foil is not used, the thickness of the sealing wax should be increased to ¼ in. (6 mm) when the required storage time is short (one week to a few months). For longer period, a thickness of ⅜ in. (9 mm) may be adequate.

NOTE 3—It is important that plastic microcrystalline wax be utilized, as it does not become brittle. Regular microcrystalline waxes are brittle and susceptible to cracking, which then severely limits their use as sealants.

7.5.2.3 Polyethylene layflat tubing is a much less effective moisture barrier than the wax-metal foil seal described in 7.5.2.2, and particular attention must be paid to the manner in which the ends of the tubing are sealed. This method of sealing may be used only if the person having testing management responsibility can prove that moisture loss can be tolerated without significantly affecting the pertinent mechanical and physical properties of the samples. The thickness of the polyethylene plastic tubing shall be at least 0.005 in. (0.13 mm). Tightly seal the ends of the tubing, for

example, by doubling the ends over at least twice and securing with adhesive tape. Make a check for seal quality; an acceptable method is to squeeze the sealed tube and observe if the air escapes easily. The effectiveness of this barrier can be improved by tightly wrapping the core with plastic film, such as vinylidene chloride, before it is placed in the polyethylene tubing. Even with the addition of this extra plastic film, the moisture barrier is still not nearly as effective as the wax-metal foil seal described in 7.5.2.2.

7.5.3 *Critical Care:*

7.5.3.1 If shock and vibration or variations in temperature, or both, may subject samples to unacceptable conditions during transport, place samples in suitable core boxes that provide cushioning or thermal insulation, or both.

7.5.3.2 Protect fragile samples from mechanical disturbance, such as vibration and shock. Completely encase each sample in the cushioning material (sawdust, rubber, polystyrene, urethane foam, or material of similar resiliency). The cushioning between the samples and walls of the core boxes should have a minimum thickness of 1 in. (25 mm). Provide a minimum thickness of 2 in. (50 mm) on the core box floor and lid. Samples must fit snugly into their assigned space.

7.5.3.3 Thermally insulate samples that are temperature sensitive. Thermal insulation is best provided by placing the core container (box or tube) inside another container that is designed specifically to provide thermal insulation. Such insulating containers are generally constructed of double or triple layers of an insulating material. Additionally, such containers are usually relatively airtight.

7.5.3.4 Seal samples that are sensitive to mechanical disturbance, fluid content, and temperature in accordance with 7.5.2.2.

7.5.4 *Soil-Like Care*—Such materials should be treated as indicated in Practices D 4220.

7.5.5 Handling of sensitive samples between borehole and transport vehicle should take place within a covered space that either provides shade against intense sunlight or can be heated to keep temperatures above freezing.

7.6 *Preparation of Storage and Shipping Containers:*

7.6.1 Core boxes must be constructed rigidly enough to prevent flexing of the core when the box is picked up by its ends. Wood is the desired construction material, and it should be ½ to ¾ in. (13 to 19 mm) thick. Partitions between core lengths shall be firmly nailed in place to increase the stiffness of the box. The lid should have sturdy hinges and a strong hasp or screw closure. Do not drive nails in the lid.

7.6.2 Packing material should be placed in the core box to support the core and prevent it from rolling around in the box.

7.6.3 Gently place the core in the core box starting with the shallowest depth at the upper left corner (nearest the hinge, if the lid is hinged) and progressing downward, as in reading a book, to the deepest depth at the lower right corner as shown in Fig. 2.

7.6.4 Core blocks should be placed at the ends of each core run.

7.6.5 Where a run of less than 100 % recovery yields a core that is too short to fill its assigned trough, the recovered core should be held secure and prevented from scattering by placement of spacer block such as a piece of wood or cardboard tube cut to the length of the missing core.

7.6.6 Unnecessary breaking of the core to fit the core box should be avoided as it may reduce the number of available test specimens. Any intentional breaks shall be recorded on the log.

7.6.7 Mark depths of the top and bottom of the core length in the box with a waterproof marker near the core ends and corresponding box corners. Intermediate depths that are accurately known should also be similarly marked.

7.6.8 Mark both the top and one edge of the core box with the following information before transportation:

7.6.8.1 Company or project name, or both,

7.6.8.2 Drill hole number or location,

7.6.8.3 Core box number in sequence down the hole, and

7.6.8.4 Depths from a specified hole datum to top and bottom of core length in box marked on appropriate corners of box.

7.6.9 The following additional information may be required to be written on the core box for specific projects:

7.6.9.1 Percent core recovery, and

7.6.9.2 Rock quality designation,

7.6.10 Mark or label tubes with the same information as core boxes. Such tubes shall have sufficient flexural rigidity to prevent core breakage due to bending. A core that requires protection from mechanical shock should have shock absorbing material packed concentrically around the core.

8. Transportation Requirements and Procedures

8.1 Damage may occur to the core if certain precautions are not taken during transportation. The mode of transportation, distance, terrain, and handling at each end are important factors. The following requirements and procedures should be considered when transporting a core from one location to another:

8.1.1 Remove the core from the drill site before it has a

chance to freeze or heat up or be damaged by activities at the drill site.

8.1.2 Handling during loading and unloading should be done gently. Never drop core boxes or tubes, but rather slide gently into position. If a box is accidentally dropped, record this fact.

8.1.2.1 A record of mishandling by commercial carriers is often difficult, if not impossible, to obtain. The use of company or agency vehicles is recommended, with someone assigned to supervise handling and storage along the transport route.

8.1.3 Provide transportation by a suitable vehicle to prevent damage by mechanical vibration, shock, freezing, and high temperatures along the entire transport route.

8.1.3.1 Rough terrain near the drill site may require the use of four-wheel drive vehicles for transport. In such cases, protect a core in the critical care category (7.5.3) by vibration insulators around the sides, bottom, and top of the core containers.

8.1.3.2 Highway transportation by passenger car, rather than van or truck, may be required for a fragile core.

8.1.4 Thermally insulate temperature sensitive cores. Thermally insulated containers are the preferred method (7.5.3.3). Such containers will maintain a thermal environment suitable for core preservation during the period of transport.

9. Cataloging and Storage Requirements and Procedures

9.1 *Cataloging*—Assign unique identification numbers to each core sample in the inventory. The identification numbers shall be easily traceable to the borehole number and depth interval from which the core was recovered. An acceptable method in many cases is to use the borehole number and depth interval as the identification numbers. It may be helpful to include the run numbers and core box numbers in the cataloging system, as well as other important subgroupings on large projects.

9.1.1 *Specimen Identification*—Core samples may be removed from inventory for a variety of tests and analyses. Preparation of specimens for these tests typically requires that the original sample be sawed and sometimes recorded. Assign core material removed for testing and analysis unique identification numbers which are easily traceable to the borehole number and depth, or which may actually be the borehole number and depth. The depth shall be taken as the depth to the midpoint of the specimen, or the depths to the top and bottom of the specimen, as required for the test program. It is often convenient to further identify the specimen with letters or numbers signifying the type of test or analysis to be performed. End pieces remaining after cutting shall be returned to their respective locations in the original core box. To hold the end pieces in their proper locations in the core box, a spacer should be placed in the location from which a test specimen has been removed.

9.2 *Storage*—Store the core so that it can be easily retrieved and in an environment that does not alter properties that are of interest.

9.2.1 *Storage Racks*—Place core boxes on racks that provide support over the entire length of the box. To reduce damage due to handling of core in the critical-care and soil-like categories, do not place more that a single layer of core boxes on a shelf. The lowest shelf in the rack should be a few inches above floor level to allow air circulation for temperature control and as a precaution against water damage.

9.2.2 *Temperature Control*—Prevent from freezing core samples that contain significant amounts of water. Also, temperature extremes may need to be limited to prevent undesirable chemical changes, such as dewatering of certain minerals. More stringent temperature control is required to prevent large temperature variations which could pump fluid from the sample.

10. Laboratory Specimen Preparation Requirements and Procedures

10.1 Samples removed from the inventory for testing and analysis in the laboratory must be handled carefully to preserve fluid content and integrity.

10.1.1 *Specimen Preparation*—Preparation techniques must be selected carefully to prevent sample alteration. Give special attention to the choice of cutting fluids that are used in machining operations such as sawing, coring, and grinding. For example, water will dissolve halite and will cause some shales to swell or slake. Specimens of water-sensitive material may be prepared with air cooling if the material is soft enough that there are no hot spots during the operation. Cutting oils, if used, should be removed from the sample using a solvent because the oil may affect certain properties of some rock types. Specimens that are moisture sensitive should be sealed between various steps of preparation if the time between steps is more than 30 min. A controlled-humidity room or chamber may be necessary for preparation of some rock types. Samples that were sealed in the inventory should be sealed after preparation until tested, unless testing occurs within 30 min after preparation.

11. Post-Test Disposition Requirements and Procedures

11.1 The requirements and procedures for disposition of test specimens vary with the nature of the project and the length of time for which there is a need to preserve the test material. The need for future re-examination may arise in the verification of engineering design or litigation.

11.2 The tests performed may be destructive or nondestructive. Reassemble and tape major fragments from destructive tests and place in a plastic bag with the fine material. Identify the bag with the specimen identification (borehole number and depth). Also place specimens from nondestructive tests in plastic bags and label.

11.3 If the specimen material is not moisture sensitive, and if easy access to the test specimens is desired, the specimens in the labeled bags may be placed in a box labeled as follows:

11.3.1 Company or project name, or both,

11.3.2 Borehole number or location,

11.3.3 Types of tests performed on specimens,

11.3.4 Name of person responsible for disposition, and

11.3.5 Date of disposition.

11.4 For moisture sensitive test specimens for which easy access is desired, the requirements of 11.3 apply except that the specimen shall be suitably sealed from the atmosphere. The sealing procedure shall be as stringent for post-test disposition as it was for the initial core preservation, as given in 7.5.

11.5 Storage requirements for boxes in 11.3 and 11.4 are the same as for the original core as given in 9.2.

11.6 If easy access to all of the test specimens is not required, return the tested specimens to their original locations in the core boxes or PVC tubes, and reseal as appropriate.

12. Records

12.1 A permanent, legible record shall be maintained by a technically qualified person for the core from each borehole. The following information shall be recorded:

12.1.1 Company and project names,

12.1.2 Drill hole number and location,

12.1.3 Orientation of borehole,

12.1.4 Elevation of hole collar, Kelly bushing, or some other datum,

12.1.5 Date(s) of coring,

12.1.6 Core box or PVC tube numbers and core depth interval contained in each,

12.1.7 Date and name of person doing initial logging,

12.1.8 Geologic log or reference to its location, and

12.1.9 Photographs of core or reference to their location.

NOTE 4—A permanent historical record needs to be maintained so that it is accessible for later use in identifying the core, ascertaining what geologic study and testing has been performed on the core, and its storage location of other disposition.

12.2 Desirable records which may also be required for some projects are as follows:

12.2.1 Each date of transportation, from where to where, mode of transportation, and name of responsible person,

12.2.2 Storage locations and source of temperature and relative humidity data,

12.2.3 Date and person removing test samples from storage, including a list of sample lengths removed and their depths,

12.2.4 Name and location of testing laboratory, nature of testing, if known, and name of responsible person,

12.2.5 Post-test disposition including date, name of responsible person, and location of tested specimens if not returned to original core boxes or PVC tubes, and

12.2.6 Date and nature of any other activity involving the core such as handling, and additional inspection and testing.

13. Keywords

13.1 identification; microcrystalline wax; sampling; transportation

 **D 5079**

2.2 VADOSE ZONE SAMPLING AND MONITORING

(See also D 5299, Section 3.2)

Standard Guide for
Soil Gas Monitoring in the Vadose Zone[1]

This standard is issued under the fixed designation D 5314; the number immediately following the designation indicates the year of original adoption or, in the case of revision, the year of last revision. A number in parentheses indicates the year of last reapproval. A superscript epsilon (ϵ) indicates an editorial change since the last revision or reapproval.

1. Scope

1.1 This guide covers information pertaining to a broad spectrum of practices and applications of soil atmosphere sampling, including sample recovery and handling, sample analysis, data interpretation, and data reporting. This guide can increase the awareness of soil gas monitoring practitioners concerning important aspects of the behavior of the soil-water-gas-contaminant system in which this monitoring is performed, as well as inform them of the variety of available techniques of each aspect of the practice. Appropriate applications of soil gas monitoring are identified, as are the purposes of the various applications. Emphasis is placed on soil gas contaminant determinations in certain application examples.

1.2 This guide suggests a variety of approaches useful to successfully monitor vadose zone contaminants with instructions that offer direction to those who generate and use soil gas data.

1.3 This guide does not recommend a standard practice to follow in all cases nor does it recommend definite courses of action. The success of any one soil gas monitoring methodology is strongly dependent upon the environment in which it is applied.

1.4 Concerns of practitioner liability or protection from or release from such liability, or both, are not addressed by this guide.

1.5 This guide is organized into the following sections and subsections that address specific segments of the practice of monitoring soil gas:

Section	
4	Summary of Practice
4.1	Basic principles, including partitioning theory, migration and emplacement processes, and contaminant degradation
4.7	Summary Procedure
5	Significance and Use
6	Approach and Procedure
6.1	Sampling Methodology
6.5	Sample Handling and Transport
6.6	Analysis of Soil Gas Samples
6.7	Data Interpretation
7	Reporting

1.6 *This guide does not purport to set standard levels of acceptable risk. Use of this guide for purposes of risk assessment is wholly the responsibility of the user.*

1.7 The values stated in either inch-pound or SI units are to be regarded separately as the standard. The values given in parentheses are for information only.

1.8 *This standard does not purport to address all of the safety problems, if any, associated with its use. It is the responsibility of the user of this standard to establish appropriate safety and health practices and determine the applicability of regulatory limitations prior to use.*

2. Referenced Documents

2.1 *ASTM Standards:*
D 653 Terminology Relating to Soil, Rock, and Contained Fluids[2]
D 1356 Terminology Relating to Atmospheric Sampling and Analysis[3]
D 1357 Practice for Planning the Sampling of the Ambient Atmosphere[3]
D 1452 Practice for Soil Investigation and Sampling by Auger Borings[2]
D 1605 Practices for Sampling Atmospheres for Analysis of Gases and Vapors[3]
D 1914 Practice for Conversion Units and Factors Relating to Atmospheric Analysis[3]
D 2652 Terminology Relating to Activated Carbon[4]
D 2820 Test Method for C_1 Through C_5 Hydrocarbons in the Atmosphere by Gas Chromatography[3]
D 3249 Practice for General Ambient Air Analyzer Procedures[3]
D 3416 Test Method for Total Hydrocarbons, Methane, and Carbon Monoxide (Gas Chromatographic Method) in the Atmosphere[3]
D 3584 Practice for Indexing Papers and Reports on Soil and Rock for Engineering Purposes[2]
D 3614 Guide for Laboratories Engaged in Sampling and Analysis of Atmospheres and Emissions[3]
D 3670 Guide for Determination of Precision and Bias of Methods of Committee D-22[3]
D 3686 Practice for Sampling Atmospheres to Collect Organic Compound Vapors (Activated Charcoal Tube Adsorption Method)[3]
D 3687 Practice for Analysis of Organic Compound Vapors Collected by the Activated Charcoal Tube Adsorption Method[3]
D 4220 Practices for Preserving and Transporting Soil Samples[2]
D 4490 Practice for Measuring the Concentration of Toxic Gases or Vapors Using Detector Tubes[3]

---

[1] This guide is under the jurisdiction of ASTM Committee D-18 on Soil and Rock and is the direct responsibility of Subcommittee D18.21 on Ground Water and Vadose Zone Investigations.
Current edition approved Nov. 15, 1992. Published January 1993.

[2] *Annual Book of ASTM Standards*, Vol 04.08.
[3] *Annual Book of ASTM Standards*, Vol 11.03.
[4] *Annual Book of ASTM Standards*, Vol 15.01.

D 4597 Practice for Sampling Workplace Atmospheres to Collect Organic Gases or Vapors with Activated Charcoal Diffusional Samplers[3]

D 4696 Guide for Pore-Liquid Sampling from the Vadose Zone[2]

D 4700 Guide for Soil Core Sampling from the Vadose Zone[2]

D 5088 Practice for the Decontamination of Field Equipment Used at Non Radioactive Waste Sites[5]

E 177 Practice for Use of the Terms Precision and Bias in ASTM Test Methods[6]

E 260 Practice for Packed Column Gas Chromatography[7]

E 355 Practice for Gas Chromatogaphy Terms and Relationships[7]

E 594 Practice for Testing Flame Ionization Detectors Used in Gas Chromatography[7]

E 697 Practice for Use of Electron-Capture Detectors in Gas Chromatography[7]

3. Terminology

3.1 *Descriptions of Terms Specific to This Standard:*

3.1.1 *capillary fringe*—the basal region of the vadose zone comprising sediments that are saturated, or nearly saturated, near the water table, gradually decreasing in water content with increasing elevation above the water table. Also see Terminology D 653.

3.1.2 *contaminant*—substances not normally found in an environment at the observed concentration.

3.1.3 *emplacement*—the establishment of contaminant residence in the vadose zone in a particular phase.

3.1.4 *free product*—liquid phase contaminants released into the environment.

3.1.5 *free vapor phase*—a condition of contaminant residence in which volatilized contaminants occur in porosity that is effective to free and open gaseous flow and exchange, such porosity generally being macroporosity.

3.1.6 *liquid phase*—contaminant residing as a liquid in vadose zone pore space, often referred to as "free product."

3.1.7 *macroporosity*—large intergranular porosity with large pore throats, including soil cracks, moldic porosity, animal burrows and other significant void space.

3.1.8 *microporosity*—intragranular porosity and microscopic intergranular porosity with submicroscopic pore throats.

3.1.9 *occluded vapor phase*—condition of contaminant residence in which volatilized contaminants occur in porosity that is ineffective to free and open gaseous flow and exchange, such porosity generally being microporosity; frequently termed dead-end pore space.

3.1.10 *partitioning*—the act of movement of contaminants from one soil residence phase to another.

3.1.11 *soil gas*—vadose zone atmosphere.

3.1.12 *solute phase*—a condition of contaminant residence in which contaminants are dissolved in ground water in either the saturated or the vadose zone.

3.1.13 *sorbed phase*—a condition of contaminant residence in which contaminants are adsorbed onto the surface of soil particles or absorbed by soil organic matter.

3.1.14 *vadose zone*—the hydrogeological region extending from the soil surface to the top of the principal water table.

4. Summary of Guide

4.1 Soil gas monitoring in the vadose zone is a method used to directly measure characteristics of the soil atmosphere that are frequently utilized as an indirect indicator of processes occurring in and below a sampling horizon. Soil gas monitoring is used as a method to suggest the presence, composition, and origin of contaminants in and below the vadose zone. Among other applications, this method is also employed in the exploration for natural resources, including petroleum, natural gas and precious metals. Soil gas monitoring is a valuable screening method for detection of volatile organic contaminants, the most abundant analytical group of ground-water contaminant compounds (1).[8]

4.2 *Basic Theoretical Principles*—The processes indicated by the soil gas monitoring method are partitioning, migration, emplacement and degradation. Partitioning represents a group of processes that control contaminant movement from one physical phase to another, these phases being liquid, free vapor (that is, through-flowing air (2)), occluded vapor (that is, locally accessible air and trapped air (2)), solute and sorbed. Migration refers to contaminant movement over distance with any vertical, horizontal or temporal component. Emplacement refers to establishment of contaminant residence in any phase within any residence opportunity. Degradation is the process whereby contaminants are attenuated by oxidation or reduction in the vadose zone, either through biogenic or abiogenic processes. Soil gas monitoring measures the result of the interaction of these processes in a dynamic equilibrium. Measurement of these processes in static equilibrium is unrealistic.

4.3 The following subsections provide detailed information on partitioning, migration, emplacement and degradation. Subsection 4.4 provides a summary procedure for soil gas sampling. Users of this guide who do not wish to study details of partitioning, migration, emplacement and degradation at this time may skip to 4.4.

4.3.1 Partitioning is the initial step by which contaminants begin to move away from their source. Partitioning occurs in water saturated and unsaturated environments. This group of processes is complex and difficult to quantify when considered in the vadose zone due to the unique makeup of the vadose matrix, i.e. air-filled porosity (microporous and macroporous), pore water, free product, solid-phase soil organic matter, clay and discrete inorganic soil particles. Important individual processes of partitioning are dissolution, volatilization, air-water partitioning, soil-water partitioning and soil-air partitioning (3).

4.3.2 Dissolution is the process whereby volatile contaminants move between the liquid phase (free product) and the solute phase (dissolved in water). At equilibrium, the product of the mole fraction of a particular compound in the liquid phase and the activity coefficient of that compound in the liquid phase is equal to the product of the mole fraction of

[5] *Annual Book of ASTM Standards*, Vol 04.09.
[6] *Annual Book of ASTM Standards*, Vol 14.02.
[7] *Annual Book of ASTM Standards*, Vol 14.01.

[8] The boldface numbers given in parentheses refer to a list of references at the end of the text.

that compound in the solute phase and the activity coefficient of that compound in the solute phase. This process is more clearly described by the following expression:

$$X^L_I \Gamma^L_I = X^W_I \Gamma^W_I \qquad (1)$$

where:
X^L_I = the mole fraction of compound (I) in the liquid (L) phase (free product),
X^W_I = the mole fraction of compound (I) in the solute (W) phase (dissolved in water),
Γ^L_I = the activity coefficient of compound (I) in the liquid (L) phase (free product), and
Γ^W_I = the activity coefficient of compound (I) in the solute (W) phase (dissolved in water).

Dissolution equilibrium is therefore influenced by concentration of the subject compound in both the free product contaminant mixture and water. The most common practical application of expression Eq (1) in soil gas monitoring is in hydrocarbon detection. Simplification of Eq (1) is achieved by the following:

assume:
$$\Gamma^W_I = 1/S_I,$$

where:
S = the solubility of compound (I) in water

and:
$$\Gamma^L_I = 1, \text{ acceptable for hydrocarbons (3)},$$

then:
$$X^W_I = X^L_I S_I \qquad (2)$$

4.3.2.1 Dissolution equilibrium is impacted by the presence of liquid phase cosolvents, such as gasoline additives, at low concentrations in liquid phase mixtures. This change in dissolution equilibrium can enhance the solubility of certain liquid phase components in water beyond what is indicated by partitioning coefficient data generated in the laboratory. This can have significant impact on downstream concentrations of the contaminant(s) in the soil atmosphere.

4.3.2.2 The effects of temperature upon dissolution equilibrium are generally insignificant for aliphatic hydrocarbons between 15 and 50°C (4), the temperature range from which most soil gas samples are recovered. However, temperature effects upon dissolution equilibrium can be significant for other common families of contaminant compounds within similar temperature ranges (5). These effects must be considered when planning or interpreting the results of a soil gas survey.

4.3.2.3 Dissolution equilibrium is altered by changes in water salinity. Modest decreases in the solubility of contaminants in water are to be expected with increases in salinity of the solution.

4.3.2.4 The rate of dissolution is strongly dependent upon the partitioning coefficient of the particular contaminant of interest and the amount of mixing of the liquid phase and water (3). For example, partitioning of a particular contaminant into ground water is accelerated by frequent water level fluctuations within a contaminated capillary fringe. The downstream implications for subsequent partitioning of the contaminant from the solute to the vapor phase for eventual soil gas recovery are obvious.

4.3.3 Volatilization is the process during which volatile contaminants move between the liquid phase (free product) or solute phase and a vapor phase, either the free vapor phase or the occluded vapor phase or both. Contaminant mixtures can contain compounds with a considerable range of vapor pressures that can contribute contaminants to the soil atmosphere by volatilization. This atmosphere will exhibit a composition similar to that of the parent contaminant but lacking in those constituents with the lowest vapor pressures. The likelihood of the presence of a particular contaminant introduced into the soil atmosphere by volatilization can be estimated by considering the partial pressure of that contaminant in a vapor phase. This partial pressure is equal to the product of the mole fraction concentration of the subject component in the liquid contaminant solution, the activity coefficient of the subject component and the vapor pressure of the pure component. This concept is more clearly expressed as follows:

$$P = X_I \Gamma_I P^o \qquad (3)$$

where:
P = the partial pressure of the subject contaminant compound in the vapor phase,
X_I = the mole fraction concentration of contaminant (I) in the liquid contaminant solution,
Γ_I = the activity coefficient of the subject contaminant in the liquid contaminant solution, and
P^o = the vapor pressure of the pure component.

4.3.3.1 The quantity of contaminant volatilized into a vapor phase and the rate of that process is strongly dependent upon temperature. Rate of volatilization is also controlled by the rate of transport of contaminant vapors from the liquid phase-vapor phase interface (3). This rate is probably higher when macroporous flow paths are available for vapor phase transport, and is promoted by a number of driving forces. These are concentration gradient, density gradient between soil atmosphere and contaminant-saturated soil atmosphere, convection currents related to temperature gradient, barometric pressure pumping and introduction of water onto the liquid phase-vapor phase interface.

4.3.4 Air-water partitioning is the process by which volatile contaminants move between the solute phase and a vapor phase, either the free vapor phase or the occluded vapor phase or both. For dilute solutions, air-water partitioning is controlled by Henry's Law, which states that the vapor pressure of a volatile compound above a dilute aqueous solution of that compound is equal to the product of the Henry's Law constant and the mole fraction of that compound in the aqueous solution. Henry's Law may be represented as:

$$P_I = k X_{I(aq)} \qquad (4)$$

where:
P_I = vapor pressure of compound (I) above a dilute aqueous solution of (I),
k = the Henry's Law constant for compound (I) at a given temperature, and
$X_{I(aq)}$ = the mole fraction of the subject contaminant compound in the aqueous solution.

Care must be exercised in using Henry's Law to approximate contaminant vapor pressures because of unknowns related to the concentration of contaminants in solution and the contribution of other partitioning processes. Some available literature pertaining to soil gas surveying places emphasis on

Henry's Law constant at 25°C and atmospheric pressure as a primary controlling factor in determining the suitability of a particular volatile contaminant to the soil gas monitoring method. Such emphasis may be inappropriate when, for example, free product is the source of contaminant vapors or when contaminants have not reached ground water. Care must also be exercised in noting the units in which Henry's Law constants are expressed, as these vary from source to source. Volatile but very highly water soluble compounds behaving according to Henry's Law may not be detectable in soil gas because of their persistence for residence in the solute phase (6).

4.3.5 Soil-water partitioning is the process by which volatile contaminants move between the sorbed phase and the solute phase. This process is generally underestimated in its importance to the success or failure of contaminant recovery by soil gas sampling, especially when utilizing the majority of active soil gas sampling techniques generally available to field personnel.[9] There is uncertainty with respect to factors controlling soil-water partitioning, creating doubt as to the reliability of soil sorption data in most applications. Problems with soil sorption data include variability in measurement protocols, the variable nature of organic matter in soils, the effect of dissolved organic matter, unusual pH effects and the effect of salinity, among others (3).

4.3.5.1 The contribution of soil-water partitioning to contaminant phase residence equilibria is strongly controlled by sorbed contaminant concentration in soil, soil makeup, vadose zone pore water content, and soil porosity configuration. Important variables in soil makeup are the quantity, type and distribution of clay in soil and the quantity, type and distribution of soil organic matter. These variables impact the surface area available to sorptive processes, that is, the storage capacity of the soil for contaminants in the sorbed phase, and the pH of the sorption environment. Variations in vadose zone pore water content directly affect the storage capacity of the soil for contaminants in the solute phase. Soil porosity configuration, principally microporosity versus macroporosity, is critical to the rate of soil-water partitioning due to the contrast in surface area between micropores and macropores and the related storage capacity of this porosity for both pore water and sorbed contaminants.

4.3.6 Soil-air partitioning is the process by which volatile contaminants move between the sorbed phase and a vapor phase, either the free vapor phase or the occluded vapor phase or both. Like soil-water partitioning, this process is underestimated in its importance to the recoverability of contaminants by many soil gas sampling techniques. In vadose zone horizons with very low pore water contents, soil-air partitioning can yield vapor phase contaminant composition that differs from free product composition. In vadose zone horizons with higher pore water content, the responsibility for this compositional inconsistency is shared, largely with soil-water partitioning. In wet soil conditions, threshold soil water content values exist for trapped soil atmosphere content to become significant (7), suggesting that

responsibility for this compositional inconsistency can be largely attributed to occluded phase residence. Additional important variables are soil clay content, type and distribution, and soil organic matter content, type and distribution. Studies have demonstrated significant impact of soil organic matter and clay content on volatile organic compound emissions from soils (8). Due to the strong control on vapor phase contaminant content by the soil-air partitioning process, it is unreasonable to expect soil contaminants with high affinity for sorption to be efficiently recovered by most soil gas sampling techniques.

4.4 Migration of contaminants in the vadose zone, that is, unsaturated flow, is highly complex and is controlled by soil characteristics, contaminant composition and contaminant phase (9). Migration through unsaturated matrix can occur through a variety of diffusion, dispersion and mass transport mechanisms which behave in a manner unique to saturated flow.

4.4.1 A major division in migratory behavior of contaminants is defined by their solubility or immiscibility in water. Contaminants are often introduced into the soil as liquid mixtures, the components of which immediately begin to partition into other phases upon soil entry. Contaminants that establish soil residence behind a migratory front change in composition with distance from their point of entry. As contaminant migration continues, pathways for individual components can become divergent, such that the composition of the liquid mixture continues to change as migration proceeds. Eventually, migration of liquid mixtures may reach ground water. This can be retarded if the contaminants partition into other phases before reaching ground water and if contaminant vapor is less dense than the uncontaminated soil atmosphere. Transport of contaminants by downward percolation of meteoric waters and upward movement of ground water accelerate the contact of contaminants with ground water. When these contaminants do reach ground water, a radically different set of migration mechanisms begins to govern contaminant transport via saturated flow. Further divergence of contaminant pathways is dependent upon the tendency of each component of the contaminant mixture to float on ground water, become dissolved in ground water or sink to an impermeable layer within the aquifer. Detailed descriptions of these phenomena are available in the literature (10).

4.4.2 The impact of migration processes on soil gas measurement is significant. Although it is impractical to estimate actual migration mechanisms by modelling prior to most soil gas monitoring efforts, a rudimentary knowledge of site characteristics can guide investigators to realistic interpretations of soil gas data expressing unusual or highly variable compositions. More thorough knowledge of relevant site characteristics, such as the presence or absence of barriers to vertical or horizontal migration, that is, foundations, buried pavement, or perched ground water, as well as preferential pathways for contaminant migration, that is, backfill rubble, utility vaults, storm sewers or soil cracks, can assist investigators to assess the migration impact on soil gas survey design.

4.5 The vadose zone is a highly complex soil-air-water-hydrocarbon system with abundant opportunity to store contaminants in all phases. Contaminants partition ac-

[9] See 6.2 for a discussion of active soil gas sampling techniques.

cording to their physical properties and the residence opportunity presented to them along their migratory path. This process has been described as an in-situ chromatographic-like separation of contaminants (11). Emplacement, or the establishment of contaminant residence, is a highly dynamic process. Contaminants move from one phase to another as changes occur in both chemical and physical equilibria. Important changes impacting phase residence change include temporal variations in moisture content, soil temperature and level of microbial activity.

4.5.1 One interesting example of disruption in equilibrium conditions is the act of sampling soil gas. Many soil gas sampling systems rely on large volume recovery of soil gas to provide a sample that is believed to be representative of the soil atmosphere in situ. Movement of this soil gas by convective flow through unsaturated soils can cause upward changes in vapor phase contaminant concentration at the expense of other phases.

4.5.2 In natural systems, temporal increases in soil moisture cause gradual increases in solute phase emplacement at the expense of other phases. It is unrealistic to attempt to characterize a static soil gas equilibrium in the vadose zone because this equilibrium is never achieved. For this reason, soil gas data sets based on specific contaminant concentrations and generated at different times are usually not comparable for the absolute values generated by each temporal sampling event. Qualitative comparison of data generated by the same soil gas method and performed at different times is permissible. Generation of a single data set by reconnaissance soil gas sampling and subsequent infilling of data to form a single data set is strongly discouraged.

4.5.3 Attempts to compensate for temporal variations in phase equilibria have been attempted by collecting samples that approximate replicates at known locations and adjusting succeeding data up or down to compensate for observed changes. This procedure is also strongly discouraged, because the number of variables affecting observed changes are too great. Moreover, the ability or willingness of most investigators to determine the most significant effects upon phase equilibria is insufficient to be of use.

4.5.4 Data sets generated by different soil gas sampling techniques may not be comparable as a direct result of differences in efficiency of recovery of contaminants from specific phases. Not only can these data sets differ in measured contaminant concentration, but they can vary substantially in composition as well.

4.6 Degradation of contaminants occurs in the vadose zone through oxidation or reduction reactions that can be biogenic or abiogenic in nature. This process can occur both aerobically and anaerobically to mitigate contaminant levels. Degradation is most often recognized in shallow, permeable soils where favorable conditions exist for oxidation of labile compounds, however other vadose environments can be conducive to degradation. Specific environmental conditions are required for degradation processes to occur. For abiogenic degradation, redox potential and soil pH can be rate controlling factors. For biodegradation, necessary environmental conditions include the presence of microorganisms capable of adaptation to the contaminant as substrate, conditions favorable to population increases of these microorganisms and migration pathways for contaminants to

come in contact with these microorganisms. Most soils contain naturally occurring populations of various microorganisms that can degrade petroleum products (12). Contaminant biodegradation is known to occur in groundwater (13) and in soils (14) prior to contaminant partitioning into a vapor phase. Contaminant biodegradation rates for some compounds are highly variable and are controlled by a number of kinetic factors influencing the distribution of microorganisms responsible for degradation. These include aerobic versus anaerobic environments, contaminant type and temperature (15, 16).

4.6.1 Degradation rate can approach, equal or periodically exceed the rate of contaminant emplacement into the vadose zone, such that contaminants are not detectable by soil gas monitoring. This mechanism can result in soil gas data which are not representative of an underlying contaminated condition (17).

4.6.2 Labile contaminants can be degraded to compounds that may or may not be detectable in soil gas. Aerobic degradation can produce carbon dioxide which can be monitored as an indirect indicator of the presence of contaminants (18), or organic acids and phenols (13) that are not routinely detectable in active whole air soil gas samples. In alternative to whole air methods, use of an appropriate adsorption medium may facilitate recovery of such compounds for analysis by desorption and gas chromatography-mass spectroscopy. Anaerobic degradation can produce compounds including methane, ethylene, propylene, acetylene, and vinyl chloride which also can be monitored as an indirect indicator of the presence of contaminants. Caution must be used in attributing elevated levels of these compounds to biodegradation, because competitive processes can confuse the interpretation of absolute concentration values and potential sources.

4.6.3 Biodegradation of contaminants in the vadose zone can proceed naturally by adaptation of indigenous microbial populations to metabolize contaminants as primary substrate, or by introduction of foreign populations which have been preconditioned to metabolize contaminants of interest. Case histories demonstrate the absence of certain compounds in soil gas contaminant suites for which biodegradation has been named as the responsible process (17, 19, 20). Such cases address the attenuation or complete absence of simple aromatic hydrocarbons, some of which are halogenated, in soil gas. This phenomenon may be controlled by the availability of oxygen as has been demonstrated in the laboratory (13). Other compound classes can exhibit similar effects.

4.6.4 Other processes may share responsibility for the actual or apparent absence or attenuation of some contaminants in soil gas sample sets. In some cases where attenuation of contaminant concentration is attributed to degradation, combinations of high soil clay, organic matter and pore water content can reduce the recovery efficiency of certain soil gas sampling techniques for certain contaminants such that contaminant concentrations fall below detection limits. Care must be exercised in attributing a lack of contaminants in soil gas samples to degradation.

4.7 Summary Procedure for Soil Gas Sampling—Vadose zone monitoring methods have a set of procedures, both general and specific, that must be consistently followed in order to provide maximum data quality and usefulness. Soil gas monitoring is no exception, with six primary procedures common to all soil gas monitoring techniques. The procedures are a planning and preparation step including definition of data quality objectives, the act of sampling soil gas in the field, handling and transporting the sample, sample analysis, interpretation of the results of analysis, and preparation of a report of findings.

4.7.1 The planning and preparation step begins with the formulation of project objectives, including purpose of the survey, appropriate application of the data to be collected and data quality objectives.

4.7.2 Data can vary in quality due to sampling methodology, sample preparation, analytical procedures, laboratory quality control, and available documentation. Quality assurance programs include all of the activities necessary to provide measurement data at a requisite precision and bias (see Practice 1357). Quality assurance objectives for soil gas monitoring are similar to those for atmospheric air monitoring. The overall quality assurance objective for measurement data is to ensure that data of known and acceptable quality are provided. In order to meet these objectives, data quality objectives should be defined for data measurements in support of the soil gas data interpretation. These are comparability, completeness, representativeness, bias and precision. The comparability of the data collected refers to the ability to interpret the results in light of previous data collection efforts. Completeness refers to the number of samples collected and analyzed compared to the planned number of samples. Representativeness is a measure of the degree to which analytical results reflect true field conditions. Field contamination and sampling intensity are two factors affecting representativeness. Bias is a generic concept of exactness related to the closeness of agreement between the average of one or more test results and an accepted reference value (see Practice E 177). The precision of a measurement process is a generic concept related to the closeness of agreement between test results obtained under prescribed like conditions from the measurement process being evaluated. Overall precision and bias targets for chemical contaminant measurements can be set at 10 % allowable deviation with 90 % confidence limits. In all of these quality assurance activities one must take into consideration that factors including geophysical conditions and definition of sampling volume in the vadose zone often have higher variability than analytical equipment calibration procedures.

4.7.3 Table 1 provides suggested quantitative limits for data quality objectives.

4.7.4 The planning and preparation step continues with the evaluation of available information already gathered for the project area. These efforts culminate in the selection of an appropriate soil gas monitoring method and a survey design which best fits the project objectives within budgetary constraints. Prior to actual field work, investigators must obtain the necessary permits and landowner permission for property access. When a survey area is pending sale, investigators should obtain written permission to conduct the survey from both the buyer and the seller. Moreover, when a soil gas survey is being performed as a service, no work should proceed on the survey without a fully executed consulting agreement between the investigator and the client

TABLE 1 Suggested Quantitative Limits for Data Quality Objectives

QA/QC Objective	Measure	Formula	Limit
Accuracy	Laboratory standard	Standard recovery	90 to 110 %
Precision	Field replicate	Relative standard deviation	< 20 %
	Laboratory replicate	Relative standard deviation	< 20 %
Representativeness	Air blank	Bias	< 10 %
	Cross contam. blank	Bias	< 10 %
Completeness	Completion (%)	Relative compl.	> 90 %
Comparability	Prof. judgment	NA	NA

for whom the survey is being conducted.

4.7.5 Actual field work consists of recovery of soil gas samples. The method selected should be based upon site specific factors and dictated by the project objectives. A detailed discussion of soil gas sampling methods is provided in 6.1.

4.7.6 As samples are being recovered, they must be handled and transported in such a way as to assure preservation prior to analysis. A detailed discussion of sampling and transport is located in 6.5.

4.7.7 The presence of contaminants is determined through analysis of the soil gas samples. This step is controlled to a large degree by the QA/QC objectives of the survey. A discussion of sample analysis is provided as 6.6.

4.7.8 Data interpretation is largely an iterative process of review of the raw soil gas data out of context, a review of the soil gas data in context of other site characteristics and the formulation of conclusions based upon all known information. A discussion of soil gas data interpretation is located in 6.7.

4.7.9 Finally, a report of findings is generated in a format that is selected to be appropriate to the requirements of the end users. Section 7 provides options that can be addressed in reporting as well as recommendations of topics that should be included in all soil gas summations.

5. Significance and Use

5.1 *Application of Soil Gas Monitoring*—Soil gas monitoring is an extremely versatile method in that it can be adapted to conform to the requirements of dissimilar industries for a wide variety of applications. A number of soil gas techniques have been utilized in the agricultural (21), petroleum (22, 23) and minerals (24) industries. Certain applications have been exercised for well over 50 years. Soil gas monitoring has been utilized in research efforts, including the monitoring of underground coal gasification retorts (25). Application to the environmental industry is comparably recent but very effective as a rapid and relatively inexpensive method of detecting volatile contaminants in the vadose zone. Field screening, of which soil gas monitoring is a basic component, has been demonstrated to be effective for selection of suitable and representative samples for other more costly and definitive monitoring methods (26). Soil gas monitoring is useful to assess the extent of ground water contamination for certain contaminants and field environments (27). Soil gas monitoring is also a viable method of monitoring subsurface contaminant discharges from underground storage tanks (28). New applications of the soil gas monitoring are periodically developed and published in the referenced literature. The method may be useful in the study of unsaturated flow. In most instances, the method can make use of very light-weight, portable and inexpensive tools made from commonly available materials. Soil gas monitoring has become a widely accepted method for locating subsequent environmental monitoring and remediation activities such as ground water monitoring wells, contaminant product recovery wells or excavations to recover contaminated soil. Soil gas monitoring has made a significant contribution to ground water monitoring and remedial planning on sites that fall under the Comprehensive Environmental Response, Compensation and Liability Act (CERCLA) (29). This method is highly useful at the initiation of Phase II environmental assessment action in determining the presence of volatile organic contamination of real property in a pending sale.

5.1.1 In any application, soil gas monitoring can be performed over a wide range of both spatial and temporal designs. Spatial designs include soil gas sampling in profiles or grid patterns at a single depth or multiple depths. Multiple depth sampling is particularly useful for contaminant determinations in cases with complex soil type distribution and multiple sources. Depth profiling can also be useful in the determination of the most appropriate depth(s) at which to monitor soil gas, as well as the demonstration of migration and degradation processes in the vadose zone. Temporal designs include the long-term monitoring of the vadose zone for the appearance of volatile organic contaminants from known potential sources such as underground storage tanks and solid waste landfills. Temporal designs are especially useful in monitoring the effectiveness of contaminant remediation efforts.

5.1.2 Soil gas monitoring in the vadose zone is an ideal reconnaissance tool and screening technique in most applications. However, site specific and contaminant specific limitations can cause this technique to be unsuccessful in meeting project objectives. Caveats exist in all soil gas monitoring procedures that can frustrate efforts to successfully apply the method to any application.

5.2 *Limitations*—The most significant limitation on soil gas monitoring is the inability to utilize the method as a stand alone technique. Soil gas monitoring does not provide repeatable quantitative information over time due primarily to the dynamic nature of phase equilibria in the vadose zone and secondarily to unavoidable inconsistencies in sampling practice. As a result of geologic variability in the vadose zone and the multitude of unique sampling devices currently being used in the field, quality assurance and quality control protocol, discussed in 6.4, cannot provide the rigor required as in a test method. For these reasons, soil gas data in itself cannot be used to provide definitive answers about the location or absence of buried contaminants. Moreover, the success of any soil gas monitoring method is strongly dependent upon effects related to geologic variation and moisture content in the sampling horizon as well as the physical properties of the target contaminants.

5.2.1 False negative results can occur as a direct result of the incompatibility of a specific procedure with the properties of the sampling horizon or the target contaminants, or both. Soil gas data cannot be used to establish bulk volume

or the commerciality of buried petroleum, natural gas, or ore bodies.

5.2.2 With the necessary analytical procedures, soil gas can be examined for compositional anomalies, a very useful technique for multiple source problems. In some instances, contaminant occurrences are limited to single species (compounds, mercury, etc.), however more often than not the contaminant source is a mixture of organic chemicals that have a unique chemical compositional character consisting of both normally evaluated priority pollutants and non-priority pollutant chemicals that may be overlooked. By identifying and using compositional information, many problematic site situations such as degradation can be minimized by targeting the more refractory compounds associated with the contaminant occurrence. This interpretive method is impossible to model for an industry wide application due to variation in methods and technique.

5.2.3 A basic limitation of the technique is that due to the ease of procurement and use of soil gas sampling devices, there is a tendency for inexperienced personnel to oversimplify any and all aspects of the method. Investigators must consider the experience level and technical ability of personnel who acquire soil gas samples and attempt to interpret the results. Certain procedural facets are not trivial, as discussed in Section 6. The results of certain techniques tend to be affected by minor variations in procedure despite apparent adherence to a "Standard Operating Procedure."

5.2.4 Atmospheric air contamination is not a trivial problem corrected by simple device-oriented field practice. Many sampling systems recover very large volumes of "soil gas" that may actually represent a mixture of soil gas and atmospheric air. This mixing occurs through the introduction of ambient air adjacent to the sampling device and through macroporous pathways in the soil which are far from the sampling device. Some environmental investigators avoid the impact of this problem by reasoning that contaminant quantities in the soil are so great that they are detected despite atmospheric mixing. For qualitative approaches with non-rigorous quality assurance/quality control (QA/QC) objectives this mixing problem can be insignificant. For detection of compounds that exhibit only marginal partitioning preference for the free vapor phase, the mixing problem can be a fatal flaw in procedure. Moreover, contaminant concentration and composition investigations can be rendered useless by variations in the magnitude of mixing at various sample locations and depths in a survey area.

5.3 *Comments on Limitations of Soil Gas Monitoring*—Many investigators believe that soil gas monitoring is not an effective vadose zone monitoring method for certain volatile organic applications, in certain geographic regions or during certain seasons of the year, or both. The applicability of soil gas monitoring is controlled by physical and chemical properties and processes in the subsurface and not by factors that are obvious at or above the surface. For example, one common misconception is that soil gas monitoring is not effective during the winter season. The impacts upon soil gas measurement of elevated soil pore water content, reduced vadose zone temperature and the presence of frost, typical of numerous regions in winter, are obvious for many facets of most soil gas monitoring methods. Modification of standard operating procedure, such as an increase in sampling depth, or selection of another soil gas monitoring method altogether can minimize the negative impacts of seasonal field conditions. It is important to understand that the responsibility for success or failure in soil gas monitoring can reside as much in the planning phase of a survey, including the method chosen, as in factors controlling the chemical and physical processes at work in the subsurface. Even with apparently ideal field conditions and with a carefully planned survey, soil gas monitoring can succeed or fail due to unknown factors controlling contaminant migration and emplacement. Soil gas monitoring is no different than any other measurement method, in that investigators must maximize effort in planning and implementation of procedure to maximize the likelihood of success.

6. Approach

6.1 *Sampling Methodology*—Soil gas sampling methodology has evolved over time and through practice in several industries. The equipment with which to perform this monitoring technique is highly varied, although it may be categorized into basic types (see 6.2.2). The literature provides numerous discussions about the design of some of this equipment (10, 30, 31, 32, 33). The selection of a soil gas sampling method involves consideration of three primary issues. These are the type of sampling system, the methodology of application of that sampling system and the rigor of the field QA/QC protocol. Each of these issues is discussed in this guide, however, no single method or procedure is recommended to the reader due to the variation in site specific factors. As many as one hundred unique soil gas sampling systems exist that arise from variations or combinations, or both, of the many facets described in this guide. Some systems are highly versatile for numerous applications. Others are functional for more limited or specific applications. Informed investigators must assume the responsibility of selecting the technique most appropriate to the subject application, whether that technique is commercially available from contractors or equipment suppliers, or reliant upon the ingenuity of the investigator in the field utilizing commonly available materials. Success in choosing an appropriate sampling device or an entire sampling system is dependent upon the investigator's level of understanding of vadose zone processes, contaminant properties and appropriate applicability of the soil gas method.

6.1.1 The application of any of these methods must be controlled by strict adherence to a standard operating procedure. Occasional deviations as dictated by unusual field conditions should be recorded in the project field notebook. Inadvertent minor deviations in field procedure can result in misinterpretation of the data acquired.

6.2 *Sampling Systems*—Six basic sampling systems exist. These are based upon the collection of soil gas by a whole-air or sorbent method in an active or passive approach, or upon the principle of collection of a soil or water sample for subsequent sampling of a contained headspace atmosphere. Contained atmosphere methods do not yield samples representative of in situ vadose zone atmospheres.

6.2.1 Whole-air methods sample the soil atmosphere as a mixture of gases, including contaminant and non-contaminant vapors. Sorbent methods sample contaminants adsorbed onto a collection medium exposed to a whole-air

sample stream. Active methods are those that obtain a soil gas sample by positioning a sampling device in the subsurface and the withdrawal of soil atmosphere through the device from the sampling horizon. Passive methods are those that obtain a soil gas sample by placing a collection device in the soil or on the soil surface, and allowing the atmosphere within the device to come into compositional equilibrium with the soil atmosphere. Four of the six basic sampling systems arise from these approaches, namely the whole air-active approach, the sorbed contaminants-active approach, the whole air-passive approach, and the sorbed contaminants-passive approach. Two additional systems exist that are based respectively upon the collection of a soil or water sample for subsequent sampling of a small volume headspace atmosphere.

6.2.2 *Whole Air-Active Approach*—This method of soil gas sample collection involves the forced movement of bulk soil atmosphere from the sampling horizon to a collection or contaminant device through a probe or other similar apparatus (10, 34). Contained samples of soil atmosphere are then transported to a laboratory for analysis, or the sampling device is directly coupled to an analytical system. Whole air-active sampling is best suited to soil gas monitoring efforts where contaminant concentrations are expected to be high and the vadose zone is highly permeable to vapor. Probes exist that must utilize pre-existing holes or that can penetrate the vadose zone by driven means. These devices can be very simple and light-weight for low cost mobilization (35), or they can be affixed to vehicle mounted drills or hammers useful for larger, more complex surveys at a higher cost of mobilization. The whole air-active technique can be combined with other monitoring methods such as soil monitoring for engineering purposes (36) in some survey environments. The success of this practice can be highly site-specific.

6.2.2.1 Ground probes can be of small to large internal volume. The development of sampling devices with smaller internal volumes equating to smaller purge volumes is a significant improvement, providing samples which are more representative of soil atmosphere, and a greater ease of equipment decontamination between usages. Sample size can vary from a few millilitres to many tens of litres depending upon the sample rate through the probe, the vapor storage capacity of the soil and the ability of the soil to deliver vapor to a probe under vacuum.

6.2.2.2 The success of the active approach is strongly dependent upon soil clay, organic matter and moisture content. Driven probes tend to destroy natural soil permeability around the body of the probe due to soil compaction concurrent with insertion. This can be a severe limitation in moist, heavy clay soils. In very dry, cemented soils, driven probes can create radial fractures that can enhance soil permeability to vapor concurrent with insertion. These fractures can communicate atmospheric air with soil atmosphere, a limiting factor for obtaining representative, large-volume soil gas samples. The effect can be so severe as to lower recovered contaminant concentrations in the soil gas sample below the limits of analytical detection. This is especially true for highly sorptive or water soluble compounds, or both. Some investigators have attributed the poor recoveries of these compounds exclusively to other processes, that is, degradation (21, 37).

6.2.2.3 Methods requiring a pre-existing hole for probe insertion (38) made with a commercially available "slam bar" can provide supportable contaminant data where contaminant concentrations and soil permeability to vapor are high, however the act of making a hole with a "slam bar" and subsequent removal of the "slam bar" can encourage soil contaminant venting and lower sample representativeness. Insertion of the sampling probe into this hole further degrades representativeness by additional venting of contaminants as the probe displaces the atmosphere in the hole upon insertion. Purging of the probe prior to sampling under conditions of low soil permeability and low contaminant concentration may lower contaminant levels below the limits of analytical detection. Methods requiring a pre-existing hole for probe insertion are not recommended for soil gas sampling from soils with high clay and moisture contents.

6.2.2.4 Excellent discussions of numerous whole air-active sampling systems may be found in the literature (10, 21, 37, 39). Investigators must consider the caveats and limitations of the whole air-active approach when selecting a certain method for a specific application.

6.2.3 *Sorbed Contaminants-Active Approach*—The sorbed contaminants-active method of soil gas sample collection also involves the forced movement of bulk soil atmosphere from the sampling horizon through a probe or other similar apparatus, but to a collection device designed to extract and trap sample stream contaminants by adsorption (40, 41). This system is well suited to sites where the soil may be highly permeable to vapor and where the contaminant concentration may be lower than required for successful whole-air surveys. Sorbent devices are designed to concentrate the components of interest and remove some of the soil gas components known to interfere with sample analysis.

6.2.3.1 Contaminant trapping is accomplished by use of an adsorbent collection medium such as charcoal or a carbonized molecular sieve adsorbent (43, 44), as well as porous polymers, silica gel and activated alumina (10). This approach is especially amenable to the detection of nonpolar volatile organic compounds. Organic compounds that are reactive, oxygenated or are gaseous at room temperature are either not adsorbed by or are not efficiently desorbed (42) from charcoal. Sorbent collection devices are commercially available or can be specially prepared with an appropriate sorbent material that concentrates desired compounds for future analysis. Colorimetric detector tubes are available which will provide an indication of the presence of target compounds at the time of sampling. These devices are limited in application by the high concentration requirements for many compounds and the compound-specific nature of these tubes.

6.2.3.2 The effectiveness of the sorbed contaminants-active approach can be limited by high vadose zone clay and water content, reducing the ability of the soil to transmit vapor through the sorbent trap. Commercially available sorbent traps come with information suggesting maximum, minimum and optimum sampling rate through the trap. Soil characteristics can limit flow rate to a point below the minimum recommended rate, affecting the performance of the trap and the reproducibility of adjacent samples. Interac-

tion of the sorption media with target compounds during desorption in the laboratory can form artifacts, restricting the interpretive value of the data. Some sorption media are prone to irreversible adsorption (see Definitions D 2652). Some may be affected by high soil gas relative humidity. Humidity greater than 60 % (very common for soil gas) can reduce the adsorptive capacity of activated charcoal to 50 % for some chemicals. Presence of condensed water in the sample tube will indicate a suspect sample (see Practice D 3686). Anticipation of these problems is recommended for all sorbent techniques, and a thorough quality control plan should be designed and implemented as is discussed in 6.4 of this guide.

6.2.3.3 Special sample preparation is required for samples adsorbed onto a trapping medium. This preparation step consists of the thermal or solvent desorption of the contaminants from the trapping medium. Proper practice will promote needed accuracy and precision in the determination of contaminant concentrations above specified values (see Practice D 3687).

6.2.4 *Whole Air-Passive Approach*—This method of soil gas sample collection involves the entry of bulk soil atmosphere or soil atmosphere components from a near-surface sampling horizon to a collection or containment device through a flux chamber or other similar apparatus (30). Enclosure devices sample vaporous emissions from a known soil surface area capped by a chamber. The volume of the chamber is continuously swept by injection of a gas of known composition, and the resultant carrier gas-contaminant mixture is collected for analysis. The rate of emission or "flux" of contaminants can be calculated if flow rate of injected gas and contaminant concentration in the sample are determined.

6.2.4.1 The whole air-passive approach is useful to some very specific applications. This method may be used, for example, to monitor contaminant emissions from soil or water to assess the health hazard risk of such emissions to the general public. Determination of the extent of contamination by volatile organic compounds has been performed with whole air-passive devices, however the application of other types of systems is far more common.

6.2.4.2 A key to successful operation of a whole air-passive system is that the system is able to recover volatile compounds as they are emitted from the vadose zone. The effects of changes in barometric pressure, soil temperature and soil moisture content are not quantifiable from site to site due to site specific variables controlling vapor phase contaminant migration and the rate of contaminant partitioning into the vapor phase. The presence of contaminants or naturally occurring organic matter floating on surface water may impact the rate of entry of certain vapor phase contaminants into the chamber.

6.2.4.3 The whole air-passive method is limited in application primarily due to the great degree of dilution of contaminants in the sample stream by injected gas. This can decrease method sensitivity by lowering contaminant concentrations to levels below the detection limits of the analysis method chosen. Further decrease in method sensitivity results from the fact that soil gas contaminant concentrations are generally lower at the surface than even at nominal depths. Soil characteristics such as high water saturation, soil cements, clay content and organic matter content will negatively impact results of these systems by restricting the rate of contaminant flux to the chamber.

6.2.4.4 Additional limitations exist. Certain devices limit flux rates into the chamber due to aspects of design. Soil macroporosity such as desiccation cracks extending beyond the collecting device will vent soil vapors to the atmosphere that will not be collected by flux chambers unless monitoring locations are biased to include these features.

6.2.5 *Sorbed Contaminants-Passive Approach*—This method of soil gas sample collection involves the passive movement of contaminants in soil to a sorbent collection device over time. Passive samplers that have been applied to sampling soil gases of environmental concern include occupational health volatile organic compound monitors (44) and a sampler originally developed for detecting the presence of hydrocarbons in petroleum exploration (33, 46). Both devices use charcoal as a sorbent; the former as a flat film and the latter coated on a wire. Passive samplers are housed in containers up to several inches in diameter, depending upon the design. They are placed open end down in holes that are usually less than 5 ft (1.5 m) deep, that are then backfilled (32). These monitors are generally left in place from two to ten days, although certain passive collectors can be left in place for a period of 30 days or more for certain applications. For at least one device, exposure efficiency can be determined.

6.2.5.1 The sorbed contaminants-passive approach can be employed in a wide range of geological conditions. Frozen ground and high water saturation may not limit the ability of the monitors to collect contaminants (46), although the composition of the contaminant suite may be impacted by related alterations in partitioning equilibria.

6.2.5.2 The sorbed contaminants-passive approach depends upon the ability of contaminants to move through the vadose zone to the passive collection device. Numerous adsorption media can be used to collect contaminants (see 6.2.4). The principle of passive-sorbent monitors relies on adsorbent reduction of the equilibrium concentration of contaminants around the monitor over time, therefore creating a concentration sink, that is, a continuous state of disequilibrium, in the vicinity of the monitor. This can encourage continued migration of contaminants toward the monitor when conditions for contaminant partitioning into the vapor phase are favorable. Migration of contaminants in the vadose zone toward a passive-sorbent device is strongly controlled by vadose zone character and the chemical and physical properties of the subject contaminants. Contaminants may move from a few feet to thousands of feet, or not at all.

6.2.5.3 Many investigators attribute the principle mechanism of contaminant migration to a passive-sorbent device to diffusion, that is, the movement of organic vapor or gas molecules from a region of high concentration to a region of low concentration as described by Fick's law (see Practice D 4597). Fick's law of diffusion states that for a constant concentration gradient, the mass of material transferred to the sampling layer can be expressed as:

$$M = \{DA(C - C_o)t\}/L \qquad (5)$$

where:

M = mass of the material, ng,
D = diffusion coefficient, cm²/min,
A = cross sectional area of diffusion cavitie(s), cm²,
L = length of diffusion path, cm,
C = concentration at face of sampler, ng/cm³,
C_o = concentration at adsorbing layer surface, ng/cm³, and
t = exposure time, min.

6.2.5.4 The cross sectional area of a diffusion cavity, the length of the diffusion path and the quantity $(C - C_o)$ are impossible to accurately measure for soil gas contaminants interacting with a passive-sorbent sampler. There is some debate as to whether passive samplers measure flux or total contaminant concentration (32) in the vicinity of the trap. Due to the fact that the mass of the material transferred to the sampler by diffusion, a key measurement, cannot be determined, the debate will no doubt continue. It is reasonable to assume that a combination of processes is responsible for contaminant migration to sorbent traps, including diffusion, dispersion and mass transfer. All migration processes are impacted by partitioning equilibria.

6.2.5.5 Ambient air represents an atmospheric contaminant concentration sink that encourages a strong vertical vector of contaminant migration. This prevailing upward movement of contaminants from sources at depth results in contaminant concentration gradients throughout the vadose zone. The sorbed contaminants-passive method makes use of this contaminant flux (see 6.2.4) to collect long-term, nondisruptive samples of volatile contaminants. The method can collect contaminants which are compositionally representative of the contaminant mixture favoring the vapor phase. The quantity of volatile organic compounds trapped by these devices is proportional to the concentration gradients of contaminants present near the collection device and the affinity of the contaminant(s) for the collection medium.

6.2.5.6 As with active sampling protocols, specific issues exist affecting the function and calibration of passive monitors. Soil gas, even in the drier climates, will be at a relatively high humidity condition. This humidity can affect the collection efficiency of the adsorbent media. In soils of low permeability, contaminants commonly move very slowly. This can create a condition of near-zero contaminant concentration in the soils immediately adjacent to the monitor if the sorptive potential of the monitor is higher than that of the soil. When soil contaminant concentrations are rapidly depleted, that is, as the result of invasion of the sampling horizon by meteoric water, the passive monitor can source contaminants back to the soil.

6.2.5.7 The sorbed contaminant-passive approach to soil gas monitoring is not immune to the migration, emplacement and degradation factors affecting all soil gas monitoring techniques. It is not possible to measure the efficiency of passive-sorbent monitoring devices because the bulk volume of soil gas affected by the sorbent trap cannot be measured. Care must be taken not to contaminate the sorbent samples during installation or by backfilling with contaminated soil. Such care is comparable to potential problems for any measurement method in which a contaminated layer is penetrated.

6.2.6 *Soil Sampling for Subsequent Headspace Atmosphere or Extraction Sampling*—This method examines contaminants that are present in a headspace atmosphere above a contained soil sample. Note well that this headspace atmosphere is not true soil gas (see 3.1.11), but is an artificial atmosphere formed above a potential contaminant source, that is, the soil sample. Contained atmosphere methods do not yield samples representative of in situ vadose zone atmospheres. Headspace atmospheres differ from in situ vadose zone atmospheres in that large percentages of vapor phase and moderate percentages of solute and sorbed phase contaminants can be lost in the act of soil sampling. This method is not generally recommended for a broad spectrum of cases due to numerous limitations and caveats. In comparison to other methods described in this guide, soil sampling for subsequent headspace atmosphere or extraction sampling can be a relatively poor method for determining many of the more volatile contaminants. Headspace atmospheres contain residual sorbed and solute phase contaminants that have partitioned to the vapor phase in the contained environment; most headspace approaches are reasonably efficient in recovery of some fraction of sorbed and solute phase contaminants. Contaminants in these phases in situ are recovered from a headspace after they have partitioned into the vapor phase. Recovery efficiency of contaminants in the vapor phase in situ ranges from moderate to poor.

6.2.6.1 Important criteria exist to consider when selecting a device that will provide suitable samples (see Guide D 4700). The equipment required is simple and readily available. Some commonly used augers are not suitable for soil sampling in support of subsequent headspace atmosphere sampling due to soil disturbance. Depths of auger investigations are limited by ground water conditions, soil characteristics and the equipment used (see Practice D 1452). Suitable procedures for some methods are described in the literature (47, 48). Current soil preservation practice may not apply (see Practice D 4220).

6.2.6.2 Limitations and special procedures exist for the application of soil sampling for subsequent headspace gas analysis. Filling head space with solvent can support a subsequent solvent extraction procedure. Some investigators minimize the effects of devolatilization by rapidly recovering small soil core plugs with polypropylene syringes which have been modified to accommodate recovery of soil plugs. Investigators also attempt to maximize partitioning of contaminants into the vapor phase by adding buffering solutions or sodium sulphate and phosphoric acid to the vial prior to sealing, in order to shift the activity coefficients of the subject contaminants to favor the vapor phase. Aqueous suspensions of solvent slurries of soil can be ineffective for the determination of high molecular weight labile compounds. Their persistence in soil is the result of physical entrapment in soil microporosity (49). Recovery efficiency of contaminants in soil headspace can be greatly enhanced by pulverization of the soil (50) in a ball mill or other similar apparatus. The method is biased toward recovery of contaminants in the sorbed, solute and occluded phases in situ due to the loss of pore space gas in preference to contaminants adsorbed onto the soil particles or trapped in soil micropores. Contaminant degradation, especially biodegradation, in the container is encouraged by the creation of an aerobic, moist environment during sample handling and transport prior to analysis. However, a simple method to minimize the effects of

biodegradation can be achieved by storing samples, when necessary, at approximately 4°C in the dark.

6.2.6.3 Acid extraction of volatile organic compounds is widely used in geochemical exploration for petroleum and natural gas. Soil samples are placed in a closed vessel, heated and evacuated to remove vapor phase contaminants. The addition of acid to the evacuated chamber causes release of hydrocarbons believed to be bound to the soils by carbonates (22). Hydrocarbons are determined by analysis of resulting vessel atmospheres. Refinements to this method have been developed (48), however the method is designed not to determine compounds in the vapor, sorbed, or solute phases. Method sensitivity is therefore greatly reduced.

6.2.7 *Soil Pore Liquid Headspace Gas Approach*—In the vadose zone, soil gas monitoring can be accomplished in combination with soil pore liquid sampling through the use of a suction lysimeter, a pan lysimeter or a free drainage glass block sampler. The suction lysimeter installed in the vadose zone is most commonly employed for this purpose. Temporally designed surveys are ideally suited to this method.

6.2.7.1 After a lysimeter has been installed for some period of time, initial aliquots of vapor sampled from a soil pore liquid sampler will be in compositional equilibrium with solute phase contaminants when pore liquid tensions are within the operating range of the lysimeter and if pore sizes are not so great as to cause loss of hydraulic contact between the soil and the porous segment of the lysimeter. Subsequent aliquots of soil gas may compositionally resemble soil vapor in situ if soil atmosphere enters the porous segment of the sampling device. When the lysimeter cannot recover a pore liquid sample, the soil gas recovered will be compositionally similar to soil vapor in situ.

6.2.7.2 The most common effort to recover soil gas from a suction lysimeter occurs when polytetrafluoroethylene (PTFE) porous segments are employed in sampling environments with high soil moisture tensions (low moisture contents). At tensions above 60 to 80 centibars, soil pore liquid samples cannot be collected (see Guide D 4696). However, soil gas can be recovered through the porous segment and collected at the surface. This alternative sampling effort can monitor soil vapor contaminants utilizing an otherwise unsuccessful procedure until soil moisture contents increase or until an alternative soil pore liquid sampler can be installed.

6.2.7.3 This technique is limited by the relative expense and complexity of installation of the sampling devices as a primary soil gas sampling method. The completeness criterion for quality assurance is difficult to satisfy due to the inability to anticipate the performance of the soil pore liquid sampler with respect to vapor recovery. Moreover, compositional bias toward solute phase contaminants and contaminants volatilized from free product is likely in soil gas samples recovered concurrently with soil pore liquid samples.

6.3 *Methodology in Application of a Sampling Technique*—The likelihood of success of the soil gas sampling technique selected is controlled in part by the methodology in application of that sampling technique. This methodology should be guided by the objectives of the subject project and the perceived spatial and temporal array of the potential sampling targets.

6.3.1 *Grids*—Many problems suitable for soil gas monitoring are best solved by obtaining data distributed over a geographic area. Sampling in grid patterns of variable design and spacing can be a very effective way to provide data coverage over a large area for a very low cost of acquisition. Common applications of soil gas grid sampling are environmental contaminant assessments, exploration for natural resources and the siting of locations for other monitoring or exploratory techniques. Compositional analyses in conjunction with properly designed grid systems are often fundamental to successful evaluation of soil gas monitoring.

6.3.1.1 Grid spacing provides for the location of soil gas samples in grid cells. The selection of grid cell size is strongly dependent upon the relationship between project confidence level requirements and cost budget. Small survey targets and complex vadose zone geology require decreased spacing between soil gas sample locations for grid methodology to be successful. Some applications, for example, defining the boundaries of contaminated soil or ground-water contaminant plumes, may require the grid cell area to be as small as 100 to 400 ft^2 (9 to 37 m^2). Most applications to natural resource exploration monitor naturally occurring volatile compounds in soil atmospheres, requiring closely spaced grids to increase the signal to noise ratio. However, a closely spaced exploratory grid equates to a broadly spaced grid for environmental application in most situations. Common petroleum exploration grid spacing utilizes a grid cell area of approximately 250 000 ft^2 (23 000 m^2), however grid cells can range from 10 000 to 1 000 000 ft^2 (9 to 90 000 m^2) depending upon perceived reservoir target area. Widely spaced grid sample arrays are useful in reconnaissance applications such as the establishment of contaminant baselines or evaluation of the exploration potential of a geologic basin. Grid cells for such purposes can be as large as a square mile or more.

6.3.1.2 The tendency exists for investigators with constrained budgets to utilize overly large grid cell spacings. This action normally results in inadequate, over-interpreted data supporting meaningless conclusions. Care must be taken to avoid this caveat.

6.3.1.3 Grid arrays can be designed as regularly spaced and predetermined locations for soil gas sampling or they can be irregularly spaced and continually field modified. Predetermined and widely spaced grid patterns are most useful for reconnaissance work, while closely spaced, irregularly situated or field modified soil gas grid sample sites, or both, are commonly used when targeting contaminant plume boundaries, contamination from underground storage tanks or other detail work.

6.3.1.4 Multiple depth sampling, discussed in 6.3.3, when coupled with a soil gas grid sampling methodology, can provide useful data in complex geologic settings and sites with multiple contaminant sources. Computer mapping of closely spaced three-dimensional soil gas grids can provide the investigator with horizontal or vertical cross sections through the subject site, making difficult observations possible.

6.3.2 *Profiling*—Profiling is a soil gas sampling methodology useful to test a linear array for the existence of contaminants. Profiling is most often performed by sampling at closely spaced intervals in a linear array and is displayed as

contaminant concentration or composition versus distance sampled on an $X - Y$ plot. Concentration data are often displayed logarithmically on the ordinant (Y) axis, while single components or ratios of compositional data are often displayed linearly on the ordinant axis.

6.3.2.1 For environmental applications such as leak detection along the length of a pipeline or monitoring of contaminant encroachment across a property boundary, soil gas samples are recovered along a profile at intervals from 25 to 100 ft (8 to 30 m) (23). Profiling for natural resource exploration can be performed at sample intervals from 50 to 500 ft (15 to 50 m), depending upon the application.

6.3.2.2 Profiling is useful as a corroborative tool for other monitoring or exploration methods. For example, a soil gas sample profile acquired coincident with a seismic profile can suggest primary contaminant migration pathways or the boundaries of confining layers in shallow, complex geologic settings. This technique has been demonstrated as highly effective in reducing exploratory risk prior to drilling for petroleum and natural gas, by suggesting the presence of hydrocarbon seepage coincident with structures with reservoir potential defined by the seismic method (51).

6.3.2.3 Soil gas profiling is also a convenient methodology effective in comparative evaluation of multiple soil gas sampling techniques. Due to variations common to the dynamic equilibrium conditions over small spatial and temporal intervals in the vadose zone (see 4.1), comparisons of multiple soil gas techniques using only one or a few soil gas samples recovered from nearly identical locations will not result in a valid comparison. However, a visual overlay of soil gas profiles resulting from the implementation of the various sampling techniques can provide a rapid and definitive comparison as to the efficiency of recovery of subject contaminants by a particular sampling system in a specific sampling environment. Similarly, comparison of profiles obtained by using the same soil gas sampling system can provide a direct measurement of system accuracy for quality control purposes.

6.3.2.4 Some investigators compare geographically coincident profiles obtained with the same sampling system at times differing by days or even years in order to generate a data correction factor in order to enhance data comparability. This practice is strongly discouraged. Factors not anticipated in this practice such as the effects of the dynamic equilibrium in the vadose zone, unavoidable changes in procedure due to personnel substitutions, contaminant movement or cultural influence on the sampling environment can have impact on results that are far more significant than the apparent correction.

6.3.3 *Multiple Depth Sampling*—Methodologies encompassing multiple depth sampling normally have one of two goals, that is, to monitor changes in soil gas contaminant fractions versus depth, and to closely follow a single sampling horizon for an entire soil gas grid or profile.

6.3.3.1 When the goal of a survey is to monitor contaminants over varying depths, some sampling systems can recover soil gas samples as probes are advanced deeper into the vadose zone. This practice is helpful in determining the optimum sampling depth for a particular site or to demonstrate the presence or absence of soil atmosphere contamination in a certain horizon. Soil gas contaminant concentrations often increase with depth as the sampling horizon approaches contaminated ground water or other source of soil gas contaminants (52). Caution must be exercised when soil gas sampling tools are advanced to increasing depths due to the fact that cross contamination of some or all of the sampling system is unavoidable. This situation limits quality control for this type of multiple depth sampling. Attempts to eliminate cross contamination in multiple depth sampling by replacement or decontamination of sampling equipment with each new sample aliquot also result in limited quality control. Tool withdrawal and tool reinsertion result in venting of the sampling environment via an open hole. The open hole behaves as a macroporous pore space, allowing enhanced partitioning into the vapor phase and convective migration to the atmosphere. The end result is a reduction in representativeness for each subsequently recovered soil gas sample.

6.3.3.2 Multiple depth sampling can also be used to focus a sampling program into a single geologic unit or suite of units without regard to depth. This practice is helpful at sites with complex lithologic changes in the vadose zone. Samples can be recovered from lithologies with greater permeability to vapor or greater storage capacity for vapor when bias in sampling depth is necessary to accomplish project goals. This practice involves greater effort and expense than most methodologies due to the necessity to establish the presence, thickness and depth of the target horizons prior to soil gas sampling. The most common application of this methodology is the sampling of soil gas at the top of the capillary fringe.

6.3.4 *Time Variant Methodologies*—Monitoring soil gas in the vadose zone over time can suggest process rates of contaminant partitioning, emplacement, migration and degradation. Practical application of this methodology includes the monitoring of the effectiveness of remedial air-injection systems, the appearance of contaminants sourced from underground storage tanks, the encroachment of contamination onto a subject property from an abutting property and the mitigation of soil and ground-water contamination by microorganisms.

6.3.4.1 Some investigators and regulators with responsibilities at more than one location delegate seemingly simple time variant soil gas monitoring tasks to local personnel. Numerous problems with time variant monitoring can arise in the field as the result of poor system maintenance and record keeping by inexperienced or unmotivated personnel (property owners or parties responsible for contamination).

6.3.4.2 Certain maintenance problems are easily corrected, that is, cleaning bacteria and other foreign matter from detectors or replacing damaged components. Other maintenance problems can be fatal flaws in the methodology. These are principally related to ice formation in the sampling system and destruction of system integrity due to soil frost heaving.

6.3.5 *Combination of Soil Gas Monitoring With Other Vadose Zone Monitoring Techniques*—Soil gas monitoring is not a stand-alone technique. Corroborative support of this reconnaissance and screening tool by other vadose zone monitoring techniques is strongly encouraged. The possible combinations of the various vadose zone techniques with soil gas surveys are numerous. Soil gas can commonly be used as a reconnaissance tool to locate other monitoring devices

such as lysimeters, neutron probes or ground water monitoring/sampling wells. Limits upon such combinations are controlled by budgetary constraints and the investigator's imagination.

6.4 *Field QA/QC*—Quality assurance and quality control procedures (QA/QC) are essential to establishing support for any interpretation of measurement data. Soil gas monitoring data requires a thorough QA/QC protocol confirming that data have been generated to satisfy the data quality objectives for the survey. This requirement is well known, however few investigators subject their soil gas data sets to the rigors of such protocol. Conclusions based upon data of unknown quality may be without merit. Justification for interpretations based upon data of unknown quality is not possible.

6.4.1 QA/QC requirements are dependent upon the data quality objectives defined in the planning phase of the survey. For example, simple contaminant audits require a less demanding QA/QC protocol than contaminant source identification. The goals of the QA/QC effort must be understood by field personnel to assure effective implementation of field QA/QC. A document control officer who is a member of the field team can provide this assurance.

6.4.2 Persons collecting descriptive data should not be varied during a soil gas survey. Soil descriptions, for example, can be somewhat subjective when estimations are made as to soil moisture or clay content. Changes in field personnel can translate into apparent changes in soil lithology that are merely functions of this subjectivity. The document control officer can review field records to discover any obvious errors related to descriptive data.

6.4.3 The results of a soil gas survey are highly sensitive to procedure. Field personnel should closely follow a standard operating procedure. This procedure should include the method(s) selected for the survey including the sampling system, means of sample collection, handling and transport of samples and field based equipment decontamination. A standard practice for equipment decontamination is essential to maximize the integrity of samples that may undergo chemical analyses (see Practice D 5088). Any deviations in the standard operating procedure should be recorded by the document control officer in a field notebook, with notes outlining the justification for the deviation. Data comparability can be severely compromised by deviations from the standard operating procedure.

6.4.4 Field based equipment decontamination can have impact on data quality. This results from the potential for cross contamination of samples due to poorly controlled field cleaning procedure or difficulties presented by the inconvenience of field decontamination. Field based equipment decontamination should not be considered a method of choice, but if unavoidable, must be performed with the data quality objectives for the survey as driving forces for procedure.

6.4.5 Bias of soil gas data describes a situation of consistently lower-than-actual or higher-than-actual soil gas contaminant concentration measurements (32). The bias of a measurement process is a generic concept related to consistent or systematic difference between a set of test results from the process and an accepted reference value of the property being measured (see Practice E 177). Bias can be imparted to the data through sample site selection, that is, exposure of a sampling device to an environment of enhanced contaminant concentration due to a preferential contaminant migration pathway, or exposure of a sampling device to an environment devoid of contaminants due to barriers to contaminant migration. Bias may also result from malfunction of the sampling system, contaminant degradation or numerous other factors. False positive or false negative values can result, lowering the value of the soil gas data set.

6.4.6 Table 2 summarizes some common problems in soil gas monitoring that can result in biased results.

6.4.7 A sampling program must be conducted during the survey to support evaluation of both the sampling system in the field and the analytical system employed. These samples are known as QA/QC samples. The type and magnitude of QA/QC sampling depends upon the purpose of the soil gas survey and the requirements for data quality attendant to it. It is the responsibility of the investigator to determine the appropriate rigor of field QA/QC protocol. The variation in QA/QC protocol from survey to survey is controlled by the purpose and magnitude of the survey, and can vary to a great degree.

6.4.8 The types of field QA/QC samples are field blanks, travel blanks, sample container blanks, sample probe blanks and sample replicates. Other types of QA/QC samples are analytical in nature and are discussed in 6.6.

6.4.9 Field blanks are samples of ambient air or nitrogen recovered from the sampling system which are recovered to determine contamination of samples by ambient atmospheric air, or, to act as system blanks to test for contamination of the sampling system. Field blanks are used to provide an indication of the probability of leakage in the sampling system or the breakthrough of atmospheric air to the sampling device through macroporous migration pathways in the vadose zone such as soil cracks or moldic porosity. If nitrogen is employed instead of atmospheric air, field blanks can have higher contaminant levels than soil gas. This is especially true for petroleum hydrocarbons in urban environments. At least one field blank should be recovered for each ten soil gas samples, or at least one field blank per sample batch or container type (53).

6.4.10 Travel blanks are the contents of a sample container handled in the same manner as those containers holding samples, except that there has been no sample inserted into the travel blank. The purpose for travel blanks

is to audit sample integrity for loss due to sample handling and transport. Travel blanks are useful when analysis is performed at an off-site laboratory. The results obtained by analysis of travel blanks can be used to indicate a potential need to modify sample handling and transport procedure. At least one travel blank should be included in each batch of samples.

6.4.11 Sample container blanks are obtained by sampling the contents of a clean sample container to ensure that residual contaminants are not present in the container prior to sample collection. If contamination is detected in the cleaned containers, the decontamination procedure must be modified to remedy the problem. Sample container blanks should be collected and analyzed prior to each use of a sample container.

6.4.12 Sample probe blanks, consisting of carrier gas or atmospheric air contrasted to atmospheric air blanks, are drawn through the sampling device and recovered in the same manner as soil gas. The purpose for sample probe blanks is to check for the presence of sample train contaminants that would impact data quality. If contaminants are detected in sample probe blanks, the decontamination procedure must be modified to remedy this condition. Sample probe blanks should be collected and analyzed prior to each use of a probe and/or other components of the sampling system.

6.4.13 Field replicates are recovered as separate soil gas samples collected from the same sample site into multiple containers. Field replicates can be used to estimate the combined precision of sampling and analysis. The recovery of field replicates is not a common practice. When field replicates are demanded by a client or as dictated by a particular situation, field replicates should be recovered as often as is economically and practically possible, however in no instance should the number of replicates fall below ten percent of the total number of soil gas samples (53).

6.4.14 Sample spiking, or the addition of a known quantity of a known compound or mixture to the soil gas sample, is sometimes performed in the field to provide internal checks of analytical quality. Sample spiking in the field is not recommended due to measurement uncertainties in the field. Moreover, caution must be exercised with this procedure because of the potential for contaminant interaction with the known compound(s).

6.4.15 A paperwork audit is recommended at the end of each working day or at the conclusion of recovery of each batch of samples recovered. The paperwork audit should be conducted by the document control officer and include evidence of an equipment inventory, sample inventory including QA/QC samples, review of field notes and chain-of-custody documentation.

6.4.16 Chain-of-custody documentation is recommended at all times, and is mandatory for soil gas surveys when samples are transmitted to an off-site laboratory. It is recommended for soil gas surveys when sample custody is transferred to someone other than the field team leader for any reason. Chain-of-custody documentation assures that samples have not been altered or mishandled prior to analysis. This procedure is mandatory for sample handling and transport in situations where there is likely to be a cost

TABLE 2 Summary of Possible Causes of False Positive and False Negative Values[A]

Result	Causes
False negatives, that is, falsely low values	Barriers to gaseous diffusion, such as perched water, clay lenses, impervious man-made debris, saturation of soil pores with water (as from rain), low subsurface temperatures.
	Biological or chemical degradation.
	Leakage or blockage in the sample train, improper purge procedure, loss of sample from sample container, problem with analytical system.
False positives, that is, falsely high values	Contamination in sampling train, sample container, or analytical system.
	Contribution of volatile organic contaminants from vegetation.
	Significant contamination in overlying soil.

[A] See Ref (32).

recovery effort or demonstration of contaminant responsibility in a court of law.

6.5 *Sample Handling and Transport*—Soil gas sampling and analysis usually involve the monitoring of contaminants at very low levels. Consideration of sample handling and transport is not trivial to this exercise.

6.5.1 The period of sample handling and transport represents the greatest opportunity for loss or gain of contaminants from or to sample containers. Loss occurs by contaminant condensation within the sampling train, sorption onto materials within the sampling train, solution into condensed water in the sampling train, chemical changes or leakage to the atmosphere through defects in the sampling apparatus or sample container. Gain of contaminants from sources other than the sampling horizon can occur through related mechanisms working in reverse. Both processes can severely limit the value of data obtained from a survey, and they must be minimized.

6.5.2 In general, the time between sample collection and analysis should be minimized. Investigators should protect samples against light and heat, and exercise precautions against leaks (see Practice D 1605).

6.5.3 *Acceptable Materials*—Investigators are responsible for selecting materials for soil gas sampling, transfer and containment that will not impact sample integrity. Containers that have parts made from porous or synthetic materials such as PTFE, rubber or many plastics are likely to retain or contribute contaminants to soil gas samples. Corrosive metals such as steel or brass become difficult to decontaminate upon corrosion due to the increased surface area of the corroded material and its enhanced sorptive capacity. Septa of any material will be responsible for measurable contaminant loss over time due to leakage. Acceptable materials can be conveniently decontaminated prior to soil gas recovery. Materials that cannot be decontaminated effectively between samples must either be replaced between samples, considered in QA/QC planning as a survey limitation or abandoned in favor of more suitable materials.

6.5.4 *Integral Systems*—Problems of sample handling and transport are minimized by integration of the sampling and analytical system. For example, a whole air-active sampling system can be coupled directly to a portable VOC (volatile organic compound) analyzer. The sample stream is fed directly to the intake port of the analyzer and passed through the detector. If there are no system malfunctions in the sample path, problems of sample degradation become trivial.

6.5.4.1 Care must be exercised with integral systems, however. The dead volume of integral systems is much higher than separate sampling and analytical systems. If the sampling system is not capable of delivering constant sample flow rates at or exceeding the requirements of the analyzer employed, data accuracy and comparability can be seriously affected. Moreover, a large sample volume is required merely to purge the sample system. In soils with moderate moisture contents or even nominal clay contents, it may not be possible to recover the volume of soil gas required to purge the system without serious negative impact to the composition of the soil gas sample recovered. Vapor phase contaminants can be lost to purge volume and atmospheric break-

through can occur, leading toward a false negative result. Although this problem may not be apparent in seriously contaminated environments, it can become a fatal flaw at low contaminant levels.

6.5.4.2 Cross-contamination is a concern with integral systems. Many integral systems employ common elements from sample to sample, namely tubing, flow meters and analyzer components. Overcoming persistent contaminants can be difficult in integral systems, especially when high soil humidity and cold weather complicate the field effort.

6.5.5 *Transfer of Samples from Sampler to Container*—The method of transfer of samples from sampling device to containers is largely dependent upon the volume of soil gas recovered.

6.5.5.1 Small volume samples are commonly recovered by syringe for immediate injection into an analyzer or small volume container. Glass gas-tight chromatography syringes are employed when rigorous QA/QC protocol is required and samples are injected into the analyzer immediately upon recovery. These syringes must be decontaminated prior to recovery of each sample aliquot. Disposable syringes are employed when samples are to be transferred to a small volume container for transport. They are inexpensive, commercially available and convenient to use. However, disposable syringes can present a disposal problem. They should be inventoried prior to use and destroyed after use, the number destroyed equalling the number inventoried and used. Destruction includes smashing the syringe cylinder and clipping the needle.

6.5.5.2 Hand pumps are also used to transfer samples into tedlar bags or glass bulbs. Hand pumps are preferably installed behind the analyzer or container in the sample train to avoid contribution from or loss of contaminants to the hand pump. Hand pumps commonly contain petroleum-based lubricants which will contribute to the hydrocarbon content of soil gas. These devices must be placed at the end of the sample train or abandoned in favor of another tool.

6.5.5.3 Large volumes of soil gas are commonly recovered by hand or mechanical pumps installed at the end of the sample train. Large volume systems can be metered for soil gas flow rate, which is controlled by the capacity of the vadose zone sampling horizon to transmit vapor to the sampling device, the volume and configuration of the sampling system and the requirements of the analyzer or sorptive trap employed.

6.5.5.4 Small volume sampling is quite sensitive to variations in sample transfer technique. Septum coring by syringe is a common problem that restricts flow of soil gas through the needle. Coring can be corrected by decreasing the needle size and using a relatively hard septum material. Coring does not occur with side-port needles, a high-cost alternative. Needles of 25 to 27 gage seldom core septa. However, flow rates through these small gage needles are slow enough to require great care in consistency of sampling rate to minimize septum bleeding during sampling. This consistency is highly subjective and must be obtained through experience. Polypropylene disposable needles may provide opportunity for contaminant loss by sorption or gain by contribution to the soil gas sample. This can be minimized by using the polypropylene syringe to purge the sampling device prior to sampling, thereby reducing the potential for loss or gain of contaminants to that of the sampling device. Luer-lock needles should be checked for tightness by twisting prior to each use.

6.5.5.5 Tubing is commonly used in large volume sampling. For low level detection, tubing can present a cross contamination problem if not replaced in the sampling train prior to sampling at a new location. Some particulate matter and condensate may be trapped in tubing prior to entry into the flow meter and analyzer by looping the tubing into three or four small diameter loops at a point near the sampling device. This can eliminate the need for water traps or particulate filters in the system that can contribute to system loss or gain of contaminants.

6.5.5.6 Vacuum can be employed to transfer soil gas from a sampler to a container. Evacuated glass bulbs, some containing adsorbents or absorbing liquids (see Practice D 1605), can be affixed to an in-place and purged sampling device and allowed to come to pressure equilibrium. Care must be exercised in recovery of the gas sample from a vacuum cylinder. Upon recovery, the sample is immediately subjected to negative pressure and atmospheric contamination of the sample is encouraged.

6.5.6 *Sample Collection: Containers*—A wide variety of sample containers is employed by field investigators. Container selection is based upon the physical properties of the contaminants sampled, the volume of the sample recovered, the physical properties of suspected contaminants, the sampling system employed, the anticipated sample holding time prior to analysis and the analytical method chosen. Container type for a soil gas survey should be held constant within the survey. A change in container type can impart bias to a portion of the data due to sorptive or desorptive processes related to container type.

6.5.6.1 Whole air samples can be contained in any device made of suitable materials (see 6.5.3) that conveniently satisfy survey, handling, transport and analytical requirements. Certain containers require special handling practice. The literature provides discourse on atmospheric sampling bags (54).

6.5.6.2 Sorbent traps are commonly self-contained. Care must be exercised to select a trapping device that is compatible with the properties of the target compounds and the technique of desorption chosen. Good practice for use of these devices, including handling and desorption procedure is required for successful implementation of sorbent traps when sampling organic compound vapors (see Practice D 3686).

6.5.6.3 Table 2 provides an inventory of sample containers, their applications, advantages and limitations (32).

6.5.6.4 Containers exist that provide for both whole-air and sorbent fractions as well as removal of sample by displacement (see Practice D 1605). Some are convenient for field use, however most are too complex or fragile to be of effective use for a field screening technique requiring rapid mobility.

6.5.6.5 Detector tubes should not be considered as a primary containment vehicle for the purpose of storage and transport of soil gas. A discussion of detector tube application is provided in 6.6.1.

6.5.6.6 Containers for soil samples to be preserved for a subsequent headspace analysis range from glass sample vials to metal cans. The choice of container for soil headspace determination is dependent upon the method of sampling chosen. For soil samples obtained by backhoe, bucket auger or other destructive technique, that is, a disturbed sample, extrusion into a sample vial is not necessary since most of the highly volatile components have already been lost through the act of soil sampling. Metal cans should be made from a material that does not rust. Coating materials and sealing waxes are likely to react with or adsorb soil contaminants, presenting limitations to the value of the data collected. Glass containers with screw threads or crimped seals are difficult to use for soil headspace methods due to the inability of investigators to consistently, thoroughly and rapidly clean the threads or crimp surfaces of all containers prior to capping.

6.5.6.7 Soil pore liquid headspace samples are whole-air or whole-air plus pore liquid samples. They may be contained in most devices suitable for whole-air containment, however investigators are cautioned to select containers from which a vapor sample can be extracted for analysis independently of the liquid present.

6.5.7 *Sample Processing*—Some investigators process soil vapor samples prior to analysis. Processing is performed in an effort to control sample degradation in containers. Efforts to check this degradation by sample processing include refrigeration, pressurization, and pasteurization. As a general practice, sample processing is strongly discouraged. Refrigeration may be somewhat effective in controlling sample degradation, however the best method is to limit or avoid soil gas sample storage whenever possible. The limited shelf life of soil gas samples is discussed in 6.5.9.

6.5.7.1 Extraction is a sample processing step used to remove soil contaminants from soil cores or other similar samples. This technique can efficiently recover contaminants from all residence phases, not just the vapor phase. As a result, the technique yields samples that are not representative of soil atmosphere contaminant suites.

6.5.8 *Sample Transport*—If samples are to be transported to an off-site laboratory for analysis, they must be properly packaged to avoid damage to sample containers. Care must be taken to keep samples from becoming overly warm or agitated during transport. Overnight air express is highly convenient if samples are properly contained, but air freight is not recommended if samples are held in containers such as gas tight syringes or tedlar bags. These containers have other limitations as discussed in 6.5.6.

6.5.9 *Sample Life*—Soil gas samples have limited shelf life even in the most effective containers. Soil gas sample life is strongly container dependent. Numerous factors limit shelf life; most involve degradation in a container. Exposure to light, heat and agitation during shipping will accelerate sample degradation. Biodegradation may occur in some sample containers if water vapor condenses in a container containing microorganisms capable of metabolizing contaminants as substrate.

6.5.9.1 The safest practice is to minimize sample storage time. This problem is greatest when off-site laboratories are engaged to analyze the samples. Prior to recovering the soil gas samples, arrangements can be made with the selected off-site testing laboratory to schedule the necessary personnel and equipment in anticipation of sample delivery.

6.5.10 *Soil Gas Archiving*—Sample archiving in anticipation of a future analytical or descriptive requirement is a common practice. Minimal effects of degradation or loss may be noted in storing certain sorbed samples. Soil gas archiving is, however, not recommended. Although dependent upon the type of container and the storage environment, the likelihood of degradation of soil gas samples is great enough to raise concern. Insertion of standard gases into an archived sample set and spiking of archived soil gas samples with standards provides a reference to determine the likelihood or extent of sample degradation.

6.6 *Analysis of Soil Gas Samples*—Soil gas analysis procedure is based upon pre-existing protocol established for the analysis of contaminants in ambient air. A common reference practice defining terms, sampling information, calibration techniques and methods for validating results may be applied to all automatic analyzers (see Practice D 3249). Basic laboratory practice common to investigators engaged in sampling and analysis of atmospheres applies to soil gas analysis. Note that air sampling protocols and soil gas sampling protocols are not equivalent; geophysical and geochemical factors as well as definition of air sample volume contribute to this lack of equivalency. This guide includes the criteria, guidelines and recommendations for analytical segments including the mode of operation of the laboratory and data validation (see Practice D 3614).

6.6.1 *Basic Analytical Approach*—Soil gas analysis is performed to identify the presence of contaminants, their type and relative concentrations. Various analytical methods are highly general, satisfying only the most rudimentary requirements of contaminant screening. Others are sophisticated, providing identification and relative concentration information for numerous chemical compounds determined to be present in a soil gas sample. The choice of basic analytical approach in soil gas analysis is driven by the purpose of the soil gas survey, quality assurance objectives and budgetary constraints placed upon investigators.

6.6.1.1 Soil gas surveying as a field screening technique can often be effective without the commitment of expenditure for highly sophisticated techniques. This survey purpose is merely to locate other, more direct, techniques. Caution is suggested when choosing highly sophisticated analytical methods for field screening by soil gas monitoring. This selection is controlled largely by the need for the analytical method chosen to be cost-effective.

6.6.1.2 Other applications of soil gas monitoring require more thorough analytical protocol. It is not possible, for example, to suggest the locations of partitioned miscible and immiscible ground-water contaminant plumes with elementary analytical systems. Moreover, the independent monitoring of multiple classes of contaminants in soil gas normally requires analytical systems with multiple detectors. Successful soil gas monitoring for petroleum exploration requires an analytical system which can separate and identify extremely similar volatile compounds occurring at very low concentration levels.

6.6.1.3 Contaminant concentrations in soil gas can vary from levels below the detection limit of the most sophisticated equipment to percent of a whole-air sample. Ideally, the analytical system chosen has enough flexibility to determine contaminants in a wide range of concentrations. Care should be taken to select an analytical system sensitive enough to avoid false negative results which can lead to invalid conclusions. Many analytical systems are not designed to perform to specifications in very high concentration environments, requiring sample dilution prior to analysis or selection of a less sensitive method.

6.6.1.4 Of primary importance to the successful analysis of soil gas is the familiarity and experience of the analyst with the analytical system chosen. The analyst must be able to independently care for and maintain the equipment as well as recognize symptoms of procedural error. The success of an analytical effort lies wholly with operator ability and experience. Excessive machine capability cannot compensate for operator inexperience.

6.6.1.5 Soil gas may be analyzed by a number of methods, including portable VOC (volatile organic compound) analyzers, gas elution chromatography, gas chromatography-mass spectroscopy, and colorimetric and color-indicating detector tubes. Infrared spectroscopy and fiber optic chemical sensors can be applied to soil gas gas analysis, however their use is currently limited and few investigators have experience with this instrumentation. In practice, gas chromatography (GC) or GC-based handheld detectors are the most widely used analytical instruments (32) for soil gas analysis. This guide uses numerous terms relating to various GC methods for soil gas analysis. Most of the terms should apply to other GC methods (see Practice E 355).

6.6.1.6 Portable VOC analyzers used for fugitive emission screening and industrial hygiene monitoring have been adopted for soil gas analytical purposes by numerous investigators. These devices are easily transported to and from the field, require minimal operator skill, provide immediate data and serve to eliminate many sample handling and transport steps which can result in uncertainty. Portable VOC analyzers are limited in application to very low level detection due to the absence of a concentration step. They exhibit limited selectivity and do not have the ability to separate contaminant compounds, leading to potential interference. These devices also are limited in accuracy due to the inability to calibrate for the wide variety of contaminant compounds encountered in soil gas, each compound having its own character of detector response. Portable VOC analyzers contain three types of detectors. These are the flame ionization detector (FID), the photoionization detector (PID) and the infrared (IR) detector. The literature contains a thorough treatment of these devices (10, 55).

6.6.1.7 Soil gas analysis by GC is by far the most versatile and the most costly soil gas analytical method. Instrumentation can be varied to accommodate field mobility, however this is not always required. The technique provides separation of compounds in a chromatographic column, tentative identification of compounds determined to be present and a relative quantitation of compound concentration based upon comparison to a known standard. Soil gas is introduced into the GC and conveyed through a chromatographic column by a carrier gas, separating the contaminants as they pass through the column. The separation is obtained when the sample mixture in the vapor phase passes through a column containing a stationary phase possessing special adsorptive properties. As the gas stream emerges from the column, it passes through a detector, providing for measurement of a specific sample property through the recording of detector electrical response. These responses, or peaks, are recorded as a function of time. Comparison of known standard compound response time with the response time of an unknown represented by a peak results in the tentative identification of the unknown. Comparison of the magnitude of detector response to the newly identified compound versus detector response to the same compound of known concentration, a laboratory standard, results in a relative quantitation of subject compound concentration in the sample.

6.6.1.8 Gas chromatography is essentially a physical separation technique. The degree of separation depends upon the differences in the distribution of volatile compounds, organic or inorganic, between a gaseous mobile phase and a selected stationary phase that is contained in a tube or GC column (see Practice E 260).

6.6.1.9 Numerous factors can impact the ability of the GC to determine contaminants in a soil gas sample. These include column characteristics, sample flow rate, sample temperature, the composition of the carrier gas and the type of detector employed. Instrumentation can be expanded to include multiple columns, multiple detectors, sample loops and temperature programming, all of which make an instrument more versatile, albeit at additional cost.

6.6.1.10 Simple GCs are portable analyzers with GC options. Field GCs are more advanced instruments with temperature programmable ovens and provide opportunity for multiple columns and detectors. They can be carried in mobile laboratories or established in a temporary base laboratory in the field. Research-grade instruments are normally based at off-site laboratories with strictly controlled environments. These are used when positive identification or very low detection limits are specified. The literature contains excellent comparisons of the advantages, limitations and applications of the various configurations of GCs, including instrument specifications (10, 32, 56, 57).

6.6.1.11 Detector tubes have been applied to safety and health atmospheric monitoring, agriculture and the chemical industry. These devices are designed to be compound specific, although this characteristic is dependent upon the contaminant compounds present in the sample drawn through the tube. Detector tubes may be used for short-term sampling (grab sampling; 1 to 10 min) or long-term sampling (dosimeter sampling; 1 to 8 h). Short-term sampling involves the movement of a given volume of gas through the tube by a mechanical pump. If the substance for which the detector tube was designed is present, the indicator chemical in the tube will change color (stain). The concentration of the gas may be estimated by either the length of the stain compared to a calibration chart or by the intensity of the color change compared to a set of standards (see Practice D 4490). Long-term sampling involves the movement of gas at a very slow rate through the tube by means of an electric pump. The use of long-term detector tube sampling for soil gas monitoring is limited to specific temporal survey designs.

6.6.1.12 Detector tubes are relatively inexpensive and provide immediate results. Their use is restricted to applications with few interfering compounds. Depending upon the contaminants present, they may be of low sensitivity and can be affected by humidity, normally high in soil gas, sample

flow rate, temperature extremes (32), storage conditions and shelf life.

6.6.1.13 The literature contains excellent discourse on the detector tube apparatus, reagents, procedure accuracy and amenable compounds (see Practice D 4490).

6.6.2 *Specific Analytical Approaches*—This subsection discusses various detectors and methods that may be integrated into soil gas analytical instrumentation. For methods providing detector alternatives, the choice of an appropriate detector should be guided by knowledge of detector properties. Key properties are as follows (after Mayer, 1989 (32)):

6.6.2.1 *Selectivity or Specificity*—Selectivity refers to the responsiveness of the detector to the compound of interest. Detectors responding to a wide range of classes of compounds are termed universal or non-selective detectors. Those that respond to only certain classes of compounds are termed selective detectors.

6.6.2.2 *Sensitivity*—Sensitivity refers to the relationship between the detector response and the quantity of the subject compound injected. It is the smallest detectable quantity of a compound; it is usually considered to be the amount that produces a response equal to twice the baseline noise of the detector.

6.6.2.3 *Linear Dynamic Range*—Linear dynamic range is the range over which the detector response to a compound is directly proportional to the amount of compound injected. Detectors vary in the range of component concentrations over which they are linear. Wide linear dynamic range is desirable because it simplifies quantitation of samples having widely varying ranges of concentrations.

6.6.2.4 *Stability*—Stability is a factor referring to detector responsivity over time. Stability is controlled by numerous factors and is seldom quantified. The required frequency of instrument calibration is determined by detector stability.

6.6.3 Specific analytical approaches are as follows:

6.6.3.1 *Flame Ionization Detectors (FID)*—Flame ionization detectors generate electric current when gases containing carbon atoms are oxidized to carbon dioxide in a hydrogen flame and potential is applied across the flame. The magnitude of the electric current generated is termed the detector response. FIDs are responsive to hydrocarbon contaminants in soil gas and are commonly employed for this purpose. These detectors are durable for field application, and have a wide linear range and nearly uniform response to organic gas species. FIDs are generally unresponsive to inorganic gases and water vapor, common constituents in soil gas. FID performance can be evaluated independently of the chromatographic column (see Practice E 594). Although highly versatile, these detectors are not selective for halogenated compounds. They require supplies of fuel gas which require careful safety practices in handling and flame ignition.

6.6.3.2 *Photoionization Detectors (PID)*—Photoionization detectors employ ultraviolet radiation to ionize contaminant molecules. Positive ions and free electrons are formed which migrate to the detector electrode(s), resulting in an electric current that is proportional to contaminant concentration at the detector. PIDs are extremely sensitive to aromatic hydrocarbons due to the great efficiency of ionization of pi bonds under ultraviolet radiation. Efficiency of ionization of sigma bonds is lower, resulting in a higher PID detection limit for aliphatic hydrocarbons. The selectivity of the method can be adjusted by selecting lamps of different energies, causing a change in response of contaminants with fixed ionization potentials to changing lamp energies. Tables exist of ionization potentials of compounds within classes common to soil gas contaminants (58). Methane has an ionization potential higher than the energies of commercially available lamps, limiting the PID to detection of compounds other than methane. PIDs are further limited by their tendency to conceal the presence of low-sensitivity compounds when high-sensitivity compounds (aromatics) are present. PID response can be impacted by condensation of water vapor in the lamp.

6.6.3.3 *Electron Capture Detectors (ECD)*—Electron capture detectors are highly sensitive to and selective for compounds with electronegative functional groups such as CFCs (chloro-fluorocarbons). The sensitivity of the detector is proportional to the number of these groups on a compound, resulting in a unique detector response to each compound. The ECD comprises a source of thermal electrons inside a reaction chamber (a radioactive source emits β radiation which ionizes the carrier gas to produce electrons). The device detects compounds with electronegative functional groups capable of reaction with thermal electrons to form negative ions. Such reactions cause a decrease in the concentration of free electrons. The detector is designed to measure changes in the concentration of these electrons inside the chamber (see Practice E 697). Calibration of the ECD is therefore linked to each compound to be determined by the detector. ECDs are also sensitive to water, oxygen and other common components of soil gas, causing potential problems in method performance. ECDs emit β radiation that should be properly vented. Operation of an ECD requires licensing under Federal regulation.

6.6.3.4 *GC/Mass Spectroscopy*—Combination of gas chromatography and mass spectroscopy results in the GC/MS method of analysis. A mass spectrometer is used to obtain a mass spectrum of each eluting compound. Positive identification of these compounds is sometimes obtained by comparison of the unknown mass spectrum to a library of known spectra. GC/MS can be extremely selective for target compounds. Use of the technique for soil gas monitoring is limited, primarily due to the cost of analyses.

6.6.3.5 *GC/Fourier Transform Infrared Spectroscopy*—This analytical method combines gas chromatography with Fourier transform infrared spectroscopy. GC/FTIR can provide a rapid identification of eluting compounds by comparison of their infrared spectra with a known spectral library. Quantitation is achieved by subsequently passing the sample through an appropriate GC detector such as the FID or ECD. This method, like GC/MS, is limited in application to soil gas monitoring by the high cost of analysis.

6.6.3.6 Other detectors are applied to soil gas analysis by GC, albeit rarely in comparison to FID, PID and ECD. They include the argon ionization detector, a non-destructive device similar in operating design to the ECD, the flame photometric detector (FPD) used to determine organic compounds containing sulfur and phosphorus, and the hot-wire (pyrolyzer) used to determine compounds containing nitrogen.

6.6.4 *Analytical QA/QC*—The validation of the analytical aspects of soil gas monitoring is fundamental to the tech-

nique. Analytical equipment and procedure must be evaluated by laboratory QA/QC, just as the sampling system, sampling plan and field procedure are evaluated by field QA/QC methods. Analytical QA/QC defines a confidence limit of performance. The utilization of well tested and uniform analytical practices is essential to the production of reliable and defensible data, the validity of which can be demonstrated at a later date through the use of written field and laboratory records (see Practice D 3614).

6.6.4.1 Most analytical QA/QC plans contain calibration steps, linearity checks, standard analyses, blank analyses, duplicate analyses and audit checks. The various analytical approaches discussed in 6.6.3 require a variety of different protocols which will satisfy the QA/QC requirements for each method. Four types of analytical QA/QC samples are required for determination of quality assurance. These are analytical reagent blanks (used to determine the potential of sample or standard contamination from a reagent), laboratory blanks (used to determine the impact potential of the laboratory atmosphere on analytical results), analytical sample replicates (used to estimate the analytical precision for samples) and analytical standard replicates (used to estimate the analytical precision for standards). Table 3 provides a summary of suggested calibration and quality control requirements for analytical systems (10).

6.6.4.2 The aspects of bias, precision, representativeness, completeness and comparability must be considered to evaluate analytical equipment performance, including the establishment of minimum detectable quantities of contaminant compounds, retention time drift and the linearity of instrument response. Bias and precision must be quantified in order to compare actual survey performance with goals established in the survey plan.

6.6.5 A data validation summary report is a common method of evaluating analytical system performance. A guide for determining parameters key to the data validation summary report is provided as follows.

6.6.5.1 *Bias*—For determination of bias, the percent recovery can be determined using the following formulas:

$$recovery\ reproducibility = (DCS/KCS)*100 \qquad (6)$$

where:
DCS = determined concentration of standard, and
KCS = known or certified concentration of standard.
The standard deviation of all standards analyzed can be determined as follows:

$$SD = \{(sum(recovery\text{-}i - recovery\text{-}ave)^2)/(n-1)\}^{0.5} \qquad (7)$$

Finally, the range of uncertainty can be determined using the following equation:

$$\pm R = \pm t*(SD)/(n^{0.5}) \qquad (8)$$

where:
t = the value of Studentized t at the 90 % confidence level and $(n-1)$ degrees of freedom.
The bias statements for data collected should be expressed as the average recovery plus or minus the range.

6.6.5.2 *Precision*—For the determination of precision, the relative standard deviation of replicates can be calculated using the following equation:

$$RSD\text{-}pair = SD/Mean \qquad (9)$$

TABLE 3 Soil Gas Sampling Containers[A]

Type	Applications	Advantages	Limitations
Stainless steel canisters	Collection of samples for delayed analysis	Durability Ease of sample handling Can be re-used Sample holding time longer than that for other whole-air sample containers Sample volume measurement not required Desorption not required Allows replicate analysis	Expense Requires vacuum pump or gage Can be difficult to decontaminate
Glass bulb	Collection of samples for delayed analysis	Glass is more inert than other sample container materials Septa possible Allows replicate analyses	Easily breakable Leakage through stopcocks or septa possible Adsorption to PTFE or other parts
Bag	Collection of samples for delayed analysis Sampling of very high vapor pressure compounds for which absorption methods are unsuitable	Bulk loss of sample is readily apparent Containers are light-weight and easy to handle Sample volume measurement not required Desorption not required Allows replicate analyses	Expense Some compounds may be lost through or adsorbed to bag walls Some container materials may contaminate samples Containers cannot be easily re-used Leaks in valves
Syringe[B]	Collection of samples for on-site analysis	Ease of sample collection Does not require special equipment to introduce sample into GC Desorption not required	PTFE plungers can adsorb sample Holding time short due to leakage or absorption Sample volume smaller than for other containers
Sorbent sampler	Allows concentration of low level samples If samples are solvent-desorbed, allows analysis of liquid sample	Ease of handling Relatively long holding time	Requires precise sample volume measurements Sorbent type must be tailored to compounds to be measured; adsorption behavior of each compound for solvent used must be accounted for Requires desorption (thermal or solvent) for analysis

[A] See Ref (32).
[B] Syringes may also be used to transfer samples from the sampling device to a container for off-site analysis.

$$RSD\text{-}ave = \{\{(sum(RSD\text{-}pair))^2\}/(n-1)\}^{0.5} \quad (10)$$

where:
$RSD\text{-}pair$ = relative standard deviation for each pair of replicates, and
$RSD\text{-}avg$ = relative standard deviation overall.
Next, the precision can be determined as follows:

$$precision = \{(t*RSD\text{-}avg)/DF\}*100 \quad (11)$$

where:
$precision$ = the percent precision,
t = the t value for $n-1$ pairs of replicates, and
DF = the degrees of freedom = $(n-1)$.
Finally, mean value is reported with associated uncertainty:

$$x \pm (x*t*SD\text{-}ave)/(DF)^{0.5} \quad (12)$$

where:
x = reported chemical concentration, and
t = the value of t at the 90 % confidence level for the appropriate degrees of freedom.

6.6.5.3 *Representativeness*—Representativeness is determined by the results of the cross contamination blanks and the air blanks. The results should be presented as a bias estimate, as follows:

$$bias\ (\%) = \{(CCC - CA)/Mean\}*100 \quad (13)$$

where:
CCC = concentration in cross contamination sample,
CA = concentration in air, and
$Mean$ = mean concentration in sample set (bias may also be expressed for a single sample by substituting sample concentration).

6.6.5.4 *Completeness*—The completeness goal is 90 % or higher. Completeness is the number of samples collected that can be validated through the procedures for bias, precision, and representativeness.

6.6.5.5 *Comparability*—Comparability is based upon professional judgment and is provided through planning steps carried out prior to initiation of field work.

6.7 *Data Interpretation*—Soil gas data interpretation is an iterative process including the examination of the raw data, selection of appropriate and useful data displays, and establishment of correlation of the data set to other vadose zone monitoring data and ground truth. Interpretation of soil gas data is not like other interpretive exercises involving measurement data, in that mathematical expressions relating soil gas contaminant concentrations to underlying soil, rock and ground-water contaminant concentrations cannot be written for most applications at a high confidence level. This is a function of a lack of site characteristics information at even the most comprehensively studied sites. Soil gas data cannot be consistently interpreted in a manner that establishes direct correlation between contaminants in a soil gas horizon and contaminants in other horizons. Processes including migration and degradation can have profound influence on the correlation of soil gas data to ground truth. Interpretive efforts excluding consideration of these influencing processes can be highly misleading. For example, the presence of contamination in an underlying horizon will not necessarily correlate to the detection of contaminants in overlying soil atmospheres, that is, the potential for a false negative result. The converse is also true, that is, the potential for a false positive result. Interpretation of GC results in the laboratory without consideration of pertinent hydrogeological information may lead to incorrect conclusions (59). However, the detection of contaminants in soil gas does suggest the existence of a contaminant source, and increases in contaminant concentration can suggest close proximity to the source or an increased quantity of the subject contaminant in the subsurface. It is the responsibility of the interpreter to examine soil gas data in context of other site characteristics, and provide an interpretation based upon sound judgment and thorough yet practical data treatment.

6.7.1 *Manipulating Data*—Soil gas data is normally interpreted as raw data. The application of correction factors is not recommended, as it is difficult if not impossible to determine if the magnitude of the correction factor is greater than that of the variance between data populations in a survey. Moreover, the need for correction factors can indicate a flaw in survey design, sampling system performance or the objectivity of the interpreter.

6.7.2 *Defining Data Subpopulations*—Soil gas monitoring seeks to define anomalous subpopulations of data that contain measurable quantities of contaminants or unusual compositions. These populations can easily be described by their contrast to normal populations, for example, contrasting populations with and without measurable contaminants. Establishment of contaminant baselines or conditions "at background" make this contrast possible. If all soil gas samples are recovered in a contaminated area, there may be no apparent contrast.

6.7.2.1 Statistical treatment of soil gas monitoring data allows the interpreter to estimate the amount of variation noted in the survey data due to errors. This practice also permits the interpreter to evaluate the data quality objectives suggested for the survey during the planning phase. Statistical treatment of soil gas data can also be of use to define anomalous data subpopulations when the boundaries of a contaminated area are not clearly defined or if the existence of multiple populations of data (that is, contaminated and uncontaminated) within a single data set is in doubt. The literature contains discourse on statistical treatment of soil gas data (10, 60).

6.7.3 *Interpreting Soil Gas Data Profiles*—Soil gas data from survey profiles displayed on an $X - Y$ plot is an effective aid to data interpretation. This display is useful to examine the overall context for soil gas measurement data potentially indicating contamination. If the profile is displayed as a cross section through a grid pattern or as a linear array of sample points, the profile display can illustrate spatially significant groupings of data subpopulations.

6.7.3.1 It is quite common for concentration data to be highly variant within a contaminated area. Soil gas profiles can be used to show variation in spatially related data. This is one method of defining subpopulations of data indicating contamination or other anomalous characteristics.

6.7.3.2 Multiple data sets can be displayed on a single profile. Comparison of one data set to another on a single profile is a simple visual method to screen for suggested data subpopulations. Comparison of concentration data and compositional data (see 6.7.5) on a single profile can further resolve this problem.

6.7.4 *Mapping Soil Gas Data*—Soil gas data obtained by sampling at a single depth is often mapped to suggest the lateral extent of subsurface contamination. Map suites of soil gas data obtained from multiple depths can sometimes aid investigators in determining the depth to the contaminant source.

6.7.4.1 Numerous algorithms can be used to interpolate between data points, including linear, inverse distance squared, inverse distance cubed, splines and kriging. The various interpolation methods will yield similar results, suggesting a general pattern of contaminant distribution in soil gas. Kriging requires a probability model for each survey site mapping application for which it is employed, the derivation of which requires data which are not normally available for a given soil gas survey area.

6.7.4.2 Caveats exist in using computer mapping programs as interpretive aids. Difficulties can arise in treatment of adjacent data points differing in contaminant concentration by an order of magnitude and more due to vapor migration barriers, preferential vapor flow paths or changes in soil moisture or porosity content. It is possible to model these characteristics and input such a model into some computer mapping programs, however this introduces bias into the mapping effort. Single point soil gas contaminant concentration highs may exist due to a sample density which is insufficient to resolve the cause for the single point anomaly. Contour mapping of such data may be meaningless without the complement of other information, especially detailed knowledge of site characteristics.

6.7.5 *Analyzing the Composition of Soil Gas Contaminants*—Certain applications of soil gas monitoring require detailed analyses available from off-site bench laboratories or mobile laboratories. Determination of a number of contaminant compounds in a soil gas sample set with either of these analytical systems enables the interpreter to make a comparative analysis of the changes in soil gas contaminant composition within that sample set.

6.7.5.1 Compositional analyses can range in scope from a simple listing of the various compounds determined in each sample to thorough data treatments. Profiles of soil gas data can be constructed to illustrate the spatial relationship between two potentially different groupings of data (see 6.7.3). Crossplots of contaminant compound concentrations are highly effective in the definition of data subpopulations, and can be used to relate contaminant types to known on-site waste streams and sources in complex settings. Known as fingerprinting, this guide compares vapor composition over a known contaminant product and the known soil atmosphere composition over that product to soil gas contaminant composition in areas being investigated on the subject site. Subtle divisions in data subpopulations can be defined by crossplots of contaminant ratios. In addition to simple ratioing, computerized multivariate pattern recognition techniques such as cluster, factor and discriminant analyses can assist in the evaluation of intra-data set compositional variations and their relationship to the physical contamination issues at a site.

6.7.5.2 Soil gas data can be examined for the appearance of target compounds determined to be present in contaminant mixtures. The success of this practice, used primarily to establish the location and extent of underlying ground water contamination, relies upon selection of appropriate target compounds and the persistence of target compounds in soil vapor.

6.7.5.3 Monitoring specific compounds in soil gas data can be utilized to determine the progress of degradation or migration of contaminants in the vadose zone and in ground water. Biodegradation has been monitored by the appearance of excessive quantities of carbon dioxide in soil gas **(61)**.

6.7.6 *Interpretation in Context of Other Vadose Zone Monitoring*—Soil gas monitoring is not a technique that can consistently support conclusions based upon interpretations of survey results. For this reason it is strongly recommended that other vadose zone monitoring methods be used to corroborate data obtained from a soil gas survey, especially when investigators are attempting to do more than simply audit a subject site for the presence of contaminants. Useful models of contaminant emplacement and transport in the vadose zone can be constructed by combining techniques. Examples of useful combinations are soil pore liquid and soil gas monitoring or neutron probe and soil gas monitoring.

6.7.7 *Correlation With Ground Truth*—Interpretation of soil gas data is difficult without establishing some form of ground truth with which to substantiate survey results. Ground truth can be in the form of monitoring well data, for purposes of determining the extent of contamination by a ground-water contaminant plume. Examples of other forms of ground truth usable in support of soil gas data interpretation are soil cores, the presence of contaminant odors in basements, observed floating contaminants in storm sewers or utility vaults, or other field observations.

7. Data Reporting Requirements

7.1 *Purpose of Reporting*—Of primary concern in a report of findings pertaining to a soil gas survey is that the report includes the information necessary to describe the results of that survey performed for a particular application. In many instances, certain interpretative methods or data reporting formats useful to end users for one particular application are not relevant to the needs of end users applying the information to a different application. Examples of these differing applications that require unique report subject matter are soil gas contaminant determinations for real property environmental assessments, soil gas monitoring of volatile organic contaminants from underground storage tanks and soil gas sampling as a tool useful in the exploration for natural resources. Certain applications require a thorough treatment of a significant number of factors impacting the meaning and usefulness of soil gas data interpretations. Examples of such applications include damage assessments, contaminant source identification or tests of the effectiveness of remediation. Other applications command minimum reporting requirements. An example of such an application is the monitoring of releases from underground storage tanks over time. Included in a discussion of the report objectives should be an identification of the end user category (for example, regulatory agency, land acquisition negotiations).

7.1.1 A decision must be made regarding the units expressed in reporting, that is, qualitative or quantitative. If quantitative, the appropriate expression of units in volume/volume or weight/volume must be determined. SI units are recommended for reporting of atmospheric measurement data (see Practice D 1914).

7.2 *Report Format*—Certain reporting requirements are commanded without regard to data application. In large part

they are related to the QA/QC objectives, and include data comparability, representativeness, bias, precision accuracy, completeness and analytical detection limits whenever possible. At a minimum, a general discussion of the reliability of results and analytical detection limits is warranted; soil gas test data may be evaluated in the same manner as is other atmosphere test data (see Practice D 3614).

7.3 *Salient Points to be Addressed in Reporting*—The report of findings of any soil gas monitoring effort can contain discussions within any number of topics that should be selected to best suit the requirements of the end user. Selection of appropriate topics is discretionary, usually based upon a scope of work determined by prior agreement between the data provider and the data end user. Efforts to limit reporting requirements for the sake of short term time and money cost savings usually result in low-confidence level treatment of the report or an ultimate time and money cost gain, or both. Discussions that should be included when appropriate and whenever possible are provided below.

7.3.1 The purpose of the soil gas study should be stated, as well as the rationale for selection of a particular soil gas monitoring technique.

7.3.2 Selection of a particular soil gas monitoring technique is typically controlled by the chemical and physical properties of the chemical compounds of interest which are known to occur or suspected to occur on site. A discussion of the sample array in three dimensions, sampling method employed and the analytical scheme chosen in context of these properties should be provided.

7.3.3 The rationale for selection of a particular soil gas monitoring technique should always be based upon the physical properties of the vadose zone as well as the chemical and physical properties of the compounds of interest. A discussion of the impact of these vadose zone properties on survey design should be included in the report. The regional and local hydrogeologic conditions within the survey area should be described. A discussion of the regional geology should include the physiographic province, a generalized geologic column, geologic structure and general ground water occurrence. The local conditions should be described with regard to soil type(s), moisture content in the vadose zone, soil/bedrock interface, stratigraphy and lithology, ground water bearing zones, flow directions and gradients, potentiometric levels, aquifer characteristics and ground water quality.

7.3.4 If known and appropriate, the characteristics of a contaminant source or spill should be addressed. Examples of such characteristics are contaminant composition, the likelihood of single or multiple contamination events or the reaction potential (above, within and beneath the vadose zone) of multiple contaminant mixtures.

7.3.5 Every subject of every vadose zone monitoring effort has unique characteristics. Those characteristics that could impact the results of the soil gas monitoring effort should be described to provide a meaningful context in which to interpret the soil gas data.

7.3.6 There are a number of topics common to most soil gas data reporting that are useful in the majority of applications. The regional and specific site location should be identified using a site plot plan. The site plot plan could include an insert showing the regional location. A discussion should be included regarding the physical structures at the site that may impact the location of sampling points and the migration of soil gas, for example, asphalt and concrete pads, buried pipelines and surface water impoundments. Site history must be considered, including the types of chemical compounds known or suspected to have been used at the site. These compounds should be listed with their chemical and physical properties as they relate to volatilization, solubility and other migration characteristics or soil gas recovery characteristics.

7.3.7 The site should be evaluated in the report of findings for the impact of the regional and local hydrogeologic conditions within the survey area on the results of the survey.

7.3.8 A detailed description should be given of the type of soil gas survey conducted. Details should include selection of active or passive method, whole air or passive sample collection method, sampling array, background sampling, equipment decontamination procedure employed prior to the survey, field or laboratory analytical methods and QA/QC procedures. Any unusual conditions should be noted, such as rainfall events during the course of the survey (especially when moveable soil gas chiefly originates from vadose zone microporosity), high pressure or low pressure front movement across the survey area during the course of the survey (especially when moveable soil gas chiefly originates from vadose zone macroporosity), or visual observations of contamination at sampling points.

7.3.9 If a subject property is found to be contaminated, a separate discussion of soil gas characterization of uncontaminated or non-anomalous contiguous property should be provided in the report of findings. This can be useful in highlighting naturally occurring petroleum hydrocarbons in soil and in establishing a regional baseline of contamination.

7.3.10 Data collected during the field sampling and field or laboratory analyses should be compiled in table form and be included in a preliminary or final report, preferably as appendices. Such data should include a listing of sampling and analysis dates, soil/rock description at each sampling point, depth and diameter of sampling point, quantity of soil gas purged prior to sampling, quantity of sample extracted, chromatogram and/or mass spectra for each sample and a tabulation of QA/QC samples recovered.

7.3.11 The report of findings should include a discussion of the results of the QA/QC efforts, establishing performance within limits set prior to the survey. Data validation involves review of the data collected for the purpose of isolating spurious values (32). Systematic errors or bias can be detected in this review. Suggestions should be made as to the origin of the errors or bias.

7.3.12 Results of analyses should be displayed on plan maps and should include sampling point locations, physical features, contours of equal concentrations of specific compounds or compound groups (for example, alkanes) and any necessary keys or other notes to guarantee map clarity. Cross-sections showing changes in contaminant concentration with depth and concentration profiles of more than one contaminant through several sample locations can be highly useful displays. The report should include text describing each map, cross-section or profile.

7.3.13 Whenever possible, discussion should be provided that correlates soil gas data to ground truth. The most common and widely accepted form of ground truth is data from ground water monitoring wells.

7.3.14 When appropriate, the report of findings should attempt to identify the source of the contaminants encountered in the soil gas survey.

7.3.15 The report should contain a section which discusses the conclusions drawn from the results of the soil gas study and any recommendations which seem appropriate to enhance the value of conducting such a soil gas study. Conclusions should include identification of the compounds detected, if any, an assessment of the appropriateness of the soil gas study method used, and any circumstances that may have significantly impacted the results of the investigation, such as weather conditions or equipment calibration. Recommendations should address need for establishing ground truth, extension of the study to adjacent areas of interest, the need for a different soil gas study method, actions to resolve questionable QA/QC results, or need for additional chemical analyses for contaminant identification.

7.4 *Disadvantages of Real-Time Reporting*—In actual practice, many end users request real-time reporting of soil gas data obtained from field-based laboratories. Presentation of such data presents the opportunity for misunderstanding by end users who are not familiar with the caveats presented by data not examined in light of the QA/QC program or site specific factors. Real-time reporting of soil gas data is therefore not recommended.

8. Keywords

8.1 contaminant; environmental monitoring; geochemistry; ground water; Henry's law; petroleum hydrocarbon; sampling; soil gas; unsaturated flow; vadose zone; vapor monitoring; volatile organic compound

TABLE 4 Summary of Suggested Calibration and Quality Control Requirements for Analytical Systems[A]

Type of Instrument	Detector Type	Type of Calibration/QC Test	Frequency	Gas Standard(s)	Acceptance Criteria	Corrective Action
Portable VOC (THC) Analyzer	FID	(1) Multipoint calibration (zero plus three upscale concentrations)	At start of program	Methane or other aliphatic compound	Correlation coefficient ≥ 0.995	Repeat multipoint calibration after checking calibration dilution system
		(2) Zero (span) calibration	Daily	UHP Air or N_2/Methane	Response factor agreement within ± 20 % of mean RF for multipoint calibration	(1) Repeat zero span calibration (2) If still unacceptable, repeat multipoint calibration
		(3) Control sample analysis	Daily, prior to testing	Methane	Measured concentration within ± 10 % of certified concentration	(1) Repeat zero span calibration (2) Repeat control sample analysis
		(4) Drift check	Daily, at conclusion of testing	Methane	Drift value ≤ 20 % of the input value	(1) Flag day's data as questionable (2) Repair or discontinue use of analyzer
	PID	(1) Multipoint calibration (zero plus three upscale concentrations)	At start of program	Benzene or other aromatic compound	Correlation coefficient ≥ 0.995	Repeat multipoint calibration after checking calibration dilution system
	PID	(2) Zero/span calibration	Daily	Benzene or other aromatic compound	Response factor agreement within ± 20 % of mean RF for multipoint calibration	(1) Repeat zero/span calibration (2) If still unacceptable, repeat multipoint calibration
		(3) Control sample analysis	Daily, prior to testing	Benzene or other aromatic compound	Measured concentration within ± 10 % of certified concentration	(1) Repeat zero/span calibration (2) Repeat control sample analysis
		(4) Drift check	Daily, at conclusion of testing	Benzene or other aromatic compound	Drift ≤ 20 % of the input value	(1) Flag day's data as questionable (2) Repair or discontinue use of analyzer
Portable Gas Chromatograph	FID	(1) Multipoint calibration (zero plus three upscale concentrations)	At start of program	Benzene or toluene	Correlation coefficient ≥ 0.995	Repeat multipoint calibration after checking calibration dilution system
		(2) Zero/span calibration	Daily	UHP air or N_2/methane	Response factor agreement within ± 20 % of mean RF for multipoint calibration	(1) Repeat zero/span calibration (2) If still unacceptable, repeat multipoint calibration
	FID	(3) Control sample analysis	Daily, prior to testing	Benzene	Measured concentration within ± 10 % of certified concentration	(1) Repeat zero/span calibration (2) Repeat control sample analysis
		(4) Drift check	Daily, at conclusion of testing	Benzene	Drift ≤ 20 % of the input value	(1) Flag day's data as questionable (2) Repair or discontinue use of analyzer
		(5) Retention time checks	Daily	Benzene or toluene	None	None
		(6) Analytical blanks	Daily	UHP air or N_2	Measured concentration ≤ 5 % of the instrument span value	Clean/replace system components until acceptable blank can be obtained
		(7) Sampling system blanks	Daily, plus after very high samples	Sample gas	Measured concentration ≤ 5 % of the instrument span value	Clean/replace system components until acceptable blank can be obtained
	FID	(8) Duplicate samples	10 % of sampling points, minimum	Sample gas	None; provides a measure of total sampling variability	None
		(9) Control point samples	After every ten samples or once per day, whichever is greater	Sample gas	None; provides a measure of temporal variability	None
		(10) Background samples	One sample per day	Sample gas	None; provides a measure of background concentration	None
	PID	(1) Multipoint calibration (zero plus three upscale concentrations)	At start of program	Benzene or toluene	Correlation coefficient ≥ 0.995	Repeat multipoint calibration after checking calibration dilution system
		(2) Zero span calibration	Daily	UHP air or N_2/methane	Response factor agreement within ± 20 % of mean RF for multipoint calibration	(1) Repeat zero/span calibration (2) If still unacceptable, repeat multipoint calibration

TABLE 4 *Continued*

Type of Instrument	Detector Type	Type of Calibration/QC Test	Frequency	Gas Standard(s)	Acceptance Criteria	Corrective Action
	PID	(3) Control sample analysis	Daily, prior to testing	Benzene	Measured concentration within ± 10 % of certified concentration	(1) Repeat zero/span calibration (2) Repeat control sample analysis
		(4) Drift check	Daily, at conclusion of testing	Benzene	Drift ≤ 20 % of the input value	(1) Flag day's data as questionable (2) Repair or discontinue use of analyzer
		(5) Retention time checks	Daily	Benzene or toluene	None	None
		(6) Analytical blanks	Daily	UHP air or N$_2$	Measured concentration ≤ 5 % of the instrument span value	Clean/replace system components until acceptable blank can be obtained
		(7) Sampling system blanks	Daily (plus after very high samples)	Sample gas	Measured concentration ≤ 5 % of the instrument span value	Clean/replace system components until acceptable blank can be obtained
		(8) Duplicate samples	10 % of sampling points, minimum	Sample gas	None; provides a measure of total sampling variability	None
	PID	(9) Control point samples	After every ten samples or once per day, whichever is greater	Sample gas	None; provides a measure of temporal variability	None
		(10) Background samples	One sample per day	Sample gas	None; provides a measure of background concentration	None
Off-site Gas Chromatograph	FID	(1) Multipoint calibration (zero plus three upscale concentrations)	1 per month	Propane/hexane	Correlation coefficient ≥ 0.995	Repeat linearity check
		(2) Single point calibration check	Daily, prior to sample analyses	Propane/hexane	Response factor agreement within ± 20 % of most recent average RFs for multipoint calibration	Repeat single point calibration
		(3) Retention time check	Daily, prior to sample analyses	Multicomponent standard	Agreement with preestablished relative retention times	Adjust GC conditions and repeat RT check
	FID	(4) Control sample analysis	Daily, prior to sample analyses	Sample gas	(1) Correct identification of 90 % of components (2) For 90 % of components, measured concentrations within ± 30 % of actual concentrations	Repeat control sample analysis
		(5) Duplicate analyses	Minimum 10 % of samples (all duplicate samples will be analyzed in duplicate)	Sample gas	CV ≤ 20 % for ten major sample components	Repeat sample analysis
		(6) Blank analysis	Daily, prior to sample analysis	UHP air or N$_2$	Total ≤ 20 ppbv-C	(1) Clean system (2) Repeat blank analysis
	PID	(1) Multipoint calibration (zero plus three upscale concentrations)	1 per month	Propane/hexane	Correlation coefficient ≥ 0.995	Repeat linearity check
	PID	(2) Single point calibration check	Daily, prior to sample analyses	Propane/hexane	Response factor agreement within ± 20 % of most recent average RFs for multipoint calibrations	Repeat single point calibration
		(3) Retention time check	Daily, prior to sample analyses	Multicomponent standard	Agreement with preestablished relative retention times	Adjust QC conditions and repeat RT check
		(4) Control sample analysis	Daily, prior to control sample analyses	Sample gas	(1) Correct identification of 90 % of components (2) For 90 % of components, measured concentrations within ± 30 % of actual concentrations	Repeat control sample analysis
		(5) Duplicate analyses	Minimum 10 % of samples. (Duplicate samples analyzed in duplicate)	Sample gas	CV ≤ 20 % for ten major sample components	Repeat sample analysis
	PID	(6) Blank analysis	Daily, prior to sample analysis	UHP air or N$_2$	Total ≤ 20 ppbv-C	(1) Clean system (2) Repeat blank analysis

TABLE 4 *Continued*

Type of Instrument	Detector Type	Type of Calibration/QC Test	Frequency	Gas Standard(s)	Acceptance Criteria	Corrective Action
	ECD	(1) Quantitative standard	Daily, prior to sample analysis	Multicomponent standard	Response factor agreement within ± 30 % of three day rolling mean RFs for all components	Repeat calibration
		(2) Retention time check	Daily, prior to sample analyses	Multicomponent standard	None; will provide basis for comparison of FID/PID results to ECD results	None
		(3) Control sample analysis	Daily, prior to sample analyses	Sample gas	(1) Correct identification of all components (2) For 90 % of components, measured concentrations within ± 30 % of actual concentrations	Repeat control sample analysis
	ECD	(4) Duplicate analyses	Minimum of 10 % of samples (all duplicate samples analyzed in duplicate)	Sample gas	CV ≤ 20 % for ten major sample components	Repeat sample analysis
		(5) Blank analysis	Daily, prior to sample analyses	UHP air or N_2	Total ≤ 20 ppbv-C	(1) Clean system (2) Repeat blank analysis

A See Ref **(10)**.

APPENDIX

(Nonmandatory Information)

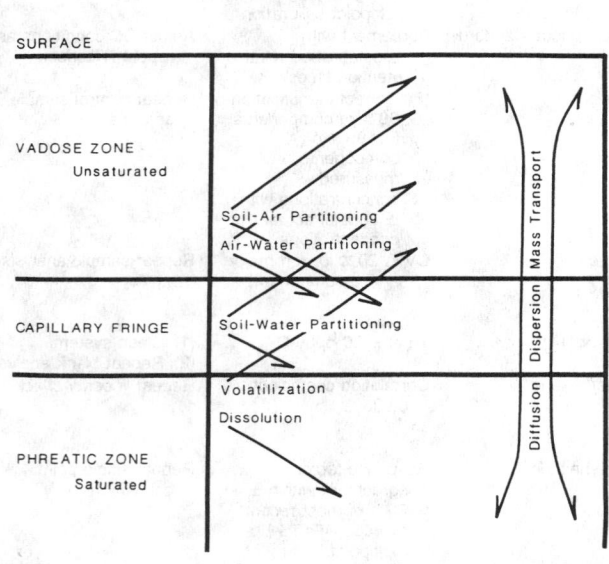

NOTE—The processes indicated by the soil gas monitoring method are partitioning, migration, emplacement and degradation. Partitioning represents a group of processes which control contaminant movement from one physical phase to another, these phases being liquid, free vapor, occluded vapor, solute and sorbed. Migration refers to contaminant movement over distance with any vertical, horizontal or temporal component. Emplacement refers to establishment of contaminant residence in any phase within any residence opportunity. Degradation is the process whereby contaminants are attenuated by oxidation or reduction in the vadose zone, either through biogenic or abiogenic processes. Soil gas monitoring measures the result of the interaction of these processes in a dynamic equilibrium.

FIG. X1.1 Arena of Soil Gas Monitoring

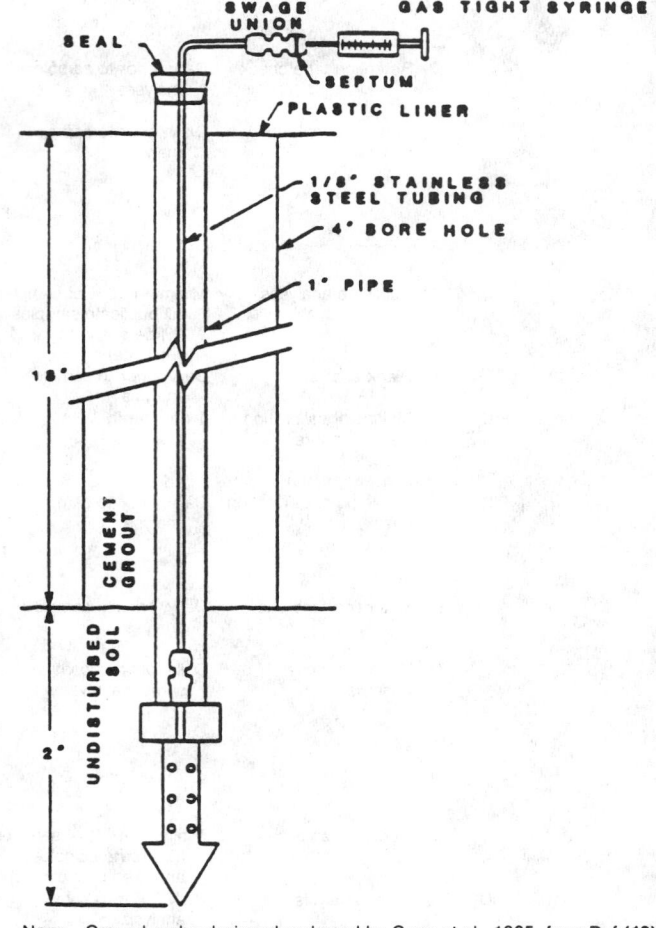

NOTE—Ground probe designed and used by Crow et al., 1985, from Ref **(10)**.

FIG. X1.2 Example of Whole-Air Active Sampling System

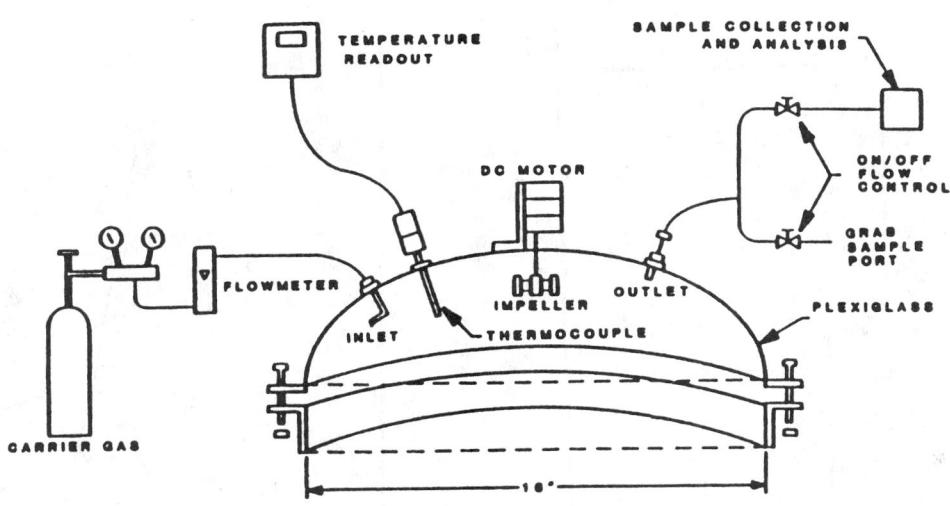

NOTE—Surface flux chamber and peripheral equipment after Eklund et al., 1984, from Ref (**10**).

FIG. X1.3   Example of Whole-Air Passive Sampling System

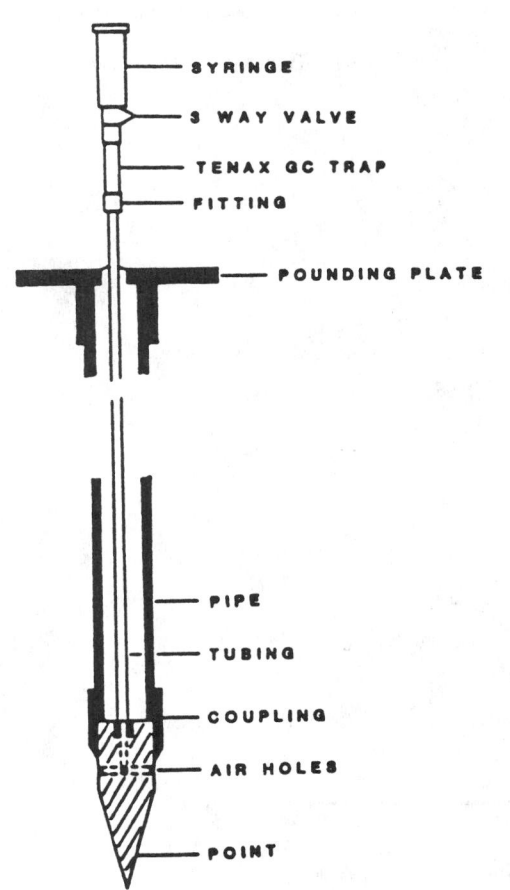

NOTE—Ground probe design used by Swallow and Gachwend, 1983, from Ref (**10**).

FIG. X1.4   Example of Sorbed Contaminant-Active System

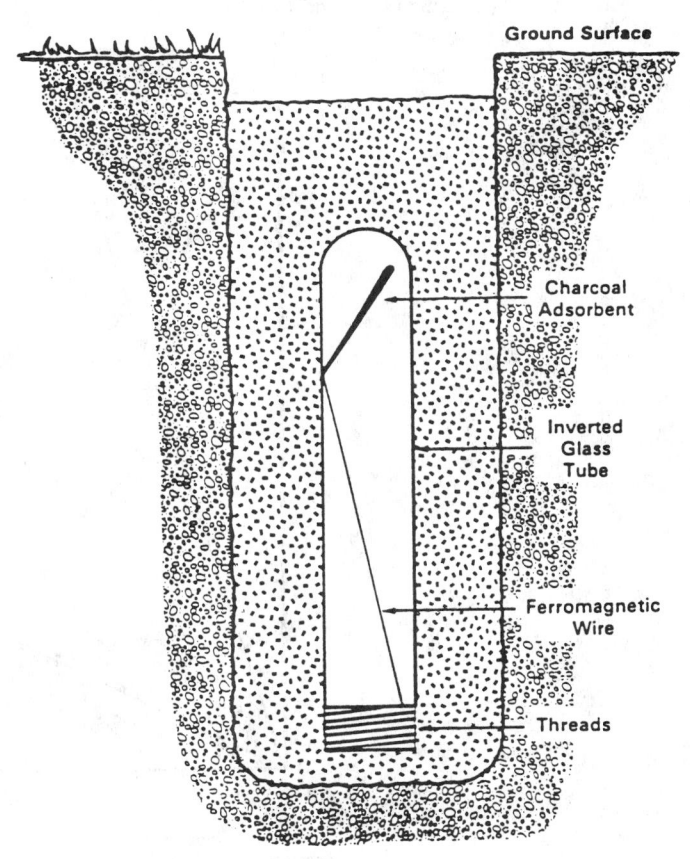

NOTE—Schematic diagram of emplacement of a sorbed contaminant-passive system (**10**).

FIG. X1.5   Example of Sorbed Contaminant-Passive System

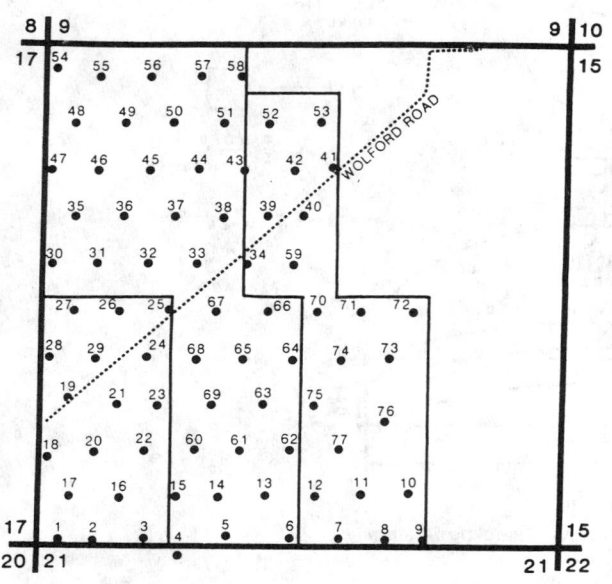

MAP SCALE: 1'' = 1,000'

NOTE—In any application, soil gas monitoring can be performed over a wide range of spatial designs, including soil gas sampling in grid patterns at a single depth or multiple depths. This example illustrates a staggered grid pattern of samples recovered at a single depth.

FIG. X1.6   Typical Soil Gas Grid Array and Map Display

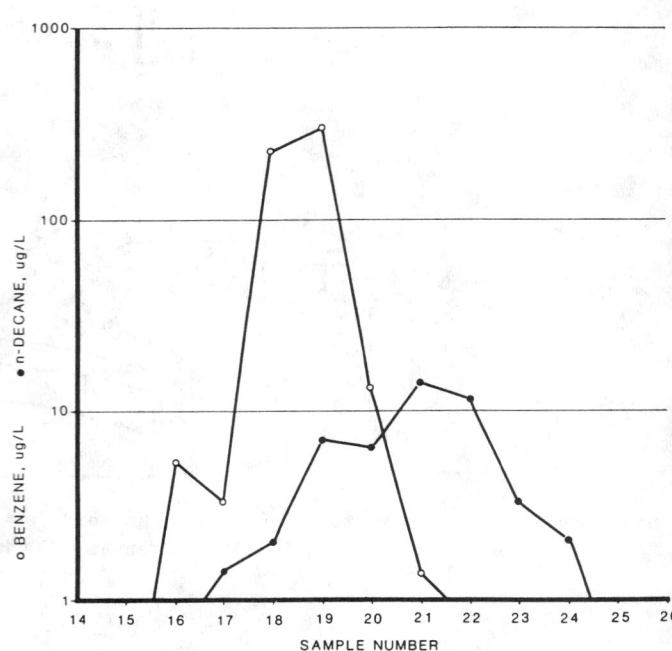

NOTE—Soil gas data from survey profiles displayed on an X – Y plot is an effective aid to data interpretation. This display is useful to examine the overall context for soil gas measurement data potentially indicating contamination. If the profile is displayed as a cross section through a grid pattern or as a linear array of sample points, the profile display can illustrate spatially significant groupings of data subpopulations.

FIG. X1.7   Typical Soil Gas Profile

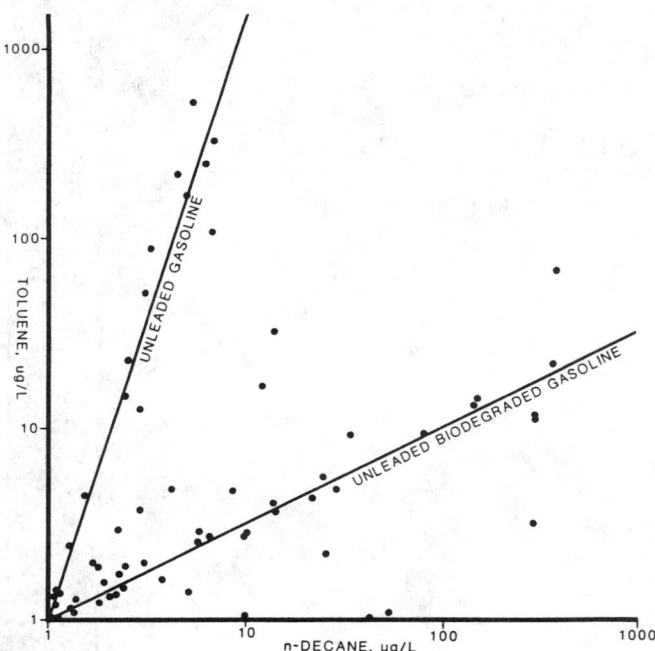

NOTE—Bimodal populations of data that represent coincident contaminant occurrences (for example, soil gas contaminant vapors sourced from converging plumes of two different fuels or mixtures of gasoline and biodegraded gasoline) can be defined using compositional analyses. One technique of compositional analysis is cross-plotting as shown.

FIG. X1.8   Soil Gas Compositional Analysis by Cross Plot

Project # _____ Sample # _____

Sampled by: _____

Date Sampled: _____, 199_____ Time: _____ (AM/PM)

Sampling System (check one):

( ) Whole air-active approach

( ) Whole air-passive approach

( ) Sorbed contaminants-active approach

( ) Sorbed contaminants-passive approach

( ) Headspace or extraction approach

( ) Soil pore liquid headspace approach

Sample Type (check one):

( ) Direct field sample

( ) Field blank

( ) Travel blank

( ) Sample container blank

( ) Sample probe blank

( ) Sample replicate

Spiked? _____ with _____ cc of _____

Potential reaction products due to spiking: _____

System purge volume: _____ Volumes purged: _____ Sample volume: _____

Sorbent Device: Installed _____ (AM/PM), _____, 199_____

Recovered _____ (AM/PM), _____, 199_____

Sample container type: _____ Sample container #_____

Integral analyzer: _____ Detector: _____

Analyzer response: _____ (units) _____

Surface conditions (pavement, wet, frost, etc.) _____

Sample depth: _____ Sampling rate: _____

Sample horizon data-visual estimates:

Vadose zone make-up: ( ) Native soil+rock ( ) Fill ( ) Rock

Soil composition: Clay, _____%

Soil organic matter, _____%

Fine granular material, _____%

Coarse granular material, _____%

100 %

Moisture content of sampling horizon (qualitative):

		( ) Dry
( ) Very	( ) Damp	
( ) Slightly	( ) Moist	
	( ) Wet	

Other characteristics of the sampling horizon:

( ) Free water present   ( ) Probable connection to surface macropores

( ) Free product present

( ) Contaminant odors   ( ) Indurated

( ) Poor perm. to vapor   ( ) Soil discoloration

( ) Near slope or vent   ( ) _____

_____

Investigator Signature/Date

_____

Investigator Affiliation

FIG. X1.9 Suggested Soil Gas Sample Data Sheet

REFERENCES

(1) Plumb, R. H., Jr., and Pitchford, A. M., "Volatile Organic Scans: Implications for Ground Water Monitoring," *Proceedings*, NWWA/API Conference on Petroleum Hydrocarbons and Organic Chemicals in Ground Water-Prevention, Detection and Restoration, Houston, Texas, November 13–15, 1985.

(2) Stonestrom, D. A., and Rubin, J., "Air Permeability and Trapped-Air Content in Two Soils," *Water Resources Research*, Vol 25, No. 9, September 1989, pp. 1959–1969.

(3) Lyman, Warren J., "Environmental Partitioning of Gasoline in Soil/Groundwater Compartments," *Seminar on the Subsurface Movement of Gasoline*, Edison, New Jersey, 1987.

(4) Price, L. C., "Aqueous Solubility of Petroleum as Applied to Its Origin and Primary Migration," *Am. Assoc. Petrol. Bull.*, Vol 60, No. 2, February 1976, pp. 213–244.

(5) Owens, J. W., Wasik, S. P., and Devoe, H., "Aqueous Solubilities and Enthalpies of Solution of n-Alkylbenzenes," *J. Chem. Eng. Data*, Vol 31, No. 47, 1986, pp. 47–51.

(6) Reisinger, H. J., Burris, D. R., Cessar, L. R., and McCleary, C. D., "Factors Affecting the Utility of Soil Vapor Assessment Data," *Proceedings*, First National Outdoor Action Conference on Aquifer Restoration, Ground Water Monitoring and Geophysical Methods, May 18–21, 1987, Las Vegas, Nevada, National Water Well Association, Dublin, Ohio, pp. 425–435.

(7) Stonestrom, D. A., and Rubin, J., "Water Content Dependence of Trapped Air in Two Soils," *Water Resources Research*, Vol 25, No. 9, September 1989, pp. 1947–1958.

(8) Manos, C. G. Jr., Williams, K. R., Balfour, W. D., and Williamson, S. J., "Effects of Clay Mineral-Organic Matter Complexes on Gaseous Hydrocarbon Emissions from Soils," *Proceedings*, NWWA/API Conference on Petroleum Hydrocarbons and Organic Chemicals in Ground Water-Prevention, Detection and Restoration, Houston, Texas, November 13–15, 1985.

(9) American Petroleum Institute, "The Migration of Petroleum Products in Soil and Ground Water: Principles and Countermeasures," *API Publication No. 4149*, Washington, D.C., 1972.

(10) Devitt, D. A., Evans, R. B., Jury, W. A., Starks, T. P., Eklund, B., Gnolson, A., and van Ee, J. J., *Soil Gas Sensing for the Detection and Mapping of Volatile Organics*, National Water Well Association, Dublin, Ohio, 1987.

(11) Kerfoot, W. B., and Sanford, W., "Four-Dimensional Perspective of an Underground Fuel Oil Tank Leakage," *Proceedings*, NWWA/API Conference on Petroleum Hydrocarbons and Organic Chemicals in Ground Water, Houston, Texas, November 12–14, 1986, National Water Well Association, Dublin, Ohio, pp. 383–403.

(12) Dragun, J., "Microbial Degradation of Petroleum Products in Soil," *Proceedings*, Conference on the Environmental and Public Health Effects of Soils Contaminated with Petroleum Products, Amherst, Massachusetts, October 30–31, 1985.

(13) Barker, J. F., Patrick, G. C., and Major, D., "Natural Attenuation of Aromatic Hydrocarbons in a Shallow Sand Aquifer," *Ground Water Monitoring Review*, Vol 7, No. 1, pp. 66–71.

(14) Davis, J. B., "Microbiology in Petroleum Exploration," *Unconventional Methods in Exploration for Petroleum and Natural Gas*, W. B. Heroy, ed., Southern Methodist University, Institute for the Study of Earth and Man, SMU Press, 1969, pp. 139–157.

(15) Jensen, B., Arvin, E., and Gundersen, A. T., "The Degradation of Aromatic Hydrocarbons with Bacteria from Oil Contaminated Aquifers," *Proceedings*, NWWA/API Conference on Petroleum Hydrocarbons and Organic Chemicals in Groundwater, Houston, Texas, November 13–15, 1985, pp. 421–435.

(16) White, K. D., Novak, J. T., Goldsmith, C. D., and Bevan, S., "Microbial Degradation Kinetics of Alcohols in Subsurface Systems," *Proceedings*, NWWA/API Conference on Petroleum Hydrocarbons and Organic Chemicals in Ground Water-Prevention,

Detection and Restoration, Houston, Texas, November 13–15, 1985.

(17) Kerfoot, H. B., and Barrows, L. J., *Soil Gas Measurement for Detection of Subsurface Organic Contamination*, U.S. Department of Commerce-National Technical Information Service, Springfield, Virginia, 1987.

(18) Diem, D., Kerfoot, H. B., and Ross, B. E., "Field Evaluation of a Soil-Gas Analysis Method for Detection of Subsurface Diesel Fuel Contamination," *Proceedings*, Second National Outdoor Action Conference on Aquifer Restoration, Ground Water Monitoring and Geophysical Methods, National Water Well Association, Dublin, Ohio, 1987.

(19) Chan, D. B., and Ford, E. A., "In-Situ Oil Biodegradation," *Military Engineer*, No. 509, 1986, pp. 447–737.

(20) Thompson, G. M., and Marrin, D. L., "Soil Gas Contaminant Investigations: A Dynamic Approach," *Ground Water Monitoring Review*, Vol 7, No. 3, pp. 88–93.

(21) Boynton, D., and Reuther, W., "A Way of Sampling Soil Gases in Dense Subsoils, and Some of Its Advantages and Limitations," *Proceedings*, Soil Science Society of America, Vol 3, 1938, pp. 37–42.

(22) Horvitz, L., "Hydrocarbon Geochemical Prospecting After Thirty Years," *Unconventional Methods in Exploration for Petroleum and Natural Gas*, W. B. Heroy, ed., Southern Methodist University, Institute for the Study of Earth and Man, SMU Press, 1969, pp. 205–218.

(23) Ullom, W. L., "Ethylene and Propylene in Soil Gas: Occurrence, Sources and Impact on Interpretation of Exploration Geochemical Data," *Bulletin*, Association of Petroleum Geochemical Explorationists, Vol 4, No. 1, December 1988, pp. 62–81.

(24) McCarthy, J. H. Jr., and Reimer, G. M., "Advances in Soil Gas Geochemical Exploration for Natural Resources: Some Current Examples and Practices," *Journal of Geophysical Research*, Vol 91, No. B12, November 1986, pp. 327–338.

(25) Jones, V. T., and Thune, H. W., "Surface Detection of Retort Gases from an Underground Coal Gasification Reactor in Steeply Dipping Beds Near Rawlins, Wyoming," SPE Paper 11050, *57th Annual Fall Technical and Exhibition of the Society of Petroleum Engineers of AIME*, New Orleans, Louisiana, September 26–29, 1982.

(26) Roffman, H. K., Neptune, M. D., Harris, J. W., Carter, A., and Thomas, T., "Field Screening for Organic Contaminants in Samples From Hazardous Waste Sites," *Proceedings*, NWWA/API Conference on Petroleum Hydrocarbons and Organic Chemicals in Ground Water, Houston, Texas, November 13–15, 1985.

(27) Wittmann, S. G., Quinn, K. J., and Lee, R. D., "Use of Soil Gas Sampling Techniques for Assessment of Ground Water Contamination," *Proceedings*, NWWA/API Conference on Petroleum Hydrocarbons and Organic Chemicals in Ground Water-Prevention, Detection and Restoration, Houston, Texas, November 13–15, 1985, pp. 291–309.

(28) Scheinfeld, R. A., and Schwendeman, T. G., "The Monitoring of Underground Storage Tanks, Current Technology," *Proceedings*, NWWA/API Conference on Petroleum Hydrocarbons and Organic Chemicals in Ground Water, Houston, Texas, November 13–15, 1985, pp. 244–264.

(29) Karably, L. S., and Babcock, K. B., "The Effects of Environmental Variables on Soil Gas Surveys," *Hazardous Materials Control*, January/February, 1989, pp. 36–43.

(30) Balfour, W. D., and Schmidt, C. E., *Sampling Approaches for Measuring Emission Rates from Hazardous Waste Disposal Facilities," U.S. EPA*, Contract No. 68-02-3171.

(31) Eklund, B., "Detection of Hydrocarbons in Groundwater by Analysis of Shallow Soil Gas/Vapor," *API Publication No. 4394*, Washington, D.C., 1985.

(32) Mayer, C. L., *Draft Interim Guidance Document for Soil-Gas Surveying*, U.S. EPA Environmental Monitoring Systems Laboratory, Office of Research and Development, Contract No. 68-03-3245, September 1989, 124 pp.

(33) Spittler, T. M., and Clifford, W. S., "A New Method for Detection of Organic Vapors in the Vadose Zone," *Proceedings*, NWWA Conference on Characterization and Monitoring of the Vadose (Unsaturated) Zone, November 19–21, 1985, Denver, Colorado, National Water Well Association, Dublin, Ohio, pp. 236–246.

(34) Kerfoot, H. B., "Shallow-Probe Soil-Gas Sampling for Indication of Ground Water Contamination by Chloroform," *International Journal of Environmental and Analytical Chemistry*, 1987, No. 30, pp. 167–181.

(35) Kerfoot, W. B., "A Portable Well Point Sampler for Plume Tracking," *Ground Water Monitoring Review*, Vol 4, No. 4, pp. 38–42.

(36) Litherland, S. T., Hoskings, T. W., and Boggess, R. T., "A New Ground Water Survey Tool: The Combined Cone Penetrometer/Vadose Zone Vapor Tool," *Proceedings*, NWWA/API Conference on Petroleum Hydrocarbons and Organic Chemicals in Ground Water, Houston, Texas, November 13–15, 1985, pp. 322–330.

(37) Nadeau, R. J., Stone, T. S., and Clinger, G. S., "Sampling Soil Vapors to Detect Subsurface Contamination: A Technique and Case Study," *Proceedings*, NWWA Conference on Characterization and Monitoring of the Vadose (Unsaturated) Zone, November 19–21, 1985, Denver, Colorado, National Water Well Association, Dublin, Ohio, pp. 215–226.

(38) U.S. EPA Environmental Response Team, Standard Operating Procedure 2149: "Soil Gas Sampling," September 30, 1988.

(39) Richers, D. M., et al., "Landsat and Soil Gas Geochemical Study of Patrick Draw Oil Field, Sweetwater County, Wyoming," *Bulletin*, American Association of Petroleum Geologists, Vol 66, No. 7, July 1982, pp. 903–922.

(40) Colenutt, B. A., and Davies, D. N., "The Sampling and Gas Chromatographic Analysis of Organic Vapors in Landfill Sites," *International Journal of Environmental and Analytical Chemistry*, No. 7, 1980, pp. 223–229.

(41) Newman, W., Armstrong, J. M., and Ettenhofer, M., "An Improved Soil Gas Survey Method Using Adsorbent Tubes for Sample Collection," *Proceedings*, Second National Outdoor Action Conference on Aquifer Restoration, Ground Water Monitoring and Geophysical Methods, May 23–26, 1988, Las Vegas, Nevada, National Water Well Association, Dublin, Ohio, pp. 1033–1049.

(42) U.S. EPA Environmental Response Team, Standard Operating Procedure 2051: "Charcoal Tube Sampling," November 7, 1988.

(43) U.S. EPA Environmental Response Team, Standard Operating Procedure 2052: "Tenax Tube Sampling," November 8, 1988.

(44) Kerfoot, H. B., and Mayer, C. L., *The Use of Industrial Hygiene Samplers For Soil-Gas Measurement*, U.S. EPA Environmental Monitoring Systems Laboratory, Advanced Monitoring Division, Contract No. 68-03-3249, September 1987.

(45) Voorhees, K. J., Hickey, J. C., and Klusman, R. W., "Analysis of Ground Water Contamination by a New Surface Static Trapping/Mass Spectrometry Technique," *Analytical Chemistry*, Vol 56, No. 13, 1984, pp. 2602–2604.

(46) Wesson, T. C., and Armstrong, F. E., *The Determination of C_1 – C_4 Hydrocarbons Adsorbed on Soils*, Bartlesville Energy Research Center Report of Investigations BERC/RI-75/13, U.S. Energy Research and Development Administration, Office of Public Affairs, Technology Information Center, Bartlesville, Oklahoma, December 1975.

(47) U.S. EPA Environmental Response Team, Standard Operating Procedure 2012: "Soil Sampling," November 8, 1988.

(48) USEPA, OWSER, SW 846, Method 5030, 1986.

(49) Sawhney, B. L., Pignatello, J. J., and Steinberg, S. M., "Determination of 1,2-Dibromoethane (EDB) in Field Soils: Implications for Volatile Organic Compounds," *Journal of Environmental Quality*, Vol 17, No. 1, January 1988, pp. 149–152.

(50) Steinberg, S. M., Pignatello, J. J., and Sawhney, B. L., "Persistence of 1,2-Dibromoethane in Soils: Entrapment in Intraparticle Micropores," *Environmental Science and Technology*, Vol 21, No. 12, December 1987, pp. 1201–1208.

(51) Rice, Gary K., "Combined Near-Surface Geochemical and Seismic Methods for Petroleum Exploration: Evidence for Vertical Migration," *Bulletin*, Association of Petroleum Geochemical Explorationists, Vol 2, No. 1, December 1986, pp. 46–62.

(52) Evans, O. D., and Thompson, G. M., "Field and Interpretation Techniques for Delineating Subsurface Petroleum Hydrocarbon Spills Using Soil Gas Analysis," *Proceedings*, NWWA/API Conference on Petroleum Hydrocarbons and Organic Chemicals in Ground Water: Prevention, Detection and Restoration, Houston, Texas, November 12–14, 1986, pp. 444–455.

(53) Riggin, R. M., *Technical Assistance Document for Sampling and Analysis of Toxic Organic Compounds in Ambient Air*, U.S. EPA Environmental Monitoring Systems Laboratory, Document No. EPA-600/X-83-025.

(54) U.S. EPA Environmental Response Team, Standard Operating Procedure 2050: "Tedlar Bag Sampling," November 8, 1988.

(55) U.S. EPA Environmental Response Team, Standard Operating Procedure 2056: "Photoionization Detector (PID) HNU," October 18, 1988.

(56) U.S. EPA Environmental Response Team, Standard Operating Procedure 2107: "Photovac 10A10 Portable Gas Chromatograph," December 29, 1988.

(57) U.S. EPA Environmental Response Team, Standard Operating Procedure 2109: "Photovac GC Analysis for Soil, Water and Air/Soil Gas," February 14, 1989.

(58) Brown, G. E., DuBose, D. A., Phillips, W. R., and Harris, G. E., "Project Summary: Response Factors of VOC Analyzers Calibrated with Methane for Selected Organic Chemicals," U.S. EPA, Industrial Environmental Research Laboratory-RTP, EPA Report No. EPA-600/52-81-002, 1981.

(59) Senn, R. B., and Johnson, M. S., "Interpretation of Gas Chromatography Data as a Tool in Subsurface Hydrocarbon Investigations," *Ground Water Monitoring Review*, Vol 7, No. 1, Winter, 1987, pp. 58–63.

(60) Smith, Rosser J. III, "Use of the Empirical Distribution Function in the Estimation of Boundaries of Anomalous Subpopulations," *Bulletin*, Association of Petroleum Geochemical Explorationists, Vol 3, No. 1, December 1987, pp. 64–87.

(61) Marrin, Donald L., "Soil Gas Analysis of Methane and Carbon Dioxide: Delineating and Monitoring Petroleum Hydrocarbons," *Proceedings*, NWWA Conference on Petroleum Hydrocarbons and Organic Chemicals in Ground Water, November 17–19, 1987, Houston, Texas, National Water Well Association, Dublin, Ohio.

Standard Guide for
Pore-Liquid Sampling from the Vadose Zone[1]

This standard is issued under the fixed designation D 4696; the number immediately following the designation indicates the year of original adoption or, in the case of revision, the year of last revision. A number in parentheses indicates the year of last reapproval. A superscript epsilon (ε) indicates an editorial change since the last revision or reapproval.

1. Scope

1.1 This guide discusses equipment and procedures used for sampling pore-liquid from the vadose zone (unsaturated zone). The guide is limited to in-situ techniques and does not include soil core collection and extraction methods for obtaining samples.

1.2 The term "pore-liquid" is applicable to any liquid from aqueous pore-liquid to oil. However, all of the samplers described in this guide were designed, and are used to sample aqueous pore-liquids only. The abilities of these samplers to collect other pore-liquids may be quite different than those described.

1.3 Some of the samplers described in this guide are not currently commercially available. These samplers are presented because they may have been available in the past, and may be encountered at sites with established vadose zone monitoring programs. In addition, some of these designs are particularly suited to specific situations. If needed, these samplers could be fabricated.

1.4 The values stated in SI units are to be regarded as the standard.

1.5 *This standard does not purport to address all of the safety problems, if any, associated with its use. It is the responsibility of the user of this standard to establish appropriate safety and health practices and determine the applicability of regulatory limitations prior to use.*

2. Referenced Documents

2.1 *ASTM Standards:*
D 653 Terminology Relating to Soil, Rock, and Contained Fluids[2]

3. Terminology

3.1 *Definitions:*

3.1.1 Where reasonable, precise terms and names have been used within this guide. However, certain terms and names with varying definitions are ubiquitous within the literature and industry of vadose zone monitoring. For purposes of recognition, these terms and names have been included in the guide with their most common usage. In these instances, the common definitions have been included in Appendix X1. Examples of such terms are soil, lysimeter, vacuum and pore-liquid tension.

3.2 *Descriptions of Terms Specific to This Standard:*

3.2.1 Appendix X1 is a compilation of those terms used in this guide. More comprehensive compilations, that were used as sources for Appendix X1, are (in decreasing order of their usage):

3.2.1.1 Terminology D 653,

3.2.1.2 *Compilation of ASTM Terminology,*[3]

3.2.1.3 *Glossary of Soil Science Terms,* Soil Science Society of America,[4] and,

3.2.1.4 *Webster's New Collegiate Dictionary.*[5]

4. Summary of Guide

4.1 Pores in the vadose zone can be saturated or unsaturated. Some samplers are designed to extract liquids from unsaturated pores; others are designed to obtain samples from saturated pores (for example, perched ground water) or saturated macropores (for example, fissures, cracks, and burrows). This guide addresses these categories. The sampler types discussed are:

4.1.1 Suction samplers (unsaturated sampling), (see Section 7),

4.1.2 Free drainage samplers (saturated sampling), (see Section 8),

4.1.3 Perched ground water samplers (saturated sampling), (see Section 9), and

4.1.4 Experimental absorption samplers (unsaturated sampling), (see Section 10).

4.2 Most samplers designed for sampling liquid from unsaturated pores may also be used to sample from saturated pores. This is useful in areas where the water table fluctuates, so that both saturated and unsaturated conditions occur at different times. However, samplers designed for sampling from saturated pores cannot be used in unsaturated conditions. This is because the liquid in unsaturated pores is held at less than atmospheric pressures (see *Richard's outflow principle,* in Appendix X1).

4.3 The discussion of each sampler is divided into specific topics that include:

4.3.1 Operating principles,

4.3.2 Description,

4.3.3 Installation,

4.3.4 Operation, and

4.3.5 Limitations.

5. Significance and Use

5.1 Sampling from the vadose zone may be an important component of some ground water monitoring strategies. It can provide information regarding contaminant transport

---

[1] This guide is under the jurisdiction of ASTM Committee D-18 on Soil and Rock and is the direct responsibility of Subcommittee D18.21 on Ground Water Vadose Zone Investigations.

Current edition approved April 15, 1992. Published June 1992.

[2] *Annual Book of ASTM Standards,* Vol 04.08.

[3] *Compilation of ASTM Terminology,* Sixth edition, ASTM, 1916 Race Street, Philadelphia, PA 19103, 1986.

[4] *Glossary of Soil Science Terms,* Soil Science Society of America, 1987.

[5] *Webster's New Collegiate Dictionary,* Fifth edition, 1977.

TABLE 1 Suction Sampler Summary

Sampler Type	Porous Section Material	Maximum[A] Pore Size (μm)	Air Entry Value (cbar)	Operational Suction Range (cbar)	Maximum Operation Depth (m)
Vacuum lysimeters	Ceramic	1.2 to 3.0 (1)[A]	>100	<60 to 80	<7.5
	PTFE	15 to 30 (2)[A]	10 to 21	<10 to 21	<7.5
	Stainless steel	NA[B]	49 to 5	49 to 5	<7.5
Pressure-vacuum lysimeters	Ceramic	1.2 to 3.0 (1)[A]	>100	<60 to 80	<15
	PTFE	15 to 30 (2)[A]	10 to 21	<10 to 21	<15
High pressure-vacuum lysimeters	Ceramic	1.2 to 3.0 (1)[A]	>100	<60 to 80	<91
	PTFE	15 to 30 (2)[A]	10 to 21	<10 to 21	<91
Filter tip samplers	Polyethylene	NA[B]	NA[B]	NA[B]	None
	Ceramic	2 to 3 (1)	>100	<60 to 80	<7.5
	Stainless steel	NA[B]	NA[B]	NA[B]	none
Cellulose-acetate hollow-fiber samplers	Cellulose Acetate Non cellulosic	<2.8	>100	<60 to 80	<7.5
	Polymer	<2.8	>100	<60 to 80	<7.5
Membrane filter samplers	Cellulose Acetate	<2.8	>100	<60 to 80	<7.5
	PTFE	2 to 5	NA[B]	NA[B]	<7.5
Vacuum plate samplers	Alundum	NA[B]	NA[B]	NA[B]	<7.5
	Ceramic	1.2 to 3.0	>100	60 to 80	<7.5
	Fritted glass	4 to 5.5	NA[B]	NA[B]	<7.5
	Stainless steel	NA[B]	49 to 5	49 to 5	<7.5

[A] Pore size determined by bubbling pressure (1) or mercury intrusion (2).

[B] NA = Not available.

and attenuation in the vadose zone. This information can be used for mitigating potential problems prior to degradation of a ground water resource (1).[6]

5.2 The choice of appropriate sampling devices for a particular location is dependent on various criteria. Specific guidelines for designing vadose zone monitoring programs have been discussed by Morrison (1), Wilson (2), Wilson (3), Everett (4), Wilson (5), Everett et al (6), Wilson (7), Everett et al (8), Everett et al (9), Robbins et al (10), Merry and Palmer (11), U.S. EPA (12), Ball (13), and Wilson (14). In general, it is prudent to combine various unsaturated and free drainage samplers into a program, so that the different flow regimes may be monitored.

5.3 This guide does not attempt to present details of installation and use of the equipment discussed. However, an effort has been made to present those references in which the specific techniques may be found.

6. Criteria for Selecting Pore-Liquid Samplers

6.1 Decisions on the types of samplers to use in a monitoring program should be based on consideration of a variety of criteria that include the following:

6.1.1 Required sampling depths,

6.1.2 Required sample volumes,

6.1.3 Soil characteristics,

6.1.4 Chemistry and biology of the liquids to be sampled,

6.1.5 Moisture flow regimes,

6.1.6 Required durability of the samplers,

6.1.7 Required reliability of the samplers,

6.1.8 Climate,

6.1.9 Installation requirements of the samplers,

6.1.10 Operational requirements of the samplers,

6.1.11 Commercial availability, and

6.1.12 Costs.

6.2 Some of these criteria are discussed in this guide.

---

[6] The boldface numbers in parentheses refer to the list of references at the end of this standard.

However, the ability to balance many of these factors against one another can only be obtained through field experience.

7. Suction Samplers

7.1 Table 1 presents the various types of suction samplers. The range of operating depths is the major criterion by which suction samplers are differentiated. Accordingly, the categories of suction samplers are as follows:

7.1.1 *Vacuum Lysimeters*—These samplers are theoretically operational at depths less than about 7.5 m. The practical operational depth is 6 m under ideal conditions.

7.1.2 *Pressure-Vacuum Lysimeters*—These samplers are operational at depths less than about 15 m.

7.1.3 *High Pressure-Vacuum Lysimeters*—(Also known as pressure-vacuum lysimeters with transfer vessels.) These samplers are normally operational down to about 46 m, although installations as deep as 91 m have been reported (15).

7.1.4 *Suction Lysimeters With Low Bubbling Pressures (Samplers With PTFE Porous Sections)*—These samplers are available in numerous designs that can be used to maximum depths varying from about 7.5 to 46 m.

NOTE 1—The samplers of 7.1.1, 7.1.2, 7.1.3, and 7.1.4 are referred to collectively as suction lysimeters. Within this standard, lysimeter is defined as a device used to collect percolating water for analyses (16).

7.1.5 *Filter Tip Samplers*—These samplers theoretically have no maximum sampling depth.

7.1.6 *Experimental Suction Samplers*—The samplers have limited field applications at the present time. They include cellulose-acetate hollow-fiber samplers, membrane filter samplers, and vacuum plate samplers. They are generally limited to depths less than about 7.5 m.

7.2 *Operating Principles:*

7.2.1 *General:*

7.2.1.1 Suction lysimeters consist of a hollow, porous section attached to a sample vessel or a body tube. Samples are obtained by applying suction to the sampler and collecting pore-liquid in the body tube. Samples are retrieved by a variety of methods.

7.2.1.2 Unsaturated portions of the vadose zone consist of interconnecting soil particles, interconnecting air spaces, and interconnecting liquid films. Liquid films in the soil provide hydraulic contact between the saturated porous section of the sampler and the soil (see Fig. 1). When suction greater than the soil pore-liquid tension is applied to the sampler, a pressure potential gradient towards the sampler is created. If the meniscuses of the liquid in the porous segment are able to withstand the applied suction (depending on the maximum pore sizes and hydrophobicity/hydrophilicity), liquid moves into the sampler. The ability of the meniscuses to withstand a suction decreases with increasing pore size and also with increasing hydrophobicity of the porous segment (see 7.6). If the maximum pore sizes are too large and hydrophobicity too great, the meniscuses are not able to withstand the applied suction. As a result, they break down, hydraulic contact is lost, and only air enters the sampler. As described in 7.6, ceramic porous segments are hydrophilic and the maximum pore sizes are small enough to allow meniscuses to withstand the entire range of sampling suctions. Presently available polytetrafluoroethylene (PTFE) porous segments are hydrophobic, the maximum pore sizes are larger, and only a very limited range of sampling suction can be applied before meniscuses break down and sampling ends (see 7.6.1.3). Therefore, samplers made with PTFE porous segments may be used only for sampling soils with low pore-liquid tensions **(12, 17)**.

7.2.1.3 The ability of a sampler to withstand applied suctions can be directly measured by its bubbling pressure. The bubbling pressure is measured by saturating the porous segment, immersing it in water, and pressurizing the inside of the porous segment with air. The pressure at which air starts bubbling through the porous segment into the surrounding water is the bubbling pressure. The magnitude of the bubbling pressure is equal to the magnitude of the maximum suction that can be applied to the sampler before air entry occurs (air entry value). Because the bubbling

pressure is a direct measure of how a sampler will perform, it is more useful than measurement of pore size distributions.

7.2.1.4 As soil pore-liquid tensions increase (low pore-liquid contents), pressure gradients towards the sampler decrease. Also, the soil hydraulic conductivity decreases exponentially. These result in lower flow rates into the sampler. At pore-liquid tensions above about 60 (for coarse grained soils) to 80 cbar (for fine grained soils), the flow rates are effectively zero and samples cannot be collected.

7.2.2 *Suction Lysimeters:*

7.2.2.1 Vacuum lysimeters directly transfer samples to the surface via a suction line. Because the maximum suction lift of water is about 7.5 m, these samplers cannot be operated below this depth. In reality, suction lifts of 6 m should be considered a practical maximum depth.

7.2.2.2 Samples may be retrieved using the same technique as for vacuum lysimeters or, for deeper applications, the sample is retrieved by pressurizing the sampler with one line; this pushes the sample up to the surface in a second line.

7.2.2.3 High pressure-vacuum lysimeters operate in the same manner as pressure-vacuum lysimeters. However, they include an inbuilt check transfer vessel or a chamber between the sampler and the surface. This prevents sample loss through the porous section during pressurization, and prevents possible cup damage due to overpressurization.

7.2.2.4 Suction lysimeters with low bubbling pressures are available in each of the three previous designs. The only difference between these samplers and the three previous designs is that these porous sections are made with PTFE. The low bubbling pressure (and hence large pore size or hydrophobicity, or both) of PTFE constrains these samplers to soils that are nearly saturated (see 7.2.1.2 and 7.6.1.3).

7.2.3 *Filter Tip Samplers*—Samples are collected from a filter tip sampler by lowering an evacuated sample vial down an access tube to a permanently emplaced porous tip. The vial is connected to the porous tip and sample flows through the porous section and into the vial. Once full, the vial is retrieved.

7.2.4 *Experimental Suction Samplers*—Experimental suction samplers generally operate on the same principle as vacuum lysimeters with different combinations of porous materials to enhance hydraulic contact. The samplers are generally fragile and difficult to install. As with vacuum lysimeters, they are generally limited to depths of less than about 7.5 m.

7.3 *Description:*

7.3.1 *Vacuum Lysimeters:*

7.3.1.1 Vacuum lysimeters generally consist of a porous cup mounted on the end of a tube, similar to a tensiometer. The cup is attached to the tube with adhesives (**18**[7]) or with "V" shaped flush threading sealed with an "O" ring. A stopper is inserted into the upper end of the body tube and fastened in the same manner as the porous cup or, in the case of rubber stoppers, inserted tightly (**12**). To recover samples, a suction line is inserted through the stopper to the base of the sampler. The suction line extends to the surface and connects to a sample bottle and suction source in series.

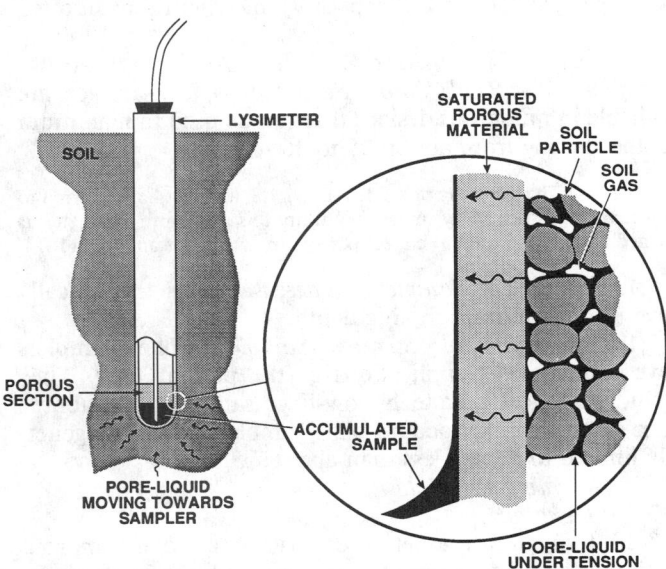

FIG. 1 Porous Section/Soil Interactions

[7] This reference is manufacturer's literature, and it has not been subjected to technical review.

Body tubes up to 1.8 m long have been reported (15) (see Fig. 2).

7.3.1.2 Harris and Hansen (19) described a vacuum lysimeter with a 6 mm by 65 mm ceramic porous cup designed for intensive sampling in small areas.

7.3.1.3 A variety of materials have been used for the porous segment including nylon mesh (20), fritted glass (21), sintered glass (22), Alundum (manufacturer name), stainless steel (23[7]), and ceramics (1.2 to 3.0 µm max pore size) (18[7]). The sampler body tube has been made with PVC, ABS, acrylic, stainless steel (24) and PTFE (18[7], 25[7]). Ceramic porous segments are attached with epoxy adhesives or with flush threading. The stopper is typically made of rubber (12), neoprene, or PTFE. The outlet lines are commonly PTFE, rubber, polyethylene, polypropylene, vinyl, nylon, and historically, copper. Fittings and valves are available in brass or stainless steel.

7.3.2 Pressure-Vacuum Lysimeters:

7.3.2.1 These samplers were developed by Parizek and Lane (26) for sampling deep moving pollutants in the vadose zone. The porous segment is usually a porous cup at the bottom of a body tube. The porous cup is attached with epoxy adhesives (18[7]) or with "V" shaped flush threading sealed with an "O" ring (25[7]). Two lines are forced through a two-hole stopper sealed into the upper end of the body tube. The discharge line extends to the base of the sampler and the pressure-vacuum line terminates a short distance below the stopper. At the surface, the discharge line connects to a sample bottle and the pressure-vacuum line connects to a pressure-vacuum pump. Designs are available that do not use a stopper but rather an "O" ring sealed, flush threaded top plug (25[7]). Tubing lines to the surface are attached to the top plug with threaded tubing fittings of appropriate materials. Body tubes are commonly available with 2.2 and 4.8 cm diameters and in a variety of lengths (see Fig. 3). The sampler and its components have been made out of the same materials used for vacuum lysimeters.

7.3.2.2 These samplers can retrieve samples from depths below 7.5 m because pressure is used for retrieval. However, during pressurization some of the sample is forced back out of the cup. At depths over about 15 m, the volume of sample lost in this fashion may be significant. In addition, at depths over about 15 m, pressures required to bring the sample to the surface may be high enough to damage the cup or to reduce its hydraulic contact with the soil (27, 28). Rapid pressurization causes similar problems. Morrison and Tsai (29) developed a tube lysimeter with the porous section located midway up the body tube instead of at the bottom (see Fig. 4). This design mitigates the problem of sample being forced back through the cup. However, it does not prevent problems with porous segment damage due to overpressurization or rapid pressurization. The sleeve lysimeter (that is no longer available) was a modification to this design for use with a monitoring well (1) (see Fig. 5). Another modification is the casing lysimeter that consists of several tube lysimeters threaded into one unit (see Fig. 6). This arrangement allows precise spacing between units (30).

7.3.2.3 Nightingale et al (31) described a design that

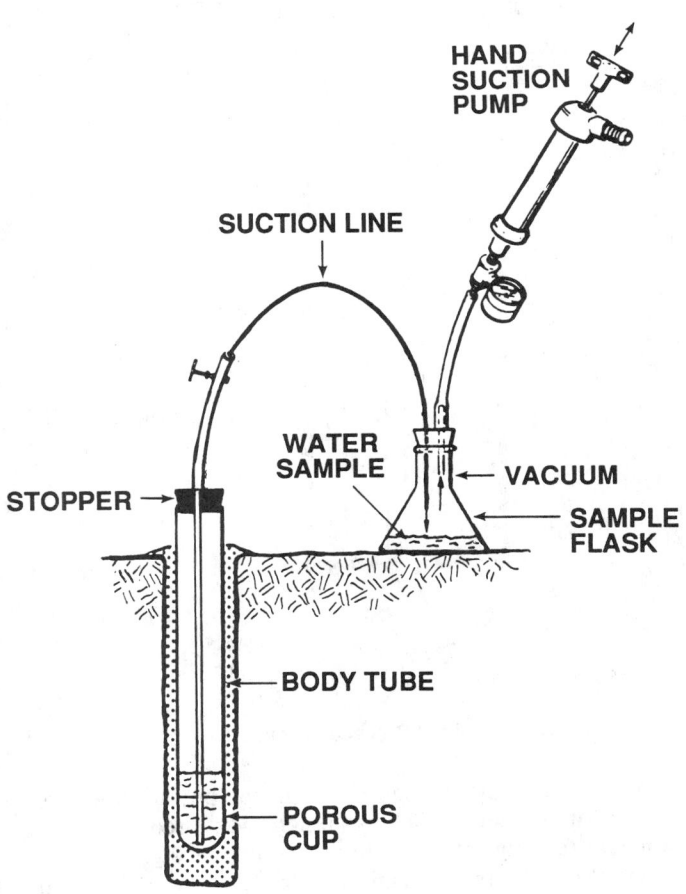

FIG. 2 Vacuum Lysimeter

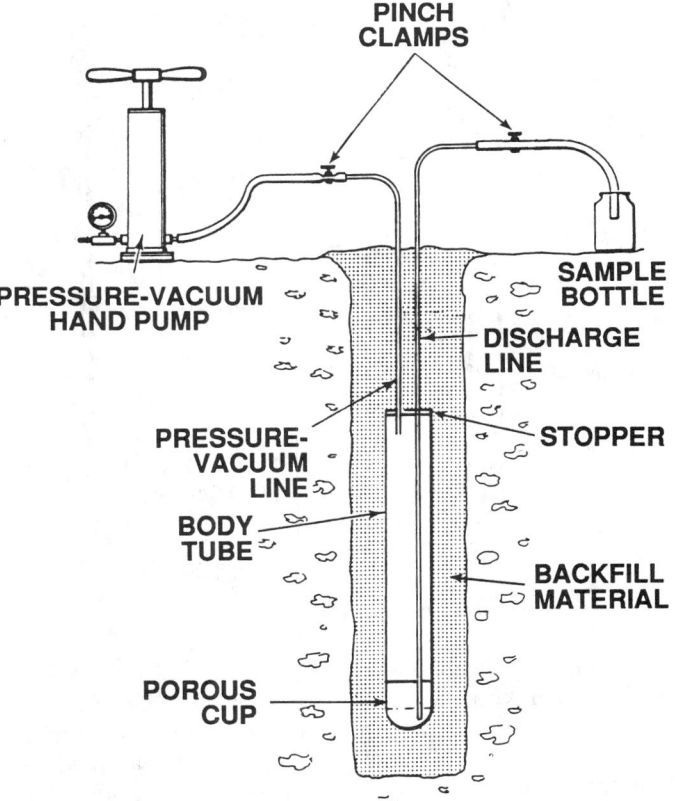

FIG. 3 Pressure-Vacuum Lysimeter

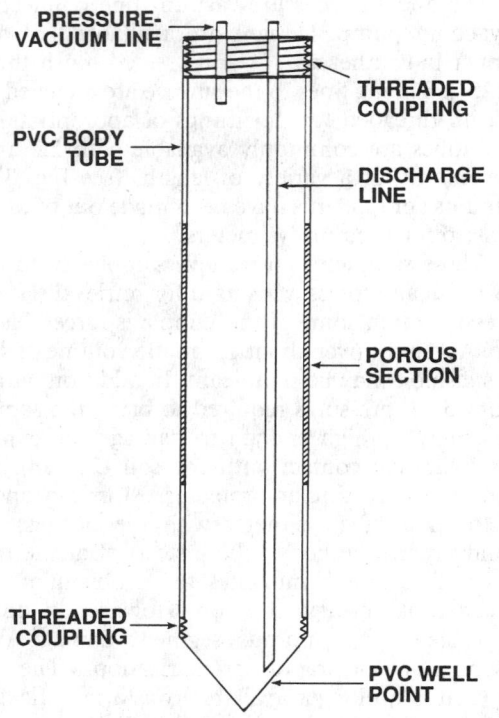

FIG. 4   Tube Pressure-Vacuum Lysimeter

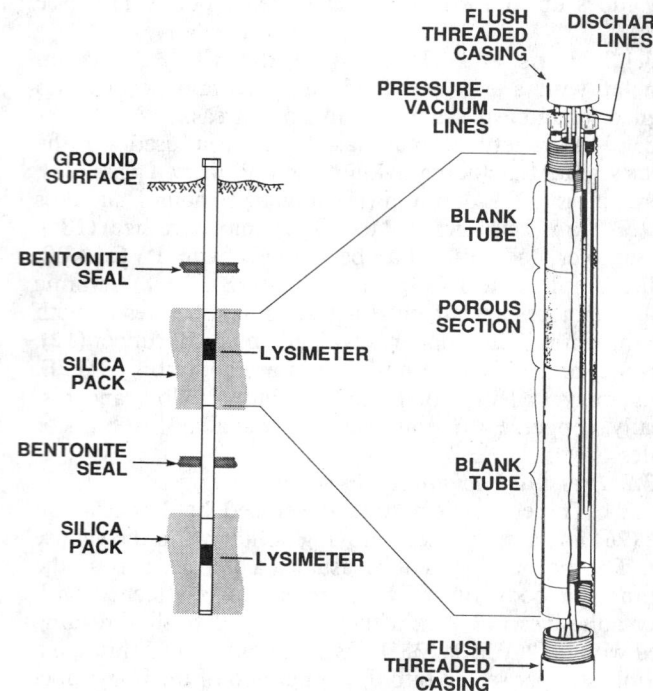

FIG. 6   Casing Lysimeter

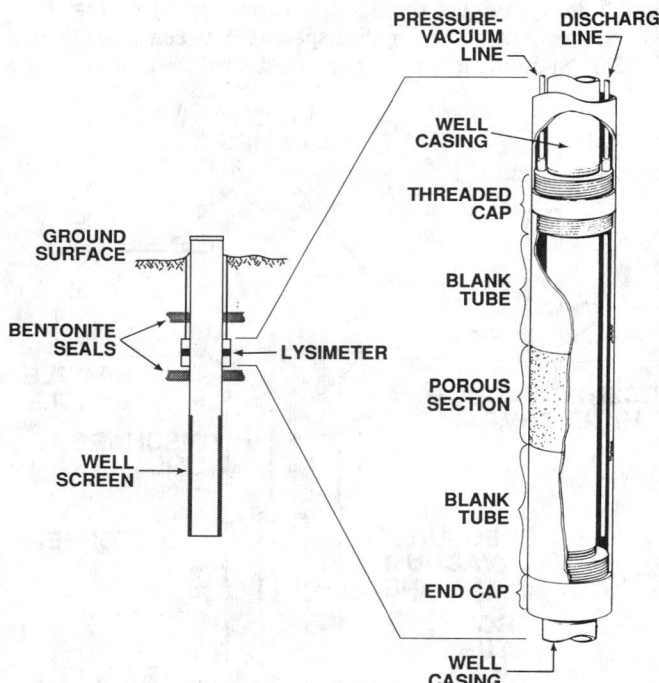

FIG. 5   Sleeve Lysimeter

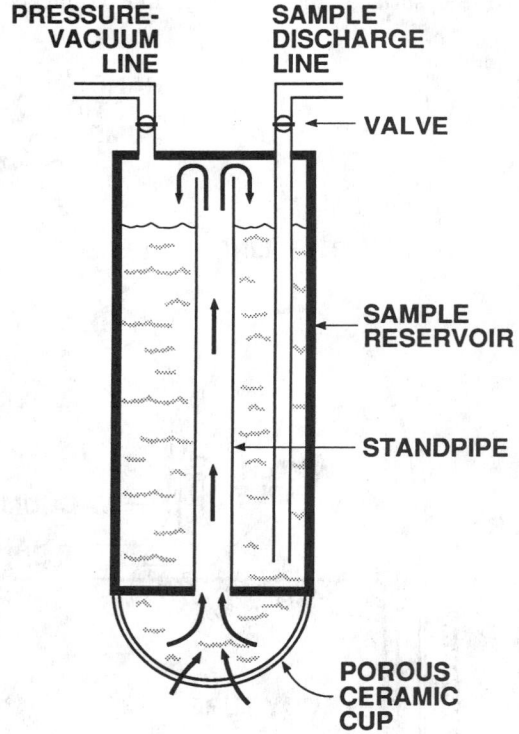

FIG. 7   Modified Pressure-Vacuum Lysimeter

allows incoming samples to flow into a portion of the sampler not in contact with the basal, porous ceramic cup (see Fig. 7). The ceramic cup is wedged into the body tube without adhesives or threading. The sampler was used to sample the vadose zone, the capillary fringe, and the fluctuating water table in a recharge area. Knighton and

Streblow (32) reported a sampler with the porous cup upon the top of a chamber. This design was used with cup diameters ranging from 7.6 to 12.7 cm (see Fig. 8). These designs also allow pressurization for sample retrieval without significant liquid loss. However, because the porous cups are

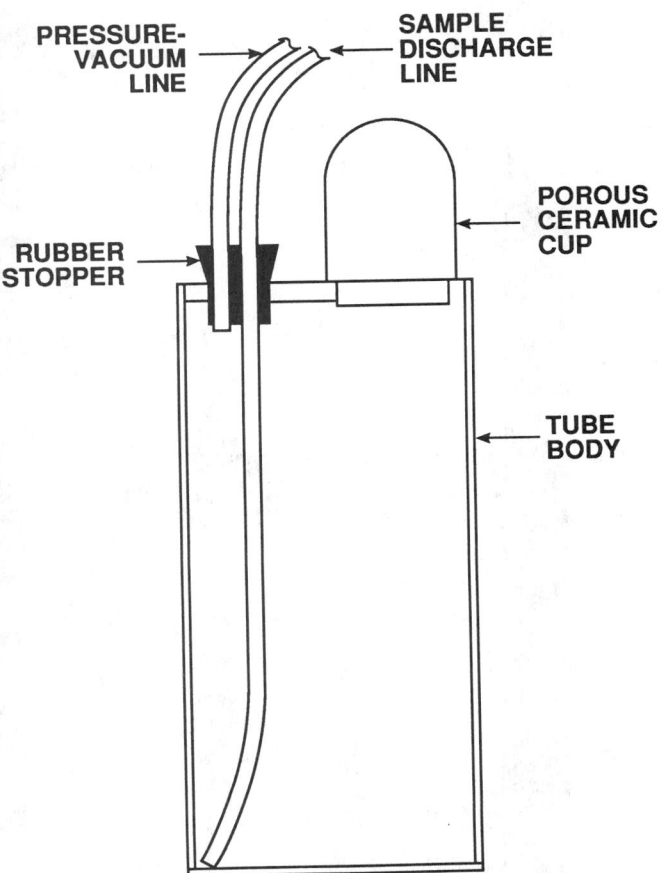

PRESSURE-VACUUM LINE

SAMPLE DISCHARGE LINE

POROUS CERAMIC CUP

RUBBER STOPPER

TUBE BODY

FIG. 8   Knighton and Streblow-Type Vacuum Lysimeter

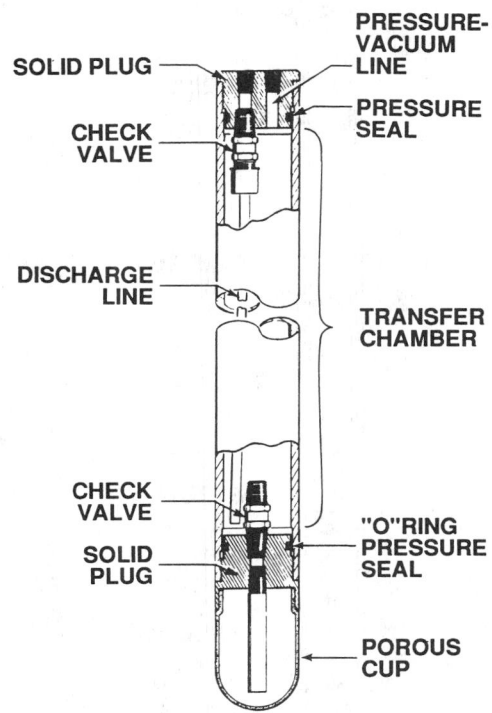

SOLID PLUG

PRESSURE-VACUUM LINE

CHECK VALVE

PRESSURE SEAL

DISCHARGE LINE

TRANSFER CHAMBER

CHECK VALVE

"O"RING PRESSURE SEAL

SOLID PLUG

POROUS CUP

FIG. 9   High Pressure-Vacuum Lysimeter

open to the rest of the samplers, possible damage due to overpressurization or rapid pressurization is still a problem.

7.3.3 *High Pressure-Vacuum Lysimeters (Lysimeters With a Transfer Vessel)*—High pressure-vacuum lysimeters overcome the problems of fluid loss and overpressurization through the use of an attached chamber or a connected transfer vessel (see Fig. 9). The porous segment is usually a porous cup at the bottom of the body tube. The cup is attached with epoxy adhesives (18[7]) or with "V" shaped flush threading sealed with an "O" ring (25[7]). In the attached chamber design, the body tube is separated into two chambers connected by a one-way check valve. A pressure-vacuum line and a discharge line enter through a two-hole plug at the top of the body tube. The pressure-vacuum line terminates below the plug. The discharge line extends to the bottom of the upper chamber. The transfer vessel design is similar. However, the vessel and body tube are integral components joined by a common double threaded, "O" ring sealed plug containing a check valve. Body tube diameters range from 2.7 to 8.9 cm outside diameter. Total sampler lengths commonly range from 15.2 to 182.9 cm. A threaded top plug allows attachment of casing to the lysimeter. This facilitates accurate placement and provides long-term protection for the tubing lines. The samplers and their components have been made out of the same materials as vacuum lysimeters.

7.3.4 *Suction Lysimeters with Low Bubbling Pressures (Samplers With PTFE Porous Sections)*—Designs are avail-

able in each of the three categories described in 7.3.1, 7.3.2, and 7.3.3. The only difference between this group of samplers and the previous three samplers is that PTFE is used for the porous sections of this group of samplers (25[7]). The porous PTFE is attached with "V" shaped flush threading sealed with an "O" ring.

7.3.5 *Filter Tip Samplers:*

7.3.5.1 Filter tip samplers consist of two components: a permanently installed filter tip, and a retrievable glass sample vial. The filter tip includes a pointed end to help with installation, a porous section, a nozzle, and a septum. The tip is threaded onto extension pipes that extend to the surface. The sample vial includes a second septum. When in use, the vial is seated in an adaptor that includes a disposable hypodermic needle to penetrate both the septa, allowing sample to flow from the porous segment into the vial (see Fig. 10). Extension pipes vary from 2.5 to 5.1 cm inside diameter. Vial volumes range from 35 to 500 mL (32[7]).

7.3.5.2 The body of the filter tip is made of thermoplastic, stainless steel, or brass. The attached porous section is available in high density polyethylene, sintered ceramic, or sintered stainless steel. The septum is made of natural rubber, nitrile rubber, or fluororubber (32[7]).

7.3.6 *Experimental Suction Samplers:*

7.3.6.1 Cellulose-acetate, hollow-fiber samplers were described by Jackson et al (33) and Wilson (3). A sampler consists of a bundle of these flexible, hollow fibers (<2.8 μm max pore size) pinched shut at one end and attached to a suction line at the other end. The suction line leads to the surface and attaches to a sample bottle and source of suction in the same manner as a vacuum lysimeter (see Fig. 11). The fibers, that are analogous to the porous sections of vacuum lysimeters, have outside diameters of up to 250 μm (33). Levin and Jackson (34) described similar fibers made from a

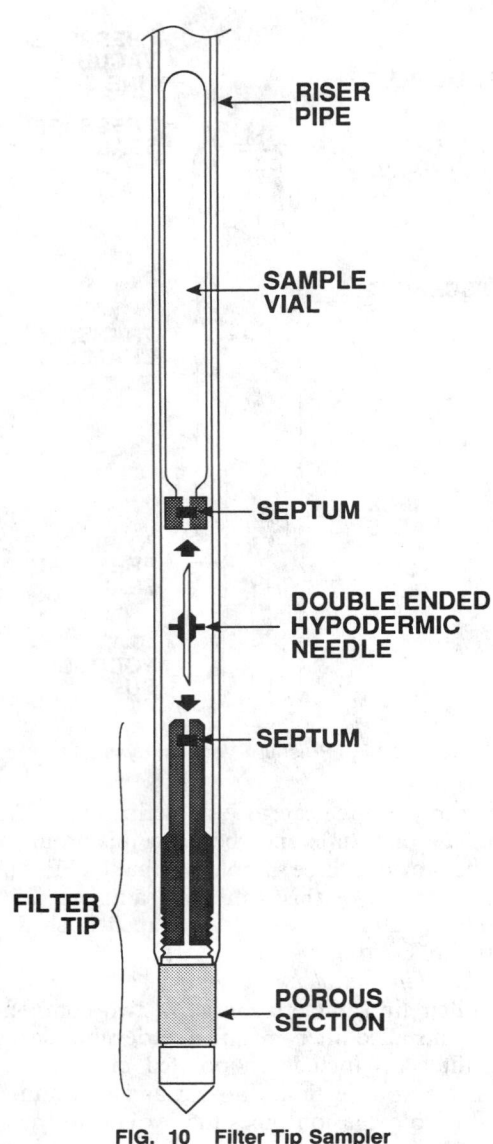

FIG. 10   Filter Tip Sampler

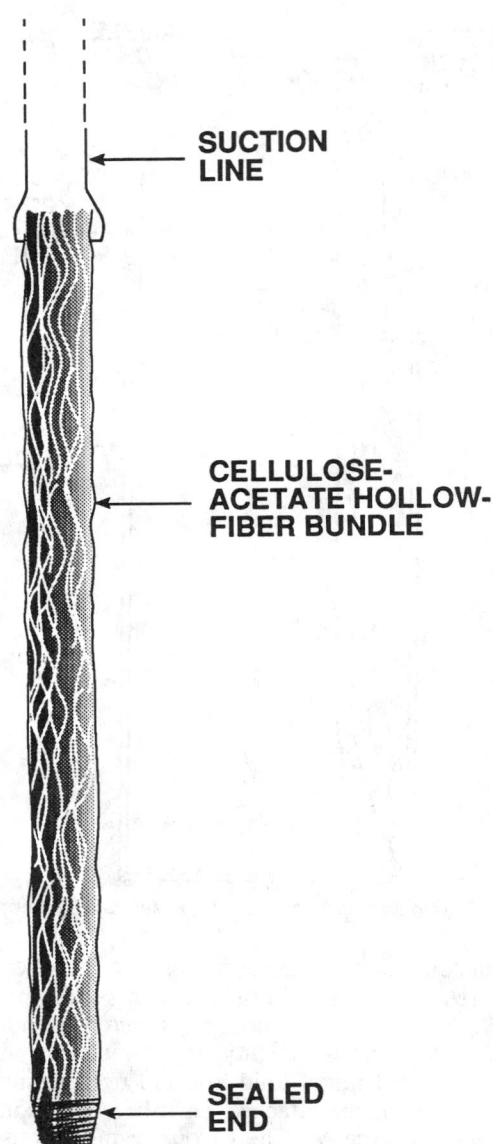

FIG. 11   Cellulose-Acetate Hollow-Fiber Sampler

noncellulosic polymer solution (max pore size <2.8 µm). Those fibers have dense inner layers surrounded by open celled, spongy layers with diameters ranging from 50 to 250 µm.

7.3.6.2 Membrane filter samplers were described by Morrison (1), Everett and Wilson (6), U.S. EPA (12) and Stevenson (35). A sampler consists of a membrane filter of polycarbonate, cellulose acetate (<2.8 µm max pore size), cellulose nitrate or PTFE (2 to 5 µm max pore size); mounted in a "swinnex" type filter holder (35, 36, 37[7]). The filter rests on a glass fiber prefilter. The prefilter rests on a glass fiber "wick" that in turn sits on a glass fiber collector. The collector is in contact with the soil and extends the sampling area of the small diameter filter (see Fig. 12 and 7.5.1.6). A suction line leads from the filter holder to the surface. At the surface, the suction line is attached to a sample bottle and suction source in a manner similar to vacuum lysimeters.

7.3.6.3 A vacuum plate sampler consists of a flat porous disk fitted with a nonporous backing attached to a suction line that leads to the surface (see Fig. 13). Plates are available in diameters ranging from 4.3 to 25.4 cm and custom designs are easily arranged (1, 18[7]). Plates are available in alundum, porous stainless steel (23[7]), ceramic (1.2 to 3.0 µm max pore size) or fritted glass (4 to 5.5 µm max pore size) (38[7], 6, 39, 40, 41, 42, 43, 44). The nonpermeable backing can be a fiberglass resin, glass, plastic, or butyl rubber.

7.3.7 *Comments:*

7.3.7.1 When some ceramic cups are glued to the inner wall of the body tube in a suction lysimeter, an inner lip is formed (45). As the discharge line is pushed through the stopper at the top of the sampler, it may catch on this lip and the operator may conclude that the line has reached the bottom of the ceramic cup (see Fig. 14). As a result, an 80 mL error can occur in sampling rate determinations. This 80 mL of fluid accumulates in the cup, is not removed during sampling, and will cause cross contamination between sampling events. Soilmoisture (18[7]) suggested that the line can be

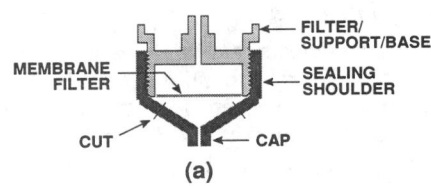

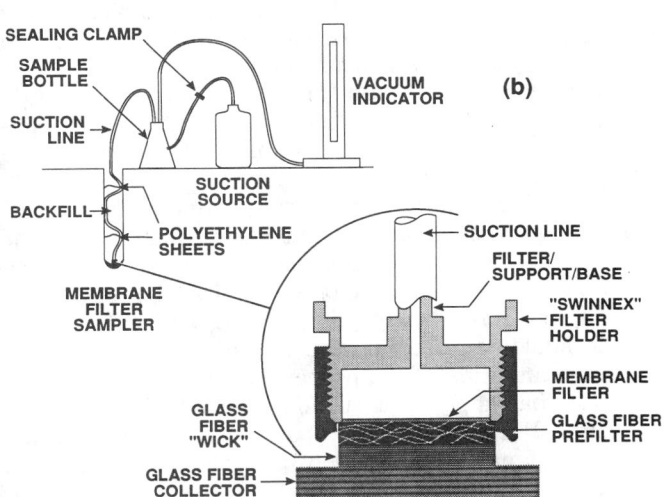

FIG. 12 Membrane Filter Sampler; (a) Preparation of Filter Sampler; and (b) Installation of Filter Sampler

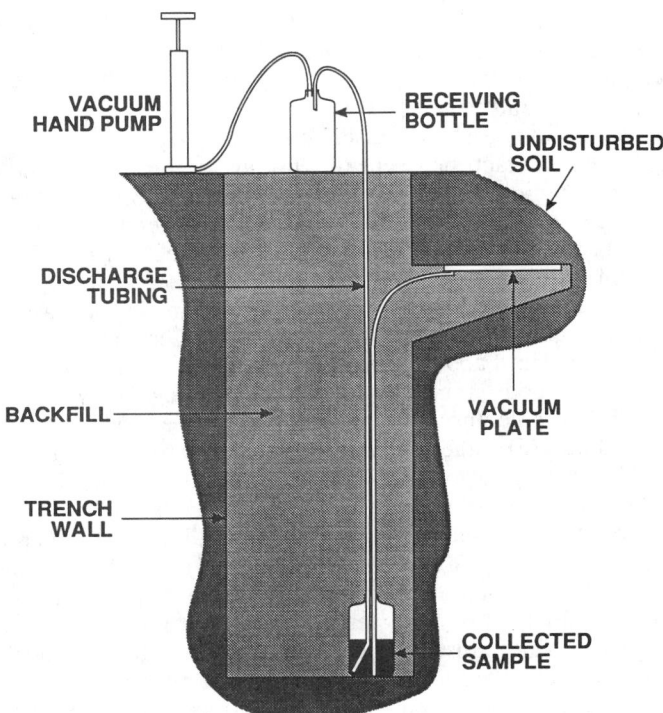

FIG. 13 Vacuum Plate Sampler Installation

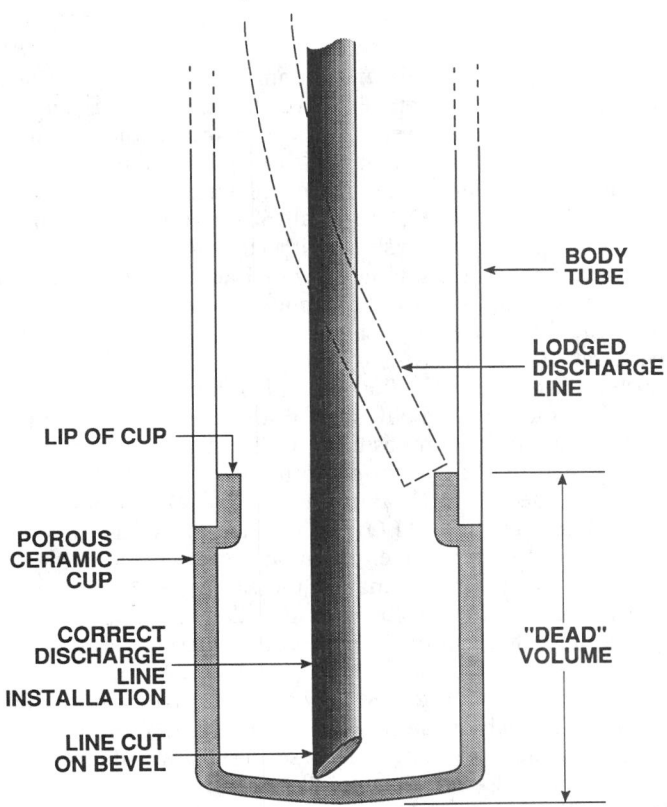

FIG. 14 Location of Potential Dead Volume in Suction Lysimeter

extending to the bottom of the cup. This results in a zero accumulation of fluid. Older samplers with PTFE porous segments and PVC body tubes have a discharge line that does not extend all the way to the bottom. This problem has been corrected in newer PTFE and PVC samplers (25[7]). This results in a 34 mL accumulation of fluid (12). Filter tip samplers develop an 8 mL accumulation of fluid. Haldorsen et al. (46) suggested collecting and discarding an initial sample to purge this accumulated fluid.

7.3.7.2 Because samplers are often handled roughly during installation, durability and ruggedness are important. It has been shown that PTFE has a higher impact strength than ceramics which need to be installed with care (25[7]). It has also been found that PTFE threads and ceramic threads (when used) are susceptible to leakage, and must be securely sealed with pipe threading tape (45). TFE-fluorocarbon (PTFE) tape is not recommended in square threaded joints since the tape is designed for tapered "V" threaded compression joints.

7.3.7.3 As described above, porous sections can be made from various materials. These materials have physical and chemical limitations that must be considered when designing a monitoring program. Physical limitations are described in 7.6.1. Chemical limitations are described in 7.6.2.

7.4 *Installation Methods:*

7.4.1 *Pre-Installation:*

7.4.1.1 As demonstrated by Neary and Tomassini (47), new samplers may be contaminated with dust during manufacturing. In order to reduce chemical interferences from substances on the porous sections, U.S. EPA (12) recom-

kept from catching by cutting its tip at an angle. In all-PTFE suction lysimeters, the discharge line is a rigid PTFE tube

mended preparation of ceramic units prior to installation following procedures originally developed by Wolff (48), modified by Wood (49) and recommended by Neary and Tomassini (47). The process involves passing hydrochloric acid (HCl) (for example, 8N) through the porous sections. This is followed by flushing with distilled water until the specific conductance of the outflowing water is within 2 % of the inflowing water. Debyle et al (50) found (in agreement with 49 and 51) that flushing with HCl strips cations off of the ceramic. This results in an initial adsorption of cations from pore-liquid onto the ceramic surface. This continues until the cation exchange capacity (CEC) of the ceramic has been satisfied. The effect is not reduced by distilled water flushing after the acid flushing. Therefore, they suggested that the sampler also be flushed, prior to installation, with a solution similar in composition to the expected soil solution. Alternately, the first sample after installation could be discarded (see 7.5.2.1). Bottcher et al (52) attributed increased adsorption of PO_4 to the acid leaching process. Therefore, they recommended a thorough flushing with a PO_4 solution of approximately the same concentration as that found in the soil solution, rather than the acid leaching procedure, when sampling for PO_4. Peters and Healy (53) used H_2SO_4 rather than HCl.

7.4.1.2 Hydrochloric acid may corrode valves within PVC and ceramic high pressure-vacuum lysimeters. Therefore, the porous segment flushing for these designs should be performed prior to attachment if possible. The maximum suction which can be applied is one atmosphere, therefore the flushing process will be slow if suction is used to draw HCl through the porous segment. The flushing can be performed more rapidly if the porous segment is filled with HCl and pressurized to force the acid out of the porous segment since more than one atmosphere of pressure can be applied. This procedure can only be used if the cups are not attached. Care must be taken to prevent overpressurization that might damage the porous section.

7.4.1.3 Corning Laboratories (38[7]) recommended washing fritted glass with hot HCl followed by a distilled water rinse. Cleaning procedures for Alundum have not been reported, although an acid and water rinse procedure similar to that for ceramic would appear to be appropriate (1). Timco (25[7]) described cleaning procedures for PTFE. The method includes passing 0.5 L of distilled water through the material. An I.P.A. bath followed by another in hydrogen peroxide or rinsing with HCl followed by a distilled water rinse.

7.4.1.4 The use of HCl to wash/flush porous segments of lysimeters, that are to be used in sanitary landfills, may cause water quality interpretation problems. Sanitary landfills are notorious generators of methane gas. Reaction of methane with free chloride ion may result in the generation of di- and trichloromethane (also known as methylene chloride and chloroform). Because of the small liquid volumes in lysimeters and the sensitivity of EPA methods (including 601), false positives for one or both of these constituents may occur.

7.4.1.5 Stevenson (35) recommended treating cellulose-acetate hollow-fibers with silver nitrate and sodium chloride to prevent biofilm growths. Morrison (1) suggested rinsing membrane filters with distilled water.

7.4.1.6 The porous section and fittings of individual samplers may have defects that could cause air entry during sampling. Therefore, prior to taking samplers to the field, each unit should be checked for its bubbling pressure, pressure tested and vacuum tested for leaks. Procedures for these tests are given in U.S. EPA (12) and Timco (25[7]). Washers or "O" rings are used to seal the plugs at the tops of body tubes. However, the accesses for pressure-vacuum and discharge lines passing through these plugs are not sealed. These accesses may leak, and should also be sealed. In the past, lubricants have been used when cutting threads into body tubes, porous segments and fittings. In addition, lubricants have been used in various pressure-vacuum pumps. The user should contact the manufacturer to determine if these lubricants are still used. If present, these lubricants should be removed.

7.4.1.7 After cleaning and testing, samplers should be bagged to prevent contamination during transport to the field. Compatibility of bag material and analytical parameters should be considered. Upon arrival at the installation location, and immediately prior to installation, the porous section should be placed in distilled water for about 30 min to ensure saturation of the porous section (1). Timco (25[7]) indicated that applying a suction of about 50 cbar to a submerged PTFE sampler for about an hour would ensure saturation. Finally, immediately prior to installation, the sampler and associated lines should be assembled and inspected for defects (for example, crimped lines).

7.4.2 *Suction Lysimeter and Filter Tip Sampler Installation:*

7.4.2.1 Suction lysimeter installation procedures have been described by U.S. EPA (12), Soilmoisture (18[7]), Timco (25[7]), Linden (54), and Rhoades and Oster (55). Filter tip sampler installation procedures were described by Torstensson and Petsonk (32).

7.4.2.2 The goals of installation are to ensure good hydraulic contact between the porous segment and the surrounding soil, and to minimize leakage of liquid along the outside of the sampler. U.S. EPA (12) recommended a silica flour/bentonite clay method to achieve these goals for suction lysimeters. A silica flour layer (installed as a slurry, see 7.4.2.6) placed around the porous segment increases hydraulic contact with the surrounding soil. Screened native backfill is placed above the silica flour, and a bentonite plug above the body tube prevents liquid leakage down the installation hole and along the body tube (see Figs. 15 and 16). Klute (56) indicated that a screened native soil slurry could be used in place of silica flour for shallow installations.

7.4.2.3 Samplers may be installed in the sidewall of an excavation or, for deeper applications, in a borehole preferably advanced with a hollow stem auger (12). U.S. EPA (12) suggested that suction lysimeters should be installed at an angle of 30 to 45° from vertical whenever possible. This ensures that an undisturbed column of soil is retained above the porous cup. Accordingly, pore liquid samples will reflect flow through pore sequences that have not been disturbed by sampler installation. This angular placement also improves the sampler's ability to collect macropore flow. When installed in the sidewall of a trench, the angled emplacement is simple (see Fig. 15). However, when installed in a borehole, angular emplacement entails angled drilling.

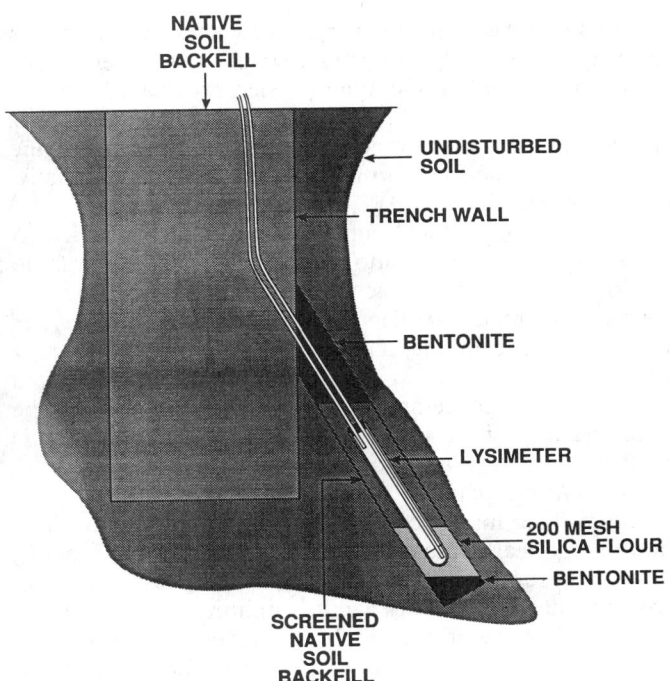

FIG. 15 Pressure-Vacuum Lysimeter Installation in the Sidewall of a Trench

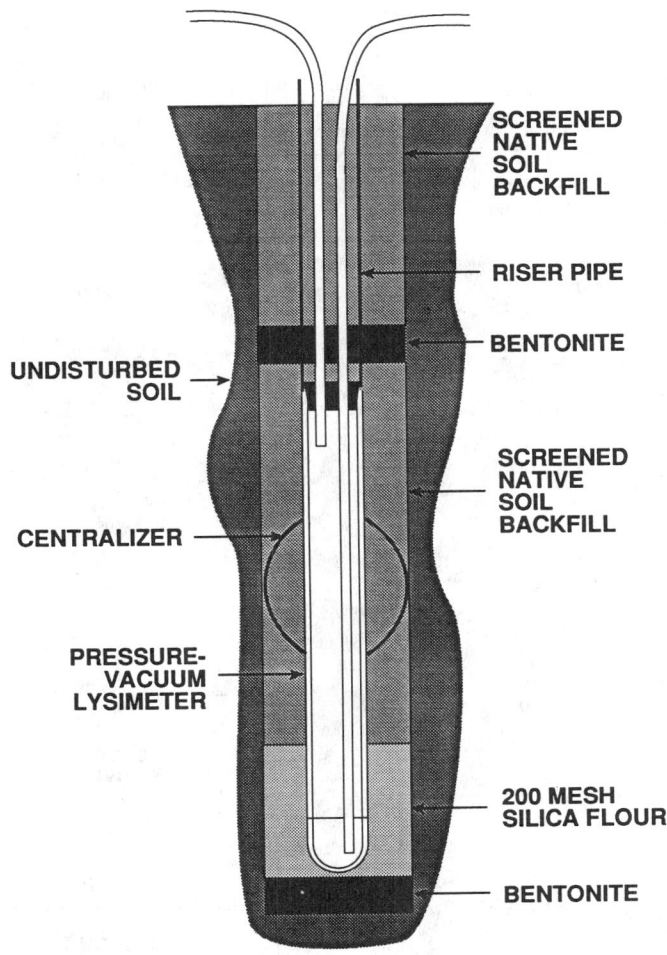

FIG. 16 Pressure-Vacuum Lysimeter Installation in a Borehole

Where soils permit, filter tip samplers can be installed by pushing the filter tip into the ground by applying a static load to the extention pipe (**32**).

7.4.2.4 When suction lysimeters are installed in a borehole advanced by a drill rig, the hole is usually advanced 15 to 20 cm below the desired location of the porous section. Morrison and Szecsody (**30**) found that the radius of sampling influence is maximized if the borehole diameter is only slightly larger than that of the sampler and if silica flour pack is used. U.S. EPA (**12**) recommended that the hole have a diameter at least 5 cm larger than the sampler. Timco (**25**[7]) recommended that the hole have a diameter at least 8 cm greater than that of the sampler to facilitate installation of the silica flour.

7.4.2.5 Suction lysimeters are preferably lowered into place attached to risers. These protect the lines and ensure exact placement at the desired depth. Centralizers are often used to center the sampler in the hole. Suction lysimeters float in the silica flour that is installed as a slurry. Therefore, the samplers should be installed full of distilled water or held in place by rigid risers.

7.4.2.6 The silica flour slurry (for example 200 to 75 µm mesh opening, silica to distilled water ratio of 0.45 kg to 150 mL) is usually installed using the tremie method (side discharge). Alternately, Brose et al (**57**) described a method for freezing the silica slurry around the sampler prior to placement. The sampler and frozen pack are then lowered to the sampling location in the borehole. They cited advantages of this technique as including ensurance of proper sampler placement in the flour pack and elimination of pack contamination by soils which slough down the borehole. U.S. EPA (**12**) recommended filling the borehole to about 30 cm above the suction lysimeter body with the silica. In addition, it was recommended that the powdered bentonite plug placed on

top of the silica be about 15 cm thick. The bentonite is also sometimes installed as a slurry, being allowed to hydrate before emplacement. Mixing the bentonite with fine sand at a 1 to 9 ratio, respectively, reduces the potential for shrinking and swelling inherent with pure bentonite (**1**). The excavated soil should be backfilled above the bentonite in the order in which it was withdrawn. An effort to compact the soil to its original bulk density should be made. When more than one suction lysimeter is installed in one borehole, these procedures are repeated at the various desired sampling depths (see Fig. 17). Care must be taken with these installations to ensure that lines from lower samplers do not interfere with the hydraulic contact of shallower samplers. Designs are available to avert these problems (**25**[7]).

7.4.2.7 U.S. EPA (**12**) recommended removal of the water within the sampler and silica slurry after installation. Litaor (**58**) recommended installation of samplers a year before sampling is to begin, in order to allow them to equilibrate with the surrounding soil. The lines at the surface should be labeled, clamped and housed in locked containers such as valve boxes or casing (**1**). Methods for cutting and splicing tubing may be found in Timco (**25**[7]). The user should be careful when using clamps and tubing provided by different manufacturers, inappropriate clamps may damage tubing. Clamps must be restricted to permanently flexible tubing

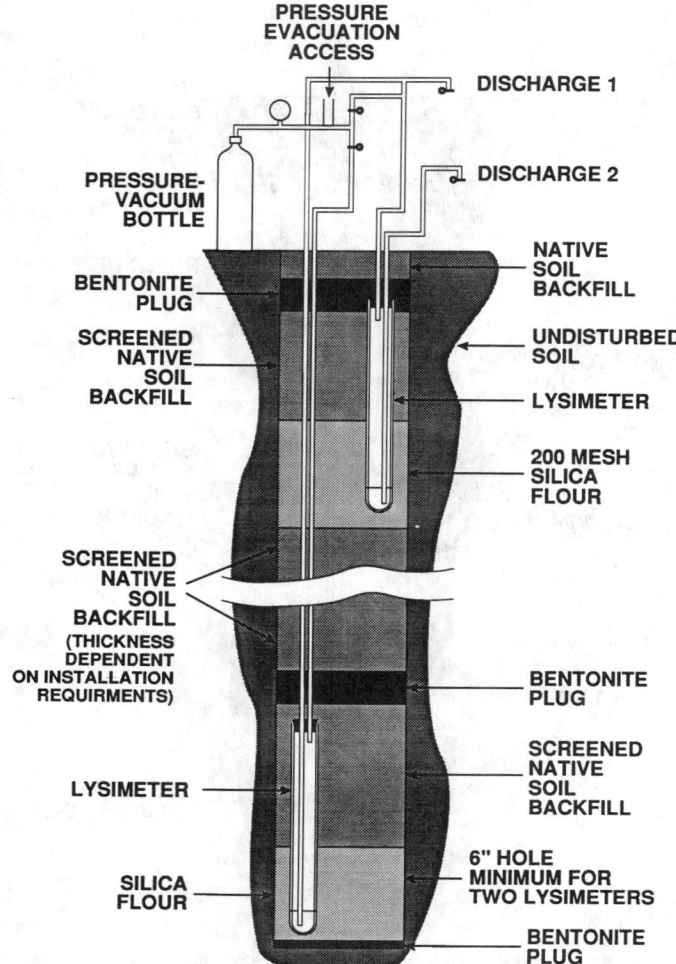

FIG. 17 Multiple Pressure-Vacuum Lysimeter Installations in a Borehole

otherwise stopcocks should be used.

7.4.3 *Experimental Suction Sampler Installation:*

7.4.3.1 Cellulose-acetate hollow-fiber sampler installation procedures were described by Everett et al (9). Membrane filter sampler installation procedures were described by Stevenson (35), Everett et al (9), and Morrison (1). Vacuum plate sampler installation procedures were described by Everett et al (9) and Morrison (1).

7.4.3.2 Cellulose-acetate hollow-fiber samplers have been used almost exclusively in laboratory studies (34). Because the samplers operate on the same principles as vacuum lysimeters, the goals and concerns of installation are similar. Good hydraulic contact between the hollow-fiber and the soil is critical. However, the fibers are too thin and fragile to be pushed into place. Therefore, the fibers must be placed in a predrilled hole (vertical or horizontal). Silkworth and Grigal (59) installed these samplers within a length of perforated, protective PVC tubing filled with soil slurry.

7.4.3.3 Membrane filter samplers are placed in a hole dug to the top of the selected sampling depth. First, sheets of the glass fiber "collectors" are placed at the bottom of the hole. These develop the necessary hydraulic contact between the sampler and the soil. In addition, the "collectors" extend the

area of sampling as they cover a larger area than the filter holder alone. Second, two or three smaller glass fiber "wick" discs that fit within the filter holder are placed on the "collectors." Third, the filter holder fitted with a glass fiber prefilter and the membrane filter is placed on top of the "wick" disks. The suction line leads to the surface. Finally, the hole is backfilled (1, 9).

7.4.3.4 Vacuum plate lysimeters are normally installed on the ceiling of a cavity cut into the side of a trench. In order to obtain the necessary contact between the porous plate and the soil, pneumatic bladders, inner tubes, or similar devices are placed beneath the sampler and are used to force it against the cavity ceiling (1). The cavity ceiling is not a smooth surface. Therefore, a layer of silica flour between the plate and the soil is sometimes used to enhance hydraulic contact.

7.4.4 *Maintenance:*

7.4.4.1 The major causes of sampler failure are line damage and leaks (caused by freezing, installation, rodents, etc.), connection leaks, and clogging of the porous material. Freeze damage to the lines can be minimized if the lines are emptied of sample prior to applying a vacuum. Care needs to be taken that the tubing line closure devices are freeze proof.

7.4.4.2 The possibility of line and connection leaks is minimized by rigorously sealing and pressure testing all connections and lines before installation. A common precaution to assist in repairing surface damage to lines is to store excess line below the surface (within the riser pipe when used) when backfilling the borehole. In the event of severed lines, an excavation to this buried length allows restoration of an operational system (1). Lines should be clamped shut when not in use to prevent foreign objects or insects from entering them. The lines should be protected from weather, sunlight exposure, and vandalism with a locked housing. The use of riser pipe around the sampler lines prevents punctures by backfill materials and prevents rodents from damaging the lines.

7.4.4.3 When shallow samplers are used, the ground surface above the sampler should be maintained in a fairly representative state. Large line housings and excessive traffic around the sampler (causing compaction of the soil) will reduce the amount of infiltration in that area. This will affect the representativeness of the pore-liquid samples. Methods to avoid these effects include angled installations, and remote operation of sampler lines.

7.4.4.4 Porous sections may clog as a function of soil composition, type of porous section material, biofilm growth, suction application, and pore-liquid content (1, 17, 20, 50). However, porous section clogging appears to be less severe than once thought (12, 17). Soils and the 200 mesh silica flour filter out fine materials before they reach the porous section (60, 61, 62). Clogging can be further reduced by periodically filling the sampler with distilled water and allowing it to drain out of the sampler. Debyle et al (50) suggested removing shallow samplers on a seasonal basis for flushing with HCl and distilled water. This process restores samplers to their original operational and chemical states. A "clogged" lysimeter may be cleaned out by filling the lysimeter with pure water and applying a pressure of 5 psi for 30 min. However, reinstallation at the same location and

736

Labels in figure:
PRESSURE EVACUATION ACCESS
DISCHARGE 1
DISCHARGE 2
PRESSURE-VACUUM BOTTLE
NATIVE SOIL BACKFILL
BENTONITE PLUG
SCREENED NATIVE SOIL BACKFILL
UNDISTURBED SOIL
LYSIMETER
200 MESH SILICA FLOUR
SCREENED NATIVE SOIL BACKFILL (THICKNESS DEPENDENT ON INSTALLATION REQUIRMENTS)
BENTONITE PLUG
SCREENED NATIVE SOIL BACKFILL
LYSIMETER
6" HOLE MINIMUM FOR TWO LYSIMETERS
SILICA FLOUR
BENTONITE PLUG

depth does not guarantee resumption of sampling from the same soil volume.

7.4.4.5 Often no sample is retrieved during a sampling attempt. The first check should be a continuity test of the lines and connections. This test can be done by applying a gentle pressure or vacuum to the V/P line and detecting air movement from the open sample line. This could be due to sampler failure or high pore-liquid tensions. Because of this, it is prudent to install a tensiometer near the sampler at a similar depth. The tensiometer that measures pore-liquid tensions allows the operator to determine if failure to obtain a sample is due to high pore-liquid tensions or due to sampler damage. The tensiometer can also be used to gage the effect of sampling on local pore-liquid flow regimes. Pore-liquid tension should be determined as an initial condition during lysimeter installation.

7.4.4.6 If a tensiometer is not available to measure pore-liquid tensions, the lysimeter can be tested to help determine reasons for failure to recover a sample. The sampler is tested by applying a suction of 80 cbar, and monitoring the decay of suction with time. Figure 18 depicts the various types of suction decay that might be found in a suction lysimeter with a 200 cbar bubbling pressure ceramic section. An almost instantaneous decay of suction is associated with lysimeter leakage. A suction decay over a period of minutes is associated with pore-liquid tensions greater than 200 cbar. Under these conditions, the porous section is desaturated and air enters the sampler. A suction decay over a period of hours reflects normal sample collection. This suggests that failure to retrieve a sample is related to damage of the sample retrieval system (for example, discharge line damage). When suction does not decay, or does so over a period of days, the pressure-vacuum line may be clogged or pore-liquid tensions may be greater than 60 cbar (but less than 200 cbar) causing liquid inflow rates that are too low for sample collection.

7.4.4.7 Morrison and Szecsody (63) described devices that could be used as tensiometers and then converted to pressure-vacuum lysimeters. However, they found that gases entering the devices prevented accurate measurement of pore-liquid tensions. Baier et al (64) described methods for converting tensiometers to pressure-vacuum lysimeters. It would also appear reasonable to convert suction lysimeters to tensiometers. However, Taylor and Ashcroft (65) found that the volume of water drawn from a converted lysimeter into the surrounding soil would significantly affect natural pore-liquid tensions. In addition, they found that the larger porous section of a lysimeter would cause more diffusion of dissolved air into the device, and that the time constant for measurement would be significantly increased over that of a tensiometer. Filter tip samplers can be converted to tensiometers with pressure transducers (32).

7.4.4.8 Operational lifetimes of suction samplers are dependent on installation, subsurface conditions, maintenance, and sampling frequency. Some samplers have been reported to be operational for as long as 25 years (64).

7.4.5 *Comments:*

7.4.5.1 Vacuum lysimeters and experimental samplers use suction to retrieve samples. Therefore, the maximum sampling depth is limited by the maximum suction lift of water (about 7.5 m) (12). In practice, these samplers are generally used to about 2 m below the surface (12). They are primarily used to monitor near-surface movement of pollutants such as those from land disposal facilities and those from irrigation return flow.

7.4.5.2 Pressure-vacuum lysimeters are generally not used at depths below about 15 m. At greater depths, sample loss and overpressurization problems are considered significant enough to warrant the use of high pressure-vacuum lysimeters that do not have these limitations. High pressure-vacuum lysimeters are not preferred at the shallower depths because they are more expensive. In addition, high pressure-vacuum units have more moving parts than pressure-vacuum units, and as a result, the possibility of failure for the former is higher.

7.4.5.3 As discussed in 7.6, two problems with suction samplers are that they may not sample from macropores (under unsaturated conditions; unless the macropores are directly intercepted) and that their results cannot be used in quantitative mass balance studies. Hornby et al (66) described an installation that could be used to surmount these problems. A barrel-sized casing (for example, 57 cm outside diameter by 85.7 cm high) is placed in a support device and gently pushed into the soil with a backhoe. As the casing is pushed, soil is excavated around it to help with insertion. The process results in an encased monolith of undisturbed soil. The monolith is then rotated and lifted, pressure-vacuum lysimeters are placed in its base, and the bottom is sealed. Subsequently the assembly is placed back into the ground at the monitoring site (see Fig. 19). All fluid draining through the monolith is collected by the samplers. Inasmuch as the boundaries of the system are sealed, the flux of liquid through the system requires maintaining a vertical hydraulic gradient by applying continual suction to the samplers.

7.5 *Operation:*

7.5.1 *Methods:*

7.5.1.1 *Vacuum Lysimeters*—Sampling methods are described by the U.S. EPA (12), by Soilmoisture (18[7]) and by Timco (25[7]). To collect a sample, suction is applied to the sampler, and the suction line is clamped shut. After sample has collected in the body tube, it is retrieved through a

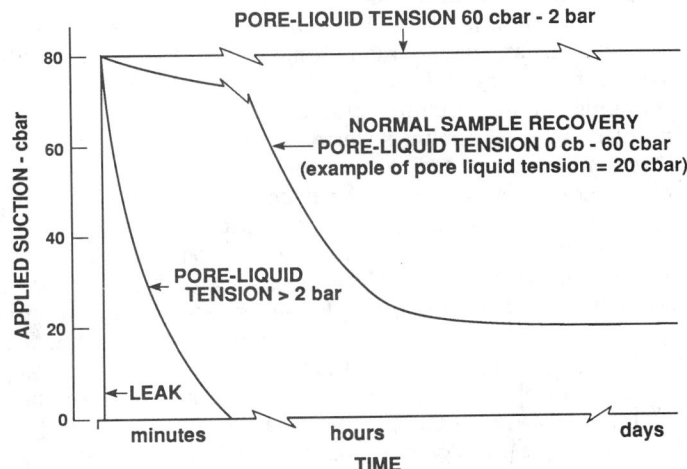

NOTE—Also shown is the almost instantaneous decay associated with an appreciable leak in the instrument.

FIG. 18 Decay Characteristics of Suction Applied to a Two Bar (Bubbling Pressure) Ceramic Cup Lysimeter in Equilibrium With Soils in Varying Ranges of Pore-Liquid Tension

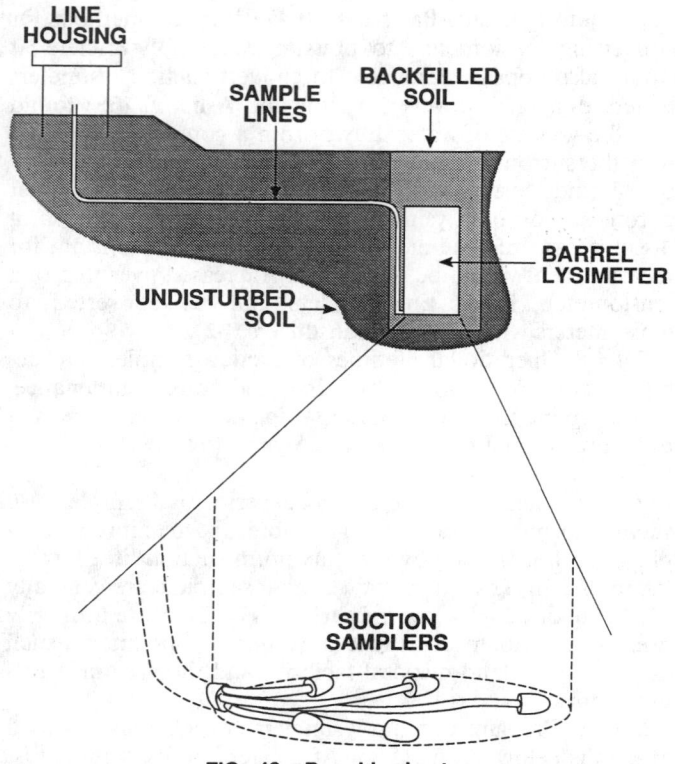

LINE HOUSING

SAMPLE LINES

BACKFILLED SOIL

UNDISTURBED SOIL

BARREL LYSIMETER

SUCTION SAMPLERS

FIG. 19 Barrel Lysimeter

discharge line extending to the base of the porous cup. In shallow installations, with the body tube extending above the soil surface, the discharge line is sometimes inserted and removed as needed. For deeper installations, the discharge line is permanently installed. At the surface, the line is connected to a sample collection flask. Suction is applied to the flask, and liquid is pulled from the sampler, up the discharge line, and into the collection flask. Cole (**42**) constructed an array of samplers that were attached to a vacuum tank connected to an electric power source. This system allowed remote operation at a constant suction. Wengel and Griffen (**67**) described methods by which samplers can be connected to a central control board and operated remotely. Brown et al (**68**) employed a solar panel to power a similar setup. Chow (**44**) described a sampler that shuts off automatically when the desired sample volume has been collected.

7.5.1.2 *Pressure-Vacuum Lysimeters*—Sampling methods are described in U.S. EPA (**12**), by Soilmoisture (**18**[7]) and by Timco (**25**[7]). To sample, suction is applied to the system via the pressure-vacuum line. The discharge line to the sample bottle is clamped shut during this time. When sufficient time has been allowed for the unit to fill with pore-liquid, suction is released and the clamp on the discharge line is opened. Gas pressure (for example, air or nitrogen; see 7.6.2) is then applied through the pressure-vacuum line. This forces the sample through the discharge line and into the collection flask at the surface (**12**). A variety of systems have been developed by which the pressure, suction, and sample volume can be controlled remotely or manually (**44, 49, 67, 69**).

7.5.1.3 *High Pressure-Vacuum Lysimeters (Lysimeters*

With a Transfer Vessel)—Sampling methods may be found in U.S. EPA (**12**), in Soilmoisture (**18**[7]) and in Timco (**25**[7]). When suction is applied to the system, it extends to the porous section through an open, one-way check valve at the bottom of the transfer vessel or chamber. A second one-way check valve in the discharge line is closed during this time. As soil solution enters the sampler it is pulled by the suction into the transfer vessel or chamber through a line attached to the open valve at its base. The sample is brought to the surface by releasing the suction and applying pressure (for example, air or nitrogen) through the pressure-vacuum line. This shuts the one-way valve to the porous segment and opens the one-way valve in the discharge line. The sample is then pushed to the surface (**12**). A variety of systems have been developed to control pressure, suction and sample volume remotely or manually (**44, 49, 67, 69**).

7.5.1.4 *Suction Lysimeters With Low Bubbling Pressures (Samplers With PTFE Porous Sections)*—Sampling methods for this group of samplers are a bit different than those for the three designs described in 7.5.1.1, 7.5.1.2, and 7.5.1.3. This system is designed to allow the soil pore water extraction process to occur separately from the movement of the collected pore water to the transfer vessel. The only difference is that maximum sampling suctions for these units are much lower (see 7.6.1.3).

7.5.1.5 *Filter Tip Samplers*—Sampling methods may be found in Torstensson and Petsonk (**32**). Samples are collected by first evacuating the sample vial. The vial is then inserted in the sampling adaptor that contains a two way hypodermic needle. The adaptor is then lowered down the extension pipe. When the adaptor connects with the nozzle of the filter tip, the needle penetrates the septa in the vial and in the filter tip. Sample then flows through the porous segment and into the sample vial due to the negative pressure in the vial. As sample is collected, the negative pressure in the vial falls towards that of the pore-liquid tension. When these negative pressures are equal, sampling ends and the sample vial is retrieved. The standard sample volume is about 35 mL. However, by connecting several vials in series, sample volumes of up to 500 mL can be obtained.

7.5.1.6 *Experimental Suction Samplers*—Cellulose-acetate hollow-fiber samplers, membrane filter samplers, and vacuum plate samplers are operated using the same general technique as for vacuum lysimeters. Jackson et al (**33**) sampled from soil columns using cellulose-acetate hollow-fiber samplers subjected to a constant suction of 81 cbar. At this suction, they were able to extract samples for chemical analyses from silty loams with moisture contents ranging from 20 to 50 %. Silkworth and Grigal (**59**) compared the performance of these samplers to suction lysimeters. They found that cellulose-acetate hollow-fiber samplers fail more often than suction lysimeters. In membrane filter samplers, the "collectors" provide hydraulic contact between the soil and the samplers. Liquid is drawn by capillarity into the "collectors." When suction is applied, liquid flows through the "wick," the prefilter, and finally the membrane filter. The prefilter reduces clogging of the membrane filter by fine soil materials (**9**). Stevenson (**35**) recommended using a suction of between 50 and 60 cbar when sampling with membrane filter samplers. A variety of constant suction methods for sampling with vacuum plates are described by Morrison (**1**).

An advantage of the larger plates is that they have large contact areas with the soil. Therefore, larger sample volumes can be collected in shorter times than with vacuum lysimeters which have porous sections with smaller surface areas.

7.5.2 *Comments:*

7.5.2.1 Nagpal (**70**) recommended several consecutive extractions of liquids during a sampling event and use of only the last one for chemical analyses. The purpose of this is to flush out cross contaminants from previous sampling periods, and to ensure that any porous segment/soil solution interactions have reached equilibrium. Debyle et al (**50**) also suggested discarding the first one or two sample volumes when sampling dilute solutions with newly flushed (HCL method) and installed samplers. The purpose of this is to allow cation exchange between the porous segment and the pore-liquid (caused by the HCL flushing) to equilibrate.

7.5.2.2 Factors which affect the volume and source of a pore-liquid sample include the amount of suction applied, the schedule of suction application, the pore-liquid content, the distribution of pore-liquid, the soil grain size distribution, the soil structure, the porous section design, and the porous section age.

7.5.2.3 Samples collected with lower suctions (about 10 cbar or less) tend to come from liquids migrating through soil macropores (**1**). Samples collected with higher suctions (greater than about 10 cbar) also include fluids held at higher tensions in micropores. The sampler may disrupt normal flow patterns due to the applied suctions. The effects may extend several meters from the sampler although the area nearest the sampler is most disturbed (**71, 72, 73**). This disturbance causes samples to be averages of the affected flow area rather than point samples (**1**). Warrick and Amoozegar-Fard (**72**) developed an approximate analytical equation which can be used to estimate the maximum radius of influence on the flow regime by a suction sampler. Narasimhan and Driess (**74**) developed a numerical technique to simulate the effects of suction samplers on the pore-liquid regime.

7.5.2.4 Sampling with falling suction produces samples with compositions that are "averages" of the liquids held at the range of tensions applied. Because suctions and therefore inflow rates decrease with time, these "averages" are weighted toward those portions of the samples obtained in early times. Samples collected over prolonged periods (due to slow inflow rates) are "averages" of the liquids fluxing past the sampling region during those times.

7.5.2.5 During wet periods, samplers affect a small volume of soil and pull liquids from a sequence of pores that may include macropores. During dry periods samplers affect a larger volume of soil, draw from micropores because the macropores have been drained, and collect less liquid (**75, 76**). The net result of this is that sampled soil solutions are "averaged" over different volumes and derived from different pores as a function of the soil moisture content and distribution.

7.5.2.6 Soil textures and pore-liquid tensions control the amount of liquid that can be removed by a sampler and its radius of influence. The slope on the pore-liquid release curve for a sand is greater than that for a clay at low pore-liquid tensions (see Fig. 20). This indicates that there

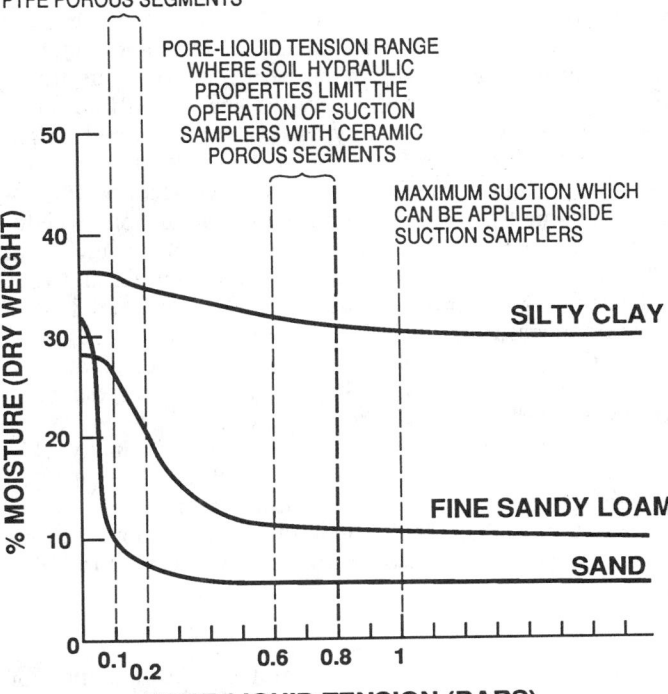

FIG. 20 Water Release Curves for Three Soils, Showing Operating Conditions for Suction Samplers

will be a larger quantity of pore-liquid released from a sand than from a clay for an equal change of pore-liquid tension at these low tensions. At higher tensions, the slope of a clay pore-liquid release curve is greater than that for a sand (see Fig. 20). This indicates that more pore-liquid will be released from a clay than from a sand for an equal change in pore-liquid tension at the higher tensions. A consequence of this is that suction samplers may not obtain samples from coarse grained soils at higher pore-liquid tensions. Morrison and Szecsody (**30**) found that (under the conditions of their study) radii of influence for suction lysimeters ranged from 10 cm in coarse soils up to 92 cm in fine grained soils.

7.5.2.7 Hansen and Harris (**20**), demonstrated that intake rates may vary substantially due to variability in the ceramic sections from one manufacturer's batch to another. As discussed in 7.4, the intake rate of a sampler is also a function of the degree of clogging. As discussed in 7.6, the range of pore-liquid tensions over which a sampler can operate is a direct function of the maximum pore size of the porous section and the surrounding silica flour pack. Finally, Morrison and Szecsody (**30**) found that the radius of influence of a sampler increases with the diameter of the porous section.

7.5.2.8 Because of these factors the following recommendations have been made for sampling with suction lysimeters. Hansen and Harris (**19**), suggested using uniform initial suctions, short sampling intervals, and uniform sampling times for different sampling events and locations to increase the uniformity of samples. Debyle et al (**50**) also recommended sampling with uniform suctions that do not significantly exceed the tension at which the percolating soil solution is being held. U.S. EPA (12) suggested sampling after infiltration events such as rain storms, spring melts, or irrigations as these periods of high pore-liquid content are accompanied by higher pore-liquid flow rates and contaminant transport. For sampling these events, it is useful to install samplers at interfaces between coarse and fine materials to take advantage of any liquid perching which might occur. Silkworth and Grigal (59) recommended using samplers with large diameter ceramic porous sections (as opposed to small diameter ceramic samplers, or hollow cellulose fiber samplers) since they showed less of a tendency to alter the pore-liquid, they had lower failure rates, and they collected larger sample volumes. These recommendations were reinforced by van der Ploeg and Beese (73) who concluded that samplers with large cross sectional area porous sections used with low extraction rates (suctions approaching those of the pore-liquid tensions) reduce the effects of sampling on compositions of samples. Finally, U.S. EPA (12) recommended that porous section material types be carefully chosen based on pore-liquid tensions expected in the sampling area. Operational ranges of various porous section types are discussed in 7.6 and are presented in Table 1.

7.6 *Limitations:*

7.6.1 *Physical Limitations:*

7.6.1.1 The most severe constraint on the operation of suction samplers involves soil around the porous sections becoming so dry (and pore-liquid tensions so high) that samples cannot be collected. The limiting factors in these conditions will be the porous segment or the soil hydraulic properties. For porous segments with bubbling pressures less than 60 cbar (for example, PTFE), the porous segment will be the limiting factor because the high suctions required to move liquids into the samplers will cause meniscuses in the porous segments to break down and air to enter. Soil hydraulic properties will be the limiting factors for porous segments with bubbling pressures greater than 60 cbar (for example, ceramics) because unsaturated hydraulic conductivity of the soil and pressures gradients across the porous segments will be so low (due to high pore-liquid tensions) that flow into the samplers will be negligible.

7.6.1.2 The maximum suction that the saturated porous section of a sampler can withstand before air enters is a function of the pore configuration and size, and its hydrophilicity/hydrophobicity (see Appendix X1 and (65)). The following variation of the capillary rise equation (8, 12, 18, 77) combines these factors:

$$P_b = \frac{-2\delta \cos\alpha}{r}$$

where:
P_b = bubbling pressure (gage), units − FLT^{-2},
δ = surface tension between pore-liquid and air, units − FT^{-1},
α = contact angle between the liquid and the material of the porous segment, D, and
r = maximum pore radius of the pore segment, units − L.

This equation shows that the bubbling pressure decreases with increasing contact angle and with increasing maximum

pore radius. The maximum sampling suction that can be applied is 100 cbar (1 atmosphere). For a hydrophilic material (that has an acute contact angle) the maximum pore size that will allow the application of 100 cbar of suction is 2.8 μm. For a hydrophobic material (that has an obtuse contact angle) a smaller pore size will be required (65). The maximum pore sizes of presently available ceramics (that are hydrophilic) used for suction lysimeters and filter tip samplers vary from 1.2 to 3 μm (as measured by the bubbling pressures) (18[7], 45, 78[7]). The maximum pore sizes of cellulose-acetate hollow-fibers and membrane filters range from less than 2.8 μm and 0.4 to 5.0 μm respectively (1, 35, 36). These pore sizes result in maximum sampling suctions near 100 cbar. Therefore, these materials will not allow air to enter during sampling, and the limiting factors will be the soil hydraulic properties. The combination of soil limiting effects result in negligible sampling rates when pore-liquid tensions are above 60 cbar (for coarse grained soils) to 80 cbar (for finer grained soils) (45). At tensions above these levels, inflow rates are too low to allow sampling.

7.6.1.3 The maximum pore sizes of presently available porous PTFE segments for suction lysimeters range from about 15 to 30 μm (calculated from bubbling pressures) (25[7]). These pore sizes allow maximum sampling suctions of about 10 to 21 cbar (25). The hydrophobicity of PTFE will further reduce the magnitude of the maximum sampling suction. Applied suctions of greater than 10 to 21 cbar (or less) will cause air to enter, and sampling to cease. Because a suction greater than 10 to 21 cbar cannot be applied to these samplers, pore-liquids held at tensions greater than 10 to 21 cbar cannot be sampled with these devices. Because of this, PTFE will be the limiting factor when it is used for the porous segment. A consequence of the small suction range available to PTFE porous sections is that only very moist soils approaching saturation may be sampled (17).

7.6.1.4 The silica flour pack, that has smaller pore sizes than PTFE, can act as an extension of the porous segment, and may extend the range of suctions that can be applied to the sampler. Everett and McMillion (45) found that the pack extended the suction range of earlier, larger pore size PTFE (70 to 90 μm) from less than 4 to 7 cbar. Timco (25[7]) suggested that the operational range of the presently available PTFE samplers (15 to 30 μm) can be extended from 10 to 21 cbar to between 61 to 71 cbar when "properly" installed within a silica flour pack (this has not been verified in peer reviewed literature). For this to be true, the silica flour pack must be able to remain saturated over the range of applied suctions. However, the results of Everett and McMillion (45) suggest that the bubbling pressure of the silica flour pack is only 7 cbar. Trainor (27) found that even if these samplers are "properly" installed, air may still enter if applied suctions exceed pore-liquid tensions by more than 30 %. Pore-liquid tensions are not always known, and technicians may not carefully control applied suctions. In addition, pressurization of pressure-vacuum lysimeters for sample retrieval appears to damage the silica flour pack (27, 28). Thus, dependency on the silica flour pack to provide the needed suction range is an extremely limited option. Because of this, suction lysimeters with PTFE porous sections are limited to near saturated sampling and have been classified separately (see Fig. 20).

7.6.1.5 Samples can be collected (using ceramic porous sections) from clays with high pore-liquid tensions (approaching 60 to 80 cbar). However, because liquid inflow rates are low at higher tensions, the amount of time required to collect sufficient sample volumes may exceed the maximum allowable holding time for many chemical analyses. Law (76) pointed out that when soils have liquid contents that allow little or no sample collection (high pore-liquid tensions), there is little or no liquid movement in the soil. Consequently, there will be little or no contaminant migration. If samples of pore-liquids held at tensions above 60 to 80 cbar are desired, soil core sampling with subsequent laboratory liquid extraction may be used (76). However, Law (76), and Brown (79) concluded that results from the two sampling methods are not comparable. Liquid from soil core samples will include constituents that are held at tensions greater than 60 to 80 cbar and that would not be picked up by suction samplers. Because of this and because samples removed from soil cores may include some of the constituents from the soil itself (for example, cations preferentially adsorbed in electrical double layers) or sorbed organics, Law (76) concluded that soil cores are more conservative estimators of cation contaminant presence in soil. Brown (79) concluded that organic contaminant concentrations derived from soil cores and pore-liquid samplers are not comparable because of preferential sorption of some compounds. Amter (80) developed an alternative to extraction of samples from soil cores. The method involves injecting a chemically blank fluid through an existing lysimeter. After a time, the fluid (now containing dilute pore-liquid) is recovered through the sampler and analyzed. The results, although qualitative, were shown to correlate well with known relative pore-liquid constituent concentrations.

7.6.1.6 Suction samplers may not intercept macropores because of the small size of their porous sections. Because of this, they may miss the majority of flow at high moisture contents in structured soils (81). The ability to intercept this flow can be enhanced by installing the samplers in large diameter silica flour packs. However, this involves drilling larger holes. Because suction samplers only sample when suction is applied, they may miss infiltration events unless a constant suction is applied. Therefore, under conditions of high moisture content in structured soils, free drainage samplers are recommended (see Section 8) (81). Pore-liquid composition changes with time. Because suction samplers sample over an extended period (especially in drier soils), the resulting sample should be considered an average of the total flux past the sampler during the sampling interval.

7.6.1.7 A major factor limiting the operation of shallow suction samplers in cold climates is that pore-liquid may freeze near the porous segments. In addition, liquid may freeze within porous segments and lines, preventing sample retrieval and perhaps fracturing the sampler during ice expansion. Because of this, lines should be emptied before the onset of cold weather. Additionally, some soils tend to heave during freezing and thawing. Consequently, the samplers may be displaced in the soil profile, resulting in a break of hydraulic contact (12).

7.6.2 *Chemical Limitations:*

7.6.2.1 The inherent heterogeneities of unsaturated pore-liquid movement and chemistry limit the degree to which samples collected with suction samplers can be considered representative. This is because the small cross sectional areas of suction samplers may not adequately integrate for spatial variability in liquid movement rates and chemistry (51, 82, 83). Biggar and Nielsen (84) suggested that results of analyses from suction lysimeter samples are good for qualitative comparisons, but that they cannot be used for quantitative analysis unless the variabilities of the parameters involved are established. Law (76) came to similar conclusions, stating that results from suction lysimeter sampling could not be used for quantitative mass balance studies.

7.6.2.2 Well structured soils have two distinct flow regions including macropores (for example, cracks, burrows, and root traces) and micropores. Under saturated conditions, liquids move more rapidly through macropores than through micropores. Because of this, the movement of liquid-borne pollutants into the finer pores may be limited. Consequently, pore-liquids in macropores may have different chemistries than those in micropores (85). This is enhanced by the fact that oxygen content in macropores can change in a matter of hours during an infiltration event, whereas micropores may remain suboxic regardless of flow conditions (75). In addition, micropores are less susceptible to leaching than macropores (1, 86, 87, 88). Because of these differences, sample chemistry can vary widely from location to location and from time to time depending on the amount of liquid drawn from these two flow systems.

7.6.2.3 Suction samplers may affect pore-liquid chemistry as it is being sampled. The major sampler related factors that can affect the sample chemistry are the porous segment material and sample storage time. The degree of chemical interaction may also be affected by the amount of porous section clogging (1). Clogging slows the rate of flow through the porous section so that contact time and chances for chemical interaction are increased (50). In addition, the types of adhesives used to attach porous segments (for example, epoxy) may alter the pore-liquid chemistry.

7.6.2.4 Interactions between porous materials and liquid can include sorption, desorption, cation exchange, precipitation (for example, ferric precipitation), and screening (20). These interactions can also occur with all other parts of the samplers that the samples contact. However, the much higher surface area of pores within porous segments makes them the most critical element chemically. Table 2 presents the results of a literature review for porous section/pore-liquid interactions. An attempt has been made to document the pertinent features of the listed studies. However, the reader should refer to the original papers to determine if experimental techniques are applicable to the situation of interest. The absence of entries for a constituent relative to a material does not infer absence of interactions. Although studies for membrane filter interactions have been performed, the results have not been included in Table 2. This is because membrane filters are made from a variety of materials that have differing chemical characteristics (36).

7.6.2.5 Suarez (89) showed that the pH of a sample may be affected by 0.28 to 0.44 pH units due to CO_2 degassing during sampling. He reduced this error by reducing the gas-liquid ratio in the sampler, and by flushing several sampler volumes of soil solution through the sampler before collecting a sample. Alternately, Suarez (89) developed a

TABLE 2 Porous Material Interactions[A,B]

	Material Sorbs Species	Material Desorbs Species[C,D]	Material Screens Species[C,D]	No Significant Interaction[C,D]	No Interaction
Al	...	C(2) A(14)	...	C(16) A(14)	...
Alkalinity	...	...	...	SF(11)	...
Ca	...	C(1,2,18) CAF(18) A(14)	...	C(3,6,10,11,25) PTFE(3) A(3) FG(18,22) CAF(10)	PTFE(13)
C	...	FG(22)	...	...	...
CO3	...	C(2)	...	...	...
HCO3	...	C(2)	...	...	...
Cd	C(11)	...	...	C(3) PTFE(3) A(3)	...
Cl	...	...	...	C(11,25) SF(11)	PTFE(13)
Cr	C(19)	C(3) PTFE(3) A(3)	...	...	...
Cu	C(11)	C(3) PTFE(3)	...	A(3)	...
Fe	C(11)	PTFE(3) A(3)	...	C(3,25) A(14) SF(11)	PTFE(13)
H	...	...	...	...	...
K	C(5,6,15)	C(18) A(14)	...	C(1,25) CAF(18)[E] FG(18,22)	...
Mg	C(6)	C(2,3,11,18) A(3,14) CAF(18)	...	C(10,25) PTFE(3) CAF(10) FG(18,22)	PTFE(13)
Mn	C(11)	A(3)	...	C(3) PTFE(3) A(14)	PTFE(13)
Na	C(6)	C(2,18) A(14) CAF(18) FG(18,22)	...	C(1,11,25)	PTFE(13)
NH4	C(4,12)	...	...	PTFE(4)	...
N	...	FG(22)	...	...	...
NO2	...	...	...	C(4,5) PTFE(4)	...
NO3	...	...	CAF(10)	C(4,8) PTFE(4)	...
NO3-N	...	...	C(10) CAF(10)	...	...
(NO2+NO3)–N	...	...	...	C(5)[F]	...
P	C(1,5,8,15,18)	...	...	CAF(18) FG(18)	...
PO4	C(4,5,7)	...	...	PTFE(4) CAF(10)	...
PO4-P	...	...	...	C(10) CAF(10)	...
Pb	...	...	...	...	PTFE(13)
SiO2	...	C(2)	...	...	...
Si	...	...	...	C(4) PTFE(4)	...
SO4	...	...	...	C(11)	...
Sr	...	C(11)	...	...	...
Zn	...	C(11)	...	...	PTFE(13)
High Molecular Weight Compounds	...		C(17,21) CAF(10)	...	
4-nitrophenol	PTFE(23)				
Chlorinated hydrocarbons	PTFE(23,24)			...	
Diethyl phthalate		...	...	PTFE(23)	...
Naphthalene	PTFE(23)	...	...	...	...
Acenaphthene	PTFE(23)	...	...	...	...

model by which pH values could be corrected. He noted that multichamber samplers had minimal pH errors and that pH corrections due to CO$_2$ degassing were not necessary. Peters and Healy (53) found that there was no significant change in

References and Notes on Experimental Techniques [G]

Reference Number in Table	Cited Reference	Porous Section was Washed	Results are a Function of Several Factors	Dilute Solutions were Tested	Experiments were Performed on Nonporous Materials
1	Ref (137)	X	...	...	...
2	Ref (48)	X	...	...	...
3	Ref (93)	X	...	...	...
4	Ref (138)	X	...	...	...
5	Ref (139)	...	X	...	...
6	Ref (50)	X	X	...	...
7	Ref (52)	X	...	...	...
8	Ref (19)	X	...	...	...
9	Ref (56)	...	X	...	...
10	Ref (34)	...	...	X	...
11	Ref (53)	X	...	...	...
12	Ref (69)	...	...	...	...
13	Ref (1)	X	...	...	...
14	Ref (47)	X	X	X	...
15	Ref (139)	X	...	...	...
16	Ref (58)	...	X	...	...
17	Ref (76)	...	...	...	...
18	Ref (59)	X	X	...	...
19	Ref (75)	...	X	...	...
20	Ref (140)	X	...	...	...
21	Ref (57)	...	...	...	...
22	Ref (36)	X	X	...	...
23	Ref (142)	...	X	X	X
24	Ref (143)	...	X	X	X
25	Ref (17)	X	X	...	...

[A] Abbreviations:
  C = porous ceramic,
  PTFE = porous PTFE,
  A = porous alundum,
  CAF = cellulose-acetate fibers,
  FG = fritted glass or glass fibers, and
  SF = silica flour.
[B] Comparisons of materials based on this table should be made cautiously. Differing experimental techniques should be considered as a source of differing conclusions. Undocumented factors often include material age and sampling history.
[C] Numbers in parenthesis refer to references cited in the second part of this table. This is indicated by (2).
[D] Valence states are often not reported in studies. This is indicated by (3).
[E] Reference (59) found that there is no significant interaction of cellulose-acetate fibers with potassium in solution. The porous section was washed prior to testing and results were found to be a function of several factors.
[F] Reference (139) found that there was no significant interaction of porous ceramic with nitrate plus nitrite nitrogen in solution. The results are a function of several factors.
[G] Absence of information on experimental techniques means that the techniques were not specified in the citation.

pH due to CO_2 degassing during long sampling times, although they recognized that pH changes could occur when the solution is originally more acidic than that which they tested. Ransom and Smeck (90) and Anderson (75) suggested purging the sampler with N_2 to preserve the subsurface redox states when sampling for redox dependent ions. Filter tip samplers do not use a purging gas, therefore, pore-liquid redox states are preserved in the samples.

7.6.2.6 Nightingale et al (31) indicated that normal suction sampling techniques are not suitable for sampling volatile organic compounds due to potential loss. Wood et al (91) devised a body tube connected to a purging chamber that is in turn connected to a trap packed with resin. Compounds that volatilize during sampling are captured in the trap. Pettyjohn et al (92) described a suitable system for sampling highly volatile organics. However, the reported system was limited to a maximum sampling depth of 6 m and a small sample volume (5 to 10 mL). Torstensson and Petsonk (32) described methods that can be used to collect samples with filter tip samplers that result in no head space in the sample vial and consequently no loss of volatile compounds.

7.6.2.7 A newly forming consensus is that the effects of suction samplers (when properly pre-treated) on sample chemistry of non-dilute solutions are generally less significant than the inherent uncertainties of sampling discussed in 7.6.2.1 and 7.6.2.2 (17, 53, 93).

7.6.3 *Microbial Limitations*—Viruses or bacteria are sometimes monitored in areas where there are livestock lots, leach fields, septic tanks, or sewage sludge spreading plots. However, it has been found that although porous ceramics will allow viruses to pass, they will screen out bacteria (for example, escherichia coli and fecal coliform) (12, 26, 94, 95).

8. Free Drainage Samplers

8.1 Free drainage samplers are classified differently by various authors, depending on the installation methods. Many free drainage samplers are installed in the side walls of trenches and are referred to as trench lysimeters. However, free drainage samplers are also installed in the walls of vertical caissons. The principle behind each of the samplers is essentially the same. However, the materials and construction differ. The general types of free drainage samplers include:

8.1.1 *Pan lysimeters,*
8.1.2 *Glass block lysimeters,*

8.1.3 *Trough lysimeters,*

8.1.4 *Vacuum trough lysimeters,*

8.1.5 *Caisson lysimeters,*

8.1.6 *Wicking soil pore-liquid samplers,* and

8.1.7 *Sand filled funnel samplers.*

8.2 *Operating Principles:*

8.2.1 A free drainage sampler consists of some sort of collection chamber that is placed at depth in the soil. Pore-liquid in excess of field capacity is free to drain through soil (usually through macropores) under the influence of gravity. This gravity drainage creates a slightly positive pressure at the soil-sampler interface causing fluid to drip into the sampler. Hence, these samplers collect liquid from those portions of the vadose zone that are intermittently saturated due to events such as rainfall, flooding, or irrigation. Some free drainage samplers apply a small suction in order to break the initial surface tension at the soil-sampler interface. Samples are retrieved either by accessing the samplers at depth or by drawing samples to the surface through a suction line.

8.2.2 As described in 4.2, suction samplers can also be used to sample free drainage flow. However, the small area of those samplers compared to the spacing of macropores limits their usefulness for this application. As described in 7.4.5.3, Hornby et al **(66)** developed an installation that includes pressure-vacuum lysimeters within an encased monolith. This enhances collection of macropore flow with these samplers.

8.3 *Description:*

8.3.1 *Pan Lysimeters:*

8.3.1.1 A pan lysimeter generally consists of a galvanized, metal pan of varying dimensions. A copper tube is soldered to a raised edge of the pan. Plastic or vinyl tubing connects the copper tube to a collection vessel. Any liquid that accumulates on the pan drains through the tubing into the vessel (see Fig. 21) **(26)**.

8.3.2 *Glass Block Lysimeters*—Barbee and Brown **(81)** developed free drainage samplers made from hollow glass bricks. These glass bricks, that are produced for ornamental masonry work, have dimensions of 30 by 30 by 10 cm and have a capacity of 5.5 L. To build a sampler, nine holes, 0.47 cm in diameter, are drilled along the perimeter of one of the square surfaces of a brick. Nylon tubing is inserted into one of the holes to allow for sample removal. The collecting surface is fitted with a fiberglass sheet to improve contact with the soil. Pore-liquid collection is enhanced by a raised lip along the edge of the surface (see Fig. 22).

8.3.3 *Trough Lysimeters:*

8.3.3.1 Trough lysimeters, also known as Ebermayer lysimeters, rely on a trough or pail to collect pore-liquid. In order for the edges of the sampler to maintain a firm contact with the soil, a fiberglass screen is suspended inside the trough. The screen is lined with glass wool and covered with soil until the soil is even with the top of the trough **(96)**.

8.3.3.2 Morrison **(1)** reported a trough lysimeter in which two parallel metal rods are inside the trough, in contact with the bottom side of the screen, and bent toward the collection tube. Liquid that enters the trough migrates along these rods towards the collection tube in response to capillary forces (see Fig. 23). A modification of this design consists of a metal trough with a length of perforated PVC pipe mounted inside.

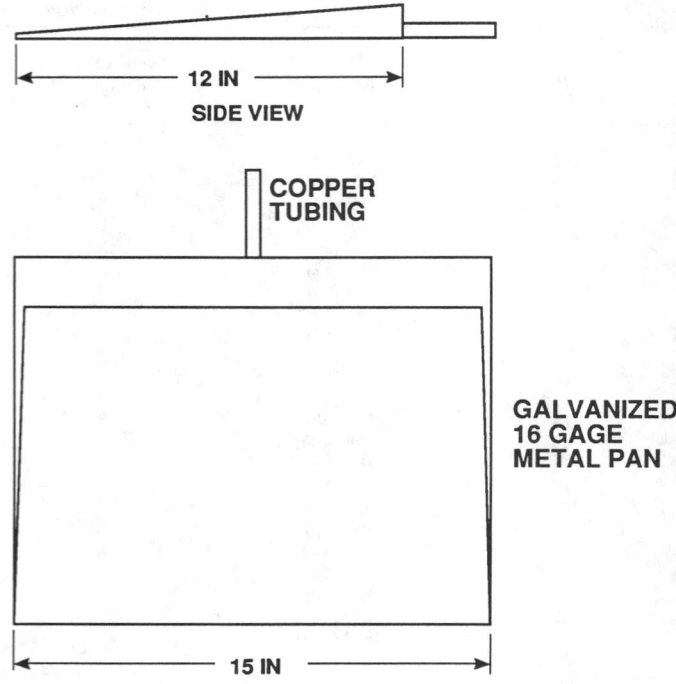

FIG. 21 Example of a Pan Lysimeter

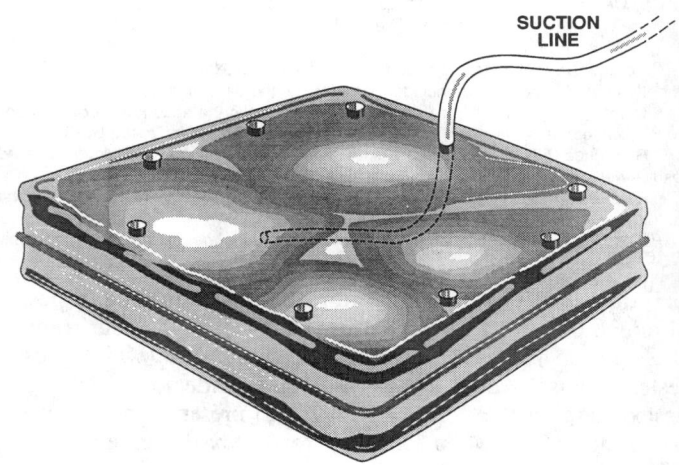

FIG. 22 Glass Block Lysimeter

The trough is filled with graded gravel so that coarse material is immediately adjacent to the PVC pipe and fine sand is at the edges and the top of the trough. The pipe is capped at one end while the other end is connected to a sample container via a drainage tube **(1)**.

8.3.4 *Vacuum Trough Lysimeters*—Montgomery et al **(97)** described a vacuum trough lysimeter consisting of a metal trough equipped with two independent strings of ceramic pipe, each 13 mm in diameter. The design, otherwise similar to trough lysimeters, allowed extraction of samples under applied suctions of up to 50 cbar. The ceramic pipes act as a vacuum system, and samples are extracted through a suction line.

8.3.5 *Caisson Lysimeters*—A caisson lysimeter consists of

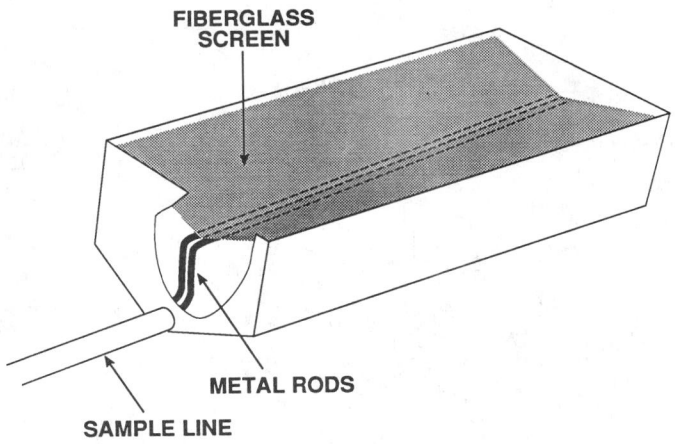

FIG. 23 Trough Lysimeter

collector pipes, radiating from a vertical chamber **(1)**. A design used by Schmidt and Clements **(98)** consists of a nearly horizontal, half-screened PVC casing (see Fig. 24). Schneider et al **(99)** designed a similar system consisting of: a 15.2 cm diameter stainless steel tube extending diagonally upward through the caisson wall into the native soil; a screened plate assembly within the tube to retain the soil; a purging system that can be used to redevelop the sampler when it becomes clogged; and an airtight cap that prevents exchange between the air in the caisson and the soil air.

8.3.6 *Wicking Soil Pore-Liquid Samplers*—Hornby et al **(66)** described a wicking sampler, alleged to combine the attributes of free drainage samplers and pressure-vacuum lysimeters. The sampler collects both free drainage liquid and liquid held at tensions to about 4 cbar. A hanging "Hurculon" fibrous column acts as a wick to exert a tension on the soil pores in contact with a geotextile fiber which serves as a plate covering a 30.5 by 30.5 by 1.3 cm pan. The terminus of the fibrous column is sealed into the cap of a tubular sample collector. The collection tube also contains an inlet pressure-vacuum line and a sample collection tube. Materials for the sample collector depend on the constituents

being sampled. Glass and PTFE were recommended materials when sampling for organics (see Fig. 25) **(66)**.

8.3.7 *Sand Filled Funnel Samplers*—K. W. Brown and Associates **(100)** discussed a sand-filled funnel for collecting freely draining liquid. The funnel is filled with clean sand and inserted into the sidewall of a trench. The funnel is connected through tubing to a collection bottle. Application of suction to a separate collection tube pulls the sample to land surface (see Fig. 26).

8.3.8 *Comments*—The dimensions of the free drainage samplers discussed are purposely left vague. Because the samplers collect fluid flowing primarily through macropores, the dimensions are often dictated on a site-by-site basis by the configurations and spacings of the macropores.

8.4 *Installation Methods:*

8.4.1 *Installation:*

8.4.1.1 Free drainage samplers are commonly installed into the side walls of trenches or caissons. The trenches can be dug either by hand or with backhoes. But they should be stabilized with timbers and siding if deeper than 1.5 m, in order to allow safe access **(26)** (see Occupational Health and Safety Administration Regulations). All wood used in the construction of permanent trenches should be treated with preservatives to protect it against degradation due to semi-saturated conditions. This may pose a problem when monitoring for organics. Any spaces between the bare trench side walls and the siding are filled with soil and peagravel to allow for free drainage. The excavations should be covered to provide positive surface drainage away from the area. Some free drainage samplers require only temporary excavations for installation. After the samplers have been installed, the excavations are backfilled with native soil.

8.4.1.2 Caissons for housing free drainage samplers are constructed with corrugated culverts or concrete drainage pipes. Schneider and Oaksford **(101)** installed caissons by excavating soil from within a concrete pipe using a crane operated shovel and manual labor. Each concrete pipe section, weighing 222.5 kN (25 tons), was set in place with a

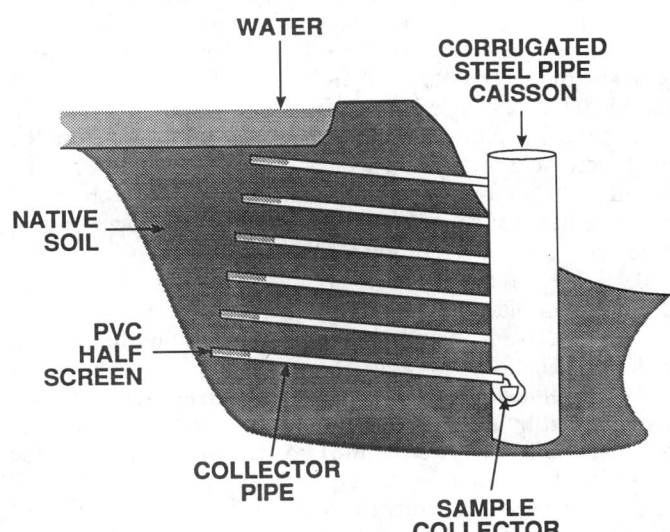

FIG. 24 Example of a Caisson Lysimeter

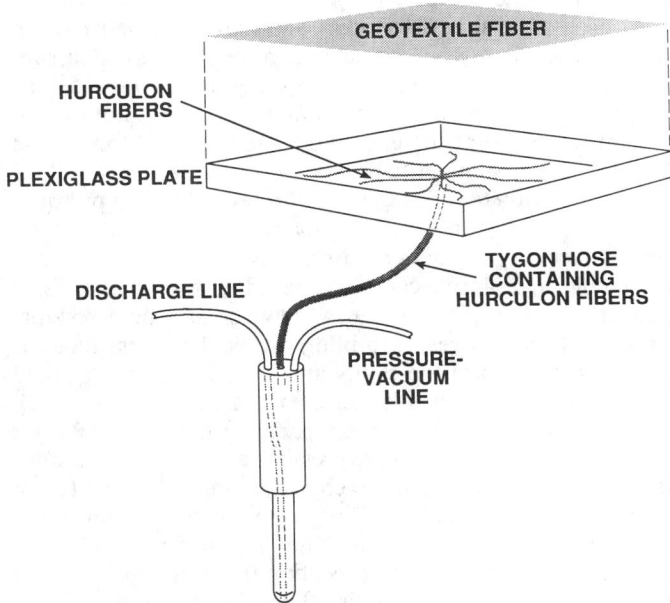

FIG. 25 Wicking Type Soil Pore-Liquid Sampler

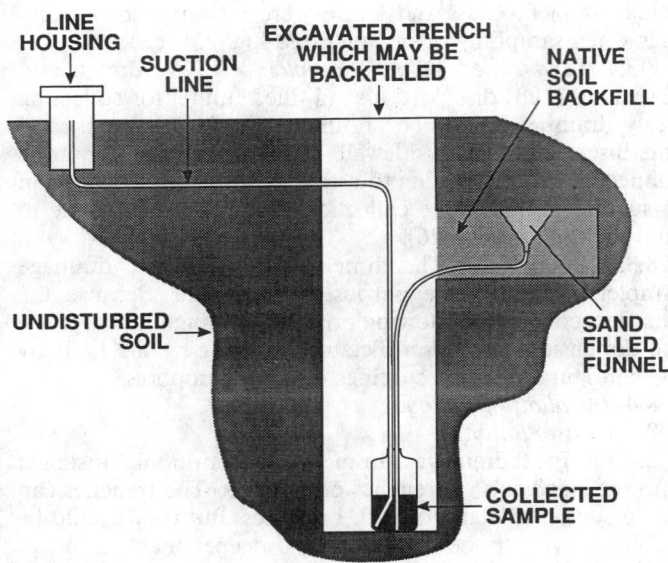

FIG. 26 Sand Filled Funnel Sampler Installation

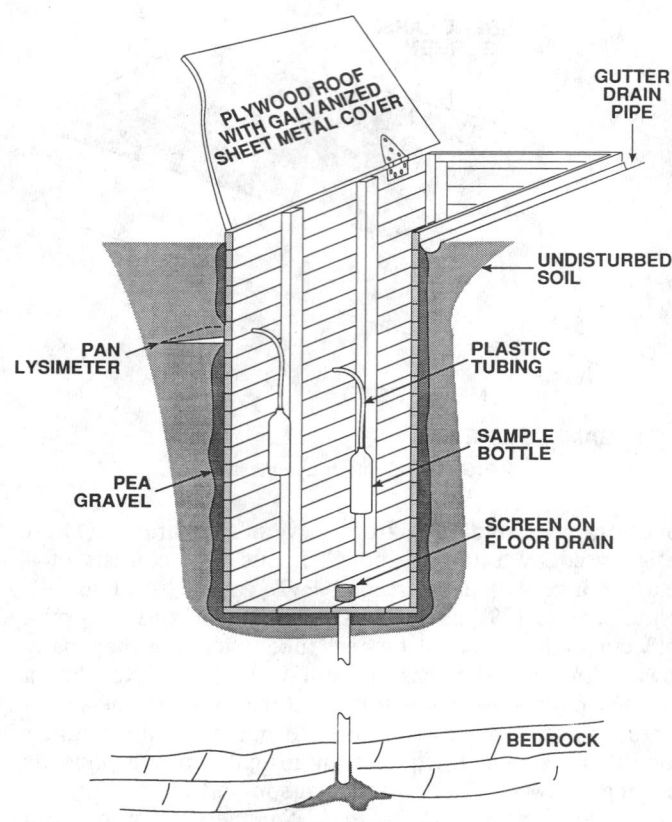

FIG. 27 Example of Pan Lysimeter Installation

crane. As excavation inside the pipe progressed, the pipe advanced downward under its own weight.

8.4.1.3 *Pan Lysimeters*—A pan lysimeter can often be pushed or driven directly into the side wall of a trench. However, if the soil is resistant, an opening for the sampler can be created by hammering a sheet metal blade into the soil profile with a sledge hammer. The pan is placed in the side wall so that it slopes gently toward the trench. Any voids above or below the pan are filled with soil (26). The end of the copper tubing is allowed to project through the trench siding and is connected to plastic tubing and a sample bottle (see Fig. 27).

8.4.1.4 *Glass Block Lysimeters*—A glass block lysimeter is installed in a cavity that is excavated in the side of a trench. Barbee and Brown (81) used a wooden model of the sampler in order to achieve the correct cavity size during excavation. They used a small knife to score the ceiling of the cavity in order to expose any pores that may have been smeared shut during excavation. Care should be taken to keep the ceiling of the cavity smooth and level so that liquids will not run off the upper surface of the glass block. Jordan (96) found that the edges of the sampler had to be in contact with the soil for the entire perimeter of the sampler in order to prevent liquid from running out through any spaces between the soil and the sampler. Level blocks are important so that the majority of the collected sample can be retrieved. However, the inside glass surface is uneven and has "low spots" where residual sample collects between sampling cycles. The glass block is pushed to the end of the cavity and wedges are used to hold its collecting surface firmly against the ceiling of the tunnel. Both the cavity and trench are partially backfilled. Barbee and Brown (81) recommend pressing a sheet of aluminum foil against the wall of the trench, extending below the top of the brick, before final backfilling in order to minimize any lateral migration of liquid from the disturbed portion of the soil profile to the undisturbed portion (see Fig. 28). It should be noted that aluminum foil is often coated with oil.

8.4.1.5 *Trough Lysimeters*—Trough lysimeters are in-

stalled in the same manner as glass block lysimeters (see Fig. 28).

8.4.1.6 *Vacuum Trough Lysimeters*—The vacuum trough lysimeter described by Montgomery et al (97) is housed in a box-like structure, with four walls and a floor, but no ceiling. The floor of the structure and the lower portions of the walls are made of steel. The upper portions of the walls are composed of fiberglass coated plywood. A slotted, plastic drain pipe is set 20 mm above the floor of the structure and is surrounded by gravel. The soil profile surrounding the trough lysimeter is reconstructed incrementally in an attempt to recreate natural conditions. The structure is filled with soil in increments of 0.5 m or less. After each increment is added, liquid is piped slowly into the structure through the drain pipe and allowed to drain back out for 24 h before the next increment is added. This working of the soil particles by liquid is believed to produce bulk densities that are fairly representative of the undisturbed conditions. However, soil macropores are not reproduced.

8.4.1.7 *Caisson Lysimeters*—Lateral collectors or free drainage samplers are installed in cavities augered by hand or by power-driven equipment through holes in the caisson walls (see Fig. 24).

8.4.1.8 *Wicking Soil Pore-Liquid Samplers*—These units are installed by the trench and cavity method similar to that for glass block lysimeters (66). A backhoe excavates the trench to the desired depth. A cavity is then dug into the wall of the trench to the dimensions of the sampler. The roof of the cavity is sometimes scarified (depending on the soil type) to remove smearing caused during excavation. The sampler

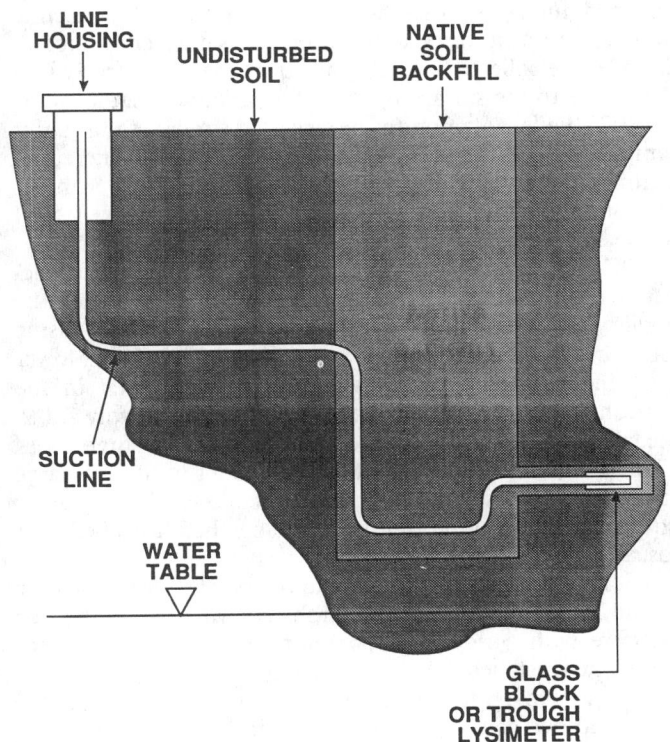

LINE HOUSING

UNDISTURBED SOIL

NATIVE SOIL BACKFILL

SUCTION LINE

WATER TABLE

GLASS BLOCK OR TROUGH LYSIMETER

FIG. 28 Example of Glass Block or Trough Lysimeter Installation

is then forced tightly into place to ensure good contact with the roof of the cavity. The cavity is large enough to accommodate the sampler, the hanging wick and the collection tube. The pressure-vacuum line and the sample collection line are extended to the surface. During backfilling of the tunnel and trench, the bulk density of the fill should be equal to or greater than the native soil.

8.4.1.9 *Sand Filled Funnel Samplers*—The installation procedure for these samplers is similar to that used for glass block lysimeters (see Fig. 26).

8.4.2 *Maintenance:*

8.4.2.1 Where samplers are accessed through permanent trenches and caissons, the sampling station must be protected against flooding due to excessive infiltration. Parizek and Lane (26) drilled a floor drain about 27 m through underlying soil into the unsaturated bedrock. This allowed drainage of excess liquid from the floor of the sampling trench, and also decreased the chances of contamination of soil surrounding the structure. Alternately, a sump pump can be used if a drain is not feasible. Parizek and Lane (26) also found that stratified soils intensified lateral flow of pore-liquid, thus aggravating any flooding problems. They concluded that flooding may be a problem in humid areas where more than about 5 cm of liquid per week is applied to the land surface.

8.4.2.2 The ground surface above the sampler should be maintained in a fairly representative state. Large housings and excessive traffic around the sampler (causing compaction of the soil) will reduce the amount of infiltration in that area. This will affect the representativeness of the pore-liquid samples.

8.4.3 *Comments*—A significant advantage of samplers

installed in the sidewalls of trenches is that in sufficiently cohesive soils, installation produces no disturbance in the overlying soil. In cohesionless, sandy soils, stable cavities may not be possible. As a result, backfilling of the fallen material may be required. This disturbs the soil profile and macropores are not preserved (97).

8.5 *Operation:*

8.5.1 *Methods:*

8.5.1.1 Since pore-liquid flows into free drainage samplers under the influence of gravity, sampling is a relatively simple procedure. Liquid accumulates in the collection device and then drains through tubing into a sample bottle. The sample can be retrieved either through access to the sampling trench or by pulling it to the surface by a suction pump. The wicking pore-liquid sampler allows the application of a slight suction (4 cbar) to improve sampling. However, this design also has a tendency to clog.

8.5.1.2 Jordan (96) found that surface tension develops in trough lysimeters at the soil-air interface and prevents some of the liquid from entering the collector. Cole (41) addressed this problem by inserting an aluminum oxide disc between the soil and the collection surface, and then applying suction to break the surface tension and draw liquid out of the soil. The problem with this approach is that it requires the soil adjacent to the aluminum oxide disc to be free of roots, cracks, and channels (96). The two parallel rods included in the trough lysimeter design overcome this problem. If one end of the metal rod touches the fiberglass screen, then the surface film of liquid surrounds the rod and the liquid moves down the rod toward the sample container. Two rods, barely touching, facilitate this migration by allowing the liquid to move in response to the capillary forces between them (96).

8.5.2 *Comments*—Under near saturated conditions with macropore flow, free drainage samplers tend to collect larger and more consistent samples than suction samplers. Since free drainage samplers are continuous samplers, they need to be emptied after each infiltration event in order to ensure sample integrity, to prevent sample container overflows, and to prevent cross contamination between hydrologic events (12).

8.6 *Advantages and Limitations:*

8.6.1 *Physical Advantages and Limitations*—A major advantage of free drainage samplers is that they are essentially passive, thus they do not alter pore-liquid flow paths. The major disadvantage of free drainage samplers is that samples can only be obtained when soil moisture conditions are in excess of field capacity. Such saturated conditions usually require constant application of surface liquid, as in the case of agricultural irrigation or at land treatment sites. Under drier conditions, free drainage samplers fail to yield any liquid and suction samplers are required.

8.6.2 *Chemical Advantages and Limitations:*

8.6.2.1 There are both advantages and disadvantages to using free drainage samplers to collect pore-liquid for chemical analysis. A major advantage is that the samplers do not distort natural flow patterns as do suction samplers. Because samples are collected over known areas, quantitative mass balance estimates are possible. Because the samplers are continuous collectors, infiltration events can be sampled without having to go to the field. The major limitation of free drainage samplers is that they cannot sample pore-liquids held at tensions greater than the field capacity. As with all samplers, analytical parameters/sampler material compatibilities should be considered. As an example, samples collected from pan lysimeters should not be analyzed for copper or zinc, particularly if the pH of the collected fluid is below seven.

8.6.2.2 Free drainage samplers tend to collect pore-liquids that drain rapidly through macropores. Since the residence time of this liquid is less than that of liquid moving under tension, the major ion chemistry appears more dilute than the fluid sampled from unsaturated pores with suction samplers. In some cases, this decrease in residence time in combination with other factors can result in an actual change in the chemical signature rather than just an overall dilution. This is because insufficient time may be available for reactions to occur with soil components that act as chemical sources or sinks (102). However, free drainage samplers have large cross-sectional areas and they are cumulative collectors. As a result, they collect samples which average soil heterogeneities and therefore give a more representative picture than suction samplers of chemical movement through wet soil, particularly through well-structured soils (81). In addition, the samplers use suction only to retrieve samples. As a result there is less potential for loss of volatile compounds than with suction samplers (12).

8.6.3 *Microbial Advantages and Limitations*—Since free drainage samplers do not have the minute openings that porous ceramic suction samplers have, they do not screen out colloidal-sized particles and soil bacteria. Consequently, they yield more representative values for suspended solids or BOD measurements (26).

9. Perched Ground Water Samplers

9.1 Perched water occurs where varying permeability layers in the vadose zone retard downward movement of liquid. Over time, liquid collects above lower permeability layers and moisture contents rise until the soil becomes saturated with liquid (9, 103). Once soil becomes saturated, wells and other devices normally installed below the water table can be used to collect samples. Separate guides are available for ground water sampling, therefore, the topics are covered briefly, with reference to appropriate documents.

9.2 Sampling perched liquid is attractive because the perching layer collects liquid over a large area. This integration allows samples to be more representative of areal conditions than suction samples (103). This also allows the sampler to potentially detect contaminants that may not be moving downward immediately adjacent to the sampler. In addition, larger sample volumes can be collected than those that can be obtained by suction samplers. Everett et al (6) and Everett et al (9) discussed the incorporation of perched ground water sampling into monitoring programs. There are a variety of systems that can be used. These include the following:

9.2.1 *Point samplers,*

9.2.2 *Wells,*

9.2.3 *Cascading water samplers,* and

9.2.4 *Drainage samplers.*

9.3 *Operating Principles:*

9.3.1 *Point Samplers*—Point samplers are open ended pipes or tubes, such as piezometers or wells with short screened intervals, installed for the purpose of collecting samples from a discrete location in saturated material. Samples are collected by bringing liquid that flows freely into the device to the surface by one of a variety of methods.

9.3.2 *Wells*—A monitoring well is similar to a point sampler except the screened interval is longer. Therefore, samples are averaged over the screened length (104). Samples are collected by bringing liquid that flows freely into the well to the surface by one of a variety of methods.

9.3.3 *Cascading Water Samplers*—Cascading water occurs when a well is screened across a perched layer and the underlying water table or when water leaks through casing joints at the perched layer. Because the water table is lower than the perched layer, water flows into the well in the portion open to the perched layer, and cascades downward to the water table. This situation is common in some areas where the practice has been to install water wells with large screened intervals (105). Samples are collected by capturing liquid flowing into the well from the perched layer before it cascades down to the water table.

9.3.4 *Drainage Samplers*—Shallow perched systems may spread contamination, cause problems with structures, or interfere with agriculture. Therefore, drainage systems are sometimes installed. These systems usually funnel liquid via gravity flow to a ditch or sump from which it is pumped out. This outflow can be sampled. Typical drainage systems include tile lines or manifold collectors. Depending on the design of the system, it may be possible to sample outflows that drain different areas.

9.4 *Description:*

9.4.1 *Point Samplers:*

9.4.1.1 Point samplers can be installed in separate boreholes or clustered together in one borehole at different depths. Fig. 29 presents different configurations that have been used. Piezometers, that are often used as point samplers, are similar to wells, in that they consist of a small diameter casing open at one end or connected to a short screened interval (106). Reeve (107), Patton and Smith (108) and Morrison (1) discussed different designs.

9.4.1.2 Point samplers can be made from a variety of materials including steel, PVC, PTFE, ABS, fiberglass, and additional materials for joints, seals and other components (106, 107).

9.4.2 *Wells:*

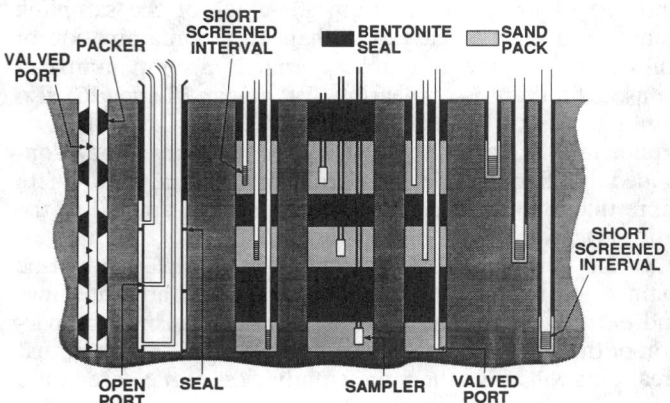

FIG. 29   Examples of Point Sampling Systems

9.4.2.1 Monitoring wells (as depth averaged samplers) are normally installed with one well in each borehole. Components of a well generally include a bottom plug, a length of screen, a length of blank casing, a cap, and a protective cover. Different monitoring well designs are presented in Figs. 30, 31, and 32. Authors who described methods for designing and installing monitoring wells include U.S. EPA (109), Driscoll (110), Gass (111), Keely (112), Minning (113), Richter and Collentine (114), Riggs (115), Riggs and Hatheway (116), Scalf et al (117), Morrison (1), Everett et al (9), Campbell and Lehr (118). Hackett (119, 120) summarized methods for designing and installing monitoring wells with hollow stem augers. Screened hollow stem augers can also be used as temporary wells for sampling (Taylor and Serafini, (121)).

9.4.2.2 Monitoring wells can be made from a variety of materials including steel, PVC, PTFE, ABS, fiberglass, and additional materials for joints, seals, and other components. Details are provided by Barcelona et al (122), U.S. EPA (109), Morrison (1) and many of the references listed above.

9.4.3 *Cascading Water Samplers:*

9.4.3.1 Cascading water is most often seen in production wells in areas with extensive ground water pumpage. Samplers simply consist of a bucket or bailer lowered to a point below the inflow of cascading water. Wilson and Schmidt (103) described methods for developing cascading water samplers (see Fig. 33).

9.4.3.2 Cascading wells differ from other wells only by the way in which water flows into them. Otherwise, the materials used for these wells are identical to those used for other types of wells. Bailers or buckets used to collect samples are also

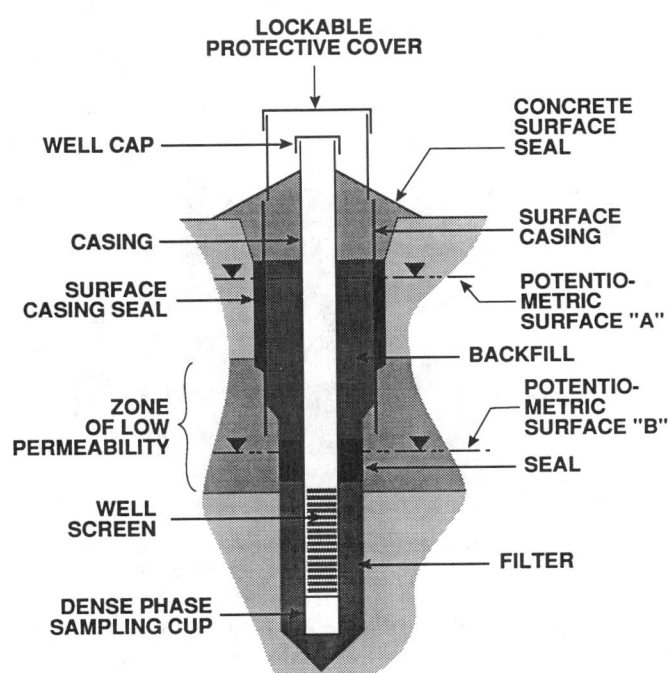

FIG. 31 A Monitoring Well Installed to Sample From the Lower of Two Ground-Water Zones

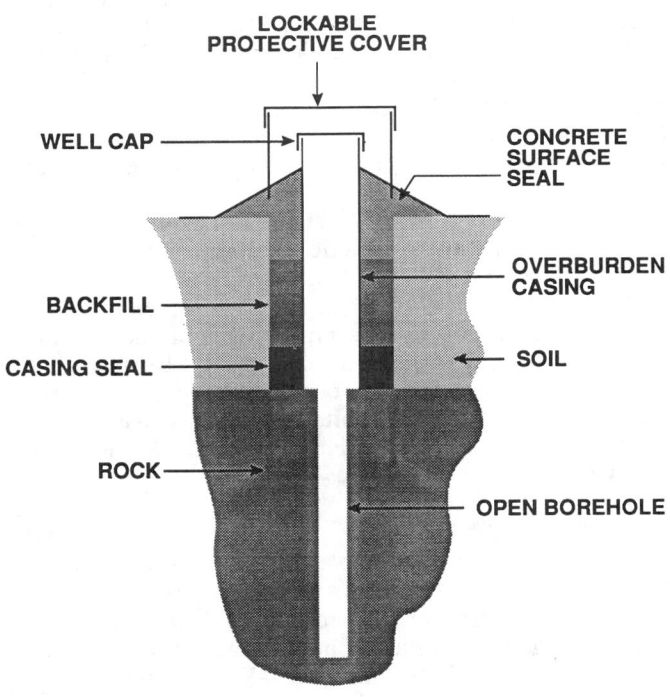

FIG. 32 An Open-Hole Ground-Water Monitoring Well in Rock

available in steel, PVC, PTFE, acrylic and other materials.

9.4.4 *Drainage Samplers:*

9.4.4.1 Drainage systems consist of conduits installed within the perched zone at sufficient slopes for water to flow to a central ditch or drain. The conduits can be tile drains, half perforated pipes, synthetic sheeting, or even layers of gravel and sand. Schilfgaarde ed. (123), contains numerous

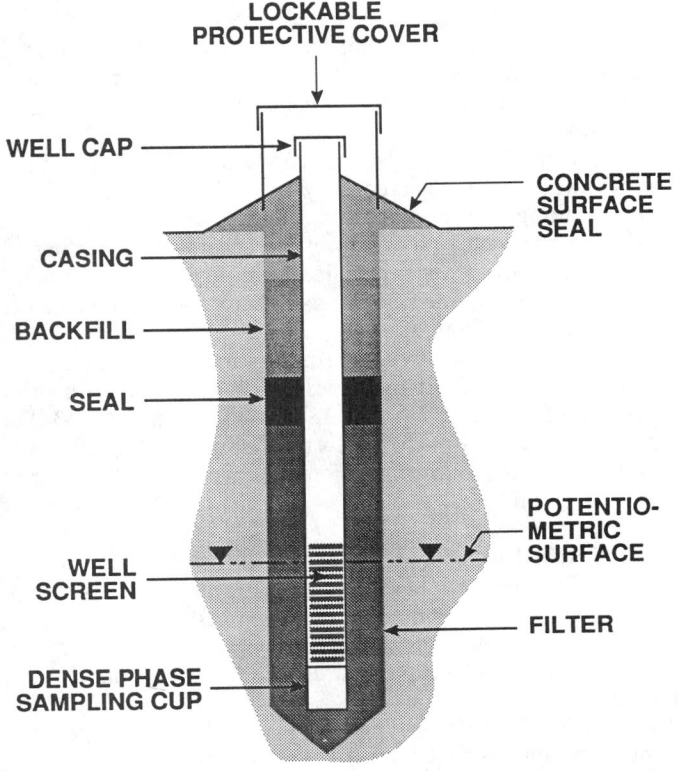

FIG. 30 A Monitoring Well With the Uppermost Ground-Water Level Intersecting the Slotted Well Screen

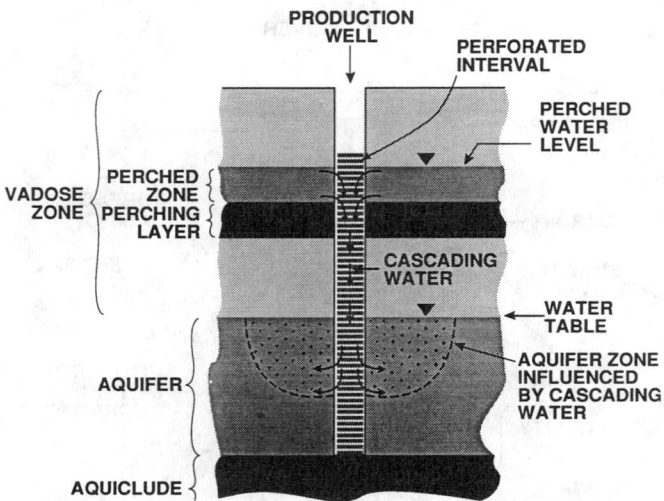

FIG. 33 Conceptualized Cross Section of a Well Showing Cascading Water from Perched Zone

papers on the design and construction of drainage systems. Donnan and Schwab (124) described sampling from agricultural drainage systems. Gilliam et al (125), Gambrell et al (126), Eccles and Gruenberg (127) and Gilliam et al (128) described sampling from tile drains. Gilliam et al (125) and Jacobs and Gilliam (129) described sampling from drainage ditches. Wilson and Small (130) described a lateral drain sampler installed beneath a new sanitary landfill. A perforated pipe collected liquid that was funneled to a sump via a drain line. In most of these systems, a thin layer of high permeability sand or gravel is installed around the drain to promote flow into the collector, and to sieve out fine materials.

9.4.4.2 Because drainage systems often require large quantities of materials, less exotic, cheaper materials such as baked clay tiles and PVC are often used.

9.4.5 *Comments*—As with all samplers, potential chemical interaction between the sampler material and the constituents of interest should be considered. Because these samplers are usually installed for other purposes, incompatibility of materials with monitoring objectives is often a problem. Everett et al (9), Dunlap (131), and U.S. EPA (109) discussed this topic.

9.5 *Installation Methods:*

9.5.1 *Point Samplers*—Reeve (107), Patton and Smith (108), and Morrison (1) discussed procedures for installing and maintaining point samplers.

9.5.2 *Wells*—Most of the references listed in 9.4.2.1 describe methods for installing monitoring wells.

9.5.3 *Cascading Water Samplers*—Wilson and Schmidt (103) discussed methods for installing cascading water samplers.

9.5.4 *Drainage Samplers*—Schilfgaarde ed. (123) contains articles that discussed installation of agricultural drainage systems. Associated hazards and costs often prohibit the installation of these systems at existing landfills. As a result, inclusion of these systems, as leachate collectors, in new landfills is more common. Everett et al (9) discussed methods for installing drainage sampling systems at hazardous waste sites.

9.6 *Operation:*

9.6.1 *Point Samplers*—Point samplers usually have diameters that are too small to allow the use of submersible pumps. As a result, suction methods are usually required (1). Sampling techniques are described in Pickens et al (106), Reeve (107), Patton and Smith (108) and Morrison (1).

9.6.2 *Wells*—Samples may be retrieved from wells in the same manner as from piezometers. However, because wells are designed for sampling or pumpage, diameters are usually large enough to accommodate most pumps. Samples can be brought to the surface by a variety of systems including bailers, suction pumps (for example, peristaltic pumps), air lift pumps, piston pumps, submersible pumps and swabbing. Each of these methods have advantages and disadvantages relating to considerations such as depth to water, required sample volume, sampling speed, alteration of the sample chemistry, equipment requirements, manpower requirements, and cost. These considerations were discussed by Everett et al (9), Fenn et al (132), Gibb et al (133), U.S. EPA (109), Dunlap et al (131), Driscoll (110), and Anderson (75). Sampling methods were described in most of the references of 9.4.2.1. As described in 9.3.2, samples from wells are averaged over the screened interval. However, samples from discrete depths along the screened interval can also be obtained using packer-pump setups such as those described by Fenn et al (132).

9.6.3 *Cascading Water Samplers:*

9.6.3.1 Wilson and Schmidt (103) described techniques for sampling from cascading wells. A bailer or bucket is decontaminated and then lowered to a position in the well below the cascading water but above the water table. When the sampler is full, it is pulled back to the surface. Alternately, as shown in Fig. 33, the chemistry of the water table immediately around a well that has been shut down will be dominated by the cascading water. Therefore, a sample can also be collected from the water table during the initial stages of pumping.

9.6.3.2 Cascading wells are usually production wells in which drawdown has lowered the water table sufficiently to cause cascading. Because of this, there is usually a pump installed in the well that will prevent access for sampling. However, the pumps are periodically removed for maintenance. Therefore, it should be possible to coordinate sampling with pump maintenance personnel.

9.6.4 *Drainage Samplers*—Samples may be collected where tile lines or drainage pipes discharge to ditches or sumps (125, 126, 127, 128, 129, 130). Willardson et al (134) described a "flow-path ground water sampler" that allows collection of water following different flowpaths along a tile drainage system.

9.7 *Limitations:*

9.7.1 Perched water systems can be difficult to find and delineate. Surface and borehole geophysical methods (for example, neutron logging) and video logging of existing wells are often used. Also, perched systems tend to be ephemeral. Therefore, suctions samplers are sometimes required as backups.

9.7.2 *Point Samplers*—The major problem with point sampling systems is that their diameters are often too small to allow adequate development after installation or to allow sampling by any method other than suction. Because the maximum suction lift of water is about 7.5 m, this is the maximum sampling depth for many of the small diameters systems. Systems such as those depicted in Fig. 29 require tight contact with the surrounding material to prevent side leakage of liquid. Depending on the material, this tight contact may not be achievable (9).

9.7.3 *Wells*—Wells provide samples that are averaged over the screened interval. As a result, when contaminants are detected, packer-pump arrangements must be used if zonation of the contaminants is to be delineated. When separate phases of water-immiscible fluid (for example, oil) are found floating in the well, it is difficult to obtain samples of the underlying water without contamination from the overlying fluid. As with all samplers, care should be taken to ensure that materials used to construct a well are compatible with the chemical analyses to be performed.

9.7.4 *Cascading Water Samplers*—Cascading water may enter a well from several distinct perched systems (103). As a result, the sample may be a mixture of water from several depths. Cascading water is most often sampled from pre-existing wells used for other purposes. As a result, materials used in the well construction may alter those chemical constituents of interest. Wells used for irrigation and water supply often have lubricant oils from the pump floating in them. With fluctuating water levels, these oils become smeared along the casing, and may even move out into the surrounding soils. Therefore, traces of these oils may appear in samples.

9.7.5 *Drainage Samplers*—Because of the limitations of excavation equipment, drainage samplers are limited to shallow depths. In addition, the systems are difficult to install in rocky or steep terrains. In areas that experience freeze-thaw cycles, they may be damaged by soil heaving (9). Drainage systems are often susceptible to clogging over time as fine particles and chemically precipitated material accumulate on the drain openings. Collected samples may or may not be representative of average conditions, depending on the distribution of soil types and contaminants in the drained area. If the area of contamination is small compared to the drained area, dilution may prevent the detection of contaminants. In addition, pollutants that are heavier than water may move below the drain if it is not located at the bottom of the perched zone. As with all samplers, there is the possibility of chemical interaction between the sampling system and the chemical constituents of interest. In the case of drainage sampling systems, this effect is amplified as contaminants may have to travel considerable distances through drains before being sampled. In addition, normally non-aerated solutions may be aerated and chemically altered as they travel through drains.

10. Experimental Absorption Samplers

10.1 *Operating Principles*—Absorbent samplers depend on the ability of the material to absorb pore-liquid (1). Samples are collected by placing the sampler in contact with soil. Liquid is allowed to absorb into the sampler material over time. The sampler is then removed, and the sample liquid is extracted for analyses.

10.2 *Description:*

10.2.1 Two designs have been described. The first design includes a cellulose-nylon sponge (0.5 by 4.8 by 30 cm)

seated in a galvanized iron trough. The trough is pressed against a soil surface with a series of lever hinges (135).

10.2.2 The second absorbent sampler design consists of tapered ceramic rods that are driven into soil (136). The rods are made from unglazed ceramic similar to that used as the porous segments of suction samplers.

10.3 *Installation:*

10.3.1 *Pre-Installation:*

10.3.1.1 Sponge samplers are prepared by soaking them for 24 h in a 1 to 5 % NaOH solution containing a washing powder, Tadros and McGarity (135). Sponges are then pressed dry using rollers, stored in a moisture tight container, and taken to the field.

10.3.1.2 Ceramic rod samplers are weighed, boiled in distilled water, oven dried, and stored in a desiccator. The rods are weighed again and then taken to the field.

10.3.2 *Installation:*

10.3.2.1 A sponge is placed in the sampling trough. The trough is then placed in a horizontal cavity cut into the side of a trench. The trough is then pressed against the cavity ceiling with the lever hinges.

10.3.2.2 A ceramic rod sampler is installed by simply driving it into the soil.

10.3.3 *Maintenance:*

10.3.3.1 The only field maintenance required for sponge samplers is the preservation of the sampling trench if future sampling is desired at that location.

10.3.3.2 There is no field maintenance for ceramic rod samplers as they are completely removed to retrieve the samples.

10.3.4 *Comments*—Theoretically, there is no maximum installation depth for sponge samplers. However, because access trenches are required for operation, installations are restricted to shallow depths dictated by excavation equipment and safety considerations. Given the NaOH treatment (without rinsing) of the sponge, measurements of pH, conductivity, TDS, metals and major cations and anions might be affected by the residual NaOH. Depending on the composition of the "washing powder", phosphate, BOD, and MBAS might also be affected.

10.3.4.1 Ceramic rod samplers will have maximum installation depths if pushed or driven from the surface. This maximum depth will generally decrease with increasing soil grain size. However, deeper installations can be achieved by drilling to the top of the interval to be sampled, lowering a rod down the hole and pushing or driving the rod into the sampling interval (136).

10.4 *Operation:*

10.4.1 *Methods:*

10.4.1.1 Sponge samplers are pressed against the soil until a sufficient volume of liquid for planned analyses has been absorbed. The sampler is then removed and the sponge is placed in a moisture proof container. The sample is extracted from the sponge with rollers.

10.4.1.2 Ceramic rod samplers are pushed or driven into the soil and left in place for a period of time. The rods are then withdrawn, and weighed. The rods are leached by boiling in a known volume of distilled water. This solution is then analyzed. The concentrations of constituents in the original pore-liquid are estimated by using the ratio of the volume of absorbed water to the volume of the boiling water.

10.4.2 *Comments*—The amount of liquid that can be sampled is dependent on time, soil type, moisture content, absorbency of the sampler material, volume of the absorbent material, and surface area of the absorbent material in contact with the soil. Generally, sponge samplers function only at higher moisture contents approaching saturation (1). Shimshi (136) used ceramic rod samplers to sample from a sandy loam with moisture contents varying from 7 to 20 %.

10.5 *Limitations:*

10.5.1 Physically, absorbent methods are limited to soils approaching saturation. Sampling requires removal of the absorbent material. Because of this, repeat sampling at the same location is difficult. Although the sampler may be placed back at its original location, identical hydraulic contact with the soil cannot be guaranteed.

10.5.2 Chemically, as with other samplers, there are problems with absorption, desorption, precipitation, cation exchange, and screening of various pore-liquid components as a function of the sampler materials. Tadros and McGarity (135) discussed these concerns in relation to sponge samplers. Shimshi (136) provided a good discussion of the limitations of sampling for nitrate with ceramic rod samplers. Specifically, he found that at lower moisture contents, sampled solutions became less representative due to vapor transfer and chromatographic separation. However, he suggested that these effects could be reduced by increasing the length of the rod insertion period. Clearly, boiling will affect analyses for organics, BOD, COD, NH_3, and some other species of nitrogen, among others.

11. Keywords

11.1 pore fluids, pressure vacuum lysimeters; soil moisture; soil water; suction lysimeters; unsaturated; vadose zone sampling

APPENDIX

(Nonmandatory Information)

X1. DESCRIPTIONS OF TERMS SPECIFIC TO THIS STANDARD

X1.1 *air entry value*—the applied suction at which water menisci of the porous segment of a suction sampler break down, and air enters.

X1.2 *bubbling pressure*—the applied air pressure at which water menisci of the porous segment of a suction sampler break down, and air exits.

X1.3 *cascading water*—perched ground water that enters a well casing via cracks or uncovered perforations, trickling, or pouring down the inside of the casing.

X1.4 *cation exchange capacity (CEC)*—the total capacity of a porous system to adsorb cations from a solution.

X1.5 *hydraulic gradient*—the change in total hydraulic head of water per unit distance of flow.

X1.6 *hydrophobicity*—the property that defines a material as being water repellent. Water exhibits an obtuse contact angle with hydrophobic materials.

X1.7 *hydrophelicity*—the property that defines a material as attracting water. Water exhibits an acute contact angle with hydrophilic materials.

X1.8 *lysimeter*—a device to measure the quantity or rate of water movement through a block of soil, usually undisturbed or in-situ; or to collect such percolated water for analyses.

X1.9 *macropore*—interaggregate cavities that serve as the principal avenues for the infiltration and drainage of water and for aeration.

X1.10 *matric potential*—the energy required to extract water from a soil against the capillary and adsorptive forces of the soil matrix.

X1.11 *matric suction*—for isothermal soil systems, matric suction is the pressure difference across a membrane separating soil solution, in-place, from the same bulk (see soil-water pressure).

X1.12 *micropore*—intraaggregate capillaries responsible for the retention of water and solutes.

X1.13 *percolation*—the movement of water through the vadose zone, in contrast to infiltration at the land surface and recharge across a water table.

X1.14 *pore-liquid*—Liquid that occupies an open space between solid soil particles. Within this guide, pore-liquid is limited to aqueous pore-liquid; that includes water and its solutes.

X1.15 *pore-liquid tension*—see *matric-suction* or *soil-water pressure.*

X1.16 *pressure head*—the head of water at a point in a porous system; negative for unsaturated systems, positive for saturated systems. Quantitatively, it is the water pressure divided by the specific weight of water.

X1.17 *Richard's outflow principle*—the principle that states that pore-liquid will not generally flow into an air-filled cavity (at atmospheric pressure) in unsaturated soil.

X1.18 *soil-water pressure*—the pressure on the water in a soil-water system, as measured by a piezometer for a saturated soil, or by a tensiometer for an unsaturated soil.

X1.19 *tensiometer*—a device for measuring soil-water matric potential (or tension or suction) of water in soil in-situ; a porous, permeable ceramic cup connected through a water filled tube to a pressure measuring device.

X1.20 *total soil-water potential*—the sum of the energy-related components of a soil-water system; for example, the sum of the gravitational, matric and osmotic potentials.

X1.21 *tremie method*—the method whereby materials are emplaced in the bottom of a borehole with a small diameter pipe.

X1.22 *vacuum*—a degree of rarefaction below atmospheric pressure: negative pressure.

X1.23 *Vadose zone*—the hydrogeological region extending from the soil surface to the top of the principle water table; commonly referred to as the "unsaturated zone" or "zone of aeration". These alternate names are inadequate as they do not take into account locally saturated regions above the principle water table (for example, perched water zones).

X1.24 *water content*—the amount of water stored within a porous matrix, expressed as either a volume (volume per unit volume) or a mass (mass per unit mass) of a given solid.

REFERENCES

(1) Morrison, Robert D., *Ground Water Monitoring Technology*, Timco Mfg., Inc., Prairie Du Sac, Wisconsin, 1983.

(2) Wilson, L. G., *Monitoring in the Vadose Zone: A Review of Technical Elements and Methods*, U.S. Environmental Protection Agency, EPA-600/7-80-134, 1980.

(3) Wilson, L. G., "The Fate of Pollutants in the Vadose Zone, Monitoring Methods and Case Studies", *Thirteenth Biennial Conference on Ground Water*, September 1981.

(4) Everett, L. G., "Monitoring in the Vadose Zone", *Ground Water Monitoring Review*, Summer 1981, pp. 44–51.

(5) Wilson, L. G., "Monitoring in the Vadose Zone: Part II", *Ground Water Monitoring Review*, Winter 1982, pp. 31–42.

(6) Everett, L. G., Wilson, L. G., and McMillion, L. G., "Vadose Zone Monitoring Concepts for Hazardous Waste Sites", *Ground Water*, Vol 20, May/June 1982, pp. 312–324.

(7) Wilson, L. G., "Monitoring in the Vadose Zone: Part III", *Ground Water Monitoring Review*, Winter 1983, pp. 155–165.

(8) Everett, Lorne G., Hoylman, Edward W., Wilson, L. Graham, and McMillion, Leslie G., "Constraints and Categories of Vadose Zone Monitoring Devices", *Ground Water Monitoring Review*, Winter 1984, pp. 26–32.

(9) Everett, L. G., Wilson, L. G., and Hoylman, E. W., *Vadose Zone Monitoring Concepts for Hazardous Waste Sites*, Noyes Data Corporation, New Jersey, 1984.

(10) Robbins, Gary A., and Gemmell, Michael M., "Factors Requiring Resolution in Installing Vadose Zone Monitoring Systems", *Ground Water Monitoring Review*, Summer 1985, pp. 75–80.

(11) Merry, W. M., and Palmer, C. M., "Installation and Performance of a Vadose Monitoring System", *Proceedings of the NWWA Conference on Characterization and Monitoring of the Vadose Zone*, NWWA, 1986, pp. 107–125.

(12) Environmental Monitoring Systems Laboratory, *Permit Guidance Manual on Unsaturated Zone Monitoring for Hazardous Waste Land Treatment Units*, U.S. Environmental Protection Agency, Office of Solid Waste and Emergency Response, EPA/530-SW-86-040, October 1986.

(13) Ball, John, and Coley, David M., "A Comparison of Vadose Monitoring Procedures", *Proceedings of the Sixth National Symposium and Exposition on Aquifer Restoration and Ground Water Monitoring*, NWWA/EPA, 1986, pp. 52–61.

(14) Wilson, L. G., "Methods for Sampling Fluids in the Vadose Zone", *Ground Water and Vadose Zone Monitoring, ASTM STP 1053*, ASTM, 1990, pp. 7–24.

(15) Bond, William R., and Rouse, Jim V., "Lysimeters Allow Quicker Monitoring of Heap Leaching and Tailing Sites", *Mining Engineering*, April 1985, pp. 314–319.

(16) Doanhue, Miller, and Shickluna, *Soils—An Introduction to Soils and Plant Growth*, 1977.

(17) Johnson, Thomas, M., and Cartwright, Keros, *Monitoring of Leachate Migration in the Unsaturated Zone in the Vicinity of Sanitary Landfills*, State Geological Survey Circular 514, Urbana, Illinois, 1980.

(18) Sales Division, *Catalog of Products*, Soilmoisture Equipment Corp., Santa Barbara, California, 1988.

(19) Hansen, Edward A., and Harris, Alfred Ray, "Validity of Soil-Water Samples Collected with Porous Ceramic Cups", *Soil Science Society of America Proceedings*, Vol 39, 1975, pp. 528–536.

(20) Quin, B. F., and Forsythe, L. J., "All-Plastic Suction Lysimeters for the Rapid Sampling of Percolating Soil Water", *New Zealand Journal of Science*, Vol 19, 1976, pp. 145–148.

(21) Long, F. Leslie, "A Glass Filter Soil Solution Sampler", *Soil Science Society of America Journal*, Vol 42, 1978, pp. 834–835.

(22) Starr, Michael R., "Variation in the Quality of Tension Lysimeter Soil Water Samples from a Finnish Forest Soil", *Soil Science*, Vol 140, December 1985, pp. 453–461.

(23) Sales Division, *Catalog of Products*, Mott Metallurgical Corp., Farmington, Conn., 1988.

(24) Smith, C. N., and Carsel, R. F., "A Stainless-Steel Soil Solution Sampler for Monitoring Pesticides in the Vadose Zone", *Soil Science Society of America Journal*, Vol 50, 1986, pp. 263–265.

(25) Caster, A., and Timmons, R., "Pourous Teflon®: Its Application in Groundwater Sampling", Timco Mfg. Inc., Prairie Du Sac, Wisc., 1988.

(26) Parizek, Richard R., and Lane, "Soil-Water Sampling Using Pan and Deep Pressure-Vacuum Lysimeters", *Journal of Hydrology*, Vol 11, 1970, pp. 1–21.

(27) Trainor, David P., *The Relationship Between Two Laboratory Leaching Procedures and Leachate Quality at Foundry Waste Landfills*, M. S. Thesis/Independent Report, Dept. of Civil and Environmental Engineering, The University of Wisconsin, Madison, Wisconsin, February, 1983.

(28) Young, Mark, "Use of Suction Lysimeters for Monitoring in the Landfill Linear Zone", *Proceedings of Monitoring Hazardous Waste Sites*, Geotechnical Engineering Division, American Society of Civil Engineers, Detroit, Mich., October, 1985.

(29) Morrison, R. D., and Tsai, T. C., *Modified Vacuum-Pressure Lysimeter for Vadose Zone Sampling*, Calscience Research Inc., Huntington Beach, California, 1981.

(30) Morrison, Robert, and Szecsody, James, "Sleeve and Casing Lysimeters for Soil Pore Water Sampling", *Soil Science*, Vol 139, May 1985, pp. 446–451.

(31) Nightingale, H. I., Harrison, Doug, and Salo, John E., "An Evaluation Technique for Ground Water Quality Beneath Urban Runoff Retention and Percolation Basins", *Ground Water Monitoring Review*, Vol 5, Winter 1985, pp. 43–50.

(32) Torstensson, B. A., and Petsonk, A. M., "A Hermetically Isolated Sampling Method for Ground Water Investigations", *Ground Water Contamination: Field Methods, ASTM STP 963*, ASTM, 1988, pp. 274–289, and Knighton, M. Dean, and Streblow, Dwight E., "A More Versatile Soil Water Sampler", *Soil Science Society of America Journal*, Vol 45, 1981, pp. 158–159.

(33) Jackson, D. R., Brinkley, F. S., and Bondietti, E. A., "Extraction of Soil Water Using Cellulose-Acetate Hollow Fibers", *Soil Science Society of America Journal*, Vol 40, 1976, pp. 327–329.

(34) Levin, M. J., and Jackson, D. R., "A Comparison of In Situ Extractors for Sampling Soil Water", *Soil Science Society of America Journal*, Vol 41, 1977, pp. 535–536.

(35) Stevenson, Craig D., "Simple Apparatus for Monitoring Land Disposal Systems by Sampling Percolating Soil Waters", *Environmental Science and Technology*, Vol 12, March 1978, pp. 329–331.

(36) Wagemann, R., and Graham, B., "Membrane and Glass Fibre Filter Contamination in Chemical Analysis of Fresh Water", *Water Research*, Vol 8, 1974, pp. 407–412.

(37) Sales Division, *Catalog of Products*, Cole-Parmer Instrument Company, 1988.

(38) Sales Division, *Catalog of Products*, Corning Glass Works, New York, 1988.

(39) Duke, H., Kruse, E., and Hutchinson, G., "An Automatic Vacuum Lysimeter for Monitoring Percolation Rates", *USDA Agricultural Research Service*, ARS, 1970, pp. 41–165.

(40) Tanner, C. B., Bourget, S. J., and Holmes, W. E., "Moisture Tension Plates Constructed from Alundum Filter Discs", *Soil Science Society of America Proceedings*, Vol 18, 1954, pp. 222–223.

(41) Cole, D. W., "Alundum Tension Lysimeter", *Soil Science*, Vol 85, June 1958, pp. 293–296.

(42) Cole, D., Gessell, S., and Held, E., "Tension Lysimeter Studies of Ion and Moisture Movement in Glacial Till and Coral Atoll Sites", *Soil Science Society of America Proceedings*, Vol 25, 1968, pp. 321–325.

(43) Nielson, D., and Phillips, R., "Small Fritted Glass Bead Plates For Determination of Moisture Retention", *Soil Science Society of America Proceedings*, Vol 22, 1958, pp. 574–575.

(44) Chow, T. L., "Fritted Glass Bead Materials as Tensiometers and Tension Plates", *Soil Science Society of America Journal*, Vol 41, 1977, pp. 19–22.

(45) Everett, Lorne, and McMillion, Leslie, G., "Operational Ranges for Suction Lysimeters", *Ground Water Monitoring Review*, Summer 1985, pp. 51–60.

(46) Haldorsen, Sylvi, Petsonk, Andrew M., and Tortensson, Bengt-Arne, "An Instrument for In-Situ Monitoring of Water Quality and Movement in the Vadose Zone", *Proceedings of the NWWA Conference on Characterization and Monitoring of the Vadose Zone*, NWWA, 1986, pp. 158–172.

(47) Neary, A. J., and Tomassini, F., "Preparation of Alundum/Ceramic Plate Tension Lysimeters for Soil Water Collection", *Canadian Journal of Soil Science*, Vol 65, February 1985, pp. 169–177.

(48) Wolff, R. G., "Weathering Woodstock Granite, near Baltimore, Maryland", *American Journal of Science*, Vol 265, 1967, pp. 106–117.

(49) Wood, Warren W., "A Technique Using Porous Cups for Water Sampling at Any Depth in the Unsaturated Zone", *Water Resources Research*, Vol 9, April 1973, pp. 486–488.

(50) Debyle, Norbert V., Hennes, Robert W., and Hart, George E., "Evaluation of Ceramic Cups for Determining Soil Solution Chemistry", *Soil Science*, Vol 146, July 1988, pp. 30–36.

(51) England, C. B., "Comments on A Technique Using Porous Cups for Water Sampling at Any Depth in the Unsaturated Zone by Warren W. Wood", *Water Resources Research*, Vol 10, October 1974, pp. 1049.

(52) Bottcher, A. B., Miller, L. W., and Campbell, K. L., "Phosphorous Adsorption in Various Soil-Water Extraction Cup Materials: Effect of Acid Wash", *Soil Science*, Vol 137, 1984, pp 239–244.

(53) Peters, Charles A., and Healy, Richard W., "The Representativeness of Pore Water Samples Collected from the Unsaturated Zone Using Pressure-Vacuum Lysimeters", *Ground Water Monitoring Review*, Spring 1988, pp. 96–101.

(54) Linden, D. R., "Design, Installation and Use of Porous Ceramic Samplers for Monitoring Soil-Water Quality", *U.S. Dep. Agric. Technical Bull.*, 1562, 1977.

(55) Rhoades, J. D., and Oster, J. D., "Solute content," Methods of Soil Analysis, Part I, Agronomy Series Monograph No 9, Second Edition, American Society of Agronomy, Inc. Madison, Wisconsin, 1986, pp. 985–1006.

(56) Klute, A., ed., *Methods of Soil Analysis, Part I*, Agronomy Series Monograph No. 9, Second Edition, American Society of Agronomy, Inc., Madison, Wisconsin, 1986.

(57) Brose, Richard J., Shatz, Richard W., and Regan, Thomas M., "An Alternate Method of Lysimeter and Flour Pack Placement in Deep Boreholes", *Proceedings of the Sixth National Symposium and Exposition on Aquifer Restoration and Ground Water Monitoring*, NWWA, 1986, pp. 88–95.

(58) Litaor, M. Iggy, "Review of Soil Solution Samplers", *Water Resources Research*, Vol 24, May 1988, pp. 727–733.

(59) Silkworth, D. R., and Grigal, D. F., "Field Comparison of Soil Solution Samplers", *Soil Science Society of America Journal*, Vol 45, 1981, pp. 440–442.

(60) Smith, J. L., and McWhorter, D. M., "Continuous Subsurface Injection of Liquid Organic Wastes", *Land as a Waste Management Alternative*, Ann Arbor Science, 1977, pp. 646–656.

(61) Grier, H. E., Burton, W., and Tiwari, C., "Overland Cycling of Animal Waste", *Land as a Waste Management Alternative*, Ann Arbor Science, 1977, pp. 693–702.

(62) Smith, J. H., Robbins, C. W., Bondurant, J. A., and Hayden, C. W., "Treatment of Potato Processing Wastewater on Agricultural Land: Water and Organic Loading, and the Fate of Applied Plant Nutrients", *Land as a Waste Management Alternative*, Ann Arbor Science, 1977, pp. 769–781.

(63) Morrison, Robert D., and Szecsody, Jim E., "A tensiometer and Pore Water Sampler for Vadose Zone Monitoring", *Soil Science*, Vol 144, November 1987, pp. 367–372.

(64) Baier, Dwight C., Aljibury, Falih K., Meyer, Jewell K., Wolfenden, Allen K., *Vadose Zone Monitoring is Effective for Evaluating the Integrity of Hazardous Waste Pond Liners*, Baier Agronomy, Inc., Woodland, California, November 1983.

(65) Taylor, S. A., and Ashcroft, G. L., *Physical Edaphology, The Physics of Irrigated and Non Irrigated Soils*, W. H. Freeman and Company, San Francisco, 1972.

(66) Hornby, W. J., Zabick, J. D., and Crawley, W., "Factors Which Affect Soil-Pore Liquid: A Comparison of Currently Available Samplers with Two New Designs", *Ground Water Monitoring Review*, Vol 6, Spring 1986, pp. 61–66.

(67) Wengel, R. W., and Griffin, G. F., "Remote Soil-Water Sampling Technique", *Soil Science Society of America Journal*, Vol 35, 1971, pp. 661–664.

(68) Brown, K. W., Thomas, J. C., and Aurelius, M. W., "Collecting and Testing Barrel Sized Undisturbed Soil Monoliths", *Soil Science Society of America Journal*, Vol 49, 1985, pp. 1067–1069.

(69) Wagner, George H., "Use of Porous Ceramic Cups to Sample Soil Water Within the Profile", Soil Science, Vol 94, 1962, pp. 379–386.

(70) Nagpal, N. K., "Comparison Among and Evaluation of Ceramic Porous Cup Soil Water Samplers for Nutrient Transport Studies", *Canadian Journal of Soil Science*, Vol 62, 1982, pp. 685–694.

(71) Bouma, J., Jongerius A., and Schoondebeek, D., "Calculation of Hydraulic Conductivity of Some Saturated Clay Soils Using Micromorpho-Meteric Data", *Soil Science Society of America Journal*, Vol 43, 1979, pp. 261–265.

(72) Warrick, A. W., and Amoozegar-Fard, A., "Soil Water Regimes Near Porous Cup Water Samplers", *Water Resources Research*, Vol 13, February 1977, pp. 203–207.

(73) Van der Ploeg, R. R., and Beese, F., "Model Calculations for the Extraction of Soil Water by Ceramic Cups and Plates", *Soil Science Society of America Journal*, Vol 41, 1977, pp. 466–470.

(74) Narasimhan, T. N., and Dreiss, Shirley, J., "A Numerical Technique for Modeling Transient Flow of Water to a Soil Water Sampler", *Soil Science*, Vol 141, March 1986, pp. 230–236.

(75) Anderson, Linda Davis, "Problems Interpreting Samples Taken with Large-Volume, Falling Suction Soil-Water Samplers", *Ground Water*, 1986.

(76) Law Engineering Testing Company, *Lysimeter Evaluation Study*, American Petroleum Institute, May 1982.

(77) Hillel, D., *Applications of Soil Physics*, Academic Press, NY, 1980.

(78) Sales Division, *Catalog of Products*, BAT Envitech Inc., Long Beach California, 1988.

(79) Brown, K. W., *Efficiency of Soil Core and Soil Pore-Liquid Sampling Systems*, U.S. EPA/600/52-86/083, Virginia, February, 1987.

(80) Amter, S., *Injection/Recovery Lysimeter Technique for Soil-Water Extraction*, M. S. Thesis, Dept. of Hydrology and Water Resources, The University of Arizona, Tucson, Arizona, 1987.

(81) Barbee, G. C., and Brown, K. W., "Comparison Between Suction and Free-Drainage Soil Solution Samplers", *Soil Science*, Vol 141, February 1986, pp. 149–154.

(82) Amoozegar-Fard, A. D., Nielsen, D. R., and Warrick, A. W., "Soil Solute Concentration Distributions for Spatially Varying Pore Water Velocities and Apparent Diffusion Coefficients" *Soil Science Society of America Journal*, Vol 46, 1982, pp. 3–9.

(83) Haines, B. L., Waide, J. B., and Todd, R. L., "Soil Solution Nutrient Concentrations Sampled with Tension and Zero-Tension Lysimeters: Report of Discrepancies", *Soil Science Society of America Journal*, Vol 46, 1982, pp. 658–660.

(84) Biggar, J. W., and Nielsen, D. R., "Spatial Variability of the

Leaching Characteristics of a Field Soil", *Water Resources Research*, Vol 12, 1976, pp. 78–84.

(85) Thomas, G. W., and Phillips, R. E., "Consequences of Water Movement in Macropores", *Journal of Environmental Quality*, Vol 8, 1979, pp. 149–152.

(86) Severson, R. C., and Grigal, D. F., "Soil Solution Concentrations: Effect of Extraction Time Using Porous Ceramic Cups Under Constant Tension", *Water Resources Bulletin*, Vol 12, December 1976, pp. 1161–1169.

(87) Shuford, J. W., Fritton, D. D., and Baker, D. E., "Nitrate-Nitrogen and Chloride Movement Through Undisturbed Field Soil", *Journal of Environmental Quality*, Vol 6, 1977, pp. 736–739.

(88) Tyler, Donald D., and Thomas, Grant W., "Lysimeter Measurements of Nitrate and Chloride Losses from Soil Under Conventional and No-Tillage Corn", *Journal of Environmental Quality*, Vol 6, 1977, pp. 63–66.

(89) Suarez, D. L., "Prediction of pH Errors in Soil Water Extractors Due to Degassing", *Soil Science Society of America Journal*, Vol 51, 1987, pp. 64–68.

(90) Ransom, M. D., and Smeck, N. E., "Water Table Characteristics and Water Chemistry of Seasonally Wet Soils of Southwestern Ohio", *Soil Science Society of America Journal*, Vol 50, 1986, pp. 1281–1290.

(91) Wood, A. L., Wilson, J. T., Cosby, R. L., Hornsby, A. G., and Baskin, L. B., "Apparatus and Procedure for Sampling Soil Profiles for Volatile Organic Compounds", *Soil Science Society of America Journal*, Vol 45, 1981, pp. 442–444.

(92) Pettyjohn, Wayne A., Dunlap, W. J., Cosby, Roger, and Keeley, Jack W., "Sampling Ground Water for Organic Contaminants", *Ground Water*, Vol 19, March-April 1981, pp. 180–189.

(93) Creasey, Carol L., and Dreiss, Shirley J., "Soil Water Sampler: Do They Significantly Bias Concentrations in Water Samples?", *Proceedings of the NWWA Conference on Characterization and Monitoring of the Vadose Zone*, NWWA, 1986, pp. 173–181.

(94) Bell, R. G., "Porous Ceramic Soil Moisture Samplers, An Application in Lysimeter Studies on Effluent Spray Irrigation", *New Zealand Journal of Experimental Agriculture*, Vol 2, June 1974, pp. 173–175.

(95) Dazzo, F., and Rothwell, D., "Evaluation of Porcelain Cup Soil Water Samplers For Bacteriological Sampling", *Applied Microbiology*, Vol 27, 1974, pp. 1172–1174.

(96) Jordan, Carl F., "A Simple, Tension-Free Lysimeter", *Soil Science*, Vol 105, 1968, pp. 81–86.

(97) Montgomery, B. R., Prunty, Lyle, and Bauder, J. W., "Vacuum Trough Extractors for Measuring Drainage and Nitrate Flux Through Sandy Soils", *Soil Science Society of America Journal*, Vol 51, 1987, pp. 271–276.

(98) Schmidt, C., and Clements, E., *Reuse of Municipal Wastewater For Groundwater Recharge*, U.S. Environmental Protection Agency, 68-03-2140, 1978, Ohio, pp. 110–125.

(99) Schneider, B. J., Oliva, J., Ku, H. F. H., and Oaksford, E. T., "Monitoring the Movement and Chemical Quality of Artificial-Recharge Water in the Unsaturated Zone on Long Island, New York", *Proceedings of the Characterization and Monitoring of the Vadose (Unsaturated) Zone*, National Water Well Association, Las Vegas, Nevada, 1984, pp. 383–410.

(100) K. W. Brown and Associates, *Hazardous Waste Land Treatment*, U.S. EPA, Office of Research and Development, SW-874, Cincinnati, Ohio, September, 1980.

(101) Schneider, B. J., and Oaksford, E. T., *Design and Monitoring Capability of an Experimental Artificial-Recharge Facility at East Meadow, Long Island, New York*, U.S. Geological Survey Open-File Report 84-070, 1986.

(102) Joslin, J. D., Mays, P. A., Wolfe, M. H., Kelly, J. M., Garber, R. W., Brewer, P. F., "Chemistry of Tension Lysimeter Water and Lateral Flow in Spruce and Hardwood Stands", *Journal of Environmental Quality*, Vol 16, 1987, pp. 152–160.

(103) Wilson, L. G., and Schmidt, K. D., "Monitoring Perched Water in the Vadose Zone", *Establishment of Water Quality Monitoring Programs, Proceedings of a Symposium*, American Water Re-

sources Association, 1978, pp. 134–149.

(104) Pickens, J. F., and Grisak, G. E., "Reply to the Preceding Discussion of Vanhof et al. at a Multilevel Device for Groundwater Sampling and Piezometric Monitoring", *Ground Water*, Vol 17(4), 1979, pp. 393–397.

(105) Smith, S. A., Small, G. S., Phillips, T. S., and Clester, M., *Water Quality in the Salt River Project, A Preliminary Report*, Salt River Project Water Resource Operations, Ground Water Planning Division, Phoenix, Arizona, August, 1982.

(106) Pickens, J. F., Cherry, J. A., Coupland, R. M., Grisak, G. E., Merritt, W. F., and Risto, G. A., "A Multi-Level Device for Ground Water Sampling", *Ground Water Monitoring Review*, Vol 1(1), 1981.

(107) Reeve, R. C., and Doering, E. J., "Sampling the Soil Solution for Salinity Appraisal", *Soil Science*, Vol 99, 1965, pp. 339–344.

(108) Patton, F. D., and Smith, H. B., "Design Considerations and the Quality of Data from Multiple-Level Groundwater Monitoring Wells", *Ground-Water Contamination: Field Methods, ASTM STP 963*, ASTM, 1988, pp. 206–217.

(109) U.S. Environmental Protection Agency, *RCRA Ground-Water Monitoring Technical Enforcement Guidance Document*, Office of Waste Programs Enforcement, Office of Solid Waste and Emergency Response, OSWER-9950.1, 1986.

(110) Driscoll, F. G., *Groundwater and Wells*, Johnson Division, St. Paul, Minnesota, 1986.

(111) Gass, T. E., "Methodology for Monitoring Wells", *Water Well Journal*, Vol 38(6), 1984, pp. 30–31.

(112) Keely, J. F., and Boateng K., "Monitoring Well Installation, Purging and Sampling Techniques", *Ground Water*, 1987, Vol 25 (3, 4).

(113) Minning, R. C., "Monitoring Well Design and Installation", *Proceedings of the Second National Symposium on Aquifer Restoration and Ground Water Monitoring*, Columbus, Ohio, 1982, pp. 194–197.

(114) Richter, H. R., and Collentine, M. G., "Will My Monitoring Well Survive Down There?: Design and Installation Techniques for Hazardous Waste Studies", *Proceedings of the Third National Symposium on Aquifer Restoration and Ground Water Monitoring*, Columbus, Ohio, 1983, pp. 223–229.

(115) Riggs, C. O., "Monitoring Well Drilling Methods", *Proceedings of the Workshop on Resource Conservation Recovery Act Ground Water Monitoring Enforcement: Use of the Technical Enforcement Guidance Document and Compliance Order Guide*, Philadelphia, Pennsylvania, 1987.

(116) Riggs, C. O., and Hatheway, A. W., "Ground Water Monitoring Field Practice—An Overview", *Ground Water Contamination: Field Methods, ASTM STP 963*, ASTM, pp. 121–136, 1988.

(117) Scalf, M. R., McNabb, J. F., Dunlap, W. J., Cosby, R. L., and Fryberger, J., *Manual of Ground Water Sampling Procedures*, National Water Well Association, Ohio, 1981.

(118) Campbell, M., and Lehr, J., *Water Well Technology*, McGraw-Hill Book Co., New York, New York, 1973.

(119) Hackett, G., "Drilling and Constructing Monitoring Wells with Hollow Stem Augers", *Ground Water Monitoring Review*, Fall 1987, pp. 51–62.

(120) Hackett, G., "Drilling and Constructing Monitoring Wells with Hollow Stem Augers", *Ground Water Monitoring Review*, Winter 1988, pp. 60–68.

(121) Taylor, T. W., and Serafini, M. C., "Screened Auger Sampling: The Technique and Two Case Studies", *Ground Water Monitoring Review*, Summer 1988, pp. 145–152.

(122) Barcelona, M. J., Gibb, J. P., and Miller, R. A., *A Guide to the Selection of Materials For Monitoring Well Construction and Ground Water Sampling*, Illinois State Water Survey, Champaign, Illinois, 1983.

(123) Schilfgaarde, J. V., ed., *Drainage for Agriculture*, Number 17 in Agronomy Series, American Society of Agronomy, Madison, Wisconsin, 1974.

(124) Donnan, W. W. and G. O. Schwab, "Current Drainage Methods in the USA," in *Drainage for Agriculture*, Agronomy Monograph Number 17, American Society of Agronomy, Madison,

Wisconsin, 1974, pp 93–114.

(125) Gilliam, J. W., Daniels, R. B., and Lutz, J. F., "Nitrogen Content of Shallow Ground Water in the North Carolina Coastal Plain", *Journal of Environmental Quality*, Vol 2, 1974, pp. 147–151.

(126) Gambrell, R. P., Gilliam, J. W., and Weed, S. B., "Denitrification in Subsoils of the North Carolina Coastal Plain as Affected by Soil Drainage", *Journal of Environmental Quality*, Vol 4, 1975, pp. 311–316.

(127) Eccles, L. A., and Gruenberg, P. A., "Monitoring Agricultural Waste Water Discharges For Pesticide Residues", *Proceedings: Establishment of Water Quality Monitoring Programs*, American Water Resources Association, 1978, pp. 319–327.

(128) Gilliam, J. W., Skaggs, R. W., and Weed, S. B., "Drainage Control to Diminish Nitrate Loss from Agricultural Fields", *Journal of Environmental Quality*, Vol 8, 1979, pp. 137–142.

(129) Jacobs, T. C., and Gilliam, J. W., "Riparian Losses of Nitrate from Agricultural Drainage Waters", *Journal of Environmental Quality*, Vol 14, 1985, pp. 472–478.

(130) Wilson, L. G., and Small, G. G., "Pollution Potential of a Sanitary Landfill Near Tucson, Hydraulic Engineering and the Environment", *Proceedings 21st Annual Hydraulics Division Specialty Conference*, ASCE, 1973.

(131) Dunlap, William J., *Some Concepts Pertaining to Investigative Methodology for Subsurface Process Research*, U.S. Environmental Protection Agency, 1977, pp. 167–172.

(132) Fenn, D. E., Hanley, K. J., Isbister, J., Briads, O., Yare, B., and Roux, P., *Procedures Manual for Ground Water Monitoring at Solid Waste Disposal Facilities*, U.S. Environmental Protection Agency, EPA/530/SW-611, 1977.

(133) Gibb, J. P., Schuller, R. M., and Griffin, R. A., *Procedures for the Collection of Representative Water Quality Data from Monitoring Wells*, Cooperative Groundwater Report 7, Illinois State Water Survey, 1981.

(134) Willardson, L. S., Meek, B. D., and Huber, M. J., "A Flow Path Ground Water Sampler", *Soil Science Society of America Proceedings*, Vol 36, 1973, pp. 965–966.

(135) Tadros, V. T., and McGarity, J. W., "A Method for Collecting Soil Percolate and Soil Solution in the Field", *Plant and Soil*, Vol 44, June 1976, pp. 655–667.

(136) Shimshi, Daniel, "Use of Ceramic Points for the Sampling of Soil Solution", *Soil Science*, Vol 101, 1966, pp. 98–103.

(137) Grover, B. L., and Lamborn, R. E., "Preparation of Porous Seramic Cups to be Used for Extraction of Soil Water Having Low Solute Concentrations", *Soil Science Society of America Proceedings*, Vol 34, 1970, pp. 706–708.

(138) Zimmermann, C. F., Price, M. T., and Montgomery, J. R., "A Comparison of Ceramic and Teflon In-Situ Samplers for Nutrient Pore Water Determinations", *Estuarine and Coastal Marine Science*, Vol 7, 1978, pp. 93–97.

(139) Faber, W. R., and Nelson, P. V., "Evaluation of Methods for Solution from Container Root Media", *Communications in Soil Science Plant Analysis*, Vol 15, No. 9, 1984, pp. 1029–1040.

(140) Barbarick, K. A., Sabey, B. R., and Klute, A., "Comparison of Various Methods of Sampling Soil Water for Determining Ionic Salts, Sodium, and Calcium Content in Soil Columns", *Soil Science Society of America Journal*, Vol 43, 1979, pp. 1053–1055.

(141) Wagemann, R., and Graham, B., "Membrane and Glass Fibre Filter Contamination in Chemical Analysis of Fresh Water", *Water Research*, Vol 8, 1974, pp. 407–412.

(142) Jones, J. N., and Miller, G. D., "Adsorption of Selected Organic Contaminants onto Possible Well Casing Materials", *Ground-Water Contamination Field Methods*, ASTM STP 963, ASTM, 1988, pp 185–198.

(143) Barcelona, M. J., Helfrich, J. A., and Garske, E. E., "Verification of Sampling Methods and Selection of Materials for Ground Water Contamination Studies", *Ground-Water Contamination Field Methods*, ASTM STP 963, ASTM, 1988.

Standard Guide for
Measuring Matric Potential in the Vadose Zone Using Tensiometers[1]

This standard is issued under the fixed designation D 3404; the number immediately following the designation indicates the year of original adoption or, in the case of revision, the year of last revision. A number in parentheses indicates the year of last reapproval. A superscript epsilon (ε) indicates an editorial change since the last revision or reapproval.

1. Scope

1.1 This guide covers the measurement of matric potential in the vadose zone using tensiometers. The theoretical and practical considerations pertaining to successful onsite use of commercial and fabricated tensiometers are described. Measurement theory and onsite objectives are used to develop guidelines for tensiometer selection, installation, and operation.

1.2 The values stated in SI units are to be regarded as the standard. The inch-pound units given in parentheses are for information only.

1.2 *This standard does not purport to address all of the safety problems, if any, associated with its use. It is the responsibility of the user of this standard to establish appropriate safety and health practices and determine the applicability of regulatory limitations prior to use.*

2. Terminology

2.1 *Descriptions of Terms Specific to This Standard:*

2.1.1 *accuracy of measurement*—the difference between the value of the measurement and the true value.

2.1.2 *hysteresis*—that part of inaccuracy attributable to the tendency of a measurement device to lag in its response to environmental changes. Parameters affecting pressure-sensor hysteresis are temperature and measured pressure.

2.1.3 *precision (repeatability)*—the variability among numerous measurements of the same quantity.

2.1.4 *resolution*—the smallest division of the scale used for a measurement and it is a factor in determining precision and accuracy.

3. Summary of Guide

3.1 The measurement of matric potential in the vadose zone can be accomplished using tensiometers that create a saturated hydraulic link between the soil water and a pressure sensor. A variety of commercial and fabricated tensiometers are commonly used. A saturated porous ceramic material that forms an interface between the soil water and bulk water inside the instrument is available in many shapes, sizes, and pore diameters. A gage, manometer, or electronic pressure transducer is connected to the porous material with small- or large-diameter tubing. Selection of these components allows the user to optimize one or more characteristics, such as accuracy, versatility, response time, durability, maintenance, extent of data collection, and cost.

4. Significance and Use

4.1 Movement of water in the unsaturated zone is of considerable interest in studies of hazardous-waste sites (1, 2, 3, 4)[2]; recharge studies (5, 6); irrigation management (7, 8, 9); and civil-engineering projects (10, 11). Matric-potential data alone can be used to determine direction of flow (11) and, in some cases, quantity of water flux can be determined using multiple tensiometer installations. In theory, this technique can be applied to almost any unsaturated-flow situation whether it is recharge, discharge, lateral flow, or combinations of these situations.

4.2 If the moisture-characteristic curve is known for a soil, matric-potential data can be used to determine the approximate water content of the soil (10). The standard tensiometer is used to measure matric potential between the values of 0 and −867 cm of water; this range includes most values of saturation for many soils (12).

4.3 Tensiometers directly and effectively measure soil-water tension, but they require care and attention to detail. In particular, installation needs to establish a continuous hydraulic connection between the porous material and soil, and minimal disturbance of the natural infiltration pattern are necessary for successful installation. Avoidance of errors caused by air invasion, nonequilibrium of the instrument, or pressure-sensor inaccuracy will produce reliable values of matric potential.

4.4 Special tensiometer designs have extended the normal capabilities of tensiometers, allowing measurement in cold or remote areas, measurement of matric potential as low as −153 m of water (−15 bars), measurement at depths as deep as 6 m (recorded at land surface), and automatic measurement using as many as 22 tensiometers connected to a single pressure transducer, but these require a substantial investment of effort and money.

4.5 Pressure sensors commonly used in tensiometers include vacuum-gages, mercury manometers, and pressure transducers. Only tensiometers equipped with pressure transducers allow for the automated collection of large quantities of data. However, the user needs to be aware of the pressure-transducer specifications, particularly temperature sensitivity and longterm drift. Onsite measurement of known zero and "full-scale" readings probably is the best calibration procedure; however, onsite temperature measurement or periodic recalibration in the laboratory may be sufficient.

---

[1] This guide is under the jurisdiction of ASTM Committee D-18 on Soil and Rock and is the direct responsibility of Subcommittee D18.21 on Ground Water and Vadose Zone Investigations.

Current edition approved May 15, 1991. Published October 1991.

[2] The boldface numbers in parentheses refer to a list of references at the end of the text.

5. Measurement Theory

5.1 In the absence of osmotic effects, unsaturated flow obeys the same laws that govern saturated flow: Darcy's Law and the Equation of Continuity, that were combined as the Richards' Equation (13). Baver et al. (14) presents Darcy's Law for unsaturated flow as follows:

$$q = -K\nabla(\psi + Z) \qquad (1)$$

where:

q = the specific flow, $\left[\dfrac{L}{T}\right]$,

K = the unsaturated hydraulic conductivity, $\left[\dfrac{L}{T}\right]$,

ψ = the matric potential of the soil water at a point, $[L]$,

Z = the elevation at the same point, relative to some datum, $[L]$, and

∇ = the gradient operator, $[L^{-1}]$.

The sum of $\psi + Z$ commonly is referred to as the hydraulic head.

5.2 Unsaturated hydraulic conductivity, K, can be expressed as a function of either matric potential, ψ, or water content, $\theta[L^3$ of water$/L^3$ of soil], although both functions are affected by hysteresis (5). If the wetting and drying limbs of the $K(\psi)$ function are known for a soil, time series of onsite matric-potential profiles can be used to determine: which limb is more appropriate to describe the onsite $K(\psi)$; the corresponding values of the hydraulic-head gradient; and an estimate of flux using Darcy's Law. If, instead, K is known as a function of θ, onsite moisture-content profiles (obtained, for example, from neutron-scattering methods) can be used to estimated K, and combined with matric-potential data to estimate flux. In either case, the accuracy of the flux estimate needs to be assessed carefully. For many porous media, $\dfrac{dK}{d\psi}$ and $\dfrac{dK}{d\theta}$ are large, within certain ranges of ψ or θ, making estimates of K particularly sensitive to onsite-measurement errors of ψ or θ. (Onsite-measurement errors of ψ also have direct effect on $\nabla(\psi + Z)$ in Darcy's Law). Other sources of error in flux estimates can result from: inaccurate data used to establish the $K(\psi)$ or $K(\theta)$ functions (accurate measurement of very small permeability values is particularly difficult) (16); use of an analytical expression for $K(\psi)$ or $K(\theta)$ that facilitates computer simulation, but only approximates the measured data; an insufficient density of onsite measurements to define adequately the θ or ψ profile, which can be markedly nonlinear; onsite soil parameters that are different from those used to establish $K(\psi)$ or $K(\theta)$; and invalid assumptions about the state of onsite hysteresis. Despite the possibility of large errors, certain flow situations occur where these errors are minimized and fairly accurate estimates of flux can be obtained (6, 17). The method has a sound theoretical basis and refinement of the theory to match measured data markedly would improve reliability of the estimates.

5.3 The concept of fluid tension refers to the difference between standard atmospheric pressure and the absolute fluid pressure. Values of tension and pressure are related as follows:

$$T_F = P_{AT} - P_F \qquad (2a)$$

where:

T_F = the tension of an elemental volume of fluid, $\left[\dfrac{M}{LT^2}\right]$,

P_{AT} = the absolute pressure of the standard atmosphere, $\left[\dfrac{M}{LT^2}\right]$, and

P_F = the absolute pressure of the same elemental volume or fluid $\left[\dfrac{M}{LT^2}\right]$.

Soil-water tension (or soil-moisture tension) similarly is equal to the difference between soil-gas pressure and soil-water pressure. Thus:

$$T_W + P_G = P_W \qquad (2b)$$

where:

T_W = the tension of an elemental volume of soil water, $\left[\dfrac{M}{LT^2}\right]$,

P_G = the absolute pressure of the surrounding soil gas, $\left[\dfrac{M}{LT^2}\right]$, and

P_W = the absolute pressure of the same elemental volume of soil water, $\left[\dfrac{M}{LT^2}\right]$.

In this guide, for simplicity, soil-gas pressure is assumed to be equal to 1 atmosphere, except as noted. Various units are used to express tension or pressure of soil water, and are related to each other by the equation:

$$1.000 \text{ bar} = 100.0 \text{ kPa} = 0.9869 \text{ atm} =$$
$$1020 \text{ cm of water at } 4°C =$$
$$1020 \text{ g per cm}^2 \text{ in a standard}$$
$$\text{gravitational field.} \qquad (3)$$

A standard gravitational field is assumed in this guide; thus, centimetres of water at 4°C are used interchangeably with grams per square centimetre.

5.4 The negative of soil-water tension is known formally as matric potential. The matric potential of water in an unsaturated soil arises from the attraction of the soil-particle surfaces for water molecules (adhesion), the attraction of water molecules for each other (cohesion), and the unbalanced forces across the air-water interface. The unbalanced forces result in the concave water films typically found in the interstices between soil particles. Baver et al. (14) present a thorough discussion of matric potential and the forces involved.

5.5 The tensiometer, formally named by Richards and Gardner (18), has undergone many modifications for use in specific problems (1, 11, 19–31). However, the basic components have remained unchanged. A tensiometer comprises a porous surface (usually a ceramic cup) connected to a pressure sensor by a water-filled conduit. The porous cup, buried in a soil, transmits the soil-water pressure to a manometer, a vacuum gage, or an electronic-pressure transducer (referred to in this guide as a pressure transducer). During normal operation, the saturated pores of the cup prevent bulk movement of soil gas into the cup.

5.6 An expanded cross-sectional view of the interface between a porous cup and soil is shown in Fig. 1. Water held by the soil particles is under tension; absolute pressure of the soil water, P_W, is less than atmospheric. This pressure is transmitted through the saturated pores of the cup to the water inside the cup. Conventional fluid statics relates the

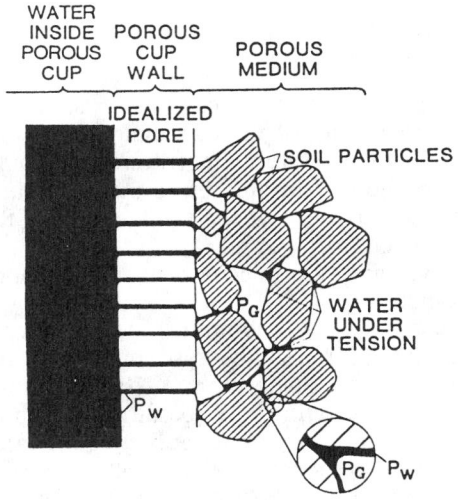

WATER INSIDE POROUS CUP | POROUS CUP WALL | POROUS MEDIUM

P_W–Absolute pressure of soil water
P_G–Absolute pressure of soil gas

FIG. 1 Enlarged Cross Section of Porous Cup-Porous Medium Interface

pressure in the cup to the reading obtained at the manometer, vacuum gage, or pressure transducer.

5.6.1 In the case of a mercury manometer (see Fig. 1(a)):

$$T_W = P_A - P_W = (\rho_{Hg} - \rho_{H_2O})r - \rho_{H_2O}(h + d) \qquad (4)$$

where:

T_W = the soil-water tension relative to atmospheric pressure, in centimetres of water at 4°C,

P_A = the atmospheric pressure, in centimetres of water at 4°C,

P_W = the average pressure in the porous cup and soil, in centimetres of water at 4°C,

ρ_{Hg} = the average density of the mercury column, in grams per cubic centimetre,

ρ_{H_2O} = the average density of the water column, in grams per cubic centimetre,

r = the reading, or height of mercury column above the mercury-reservoir surface, in centimetres,

h = the height of the mercury-reservoir surface above land surface, in centimetres, and

d = the depth of the center of the cup below land surface, in centimetres.

5.7 Although the density of mercury and water both vary about 1 % between 0 and 45°C, Eq. 4 commonly is used with ρ_{Hg} and ρ_{H_2O} constant.

5.7.1 Using $\rho_{Hg} = 13.54$ and $\rho_{H_2O} = 0.995$ (the median values for this temperature range) yields about a 0.25 % error (1.5 cm H_2O) at 45°C, for $T_W \approx 520$ cm H_2O. This small, but needless, error can be removed by using the following density functions:

$$\rho_{Hg} = 13.595 - 2.458 \times 10^{-3} (T) \qquad (5)$$

and

$$\rho_{H_2O} = 0.9997 + 4.879 \times 10^{-5} (T) - 5.909 \times 10^{-6} (T)^2 \qquad (6)$$

where: ρ_{Hg} and ρ_{H_2O} are as defined above, and
T = average temperature of the column, in °C.

5.7.2 Average temperature of the buried segment of water

column can be estimated with a thermocouple or thermistor in contact with the tubing, buried at about 45 % of the depth of the porous cup. Air temperature is an adequate estimate for exposed segments.

5.8 Most vacuum gages used with tensiometers are graduated in bars (and centibars) and have an adjustable zero-reading. The zero adjustment is used to offset the effects of altitude, the height of the gage above the porous cup (see Fig. 3(b)), and changes in the internal characteristics of the gage with time. The adjustment is set by filling the tensiometer with water and then setting the gage to zero while immersing the porous cup to its midpoint in a container of water. This setting is done at the altitude at which the tensiometer will be used and it needs to be repeated periodically after installation either by removing the tensiometer from the soil or by unscrewing the gage and measuring a tension equal to that used in the original calibration. The gage then reads directly the tension in the porous cup. Use of a vacuum gage without an adjustable zero reading could result in inaccurate measurements because the zero-reading could become negative and, therefore, would be indeterminate.

5.9 Pressure transducers convert pressure, or pressure difference, into a voltage (or current) signal. The pressure transducer can be connected remotely to the porous cup with tubing (22, 24) attached directly to the cup (19, 32), or transported between sites (24). An absolute pressure transducer measures the absolute pressure (P_P) in its port. A gage pressure transducer measures the difference between ambient-atmospheric pressure (P_A) and the pressure in its port (P_P), known as gage pressure. When $P_P < P_A$, gage pressure is identical to tension. A differential pressure transducer measures the difference between two pressures; one in each of its two ports. When used with tensiometers, the second port usually is connected to the atmosphere; the unit is used as a gage pressure transducer and it measures tension.

5.10 A calibration equation supplied by the manufacturer, or determined by the user, is used to convert the measured signal into pressure or tension at the pressure-transducer port. The tension in the porous cup and soil is then (see Fig. 3(c)):

$$T_W = T_P - t (\rho_{H_2O}) \qquad (7)$$

where:

T_W = the average tension in the porous cup and soil, in centimetres of water at 4°C,

T_P = the tension in the pressure-transducer port, in centimetres of water at 4°C,

t = the difference in elevation between the pressure-transducer port and the center of the porous cup, in centimetres, and

ρ_{H_2O} = the average density of the water column connecting the porous cup and transducer, in grams per cubic centimetre.

5.11 At 15°C, pure liquid water begins to cavitate (vaporize) if its tension exceeds 969 cm H_2O. If cavitation happens in a tensiometer, liquid continuity is interrupted and tension readings are invalid. Water used in tensiometers is deaerated as completely as practicable, but some impurities and dissolved gases remain that decrease the tension sustainable by liquid water to about 867 cm H_2O (33). Thus, the operating range of tensiometers is described by the following equations:

$$T_C + \Delta h < 867 \text{ cm} \qquad (8)$$

and

$$T_C < 867 \text{ cm} \qquad (9)$$

where:
T_C = the tension in the porous cup, in centimetres of water at 4°C, and
Δh = the elevation of the highest point in the hydraulic connection between the porous cup and the pressure sensor, minus the elevation of the porous cup, in centimetres.

Equation 8 indicates that a "trade-off" occurs between depth of installation of the porous cup and the maximum tension measurable; Eq. 9 sets the upper limit of that tension. Eqs. 8 and 9 are approximate; if the water is insufficiently deaerated, the value 867 would be replaced with a smaller value.

5.12 The only tensiometer described thus far that measures absolute soil-water pressure (P_W) directly is the absolute-pressure-transducer type. The others, differential tensiometers, measure the quantity $P_A - P_W$, where P_A is ambient atmospheric pressure. The driving forces for liquid water in the unsaturated zone (ignoring osmotic potential) are the absolute pressure gradient in the liquid-water phase and gravity (see Eq. 1). If the pressure wave propagates easily through the unsaturated zone, then differential tensiometers can be used directly to determine pressure gradients. However, if a barometric-pressure change is transmitted readily to one differential tensiometer porous cup and not to another (because of an intermediate confining layer), the calculated gradient between the two porous cups would be in error. If a porous cup is isolated from the atmosphere by a confining layer, then a time series of soil-water pressure at the porous cup, calculated with P_A constant, will indicate fluctuations that correlate well with barometric fluctuations. In this case, a recording barometer will provide a record of ambient atmospheric pressure from which absolute soil-water pressure and pressure gradients can be determined. The resulting series of absolute soil-water pressure at the isolated porous cup will be a smoother curve, that will indicate real pressure changes in the water phase.

5.13 Richards **(12)** defined the time constant of a tensiometer as follows:

$$\tau = \frac{1}{K_c S} \qquad (10)$$

where:
τ = the time constant, or time required for 63.2 % of a step change in pressure to be recorded by a tensiometer, when the cup is surrounded by water, in seconds,
K_c = the conductance of the saturated porous cup, or the volume of water passing through the cup wall per unit of time per unit of hydraulic-head difference, in centimetre^2 second^{-1}, and
S = the tensiometer sensitivity, or change in pressure reading per unit volume of water passing through the porous-cup wall, in centimetre^{-2}.
Also, the porous cup conductance may be expressed as:

$$K_c = \frac{kA}{W}; \qquad (11)$$

where:
K_c = the cup conductance, in centimetre^2 second^{-1},
k = the permeability of the cup material to water at the prevailing temperature, in centimetre second^{-1},
A = the average surface area of porous-cup material, estimated as the mean of the inside area and the outside area, in centimetre^2, and
W = the average wall thickness of the porous cup, in centimetres.

5.14 Richards' **(12)** definition does not apply to a tensiometer buried in a soil because soil conductance (K_s) is in series with K_c and usually $K_s \ll K_c$. In fact, an onsite time constant cannot be defined **(19)** because the response is not logarithmic due to a varying K_s during equilibration. However, the phrase "response time" is used to describe the rate of onsite response to pressure changes **(33)**. The term is not to be confused with the time constant because two tensiometers with equal time constants emplaced in the same soil can have different response times. For example, if $K_{c_1} = 10 K_{c_2}$ and $S_2 = 10 S_1$, then $\tau_1 = \tau_2$; but if $K_s \approx K_{c_2}$, then *response time$_1$ > response time$_2$*. Nonetheless, τ as defined here can be used comparatively to help evaluate tensiometer design. Greater sensitivity, large porous-cup surface area and permeability, and thin porous-cup walls are characteristics of a tensiometer with a short response time. Use of a sensitive pressure transducer is the most effective way to decrease response time in a soil of low hydraulic conductivity.

5.15 A bubble that interrupts hydraulic continuity between the porous cup and the pressure sensor will cause a change in the calculated value of P_W as follows:

$$\Delta = (E_P - E_C) \rho_{H_2O} \qquad (12)$$

where:
Δ = the change in the calculated value of P_W, in centimetres of water at 4°C,
E_P = the elevation of the end of the bubble nearest the pressure sensor, in centimetres,
E_C = is the elevation of the end of the bubble nearest the cup, in centimetres, and
ρ_{H_2O} = the density of water adjacent to the air bubble, in grams centimetre^{-3}.
If bubbles are detected and measured, the above correction(s) can be made to P_W as calculated in Eqs. 4 or 7. Small bubbles that cling to the wall of the tubing and do not block the entire cross-section do not affect the calculated value of P_W.

6. Procedure

6.1 *Construction and Applications:*

6.1.1 The definitions used to describe the quality of a measurement and used in Table 1 to compare types of tensiometers are given in Section 2.

6.1.2 The operating characteristics of commonly available tensiometers vary (see Table 1) and they need to be matched to the specific installation, cost constraints, and the desired quality of data collection. Complete tensiometers may be purchased from soils and agricultural research companies, made entirely from parts, or made from parts of commercial units modified to suit the user's needs. The advantages and

TABLE 1 Tensiometer Characteristics

Characteristic	Commercial		Constructed			
	Vacuum Gage	Manometer (Hybrid)	Manometer		Pressure Transducer	
			Small Diameter	Hybrid	Small Diameter	Hybrid
Accuracy	Poor	Excellent	Excellent	Excellent	Good to excellent	Good to excellent
Precision[A]	Poor	Good	Good	Good	Excellent	Excellent
Hysteresis	Poor	Excellent	Excellent	Excellent	Fair to excellent	Fair to excellent
Response time	Poor to excellent	Fair	Fair	Fair	Excellent	Excellent
Versatility of application	Fair	Fair	Excellent	Fair	Excellent	Fair
Durability	Good	Good	Good to excellent	Good	Good	Good
Purging	Seldom	Occasionally	Often	Occasionally	Often	Occasionally
Recalibration	Occasionally	Never	Never	Never	Often	Often
Data-collection method	Manual	Manual	Manual	Manual	Manual or automatic	Manual or automatic
Cost of Five[B]	$260.00	$200.00	$120.00[C]	$150.00	$410.00[D]	$440.00[D]

[A] Precision (repeatability) is rated for either a wetting or drying cycle to distinguish from hysteresis effects.
[B] Estimated for five 0.914 m (3-ft) deep tensiometers.
[C] Does not include cost of deaerating water.
[D] Does not include cost of deaerating water or recording equipment.

disadvantages of some of the different types are discussed in the following sub-sections and in Table 1.

6.1.3 Commercially available vacuum-gage-type units (see Fig. 2(b)) usually have a large diameter porous cup cemented to a rigid plastic tube of equal diameter (19 or 22 mm). A vacuum gage that indicates from 0 to 100 centibars of tension is screwed into the side of the tube, several centimetres below the top. The space between the vacuum gage and the top of the tube is a reservoir for air (the water may or may not be deaerated beforehand) to collect. When the water level inside the tube approaches the vacuum-gage inlet, the tube cap is unscrewed and the air space is refilled with water. Some vacuum-gage tensiometers have a large water reservoir connected to the top of the tube with a spring-loaded valve to simplify refilling.

6.1.4 The advantages of vacuum-gage tensiometers include simplicity of use, relatively low cost, and the maintenance of a hydraulic connection between the porous cup and gage, even with large quantities of air present. However, this last advantage is typically offset by the use of a vacuum gage with a resolution of 0.5 centibar (5 cm H_2O) and an overall accuracy of 3 centibars (31 cm H_2O). Response time is excellent immediately after removing all air, but it slows rapidly as the air reservoir fills up. Efforts can be made to minimize thermal effects on the air column by shielding it from the sun. The construction is fairly durable, but its rigidity can transfer shock and actually damage the porous cup, cup-tube bond, or hydraulic connection with the soil if the top is impacted after installation. Although the tube usually is installed vertically, it can be inclined to a nearly horizontal orientation as long as the zero adjustment of the vacuum gage is made at the same inclination. Installations greater than 45° from the vertical are more likely to have air accumulation problems.

6.1.5 A vacuum-gage tensiometer is used predominantly for irrigation scheduling where extreme accuracy is not necessary. It is not recommended for measurement of unsaturated hydraulic gradients (33). However, replacement of a standard vacuum gage with a more accurate, higher-resolution gage, or with an accurate pressure transducer, would improve the usability of the tensiometer.

6.1.6 In this guide, a tensiometer with a large diameter cup-tube assembly connected to the pressure sensor with small-diameter (3.2 mm, for example) tubing is referred to as a hybrid tensiometer (see Fig. 2(a)). Hybrid tensiometers, like vacuum-gage tensiometers, have a space at the top of the large tube to collect air. Hydraulic continuity is not broken, unless air bubbles block an entire cross-section of the small-diameter tubing.

6.1.7 Commercial manometer-type tensiometers commonly are hybrid types. Almost all of the air that enters the tensiometer through the porous cup collects harmlessly at the top. However, air also tends to be liberated from solution near the top of the manometer, where the maximum tension occurs; use of deaerated water minimizes air production.

6.1.8 A mercury manometer is probably the most accurate pressure-sensor commonly used in tensiometers and it never needs calibrating (a water manometer, usable only in special cases, is more accurate because of a better resolution). Hysterisis in a manometer tensiometer (from surface tension at the interface) or pressure-transducer tensiometer is usually much less than that in a vacuum-gage tensiometer. Thus, the hybrid-manometer tensiometer and pressure-transducer tensiometer combine fairly maintenance-free operation with excellent accuracy.

6.1.9 The major advantages of constructing a manometer tensiometer with small-diameter tubing are the versatility of onsite application, accuracy, and low cost. Flexible nylon tubing can be routed around obstructions, connecting a porous cup to a gage hundreds of feet away. Installation orientation is limited only by backfilling capabilities. A typical design (30) employs two 3.2 mm-diameter (nominal 1/8-in.-diameter) nylon tubes cemented with epoxy directly to a 9.5 mm-diameter (nominal 3/8-in.-diameter) porous cup (see Fig. 3(a)). The water-supply tube is connected via a shutoff valve to a deaerated water supply and the measurement tube is routed directly to the mercury reservoir. Manometer sensitivity (S, see Eq. 10) is the reciprocal of the cross-sectional area of the manometer tubing; response time is minimized by using small-diameter tubing. The design is simple, but the epoxy-nylon-tube bond is somewhat susceptible to rupture, from differential movement of the nylon tubes. A more robust design (see Fig. 3(b)) uses a larger porous cup and a metal plate to separate the nylon tubes,

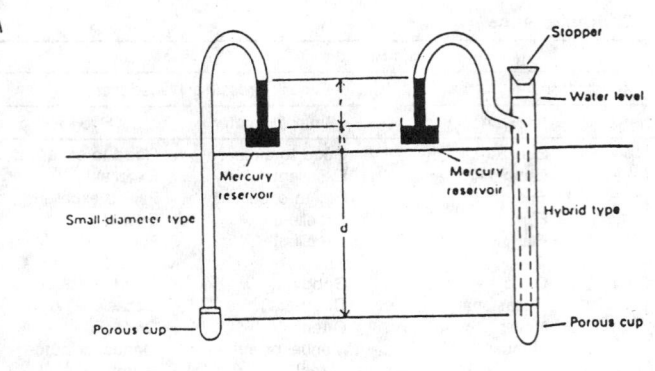

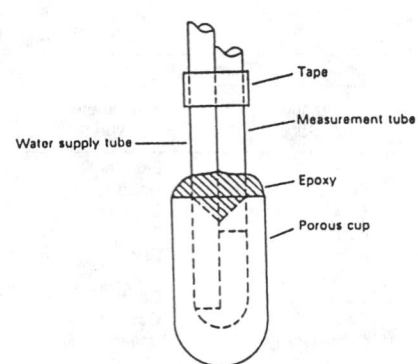

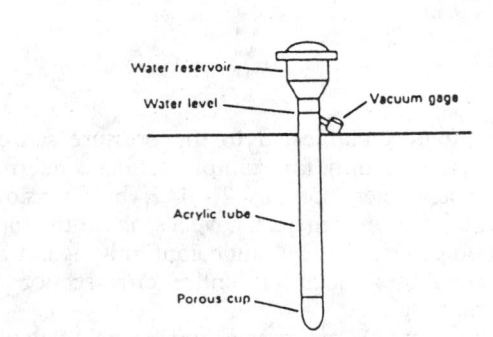

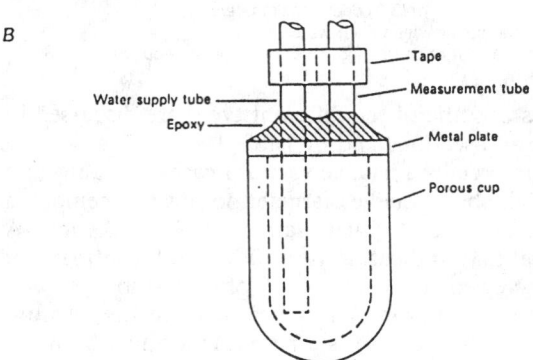

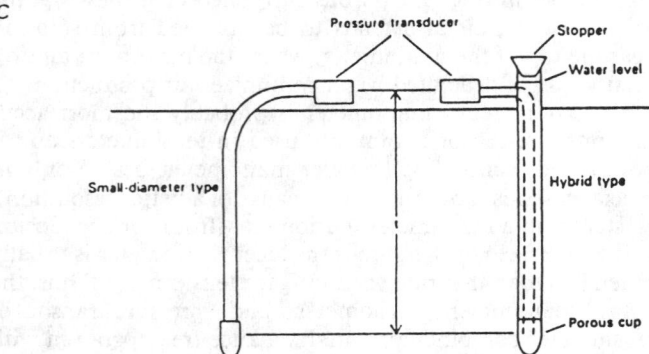

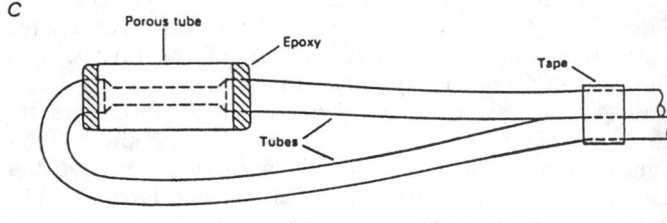

FIG. 3 Porous-Cup and Tube Designs

FIG. 2 Three Common Types of Tensiometers: (a) Manometer;
(b) Vacuum Gage; and (c) Pressure Transducer

allowing the epoxy to form a stronger bead around the tubes. A third design (see Fig. 3(c)) uses a porous tube, made by cutting the rounded end off a 6.4 mm-diameter (nominal ¼-in.-diameter) porous cup. One of the nylon tubes is molded into a "U" shape with heat, to produce the design in Fig. 3(c). The third design is extremely durable and versatile. The porous tube can be purged of air in any orientation, which is not entirely true of designs (a) and (b).

6.1.10 Tensiometers tend to collect air bubbles onsite that originate from the following: insufficiently deaerated water; air that diffuses through the water-filled pores of the porous cup; dissolved gases in the soil moisture that flow into the porous cup during a wetting cycle; and air that diffuses through the tubing material.

6.1.11 The major disadvantage of the small-diameter

designs is that air bubbles can easily block the entire cross-section of the tubing, interrupting hydraulic continuity. Thus, small-diameter designs require frequent purging of air—especially at large tensions. Use of a thick-walled tubing decreases diffusion through the tube wall. Some plastics (such as polyethylene) are relatively permeable to air and are unsatisfactory tubing material. Use of metal tubing almost eliminates air diffusion through the tube wall, however, other adverse affects can occur from the use of materials that have a high thermal conductivity.

6.1.12 A hybrid-manometer tensiometer can be constructed for two-thirds to three-fourths the price of a commercial unit. The cost in Table 1 was determined for five "cup-tube kits" purchased at a supply company and outfitted with a manometer made from parts. A cup-tube kit consists of a porous cup cemented to a stoppered acrylic tube with a fitting to accept 1.6 mm-diameter (nominal ⅟₁₆-in.-diameter) manometer tubing. The manometer tubing extends to the bottom of the acrylic tube. The cost could be decreased further by assembling the entire unit from basic parts.

6.1.13 Tensiometers equipped with pressure transducers are well-suited to collect large quantities of data. Measurements can be made often and recorded automatically by a data logger or a strip-chart recorder. The extreme sensitivity of the pressure transducer results in the shortest time constant obtainable, making this type of tensiometer ideal for tracking wetting fronts. However, the extreme sensitivity also results in the transducer becoming susceptible to measuring transient-temperature effects (34), caused by thermal expansion and contraction of water or air, or both, in the tensiometer. Water that freezes inside a pressure-transducer cavity will affect the calibration, and it can rupture the unit. Pressure transducers (and above-ground connecting tubing) can be enclosed in an insulated (and, if need be, heated) shelter or surface pit to minimize temperature fluctuations. The shelter or pit needs to be located so that it does not disturb the natural flow field at the porous cup(s).

6.1.14 Sensing elements of a typical pressure transducer are semiconductive resistors, embedded in a diaphragm that moves from applied pressure(s). As the resistors shorten or lengthen, their resistances change. Inherently, the resistance is a nonlinear function of pressure (or pressure difference) and temperature. The resistors are included in a modified Wheatstone bridge, excited by a regulated voltage (or current) source. The output of the bridge is a nearly linear function of pressure that is independent of temperature; however, all pressure transducers retain some nonlinearity and a slight temperature dependence (35). In addition, the zero offset and, to a lesser extent, the sensitivity may change with time (known as drift), possibly requiring recalibration at regular intervals (35). Other sources of error are repeatability and pressure and temperature hysteresis effects (35). Onsite application determines required pressure-transducer specifications; for example, gradient-measurement normally warrants a more accurate transducer than simple pressure measurement does.

6.1.15 A large degree of accuracy may be achieved in a variety of ways. A sophisticated pressure transducer costing seven to eight hundred dollars typically has an overall accuracy of 1 cm H_2O, at 23°C, and a temperature coefficient of 0.07 cm H_2O °C^{-1}. Without temperature correction, a worst-case error (at 0°C) of 2.6 cm H_2O could result. At the other extreme, a simple pressure transducer may be purchased for about fifty dollars and its output may be corrected for nonlinearity and temperature dependence by measuring temperature and applying a second-order polynomial fit to the measured data (36). The decreased linearity and temperature errors, combined with a typical repeatability and hysteresis error of 1.5 cm H_2O, produce a root-sum-square error of 1.8 cm H_2O. Such a transducer is listed in Table 1. Of course, the same curve-fitting procedure can be applied to a pressure transducer with better repeatability and hysteresis to achieve greater accuracy.

6.1.16 Accuracies determined in the two examples discussed in 6.1.15 further are degraded by drift or lack of long-term-stability. If a long-term-stability specification cannot be supplied by the pressure-transducer manufacturer, much of the data collected may be inaccurate. A long-term-stability specification is used to determine how often a transducer needs to be recalibrated to maintain desired accuracy. A hanging column of water or mercury (or a

calibrated vacuum source) is used for initial and then periodic recalibration.

6.1.17 Air that collects in an automated-tensiometer system can be purged manually or automatically at regular intervals using solenoid valves triggered by a data logger. Air collects more rapidly when tension is greater; onsite experience will determine the necessary time interval to maintain desired accuracy. The time interval can be maximized by using horizontal sections of tubing at high points in the system. Bubbles that collect in these horizontal sections do not cause errors but they do increase the response time.

6.1.18 Porous cups used in tensiometers remain saturated during normal operation. If the difference between air pressure outside the porous cup and water pressure inside the cup exceeds the bubbling pressure of the cup, air will displace the water in the largest pores and eventually will enter the interior of the cup. The more commonly used tensiometer cups have a bubbling pressure of 1 bar (1020 cm H_2O). Porous cups with bubbling pressures of 2, 3, and 5 bars are available, but they only have applications in the laboratory. If onsite tensions are known to not exceed 0.5 bar, a porous cup with a 0.5-bar bubbling pressure can be used to decrease response time. However, a better way to decrease response time is by the use of commonly available, "high-flow" porous cups, made with a pore-size distribution that emphasizes the larger pores. (Cost of a porous cup is between two and fifteen dollars. Larger porous cups are more expensive than smaller ones, and high-flow-cups are more expensive than standard-flow-cups.)

6.1.19 Specialized tensiometers have been developed to address specific onsite problems. Peck and Rabbidge (35) extended the upper limit of measurable tension by using a reference solution with a low osmotic potential and a porous cup coated with a semipermeable membrane. Using a pressure transducer, soil-water tensions as large as 153 m H_2O (15 bars) can be measured. The solution in the tensiometer usually is under positive pressure, thus decreasing the problem of air invasion. However, the pressure can be quite large and it could cause permanent "creep" of the transducer diaphragm. Depolymerization of the solute and subsequent loss through the membrane also could cause creep. Ambient temperature affects the osmotic potential directly (as qualitatively predicted by van't Hoff's law) and indirectly by causing flow of water through the membrane. Thus, an osmotic tensiometer is valuable for measuring extremely low matric potentials, but some sources of measurement error are unique to it.

6.1.20 The U.S. Forest Service developed an inexpensive recording tensiometer for use in remote areas where electric power is unavailable (31). This instrument can record as much as 1 month of continuous data on a battery-driven raingage chart. Oaksford (27) designed a unit based on a coaxial water manometer that provides maximum sensitivity at depths as deep as about 6 m. This unit uses a calibrated wire and ohmmeter to sense a below-ground, free-water surface inside the unit.

6.1.21 Fluid-scanning switches have been used successfully (20, 34, 37, 38) to connect as many as 22 tensiometers sequentially to a single pressure transducer. The approach minimizes cost and removes the bias between tensiometers caused by using different pressure transducers. Equally relevant, this network has the capability of measuring a zero and a full-scale calibration tension before each scan. This capability removes measurement errors from hysteresis, temperature dependence, and long-term drift. However, if the pressure transducer fails, data from all its tensiometers are lost. Also, the scanning switch is made with precise tolerances and it may develop leaks over time.

6.1.22 Another approach to efficient data collection with large numbers of tensiometers (25, 39) is to connect a portable pressure transducer to each tensiometer using a hypodermic needle and septum. The needle tip is inserted in an air space above the tensiometer fluid. A small change in tensiometer pressure is caused by the connection, probably affecting measurements by a few cm H_2O. Also, changes in the fluid-surface elevation, although small, affect measurements directly unless the changes are accounted for.

6.1.23 A small proportion of water in a frozen soil may remain liquid and, therefore, mobile. Measurement of tension in frozen soils can be accomplished using a pressure transducer and ethylene glycol (26) as a tensiometer fluid. The pressure transducer minimizes exchange of fluid between the soil and tensiometer, but slight bulk flow and diffusive flux do occur that decrease the freezing point of the soil fluid. The osmotic tensiometer probably will work well in frozen soils.

6.1.24 Porous cups have been connected directly to pressure transducers; the internal cavity has been filled with deaerated fluid and the entire assembly has been buried in a soil, with no provisions for purging of the fluid (19, 32). Although this approach provides a stable temperature environment, the pressure transducer cannot be recalibrated readily for drift. Also, soil gas diffusing through the porous cup creates an air pocket in the cavity. Because this pocket is at a pressure less than that of the soil gas, air will continue to diffuse through the porous-cup wall, eventually emptying the cup of water. A purging system is needed for extended undisturbed operation.

6.1.25 The specialized tensiometers developed thus far have resulted from insight and persistence of researchers faced with particular problems. Most of these solutions require extra effort, care, and expense; review of the cited report(s) is needed before implementation.

6.2 *Installation:*

6.2.1 Continuous hydraulic connection between the porous cup and soil (18), and minimal disturbance of the natural flow field are essential to collection of accurate tensiometric data. When a hole is made to accept the tensiometer, the cuttings need to be preserved in the order they were removed if the hole is to be backfilled.

6.2.2 Hydraulic connection can be established in several ways. Commercial tensiometers fit snugly into holes made with a coring tool (available as a tensiometer accessory) or made with standard iron pipe. The porous cup is forced against the hole bottom and no backfill is used. When the soil is rocky, or when a small-diameter tensiometer is installed, a hole larger in diameter than the porous cup is excavated. If the soil at the bottom of the hole is soft, the porous cup may be forced into the soil. A hard soil sometimes can be softened with water. If the porous cup will not penetrate the moistened soil, the last cuttings from the hole need to be used to backfill around the cup, either by

making them into a slurry or by careful tamping. A tremie pipe ensures clean delivery of cuttings or slurry to the bottom of the hole. Gaps between the porous cup and soil increase the tensiometer response time by reducing the effective area of the porous cup. In the worst case, no hydraulic connection occurs and the tensiometer will not indicate the soil-water tension.

6.2.3 If water is used to establish hydraulic connection, the tension adjacent to the porous cup will be reduced, and it will recover asymptotically as the added water is dispersed in the soil. The rate of dispersal will depend on the K (see Eq. 1) of the soil and a time series of tension will indicate when natural conditions are restored sufficiently.

6.2.4 If vertical profiles or gradients of pressure are to be measured, multiple small-diameter tensiometers can be installed in a single hole. Ideally, the original lithology is duplicated (except that gravel larger than 6 mm in diameter and cobbles and boulders can be removed) by using the cuttings in reverse order for backfilling. A backfill that is more compacted than the undisturbed soil will tend to shed infiltrating water, but, after a short time, the tension in the undisturbed soil and in the backfill will be in equilibrium. A backfill that is less compacted than the surrounding soil, or one with excessive gaps, will be a conduit for infiltrating water, resulting in abnormally low tensions in the backfill and in the undisturbed soil. Therefore, compaction of the backfill to a slightly greater bulk density than that of the undisturbed soil is desirable. Less permeable layers (such as clay lenses) need to be reproduced or even exaggerated by importing a fine-grained material.

6.2.5 Deep installation of tensiometers can be accomplished by drilling horizontal holes radially from a central caisson hole. This method also preserves undisturbed conditions above and below each porous cup. Backfilling horizontal holes with a tamping rod is painstaking. An alternative is to backfill in the vicinity of the porous cup and to fill the remainder of the hole with an expanding insulation foam. Pressure transducers work particularly well in caisson holes. If the entire length of tubing connecting the porous cup to the pressure transducer is horizontal, air bubbles do not cause errors. Also, the problem of transducer sensitivity to temperature change is minimized, because temperature in the caisson hole remains relatively constant. A caisson hole allows use of water manometers for improved precision because the entire manometer can be placed below the level of the porous cup.

6.3 *Operation:*

6.3.1 Testing the porous cup, the porous cup-tubing interface, all fittings, and the measurement and recording device(s), before installation, is desirable. After saturating the porous cup, apply air pressure to the interior of the tensiometer while the parts to be tested are submersed. If bubbles appear at a gage pressure substantially less than the bubbling pressure of the porous cup, the unit is faulty and the appropriate parts need to be repaired or replaced.

6.3.2 Water used in tensiometers is deaerated in a carboy by applying a vacuum or heat, or both. Excellent results have been obtained using a pump that generates 970 cm H_2O vacuum with a heated magnetic stirrer for 48 h. Insufficiently deaerated water requires frequent onsite purging, or, in the worst case, allows bubbles to form before a tensiometer has reached equilibrium, preventing accurate data collection. The deaerated water is siphoned (to minimize reaeration) from the carboy to a collapsible plastic container (available at a sporting goods store) for onsite use. Although the air above the water is forced out of the spout immediately, the water reaerates slowly by diffusion of air through the container wall. Proper fittings should be used to connect the container spout to the tensiometer supply tube. Use of water native to the soil being measured will minimize the effect of osmotic potential across the cup on the measured matric potential.

6.3.3 Purge a small-diameter tensiometer by connecting the water supply container to the water supply tube and raising it above the top of the tensiometer, while opening the supply valve. Several tensiometers can be connected in a "tee" network to simplify multiple purging. Purging instructions for vacuum-gage and hybrid tensiometers are supplied by the manufacturer. These instructions can be modified if use of deaerated water is desired. Purging time needs to be short to minimize wetting of the soil immediately surrounding the porous cup. When purging is complete, the system is closed and the soil draws water through the porous cup until equilibrium is established. The pressure inside the porous cup approaches the soil-water pressure asymptotically at a rate determined by the time constant and the unsaturated hydraulic conductivity of the soil. When equilibrium is reached, make the measurement. Record a single value of pressure when using a vacuum-gage or pressure transducer; a manometer requires measurement of the mercury column and reservoir elevations.

6.3.4 The most reliable data are obtained by purging a tensiometer and allowing it to equilibrate before recording the measurement. However, wet soils, lack of wetting fronts, low permeability tubing, or thoroughly deaerated water tend to prevent air accumulation for long periods; these conditions, either singly or in combination permit reliable data collection without purging.

6.3.5 Dry soils or inadequate manometer design, or both, occasionally result in mercury being pulled over the top of the manometer and into the porous cup. The porous cups shown in Fig. 3 may be purged of mercury by applying pressure to the measurement tube, forcing mercury out the supply tube. Pressure applied to the top of a hybrid tensiometer will force the mercury out the measurement tube.

6.3.6 Porous cups that are removed from a soil to be reused need to be washed with warm water to prevent plugging of the pores. A porous cup with plugged pores possibly can be restored by sanding or rinsing in a weak HCl solution (33).

REFERENCES

(1) Healy, R. W., Peters, C. A., DeVries, M. P., Mills, P. C., and Moffett, D. L., "Study of the Unsaturated Zone at a Low-Level Radioactive-Waste Disposal Site near Sheffield, Ill.," *Proceedings*, National Water Well Association Conference on Characterization and Monitoring of the Vadose Zone, Los Vegas, NV, Dec. 1983, pp. 820–831.

(2) McMahon, P. B., and Dennehy, K. F., "Water Movement in the Vadose Zone at Two Experimental Waste-Burial Trenches in South Carolina," *Proceedings*, National Water Well Association Conference on Characterization and Monitoring of the Unsaturated (Vadose) Zone, Denver, CO, Nov. 1985, pp. 34–54.

(3) Ripp, J. A., and Villaume, J. F., "A Vadose Zone Monitoring System for a Flyash Landfill," *Proceedings*, National Water Well Association Conference on Characterization and Monitoring of the Unsaturated (Vadose) Zone, Denver, CO, Nov. 1985, pp. 73–96.

(4) Ryan, B. J., DeVries, M. P., Garklavs, George, Gray, J. R., Healy, R. W., Mills, P. C., Peters, C. A., and Striegl, R. G., "Results of Hydrologic Research at a Low-Level Radioactive-Waste Disposal Site Near Sheffield, Illinois," *U.S. Geological Survey Water-Supply Paper 2367*, 1991.

(5) Lichtler, W. F., Stannard, D. I., and Kouma, E., "Investigation of Artificial Recharge of Aquifers in Nebraska," *U.S. Geological Survey Water-Resources Investigations Report 80-93*, 1980.

(6) Sophocleous, Marios, and Perry, C. A., "Experimental Studies in Natural Groundwater Recharge Dynamics: Analysis of Observed Recharge Events," *Journal of Hydrology*, Vol 81, 1985, pp. 297–332.

(7) Anderson, E. M., "Tipburn of Lettuce. Effect of Maturity, Air and Soil Temperature, and Soil Moisture Tension," Cornell Agricultural Experimental Station, *Bulletin*, No. 829, Cornell University, New York, 1946.

(8) Richards, L. A., and Neal, O. R., "Some Field Observations with Tensiometers," *Proceedings*, Soil Science Society of America (1936), Vol 1, 1937, pp. 71–91.

(9) Richards, S. J., Willardson, L. S., Davis, Sterling, and Spencer, J. R., "Tensiometer Use in Shallow Ground-Water Studies," *Proceed-*

ings, American Society of Civil Engineers, Vol 99, No. IR4, 1973, pp. 457–464.

(10) McKim, H. L., Walsh, J. E., and Arion, D. N., "Review of Techniques for Measuring Soil Moisture In Situ," *U.S. Army Corps of Engineers, Cold Regions Research and Engineering Lab Special Report 80-31*, 1980.

(11) Richards, L. A., Russell, M. B., and Neal, O. R., "Further Developments on Apparatus for Field Moisture Studies," *Proceedings*, Soil Science Society of America, Vol 2, 1938, pp. 55–64.

(12) Richards, L. A., "Methods of Measuring Soil Moisture Tension," *Soil Science*, Vol 68, 1949, pp. 95–112.

(13) Richards, L. A., "Capillary Conduction of Liquids through Porous Mediums," *Physics 1*, 1931, pp. 318–333.

(14) Baver, L. D., Gardner, W. H., and Gardner, W. R., *Soil Physics*, Wiley, New York, 1972, pp. 291–296.

(15) Brooks, R. H., and Corey, A. T., "Hydraulic Properties of Porous Media," *Hydrology Papers*, No. 3, Colorado State University, Fort Collins, CO, 1964.

(16) Marshall, T. J., "Relations Between Water and Soil," Commonwealth Bureau of Soil Science, England, *Technical Communication* No. 50, 1960.

(17) Haverkamp, R., Vauclin, M., Touma, J., Wierenga, P. J., and Vachaud, G., "A Comparison of Numerical Simulation Models for One-Dimensional Infiltration," *Soil Science Society of America Journal*, Vol 41, 1977, pp. 285–294.

(18) Richards, L. A., and Gardner, W., "Tensiometers for Measuring the Capillary Tension of Soil Water," *Journal of American Society of Agronomy*, Vol 28, 1936, pp. 352–358.

(19) Bianchi, W. C., "Measuring Soil Moisture Tension Changes," *Agricultural Engineering*, Vol 43, 1962, pp. 398–404.

(20) Bianchi, W. C., and Tovey, Rhys, "Continuous Monitoring of Soil Moisture Tension Profiles," *Transactions*, American Society of Agricultural Engineers, Vol 11, No. 3, 1968, pp. 441–447.

(21) Colman, E. A., Hanawalt, W. B., and Burck, C. R., "Some Improvements in Tensiometer Design," *Journal of the American Society of Agronomy*, Vol 38, 1946, pp. 455–458.

(22) Healy, R. W., deVries, M. P., and Striegl, R. G., "Concepts and Data-Collection Techniques Used in a Study of the Unsaturated Zone at a Low-Level Radioactive-Waste Disposal Site Near Sheffield, Illinois," *U.S. Geological Survey Water-Resources Investigations Report 85-4228*, 1986.

(23) Hunter, A. S., and Kelley, O. J., "Changes in the Construction of Soil Moisture Tensiometers for Field Use," *Soil Science*, Vol 61, 1946, pp. 215–218.

(24) Klute, A., and Peters, D. B., "A Recording Tensiometer with a Short Response Time," *Proceedings*, Soil Science Society of America, Vol 26, 1962, pp. 87–88.

(25) Marthaler, H. P., Vogelsanger, W., Richard, F., and Wierenga, P. J., "A Pressure Transducer for Field Tensiometers," *Soil Science Society of America Journal*, Vol 47, 1983, pp. 624–627.

(26) McKim, H. L., Berg, R. L., McGraw, R. W., Atkins, R. T., and Ingersoll, J., "Development of a Remote-Reading Tensiometer Transducer System for Use in Subfreezing Temperatures," *Proceedings*, American Geophysical Union Second Conference on Soil-Water Problems in Cold Regions, Edmonton, Alberta, 1976, pp. 31–45.

(27) Oaksford, E. T., "Water-Manometer Tensiometers Installed and Read from the Land Surface," *Geotechnical Testing Journal*, Vol 1, No. 4, 1978, pp. 199–202.

(28) Peck, A. J., and Rabbidge, R. M., "Design and Performance of an Osmotic Tensiometer for Measuring Capillary Potential," *Proceedings*, Soil Science Society of America, Vol 33, 1969, pp. 196–202.

(29) Richards, L. A., "Soil Moisture Tensiometer Materials and Construction," *Soil Science*, Vol 53, 1942, pp. 241–248.

(30) Stannard, D. I., "Theory, Construction, and Operation of Simple Tensiometers," *Ground Water Monitoring Review*, Vol 6, No. 3, 1986, pp. 70–78.

(31) Walkotten, W. J., "A Recording Soil Moisture Tensiometer," *U.S.D.A. Forest Service Research Note PNW-180*, 1972.

(32) Watson, K. K., "A Recording Field Tensiometer with Rapid Response Characteristics," *Journal of Hydrology*, Vol 5, 1967, pp. 33–39.

(33) Cassel, D. K., and Klute, A., "Water Potential: Tensiometry," American Society of Agronomy—Soil Science Society of America, *Agronomy Monograph* No. 9 (Second Edition). *Methods of Soil Analysis, Part 1. Physical and Mineralogical Methods*, 1986, pp. 563–596.

(34) Rice, Robert, "A Fast-Response, Field Tensiometer System," *Transactions*, American Society of Agricultural Engineers, Vol 12, 1969, pp. 48–50.

(35) Omega Engineering, Inc., *Omega 1987 Complete Pressure, Strain and Force Measurement Handbook and Encyclopedia*, Dec 1986.

(36) Micro Switch, Freeport, Illinois, *Pressure Sensors, Catalog 15*, 1985.

(37) Fitzsimmons, D. W., and Young, N. C., "Tensiometer-Pressure Transducer System for Studying Unsteady Flow Through Soils," *Transactions*, American Society of Agricultural Engineers, Vol 15, No. 2, 1972, pp. 272–275.

(38) Williams, T. H. Lee, "An Automatic Scanning and Recording Tensiometer System," *Journal of Hydrology*, Vol 39, 1978, pp. 175–183.

(39) Haldorsen, Sylvi, Petsonk, A. M., and Torstensson, B. A., "An Instrument for In Situ Monitoring of Water Quality and Movement in the Vadose Zone," *Proceedings*, National Water Well Association Conference on Characterization and Monitoring of the Unsaturated (Vadose) Zone, Denver, CO, Nov. 1985, pp. 158–172.

(40) Mualem, Yechezkel, "A New Model for Predicting the Hydraulic Conductivity of Unsaturated Porous Media," *Water Resources Research*, Vol 12, No. 3, 1976, pp. 513–522.

Standard Test Method for
Field Determination of Water (Moisture) Content of Soil by the Calcium Carbide Gas Pressure Tester Method[1]

This standard is issued under the fixed designation D 4944; the number immediately following the designation indicates the year of original adoption or, in the case of revision, the year of last revision. A number in parentheses indicates the year of last reapproval. A superscript epsilon (ϵ) indicates an editorial change since the last revision or reapproval.

1. Scope

1.1 This test method outlines procedures for determining the water (moisture) content of soil by chemical reaction using calcium carbide as a reagent to react with the available water in the soil producing a gas. A measurement is made of the gas pressure produced when a specified mass of wet or moist soil is placed in a testing device with an appropriate volume of reagent and mixed.

1.2 This test method is not intended as a replacement for Test Method D 2216; but as a supplement when rapid results are required, when testing is done in field locations, or where an oven is not practical for use. Test Method D 2216 is to be used as the test method to compare for accuracy checks and correction.

1.3 This test method is applicable for most soils. Calcium carbide, used as a reagent, reacts with water as it is mixed with the soil by shaking and agitating with the aid of steel balls in the apparatus. To produce accurate results, the reagent must react with all the water which is not chemically hydrated with soil minerals or compounds in the soil. Some highly plastic clay soils or other soils not friable enough to break up may not produce representative results because some of the water may be trapped inside soil clods or clumps which cannot come in contact with the reagent. There may be some soils containing certain compounds or chemicals that will react unpredictably with the reagent and give erroneous results. Any such problem will become evident as calibration or check tests with Test Method D 2216 are made. Some soils containing compounds or minerals that dehydrate with heat (such as gypsum) which are to have special temperature control with Test Method D 2216 may not be affected (dehydrated) in this test method.

1.4 This test method is limited to using calcium carbide moisture test equipment made for 20 g, or larger, soil specimens and to testing soil which contains particles no larger than the No. 4 Standard sieve size.

1.5 *This standard does not purport to address all of the safety problems, if any, associated with its use. It is the responsibility of the user of this standard to establish appropriate safety and health practices and determine the applicability of regulatory limitations prior to use. For specific hazards statements, see Section 7.*

2. Referenced Documents

2.1 *ASTM Standards:*
D 653 Terminology Relating to Soil, Rock, and Contained Fluids[2]
D 2216 Test Method for Laboratory Determination of Water (Moisture) Content of Soil, Rock, and Soil-Aggregate Mixtures[2]
E 11 Specification for Wire-Cloth Sieves for Testing Purposes[3]

3. Terminology

3.1 Definitions of terms used in this test method can be found in Terminology D 653.

4. Summary of Test Method

4.1 A measured volume of calcium carbide is placed in the testing apparatus along with two steel balls and a representative specimen of soil having all particles smaller than the No. 4 sieve size and having a mass equal to that specified by the manufacturer of the instrument or equipment. The apparatus is shaken vigorously in a rotating motion so the calcium carbide reagent can contact all the available water in the soil. Acetylene gas is produced proportionally to the amount of available water present. The apparent water content is read from a pressure gage on the apparatus calibrated to read in percent water content for the mass of soil specified.

4.2 A calibration curve is developed for each instrument and each soil type by plotting the pressure gage reading and the water content determined from Test Method D 2216 using representative specimens of the soil. The calibration curve is used to determine a corrected water content value for subsequent tests on the same type of soil.

5. Significance and Use

5.1 The water content of soil is used throughout geotechnical engineering practice, both in the laboratory and in the field. Results are sometimes needed within a short time period and in locations where it is not practical to install an oven or to transport samples to an oven. This test method is used for these occasions.

5.2 The results of this test have been used for field control of compacted embankments or other earth structures such as in the determination of water content for control of soil moisture and dry density within a specified range.

---

[1] This test method is under the jurisdiction of ASTM Committee D-18 on Soil and Rock and is the direct responsibility of Subcommittee D18.08 on Special and Construction Control Tests.

Current edition approved April 28, 1989. Published June 1989.

[2] *Annual Book of ASTM Standards*, Vol 04.08.
[3] *Annual Book of ASTM Standards*, Vol 14.02.

5.3 This test method requires specimens consisting of soil having all particles smaller than the No. 4 sieve size.

5.4 This test method may not be as accurate as other accepted methods such as Test Method D 2216. Inaccuracies may result because specimens are too small to properly represent the total soil, from clumps of soil not breaking up to expose all the available water to the reagent and from other inherent procedural, equipment or process inaccuracies. Therefore, other methods may be more appropriate when highly accurate results are required, or when the use of test results is sensitive to minor variations in the values obtained.

6. Apparatus

6.1 *Calcium Carbide Pressure Tester Set* (including testing chamber with attached pressure gage and a set of tared balances), for water content testing of specimens having a mass of at least 20 g, (10 g for the half measure required for wetter specimens). Testers that use a smaller mass are available, but are considered too inaccurate for this standard. The testing chamber with pressure gage and the balances are calibrated as a set (see Section 8). A typical apparatus is shown in Fig. 1.

6.2 *Small Scoop*, for measuring reagent.

6.3 *Two Steel Balls*, (manufacturer supplied).

6.4 *Brush and Cloth*, for cleaning and other incidental items.

6.5 *Sieve*, No. 4 (4.75 mm), conforming to the requirements of Specification E 11.

6.6 *Supply of Calcium Carbide*, meeting the requirements of the manufacturer for use in the equipment. The particle size of the calcium carbide shall be no coarser than the No.

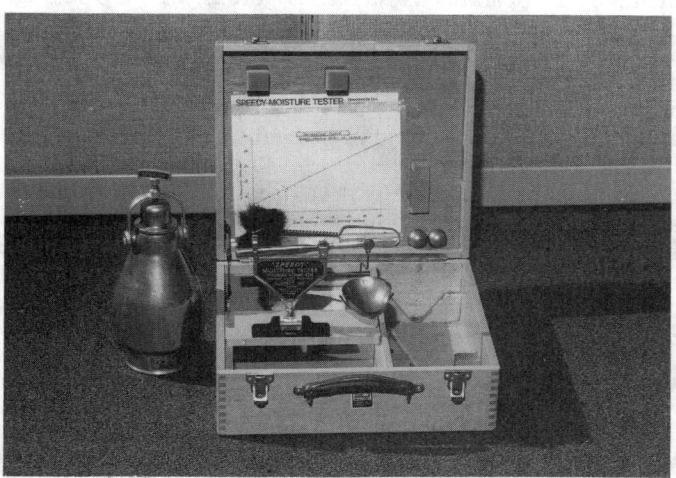

FIG. 1 Typical Calcium Carbide Gas Pressure Test Apparatus for Water Content of Soil

50 sieve (300 μm) and no more than 10 % shall pass the No. 140 sieve (106 μm). It is best to purchase calcium carbide in small containers with air tight replaceable lids, to store it in a dry place, to keep the lid on the container at all times except

when measuring out a portion for use in a test, and to use a complete container before opening a new one. Calcium carbide quality will deteriorate with time after it becomes exposed to the atmosphere or any source of moisture. Periodic purchase of a new supply is recommended.

6.7 *Miscellaneous Clothing or Safety Equipment*, such as goggles to protect the operator (see 7.2).

6.8 *Equipment*, as listed in Test Method D 2216, for performing comparison tests to make calibration curves.

NOTE 1—Calibration kits are available from manufacturers for testing gasket leakage and for calibrating the gage. Periodic checks for gasket leakage are recommended. The gasket should be changed when leakage is suspected. Gage calibration problems can usually be detected as the instrument calibration curves are made (see Section 8). When the gage needs adjusting, any good quality calibrating gage can be used.

7. Safety Hazards

7.1 When combined with water, the calcium carbide reagent produces a highly flammable or explosive acetylene gas. Testing should not be carried out in confined spaces or in the vicinity of an open flame, embers or other source of heat that can cause combustion. Care should be exercised when releasing the gas from the apparatus to direct it away from the body. Lighted cigarettes, hot objects or open flames are extremely dangerous in the area of testing.

7.2 As an added precaution, the operator should use a dust mask, clothing with long sleeves, gloves and goggles to keep the reagent from irritating the eyes, respiratory system, or hands and arms.

7.3 Attempts to test excessively wet soils or improper use of the equipment, such as adding water to the testing chamber, could cause pressures to exceed the safe level for the apparatus. This may cause damage to the equipment and an unsafe condition for the operator.

7.4 Care should be taken not to dispose or place a significant amount of the calcium carbide reagent where it may contact water because it will produce an explosive gas.

8. Calibration

8.1 The manufacturer-supplied equipment set, including the testing chamber with attached gage and the balance scales, are calibrated as a unit and paired together for the testing procedure.

8.2 Calibration curves must be developed for each equipment set using the general soil types to be tested and the expected water content range of the soil. As new materials are introduced, further calibration is needed to extend the curve data for the specific instrument. If tests are made over a long period of time on the same soil, a new calibration curve should be made periodically, not exceeding 12 months.

8.3 Calibration curves are produced by selecting several samples representing the range of soil materials to be tested and having a relatively wide range of water content. Each sample is carefully divided into two specimens by quartering procedures or use of a sample splitter. Taking care to not lose any moisture, one specimen is tested in accordance with the procedure of this test method (see 10.1 through 10.6) without

CALIBRATION CURVE

CALCIUM CARBIDE TESTER NO. 721915

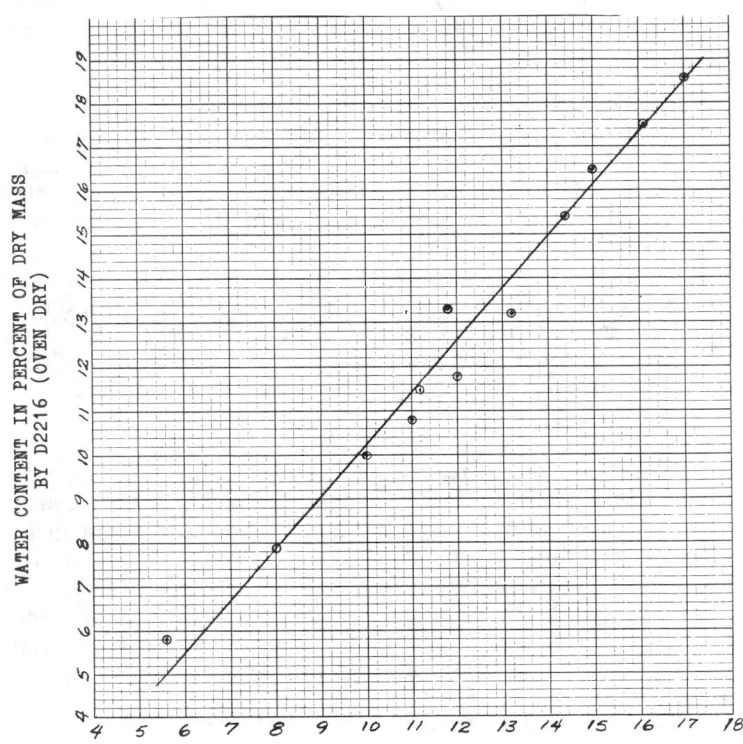

FIG. 2 Typical Calibration Curve

using a calibration curve, and the other specimen is tested in accordance with Test Method D 2216.

8.4 The results of the oven dry water content determined by Test Method D 2216 from all the selected samples are plotted versus the gage reading from the calcium carbide tester for the corresponding test specimen pair. A best fit curve is plotted through the points to form a calibration curve for each soil type. Comparisons should be relatively consistent. A wide scatter in data indicates that either this test method or Test Method D 2216 is not applicable to the soil or conditions. Figure 2 shows a typical calibration curve.

8.5 A comparison of this test method with Test Method D 2216 for a given soil can be made by using the calibration curve. Points that plot off the curve indicate deviations. Standard and maximum deviations can be determined if desired.

9. Sampling

9.1 For water content testing being done in conjunction with another method (such as Test Method D 2216), the requirements for sample and test specimen selection and handling in the other standard shall govern.

9.2 Equipment limitations require the use of specimens smaller than is recommended to properly represent the total soil. Extra care must be exercised to select specimens that are representative of the soil.

9.3 Specimens are to contain only soil particles smaller than the No. 4 Standard sieve size.

10. Procedure

10.1 Remove the cap from the testing chamber of the apparatus and place the recommended amount of calcium carbide reagent along with the two steel balls into the testing chamber. Most equipment built to test 20-g samples requires approximately 22 g of reagent (measured using the supplied scoop, which is filled two times).

10.2 Use the balance to obtain a specimen of soil that has a mass recommended for the equipment and contains particles smaller than the No. 4 sieve size. One-half specimen size should be used when the water content is expected to exceed the limits of the gage on the gas pressure chamber or when it actually reaches or exceeds the gage limit in any test (see 10.6).

10.3 Place the soil specimen in the testing chamber cap; then, with the apparatus in the horizontal position, insert the cap in the testing chamber and tighten the clamp to seal the cap to the unit. Take care that no calcium carbide comes in contact with the soil until a complete seal is achieved.

NOTE 2—The soil specimen may be placed in the chamber with the calcium carbide in the cap if desired.

10.4 Raise the apparatus to the vertical (upright) position so that the contents of the cap fall into the testing chamber.

Strike the side of the apparatus with an open hand to assure that all the material falls out of the cap.

10.5 Shake the apparatus vigorously with a rotating motion so that the steel balls roll around the inside circumference and impact a grinding effect on the soil and reagent. This motion also prevents the steel balls from striking the orifice that leads to the pressure gage. Shake the apparatus for at least 1 min for sands, increasing the time for silts, and up to 3 min for clays. Some highly plastic clay soils may take more than 3 min. Periodically check the progress of the needle on the pressure gage dial. Allow time for the needle to stabilize as the heat from the chemical reaction is dissipated.

10.6 When the pressure gage dial needle stops moving, read the dial while holding the apparatus in the horizontal position. If the dial goes to the limit of the gage, 10.1 through 10.6 should be repeated using a new specimen having a mass half as large as the recommended specimen. When a half size specimen is used, the final dial reading is multiplied by two for use with the calibration curve.

10.7 Record the final pressure gage dial reading and use the appropriate calibration curve to determine the corrected water content in percent of dry mass of soil and record.

10.8 With the cap of the testing chamber pointed away from the operator, slowly release the gas pressure (see Section 7). Empty the chamber and examine the specimen for lumps. If the material is not completely pulverized, the test should be repeated using a new specimen.

10.9 Clean the testing chamber and cap with a brush or cloth and allow the apparatus to cool before performing another test. Repeated tests can cause the apparatus to heat up which will affect the results of the test. The apparatus should be at about the same temperature as it was during calibration (determined by touch). This may require warming the instrument up to calibration temperature before use when the temperature is cold.

10.10 Discard the specimen where it will not contact water and produce an explosive gas. It is recommended that the specimen soil not be used for further testing as it is contaminated with the reagent.

11. Report

11.1 Report the following information:

11.1.1 Test number assigned and identification of the sample by location (segment of the project, station, elevation, zone or feature) and by classification or description of the material,

11.1.2 Apparatus identification by number,

11.1.3 Specimen mass and final pressure gage dial reading from the apparatus, and

11.1.4 Water content of the sample (from the calibration curve) to the nearest 1 %.

12. Precision and Bias

12.1 The precision of this test method has not been determined. Limited data are being evaluated to determine the precision of this test method. Subcommittee D18.08 is seeking pertinent data from users of this test method.

12.2 There is no accepted reference value for this test method; therefore, bias cannot be determined. Deviations from Test Method D 2216 can be determined from calibration curves (see 8.5).

13. Keywords

13.1 acceptance test; gas pressure; moisture content; pressure-measuring instrument; quick test; soil moisture; water content

Standard Test Method for
Water Content of Soil and Rock in Place by Nuclear Methods (Shallow Depth)[1]

This standard is issued under the fixed designation D 3017; the number immediately following the designation indicates the year of original adoption or, in the case of revision, the year of last revision. A number in parentheses indicates the year of last reapproval. A superscript epsilon (ϵ) indicates an editorial change since the last revision or reapproval.

This standard has been approved for use by agencies of the Department of Defense. Consult the DoD Index of Specifications and Standards for the specific year of issue which has been adopted by the Department of Defense.

1. Scope

1.1 This test method covers the determination of water content of soil and rock by the thermalization or slowing of fast neutrons where the neutron source and the thermal neutron detector both remain at the surface.

1.2 The water content in mass per unit volume of the material under test is determined by comparing the detection rate of thermalized or slow neutrons with previously established calibration data.

1.3 The values stated in SI units are to be regarded as the standard. The inch-pound equivalents may be approximate.

1.3.1 It is common practice in the engineering profession to concurrently use pounds to represent both a unit of mass (lbm) and of force (lbf). This implicitly combines two systems of units, that is, the absolute system and the gravitational system. This standard has been written using the absolute system for water content (kilograms per cubic metre) in SI units. Conversion to the gravitational system of unit weight in lbf/ft³ may be made by multiplying by 0.06243 or in kN/m³ by multiplying by 9.807. The recording of water content in pound-force per cubic foot should not be regarded as non-conformance with this standard although the use is scientifically incorrect.

1.4 *This standard does not purport to address all of the safety concerns, if any, associated with its use. It is the responsibility of the user of this standard to establish appropriate safety and health practices and determine the applicability of regulatory limitations prior to use.*

2. Referenced Documents

2.1 *ASTM Standards:*
D 1556 Test Method for Density of Soil in Place by the Sand-Cone Method[2]
D 2167 Test Method for Density and Unit Weight of Soil in Place by the Rubber Balloon Method[2]
D 2216 Test Method for Laboratory Determination of Water (Moisture) Content of Soil, Rock, and Soil-Aggregate Mixtures[2]
D 2922 Test Methods for Density of Soil and Soil Aggregate and Rock in Place by Nuclear Methods (Shallow Depth)[2]

D 2937 Test Method for Density of Soil in Place by the Drive-Cylinder Method[2]
D 4643 Test Method for Determination of Water (Moisture) Content of Soil by the Microwave Oven Method[2]
D 4718 Practice for Correction of Unit Weight and Water Content for Soils Containing Oversize Particles[2]

3. Significance and Use

3.1 The test method described is useful as a rapid, nondestructive technique for the in-place determination of water content of soil and rock.

3.2 The test method is used for quality control and acceptance testing of compacted soil and rock for construction and for research and development. The non-destructive nature allows repetitive measurements at a single test location and statistical analysis of the results.

3.3 The fundamental assumptions inherent in the test method are that the hydrogen present is in the form of water as defined by Test Method D 2216, and that the material under test is homogeneous.

3.4 Test results may be affected by chemical composition, sample heterogeneity, and, to a lesser degree, material density and the surface texture of the material being tested. The technique also exhibits spatial bias in that the apparatus is more sensitive to water contained in the material in close proximity to the surface and less sensitive to water at deeper levels.

4. Interferences

4.1 The chemical composition of the sample may dramatically affect the measurement and adjustments may be necessary. Hydrogen in forms other than water, as defined by Test Method D 2216, and carbon will cause measurements in excess of the true value. Some chemical elements such as boron, chlorine, and minute quantities of cadmium will cause measurements lower than the true value.

4.2 The water content determined by this test method is not necessarily the average water within the volume of the sample involved in the measurement. The measurement is heavily influenced by the water content of the material closest to the surface. The volume of soil and rock represented in the measurement is indeterminate and will vary with the water content of the material. In general, the greater the water content of the material, the smaller the volume involved in the measurement. At 160 kg/m³ (10 lbf/ft³), approximately 50 % of the typical measurement results from

---

[1] This test method is under the jurisdiction of ASTM Committee D-18 on Soil and Rock and is the direct responsibility of Subcommittee D18.08 on Special and Construction Control Tests.
Current edition approved Oct. 10, 1996. Published February 1997. Originally published as D 3017 – 72. Last previous edition D 3017 – 88 (1993)$^{\epsilon 1}$.
[2] *Annual Book of ASTM Standards*, Vol 04.08.

the water content of the upper 50 to 75 mm (2 to 3 in.).

4.2.1 If samples of the measured material are to be taken for purposes of correlation with other test methods or rock correction, the volume measured can be approximated by a 200-mm (8 in.) diameter cylinder located directly under the center line of the fast neutron source and thermal neutron detector. The height of the cylinder to be excavated is approximated by:

Moisture Content		Cylinder Height		Volume	
kg/m³	lbf/ft³	mm	in.	m³	ft³
80	5	250	10	0.0079	0.29
160	10	200	8	0.0063	0.23
240	15	150	6	0.0047	0.17
320	20	125	5	0.0039	0.15
400	25	112	4.5	0.0035	0.13
480	30	100	4	0.0031	0.12

NOTE 1—The volume of field compacted material sampled by the test can effectively be increased by repeating the test at immediately adjacent (vertically or horizontally) locations and averaging the results.

4.3 Other neutron sources must not be within 8 m (25 ft) of equipment in operation.

5. Apparatus

5.1 While exact details of construction of the apparatus may vary, the system shall consist of:

5.1.1 *Fast Neutron Source*—A sealed mixture of a radioactive material such as americium or radium and a target material such as beryllium.

5.1.2 *Slow Neutron Detector*—Any type of slow neutron detector such as boron trifluoride or helium-3 proportional counter.

5.1.3 *Readout Device*—A suitably timed scaler(s). Usually the readout device will contain the high-voltage supply necessary to operate the detector, and low-voltage power supply to operate the readout and accessory devices.

5.1.4 *Housing*—The source, detector, readout device, and power supply shall be in housings of rugged construction which shall be water and dust resistant.

5.1.5 *Reference Standard*—A block of hydrogeneous material for checking equipment operation and to establish conditions for a reproducible count rate.

5.1.6 *Site Preparation Device*—A steel plate, straightedge, or other suitable leveling tools which may be used to plane the test site to the required smoothness.

5.2 Calibrate apparatus in accordance with Annex A1.

5.3 Determine the precision of the apparatus in accordance with Annex A2.

6. Hazards

6.1 This equipment utilizes radioactive materials which may be hazardous to the health of the users unless proper precautions are taken.

6.2 Effective operator instruction together with routine safety procedures such as source leak tests, recording and evaluation of film badge data, use of survey meters, etc., are a recommended part of the operation of equipment of this type.

7. Standardization

7.1 All nuclear water content instruments are subject to long-term aging of the radioactive source, detectors, and electronic systems, which may change the relationship between count rate and water content. To offset this aging instruments are calibrated as a ratio of the measurement count rate to a count rate made on a reference standard. The reference count rate should be in the same or higher order of magnitude than the range of measurement count rates over the useful water range of the equipment.

7.2 Standardization of equipment on the reference standard is required at the start of each day's use and a permanent record of these data shall be retained. The standardization shall be performed with the equipment located at least 8 m (25 ft) away from other gages and clear of large masses of water or other items which may affect the gage readings.

7.2.1 Turn on the instrument and allow for stabilization in accordance with the manufacturer's recommendations. If the instrument is to be used either continuously or intermittently during the day, it is generally best to leave it in the "power on" condition to prevent having to repeat the stabilization. This will provide more stable, consistent results.

7.2.2 Using the reference standard take at least four repetitive readings at the normal measurement period and obtain the mean. If available on the instrument, one measurement at a period of four or more times the normal period is acceptable. This constitutes one standardization check.

7.2.3 If the value obtained above is within the limits stated below, the equipment is considered to be in satisfactory condition and the value may be used to determine the count ratios for the day of use. If the value obtained is outside these limits, another standardization check should be made. If the second standardization check is within the limits, the equipment may be used, but if it also fails the test, the equipment shall be adjusted or repaired as recommended by the manufacturer.

$$N_s \leqq N_o + \frac{2.0 \sqrt{N_o}}{\sqrt{F}} \qquad (1)$$

and

$$N_s \geqq N_o - \frac{2.0 \sqrt{N_o}}{\sqrt{F}} \qquad (2)$$

where:

N_s = value of current standardization check (7.2.2) on the reference standard,

N_o = average of the past four values of N_s taken for prior usage, and

F = value of prescale (A2.2.1).

7.3 The value of N_s will be used to determine the count ratios for the current day's use of the equipment. If, for any reason, measured water content becomes suspect during the day's use, perform another standardization.

8. Procedure

8.1 Standardize the instrument (Section 7).

8.2 Select a location for test where the instrument in test position will be at least 250 mm (10 in.) away from any vertical projection.

8.3 Prepare the test site in the following manner:

8.3.1 Remove all loose and disturbed material, and remove additional material as necessary to reach the top of the vertical interval to be tested. Surface drying and the spatial bias should be considered in determining the depth at which the instrument is to be seated.

8.3.2 Prepare a horizontal area, sufficient in size to accommodate the instrument, by planing to a smooth condition so as to obtain maximum contact between the instrument and material being tested. If the instrument base is to be placed below the surrounding surface, the horizontal area shall be at least twice the area of the base of the instrument. If the depression is greater than 25 mm (1 in.), the condition in 8.2 must be met by clearing a larger area.

8.3.3 The placement of the instrument on the surface of the material to be tested is critical to the successful determination of water. The optimum condition is total contact between the bottom surface of the instrument and the surface of the material being tested. The maximum void beneath the instrument shall not exceed approximately 3 mm (⅛ in.). Use native fines of similar water content or dry quartz sand to fill voids and level the excess with a rigid plate or other suitable tool. The total area filled shall not exceed 10 % of the bottom area of the instrument.

8.4 Proceed with the test in the following manner:

8.4.1 Seat the instrument firmly, place the source in the proper position and take a count for the normal measurement period.

8.4.2 Determine the ratio of the reading to the standard count (Section 7). From this ratio and the calibration and adjustment data, determine the in-place water content per unit volume (Note 2).

NOTE 2—Some instruments have built-in provisions to compute the ratio, the water content per unit volume with adjustments, the dry density, and the water content in percent of dry density (or dry unit weight).

8.5 If the volume tested as defined in 4.2.1 is insufficient for the size of rock contained in the soil (refer to Practice D 4718), take additional tests at adjacent locations and average the results (Note 3).

NOTE 3—The water content value obtained should be compared to other water contents obtained for similar soils and conditions. The presence of a large rock particle or void in the soil being tested may give an unrepresentative value of water content. If the value is unusually high or low, another determination of water content should be performed. To avoid preparation of another test site, the gage may be repositioned (such as rotating the gage 90°) at the original site.

9. Calculation

9.1 Calculate the water content, w, in percent of dry density (or dry unit weight) of soil as follows:

$$w = \frac{M_m \times 100}{\rho_d} \qquad (3)$$

or

$$w = \frac{M_m \times 100}{\rho - M_m} \qquad (4)$$

where:

w = water content, percent of dry density,

TABLE 1 Results of Statistical Analysis

Precision and Soil Type	Average lb/ft³ (kg/m³)	Standard Deviation, lb/ft³ (kg/m³)	Acceptable Range of Two Results lb/ft³ (kg/m³)
Single Operator Precision:			
CL	12.1 (193.8)	0.35 (5.6)	0.97 (15.5)
SP	18.7 (299.5)	0.46 (7.4)	1.29 (20.7)
ML	19.6 (314.0)	0.35 (5.6)	0.99 (15.8)
Multilaboratory Precision:			
CL	12.1 (193.8)	0.52 (8.3)	1.44 (23.1)
SP	18.7 (299.5)	0.75 (12.0)	2.10 (33.6)
ML	19.6 (314.0)	0.58 (9.3)	1.63 (26.1)

M_m = water content, kg/m³ (lbf/ft³),

ρ_d = dry density of soil (kg/m³) or dry unit weight (lbf/ft³), and

ρ = wet (total) density of soil (kg/m³) or wet unit weight (lbf/ft³).

10. Report

10.1 Report the following information:

10.1.1 Make, model, and serial number of the test device,

10.1.2 Standard count and adjustment data for the date of the tests,

10.1.3 Name of the operator,

10.1.4 Test site identification,

10.1.5 Visual description of material tested,

10.1.6 Count rate for each reading, if applicable,

10.1.7 Water content in kg/m³ or lbf/ft³,

10.1.8 Wet and dry densities in kg/m³ or unit weights in lbf/ft³,

10.1.9 Water content in percent of dry density or dry unit weight.

11. Precision and Bias

11.1 *Precision*—Criteria for judging the acceptability of the water content results obtained by this test method are given in Table 1. The value in column two is in the units actually measured by the nuclear gage. The figures in column three represent the standard deviations that have been found to be appropriate for the materials of test given in column one. The figures given in column four are the limits that should not be exceeded by the difference between the results of two properly conducted tests. The figures given are based upon an interlaboratory study in which five test sites containing soils, with water content as shown in column two, were tested by eight different devices and operators. The water content of each test site was determined three times by each device.[3]

12. Keywords

12.1 compaction test; construction control; density; field control; inspection; moisture content; moisture control; nuclear methods; nuclear moisture; quality control; soil moisture; test procedure; water content

---

[3] Details of the study are contained in a Research Report available from ASTM Headquarters. Request RR:D18-1005.

ANNEXES

(Mandatory Information)

A1. CALIBRATION

A1.1 *Calibration Curves*—Calibration curves, tables, or equations shall be established or verified once each year by determining by test the nuclear count rate of at least two samples of different known water content. This data may be plotted or equations determined to create tables or computer programs for conversion of count data to water content. The method and test procedures used in obtaining the count rate (or ratio as defined in Section 7) to establish the calibration relationship must be the same as those used for measuring the water content of the material to be tested. The water content of materials used to establish the calibration must vary through a range to include the water content of materials to be tested and be in the density range of 1600 to 2240 kg/m³ (100 to 140 lbf/ft³). Due to the effect of chemical composition, the calibration supplied by the manufacturer with the apparatus will not be applicable to all materials. It shall be accurate for mixes of silica sand and water; therefore, the calibration must be checked and adjusted, if necessary, in accordance with A1.3.

A1.2 *Calibration Standards*—Calibration standards may be established using any of the following methods. Prepared containers or standards must be large enough to not change the observed count rate (or ratio as defined in Section 7) if made larger in any dimension.

NOTE A1.1—Dimensions of approximately 610 mm long by 460 mm wide by 360 mm deep (approximately 24 by 18 by 14 in.) have proven satisfactory.

A1.2.1 Prepare a homogeneous standard of hydrogenous materials having an equivalent water content determined by comparison (using a nuclear instrument) with a saturated silica sand standard prepared in accordance with A1.2.2. A metallic density standard as defined in Test Methods D 2922 is a convenient zero water content standard.

A1.2.2 Prepare containers of compacted material with a percent water content determined by oven dry (Test Method D 2216) and a wet density calculated from the mass of the material and the inside dimensions of the container. The water content may be calculated as follows:

$$M_m = \frac{\rho \times w}{100 + w} \qquad (A1.1)$$

where:
M_m = water content, kg/m³ or lbf/ft³,
w = water content, percent of dry mass, and
ρ = wet (total) density, kg/m³ or lbf/ft³.

A1.2.3 Where neither of the previous calibration standards are available, the instrument may be calibrated by using a minimum of four selected test sites in an area of a compaction project where a homogenous material has been placed at several different water contents. The test sites shall represent the range of water contents over which the calibration is to be used. At least four replicate nuclear measurements shall be made at each test site. The density at each site shall be measured by making four closely spaced determinations with calibrated equipment in accordance with Test Methods D 2922, Test Method D 1566, Test Method D 2167, or Test Method D 2937. The water content of each of the density tests shall be determined by Test Method D 2216. Use the mean value of the replicate readings as the calibration point value for each site.

A1.3 *Calibration Adjustments:*

A1.3.1 The calibration of newly acquired or repaired instruments shall be verified and adjusted prior to use. Calibration curves shall be checked prior to performing tests on materials that are distinctly different from material types previously used in obtaining or adjusting the calibration. Sample materials may be selected by either A1.3.1.1 or A1.3.1.2. The amount of water shall be within ±2 % of the water content established as optimum for compaction. Determine the water content in kg/m³ or lbf/ft³ by Eq A1.1. A microwave oven or direct heater may be utilized for drying materials which are not sensitive to temperature, in addition to the methods listed in A1.2.3. A minimum of four comparisons is required and the mean of the observed differences used as the correction factor.

A1.3.1.1 Container(s) of compacted material taken from the test site may be prepared in accordance with A1.2.2.

A1.3.1.2 Test site(s) on the compacted material may be selected in accordance with A1.2.3.

A1.3.2 The method and test procedures used in obtaining the count rate (or ratio as defined in Section 7) to establish the error must be the same as those used for measuring the water content of the material to be tested.

A1.3.3 The mean value of the difference between the moisture content of the test samples as determined in A1.3.1.1 or A1.3.1.2 and the values measured with the instrument shall be used as a correction to measurements made in the field. Some instruments utilizing a microprocessor may have provision to input a correction factor that is established by the relative values of water content as a percentage of dry density, thus eliminating the need to determine the difference in mass units of water.

A2. DETERMINING PRECISION OF APPARATUS

A2.1 The precision of the apparatus at a water content of 160 kg/m³ (10 lbf/ft³) shall be better than 4 kg/m³ (0.25 lbf/ft³) at the manufacturer's stated period of time for the measurement (Note A2.1). Other timing periods may be available (usually multiples of four of the normal period) which may be used where higher or lower precisions are desired for statistical purposes. The precision shall be determined by the procedure defined in 5.2.1 or 5.2.2.

NOTE A2.1—While 1 min is the usual timing period and may be used for the comparison of various apparatus, the intent of the test method is to require a measurement period that produces the stated precision for all acceptance testing.

A2.2 The precision of the apparatus is determined from the slope of the calibration response and the statistical deviation of the count (detected neutrons) for the period of measurement, as follows:

$$P = \frac{\sigma}{S} \qquad (A2.1)$$

where:

P = apparatus precision in water content (kg/m³ or lbf/ft³),

σ = standard deviation in counts per measurement period, and

S = slope in change in counts per measurement period divided by the change in water content (kg/m³ or lbf/ft³).

A2.2.1 The count per measurement period shall be the total number of thermal neutrons detected during the timed period. The displayed value must be corrected for any prescaling which is built into the apparatus. The prescale value (F) is a divisor which reduces the actual value for the purpose of display. The manufacturer will supply this value if other than 1.0.

A2.2.2 The standard deviation in counts per measurement period shall be obtained as follows:

$$\sigma = \frac{\sqrt{C}}{\sqrt{F}} \qquad (A2.2)$$

where:

σ = standard deviation in counts per measurement period,

C = counts per measurement period (before prescale correction) at a water content of 160 kg/m³ (10 lbf/ft³), and

F = value of prescale (A2.2.1).

A2.2.3 The counts per measurement period (before prescale correction) shall be obtained form the calibration curve, tables or equation.

A2.2.4 The slope of calibration response in counts per measurement period (before prescale correction) at a water content of 160 kg/m³ (10 lbf/ft³) shall be determined from the calibration curve, tables, or equation.

A2.3 The precision shall be computed by determining the standard deviation of at least 20 repetitive measurements (instrument not moved after the first measurement) on material having a water content of 130 to 190 kg/m³ (8.1 to 11.9 lbf/ft³). In order to perform this procedure, the resolution of the count display, calibration response, or other method of displaying water content must be equal to or better than ±1 kg/m³ or 0.1 lbf/ft³.

Standard Test Method for
Water Content of Soil and Rock In-Place by the Neutron Depth Probe Method[1]

This standard is issued under the fixed designation D 5220; the number immediately following the designation indicates the year of original adoption or, in the case of revision, the year of last revision. A number in parentheses indicates the year of last reapproval. A superscript epsilon (ε) indicates an editorial change since the last revision or reapproval.

1. Scope

1.1 This test method covers the calculation of the water content of soil and rock by thermalization or slowing of fast neutrons where the neutron source and the thermal neutron detector are placed at the desired depth in the bored hole lined by an access tube (see Note 1).

1.2 The water content, in mass per unit volume of the material under test, is calculated by comparing the thermal neutron count rate with previously established calibration data (see Annex A1).

1.3 The values expressed in SI units are regarded as the standard. The inch-pound units given in parentheses may be approximate and are provided for information only.

1.4 *This standard does not purport to address all of the safety problems, if any, associated with its use. It is the responsibility of the user of this standard to establish appropriate safety and health practices and determine the applicability of regulatory limitations prior to use.* Specific hazards are given in Section 7.

2. Referenced Documents

2.1 *ASTM Standards:*
D 1452 Practice for Soil Investigation and Sampling by Auger Borings[2]
D 1586 Method for Penetration Test and Split-Barrel Sampling of Soils[2]
D 1587 Practice for Thin Walled Tube Sampling of Soils[2]
D 2113 Practice for Diamond Core Drilling for Site Investigation[2]
D 2216 Test Method for Laboratory Determination of Water (Moisture) Content of Soil, Rock, and Soil-Aggregate Mixtures[2]
D 2937 Test Method for Density of Soil in Place by the Drive-Cylinder Method[2]
D 3017 Test Method for Moisture Content of Soil and Soil-Aggregate in Place by Nuclear Methods (Shallow Depth)[2]
D 3550 Practice for Ring-Lined Barrel Sampling of Soils[2]
D 4428/D 4228M Test Method for Crosshole Seismic Testing[2]
D 5195 Test Method for Density of Soil and Rock in Place at Depths Below the Surface by Nuclear Methods

3. Summary of Test Method

3.1 This test method uses thermalization of neutron radiation to calculate the in-place water content of soil and rock at various depths by placing a probe containing a neutron source and a thermal neutron detector at desired depths in a bored hole lined by an access tube as opposed to surface measurements in accordance with Test Method D 3017.

3.2 Neutrons emitted by the source are thermalized (slowed) by collisions with materials of low atomic numbers. Hydrogenous materials, such as water and other compounds containing hydrogen, are most effective in thermalizing neutrons. In this apparatus the neutrons thermalized by the material under test are detected by the thermal neutron detector.

3.3 In the absence of interference elements as discussed in Section 5, the number of thermalized neutrons is a function of the hydrogen content of the material under test and the water content is proportional to the hydrogen content.

3.4 By the use of a calibration process the water content is calculated by correlating the count rate to known water contents.

4. Significance and Use

4.1 This test method is useful as a rapid, nondestructive technique for the calculation of the in-place water content of soil and rock at desired depths below the surface.

4.2 With proper calibration in accordance with Annex A1, this test method can be used for quality control and acceptance testing for construction and for research and development applications.

4.3 The non-destructive nature of this test method allows repetitive measurements to be made at a single test location for statistical analysis and to monitor changes over time.

4.4 The fundamental assumptions inherent in this test method are that the material under test is homogeneous and hydrogen present is in the form of water as defined by Test Method D 2216.

5. Interferences

5.1 The sample heterogeneity, density, and chemical composition of the material under test will affect the measurements. The apparatus must be calibrated to the material under test or adjustments made in accordance with Annex A2.

5.1.1 Hydrogen, in forms other than water, as defined by Test Method D 2216 and carbon, present in organic soils, will cause measurements in excess of the true water value. Some elements such as boron, chlorine, and minute quanti-

---

[1] This test method is under the jurisdiction of ASTM Committee D-18 on Soil and Rock and is the direct responsibility of Subcommittee D18.02 on Sampling and Related Field Testing for Soil Investigations.
Current edition approved Jan. 15, 1992. Published March 1992.
[2] *Annual Book of ASTM Standards*, Vol 04.08.

ties of cadmium, if present in the material under test, will cause measurements lower than the true water value.

5.2 This test method exhibits spatial bias in that it is more sensitive to water contained in the material closest to the access tube. The measurement is not necessarily an average water content of the total sample involved.

5.2.1 Voids around the access tube can affect the measurement (see 11.1.2).

5.3 The sample volume is approximately 0.048 m³ (1.7 ft³) with a water content of 200 kg/m³ (12.5 lbf/ft³). The actual sample volume is indeterminate and varies with the apparatus and the water content of the material. In general, the greater the water content of the material, the smaller the volume involved in the measurement.

6. Apparatus (See Fig. 1)

6.1 The apparatus shall consist of a nuclear instrument capable of measuring water content at various depths below the surface containing the following:

6.1.1 A sealed mixture of a radioactive material such as americium or radium with a target element such as beryllium, and a suitable thermal neutron detector, and

6.1.2 A suitable timed scaler and power source.

6.2 The apparatus shall be equipped with a cylindrical probe containing the neutron source and detector, connected by a cable of sufficient design and length, that is capable of being lowered down the cased hole to desired test depths.

6.3 The apparatus shall be equipped with a reference standard, a fixed shape of hydrogenous material used for checking apparatus operation and to establish conditions for a reproducible reference count rate. It may also serve as a radiation shield.

6.4 *Apparatus Precision*—See Annex A3 for the precision of the apparatus.

6.5 *Accessories:*

6.5.1 *Access Tubing*—The access tubing (casing) is required for all access holes in nonlithified materials (soils and poorly consolidated rock) that cannot maintain constant borehole diameter with repeated measurements. If access tubing is required the tubing shall be of a material such as aluminum, steel, or polyvinyl chloride, having an interior diameter large enough to permit probe access without binding. The tubing shall be as thin-walled as possible to provide close proximity of the probe to the material under test. The same type of tubing shall be used in the field as is used in calibration.

6.5.2 Hand auger or power drilling equipment that can be used to establish the access hole. Any drilling equipment that provides a suitable clean open hole for installation of access tubing and insertion of the probe shall be acceptable. The equipment used shall be capable of maintaining constant borehole diameter to ensure that the measurements are performed on undisturbed soil and rock. The type of equipment and methods of advancing the access hole should be reported.

6.5.3 *Dummy Probe*—A cylindrical probe the same size as the probe containing the neutron source and a chain or cable of sufficient design and length to permit lowering the dummy probe down the cased hole to desired test depths.

7. Hazards

7.1 This equipment utilizes radioactive materials that may be hazardous to the health of the users unless proper precautions are taken. Users of this equipment must become completely familiar with possible safety hazards and with all applicable regulations concerning the handling and use of radioactive materials. Effective user instructions together with routine safety procedures are a recommended part of the operation of this apparatus.

8. Calibration, Standardization, and Reference Check

8.1 Calibrate the instrument in accordance with Annex A1.

8.2 Adjust the calibration in accordance to Annex A2 if adjustments are necessary.

8.3 *Standardization and Reference Check:*

8.3.1 Nuclear apparatus are subject to the long-term decay of the radioactive source and aging of detectors and electronic system, that may change the relationship between count rate and water content. To offset these changes, the apparatus may be calibrated as the ratio of the measurement count rate to a count rate made on a reference standard. The reference count rate should be in the same or a higher order of magnitude than the range of measurement count rates over the useful water content range of the apparatus.

8.3.2 Standardization of equipment should be performed at the start of each day's work and a permanent record of these data retained. Perform the standardization with the apparatus located at least 10 m (30 ft) away from other apparatus containing neutron emitting radioactive sources and clear of large masses of water or other items which may

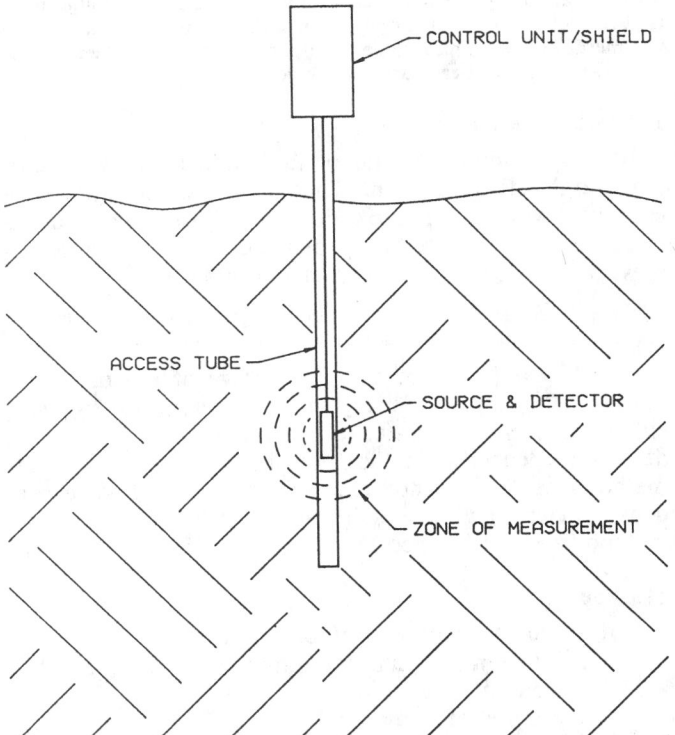

FIG. 1 Schematic Diagram; Water Content by Neutron Depth Probe Method

CONTROL UNIT/SHIELD

ACCESS TUBE

SOURCE & DETECTOR

ZONE OF MEASUREMENT

affect the reference count rate.

8.3.2.1 If recommended by the apparatus manufacturer to provide more stable and consistent results: turn on the apparatus prior to use to allow it to stabilize; and leave the power on during the day's testing.

8.3.2.2 Using the reference standard, take at least four repetitive readings at the manufacturer's recommended measurement period and determine the mean. If available on the apparatus, one measurement at a period of four or more times the recommended period is acceptable. These measurements constitute one standardization check.

8.3.2.3 If the value obtained above is within the limits stated below, the equipment is considered to be in satisfactory condition and the value may be used to determine the count ratios for the day of use. If the value is outside these limits, allow additional time for the apparatus to stabilize, make sure the area is clear of sources of interference and then conduct another standardization check. If the second standardization check is within the limits, the apparatus may be used, but if it also fails the test, the apparatus shall be adjusted or repaired as recommended by the manufacturer. The limits are as follows:

$$N_s <= N_o + \frac{2.0\sqrt{N_o}}{\sqrt{F}}$$

and

$$N_s >= N_o - \frac{2.0\sqrt{N_o}}{\sqrt{F}}$$

where:
N_s = value of current standardization check,
N_o = average of the past four values of N_s taken previously, and
F = value of prescale

NOTE 1—Some instruments have built-in provisions to compute and display the results of a statistical test of the standardization counts and to indicate if the apparatus is in satisfactory condition.

8.3.3 Use the value of N_s to determine the count ratios for the current day's use of the apparatus. If, for any reason, the measured water content becomes suspect during the day's use, perform another standardization check.

8.3.4 If the instrument was calibrated in the field using methods described in Annex A1.2.3 the count rate on any trial reading may be adjusted by a correction factor determined in initial calibration.

9. Procedure

9.1 *Installation of Access Tubing (Casing):*

9.1.1 Drill the access tube hole and install access tubing in a manner dependent upon the material to be tested, the depth to be tested, and the available drilling equipment.

9.1.2 The access hole must be clear enough to allow installing the tubing yet must provide a snug fit. Voids along the sides of the tubing may cause erroneous readings.

9.1.2.1 If voids are suspected to be caused by the drilling process they can be grouted using procedures in Test Method D 4428.

9.1.2.2 The only method to determine the presence of voids is to perform field calibrations provided in Annex A2.2.3.

9.1.3 Record and note the position of the ground water table, perched water tables, and changes in strata as drilling progresses.

9.1.3.1 If ground water is encountered or saturated conditions are expected to develop, seal the tube at the bottom to prevent water seepage into the tube using procedures given in Test Method D 4428 or the manufacturer's recommended procedures. This will prevent erroneous readings and possible damage to the probe.

9.1.4 The tubing should project above the ground and be capped to prevent foreign material from entering. The access tube should not project above the ground so high that it might be damaged by equipment passing over it.

9.1.4.1 Install all tubes at the same height above the ground as this enables marking the cable to indicate the measured depth to be used for all tubes.

9.2 Lower a dummy probe down the access tube to verify proper clearance before lowering the probe containing the radioactive source.

9.3 Standardize the apparatus.

9.4 Proceed with the test as follows:

9.4.1 Seat the apparatus firmly over the access tube, then lower the probe into the tube to the desired depth. Secure the probe by cable clamps (usually provided by the apparatus manufacturer).

9.4.2 Take a measurement count at the selected timing period.

NOTE 2—The above procedure is performed in an installed access tube that will allow repeated in-place measurements. In some field situations it may be more appropriate to use a drilling technique involving alternating between access tubing and one of the following: a large diameter hollow stem auger, a split-spoon sampler, or a thin-walled volumetric sampler. This technique is destructive and only one measurement can be made at each depth per hole.

10. Interpretation of Results

10.1 Determine the ratio of the reading obtained compared to the standard count. Then using the calibration data combined with appropriate calibration adjustments, or apparatus direct readout features, calculate the water content in mass per unit volume of the material under test.

NOTE 3—Some instruments have built-in provisions to compute and display the ratio and corrected water content per unit volume.

10.2 If water content as a percentage of dry density is required, the in-place density may be determined by using either the same apparatus or a different apparatus that determines density at depths below the surface by the nuclear method (see Test Method D 5195) or by a method such as density determination of soil in-place by the drive-cylinder method (see Test Method D 2937).

11. Report

11.1 Report the following information:

11.1.1 Make, model, and serial number of the apparatus,

11.1.2 Date of calibration,

11.1.3 Method of calibration, such as field, factory, etc.

11.1.4 Calibration adjustments,

11.1.5 Date of test,

11.1.6 Standard count(s) for the day of the test,

11.1.7 Any adjustment data for the day of the test,

11.1.8 Test site identification including tube location(s), tube number(s),

11.1.9 Tube type and tube installation methods (method of drilling, installing and any initial gravimetric and count data),

11.1.10 Geologic log of the borehole, and

11.1.11 Depth, measurement count data, and calculated water content of each measurement.

12. Precision and Bias

12.1 *Precision*—The precision of this test method has not been determined. While the apparatus precision (repeatability on the same sample) can be defined (see Annex A3), no data are presently available to determine true test precision.

12.2 *Bias*—No methods are presently available that provide sufficiently accurate values of water content of soil and rock in-place against which these methods can be compared.

13. Keywords

13.1 depth probe; in-place water content; in situ water content; neutron probe; nuclear methods

ANNEXES

(Mandatory Information)

A1. APPARATUS CALIBRATION

A1.1 At least once each year, establish or verify calibration curves, tables, or equations by determining the count rate of at least three samples of different known water contents. These data may be presented in the form of a graph, table, equation coefficients, or stored in the apparatus, to allow converting the count rate data to water content. The method and test procedures used in establishing these count ratios must be the same as those used for obtaining the count ratios for in-place material. The water content of materials used to establish the calibration must vary through a range to include the water content of materials to be tested and be of a similar density.

A1.2 Calibration standards may be established using one of the following methods. The standards must be verified to be of sufficient size to not change the count rate if enlarged in any dimension. Access tubing used in the standards must be the same type and size as that to be used for in-place measurements.

A1.2.1 Prepare homogeneous standards of hydrogenous materials having a water content determined by comparison (using a nuclear instrument gage) to a saturated silica sand standard with a known water content. As an alternate, determine the equivalent water content by calculation if the hydrogen, carbon, and oxygen content is known or can be calculated from the specific gravity and chemical composition. A zero water content standard can be prepared by using a non-hydrogenous material, such as magnesium, as the standard.

A1.2.2 Prepare containers of soil and rock compacted to uniform densities with a range of water contents. Determine the percent water content of the materials by oven drying (see Test Method D 2216) and a wet density calculated from the mass of the material and the inside dimensions of the container. Whenever possible, use soil and rock obtained from the test site for this calibration.

A1.2.3 Where neither of the previous calibration standards are available or a higher accuracy of calibration is required, the apparatus may be calibrated in the field by using the following methods:

A1.2.3.1 Use a minimum of three selected test sites containing a homogeneous material with as wide a range of water contents as possible. As the access hole is drilled, take gravimetric samples from the soil or rock samples taken by any suitable drilling and sampling method appropriate for the material (see Method D 1586 and Practices D 1452, D 1587, D 2113, and D 3550) and determine the percent water content by oven drying (see Test Method D 2216). Note the sampling intervals for the gravimetric samples. (See Note A1.1.)

NOTE A1.1—For agricultural purposes it is highly practical to obtain the gravimetric samples (see Practice D 1452) above the water table in shallow installations. The gravimetric sample will represent a mixture of materials over the interval samples. For a higher level of calibration accuracy it is recommended to obtain samples directly in the measurement interval by other referenced methods.

A1.2.3.2 As soon as possible after the access tubing has been installed, take sufficient measurements at the desired depths in accordance with Section 8 and calculate the count ratio and measured volumetric water content.

A1.2.3.3 The test measurement counts are to be taken at approximate depths that will correspond to the depth location of the gravimetric samples.

A1.2.4 Report any anomalous values, such as voids, grout plugs, changes in strata, or perched water layers, and their probable causes obtained from this calibration.

A1.3 For the highest level of calibration, obtain gravimetric samples from each access hole over the measurement intervals to be tested using the methods given in A1.2.3. At a minimum, obtain gravimetric samples at 2 m intervals and at changes in strata.

A1.3.1 As soon as possible after the access tubing has been installed, take sufficient measurements at the desired depths in accordance with Section 8 and calculate the count ratio and measured water content. The test measurement counts are to be taken at approximate depths that will correspond to the depth location of the gravimetric samples. The initial count profile and adjusted water content data should be reported with later readings to review changes in water content with subsequent readings.

A2. CALIBRATION ADJUSTMENTS

A2.1 Check the calibration response prior to performing tests on materials that are distinctly different from the material types used in establishing the calibration. Also check the calibration response on newly acquired or repaired apparatus.

A2.2 Take sufficient measurements and compare them to samples obtained by other accepted methods such as a volumetric sampling (see Test Method D 2937) to establish a correlation between the apparatus calibration and the other method.

A2.2.1 Adjust the existing calibration to correct for the difference or establish a new calibration in accordance with Annex A1.

NOTE A2.1—Some apparatus utilizing a microprocessor may have provision to input a correction factor that is established by determining the correlation between the apparatus measurement and oven drying (see Test Method D 2216)

A3. PRECISION OR APPARATUS

A3.1 The precision of the apparatus at a water content of 200 kg/m^3 (12.5 lbf/ft^3) shall be better than 5 kg/m^3 (0.3 lbf/ft^3) at the manufacturer's stated period of time for the measurement (see Note A3.1). Other timing periods may be available that may be used where higher or lower precisions are desired for statistical purposes. The precision shall be determined by the procedure defined in A3.2 or A3.3.

NOTE A3.1—While 30 s is the usual timing period and may be used for comparison of various apparatus, the intent of this test method is to require a measurement period that produces the stated precision for all acceptance testing.

A3.2 The precision of the apparatus is determined from the slope of the calibration response and the statistical deviation of the count (detected neutrons) for the period of measurement:

$$P = \sigma/S \qquad (A3.1)$$

where:
P = apparatus precision in water content (kg/m^3 or lbf/ft^3),
σ = standard deviation in counts per measurement period, and
S = slope in change in counts per measurement period divided by the change in water content (kg/m^3 or lbf/ft^3).

A3.2.1 The counts per measurement period shall be the total number of thermal neutrons detected during the timed period. The displayed value must be corrected for any prescaling that is built into the apparatus. The prescale value (F) is a divisor that reduces the actual value for the purpose of display. The manufacturer will supply this value if other than 1.0.

A3.2.2 The standard deviation in counts per measurement period shall be obtained by:

$$\sigma = \sqrt{C/F} \qquad (A3.2)$$

where:
σ = standard deviation in counts per measurement period,
C = counts per measurement period (before prescale correction) at a water content of 200 kg/m^3 (12.5 lbf/ft^3); and
F = value of prescale (see A3.2.1).

A3.2.3 The counts per measurement period (before prescale correction) may be obtained from the calibration curve, tables, or equation by multiplying the count ratio by the apparatus standard count.

A3.2.4 The slope of calibration response in counts per measurement period (before prescale correction) at a water content of 200 kg/m^3 (12.5 lbf/ft^3) shall be determined from the calibration curve, tables, or equation.

A3.3 Compute the precision by determining the standard deviation of at least 20 repetitive measurements (apparatus not moved after the first measurement) on material having a water content of 160 to 240 kg/m^3 (10 to 15 lbf/ft^3). In order to perform this procedure, the resolution of the count display, calibration response, or other method of displaying water content must be equal to or better than ± 1 kg/m^3 (± 0.1 lbf/ft^3).

Standard Test Method for
Logging In Situ Moisture Content and Density of Soil and Rock by the Nuclear Method in Horizontal, Slanted, and Vertical Access Tubes[1]

This standard is issued under the fixed designation D 6031; the number immediately following the designation indicates the year of original adoption or, in the case of revision, the year of last revision. A number in parentheses indicates the year of last reapproval. A superscript epsilon (ϵ) indicates an editorial change since the last revision or reapproval.

1. Scope

1.1 This test method covers collection and comparison of logs of thermalized-neutron counts and back-scattered gamma counts along horizontal or vertical air-filled access tubes.

1.2 The in situ water content in mass per unit volume and the density in mass per unit volume of soil and rock at positions or in intervals along the length of an access tube are calculated by comparing the thermal neutron count rate and gamma count rates respectively to previously established calibration data.

1.3 The values stated in SI units are regarded as the standard. The inch-pound units given in parentheses may be approximate and are provided for information only.

1.4 *This standard does not purport to address all of the safety concerns, if any, associated with its use. It is the responsibility of the user of this standard to establish appropriate safety and health practices and determine the applicability of regulatory limitations prior to use. For specific hazards, see Section 6.*

2. Referenced Documents

2.1 *ASTM Standards:*
D 1452 Practice for Soil Investigation and Sampling by Auger Borings[2]
D 1586 Test Method for Penetration Test and Split/Barrel Sampling of Soils[2]
D 1587 Practice for Thin-Walled Tube Sampling of Soils[2]
D 2113 Practice for Diamond Core Drilling for Site Investigation[2]
D 2216 Test Method for Laboratory Determination of Water (Moisture) Content of Soil and Rock[2]
D 2922 Test Methods for Density of Soil and Soil-Aggregate in Place by Nuclear Methods (Shallow Depth)[2]
D 2937 Test Method for Density of Soil in Place by the Drive-Cylinder Method[2]
D 3017 Test Method for Water Content of Soil and Rock in Place by Nuclear Methods (Shallow Depth)[2]
D 4428/D 4428M Test Methods of Crosshole Seismic Testing[2]

D 4564 Test Method for Density of Soil in Place by the Sleeve Method[2]
D 5195 Test Method for Density of Soil and Rock In-Place at Depths Below the Surface by Nuclear Methods[3]
D 5220 Test Method for Water Content of Soil and Rock In-Place by the Neutron Depth Probe Method[3]

3. Significance and Use

3.1 This test method is useful as a repeatable, nondestructive technique to monitor in-place density and moisture of soil and rock along lengthy sections of horizontal, slanted, and vertical access holes or tubes. With proper calibration in accordance with Annex A1, this test method can be used to quantify changes in density and moisture content of soil and rock.

3.2 This test method is used in vadose zone monitoring, for performance assessment of engineered barriers at waste facilities, and for research related to monitoring the movement of liquids (water solutions and hydrocarbons) through soil and rock. The nondestructive nature of the test allows repetitive measurements at a site and statistical analysis of results.

3.3 The fundamental assumptions inherent in this test method are that the dry bulk density of the test material is constant and that the response to fast neutrons and gamma-ray energy associated with soil and liquid chemistry is constant.

4. Interferences

4.1 The sample heterogeneity and chemical composition of the material under test will affect the measurement of both moisture and density. The apparatus should be calibrated to the material under test at a similar density of dry soil or rock and in the similar type and orientation of access tube, or adjustments must be made in accordance with Annex A2.

4.2 Hydrogen, in forms other than water, as defined by Test Method D 2216, will cause measurements in excess of the true moisture content. Some elements such as boron, chlorine, and minute quantities of cadmium, if present in the material under test, will cause measurements lower than the true moisture content. Some elements with atomic numbers greater than 20 such as iron or other heavy metals may cause measurements higher than the true density value.

4.3 The measurement of moisture and density using this test method exhibits spatial bias in that it is more sensitive to

[1] This test method is under the jurisdiction of ASTM Committee D-18 on Soil and Rock and is the direct responsibility of Subcommittee D18.21 on Ground Water and Vadose Zone Investigations.
Current edition approved Oct. 10, 1996. Published February 1997.
[2] *Annual Book of ASTM Standards*, Vol 04.08.

[3] *Annual Book of ASTM Standards*, Vol 04.09.

the material closest to the access tube. The density and moisture measurements are necessarily an average of the total sample involved.

4.4 The sample volume for a moisture measurement is approximately 0.11 m³ (3.8 ft³) at a moisture content of 200 kg/m³ (12.5 lbf/ft³). The actual sample volume for moisture is indeterminate and varies with the apparatus and the moisture content of the material. In general the greater the moisture content of the material, the smaller the measurement volume.

4.5 A density measurement has a sample volume of approximately 0.028 m³ (0.8 ft³). The actual sample volume for density is indeterminate and varies with the apparatus and the density of the material. In general, the greater the density of the material, the smaller the measurement volume.

4.6 Air gaps between the probe and the access tube or voids around the access tube will cause the indicated moisture content and density to be less than the calibrated values.

4.7 Condensed moisture inside the access tube may cause the indicated moisture content to be greater than the true moisture content of material outside the access tube.

5. Apparatus

5.1 While exact details of construction of the apparatus may vary, the system shall consist of:

5.1.1 *Fast Neutron Source*—A sealed mixture of a radioactive material such as americium or radium and a target material such as beryllium, or other fast neutron sources such as californium that do not require a target.

5.1.2 *Slow Neutron Detector*—Any type of slow neutron detector, such as boron trifluoride or helium-3 proportional counters.

5.1.3 *High-Energy Gamma-Radiation Source*—A sealed source of radioactive material, such as cesium-137, cobalt-60, or radium-226.

5.1.4 *Gamma Detector*—Any type of gamma detector, such as a Geiger-Mueller tube.

5.1.5 *Suitable Readout Device:*

5.1.6 *Cylindrical Probe*—The apparatus shall be equipped with a cylindrical probe, containing the neutron and gamma sources and the detectors, connected by a cable or cables of sufficient design and length, that are capable of raising and lowering the probe in vertical applications and pulling it in horizontal applications, to the desired measurement location.

5.1.7 *Reference Standard*—A device containing dense, hydrogenous material for checking equipment operation and to establish conditions for a reproducible reference count rate. It also may serve as a radiation shield.

5.2 Accessories shall include:

5.2.1 *Access Tubing*—The access tubing (casing) is required for all access holes in nonlithified materials (soils and poorly consolidated rock) that cannot maintain constant borehole diameter with repeated measurements. If access tubing is required it must be of a material, such as aluminum, steel, or plastic, having an interior diameter large enough to permit probe access without binding, and an exterior diameter as small as possible to provide close proximity of the material under test. The same type of tubing must be used in the field as is used in calibration.

5.2.2 *Hand Auger or Power Drilling/Trenching Equipment*—Equipment that can be used to establish the access hole or position the access tube when required (see 5.2.1). Any equipment that provides a suitable clean open hole for installation of access tubing and insertion of the probe that ensures the measurements are performed on undisturbed soil and rock while maintaining a constant diameter per width shall be acceptable. The type of equipment and methods of advancing the access hole should be reported.

5.2.3 *Winching Equipment or Other Motive Devices*—Equipment that can be used to move the probe through the access tubing. The type of such equipment is dependent upon the orientation of the access tubing and the distance over which the probe must be moved.

6. Hazards

NOTE: **WARNING**—This equipment utilizes radioactive materials that may be hazardous to the health of the users unless proper precautions are taken. Users of this equipment must become completely familiar with all possible safety hazards and with all applicable regulations concerning the handling and use of radioactive materials. Effective user instructions together with routine safety procedures are a recommended part of the operation of this apparatus.

NOTE: **CAUTION**—When using winching or other motive equipment, the user should take additional care to learn its proper use in conjunction with measurement apparatus. Known safety hazards such as cutting and pinching exist when using such equipment.

NOTE—This test method does not cover all safety precautions. It is the responsibility of the users to familiarize themselves with all safety precautions.

7. Calibration, Standardization, and Reference Check

7.1 Calibrate the instrument in accordance with Annex A1.

7.2 Adjust the calibration in accordance with Annex A2 if adjustments are necessary.

7.3 *Standardization and Reference Check:*

7.3.1 Nuclear apparatus are subject to the long-term decay of the radioactive source and aging of detectors and electronic systems that may change the relationship between count rate and either the material density or the moisture content of the material, or both. To correct for these changes, the apparatus may be calibrated periodically. To minimize error, moisture and density measurements commonly are reported as count ratios, the ratio of the measured count rate to a count rate made in a reference standard. The reference count rate should be similar or higher than the count rates over the useful measurement range of the apparatus.

7.3.2 Standardization of equipment on the reference standard is required at the start of each day's use and a permanent record of these data shall be retained. The standardization shall be performed with the equipment located at least 10 m (33 ft) away from other radioactive sources and large masses or other items that may affect the reference count rate.

7.3.3 If recommended by the apparatus manufacturer to provide more stable and consistent results, turn on the apparatus prior to use to allow it to stabilize and leave the power on during the day's testing.

7.3.4 Using the reference standard, take at least four repetitive readings at the manufacturer's recommended measurement period of 20 or more at some shorter period and obtain the mean. If available on the instrument, one measurement at a period of four or more times the normal test measurement period is acceptable. This constitutes one standardization check.

7.3.5 If the value obtained in 7.3.4 is within the following limits, the equipment is considered to be in satisfactory condition and the value may be used to determine the count ratios for the day of use. If the value obtained is outside these limits, another standardization check should be made. If the second standardization check is within the limits, the equipment may be used. If it also fails the test, however, the equipment shall be adjusted or repaired as recommended by the manufacturer.

$$No + 2F \sqrt{\frac{No}{F}} > Ns > No - 2F \sqrt{\frac{No}{F}}$$

where:

Ns = value of current standardization check (7.3.4) on the reference standard,

No = average of the past values of Ns taken for prior usage, and

F = value of prescale, a multiplier that alters the count value for the purpose of display (see A3.1.1.1).

7.3.6 If the apparatus standardization has not been checked within the previous three months, perform at lest four new standardization checks and use the mean as the value for No.

7.3.7 The value of Ns will be used to determine the count ratios for the current day's use of the equipment. If, for any reason, either the measured density or moisture content become suspect during the day's use, perform another standardization to ensure that the equipment is stable.

8. Procedure

8.1 *Installation of Access Tubing (Casing):*

8.1.1 Drill the access hole or excavate a trench at the desired location and install the access tube in a manner to maximize contact with test material and minimize voids. The access tubes should fit snugly into the access hole or trench. Unstable conditions in fill material around the access tube may result in redistribution of solids over time, piping, or other phenomena that will degrade precision. Voids caused during drilling, tube installation, or backfilling, or a combination thereof, may cause erroneously low results. Excessive compaction of clay-rich backfill material will limit the effectiveness of moisture monitoring for leak detection. Backfill should approximate the composition, water content, and bulk density of test material as nearly as possible.

8.1.2 Grouting of annular spaces, if required, should be of minimum functional thickness, and grout mixtures should not contain excessive water. Grouts thicker than 5 cm (2 in.) create high background counts that will obscure moisture content changes in fine-textured soils and severely limit meaningful density measurements in all soil types. Grouting should not be used unless it is required to seal off flow pathways along the access tube, such as in some vertical borings and where trenches cross engineered barriers.

Grouting can be accomplished using procedures described in Test Methods D 4428/D 4428M.

8.1.3 Record and note the position of the ground water table, perched water tables, and changes in soil texture as drilling or trenching progresses.

8.1.4 If ground water is encountered or saturated conditions are expected to develop, seal the tube at seams and open ends to prevent water seepage into the tube. This will prevent erroneous measurements and possible damage to the probe.

8.1.5 The access tube should project above the ground and be capped to prevent foreign material from entering. The access tube should not project out of the test material far enough to be damaged by equipment traffic.

8.2 Pass a dummy probe through the access tube to verify proper clearance before deploying the radioactive sources.

8.3 Standardize the apparatus (see 7.3).

8.4 Proceed with the test run in a continuous logging mode or in a noncontinuous logging mode as follows:

8.4.1 Set up the winching equipment or other motive devices (see 5.2.3) to begin a logging run by stationing the probe at one end of the access tube to be logged.

8.4.2 Select a timing period for collecting measurement counts based on desired precision (see Annex A3), anticipated measurement response, or site-specific logistical criteria.

8.4.3 For testing in continuous logging mode, advance the probe continuously through the access tube while recording data that relate gamma counts and thermal neutron counts to position intervals or time (for constant logging speed), or both.

8.4.4 For testing in noncontinuous logging mode, advance the probe through the access tube to the desired position and stop, record counts while probe is stationary, advance the probe to the next desired position, and repeat. Record data relating gamma counts and thermal neutron counts to discrete positions along the access tube.

9. Calculation

9.1 Calculations related to reporting density as calibrated units are provided in Test Method D 5195. For moisture content, these same calculations are provided in Test Method D 3017.

9.2 Data can be used in a comparative mode, as in graphs or charts. For example, measurements from repeated logging events can be compared directly at each position (or interval) and analyzed to detect statistically significant changes from background.

9.2.1 For data reported as uncalibrated counts, the accepted estimator of the standard deviation of a population of nuclear count measurements is equal to the square root of the mean.[4] Standard deviation estimated from more than one background measurement at any given position (or over any specific interval) can be used to define tolerance levels. The tolerance level defines a threshold neutron count above which there is a defined probability that the count is higher than background.

---

[4] Kramer, J. H., Everett, L. G., and Cullen, S. J., 1992. "Vadose Zone Monitoring with Neutron Moisture Probe," *Ground Water Monitoring Review*, Vol 12, No. 2, 1992, pp. 177–187.

10. Report

10.1 Report the following information:

10.1.1 Make, model, and serial number of the apparatus.

10.1.2 Date of test.

10.1.3 Standard count for day of the test.

10.1.4 Test site identification including tube location(s) and tube number(s).

10.1.5 Distance (depth), measurement count data, and count ratios or calculated density and moisture content.

10.1.6 Optional graphical display of the magnitude of count measurements along the access tube transect.

10.1.7 Report results in both SI and inch-pound units.

11. Precision and Bias

11.1 *Precision*—The precision of the procedure in Test Method D 6031 must be determined using site-specific samples. Annex A3 is the precision of the instrument and should not be confused with the precision of the test method.

11.2 *Bias*—Since there is no accepted reference material suitable for determining the bias for Test Method D 6031 for measuring the moisture or density, or both, of soil, bias cannot be determined.

ANNEXES

(Mandatory Information)

A1. CALIBRATION

A1.1 *Calibration Curves*—Calibration curves, tables, or equations shall be established or verified once each year or as recommended by the manufacturer, by determining the nuclear count rate of at least two samples of different known moisture content and at least three samples of different known density. This data may be presented in the form of a graph, table, equation coefficients, or stored in the apparatus to allow converting the count rate data to material moisture content or density. The method and test procedures used in establishing these count rate data must be the same as those used for obtaining the count rate data for in-place material.

A1.2 *Density*—Calibration standards may be established using one of the following methods, or as recommended by the manufacturer. The standards must be of sufficient size to not change the count rate if enlarged in any dimension. Access tubing used in the standards must be the same type and size as that to be used for in-place measurements.

A1.2.1 Prepare containers of soil and rock of a range of different densities. Place the material in lifts of thickness that depends upon the compaction method being used. Each lift is to receive equal compactive effort. Calculate the density of each container of material based on the measured volume and mass (weight) of the material.

A1.2.2 Prepare containers of cured concrete using different aggregate to sand ratio mixes to obtain a range of densities. Place the concrete in the containers in a way that will ensure a uniform mixture and uniform densities.

A1.2.3 Prepare containers of non-soil materials. Calculate the soil and rock equivalent density of each container of material based on the measured volume and mass (weight) of the material.

A1.2.4 Take sufficient measurements in each prepared container to establish a correlation between the apparatus measurements and the densities of the material in the containers.

A1.3 *Field Calibration for Density*—The apparatus may be calibrated in the field by using the following method when a verification of laboratory calibration accuracy to field materials is required, or in instances where neither of the previous calibration standards are available, or a more accurate calibration is required.

A1.3.1 During placement of access tubing, obtain undisturbed samples of the material around the tubing from points along it that are representative of the material to be tested. Take undisturbed samples from the soil or rock by any suitable drilling and sampling method appropriate for the material (see Practices D 1452, D 1587, and D 2113, double-tube or triple-tube core samplers, piston samplers, or double-tube hollow stem samplers), and determine the average sample density by trimming excess material and measuring the mass and volume of the samples. Samples should be taken over the length of the access tube in which the probe will be used. At a minimum, obtain undisturbed samples at 2-m (6.6-ft) intervals and at all locations where the material around the access tube changes composition or texture.

A1.3.2 As soon as possible after the access tubing has been installed, take measurements in accordance with Section 8 using the appropriate type of winching equipment detailed in Section 5. The winching speed for continuous logging mode shall be determined by the user, but generally it will fall within the range from 0.6 to 3.0 m (2.0 to 10.0 ft)/min. Based upon laboratory calibrations, calculate the gage density measurement for each reading taken. Take the test measurement counts so that they will include or be adjacent to the location of the undisturbed samples. Compare the sample densities to the gage measurement(s) closest to it (with respect to length along the tubing), and make any needed adjustments to the laboratory calibrations (see Annex A2). Follow the manufacturer's recommendations for any such adjustments. The sample density and measurement count ratios may be presented in the form of a graph, table, equation coefficients, or stored in the gage to allow converting future instrument count ratios to material densities.

A1.3.3 Report all sample data including changes in strata and all anomalous data obtained, such as voids. The initial count profile and adjusted density data should be reported with later readings to review changes in density with subsequent readings.

A1.4 *Moisture Content*—Calibration standards may be

established using one of the following methods or as recommended by the manufacturer. The standards must be verified to be large enough to not change the observed count rate (or ratio as defined in 7.3.1) if made larger in any dimension. Access tubing used in the standards must be the same type and size as that to be used for in-place measurements.

A1.4.1 Prepare homogenous standards of hydrogenous materials having moisture contents determined by comparison (using a nuclear instrument) to saturated silica sand standards with known moisture contents. As an alternative, determine the equivalent moisture content by calculation if the hydrogen, carbon, and oxygen content is known or can be calculated from the specific gravity and chemical composition. A zero moisture content standard can be prepared by using an non-hydrogenous material, such as a magnesium alloy, as the standard.

A1.4.2 Prepare containers of soil and rock compacted to uniform densities with a range of moisture contents. Determine the moisture content of the materials by oven drying (see Test Method D 2216). If desired, calculate volumetric moisture content θ_v using Test Methods D 2937 or D 4564 and Eq A1.1. Whenever possible, use soil and rock obtained from the test site for this calibration.

$$\theta_v = \theta_g \times \frac{\rho_d}{\rho_w} \qquad (A1.1)$$

where:
θ_v = volumetric moisture content, cm^3/cm^3,
θ_g = gravimetric moisture content, g water/g soil,
ρ_d = in-place dry density of soil, g/cm^3, and
ρ_w = density of water, 1 g/cm^3.

A1.4.3 Take sufficient measurements in each prepared container to establish a correlation between the apparatus measurements and the moisture contents of the material in the containers.

A1.5 *Field Calibration for Moisture Content*—The instrument may be calibrated in the field using the following method when a verification of laboratory calibration accuracy to field materials is required, or in instances where neither of the previous calibration standards are available or a more accurate calibration is required.

A1.5.1 During placement of access tubing obtain undisturbed samples of the material from around the tubing. Take volumetric or gravimetric samples from the soil or rock by any suitable drilling and sampling method appropriate for the material (see Test Method D 1586 and Practices D 1452, D 1587, D 2113, and D 3550) and determine the percent moisture content by oven drying (see Test Method D 2216). Note the sampling intervals for the samples. Samples should be taken over the length of the access tube that the probe will be taking measurements. At a minimum obtain samples at 2-m (6.6-ft) intervals and at all locations where the material around the access tube changes composition.

A1.5.2 As soon as possible after the access tubing has been installed, take measurements in accordance with Section 8 using the appropriate type of winching equipment detailed in Section 5. The winching speed for continuous logging mode shall be determined by the user. Generally, it will fall within the range from 0.6 to 3.0 m (2.0 to 10.0 ft)/min. In addition to these initial measurements, measurements also should be taken when periodic samples are taken. Take the test measurement counts so that they will include or be adjacent to the location of the gravimetric or volumetric samples. Compare the sample moisture contents to the gage measurement(s) closest to it (with respect to length along the tubing), and make any needed adjustments to the laboratory calibrations (see Annex A2). Follow the manufacturer's recommendations for any such adjustments. The sample moisture content and measurement count ratios may be presented in the form of a graph, table, equation coefficients, or stored in the gage to allow converting future instrument count ratios to material moisture contents.

A1.5.3 Report all sample data including changes in strata and all anomalous data obtained, such as voids. The initial count profile and adjusted moisture content data should be reported with later readings to review changes in moisture content with subsequent readings.

A2. CALIBRATION ADJUSTMENTS

A2.1 Check the calibration response prior to performing tests on materials that are distinctly different from the material types used in establishing the apparatus calibration. The calibration response also shall be checked on newly acquired or repaired apparatuses.

NOTE A2.1—Some apparatus utilizing a microprocessor may have provision to input a correction factor that is established by determining the correlation between the apparatus measurement and gravimetric measurements.

A2.2 Take sufficient measurements and compare them to other accepted methods, such as volumetric sampling (see Test Methods D 2937 or D 4564), to establish a correlation between the apparatus calibration and the other method.

A2.2.1 Adjust the existing calibration to correct for the difference or establish a new calibration in accordance with Annex A1.

A3. PRECISION OF APPARATUS

A3.1 *Density*—The precision of the apparatus on a sample of approximately 2000 kg/m^3 (125 lbf/ft^3) shall be better than 8 kg/m^3 (0.5 lbf/ft^3) at the manufacturer's stated period of time for the measurement. Other timing periods may be available that may be used where higher or lower precision is desired for statistical purposes. The precision shall be determined by the procedure defined in A3.1.1 and A3.1.2.

A3.1.1 The precision of the apparatus is determined from the slope of the calibration response and the statistical deviation of the count (detected gamma radiation) for the period of measurement as follows:

$$P = \sigma/S$$

where:

P = apparatus precision in density, kg/m³ or lbf/ft³,
σ = standard deviation in counts/measurement period, and
S = slope of change in counts/measurement period at a density of 2000 kg/m³ (125 lbf/ft³) divided by the change in density, kg/m³ or lbf/ft³.

A3.1.1.1 The count per measurement period shall be the total number of photons detected during the time period. The displayed value must be corrected for any prescaling which is built into the apparatus. The prescale value (F) is a factor that changes the actual value for the purpose of display. The manufacturer will supply this value if other than 1.0.

A3.1.1.2 The standard deviation in counts/measurement period shall be obtained as follows:

$$\sigma = \sqrt{(C/F)}$$

where:

σ = standard deviation in counts per measurement period,
C = reported counts/measurement period (before prescale correction) at a density of 2000 kg/m³ (125 lbf/ft³), and
F = value of prescale (see A3.1.1.1).

A3.1.1.3 The counts/measurement period (before prescale correction) may be obtained from the calibration curve, tables, or equation by multiplying the count ratio by the instrument standard count.

A3.1.1.4 The slope of calibration response in counts/measurement period (before prescale correction) at a density of 2000 kg/m³ (125 lbf/ft³) shall be determined from the calibration curve, tables, or equation.

A3.1.2 Compute the precision by determining the standard deviation of at least 20 repetitive measurements (apparatus not moved after the first measurement) on material having a density of 1600 to 2400 kg/m³ (100 to 150 lbf/ft³). In order to perform this procedure, the resolution of the count display, calibration response, or other method of displaying density must be equal to or better than 1.6 kg/m³ (±0.1 lbf/ft³).

A3.2 *Moisture Content*—The precision of the apparatus at a moisture content of 200 kg/m³ (12.5 lbf/ft³) shall be better than 3 kg/m³ (0.2 lbf/ft³) at the manufacturer's stated period of time for the measurement. Other timing periods may be available that may be used where higher or lower precisions are desired for statistical purposes. The precision shall be determined by the procedure defined in A3.2.1 or A3.2.2.

A3.2.1 The precision of the apparatus is determined from the slope of the calibration response and the statistical deviation of the count (detected thermal neutrons) for the period of measurement:

$$P = \sigma/S$$

where:

P = apparatus precision in moisture content, kg/m³ or lbf/ft³,
σ = standard deviation, counts per measurement period, and
S = slope in change in counts/measurement period divided by the change in moisture content, kg/m³ or lbf/ft³.

A3.2.1.1 The counts per measurement period shall be the total number of thermal neutrons detected during the timed period. The displayed value must be corrected for any prescaling that is built into the apparatus. The prescale value, F, is a factor that changes the actual value for the purpose of display. The manufacturer will supply this value if other than 1.0.

A3.2.1.2 The standard deviation in counts/measurement period shall be obtained by:

$$\sigma = \sqrt{(C/F)}$$

where:

σ = standard deviation in counts per measurement period,
C = reported counts per measurement period (before prescale correction) at a water content of 200 kg/m³ (12.5 lbf/ft³), and
F = value of prescale (see A3.2.1.1).

A3.2.1.3 The counts per measurement period (before prescale correction) may be obtained from the calibration curve, tables, or equation by multiplying the count ratio by the apparatus standard count.

A3.2.1.4 The slope of calibration response in counts per measurement period (before prescale correction) at a moisture content of 200 kg/m³ (12.5 lbf/ft) shall be determined from the calibration curve, tables, or equation.

A3.2.2 Compute the precision by determining the standard deviation of at least 20 repetitive measurements (apparatus not moved after the first measurement) on material having a moisture content of 160 to 240 kg/m³ (10 to 15 lbf/ft³). In order to perform this procedure, the resolution of the count display, calibration response, or other method of displaying moisture content must be equal to or better than ±1.6 kg/m³ (±0.1 lbf/ft³).

2.3 SEDIMENT SAMPLING

Standard Guide for
Sampling Fluvial Sediment in Motion[1]

This standard is issued under the fixed designation D 4411; the number immediately following the designation indicates the year of original adoption or, in the case of revision, the year of last revision. A number in parentheses indicates the year of last reapproval. A superscript epsilon (ϵ) indicates an editorial change since the last revision or reapproval.

1. Scope

1.1 This guide covers the equipment and basic procedures for sampling to determine discharge of sediment transported by moving liquids. Equipment and procedures were originally developed to sample mineral sediments transported by rivers but they are applicable to sampling a variety of sediments transported in open channels or closed conduits. Procedures do not apply to sediments transported by flotation.

1.2 This guide does not pertain directly to sampling to determine nondischarge-weighted concentrations, which in special instances are of interest. However, much of the descriptive information on sampler requirements and sediment transport phenomena is applicable in sampling for these concentrations, and 9.2.8 and 13.1.3 briefly specify suitable equipment. Additional information on this subject will be added in the future.

1.3 The cited references are not compiled as standards; however they do contain information that helps ensure standard design of equipment and procedures.

1.4 Information given in this guide on sampling to determine bedload discharge is solely descriptive because no specific sampling equipment or procedures are presently accepted as representative of the state-of-the-art. As this situation changes, details will be added to this guide.

1.5 *This standard does not purport to address all of the safety problems, if any, associated with its use. It is the responsibility of the user of this standard to establish appropriate safety and health practices and determine the applicability of regulatory limitations prior to use.* Specific precautionary statements are given in Section 12.

2. Referenced Documents

2.1 *ASTM Standards:*
D 1129 Terminology Relating to Water[2]
D 3977 Practice for Determining Suspended Sediment Concentration in Water Samples[3]

3. Terminology

3.1 *Definitions:*

3.1.1 *isokinetic*—a condition of sampling, whereby liquid moves with no acceleration as it leaves the ambient flow and enters the sampler nozzle.

3.1.2 *sampling vertical*—an approximately vertical path from water surface to the streambed. Along this path, samples are taken to define various properties of the flow such as sediment concentration or particle-size distribution.

3.1.3 *sediment discharge*—mass of sediment transported per unit of time.

3.1.4 *suspended sediment*—sediment that is carried in suspension in the flow of a stream for appreciable lengths of time, being kept in this state by the upward components of flow turbulence or by Brownian motion.

3.1.5 For definitions of other terms used in this guide, see Terminology D 1129.

3.2 *Descriptions of Terms Specific to This Standard:*

3.2.1 *concentration, sediment*—the ratio of the mass of dry sediment in a water-sediment mixture to the volume of the water-sediment mixture. Refer to Practice D 3977.

3.2.2 *depth-integrating suspended sediment sampler*—an instrument capable of collecting a water-sediment mixture isokinetically as the instrument is traversed across the flow; hence, a sampler suitable for performing depth integration.

3.2.3 *depth-integration*—a method of sampling at every point throughout a sampled depth whereby the water-sediment mixture is collected isokinetically to ensure the contribution from each point is proportional to the stream velocity at the point. This method yields a sample that is discharge-weighted over the sampled depth. Ordinarily, depth integration is performed by traversing either a depth- or point-integrating sampler vertically at an acceptably slow and constant rate; however, depth integration can also be accomplished with vertical slot samplers.

3.2.4 *point-integrating suspended-sediment sampler*—an instrument capable of collecting water-sediment mixtures isokinetically. The sampling action can be turned on and off while the sampler intake is submerged so as to permit sampling for a specified period of time; hence, an instrument suitable for performing point or depth integration.

3.2.5 *point-integration*—a method of sampling at a fixed point whereby a water-sediment mixture is withdrawn isokinetically for a specified period of time.

3.2.6 *stream discharge*—the quantity of flow passing a given cross section in a given time. The flow includes the mixture of liquid (usually water), dissolved solids, and sediment.

4. Significance and Use

4.1 This guide is general and is intended as a planning guide. To satisfactorily sample a specific site, an investigator must sometimes design new sampling equipment or modify existing equipment. Because of the dynamic nature of the transport process, the extent to which characteristics such as mass concentration and particle-size distribution are accurately represented in samples depends upon the method of

---

[1] This guide is under the jurisdiction of ASTM Committee D-19 on Water and is the direct responsibility of Subcommittee D19.07 on Sediments.

Current edition approved April 15, 1993. Published June 1993. Originally published as D 4411 – 84. Last previous edition D 4411 – 84 (1991).

[2] *Annual Book of ASTM Standards*, Vol 11.01.

[3] *Annual Book of ASTM Standards*, Vol 11.02.

collection. Sediment discharge is highly variable both in time and space so numerous samples properly collected with correctly designed equipment are necessary to provide data for discharge calculations. General properties of both temporal and spatial variations are discussed.

5. Design of the Sampling Program

5.1 The design of a sampling program requires an evaluation of several factors. The objectives of the program and the tolerable degree of measurement accuracy must be stated in concise terms. To achieve the objectives with minimum cost, care must be exercised in selecting the site, the sampling frequency, the spatial distribution of sampling, the sampling equipment, and the operating procedures.

5.2 A suitable site must meet requirements for both stream discharge measurements and sediment sampling (1).[4] The accuracy of sediment discharge measurements are directly dependent on the accuracy of stream discharge measurements. Stream discharge usually is obtained from correlations between stream discharge, computed from flow velocity measurements, the stream cross-section geometry, and the water-surface elevation (stage). The correlation must span the entire range of discharges which, for a river, includes flood and low flows. Therefore, it is advantageous to select a site that affords a stable stage-discharge relationship. In small rivers and man-made channels, artificial controls as weirs can be installed. These will produce exceptionally stable and well defined stage-discharge relationships. In large rivers, only natural controls ordinarily exist. Riffles and points where the bottom slope changes abruptly, such as immediately upstream from a natural fall, serve as excellent controls. A straight uniform reach is satisfactory, but the reach must be removed from bridge piers and other obstructions that create backwater effects.

5.3 A sampling site should not be located immediately downstream from a confluence because poor lateral mixing of the sediment will require an excessive number of samples. Gaging and sampling stations should not be located at sites where there is inflow or outflow. In rivers, sampling during floods is essential so access to the site must be considered. Periods of high discharge may occur at night and during inclement weather when visibility is poor. In many instances, bridges afford the only practical sampling site.

5.4 Sampling frequency can be optimized after a review of the data collected during an initial period of intensive sampling. Continuous records of water discharge and gage height (stage) should be maintained in an effort to discover parameters that correlate with sediment discharge, and, therefore, can be used to indirectly estimate sediment discharge. During periods of low-water discharge in rivers, the sampling frequency can usually be decreased without loss of essential data. If the sediment discharge originates with a periodic activity, such as manufacturing, then periodic sampling may be very efficient.

5.5 The location and number of sampling verticals required at a sampling site is dependent primarily upon the degree of mixing in the cross section. If mixing is nearly complete, that is the sediment is evenly and uniformly distributed in the cross section, a single sample collected at one vertical and the water discharge at the time of sampling will provide the necessary data to compute instantaneous sediment-discharge. Complete mixing rarely occurs and only if all sediment particles in motion have low fall velocities. Initially, poor mixing should be assumed and, as with sampling any heterogeneous population, the number of sampling verticals should be large.

5.6 If used properly, the equipment and procedures described in the following sections will ensure samples with a high degree of accuracy. The procedures are laborious but many samples should be collected initially. If acceptably stable coefficients can be demonstrated for all anticipated flow conditions, then a simplified sampling method, such as pumping, may be adopted for some or all subsequent sampling.

6. Hydraulic Factors

6.1 *Modes of Sediment Movement:*

6.1.1 Sediment particles are subject to several forces that determine their mode of movement. In most instances where sediment is transported, flow is turbulent so each sediment particle is acted upon by both steady and fluctuating forces. The steady force of gravity and the downward component of turbulent currents accelerate a particle toward the bed. The force of buoyancy and the upward components of turbulent currents accelerate a particle toward the surface. Relative motion between the liquid and the particle is opposed by a drag force related to the fluid properties and the shape and size of the particle.

6.1.2 Electrical charges on the surface of particles create forces that may cause the particles to either disperse or flocculate. For particles in the submicron range, electrical forces may dominate over the forces of gravity and buoyancy.

6.1.3 Transport mode is determined by the character of a particle's movement. Clay and silt-size particles are relatively unaffected by gravity and buoyant forces; hence, once the particles are entrained, they remain suspended within the body of the flow for long periods of time and are transported in the suspended mode.

6.1.4 Somewhat larger particles are affected more by gravity. They travel in suspension but their excursions into the flow are less protracted and they readily return to the bed where they become a part of the bed material until they are resuspended.

6.1.5 Still larger particles remain in almost continuous contact with the bed. These particles, termed bedload, travel in a series of alternating steps interrupted by periods of no motion when the particles are part of the streambed. The movement of bedload particles invariably deforms the bed and produces a bed form (that is, ripples, dunes, plane bed, antidunes, etc.), that in turn affects the flow and the bedload movement. A bedload particle moves when lift and drag forces or impact of another moving particle overcomes resisting forces and dislodges the particle from its resting place. The magnitudes of the forces vary according to the fluid properties, the mean motion and the turbulence of the flow, the physical character of the particle, and the degree of exposure of the particle. The degree of exposure depends largely on the size and shape of the particle relative to other particles in the bed-material mixture and on the position of the particle relative to the bed form and other relief features on the bed. Because of these factors, even in steady flow, the bedload discharge at a point fluctuates significantly with time. Also, the discharge varies substantially from one point to another.

6.1.6 Within a river or channel, the sizes of the particles in transport span a wide range and the flow condition determines the mode by which individual particles travel. A change in flow conditions may cause particles to shift from one mode to the other.

6.1.7 For transport purposes, the size of a particle is best characterized by its fall diameter because this describes the particle's response to the steady forces in the transport process.

6.2 *Dispersion of Suspended Sediment:*

6.2.1 The various forces acting on suspended-sediment particles cause them to disperse vertically in the flow. A particle's upward velocity is essentially equal to the *difference* between the mean velocity of the upward currents and the particle's fall velocity. A particle's downward velocity is essentially equal to the *sum* of the mean velocity of the downward currents and the particle's fall velocity. As a result, there is a tendency for the flux of sediment through any horizontal plane to be greater in the downward direction. However, this tendency is naturally counteracted by the establishment of a vertical concentration gradient. Because of the gradient, the sediment concentration in a parcel of water-sediment mixture moving upward through the plane is higher than the sediment concentration in a parcel moving downward through the plane. This difference in concentration produces a net upward flux that balances the net downward flux caused by settling. Because of their high fall velocities, large particles have a steeper gradient than smaller

particles. Figure 1 (2) shows (for a particular flow condition) the gradients for several particle-size ranges. Usually, the concentration of particles smaller than approximately 60 μm will be uniform throughout the entire depth.

6.2.2 Turbulent flow disperses particles laterally from one bank to the other. Within a long straight channel of uniform cross section, lateral concentration gradients will be nearly symmetrical and vertical concentration gradients will be similar across the section. However, within a channel of irregular cross section, lateral gradients will lack symmetry and vertical gradients may differ significantly. Figure 2 (3) illustrates the variability within one cross section of the Rio Grande.

6.2.3 Sediment entering from the side of a channel slowly disperses as it moves downstream and lateral gradients may exist for several hundred channel widths downstream. In or near a channel bend, secondary flow accentuates both horizontal and vertical gradients. Until data have been collected to prove the contrary, one must assume both gradients exist and design sampling procedures accordingly.

6.2.4 At sections where spatial variability exists, samples must be collected from many regions within a cross section. Only for special conditions will samples from one or two points be adequate.

6.2.5 Despite turbulent currents that disperse particles along the direction of flow, the concentration at a fixed point will vary with time even if flow conditions are steady. Temporal variability depends upon many factors. Within a group of samples collected during a short period of time, the concentration of any sample generally will not deviate from the mean by more than approximately 20 %; however, every sample must be composed of a stream filament at least 50 ft long.

7. Spatial and Temporal Variations in Bedload Discharge

7.1 Bedload discharges vary both within a section and

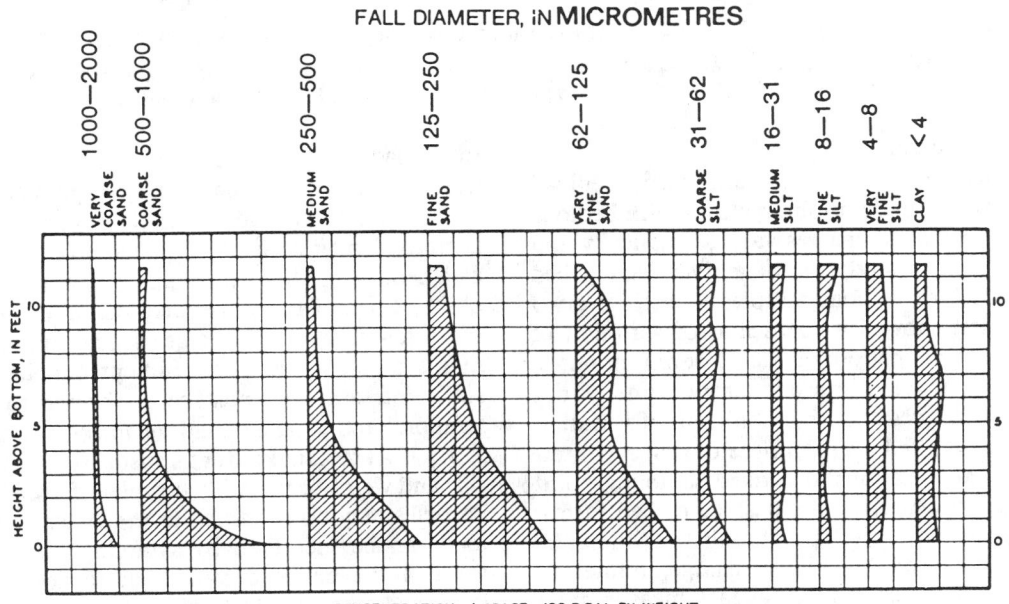

FIG. 1 (2) Vertical Distribution of Sediment in the Missouri River at Kansas City, MO

along the channel due to variations in the sediment and mean flow properties, turbulence, patterns of secondary circulation and position relative to the bed relief. (See 13.1, also 7.2.) Also, because of the intimate relationship between bedload discharge and the flow forces, particles that move as bedload at one section may be immobile or may move as suspended load at another cross section. As a result, the proportion of bedload discharge to total sediment transport may vary longitudinally and bedload discharge observed at one section may not be representative of the bedload

the sampling position must be moved longitudinally so that samples are obtained randomly over parts of several bed-form wave-lengths.

8. Spatial and Temporal Variations in Total-Sediment Discharge

8.1 Temporal and spatial variations in the total sediment discharge result from the combined effects of variations in the suspended-sediment discharge and the bedload discharge. Detailed information on the extent of temporal variations in

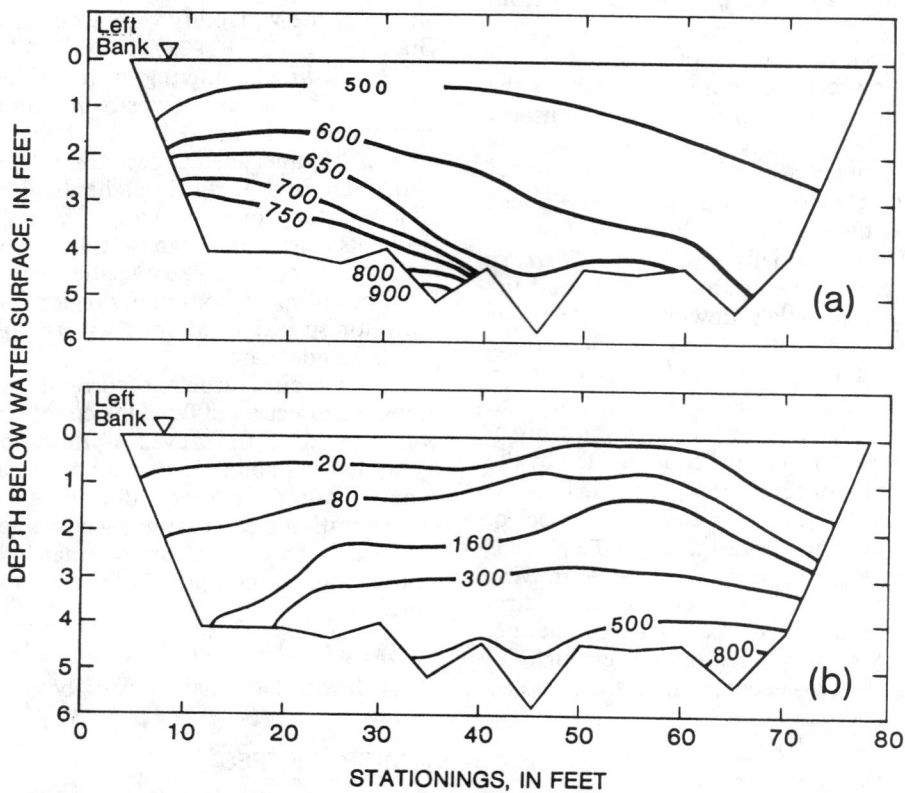

FIG. 2 (3) Cross-Sectional Variability of Suspended Material in Two Different Size Ranges, Rio Grande, near Bernardo, NM (a) Contours in mg/L for Material Between 0.0625 and 0.125 mm; (b) Contours in mg/L for Material Between 0.25 and 0.5 mm

discharge at another section.

7.2 Although data on the temporal variation in bedload discharge are far from abundant, observations with bedload samplers have shown that discharges vary dramatically and tend to be cyclic. In one study (4) of a river having bed material of coarse cobbles, bedload samples collected every 3 min during a 3-h test showed a coefficient of variation of 41 % and an oscillation period of about seven minutes. Another study (5), conducted in a laboratory flume with a bed of coarse gravel, showed that the coefficient of variation of bedload samples collected every minute during a 1-h test was 100 %. Temporal variations at a fixed sampling point are caused, in large measure, by the passage of bed forms. Because a single measurement at a point probably will not be representative of the mean bedload discharge, numerous repetitive measurements must be made at each measurement point during a time interval that is sufficiently long to allow a number of bed-form wave-lengths to pass. Alternatively,

total load are scarce; however, as with variations in suspended sediment discharge, the variations in total load can be expected to change according to particle size. Ordinarily, at normal river sections, the total load cannot be measured as a separate entity; therefore, it is obtained by combining observations of the suspended load and the bedload. When the total-sediment discharge is determined from measurements of the suspended-sediment and bedload discharges, sufficient sampling must be performed to account for the temporal and spatial variations in both quantities.

8.1.1 At certain kinds of unusual sections, such as outfalls, sills and weirs, or in highly turbulent flow, all of the sediment particles may be entrained in the water; consequently, total load can be measured by sampling through the nappe or through the entire depth. Such sections are often called total-load sections. At total-load sections, spatial variations in the total sediment discharge can be significant and are functions of the lateral variations in flow properties, suspended-sediment concentration, and bedload discharge. At total-load sections, sampling must be carried out in accordance with the principles of suspended-sediment sampling and replicate samples must be collected at a sufficient number of lateral locations to account for variations in the discharge of entrained bedload particles.

9. Selection and Design of Sampling Apparatus

9.1 Apparatus selection depends upon the object of the sampling program and the physical and hydraulic characteristics of the site. To sample for total sediment discharge within a straight section of open channel, use a suspended-sediment sampler in conjunction with a bedload sampler. If initial measurements show that nearly all of the total load is transported in suspension, routine sampling can be simplified by eliminating bedload measurements. At an outfall, total load may be measured by sampling through the nappe with a depth-integrating sampler. Because these samplers are calibrated when fully submerged, the depth of the nappe should be great enough to ensure the flow contacts the region downstream of the air exhaust port. For continuous sampling of total load, a traveling-slot or a stationary-slot sampler may be used.

9.2 *Suspended Sediment Samplers:*

9.2.1 Whenever the fluid within a streamtube accelerates by changing either its direction or speed, sediment particles tend to migrate across the streamtube boundaries. This migration causes a local enrichment or depletion in the sediment concentration. To avoid such changes at a sampling nozzle, suspended-sediment samplers must operate isokinetically (or nearly isokinetically). If the velocity at the entrance of the sampler nozzle deviates from ambient velocity by less than ±15 %, the error in concentration will seldom exceed ±5 %. The angle between the axis of the nozzle and the approaching flow should not exceed 20°.

9.2.2 Two basic types of isokinetic instruments are commonly used to sample suspended sediment. One type (integrating) accumulates the liquid-sediment mixture by withdrawing it during a long period of time. The other type (trap) instantaneously traps a volume of the mixture by simultaneously closing off the ends of a flow-through chamber. The integrating type collects a long filament of flow, hence, the sample concentration is only slightly affected by short-term fluctuations in the concentration within the approaching flow. For this reason, integrating types are recommended over trap types.

9.2.3 For integrating-type samplers it is recommended that the nozzle entrance be circular in cross section and have an inside diameter of 3.2 mm (⅛ in.) or larger. A nozzle with a diameter of 4.8 mm or 6.3 mm (³⁄₁₆ in. or ¼ in.) is preferred. At the nozzle entrance, the wall thickness should not exceed 1.6 mm (¹⁄₁₆ in.) and the outside edge should be gently rounded.

9.2.4 To ensure an undisturbed flow pattern, the nozzle must extend upstream from its support which may be a tethered body or a fixed support strut. An upstream distance of 25.4 mm (1 in.) is adequate provided the support is well streamlined and its largest dimension lateral to the flow is not more than 40 nozzle diameters.

9.2.5 After entering the nozzle, the sample must be conveyed, without a change in concentration, to a container.

If the volume of the conduit is more than approximately 5 % of the sample volume, the velocity within the conduit must be adequate to ensure transport as a homogeneous suspension. A velocity exceeding 17 W is recommended where W equals settling velocity of the largest particle in suspension.

9.2.6 Integrating samplers that meet the above requirements are fabricated commercially in the United States. The samplers, which are listed in Table 1 (6), belong to the "US series" designed by the Federal Interagency Sedimentation Project. The samplers are of two types, depth-integrating and point-integrating.

9.2.7 *Depth-Integrating Samplers*—US series depth-integrating samplers have an intake nozzle and exhaust port but they do not have a valve; therefore, they sample the water-sediment mixture continuously when submerged. They are highly reliable because they do not contain moving parts; furthermore, they are suitable for use in a sampling technique termed "depth integration" (see 13.1.4). Depth-integrating samplers have a maximum operating depth (see Table 1) (6). Figure 3 (7) shows the shape of one member of the US series of depth-integrators. Auxiliary equipment includes a cable-and-reel suspension system, or for the DH-48 (8) and DH-75, a wading-rod suspension. During the depth-integration process, a sampler must be lowered and raised at a uniform rate so cable-speed indicators or timing devices are used whenever possible.

9.2.8 *Point-Integrating Samplers*—US series point-integrating samplers have an intake nozzle and exhaust port that can be opened and closed while the samplers are submerged. They also contain a pressure-equalization system to ensure that the pressure within the sample container equals the hydrostatic pressure whenever the intake-exhaust valve is opened. These features allow the samplers to be used for sampling by either the depth integration or point integration (see 13.1.3) techniques. Maximum allowable depths listed for these samplers in Table 1 (6) apply when they are used for point integration. When the samplers are used for depth integration starting at the water surface, the depth limitations given in footnote B of Table 1 (6) specify the length of the allowable two-way vertical sampling path for any single-sample container; segments of an allowable path length can be sampled throughout all or any part of the maximum allowable depth by using multiple containers and opening and closing the intake-exhaust valve appropriately. If sampling is done by one-way integration, the allowable path length is twice the listed value. In addition to a suspension and speed indicating system, the samplers also require a source of electrical power.

10. Bedload Samplers

10.1 Both in Europe and the United States many different kinds of bedload monitoring apparatus (9) have been developed to measure the transport of a wide variety of bed-material particles that occur in nature. In general, each kind of apparatus was designed to monitor a particular range of bedload sizes and transport rates. Two broad classifications exist, direct-measuring apparatus and indirect-measuring apparatus. Direct-measuring apparatus collect and accumulate bedload particles for a given period of time. Indirect-measuring apparatus monitor some property of the bedload or some phenomena that occurs as a result of bedload

TABLE 1 (6) Physical Characteristics of US-Series Depth-Integrating and Point-Integrating Samplers for Collecting Samples of Water-Suspended Sediment Mixtures (after Table 3-3, National Handbook of Recommended Methods for Water-Data Acquisition)

NOTE—[Type: DI, depth-integrating; PI, point-integrating. Available nozzle size: A, 6.4 mm; B, 4.8 mm; C, 3.2 mm; D, 7.9 mm. Body material: AL, aluminum; BR, bronze; PS, plated steel].

Name	Type of Sampler	Method of Suspension	Mass, kg	Overall Length, m	Available Nozzle Size	Sample Container Size, mL	Maximum Allowable Depth, m	Maximum Calibrated Velocity, m/s	Distance Between Nozzle and Sampler Bottom, mm	Body Material	Remarks
US DH-48	DI	rod	2.0	0.33	A, B[A]	473	[B]	2.7	90	AL[C]	for wading.
US DH-59	DI	cable	10.2	0.42	A, B, C	473	[B]	1.5	114	BR[C]	for hand-line operation.
US DH-75P	DI	rod	0.4	0.26	B	500	[B]	2.0	83	PS[C]	for sampling only in sub-freezing temperatures.
US DH-75Q	DI	rod	0.4	0.29	B	1000	[B]	2.0	114	PS[C]	similar to US DH-75P.
US DH-76	DI	cable	10.9	0.47	A, B, C	946	[B]	2.0	80	BR[C]	similar to US DH-59.
US D-43	DI	cable	22.6	0.52	A, B, C	473	[B]	2.1	105	BR	no longer available.
US D-49	DI	cable	28.0	0.61	A, B, C	473	[B]	2.1	103	BR[C]	
US D-49AL	DI	cable	18.0	0.61	A, B, C	473	[B]	2.0	103	AL[C]	similar to US D-49.
US D-74	DI	cable	28.2	0.66	A, B, C	473 or 946	[B]	2.0	103	BR[C]	similar to US D-49.
US D-74AL	DI	cable	11.4	0.66	A, B, C	473 or 946	[B]	1.8	111	[C,D]	similar to US D-74.
US D-77	DI	cable	34.0	0.75	D	3000	4.72	2.4	177	BR	can be converted to a bag-type sampler.
US P-46	PI	cable	45.2	0.66	B	473	42.7	3.0	122	BR	no longer available.
US P-50	PI	cable	135.6	1.12	B	473 or 946	61.0[E] 41.0[F]	3.0	140	BR	
US P-61-A1	PI	cable	47.5	0.71	B	473 or 946 473 or 946	54.9[E] 36.6[F]	2.0	109	BR[C]	
US P-63	PI	cable	90.4	0.85	B		54.9[E] 36.6[F]	2.0	150	BR[C]	
US P-72	PI	cable	17.7	0.71	B	473 or 946	22.0[E] 15.5[F]	1.6	109	AL[C]	similar to P-61-A1 but for hand-line operation.

[A] 4.8-mm nozzle available by special order.
[B] Varies with nozzle and container sizes as follows:

Nozzle Size	Container Size	
	473 mL	946 mL
C	5.8 m	4.9 m
B	4.9 m	4.9 m
A	2.7 m	4.9 m

[C] Available with epoxy-painted body, nylon nozzles, and silicone-rubber gaskets for trace metals.
[D] Aluminum body, bronze head.
[E] With 473-mL container.
[F] With 946-mL container.

movement. In addition, bedload discharge can be determined from measurements of the rate of (1) migration of bedforms, (2) movement of tracer particles, (3) deposition or erosion in a given area, and (4) change with distance in the concentration of some nonconservative property associated with the bedload particles. This nonconservative property, such as radioactivity, must have a known time rate of decay.

10.1.1 No portable direct-measuring apparatus nor indirect technique is generally accepted at this time as being entirely suitable for determining bedload discharge.

10.2 *Direct Measuring Apparatus:* Direct-measuring apparatus can be classified into four general categories; box or basket samplers, pan or tray samplers, pressure-difference samplers, and slot or pit samplers.

10.2.1 *Box or Basket Samplers*—Enclosures are open at the upstream end and possibly at the top, and have either solid sides, mesh sides, or a combination of both. Particles are retained within the sampler either by being screened from the flow or by settling in regions of reduced flow velocities within the sampler.

10.2.2 *Pan or Tray Samplers*—These samplers collect particles that drop into one or more sections or slots after the particles have been transported up an entrance ramp.

10.2.3 *Pressure-Difference Samplers*—Essentially box or basket samplers that have entrances or other features that create a pressure drop that overcomes the flow resistance within the sampler and thereby keeps flow velocities at the entrance about the same as the stream velocity.

10.2.4 *Slot or Pit Samplers*—These samplers consist of collection chambers that accumulate particles as they drop over the forward edge of a chamber that is buried in the stream bed.

10.3 *Indirect-Measuring Apparatus:* Most indirect-measuring apparatus are acoustical devices that measure (1) the magnitude and frequency of particle-sampler or particle-particle collisions or (2) the attenuation of energy. Apparatus of this type ordinarily give only qualitative information and their outputs must be correlated with known discharges to provide quantitative results. Acoustic devices are seldom used in routine data collection programs.

11. Total-Sediment Discharge Samplers

11.1 Because the total sediment discharge is composed of suspended-sediment particles, which moves along within the body of the flow essentially at stream velocity, and bedload particles, which moves in an interrupted fashion essentially in continuous contact with the bed, no practical sampler has been designed for sampling total-sediment discharge at regular river sections. Normally, the total sediment discharge is determined from separate measurements of the suspended

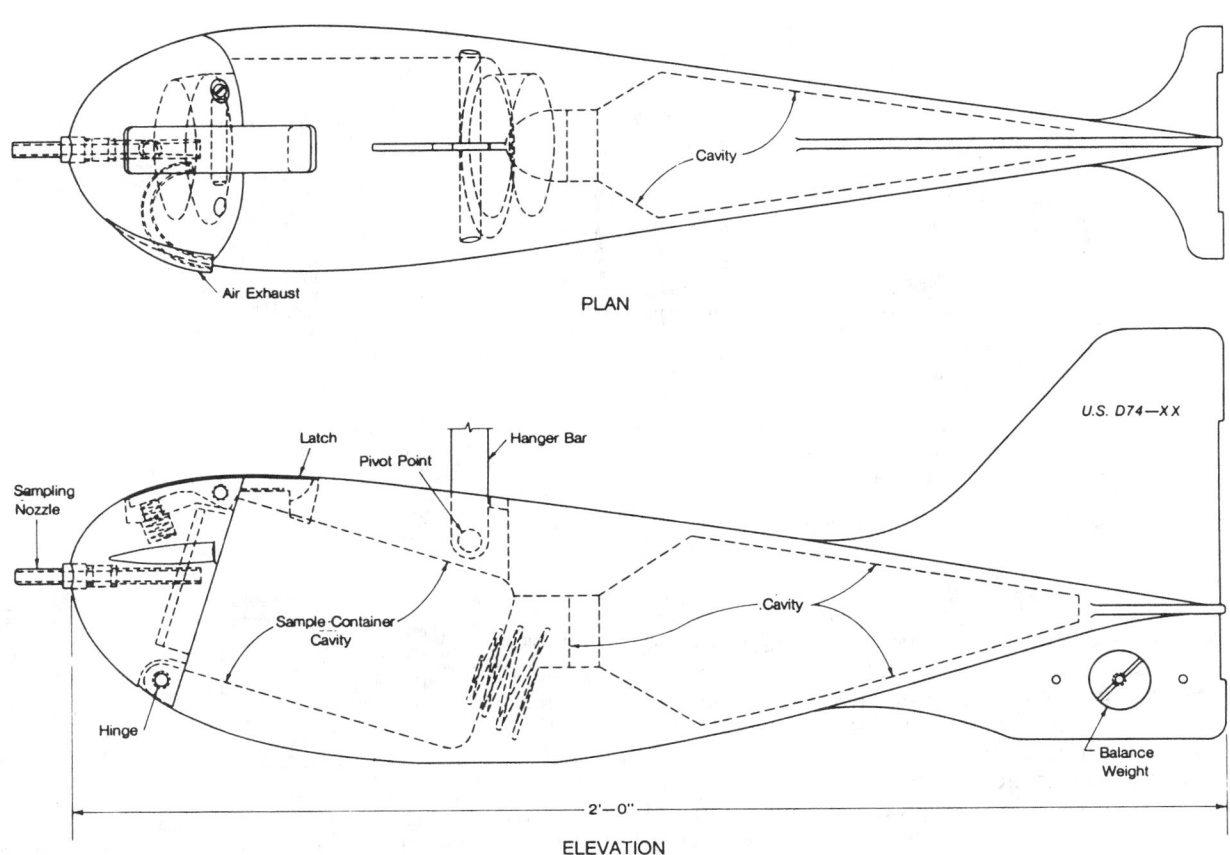

FIG. 3 (7)   US D-74 Suspended Sediment Sampler

sediment discharge and the bedload discharge. Conventional sampling equipment can be used to measure the total sediment discharge at certain sections termed total-load sections. At an outfall, a sill, a weir, or a section where flow turbulence is sufficient to entrain all the sediment within the flow, suspended-sediment sampling equipment and techniques can be used to determine the discharge of particles finer than 2 mm. For particles coarser than 2 mm, use equipment that is capable of collecting and retaining coarse particles, and that is based on the isokinetic principles of suspended-sediment sampling. Such equipment includes slot samplers. Economic considerations usually preclude the construction of artificial total-load sections except on small streams. The ASCE Manual (10) illustrates a large but expensive turbulence flume.

11.2 If sampling can be conducted at a free outfall, slot samplers can be installed. By means of a slotted conduit positioned in the outfall, the slot sampler diverts a fraction of the water-sediment mixture into a suitable container. Slot samplers have been used extensively in erosion research and in laboratory flumes but standard designs have not been perfected for sampling sediment or industrial waste water in open channels or streams. Slot samplers are normally used in conjuction with a flume, weir, or other flow measuring devices.

11.3 The slot width must be adequate to permit free passage of the largest sediment particle; the conduit must be streamlined to minimize disturbance to the flow. Sides of the slot may be formed from rigid-metal sheets that are supported so that the slot opening faces the flow. The slot edges should be knife sharp and true to line. A tube or flexible pipe connected to the bottom of the slot carries the sample to a suitable storage container. The sampler is mounted on the downstream end of a flow measuring device with a free overfall. The height of the sampler slot must exceed the depth of flow. Some slot samplers will not function properly if the flow transports debris capable of clogging the slot. The slot may be located at a fixed point in the flow or it may be propelled across the flow. Accordingly, slot samplers may be divided into two broad categories, stationary or transversing.

11.4 *Stationary Slot Samplers*—Stationary slot samplers are simple to build and operate. They require no external source of power. To enhance self-clearing of debris, incline the slot at a slight downward angle. Figure 4 (11) illustrates several types that have been tested. Samples are extracted along one fixed line so they are less representative than those collected with a traversing slot.

11.5 *Traversing Slots*—Traversing slots collect a sample representative of the entire cross section. The vertical slot sampler requires electric power, and is relatively insensitive to approach conditions. In situations where only infrequent clogging is anticipated, satisfactory performance may be obtained by using brushes or other equipment to periodically clean the slot. Fiber brushes mounted so that the slot is brushed before each pass through the flow nappe will usually assure satisfactory performance. Figure 5 (12) illustrates one

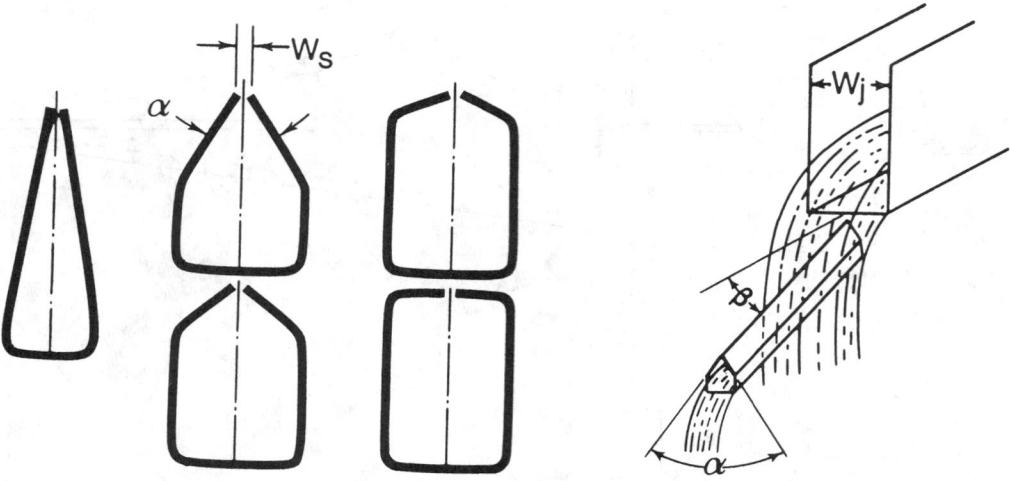

FIG. 4 (11)   Cross Sections and Installation of Slot-Type Sampler

type which has been tested.

11.6 *Rotating (Coshocton-Type) Sampler*—The rotating (Coshocton-type) sampler, (13) which is in the traversing category, consists of an elevated slot affixed to a revolving water wheel that is mounted on the downstream end of a small H-flume. Discharge from the flume falls on the water wheel and causes it to rotate. With each revolution the sampling slot cuts across the flow jet and extracts a sample. The sample falls into a collecting pan beneath the wheel and is routed through a closed conduit to a storage tank. A typical Coshocton-type sampler is pictured in Fig. 6 (13) and Fig. 7 (13). Sampler size, maximum discharge rate, sampling ratio, and other pertinent data are given in Table 2 (14).

11.6.1 The Coshocton sampler requires no external power source, however it is sensitive to upstream approach conditions. Rotation of the wheel may become erratic at stream discharges that exceed 80 % of the flume capacity.

12. Hazards

12.1 *Personal Clothing, Equipment, and Training:* **Precautions**—Operators should wear protective footgear and protective headgear, safety glasses, and leather gloves in addition to high-visibility clothing that is warm enough to prevent hypothermia. Where drowning is a hazard, perform sampling by teams that are equipped with Coast Guard-approved personal-flotation devices and that are proficient in both swimming and first aid.

12.2 *Electrical Hazard:* **Warning**—Equipment powered from low-voltage batteries is safer than equipment powered from 120-V, a-c distribution circuits. Regardless of the power source, ground the frames of hoists and other equipment to nearby metal objects such as bridge railings, bridge decks, or boat hulls. Use ground-fault detectors where applicable. During electrical storms, operators should retreat to low ground or take cover in a building or metal-topped vehicle.

12.3 *Vehicles:* **Precautions**—Equip vehicles that must be parked on road shoulders with warning lights and flares in compliance with local regulations. Isolate the cargo area from the driver-passenger area; lash the cargo to prevent tipping or sliding.

TABLE 2   (14) Size Schedule for Coshocton Samplers

Sampler No.	Wheel Diameter	Capacity	Headroom Requirement	Aliquot	Approximate Weight
	Ft	Ft³/s	Ft	Pct	Lb
N-1	1	⅓	1½	1	26
N-2	2	2	2½	½	85
N-3	3	5½	3¾	⅓	270

12.4 *Sampling Wadable or Ice-Covered Streams:* **Precautions**—At crossings that appear marginal from safety aspects, test the surface with a rod or ice chisel, and wear a safety line anchored to a firm object on the shore.

12.5 *Sampling from Overhead Cableways and Bridges:* **Precautions**—Inspect supports and safety railings regularly for loose, worn, or weak components. At sites where trees or other heavy debris may snag a submerged sampler, the operator should be prepared to sever the suspension cable if the sample cannot be retrieved.

12.6 *Reports and Medical Treatment*—Report all accidents and potentially dangerous situations promptly to the local safety officer. To save valuable time when an accident occurs, procedures for obtaining professional emergency treatment should be clearly understood by all operators.

13. Sampling Techniques

13.1 *Techniques for Sampling Suspended Sediment:*

13.1.1 Because of spatial variations in suspended sediment concentrations and in flow velocity, the discharge of suspended sediment through an area at any given instant is defined by Eq 1 (15) as follows:

$$G_{ss} = \int_A C\, U\, \mathrm{d}A \tag{1}$$

where:

G_{ss} = "instantaneous" suspended sediment discharge through a section of area A,

U = velocity of sediment particles through an elemental area $\mathrm{d}A$,

C = suspended sediment concentration in the elemental spatial volume $U\,t'\,\mathrm{d}A$,

for which:

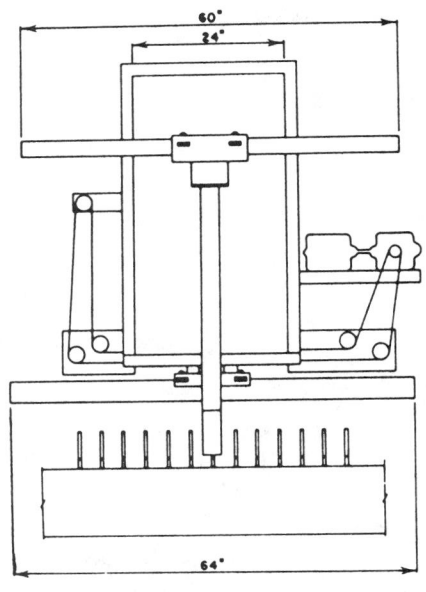
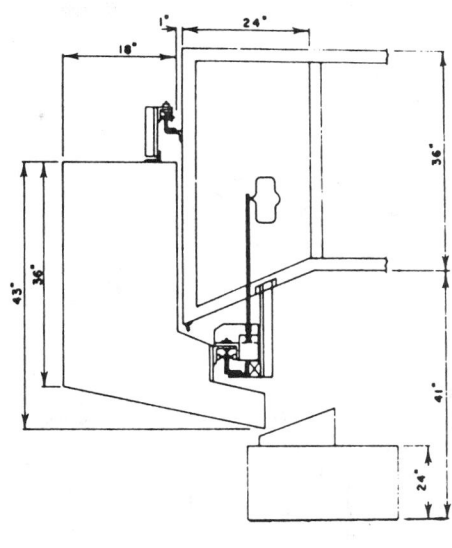

FRONT ELEVATION   SIDE ELEVATION

FIG. 5 (12)   Space Required for the Traversing-Slot Sampler Mounted on a 2-ft Parshall Flume

t' = unit of the time used to express U.

In the practical application of Eq 1, U is considered to equal the flow velocity and C is considered to be constant during any given sampling period.

13.1.2 Three different techniques are commonly used to evaluate Eq 1; point integration, depth integration, and area integration. In point integration, samplers and sampling procedures are designed to yield spatial concentrations at a series of points throughout an area. These concentrations together with flow velocities from individual points are used to define concentration and velocity gradients, which, in turn are integrated according to Eq 1 to give the instantaneous suspended-sediment discharge through the area.

13.1.3 To sample by point integration, divide the flow area laterally into increments and collect samples at several depths along a vertical in each increment. Select increment widths and sampling depths so that between adjacent sampling points the difference in concentration and difference in velocity are small enough to conform with desired accuracy. Use a P-61 or any other sampler that meets requirements of Section 9.

13.1.4 In depth integration and area integration, the sampling equipment and procedures are designed to mechanically and hydraulically perform the integration over the flow area. With both depth integration and area integration, an isokinetic sampler is traversed through the flow so that each incremental volume of mixture collected from the corresponding element of traversed area is in the same proportion to the sample volume as the stream discharge in each corresponding element is to the stream discharge in the sampled area. This procedure yields samples having "discharge-weighted" concentrations that can be multiplied directly with the stream discharge through the sampled area to yield the instantaneous suspended-sediment discharge

through the area. The following derivation, which uses Eq 1 in discrete form, mathematically explains the concept. Consider a sampled area divided into N elemental areas of size $\Delta x \Delta y$. Let Q_i be the water discharge and C_i be the suspended-sediment concentration of mixture flowing through the ith element. The suspended-sediment discharge, G_{ss}, through the sampled area then is defined by Eq 2 as follows:

$$G_{ss} = \sum_{i=0}^{N} C_i Q_i \qquad (2)$$

Since by definition, $C_i = w_i/v_i$, $w = \sum_{i=0}^{N} w_i$, and $C_m = w/v$, and by virtue of the sampling technique, $v_i/v = Q_i/Q$.

where:

w_i and v_i = mass of sediment and volume of mixture, respectively, collected from the ith element,

w, v, and C_m = mass of sediment, volume of mixture, and the "discharge-weighted" concentration, respectively, in and of the sample collected from the sampled area, and

Q = stream discharge through the sampled area.

Substituting the defining equations and the "sampling technique" equation in Eq 2 we obtain:

$$G_{ss} = \sum_{i=0}^{N} \left(\frac{w_i}{v_i}\right)\left(\frac{v_i Q}{v}\right) = \frac{Q}{v}\sum_{i=0}^{N} w_i = \frac{Q_w}{v} = C_m Q \qquad (3)$$

13.1.5 In area integration, the entire flow cross section is sampled, hence, C_m and Q are the discharge-weighted sediment concentration and the stream discharge for the entire cross section. In depth integration, the sampled area is only that part of the stream cross section transversed by the intake nozzle at a single vertical. To determine the suspended-sediment discharged through an entire cross section, a series of verticals must be sampled by depth integration. By

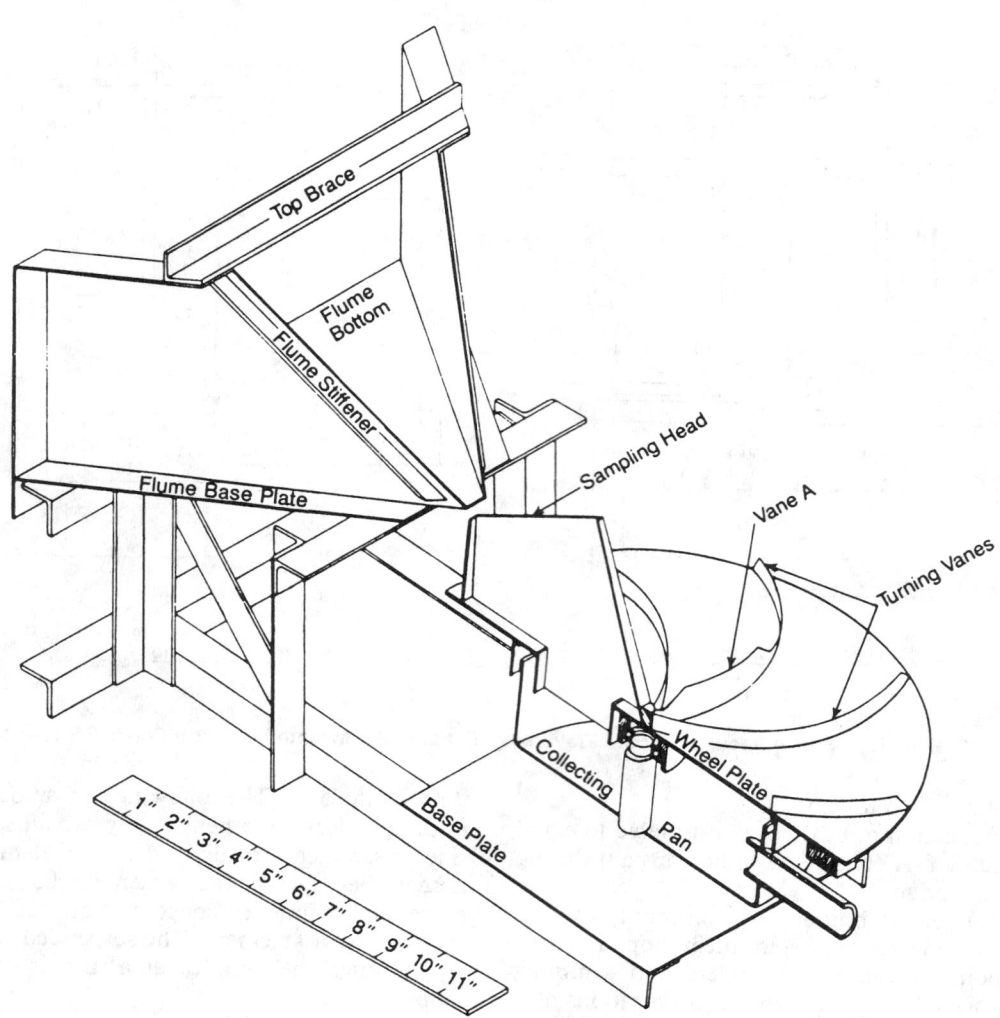

FIG. 6 (13)   The N-1 Coshocton-Type Runoff Sampler

assuming that the discharge-weighted sediment concentration from each vertical represents a certain proportion of the total stream discharge, a discharge-weighted sediment concentration for the entire cross section can be obtained and combined with the total stream discharge to give the suspended-sediment discharge through the cross section. The accuracy of the samples improves with an increase in the number of sampling verticals.

13.1.6 In principle, the fixed slot performs depth integration at a single vertical. Instead of sampling along the vertical with a moving nozzle, the slot instantaneously samples the whole depth.

13.2 *Depth Integration:*

13.2.1 To sample by depth integration, collect water-sediment mixture along a vertical line throughout the entire depth by using a US D- or P-series sampler. While a sample is being collected, the sampler must be moved vertically at a uniform velocity termed the transit rate (16). With the D series, sampling is continuous through the entire depth. P series samplers are equipped with a valve so sampling may be continuous or interrupted in a series of segments through the depth. The only basic requirements are that at each sampling vertical, (1) the entire depth is sampled isokinetically, (2) the

vertical transit rate in any given direction is the same over all parts of the depth and the rate never exceeds the product of K, (Figs. 8 **(17)** and 9 **(18)**) multiplied by the mean stream velocity at the vertical, and (3) sampling does not extend in one-way integration over a distance greater than $K_e V_i/A_n$, or half this distance in two-way integration (K_e must be ≥ 0.4 and is the largest K value possible for a given nozzle area, A_n, and sample volume, V_i). Figure 8 **(17)** and Figure 9 **(18)** show graphs of K versus stream depth for standard nozzle sizes and for 1- and 2-pint containers. K is a dimensionless number.

13.3 *Depth Integration Throughout a Flow Cross Section:*

13.3.1 Sampling throughout a cross section requires depth integration at several verticals spaced across the section. Two methods are commonly used: the equal-width-increment (EWI) method and the equal-discharge-increment (EDI) method. Either method is applicable, but depending on the situation, one method may be more convenient to apply than the other.

13.4 *EWI (Equal-Width Increment) Method:*

13.4.1 The EWI method produces for each increment a sample volume proportional to the stream discharge through the increment. All increments are of equal width and

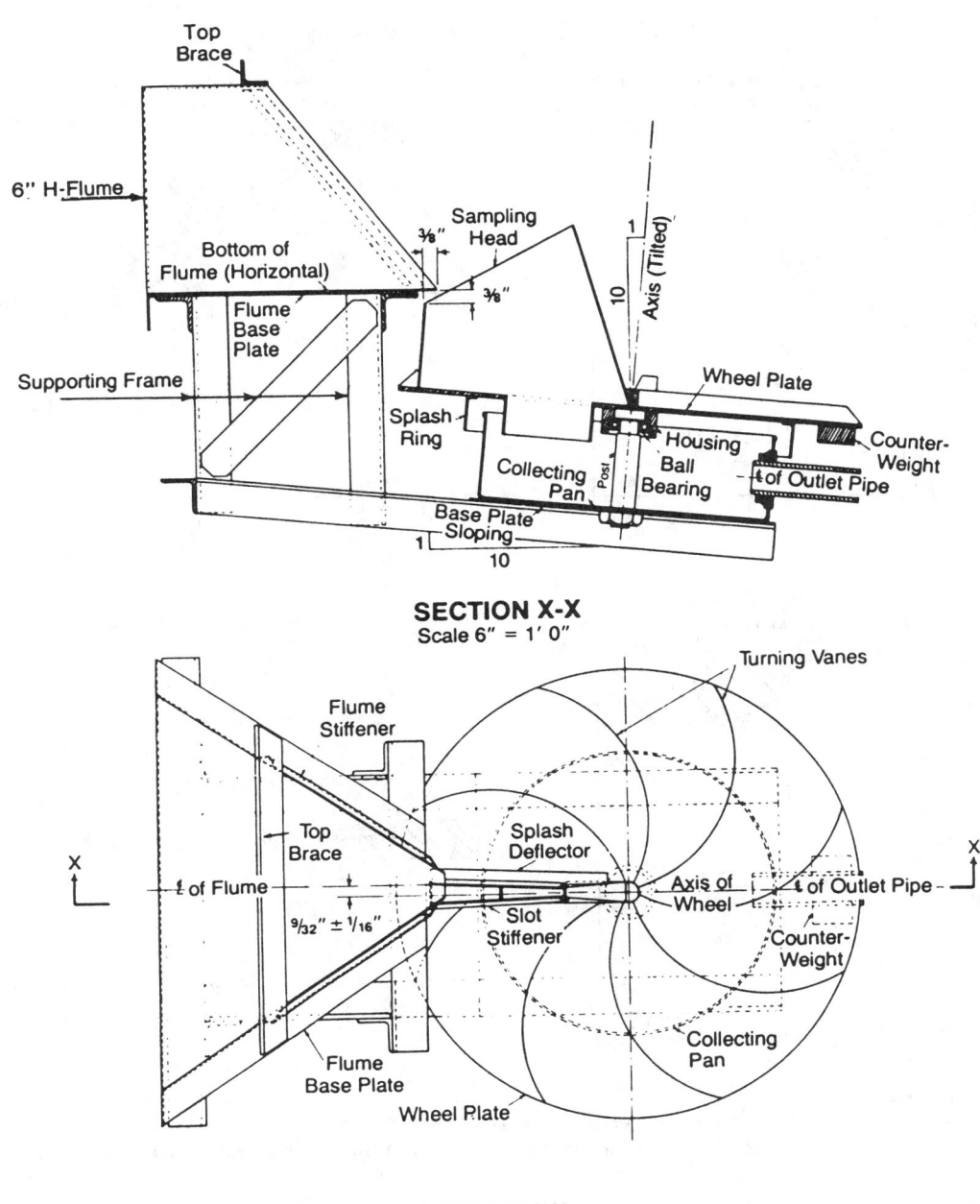

SECTION X-X
Scale 6″ = 1′ 0″

TOP VIEW
Scale 6″ = 1′-0″

NOTE—1 in. = 25.4 mm, 1 ft = 0.3 m.

FIG. 7 (13)   Details of the N-1 Coshocton-Type Runoff Sampler

sampling verticals are located at the center of each increment. Furthermore, the transit rate is the same at all verticals. This method produces a gross sample in which the concentration of the sediment is weighted both vertically and laterally according to stream discharge. The method is often used in shallow, sandbed streams where the discharge rating is unstable. The number of verticals depends upon the streamflow, the sediment characteristics of the stream, and the sampling accuracy desired. In general, not more than 25 verticals will be necessary.

13.4.2 Initially, select a sampling interval such that at least several sampling verticals are located within each part

of the cross section where the suspended-sediment discharge is substantially different from the other parts. In sections that have a uniform shape and a relatively uniform lateral distribution of suspended-sediment discharge, 10 or more sampling verticals are ordinarily sufficient. In sections that have a nonuniform lateral distribution of suspended-sediment discharge, 20 or more verticals are required. Make selection of intervals on the basis of detailed information on the lateral distribution of stream discharge. If such information is unavailable make the selection only after a visual survey.

13.4.3 Establish the distance, S, of the first sampling

350 ml<Sample Volume<440 ml

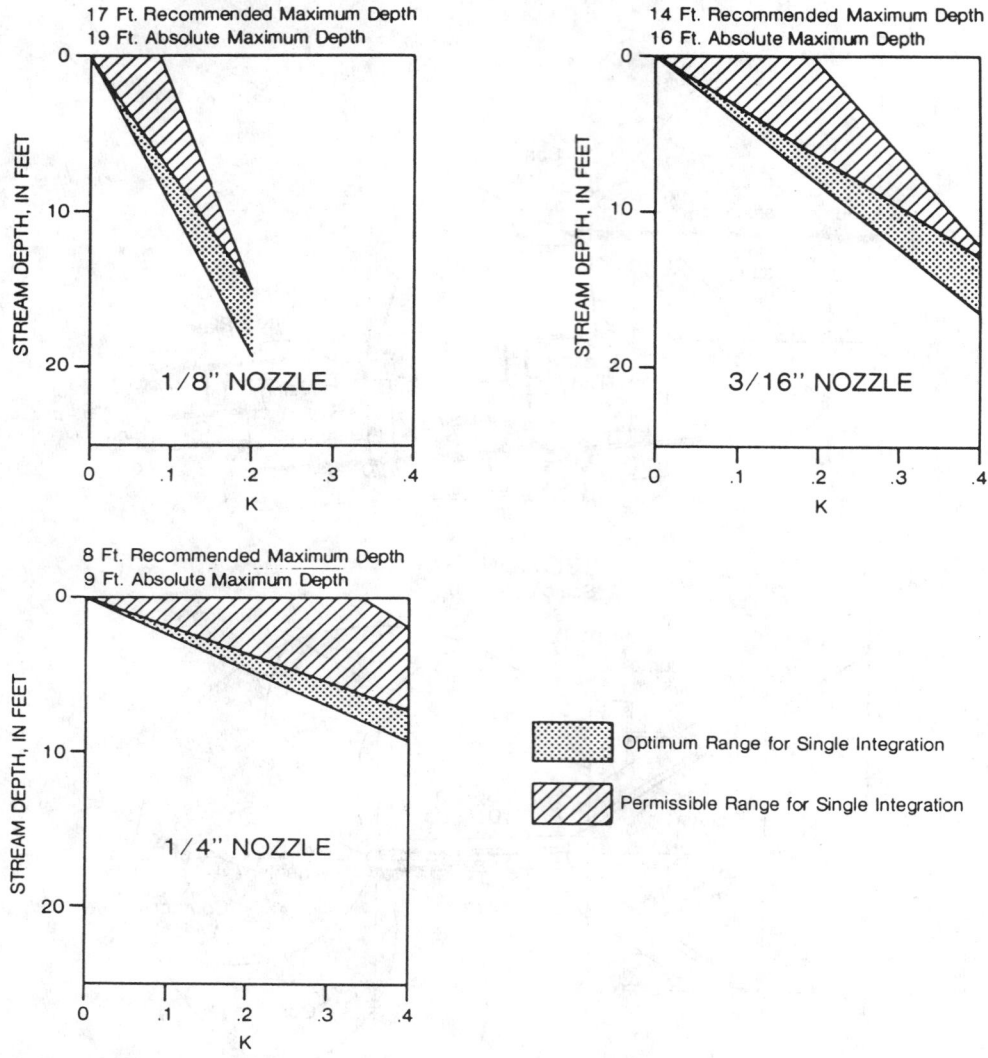

NOTE 1—1 in. = 25.4 mm, 1 ft = 0.3 m.
NOTE 2—Sampler's transit rate equals K multiplied by mean flow velocity at the sampling vertical.

FIG. 8 (17)   Transit Rates for Depth-Integrating Sampler Type US D-74 with Pint Sample Container

vertical from the edge of the water, by first dividing the selected sampling interval, I, into the surface width, W, to ascertain the number of times, N, the interval will divide into the width evenly; then, compute the distance E from $E = \frac{1}{2}(W - NI)$; and finally, determine S from (a) $S = E + I/2$ when $E < I/4$, (b) $S = I/2$ when $E = 0$, or (c) $S = E$ when $E \geq I/4$. Subsequent sampling verticals are spaced according to the selected sampling interval. The last vertical is located at a distance of S units from the other edge of the water.

13.4.4 By experimentation, determine a transit rate at the vertical with the greatest water discharge per foot of width. The rate should be the slowest rate possible that falls within the optimum range designated in the appropriate graph in Figs. 8 **(17)** and 9 **(18)**; this rate is also used at all other sampling verticals.

13.4.5 Lower the sampler to the water surface then, while keeping the nozzle out of the water, allow the sampler to

become oriented with the flow. To sample, lower the sampler to the streambed and then raise the sampler to the surface at the transit rate determined from 13.4.4. Once established, maintain the transit rate while lowering and raising; furthermore, maintain the same transit rate at all verticals.

13.4.6 Move to the next vertical and repeat the sampling process. Water-sediment mixture from more than one vertical can be accumulated in a single bottle; however, care must be taken not to sample a vertical if the additional mixture will cause the bottle to overfill.

13.4.7 Mark each bottle with appropriate identification information. Prior to analysis, samples from all verticals may be composited.

13.4.8 The number of verticals initially selected may be reduced if it can be established that fewer verticals will provide the required accuracy.

13.5 *EDI (Equal-Discharge Increment) Method:*

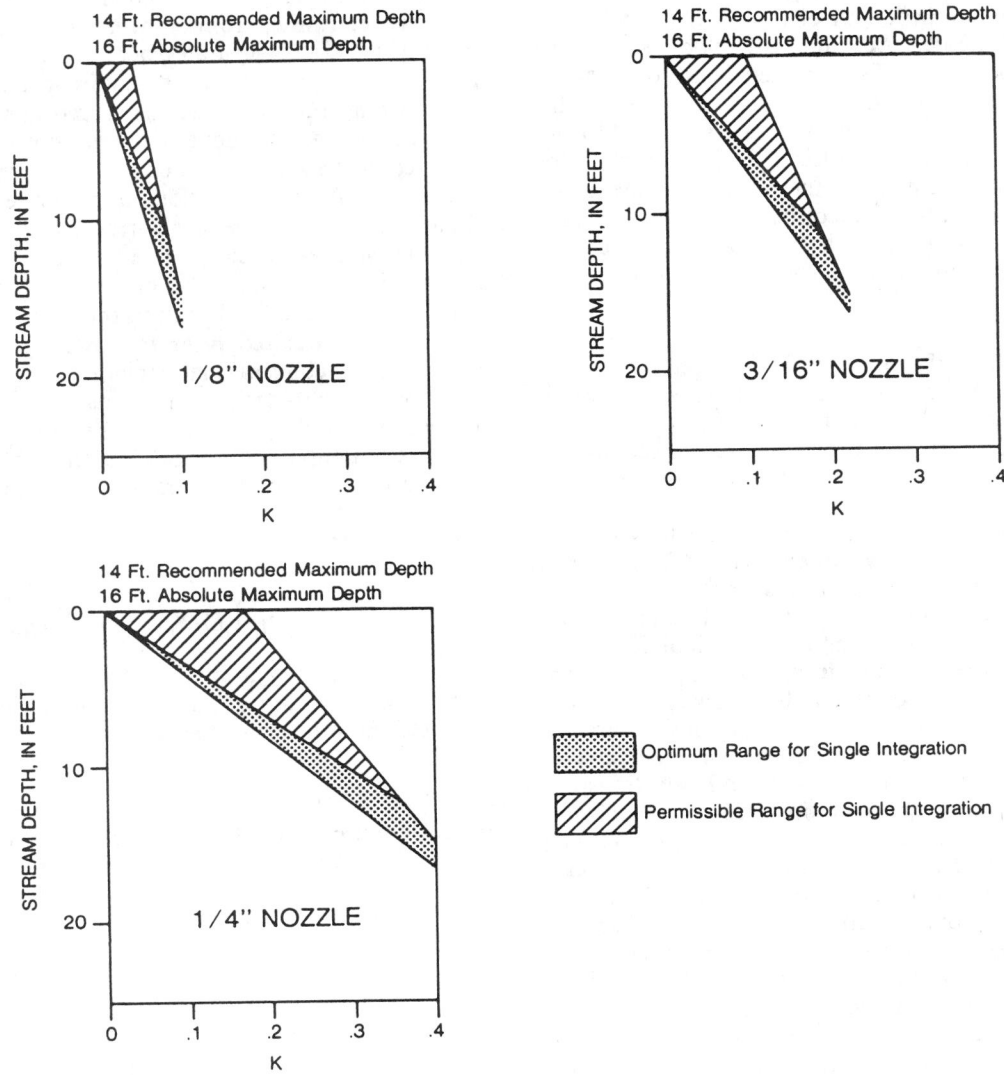

NOTE 1—1 in. = 25.4 mm. 1 ft = 0.3 m.
NOTE 2—Sampler's transit rate equals K multiplied by mean flow velocity at the sampling vertical.

FIG. 9 (18)   Transit Rates for Depth-Integrating Sampler Type US D-74 with Quart Sample Container

13.5.1 The EDI method produces a group of samples each of which represents the same proportion of the total water discharge. If the samples are of equal volume (within 10 % of each other), they can be composited to form a single gross sample representative of the cross section. Otherwise, each sample must be analyzed separately then concentration values must be averaged numerically to obtain the mean discharge-weighted concentration for the cross section. The method is commonly used to sample rivers that have stable channels and stable discharge rating.

13.5.2 Use stream-gaging equipment or other appropriate apparatus to accurately measure the lateral distribution of water discharge in the cross section. Arrange the results to express accumulated discharge as a function of distance from one edge of the water. Select a sufficient number of verticals to be sampled so as to assure that at least one vertical is located in each part of the cross section where the stream discharge per foot of width is distinctly different from the other parts. For uniformly shaped sections where the lateral distribution of suspended sediment discharge is relatively uniform, seven sampling verticals will usually be satisfactory. For irregular sections more verticals are required; 14 verticals will usually be sufficient. Divide the number of verticals, N, into the total water discharge to determine the discharge to be represented by each sampling vertical, and then partition the width into N increments of equal discharge (19). Within each increment, sample at one vertical that is located so half of the increment's discharge is on one side of the vertical. The transit rate at a vertical must be uniform; but transit rates may differ among the verticals. Care must be taken not to overfill the sample container nor to exceed the maximum transit rate designated in the appropriate graph in Figs. 8 (17) and 9 (18).

13.6 *Area Integration:*

13.6.1 To sample by area integration, a traversing slot must be installed in accordance with 11.3. The slot must sample through the entire depth of flow and must sample isokinetically at all depths. At a uniform velocity, move the slot laterally across the flow. The sample consists of all liquid and sediment that enters the slot during one traverse. By this procedure an element of sampled volume is collected from every area element in the cross section. The ratio of the volume of a sample element to the volume of the entire sample is equal to the ratio of the discharge through the area element to the discharge through the entire cross section; hence, the concentration of the sample is "discharge weighted."

14. Techniques for Sampling Bedload

14.1 With direct-measuring bedload samplers, information on the bedload discharge is obtained by accumulating particles for a given period of time. To average spatial and temporal variations in the transport rate, a number of individual samples must be collected at each of several different lateral positions across the width of the flow section. Whenever it is possible, many samples should be collected and analyzed statistically to determine the optimum number necessary for a given accuracy. Because the rate of bedload transport may vary over the length of the individual bed forms, samples must be collected from a number of random positions along the lengths of several bed forms.

14.2 The number of lateral sampling positions required to adequately represent the discharge depends on the character of the lateral variations in transport. If both the transport rate and the bed form configuration are fairly uniform laterally, sampling sequentially at numerous points across a section provides samples from a number of random positions along the length of the bed forms. Hence, if traverses are made repetitively over a long enough time for several bed-form wave lengths to pass by the measuring section, the necessity to sample both randomly along the length of the bed forms and at different lateral positions is satisfied.

14.3 If the sampler is supported on a suspension line and if the velocities or bed-form relief is high, extreme care must be taken to avoid inadvertently collecting stationary bed material. As the sampler is lowered into layers of progressively lower velocity, the drag force on the sampler continually decreases so the unit achieves an upstream motion which must be arrested before it strikes the bed and scoops bed material. Also, when the sampler is retrieved, there is a tendency for it to slide upstream prior to lifting off the bed; during this process scooping is possible. Scooping can sometimes be reduced by using a front stay-line to limit the downstream motion. Even with a stay-line, extreme turbulent pulsations combined with line elasticity may cause the sampler to oscillate on the bed and thereby scoop some material. Scooping can be minimized by installing devices on the sampler that cause its entrance to lift off the bed immediately upon retrieval.

15. Calculation

15.1 *Calculation of Suspended Sediment Discharge:*

15.1.1 Instantaneous suspended-sediment discharge can be computed from either Eq 1 (see 13.1.1) or Eq 3 (see 13.1.4). In either case, the accuracy of the result depends upon the representativeness of the samples and the accuracy of the stream-discharge measurements. When sediment discharge is determined from point-integrated samples in accordance with Eq 1, accurate measurements of vertical and lateral velocity gradients are required. These are best obtained from point velocities measured at the time of sampling. Suspended-sediment discharge computed according to Eq 3 requires accurate data on stream discharge at the time of sampling. At many locations, an accurate discharge can be obtained from the gage-height record and a stage-discharge rating curve. If an accurate rated discharge is not available, stream discharge should be measured immediately before sampling or immediately after sampling. When the stream discharge is measured prior to sampling and is measured according to conventional stream-gaging techniques, the measurement also provides the data necessary to locate the position of sampling verticals for the EDI sampling method.

15.1.2 If temporal variations in both stream discharge and suspended-sediment concentrations are significant, the quantity of suspended sediment, S, discharged through an area during a specified duration, T, must be determined as follows:

$$S = \int_T \int_A C\, U\, \mathrm{d}A\, \mathrm{d}T = \int_T Q_s \mathrm{d}T \tag{4}$$

Whenever sampling is in accordance with depth- or area-integration procedures and concentrations are discharge weighted, Eq 4 becomes the following:

$$S = \int_T C_m Q \mathrm{d}T \tag{5}$$

If, during time period T, C_m, or Q or both, are nearly constant, integration of Eq 5 can be simplified as follows:

$$\text{if } C_m = \text{constant} = \overline{C}, \text{ then } S = \overline{C} \int_t Q \mathrm{d}T = \overline{C}\overline{Q}T \tag{6}$$

$$\text{if } Q = \text{constant} = \overline{Q}, \text{ then } S = \overline{Q} \int_T C_m \mathrm{d}T = \overline{C}_m \overline{Q}T \tag{7}$$

if C_m = constant = $\overline{C}_m$ and Q = constant = $\overline{Q}$,

$$\text{then } S = \overline{C}_m \overline{Q} \int_T \mathrm{d}T = \overline{C}_m \overline{Q}T \tag{8}$$

15.1.3 If the sample was collected by a slot sampler, the sample is a time-integrated composite. Measure both the mass of sediment and volume of the mixture in the sample container. Divide each quantity by R, the sampling ratio, to obtain the mass of sediment and volume of mixture discharged by the channel.

15.1.4 For a fixed slot

$$R = W_s/W_j \tag{9}$$

15.1.5 For the vertical traversing slot,

$$R = W_s T_i/W_j T_2 \tag{10}$$

where:
W_s = slot width,
W_j = width of the flow jet,
T_i = time the sampler is in the flow jet per traverse, and
T_2 = time for one complete traverse.

15.1.6 for the Coshocton-type sampler [20]

$$R = W_s/2\pi r \tag{11}$$

where:

r = radial distance from the center of rotation.

15.1.7 Equations 1 through 6 are written without specification of units. In addition to the preferred units for concentration, mass of sediment per unit volume of mixture, concentration may be expressed in terms of volume of sediment per unit volume of water-sediment mixture or in terms of mass of sediment per unit mass of water sediment mixture. Parts per million by mass and milligrams per litre are commonly used. Stream discharge is ordinarily expressed as volume per unit time and sediment discharge is usually expressed as mass per unit time. Because of the disparity in units, each of the equations requires a units conversion constant to make it dimensionally correct. The constant may involve the density of either the solid sediment or the water-sediment mixture in addition to mass, volume and time conversions. Reported values for concentration, stream discharge, and sediment discharge should be specified in metric units.

15.1.8 A sampler in the US series will not sample from a zone close to the streambed. This zone extends upward from the bed to a height equal to the distance from the sampler nozzle to the sampler bottom (see Table 1). As a result, depth-integrated samples collected with these samplers do not contain a contribution from this unsampled zone. When the suspended-sediment discharge is computed from Eq 3 by using the discharge-weighted concentration from such samples, there is an unaccounted portion of the total suspended-sediment discharge that is called the "unmeasured suspended-sediment discharge." This discharge equals $Q''(C_m'' - C_m)$ in which Q'' is the water discharge in the unsampled zone and C_m'' is the discharge-weighted concentration in the unsampled zone. If the suspended sediment is distributed uniformly with depth so that $C_m'' = C_m$ or if the unsampled depth is very small relative to the total stream depth, then the unmeasured suspended-sediment discharge is usually considered to be inconsequential. However, if the suspended sediment concentration gradient increases significantly near the streambed or if the unsampled zone is a large fraction of the sampled depth, then the unmeasured discharge can be significant.

16. Precision and Bias

16.1 *General Comments:*

16.1.1 The selection of a sampling procedure and the evaluation of probable measurement error must be based on statistical parameters that are site dependent. These parameters can be estimated by analyzing results from an initial sampling program and by studying the nature of the sediment source. This initial sampling should be intensive, so that spatial variability and temporal variability can be determined. The basic data can also be used to estimate errors associated with more abbreviated programs or with more simplified sample-collecting procedures.

16.1.2 In a large river, sampling errors that occur because of a change in equipment or procedure can be estimated by comparing the data sets; however, this comparison will not reveal sampling errors that are common to all data sets. In some small streams and flumes, more accurate error analysis can be performed by comparing sediment discharge computed from samples with sediment discharge obtained from an independent measurement. This comparison will reveal all sampling errors regardless of their source. One experiment technique involves the use of a vibrating-type of material feeder. With a feeder, inject dry sediment into the flow at a suitable location upstream of the sampling site. Sediment discharge determined by sampling can be compared with the sediment injection rate. A correction must be made for sediment stored within the channel between the sampling site and the feeder.

16.1.3 Particles transported in suspension can be sampled more accurately than particles transported near or on the bed; consequently, for a given flow condition, sampling errors increase with an increase in the size of the particles in transport.

16.2 *Suspended Sediment Samplers:*

16.2.1 With the U.S. Series of samplers the largest errors occur while sampling close to the bed. Each sampler is constructed with its nozzle several centimeters above the bottom of the sampler; therefore, in a channel with a firm bed, the flow between the bed and the nozzle entrance is unsampled. Samplers should be chosen or designed to minimize this unsampled depth. In a channel having a bed composed of soft deposits or dunes, oversampling may result if the sampler is allowed to settle into the bed or into a trough. In deep streams the long length of submerged cable and the heavy weight of the sampler act to increase the time required to reverse the direction of sampler motion. This reversal time combined with the high concentration near the bed result in oversampling.

16.2.2 A point-integrating or a depth integrating sampler should be checked to ensure it samples isokinetically. The first step is to adjust the balance of the sampler. The nozzle must be level to keep the entrance facing directly into the approaching flow. Adjustments can be made by adding or removing weights from the lower tail vane (See Fig. 3).

16.2.3 The hydraulic tests can be performed either in a flume or in a tow tank (21). If a flume is used, the breadth and depth must be sufficient to eliminate blockage effects between the walls of the flume and the body of the sampler. Tests indicate a flume width of 3.0 ft and a water depth of 6.0 ft are sufficient for testing P-61's, which are the largest of the commonly used suspended-sediment samplers. The test section of the flume must be longer than about 18 ft.

16.2.4 The first step in performing the hydraulic tests is to set the desired flow speed at the test point, which is usually taken midway between the flume walls and about two feet below the water surface. Measure the speed V_f with a rotating-bucket current meter. Remove the meter before proceeding to the next step.

16.2.5 If the sampler is a point integrator, close the sampling valve before positioning the nozzle at the test point. When all is ready, switch the valve open and at the same instant, start a timer. After a few seconds, close the valve and stop the timer.

16.2.6 Hoist the sampler above the flow and measure the volume collected in the sample container. The optimum volume is about 80 % of the bottle's rated capacity. If the volume is too small, discard the sample and collect another during a longer interval; if the volume is too large, shorten the interval.

16.2.7 The flow speed inside the nozzle, V_n (in feet per second), is given by the equation $V_n = KQ/T$ where Q is the sample volume in millilitres, T is the sampling interval in seconds, and K is a constant that depends on the diameter of the hole at the nozzle's entrance. K is 0.4143 for a ⅛ in. hole, 0.1841 for a 3/16 in. hole, 0.1036 for a ¼ in. hole, and 0.0663 for a 5/16 in. hole.

16.2.8 For isokinetic sampling, V_n must match V_f; however, in practice a 10 % deviation from this ideal condition is permissible. In other words, V_n should be between $0.9 V_f$ and $1.1 V_f$. At flow speeds slower that about 1.5 ft/s, a larger deviation is acceptable.

16.2.9 V_n can be increased by enlarging the hole where the sample water emerges from the nozzle. Use a reamer with a taper of ¼ in./ft to cut away a small amount of material. Proceed cautiously: alternative between reaming and hydraulic testing. Once removed, the material cannot be replaced. Normally, this reaming operation is performed only once when a sampler is new.

16.2.10 If the sampler under test is a depth integrator, set V_f as explained in 16.2.3. Then lower the sampler and start the timer at the instant the nozzle penetrates the water surface. Hold the sampler at the test point for a few seconds, then hoist the sampler above the flow. Stop the timer when the nozzle breaks through the water surface.

16.2.11 Refer to 16.2.6 through 16.2.8 for computations and reaming.

16.2.12 If a tow tank is used for the hydraulic tests, the breadth and depth must be sufficient to eliminate blockage effects between the walls and sampler. Satisfactory results have been obtained in a tank 12 ft wide and 12 ft deep. The minimum length of the tank must be about 450 ft if the tests are to be run at speeds ranging up to 10 ft/s.

16.2.13 If the sampler is a point integrator, close the sampling valve before positioning the sampler at the test point. When the carriage has stabilized at the desired speed, open the sampling valve and at the same instant start the timer.

16.2.14 After a few seconds, close the valve and, at the same instant, stop the timer. Refer to 16.2.5 for the optimum sample volume. It may be necessary to collect samples smaller than the optimum in order to stop the carriage safely at the end of the run.

16.2.15 Refer to 16.2.6 through 16.2.8 for computations and adjustments. In the towing tests, V_f corresponds to the towing speed that is usually read from a meter aboard the towing carriage.

16.2.16 If the sampler is a depth integrator, the tow tank carriage must be moving at a steady speed before the sampler is lowered into the water. Start the timer at the instant the nozzle penetrates the water surface. Wait a few seconds, then hoist the sampler. Stop the timer at the instant the nozzle clears the water. Refer to 16.2.6 through 16.2.9 for computations and adjustments.

16.3 *Slot Samplers:*

16.3.1 To assure acceptable performance, slot samplers must be calibrated under conditions for which they were designed and are to be used. Their liquid sampling ratio and sediment sampling ratio should be calibrated for a wide range in discharge, sediment concentration, and sediment particle size.

16.3.2 The volumetric or weight method of calibration is recommended. If the stream discharge is known from the flow measuring device used in conjunction with the sampler, the liquid sampling ratio is readily determined from timed volumetric or weight measurements of discharge from the slot sampler. A calibrated tank will greatly facilitate the measurement of sample volume.

16.3.3 Determining the sediment sampling ratio is more difficult. Both the liquid and sediment discharge must be known. Thorough mixing of a known quantity of sediment in the flow upstream from the sampler and accurate measurements of sample volume and weight are required. When the sediment concentration in the sample is the same as the sediment concentration in the flow, the sediment sampling ratio is assumed to equal the liquid sampling ratio.

16.3.4 If either sampling ratio varies with stream discharge, a composite sample collected during a period of variable flow may not be representative of the total flow. Samples composited under such conditions must be discharge weighted to be truly representative.

16.3.5 Normally the assumption is made that the sediment sampling ratio is the same as the liquid sampling ratio. This may not be true if the sediment is composed of heavy particles. Calibration to determine the sediment sampling ratio is required.

16.4 *Vertical Slot Samplers:*

16.4.1 The liquid sampling ratio of the vertical slot, Fig. 5, **(12)** increases with increasing flow depth (discharge) and increases as W_s/W_j increases.

16.4.2 The accuracy of vertical slot samplers is dependent on the precision used in construction. The slot must be straight and must have knife-sharp edges. Flow within the sampler and from the sampler to the storage tank must be unrestricted to prevent sediment deposition.

16.5 *Inclined Slot:*

16.5.1 The accuracy of inclined slot samplers is dependent on the precision used in construction. The slot must be straight and have knife-sharp edges. Flow within the sampler and from the sampler to the storage tank must be unrestricted to prevent sediment deposition.

16.5.2 Limited test data **(11)** for the inclined slot (Fig. 4) show that the sampling ratio (S): (*a*) does not change with moderate changes in Reynolds number (from 1600 to 3600); (*b*) is relatively constant for angles of inclination (β) between 5° and 20°; (*c*) increases as the ratio of slot width (W_s) to jet flow width (W_j) increases; (*d*) decreases as the angle (α) formed by the sides of the sampler increases; and (*e*) decreases as flow depth (discharge) increases. A relatively constant sampling ratio may be obtained for a range of flow depths (discharge) by using a slightly diverging slot that increases in width toward the downstream end. The width of the slot at any distance from the end of the flume will depend upon the shape (width) of the nappe at that point **(22)**.

16.6 *Coshocton Samplers:*

16.6.1 Coshocton samplers will not function precisely as indicated by Eq 11. The liquid sampling ratio is influenced by the following: (*a*) the speed and uniformity of rotation of the sampling wheel and slot; (*b*) size and geometry of the slot; (*c*) position of the slot relative to the jet; (*d*) flow approach conditions upstream from the sampler; (*e*) splashing; and (*f*) velocity and direction of the jet.

16.6.2 The width of the diverging slot at any radius (r) is approximately the following:

$$W_s = \frac{2\pi r\, q/Q}{1 + 1.6\, T}$$

where:

T = thickness of metal at the slot edge (in.), and

q/Q = desired proportion of stream discharge to be extracted; typically 0.01, 0.005, and 0.0033 for the 1, 2, and 3 ft samplers, respectively.

16.6.3 At high stream discharges the liquid sampling ratio tends to decrease rapidly with an increase in the stream discharge; therefore, the sampler should not be used if the discharge exceeds 80 % of flume capacity (23).

16.7 *Bedload Samplers:*

16.7.1 A bedload sampler must be calibrated to determine its sampling efficiency, which is defined as the ratio of the mass of bedload collected during any single sampling time to the mass of bedload that would have passed through the width occupied by the sampler entrance during the same time if the sampler had been absent. Many samplers have a sampling efficiency that varies from one particle size to another. Except for slot or pit samplers, which ordinarily have efficiencies close to 100 % for all conditions, the overall sampling efficiency of a specific sampler is not constant but varies with the size distribution of the bedload particles, the flow conditions, the rate of bedload transport, and the degree of filling of the sampler. The sampling efficiency is influenced, but is not solely determined by the sampler's hydraulic efficiency, which is the ratio of the volume of water that passed through the sampler entrance in a unit of time to the volume of water that would have passed through this entrance area if the sampler had been absent. Hydraulic efficiency and sampling efficiency can only be determined accurately through comprehensive and detailed procedures.

17. Keywords

17.1 fluvial sediment; samplers; sampler calibration; sediment discharge; sediment transport

REFERENCES

(1) Guy, H. P., and Norman, W. W., "Field Methods for Measurement of Fluvial Sediments," *Techniques of Water-Resources Investigations of the United States Geological Survey*, Book 3, Chapter C2, 1970, pp. 23–24.

(2) Inter-Agency Committee on Water Resources, "Determination of Fluvial Sediment Discharge," *Rept. 14, A Study of Methods Used in Measurement and Analysis of Sediment Loads in Streams*; Subcommittee on Sedimentation, Minneapolis, MN, 1963, p. 28.

(3) Bennet, J. P., and Nordin, C. F., Jr., "Suspended Sediment Sampling Variability," Chap. 17 in *Environmental Impact on Rivers (River Mechanics III)*, ed. and pub. by H. W. Shen, Fort Collins, CO, 1973.

(4) Ehrenberger, R., "Direkte Geschiebemessungen an der Donau bei Wein und deren bisherige Ergebnisse" (Direct Bedload Measurements on the Danube at Vienna and Their Results to Date): *Vienna, Die Wasserwirtschaft*, Issue 34, 1931, pp. 1–9. Translation No. 39–20, U.S. Waterways Experiment Station, Vicksburg, MS.

(5) Einstein, A. H., 1937, "Die Eichung des im Rhein verwendeten Geschiebefangers" (Calibrating the Bedload Trap as Used in the (Rhine): *Schweizer. Bauzeitung*, Vol 110, No. 12. Translation, Soil Conservation Service, California Institute of Technology, Pasadena, CA.

(6) Federal Interagency Work Group III 1978, Chapter 3—Sediment, in *National Handbook of Recommended Methods for Water Data Acquisition*, Office of Water Data Coordination, Geological Survey, U.S. Dept. of Interior, Reston, VA, pp. 3–20.

(7) Skinner, J. V., and Beverage, J. P., 1976, Instructions for Sampling with Depth-Integrating, Suspended-Sediment Samplers D-74, D-74AL. D-74-TM, and D-74AL-TM, *Federal Inter-Agency Sedimentation Project*, St. Anthony Falls Hydraulic Laboratory, Minneapolis, MN, Fig. 1.

(8) Inter-Agency Committee on Water Resources, "The Design of Improved Types of Suspended Sediment Samplers," Rept 6, *A Study of Methods Used in Measurement and Analysis of Sediment Loads in Streams*; Subcommittee on Sedimentation, Minneapolis, MN, 1952, pp. 60–63.

(9) Hubbell, D. W., "Apparatus and Techniques for Measuring Bedload," *U.S. Geological Survey Water-Supply Paper* 1748, Washington, DC, 1964.

(10) American Society of Civil Engineers, *Sedimentation Engineering*. by Task Committee. V. A. Vanoni, ed., ASCE, New York, NY, 1975, p. 342.

(11) Barnes, L. L., and Frevert, R. K.,"A Runoff Sampler for Large Watersheds, Part 1, Laboratory Investigations," *Agricultural Engineering*, Vol 35, No. 2, 1954, pp. 84–90.

(12) Dendy, F. E., "Traversing Slot Runoff Sampler for Small Watershed," *ARS-S-15*, U.S. Dept. of Agriculture, Southern Region, 1973.

(13) Parsons, D. A., "Coshocton-Type Runoff Samplers—Laboratory Investigations," *TP-124*, U.S. Dept. of Agriculture, Soil Conservation Service, 1954, Sheet 1.

(14) Dendy, F. E., Allen, P. B., and Piest, R. F., Chapter 4, Sedimentation, in "Field Manual for Research in Agricultural Hydrology," *Agricultural Handbook No. 224*; Science and Education Administration, U.S. Dept. of Agriculture, Beltsville, IN 20705.

(15) ASCE, *ibid*, p. 338.

(16) Inter-Agency Committee on Water Resources, Report No. 6, *ibid*, p. 28.

(17) Skinner, J. V., and Beverage, J. P., 1976, *ibid*, Fig. 2.

(18) Skinner, J. V., and Beverage, J. P., 1976, *ibid*, Fig. 3.

(19) American Society of Civil Engineers, *Sedimentation Engineering*, by Task Committee, V. A. Vanoni, ed., ASCE, New York, NY, 1975, p. 339.

(20) Carter, C. E., and Parsons, D. A., "Field Test on the Coshocton-Type Wheel Runoff Sampler," *ASAE Transactions*, 1967, Vol 10, No. 1, pp. 133–135.

(21) Beverage, Joseph P., and Futrell II, James C., "Comparison of Flume and Towing Methods for Verifying the Calibration of a Suspended-Sediment Sampler," Water Resources Investigations Report 86-4193 of the U.S. Geological Survey, 1986.

(22) Barnes, K. K., and Johnson, H. P. "A Runoff Sampler for Large Watersheds, Part II. Design of Field Installation," *Agricultural Engineering*, Vol 37, No. 12, pp. 813–815, 824.

(23) Dendy, F. E., Allen, P. B., and Piest, R. F., *ibid*, pp. 248–50, 378–87.

Standard Guide for
Core Sampling Submerged, Unconsolidated Sediments[1]

This standard is issued under the fixed designation D 4823; the number immediately following the designation indicates the year of original adoption or, in the case of revision, the year of last revision. A number in parentheses indicates the year of last reapproval. A superscript epsilon (ϵ) indicates an editorial change since the last revision or reapproval.

1. Scope

1.1 This guide covers core-sampling terminology, advantages and disadvantages of different types of core samplers, core-distortions that may occur during sampling, techniques for detecting and minimizing core distortions, and methods for dissecting and preserving sediment cores.

1.2 In this guide, sampling procedures and equipment are divided into the following categories based on water depth: sampling in depths shallower than 0.5 m, sampling in depths between 0.5 m and 10 m, and sampling in depths exceeding 10 m. Each category is divided into two sections: equipment for collecting short cores and equipment for collecting long cores.

1.3 This guide emphasizes general principles. Only in a few instances are step-by-step instructions given. Because core sampling is a field-based operation, methods and equipment must usually be modified to suit local conditions. This modification process requires two essential ingredients: operator skill and judgment. Neither can be replaced by written rules.

1.4 Drawings of samplers are included to show sizes and proportions. These samplers are offered primarily as examples (or generic representations) of equipment that can be purchased commercially or built from plans in technical journals.

1.5 This guide is a brief summary of published scientific articles and engineering reports. These references are listed in this guide. These documents provide operational details that are not given in this guide but are nevertheless essential to the successful planning and completion of core sampling projects.

1.6 *This standard does not purport to address all of the safety concerns, if any, associated with its use. It is the responsibility of the user of this standard to establish appropriate safety and health practices and determine the applicability of regulatory limitations prior to use. For specific hazard statements, see Notes 1 and 2.*

2. Referenced Documents

2.1 *ASTM Standards:*
D 420 Practice for Investigating and Sampling Soil and Rock[2]
D 1129 Terminology Relating to Water[3]

D 1452 Practice for Soil Investigation and Sampling by Auger Borings[2]
D 1586 Method for Penetration Test and Split-Barrel Sampling of Soils[2]
D 1587 Practice for Thin-Walled Tube Sampling of Soils[2]
D 4220 Practice for Preserving and Transporting Soil Samples[2]
D 4410 Terminology for Fluvial Sediment[3]

3. Terminology

3.1 *Definitions*—For definitions of terms used in this guide, refer to Terminology D 1129 and Terminology D 4410.

3.2 *Descriptions of Terms Specific to This Standard:*

3.2.1 *check valve*—a device (see Fig. 1) mounted atop an open-barrel core sampler. As the sampler moves down through water and sediment, the valve remains open to allow water to flow up through the barrel. When downward motion stops, the valve closes. During retrieval, the valve remains closed and creates suction that holds the core inside the barrel.

3.2.2 *piston sampler*—a core sampler (see Fig. 2) with a solid cylinder (piston) that seals against the inside walls of the core barrel. The piston remains fixed at the bed-surface elevation while the core barrel cuts down through the sediment.

3.2.3 *core*—a vertical column of sediment cut from a parent deposit.

3.2.4 *core catcher*—a device (see Fig. 3) that grips and supports the core while the sampler is being pulled from the sediment and hoisted to the water surface.

3.2.5 *core conveyor*—a device (see Fig. 4) for reducing friction between a core and the inside surface of a core barrel.

3.2.6 *core-barrel liner*—a rigid, thin-wall tube mounted inside the barrel of a core sampler. During the core-cutting process, sediment moves up inside the liner.

3.2.7 *core sampler*—an instrument for collecting cores.

3.2.8 *extrude*—The act of pushing a core from a core barrel or a core-barrel liner.

3.2.9 *open-barrel sampler*—in simplest form, a straight tube open at both ends. More elaborate open-barrel samplers have core catchers and check valves.

3.2.10 *trip release*—a mechanism (see Figs. 2 and 5(*b*)) that releases a core sampler from its suspension cable and allows the sampler to freely fall a predetermined distance before striking the bed.

3.2.11 *piston immobilizer*—a special coupling (see Fig. 6) that protects a core from disruptive forces that arise during sampler pull-out. Piston immobilizers are also called *split pistons* or *break-away pistons*.

3.2.12 *recovery ratio*—the ratio A/B where "A" (see Fig.

[1] This guide is under the jurisdiction of ASTM Committee D-19 on Water and is the direct responsibility of Subcommittee D19.07 on Sediments, Geomorphology, and Open Channel Flow.
Current edition approved Sept. 10, 1995. Published November 1995. Originally published as D 4823 – 88. Last previous edition D 4823 – 88.
[2] *Annual Book of ASTM Standards*, Vol 04.08.
[3] *Annual Book of ASTM Standards*, Vol 11.01.

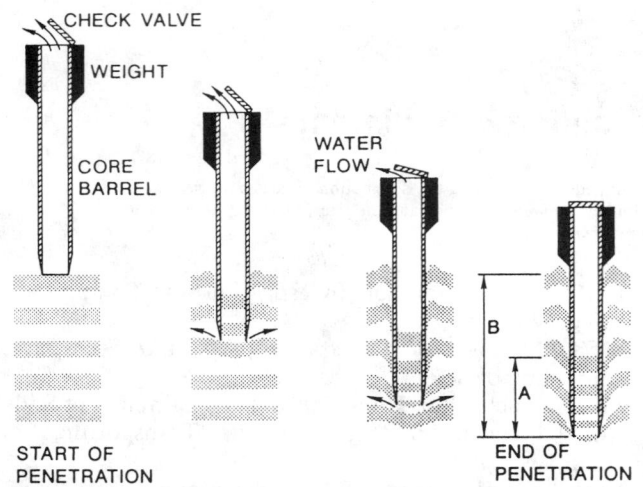

START OF PENETRATION

END OF PENETRATION

NOTE—Dark bands represent stiff sediments; light bands represent plastic sediments. As coring proceeds, sediment below the barrel moves laterally away from the cutting edge and plastic sediments inside the barrel are compressed. "A" is the core's length and "B" is the barrel's penetration depth.

FIG. 1   Deformations Caused by Open-Barrel Core Samplers (1)[4]

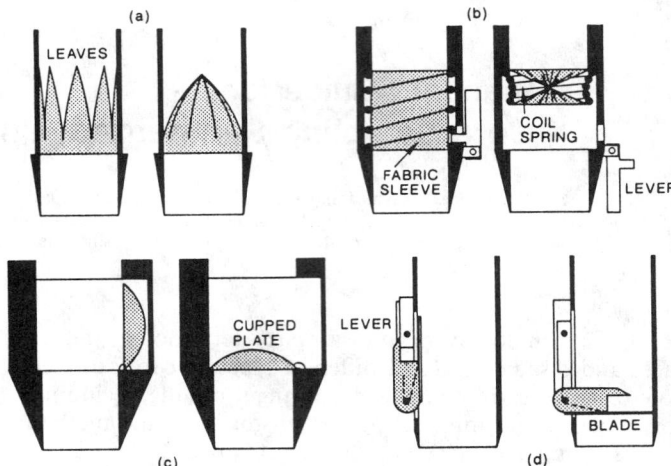

NOTE—(a) The leaves separate during penetration and then close during retrieval. Strips of gauze can be woven around the leaves to provide additional support. (3) (b) The lever trips down during retrieval to release the spring and twist the fabric sleeve shut. (4) (c) The cupped plate drops during retrieval to block the entrance and support the core. (4) (d) The lever releases the spring-loaded blade which pivots downward to hold the core. (4)

FIG. 3   Core Catchers

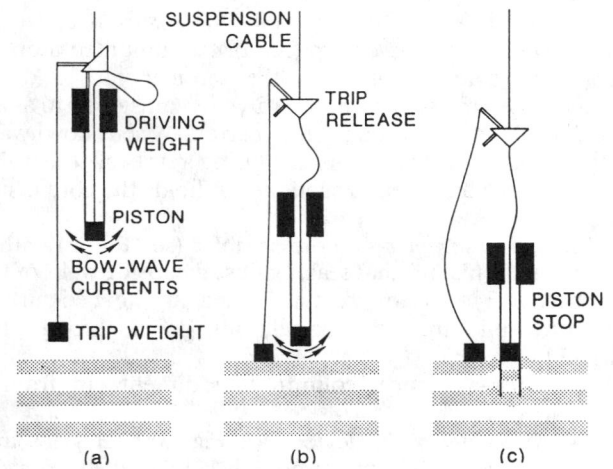

NOTE—(a) The sampler is lowered slowly through the water. (b) The sampler falls free when the trip weight contacts the bed. (c) The core barrel cuts downward but the piston remains stationary.

FIG. 2   Operation of a Piston-Type Core Sampler (2)

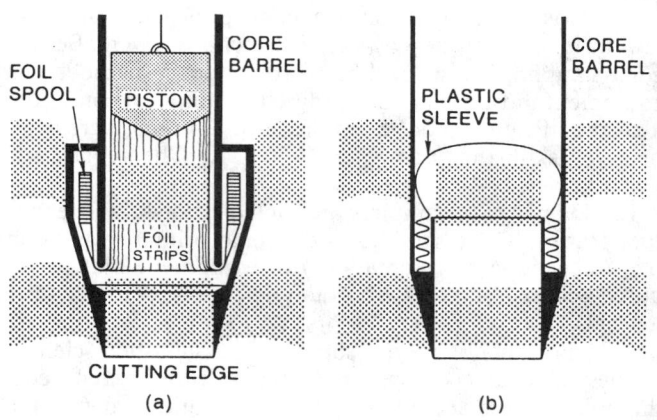

NOTE—(a) Strips of metal foil slide up through the core barrel as the cutting edge advances downward. (5) (b) The plastic sleeve unfolds from pleats stored near the cutting edge. This sleeve surrounds the core as the barrel moves down. (4)

FIG. 4   Core Conveyors

1) is the distance from the top of the sediment core to the bottom of the cutting bit and "B" is the distance from the surface of the parent deposit to the bottom of the cutting bit.

3.2.13 *repenetration*—a mishap that occurs when a core sampler collects two or more cores during one pass.

3.2.14 *surface sampler*—a device for collecting sediment from the surface of a submerged deposit. Surface samplers are sometimes referred to as grab samplers.

3.2.15 *undisturbed sample*—sediment particles that have not been rearranged relative to one another by the process used to cut and isolate the particles from their parent deposit. All core samples are disturbed to some degree because raising the cores to the water surface causes pore water and trapped gases to expand (10). In common usage, the term "undisturbed sample" describes particles that have been rearranged but only to a slight degree.

.4. Critical Dimensions of Open-Barrel and Piston Samplers

4.1 Dimensions of a sampler's cutting bit, core tube, and core-tube liner (see Fig. 7) are critical in applications requiring undisturbed samples. These dimensions control the amount of distortion in recovered cores. The recommendations in this section were developed from tests on open-barrel core samplers (11); however, the recommendations are usually extended to cover piston-type core samplers.

4.2 *Cutting-Bit Angle*—The angle "b" on the cutting bit (see Fig. 7) should be less than about 10°; the optimum angle is about 5°. If the angle is smaller than about 2°, the bit cuts efficiently but its edge chips and dulls easily.

---

[4] The boldface numbers in parentheses refer to the list of references at the end of this guide.

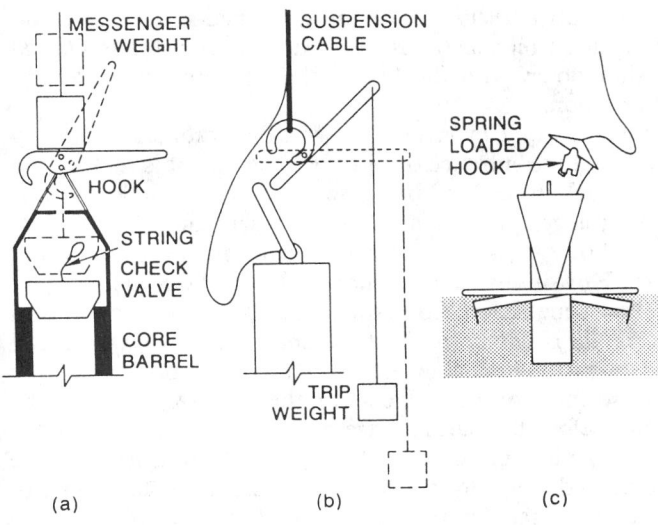

NOTE—(a) The messenger weight strikes the hook and releases the string holding the check valve. **(6)** (b) The trip weight strikes the sediment and unhooks the sampler. **(7)** (c) The cable slackens and allows the spring-loaded hook to open. **(8)**

FIG. 5   Release Mechanism

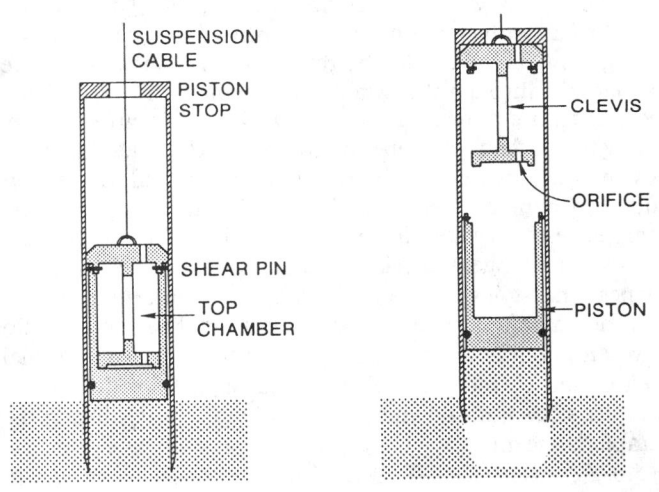

NOTE—During penetration the shear pins break but the flow-restricting orifice holds the clevis and piston together. During retrieval, water in the top chamber flows through the orifice and allows the piston and clevis to separate. Cable tension pulls the clevis up against the stop but friction locks the piston and core barrel together.

FIG. 6   Piston Immobilizer (9)

4.3 *Core-Liner Diameter, D_s* (see Fig. 7)—D_s should be larger than about 5 cm; however, the upper limit for D_s is difficult to establish. As D_s increases, the amount of core compaction decreases but the sampler becomes heavier and larger. A survey of existing samplers shows that 10 cm is a practical upper limit. A few samplers have barrels larger than 10 cm but these are used only for special applications **(12)**.

4.4 *Inside Friction Factor*—The dimensions D_s and D_e (see Fig. 7) set the inside friction factor defined as $C_i = (D_s - D_e)100/D_e$. For a barrel without a core conveyor, the optimum C_i value depends mainly on the barrel's length. C_i should be smaller than 0.5 if the barrel is shorter than about 2 m. If the barrel is longer than about 2 m, C_i should fall

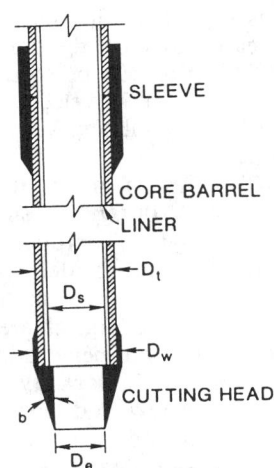

FIG. 7   Critical Dimensions for Cutting Bits and Core Barrels (11)

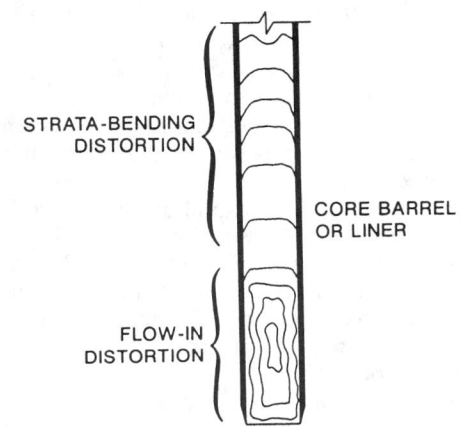

FIG. 8   Flow in and Strata-Bending Distortions Inside a Core Barrel (13)

between 0.75 and 1.5. For a barrel with a core conveyor, C_i should be smaller than 0.5 regardless of the barrel's length. Notice that in all instances D_s is lightly greater than D_e. The small expansion above the cutting bit minimizes friction where the outside of the core contacts the inside of the barrel or liner. Friction distorts the core's strata by bending horizontal layers into curved, bowl-shaped surfaces shown on the upper part of Fig. 8. Friction also causes overall end-to-end compaction of the core and thereby reduces recovery ratios. If friction becomes very large, sediment fails to enter the cutting bit. Instead, sediment moves aside as the bit penetrates downward. This lateral motion, commonly referred to as "staking," prevents deep-lying strata from being sampled. It is important to observe upper limits on C_i because too large an expansion causes another form of distortion, the core slumps against the walls as the sediment slides up into the barrel.

4.5 *Outside Friction Factor*—The dimensions D_w and D_t (see Fig. 7) set the outside friction factor defined as $C_o = (D_w - D_t)100/D_t$. C_o should be zero for barrels used in cohesionless sediments; but C_o should be between 1.0 and about 3.0 for barrels used in cohesive sediments. Notice that in all instances D_w is larger than D_t. The small contraction above the bit reduces friction at the outside surface of the barrel and makes it easier to push the core barrel into the bed. On a long barrel, friction can be reduced by installing one or more sleeves (see Fig. 7). The sleeves not only plough a path for the barrel but they also serve as clamps to hold barrel sections together.

4.6 *Area Factor*—The dimensions D_w and D_e set the area factor defined as $C_a = (D_w^2)\,100/D_e^2$. C_a should be less than 10 or possibly 15. Notice that C_a is proportional to the area of sediment displaced by the bit divided by the area of the bit's entrance; therefore, C_a is an index of disturbance at the cutting edge. A sampler with too large an area factor tends to oversample during early stages of penetration when friction along the inner wall of the barrel is low. Oversampling occurs because sediment laying below and outside the bit shift inward as the bit cuts downward.

4.7 *Core-Barrel Length*—A sampler's core barrel should be slightly longer than L, the longest core that can be collected without causing significant compaction. L and D_s (see Fig. 7) set the core-length factor defined as $L_f = L/D_s$. L_f should be less than 5.0 (or possibly 10) for a sampler used in cohesive sediments, but L_f should be less than 10 (or possibly 20) for a sampler used in cohesionless sediments. The constant factors 5, 10, and 20 apply to slow-penetrating, open-barrel samplers. Studies suggest that all of these factors can be increased by raising the sampler's penetration speed or using a piston sampler instead of an open-barrel sampler.

4.8 *Barrel Surfaces*—All surfaces contacting the core should be smooth and free of protruding edges to reduce internal friction and minimize core distortion. The surfaces should also be clean and chemically inert if the core is to be analyzed for contaminants or if the core is to be stored in its liner for long periods of time.

4.9 *Chemical Composition of Sampler Parts*—Sampler parts must not contain substances that interfere with chemical analysis of the cores. For example, barrels, pistons, and core catchers made of plastic should not be used if tests include phthalate concentrations. Misleading data will result from plasticizer contamination of the sediments.

5. Open-Barrel Samplers Versus Piston Samplers

5.1 Users sometimes face difficult decisions in choosing between an open-barrel sampler and a piston sampler. The decision frequently depends not only upon characteristics of the two samplers but also upon other factors such as hoisting-equipment capabilities, working platform stability, water depth, operator experience, and the purpose for collecting the cores. This section covers factors to consider before making the final choice.

5.2 *Depth of Penetration*—Most open-barrel samplers and most piston samplers rely on momentum to drive their barrels into sediment deposits. Momentum-driven samplers are released at a predetermined point so as to acquire momentum while falling toward the bed. A momentum-driven piston sampler generally penetrates deeper than a momentum-driven open-barrel sampler provided the two samplers have equal weights, equal barrel-diameters, and equal fall-distances (2).

5.3 *Core Compaction*—When compared under equal test conditions (see 5.2), a piston sampler causes less core compaction than an open-barrel sampler. However, the piston must be held motionless at the bed-surface elevation while the barrel penetrates downward. If the piston is allowed to shift down with the barrel, the core undergoes serious compaction.

5.4 *Flow-in Distortion*—Flow-in distortion is caused by suction at the entrance of a sampler. Sediment is sucked into the barrel instead of being severed and encircles by the cutting edge. Flow-in rarely occurs with open-barrel samplers; however, it can be a problem with piston samplers (14). Flow-in usually occurs during pull-out following a shallow penetration. Conditions leading to flow-in are shown in Fig. 2(c). The barrel is at the end of its downward travel but the piston lies below the piston stop. During pull-out, the upward force on the cable slides the piston up through the barrel before the cutting edge clears the bed. As the piston slides, it pulls the core up through the barrel. As the core moves, sediment flows in to fill the void at the lower end of the barrel. Strata lines at the bottom of the recovered core are distorted and resemble those in Fig. 8. A piston immobilizer helps prevent flow-in distortion by breaking the connection between the cable and the piston during the pull-out process.

5.5 *Surface Disturbance*—Surface sediment, material lying at the interface between water and bed, is easily disturbed by bow-wave currents (see Fig. 2(a)) that travel ahead of a sampler's cutting bit. A piston sampler creates a strong bow wave as the barrel, which is blocked by the piston, falls through the water. Fine-grained, unconsolidated sediments are blown aside just before the cutting edge contacts the bed. An open-barrel sampler creates a weak bow-wave because the barrel is unobstructed. However, adding a core catcher or a check valve to an open-barrel sampler restricts water flow through the barrel and makes the bow-wave stronger. Check valves come in a variety of sizes, shapes, and styles. These characteristics should be carefully considered before making a final selection. The valve should have an opening approximately equal to the cross-sectional area of the barrel. The valve should open fully during the sampler's descent and then close and seal tightly during the sampler's ascent.

5.6 *Repenetration*—Repenetration occasionally occurs in shallow-water sampling if the working platform (boat or barge) rolls and heaves; however, repenetration usually occurs in deep-water sampling that requires a long cable (1). During the initial stage of pull out, the cable stretches as tension gradually increases. Suddenly, the sediment relaxes its grip on the barrel and the lower section of cable contracts as the sampler springs upward. A rapid sequence of events follow. A shock wave races up the cable, reflects off the hoist drum, and then travels back down to the sampler (15). Upon reaching the bottom, the shock wave abruptly lowers the sampler and the cutting bit cores the top layer of sediment a second time. This up and down bobbing action, may occur several times before the sampler can be hoisted to a safe level above the bed. The severity of repenetration depends on the type of sampler used. With an open-barrel sampler, the first core that is cut can shift up the barrel and easily escape through the check valve as additional cores enter the bit. With a piston sampler, the first core fills the barrel if the sampler cuts to full penetration. Since the first core cannot move past the piston, the sampler offers high resistance to repenetration.

6. Driving Core Samplers into Sediment Deposits

6.1 Two techniques are frequently used to drive core samplers into sediment deposits. One technique depends entirely on weight. A weight-driven sampler is lowered slowly until friction along the barrel wall stops downward penetration. The other technique is based on momentum. A momentum-driven sampler is dropped from a specified height by a trip-release mechanism. As the sampler falls, it gains momentum that drives the barrel into the deposit. Paragraphs 6.2 through 6.6 cover other driving techniques that are occasionally used in special situations.

6.2 *"Free" Core Samplers*—Operation of a "free" core sampler, sometimes referred to as a "boomerang-core sampler" or a "free-fall corer" is shown in Fig. 9. The sampler is dropped into the water and then gains speed and momentum by falling through the entire water column (see Fig. 9(*a*)). After the core barrel has reached full penetration (see Fig. 9(*b*)), a latch (not shown) disconnects the core barrel and float from the heavily weighted lower section (Fig. 9(*c*)). The lower section remains on the bed; however, the float, core barrel, and core sample rise to the water surface (Fig. 9(*d*)) where they are retrieved. Free corers are useful if many samples must be collected rapidly. Free cores are costly to operate because the lower sections must be replaced and because the latches sometimes fail.

6.3 *Implosive and Explosive Samplers*—These samplers are driven by high pressures developed by either implosions or explosions (16). An implosive-driven sampler has an electrically operated valve and a cylindrical cavity fitted with a piston. The sampler is lowered to the bed, then the valve is opened so that high-pressure water around the sampler can rush into the cavity and push against the piston. As the piston slides, it pulls against cables (or rods) which exert a downward thrust on the barrel and upward thrust on the sampler's frame. Implosive samplers are complicated, expensive to purchase, and restricted to deep-water applications. However, the samplers have the advantage of being lighter

than momentum-driven samplers. An explosive-driven sampler has a charge that detonates when the sampler touches bottom. The expanding gas produces a strong downward force on the core barrel. Using explosive-driven samplers has a redeeming feature in that they are lighter than momentum-driven samplers.

NOTE 1: **Warning**—Because of the possibility of injury when using explosive samplers, it is suggested that specially trained personnel handle this apparatus.

6.4 *Punch-Corer Samplers*—Punch corers are pushed downward by using a stiff rod connected to a jack, drill rig, or heavy weight. The samplers may be either open-barrel or piston types. Punch corers are commonly used in shallow water. Maximum operating depths are set primarily by the rigidity and length of the push rod. A sampling spud (Fig. 10) is a form of punch corer since the spud is pushed with a rod; however, the spud does not collect a true core sample. Instead, small specimens of sediment are trapped in the cup-shaped cavities. Color, softness, and grain-size profiles along the spud are approximate indexes of profiles in the sampled deposit.

6.5 *Vibratory-Driven Samplers*—High-frequency vibration helps to reduce friction on a core barrel. Sediment is pulsed away from the barrel, then the sampler advances downward a short distance. Pulsing and advancing alternates rapidly so that the barrel cuts downward at a nearly uniform rate. The vibrator (see Fig. 11) which is fastened to the top end of the core barrel, receives power through an electric cable or compressed-air tube. Sediment grains inside the core are realigned by the vibration; however, compaction and strata-bending are nearly eliminated. Vibratory-driven samplers can be used through a broad range of water depths. According to Hubbell and Glenn (18), the samplers work especially well in sandy sediments that are difficult to penetrate with other types of core samplers.

6.6 *Impact-Driven Samplers*—Some gravity deposits that cannot be penetrated with open-ended barrels, can be pierced with pointed pipes (see Fig. 12(*a*)) driven with a heavy hammer. The pipes are filled with carbon-dioxide gas which slowly freezes the surrounding sediment. When freezing is complete, the pipes along with their load of frozen

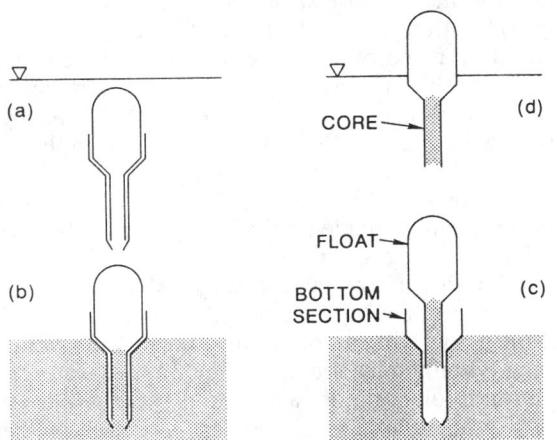

NOTE—(*a*) The sampler falls toward the bed. (*b*) The bottom section drives the core barrel into the sediment. (*c*) The bottom section unlatches and releases the float. (*d*) The float and core rise to the surface.

FIG. 9 Operation of a Free Corer (2)

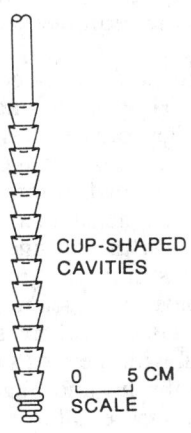

NOTE—The spud is pushed or driven into the sediment deposit. As the spud is pulled up, sediment becomes trapped in the cup-shaped cavities.

FIG. 10 Sampling Spud (17)

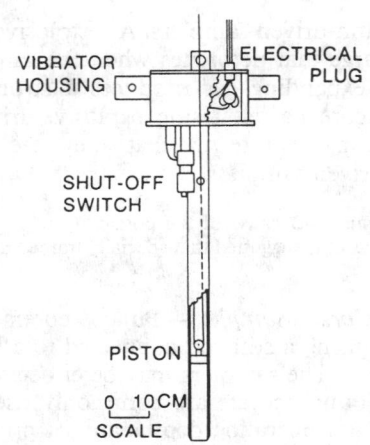

FIG. 11 Vibratory-Type Core Sampler (19)

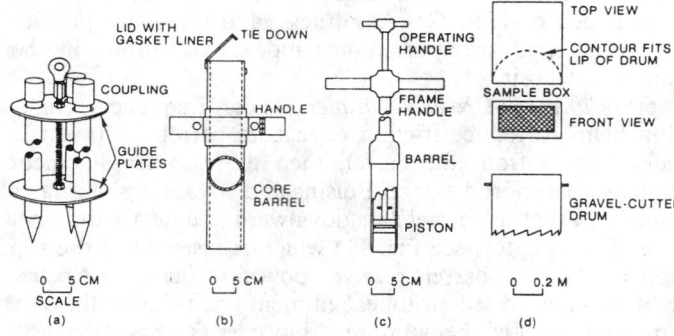

NOTE 1—(a) The cryogenic-gravel sampler, a freeze-type sampler. (20) (b) The Van Stratten, an open-barrel sampler. (4) (c) The BMH-53, a piston sampler. (21) (d) The gravel-cutting sampler. (22)

NOTE 2—Fig. 12(b) has been reprinted from Bouma, A. H., Methods for the Study of Sedimentary Structures, 1969, with the permission of John Wiley and Sons, Inc., New York, NY.

FIG. 12 Core Samplers for Water Depths Less Than About ½ m

sediment are pulled free with a hoist suspended from a portable tripod.

7. Samplers for Specific Field Conditions

7.1 *Collecting Short Cores in Shallow Water:*

7.1.1 The Van Stratten sampler shown in Fig. 12(b) has been found satisfactory for the purpose of coring soft, cohesive sediments covered by water shallower than about 50 cm. This sampler is easy to make with a lathe and ordinary hand tools. The core barrel is a pipe with a diameter of about 10 cm and a length of about 60 cm. On thick-walled pipe, one end must be turned to form a sharp cutting edge: on thin-walled stove pipe, no sharpening is required. Scribe a vertical reference line on the outside surface if north-south alignment of the core is important in subsequent laboratory analysis. The stove-pipe has a seam that serves as a ready-made reference line. Glue a rubber sheet under the lid to form a water-tight seal with the pipe's upper edge. To use the sampler, first loosen the lid and align the reference mark. Then apply a steady pressure to force the sampler down into the sediment. Avoid hammering; it disturbs the core and usually fails to increase penetration. Holding the core inside the barrel during pullout is sometimes difficult. One solution

is to excavate sediment from around the pipe and push a flat plate under the core. The pipe, plate, and core can then be lifted as a unit. Another solution is to fill the pipe brimful with water and then close and seal the lid. The suction formed during pullout helps to support the core.

7.1.2 The BMH-53 sampler shown in Fig. 12(c) has been found satisfactory for the purpose of coring sandy sediments that are difficult to penetrate with the sampler shown in Fig. 12(b). A BMH-53 sampler is frequently used for sampling beds of wadeable rivers. The operating handle is connected to the piston and the frame handle is connected to the core barrel. Before collecting a sample, push the two handles together to set the piston flush with the barrel's cutting edge. Set the cutting edge against the bed and then cut the core by pressing down on the frame handle while holding the operating handle stationary. A slight rocking motion may be necessary to achieve full penetration and break the core loose from the bed. To retrieve the core, first grip the stem of the operating handle so that the piston cannot shift inside the barrel and then quickly lift the sampler above the water. To eject the core, push the handles together and catch the sediment in a clear carton. If the core slumps, it must be regarded as a highly disturbed sample.

7.1.3 Gravel beds can be sampled with the cryogenic sampler shown in Fig. 12(a). The pointed stainless-steel stakes are 1.3-m long sections of 2.5-cm pipe. Using a sledge, drive the stakes through holes in the guide plates and down into the bed. The flat guides hold the stakes upright and maintain a 7.6 cm spacing between centers. After all three stakes have been seated, connect the couplings to the manifold on a 9 kg CO_2 fire extinguisher fitted with a hand-wheel valve. When the valve is opened, cold CO_2 fills the pipes and freezes the sediment to the stakes. Lift the entire unit out of the bed with a hoist suspended from a portable tripod erected over the sampler. Collect subsamples of the stratified-sediment layers by first laying the stakes and frozen core across metal boxes placed side-by-side. About seven boxes, each 10 cm wide, are usually required. As the sediment thaws, particles fall into the boxes and are segregated according to position along the core. A blowtorch helps to speed the thawing process.

7.1.4 Another gravel-bed sampler is shown in Fig. 12(d). The gravel cutter, a ½-m diameter cylinder, is turned and pushed into the bed by hand. Serrations on the cutting edge help to plow through sand and gravel-size particles. When the cutter is in place, sediment is excavated layer-by-layer and placed in the sample box. If desired, samples can be sieved through the screened opening.

7.1.5 A lightweight sampler for use in shallow, cobblebed streams (36) can be made by removing both ends from a 30 gallon barrel. By pressing down on the cylinder formed by the barrel walls, a circular section of stream bed and the water column above it are isolated and shielded from the flow. Fine grains lying on or between the cobbles can then be lifted and entrained by circulating the trapped water through a small pump or, if milder agitation is preferred, by stirring with a large paddle. The suspended particles are then dip sampled with quart bottles. If the fine-grain deposits inside the barrel's footprint are thicker than about 10 mm, they can be sampled with a scoop. Next, the armoured surface layer is sampled by manually removing the particles within the circle

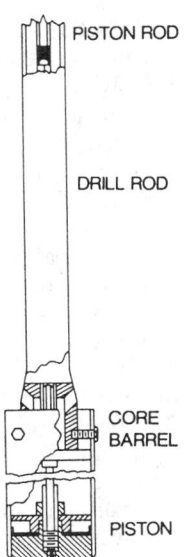

NOTE—The center section houses complex mechanical linkages.

FIG. 13 Hvorslev's Thin-Wall Fixed-Piston Sampler (23)

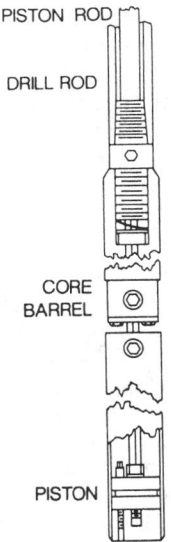

FIG. 14 Butter's Thin-Wall Fixed-Piston Sampler (23)

formed by the drum's rim. To avoid sampling the subsurface layer, the operator starts at one point and makes only one traverse around the area. Particles lying under the rim are collected only if more than half their surface lies inside the circle. In the last phase, subsurface material is sampled to a depth of 0.15 to 0.30 m by using a stainless steel bowl. A prybar may be required to loosen large, tightly wedged particles.

7.1.6 To avoid loosing fines when sampling flows that overtop the barrel, a bag (0.81 m wide by 1.5 m long) made from filter mesh having about 0.149 mm openings should be placed over the barrel's upper rim and secured with an elastic cord. A slit in the bag about 0.15 m long allows the operator to insert an arm and the necessary tools. Anchors are needed in flows that overtop the barrel and exceed speeds of about 0.8 m/s. Snorkeling or SCUBA equipment is required when

sampling depths exceed about 1.3 m.

7.2 *Collecting Long Cores in Shallow Water:*

7.2.1 Collecting long cores requires a well-drilling rig modified for soil-sampling applications. Because the rig must be towed or winched to the sampling site, water depths must be shallower than about 0.5 m and the underlying sediment must be strong enough to support the rig's weight.

7.2.2 Hvorslev's sampler (Fig. 13), or Butter's sampler (Fig. 14), have been found satisfactory for this purpose. Trained operators are needed because these samplers are easily damaged by improper use, incorrect assembly, or poor lubrication.

7.2.3 Cores are collected in segments that, depending on soil firmness, range in length from about 0.4 to about 0.6 m.

7.2.4 To begin a coring operation, set the piston flush with the barrel's cutting edge and then set the bottom of the barrel against the sediment. While holding the sampler in this position, lower a section of piston rod (1.3 cm pipe fitted with flush couplings) through the top end of the drill rod then screw the piston rod into the coupling atop the sampler. When the threads are fully engaged, unlock the piston by rotating the rod through the proper angle (five clockwise revolutions for Hvorslev's sampler and ¼-clockwise revolution for Butter's sampler). Clamp the piston rod to the drill-rig frame or, to a stationary frame independent of the drill rig. The first section of core is cut by pushing down on the drill rod. This drive should be made in one continuous stroke and the barrel should not be allowed to rotate as it moves down. Avoid overdriving. Stop the motion when the barrel has advanced to within a few centimetres of its rated maximum travel. Remove the piston rod, break the core from the parent material by turning the drill rod a few degrees, and then carefully lift the sampler along with the core back to the surface. A mechanism inside the sampler automatically locks the piston to the barrel and thereby supports the core during the lifting operation. As an added precaution against losing the core, slide a plate (or hand) under the barrel before the cutting edge clears the water.

7.2.5 After extruding the core, enlarge and then clean the sampling hole (see Fig. 15) by augering or wash boring. If the sampling walls are too weak to stand, they must be cased. When the hole is ready, collect another core section by repeating the procedure described in 7.2.4. Repeat the cycle of coring, enlarging, cleaning, and casing until the desired

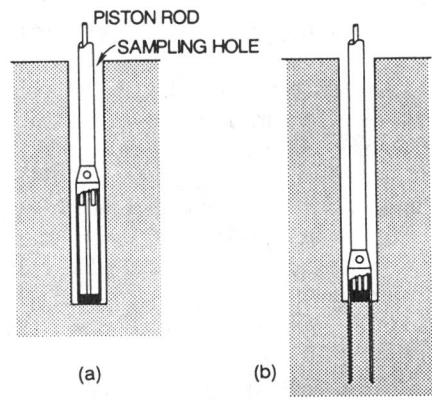

FIG. 15 Sampling with Hvorslev's or Butter's Fixed-Piston Sampler (23)

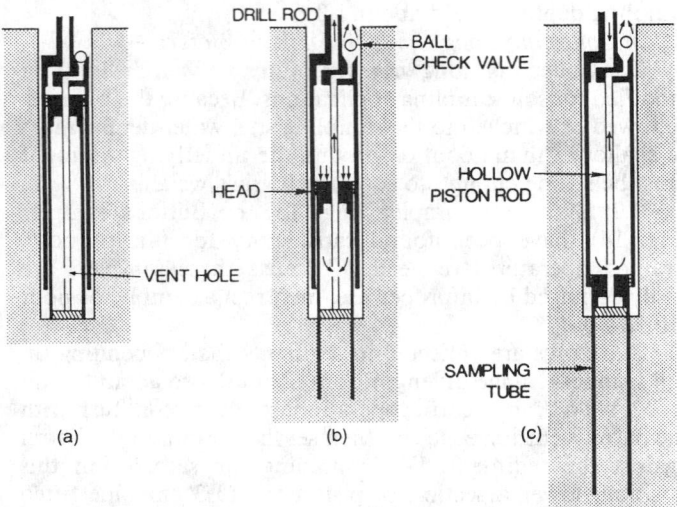

NOTE—(a) Retracted sampling tube, (b) partially extended sampling tube, and (c) fully extended sampling tube. (Arrows indicate water flow.)

FIG. 16 Sampling with Osterberg's Thin-Wall Fixed-Piston Sampler (23)

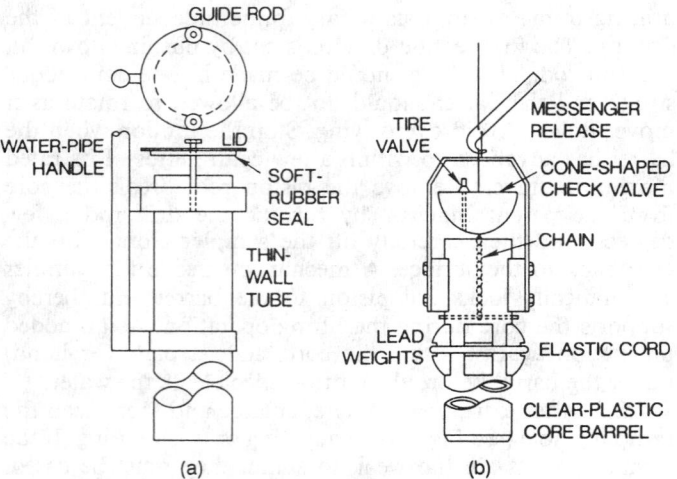

NOTE 1—(a) Modified Van-Stratten sampler. (4) (b) Milbrink's bottom sampler. (6)

NOTE 2—Fig. 17(a) has been reprinted from Bouma, A. H., *Methods for the Study of Sedimentary Structures*, 1969, with the permission of John Wiley and Sons, Inc., New York, NY.

FIG. 17 Core Samplers for Use in Depths from 1 to 10 m

sampling depth has been reached.

7.2.6 The Osterberg sample shown in Fig. 16 has been found satisfactory for the purpose of sampling without using a piston rod. Instead of using a piston rod, the sampler's top section is held with a drill rod while the core barrel is driven downward with water pressure. Overdrives are prevented because the head automatically vents the water pressure at the proper point in the stroke.

7.3 *Collecting Short Cores in Water Depths Ranging From About 0.5 m to About 10 m:*

7.3.1 A boat is required to sample in water deeper than about 0.5 m. However, a boat's propeller creates backwash currents that can easily disturb loose, fluffy deposits (24).

This backwash problem is most severe in depths of 1 to 6 m. When working in this critical-depth range, approach sampling points slowly and apply engine power gradually.

7.3.2 Operators sometimes prefer a sampler that can be pushed down with a stiff rod. The modified Van Stratten sampler (Fig. 17(a)), a simple, lightweight sampler that can be coupled to a water-pipe handle up to 6 m long has been found satisfactory for this purpose. During penetration, the lid remains open. During retrieval, the lid slides down the guide rods and seals against the core-barrel's rim. Schneider (25) gives plans for a lightweight piston sampler that penetrates 1.2 m into loose sediments commonly found on lake and pond bottoms.

7.3.3 Milbrink's sampler (see Fig. 17(b)) can be operated with a cable or wire. The core barrel is a 7 by 30-cm plastic tube sharpened at one end and banded with lead weights. The check valve, which is a cone-shaped rubber stopper, has a vent hole capped with an automobile tire valve. The procedure for collecting samples is simple. Slowly lower the sampler to a point about 1 m above the bed, then allow the sampler to fall the remaining distance to gain momentum. After the core has been cut, close the check valve by sliding a messenger weight down the suspension cable to active the release mechanism, and seat the check valve. The sampler does not have a core catcher so the entire apparatus must be lifted slowly and carefully to avoid losing the core. Seal the barrel's bottom end with a flat plate before the sampler is lifted above the water. When the sampler is safely on board, open the tire valve to break the vacuum and facilitate removing the check valve and core.

7.3.4 If a core must be longer than about 0.5 m, a simple-drive apparatus will sometimes suffice. Figure 18 shows a lightweight sampling platform and pulley arrangement that can be used with a Swedish-foil sampler or basic piston samplers. The line on the top hook holds the piston stationary while the line on the bottom hook pulls the core barrel downward. When the drive is complete, the lines are restrung for lifting. The top-hook line is clamped to the core barrel and the bottom-hook line is removed from the bottom pulley and tied to the core barrel.

7.3.5 Fuller and Meisburger (27) give plans for an air-driven vibratory corer for use in lakes and estuaries. The vibrator assembly is commercially available and the remaining parts are easy to build. Core lengths range from about 1 m in sand to about 2 m in silt.

7.3.6 Coring the beds of estuaries and large rivers requires

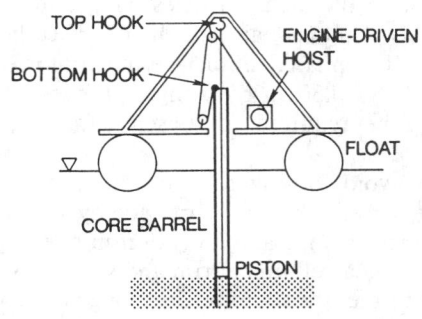

NOTE—Cables strung for coring phase.

FIG. 18 Sampling Platform for Swedish Foil and Piston-Type Samplers (26)

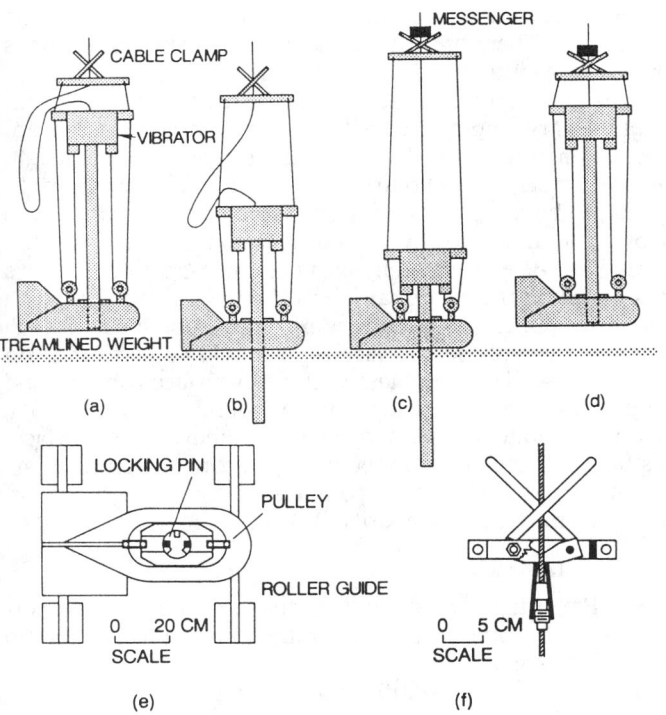

NOTE—(a) lowering, (b) coring, (c) extracting, (d) ascending, (e) top view of streamlined weight, and (f) detail of cable clamp.

FIG. 19 Prych-Hubbell Core Sampler (28)

a sampler light enough to deploy from a tugboat-size craft yet heavy enough to stand upright and resist the tipping forces created by flowing water. The Prych-Hubbell sampler (see Fig. 19) has been found satisfactory for the purpose of withstanding flow velocities up to 1.5 m/s and coring sandy bottoms to a depth of about 1.8 m. Figure 19(a) through (d) shows cable configurations during each of the four operational phases. During the lowering phase, spring-loaded

locking pins (Fig. 19(e)) hold the barrel in a retracted position. These pins release when the sampler strikes bottom and the suspension cable momentarily slackens. During the coring phase, (Fig. 19(b)) three forces, tension in the side cables, vibration, and weight, push the barrel downward. Coring continues until an electrical switch (see Fig. 11) hits the streamlined weight and disconnects power to the vibrator. The "extraction" phase (Fig. 19(c)) begins when a messenger weight releases the cable clamp (Fig. 19(f)) holding the center cable. The barrel is then pulled upward by reeling in the suspension cable. The "ascending" phase (Fig. 19(d)) begins when the cutting edge passes a spring-loaded plate (not shown) that rotates under the barrel to support the core.

7.4 *Collecting Long Cores in Water Depths Ranging From About 0.5 m to About 10 m:*

7.4.1 The Butter sampler shown in Fig. 14 or the Osterberg sampler shown in Fig. 16, have been found satisfactory for the purpose of collecting cores longer than about 3 m. The drill rig for the sampler is loaded onto a

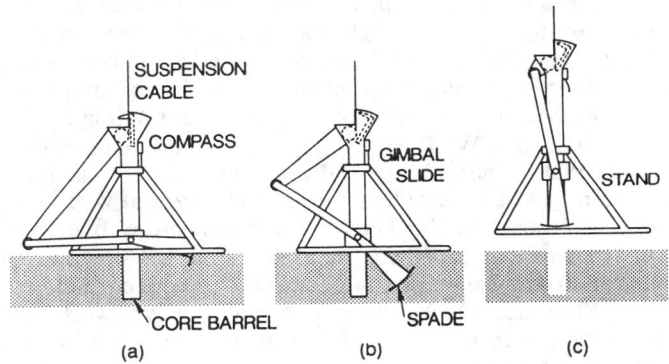

NOTE—(a) The core barrel cuts downward. (b) Cable tension rotates the spade toward the cutting edge. (c) The core sample rests on the spade during pull out and ascent.

FIG. 21 USNEL Spade Corer (34)

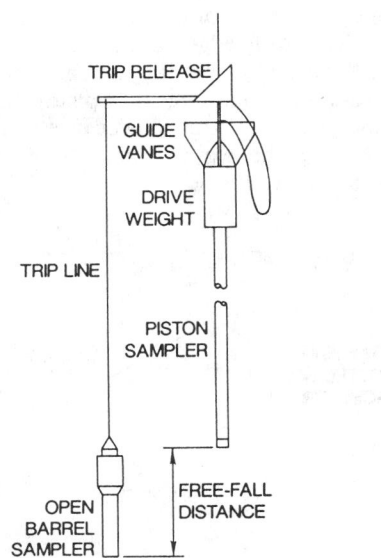

FIG. 20 A Deep-Water Piston Sampler Coupled to an Open-Barrel Sampler

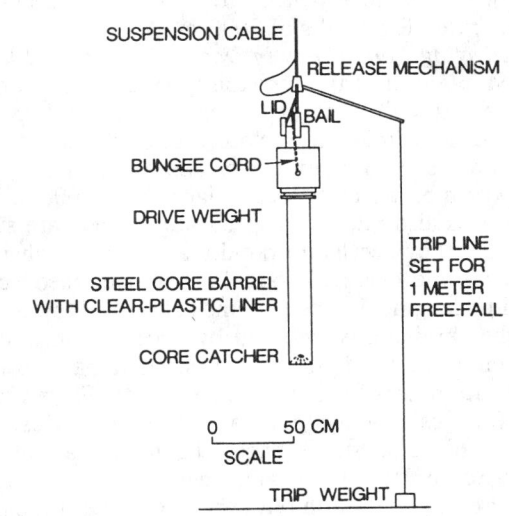

NOTE—The bungee cord closes the check valve when the bail rotates to the right.

FIG. 22 SCCWRP Corer for Soft Sediments (7)

barge, towed to the sampling site, and then anchored in position. Basically, the sampling techniques are identical to those described in 7.2.

7.4.2 Long cores of soft sediment can be collected with large, momentum-driven piston samplers similar to the one in Fig. 20. These samplers can be used only if water depths exceed about 8 m. The samplers are several metres long and must fall about 3 m before striking the bed.

7.5 *Collecting Short Cores in Depths Exceeding 10 m:*

7.5.1 Box corers, sometimes referred to as spade corers, can be used to collect undisturbed samples of soft sediments (29). Figure 21 shows a USNEL (U.S. Navy Electronics Laboratory) box corer that penetrates 20 to 30 cm. Since the barrel is completely open during descent, particles at the sediment-water interface are not disturbed as the sampler approaches. During penetration (Fig. 21(*a*)), the stand rests on the sediment as the rectangular-shaped core barrel slides down through the gimbals. The sampler's north-south orientation registers on the compass which locks when the compass wire trips. The closing stage (Fig. 21(*b*)) begins when the suspension-cable slackens and releases the hook. As the operator reels the cable, the spade rotates downward toward the barrel's cutting edge. The ascent stage (Fig. 21(*c*)) begins when the spade reaches the barrel and closes under the core. Box corers collect high-quality samples (30), but the equipment is bulky and heavy (most box corers weigh about 400 kg).

7.5.2 The SCCWRP corer (Fig. 22) is an open-barrel sampler that weighs 130 kg and penetrates about 1 m into soft sediments. The sampler is lowered slowly. But when the trip-weight contacts the bed, the sampler drops free and quickly gains speed. When the barrel stops its downward penetration, the bail trips a latch and the bungee cord closes the lid. The lid, which serves as a check valve, works in conjunction with the core catcher to hold the core in the barrel during retrieval.

7.5.3 Free corers (Fig. 9) are designed to collect cores about 1 m long (3). The samplers are costly to operate and their impact speed is difficult to control; however, these disadvantages are offset by compactness and ease of deployment (32). Commercially-made free corers weigh about 90 kg and are rated for depths of 9000 m.

7.6 *Collecting Long Cores in Depths Exceeding 10 m:*

7.6.1 Momentum-driven piston-type samplers are frequently used to collect cores longer than about 1 m. Since momentum and mass are closely related, all long-core samplers are heavy. For example, one commercially made sampler with a barrel 15 m long weighs about 1000 kg. Since momentum is also related to speed, long corers are streamlined so that they accelerate rapidly and reach high impact speeds. To some degree, a sampler's impact speed can be increased by lengthening its trip line (Fig. 21). However, this practice has limitations imposed by acceleration and rotation. When a sampler is released, it starts to gain speed but the acceleration lasts for only a few seconds. Thereafter, the sampler plunges downward at a steady rate. Most heavy samplers achieve a stable speed in falling about 3 m. Dropping a sampler from heights greater than 3 m not only fails to increase penetration but also accentuates problems of rotation. As a sampler falls, it rotates out of plumb. Penetration depths become shallow, flow-in distortion becomes severe, and the danger of breaking or bending the

barrel increases. Optimum free-fall distances vary among samplers. Therefore, manufacturers' recommendations should be carefully reviewed before setting trip-line lengths.

7.6.2 Piston immobilizers should be used if depths of penetration vary unpredictably from one sampling site to another. Penetration depth can usually be gaged by the length of mud smears on the outside walls of the sampler's barrel.

7.6.3 The top portion of most piston-sampler cores fail to show true interface-zone conditions. This deficiency can sometimes be eliminated by using an open-barrel corer (or a box corer) for the piston sampler's trip weight (Fig. 20). With this arrangement, two core samples are obtained in each pass. Analysis of the interface zone should be performed on the open-barrel core and analysis of sub-surface zones should be performed on the piston-sampler core.

7.6.4 Vibratory corers overcome stability and surface-disturbance problems inherent to piston samplers. Long vibratory corers driven by compressed air cut to depths of about 12 m in sandy deposits (33).

8. Field Records

8.1 Preparing accurate and complete records is an important part of every sampling operation. Attach a log sheet to each core giving:

8.1.1 The core's identification number,

8.1.2 The operator's name or initials,

8.1.3 The purpose for collecting the core,

8.1.4 The type and serial number of the sampler used,

8.1.5 The exact location of the sampling site, and

8.1.6 Descriptive information on the sampling procedure.

8.2 Also record all information bearing on the core's overall quality (representativeness). For example, if the bottom section of the core slipped through the catcher, or if the barrel failed to make full penetration, these facts should be noted. If a core of substandard quality must be analyzed, an accurate record is helpful in interpreting laboratory data.

8.3 Inspect and measure each core before removing it from the barrel. Determine the length of the core (designated "A" in Fig. 1) by first drawing a line on the outside of the barrel opposite the top of the core and then measuring from the line to the cutting edge. In an open-barrel sampler, locate the top of the core by opening the check valve and inserting a ruler through the opening. In a piston sampler, the top of the core touches the piston's lower face which can be located after measuring the exposed length of piston wire. In a cryogenic sampler, the entire core is fully exposed so its

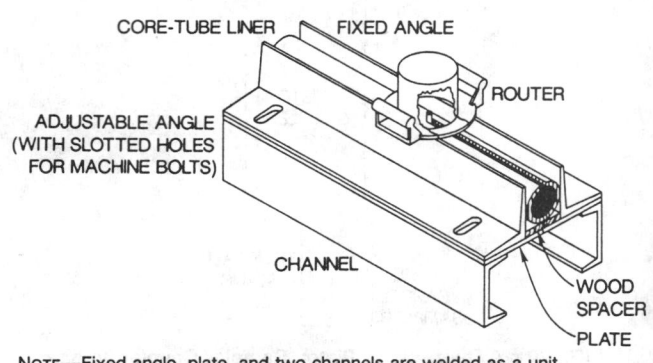

NOTE—Fixed angle, plate, and two channels are welded as a unit.

FIG. 23 Core-Liner Slitting Apparatus (34)

length can be measured directly.

8.4 Determine the barrel's penetration (designated "B" in Fig. 1) by marking the interface line on the barrel and then measuring from this line down to the cutting edge. The method for locating the interface depends on the water depth. In shallow water, mark the interface on the barrel before retrieving the sampler. In deep water, the interface usually lies at the uppermost extremity of mud smears on the barrel's outer surface. However, mud smears are sometimes deceptive. If the sampler tips over, smears may appear along the full length of the barrel. Smears that girdle only a fraction of the barrel's perimeter are poor penetration-depth indicators.

8.5 Check the lower portion of each core for missing sediment. If the bottom of a core lies above the catcher, measure and record the length of the intervening gap.

8.6 It is helpful to record information on printed forms that include headings for laboratory test results in addition to the field observations. Fig. 2–23 of the Earth Manual (23) shows a form for punch-core samples intended for use in moisture content and density tests.

9. Extruding Cores

9.1 Some samplers have no liners so cores must be extruded (pushed from the barrels) before the samplers can be reused. Start the extrusion process immediately after finishing the field records. Speed is essential because mechanical bonds between a core and a core-barrel strengthen quickly. Delays make the extrusion process more difficult.

9.2 Two special tools are required; a stiff rod with a piston at one end and a T handle at the other and a round-bottom trough with a smooth inner surface. The diameter of the piston and trough must match the inside diameter of the core barrel. The rod and trough must both be longer than the barrel. However, beyond this, lengths are not critical.

9.3 To begin the extrusion process, lay the sampler on a level surface, then clear the ends of the barrel by removing the core catcher and check valve. On some samplers, the barrel must be unscrewed from suspension hardware and driving weights. Insert the piston on the T-handle rod inside the cutting end of the barrel. It is important to work from the cutting end (4) because sediments in this region are tightly compacted by the core cutting operation. By comparison, sediments at the top end of the barrel are only loosely compacted and therefore are easily compressed by forces on the piston. When the piston is in place, align the trough with the top end of the barrel.

9.4 Equipment for pushing a piston through the barrel depends on the sampler's size. On small samplers, hand pressure is sufficient to dislodge the cores but on large samplers block-and-tackle pullers are required. To use a puller, anchor the T-handle and one end of the puller to a sturdy support and then fasten the other end of the puller to the bottom section of the barrel. Slowly increase the pressure to break the core free. Then maintain a steady pressure to slide the core into the trough.

10. Slitting Core Liners

10.1 A core encased in a liner can be extruded. However, slitting and removing the liner causes less distortion within the core itself. Power tools and bulky supports are required, so the slitting operation must usually be performed in a laboratory.

10.2 Freeze loosely packed cores to prevent slumping during and after the slitting process.

10.3 Figure 23 shows a clamp for holding the liner during the slitting operation. Fasten two angle irons to a steel plate which, in turn, is fastened to two 102-mm channels. The rear angle iron is welded but the front angle iron is bolted through slotted holes that permit adjustment.

10.4 The liner, which rests on the wooden spacer, is clamped with its top surface about 1 cm below the angle iron edges.

10.5 Cuts are made with a 6-mm carbide-tipped bit driven by a heavy-duty router. Set the router base across the irons then, while holding the router fence against one of the irons, center the bit over the liner. Lower the bit to cut about halfway through the liner wall. Then tighten all adjustments and make an end-to-end cutting pass. Return to the starting point, lower the bit to cut about 1 mm above the core, and then make a second cutting pass.

10.6 After one side has been cut, rotate the liner 180° and repeat the procedure in 10.5.

10.7 Carefully move the liner and core to a work table. Break away all uncut liner fragments from the grooves then lift the top half of the liner to expose the core.

11. Slitting Cores

11.1 An exposed core can be slit (cut lengthwise) to reveal internal grain structures. The slitting process works best if the core is resting in its extrusion trough or in the bottom half of its core liner.

11.2 A core containing a high percentage of sand can be slit with a slender wire strung on a cheese cutter or coping-saw frame. Guide the wire along the edges of the liner (or trough) and, at the same time, pull the wire through the sediment. Insert thin sheet-metal strips behind the wire to prevent the freshly cut surfaces from adhering to one another.

11.3 A core that has dried and hardened can be cut with a hacksaw blade driven by hand pressure. A gentle sawing action usually eases the cutting process.

11.4 A core containing a high percentage of clay can be slit with a thin wire, a sharp knife, or an electro-osmotic cutter.

11.5 An electro-osmotic cutter (see Fig. 24), is an ordinary knife (or wire) charged with direct current. The cutter cleanly severs sediments (4).

NOTE 2: Precaution—To minimize harm from shocks delivered by electro-osmotic cutters, operators should wear insulated gloves and boots.

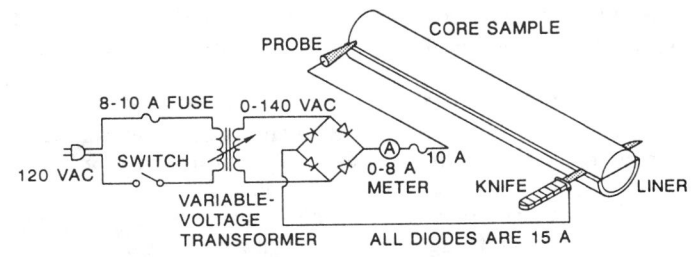

FIG. 24 Schematic of Electro-Osmotic Cutter (4)

11.6 To make the setup, imbed the probe in the core. The probe, which can be a nail or awl, serves as a ground return for the electric current. Polarity is important. The probe must be connected to the positive lead and the knife to the negative lead. To start a cut, moisten the knife, press it to the core, and then adjust the transformer for the best cutting action. For most cores, ammeter currents should be 1 to 4 A. Rewet the knife frequently as cutting proceeds. To prevent blowing fuses, avoid touching the probe with the knife.

11.7 After the core has been slit, separate the halves and clean the cut surfaces. Particles dislodged by the cutter can be removed with tweezers or a soft brush. Surface smears can be removed with a brush or scraped away with a razor blade.

11.8 Inspect the core, particularly its bottom end, for sections disturbed by flow-in. Centers of these sections are marked by rectangular grain patterns as shown in Fig. 8. Ends of flow-in sections fall where the strata lines shift abruptly from horizontal alignment to vertical alignment (13). All disturbed sections should be clearly marked because they do not represent bed-deposit structures.

12. Sectioning Cores

12.1 Sectioning (cutting across a core's axis) is the first step in performing certain types of laboratory analysis. Perform the sectioning with the core inside its liner. Because the core will be cut into several pieces, it is important to label several points along the core. Lay the liner on a table, mark the desired cut points, then start at the bottom of the core and number the sections in sequence. Also, label the top and bottom of each section.

12.2 To prevent slumping, freeze points along the core before the cuts are made. Set the lower halves of dry-ice boxes (Fig. 25) under the liner at all cut-point marks, then set the upper halves of the boxes in place. Fill all boxes with dry ice, then cover them to speed the freezing process. Allow only enough time to freeze sediment near each cut point (one 4-cm core takes about 25 min). At the proper time, remove the boxes and cut through the liner and core with an ordinary carpenter saw. Cut rapidly to avoid premature thawing.

12.3 After completing each cut, support the exposed sediment by taping a disk of polystryene over the section ends.

12.4 Refer to 11.8 regarding contamination.

13. Sampling Through Liner Walls

13.1 Small specimens of a core may be collected after cutting an opening in the core-liner wall. Use an electric drill and carpenter's hole saw to cut a circular disk from the liner. Exert only gentle pressures to prevent the saw from breaking through the liner and plunging into the core. Remove the specimen, then replace the disk, and seal it in place with adhesive-backed cloth tape (commonly known as duct tape).

14. Preserving and Displaying Cores

14.1 Cores to be preserved in their natural state must be hermetically sealed to prevent evaporation and condensation.

14.2 An extruded core or slit core should be sealed, along with its extrusion trough or the bottom half of its liner. Wrap the core and its liner with several layers of aluminum foil,

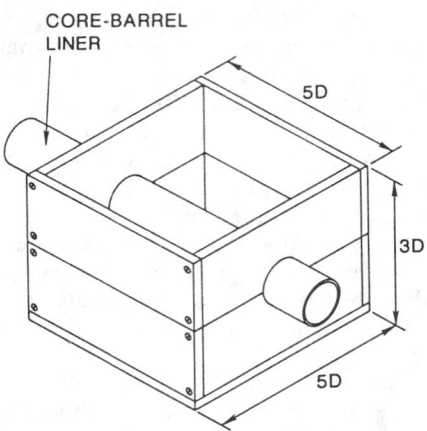

NOTE—Dimension "D" matches outside diameter of core-barrel liner.
FIG. 25   Dry-Ice Freezer Box (35)

and then seal all edges with molten wax. An alternative procedure is to slide the core and its liner into heat-shrinkable tubing, and then shrink the tubing by fanning it with warm air from an electric hair dryer. Ends of a core are particularly vulnerable; they should be sealed with several layers of wax and foil.

14.3 Sealed cores should be stored in a temperature-controlled chamber to retard the growth of bacteria. Set the temperature at approximately 2°C.

14.4 Polishing core surface is effective for highlighting grain-structure details. A core scheduled for polishing must first be impregnated—pores between sediment grains are filled with liquid resin that hardens and bonds all particles together. The impregnation process (4) involves several operations. First, section the core into pieces approximately 0.15 m long. Place the pieces in an aluminum pan, and then set the pan and its contents in an oven at 93°C. After a few hours, raise the temperature to about 170°C. Maintain this temperature until all traces of moisture have disappeared, and let the sections cool to room temperature. Next, fill the pan with a clear casting resin, and place the pan and its contents in a vacuum desiccator. Release the vacuum after 30 min, and place the pan and its contents in a dry-nitrogen pressure chamber. Hold the pressure at 1 atm for about 30 min, and slowly release the pressure. After the resin has completely cured and hardened, saw and polish the sections. Bouma (4) gives several other impregnation techniques.

14.5 Making a lacquer peel (4) is another way to preserve and display grain-structure details. This technique works best on flat surfaces made with a slitting knife or osmotic cutter. Carefully scrape the surface to remove smears left by the knife, and allow the surface to dry completely. Apply one coat of lacquer with a sprayer, and after several minutes, apply one or two more coats with a brush. When the last coat is firm but still tacky, cover it with a layer of cheesecloth, and paint the cloth with another layer of lacquer. After the surface and coatings have hardened, lift and peel away the cloth-lacquer laminate to expose the thin layer of sediment adhering to the first coat. Protect this sediment by spraying it with a thin coat of lacquer, cut away the excess cloth, and trim the edges. The finished lacquer peel can be mounted on a stiff board for display.

15. Keywords

15.1 core distortion; core samplers; freeze core samplers; grab samplers; piston samplers; sampling errors; sediment sampling; surface samplers; vibratory samplers

REFERENCES

(1) Weaver, P. P. E., and Schultheiss, P. J., "Detection of Repenetration and Sediment Disturbance in Open-Barrel Gravity Cores," *Journal of Sedimentary Petrology*, Vol 53, No. 2, June 1983, pp. 649–654.

(2) Schultheiss, P. J., "Geotechnical Properties of Deep Sea Sediments: A Critical Review of Measurement Techniques," Institute of Oceanographic Sciences, Wormley, U. K., Report No. 134, 1982.

(3) Sachs, P. L., and Raymond, S. O., "A New Unattached Sediment Sampler," *Journal of Marine Research*, Vol 23, 1965, pp. 44–53.

(4) Bouma, A. H., *Methods for the Study of Sedimentary Structures*, Wiley-Interscience, New York, NY, 1969, pp. 301–378.

(5) Carrigan, P. H., Jr., "Inventory of Radionuclides in Bottom Sediment of the Clinch River Eastern Tennessee," U.S. Geological Survey Professional Paper 433-I, U.S. Government Printing Office, Washington, DC, 1969.

(6) Milbrink, G., "A Simplified Tube Bottom Sampler," Oikos:Acta-Oecologica-Scandinavica. No. 22, 1971, pp. 260–263.

(7) Bascom, W., Mardesich, J., and Stubbs, H., "An Improved Corer for Soft Sediments," Coastal Water Research Project, Biennial Report for 1981–1982, Willard Bascom, ed., Southern California Coastal Water Research Project, Long Beach, CA, 1982, pp. 267–271.

(8) Peters, R. D., Timmins, N. T., Calvert, S. E., and Morris, R. J., "The IOS Box Corer: Its Design, Development, Operation and Sampling," Institute of Oceanographic Sciences, Wormley, U. K., Report No. 106, 1980.

(9) Kermabon, A., Blavier, P., Cortis, V., and Delauze, H., "The "Sphincter" Corer: A Wide-Diameter Corer with Watertight Core-Catcher," *Marine Geology*, Vol 4, 1966, pp. 149–162.

(10) Kallstenius, T., "Mechanical Disturbances in Clay Samples Taken with Piston Samplers," *Royal Swedish Geotechnical Institute Proceedings*, No. 16, Ivar Haeggstroms Baktryckeri AB, Stockholm, 1958.

(11) Hvorslev, M. J., *Subsurface Exploration and Sampling of Soils for Civil Engineering Purposes*, U.S. Army, Corps of Engineers, Waterways Experiment Station, Vicksburg, MI, 1949.

(12) Silva, A. J., and Hollister, C. D., "Geotechnical Properties of Ocean Sediments Recovered with Giant Piston Corer," *Journal of Geophysical Research*, Vol 78, No. 18, June 20, 1973, pp. 3597–3616.

(13) Bouma, A. H., and Boerma, J. A. K., "Vertical Disturbances in Piston Cores," *Marine Geology*, Vol 6, 1968, pp. 231–241.

(14) Burns, R. E., "A Note on Some Possible Misinformation From Cores Obtained by Piston-Type Coring Devices," *Journal of Sedimentary Petrology*, Vol 33, 1963, pp. 950–952.

(15) Kermabon, A., and Cortis, V., "A New "Sphincter" Corer with a Recoilless Piston," *Marine Geology*, Vol 7, 1969, 147–159.

(16) Hopkins, T. L., "A Survey of Marine Bottom Samplers," *Progress in Oceanography*, Vol 2, Pergamon Press, Inc., Elmsford, NY, 1964, pp. 215–256.

(17) Brakensiek, D. L., Osborn, H. B., and Rawls, W. J., "Field Manual for Research in Agricultural Hydrology," Agricultural Handbook No. 224, U.S. Dept. of Agriculture, Science and Education Administration, 1979.

(18) Hubbell, D. W., and Glenn, J. L., "Distribution of Radionuclides in Bottom Sediments of the Columbia River Estuary," U.S. Geological Survey Professional Paper 433-L, U.S. Government Printing Office, Washington, DC, 1973, pp. L6–L7.

(19) Prych, E. A., Hubbell, D. W., and Glenn, J. L., "New Estuarine Measurement Equipment and Techniques," *Journal of the Waterways and Harbors Division*, *Proceedings* of the American Society of Civil Engineers, Vol 93, No. WW 2, May, 1967, pp. 41–58.

(20) Everest, F. H., McLemore, C. E., and Ward, J. F., "An Improved Tri-Tube Cryogenic Gravel Sampler," U.S. Department of Agriculture, Forest Service, 809 NE 6th Ave., Portland, OR, PNW-350, March 1980.

(21) Inter-Agency Committee on Water Resources, Subcommittee on Sedimentation, "Determination of Fluvial Sediment Discharge," *A Study of Methods Used in Measurement and Analysis of Sediment Loads in Streams*, Federal Inter-Agency Sedimentation Project, Minneapolis, MN, Report No. 14, 1963.

(22) Yuzyk, T. R., "Bed Material Sampling in Gravel-Bed Streams," *Environment Canada*, IWS-HQ-WRB-SS-86-8, 1986.

(23) Earth Manual, A Water Resources Technical Publication, 2nd Ed., U.S. Dept. of Interior, Water and Power Resources Services, U.S. Government Printing Office, Washington, DC, 1974, pp. 361–383.

(24) Sly, P. G., "Bottom Sediment Sampling," Center for Great Lakes Research, Canada Centre for Inland Waters, Burlington, Ontario, 1969, pp. 883–898.

(25) Schneider, R. F., "A Coring Device for Unconsolidated Lake Sediments," Water Resources Research, U.S. Department of Interior, Federal Water Pollution Control Administration, Southeast Water Laboratory, Athens, GA, Vol 5, No. 2, 1969, pp. 524–526.

(26) Pickering, R. J., "River-Bottom Sediment Sampling With a Swedish Foil Sampler," Selected Techniques of Water-Resources Investigations, 1965, U.S. Geological Survey Water-Supply Paper 1822, U.S. Government Printing Office, Washington, DC, pp. 94–99.

(27) Fuller, J. A., and Meisburger, E. P., "A Lightweight Pneumatic Coring Device: Design and Field Test," U.S. Army, Corps of Engineers, Coastal Engineering Research Center, Fort Belvoir, VA, Miscellaneous Report No. 82-8, September 1982.

(28) Prych, E. A., and Hubbell, D. W., "A Sampler for Coring Sediments in Rivers and Estuaries," *Geological Society of Amer-ica Bulletin.* Geological Society of America Publications, Boulder, CO, Vol 77, May 1966, pp. 549–557.

(29) Hagerty, Royal, *Deep Sea Sediments: Physical and Mechanical Properties,*, Marine Series, Plenum Press, New York, NY, 1974, pp. 169–186.

(30) Bouma, A. H., Marshall, N. F., "A Method for Obtaining and Analyzing Undisturbed Oceanic Sediment Samples," *Marine Geology*, Vol 2, 1964, pp. 81–99.

(31) Rosfelder, A. M., and Marshall, N. F., "Obtaining Large, Undisturbed, and Oriented Samples in Deep Water," *Marine Geotechnique*, A. F. Richards, ed., (*Proceedings*, 1966 International Conference on Marine Geotechnique), University of Illinois Press, Urbana, IL, 1967, pp. 243–263.

(32) Moore, D. G., "The Free Corer: Sediment Sampling Without Wire and Winch," *Journal of Sedimentary Petrology*, Vol 31, 1961, pp. 627–630.

(33) Meisburger, E. P., and Williams, S. J., "Use of Vibratory Coring Samplers for Sediment Surveys," *Coastal Engineering Technical Aid* No. 81-9, U.S. Army, Coastal Engineering Research Center, Fort Belvoir, VA, No. 81-9, July 1981.

(34) Meisburger, E. P., Williams, S. J., and Prins, D. A., "An Apparatus for Cutting Core Liners," *Journal of Sedimentary Petrology*, Vol 50, No. 2, June 1980, pp. 641–645.

(35) Lara, J. M., and Sanders, J. I., "Investigations of Sediment Properties," The 1963–64 Lake Mead Survey, U.S. Department of Interior, Bureau of Reclamation, REC-OCE-70-21, Part 4, August 1970, pp. 75–140.

(36) Hogan, Scott A., Abt, Steven R., and Watson, Chester C., "Development and Testing of a Bed Material Sampling Method for Gravel and Cobble Bed Streams," Department of Civil Engineering, Colorado State University, Fort Collins, CO, 80523, 1993.

Standard Practices for
Handling, Storing, and Preparing Soft Undisturbed Marine Soil[1]

This standard is issued under the fixed designation D 3213; the number immediately following the designation indicates the year of original adoption or, in the case of revision, the year of last revision. A number in parentheses indicates the year of last reapproval. A superscript epsilon (ε) indicates an editorial change since the last revision or reapproval.

1. Scope

1.1 These practices cover methods for project/cruise reporting, and handling, transporting and storing soft cohesive undisturbed marine soil. Procedures for preparing soil specimens for triaxial strength, and consolidation testing are also presented.

1.2 These practices may include the handling and transporting of sediment specimens contaminated with hazardous materials and samples subject to quarantine regulations.

1.3 *This standard does not purport to address all of the safety problems, if any, associated with its use. It is the responsibility of the user of this standard to establish appropriate safety and health practices and determine the applicability of regulatory limitations prior to use.* Specific precautionary statements are given in Sections 1, 2 and 7.

1.4 The values in acceptable SI units are to be regarded as the standard. The values given in parentheses are for information only.

2. Referenced Documents

2.1 *ASTM Standards:*
D 653 Terminology Relating to Soil, Rock, and Contained Fluids[2]
D 1587 Practice for Thin-Walled Tube Sampling of Soils[2]
D 2435 Test Method for One-Dimensional Consolidation Properties of Soils[2]
D 2488 Practice for Description and Identification of Soils (Visual Manual Procedure)[2]
D 2850 Test Method for Unconsolidated, Undrained Compressive Strength of Cohesive Soils in Triaxial Compression[2]
D 4186 Test Method for One-Dimensional Consolidation Properties of Soils Using Controlled-Strain Loading[2]
D 4220 Practices for Preserving and Transporting Soil Samples[2]
D 4452 Methods for X-Ray Radiography of Soil Samples[2]

3. Terminology

3.1 *Definitions*—The definitions of terms used in these practices shall be in accordance with Terminology D 653.

4. Summary of Practice

4.1 Procedures are presented for handling, transporting, storing, and preparing very soft and soft, fine-grained marine sediment specimens that minimize disturbance to the test specimen from the time it is initially sampled at sea to the time it is placed in a testing device in the laboratory.

5. Significance and Use

5.1 Disturbance imparted to sediments after sampling can significantly affect some geotechnical properties. Careful practices need to be followed to minimize soil fabric changes caused from handling, storing, and preparing sediment specimens for testing.

5.2 The practices presented in this document should be used with soil that has a very soft or soft shear strength (undrained shear strength less than 25 kPa (3.6 psi)) consistency.

NOTE 1—Some soils that are obtained at or just below the seafloor quickly deform under their own weight if left unsupported. This type of behavior presents special problems for some types of testing. Special handling and preparation procedures are required under those circumstances. Test are sometimes performed at sea to minimize the effect of storage time and handling on soil properties. An undrained shear strength of less than 25 kPa was selected based on Terzaghi and Peck.[3] They defined a very soft saturated clay as having undrained shear strength less than 25 kPa.

5.3 These practices shall apply to specimens of naturally formed marine soil (that may or may not be fragile or highly sensitive) that will be used for density determination, consolidation, permeability testing or shear strength testing with or without stress-strain properties and volume change measurements (see Note 2). In addition, dynamic and cyclic testing can also be performed on the sample.

NOTE 2—To help evaluate disturbance, X-Ray Radiography has proven helpful, refer to Methods D 4452.

5.4 These practices apply to fine-grained soils that do not allow the rapid drainage of pore water. Although many of the procedures can apply to coarser-grained soils, drainage may occur rapidly enough to warrant special handling procedures not covered in these practices.

5.5 These practices apply primarily to soil specimens that are obtained in thin-walled or similar coring devices that produce high-quality cores or that are obtained by pushing a thin-walled tube into cores taken with another sampling device.

5.6 These practices can be used in conjunction with soils containing gas, however, more specialized procedures and equipment that are not covered in these practices have been

---

[1] These practices are under the jurisdiction of ASTM Committee D-18 on Soil and Rock and are the direct responsibility of Subcommittee D18.13 on Marine Geotechnics.
Current edition approved May 15, 1991. Published July 1991.
[2] *Annual Book of ASTM Standards*, Vol 04.08.

[3] Terzaghi, K. and Peck, R. B., *Soil Mechanics in Engineering Practice*, 2nd ed., Wiley, 1967, p. 729.

developed for use with such materials.

NOTE 3—For information on handling gas charged sediments, the reader is referred to papers by Johns, et al.,[4] and Lee.[5]

6. Apparatus

6.1 *Coring Device*, capable of obtaining high-quality soil specimens, including related shipboard equipment such as cable and winch. Typical coring devices used in industry are the wireline push or piston samplers.

NOTE 4—Some sampling devices, for example, box corers, obtain samples of a size or shape that are difficult to preserve. Such cores can be subsampled aboard ship by pushing a thin-walled sampler into the larger size core. This method can produce samples from soils obtained near the seafloor. The subsamples can then be handled and stored according to these practices.

6.1.1 *Metal or Plastic Liners or Barrels (Pipe or Thin-Walled Tubes)*, the soil will be obtained or stored within, or both. Short sections of the liner, sharpened on one end, may also be used to subsample larger sized cores (see Note 4). It is important to note that liners constructed of cellulose acetate butyrate (CAB) plastic are pervious to water. Polycarbonate is nearly impervious and polyvinyl chloride (PVC) is impervious to water migration.

6.2 *Equipment Required on Board Ship to Seal and Store Soil Samples:*

6.2.1 *Identification Material*—This includes the necessary writing pens, tags, and labels to properly identify the sample(s).

6.2.2 *Caps*, either plastic, rubber, or metal, to be placed over the end of thin-walled tubes, liners and rings, and sealed with tape or wax, or both.

6.2.3 *Packers*, or add wax to top and bottom of core to seal the ends of samples within thin-walled tubes.

NOTE 5—Plastic expandable packers are preferred. Metal expandable packers seal equally well; however, long-term storage using metal expandable packers may cause corrosion problems.

6.2.4 *Filler Material*, used to occupy the voids at the top and bottom of the sediment container. The material must be slightly smaller than the inside dimensions of the container and must be a light-weight, nonabsorbing, nearly incompressible substance. For example, wooden disks of various thicknesses that have been coated with a waterproofing material can be used.

6.2.5 *Tape*, either waterproof electrical or duct tape.

6.2.6 *Cheesecloth or Aluminum Foil*, to be used in conjunction with wax for block sample.

6.2.7 *Sealing Wax*, non-shrinking, non-cracking wax, includes microcrystalline wax, beeswax, ceresine, carnaubawax, or combination thereof.

NOTE 6—The wax must be able to adhere to the container and be ductile enough not to chip or flake off during handling at cold temperatures. Microcrystalline wax alone or in combination with other waxes has been shown to be satisfactory in sealing the ends of cores stored at low temperatures.

6.2.8 *Plastic Wrap*, used to prevent the wax from adhering to other objects and providing additional protection against soil moisture loss.

6.2.9 *Core Storage Boxes*.

6.2.10 *Rope, Cord, or Chains*, used to immobilize containers, boxes, or other core storage fixtures aboard ship.

6.2.11 *Shipboard Refrigeration Equipment*, when geochemical, or gas charged sediments are present or other special use. Refrigeration may not be needed under some circumstances, such as coring in shallow water in the tropics.

6.3 *Equipment for Transporting Cores*, used from the ship to a shore-based laboratory facility.

6.3.1 *Packing*—Material to protect against vibration and shock, includes foam rubber.

6.3.2 *Insulation*, if refrigeration is not used, either granule (bead) sheet, or foam type, to resist temperature change of soil or to prevent freezing.

6.3.3 *Shipping Containers*, either box or cylindrical type and of proper construction to protect against vibration, shock, and the elements. Refer to Practices D 4220.

NOTE 7—The length, girth, and weight restrictions for commercial transportation must be considered.

6.4 *Equipment for Storing Cores*, used at the shore-based laboratory facility.

6.4.1 *Refrigeration Unit*, capable of maintaining a temperature close to the in situ condition, see section 6.2.11.

6.4.2 *Core Storage Boxes or Racks*, capable of supporting all cores in the vertical orientation in which they were obtained.

NOTE 8—An environment that is close to 100 % relative humidity may be required to minimize sediment water loss during storage of samples obtained within cellulose acetate butyrate (CAB) liners unless they are totally coated with impervious wax and plastic wrap. Other liner materials, such as polycarbonate or polyvinyl chloride (PVC) may be more suitable for sample storage because of their low water transmissibility.

6.5 *Equipment for Preparing Specimens*, used for laboratory testing.

6.5.1 *Thin-Walled Rings*, made of stainless steel or other noncorrosive metal or material, used to obtain samples for consolidation or permeability testing.

NOTE 9—The sampling ring may also be used as the test confining ring. For size and deformation requirements of consolidation test rings refer to Test Methods D 2435 and D 4186. Because of the small height to diameter ratio of consolidation samples and due to the nature of consolidation testing, the inside clearance ratio as specified by Practice D 1587 can be reduced from 1 % to zero. The ring area ratio, A_r, equal to $[(D_o^2 - D_i^2)/D_i^2] \times 100$ (terms are defined in Practice D 1587) should be less than 13 % to minimize subsampling disturbance.[6]

6.5.2 *Thin-Walled Piston Subsampler*, used to obtain triaxial test specimens from soil that quickly deforms under its own weight if left unsupported (see Fig. 1).

NOTE 10—To minimize soil disturbance, the sampler wall thickness should be the thinnest possible that will adequately obtain a test specimen. The area ratio (see Note 9) should be less than 10 % and the inside clearance ratio (refer to Practice D 1587) should be zero.

[4] Johns, M. W., Taylor, E., and Bryant, W. R., "Geotechnical Sampling and Testing of Gas-Charged Marine Sediments at In Situ Pressures," *Geo-Marine Letters*, Vol 2, 1982, pp. 231–236.

[5] Lee, H. J., "State of the Art: Laboratory Determination of the Strength of Marine Soils," *Strength Testing of Marine Sediments, ASTM STP 883*, ASTM, 1985, pp. 181–250.

[6] International Society for Soil Mechanics and Foundation Engineering. *International Manual for the Sampling of Soft Cohesive Soils*, Tokai University Press, Tokyo, 1981, p. 129.

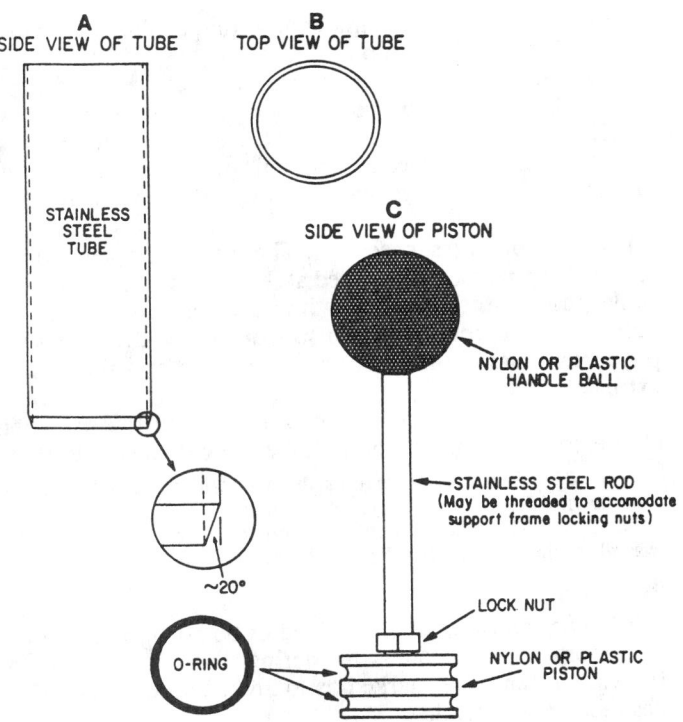

A SIDE VIEW OF TUBE

B TOP VIEW OF TUBE

STAINLESS STEEL TUBE

C SIDE VIEW OF PISTON

NYLON OR PLASTIC HANDLE BALL

~20°

STAINLESS STEEL ROD (May be threaded to accomodate support frame locking nuts)

O-RING

LOCK NUT

NYLON OR PLASTIC PISTON

FIG. 1 Thin-Walled Piston Sampler

7. Hazards

7.1 Preserving and transporting soil samples may involve personnel contact with hazardous materials, operations, and equipment. It is the responsibility of whoever uses these practices to consult and establish appropriate safety and health practices and to determine the applicability of regulatory limitations and requirements prior to use.

7.2 Special instructions, descriptions, and marking of containers must accompany and be affixed to any sample container that may include radioactive material, toxic chemicals, or other hazardous materials.

7.3 Interstate transportation, containment, storage, and disposal of soil samples obtained from certain areas within the United States and the transportation of foreign soils into or through the United States are subject to regulations established by the U.S. Department of Agriculture, Animal and Plant Health Service, Plant Protection, and Quarantine Programs, and possibly to regulations of other federal, state, or local agencies.

8. Procedure

8.1 *Shipboard Handling of Soil Cores not Requiring Subsampling:*

8.1.1 Carefully bring soil sampling or coring device aboard ship, avoid contact with either the side of the ship or moon pole, or dropping the device onto the deck during this process. For drop corers, have an end cap available to prevent material from dropping out.

NOTE 11—Proper coring and sampling operations may not be possible during adverse weather conditions or sea states.

8.1.2 Remove liner or core tube from soil sampling or coring device.

8.1.3 *Sealing the Bottom of the Sample Liner:*

8.1.3.1 Either insert expandable packer and tighten (some sediment may have to be removed) or add wax at top and bottom of core in its tube.

8.1.3.2 Apply an end cap and securely tape in place with waterproof electrical tape or duct tape. If the sample is to be stored for over 2 weeks prior to testing, insure that the tape is completely covered with wax by dipping the liner and end cap into a container of melted wax. Cover with plastic wrap prior to storage. Leakage or evaporation of pore water during storage is not acceptable.

NOTE 12—If an air void is present between the end of the liner and the soil surface, cut the liner level with the soil surface prior to applying the end cap or fill the void with a nearly incompressible, nonabsorbing inert material, or add wax (see Note 5). Free water accumulating above the sample when held vertically can be drained by either a small cut in the liner, drilled hole, or decanting. The liner is then cut off level with the soil surface. When cutting the liner, be sure that the method used does not impart significant vibrations to the soil or distort the liner.

NOTE 13—A rotary pipe cutter, fine-toothed crosscut hand or power saw, or custom-made device may be used to cut liner with a circular cross section. Snug-fitting metal sleeves applied around the liner perimeter on each side of the cut can be used to minimize liner distortion during the cutting process.

NOTE 14—Cheesecloth or aluminum foil taped to the end cap and liner allows better adhesion of the wax and reduces the potential of wax to chip or crack during handling. Cheesecloth or aluminum foil and wax may be applied to entire core sections to reduce leakage and evaporation.

8.1.4 *Recording Information on Core Liner:*

8.1.4.1 Mark a series of arrows or other appropriate symbols or text on the liner indicating the top of the core.

8.1.4.3 Mark the core liner in accordance with the project requirements with information such as cruise, station, and boring number/core identifiers and subbottom depth intervals on the liner at several locations. Adhesive labels can be used if they are securely fastened in place with tape or a coating of wax.

8.1.5 *Sealing the Top of the Sample Liner:*

8.1.5.1 Refer to 8.1.2.1 or 8.1.3.2 and Notes 12 and 14.

8.1.5.2 Cores that are stored vertically for less than two weeks do not need the top cap waxed.

8.1.5.3 Cores that are stored horizontally for longer than 2 weeks require either waxing of both ends if packers are not used. Either brush wax on or dip ends into wax.

8.1.6 If the recovered core is larger than 1.5 m (4.9 ft), cut the liner into lengths of 1 m (3.3 ft) or less. Seal the ends in accordance with 8.1.3.1 or 8.1.3.2. It is desirable for ease of storage and handling that cores are cut to the same length as far as practical.

8.1.7 Securely store the core in the same orientation in which it was obtained away from sources of vibration in an area typically along the vessel centerline and amidships, where the ship motion will be minimized.

NOTE 15—Cores are typically obtained and stored vertically. Some circumstances may require horizontal storage; for example, if significant gravitational compaction is anticipated, then the cores can be stored horizontally. However, horizontal storage can cause problems. For example, a smear zone can develop along one side of the core by gravitational compaction while oriented horizontally. If the cores are stored horizontally, make a note to that effect on the core sample information report. Better sample quality will be obtained by storing short cores sections vertically than by storing long cores horizontally.

NOTE 16—If refrigeration equipment is available aboard ship the soil samples should ideally be stored at the in situ, seafloor temperature. Deep sea soil is often stored at 5 ± 2°C. Under no circumstances should cores or subsamples be frozen because of possible water migration and volume change if index property, compressibility, and strength testing are to be performed, unless they contain permafrost or were naturally frozen.

NOTE 17—Where applicable, follow procedures for soil Groups C and D in Practices D 4220.

8.2 *Shipboard Handling of Soil Requiring Subsampling:*

8.2.1 Carefully bring coring device aboard ship; avoid contact with the side of the vessel or dropping the device onto the deck during this process.

8.2.2 Expose sediment surface to be subsampled.

8.2.3 Insert subsampling device into core.

NOTE 18—The subsampling device may be a piece of liner sharpened on one end, a thin-walled stainless steel ring, or a thin-walled stainless steel piston sampler (Fig. 1). The inside diameter of the thin-walled subsampler may be the same as the outside diameter of the laboratory test specimen.

8.2.4 Remove the subsampler from the large-size core and clean the excess soil from the exterior of the subsampler with a spatula and wipe with either a cloth or paper towel.

8.2.5 Seal the ends of the subsampling device to prevent drainage. Refer to 8.1.3.1 or 8.1.3.2.

8.2.6 Refer to 8.1.4 and 8.1.7.

NOTE 19—Sealed subsampled sections for long term storage can be stored in tanks of seawater to reduce the potential for moisture loss. The salinity of the seawater should be similar to the salinity of the sediment's pore water.

8.3 *Transportation of Samples from the Ship to the Shore-Based Laboratory:*

8.3.1 Transport cores or subsamples in accordance with Practice D 4220 for soil Groups C and D where applicable.

NOTE 20—If cores are stored at in situ temperature aboard ship, then provision for maintaining that temperature during transportation to the shore-based laboratory would be optimal. However, if transit time is short and containers are adequately insulated, special provisions are generally unnecessary.

NOTE 21—Tilt indicators can be applied to the interior and exterior of shipping containers to record mishandling. Any mishandling must be noted and considered when analyzing data from the resulting test program.

8.4 *Storage of Samples at the Shore-Based Laboratory:*

8.4.1 Securely store cores and subsamples in the orientation in which they were obtained in situ. Refer to Note 15.

8.4.2 Check the cores as appropriate for signs of leakage. Report and correct any leakage.

8.4.3 Store cores in the dark at their in situ temperature at high humidity.

8.4.4 Test cores as soon as possible after sampling.

NOTE 22—Storage time should be minimal. Some properties may change for some types of soil within hours or days of sampling. Storage should not be allowed to adversely affect the soil properties to be measured.

8.5 *Specimen Preparation:*

8.5.1 Follow specimen handling procedures for each respective test.

8.5.2 If any test, for example, a consolidation test, requires that a top cap be placed on the soil make the top cap

out of a light-weight inert material so that applied consolidation or shear stresses are minimized.

8.5.3 For the triaxial compression test, refer to Test Method D 2850, and the following special provisions.

8.5.3.1 Specimens that cannot be trimmed with a common soil lathe because the soil quickly deforms if left unsupported can be prepared by using a thin-walled piston sampler (Fig. 1).

8.5.3.2 Obtain test specimen. The piston sampler is operated by keeping the piston fixed at the level of the soil surface while pushing the tube down into the soil. Use the piston sampler in conjunction with a support device that keeps the piston stationary, during both subsampling and subsequent extrusion.

8.5.3.3 Quickly place a membrane around the specimen after extrusion, assemble and fill the test chamber with fluid.

NOTE 23—Whenever water must be added it is preferable to add water having the same electrolyte as the natural water.

NOTE 24—The above procedure should provide an acceptable triaxial specimen for all but the weakest marine soil.

9. Report

9.1 The following data obtained onboard ship and in the laboratory for each core or boring should be reported. Individual items are marked as follows: E — essential, D — desirable, U — useful.

9.1.1 Cruise or Project identification (E),

9.1.2 Station/boring number and core identification (E),

9.1.3 Date and time of sampling (D),

9.1.4 Location, including latitude and longitude (E),

9.1.5 Water depth (D),

9.1.6 Sea state, weather conditions (U),

9.1.7 Corer or sampler type (E),

9.1.8 Core barrel or sampler length, and cross section dimension(s) (D),

9.1.9 Free-fall height or rate of penetration (U),

9.1.10 Amount of weight added and total weight of coring device (U),

9.1.11 Bottom elevation (D),

9.1.12 Subbottom penetration depth, depth of sampling interval (E),

9.1.13 Total recovered length and length of individual subsections (D),

9.1.14 Soil description (see Practice D 2488) (D),

9.1.15 Storage orientation (U),

9.1.16 Storage temperature (U), and

9.1.17 Names of party members (U),

9.2 Note if any of the following conditions were present or occurred:

9.2.1 Visible degassing (D),

9.2.2 Pore water leakage, when it occurred, how it was corrected (D),

9.2.3 Soil disturbance, at what subbottom depths in the core (D),

9.2.4 Mishandling or problems during recovery, transportation and storage (E), and

9.2.5 Potential hazardous substances (E).

10. Keywords

10.1 laboratory fine-grained soils; marine; soft; storage; undisturbed samples

Standard Practice for
Preparation of Sediment Samples for Chemical Analysis[1]

This standard is issued under the fixed designation D 3976; the number immediately following the designation indicates the year of original adoption or, in the case of revision, the year of last revision. A number in parentheses indicates the year of last reapproval. A superscript epsilon (ε) indicates an editorial change since the last revision or reapproval.

1. Scope

1.1 This practice describes standard procedures for preparation of test samples (including the removal of occluded water and moisture) of field samples collected from locations such as streams, rivers, ponds, lakes, and oceans.

1.2 These procedures are applicable to the determination of volatile, semivolatile, and nonvolatile constituents of sediments.

1.3 *This standard does not purport to address all of the safety concerns, if any, associated with its use. It is the responsibility of the user of this standard to establish appropriate safety and health practices and determine the applicability of regulatory limitations prior to use. For a specific precautionary statement, see Note 3.*

2. Referenced Documents

2.1 *ASTM Standards:*
D 596 Practice of Reporting Results of Analysis of Water[2]
D 1129 Terminology Relating to Water[2]
D 1192 Specification for Equipment for Sampling Water and Steam in Closed Conduits[2]
D 3370 Practices for Sampling Water from Closed Conduits[2]
D 4410 Terminology for Fluvial Sediment[2]

3. Terminology

3.1 *Definitions*—For definitions of terms used in this practice, refer to Terminologies D 1129 and D 4410.

4. Summary of Practice

4.1 Samples collected (see Practices D 3370 and Specification D 1192) in the field are screened to remove foreign objects prior to homogenization for chemical examination and analysis. Large objects are mechanically removed and small ones are eliminated by sieving the sample through a 10-mesh (2-mm openings) sieve.

4.2 Wet, sieved samples are mixed for preliminary homogenization, then allowed to settle to remove most of the occluded water.

4.3 Moisture determinations are made on separate samples from those analyzed for volatile or semivolatile constituents.

4.4 Analyses for volatile constituents are made using wet, settled samples from which supernatant liquid has been removed by decantation. The results are corrected to those that would have been obtained on samples dried to constant weight at 105 ± 2°C, on the basis of a moisture determination using a separate sample.

4.5 Analyses for semivolatile constituents (for example, mercury) are made on samples previously dried at a temperature found to be adequate for the purpose, and specified in the corresponding analytical procedure.

4.6 Analyses for nonvolatile constituents are made on samples previously dried to constant weight at 105 ± 2°C.

4.7 A flow diagram, outlining typical procedures, is shown in Fig. 1.

5. Significance and Use

5.1 The chemical analysis of sediments, collected from such locations as streams, rivers, ponds, lakes, and oceans can provide information of environmental significance.

5.2 Sediment samples are inherently heterogeneous in that they contain occluded water in varying and unpredictable amounts and may contain foreign objects or material not ordinarily considered as sediment, the inclusion of which would result in inaccurate analysis.

5.3 Standard methods for separating foreign objects to facilitate homogenization will minimize errors due to poor mixing and inclusion of extraneous material.

5.4 Standardized procedures for drying provide a means for reporting analytical values to a common dry weight basis.

6. Preliminary Treatment of Field Samples

6.1 The analytical sample is arbitrarily defined as that which passes a 10-mesh (approximately 2-mm openings) sieve. The purpose of this is to provide a basis for discrimination of sediment and foreign objects or materials. Stainless steel or nylon sieves may be used when inorganic constituents are to be determined. Stainless steel or brass sieves are suitable for use when organic substances are to be determined.

NOTE 1—For inorganic analyses, stainless steel sieves are acceptable provided the mesh is not soldered or welded to the frame. For organic analyses, organic materials such as rubber or plastics should not be used in the storage or handling of samples.

6.2 Sieve dry samples without further pretreatment and follow the procedures given in 7.3, 7.4, or 7.5, or a combination thereof, as appropriate.

6.3 Vigorously stir wet field samples, which may have settled during transit, to incorporate as much field water as possible, thereby facilitating subsequent wet sieving.

NOTE 2—Do not add additional laboratory water since this may extract constituents or otherwise change the composition of the sedi-

[1] This practice is under the jurisdiction of ASTM Committee D19 on Water and is the direct responsibility of Subcommittee D19.07 on Sediments, Geomorphology, and Open-Channel Flow.
Current edition approved Oct. 15, 1992. Published December 1992. Originally published as D 3976 – 80. Last previous edition D 3976 – 88.
[2] *Annual Book of ASTM Standards*, Vol 11.01.

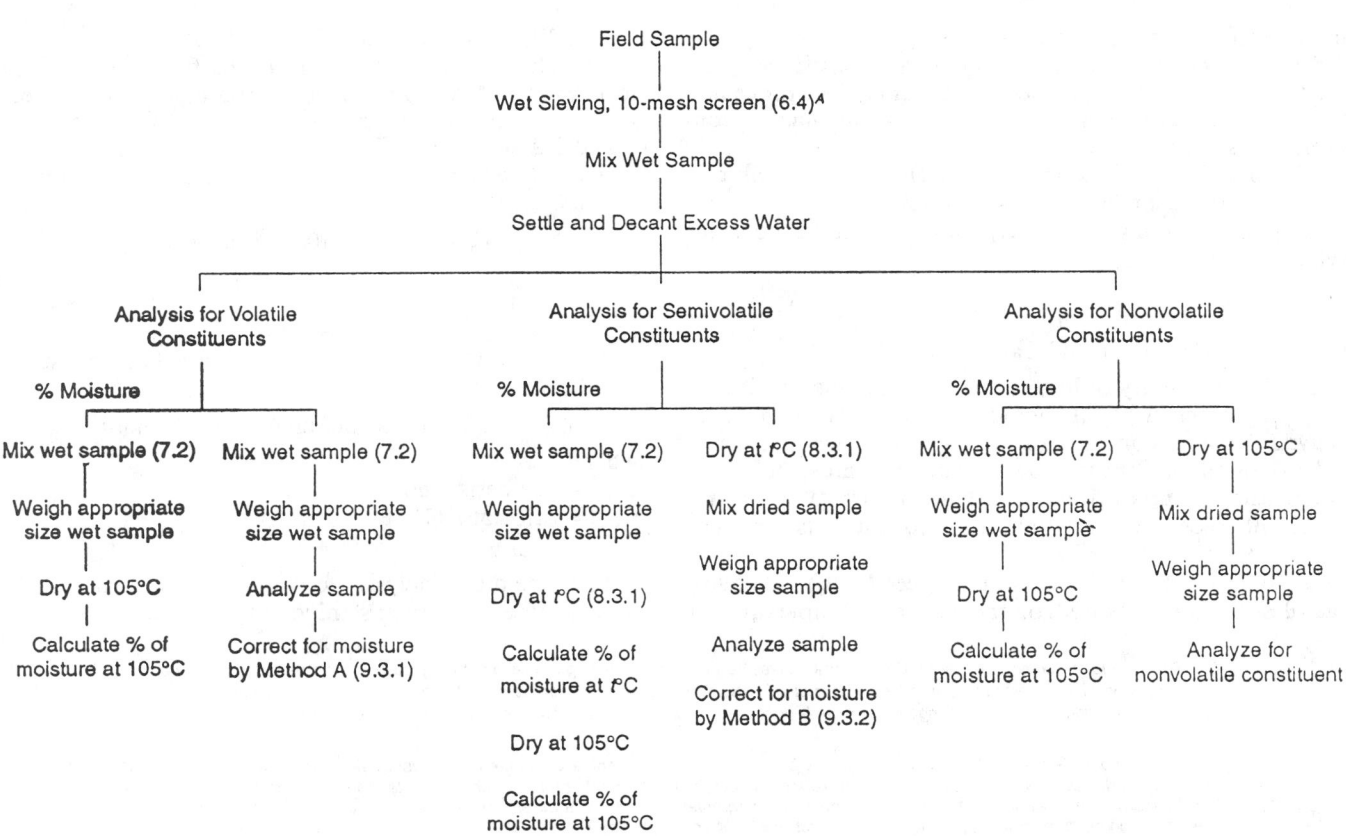

Field Sample
|
Wet Sieving, 10-mesh screen (6.4)[A]
|
Mix Wet Sample
|
Settle and Decant Excess Water

Analysis for Volatile Constituents — % Moisture
- Mix wet sample (7.2) → Weigh appropriate size wet sample → Dry at 105°C → Calculate % of moisture at 105°C
- Mix wet sample (7.2) → Weigh appropriate size wet sample → Analyze sample → Correct for moisture by Method A (9.3.1)

Analysis for Semivolatile Constituents — % Moisture
- Mix wet sample (7.2) → Weigh appropriate size wet sample → Dry at t°C (8.3.1) → Calculate % of moisture at t°C → Dry at 105°C → Calculate % of moisture at 105°C
- Dry at t°C (8.3.1) → Mix dried sample → Weigh appropriate size sample → Analyze sample → Correct for moisture by Method B (9.3.2)

Analysis for Nonvolatile Constituents — % Moisture
- Mix wet sample (7.2) → Weigh appropriate size wet sample → Dry at 105°C → Calculate % of moisture at 105°C
- Dry at 105°C → Mix dried sample → Weigh appropriate size sample → Analyze for nonvolatile constituent

FIG. 1 Flow diagram for Sediment-Sample Treatment

ment. However, it is permissible to slurry the sediment with a minimum quantity of field water collected with the sample, when necessary, to facilitate wet-sieving.

6.4 Pass the wet sample, preferably as a slurry, through the sieve (plastic or stainless steel). The bottom of an appropriate size Erlenmeyer flask may be used to gently press the sediment through the sieve, as necessary.

6.4.1 Manually remove foreign objects such as stones, twigs, leaves, trash, etc., which would obviously not pass through the sieve and which may interfere with the sieving operation.

6.5 Mix the sieved material by stirring and allow it to settle for subsequent removal of supernatant liquid.

6.5.1 Store the material, as prepared above, in contact with its supernatant liquid, until time of use for chemical examination.

NOTE 3: Precaution—Samples intended for both organic and inorganic compound analysis may undergo changes in composition during storage. The analytical method should specify the conditions necessary to assure requisite stability. In the absence of specific instructions, storage at a temperature of 4°C or lower for a period of time not to exceed 1 week is recommended, although it is known that microbiological activity does not cease under these conditions.

7. Preparation of Analytical Samples

7.1 Decant the supernatant liquid from the settled sediment prepared in accordance with Section 6. Save the supernatant liquid for separate analysis of any suspended material, as necessary.

7.2 Mix the sediment, using a glass rod or porcelain spatula, to minimize stratification effects due to differential rates of settling.

7.3 Remove a number of small portions (at least ten) from random locations in the sample container and composite them to obtain a representative sample of size suitable for determination of moisture (see Section 9), or for drying to prepare material for analysis of semivolatile or nonvolatile constituents.

7.4 Determine volatile constituents, using wet samples prepared in accordance with 7.3.

7.5 Determine semivolatile and nonvolatile constituents, using samples prepared in accordance with 7.3 and Section 8.

7.5.1 Disaggregate the dried material by gently crushing any lumps or clumps in a mortar. Mix the disaggregated material and prepare a composite sample resulting from removal of a number of smaller portions as indicated in 7.3. The use of freeze drying facilitates sample disaggregation.

8. Drying Procedure

8.1 Use a sample prepared in accordance with directions given in Section 7.

8.2 Accurately weigh 5 to 10 g (±1 mg) or 10 to 25 g (±10 mg) of the sediment into a previously tared porcelain dish, weighed with the same accuracy.

8.2.1 When a limited amount of sample is available, the moisture may be determined on 1- to 2-g samples, weighed with an accuracy of ±0.1 mg. The use of samples smaller than 1 g is not recommended for moisture determination.

8.3 Transfer the dishes containing the weighed sediment

to an oven and dry for 2 h as follows:

8.3.1 For determination of semivolatile constituents, use the temperature (t°C) specified in the analytical procedure for that constituent. For determination of nonvolatile constituents use 105 ± 2°C.

8.4 Cool in a desiccator, then weigh the dried samples with the same accuracy as the wet sample.

8.5 Repeat drying at hourly intervals, to a constant weight.

9. Moisture Correction and Calculations

9.1 The possibility of loss of volatile constituents dictates the drying procedure that should be used, prior to chemical analysis. Volatile constituents are usually determined using undried samples. Semivolatile constituents must be determined using samples dried at a temperature at which no significant losses occur. Nonvolatile constituents are analyzed using samples dried at 105 ± 2°C.

9.2 Report all analytical values (see Method D 596), regardless of how measured, on the basis of a sample dried at 105 ± 2°C to facilitate correlation of data.

9.3 The following equations are useful to correct analytical results, C (weight percent, µg/g, etc.), obtained on an undried sample or one partially dried at t°C, to the basis of one dried at 105 ± 2°C.

9.3.1 *Test Method A*—Calculate volatile constituents as follows:

$$C_v = C_{vl} (100)/(100 - \% \text{ moisture}_{105})$$

where:
C_v = dry basis, and
C_{vl} = wet basis.

9.3.2 *Test Method B*—Calculate semivolatile constituents as follows:

$$C_{sv} = C_{svl} (100 - \% \text{ moisture})/(100 - \% \text{ moisture}_{105})$$

where:
C_{sv} = dry basis, and
C_{svl} = dried at t°C.

10. Keywords

10.1 chemical analysis; nonvolatile constituents; preparation; sediment samples; semivolatile; volatile

Standard Guide for
Collection, Storage, Characterization, and Manipulation of Sediments for Toxicological Testing[1]

This standard is issued under the fixed designation E 1391; the number immediately following the designation indicates the year of original adoption or, in the case of revision, the year of last revision. A number in parentheses indicates the year of last reapproval. A superscript epsilon (ϵ) indicates an editorial change since the last revision or reapproval.

1. Scope

1.1 This guide covers procedures for obtaining, storing, characterizing, and manipulating saltwater and freshwater sediments, for use in laboratory sediment toxicity evaluations. It is not meant to provide guidance for all aspects of sediment assessments, such as chemical analyses or monitoring, geophysical characterization, or extractable phase and fractionation analyses. However, some of this information might have applications for some of these activities. A variety of test methods are reviewed in this guide. A statement on the consensus approach then follows this review of the test methods. This consensus approach has been included in order to foster consistency among studies. The state-of-the-art is currently in its infancy, and the development of standard test methods is not feasible; however, it is crucial that there be an understanding of the significant effects that these test methods have on sediment quality evaluations. It is anticipated that recommended test methods and this guide will be updated routinely to reflect progress in our understanding of sediments and how to best study them.

1.2 There are several regulatory guidance documents concerned with sediment collection and characterization procedures that might be important for individuals performing federal or state agency-related work. Discussion of some of the principles and current thoughts on these approaches can be found in Dickson, et al **(1)**.[2]

1.3 This guide is arranged as follows:

1.4 Field-collected sediments might contain potentially toxic materials and should thus be treated with caution to minimize occupational exposure to workers. Worker safety must also be considered when working with spiked sediments containing various organic, inorganic, or radiolabeled contaminants, or some combination thereof. Careful consideration should be given to those chemicals that might biodegrade, volatilize, oxidize, or photolyze during the exposure.

1.5 The values stated in either SI or inch-pound units are to be regarded as the standard. The values given in parentheses are for information only.

1.6 *This standard does not purport to address all of the safety concerns, if any, associated with its use. It is the responsibility of the user of this standard to establish appropriate safety and health practices and determine the applicability of regulatory limitations prior to use. Specific hazards statements are given in Section 8.*

2. Referenced Documents

2.1 *ASTM Standards:*
D 1129 Terminology Relating to Water[3]
D 4387 Classification of Grab Sampling Devices for Collecting Benthic Macroinvertebrates[4]
D 4822 Guide for Selection of Methods of Particle Size Analysis of Fluvial Sediments (Manual Methods)[5]
D 4823 Guide for Core-Sampling Submerged, Unconsolidated Sediments[5]
E 380 Practice for Use of the International System of Units (SI) (the Modernized Metric System)[6]
E 729 Guide for Conducting Acute Toxicity Tests with Fishes, Macroinvertebrates, and Amphibians[4]
E 943 Terminology Relating to Biological Effects and Environmental Fate[4]
E 1367 Guide for Conducting 10-Day Static Sediment Toxicity Tests with Marine and Estuarine Amphipods[4]
E 1383 Guide for Conducting Sediment Toxicity Tests with Freshwater Invertebrates[4]

3. Terminology

3.1 *Definitions:*
3.1.1 The words "must," "should," "may," "can," and "might" have very specific meanings in this guide. "Must" is used to express an absolute requirement, that is, to state that the test ought to be designed to satisfy the specified condi-

---

[1] This guide is under the jurisdiction of ASTM Committee E-47 on Biological Effects and Environmental Fate and is the direct responsibility of Subcommittee E47.03 on Sediment Toxicology.
Current edition approved July 15, 1994. Published September 1994. Originally published as E 1391 – 90. Last previous edition E 1391 – 90.
[2] The boldface numbers in parentheses refer to the list of references at the end of this standard.

[3] *Annual Book of ASTM Standards*, Vol 11.01.
[4] *Annual Book of ASTM Standards*, Vol 11.04.
[5] *Annual Book of ASTM Standards*, Vol 11.02.
[6] *Annual Book of ASTM Standards*, Vol 14.02.

tion, unless the purpose of the test requires a different design. "Must" is used only in connection with the factors that relate directly to the acceptability of the test. "Should" is used to state that the specified condition is recommended and ought to be met in most tests. Although the violation of one "should" is rarely a serious matter, the violation of several will often render the results questionable. Terms such as "is desirable," "is often desirable," and "might be desirable" are used in connection with less important factors. "May" is used to mean "is (are) allowed to," "can" is used to mean "is (are) able to," and "might" is used to mean "could possibly." Thus, the classic distinction between "may" and "can" is preserved, and "might" is never used as a synonym for either "may" or "can."

3.1.2 For definitions of terms used in this guide, refer to Guide E 729, Terminologies D 1129 and E 943, and Classification D 4387; for an explanation of units and symbols, refer to Practice E 380.

4. Summary to Guide

4.1 This guide provides a review of widely used test methods for collecting, storing, characterizing, and manipulating sediments for toxicity testing. Where the science permits, recommendations are provided on which procedures are appropriate, while identifying their limitations.

5. Significance and Use

5.1 Sediment toxicity evaluations are a critical component of environmental quality and ecosystem impact assessments, used to meet a variety of research and regulatory objectives. The manner in which the sediments are collected, stored, characterized, and manipulated can influence the results of any sediment quality or process evaluation greatly. Addressing these variables in a systematic and uniform manner will aid the interpretations of sediment toxicity or bioaccumulation results and may allow comparisons between studies.

6. Interferences

6.1 Maintaining the integrity of a sediment sample relative to ambient environmental conditions during its removal, transport, and testing in the laboratory is extremely difficult. The sediment environment is composed of a myriad of microenvironments, redox gradients, and other interacting physicochemical and biological processes. Many of these characteristics influence sediment toxicity and bioavailability to benthic and planktonic organisms, microbial degradation, and chemical sorption. Any disruption of this environment complicates interpretations of treatment effects, causative factors, and in situ comparisons. See Section 9 for additional information.

7. Apparatus

7.1 A variety of sampling, characterization, and manipulation methods exist using different equipment. These are reviewed in Sections 9 and 14.

7.2 *Cleaning*—Test chambers and equipment used to collect and store sediment samples, prepare and store dilution water and stock solutions, and expose test organisms should be cleaned before use. New glassware and plasticware should be soaked in 1:1 concentrated acid prior to use.

Soaking overnight is adequate for glassware. Soaking for seven days in HCl, followed by seven days in HNO_3, followed by seven days in deionized water is recommended for plasticware. Used sample containers should be washed following these steps: (1) non-phosphate detergent wash, (2) triple water rinse, (3) water-miscible organic solvent wash (acetone followed by pesticide-grade hexane (2, 3), (4) water rinse, (5) acid wash (such as 5 % concentrated hydrochloric acid), and (6) triple rinse with deionized-distilled water. Altering this cleaning procedure might result in problems. Many organic solvents might leave a film that is insoluble in water (Step 3). A dichromate-sulfuric acid cleaning solution can generally be used in place of both the organic solvent and the acid (Steps 3 through 5), but it might attack silicone adhesive. (See 9.10 for cleaning during sample collection.)

8. Safety Hazards

8.1 Many substances can affect humans adversely if adequate precautions are not taken. Information on the toxicity to humans (4) and recommended handling procedures of toxicants (5) should be studied before tests are begun with any contaminant or sediment. Health and safety precautions should be incorporated into any study plan prior to initiating any work with contaminants or sediments.

8.2 Field-collected sediments might contain a mixture of hazardous contaminants or disease-causing agents such that proper handling to avoid human exposure is critical. Skin contact with all test materials and solutions should therefore be minimized by such means as wearing appropriate protective gloves, especially when putting hands into sediments, overlying water, or washing equipment. Proper handling procedures might include the following: (1) sieving and distributing sediments under a ventilated hood or enclosed glove box; (2) enclosing and ventilating the toxicity test water bath; and (3) using respirators, aprons, safety glasses, and gloves when handling potentially hazardous sediments. Special procedures might be necessary with radiolabeled test materials (6) and materials that are, or are suspected of being, carcinogenic (5).

8.3 The disposal of sediments, dilution water over sediments, and test organisms containing hazardous compounds might pose special problems. Removal or degradation of the toxicant(s) before disposal is sometimes desirable for tests involving spiking sediments with known toxicants. Disposal of all hazardous wastes should adhere to the requirements and regulations of the Resource Conservation and Recovery Act and any relevant state or local regulations.

9. Sampling and Transport

9.1 Sediments have been collected for a variety of chemical, physical, toxicological, and biological investigations. The sediments should be collected from depositional zones in which fine-grained sediments accumulate. Site selection should also consider the location of pollutant loadings and hydrological flow patterns. The site selection may also need to be of a random or stratified random nature, depending on the study objectives. Sediment variability must be considered since most sediments are very heterogeneous (both vertically and horizontally) in nature. A preliminary survey or review of background data may therefore be required to determine accurately the appropriate number of sediment replicates to collect.

9.2 Sediment collections have been made with grab and dredge sampling devices and core samplers (see Table 2 and Guide D 4823). The advantages and disadvantages of the various collection methods have been reported previously (7, 8) and are summarized in Table 1. All sampling methods disturb the sediment integrity to a degree. It is important to obtain sediments with as little disruption as possible when using sediment toxicity evaluations for realistic laboratory evaluations of in situ conditions. Core sampling is preferred above other methods for this reason. Choosing the most appropriate sediment sampler for a study will depend on the sediment's characteristics, efficiency required, and study objectives. Several references are available that discuss the various collection devices (7–11). Grab samplers can penetrate sediments to depths of 10 to 50 cm. Dredge samplers collect to a depth of 10 cm and disrupt sediment integrity. Core samplers collect up to 1 or 2 m when collected by hand or gravity. However, vibratory or piston corers can reach depths of 10 m. The depth of penetration is limited to 10 core diameters in sandy substrates and 20 diameters in predominately clay sediments. The efficiency of these samplers for benthic collections has been compared, and the grab samplers are less efficient collectors than the corers in general, but they are easier to handle in rough water, often require fewer personnel, and are obtained more easily (9, 11, 12). Most of the reported studies used grab samplers, although box corers (13–15), gravity corers (16), and hand collection (17–19) test methods are reported with increasing frequency.

9.3 The disadvantages of grab and dredge samplers (Table 1) include a shallow depth of penetration and the presence of a shock wave that results in loss of the fine surface sediments. Murray and Murray (20), however, described a grab sampler usable in rough water that samples the top 1 cm of sediment quantitatively and retains fine materials. Other grab samplers that sample surface sediments quantitatively have been described by Grizzle (21). The depth profile of the sample may be lost in removal of the sample from the sampler. Grab sampling promotes the loss of not only fine sediments (Table 1), but also water-soluble compounds and volatile organic compounds present in the sediment. Dredge samplers are appropriate only for collecting sediments that are to be dredged because they disrupt sediment integrity severely and lose surficial fines.

9.4 Studies of macroinvertebrate sampling efficiency with various grab samplers have provided useful information for sampling in sediment toxicity and sediment quality evaluations. These data provide information that would indicate sampler efficiency at retaining surficial sediment layers. The modified van Veen is used commonly in coastal sampling (22). The Ekman grab is a commonly used sampler for benthic investigations (21). The Ekman's efficiency is limited to less compacted, fine-grained sediments, as are the corer samplers. Blomqvist (23) reviewed the various Ekman modifications and their associated problems and concluded that the Ekman grab could be used reliably if caution was used during operation. The most commonly used corer is the Kajak-Brinkhurst corer. The Petersen, PONAR, and Smith-McIntyre grabs are used most often (9) in more resistant

sediments. Based on studies of benthic macroinvertebrate populations the sediment corers are the most accurate samplers, in most cases followed by the Ekman grab (9). The PONAR grab was the most accurate and the Petersen the least for compacted sediments (9). A comparison of sampler precision indicated the van Veen sampler to be the least precise; the most precise were the corers and Ekman grab (9).

9.5 Many of the problems associated with grab and dredge samplers are largely overcome with the corers. The best corers for most sediment studies are hand-held polytetrafluoroethylene (PTFE) plastic, high-density polyethylene, glass corers (liners), or large box-corers. The corers can maintain the integrity of the sediment surface while collecting a sufficient depth. Furthermore, the box core can be sub-cored or sectioned at specific depth intervals, as required by the study. Unfortunately, the box corer is large and cumbersome; it is thus difficult to use. Freefall or gravity cores tend to cause compaction, disrupting the vertical gradients in the sediment. Compaction is reduced using the piston corer. Other coring devices that have been used successfully include the percussion corer (24) and vibratory corers (25–27).

9.6 Corer samplers have several limitations. Most corers do not work well in sandy sediments; grab samplers or diver-collected material remain the only current alternatives. In general, corers collect less sediment than grab samplers, which may provide inadequate quantities for some studies. Small cores tend to increase bow waves (that is, disturbance of surface sediments) and compaction, thus altering the vertical profile. However, these corers provide better information on spatial variation when multiple cores are obtained (9, 28–32). As shown by Rutledge and Fleeger (33) and others, care must be taken in subsampling from core samples since surface sediments might be disrupted in even hand-held core collection. They recommend subsampling in situ or homogenizing core sections before subsampling. See Ref (8) for additional information of various core types.

9.7 Core sampling should be used to best maintain the complex integrity of the sediment for studies of sediment toxicity, interstitial waters, microbiological processes, or chemical fate. When obtaining cores from shallow waters, one must ensure that the vessel does not disturb the sediments before sampling (34). If core sampling is not possible due to an inability of the core to penetrate the sediment (for example, highly compacted sediment) or retain the sample (for example, primarily sand composition), grab samplers should be used that reduce the loss of fine-grained surficial sediments.

9.8 Subsampling, compositing, or homogenization of sediment samples is often necessary, and the optimal methods will depend on the study objectives. Important considerations include the following: loss of sediment integrity and depth profile; changes in chemical speciation by means of oxidation and reduction or other chemical interactions; chemical equilibrium disruption resulting in volatilization, sorption, or desorption; changes in biological activity; completeness of mixing; and sampling container contamination. It is advantageous in most studies of sediment toxicity to subsample the inner core area (not contacting the sampler) since this area is most likely to have maintained its integrity and depth profile and not be contaminated by the sampler.

TABLE 1 Summary of Bottom Sampling Equipment[A]

Device	Use	Advantages	Disadvantages
PTFE or glass tube	Shallow wadeable waters or deep waters if SCUBA available. Soft or semi-consolidated deposits.	Preserves layering and permits historical study of sediment deposition. Rapid—Samples immediately ready for laboratory shipment. Minimal risk of contamination.	Small sample size requires repetitive sampling.
Hand corer with removable PTFE or glass liners	Same as above except more consolidated sediments can be obtained.	Handles provide for greater ease of substrate penetration. Above advantages.	Careful handling necessary to prevent spillage. Requires removal of liners before repetitive sampling. Slight risk of metal contamination from barrel and core cutter.
Box corer	Same as above.	Collection of large sample undisturbed, allowing for subsampling.	Hard to handle.
Gravity corers, that is, Phleger Corer	Deep lakes and rivers. Semi-consolidated sediments.	Low risk of sample contamination. Maintains sediment integrity relatively well.	Careful handling necessary to avoid sediment spillage. Small sample, requires repetitive operation and removal of liners. Time consuming.
Young Grab (PTFE- or kynar-lined modified 0.1-m² van Veen)	Lakes and marine areas.	Eliminates metal contamination. Reduced bow wake.	Expensive. Requires winch.
Ekman or box dredge	Soft to semisoft sediments. Can be used from boat, bridge, or pier in waters of various depths.	Obtains a larger sample than coring tubes. Can be subsampled through box lid.	Possible incomplete jaw closure and sample loss. Possible shock wave, which may disturb the fines. Metal construction may introduce contaminants. Possible loss of "fines" on retrieval.
PONAR Grab Sampler	Deep lakes, rivers, and estuaries. Useful on sand, silt, or clay.	Most universal grab sampler. Adequate on most substrates. Large sample obtained intact, permitting subsampling.	Shock wave from descent may disturb "fines." Possible incomplete closure of jaws results in sample loss. Possible contamination from metal frame construction. Sample must be further prepared for analysis.
BMH-53 Piston Corer	Waters of 4 to 6 ft deep when used with extension rod. Soft to semi-consolidated deposits.	Piston provides for greater sample retention.	Cores must be extruded on site to other containers. Metal barrels introduce risk of metal contamination.
Van Veen	Deep lakes, rivers, and estuaries. Useful on sand, silt, or clay.	Adequate on most substrates. Large sample obtained intact, permitting subsampling.	Shock wave from descent may disturb "fines." Possible incomplete closure of jaws results in sample loss. Possible contamination from metal frame construction. Sample must be further prepared for analysis.
BMH-60	Sampling moving waters from a fixed platform.	Streamlined configuration allows sampling where other devices could not achieve proper orientation.	Possible contamination from metal construction. Subsampling difficult. Not effective for sampling fine sediments.
Petersen Grab Sampler	Deep lakes, rivers, and estuaries. Useful on most substrates.	Large sample; can penetrate most substrates.	Heavy. May require winch. No cover lid to permit subsampling. All other disadvantages of Ekman and Ponar.
Shipek Grab Sampler	Used primarily in marine waters and large inland lakes and reservoirs.	Sample bucket may be opened to permit subsampling. Retains fine-grained sediments effectively.	Possible contamination from metal construction. Heavy. May require winch.
Orange-Peel Grab Smith-McIntyre Grab	Deep lakes, rivers, and estuaries. Useful on most substrates.	Designed for sampling hard substrates.	Loss of fines. Heavy. May require winch. Possible metal contamination.
Scoops, Drag Buckets	Various environments, depending on depth and substrate.	Inexpensive, easy to handle.	Loss of fines on retrieval through water column.

[A] Comments represent subjective evaluations.

Subsamples from the depositional layer of concern, for example, the top 1 or 2 cm, should be collected with a nonreactive sampling tool such as a PTFE-lined calibration scoop (35). Samples are frequently of a mixed depth; however, a 2-cm sample (36) is the most common depth obtained, although depths up to 12 m have been used in some dredging studies. It is advantageous or necessary for some studies to composite or mix single sediment samples (35, 37, 38). Composites usually consist of three to five grab samples. An advantage of composited samples is that they reduce the likelihood of missing a "hot spot" due to site heterogeneity. However, a disadvantage is the loss of information on the spatial variability at the site and reduction of the toxicity of hot spot samples when they are diluted with cleaner samples. This is a more critical issue in the boundary areas of the site contamination, also known as the grey zone. Subsamples are collected with a nonreactive sampling scoop and placed in a nonreactive bowl or pan. The composite sample should be mixed until the texture and color appear uniform.

9.9 The assessment of in situ sediment toxicity or bioaccumulation is aided by the collection and testing of reference and control samples. For the purposes of this guide, a reference sediment is defined as a whole sediment near an area of contamination used to assess sediment condition exclusive of the material(s) of interest. It should contain characteristics similar to those of the test sediment. Sediment characteristics, such as particle size distribution and percent organic carbon, should bracket that of the test sediment. If a wide range of test sediment types exists, the reference sediment characteristics should be in an intermediate range unless the test species is affected by particle size. The appropriate ASTM guides for marine (Guide E 1367) and freshwater (Guide E 1383) invertebrates should then be consulted to determine the particle size requirements of the test species. It is preferable that reference sediments be collected from the same aquatic system and be located close to and have physical, chemical, and biological characteristics similar to those of the test sediment. The reference sediment test results might be analyzed as either a treatment or a control variable, depending on the study objectives. The reference sediment might be toxic in some situations due to naturally occurring chemical, physical, or biological properties. It is important for this reason to also use control sediments in the evaluation of test sediments. A control sediment is defined as a sediment that is essentially free of contamination and is used routinely to assess the acceptability of a test. Control sediments have been used successfully in toxicity evaluations (39).

9.10 When collecting sediment grab samples, it is important to clean the sampling device, scoop, spatula, and mixing bowls between sample sites. The cleaning procedure can follow that outlined in Section 7 or the following (40): (1) soap and water wash, (2) distilled water rinse, (3) methanol rinse, (4) methylene chloride rinse, and (5) site water rinse. Waste solvents should be collected in labeled hazardous waste containers.

9.11 The transport conditions for the samples were not specified in the references reviewed in most cases. Where conditions were specified, the sediments were usually transported whole, in both plastic, polyethylene (41–43), and glass (18, 19, 44) containers, and transported under refrigeration or on ice (18, 19, 36, 44–49).

9.12 The collection, transport, storage, and test chamber material composition should be chosen based on a consideration of sorption effects, sample composition, and contact time (Table 2). For example, in sediments in which organics are of concern, brown borosilicate glass containers with PTFE lid liners are optimal, while plastic containers are recommended for metal samples. PTFE or high-density polyethylene containers are relatively inert and optimal for samples contaminated with multiple chemical types. Additionally, polycarbonate containers have been shown not to sorb metal species (50). However, Moody and Lindstrom (51) found that all plastics (including PTFE) leached elements and should be preconditioned with a seven-day soaking in 1:1 HCl, HNO$_3$, and deionized water. Shipping containers with insulation 1 in. (25.4 mm) in thickness kept samples at 4°C for 21 h, while insulation of 2-in. (51-mm) thickness maintained 4°C for 60 to 82 h (52). Additional information regarding chemical analyses on sample containers, preservation, storage times, and volume requirements is available in other guidance documents (7, 8, 37, 53–55). These criteria are applicable to toxicity test requirements in many cases.

10. Storage

10.1 Containers for storage were generally not specified, although it was assumed that the containers were the same as the transport containers, where specified, and were generally high-density polyethylene (see 9.12). Where sediments con-

tain volatile compounds, transport and storage should be in airtight PTFE or glass containers with PTFE-lined screw caps. Volatile and semi-volatile compounds must be stored at 4°C and are lost in seven or eight days, respectively (52). See Table 2 for further information on the storage requirements for chemical analyses.

10.2 Drying, freezing, and cold storage conditions all affect toxicity and bioavailability (56–61). The storage time of sediments used in toxicity tests was often not specified and, where specified, ranged from a few days (62) to one year (42). The storage of sediments after arrival at the laboratory was generally by refrigeration at 4°C (41–43, 45–49, 59, 62–65). Significant changes in metal toxicity to cladocerans and microbial activity have been observed in stored sediments (60, 66). Recommended limits for the storage of metal-spiked sediments have ranged from within two days (56) to five days (62) and seven days (67, 68). Cadmium toxicity in sediments has been shown to be related to acid volatile sulfide (AVS) complexation (69). When anoxic sediments were exposed to air, AVS was volatilized rapidly. AVS is apparently the reactive solid phase sulfide pool that binds metal, thus reducing toxicity. If a study objective is to investigate metal toxicity and the sediment environment is anoxic, exposure to air might reduce or increase toxicity due to the oxidation and precipitation of the metal species or loss of acid volatile sulfide complexation. A study of sediments contaminated with nonpolar organics found that the interstitial water storage time did not affect toxicity to polychaetes when samples were frozen (70); however, it is generally agreed that sediments to be used for toxicity testing must not be frozen (59, 61, 62, 67, 71).

10.3 Although risking changes in sediment composition, several studies elected to freeze samples (36, 59, 72–76). Fast-freezing of sediment cores has been recommended for some metal and organic chemical analyses (22, 63, 77); however, this alters the sediment structure and profile distortion occurs (33). Freezing has been reported to inhibit the oxidation of reduced iron and manganese compounds (73). It has also been recommended for stored sediments that are to be analyzed for organics and nutrients (78). Thomson, et al (77) found that no storage method for sediments preserved the initial chemical and physical characteristics of the sediment. Freezing was adequate for the chemical analyses of several metals and organic material (63, 77). Changes were observed at 15 days in sediments stored at 4°C. Oxidation was greater than reduction during storage (77). Carr and Wilkniss (79) showed no mercury loss in sediments acidified to pH 1 for up to eight days, but the sorbed fraction decreased from 80 to 15 % of the total concentration. If sediments are to be frozen for chemical analyses, they should be a split sample from those used for toxicity testing that are kept at 4°C.

10.4 Interstitial water chemistry can change significantly after 24-h storage (80, 81), even when stored at in situ temperatures (81). The coagulation and precipitation of humic material was noted when interstitial water was stored at 4°C for more than one week (82). Oxidation of reduced arsenic species in the pore water of stored sediments was unaffected for up to six weeks when samples were acidified and kept near 0°C, without deoxygenation. Deoxygenation

**TABLE 2   Sampling Containers, Preservation Requirements, and Holding Times for Sediment Samples[A] (EPA, 196, 197).
See also Rochon and Chevalier (160)**

Contaminant	Container[B]	Preservation	Holding Time
Acidity	P, G	Cool, 4°C	14 days
Alkalinity	P, G	Cool, 4°C	14 days
Ammonia	P, G	Cool, 4°C	28 days
Sulfate	P, G	Cool, 4°C	28 days
Sulfide	P, G	Cool, 4°C	28 days
Sulfite	P, G	Cool, 4°C	48 h
Nitrate	P, G	Cool, 4°C	48 h
Nitrate-nitrite	P, G	Cool, 4°C	28 days
Nitrite	P, G	Cool, 4°C	48 h
Oil and grease	G	Cool, 4°C	28 days
Organic	P, G	Cool, 4°C	28 days
Metals[C]			
Chromium VI	P, G	Cool, 4°C	48 h
Mercury	P, G		8 days
Metals (except Cr or Hg)	P, G		6 months
Organic Compounds[C]			
Extractables (including phthalates, atrosamines organochlorine pesticides, PCB's artroaromatics, isophorone, Polynuclear aromatic hydrocarbons, haloethers, chlorinated hydrocarbons, and TCDD)	G, PTFE-lined cap	Cool, 4°C	7 days (until extraction) 30 days (after extraction)
Extractractables (phenols)	G, PTFE-lined cap	Cool, 4°C	7 days (until extraction) 30 days (after extraction)
Purgables (halocarbons and aromatics)	G, PTFE-lined septum	Cool, 4°C	14 days
Purgables (acrolein and acrylonitrate)	G, PTFE-lined septum	Cool, 4°C	3 days
Orthophosphate	P, G	Cool, 4°C	48 h
Pesticides	G, PTFE-lined cap	Cool, 4°C	7 days (until extraction) 30 days (after extraction)
Phenols	P, G	Cool, 4°C	28 days
Phosphorus (elemental)	G	Cool, 4°C	48 h
Phosphorus, total	P, G	Cool, 4°C	28 days
Chlorinated organic compounds	G, PTFE-lined cap	Cool, 4°C	7 days (until extraction) 30 days (after extraction)

[A] Taken from EPA 600-4-84-075 and EPA 600-4-85-048. See also Ref (85) and USEPA/COE 1991.
[B] Polyethylene (P) or Glass (G).
[C] Freezing is recommended by some for metals and organics with holding times of 30 and 10 days, respectively (USEPA/COE 1991).

was necessary when samples were not acidified (83). See also Section 11.

10.5 In summary, it is recommended that sediments for toxicity tests and chemical analyses be refrigerated or placed on ice in polyethylene containers during transport. In addition, if samples are to be used for chemical analyses, the appropriate container and holding time should be used as previously described and in Table 2. The storage conditions should be refrigeration at 4°C and under anoxic conditions, if appropriate (37, 53, 84). It has been shown that some contaminated sediments can be stored at 4°C for up to 12 months without significant alterations in toxicity (85). Limits to storage time before testing therefore appear to be a function of both the sediment and contaminant characteristics. Storage should be limited to a two-week period at 4°C unless previous data exist that indicate that the study site sediments can be stored without affecting toxicity.

11. Collection of Interstitial Water

11.1 Interstitial water (pore water), defined as the water occupying the space between sediment or soil particles, is often isolated to provide either a matrix for toxicity testing or an indication of the concentration and partitioning of contaminants within the sediment matrix. There is some indication that the interstitial water may be as useful as whole sediment for evaluating the toxicity of some sediment-associated compounds, for example, those that are not sorbed strongly to particles and where the ingestion of contaminated particles is not a major route of accumulation. The isolation of sediment interstitial water can be accomplished by several methods: centrifugation, squeezing, gas pressurization, suction, and equilibrium dialysis. These techniques have been reviewed recently by Adams (86) and Burton (87). In general, where relatively large volumes of water are required only centrifugation (for example, 44, 82, 88–96) and sediment squeezing (70, 97) can provide large quantities. Other methods, such as suction (98–101) and in situ samplers (86), do not produce sufficient volumes from most sediments easily.

11.2 Most collection methods have been shown to alter interstitial water chemistry and therefore may alter toxicity. There are a number of precautions that one should take to reduce the likelihood of causing significant sample change from in situ conditions. Some interstitial water constituents, for example, dissolved organic carbon, dimethylsulfide, ammonia, and major cations, can be altered significantly by the collection method (102–105). Increased sample handling by means of methods such as centrifugation or squeezing, compared to in situ "peepers" or core-port suction, may cause increased ammonia and decreased sulfide concentrations (103). Other constituents, such as salinity, dissolved inorganic carbon, ammonia, sulfide, and sulfate, might not be affected by collection, providing that oxidation is prevented (103). If the sediments are anoxic, all of the steps involved in sample processing should be conducted in inert atmospheres or by limited contact with the atmosphere in order to prevent oxidation (and the subsequent sorption and precipitation) of reduced species (102, 103, 106). Immediate collection of the interstitial water is recommended since chemical changes might occur even when the sediments are stored for short periods of time (for example, 24 h) at in situ temperatures (80, 81). Toxicity changes have been observed in interstitial water stored for less than 24 h (107). The coagulation and precipitation of humic material was noted when interstitial water was stored at 4°C for more than one week (82). The oxidation of reduced arsenic species in the interstitial water of stored sediments was unaffected for up to six weeks when the samples were acidified and kept near 0°C, without deoxygenation. Deoxygenation was necessary when the samples were not acidified, (83). Others have recommended that interstitial waters be frozen after extraction, prior to toxicity testing, to prevent changes (70). The optimal collection method will depend on the intended use of the sample (for example, acidification for metal analysis and not toxicity testing), characteristics of the sediment, and contaminants of concern.

11.3 The conditions for isolation of interstitial waters by centrifugation have varied considerably. Interstitial waters have been isolated for toxicity testing over a range of centrifugal forces and temperatures (44, 82, 88–93) with centrifuge bottles of various compositions. When centrifugation followed by filtration has been compared with in situ dialysis, higher speed centrifugation followed by filtration with 0.2 membrane filters has produced results that were more comparable for metals and organic carbon (73, 107, 108). Centrifuging at low speeds or the use of a 0.45-µm pore size membrane will result in the collection of both dissolved contaminants, colloidal materials, and aquatic bacteria (73). High-speed centrifugation (for example, 10 000 × g) is necessary to remove colloids and dispersible clays (86, 109, 110). The duration of the centrifugation has varied in the literature, but 30 min is relatively common and is the recommended time. The temperature for the centrifugation should reflect the ambient temperature of collection to ensure that the equilibrium between the particles and interstitial water is not shifted. Since trace metals and organics concentrate on solids, their removal is important in sorption and partitioning studies (73, 109). However, filtration through a wide range of filter types such as glass fiber or polycarbonate membranes may be inappropriate since they sorb some dissolved metals and organics (111). If filtration is used, a nonfiltered sample should also be tested for toxicity and contaminant concentrations. The effects of centrifugation speed, filtration, and oxic conditions on some chemical concentrations in interstitial waters have been well documented (for example, 86, 112, 113). It is recommended that sediments should be centrifuged at 10 000 × g for a 30-min period for routine toxicity testing of interstitial waters.

11.4 It is difficult to collect interstitial water from sediments that are predominately coarse sand. A modified centrifuge bottle has been developed, with an internal filter that can recover 75 % of the interstitial water, compared to 25 to 30 % from squeezing (114).

11.5 If sorptive organic compounds or mixtures of inorganic and organic compounds are to be isolated, PTFE centrifuge bottles should be used. Polytetrafluorethylene bottles will collapse at 3000 × g but have been used successfully in the range of 2500 g when filled to 80 % of capacity (88). So, in this case, the isolation of interstitial water should be at the temperature of collection, at a slower speed of 2500 g for 30-min duration. This material will contain colloidal material as well as dissolved compounds. Removal of the colloids may not be possible at low centrifugation speeds, without filtration. The influence of dissolved and colloidal organic carbon may be estimated by measuring the organic carbon content. Centrifugation can be performed with glass tubes (up to 10 000 × g) (111) if small volumes of water are required, for example, 50 mL, for testing higher speed. High-speed centrifugation in stainless steel centrifuge tubes can be performed if metals are not an issue.

11.6 The isolation of interstitial water by squeezing has been performed by means of a variety of practices (70, 86, 94–97, 115). In all cases, the interstitial water is passed through a filter that is a part of the apparatus. Filters have different sorptive capacities for different compounds. The characteristics of filters and the filtering apparatus should be considered carefully based on the types of contaminants expected. Squeezing has been demonstrated to yield results equivalent to those for other methods for silica (116) but not for sulfide (117). However, squeezing has been shown to produce a number of artifacts due to shifts in equilibrium from pressure, temperature, and gradient changes (for example, 95, 104, 105, 118–120). Squeezing can affect the electrolyte concentration in the interstitial water with a drop near the end of the squeezing process. It is therefore recommended that moderate pressures be used with electrolyte (conductivity) monitoring during extraction (119). Several studies revealed significant alterations to the interstitial water composition when squeezing was at temperatures that differed from ambient temperatures (for example, 104, 105). The major sources of alteration of the interstitial water, when using the squeezing method, are as follows: contamination from overlying water, internal mixing of the interstitial water during extrusion, and solid-solution reactions as the interstitial water is expressed through the overlying sediment. As interstitial waters are displaced into upper sediment zones during squeezing, they come into contact with solids that they are not in equilibrium with. This inter-mixing causes solid-solution reactions to occur. These reactions will generally reflect an approach to saturation, adsorption or desorption, and ion exchange. The chemistry of the sample may be altered due to the fast kinetics (minutes to hours) of these reactions. Most interstitial water species are out of metastable equilibrium with overlying sediments and are transformed rapidly, such as the case observed with ammonia and trace metals (121, 122). Bollinger, et al (118) found elevated levels of several ions and dissolved organic carbon in squeezed samples compared to samples collected by peepers. The degree of artifact will depend on the element, sediment characteristics, and redox potential. It is unlikely that reactive species gradients can be established by means of the squeezing of sediment cores (116).

11.7 Small-volume isolation of interstitial water, generally for chemical analysis, can also be performed by vacuum filtration (73, 101, 123), gas pressurization (35, 96), or displacement after removing the sediment from the aquatic environment (86). When preparing the sediments for interstitial water isolation of metals, care must be taken to maintain the anoxic conditions of deeper sediments by performing the procedures under an inert atmosphere (86). When core suction was compared to centrifugation and squeezing, it was found that the recovery of spiked tritium was similar; however, chlorobenzene differed significantly among methods, with suction exhibiting the highest recovery, followed by squeezing and centrifugation (101). Suction using an aquarium air stone recovered up to 1500 mL from sediment (4 L) and suctioned in an anoxic environment (122). Problems common to suction methods are a loss of equilibration between the interstitial water and the solids, filter clogging, and oxidation (124). However, in situ suction or suction by means of core ports has been shown to define small gradients of some sediment-associate compounds accurately, including ammonia, which can change an order of magnitude over a 1-cm depth (113). However, these definitive suction methods do not provide an adequate volume for conducting most toxicity test procedures.

11.8 Perhaps the optimal method of pore water collection is by the use of equilibrium dialysis (86, 125–128) or in situ suction techniques (98–101, 103). These methods have the greatest likelihood of maintaining in situ conditions and have been used to sample dissolved gases (129) and volatile organic compounds (101). However, these techniques isolate only relatively small volumes of interstitial water and must be placed by divers in deeper waters which limits the depth and conditions at which the devices can be deployed. Suction of undisturbed sediments is also possible from intact box core-collected sediments. The duration of equilibration for dialysis has ranged from hours to a month, but one to two weeks is most often used (86). The optimal equilibration time is a function of the sediment type, contaminants of concern, and temperature (for example, 103, 113, 127, 130–132). Many of the artifact problems associated with dialysis samplers have been discussed (108). The total organic carbon may be elevated in peepers (4 to 8-μm pore size) due to biogenic production; however, colloidal concentrations are lower than in centrifuged samples (109). When ionizable compounds, for example, metals, are to be collected, it is important to pre-equilibrate the samplers with an inert atmosphere in order to avoid introducing oxygen into the sediments and thereby changing the equilibrium. Plastic samplers can contaminate anoxic sediments with diffusable oxygen and should be stored before testing in inert atmospheres (130). In addition, samples should also be kept under an inert atmosphere and processed quickly when they are collected and processed. Cellulose membranes are unsuitable because they decompose too quickly. A variety of polymer materials have been used, some of which may be inappropriate for studies of certain nonpolar compounds. However, efforts to use semipermeable membrane devices filled with a nonpolar sorbant show some promise for use in dialysis systems for organic compounds (133). Test organisms have recently been exposed within peeper chambers in which larger mesh sizes of 149 μm were used successfully in oxic sediments (132, 134, 135). Equilibration of conductivity was observed within hours of peeper insertion into the sediment.

Replicate peepers revealed extreme heterogeneity in sediment interstitial water concentrations of ammonia and dissolved oxygen. Sediments that were high in clay and silt fractions were usually anoxic and did not allow for organism exposure in situ.

11.9 Based on the literature previously discussed, no clear superior method exists for isolating interstitial water for toxicity testing purposes. Each approach has unique strengths and weaknesses that vary with the sediment's characteristics, contaminants of concern, toxicity test methods to be used, and resolution necessary (that is, the data quality objectives). For most toxicity test procedures, relatively large volumes of interstitial water (for example, litres) are frequently needed for static or static renewal exposures with the associated water chemistry analyses. The use of in situ methods are preferred if smaller volumes are adequate and logistics allow because they are less likely to produce sample artifacts. The collection of core samples that are then subjected to immediate side port suctioning or centrifugation at ambient bottom water temperatures is recommended if logistics do not permit the placement of in situ samplers. However, it will be necessary for most studies to collect larger quantities of samples, preferably multiple cores, that are processed in an inert environment and centrifuged at ambient temperatures as rapidly as possible. If other methods and procedures are used for interstitial water collection (such as grab samplers, exposure to oxygen, extraction at room temperature, delayed extraction, squeezing, and filtration), the investigator should realize that the interstitial water sample has been altered from in situ conditions.

12. Characterization

12.1 Sediments that are to be analyzed for toxicity should be characterized physically and chemically. At a minimum, this characterization should include moisture content (total solids and specific gravity), organic carbon or volatile matter content, and particle size. More extensive characterization may be necessary to meet the study objectives. The degree of precision and accuracy necessary for these analyses will depend on the study's data quality objectives. By their nature, sediments are very heterogenous; they exhibit significant temporal and spatial heterogeneity in the laboratory and in situ. Lappalainen (136) demonstrated seasonal effects on interstitial water chemistry due to differences between sediment and overlying water temperature. Convectional heat transfer, interstitial water currents, and the transfer of soluble and gaseous materials was observed in the spring and autumn to sediment depths of tens of centimetres. Replicate samples should be analyzed to determine the variance in sediment characteristics and analytical methods. Sediment characterization will depend on the study objectives and contaminants of concern. Several additional characteristics that may assist in data interpretation and the quality assurance (QA)/quality control (QC) process (that is, assessing sediment integrity, artifact production, optimal extraction, and test procedures) include the following: in situ temperature, ash-free weight (total volatile solids), total and dissolved organic carbon (determined by titration or combustion), pore water salinity (for estuarine and marine sediments), pH, Eh, ammonia, and cation-exchange capacity. Many of the characterization methods have been based on analytical techniques for soils and waters, and the literature should be consulted for further information (**11, 137–139**).

12.2 The moisture content of sediments is measured by drying the sediments at 50 to 105°C to a consistent weight (**11**).

12.3 Volatile matter content is often measured instead of, and in some cases in addition to, organic carbon content as a measure of the total amount of organic matter in a sample. This measurement is made by ashing the sediments at high temperature and reporting the percent ash-free dry weight (**140–142**). Although the exact method for ashing the sample is often not specified, the normally accepted temperature is 550 ± 50°C (**11, 37**) for 2 to 24 h.

12.4 Carbon fractions that may be of importance in determining toxicant fate and bioavailability include the following: total organic carbon (**37, 143–145**), dissolved organic carbon (**82**), dissolved inorganic carbon, sediment carbonates, and reactive particulate carbon (**146, 147**). Reactive particulate carbon is that portion which equilibrates with the aqueous phase. Sediment organic carbon content has been measured by wet oxidation, which is also useful for determination of the organic carbon content of water (**148**). Organic carbon analyses have also been conducted by titration (**149**), modification of the titration method (**150**), or combustion after the removal of carbonate by the addition of HCl and subsequent drying (**65**).

12.5 Sediment particle size can be measured by numerous methods (**137, 151,** and see Guide D 4822), depending on the particle properties of the sample (**152**). Greater agreement exists between sizing and settling methods when the clay fractions are greater than 15 % (**153**). Particle size distribution is often determined by wet sieving (**2, 11, 37, 137, 153**). Particle size classes might also be determined by the hydrometer method (**154, 155**), pipet method (**137, 156**), settling techniques (**157**), X-ray absorption (**153, 156**), and laser light scattering (**158**). The pipet method may be superior to the hydrometer method (**159**). A method using a Coulter (particle size) counter might be used (**160, 161**) to obtain definite particle sizes for the fine material. This device gives the fraction of particles with an apparent spherical diameter. The Coulter was found to be the most versatile method overall in a review by Swift, et al (**162**); however, this method does not provide settling information. Another method for determining the particle size distribution of a very fine fraction is through the use of electron microscopy (**163**). The collection technique for the very fine materials can result in aggregation to larger colloidal structures (**163–166**). Comparisons of particle sizing methods have shown that some produce similar results and others do not. These differences might be attributed to differences in the particle property being measured. That is, the Malvern Laser Sizer and Electrozone Particle Counter are sizing techniques, and the hydrophotometer and SediGraph determine sedimentation diameter based on particle settling (**152, 167–169**). It is preferable to use a method that incorporates particle settling as a measure, as opposed to strictly sediment sizing.

12.6 Various methods have been recommended for determining the bioavailable fractions of metals in sediments (**69,** 170–172). One extraction procedure, cation-exchange capacity, provides information relevant to metal bioavailability studies (**138**). Amorphic oxides of iron and manganese, and reactive particulate carbon, have been implicated as the primary influences on the metal sorption potential in sediments (**73, 171, 173–175**). The measurement of acid volatile sulfide (AVS) and divalent metal concentrations associated with AVS extraction provides insight into metals availability in anaerobic sediments (**69**). Easily extractable fractions are usually removed with cation displacing solutions, for example, neutral ammonium acetate, chloride, sodium acetate, or nitrate salts (**176**). However, the extraction of saltwater or calcareous sediments, is often complicated by complexation effects or the dissolution of other sediment components (**172, 177**). Other extractants and associated advantages and disadvantages have been discussed (**172, 175, 178, 179**). Some extractants that have been used successfully in evaluations of trace metals in nondetrital fractions of sediments are EDTA or HCl (**172, 180, 181**). Metal partitioning in sediments might be determined by using sequential extraction procedures that fractionate the sediments into several components such as interstitial water, ion exchangeable, easily reducible organic, and residual sediment components (**93, 178, 182, 183**). Unfortunately, no one method is clearly superior to the others at this time (**177**). This might partly be due to site-specific characteristics that influence bioavailability, for example, desorption and equilibration processes.

12.7 pH is important for many chemicals and can be measured directly (**11**) or in a 1 to 1 mixture of sediment and soil to water (**184**).

12.8 Eh measures are particularly important for metal speciation and for determining the extent of sediment oxidation. Redox gradients in sediments often change rapidly over a small depth and are disturbed easily. Care must be taken in probe insertion to allow equilibration to occur when measuring Eh. These measurements are potentiometric and measured with a platinum electrode relative to a standard hydrogen electrode (**11**).

12.9 Biochemical oxygen demand and chemical oxygen demand might provide useful information in some cases (**11**). Sediment oxygen demand might also be a useful descriptor; however, a wide variety of methods exists (**84, 135–188**).

12.10 The analysis of toxicants in sediments is generally performed by standard methods such as those of the Environmental Protection Agency (EPA) (**2, 11, 189**). Acid digests are necessary for bound metal extraction. Soxhlet extraction is generally best for organics but depends on the extraction parameters (**190, 191**). Concentrations are generally reported on a dry weight or organic carbon basis. The sample size requirements for chemical and physical analysis are generally as follows: organics, 250 g (wet); metals, 100 g; ammonia, 100 g; grain size, 500 g; total organic carbon, 50 g; and total solids, 50 g.

13. Manipulations

13.1 Manipulation of sediments is often required to yield consistent material for toxicity testing and laboratory experiments. The manipulations reviewed in this section are as follows: (*1*) mixing, (*2*) spiking, (*3*) sieving, (*4*) dilutions for concentration-effect determinations, (*5*) elutriates, and (*6*) capping. See 9.7 for discussions of subsampling, compositing, or homogenization.

13.2 Mixing of sediments is conducted to produce an homogeneous sample that is uniform in color, texture, and moisture and that yields precise results in replicate determination of toxicity. For field-collected sediments, the sediment quality will be influenced by the depth of sampling, depth of biological activity, contaminant solubility and partitioning characteristics, and depth of the contaminant concentration peak, which is dependent on the historical contamination and sedimentation rates for the study site. As a result, mixing of various layers of sediments might result in either the dilution or enhancement of concentrations (see Section 10 for additional relevant discussions). Hand mixing can be accomplished by blending with a spatula (43, 59, 192–196), rolling the sediment out flat on a sheet of plastic or pre-combusted foil and tumbling by raising each corner of the sheet in succession, or by coning (mounding the sediment) followed by quartering and remixing (197, 198). A variety of mechanical mixers, such as a hand-held drill equipped with a polypropylene stirrer (for example, 60, 199), a rolling mill (199–201), or gyro-rotary and Eberbach shakers (60), have also been used. The mixing time for sediments that differ in color, texture, moisture, volume, and layering will vary but will generally be in the range from one to several minutes (199, 202). Mechanical mixing may alter the particle size distribution. It is therefore recommended that the particle size be determined prior to and following the mixing process in order to monitor potential changes in grain size due to the mixing process. Regardless of the mixing method, the efficiency of mixing must also be demonstrated by determining the coefficients of variation (203) for chemical or physical analyses from replicated samples (see 13.5 for further discussion).

13.3 Spiking—Whole sediments may be spiked with specific chemicals in order to determine the effects of single toxicants or mixtures of toxicants on biota (43, 72, 74, 140, 198, 204-208). The primary methods used to spike sediments with contaminants involve dry- and wet-spiking techniques. Air-dried sediments have been spiked successfully with organic compounds in dose-response toxicity tests (43, 141, 206, 208, 209). However, air drying may result in losses of volatile compounds as well as changes in sediment characteristics, especially particle size. The presence of air and air drying have also been shown to change metal availability and complexation, and dry-spiking is therefore not recommended. Wet-spiking techniques are currently the most acceptable for the preparation of a spiked sediment, and several techniques have been used, depending on the chemical used in spiking (72, 198, 199, 204, 205, 208, 210, 211).

13.4 Wet-spiking methodologies differ mainly in the amount of water present in the mixture during spiking, solvent used to apply the toxicant, and method of mixing. In many cases, the compound is either coated on the walls of the flask, and an aqueous slurry (sediment and water in various proportions) added, or the carrier-containing mixture is added directly to the slurry. When the sediment-to-water ratio is adjusted for optimal mixing, sediments that are too dense to mix by slurrying in water have been mixed successfully using the rolling mill (199–201). In addition to the rolling mill technique, thorough mixing of spiked sedi-

ments has been accomplished using Eberback and gyro-rotary shakers (60). A chemical can also be added to the water overlying the sediment and allowed to sorb with no mixing (63, 212–218). A carrier has occasionally been added directly to sediment (39, 74–76, 167, 209, 211, 219–222) and the carrier evaporated before the addition of water, leaving the chemical in a crystalline form. This approach does not seem to result in compounds being sorbed to sediment at the same sites as dosing under aqueous conditions (203). Care should be taken to ensure complete and homogenous mixing (see 13.2) no matter what technique is used for spiking. In addition, chemical analyses should be conducted to ensure that spiking is uniform in the mixed material (see 13.5). The mixing time following spiking should be limited to a few minutes or hours (1–24), and temperatures should be kept to a minimum (for example, 4°C) due to the rapid alterations that may occur in the sediment's physicochemical and microbiological characteristics that could alter bioavailability and toxicity. The mixing time might be extended for recalcitrant organics and some metals (for example, cadmium and copper) without adverse effects (see Sections 9 through 12 for additional discussion).

13.5 One of the most important criteria for the choice of both the mixing methodology and chemical used in the preparation of a spiked sediment is that homogeneous mixing occurs within the substrate. Ditsworth, et al (199) found that coefficients of variation (CVs) ranged from 2.2 to 10.9 % (mean of 4.8 %) for cadmium levels in cadmium-spiked sediment samples collected along a longitudinal axis of an horizontally lying mixing jar (199). The CVs did not increase with nominal cadmium levels (as $CdCl_2$, range from 3.5 to 14 mg/kg) added to the sediment. Significant differences in cadmium concentration existed among sampling locations within jars in some cases. Regarding organics, Ditsworth, et al (199) reported that mixing fluoranthene into one jar of sediment using the rolling-jar technique provided a CV of 11.5 % between sample locations within the jar, and no significant effect of the sample location was found. Good mixing efficiency for fluoranthene was also shown by Suedel, et al (211) with a CV of 10.3 % when the chemical was added directly to sediment, the carrier evaporated, and the sediment mixed by hand for 60 s before the addition of test water. Landrum and coworkers have found the following CVs for sediments at various concentrations in different experiments and using the slurry technique: pyrene, 4.8 to 6.9 %; phenanthrene, 4.7 to 9.3 %; BaP, 5.8 ± 3.2 %; hexachlorobiphenyl, 7.8 ± 4.5 %; and tetrachlorobiphenyl, 9.1 ± 5.0 % (198, 208, unpublished data). CVs should be ≤20 % for the homogeneity of mixing to be considered valid (223). However, it should be noted that the concentrations of total chemical determined in the sediment matrix do not reflect the bioavailable fraction of the chemical.

13.6 The spiking method to be used is contingent on the study objectives, sediment type, and compound(s) of interest. For example, when attempting to mimic in situ conditions, sediment cores should be spiked by adding an aqueous or suspended sediment solution of toxicants to the overlying water column, as would occur in the natural environment; or, when investigating the dredging effects or conditions of sediment perturbation where toxicant sorption processes are accelerated, mixing toxicants into sediment slurries may be advantageous. When investigating the source of sediment toxicity or interactive effects of sediment toxicants, it is useful to spike both the reference and control sediments with the toxicant of concern present in the test sediment.

13.7 Organic compounds are generally added by means of a carrier solvent, such as acetone or methanol, to ensure that they are soluble and remain in solution during mixing. Word, et al (112) compared several sediment-labeling techniques using methylene chloride, ethanol, and glycine as carriers. They found that glycine was superior when mixed with sediment for seven days. The use of a polar water-soluble carrier such as methanol has little effect on the partitioning of nonpolar compounds to dissolved organic matter at concentrations up to 15 % carrier by volume (224). However, another study shows that changes in partitioning by a factor of approximately two might well occur with 10 % methanol as a cosolvent for anthracene sorption (225). Caution should thus be taken to minimize the amount of carrier used. Metals are added in aqueous solutions while organic compounds are generally added in an organic carrier.

13.8 A variety of methods have been used to spike sediments with metals, but the two principal categories of methods are as follows: metal addition directly to the sediment, which is mixed and then water added (60, 85, 205, 210, 226); and addition of the metal to the overlying waters (72, 227–229).

13.9 Highly volatile compounds have been spiked into sediments using cosolvents followed by shaking in an aqueous slurry. Immediate testing in covered flow-through systems is recommended (230) when highly volatile compounds are used.

13.10 If a solvent other than water is used, both a sediment solvent control and sediment negative control or reference sediment, or both, must be included in the test. The solvent control must contain the highest concentration of solvent present and must use solvent from the same batch used to make the stock solution.

13.11 Once a sediment has been spiked with the toxicant of choice, it is necessary to allow the mixture to reach equilibrium before commencing a whole-sediment toxicity test. Equilibrium is defined as in equilibrium partitioning and refers to the assumption that an equilibrium exists between the chemical sorbed to the particulate sediment components and the pore water (231). The equilibration times and storage procedures for spiked sediments vary widely among studies (232), and there has been no attempt to standardize them. This is partly because accurate methods for measuring true equilibrium scientifically (that is, accurately isolating interstitial water and measuring the freely dissolved fraction of the compound of interest) are currently lacking, and little information exists on how long it will take for equilibrium to be established for any compound. In addition, the time to reach equilibrium will differ for compounds and sediments of differing characteristics. For metals, the time could be as short as 24 h (228, 233) or as long as 120 days. Similarly, for organics, the time allowed for the sediment and water to equilibrate has been as short as 24 h (234) or as long as 5 weeks (201). The duration of contact between the toxicant and sediment particles can affect both the partitioning and bioavailability of the toxicant. For

example, Landrum, et al (198) found that the partitioning of pyrene and phenanthrene between sediment particles and interstitial water increased significantly, whereas the uptake rate coefficients for the amphipod, *Diporeia sp.*, decreased significantly for both chemicals as the contact time increased. This effect occurs apparently because of an initial rapid labile sorption followed by movement of the toxicant into resistant sorption sites or in the particle (235–237). The contact time can be important when spiking sediments because of the kinetically controlled changes in the partitioning that results in changes in bioavailability (198, 208, 225, 238). Bounds on the sorption time can be estimated from the partition coefficient for the sediment following the calculations in Karickhoff and Morris (236). In addition, it is important to recognize that the quantity of toxicant spiked might exceed the complexation capacity of the test sediment system and not allow reactions to attain equilibrium. These phenomena will complicate the interpretation of test results (60, 178). Until more definitive information is generated, it is recommended that a standard equilibration time (for example, 2 weeks at 4°C) be established between the initial contact of the contaminant with the sediment and the initiation of toxicity tests.

13.12 The organic carbon content of sediments may be one of the most important characteristics affecting the biological availability of contaminants. Modifications of the carbon content have therefore been made in many studies. Methods for modification include dilution with clean sand (42, 43, 49, 230) or humics (222), and other organics such as sheep manure (39, 62), or the addition of organic detritus such as feces of *Crassostrea gigas* or *Callianassa californiensis* (201). Such dilutions also change the particle composition and size distribution of the particles; results from such experiments should thus be interpreted with care. The organic carbon content has also been altered by the use of combustion (39, 239). Combustion may alter the type of carbon as well as oxidize some of the inorganic components, thus altering the characteristics of the sediment greatly.

13.13 Although the sieving of field-collected sediments is known to disrupt chemical equilibrium, such manipulations may be necessary before toxicity tests are performed (16, 39, 45–47, 59, 62, 141, 142, 204, 206, 207, 209, 215, 221, 240). Justifications for sieving include the removal of large stones and other debris; removal of endemic species; improved sample homogeneity and replication; improved counting efficiency of organisms; increased ease of sediment handling and subsampling; and ability to study the influence of particle size on toxicity, bioavailability, or contaminant partitioning. Sediments can be either wet sieved (60, 192, 196, 197, 241) or pressure-sieved (242). Wet sieving involves agitating or swirling the sieve containing sediment in water so that particles smaller than the selected mesh size are washed through the sieve into a container. The sieve may be placed on a mechanical shaker, or the sediments on the screen can be stirred with a nylon brush (197), to facilitate the process. Alternatively, the particles may be washed through the sieve with a small volume of running water (243). Particles retained in the sieve (the coarse fraction) are examined and retained if they are of interest to the study. Pressure sieving involves the pressing of sediment particles through a sieve having an appropriate mesh size with a mechanical, piston-type arrangement, or with a flat-surfaced, hand-held tool. This technique works well with sediments containing few stones or other large objects and with a low to moderate clay content. Also, the method is applied best using sieves with mesh sizes >0.50 mm. Sieves used in toxicity tests can be constructed of stainless steel or plastic (for example, polyethylene, polypropylene, nylon, and PTFE), with mesh sizes varying from 0.25 to 2.0 mm (60, 112, 193, 196, 206, 207, 242, 244). The mesh size used most frequently for sieving is 1.0 mm, but the choice of mesh size is dependent on the objectives of the study and whether indigenous organisms must be removed from the test sediment (see 13.14 and 13.15).

13.14 The sieving of sediments may also (increase or decrease) the concentrations of contaminants contained in test sediments. Particles and their attendant contaminant loads may be either concentrated or removed. Also, sieving may disrupt chemical equilibria through the volatilization or modification of sorption and desorption characteristics. For example, Day, et al (245) found that sieving contaminated sediment through 250-μm mesh decreased concentrations of PCBs and PAHs as much as four-fold. Surface areas (in relation to the weight of the sample) and sorptive capacities are higher in fine-grained sediments (that is, clay and silt), and organic carbon concentrations as well as toxic chemicals thus tend to be higher in these sediments. Measuring size fractions of less than 63 μm has been recommended in contaminant studies with sediments, particularly for metals (223, 246). In studies of metal concentrations in sediments, normalizing to the <63-μm size fractions was superior for describing metal binding in sediments, compared to sediment concentrations normalized to dry weight, by organic carbon content, or corrected by a centrifugation procedure (223). Small-size fractions are characteristic of depositional areas in aquatic systems; however, the sieving of sediments from non-depositional sites to obtain the fine fraction might alter the sediment characteristics significantly. It is recommended that sediments not be sieved, unless the sediments contain excessive quantities of large organic debris or if the contamination is being normalized to a specific grain size.

13.15 The suspected presence of endemic organisms that will interfere with the results of chronic toxicity tests (for example, oligochaete worms, leeches, chironomids, etc.) will also necessitate the sieving of some field-collected sediments.

13.16 The presence of endemic species in sediments used in toxicity tests has been shown to complete the interpretation of acute and chronic endpoints (194, 245, 247, 248). Swartz, et al (200) demonstrated that the optimum mesh size for the removal of endemics was 0.50 mm for marine sediments. In freshwater sediments, the removal of large predators such as leeches can be accomplished by hand-picking with tweezers, but species of invertebrates that are morphologically similar to or in competition for space and food with species used in toxicity tests, or both, can be eliminated only by sieving with a mesh size of ≤0.25 mm (244, 246). In order to eliminate potential interferences from endemic species in freshwater samples, but limit the unnecessary sieving of sediments, it is recommended that a subsample of field-collected sediment be examined under low magnification using a stereomicroscope and, if coccoons, juvenile instars, or adults of endemic species such as

oligochaete worms or chironomids are noted, that sieving of test sediments be conducted.

13.17 Methods other than sieving to inhibit endemic biological activity in field-collected sediments include autoclaving, freezing, and gamma irradiation of sediments (245, 249). Caution is required in the use of these techniques, depending on the objectives of the study and test species to be used in the subsequent toxicity test. For example, Day, et al (245) found that survival of the amphipod, *Hyalella azteca*, was reduced significantly in any sediment that was frozen, autoclaved, or gamma irradiated. The reasons for this response are unknown but may relate to changes in the physical structure of sediments during these manipulations, an increased bioavailability of toxic compounds within the sediment matrice due to changes in chemical equilibria, or a reductions in sources of food for *H. azteca* due to sterilization. Malueg, et al (59) found that freezing sediment attenuated the release of total and soluble copper from the sediment into the overlying water. In contrast to the studies with *H. azteca*, growth of the chironomid, *Chironomus riparius*, and reproduction of the tubificid worm, *Tubifex tubifex*, were enhanced in sediments that had been sterilized by autoclaving or gamma irradiation (245). Tubificid worms and chrionomids feed on organic material as well as particles of sediment within the benthos, and the sterilization of sediments may increase organic material (250), thereby providing more food for the test organisms and thus better growth. Other sterilization techniques have included the use of antibiotics such as streptomycin and ampicillin (Danso, et al, 1973; Burton, et al, 1987) or the addition of chemical inhibitors such as $HgCl_2$ or sodium azide. Information on the effects of sediments that have received these treatments to toxicity test responses is not available. Some antibiotics are labile and light sensitive or bind readily to organic matter, so their use in all situations may not be appropriate. Mercuric chloride appears to be superior to sodium azide as a bacteriocide. It is crucial that a sterility control be incorporated in studies requiring sterility.

13.18 Diluting a test sediment with a clean, non-contaminated sediment has been suggested as an approach in order to obtain concentration-effects information in solid phase toxicity testing (232, 251, 252). Such dilutions have been performed with reference sediments (234, 242, 253–256) or clean sand (256, 257). Dilutions with test sediment have generally led to reductions in the toxicity of the diluted material relative to the test sediment. However, the toxicity decreased and then increased subsequently for some sediments (256) when sand was used as a diluent, although the sand alone was not toxic compared to controls. The mechanism for this effect is not known. The dilutions were generally mixed to visual homogeneity where described, and the only report of a definitive storage time after mixing was for 10 days (242, 257) and the temperature for storing diluted sediments was 4°C (242). No definitive testing has been performed on the appropriate length of storing dilutions. The actual amount of dilution can be estimated by determining the fraction of fine material and organic carbon content in the reference sediment, test sediment, and diluted material (256). Little information remains on the most appropriate method for diluting test sediments to obtain graded contaminant concentrations. Little is known concerning the role of sediment composition, equilibrium time, and alteration of chemistry during mixing on the exposure to the test sediment contaminants in the diluted material. A clean, noncontaminated sediment should be used as the diluent. This sediment should optimally have characteristics similar to the test sediment, such as organic matter and carbon concentration and particle size distribution, and should not contain elevated levels of the toxicants. Pure sand does not appear to be an appropriate dilution material because of the changes in toxicity with differing dilutions.

13.19 Elutriate tests, or aqueous extractions of resuspended sediments, have been conducted routinely (90, 258, 259). The method of elutration was originally developed (260) to simulate processes that might disturb the sediment and thus bring contaminants into the water column, that is, dredging activities, but the method has been adapted further to evaluate the effects of other common events that disrupt sediments and affect water quality, such as bioturbation and storms (22, 88). Elutriates are generally prepared by combining various mixtures of water and sediment (usually 4:1 ratio, v/v) and shaking, bubbling, or stirring the mixture for 1 h (88, 90, 259, 261). The water phase is then separated from the sediment by centrifugation, and the supernatant is used in various toxicity tests (for example, fathead minnow, *Pimephales promelas*; bioluminescence assay, *Photobacterium phosphoreum*; and sea urchin (*Arbacia punctulata*), fertilization test). Filtration of the supernatant through filters (0.45 to 1.2 µm) may be necessary when the elutriate is used in some toxicity tests such as the algal growth assay with *Selenastrum capricornutum*. However, as discussed in previous sections, filtration can remove toxicity due to the sorption of dissolved chemicals to the filtration membrane and retention of colloids. Elutriates have generally been found to be less toxic than bulk sediments or interstitial water fractions (88, 90) to various biota, but there have been isolated cases in which resuspension increased the bioavailability of toxicants in the water column. Partitioning to organic colloids in the interstitial water has been suggested as a possible explanation for the discrepancies between suspended-phase and interstitial water exposures (88). Toxicity may be affected significantly by the method of elutriation; data comparisons should therefore be made only where standardized elutriate methods were used.

13.20 The remediation of sediment might include capping the contaminated sediments with clean sediments. The laboratory design of such experiments should vary the depth of both the contaminated sediments and the capping sediment layers to evaluate contaminant transport by means of physiochemical and biological (bioturbation) processes.

14. Quality Assurance

14.1 The QA guidelines (7, 8, 37, 55, 262) should be followed. The QA considerations for sediment modeling, QA-QC plans, statistical analyses (for example, sample number and location), and sample handling have been addressed in depth (55).

14.2 Sediment heterogeneity significantly influences studies of sediment quality, contaminant distribution, and both benthic invertebrate and microbial community effects. Spatial heterogeneity might result from numerous biological, chemical, and physical factors and should be considered both horizontally (such as on the sediment surface) and vertically (that is, depth). Accumulation areas with similar particle size distributions might yield significantly different toxicity patterns when subsampled (71, 263); an adequate number of replicates should therefore be processed to determine site variance. When determining site variance, one should consider within sample (that is, subsample) variance, analytical variance (for example, chemical or toxicological), and the sampling instruments' accuracy and precision. A sampling design can be constructed after these considerations that addresses the resource limitations and study objectives.

14.3 As stated in previous sections, the methodological approach used, such as number of samples, will depend on the study objectives and sample characteristics. There are a number of references available for information on sediment heterogeneity; splitting; compositing; controls; or determining sample numbers, sampler accuracy and precision, and resource requirements (8, 9, 55, 78, 223, 264, 265).

14.4 Quality assurance is an integrated system of management activities involving planning, implementation, assessment, reporting, and quality improvement to ensure that an environmental assessment is of the type and quality necessary. Quality control is the overall system of technical activities that measures and controls the quality of the assessment. The primary mechanism for ensuring that there is an adequate QA-QC program is through a Quality Assurance Project Plan (QAPP). This formal document describes, in detail, the necessary QA and QC procedures that are implemented to ensure that the results of the assessment will satisfy the stated performance criteria. This process is described in detail in USEPA (3). The QAPP describes the following: project description; project organization and responsibilities; QA objectives for the measurement data (including data quality objectives, precision, accuracy, test acceptability, representativeness, completeness, and comparability; sampling; analytical or test procedures (standard operating procedures); sample custody procedures; calibration procedures and frequency; internal QC checks and frequency; performance and system audits; analytical procedures; data reduction, validation, assessment, and reporting procedures; preventive maintenance procedures and schedules; corrective action; and QA reports to management. Refer to the appropriate standard test method guidance (for example, ASTM, USEPA, and APHA) for acceptable quality control limits for test measurements (for example, toxicity assay performance criteria, analytical precision, accuracy, completeness, and method detection limit).

15. Report

15.1 *Documentation*—Include the following information, either directly or by reference to existing documents, in the record of sediment collection, storage, handling, and manipulation. Published reports should contain enough information to identify the methodology used and quality of the results clearly. Specific information should include the following:

15.1.1 Name of the test and investigator(s); name and location of the sample station and test laboratory; field conditions (for example, water depth, sampler penetration depth in sediment, sediment characteristics, collection and storage methods, and dates of starting and ending of sampling and sediment manipulation;

15.1.2 Source of the control, reference, or test sediment; method for handling, storage, and disposal of the sediment;

15.1.3 Source of the water; its chemical characteristics; a description of any pretreatment;

15.1.4 Methods used for, and results (with confidence limits) of, physical and chemical analyses of the sediment; and

15.1.5 Anything unusual concerning the study, any deviation from these procedures, manipulations, and any other relevant information.

16. Keywords

16.1 characterization; collection; manipulation; sediment; storage; toxicity; transport

REFERENCES

(1) Dickson, K. L., Maki, A. W., and Brungs, W. A., eds., *Fate and Effects of Sediment-Bound Chemicals in Aquatic Systems*, Lewis Publishers, Boca Raton, FL, 1987.

(2) U.S. Environmental Protection Agency, "Chemistry Laboratory Manual for Bottom Sediments and Elutriate Testing," EPA-905-4-79-014 (NTIS PB 294596), Region V, Chicago, IL, 1979.

(3) U.S. Environmental Protection Agency, "Samplers and Sampling Procedures for Hazardous Waste Streams," EPA 600/2-80-018, 1980.

(4) International Technical Information Institute, *Toxic and Hazardous Industrial Chemicals Safety Manual*, Tokyo, Japan, 1977; for example, see: Sax, N. I., *Dangerous Properties of Industrial Materials*, 5th Ed., Van Nostrand Reinhold Co., New York, NY, 1979; Patty, F. A., ed., Industrial Hygiene and Toxicology, Vol. II, 2nd Ed., Interscience, New York, NY, 1963; Hamilton, A., and Hardy, H. L., *Industrial Toxicology*, 3rd Ed., Publishing Sciences Group, Inc., Acton, MA, 1974; and Goselin, R. E., Hodge, H. C., Smith, R. P., and Gleason, M. H., *Clinical Toxicology of Commercial Products*, 4th Ed., Williams and Wilkins Co., Baltimore, MD, 1976.

(5) Green, N. E., and Turk, A., *Safety in Working with Chemicals*, MacMillan, New York, NY, 1978; for example, see: National Research Council, *Prudent Practices for Handling Hazardous Chemicals in Laboratories*, National Academy Press, Washington, DC, 1981; Walters, D. B., ed., *Safety Handling of Chemical Carcinogens, Mutagens, Teratogens and Highly Toxic Substances*, Ann Arbor Science, Ann Arbor, MI, 1980; and Fawcett, H. H., and Wood, W. S., eds., *Safety and Accident Prevention in Chemical Operations*, 2nd Ed., Wiley-Interscience, New York, NY, 1982.

(6) National Council on Radiation Protection and Measurement, "Basic Radiation Protection Criteria," NCRP Report No. 39, Washington, DC, 1971; for example, see: Shapiro, J., *Radiation Protection*, 2nd Ed., Harvard University Press, Cambridge, MA, 1981.

(7) U.S. Environmental Protection Agency, "Sampling Protocols for Collecting Surface Water, Bed Sediment, Bivalves, and Fish for Priority Pollutant Analysis," Final Draft Report, Office of Water Regulations and Standards Monitoring and Data Support Division, Washington, DC, 1982a.

(8) U.S. Environmental Protection Agency, *Handbook for Sampling and Sample Preservation of Water and Wastewater*, EPA-600/4-82-029, Environment Monitoring and Support Laboratory, Cincinnati, OH, 1982b.

(9) Downing, J. A., *Sampling the Benthos of Standing Waters, A Manual on Methods for the Assessment of Secondary Productivity in Freshwaters*, 2nd Ed., IBP Handbook 12, Blackwell Scientific Publications, Boston, MA, 1984, pp. 87–130.

(10) Hopkins, T. L., "A Survey of Marine Bottom Samples," *Progress in Oceanography*, Vol 2, Pergamon Press, New York, NY, 1964, pp. 215–254.

(11) Plumb, R. H., "Procedures for Handling and Chemical Analysis of Sediment and Water Samples," Environmental Protection Agency/Corps of Engineers Technical Committee on Criteria for Dredged and Fill Material, Contract EPA-4805572010, EPA/CE-81-1, 1981, 478 pp.

(12) Word, J. Q., "An Evaluation of Benthic Invertebrate Sampling Devices for Investigating Habitats of Fish," *Proceedings of First Pacific Northwest Technical Workshop on Fish Food Habits Studies*, Washington University Sea Grant, Seattle, WA, June 1977, pp. 43–55.

(13) Carlton, R. G., and Wetzel, R., "A Box Corer for Studying Metabolism of Epipelic Microorganisms in Sediment Under In Situ Conditions," *Limnology and Oceanography*, Vol 30, 1985, pp. 422–426.

(14) Oliver, B. G., "Uptake of Chlorinated Organics from Anthropogenically Contaminated Sediments by Oligochaete Worms," *Canadian Journal of Fisheries and Aquatic Science*, Vol 41, 1984, pp. 878–883.

(15) Oliver, G. B., "Biouptake of Chlorinated Hydrocarbons from Laboratory-Spiked and Field Sediments by Oligochaets Worms," *Environmental Science and Technology*, Vol 21, 1987, pp. 785–790.

(16) Klump, J. V., Krezoski, J. R., Smith, M. E., and Kaster, J. L., "Dual Tracer Studies of the Assimilation of an Organic Contaminant from Sediments by Deposit Feeding Oligochaetes," *Canadian Journal of Fisheries and Aquatic Science*, Vol 44, 1987, pp. 1574–1583.

(17) Breteler, R. J., and Saksa, F. I., "The Role of Sediment Organic Matter on Sorption Desorption Reactions and Bioavailability of Mercury and Cadmium in an Intertidal Ecosystem," *Aquatic Toxicology and Hazard Assessment, Seventh Symposium, ASTM STP 854*, ASTM, Philadelphia, PA, 1985, pp. 454–468.

(18) Jafvert, C. T., and Wolfe, N. L., "Degradation of Selected Halogenated Ethanes in Anoxic Sediment-Water Systems," *Environmental Toxicology and Chemistry*, Vol 6, 1987, pp. 827–837.

(19) Wolfe, N. L., Kitchen, B. E., Maclady, D. L., and Grundl, T. J., "Physical and Chemical Factors that Influence the Anaerobic Degradation of Methylparathione in Sediment Systems," *Environmental Toxicology and Chemistry*, Vol 5, 1986, pp. 1019–1026.

(20) Murray, W. G., and Murray, J., "A Device for Obtaining Representative Samples from the Sediment-Water Interface," *Marine Geology*, Vol 76, 1987, pp. 313–317.

(21) Grizzle, R. E., and Stegner, W. E., "A New Quantitative Grab for Sampling Benthos," *Hydrobiologia*, Vol 126, 1985, pp. 91–95.

(22) U.S. Environmental Protection Agency/U.S. Army Corps of Engineers, *Evaluation of Dredged Material Proposed for Ocean Disposal (Testing Manual)*, EPA 503/8-91-001, Washington, DC, 1991.

(23) Blomqvist, S., "Sampling Performance of Ekman Grabs—*In Situ* Observations and Design Improvements," *Hydrobiologia*, Vol 206, 1990, pp. 245–254.

(24) Gilbert, R., and Glew, J., "A Portable Percussion Coring Device for Lacustrine and Marine Sediments," *Journal of Sediment Petrology*, Vol 55, 1985, pp. 607–608.

(25) Fuller, J. A., and Meisburger, E. P., "A Simple, Ship-Based Vibratory Corer," *Journal of Sediment Petrology*, Vol 52, 1982, pp. 642–644.

(26) Imperato, D., "A Modification of the Vibracoring Technique for Sandy Sediment," *Journal of Sediment Petrology*, Vol 57, 1987, pp. 788–789.

(27) Lanesky, D. E., Logan, B. W., Brown, R. G., and Hine, A. C., "A New Approach to Portable Vibracoring Underwater and on Land," *Journal of Sediment Petrology*, Vol 49, 1979, pp. 654–657.

(28) Elliott, J. M., "Some Methods for the Statistical Analysis of Samples of Benthic Invertebrates," *Freshwater Biological Association of Science Publications*, Vol 25, 1977, 156 pp.

(29) Blomqvist, S., "Reliability of Core Sampling of Soft Bottom Sediment—An *In Situ* Study," *Sedimentology*, Vol 32, 1985, pp. 605-612.

(30) Findlay, S. E. G., "Influence of Sampling Scale on Apparent Distribution of Meiofauna on a Sandflat," *Estuaries*, Vol 5, 1982, pp. 322–324.

(31) Fleeger, J. W., Sikora, W., and Sikora, J., "Spatial and Long-Term Variation of Meiobenthic-Hyperbentic Copepods in Lake Pontchartrain, Louisiana," *Estuarine Coastal Shelf Science*, Vol 16, 1983, pp. 441–453.

(32) McIntyre, A. D., and Warwick, R., "Meiofauna Techniques," *Methods for the Study of Marine Benthos*, 2nd ed., N. Holme and A. McIntyre, eds., 1984, pp. 217–244.

(33) Rutledge, P. A., and Fleeger, J., "Laboratory Studies on Core Sampling with Application to Subtidal Meiobenthos Collection," *Limnology and Oceanography*, Vol 33, 1988, pp. 274–280.

(34) Sly, P. G., "Bottom Sediment Sampling," *Proceedings of 12th Conference Great Lakes Research*, 1969, pp. 883–898.

(35) Long, E. R., and Buchman, M. F., "An Evaluation of Candidate Measures of Biological Effects for the National Status and Trends Program," NOAA Technical Memorandum NOS OMA 45, Seattle, Washington, 1989.

(36) Reichert, W. L., Le Eberhart, B., and Varanasi, U., "Exposure of Two Species of Deposit-Feeding Amphipods to Sediment-Associated [3H]benzoapyrene: Uptake, Metabolism and Covalent Binding to Tissue Macromolecules," *Aquatic Toxicology*, Vol 6, 1985, pp. 45–56.

(37) U.S. Environmental Protection Agency, Region 10, "Recommended Protocols for Conducting Laboratory Bioassays on Puget Sound Sediments," Final Report TC-3991-04, Puget Sound Estuary Program, Prepared by TetraTech, Inc., and E.V.S. Consultants, Inc., USEPA, Seattle, WA, 1986b.

(38) Garner, F. C., et. al., "Composite Sampling for Environmental Monitoring," *American Chemical Society*, 1988, pp. 363–374.

(39) Adams, W. U., Kimerle, R. A., and Mosher, R. G., "Aquatic Safety Assessment of Chemicals Sorbed to Sediments," *Aquatic Toxicology and Hazard Assessment, Seventh Symposium, ASTM STP 854*, ASTM, Philadelphia, PA, 1985, pp. 429–453.

(40) NOAA, "Collection of Bivalve Molluscs and Surficial Sediments, and Performance of Analysis for Organic Chemicals and Toxic Trace Elements," Phase 1 Final Report, National Statistics and Trends Mussel Watch Program, Prepared by Battelle Ocean Sciences for the National Oceanic and Atmospheric Administration, Rockville, MD, 1987.

(41) Alden, R. W., and Butt, A. J., "Statistical Classification of the Toxicity and Polynuclear Aromatic Hydrocarbon Contamination of Sediments from a Highly Industrialized Seaport," *Environmental Toxicology and Chemistry*, Vol 6, 1987, pp. 673–684.

(42) Clark, J. R., Patrick, J. M., Moore, J. C., and Forester, J., "Accumulation of Sediment-Bound PCBs by Fiddler Crabs." *Bulletin of Environmental Contamination and Toxicology*, Vol 36, 1986, pp. 571–578.

(43) Clark, J. R., Patrick, J. M., Moore, J. C., and Lores, E. M., "Waterborn and Sediment-Source Toxicities of Six Organic Chemicals to Grass Shrimp (*Palaemonetes pugio* and *Amphioxus* (*Branchiostoma caribaeum*)," *Archives of Environmental Contamination and Toxicology*, Vol 16, 1987, pp. 401–407.

(44) Giesy, J. P., Graney, R. L., Newsted, J. L., Rosiu, C. J., Benda, A., Kreis, R. G., and Horvath, F. J., "Comparison of Three Sediment Bioassay Methods Using Detroit River Sediments," *Environmental Toxicology and Chemistry*, Vol 7, 1988, pp. 483–498.

(45) Malueg, K. W., Schuytema, G. S., Gakstatter, J. H., and Krawczyk, D. F., "Effect of *Hexagenia* on *Daphnia* Response in Sediment Toxicity Tests," *Environmental Toxicology and Chemistry*, Vol 2, 1983, 73–82.

(46) Malueg, K. W., Schuytema, G. S., Gakstatter, J. H., and Krawczyk, D. F., "Toxicity of Sediments from Three Metal Contaminated Areas," *Environmental Toxicology and Chemistry*, Vol 3, 1984, pp. 279–291.

(47) Malueg, K. W., Schuytema, G. S., Krawczyk, D. F., and Gakstatter, J. H., "Laboratory Sediment Toxicity Tests, Sediment Chemistry and Distribution of Benthic Macroinvertebrates in Sediments from the Keweenaw Waterway, Michigan," *Environmental Toxicology and Chemistry*, Vol 3, 1984, pp. 233–242.

(48) Rubinstein, N. I., Lores, E., and Gregory, N. R., "Accumulation of PCBs, Mercury and Cadmium by *Nereis virens, Mercenaria mercenaria* and *Palaemonetes pugio* from Contaminated Harbor Sediments," *Aquatic Toxicology*, Vol 3, 1983, pp. 249–260.

(49) Tatem, H. E., "Bioaccumulation of Polychlorinated Biphenyls and Metals from Contaminated Sediment by Freshwater Prawns, *Macrobracium rosenbergii*, and Clams, *Corbicula fluminea*," *Archives of Environmental Contamination and Toxicology*, Vol 15, 1986, pp. 171–183.

(50) Fitzwater, S. E., Knauer, G. A., and Martin, J. H., "Metal Contamination and Its Effect on Primary Production Measurements," *Limnology and Oceanography*, Vol 27, 1982, 544–551.

(51) Lindstrom, R. M., and Moody, J. R., "Selection and Cleaning of Plastic Containers for Storage of Trace Element Samples," *Analytical Chemistry*, Vol 49, 1977, pp. 2264–2267.

(52) *The Hazardous Waste Consultant*, "The Impact of Transport and Storage Temperature on Sample Integrity," 1992, pp. 13–15.

(53) U.S. Environmental Protection Agency, "A Compendium of Superfund Field Operation Methods," EPA/540-P-87/001, Office of Emergency and Remedial Response, Washington, DC, 1987.

(54) U.S. Environmental Protection Agency, "Sampling Guidance Manual for the National Dioxin Manual," Draft, Office of Water Regulations and Standards, Monitoring and Data Support Division, Washington, DC, 1984.

(55) U.S. Environmental Protection Agency, "Sediment Sampling Quality Assurance User's Guide," EPA 600/4/85/048 (NTIS PB 85-233542), Environmental Monitoring and Support Laboratory, Las Vegas, NV, 1985b.

(56) Burton, G. A., and Stemmer, B. L., "Factors Affecting Effluent and Sediment Toxicity Using Cladoceran, Algae and Microbial Indicator Assays," *Abstract of the Annual Meeting of the Society of Environmental Toxicology and Chemistry*, Pensacola, FL, 1987.

(57) Lee, G. F., and Jones, R. A., "Discussion of: Dredged Material Evaluations: Correlation Between Chemical and Biological Evaluation Procedures," *Journal of Water Pollution Control Federation*, Vol 54, 1982, pp. 406–409.

(58) Lee, G. F., and Plumb, R., "Literature Review on Research Study for the Development of Dredged Material Disposal Criteria," Contract Report D-74-1, U.S. Army Engineer Waterways Experiment Station, Vicksburg, MS, 1974.

(59) Malueg, K. W., Schuytema, G. S., and Krawczyk, D. F. "Effects of Sample Storage on a Copper-Spiked Freshwater Sediment," *Environmental Toxicology and Chemistry*, Vol 5, 1986, pp. 245–253.

(60) Stemmer, B. L., Burton, G. A., Jr., and Leibfritz-Frederick, S., "Effect of Sediment Test Variables on Selenium Toxicity to *Daphnia magnia*," *Environmental Toxicology and Chemistry*, Vol 9, in press, 1990.

(61) Swartz, R. C., "Toxicological Methods for Determining the Effects of Contaminated Sediment on Marine Organisms," *Fate and Effects of Sediment-Bound Chemicals in Aquatic Systems*, Pergamon Press, New York, NY, 1987, pp. 183–198.

(62) Swartz, R. C., DeBen, W. A., Jones, J. K., Lamberson, J. O., and Cole, F. A., "Phoxocephalid Amphipod Bioassay for Marine Sediment Toxicity," *Aquatic Toxicology and Hazard Assessment, Seventh Symposium, ASTM STP 854*, ASTM, Philadelphia, PA, 1985, pp. 284–307.

(63) Silver, M. L., "Handling and Storage of Chemically and Biologically Active Ocean Sediments," *MTS Journal*, Vol 6, 1972, pp. 32–36.

(64) Swartz, R. C., Ditsworth, G. R., Schults, D. W., and Lamberson, J. O., "Sediment Toxicity to a Marine Infaunal Amphipod: Cadmium and Its Interaction with Sewage Sludge," *Marine and Environment Research*, Vol 18, 1985, pp. 133–153.

(65) Wood, L. W., Rhee, G. Y., Bush, B., and Barnard, E., "Sediment Desorption of PCB Congeners and Their Bio-Uptake by Dipteran Larvae," *Water Research*, Vol 21, 1987, pp. 873–884.

(66) Burton, G. A., Jr., and Stemmer, B. L., "Spiking Method, Sample Storage and Spatial Heterogeneity Effects on Sediment Toxicity to *Daphnia magna*," *Abstract of the Annual Meeting of the Society of Environmental Toxicology and Chemistry,*, Arlington, VA, 1988.

(67) Anderson, J., Birge, W., Gentile, J., Lake, J., Rodgers, J., Jr., and Swartz, R., "Biological Effects, Bioaccumulation, and Ecotoxicology of Sediment-Associated Chemicals," *Fate and Effects of Sediment-Bound Chemicals in Aquatic Systems*, Pergamon Press, New York, NY, 1987, pp. 267–296.

(68) Plumb, R. H., Jr., "Sampling, Preservation, and Analysis of Sediment Samples: State-of-the-Art Limitations," *Proceedings Fifth United States-Japan Experts Meeting on Management of Bottom Sediments Containing Toxic Substances*, EPA/60019-80-044, New Orleans, LA, 1980, pp. 259–272.

(69) Di Toro, D. M., Mahony, J. D., Hansen, D. J., Scott, K. J., Hicks, M. B., Mayr, S. M., and Redmond, M. S., "Toxicity of Cadmium in Sediments: The Role of Acid Volatile Sulfide," Draft Report to U.S. Environmental Protection Agency, Criteria and Standards Division, Washington, DC, 1989.

(70) Carr, R. S., Williams, J. W., and Fragata, C. T. B., "Development and Evaluation of a Novel Marine Sediment Pore Water Toxicity Test with the Polychaete *Dinophilus gyrociliatus*," *Environmental Toxicology and Chemistry*, Vol 8, 1989, pp. 533–543.

(71) Stemmer, B. L., "An Evaluation of Various Effluent and Sediment Toxicity Tests," MS Thesis, Wright State University, Dayton, OH, 1988.

(72) Cairns, M. A., Nebeker, V., Gakstatter, J. H., and Griffis, W. L., "Toxicity of Copper-Spiked Sediments to Freshwater Invertebrates," *Environmental Toxicology and Chemistry*, Vol 3, 1984, pp. 435–445.

(73) Jenne, E. A., and Zachara, J. M., "Factors Influencing the Sorption of Metals," *Fate and Effects of Sediment-Bound Chemicals in Aquatic Systems*, Pergamon Press, New York, NY, 1987, pp. 83–98.

(74) Muir, D. C. G., Griff, N. D., Townsend, B. E., Matner, D. A., and Lockhart, W. L., "Comparison of the Uptake and Bioconcentration of Fluridone and Terbutryn by Rainbow Trout and *Chironomus tentans* in Sediment and Water Systems," *Archives of Environmental Contamination and Toxicology*, Vol 11, 1982, pp. 595–602.

(75) Muir, D. C. G., Townsend, B. E., and Lockhart, W. L., "Bioavailability of Six Organic Chemicals to *Chironomus tentans* Larvae in Sediment and Water," *Environmental Toxicology and Chemistry*, Vol 2, 1983, pp. 269–281.

(76) Muir, D. C. G., Rawn, G. P., Townsend, B. E., Lockhart, W. L., and Greenhalgh, R., "Bioconcentration of Cypermethrin, Deltamethrin, Fenvalerate and Permethrin by *Chironomus tentans* Larvae in Sediment and Water," *Environmental Toxicology and Chemistry*, Vol 4, 1985, pp. 51–61.

(77) Thomson, E. A., et al., "The Effect of Sample Storage on the Extraction of Cu, Zn, Fe, Mn and Organic Material from Oxidized Estuarine Sediments," *Water, Air, and Soil Pollution*, Vol 14, 1980, pp. 215–233.

(78) Rochon, R., and Chevalier, M., "Sediment Sampling and Preservation Methods for Dredging Projects," Conservation and Protection-Environment Canada, Quebec Region, 1987.

(79) Wilkness, P. E. and Carr, R. A., "Mercury: A Short-Term Storage of Natural Waters," *Environmental Science and Technology*, Vol 7, No. 1, 1973, pp. 62–63.

(80) Watson, P. G., Frickers, P., and Goodchild, C., "Spatial and Seasonal Variations in the Chemistry of Sediment Interstitial Waters in the Tamar Estuary," *Estuaries and Coastal Shelf Science*, Vol 21, 1985, pp. 105–119.

(81) Hulbert, M. H. and Brindle, M. P., "Effects of Sample Handling on the Composition of Marine Sedimentary Pore Water," *Geological Society of America Bulletin*, Vol 86, 1975, pp. 109–110.

(82) Landrum, P. F., Nihart, S. R., Eadie, B. J., and Herche, L. R., "Reduction in Bioavailability of Organic Contaminants to the Amphipod *Pontoporeia hoyi* by Dissolved Organic Matter of Sediment Interstitial Waters," *Environmental Toxicology and Chemistry*, Vol 6, 1987, pp. 11–20.

(83) Aggett, J., and Kriegman, M. R., "Preservation of Arsenic (III) and Arsenic (V) in Samples of Sediment Interstitial Water," *Analyst*, Vol 112, 1987, pp. 153–157.

(84) Andersen, F. O., and Helder, W., "Comparison of Oxygen Microgradients, Oxygen Flux Rates and Electron Transport System Activity in Coastal Marine Sediments," *Marine Ecology Progress Series*, Vol 37, 1987, pp. 259–264.

(85) Tatem, H. E., "Use of *Daphnia magna* and *Mysidopsis almyra* to Assess Sediment Toxicity," *Water Quality '88,* Seminar Proceedings, Charleston, SC, U.S. Army Corps of Engineers Committee on Water Quality, Washington, DC, 1988.

(86) Adams, D. D., "Sediment Pore Water Sampling," *Handbook of Techniques for Aquatic Sediments Sampling*, A. Mudroch and S. D. MacKnight, eds., CRC Press, Boca Raton, FL, 1991, pp. 171–202.

(87) Burton, G. A., Jr., "Sediment Collection and Processing: Factors Affecting Realism, G. A. Burton, Jr., ed., *Sediment Toxicity Assessment, Lewis Publishers, Boca Raton, FL.*

(88) Burgess, R. M., Schweitzer, K. A., McKinney, R. A., and Phelps, D. K., *Environmental Toxicology and Chemistry*, Vol 12, 1993, pp. 127–138.

(89) Ankley, G. T., Katko, A., and Arthur, J. W., "Identification of Ammonia as an Important Sediment-Associated Toxicant in the Lower Fox River and Green Bay, Wisconsin," *Environmental Toxicology and Chemistry*, Vol 9, 1990, pp. 313–322.

(90) Ankley, G. T., Schubauer-Berigan, M. K., and Dierkes, J. R., "Predicting the Toxicity of Bulk Sediments to Aquatic Organisms with Aqueous Test Fractions: Pore Water vs. Elutriate," *Environmental Toxicology and Chemistry*, Vol 10, 1991, pp. 1359–1366.

(91) Schubauer-Berigan, M. K., and Ankley, G. T., "The Contribution of Ammonia, Metals and Nonpolar Organic Compounds to the Toxicity of Sediment Interstitial Water from an Illinois River Tributary," *Environmental Toxicology and Chemistry*, Vol 10, 1991, pp. 925–939.

(92) Edmunds, W. M., and Bath, A. H., "Centrifuge Extraction and Chemical Analysis of Interstitial Waters," *Environmental Science and Technology*, Vol 10, 1976, pp. 467–472.

(93) Engler, R. M., Brannon, J. M., Rose, J., and Bigham, Cr., "A Practical Selective Extraction Procedure for Sediment Characterization," *Chemistry of Marine Sediments*, Ann Arbor Science, Ann Arbor, MI, 1977, p. 163–171.

(94) Jahnke, R. A., "A Simple, Reliable, and Inexpensive Pore-Water Sampler," *Limnology and Oceanography*, Vol 33, 1988, pp. 483–487.

(95) Kalil, E. K., and Goldhaker, M., "A Sediment Squeezer for Removal of Pore Waters Without Air Contact," *Journal of Sediment Petrology*, Vol 43, 1973, pp. 554–557.

(96) Reeburgh, W. S., "An Improved Interstitial Water Sampler," *Limnology and Oceanography*, Vol 12, 1967, pp. 163–165.

(97) Long, E. R., Buchman, M. F., Bay, S. M., Breteler, R. J., Carr, R. S., Chapman, P. M., Hose, J. E., Lissner, A. L., Scott, J., and Wolfe, D. A., "Comparative Evaluation of Five Toxicity Tests with Sediments from San Francisco Bay and Tomales Bay, California," *Environmental Toxicology and Chemistry*, Vol 9, 1990, pp. 1193–1214.

(98) Watson, P. G., and Frickers, T. E., "A Multilevel, In Situ Pore-Water Sampler for Use in Intertidal Sediments and Laboratory Microcosms, *Limnology and Oceanography*, Vol 35, 1990, pp. 1381–1389.

(99) Whitman, R. L., "A New Sampler for Collection of Interstitial Water from Sandy Sediments," *Hydrobiologia*, Vol 176/177, 1989, pp. 531–533.

(100) Pittinger, C. A., Hand, V. C., Masters, J. A., and Davidson, L. F., "Interstitial Water Sampling in Ecotoxicological Testing: Partitioning of a Cationic Surfactant," *Toxicology and Hazard Assessment*, Vol 10, 1988, pp. 138–148.

(101) Knezovich, J. P., and Harrison, F. L., "A New Method for Determining the Concentrations of Volatile Organic Compounds in Sediment Interstitial Water," *Bulletin of Environmental Contamination and Toxicology*, Vol 38, 1987, pp. 937–940.

(102) Lyons, W. B., Gaudette, J., and Smith, G., "Pore Water Sampling in Anoxic Carbonate Sediments: Oxidation Artifacts," *Nature*, Vol 277, 1979, pp. 48–49.

(103) Howes, B. L., Dacey, J. W. H., and Wakeham, S. G., "Effects of Sampling Technique on Measurements of Porewater Constituents in Salt Marsh Sediments," *Limnology and Oceanography*, Vol 30, 1985, pp. 221–227.

(104) Sayles, F. L., Wilson, T. R. S., Hume, D. N., and Mangelsdorf, P. C., Jr., "In Situ Sampler for Marine Sedimentary Pore Waters: Evidence for Potassium Depletion and Calcium Enrichment," *Science*, Vol 180, 1973, pp. 154–156.

(105) Bischoff, J. L., Greer, R. E., and Luistro, A. O., "Composition of Interstitial Waters of Marine Sediments: Temperature of Squeezing Effect," *Science*, Vol 167, 1970, pp. 1245–1246.

(106) Bray, J. T., Bricker, O. P., and Troup, B. N., "Phosphate in Interstitial Waters of Anoxic Sediments: Oxidation Effects During Sampling Procedure," *Science*, Vol 180, 1973, pp. 1362–1364.

(107) Kemble, N. E., Brumbaugh, W. G., Brenson, E. L., Dwyer, F. J., Ingersoll, G., Monda, D. P., and Woodward, D. F., "Toxicity of Metal Contaminated Sediments from the Upper Clark Fork River Mountain, to Aquatic Invertabrates in Laboratory Exposures," *Environmental Toxicology and Chemistry*, 1994.

(108) Carignan, R., Rapin, F., and Tessier, A., "Sediment Porewater Sampling for Metal Analysis: A Comparison of Techniques," *Geochimica et Cosmochimica Acta*, Vol 49, 1985, pp. 2493–2497.

(109) Chin, Y., and Gschwend, P. M., "The Abundance, Distribution, and Configuration of Porewater Organic Colloids in Recent Sediment," *Geochimica et Cosmochimica Acta*, Vol 55, 1991, pp. 1309–1317.

(110) Brownawell, B. J., and Farrington, J. W., "Biogeochemistry of PCBs in Interstitial Waters of a Coastal Marine Sediment," *Geochimica et Cosmochimica Acta*, Vol 50, 1986, pp. 157–169.

(111) Word, J. Q., Ward, J. A., Franklin, L. M., Cullinan, V. I., and Kiesser, S. L., "Evaluation of the Equilibrium Partitioning Theory for Estimating the Toxicity of the Nonpolar Organic Compound DDT to the Sediment Dwelling Amphipod Rhepoxynius abronius," *Battelle/Marine Research Laboratory Report*, Task 1, WA56, Sequim, WA, 1987, 60 pp.

(112) Klinkhammer, G. P., "Early Diagenesis in Sediments from the Eastern Equatorial Pacific, II: Pore Water Metal Results," *Earth and Planetary Science Letters*, Vol 49, 1980, pp. 81–101.

(113) Simon, N. S., Kennedy, M. M., and Massoni, C. S., "Evaluation and Use of a Diffusion Controlled Sampler for Determining Chemical and Dissolved Oxygen Gradients at the Sediment-Water Interface," *Hydrobiologia*, Vol 126, 1985, pp. 135–141.

(114) Saager, P. M., Sweerts, J-P, and Ellermeijer, H. J., "A Simple Pore-Water Sampler for Coarse, Sandy Sediments of Low Porosity," *Limnology and Oceanography*, Vol 35, 1990, pp. 747–751.

(115) Boulegue, J., Lord, C. J., and Church, T. M., "Sulfur Speciation and Associated Trace Metals (Fe, Cu) in the Pore Waters of Great March, Delaware, *Geochimica et Cosmochimica Acta*, Vol 46, 1982, pp. 453–464.

(116) Bender, M., Martin, W., Hess, J., Sayles, F., Ball, L., and Lambert, C., "A Whole-Core Squeezer for Interfacial Pore-Water Sampling," *Limnology and Oceanography*, Vol 32, 1987, pp. 1214–1225.

(117) Hines, M. E., Knollmeyer, S. L., and Tugel, J. B., "Sulfate Reduction and Other Sedimentary Biogeochemistry in a Northern New England Salt Marsh, *Limnology and Oceanography*, Vol 34, 1989, pp. 578–590.

(118) Bolliger, R., Brandl, H., Hohener, P., Hanselmann, K. W., and Bachofen, R., "Squeeze-Water Analysis for the Determination of Microbial Metabolites in Lake Sediments—Comparison of Methods," *Limnology and Oceanography*, Vol 37, 1992, pp. 448–455.

(119) Kriukov, P. A., and Manheim, F. T., "Extraction and Investigative Techniques for Study of Interstitial Waters of Unconsolidated Sediments: A Review, *The Dynamic Environment of the Ocean Floor*, K. A. Fanning and F. T. Manheim, eds., Lexington Books, Washington, DC, 1982, pp. 3–26.

(120) Froelich, P. M., Klinkhammer, G. P., Bender, M. L., Luedtke, N. A., Heath, G. R., Cullen, D., Dauphin, P., Hammond, D., Hartmann, B., and Maynard, V., "Early Oxidation of Organic Matter in Pelagic Sediments of the Eastern Equatorial Atlantic: Suboxic Diagenesis, *Geochimica et Cosmochimica Acta*, Vol 43, 1979, pp. 1075–1090.

(121) Rosenfeld, J. K., "Ammonia Absorption in Nearshore Anoxic Sediments, *Limnology and Oceanography*, Vol 24, 1979, pp. 356–364.

(122) Santschi, P. H., Nyffeler, V. P., O'Hara, P., Buchholtz, M., and Broecker, W. S., "Radiotracer Uptake on the Seafloor: Results from the MANOP Chamber Deployments in the Eastern Pacific, *Deep-Sea Research*, Vol 31, pp. 451–468.

(123) Winger, P. V., and Lasier, P. J., "A Vacuum-Operated Pore-Water Extractor for Estuarine and Freshwater Sediments," *Archives of Environmental Contamination and Toxicology*, Vol 21, 1991, pp. 321–324.

(124) Brinkman, A. G., van Raaphorst, W., and Lijklema, L., "In Situ Sampling of Interstitial Water from Lake Sediments," *Hydrobiologia*, Vol 92, 1982, pp. 659–663.

(125) Carignan, R., and Lean, D. R. S., "Regeneration of Dissolved Substances in a Seasonally Anoxic Lake: The Relative Importance of Processes Occurring in the Water Column and in the Sediments," *Limnology and Oceanography*, Vol 36, 1991, pp. 683–707.

(126) Mayer, L. M., "Chemical Water Sampling in Lakes and Sediments with Dialysis Bags," *Limnology and Oceanography*, Vol 21, 1976, pp. 909–911.

(127) Bottomley, E. Z., and Bayly, I. L., "A Sediment Porewater Sampler Used in Root Zone Studies of the Submerged Macrophyte, *Myriophyllum spicatum*," *Limnology and Oceanography*, Vol 29, 1984, pp. 671–673.

(128) Hesslin, R. H., "An In Situ Sampler for Close Interval Pore Water Studies," *Limnology and Oceanography*, Vol 21, 1976, pp. 912–914.

(129) Barnes, R., "Interstitial Water Sampling by Dialysis: Methodological Notes," *Limnology and Oceanography*, Vol 29, 1984, p. 667.

(130) Carr, R. S., Williams, J. W., and Fragata, C. T. B., "Development and Evaluation of a Novel Marine Sediment Pore Water Toxicity Test with the Polychaete *Dinophilus Gyrociliatus*," *Environmental Toxicology and Chemistry*, Vol 8, 1989, pp. 533–543.

(131) Skalski, C., "Laboratory and In Situ Sediment Toxicity Evaluations Using Early Life Stages of *Pimephales promelas*," MS Thesis, Wright State University, Dayton, OH, 1991.

(132) Fisher, R., "Sediment Interstitial Water Toxicity Evaluations Using *Daphnia magna*," MS Thesis, Wright State University, Dayton, OH, 1991.

(133) Huckins, J. N., Tubergen, M. W., and Manuweera, G. K., "Semipermeable Membrane Devices Containing Model Lipid: A New Approach to Monitoring the Bioavailability of Lipophilic Contaminants and Estimating Their Bioconcentration Potential," *Chemosphere*, Vol 20, 1990, pp. 533–552.

(134) Burton, G. A., Jr., "Sediment Toxicity Assessments Using In Situ Assays", *Abstract of the First Society of Environmental Toxicology and Chemistry World Congress*, Lisbon, Portugal, 1993.

(135) Skalski, C., and Burton, G. A., Jr., "Comparison of Laboratory and In Situ Sediment Toxicity Responses of *Pimephales promelas* Larvae to Sediment," *Abstract of the Annual Meeting of the Society of Environmental Toxicology and Chemistry*, Seattle, WA.

(136) Lappalainen, K. M., "Convection in Bottom Sediments and its Role in Material Exchange Between Water and Sediments," *Hydrobiologia*, Vol 86, 1982, pp. 105–108.

(137) U.S. Geological Survey, "Techniques of Water-Resources Investigations of the U.S.G.S.," Chapter C1, Harold P. Guy, p. 58, *Laboratory Theory and Methods for Sediment Analysis: Book 5*, Laboratory Analysis, Geological Survey, Arlington, VA, 1969.

(138) Black, C. A., ed., "Methods of Soil Analysis," Agronomy Monograph No. 9, American Society of Agronomy, Madison, WI, 1965.

(139) Page, A. L., Miller, R. H., and Keeney, D. R. eds., "Methods of Soil Analysis," Parts 1 and 2, American Society of Agronomy, Madison, WI, 1982.

(140) American Public Health Association, "Standard Methods for the Examination of Water and Wastewater," 16th ed., Washington, DC, 1985.

(141) Keilty, T. J., White, D. S., and Landrum, P. F., "Sublethal Responses to Endrin in Sediment by *Limnodrilus hoffmeisteri* (Tubificidae) and in Mixed-Culture with *Stylodrilius heringianus* (Lumbriculidae)," *Aquatic Toxicology*, Vol 13, 1988, pp. 251–270.

(142) McLeese, D. W., Metcalfe, C. D., and Pezzack, D. S., "Uptake of PCB's from Sediment by *Nereis virens* and *Crangon septemspinosa*," *Archives of Environmental Contamination and Toxicology*, Vol 9, 1980, pp. 507–518.

(143) Kadeg, R. D., and Pavlou, S., "Reconnaissance Field Study Verification of Equilibrium Partitioning: Nonpolar Hydrophobic Organic Chemicals," submitted by Battelle Washington Environmental Program Office, Washington, DC, for U.S. Environmental Protection Agency, Criteria and Standards Division, Washington, DC, 1987.

(144) Kadeg, R. D., Pavlou, S., and Duxbury, A. S., "Sediment Criteria Methodology Validation, Work Assignment 37, Task II: Elaboration of Sediment Normalization Theory for Nonpolar Hydrophobic Organic Chemicals," prepared by Envirosphere Co., Inc., for Battelle Pacific Northwest Laboratories and U.S. Environmental Protection Agency, Criteria and Standards Division, Washington DC, 1986.

(145) Pavlou, S., Kadeg, R., Turner, A., and Marchlik, M., "Sediment Quality Criteria Methodology Validation: Uncertainty Analysis of Sediment Normalization Theory for Nonpolar Organic Contaminants," Envirosphere Co., Inc., for Environmental Protection Agency, Criteria and Standards Division, Washington DC, submitted by Battelle, Washington Environmental Program Office, Washington, DC, 1987.

(146) Burford, J. R., and Bremmer, J. M., "Relationships Between the Dentrification Capacities of Soils and Total Water-Soluble and Readily Decomposable Soil Organic Matter," Soil Biology and Biochemistry, Vol 7, pp. 389–394.

(147) Cahill, R. A., and Autrey, A. D., "Improved Measurement of the Organic Carbon Content of Various River Components," *Journal of Freshwater Ecology*, Vol 4, 1987, pp 219–223.

(148) Menzel, D. W., and Vaccaro, R. F., "The Measurement of Dissolved Organic and Particulate Carbon in Sea Water," *Limnology and Oceanography*, Vol 9, 1964, pp. 138–142.

(149) Walkley, A., and Black, I. A., "An Examination of the Degtjareff Method for Determining Soil Organic Matter and a Proposed Modification of the Chromic Acid Titration Method," *Soil Science*, Vol 37, 1934, pp. 29–38.

(150) Yeomans, J. C., and Bremmer, J. M., "A Rapid and Precise Method for Routine Determination of Organic Carbon in Soil," *Communications in Soil Science and Plant Analysis*, Vol 19, 1985, pp. 1467–1476.

(151) Allen, T., *Particle Size Measurement*, John Wiley and Sons, New York, NY 1975, 452 pp.

(152) Singer, J. K., Anderson, J. B., Ledbetter, M. T., McCave, I. N., Jones, K. P. N., and Wright, R., "An Assessment of Analytical Techniques for the Size Analysis of Fine-Grained Sediments," *Journal of Sediment Petrology*, Vol 58, 1988, pp. 534–543.

(153) Duncan, G. A., and Lattaie, G. G., "Size Analysis Procedures Used in the Sedimentology Laboratory, NWRI Manual," National Water Research Institute, Canada Centre for Inland Waters, 1979.

(154) Day, P. R., "Particle Fractionation and Particle-Size Analysis, Section 43-5: Hydrometer Method of Particle Size Analysis," Monograph No. 9, American Society of Agronomy, Madison, WI, 1965, pp. 562–566.

(155) Patrick, W. H., Jr., "Modification of Method Particle Size Analyses," Proceedings of the Soil Science Society of America, Vol 4, 1958, pp. 366–367.

(156) Rukavina, N. A., and Duncan, G. A., "F.A.S.T.—Fast Analysis of Sediment Texture," Proceedings 13th Conference of Great Lakes Research, 1970, pp. 274–281.

(157) Sanford, R. B., and Swift, D. J. P., "Comparisons of Sieving and Settling Techniques for Size Analysis, Using a Benthos Rapid Sediment Analyzer," *Sedimentology*, Vol 17, 1971, pp. 257–264.

(158) Cooper, L. R., Haverland, R., Hendricks, D., and Knisel, W., "Microtrac Particle Size Analyzer: An Alternative Particle Size Determination Method for Sediment and Soils," *Soil Science*, Vol 138, pp. 138–146.

(159) Sternberg, R. W., and Creager, J. S., "Comparative Efficiencies of Size Analysis by Hydrometer and Pipette Methods," *Journal of Sediment Petrology*, Vol 31, 1961, pp. 96–100.

(160) McCave, I. M., and Jarvis, J., "Use of the Model T Coulter® Counter in Size Analysis of Fine to Coarse Sand," *Sedimentology*, Vol 20, 1973, pp. 305–315.

(161) Vanderpleog, H. A., "Effect of the Algal Length/Aperture Length Ratio on Coulter Analyses of Lake Seston," *Canadian Journal of Fisheries and Aquatic Science*, Vol 38, 1981, pp. 912–916.

(162) Swift, D. J. P., Schubel, J. R., and Sheldon, R. W., "Size Analysis of Fine-Grained Suspended Sediments: A Review, *Journal of Sediment Petrology*, Vol 42, 1972, pp. 122–134.

(163) Leppard, G. G., Buffle, J., De Vitore, R. R., and Perseet, D., "The Ultrastructure and Physical Characteristics of a Distinctive Colloidal Iron Particulate Isolated from a Small Eutrophic Lake," *Archives Hydrobiology*, Vol 113, 1988, pp. 405–424.

(164) Burnison, B. K., and Leppard, G. G., "Isolation of Colloidal Fibrils from Lake Water by Physical Separation Techniques," *Canadian Journal of Fisheries and Aquatic Science*, Vol 40, pp. 373–381.

(165) Leppard, G. G., "The Fibrillar Matrix Component of Lacustrine Biofilms," *Water Research*, Vol 20, 1986, pp. 697–702.

(166) Leppard, G. G., Massalski, A., and Lean, D. S. R., "Electron-Opaque Microscopic Fibrils in Lakes: Their Demonstration, Their Biological Derivation and Their Potential Significance in the Redistribution of Cations," *Protoplasma*, Vol 92, 1977, pp. 289–309.

(167) Kaddah, M. T., "The Hydrometer Method for Detailed Particle-Size Analysis, 1: Graphical Interpretation of Hydrometer Readings and Test of Method," *Soil Science*, Vol 118, 1974, pp. 102–108.

(168) Stein, R., "Rapid Grain-Size Analyses of Clay and Silt Fraction by Sedigraph 5000D: Comparison with Coulter Counter and Atterberg Methods," *Journal of Sediment Petrology*, Vol 55, 1985, pp. 590–615.

(169) Welch, N. H., Allen, P. B., and Galinds, D. J., "Particle-Size Analysis by Pipette and Sedigraph," *Journal of Environmental Quality*, Vol 8, 1979, pp. 543–546.

(170) Chao, T. T., and Zhou, L., "Extraction Techniques for Selective Dissolution of Amorphous Iron Oxides from Soils and Sediments," *Soil Science Society of America Journal*, Vol 47, 1983, pp. 225–232.

(171) Crecelius, E. A., Jenne, E. A., and Anthony, J. S., "Sediment Quality Criteria for Metals: Optimization of Extraction Methods for Determining the Quantity of Sorbents and Adsorbed Metals in Sediments," Battelle Report to USEPA, Criteria and Standards Division, Washington, DC, 1987.

(172) Kersten, M., and Forstner, U., "Cadmium Associations in Freshwater and Marine Sediment," *Cadmium in the Aquatic Environment*, John Wiley and Sons, New York, NY, 1987, pp. 51–88.

(173) Jenne, E. A., "Controls on Mn, Fe, Co, Ni, Ca, and Zn Concentrations in Soils and Water: The Significant Role of

Hydrous Mn and Fe Oxides," *Advanced Chemistry*, Vol 73, 1968, pp. 337–387.

(174) Jenne, E. A., "Trace Element Sorption by Sediments and Soil-Sites and Processes," *Symposium on Molybdenum in the Environment*, Vol 2, M. Dekker, Inc., New York, NY, 1977, pp. 415–553.

(175) Jenne, E. A., "Sediment Quality Criteria for Metals, II: Review of Methods for Quantitative Determination of Important Adsorbants and Sorbed Metals in Sediments," Battelle Report to USEPA, Criteria and Standards Division, Washington, DC, 1987.

(176) Lake, D. L., Kirk, P., and Lester, J., "Fractionation, Characterization, and Speciation of Heavy Metals in Sewage Sludge and Sludge-Amended Soils: A Review," *Journal of Environmental Quality*, Vol 13, 1984, pp. 175–183.

(177) Maher, W. A., "Evaluation of a Sequential Extraction Scheme to Study Associations of Trace Elements in Estuarine and Oceanic Sediments," *Bulletin of Environmental Contamination and Toxicology*, Vol 32, 1984, p. 339.

(178) O'Donnel, J. R., Kaplan, B. M., and Allen, H. E., "Bioavailability of Trace Metals in Natural Waters," *Aquatic Toxicology and Hazard Assessment: Seventh Symposium, ASTM STP 854*, ASTM, Philadelphia, PA, 1985, pp. 485–501.

(179) Salomons, W., and Forstner, U., "Trace Metal Analysis on Polluted Sediment, II: Evaluation of Environmental Impact," *Environmental Technology Letters*, Vol 1, 1980, pp. 506–517.

(180) Fiszman, M., Pfeiffer, W. C., and Drude de Lacerda, L., "Comparison of Methods Used for Extraction and Geochemical Distribution of Heavy Metals in Bottom Sediments from Sepetiba Bay," *Environmental Technology Letters*, Vol 5, 1984, pp. 567–575.

(181) Malo, B. A., "Partial Extraction of Metals from Aquatic Sediments," *Environmental Science and Technology*, Vol 11, 1977, pp. 277–282.

(182) Khalid, R. A., Gambrell, R., and Patrick, W., Jr., "Chemical Availability of Cadmium in Mississippi River Sediment," *Journal of Environmental Quality*, Vol 10, 1981, pp. 523–528.

(183) Tessier, A., Campbell, D., and Bisson, M., "Sequential Extraction Procedure for the Speciation of Particulate Trace Metals," *Analytical Chemistry*, Vol 51, 1979, pp. 844–851.

(184) Jackson, M. L., *Soil Chemical Analysis*, Prentice-Hall, Inc., Englewood Cliffs, NJ, 1958.

(185) Bott, T. L., "Benthic Community Metabolism in Four Temperate Stream Systems: An Inter-Biome Comparison and Evaluation of the River Continuum Concept," *Hydrobiologia*, Vol 123, 1985, pp. 3–45.

(186) Davis, W. S., Fay, L. A., and Herdendort, C. E., "Overview of USEPA/CLEAR Lake Erie Sediment Oxygen Demand Investigations During 1979," *Journal of Great Lakes Research*, Vol 13, 1987, pp. 731–737.

(187) Davis, W. S., Brosnan, T. M., and Sykes, R. M., "Use of Benthic Oxygen Flux Measurements in Wasteload Allocation Studies," *Chemical and Biological Characterization of Sludges, Sediments, Dredge Soils, and Drilling Muds, ASTM STP 976*, ASTM, Philadelphia, PA, 1988, pp. 450–462.

(188) Uchrin, G. G., and Ahlert, W. K., "In Situ Sediment Oxygen Demand Determination in the Passaic River (NJ) During the Late Summer/Early Fall 1983," *Water Research*, Vol 19, 1985, pp. 1141–1144.

(189) U.S. Environmental Protection Agency, "Test Methods for Evaluating Solid Waste," USEPA, Office of Solid Waste and Emergency Response, Wshington, DC, 1986.

(190) Haddock, J. D., Landrum, P. F., and Giesy, J. P., "Extraction Efficiency of Anthracene from Sediments," *Analytical Chemistry*, Vol 55, 1983, pp. 1197–1200.

(191) Sporstoel, S., Gjos, N., and Carlberg, G. E., "Extraction Efficiencies for Organic Compounds Found in Aquatic Sediments," *Analytical Cimmica Acta*, Vol 151, 1983, pp. 231–235.

(192) Burton, G. A., Jr., Stemmer, B. L., Winks, K. L., Ross, P. E., and Burnett, L. C., "A Multitrophic Level Evaluation of Sediment Toxicity in Waukegan and Indiana Harbors," *Environmental Toxicology and Chemistry*, Vol 8, 1989, pp. 1057–1066.

(193) Pastorok, R. A., and Becker, D. S., "Comparative Sensitivity of Sediment Toxicity Bioassays at Three Superfund Sites in Puget Sound," *Aquatic Toxicology and Risk Assessment, ASTM STP 1096*, Vol 13, W. G. Landis and W. H. van der Schalie, eds., ASTM, Philadelphia, PA, 1990, pp. 123–139.

(194) Ingersoll, C. G., and Nelson, M. K., "Testing Sediment Toxicity with *Hyalella azteca* (Amphipoda) and *Chironomus riparius* (Diptera)," *Aquatic Toxicology and Risk Assessment, ASTM STP 1096*, Vol 13, W. G. Landis and W. H. van der Schalie, eds., ASTM, Philadelphia, PA, 1990, pp. 93–109.

(195) Carr, R. S., and Chapman, D. C., "Comparison of Solid-Phase and Pore-Water Approaches for Assessing the Quality of Marine and Estuarine Sediments," *Chemical Ecology*, Vol 7, 1992, pp. 19–30.

(196) Johns, D. M., Pastorak, R. A., and Ginn, T. C., "A Sublethal Sediment Toxicity Test Using Juvenile *Neanthes sp.* (Polychaeta; Nereidae)," *Aquatic Toxicology and Risk Assessment, ASTM STP 1124*, Vol 14, M. A. Mayes and M. G. Barron, eds., ASTM, Philadelphia, PA, 1991, pp. 280–293.

(197) Mudroch, A., and MacKnight, S. D., *CRC Handbook of Techniques for Aquatic Sediments Sampling*, CRC Press, Boca Raton, FL, 1991, 210 pp.

(198) Landrum, P. F., Eadie, B. J., and Faust, W. R., "Variation in the Bioavailability of Polycyclic Aromatic Hydrocarbons to the Amphipod *Diporeia* (spp.) with Sediment Aging," *Environmental Toxicology and Chemistry*, Vol 11, 1992, pp. 1197–1208.

(199) Ditsworth, G. R., Schults, D. W., and Jones, J. K. P., "Preparation of Benthic Substrates for Sediment Toxicity Testing," *Environmental Toxicology and Chemistry*, Vol 9, 1990, pp. 1523–1529.

(200) Swartz, R. C., Schults, D. W., DeWitt, T. H., Ditsworth, G. R., and Lambertson, J. O., "Toxicity of Fluoranthene in Sediment to Marine Amphipods: A Test of the Equilibrium Partitioning Approach to Sediment Quality Criteria," *Environmental Toxicology and Chemistry*, Vol 9, 1990, pp. 1071–1080.

(201) Dewitt, T. H., Ozrethich, R. J., Swartz, R. C., Lamberson, J. O., Schults, D. W., Ditsworth, G. R., Jones, J. K. P., Hoselton, L., and Smith, L. M., "The Influence of Organic Matter Quality on the Toxicity and Partitioning of the Sediment-Associated Fluoranthene," *Environmental Toxicology and Chemistry*, Vol 16, 1992, pp. 401–407.

(202) Sasson-Brickson, G., and Burton, G. A., "In Situ Laboratory Sediment Toxicity Testing with *Ceriodaphnia dubia*," *Environmental Toxicology and Chemistry*, Vol 10, 1991, pp. 201–207.

(203) Zar, J. H., *Biostatistical Analysis*, 2nd Ed., Prentice-Hall, Inc., NJ, 1984, 718 p.

(204) Schuytema, G. S., Nelson, P. O., Malueg, K. W., Nebeker, A. V., Krawczyk, D. F., Ratcliff, A. K., and Gakstatter, J. H., "Toxicity of Cadmium in Water and Sediment Slurries to *Daphnia magna*," *Environmental Toxicology and Chemistry*, Vol 3, 1984, pp. 293–308.

(205) Francis, P. C., Birge, W., and Black, J., "Effects of Cadmium-Enriched Sediment on Fish and Amphibian Embryo-Larval Stages," *Ecotoxicology and Environmental Safety*, Vol 8, 1984, pp. 378–387.

(206) Keilty, T. J., White, D. S., and Landrum, P. F., "Sublethal Responses to Endrin in Sediment by *Stylodrilius heringianus* (Lumbriculidae) as Measured by a 137 Cesium Marker Layer Technique," *Aquatic Toxicology*, Vol 13, 1988, pp. 227–250.

(207) Keilty, T. J., White, D. S., and Landrum, P. F., "Short-Term Lethality and Sediment Avoidance Assays with Endrin-Contaminated Sediment and Two Oligochaetes from Lake Michigan," *Archives of Environmental Contamination and Toxicology*, Vol 17, 1988, pp. 95–101.

(208) Landrum, P. F., and Faust, W. R., "Effect of Variation in Sediment Composition on the Uptake Rate Coefficient for Selected PCB and PAH Cogeners by the Amphipod, *Diporeia sp.*," *Aquatic Toxicology and Risk Assessment, ASTM STP 1124*, M. A. Mayes and M. G. Barron, eds., ASTM, Philadelphia, PA, 1991, pp. 263–279.

(209) Foster, G. D., Baksi, S. M., and Means, J. C., "Bioaccumulation of Trace Organic Contaminants From Sediment by Baltic Clams (*Macoma balthica*) and Soft-Shell Clams (*Mya arenaria*)," *Environmental Toxicology and Chemistry*, Vol 6, 1987, pp. 969–976.

(210) Birge, W. J., Black, J., Westerman, S., and Francis, P., "Toxicity of Sediment-Associated Metals to Freshwater Organisms: Biomonitoring Procedures," *Fate and Effects of Sediment-Bound Chemicals in Aquatic Systems*, Pergamon Press, New York, NY, 1987, pp. 199–218.

(211) Suedel, B. C., Rodgers, J. H., Jr., and Clifford, P. A., "Bioavailability of Fluoranthene in Freshwater Sediment Toxicity Tests," *Environmental Toxicology and Chemistry*, Vol 12, 1993, pp. 155–165.

(212) Tsushimoto, G., Matsumura, F., and Sago, R., "Fate of 2,3,7,8-tetrachlorodebenzo-p-dioxin (TCDD) in an Outdoor Pond and in Model Aquatic Ecosystems," *Environmental Toxicology and Chemistry*, Vol 1, 1982, pp. 61–68.

(213) Lay, J. P., Schauerte, W., Klein, W., and Korte, F., "Influence of Tetrachloroethylene on the Biota of Aquatic Systems: Toxicity to Phyto- and Zooplankton Species in Compartments of a Natural Pond," *Archives of Environmental Contamination and Toxicology*, Vol 13, 1984, pp. 135–142.

(214) Stephenson, R. R., and Kane, D. F., "Persistence and Effects of Chemicals in Small Enclosures in Ponds," *Archives of Environmental Contamination and Toxicology*, Vol 13, 1984, pp. 313–326.

(215) O'Neill, E. J., Monti, C. A., Prichard, P. H., Bourquin, A. W., and Ahearn, D. G., "Effects of Lugworms and Seagrass on Kepone (Chlordecone) Distribution in Sediment-Water Laboratory Systems," *Environmental Toxicology and Chemistry*, Vol 4, 1985, pp. 453–458.

(216) Crossland, N. O., and Wolff, C. J. M., "Fate and Biological Effects of Pentachlorophenol in Outdoor Ponds," *Environmental Toxicology and Chemistry*, Vol 4, 1985, pp. 73–86.

(217) Pritchard, P. H., Monti, C. A., O'Neill, E. J., Connolly, J. P., and Ahearn, D. G., "Movement of Kepone (Chlordecone) Across an Undisturbed Sediment-Water Interface in Laboratory Systems," *Environmental Toxicology and Chemistry*, Vol 5, 1986, pp. 647–657.

(218) Gerould, S., and Gloss, S. P., "Mayfly-Mediated Sorption of Toxicants into Sediments," *Environmental Toxicology and Chemistry*, Vol 5, 1986, pp. 667–673.

(219) McLeese, D. W., Metcalfe, C. D., and Pezzack, D. S., "Uptake of PCBs from Sediment by *Nereis virens* and *Crangon septemspinosa*," *Archives of Environmental Contamination and Toxicology*, Vol 9, 1980, pp. 507–518.

(220) Greaves, M. P., Davies, H. A., Marsh, J. A. P., and Wingfield, G. I., "Effects of Pesticides on Soil Microflora Using Dalapon as an Example," *Archives of Environmental Contamination and Toxicology*, Vol 10, 1981, pp. 437–439.

(221) Tagatz, M. E., Plaia, G. R., and Deans, C. H., "Toxicity of Dibutyl Phthalate-Contaminated Sediment to Laboratory and Field Colonized Estuarine Benthic Communities," *Bulletin of Environmental Contamination and Toxicology*, Vol 37, 1986, pp. 141–150.

(222) Swindoll, C. M., and Applehans, F. M., "Factors Influencing the Accumulation of Sediment-Sorbed Hexachlorobiphenyl by Midge Larvae," *Bulletin of Environmental Contamination and Toxicology*, Vol 39, 1987, pp. 1055–1062.

(223) Häkanson, L., "Sediment Sampling in Different Aquatic Environments: Statistical Aspects," *Water Resource Research*, Vol 20, 1984, pp. 41–46.

(224) Webster, G. R. B., Servos, M. R., Choudhry, G. G., Sarna, L. P., and Muir, G. C. G., "Methods for Dissolving Hydrophobics in Water for Studies of Their Interactions with Dissolved Organic Matter," Presented at 193rd National Meeting of the American Chemical Society, Division of Environmental Chemistry, *Advances in Chemistry Series*, Entended Abstracts, Vol 27, 1990, pp. 191–192.

(225) Nkedi-Kizza, P., Rao, P. S. C., and Hornsby, A. G., "Influence of Organic Cosolvents on Sorption of Hydrophobic Organic Chemicals by Soils," *Environmental Science and Technology*, Vol 19, 1985, pp. 975–979.

(226) Ziegenfuss, P. S., and Adams, W., "A Method for Assessing the Acute Toxicity of Contaminated Sediments and Soils with *Daphnia magna* and *Chironomus tentans*," Report No. MSL-4549, ESC-EAG-M-85-01, Monsanto Corp., St. Louis, MO. 1985.

(227) Titus, J. A., and Pfister, R. M., "Bacterial and Cadmium Interactions in Natural and Laboratory Model Aquatic Systems," *Archives of Environmental Contamination and Toxicology*, Vol 13, 1984, pp. 271–277.

(228) Nebeker, A. V., Onjukka, S. T., Cairns, M. A., and Krawczyk, D. F., "Survival of *Daphnia magna* and *Hyalella azteca* in Cadmium Spiked Water," *Environmental Toxicology and Chemistry*, Vol 5, 1986, pp. 933–938.

(229) Burton, G. A., Lazorchak, J. M., Waller, W. T., and Lanza, G. R., "Arsenic Toxicity Changes in the Presence of Sediment," *Bulletin of Environmental Contamination and Toxicology*, Vol 38, 1987, pp. 491–499.

(230) Knezovich, J. P., and Harrison, F. L., "The Bioavailability of Sediment-Sorbed Chlorobenzenes to Larvae of the Midge Chironomus decorus," *Ecotoxicology and Environmental Safety*, Vol 15, 1988, pp. 226–241.

(231) Di Toro, D. M., Zarba, C. J., Hansen, D. J., Berry, W. J., Swartz, R. W., Cowna, C. E., Pavlou, S. P., Allen, H. E., Thomas, N. A., and Paquin, P. R., "Technical Basis for Establishing Sediment Quality Criteria for Nonionic Organic Chemicals Using Equilibrium Partitioning," *Environmental Toxicology and Chemistry*, Vol 10, 1991, pp. 1541–1583.

(232) Burton, G. A., Jr., "Assessing the Toxicity of Freshwater Sediments," *Environmental Toxicology and Chemistry*, Vol 10, 1991, pp. 1585–1627.

(233) Jenne, E. A., and Zachara, J. M., "Factors Influencing the Sorption of Metals," *Fate and Effects of Sediment-Bound Chemicals in Aquatic Systems*, K. L. Dickson, A. W. Maki, and W. A. Brungs, eds., Pergamon Press, Elmsford, NY, 1984, pp. 83–98.

(234) DeWitt, T. H., Swartz, R. C., and Lamberson, J. O., "Measuring the Acute Toxicity of Estuarine Sediments," *Environmental Toxicology and Chemistry*, Vol 8, 1989, pp. 1035–1048.

(235) Di Toro, D. M., Horzempa, L. M., Casey, M. M., and Richardson, W., "Reversible and Resistant Components of PCB Adsorption-Desorption: Adsorbent Concentration Effects," *Journal of Great Lakes Research*, Vol 8, 1982, pp. 336–349.

(236) Karickhoff, S. W., and Morris, K. R., "Sorption Dynamics of Hydrophobic Pollutants in Sediment Suspensions," *Environmental Toxicology and Chemistry*, Vol 4, 1985, pp. 469–479.

(237) Karickhoff, S. W., "Sorption of Hydrophobic Pollutants in Natural Sediments," *Contaminants and Sediments, Vol 2, Analysis, Chemistry, Biology*, Ann Arbor Science, Ann Arbor, MI, 1986, pp. 193–205.

(238) Landrum, P. F., "Bioavailability and Toxicokinetics of Polycyclic Aromatic Hydrocarbons Sorbed to Sediments for the Amphipod, *Pontoporeia hoyi*," *Environmental Science and Technology*, Vol 23, 1989, pp. 585–588.

(239) International Joint Commission, "Procedures for the Assessment of Contaminated Sediment Problems in the Great Lakes," IJC, Windsor, Ont., Canada, 1988, p. 140.

(240) Landrum, P. F., and Poore, R., "Toxicokinetics of Selected Xenobiotics in *Hexagenia limbata*," *Journal of Great Lakes Research*, Vol 14, 1988, pp. 427–437.

(241) American Public Health Association, American Water Works Association, and Water Pollution Control Federation, *Standard Methods for the Examination of Water and Wastewater*, 15th Ed., American Public Health Association, Washington, DC, 1980, pp. 615–714.

(242) Giesy, J. P., Rosiu, C. J., Graney, R. L., and Henry, M. G., "Benthic Invertebrate Bioassays with Toxic Sediment and Pore Water," *Environmental Toxicology and Chemistry*, Vol 9, 1990, pp. 233–248.

(243) *American Standards for Testing and Materials, Section 14: General Methods and Instrumentation*, ASTM, Philadelphia, PA, 1986.

(244) Lydy, M. J., Bruner, K. A., Fry, K. A., and Fisher, S. W., "Effects of Sediment and the Route of Exposure on the Toxicity and Accumulation of Neutral Lipophilic and Moderately Water Soluble Metabolizable Compounds in the Midge, *Chironomus riparius*, *Aquatic Toxicology and Risk Assessment, ASTM STP 1096*, Vol 13, W. G. Landis and W. H. van der Schalis, eds., ASTM, Philadelphia, PA, 1990, pp. 140–164.

(245) Day, K. E., Kirby, R. S., and Reynoldson, T. B., "The Effects of Sediment Manipulations on Chronic Sediment Bioassays with Three Species of Benthic Invertebrates," Abstract, Presented at the 13th Annual Meeting, Society of Environmental Toxicology and Chemistry, Cinncinati, OH, Nov. 8–12, 1992.

(246) Förstner, U., and Salomons, W., "Trace Metal Analysis on Polluted Sediments," Publication 248, Delft Hydraulic Laboratory, Delft, The Netherlands, 1981, pp. 1–13.

(247) Moore, D. W., and Dillon, T. M., "Chronic Sublethal Effects of San Francisco Bay Sediments on *Nereis (Neanthes) arenaceodentata*: Nontreatment Factors," U.S. Army Corps of Engineers, Miscellaneous Paper D-92-4, 1992.

(248) Reynoldson, T. B., Day, K. E., Kirby, S. K., Clarke, C., and Millani, D., "Effect of Indigenous Animals on Chronic Endpoints in Freshwater Sediment Toxicity Tests," *Environmental Toxicology and Chemistry*, submitted.

(249) Powlson, D. S., and Jenkinson, D. S., "The Effects of Biocidal Treatments on Metabolism in Soil—II: Gamma Irradiation, Autoclaving, Air-Drying and Fumigation," *Soil Biology and Biochemistry.*, Vol 8, 1976, pp. 179–188.

(250) Jenkinson, D. S., "The Effects of Biocidal Treatments on Metabolism in Soil—IV: The Decomposition of Fumigated Organisms in Soil," *Soil Biology and Biochemistry*, Vol 8, 1976, pp. 203–208.

(251) Chapman, P. M., "Marine Sediment Toxicity Tests," *Symposium on Chemical and Biological Characterization of Sludges, Sediments, Dredge Spoils and Drilling Muds, ASTM STP 976*, ASTM, Philadelphia, PA, 1987, pp. 391–402.

(252) Giesy, J. P., and Hoke, R. A., "Freshwater Sediment Toxicity Bioassessment: Rationale for Species Selection and Test Design," *Journal of Great Lakes Research*, Vol 15, 1989, pp. 539–569.

(253) Pastorok, R. A., and Becker, D. S., "Comparative Sensitivity of Sediment Toxicity Bioassays at Three Superfund Sites in Puget Sound," *Aquatic Toxicology and Risk Assessment, ASTM STP 1096*, W. G. Landis and W. H., eds., ASTM, Philadelphia, PA, 1990, pp. 123–139.

(254) Swartz, R. C., Kemp, P. F., Schults, D. W., Ditsworth, G. R., and Ozretich, R. J., "Acute Toxicity of Sediments from Eagle Harbor, Washington, to the Infaunal Amphipod, *Rhepoxynius abronius*,"

(255) McGee, D. J., Thomulka, K. W., and Lange, J. H., "Use of Bioluminescent Bacterium *Photobacterium phosphoreum* to Detect Potentially Biohozardous Materials in Water," *Bulletin of Environmental Contamination and Toxicology*, Vol 51, 1993, pp. 538–544.

(256) Nelson, M. K., Landrum, P. F., Burton, G. A., Klaine, S. J., Crecelius, E. A., Byl, T. D., Gossiaux, D. C., Tysmbal, V. N., Cleveland, L., Ingersoll, C. G., and Sasson-Brickson, G., "Toxicity of Contaminated Sediments in Dilution Series with Control Sediments," *Chemosphere*, in press.

(257) Tay, K. L., Doe, K. G., Wade, S. J., Vaughan, D. A., Berrigan, R. E., and Moore, M. J., "Sediment Bioassessment in Halifax Harbor," *Environmental Toxicology and Chemistry*, Vol 11, 1992, pp. 1567–1581.

(258) Chapman, P. M., and Fink, R., "Effects of Puget Sound Sediments and Their Elutriates on the Life Cycle of *Capitella capitata*," *Bulletin of Environmental Contamination and Toxicology*, Vol 33, 1984, pp. 451–459.

(259) Ross, P. E., and Henebry, M. S., "Use of Four Microbial Tests to Assess the Ecotoxicological Hazard of Contaminated Sediments," *Toxicity Assessment*, Vol 4, 1989, pp. 1–21.

(260) U.S. Army Corps of Engineers, "Ecological Evaluation of Proposed Discharge of Dredged or Fill Material into Navigable Waters," Miscellaneous Paper D-76-17, Waterways Experiment Station, Vicksburg, MS, 1976.

(261) Daniels, S. A., Munawar, M., and Mayfield, C. I., "An Improved Elutriation Technique for the Bioassessment of Sediment Contaminants," *Hydrobiologia*, Vol 188/189, 1989, pp. 619–631.

(262) Shumacker, B., Fox, R., Filkins, J. C., and Barrick, B., "Quality Assurance and Quality Control," *Assessment and Remediation of Contaminated Sediments (ARCS) Program*, U.S. EPA Assessment Guidance Document, 1994, pp. 10–32.

(263) Stemmer, B. L., Burton, G. A., Jr., and Sasson-Brickson, G., "Effect of Sediment Spatial Variance and Collection Method in Cladoceran Toxicity and Indigenous Microbial Activity Determinations," *Environmental Toxicology and Chemistry*, Vol 9, 1990, pp. 1035–1044.

(264) Downing, J. A., and Rath, L. C., "Spatial Patchiness in the Lacustrine Sedimentary Environment," *Limnology and Oceanography*, Vol 33, 1988, pp. 447–458.

(265) Morin, A., "Variability of Density Estimates and the Optimization of Sampling Programs for Stream Benthos," *Canadian Journal of Fisheries and Aquatic Science*, Vol 42, 1985, pp. 1530–1534.

PART 3.   WATER SAMPLING AND MONITORING

3.1 GENERAL

Standard Guide for
Quality Planning and Field Implementation of a Water Quality Measurement Program[1]

This standard is issued under the fixed designation D 5612; the number immediately following the designation indicates the year of original adoption or, in the case of revision, the year of last revision. A number in parentheses indicates the year of last reapproval. A superscript epsilon (ϵ) indicates an editorial change since the last revision or reapproval.

1. Scope

1.1 This guide covers planning and implementation of the sampling aspects of environmental data generation activities. Environmental data generation efforts are comprised of four parts: (*1*) establishment of data quality objectives (DQOs); (*2*) design of field sampling and measurement strategies and specification of laboratory analyses and data acceptance criteria; (*3*) implementation of sampling and analysis strategies; and (*4*) data quality assessment.

1.2 This guide defines the criteria that must be considered to ensure the quality of the field aspects of environmental data and sample generation activities.

1.3 DQOs should be adopted prior to the application of this guide. The data generated in accordance with this guide are subject to a final assessment to determine whether the DQOs were met. For example, many screening activities do not require all of the quality assurance (QA) and quality control (QC) steps found in this guide to generate data adequate to meet the project needs. The extent to which all of the requirements must be met remains a matter of technical judgement as it relates to the established DQOs.

1.4 This guide presents extensive management requirements designed to ensure high-quality samples and data. The words "must," "shall," "may," and "should" have been selected carefully to reflect the importance placed on many of the statements made in this guide.

1.5 *This standard does not purport to address all of the safety concerns, if any, associated with its use. It is the responsibility of the user of this standard to establish appropriate safety and health practices and determine the applicability of regulatory limitations prior to use.*

2. Referenced Documents

2.1 *ASTM Standards:*
D 596 Practice for Reporting Results of Analysis of Water[2]
D 1129 Terminology Relating to Water[2]
D 2777 Practice for Determination of Precision and Bias of Applicable Methods of Committee D-19 on Water[2]
D 3370 Practices for Sampling Water[2]
D 3856 Guide for Good Laboratory Practices in Laboratories Engaged in Sampling and Analysis of Water[2]
D 4210 Practice for Interlaboratory Quality Control Procedures and a Discussion on Reporting Low-Level Data[2]

D 4447 Guide for the Disposal of Laboratory Chemicals and Samples[3]
D 4448 Guide for Sampling Groundwater Monitoring Wells[3]
D 4840 Guide for Sampling Chain of Custody Procedures[2]
D 4841 Practice for Estimation of Holding Time for Water Samples Containing Organic and Inorganic Constituents[2]
D 5172 Guide for Documenting the Standard Operating Procedures Used in a Specific Laboratory[2]
D 5283 Practice for Generation of Environmental Data Related to Waste Management Activities: Quality Assurance and Quality Control Planning and Implementation[3]
E 29 Practice for Using Significant Digits in Test Data to Determine Conformance with Specifications[4]
E 178 Practice for Dealing with Outlying Observations[4]
E 1187 Terminology Relating to Laboratory Accreditation[4]

2.2 *U.S. Environmental Protection Agency Documents:*[5]
QAMS-005/80 (NTIS No. PB83170514/LL), Interim Guidelines and Specifications for Preparing Quality Assurance Project Plans, Office of Monitoring Systems and Quality Assurance, Dec. 29, 1980
QAMS-500/80, Development of Data Quality Objectives, Description of Stages I and II, July 16, 1986
QAMS-004/80 (NTIS No. PB83219667/LL), Guidelines and Specifications for Preparing Quality Assurance Program Plans, Office of Monitoring Systems and Quality Assurance, Sept. 20, 1980

3. Terminology

3.1 *Definitions*—The terms that are most applicable to this guide have been defined in Terminologies D 1129 and E 1187.

3.2 *Descriptions of Terms Specific to This Standard:*

3.2.1 *background sample*—a sample taken from a location on or proximate to the site of interest. This sample is taken to document baseline or historical information.

3.2.2 *collocated samples*—independent samples collected as close as possible to the same point in space and time and intended to be identical.

3.2.3 *data quality objectives (DQOs)*—statements on the level of uncertainty that a decision maker is willing to accept

---

[1] This guide is under the jurisdiction of ASTM Committee D-19 on Water and is the direct responsibility of Subcommittee D19.02 on General Specifications, Technical Resources, and Statistical Methods.
Current edition approved Sept. 15, 1994. Published November 1994.
[2] *Annual Book of ASTM Standards*, Vol 11.01.

[3] *Annual Book of ASTM Standards*, Vol 11.04.
[4] *Annual Book of ASTM Standards*, Vol 14.02.
[5] Available from Standardization Documents Order Desk, Bldg. 4 Section D, 700 Robbins Ave., Philadelphia, PA 19111-5094, Attn. NPODS.

in the results derived from environmental data (see QAMS-500/80).

3.2.4 *material blank*—a sample composed of construction materials such as those used in well installation, well development, pump and flow testing, and slurry wall construction. Examples of these materials are bentonite, sand, drilling fluids, and source and purge water. This blank documents the contamination resulting from usage of the construction materials.

3.2.5 *quality assurance program plan (QAPP)*—an orderly assemblage of management policies, objectives, principles, and general procedures by which an organization involved in environmental data generation activities outlines how it intends to produce data of known quality.

3.2.6 *quality assurance project plan (QAPjP)*—an orderly assemblage of detailed procedures designed to produce data of sufficient quality to meet the DQOs for a specific data collection activity.

4. Summary of Guide

4.1 This guide describes the criteria and activities for organizations involved in obtaining water samples and generating field data in terms of human and physical resources and QC procedures and documentation requirements depending on the DQOs or agreed upon project plan.

5. Significance and Use

5.1 Environmental data are often required for making regulatory and programmatic decisions. These data must be of known quality commensurate with their intended use.

5.2 Certain minimal criteria must be met by the field organizations in order to meet the objectives of the water monitoring activities.

5.3 This guide defines the criteria for organizations taking water samples and generating environmental data and identifies other activities that may be required based on the DQOs.

5.4 This guide emphasizes the importance of communication among those involved in establishing the DQOs, planning, and implementing the sampling and analysis aspects of environmental data generation activities, and assessing data quality.

6. Project Specification

6.1 *Overall Project Objectives*—The overall objectives of the project must be defined prior to the start of any field and laboratory activities.

6.2 *Data Quality Objectives*—DQOs for the data generation activity should be defined prior to the initiation of field and laboratory work, and they must be compatible with project objectives. It is desirable that the field and laboratory organizations be aware of the DQOs so that the personnel conducting the work are able to make informed decisions during the course of the project.

6.3 *Project Plan*—The project plan should be designed to meet the project objectives and DQOs. The project plan should define the following:

6.3.1 *Specific Project Objectives*—The objectives of the field and laboratory work must be defined clearly, define specific objectives for the sampling location, and describe the intended uses for the data. The project objective may need to

be reviewed as information is gathered. Any changes in the project objective affecting field and laboratory activities should be communicated to the field and laboratory personnel.

6.3.2 *Background Information*—Any background information that could affect meeting the project objective or DQOs should be provided. For example, the identification of any regulatory programs governing data collection and analysis and the reason for conducting the sample collection work should be included in the background information.

6.3.3 Project management shall have individuals designated as having responsibility and authority for the following: (*1*) developing project documents that implement the DQOs; (*2*) selecting field and laboratory organizations to conduct the work; (*3*) coordinating communication among the field and laboratory organizations and government agencies, as required; and (*4*) reviewing and assessing the final data.

6.3.4 Sampling requirements shall be specified, including sampling locations, equipment and procedures, and sample preservation and handling.

6.3.5 Analytical requirements shall be specified, including the analytical procedures, analyte list, required detection limits, and required precision and bias values. Regulatory requirements and DQOs shall be considered when developing the specifications.

NOTE 1—The above does not imply that the specified analytical requirements can be met.

6.3.6 The QA and QC requirements shall address both field and laboratory activities. The means for controlling false positives and false negatives shall be specified.

6.3.6.1 The types and frequency of field QC samples to be collected, including field blanks, duplicates, and spikes, trip blanks, equipment rinsates, background samples, reference materials, material blanks, and split samples, should be specified. Control parameters for field activities shall be described (see 7.6.3).

6.3.6.2 The types and frequency of laboratory QC samples, such as laboratory control samples, laboratory blanks, matrix spikes, matrix duplicates, and matrix spike duplicates, shall be specified. Any specific performance criteria shall be specified. Data validation criteria shall be defined.

6.4 *Project Documentation*—All documents required for planning, implementing, and evaluating the data collection effort shall be specified. These may include, although are not limited to, a statement of work, technical and cost proposals, work plan, sampling and analysis plan, QAPjP, health and safety plan, community relations plan, documents required by regulatory agencies, requirements for raw field and analytical records, technical reports assessing the environmental data, and records retention policy. Planning documents shall specify the required level of document control and identify the personnel having access. Document formats that may be required to ensure that all data needs are satisfied shall be specified. In addition, a project schedule that identifies critical milestones and completion dates should be available.

7. Standard Guide for Environmental Field Operations

7.1 *Purpose*—the field organization must conduct its operations in such a manner as to provide reliable information that meets the DQOs. To achieve this goal, certain minimum policies and procedures must be implemented in order to meet the DQOs.

7.2 *Organization*—The field organization shall be structured such that each member of the organization has a clear understanding of his or her duties and responsibilities and the relationship of those responsibilities to the total effort. The organizational structure, functional responsibilities, levels of authority, job descriptions, and lines of communication for activities shall be established and documented. One person may cover more than one organizational function.

7.2.1 *Management*—The management personnel of the field organization is responsible for establishing organizational, operational, health and safety, and QA policies. Management shall ensure that the following requirements are met: (*1*) the appropriate methodologies are followed, as documented in the standard operating procedures (SOPs); (*2*) personnel understand clearly their duties and responsibilities; (*3*) each staff member has access to appropriate project documents; (*4*) any deviations from the project plan are communicated to project management; and (*5*) communication occurs between the field, laboratory, and project managements, as specified in the project plan. Management shall foster an attitude within the organization that emphasizes the importance of quality and supports implementation of the QAPjP.

7.2.2 *Quality Assurance Function*—The organization shall appoint an individual(s) to be responsible for monitoring field operations in order to ensure that the site facilities, equipment, personnel, procedures, practices, and documentation are in conformance with the organization's QAPP and any applicable QAPjP. The QA monitoring function should be entirely separate from and independent of personnel engaged in the work being monitored. The QA function shall be responsible for the QA review in accordance with 7.7.

7.2.3 *Personnel*—It is the responsibility of the organization to establish personnel qualifications and training requirements for all positions. Each member of the organization shall possess the education, training, technical knowledge, and experience, or a combination thereof, to enable that individual to perform his or her assigned functions. Personnel qualifications shall be documented in terms of education, experience, and training. Training shall be provided for all staff members, as necessary, so that they can perform their functions properly.

7.2.4 *Subcontractors*—The use of subcontractors shall not jeopardize data quality. The field organization is therefore responsible for ensuring that its subcontractors are in compliance with the requirements of this section as is appropriate to the specific task(s) they are performing.

7.3 *Field Logistics:*

7.3.1 *General*—Sampling site facilities shall be examined prior to the start of work in order to ensure that all required items are available. The actual sampling area shall be examined to ensure that trucks, drilling equipment, and personnel have access to the site. Security, health and safety, and protection of the environment shall be controlled at the site support areas and sampling site.

7.3.2 *Field Measurements*—Project planning documents shall both address the type of field measurements to be performed and plan for the appropriate area to perform the work. Planning documents shall address ventilation, protection from extreme weather and temperatures, access to stable power, and provisions for water and gases of required purity. Plans shall be made to identify and supply applicable safety equipment, as specified in the project health and safety plan.

7.3.3 *Sample Handling, Shipping, and Storage Area*—The determination of whether sample shipping is necessary shall be made during project planning. This need is established by evaluating the analyses required, holding times (see Practice D 4841), and location of the site and laboratory. Shipping or transporting of the samples to a laboratory shall be completed in a timely manner, ensuring that the laboratory is allowed sufficient time to perform its analysis within any required holding times.

7.3.3.1 Samples shall be packaged, labeled, and documented in an area that minimizes sample contamination and provides for safe storage. The level of custody and whether sample storage is required shall be outlined in the planning documents.

7.3.4 *Chemical Storage*—Safe storage areas for solvents, reagents, standards, and reference materials shall be adequate to preserve their identity, concentration, purity, and stability prior to use.

7.3.5 *Decontamination*—Decontamination of sampling equipment may be performed at the location at which sampling occurs, prior to transfer to the sampling site, or in designated areas near the sampling site. Project documentation shall specify where this work will be performed and how it will be accomplished. Water and solvents of appropriate purity shall be available if decontamination is to be conducted at the site. This method of accomplishing decontamination of materials, solvents, and water purity shall be specified in the planning documents or SOPs.

7.3.6 *Waste Storage Area*—Waste materials may be generated during both the sampling process and on site or in situ analysis. Planning documents and SOPs shall outline the method for storage and disposal of these waste materials. Adequate facilities shall be provided for the collection and storage of all wastes. These facilities shall be operated so as to minimize environmental contamination. Waste storage and disposal facilities shall comply with applicable federal, state, and local regulations.

7.4 *Equipment and Instrumentation:*

7.4.1 *Equipment and Instrumentation*—The equipment, instrumentation, and supplies required at the sampling site shall be appropriate to accomplish the activities planned. The equipment and instrumentation shall meet the requirements of pertinent specifications, methods, and SOPs. Before the field staff arrives at the site, a list of required items shall be prepared and checked to ensure availability at the site.

7.4.2 *Maintenance and Calibration of Equipment and Instrumentation*—An SOP or operation and maintenance manual shall set forth the methods, materials, and schedules to be used in the routine inspection, cleaning, maintenance testing, and calibration of the equipment and instrumentation used in performing geophysical, analytical, or in situ measurements. Procedures or manuals may outline typical problems for common malfunctions. Procedures shall designate a person(s) or organizations responsible for maintenance and calibration. Records of all inspections, maintenance, repairs, testing, and calibration shall be maintained.

7.5 *Standard Operating Procedures*—The organization shall have written SOPs for all procedures performed routinely that affect data quality. Guide D 5172 contains information for documenting standard operating procedures. SOPs shall be available for the following areas and shall contain the information described:

7.5.1 *Sample Management*—These SOPs describe the numbering and labeling systems, chain-of-custody procedures, and tracking of samples from collection to shipment or relinquishment to the laboratory. Sample management includes the specification of holding times, volume of sample required by the laboratory, preservatives, and shipping requirements.

7.5.2 *Reagent and Standard Preparation*—These SOPs describe the procedures used to prepare standards and reagents. Information should include the specific grades of materials used in reagent and standard preparation, appropriate glassware and containers for preparation and storage, labeling and record keeping for stocks and dilutions, and safety precautions to be taken.

7.5.3 *Decontamination*—These SOPs describe the procedures used to clean field equipment before and during the sample collection process. The SOPs should include the cleaning materials used, order of washing and rinsing with the cleaning materials, requirements for protecting or covering cleaned equipment, procedures for disposing of cleaning materials, and safety considerations.

7.5.4 *Sample Collection Procedures*—SOPs for sample collection procedures shall describe how the procedures are actually performed in the field and shall not be a simple reference to a standard sampling method, unless the procedure is performed exactly as described in the published sampling method. If possible, industry-recognized sample collection methods from source documents published by the U.S. Environmental Protection Agency, ASTM, U.S. Department of the Interior, National Water Well Association, American Petroleum Institute, or other recognized organizations should be used. The SOP for sample collection procedures should include the following information:

7.5.4.1 Applicability of the procedure.

7.5.4.2 Equipment and reagents required.

7.5.4.3 Detailed description of the procedures to be followed when collecting the samples (see Guide D 4448 and Practices D 3370 for sampling guidance and common practices).

7.5.4.4 Common problems encountered.

7.5.4.5 Precautions to be taken.

7.5.4.6 Health and safety considerations.

7.5.5 *Equipment Calibration and Maintenance*—These SOPs describe the procedures used to ensure that field equipment and instrumentation are in working order. The SOPs describe calibration and maintenance procedures and schedules, maintenance logs, service contracts or service arrangements for equipment, and spare parts available in-house. The calibration and maintenance of field equipment and instrumentation should be in accordance with the manufacturer's specifications and shall be documented.

7.5.6 *Field Measurements*—These SOPs describe all methods used in the field to determine a chemical or physical parameter.

7.5.7 *Corrective Action*—These SOPs describe procedures used to identify and correct deficiencies in the sample collection process. These should include specific steps to take when correcting deficiencies such as performing additional decontamination of equipment, resampling, or additional training or field personnel in methods procedures. The SOP shall specify that each corrective action must be documented with a description of the deficiency, corrective action taken, and person(s) responsible for implementing the corrective action.

7.5.8 *Data Reduction and Validation*—These SOPs describe procedures used to compute the results from field measurements and to review and validate these data. They should include all formulas used to calculate the results and procedures used to verify independently that the field measurement results are correct.

7.5.9 *Reporting*—These SOPs describe the process for reporting the results of field activities (see Practices E 29 and D 4210 for additional information).

7.5.10 *Records Management*—These SOPs describe the procedures for generating, controlling, and archiving field records. The SOPs should describe the responsibilities for record generation and control and the policies for record retention, including type, time, security, and retrieval and disposal authorities. Records should include project-specific and field operations records.

7.5.10.1 Project-specific records relate to field work performed for a group of samples. Project records may include correspondence, chain-of-custody, field notes, all reports issued as a result of the work, project planning documents, and procedural SOPs used.

7.5.10.2 Field operations records document overall field operations. These records may include equipment performance and maintenance logs, personnel files, general field SOPs, and corrective action reports.

7.5.11 *Waste Disposal*—These SOPs describe policies and procedures for the disposal of waste materials resulting from field operations (see Guide D 4447). The disposal of all wastes must conform to federal, state, and local regulations, including those associated with the Resource Conservation and Recovery Act, Superfund Act Reauthorization and Amendments, Department of Transportation, and Occupational Safety and Health Administration.

7.5.12 *Health and Safety*—These SOPs describe policies and procedures designed both to provide a safe and healthy working environment for field personnel and to comply with federal and state regulations.

7.6 *Field Quality Assurance and Quality Control Requirements:*

7.6.1 *Quality Assurance Program Plan*—The field organization shall have a written QAPP that describes the organization's QA policy. The plan shall specify the responsibilities of the field management and field staff and the QA function in the areas of QA and QC, and it shall also describe the QC procedures followed by the organization (see QAMS-004/80 for an example).

7.6.2 *Quality Assurance Project Plan*—Some projects, particularly those that are large or complex, require a QAPjP. The QAPjP details the QA and QC goals and protocol for a specific data collection activity to ensure that the data generated by sampling and analysis activities are of quality commensurate with their intended use. The QAPjP elements should include a discussion of the quality objectives of the project, identification of those involved in the data collection and their responsibilities and authorities, enumeration of the QC procedures to be followed, and reference to the specific SOPs that will be followed for all aspects of the project. Elements may be added or removed, as required, by the project or the end-user of the data (see QAMS-005/80 for an example).

7.6.3 *Control Samples*—Control samples are QC samples that are introduced into a process to monitor the performance of the system. Control samples, which may include blanks, duplicates, spikes, analytical standards, and reference materials, can be used in different phases of the overall process, beginning with sampling and continuing through transportation, storage, and analysis. The types of control samples used, and the frequency of usage, are dependent on the DQOs of the data collection effort and must be specified for each project.

7.6.4 *Procedures for Establishing Acceptance Criteria*—Procedures shall be in place for establishing acceptance criteria for field activities, as required, in the project planning documents. Acceptance criteria may be qualitative or quantitative. Field events or data that fall outside of the established acceptance criteria may indicate a problem with the sampling process that must be investigated.

7.6.5 *Deviations*—Any activity not performed in accordance with the SOPs or project planning documents is considered a deviation from the plan. Deviations from the plan may or may not effect data quality. All deviations from the plan shall be documented as to the extent of or the reason for the deviation, or both.

7.6.6 *Corrective Action*—Errors, deficiencies, deviations, or field events or data that fall outside the established acceptance criteria require investigation. Corrective action may be necessary to resolve the problem and restore proper functioning to the system in some instances. Investigation of the problem and any subsequent corrective action taken shall be documented.

7.6.7 *Data Handling Procedures:*

7.6.7.1 *Data Reduction*—All field measurement data are reduced in accordance with protocol described in the appropriate SOP. Computer programs used for data reduction shall be validated before use and verified on a regular basis. All information used in the calculations shall be recorded to enable reconstruction of the final result at a later date.

7.6.7.2 *Data Review*—All data are reviewed in accordance with SOPs to ensure that the calculations are correct and to detect transcription errors. Spot checks are performed on computer calculations to verify program validity.

7.6.7.3 *Data Reporting*—Data are reported in accordance with the requirements of the end-user.

7.7 *Quality Assurance Review:*

7.7.1 *General*—The QA review consists of internal and external assessments to ensure that both QA and QC procedures are in use and field staff conform to these procedures. Planning documents shall specify the requirements for internal, external, and on-site assessment. These documents shall specify the frequency and documentation of these assessments.

7.7.1.1 *Internal Assessment*—Personnel responsible for performing field activities are responsible for continually monitoring individual compliance with the QA and QC programs and planning documents. A QA officer or an appropriate management designee shall review the field results and findings for compliance to the QA and QC programs and planning documents. The results of this internal assessment should be reported to management with requirements for a plan to correct the observed deficiencies.

7.7.1.2 *External Assessment*—The field staff may be reviewed by personnel external to the organization. The results of the external assessment should be submitted to management with requirements for a plan to correct the observed deficiencies.

7.7.1.3 *On-Site Evaluation*—On-site evaluations may be conducted as part of both internal and external assessments. On-site evaluations may include, but are not limited to, a complete review of the facilities, staff, training, instrumentation, SOPs, methods, field analysis, sample collection, QA and QC policies, and procedures related to the generation of environmental data. Records of each evaluation shall be maintained in accordance with regulation or the organization's policy. These records should include the date of the evaluation, area or site, areas reviewed, individual performing the evaluation, findings and problems, actions recommended and taken to resolve the problems, and scheduled date for re-inspection. Any problems identified that are likely to affect data integrity shall be brought to the attention of management immediately.

7.7.2 *Evaluation of Field Records*—The review of field records shall be conducted by one or more individuals knowledgeable in the field activities, evaluating the following subjects at a minimum:

7.7.2.1 *Completeness of Field Records*—This review ensures that all requirements for field activities in the planning documents have been fulfilled, that complete records exist for each field activity, and that the procedures specified in the planning documents have been implemented. Emphasis on documentation will help ensure sample integrity and that sufficient technical information is available to recreate each field event. The results of this completeness check shall be documented, and environmental data affected by incomplete records shall be identified.

7.7.2.2 *Identification of Valid Samples*—This review involves interpretation and evaluation of the field records to detect problems affecting the representativeness of environmental samples. Examples of items that could indicate invalid samples include improper well development, improperly screened wells, instability of pH or conductivity, and collection of volatiles near combustion engines. The field records shall be evaluated against planning documents and SOPs. The reviewer shall document the sample validity and identify the environmental data associated with poor or incorrect field work.

7.7.2.3 *Correlation of Field Test Data*—The results of field measurements obtained by more than one method shall be compared. For example, surface geophysics may be surveyed using both ground-penetrating radar and a resistivity survey.

7.7.2.4 *Identification of Anomalous Field Test Data*—Anomalous field test data should be identified. For example, a water temperature for one well that is 5° higher than any other well temperature in the same aquifer should be noted. The impact of anomalous field measurement results on the associated environmental data shall be evaluated.

7.7.2.5 *Validation of Field Analysis*—All data from field analysis that are generated in situ or from mobile laboratory shall be validated by one or more individuals knowledgeable in the analysis. The results of the validation shall be reported. The report shall discuss whether the QC checks meet the acceptance criteria and whether corrective actions were taken for any analysis performed when the acceptance criteria were not met.

7.7.3 *Quality Assurance Reports to Management*—The QA program shall provide for the periodic reporting of pertinent QA and QC information to management to allow assessment of the overall effectiveness of the QA program.

7.7.3.1 *Report on Measurement Quality Indications*—This report shall include the assessment of QC data (such as that generated in accordance with 7.6.3) gathered over the period, frequency of repeating work due to unacceptable performance, and corrective action taken.

7.7.3.2 *Report on Quality Assurance Assessments*—This report shall be submitted immediately following any internal or external on-site evaluations or upon receipt of the results of any performance evaluation studies. The report shall include the results of the assessment and the plan for correcting identified deficiencies.

7.7.3.3 *Report on Key Quality Assurance Activities During the Period*—A report shall be delivered to management summarizing key QA activities during the period. The report shall stress measures that are being taken to improve data quality and shall include a summary of the significant quality problems observed and corrective actions taken. The report shall also include a summary of involvements in resolution of quality issues with clients or agencies, QA organizational changes, and notice of the distribution of any revised documents controlled by the QA function.

7.8 *Field Records:*

7.8.1 Records provide direct evidence and support for the necessary technical interpretations, judgements, and discussions concerning project activities. These records, particularly those that are anticipated for use as evidential data, must support current or ongoing technical studies and activities directly and must provide the historical evidence necessary for later reviews and analyses. Records shall be legible, identifiable, and retrievable and protected from damage, deterioration, or loss. Field records generally consist of bound field notebooks with prenumbered pages, sample collection forms, personnel qualifications and training forms, sample location maps, equipment maintenance and calibration forms, chain-of-custody forms, sample analysis request forms, and field change request forms. All records shall be completed with black, waterproof ink.

7.8.2 Procedures for reviewing, approving, and revising field records must be defined clearly, with the lines of authority included. At a minimum, all documentation errors shall be corrected by drawing a single line through the error and initialing by the responsible individual, along with the date of change. The correction is written adjacent to the

error. Deviations from field SOPs shall be documented.

7.8.3 *Personnel Training and Qualification Records*—It is the responsibility of the organization to establish personnel qualifications and training requirements. Each staff member shall have the education, training, technical knowledge, and experience, or a combination thereof, to enable that individual to perform his or her assigned functions. Personnel qualifications shall be documented in terms of education, experience, and training. Training shall be provided for all staff members so that they can perform their functions properly.

7.8.4 SOPs shall be available to those performing the task outlined, and revisions to field SOPs shall be written and distributed to all affected individuals to ensure the implementation of changes. The areas covered by SOPs are given in 7.5.

7.8.5 *Quality Assurance Plans*—The QAPP and all applicable QAPjPs shall be on file.

7.8.6 *Equipment Maintenance*—Maintenance procedures shall be defined clearly and written for each measurement system and required support equipment. When maintenance is necessary, it shall be documented in either standard forms or in logbooks. A history of the maintenance record of each system serves as an indication of the adequacy of maintenance schedules and parts inventory.

7.8.7 *Calibration and Traceability of Standards and Reagents*—Calibration is a reproducible reference base to which all sample measurements can be correlated. A sound calibration program shall include provisions for documentation of the frequency, conditions, standards, and records reflecting the calibration history of a measurement system. The accuracy of calibration standards is an important point to consider because all data will be in reference to the standards used. A program for verifying and documenting the accuracy of all working standards against primary grade standards shall be followed routinely.

7.8.8 *Sample Collection and Tracking Records*—To ensure maximum utility of the sampling effort and resulting data, documentation of the sampling protocol, as performed in the field, is essential. Sample collection records shall contain the persons conducting the activity, sample number, sample location, equipment used, climatic conditions, documentation of adherence to protocol, and unusual observations as a minimum. The actual sample collection record is usually one of the following: a bound field notebook with prenumbered pages, a preprinted form, or digitized information on a computer tape or disc.

7.8.8.1 Sample tracking records (chain of custody) involving the possession of samples from the time at which they are obtained until they are relinquished shall be documented with the following minimum information: (*1*) project name; (*2*) signatures of the samplers; (*3*) sample number, date and time of collection, and grab or composite sample designation; (*4*) signatures of the individuals involved in sample transfer; and (*5*) the air bill or other shipping number, if applicable. Additional chain of custody information may be found in Practice D 4840.

7.8.9 *Maps and Drawings*—Project planning documents and reports often contain maps. The maps are used to document the location of sample collection points and monitoring wells, and as a means of presenting environmental data. Information used to prepare maps and drawings is normally obtained through field surveys, property surveys, surveys of monitoring wells, serial photography, or photogrammetric mapping. The final, approved maps shall have a revision number and date and shall be subject to the same controls as other project records.

7.8.10 *Results from Control Samples*—Documentation for the collection of QC samples, such as field, trip, and equipment rinsate blanks, duplicate samples, spikes, and reference materials, shall be maintained.

7.8.11 *Correspondence*—Project correspondence can provide evidence supporting technical interpretations. Correspondence pertinent to the project shall be kept and placed in the project files.

7.8.12 *Deviations*—Field changes and deviations from the planning documents shall be reviewed and approved by either the authorized personnel who performed the original technical review or their designees. All deviations from the procedural and planning documents shall be recorded in the site log.

7.8.13 *Final Report*—The final report shall summarize the field activities, data, results of deviations from the planning documents, and interpretation of the data. The planning documents shall outline the items to be included in the report, which may include any special formats required, QC reporting requirements, conclusions, and recommendations.

8. Data Quality Assessment

8.1 The assessment of environmental data occurs in two phases. Field records and analytical data are reviewed during the first phase to identify whether the data are accurate and defensible. The data are interpreted in the second phase with respect to meeting DQOs or the project plan.

8.2 Technical reports of environmental data collection efforts should summarize the information contained in the field records and the results of the laboratory data review, in accordance with 7.7.2. This information should be used to identify clearly the data that are not representative of environmental conditions or that have been generated using poor field or laboratory practices.

8.3 The combined field and laboratory data are then subject to a final assessment to determine whether the DQOs or project plan have been met. Practices D 2777, E 29, and E 178 will be of help in the final assessment process.

9. Standard Practice for Analytical Operations

9.1 Analytical operations are an integral part of the water sampling and field analysis process. It is not the intent of this guide to cover all aspects of analytical operations. The following documents will help both the field and analytical personnel to determine that which is needed to meet the DQOs: Practices D 596 and D 5283 (section titled Standard Practices for Environmental Laboratory Operations) and Guides D 3856 and D 5172.

10. Documentation Storage

10.1 *Documentation Archive*—Procedures shall be established to ensure that the documents required to recreate the sampling, analysis, and reporting of information are stored. These documents may include, but are not limited to, planning documents, SOPs, logbooks, field data records, sample tags and labels, chain-of-custody records, photographs, and any other information noted in 7.8.

10.2 *Storage Time*—The length of storage time for field records shall comply with regulatory requirements, organizational policy, or project requirements, whichever is or are more stringent.

10.3 *Filing System*—The control of records is essential for providing evidence of technical adequacy and quality for all project activities. These records shall be identified, retrievable, and organized to prevent loss.

10.4 *Personnel Authorized to Enter Archive*—Access to project files shall be controlled to restrict unauthorized personnel from having free and open access. An authorized access list shall be prepared for the project files and shall name the personnel who have unrestricted access to the files.

11. Keywords

11.1 field; quality assurance; quality control; sampling

Standard Guide for
Planning and Implementing a Water Monitoring Program[1]

This standard is issued under the fixed designation D 5851; the number immediately following the designation indicates the year of original adoption or, in the case of revision, the year of last revision. A number in parentheses indicates the year of last reapproval. A superscript epsilon (ε) indicates an editorial change since the last revision or reapproval.

INTRODUCTION

Water resource monitoring has taken place in many forms for scores of years. This monitoring has been sponsored and performed by a variety of federal, state, and local public agencies; and perhaps by an even wider variety of private, quasi-public and industrial entities. Historically, much of the early data dealt with quantities of flow, and drinking water quality was judged by the standards of the period.

During the past several years the problems related to point and nonpoint sources of pollution of water resources have become increasingly apparent. Technology has improved dramatically, as the need for monitoring data has improved. There is a necessity for information on marine beaches and estuarine areas, fresh water swamps, ground water, wetlands, streams, and sediment deposits, and to better understand the entire hydrologic cycle.

The need for more and varied water quality information has expanded as rapidly as our technological ability to generate the information. Further, it has become increasingly difficult and sometimes impossible to understand and resolve conflicts among the different data sets available. Much of the data have been collected at different times, in different geographic areas, and for different purposes. The data have been collected by persons with varied training, using different methods, and with vastly different analytical capabilities. As a consequence, we presently are at the stage where we may know more about a given situation than we understand and workers in the field who receive the data are unable to integrate the data available into a useful solution. The need for standardization of monitoring programs is evident. Standardization does not herein mean everyone doing everything exactly the same way. It does mean the use of methods and procedures, where applicable, that follow recognized and documented protocols as well as the accurate recording and storage of the data in accessible formats.

Realizing the difficulties in water monitoring, the Office of Management and Budget (OMB) of the federal government charged the Water Information Program (WICP), a program of the U.S. Geological Survey's Office of Water Data Coordination, with studying water quality monitoring in the United States and recommending improvements. The Intergovernmental Task Force on Monitoring Water Quality (ITFM), a federal, state, and tribal partnership, was established under the WICP's Interagency Advisory Committee on Water Data to carry out this study. The results of three years of work by about 200 contributors have been captured in a series of three annual reports (**1, 2, 3**).[2]

The following summarizes the conclusions from those reports:

(*1*) Monitoring programs shall keep pace with changing water-management programs.

(*2*) A collaborative strategy is needed to link the many separate monitoring programs.

(*3*) A genuine appreciation of the need for cooperation currently exists among monitoring agencies.

(*4*) Recent advances in technology provide opportunities for interaction and cooperation that previously were impossible.

Based upon those conclusions, the following recommendations were made:

(*1*) Implement an integrated, voluntary, nationwide strategy to improve water quality monitoring.

(*2*) Charter a permanent national body to guide the implementation of ITFM recommendations.

---

[1] This guide is under the jurisdiction of ASTM Committee D-19 on Water and is the direct responsibility of Subcommittee D19.02 on General Specifications, Technical Resources, and Statistical Methods.

Current edition approved Oct. 10, 1995. Published December 1995.

[2] The boldface numbers in parentheses refer to a list of references at the end of this guide.

(3) Develop a framework for monitoring water quality that defines the components of a monitoring program.

(4) Develop criteria with which to select parameters that measure progress in achieving water quality goals.

(5) Recommend indicators to measure whether water quality uses designated by the state are being met.

(6) Charter a Methods and Data Comparability Council to foster the development and use of performance-based methods of collection and analysis.

(7) Use the ecoregions concept, reference conditions, and index calibration.

1. Scope

1.1 *Purpose*—This guide is generic in its application to surface or ground water, rivers, lakes, or estuaries (quantity and quality). It proposes a series of options that offer direction without recommending a definite course of action and discusses the major elements that are common to all purposes of water monitoring.

1.2 The elements described are applicable whether the monitoring is only for one location or integrates multiple measurement sites for the purpose of assessing a whole watershed, estuary, or aquifer system.

1.3 This guide is intended to outline for planners and administrators the components, process, and procedures which should be considered when proposing, planning, or implementing a monitoring program. The guide is not a substitute for obtaining specific technical advice. The reader is not assumed to be a technical practitioner in the water field; however, practitioners will find it a good summary of practice and a handy checklist. Other standard guides have or will be prepared that address the necessary detail.

1.4 *Monitoring Components*—A water monitoring program is composed of a set of activities, practices, and procedures designed to collect reliable information of known accuracy and precision concerning a particular water resource in order to achieve a specific goal or purpose. The purposes may range in scope from tracking status and trends on a regional or national basis to gathering data to determine the effects of a specific management practice or pollution incident such as a spill. This guide suggests and discusses the following process and components:

1.4.1 Establishment of program goals and objectives and recording of decisions in a written plan (see 6.1),

1.4.2 Developing background data and a conceptual model (see 6.1.12),

1.4.3 Establishment of data (quality, quantity, type) objectives (see 6.2),

1.4.4 Design of field measurement and sampling strategies and specification of laboratory analyses and data acceptance criteria (see 6.3),

1.4.5 Data storage and transfer (see 6.6),

1.4.6 Implementation of sampling and analysis strategies (see 6.4),

1.4.7 Data quality assessment (see 6.5),

1.4.8 Assessment of data (see 6.7),

1.4.9 Program evaluation (see 6.8), and

1.4.10 Reporting (see 6.9).

See also Fig. X1 in Appendix X1 and the condensed list of headings in Appendix X2.

1.5 *Monitoring Purposes*—Establishing goals defines the purpose for monitoring. Each purpose has some monitoring design needs specific to itself. There are six major purposes for water monitoring. They are as follows:

1.5.1 *Determining the Status and Trends of Water Conditions*—This can require long term, regular monitoring to determine how parameters change over time.

1.5.2 *Detecting Existing and Emerging Problems*—Determining if, how, or where a substance may move through an aquatic system, or if water quantities are changing.

1.5.3 *Developing and Implementing Management and Regulatory Programs*—Includes baseline and reconnaissance monitoring to characterize existing conditions such as to identify critical areas or hot spots; implementation monitoring to assess whether activities were carried out as planned; and compliance monitoring to determine if specific water quality or water use criteria were met.

1.5.4 *Responding to an Emergency*—Performed to provide information in the near term.

1.5.5 *Evaluating the Effectiveness of Water Monitoring Programs*—Is the monitoring able to achieve the stated goals? Also, monitoring to check on monitoring.

1.5.6 Supporting research objectives or validating of simulation models.

1.6 This guide is applicable to these purposes and provides guidance on some of the specific needs of each. After goals and objectives have been established, a specialist can define the type, frequency, and duration of sampling and measurements. The specialist also will be able to forecast the data analysis needed to meet the objectives.

1.7 There are related standards currently available or under development and several documents that prescribe protocols for water monitoring (4–9). See also Section 2.

1.8 This guide suggests that water monitoring programs use standardized documented protocols for all aspects of the program. Where they are not available or appropriate, the methods used should be documented.

1.9 *This standard does not purport to address all of the safety concerns, if any, associated with its use. It is the responsibility of the user of this standard to establish appropriate safety and health practices and determine the applicability of regulatory limitations prior to use.*

2. Referenced Documents

2.1 *ASTM Standards:*

D 1129 Terminology Relating to Water[3]

D 4840 Practice for Sampling Chain of Custody Procedures[3]

---

[3] *Annual Book of ASTM Standards*, Vol 11.01.

D 5847 Practice for Writing Quality Control Specifications for Standard Test Methods for Water Analysis[4]

2.2 *Other Documents:*

Compilation of Scopes of ASTM Standards Relating to Environmental Monitoring, 1994, ASTM, Philadelphia, PA. PCN: 13-600003-16 (700 standards)[5]

ASTM Standards on Ground Water and Vadose Zone Investigations. PCN: 03-418094-38[5]

2.3 *EPA Documents:*

U.S. EPA 813/B-92-002 Definitions for the Minimum Set of Data Elements for Ground Water Quality[6]

U.S. EPA 910/9-91/001 Monitoring Guidelines to Evaluate Effects of Forestry Activities on Streams in the Pacific Northwest and Alaska[6]

3. Terminology

3.1 For definitions of terms used in this guide, refer to Terminology D 1129.

3.2 *Descriptions of Terms Specific to This Standard:*

3.2.1 *analyze*—to determine the relationship of parts or the value of a particular parameter.

3.2.2 *assess*—to determine importance of data.

3.2.3 *evaluate*—to determine significance or worth.

3.2.4 *measurement*—determining the values of a characteristic within a sample or in situ.

3.2.5 *metadata*—ancillary data that describe the natural conditions under which an environmental data value is measured, the purpose for collection, the methods and standards employed, and the organization responsible.

3.2.6 *sampling*—the removal of a portion of the water which may or may not be representative of the whole. This is not monitoring.

3.2.7 *water monitoring*—water monitoring consists of systematic activities conducted to characterize the quantity or quality, or both, of water.

4. Significance and Use

4.1 The user of this guide is not assumed to be a technical practitioner in the water field. This guide is an assembly of the components common to all aspects of water monitoring and fulfills a need in the development of a common framework for a better coordinated and more unified approach to monitoring water.

4.2 *Limitations*—This guide does not establish a standard procedure to follow in all cases and it does not cover the details necessary to meet a particular monitoring objective.

5. A Primer on Water Monitoring Programs

5.1 *The Problem*—Why is water monitoring difficult?

5.1.1 The movement of water through the hydrologic cycle dwarfs other material cycles operating on the earth's surface, such as the carbon and oxygen cycles. Water's chemical and physical properties enable it to dissolve many substances and physically remove and suspend others. Consequently, as water encounters various substances in the atmosphere, on land surfaces, and below ground, the water's chemical composition changes, and the composition of materials suspended in the water changes. Physical and chemical processes further change its composition as water moves through the hydrologic cycle.

5.1.2 Human activities using land and water have greatly altered the kind and amount of substances that water encounters as it moves through the hydrological cycle. Often, some substances, including those biological communities living within water, are present at concentrations that impair various water uses. These substances are regarded as pollutants. Much of our effort to manage water resources is directed at reducing the addition of pollutants in water. Other management efforts are directed toward altering local pathways of water flow and maintaining or enhancing aquatic and marine habitats.

5.1.3 Across the globe or across a county there are large spatial and temporal variations in water flows and volumes, in the natural features, which impact water resources, and in the nature and extent of human land and water uses. Consequently, there can be large spatial and temporal variations in the composition of water. The problem that must be addressed in water monitoring is how to spatially and temporally characterize the composition of water and the source of this composition with sufficient accuracy and precision to support local and regional water uses and management efforts. Monitoring water as it flows through the hydrologic cycle is not easy.

5.1.4 Reading through the following list of procedures and considerations will provide the administrator or planner with insight into the details of needed expertise, complexity, and work tasks in the design, implementation, and evaluation of a monitoring project.

6. Procedure

6.1 *Establish Program Goals and Objectives:*

6.1.1 Define goals and objectives using a multidisciplinary team. This requires a variety of professionals with special insights in order to prepare a sensible plan.

6.1.1.1 Review existing data.

6.1.2 Prepare a plan of work from the goals, objectives, and decisions. This will be an iterating process as progress is made developing the components. The plan can use a pilot approach or phased-in approach.

6.1.2.1 Coordinate activities with other relevant agencies, groups, and persons.

6.1.3 Develop a project schedule and budget.

6.1.3.1 Establish budgetary and personnel requirements.

6.1.4 Set timelines.

6.1.5 Set interim goals, checkpoints, and review periods.

6.1.6 Identify adaptive management parameters in accordance with the project's objectives; these are project specific parameters, such as ground water flow direction and concentration, that are selected to be observed on a regular basis in order to determine the need for change of monitoring procedures.

6.1.7 Establish feedback loops related to review inputs. As data are collected they shall be reviewed in light of quality standards and in meeting program objectives.

6.1.8 Schedule flexibility for program adjustment.

6.1.9 Determine program costs and sources of funding.

[4] *Annual Book of ASTM Standards*, Vol 11.02.

[5] Available from ASTM Headquarters, 100 Barr Harbor Drive, West Conshohocken, PA 19428-2959.

[6] Available from Superintendent of Documents, Government Printing Office, Washington, DC 20402.

6.1.10 Identify who will need or use the data and who will benefit from the project.

6.1.11 Identify and describe the existing environmental setting including its surface and ground water hydrology, physiography, climate, biology, and ancillary information such as population, land use, and water use.

6.1.12 Develop a conceptual model of the project area that relates the known water data and the surroundings that influence water conditions. The model will aid in predicting influences and selecting sampling sites.

6.1.13 Collaborate with others who can contribute information and support.

6.2 *Establish Data Objectives:*

6.2.1 The what, how, how many, and how good of measurements depends on many factors, especially why the monitoring is being done. The needs of the end users have to be clearly identified. Data shall be collected and measured in accordance with established norms and standards. Measurements of physical parameters and environmental indicators are made to determine the following:

6.2.1.1 Concentrations of both natural and synthetic constituents dissolved or suspended in water,

6.2.1.2 Physical characteristics of water (temperature, turbidity, color, density, and conductivity),

6.2.1.3 The volumes of water present in various compartments of the hydrologic cycle,

6.2.1.4 The flow rates of water between various compartments of the hydrological cycle,

6.2.1.5 The loading of dissolved and suspended constituents between compartments of the hydrologic cycle,

6.2.1.6 The rates of chemical and physical processes,

6.2.1.7 The status of biological communities living within or adjacent to surface and ground waters,

6.2.1.8 The quality of aquatic habitats,

6.2.1.9 The factors that influence any of the above, and

6.2.1.10 The suitability of water for a particular use.

6.2.2 Define requirements for data analyses. For example, what is the supporting information, such as land use, that is needed to analyze the data?

6.2.3 Define interaction of various professional skills, for example, field worker taking samples, chemist, hydrologist, data manager, data analyst, and the person who interprets the data, to ensure that all work to be done and who is to do it are identified.

6.2.4 Based upon the stated program objectives, determine the scope of the monitoring program by doing the following:

6.2.4.1 Determining the areal extent needed to meet objectives,

6.2.4.2 Determining the analysis or parameters needed to meet objectives,

6.2.4.3 Determining what is known,

6.2.4.4 Investigating related prior work,

6.2.4.5 Correlating objectives and scope with objectives attained by prior work,

6.2.4.6 Evaluating existing information to depict the known or suspected surface and ground-water quality conditions, problems, or information gaps,

6.2.4.7 Providing a current conceptual understanding,

6.2.4.8 Identifying management concerns and alternatives,

6.2.4.9 Analyzing prior data for integration with new data,

6.2.4.10 Determining whether the work can be used,

6.2.4.11 Determining impact of locations of monitoring sites on data from prior work and upon proposed work,

6.2.4.12 Determining impact of access to prior and future sites upon prior data and data to be collected, and

6.2.4.13 Evaluating impact of past and present sampling methodology upon past and proposed data, including equipment variations, for example, manual, automatic, remote, and experience of personnel, with respect to environmental requirements and data needs.

6.2.5 Identify null hypothesis option, that is, what happens if monitoring is not performed.

6.2.6 Establish reference conditions for environmental indicators that can be monitored to provide a baseline water-quality assessment.

6.2.7 Define data management needs.

6.2.8 Evaluate monitoring program, that is, can goals be achieved?

6.3 *Design of Field Measurements*—All measurements should conform to standard methods, documented protocols, or at least documented to provide a clear description of the methods used. The use of nonstandard methods is appropriate where the use of standard methods would not be suitable for the successful implementation of the work.

6.3.1 Evaluate spatial aspects of monitoring activities, for example, where to sample or measure. Will locations be acceptable and accessible? Are the samples representative? What will be the exact location for the measurement point in reference to location in a stream, depth in a well, location in a lake, or depth in a bay in order to get representative samples within the scope of the project.

6.3.2 Evaluate temporal aspects of monitoring activities, for example, how frequently and how long to sample. Factors to be considered include tidal effects, climatic effects, seasonal effects, scale effects, daily effects, and annual effects.

6.3.3 Select monitoring sites. Consider access, long term use, physical hardships, and special equipment.

6.3.4 Select environmental indicators and data parameters. Evaluate the environmental indicators and habitat and related chemical, physical, biological, and ancillary data parameters to be monitored. Select monitoring mechanisms and methods.

6.3.5 Identify specific factors that impact sampling sites. Consider the following:

6.3.5.1 Relationship between site monitored and point sources,

6.3.5.2 Relationship between site monitored and non-point sources,

6.3.5.3 Relationship between site-monitored and environmental variations (influencing conditions), for example, the edge of a swamp, thalweg of river, intertidal position, depth of well screen, thickness of strata, riffle, rapid, pool, aquatic zone in a lake, or depth in ocean, that is, any and all significant variations,

6.3.6 Define probable water level, stage and frequency variations, and relationship to monitoring methodology,

6.3.7 Select sampling approach to achieve objectives. Consider the following:

(*1*) Fixed station,

(2) Synoptic,
(3) Intensive,
(4) Event sampling,
(5) Sample type (grab/composite point/spatially integrated/continuous),
(6) Frequency and duration, and
(7) Scale.

6.3.8 Define monitoring and sampling program; develop sampling plans and identify standardized protocols and methods (performance based if possible); and document data to enable comparison with other monitoring programs. Considerations are as follows:

6.3.8.1 Published consensus standards,
6.3.8.2 Methodology,
6.3.8.3 Equipment,
6.3.8.4 Procedures,
6.3.8.5 Frequency,
6.3.8.6 Ancillary events (temperature, rain, relationship to stage, releases, current),
6.3.8.7 Seasonal and other temporal variations,
6.3.8.8 Constituents,
6.3.8.9 Environmental Indicators (chemical, physical, biological, surrogate indicators),
6.3.8.10 In situ analysis,
6.3.8.11 Sample transport, and
6.3.8.12 Qualifications of personnel.

6.3.9 Define analytical methodology.

6.3.9.1 Define data-quality objectives, including the level of confidence needed to meet the management objectives.

6.3.9.2 Determine basis for monitoring design that will allow successful interpretation of the data at a resolution that meets project purposes.

6.3.9.3 Identify field and laboratory protocols, or performance-based methods including sample collection, preservation, and transportation and including detection level accuracy, precision, and turnaround time. (See *Compilation of Scopes of ASTM Standards Relating to Environmental Monitoring* and *ASTM Standards on Ground Water and Vadose Zone Investigations.*)

6.3.10 Establish a Quality Assurance/Quality Control (QA/QC) Program (Quality Assurance Program Plan (QAPP)).

6.3.10.1 Identify and designate requirements for QA/QC procedures for field sampling methodology. Evaluate the following:

(1) Replicability of field measurements,
(2) Representativeness of field measurements,
(3) Natural variation of data,
(4) Limitations of data collected as a consequence of unavoidable variations, and
(5) Impact of field methods upon project objectives, that is, if field sampling has a probable variability of two orders of magnitude, the project objectives cannot be reported with less.

6.3.10.2 Identify and designate requirements for QA/QC for laboratory procedures.

6.3.10.3 Develop a Quality Assurance Project Plan documenting accuracy and precision, representativeness of the data set, and comparability of data to prior data or data collected by others, or both.

6.3.10.4 Define laboratory analytical requirements, qual-

ity assurance protocols, chain of custody procedures, and quality assurance procedures for all testing, for example, suspended solids, particulate gradation.

6.3.10.5 Select confidence limits that shall be obtained for the data to be meaningful.

6.3.11 Define qualifications of personnel.

6.3.11.1 Prepare safety plan for project.

6.3.11.2 Define skill levels that are required for various portions of the work, including training requirements, and certification and registration.

6.3.11.3 Define training program necessary to obtain needed skill levels.

6.3.11.4 Define test for personnel as a function of the field QA/QC. Can personnel make reliable measurements for the specified precision?

6.3.12 Evaluate field measurement program.

6.4 *Implementation of Sampling and Analysis Strategies:*

6.4.1 Collect data according to monitoring design and protocols.

6.4.2 Review data collection activities relative to protocols and monitoring design.

6.4.3 Follow QAPP plan and document deviations from plan.

6.4.4 Regularly review data for completeness and for accuracy and precision from QC data.

6.4.5 Regularly review data relative to changing monitoring design to accomplish objectives.

6.5 *Data Quality Assessment*—Base data QA on results of QAPP.

6.5.1 Review custody records.
6.5.2 Compare laboratory spikes.
6.5.3 Consider collection conditions.
6.5.4 Meet statistical protocol.
6.5.5 Meet confidence criteria.
6.5.6 Determine statistical reliability of data using QA information.
6.5.7 Evaluate data, that is, does it meet goals?

6.6 *Data Management:*

6.6.1 Identify and use standard protocols, if appropriate, for reporting and storing of data (see EPA 813).
6.6.2 Identify storage methods for ancillary data.
6.6.3 Define metadata storage needs. This is information that can be used to support the measured data (for example, date, time, location, collection methodology, QA data).
6.6.4 Define georeferencing standards.
6.6.5 Describe the data management protocol.
6.6.6 Transmit data to other users.
6.6.7 Archive data.
6.6.8 Secure data.

6.7 *Data Interpretation and Assessment (Evaluate, Analyze, and Assess Data):*

6.7.1 Evaluate data to determine if objectives can be achieved.
6.7.1.1 Test and quantify hypotheses.
6.7.2 Analyze data to determine statistical significance.
6.7.2.1 Determine temporal variability and spatial distribution.
6.7.2.2 Identify statistical and deterministic significance of data.

NOTE—Nonparametric and other statistical packages fit here.

6.7.2.3 Quantify source, cause, transport, and fate of constituents.

6.7.3 Assess data to measure impact of results.

6.7.3.1 Interpret data to meet objectives.

6.7.3.2 Show relationship between ancillary and water quality data.

6.7.4 Coordinate interpretation with collaborators and customers.

6.7.4.1 Review interpretation with collaborators and customers and provide feedback.

6.7.4.2 Address management alternatives.

6.7.5 Use deterministic models for testing management scenarios.

6.8 *Program Evaluation:*

6.8.1 Identify problems.

6.8.1.1 Identify problems associated with collecting and analyzing the data and with storing and disseminating the data to intended users.

6.8.1.2 Identify problems associated with interpreting the data and with reporting the information to water resource managers and the public.

6.8.2 Evaluate costs.

6.8.2.1 Evaluate costs as compared to program objectives to determine need to modify program.

6.8.2.2 Make recommendations for changes in program design, if needed.

6.8.3 Evaluate water monitoring program.

6.8.3.1 Determine if objectives were met.

6.8.3.2 Identify new purposes or revised objectives for a second phase of sampling.

6.8.3.3 Identify changes in the monitoring design.

6.8.4 Determine if information satisfies the stated objectives.

6.8.4.1 Determine if land and water use activity laws, standards, or guidelines or any combination thereof, need to change.

6.8.5 Identify current and future needs not addressed in the current program.

6.9 *Communication and Report:*

6.9.1 Prepare information for identified audiences. Consider the following items:

6.9.1.1 Write the report with regard to stated objectives.

6.9.1.2 Write concise and timely reports.

6.9.1.3 Use graphical and visual techniques to assist reader and listener.

6.9.1.4 Describe current water related conditions, spatial distribution, temporal variability, source, cause, transport, fate, and effects to human, and wildlife as appropriate.

6.9.1.5 Write lay reports or executive summaries for nontechnical audiences.

6.9.1.6 Write reports subject to peer review.

6.9.1.7 Place large data sets and statistical tables in appendices or in separate data reports.

6.9.1.8 Distribute reports to identified users.

6.9.1.9 Make presentations to assist management and the public in understanding the significance of results.

6.9.1.10 Use electronic word processing formats of accepted and widespread use.

6.9.1.11 Make reports available in electronic format.

6.9.2 Prepare data for data users.

6.9.2.1 Make data readily available in standardized computer format.

6.9.2.2 Make data, reports, and presentations in a timely manner (real time data may require special arrangements).

6.9.3 Determine how, whether, and when (by schedule) data and its interpretation will be communicated, for example, through press releases, public meetings, agency meetings, conferences, agency reports, journal articles, etc.

6.9.4 Report the following information:

6.9.4.1 Data citation and explanation,

6.9.4.2 Limitations, variability for other problems with data specifically including limitations with field and laboratory methodology and procedures,

6.9.4.3 Specific discussion of the data and the conclusions indicated by the data, and the precise way these relate to the defined objectives of the project,

6.9.4.4 Determine and discuss how the data, the conclusions, and the project in its entirety relates to other work, and how better coordination and assimilation can be attained, and

6.9.4.5 A summarization of conclusions and recommendations.

6.10 *Cooperation and Coordination*—A water monitoring program seldom can be implemented without coordination of several individual entities. Planning, field work, transport, laboratory work, data management, and data analysis are seldom handled by one group. Formal procedures shall be written to ensure that all understand their role in the process. All levels should understand the program objectives.

7. Keywords

7.1 ground water; guide for administrators; metadata; monitoring components; monitoring goals and objectives; monitoring purposes; surface water; water monitoring; water quality; water quantity

APPENDIXES

(Nonmandatory Information)

X1. PROCEDURAL FLOW CHART

X1.1 A procedural flow chart is shown in Fig. X1.

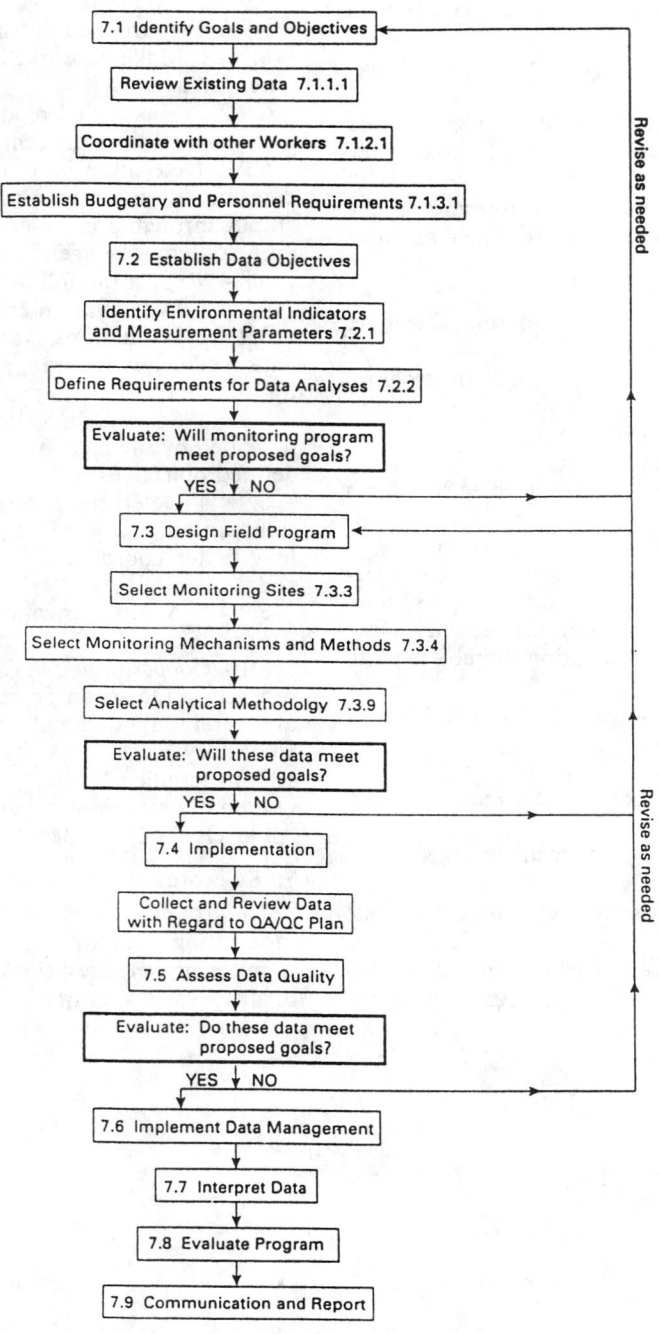

FIG. X1 Procedural Flow Chart

X2. CONDENSED LIST OF HEADINGS OF SECTION 7, PROCEDURE

X2.1 Use as a checklist for developing a water monitoring program.

X2.1.1 *Establish Program Goals and Objectives (6.1):*

X2.1.1.1 Use a multidisciplinary team to define goals and objectives (6.1.1).

X2.1.2 Document goals and objectives and decisions into a plan of work (6.1.2).

X2.1.3 Develop project schedule and budget (6.1.3).

X2.1.4 Set timelines (6.1.4).

X2.1.5 Set interim goals, checkpoints, and review periods (6.1.5).

X2.1.6 Identify adaptive management parameters in accordance with the projects objectives (6.1.6).

X2.1.7 Establish feedback loops related to review inputs (6.1.7).

X2.1.8 Schedule flexibility for program adjustment (6.1.8).

X2.1.9 Determine program costs and sources of funding (6.1.9).

X2.1.10 Identify who will need or use the data and who will benefit from the project (6.1.10).

X2.1.11 Identify and describe the existing environmental setting (6.1.11).

X2.1.12 Develop a conceptual model of the project area (6.1.12).

X2.1.13 Collaborate with others who can contribute information and support (6.1.13).

X2.2 *Establish Data Objectives*—The what, how, and how many of measurements depends on many factors, especially why the monitoring is being done. The needs of the end user have to be clearly identified (6.2).

X2.2.1 Determine measurements of physical parameters and environmental indicators (6.2.1).

X2.2.2 Define requirements for data analyses, for example, what is the supporting information? (6.2.2).

X2.2.3 Define interaction of various professional skills (6.2.3).

X2.2.4 Based upon the stated program objectives, determine the scope of the monitoring program (6.2.4).

X2.2.5 Identify null hypothesis option, that is, what happens if no monitoring is performed? (6.2.5).

X2.2.6 Establish reference conditions for environmental indicators that can be monitored to provide a baseline water-quality assessment (6.2.6).

X2.2.7 Define data management needs (6.2.7).

X2.3 *Design of Field Measurements*—All measurements should conform to standard methods, documented protocols, or at least documented to provide a clear description of the methods used. The use of nonstandard methods is appropriate where the use of standard methods would place a constraint on the successful implementation of the work (6.3).

X2.3.1 Evaluate spatial aspects of monitoring activities, for example, where to sample or measure (6.3.1).

X2.3.2 Evaluate temporal aspects of monitoring activities, such as how frequently and how long to sample (6.3.2).

X2.3.3 Select monitoring sites (6.3.3).

X2.3.4 Select environmental indicators and data parameters (6.3.4).

X2.3.5 Identify specific factors impacting sampling sites (6.3.5).

X2.3.6 Define probable water level, stage and frequency variations, and relationship to monitoring methodology (6.3.6).

X2.3.7 Select sampling approach to achieve objectives (6.3.7).

X2.3.8 Define monitoring and sampling program; develop sampling plans and identify standardized protocols and methods (performance based if possible) and document data to enable data comparison with other monitoring programs (6.3.8).

X2.3.9 Define analytical methodology (6.3.9).

X2.3.10 Establish QA/QC program (Quality Assurance Program Plan (QAPP)) (6.3.10).

X2.3.11 Define skill requirements (6.3.11).

X2.4 *Implementation of Sampling and Analysis Strategies (6.4):*

X2.4.1 Collect data according to monitoring design and protocols (6.4.1).

X2.4.2 Review data collection activities relative to protocols and monitoring design (6.4.2).

X2.4.3 Follow QAPP plan and document deviations from plan (6.4.3).

X2.4.4 Regularly review data for completeness and for accuracy and precision from QC data (6.4.4).

X2.4.5 Regularly review data relative to changing monitoring design to accomplish objectives (6.4.5).

X2.5 *Data Quality Assessment*, based on results of QAPP (6.5).

X2.5.1 Review custody records (6.5.1).

X2.5.2 Compare laboratory spikes (6.5.2).

X2.5.3 Consider collection conditions (6.5.3).

X2.5.4 Meet statistical protocol (6.5.4).

X2.5.5 Meet confidence criteria (6.5.5).

X2.5.6 Define statistical reliability of data using quality assurance information (6.5.6).

X2.6 *Data Management (6.6):*

X2.6.1 Identify and use standard protocols, if appropriate, for reporting and storage of data (EPA 813) (6.6.1).

X2.6.2 Identify ancillary data storage needs (6.6.2).

X2.6.3 Define metadata storage needs. This is information that can be used to support the measured data (for example, date, time, location, collection methodology, QA data) (6.6.3).

X2.6.4 Define georeferencing standards (6.6.4).

X2.6.5 Describe the data management protocol (6.6.5).

X2.6.6 Share data; ensure that data can and is being transmitted to other users (6.6.6).

X2.6.7 Archive data (6.6.7).

X2.6.8 Secure data (6.6.8).

X2.7 *Data Interpretation and Assessment (Evaluate, Analyze, and Assess Data) (6.7):*

X2.7.1 Evaluate data to determine if objectives can be achieved (6.7.1).

X2.7.2 Analyze data to determine statistical significance (6.7.2).

X2.7.3 Assess data to measure impact of results (6.7.3).

X2.7.4 Coordinate interpretation with collaborators and customers (6.7.4).

X2.7.5 Use deterministic models for testing management scenarios (6.7.5).

X2.8 *Program Evaluation (6.8):*

X2.8.1 Identify problems (6.8.1).

X2.8.2 Evaluate costs (6.8.2).

X2.8.3 Evaluate water monitoring program (6.8.3).

X2.8.4 Determine if information satisfies the stated objectives (6.8.4).

X2.8.5 Identify current and future needs not addressed in the current program (6.8.5).

X2.9 *Communication and Report (6.9):*

X2.9.1 Prepare information for identified audiences (6.9.1).

X2.9.2 Prepare data for data users (6.9.2).

X2.9.3 Determine how, whether, and when (by schedule) data and interpretive information can be communicated; for example, through press releases, public meetings, agency meetings, conferences, agency reports, journal articles etc. (6.9.3).

X2.9.4 Report format (6.9.4).

X2.10 *Cooperation and Coordination (6.10):*

X2.10.1 A water monitoring program seldom can be implemented without coordination of several individual entities. Planning, field work, transport, lab work, data management, and data analysis are seldom handled by one group. Formal intergroup or agency procedures shall be written to ensure that all understand their role in the process. All levels should understand the program objectives (6.10.1).

REFERENCES

(1) *Ambient Water-Quality Monitoring in the United States,* First Year Review, Evaluation, and Recommendations, Intergovernmental Task Force on Monitoring Water Quality, *U.S. Geological Survey,* Reston, VA, 1992.

(2) *Water-Quality Monitoring in the United States,* 1993 Report of the Intergovernmental Task Force on Monitoring Water Quality. *U.S. Geological Survey,* Reston, VA, 1994.

(3) *Water-Quality Monitoring in the United States,* 1993 Report of the Intergovernmental Task Force on Monitoring Water Quality, Technical Appendices, *U.S. Geological Survey,* Reston, VA, 1994.

(4) IHD-WHO Working Group on the Quality of Water, "Water Quality Surveys," *Studies and Reports in Hydrology,* No. 23, World Health Organization, Geneva, Switzerland, 1978.

(5) *The Strategy for Improving Water-Quality Monitoring in the United States,* Final Report of the Intergovernmental Task Force on Monitoring Water Quality, Office of Water Data Coordination, *U.S. Geological Survey,* Reston, VA, 1994.

(6) U.S. Army Corps of Engineers, *Sampling Design for Reservoir Water Quality Investigations,* 1987.

(7) United States Department of Agriculture Natural Resources Conservation Service, *National Handbook of Water Quality Monitoring* (J. C. Clausen, in press), Washington, DC.

(8) United States Department of Agriculture *Field Manual for Research Agricultural Hydrology,* Agricultural Handbook 224, Washington, DC, 1979.

(9) *National Handbook of Recommended Methods for Water Data Acquisition,* Office of Water Data Coordination, *U.S. Geological Survey,* 1977.

Standard Guide for
Design of Ground-Water Monitoring Systems in Karst and Fractured-Rock Aquifers[1]

This standard is issued under the fixed designation D 5717; the number immediately following the designation indicates the year of original adoption or, in the case of revision, the year of last revision. A number in parentheses indicates the year of last reapproval. A superscript epsilon (ε) indicates an editorial change since the last revision or reapproval.

INTRODUCTION

This guide for the design of ground-water monitoring systems in karst and fractured-rock aquifers promotes the design and implementation of accurate and reliable monitoring systems in those settings where the hydrogeologic characteristics depart significantly from the characteristics of porous media. Variances from government regulations that require on-site monitoring wells may often be necessary in karst or fractured-rock terranes (see 7.3) because such settings have hydrogeologic features that cannot be characterized by the porous-media approximation. This guide will promote the development of a conceptual hydrogeologic model that supports the need for the variances and aids the designer or governmental reviewer in establishing the most reliable and efficient monitoring system for such aquifers.

Many of the approaches contained in this guide may also have value in designing ground-water monitoring systems in heterogeneous and anisotropic unconsolidated and consolidated granular aquifers. The focus of this guide, however, is on unconfined karst systems where dissolution has increased secondary porosity and on other geologic settings where unconfined ground-water flow in fractures is a significant component of total ground-water flow.

1. Scope

1.1 *Justification*—This guide considers the characterization of karst and fractured-rock aquifers as an integral component of monitoring-system design. Hence, the development of a conceptual hydrogeologic model that identifies and defines the various components of the flow system is recommended prior to the design and implementation of a monitoring system.

1.2 *Methodology and Applicability*—This guide is based on recognized methods of monitoring-system design and implementation for the purpose of collecting representative ground-water data. The design guidelines are applicable to the determination of ground-water flow and contaminant transport from existing sites, assessment of proposed sites, and determination of wellhead or springhead protection areas.

1.3 *Objectives*—The objectives of this guide are to outline procedures for obtaining information on hydrogeologic characteristics and water-quality data representative of karst and fractured-rock aquifers.

1.4 *This standard does not purport to address all of the safety concerns, if any, associated with its use. It is the responsibility of the user of this standard to establish appropriate safety and health practices and determine the applicability of regulatory limitations prior to use.*

2. Referenced Documents

2.1 *ASTM Standards:*
D 653 Terminology Relating to Soil, Rock, and Contained Fluids[2]
D 5092 Practice for Design and Installation of Ground Water Monitoring Wells in Aquifers[3]
D 5254 Practice for Minimum Set of Data Elements to Identify a Ground-Water Site[3]

3. Terminology

3.1 *Definitions:*
3.1.1 For terms not defined below, see Terminology D 653.
3.2 *Descriptions of Terms Specific To This Standard:*
3.2.1 *aliasing*—the phenomenon in which a high-frequency signal can be interpreted as a low-frequency signal or trend because the sampling was too infrequent to characterize the signal.
3.2.2 *conduit*—pipe-like opening formed and enlarged by dissolution of bedrock and that has dimensions sufficient to sustain turbulent flow under ordinary hydraulic gradients.
3.2.3 *dissolution zone*—a zone where extensive dissolution of bedrock has occurred; void size may range over several orders of magnitude.
3.2.4 *epikarst*—a zone of enhanced bedrock-dissolution immediately beneath the soil zone; characterized by storage of water in dissolutionally enlarged fractures and bedding planes, and that may be separated from the phreatic zone by

---

[1] This guide is under the jurisdiction of ASTM Committee D-18 on Soil and Rock and is the direct responsibility of Subcommittee D18.21 on Ground Water and Vadose Zone Investigations.
Current edition approved April 15, 1995. Published June 1995.

[2] *Annual Book of ASTM Standards*, Vol 04.08.
[3] *Annual Book of ASTM Standards*, Vol 04.09.

a relatively waterless interval locally breached by vertical vadose flow.

3.2.5 *fractured-rock aquifer*—an aquifer in which flow of water is primarily through fractures, joints, faults, or bedding planes that have not been significantly enlarged by dissolution.

3.2.6 *karst aquifer*—an aquifer in which all or most flow of water is through one or more of the following: joints, faults, bedding planes, pores, cavities, conduits, and caves, any or all of which have been significantly enlarged by dissolution of bedrock.

3.2.7 *karst terrane*—a landscape and its subsurface characterized by flow through dissolutionally modified bedrock and characterized by a variable suite of surface landforms and subsurface features, not all of which may be present or obvious. These include: sinkholes, springs, caves, sinking streams, dissolutionally enlarged joints or bedding planes, or both, and other dissolution features. Most karsts develop in limestone or dolomite, or both, but they may also develop in gypsum, salt, carbonate-cemented sandstones, and other soluble rocks.

3.2.8 *overflow spring*—a spring that discharges generally intermittently at a ground-water stage above base flow (compare with underflow spring).

3.2.9 *rapid flow*—ground-water flow with a velocity >0.001 m/s.

3.2.10 *secondary porosity*—joints, fissures, faults, that develop after the rock was originally lithified; these features have not been modified by dissolution.

3.2.11 *sinkhole*—a topographic depression formed as a result of karst-related processes such as dissolution of bedrock, collapse of a cave roof, or flushing or collapse, or both, of soil and other sediment into a subjacent void.

3.2.12 *slow flow*—ground-water flow with a velocity <0.001 m/s.

3.2.13 *swallet*—the hole into which a surface stream sinks.

3.2.14 *tertiary porosity*—porosity caused by dissolutional enlargement of secondary porosity.

3.2.15 *tracer*—a substance added to a medium, typically water, to give it a distinctive signature that makes the medium recognizable elsewhere.

3.2.16 *underflow spring*—a spring that is at or near the lowest discharge point of a ground-water basin and that usually flows perennially (compare with overflow spring).

4. Significance and Use

4.1 *Users*—This guide will be useful to the following groups of people:

4.1.1 Designers of ground-water monitoring networks who may or may not have experience in karst or fractured-rock terranes;

4.1.2 The experienced ground-water professional who is familiar with the hydrology and geomorphology of karst terranes but has minimal familiarity with monitoring problems; and

4.1.3 Regulators who must evaluate existing or proposed monitoring for karst or fractured-rock aquifers.

4.2 *Reliable and Efficient Monitoring Systems*—A reliable and efficient monitoring system provides information relevant to one or more of the following subjects:

4.2.1 Geologic and hydrologic properties of an aquifer;

4.2.2 Distribution of hydraulic head in time and space;

4.2.3 Ground-water flow directions and rates;

4.2.4 Water quality with respect to relevant parameters; and

4.2.5 Migration direction, rate, and characteristics of a contaminant release.

4.3 *Limitations:*

4.3.1 This guide provides an overview of the methods used to characterize and monitor karst and fractured-rock aquifers. It does not address the details of these methods, field procedures, or interpretation of the data. Numerous references are included for that purpose and are considered an essential part of this guide. It is recommended that the user of this guide be familiar with the relevant material within this guide and the references cited. This guide does not address the application of ground-water flow models in the design of monitoring systems in karst or fractured-rock aquifers. The use of flow and transport mode at fractured-rock sites summarized in Ref (1)[4] provide a more recent comparison of fracturent and transport modeling.

4.3.2 The approaches to the design of ground-water monitoring systems suggested within this guide are the most appropriate methods for karst and fractured-rock aquifers. These methods are commonly used and are widely accepted and proven. However, other approaches or methods of ground-water monitoring which are technically sound may be substituted if justified and documented.

5. Special Characteristics of Karst and Fractured-Rock Aquifers

5.1 Karst and fractured-rock aquifers differ from granular aquifers in several ways; these differences are outlined in 5.2. Designing reliable and efficient monitoring systems requires the early development of a conceptual hydrogeologic model that adequately describes the flow and transmission characteristics of the site under investigation. Section 5.3 outlines various approaches to conceptualizing these systems and 5.4 contains subjective guidelines for determining which conceptual approach is appropriate for various settings.

5.2 *Comparison of Granular, Fractured-Rock, and Karst Aquifers*—Table 1 lists aquifer characteristics and compares the qualitative differences between granular, fractured-rock, and karst aquifers. This table represents points along a continuum. For this guide a karst aquifer is defined as an aquifer in which most flow of water is through one or more of the following: joints, faults, bedding planes, pores, cavities, conduits, and caves, any or all of which have been significantly enlarged by dissolution of bedrock (2). For this guide a fractured-rock aquifer is defined as an aquifer in which the flow is primarily through fractures that have not been significantly enlarged by dissolution. Fracture is "a general term for any break in rock, whether or not it causes displacement, due to mechanical failure by stress. Fractures include cracks, joints, and faults" (3). The following factors must be evaluated to properly characterize an aquifer's position in the continuum.

---

[4] The boldface numbers given in parentheses refer to a list of references at the end of the text.

TABLE 1 Comparison of Granular, Fractured-Rock, and Karst Aquifers (3)

Aquifer Characteristics	Aquifer Type		
	Granular	Fractured Rock	Karst
Effective Porosity	Mostly primary, through intergranular pores	Mostly secondary, through joints, fractures, and bedding plane partings	Mostly tertiary (secondary porosity modified by dissolution); through pores, bedding planes, fractures, conduits, and caves
Isotropy	More isotropic	Probably anisotropic	Highly anisotropic
Homogeneity	More homogeneous	Less homogeneous	Non-homogeneous
Flow	Slow, laminar	Possibly rapid and possibly turbulent	Likely rapid and likely turbulent
Flow Predictions	Darcy's law usually applies	Darcy's law may not apply	Darcy's law rarely applies
Storage	Within saturated zone	Within saturated zone	Within both saturated zone and epikarst
Recharge	Dispersed	Primarily dispersed, with some point recharge	Ranges from almost completely dispersed- to almost completely point-recharge
Temporal Head Variation	Minimal variation	Moderate variation	Moderate to extreme variation
Temporal Water Chemistry Variation	Minimal variation	Minimal to moderate variation	Moderate to extreme variation

5.2.1 *Porosity*—The type of porosity is the most important difference between these three types of aquifers. All other differences in characteristics are a function of porosity. In a granular aquifer, effective porosity is primarily a consequence of depositional setting, diagenetic processes, texture, and mineral composition while in fractured-rock and karst aquifers, effective porosity is a secondary result of fractures, faults, and bedding planes. Secondary features modified by dissolution comprise tertiary porosity.

5.2.2 *Isotropy*—Fractured-rock and karst aquifers are typically anisotropic in three dimensions. Hydraulic conductivity can frequently range over several orders of magnitude, depending upon the direction of measurement. Ground water in anisotropic media does not usually move perpendicular to the hydraulic gradient, but at some angle to it (4, 5).

5.2.3 *Homogeneity*—The variation of aquifer characteristics within the spatial limits of the aquifer is frequently large in fractured-rock and karst aquifers. Hydraulic conductivity differences of several orders of magnitude can occur over very short horizontal and vertical distances.

5.2.4 *Flow*—Flow in fractured rocks that are not significantly soluble is dependent upon the number of fractures per unit volume, their apertures, their distribution, and their degree of interconnection. Aquifers with a large number of well-connected and uniformly distributed fractures may approximate porous media. In these settings, the equations describing flow in granular media, based on Darcy's law, are sometimes applicable. Fractured-rock aquifers that have a few localized highly transmissive fractures, or fracture zones that exert a dominant control on ground-water occurrence and movement, are not accurately characterized by the porous-media approximation; they more closely resemble karst aquifers. Ground water moves through most karst aquifers predominantly through conduits formed by dissolution and fractures enlarged by dissolution that occupy a small percentage of the total rock mass. Ground-water flow in the rock mass is both intergranular and through fractures that have not been significantly modified by dissolution. Such flow is usually only a small percentage of the volume of water discharging from the aquifer, though it provides most of the storage (6).

5.2.4.1 It was formerly thought, after the work of Shuster and White (7), that conduit flow was dominant in some aquifers, and diffuse flow was dominant in others. The diffuse-flow dominated regime was thought to be characterized by low variation in hardness, turbidity, and discharge—as measured at a spring. It is now recognized that the variations of these parameters are due to the aquifer boundary conditions, such as the number of sinking stream inputs or whether the spring is an underflow or overflow spring (8, 9, 10).

5.2.4.2 The terms *rapid flow* and *slow flow* should be used rather than *conduit flow* and *diffuse flow*. The latter terms are ambiguous when used in reference to karst aquifers because they have been used to describe types of flow within an aquifer, types of recharge, and types of spring-flow as affected by recharge events, as well as flow hydraulics, and water chemistry. Rapid flow takes place in conduits >5 to 10 mm in diameter (11) where velocities generally exceed 0.001 m/s. The swallet-flow component of karst aquifers typically yields flow in conduits >0.001 m/s (10). Such rapid flow can also occur in open fractures. Flow in the rock matrix and through fractures that have not been significantly modified by dissolution is typically slow (<0.001 m/s). However, flow in conduits and fractures can also be slow.

5.2.5 *Storage*—In most aquifers, ground water is stored within the zone of saturation (phreatic zone); however, karst aquifers can store large volumes of ground water in a part of the unsaturated (vadose) zone known as the epikarst (subcutaneous zone) (12, 13, 14). The epikarst, the uppermost portion of carbonate bedrock, commonly about 10 to 15 m thick, consists of highly-fractured and dissolved bedrock (see Fig. 1). Highly permeable vertical pathways are formed along intersections of isolated vertical fractures. The epikarst behaves as a locally saturated, sometimes perennial, storage zone that functions similarly to a leaky capillary barrier or a perched aquifer, but it is commonly not perched on a lithologic discontinuity. Flow into this zone is more rapid than flow out of it, as only limited vertical pathways transmit water downward.

5.2.6 *Recharge*—In granular aquifers, recharge tends to be areally distributed and an aquifer's response to a given recharge event tends to be damped by movement of the recharging water through the unsaturated zone. Generally there is some temporal lag between a recharge event and a resultant rise in water-table; water-table fluctuations in granular aquifers rarely range more than a few meters. By contrast, in karst and fractured-rock aquifers with minimal

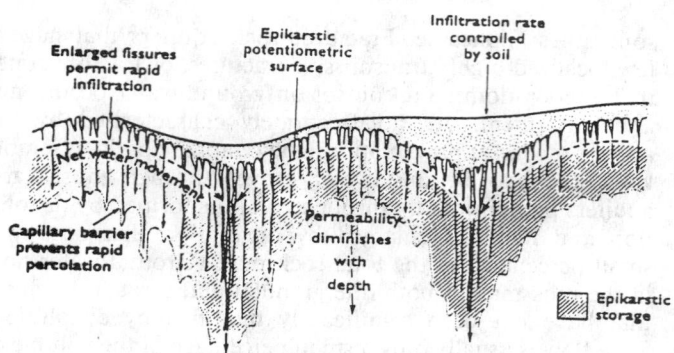

FIG. 1 Cross-Section Illustrating Epikarstic Zone in Carbonate Terrane (14)

unlithified overburden, recharge tends to be rapid; water-levels may rise within minutes of the onset of the storm and water-table fluctuations may range up to many tens of meters. Karst and fractured-rock aquifers with thick unlithified overburden may have a long temporal lag similar to that of granular aquifers. Recharge may be distributed through an areally extensive network of fractures or through soil (*dispersed recharge*), or it may be concentrated at points that connect directly to the aquifer (*point recharge*). The percentage of point recharge of an aquifer strongly influences the character and variability of its discharge and water quality **(10, 14)**.

5.3 *Conceptual Models of Ground-Water Flow in Fractured-Rock and Karst Aquifers:*

5.3.1 Three conceptual models of ground-water flow can be used to characterize fractured-rock and karst aquifers: continuum, discrete, and dual porosity. A hydrogeologic investigation must be conducted to determine which model applies to the site of interest.

5.3.2 The continuum model assumes that the aquifer approximates a porous medium at some working scale (sometimes called the "equivalent porous-media" approach). In this approach, the properties of individual fractures or conduits are not as important as the properties of large regions or large volumes of aquifer material. The porous-medium approximation implies that the classical equations of ground-water movement hold at the problem scale, that knowledge of the hydraulic properties of individual fractures is not important, and that aquifer properties can be characterized by field and laboratory techniques developed for porous media. The discrete model assumes that the majority of the ground water moves through discrete fractures or conduits and that the hydraulic properties of the matrix portion of the aquifer are unimportant. Measurement of the hydraulic characteristics of individual fractures or conduits are used to characterize ground-water movement. The dual-porosity model of ground-water flow lies somewhere between that of the continuum and discrete models. A dual-porosity approach attempts to characterize ground-water flow in individual conduits or fractures as well as in the matrix portion of the aquifer.

5.3.3 These theoretical models are useful tools for conceptualizing ground-water flow in fractured-rock and karst aquifers. However, the design of a ground-water monitoring system must be based on empirical data from the site to be

monitored. It is important to realize that standard hydrogeologic field techniques may not be valid in fractured-rock and karst aquifers, because many of these techniques are based on the continuum model. The following section provides subjective guidelines for determining which conceptual approach will best characterize ground-water flow in the aquifer under investigation.

5.4 *Subjective Guidelines for Determining the Appropriate Conceptual Model:*

5.4.1 The question of which conceptual approach is most suitable for a given aquifer is somewhat a question of scale. Implicit in the porous-medium approximation is the idea that aquifer properties, such as hydraulic conductivity, porosity, and storativity, can be measured for some representative elementary volume (REV) of aquifer material and that these values are representative over a given portion of the aquifer. For granular aquifers and some densely-fractured aquifers, the REV is likely to be encompassed by standard field-monitoring devices such as monitoring wells. In such aquifers, the continuum approach is appropriate for site-specific investigations provided aquifer heterogeneity is adequately characterized. The porous-medium approximation is not a valid conceptual model for those fractured-rock and karst aquifers where flow is primarily through widely-spaced discrete fractures or conduits, (14, 15, 16).

5.4.2 The discrete approach is most appropriate for those aquifers where there is a great contrast between matrix and fracture or conduit hydraulic conductivity. The dual-porosity approach is most appropriate for those aquifers where the matrix is relatively permeable and yet there are discrete zones of higher conductivity such as dissolution zones, fractures, or conduits.

5.4.3 Determining which conceptual model is appropriate for a given aquifer requires that an investigator determine the influence of fractures and conduits on the flow system. Existing data may provide valuable information. However, relevant and appropriate site-specific field investigations are necessary to fully characterize the flow system.

5.4.4 Below is a list of subjective criteria that can be used to help determine which conceptual ground-water flow model is appropriate for use at a given site. Reference (3) lists several criteria for determining whether the continuum approach is appropriate for a fractured-rock aquifer; these are summarized in 5.4.1 to 5.4.5. Additional criteria for determining the applicability of the porous-medium approximation in karst aquifers (5.4.8) are provided by Ref (2). All of these guidelines are subjective because fractured-rock and karst aquifers range from porous-medium-equivalent to discrete fracture or conduit-dominated systems. The decision as to which conceptual model is most appropriate will always require professional judgment and experience.

5.4.5 *Ratio of Fracture Scale to Site Scale*—For porous-medium-equivalent aquifers, the observed vertical and horizontal fractures should be numerous, the distance between the fractures should be orders of magnitude smaller than the size of the site under investigation, and the fractures should show appreciable interconnection.

5.4.6 *Hydraulic Conductivity Distribution*—In porous-medium-equivalent settings, the distribution of hydraulic conductivity, as estimated from piezometer slug tests or from specific capacity analyses, tends to be approximately lognormal. In aquifers where the hydraulic conductivity distribution is strongly bimodal or polymodal, the porous-medium approximation is probably not valid. It is also possible to obtain a log-normal distribution of hydraulic conductivity for wells in those aquifers that do not fit the porous-medium approximation (see 6.5) because most wells are preferentially completed in high-yielding zones. In addition, hydraulic conductivity values vary with the scale of measurement (16, 17, 18, 19) and slug tests completed in open boreholes will yield averaged hydraulic conductivities that do not represent the full variability in hydraulic conductivity.

5.4.7 *Water-Table Configuration*—For porous-medium-equivalent aquifers, a water-table map should show a smooth and continuous surface without areas of rapidly changing or anomalous water levels. In particular, the water table should not have the "stair-step" appearance that can occur in sparsely fractured rocks with large contrasts in hydraulic conductivity between blocks and fractures, nor should the map exhibit contours that appear to "V" upgradient, where no topographic valley exists. In such settings, flow within a conduit may be affecting the configuration of the water table. Although the "stair-step" or "V-shaped" anomalies (for an example, see Ref (20) clearly indicate a failure of the porous-medium approximation, a smooth water table does not prove a porous-medium-equivalent setting because the density of measuring points may not be sufficient to detect irregularities in the water-table configuration (see 6.3.1.1).

5.4.8 *Pumping Test Responses*—There are several criteria for determining how closely a fractured-rock aquifer approximates a porous medium by using an aquifer pumping test.

5.4.8.1 The drawdown in observation wells should increase linearly with increases in the discharge rate of the pumping well.

5.4.8.2 Time-drawdown curves for observation wells located in two or more different directions from the pumped well should be similar in shape and should not show sharp inflections, which could indicate hydraulic boundaries.

5.4.8.3 Distance-drawdown profiles that are highly variable (for example, distant points respond more strongly while nearby points have little or no response) indicate that the porous-media approximation is not valid.

5.4.8.4 A plotted drawdown cone from a pumping test using multiple observation wells should be either circular or near-circular (elliptical). Linear, highly elongated, or very irregular cones, in areas where no obvious hydraulic boundaries are present, indicate that the assumption of a porous medium is invalid.

5.4.9 *Variations in Water Chemistry*—Large spatial and temporal variations in the chemistry of natural waters can be observed in fractured-rock and karst aquifers because of the rapid movement of water through discrete fractures or solution conduits. The coefficient of variation of specific conductance (or hardness) of spring and well water is a function of the percentage of rapid versus slow recharge to an aquifer and can be used to infer that percentage except where anthropogenic influences will impact the conductivity of the recharging water (8, 9, 10).

5.4.9.1 Many wells and springs, particularly those used for public water supply, are sampled on a regular basis for such parameters as temperature, pH, specific conductance, hardness, turbidity, and bacteria. If sampling results indicate large, short-term fluctuations in any of these parameters, the porous-medium approximation cannot be assumed.

NOTE 1—The last sentence of the preceding paragraph assumes that the short-term fluctuations (on the order of hours or days) are not a consequence of initiation of pumping or other withdrawal methods.

5.4.9.2 Water-supply wells and springs are often sampled on a monthly basis and while monthly variation in water-quality parameters may provide a general indication of whether the aquifer behaves as a porous medium, water-quality variations in response to recharge events are frequently a better test of the porous-medium approximation. In order to determine the validity of the porous-medium approximation at a monitoring point, observe and record at least two, and preferably all, of the following: spring discharge or hydraulic head, turbidity, specific conductance, and temperature, preferably a day before, during, and for several days or weeks after several major recharge events. If the water becomes turbid and the other parameters show rapid and flashy responses to the recharge event, the porous-medium approximation is most likely not valid. A bimodal or polymodal distribution of daily or continuous measurements of specific conductance (14, 21) also indicates that the porous-medium approximation may not be valid.

5.4.10 *Presence of Karst Features*—The presence in the same contiguous formation within several kilometres of a site of landforms such as sinkholes, sinking streams, blind valleys, and subsurface features such as caves and dissolutionally enlarged joints, indicates a degree of dissolutional modification that probably invalidates the porous-medium approximation and denotes a karst terrane. As a generalization, if there is carbonate rock, it is highly probable that there is both a karst terrane and a karst aquifer. If a carbonate aquifer has been or is presently subaerially exposed, and if total hardness is less than 500 mg/L, then a rapid-flow component and a karst aquifer are present (10).

5.4.11 *Variations in Hydraulic Head*—Monitoring wells in granular media tend to exhibit predictable and minor changes in hydraulic head in response to recharge events. In fractured-rock and karst aquifers it is not uncommon to see large variations in head in immediate response to recharge events. The degree of response of hydraulic head in a given well is dependent upon the size of fractures or conduits encountered by the well and the directness of their connections to surface inputs.

5.4.11.1 Aquifers with a high contrast in hydraulic conductivity over short distances can exhibit non-coincident water levels in closely spaced wells that are screened or open over the same vertical interval. In Karst and fractured-rock terrane such non-coincident water levels indicate that the porous-medium approximation is probably not valid.

5.4.12 *Borehole Logging*—Several borehole logging techniques can help determine if high-permeability zones are present within a borehole. The presence of such zones suggests that the aquifer is not a porous-medium equivalent. Zones of high permeability are indicated by the following:

5.4.12.1 Presence of open fractures or dissolution features as indicated by a caliper log, borehole television logs (for example, Ref (22)), or acoustic televiewer (23).

5.4.12.2 Significant variation in specific conductance or

temperature as interpreted from borehole logs (for example, Ref (**24**)).

5.4.12.3 Significant variations in borehole fluid movement as measured by a flow meter in a pumped or unpumped well (for example, Refs (**25, 26, 27, 28**).

5.4.12.4 Significant increase in porosity within a rock unit that otherwise has a constant porosity as measured by a porosity (neutron-neutron) log; and

5.4.12.5 Significant decrease in density within a rock unit that otherwise has a consistent density as measured in a density (gamma-gamma) log.

6. Hydrogeologic Setting

6.1 Hydrogeologic characterization of fractured-rock and karst aquifers is complicated by the presence of high-permeability fractures, conduits, and dissolution zones that exert a controlling influence on ground-water flow systems. Locating and characterizing these high-permeability zones can be logistically difficult if not impossible, because conduits, dissolution zones, or subsurface fractures that transmit a large percentage of the flow may be as small as a few millimetres in size. Benson and Yuhr (**29**) note that borings alone are inadequate for subsurface characterization in karst settings. They provide some insights into the number of borings required for locating a subsurface cavity by noting the detection probabilities. The example they provide is that "if a 1 acre site contains a spherical cavity with a projected surface area of 1/10 acre (a site to target ratio of 10), 10 borings spaced over a regular grid will be required to provide a detection probability of 90 %. Sixteen borings will be required to provide a detection probability of 100 %...for smaller targets, such as widely spaced fractures, the site-to-target ratio can increase significantly to 100 or 1000, thus requiring 100 to 1000 borings to achieve a 90 % detection confidence level" (**29**).

6.1.1 In granular media, the monitoring well is the standard measuring point for both obtaining representative ground-water samples and determining aquifer properties. However, the discrete and dual-porosity conceptual models require an investigator to identify sampling points and perform aquifer tests or tracer tests, or both, that do not rely on the porous-medium approximation (continuum approach). In karst and fractured-rock settings, an investigator cannot assume that a monitoring well will provide representative data either for water-quality or aquifer characteristics (**14, 30, 31**). Tracer tests (see 6.7) are one of the more valuable tools for determining ground-water flow directions and velocities because the interpretation of these tests does not require the porous-medium approximation (continuum approach).

6.1.2 This section discusses the importance of understanding stratigraphic and structural influences on ground-water flow systems (see 6.2); location and characterization of fracture patterns and karst features (see 6.3); delineation of ground-water basin boundaries and flow directions (see 6.4); applicability of geophysical techniques (see 6.5); and measurement of aquifer characteristics (see 6.6).

6.2 *Regional Geology and Structure*—The design of a ground-water monitoring network should include a determination of how the site fits into the regional geologic setting because regional stratigraphic and structural patterns provide the constraints within which the local ground-water flow system is developed.

6.2.1 *Sources of Data*—Information on regional geology and hydrogeology, (that is, geologic maps, stratigraphic cross-sections, geophysical logs from nearby sites, cave maps, water-table or potentiometric-surface maps, long-term records of water levels or water quality in monitoring wells) can be obtained from both published and unpublished sources including federal and state publications, academic theses and dissertations, journal articles, and available consultants' reports. Additional information can be obtained from local land owners, quarry operators, highway departments, local construction firms, as well as geologic logs, drillers' logs, and well-construction reports from domestic wells. Data on the number, distribution, and construction of domestic wells are best obtained by house-to-house survey; state and federal files for most areas rarely include more than a small percentage of the wells that exist. The most information about caves can be obtained from consultation with the National Speleological Society[5] whose members compile information on a state-by state basis.

6.2.2 *Integrating Geologic Information With Flow-System Characteristics*—When reviewing the existing data, an investigator should take extra note of any information that indicates the presence of conduits or high permeability dissolution or fracture zones (see guidelines outlined in 5.4). The initial hydrogeologic characterization should include a survey of bedrock outcrops in the area. Special attention should be paid to the relationships between stratigraphy and structure and the distribution of lineaments, fracture patterns, karst landforms, sinkhole alignments, and hydrologic features such as seeps or springs.

6.2.3 *Stratigraphy:*

6.2.3.1 In any layered rock sequence, either sedimentary rocks or layered volcanics, stratigraphy can be a controlling factor in the development of zones of enhanced flow of ground water. Bedrock outcrops, including quarries and caves, should be examined in order to determine the stratigraphic position of springs, seeps, caves, zones of dissolution, or zones of intense fracturing.

6.2.3.2 In fractured carbonate terranes, the development of conduits and dissolution zones is most commonly controlled by bedding plane partings rather than vertical fractures. Dissolution preferentially develops along bedding planes with substantial depositional unconformities, planes with shale laminae or thicker partings, and planes with nodules or beds of chert (**14**). The relationship of shale beds or other low-permeability units to hydrologic features should be noted. These units cannot be assumed to provide effective barriers to ground-water flow because in fractured-rock and karst terranes they are frequently breached by fractures or shafts (**32**). In carbonate terranes, interbedded shales are frequently calcareous and hence subject to dissolution; in addition, shale beds may enhance dissolution, because of oxidation of included sulfides and production of sulfuric acid (**9**).

6.2.3.3 In layered volcanic terranes, interbedded basalts and pyroclastic deposits have different hydrologic properties.

---

[5] National Speleological Society, Cave Ave., Huntsville, AL 35810.

Pyroclastic deposits can range greatly in terms of primary porosity and hydraulic conductivity due to differences in welding and the development of secondary fractures. In general, ash-flow tuffs in the upper portions of flows exhibit high values of porosity and permeability (33). In flood basalts, the interflow zones (top of one flow and bottom of overlying flow) tend to be zones of high porosity and high conductivity due to primary depositional features such as "flow breccias, clinkers, shrinkage cracks, flow-top rubble, and gas vesicles" (33). In flood basalt terranes, tubes or conduits (lava caves) should be suspected, and although these features will usually be influential only in the shallow zone, they could cause preferential flow similar to a conduit system in a karst terrane. Springs are also common in volcanic rocks; their location is determined by topography, structure, and depth to ground water.

6.2.3.4 In regions with no quarries, accessible caves, and few bedrock outcrops, geologic characterization will have to be based on information obtained from drilling. Core drilling provides a good record of both subsurface stratigraphy and fracture distribution in areas of good core recovery. However, core recovery often fails in zones of poor rock quality or in areas with extensive voids. An alternative approach for such situations is to drill destructively (without coring) and then log the hole with applicable geophysical techniques (that is, gamma, resistivity, or conductivity for stratigraphy; and caliper and television for fractures). While vertical boreholes provide useful information about horizontal fractures and dissolution zones, angle drilling with collection of oriented core can be used to better characterize vertical and near-vertical fracture systems, and steeply dipping beds (34). A review of existing well logs, including geophysical logs, should note stratigraphic zones where circulation was lost during drilling, where enhanced yields were obtained during well development or aquifer tests, and where open or mud-filled cavities or fractures were encountered.

6.2.4 *Structure:*

6.2.4.1 Structural features commonly associated with concentration of ground-water flow include anticlines, synclines, and faults. Anticlines are important because extension of joints along their crests can favor development of joint-controlled conduits. Synclines tend to concentrate flow, usually with down-dip inputs to a conduit located close to the base of the trough (14). Faults, especially faults formed by extension, can concentrate ground-water flow, provided that they have not been filled by secondary mineralization (14). Faults can also provide barriers to ground-water flow if secondary mineralization or fault gouge is extensive or if a low-permeability fault block truncates an aquifer.

6.2.4.2 In dipping carbonate rocks (that is, 2 to 5° or more), initial ground-water flow is commonly downdip with eventual discharge along the strike of the beds. In these settings there is substantial evidence that strike-aligned flow is common, and can extend up to tens of kilometres. When designing monitoring systems in these settings, discharge points along the strike must be located even if they are several kilometres away from the site to be monitored. In dipping carbonate strata, depth of ground-water circulation is influenced by fissure frequency, down-dip resistance to flow, ground-water basin length, and angle of dip (9, 14).

6.2.4.3 In crystalline rocks, fractures are typically most abundant near the land surface; fracture density diminishes with depth. However, high-permeability fractures have been found at depths greater than 1500 m (35). Water-table configuration and hence ground-water flow direction in these settings appears to be topographically controlled (36). Enhanced well-yields indicate that the zones of enhanced ground-water flow occur along fracture traces, at the intersection of fracture traces and in valley bottoms which are probably fracture-controlled (36, 37). Reference (37) also notes that "sheet joints", subparallel to the land surface at shallow depths and horizontal at greater depths, may play an important role in ground-water movement in plutonic rocks.

6.3 *Field Mapping and Site Reconnaissance*—In areas where the surficial materials are thin or absent, high-angle fractures and the location of large karst features can sometimes be mapped from topographic maps and aerial photographs. Most fractures and many karst features are not recognizable on topographic maps or air photos and field mapping will be necessary to locate them. Field reconnaissance, completed early in the project, is an important component of site investigation and is essential for the identification of open fractures, swallets, small sinkholes, springs, and cave entrances. (A detailed discussion of fracture-mapping methods can be found in Ref (38). Fractures and karst features will have a large impact on the subsurface hydrology, even if their surface expression is slight. Field mapping can provide detail on the distribution of karst features and on fracture orientation and density. However, it gives little information about the distribution of fractures or conduits at depth.

6.4 *Determination of Ground-water Flow Directions, Velocities, and Basin Boundaries*—Water-table or potentiometric-surface maps, or both, are used to estimate the direction and rate of ground-water and contaminant movement in granular aquifers. Such estimates are complicated in fractured-rock and karst aquifers. Even porous-medium-equivalent fractured-rock aquifers frequently exhibit significant horizontal anisotropy, which can make prediction of ground-water flow directions difficult. Some fractured-rock aquifers respond rapidly enough to recharge events that temporary ground-water mounding may develop and lead to reversals of flow directions. The concept of a "water-table" becomes less clearly defined in those fractured-rock and karst systems where there are discrete high-permeability zones in a much lower-permeability matrix. In these settings, the fractures and conduits respond quickly to recharge events and may spill over into empty, higher-lying conduits or fractures while the lower-permeability portion of the aquifer remains unsaturated.

6.5 *Variation of Hydraulic Head:*

6.5.1 *Potentiometric-Surface Mapping*—Constructing a potentiometric-surface map with water levels from existing wells assumes the following: vertical hydraulic gradients are not significant, and the well intersects enough fractures or conduits to provide a representative water level for the aquifer. If significant vertical gradients are present, construction of a potentiometric-surface map will require screening out of apparent anomalies in water levels resulting by measuring water levels from wells cased at different depths in an aquifer's recharge and discharge zones (39). The elevation of base-level springs, lakes, and streams should be regarded as possible data points for a potentiometric map, provided it can be established that they are not perched.

6.5.1.1 When constructing a potentiometric map based on field-measured water levels, it is necessary to measure water levels in all representative accessible wells over a short time-interval. Data from existing wells may be adequate, provided well-construction information is available, and the water-levels are evaluated with respect to the length and depth of the open interval. Depending on the nature of the investigation and the level of detail required, it may be necessary to install additional wells.

6.5.1.2 It is difficult to state a universal rule that unambiguously specifies the appropriate contour interval or the density and distribution of data points that are needed for construction of a potentiometric map. The steepness of the hydraulic gradient guides the choice of contour interval and the data density should be such that, on average, no more than two to three contour lines are interpolated between data points.

6.5.1.3 Contaminants in karst terranes can quickly travel several kilometres or more. Therefore, it is necessary to extend potentiometric maps significantly beyond property boundaries in order to determine the likely extent and direction of contaminant travel, and to increase the accuracy of the map.

6.5.2 *Vertical Distribution of Hydraulic Head:*

6.5.2.1 The vertical component of flow should be considered in the delineation of ground-water flow direction. Ground-water flow systems typically have a downward flow component in recharge areas that gradually becomes horizontal before changing to upward flow in discharge areas. In granular and porous-media equivalent aquifers, vertically nested or closely spaced piezometers along the flow path are sufficient to describe these gradients.

6.5.2.2 In karst aquifers, the matrix, fractures, and conduits each have very different vertical flow regimes which can be difficult to characterize. For some karst aquifers, recharge is concentrated at very specific points (for example, at sinkholes and swallets of sinking streams) that feed a complex network of conduits. Whether flow is predominantly horizontal or vertical at various points along the flow path is controlled by hydraulic head, the geometry of the conduit system, and location of the discharge point. The degree of connection between the fractured and matrix portions of the aquifer and the conduits will be a function of fracture density, primary porosity, extent of dissolution, and hydraulic gradient. Characterization of the vertical component of flow in conduit-dominated aquifers requires locating point inputs and point discharges, determining the vertical component of flow in the fractured and matrix portions of the aquifer, and evaluating the degree of connection between the fractured and matrix portions of the aquifer and the conduits.

6.5.3 *Temporal Changes in Hydraulic Head:*

6.5.3.1 The response of springs or wells to recharge events is useful for characterizing an aquifer. On the continuum from porous-media-equivalent aquifers to discrete fracture or conduit-dominated aquifers, head variations in the latter tend to increase in magnitude; lag times to the hydrograph peak after the recharge event tend to decrease. In brief, flow in aquifers with numerous direct surface inputs (point

recharge) and discrete fractures or conduits is more flashy than in those aquifers where direct surface inputs are minimal. Individual well-responses to recharge events can be used to indicate the degree of connection between the well and the fracture or conduit system.

6.5.3.2 Complex responses to recharge events commonly occur in fractured-rock and karst aquifers. Flow-system configurations can change dramatically in response to recharge events. In karst aquifers, as deeper conduits fill, ground water may spill over to higher conduits and discharge to a different ground-water basin than it does during low-flow conditions (30, 40). Moderately permeable fractured-rock aquifers may also exhibit such ground-water flow reversals if temporary ground-water mounds develop (40).

6.5.3.3 Investigators working in fractured-rock and karst aquifers need to assess whether temporal changes in hydraulic head can lead to changes in ground-water flow direction or the position of ground-water basin boundaries. The frequency of water-level measurements needs to be determined by the variability of the system rather than by reporting requirements. Continuous water-level records on representative wells are recommended in the early phase of the investigation; after monitoring the response of an aquifer to several recharge events, the measuring frequency can then be adjusted.

6.6 *Determination of the Directions and Rates of Ground-Water Flow:*

6.6.1 *Flow Directions*—Water-table and potentiometric-surface maps are valuable guides for predicting ground-water flow directions. However, the predicted flow directions will be correct only if the assumption of two-dimensional flow is valid and anisotropic aquifer characteristics, if present, are taken into account. In some fractured-rock aquifers and most karst aquifers, the assumption of two-dimensional flow is probably not valid and anisotropy ratios are frequently unavailable for site-specific scales (15, 18). In some settings the potentiometric surface can provide a reasonable first approximation for the delineation of ground-water flow directions and basin boundaries, but this approximation must be confirmed with tests that are not dependent on the assumption of two-dimensional flow. Such confirmation can be provided by properly conducted tracer tests performed on both sides of a proposed boundary, as shown by Quinlan and Ewers (40) and discussed by Quinlan (30, 31).

6.6.2 *Flow Rates*—It is usually inappropriate to use water-table or potentiometric surface-maps to predict regional or local ground-water flow rates in fractured-rock and karst aquifers. Such calculations assume that the porous-media approximation is valid, flow is two-dimensional, and the hydraulic conductivity distribution is relatively homogeneous. While these conditions might be met over very short distances, they are rarely, if ever, met for site-specific or larger areas. See 6.9 for a discussion of aquifer characteristics. Flow rates are directly determined from the results of aquifer-scale or site-scale tracer tests.

6.7 *Use of Tracer Tests:*

6.7.1 Tracer tests are a valuable tool for characterization of fractured-rock and karst aquifers. They can yield empirical determinations of ground-water flow directions, flow rates, flow destinations, and basin boundaries. The results of these tests depend on the conservative nature of the tracer, its unambiguous detectability, proper test design and execution, and correct interpretation.

6.7.2 Two broad classes of tracers have been used: labels and pulses, both of which can be usefully subdivided into natural and artificial tracers. The purpose of the labels is to enable identification of the investigator's water which serves as a surrogate for a pollutant. The purpose of pulse-tracing is to be able to send an identifiable signal through the ground-water system. A partial outline of tracers that has been used can be found in Table 2.

6.7.3 The various types of tracers have different advantages and disadvantages and they yield different types of information about a hydrogeologic system. As the level of sophistication of an investigation increases, comparison of the results obtained with different tracers can often yield additional information about the system properties.

6.7.4 An ideal tracer has the following properties. It is:

6.7.4.1 Non-toxic to people and the ecosystem;

6.7.4.2 Either not naturally present in the system or present at very low, near-constant levels;

6.7.4.3 In the case of chemical substances, soluble in water with the resulting solution having approximately the same density as water; (care should be taken, in the design of tracer tests, to address concerns that may arise when the pollutants of concern are light or dense non-aqueous phase liquids);

6.7.4.4 Neutral in buoyancy and, in the case of particulate tracers, with a sufficiently small diameter to avoid significant losses by natural filtration;

6.7.4.5 Unambiguously detectable in very small concentrations;

6.7.4.6 Resistant to adsorptive loss or to chemical, physical, or biological degradation, or all of the aforementioned;

6.7.4.7 Capable of being analyzed quickly, economically, and quantitatively;

6.7.4.8 Easy to introduce to the flow system; and

6.7.4.9 Inexpensive and readily available.

6.7.5 Non-toxicity is the most important tracer characteristic. Few tracers satisfy all of these criteria, but several of the fluorescent dyes meet most in many situations. For most settings, dyes are the most practical tracers. Toxicity studies indicate that most fluorescent dyes are not harmful in the concentrations conventionally employed in tracer tests. It has been determined that a concentration of one part per million of the most commonly used fluorescent dyes, over an exposure period of 24 h, poses no threat to human or ecosystem health (41).

6.7.6 Tracer tests are appropriate when:

6.7.6.1 Flow velocities are likely to be such that results will be obtained within a reasonable period of time, usually less than a year;

6.7.6.2 The consequences of existing or possible future ground-water contamination must be determined;

6.7.6.3 It is necessary to delineate recharge areas or ground-water basin boundaries; or

6.7.6.4 It is necessary to design or test a ground-water monitoring system, or both.

6.7.7 Tracer tests can be classified in several ways which are outlined in Table 3. Techniques for tracing ground water, with emphasis on the use of fluorescent dyes, are described and discussed in Ref (43).

6.7.8 Tracing techniques and approaches that an investigator might use vary greatly in levels of sophistication. For example, the question "Is the septic system of this house connected to the nearest sewer main?" can sometimes be adequately answered with a few pennies worth of dye and a few minutes of someone's time. Similarly, questions about the internal connections in some caves can often be answered with about the same level of resources and time. In contrast, questions about the regional-scale dispersal of pollutants from a major Superfund site in a densely populated karst terrane require a considerably greater investment of time, resources, and effort.

6.7.9 Dye-detection techniques range from visual detection, to detection by fluorometer, to instrumental analysis of water samples using a scanning spectrofluorophotometer or (rarely) high performance liquid chromatography (HPLC).

TABLE 2 Types of Tracers (43)

Labels:
  Natural
    Flora and fauna (chiefly, but not exclusively microorganisms)
    Ions in solution
    Environmental isotopes
    Temperature
    Specific conductance
  Introduced
    Dyes and dye-intermediates
    Radiometrically detected substances
    Salts and other inorganic compounds
    Spores
    Fluorocarbons
    Gases
    A wide variety of organic compounds
    Biological entities (bacteria, viruses, yeasts, phages)
    Effluent and spilled substances
    Organic particles, microspheres
    Inorganic particles (including sediment)
    Temperature
    Specific conductance
    Exotica (eels, ducks, marked fish, etc.)

Pulses Significantly Above Background or Base-Flow Levels:
  Natural
    Discharge (change in stage or flow)
    Temperature
    Turbidity
  Introduced
    Discharge
    Temperature

TABLE 3 Classifications of Tracer Tests (43)

A. Degree of Quantification:
    Qualitative
    Semi-quantitative
    Quantitative
B. Degree of Alteration of Hydraulic Gradient:
    Natural gradient
    Forced gradient, accomplished by:
      Injection (input raises potentiometric surface)
      Discharge (pumping lowers potentiometric surface)
C. Type of Injection Site:
    Natural
      Sinkhole or swallet
      Cave stream
    Artificial
      From a well or other man-made contrivance
D. Type of Recovery Site:
    Natural discharge site
      Spring, cave stream, etc.
    Artificial discharge site
      Monitoring well
      One or more domestic wells

The use of simple visual detection of dyes is now considered usually unacceptable in a major project, but it can still be a very effective demonstration of a connection in some settings.

6.8 *Geophysical Techniques:*

6.8.1 Geophysical techniques can be used in a number of ways to aid subsurface investigations and to characterize some subsurface features of karst or fractured-rock aquifers (29, 44). Surface geophysics can provide general information over a large area and can also be used to provide detailed, site-specific data. Borehole logging techniques provide localized information within and immediately around a borehole or well. Some borehole or hole-to-hole techniques can be used to detect fractures and karst features. The method or methods to be used must be selected to meet both project objectives and site conditions. Interpretations based upon surface and borehole geophysical data should be verified by other data and require experienced field crew and interpretation.

6.8.2 *Surface Geophysical Techniques*—Surface geophysics provides a means of characterizing subsurface conditions by making measurements of some physical parameter (acoustical properties, electrical properties, etc.) at the surface. Surface geophysical methods can help characterize subsurface features such as depth to rock, depth to water table, or to locate buried channels. Large structural features such as dip, folds, and faults can be located and mapped. Fracture orientation and areal variations in water quality can also be determined. Surface geophysical methods can sometimes be used to detect conduits directly if they are shallow and large enough (29, 44). Effectiveness of surface geophysical methods diminishes as the feature of interest occurs at an increasing depth and with decreasing size of the feature. Fractures or conduits that are deeper than can be detected by surface geophysical methods, can sometimes be located indirectly by using near-surface indicators (29, 45). Geophysical methods are often used to indicate anomalous conditions caused by a fracture or conduit. The anomalous conditions can then be investigated further by boreholes where the borings are focused into the anomalous area(s) and have a much better probability of encountering the fracture or conduit than a randomly placed borehole.

6.8.2.1 Surface geophysics may be a good reconnaissance tool that can be used to determine areas in need of further study. Several of the following methods may be applicable in fractured-rock and karst settings, including ground-penetrating radar, electromagnetic or electrical resistivity surveys, natural potential (SP), and or microgravity. Such methods as electromagnetics (46, 47, 48) azimuthal resistivity (48, 49), and azimuthal seismic measurements (50) can be applied to determine dominant fracture orientations.

6.8.3 *Borehole Logging Methods*—Borehole logging can be used to identify strata (for example, shale versus limestone) and to correlate stratigraphy between boreholes. These methods are particularly useful for investigation of fractured-rock aquifers because they provide detailed information about rock properties in the immediate vicinity of borehole walls. They are useful for determining water-bearing zones within a borehole and for determining hydraulic properties of inclined and horizontal fractures (see Ref (51) for general borehole logging techniques applied to ground water investigations).

6.8.3.1 Borehole logs most commonly used to correlate stratigraphy include natural gamma, gamma-gamma, resistivity (or conductivity), and spontaneous potential. Borehole methods particularly useful for locating and characterizing fractures and conduits include video, temperature, caliper, acoustic televiewer, flow meter, borehole fluid logging, and cross-hole tomography (29, 44). When budget limitations preclude the use of multiple logging techniques, it is recommended that video logging be used to determine the location and orientation of fractures and conduits to aid in the placement of monitoring well screen. Tomography carried out by radar and attenuation of higher frequency acoustic signals can be used to detect fractures and conduits.

6.8.3.2 Borehole methods are often used in conjunction with each other. Borehole diameter, well construction, and proper well development can affect the results and usefulness of borehole logs.

6.9 *Aquifer Characteristics*—One of the special problems of monitoring in fractured-rock or karst aquifers is that aquifer characteristics such as aquifer thickness, porosity, hydraulic conductivity, and storativity can be difficult to quantify.

6.9.1 *Aquifer Thickness*—In aquifers with mainly primary porosity, the thickness of the aquifer can frequently be defined by lithologic or stratigraphic boundaries. In fractured-rock aquifers, fracture density and aperture frequently decrease with depth and it is difficult to determine at what depth the fractures are no longer capable of transmitting significant amounts of water. Examination of cores and borehole logging data may be helpful in identifying the "productive" portion of the aquifer. Karst aquifers present similar problems in that karstification may decrease with depth or be confined to very specific zones or beds within the carbonate rock. While it is often difficult to determine the base of karstification, Worthington (9) suggests that stratal dip and length of ground-water basin can be used to estimate the mean depth of flow. Reference (52) suggests that packer tests at successively lower depths can be used for estimates of depth of karstification.

6.9.2 *Porosity*—Primary porosity can be measured on the scale of a hand sample or a core sample; secondary and tertiary porosity need to be measured at a scale that statistically represents the distribution of heterogeneities in the aquifer. For densely-fractured rock, the sample volume may be relatively small and encompassed by borehole geophysical measurement techniques, while for aquifers with widely spaced heterogeneities (that is, sparsely fractured rock or conduit systems) the huge volume of rock needed prohibits meaningful evaluations of porosity.

6.9.3 *Hydraulic Conductivity and Storativity*—Hydraulic conductivity values vary with the scale of measurement (16, 17, 53). The range of hydraulic conductivities and associated ground-water velocities for karst aquifers is illustrated in Fig. 2. Hydraulic conductivity values from lab and field tests (Methods A through D) are compared to velocities of ground-water flow in conduits (Method E). The presence of conduits in a karst aquifer requires a dual-porosity approach to aquifer characterization, or at least a discrete-porosity approach, rather than a porous-medium approximation because the hydraulic conductivity would be grossly underestimated with the porous-medium approach.

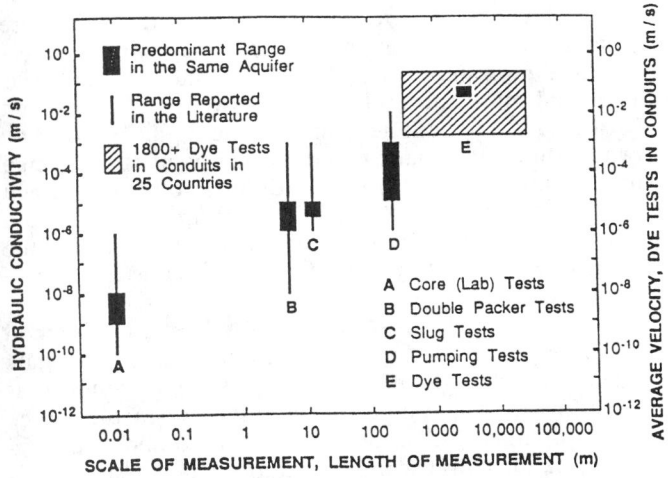

NOTE—The data represented by heavy bars are from a Jurassic karst aquifer in the Swabian Alb of Germany, as described by Sauter (69). The hatchured box represents velocity data from more than 1800 dye traces from sinking streams to springs (that is, in conduits) from 25 countries (modified after (16, 69)).

FIG. 2   Range in Hydraulic Conductivity and Ground-Water Velocity in Karst Aquifers as a Function of Scale of Measurement

6.9.3.1 Hydraulic conductivity in granular media is frequently evaluated by single-well or multiple-well pumping tests. Results of such tests performed in fractured-rock and karst aquifers should be interpreted with respect to the portion of the aquifer that responds and the measurement-scale effects, illustrated in Fig. 2 should be recognized. The discrete nature of high-conductivity zones in fractured-rock and karst aquifers can yield hydraulic conductivity values ranging over several orders of magnitude at a specific measurement scale. Figure 3 illustrates the range of hydraulic conductivity values measured from slug tests (borehole-scale) at a small site in horizontally-bedded fractured dolomite. Significant errors can occur when aquifer characterization tests are designed, conducted, or interpreted without regard to the portion of the aquifer being tested.

6.9.3.2 Site-specific investigations may require detailed information on the transmissivity of specific zones within the aquifer. In fractured-rock and karst aquifers, borehole packers can be used to segregate specific zones within the borehole. Slug tests and single-well pumping tests can then be performed to determine transmission characteristics of different portions of the aquifer. Borehole-fluid logging in a pumping well (24) can also help to characterize the producing zones within fractured-rock aquifers.

6.9.3.3 Measurements of transmissivity and storativity averaged over a ground-water basin in karst aquifers can be estimated from discharge rates at springs (32, 52). Such averaged parameters may be appropriate for regional assessments of ground-water resources but they are less appropriate for site-specific investigations.

7. Developing a Reliable Monitoring System

7.1 *Applicable Monitoring Points:*

7.1.1 Determination of applicable monitoring points will depend on which conceptual approach most accurately describes the setting under investigation. If the aquifer deviates significantly from the porous-medium approxima-

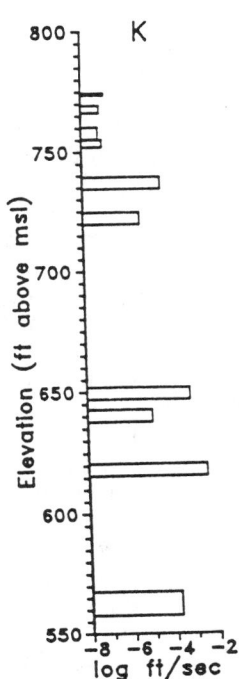

NOTE—Elevation is on the vertical axis (land surface is 800 ft (243.84 m) above msl), thickness of bar indicates length of open interval, length of the bar indicates measured hydraulic conductivity. Hydraulic conductivity values range over five orders of magnitude (compiled from data in Ref (17)).

FIG. 3   Range of Hydraulic Conductivity Values from Slug Tests (Borehole-Scale) at a Site in Fractured Dolomite

tion, monitoring wells probably will not yield representative ground-water samples unless it is demonstrated by properly designed tracer studies and hydraulic tests that the monitoring points are connected to the site to be monitored. Alternative monitoring points (such as springs, cave streams, and seeps) are usually more appropriate in karst terranes. These natural discharge points intercept flow from a larger area than a monitoring well and, as a result, they are more likely to capture drainage from a site. Monitoring sites that integrate drainage from a large area are likely to show more dilute concentrations of contaminants than monitoring sites that intercept drainage from a small area. Monitoring of alternative sampling points requires evaluation of the significance of dilution of contaminants. Designers of a monitoring system must weigh the desirability of analysis of diluted waters that are known to drain from a site versus analysis of waters that are not demonstrably derived from the site.

7.1.2 Current ground-water monitoring practices utilize both upgradient and downgradient-monitoring points in order to meet regulatory requirements. In fractured-rock and karst aquifers, rapid variations in hydraulic head can lead to changes and even reversals in ground-water flow directions (see 6.5). In these settings, determination of flow-directions from water-table or potentiometric maps may not be adequate to determine placement of monitoring points. Samples for background water-quality should be collected at springs, cave streams, and wells that yield water that is geochemically representative of the aquifer. These monitoring points might be located in an adjacent ground-water basin (30, 31, 54).

7.2 *Methods of Testing Applicability of Monitoring*

Points—Tracing studies and hydraulic tests should be used to demonstrate whether or not sampling sites are connected to the site being monitored. "Downgradient" monitoring-points cannot be assumed to intercept drainage from a site unless a positive connection from the site to the monitoring point is demonstrated.

7.2.1 *Tracer Tests:*

7.2.1.1 Tracer tests that monitor the presence or absence of tracer at monitoring points are usually sufficient for determining flow directions and validating monitoring points. Tracer tests in which tracer concentration is determined on samples collected at short-time intervals (that is, minutes to hours) can be used to determine optimum monitoring frequency that avoids aliasing; without this knowledge a large number of monitoring points might be sampled for contaminants more frequently or less frequently than is necessary for accurate characterization.

7.2.1.2 Measuring tracer concentrations and discharges at monitoring points can provide additional mass-balance data that will make the design or modification of a monitoring system more efficient. At sites where there are multiple flow-directions and discharges to numerous monitoring points, it is useful to know whether a majority of the site's drainage is to one or just a few of the monitoring points, as contrasted with nearly equal discharge to all of them. This mass-balance data can also be used to assess the significance of contaminant dilution.

7.2.1.3 As a general principle, the cost-benefit ratio of measuring both tracer concentration and discharge rises as the number of potential monitoring points increases. If mass-balance tracing results are deemed necessary, qualitative traces should be performed first. This may eliminate the cost of sampling and analyzing monitoring points which do not receive tracers. For a discussion of mass-balance tracing techniques as applied to the design of ground-water monitoring plans, see Refs (30, 43, 55, 56).

7.2.1.4 The mass-balance tracing technique described by Mull et al. (55), would be useful (in some settings) for evaluating the possible consequences of a spill into an open sinkhole draining to a cave stream. However, this technique is incapable of evaluating leakage from a waste disposal site. Therefore, use of this method should be limited to settings where there is point recharge directly into a cave stream.

7.2.2 *Hydraulic Testing Methods:*

7.2.2.1 A variety of hydraulic tests can also be used to determine the relative "connectedness" of an individual monitoring point to the fracture-flow system, connections between monitoring wells, and connection to the site being monitored. Any hydraulic testing program requires careful design because of the discrete nature of high-conductivity zones in fractured-rock and karst aquifers. Ideally, monitoring wells should intersect the producing zones that are more likely to carry contaminants from the site. However, in some cases it may also be necessary to monitor the matrix portion of the aquifer.

7.2.2.2 Packer tests and borehole logging techniques can help locate both high-conductivity and low conductivity zones within the aquifer (see 6.8.3). Pumping tests can then be designed to test the connections between various parts of the system (57). If possible, a pumping well could be placed at the source of contamination and the response of indi-

vidual monitoring wells to pumping (both rate of response and overall drawdown) could be used to determine connection to the monitoring site. Sometimes more distant wells will respond more quickly than nearby wells, indicating that they are better connected to the pumping well. Any drawdown indicates that the monitoring point is connected to the pumping well; however, it is difficult to use drawdown to assess the degree of connection. Small drawdowns could indicate a weak connection or they could indicate that the connected zone is highly transmissive (and thus difficult to draw down).

7.3 *Monitoring Wells*—Monitoring wells are the method of ground-water monitoring required by federal and state regulatory agencies and should always be considered as possible monitoring points in a karst or fractured-rock aquifer. Boreholes drilled onsite or offsite to obtain geological information can be converted to piezometers since most ground-water monitoring plans include the installation of piezometers in order to determine the variation in hydraulic head. The piezometers can then be considered as temporary or surrogate monitoring wells. If any of the piezometers later prove to be capable of providing ground-water samples representative of the water draining from the site, they can be converted to, or replaced by, true monitoring wells that meet regulatory standards.

7.3.1 *Placement of Monitoring Wells*—Placement of monitoring wells should be guided by interpretation of the data gathered in the site characterization (see 6.3 through 6.9). If the aquifer is uniformly and densely fractured, monitoring well placement, construction, and development are similar to that for granular aquifers (see Practice D 5092 and Refs (58 and 60). Different placement and construction techniques are necessary for aquifers characterized by discrete high-permeability zones (enlarged fractures, dissolution zones, and conduits) that carry the majority of the water. Wells placed in these high-permeability zones are more likely to intercept drainage from a site than randomly placed wells or wells completed in low-permeability zones. In settings where the matrix blocks have appreciable porosity, it may also be important to monitor the blocks as well as the high-permeability zones because the blocks may function as storage reservoirs for pollutants.

7.3.1.1 Fracture lineaments and the intersection of vertical fractures are potential sites for monitoring wells, especially in crystalline rocks. However, in carbonate rocks, most conduits and high-permeability zones are developed along bedding planes; monitoring wells located on the basis of fracture-trace and lineament analysis are not likely to intercept major conduits. Horizontal zones of high permeability are important in determining placement of monitoring wells. If the site characterization has identified zones of enhanced permeability (that is, noted by borehole geophysical logs, loss of circulation when drilling, etc.), monitoring wells should be constructed so as to intersect these zones.

7.3.1.2 In most karst terranes, substantial flow occurs at the soil-bedrock interface and within the subjacent epikarst. Wells placed across this interface or within the epikarst may only be intermittently saturated. However, these wells are likely to intercept the early movement of contaminants from an overlying source.

7.3.1.3 Wells drilled to intersect cave streams may also function as good monitoring points. While most geophysical techniques are incapable of detecting the flow of water within conduits, the natural potential method (a type of spontaneous potential measurement) is the only geophysical method that can detect flowing water and it can sometimes be used effectively (50, 60, 61, 62).

7.3.1.4 Monitoring wells are typically constructed within the boundaries of the site to be monitored. In fractured-rock and karst aquifers, the location of high-permeability zones should guide the placement of monitoring wells even if they are located offsite.

7.3.2 *Construction of Monitoring Wells:*

7.3.2.1 The presence of zones of enhanced dissolution can complicate construction of monitoring wells in karst and fractured-rock aquifers. Drilling methods and well construction techniques should be chosen so as to minimize loss of drilling fluids, cuttings, or construction materials to the formation. Air-rotary drilling is one possibility if circulation can be maintained and risk of partial plugging of fissures can be tolerated; rotary drilling with over-shot casing can effectively reduce loss of fluids to a formation.

7.3.2.2 The open interval of the monitoring well should be designed to intercept zones of high-permeability. If no such zones are present, the well should be cased to the depth where competent rock is encountered and left open below that. The annular space between the casing and the borehole wall should be sealed in such a way as to minimize loss of materials to the aquifer. It may be possible to use standard well construction techniques such as a bentonite slurry or grouting to set the casing if no high-permeability zones are present.

7.3.2.3 If discrete high-permeability zones are encountered, the wells should be constructed so as to be open to those zones. When smaller-diameter monitoring wells are placed within a larger-diameter borehole, a gravel pack should be installed around the screen and an annular seal placed above the gravel pack. The gravel pack should be constructed of materials that will minimize any chemical reactions with the ground-water. Bentonite chips and pellets are recommended for the annular seal because these materials are not as easily lost to the formation as are slurries of cement or bentonite.

7.3.2.4 All materials used in monitoring-well construction should meet federal regulations (63) and state guidelines. Practice D 5092 provides general recommendations for monitoring-well construction.

7.3.2.5 *Development and Maintenance*—Wells that intersect high-permeability zones frequently exhibit high turbidity if finer particles from the fractures, conduits, and other dissolution zones are drawn into the well. These wells will require more extensive development than most monitoring wells. In many wells, turbidity may be a persistent problem, particularly during and after storm events. If siltation is a persistent problem, routine maintenance to remove the accumulated sediment may be necessary.

7.3.3 *Alternative Monitoring Points*—When tracer studies and hydraulic tests do not indicate a connection between a monitoring well and the site being monitored, the well should be considered inadequate for its intended purpose; alternative monitoring points must be used. Monitoring

water quality at seeps, springs, or cave streams shown to be connected to the site by tracing studies is one alternative (31, 40, 54) provided a waiver from existing federal or state regulations can be obtained (see 7.5). Regulators are increasingly recognizing springs and cave streams as viable, efficient, and reliable monitoring points that meet the intent of the laws, even though these features may be found offsite (see 7.5.3). However, whenever possible, the protocols of locating monitoring points at the site should be followed. Detection of contaminants prior to migration offsite, the desired monitoring goal, may not be possible if the only relevant monitoring sites are offsite. However, because cave streams and seeps are natural discharge points for ground water flowing through discrete, difficult-to-locate, high-permeability zones, monitoring at these offsite sampling points may be the only appropriate and practicable monitoring strategy.

7.3.3.1 When documenting monitoring points, horizontal and vertical coordinates should be noted (see Practice D 5254), and pertinent geologic information should be recorded. Pertinent geologic information would include such things as identifying the formation from which a spring is discharging and noting particular lithologic and structural descriptors (for example, spring issues at intersection of vertical joint in limestone and bedding plane of a shale bed).

7.3.3.2 In carbonate terranes, the ratio between maximum and minimum discharge of a spring and the shape of the hydrographs are indications of whether a spring is classified as an overflow or underflow spring (64). In addition, Worthington (9) used the coefficient of variation of bicarbonate and sulfate to determine overflow/underflow springs. This classification is important in assessing the number of potential discharge points and monitoring points for a karst ground-water basin. If a spring is recognized as an overflow spring, it indicates the presence of underflow springs that carry some of the ground-water discharge, sometimes all of it when the overflow spring is not discharging. Both underflow and overflow springs must be included in a comprehensive ground-water monitoring network in karst terranes.

7.3.3.3 When collecting samples from alternative monitoring points, it is best to sample as close to a spring orifice or seep discharge as possible. Where possible, spring discharge should be measured and recorded whenever samples are taken. If the discharge cannot be accurately measured, stage height is an acceptable alternative, and even a visual estimate of discharge is better than no record at all. As in sampling a well, a visual description of the water sample should be recorded (for example, level of turbidity, coloration, presence of iron staining, presence of oil sheen, noticeable odors, etc.) and standard field parameters (for example, specific conductance, temperature, and pH) should be measured and recorded.

7.4 *Sampling Frequency:*

7.4.1 Water-quality parameters can be extremely variable in karst and fractured-rock aquifers. This is particularly true during and after recharge events that cause rapid changes in discharge at springs or rapid changes in hydraulic head at wells. Such events typically cause high-frequency, high-amplitude changes in water quality. These water-quality changes may be either in-phase or out-of-phase with discharge peaks or with each other. In order for samples to be representative of conditions in the aquifer, frequency of sampling should be selected to reflect this inherent variability rather than at pre-specified, fixed intervals as is often dictated by regulatory programs. A general discussion of sampling frequency is given in Ref (65).

7.4.1.1 The correct interpretation of the variation of water-quality data for determining proper sampling frequency cannot be done with confidence unless it is known that the results were not subject to aliasing, a phenomenon in which a high-frequency signal can be interpreted as a low-frequency signal or trend because the sampling was too infrequent to accurately characterize the signal (66).

7.4.1.2 The following discussion of sampling frequency assumes that properly designed and conducted tracer tests and hydraulic tests have identified representative down-gradient monitoring points that are connected to the site and representative background monitoring sites that are not connected to the monitored site (see 7.2).

7.4.2 *Hydrographs and Chemographs*—The determination of an appropriate sampling frequency should be based on interpretation of the behavior of several physical and chemical parameters at springs and wells. Plots of spring discharge (or stage) and water quality as functions of time (hydrograph and chemograph analysis, respectively) have been used extensively in karst-aquifer studies (14) as tools for obtaining information about the ground-water flow dynamics of the karst system. (In aquifers where discharge is to subaqueous springs or seeps, monitoring is difficult, but possible—if they can be found.) Monitoring wells completed in fractured dolomite may also exhibit extreme temporal variations in hydraulic head and water-quality parameters and these variations can be used to characterize ground-water flow dynamics in similar aquifers (17).

7.4.3 *Conventional Parameters*—A suite of easily measured parameters has commonly been used to characterize the variability of water-quality in karst aquifers. These parameters include discharge or head, specific conductance, temperature, and turbidity and are recommended as a minimum set of data to be collected. They should be measured at representative monitoring points continuously, or near-continuously, for a period of several weeks to several months and at least until several major recharge events have occurred.

7.4.4 *Determining Sampling Frequency for Target Compounds*—When monitoring pollutant releases from a site in karst or fractured-rock aquifers, the inherent variability of the system must be considered. The natural variation of head or discharge, temperature, specific conductance, and turbidity can be used to select the appropriate sampling frequency for contaminant or target compounds. For a monitoring system, the most important question to be answered is whether the maximum concentration of the target compound exceeds an established background or regulatory-action value at the point of compliance.

7.4.5 A general procedure for determining the sampling frequency of target compounds is outlined below:

7.4.5.1 Plot discharge (stage) for a spring or head for a well, specific conductance, temperature, and turbidity against time (plot all of them on the same graph, using different vertical scales, so that they may be compared). The

continuous or near-continuous measurement of these parameters is necessary in order to prevent aliasing of the data.

7.4.5.2 Determine which parameter varies the most.

7.4.5.3 Establish the correlation and time-lag (if any) between maxima and minima of discharge (stage) or head, specific conductance, temperature, and turbidity.

7.4.5.4 Determine a sampling frequency that will capture the variability of the most-variable parameter.

7.4.5.5 Sample for the target compound(s) at this sampling frequency both at the background and at the downgradient-monitoring points. Samples should be collected through at least one major recharge event. This recharge event should be near to or greater than the average annual maximum recharge event. Samples must also be collected during baseflow conditions.

7.4.5.6 Plot the concentration of the contaminant compound(s), discharge (stage) or head, specific conductance, temperature, and turbidity against time for both high-flow and baseflow conditions. Establish the correlation and lag-time between maximum target compound concentration and the maxima and minima of discharge (stage) or head, specific conductance, temperature, and turbidity. These correlations determine the subsequent sampling frequencies for the target compounds.

7.4.5.7 If the maximum target compound concentrations are measured under baseflow conditions, periodic samples collected during lowest flow conditions may achieve the monitoring goals. If maximum target compound concentrations occur during high-flow conditions, then subsequent samples for those compounds must be collected during high flow. High-flow sampling frequency should initially be based on 7.4.5.1 through 7.4.5.4 and modified as data are collected and interpreted.

7.4.6 This procedure may indicate an optimum storm-related sampling frequency ranging from minutes to hours for some systems; sampling at this frequency is necessary through at least one major recharge event. Analytical costs can be lessened by analyzing every third sample. If the contaminant concentration data plot smoothly, it may not be necessary to analyze the stored samples; if they do not, it will.

7.4.7 At the start of a recharge event, it is impossible to know how significant it will be. At its middle or end, it is too late to collect samples that will characterize its beginning. Accordingly, it is always necessary to commence sampling at the start of an event. After the event, the decision to analyze or not to analyze the samples should be based on professional judgment and evaluation of the significance of the event.

7.5 *Meeting Regulatory Goals:*

7.5.1 Regulatory agencies require ground-water monitoring wells as a means of detecting statistically significant changes in water quality resulting from releases to ground water by various operations. However, alternative monitoring points such as springs and cave streams, shown to be draining from the site, may provide the most representative ground-water samples in some settings, and have been required in some states (see 7.3.3).

7.5.2 *Current Federal Regulations*—The Code of Federal Regulations (CFR) that addresses ground-water monitoring and corrective action at hazardous-waste-disposal sites can be found in Ref **(67)** (40 CFR Subpart F §§ 264.90 to 264.101 and 40 CFR Subpart B § 270.14(c)). These regulations contain specific information on monitoring ground-water quality (§ 264) and reporting requirements based on a comprehensive site investigation (§ 270). Section 264.97 specifically lists ground-water monitoring requirements that must be met in order to satisfy the requirements of § 264.98, Detection Monitoring; § 264.99, Compliance Monitoring; and § 264.100, Corrective Action. Section 270.14(c), *Additional Information Requirements*, provides for specific information to be compiled and submitted in order to meet the provisions required under §§ 264.90 through 264.101 **(66)**.

7.5.3 *Modifications of Current Regulations*—Recent program evaluations have shown that specifically requiring certain items (for example, monitoring wells) has led to the development of inadequate ground-water monitoring systems that are designed solely to meet regulatory guidelines. The U.S. EPA is currently considering new, more flexible guidelines/regulations that allow the use of seeps, springs, and cave streams as monitoring points to supplement a monitoring-well network. Alternative monitoring points would have to meet the performance criteria described in Ref **(67)** § 264.97 [for example, monitoring wells at the point of compliance Ref **(66)** § 264.97(a)(2)] and the use of such alternative monitoring points would be based on a site investigation designed to assess the hydrogeologic conditions of the site. These new procedures have been outlined in Refs **(63, 67, 68)**.

7.5.3.1 Alternative monitoring schemes have been proposed and implemented at some facilities regulated under the Resource Conservation and Recovery Act (RCRA), where hydrogeologic conditions did not conform to the porous-medium approximation. The reasoning behind such variances was that a monitoring system consisting solely of monitoring wells that are unable to provide ground-water samples representative of a given site would not meet the intent or spirit of the law. However, monitoring ground-water quality at alternative, offsite monitoring points, while violating the letter of the law (because the points are offsite), would meet the intent and spirit of the law. Such variances from RCRA regulations are considered acceptable for existing and interim status facilities. Variances have not been allowed for a new facility because § 264.97(a)(2) **(66)** stipulates that monitoring must be conducted at the point of compliance (for example, the unit boundary or facility boundary).

8. Keywords

8.1 carbonate aquifers; fractured-rock; ground water; ground-water monitoring; ground-water sampling; karst; springs

REFERENCES

(1) Schmelling, S. G., and Ross, R. R., "Superfund Ground Water Issue. Contaminant Transport in Fractured Media: Models for Decision Makers," U.S. Environmental Protection Agency, Office of Emergency and Remedial Response, Washington, DC, *EPA/540/4-89/004*, 1989, (Available from NTIS as document PB90-268517).

(2) Quinlan, J. F., Smart, P. L., Schindel, G. M., Alexander, E. C., Jr., Edwards, A. J., and Smith, A. R., "Recommended Administrative/Regulatory Definition of Karst Aquifers, Principles for Classification of Carbonate Aquifers, Practical Evaluation of Vulnerability of Karst Aquifers, and Determination of Optimum Sampling Frequency at Springs," *Proceedings*, Hydrology, Ecology, Monitoring, and Management of Ground Water in Karst Terranes Conference, National Ground Water Association, Dublin, Ohio, 1992, pp. 573–635.

(3) Bradbury, K. R., Muldoon, M. A., Zaporozec, A., and Levy, J., *Delineation of Wellhead Protection Areas in Fractured Rocks*, U.S. Environmental Protection Agency, Office of Ground Water and Drinking Water, Washington, DC, EPA 570/9-91-009, 1991.

(4) Fetter, Jr., C. W., "Determination of the Direction of Groundwater Flow," *Ground Water Monitoring Review*, Vol 1, No. 4, 1981, pp. 28–31.

(5) Cleary, T. C. B. F., and Cleary, R. W., "Delineation of Wellhead Protection Areas: Theory and Practice," *Water Science and Technology*, Vol 24, 1991, pp. 239–250.

(6) Atkinson, T. C., "Diffuse Flow and Conduit Flow in Limestone Terranes in the Mendip Hills, Somerset (Great Britain)," *Journal of Hydrology*, Vol 35, 1977, pp. 93–110.

(7) Shuster, E. T., and White, W. B., "Seasonal Fluctuations in the Chemistry of Limestone Springs: A Possible Means for Characterizing Carbonate Aquifers," *Journal of Hydrology*, Vol 14, 1971, pp. 93–128.

(8) Newson, M. D., "A Model of Subterranean Limestone Erosion in the British Isles Based on Hydrology," *Transactions*, Institute of British Geographers, Vol 54, 1971, pp. 51–70.

(9) Worthington, S. R. H., *Karst Hydrogeology of the Canadian Rocky Mountains*, Ph.D. thesis, Geography Department, McMaster University, Hamilton, Ontario, 1991. (Available from University Microfilms, Ann Arbor, MI.)

(10) Worthington, S. R. H., Davies, G. J., and Quinlan, J. F., "Geochemistry of Springs in Temperate Carbonate Aquifers: Recharge Type Explains Most of the Variation," *Proceedings*, Colloque d'Hydrologie en Pays Calcaire et en Milieu Fissuré (5th Neuchâtel, Switzerland), *Annales Scientifiques de l'Université de Bescançon, Geologie—Mémoires Hors Série*, No. 11, 1992, pp. 341–347.

(11) White, W. B., and Longyear, J., "Some Limitations on Speleo-Genetic Speculation Imposed by the Hydraulics of Groundwater Flow in Limestone," *Nittany Grotto Newsletter*, Vol 10, 1962, pp. 155–167.

(12) Williams, P. W., "Role of the Subcutaneous Zone in Karst Hydrology," *Journal of Hydrology*, Vol 61, 1983, pp. 45–67.

(13) Smart, P. L., and Freiderich, H., "Water Movement and Storage in the Unsaturated Zone of a Maturely Karstified Carbonate Aquifer, Mendip Hills, England," *Proceedings*, Environmental Problems in Karst Terranes and Their Solutions Conference, National Water Well Association, Dublin, Ohio, 1986, pp. 59–87.

(14) Ford, D. C., and Williams, P. W., *Karst Geomorphology and Hydrology*, Unwin Hyman, Boston, Massachusetts, 1989.

(15) Dreiss, S. J., "Regional Scale Transport in a Karst Aquifer, 2. Linear Systems and Time Moment Analysis," *Water Resources Research*, Vol 25, 1989, pp. 126–134.

(16) Quinlan, J. F., Davies, G. J., and Worthington, S. R. H., "Rationale for the Design of Cost-Effective Groundwater Monitoring Systems in Limestone and Dolomite Terranes: Cost Effective as Conceived is not Cost Effective as Built if the System Design and Sampling Frequency Inadequately Consider Site Hydrogeology," *Proceedings*, Annual Waste Testing and Water Quality Assurance Symposium, U.S. Environmental Protection Agency, Washington, DC, 1992, pp. 552–570.

(17) Bradbury, K. R., and Muldoon, M. A., "Hydraulic Conductivity Determination in Unlithified Glacial and Fluvial Materials," *Ground Water and Vadose Zone Monitoring, ASTM STP 1053*, D. M. Nielsen and A. I. Johnson, Eds., ASTM, Philadelphia, 1990, pp. 138–151.

(18) Clauser, C., "Permeability of Crystalline Rocks," *Eos*, Vol 73, No. 21, May 26, 1992, pp. 233, 237–238.

(19) Smart, P. L., Edwards, A. J., and Hobbs, S. L., "Heterogeneity in Carbonate Aquifers: Effect of Scale, Fissuration, and Karstification," *Proceedings*, Hydrology, Ecology, Monitoring, and Management of Ground Water in Karst Terranes Conference, National Ground Water Association, Dublin, Ohio, 1992, pp. 39–57.

(20) Smith, E. D., and Vaughan, N. D., "Experience with Aquifer Testing and Analysis in Fractured Low-Permeability Sedimentary Rocks Exhibiting Nonradial Pumping Response," *Proceedings*, Hydrogeology of Rocks of Low Permeability, International Association of Hydrogeologists, 17th Congress, Memoirs, Vol 27, 1985, pp. 137–149.

(21) Bakalowicz, M., and Mangin, A., "L'aquifère Karstique. Sa Définition, Ses Charactéristiques et son Identification," *Hydrogéologie: Interactions Entre l'eau Souterrain et son Milieu*, Société Géologique de France, Mémoire Hors Série No. 11, 1980, pp. 71–79.

(22) Safko, P. S., and Hickey, J. J., "A Preliminary Approach to the Use of Borehole Data, Including Television Surveys, for Characterizing Secondary Porosity of Carbonate Rocks in the Floridan Aquifer System," *U.S. Geological Survey, Water Resources Investigation Report*, No. 91-4168, 1992.

(23) Williams, J. H., and Conger, R. W., "Preliminary Delineation of Contaminated Water-Bearing Fractures Intersected by Open-Hole Bedrock Wells," *Ground Water Monitoring Review*, Vol 10, No. 4, 1990, pp. 118–126.

(24) Pedler, W. H., Barvenik, M. J., Tsang, C. F., and Hale, F. V., "Determination of Bedrock Hydraulic Conductivity and Hydrochemistry Using a Wellbore Fluid Logging Method," *Proceedings*, National Outdoor Action Conference on Aquifer Restoration, Ground Water Monitoring, and Geophysical Methods, National Well Water Association, Dublin, Ohio, 1990, pp. 39–53.

(25) Molz, F. J., Morin, R. H., Hess, A. E., Melville, J. G., and Güven, O., "The Impeller Meter for Measuring Aquifer Permeability Variations: Evaluation and Comparison With Other Tests," *Water Resources Research*, Vol 25, 1989, pp. 1677–1683.

(26) Hess, A. E., and Paillet, F. L., "Application of the Thermal-Probe Flowmeter in the Hydraulic Characterization of Fractured Rocks," *Geophysical Investigations for Geotechnical Investigations, ASTM STP 1101*, ASTM, 1990, pp. 99–112.

(27) Young, S. C., and Pearson, J. S., "Characterization of Three-Dimensional Hydraulic Conductivity Field with an Electromagnetic Borehole Flowmeter," *Proceedings*, National Outdoor Action Conference on Aquifer Restoration, Ground-Water Monitoring, and Geophysical Methods, National Well Water Association, Dublin, Ohio, 1990, pp. 83–97.

(28) Kerfoot, W. B., Beaulieu, G., and Kieley, L., "Direct-Reading Borehole Flowmeter Results in Field Applications," *Proceedings*, National Outdoor Action Conference on Aquifer Restoration, Ground-Water Monitoring, and Geophysical Methods, National Well Water Association, Dublin, Ohio, 1991, pp. 1073–1084.

(29) Benson, R. C., and Yuhr, L., "Spatial Sampling Considerations and Their Applications to Characterizing Fracture and Cavity Systems," *Proceedings*, Multidisciplinary Conference on Sinkholes and the Environmental Impacts of Karst, Balkema, Rotterdam, 1993, pp. 99–113.

(30) Quinlan, J. F., *Ground-Water Monitoring in Karst Terranes—Recommended Protocols and Implicit Assumptions*, U.S. Environmental Protection Agency, Environmental Monitoring Systems Laboratory, Las Vegas, Nevada, EPA 600/X-89/050 1989. (Draft; final version to be published 1993.)

(31) Quinlan, J. F., "Special Problems of Ground-Water Monitoring in Karst Terranes," *Ground Water and Vadose Zone Monitoring, ASTM STP 1053*, D. M. Nielsen and A. I. Johnson, Eds., American Society for Testing and Materials, ASTM, Philadelphia, 1990, pp. 275–304.

(32) White, W. B., *Geomorphology and Hydrology of Karst Terrains*, Oxford, New York, 1988.

(33) Fernandez, L. A., and Wood, W. W., "Volcanic Rocks," *Hydrogeology, The Geology of North America*, Vol O-2, W. Back, J. S. Rosenshein, and P. R. Seaber, Eds., Geological Society of America, Boulder, Colorado, 1988, pp. 353–365.

(34) Banks, David, "Optimal Orientation of Water-Supply Boreholes in Fractured Aquifers," *Ground Water*, Vol 30, 1992, pp. 895–900.

(35) Fetter, Jr., C. W., *Applied Hydrogeology*, 2nd ed., Macmillan, New York, New York, 1988.

(36) Farvolden, R. N., Pfannkuch, O., Pearson, R., and Fritz, P., "Region 12, Precambrian Shield," *Hydrogeology, The Geology of North America*, Vol O-2, W. Back, J. S. Rosenshein, and P. R. Seaber, Eds., Geological Society of America, Boulder, Colorado, 1988, pp. 101–114.

(37) Trainer, F. W., "Plutonic and Metamorphic Rocks," *Hydrogeology, The Geology of North America*, Vol O-2, W. Back, J. S. Rosenshein, and P. R. Seaber, Eds., Geological Society of America, Boulder, Colorado, 1988, pp. 367–380.

(38) LaPointe, P. R., and Hudson, J. A., *Characterization and Interpretation of Rock Mass Joint Patterns*, Geological Society of America, Special Paper 199, 1985.

(39) Saines, M., "Errors in the Interpretation of Ground-Water Level Data," *Ground Water Monitoring Review*, Vol 2, No. 1, 1981, pp. 56–61.

(40) Quinlan, J. F., and Ewers, R. O., "Subsurface Drainage in the Mammoth Cave Area: In *Karst Hydrology—Concepts from the Mammoth Cave Area*, White, W. B., and White, E. L. (eds.), Van Nostrand Reinhold, 1989.

(41) Bradbury, K. R., and Muldoon, M. A., "Hydrogeology and Ground-Water Monitoring of Fractured Dolomite in the Upper Door County Priority Watershed, Door County, Wisconsin," Wisconsin Geological and Natural History Survey, Open File Report (*WOFR 92-2*), 1992.

(42) Smart, P. L., "A Review of the Toxicity of Twelve Fluorescent Dyes Used in Water Tracing," *National Speleological Society Bulletin*, Vol 46, No. 2, 1984, pp. 21–33.

(43) Alexander, E. C., Jr., and Quinlan, J. F., *Practical Tracing of Groundwater with Emphasis on Karst Terranes*, Short Course Manual. Geological Society of America, Boulder, Colorado, 1992.

(44) Franklin, A. G., Patrick, D. M., Butler, D. K., Strohm, W. E., and Hayes-Griffin, M. E., *Foundation Considerations in Siting of Nuclear Facilities in Karst Terrains and Other Areas Susceptible to Ground Collapse*, NUREG/CR-2062, R6, RA, CA, CG. U.S. Army Waterways Experiment Station, Vicksburg, Mississippi, 1981.

(45) Benson, R. C., and LaFountain, L. J., "Evaluation of Subsidence or Collapse Potential Due to Subsurface Cavities," *Proceedings*, Multidisciplinary Conference on Sinkholes, Balkema, Rotterdam, 1984, pp. 161–169.

(46) Morgenstern, K. A., and Syverson, T. L., "Determination of Contaminant Migration in Vertical Faults and Basalt Flows With Electromagnetic Conductivity Techniques," *Proceedings*, National Outdoor Action Conference on Aquifer Restoration, Ground Water Monitoring, and Geophysical Methods, 1988a, National Well Water Association, Dublin, Ohio, pp. 597–615.

(47) Morgenstern, K. A., and Syverson, T. L., "Utilization of Vertical and Horizontal Dipole Configurations of the EM 34-3 for Contaminant Mapping in Faulted Terrain" *Proceedings*, Superfund '88, Hazardous Materials Control Research Institute, Silver Spring, Maryland, 1988b, pp. 84–92.

(48) Jansen, J., "Surficial Geophysical Techniques for the Detection of Bedrock Fracture Systems," *Proceedings*, Eastern Ground Water Conference, National Water Well Association, Dublin, Ohio, 1990, pp. 239–253.

(49) Taylor, R. W., and Fleming, A., "Characterization of Jointed Systems by Azimuthal Resistivity Survey," *Ground Water*, Vol 26, 1988, pp. 468–474.

(50) Karous, M., and Mareš, S. *Geophysical Methods in Studying Fracture Aquifers*, Charles University, Prague, 1988.

(51) Keys, W. S., *Borehole Geophysics Applied to Ground-Water Investigations.* National Water Well Association, 1989. (Also published as: *Techniques of Water-Resources Investigations, United States Geological Survey,* Book 2, Chapter E2, USGS, Denver, Colorado, 1990.)

(52) Milanović, P. T., *Karst Hydrogeology,* Water Resources Publications, Littleton, Colorado, 1981.

(53) Kiraly, L., "Rapport Sur l'état Actuel des Connaissances dans le Domaine Des Charactères Physiques des Roches Karstiques," *Hydrogeology of Karstic Terrains,* A. Burger, and L. Dubertret, Eds., International Union Geological Sciences, Series B, No. 3, Paris, 1975, pp. 53–67.

(54) Quinlan, J. F., and Ewers, R. O., "Ground Water Flow in Limestone Terranes: Strategy Rationales and Procedure for Reliable, Efficient Monitoring of Ground-Water Quality in Karst Areas," *Proceedings,* National Symposium and Exposition on Aquifer Restoration and Ground Water Monitoring, National Water Well Association, Worthington, Ohio, 1985, pp. 197–234.

(55) Mull, D. S., Liebermann, T. D., Smoot, J. L., and Woosley, L. H., Jr., *Application of Dye-Tracing Techniques for Determining Solute-Transport Characteristics of Ground Water in Karst Terranes,* U.S. Environmental Protection Agency, Region IV, Atlanta, Georgia, EPA 904/6-88-001, 1988.

(56) McCann, M. R., and Krothe, N. C., "Development of a Monitoring Program at a Superfund Site in a Karst Terrane Near Bloomington, Indiana," *Proceedings,* Hydrology, Ecology, Monitoring, and Management of Ground Water in Karst Terranes Conference, National Ground Water Association, Dublin, Ohio, 1992, pp. 349–371.

(57) Robinson, J. L., and Hutchinson, C. B., "Ground-Water Tracer Tests in West-Central Florida," *American Institute of Hydrology Annual Meeting, Abstracts,* 1991.

(58) U.S. Environmental Protection Agency, "Revision of Chapter Eleven of SW-846: Ground-Water Monitoring System Design, Installation, and Operating Practices, Final Draft," 1991c, pp. 71–85.

(59) Aller, L., Bennett, T. W., Hackett, G., Petty, R. J., Lehr, J. H., Sedoris, H., Nielsen, D. M., and Denne, J. E., *Handbook of Suggested Practices for the Design and Installation of Ground-Water Monitoring Wells,* U.S. Environmental Protection Agency, Environmental Monitoring Systems Laboratory, Las Vegas, NV, *EPA 600/4-89/034,* 1991.

(60) Kilty, K. T., and Lange, A. L., "Electrochemistry of Natural Potential Processes in Karst," *Proceedings,* Hydrology, Ecology, Monitoring, and Management of Ground Water in Karst Terranes Conference, National Ground Water Association, Dublin, Ohio, 1992, pp. 163–177.

(61) Lange, A. L., and Kilty, K. T., "Natural Potential Responses of Karst Systems at the Ground Surface," *Proceedings,* Hydrology, Ecology, Monitoring, and Management of Ground Water in Karst Terranes Conference, National Ground Water Association, Dublin, Ohio, 1992, pp. 179–196.

(62) Merkler, G. P., Miltzer, H., Hötzl, H., Armbruster, H., and Brauns, J., Eds., *Detection of Subsurface Flow Phenomena,* Lecture Notes in Earth Sciences, No. 27. Springer-Verlag, Berlin, 1989.

(63) U.S. Environmental Protection Agency, *Test Methods for Evaluating Solid Wastes: SW-846,* Third ed., Office of Solid Waste and Emergency Response, U.S. Environmental Protection Agency, Washington, DC, Vols IA, IB, IC, and II, 1986.

(64) Smart, C. C., "Hydrology of a Glacierised Alpine Karst," Ph.D. Thesis, McMaster University, Hamilton, Ontario, Canada, 1983. (Available from University Microfilms, Ann Arbor, MI.)

(65) Barcelona, M. J., Wehrmann, H. A., Schock, M. R., Sievers, M. E., and Karny, J. R., Sampling Frequency for Ground-Water Quality Monitoring, U.S. Environmental Agency, Environmental Monitoring Systems Laboratory, Las Vegas, NV, *EPA/600/4-89/032,* 1989.

(66) Gottman, J. M., *Time Series Analysis: A Comprehensive Introduction for Social Scientists,* Cambridge University Press, Cambridge, 1981.

(67) U.S. Environmental Protection Agency, "Proposed Modifications to Title 40 CFR Part 264—Standards for Owners and Operators of Hazardous Waste Treatment, Storage, and Disposal Facilities," 1991b, pp. 2, 9, 40–44, and 51–52.

(68) U.S. Environmental Protection Agency, "Notice of Proposed Rulemaking for Ground-Water Monitoring Constituents (Phase II) and Methods Under Subtitle C of the Resource Conservation and Recovery Act—ACTION MEMORANDUM," from Don Clay, Assistant Administrator for Solid Waste and Emergency Response to William K. Riley, Administrator, EPA, 1991.

(69) Sauter, M., "Assessment of Hydraulic Conductivity in a Karst Aquifer at Local and Regional Scale," *Proceedings,* Hydrology, Ecology, Monitoring, and Management of Ground Water in Karst Terranes Conference, National Ground Water Association, Dublin, Ohio, 1992, pp. 39–57.

3.2 WATER SAMPLING

Standard Practice for
Sampling with a Dipper or Pond Sampler[1]

This standard is issued under the fixed designation D 5358; the number immediately following the designation indicates the year of original adoption or, in the case of revision, the year of last revision. A number in parentheses indicates the year of last reapproval. A superscript epsilon (ε) indicates an editorial change since the last revision or reapproval.

[ε1] NOTE—Title was corrected editorially in February 1995.

1. Scope

1.1 This practice describes the procedure and equipment for taking surface samples of water or other liquids using a dipper. A pond sampler or dipper with extension handle allows the operator to sample streams, ponds, waste pits, and lagoons as far as 15 ft from the bank or other secure footing. The dipper is useful in filling a sample bottle without contaminating the outside of the bottle.

1.2 *This standard does not purport to address all of the safety concerns, if any, associated with its use. It is the responsibility of the user of this standard to establish appropriate safety and health practices and determine the applicability of regulatory limitations prior to use.*

2. Referenced Documents

2.1 *ASTM Standards:*
D 4687 Guide for General Planning of Waste Sampling[2]
D 5088 Practice for Decontamination of Field Equipment Used at Nonradioactive Waste Sites[3]
2.2 *Other Documents:*
EPA-600/2-80-018 Samplers and Sampling Procedures for Hazardous Waste Streams[4]
EPA-600/4-84-076 Characterization of Hazardous Waste Sites-A Methods Manual: Volume II. Available Sampling Methods, Second Edition[4]

3. Summary of Practice

3.1 The dipper is lowered into the liquid and lifted out while avoiding splashing or otherwise disturbing the surface layer. The sample is then poured into a sample container.

4. Significance and Use

4.1 This practice is intended for use in the sampling of surface waters and other liquids.

4.2 Dipper equipment is uncomplicated in construction, simple to use, and relatively easy to decontaminate; however, this practice will not provide accurate results with multiphase liquids.

4.3 This practice is to be used by personnel acquiring samples.

4.4 The dipper is best used to take a *surface* sample of liquid. No attempt should be made to take subsurface samples with a dipper because mixing and dilution will occur as the dipper is brought to the surface. Subsurface layers must be sampled using a device that can be closed prior to bringing it to the surface (for example, a COLIWASA, tube sampler, or bottle sampler).

4.5 The dipper is not appropriate for sampling multiphase liquids if quantitative characterization is needed.

4.6 When volatile organic analysis (VOA) will be performed, samples obtained with a dipper should be poured into an appropriate container (VOA vial) with minimal air contact and agitation.

4.7 This practice should be used in conjunction with Guide D 4687, which covers sampling plans, safety, Quality Assurance (QA), preservation, decontamination, labeling, and chain-of-custody procedures, also Practice D 5088, which covers decontamination of field equipment used at waste sites. Other documents pertinent to this practice are EPA-600/2-80-018 and EPA-600/4-84-076.

5. Sampling Equipment

5.1 Dippers may be fabricated as shown in Fig. 1 and are also available commercially (see Fig. 2). Disposable dippers are convenient for use with hazardous materials. Dippers selected must be constructed of materials compatible with the liquid being sampled and with the tests or analyses to be performed. Light weight and rigidity are important characteristics of the extension handle.

6. Sample Containers

6.1 Plastic, glass, or other nonreactive containers should be used. Refer to Guide D 4687 for further information on containers.

7. Procedure

7.1 Have sample containers ready for use.

7.2 Assemble the dipper and, if desired, the extension handle for the pond sampler.

7.3 Lower the dipper slowly into the liquid. The dipper should be on its side so that liquid runs into the container without bubbling or swirling. Rotate the dipper so that the lip of the container is up, preventing the sample from running out. Lift the dipper from the liquid.

7.4 Remove the cap from the sample container, and tilt the container slightly.

7.5 Pour the sample from the dipper down the inside of the sample container. Avoid splashing of the sample.

---
[1] This practice is under the jurisdiction of ASTM Committee D-34 on Waste Management and is the direct responsibility of Subcommittee D34.01 on Sampling and Monitoring.
Current edition approved Jan. 15, 1993. Published March 1993.
[2] *Annual Book of ASTM Standards*, Vol 11.04.
[3] *Annual Book of ASTM Standards*, Vol 04.08.
[4] Available from Standardization Documents Order Desk, Bldg. 4 Section D, 700 Robbins Ave., Philadelphia, PA 19111-5094, Attn: NPODS.

7.6 Leave adequate air space in the container to allow for expansion, except for VOA vials. Fill VOA vials to the lip, and seal with no headspace to prevent loss of volatiles.

7.7 Complete the field log book and chain-of-custody forms.

7.8 Decontaminate the used equipment in accordance with Practice D 5088.

8. Keywords

8.1 dipper; liquid sampling; pond sampler; sampling; waste

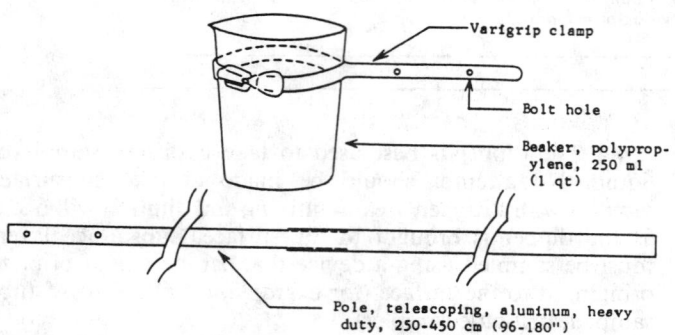

FIG. 1 Dipper with Extension Handle (also known as a Pond Sampler)

FIG. 2 Typical Commercially Available Dipper

Standard Guide for
Continual On-Line Monitoring Systems for Water Analysis[1]

This standard is issued under the fixed designation D 3864; the number immediately following the designation indicates the year of original adoption or, in the case of revision, the year of last revision. A number in parentheses indicates the year of last reapproval. A superscript epsilon (ε) indicates an editorial change since the last revision or reapproval.

1. Scope

1.1 This guide covers the selection, establishment, application, and validation and verification of monitoring systems for determining water characteristics by continual sampling, automatic analysis, and recording or otherwise signaling of output data. The system chosen will depend on the purpose for which it is intended: whether it is for regulatory compliance, process monitoring, or to alert the user of adverse trends. If it is to be used for regulatory compliance, the method published or referenced in the regulations should be used in conjunction with this guide and other ASTM methods.

1.2 *This standard does not purport to address all of the safety concerns, if any, associated with its use. It is the responsibility of the user of this standard to establish appropriate safety and health practices and determine the applicability of regulatory limitations prior to use.* Specific hazard statements are given in Section 7.

2. Referenced Documents

2.1 *ASTM Standards:*
D 1129 Terminology Relating to Water[2]
D 1193 Specification for Reagent Water[2]
D 2579 Test Methods for Total and Organic Carbon in Water[2]
D 3370 Practices for Sampling Water from Closed Conduits[2]
D 4210 Practice for Intralaboratory Quality Control Procedures and a Discussion on Reporting Low-Level Data[2]
D 5540 Practice for Flow Control and Temperature Control for On-line Water Sampling and Analysis[2]
E 178 Practice for Dealing with Outlying Observations[3]
2.2 *ASTM Special Technical Publication:*
STP 442 Manual on Water[4]

3. Terminology

3.1 *Definitions*—For definitions of terms used in this guide refer to Terminology D 1129.
3.2 *Descriptions of Terms Specific to This Standard:*
3.2.1 *Calibrations:*

3.2.1.1 *laboratory check sample for flow-through systems*—calibration curve calculated from withdrawn samples or additional standards that may be spiked or diluted and analyzed using the appropriate laboratory analyzer.

3.2.1.2 *line sample calibration*—coincidental comparison of a line sample and adjustment of a continuous analyzer to the compared laboratory analyzer or a second continuous analyzer.

3.2.1.3 *multiple standard calibration*—where the calibration curve is calculated from a series of calibration standards covering the range of the measurements of the sample being analyzed.

3.2.1.4 *probe calibration*—where the probe is removed from the sample stream and exposed to a calibration solution and the analyzer is adjusted to indicate the appropriate value. Alternately, two probes are exposed to the same solution and the on-line analyzer is adjusted to coincide with the pre-calibrated laboratory instrument.

3.2.1.5 *reference sample calibration*—coincidental comparison of a reference sample and adjustment of a continuous analyzer to the compared laboratory analyzer results.

3.2.2 *cycle time*—the interval between repetitive sample introductions in a monitoring system with discrete sampling.

3.2.3 *drift*—the change in system output, with constant input over a stated time period of unadjusted, continuous operation; usually expressed as percentage of full scale over a 24-h period.

3.2.3.1 *span drift*—drift when the input is at a constant, stated upscale value.

3.2.3.2 *zero drift*—drift when the input is at zero.

3.2.4 *full scale*—the maximum measuring limit of the system for a given range.

3.2.5 *input*—the value of the parameter being measured at the inlet to the analyzer.

3.2.6 *interference*—an undesired output caused by a substance or substances other than the one being measured. The effect of interfering substance(s) on the measured parameter of interest shall be expressed as a percentage change (±) in the measured component as the interference varies from 0 to 100 % of the measuring scale. If the interference is nonlinear, an algebraic expression should be developed (or curve plotted) to show the varying effect.

3.2.7 *laboratory analyzer*—a device that measures the chemical composition or a specific physical, chemical, or biological property of a sample.

3.2.8 *limit of detection*—a concentration of twice the criterion of detection when it has been decided that the risk of making a Type II error is equal to a Type I error as described in Practice D 4210.

3.2.9 *linearity*—the extent to which an actual analyzer reading agrees with the reading predicted by a straight line

[1] This guide is under the jurisdiction of ASTM Committee D-19 on Water and is the direct responsibility of Subcommittee D19.03 on Sampling of Water and Water-Formed Deposits, Surveillance of Water, and Flow Measurement of Water.
Current edition approved Feb. 10, 1996. Published May 1996. Originally published as D 3864 – 79. Last previous edition D 3864 – 79 (1990).
[2] *Annual Book of ASTM Standards*, Vol 11.01.
[3] *Annual Book of ASTM Standards*, Vol 14.02.
[4] Available from ASTM Headquarters. Contact Customer Service, 100 Barr Harbor Drive, West Conshohocken, PA 19428-2959.

drawn between upper and lower calibration points—generally zero and full-scale. (The maximum deviation from linearity is frequently expressed as a percentage of full-scale.)

3.2.10 *monitoring system*—the integrated equipment package comprising sampling system, analyzer, and data output equipment, required to perform water quality analysis automatically.

3.2.10.1 *analyzer*—a device that continually measures the specific physical, chemical, or biological property of a sample.

3.2.10.2 *data acquisition equipment*—analog or digital devices for acquiring, processing, or recording, or a combination thereof, the output signals from the analyzer.

3.2.10.3 *sampling system*—equipment necessary to deliver a continual representative sample to the analyzer.

3.2.11 *output*—a signal, usually electrical, that is related to the parametric measurement and is the intended input to data acquisition equipment.

3.2.12 *range*—the region defined by the minimum and maximum measurable limits.

3.2.13 *repeatability*—a measure of the precision of one analyzer to repeat its results on independent introduction of the same sample at different time intervals.

3.2.14 *reproducibility*—a measure of the precision of different analyzers to repeat results on the same sample.

3.2.15 *response time*—the time interval from a step change in the input or output reading to 90 % of the ultimate reading.

3.2.15.1 *lag time*—the time interval from a step change in input to the first corresponding change in output.

3.2.15.2 *total time*—the time interval from a step change in the input to a constant analyzer signal output.

3.2.16 *sample port*—that point in the sampling system located between the sample conditioning unit and the analyzer or at the outlet of the analyzer from which samples for laboratory analysis are taken.

3.2.17 *samples:*

3.2.17.1 *line sample*—a process sample withdrawn from the sample port (3.2.16) during a period when the process stream flowing through the continuous analyzer is of uniform quality and the analyzer result displayed is essentially constant. Laboratory tests or results from a second continuous analyzer are obtained from each sample and compared with the continuous analyzer results obtained at the time of sampling.

3.2.17.2 *reference sample*—can be a primary standard or a dilution of a primary standard of known reference value. The reference value must be established through multiple testing using an appropriate ASTM or other standard laboratory test method. Bulk quantities of the reference sample must be stored and handled to avoid contamination or degradation. One or more reference samples encompassing the range of the analyzer may be required.

NOTE 1—It is essential that the laboratory analyzer be checked carefully before these tests are performed to ensure compliance with the requirements of the standard test procedure. To further ensure proper operation it is recommended that a previously calibrated reference sample or an in-house control standard of known concentration be tested to validate the operations of the laboratory analyzer.

3.2.18 *validations*—a one-time comprehensive examination of analytical results.

3.2.18.1 *reference sample validations*—a reference sample is analyzed a minimum of seven times by an appropriate continuous analyzer and by an appropriate laboratory analyzer. A comparison is made between the average continuous analyzer results and the average laboratory results using the Student's t test at 95 % confidence coefficient, two-tailed test as described in 14.1. Passing the Student's t test signifies the continuous analyzer's average analysis of the reference sample is not statistically significantly different from the laboratory analyzer's average analysis of the same reference sample (validation test acceptable). Failing the "t" test signifies a statistically significant difference exists (validation test not acceptable).

3.2.18.2 *line sample validations*—a line sample is analyzed coincidentally a minimum of seven times by an appropriate continuous analyzer and an appropriate laboratory analyzer or a second continuous analyzer. A comparison is made on the differences between the coincidental results using the Student's t test at 95 % confidence coefficient, two-tailed test, to evaluate whether the average difference is statistically significantly different from zero difference as described in 14.2.

3.2.19 *verification*—a periodic or routine procedure to ensure reliability of analytical results.

3.2.19.1 *line sample verification*—a line sample is analyzed as described in 3.2.18.2, and the results of the difference between the continuous analyzer and the laboratory analyzer or a second continuous analyzer is plotted on a control chart. If the calculated difference between the continuous analyzer and the laboratory analyzer or a second continuous analyzer is within $\pm 3\ S_d$, the continuous analyzer is considered verified. If the calculated difference is outside $\pm 3\ S_d$ the continuous analyzer is considered out of control (not verified).

3.2.19.2 *reference sample verification*—a reference sample is analyzed as described in 3.2.18.1 and the results of the differences between the continuous analyzer and the laboratory analyzer are plotted on a control chart. If the calculated difference between the continuous analyzer and the laboratory analyzer is within $\pm 3\ S_d$ the continuous analyzer is considered verified. If the calculated difference is outside $\pm 3\ S_d$ the continuous analyzer is considered out of control (not verified).

4. Summary of Guide

4.1 This guide provides a unified approach to the use of on-line monitoring systems for water quality analysis. It presents definitions of terms, safety precautions, system design and installation considerations, calibration techniques, general operating procedures, and comments relating to validation and verification procedures.

5. Significance and Use

5.1 Many of the manual and automated laboratory methods for measurement of physical, chemical, and biological parameters in water and waste water are adaptable to on-line sampling and analysis. The resulting real-time data output can have a variety of uses, including confirming regulatory compliance, controlling process operations, or detecting leaks or spills.

5.2 This guide is intended to be a common reference that

can be applied to all water quality monitoring systems. However, calibration, validation, and verification sections may be inappropriate for certain tests since the act of removing a sample from a flowing stream may change the sample.

5.3 Technical details of the specific methodology are contained in the pertinent ASTM standard test methods, which will reference this practice for guidance in selection of systems and their proper implementation.

5.4 This guide complements descriptive information on this subject found in the *ASTM Manual on Water*.[4]

6. Reagents

6.1 *Purity of Reagents*—Reagent grade chemicals shall be used in all tests. Unless otherwise indicated, it is intended that all reagents shall conform to the specifications of the Committee on Analytical Reagents of the American Chemical Society.[5] Other grades may be used, provided it is first ascertained that the reagent is of sufficiently high purity to permit its use without lessening the accuracy of the determination.

6.2 *Purity of Water*—Unless otherwise indicated, references to water shall be understood to mean reagent water conforming to Specification D 1193, Type II.

7. Hazards

7.1 Each analyzer installation shall be given a thorough safety engineering study.[6]

7.2 Electrically, the monitoring system as well as the individual components, shall meet all code requirements for the particular area classification.

7.2.1 All analyzers using 120 V, alternating current, 60 Hz, 3-wire systems shall observe polarity and shall not use mechanical adapters for 2-wire outlets.

7.2.2 Check the neutral side of the power supply at the analyzer to see that it is at ground potential.

7.2.3 Connect the analyzer's ground connection to earth ground and check for proper continuity.

7.2.4 The metallic framework of the analyzer shall be at ground potential.

7.2.5 Consider additional protection in the form of properly sized ground fault interrupters for each individual application.

7.2.6 Analyzers containing electrically heated sections shall have a temperature-limit device.

7.2.7 The analyzer, and any related electrical equipment (the system), shall have a properly sized power cutoff switch and a fuse or breaker on the "hot" side of the line(s) of each device.

7.3 Give full consideration to safe disposal of the analyzer's spent samples and reagents.

7.4 Provide pressure relief valves, if applicable, to protect both the analyzer and monitoring system.

7.5 Take precautions when using cylinders containing gases or liquids under pressure. Helpful guidance may be obtained from Refs (1–4).[7]

7.5.1 Gas cylinders must be handled by trained personnel only.

7.5.2 Fasten gas cylinders to a rigid structure.

7.5.3 Take special safety precautions when using or storing combustible or toxic gases to ensure that the system is safe and free from leaks.

7.6 Gas piping, where possible, shall be metallic, especially inside the analyzer housing.

8. Measurement Objectives

8.1 Carefully define the measurement objective for the monitoring system before selecting components of the system and set specifications realistically, to meet the objective. Terms used as specifications shall be consistent with the terminology in Section 3.

8.2 If the monitoring system is intended primarily to determine compliance with regulatory standards, the accuracy, precision, frequency of sampling, and response time may be dictated by the requirements of the regulations. A high degree of stability and on-line reliability is generally required. The analyzer response for a specific parameter must be referenced to a recognized or specified laboratory method approved by the regulatory agency.

8.3 Monitoring systems intended to detect leaks and uncontrolled discharges, that is, spills, to protect treatment plants or receiving waters, require short sampling cycles and rapid response. Typically, these will activate alarms to alert operating personnel. They then may cause flow to be diverted from normal channels until the upset has passed or has been corrected. Frequently, the monitoring system is used in some way to locate and identify the source of the spill.

8.4 Systems that monitor the performance of process operations such as waste treatment, may have varying degrees of sophistication and complexity, depending on the specific nature of the application.

8.4.1 Simple, inexpensive, and low-precision analyzers with indicating or recording devices and alarms are acceptable for monitoring trends in operating parameters and for alerting operating personnel to off-standard performance.

8.4.2 Monitoring systems that provide data to be used to manually control process operations or to manually set automatic controllers are generally more complex and frequently require that outputs be transmitted long distances.

8.4.3 Monitoring systems intended to process data for operating guidance or management presentation and to provide varying degrees of automatic process control must be compatible with digital computers or telemetering systems. The reliability and stability of such systems, particularly the data output equipment, shall be high.

9. Sample System Design Considerations

9.1 Carefully examine the measurement objectives of the

---

[5] *Reagent Chemicals, American Chemical Society Specifications*, American Chemical Society, Washington, DC. For suggestions on the testing of reagents not listed by the American Chemical Society, see *Analar Standards for Laboratory Chemicals*, BDH Ltd., Poole, Dorset, U.K., and the *United States Pharmacopeia and National Formulary*, U.S. Pharmacopeial Convention, Inc. (USPC), Rockville, MD.

[6] The user, equipment, supplier, and installer should be familiar with requirements of the National Electrical Code, any local applicable electrical code, U.L. Safety Codes, and the Occupational Safety and Health Standards (*Federal Register*, Vol 36, No. 105, Part II, May 29, 1971).

[7] The boldface numbers in parentheses refer to the list of references at the end of this standard.

monitoring system and select a sampling system that matches these requirements.

9.2 Review all sample requirements with the equipment supplier. Be sure to define accurately all conditions of intended operation, the components in the sample and expected variations in the measured parameters.

9.3 Choose materials of construction for the parts that will be in contact with the sample, that do not react with the sample to cause subsequent contamination, corrosion, or other damage to critical parts or sorption of measurable components and maintain sample integrity.

9.4 Select the sampling point(s) so as to provide a representative and measurable sample as close as possible to the sample system and analyzer, and as outlined in Practices D 3370.

9.5 Design the sample probe to be consistent with the measurement objective and to require a minimum of maintenance.

9.6 Select the sample transfer system, including pumps and transfer lines, so that the integrity of the sample is maintained from sampling point to analyzer, especially with respect to suspension of solids and biological growth.

9.7 Provide necessary sample conditioning equipment (for example, filters, diluters, homogenizers, stream splitters), that is consistent with the defined measurement objective.

9.8 Provide a connection, when necessary, for introducing standard samples or withdrawing check samples immediately upstream of the analyzer.

9.9 Keep single- or multiple-sample streams that interface a single analyzer flowing all the time. Keep the manifold close to the analyzer to minimize cross-contamination.

9.10 Always keep sample lines as short as possible.

9.11 Provide appropriate protection of sample lines from extremely hot or freezing temperatures.

10. Considerations for Analyzer Selection

10.1 The analyzer selected must meet the measurement objective of the system over the complete range of application.

10.1.1 Precision and accuracy of measurement and response time for the parameter of interest shall coincide with system specifications at all levels of measurement.

10.1.2 Interference shall be insignificant relative to the measured component or shall be controllable. When used for regulatory compliance, known interferences shall not affect the reading more than 5 % from the true value.

10.1.3 If required for compliance, the analyzer shall be capable of validation by calibration with approved and certified standard reference materials using standard ASTM (or equivalent) tests.

10.2 In choosing a specific analyzer for a specific application, on line reliability of the instrument is of prime concern.

10.2.1 Downtime for maintenance because of component failures or other malfunction shall be minimal. Ease, promptness, minimal cost of repair or replacement are essential.

10.2.2 The analyzer shall be stable. Drift and changes in response with changes in conditions such as flow and temperature shall be insignificant or means for compensation shall be provided. Sample flow variations may have a significant effect on measured analyte concentrations. Flow

rate control shall be established as specified in Practice D 5540. Sample flow rate shall be maintained within limits to maintain the necessary precision of the continuous on line monitor.

10.2.3 The analyzer shall be relatively simple and easy to operate and maintain at a satisfactory level of performance.

11. Data Output Equipment Considerations

11.1 Equipment for the acquisition of output data from the analyzer shall meet the requirements of the measuring objectives for the monitoring system.

11.2 Visual or audible alarms and simple output meters are acceptable and desirable in many applications.

11.3 The analyzer output can be recorded locally at the field location. The digital or analog signal is frequently transmitted to a centralized location, such as a control room, often by a data line shared with other instruments.

11.4 Records or real-time data can be transferred to computers for storage, process control, or report generation.

11.5 Process equipment such as valves and pumps can be actuated by output generated by analyzers in a number of ways:

11.5.1 Recorded and output meters can have set points as integral parts of their design which actuate the equipment directly for either on-off or proportional control.

11.5.2 Controllers can be manually adjusted in response to analyzer signals read from a recorder or from output presented in a data report, typed or displayed on a cathode ray tube.

11.5.3 Direct digital process control is possible in more complicated and sophisticated systems, where real-time analyzer output is integrated with other process data and used to maintain desirable process conditions.

12. Installation of Monitoring System

12.1 Obtain information required for installation and operation of the monitoring system from the supplier.

12.2 Study operational data and design parameters furnished by the supplier before installation.

12.3 Choose materials of construction and components of the monitoring system to withstand the environment in which it is installed.

12.4 Select a location for the analyzer that is as close as possible to the sample intake and which provides adequate protection from extremes of temperature and humidity, where this is essential for proper performance.

12.5 Provide a convenient access to the entire monitoring system.

12.6 Provide proper outlets for the analyzer's exit streams so that no liquid or gas pressure buildup occurs (see 7.4).

12.7 After the installation has been completed, allow the analyzer to stabilize and calibrate before testing performance specifications.

13. Calibration

13.1 Establish a written calibration procedure and frequency consistent with the parameter being measured and the accuracy and reliability demanded by the measurement or control objectives based on the following:

13.1.1 Consult the analyzer supplier to determine the best

calibration procedure to use with the specific analyzer in a particular application.

13.1.2 When required for regulatory compliance, use calibration procedures specified by the appropriate agency.

13.1.3 Refer to ASTM standards, where applicable, to determine appropriate calibration standards.

13.1.4 Provide calibration standards at concentrations and compositions as close as possible to those of the sample stream being analyzed.

13.1.5 Before calibration, ensure that the sampling system and output instrumentation are functioning properly and that all preliminary adjustments to the analyzer required by the procedure have been made.

NOTE 2—Flow rate changes may affect continuous on line analyzer measured analyte concentration. If flow rates cannot be maintained constant, the effect of flow rate variation on measured analyte concentration shall be evaluated. Limits for flow rate variation shall be established to maintain the necessary precision of the continuous on line monitor.

13.2 *Reference Sample Calibration:*

13.2.1 With the reference sample flowing uniformly through the analyzer sampling line, allow the continuous analyzer readout to equilibrate.

13.2.2 Record time, sample number, date, and the corresponding continuous analyzer readout, and immediately analyze the reference sample using the appropriate laboratory analysis test method.

13.2.3 Determine the continuous analyzer calibration adjustment required so that results of laboratory analysis and the continuous analyzer readout coincide. Adjust the analyzer controls accordingly.

13.2.4 Repeat this procedure until no further change is needed, consistent with the quality of data required.

13.3 *Line Sample Calibration:*

13.3.1 With the sample flowing through the continuous analyzer sampling line uniformly and the continuous analyzer readout as close as possible to an equilibrium value, connect a second on line analyzer either downstream or on a parallel sample line, or withdraw a sample from the inlet stream as described in Practices D 3370.

NOTE 3—The connection should be made in such a way so as not to contaminate the flowing sample.

13.3.2 Record time, date, continuous analyzer results and the second on line analyzer results, or immediately analyze the withdrawn sample using the appropriate laboratory analysis test method.

13.3.3 Determine the continuous analyzer calibration adjustment required so that the results of the on line continuous analyzers agree with the second on line analyzer or the laboratory analysis.

NOTE 4—It is essential that the second on line continuous analyzer be checked carefully before this calibration is performed to ensure compliance with the requirements of the standard test procedure. To further ensure proper operation it is recommended that a reference sample or in-house control standard of known quality be tested to validate the operation of the second on line continuous analyzer.

13.3.4 Adjust the continuous analyzer with the analyzer controls accordingly.

13.4 *Multiple Standard Calibration:*

13.4.1 Prepare a series of calibration standards covering the range of measurements for the sample being analyzed,

following instructions in the test method or in the analyzer supplier's instructions.

13.4.2 Check all operating conditions of the system in accordance with the analyzer specifications, and allow sufficient time for instrument equilibrium.

13.4.3 Introduce a calibration standard of a concentration level recommended by the instrument supplier into the analyzer using the recommended instrument operating procedure. Activate the readout equipment.

13.4.4 After sufficient sample has been allowed to flow through the analyzer, adjust the readout to conform to the desired value.

13.4.5 Repeat 13.3.3 for the remaining standards from the calibration series, recording the equilibrium readout value each time.

13.4.6 Plot a calibration curve of standard value versus readout response from the above data.

13.4.7 Discard any standard when any change of composition is detected.

13.5 *Laboratory Check Sample for Flow-Through System:*

13.5.1 Withdraw from the spot sampling line or otherwise obtain directly from the sample stream sufficient sample for calibration, representative of one concentration within the range of measurement of the analyzer (see Practices D 3370).

13.5.2 Analyze the sample for the parameter of interest using the appropriate laboratory analysis test method.

13.5.3 If necessary, prepare additional standards to cover the range of interest by dilution with reagent water or by "spiking" with known amounts of an appropriate standard.

13.5.4 Serially, introduce the standards into the continuous analyzer, using the recommended instrument operating procedures. Allow the continuous analyzer readout to reach equilibrium, and record the equilibrium readout value each time.

13.5.5 Plot a calibration curve of concentration of parameter being determined versus readout response from the readout data.

13.6 *Probe Calibration:*

13.6.1 Provide special calibration procedure for continuous analyzers for which the instrumental measuring technique utilizes a sensor that is inserted directly into the sample, for example, pH, dissolved oxygen, conductivity.

13.6.2 Prepare two calibration solutions in accordance with the appropriate test method, selecting them to bracket the anticipated value of measurement.

13.6.3 Remove the probe from the sample stream, clean if appropriate and perform any necessary maintenance.

13.6.4 Fill a test container with the first calibration solution. The container shall have the means for monitoring temperature and, where appropriate, provide and maintain an adequate flow of sample past the sensor.

13.6.5 Insert the probe in the container containing the calibration solution and, using the procedure provided by the suppliers, adjust controls so that the analyzer output coincides with the accepted value of the standard. Make necessary adjustments for temperature compensation.

13.6.6 Rinse the probe thoroughly, place it in a second container containing the other calibration solution and readjust the controls, if necessary, so that the output agrees with the value of this guide.

13.6.7 Recheck with both solutions at least once. If either

point differs from the true value by a significant amount, as determined by the quality of measurement required, perform necessary maintenance, and recalibrate.

13.6.8 Alternatively, insert a second probe, with independent readout equipment and previously calibrated, into the sample alongside the probe and calibrate in situ, by adjusting its controls until the outputs of the two probes coincide.

13.7 After initial calibration with standard solutions or actual samples, as in 13.2 through 13.5, analyzer calibrations can be rechecked with secondary standards.

13.7.1 An electrical signal may be imposed to produce an analyzer output corresponding to a specific value produced by the parameters being analyzed.

13.7.2 A solution containing material other than the component of interest, but producing the same analyzer output as that component, may be used in place of the standard solution.

13.7.3 An optical filter may be placed in the beam of a photometric analyzer to produce an output equivalent to that produced by the component of interest.

14. Validation Procedures

14.1 *Reference Sample Validation Procedure:*

14.1.1 Obtain the reference sample and determine the reference value in accordance with 3.2.18.2.

14.1.2 Store the reference sample under conditions that will not cause contamination or degradation of the reference sample concentration. Because storage conditions and factors that affect sample stability change with time, confirm the reference value at periodic intervals. The frequency of confirmation can best be determined by the user of the analyzer.

14.1.3 Obtain a minimum of seven coincidental laboratory and continuous analyzer results of the reference sample, by introducing the reference sample into the continuous analyzer or laboratory analyzer and recording the results. Preferably use different qualified operators to make the multiple determinations over a period of time, with routine testing in the interim, until sufficient data have been obtained for analysis.

14.1.4 More than seven test results on the reference sample are often necessary to attain an average value with acceptable confidence limits. This will vary significantly for different laboratory procedures and reference sample concentrations. This applies for both laboratory and continuous analysis.

14.1.5 Tabulate the laboratory and continuous analyzer results and their differences. Check for outliers using the Grubbs test criterion in Annex A1.

14.1.6 Calculate the laboratory analyzer variance from the individual test results, excluding any outliers found in 14.1.5, as follows:

$$S_L^2 = \frac{\left[\sum\limits_{i=1}^{n} X_L^2 - \frac{(\Sigma X_L)^2}{n_L}\right]}{(n_L - 1)} \qquad (1)$$

where:
S_L^2 = variance of the laboratory test results,
X_L = individual laboratory analyzer test results on the reference sample,
n_L = number of laboratory analyzer test results, and

$\bar{X}_L = \dfrac{\Sigma X_L}{n_L}$ = arithmetic average of the laboratory analyzer test results.

14.1.7 Determine whether the precision of the laboratory test results on the reference sample is statistically significantly different from the historical precision of the laboratory test method. The statistical criterion for this purpose is the F test as follows:

$$F = \frac{S_B^2}{S_s^2} \qquad (2)$$

where:
S_B^2 = larger variance, either S_L^2 or S_h^2,
S_s^2 = smaller variance, either S_L^2 or S_h^2,
S_L^2 = variance of the laboratory test results on the reference sample as determined in 14.1.6,
v_L = degrees of freedom for laboratory analysis of reference sample ($n_L - 1$),
S_h^2 = historical variance for the laboratory analysis with n_h determinations, and
v_h = degrees of freedom for historical laboratory analyzer tests ($n_L - 1$).

14.1.8 Compare the calculated F value with the critical F value given in Table A1.2 for the appropriate degrees of freedom in the numerator (v_L or v_h) and appropriate degrees of freedom in the denominator (v_h or v_L).

14.1.8.1 If the calculated F value exceeds the critical F value obtained from Table A1.2, there is at least a 95 % probability that the reference sample laboratory analyzer data precision is statistically significantly different from the historical precision for that laboratory analyzer. In this event, the reasons for the substandard test precision should be determined, appropriate corrective actions to the procedure or laboratory analyzer, or both, and a minimum of seven new tests on the reference sample repeated in accordance with 14.1.3 through 14.1.8 until acceptable laboratory test precision is obtained.

14.1.9 Calculate the variance of the continuous analyzer excluding outliers rejected in 14.1.5 as follows:

$$S_c^2 = \frac{\left[\sum\limits_{i=1}^{n} X_c^2 - \frac{(\Sigma X_c)^2}{n_c}\right]}{(n_c - 1)} \qquad (3)$$

or

$$\frac{\sum\limits_{i=1}^{n} (X_c - \bar{X}_c)^2}{n_c - 1} \qquad (4)$$

where:
X_c = individual continuous analyzer results on the reference sample,
n_c = number of continuous analyzer test results, and
$\bar{X}_c = \dfrac{\Sigma X_c}{n_c}$ = arithmetic average of the continuous analyzer test results.

14.1.10 Apply the F test as follows to determine whether the variance of the laboratory analyzer (S_L^2) and the variance of the continuous analyzer (S_c^2) are statistically significantly different:

$$F = \frac{S_B^2}{S_s^2} \qquad (5)$$

where:

S_B^2 = larger variance, either S_L^2 or S_c^2,
S_s^2 = smaller variance, either S_L^2 or S_c^2,
S_L^2 = variance of the laboratory test results on the reference sample as determined in 14.1.6,
v_L = degrees of freedom for laboratory analysis of reference sample ($n_L - 1$),
S_c^2 = variance of the continuous analyzer test results on the reference sample in 14.1.9, and
v_c = degrees of freedom for the continuous analyzer, ($n_c - 1$).

14.1.11 Compare the calculated F value with the critical F value given in Table A1.2 for the appropriate degrees of freedom in the numerator (v_L or v_h) and the appropriate degrees of freedom in the denominator (v_h or v_L).

14.1.12 If the calculated F value is equal to or less than the critical F value obtained from Table A1.2, apply the Student's t test to determine if there is a statistically significant difference between the average continuous analyzer results and the average laboratory analyzer result. If the computed F is greater than the critical F value proceed to 14.1.16.

$$t = \frac{|\overline{X}_c - \overline{X}_L|}{S_p} \qquad (6)$$

$$S_p = \sqrt{\frac{(v_L) S_L^2 + (v_c) S_c^2}{n_L + n_c - 2} \times \left(\frac{1}{n_L} + \frac{1}{n_c}\right)} \qquad (7)$$

where:

S_p = pooled standard deviation for the difference between $\overline{X}_L$ and $\overline{X}_c$,
$\overline{X}_L$ = arithmetic average laboratory analyzer results,
$\overline{X}_c$ = arithmetic average continuous analyzer results,
v_L = degrees of freedom for laboratory analysis ($n_L - 1$),
v_c = degrees of freedom for the continuous analyzer results ($n_c - 1$),
S_L^2 = variance of laboratory analyzer results, and
S_c^2 = variance of the continuous analyzer results.

14.1.13 Compare the calculated t value from Eq 6 with the critical t value from Table A1.3 for the ($n_L + n_c - 2$).

14.1.14 If the calculated t value is equal to or less than the critical t value, the continuous analyzer can be expected to give essentially the same average results as the laboratory analyzer.

14.1.15 If the calculated t value exceeds the critical t value there is at least a 95 % probability that the continuous analyzer and the laboratory analyzer are not giving the same average test results. Continuous analyzer validity is therefore suspect. Further investigation of the continuous analyzer function and operation should be made to correct the probable bias.

14.1.16 Calculate the t value as follows:

$$t = \frac{|X_c - X_L|}{\sqrt{\frac{S_L^2}{n_L} + \frac{S_c^2}{n_c}}} \qquad (8)$$

14.1.16.1 Compute the degrees of freedom for the t test as follows:

$$\text{degrees of freedom} = \frac{\left[\left(\frac{S_L^2}{n_L} + \frac{S_c^2}{n_c}\right)\right]^2}{\frac{\left(\frac{S_L^2}{n_L}\right)^2}{n_L + 1} + \frac{\left(\frac{S_c^2}{n_c}\right)^2}{n_c + 1}} - 2 \qquad (9)$$

round the computed value to the nearest whole number.

14.1.17 Compare the calculated t value from above with the critical t value from Table A1.3 for the degrees of freedom computed above.

14.1.18 Refer to 14.1.14 and 14.1.15 for the proper interpretation of the comparison in 14.1.17.

14.1.19 Calculate the t value for the differences as follows:

$$t = \frac{\overline{d} \sqrt{n_d}}{S_d} \qquad (10)$$

where:

$\overline{d}$ = average difference,
n_d = number of differences, and
S_d = standard deviation of the differences.

14.1.20 Compare the calculated t value from above with the critical t value from Table A1.3 for $n_d - 1$ degrees of freedom.

14.1.21 Refer to 14.1.14 and 14.1.15 for the proper interpretation of the comparison in 14.1.20.

14.2 *Line Sample Validation Procedure:*

14.2.1 The line sample method is used primarily for the validation of the continuous analyzer operation where the process stream is in service and available. The line sample method therefore is not applicable for predelivery validation of the continuous analyzer or for calibration before start-up and is not a viable alternative to the reference sample method under these conditions.

14.2.2 For continuous analyzer applications or process stream conditions, or both, that negate the practical use of the reference sample method for predelivery or initial validation, the line sample method is used for the analyzer validation and is applied at appropriate times in the process when the process stream corresponds to low, midscale, and high concentrations within the range of continuous analyzer operation.

14.2.3 Obtain a minimum of seven line samples, preferably during times of stable continuous analyzer results, using different qualified operators, over a period of time, with routine testing in the interim, until sufficient samples have been obtained.

14.2.4 Record the continuous analyzer results at the time each sample is withdrawn.

14.2.5 Determine the laboratory analyzer results using an appropriate ASTM or standard test method.

14.2.6 Tabulate the difference between each continuous analyzer result and its corresponding laboratory analyzer result as follows:

$$d_i = X_c - X_L \qquad (11)$$

where:

d_i = individual difference,

X = continuous analyzer result, and
X_c = laboratory analyzer result.

14.2.7 Check the set of differences for outliers by the Grubbs test as described in Annex A1.

14.2.8 Compute the average difference and the standard deviation of the individual differences excluding outliers rejected in 14.2.6 as follows:

$$\bar{d} = \frac{\Sigma d_i}{n_d} \qquad (12)$$

$$S_d = \sqrt{\frac{\sum\limits_{i=1}^{n} d_i{}^2 - \frac{(\Sigma d_i)^2}{n_d}}{(n_d - 1)}} \qquad (13)$$

where:
$\bar{d}$ = average differences,
d_i = individual differences,
n_d = number of differences, and
S_d = standard deviation of the differences.

14.2.9 Apply the t test as follows to check for a possible systematic difference (bias) between the continuous analyzer results and the laboratory analyzer results:

$$t = \frac{\bar{d}\ \sqrt{n_d}}{S_d} \qquad (14)$$

14.2.10 Compare the calculated t value to the critical t value from Table A1.3 for $(n_d - 1)$ degrees of freedom.

14.2.11 If the calculated t value is equal to or less than the critical t value, the continuous analyzer can be expected to give essentially the same average results as the laboratory analyzer.

14.2.12 If the t value is greater than the critical t value, there is at least a 95 % probability that the continuous analyzer and the laboratory analyzer are not giving the same average results. Therefore, the continuous analyzer validity is suspect. Make further investigations of the continuous analyzer function and operation to resolve the probable bias indicated.

15. Verification Procedures

15.1 *Reference Sample Verification Procedure:*

15.1.1 Tabulate the differences between the continuous analyzer results and the laboratory analyzer results to the reference sample in 14.1 as follows:

$$d_i = X_c - X_L \qquad (15)$$

where:
d_i = individual difference,
X_c = continuous analyzer result, and
X_L = laboratory analyzer result.

15.1.2 Apply the Grubbs test for outliers to the tabulated differences in accordance with Annex A1.

15.1.3 Calculate the average difference and the standard deviation of the individual differences excluding outliers rejected in 15.1.2 as follows:

$$\bar{d} = \frac{\Sigma d_i}{n_d} \qquad (16)$$

$$S_d = \sqrt{\frac{\sum\limits_{i=1}^{n} d_i{}^2 - \frac{(\Sigma d_i)^2}{n_d}}{(n_d - 1)}} \qquad (17)$$

where:
$\bar{d}$ = average of differences,
d_i = individual differences,
n_d = number of differences, and
S_d = standard deviation of the differences.

15.1.4 Apply the Student's t test to determine if a statistical significant bias exists between the average difference and zero difference as follows:

$$t = \frac{\bar{d}\ \sqrt{n_d}}{S_d} \qquad (18)$$

15.1.5 Compare the calculated t value to the critical t value from Table A1.3 for $(n - 1)$ degrees freedom.

15.1.6 If the calculated t value is equal to or less than the critical t value, center the control chart around zero difference.

15.1.7 If the calculated t value is greater than the critical t value, center the control chart around the average difference ($\bar{d}$). Make further investigations of the continuous analyzer function and operation to resolve the probable bias indicated.

15.1.8 Calculate the upper and lower control chart limits as follows:

$$UCL = 0 \text{ or } \bar{d} + 3 \times S_d \qquad (19)$$

$$LCL = 0 \text{ or } \bar{d} - 3 \times S_d \qquad (20)$$

where:
UCL = upper control limit,
LCL = lower control limit,
0 = center line if pass t test,
$\bar{d}$ = center line if fail t test, and
$3 \times S_d$ = 99 % confidence interval, where S_d is the standard deviation of the differences.

15.1.9 Periodically, by introducing the reference sample, compare the calculated differences between the continuous analyzer and the coincidental analysis results from the laboratory analyzer. The frequency for the periodic check will depend on the stability of the continuous analyzer. The frequency should be short enough to detect instrument drift or malfunction but long enough so as to not be a nuisance to the operator of the continuous analyzer.

15.1.10 If the calculated difference is within the UCL and LCL, the continuous analyzer is considered verified.

15.1.11 If the calculated difference is outside the UCL or LCL, the continuous analyzer is considered out of control. Further investigation of the continuous analyzer function and operation should be made to correct the problem.

15.2 *Line Sample Verification Procedure:*

15.2.1 Calculate the upper and lower control chart limits as follows:

$$UCL = 0 \text{ or } \bar{d} + 3 \times S_d \qquad (21)$$

$$LCL = 0 \text{ or } \bar{d} - 3 \times S_d \qquad (22)$$

where:
UCL = upper control limit,
LCL = lower control limit,

0 = center line if pass t test,
$\bar{d}$ = center line if fail t test, and
$3 \times S_d$ = 99 % confidence interval, where S_d is the standard deviation of the differences.

15.2.2 Periodically, compare the calculated differences between the continuous analyzer and a coincidental laboratory analysis or a second continuous analyzer. The frequency for the periodic check depends on the stability of the continuous analyzer. The frequency should be short enough to detect instrument drift or malfunction but not long enough to be a nuisance to the operator of the continuous analyzer.

15.2.3 If the calculated difference is within the UCL and LCL, the continuous analyzer is considered verified.

15.2.4 If the calculated difference is outside the UCL or LCL, the continuous analyzer is considered out of control. Further investigation of the continuous analyzer function and operation should be made to correct the problem.

16. Calculation

16.1 Each individual monitoring system and ASTM test method chosen determines the calculations necessary to perform on the output signal. Most analyses are recorded as direct readouts based on instrument calibration.

17. Precision

17.1 Preferably, each laboratory standard test method that is applied to on line monitoring shall include its own precision section based on cooperative test program results.

17.2 If it is desirable to validate the monitoring system results relative to the laboratory method, statistical equivalency of the data generated by the two techniques shall be demonstrated by applying the "t" test for mean differences of the paired observations at the 95 % ($P \leq 0.05$) confidence level ($t_{0.025}$ — two-tailed). This shows whether any significant difference exists between the mean values of the differences and zero for paired observations having different values. The procedure for this test is given in the annex.

18. Keywords

18.1 automatic analysis; continuous sampling; monitoring systems; on line; validation; verification; water analysis

ANNEX

(Mandatory Information)

A1. REJECTION OF INDIVIDUAL OUTLIERS

A1.1 *Rejection of Individual Outliers*—Absolutely no data should be discarded unless valid statistical criteria show them clearly to be erroneous or aberrant. Control charts and variance analyses may be misleading in some cases. Statistical references provide valid criteria for excluding data from precision evaluations. When the experimenter is aware that a gross deviation from prescribed experimental procedure has occurred, the resultant observation should be discarded. Otherwise the most extreme value among the data at each concentration of material may be tested by calculating its T value.

A1.1.1 If the T value is greater than the critical value recorded in Table A1.1 at the selected significance level (see Practice E 178), the outlier may be rejected (the selection of the significance level is left up to the collaborative study chairman). The test criterion for the suspected outlier, x_l or x_n, is as follows:

$$T_n = (x_n - \bar{x})/s_t \tag{A1.1}$$

$$T_l = (\bar{x} - x_l)/s_t \tag{A1.2}$$

where: $\bar{x}$ and s_t are the current estimates of the mean and overall standard deviation for all retained data at the concentration of material associated with x_n (the highest data) and x_l (the lowest data).

A1.1.1.1 Because the direction of the value being tested is not known beforehand and because the experimenter is interested in detecting values that could be either high or low, a two-sided test is being performed (see Grubbs (5) for detailed explanation).

A1.1.2 If the suspected outlier fails the test in 10.5.1, recalculate the $\bar{x}$ and s_t from the remaining data at this concentration of material.

A1.1.3 If there is another suspected value among the remaining data for a specific concentration level, a second iteration of 10.5.1 may be justified but is generally not recommended.

A1.1.4 A completed example is shown in X1.3.

TABLE A1.1 Grubbs Distribution

Critical Values for T (Two-Sided Test) When Standard Deviation is Calculated from the Same Samples (for Outliers)[A]

Number of Observations, n	10 % Significance Level	5 % Significance Level	2 % Significance Level	1 % Significance Level
3	1.15	1.15	1.15	1.15
4	1.46	1.48	1.49	1.50
5	1.67	1.71	1.75	1.76
6	1.82	1.89	1.94	1.97
7	1.94	2.02	2.10	2.14
8	2.03	2.13	2.22	2.27
9	2.11	2.21	2.32	2.39
10	2.18	2.29	2.41	2.48
11	2.23	2.36	2.48	2.56
12	2.29	2.41	2.55	2.64
13	2.33	2.46	2.61	2.70
14	2.37	2.51	2.66	2.75
15	2.41	2.55	2.70	2.81
16	2.44	2.58	2.75	2.85
17	2.47	2.62	2.78	2.89
18	2.50	2.65	2.82	2.93
19	2.53	2.68	2.85	2.97
20	2.56	2.71	2.88	3.00
21	2.58	2.73	2.91	3.03
22	2.60	2.76	2.94	3.06
23	2.62	2.78	2.96	3.09
24	2.64	2.80	2.99	3.11
25	2.66	2.82	3.01	3.13
30	2.75	2.91	3.10	3.24
35	2.82	2.98	3.18	3.32
40	2.87	3.04	3.24	3.38
45	2.92	3.08	3.29	3.43
50	2.96	3.13	3.34	3.48
60	3.03	3.20	3.41	3.56
70	3.09	3.26	3.47	3.62
80	3.14	3.30	3.52	3.67
90	3.18	3.35	3.56	3.72
100	3.21	3.38	3.60	3.75

[A] Values of T for N ≤ 25 are based on those given in Ref (5). For n > 25, the values of T are approximated. All values have been adjusted for division by n − 1 instead of n in calculating s. Tabulated values come from Practice E 178 and may also be found in Grubbs (5). Levels of significance shown in Practice E 178 were doubled, since this is a two-sided test for significance instead of a one-sided test.

TABLE A1.3 Table of t at 5 % Probability Level

Degrees of Freedom (N − 1)	t
1	12.706
2	4.303
3	3.182
4	2.776
5	2.571
6	2.447
7	2.365
8	2.306
9	2.262
10	2.228
11	2.201
12	2.179
13	2.160
14	2.145
15	2.131
16	2.120
17	2.110
18	2.101
19	2.093
20	2.086

TABLE A1.2 F-Distribution: Degrees of Freedom for Numerator

d/n[A]	1	2	3	4	5	6	7	8	9	10	12	15	20[B]
1	161	200	216	225	230	234	237	239	241	242	244	246	248
2	18.5	19.0	19.2	19.2	19.3	19.3	19.4	19.4	19.4	19.4	19.4	19.4	19.4
3	10.1	9.55	9.28	9.12	9.01	8.94	8.87	8.85	8.81	8.79	8.74	8.70	8.66
4	7.71	6.94	6.59	6.39	6.26	6.16	6.09	6.04	6.00	5.96	5.91	5.86	5.80
5	6.61	5.79	5.41	5.19	5.05	4.95	4.88	4.81	4.77	4.74	4.68	4.62	4.56
6	5.99	5.14	4.76	4.53	4.39	4.28	4.21	4.15	4.10	4.06	4.00	3.94	3.87
7	5.59	4.74	4.35	4.12	3.97	3.87	3.79	3.73	3.68	3.64	3.57	3.61	3.44
8	5.32	4.46	4.07	3.84	3.69	3.58	3.50	3.44	3.39	3.35	3.28	3.22	3.15
9	5.12	4.26	3.86	3.63	3.48	3.37	3.29	3.23	3.18	3.14	3.07	3.01	2.94
10	4.96	4.10	3.70	3.48	3.33	3.22	3.14	3.07	3.02	2.98	2.91	2.85	2.77
11	4.84	3.98	3.59	3.36	3.20	3.09	3.01	2.95	2.90	2.85	2.79	2.72	2.65
12	4.75	3.89	3.49	3.26	3.11	3.00	2.91	2.85	2.80	2.75	2.69	2.62	2.54
13	4.67	3.81	3.41	3.18	3.03	2.92	2.83	2.77	2.71	2.67	2.60	2.53	2.46
14	4.60	3.74	3.34	3.11	2.96	2.85	2.76	2.70	2.65	2.60	2.53	2.46	2.39
15	4.54	3.68	3.29	3.06	2.90	2.79	2.71	2.64	2.59	2.54	2.48	2.40	2.33
16	4.49	3.63	3.24	3.01	2.85	2.74	2.66	2.59	2.54	2.49	2.42	2.35	2.28
17	4.45	3.59	3.20	2.96	2.81	2.70	2.61	2.55	2.49	2.45	2.38	2.31	2.33
18	4.41	3.55	3.16	2.93	2.77	2.66	2.58	2.51	2.46	2.41	2.34	2.27	2.19
19	4.38	3.52	3.13	2.90	2.74	2.63	2.54	2.48	2.42	2.38	2.31	2.23	2.16
20	4.35	3.49	3.10	2.87	2.71	2.60	2.51	2.45	2.39	2.35	2.28	2.20	2.12
∞	3.84	3.00	2.60	2.37	2.21	2.10	2.01	1.94	1.88	1.83	1.75	1.67	1.57

[A] Where: n = degrees of freedom in the numerator and d = degrees of freedom in the denominator (for example, if n = 6 and d = 15 the critical F value is 2.79).
[B] Expanded tables may be found in statistical reference books, see also "Standard Probability and Statistics," CRC Press, 1991.

APPENDIX

(Nonmandatory Information)

X1. REFERENCE SAMPLE VALIDATION DETERMINATION

X1.1 The following is a sample calculation for determining the validity of continuous analyzer system results relative to a reference sample analyzed on both a continuous and laboratory analyzer or a second continuous analyzer in accordance with 14.1.

X1.2 Tabulate continuous and laboratory analyzer or a second continuous analyzer results for at least seven matched paired analysis of the reference sample, as in Table X1.1.

X1.3 Check the extreme outliers using the Grubbs rejection criterion in A1.1 for continuous, laboratory, and difference values.

X1.3.1 *Laboratory:*

$$T_N = \frac{X_{nr} - \overline{X}_r}{S_r} = \frac{31 - 21.09}{4.826} = 2.053 \qquad (X1.1)$$

$$T_l = \frac{\overline{X}_r - X_{lr}}{S_r} = \frac{21.09 - 15}{4.826} = 1.262 \qquad (X1.2)$$

$$T_{(\alpha/2=0.05,11)} = 2.36, \text{ conclude data are not outliers} \qquad (X1.3)$$

where:
X_{nr} = highest laboratory data,
$\overline{X}_r$ = average of the laboratory data,
S_r = standard deviation of the laboratory data, and
X_{lr} = lowest laboratory data.

Continuous:

$$T_N = \frac{X_{nc} - \overline{X}_c}{S_c} = \frac{26.8 - 21.69}{3.269} = 1.563 \qquad (X1.4)$$

$$T_l = \frac{\overline{X}_c - X_{lc}}{S_c} = \frac{21.69 - 15}{3.269} = 2.046 \qquad (X1.5)$$

$$T_{(\alpha/2=0.05,11)} = 2.36, \text{ conclude data are not outliers} \qquad (X1.6)$$

where:
X_{nc} = highest continuous data,
$\overline{X}_c$ = average of the continuous data,
S_c = standard deviation of the continuous data, and
X_{lc} = lowest continuous data.

Difference:

$$T_N = \frac{X_{nd} - \overline{X}_d}{S_d} = \frac{2.9 - 0.60}{2.244} = 1.025 \qquad (X1.7)$$

$$T_l = \frac{\overline{X}_d - X_{ld}}{S_d} = \frac{0.60 - (-5)}{2.244} = 2.495 \qquad (X1.8)$$

$$T_{(\alpha=0.05,11)} = 2.36, \text{ conclude data are outlier} \qquad (X1.9)$$

where:
X_{nd} = highest difference data,
$\overline{X}_d$ = average of the difference data,
S_d = standard deviation of the difference data, and
X_{ld} = lowest difference data.

Reject this data pair, and recalculate Grubbs values from data shown in Table X1.2.

Laboratory:

$$T_N = \frac{27 - 20.10}{3.725} = 1.852 \qquad (X1.10)$$

$$T_l = \frac{20.10 - 15}{3.725} = 1.369 \qquad (X1.11)$$

$$T_{(\alpha/2=0.05,10)} = 2.29, \text{ conclude data are not outliers} \qquad (X1.12)$$

Continuous:

$$T_N = \frac{26.8 - 21.26}{3.099} = 1.788 \qquad (X1.13)$$

$$T_l = \frac{21.26 - 15}{3.099} = 2.020 \qquad (X1.14)$$

$$T_{(\alpha/2=0.05,10)} = 2.29, \text{ conclude data are not outliers} \qquad (X1.15)$$

Difference:

$$T_N = \frac{3 - 1.160}{1.327} = 1.386 \qquad (X1.16)$$

$$T_l = \frac{1.160 - (-1)}{1.327} = 1.628 \qquad (X1.17)$$

$$T_{(\alpha/2=0.05,10)} = 2.29, \text{ conclude data are not outlier} \qquad (X1.18)$$

X1.4 Determine whether the precision of the laboratory test results on the reference sample are statistically significantly different from the historical precision of the laboratory of the analysis as follows:

TABLE X1.1 Reference Sample Tabulated On Line and Laboratory Results

On line Response	Laboratory Response	Difference
26.8	27	−0.2
20	17	3
26	31	−5
25	26	−1
21.3	20	1.3
21	20	1
20.9	18	2.9
20.8	20	0.8
20.5	19	1.5
15	15	0
21.3	19	2.3
$\overline{X}$ = 21.69	21.09	0.60
S = 3.269	4.826	2.244

TABLE X1.2 Reference Sample Tabulated On Line and Laboratory Results without Outliers

On line Response	Laboratory Response	Difference
26.8	27	−0.2
20	17	3
25	26	−1
21.3	20	1.3
21	20	1
20.9	18	2.9
20.8	20	0.8
20.5	19	1.5
15	15	0
21.3	19	2.3
$\overline{X}$ = 21.26	20.10	1.160
S = 3.099	3.725	1.327

$$F = \frac{S_r^2}{\sigma_r^2} \tag{X1.19}$$

$$F = \frac{(4.826)^2}{(3.575)^2} = 1.822 \tag{X1.20}$$

$$F_{(\alpha=0.05,9,9)} = 3.18, \text{ conclude variances are not} \\ \text{statistically significantly different} \tag{X1.21}$$

where:

S_r^2 = variance of laboratory results from the reference sample validation $(S_r)^2$, and

σ_r^2 = historical variance for this laboratory method, from previous quality control data.

If the F test is failed, evaluate, investigate, and eliminate the reason for failure.

X1.5 Determine whether the precision of the laboratory analysis of the reference sample are statistically significantly different from the precision of the continuous analysis as follows:

$$F = \frac{S_L^2}{S_S^2} \tag{X1.22}$$

where:

S_L^2 = larger variance, and

S_S^2 = smaller variance.

$$F = \frac{(3.725)^2}{(3.099)^2} = 1.445 \tag{X1.23}$$

$$F_{(\alpha=0.05,9,9)} = 3.18, \text{ conclude variances are not} \\ \text{statistically significantly different} \tag{X1.24}$$

X1.6 Calculate the Student's t test for the averages as follows:

$$t = \frac{\overline{X}_c - \overline{X}_l}{S_p} \tag{X1.25}$$

$$S_p = \sqrt{\frac{\nu_l(S_l^2) + \nu_c(S_c^2)}{n_l + n_c - 2} \times \left(\frac{1}{n_l} + \frac{1}{n_c}\right)} \tag{X1.26}$$

$$\nu = n_l + n_c - 2 \tag{X1.27}$$

$$t = \frac{21.26 - 20.10}{1.532} = 0.757 \tag{X1.28}$$

$$S_p = \sqrt{\frac{9(3.099)^2 + 9(3.725)^2}{10 + 10 - 2} \times \left(\frac{1}{10} + \frac{1}{10}\right)} = 1.532 \tag{X1.29}$$

$$t_{(\alpha/2=0.05,9)} = 2.262, \text{ conclude no statistically} \\ \text{significantly difference. Instrument pairs} \\ \text{are not statistically significantly different.} \tag{X1.30}$$

X1.7 Calculate the Student's t test for the difference as follows:

$$t = \frac{\overline{d}\sqrt{n_d}}{S_d} = \frac{1.160\sqrt{10}}{1.327} = 2.762 \tag{X1.31}$$

$$t_{(\alpha/2=0.05,9)} = 2.262, \text{ conclude there is a statistical} \\ \text{significant bias in the difference values.} \\ \text{Instrument pair can not be used for validation.} \tag{X1.32}$$

NOTE X1.1—If the Student's t test fails in either X1.6 or X1.7, the instrument pair can not be validated.

X2. LINE SAMPLE VALIDATION DETERMINATION

X2.1 The following is a sample calculation for determining the validity of continuous analyzer system results relative to a laboratory or second continuous analyzer in accordance with 14.2.

X2.2 Tabulate continuous and laboratory analyzer or a second continuous analyzer results for at least seven matched paired analysis of the process stream, as in Table X2.1.

X2.3 Check for extreme outliers using the Grubbs rejection criterion in A1.1 for difference values.

Differences:

$$T_n = \frac{X_{nd} - \overline{X}_d}{S_d} = \frac{0.70 - 0.155}{0.299} = 1.822 \tag{X2.1}$$

$$T_l = \frac{\overline{X}_d - X_{ld}}{S_d} = \frac{0.155 - (-0.19)}{0.299} = 1.154 \tag{X2.2}$$

$$T_{(\alpha=0.05,7)} = 2.02, \text{ conclude data are not outliers} \tag{X2.3}$$

X2.4 Calculate the Student's t test for the difference as follows:

TABLE X2.1 Line Sample Tabulated On Line and Laboratory Results

Continuous Analyzer Response	Second Continuous Analyzer Response	Difference
8.37	8.33	0.04
4.61	4.52	0.09
5.87	6.06	−0.19
5.80	5.81	−0.01
6.86	6.81	0.05
6.03	5.33	0.70
5.45	5.04	0.41
$\overline{X}$ = ...	...	0.155
S = ...	...	0.299

$$t = \frac{\overline{d}\sqrt{n_d}}{S_d} = \frac{0.155\sqrt{7}}{0.299} = 1.372 \tag{X2.4}$$

$$t_{(\alpha/2=0.05,6)} = 2.447, \text{ conclude no statistical significant} \\ \text{difference. Instrument pair can be used} \\ \text{for validation.} \tag{X2.5}$$

NOTE X2.1—If the t test fails, the instrument pair cannot be used for validation.

X3. REFERENCE SAMPLE VERIFICATION

X3.1 Develop a control chart centered at zero difference $\pm 3 S_d$

where: S_d = standard deviation of the differences.

X3.2 Periodically compare, calculate, and plot the difference between the continuous analyzer and the coincidental laboratory analysis or a second continuous analyzer to the reference sample.

X3.3 If the difference value is within $\pm 3\ S_d$, the continuous analyzer is considered verified. If not, the continuous analyzer should be considered out-of-service until the difference is resolved.

X4. LINE SAMPLE VERIFICATION

X4.1 Develop a control chart centered at zero differences $\pm 3\ S_d$ where:

S_d = standard deviation of the differences

X4.2 Periodically compare, calculate, and plot the difference between the continuous analyzer and the coincident laboratory or second continuous to the process stream.

X4.3 If the difference value is within $\pm 3\ S_d$, the continuous analyzer is considered verified. If not, the continuous analyzer should be considered out-of-service until the difference is resolved.

REFERENCES

(1) "Safe Handling of Compressed Gases," Pamphlet P-1, Compressed Gas Association, Inc., New York, NY.

(2) "Compressed Gases, Safe Practices," Pamphlet No. 95, National Safety Council, Chicago, IL.

(3) "Chemical Safety Data Sheets," Manufacturing Chemists Association, 1825 Connecticut Ave., N.W., Washington, DC 20009.

(4) Sax, N. I., *Dangerous Properties of Industrial Materials*, 3rd ed., 1968, Reinhold Book Corp., New York, NY.

(5) Grubbs, F. E., and Beck, G., "Extension of Sample Sizes and Percentage Points for Significance Tests of Outlying Observations," *Technometrics*, Vol 14, No. 4, November 1972, pp. 847–854.

Standard Practices for
Sampling Water-Formed Deposits[1]

This standard is issued under the fixed designation D 887; the number immediately following the designation indicates the year of original adoption or, in the case of revision, the year of last revision. A number in parentheses indicates the year of last reapproval. A superscript epsilon (ϵ) indicates an editorial change since the last revision or reapproval.

1. Scope

1.1 These practices cover the sampling of water-formed deposits for chemical, physical, biological, or radiological analysis. The practices cover both field and laboratory sampling. It also defines the various types of deposits. The following practices are included:

	Sections
Practice A—Sampling Water-Formed Deposits From Tubing of Steam Generators and Heat Exchangers	8 to 10
Practice B—Sampling Water-Formed Deposits From Steam Turbines	11 to 14

1.2 The general procedures of selection and removal of deposits given here can be applied to a variety of surfaces that are subject to water-formed deposits. However, the investigator must resort to his individual experience and judgment in applying these procedures to his specific problem.

1.3 *This standard does not purport to address all of the safety concerns, if any, associated with its use. It is the responsibility of the user of this standard to establish appropriate safety and health practices and determine the applicability of regulatory limitations prior to use.* See Section 7 and Notes 1 through 4 for specific hazards statements.

2. Referenced Documents

2.1 *ASTM Standards:*

D 512 Test Methods for Chloride Ion in Water[2]
D 934 Practices for Identification of Crystalline Compounds in Water-Formed Deposits by X-Ray Diffraction[3]
D 993 Test Methods for Sulfate-Reducing Bacteria in Water and Water-Formed Deposits[4]
D 1129 Terminology Relating to Water[2]
D 1193 Specification for Reagent Water[2]
D 1245 Practice for Examination of Water-Formed Deposits by Chemical Microscopy[3]
D 1293 Test Methods for pH of Water[2]
D 1428 Test Methods for Sodium and Potassium in Water and Water-Formed Deposits by Flame Photometry[2]
D 2331 Practices for Preparation and Preliminary Testing of Water-Formed Deposits[3]
D 2332 Practice for Analysis of Water-Formed Deposits by Wavelength-Dispersive X-Ray Fluorescence[3]

D 2579 Test Methods for Total and Organic Carbon in Water[3]
D 3483 Test Methods for Accumulated Deposition in a Steam Generator Tube[3]

3. Terminology

3.1 *Descriptions of Terms Specific to This Standard:*

3.1.1 *water-formed deposits*—any accumulation of insoluble material derived from water or formed by the reaction of water upon surfaces in contact with the water.

3.1.1.1 Deposits formed from or by water in all its phases may be further classified as scale, sludge, corrosion products, or biological deposit. The overall composition of a deposit or some part of a deposit may be determined by chemical or spectrographic analysis; the constituents actually present as chemical substances may be identified by microscope or x-ray diffraction studies. Organisms may be identified by microscopic or biological methods.

3.1.2 *scale*—a deposit formed from solution directly in place upon a surface.

3.1.2.1 Scale is a deposit that usually will retain its physical shape when mechanical means are used to remove it from the surface on which it is deposited. Scale, which may or may not adhere to the underlying surface, is usually crystalline and dense, frequently laminated, and occasionally columnar in structure.

3.1.3 *sludge*—a water-formed sedimentary deposit.

3.1.3.1 The water-formed sedimentary deposits may include all suspended solids carried by the water and trace elements which were in solution in the water. Sludge usually does not cohere sufficiently to retain its physical shape when mechanical means are used to remove it from the surface on which it deposits, but it may be baked in place and be hard and adherent.

3.1.4 *corrosion products*—a result of chemical or electrochemical reaction between a metal and its environment.

3.1.4.1 A corrosion deposit resulting from the action of water, such as rust, usually consists of insoluble material deposited on or near the corroded area; corrosion products may, however, be deposited a considerable distance from the point at which the metal is undergoing attack.

3.1.5 *biological deposits*—water-formed deposits of organisms or the products of their life processes.

3.1.5.1 The biological deposits may be composed of microscopic organisms, as in slimes, or of macroscopic types such as barnacles or mussels. Slimes are usually composed of deposits of a gelatinous or filamentous nature.

3.2 *Definitions*—For definitions of other terms used in these practices, refer to Definitions D 1129.

---

[1] These practices are under the jurisdiction of ASTM Committee D-19 on Water, and is the direct responsibility of Subcommittee D19.03 on Sampling of Water and Water-Formed Deposits, Surveillance of Water, and Flow Measurement of Water.

Current edition approved Oct. 29, 1982. Published March 1983. Originally published as D 887 – 46 T. Last previous edition D 887 – 77.

[2] *Annual Book of ASTM Standards*, Vol 11.01.

[3] *Annual Book of ASTM Standards*, Vol 11.02.

[4] *Discontinued*—See 1987 *Annual Book of ASTM Standards*, Vol 11.02.

4. Summary of Practices

4.1 These practices describe the procedures to be used for sampling water-formed deposits in both the field and laboratory from boiler tubes and turbine components. They give guidelines on selecting tube and deposit samples for removal and specify the procedures for removing, handling, and shipping of samples.

5. Significance and Use

5.1 The goal of sampling is to obtain for analysis a portion of the whole that is representative. The most critical factors are the selection of sampling areas and number of samples, the method used for sampling, and the maintenance of the integrity of the sample prior to analysis. Analysis of water-formed deposits should give valuable information concerning cycle system chemistry, component corrosion, erosion, the failure mechanism, the need for chemical cleaning, the method of chemical cleaning, localized cycle corrosion, boiler carryover, flow patterns in a turbine, and the rate of radiation build-up. Some sources of water-formed deposits are cycle corrosion products, make-up water contaminants, and condenser cooling water contaminants.

6. Reagents and Materials

6.1 *Purity of Reagents*—Reagent grade chemicals shall be used in all cases. Unless otherwise indicated, it is intended that all reagents shall conform to the specifications of the Committee on Analytical Reagents of the American Chemical Society, where such specifications are available.[5] Other grades may be used, provided it is first ascertained that the reagent is of sufficiently high purity to permit its use without lessening the accuracy of analysis.

6.1.2 *Purity of Water*—Unless otherwise indicated, references to water should be understood to mean Type III reagent water, Specification D 1193.

6.2 *Materials:*

6.2.1 The highest purity material available should be used for removing the deposit samples.

6.2.2 *Filter Paper* may contain water leachable contaminants (chloride, fluoride, and sulfur) which can be removed by pretreatment prior to sampling.

6.2.3 *Polyester Tape* may contain impurities of antimony and cadmium which must be considered during analysis.

7. Hazards

7.1 *Warnings:*

7.1.1 Special safety precautions are necessary in using acetone on a wipe material for removing water-formed deposits (see Note 2).

7.1.2 Special handling precautions may be required for working with water-formed deposits containing radioactive nuclides (see 9.14).

7.2 *Cautions:*

7.2.1 Extreme care must be taken not to damage the underlying surface when removing water-formed deposit samples from equipment in the field (see Note 1).

7.2.2 The selection of samples necessarily depends on the experience and judgment of the investigator. The intended use of the sample, the accessibility and type of the deposit, and the problem to be solved will influence the selection of the samples and the sampling method.

7.2.3 The most desirable amount of deposit to be submitted as a sample is not specific. The amount of deposit should be consistent with the type of analysis to be performed.

7.2.4 The samples must be collected, packed, shipped, and manipulated prior to analysis in a manner that safeguards against change in the particular constituents or properties to be examined.

7.2.5 The selection of sampling areas and number of samples is best guided by a thorough investigation of the problem. Very often the removal of a number of samples will result in more informative analytical data than would be obtained from one composite sample representing the entire mass of deposit. A typical example is the sampling of deposits from a steam turbine. Conversely, in the case of a tube failure in a steam generator, a single sample from the affected area may suffice.

7.2.6 Most deposits are sampled at least twice before being submitted to chemical or physical tests. The gross sample is first collected from its point of formation in the field and then this sample is prepared for final examination in the laboratory.

7.2.7 A representative sample is not an absolute prerequisite. The quantity of deposit that can be removed is often limited. In such cases, it is better to submit a single mixed sample (composite) and to describe how the sample was obtained. For radiological analysis all samples should be checked for activity levels before preparing a composite since wide variations in radioactive content may occur in samples of similar appearance and chemical composition.

7.2.8 It is good practice for deposits to be taken and analyzed every time a turbine is opened for repairs or inspection. Deposit history can then supplement chemical records of a unit, and deposit chemistry of units with and without corrosion and other problems can be compared. Enough information on deposits has been published **(1, 2)**[6] that a comparison between different types of boilers and different water treatments, as well as an assessment of deposit corrosiveness, are possible. It has been a general experience that about 0.2 % of a corrosive impurity, such as chloride, in a deposit, is a division between corrosive and noncorrosive deposits.

7.2.9 Deposits taken after a turbine is open do not exactly represent chemical composition of deposits in an operating turbine. Chemical thermodynamic data on steam additives and impurities, such as vapor pressures of solutions, ionization, and volatility data are needed to reconstruct chemistry of environment during operation.

7.2.10 Typical changes which occur after the hot turbine is shut down and air is admitted are: (*1*) reactions with

[5] *Reagent Chemicals, American Chemical Society Specifications,* American Chemical Society, Washington, DC. For suggestions on the testing of reagents not listed by the American Chemical Society, see *Analar Standards for Laboratory Chemicals,* BDH Ltd., Poole, Dorset, U.K., and the *United States Pharmacopeia and National Formulary,* U.S. Pharmaceutical Convention, Inc. (USPC), Rockville, MD.

[6] The boldface numbers in parentheses refer to the references at the end of these practices.

TABLE 1 Selection of Samples

	Tube 1		Tube 2
	Principal Area	Adjacent Area	Adjacent Tube 1 or Related Tube
Preferred selection procedure	X	X	X
Alternative selection procedure 1	X	X	
Alternative selection procedure 2	X		
Alternative selection procedure 3		X	
Alternative selection procedure 4			X

oxygen and carbon dioxide, (*2*) drying of some deposits and water absorption by others, (*3*) leaching and recrystallization where moisture is allowed to condense, and (*4*) formation of iron hydroxide and hematite.

PRACTICE A—SAMPLING WATER-FORMED DEPOSITS FROM TUBING OF STEAM GENERATORS AND HEAT EXCHANGERS

8. Scope

8.1 This practice covers the sampling of water-formed deposits from tubing of steam generators and heat exchangers. It covers both field and laboratory sampling of water-formed deposits. It gives guidelines on selecting tube samples for removal and specifies the procedure for removing tube samples from the unit.

9. Field Sampling

9.1 *Selection of Tube Samples*—Whenever feasible, remove the tube containing the water-formed deposit. The length of tubing removed depends on the amount of deposits present and the type of analyses to be performed. As a guideline, 3 ft (0.9 m) of tubing is suggested. Table 1 contains a summary of the various procedures for selection of samples in the order of preference.

9.1.1 *Preferred Selection Procedure*—Select one or more separate tube samples containing the area of failure, heaviest deposition, or principle concern (primary area) and include any adjacent or closely related areas of these tube samples that might contain deposits significantly different from the primary area. Also, one or more tube samples is selected from adjacent rows or other related areas that might contain deposits significantly different from the primary area.

9.1.2 *Alternative Selection Procedure 1*—Select one or more separate tube samples containing the area of failure, heaviest deposition, or principle concern (primary area) and include any adjacent or closely related areas of these tube samples that might contain deposits significantly different from the primary area. Use this procedure when it is impractical to remove the samples from adjacent rows or other related areas or when it is improbable that the information gained by such sampling will justify the additional work involved.

9.1.3 *Alternative Selection Procedure 2*—Select one or more separate tube samples containing the area of failure, heaviest deposition, or principle concern (primary area). Use this procedure when only the tube section containing the primary area can be removed or when it is impractical to remove adjacent or closely related areas, or tube samples from adjacent rows or other related areas, or when it is

improbable that the information gained by such sampling will justify the additional work involved.

9.1.4 *Alternative Selection Procedure 3*—Select one or more tube samples containing an area adjacent or closely related to the primary area. Use this procedure only when it is not possible to obtain a tube section containing the primary area.

9.1.5 *Alternative Selection Procedure 4*—Select one or more separate tube samples from adjacent rows or other related areas. Use this procedure only when it is not possible to remove a tube section from the primary area, adjacent to the primary area, or closely related to the primary area.

9.2 *Taking the Tube Sample*—Mark the tube that is to be removed (sampled) with a crayon. A long arrow can be used to show: (*1*) the ligament that is facing into the furnace, and (*2*) which end of the tube is up. Mark the tube before it is removed. The marking should not involve the use of a hammer and die or paint.

9.2.1 Whenever possible, remove the tube samples by sawing. The tube should be dry-cut (no oil). Grinding wheels and cutting torches can be used to obtain tube samples which cannot be sawed. Grinding wheels and cutting torches can produce sufficient heat to alter the composition of the deposit near the cutting point. If a grinding wheel or cutting torch is used, make the cut a minimum of 6 in. (152 mm) from the area of concern.

9.2.2 Usually it is impractical and inconvenient to remove short sections of tubes from a water-cooled furnace wall in order to obtain deposits from the waterside surfaces. Several convenient ways can be used to remove such surfaces. Avoid torch burning whenever possible, since original sample environment is often destroyed and a valid metallographic examination cannot be made of the specimen because of the effects of burning.

9.2.3 If a tube cannot be removed, trepan (hole-saw cutting) above and below the affected area harboring the deposits to remove "window section" from the tube. The window section is removed by connecting the holes with longitudinal cuts using an abrasive wheel. In most cases, the original deposit can be retained or collected from the affected area on the waterside surfaces of the tube. Also, a metallographic examination can be made because the original area is not altered or destroyed.

9.2.4 Penetrate the tube on the casing side or cold side of the tube if entrance on the furnace side is impractical from a time and cost standpoint. The deposit can be removed readily and a local examination of the waterside surfaces of the tube can be made.

9.2.5 Remove "window sections" only from tubes which have not failed. It is difficult to remove "window sections" from small diameter, thick-walled tubes.

9.3 *Sealing the Sample*—After removing the tube sample, allow the ends to cool. Dry the tube sample as soon as possible. Seal the tube ends with rubber stoppers or cardboard and secure the seal with tape.

9.4 *Sample Label*—Affix a label or a cardboard or linen tag to the sample.

9.4.1 Note the following information on the label or tag as soon as it becomes available. If this information is too voluminous for inclusion on the tag, it can be forwarded in a separate letter or report, properly identified with the samples concerned.

9.4.1.1 Name of organization supplying sample.

9.4.1.2 Name and location of plant.

9.4.1.3 Name and other designation of unit from which sample was removed.

9.4.1.4 Number of sample.

9.4.1.5 Date and time of sampling.

9.4.1.6 Precise location occupied in service.

9.4.1.7 Appearance of sample (note failures, bulges, pits, cracks, etc.).

9.4.1.8 Type of deposit (whether scale, sludge, or corrosion products).

9.4.1.9 Appearance of deposit (note the color, uniformity, texture, odor, and oily matter).

9.4.1.10 Exact procedure that was used in removing the sample and notes concerning any contamination that might have occurred during the process.

9.4.1.11 Identification of opposite walls of the tube sample that might contain different deposits.

9.4.1.12 Statement of whether liquid or vapor was present in the tube sample during operation.

9.4.1.13 Operating temperature, pressure, and rate of flow of liquid or vapor in the tube sample.

9.4.1.14 Type of treatment applied to the water and a chemical analysis of the water that formed the deposit or furnished steam to the affected zone.

9.4.1.15 Description of why the sample was taken; that is, the problem involved.

9.4.1.16 Description of discrepancies in operating condition that could have contributed to the problem.

9.4.1.17 Results of field tests made on the sample or related equipment.

9.4.1.18 Type of analysis necessary to solve the problem.

9.4.1.19 Signature of sampler.

9.4.2 Provide other background information that may be necessary for the specific problem. A few of these are date of unit startup, dates of chemical cleanings, type and location of water purification equipment, and direction of flow of fluid in the sample.

9.4.3 When numerous samples are collected, forward a diagram of the unit or affected zone to show the precise location of where the samples were removed.

9.5 *Sample Shipping Container*—Use a clean wooden box having a separate compartment for each sample as the shipping container. Line the compartment with corrugated paper, an elastic packing material, or other suitable material. Place each sample in a plastic bag or wrap in heavy paper and place in a compartment in the container. Seal the container properly to protect the samples during transit.

9.6 *Shipping Label*—Print the addresses of consignee and consignor plainly upon two sides of the outer container, or attach firmly thereon by cards or labels. Attach warning and descriptive labels to the outer container.

9.7 *Selection of Deposit Samples*—When it is not possible to remove the tube containing the water-formed deposit, remove the deposit directly from the tube surface in the field. The selection of sampling points will be somewhat limited and depend mainly on accessibility. Representative samples of water-formed deposits on tubes usually can be obtained near steam and mud drums, handholes, and manways. These samples can provide useful information upon analysis, depending on the specific problem involved. A photograph of the area, before and after removal of the deposit, could be a valuable aid in studying the problem and planning the analysis of the sample. The composition next to the underlying surface may be different from that which was in contact with the water. If possible, separation of these two different surfaces should be made to ascertain this possibility. At times, the quantity of deposit that can be removed is limited. In such cases, it is better to submit a single mixed sample (composite) and to describe how the sample was obtained, than to collect no sample. If the deposit weight per unit area is to be determined, Test Methods D 3483 should be reviewed prior to removal of the deposit.

9.8 *Taking the Deposit Sample*—The method of deposit removal will depend on accessibility and the type and amount of deposit present.

NOTE 1: **Caution**—in all cases, extreme care must be taken to minimize damage to the underlying surface or contaminate the sample. To avoid chloride and sodium contamination, deposits shall not be contacted with the hands.

9.8.1 *Tightly Adherent Deposits*—Remove hard, adherent deposits by using a sharp penknife, steel scraper, or scalpel. A vibrating mechanical power device may be required for extremely tenacious deposits.

9.8.2 *Loosely Adherent Deposits*—Remove deposits that adhere loosely to the surface with a knife, spatula, or spoon.

9.8.3 *Thin Hard Films*—Remove thin film deposits on rough or irregular surfaces by using a stainless steel brush, knife, or spatula.

9.8.4 *Thin Soft Films*—Remove thin, soft film deposits by using polyester tape or wiping with filter paper or lint-free cloth. Scotch tape or a comparable polyester tape has been found to be suitable for this application. When using filter paper or lint-free cloth, use the following procedure.

9.8.4.1 Wet a filter paper or lint-free cloth with water. (Whatman No. 40 or comparable filter paper, 12.5 cm has been found to be a suitable wiping material.) Wear rubber gloves to reduce the possibility of contaminating the wipe material and the sample.

9.8.4.2 Fold the wipe material to make a compact wad.

9.8.4.3 Wipe the surface to remove the deposit.

9.8.4.4 Continue wiping until it appears that no additional deposit is being removed. Unfold, turn, and refold the wipe material, if necessary. Keep the wipe material moist. Use more than one wipe material if required, but do not use more than is necessary.

9.8.4.5 Measure and record the area wiped.

9.8.4.6 Wet another wipe material with acetone and repeat 9.8.4.2, 9.8.4.3, and 9.8.4.4 using the same area.

NOTE 2: **Warning**—Acetone is an extremely flammable liquid and precautions are necessary when it is used. Methods for shipping, handling, storage, and disposal of acetone have been published by the National Safety Council (3).

9.8.4.7 Place the wipe materials (water and acetone wipes) containing the sample onto a sample container. Allow the acetone soaked wipe materials to air dry before placing into the sample container.

9.8.4.8 Select six clean filter papers or three clean cloths from the same lot as was used for removing the deposit and wet with water.

9.8.4.9 Select another six clean filter papers or three clean cloths from the same lot as was used for removing the deposit and wet with acetone. Allow to air dry.

9.8.4.10 Place the clean wipe materials (water and acetone soaked) into a sample container. These will serve as "blanks."

9.8.5 *Heterogeneous Layered Deposits*—Remove these deposits as near to the underlying surface as possible. The separation of the individual layers for examination is best done in the laboratory.

9.8.6 *Deposits in Pits*—Remove deposits in pits with a sharp penknife, steel scraper, scalpel, icepick, or sturdy needle-like probe.

9.9 *Quantity of Samples*—See 7.2.3 and 7.2.7.

9.10 *Sample Container*—Place each sample, as soon as possible after removal from the surface, in a container that will protect it from contamination or chemical change. Sample containers shall be made of materials that will not contaminate the sample and shall be clean and dry. A wide-mouth plastic bottle, small plastic vial, plastic bag, or noncorrodible metal container are suitable containers. Do not use corrodible metal containers for wet samples. Glass bottles and vials are also suitable, but are fragile and subject to breakage. Paper containers are not recommended. Fix the closures for the sample containers in place by wire, tape, or cord to prevent loss of the sample in transit if they are to be shipped.

9.11 *Sample Labels*—See 9.4.

9.12 *Sample Shipping Container*—Use a wooden box, cardboard box, or envelope as the shipping container, depending on the number and size of the samples. Line the shipping container with corrugated paper, an elastic packing material, or other suitable material and properly seal.

9.13 *Shipping Labels*—See 9.6.

9.14 *Radioactive Material*—In sampling water-formed deposits for the determination of its radioactivity, apply the normal principles of sampling. However, it may be impossible to remove deposit samples by the procedures given in 9.8. In these cases, chemical removal of the deposit may be the only alternative.

NOTE 3: **Warning**—Because of the potential hazards related to working with water-formed deposits containing radioactive nuclides, special handling precautions may be required. Such handling will depend on the amount and type of radioactive substances contained in the water-formed deposits and whether determination is to be made for gross radioactivity, for a specific nuclide, or for nonradioactive constituents by established chemical or physical tests.

NOTE 4: **Warning**—The radiation hazard associated with deposit samples may increase significantly when the sample is removed from a system which shields it or when suspended solids are filtered from water.

9.14.1 *Handling Precautions*—When sampling water-formed deposits where radiation levels or radioactive contamination may be high, such as in nuclear steam systems, follow the applicable health physics regulations. Anyone who is not experienced in the handling of radioactive materials should consult a health physicist or other person experienced in such activities. Personnel monitoring devices may be necessary where external radiation levels present a potential hazard. Where the level of radioactivity in the sampling area is sufficiently great to introduce a radiation hazard, shielding may be required to minimize exposure to radiation. Also, the use of suitable protective clothing may be required and

precautionary measures should be exercised to prevent contamination.

9.14.2 For information on radiological hazards and recommendations on radiation protection, refer to the publications by the National Committee on Radiation Protection (4, 5). The recommendations of the Federal Radiation Council on radiation protection guides have been adopted by all federal agencies by executive order of the President (6). The existing standards for protection against radiation are set fourth in the Code of Federal Regulations (7). Special methods and precautionary measures for handling radioactive samples are described in the literature (8, 9).

9.14.3 *Sample Containers*—Select the sample container by considering the type of sample involved and its level of radioactivity. Plastic containers usually are preferable when sampling for radioactivity determinations. The level of radioactivity in the sampling area and of the sample itself must be considered. A special apparatus may be required for shielding sample containers.

9.14.4 *Time Interval Between Collection and Analysis of Samples*—When sampling for radioactivity determinations, note the exact time of sample collection. If short-lived activity is of interest, the analysis should be made as rapidly as practical to minimize loss of activity by radioactive decay. If only long-lived radioactivity is of interest, measurement sometimes can be simplified by allowing sufficient time for the decay of the short-lived radionuclides before analysis.

9.14.5 *Shipping Containers*—Shield and pack the samples to comply with regulations for shipping radioactive materials (10, 11).

10. Laboratory Sampling

10.1 *Preliminary Examination of Tube Samples*—Carefully unpack the tube sample and remove seals and tape from the tube end. Burrs and sharp edges can be removed with a small file. Excessive handling of the sample, which can dislodge or contaminate the deposits, should be avoided. A photograph of the deposit in its original state and in different stages of removal can be taken to provide a valuable record of its appearance. The sample should be examined visually and its length and outside diameter plus any failures, cracks, or pits on the tube noted and recorded. Also, the characteristics of the water-formed deposit on the tube should be noted. The history of the sample should be reviewed and a definite plan for examination and analysis of the sample determined. If a metallurgical analysis (destructive or microchemical, or both) is to be performed, the examining metallurgist should remove specimens prior to removal of the water-formed deposit. The procedure of deposit removal and the amount of deposit removed must be consistent with the type of examination and analyses planned.

10.2 *Removal of Deposits from a Concave Surface (Interior Surface):*

10.2.1 *Preferred Procedure*—Apply this procedure to remove water-formed deposits from a specific location on a tube sample. The procedure can be used when the underlying metal surface is required for metallurgical examination. A cylindrical section that contains adequate deposits for the planned analyses is removed from each tube (or portion thereof) to be sampled. The section is split longitudinally with a shaper or dry saw (no oil). If opposite walls of the section might contain different types or amounts of deposits, as in a furnace wall boiler tube, the longitudinal split is made so as to separate the dissimilar portions. Care should be taken not to contaminate the sample with deposits from the exterior (convex) surface of the section. The deposits are separated from the tube by a suitable mechanical means that will not score the tube metal or otherwise contaminate the sample. See 9.8. If possible, deposits near the tube ends where saw filings and slag may have contaminated the deposit are avoided. If the tube sample was removed with a cutting torch in the field, the deposit near the cutting point should be avoided. The exact location of the deposit removed is noted. The surface area from which the deposit was removed is measured and recorded.

10.2.2 *Alternative Removal Procedure 1*—Apply this procedure when only a bulk, random sample of deposit from the tube surface is desired. Do not use this procedure when the underlying metal is required for metallurgical tests. Not all types of deposits can be dislodged by this method. A cylindrical section is removed from each tube and sectioned longitudinally as in 9.2.1. The deposit is dislodged by squeezing the freshly cut, open edges of the two resulting hemicylindrical sections in a vise.

10.2.3 *Alternative Removal Procedure 2*—When it is not permissible to destroy the tube sample by sectioning, remove superficial deposits by brushing the interior surface and dislodge the principal deposit with a turbine, vibrating head, or rotary cutter. The action of the cutter shall be vigorous enough to remove the deposit, instead of simply polishing it, but shall not be so severe that it causes excessive abrasion of the underlying surface and contamination of the sample.

10.2.4 *Alternative Removal Procedure 3*—Use inhibited chemical solvents when it is desirable to strip the tightly adherent water-formed deposits or protective film, or both from the tube surface after the overlaying loose deposits have been removed. Extreme care is necessary to ensure that only the desired deposits are removed and contamination from other surfaces of the tube sample is avoided.

10.3 *Removal of Deposits from a Convex Surface (Exterior Surface):*

10.3.1 *Preferred Removal Procedure*—See 10.2.1.

10.3.2 *Alternative Removal Procedure 4*—See 10.2.2.

10.3.3 *Alternative Removal Procedure 5*—See 10.2.4.

10.4 *Storage of Samples*—See 9.10.

10.5 *Sample Label*—See 9.4.

10.6 The final preparation of water-formed deposit samples received from the field or obtained from tube samples in the laboratory is a part of the specific analysis to be performed. Various physical and chemical treatments may be given to portions of the sample undergoing identification. Care must be exercised in selecting a representative sample for analysis, since the value of an analysis is related to the sampling technique employed. The following ASTM standards should be reviewed for specific instructions in selecting and preparing samples: Practices D 934, Test Methods D 993, Practices D 1245, Test Methods D 1428, Practices D 2331, Practice D 2332, and Test Methods D 3483.

10.7 *Radioactive Material*—See 9.14.

PRACTICE B—SAMPLING WATER-FORMED DEPOSITS FROM STEAM TURBINES

11. Scope

11.1 This practice covers the sampling of water-formed deposits from steam turbines. It covers both field and laboratory sampling of water-formed deposits. It gives guidelines for removal of turbine components.

12. Field Sampling

12.1 Field sampling immediately after a turbine is open is preferred since the deposits are easily contaminated by air-borne impurities, lubricants and solvents used in disassembly, and by nondestructive test (NDT) fluids. Most of these fluids contain chloride and sulfur as contaminants, some are chlorinated hydrocarbons.

12.2 *Selection of Deposit Samples*—Significance of a sample location differs for different types of turbine cycles (superheated, wet) and depends upon the reasons for deposit analysis (routine, corrosion failure analysis, component malfunction due to deposits, loss of thermodynamic efficiency due to deposits).

12.2.1 For a routine analysis in turbines using superheated steam, deposits from rotating blades at the inlet and exit of each turbine of a turbine set (high, intermediate and low pressure) as well as from the blades at the dry-wet transition region (Wilson line) should be sampled. Both, convex and concave surfaces and the whole length of the blades should be scraped. If there is anything indicating a flow or washing pattern or localized deposition anywhere on the rotating and stationary blades, disks, seals, pipes, valves, etc., a picture or sketch and a localized deposit should be taken.

12.2.2 For a corrosion failure analysis and any time there is a possibility of corrosion influence, deposits from the failed or cracked components and its vicinity should be taken in addition to all the deposits identified in 12.2.3. Frequently, there is a problem with deposit collection from cracked turbine components because they are cleaned (usually by alumina blasting) before the cracks are discovered by NDT. For cases like that, deposits from adjacent stationary blades and inlet pipes should be taken.

12.2.4 When corrosion and corrosion cracking (stress corrosion, corrosion fatigue) is found before component cleaning, a part of the surface should be masked by taping a plastic pad over a deposited surface to allow later microscopic analysis of the corroded surface.

12.2.5 Valve and seal malfunction due to buildup of deposits has been experienced. When solvents are needed in these cases for a disassembly, their chemistry should be known and accounted for in the deposit analysis.

12.2.6 Where deposition of rotating blades results in loss of thermodynamic efficiency, flow pressure changes and increase of thrust load, a deposit profile along all the blade rows is useful. In these cases, deposits may be taken from every rotating and stationary row and localized deposition of individual chemicals, thickness and pH profile established.

12.3 *Taking Deposit Samples*—Refer to 9.8.

12.4 Deposits within pits, cracks, and crevices on components which cannot be brought into a laboratory can be stripped by an acetate replicating tape used in fractography. Such a deposit is suitable for a microscopic, energy dispersive

X-ray (EDX) and X-ray diffraction analysis.

12.5 *Quantity of Samples*—See 7.2.3 and 7.2.7.

12.6 *Sample Container*—See 9.10.

12.7 *Sample Label*—See 9.4.

12.8 *Sample Shipping Container*—See 9.12.

12.9 *Shipping Label*—See 9.6.

12.10 *Radioactive Material*—See 9.14.

13. Removal of Turbine Components for Analysis

13.1 Turbine components are usually removed from a disassembled turbine for replacement or laboratory failure analysis. They are rarely removed for analysis of deposits. In analysis of a corrosion problem, analysis of surface morphology and deposition is important.

13.2 *Removal of a Component*—Critical parts of a turbine component destined for further analysis should be protected against contamination during removal, handling, and shipment. Use of solvents and other fluids should be avoided or their chemistry should be known and accounted for.

13.3 *Shipping*—The whole component or a part of its surface should be reliably protected against rain and airborne contamination, particularly by sea salt and road salts. For labeling and containers, refer to 9.4, 9.6, 9.10, 9.12, and 9.14.

14. Laboratory Sampling

14.1 Laboratory sampling of deposits from turbine components is done similar to the field sampling on a section of a component. Dry cutting of such a section is recommended to avoid deposit washing and contamination.

14.2 Such samples are suitable for microscopic and EDX analysis, studies of deposits within cracks and pits, and for secondary ion mass spectroscopy (SIMS) analysis. Since the quantity of a deposit is usually low, highly sensitive analytical techniques such as liquid ion chromatography and inductively coupled plasma emission of washed deposits are very useful.

14.3 Turbine deposits are usually analyzed semi-quantitatively (by spectroscopy or EDX); quantitatively for anions, sodium, total carbon and total sulfur; and by X-ray diffraction.

14.4 The analyses considered most important are: chloride, sulfate, total sulfur, phosphate, carbonate, hydroxide, silicon, copper, total carbon, iron, pH, and X-ray diffraction identification of chemical compounds. In special cases, additional analyses, such as for chromium, manganese, molybdenum, calcium, potassium, aluminum, lead, fluoride, etc., may be needed. Deposit thickness should be measured or estimated. A list of commonly used procedures is shown in Table 2.

TABLE 2   Procedures for Analysis of Turbine Deposits

ASTM Designation	Measurement
Test Methods D 1428	sodium, potassium
Practices D 2331	phosphates, sulfates, and carbonates
Test Methods D 2579	total carbon
	total sulfur
Test Methods D 512	chloride
Practices D 934	crystalline compounds
Specification D 1293	pH
Semiquantitative spectroscopy	Fe, Si, Ni, Cr, Mo, Cu, Mn, Al, Co, Ti, V, Ca, Cb, Pb, Mg, Zr
Secondary ion mass spectroscopy (SIMS)	compounds in thin layers
Energy dispersive x-ray (EDX)	elements with atomic number 11 and higher
Liquid ion chromatography (12, 13)	Na^+, NH_4^+, K^+, F^-, Cl^-, NO_2^-, NO_3^-, PO_4^{-3}, Br^-, SO_4^{-2}
Inductively coupled plasma (ICP) (14)	metals

REFERENCES

(1) Jonas, O., "Survey of Steam Turbine Deposits," Proceedings of the 34th International Water Conference, Pittsburgh, PA, 1973, pp. 73–81.

(2) Jonas, O., "Transfer of Chemicals in Steam Power Systems," Presented at EEE-ASME Joint Power Generation Conference, Buffalo, NY, September 1976.

(3) Acetone, Data Sheet 398 (revised), Published by National Safety Council, 425 North Michigan Ave., Chicago, IL, 1963.

(4) "Basic Radiation Protection Criteria," N.R.C.P. Report No. 39, January 15, 1971.

(5) Code of Federal Regulations, Title 10, Part 20. Standards for Protection Against Radiation. Appendix B and Addenda 1 to 69, issued August 1963, or latest applicable references.

(6) *Federal Register*, May 18, 1969, p. 4402.

(7) Code of Federal Regulations, Title 10-Atomic Energy, Part 20, Standards for Protection Against Radiation.

(8) Friedlander, G., and Kennedy, J. W., *Nuclear and Radiochemistry*, 2nd Edition Revised, John Wiley and Sons, Inc., New York, NY, 1965.

(9) Overman, R. T., and Clark, H. M., *Radioisotope Techniques*, McGraw-Hill Book Company, Inc., New York, NY, 1960.

(10) Hazardous Material Regulations of the DOT, Title 49, Parts 170–190, revised Dec. 31, 1968, or latest applicable references.

(11) Current CAB Regulations for Air Shipments.

(12) Small, H., Stevens, T., and Bauman, W. "Novel Ion Exchange Chromatographic Method Using Conductimetric Detection," *Analytical Chemistry*, Vol 47 (11) 1975, p. 1801.

(13) Stevens, T., and Turkelson, V. T., "Determination of Anions in Boiler Blow-Down Water with Ion Chromatography," *Analytical Chemistry*, Vol 49 (8) 1977, p. 1176.

(14) Garbarino, J. R., and Taylor, H. E., *Applied Spectroscopy*, Vol 33/3, 1979, p. 220.

Standard Practices for
Sampling of Waterborne Oils[1]

This standard is issued under the fixed designation D 4489; the number immediately following the designation indicates the year of original adoption or, in the case of revision, the year of last revision. A number in parentheses indicates the year of last reapproval. A superscript epsilon (ε) indicates an editorial change since the last revision or reapproval.

1. Scope

1.1 These practices describe the procedures to be used in collecting samples of waterborne oils (see Practice D 3415), oil found on adjoining shorelines, or oil-soaked debris, for comparison of oils by spectroscopic and chromatographic techniques, and for elemental analyses.

1.2 Two practices are described. Practice A involves "grab sampling" macro oil samples. Practice B can be used to sample most types of waterborne oils and is particularly applicable in sampling thin oil films or slicks. Practice selection will be dictated by the physical characteristics and the location of the spilled oil. These two practices are:

	Sections
Practice A (for grab sampling thick layers of oil, viscous oils or oil soaked debris, oil globules, tar balls, or stranded oil)	9 to 13
Practice B (for TFE–fluorocarbon polymer strip samplers)	14 to 17

1.3 Each of the two practices is designed to collect oil samples with a minimum of water, thereby reducing the possibility of chemical, physical, or biological alteration by prolonged contact with water between the time of collection and analysis.

1.4 *This standard does not purport to address all of the safety concerns, if any, associated with its use. It is the responsibility of the user of this standard to establish appropriate safety and health practices and determine the applicability of regulatory limitations prior to use.* For specific hazards statements, see Section 7.

2. Referenced Documents

2.1 *ASTM Standards:*
D 1129 Terminology Relating to Water[2]
D 3415 Practice for Identification of Waterborne Oils[3]

3. Terminology

3.1 *Definitions*—For the definitions of terms used in these practices, refer to Terminology D 1129.

3.2 *Descriptions of Terms Specific to This Standard:*

3.2.1 *chain of custody*—a documented accountability of each sample, that is, date, time, and signature of each recipient when the sample changes hands, from the time of collection until the requirement for each sample is terminated.

3.2.2 *waterborne oil*—refer to Practice D 3415.

4. Significance and Use

4.1 Identification of the source of a spilled oil is established by comparison with known oils selected because of their possible relationship to the spill, that is, potential sources. Generally, the suspected source oils are from pipelines, tanks, etc., and therefore pose little problems in sampling compared to the spilled oil. This practice addresses the sampling of spilled oils in particular, but could be applied to appropriate source situations, for example, a ship's bilge.

5. Apparatus

5.1 *Sample Containers,* 100 to 125-mL wide-mouth glass jars that have been thoroughly cleaned. When field expedients must be employed, an empty container of each type used should be included in the shipment to the laboratory, to be used as a blank to measure inadvertent contamination.

5.2 *Closures*—Lids for the glass jars should have TFE-fluorocarbon polymer film or aluminum-coated insert.

5.3 *Strip Samplers,* 5 by 7.5 cm pieces of TFE-fluorocarbon polymer sheets (0.25 mm thickness, or screen or fabric (50–70 mesh)).

5.4 *Wooden Tongue Depressor.*

5.5 *TFE-Fluorocarbon Polymer Net Sampling Kit.*[4]

6. Reagents

6.1 *High Purity Solvents,*[5] that must be used for rinsing samplers and sample containers. The solvents which may be used are *n*-hexane, mixed hexanes, cyclohexane, pentane, or dichloromethane, acetone, or chloroform.

7. Hazards

7.1 **Precaution:** Extreme care should be exercised so as not to contaminate the samples or cause their integrity to be questioned.

7.2 **Warning:** The rinsing solvents are volatile and, except for dichloromethane, are flammable, and therefore should be handled with appropriate care. Dichloromethane will release toxic vapors when heated.

7.3 Minimize contact with oil even when wearing gloves.

8. General Sampling Guidelines

8.1 The objective is to obtain a sample for analysis that is representative of the spilled oil. The most critical factors in sampling are selecting a suitable location, collecting a sample of oil with the least water possible (to minimize possible

---

[1] These practices are under the jurisdiction of ASTM Committee D-19 on Water and are the direct responsibility of Subcommittee D19.31 on Identification of Waterborne Oils.
Current edition approved Sept. 10, 1995. Published November 1995. Originally published as D 4489 – 85. Last previous edition D 4489 – 85 (1990)ε1.
[2] *Annual Book of ASTM Standards,* Vol 11.01.
[3] *Annual Book of ASTM Standards,* Vol 11.02.

[4] Sampling kit available from General Oceanics, Miami, FL, or equivalent, is suitable.
[5] MCB Spectroquality solvents, available from MCB Manufacturing Chemists, Inc. (Associate of E. Merck, Darmstadt, Germany), 480 Democrat Rd., Gibbstown, NJ 08027, or equivalent are suitable.

sample alteration), and maintaining the sample integrity.

8.2 It is recommended that at least three samples be taken of each waterborne oil in order to demonstrate the homogeneity of the spill. These samples should be taken in different regions of the oil slick at points where the accumulation is heaviest. This will increase the volume of oil available for analysis. In the event that multiple samples cannot be collected, then a single sample should be collected from the area where the accumulation of oil visually appears to be the heaviest.

8.3 The following general rules are applicable to sampling of waterborne oils:

8.3.1 Take a sample that contains sufficient oil for the method or methods of analysis to be employed and for any replicate analyses that may be required.

8.3.2 Affix a label or tag to the sample jar in such a manner that it becomes an integral part of the container. The label or tag should contain the following information: sample identification, date and time of collection, location of collection, signature of person collecting the sample, and at least one witness to the collection.

8.3.3 Pack the samples, ship, and manipulate prior to analysis in a manner that maintains a continuous chain of custody and safeguards against tampering or changes in the properties of the samples.

8.4 Store collected samples at refrigerator temperatures (4 to 5°C).

NOTE 1—Storage at lower temperatures (−10°C or lower) may cause irreversible crystallization of waxes. Storage at 4 to 5°C obviates this problem; biological degradation at 4 to 5°C has been found negligible over a 3 to 5 year storage with respect to qualitative identification of oil.

PRACTICE A—GRAB SAMPLING

9. Scope

9.1 This practice is applicable to thick layers of waterborne oil films, viscous oils, oil globules, and tar balls.

9.2 This practice is also applicable to sampling oil stranded on shorelines or oil-soaked debris.

10. Summary of Practice

10.1 The sampling consists of collecting the sample directly with the sample container, that is, scooping the sample up in the sample jar and sealing.

11. Apparatus

11.1 The sample container serves as the sampling device (see 5.1). The glass jars and lid liners should be rinsed three times with a high purity solvent (see 6.1), allowed to air dry, and assembled prior to use. Sample jars that are precleaned using EPA recommended wash procedures for organics are acceptable.

NOTE 2—To avoid possible sample contamination, do not reuse sample containers, lids, or liners.

11.2 Nitrile gloves are to be worn during sampling.

11.3 A detachable ring for the sample jar and sampling pole may be useful to extend sampling range.

12. Procedure for Floating Samples

12.1 Select the sampling site.

12.2 Unscrew the lid from the sample jar. Hold the jar in

position for sampling; hold the lid in a free hand or place the lid in a safe position. Gently lower the sample jar into the water and gently skim the oil layer or oil globules from the water surface into the sample container. Continue the process until the sample container is approximately three-quarters full.

12.3 Remove the sample container from the water surface, replace and tighten the lid. Invert the jar and allow the container to stand in this position for 2 to 3 min.

12.4 Gently unscrew the sample jar lid and allow the water layer to drain out of the inverted container. Seal the lid and return the jar to the upright position.

12.5 Repeat 12.2 to 12.4, if necessary, until approximately 60 mL of oil is collected, or until there is no increase in the amount of recovered oil.

12.6 When the collection is complete, invert the jar and allow to stand for 10 min. Gently unscrew the lid to drain off excess water a final time. Tighten the lid and return the jar to the upright position. Wipe excess water and oil from the outside surface of the sample container.

12.7 Attach a sample label or tag to the container, bearing the information cited in 8.3.2.

13. Procedure for Shoreline Sampling (Oil on Sand and Debris)

13.1 Select a sampling site where oil accumulation is largest.

13.2 Open the sample jar; hold the jar in one hand and lid in the other. Using either the sample jar or the lid as a scoop, fill the jar three-quarters full with oil-saturated material. Use a wooden tongue depressor to maneuver the sample into the jar, if necessary.

13.3 Replace and tighten the lid. Wipe excess material from the outside surfaces of the sample container and lid.

13.4 Attach the sample label or tag to the container, bearing the information cited in 8.3.2.

PRACTICE B—TFE-FLUOROCARBON POLYMER SAMPLERS

14. Scope

14.1 This practice is applicable to sampling all types of oil by preferential adherence to a film or sheet of TFE-fluorocarbon material. It depends in principle on the lipophilic properties of TFE-fluorocarbon polymer, that is, the preferential adhesion of oil rather than of water to TFE-fluorocarbon polymer.

14.2 In general, the use of TFE-fluorocarbon polymer screening (approximately 50 to 70 mesh) will collect significantly more oil than strips of sheet material. The screen reduces the rate at which oil rolls off the surface and presents more surface openings to trap oil droplets, thereby collecting more oil per unit area of film.

14.3 TFE-fluorocarbon polymer mesh fabric sheets are commercially available. These are the most efficient form of TFE-fluorocarbon polymer for sampling oil. Although they can be used directly in sheet form, this material can be fabricated into nets of design similar to a miniature plankton net, further increasing the ease with which the collection device can be brought into contact with the oil sheen. There is little difference in the performance of the nets ranging from 100 to 200 micron mesh size.

14.4 The efficiency of collecting oil increases as the viscosity of the oil increases. This practice is particularly useful in sampling highly weathered oils.

15. Summary of Practice

15.1 Sampling is accomplished by slowly dragging the TFE-fluorocarbon polymer through the slick and using its natural affinity to collect the oil.

15.2 The procedure is for the use of strips of TFE-fluorocarbon polymer sheet, TFE-fluorocarbon screen or fabric screen, or commercially available prefabricated nets with support rings.

16. Materials

16.1 TFE-fluorocarbon polymer can be obtained in sheets of 0.25 mm thickness. To increase the efficiency of oil collection, sheets can be obtained containing five 1.5-mm holes per square centimetre. Higher oil collecting efficiency can be obtained by using the previously mentioned 50 to 70-mesh screen (see 14.2) or fabric (see 14.3).

16.2 TFE-fluorocarbon polymer is cut into 5 by 7.5 cm strips. Carefully rinse the strips with high purity solvent (see 6.1) and air dry. Place eight strips in each sample jar that has been precleaned as described in 11.1, and tightly seal the lids.

16.3 Each jar, containing eight strips, is used for taking a single sample. One jar is set aside in each sampling situation for use as a blank in subsequent analyses.

16.4 Clean tweezers, hemostats, or pliers are required for handling the TFE-fluorocarbon strips to avoid contamination.

16.5 TFE-fluorocarbon polymer mesh fabric, with 150 µm distance between the polymer threads, is available prefabricated into nets that are detachable from support rings. The nets and rings are cleaned and prepackaged with nitrile gloves in plastic bags. To avoid contaminating the nets with finger oils, it is critical to handle the nets only with the nitrile gloves.

17. Sampling Waterborne Oils

17.1 Select the sampling site where oil accumulation is heaviest.

17.2 For the strip technique, remove the lid from the sample jar. Using precleaned tweezers, hemostats, or pliers as holders, carefully lay the TFE-fluorocarbon strips on the inverted lid. Using the holder, take the strips one at a time, and gently drag them through the slick. Expose both sides of the TFE-fluorocarbon polymer strip to the slick by turning the strip over and dragging it again through the slick, taking care not to get oil on the tweezers, hemostats, or pliers. Place the strip in the empty sample jar as quickly as possible to prevent loss of oil.

17.3 Repeat 17.2 until all eight strips have been used.

NOTE 3—For extremely thin sheets, it is recommended that 16 strips be used, rather than 8.

17.4 For the net technique, remove the nitrile gloves from the kit. Do not handle anything other than the sample kit. Remove the net from its sealed bag using the handle of the support ring to which it is attached. If necessary, attach the net support ring to the extension pole. Do not allow the net to come into contact with anything other than the spill.

17.5 Take the sample by skimming through the sheen and straining the oily water through the net. Make sure that the sheen is entering through the mouth of the net and straining through the fine mesh of the net. Slowly skim the surface with the net back and forth through the full length of the sheen at least eight times.

17.6 While still wearing the gloves, unclip the net from the support ring and place the net into a clean sample jar. Expect some water to remain in the jar. Touch the net material as little as possible. Discard the gloves.

17.7 Attach each jar lid and tighten.

17.8 Attach a sample label or tag to each container, bearing the information cited in 8.3.2.

18. Keywords

18.1 oil identification; sampling; spilled oil; TFE-fluorocarbon polymer sampler; waterborne petroleum oils

Standard Practice for
Preservation of Waterborne Oil Samples[1]

This standard is issued under the fixed designation D 3325; the number immediately following the designation indicates the year of original adoption or, in the case of revision, the year of last revision. A number in parentheses indicates the year of last reapproval. A superscript epsilon (ε) indicates an editorial change since the last revision or reapproval.

ε1 NOTE—Keywords were added editorially in December 1996.

1. Scope*

1.1 This practice covers the preservation of waterborne oil samples from the time of collection to the time of analysis. Information is provided to ensure sample integrity and to avoid contamination and to minimize microbial degradation.

1.2 The practice is for controlled field or laboratory conditions and specifies thorough preparation of equipment and precise operation. Where these details must be compromised in a field emergency, nonstandard simplifications are recommended that will minimize or eliminate consequent errors.

NOTE 1—Procedures for the analysis of oil spill samples are Practices D 3326, D 3415, D 3650, and D 4489, and Test Methods D 3327, D 3328, D 3414, and D 5037. A guide to the use of ASTM test methods for the analysis of oil spill samples is found in Practice D 3415.

1.3 *This standard does not purport to address all of the safety concerns, if any, associated with its use. It is the responsibility of the user of this standard to establish appropriate safety and health practices and determine the applicability of regulatory limitations prior to use.* For a specific hazard statement, see Note 3.

2. Referenced Documents

2.1 *ASTM Standards:*
D 1129 Terminology Relating to Water[2]
D 3326 Practices for Preparation of Samples for Identification of Waterborne Oils[3]
D 3327 Test Methods for Analysis of Selected Elements in Waterborne Oils[4]
D 3328 Test Methods for Comparison of Waterborne Petroleum Oils by Gas Chromatography[3]
D 3414 Test Method for Comparison of Waterborne Petroleum Oils by Infrared Spectroscopy[3]
D 3415 Practice for Identification of Waterborne Oils[3]
D 3650 Test Method for Comparison of Waterborne Petroleum Oils by Fluorescence Analysis[3]
D 4489 Practices for Sampling Waterborne Oils[3]
D 5037 Test Method for Comparison of Waterborne Petroleum Oils by High Performance Liquid Chromatography[3]

3. Terminology

3.1 *Definitions*—For definitions of terms used in this practice, refer to Terminology D 1129.

4. Summary of Practice

4.1 Special types of sample containers and shipping containers are recommended. Samples may be of several types: tar balls, collected oil, oil-water mixtures, emulsions, and oil and water on collecting devices such as silanized glass cloth, TFE-fluorocarbon polymer, or other materials. Instructions are given for the care of samples to minimize changes due to autoxidation and microbial attack between the time of sampling and the time of analysis. Services available for transportation of samples are described.

5. Apparatus

5.1 *Sample Containers*—Borosilicate glass containers that have been thoroughly cleaned are preferable. All glass containers, new or used, must be thoroughly cleaned and washed prior to use. The cleaning steps consist of an initial wash with a warm aqueous detergent mixture followed by six hot tap water rinses, two rinses with reagent water, a rinse with reagent-grade acetone, and a final rinse with a solvent such as pentane, hexane, cyclohexane, dichloromethane, or chloroform followed by drying in a clean oven at 105°C or hotter for 30 min. If the glassware requires cleaning under field conditions, it should be washed with warm aqueous detergent followed by extensive water rinsing. A solvent rinse with acetone should be made, if possible, followed by lengthy air drying to remove residual solvent.

NOTE 2—Hot reagent water rinses are advisable where hot tap water might reintroduce contamination.
NOTE 3—**Caution:** For safety reasons, the use of pentane, hexane, or cylcohexane is recommended over use of dichloromethane or carbon tetrachloride.

5.1.1 Plastic containers are not acceptable since volatile hydrocarbons diffuse readily through many commercial plastic containers or may be absorbed into the plastic. In addition, the plasticizer may dissolve in the sample causing misleading results.

5.1.2 Metal containers usually should be avoided because

[1] This practice is under the jurisdiction of ASTM Committee D-19 on Water and is the direct responsibility of Subcommittee D19.31 on Identification of Waterborne Oils.
Current edition approved May 25, 1990. Published February 1991. Originally published as D 3325 – 74 T. Last previous edition D 3325 – 85ε1.
[2] *Annual Book of ASTM Standards*, Vol 11.01.
[3] *Annual Book of ASTM Standards*, Vol 11.02.
[4] Discontinued; see *1993 Annual Book of ASTM Standards*, Vol 11.02.

* A Summary of Changes section appears at the end of this practice.

TABLE 1 Procedures for the Preservation of Waterborne Oil Samples

Recommended Operation	Procedure for Emergency Conditions	Procedures for Controlled Field or Laboratory
Sample containers	Borosilicate glass jars for high boiling samples.	Borosilicate glass jars for high boiling samples.
Cleaning containers	Wash with warm aqueous detergent followed by water rinsing. Rinse with acetone, if available, followed by air drying.	Wash with warm aqueous detergent followed with hot tap water and reagent water rinses. Rinse with acetone, chloroform, and oven dry.
Closures	TFE-fluorocarbon sheet or TFE-fluorocarbon or aluminum-coated cardboard inserts.	TFE-fluorocarbon sheet or TFE-fluorocarbon or aluminum-coated cardboard inserts.
Protection against autoxidation	Store in dark.	Remove air with nitrogen or carbon dioxide. Store in dark in refrigerator.
Protection against microbial attack	Refrigerate to 4 to 5°C, if possible.	Samples should be stored in laboratory refrigerator at about 4 to 5°C.
Shipment of samples	Pack in ice, if possible, and label appropriately. Notify recipient.	Pack in ice, if possible, and label appropriately. Notify recipient.

the nickel and vanadium determinations could be invalidated by introduction of metal from the can.

5.1.3 When field expedients must be employed, an empty container of each type used should be included in the shipment to the laboratory to be used as a blank to measure inadvertent contamination.

5.2 *Closures*—Proper choice of closures is critical to avoid contamination and to preserve sample. Use caps with aluminum-coated or TFE-fluorocarbon-coated cardboard inserts. Aluminum foil sheet should not be used. Inserts of TFE-fluorocarbon sheet, however, are acceptable.

5.3 *Refrigerator*, explosion-proof at about 4 to 5°C.

5.4 *Shipping Containers*—Sturdy cartons or wooden boxes should be used. These should be sufficiently large so the sample containers are adequately surrounded by absorbent packing material, such as vermiculite, sufficient to absorb the entire contents should breakage occur and be lined with a grease proof plastic bag.

NOTE 4—After Dec. 31, 1990, packagings used for shipment of dangerous goods such as oil must be tested for compliance with UN Performance Packaging Standards and certified by a marking applied by the packaging manufacturer.

6. Reagents and Materials

6.1 *Purity of Reagents*—Reagent grade chemicals shall be used in all tests. Unless otherwise indicated, it is intended that all reagents shall conform to the specifications of the Committee on Analytical Reagents of the American Chemical Society.[5] Other grades may be used, provided it is first ascertained that the reagent is of sufficiently high purity to permit its use without lessening the accuracy of the determination.

6.2 *Acetone.*

6.3 *N-hexane, Mixed Hexanes, Cyclohexane, Pentane, Dichloromethane, or Chloroform*, spectroquality or equivalent high purity.

7. Sampling

7.1 Collect a representative sample in accordance with Practices D 4489.

---

[5] *Reagent Chemicals, American Chemical Society Specifications*, American Chemical Society, Washington, DC. For suggestions on the testing of reagents not listed by the American Chemical Society, see *Analar Standards for Laboratory Chemicals*, BDH Ltd., Poole, Dorset, U.K., and the *United States Pharmacopeia and National Formulary*, U.S. Pharmacopeial Convention, Inc. (USPC), Rockville, MD.

7.2 Sample containers should be carefully prepared as described in 5.1.

8. Preservation of Samples

8.1 *Protection Against Autoxidation*—Treat the sample container to displace air and store in a dark area in a refrigerator. Nitrogen or carbon dioxide can be used as inert gases to displace air.

8.2 *Protection Against Microbial Attack*—Maintain refrigeration (4 to 5°C) once samples are received in the laboratory.

9. Shipment of Samples

9.1 The shipping of oil samples is regulated by both Department of Transportation (DOT) and United States Postal Service regulations.

NOTE 5—As of Sept. 18, 1988, the U.S. Postal Service regulations were amended to reflect implementation of the International Civil Aviation Organization (ICAO) Technical Instructions for the Safe Transport of Dangerous Goods by Air. Under these international regulations, the definition of flammable liquid is different from the U.S. DOT definition. Under ICAO rules, a Flammable Liquid—Class 3 is defined as a liquid having a flashpoint of 141°F (60.5°C) or less. In addition, the limited quantity provisions for flammable liquids under U.S. DOT regulations are not applicable under the ICAO rules.

NOTE 6—For more information about the legal requirements for packaging and shipping petroleum oils and other hazardous materials, refer to U.S. Postal Service Publication 52, "Acceptance of Hazardous, Restricted, or Perishable Matter," the Domestic Mail Manual, Part 124, "Nonmailable Matter—Articles and Substances; Special Mailing Rules," and the packaging requirements listed in the Domestic Mail Manual, Part 121.

9.2 Common carriers such as air express or air freight are often employed for transportation of oil spill samples. Consult the carrier for current packing and labelling requirements.

9.3 It is good practice to notify the receiving laboratory of shipment arrangements.

10. Procedure

10.1 The prescribed standard procedures for controlled field or laboratory conditions and recommended modification for emergency conditions are summarized in Table 1. Use this table with the specific sections of this practice to choose the proper measures for preservation of the specific waterborne oils under consideration.

11. Keywords

11.1 identification; oil spill; preservation; sample; shipment; storage; waterborne oil

SUMMARY OF CHANGES

This section identifies the location of selection changes to this practice that have been incorporated since the last issue. For the convenience of the user, Committee D-19 has highlighted those changes that may impact the use of this practice. This section may also include descriptions of the changes or reasons for the changes or both.

(*1*) Paragraph 5.1 includes the addition of alternative solvents for cleaning sample containers.

(*2*) Notes 4, 5, and 6 contain information on changes to shipping regulations and references useful for determining current regulations in effect by the date of this revision.

Standard Practice for
Preparation of Samples for Identification of Waterborne Oils[1]

This standard is issued under the fixed designation D 3326; the number immediately following the designation indicates the year of original adoption or, in the case of revision, the year of last revision. A number in parentheses indicates the year of last reapproval. A superscript epsilon (ε) indicates an editorial change since the last revision or reapproval.

[ε1] NOTE—Keywords were added editorially in December 1996.

1. Scope

1.1 This practice covers the preparation for analysis of waterborne oils recovered from water. The identification is based upon the comparison of physical and chemical characteristics of the waterborne oils with oils from suspect sources. These oils may be of petroleum or vegetable/animal origin, or both. Seven procedures are given as follows:

	Sections
Procedure A (for samples of more than 50-mL volume containing significant quantities of hydrocarbons with boiling points above 280°C)	8 to 12
Procedure B (for samples containing significant quantities of hydrocarbons with boiling points above 280°C)	13 to 17
Procedure C (for waterborne oils containing significant amounts of components boiling below 280°C and to mixtures of these and higher boiling components)	18 to 22
Procedure D (for samples containing both petroleum and vegetable/animal derived oils)	23 to 27
Procedure E (for samples of light crudes and medium distillate fuels)	28 to 34
Procedure F (for thin films of oil-on-water)	35 to 39
Procedure G (for oil-soaked samples)	40 to 44

1.2 Procedures for the analytical examination of the waterborne oil samples are described in Practice D 3415 and Test Methods D 3327, D 3328, D 3414, and D 3650. Refer to the individual oil identification test methods for the sample preparation method of choice. The deasphalting effects of the sample preparation method should be considered in selecting the best methods.

1.3 *This standard does not purport to address all of the safety concerns, if any, associated with its use. It is the responsibility of the user of this standard to establish appropriate safety and health practices and determine the applicability of regulatory limitations prior to use.* Specific caution statements are given in Sections 6 and 32.

2. Referenced Documents

2.1 *ASTM Standards:*
D 95 Test Method for Water in Petroleum Products and Bituminous Materials by Distillation[2]
D 96 Test Method for Water and Sediment in Crude Oil by Centrifuge Method (Field Procedure)[2]
D 1129 Terminology Relating to Water[3]

D 1193 Specification for Reagent Water[3]
D 1959 Test Method for Iodine Value of Drying Oils and Fatty Acids[4]
D 1983 Test Method for Fatty Acid Composition by Gas-Liquid Chromatography of Methyl Esters[4]
D 2800 Test Method for Preparation of Methyl Esters from Oils for Determination of Fatty Acid Composition by Gas Chromatography[4]
D 3325 Practice for Preservation of Waterborne Oil Samples[5]
D 3327 Test Methods for Analysis of Selected Elements in Waterborne Oils[6]
D 3328 Test Methods for Comparison of Waterborne Petroleum Oils by Gas Chromatography[5]
D 3414 Test Method for Comparison of Waterborne Petroleum Oils by Infrared Spectroscopy[5]
D 3415 Practice for Identification of Waterborne Oils[5]
D 3650 Test Method for Comparison of Waterborne Petroleum Oils by Fluorescence Analysis[5]
D 4489 Practices for Sampling of Waterborne Oils[5]
E 1 Specification for ASTM Thermometers[7]
E 133 Specification for Distillation Equipment[8]

3. Terminology

3.1 *Definitions*—For definitions of terms used in this practice, refer to Terminology D 1129.

3.2 *Definitions of Terms Specific to This Standard:*

3.2.1 *animal/vegetable-derived oils*—a mixture made of mono-, di-, and triglyceride esters of fatty acids and other substances of animal or vegetable origin, or both.

3.2.2 *Simulated weathering of waterborne oils by distillation* considers only the effect of evaporation, which likely is the most significant short-term weathering effect in the environment.

3.2.3 *Simulated weathering of waterborne oils by evaporation* under ultraviolet light simulates the loss of light components on weathering, as well as some oxidative weathering.

4. Significance and Use

4.1 Identification of a recovered oil is determined by comparison with known oils selected because of their possible relationship to the particular recovered oil, for example,

---

[1] This practice is under the jurisdiction of ASTM Committee D-19 on Water and is the direct responsibility of Subcommittee D19.31 on Identification of Waterborne Oils.
Current edition approved May 25, 1990. Published October 1990. Originally published as D 3326 – 74 T. Last previous edition D 3326 – 84.
[2] *Annual Book of ASTM Standards,* Vol 05.01.
[3] *Annual Book of ASTM Standards,* Vol 11.01.

[4] *Annual Book of ASTM Standards,* Vol 06.03.
[5] *Annual Book of ASTM Standards,* Vol 11.02.
[6] Discontinued; see *1993 Annual Book of ASTM Standards,* Vol 11.02.
[7] *Annual Book of ASTM Standards,* Vol 14.03.
[8] *Annual Book of ASTM Standards,* Vol 14.02.

suspected or questioned sources. Thus, samples of such known oils must be collected and submitted along with the unknown for analysis. It is unlikely that identification of the sources of an unknown oil by itself can be made without direct matching, that is, solely with a library of analyses.

5. Reagents and Materials

5.1 *Purity of Reagents*—Reagent grade chemicals shall be used in all tests. Unless otherwise indicated, it is intended that all reagents shall conform to the specifications of the Committee on Analytical Reagents of the American Chemical Society.[9] Special ancillary procedures such as fluorescence may require higher purity grades of solvents. Other grades may be used provided it is first ascertained that the reagent is of sufficiently high purity to permit its use without lessening the accuracy of the determination.

5.2 *Purity of Water*—Unless otherwise indicated, references to water shall be understood to mean reagent Type II water conforming to Specification D 1193.

6. Caution

6.1 Solvents used in this practice are volatile, flammable, or may cause the harm to the health of the user. Specifically, benzene is a known carcinogen, while chloroform and carbon tetrachloride are suspected carcinogens. Consequently, it is important that extractions and separations utilizing these substances must be carried out in a laboratory hood with a minimum linear face velocity of 38 to 45 m/min (125 to 150 ft/min) located in a regulated area posted with signs bearing the legends: NO SMOKING or (if appropriate) DANGER-CHEMICAL CARCINOGEN-AUTHORIZED PERSONNEL ONLY, or both.

7. Sampling

7.1 Collect representative samples in accordance with Practices D 4489.

7.2 Preserve the waterborne oil samples in accordance with Practice D 3325.

7.3 The portion of the sample used must be representative of the total sample. If the material is liquid, thoroughly stir the sample as received, warming if necessary to ensure uniformity.

PROCEDURE A—LARGE SAMPLES

8. Scope

8.1 This procedure covers the preparation for analysis of samples in which the volumes of waterborne oil in the environmental and suspect source samples equal or exceed 50 mL and in which the oil portion contains significant amounts of hydrocarbons with boiling points above 280°C.

NOTE 1—The boiling point may be ascertained by injecting the neat samples into the gas chromatograph and checking the elution times above that of pentadecane on a nonpolar column.

9 *Reagent Chemicals, American Chemical Society Specifications*, American Chemical Society, Washington, DC. For suggestions on the testing of reagents not listed by the American Chemical Society, see *Analar Standards for Laboratory Chemicals*, BDH Ltd., Poole, Dorset, U.K., and the *United States Pharmacopeia and National Formulary*, U.S. Pharmacopeial Convention, Inc. (USPC), Rockville, MD.

8.2 The preparation of samples containing mostly hydrocarbons of boiling points below 280°C, such as petroleum distillate fuels, is beyond the scope of this procedure (see Procedure C or E).

9. Summary of Procedure

9.1 A neat portion of the waterborne oil is retained. If not possible to obtain a neat portion, then retain a portion of the waterborne oil as received. This is to be used in those analyses performed on samples containing significant quantities of hydrocarbons with boiling points below 280°C. Preparation of these samples is beyond the scope of this procedure, but are covered in Procedure C.

NOTE 2—Waterborne oil samples containing significant quantities of hydrocarbons with boiling points below 280°C (see Note 1), such as gasoline and kerosene, can usually be obtained as neat samples without any sample preparation.

9.2 The waterborne oil sample is dissolved in an equal volume of chloroform or dichloromethane and centrifuged to remove the free water, solids, and debris in accordance with Test Method D 96. The water layer, if present, is separated from the organic layer. Other debris, if present, is removed by filtration through glass wool.

NOTE 3—The use of spectrograde cyclohexane is required for the extraction of samples to be analyzed by fluorescence spectrometry by Test Method D 3650. Separation of water may be accomplished by centrifugation or dying, or both, with anhydrous sodium sulfate.

9.3 When centrifugation will not separate the water from the chloroform solution of the sample, it is refluxed with an aromatic or petroleum distillate solvent in accordance with Test Method D 95.

NOTE 4—Pressure filtration has also been found useful for breaking emulsions.

9.4 A portion of the solvent/sample solution is retained. The solvent may be removed by evaporation. This portion of the sample may be used in the preliminary gas chromatographic analysis, Test Methods D 3328 (Test Method A), and other analyses in which the results are unaffected by weathering.

9.5 The remainder of the solvent/sample solution is distilled using nitrogen purge to a liquid temperature of 280°C to remove the solvent and simulate weathering conditions as nearly as possible. The distillate may be discarded or saved for characterization by gas chromatography (Test Methods D 3328). This simulated weathering treatment is necessary to bring the unweathered suspect samples and the waterborne oil sample to as nearly comparable physical condition for subsequent analysis as possible. Analyses requiring the use of this treated residue include elemental analysis (Test Methods D 3327); gas chromatographic analysis (Test Methods D 3328, Test Methods A and B); an infrared procedure (Test Method D 3414); a fluorescence test method (Test Method D 3650); and any applicable test method or practice described in Practice D 3415.

NOTE 5—The distillate might yield useful information but is discarded in this practice.

10. Apparatus

10.1 *Centrifuge*, capable of whirling two or more filled 100-mL centrifuge tubes at a speed that is controlled to give a relative centrifugal force (rcf) between 500 and 800 at the tip of the tubes, as specified in Test Method D 96.

10.2 *Centrifuge Tubes*, cone shaped, 100 mL, as specified in Test Method D 96.

10.3 *Distillation Apparatus for Water Determination*, as specified in Test Method D 95.

10.4 *Distillation Apparatus for Simulated Weathering*, as described in Specification E 133 except fitted with nitrogen-stripping tubulation as illustrated in Fig. 1.

10.5 *Distillation Flask*, 200 mL, as described in Specification E 133.

10.6 *Thermometer*, ASTM high distillation, having a range from −2 to +400°C and conforming to the requirements for thermometer 8C as prescribed in Specification E 1.

10.7 *Flowmeter*, to regulate flow of nitrogen to distillation flask. It should be calibrated and graduated for the range 10 to 15 mL/min.

11. Reagents and Materials

11.1 *Filter Paper*, medium retention, medium fast speed, prewashed with solvent used.

11.2 *Glass Wool*, prewashed with solvent used.

11.3 *Solvent*—Chloroform (stabilized with ethanol) or dichloromethane is used for dissolution of the waterborne oil samples. If water is to be removed by distillation, an aromatic, petroleum distillate, or volatile spirits solvent is required as specified in Test Method D 95. The safety precautions associated with the use of the solvent selected should be considered before it is used (see Note 3).

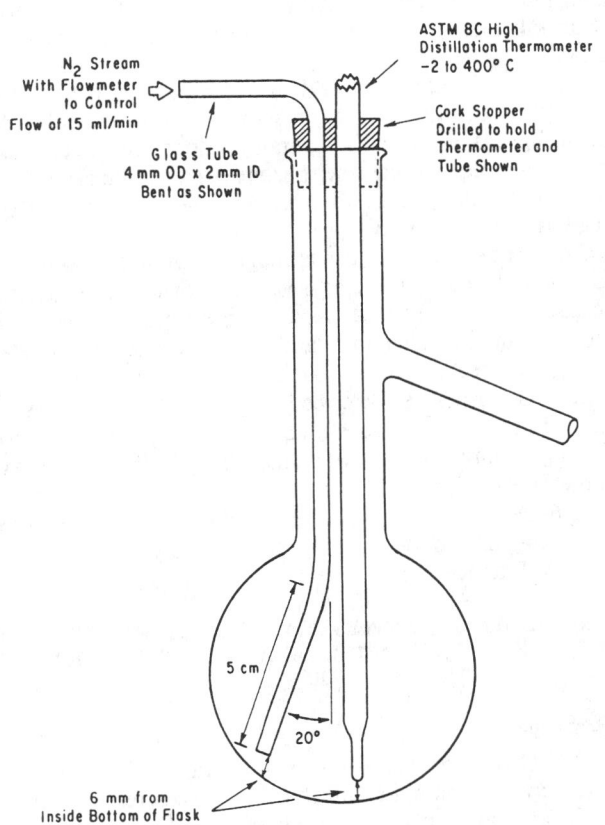

N₂ Stream
With Flowmeter
to Control
Flow of 15 ml/min

ASTM 8C High
Distillation Thermometer
−2 to 400° C

Cork Stopper
Drilled to hold
Thermometer and
Tube Shown

Glass Tube
4 mm OD x 2 mm ID
Bent as Shown

5 cm

20°

6 mm from
Inside Bottom of Flask

FIG. 1 Adaptation of ASTM Distillation Flask for Topping Chloroform Solutions of Oil to Simulate Weathering

12. Procedure

12.1 *Retention of Neat Samples:*

12.1.1 Decant or siphon off a portion of the neat waterborne oil if possible.

12.1.2 If not possible to obtain a neat sample, retain a portion of the original oil.

12.2 *Removal of Water, Sediment, and Debris:*

12.2.1 Transfer about 50 mL of original waterborne oil to a 100-mL centrifuge tube. Add about 50 mL of chloroform or dichloromethane to the tube and mix thoroughly. For waxy samples, use chloroform. Warm solutions to 50°C to prevent precipitation (see Note 3).

12.2.1.1 Centrifuge the mixture at 500 to 800 rcf (relative centrifugal force) for 10 min to separate free water and solids as specified in Test Method D 96. For waxy samples, use chloroform. Warm solutions to 50°C to prevent precipitation (see Note 3).

12.2.1.2 Withdraw the water layer if present. Decant the chloroform or dichloromethane solution to a sample bottle. Filter through a glass wool plug, if necessary, to afford a clean separation.

12.2.2 Process those samples from which water cannot be separated by centrifugation by Test Method D 95 distillation procedure. Filter the dry solution through medium retention filter paper. Rinse filter paper with solvent to remove oil. (For waxy samples, use chloroform and keep filter funnel and contents at 50°C during filtration.) (See Note 3.)

12.2.3 Starting at 12.1, treat all reference or suspect samples in an identical fashion. If it is apparent that the reference or suspect samples contain less than 1 % water and sediment, centrifugation may be eliminated and the reference or suspect samples should be diluted with an equal volume of chloroform or dichloromethane before proceeding.

12.3 *Removal of Solvent and Simulated Weathering:*

12.3.1 Transfer approximately 100 mL of the solution to a chemically clean 200-mL flask. Assemble apparatus so the ASTM high distillation thermometer (8C) and nitrogen stripping tubulation are about 6 mm from the bottom of the flask. Direct flow away from thermometer bulb to prevent local cooling of thermometer (see Fig. 1).

12.3.2 Perform distillation using a nitrogen flow of 10 to 15 mL/min. Terminate distillation at a liquid temperature of 280°C. Shut off the nitrogen flow when the temperature of the liquid in the distillation flask cools below 175°C. Pour the hot residue into a suitable container.

12.3.3 Treat all reference and suspect oils in the same manner as the waterborne oil samples. Repeat 12.2.1 through 12.3.2.

PROCEDURE B—LIMITED SAMPLE VOLUMES OF HEAVY OILS

13. Scope

13.1 This procedure covers the preparation for analysis of waterborne oil samples of petroleum derived origin in which the volumes equal or are less than 1 mL. An aliquot of larger oil samples may also be used.

13.2 The procedure is applicable to oils containing significant amounts of hydrocarbons boiling above 280°C.

13.3 The preparation of samples containing lower boiling

hydrocarbon is beyond the scope of this procedure, but is covered by Procedures C and E.

14. Summary of Procedure

14.1 The sample is dissolved in pentane or hexane, and the water and insolubles are removed by centrifugation. The organic solvent phase is dried with anhydrous magnesium sulfate, filtered, and the volatile components and solvents are removed by evaporation under a nitrogen stream (see Note 3).

15. Apparatus

15.1 *Centrifuge*, see 10.1.

15.2 *Centrifuge Tubes*, see 10.2.

15.3 *Flow Control on Nitrogen Cylinder*, to control nitrogen flow over sample surface.

15.4 *Steam Bath*, or commercial temperature controlled solvent evaporator, maintained between 40 and 50°C.

16. Reagents and Materials

16.1 *Magnesium Sulfate*, anhydrous.

16.2 *Nitrogen*, a high purity grade.

16.3 *Pentane or Hexane*, chromatographic grade.

17. Procedure

17.1 Remove approximately 1 mL of the oil phase from the water-oil sample if possible and place it in a 100-mL centrifuge tube.

17.1.1 Add 40 mL of pentane or hexane and 1 g of anhydrous magnesium sulfate. Mix to remove water. If the sample tube is warm, additional magnesium sulfate may be required. Add magnesium sulfate in 1-g aliquots, mixing after each addition until no temperature change is detectable to the touch.

17.2 Alternatively, estimate the volume of oil in the sample and add approximately 40 vol of pentane per 1 vol of oil.

17.2.1 Shake or rapidly mix the oil and solvent.

17.2.2 Allow phases to separate, withdraw the solvent phase with a pipet, and place it in a 100-mL centrifuge tube.

17.3 Centrifuge as described in 12.2.1.1 for 5 min.

17.4 Decant supernatant liquid into a 250-mL beaker and evaporate the solvent and volatiles initially at 25 to 35°C and then at 40 to 50°C for 2 h in the presence of a stream of nitrogen. Transfer the sample to a sample vial when there is approximately 4 mL sample remaining and continue the solvent removal. The samples can then be used for analysis in accordance with Practice D 3415.

NOTE 6—This treatment with 70 mg of oil, evaporated at 40°C for 15 min in the presence of an airstream, yielded gas chromatograms resembling those of the distillation test method in 12.3.[10]

PROCEDURE C—OILS BOILING BELOW 280°C

18. Scope

18.1 This procedure covers the preparation for analysis of waterborne oil samples containing significant amounts of

components boiling below 280°C.

18.2 The procedure is applicable to samples of distillate fuel oils, light and heavy naphthas, and other petroleum solvents.

19. Summary of Procedure

19.1 The oil and water phases are separated by centrifugation, and the oil phase is dried with anhydrous magnesium sulfate.

20. Apparatus

20.1 *Centrifuge*, see 10.1.

20.2 *Centrifuge Tubes*, see 10.2.

20.3 *Separatory Funnel*, glove or pearshaped, 100 mL, with TFE-fluorocarbon stopcock.

20.4 *Pipets*, disposable glass.

21. Reagents and Materials

21.1 *Magnesium Sulfate*, anhydrous.

22. Procedure

22.1 Transfer up to 10 mL of sample into a 100-mL separatory funnel. If phases separate, withdraw and discard aqueous (lower) phase. Transfer the organic phase into a 12.5-mL centrifuge tube. Alternatively, if there is enough oil on the water, the oil may be transferred directly with a pipet. Proceed to 22.4.

22.2 Prepare emulsified samples in the following manner: Transfer 10 mL of the sample to a centrifuge tube and centrifuge for 30 min at 1000 rcf (relative centrifugal force). If an oil layer appears, remove and proceed as directed in 22.4.

22.3 If a distinct oil layer does not appear, add to the test tube a maximum of 1 g of sodium chloride, mix, and centrifuge as in 22.2. If separation does not occur after centrifugation, add pentane, hexane, or cyclohexane, up to one quarter the sample volume, mix thoroughly, and proceed as in Procedure B.

22.4 Add 1 g of anhydrous magnesium sulfate, and mix for 1 min. If the sample tube is warm, additional magnesium sulfate may be required. Add magnesium sulfate in 1-g aliquots, mixing after each addition until no temperature change is detectable to the touch.

22.5 Centrifuge as described in 12.2.1.1 for at least 10 min. If magnesium sulfate is not completely removed from the oil, it may interfere with analysis by infrared spectroscopy, Test Method D 3414.

22.6 Decant the supernatant, leaving some oil to avoid disturbing the solids, and use for analysis by procedures given in Practice D 3415.

PROCEDURE D—SAMPLES COMPOSED OF MIXTURES OF PETROLEUM-BASED AND ANIMAL-/VEGETABLE-DERIVED OILS

23. Scope

23.1 This procedure covers the preparation for analysis of waterborne oil samples composed of mixtures containing significant amounts of petroleum-based and animal-/vegetable-derived oils.

23.2 The procedure incorporates a column chromato-

---

[10] Gruenfeld, M., and Frederick, R., "The Ultrasonic Dispersion, Source Identification, and Quantitative Analysis of Petroleum Oils in Water," Rapp. P-V, Reun. Cons. int. Explor. Mer. 171:33, 1977.

graphic procedure to separate the animal-/vegetable-derived oil fraction from the mixture.

24. Summary of Procedure

24.1 The waterborne oil phase is separated from the sample and dried with anhydrous magnesium sulfate. The petroleum hydrocarbon phase is separated from the animal vegetable oil phase by dissolving in carbon tetrachloride, followed by column chromatography using a silica gel-alumina column. The animal/vegetable oil fraction may be recovered from the column by elution with methanol.[11]

25. Apparatus

25.1 *Centrifuge*, see 10.1.

25.2 *Centrifuge Tubes*, see 10.2.

25.3 *Separatory Funnel*, glove- or pear-shaped, 100 mL, with TFE-fluorocarbon stopcock.

25.4 *Chromatographic Column*, 10 mm diameter by 250 mm high, loaded as shown in Fig. 2. The column is preconditioned by eluting with 100 mL of carbon tetrachloride and is kept saturated with solvent.

25.5 *Evaporating Dish*, porcelain or borosilicate glass, 100 mL.

26. Reagents and Materials

26.1 *Alumina*, neutral, Brockman activity 1, activated, 80 to 200 mesh, deactivated to 3 % water (wt/wt) vol per weight of water.

26.2 *Carbon Tetrachloride*, spectral or chromatographic grade.

26.3 *Magnesium Sulfate*, anhydrous, prewashed with carbon tetrachloride and dried at 103°C for 1 h.

26.4 *Methanol*, spectral or chromatographic grade.

26.5 *Silica Gel*, activated, 100 to 200 mesh, deactivated to 3 % water (wt/wt).

26.6 *Sand*, sea, washed and ignited, 20 to 30 mesh.

27. Procedure

27.1 Transfer the sample into a 1-L separatory funnel.

27.2 Allow phases to separate.

27.3 Withdraw and discard aqueous (lower) phase.

27.4 Transfer 1 to 10 mL of the organic phase into a 100-mL centrifuge tube.

27.5 Dilute with carbon tetrachloride using a 10:1 volume per volume carbon tetrachloride to sample ratio (see 5.1 regarding purity of solvent).

27.6 Add 5 g of anhydrous magnesium sulfate and shake for 1 min.

27.7 Centrifuge as indicated in 12.2.1.1 for 10 min.

NOTE 7—Centrifugation may not be necessary if a clean supernatant is obtained upon addition of the magnesium sulfate.

27.8 Place lower phase into a 100-mL evaporating dish and evaporate to 2 mL or the original sample volume, whichever is larger.

27.9 Place 2 mL of the residue from 27.8 into the chromatographic column (see 25.4 and Fig. 2).

---

[11] Kahn, L., Dudenbostel, B., Speis, D. N., and Karras, G., "Determination of Mineral Oils and Animal/Vegetable Oils in the Presence of Each Other," *American Laboratory*, Vol 9, No. 3, 1977, pp. 61 to 66.

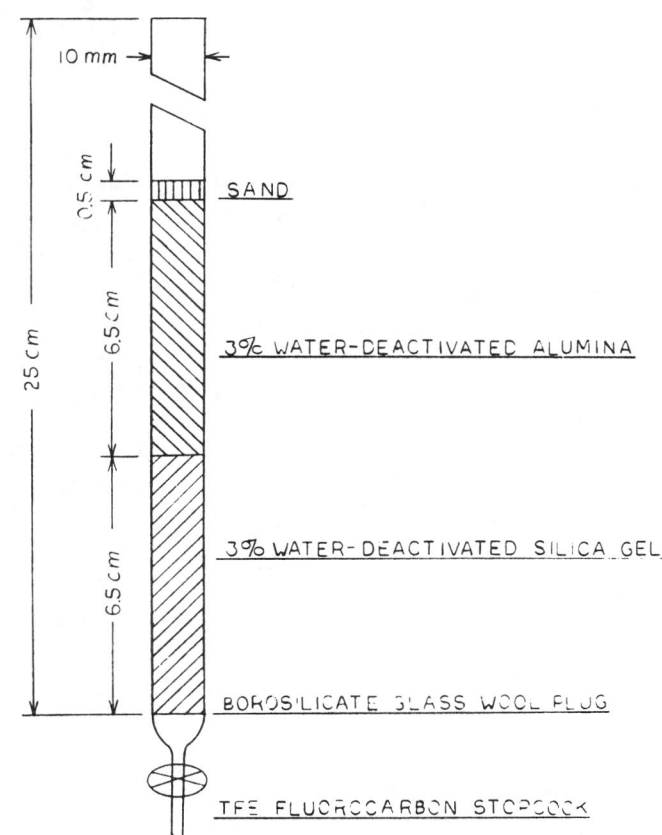

FIG. 2   Chromatographic Column

27.10 Elute with carbon tetrachloride at a rate of 3.0 mL/min and collect 30 mL of eluate.

NOTE 8—This eluate contains the hydrocarbons fraction and includes all the petroleum-based oils as well as a small fraction of the hydrocarbons in the liquids from animal/vegetable wastes and oils. This fraction may be analyzed after solvent evaporation by procedures given in Practice D 3415.

27.11 Elute the residual material in the column with 100 mL of methanol into a separate container.

NOTE 9—This eluate contains the mono-, di-, and triglycerides, other fatty esters, and polar components of animal-/vegetable-derived oils.

27.12 Evaporate the solvent from 27.10 and 27.11, respectively, at 40 to 50°C by gently blowing nitrogen over the surface of the liquid.

NOTE 10—This fraction may be analyzed by a variety of adjunct procedures such as infrared analysis (Test Method D 3414), iodine number (Test Method D 1959), gas chromatography (Test Methods D 2800 and D 1983), etc.

PROCEDURE E—LIMITED SAMPLE VOLUMES OF LIGHT OILS

28. Scope

28.1 This procedure covers the preparation for analysis of weathered light crudes and weathered oils of petroleum derived origin having significant amounts of hydrocarbons

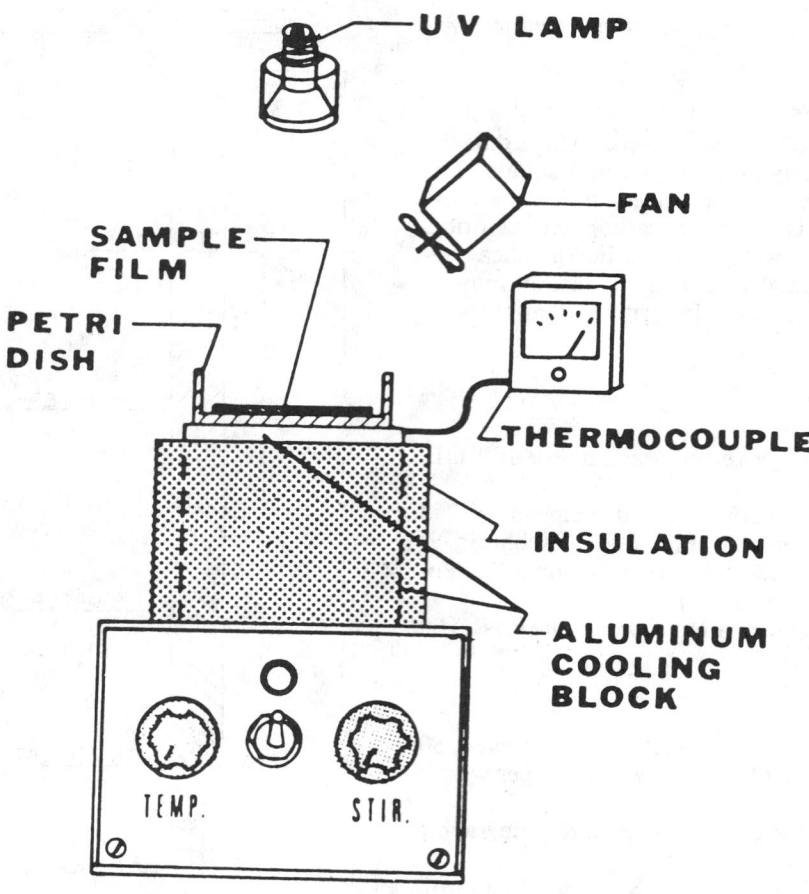

FIG. 3 Weathering Apparatus

with boiling points below 280°C.

28.2 The procedure provides a means to simulate the effects of environmental weathering on petroleum oils, thus simplifying comparison of spilled oils to suspected sources.

28.3 The procedure is applicable to simulation of light to moderate weathering, equivalent to one to three days exposure to the marine environment.

29. Summary of Procedure

29.1 Neat samples of unweathered oils (suspected sources), intended to be compared with untreated samples of the weathered waterborne oil, are irradiated as thin films with long wavelength UV light while exposed to an accelerated air flow.

29.2 Exposure of the oil for 3 to 6 h in this manner simulates the effects of one to three days of weathering of light fuel oils in the marine environment.

30. Apparatus

30.1 *Cooling Block*, capable of maintaining a surface temperature of 20°C at room temperature.

30.2 *Fan*, to provide air current on the order of 2.6 m/s.

30.3 *Ultraviolet Lamp*, long wave high intensity, using a high pressure 100 W, sealed beam bulb.

30.4 *Thermometer* (0 to 50°C), or thermocouple.

31. Reagents and Materials

31.1 *Glass Petri Dish*, 7.5 cm, for weathering up to 0.75 mL of oil.

31.2 *Infrared Salt Plate*, (preferably KBr) can be used as the support for the thin film of oil. This is used for infrared analysis of small samples as little as 30 mg.

32. Caution

32.1 Apparatus should be used in a well ventilated area due to hazardous vapors created.

33. Preparation of Apparatus

33.1 Assemble apparatus as shown in Fig. 3.

33.2 Set fan speed and distance to provide moderate air flow on the order of 2.6 m/s at cooling block.

33.3 Place UV lamp 10 cm from the cooling block (this provides 15.6 mm/cm² for the lamp specified).

33.4 With the lamp on and air flow across the cooling block, adjust the cooling block surface temperature to approximately 22°C (20 to 25°C).

34. Procedure

34.1 Transfer 0.25 to 0.75 mL of oil sample to be weathered onto the glass petri dish placed on apparatus thermally equilibrated as described in 33.4. (For infrared analysis, if only small amounts of sample are available, 30 to

100 mg will uniformly coat one side of a 19 by 34 mm KBr salt window.)

34.2 Place the sample on the cooling block. (If salt window is used place a nonabrasive, noncontaminating paper between the window and the cooling block to avoid scratching the salt window).

34.3 Remove the sample after 3 to 6 h of continuous exposure in this apparatus. The sample is now ready for analysis by procedures given in Practice D 3415.

PROCEDURE F—THIN FILMS OF OIL-ON-WATER

35. Scope

35.1 This procedure covers the preparation for analysis of waterborne oil samples having an insufficient sample volume for preparation by Procedures A through E.

35.2 The procedure is applicable to both petroleum and nonpetroleum derived oils. It is also applicable to water samples that may or may not contain spilled oil. These samples may be encountered in cases where water is sampled to assess the extent to which oil has dispersed.

35.3 Traces of oil recovered by the techniques in this procedure have usually been subjected to environmental weathering; analysis of such samples may be inconclusive.

36. Summary of Procedure

36.1 Thin films of oil-on-water are removed with a TFE-fluorocarbon strip.

36.2 Water samples are extracted with cyclohexane and either evaporated to a neat oil or used directly for analysis by fluorescence spectroscopy (Test Method D 3650) or gas chromatography (Test Methods D 3328), or both.

37. Apparatus

37.1 *Centrifuge*, see 10.1.

37.2 *Centrifuge Tubes*, see 10.2.

37.3 *Flow Control on Nitrogen Cylinder*, see 15.3.

37.4 *Steam Bath, Hot Plate*, or commercial temperature controlled solvent evaporator, see 15.4.

37.5 *Test Tubes*, disposable, 16 by 125 mm, 15 mL, borosilicate glass culture tubes.

37.6 *Weighing Pans*, 5 to 7 cm diameter, 18 mm deep, made of aluminum or equivalent.

38. Reagents and Materials

38.1 *Cyclohexane*, spectroquality grade, with a fluorescence solvent blank less than 2 % of the intensity of the major peak of the sample fluorescence generated with the same instrumental settings over the emission range used.

38.2 *TFE-Fluorocarbon Strips*, 25 by 75 mm, 0.25 mm thickness.

38.3 *Magnesium Sulfate*, anhydrous.

38.4 *Nitrogen*, compressed gas cylinder.

39. Procedure

39.1 Samples may be concentrated in the following manner: Dip or pass a TFE-fluorocarbon strip through the oil layer, then allow the oil to drip into a clean aluminum weighing pan. Continue until enough oil has been recovered to use a micropipet. Continue with analysis by fluorescence spectroscopy (Test Method D 3650) or gas chromatography (Test Methods D 3328), or both.

39.2 To recover oil remaining after the procedure in 39.1 is used, or if insufficient oil is recovered by the TFE-fluorocarbon strip procedure, add 10 mL of cyclohexane to the sample container. (Appropriate sample containers with TFE-fluorocarbon-lined lids are described in Practices D 4489.)

39.3 Replace the lid, then gently shake or swirl for approximately 1 min. Allow the sample container to stand undisturbed until the phases have separated.

39.4 Transfer the cyclohexane (upper) phase to a disposable glass test tube or a centrifuge tube. Centrifuge as in 12.2.1.1.

39.5 Transfer the organic phase to a new test tube, add 1 g of magnesium sulfate, and centrifuge as in 22.2.

39.6 Remove solvent under a stream of nitrogen until either all the solvent has evaporated or only a few drops of liquid remain. If the remaining solution appears to contain no visible oil, add 2 mL of cyclohexane. This solution may be used directly for analysis by fluorescence spectroscopy (Test Method D 3650) or gas chromatography (Test Methods D 3328), or both.

PROCEDURE G—OIL-SOAKED SAMPLES

40. Scope

40.1 This procedure covers the preparation for analysis of oil-soaked samples, such as sand, debris, sorbent pads, or any other substrate having a limited quantity of free flowing oil.

40.2 Oil recovered by the techniques in this procedure is subject to potential interferences from environmental weathering, as well as contamination by organic substances derived from the substrate material.

41. Summary of Procedure

41.1 The oil is separated from the substrate by squeezing or centrifugation.

42. Apparatus

42.1 *Centrifuge*, see 10.1.

42.2 *Centrifuge Tubes*, see 10.2.

42.3 *Test Tubes*, disposable, 16 by 125 mm, 15 mL, borosilicate glass culture tubes.

42.4 *Weighing Pans*, 5 to 7 cm diameter, 18 mm deep, made of aluminum or equivalent. (Double-weight aluminum foil may be substituted.)

43. Reagents and Materials

43.1 *Purity of Water*—References to water in this procedure shall be understood to mean Type IV reagent water conforming to Specification D 1193. However, because fluorescent organic impurities in the water may constitute an interference, the purity of the water should be checked by running a water blank.

44. Procedure

44.1 For light oil collected on soft absorbent material, place a portion of the oil-soaked material in the center of a clean aluminum weighing pan or piece of aluminum foil. Fold the sides of the aluminum over the material and squeeze out the oil into a clean aluminum pan. If sufficient oil is recovered, Procedure C may then be followed for sample preparation. Otherwise, microlitre amounts of the oil may be removed with a micropipet and analyzed directly by fluorescence spectroscopy (Test Method D 3650) or gas chromatography (Test Methods D 3328), or both.

44.2 For oil collected on substrates heavier than the oil, fill a centrifuge tube one-fourth to one-third full with the oil soaked material, plus an equal amount of water. Centrifuge as stated in 12.2.1.1 to free the oil from the material. After this is done, prepare the sample as in Procedure B or C.

44.3 For medium to heavy weight oil collected on light substrates, (grass, feathers, sticks), the oil may be physically removed by transfer with a clean inert material, such as a clean TFE-fluorocarbon strip or spatula, and prepared as in Procedure B or C.

44.4 If none of the above techniques are successful in removing sufficient oil for analysis, a solvent extraction should be used as described in Procedure B.

45. Keywords

45.1 identification; oil spill; preparation; sample; water-borne oil

Standard Guide for
Use of Test Kits to Measure Inorganic Constituents in Water[1]

This standard is issued under the fixed designation D 5463; the number immediately following the designation indicates the year of original adoption or, in the case of revision, the year of last revision. A number in parentheses indicates the year of last reapproval. A superscript epsilon (ε) indicates an editorial change since the last revision or reapproval.

1. Scope

1.1 This guide covers general considerations for the use of test kits for quantitative determination of analytes in water and wastewater. Test kits are available from various manufacturers for the determination of a wide variety of analytes in drinking water, surface or ground waters, domestic and industrial feedwaters and wastes, and water used in power generation and steam raising. See Table 1 for a listing of some of the types of kits that are available for various inorganic analytes in water.[2]

1.2 Ranges, detection limits, sensitivity, accuracy, and susceptibility to interferences vary from kit to kit, depending on the methodology selected by the manufacturer. In some cases, kits are designed to replicate exactly an official test method of a standard-setting organization such as the Association of Official Analytical Chemists (AOAC), American Public Health Association (APHA), ASTM, or the U.S. Environmental Protection Agency (USEPA). In other cases, minor modifications of official test methods are made for various reasons, such as to improve performance, operator convenience, or ease of use. Adjustments may be made to sample size, reagent volumes and concentrations, timing, and details of the analytical finish. In yet other cases, major changes may be made to the official test method, such as the omission of analytical steps, change of the analytical finish, omission of reagents, or substitution of one reagent for another. Reagents in test kits are often combined to obtain a fewer number and make the test easier to use. Additives may also be used to minimize interferences and to make the reagent more stable with time. A kit test method may be based on a completely different technology, not approved by any official or standard-setting organization. Combinations of test kits—multi-parameter test kits—may be packaged to satisfy the requirements of a particular application conveniently. The test kits in such combination products may be used to make dozens of determinations of several parameters.

1.3 Test kit reagent refills are commonly available from manufacturers. Refills permit cost savings through reuse of the major test kit components.

1.4 Because of the wide differences among kits and methodologies for different analytes, universal instructions cannot be provided. Instead, the user should follow the instructions provided by the manufacturer of a particular kit.

1.5 A test kit or kit component should not be used after the manufacturer's expiration date; it is the user's responsibility to determine that the performance is satisfactory.

1.6 *This standard does not purport to address all of the safety problems, if any, associated with its use. It is the responsibility of the user of this standard to establish appropriate safety and health practices and determine the applicability of regulatory limitations prior to use. For specific precautionary statements, see Section 10.*

TABLE 1 Availability and Types of Test Kits

Analyte	Kit Methodology[A]
Acidity	T
Alkalinity	C, P, T
Aluminum	C, P
Ammonia	C, P
Boron	C, P
Bromine	C, P, T
Cadmium	C
Calcium	P, T
Carbon dioxide	T
Chloride	A, C, P, T
Chlorine	C, P, T
Chlorine dioxide	C, P, T
Chromium (III)	C
Chromium (VI)	C, P, T
Cobalt	C
Copper	C, P, T
Cyanide	C, P, T
Fluoride	P
Hardness	C, GNG, P, T
Hydrazine	C, P
Hydrogen peroxide	C, P, T
Iodine	C, P, T
Iron	C, P
Lead	C, P
Manganese	C, P
Magnesium	C, T
Molybdate	C, P, T
Nickel	C, P
Nitrate	C, P
Nitrite	C, P, T
Oxygen (dissolved)	C, P, T
Ozone	C, P
Permanganate	C, T
pH	C, P
Phosphate	C, P
Silica	C, P
Silver	P
Sulfate	A, C, P, T
Sulfide	C, P, T
Sulfite	C, P, T
Thiocyanate	C
Tin	C
Vanadium	C
Zinc	C, P, T

[A] Kit Methodology: A = appearance/turbidity, C = visual colorimetric, GNG = go no go, P = photometric, and T = titrimetric.

[1] This guide is under the jurisdiction of ASTM Committee D-19 on Water and is the direct responsibility of Subcommittee D19.05 on Inorganic Constituents in Water.
Current edition approved Sept. 15, 1993. Published November 1993.
[2] Test kits for determining inorganic analytes in water are available from various United States and foreign manufacturers, as well as from laboratory supply companies.

2. Referenced Documents

2.1 *ASTM Standards:*
D 1129 Terminology Relating to Water[3]
D 1192 Specification for Equipment for Sampling Water and Steam[3]
D 1193 Specification for Reagent Water[3]
D 3370 Practices for Sampling Water[3]
D 4453 Practice for Handling of Ultra-Pure Water Samples[3]
D 4691 Practice for Measuring Elements in Water by Flame Atomic Absorption Spectrophotometry[3]
E 178 Practice for Dealing with Outlying Observations[4]
E 275 Practice for Describing and Measuring Performance of Ultraviolet, Visible, and Near Infrared Spectrophotometers[5]
E 958 Practice for Measuring Practical Spectral Bandwidth of Ultraviolet-Visible Spectrophotometers[5]

3. Terminology

3.1 *Definitions*—For definitions of terms used in this guide, refer to Terminology D 1129 and Practice D 4691.

3.2 *Descriptions of Terms Specific to This Standard:*

3.2.1 *analyte*—the chemical or constituent being determined.

3.2.2 *carryover*—the contamination of a subsequent sample by a previous sample, typically due to incomplete cleaning of a reused test kit component.

3.2.3 *expiration date*—a date applied by the manufacturer after which an accurate result is not ensured by the manufacturer.

3.2.4 *finish (usually analytical finish)*—the analytical methodology used for the measuring step of the analysis.

3.2.5 *kit (or test kit)*—a commercially packaged collection of components that is intended to simplify the analytical testing function.

3.2.6 *interference*—an effect of a matrix component that might cause an analytical bias or that might prevent a successful analysis.

3.2.7 *material safety data sheet*—a federally-mandated, safety-related document that must be made available to kit chemistry users.

3.2.8 *matrix*—sample contents other than the target analyte.

3.2.9 *official method*—an analytical test method officially approved by an industry consensus organization such as ASTM, AOAC, or APHA or by a government entity such as the USEPA.[6]

3.2.10 *refill*—a replacement package of test kit components used in testing.

3.2.11 *spike*—a small volume, high relative concentration aliquot of analyte added quantitatively to a split sample as a quality check.

3.2.12 *split sample*—a sample that is split into subsamples that are intended to have the same composition as the original sample.

4. Summary of Guide

4.1 Analytical test kits simplify the operational procedures necessary to perform an analysis. This guide includes general considerations relating to the procedures to be followed in order to ensure an accurate determination. This guide also describes, in general terms, the characteristics of some kit types and kit components and includes some comments on their capabilities, benefits and, where appropriate, their limitations.

5. Significance and Use

5.1 Inorganic constituents in water and wastewater must be identified and measured to support effective water quality monitoring and control programs. Currently, one of the simplest, most practical and cost effective means of accomplishing this is through the use of chemical test kits and refills. A more detailed discussion is presented in ASTM STP 1102.[7]

5.2 Test kits have been accepted for many applications, including routine monitoring, compliance reporting, rapid screening, trouble investigation, and tracking contaminant source.

5.3 Test kits offer time-saving advantages to the user. They are particularly appropriate for field use and usually are easy to use. Users do not need to have a high level of technical expertise. Relatively unskilled staff can be trained to make accurate determinations using kits that include a premixed liquid reagent, premeasured reagent (tablets, powders, or glass ampoules), and premeasured sample (evacuated glass ampoules).

6. General Considerations

6.1 *Personnel*—The selection of a test kit and determination that the test kit analysis is appropriate should be conducted by a responsible chemist. The development of suitable protocols and conditions for safe use should be conducted by the responsible chemist with the assistance of an industrial hygienist. The kit user may be a relatively unskilled staff person but must be trained to an appropriate level of proficiency.

6.2 *Completeness of Kits*—The kit's components may or may not be complete for the required determination. The user must assemble all instruments and materials necessary for the determination. For example, if the test kit is used for field screening to indicate the need for samples requiring a high accuracy measurement, the user may need to provide a means of preserving a sample for later measurements at a laboratory.

6.3 *Protocol Established by a Responsible Chemist*—A responsible chemist must determine whether the sample can be analyzed correctly by a particular kit chemistry. The responsible chemist should determine whether matrix factors, interferences, and temperature are handled correctly by

---

[3] *Annual Book of ASTM Standards*, Vol 11.01.
[4] *Annual Book of ASTM Standards*, Vol 14.02.
[5] *Annual Book of ASTM Standards*, Vol 14.01.
[6] Other documents: *Official Methods of Analysis of the Association of Official Analytical Chemists*, 15th Ed., AOAC, Arlington, VA, 1990. Changes are published in annual supplements. *Standard Methods for the Examination of Water and Wastewater*, 17th Ed., APHA, AWWA, and WPCF. Washington, DC, 1989. *Methods for the Chemical Analysis of Water and Wastes*, USEPA, Cincinnati, OH, March 1983.

[7] Spokes, G. Neil, and Bradley, Julie A., "Performance Testing of Selected Test Kits for Analysis of Water Samples," *ASTM STP* 1102, ASTM, Philadelphia, PA, 1991.

the kit chemistry. Questions to be answered include the following: Has the kit chemistry previously given satisfactory results under the proposed conditions? What changes have occurred that must be taken into account? For example, the chemist should consider seasonal changes, new interferences, sample pH changes, new dischargers upstream, and new process wastes in the sample. The responsible chemist must also decide whether the proposed kit chemistry is applicable to the particular circumstances. For example, it is necessary to determine whether the test range is appropriate, ensure that a colorimetric test kit that compensates for color is used with a highly colored sample, and ensure that a colorblind user is able to run a test requiring visual color comparisons accurately. The chemist must also ensure that an officially approved kit chemistry is used when an official method is required.

6.4 *Technical Support*—In case of difficulties, many kit manufacturers may provide technical assistance.

7. Interferences

7.1 Kit chemistries that are based on an official test method are subject to the same interferences as that test method. If the kit manufacturer uses a revised version of the official test method, the revision may increase or decrease interference effects.

7.2 Sample carryover effects may occur if a common sampling cup or tube is used. Appropriate care is necessary under such conditions in order to prevent sample carryover. The carryover may be prevented or reduced by either cleaning the reused item or rinsing with fresh sample several times. Aggressive cleaning action may be necessary after a sample containing a high concentration is tested.

7.3 Careful note should be made of the manufacturer's comments concerning interferences, and appropriate action should be taken.

7.4 Temperature may affect kit performance.

8. Apparatus

8.1 *Colorimetric Determinations*—Many procedures depend on color determination with a color comparator, photometer, or spectrophotometer. The manufacturer may offer a color comparator for visual comparisons based on liquid, glass, plastic, or printed color standards. The manufacturer may offer a photometer or may recommend the use of a spectrophotometer for photo-electric color determinations. The manufacturer's photometer may be based on optical filters using either colored glass or plastic, or on interference filters or LEDs. The filter bandwidth may be wide (up to 100-nm full width half maximum height) for colored glass or plastic filters and LEDs or quite narrow (10 nm) with interference filters. The laboratory spectrophotometer may have a 1- to 20-nm bandwidth and is typically more accurate than a kit photometer or colorimeter. Refer to Practices E 275 and E 958 for additional discussion of colorimetry.

NOTE 1—Visual comparator kits may require the use of a particular type of background illumination. The user should use the light source that produces the correct color or spectrum of background illumination, as specified by the manufacturer.

NOTE 2—Color standards may not be permanent; reference should be made to the manufacturer's recommendations.

8.2 *Titrimetric Determinations*—Many procedures depend on measuring the volume of a standard solution required to react with an analyte completely. The manufacturer may offer a buret, digital titrator, drop-test, or calibrated sample container to dispense and measure the volume of a standard solution. A buret or digital titrator typically provides more accuracy than a drop-test or calibrated sample container.

9. Reagents and Materials

9.1 *Purity of Reagents*—Reagent grade or better chemicals shall be used in all tests. Unless otherwise indicated, it is intended that all reagents shall conform to the specifications of the Committee on Analytical Reagents of the American Chemical Society, where such specifications are available.[8] However, these reagents may not be of sufficient purity in some cases due to the sensitivity of the technique. It is the responsibility of the manufacturer to provide reagents and accessory solutions of sufficient quality to meet the performance specification claims of the test kit. In addition, the manufacturer should specify acceptable conditions of storage and provide expiration dates, where appropriate. It is the responsibility of the kit user to ensure that no unacceptable deterioration has occurred in transit or due to improper storage conditions and that the kits are not used improperly after their expiration dates (see 9.4).

9.2 *Purity of Water*—Water must be of sufficient purity that it does not interfere with the test. Manufacturer's instructions should be followed. Unless otherwise indicated, references to water shall be understood to mean reagent water as defined by Type III or better of Specification D 1193.

9.3 *Kit Components and Packaging*—The test kit components and packaging are usually designed carefully by the manufacturer to facilitate quick and easy determination of the analyte. Pre-mixed liquid reagents eliminate the need for making up reagents and offer the benefits of simplicity of use, reduced need for operator measurements, and good immunity to environmental effects. Test kits with unit dose disposable reagent packs offer the further benefits of simplicity of use and reduced need for operator measurements. In particular, tablets, foil packs, powders, and glass ampoule reagent packaging techniques eliminate the need for making up reagents and then measuring the reagent volume. Glass ampoules with reagents packaged under vacuum also offer immunity to reagent oxidation and make the measurement of sample volume unnecessary. Test kit reagent refills may often be obtained from manufacturers. Refills save the expense of purchasing an entire kit by permitting the reuse of major kit components.

9.4 *Storage*—Users should follow the manufacturer's instructions for acceptable storage conditions. Some kit components can be refrigerated to prolong usability. If the user is uncertain about past storage, a quality control check sample

[8] *Reagent Chemicals, American Chemical Society Specifications*, American Chemical Society, Washington, DC. For suggestions on the testing of reagents not listed by the American Chemical Society, see *Analar Standards for Laboratory Chemicals*, BDH Ltd., Poole, Dorset, U.K., and the *United States Pharmacopeia and National Formulary*, U.S. Pharmaceutical Convention, Inc. (USPC), Rockville, MD.

(see 13.1.1) should be used to determine the acceptability of a test component.

9.5 *Kit Expiration*—The user should not use kit components after their expiration date.

10. Safety Precautions

10.1 The majority of kit test methods use chemicals that have some type of hazard associated with them. The responsible chemist/industrial hygienist should ensure that proper heed is taken of warnings in the instructions. Material safety data sheets must be obtained for each kit, and all appropriate care should be taken based on the manufacturer's warnings. Protective clothing and safety eyewear may be required, depending on the hazards posed by sample and kit. A ventilation hood may be required in the laboratory. Ventilation of the work area is rarely a problem in the field, except when working in confined spaces such as in manholes. Kits involving instruments should not be used where flammables are in use unless previously certified as acceptable by the responsible safety officer.

10.1.1 Test kits containing premixed reagents, including the unit dose type of test kit, greatly reduce the need for operator contact with chemicals and therefore improve safety compared with the officially approved laboratory test methods.

10.1.2 The responsible chemist/industrial hygienist must take all necessary steps to ensure that the operator performs tests using the kits in a safe manner. Protective clothing and safety eyewear should be used as necessary. Used test kit components must be disposed of in accord with federal, state, and local laws. The responsible chemist/industrial hygienist should provide written instructions where special disposal techniques are required.

10.2 *Safety Training*—The responsible chemist/industrial hygienist must ensure that the operator is trained properly in the use of the kit, with due regard to the kit's safety hazards.

10.3 *Right-to-Know and Other Laws*—The responsible chemist/industrial hygienist must ensure that the test user is trained in safe use of the test kit and is informed properly regarding the various hazards associated with the kit chemicals. The manufacturer will provide the required information through a material safety data sheet.

11. Samples and Sampling Procedures

11.1 *Sample Collection*—Unless otherwise specified, collect all samples in accordance with Specification D 1192 and Practices D 3370.

11.2 *Sample Handling*—Follow the kit manufacturer's recommended sample handling procedures, when available. Contamination and loss are of prime concern for the determination of trace metals. Environmental dust and impurities in apparatus components that contact the sample are potential sources of contamination. Containers can introduce errors in the measurement of trace metals by contributing contaminants through leaching or surface desorption or by depleting the concentration of the analyte through adsorption. Refer to Practice D 4453 for the handling of ultra-pure water samples.

11.3 *On-Site Sampling*—Test kits that permit on-site analysis offer great advantages to their users, inasmuch as error-prone sample preservation techniques are often unnec-

essary. However, the value of an analytical result is only as good as the sample that is obtained, and the usual care is required to ensure the representativeness of the sample.

11.4 *Sample Containers*—Take care to collect a representative sample in clean containers that will not cause contamination in any significant way.

11.5 *Sample Size*—The sample size shall be sufficient to complete the determinations. A larger sample may be necessary when processing or multiple determinations, or both, are required.

11.6 *Sample Preservation and Storage*—The kit method ordinarily permits an immediate measurement. If the test result is equivocal or the responsible chemist directs that a preserved sample be obtained, preserve a sample in accord with the applicable approved procedure.

12. Calibration and Standardization

12.1 Kit chemistry manufacturers may provide a precalibrated color comparator or colorimeter. Alternatively, a spectrophotometric analytical finish may be recommended. A titration kit test method may require a drop count determination or titration scale reading. The drop count or scale reading may be multiplied by one factor to generate an analytical result. The calibration accuracy depends on the accuracy of the system design and ensuing manufacturing processes. All kits that lead directly to an analytical result rely on an internal calibration. Manufacturers will offer to provide a calibration chart in some cases. Wherever possible, known reference samples or standards should be analyzed to validate the results from test kits.

13. Quality Control

13.1 *Quality Control*—Test kit quality control is the responsibility of the manufacturer. The manufacturer may be requested to furnish supporting quality control data. The user should verify that the final color resulting when using a colorimetric test kit is the same as that of a calibration sample.

13.1.1 *Quality Control Check Samples*—Kit quality may be determined by occasionally running a blank sample, a known mid-range sample, a known upper mid-range sample, and a known higher than a full scale sample. The user should verify independently that the kit calibration and recovery are satisfactory for the intended application, with due regard for interferences typically found in the user's samples.

NOTE 3—In some cases, it is not possible to establish an analytical standard solution that can be routed to a user laboratory (for example, dissolved oxygen in water). In that event, it may be possible to use a surrogate compound instead of the analyte. In the event of difficulty in obtaining a standard, the manufacturer should provide the necessary product support information.

13.1.2 *Split Samples*—Split samples should be analyzed frequently. As a recommended minimum, analyze one set of split samples every 10 samples (see 14.4).

13.1.3 *Laboratory Spikes*—Prepare and analyze laboratory spikes. The spiked sample should contain approximately double the anticipated quantity of target analyte in the test sample, and the spike should be of comparatively small volume. The spike should not be so great as to exceed the range of the test kit method.

13.1.4 *Field Spikes*—Field spikes may be prepared in the

laboratory and added to split samples in the field.

14. Procedure

14.1 *Sample Pretreatment*—Sample pretreatments specified by the kit manufacturer must be conducted prior to the use of the kit. The responsible chemist must ensure that any necessary sample pretreatments are conducted prior to use of the kit. Failure to conduct the sample pretreatments can result in the generation of meaningless test results. The manufacturer should be able to provide the necessary information, but such procedures are often not addressed adequately in manufacturer's test kit instructions.

14.2 Prepare quality control check samples in the laboratory.

14.3 Perform sample analysis procedures in accord with the kit manufacturer's instructions.

14.4 The frequency of quality control check samples should be based on a statistical plan established to achieve a desired level of confidence. An alternative plan is to check at least one calibration standard after every 10 samples and at the end of the analyses. Analyze the quality control check samples, split samples, and spiked samples as directed by quality control procedures.

15. Calculation

15.1 In most cases, kit test methods provide instructions that lead to direct numeric results. The user should follow the manufacturer's instructions.

15.2 *Units*—The numeric results obtained from the test kit should be stated clearly in commonly used units, for example, mg/L, µg/L, ppm (parts per million), or ppb (parts per billion). If a result is presented as ppm or ppb, it should specify whether this is a weight or volume specification. Note that ppm in gas analyses is almost always volume per volume, whereas ppm in dissolved gas results usually means mg/L, for example, dissolved oxygen in water. The numeric results should specify the measurement units where chemical species may vary, for example, nitrate as nitrogen versus nitrate as nitrate. In the use of test kits for process control, the analyte may be measured in terms of a treating chemical, not in mg/L of a particular ion.

15.3 The user should refer to Practice E 178 to deal with outlying observations.

16. Precision and Bias

16.1 The accuracy and precision of a test kit method should be determined by the user for the specific application. The reference method or official method may be used to confirm the accuracy of a test kit method.

17. Keywords

17.1 inorganic analysis; kits; test kits; wastewater; water

Standard Practice for
Estimation of Holding Time for Water Samples Containing Organic Constituents[1]

This standard is issued under the fixed designation D 4515; the number immediately following the designation indicates the year of original adoption or, in the case of revision, the year of last revision. A number in parentheses indicates the year of last reapproval. A superscript epsilon (ε) indicates an editorial change since the last revision or reapproval.

[ε1] NOTE—Section 12 was added editorially in June 1995.

1. Scope

1.1 This practice describes the means of estimating the period of time during which a water sample can be stored after collection and preservation without significantly affecting the accuracy of analysis.

1.2 The maximum holding time is highly matrix-dependent and is also dependent on the specific analyte of interest. Therefore, water samples from a specific source must be tested to determine the period of time that sample integrity is maintained by standard preservation practices.

1.3 In those cases where it is not possible to analyze the sample immediately at the time of collection, this practice does not provide information regarding degradation of the constituent of interest or changes in matrix that may occur from the time of sample collection to the time of the initial analysis.

1.4 This practice does not provide information regarding holding time for concentration of analyte less than one order of magnitude above the criterion of detection.

1.5 *This standard does not purport to address all of the safety concerns, if any, associated with its use. It is the responsibility of the user of this standard to establish appropriate safety and health practices and determine the applicability of regulatory limitations prior to use.*

2. Referenced Documents

2.1 *ASTM Standards:*
D 1129 Terminology Relating to Water[2]
D 1192 Specification for Equipment for Sampling Water and Steam[2]
D 1193 Specification for Reagent Water[2]
D 2777 Practice for Determination of Precision and Bias of Applicable Methods of Committee D-19 on Water[2]
D 3694 Practices for Preparation of Sample Containers and for Preservation of Organic Constituents[3]
D 4210 Practice for Intralaboratory Quality Control Procedures and a Discussion on Reporting Low-Level Data[2]

3. Terminology

3.1 *Definitions*—For definitions of terms used in this practice, refer to Terminology D 1129.

3.2 *Descriptions of Terms Specific to this Standard:*

3.2.1 *acceptable holding time*—acceptable holding time is any period of time less than or equal to the maximum holding time.

3.2.2 *maximum holding time*—maximum holding time is the maximum period of time during which a properly preserved sample can be stored before such degradation of the constituent of interest occurs or change in sample matrix occurs that the systematic error exceeds the 99 % confidence interval (not to exceed 15 %) of the test about the mean concentration found at zero time.

4. Summary of Practice

4.1 Holding time is estimated by means of replicate analysis at discrete time intervals of a large volume of a water sample that has been properly collected and preserved. Concentration of the constituent of interest is plotted versus time. The maximum holding time is the period of time from sample collection to such time that degradation of the constituent of interest occurs or change in sample matrix occurs that the systematic error exceeds the 99 % confidence interval (not to exceed 15 %) of the test about the mean concentration at zero time. Prior to determination of holding time, each laboratory must generate its own precision data for use in the calculation. For those tests which are relatively imprecise, replicate determinations are performed at each time interval to maintain the 99 % confidence interval within 15 % of the concentration found at zero time.

NOTE 1—This practice generates only limited data that may not lead to consistent conclusions each time the test is applied. In cases where the concentration of the constituent of interest changes very gradually over an extended period of time, the inherent variability in test results may lead to somewhat different conclusions each time that the practice is applied.

5. Significance and Use

5.1 In order to obtain meaningful analytical data, sample preservation techniques must be effective from the time of sample collection to the time of analysis. This period of time must be defined in order that the analyst may know how long samples may be stored prior to analysis.

6. Reagents

6.1 *Purity of Reagents*—Reagent grade chemicals shall be

---

[1] This practice is under the jurisdiction of ASTM Committee D-19 on Water and is the direct responsibility of Subcommittee D19.06 on Methods for Analysis for Organic Substances in Water.
Current edition approved Aug. 30, 1985. Published October 1985.
[2] Annual Book of ASTM Standards, Vol 11.01.
[3] Annual Book of ASTM Standards, Vol 11.02.

TABLE 1 Estimated Number of Replicate Determinations Required at Each Interval in the Holding Time Study Based on the Estimated Relative Standard Deviation of the Test in the Matrix Under Study

Estimated RSD, %	Approximate Number of Replicates
1 to 4	1
5 to 6	2
7 to 8	3
9	4
10	5
11	6
12	7
13	8
14	10
15	11

used in all tests. Unless otherwise indicated, it is intended that all reagents shall conform to the specifications of the committee on Analytical Reagents of the American Chemical Society, where such specifications are available.[4] Other grades may be used, provided it is first ascertained that the reagent is of sufficiently high purity to permit its use without lessening the accuracy of the determination.

6.1.1 Refer to the specific test method and to Practices D 3694 for information regarding necessary equipment and preparation of reagents.

6.2 *Purity of Water*—Reference to water shall be understood to mean reagent water conforming to Specification D 1193, Type II and demonstrated to be free of specific interference for the test being performed.

7. Determination of Holding Time

7.1 *Collection of Sample:*

NOTE 2—In some instances, it may be of interest to determine the holding time of standard solutions prepared in water. In such cases, a large volume of properly preserved standard solution should be prepared and carried through the steps of the practice in the same manner as a sample. The volume of solution required can be estimated using the equation in 7.1.1.

7.1.1 Based on the estimated precision of the test in the matrix to be tested, calculate the estimated total volume of sample required to perform the holding time determination plus a precision study. The following formula may be used to estimate this volume.

$$V = (A \times B \times C) + 2 (A \times D)$$

where:
V = estimated volume of sample required, mL,
A = volume of sample required to perform each separate analysis, mL,
B = estimated number of replicate analyses required at each interval in the holding time study (see Table 1),
C = estimated number of time intervals required for the holding time study (excluding the initial time zero precision study), and

D = number of replicate determinations performed in initial precision study (usually 10).

7.1.2 Based on the volume calculated in 7.1.1, collect a sufficient volume of the specific matrix to be tested to perform the holding time study and a precision study. The sample must be collected in a properly prepared sample container or series of containers. Refer to Practices D 3694 and the procedure for the constituent of interest for specific instructions on sample collection procedures.

NOTE 3—The total volume of sample calculated in 7.1.1 is only an estimate. Depending upon the degree of certainty with which the precision can be estimated, it is recommended that a volume somewhat in excess of that calculated in 7.1.1 be collected in order to make certain that sufficient sample will be available to complete the holding time study. The analyst may want to consider performing a preliminary precision study prior to sample collection in order to be certain that the estimate of precision made in 7.1.1 is reasonably accurate.

7.1.3 Add the appropriate preservation reagents to the sample. Immediately proceed to 7.2.

7.2 *Determination of Single Operator Precision:*

7.2.1 *General Organic Constituent Methods:*

7.2.1.1 Immediately after sample collection, analyze an appropriate number (usually 10) of measured volumes of sample as described in the appropriate procedure. If a sufficiently high concentration of the constituent of interest is found (concentration must be at least one order of magnitude higher than the criterion of detection) proceed to 7.2.1.2. If not, collect another sample and repeat the analysis until a sample containing a sufficiently high concentration is obtained.

NOTE 4—Since there is no way of positively identifying all of the compounds which may be contributing to the values found in the General Organic Constituent Methods, the sample cannot be fortified. In order to carry out the holding time determination, a sample must be obtained which contains a sufficiently high concentration to carry out the study.

7.2.1.2 Calculate the mean concentration, the standard deviation, and the relative standard deviation of these replicate determinations. (See Practice D 2777.) Proceed to 8.1.

7.2.2 *Specific Organic Constituent Methods* (Applicable to methods that do not require extraction of the sample container.):

7.2.2.1 Immediately after sample collection, analyze an appropriate number (usually 10) of measured volumes of sample as described in the appropriate procedure. If a sufficiently high concentration of the constituent of interest is found (mean concentration must be at least one order of magnitude higher than the criterion of detection), proceed to 7.2.2.4. If not, fortify the sample as described in 7.2.2.2 and reanalyze.

7.2.2.2 Accurately measure the volume of the remainder of the sample and fortify with a known concentration of the constituent of interest. The fortified sample must contain a concentration of the constituent of interest which is at least one order of magnitude higher than the criterion of detection of the method.

7.2.2.3 Immediately perform an appropriate number (usually 10) of replicate analyses of the fortified sample as described in the appropriate procedure.

7.2.2.4 Calculate the mean concentration, the standard deviation and relative standard deviation of these replicate

---

[4] *Reagent Chemicals, American Chemical Society Specifications*, American Chemical Society, Washington, DC. For suggestions on the testing of reagents not listed by the American Chemical Society, see *Analar Standards for Laboratory Chemicals*, BDH Ltd., Poole, Dorset, U.K., and the *United States Pharmacopeia and National Formulary*, U.S. Pharmaceutical Convention, Inc. (USPC), Rockville, MD.

determinations. (See Practice D 2777.) Proceed to 8.1.

7.2.3 *Specific Organic Constituent Methods* (Applicable to methods that require extraction of the sample container.):

7.2.3.1 If the sample was collected in a container other than litre glass bottles, immediately transfer shaken 1-L portions of the sample to separate properly prepared (see Practices D 3694) litre glass bottles which have had the litre mark placed on the neck of the container.

7.2.3.2 Immediately perform an appropriate number (usually 10) replicate determinations of the constituent of interest by analyzing the sample in the containers. If a sufficiently high concentration of the constituent of interest is found (mean concentration must be at least one order of magnitude higher than the criterion of detection), proceed to 7.2.3.5. If not, fortify the sample as described in 7.2.3.3 and reanalyze.

7.2.3.3 Fortify the sample in all of the remaining glass bottles with a known concentration of the constituent of interest by adding an accurately measured small volume of a concentrated standard solution of the analyte. The fortified sample must contain a concentration of the constituent of interest which is at least one order of magnitude higher than the criterion of detection of the method.

7.2.3.4 Immediately perform an appropriate number (usually 10) of replicate analyses of the fortified sample as described in the appropriate procedure.

7.2.3.5 Calculate the mean concentration, the standard deviation, and the relative standard deviation of these replicate determinations. (See Practice D 2777.) Proceed to 8.1.

7.2.4 *Purgeable Organic Compounds:*

7.2.4.1 Immediately after collection, perform an appropriate number (usually 10) of replicate determinations of the constituent of interest by analyzing separate aliquots of the sample that have been collected in hermetically sealed containers. If the concentration is sufficiently high (concentration must be at least one order of magnitude higher than the criterion of detection), proceed to 7.2.4.5.

7.2.4.2 If the concentration found in 7.2.4.1 is not sufficiently high to accurately determine holding time (concentration must be at least one order of magnitude higher than the criterion of detection of the method), collect another sample and repeat the analysis or fortify the sample as described in 7.2.4.3.

7.2.4.3 If the sample requires fortification, open all of the remaining containers and transfer the contents to a graduated cylinder to measure the total volume and the remaining sample. Then transfer the sample to an aspirator bottle fitted with a stopcock at the bottom. Transfer, by means of a syringe, a measured volume of stock solution containing a known concentration of the constituent of interest into the sample. The syringe needle should be below the surface of the liquid during the transfer. Stopper the bottle and mix well. Carefully transfer (by draining through the stopcock) the sample to separate small glass sample vials. Great care must be exercised to carry out the sample transfer with a minimum of sample agitation and aeration. Each sample vial must be filled to overflowing so that a convex miniscus forms at the top. Seal each vial as described in Practices D 3694.

NOTE 5—It is recommended that the operator test his or her technique in transferring solutions of purgeable organic compounds by

preparation and analysis of replicates prepared from a standard solution. This should be done to make certain that no loss of purgeable organic compounds is occurring during transfer. Such loss can seriously bias the results of this test.

7.2.4.4 Perform an appropriate number (usually ten) replicate analyses of the fortified sample as described in the appropriate procedure.

7.2.4.5 Calculate mean concentration, the standard deviation and relative standard deviation of the values found in either 7.2.4.1 or 7.2.4.4. (See Practice D 2777.) Proceed to 8.1.

8. Calculation of Replicates Required for Holding Time Study

8.1 Based on the relative standard deviation found in 7.2, calculate the number of replicate determinations that will be required at each time interval in the holding time study. The following formula is used for the calculations:

$$n = \left(\frac{ts_0}{D}\right)^2$$

where:
n = number of replicates required in the holding time determination,
t = student's t (based on number of replicates used in precision study. See Table 2.),
s_0 = relative standard deviation expressed as percent (Determined in 7.2.), and
D = 15 % (maximum variation from mean concentration to be tolerated).

NOTE 6—The number of replicate determinations calculated using this formula is rounded off to the next highest whole number. For example, a value of 1.09 would be rounded to 2.

9. Analyses at Specified Time Intervals

9.1 At appropriate intervals following the initial analysis, perform the appropriate number of replicate analyses as calculated in 8.1. The intervals at which the subsequent analyses are carried out are left to the judgment of the analyst and are somewhat dependent on whether a measure of maximum or acceptable holding time is desired. For example, days 1, 5, 10, and 14 would be appropriate for a two week study. In some cases, shorter or longer time intervals may be appropriate. During this period, the sample

TABLE 2 Values of Student *t* at 99 % Confidence Interval[A]

Number of Replicates	t Value
2	63.657
3	9.925
4	5.841
5	4.604
6	4.032
7	3.707
8	3.499
9	3.355
10	3.250
11	3.169
12	3.106
13	3.055
14	3.012
15	2.977

[A] University of Kentucky College of Engineering, "Design of Experiments Course", Vol 7, p. 146.

must be stored under the conditions defined for sample preservation.

NOTE 7—In some cases, degradation of the analyte may occur more rapidly than anticipated and acceptable range of variation is exceeded after the first or second chosen interval. In such cases, the holding time study should be repeated using shorter time intervals if an accurate estimation of maximum holding time is required.

NOTE 8—If it is desired to know only whether a specific time interval is an acceptable holding time, a single time interval may suffice.

10. Calculation and Evaluation of Data

10.1 Calculate the average concentration found at each time interval in the holding time study.

10.2 Calculate the tolerable range of variation (99 % confidence interval) from the initial mean concentration that will be used as the criterion for the holding time evaluation. Use the following equation:

$$d = \pm(ts/\sqrt{n})$$

where:

d = range of tolerable variation from the initial mean concentration (in concentration terms),

t = student's t (based on the number of replicates used in the precision study),

s = standard deviation (in concentration terms) calculated in 7.2, and

n = number of replicate determinations used at each time interval in the holding time determination (calculated in 8.1).

10.3 Plot the average concentration found at each time interval versus time on linear graph paper. Indicate on the plot the range of variation from the initial mean concentration that can be tolerated before the holding time is exceeded.

10.4 Draw the best graphical fit of the data points. Evaluate the changes in concentration as a function of time to determine whether the changes represent a significant systematic error in analysis due to increase or decrease in analyte concentration. The maximum holding time is the maximum period of time during which a properly preserved sample can be stored before the systematic error exceeds the tolerable range of variation calculated in 10.2. See Note 1.

11. Example of Holding Time Evaluation

11.1 Assume that a laboratory is planning on determining the holding time for a specific organic constituent in a specific water. Historically, the concentration of the constituent of interest has ranged from below the criterion of detection (<1 mg/L) to as high as 80 mg/L. Based on limited precision studies performed in the past and experience with the method, the single operator precision is estimated to be in the range of 3 to 8 % (RSD) over the concentration range of 10 to 50 mg/L. The laboratory is interested in determining whether the analyte is stable in the water for a period of up to 30 days. The time intervals chosen for the study are 0, 6, 12, 18, 24, and 30 days. The volume required to perform each individual test is 100 mL.

11.2 The total amount of sample required for the study is calculated using the equation in 7.1.1.

$$V = (100 \times 3 \times 5) + 2(100 \times 10) = 3500 \text{ mL}$$

The laboratory decides to collect a total of 5000 mL of

sample in case the estimate of precision is somewhat low.

11.3 Immediately after sample collection and preservation, ten measured aliquots of sample are analyzed according to the prescribed procedure. The mean concentration found is 8.5 mg/L. This value is less than one order of magnitude above the criterion of detection. The remaining sample is fortified with 40 mg/L of the constituent of interest. Ten measured aliquots of the fortified sample are then immediately analyzed. These data are tabulated and the mean, standard deviation, and relative standard deviation of the fortified values are calculated.

Replicate Number	Concentration, mg/L
1	44.8
2	46.5
3	52.2
4	46.2
5	46.6
6	49.5
7	47.6
8	51.1
9	55.2
10	46.3

The mean of the above values is calculated by summing the concentrations and dividing by the number of replicate determinations.

Sum of concentrations = 486.0

$$\text{Mean concentration}, \bar{X} = \frac{486.0}{10} = 48.6 \text{ mg/L}$$

Calculate the standard deviation of the concentration values using the following equation:

$$s = \sqrt{\Sigma(Xi - \bar{X})^2/(n - 1)}$$

where:

s = estimated standard deviation of the series of results,

Xi = each individual concentration value,

$\bar{X}$ = the mean concentration (calculated above), and

n = number of replicate determinations.

Replicate Number	$(Xi - \bar{X})$	$(Xi - \bar{X})^2$
1	−3.8	14.44
2	−2.1	4.41
3	3.6	12.96
4	−2.4	5.76
5	−2.0	4.00
6	0.9	0.81
7	−1.0	1.00
8	2.5	6.25
9	6.6	43.56
10	−2.3	5.29
		98.48

$$\Sigma(Xi - \bar{X})^2 = 98.48$$
$$s = \sqrt{98.48/9} = 3.3079 = 3.31 \text{ mg/L}$$

Replicate No. 9 is tested to determine whether it is an outlier (See Practice D 2777) and found not to be an outlier.

Calculate the relative standard deviation (RSD):

$$\text{RSD, \%} = \frac{s}{\bar{X}} \times 100 = \frac{3.31}{48.6} \times 100 = 6.8 \text{ \%}$$

The final tabulation of the data is as follows:

Number of Replicates	Mean, mg/L	Standard Deviation, mg/L	Relative Standard Deviation, %
10	48.6	3.31	6.8

11.4 Calculate the number of replicates required in the

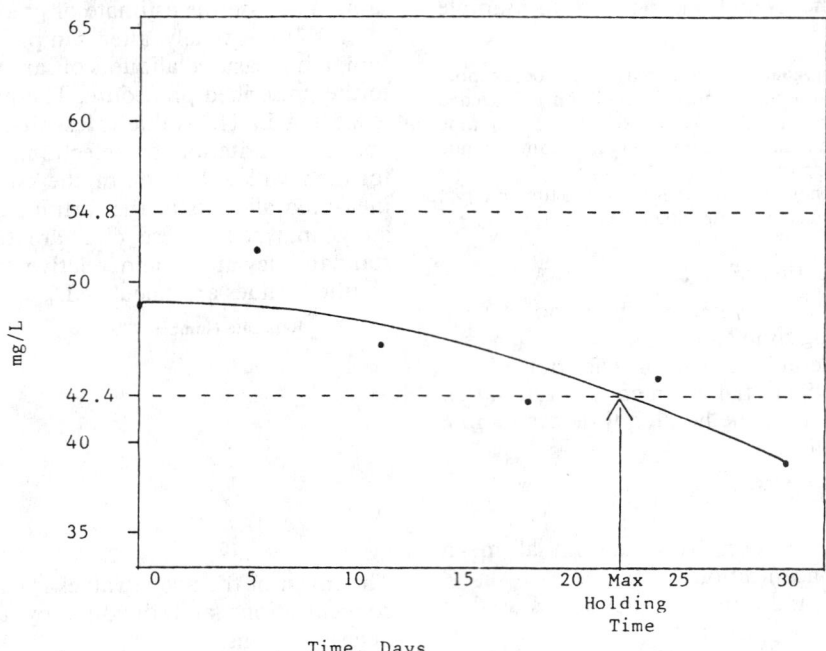

FIG. 1 Plot of Data for Holding Time Estimation

holding time study using the equation in 8.1.

$$n = \left(\frac{3.25 \times 6.8}{15}\right)^2 = 2.17$$

The calculated value of 2.17 is rounded to 3. Three replicate determinations will be required at each time interval in the holding time study.

11.5 All of the tests are carried out at the appropriate time intervals. The average concentration found at each time interval is calculated. The tolerable range of variation from the mean concentration (99 % confidence interval) is calculated using the equation in 10.2.

$$d = \pm \frac{3.25 \times 3.31}{\sqrt{3}} = \pm 6.2 \text{ mg/L}$$

The tolerable interval of variation is therefore, 48.6 ± 6.2 = 42.4 to 54.8 mg/L.

11.6 A plot of the data is prepared and the best graphical fit of the data is drawn (see Fig. 1). The point at which this line crosses the tolerable range of variation is the estimated maximum holding time.

Evaluation of Data for Holding Time Determination

Day	Concentration Found, mg/L
0	48.6
6	51.9
12	45.6
18	42.1
24	43.2
30	37.9

12. Keywords

12.1 acceptable holding time; maximum holding time; preserved samples; purgeable organic compounds; specific organic constituents

ASTM Designation: D 4841 – 88 (Reapproved 1993)ε[1]

Standard Practice for
Estimation of Holding Time for Water Samples Containing Organic and Inorganic Constituents[1]

This standard is issued under the fixed designation D 4841; the number immediately following the designation indicates the year of original adoption or, in the case of revision, the year of last revision. A number in parentheses indicates the year of last reapproval. A superscript epsilon (ε) indicates an editorial change since the last revision or reapproval.

ε[1] NOTE—Section 12 was added editorially in May 1993.

1. Scope

1.1 This practice covers the means of estimating the period of time during which a water sample can be stored after collection and preservation without significantly affecting the accuracy of analysis.

1.2 The maximum holding time is dependent upon the matrix used and the specific analyte of interest. Therefore, water samples from a specific source must be tested to determine the period of time that sample integrity is maintained by standard preservation practices.

1.3 In the event that it is not possible to analyze the sample immediately at the time of collection, this practice does not provide information regarding degradation of the constituent of interest or changes in the matrix that may occur from the time of sample collection to the time of the initial analysis.

1.4 *This standard does not purport to address all of the safety problems, if any, associated with its use. It is the responsibility of the user of this standard to establish appropriate safety and health practices and determine the applicability of regulatory limitations prior to use.*

2. Referenced Documents

2.1 *ASTM Standards:*
D 1129 Terminology Relating to Water[2]
D 1192 Specification for Equipment for Sampling Water and Steam[2]
D 1193 Specification for Reagent Water[2]
D 2777 Practice for Determination of Precision and Bias of Applicable Methods of Committee D-19 on Water[2]
D 3694 Practices for Preparation of Sample Containers and for Preservation of Organic Constituents[3]
D 4210 Practice for Intralaboratory Quality Control Procedures and a Discussion on Reporting Low-Level Data[2]
D 4375 Terminology for Basic Statistics in Committee D-19 on Water[2]
E 178 Practice for Dealing with Outlying Observations[4]

3. Terminology

3.1 *Definitions:*
3.1.1 For definitions of terms used in this practice, refer to Terminology D 1129.
3.1.2 *criterion of detection*—the minimum quantity that must be observed before it can be stated that a substance has been discerned with an acceptable probability that the statement is true (see Practice D 4210).
3.2 *Description of Terms Specific to This Standard:*
3.2.1 *maximum holding time*—the maximum period of time during which a properly preserved sample can be stored before such degradation of the constituent of interest or change in sample matrix occurs that the systematic error exceeds the 99 % confidence interval (not to exceed 15 %) of the test calculated around the mean concentration found at zero time.
3.2.2 *acceptable holding time*—any period of time less than or equal to the maximum holding time.

4. Summary of Practice

4.1 Holding time is estimated by means of replicate analyses at discrete time intervals using a large volume of a water sample that has been properly collected and preserved. A sufficient number of replicate analyses are performed to maintain the 99 % confidence interval within 15 % of the concentration found at zero time. Concentration of the constituent of interest is plotted versus time. The maximum holding time is the period of time from sample collection to such time that degradation of the constituent of interest or change in sample matrix occurs and the systematic error exceeds the 99 % confidence interval (not to exceed 15 %) of the test calculated around the mean concentration at zero time. Prior to the determination of holding time, each laboratory must generate its own precision data in matrix water. These data are compared to the pooled single-operator precision data on reagent water reported in the test method and, the less precise of the two sets of data are used in the calculation.

NOTE 1—This practice generates only limited data which may not lead to consistent conclusions each time that the test is applied. In cases where the concentration of the constituent of interest changes gradually over an extended period of time, the inherent variability in test results may lead to somewhat different conclusions each time that this practice is applied.

5. Significance and Use

5.1 In order to obtain meaningful analytical data, sample

[1] This practice is under the jurisdiction of ASTM Committee D-19 on Water and is the direct responsibility of Subcommittee D19.02 on General Specifications, Technical Resources, and Statistical Methods.
Current edition approved June 24, 1988. Published September 1988.
[2] *Annual Book of ASTM Standards*, Vol 11.01.
[3] *Annual Book of ASTM Standards*, Vol 11.02.
[4] *Annual Book of ASTM Standards*, Vol 14.02.

preservation techniques must be effective from the time of sample collection to the time of analysis. A laboratory must confirm that sample integrity is maintained throughout maximum time periods between sample collection and analysis. In many cases, it is useful to know the maximum holding time. An evaluation of holding time is useful also in judging the efficacy of various preservation techniques.

6. Reagents

6.1 *Purity of Reagents*—Reagent grade chemicals shall be used in all tests. Unless otherwise indicated, it is intended that all reagents shall conform to the specifications of the Committee on Analytical Reagents of the American Chemical Society, where such specifications are available.[5] Other grades may be used provided it is first ascertained that the reagent is of sufficiently high purity to permit its use without lowering the accuracy of the determination.

6.1.1 Refer to the specific test method and to Practices D 3694 for information regarding necessary equipment and preparation of reagents.

6.2 *Purity of Water*—Reference to water shall be understood to mean reagent water conforming to Specification D 1193, Type II, and demonstrated to be free of specific interference for the test being performed.

7. Determination of Holding Time

7.1 *Collection of Sample:*

NOTE 2—In some instances, it may be of interest to determine the holding time of standard solutions prepared in water. In such cases, a large volume of properly preserved, standard solution should be prepared and carried through the steps of the practice in the same manner as a sample. The volume of solution required can be estimated using the equation in 7.1.1.

7.1.1 Based on the estimated precision of the test (determined from past experience or from precision data reported in the test method), calculate the estimated total volume of sample required to perform the holding time determination plus a precision study. Estimate this volume as follows:

$$V = (A \times B \times C) + 2 (A \times D) \qquad (1)$$

where:

V = estimated volume of sample required, mL,

A = volume of sample required to perform each separate analysis, mL,

B = estimated number of replicate determinations required at each interval in the holding time study (see Table 1),

C = estimated number of time intervals required for the holding time study (excluding the initial time zero precision study), and

D = number of replicate determinations performed in initial precision study (usually 10).

7.1.2 Based on the volume calculated in 7.1.1, collect a sufficient volume of the specific matrix to be tested to perform a precision study and the holding time study. Collect the sample in a properly prepared sample container

[5] "Reagent Chemicals, American Chemical Society Specifications," Am. Chemical Soc., Washington, DC. For suggestions on the testing of reagents not listed by the American Chemical Society, see "Analar Standards for Laboratory Chemicals," BDH Ltd., Poole, Dorset, U. K., and the "United States Pharmacopeia."

TABLE 1   Approximate Number of Replicate Determinations Required at Each Interval in the Holding Time Study Based on the Estimated Relative Standard Deviation of the Test in the Matrix Under Study

Estimated RSD, %	Approximate Number of Replicates
1–4	1
5–6	2
7–8	3
9	4
10	5
11	6
12	7
13	8
14	10
15	11

or series of containers. Refer to the procedure for the constituent of interest for specific instructions on sample collection procedures.

NOTE 3—The total volume of sample calculated in 7.1.1 is only an estimate. Depending upon the degree of certainty with which the precision can be estimated, it is recommended that a volume somewhat in excess of that calculated in 7.1.1 be collected in order to make certain that sufficient sample will be available to complete the holding time study. The analyst may want to consider performing a preliminary precision study prior to sample collection in order to be certain that the estimate of precision used in 7.1.1 is reasonably accurate.

7.1.3 Add the appropriate preservation reagents to the sample immediately after collection. Immediately proceed to 7.2 or 7.3 depending upon whether inorganic or organic compounds are being determined.

7.2 *Determination of Single Operator Precision—Inorganic Methods:*

7.2.1 Immediately after sample collection, analyze an appropriate number (usually 10) of measured volumes of sample as described in the appropriate procedure. If a measurable concentration of the constituent of interest is found, proceed to 7.2.4. If the concentration of the constituent of interest is below the criterion of detection at a P level of ≤ 0.05, fortify the sample as described in 7.2.2 and reanalyze or collect another sample.

NOTE 4—If the concentration of the constituent of interest is very low such that it approaches the criterion of detection at a P level of ≤ 0.05, the precision will be very poor. At such very low concentrations, a fairly large number of replicate determinations will be required to bring the 99 % confidence interval to within 15 % of the concentration found. Under these circumstances, it may be desirable to fortify the sample with the constituent of interest to increase the concentration to a point where the precision will be improved and fewer replicates will be required for the holding time determination. However, the holding time may be different at the higher concentration than it would be at the lower concentration. This decision is left to the judgement of the analyst.

7.2.2 Accurately measure the volume of the remainder of the sample and fortify with a known concentration of the constituent of interest.

7.2.3 Immediately perform an appropriate number (usually 10) of replicate analyses of the sample as described in the appropriate procedure.

7.2.4 Calculate the mean concentration, the standard deviation, and relative standard deviation of these replicate determinations (see Practice D 4375). Proceed to 8.1.

7.3 *Determination of Single-Operator Precision—Organic Methods:*

7.3.1 *General Organic Constituent Methods*—Immediately after sample collection, analyze an appropriate number (usually 10) of measured volumes of sample as described in the appropriate procedure. If a measurable concentration of organics is found, proceed to 7.3.1.1. If the concentration of the organic compounds is below the criterion of detection at a P level of ≤ 0.05, collect another sample and repeat the analysis until a sample containing a measurable concentration is obtained (see Note 4).

NOTE 5—Since there is no way of positively identifying all of the compounds that may be contributing to the values found in the general organic constituent methods, the sample cannot be fortified. To carry out the holding time determination, a sample must be obtained that contains a measurable concentration of organics in order to carry out the study.

7.3.1.1 Calculate the mean concentration, the standard deviation, and the relative standard deviation of these replicate determinations (see Practice D 4375). Proceed to 8.1.

7.3.2 *Specific Organic Constituent Methods* (Applicable to methods that do not require extraction of the sample container):

7.3.2.1 Immediately after sample collection, analyze an appropriate number (usually 10) of measured volumes of sample as determined in the appropriate procedure. If a measurable concentration of the constituent of interest is found, proceed to 7.3.2.4. If not, either collect another sample or fortify the sample as described in 7.3.2.2 and reanalyze (see Note 4).

7.3.2.2 Accurately measure the volume of the remainder of the sample and fortify it with a known concentration of the constituent of interest.

7.3.2.3 Immediately perform an appropriate number (usually 10) of replicate analyses of the fortified sample as described in the appropriate procedure.

7.3.2.4 Calculate the mean concentration, the standard deviation, and the relative standard deviation of these replicate determinations (see Practice D 4375). Proceed to 8.1.

7.3.3 *Specific Organic Constituent Methods* (Applicable to methods that require extraction of the sample container):

7.3.3.1 If the sample was collected in a container other than litre glass bottles, immediately transfer shaken, 1-L portions of the sample to separate properly prepared (see Practices D 3694) litre glass bottles which have had the litre mark placed on the neck of the container.

7.3.3.2 Immediately perform an appropriate number (usually 10) of replicate determinations of the constituent of interest by analyzing the sample in the containers. If a measurable concentration of the constituent of interest is found, proceed to 7.3.3.5. If not, fortify the sample as described in 7.3.3.3 and reanalyze (see Note 4).

7.3.3.3 Fortify the sample in all of the remaining glass bottles with a known concentration of the constituent of interest by adding an accurately measured small volume of a concentrated standard solution of the analyte.

7.3.3.4 Immediately perform an appropriate number of replicate analyses of the fortified sample as described in the appropriate procedure.

7.3.3.5 Calculate the mean concentration, the standard deviation, and the relative standard deviation of these

replicate determinations (see Practice D 4375). Proceed to 8.1.

7.3.4 *Purgeable Organic Compounds:*

7.3.4.1 Immediately after collection, perform an appropriate number (usually 10) of replicate determinations of the constituent of interest by analyzing separate aliquots of sample that have been collected in hermetically sealed containers. If a measurable concentration is found, proceed to 7.3.4.3. If the concentration is below the criterion of detection at a P level of ≤ 0.05, either fortify the sample as described in 7.3.4.2 or collect another sample and repeat the analysis (see Note 4).

7.3.4.2 If the sample requires fortification, open all of the remaining containers and transfer the contents to a graduated cylinder to measure the total volume of the remaining sample. Then transfer the sample to an aspirator bottle fitted with a stopcock at the bottom. Transfer, by means of a syringe, a measured volume of stock solution containing a known concentration of the constituent of interest into the sample. The syringe needle should be below the surface of the liquid during the transfer. Stopper the bottle and mix well. Carefully transfer (by draining through the stopcock) the sample to separate small glass vials. Take care to carry out the sample transfer with a minimum of sample agitation and aeration. Fill each sample vial to overflowing so that a convex meniscus forms at the top. Seal each vial as described in Practices D 3694.

NOTE 6—It is recommended that the operator's technique used in transferring solutions of purgeable organic compounds be tested by preparation and analysis of replicates prepared from a standard solution. This should be done to make certain that no loss of purgeable organic compounds is occurring during transfer. Such loss can seriously bias the results of this test.

7.3.4.3 Perform an appropriate number (usually 10) of replicate analyses of the fortified sample as described in the appropriate procedure.

7.3.4.4 Calculate the mean concentration, standard deviation, and relative standard deviation of the values found in either 7.3.4.1 or 7.3.4.3 (see Practice D 4374). Proceed to 8.1.

8. Calculation of Replicates Required for Holding Time Study

NOTE 7—Since some analytical methods are very precise (especially those used in determination of inorganic constituents), it is possible that the single operator precision as generated by the laboratory on a single day may be significantly better than the day-to-day variation caused by random errors. If so, this would significantly bias the results of the test. Consequently, the pooled single-operator precision on reagent water generated in the round-robin testing of the method should be used as the basis for calculation of the 99 % confidence interval if these data show poorer precision than the data generated in 7.2 or 7.3. It is recognized that such data do not include the variability caused by the matrix. However, it is assumed that if the single-operator precision as determined in matrix water is better than the pooled single-operator precision found in reagent water, the contribution of the matrix to the variability is negligible.

8.1 Based on the relative standard deviation found in 7.2 or 7.3 or the pooled single-operator precision in reagent water (see Note 7), calculate the number of replicate determinations that will be required at each time interval in the holding time study. Calculate the number of replicate determinations as follows:

TABLE 2   Values of Student's t for a Two-Tailed 99 % Confidence Interval[6]

No. of Replicates	t Value
2	63.657
3	9.925
4	5.841
5	4.604
6	4.032
7	3.707
8	3.499
9	3.355
10	3.250
11	3.169
12	3.106
13	3.055
14	3.012
15	2.977

$$n = \left(\frac{t\,RSD_o}{D}\right)^2 \qquad (2)$$

where:

n = number of replicates required in the holding time determination,

t = Student's t (Based on the number of replicates used in the precision study. See Table 2 and Note 8),

RSD_o = relative standard deviation, %, (Determined in 7.2 or 7.3 or use pooled single operator precision in reagent water), and

D = 15 % (maximum variation from mean concentration to be tolerated).

Note 8—If the pooled single-operator precision in reagent water reported in the test method is used in this calculation, information on the number of replicates used in the precision study may not be available. Under these circumstances, use $t = 3.00$ to obtain a reasonably accurate estimate of the 99 % confidence interval.

Note 9—The number of replicate determinations calculated using this formula is rounded to the next highest whole number. For example, a value of 1.09 would be rounded to 2.

Note 10—The value of 15 % was chosen as the maximum variation from the mean concentration to avoid the need to run an unrealistic number of replicates on tests that are very imprecise. Note that only one determination will be required on tests with a relative standard deviation (RSD) of about 4.5 % or less.

9. Analysis at Specified Time Intervals

9.1 At appropriate intervals following the initial analysis, perform the appropriate number of replicate analyses as calculated in 8.1. The intervals at which the subsequent analyses are carried out are left to the judgment of the analyst and are somewhat dependent on whether a measure of maximum or acceptable holding time is desired. For example, days 1, 5, 10, and 14 would be appropriate for a 2-week study. In some cases, shorter or longer time intervals may be appropriate. During this period, the sample must be stored under the conditions defined for sample preservation.

Note 11—In some cases, degradation of the analyte may occur more rapidly than anticipated and the acceptable range of variation is exceeded after the first or second chosen interval. In such cases, the holding time study should be repeated using shorter time intervals if an accurate estimation of maximum holding time is required.

---
[6] "Design of Experiments Course," University of Kentucky College of Engineering, Vol 7, p. 146.

Note 12—If it is desired to know only whether a specific time interval is an acceptable holding time, a single time interval may suffice.

10. Calculations and Evaluation of Data

10.1 Calculate the average concentration found at each time interval in the holding time study.

10.2 Calculate the tolerable range of variation (99 % confidence interval) from the initial mean concentration that will be used as the criterion for the holding time evaluation as follows:

$$d = \pm \frac{t\,s}{\sqrt{n}} \qquad (3)$$

where:

d = range of tolerable variation from the initial mean concentration (in concentration terms),

t = Student's t (based on the number of replicates used in the precision study or use 3.00 if the pooled single-operator precision in reagent water is used),

s = standard deviation (in concentration terms) calculated in 7.2 or 7.3 or based on pooled single-operator precision in reagent water, and

n = number of replicate determinations used at each time interval in the holding time determination (calculated in 8.1).

10.3 Plot the average concentration found at each time interval versus time on linear graph paper. Indicate on the plot the range of variation from the initial mean concentration that can be tolerated before the holding time is exceeded.

10.4 If the loss of analyte versus time appears to be a linear relationship, calculate and plot the best straight line through the points using the method of least squares. Otherwise, draw the best graphical fit of the data points. Evaluate the changes in concentration as a function of time to determine whether the changes represent a significant systematic error in analysis due to increase or decrease in analyte concentration. The maximum holding time is the maximum period of time during which a properly preserved sample can be stored before the systematic error exceeds the tolerable range of variation calculated in 10.2 (see Note 1).

11. Example of Holding Time Evaluation

11.1 Assume a laboratory is planning on determining the holding time for a specific organic constituent in a specific water. Historically, the concentration of the constituent of interest has ranged from below the criterion of detection (< 1 mg/L) to as high as 80 mg/L. Based on limited precision studies performed in the past and experience with the method, the single-operator precision is estimated to be in the range from 3 to 8 % RSD over the concentration range from 10 to 50 mg/L. The laboratory is interested in determining whether the analyte is stable in the water for a period of up to 30 days. The time intervals chosen for the study are 0, 6, 12, 18, 24, and 30 days. The volume required to perform each individual test is 100 mL.

11.2 The total amount of sample required for the study is calculated using the equation in 7.1.1.

$$V = (100 \times 3 \times 5) + 2\,(100 \times 10) = 3500 \text{ mL} \qquad (4)$$

The laboratory decides to collect a total of 5000 mL of sample in case the estimate of precision is somewhat low.

TABLE 3 Example Data

Replicate No.	Concentration, mg/L
1	44.8
2	46.5
3	52.2
4	46.2
5	46.6
6	49.5
7	47.6
8	51.1
9	55.2
10	46.3

TABLE 4 Standard Deviation of Concentration Values

Replicate No.	$(X_i - \bar{X})$	$(X_i - \bar{X})^2$
1	−3.8	14.44
2	−2.1	4.41
3	3.6	12.96
4	−2.4	5.76
5	−2.0	4.00
6	0.9	0.81
7	−1.0	1.00
8	2.5	6.25
9	6.6	43.56
10	−2.3	5.29
Total		98.48

TABLE 5 Tabulation of Statistics

Number of Replicates	Mean, mg/L	Standard Deviation, mg/L	Relative Standard Deviation, %
10	48.6	3.31	6.8

11.3 Immediately after sample collection and preservation, ten measured aliquots of sample are analyzed according to the prescribed procedure. The mean concentration found is 8.5 mg/L. To improve the precision of the measurement, the remaining sample is fortified with 40 mg/L of the constituent of interest. Ten measured aliquots of the fortified sample are then immediately analyzed. These data are tabulated (see Table 3) and the mean, standard deviation, and relative standard deviation of the fortified values are calculated.

11.3.1 The mean of the values in Table 3 are calculated by summing the concentrations and dividing by the number of replicate determinations as follows:

$$\text{Sum of concentrations} = 486.0$$

$$\text{Mean Concentration } (\bar{X}) = \frac{486.0}{10} = 48.6 \text{ mg/L}$$

11.3.2 The standard deviation of the concentration values (see Table 4) is then calculated as follows:

$$s = \sqrt{\Sigma(X_i - \bar{X})^2/(n-1)} \tag{5}$$

where:

s = estimated standard deviation of the series of results,
X_i = each individual concentration value,

TABLE 6 Evaluation of Data for Holding Time Determination

Day	Concentration Found, mg/L
0	48.6
6	51.9
12	45.6
18	42.1
24	43.2
30	37.9

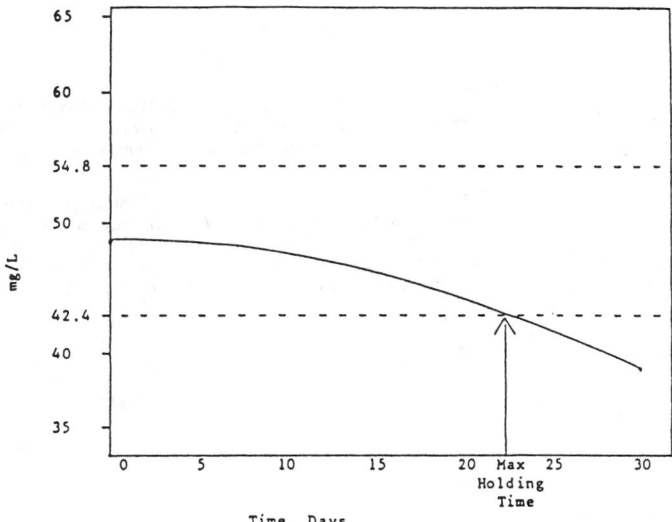

FIG. 1 Plot of Data for Holding Time Determination

$\bar{X}$ = the mean concentration (calculated in Eq. 5), and
n = number of replicate determinations.

$$\Sigma(X_i - \bar{X})^2 = 98.48 \tag{6}$$

$$s = \sqrt{98.48/9} = 3.3079 = 3.31 \text{ mg/L} \tag{7}$$

11.3.3 Replicate No. 9 is tested to determine whether it is an outlier (see Recommended Practice E 178 and Practice D 2777) and found not to be an outlier.

11.3.4 The RSD is then calculated as follows:

$$\text{RSD } (\%) = \frac{s}{\bar{X}}(100) = \frac{3.31}{48.6}(100) = 6.8 \% \tag{8}$$

11.3.5 The final tabulation of the statistics is shown in Table 5.

11.4 Calculate the number of replicates required in the holding time study using Eq. 2 in 8.1.

$$n = \left(\frac{3.25 \ (6.8)}{15}\right)^2 = 2.17 \tag{9}$$

11.4.1 The calculated value of 2.17 is rounded to 3. Three replicate determinations will be required at each time interval in the holding time study.

11.5 All of the tests are then carried out at the appropriate time intervals. The average concentration found at each time interval is calculated. The tolerable range of variation from the mean concentration (99 % confidence interval) is then calculated using Eq. 3 in 10.2.

$$d = \frac{\pm 3.25\ (3.31)}{\sqrt{3}} = \pm 6.2\ \text{mg/L} \tag{10}$$

The tolerable interval of variation is therefore, $48.6 \pm 6.2 = 42.4$ to 54.8 mg/L.

11.6 A plot of the data is then prepared as shown in Table 6 and Fig. 1. Since the loss of analyte does not appear to be linear with time, the best graphical fit of the data is drawn. The point at which this line crosses the tolerable range of variation is the estimated maximum holding time.

12. Keywords

12.1 degredation; estimation; holding time; storage limit; water

3.3 GROUND WATER MONITORING WELLS
(See also Drilling Methods, Section 1.6)

Designation: D 5092 – 90 (Reapproved 1995)

Standard Practice for
Design and Installation of Ground Water Monitoring Wells in Aquifers[1]

This standard is issued under the fixed designation D 5092; the number immediately following the designation indicates the year of original adoption or, in the case of revision, the year of last revision. A number in parentheses indicates the year of last reapproval. A superscript epsilon (ε) indicates an editorial change since the last revision or reapproval.

INTRODUCTION

This practice for the design and installation of ground water monitoring wells in aquifers will promote (*1*) durable and reliable construction, (*2*) extraction of representative ground water quality samples, and (*3*) efficient and site hydrogeological characterizations. The guidelines established herein are affected by governmental regulations and by site specific geological, hydrogeological, climatological, topographical, and subsurface chemistry conditions. To meet these geoenvironmental challenges, this guidance promotes the development of a conceptual hydrogeologic model prior to monitoring well design and installation.

1. Scope

1.1 This practice considers the selection and characterization (that is, defining soil, rock types, and hydraulic gradients) of the target monitoring zone as an integral component of monitoring well design and installation. Hence, the development of a conceptual hydrogeologic model for the intended monitoring zone(s) is recommended prior to the design and installation of a monitoring well.

1.2 These guidelines are based on recognized methods by which monitoring wells may be designed and installed for the purpose of detecting the presence or absence of a contaminant, and collecting representative ground water quality data. The design standards and installation procedures herein are applicable to both detection and assessment monitoring programs for facilities.

1.3 The recommended monitoring well design, as presented in this practice, is based on the assumption that the objective of the program is to obtain representative ground water information and water quality samples from aquifers. Monitoring wells constructed following this practice should produce relatively turbidity-free samples for granular aquifer materials ranging from gravels to silty sand and sufficiently permeable consolidated and fractured strata. Strata having grain sizes smaller than the recommended design for the smallest diameter filter pack materials should be monitored by alternative monitoring well designs which are not addressed in this practice.

1.4 The values stated in inch-pound units are to be regarded as standard. The values in parentheses are for information only.

1.5 *This standard does not purport to address all of the safety concerns, if any, associated with its use. It is the responsibility of the user of this standard to establish appropriate safety and health practices and determine the applicability of regulatory limitations prior to use.*

2. Referenced Documents

2.1 *ASTM Standards:*
C 150 Specification for Portland Cement[2]
C 294 Descriptive Nomenclature of Constituents of Natural Mineral Aggregates[3]
D 653 Terminology Relating to Soil, Rock, and Contained Fluids[4]
D 1452 Practice for Soil Investigation and Sampling by Auger Borings[4]
D 1586 Method for Penetration Test and Split-Barrel Sampling of Soils[4]
D 1587 Practice for Thin-Walled Tube Sampling of Soils[4]
D 2113 Practice for Diamond Core Drilling for Site Investigation[4]
D 2487 Classification of Soils for Engineering Purposes (Unified Soil Classification System)[4]
D 2488 Practice for Description and Identification of Soils (Visual-Manual Procedure)[4]
D 3282 Classification of Soils and Soil Aggregate Mixtures for Highway Construction Purposes[4]
D 3550 Practice for Ring Lined Barrel Sampling of Soils[4]
D 4220 Practice for Preserving and Transporting Soil Samples[4]

3. Significance and Use

3.1 An adequately designed and installed ground water monitoring well system for aqueousphase liquids provides essential information for decisions pertaining to one or more of the following subjects:

3.1.1 Aquifer and aquitard properties, both geologic and hydraulic;

---

[1] This practice is under the jurisdiction of ASTM Committee D-18 on Soil and Rock and is the direct responsibility of Subcommittee D18.21.05 on Design and Installation of Ground-Water Monitoring Wells.
Current edition approved June 29, 1990. Published October 1990.

[2] *Annual Book of ASTM Standards*, Vol 04.01.
[3] *Annual Book of ASTM Standards*, Vol 04.02.
[4] *Annual Book of ASTM Standards*, Vol 04.08.

3.1.2 Potentiometric surface of a particular hydrologic unit(s);

3.1.3 Water quality with respect to various indicator parameters;

3.1.4 Migration characteristics of a contaminant release;

3.1.5 Additional installations or decommissioning of installations, or both, no longer needed.

4. Terminology

4.1 *Definitions:*

4.1.1 *annular space; annulus*—the space between two concentric tubes or casings, or between the casing and the borehole wall. This would include the space(s) between multiple strings of tubing/casings in a borehole installed either concentrically or multi-cased adjacent to each other.

4.1.2 *assessment monitoring*—an investigative monitoring program that is initiated after the presence of a contaminant in ground water has been detected. The objective of this program is to determine the concentration of constituents that have contaminated the ground water and to quantify the rate and extent of migration of these constituents.

4.1.3 *ASTM cement types*—Portland cements meeting the requirements of Specifications C 150. Cement types have slightly different formulations that result in various characteristics which address different construction conditions and different physical and chemical environments. They are as follows:

4.1.3.1 *Type I (Portland)*—a general-purpose construction cement with no special properties.

4.1.3.2 *Type II (Portland)*—a construction cement that is moderately resistant to sulfates and generates a lower head of hydration at a slower rate than Type I.

4.1.3.3 *Type III (Portland; high early strength)*—a construction cement that produces a high early strength. This cement reduces the curing time required when used in cold environments, and produces a higher heat of hydration than Type I.

4.1.3.4 *Type IV (Portland)*—a construction cement that produces a low head of hydration (lower than Types I and II) and develops strength at a slower rate.

4.1.3.5 *Type V (Portland)*—a construction cement that is a high sulfate resistant formulation. Used when there is severe sulfate action from soils and ground water.

4.1.4 *bailer*—a hollow tubular receptacle used to facilitate withdrawal of fluid from a well or borehole.

4.1.5 *ballast*—materials used to provide stability to a buoyant object (such as casing within a borehole filled with water).

4.1.6 *blow-in*—the inflow of ground water and unconsolidated material into a borehole or casing caused by differential hydraulic heads; that is, caused by the presence of a greater hydraulic head outside of a borehole/casing than inside.

4.1.7 *borehole* a circular open or uncased subsurface hole created by drilling.

4.1.8 *borehole log*—the record of geologic units penetrated, drilling progress, depth, water level, sample recovery, volumes, and types of materials used, and other significant facts regarding the drilling of an exploratory borehole or well.

DISCUSSION—The definition of aquifer as currently included in Terminology D 653 varies from the definition as prescribed by US federal regulations. Since this federal definition is associated with the installation of many monitoring wells it is provided herein as a technical note:

aquifer—a geologic formation, group of formation, or part of a formation that is saturated, and is capable of providing a significant quantity of water.

4.1.9 *bridge*—an obstruction within the annulus which may prevent circulation or proper emplacement of annular materials.

4.1.10 *casing*—pipe, finished in sections with either threaded connections or bevelled edges to be field welded, which is installed temporarily or permanently to counteract caving, to advance the borehole, or to isolate the zone being monitored, or combination thereof.

4.1.11 *casing, protective*—a section of larger diameter pipe that is emplaced over the upper end of a smaller diameter monitoring well riser or casing to provide structural protection to the well and restrict unauthorized access into the well.

4.1.12 *casing, surface*—pipe used to stabilize a borehole near the surface during the drilling of a borehole that may be left in place or removed once drilling is completed.

4.1.13 *caving; sloughing*—the inflow of unconsolidated material into a borehole which occurs when the borehole walls lose their cohesive strength.

4.1.14 *cement; Portland cement*—commonly known as Portland cement. A mixture that consists of a calcareous, argillaceous, or other silica-, alumina-, and iron-oxide-bearing materials that is manufactured and formulated to produce various types which are defined in Specification C 150. Portland cement is also considered a hydraulic cement because it must be mixed with water to form a cement-water paste that has the ability to harden and develop strength even if cured under water (see *ASTM cement types*).

4.1.15 *centralizer*—a device that assists in the centering of a casing or riser within a borehole or another casing.

4.1.16 *circulation*—applies to the fluid rotary drilling method; drilling fluid movement from the mud pit, through the pump, hose and swivel, drill pipe, annular space in the hole and returning to the mud pit.

4.1.17 *conductance (specific)*—a measure of the ability of the water to conduct an electric current at 77°F (25°C). It is related to the total concentration of ionizable solids in the water. It is inversely proportional to electrical resistance.

4.1.18 *confining unit*—a term that is synonymous with "aquiclude," "aquitard," and "aquifuge;" defined as a body of relatively low permeable material stratigraphically adjacent to one or more aquifers.

4.1.19 *contaminant*—an undesirable substance not normally present in water or soil.

4.1.20 *detection monitoring*—a program of monitoring for the express purpose of determining whether or not there has been a contaminant release to ground water.

4.1.21 *drill cuttings*—fragments or particles of soil or rock, with or without free water, created by the drilling process.

4.1.22 *drilling fluid*—a fluid (liquid or gas) that may be used in drilling operations to remove cuttings from the borehole, to clean and cool the drill bit, and to maintain the integrity of the borehole during drilling.

4.1.23 *d-10*—the diameter of a soil particle (preferably in millimetres) at which 10 % by weight (dry) of the particles of a particular sample are finer. Synonymous with the effective size or effective grain size.

4.1.24 *d-60*—the diameter of a soil particle (preferably in millimetres) at which 60 % by weight (dry) of the particles of a particular sample are finer.

4.1.25 *flow path*—represents the area between two flow lines along which ground water can flow.

4.1.26 *flush joint or flush coupled*—casing or riser with ends threaded such that a consistent inside and outside diameter is maintained across the threaded joints or couplings.

4.1.27 *gravel pack*—common nomenclature for the terminology, primary filter of a well (see *primary filter pack*).

4.1.28 *grout (monitoring wells)*—a low permeability material placed in the annulus between the well casing or riser pipe and the borehole wall (that is, in a single-cased monitoring well), or between the riser and casing (that is, in a multi-cased monitoring well), to maintain the alignment of the casing and riser and to prevent movement of ground water or surface water within the annular space.

4.1.29 *grout shoe*—a *plug* fabricated of relatively inert materials that is positioned within the lowermost section of a permanent casing and fitted with a passageway, often with a flow check device, through which grout is injected under pressure to fill the annular space. After the grout has set, the grout shoe is usually drilled out.

4.1.30 *head (static)*—the height above a standard datum of the surface of a column of water (or other liquid) that can be supported by the static pressure at a given point. The static head is the sum of the elevation head and the pressure head.

4.1.31 *head (total)*—the sum of three components at a point: (*1*) elevation head, h_e, which is equal to the elevation of the point above a datum; (*2*) pressure head, hp, which is the height of a column of static water than can be supported by the static pressure at the point; and (*3*) velocity head, hv, which is the height the kinetic energy of the liquid is capable of lifting the liquid.

4.1.32 *hydrologic unit*—geologic strata that can be distinguished on the basis of capacity to yield and transmit fluids. Aquifers and confining units are types of hydrologic units. Boundaries of a hydrologic unit may not necessarily correspond either laterally or vertically to lithostratigraphic formations.

4.1.33 *jetting*—when applied as a drilling method, water is forced down through the drill rods or casings and out through the end aperture. The jetting water then transports the generated cuttings to the ground surface in the annulus of the drill rods or casing and the borehole. The term jetting may also refer to a development technique (see well screen jetting).

4.1.34 *loss of circulation*—the loss of drilling fluid into strata to the extent that circulation does not return to the surface.

4.1.35 *mud pit*—usually a shallow, rectangular, open, portable container with baffles into which drilling fluid and cuttings are discharged from a borehole and that serves as a reservoir and settling tank during recirculation of the drilling fluids. Under some circumstances, an excavated pit with a lining material may be used.

4.1.36 *multi-cased well*—a well constructed by using successively smaller diameter casings with depth.

4.1.37 *neat cement*—a mixture of Portland cement (Specification 150) and water.

4.1.38 *observation well*—typically, a small diameter well used to measure changes in hydraulic heads, usually in response to a nearby pumping well.

4.1.39 *oil air filter*—a filter or series of filters placed in the air flow line from an air compressor to reduce the oil content of the air.

4.1.40 *oil trap*—a device used to remove oil from the compressed air discharged from an air compressor.

4.1.41 *packer (monitoring wells)*—a transient or dedicated device placed in a well that isolates or seals a portion of the well, well annulus, or borehole at a specific level.

4.1.42 *potentiometric surface*—an imaginary surface representing the static head of ground water. The water table is a particular potentiometric surface.

DISCUSSION—Where the head varies with depth in the aquifer, a potentiometric surface is meaningful only if it describes the static head along a particular specified surface or stratum in that aquifer. More than one potentiometric surface is required to describe the distribution of head in this case.

4.1.43 *primary filter pack*—a clean silica sand or sand and gravel mixture of selected grain size and gradation that is installed in the annular space between the borehole wall and the well screen, extending an appropriate distance above the screen, for the purpose of retaining and stabilizing the particles from the adjacent strata. The term is used in place of *gravel pack*.

4.1.44 *PTFE tape*—joint sealing tape composed of polytetrafluoroethylene.

4.1.45 *riser*—the pipe extending from the well screen to or above the ground surface.

4.1.46 *secondary filter pack*—a clean, uniformly graded sand that is placed in the annulus between the primary filter pack and the over-lying seal, or between the seal and overlying grout backfill, or both, to prevent movement of seal or grout, or both, into the primary filter pack.

4.1.47 *sediment sump*—a blank extension beneath the well screen used to collect fine-grained material from the filter pack and adjacent strata. The term is synonymous with rat trap or tail pipe.

4.1.48 *shear strength (monitoring wells)*—a measure of the shear or gel properties of a drilling fluid or grout.

4.1.49 *single-cased well*—a monitoring well constructed with a riser but without an exterior casing.

4.1.50 *static water level*—the elevation of the top of a column of water in a monitoring well or piezometer that is not influenced by pumping or conditions related to well installation, hydrologic testing, or nearby pumpage.

4.1.51 *tamper*—a heavy cylindrical metal section of tubing that is operated on a wire rope or cable. It slips over the riser and fits inside the casing or borehole annulus. It is generally used to tamp annular sealants or filter pack materials into place and prevent bridging.

4.1.52 *target monitoring zone*—the ground water flow path from a particular area or facility in which monitoring wells will be screened. The target monitoring zone should be a stratum (strata) in which there is a reasonable expectation that a vertically placed well will intercept migrating contaminants.

4.1.53 *test pit*—a shallow excavation made to characterize the subsurface.

4.1.54 *transmissivity*—the rate at which water of the prevailing kinematic viscosity is transmitted through a unit width of the aquifer under a unit hydraulic gradient.

DISCUSSION—It is equal to an integration of the hydraulic conductivities across the saturated part of the aquifer perpendicular to the flow paths.

4.1.55 *tremie pipe*—a pipe or tube that is used to transport filter pack materials and annular sealant materials from the ground surface into the borehole annulus or between casings and casings or riser pipe of a monitoring well.

4.1.56 *uniformly graded*—a quantitative definition of the particle size distribution of a soil which consists of a majority of particles being of the same approximate diameter. A granular material is considered uniformly graded when the uniformity coefficient is less than about five (Test Method D 2487). Comparable to the geologic term *well sorted*.

4.1.57 *vented cap*—a cap with a small hole that is installed on top of the riser.

4.1.58 *washout nozzle*—a tubular extension with a check valve utilized at the end of a string of casing through which water can be injected to displace drilling fluids and cuttings from the annular space of a borehole.

4.1.59 *weep hole*—a small diameter hole (usually ¼ in.) drilled into the protective casing above the ground surface that serves as a drain hole for water that may enter the protective casing annulus.

4.1.60 *well completion diagram*—a record that illustrates the details of a well installation.

4.1.61 *well screen*—a filtering device used to retain the primary or natural filter pack; usually a cylindrical pipe with openings of a uniform width, orientation, and spacing.

4.1.62 *well screen jetting (hydraulic jetting)*—when jetting is used for development, a jetting tool with nozzles and a high-pressure pump is used to force water outwardly through the screen, the filter pack, and sometimes into the adjacent geologic unit.

4.1.63 *zone of saturation*—a hydrologic zone in which all the interstices between particles of geologic material or all of the joints, fractures, or solution channels in a consolidated rock unit are filled with water under pressure greater than that of the atmosphere.

5. Site Characterization

5.1 *General*—Soil mechanics, geomorphological concepts, geologic structure, stratigraphy, and sedimentary concepts, as well as the nature and behavior of the solutes of interest, must be combined with a knowledge of ground water movement to make a complete application of the results of the monitoring well design and installation guidance. Therefore, development of a conceptual hydrogeologic model that identifies potential flow paths and the target monitoring zone(s) is recommended prior to monitoring well design and installation. Development of the conceptual model is accomplished in two phases—an initial reconnaissance and a field investigation. When the hydrogeology of a project area is relatively uncomplicated and well docu-

mented in the literature, the initial reconnaissance may provide sufficient information to identify flow paths and the target monitoring zone(s). However, where little background data is available or the geology is complicated, a field investigation will generally be necessary to completely develop a conceptual hydrogeologic model.

5.2 *Initial Reconnaissance of Project Area*—The goal of the initial reconnaissance of the project area is to identify and locate those zones with the greatest potential to transmit a fluid from the project area. Identifying these flow paths is the first step in selecting the target ground water monitoring zone(s).

5.2.1 *Literature Search*—Every effort should be made to collect and review all applicable field and laboratory data from previous investigations of the project area. Data such as, but not limited to, topographic maps, aerial imagery, site ownership and utilization records, geologic and hydrogeologic maps and reports, mineral resource surveys, water well logs, personal information from local well drillers, agricultural soil reports, geotechnical engineering reports, and other engineering maps and report related to the project area should be reviewed.

5.2.2 *Field Reconnaissance*—Early in the investigation, the soil and rocks in open cut areas in the vicinity of the project should be studied, and various soil and rock profiles noted. Special consideration should be given to soil color and textural changes, landslides, seeps, and springs within or near the project area.

5.2.3 *Preliminary Conceptual Model*—The distribution of the predominant soil and rock units likely to be found during subsurface exploration may be hypothesized at this time in a preliminary hydrogeologic conceptual model using data obtained in the literature search and field reconnaissance. In areas where the geology is relatively uniform, well documented in the literature, and substantiated by the field reconnaissance, further refinement of the conceptual model may not be necessary unless anomalies are discovered in the well drilling stage.

5.3 *Field Investigation*—The goal of the field investigation is to refine the preliminary conceptual hydrogeologic model so that the target monitoring zone(s) is selected prior to monitoring well installation.

5.3.1 *Exploratory Borings and Test Pits*—Characterization of the flow paths conceptualized in the initial reconnaissance involves defining the porosity, hydraulic conductivity, gradation, stratigraphy, lithology, and structure of each hydrologic unit. The characteristics are defined by conducting an exploratory boring program which may include test pits. Exploratory borings and test pits should be deep enough to develop the required engineering and hydrogeologic data for determining the flow path(s), target monitoring zone, or both.

5.3.1.1 *Sampling*—Soil and rock properties should not be predicted wholly on field identification or classification, but should be checked by laboratory and field tests made on samples. Representative soil or rock samples, or both, of each material that is significant to the analysis and design of the monitoring system should be obtained and evaluated by a geologist, hydrogeologist, or engineer trained and experienced in soil and rock analysis. Soil sample extraction should be conducted according to Practice D 1452, Method D 1586,

Practice D 3550, or Practice D 1587, whichever is appropriate given the anticipated characteristics of the soil samples. Rock samples should be extracted according to Practice D 2113. Soil samples obtained for evaluation of hydraulic properties should be containerized and identified for shipment to a laboratory. Special measures to preserve either the continuity of the sample or the natural moisture are not usually required. However, soil and rock samples obtained for evaluation of chemical properties often require special field preparation and preservation to prevent significant alteration of the chemical constituents during transportation to a laboratory (see Practice D 4220). Rock samples for evaluation of hydraulic properties are usually obtained using a split-inner-tube core barrel. Evaluation and logging of the core samples is usually made in the field before the core is removed from half of the split inner tube core barrel.

5.3.1.2 *Boring Logs*—Care should be taken to prepare and retain a complete boring log and sampling record for each exploratory borehole and test pit.

NOTE 1—Site investigations for the installation of ground-water monitoring wells can vary greatly due to the availability of reliable site data or the lack thereof. The general procedure would however be as follows: (*1*) gather factual data regarding the surficial and subsurface conditions, (*2*) analyze the data, (*3*) develop a conceptual model of the site conditions, (*4*) locate the monitoring wells based on the first three steps. Monitoring wells should only be installed with sufficient understanding of the geologic and hydrogeologic conditions present on site. Monitoring wells often serve as part of an overall site investigation for a specific purpose, such as determining the extent of contamination present, or for prediction of the effectiveness of aquifer remediations. In these cases extensive additional geotechnical and hydrogeologic information may be required that would go beyond the Section 5 Site Characterization description.

Boring logs should include the location, geotechnical (that is, penetration rates or blow counts), and sampling information for each material identified in the borehole either by symbol or word description, or both. Identification of all soils should be in accordance with Practice D 2488 or Practice D 3282. Identification of rock material should be based on Nomenclature C 294 or by an appropriate geologic classification system. Observations of seepage, free water, and water levels should also be noted. The boring logs should be accompanied by a report that includes a description of the area investigated; a map illustrating the vertical and horizontal location (with reference to nearest National Geodetic Vertical Datum [NGVD] and to a standardized survey grid, respectively) of each exploratory borehole or test pit, or both; and color photographs of rock cores, soil samples, and exposed strata labeled with a date and identification.

5.3.2 *Geophysical Exploration*—Geophysical surveys may be used to supplement borehole and outcrop data and to aid in interpretation between boreholes. Surface geophysical methods such as seismic surveys, and electrical-resistivity and electromagnetic conductance surveys can be particularly valuable when distinct differences in the properties of contiguous subsurface materials are indicated. Borehole methods such as resistivity, gamma, gamma-gamma, neutron, and caliper logs can be useful to confirm specific subsurface geologic conditions. Gamma logs are particularly useful in existing cased wells.

5.3.3 *Ground Water Flow Direction*—Ground water flow direction is generally determined by measuring the vertical and horizontal hydraulic gradient within each conceptualized flow path. However, because water will flow along the path of least resistance, flow direction may be oblique to the hydraulic gradient (buried stream channels or glacial valleys, for example). Flow direction is determined by first installing piezometers in the exploratory boreholes. The depth and location of the piezometers will depend upon anticipated hydraulic connections between conceptualized flow paths and their respective lateral direction of flow. Following careful evaluation, it may be possible to utilize existing private or public wells to obtain water level data. The construction integrity of such wells should be verified to ensure that the water levels obtained from the wells are representative only of the zones of interest. Following water level data acquisition, a potentiometric surface map should be prepared. Flow paths are ordinarily determined to be at right angles, or nearly so, to the equipotential lines.

5.4 *Completing the Conceptual Model*—A series of hydrogeologic cross sections should be developed to refine the conceptual model. This is accomplished by first plotting logs of soil and rock observed in the exploratory borings or test pits, and interpreting between these logs using the geologic and engineering interrelationships between other soil and rock data observed in the initial reconnaissance or with geophysical techniques. Extrapolation of data into adjacent areas should be done only where geologically uniform subsurface conditions are known to exist. The next step is to integrate the profile data with the piezometer data for both vertical and horizontal hydraulic gradients. Plan view and cross-sectional flow nets may need to be constructed. Following the analysis of these data, conclusions can be made as to which flow path(s) is the appropriate target monitoring zone(s).

NOTE 2—Ground water monitoring is difficult and may not be a reliable technology in fine-grain, low hydraulic conductivity, primary porosity strata because of (*1*) the disproportionate influence that microstratigraphy has on ground water flow in fine-grain strata; (*2*) flow lines proportionally higher for the vertical flow component in low hydraulic conductivity strata; and (*3*) the presence of indigenous metallic and inorganic constituents that make water quality data evaluation difficult.

6. Monitoring Well Construction Materials

6.1 *General*—The materials that are used in the construction of a monitoring well and that come in contact with the water sample should not measurably alter the chemical quality of the sample for the constituents being examined using the appropriate sampling protocols. Furthermore, the riser, well screen, and annular sealant injection equipment should be steam cleaned or high-pressure water cleaned (if appropriate for the selected riser material) immediately prior to well installation or certified clean from the manufacturer and delivered to site in a protective wrapping. Samples of the cleaning water, filter pack, annular seal, and mixed grout should be retained to serve as quality control until the completion of at least one round of ground water quality sampling and analysis.

6.2 *Water*—Water used in the drilling process, to prepare grout mixtures and to decontaminate the well screen, riser, and annular sealant injection equipment, should be obtained from a source of known chemistry that does not contain constituents that could compromise the integrity of the well installation.

6.3 *Primary Filter Pack:*

6.3.1 *Materials*—The primary filter pack (gravel pack) consists of a granular material of known chemistry and selected grain size and gradation that is installed in the annulus between the screen and the borehole wall. The filter pack is usually selected to have a 30 % finer (d-30) grain size that is about 4 to 10 times greater than the 30 % finer (d-30) grain size of the hydrologic unit being filtered (see Fig. 1). Usually, the filter is selected to have a low (that is, less than 2.5) uniformity coefficient. The grain size and gradation of the filter are selected to stabilize the hydrologic unit adjacent to the screen and permit only the finest soil grains to enter the screen during development. Thus, after development, a correctly filtered monitoring well is relatively turbid-free.

NOTE 3—When installing a monitoring well in Karst or highly fractured bedrock, the borehole configuration of void spaces within the formation surrounding the borehole is often unknown. Therefore, the installation of a filter pack becomes difficult and may not be possible.

6.3.2 *Gradation*—The filter pack should be uniformly graded and comprised of hard durable siliceous particles washed and screened with a particle size distribution derived by multiplying the d-30 size of the finest-grained screened stratum by a factor between 4 and 10. Use a number between four and six as the multiplier if the stratum is fine and uniform; use a factor between six and ten where the material has highly nonuniform gradation and includes silt-sized particles. The grain-size distribution of the filter pack is then plotted using the d-30 size as the control point on the graph. The selected filter pack should have a uniformity coefficient of approximately 2.5 or less.

NOTE 4—This practice presents a design for monitoring wells that will be effective in the majority of aquifers. Applicable state guidance may differ from the designs contained in this practice.

NOTE 5—Because the well screen slots have uniform openings, the filter pack should be composed of particles that are as uniform in size as is practical. Ideally, the uniformity coefficient (the quotient of the 60 % passing, D-60 size divided by the 10 % passing D-10 size [effective size]) of the filter pack should be 1.0 (that is, the D-60 % and the D-10 % sizes should be identical). However, a more practical and consistently achievable uniformity coefficient for all ranges of filter pack sizes is 2.5. This value of 2.5 should represent a maximum value, not an ideal.

NOTE 6—Although not recommended as standard practice, often a project requires drilling and installing the well in one phase of work.

Therefore, the filter pack materials must be ordered and delivered to the drill site before soil samples can be collected. In these cases, the suggested well screen slot size and filter pack materials are presented in Table 1.

6.4 *Well Screen:*

6.4.1 *Materials*—The well screen should be new, machine-slotted or continuous wrapped wire-wound and composed of materials most suited for the monitoring environment and site characterization findings. The screen should be plugged at the bottom. The plug should be of the same material as the well screen. This assembly must have the capability to withstand installation and development stresses without becoming dislodged or damaged. The length of the slotted area should reflect the interval to be monitored. Immediately prior to installation, the well screen should be steam cleaned or high-pressure water cleaned (if appropriate for the selected well screen materials) with water from a source of known chemistry if not certified by the manufacturer, delivered, and maintained clean at the site.

NOTE 7—Well screens are most commonly composed of PVC, stainless steel, fiberglass, or fluoropolymer materials.

6.4.2 *Diameter*—The minimum nominal internal diameter of the well screen should be chosen based on the particular application. However, in most instances, a minimum of 2 in. (50 mm) is needed to allow for the introduction and withdrawal of sampling devices.

6.4.3 *Slot Size*—The slot size of the well screen should be determined relative to the grain size analysis of the stratum interval to be monitored and the gradation of the filter pack material. In granular non-cohesive strata that will fall in easily around the screen, filter packs are not necessary. In these cases of natural development, the slot size of the well screen is to be determined using the grain size of the materials in the surrounding strata. The slot size and arrangement should retain at least 90 % and preferably 99 % of the filter pack. The method for determining the correct gradation of filter pack material is described in 6.3.2.

6.5 *Riser:*

6.5.1 *Materials*—The riser should be new and composed of materials that will not alter the quality of water samples for the constituents of concern and that are appropriate for the monitoring environment. The riser should have adequate wall thickness and coupling strength to withstand installation and development stresses. Each section of riser should be steam cleaned or high-pressure water cleaned (if appropriate for the selected material) using water from a source of known chemistry immediately prior to installation.

NOTE 8—Risers are generally constructed of PVC, stainless steel, fiberglass, or fluoropolymer materials.

6.5.2 *Diameter*—The minimum nominal internal diameter of the riser should be chosen based on the particular application. However, in most instances, a minimum of 2 in. (50 mm) is needed to accommodate sampling devices.

6.5.3 *Joints (Couplings)*—Threaded joints are recommended. Glued or solvent welded joints of any type are *not* recommended since glues and solvents may alter the chemistry of the water samples. In most cases, square profile flush joint threads do not require PTFE taping, however, tapered thread joints should be PTFE taped to prevent leakage of water into the riser. Alternatively, O-rings composed of

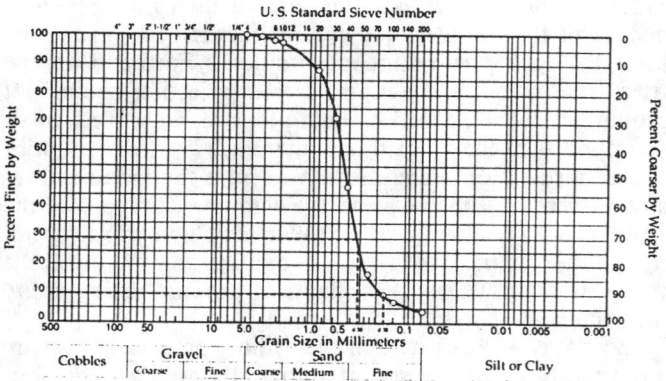

FIG. 1 Example Grading Curve for Design of Monitoring Well Screens

TABLE 1 Recommended (Achievable) Filter Pack Characteristics for Common Screen Slot Sizes

Size of Screen Opening, mm (in.)	Slot No.	Sand Pack Mesh Size Name(s)	1 % Passing Size (D-1), mm	Effective Size, (D-10), mm	30 % Passing Size (D-30), mm	Range of Uniformity Coefficient	Roundness (Powers Scale)
0.125 (0.005)	5[A]	100	0.09 to 0.12	0.14 to 0.17	0.17 to 0.21	1.3 to 2.0	2 to 5
0.25 (0.010)	10	20 to 40	0.25 to 0.35	0.4 to 0.5	0.5 to 0.6	1.1 to 1.6	3 to 5
0.50 (0.020)	20	10 to 20	0.7 to 0.9	1.0 to 1.2	1.2 to 1.5	1.1 to 1.6	3 to 6
0.75 (0.030)	30	10 to 20	0.7 to 0.9	1.0 to 1.2	1.2 to 1.5	1.1 to 1.6	3 to 6
1.0 (0.040)	40	8 to 12	1.2 to 1.4	1.6 to 1.8	1.7 to 2.0	1.1 to 1.6	4 to 6
1.5 (0.060)	60	6 to 9	1.5 to 1.8	2.3 to 2.8	2.5 to 3.0	1.1 to 1.7	4 to 6
2.0 (0.080)	80	4 to 8	2.0 to 2.4	2.4 to 3.0	2.6 to 3.1	1.1 to 1.7	4 to 6

[A] A 5-slot (0.152-mm) opening is not currently available in slotted PVC but is available in Vee wire PVC and Stainless; 6-slot opening may be substituted in these cases.

materials that would not impact the water sample for the constituents of concern may be selected for use on flush joint threads.

6.6 *Casing*—Where conditions warrant, the use of permanent casing installed to prevent communication between water-bearing zones is encouraged. The following subsections address both temporary and permanent casings.

6.6.1 *Materials*—The material type and minimum wall thickness of the casing should be adequate to withstand the forces of installation. All casing that is to remain as a permanent part of the installation (that is, multi-cased wells) should be new and cleaned to be free of interior and exterior protective coatings.

NOTE 9—The exterior casing (temporary or permanent multi-cased) is generally composed of steel, although other appropriate materials may be used.

6.6.2 *Diameter*—Several different casing sizes may be required depending on the subsurface geologic conditions penetrated. The diameter of the casing for filter packed wells should be selected so that a minimum annular space of 2 in. (50 mm) is maintained between the inside diameter of the casing and outside diameter of the riser. In addition, the diameter of the casings in multi-cased wells should be selected so that a minimum annular space of 2 in. is maintained between the casing and the borehole (that is, a 2-in. diameter screen will require first setting a 6-in. (152-mm) diameter casing in a 10-in. (254-mm) diameter boring).

NOTE 10—Under difficult drilling conditions (collapsing soils, rock, or cobbles), it may be necessary to advance temporary casing, under these conditions a smaller annular space may be maintained.

6.6.3 *Joints (Couplings)*—The ends of each casing section should be either flush-threaded or bevelled for welding.

6.7 *Protective Casing:*

6.7.1 *Materials*—Protective casings may be made of aluminum, steel, stainless steel, cast iron, or a structural plastic. The protective casing should have a lid capable of being locked shut by a locking device.

6.7.2 *Diameter*—The inside dimensions of the protective casing should be a minimum of 2 in. (50 mm) and preferably 4 in. (101 mm) larger than the nominal diameter of the riser to facilitate the installation and operation of sampling equipment.

6.8 *Annular Sealants*—The materials used to seal the annulus may be prepared as a slurry or used un-mixed in a dry pellet, granular, or chip form. Sealants should be selected to be compatible with ambient geologic, hydrogeologic, and climatic conditions and any man-induced conditions anticipated to occur during the life of the well.

6.8.1 *Bentonite*—Bentonite should be powdered, granular, pelletized, or chipped sodium montmorillonite furnished in sacks or buckets from a commercial source and free of impurities which adversely impact the water quality in the well. Pellets consist of roughly spherical or disk shaped units of compressed bentonite powder. Chips are large, irregularly shaped, and coarse granular units of bentonite free of additives. The diameter of pellets or chips selected for monitoring well construction should be less than one fifth the width of the annular space into which they are placed to reduce the potential for bridging. Granules consist of coarse particles of unaltered bentonite, typically smaller than 0.2 in. (50 mm).

6.8.2 *Cement*—Each type of cement has slightly different characteristics that may be appropriate under various physical and chemical conditions. Cement should be one of the five Portland cement types that are specified in Specification C 150. The use of quick-setting cements containing additives is not recommended for use in monitoring well installation. Additives may leach from the cement and influence the chemistry of the water samples.

6.8.3 *Grout*—The grout backfill that is placed above the bentonite annular seal and secondary filters (see Fig. 2) is ordinarily a liquid slurry consisting of either a bentonite (powder or granules, or both) base and water, or a Portland cement base and water. Often, bentonite-based grouts are used when it is desired that the grout remain flexible (that is, to accommodate freeze-thaw) during the life of the installation. Cement or bentonite-based grouts are often used when the filling in of cracks in the surrounding geologic material, adherence to rock units, or a rigid setting is desired.

6.8.3.1 *Mixing*—The mixing (and placing) of a grout backfill should be performed with precisely recorded weights and volumes of materials, and according to procedures stipulated by the manufacturer that often include the order of component mixing. The grout should be thoroughly mixed with a paddle type mechanical mixer or by recirculating the mix through a pump until all lumps are disintegrated. Lumpy grout should not be used in the construction of a monitoring well to prevent bridging within the tremie.

NOTE 11—Lumps do not include lost circulation materials that may be added to the grout if excessive grout losses occur.

6.8.3.2 *Typical Bentonite Base Grout*—When a bentonite base grout is used, bentonite, usually unaltered, *must* be the first additive placed in the water through a venturi device. A typical unbeneficiated bentonite base grout consists of about 1 to 1.25 lb (0.57 kg) of unaltered bentonite to each 1 gal (3.8 L) of water. After the bentonite is mixed and allowed to "yield or hydrate," up to 2 lb (0.9 kg) of Type I Portland cement (per gallon of water) is often added to stiffen the mix.

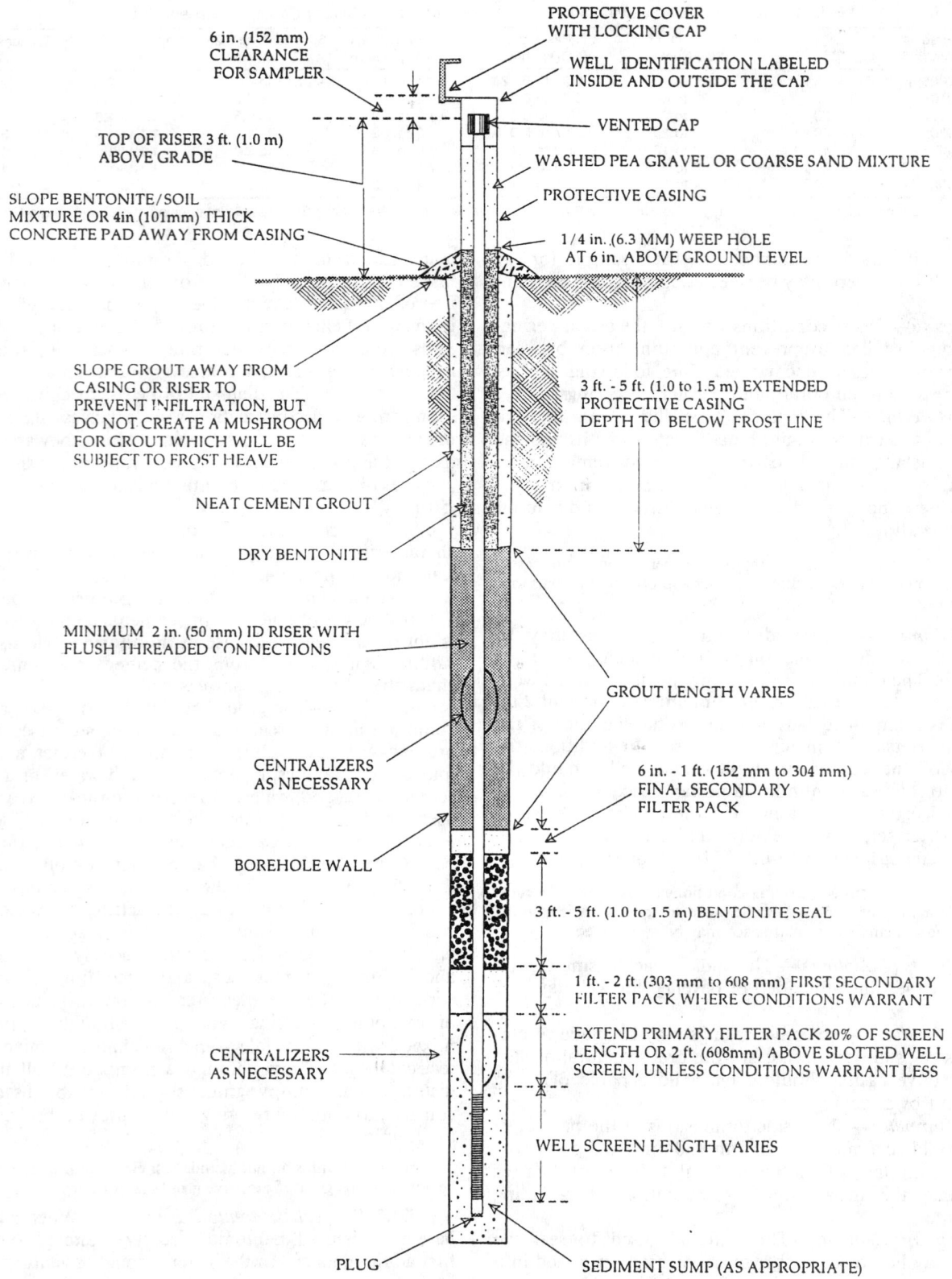

PROTECTIVE COVER
WITH LOCKING CAP

WELL IDENTIFICATION LABELED
INSIDE AND OUTSIDE THE CAP

6 in. (152 mm)
CLEARANCE
FOR SAMPLER

VENTED CAP

TOP OF RISER 3 ft. (1.0 m)
ABOVE GRADE

WASHED PEA GRAVEL OR COARSE SAND MIXTURE

PROTECTIVE CASING

SLOPE BENTONITE/SOIL
MIXTURE OR 4in (101mm) THICK
CONCRETE PAD AWAY FROM CASING

1/4 in. (6.3 MM) WEEP HOLE
AT 6 in. ABOVE GROUND LEVEL

SLOPE GROUT AWAY FROM
CASING OR RISER TO
PREVENT INFILTRATION, BUT
DO NOT CREATE A MUSHROOM
FOR GROUT WHICH WILL BE
SUBJECT TO FROST HEAVE

3 ft. - 5 ft. (1.0 to 1.5 m) EXTENDED
PROTECTIVE CASING
DEPTH TO BELOW FROST LINE

NEAT CEMENT GROUT

DRY BENTONITE

MINIMUM 2 in. (50 mm) ID RISER WITH
FLUSH THREADED CONNECTIONS

GROUT LENGTH VARIES

6 in. - 1 ft. (152 mm to 304 mm)
FINAL SECONDARY
FILTER PACK

CENTRALIZERS
AS NECESSARY

BOREHOLE WALL

3 ft. - 5 ft. (1.0 to 1.5 m) BENTONITE SEAL

1 ft. - 2 ft. (303 mm to 608 mm) FIRST SECONDARY
FILTER PACK WHERE CONDITIONS WARRANT

EXTEND PRIMARY FILTER PACK 20% OF SCREEN
LENGTH OR 2 ft. (608mm) ABOVE SLOTTED WELL
SCREEN, UNLESS CONDITIONS WARRANT LESS

CENTRALIZERS
AS NECESSARY

WELL SCREEN LENGTH VARIES

PLUG

SEDIMENT SUMP (AS APPROPRIATE)

FIG. 2 Monitoring Well Design—Single-Cased Well

100 % Bentonite grouts should not be used solely for monitoring well annular sealants in the vadose zone of arid regions because of their propensity to desiccate. This could result in non-representative waters affecting the target monitoring zone.

NOTE 12—High solids bentonite grouts (minimum 20 % by weight with water) and other bentonite-based grouts may contain granular bentonite to increase the solids content and other components added under manufacturer's directions to either stiffen or retard stiffening of the mix.

All additives to grouts should be evaluated for their effects on subsequent water samples.

6.8.3.3 *Typical Cement Base Grout*—When a cement-based grout is used, cement is usually the first additive placed in the water. A typical cement-based grout consists of about 6 to 7 gal (23 to 26 L) of water per 94-lb (43-kg) bag of Type I Portland cement. From 0 to 10 % (by dry weight of cement) of unaltered bentonite powder is often added after the initial mixing of cement and water to retard shrinkage and provide plasticity. The bentonite is added *dry* to the cement-water slurry without first mixing it with water.

6.9 *Secondary Filter Packs:*

6.9.1 *Materials*—A secondary filter pack is a layer of material placed in the annulus between the primary filter pack and the bentonite seal, and between the bentonite seal and the grout backfill (see Figs. 2 and 3).

6.9.2 *Gradation*—The secondary filter pack should be uniformly graded fine sand with a 100 % by weight passing the No. 30 U.S. Standard sieve, and less than 2 % by weight passing the 200 U.S. Standard sieve.

6.10 *Annular Seal Equipment*—The equipment used to inject the annular seals and filter pack should be steam cleaned or high-pressure water cleaned (if appropriate for the selected material) using water from a source or known quality prior to use. This procedure is performed to prevent the introduction of materials that may ultimately alter the water sample quality.

7. Drilling Methods

7.1 The type of equipment required to create a stable, open, vertical borehole for installation of a monitoring well depends upon the site geology, hydrology, and the intended use of the data. Engineering and geological judgment is required for the selection of the drilling methods utilized for drilling the exploratory boreholes and monitoring wells. Whenever feasible, drilling procedures should be utilized that do not require the introduction of water or liquid fluids into the borehole, and that optimize cuttings control at ground surface. Where the use of drilling fluid is unavoidable, the selected fluid should have as little impact as possible on the water samples for the constituents of interest. In addition, care should be taken to remove as much drilling fluid as possible from the well and the aquifer during the well development process. It is recommended that if an air compressor is used, it is equipped with an oil air filter or oil trap.

8. Monitoring Well Installation

8.1 *Stable Borehole*—A stable borehole must be constructed prior to attempting to install the monitoring well screen and riser. Steps must be taken to stabilize the borehole before attempting installation if the borehole tends to cave or blow-in, or both. Boreholes that are not straight or are partially obstructed should be corrected prior to attempting the installations described herein.

8.2 *Assembly of Well Screen and Riser:*

8.2.1 *Handling*—The well screen, bottom plug, riser, should be either certified clean from the manufacturer or steam cleaned or high-pressure water cleaned (if appropriate for the selected material) using water from a source of known chemistry immediately prior to assembly. Personnel should take precautions to assure that grease, oil, or other contaminants that may ultimately alter the water sample do not contact any portion of the well screen and riser assembly. As one precaution, for example, personnel should wear a clean pair of cotton or surgical (or equivalent) gloves while handling the assembly.

8.2.2 *Riser Joints (Couplings)*—Flush joint risers with square profile threads normally do not require additional PTFE taping to obtain a water tight seal. In addition, O-rings of known chemistry, selected on the basis of prevailing environmental or physical conditions, may be used to assure a tight seal of flush-joint couplings. Couplings are often tightened by hand; however, if necessary, steam cleaned or high-pressure water cleaned wrenches may be utilized. Precautions should be taken to prevent damage to the threaded joints during installation.

8.3 *Setting the Well Screen and Riser Assembly*—When the well screen and riser assembly is lowered to the predetermined level and held into position, the assembly may require ballast to counteract the tendency to float in the borehole. Ballasting may be accomplished by continuously filling the riser with water from a source of known chemistry or, preferably, water which was previously removed from the borehole. Alternatively, the riser may be slowly pushed into the fluid in the borehole with the aid of hydraulic rams on the drill rig and held in place as additional sections of riser are added to the column. Care must be taken to secure the riser assembly so that personnel safety is assured during the installation. The assembly must be installed straight with the appropriate centralizers to allow for the introduction and withdrawal of sampling devices. Difficulty in maintaining a straight installation may be encountered where the weight of the well screen and riser assembly is significantly less than the buoyant force of the fluid in the borehole. The riser should extend above grade and be capped temporarily to deter entrance of foreign materials during completion operations.

8.4 *Installation of the Primary Filter Pack:*

8.4.1 *Volume of Filter Pack*—The volume of filter pack required to fill the annular space between the well screen and borehole should be computed, measured, and recorded on the well completion diagram during installation. To be effective, the filter pack should extend above the screen for a distance of about 20 % of the length of the well screen but not less than 2 ft (600 mm) (see Figs. 2 and 3). Where there is hydraulic connection between the zone to be monitored and the overlying strata, this upward extension should be gauged to prevent seepage from overlying hydrologic units into the filter pack. Seepage from other units may alter the water sample.

8.4.2 *Placement of Primary Filter Pack*—Placement of

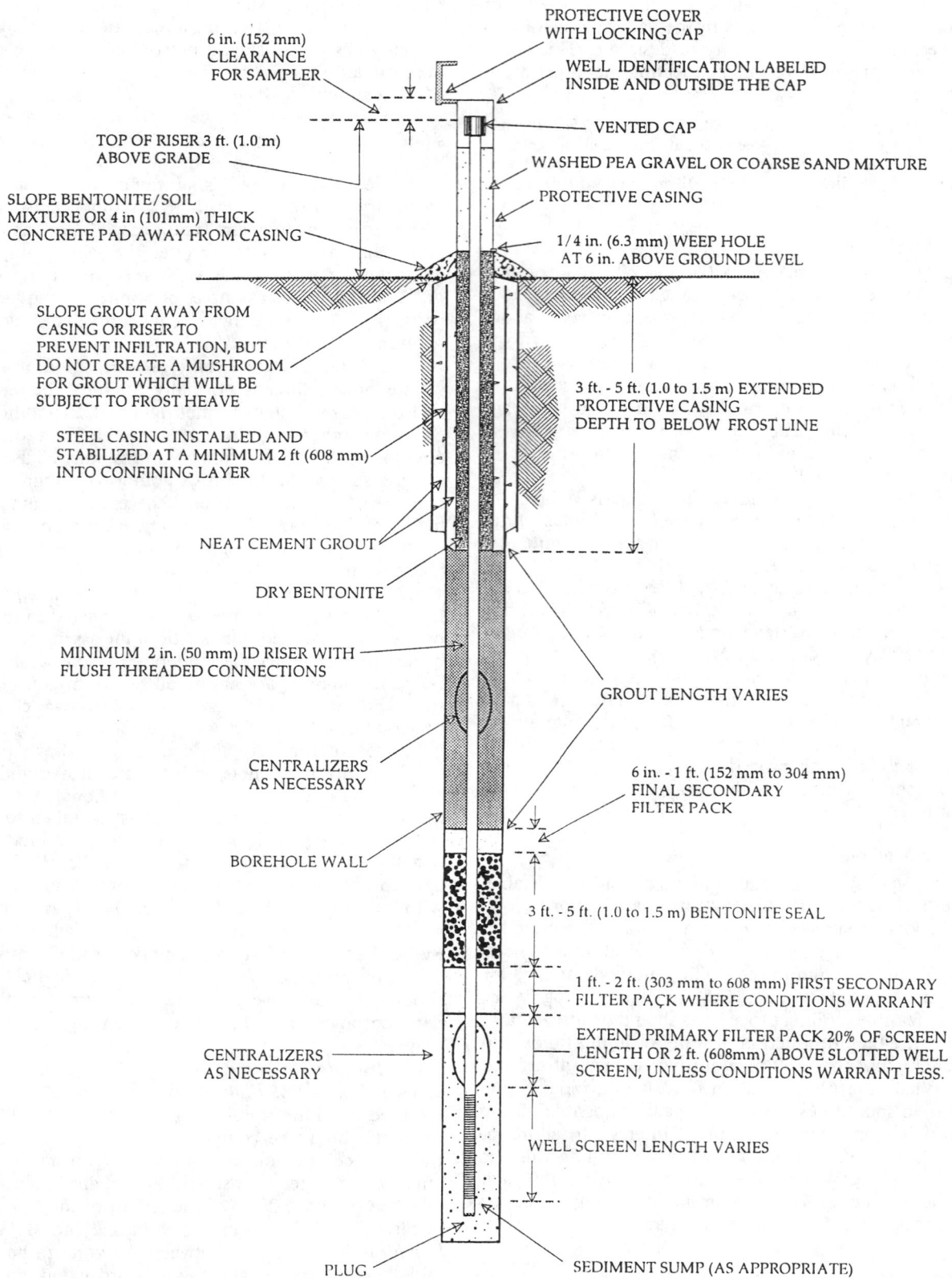

6 in. (152 mm)
CLEARANCE
FOR SAMPLER

PROTECTIVE COVER
WITH LOCKING CAP

WELL IDENTIFICATION LABELED
INSIDE AND OUTSIDE THE CAP

TOP OF RISER 3 ft. (1.0 m)
ABOVE GRADE

VENTED CAP

WASHED PEA GRAVEL OR COARSE SAND MIXTURE

SLOPE BENTONITE/SOIL
MIXTURE OR 4 in (101mm) THICK
CONCRETE PAD AWAY FROM CASING

PROTECTIVE CASING

1/4 in. (6.3 mm) WEEP HOLE
AT 6 in. ABOVE GROUND LEVEL

SLOPE GROUT AWAY FROM
CASING OR RISER TO
PREVENT INFILTRATION, BUT
DO NOT CREATE A MUSHROOM
FOR GROUT WHICH WILL BE
SUBJECT TO FROST HEAVE

3 ft. - 5 ft. (1.0 to 1.5 m) EXTENDED
PROTECTIVE CASING
DEPTH TO BELOW FROST LINE

STEEL CASING INSTALLED AND
STABILIZED AT A MINIMUM 2 ft (608 mm)
INTO CONFINING LAYER

NEAT CEMENT GROUT

DRY BENTONITE

MINIMUM 2 in. (50 mm) ID RISER WITH
FLUSH THREADED CONNECTIONS

GROUT LENGTH VARIES

6 in. - 1 ft. (152 mm to 304 mm)
FINAL SECONDARY
FILTER PACK

CENTRALIZERS
AS NECESSARY

BOREHOLE WALL

3 ft. - 5 ft. (1.0 to 1.5 m) BENTONITE SEAL

1 ft. - 2 ft. (303 mm to 608 mm) FIRST SECONDARY
FILTER PACK WHERE CONDITIONS WARRANT

EXTEND PRIMARY FILTER PACK 20% OF SCREEN
LENGTH OR 2 ft. (608mm) ABOVE SLOTTED WELL
SCREEN, UNLESS CONDITIONS WARRANT LESS.

CENTRALIZERS
AS NECESSARY

WELL SCREEN LENGTH VARIES

PLUG

SEDIMENT SUMP (AS APPROPRIATE)

FIG. 3 Monitoring Well Design—Multi–Cased Well

the well screen is preceded by placing no less than 2 % and no more than 10 % of the primary filter pack into the bottom of the borehole using a decontaminated, flush threaded, 1-in. (25-mm) minimum internal diameter tremie pipe. Alternatively, the filter pack may be added directly between the riser pipe and the auger or borehole or casing and the top of the filter pack located using a tamper or a weighted line. The well screen and riser assembly is then centered in the borehole using one or more centralizer(s) or alternative centering device located not more than 10 ft (3 m) above the bottom of the well screen (see Figs. 2 and 3). The centralizer should not be located in the bentonite seal. The remaining primary filter pack is then placed in increments as the tremie is gradually raised. As primary filter pack material is poured into the tremie pipe, water from a source of known chemistry may be added to help move the filter pack. The tremie pipe or a weighed line inserted through the tremie pipe can be used to measure the top of the primary filter pack as work progresses. If bridging of the primary filter pack occurs, the bridged material should be broken mechanically prior to proceeding with the addition of more filter pack material. The elevation, volume, and gradation of primary filter pack is recorded on the well completion diagram.

8.4.3 *Withdrawal of the Temporary Casing/Augers*—If used, the temporary casing or hollow stem auger is withdrawn, usually in stipulated increments. Care should be taken to minimize lifting the riser with the withdrawal of the temporary casing/augers. To limit borehole collapse, the temporary casing or hollow stem auger is usually withdrawn until the lower most point on the temporary casing or hollow stem auger is at least 2 ft (608 mm), but no more than 5 ft (1.5 m), above the filter pack for unconsolidated materials; or at least 5 ft, but no more than 10 ft (3.0 m), for consolidated materials. In highly unstable formations, withdrawal intervals may be much less. After each increment, it should be ascertained that the primary filter pack has not been displaced during the withdrawal operation (that is, a weighed measuring device).

8.5 *Placement of First Secondary Filter*—A secondary filter pack may be installed above the primary filter pack to prevent the intrusion of the bentonite grout seal into the primary filter pack (see Figs. 2 and 3). To be effective, measured and recorded volume of secondary filter material should be added to extend 1 to 2 ft (304 to 608 mm) above the primary filter pack. As with the primary filter, a secondary filter must not extend into an overlying hydrologic unit (see 8.4.1). The well designer should evaluate the need for this filter pack by considering the gradation of the primary filter pack, the hydraulic heads between adjacent units, and the potential for grout intrusion into the primary filter pack. The secondary filter material is poured into the annular space through a decontaminated, flush threaded, 1-in. (25-mm) minimum internal diameter tremie pipe lowered to within 3 ft (1.0 m) of the placement interval. Water from a source of known chemistry may be added to help move the filter pack into its proper location. The tremie pipe or weighed line inserted through the tremie pipe can be used to measure the top of the secondary filter pack as work progresses. The elevation, volume, and gradation of the

secondary filter pack is recorded on the well completion diagram.

8.6 *Installation of the Bentonite Seal*—A bentonite pellet or a slurry seal is placed in the annulus between the borehole and the riser pipe on top of the secondary or primary filter pack (see Figs. 2 and 3). This seal retards the movement of cement-based grout backfill into the primary or secondary filter packs. To be effective, the bentonite seal should extend above the filter packs approximately 3 to 5 ft (1.0 to 1.5 m)—depending on local conditions. The bentonite seal should be installed using a tremie pipe lowered to the top of the filter packs and slowly raised as the bentonite pellets or the slurry fill the annular space. Bentonite pellets may bridge and block the tremie pipe in deep wells. In these cases, pellets may be allowed to free-fall into the borehole. As a bentonite pellet seal is poured into the tremie pipe or allowed to free-fall into the borehole, a tamper or weighed line may be necessary to tamp pellets into place. If the seal is installed above the water level, water from a source of known chemistry would be added to allow proper hydration of the annular seal. The tremie pipe or a weighed line inserted through the tremie pipe can be used to measure the top of the bentonite seal as the work progresses. If a bentonite pellet seal is being constructed above the water level, approximately 5 gal (20 L) of water from a source of known chemistry can be poured into the annulus to ensure that the pellets hydrate. Sufficient time should be allowed for the bentonite pellet seal to hydrate or the slurry annular seal to expand prior to grouting the remaining annulus. The volume and elevation of the bentonite seal material should be measured and recorded on the well completion diagram.

8.7 *Final Secondary Filter Pack*—A 6-in. to 1-ft (152 to 304-mm) secondary filter may be placed above the bentonite seal in the same manner described in 8.5 (see Figs. 2 and 3). This secondary filter pack will provide a confining layer over the bentonite seal to limit the downward movement of cement-based grout backfill into the bentonite seal. The volume, elevation, and gradation of this final secondary filter pack should be documented on the well completion diagram.

8.8 *Grouting the Annular Space:*

8.8.1 *General*—Grouting procedures vary with the type of well design. The following procedures will apply to both single- and multi-cased monitoring wells. Paragraphs 8.8.2 and 8.8.3 detail those procedures unique to single- and multi-cased installations, respectively.

8.8.1.1 *Volume of Grout*—The volume and location of grout used to backfill the remaining annular space is recorded on the well completion diagram. An ample volume of grout should be premixed on site to compensate for unexpected losses. The use of alternate grout materials, including grouts containing gravel, may be necessary to control zones of high grout loss.

8.8.1.2 *Injection Procedures*—The grout backfill should be injected under pressure to reduce the chance of leaving voids in the grout, and to displace any liquids and drill cuttings that may remain in the annulus. Depending upon the well design, grouting may be accomplished using a pressure grouting technique or by gravity feed through a tremie pipe. With either method, grout is introduced in one continuous operation until full strength grout flows out at the ground surface without evidence of drill cuttings or fluid.

The grout should slope away from the riser or casing at the surface, but care should be taken not to create a grout mushroom that would be subjected to frost heave.

8.8.1.3 *Grout Setting and Curing*—The riser or casing or both should not be disturbed until the grout sets and cures for the amount of time necessary to prevent a break in the seal between the grout and riser or grout and casing or both. The amount of time required will vary with grout content and climatic conditions and should be documented on the well completion diagram.

8.8.2 *Specific Procedures for Single-Cased Wells*—Grouting should begin at a level directly above the final secondary filter pack (see Fig. 2). Grout should be injected using a tremie pipe equipped with a side discharge; this dissipates the fluid-pumping energy against the borehole wall and riser, reducing the potential for infiltration of grout into the primary filter pack. The tremie pipe should be kept full of grout from start to finish with the discharge end of the pipe completely submerged as it is slowly and continuously lifted. Approximately 5 to 10 ft (1.5 to 3.0 m) of tremie pipe should remained submerged until grouting is complete. For deep installations or where the joints or couplings of the selected riser cannot withstand the shear or collapse stress exerted by a full column of grout as it sets, a staged grouting procedure may be considered. If used, the temporary casing or hollow stem auger should be removed in increments immediately following each increment of grout installation and in advance of the time when the grout begins to set. If casing removal does not commence until grout injection is completed, then, after the casing is removed, additional grout may be periodically injected into the annular space to maintain a continuous column of grout up to the ground surface.

8.8.3 *Specific Procedures for Multi-Cased Wells*—If the outer casing of a multi-cased well cannot be driven to form a tight seal between the surrounding stratum (strata) and the casing, it should be installed in a predrilled borehole. After the borehole has penetrated not less than 2 ft (608 mm) of the first targeted confining stratum, the outer casing is lowered to the bottom of the boring and the annular space is filled with grout. Grouting may be accomplished using a pressure grouting method or gravity feed through a tremie pipe. Pressure grouting will require the use of a grout shoe or packer installed at the end of the outer casing to prevent grout from moving up into the casing. If a tremie pipe is used to inject grout into the annular space, it should be equipped with a side discharge. With each alternative, the grout must be allowed to cure and form a seal between the casing and the grout prior to advancing the hole to the next hydrologic unit. This procedure is repeated as necessary to advance the borehole to the desired depth. Upon reaching the final target depth, the riser and screen is set through the inner casing. Subsequent to the placement of the filter packs and bentonite seal, the remaining annular space is grouted as described in 8.8.2 (see Fig. 3).

NOTE 13—When using a packer, pressure may build up during grout injection and force grout up the sides of the packer and into the casing.

8.9 *Well Protection*—Well protection refers specifically to installations made at the ground surface to deter unauthorized entry to the monitoring well and to prevent surface water from entering the annulus.

8.9.1 *Protective Casing*—The protective casing should extend from below the frost line (3 to 5 ft [1.0 to 1.5 m]) below the grade depending on local conditions to slightly above the well casing tip. The protective casing should be initially placed before final set of the grout backfill. The protective casing should be sealed and immobilized in concrete placed around the outside of the protective casing above the set grout backfill. The casing should be positioned and stabilized in a position concentric with the riser (see Figs. 1 and 2). Sufficient clearance, usually 6 in. (152 mm) should be maintained between the lid of the protective casing and the top of the riser to accommodate sampling equipment. A ¼-in. (6.3-mm) diameter weep hole should be drilled in the casing 6 in. above the ground surface to permit water to drain out of the annular space. In cold climates, this hole will also prevent water freezing between the well protector and the well casing. Dry bentonite pellets, granules, or chips should then be placed in the annular space below ground level within the protective casing. Coarse sand or pea gravel or both is placed in the annular space above the dry bentonite pellets and above the weep hole to prevent entry of insects. All materials chosen should be documented on the well completion diagram. The monitoring well identification number should be clearly visible on the inside and outside of the lid of the protective casing.

8.9.2 *Completion of Surface Installation*—The well protection installation may be completed in one of three ways:

8.9.2.1 In areas subject to frost heave, place a soil or bentonite/sand layer adjacent to the protective casing sloped to direct water drainage away from the well.

8.9.2.2 In regions *not* subject to frost heave, a 4-in. (101-mm) thick concrete pad sloped to provide water drainage away from the well may be placed around the installation. Care must be taken not to lock the concrete pad onto the protective casing if subsidence of the surface may occur in the future.

8.9.2.3 Where monitoring well protection must be flushed with the ground, an internal cap should be fitted on top of the riser within the manhole or vault. This cap should be leak-proof so that if the vault or manhole should fill with water, the water will not enter the well casing. Ideally, the manhole cover cap should also be leak-proof.

8.9.3 *Additional Protection*—In areas where there is a high probability of damaging the well (high traffic, heavy equipment, poor visibility), it may be necessary to enhance the normal protection of the monitoring well through the use of posts, markers, signs, etc. The level of protection should meet the damage threat posed by the location of the well.

9. Well Development

9.1 *General*—The development serves to remove the finer grained material from the well screen and filter pack that may otherwise interfere with water quality analyses, restore the ground-water properties disturbed during the drilling process and to improve the hydraulic characteristics of the filter pack and hydraulic communication between the well and the hydrologic unit adjacent to the well screen. Methods of well development vary with the physical characteristics of hydrologic units in which the monitoring well is screened and with the drilling method used.

9.2 *Development Methods*—Methods of development

most often used include mechanical surging and bailing or pumping, over-pumping, air-lift pumping, and jetting. An important factor in any method is that the development work be stated slowly and gently and be increased in vigor as the well is developed. Most methods of well development require the application of sufficient energy to disturb the filter pack, thereby freeing the fines and allowing them to be drawn into the well. The coarser fractions then settle around and stabilize the screen. The well development method chosen should be documented on the well completion diagram.

NOTE 14—Any time an air compressor is used, it should be equipped with an oil air filter or oil trap to minimize the introduction of oil into the screen area. The presence of oil would impact the organic constituent concentrations of the water samples.

NOTE 15—Development procedures for wells completed in fine sand and silt strata should involve methods that are relatively gentle so that the strain material will not be incorporated into the filter pack. Vigorous surging for development can produce mixing of the fine strata and filter pack and produce turbid samples from the installation. Also, development methods should be carefully selected based upon the potential contaminant(s) present, quality of waste water generated, and requirements for containerization or treatment of waste water.

9.2.1 *Mechanical Surging*—In this method, water is forced to flow into and out of the well screen by operating a plunger (or surge block) or bailer up and down in the riser. A pump or bailer should then be used to remove the dislodged sediments following surging.

9.2.2 *Over Pumping*—With this method, the monitoring well is pumped at a rate considerably higher than it would be during normal operation. The fine-grain materials would be dislodged from the filter pack and surrounding strata influenced by the higher pumping rate. This method is usually conducted in conjunction with mechanical surging.

9.2.3 *Air Lift Pumping*—In this method, an air lift pump is operated by cycling the air pressure on and off for short periods of time. This operation will provide a surging action that will dislodge fine-grained particles. Applying a steady, low pressure will remove the fines that have been drawn into the well by the surging action. Efforts should be made (that is, through the use of a foot valve) to avoid pumping air into the filter pack and adjacent hydrologic unit because the air may lodge there and inhibit future sampling efforts and may alter ambient water chemistry. Furthermore, application of high air pressures should be avoided to prevent damage to small diameter PVC risers, screens, and filter packs.

9.2.4 *Well Jetting*—Another method of development involves jetting the well screen area with water while simultaneously air-lift pumping the well. However, the water added during this development procedure will alter the natural, ambient water quality and may be difficult to remove. Therefore, the water added should be obtained from a source of known chemistry. Water from the monitoring well being developed may also be used if the suspended sediments are first removed.

9.3 *Duration of Well Development*—Well development should begin after the monitoring well is completely installed and prior to water sampling. Development should be continued until representative water, free of the drilling fluids, cuttings, or other materials introduced during well construction is obtained. Representative water is assumed to have been obtained when pH, temperature, and specific conduc-

tivity readings stabilize and the water is visually clear of suspended solids. The minimum duration of well development should vary in accordance with the method used to develop the well. For example, surging and pumping the well may provide a stable, sediment-free sample in a matter of minutes; whereas, bailing the well may require several hours of continuous effort to obtain a clear sample. The duration of well development and the pH, temperature, and specific conductivity readings should be recorded on the well completion diagram.

9.4 *Well Recovery Test*—A well recovery test should be performed immediately after and in conjunction with well development. The well recovery test not only provides an indication of well performance but also provides data for determining the transmissivity of the screened hydrologic unit. Estimates of the hydraulic conductivity of the unit can then be determined. Readings should be taken at intervals suggested in the table below until the well has recovered to 90 % of its static water level.

NOTE 16—If a monitoring well does not recover sufficiently for sampling within a 24-h period and the well has been properly developed, the installation should not generally be used as a monitoring well for detecting or assessing low level organic constituents. The installation may, however, be used for long-term water level monitoring if measurements of shorter frequency water level changes are not required.

10. Installation Survey

10.1 *General*—The vertical and horizontal position of each monitoring well in the monitoring system should be surveyed and subsequently mapped by a licensed surveyor. The well location map should include the location of all monitoring wells in the system and their respective identification numbers, elevations of the top of riser position to be used as the reference point for water level measurements, and the elevations of the ground surface protective installations. The locations and elevations of all permanent benchmark(s) and pertinent boundary marker(s) located on-site or used in the survey should also be noted on the map.

10.2 *Water Level Measurement Reference*—The water level measurement reference point should be permanently marked, for instance, by cutting a V-notch into the top edge of the riser pipe. This reference point should be surveyed in reference to the nearest NGVD reference point.

10.3 *Location Coordinates*—The horizontal location of all monitoring wells (active or decommissioned) should be surveyed by reference to a standardized survey grid or by metes and bounds.

11. Monitoring Well Network Report

11.1 To demonstrate that the goals as set forth in Section 1, the Scope, have been met, a monitoring well network report should be prepared. This report should:

11.1.1 Locate the area investigated in terms pertinent to the project. This should include sketch maps or aerial photos

TABLE 2 Suggested Recording Intervals for Well Recovery Tests

Time Since Starting Test	Time Interval
0 to 15 min	1 min
15 to 50 min	5 min
50 to 100 min	10 min
100 to 300 min (5 h)	30 min
300 to 1440 min (24 h)	60 min

on which the exploratory borings, piezometers, sample areas, and monitoring wells are located, as well as topographic items relevant to the determination of the various soil and rock types, such as contours, streambeds, etc. Where feasible, include a geologic map and geologic cross sections of the area being investigated.

11.1.2 Include copies of all well boring test pits and exploratory borehole logs, initial and post-completion water levels, all laboratory test results, and all well completion diagrams.

11.1.3 Include the well installation survey.

11.1.4 Describe and relate the findings obtained in the initial reconnaissance and field investigation (Section 5) to the design and installation procedures selected (Sections 7 to 9) and the surveyed locations (Section 10).

11.1.5 This report should include a recommended decommission procedure that is consistent with the well construction and local regulatory requirements.

12. Keywords

12.1 aquifer; borehole drilling; geophysical exploration; ground water; monitoring well; site investigation

Standard Practice for
Monitoring Well Protection[1]

This standard is issued under the fixed designation D 5787; the number immediately following the designation indicates the year of original adoption or, in the case of revision, the year of last revision. A number in parentheses indicates the year of last reapproval. A superscript epsilon (ϵ) indicates an editorial change since the last revision or reapproval.

INTRODUCTION

This practice for monitoring well protection is provided to promote durable and reliable protection of installed monitoring wells against natural and man caused damage. The practices contained promote the development and planning of monitoring well protection during the design and installation stage.

1. Scope

1.1 This practice identifies design and construction considerations to be applied to monitoring wells for protection from natural and man caused damage or impacts.

1.2 The installation and development of a well is a costly and detailed activity with the goal of providing representative samples and data throughout the design life of the well. Damages to the well at the surface frequently result in loss of the well or changes in the data. This standard provides for access control so that tampering with the installation should be evident. The design and installation of appropriate surface protection will mitigate the likelihood of damage or loss.

1.3 This practice may be applied to other surface or subsurface monitoring device locations, such as piezometers, permeameters, temperature or moisture monitors, or seismic devices to provide protection.

2. Referenced Documents

2.1 *ASTM Standards:*
C 150 Specification for Portland Cement
C 294 Descriptive Nomenclature of Constituents of Natural Mineral Aggregates
D 5092 Design and Installation of Ground Water Monitoring Wells in Aquifers

3. Terminology

3.1 *Definitions:*

3.1.1 *barrier*—any device that physically prevents access or damage to an area.

3.1.2 *barrier markers*—plastic, or metal posts, often in bright colors, placed around a monitoring well to aid in identifying or locating the well.

3.1.3 *barrier posts*—steel pipe, typically from 4 to 12 inches in diameter and normally filled with concrete or grout that are placed around a well location to protect the well from physical damage, such as from vehicles.

3.1.4 *borehole*—a circular open or uncased subsurface hole created by drilling.

3.1.5 *casing*—pipe, finished in sections with either threaded connections or bevelled edges to be field welded, which is installed temporarily or permanently to counteract caving, to advance the borehole, or to isolate the zone being monitored, or a combination thereof.

3.1.6 *casing, protective*—a section of larger diameter pipe that is emplaced over the upper end of a smaller diameter monitoring well riser or casing to provide structural protection to the well and restrict unauthorized access into the well.

3.1.7 *riser*—the pipe extending from the well screen to or above the ground surface.

3.1.8 *sealed cap*—a sealable riser cap, normally gasketed or sealed, that is designed to prevent water or other substances from entering into, or out of the well riser.

3.1.9 *vented cap*—a cap with a small hole that is installed on top of the riser.

4. Significance and Use

4.1 An adequately designed and installed surface protection system will mitigate the consequences of naturally or man caused damages which could otherwise occur and result in either changes to the data, or complete loss of the monitoring well.

4.2 The extent of application of this practice may depend upon the importance of the monitoring data, cost of monitoring well replacement, expected or design life of the monitoring well, the presence or absence of potential risks, and setting or location of the well.

4.3 Monitoring well surface protection should be a part of the well design process, and installation of the protective system should be completed at the time of monitoring well installation and development.

4.4 Information determined at the time of installation of the protective system will form a baseline for future monitoring well inspection and maintenance. Additionally, elements of the protection system will satisfy some regulatory requirements such as for protection of near surface ground water and well identification.

5. Design Considerations

5.1 The design of a monitoring well protective system is like other design processes, where the input considerations are determined and the design output seeks to remedy or

---

[1] This practice is under the jurisdiction of ASTM Committee D-18 on Soil and Rock and is the direct responsibility of Subcommittee D18.21 on Ground Water and Vadose Zone Investigations.

Current edition approved Sept. 10, 1995. Published January 1996.

mitigate the negative possibilities, while taking advantage of the site characteristics.

5.2 The factors identified in this practice should be considered during the design of the monitoring well protective system. The final design should be included in the monitoring well design and installation documentation and be completed and verified during the final completion and development of the well.

5.3 In determining the level or degree of protection required, the costs and consequences, such as loss of data or replacement of the well, must be weighed against the probability of occurrence and the desired life of the well. For monitoring wells which will be used to obtain data over a short time period, the protection system may be minimal. For wells which are expected to be used for an indefinite period, are in a vulnerable location, and for which the costs of lost data could be high, the protective system should be extensive. Factors to consider and methods of mitigating them are presented in the following sections.

5.3.1 *Impact Damages*—Physical damages resulting from construction equipment, livestock, or vehicles striking the monitoring well casing frequently occur. Protective devices and approaches include:

5.3.1.1 Extra heavy protective casings with a reinforced concrete apron extending several feet around the casing may be an acceptable design in those areas where frost heave is not a problem. The principle behind this is to design the protective casing so that it will be able to withstand the impact of vehicles without damage to the riser within.

5.3.1.2 Barrier Posts placed in an array such that any anticipated vehicle can not pass between them to strike the protective casing. Barrier posts are typically filled with concrete and set in post holes several feet deep which are backfilled with concrete. Barrier posts typically extend from 3 to 5 feet above the ground surface. Barrier posts are frequently used in and around industrial or high vehicle traffic areas. Costs for installation can be substantial however they provide a high degree of protection for exposed wells. Cost of removal at decommissioning can also be substantial.

NOTE 1—Cattle frequently rub against above ground completions leading to damage of unprotected casings. Concrete filled posts or driven T-posts, wrapped with barbed wire, are frequently used.

5.3.1.3 Barrier Markers are relatively lightweight metal or often plastic posts which provide minimal impact resistance but which by their color, location, and height, warn individuals of the well presence. The use of barrier markers is effective in areas that are well protected from impact type damage by other features, such as surrounding structures or fences. They are relatively inexpensive to install.

5.3.1.4 *Signs*—An inexpensive means of identifying the presence of a monitoring well. Signs provide protection only by warning of the well presence. Signs may be required in some circumstances and appropriate in others. Wells known to contain hazardous, radioactive, or explosive compounds should be marked to warn sampling personnel of potential dangers. When a potential exists for water usage, signage indicating that the water is non-potable and is utilized strictly as a monitoring well, and not for any other purpose, may be appropriate. Disadvantages of signs are that they may be ignored, are often difficult to maintain, and may invite vandalism to the well.

5.3.1.5 Recessed or Subsurface casings may be used to mitigate impact damage by allowing the vehicles to pass over. Frequently used techniques include recessing the casing below ground level, using commercially available covers. These may take the form of valve pits or manholes, as examples. Advantages include both protecting the well while minimizing the interference to surface traffic, such as in parking lots or urban areas and screening the well from view. Using this technique, wells may be located in the most desired locations from a ground-water monitoring perspective. Disadvantages include the need to assure surface drainage does not enter the well riser, either by maintaining positive drainage or by using a sealed riser cap (or both). When the risk is from the influx of surface water, drains below the level of the riser should be installed. In extreme cases, such as in location with high ground-water levels or potential drainage from surrounding areas, automatic sump pumps may be required. Consideration should be given to the sampling personnel who will require adequate space to perform sampling, particularly in manhole situations. Additionally, personnel protection requirements from working in a confined space should be considered.

5.3.1.6 Fencing, such as commercial chainlink type fences may provide adequate protection in areas with light risk from vehicles, but where people or animals may interfere or affect the well. Advantages are relative minimal costs, ease of removal or opening. Disadvantages include maintenance, adequacy of protection from hard vehicle impacts, and visual and traffic interference.

5.3.2 *Vandalism*—Damage from vandals can take two forms, those which seek to damage or destroy the well itself, and those which intend to damage the data that the well may provide. Theft of sampling pumps, loss of access to the riser, plugging of the well with foreign debris, or injection of foreign materials or chemicals are potential results of vandalism.

5.3.2.1 Physical damage to the well can be minimized with many of the same techniques as used to protect the well

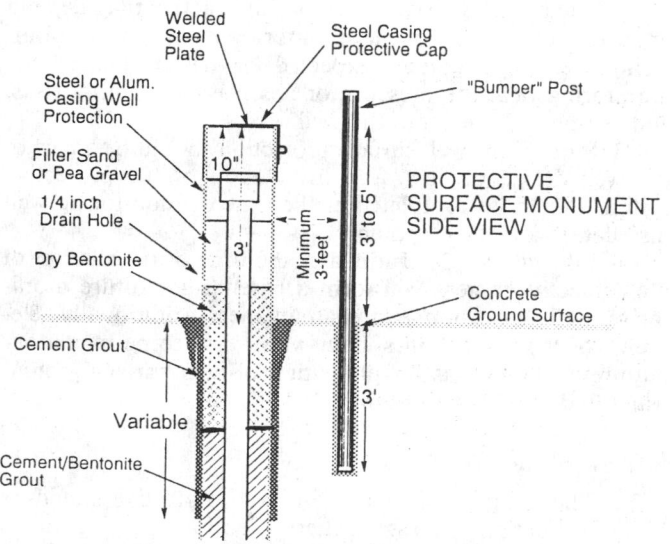

FIG. 1 Example of Protective Design

from impact damages. Generally two techniques can be used to protect a well from physical damage, one, by hiding or camouflaging the well, the other by constructing the surface protection of the well with multiple physical barriers. Hiding or camouflaging the well utilizes the philosophy that what can't be found can't be damaged. Camouflage techniques include enclosing the well in manholes or sumps, planting shrubs or vegetation to shield the well from view, enclosing the well in another structure, such as inside a raised planter or a small shed. Color characteristics of the above ground can be used to disguise the well or to assist in making it blend into the surroundings. Costs for camouflage can vary widely, but are generally minimal when included with other protections. Disadvantages are that if found, the well is still susceptible to damage by vandals, that damage may be undetected, and that sampling personnel not familiar with the well may have difficulty locating it.

5.3.2.2 Protection from vandalism is generally achieved by constructing multiple physical barriers. The first barrier should always include a rugged protective casing with a locking cap or lid. The lock quality can vary from relatively inexpensive and easily broken types to more costly high security type locks. Locks used on wells are subject to weather, dirt and deterioration. Frequently locks must be cut if not regularly maintained and the design and selection of the cap and lock should include this consideration. Construction of the hasps, locking lugs, or other mechanisms should be rugged, made of metal and welded to prevent access to the casing by prying, hammering or other typical vandalism. The casing should be heavy enough to resist penetration by bullets in areas where shooting may occur. A concrete apron or grout collar around the casing will provide mass to defeat attempts to pull the casing upwards, or sideways. Additional physical barriers should be added in consideration of the location and likelihood of vandalism. These include locked chainlink fences, use of barbed or concertina wire, concrete walls, or enclosure inside of buildings or other fenced or enclosed areas. When placed in below ground level structures, such as sumps or manholes, the access covers can be equipped with a lock. Access to keys should be controlled to prevent unauthorized use and entry.

5.3.2.3 Protection of the well and the data, (for example, ground-water level elevations), that the well will provide can be generally achieved by the physical barriers previously described. Detection of access to a well should also be considered. While not protecting the well and the sample data directly, it will be valuable in evaluating the data derived from the well samples. Sampling personnel should be alert and inspect the well and the protective devices for signs of vandalism. Foil or paper seals can be applied to the riser and cap at the end of each sampling to allow visual verification that the riser cap has not been disturbed between samplings. Seals are inexpensive and provide assurance of the well integrity and should be considered for use on all wells.

5.3.3 *Landslides*—Movement of the surface layers of soil due to seismic activity or other changes can result in lateral movement with the riser being bent or ultimately sheared. The primary protection against this type of damage is location. Whenever possible, the well should be located outside of the slide area. When relocation is not possible and

the moving soil layer is relatively thin, limited protection may be achieved by extending the protective casing several feet below the shear line. Additional protection may be gained by driving piling or posts through the surface layer and below the shear line to anchor the surface. Protection and maintenance of wells in slide areas can be expensive and may result in only delaying the loss of the well.

5.3.4 *Freeze Damage*—Freezing of the ground surrounding a well riser can result in heaving which can sever the riser resulting in the loss of the well. In areas where extended freezing temperatures are expected, the well protective casing should be constructed to minimize the possibility of damage. The protective casing should extend several feet below the frost line and the space between the well riser and the protective casing should be filled with a granular, free draining fine gravel down to the ground surface elevation and the bentonite below the gravel. Alternative designs in frost heave problem areas use a tapered concrete collar preferable to a 4-in. concrete pad. This will allow vertical movement of the protective casing and apron or collar without placing stress on the riser. The casing should have drain holes at several locations and heights to allow any water that may accumulate to drain freely. In areas where freezing occurs, the top of riser elevation and casing or concrete apron should be periodically checked to verify if movement has occurred. This will also allow for correction of ground water levels measured from the surface reference points.

5.3.5 *Floods*—Flood waters provide opportunities for physical damage to the well and to the integrity of the data that the well may provide. Wells located in low areas, floodplains, or areas where there is a potential for ponding of water should consider protection from physical damage and infiltration. Physical damage is generally mitigated by the protective devices described previously. Infiltration of water into the well from flooding can affect the data that will be derived from the well samples. It can also serve as a pathway for surface contaminants to reach the ground water. When flood or ponding surface waters are a possibility, the riser should be extended above an anticipated water level, such as the 100 year flood level. Additionally, a sealed cap capable of preventing leakage should be considered. If extending the riser is not possible or desirable, the riser cap should be sealable and capable of withstanding the anticipated head pressures that the site may experience.

NOTE 2—Where a sealed cap is used to prevent surface water from entering a well, it can also impede the vertical movement of water in the well casing, thus affecting the accuracy of the water level measurements.

5.3.6 *Elevation Changes*—Frequently wells are placed in locations where the surface elevations are expected to change, such as in landfills, borrow areas, or construction fill areas. Design of the well surface protection should take these changes into consideration. In areas where the surface elevation will be raised to a known level, the riser and casing should be extended to above the expected level. A second concrete collar and surface protection system can be put into place at that time. When the elevation change was not anticipated or known, the protection of the well and casing during the change should be carefully planned to protect the well riser from damage. Damage can occur due to equipment impacts, shearing of slopes, having the riser and casing buried, or falling over when surrounding soils are removed. Elevations before the change should be known. Elevations should be re-established at the completion and included in the well history records. Costs for protecting a well during elevation changes are minimal and less than the cost of replacing a well that was badly damaged or lost.

5.4 *Signage*—External signs provide a means of economical administrative control. Signs can protect the well from damage or accidental extraction by informing personnel of the well's purpose for providing ground-water data. Signs also have the negative aspect of informing vandals and others of the well's presence. The use of signs should consider the benefits and disadvantages of identifying the well. Signs may be required by regulation, to identify the owner, permit identifications, location identification, date of installation, and other information.

5.4.1 Internal signs or tagging may also be placed inside the protective casing, an attached locking cap, or other structures to provide information to the sampling personnel and to prevent inadvertent errors, such as sampling the wrong wells. Signs, or tags may also inform the samplers of relevant information or requirements, such as recording the total volume of water extracted, or other information such as the well elevation, chemical or other hazards, explosive potential, or required safety precautions. Signs or tags used for identification should be positively attached inside the individual well protective casing. When several well risers are clustered inside a single protective casing, the identification should be affixed to the riser, rather than the cap to prevent inadvertent misidentification.

5.5 *Decommissioning*—If the well is to be used only for a limited period, the ease or difficulty and costs of removing the surface protective devices should be considered along with the need to protect the well while in use. In cases where the surface or surrounding soils may become contaminated by materials extracted from the well, preventative protection should be considered, such as placing liners below the soil surface, and concrete cap, using raised berms, and protective coatings on the metal and concrete surfaces. These will prevent the spread of contamination, should it occur, and will minimize the amount of material that must be decontaminated or removed at the time of cleanup or decommissioning.

6. Records and Reports

6.1 *Well Design Considerations*—Surface protection, including signs, labeling, barriers, and other details should be included in the well design documentation. Other information, such as the initial survey elevations at the time of well completion and asbuilt configuration should also be included as baseline information that can be referred to during the life of the well and for use during the decommissioning. The design information should be updated as new information becomes available. Such information may include periodic elevation surveys, records or changes to the surface devices, changes in surrounding grade or land uses, and the like.

6.2 *Well Condition Checklists*—Checklists for use during the life of the well should be developed during the design stage to insure that the well protective devices are maintained throughout the life of the well. These checklists should be completed whenever the well is sampled or in a predefined inspection schedule. Needed repairs to the surface protective devices should be recorded and accomplished to maintain the well protection. Any damage, or suspected intrusions into the well should be recorded. This information may be useful during evaluation of the well sample data, particularly when unauthorized spiking or tampering is suspected.

6.3 *Repair/Maintenance History*—Any maintenance performed, including the costs, should be included in the well documentation.

7. Keywords

7.1 ground water; surface protection; well damage; well protection; well vandalism

Standard Guide for
Development of Ground-Water Monitoring Wells in Granular Aquifers[1]

This standard is issued under the fixed designation D 5521; the number immediately following the designation indicates the year of original adoption or, in the case of revision, the year of last revision. A number in parentheses indicates the year of last reapproval. A superscript epsilon (ϵ) indicates an editorial change since the last revision or reapproval.

1. Scope

1.1 This guide covers the development of screened wells installed for the purpose of obtaining representative ground-water information and water quality samples from granular aquifers, though the methods described herein could also be applied to wells used for other purposes. Other well-development methods that are used exclusively in open-borehole bedrock wells are not described in this guide.

1.2 The applications and limitations of the methods described in this guide are based on the assumption that the primary objective of the monitoring wells to which the methods are applied is to obtain representative water quality samples from aquifers. Screened monitoring wells developed using the methods described in this guide should yield relatively sediment-free samples from granular aquifer materials, ranging from gravels to silty sands. While many monitoring wells are considered "small-diameter" wells (that is, less than four inches in inside diameter), some of the techniques described in this guide will be more easily applied to large-diameter wells (that is, four-inches or greater in inside diameter).

1.3 The values stated in inch-pound units are to be regarded as standard. All other units in parentheses are provided for information only.

1.4 *This standard does not purport to address all of the safety concerns, if any, associated with its use. It is the responsibility of the user of this standard to establish appropriate safety and health practices and determine the applicability of regulatory limitations prior to use.*

2. Referenced Documents

2.1 *ASTM Standards:*
D 653 Terminology Relating to Soil, Rock, and Contained Fluids[2]
D 5088 Practice for Decontamination of Field Equipment Used at Nonradioactive Waste Sites[2]
D 5092 Practice for Design and Installation of Ground-Water Monitoring Wells in Aquifers[2]

3. Terminology

3.1 *Definitions:*
3.1.1 Many of the terms discussed in this guide are contained in Terminology D 653. The reader should refer to this for definitions of selected terms.

3.2 *Descriptions of Terms Specific to This Standard:*

3.2.1 *air entrapment*—trapping of air or other gas in pore spaces of the formation or filter pack during development with compressed air.

3.2.2 *air lift pump*—a device consisting of two pipes, with one (the air line) inside the other (the eductor pipe), used to withdraw water from a well. The lower ends of the pipes are submerged, and compressed air is delivered through the inner pipe to form a mixture of air and water. This mixture rises in the outer pipe to the surface because the specific gravity of this mixture is less than that of the water column.

3.2.3 *air line*—a small vertical air pipe used in air-lift pumping. It usually extends from the ground surface to near the submerged lower end of the eductor pipe. The length of the air line below the static water level is used in calculating the air pressure required to start air-lift pumping.

3.2.4 *annular seal*—material used to provide a seal between the borehole and the casing of a well. The annular seal should have a hydraulic conductivity less than that of the surrounding geologic materials and be resistant to chemical or physical deterioration.

3.2.5 *backwashing*—the reversal of water flow caused by the addition of water to a well that is designed to loosen bridges and facilitate the removal of fine-grained materials from the formation surrounding the borehole.

3.2.6 *bailer*—a long, narrow tubular device with an open top and a check valve at the bottom that is used to remove water and sediment from a borehole or well.

3.2.7 *bailing (development)*—a development technique using a bailer which is raised and lowered in the well to create a strong inward and outward movement of water from the formation to break sand bridges and to remove fine materials from the well.

3.2.8 *borehole wall*—the face of an open borehole.

3.2.9 *bridge*—an obstruction to fluid and sediment movement in the filter pack or formation adjacent to the well due to the arching of fine sand grains across pore spaces. This condition is caused by one-directional movement of water into the well during development, as might occur during overpumping.

3.2.10 *bridging (development)*—the creation of obstructions to fluid and sediment movement in filter pack or formation materials during well development.

3.2.11 *cable tool drilling*—a drilling technique in which a drill bit attached to the bottom of a weighted drill stem is raised and dropped to crush and grind formation materials. In unconsolidated formations, casing is usually driven as drilling proceeds to prevent collapse of noncohesive materials into the borehole.

[1] This guide is under the jurisdiction of ASTM Committee D-18 on Soil and Rock and is the direct responsibility of Subcommittee D18.21 on Ground Water and Vadose Zone Investigations.
Current edition approved March 15, 1994. Published September 1994.
[2] *Annual Book of ASTM Standards*, Vol 04.08.

3.2.12 *cathead*—a rotating power unit on a drilling rig employing either a plain spool or an automatic spool used to hoist drill pipe, casing, small tools, or other drilling equipment.

3.2.13 *centrifugal pump (submersible)*—a downhole pump consisting of a sealed electric motor that drives impellers through a rotating shaft and seal arrangement at high revolutions per minute.

3.2.14 *centrifugal pump (surface)*—a pump that moves a liquid by accelerating it radially outward from within a rotating impeller to a surrounding circular-shaped chamber.

3.2.15 *development*—see *well development*.

3.2.16 *drilling fluid*—a water- or air-based fluid used in the well drilling operation to remove cuttings from the borehole, to clean and cool the bit, to reduce friction between the drill string and the sides of the borehole and to hold the borehole open during the drilling operation.

3.2.17 *eductor pipe*—the vertical discharge pipe used in air-lift pumping, submerged at least one third but usually two thirds of its length below the pumping water level in the well.

3.2.18 *filter cake*—the solids from a drilling fluid that are deposited on the walls of a borehole in a geologic formation during the process of drilling. Also called mudcake.

3.2.19 *filter-packed well*—a well in which the natural formation materials adjacent to the well screen has been replaced by a filter pack material.

3.2.20 *formation damage*—reduction of formation hydraulic conductivity at the borehole wall caused by the drilling process. May consist of compaction, clay smearing, clogging of pores with drilling mud filtrate, or other drilling-related damage.

3.2.21 *hydraulic jetting*—a well-development method that employs a jetting tool with nozzles and a high-pressure pump to force water outwardly through the well screen, the filter pack, and sometimes into the adjacent geologic unit, for the purpose of dislodging fine sediment and correcting formation damage done during drilling.

3.2.22 *indicator parameters*—chemical parameters, including pH, specific conductance, temperature and dissolved oxygen content, which are used to determine when formation water is entering a monitoring well.

3.2.23 *jetting*—see *hydraulic jetting*.

3.2.24 *monitoring well*—a well that is constructed by one of a variety of techniques that may serve a variety of purposes: (*1*) extracting ground water for physical, chemical, or biological testing; (*2*) measuring water levels; (*3*) measuring formation hydraulic parameters; or (*4*) measuring formation fluid chemical or physical parameters.

3.2.25 *naturally developed well*—a well in which the formation materials collapse around the well screen, and fine formation materials are removed using standard development techniques.

3.2.26 *overpumping*—a well-development technique that involves pumping the well at a rate that exceeds the design capacity of the well.

3.2.27 *rawhiding*—starting and stopping a pump intermittently to produce rapid changes in the pressure head in the well.

3.2.28 *sandlocking*—refers to the accumulation of sand and other sediment on development tools while they are working in the well screen, resulting in the tools becoming lodged in the screen. Also refers to the accumulation of sand and other sediment in the impeller section of a submersible pump, resulting in the impellers binding.

3.2.29 *sloughing*—caving of formation materials into an unstabilized open borehole.

3.2.30 *spudding*—the operation, in cable-tool drilling, of drilling a collar hole and advancing a casing through overburden. Also a general term in rotary or diamond core drilling applied to drilling through overburden.

3.2.31 *sump*—a blank extension of easing beneath the well screen that provides a space for sediment brought into the well during development to accumulate.

3.2.32 *surge block*—a plunger-like tool consisting of disks of flexible material (for example, neoprene) sandwiched between rigid (for example, metal) disks that may be solid or valved, and that is used in well development. See *surging*.

3.2.33 *surging*—a well-development technique in which a surge block is alternately raised and lowered within the well casing or screen, or both, to create a strong inward and outward movement of water through the well screen.

3.2.34 *tool string*—the drill pipe or drill rod and all attached drilling or development tools used in the borehole or well.

3.2.35 *turbidity*—cloudiness in water due to suspended and colloidal material.

3.2.36 *well casing*—a durable pipe placed in a borehole to prevent the walls of the borehole from caving, and to seal off surface drainage or undesirable water, gas, or other fluids and prevent their entrance into the well.

3.2.37 *well development*—the act of repairing damage to the borehole caused by the drilling process and removing fine-grained materials or drilling fluids, or both, from formation materials so that natural hydraulic conditions are restored and well yields are enhanced.

3.2.38 *well screen*—a filtering device that allows ground water to flow freely into a well from the adjacent formation, while minimizing or eliminating the entrance of fine-grained material into the well.

4. Significance and Use

4.1 A properly designed, installed, and developed ground-water monitoring well, constructed in accordance with Practice D 5092 should provide the following: representative samples of ground water that can be analyzed to determine physical properties and water-quality parameters of the sample or potentiometric levels that are representative of the total hydraulic head of that portion of the aquifer screened by the well, or both. Such a well may also be utilized for conducting aquifer tests used for the purpose of determining the hydraulic properties of the geologic materials in which the well has been completed.

4.2 Well development is an important component of monitoring well completion. Monitoring wells installed in aquifers should be sufficiently developed to ensure that they serve their intended objectives. Well development methods vary with the physical characteristics of the geologic formation in which the monitoring well is screened, the construction details of the well, the drilling method used during the construction of the borehole in which the well is installed, and the quality of the water. The development method for

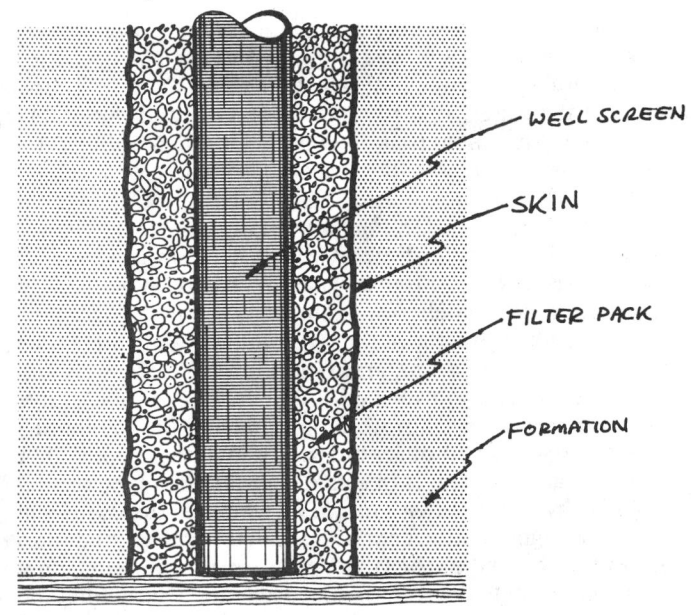

NOTE—One of the purposes of development is to rectify the damage done to the borehole wall during drilling, such as the "skin" of fine-grained materials that accumulates on the borehole wall during mud-rotary drilling.

FIG. 1 Example of Rectifying Damage Done During Drilling

each individual monitoring well should be selected from among the several methods described in this guide and should be employed by the well construction contractor or the person responsible for monitoring well completion.

4.3 The importance of well development in monitoring wells cannot be overestimated; all too often development is not performed or is carried out inadequately. Proper and careful well development will improve the ability of most monitoring wells to provide representative, unbiased chemical and hydraulic data. The additional time and money spent performing this important step in monitoring well completion will minimize the potential for damaging pumping equipment and in-situ sensors, and increase the probability that ground-water samples are representative of water contained in the monitored formation.

_____

[3] Figure adapted from _Ground Water and Wells,_ Second Edition, 1986.

5. Purposes of Monitoring Well Development

5.1 Monitoring wells are developed primarily for the following reasons:

5.1.1 To rectify damage done during drilling to the borehole wall and the adjacent formation (that is, clogging, smearing, or compaction of formation materials) that may result in a localized reduction in hydraulic conductivity of the formation near the borehole (see Fig. 1);

5.1.2 To remove fine-grained materials from the formation and filter pack (where applicable) that may result in the acquisition of turbid, sediment-laden samples;

5.1.3 To stabilize formation and artificial filter pack materials (where applicable) adjacent to the well screen (see Fig. 2);

5.1.4 To retrieve lost drilling fluid (if drilling fluid was used in the borehole installation process) that may alter the quality of water in the vicinity of the well and interfere with water quality analysis (see Fig. 3); and

5.1.5 To maximize well efficiency and hydraulic communication between the well and the adjacent formation to provide for the acquisition of representative ground-water samples and formation hydraulic test data.

6. Conducting a Monitoring Well-Development Program

6.1 _Well Development Process_—The well development process consists of three phases: predevelopment, preliminary development, and final development.

6.1.1 Predevelopment refers to techniques used to mitigate formation damage during well construction. This is particularly important when using direct or reverse rotary drilling systems that depend on drilling fluid to carry cuttings to the surface and support an open borehole. Control of drilling fluid properties, during the drilling operation and immediately prior to the installation of screen, casing, and filter pack, is very important.

6.1.2 Preliminary development takes place after the screen, casing, and filter pack have been installed. Methods used to accomplish this task include surging, bailing, hydraulic jetting, and air lifting. The primary purpose of this operation is to apply sufficient energy in the well to facilitate rectification of formation damage due to drilling; removal of fine-grained materials from the screen, filter pack, and formation; stabilization and consolidation of the filter pack; retrieval of drilling fluid (if used); and creation of an effective

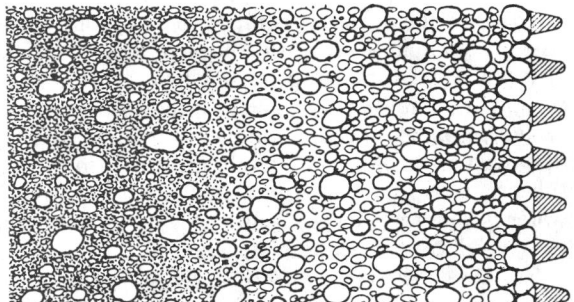
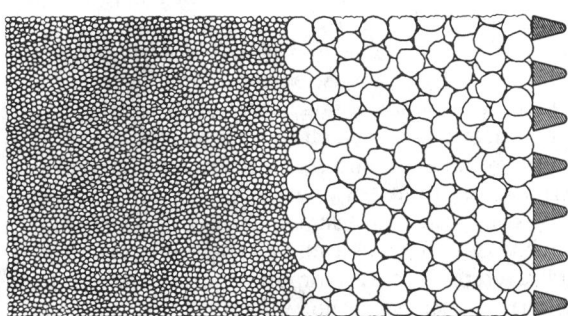

NOTE—After well development, formation materials in "naturally developed" wells (left) and filter packed wells (right) should be stabilized so that entry of fine-grained materials into the well is minimized and no settlement occurs.[3]

FIG. 2 Formation Materials in Wells

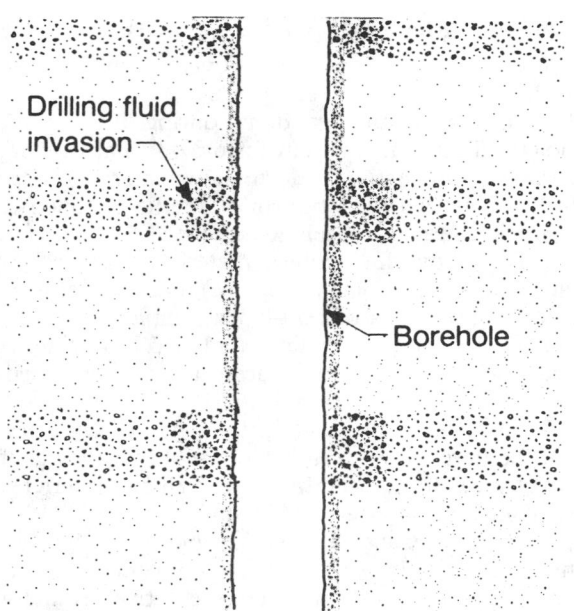

Drilling fluid invasion

Borehole

NOTE—When drilling with water-based drilling fluids, some fluid will infiltrate beyond the borehole into the most permeable zones. One of the purposes of development is to remove lost drilling fluid from the formation adjacent to the open interval of the well.[3]

FIG. 3 Removal of Lost Drilling Fluid

hydraulic interface between the filter pack and the formation.

6.1.3 During this phase of well development, the preferred technique is to gradually apply the selected method, increasing intensity as long as the well responds to treatment. Response generally is indicated by increased yields of water and sediment. Intensive development of a well that appears to be plugged should not be attempted because damage and destruction of the well may result.

6.1.4 Final development refers to procedures performed with a pump, such as pumping and surging, and backwashing. These techniques are used as the final step in achieving the objectives of well development. If preliminary development methods have been effective, the time required for final development should be relatively short. However, if the preliminary methods have not been successful, or if conditions preclude the use of the preliminary techniques listed, the final development phase should be continued until the development completion criteria (described below) are satisfied.

6.2 *Factors Affecting the Selection of a Well-Development Method*—A variety of factors must be considered in selecting the method(s) used for developing any given monitoring well; these include: the construction of the well (that is, material used for well casing and screen, type and open area of well screen, type of joint between casing sections, screen length and slot size, casing and screen diameter, whether or not a filter pack was used in the well and the thickness of the filter pack); characteristics and hydraulic conductivity of the formation materials adjacent to the well screen; water quality in the aquifer in which the well is installed (that is, whether or not it may be contaminated, requiring special safety or handling considerations, or both, such as containment or treatment upon removal from the well); consequences of

introducing foreign fluids (that is, air, water, or chemical solutions) into the well and aquifer; drilling method used during borehole installation; depth to static water level and height of the water column in the well; type and portability of available equipment (that is, whether or not a drilling rig is required); time available for development; and cost effectiveness of the method.

6.3 *Timing of Well Development: When and How Long to Develop*—The point in time at which a monitoring well is developed is a decision that is generally based on design and construction of the well. For example, if the well is installed with the intent of using natural formation material as the filter pack (that is, a "naturally developed" well), development is generally performed after the screen and casing have been installed and the formation material has collapsed against the screen (to at least 5 feet above the screen), but before the annular seal is installed. Because this type of well design is based on the assumption that well development will remove a significant fraction of the formation materials adjacent to the well screen (therefore causing some sloughing in the borehole), developing the well after installing the annular seal may result in portions of the annular seal collapsing into the vicinity of the well screen. On the other hand, properly designed and constructed filter-packed wells may be developed after the annular seal materials have been installed and given sufficient time to set or cure, because the well screen is designed to retain at least 90 % and preferably 99 % of filter pack materials and little or no sloughing should occur.

6.3.1 The duration of well development is based on the primary purpose(s) of the development process. For example, if the primary purpose for development is to remove drilling fluid lost to the formation during borehole installation, the time required for completion of development may be based on the time it takes to remove from the well some multiple of the estimated volume lost. If the primary purpose of development is to rectify damage done during drilling to the borehole wall and the adjacent formation, the time for development may be based on the response of the well to pumping. An improvement in recovery rate of the well indicates that the localized reduction in hydraulic conductivity has been effectively rectified by development. If the primary purpose of development is to remove fine-grained materials, development may continue until visibly clear water is discharged from the well, or until the turbidity of water removed from the well is at some specified level. These criteria may be difficult or impossible to satisfy in formations with a significant fraction of fine-grained material. Another criterion used for determining when development is complete is the stabilization of certain indicator parameters (that is, temperature, specific conductance, pH, redox potential, dissolved oxygen) that are easily measured in the field. While this criterion may be an indicator of when native formation water is being produced, it does not necessarily indicate that well development is complete.

6.4 *Decontamination of Well Development Equipment*—Any equipment or materials used to develop a monitoring well should be thoroughly cleaned in accordance with Practice D 5088. Cleaning should take place prior to the use of any equipment in any monitoring well, and between uses in either the same well or in other wells.

7. Limitations of Well Development

7.1 Well development should be applied with great care to wells installed in predominantly fine-grained formation materials (that is, in formations dominated by fine sand, silt or clay). If vigorous development is attempted in such wells, the turbidity of water removed from the well may actually increase many times over. In some fine-grained formation materials, no amount of development will measurably improve formation hydraulic conductivity or the hydraulic efficiency of the well.

7.2 While development methods which require the addition of a foreign fluid to a well may be applied to groundwater monitoring wells, such methods should be used with an understanding of the negative effects that added fluids may have on the ability of the well to yield representative ground-water quality samples. Only in very extreme or special cases should fluids other than clean water or filtered air be considered for use in a well during development. Fluids other than water, including deflocculating or dispersing agents (that is, polyphosphates), acids (that is, hydrochloric or hydrofluoric acid), surfactants, and disinfectants (that is, sodium hypochlorite), may produce severe and persistent chemical alterations of water quality in the immediate vicinity of the well. The use of chemicals for well development is not discussed further for these reasons.

7.2.1 Any water added to a monitoring well for the purpose of development should be of known and acceptable chemistry. The impact of added water on in situ water quality should be evaluated and, to the extent possible, this water should be removed by pumping after development is complete. One possible means of reducing potential problems related to the addition of water to the well is to obtain water-quality samples from the well only after natural ground-water flow in the aquifer has had time to flush the remnants of well-development fluids beyond the well. Another means may be to use water that has been taken from the formation itself (that is, water pumped from the formation either prior to or during development) for the development process.

7.3 Development methods using compressed air (that is, air-lift pumping) should be attempted only after great care has been taken to remove any compressor oil or other foreign substances from the air stream prior to introduction into the well. Air should not be forced into the formation or allowed to be released directly into the well without the use of a containment device (that is, an eductor pipe). The injection of air into the formation may cause air entrapment and result in a dramatic reduction in formation hydraulic conductivity. An uncontrolled release of air into the well may cause significant chemical changes in the water in the well and the adjacent formation.

7.4 Development methods that rely only on pumping ("passive" development), especially at low-flow rates, do not sufficiently stabilize formation or filter pack material and do not effectively remove fine-grained material or rectify formation damage done during drilling (see Fig. 4). Effective development action requires movement of water in both directions through the well screen openings (see Fig. 5). Although visibly clear water may eventually be discharged as a result of such pumping, any subsequent activity that agitates the water column in the well (that is, conducting a

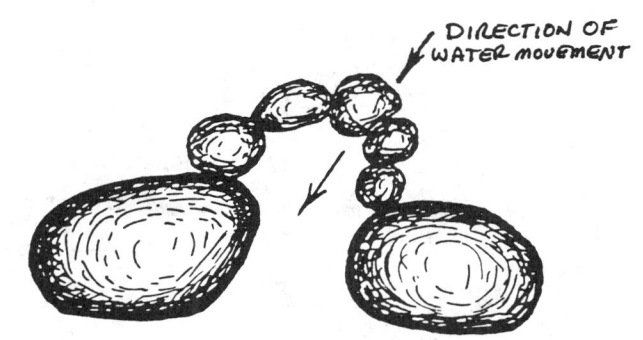

Note—Bridging in formation and filter pack materials is caused by movement of water in one direction only during well development.[3]

FIG. 4   Bridging in Formation and Filter Pack Materials

formation hydraulic test, purging prior to sampling, or sampling, especially with bailers) can cause considerable turbidity in the well.

7.5 Development should be applied very cautiously to wells that are known or suspected to be contaminated with hazardous chemical constituents, particularly constituents which pose a hazard through inhalation or dermal contact. Appropriate safety precautions should be taken to protect field personnel. It should be noted that contaminated water and sediment removed from monitoring wells during development may also have to be contained in drums, tanks, or other storage vessels until the water and sediment have been tested and evaluated to determine an appropriate disposal or treatment method. This could significantly increase the cost of development.

8. Methods and Processes Available for Monitoring Well Development

8.1 *General*—Of the various methods available for use in developing wells in general, mechanical surging, overpumping and backwashing, and high-velocity hydraulic jetting with pumping (or combinations of two or more of these methods) are best suited for use in developing ground-water monitoring wells. The method most appropriate for use in a given situation depends on a variety of factors discussed in 6.2. The user should evaluate the methods described herein and select the method that is most appropriate for the situation at hand.

8.2 *Mechanical Surging*—Mechanical surging is accomplished by using a close-fitting surge block (sometimes referred to as a surge plunger or swab) affixed to the end of a length of drill pipe, a solid rod, or a cable, operating like a piston in the well casing or screen. The up-and-down plunging action alternately forces water to flow into (on the upstroke) and out of (on the downstroke) the well, similar to a piston in a cylinder (see Fig. 6). The downstroke causes a backwash action to loosen bridges in the formation or filter pack and the upstroke then pulls dislodged fine-grained material into the well. This method is equally applicable to small-diameter and large-diameter wells and is the most effective method for small-diameter wells.

8.2.1 Several designs for surge blocks, including a solid surge block, a valved or vented surge block, a spring-loaded surge block, and a multiple-flange surge block (see Fig. 7) can

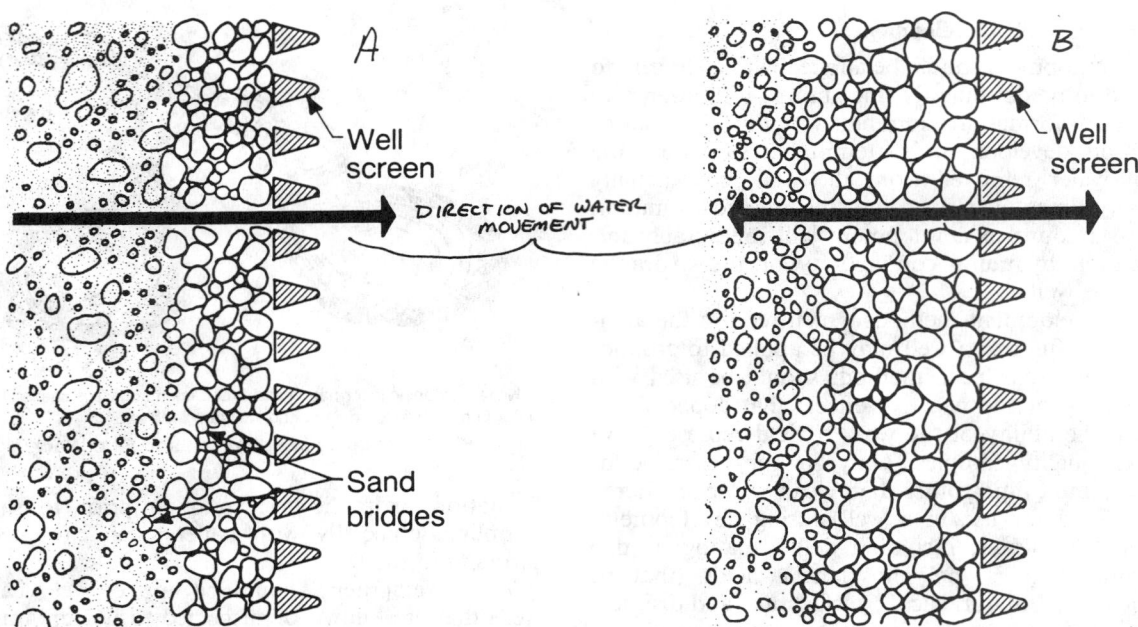

NOTE—Effective development action requires movement of water in both directions through the well screen openings. (*B*) Reversing flow helps to minimize bridging in the formation and filter pack (if used). (*A*) Movement of water in only one direction, as when overpumping the well, does not produce the proper development effect.[3]

FIG. 5 Movement of Water in Both Directions

be utilized. A heavy bailer or a pump (such as a gas-drive pump or an inertial lift pump) fitted with flexible disks similar to those on a surge block (see Fig. 8) may also be used to produce the surging action, but these are not as effective as a close-fitting surge block.

8.2.2 The proper procedure for mechanical surging is to bail or pump the well first to make sure that the well will yield water. If the screen is completely plugged and water does not enter the well upon bailing or pumping, surging should not be attempted, as the strong negative pressure created on the upstroke of the surge block may cause the screen to collapse. When it is determined that the well will yield water, the surge block is lowered until it is below the static water level but above the screen, and a relatively slow, gentle surging action is started. This surging action should allow any material blocking the screen to break up, go into suspension, and move into the well. The surge block should be operated with particular care if the formation above the screen consists mainly of fine sand, silt or clay which may slump into the screened interval. The water column should effectively transmit the action of the surge block to the screened section of the well. As water begins to move easily in and out of the well, the surge block is lowered (in steps) farther into the well and the speed (and, therefore, the force) of the surging movement is increased. If initial development is too vigorous, particularly in fine-grained formations, surging can harm a well rather than improve it. Because significant pressure differentials can occur during mechanical surging, great care must be taken to avoid damaging (that is, collapsing) the casing or screen by overzealous development.

8.2.3 In wells with short (that is, less than five feet) screens, it may not be necessary to operate the surge block within the screen to develop the entire screened interval; in wells with longer (that is, ten feet or more) screens, it may

prove more effective to operate the surge block within the screen to concentrate its action at various levels. Surging should always begin above the screen and move progressively downward to prevent the surge block from becoming sand locked and to prevent damage to the screen. The surge block should be lowered in intervals equal to the length of the stroke until the entire screen has been surged. If surging of long screened wells is done exclusively in the casing, especially in situations in which the formation adjacent to the screen is highly variable, surging may preferentially develop only the material adjacent to the top of the screen or the most permeable zones of material adjacent to the screen.

8.2.4 The force exerted on the formation depends in part on the length of the stroke and the vertical velocity of the surge block. The length of the stroke depends on the mechanism used to operate the surge block. For cable-tool rigs, that are ideally suited to the surging operation, the length of the stroke is determined by the spudding motion. For rigs using a cathead to surge, the length of the stroke can be varied by varying the length of time that the rope is tightened on the spool. For manual surging, the length of the stroke is generally limited to the range of motion of the operator's arms. Generally, a 2 to 3 ft (0.61 to 0.91 m) stroke is sufficient to achieve proper well development with mechanical surging.

8.2.5 The vertical velocity of the surge block depends on the weight exerted on the block and the retraction speed. Generally, a vertical velocity of between three and five feet per second (0.91 to 1.52 m/s) is most effective. On the downward stroke, the surge block assembly must be of sufficient weight to free-fall through the water column in the well and create a vigorous outward surge. The surge block assembly should not be permitted to fall out of plumb because of the hazard of the assembly falling against and possibly damaging the screen. When used with a cable-tool

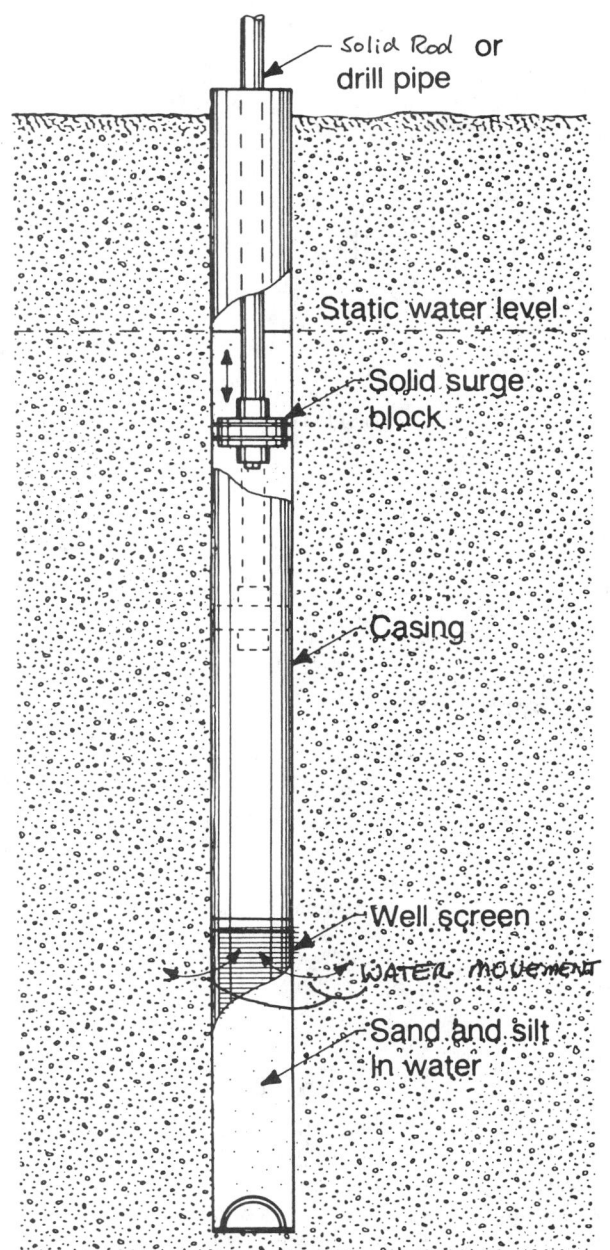

NOTE—For certain types of formations, a surge block is an effective tool for well development. On the downstroke, water is forced outward into the formation; water, silt, and fine sand are then pulled into the well screen during the upstroke.[3]

FIG. 6 Mechanical Surging

rig, a surge block may be weighted using the drilling tool string so it will fall at the desired rate, with the spudding motion controlled to vary the retraction speed. If a rotary rig is used, the weight on the surge block can be provided by drill pipe, with the retraction speed controlled by the rate of response of the hydraulic system. For rigs using a cathead to assist in surging, weight must be supplied to the surge block through either drill rod or the addition of weights above or

below the surge block; the speed of retraction can be varied by the tension of the rope on the spool. For manual surging, only a limited amount of weight can be added to the surge block assembly because of the difficulty of working by hand with a heavy tool string in a well. Down force may be applied more easily manually by pushing the surge block into the well. The speed of manual retraction is controlled by the rate at which the individual is capable of pulling the surge block assembly back out of the well.

NOTE 1—Manual surging is a very tiring and laborious procedure that commonly exhausts field personnel long before development is complete.

8.2.6 The effectiveness of mechanical surging is also governed in part by how tightly the surge block fits into the well casing or screen. If surging is to be performed only in the casing, the outside diameter of the surge block should be sized to be within ⅛ to ¼ in. (0.32 to 0.64 mm) of the inside diameter of the casing. (Care should be taken during casing assembly to ensure a smooth inner surface, especially at joints; shoddy assembly or irregular surfaces at casing joints could result in damage to the surge block or the casing, or both.) If surging is to be performed within the screen, care should be taken to avoid "sandlocking" the surge block. If sandlocking is a concern, the surge block can be sized to be slightly smaller. This reduction in size reduces the pressure exerted on the screen by allowing some water to flow past the surge block on both the upstroke and the downstroke, and it allows fine sediment to flow around the block rather than lodging between the block and the screen. A valved or vented surge block creates the same effect. Prior to the surging operation, the surge block or a dummy pipe of equivalent diameter and length should be tested to be certain that it will fit into the casing or screen, or both, without becoming lodged.

8.2.7 The first surging run should be very short so that the operator can judge the amount of sediment entering the screen. Surging should continue for several minutes, and the surge block should be removed from the well and the accumulation of sediment measured. Subsequent surging runs should be done over short to increasingly longer periods, keeping records of how quickly sediment enters the screen. When the sediment accumulated at the bottom of the screen begins to block off a portion of the screen, the sediment should be removed by pumping or bailing. Alternately, it may be possible to surge and pump simultaneously, using a specially designed surging tool (see Fig. 9) or a surge block affixed to an open pipe with a swivel for attachment to a pump and alternating surging without pumping with surging and pumping. Because development is more effective if the amount of sediment in the screen is kept to a minimum, accumulated sediment should be removed as often as possible. A "sump" or length of blank casing installed beneath the screen may help to keep the screen free of sediment. Surging should not be attempted when the screen is full of sediment because the force of surging may cause the casing above the screen to collapse. The rate and volume of sediment accumulation should be recorded to provide data on the progress of development. Surging and

---

[4] Figures *a* and *b* adapted from *Ground Water Wells*, Second Edition, 1986, and Figs. *c* and *d* adapted from *Handbook of Ground Water Development*, 1990.

[5] Figure adapted from *Handbook of Ground Water Development*, 1990.

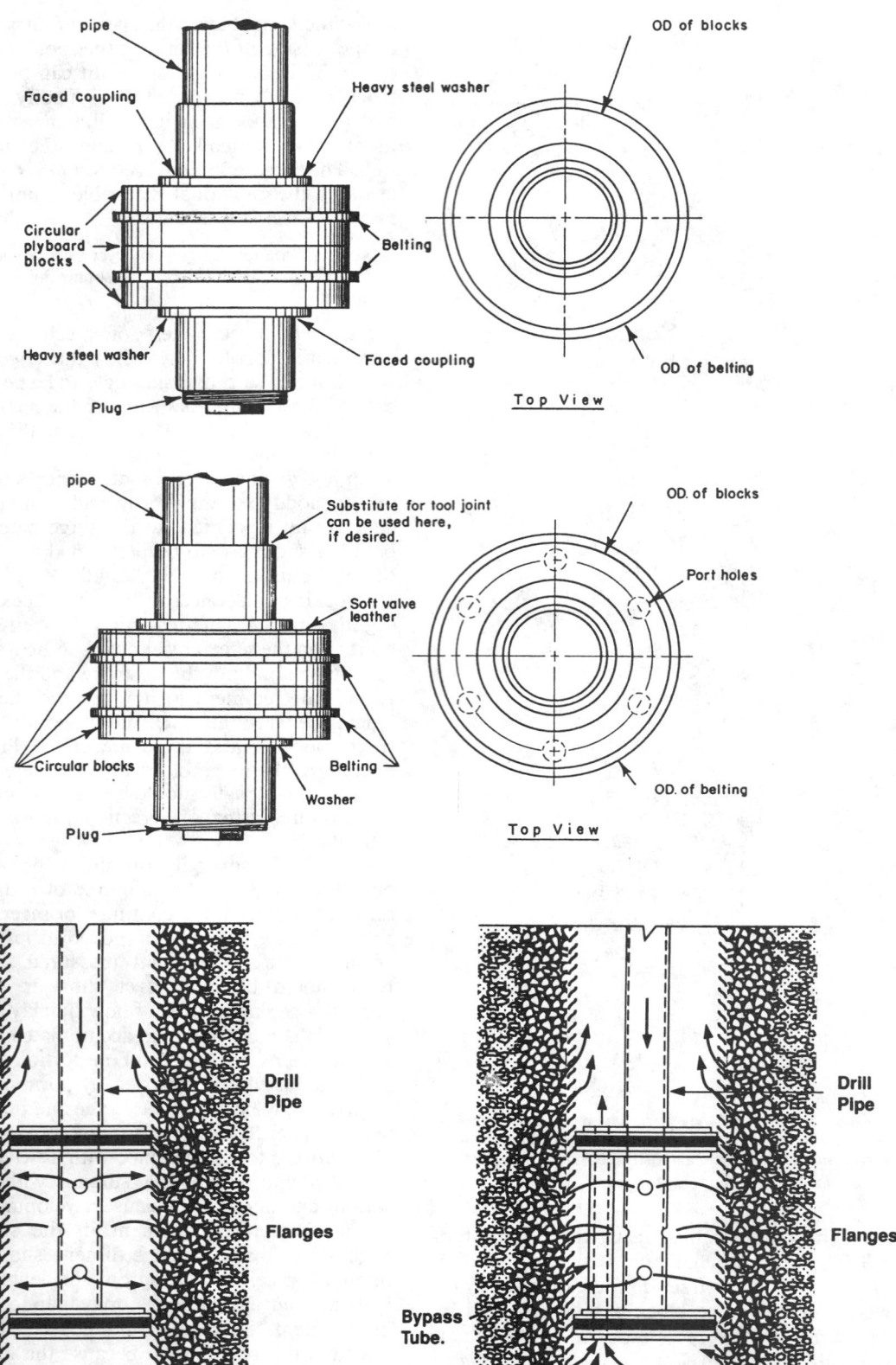

NOTE—Various configurations of surge blocks: (*a*) solid surge block; (*b*) valved surge block; (*c*) double-flanged surge block; and (*d*) valved double-flanged surge block.[3,4]

FIG. 7 Various Configurations of Surge Blocks

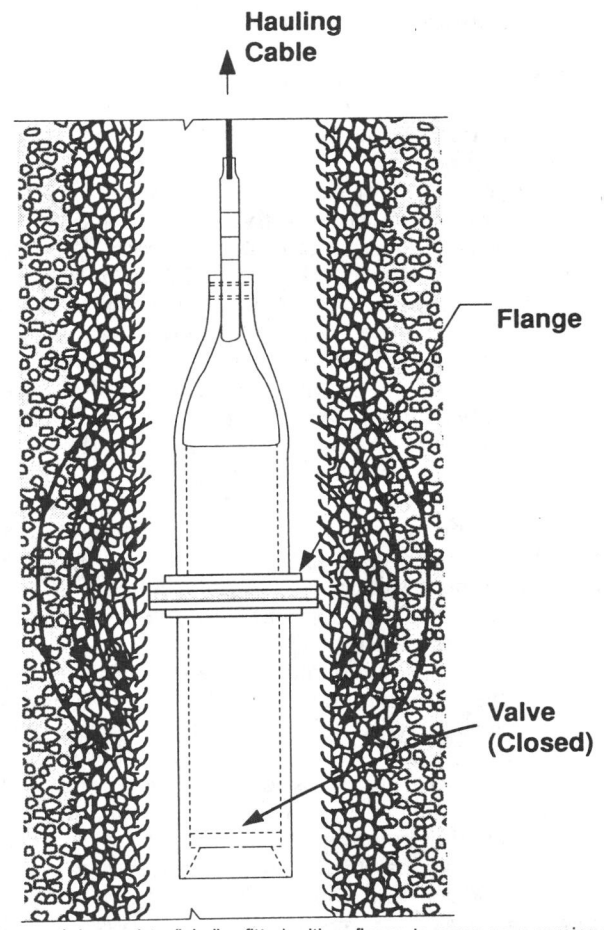

NOTE—A heavy (steel) bailer fitted with a flange to serve as a surging tool. Arrows indicate the direction of water movement during retraction of the bailer.[5]

FIG. 8 Bailer

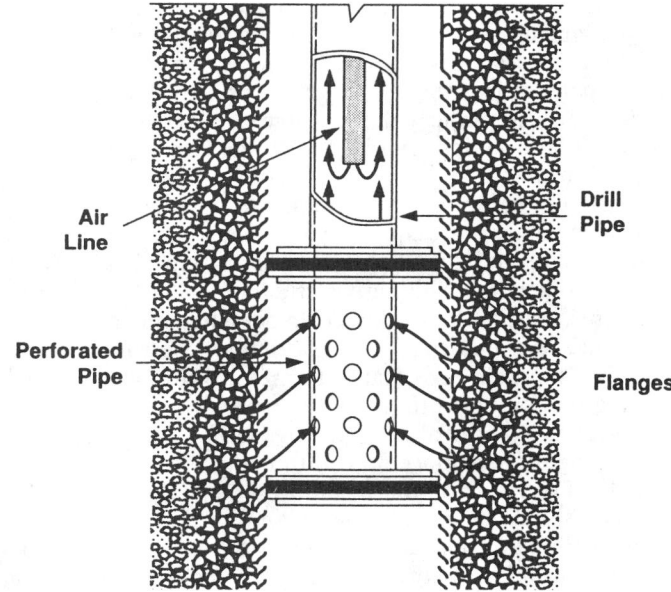

NOTE—A specially designed surging tool combining a double-flanged surge block with an air-lift pumping system to allow simultaneous surging and pumping. Arrows indicate the direction of water and sediment movement during pumping.[5]

FIG. 9 Specially Designed Surging Tool

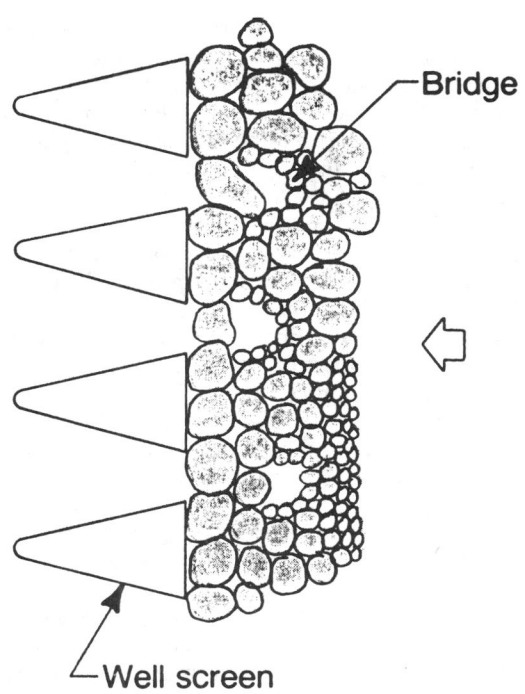

NOTE—During development by overpumping, sand grains can bridge openings because flow occurs in only one direction. Once the well is placed into service, agitation by normal pump cycling can break down the bridges, causing sand pumping.[3]

FIG. 10 Overpumping

cleaning should be continued until little or no sediment is measured after surging. The time required to properly surge a well depends on the character of the aquifer material and its apparent response to development, and may vary widely from well to well.

8.2.8 Manual mechanical surging can be accomplished effectively only in relatively shallow small-diameter wells (that is, wells less than 4 in. in diameter that are less than 50 ft (15.2.4 m) deep). Development of larger diameter or (10.16 cm) deeper wells will require either mechanical assistance (that is, by means of a block and tackle or pulley system used with a tripod assembled atop the well) or the use of a drilling rig or pump-pulling rig.

8.3 *Overpumping and Backwashing*—The simplest of method removing formation fines is by overpumping, or pumping at a higher rate then the well will be pumped when it is purged and sampled. Theoretically, increasing the drawdown to the lowest possible level results in increased flow velocities toward the well, resulting in the movement of fine-grained materials into the well (that is, any monitoring well that can be pumped sediment-free at a high pumping rate can then be pumped sediment-free at a lower rate). However, five important limitations to overpumping include: overpumping by itself will not adequately develop a well because water flow is in only one direction; it may cause bridges to form in the formation or filter pack, or both, resulting in only partial stabilization of these materials (see Fig. 10); it often requires the use of larger pumping equipment than will fit into the small-diameter casings that are used in many monitoring wells; it subjects the pump used in the operation to abrasion, excessive wear, and loss of efficiency, as well as the possibility of sandlocking; and it results in the production of potentially large volumes of water that may require containment or treatment. Furthermore, because the pump is normally set above the top of the well screen, most of the development that takes place during overpumping occurs in the zones of highest hydraulic conductivity closest to the top of the screen. For a given pumping rate, the longer the screen, the less development will take place in the lower part of the screen. After fine-grained material has been removed from the high hydraulic conductivity zones near the top of the screen, water entering the screen moves preferentially through these zones, leaving the rest of the well poorly developed. Overpumping may be effective in filter-packed wells or in non-stratified sands in which flow toward the well is more or less uniform. Overpumping is also useful for removing drilling fluid lost to the formation, but it is not an adequate development method if used alone. Overpumping is best used in combination with backwashing. Backwashing is the term applied to the method of well development in which water is added to the well to create a flow reversal.

8.3.1 A commonly used backwashing procedure is to pump water into the well in a sufficient volume to maintain a head greater than that in the formation. This requires a high-capacity water source. Water can be obtained by diverting part of the water pumped from the well during overpumping into a large tank. After the well is pumped, the stored water is pumped back down the well, either through the pump column or through a separate pipe. This method of backwashing should not be used in cases in which the water pumped from the well is potentially contaminated.

NOTE 2—There may be regulatory or legal constraints on returning potentially contaminated fluids to a monitoring well.

8.3.2 "Rawhiding," consists of starting and stopping a pump intermittently to produce rapid changes in the pressure head within the well. The alternate lifting and dropping of a column of water in the pump discharge pipe creates a surging action in the well. During a surge, the amount of water contained in the discharge pipe, combined with the well's natural recovery, may not be sufficient to cause the water level in the well to rise above the static water level. In this situation, supplemental water is generally added to the well when pumping is stopped for a surge. Judgment should be exercised concerning the duration of the backwash and the quantity of water added, because to some extent the success of this method depends on the frequency of flow reversals rather than the magnitude of water level change in the well.

8.3.3 Before beginning the rawhiding procedure, the pump should be started at reduced capacity and gradually increased to full capacity to minimize the danger of sandlocking the pump. When the pump discharge is clear of sediment, the pump is shut off and the water in the pump discharge or column pipe falls back into the well; the pump is then repeatedly started and stopped as rapidly as the power unit and starting equipment will permit. The pump used must not be equipped with a check valve or other backflow prevention device. To avoid damaging submersible pumps, the control box should be equipped with a starter lockout so the pump cannot be started when it is back spinning. During the rawhiding procedure, the well should be pumped occa-

sionally to remove the sediment that has been brought in by the surging action.

8.3.4 Some wells respond satisfactorily to rawhiding, but in some cases the surging effect is not vigorous enough to obtain optimum results. Also, rawhiding is very hard on pumping equipment.

8.3.5 Various types of pumps can be used to rawhide a monitoring well. To be effective for development by rawhiding, a pump must be capable of pumping at a rate of at least 5 to 10 gal/min (18.93 to 37.85 L/mm) (gpm). Surface centrifugal or diaphragm pumps are capable of these rates, but can only be used if the depth to static water level in the well is less than about 20 ft (6.1 m) below ground surface. Submersible centrifugal pumps with these pumping capacities are available for small-diameter monitoring wells, but the impellers in these pumps are relatively easily damaged by sand because of the very high speed at which they operate. Air-lift pumps can be constructed to easily fit into a small-diameter monitoring well and these pumps are capable of pumping in excess of 10 gpm. However, air-lift pumping and backwashing results in mixing of aerated water with water in the well and adjacent formation, temporarily altering groundwater quality.

8.4 *High-Velocity Hydraulic Jetting*—Where conditions permit, another effective method available for use in developing monitoring wells, is high-velocity hydraulic jetting. Because of the size of the equipment required, this method is more easily applied to wells of four inches or greater diameter. Development by high-velocity hydraulic jetting employs several horizontal jets of water operated from inside the well screen so that high-velocity streams of water exit through the screen and loosen fine-grained material and drilling mud residue from the formation (Fig. 11). The loosened material moves inside the well screen and can be removed from the well by concurrent pumping or by bailing. Jetting is particularly successful in developing highly stratified unconsolidated formations, consolidated bedrock wells, large-diameter wells, and naturally developed wells. Jetting can also be useful in filter-packed wells provided that filter pack material is carefully chosen to provide some, but not excessive, loss through the well screen, and that the filter pack is not too thick (that is, not greater than 2 or 3 in. (5.1 or 7.6 cm)).

8.4.1 The equipment required for jetting includes a jetting tool with two or more equally spaced nozzles; a high-pressure pump, hose, and connectors; a string of pipe; and a water tank or other high-volume water supply. The high-velocity jets direct water through the screen openings, and agitate and rearrange the particles of the formation surrounding the screen. The filter cake that is formed on the borehole wall during mud rotary drilling is broken down and dispersed, allowing easy removal by pumping. Jetting is also useful in correcting other damage done during drilling.

8.4.2 Figure 12 shows a jetting tool with four nozzles. Nozzles should be equally spaced around the circumference of the jetting tool, and generally range from 5/32- to 3/8-in. (0.4 to 0.95 cm). The jetting tool should be constructed so that the nozzle outlets or holes are as close to the inside diameter of the screen as practical (generally within 1/2 in (1.27 cm)), with the bottom end of the tool capped and the top end threaded so it can be screwed on to the lower end of the pipe

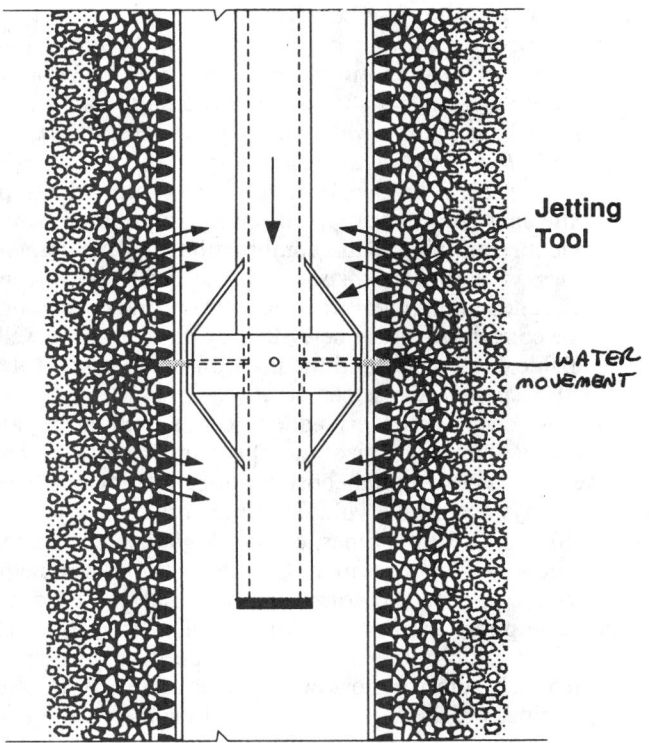

NOTE—Development by high-velocity hydraulic jetting employs a tool using several horizontal jets of water operated from inside the well screen. High-velocity streams of water are directed through the screen into the formation or filter pack to loosed fine-grained material and drilling mud residue and bring it into the well, where it can be pumped out.[5]

FIG. 11 High-Velocity Hydraulic Jetting

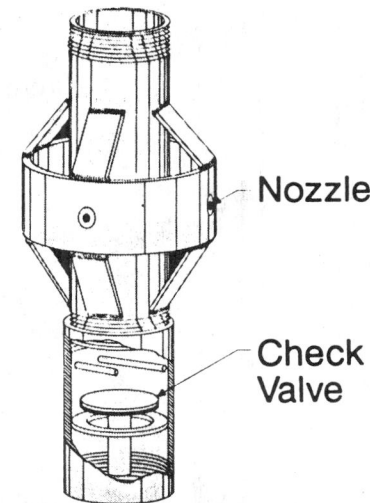

NOTE—A jetting tool with four nozzles spaced 90° apart around the circumference of the tool. The check valve at the bottom of the tool closes when water pressure is introduced from the surface, and opens to allow pumping through the drop pipe after jetting is completed.[3]

FIG. 12 Jetting Tool With Four Nozzles

string. If the outside diameter of the jetting tool is too small in relation to the inside diameter of the screen, much of the energy of the jet is dissipated by the turbulence created within the well.

8.4.3 Every effort should be made to limit sediment concentrations in the water used for jetting. Sediment-laden water that is recirculated through the jetting tool, causes erosion of the nozzle bores and may produce a pronounced pressure reduction at the nozzle face. Sediment in water used for jetting can also damage screens if the jets are directed at one area for a prolonged period of time.

8.4.4 A nozzle velocity of as low as 100 ft/s (30.5 m/s) can produce effective jetting results, though much better results can be expected when the nozzle velocities are between 150 and 300 ft/s (m/s). Velocities greater than 300 ft/s may not result in sufficient additional benefit to justify the additional cost. In general, 200 psi (45.7 and 91.4 kPa) at the nozzle is the preferred operating pressure for jetting in wells constructed with metallic screens. However, great care must be exercised in jetting screens constructed of PVC or other less-abrasion-resistant materials. All jetting of PVC screens should be done only with clean, sediment-free water to minimize abrasion, and the pressure used should not exceed 100 psi. In general, pressures higher than 100 psi are required when working in predominantly fine-grained formations or filter-packed wells.

8.4.5 The energy that agitates formation particles outside the well screen is a function of both the velocity and the diameter of the water stream. The energy of the jetting stream will depend on the capacity of the pump (maximum pressure output and flow rate at that pressure) and the orifice size of the discharge nozzle. The nozzle size should be the largest possible diameter that will maintain a minimum line pressure of 100 psi at the minimum anticipated flow rate of the pump. For a given nozzle size (water stream diameter), greater pressures will result in greater energy to penetrate formation materials. The pipe that is attached to the jetting tool should be large enough to minimize friction losses so that the velocity at the nozzle is as high as possible.

8.4.6 Although jetting can be accomplished without a drilling rig, it is usually done with a rig such as a rotary rig or another type of rig with rotating capability that is equipped with a mud pump that can supply the required down-hole pressure. When using a drilling rig, the jetting tool is attached to the lower end of the drill string and tool rotation is controlled by the rig. The jetting tool is placed near the bottom of the well screen and rotated slowly while being pulled upward at a rate of 5 to 15 min/ft (1.57 to 4.57 min/m) of screen. Material loosened from the formation is brought into the well by the turbulence created above and below the jets, and accumulates at the bottom of the screen (or in a sump if one is used) as the jetting tool is raised. This material must be removed concurrently or periodically. Slowly rotating and gradually raising the tool exposes the entire surface of the screen to the vigorous action of the jets. Several passes up and down the screen are made until the amount of additional material removed from the formation becomes negligible. To avoid erosion of the screen and to expedite development, the jetting tool should never be operated in a stationary position; it should always be slowly rotated and slowly raised or lowered.

8.4.7 In general, the effectiveness of the jetting process is controlled by the ratio of the filter-pack thickness to the jet radius, the velocity of the jet, the distance from the jet to its impact point on the screen, and the ratio of the hydraulic

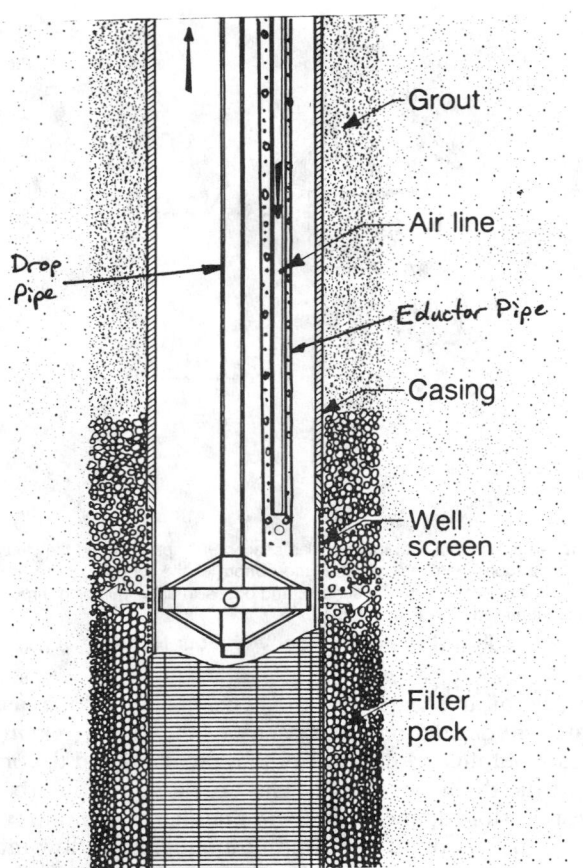

NOTE—In this configuration, the jetting tool and drop pipe are separate from an air-lift pumping system so that jetting and air-lift pumping can be done simultaneously.[3]

FIG. 13   Air-Lift Pump

conductivity of formation materials to that of the filter pack materials.

8.4.8 Optimal removal of sediment by jetting will depend on the time allotted to the process. Because the jetting energy can focus on only a small part of the formation at a given moment, more time may be necessary for jetting than for other methods that affect a larger portion of the formation. Less satisfactory results from jetting almost inevitably occur when not enough time is allowed for a thorough job.

8.5 *High-Velocity Hydraulic Jetting Combined with Simultaneous Pumping*—Although jetting is effective in dislodging material from the formation, maximum development efficiency is achieved when jetting is combined with simultaneous pumping. This combination of development techniques is particularly successful for wells in unconsolidated sands and gravels. It is not always practicable, but should be considered where permitted by the size of the well, the available equipment, and the position of the static water level with respect to the screen.

8.5.1 In jetting, water is added to the well at a rate governed by the nozzle size and the pump pressure. The volume of water simultaneously pumped from the well should always exceed the volume pumped in during jetting, by as much as 1.5 to 2 times, so that a gradient is created toward the well. The movement of water into the well helps to remove some of the formation material loosened by jetting. The pump then delivers the sediment from the well before it can settle in the screen. This procedure also serves to immediately remove most of the water added to the well during jetting, which may be a concern in some monitoring wells.

8.5.2 An added advantage of pumping is that water removed from the well provides a continuous supply that can be recirculated through the pump and jetting equipment after the sediment has settled out in a tank (to avoid damaging the high-pressure pump, jetting nozzles, or screen). This partially alleviates concerns regarding the quality of water added to the well during jetting, because the water used in this situation is water from the well being pumped. This water may be altered chemically during pumping, particularly if the equipment is not cleaned prior to use. Pumping while developing also permits an appraisal of the effectiveness of the jetting, based on the size and volume of the sediment collected at the bottom of the holding tank.

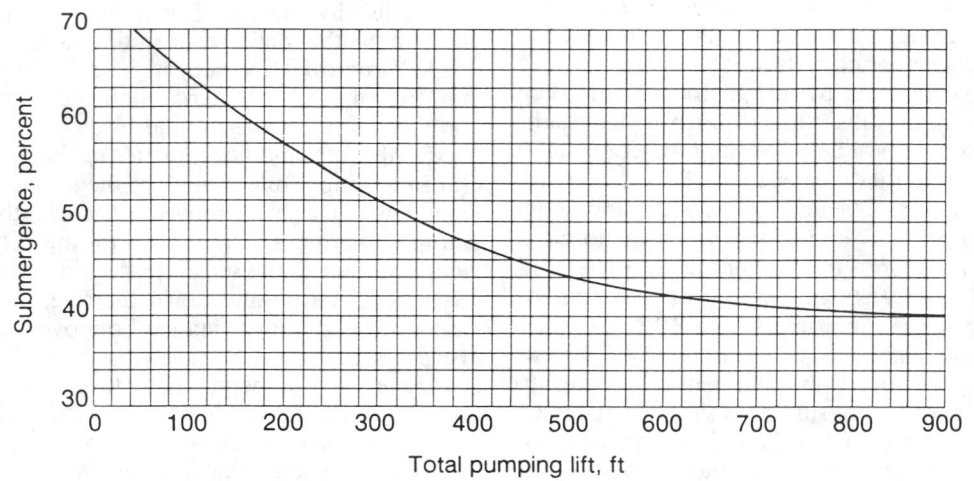

NOTE—Approximate percent pumping submergence for optimum air-lift efficiency. In general, development proceeds most efficiently when the discharge is maximized. Therefore, the submergence should always be as great as possible within practical limits.[3]

FIG. 14   Submerged Air Line

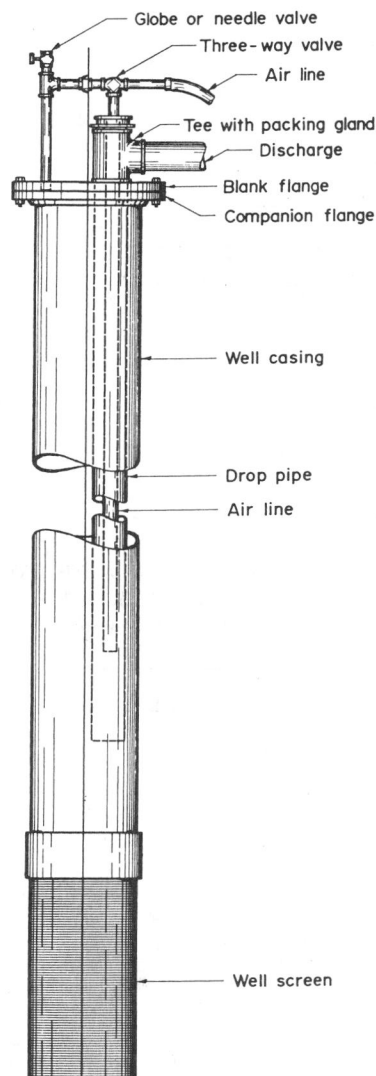

- Globe or needle valve
- Three-way valve
- Air line
- Tee with packing gland
- Discharge
- Blank flange
- Companion flange
- Well casing
- Drop pipe
- Air line
- Well screen

NOTE—See Footnote 3.

FIG. 15 A Typical Air-Lift Pumping Configuration

8.5.3 An air-lift pump (see Fig. 13) is the most common type of pump employed in this development operation, because of its small diameter and efficiency in removing sediment without damage to the pumping mechanism. Occasionally, the size of the air line and eductor pipe may have to be decreased somewhat so that the air line and eductor pipe will both fit into the annulus between the jetting pipe and the well casing. The air-lift system operates best when the air line is at least 40 % submerged (see Fig. 14) and when the bottom of the eductor pipe is placed just above the jetting tool allowing more suspended sediment to be pumped out of the well.

8.5.4 Compressed air used for development must be filtered to ensure that oil from the air compressor is not introduced into the well. In theory, this could be accomplished relatively easily with high-volume carbon filters. In

practice, however, the effectiveness of air/oil filters in removing entrained oil is questionable. It is difficult to ensure that the air used has a non-detectable oil content. Thus, development with compressed air may actually introduce oil into the well, albeit at very low levels. This oil may have the effect of confusing water quality analytical results. Alternately, oilless or oil-free compressors may be used to obviate the problems inherent with common oil-lubricated air compressors.

8.5.5 When air-lift pumping is impractical or undesirable, a submersible pump can be used, but this is only possible in large-diameter wells. Usually, the pump must be placed above the jetting tool so that the amount of sediment passing through the pump is minimized to avoid damaging the pump. Thus, the pump causes material temporarily placed in suspension by the jetting action to move into the well, but much of the sediment falls to the bottom. This sediment must be removed periodically during the jetting and pumping operation so that the entire screen can be effectively developed.

8.6 *Developing with Air*—Developing with air is not recommended for monitoring wells. Air development may force air into contact with the formation, which may alter the oxidation-reduction potential of the formation water and change the chemistry of the water in the vicinity of the well. The effects of this type of chemical disturbance may persist for several weeks or more after well development.

8.6.1 Blowing air into the well during the process of air development may cause air to become entrapped in the narrow slots of some monitoring well screens, the pores of the filter pack or formation materials immediately adjacent to the borehole. Entrapped air is difficult to remove and it may significantly reduce formation hydraulic conductivity and effectively reduce the amount of open area of the well screen.

8.6.2 In situations in which the well is installed in an area of contaminated groundwater, development with air may potentially result in the exposure of field personnel to hazardous materials. Though precautions can be taken to minimize personnel exposure, other development methods, which may also be more effective, can be used without such exposure problems and are, therefore, generally preferable.

8.6.3 For these reasons, compressed air alone should not be used to develop monitoring wells. Air-lift pumping, in which the air stream is not released directly into the well but instead is confined within an eductor pipe (see Fig. 15), should not produce the deleterious effects of air surging, and can thus be used in conjunction with other development processes.

9. Report

9.1 Keep records of the various operations performed during development and the progress and results of the development procedure. Report the following information: recorded:

9.1.1 Baseline turbidity (to be compared with turbidity levels obtained later in the life of the well);

9.1.2 Chemical quality, physical characteristics (that is, color, odor, etc.) volume, and sediment content of water added to the well or removed from the well;

 **D 5521**

WELL DEVELOPMENT DATA

SITE TYPE   SITE ID

| WELL | |

DEPTH TO BOTTOM   (INITIAL) _____   PROJECT NO. _____

                 (FINAL) _____   DATE(S) INSTALLED _____

STATIC WATER LEVEL   (INITIAL) _____   DATE(S) DEVELOPED _____

                 (FINAL) _____   PUMP (TYPE) _____

MEASURING POINT _____   (CAPACITY) _____

CASING I.D. _____   BAILER (TYPE) _____

Responsible Professional _____   (CAPACITY) _____

DRILLER _____

TIME	VOLUME OF WATER REMOVED	pH	SPECIFIC CONDUCTANCE AT 25°C	TEMP	SAND CONTENT	OTHER PHYSICAL CHARACTERISTICS (CLARITY, ODOR, PARTICULATES, COLOR)

FOPM 14 / SEPT 87

FIG. 16   Example of a Well Development Data Form

9.1.3 Amount and particle size of sediment that accumulates in the well between phases of development;

9.1.4 Type(s) of equipment used during development;

9.1.5 Well construction details;

9.1.6 Static water level prior to, during, and after development;

9.1.7 Time spent during development and

9.1.8 Other details pertinent to the development process and the objectives of development. A well development data form, such as that illustrated in Fig. 16, can be used for this purpose.

10. Keywords

10.1 aquifer; ground water; high-velocity; hydraulic jetting; jetting; mechanical surging; monitoring well; overpumping and backwashing; representative sample; well development

Standard Test Method for
Determining Subsurface Liquid Levels in a Borehole or Monitoring Well (Observation Well)[1]

This standard is issued under the fixed designation D 4750; the number immediately following the designation indicates the year of original adoption or, in the case of revision, the year of last revision. A number in parentheses indicates the year of last reapproval. A superscript epsilon (ε) indicates an editorial change since the last revision or reapproval.

[ε1] NOTE—Section 11 was added editorially in September 1993.

1. Scope

1.1 This test method describes the procedures for measuring the level of liquid in a borehole or well and determining the stabilized level of liquid in a borehole.

1.2 The test method applies to boreholes (cased or uncased) and monitoring wells (observation wells) that are vertical or sufficiently vertical so a flexible measuring device can be lowered into the hole.

1.3 Borehole liquid-level measurements obtained using this test method will not necessarily correspond to the level of the liquid in the vicinity of the borehole unless sufficient time has been allowed for the level to reach equilibrium position.

1.4 This test method generally is not applicable for the determination of pore-pressure changes due to changes in stress conditions of the earth material.

1.5 This test method is not applicable for the concurrent determination of multiple liquid levels in a borehole.

1.6 The values stated in inch-pound units are to be regarded as the standard.

1.7 *This standard does not purport to address all of the safety problems, if any, associated with its use. It is the responsibility of the user of this standard to establish appropriate safety and health practices and determine the applicability of regulatory limitations prior to use.*

2. Referenced Document

2.1 *ASTM Standard:*
D 653 Terminology Relating to Soil, Rock, and Contained Fluids[2]

3. Terminology

3.1 *Descriptions of Terms Specific to This Standard:*

3.1.1 *borehole*—a hole of circular cross-section made in soil or rock to ascertain the nature of the subsurface materials. Normally, a borehole is advanced using an auger, a drill, or casing with or without drilling fluid.

3.1.2 *earth material*—soil, bedrock, or fill.

3.1.3 *ground-water level*—the level of the water table surrounding a borehole or well. The ground-water level can

be represented as an elevation or as a depth below the ground surface.

3.1.4 *liquid level*—the level of liquid in a borehole or well at a particular time. The liquid level can be reported as an elevation or as a depth below the top of the land surface. If the liquid is ground water it is known as water level.

3.1.5 *monitoring well* (*observation well*)—a special well drilled in a selected location for observing parameters such as liquid level or pressure changes or for collecting liquid samples. The well may be cased or uncased, but if cased the casing should have openings to allow flow of borehole liquid into or out of the casing.

3.1.6 *stabilized borehole liquid level*—the borehole liquid level which remains essentially constant with time, that is, liquid does not flow into or out of the borehole.

3.1.7 *top of borehole*—the surface of the ground surrounding the borehole.

3.1.8 *water table* (*ground-water table*)—the surface of a ground-water body at which the water pressure equals atmospheric pressure. Earth material below the ground-water table is saturated with water.

3.2 *Definitions:*

3.2.1 For definitions of other terms used in this test method, see Terminology D 653.

4. Significance and Use

4.1 In geotechnical, hydrologic, and waste-management investigations, it is frequently desirable, or required, to obtain information concerning the presence of ground water or other liquids and the depths to the ground-water table or other liquid surface. Such investigations typically include drilling of exploratory boreholes, performing aquifer tests, and possibly completion as a monitoring or observation well. The opportunity exists to record the level of liquid in such boreholes or wells, as the boreholes are being advanced and after their completion.

4.2 Conceptually, a stabilized borehole liquid level reflects the pressure of ground water or other liquid in the earth material exposed along the sides of the borehole or well. Under suitable conditions, the borehole liquid level and the ground-water, or other liquid, level will be the same, and the former can be used to determine the latter. However, when earth materials are not exposed to a borehole, such as material which is sealed off with casing or drilling mud, the borehole water levels may not accurately reflect the ground-water level. Consequently, the user is cautioned that the liquid level in a borehole does not necessarily bear a

---

[1] This test method is under the jurisdiction of ASTM Committee D-18 on Soil and Rock and is the direct responsibility of Subcommittee D18.21 on Ground Water and Vadose Zone Investigations.
Current edition approved Nov. 27, 1987. Published January 1988.
[2] *Annual Book of ASTM Standards*, Vol 04.08.

relationship to the ground-water level at the site.

4.3 The user is cautioned that there are many factors which can influence borehole liquid levels and the interpretation of borehole liquid-level measurements. These factors are not described or discussed in this test method. The interpretation and application of borehole liquid-level information should be done by a trained specialist.

4.4 Installation of piezometers should be considered where complex ground-water conditions prevail or where changes in intergranular stress, other than those associated with fluctuation in water level, have occurred or are anticipated.

5. Apparatus

5.1 Apparatus conforming to one of the following shall be used for measuring borehole liquid levels:

5.1.1 *Weighted Measuring Tape*—A measuring tape with a weight attached to the end. The tape shall have graduations that can be read to the nearest 0.01 ft. The tape shall not stretch more than 0.05 % under normal use. Steel surveying tapes in lengths of 50, 100, 200, 300, and 500 ft (20, 30, 50 or 100 m) and widths of ¼ in. (6 mm) are commonly used. A black metal tape is better than a chromium-plated tape. Tapes are mounted on hand-cranked reels up to 500 ft (100 m) lengths. Mount a slender weight, made of lead, to the end of the tape to ensure plumbness and to permit some feel for obstructions. Attach the weight to the tape with wire strong enough to hold the weight but not as strong as the tape. This permits saving the tape in the event the weight becomes lodged in the well or borehole. The size of the weight shall be such that its displacement of water causes less than a 0.05-ft (15-mm) rise in the borehole water level, or a correction shall be made for the displacement. If the weight extends beyond the end of the tape, a length correction will be needed in measurement Procedure C (see 7.2.3).

5.1.2 *Electrical Measuring Device*—A cable or tape with electrical wire encased, equipped with a weighted sensing tip on one end and an electric meter at the other end. An electric circuit is completed when the tip contacts water; this is registered on the meter. The cable may be marked with graduations similar to a measuring tape (as described in 5.1.1).

5.1.3 *Other Measuring Devices*—A number of other recording and non-recording devices may be used. See Ref. (1) for more details.[3]

6. Calibration and Standardization

6.1 Calibrate measuring apparatus in accordance with the manufacturers' directions.

7. Procedure

7.1 Liquid-level measurements are made relative to a reference point. Establish and identify a reference point at or near the top of the borehole or a well casing. Determine and record the distance from the reference point to the top of the borehole (land surface). If the borehole liquid level is to be reported as an elevation, determine the elevation of the

---

[3] The boldface numbers in parentheses refer to the list of references at the end of this standard.

reference point or the top of borehole (land surface). Three alternative measurement procedures (A, B, and C) are described.

NOTE 1—In general, Procedure A allows for greater accuracy than B or C, and B allows for greater accuracy than C; other procedures have a variety of accuracies that must be determined from the referenced literature (2–5).

7.2 *Procedure A—Measuring Tape:*

7.2.1 Chalk the lower few feet of tape by drawing the tape across a piece of colored carpenter's chalk.

7.2.2 Lower a weighted measuring tape slowly into the borehole or well until the liquid surface is penetrated. Observe and record the reading on the tape at the reference point. Withdraw the tape from the borehole and observe the lower end of the tape. The demarcation between the wetted and unwetted portions of the chalked tape should be apparent. Observe and record the reading on the tape at that point. The difference between the two readings is the depth from the reference point to the liquid level.

NOTE 2—Submergence of the weight and tape may temporarily cause a liquid-level rise in wells or boreholes having very small diameters. This effect can be significant if the well is in materials of very low hydraulic conductivity.

NOTE 3—Under dry surface conditions, it may be desirable to pull the tape from the well or borehole by hand, being careful not to allow it to become kinked, and reading the liquid mark before rewinding the tape onto the reel. In this way, the liquid mark on the chalked part of the tape is rapidly brought to the surface before the wetted part of the tape dries. In cold regions, rapid withdrawal of the tape from the well is necessary before the wet part freezes and becomes difficult to read. The tape must be protected if rain is falling during measurements.

NOTE 4—In some pumped wells, or in contaminated wells, a layer of oil may float on the water. If the oil layer is only a foot or less thick, read the tape at the top of the oil mark and use this reading for the water-level measurement. The measurement will not be greatly in error because the level of the oil surface in this case will differ only slightly from the level of the water surface that would be measured if no oil was present. If several feet of oil are present in the well, or if it is necessary to know the thickness of the oil layer, a water-detector paste for detecting water in oil and gasoline storage tanks is available commercially. The paste is applied to the lower end of the tape that is submerged in the well. It will show the top of the oil as a wet line and the top of the water as a distinct color change.

7.2.3 As a standard of good practice, the observer should make two measurements. If two measurements of static liquid level made within a few minutes do not agree within about 0.01 or 0.02 ft (generally regarded as the practical limit of precision) in boreholes or wells having a depth to liquid of less than a couple of hundred feet, continue to measure until the reason for the lack of agreement is determined or until the results are shown to be reliable. Where water is dripping into the hole or covering its wall, it may be impossible to get a good water mark on the chalked tape.

7.2.4 After each well measurement, in areas where polluted liquids or ground water is suspected, decontaminate that part of the tape measure that was wetted to avoid contamination of other wells.

7.3 *Procedure B—Electrical Measuring Device:*

7.3.1 Check proper operation of the instrument by inserting the tip into water and noting if the contact between the tip and the water surface is registered clearly.

NOTE 5—In pumped wells having a layer of oil floating on the water, the electric tape will not respond to the oil surface and, thus, the liquid

BOREHOLE OR WELL SCHEDULE FORM

Date ...., 19...... Field No. ....

Record by .... Office No. ....

Source of data ....

1. *Location:* State .... County ....

   Map ....

   .... ¼ .... ¼ sec. .... T .... $\frac{N}{S}$ R .... $\frac{E}{W}$

2. *Owner:* .... Address ....

   Tenant .... Address ....

   Driller .... Address ....

3. *Topography* ....

4. *Elevation* .... ft. $\frac{above}{below}$ ....

5. *Type:* Dug, drilled, driven, bored, jetted .... 19......

6. *Depth:* Rept. .... ft. Meas. .... ft.

7. *Casing:* Diam. .... in., to .... in., Type ....

   Depth .... ft., Finish ....

8. *Chief Aquifer* .... From .... ft. to .... ft.

   Others ....

9. *Water level* .... ft. $\frac{rept.}{meas.}$ .... 19...... $\frac{above}{below}$ ....

   .... which is .... ft. $\frac{above}{below}$ surface

10. *Pump:* Type .... Capacity .... G. M.

    Power: Kind .... Horsepower ....

11. *Yield:* Flow .... G. M., Pump .... G. M., Meas., Rept. Est.

    Drawdown .... ft. after .... hours pumping .... G. M.

12. *Use:* Dom., Stock, PS., RR., Ind., Irr., Obs. ....

    Adequacy, permanence ....

13. *Quality* .... Temp .... °F.

    Taste, odor, color .... Sample $\frac{Yes}{No}$ ....

    Unfit for ....

14. *Remarks:* (Log, Analyses, etc.) ....

    ....

    ....

FIG. 1   Example of a Borehole or Well Schedule Form

LIQUID LEVEL MEASUREMENT FORM

Field No. ....

Owner .... Office No. ....

Location .... Project ....

Measuring Point ....

Elevation of Measuring Point ....

Date	Hour	Depth to Water	Elev. of Water Surface	Meas. by	Remarks (Nearby wells pumping, etc.)

FIG. 2   Example of a Liquid Level Measurement Form

level determined will be different than would be determined by a steel tape. The difference depends on how much oil is floating on the water. A miniature float-driven switch can be put on a two-conductor electric tape that permits detection of the surface of the uppermost fluid.

7.3.2 Dry the tip. Slowly lower the tip into the borehole or well until the meter indicates that the tip has contacted the surface of the liquid.

7.3.3 For devices with measurement graduations on the cable, note the reading at the reference point. This is the liquid-level depth below the reference point of the borehole or well.

7.3.4 For measuring devices without graduations on the cable, mark the cable at the reference point. Withdraw the cable from the borehole or well. Stretch out the cable and measure and record the distance between the tip and the mark on the cable by use of a tape. This distance is the liquid-level depth below the reference point.

7.3.5 A second or third check reading should be taken before withdrawing the electric tape from the borehole or well.

7.3.6 Decontaminate the submerged end of the electric tape or cable after measurements in each well.

NOTE 6—The length of the electric line should be checked by measuring with a steel tape after the line has been used for a long time or after it has been pulled hard in attempting to free the line. Some electric

lines, especially the single line wire, are subject to considerable permanent stretch. In addition, because the probe is usually larger in diameter than the wire, the probe can become lodged in a well. Sometimes the probe can be attached by twisting the wires together by hand and using only enough electrical tape to support the weight of the probe. In this manner, the point of probe attachment is the weakest point of the entire line. Should the probe become "hung in the hole," the line may be pulled and breakage will occur at the probe attachment point, allowing the line to be withdrawn.

7.4 *Procedure C—Measuring Tape and Sounding Weight:*

7.4.1 Lower a weighted measuring tape into the borehole or well until the liquid surface is reached. This is indicated by an audible splash and a noticeable decrease in the downward force on the tape. Observe and note the reading on the tape at the reference point. Repeat this process until the readings are consistent to the accuracy desired. Record the result as the liquid-level depth below the reference point.

NOTE 7—The splash can be made more audible by using a "plopper," a lead weight with a concave bottom surface.

7.4.2 If the liquid level is deep, or if the measuring tape adheres to the side of the borehole, or for other reasons, it may not be possible to detect the liquid surface using this method. If so, use Procedure A or Procedure B.

8. Determination of a Stabilized Liquid Level

8.1 As liquid flows into or out of the borehole or well, the

SITE NO.

BOREHOLE OR WELL SCHEDULE FORM

Recorded by _____

Date _____

Check One _____ English _____ Metric Units

GENERAL SITE DATA (0)

Site Ident No | | | | | | | | | | | |

RG Number `R - 0 ·`

Transaction `T · A D M V ·`
add delete modify verified

Site Type `2 · C D E H I M Ø P S T W X ·`
collector, drain, engine, sink- connector, multiple, outcrop, pond, spring, tunne- well, test
tion, hole, well, or shaft, hole

Data Reliability `3 · C U ·`
field checked, unchecked

Reporting Agency `4 · | | |`

Project No. `5 · | | | | |`

District `6 · | | ·`

State `7 · | | ·`

County (or town) _____ `8 · | ·`

Latitude `9 · | | | | ·`
deg min sec

Longitude `10 · | | | | ·`
deg min sec

Lat-Long Accuracy `11 · S F T M ·`
sec. 5 sec. 10 sec. Min

Local Number `12 · | | | | | | | | | | | |`

Land Net Loc `13 · S T R ·`
1/4 1/4 1/4 section township range mns

Location Map `14 · | | | | | | | ·`

Scale `15 · | | | ·`

Altitude `16 · | | | | ·`

Method of Measurement `17 · A L M ·`
altimeter, level, map

Accuracy `18 · | | ·`

Topo Setting `19 · A B C D E F G H K L M Ø P S T U V W ·`
alluv playa, stream, depres dunes, flat, flood hill sink swamp mangrove off- ped- hill- terrace, undulating, valley, upland
al fan, channel sion, plain, top, swamp shore, ment side, flat draw

Hydrologic Unit (OWDC) `20 · | | ·`

Use of Site `23 · A C D E G H M Ø P R S T U W X Z ·`
anadc, standby, drain, gen seismic, heat, mine, observa oil or recharge, repress, test, unused, with- waste, destroyed
emer thermal, reser, tion gas drawal,
supply

Secondary Site Use `301 · | ·`

Tertiary Site Use `302 · | |`

Use of Water `24 · A B C D E F H I J K M N P O R S T U Y Z ·`
air bot com power, fire, do irri industrial mini medi indus public aqua recreation, stock, institution, unused, desa other
cond, tling, mercial, water, mestic, gation (cooling) cinal, trial, supply culture

Secondary Water Use `25 · | ·`

Tertiary Use of Water `26 · | ·`

Depth of Hole `27 · | | | · |`

Depth of Well `28 · | | | · |`

Source of Depth Data `29 · | ·`

Water Level `30 · | | | · | |`

Data Measured `31 · | / | / | |`
month day year

Source `33 · | ·`

Method of Measurement `34 · A B C E G H L M N R S T V Z ·`
airline, analog, calibrated, estimated, pressure, calibrated, geophysical, manometer, non-rec, reported, steel, electric, calibrated, other
airline gage pressure gage logs gage tape tape electric tape

Site Status `37 · D E F G H I J N Ø P R S T U V W X Z ·`
dry, recently, flowing, nearby, nearby, injector, inj site, meas, obstruction, pumping recently, nearby, nearby, foreign well, surface water other
flowing flowing recently site monitor discon. pumped pumping recently substance destroyed effects
flowing pumped

Source of Geohydrologic Data `36 · | ·`

Pump Used `35 · | ·`
no

Date of First Construction Completion `21 · | / | / | |`
month day year

OWNER IDENTIFICATION (1)

`R · 158 ·`   `T · A D M ·`
add, delete, modify

Date of Ownership `159 · | / | / | ·`
month day year

Name: Last `161 = | | | | | | | | | | | ·`

First `162 · | | | | | | | | | | | ·`

Middle Initial `163 · | ·`

OTHER SITE IDENTIFICATION NUMBERS (1)

`R · 189 ·`   `T · A D M ·`
add, delete, modify

Ident `190 = | | | | | | |`

Assigner `191 · | | | |`

New Card Same R & T

Ident `190 = | | | | | | |`

Assigner `191 · | | | |`

SITE VISIT DATA (1)

`R · 186 ·`   `T · A D M ·`
add, delete, modify

Date of Visit `187 = | / | / | ·`
month day year

Name of Person `188 · | | | | | | | | | |`

FIELD WATER QUALITY MEASUREMENTS (1)

`R · 192 ·`   `T · A D M ·`
add, delete, modify

Date `193 = | / | / | |`
month day year

Geohydrologic Unit `195 = | | | | | ·`

New Card Same R thru 195

Temperature `196 = 0 , 0 , 0 , 1 , 0 ·`

Degrees C `197 · | | | | · | ·`

Conductance `196 = 0 , 0 , 0 , 9 , 5 ·`

μ Mhos `197 · | | | | · | ·`

Other (STORET) Parameter `196 = | | | | ·`

Value `197 · | | | | · | |`

Other (STORET) Parameter `196 = | | | | ·`

Value `197 · | | | | · | |`

FOOT NOTES

① Source of Data Codes

| A | D | G | L | M | O | R | S | Z |

other driller geologist logs, memory owner other reporting other
gov't reported agency

FIG. 3   Example of a Borehole or Well Schedule Form

liquid level will approach, and may reach, a stabilized level. The liquid level then will remain essentially constant with time.

NOTE 8—The time required to reach equilibrium can be reduced by removing or adding liquid until the liquid level is close to the estimated stabilized level.

8.2 Use one of the following two procedures to determine the stabilized liquid level.

8.2.1 *Procedure 1*—Take a series of liquid-level measurements until the liquid level remains constant with time. As a minimum, two such constant readings are needed (more readings are preferred). The constant reading is the stabilized liquid level for the borehole or well.

NOTE 9—If desired, the time and level data could be plotted on graph paper in order to show when equilibrium is reached.

8.2.2 *Procedure 2*—Take at least three liquid-level measurements at approximately equal time intervals as the liquid level changes during the approach to a stabilized liquid level.

8.2.2.1 The approximate position of the stabilized liquid level in the well or borehole is calculated using the following equation:

$$h_o = \frac{y_1^2}{y_1 - y_2}$$

where:

h_o = distance the liquid level must change to reach the stabilized liquid level,

y_1 = distance the liquid level changed during the time interval between the first two liquid-level readings, and

y_2 = distance the liquid level changed during the time interval between the second and the third liquid level readings.

8.2.2.2 Repeat the above process using successive sets of three measurements until the h_o computed is consistent to the accuracy desired. Compute the stabilized liquid level in the well or borehole.

NOTE 10—The time span required between readings for Procedures 1 and 2 depends on the permeability of the earth material. In material with comparatively high permeability (such as sand), a few minutes may be sufficient. In materials with comparatively low permeability (such as clay), many hours or days may be needed. The user is cautioned that in clayey soils the liquid in the borehole or well may never reach a stabilized level equivalent to the liquid level in the earth materials surrounding the borehole or well.

9. Report

9.1 For borehole or well liquid-level measurements, report, as a minimum, the following information:

9.1.1 Borehole or well identification.

9.1.2 Description of reference point.

9.1.3 Distance between reference point and top of borehole or land surface.

9.1.4 Elevation of top of borehole or reference point (if the borehole or well liquid level is reported as an elevation).

9.1.5 Description of measuring device used, and graduation.

9.1.6 Procedure of measurement.

9.1.7 Date and time of reading.

9.1.8 Borehole or well liquid level.

9.1.9 Description of liquid in borehole or well.

9.1.10 State whether borehole is cased, uncased, or contains a monitoring (observation) well standpipe and give description of, and length below top of borehole of, casing or standpipe.

9.1.11 Drilled depth of borehole, if known.

9.2 For determination of stabilized liquid level, report:

9.2.1 All pertinent data and computations.

9.2.2 Procedure of determination.

9.2.3 The stabilized liquid level.

9.3 *Report Forms*—An example of a borehole or well-schedule form is shown in Fig. 1. An example of a liquid-level measurement form, for recording continuing measurements for a borehole or well, is shown in Fig. 2. An example of a borehole or well schedule form designed to facilitate computer data storage is shown in Fig. 3.

10. Precision and Bias

10.1 Borehole liquid levels shall be measured and recorded to the accuracy desired and consistent with the accuracy of the measuring device and procedures used. Procedure A multiple measurements by wetted tape should agree within 0.02 ft (6 mm). Procedure B multiple measurements by electrical tape should agree within 0.04 ft (12 mm). Procedure C multiple measurements by tape and sounding weight should agree within 0.04 ft (12 mm). Garber and Koopman (2) describe corrections that can be made for effects of thermal expansion of tapes or cables and of stretch due to the suspended weight of tape or cable and plumb weight when measuring liquid levels at depths greater than 500 ft (150 m).

11. Keywords

11.1 borehole; electrical measuring device; ground water; liquid level; measuring tape; well

REFERENCES

(1) "*National Handbook of Recommended Methods for Water Data Acquisition—Chapter 2—Ground Water*", Office of Water Data Coordination, Washington, DC, 1980.

(2) Garber, M. S., and Koopman, F. C., "Methods of Measuring Water Levels in Deep Wells," *U.S. Geologic Survey Techniques for Water Resources Investigations*, Book 8, Chapter A-1, 1968.

(3) Hvorslev, M. J., "Ground Water Observations," in *Subsurface Exploration and Sampling of Soils for Civil Engineering Purposes*, American Society Civil Engineers, New York, NY, 1949.

(4) Zegarra, E. J., "Suggested Method for Measuring Water Level in Boreholes," *Special Procedures for Testing Soil and Rock for Engineering Purposes, ASTM STP 479*, ASTM, 1970.

(5) "Determination of Water Level in a Borehole," CSA Standard A 119.6 – 1971, Canadian Standards Association, 1971.

ASTM Designation: D 5978 – 96

Standard Guide for
Maintenance and Rehabilitation of Ground-Water Monitoring Wells[1]

This standard is issued under the fixed designation D 5978; the number immediately following the designation indicates the year of original adoption or, in the case of revision, the year of last revision. A number in parentheses indicates the year of last reapproval. A superscript epsilon (ε) indicates an editorial change since the last revision or reapproval.

INTRODUCTION

This guide for maintenance and rehabilitation promotes procedures appropriate to ground-water monitoring wells installed to evaluate the extent and nature of contamination, progress of remediation, and for long-term monitoring of either water quality or water level.

1. Scope

1.1 This guide covers an approach to selecting and implementing a well maintenance and rehabilitation program for ground-water monitoring wells. It provides information on symptoms of problems or deficiencies that indicate the need for maintenance and rehabilitation. It is limited to monitoring wells, that are designed and operated to provide access to, representative water samples from, and information about the hydraulic properties of the saturated subsurface while minimizing impact on the monitored zone. Some methods described herein may apply to other types of wells although the range of maintenance and rehabilitation treatment methods suitable for monitoring wells is more restricted than for other types of wells. Monitoring wells include their associated pumps and surface equipment.

1.2 This guide is affected by governmental regulations and by site specific geological, hydrogeological, geochemical, climatological, and biological conditions.

1.3 The values stated in SI units are to be regarded as the standard.

1.4 *This standard does not purport to address all of the safety concerns, if any, associated with its use. It is the responsibility of the user of this standard to establish appropriate safety and health practices and determine the applicability of regulatory limitations prior to use.*

2. Referenced Documents

2.1 *ASTM Standards:*
D 653 Terminology Relating to Soil, Rock, and Contained Fluids[2]
D 1889 Test Method for Turbidity of Water[3]
D 4044 Test Method for (Field Procedures) Determining Instantaneous Change in Head (Slug Tests) for Determining Hydraulic Properties of Aquifers[2]

D 4412 Test Methods for Sulfide Reducing Bacteria in Water and Water-Formed Deposits[4]
D 4448 Guide for Sampling Ground Water Monitoring Wells[5]
D 4750 Test Method for Determining Subsurface Liquid Levels in a Borehole or Monitoring Well (Observation Well)[2]
D 5088 Practice for Decontamination of Field Equipment Used at Nonradioactive Waste Sites[6]
D 5092 Practice for Design and Installation of Ground Water Monitoring Wells in Aquifers[6]
D 5254 Practice for the Minimum Set of Data Elements to Identify a Ground-Water Site[6]
D 5299 Guide for the Decommissioning of Ground Water Wells, Vadose Zone Monitoring Devices, Boreholes, and Other Devices for Environmental Activities[6]
D 5408 Guide for the Set of Data Elements to Describe a Ground-Water Site; Part 1—Additional Identification Descriptors[6]
D 5409 Guide for the Set of Data Elements to Describe a Ground-Water Site; Part 2—Physical Descriptors[6]
D 5410 Guide for the Set of Data Elements to Describe a Ground-Water Site; Part 3—Usage Descriptors[6]
D 5472 Test Method for Determining Specific Capacity and Estimating Transmissivity at the Control Well[6]
D 5474 Guide for Selection of Data Elements for Ground-Water Investigations[6]
D 5521 Guide for Development of Ground Water Monitoring Wells in Granular Aquifers[6]
2.1.1 In addition, ASTM Volume 11.01 on Water (I) and Volume 11.02 on Water (II) contain numerous test methods and standards that may be of value to the user of this guide.

3. Terminology

3.1 *Definitions:*

3.1.1 Except where noted, all terms and symbols in this guide are in accordance with the following publications in their order of consideration:

[1] This guide is under the jurisdiction of ASTM Committee D-18 on Soil and Rock and is the direct responsibility of Subcommittee D18.21 on Ground Water and Vadose Zone Investigations.
Current edition approved July 10, 1996. Published November 1996.
[2] *Annual Book of ASTM Standards*, Vol 04.08.
[3] *Annual Book of ASTM Standards*, Vol 11.01.

[4] *Annual Book of ASTM Standards*, Vol 11.02.
[5] *Annual Book of ASTM Standards*, Vol 11.04.
[6] *Annual Book of ASTM Standards*, Vol 04.09.

3.1.1.1 Terminology D 653,

3.1.1.2 Guide D 5521,

3.1.1.3 *Compilation of ASTM Standard Terminology*, 7th Edition, 1990, and

3.1.1.4 *Webster's Ninth New Collegiate Dictionary*, 1989.

3.2 *Definitions of Terms Specific to This Standard:*

3.2.1 *well development*—actions taken during the installation and start-up of a well for the purpose of mitigating or correcting damage done to the adjacent geologic formations and filter materials that might affect the well's ability to produce representative samples.

3.2.2 *well maintenance*—any action that is taken for the purpose of maintaining well performance (see Discussion) and extending the life of the well to provide samples that are representative of the ground water surrounding it. Maintenance includes both physical actions taken at the well and the documentation of those actions and all operating data in order to provide benchmarks for comparisons at later times.

3.2.2.1 *Discussion*—Desired level of well performance can vary depending on the design objectives.

3.2.3 *well preventive maintenance*—any well maintenance action that is initiated for the purpose of meeting some preestablished rule or schedule that applies while well performance is still within preestablished ranges.

3.2.4 *well reconstructive maintenance*—any preventive or rehabilitative well maintenance action involving the replacement of a major component (for example, pump, surface protection).

3.2.5 *well redevelopment*—any preventive or rehabilitative well maintenance action, taken after start-up, for the purpose of mitigating or correcting deterioration of the filter pack or adjacent geologic formations, or both, due to the well's presence and operation over time, usually involving physical development procedures, applied in reaction to deterioration.

3.2.6 *well rehabilitation*—for the purposes of this guide, synonymous with well rehabilitative or restorative maintenance.

3.2.7 *well rehabilitative or restorative maintenance*—any well maintenance action that is initiated for the purpose of correcting well performance that has moved outside of preestablished ranges.

4. Significance and Use

4.1 The process of operating any engineered system, such as monitoring wells, includes active maintenance to prevent, mitigate, or reverse deterioration. Lack of or improper maintenance can lead to well performance deficiencies (physical problems) or sample quality degradation (chemical problems). These problems are intrinsic to monitoring wells, which are often left idle for long periods of time (as long as a year), installed in non-aquifer materials, and installed to evaluate contamination that can cause locally anomalous hydrogeochemical conditions. The typical solutions for these physical and chemical problems that would be applied by owners and operators of water supply, dewatering, recharge, and other wells may not be appropriate for monitoring wells because of the need to minimize their impact on the conditions that monitoring wells were installed to evaluate.

4.2 This guide covers actions and procedures, but is not an encyclopedic guide to well maintenance. Well mainte-

nance planning and execution is highly site and well specific.

4.3 The design of maintenance and rehabilitation programs and the identification of the need for rehabilitation should be based on objective observation and testing, and by individuals knowledgeable and experienced in well maintenance and rehabilitation. Users of this guide are encouraged to consult the references provided.

4.4 For additional information see Test Methods D 1889, D 4412, D 5472, and Guides D 4448, D 5409, D 5410 and D 5474.

5. Well Performance Deficiencies

5.1 Proper well design, installation, and development can minimize well performance deficiencies that result in the need for maintenance and rehabilitation. Practice D 5092 and Guide D 5521 should be consulted. Performance deficiencies include: sand, silt, and clay infiltration; low yield; slow responses to changes in ground-water elevations; and loss of production.

5.2 *Preventable Causes of Poor Well Performance:*

5.2.1 Inappropriate well location or screened interval. These may be unavoidable if a requirement for site characterization or monitoring exists,

5.2.2 Inappropriate drilling technique or methodology for materials screened,

5.2.3 Inadequate intake structure design (screen, filter material, and so forth),

5.2.4 Inappropriate well construction materials. This may lead to corrosion or collapse,

5.2.5 Improper construction, operation, or maintenance, or combination thereof, of borehole or well, wellhead protection, well cap, and locking device,

5.2.6 Ineffective development,

5.2.7 Inappropriate pump selection, and

5.2.8 Introduction of foreign substances.

5.3 *Physical Indicators of Well Performance Deficiencies Include:*

5.3.1 *Sand, Silt, and Clay Infiltration*—Causes include inappropriate and inadequate well drilling (for example, auger flight smearing), improper screen and filter pack, improper casing design or installation, incomplete development, screen corrosion, or collapse of filterpack. In rock wells, causes include the presence of fine material in fractures. The presence of sand, silt, or clay can result in pump and equipment wear and plugging, turbid samples, filterpack plugging, or combination thereof.

5.3.2 *Low Yield*—Causes include dewatering, collapse or consolidation of fracture or water-bearing zone, pump malfunction or plugging, screen encrustation or plugging, and pump tubing corrosion or perforation.

5.3.3 *Water Level Decline*—Causes include area or regional water level decline, well interference, and chemical or microbial plugging or encrustation of the borehole, screen, or filterpack.

5.3.4 *Loss of Production*—Usually caused by pump failure, but can also be caused by dewatering, plugging, or well collapse.

5.3.4.1 *Well Collapse*—Can be caused by tectonism, ground subsidence, failure of unsupported casing (that is, in caves or because of faulty grout), corrosion and subsequent failure of screen and casing, improper casing design, local site operations, freeze-thaw, or improper chemical or mechanical rehabilitation.

5.3.5 Observation of physical damage or other indicator.

6. Sample Quality Degradation

6.1 All of the preceding physical well performance deficiencies can result in sample quality degradation by dilution, cross-contamination, or entrainment of solid material in water samples. In addition, chemical and biological activity can both degrade well performance and sample quality. Any change in well or aquifer chemistry that results from the presence of the well can interfere with accurate characterization of a site.

6.2 *Physical Indicators*—Chemical and biological activity that can lead to sample quality degradation include:

6.2.1 *Chemical Encrustation*—Precipitation of calcium or magnesium carbonate or sulfate, iron, or sulfide compounds can reduce well yield and specific capacity.

6.2.2 *Biofouling (Biological Fouling)*—Microbial activity can result in slime production and the precipitation of iron, manganese, or sulfur compounds and occasionally other materials such as aluminum oxides. Biofouling may be accompanied by corrosion or encrustation, or both, and can result in reduced specific capacity and well yield. Biochemical deposits can interfere with sample quality by acting as chemical sieves.

6.2.3 *Corrosion*—Corrosion of metal well and pump components (that is, stainless steel, galvanized steel, carbon steel, and low carbon steel) can result from naturally aggressive waters (containing H_2S, $NaCl$) or electrolysis. The presence of contaminants contributes to corrosion through contributions to microbial corrosion processes and formation of redox gradients. Nonaqueous phase solvents may degrade PVC and other plastics. Other environmental conditions such as heat or radiation may contribute to material deterioration (such as enhanced embrittlement). Metals such as nickel or chromium may be leached from corroding metals. Degradation of plastic well components may result in a release of monomers (such as vinyl chloride) to the environment (see Note 1).

NOTE 1—Naturally aggressive (for metals) waters have been defined as low pH (<7.0), high DO (>2 mg/L), high H_2S (>1 mg/L), high dissolved solids (>1000 mg/L), high CO_2 (>50 mg/L), and high Cl^- content (>500 mg/L). However, local conditions may result in corrosion at less extreme values. Expression of corrosion is also dependent on materials load.

6.2.4 *Change in Turbidity*—Causes include biofouling and intake structure, screen or filter pack clogging or collapse. Increase in turbidity may not always be the result of a problem with the well. Changes in the purging and sampling procedures and devices used can affect the turbidity of water from a monitoring well. For example, using a bailer where a pump was previously utilized, or pumping at a higher rate than previously used could increase turbidity; likewise, pumping a well that was previously bailed could increase turbidity.

6.2.5 *Change in Sand/Silt Content or Particle Counts*— Causes include biofouling (resulting in clogging or sloughing) and intake structure clogging or collapse. Increase in the sand/silt content may not always be the result of a problem with the well. Changes in the purging and sampling proce-

dures and devices used can affect the sand/silt content of water from a monitoring well. For example, using a bailer where a pump was previously utilized, or pumping at a higher rate than previously used could increase the sand/silt content; likewise, pumping a well that was previously bailed could increase the sand/silt content.

6.3 *Chemical Indicators (Observed in Ground Water Samples)*—Chemical and biological activity that can lead to sample quality degradation include (see Note 2):

NOTE 2—Changes in chemical indicators can also be a result of site-wide changes in hydro-geochemistry.

6.3.1 *Iron (Changes in Total Fe, Fe^{2+}/Fe^{3+}, Iron Minerals and Complexes)*—Causes include corrosion, changes in redox potential, and biofouling.

6.3.2 *Manganese (Changes in Total Mn, Mn^{2+}/Mn^{4+}, Manganese Minerals and Complexes)*—Causes include changes in redox potential and biofouling.

6.3.3 *Sulfur (Changes in Total $S^{2-}/S^0/SO_4^{2-}$, Sulfur Minerals and Complexes)*—Causes include changes in redox potential and biofouling.

6.3.4 *Changes in Redox Potential (Eh)*—Causes include microbial activity and changes in O_2, CH_4, CO_2, N, S, Fe, and Mn species present in the system.

6.3.5 *Changes in pH*—Causes include corrosion; microbial activity; dissolved gases such as oxygen, carbon dioxide, and hydrogen sulfide; and encrustation.

6.3.6 *Changes in Conductivity*—Causes include changes in total solids content, microbial activity, and corrosion.

6.3.7 *Changes in the Type and Concentration of Gases*— Dissolved oxygen, carbon dioxide, nitrogen, hydrogen sulfide, and methane are indicators of redox status and microbial activity.

7. Maintenance Planning, Monitoring, and Treatment

7.1 The purpose of maintenance is to detect and control deterioration in well performance. Maintenance should be based on objective observation and testing of the well and aquifer to determine the factors that can cause clogging, turbidity, and corrosion. Monitoring well maintenance must not alter the chemistry of the ground water being monitored. Maintenance is best implemented routinely, from installation through the life of the well, but can be implemented after deteriorated wells have been rehabilitated.

7.2 *Goals for Maintenance:*

7.2.1 Maintenance is intended, to the degree possible, to prevent or slow deterioration of the well system's structure, prevent contamination of ground water, or to ensure hydraulic performance. To address these goals, a maintenance plan should be developed and followed with adjustments to meet changing conditions.

7.2.2 A maintenance plan includes those practices, including preventive design and construction practices (see 5.1), an assessment of identified and potential problems (see 5.2, 6.1, 6.2), procedures for how these potential problems will be monitored and evaluated (see Sections 6 and 7), and a decision-making process on how to proceed to address problems as they occur. The decision-making process should include, as a minimum, who will make the decisions based on what criteria, a set of alternatives such as establishing a program of preventive treatment, replacing components on an as-needed basis, and how to proceed if more intrusive rehabilitation or decommissioning is needed. This decision-making process should be triggered if there are changes in condition or performance detected in routine monitoring that show deterioration or the potential to affect the well's ability to provide acceptable information. The decision-maker must decide what the standards are and the importance of detected changes. It is understood that there is no single level of performance or maintenance standards that exists or is possible due to the individual character of wells and site conditions.

7.2.3 In setting the goal(s) for an acceptable level of performance, the users of this guide should keep in mind what is possible in a given situation and evaluate whether desired standards can be met. The decision process should include personnel with special knowledge or skill in well maintenance and rehabilitation, especially field or contractor personnel with direct experience in these activities.

7.3 *Maintenance Program Design*—The design of a maintenance program should incorporate all available information about site-specific factors that could cause sand, silt, or clay infiltration, sample turbidity or alteration, corrosion, or clogging. Such information can include biological activity, redox potential, pH, conductivity, alkalinity, and major ions present in the ground water. Hydraulic performance and water chemistry should be benchmarked at installation and periodically during operations so that changes in performance can be detected. The frequency of maintenance is typically site specific and may be dependent on the proposed sampling schedule. Quantities of sediment in samples should be recorded and compared through the life of the well.

7.4 *Maintenance Monitoring*—Monitoring well maintenance includes routine physical inspection and analyses of hydraulic performance and sample quality. Personnel should first review records for as-built and previous conditions and compare the current conditions and measurements to those recorded previously. Any deviation, for example in total depth, should trigger a repair or rehabilitation decision.

7.4.1 *Methods of Physical Inspection Include:*

7.4.1.1 Surface facility inspection, including check of location, coordinates, elevation, and unique well identification,

7.4.1.2 Borehole mirror survey (above the water surface), camera, or televiewer,

7.4.1.3 Geophysical logs as appropriate to evaluate well construction,

7.4.1.4 Measurement of total depth, and

7.4.1.5 Inspection of pulled components.

7.4.2 *Methods of Analysis of Hydraulic Performance Include:*

7.4.2.1 Geophysical logs as appropriate to evaluate geology/hydrologic conditions, and

7.4.2.2 Drawdown/recovery measurements (in response to pumping).

7.4.2.3 *Flow Measurements*—Both temporary and permanent methods are used. Temporary methods such as bucket or weir are used to test new pumps or retest existing pumps. Permanent wellhead methods such as turbine or Doppler flow meters are more appropriate for extraction well arrays, but may be used for monitoring wells in some circumstances.

7.4.2.4 *Slug Testing*—If slug test data is available from an earlier test, the change in hydraulic performance can be inferred by performing another slug test. Slug tests are especially useful with low flow conditions or in contaminated settings. The reader should refer to Test Method D 4044.

7.4.3 *Methods of Analysis of Sample Quality Include:*

7.4.3.1 Time-series monitoring of site-specific chemical parameters of maintenance concern.

7.4.3.2 Pumped grab samples or biofilm collection for biofouling indicators such as Biological Activity Reaction Test (BART) analysis, heterotrophic iron and sulfur bacteria, sulfate reducing bacteria (SRB), microscopy, and biofilm mineralogical and elemental analyses (see Note 3).

NOTE 3—Biofilm indicator methods can only be considered qualitative at the present time.

7.5 *Rehabilitative Maintenance:*

7.5.1 Rehabilitation for removal of entrapped pollutants should be the last phase in the life cycle of a working well. If rehabilitation is unsuccessful, decommissioning may be required. Rehabilitation of a viable well is not a permanent solution for performance problems and should be followed by maintenance. Methods of rehabilitation must not, more than transiently, change the chemistry of the ground water being monitored. Methods are also limited by the typically small size and relative fragility of monitoring wells.

7.5.2 When determining whether rehabilitation or decommissioning is appropriate, decision criteria should include: planned life length of well, cost, and effectiveness of rehabilitation. In the event that well replacement is chosen, Guide D 5299 should also be consulted.

7.5.3 The appendix contains a list of references for detailed information on maintenance and rehabilitation.

8. Equipment and Materials

8.1 Selection of equipment and materials for maintenance and rehabilitation depends on well construction and site-specific geological, hydrogeological, geochemical, climatological, and biological conditions. Practice D 5088 should be consulted.

8.2 *Equipment for Physical and Chemical Measurements:*

8.2.1 Drawdown (water depth) equipment includes measuring (tape) devices, airline, electric or acoustic sounder, and recording transducer. See Test Method D 4750.

8.2.2 Flow meters include calibrated bucket (<10 gpm or 0.6 L/s), and orifice weir (>10 gpm) or any other appropriate, accurate device.

8.2.3 Other instruments, such as electronic colorimetric instruments, spectrophotometers, electronic pH and mV meters, turbidometers, particle counters, multi-probes, flow-through cells, multiparameter meters and other types of probes (dissolved oxygen, temperature, TDS, specific and ion electrodes, and so forth), and geophysical logging tools (see Note 4).

NOTE 4—Calibrated portable instruments may be used for maintenance monitoring to encourage frequent monitoring without significant loss of accuracy. Some redox-sensitive parameters are preferably analyzed at the well head using flow-through cells.

8.3 Equipment for analysis of microbial components includes light microscope and biofilm sample collection apparatus.

8.4 Equipment for redevelopment and rehabilitative maintenance of wells will depend on the action needed. Routine hand tools would be needed for a variety of purposes, and special tools may be required for pump service. Spare parts and major components for pumps used should be readily available to maintenance personnel. Devices used for well redevelopment are identical to those used in development, and described in Guide D 5521 and references. If chemicals, flushing, or specialized procedures such as cryogenic CO_2 treatments are employed, the necessary mixing and pumping equipment should be onsite in working order.

9. Maintenance

9.1 Selection of procedures for both maintenance and rehabilitation is limited by the need (and often regulatory requirements) to minimize their impact on the conditions that monitoring wells were installed to evaluate. Usually only physical, not chemical, methods are acceptable. If chemicals are used, chemical purity, alteration of existing conditions, and regulations must be considered. Maintenance includes routine preventive practices to avoid damage to the physical structure and access to the well, including nonchemical weed removal (to avoid concrete splitting) or changing or protecting locks (if they are subject to corrosion or freezing).

9.2 *Maintenance Evaluation*—Methods by which the need for maintenance is identified include collection and analysis of physical and chemical data on a routine basis. Some methods include:

9.2.1 Visual inspection of surface facility, borehole, and pulled components. Concrete pads should be inspected for cracks, separation from well, and heaving. Surface casing should be inspected for cracks or damage. Traffic cover (for flush-mounted wells) should be inspected for fit, cracks, and leaks. Locks should be serviceable and prevent unauthorized entry into the well. (See Practice D 5092.)

9.2.2 Borehole geophysical logging using televisions, flowmeters, and calipers can be useful to identify water movement, casing breaks and damage, clogging, and biofouling.

9.2.3 *Water Level and Well Depth Measurement.* Well depth measurement may indicate that materials may be filling up the well or that other obstructions may be present. A weighted measuring tape is typically used for bottom depth measurements. (See Test Method D 4750.) Bottom sounding in wells with dedicated pumping systems may be difficult or impossible without removing the system. Dedicated bottom sounders, consisting of a dedicated weight and cable that extends from the well bottom to the well cap have been used to eliminate this concern.

9.2.4 *Pump Performance*—Manufacturer's specifications for maintenance should be met, and there should be a visual inspection for clogging. While visual inspection of a pump or associated hardware could reveal the cause of diminished pump performance, removal and reinstallation of the equipment may introduce contaminants to the well or sampling system. Some manufacturers publish performance testing and trouble-shooting procedures to assess pump performance without pump removal. Also, some manufacturers do not require or recommend routine maintenance for their sampling pumps, only repairs when needed, using perfor-

mance testing instead of routine maintenance.

9.2.5 *Drawdown Measurement*—See Practice D 5092.

9.2.6 *Flow Measurement*—Obvious increases or decreases in flow capacity may indicate the need for rehabilitation. The pump flow output should be checked against nominal performance to evaluate pump performance.

9.2.7 *Evaluation of Chemical Data Trends*—Obvious deviations from established trends not attributable to other causes may indicate the need for rehabilitation.

10. Rehabilitation

10.1 Rehabilitation is the repair and replacement of surface and downhole components of the well found to be deficient by visual inspection.

10.1.1 The goals for rehabilitation are by nature site-specific. What is possible in a given formation, or with feasible means are dependent on site-specific factors.

10.1.2 Standards for rehabilitate should be flexible. There are limits to effectively rehabilitate monitoring wells. Some wells cannot be rehabilitated to set standards due to conditions external to the well or due to well deficiencies.

10.2 *Rehabilitation Procedures:*

10.2.1 Methods include redevelopment to remove fine-grained materials from the well and to remove materials clogging the well screen. An economic evaluation comparing well replacement cost with the cost of the time and materials for rehabilitation should be performed. Monitoring wells are usually replaced rather than rehabilitated if redevelopment is not effective.

10.2.2 Rehabilitation should continue by the means selected until an irreducible minimum in the condition is reached. At that point, a decision must be made to employ another method, accept the condition, reevaluate and make repairs indicated, or decommission the well. If a more effective method of rehabilitation is employed, the process should be repeated until another irreducible minimum is reached, then evaluated again. If a condition is uncovered that cannot be rehabilitated, then decommissioning and new construction are indicated.

10.3 *Redevelopment*—Redevelopment can be accomplished using pumps, surge blocks, compressed air (for example, air lift method), or water jetting, or combination thereof. In certain conditions, chemicals and steam may be used for redevelopment for bacterial problems. The reader should consult Guide D 5521 for appropriate methods. It is noted that the goals for redevelopment are the same as the goals identified in Guide D 5521.

11. Reporting and Record Keeping

11.1 Reporting and record keeping are important components of both maintenance and rehabilitation.

11.2 *Maintenance*—Because monitoring well maintenance includes routine physical inspection and analysis of hydraulic performance and sample quality in order to detect and control deterioration in well performance, data must be compared for each well through time. An organized record-keeping system that permits data storage and retrieval is necessary to document and analyze changes through time. Such systems include paper files and computer databases.

11.3 *Rehabilitation*—Records of test method's results and observations that led to the decision for rehabilitation should be kept. Records of the rehabilitation and subsequent test methods and results should also be retained.

11.4 *General Information That Should Be Recorded:*
11.4.1 Location of well,
11.4.2 Well name/number,
11.4.3 Method/materials/date of construction,
11.4.4 As-built diagram,
11.4.5 Drill log,
11.4.6 Purpose of well,
11.4.7 Historical trends,
11.4.8 Water quality data,
11.4.9 Observations leading to maintenance or rehabilitation,
11.4.10 Test methods and results prior to maintenance or rehabilitation,
11.4.11 Dates of observation/testing/maintenance/rehabilitation,
11.4.12 Work performed, and
11.4.13 Post-maintenance or rehabilitation test methods and results.

11.5 Additional information related to the other data needs can be found in Practice D 5254. If additional information is needed, the users of this guide are referred to Guide D 5408.

12. Keywords

12.1 biofouling; development; encrustation; ground water; maintenance; monitoring well; rehabilitation

APPENDIX

(Nonmandatory Information)

X1. ADDITIONAL REFERENCES

NOTE X1.1—Not all methods referenced for rehabilitation of water wells are directly appropriate for monitoring wells, but may be adapted for some purposes.

Alford, G., Mansuy, N., and Cullimore, D. R., "The Utilization of the Blended Chemical Heat Treatment (BCHT) Process to Restore Production Capacities to Biofouled Water Wells," *Proceedings of the Third Annual Outdoor Action Conference, NWWA*, 1989, pp. 229–237.

Aller, L., et al, *Handbook of Suggested Practices for the Design and Installation of Ground-Water Monitoring Wells*, EPA 600/4-89/034, Published by National Water Well Association, Dublin, OH, 1989.

Borch, M. A., Smith, S. A., and Noble, L. N., *Evaluation and Restoration of Water Supply Wells*, AWWA Research Foundation, Denver, CO, 1993.

Cullimore, D. R., *Practical Manual of Groundwater Microbiology*, Lewis Publishers, Chelsea, MI, 1993.

Driscoll, F. G., *Groundwater and Wells*, 2nd ed., Johnson Division, 1986.

Fountain, J., and Howsam, P., "The Use of High Pressure Water Jetting as a Rehabilitation Technique," *Water Wells Monitoring, Maintenance, and Rehabilitation*, P. Howsam, ed., E.&F.N. Spon., London, 1990, pp. 180–194.

Gass, T. E., Bennett, T. W., and Miller, J., *Manual of Water Well Maintenance and Rehabilitation Technology*, National Water Well Association, Worthington, OH, 1980.

Gates, W. C. B., "Protection of Ground-Water Monitoring Wells Against Frost Heave," *Bulletin of the Association of Engineering Geologist, 26(2)*, 1989, pp. 241–251.

Hem, J. D., "Study and Interpretation of the Chemical Characteristics of Natural Water," 3rd ed., *U.S. Geological Survey Water Supply Paper 2254*, 1985.

Kraemer, C. A., Schultz, J. A., and Ashley, J. W., "Monitoring Well Post-Installation Considerations," *Practical Handbook of Ground Water Monitoring*, Nielsen, D. M., ed., Lewis Publishers, 1991, pp. 333–365.

Leach, R., Mikell, A., Richardson, C., and Alford, G., "Rehabilitation of Monitoring, Production, and Recharge Wells," *Proceedings of 15th Annual Army Environmental R & D Symposium (1990), CETHA-TS-CR-91077*, U.S. Army Toxic and Hazardous Materials Agency, Aberdeen Proving Grounds, MD, 1991, pp. 623–646.

Macaulay, D., "Maintenance for Monitoring Wells: How to Extend the Usefulness of Monitoring Wells," in *Ground Water Age*, Vol 22, No. 8, 1988, pp. 24–27.

Mansuy, N., Nuzman, C., and Cullimore, D. R., "Well Problem Identification and Its Importance in Well Rehabilitation," *Water Wells Monitoring, Maintenance, and Rehabilitation*, P. Howsam, ed., E.&F.N. Spon., London, 1990, pp. 87–88.

McLaughlan, R. G., Knight, M. J., and Steutz, R. M., *Fouling and Corrosion of Groundwater Wells, A Research Study*, National Centre for Groundwater Management, The University of Technology, Sydney, Australia, 1993.

McCullom, K. M., and Cronin, J. E., "Installation and Maintenance of Ground Water Monitoring Wells Located in Permafrost Soils," *Ground Water Management 11: 31–44 (6th NOAC)*, 1992.

McTique, W. H., and Kunzel, R. G., "A Technique for Renovating Clogged Monitor Wells," *Proceedings of the Third National Symposium on Aquifer Restoration and Ground Water Monitoring*, NGWA, Worthington, OH, 1983, pp. 247–252.

Nuckols, T. E., "Development of Small Diameter Wells," *Proceedings of the Fourth Annual Conference on Aquifer Restoration, Ground Water Monitoring and Geophysical Methods, NGWA*, Dublin, OH, 1992.

Powers, J. P., *Construction Dewatering*, Wiley-Interscience, New York, 1992.

Roscoe Moss Company, *Handbook of Ground Water Development*, Wiley-Interscience, New York.

Sevee, J. E., and Maher, P. M., "Monitoring Well Rehabilitation Using the Surge Block Technique," *Ground Water and Vadose Zone Monitoring*, D. M. Nielsen, and A. I. Johnson, eds., *ASTM STP 1053*, 1990, pp. 91–97.

Smith, S. A., "Maintenance Problems of Environmental Site Wells: A Thumbnail Sketch," *National Drillers Buyers Guide*, Vol 12, No. 5, 1991.

Smith, S. A., *Methods for Monitoring Iron and Manganese Biofouling in Water Supply Wells*, AWWA, 1992.

Smith, S. A., *Monitoring and Remediation Wells—Problem Prevention, Maintenance, and Remediation*, CRC Press/Lewis Publishers, Boca Raton, FL, 1995.

Winegardner, D. L., "Monitoring Wells: Maintenance, Rehabilitation, and Abandonment," *Ground Water and Vadose Zone Monitoring*, D. M. Nielsen, and A. I. Johnson, eds., *ASTM STP 1053*, 1990, pp. 98–107.

 **D 5978**

Standard Guide for
Decommissioning of Ground Water Wells, Vadose Zone Monitoring Devices, Boreholes, and Other Devices for Environmental Activities[1]

This standard is issued under the fixed designation D 5299; the number immediately following the designation indicates the year of original adoption or, in the case of revision, the year of last revision. A number in parentheses indicates the year of last reapproval. A superscript epsilon (ε) indicates an editorial change since the last revision or reapproval.

1. Scope

1.1 This guide covers procedures that are specifically related to permanent decommissioning (closure) of the following as applied to environmental activities. It is intended for use where solid or hazardous materials or wastes are found, or where conditions occur requiring the need for decommissioning. The following devices are considered in this guide:

1.1.1 A borehole used for geoenvironmental purposes (see Note 1),

1.1.2 Monitoring wells,

1.1.3 Observation wells,

1.1.4 Injection wells (see Note 2),

1.1.5 Piezometers,

1.1.6 Wells used for the extraction of contaminated ground water, the removal of floating or submerged materials other than water such as gasoline or tetrachloroethylene, or other devices used for the extraction of soil gas,

1.1.7 A borehole used to construct a monitoring well, and

1.1.8 Any other vadose zone monitoring device.

1.2 Temporary decommissioning of the above is not covered in this guide.

NOTE 1—This guide may be used to decommission boreholes where no contamination is observed at a site (see Practice D 420 for details); however, the primary use of the guide is to decommission boreholes and wells where solid or hazardous waste have been identified. Methods identified in this guide can also be used in other situations such as the decommissioning of water supply wells and boreholes where water contaminated with nonhazardous pollutants (such as nitrates or sulfates) are present. This guide should be consulted in the event that a routine geotechnical investigation indicates the presence of contamination at a site.

NOTE 2—The term "well" is used in this guide to denote monitoring wells, piezometers, or other devices constructed in a manner similar to a well. Some of the devices listed such as injection and extraction wells can be decommissioned using this guide for information, but are not specifically covered in the text.

NOTE 3—Details on the decommissioning of multiple-screened wells are not provided in this guide due to the many methods used to construct these types of wells and the numerous types of commercially available multiple-screened well systems. However, in some instances, the methods presented in this guide may be used with few changes. An example of how this guide may be used is the complete removal of the multiple-screened wells by overdrilling.

1.3 Most monitoring wells and piezometers are intended primarily for water quality sampling, water level observation, or soil gas sampling, or combination thereof, to determine quality. Many wells are relatively small in diameter (<4-in. (10.1 cm) inside diameter) and are used to monitor for hazardous chemicals in ground water. Decommissioning of monitoring wells is necessary to:

1.3.1 Eliminate the possibility that the well is used for purposes other than intended,

1.3.2 Prevent migration of contaminants into an aquifer or between aquifers,

1.3.3 Prevent migration of contaminants in the vadose zone,

1.3.4 Reduce the potential for vertical or horizontal migration of fluids in the well or adjacent to the well, and

1.3.5 Remove the well from active use when the well is no longer capable of rehabilitation, or has failed structurally; no longer required for monitoring; no longer capable of providing representative samples or is providing unreliable samples; or required to be decommissioned; or to meet regulatory requirements.

NOTE 4—The determination of whether a well is providing a representative water quality sample is not defined in this guide. Examples of when a representative water quality sample may not be collected include the biological or chemical clogging of well screens, a drop in water level to below the base of the well screen, or complete silting of a tail pipe. These conditions may indicate that a well is not functioning properly.

1.4 This guide is intended to provide information for effective permanent closure of wells so that the physical structure of the well does not provide a means of hydraulic communication between aquifers or react chemically in a detrimental way with the environment.

1.5 The intent of this guide is to provide procedures that when followed result in a reasonable level of confidence in the integrity of the decommissioning activity. However, it may not be possible to verify the integrity of the decommissioning procedure. At this time, methods are not available to substantially determine the integrity of the decommissioning activity.

1.6 The values stated in inch-pound units are to be regarded as the standard. The SI units given in parentheses are for information only.

1.7 *This standard does not purport to address all of the safety problems, if any, associated with its use. It is the responsibility of the user of this standard to establish appropriate safety and health practices and determine the applicability of regulatory limitations prior to use.*

[1] This guide is under the jurisdiction of ASTM Committee D-18 on Soil and Rock and is the direct responsibility of Subcommittee D18.21 on Ground Water and Vadose Zone Investigations.

Current edition approved Sept. 15, 1992. Published January 1993.

NOTE 5—If state and local regulations are in effect where the decommissioning is to occur, the regulations take precedence over this guide.

2. Referenced Documents

2.1 *ASTM Standards:*

C 150 Specification for Portland Cement[2]

C 618 Specification for Fly Ash and Raw or Calcined Pozzolan for Use as a Mineral Admixture in Portland Concrete Cement[2]

D 420 Practice for Investigating and Sampling Soil and Rock for Engineering Purposes[3]

D 4380 Test Method for Density of Bentonitic Slurries[3]

D 5088 Practice or the Decontamination of Field Equipment Used at Non-Radioactive Waste Sites[3]

D 5092 Practice for Design and Installation of Ground Water Monitoring Wells in Aquifers[3]

3. Terminology

3.1 *Descriptions of Terms Specific to This Standard:*

3.1.1 **abandonment**—see *decommissioning.*

3.1.2 *attapulgite clay*—a chain-lattice clay mineral. The term also applies to a group of clay minerals that are lightweight, tough, matted, and fibrous.

3.1.3 *borehole television log*—a borehole or well video record produced by lowering a television camera into the borehole or well. This record is useful in visually observing downhole conditions such as collapsed casing or a blocked screen.

3.1.4 *blowout*—a sudden or violent uncontrolled escape of fluids or gas, or both, from a borehole.

3.1.5 *caliper log*—a geophysical borehole log that shows to scale the variations with depth in the mean diameter of a cased or uncased borehole.

3.1.6 *cement, API, Class A*—a cement intended for use from the surface to a depth of 6000 ft (1828 m). This cement is similar to ASTM Type I cement.

3.1.7 *cement, API, Class B*—a cement intended for use from the surface to a depth of 6000 ft (1828 m) when conditions require moderate- to high-sulfate resistance. This cement is similar to ASTM Type II cement.

3.1.8 *cement, API, Class C*—this cement is intended for use from the surface to a depth of 6000 ft (1828 m) when conditions require high early strength. This cement is similar to ASTM Type III cement. Also available as a high sulfate resistant type.

3.1.9 *cement, API, Class G*—this cement is intended for use from the surface to a depth of 8000 ft (2438 m). It can be used with accelerators or retarders to cover a wide range of well depths and temperatures. No additions other than calcium sulfate or water, or both, can be interground or blended with the clinker during manufacture of the cement. Also available as several sulfate-resistant types.

3.1.10 *cement, API, Class H*—this cement is intended for use from the surface to a depth of 8000 ft (2438 m). It can be used with accelerators or retarders to cover a wide range of well depths and temperatures. No additions other than calcium sulfate or water, or both, can be interground or

blended with the clinker during manufacture of the cement. Also available as a sulfate-resistant type.

3.1.11 *cement, API, Class J*—this cement is intended for use from depths of 12 000 to 16 000 ft (3658 to 4877 m) under conditions of extremely high temperatures and pressures. It can be used with accelerators and retarders to cover a range of well depths and temperatures. No additions of retarders other than calcium sulfate, or water, or both, can be interground or blended with the clinker during manufacture of the cement.

3.1.12 *cement bond (sonic) log*—a borehole geophysical log that can be used to determine the effectiveness of a cement seal of the annular space of a well.

3.1.13 *channeling*—the process of forming a vertical cavity resulting from a faulty cement job in the annular space.

3.1.14 *curing accelerator*—a material added to cement to decrease the time for curing. Examples are sodium chloride, calcium sulfate (gypsum), and aluminum powder.

3.1.15 *curing retarder*—a material added to cement to increase the time for curing. Sodium chloride in high concentrations is an example.

3.1.16 *decommissioning (closure)*—the engineered closure of a well, borehole, or other subsurface monitoring device sealed with plugging materials. Decommissioning also includes the planning and documenting of all associated activities. A synonym is abandonment.

3.1.17 *decontamination*—the process of removing undesirable physical or chemical constituents, or both, from equipment to reduce the potential for cross-contamination.

3.1.18 *fallback*—shrinkage, settlement, or loss of plugging material placed in a borehole or well.

3.1.19 *fire clay*—a silicious clay rich in hydrous aluminum silicates.

3.1.20 *flow log*—a borehole geophysical log used to record vertical movement of ground water and movement of water into or out of a well or borehole and between formations within a well.

3.1.21 *geophysical borehole log*—a log obtained by lowering an instrument into a borehole and continuously recording a physical property of native or backfill material and contained fluids. Examples include resistivity, induction, caliper, sonic, and natural gamma logs.

3.1.22 *grout*—material consisting of bentonite, cement, or a cement-bentonite mixture.

3.1.23 *grout pipe*—a pipe or tube that is used to transport cement, bentonite, or other plugging materials from the ground surface to a specified depth in a well or borehole. The material may be allowed to flow freely or it may be injected under pressure. The term tremie pipe is frequently used interchangeably.

3.1.24 *hydraulic communication*—the migration of fluids from one zone to another, with reference to this guide; especially along a casing, grout plug, or through backfill materials.

3.1.25 *multiple-screened wells*—two or more monitoring wells situated in the same borehole. These devices can be either individual casing strings and screen set at a specific depth, a well with screens in more than one zone, or can consist of devices with screens with tubing or other collecting devices attached that can collect a discrete sample.

[2] *Annual Book of ASTM Standards,* Vol 04.02.
[3] *Annual Book of ASTM Standards,* Vol 04.08.

3.1.26 *native material*—in place geologic (or soil) materials encountered at a site.

3.1.27 *overdrilling*—the process of drilling out a well casing and any material placed in the annular space.

3.1.28 *perforation*—a slot or hole made in well casing to allow for communication of fluids between the well and the annular space.

3.1.29 *permanent plugging*—a seal that has a hydraulic conductivity that is equivalent or less than the hydraulic conductivity of the geologic formation. This term is often used with uncased boreholes.

3.1.30 *plow layer*—the depth typically reached by a plow or other commonly used earth turning device used in agriculture. This depth is commonly one to two feet (.3 m to .61 m) below land surface.

3.1.31 *plugging material*—a material that has a hydraulic conductivity equal to or less than that of the geologic formation(s) to be sealed. Typical materials include portland cement and bentonite.

3.1.32 *pre-conditioning*—an activity conducted prior to placing plugging material into a borehole in order to stabilize the hole.

3.1.33 *temporary decommissioning*—the engineered closure of a well intended to be returned to service at some later date (generally no more than six months). Temporary plugging should not damage the structural integrity of the well. Plugging materials consist of sand, bentonite, or other easily removed materials.

4. Summary of Guide

4.1 Information is provided on the significance of properly decommissioning boreholes and wells at sites containing or formerly containing solid or hazardous waste or hazardous materials or their byproducts, or that may be affected by solid or hazardous waste materials or their byproducts in the future. This guide may be used in situations where water quality in one aquifer may be detrimental to another aquifer either above or below the aquifer. The primary purpose of decommissioning activities is to permanently decommission the borehole or monitoring device so that the natural migration of ground water or soil vapor is not significantly influenced. Decommissioned boreholes and wells should have no adverse influence on the local environment than the original geologic setting.

4.2 It is important to have a good understanding of the geology, hydrogeology, well construction, historic and future land use, chemicals encountered, and the regulatory environment for successful decommissioning to occur.

4.3 Various materials suitable for decommissioning boreholes and wells are discussed, including their positive and negative attributes for decommissioning. A generalized procedure is provided that discusses the process from planning through implementation and documentation. Examples of typical practices are provided in the appendix.

5. Significance and Use

5.1 Decommissioning of boreholes and monitoring wells, and other devices requires that the specific characteristics of each site be considered. The wide variety of geological, biological, and physical conditions, construction practices, and chemical composition of the surrounding soil, rock,

waste, and ground water precludes the use of a single decommissioning practice. The procedures discussed in this guide are intended to aid the geologist or engineer in selecting the tasks required to plan, choose materials for, and carry out an effective permanent decommissioning operation. Each individual situation should be evaluated separately and the appropriate technology applied to best meet site conditions. Considerations for selection of appropriate procedures are presented in this guide, but other considerations based on site specific conditions should also be taken into account.

NOTE 6—Ideally, decommissioning should be considered as an integral part of the design of the monitoring well. Planning at this early stage can make the decommissioning activity easier to accomplish. See Practice D 5092 for details on monitoring well construction.

5.2 This guide is intended to provide technical information and is not intended to supplant statutes or regulations. Approval of the appropriate regulatory authorities should be an important consideration during the decommissioning process.

6. Materials

6.1 The materials used for construction of a monitoring well or other monitoring device to be decommissioned in part determines how it is decommissioned. Various materials are available for use in plugging boreholes and monitoring wells. This section provides information on these materials.

6.2 *Casing and Screen Materials:*

6.2.1 Various materials are used for well casing and screen. The most common materials used are: PVC, PTFE, fiberglass, carbon steel, stainless steel, and aluminum. Typically, the same material is used for casing and screen in a well, however, in some instances different materials may be used in a well to achieve a particular purpose such as corrosion protection, reduction of material costs, or improving the integrity of ground water or soil vapor samples. This guide does not specifically address the use of more than one type of casing or screen material used in a well, however, the same decommissioning methods can frequently be used when more than one material is used (for example, PVC and PTFE, or stainless steel and carbon steel) in a well.

6.2.2 In selecting a well decommissioning method, PVC, PTFE, and fiberglass wells can be decommissioned using similar methods as all three types of materials tend to be low in tensile strength and easy to drill out or perforate. Appendix X1 provides a discussion on various procedures that can be used for the decommissioning of PVC wells and by reference PTFE and fiberglass wells.

6.2.3 Wells constructed of carbon steel, stainless steel, and heavy walled aluminum can be decommissioned using similar methods as these materials tend to have a higher tensile strength that allows for the casing to be removed. Appendix X1 provides a discussion on various procedures that can be used for the decommissioning of steel wells and by reference stainless steel and aluminum wells.

6.3 *Plugging Materials:*

6.3.1 Plugging materials should be carefully chosen for well closure to be permanent. Basic material characteristics are listed as follows:

6.3.1.1 Plugging materials should not react with contaminants or adversely react with ground water or geologic materials.

6.3.1.2 Plugging materials used in decommissioning wells, borings, etc. should have hydraulic conductivity (saturated condition) that is comparable to or lower than that of the lowest hydraulic conductivity of the geologic material being sealed.

6.3.1.3 Plugging materials must have sufficient structural strength to withstand pressures expected from native conditions.

6.3.1.4 Plugging materials must maintain sealing capabilities and not degrade due to chemical interaction, corrosion, dehydration, or other physical or chemical processes. Materials should maintain their design characteristics for the length of time contamination is present at the site.

6.3.1.5 Plugging materials should not be readily susceptible to cracking or shrinkage, or both.

6.3.1.6 Plugging materials must be capable of being placed at the position in the well or borehole in which they are needed and must have properties that reduce their unintended movement vertically and horizontally.

6.3.1.7 Plugging materials must be capable of forming a tight bond and seal with well casing and the formation.

6.3.18 Plugging materials must have properties that eliminate leaching or erosion of the material, under the conditions the material will be subjected. These include vertical or horizontal movement, or both, or contact with ground water or other existing conditions.

NOTE 7—The grain size of plugging material used in decommissioning operations conducted in areas where thick vadose zones occur should be coarser than materials used in areas where thin vadose zones or shallow saturated conditions occur. This is necessary as water is not transported effectively in coarse-grained materials under negative pore pressures. Coarse-grained materials should not be used where saturated conditions are likely to exist during the period of time that hazardous materials can be expected to occur at the site. It is important to determine the lithology and grain size distribution of materials adjacent to the borehole or well prior to selection of plugging materials.

NOTE 8—If coarse-grained materials are used to decommission the borehole or well, a layer of fine-grained material (such as cement or bentonite, or both) 1 or 2 ft (.3 or .61 m) thick should be placed at 10 ft (3 m) intervals in the borehole in the saturated zone. This layer should extend 2 to 3 ft (.61 to .91 m) above the highest expected level saturation is expected based on historical information on the water table for unconfined aquifers. A similar thickness of these materials should be used for confined aquifers. A similar 5-ft (1.5-m) seal of a low-permeability material should be placed near the ground surface to reduce the potential for entrance of fluids at the ground surface.

6.4 *Commonly Used Materials*—Subsections 6.2 and 6.3 introduced the general criteria that must be evaluated during the process of selecting the appropriate procedure and material for plugging a specific well. Because well construction and local geological conditions are site specific, a wide variety of materials and procedures may be used to complete the closure.

6.4.1 Section 6.4 presents a review of the plugging materials most commonly used to decommission monitoring wells. Table 1 summarizes these materials and lists the most important considerations (positive and negative) for their use. A detailed discussion of each material is presented in the following subsections.

6.4.2 *Portland Cement*—Portland cement may be used in any of its various forms to meet placement, strength, and durability criteria listed in 6.1. The amount of shrinkage or settling of neat cement is dependent on the amount of water used. Higher water to cement ratios tend to increase shrinkage (1).[4]

6.5 *Specification C 150:*

6.5.1 *Type I*—Type 1 cement, a general-purpose material, is the most commonly used cement. This material has a tendency to develop a relatively high heat of hydration when used in confined situations and has relatively low-sulfate resistance.

6.5.2 *Type II*—Somewhat slower strength development than Type I; however, Type II cement has moderate heat of hydration and moderate sulfate resistance.

6.5.3 *Type III*—Type III cement is used when high early strength is desired. This material is not commonly used in decommissioning activities because of its ability to quickly set. Care must be used in working with this material.

6.5.4 *Type IV*—Type IV cement is used where a low heat of hydration is desired. It is not commonly used in decommissioning activities.

6.5.5 *Type V*—Type V cement has high resistance to sulfate, and brine solutions. This material has ultimate strength development somewhat less than either Types I or II.

6.5.6 Type K cement is expansive and can be used to compensate for shrinkage. This cement is essentially Type I or more commonly Type II Portland Cement with additives to produce expansion. It can be of use in plugging situations where water-tightness is important. Type K cement contains calcium sulfoaluminate. When mixed with water, the hydration causes an expansion ranging from approximately 0.05 to 0.20 % (2).

6.6 *API Cements* (3):

6.6.1 *Class A*—Class A cement corresponds closely to ASTM Type 1. This cement is intended to be used from the surface to a depth of 6000 ft (1828 m).

6.6.2 *Class B*—Class B cement corresponds closely to ASTM Type II. It is intended for use from the surface to a depth of 6000 ft (1828 m) and is also available as a high-sulfate resistant variety.

6.6.3 *Class C*—Class C cement corresponds closely to ASTM Type III. It is intended for use from the surface to a depth of 6000 ft (1828 m). It is also available as a high-sulfate resistant variety.

6.6.4 *Class G*—Class G cement is intended for use from the surface to a depth of 8000 ft (2438 m) and can be used with accelerators or retarders to cover a wide range of depths and temperatures. The cement is also available as a high-sulfate resistant variety.

6.6.5 *Class H*—Class H cement is intended for use from the surface to a depth of 8000 ft (2438 m). It can be used with a wide variety of accelerators and retarders to cover a wide range of depths and temperatures. It is available only as a moderate-sulfate resistant type.

6.6.6 *Class J*—This cement is intended for use from a depth of 12 000 to 16 000 ft (3658 to 4877 m) where

---

[4] The boldface numbers in parentheses refer to a list of references at the end of the text.

TABLE 1 Properties of Common Plugging Materials

Plugging Material	Description	Positive Attributes	Negative Attributes
ASTM C-50 Portland Cement			
Type I	Most commonly used type of cement for plugging	Forms a good seal when used with bentonite in 3 to 5 % concentration. Commonly available and can be purchased premixed on-site.	High heats of hydration may be a problem in PVC-cased wells. Can shrink and crack; low-sulfate resistance. Should not be used in the presence of strong acids or in low-pH environments.
Type II	Similar to Type I, but with a moderate heat of hydration.	Moderate heat of hydration. Moderate resistance to sulfate.	Somewhat slower strength development than Type I; expensive. Can shrink and crack. Can be difficult to use. Should not be used in the presence of strong acids or in low-pH environments.
Type III	High early strength.	May prove useful in situations where high early strength is needed, such as borehole walls that have a tendency to collapse.	Not a common cement. Can set very quickly before decommissioning is completed. Should not be used in the presence of strong acids or in low-pH environments.
Type IV	Low heat of hydration.	May prove useful in situations where a low heat of hydration is required	Not a common cement. Should not be used in the presence of strong acids or in low-pH environments.
Type V	Similar to Type I, with high resistance to sulfate and brine.	High resistance to sulfate and brine. Low heat of hydration.	Ultimate strength is less than Types I and III. Expensive; should not be used in the presence of strong acids or in low-pH environments. Can be difficult to use. Can shrink or crack.
K	Expansive cement.	Basically Type I or Type II Portland Cement with additions (tricalcium sulfo aluminate for example) to provide for expansion. Expansion is generally in the range from 0.05 to 0.20 % Good resistance to sulfate attack.	
API 10			
Class A	Similar to ASTM Type I.	Can be used to a depth of 6000 ft (1828 m). Forms a good seal when used with bentonite in 3 to 5 % concentration. Commonly available and can be purchased premixed on-site.	High heats of hydration may be a problem in PVC-cased wells. Can shrink and crack; low-sulfate resistance. Should not be used in the presence of strong acids or in low-pH environments.
Class B	Similar to ASTM Type II.	Can be used to depth of 6000 ft (1828 m). Moderate heat of hydration. Moderate resistance to sulfate. Available as a high-sulfate resistant variety.	Somewhat slower strength development than Type I; expensive. Can shrink and crack. Can be difficult to use. Should not be used in the presence of strong acids or in low-pH environments.
Class C	Similar to ASTM Type III.	Can be used to a depth of 6000 ft (1828 m).	Can set very quickly before decommissioning is completed. Should not be used in the presence of strong acids or in low-pH environments. Can shrink and crack.
Class G	Useful in a wide range of temperatures and depths through the use of accelerators or retarders.	Can be used to a depth of 8000 ft (2438 m). Available as a sulfate-resistant variety.	Should not be used in the presence of strong acids or in low-pH environments. Can shrink and crack.
Class H	Useful in a wide range of depths and temperatures through the use of accelerators or retarders.	Can be used to a depth of 8000 ft (2438 m). Available only as a moderate sulfate type.	Should not be used in the presence of strong acids or in low-pH environments. Can shrink and crack.
Class J	Intended for use from a depth 12 000 to 16 000 ft (3658 to 4877 m).	Has use where extremely high temperatures and pressures occur.	Should not be used in the presence of strong acids or in low-pH environments. Can shrink and crack.
Pozzolanic cement	Addition of silicious materials to ASTM Type V or API Class A cement.	Good resistance to corrosive conditions and in reducing the permeability of cement.	Many types of materials can be used that can result in variable results.
Epoxy cements	Vinyl ester resins.	Good chemical resistance to acids and bases. Can use available equipment to place cement.	Very expensive. Poor chemical resistance to chlorinated hydrocarbons and acetic aid. Should be used only by experienced personnel. Water accelerates curing, must use diesel oil to precondition hole (diesel may increase contamination of site if hydrocarbons are a concern).
Bentonite			
Pellets	Granular bentonite compressed into a tablet	Uniform in size. Easy to use. Produces a low permeability seal.	Must be hydrated after placement. Shrinkage may occur when desiccated or when in contact with high concentrations of organic compounds (greater than 2 %) or materials that are strongly acidic or alkaline. Expensive.
Chips	Raw mined montmorillonite in the form of chunks ¼ to ¾ in. (.64 to 1.91 m) in size.	Inexpensive. No mixing equipment required. Forms a low-permeability seal.	Difficult to place. Must be hydrated after placement. Less swelling than beneficiated bentonite. Shrinkage may occur when desiccated when in contact with high concentrations of organic compounds (greater than 2 %) or materials that are strongly acidic or alkaline.
Granular	Raw mined montmorillonite crushed and seared to an 8 to 20-mesh size.	Can be placed at depth in dry holes. Forms a low- permeability seal.	Difficult to place in holes containing water as it quickly hydrates. Can bridge in hole. May desiccate when in contact with high concentrations of organic compounds (greater than 2 %) or materials that are strongly acidic or alkaline causing shrinkage.

TABLE 1 *Continued*

Plugging Material	Description	Positive Attributes	Negative Attributes
Powdered	Pulverized and seared bentonite that passes a 200- mesh screen. Used as drilling mud or as an additive to cement.	Used with cement to compensate for shrinkage (under saturated conditions). Other additives can be used to inhibit swelling, etc. Retards cement set; lowers heat of hydration.	May not be a desirable plugging material in deep vadose zones due to the drying out of the material, resulting in cracking. Difficult to place in holes containing water, as it quickly hydrates. Can bridge in hole. May desiccate when in contact with high concentrations of organic compounds (greater than 2 %) or materials that are strongly acidic or alkaline causing shrinkage.
High solids clay grout	Powdered bentonite (200 mesh) mixed with fresh water to form a slurry with a minimum of 20 % solids and a density of 9.4 lb/gal (1126 Kg/m³ g/L).	Does not shrink during curing. Low density reduces formation losses. Forms a low-permeability seal that stays flexible as long as it is hydrated.	May not be a desirable plugging material in deep vadose zones due to the drying out of the material, resulting in cracking. May desiccate when in contact with high concentrations of organic compounds (greater than 2 %) or materials that are strongly acidic or alkaline causing shrinkage. A low-strength material subject to expansion under low-pressure differentials such as artesian conditions.

extremely high temperatures and pressures can be expected to occur.

6.6.7 *Other Cements*—Other cements have been developed that may have applicability in decommissioning activities. These include the following:

6.6.7.1 Ultralight cements with a slurry density that can be as low as 6 lb/gal (719 cm Kg/L). This material can be made by foaming the cement with nitrogen or through the addition of hollow glass microspheres between 60 and 315 μm in diameter. The latter forms a slurry of between 9 and 12 lb/gal (1078 and 1438 Kg/L). Ultralight cements and microspheres have been reported **(4)** for cement unconsolidated sands and for plugging cavernous formations and lost circulation zones. Reference **(5)** provided similar information on microspheres. Microspheres can also be used in high-pressure applications when it may be desirable to limit density increases. Another advantage is the low water/cement ratio due to the low water absorbency and low density **(5)**.

6.6.7.2 *Pozzolanic-Portland Cements*—These cements consist of silicious materials that develop into a cement in the presence of lime and water. Both natural materials of volcanic origin such as perlites (volcanic ashes), heat-treated clays, shales, tuffs, opaline cherts, diatomaceous earth and artificial materials consisting of byproducts from glass factories, furnace slag, and fly ash have been used **(2, 4, 6)**. The large variety of materials that can be used as a source for pozzolans may result in variable results.

6.6.7.3 Pozzolans act to extend cement and decrease density. The specific gravity of fly ash ranges from 2.3 to 2.7 (depending upon the source) while portland cement is 3.1 to 3.2 **(2)**. These materials can also provide improved resistance to corrosive fluids. Table 2 provides a comparison of sulfate resistance between ASTM Type V cement with and without pozzolans.

6.6.7.4 The improved resistance to corrosive materials is accomplished in part as many pozzolans contain zeolites which have the ability for ion exchange between the corrosive material and the alkaline component in the cement **(7)**. Secondly, the use of pozzolans also decreases cement permeability over time. This occurs as a result of the increased percentage in hydrated cement containing materials resulting from the release of calcium hydroxide and the silica combining with lime from the cement to form a stable material **(8)**.

6.6.7.5 Pozzolans are added to portland cement by adding 74 lb (33.6 Kg) (as fly ash) per sack of cement. If perlites are used, 2 to 6 % of bentonite by weight is needed to keep the perlite from floating **(9)**.

6.6.8 Gypsum cements can be used for high early strength development and their ability to set rapidly. These cements expand approximately 0.3 %. This cement may have use in plugging highly permeable formations. Care should be taken in using gypsum cements due to the solubility of gypsum.

6.6.9 *Epoxy Resin Cements*—Epoxy (vinyl ester resins) cements may have applicability in decommissioning wells and boreholes where corrosive materials may be present **(5)**.

6.6.9.1 Epoxy cement consists of an epoxy base, hardener, accelerator, and inert filler. Resin viscosity is reduced by the addition of a nonreactive liquid diluent that also controls exothermic heat during polymerization. An inert solid filler such as very fine silica or barite is added to further reduce reaction heat and increase strength **(10)**.

6.6.9.2 There are several advantages of using epoxy cements. Reference **(11)** reports that a stronger bond occurred between the casing and the formation. Cole also reported resistance to various chemicals (see Table 3). The cement is highly resistant to high concentrations of hydrochloric and sulfuric acid, but is not suitable for use in environments where acetic acid, chlorinated hydrocarbons, or toluene are present.

TABLE 2 **Comparison of ASTM Type V Cement With and Without Pozzolan Materials**[A]

Cement Type	Relative Degree of Sulfate Attack	Percentage of Water Soluble Sulfate (as SO_4) in Soil, ppm	Sulfate (as SO_4) in Water Samples, ppm
V	Severe	0.20 to 2.00	1 500 to 10 000
V (plus pozzolan)	Very severe	2.00 or more	10 000 or more

[A] See Ref **(4)**.

TABLE 3 **List of Chemicals Reported Not to Affect Epoxy Cement**[A]

Chemical	Concentration, %
Hydrochloric acid	30
Sulfuric acid	25
Chromic acid	up to 10
Nitric acid	5
Hydroxide	up to 20
Hypochlorite	up to 6

[A] See Ref **(11)**.

6.6.9.3 This cement is expensive and requires removal of water (that reduces settling time) through the use of gelled and weighted diesel oil to precondition the hole. The use of this material may increase contamination at the site, if diesel oil (hydrocarbons) are a concern at the site.

6.6.10 *Cement Additives*—A number of materials can be added to cement to modify properties to meet a specific need. Cement additives can be used to extend, accelerate, retard, increase density, control fluid losses, control circulation losses, or reduce friction (4). Several of these materials have more than one use; for example, sodium chloride can be used to accelerate or retard cement. The most common cement additives are discussed in the following subsections. Figure 1 lists these additives and presents the relative impact of their use on selected performance criteria.

6.7 *Extenders:*

6.7.1 Bentonite is the most commonly used material in modifying cement properties. It can be added to most ASTM and API cements. In decommissioning activities, the percentage of bentonite added is generally no more than 4 %. It has the following effects when added to cement (4):

6.7.1.1 Lowers the hydraulic conductivity of the cement;

6.7.1.2 Increases slurry viscosity;

6.7.1.3 Reduces fluid loss to the formation;

6.7.1.4 Provides for a longer pumpability at normal pressures as a result of delaying strength development;

6.7.1.5 Reduces compressive strength; and

6.7.1.6 Lowers resistance to chemical attack.

6.7.2 Bentonite increases shrinkage as it ties up large volumes of water that would normally be in the cement. A second common extender are pozzolans which have already been addressed in 6.6.7.2.

6.8 *Accelerators:*

6.8.1 Accelerators hasten the settling of cement and are useful when voids occur, or when cement plugs are to be used in the first pour. Two common materials are used; calcium chloride and sodium chloride.

6.8.2 Calcium chloride is available as a powder or flake. Flakes are the most commonly used form, as it is easy to store and can absorb some moisture without becoming lumpy (2). Two to four % of calcium chloride by weight is used to achieve maximum acceleration. The use of calcium chloride should be considered when a rapid set, a decrease in viscosity, and early strength are desired.

6.8.3 Sodium chloride can be added between 1.5 to 5 % by weight of cement to reduce setting time. Maximum acceleration occurs at a concentration of 2 to 2.5 % except when higher water ratios are used (2).

6.9 *Retarders:*

6.9.1 Sodium chloride can be used to retard the settling of cement as well as accelerate the setting of cement. Fifteen to seventeen % salt (14 to 16 lb (6.35 to 7.26 Kg) of salt per sack) by weight is added to retard cement (2).

6.9.2 Other chemicals (cellulose, lignosulfates) have been used as retarders, but are not appropriate for decommissioning activities without additional information on their compatibility with waste and their effect on water quality. Reference (12) indicates that sugar-derived retarders such as cellulose lignosulfates are destructive to cement strength and should not be used where strength is important. Organic

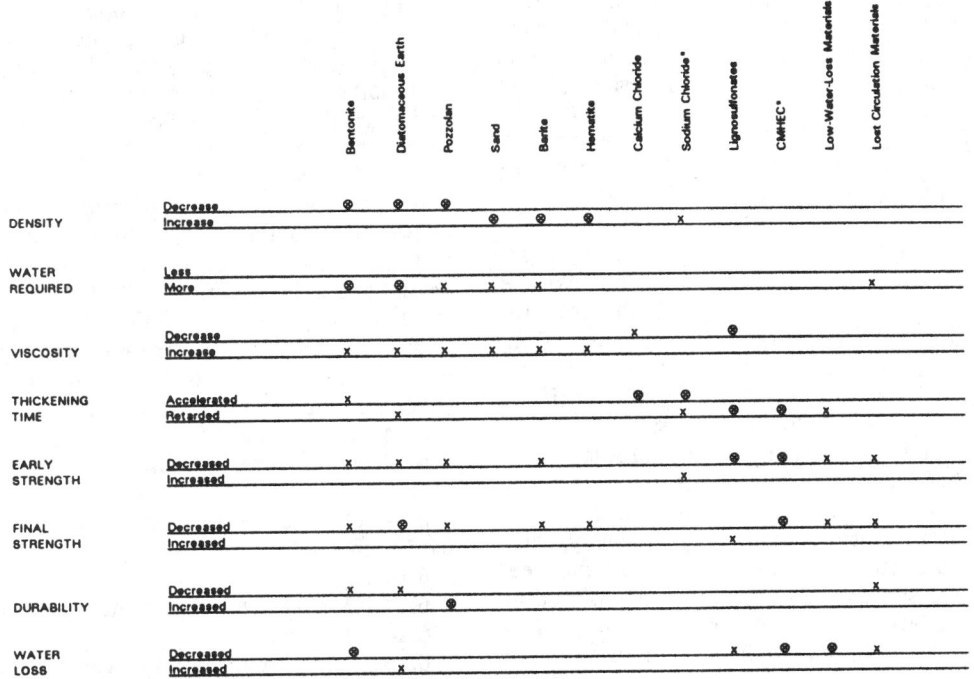

NOTE—See Ref (3).

FIG. 1 Effects of Some Additives on the Physical Properties of Cement

retarders should not be used for decommissioning activities.

6.10 *Density Improvers*—The density of cement can be improved to increase hydrostatic pressure. Sand can be used to increase density without affecting the cement chemically although additional water is required. Barite has been used, but may interact with waste and should not be used.

6.11 *Fluid Loss Controllers:*

6.11.1 Various organic materials such as cellulose can be used to produce a constant water to solids ratio that may have applicability when a grout is placed under pressure and water loss can occur. However, these materials may not be suitable for decommissioning activities, as they may contribute to contamination.

6.11.2 These organic materials (fibrous materials, cellophane flakes) act to block the movement of the grout into the formation. It is not desirable to use these materials in decommissioning activities due to their organic content that may adversely affect water quality and also may not result in a good plug.

6.12 *Friction Reducers (Dispersants):*

6.12.1 These materials reduce friction to improve flow and can be effective when the water cement ratio is reduced. Reduction of the water cement ratio is a method to decrease cement friction. (It is possible to reduce the amount of water added by using a dispersant (5). These materials (sodium chloride, polymers, and calcium lignosulfonate) also help to reduce the energy required to pump the grout. Polymers and calcium lignosulfonate may not be appropriate materials for decommissioning activity as they may affect water quality.

6.13 *Bentonite*—Bentonite is predominantly composed of the clay mineral sodium montmorillonite. It has the ability to absorb large quantities of water and swell to many times its original size when hydrated, and the material remains flexible. Bentonite clay may be used in any of its various forms to meet placement, strength, and sealing criteria listed in 6.3. The amount of shrinkage or settling of a bentonite seal is dependent on the percent solids of bentonite, composition of surrounding formation and its soil moisture. Higher water to bentonite ratios increase the likelihood of dehydration.

6.13.1 The permeability of bentonite is very low; hydraulic conductivities of 1×10^{-6} cm/s or less can be achieved. However, bentonite may desiccate in the presence of high concentrations of some organic chemicals, strong acids or bases, saline ground water, or when allowed to dry, thereby increasing its hydraulic conductivity. Bentonite is commercially available in the following forms:

6.13.1.1 *Pellets*—Pellets are made from granular powdered bentonite that has been compressed into tablets, commonly ¼ to ¾ in. (.64 cm to 1.91 cm) in diameter. Pellets have a low-moisture content, high density, and uniform size. Pellets should be composed of additive-free, high-swelling granular sodium bentonite. Properly placed in a well or borehole, pellets hydrate and expand creating a low permeability (1×10^{-6} cm/s) plug. Pellets can be used in the saturated zone provided the length of the water column is short. The rate of pour into the hole should not be more than 50 lb (22.7 Kg) of bentonite in 5 min (1).

6.13.1.2 *Preformed Donuts*—Commercial preformed donuts consist of compressed bentonite and may have use in decommissioning activities.

6.13.2 *Chips*—Raw mined sodium montmorillonite in the form of chunks that are ¼ to ¾ in. (6.4 to 1.91 cm) in diameter. Their angular shape can make it difficult to place chips to the desired depth in a small-diameter well or borehole without bridging.

6.13.2.1 Fine-grained material resulting from the mechanical breakdown during shipping may cause a problem in the placement of chips due to clumping. Fines should be screened through a ¼-in. (6.4-mm) mesh screen before use.

6.13.2.2 The lower affinity for water that chips have allow them to fall through a water column without rapid hydration.

6.13.2.3 Chips have applicability in large-diameter boreholes and when carefully dropped into the hole to reduce bridging.

6.13.3 *Granular*—Raw-mined sodium montmorillonite without any additives that has been crushed and seared to an 8 to 20-mesh size. This material can be placed at depth in dry holes but hydrates quickly when placed into water. It often sticks to wet borehole walls and bridges when placed through water. Granular material is best suited for use in the unsaturated zone with enough water added to provide adequate hydration.

6.13.3.1 Fines can clump when in contact with water (1). Fines result from mechanical breakdown of the material during shipping. Granular bentonite should be poured slowly to reduce the potential for bridging. In some situations, a pour rate not exceeding 50 lb (227 Kg) in 5 min has been used successfully (1).

6.13.4 *Powdered*—Untreated, seared, and ground bentonite that passes through a 200-mesh screen. It is designed to be used in drilling fluids (muds) and as an additive to other plugging materials such as cement. Bentonite powder slurry can become an effective grout material when combined with density-increasing additives and swelling inhibitors. Powdered bentonite should not be placed in dry form through water as it can bridge and stick to the borehole walls.

6.13.5 *High Solids Clay Grout*—This material is a blend of powdered polymer-free bentonite clays mixed with fresh water that forms a slurry with a minimum 20 % solids by weight and a density of 9.4 lb/gal (1126m³ Kg/L). The slurry sets to a low-permeable plastic grout that generates no heat of hydration and does not shrink during curing in the presence of moisture. High solids clay grouts are commonly used for borehole plugging.

6.14 *Other Materials:*

6.14.1 A number of other materials have been used for plugging:

6.14.1.1 Attapulgite clay (may have applicability when used with a salt cement grout)[5]

6.14.1.2 Fire clay

6.14.1.3 Commercial packing materials

6.14.1.4 Packers

6.14.2 These materials are either inappropriate for use in decommissioning wells or boreholes where hazardous waste are encountered, or are not well studied for decommissioning wells in hazardous waste situations. Therefore, they

[5] Sutton, Fred, Personal Communication, 1990.

are not discussed in this guide.

6.14.3 Other materials have been used in the past for plugging wells and boreholes. Such materials as wooden or lead plugs should not be used because wood plugs may decay and lead is a potentially hazardous material. Mechanical packers composed of steel, plastic, or other materials can be used to assist in plugging.

7. Procedure

7.1 The primary purpose of most boreholes and monitoring wells is to monitor chemical compounds in the soil and ground water; however, there are other uses for subsurface monitoring including the measurement of temperature, soil gas sampling, or measurement of geophysical parameters. Use significant care in planning and implementing the decommissioning activity. It is important to obtain any required approvals from regulatory agencies, land owners, responsible parties, and other parties involved with the site. The following subsections present a recommended list of tasks in order that the decommissioning activity is successfully completed. Several of the steps outlined below do not pertain to boreholes and may be omitted.

7.2 *Planning:*

7.2.1 *Records Review*—Carefully review all available records and information relating to use of the monitoring well, borehole, etc. This review may include the following information:

7.2.1.1 Review applicable Federal, state, and local regulations relating to decommissioning activities. This may include contacting the applicable state or local agency having jurisdiction over drilling activities and preparation of the necessary documentation to drill (start card),

7.2.1.2 Collection of drillers' logs, geophysical logs, well construction, or geologic logs, including stratigraphy, structural geology, subsurface information, construction materials, screened interval, depth, hydraulic gradients (if water levels are available from other wells for its determination), legal location, date of installation, and photographs of the well,

7.2.1.3 Review of analytical chemical data for soil and ground water over the life of the well, and variations in water levels over time,

7.2.1.4 Review of records of the repairs, modifications, or other changes made to the well during the lifetime of the well,

7.2.1.5 Evaluation of historic, current, and planned land use,

7.2.1.6 Interviews with local workers and collection of other pertinent data such as discussing site conditions with local drillers, and

7.2.1.7 While not directly part of the decommissioning activity, proper disposal of displaced fluids and other materials (such as pulled or drilled out casing and cement seals) should be considered. Some of these materials may be classified as a hazardous waste under Federal, state, or local regulations. Conduct a review of these regulations and appropriate analytical documentation prior to classifying a material as a hazardous waste.

NOTE 9—This information may be summarized in a work plan. A work plan would also include a description of the site geology and hydrogeology and the decommissioning method to be used.

7.2.2 *Verification of Field Data*—The variety and quality of field practices and reporting require that the well be inspected to verify the actual field situation prior to decommissioning the well. The following list of procedures is recommended so that the actual condition of the well is known. Some of the borehole geophysical logs may not be applicable or may not be available for small diameter (2 in. (5.08 cm) or less) holes or wells.

7.2.2.1 Inspection of well head installation for integrity,

7.2.2.2 Current depth measurement of the casing and well. (The original depth of the borehole may be different than the well.),

7.2.2.3 *Water Quality Sampling and Analysis*—A final water quality sample taken from the well may be required for regulatory purposes;

7.2.2.4 *Downhole Inspections*—Including caliper logs to measure inside diameter; television logs to determine in-well conditions such as casing breaks, screen size, etc.; gamma logs to verify geologic information, if not already available; cement bond logs (sonic) to determine if the casing is firmly attached to grout (presently available for holes 2½ in. (6.35 cm) or larger in diameter); flow logs (flow meter or spinners) to determine if vertical flow occurs within the casing; and hydraulic integrity test to determine if the well casing is intact.

NOTE 10—Care should be taken in running any of these tools in a well with a collapsed or broken casing, or in boreholes that may collapse on the tool. Tools with active radioactive sources should not be used under these circumstances. Conduct downhole inspections only after obstructions are removed from the well casing or borehole.

7.2.2.5 Contact local owner, resident operator/observer to verify operations at the site.

7.2.2.6 Verification of field data is an ongoing responsibility. Use verified information to modify plans in order that the decommissioning activity is correctly conducted. Continue this activity during the field phase and change specifications as needed.

7.2.3 *Review of Decommissioning Options*—After the records have been thoroughly reviewed and verified in the field, select an appropriate decommissioning procedure. Evaluate each possible option to determine the most appropriate method for the selection. The following list of evaluation criteria is recommended:

7.2.3.1 The potential for fluid movement from one aquifer into another by means of the borehole or well should be eliminated.

7.2.3.2 Materials to be used in plugging must be compatible with well casing and screen (if left in place), and with subsurface formation and ground water, etc. over the period of time hazardous materials are found at these sites.

7.2.3.3 Future land use (as is known at the time of decommissioning) should be compatible with decommissioning plans.

7.2.3.4 Closure options should be compatible with applicable federal, state, and local requirements.

7.3 *Implementation:*

7.3.1 *Field Procedure:*

7.3.1.1 Satisfactory completion of decommissioning is the primary purpose of this guide. All work performed on the borehole or well should be completed by competently trained drillers, equipped with appropriate tools, under the direction of a geological or engineering professional who is qualified to certify that the decommissioning is completed according to the planned procedures and is consistent with applicable regulations.

7.3.1.2 Approve any modifications to the proposed work plan and record in writing by the on-site geologist or engineer (or their representatives) prior to implementation.

7.3.1.3 The geologist or engineer should be on-site during the field activities to verify that the activities are completed as planned. Decommissioning operations can be successfully accomplished by careful planning and documentation (see 7.5). Maintain documentation of decommissioning activities for the post-closure period or period required by regulations (if specified). While regulations may require documentation for a period of 30 years, it is advisable to continue this activity for a period lasting as long as hazardous materials occur at the site.

7.3.1.4 Remove casing from the ground by either pulling or overdrilling (see 7.3.7). Depending upon construction, it may be necessary to leave the casing in place and produce suitable perforations in the screen and blank casing to allow for the plugging material to penetrate the annular space and formation (see 7.3.7). If grout in the annular space can be verified to be in good condition, the well can be decommissioned by cutting the blank casing and filling the screened interval with grout. Verifying the integrity of grout may be difficult to impossible. If a filter pack is present, it may be necessary to remove the filter pack after perforating the casing by washing or overdrilling. Several of the methods identified in this subsection are briefly discussed in the Appendix X1.

7.3.1.5 If well construction conditions are not adequately known and the well site contains hazardous materials, it may not be appropriate to remove the casing and screen, as this may increase the mobility of hazardous materials.

7.3.1.6 The borehole or well, or both, may require preconditioning for decommissioning to be successful. Preconditioning can reduce the potential for sloughing of the borehole wall if for example, a sodium montmorillonite clay occurs naturally in the formation and cement is used as the plugging material. The calcium contained in Portland cement exchanges with the sodium cation in bentonite clay decreasing the water contained in the clay and inducing sloughing. This problem can be significant in sediments or rocks that are under considerable pressure, causing a loss in part of the hole, thereby not completely plugging the borehole and possibly causing loss of all downhole equipment. These conditions are usually known by local drilling and well servicing contractors who should be contacted prior to the start of field operations.

7.3.1.7 Preconditioning consists of removing mud from borehole walls (when mud is used for drilling the borehole), or stabilizing a borehole prior to placement of the plugging material. If a drilling mud has been used to drill a borehole, preconditioning can involve the circulation of a high-quality, low-solids drilling fluid to remove gelled mud from the borehole and borehole walls prior to plugging. For the above situation, a high-quality bentonite or drilling fluid can be used. If a drilling mud is used as the plugging material, prepare fresh mud. The mud used in drilling contains cuttings and may also not have suitable properties (13). The

selection and use of material(s) should meet the requirements specified in Section 6.

7.3.2 *Volume of Plugging Material Required:*

7.3.2.1 The geologist or engineer should calculate the volume of plugging material required for the borehole or well after first taking into consideration applicable loss of material to the formation, voids intersecting the borehole, changes in borehole diameter, washout zones, and swelling or shrinkage of material. An approximation of the volume of material that may be required can frequently be provided by contacting local drillers or professionals.

7.3.2.2 Loss of plugging material into the formation may occur rapidly (within minutes) or after several hours or days. The volume of plugging material required to be on-site should be at a minimum, enough to fill one borehole volume, however, it is advisable to have available a minimum of 25 to 50 % in excess of the calculated borehole volume (1, 2). Additional plugging material should be readily available to site personnel under short notice, especially if it is common to lose plugging material into the formation or the material is required for clearing a hole (see 7.3.8.2).

7.3.2.3 A caliper log is helpful in boreholes or overdrilled holes to define hole diameter. If this information is not available, use an estimate of borehole diameter from the available well construction specifications. Calculate the volume of the borehole using the following:

$$v = \pi r^2 L$$

where:
v = volume,
L = length of borehole or well to be plugged, and
r = radius of hole.

7.3.2.4 Manuals listing the volume of a hole per linear foot (metre) such as Ref (3) can also be used. These manuals are available through major deep well cementing service companies. Assume in the calculations or table used that the derived volume is the minimum required for actual conditions due to possible loss of plugging material into the formation. Verify grout emplaced through estimating the volume of material leaving the hole for holes filled with mud or water. For all boreholes, measure the volume of material emplaced and check for the appearance of the material at the top of the interval grouted or land's surface, whichever is applicable.

7.3.2.5 Location of the grout at depth can be difficult because the top of the grout may not be able to be distinguished from water or other fluids in the borehole. One field procedure that has been applied to measure the level of the grout is to use a wooden sounding block with a weight attached to the block. The weight should be slightly denser than the grout.

NOTE 11—Location of the depth of plugging material in a borehole or well does not confirm that the plug has performed its intended purpose. Plugging materials may not have adhered to the walls of the borehole or well or may have been bypassed from the target zone. Therefore, placement of plugging materials is highly dependent upon proper preparation of the hole.

7.3.2.6 In areas where coarse materials are encountered, or considerable fracturing or solution openings occur, two or three times the calculated volume of material may be required to fill the hole. In extreme cases such as in some karst terrains where large conduits may exist, it may be difficult to plug a well or borehole. In these situations, packers can be used to isolate critical intervals for filling with plugged materials above or below the karstic formations. Packers may also be used to isolate critical intervals within a well or borehole throughout the entire depth.

7.3.3 *Quality of Water Used for Grout*—Grout must be carefully mixed using water of a known chemical quality. The quality of the water must be compatible with the grouting material and not introduce contamination into the subsurface. For cement, water used should be free of silt, organic matter, alkali compounds, and have a total solids content of 500 mg/L or less (14). The grout should be weighed using a mud balance.

NOTE 12—Thicker grout mixtures may have greater effectiveness in plugging highly permeable materials or materials with large voids, but may be very difficult to pump through a grout pipe. Less viscous mixtures are easy to pump, but may be too mobile, have greater shrinkage, and take longer to set if additives to compensate these problems are not added.

7.3.4 *Mixing of Cement and Bentonite On-Site*—Both cement and bentonite can be mixed on site. Cement can also be purchased from a contractor and delivered premixed on site, provided the water used is of known and acceptable quality.

7.3.5 *Use of Curing Accelerators*—Curing accelerators can be used to hasten the initial set time for cement from 6 to 8 h to perhaps 2 or 3 h, provided the accelerator (typically calcium chloride, sodium chloride, aluminum powder, or gypsum) does not degrade the cement or react with the environment. The use of these materials may also require approval by regulatory authorities.

7.3.6 *Decommissioning of Boreholes:*

7.3.6.1 To achieve an effective seal, the borehole should be free of debris and foreign matter that may restrict the adhesion of the plugging materials to the borehole wall. Clear boreholes of excessive mud filtercake, or gelled mud (if used) and any bridges resulting from the removal of temporary casing, or when noncohesive materials (such as sand and gravel) are encountered that can lead to a collapsed borehole during decommissioning activities.

7.3.6.2 One method commonly used for effective removal of these materials is to advance a small grout pipe to the bottom of the borehole by use of water or a high-quality bentonite slurry and flush the hole. Flushing is continued until the blockage is removed, or noncohesive formations are stabilized. Use at least two borehole volume of materials. If the flushing fluid returning to the surface contains potentially hazardous materials, it may be necessary to place this material into containers for proper disposal.

7.3.6.3 As soon as the borehole is prepared, pump plugging material slowly through a grout pipe to displace the flushing fluid. Inject grout starting from the bottom of the hole, forcing other fluids upward. Complete grouting slowly to prevent channeling of the grout around any undesirable material remaining in the hole. Complete this operation in one continuous operation. Raise the grout pipe when pumping pressure increases significantly, or when undiluted grout reaches the surface. Regularly sample and evaluate overflowing grout for weight, presence of foreign material, or other changes. When the overflowing grout is similar to that

being pumped down the hole, the plugging is considered complete.

7.3.6.4 Grout pipes should be the largest diameter that is practical for field operations to reduce the required pumping pressure. Cut the lower end of the pipe at an angle to allow for the side discharge of the grout. Side discharge reduces the potential for erosion of the borehole wall. Typically, heavy-walled PVC (Schedule 40 or greater) or thin-walled steel pipe is used for grouting purposes. Use caution while grouting deeper holes in order that downhole pressure does not exceed the rupture strength of the grout pipe.

7.3.6.5 Grouting of shallow auger holes or other boreholes where grout pumping equipment is not readily available may be accomplished by placing grout through a side discharge grout pipe that has a funnel attached to the top. As the grouting progresses, slowly raise the pipe. Take care when using this procedure, as the low placement pressure may not completely fill or flush undesirable materials from the borehole. This procedure is only recommended when the entire borehole depth is 10 to 15 ft (3 to 4.6 m) or less.

7.3.6.6 Small diameter (<2-in. (5.08 cm-mm)) holes are difficult to plug. A small diameter (¾-in. (1.9 cm-mm)) grout pipe can be used, however, high-pumping pressures may be required or less viscous plugging materials may be necessary.

NOTE 13—It is important to avoid hydrofracturing or blowout in shallow holes in unconsolidated materials when pressure grouting. A general rule to avoid hydrofracturing is to restrict pumping pressure to about 0.6 lb/in.^{2} (0.42 Kg/cm^2) of hydraulic pressure for each foot (metre) of overburden. In some instances, the pressure used must be reduced further.

7.3.6.7 A conservative method (15) of calculating maximum pressure is as follows:

$$Pm = (0.733 - 0.433\ Sg)d$$

where:

Pm = pressure of fluid injected at wellhead, lb/in.^{2} (Kg/cm^2),

Sg = density (specific gravity) of the plugging material (grout) (unitless), and

d = depth measured for the surface to the opening of the grout pipe (metres) feet.

Hydrofracturing typically occurs during start-up or restart of grouting.

7.3.6.8 Inject grout at the bottom of the hole, displacing grout, loose formation materials, and borehole fluids upward. Slowly pump grout to avoid channeling. Do this in one continuous operation. Raise the grout pipe when grout can no longer be easily forced from the pipe into the hole or when undiluted grout reaches the top of the hole and flows out. Sample and weigh this material. The weight of the grout returning to the surface should be the same as the grout pumped into the hole.

7.3.7 *Control of Elevated Formation Pressures*—Occasionally, a borehole or well may penetrate a formation that is under confined conditions (artesian head), or from which a gas is being released, under pressure. (Gas bubbling upward through the grout may result in open channels, or increase the permeability of the grout.) When this condition is encountered, sealing of the borehole requires that the grout pressure be maintained greater than the formation pressure until initial grout set occurs. The "shut-in-pressure" of the formation can usually be determined by use of pressure gages attached to a casing that has been pushed or driven into the confining unit. This casing must be tightly sealed to prevent leakage around its annulus. (Separate packers or casing grout, or both, may be required to prevent leakage around the casing.)

7.3.7.1 Several procedures may be used to balance the formation pressure until the initial set of the grout has occurred. (See Fig. 2.) The procedure most often used to contain the pressure is to use a sufficient column of grout. If additional head pressure is required, increase the unit weight of the grout by the addition of sand, densifying additives (barite, or hematite) or decreasing the water cement ratio. Another alternative commonly used is to pump the grout through a check valve into the casing until the top of grout reaches the desired elevation. Maintain pressure above the grout by use of compressed air, or by use of a simple standpipe filled with water. If the standpipe procedure is used, pump grout through a tremie pipe extending through the water to prevent dilution.

7.3.7.2 After the initial grout set has occurred (minimum 24 h), the sustaining pressure (above atmospheric pressure) may be released. However, record the air pressure (or fluid level) in the air filled casing above the grout frequently for several days to verify that no leakage is occurring. If gas (of any variety) is present in the formation, analyze a sample of the air from the casing above the grout. Care should be taken under these conditions, as explosive or hazardous conditions may be present. Methane, the most commonly produced gas, is odorless, colorless, toxic, and extremely flammable. Safety precautions are advised.

7.3.7.3 Only after testing results confirm that the seal is effective, is it advisable to complete the permanent closure or removal of the near (or above) surface casing.

7.3.7.4 Measure any fallback or settlement of the grout surface (whether by gravity or under pressure) that occurs after grout emplacement to the nearest foot and record. Correct any fallback with native or imported materials to grade, or to a specified depth below grade, such as below the plow layer. The depth is based on existing and proposed land use and regulatory requirements.

7.3.8 *Decommissioning of Wells and Other Monitoring Devices:*

7.3.8.1 This discussion has applicability in decommissioning ground water monitoring and injection wells and soil gas monitoring wells, neutron probe access tubes, lysimeter and tensiometer and installations, and similar devices. Appendix X1 should be consulted for additional information.

7.3.8.2 It is desirable to remove all existing well construction materials such as screen, casing, filter pack, seal, and grout from the hole to reduce the potential for the formation of a vertical conduit to occur at the contact between casing and annular seal; reduce the potential of these materials interfering with the decommissioning operation; decrease the potential of a reaction between the materials used; or to minimize the reaction with the native materials or ground water. In situations where well materials are removed and borehole collapse occurs, redrill the borehole following the guidance provided in 7.3.6.

7.3.8.3 Steel casing may be removed using jacks to free casing from the hole followed by lifting the casing out by using a drill rig, backhoe, cranes, etc. of sufficient capacity. If

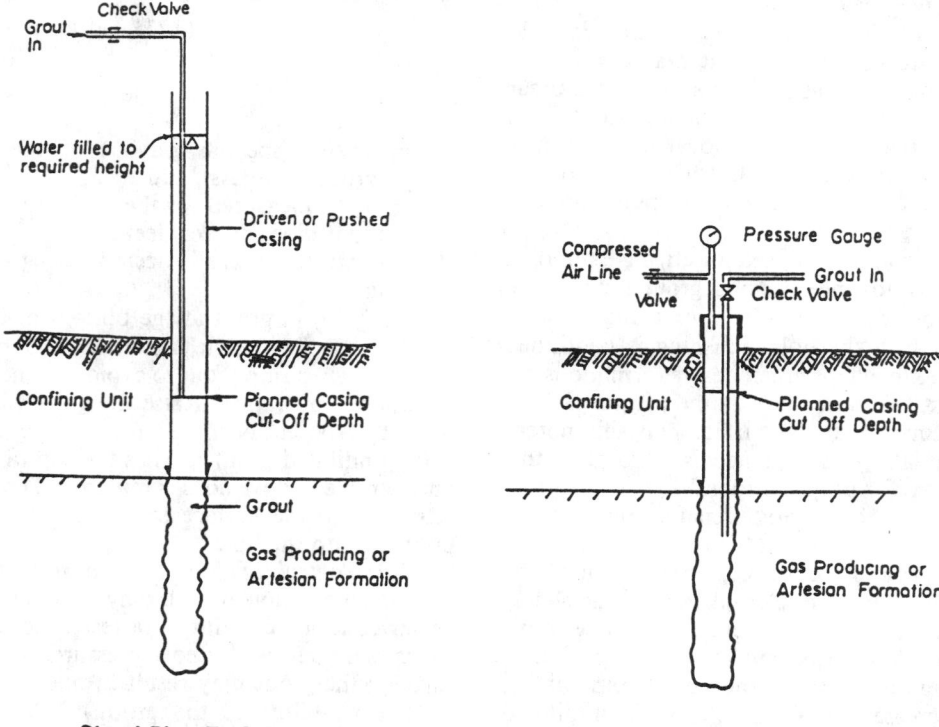

Stand Pipe To Contain Excess Formation Pressure

Compressed Air To Contain Excess Formation Pressure

FIG. 2   Two Procedures That Can Be Used to Overcome Elevated Formation Pressure

the annular space has been cemented over a long distance, this method may not be readily used unless a poor contact occurs between the casing and borehole. Small lengths of cement (typically less than 10 ft (3 m)) can be removed along with the casing if the drill rig has sufficient lifting capacity. When the casing cannot be removed, perforate the casing and screen using a suitable tool. This is necessary as encrustation or corrosion of the screen can block or completely close openings. A wide variety of commercial equipment is available for perforating casings and screens. Due to the diversity of application, consult an experienced contractor prior to selection of the technique. A minimum of four rows of perforations several inches (millimetres) long and a minimum of five perforations per linear foot (metre) of casing or screen is recommended.

7.3.8.4 Remove steel casing by overdrilling (overreaming) the casing. Overreaming can be done using an overreaming tool (see Fig. 3). Select a pilot bit that is nearly the same size as the inside diameter of the casing. The reaming bit should be slightly larger than the borehole diameter to remove all well construction materials and a small amount of native material. As drilling proceeds, the casing, grout, bentonite seal, filter pack, and other well materials are destroyed and returned to the surface. In situations when the grout in the annular space can be verified to be in good condition, it may be very difficult to remove casing and grout and grout can be left in place by pressure grouting the screened interval and casing. Verifying the integrity of the grout can be difficult.

7.3.8.5 PVC and other low tensile strength materials generally cannot be removed by pulling if they have been properly cemented into place. Overdrilling is necessary to remove these casing and screen materials.

7.3.8.6 A hollow stem auger equipped with outward facing carbide cutting teeth with a diameter 2 to 4 in. (50.8 to 1016 cm) larger than the casing may be used for overdrilling. Place the lead auger over the casing and rotate downward (see Fig. 4). The casing guides the cutting head and remains inside the auger. When the full diameter and length of the well has been penetrated, the casing and screen can be retrieved from the center of the auger. It is important to use outward facing cutting teeth in order that the cutting tool does not sever the casing and drift off center. An alternative is to install a steel guide pipe inside the well casing so that the augers can be centered.[6] Firmly attach this temporary working pipe to the inside of the casing by use of a packer, or other type of expansion or friction device. When the auger reaches full depth and the well materials have been removed, pump plugging materials through the hollow stem as the augers are withdrawn.

NOTE 14—Local regulations may allow leaving the PVC casing in place, if a good annular seal exists, or just removing blank casing if the integrity of the annular seal can be documented. It is generally difficult to document a good annular seal (tight fit, no fractures or channels) without the use of a cement bond log or similar method. A cement bond log cannot be run in small diameter wells (less than 2.5 in. (6.35 cm) in diameter) because a small diameter tool is not currently available. Also, this method is not reliable for PVC casings.

7.3.8.7 After removal of the casing, decommissioning can be completed in accordance with 7.3.6.

---

[6] Baker, Robert, Personal Communication, Layne Environmental, 1990.

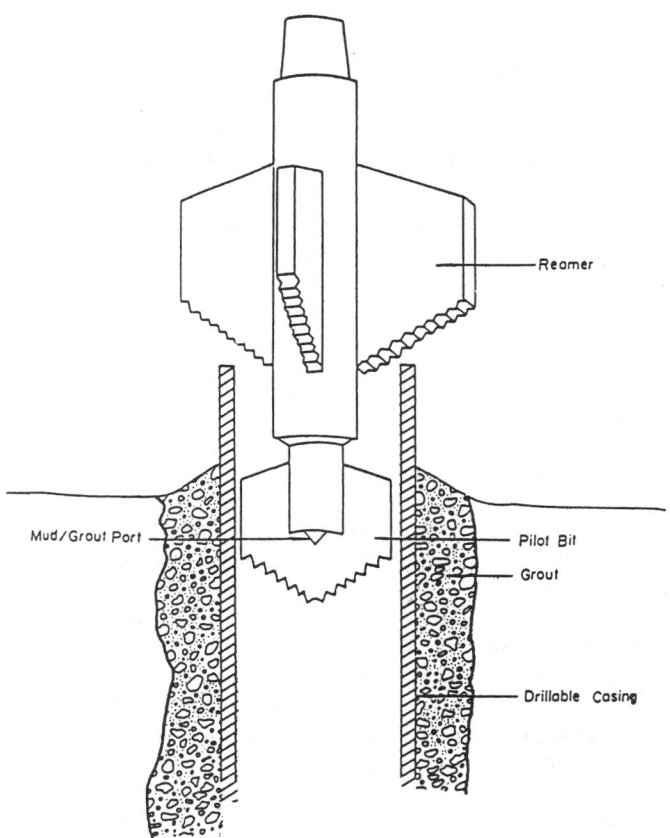

FIG. 3 Over Drilling by Mud Rotary Procedure

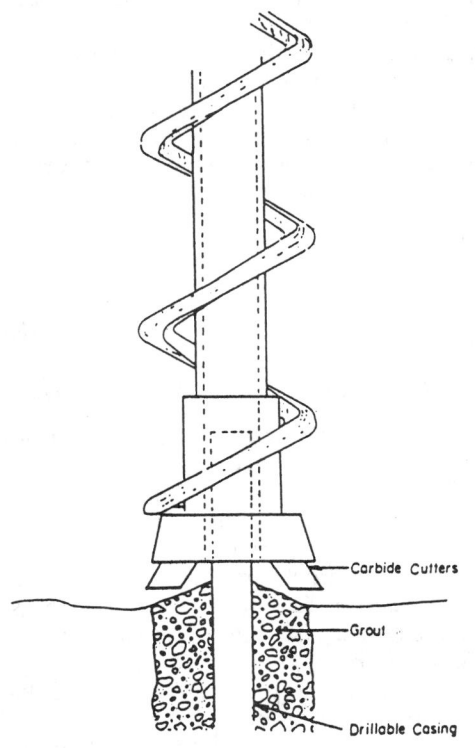

FIG. 4 Over Drilling by Hollow Stem Auger Procedure

NOTE 15—Wells with a maximum depth of 10 to 15 ft (3 to 4.6 m) can be removed by overdrilling, or when possible by pulling the casing out of the ground. The open hole is then filled with grout or other suitable plugging materials. A tamping rod can be used to tamp the material into place.

7.4 Decontamination of Downhole Equipment:

7.4.1 If hazardous waste or other contaminants occur at the site, decontaminate all tools placed into the borehole or well prior to entry (see Practice D 5088). Follow state and local regulations, if applicable. A hot water pressure wash or steam cleaning are examples of two typical decontamination methods. Other methods such as the use of solvents followed by rinsing with clean tap water and allowing equipment to air dry may also be acceptable. Also contain water and other materials used for decontamination for chemical analysis to determine proper disposal methods.

7.4.2 Do not flush grout pipes out with water while in or above the hole. Conduct all cleaning operations aboveground and manage appropriately as a hazardous or nonhazardous situation depending upon chemical analytical data on the site.

7.5 Documentation:

7.5.1 *Inspection Records*—The primary purpose of records of field work is to provide that appropriate measures have been taken so that the borehole or monitoring well is permanently decommissioned in a manner that minimizes it from being a conduit for fluid, water, or vapor migration. Properly decommissioned boreholes and wells should have no adverse influence on the local environment than the original geological setting.

7.5.2 *Narrative Report*—Maintain the narrative report of activities as a permanent record of the decommissioning activities and include the full set of field activities including:

7.5.2.1 Decommissioning date,

7.5.2.2 Personnel (with listing of company representation, including phone number and address),

7.5.2.3 Source of decommissioning method,

7.5.2.4 Step-by-step procedures used in the field,

7.5.2.5 Record of all measurements made, depths encoun-

tered, types and volume of fluids pumped and photographs before and after, and

7.5.2.6 All other pertinent information that is required based on site conditions or regulatory requirements. Any problems encountered should also be documented in detail.

8. Report

8.1 At the completion of field work, present a formal report to document the entire procedure used in the decommissioning process for the borehole or well. Additional information or requirements may be specified in government regulations and should also be included in this report. Report in detail the following items as a minimum:

8.1.1 *Background Information:*

8.1.1.1 Location of the borehole/well,

8.1.1.2 Purpose of the borehole/well,

8.1.1.3 History of the borehole/well,

8.1.1.4 Chemical parameters and concentrations present during the active life of the well and at the time of plugging; and

8.1.1.5 Information collected prior to the inspection record.

8.1.2 *Testing Prior to Decommissioning:*

8.1.2.1 Site characterization (if applicable),

8.1.2.2 Physical testing (depth, structural condition, etc.),

8.1.2.3 Chemical testing results (last chemical analysis for record), and

8.1.2.4 Geophysical logging results (if any).

8.1.3 *Decommissioning Design Procedure:*

8.1.3.1 Rationale for selection of method used, and

8.1.3.2 Presentation of a decommissioning plan.

8.1.4 *Field Implementation:*

8.1.4.1 Description of activities,

8.1.4.2 Variance from plan, and

8.1.4.3 Result of testing and measurements.

8.1.5 *Confirmation*—Statement of regulatory compliance.

9. Keywords

9.1 abandonment; decommissioning; ground water monitoring wells; hazardous waste; plugging

APPENDIX

(Nonmandatory Information)

X1. EXAMPLES OF SUCCESSFUL MONITORING WELL DECOMMISSIONING PRACTICES

X1.1 This appendix covers a general discussion of field procedures that have been used to decommission monitoring wells. These procedures are presented to provide technical guidance to assist in the development of decommissioning wells or boreholes. Local regulations may not allow implementation of some procedures. Two subsets of procedures are outlined, one for rigid steel casing and the second for lower tensile strength casing materials (plastic).

X1.2 *Steel Casing:*

X1.2.1 *Condition*—Properly grouted steel casing and screen, difficult to pull casing:

X1.2.1.1 *Suggestion*—Pressure pump plugging materials into well to above aquifer, place 3-ft (.91-mm) plug of plastic

material (bentonite plug) above rigid plug then complete with rigid plugging material to surface. Place a PVC or neoprene plug or packer, located at the base of the rigid (cement) material to reduce the co-mingling of fluids due to different specific gravities. Complete decommissioning at surface according to local regulations.

X1.2.1.2 *Alternative*—Perforate steel casing and screen to allow plugging material to come into contact with the annular space and formation.

X1.2.1.3 *Advantage*—If (when) casing material corrodes, the continually plastic bentonite plug will continue to provide a suitable sealing zone.

X1.2.2 *Condition*—Steel well materials, that are removable:

X1.2.2.1 *Suggestion*—Pull screen and casing, then redrill borehole to original depth using slightly larger diameter drill bit than the borehole. Complete closure by pressure sealing the borehole by grout pipe from bottom to top.

X1.2.2.2 *Advantage*—Casing and annular materials are completely removed and replaced with sealing material of equal or lower permeability (saturated zone) than the native geologic materials.

X1.2.3 *Condition*—Shallow screened well completed into permeable sand or gravel aquifer:

X1.2.3.1 *Suggestion*—Pull the casing and screen (grout also, if possible) then redrill the borehole to original depth. Completion should be by pressure sealing to the surface.

X1.2.3.2 *Advantage*—The possibility of hydraulic connections with the surface is greatly reduced.

X1.2.4 *Condition*—Steel casing and screen set into highly permeable formation (such as fractured limestone) that results in lost fluid circulation during drilling or decommissioning procedures:

X1.2.4.1 *Suggestion*—Fill highly permeable section with quick setting cement grout (addition of calcium chloride), or coarse gravel to base of casing, or hang "cement basket" at base of casing. Complete decommissioning with plugging materials pumped into place, including at least one section of permanent plastic seal.

X1.2.4.2 *Advantage*—Extremely permeable or highly cavernous materials may not be able to be filled sufficiently to allow closure similar to other cases. In such situations, prevention of vertical migration of fluids to or from that zone is the primary purpose of decommissioning.

X1.3 *Low-Tensile Strength Materials (that is, PVC, ABS, PTFE):*

X1.3.1 *Condition*—Properly constructed well that has a sealed annulus from the surface to the top of screen:

X1.3.1.1 *Suggestions*—Pressure grout in place after perforating the entire casing so that the screen and gravel pack are filled with grout; or over-ream the well using a drill bit diameter that is a minimum or slightly larger than the diameter of the annular space.

X1.3.2 *Condition*—Annulus is not permanently sealed and casing/screen cannot be pulled:

X1.3.2.1 *Suggestion*—Overdrill casing, screen and filter pack with a hollow stemmed auger or other suitable drilling method. Remove casing from bottom to top as augers are removed.

X1.3.2.2 *Advantage*—Complete well structure is removed. Grouting materials provide no avenue of fluid or gas migration after decommissioning.

X1.3.3 *Condition*—Annulus seal questionable; casing screen cannot be pulled:

X1.3.3.1 *Suggestion*—Use rotary tools equipped with pilot bit. Large portion of bit must be equal to or larger than diameter of original borehole. Drill out casing, annular material, to depth of well. Pressure seal after tools are removed.

X1.3.3.2 *Advantage*—Complete well structure is removed. Grouting materials do not provide pathways for fluid or gas migration after decommissioning.

X1.3.4 *Condition*—Annulus seal questionable, casing and screen may be pulled from firm clay soils:

X1.3.4.1 *Suggestion*—Use pulling rig equipped with down hole casing latch tool. Anchor tool at bottom of well and lift casing and screen. Pressure grout open hole from bottom to top. (This method may have applicability for shallow wells.)

X1.3.4.2 *Advantage*—This method is often possible in firm clay soils. Requires less expensive equipment than other procedures.

X1.3.5 *Condition*—Vadose zone monitoring devices such as lysimeters, tensiometers, neutron access holes, and similar devices installed vertically:

X1.3.5.1 *Suggestion*—Drill out the device and backfill as suggested in 7.3.8.5 and 7.3.8.6. (See also Note 15 in the text.)

X1.3.5.2 *Advantage*—The device and materials used to construct the device are removed reducing the potential for fluid movement from the surface to underlying materials and ground water.

REFERENCES

(1) Gaber, M. S., and Fisher, B., *Michigan Water Well Grouting Manual*, Michigan Department of Public Health, Lansing, MI, 1988, pp. 83.

(2) Roscoe Moss Company, *Handbook of Ground Water Development, Chapter 12, Water Well Cementing*, John Wiley, New York, NY 1990, pp. 232–252.

(3) Dowell Division of Dow Chemical, U.S.A., *Engineer's Handbook*, Houston, 1983, p. 77.027.

(4) Williams, Camila, and Evans, L., "Guide to the Selection of Cement, Bentonite, and Other Additives for Use in Monitor Well Construction," *First National Outdoor Action Conference on Aquifer Restoration, Monitoring and Geophysical Methods*, NWWA, Las Vegas, NV, 1987, pp. 325–343.

(5) George, Charles, and Bruce, T., "Cementing to Achieve Zone Isolation in Disposal Wells," in *International Symposium Subsurface Inspection of Liquid Wastes*, NWWA, New Orleans, LA, March 3–6, 1986, pp. 77–89.

(6) Anonymous, "Special Cements and Their Uses," *Concrete Construction*, Vol 18, No. 1, January, 1973, pp. 3–7.

(7) Berry, E., and Malhortra, V., "Fly Ash for Use in Concrete—A Critical Review," *Journal of the American Concrete Institute*, Vol 77, No. 2, 1980, pp. 57–73.

(8) Coleman, R. J., and Corrigan, G., "Fineness and Water-to-Cement Ratio in Relation to Volume and Permeability of Cement," *Petroleum Technology, Technical Publication No. 1266*, 1941, pp. 1–11.

(9) Dowell Schumberger, *Cementing Technology*, Nova Communications Ltd., New York, 1984.

(10) Creech, John, 1986, "Class I Injection Well Design Considerations Using Fiberglass Tubular and Epoxy Cement," *Proceedings of International Symposium Subsurface Injections of Liquid Wastes*, New Orleans, LA, March 3–5, 1986.

(11) Cole, Robert C., "Epoxy Sealant for Combating Well Corrosion," *International Symposium on Oilfield and Geothermal Chemistry*, SPE of AIME, Houston, 1979.

(12) Jeffries, S. A., "Bentonite-Cement Slurries for Hydraulic Cut-Offs," *Soil Mechanics and Foundation Engineering*, Tenth International Congress, Vol 1, Stockholm, 1981, pp. 435–440.

(13) Herndon, Joe, and Smith, Dwight K., "Settling Downhole Plugs: A State-of-the-Art," *Petroleum Engineer International*, Vol 50, No. 4, April 1978, p. 56.

(14) Smith, D. K., Cementing, "Society of Petroleum Engineers Monograp Series," Vol 4, *Society of Petroleum Engineers of AIME*, 1976.

(15) U.S. EPA, "Construction Requirements for Wells Authorized by Rule," 40 CRF 147.2912b(1), (*Subpart GGG, Osage Mineral Reserve, Class II Wells*), 1988.

(16) Portland Cement Association, *Design and Control of Concrete Mixtures*, Skokie, IL, 1979.

3.4 GROUND WATER SAMPLING

Standard Guide for
Planning and Preparing for a Groundwater Sampling Event[1]

This standard is issued under the fixed designation D 5903; the number immediately following the designation indicates the year of original adoption or, in the case of revision, the year of last revision. A number in parentheses indicates the year of last reapproval. A superscript epsilon (ε) indicates an editorial change since the last revision or reapproval.

1. Scope

1.1 This guide covers planning and preparing for a ground-water sampling event. It includes technical and administrative considerations and procedures. Example checklists are also provided as Appendices.

1.2 This guide may not cover every consideration procedure, or both, that is necessary before all ground-water sampling projects. In karst or fractured rock terranes, it may be appropriate to collect ground water samples from springs (see Guide D 5717). This guide focuses on sampling of ground water from monitoring wells; however, most of the guidance herein can apply to the sampling of springs as well.

1.3 *This standard does not purport to address all of the safety concerns, if any, associated with its use. It is the responsibility of the user of this standard to establish appropriate safety and health practices and determine the applicability of regulatory limitations prior to use.*

2. Referenced Documents

2.1 *ASTM Standards:*
D 653 Terminology Relating to Soil, Rock, and Contained Fluids[2]
D 5717 Guide to the Design of Ground-Water Monitoring Systems in Karst and Fractured-Rock Aquifers

3. Significance and Use

3.1 The success of a sampling event is influenced by adequate planning and preparation. Use of this guide will help the ground-water sampler to methodically execute the planning and preparation.

3.2 This guide should be used by a professional or technician that has training or experience in ground-water sampling.

4. Considerations and Procedures

4.1 Evaluate the scope of the sampling and analysis program.

4.1.1 Review plans, protocols, and objectives of the sampling program and event. The sampler should review the sampling and analysis plan, site health and safety plan, sampling protocol, and quality assurance/quality control plan, when available. These documents will provide information on required sampling procedures and also should provide the information in the following paragraphs.

4.1.2 Determine which wells will be sampled. The sampler should have a map or diagram showing the locations of the wells to be sampled. Determine if there is a preferred well sampling sequence specified in the sampling and analysis plan.

4.1.3 Identify the laboratory analyses to be performed on samples from each well. The analytical requirements are often, but not always, the same for each well. Determine if there is a preferred order in filling containers based on analytes.

4.1.4 Identify data to be collected in the field. The sampler must know in advance what types of data must be collected in the field (that is, chemical measurements, water level measurements, etc.) Many samplers use a form to record field data and other observations. The use of a form can help the sampler to collect and record information in a consistent manner and can reduce the chance of failure to collect needed data.

4.1.5 Determine from what depth range within the well the samples will be collected.

4.1.6 Evaluate the need for specialized handling of purged water and decontamination wastewater. The waters may be released to the ground surface, discharged to a sanitary or industrial sewer, or containerized and handled as a potentially hazardous waste. Hazardous wastes will require specialized labeling, storage, transportation, and disposal.

4.1.7 Identify all documentation and field quality control procedures stipulated in the sampling and analysis plan or quality control plan.

4.2 Review available information.

4.2.1 Review well construction details. The sampler should know the material of construction, the inside diameter, the completion depth, the screened interval, and the cap type and lock type (if locked). This information is needed to select purging and sampling equipment, and may be needed to select other tools (for example, a strap wrench to remove the cap, boltcutters or a hack saw to remove damaged locks, or keys for locks).

4.2.2 Evaluate historical well performance and chemical characteristics of the water from each well, if available. The behavior of the well during past sampling events is useful information in the planning process. This may include the flow rate in the screened interval, the maximum pumping rate, the time required to purge the well, whether the well is easily bailed or pumped dry, etc. Knowledge of the past ground-water chemistry and non-aqueous phase liquids in the well also can be useful. The turbidity of the water may influence sampling methods and the need for or approach to filtration of samples. Use of personal protective equipment also may be dictated by known contamination of the water from a well.

4.2.3 Evaluate the physical setting of the well locations.

---

[1] This guide is under the jurisdiction of ASTM Committee D-18 on Soil and Rock and is the direct responsibility of Subcommittee D18.21 on Ground Water and Vadose Zone Investigations.
Current edition approved March 10, 1996. Published May 1996.
[2] *Annual Book of ASTM Standards*, Vol 04.08.

This is necessary to determine the accessibility of the wells. Access could be impeded or difficult due to mud, snow, trees, fences, steep hills, secured areas, etc. This information will help the sampler determine what type of vehicle is needed, whether special tools are needed, or whether administrative clearances are required, or both.

4.2.4 For wells with dedicated sampling equipment find out the type of equipment, pump depth, whether there are any packers in the well, where packers are set, and the power source for equipment.

4.3 Estimate the time required to complete the sample collection and associated field work. The amount of time required will affect equipment needs and possibly lodging or other administrative arrangements. It is usually necessary to inform the laboratory when samples should arrive at the laboratory.

4.4 Coordinate with the analytical laboratory.

4.4.1 Notify the laboratory in advance of the number of analyses of each type to include quality control sample analyses. This notification allows the laboratory to plan for adequate equipment and personnel resources to complete the analyses.

4.4.2 Determine the volume of sample needed for each analysis.

4.4.3 Coordinate the preparation or shipment, or both, of sample containers, preservatives, and shipping containers to the site and to the laboratory. The analytical laboratory often supplies the sample containers and preservatives, and sometimes the shipping containers for the return of samples. The project manager or sampler will need to provide the details needed to accomplish this. The laboratory will need to know the number of containers and preservatives for each analyte, when the containers are needed, whether containers will be picked up or shipped, and the address of the location to which containers/preservatives must be shipped. The laboratory should specify any related administrative requirements. The return of samples to the laboratory also must be coordinated. The sampler will need to be aware of any special instructions regarding shipment or receipt of the samples (that is, times when samples cannot be received, unacceptable shipping containers, Department of Transportation restrictions, and documentation requirements). The sampler also must have the address of the laboratory if samples will be shipped.

4.4.4 When the sampler is also the project manager, the methods of analyses and lower reporting limits also must be coordinated with the laboratory. These are chosen based on the data quality objectives.

4.4.5 Identify the sample volumes, preparation, and holding time requirements. The sampler should be aware of the total volume of water that must be collected from each well. This may influence the selection of sampling equipment. The sampler also should know what will be involved in the preparation of samples (that is, chemical and physical preservation). This knowledge is needed to make logistical arrangements. For example, the sampler may need to use an area near the site that has an electrical outlet and a sink if filtration is required. Lastly, the sampler must know if any of the samples have a short holding time (maximum allowable time between sample collection and preparation or analysis). Collecting samples with short holding times could influence the timing or method of sample shipment.

4.4.6 Inform the laboratory of any special requirements that are different than normal laboratory procedures.

4.4.7 Notify the laboratory of the types and numbers of field quality control samples that will be submitted. Some quality control samples will be prepared or collected in the field; others will be prepared in the laboratory. The sampler must know how to collect and prepare the field quality control samples.

4.4.8 Identify laboratory documentation needs. The laboratory may have certain project identifiers, sample identifiers, or forms that they use for sample tracking or data reporting, or both. It is important that the sampler and the laboratory agree on all means of documentation that will be used by the laboratory.

4.4.9 Determine when the laboratory must be notified regarding sample arrival times and how accurate the time estimates must be (that is, within a day, a half a day, a week). The sampler should discuss this notification process with the laboratory.

4.4.10 Provide information to the laboratory on when data are needed. This is the responsibility of the project manager; however, the sampler and the project manager may be the same person.

4.5 Coordinate with the client or site-related personnel. Coordination with the client is necessary when sampling at a site not owned by you or your company. It also may be necessary to coordinate with people at your own site if they should be notified or have some involvement in your project.

4.5.1 Notify the client or site workers, or both, of when the sampling event will take place.

4.5.2 Request logistical support as needed. This may be as simple as requesting use of the phone. In some cases, logistical support needs may be more extensive. Other logistical support items could include an area for sample preparation and storage, a potable water source, a vehicle, fuel, maintenance support, tools, etc. The sampler should ensure that all support needed from outside sources is prearranged.

4.5.3 Obtain necessary site and well access. It may be necessary to get a pass to enter a site or to have a permit to sample the wells. It also may be necessary to obtain keys to gates or wells, or both. All possible access restrictions should be identified in advance to prevent a delay in the sampling event.

4.5.4 Address site-specific safety concerns. This information should be available in a site health and safety plan. If no such plan exists, at a minimum the sampler should obtain emergency phone numbers and a map showing the location of the nearest health care facility, and identify any safety hazards or weather conditions unique to the site.

4.6 Identify equipment needs. This identification will include selecting purging and sampling devices; field measurement equipment; sample handling, filtration, preservation, and shipping supplies; documentation; personal protective equipment, and other incidental equipment. Appendix X1 is an example checklist of supplies and equipment. Using a comprehensive checklist will reduce the chance of overlooking a needed item.

4.7 Make provisions to keep sample containers separated

from potential sources of contamination such as decontamination reagents and fuel.

4.8 Prepare sampling equipment and supplies for use. It is important that sampling equipment be in good operating condition before going into the field. The sampler should pack necessary and contingency supplies. Appendix X2 is an example checklist.

4.9 Prepare field measurement equipment for use. The sampler should check all field measurement devices to ensure that they are operational. This should include calibration of test instruments.

4.10 Make lodging and transportation arrangements if necessary.

5. Keywords

5.1 ground-water sampling; laboratory coordination; monitoring well; sampling and analysis plan

APPENDIXES

(Nonmandatory Information)

X1. SAMPLING EQUIPMENT CHECKLIST

X1.1 *Personal Protection:*
X1.1.1 Gloves,
X1.1.2 Coveralls,
X1.1.3 Respirators (with appropriate filters),
X1.1.4 Protective eyewear and footwear, and
X1.1.5 *Comfort Items*—Sunscreen, water, insect repellant, rain/snow gear, space heater.
X1.2 *Measurement:*
X1.2.1 Water level measuring device,
X1.2.2 Hydrocarbon/water interface probe,
X1.2.3 Thermometer,
X1.2.4 Ph meter and probes,
X1.2.5 Conductivity meter and probe,
X1.2.6 Dissolved oxygen meter and probe,
X1.2.7 Organic vapor analyzer,
X1.2.8 Turbidity meter,
X1.2.9 Oxidation reduction potential meter and probe,
X1.2.10 Flow-through cell/beakers,
X1.2.11 Calibration standards for all meters,
X1.2.12 Maintenance supplies and spare batteries for meters/probes,
X1.2.13 Deionized water and squeeze bottle,
X1.2.14 Timekeeping device, and
X1.2.15 Explosimeter.
X1.3 *Incidentals:*
X1.3.1 Plastic ground cover,
X1.3.2 Paper towels,
X1.3.3 Scissors,
X1.3.4 Miscellaneous tools,
X1.3.5 Duct tape,
X1.3.6 Trash bags,
X1.3.7 Keys for site or well access,
X1.3.8 Calculator,
X1.3.9 Funnel, and
X1.3.10 Extension cord.
X1.4 *Portable Sampling:*

X1.4.1 Bailer,
X1.4.2 Disposable haul line,
X1.4.3 Pump, cables, hoses, reel,
X1.4.4 Pump control box,
X1.4.5 Pump power supply,
X1.4.6 Fuel for pump or generator,
X1.4.7 Discharge tubing,
X1.4.8 Maintenance supplies and spare parts,
X1.4.9 Graduated cylinders or buckets for measuring discharge rate,
X1.4.10 Container for purged water, and
X1.4.11 *Decontamination Supplies*—Solutions, brushes, drums, buckets, spray bottles.
X1.5 *Sample Preparation and Shipment:*
X1.5.1 Filtration system,
X1.5.2 Chemical preservatives,
X1.5.3 Material Safety Data Sheets,
X1.5.4 Pipettes,
X1.5.5 Sample containers,
X1.5.6 Plastic bags (to keep containers dry),
X1.5.7 Shipping containers,
X1.5.8 Trash bags to line shipping containers,
X1.5.9 Packing material,
X1.5.10 Ice, and
X1.5.11 Packing tape.
X1.6 *Documentation:*
X1.6.1 Sampling and analysis plan,
X1.6.2 Well completion data,
X1.6.3 Sample container labels,
X1.6.4 Address labels,
X1.6.5 Chain of custody forms,
X1.6.6 Field data sheet or logbook,
X1.6.7 Calibration sheets,
X1.6.8 Custody seals, and
X1.6.9 Permanent marker.

X2. CHECKLIST FOR PREPARING SAMPLING EQUIPMENT AND SUPPLIES

X2.1 *Check Performance of Power Supplies and Controls:*

X2.1.1 Visually inspect power sources for damage or wear (hoses; cables, etc.)

X2.1.2 Check fluid levels, and fill to proper levels as needed.

X2.1.3 Check/tighten drive belts, shafts or gears, or both.

X2.1.4 Inspect for presence and condition of safety shrouds and guards.

X2.1.5 If electric start, check battery condition; if manual start, check pull cord condition.

X2.1.6 Perform maintenance per manufacturer's guidelines (for example, oil change).

X2.1.7 Operate to check performance and output if possible.

X2.2 *Check Condition and Operation of Purging and Sampling Devices:*

X2.2.1 Visually inspect tubing, hoses, electrical cable, support cable, etc. for damage or wear.

X2.2.2 Check condition of fittings, electrical connectors, and support cable attachments.

X2.2.3 Operate pumps to check performance and output if possible.

X2.3 *Prepare Spare Parts, Fuels and Lubricants for Equipment and Power Sources:*

X2.3.1 *Power Sources:*

X2.3.1.1 Lubricating oil, gasoline, etc.

X2.3.1.2 Spare spark plug and plug wrench.

X2.3.1.3 Funnel for refueling.

X2.3.2 *Pumps and Samplers:*

X2.3.2.1 Spare fittings or ferrules, or both.

X2.3.2.2 Check valves or valve components, or both.

X2.3.2.3 O-rings/seals.

X2.3.2.4 Retaining pins/clips.

X2.3.2.5 Polytetrafluoroethylene thread tape.

X2.3.2.6 Tools for service or disassembly, especially special tools for specific devices.

X2.3.2.7 Batteries/charger/extension cord.

Standard Guide for
Sampling Groundwater Monitoring Wells[1]

This standard is issued under the fixed designation D 4448; the number immediately following the designation indicates the year of original adoption or, in the case of revision, the year of last revision. A number in parentheses indicates the year of last reapproval. A superscript epsilon (ε) indicates an editorial change since the last revision or reapproval.

1. Scope

1.1 This guide covers procedures for obtaining valid, representative samples from groundwater monitoring wells. The scope is limited to sampling and "in the field" preservation and does not include well location, depth, well development, design and construction, screening, or analytical procedures.

1.2 This guide is only intended to provide a review of many of the most commonly used methods for sampling groundwater quality monitoring wells and is not intended to serve as a groundwater monitoring plan for any specific application. Because of the large and ever increasing number of options available, no single guide can be viewed as comprehensive. The practitioner must make every effort to ensure that the methods used, whether or not they are addressed in this guide, are adequate to satisfy the monitoring objectives at each site.

1.3 *This standard does not purport to address all of the safety problems, if any, associated with its use. It is the responsibility of the user of this standard to establish appropriate safety and health practices and determine the applicability of regulatory limitations prior to use.*

2. Summary of Guide

2.1 The equipment and procedures used for sampling a monitoring well depend on many factors. These include, but are not limited to, the design and construction of the well, rate of groundwater flow, and the chemical species of interest. Sampling procedures will be different if analyzing for trace organics, volatiles, oxidizable species, or trace metals is needed. This guide considers all of these factors by discussing equipment and procedure options at each stage of the sampling sequence. For ease of organization, the sampling process can be divided into three steps: well flushing, sample withdrawal, and field preparation of samples.

2.2 Monitoring wells must be flushed prior to sampling so that the groundwater is sampled, not the stagnant water in the well casing. If the well casing can be emptied, this may be done although it may be necessary to avoid oxygen contact with the groundwater. If the well cannot be emptied, procedures must be established to demonstrate that the sample represents groundwater. Monitoring an indicative parameter such as pH during flushing is desirable if such a parameter can be identified.

2.3 The types of species that are to be monitored as well as the concentration levels are prime factors for selecting sampling devices (1, 2).[2] The sampling device and all materials and devices the water contacts must be constructed of materials that will not introduce contaminants or alter the analyte chemically in any way.

2.4 The method of sample withdrawal can vary with the parameters of interest. The ideal sampling scheme would employ a completely inert material, would not subject the sample to negative pressure and only moderate positive pressure, would not expose the sample to the atmosphere, or preferably, any other gaseous atmosphere before conveying it to the sample container or flow cell for on-site analysis.

2.5 The degree and type of effort and care that goes into a sampling program is always dependent on the chemical species of interest and the concentration levels of interest. As the concentration level of the chemical species of analytical interest decreases, the work and precautions necessary for sampling are increased. Therefore, the sampling objective must clearly be defined ahead of time. For example, to prepare equipment for sampling for mg/L (ppm) levels of Total Organic Carbon (TOC) in water is about an order of magnitude easier than preparing to sample for μg/L (ppb) levels of a trace organic like benzene. The specific precautions to be taken in preparing to sample for trace organics are different from those to be taken in sampling for trace metals. No final Environmental Protection Agency (EPA) protocol is available for sampling of trace organics. A short guidance manual, (3) and an EPA document (4) concerning monitoring well sampling, including considerations for trace organics are available.

2.6 Care must be taken not to cross contaminate samples or monitoring wells with sampling or pumping devices or materials. All samples, sampling devices, and containers must be protected from the environment when not in use. Water level measurements should be made before the well is flushed. Oxidation-reduction potential, pH, dissolved oxygen, and temperature measurements and filtration should all be performed on the sample in the field, if possible. All but temperature measurement must be done prior to any significant atmospheric exposure, if possible.

2.7 The sampling procedures must be well planned and all sample containers must be prepared and labeled prior to going to the field.

3. Significance and Use

3.1 The quality of groundwater has become an issue of national concern. Groundwater monitoring wells are one of

[1] This guide is under the jurisdiction of ASTM Committee D-34 on Waste Management and is the direct responsibility of Subcommittee D34.01 on Sampling and Monitoring.

Current edition approved Aug. 23 and Oct. 25, 1985. Published May 1986.

[2] The boldface numbers in parentheses refer to a list of references at the end of this guide.

TABLE 1 Typical Container and Preservation Requirements for a Ground-Water Monitoring Program

Sample and Measurement	Volume Required (mL)	Container P—Polyethylene G—Glass	Preservative	Maximum Holding Time
Metals As/Ba/Cd/Cr/Fe Pb/Se/ Ag/Mn/Na	1000–2000	P/G (special acid cleaning)	high purity nitric acid to pH <2	6 months
Mercury	200–300	P/G (special acid cleaning)	high purity nitric acid to pH <2 +0.05 % $K_2Cr_2O_7$	28 days
Radioactivity alpha/beta/radium	4000	P/G (special acid cleaning)	high purity nitric acid to pH <2	6 months
Phenolics	500–1000	G	cool, 4°C H_2SO_4 to pH <2	28 days
Miscellaneous	1000–2000	P	cool, 4°C	28 days
Fluoride	300–500	P		28 days
Chloride	50–200	P/G		28 days
Sulfate	100–500	P/G		48 hours
Nitrate	100–250	P/G		6 h
Coliform	100	P/G		on site/24 h
Conductivity	100	P/G		on site/6 h
pH	100	P/G		48 h
Turbidity	100	P/G		
Total organic carbon (TOC)	25–100	P/G	cool, 4°C or cool, 4°C HCl or H_2SO_4 to pH <2	24 h / 28 days
Pesticides, herbicides and total organic halogen (TOX)	1000–4000	G/TFE-fluoro-carbon lined cap solvent rinsed	cool, 4°C	7 days/extraction +30 days/analysis
Extractable organics	1000–2000	G/TFE-fluoro-carbon-lined cap solvent rinsed	cool, 4°C	7 days/extraction +30 days/analysis
Organic purgeables acrolein/acrylonitrile	25–120	G/vial TFE-fluorocar-bon-lined sep-tum	cool, 4°C	14 days / 3 days

the more important tools for evaluating the quality of groundwater, delineating contamination plumes, and establishing the integrity of hazardous material management facilities.

3.2 The goal in sampling groundwater monitoring wells is to obtain samples that are truly representative of the aquifer or groundwater in question. This guide discusses the advantages and disadvantages of various well flushing, sample withdrawal, and sample preservation techniques. It reviews the parameters that need to be considered in developing a valid sampling plan.

4. Well Flushing (Purging)

4.1 Water that stands within a monitoring well for a long period of time may become unrepresentative of formation water because chemical or biochemical change may cause water quality alterations and even if it is unchanged from the time it entered the well, the stored water may not be representative of formation water at the time of sampling, or both. Because the representativeness of stored water is questionable, it should be excluded from samples collected from a monitoring well.

4.2 The surest way of accomplishing this objective is to remove all stored water from the casing prior to sampling. Research with a tracer in a full scale model 2 in. PVC well (5) indicates that pumping 5 to 10 times the volume of the well via an inlet near the free water surface is sufficient to remove all the stored water in the casing. The volume of the well may

be calculated to include the well screen and any gravel pack if natural flow through these is deemed insufficient to keep them flushed out.

4.3 In deep or large diameter wells having a volume of water so large as to make removal of all the water impractical, it may be feasible to lower a pump or pump inlet to some point well below the water surface, purge only the volume below that point then withdraw the sample from a deeper level. Research indicates this approach should avoid most contamination associated with stored water (5, 6, 7). Sealing the casing above the purge point with a packer may make this approach more dependable by preventing migration of stored water from above. But the packer must be above the top of the screened zone, or stagnant water from above the packer will flow into the purged zone through the well's gravel/sand pack.

4.4 In low yielding wells, the only practical way to remove all standing water may be to empty the casing. Since it is not always possible to remove all water, it may be advisable to let the well recover (refill) and empty it again at least once. If introduction of oxygen into the aquifer may be of concern, it would be best not to uncover the screen when performing the above procedures. The main disadvantage of methods designed to remove all the stored water is that large volumes may need to be pumped in certain instances. The main advantage is that the potential for contamination of samples with stored water is minimized.

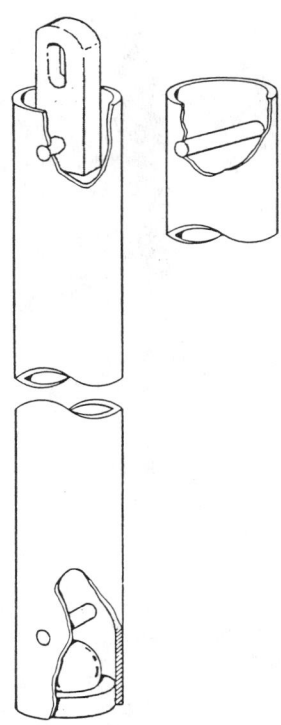

NOTE—Taken from Ref (15).

FIG. 1   Single Check Valve Bailer

4.5 Another approach to well flushing is to monitor one or more indicator parameters such as pH, temperature, or conductivity and consider the well to be flushed when the indicator(s) no longer change. The advantage of this method is that pumping can be done from any location within the casing and the volume of stored water present has no direct bearing on the volume of water that must be pumped. Obviously, in a low yielding well, the well may be emptied before the parameters stabilize. A disadvantage of this approach is that there is no assurance in all situations that the stabilized parameters represent formation water. If significant drawdown has occurred, water from some distance away may be pulled into the screen causing a steady parameter reading but not a representative reading. Also, a suitable indicator parameter and means of continuously measuring it in the field must be available.

4.6 Gibb (4, 8) has described a time-drawdown approach using a knowledge of the well hydraulics to predict the percentage of stored water entering a pump inlet near the top of the screen at any time after flushing begins. Samples are taken when the percentage is acceptably low. As before, the advantage is that well volume has no direct effect in the duration of pumping. A current knowledge of the well's hydraulic characteristics is necessary to employ this approach. Downward migration of stored water due to effects other than drawdown (for example density differences) is not accounted for in this approach.

4.7 In any flushing approach, a withdrawal rate that minimizes drawdown while satisfying time constraints should be used. Excessive drawdown distorts the natural flow patterns around a well and can cause contaminants that were not present originally to be drawn into the well.

5. Materials and Manufacture

5.1 The choice of materials used in the construction of sampling devices should be based upon a knowledge of what compounds may be present in the sampling environment and how the sample materials may interact via leaching, adsorption, or catalysis. In some situations, PVC or some other plastic may be sufficient. In others, an all glass apparatus may be necessary.

5.2 Most analytical protocols suggest that the devices used in sampling and storing samples for trace organics analysis (μg/L levels) must be constructed of glass or TFE–fluorocarbon resin, or both. One suggestion advanced by the EPA is that the monitoring well be constructed so that only TFE–fluorocarbon tubing be used in that portion of the sampling well that extends from a few feet above the water table to the bottom of the borehole. (3, 5) Although this type of well casing is now commercially available, PVC well casings are currently the most popular. If adhesives are avoided, PVC well casings are acceptable in many cases although their use may still lead to some problems if trace organics are of concern. At present, the type of background presented by PVC and interactions occurring between PVC and groundwater are not well understood. Tin, in the form of an organotin stabilizer added to PVC, may enter samples taken from PVC casing. (9)

5.3 Since the most significant problem encountered in trace organics sampling, results from the use of PVC adhesives in monitoring well construction, threaded joints might avoid the problem (3, 5). Milligram per litre (parts per million) levels of compounds such as tetrahydrofuran, methyl-ethyl-ketone, and toluene are found to leach into

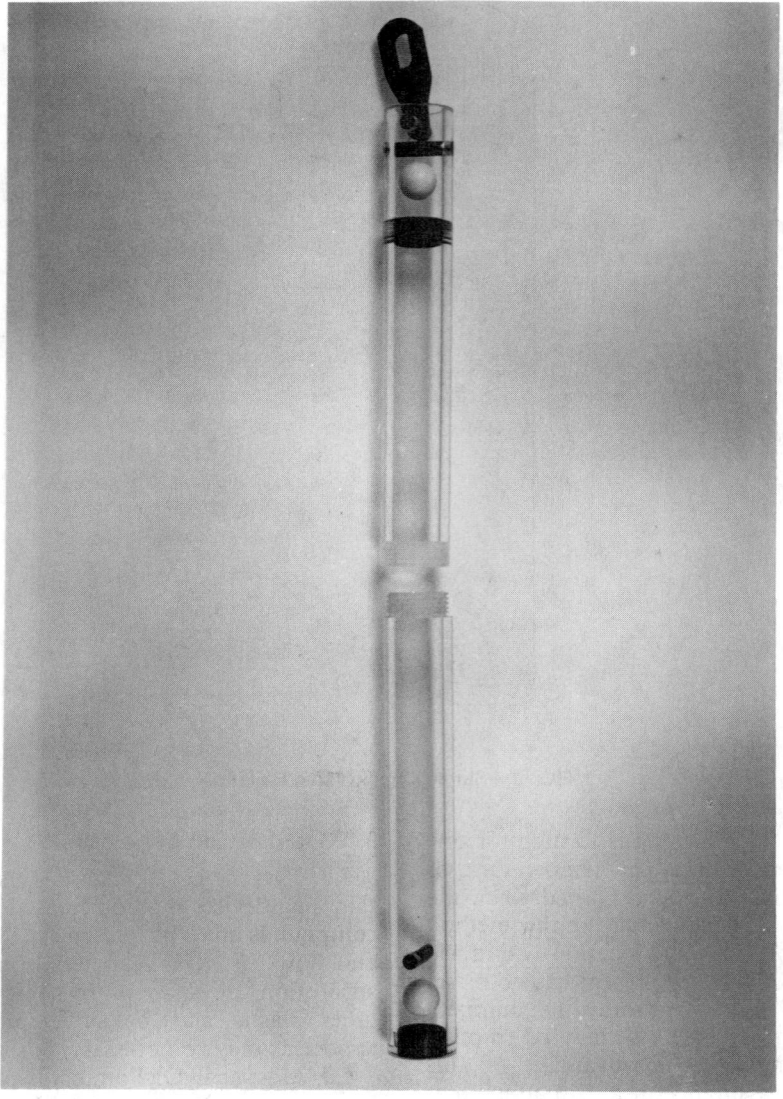

NOTE—Taken from Ref **(17)**.

FIG. 2 Acrylic Point Source Bailer

groundwater samples from monitoring well casings sealed with PVC solvent cement. Pollutant phthalate esters **(8, 10)** are often found in water samples at ppb levels; the EPA has found them on occasion at ppm levels in their samples. The ubiquitous presence of these phthalate esters is unexplained, except to say that they may be leached from plastic pipes, sampling devices, and containers.

5.4 TFE–fluorocarbon resins are highly inert and have sufficient mechanical strength to permit fabrication of sampling devices and well casings. Molded parts are exposed to high temperature during fabrication which destroys any organic contaminants. The evolution of fluorinated compounds can occur during fabrication, will cease rapidly, and does not occur afterwards unless the resin is heated to its melting point.

5.5 Extruded tubing of TFE-fluorocarbon for sampling may contain surface traces of an organic solvent extrusion aid. This can be removed easily by the fabricator and, once

removed by flushing, should not affect the sample. TFE-fluorocarbon FEP and TFE-fluorocarbon PFA resins do not require this extrusion aid and may be suitable for sample tubing as well. Unsintered thread-sealant tape of TFE-fluorocarbon is available in an "oxygen service" grade and contains no extrusion aid and lubricant.

5.6 Louneman, et al. **(11)** alludes to problems caused by a lubricating oil used during TFE-fluorocarbon tubing extrusion. This reference also presents evidence that a fluorinated ethylene-propylene copolymer adsorbed acetone to a degree that later caused contamination of a gas sample.

5.7 Glass and stainless steel are two other materials generally considered inert in aqueous environments. Glass is probably among the best choices though it is not inconceivable it could adsorb some constituents as well as release other contaminants (for example, Na, silicate, and Fe). Of course, glass sampling equipment must be handled carefully in the field. Stainless steel is strongly and easily machined to

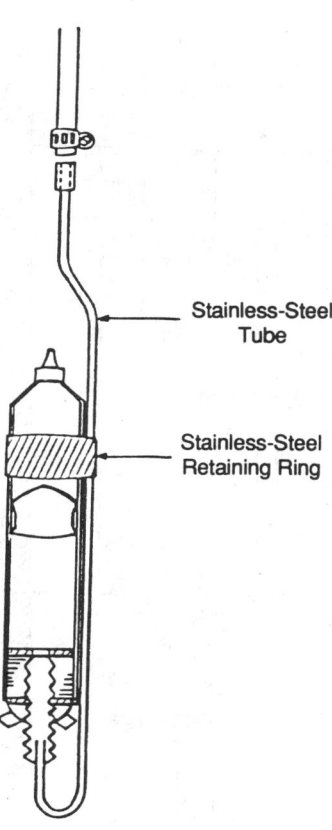

Stainless-Steel
Tube

Stainless-Steel
Retaining Ring

NOTE—Taken from Ref (21).

FIG. 3   Schematic of the Inverted Syringe Sampler

fabricate equipment. Unfortunately, it is not totally immune to corrosion that could release metallic contaminants. Stainless steel contains various alloying metals, some of these (for example Ni) are commonly used as catalysts for various reactions. The alloyed constituents of some stainless steels can be solubilized by the pitting action of nonoxidizing anions such as chloride, fluoride, and in some instances sulfate, over a range of pH conditions. Aluminum, titanium, polyethylene, and other corrosion resistant materials have been proposed by some as acceptable materials, depending on groundwater quality and the constituents of interest.

5.8 Where temporarily installed sampling equipment is used, the sampling device that is chosen should be non-plastic (unless TFE-fluorocarbon), cleanable of trace organics, and must be cleaned between each monitoring well use in order to avoid cross-contamination of wells and samples. The only way to ensure that the device is indeed "clean" and acceptable is to analyze laboratory water blanks and field water blanks that have been soaked in and passed through the sampling device to check for the background levels that may result from the sampling materials or from field conditions. Thus, all samplings for trace materials should be accompanied by samples which represent the field background (if possible), the sampling equipment background, and the laboratory background.

5.9 Additional samples are often taken in the field and spiked (spiked-field samples) in order to verify that the sample handling procedures are valid. The American Chem-ical Society's committee on environmental improvement has published guidelines for data acquisition and data evaluation which should be useful in such environmental evaluations **(10, 12)**.

6. Sampling Equipment

6.1 There is a fairly large choice of equipment presently available for groundwater sampling from single screened wells and well clusters. The sampling devices can be categorized into the following eight basic types.

6.1.1 *Down-Hole Collection Devices:*

6.1.1.1 Bailers, messenger bailers, or thief samplers **(13, 14)** are examples of down-hole devices that probably provide valid samples once the well has been flushed. They are not practical for removal of large volumes of water. These devices can be constructed in various shapes and sizes from a variety of materials. They do not subject the sample to pressure extremes.

6.1.1.2 Bailers do expose part of the sample to the atmosphere during withdrawal. Bailers used for sampling of volatile organic compounds should have a sample cock or draft valve in or near the bottom of the sampler allowing withdrawal of a sample from the well below the exposed surface of the water or the first few inches of the sample should be discarded. Suspension lines for bailers and other samplers should be kept off the ground and free of other contaminating materials that could be carried into the well. Down-hole devices are not very practical for use in deep

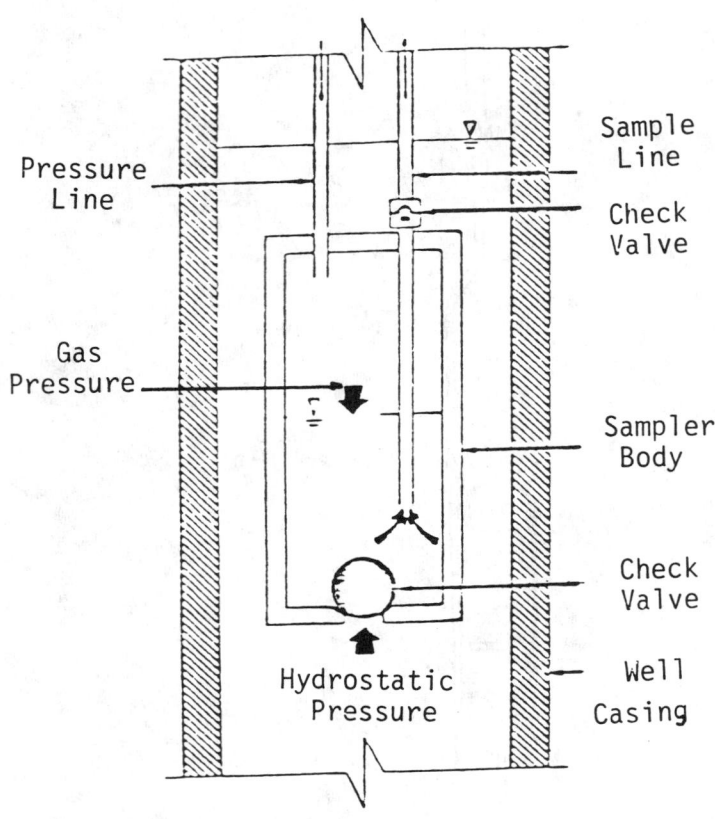

NOTE—Taken from Ref (5).

FIG. 4 The Principal of Gas Displacement Pumping

wells. However, potential sample oxidation during transfer of the sample into a collection vessel and time constraints for lowering and retrieval for deep sampling are the primary disadvantages.

6.1.1.3 Three down-hole devices are the single and double check valve bailers and thief samplers. A schematic of a single check valve unit is illustrated in Fig. 1. The bailer may be threaded in the middle so that additional lengths of blank casing may be added to increase the sampling volume. TFE-fluorocarbon or PVC are the most common materials used for construction (15).

6.1.1.4 In operation, the single check valve bailer is lowered into the well, water enters the chamber through the bottom, and the weight of the water column closes the check

valve upon bailer retrieval. The specific gravity of the ball should be about 1.4 to 2.0 so that the ball almost sits on the check valve seat during chamber filling. Upon bailer withdrawal, the ball will immediately seat without any samples loss through the check valve. A similar technique involves lowering a sealed sample container within a weighted bottle into the well. The stopper is then pulled from the bottle via a line and the entire assembly is retrieved upon filling of the container (14, 16).

6.1.1.5 A double check valve bailer allows point source sampling at a specific depth (15, 17). An example is shown in Fig. 2. In this double check valve design, water flows through the sample chamber as the unit is lowered. A venturi tapered inlet and outlet ensures that water passes freely through the

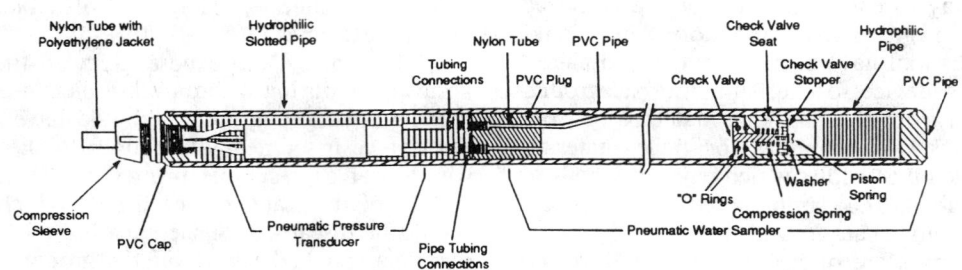

NOTE—Taken from Ref (41).

FIG. 5 Pneumatic Water Sampler With Internal Transducer

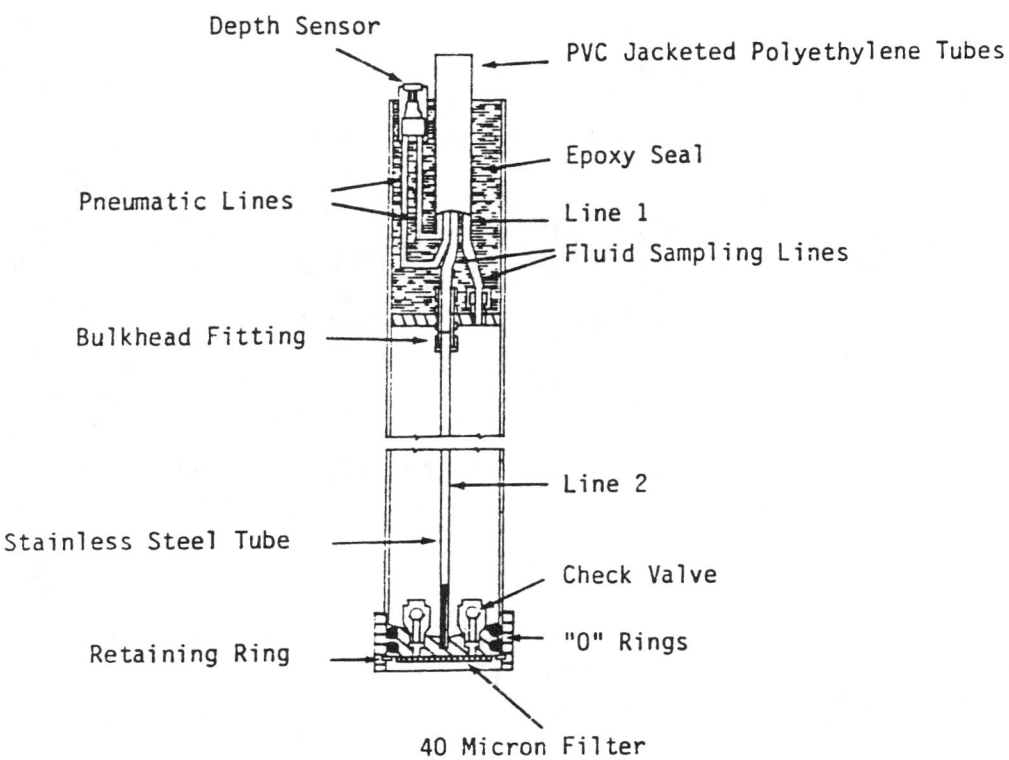

Depth Sensor

PVC Jacketed Polyethylene Tubes

Pneumatic Lines

Epoxy Seal

Line 1

Fluid Sampling Lines

Bulkhead Fitting

Line 2

Stainless Steel Tube

Check Valve

Retaining Ring

"O" Rings

40 Micron Filter

NOTE—Taken from Ref (42).

FIG. 6 Pneumatic Sampler With Externally Mounted Transducer

unit. When a depth where the sample is to be collected is reached, the unit is retrieved. Because the difference between each ball and check valve seat is maintained by a pin that blocks vertical movement of the check ball, both check valves close simultaneously upon retrieval. A drainage pin is placed into the bottom of the bailer to drain the sample directly into a collection vessel to reduce the possibility of air oxidation. The acrylic model in Fig. 2 is threaded at the midsection allowing the addition of threaded casing to increase the sampling volume.

6.1.1.6 Another approach for obtaining point source samples employs a weighted messenger or pneumatic change to "trip" plugs at either end of an open tube (for example, tube water sampler or thief sampler) to close the chamber (18). Foerst, Kemmerer, and Bacon samplers are of this variety (14, 17, 19). A simple and inexpensive pneumatic sampler was recently described by Gillham (20). The device (Fig. 3) consists of a disposable 50 mL plastic syringe modified by sawing off the plunger and the finger grips. The syringe is then attached to a gas-line by means of a rubber stopper assembly. The gas-line extends to the surface, and is used to drive the stem-less plunger, and to raise and lower the syringe into the hole. When the gas-line is pressurized, the rubber plunger is held at the tip of the syringe. The sampler is then lowered into the installation, and when the desired depth is reached, the pressure in the gas-line is reduced to atmospheric (or slightly less) and water enters the syringe. The sampler is then retrieved from the installation and the syringe detached from the gas-line. After the tip is sealed, the syringe is used as a short-term storage container. A number

of thief or messenger devices are available in various materials and shapes.

6.1.2 *Suction Lift Pumps:*

6.1.2.1 Three types of suction lift pumps are the direct line, centrifugal, and peristaltic. A major disadvantage of any suction pump is that it is limited in its ability to raise water by the head available from atmospheric pressure. Thus, if the surface of the water is more than about 25 ft below the pump, water may not be withdrawn. The theoretical suction limit is about 34 ft, but most suction pumps are capable of maintaining a water lift of only 25 ft or less.

6.1.2.2 Many suction pumps draw the water through some sort of volute in which impellers, pistons, or other devices operate to induce a vacuum. Such pumps are probably unacceptable for most sampling purposes because they are usually constructed of common materials such as brass or mild steel and may expose samples to lubricants. They often induce very low pressures around rotating vanes or other such parts such that degassing or even cavitation may occur. They can mix air with the sample via small leaks in the casing, and they are difficult to adequately clean between uses. Such pumps are acceptable for purging of wells, but should not generally be used for sampling.

6.1.2.3 One exception to the above statements is a peristaltic pump. A peristaltic pump is a self-priming, low volume suction pump which consists of a rotor with ball bearing rollers (21). Flexible tubing is inserted around the pump rotor and squeezed by heads as they revolve in a circular pattern around the rotor. One end of the tubing is placed into the well while the other end can be connected

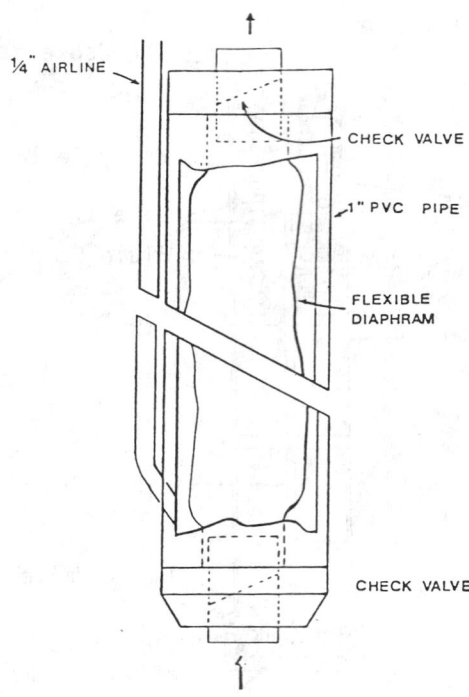

¼" AIRLINE

CHECK VALVE

1" PVC PIPE

FLEXIBLE DIAPHRAM

CHECK VALVE

NOTE—Taken from Ref (4).

FIG. 7 Bladder Pump

directly to a receiving vessel. As the rotor moves, a reduced pressure is created in the well tubing and an increased pressure (<40 psi) on the tube leaving the rotor head. A drive shaft connected to the rotor head can be extended so that multiple rotor heads can be attached to a single drive shaft.

6.1.2.4 The peristaltic pump moves the liquid totally within the sample tube. No part of the pump contacts the liquid. The sample may still be degassed (cavitation is unlikely) but the problems due to contact with the pump mechanism are eliminated. Peristaltic pumps do require a fairly flexible section of tubing within the pumphead itself. A section of silicone tubing is commonly used within the peristaltic pumphead, but other types of tubing can be used particularly for the sections extending into the well or from the pump to the receiving container. The National Council of the Paper Industry for Air and Stream Improvement (22) recommends using medical grade silicone tubing for organic sampling purposes as the standard grade uses an organic vulcanizing agent which has been shown to leach into samples. Medical grade silicone tube is, however, limited to use over a restricted range of ambient temperatures. Various manufacturers offer tubing lined with TFE-fluorocarbon or Viton[3] for use with their pumps. Gibb (1, 8) found little difference between samples withdrawn by a peristaltic pump and those taken by a bailer.

6.1.2.5 A direct method of collecting a sample by suction consists of lowering one end of a length of plastic tubing into the well or piezometer. The opposite end of the tubing is connected to a two way stopper bottle and a hand held or

mechanical vacuum pump is attached to a second tubing leaving the bottle. A check valve is attached between the two lines to maintain a constant vacuum control. A sample can then be drawn directly into the collection vessel without contacting the pump mechanism (5, 23, 24).

6.1.2.6 A centrifugal pump can be attached to a length of plastic tubing that is lowered into the well. A foot valve is usually attached to the end of the well tubing to assist in priming the tube. The maximum lift is about 4.6 m (15 ft) for such an arrangement (23, 25, 26).

6.1.2.7 Suction pump approaches offer a simple sample retrieval method for shallow monitoring. The direct line method is extremely portable though considerable oxidation and mixing may occur during collection. A centrifugal pump will agitate the sample to an even greater degree although pumping rates of 19 to 151 Lpm (5 to 40 gpm) can be attained. A peristaltic pump provides a lower sampling rate with less agitation than the other two pumps. The withdrawal rate of peristaltic pumps can be carefully regulated by adjustment of the rotor head revolution.

6.1.2.8 All three systems can be specially designed so that the water sample contacts only the TFE flourocarbon or silicone tubing prior to sample bottle entry. Separate tubing is recommended for each well or piezometer sampled.

6.1.3 *Electric Submersible Pumps:*

6.1.3.1 A submersible pump consists of a sealed electric motor that powers a piston or helical single thread worm at a high rpm. Water is brought to the surface through an access tube. Such pumps have been used in the water well industry for years and many designs exist (5, 26).

6.1.3.2 Submersible pumps provide relatively high discharge rates for water withdrawal at depths beyond suction

---

[3] Viton is a trademark of E. I. du Pont de Nemours & Co., Wilmington, DE 19898 and has been found suitable for this purpose.

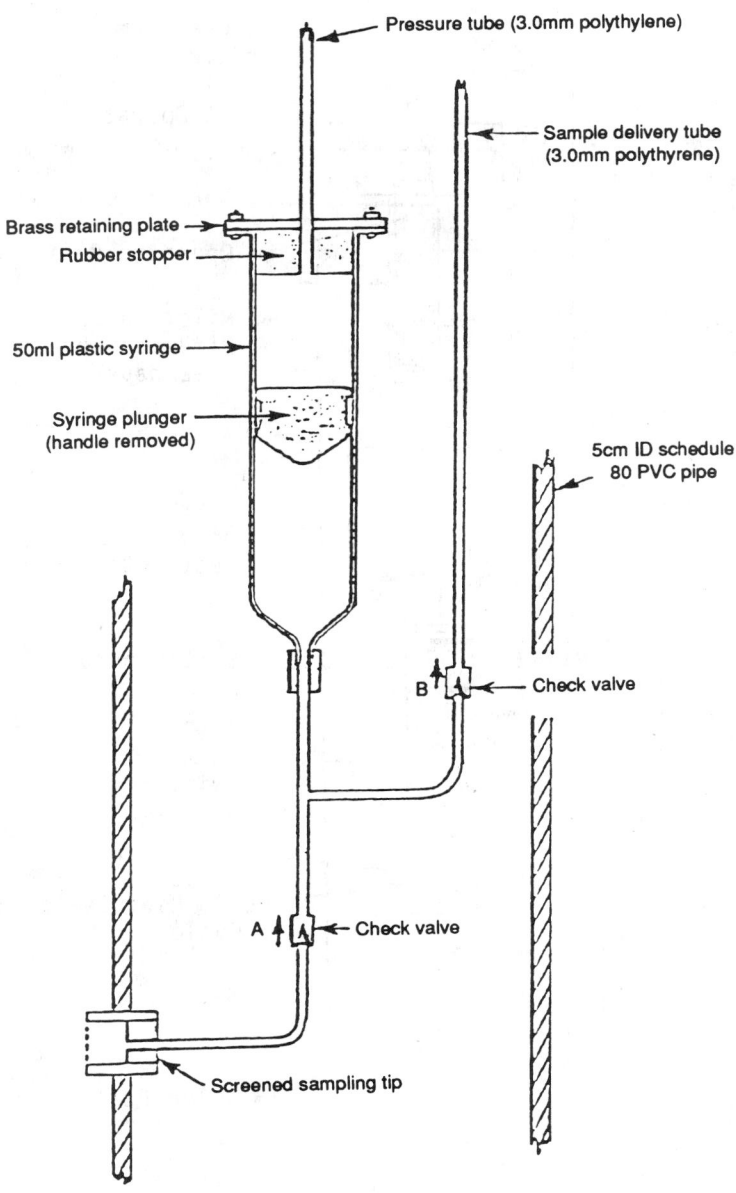

Pressure tube (3.0mm polythylene)

Sample delivery tube
(3.0mm polythyrene)

Brass retaining plate

Rubber stopper

50ml plastic syringe

Syringe plunger
(handle removed)

5cm ID schedule
80 PVC pipe

B — Check valve

A — Check valve

Screened sampling tip

NOTE—Taken from Ref **(48)**.

FIG. 8 Positive Displacement Syringe Pump

lift capabilities. A battery operated unit 3.6 cm (1.4 in.) in diameter and with a 4.5 Lpm (1.2 gpm) flow rate at 33.5 m (110 ft) has been developed **(27)**. Another submersible pump has an outer diameter of 11.4 cm (4.5 in.) and can pump water from 91 m (300 ft). Pumping rates vary up to 53.0 Lpm (14 gpm) depending upon the depth of the pump **(28)**.

6.1.3.3 A submersible pump provides higher extraction rates than many other methods. Considerable sample agitation results, however, in the well and in the collection tube during transport. The possibility of introducing trace metals into the sample from pump materials also exists. Steam cleaning of the unit followed by rinsing with unchlorinated, deionized water is suggested between sampling when analysis for organics in the parts per million (ppm) or parts per billion (ppb) range is required **(29)**.

6.1.4 *Gas-Lift Pumps:*

6.1.4.1 Gas-lift pumps use compressed air to bring a water sample to the surface. Water is forced up an eductor pipe that may be the outer casing or a smaller diameter pipe inserted into the well annulus below the water level **(30, 31)**.

6.1.4.2 A similar principle is used for a unit that consists of a small diameter plastic tube perforated in the lower end. This tube is placed within another tube of slightly larger diameter. Compressed air is injected into the inner tube; the air bubbles through the perforations, thereby lifting the water sample via the annulus between the outer and inner tubing **(32)**. In practice, the eductor line should be submerged to a depth equal to 60 % of the total submerged eductor length during pumping **(26)**. A 60 % ratio is considered optimal although a 30 % submergence ratio is adequate.

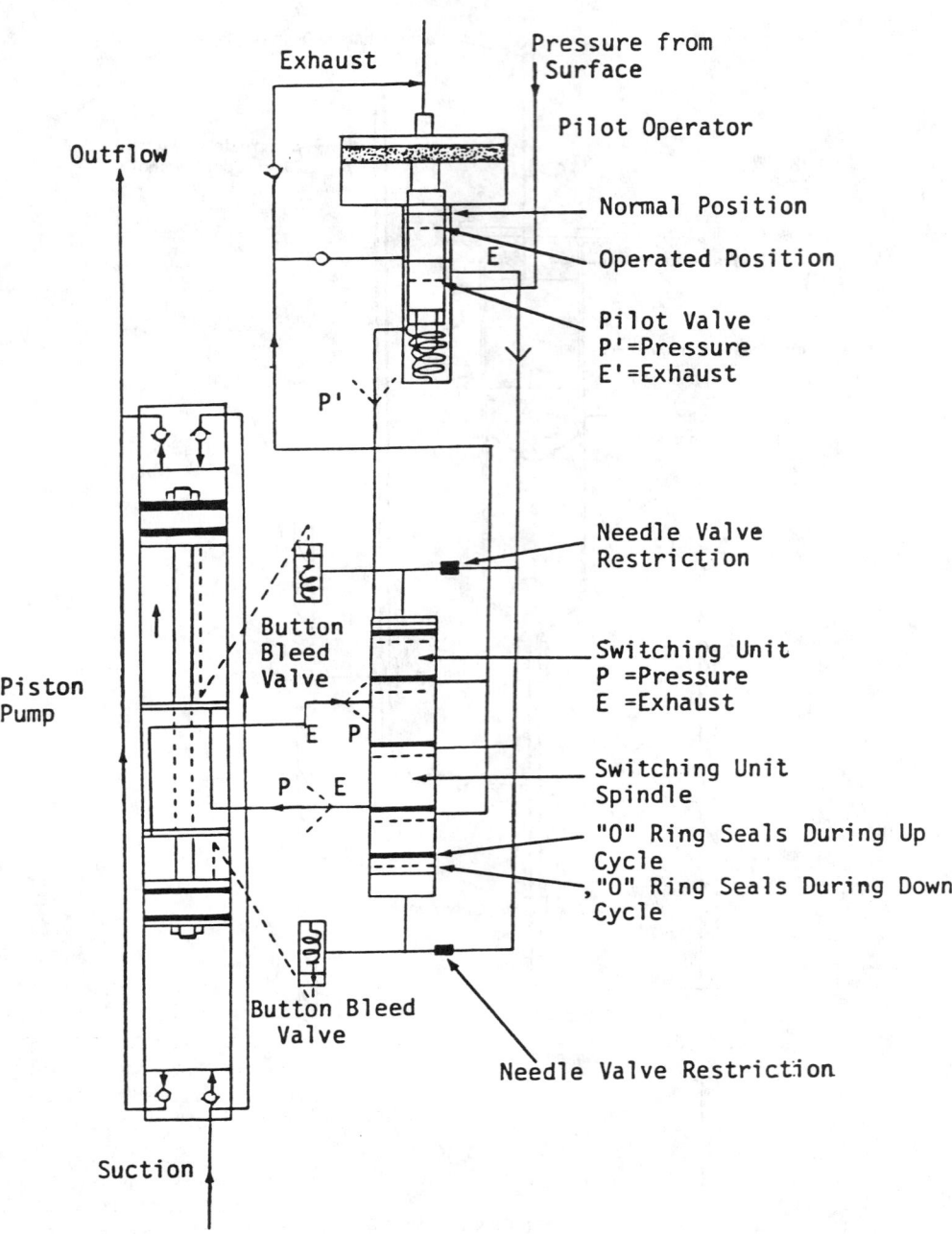

Exhaust

Outflow

Pressure from Surface

Pilot Operator

Normal Position

Operated Position

E

Pilot Valve
P'=Pressure
E'=Exhaust

P'

Piston
Pump

Button
Bleed
Valve

E   P

P   E

Needle Valve
Restriction

Switching Unit
P =Pressure
E =Exhaust

Switching Unit
Spindle

"O" Ring Seals During Up
Cycle

"O" Ring Seals During Down
Cycle

Button Bleed
Valve

Needle Valve Restriction

Suction

Note—Taken from Ref (49).

FIG. 9   Gas Driven Piston Pump

6.1.4.3 The source of compressed gas may be a hand pump for depths generally less than 7.6 m (25 ft). For greater depths, air compressors, pressurized air bottles, and air compressed from an automobile engine have been used.

6.1.4.4 As already mentioned, gas-lift methods result in considerable sample agitation and mixing within the well, and cannot be used for samples which will be tested for volatile organics. The eductor pipe or weighted plastic tubing is a potential source of sample contamination. In addition, Gibb (8) uncovered difficulties in sampling for inorganics. These difficulties were attributed to changes in redox, pH,

and species transformation due to solubility constant changes resulting from stripping, oxidation, and pressure changes.

6.1.5 *Gas Displacement Pumps:*

6.1.5.1 Gas displacement or gas drive pumps are distinguished from gas-lift pumps by the method of sample transport. Gas displacement pumps force a discrete column of water to the surface via mechanical lift without extensive mixing of the pressurized gas and water as occurs with air-lift equipment. The principle is shown schematically in Fig. 4. Water fills the chamber. A positive pressure is applied to the

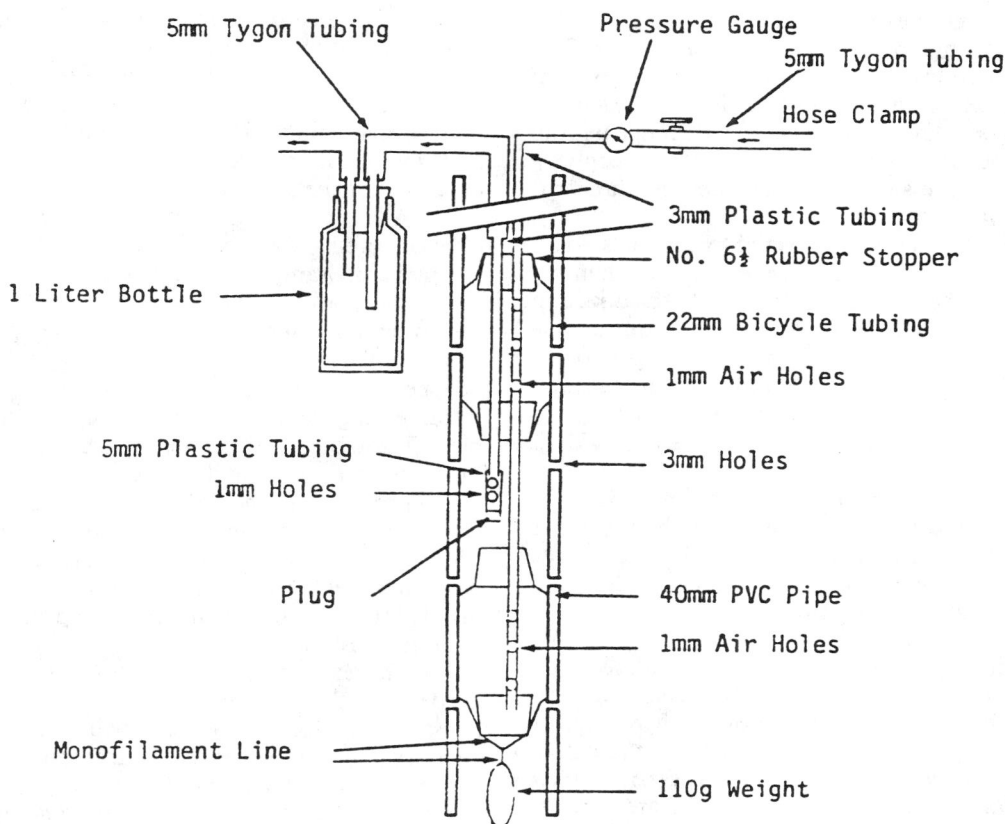

5mm Tygon Tubing

Pressure Gauge

5mm Tygon Tubing

Hose Clamp

3mm Plastic Tubing

No. 6½ Rubber Stopper

22mm Bicycle Tubing

1mm Air Holes

3mm Holes

1 Liter Bottle

5mm Plastic Tubing

1mm Holes

Plug

40mm PVC Pipe

1mm Air Holes

Monofilament Line

110g Weight

NOTE—Taken from Ref (53).

FIG. 10 Packer Pump Arrangement

gas line closing the sampler check valve and forcing water up the sample line. By removing the pressure the cycle can be repeated. Vacuum can also be used in conjunction with the gas (30). The device can be permanently installed in the well (33, 34, 35) or lowered into the well (36, 37).

6.1.5.2 A more complicated two stage design constructed of glass with check valves made of TFE-fluorocarbon has been constructed (38, 39). The unit was designed specifically for sample testing for trace level organics. Continuous flow rates up to 2.3 Lpm (0.6 gpm) are possible with a 5.1 cm (2 in.) diameter unit.

6.1.5.3 Gas displacement pumps have also been developed with multiple functions. The water sample in Fig. 5 provides piezometric data measurements with an internally mounted transducer (40). A sample with its transducer exposed externally for piezometric measurements is illustrated in Fig. 6 (41). The sensor can activate the gas source at the surface to cause sample chamber pressurization at the predetermined depth. Another design can be used as a water sampler or as a tool for injecting brine or other tracers into a well (42).

6.1.5.4 Gas displacement pumps offer reasonable potential for preserving sample integrity because little of the driving gas comes in contact with the sample as the sample is conveyed to the surface by a positive pressure. There is, however, a potential loss of dissolved gasses or contamination from the driving gas and the housing materials.

6.1.6 *Bladder Pumps:*

6.1.6.1 Bladder pumps, also referred to as gas-operated squeeze pumps, consist of a flexible membrane enclosed by a rigid housing. Water enters the membrane through a check valve in the vessel bottom; compressed gas injected into the cavity between the housing and bladder forces the sample through a check valve at the top of the membrane and into a discharge line (Fig. 7). Water is prevented from re-entering the bladder by the top check valve. The process is repeated to cycle the water to the surface. Samples taken from depths of 30.5 m (100 ft) have been reported.

6.1.6.2 A variety of design modifications and materials are available (43, 44). Bladder materials include neoprene, rubber, ethylene propylene terpolymer (E.P.T.), nitrile, and the fluorocarbon Viton.[3] A bladder made of TFE-fluorocarbon is also under development (45). Automated sampling systems have been developed to control the time between pressurization cycles (46).

6.1.6.3 Bladder pumps provide an adaptable sampling tool due primarily to the number of bladder shapes that are feasible. These devices have a distinct advantage over gas displacement pumps in that there is no contact with the driving gas. Disadvantages include the large gas volumes required, low pumping rates, and potential contamination from many of the bladder materials, the rigid housing, or both.

6.1.7 *Gas Driven Piston Pumps:*

1023

6.1.7.1 A simple and inexpensive example of a gas driven piston pump is a syringe pump **(47)**. The pump (Fig. 8) is constructed from a 50 mL plastic syringe with plunger stem removed. The device is connected to a gas line to the surface and the sample passes through a check valve arrangement to a sampling container at the surface. By successively applying positive and negative pressure to the gas-line, the plunger is activated driving water to the surface.

6.1.7.2 A double piston pump powered by compressed air is illustrated in Fig. 9. Pressurized gas enters the chamber between the pistons; the alternating chamber pressurization activates the piston which allows water entry during the suction stroke of the piston and forces the sample to the surface during the pressure stroke **(48)**. Pumping rates between 9.5 and 30.3 L/hr (2.5 to 8 gal/hr) have been reported from 30.5 m (100 ft). Depths in excess of 457 m (1500 ft) are possible.

6.1.7.3 The gas piston pump provides continuous sample withdrawal at depths greater than is possible with most other approaches. Nevertheless, contribution of trace elements from the stainless steel and brass is a potential problem and the quantity of gas used is significant.

6.1.8 *Packer Pump Arrangement:*

6.1.8.1 A packer pump arrangement provides a means by which two expandable "packers" isolate a sampling unit between two packers within a well. Since the hydraulic or pneumatic activated packers are wedged against the casing wall or screen, the sampling unit will obtain water samples only from the isolated well portion. The packers are deflated for vertical movement within the well and inflated when the desired depth is attained. Submersible, gas lift, and suction pumps can be used for sampling. The packers are usually constructed from some type of rubber or rubber compound **(48, 49, 50, 51)**. A packer pump unit consisting of a vacuum sampler positioned between two packers is illustrated in Fig. 10 **(52)**.

6.1.8.2 A packer assembly allows the isolation of discrete sampling points within a well. A number of different samplers can be situated between the packers depending upon the analytical specifications for sample testing. Vertical movement of water outside the well casing during sampling is possible with packer pumps but depends upon the pumping rate and subsequent disturbance. Deterioration of the expandable materials will occur with time with the increased possibility of undesirable organic contaminants contributing to the water sample.

7. Sample Containers and Preservation

7.1 Complete and unequivocal preservation of samples, whether domestic wastewater, industrial wastes, or natural waters, is practically impossible. At best, preservation techniques only retard the chemical and biological changes that inevitably continue after the sample is removed from the source. Therefore, insuring the timely analysis of a sample should be one of the foremost considerations in the sampling plan schedule. Methods of preservation are somewhat limited and are intended to retard biological action, retard hydrolysis of chemical compounds and complexes, and reduce the volatility of constituents. Preservation methods are generally limited to pH control, chemical addition, refrigeration and freezing. For water samples, immediate refrigeration just above freezing (4°C in wet ice) is often the best preservation technique available, but it is not the only measure nor is it applicable in all cases. There may be special cases where it might be prudent to include a recording thermometer in the sample shipment to verify the maximum and minimum temperature to which the samples were exposed. Inexpensive devices for this purpose are available.

7.2 All bottles and containers must be specially pre-cleaned, pre-labelled, and organized in ice-chests (isolating samples and sampling equipment from the environment) before one goes into the field. Otherwise, in any comprehensive program utter chaos usually develops in the field or laboratory. The time in the field is very valuable and should be spent on taking field notes, measurements, and in documenting samples, not on labelling and organizing samples. Therefore, the sampling plan should include clear instructions to the sampling personnel concerning the information required in the field data record logbook (notebook), the information needed on container labels for identification, the chain-of-custody protocols, and the methods for preparing field blanks and spiked samples. Example of detailed plans and documentation procedures have been published **(14, 53)**.

7.3 The exact requirements for the volumes of sample needed and the number of containers to use may vary from laboratory to laboratory. This will depend on the specific analyses to be performed, the concentration levels of interest, and the individual laboratory protocols. The manager of the sampling program should make no assumptions about the laboratory analyses. He should discuss the analytical requirements of the sampling program in detail with the laboratory coordinator beforehand. This is especially the case since some analyses and preservation measures must be performed at the laboratory as soon as possible after the samples arrive. Thus, appropriate arrangements must be made.

7.4 There are a number of excellent references available which list the containers and preservation techniques appropriate for water and soils **(13, 14, 50, 54, 55, 56)**. The "Handbook for Sampling and Sample Preservation of Water and Wastewater" is an excellent reference and perhaps the most comprehensive one **(14)**. Some of this information is summarized in Table 1.

7.5 Sample containers for trace organic samples require special cleaning and handling considerations **(57)**. The sample container for purgeable organics consist of a screw-cap vial (25 to 125 mL) fitted with a TFE-flourocarbon faced silicone septum. The vial is sealed in the laboratory immediately after cleaning and is only opened in the field just prior to pouring sample into it. The water sample then must be sealed into the vial headspace free (no air bubbles) and immediately cooled (4°C) for shipment. Multiple samples (usually about four taken from one large sample container) are taken because leakage of containers may cause losses, may allow air to enter the containers, and may cause erroneous analysis of some constituents. Also, some analyses are best conducted on independent protected samples.

7.6 The purgeable samples must be analyzed by the laboratory within 14 days after collection, unless they are to be analyzed for acrolein or acrylonitrile (in which case they are to be analyzed within 3 days). For samples for solvent extractions (extractable organics-base neutrals, acids and pesticides), the sample bottles are narrow mouth, screw cap quart bottles or half-gallon bottles that have been precleaned, rinsed with the extracting organic solvent and oven dried at 105°C for at least 1 h. These bottles must be sealed with TFE-fluorocarbon lined caps (Note). Samples for organic extraction must be extracted within 7 days and analyzed within 30 days after extraction. Special pre-cleaned, solvent rinsed and oven-dried stainless steel beakers (one for each monitoring well) may be used for transferring samples from the sampling device to the sample containers.

NOTE—When collecting samples, the bottles should not be overfilled or prerinsed with sample before filling because oil and other materials may remain in the bottle. This can cause erroneously high results.

7.7 For a number of groundwater parameters, the most meaningful measurements are those made in the field at the time of sample collection or at least at an on-site laboratory. These include the water level in the well and parameters that sometimes can change rapidly with storage. A discussion of the various techniques for measuring the water level in the well is contained in a NCASI publication (5) and detailed procedures are outlined in a U.S. Geological Survey publication (58). Although a discussion of these techniques is beyond the scope of this guide, it is important to point out that accurate measurements must be made before a well is flushed or only after it has had sufficient time to recover. Parameters that can change rapidly with storage include specific conductance, pH, turbidity, redox potential, dissolved oxygen, and temperature. For some of the other parameters, the emphasis in groundwater monitoring is on the concentration of each specific dissolved component, not the total concentration of each. Samples for these types of measurements should be filtered through 0.45 μm membrane filters ideally in the field or possibly at an on-site laboratory as soon as possible. Analyses often requiring filtered samples include all metals, radioactivity parameters, total organic carbon, dissolved orthophosphate (if needed), and total dissolved phosphorous (if needed) (13, 14). If metals are to be analyzed, filter the sample prior to acid preservation. For TOC organics, the filter material should be tested to assure that it does not contribute to the TOC. The type or size of the filter to be used is not well understood. However, if results of metal, TOC or other parameters that could be effected by solids are to be compared, the same filtering procedure must be used in each case. Repeated analytical results should state whether the samples were filtered and how they were filtered.

7.8 Shipment and receipt of samples must be coordinated with the laboratory to minimize time in transit. All samples for organic analysis (and many other parameters), should arrive at the laboratory within one day after it is shipped and be maintained at about 4°C with wet ice. The best way to get them to the laboratory in good condition is to send them in sturdy insulated ice chests (coolers) equipped with bottle dividers. 24-h courier service is recommended, if personal delivery service is not practical.

REFERENCES

(1) Gibb, J. P., Schuller, R. M., Griffin, R. A., *Monitoring Well Sampling and Preservation Techniques*, EPA-600/9-80-101, 1980.

(2) Pettyjohn, W. A., Dunlap, W. J., Cosby, R. L., Keeley, J. W., "Sampling Ground Water for Organic Contaminants," *Ground Water*, Vol 19, (2), March/April 1981, pp. 180–189.

(3) Dunlap, W. J., McNabb, J. F., Scalf, M. R., Cosby, R. L., *Sampling for Organic Chemicals and Microorganisms in the Subsurface*, EPA-600/2-77-176, NTIS PB 276 679, August 1977, 35 pp.

(4) Scalf, M. R., McNabb, J. F., Dunlap, W. J., and Cosby, R. L., *Manual of Ground Water Quality Sampling Procedures*, National Water Well Association, NTIS PB-82 103 045, 1981.

(5) "A Guide to Groundwater Sampling," *NCASI Technical Bulletin*, No. 362, January 1982.

(6) Humenick, M. J., Turk, L. J., Coldrin, M., "Methodology for Monitoring Ground Water at Uranium Solution Mines," *Ground Water*, Vol 18 (3), May–June 1980, p. 262.

(7) Marsh, J. M., and Lloyd, J. W., "Details of Hydrochemical Variations in Flowing Wells," *Ground Water*, Volume 18 (4), July–August 1980, p. 366.

(8) Gibb, J. P., Schuller, R. M., Griffin, R. A., "Collection of Representative Water Quality Data from Monitoring Wells," *Proceeding of the Municipal Solid Waste Resource Recovery Symposium*, EPA-600/9-81-002A, March 1981.

(9) Boettner, E. A., Gwendolyn, L. B., Zand, H., Aquino, R., *Organic and Organotin Compounds Leached from PVC and CPVC Pipe*, NTIS P8 82-108 333, 1982.

(10) Junk, G. A., Svec, H. J., Vick, R. D., Avery, M. J., "Contamination of Water by Synthetic Polymer Tubes," *Environmental Science and Technology*, Vol 8 (13) 1100, December 1974.

(11) Louneman, W. A., Bufalini, J. J., Kuntz, R. L., and Meeks, S. A., "Contamination from Fluorocarbon Films," *Environmental Science and Technology*, Vol 15 (1), January 1981.

(12) ASC Committee on Environmental Improvement, "Guidelines for Data Acquisition and Data Quality Evaluation in Environmental Chemistry." *Analytical Chemistry*, Vol 52, 1980, pp. 2242–2249.

(13) *Procedures Manual for Ground Water Monitoring at Solid Waste Disposal Facilities*, EPA/530/SW-611, August 1977.

(14) *Handbook for Sampling and Sample Preservation of Water and Wastewater*, U.S. Dept. of Commerce NTIS PB-259 946, September 1976.

(15) Timco Manufacturing Co., Inc., "Variable Capacity Bailer," *Timco Geotechnical Catalogue*, Prairie du Sac, WI, 1982.

(16) deVera, E., Simmons, B., Stephens, R., Storm, D., *Samplers and Sampling Procedures for Hazardous Waste Streams*, Environmental Protection Agency, EPA-600/2-80-018, 1980, p. 51.

(17) Morrison, R., *Ground Water Monitoring Technology*, Timco Manufacturing Co., 1982, p. 276.

(18) Eijelkamp, "Equipment for Soil Research," *General Catalogue*, Geisbeek, The Netherlands, 1979, pp. 82–83.

(19) Wood, W., "Guidelines for Collection and Field Analysis of Ground-Water Samples for Selected Unstable Constituents," *Techniques of Water-Resources Investigations of the United States Geological Survey*, Chapter D2, 1976, p. 24.

(20) Gilham, R. W., "Syringe Devices for Groundwater Monitoring," *Ground Water Monitoring Review*, Vol 2 (2), Spring 1982, p. 36.

(21) Masterflex, *Masterflex Pump Catalogue*, Barnant Corp., Barrington, IL, 1981.

(22) "Guidelines for Contracting Sampling and Analyses for Priority Pollutants in Pulp and Paper Industry Effluents," *NCASI Stream Improvement Technical Bulletin*, No. 335, August 1980.

(23) Allison, L., "A Simple Device for Sampling Ground Water in Auger Holes," *Soil Science Society of America Proceedings* 35: 844–45, 1971.

(24) Willardson, L., Meek, B., Huber, M., "A Flow Path Ground Water

Sampler," *Soil Science Society of America Proceedings* 36: 965–66, 1972.

(25) Wilson, L., *Monitoring in the Vadose Zone: A Review of Technical Elements and Methods*, U.S. Environmental Protection Agency, EPA-60017-80-134, 1980, p. 180.

(26) *Ground Water and Wells*, Johnson, E. E., Inc., St. Paul, MN, 1980, p. 440.

(27) Keck, W. G. and Associates, *New "Keck" Submersible Water Sampling Pump for Groundwater Monitoring*, Keck, W. G. and Associates, East Lansing, MI, 1981.

(28) McMillion, L., and Keeley, J. W., "Sampling Equipment for Ground-Water Investigation," *Ground Water*, Vol 6, 1968, pp. 9–11.

(29) Industrial and Environmental Analysts, Inc., *Procedures and Equipment for Groundwater Monitoring*, Industrial and Environmental Analysts, Inc., Essex Junction, VT, 1981.

(30) Trescott, P., and Pinder, G., "Air Pump for Small-Diameter Piezometers," *Ground Water*, Vol 8, 1970, pp. 10–15.

(31) Sommerfeldt, T., and Campbell, D., "A Pneumatic System to Pump Water From Piezometers," *Ground Water*, Vol 13, p. 293.

(32) Smith, A., "Water Sampling Made Easier with New Device," *The Johnson Drillers Journal*, July–August 1976, pp. 1–2.

(33) Morrison, R., and Ross, D., "Monitoring for Groundwater Contamination at Hazardous Waste Disposal Sites," *Proceedings of 1978 National Conference on Control of Hazardous Material Spills*, April 13, Miami Beach, FL, 1968, pp. 281–286.

(34) Morrison, R., and Brewer, P. "Air-Lift Samplers for Zone-of-Saturation Monitoring," *Ground Water Monitoring Review*, Spring 1981, pp. 52–54.

(35) Morrison, R., and Timmons, R., "Groundwater Monitoring II," *Groundwater Digest*, Vol 4, 1981, pp. 21–24.

(36) Bianchi, W. C., Johnson, C., Haskell, E., "A Positive Action Pump for Sampling Small Bore Holes," *Soil Science Society of America Proceedings*, Vol 26, 1961, pp. 86–87.

(37) Timmons, R., Discussion of "An All-Teflon Bailer and An Air-Driven Pump for Evacuating Small-Diameter Ground-Water Wells" by D. Buss and K. Bandt, *Ground Water*, Vol 19, 1981, pp. 666–667.

(38) Timco Manufacturing Co., Inc., "Gas Lift Teflon Pump," *Timco Geotechnical Catalogue*, Prairie du Sac, WI, 1982.

(39) Tomson, M., King, K., Ward, C., "A Nitrogen Powered Continuous Delivery, All Glass Teflon Pumping System for Groundwater Sampling from Below 10 Meters," *Ground Water*, Vol 18, 1980, pp. 444–446.

(40) *Pneumatic Water Sampler*, Slope Indicator Co., Seattle, WA, 1982.

(41) Petur Instrument Co., Inc., *Petur Liquid Sampler*, Petur Instrument Co., Inc., Seattle, WA, 1982.

(42) Idler, G., "Modification of an Electronic Downhole Water Sampler," *Ground Water*, Vol 18, 1980, pp. 532–535.

(43) *Remote Sampler Model 200*, Markland Specialty Engineering, Ltd., Etobicoke, Ontario, Bulletin 200/78, 1978.

(44) Middleburg, R., "Methods for Sampling Small Diameter Wells for Chemical Quality Analysis," *Presented at the National Conference on Quality Assurance of Environmental Measurements*, Nov. 27–29, Denver, CO, 1978.

(45) *Air Squeeze Pump*, Leonard Mold and Die Works, Denver, CO, 1982.

(46) *Automatic Sampler Controller: Markland Model 105 and 2105*, Markland Specialty Engineering, Ltd., Etobicoke, Ontario, Bulletin 105/78, 1981.

(47) Gillham, R. W., and Johnson, P. E., "A Positive Displacement Ground-Water Sampling Device," *Ground Water Monitoring Review*, Vol 1 (2), Summer 1981, p. 33.

(48) Signor, D., "Gas-Driven Pump for Ground-Water Samples," *U.S. Geological Survey, Water Resources Investigation 78–72*, Open File Report, 1978.

(49) *Tigre Tierra HX Pneumatic Packer*, Tigre Tierra, Inc., Puyallup, WA, 1981.

(50) Cherry, R., "A Portable Sampler for Collecting Water Samples from Specific Zones in Uncased or Screened Wells," *U.S. Geological Survey*, Prof. Paper 25-C, 1965, pp. 214–216.

(51) Grisak, G., Merritt, W., Williams, D., "Fluoride Borehole Dilution Apparatus for Groundwater Velocity Measurements," *Canadian Geotechnical Journal*, Vol 14, 1977, pp. 554–561.

(52) Galgowski, C., Wright, W., "A Variable-Depth Ground-Water Sampler," *Soil Science Society of America Proceedings*, Vol 44, 1980, pp. 1120–1121.

(53) *Samplers and Sampling Procedures for Hazardous Waste Streams*, USEPA MERL Laboratory, Cincinnati, OH, EPA-600/2-80-018, January 1980.

(54) *Methods for Chemical Analysis of Water and Wastes*, EPA-600/4-79-020, USEPA EMSL Laboratory, Cincinnati, OH, March 1979.

(55) *Federal Register*, Vol 44, No. 244, Dec. 18, 1979, pp. 75050–75052.

(56) *Standard Methods for the Examination of Water and Wastewater*, APAA, 14th ed., Washington, DC, 1976, pp. 38–45.

(57) *Handbook for Analytical Quality Control in Water and Wastewater Laboratories*, EPA-600/4-79-019, USEPA EMSL Laboratory, Cincinnati, OH, March 1979.

(58) U.S. Department of Interior, "Groundwater," Chapter II, *National Handbook of Recommended Methods for Water Data Acquisition*, 1980.

Standard Guide for
Direct-Push Water Sampling for Geoenvironmental Investigations[1]

This standard is issued under the fixed designation D 6001; the number immediately following the designation indicates the year of original adoption or, in the case of revision, the year of last revision. A number in parentheses indicates the year of last reapproval. A superscript epsilon (ϵ) indicates an editorial change since the last revision or reapproval.

1. Scope

1.1 This guide covers a review of methods for sampling ground water at discrete points or in increments by insertion of sampling devices by static force or impact without drilling and removal of cuttings. By directly pushing the sampler, the soil is displaced and helps to form an annular seal above the sampling zone. Direct-push water sampling can be one time, or multiple sampling events. Methods for obtaining water samples for water quality analysis and detection of contaminants are presented.

1.2 Direct-push methods of water sampling are used for ground-water quality studies. Water quality may vary at different depths below the surface depending on geohydrologic conditions. Incremental sampling or sampling at discrete depths is used to determine the distribution of contaminants and to more completely characterize geohydrologic environments. These investigations are frequently required in characterization of hazardous and toxic waste sites.

1.3 Direct-push methods can provide accurate information on the distribution of water quality if provisions are made to ensure that cross-contamination or linkage between water bearing strata are not made. Discrete point sampling with a sealed (protected) screen sampler, combined with on-site analysis of water samples, can provide the most accurate depiction of water quality conditions at the time of sampling. Direct-push water sampling with exposed-screen sampling devices may be useful and are considered as screening tools depending on precautions taken during testing. Exposed screen samplers may require development or purging depending on sampling and quality assurance plans. Results from direct-push investigations can be used to guide placement of permanent ground-water monitoring wells and direct remediation efforts. Multiple sampling events can be performed to depict conditions over time. Use of double tube tooling, where the outer push tube seals the hole, prevents the sampling tools from coming in contact with the formation, except at the sampling point.

1.4 Field test methods described in this guide include installation of temporary well points, and insertion of water samplers using a variety of insertion methods. Insertion methods include: (1) soil probing using combinations of impact, percussion, or vibratory driving with or without additions of smooth static force; (2) smooth static force from the surface using hydraulic penetrometer or drilling equip-

ment, and incremental drilling combined with direct-push water sampling events. Under typical incremental drilling operations, samplers are advanced with assistance of drilling equipment by smooth hydraulic push, or mechanical impacts from hammers or other vibratory equipment. Methods for borehole abandonment by grouting are also addressed.

1.5 Direct-push water sampling is limited to soils that can be penetrated with available equipment. In strong soils damage may result during insertion of the sampler from rod bending or assembly buckling. Penetration may be limited, or damage to samplers or rods can occur in certain ground conditions, some of which are discussed in 4.6. Information in this procedure is limited to sampling of saturated soils in perched or saturated ground-water conditions.

1.6 This guide does not address installation of permanent water sampling systems such as those presented in Practice D 5092.

1.7 Direct-push water sampling for geoenvironmental exploration will often involve safety planning, administration, and documentation.

1.8 *This guide does not purport to address all aspects of exploration and site safety. It is the responsibility of the user of this guide to establish appropriate safety and health practices and determine the applicability of regulatory limitations before its use.*

2. Referenced Documents

2.1 *ASTM Standards:*
D 653 Terminology Relating to Soil, Rock, and Contained Fluids[2]
D 2488 Practice for Description and Identification of Soils (Visual-Manual Procedure)[2]
D 3441 Test Method for Deep, Quasi-Static, Cone and Friction-Cone Penetration Tests of Soil[2]
D 4448 Guide for Sampling Groundwater Monitoring Wells[2]
D 4750 Test Method for Determining Subsurface Liquid Levels in a Borehole or Monitoring Well (Observation Well)[2]
D 5088 Practice for Decontamination of Field Equipment Used at Nonradioactive Waste Sites[3]
D 5092 Practice for Design and Installation of Ground Water Monitoring Wells in Aquifers[3]
D 5229 Practice for Decommissioning Monitoring Wells[3]
D 5254 Guide for Minimum Set of Data Elements to Identify a Ground Water Site[3]

---

[1] This guide is under the jurisdiction of ASTM Committee D-18 on Soil and Rock and is the direct responsibility of Subcommittee D18.21 on Ground-Water and Vadose-Zone Investigations.
Current edition approved Aug. 10, 1996. Published January 1997.

[2] *Annual Book of ASTM Standards*, Vol 04.08.
[3] *Annual Book of ASTM Standards*, Vol 04.09.

D 5314 Guide for Soil Gas Sampling in the Vadose Zone[3]

D 5434 Guide for Field Logging of Subsurface Explorations of Soil and Rock[3]

D 5474 Guide for Selection of Data Elements for Groundwater Investigation[3]

D 5521 Guide for Development of Ground Water Monitoring Wells in Granular Aquifers[3]

D 5778 Test Method for Performing Electronic Friction Cone and Piezocone Penetration Tests[3]

D 5730 Guide to Site Characterization for Environmental Purposes[3]

2.1.1 *Drilling Methods:*

D 5781 Guide for the Use of Dual-Wall Reverse Circulation Drilling for Geoenvironmental Exploration and Installation of Subsurface Water Quality Monitoring Devices[3]

D 5782 Guide for the Use of Direct Air-Rotary Drilling for Geoenvironmental Exploration and Installation of Subsurface Water Quality Monitoring Devices[3]

D 5783 Guide for the Use of Direct Rotary Drilling with Water-Based Drilling Fluid for Geoenvironmental Exploration and Installation of Subsurface Water Quality Monitoring Devices[3]

D 5784 Guide for the Use of Hollow-Stem Augers for Geoenvironmental Exploration and Installation of Subsurface Water Quality Monitoring Devices[3]

D 5786 Guide for the Use of Direct Rotary Wireline Casing Advancement Drilling Methods for Geoenvironmental Exploration and the Installation of Subsurface Water-Quality Monitoring Devices

D 5785 Guide for the Use of Cable Tool Drilling and Sampling Methods for Geoenvironmental Explorations and Installation of Subsurface Water Quality Monitoring Devices

2.1.2 *Soil Sampling:*

D 1586 Method for Penetration Test and Split-Barrel Sampling of Soils[2]

D 1587 Practice for Thin-Walled Tube Sampling of Soils[2]

D 3550 Practice for Ring-Lined Barrel Sampling of Soils[2]

D 4700 Guide for Soil Sampling in the Vadose Zone[2]

3. Terminology

3.1 Terminology used within this guide is in accordance with Terminology D 653 with the addition of the following:

3.2 Definitions in accordance with Practice D 5092.

3.2.1 *bailer*—a hollow tubular receptacle used to facilitate removal of fluid from a well or borehole.

3.2.2 *borehole*—a circular open or uncased subsurface hole created by drilling.

3.2.3 *casing*—pipe, finished in sections with either threaded connections or beveled edges to be field welded, which is installed temporarily or permanently to counteract caving, to advance the borehole, or to isolate the interval being monitored, or combination thereof.

3.2.4 *caving; sloughing*—the inflow of unconsolidated material into a borehole that occurs when the borehole walls lose their cohesive strength.

3.2.5 *centralizer*—a device that helps in the centering of a casing or riser within a borehole or another casing.

3.2.6 *jetting*—when applied as a drilling method, water is

forced down through the drill rods or riser pipe and out through the end openings. The jetting water then transports the generated cuttings to the ground surface in the annulus of the drill rods or casing and the borehole. The term jetting may also refer to a well development technique.

3.2.7 *PTFE tape*—joint sealing tape composed of polytetrafluorethylene.

3.2.8 *well screen*—a filtering device used to retain the primary or natural filter pack; usually a cylindrical pipe with openings of uniform width, orientation, and spacing.

3.3 *Descriptions of Terms Specific to This Standard:*

3.3.1 *assembly length*—length of sampler body and riser pipes.

3.3.2 *bentonite*—the common name for drilling fluid additives and well construction products consisting mostly of naturally occurring sodium montmorillonite. Some bentonite products have chemical additives that may affect water quality analyses (see 9.3.3).

3.3.3 *direct-push sampling*—sampling devices that are directly inserted into the soil to be sampled without drilling or borehole excavation.

3.3.4 *drill hole*—a cylindrical hole advanced into the subsurface by mechanical means; also, known as borehole or boring.

3.3.5 *effective screen length*—the length of a screen open or exposed to water bearing strata.

3.3.6 *effective seal length*—the length of soil above the well screen that is in intimate contact with the riser pipe and prevents connection of the well screen with ground water from other zones.

3.3.7 *grab sampling*—the process of collecting a sample of fluid exposed to atmospheric pressure through the riser pipe with bailers or other methods that may include pumping; also know as batch sampling.

3.3.8 *incremental drilling and sampling*—insertion method where rotary drilling and sampling events are alternated for incremental sampling. Incremental drilling is often needed to penetrate harder or deeper formations.

3.3.9 *in situ testing devices*—sensors or samplers, used for obtaining mechanical or chemical test data, that are typically pushed, rotated, or driven from the surface or below the bottom of a borehole following completion of an increment of drilling.

3.3.10 *intermittent sampling devices*—usually barrel-type samplers driven or pushed below the bottom of a borehole following completion of an increment of drilling.

3.3.11 *percussion driving*—insertion method where rapid hammer impacts are performed to insert the sampling device. The percussion is normally accompanied with application of static down force.

3.3.12 *push depth*—the depth below a ground surface datum that the end or tip of the direct-push water sampling device is inserted.

4. Summary of Guide

4.1 Direct-push water sampling consists of pushing a protected well screen to a known depth, opening the well screen over a known interval, and sampling water from the interval. A well point with an exposed screen can also be pushed with understanding of potential cross-contamination effects and purging requirements considered. A sampler with constant outside diameter is inserted directly into the soil by hydraulic jacking or hammering until sufficient riser pipe is seated into the soil to ensure a seal. Protected well screens can be exposed by retraction of riser pipes. While the riser is seated in the soil, water samples can be taken, and water injection or pressure measurements may be performed.

5. Significance and Use

5.1 Direct-push water sampling is an economical method for obtaining discrete ground-water samples without the expense of permanent monitoring well installation (1-4).[4] This guide can be used to profile potential ground-water contamination with depth by performing repetitive sampling events. Soils to be sampled must be permeable to allow filling of the sample in a relatively short time. The zone to be sampled can be isolated by matching well screen length to obtain discrete samples of thin aquifers. Use of these sampling techniques will result in more detailed site characterization of sites containing multiple aquifers. By inserting a protected sampling screen in direct contact with soil and with watertight risers, initial well development (Guide D 5521) and purging of wells may not be required for the first sampling event. Discrete water sampling, combined with knowledge of location and thickness of target aquifers, may better define conditions in thin multiple aquifers than monitoring wells with screened intervals that can intersect and allow for intercommunication of multiple aquifers (2,4,5,7,8,11). Direct-push sampling performed without knowledge of the location and thickness of target aquifers can result in sampling of the wrong aquifer or penetration through confining beds.

5.2 For sites that allow surface push of the sampling device, discrete water sampling is often performed in conjunction with the cone penetration test (Test Method D 5778) (2-9), which is often used for stratigraphic mapping of aquifers, and to delineate high-permeability zones. In such cases, direct-push water sampling is normally performed close to cone holes. In complex alluvial environments, thin aquifers may vary in continuity such that water sampling devices may not intersect the same layer at equivalent depths as companion cone penetrometer holes.

5.3 Water sampling chambers may be sealed to maintain in situ pressures and to allow for pressure measurements and permeability testing (4,7,10). Sealing of samples under pressure may reduce the possible volatilization of some organic compounds. Field comparisons may be used to evaluate any systematic errors in sampling equipments and methods. Comparison studies may include the need for pressurizing samples, or the use of vacuum to extract fluids more rapidly from low hydraulic conductivity soils (8.1.5.3).

5.4 Degradation of water samples during handling and transport can be reduced if discrete water sampling events with protected screen samplers are combined with real time field analysis of potential contaminants. In limited studies, researchers have found that the combination of discrete protected screen sampling with onsite field analytical testing provide accurate data of aquifer water quality conditions at the time of testing (2,4). Direct-push water sampling with exposed screen sampling devices, which may require development or purging, are considered as screening tools depending on precautions that are taken during testing.

5.5 A well screen may be pushed into undisturbed soils at the base of a drill hole and backfilled to make permanent installed monitoring wells. Procedures to complete direct-push wells as permanent installations are similar to those given in Practice D 5092. These procedures allow for permanent sealing of riser pipe in the borehole. Some state or local regulations may not allow for certain types of direct-push installations as permanent monitoring wells depending on the application. Sometimes, where temporary well screens are inserted at the top of the ground water table, sealing an annulus may not be necessary.

5.6 In difficult driving conditions, penetrating to the required depth to ensure sealing of the sampler well screen may not be possible. If the well screen cannot be inserted into the soil with an adequate seal, the water-sampling event would require sealing in accordance with Practice D 5092 to isolate the required aquifer. Selection of the appropriate equipment and methods to reach required depth at the site of concern should be made in consultation with experienced operators or manufacturers. If there is no information as to the subsurface conditions, initial explorations consisting of penetration-resistance tests, such as Method D 1586, or actual direct-push testing trials can be performed to select the appropriate testing system.

5.6.1 Typical penetration depths for a specific equipment configuration depend on many variables. Some of the variables are the driving system, the diameter of the sampler and riser pipes, and the resistance of the materials.

5.6.2 Certain subsurface conditions may prevent sampler insertion. Penetration is not possible in hard rock and usually not possible in softer rocks such as claystones and shales. Coarse particles such as gravels, cobbles, and boulders may be difficult to penetrate or cause damage to the sampler or riser pipes. Cemented soil zones may be difficult to penetrate depending on the strength and thickness of the layers. If layers are present that prevent direct push from the surface, the rotary or percussion drilling methods (Guides D 5781, D 5782, D 5783, and D 5784, D 5785, D 5786, and see 2.1.1) can be employed to advance a boring through impeding layers to reach testing zones.

5.6.3 Driving systems are generally selected based on required testing depths and the materials to be penetrated. For systems using primarily static reaction force to insert the sampler, depth will be limited by the reaction weight of the equipment and penetration resistance of the material. The ability to pull back the rod string is also a consideration. Impact or percussion soil probing has an advantage of reducing the reaction weight required for penetration. Penetration capability in clays may be increased by reducing rod friction by enlarging tips or friction reducers. However, over reaming of the hole may increase the possibility of rod buckling and may allow for communication of differing ground-water tables. Hand-held equipment is generally used on very shallow investigations, typically less than 5-m depth, but depths on the order of 10^1 m have been reached in very soft lacustrine clays. Intermediate size driving systems, such as small truck-mounted hydraulic-powered push and impact

---

[4] The boldface numbers in parentheses refer to a list of references at the end of this guide.

drivers, typically work within depth ranges from 5 to 30 m, but can reach depths on the order of 10^2 m. Heavy static-push cone penetrometer vehicles, such as 20-ton trucks, typically work within depth ranges from 15 to 45 m, and also reach depth ranges on the order of 10^2 m in soft ground conditions. Drilling methods (Guides D 5781, D 5782, D 5783, D 5784, D 5785, D 5786, and also see 2.1.1) using drilling and incremental sampling are frequently used in all depth ranges and can be used to reach depths on the order of 10^3 m.

NOTE 1—Users and manufacturers cannot agree on depth ranges for different soil types. Users should consult with experienced producers and manufacturers to determine depth capability for their site conditions.

5.7 Combining multiple-sampling events in a single-sample chamber without decontamination (Practice D 5088) is generally unacceptable. In this application, purging of the chamber should be performed to ensure isolation of the sampling event. Purging should be performed by removing several volumes of fluid until new chemical properties have been stabilized or elements are flushed with fluid of known chemistry. Purging requirements may depend upon the materials used in the sampler and the sampler design.

6. Apparatus

6.1 *General*—A direct-push sampling system consists of a tip; well screen; chambers, if present; and riser pipes extending to the surface. Direct-push water sampling equipment can be grouped into two classes, either with a sealed protected screen or exposed screen. Samplers with sealed screens depend on the seal to avoid exposure of the sampling interval to soil or water from other layers. They can be considered as accurate point-source detectors. They are normally decontaminated between sampling events. Exposed-screen samplers may require purging and development and as such are considered as screening devices for profiling relative degrees of contamination.

6.1.1 *Exposed-Screen Samplers*—Some direct-push samplers may consist of a simple exposed well screen and riser pipe that allows grab sampling with bailers or pumps. An example of this arrangement is the simple push or well point shown in Fig. 1. **(12)**. The practice of jetting well points is often not acceptable due to the large quantities of water used for insertion and the resulting potential for disturbance and dilution in the aquifer. If water is used for insertion, knowing the chemical constituents in the water may be necessary. Bias may be possible if an exposed-screen sampler is pushed through multiple contaminated layers. If exposed-screen well points are pushed through predrilled holes the screen and riser may fill with water present in the drill hole and require purging before sampling.

6.1.1.1 Another form of an exposed-screen sampler has been incorporated into cone penetrometer bodies **(6)**. The cone penetrometers have sample chambers with measurement devices such as temperature and conductivity. Some cone penetrometers have been equipped with pumps for drawing in water samples into sample chambers or to the surface. Samplers equipped with chambers and subjected to multiple sampling events may require purging between sampling events. Although several of these designs have been proposed, they have not been successful in production

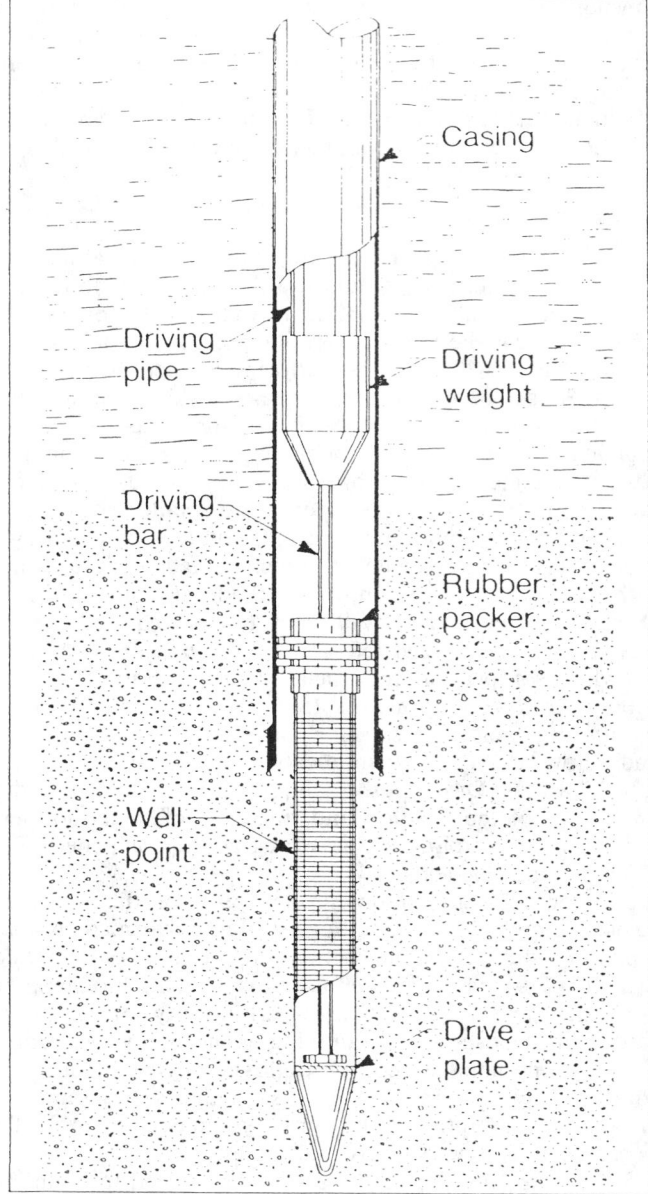

FIG. 1 Exposed-Screen Sampler—Well Point Driven Below the Base of a Borehole (12)

practice. This is because of lengthy and time consuming purging requirements. In most cases, purging requirements and the depths of testing may be such that single-sampling events without cone penetrometers may be more economical than multiple-sampling events requiring purging.

6.1.2 *Sealed-Screen Samplers*—Protected well screen and simple riser pipes for grab sampling are also deployed. An example is shown in Fig. 2 **(13)**. This simple well screen arrangement allows for grab sampling through the riser pipe without purging or development if there is no leakage at the screen seals and riser pipes. Figure 3 shows a schematic of a direct-push water sampler with a protected screen and with the ability to work in the grab sampling mode or by allowing water to enter a sample chamber in the sampler body **(1)**. Most simple sample chambers allow for flow through the

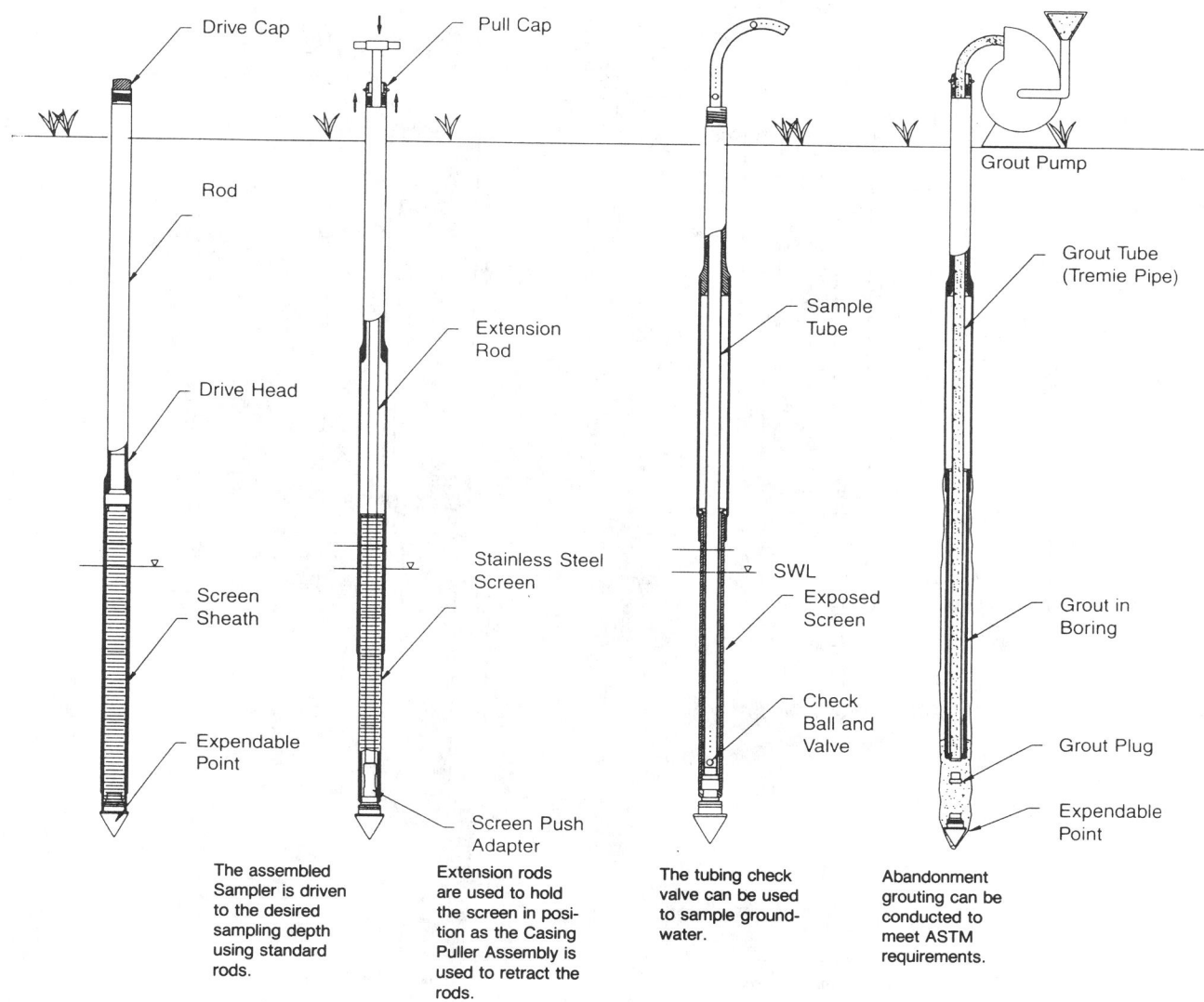

Drive Cap

Pull Cap

Rod

Drive Head

Extension
Rod

Screen
Sheath

Stainless Steel
Screen

Expendable
Point

Screen Push
Adapter

Sample
Tube

Grout Pump

Grout Tube
(Tremie Pipe)

SWL

Exposed
Screen

Grout in
Boring

Check
Ball and
Valve

Grout Plug

Expendable
Point

The assembled
Sampler is driven
to the desired
sampling depth
using standard
rods.

Extension rods
are used to hold
the screen in posi-
tion as the Casing
Puller Assembly is
used to retract the
rods.

The tubing check
valve can be used
to sample ground-
water.

Abandonment
grouting can be
conducted to
meet ASTM
requirements.

FIG. 2 Simple Protected Screen Sampler (13)

chamber. When flow through chambered samplers is opened, it is possible that the ground water from the test interval can fill into the rods above the chamber. In those cases, it may be advisable to add water of known chemistry into the rods prior to opening the screen. Some protected-screen samplers have sample chambers designed to reduce volume and pressure changes in the sample to avoid possible volatilization of volatile compounds (4,7,10). The need for pressurization is dependent on the requirements of the investigation program and should be evaluated by comparison studies in the field with simpler systems allowing the sample to equalize at atmospheric pressure. There are different approaches to pressurizing the sample chamber including use of inert gas pressure or using sealed systems. An example of a sealed vial-septum system is shown in Fig. 4 (4). In the sealed vial system, a septum is punctured with a hypodermic needle connected to a sealed vial. With this approach the vial will contain both a liquid and gas at aquifer pressure. The sealed vial-septum system has been used in an exposed-screen mode.

6.1.3 *Materials of Manufacture*—The choice of materials used in the construction of direct-push water sampling devices should be based on the knowledge of the geochemical environment to be sampled and how the materials may interact with the sample by means of physical, chemical, or biological processes. Due to the nature of insertion of these devices, the sampler body is typically comprised of steel, stainless steel, or metals of other alloys. The type of metal should be selected based on possible interaction effects with the fluid to be sampled. Well-screen materials can be selected from a variety of materials. Materials commonly used for well-screen elements include steel, stainless steel, rigid poly-vinyl chloride (PVC), polytetrafluorethylene (PTFE), poly-ethylene (PE), polypropylene (PP), and brass. Sample chambers, pumps, and connector lines are also constructed with a variety of materials. Evaluating the possible interaction of materials that will be exposed to the water during the sampling event is important.

6.2 *Sampler Body*—The sampler body consists of a tip, and a barrel that consists of well screen, a protective sleeve if used, and a sampling chamber if used, with a connector

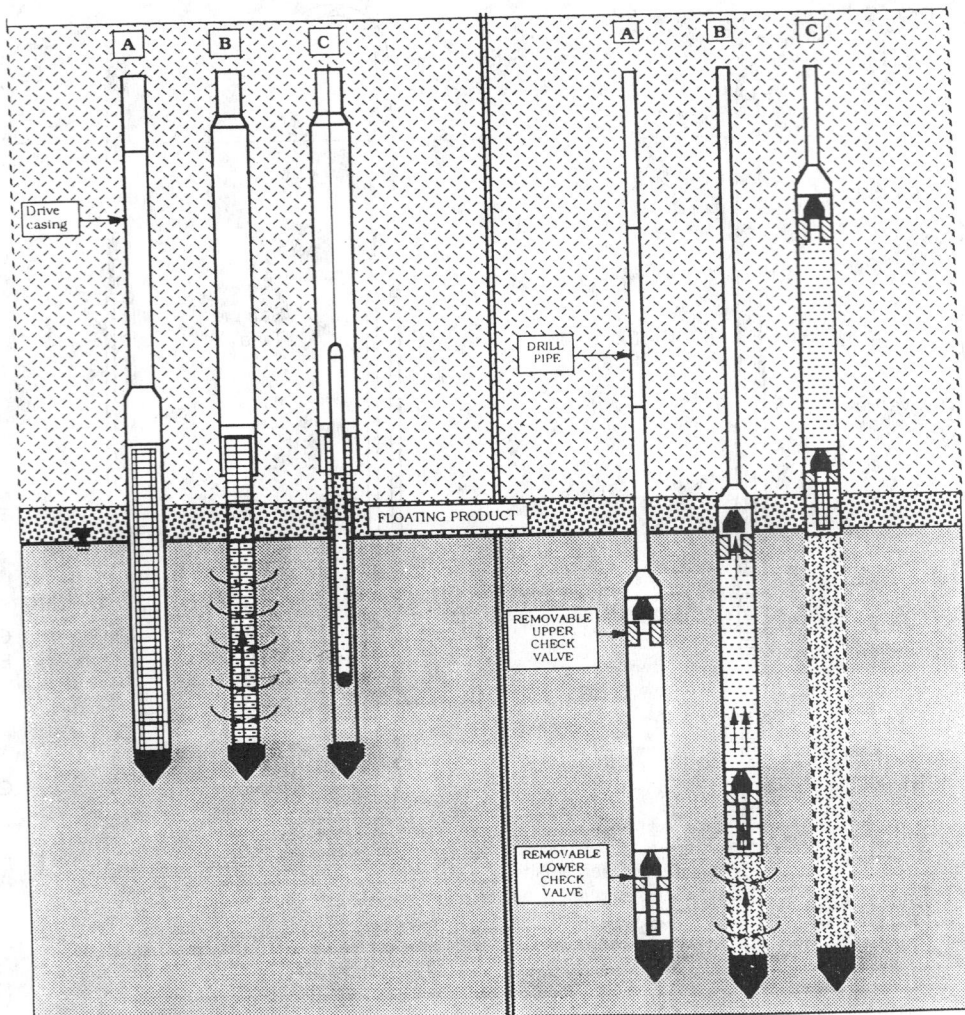

Legend: Grab Sampling

A Penetrometer closed while being driven into position.
B Tool opened and 5 foot screen telescopes into position for collection of hydrocarbon or water sample at the very top of the aquifer.
C Hydrocarbon sample being collected using bailer lowered through drive casing.

Legend: Water Sampling in Chamber

A Penetrometer closed while being driven into position.
B Cone separated and tool open to collect sample.
C Check valves closed as sample is retrieved within body of the tool.

FIG. 3 Protected Screen Sampler Capable of Working in Grab or Chamber Sampling Modes (1)

assembly to attach to riser pipes. The sampler is normally constructed of steel to withstand insertion forces. The sampler barrel should be of constant outside diameter to ensure intimate contact with the soil to be tested. Protective sleeves shall be equipped with O-rings to prevent the ingress of water before the sampling event.

6.2.1 *Expendable Sampler Tips*—Some sampler tips are expendable and are left in the ground after the sampling event. The tip should be equipped with an O-ring seal to the sampler sleeve to prevent leakage into the riser pipe until the sampling depth is reached.

6.2.1.1 Sampler tips are designed so that upon pull back of the sampler body and riser pipe, the tip is disconnected from the sampler. The required diameter, and the ability to expend the tip successfully, depends on the soils to be penetrated. The tip diameter can be set equal to, or slightly less than, the sampler body. If there are problems with tip

retraction, tips can be designed with a diameter of 1 to 3 mm (1/8 to 1/16 in.) larger than the sampler body. The use of an enlarged diameter with a larger shoulder or tip may help in reaching greater depths because it acts as a friction reducer. An enlarged tip should not leave too large an annulus above the sampler body and riser pipes as to maintain a seal above the well screen and to prevent potential cross contamination.

6.2.1.2 Most sampler tips are made of steel to withstand pushing forces. With some samplers, after the sampling event, the tip may remain in the ground and the hole may be grouted. The user should consider if leaving the tips below the ground will adversely affect surrounding ground-water chemistry depending on site conditions.

6.2.2 *Well Screen*—Many materials for well screens are available for direct-push samplers. The material of manufacture should be selected with consideration of chemical composition of the ground water to be sampled and possible

1032

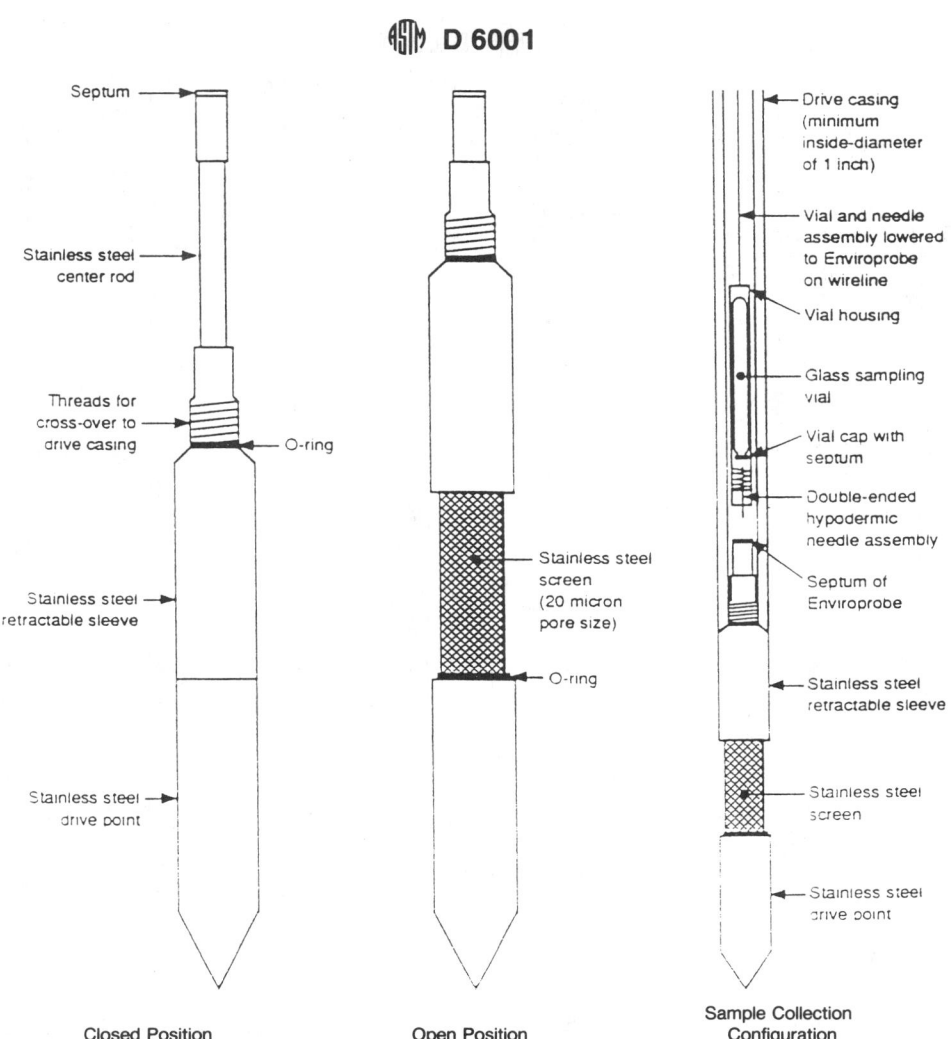

FIG. 4 Protected Screen Sampler with Sealed Vial System (4)

interactive effects (see 6.1.3). Some samplers use simple mill slotted steel, or PVC tube. Steel or brass screen formed into a cylinder can be used to cover inlets. Continuous-wrapped, wire-wound well points are also commonly used. The effective opening size of the well screen material should be selected based on the material to be sampled, the time required to sample, and soil sediment that can be tolerated in the water sample. Methods to size well-screen and filter-pack materials are given in Practice D 5092. Clean sands and gravels can be sampled with a screen with larger openings without producing excessive sediment. Clayey and silty soils containing fines may require finer openings. Typical openings of 10 to 60 µm are used. Finer openings will reduce sediment but may also slow ingress of fluid.

6.2.3 Some sampler inlets are not protected by well screen or slotting. The simplest form of sampler can be an open riser pipe with an expendable tip. The use of unprotected inlets has sometimes been useful to sample ground water at soil/bedrock interface. If unprotected inlets are used, one must consider the amount of soil sediment that can be tolerated in the sample.

6.3 *Riser Pipes*—Also commonly referred to as "push rods" or "extension rods", riser pipes are normally constructed of steel to withstand pushing forces. Some tempo-

rary well-point installations may use a double-tube system such as a small-diameter PVC riser pushed by the steel tube (Fig. 5) **(14)**. Double-tube systems are advantageous if multiple sampling events are required in a single push. Other temporary systems may use a flexible tubing system connected to the well point (Fig. 6) **(14)**. For PVC riser pushed with outside steel tubing, the withdrawal of steel push rods will leave a small annulus between the soil and PVC riser or tubing. This annulus may require grouting depending on the effective seal above the well screen and the possibility of cross contamination of overlying layers. Cone penetrometer rods as specified in Test Method D 5778 are sometimes used in sampling systems deployed with cone penetrometer equipment. Larger diameter rods, typically 45 mm (1.75 in.), are sometimes used with cone penetrometer equipment. The maximum rod diameter that can be used depends on the material to be penetrated and the driving system. Increased rod diameter causes increase in the required driving force required to penetrate a sufficient distance. Most surface direct-push riser pipes are less than 50 mm (2 in.) in diameter.

6.4 Standard drilling rods used for rotary drilling are normally used when sampling is done at the base of drill holes. Many drill rods are available (see Guides D 5781,

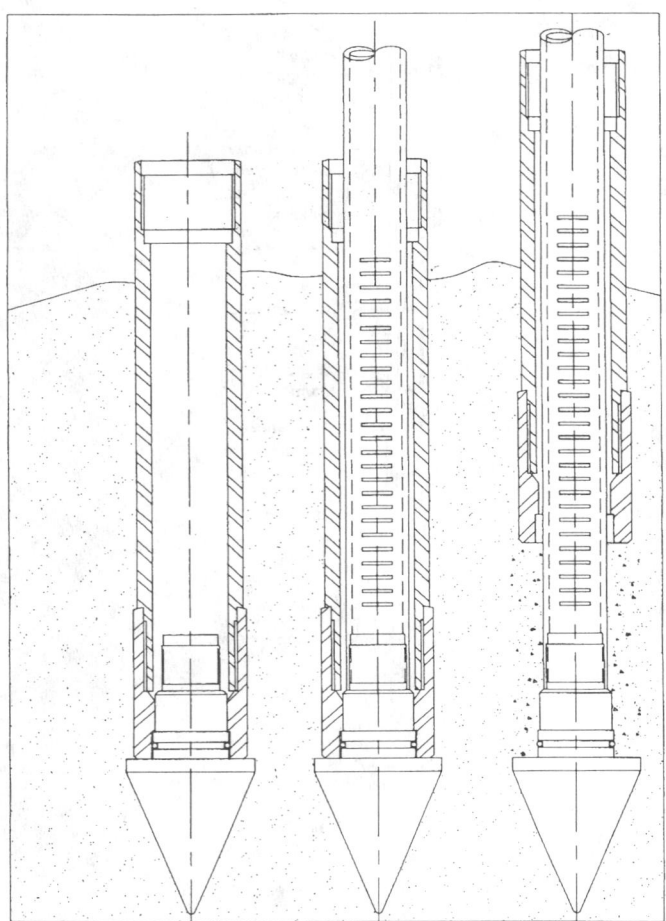

FIG. 5   Double-Tube Temporary Well Point System (14)

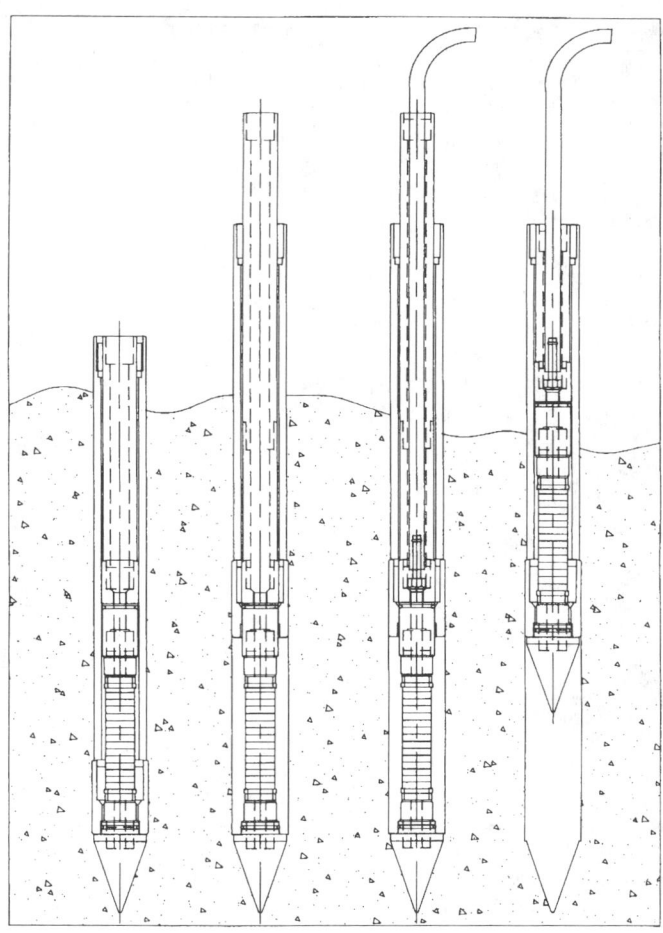

FIG. 6   Protected Screen Sampler with Sample Tubing (14)

D 5782, D 5783, D 5784, D 5785, D 5786, and also see 2.1.1). For direct-push sampling systems that depend on the riser pipe for grab sampling within the riser, ensuring that joints are watertight will be necessary such that water enters through the well screen interval to be sampled. Rods should be wrench-tightened, and PFTE tape can be used on the threads to stop leakage. The quality checks discussed in Section 8 can be performed to evaluate possible leakage. Sometimes it may be necessary to equip rod joint shoulders with O-rings to prevent leakage. Cone penetrometer rods with precision tapered threads are normally watertight during short sampling events lasting up to 1 h if they are not damaged.

6.4.1 *Friction Reducers*—Friction reducers that have enlarged outside diameters of the riser pipe are sometimes employed to reduce thrust capacity needed to advance the well point or sampler. If friction reducers are used, they must be a sufficient distance above the sampling location to ensure that fluids from overlying layers can enter the sampling zone. If cross-contamination is possible, use of friction reducers should be avoided. In some cases the use of friction reducers can help in forming an annular seal. Donut-type reducers ream the hole smoothly. Lug-type reducers rip and remold the soil and may provide a better annular seal. The type and location of friction reducers should be documented in the project report.

6.4.2 *Mud Injection*—Some direct-push systems inject betonite drill fluid along the drill rods to reduce friction. These systems normally inject the fluid behind friction reducers. These systems may provide better sealing above the sampler for the sampling process but are also more difficult to operate.

6.5 *Sampling Devices*—Methods to obtain water samples vary widely. Examples are given in 6.1.1 and 6.1.2. Simple grab samplers, most often bailers, are used with simpler systems. Other systems draw water into chambers or sealed vials for retrieval to the surface. Some systems may have pumps and circulation systems to retrieve samples to the surface. The materials of manufacture of samplers, sample containers, pumps, and circulation lines should be selected considering possible interaction effects discussed in 6.1.3. Selection of devices for sampling ground water is presented in Guide D 4448. Sampling methods and devices should be selected based on the potential impact on sample integrity as addressed in 6.1.3 and other areas in Guide D 4448.

6.6 *Sample Containers*—Sample containers for sampling ground water are addressed in Guide D 4448.

6.7 *Driving or Pushing Equipment*—Soil probing (percussion driving) systems, penetrometer systems, and rotary drilling equipment are used for inserting direct-push water sampling devices. The equipment should be capable of applying sufficient mechanical force or have sufficient reaction weight, or both, to advance the sampler or screen to a sufficient depth to ensure an effective seal above the area to be sampled. The advancement system must also have sufficient retraction force to remove the rods, which is often a more difficult task than advancing the rods. Simple advancement systems include hand-held rotary-impact hammers with mechanical-extraction jacks. Many systems use hydraulic- or vibratory-impact hammers operating at high frequency to drive rods into the sampling interval. Reaction force can be reduced if impact hammers are employed. Multipurpose driving systems such as those commonly deployed for soil gas sampling (Guide D 5314) are frequently used in shallow explorations. Some vibratory drilling systems can provide vibration to the rods and easily penetrate cohesionless soils. On soft ground sites, cone penetrometer systems use hydraulic rams to push the sampler and riser pipe into the ground. Conventional rotary drilling rigs can use either hydraulic pull-down capability or hammers to drive the sampler to the required depth. Rotary drilling rigs are often used with the incremental drilling and sampling method. A 140-lb SPT hammer (Method D 1586) is available on most rotary drilling rigs and can be used to advance the sampler. Use of impact or vibration may allow for penetration of harder soils. If a significant length of rods whip during driving, they should be restrained to prevent damaging of the annular seal at the base of a borehole from lateral movement.

7. Conditioning

7.1 *Decontamination*—Sampling equipment that contacts ground water to be sampled before and after the sampling event may require decontamination. Decontamination should be performed following the procedures outlined in Practice D 5088 and the site-sampling plan. The sampler body normally requires complete decontamination before sampling. Well-screen components are sometimes expendable. Newly manufactured screens and sampler components may contain residues from manufacture and should be cleaned before the sampling event. Riser pipes should be decontaminated if grab sampling will be performed within the tube.

7.2 *Purging*—For exposed-screen sampling devices and sampling systems open to overlying ground water, purging may be required before the sampling event. With both protected- and exposed-screen samplers, purging may be required if ground water from overlying sources infiltrates into the riser pipes into the sampling area. Purging should consist of removal of overlying ground water from the sampling system prior to the sampling event. Purging requirements are outlined in Guide D 4448.

8. Procedure

8.1 Two procedures are outlined depending on whether the sampling device is pushed directly from the surface or whether drilling is used to advance an open hole close to the sampling interval. In either event, the sampling screen should be advanced into undisturbed soil a sufficient distance to ensure that the sampling depth cannot be exposed to overlying ground water, if present.

8.1.1 *Incremental Drilling and Sampling*—In this method, advance a drill hole close to the sampling interval

using drilling methods listed in 2.1.1. Of the drilling methods listed, the most commonly employed is rotary hollow-stem auger drilling because fluids are not introduced during the drilling process. If a rotary drilling method using drilling fluid or air is employed, the impact of the fluid or air to the sample quality and quality of the surrounding aquifer should be considered. If caving or sloughing occurs the use of protective casings may be required.

8.1.1.1 Stabilize the drill rig and erect the drill rig mast. Establish and document a datum for measuring hole depth. This datum may consist of a stake driven into a stable ground surface, the top of the surface casing, or the drilling deck. Do not use surface casing as a datum if it is subject to movement. If the hole is to be later surveyed for elevation, record and report the elevation difference between the datum and the ground surface. Proceed with drilling until a depth is reached above the target sampling interval. Check and document the depth of the borehole and condition of the base of the hole. Establish the depth and condition of the base of the boring by resting the sampler at the base of the boring and checking depth to the sampler tip. If casing is used and heave occurs into the casing, remove this material and advance the hole deeper. Heave of soil into the casing may make it impossible to drive the well point without it carrying the casing along with the well point or sampler. If excessive heave, caving, or sloughing of soil occurs, consider using an alternative drilling method capable of maintaining stable soil conditions.

8.1.2 If the sampling event is to occur at the ground-water table and equipment depends on a dry-hole condition, that is, an exposed screen sampler with no purging requirements, test the drill hole to confirm that ground water has not entered the hole. Water levels can be determined using Test Method D 4750.

8.1.3 Attach the well point or sampler to riser pipes and lower into the borehole. Carefully record the assembly length as rod sections are added to the assembly. Centralizers may be used to maintain verticality of the assembly and to reduce rod whip. Rest the assembly on the base of the borehole. Determine and record the depth to the tip of the assembly.

8.1.4 Either push or drive the well point or sampler a sufficient distance below the base of the boring. This distance should be at least 1 m (3 ft), or the minimum to ensure an effective seal. For protected-screen samplers where a protective screen is exposed by pulling back the riser pipe, the withdrawal action may shear or crack soil, allowing connection to the base of the borehole. In these cases, adjust the insertion and retraction lengths according to soil conditions. In general, the sampler should be inserted at least three times the effective screen length from retraction. To check the seal in fluid filled holes, tracers can be introduced into the fluid in the base of the borehole. Document the final depth of insertion to the tip of the sampler and midpoint of the well screen. If the sampler is driven with hammer blows, accomplish the penetration without excessive vibrations that could reduce the effective seal of the riser pipe above the well screen. Normally, if smooth penetration is accomplished with each hammer blow, the seal should be intact.

8.1.4.1 The process of jetting well points is not preferred because of the addition of water, disturbance to the sampling zone, and lack of an effective seal above the screen. These installations are usually intended for permanent installations with the drill hole completed as a monitoring well. If jetting is used, document the approximate volume and chemical quality of water.

8.1.5 *Sampling*—The sampling process depends on the type of the sampling equipment used, that is, exposed- or protected-screen samplers.

8.1.5.1 *Sampling of Exposed-Screen Samplers*—Exposed-screen samplers can be sampled after fluids have been purged from the screen and riser pipes. Purge these systems in accordance with Guide D 4448.

8.1.5.2 *Sampling of Protected-Screen Samplers*—Test protected-screen samplers that are open to the surface through the riser for grab sampling for system leakage before exposing the screen for sampling. Before screen exposure, test the riser for presence of water that may have leaked through joints and connections using Test Method D 4750. If water is present from unknown sources, this should be noted and either purging or abandoning of the test should be considered. After quality checks for leakage, the riser pipes may be pulled or twisted to expose the well screen to the aquifer.

8.1.5.3 Several methods for sampling water are available. If the sampling device uses head pressure available in the aquifer, sufficient time should be allowed for water to fill the sampling chamber or riser pipes. Some systems allow for connection of a sealed sampling chamber, or tubing, to a port in the sampler body after the screen is opened, allowing direct connections to the screened sampling area. By using these systems, one may avoid the necessity to check inside the riser pipes for leakage water. Use of sampling pumps to draw in the sample may be allowed, but consideration should be given the changes in ambient pressures and temperatures that may change chemical compositions. With an open tube well screen using grab sampling in low permeability soils, a vacuum is sometimes applied to the top of the riser pipe to accelerate ground-water inflow. The use of a vacuum and its effect on chemical composition should be considered and evaluated if site requirements dictate.

8.1.5.4 After a sufficient volume of the sample is obtained, place the samples in suitable containers for analysis. The volume of a sample to obtain depends on the chemical composition of ground water, testing protocols, and the data-quality objectives. Depending on the screen used, samples may contain sediment and may require filtering before placement of samples in containers. Certain testing procedures or regulations may require filtration of water samples.

8.1.6 After sampling, either retrieve the sampler or leave it in place for permanent installation in accordance with Practice D 5092. Some retrievable samplers leave a tip or a well screen element, or both, below the bottom of the boring. If repeated sampling events are to be performed in the same drill hole, drilling it through these pieces if present will be necessary. Depending on the drilling method, a pilot bit should be reinserted in the drill string and drilling continued to a depth exceeding the depth of the previous sampling event. Normally tips or screens, or both, will be moved to the side of the drill hole before the next sampling event. Sometimes the presence of a tip or element, or both, can be detected by drilling action. If drilling action detects these pieces, note the location. Drilling continues to the next depth

of concern and sampling may be repeated. The depth of the extended drill hole should equal or exceed the depth to the sampling tip of the previous interval.

8.1.7 After the drilling is completed, the drill hole should be completed following guidelines in drilling methods (Guides D 5781, D 5782, D 5783, D 5784, D 5785, D 5786, and also see 2.1.1) or those given in Section 9.

8.2 *Direct Push from the Surface*—Well points and samplers may be advanced directly from the surface with multipurpose percussion driving systems, hand-held rotary percussion drills, cone penetrometer systems, or any other systems capable of supplying sufficient force to reach the depths of concern.

8.2.1 Stabilize and level the rig for testing. For some tire-mounted equipment, the rig can be raised off the ground and leveled with hydraulic rams to lift the rig from the tires to avoid shifting during difficult driving conditions. Establish and document a datum for measuring hole depth. If the hole is to be later surveyed for elevation, record and report the height of the datum to the ground surface.

8.2.2 The sampler body is connected to riser pipes along with any subassemblies such as friction reducers. Prior to driving, measure the length of the sampler assembly and riser pipes to determine the depth of sampling. Some temporary well systems drive a double tube or cased system, where riser pipe and casing are added as it is advanced. This allows for easy annulus grouting as the casing is retracted. The rods are then pushed using smooth quasi static push or impacts, or both. Additional riser pipes are added as pushing progresses. As driving progresses, operators should carefully record the rods added to ensure that sampling occurs at the correct depth.

8.2.3 *Sampling of Exposed-Screen Samplers*—Use the same procedures in accordance with 8.1.5.1.

8.2.3.1 *Sampling of Protected-Screen Samplers*—Use the procedure in accordance with 8.1.5.2 with the addition that the riser pipes should be periodically checked for leakage using Test Method D 4750.

8.2.4 After sufficient volume of a sample is procured, place the samples in suitable containers for analysis. The volume of the sample to obtain depends on the chemical composition of ground water, testing protocols, and the data-quality objectives. Depending on the screen used, samples may contain sediment and may require filtering before placement of samples in containers.

8.3 After sampling, the sampler is either retrieved or left in place for permanent installation (Section 9). Some retrievable samplers leave a tip or a well-screen element, or both, at the bottom of the sounding. If repeated sampling events are to be done in the same hole, they must be done with samplers pushed to greater depths.

8.4 After the testing is finished, complete the borehole following the guidelines in Section 9.

9. Completion and Abandonment

9.1 *Permanent or Temporary Well Installations*—Wells inserted by either drilling methods or direct push from the surface may be left in the ground as permanent or temporary installations. Some state or local regulations may not allow for certain types of direct-push installations as permanent monitoring wells depending on the application. If there are questions as to the performance of direct-push wells, they can be compared to wells installed using rotary drilling methods (Guides D 5781, D 5782, D 5783, and D 5784, D 5785, D 5786, and also see 2.1.1) in accordance with Practice D 5092. For wells inserted in drill holes, the drill hole will require completion with sealing materials to ensure a seal between the hole wall and riser pipes. Sealing procedures are given Practice D 5092.

9.1.1 For wells installed by direct push from the surface, the need for sealing depends on the size of the annulus, ground-water quality, and the ability for cross-contaminating or accelerating contamination movements among aquifer(s). Temporary well points installed into the top of the first ground-water layer may only require surface sealing. If the annulus is very small, soil cave and squeeze may reduce effective vertical hydraulic conductivity. If the well riser intersects perched aquifers, cross-communication of aquifers may be possible if too large an annulus is left open. Communication can be evaluated by performing tracer tests, if necessary. Friction reducers used on cone penetrometer equipments may only increase hole diameters by 6 to 13 mm (¼ to ½ in.) of that of the steel pipes for pushing.

9.2 *Other Completion Methods*—Depending on the requirements of the investigation, performing special completions with protective casings or other sealing may be necessary. For holes using rotary drilling methods and incremental sampling, the hole could be completed as a monitoring well (Practice D 5092) or with grouted casings for other testing such as geophysical tests. Several methods are available for grouting of casings. Using injection grouting where injection is done at the base of the boring is most desirable and grouts are pumped up the annulus until they reach the surface showing a continuous seal.

9.3 *Hole Abandonment*—For test holes where there are no installations or other completion methods, the hole should be abandoned following program requirements. The need for and the method of sealing for abandonment depends on state and local regulations, site conditions, ground-water quality, and the ability for cross-contaminating or accelerating contamination movements among aquifer(s).

9.3.1 Large-diameter drill holes from rotary drill operation often require sealing. State, federal, and local regulations may dictate abandonment requirements for boreholes intersecting the water table.

9.3.1.1 The need for sealing of holes is also dependent on geohydrologic conditions. If the hole intersects the top of the first ground-water table, complete sealing may not be required. Under a homogeneous single aquifer system, where there are no perched water table or artesian conditions, there will be little hydraulic gradient to move potential contaminants at differing elevations. The worst case for possible cross-communication of aquifers occurs under perched or confined ground-water conditions.

9.3.1.2 In most cases, direct-push holes intersecting ground-water tables will require complete sealing. In cases where the hole is to be backfilled completely, the condition of the hole should be evaluated and documented. Any zones of caving or blocking which preclude complete sealing should be documented. Displacement grouting may displace ground water from the hole to the surface. If this water is considered contaminated then provisions must be made to collect these fluids at the surface. A minimum requirement for sealing should be that the surface of the hole is sealed to prevent hazards to those at the surface and to eliminate direct movement of surface contaminants to the water table through the hole.

9.3.2 *Completion of Drill Holes*—Completion of boreholes using drilling methods are addressed in Guides D 5781, D 5782, D 5783, D 5784, D 5785, D 5786, and also see 2.1.1.

9.3.3 *Completion of Surface Direct-Push Holes*—Several methods have been used successfully for sealing or grouting of surface direct-push holes (15). The method of grouting depends on the types of equipment deployed and the subsurface conditions encountered.

9.3.3.1 One method of grouting is retraction grouting directly through the sampler tip or friction reducer as the sampler is withdrawn after the sampling event. Tip retraction grouting is normally performed through small diameter tubes and a knockoff tip. Tip retraction grouting is the least frequently used due to difficulty in pumping grout mixtures without significant head loss through the tubing. Cement grouts for tip retraction grouting may require higher water content or additives to reduce viscosity.

9.3.3.2 Retraction grouting is sometimes performed through grouting points above the sampler tip. This is normally accomplished using an enlarged diameter grouting port above the sampler as shown in Fig. 7.

9.3.3.3 Reentry grouting may have an advantage of freeing pushing equipments for production while grouting

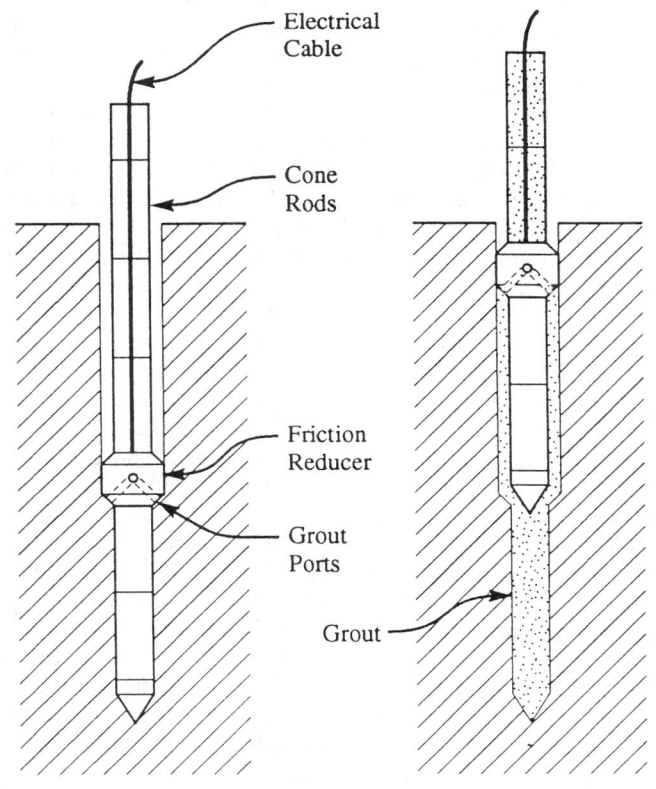

(a) Installation   (b) Grouting

FIG. 7 Grouting Through Ports in Friction Reducers (15)

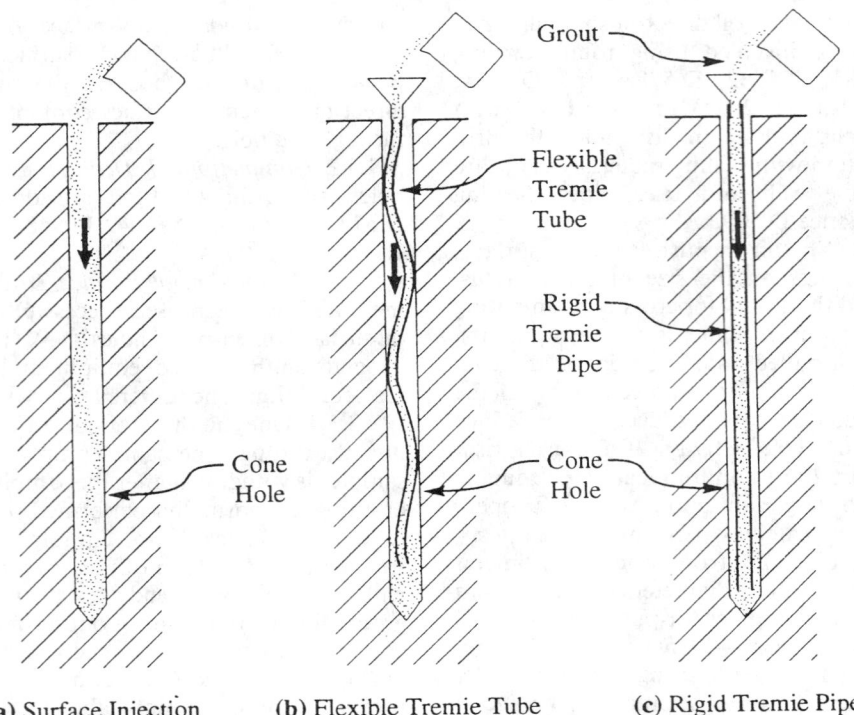

(a) Surface Injection　　(b) Flexible Tremie Tube　　(c) Rigid Tremie Pipe

FIG. 8　Rigid Pipe with Internal Flexible Tremie Tube (15)

operations follow. Reentry grouting allows temporary connection of aquifers between the removal and reinsertion process but is normally acceptable if grouting follows promptly minimizing exposure. The selection of retraction or reentry grouting is an economic decision and it depends on site conditions and depth of soundings.

9.3.3.4 In reentry grouting, Figs. 8 and 9, the test string is completely withdrawn from the hole and a secondary grouting tube or tubing is reinserted to the complete depth of the hole. If the hole remains open after retraction of the test string, inserting flexible tubing or small-diameter PVC into the hole by hand directly after testing may be possible. In this case, reinserting the grout line is desirable close to the original depth of the hole. In some cases, depending on project needs, locations of water bearing strata, and soil stratigraphy, it may be acceptable if the grout line does not reach the bottom of the hole.

9.3.3.5 Usually, with squeezing clays or caving sands, reaction equipment may be required to push rigid tubing of steel or plastic with a sacrificial or grouting tip to the complete depth of the hole (Figs. 8 and 9). The reentry string should follow the original hole alignment because it is the path of least resistance. If deviation is suspected, it should be reported. If a knockoff tip is to be retracted in high hydraulic conductivity sands it may be necessary to add grout into rods prior to tip retraction to avoid water filling the rods. Grout is then pumped through the hole until it rises to the surface, or tremie grouting is performed by maintaining a grout column in the rods as they are removed. Grouting is continued to maintain a full hole as tubing is withdrawn. The simplest method of sealing a direct-push hole in stable materials is to place dry materials by pouring or placing directly into the open hole after testing. This method is normally only

acceptable in stable clay soils where the hole remains open after testing. This method is not acceptable if there are zones of hole caving or squeezing or there is appreciable presence of ground water in the hole. The holes can be probed with small-diameter rods to evaluate these conditions. Small-diameter granular bentonite is normally used in this application.

9.3.3.6 Direct-push water sampling holes can be grouted

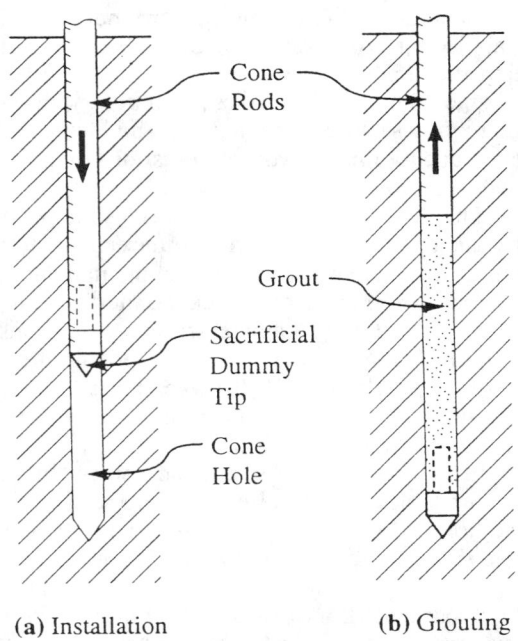

(a) Installation　　　　　(b) Grouting

FIG. 9　Reentry with CPT Rods and Sacrificial Tip (15)

with either cement or bentonite grouts. The grout consistency may have to be wetter than standard mixes used for sealing boreholes (Practice D 5092). There has been no research to confirm the best proportions. A typical mixture is 1 sack of Portland cement to 19 to 22 L (5 to 8 gal) of water. Bentonite is added in a small percentage, 2 to 5 %, to reduce shrinkage. Typical bentonite-based mixtures consist of 22.7 kg of dry powered bentonite to 50 to 200 L (24 to 55 gal) of water. It is difficult to mix dry high-yield bentonite without good circulation equipment and time to allow for mixing and hydration. Pre-hydrated bentonite is easier to mix. Some bentonites contain additives that may not be acceptable for grouting use and the user should check with regulators to ensure sealing products are acceptable.

9.3.3.7 Record the volumes of grout injected and compare them with theoretical hole volumes. Often the grouting pressure at depth is unknown due to head losses through pipes, grout tubing, and connections. Pressure grouting equipments should at a minimum include a pressure gage at the surface. To avoid excessive hydraulic fracturing of the units, downhole pressures should be restricted to ½ psi per foot of hole depth. Record any unusual changes in grouting pressures that may suggest the presence of obstructions, caved zones, or occurrence of fracturing.

10. Field Report and Project Control

10.1 Report information recommended in Guide D 5434 and identified as necessary and pertinent to the needs of the exploration program. Information is normally required for the project, exploration type and execution, drilling equipment and methods, subsurface conditions encountered, ground-water conditions, sampling events, and installations. Some of the data collected during these investigations may be reported as data elements for describing ground-water sites (Guides D 5254, and D 5474).

10.2 Other information besides that mentioned in Guide D 5434 should be considered if deemed appropriate and necessary to the needs of the exploration program. Additional information should be considered as follows:

10.2.1 *Drilling Methods*—If rotary drilling methods are used for predrilling holes, report information particular to the drilling methods as outlined in Guides D 5781, D 5782, D 5783, D 5784, D 5785, D 5786, and also see 2.1.1.

10.2.2 *Percussion Driving and Penetrometer Equipment*—For equipment used for surface direct push, report the equipment type, make, model, and manufacturers. Report conditions during push of the sampler such as the occurrence of hard layers. Report datums established for monitoring depth of penetration. For combined cone penetrometers and water-sampling devices, report cone-penetration information in accordance with Test Methods D 3441 and D 5778.

10.3 *Sampling:*

10.3.1 *Equipment*—Report the types of sampling equipment used including materials of manufacture of the components. Provide dimensions of the equipment including outside diameter, screen length and diameter, and friction reducers. Report methods for cleaning of the equipment before and after sampling. Note materials left in the hole or discarded between sampling events. Report any purging or development actions taken before the sampling event.

10.3.2 When water sampling is performed at the base of the borehole, report the condition of the base of the hole before sampling, and report any slough or cuttings present in the recovered sample.

10.3.3 During insertion of the sampler or well point, note any difficulties in advancing the point and retraction of a protective sleeve. Report the retraction distance for protected-screen samplers. If the sampler cannot be advanced more than the minimum required distance of the sampler given in 8.1.4, report the distance driven. Note and record sampling depths including depths to the tip and midpoint of the well screen. Note any unusual occurrence during sampling such as fluid exposure, or evidence of cross-contamination contained in the samples recovered. Note and record the volume of the sample taken and other sample handling and preservation methods taken.

10.3.4 Report any measurements of water samples routinely performed in the field. These measurements may include temperature, PH, and conductivity. Report methods of testing, calibrations, and equipment used.

10.4 *Completion and Installations*—A description of completion materials and methods of placement, approximate volumes placed, intervals of placement, methods of confirming placement, and areas of difficulty or unusual occurrences.

11. Precision and Bias

11.1 The precision and bias of this method have not been established. Due to variability of subsurface conditions, comparative studies of differing approaches to direct-push sampling have not been statistically significant, because site spatial variability exceeded differences between methods (2). Comparisons between water samples obtained from direct-push samples and standard-monitoring wells have been favorable (11). Additional studies are needed and are actively pursued by Subcommittee D18.21.

12. Keywords

12.1 direct-push; water sampling; well point

REFERENCES

(1) Cordry, K., "Ground Water Sampling Without Wells," *Proceedings of the National Water Well Association, Sixth National Symposium on Aquifer Restoration and Ground Water Monitoring,* May 19–22, 1986, Columbus, OH, pp. 262–271.

(2) Zemo, D., et al, "Cone Penetrometer Testing and Discrete-Depth Groundwater Sampling Techniques: A Cost Effective Method of Site Characterization in a Multiple Aquifer Setting," *Proceedings of the National Ground Water Association,* Outdoor Action Conference, May 11–13, 1992, Las Vegas, NV, pp. 299–313.

(3) Smolley, M., and Kappemyer, J., "Cone Penetrometer Tests and Hydropunch™ Sampling—An Alternative to Monitoring Wells for Plume Detection," *Proceedings of the Hazmacon 1989 Conference,* April 18–20, 1989, Santa Clara, CA, pp. 71–80.

(4) Berzins, N. A., "Use of the Cone Penetration Test and BAT Groundwater Monitoring System to Assess Deficiencies in Monitoring Well Data," *Proceedings of the National Ground Water Association Outdoor Action Conference,* 1992, pp. 327–341.

(5) Strutynsky, A., and Bergen, C., "Use of Piezometric Cone Penetra-

tion Testing and Penetrometer Groundwater Sampling for Volatile Organic Contaminant Plume Detection," *Proceedings of the National Water Well Association Petroleum Hydrocarbons Conference*, Oct. 31–Nov. 2, 1990, Houston, TX, pp. 71–84.

(6) Woeller, D. J., Weemees, I., Kokan, M., Jolly, G., and Robertson, P. K., "Penetration Testing for Groundwater Contaminants," *ASCE Geotechnical Engineering Congress, Special Technical Publication*, June 1991.

(7) Klopp, R. A., Petsonk, A. M., and Torstensson, "In-Situ Penetration Testing for Delineation of Ground Water Contaminant Plumes," *Proceedings of the Third National Outdoor Action Conference on Aquifer Restoration, Ground Water Monitoring and Geophysical Methods*, May 22–25, 1989, National Ground Water Association, Dublin, OH, pp. 329–343.

(8) Campanella, R. G., Davies, M. P., Boyd, T. J., and Everard, J. L., "In-Situ Teating Methods for Groundwater Contamination Studies," *Developments in Geotechnical Engineering*, Balasubramaniam et al (eds), Balkema, Rotterdam, pp. 371–379.

(9) Zemo, D. A., Pierce, Y. G., Gallinatti, J. D., "Cone Penetrometer Testing and Discrete-Depth Ground Water Sampling Techniques: A Cost Effective Method of Site Characterization in a Multiple Aquifer Setting," *Ground Water Monitoring Review*, Fall 1994,

National Ground Water Association, Dublin, OH, pp. 176–182.

(10) Eckard, T. L., Millison, D., Muller, J., Vander Velde, E., and Bowallius, R. U., "Vertical Ground Water Monitoring Using the BAT Groundwater Monitoring System," *Proceedings of the Third National Outdoor Action Conference on Aquifer Restoration, Ground Water Monitoring and Geophysical Methods*, May 22–25, 1989, National Ground Water Association, Dublin, OH, pp. 313–327.

(11) Church, P. E., and Gvanato, G. E., "Bias in Ground-Water Data Caused by Well Bore Flow in Long Screen Wells," *Ground Water*, March–April 1996, Ground Water Publishing, Columbus, OH, 1996, pp. 262–273.

(12) Driscoll, F. G., *Groundwater and Wells*, Johnson Filtration Systems Inc., St. Paul, MN, 55112, 1986.

(13) *Geoprobe Screen Point Ground Water Sampler—Standard Operating Procedure, Technical Bulletin No. 94-440*, Geoprobe Systems, April 1994.

(14) Schematic Drawings Courtesy of Diedrich Drilling Inc., LaPorte, IN.

(15) Lutennegger, A. J., and DeGroot, D. J., "Techniques for Sealing Cone Penetrometer Holes," *Canadian Geotechnical Journal*, Vol 32, No. 5, pp. 880–891.

Provisional Standard Guide for
Developing Appropriate Statistical Approaches for Ground-Water Detection Monitoring Programs[1]

This standard is issued under the fixed designation PS 64; the number immediately following the designation indicates the year of original adoption.

Scope

1.1 This provisional guide covers the context of ground-water monitoring at waste disposal facilities, regulations have required statistical methods as the basis for investigating potential environmental impact due to waste disposal facility operation. Owner/operators must perform a statistical analysis on a quarterly or semiannual basis. A statistical test is performed on each of many constituents (for example, 10 to 50 or more) for each of many wells (5 to 100 or more). The result is potentially hundreds, and in some cases, a thousand or more statistical comparisons performed on each monitoring event. Even if the false positive rate for a single test is small (for example, 1 %), the possibility of failing at least one test on any monitoring event is virtually guaranteed. This assumes you have done the correct statistic in the first place.

1.2 This guide is intended to assist regulators and industry in developing statistically powerful ground-water monitoring programs for waste disposal facilities. The purpose of these methods is to detect a potential ground-water impact from the facility at the earliest possible time while simultaneously minimizing the probability of falsely concluding that the facility has impacted ground water when it has not.

1.3 When applied inappropriately existing regulation and guidance on statistical approaches to ground-water monitoring often suffer from a lack of statistical clarity and often implement methods that will either fail to detect contamination when it is present (a false negative result) or conclude that the facility has impacted ground water when it has not (a false positive). Historical approaches to this problem have often sacrificed one type of error to maintain control over the other. For example, some regulatory approaches err on the side of conservatism, keeping false negative rates near zero while false positive rates approach 100 %.

1.4 The purpose of this provisional guide is to illustrate a statistical ground-water monitoring strategy that minimizes both false negative and false positive rates without sacrificing one for the other.

1.5 This provisional guide is applicable to statistical aspects of ground-water detection monitoring for hazardous and municipal solid waste disposal facilities.

1.6 It is of critical importance to realize that on the basis of a statistical analysis alone, it can never be concluded that a waste disposal facility has impacted ground water. A statistically significant exceedance over background levels indicates that the new measurement in a particular monitoring well for a particular constituent is inconsistent with chance expectations based on the available sample of background measurements.

1.7 Similarly, statistical methods can never overcome limitations of a groundwater monitoring network that might arise due to poor site characterization, well installation and location, sampling, or analysis.

1.8 It is noted that when justified, intra-well comparisons are generally preferable to their inter-well counterparts because they completely eliminate the spatial component of variability. Due to the absence of spatial variability, the uncertainty in measured concentrations is decreased making intra-well comparisons more sensitive to real releases (that is, false negatives) and false positive results due to spatial variability are completely eliminated.

1.9 Finally, it should be noted that the statistical methods described here are not the only valid methods for analysis of ground-water monitoring data. They are, however, currently the most useful from the perspective of balancing site-wide false positive and false negative rates at nominal levels. A more complete review of this topic and the associated literature is presented by Gibbons (1).[2]

1.10 The values stated in both inch-pound and SI units are to be regarded as the standard. The values given in parentheses are for information only.

1.11 *This standard does not purport to address all of the safety concerns, if any, associated with its use. It is the responsibility of the user of this standard to establish appropriate safety and health practices and determine the applicability of regulatory limitations prior to use.*

2. Terminology

2.1 *Definitions:*

2.1.1 *assessment monitoring program, n*—ground-water monitoring that is intended to determine the nature and extent of a potential site impact following a verified statistically significant exceedance of the detection monitoring program.

2.1.2 *combined Shewart (CUSUM) control chart, n*—a statistical method for intra-well comparisons that is sensitive to both immediate and gradual releases.

2.1.3 *detection limit (DL), n*—the true concentration at which there is a specified level of confidence (for example, 99 % confidence) that the analyte is present in the sample (2).

2.1.4 *detection monitoring program, n*—ground-water monitoring that is intended to detect a potential impact from a facility by testing for statistically significant changes in

---

[1] This provisional guide is under the jurisdiction of ASTM Committee D-18 on Soil and Rock and is the direct responsibility of Subcommittee D18.21 on Ground Water and Vadose Zone Investigations.
Current edition approved May 7, 1996. Published June 1996.

---

[2] The boldface numbers given in parentheses refer to a list of references at the end of the text.

geochemistry in a downgradient monitoring well relative to background levels.

2.1.5 *intra-well comparisons, n*—a comparison of one or more new monitoring measurements to statistics computed from a sample of historical measurements from that same well.

2.1.6 *inter-well comparisons, n*—a comparison of a new monitoring measurement to statistics computed from a sample of background measurements (for example, upgradient versus downgradient comparisons).

2.1.7 *prediction interval or limit, n*—a statistical estimate of the minimum or maximum concentration, or both, that will contain the next series of k measurements with a specified level of confidence (for example, 99 % confidence) based on a sample of n background measurements.

2.1.8 *quantification limit (QL), n*—the concentration at which quantitative determinations of an analyte's concentration in the sample can be reliably made during routine laboratory operating conditions (3).

2.2 *Definitions of Terms Specific to This Standard:*

2.2.1 *false negative rate, n—in detection monitoring*, the rate at which the statistical procedure does not indicate possible contamination when contamination is present.

2.2.2 *false positive rate, n—in detection monitoring*, the rate at which the statistical procedure indicates possible contamination when none is present.

2.2.3 *nonparametric, adj*—a term referring to a statistical technique in which the distribution of the constituent in the population is unknown and is not restricted to be of a specified form.

2.2.4 *nonparametric prediction limit, n*—the largest (or second largest) of n background samples. The confidence level associated with the nonparametric prediction limit is a function of n and k.

2.2.5 *parametric, adj*—a term referring to a statistical technique in which the distribution of the constituent in the population is assumed to be known.

2.2.6 *verification resample, n*—in the event of an initial statistical exceedance, one (or more) new independent sample is collected and analyzed for that well and constituent which exceeded the original limit.

2.3 *Symbols:*

2.3.1 α—the false positive rate for an individual comparison (that is, one well and constituent).

2.3.2 α^*—the site-wide false positive rate covering all wells and constituents.

2.3.3 k—the number of future comparisons for a single monitoring event (for example, the number of downgradient monitoring wells multiplied by the number of constituents to be monitored) for which statistics are to be computed.

2.3.4 n—the number of background measurements.

2.3.5 σ^2—the true population variance of a constituent.

2.3.6 s—the sample-based standard deviation of a constituent computed from n background measurements.

2.3.7 s^2—the sample-based variance of a constituent computed from n background measurements.

2.3.8 μ—the true population mean of a constituent.

2.3.9 $\bar{x}$—the sample-based mean or average concentration of a constituent computed from n background measurements.

3. Summary of Guide

3.1 This guide is summarized in Fig. 1, that provides a flowchart illustrating the steps in developing a statistical monitoring plan. The monitoring plan is based either on background versus monitoring well comparisons (for example, upgradient versus downgradient comparisons or intra-well comparisons, or a combination of both). Figure 1 illustrates the various decision points at which the general comparative strategy is selected (that is, upgradient background versus intra-well background) and how the statistical methods are to be selected based on site-specific considerations. The statistical methods include parametric and nonparametric prediction limits for background versus monitoring well comparisons and combined Shewart-CUSUM control charts for intra-well comparisons. Note that the background database is intended to expand as new data become available during the course of monitoring.

4. Significance and Use

4.1 The principal use of this provisional guide is in ground-water detection monitoring of hazardous and municipal solid waste disposal facilities. There is considerable variability in the way in which existing Guide USEPA regulation and guidance are interpreted and practiced. Often, much of current practice leads to statistical decision rules that lead to excessive false positive or false negative rates, or both. The significance of this proposed guide is that it jointly minimizes false positive and false negative rates at nominal levels without sacrificing one error for another (while maintaining acceptable statistical power to detect actual impacts to ground-water quality (4)).

Development of a Statistical Detection Monitoring Plan

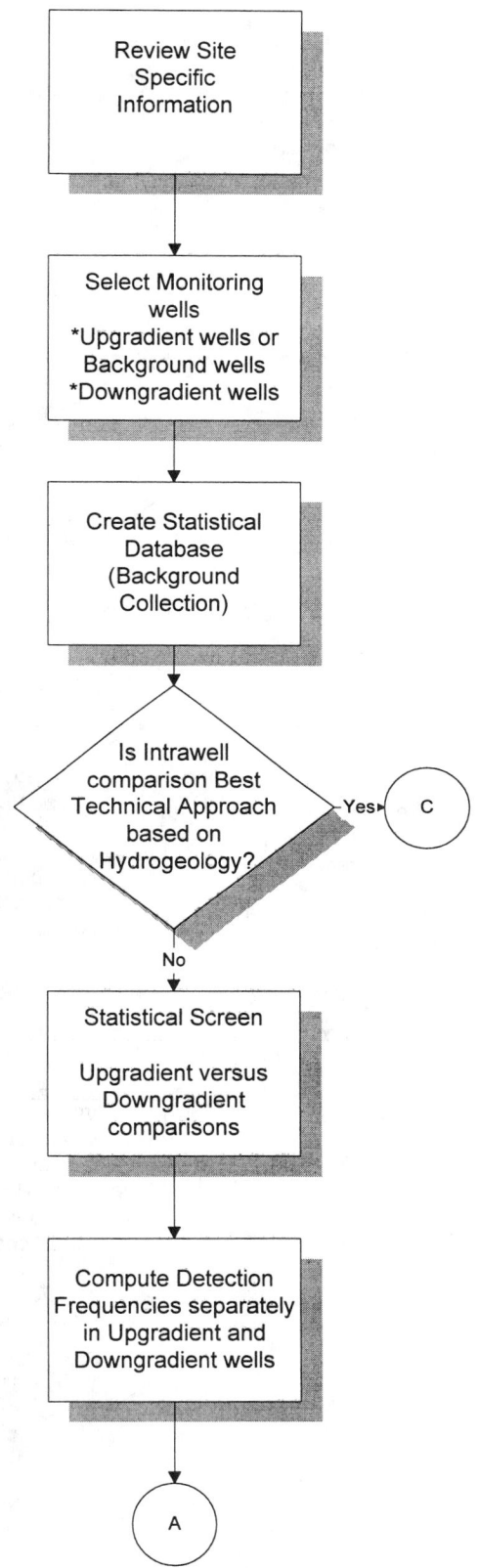

FIG. 1 Development of a Statistical Detection Monitoring Plan

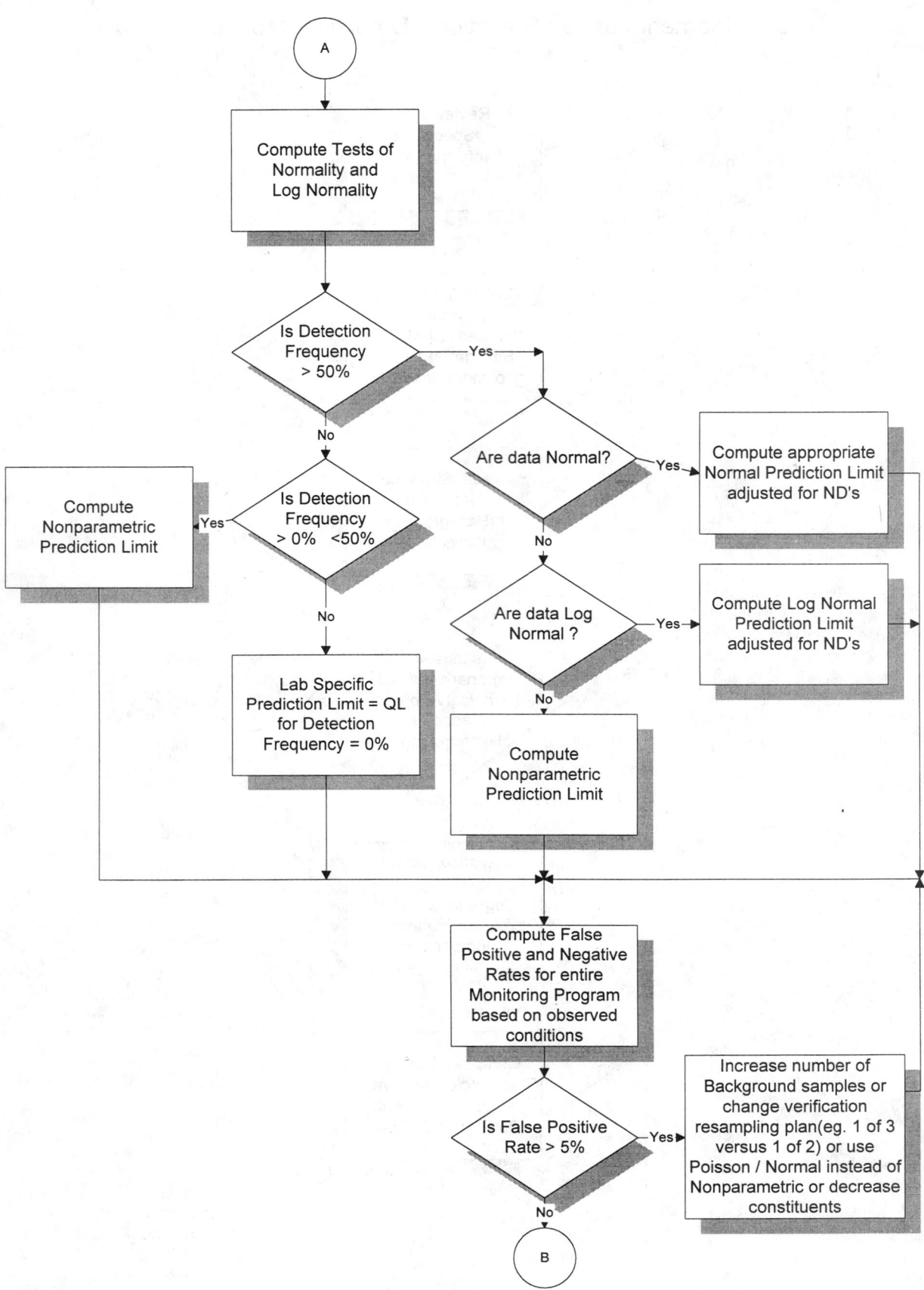

FIG. 1 (*Continued*)

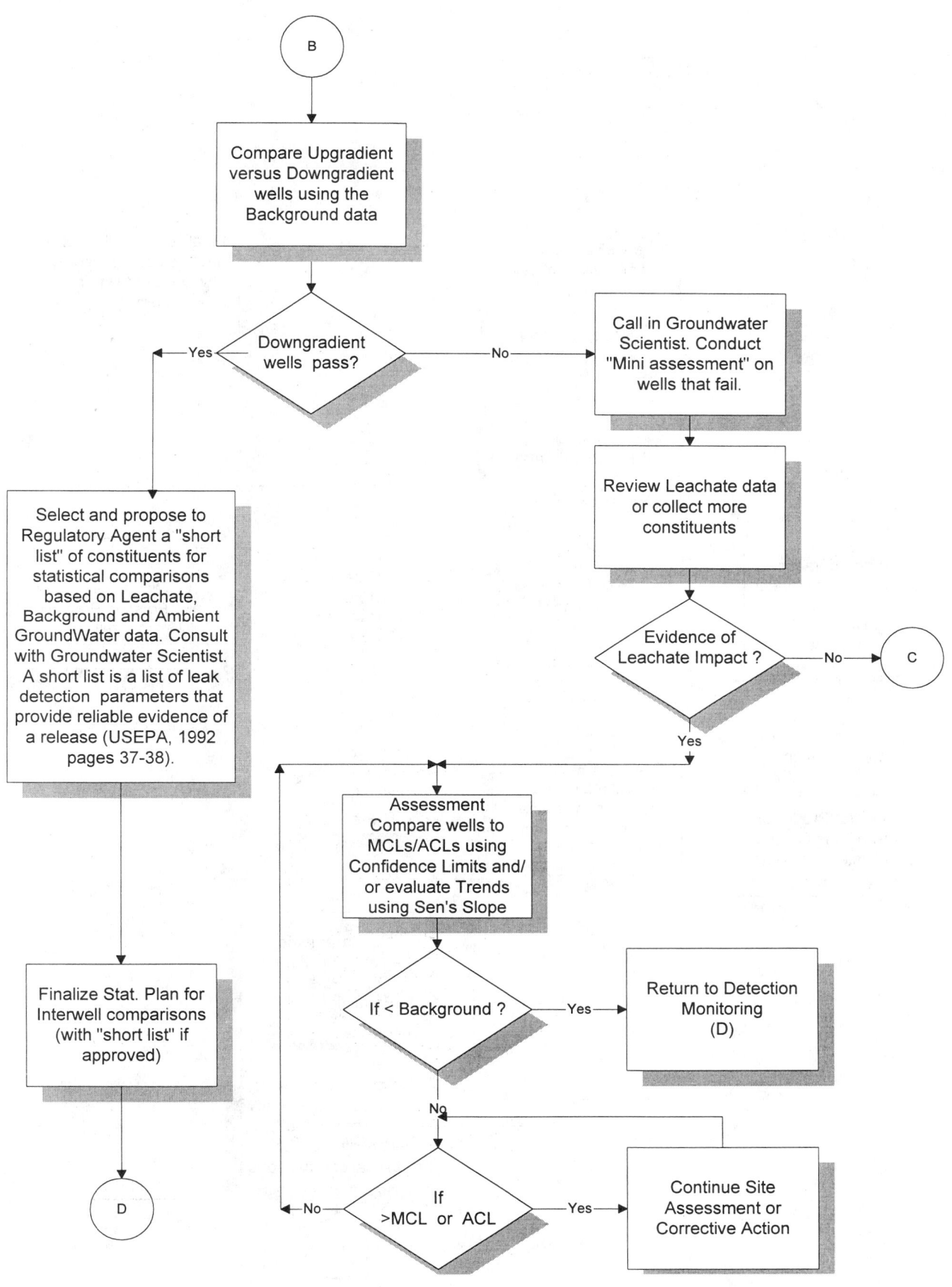

FIG. 1 (*Continued*)

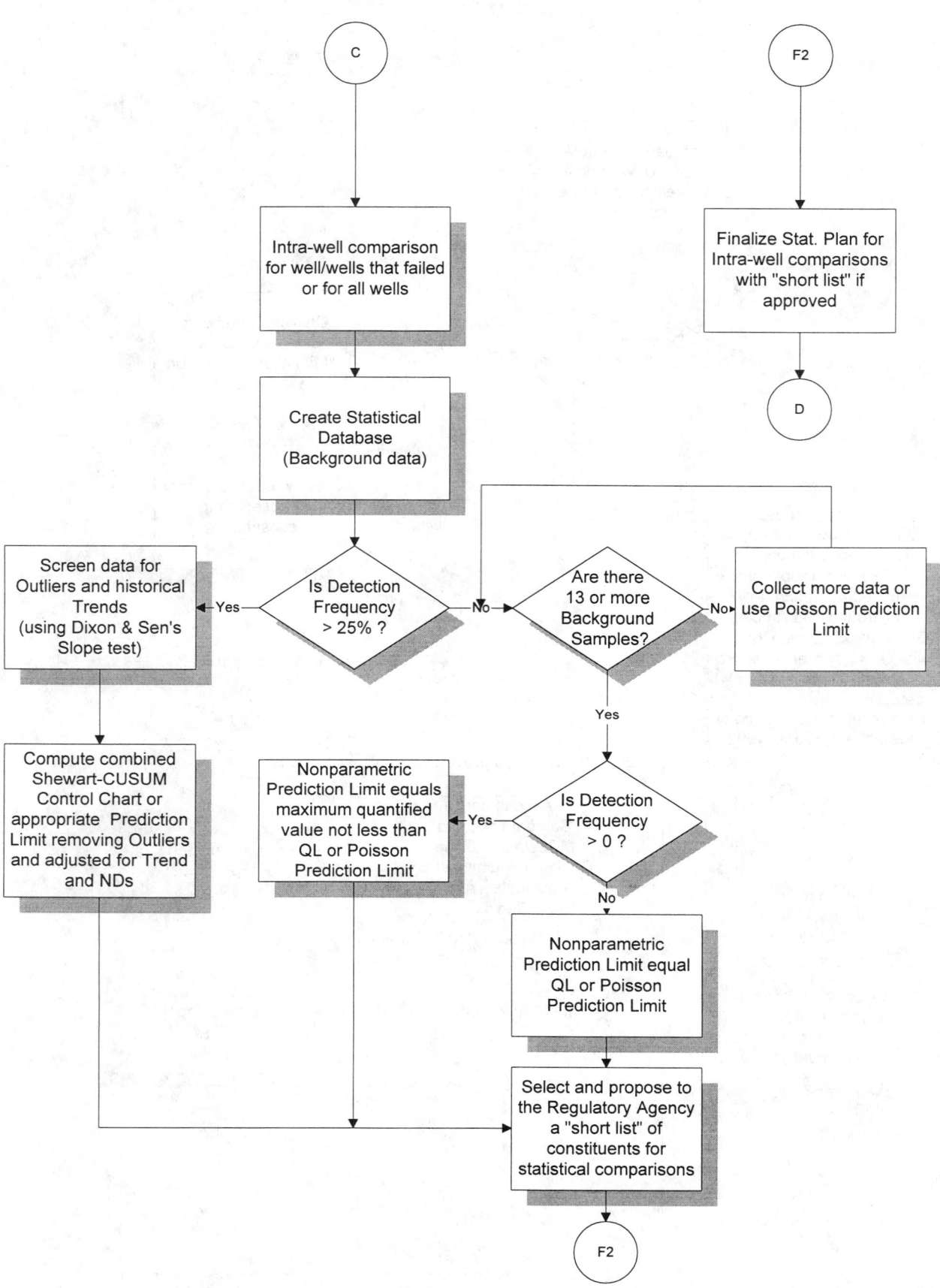

FIG. 1 (*Continued*)

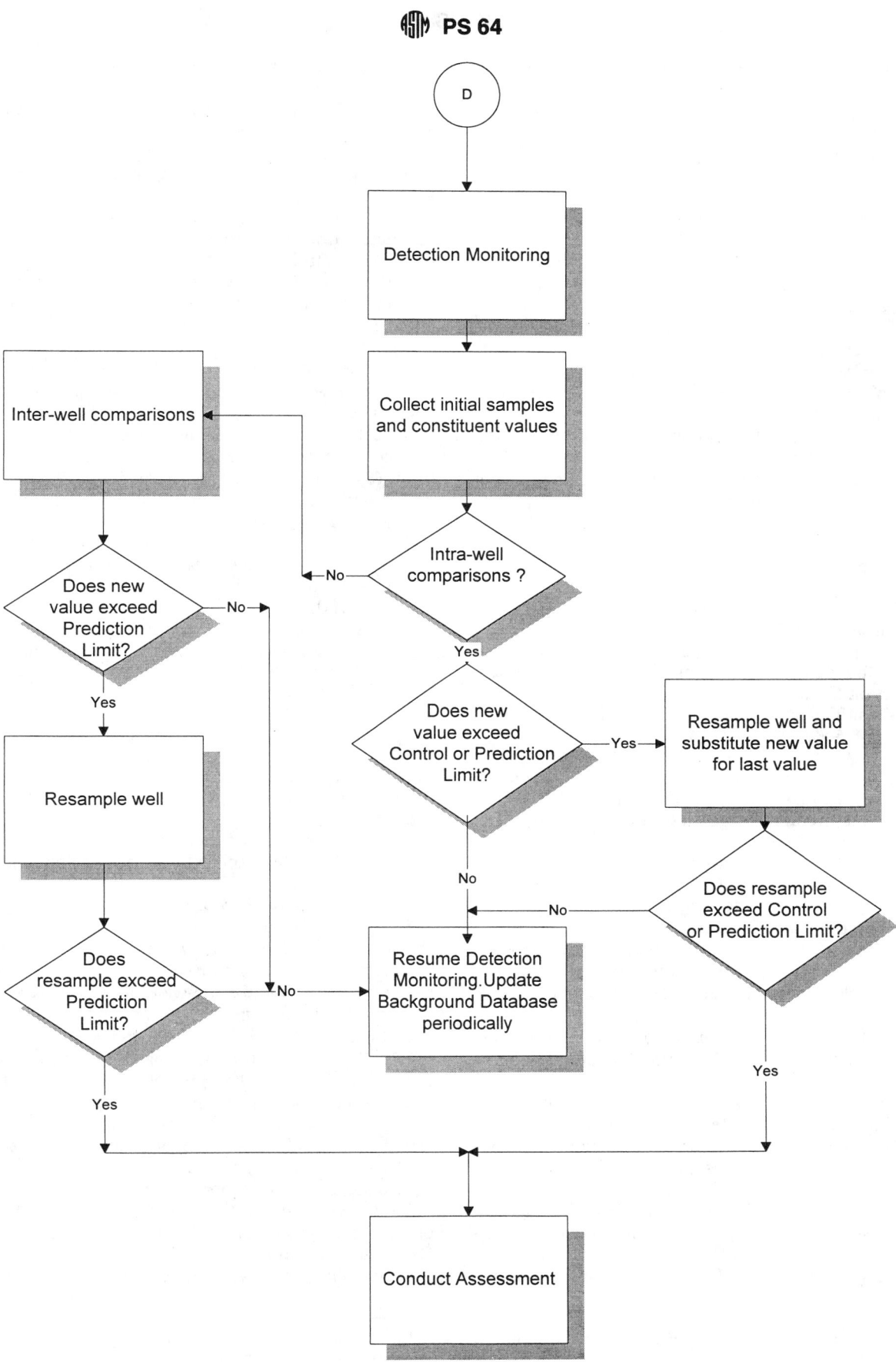

FIG. 1 (*Continued*)

4.2 Using this provisional guide, an owner/operator or regulatory agency should be able to develop a statistical detection monitoring program that will not falsely detect contamination when it is absent and will not fail to detect contamination when it is present.

5. Procedure

NOTE 1—In the following, an overview of the general procedure is described with specific technical details described in Section 6.

5.1 *Detection Monitoring:*

5.1.1 *Upgradient Versus Downgradient Comparisons:*

5.1.1.1 Detection frequency ≥50 %.

5.1.1.2 If the constituent is normally distributed, compute a normal prediction limit (5) selecting the false positive rate based on number of wells, constituents, and verification resamples (6) adjusting estimates of sample mean and variance for nondetects.

5.1.1.3 If the constituent is lognormally distributed, compute a lognormal prediction limit (7).

5.1.1.4 If the constituent is neither normally nor lognormally distributed, compute a nonparametric prediction limit (7) unless background is insufficient to achieve a 5 % site-wide false positive rate. In this case, use a normal distribution until sufficient background data are available (7).

5.1.1.5 If the background detection frequency is greater than zero but less than 50 %.

5.1.1.6 Compute a nonparametric prediction limit and determine if the background sample size will provide adequate protection from false positives.

5.1.1.7 If insufficient data exist to provide a site-wide false positive rate of 5 %, more background data must be collected.

5.1.1.8 As an alternative to 5.1.1.7 use a Poisson prediction limit which can be computed from any available set of background measurements regardless of the detection frequency (see 2.2.4 of Ref (4)).

5.1.1.9 If the background detection frequency equals zero, use the laboratory-specific QL (recommended) or limits required by applicable regulatory agency (8).[1]

5.1.1.10 This only applies for those wells and constituents that have at least 13 background samples. Thirteen samples provides a 99 % confidence nonparametric prediction limit with one resample for a single well and constituent (see Table 1).

5.1.1.11 If less than 13 samples are available more background data must be collected to use the nonparametric prediction limit.

5.1.1.12 An alternative would be to use a Poisson prediction limit that can be computed from four or more background measurements regardless of the detection frequency and can adjust for multiple wells and constituents.

5.1.1.13 If downgradient wells fail, determine cause.

5.1.1.14 If the downgradient wells fail because of natural or off-site causes, select constituents for intra-well comparisons (9).

5.1.1.15 If site impacts are found, a site plan for assessment monitoring may be necessary (10).

5.1.2 *Intra-well Comparisons:*

5.1.2.1 For those facilities that either have no definable hydraulic gradient, have no existing contamination, have too few background wells to meaningfully characterize spatial variability (for example, a site with one upgradient well or a facility in which upgradient water quality is either inaccessible or not representative of downgradient water quality), compute intra-well comparisons using combined Shewart-CUSUM control charts (9).[2]

5.1.2.2 For those wells and constituents that fail upgradient versus downgradient comparisons, compute combined Shewart-CUSUM control charts. If no volatile organic compounds (VOCs) or hazardous metals are detected and no trend is detected in other indicator constituents, use intra-well comparisons for detection monitoring of those wells and constituents.

5.1.2.3 If data are all non-detects after 13 quarterly sampling events, use the QL as the nonparametric prediction limit (8). Thirteen samples provides a 99 % confidence nonparametric prediction limit with one resample (1). Note that 99 % confidence is equivalent to a 1 % false positive rate, and pertains to a single comparison (that is, well and constituent) and not the site-wide error rate (that is, all wells and constituents) that is set to 5 %.

5.1.2.4 If detection frequency is greater than zero (that is, the constituent is detected in at least one background sample) but less than 25 %, use the nonparametric prediction limit that is the largest (or second largest) of at least 13 background samples.

5.1.2.5 As an alternative to 5.1.2.3 and 5.1.2.4 compute a Poisson prediction limit following collection of at least four background samples. Since the mean and variance of the Poisson distribution are the same, the Poisson prediction limit is defined even if there is no variability (for example, even if the constituent is never detected in background). In this case, one half of the quantification limit is used in place of the measurements, and the Poisson prediction limit can be computed directly.

5.1.3 *Verification Resampling:*

5.1.3.1 Verification resampling is an integral part of the statistical methodology (see Section 5 of Ref (4)). Without verification resampling much larger prediction limits would be required to obtain a site-wide false positive rate of 5 %. The resulting false negative rate would be dramatically increased.

5.1.3.2 Verification resampling allows sequential application of a much smaller prediction limit, therefore minimizing both false positive and false negative rates.

5.1.3.3 A statistically significant exceedance is not declared and should not be reported until the results of the verification resample are known. The probability of an initial exceedance is much higher than 5 % for the site as a whole.

5.1.3.4 Note that in the parametric case requiring passage of two verification resamples (for example, in the state of California regulation) will lead to higher false negative rates (for a fixed false positive rate) because larger prediction limits

---

[1] Note, if background detection frequency is zero, one should question whether the analyte is a useful indicator of contamination. If it is not, statistical testing of the constituent should not be performed.

[2] Some examples of inaccessible or nonrepresentative background upgradient wells may include slow moving ground water, radial or convergent flow, or sites that straddle ground-water divides.

TABLE 1 Probability That the First Sample or the Verification Resample Will Be Below the Maximum of *n* Background Measurements at Each of *k* Monitoring Wells for a Single Constituent

Previous *n*	Number of Monitoring Wells (*k*)														
	1	2	3	4	5	6	7	8	9	10	11	12	13	14	15
4	0.933	0.881	0.838	0.802	0.771	0.744	0.720	0.698	0.679	0.661	0.645	0.630	0.617	0.604	0.592
5	0.952	0.913	0.879	0.849	0.823	0.800	0.779	0.760	0.742	0.726	0.711	0.697	0.684	0.672	0.661
6	0.964	0.933	0.906	0.882	0.860	0.840	0.822	0.805	0.789	0.774	0.761	0.748	0.736	0.725	0.714
7	0.972	0.947	0.925	0.905	0.886	0.869	0.853	0.838	0.825	0.812	0.799	0.788	0.777	0.766	0.757
8	0.978	0.958	0.939	0.922	0.906	0.891	0.878	0.864	0.852	0.841	0.830	0.819	0.809	0.800	0.791
9	0.982	0.965	0.949	0.935	0.921	0.908	0.896	0.885	0.874	0.864	0.854	0.844	0.835	0.827	0.818
10	0.985	0.971	0.957	0.945	0.933	0.922	0.911	0.901	0.891	0.882	0.873	0.865	0.857	0.849	0.841
11	0.987	0.975	0.964	0.953	0.942	0.933	0.923	0.914	0.906	0.897	0.889	0.882	0.874	0.867	0.860
12	0.989	0.979	0.969	0.959	0.950	0.941	0.933	0.925	0.917	0.910	0.902	0.896	0.889	0.882	0.876
13	0.990	0.981	0.973	0.964	0.956	0.948	0.941	0.934	0.927	0.920	0.914	0.907	0.901	0.895	0.889
14	0.992	0.984	0.976	0.969	0.961	0.954	0.948	0.941	0.935	0.929	0.923	0.917	0.912	0.906	0.901
15	0.993	0.986	0.979	0.972	0.966	0.959	0.953	0.947	0.942	0.936	0.931	0.926	0.920	0.915	0.910
16	0.993	0.987	0.981	0.975	0.969	0.964	0.958	0.953	0.948	0.943	0.938	0.933	0.928	0.923	0.919
17	0.994	0.988	0.983	0.978	0.972	0.967	0.962	0.957	0.953	0.948	0.943	0.939	0.935	0.930	0.926
18	0.995	0.990	0.985	0.980	0.975	0.970	0.966	0.961	0.957	0.953	0.949	0.944	0.940	0.937	0.933
19	0.995	0.991	0.986	0.982	0.977	0.973	0.969	0.965	0.961	0.957	0.953	0.949	0.946	0.942	0.938
20	0.996	0.991	0.987	0.983	0.979	0.975	0.972	0.968	0.964	0.960	0.957	0.953	0.950	0.947	0.943
25	0.997	0.994	0.992	0.989	0.986	0.984	0.981	0.978	0.976	0.973	0.971	0.968	0.966	0.964	0.961
30	0.998	0.996	0.994	0.992	0.990	0.988	0.986	0.984	0.983	0.981	0.979	0.977	0.975	0.974	0.972
35	0.998	0.997	0.996	0.994	0.993	0.991	0.990	0.988	0.987	0.986	0.984	0.983	0.981	0.980	0.979
40	0.999	0.998	0.997	0.995	0.994	0.993	0.992	0.991	0.990	0.989	0.988	0.987	0.985	0.984	0.983
45	0.999	0.998	0.997	0.996	0.995	0.995	0.994	0.993	0.992	0.991	0.990	0.989	0.988	0.987	0.987
50	0.999	0.998	0.998	0.997	0.996	0.996	0.995	0.994	0.993	0.993	0.992	0.991	0.990	0.990	0.989
60	0.999	0.999	0.998	0.998	0.997	0.997	0.996	0.996	0.995	0.995	0.994	0.994	0.993	0.993	0.992
70	1.00	0.999	0.999	0.998	0.998	0.998	0.997	0.997	0.997	0.996	0.996	0.995	0.995	0.995	0.994
80	1.00	0.999	0.999	0.999	0.998	0.998	0.998	0.998	0.997	0.997	0.997	0.996	0.996	0.996	0.996
90	1.00	1.00	0.999	0.999	0.999	0.999	0.998	0.998	0.998	0.998	0.997	0.997	0.997	0.997	0.996
100	1.00	1.00	0.999	0.999	0.999	0.999	0.999	0.998	0.998	0.998	0.998	0.998	0.998	0.997	0.997

Previous *n*	Number of Monitoring Wells (*k*)													
	20	25	30	35	40	45	50	55	60	65	70	75	80	90	100
4	0.542	0.504	0.474	0.449	0.428	0.410	0.394	0.380	0.367	0.356	0.345	0.336	0.327	0.312	0.299
5	0.612	0.574	0.543	0.517	0.495	0.476	0.459	0.443	0.430	0.417	0.406	0.396	0.386	0.369	0.355
6	0.668	0.631	0.600	0.574	0.552	0.532	0.514	0.499	0.484	0.472	0.460	0.449	0.439	0.420	0.405
7	0.713	0.678	0.648	0.623	0.600	0.580	0.563	0.547	0.532	0.519	0.507	0.496	0.485	0.466	0.450
8	0.750	0.717	0.688	0.664	0.642	0.622	0.605	0.589	0.574	0.561	0.549	0.537	0.527	0.507	0.490
9	0.781	0.750	0.723	0.699	0.678	0.659	0.642	0.626	0.612	0.598	0.586	0.574	0.564	0.544	0.527
10	0.807	0.777	0.752	0.729	0.709	0.691	0.674	0.659	0.644	0.631	0.619	0.608	0.597	0.578	0.560
11	0.828	0.801	0.777	0.755	0.736	0.718	0.702	0.687	0.674	0.661	0.649	0.638	0.627	0.608	0.590
12	0.847	0.821	0.799	0.778	0.760	0.743	0.727	0.713	0.700	0.687	0.675	0.664	0.654	0.635	0.618
13	0.862	0.839	0.817	0.798	0.781	0.764	0.750	0.736	0.723	0.711	0.699	0.689	0.678	0.660	0.643
14	0.876	0.854	0.834	0.816	0.799	0.784	0.769	0.756	0.744	0.732	0.721	0.710	0.701	0.682	0.666
15	0.888	0.867	0.848	0.831	0.815	0.801	0.787	0.774	0.762	0.751	0.740	0.730	0.721	0.703	0.686
16	0.898	0.879	0.861	0.845	0.830	0.816	0.803	0.791	0.779	0.768	0.758	0.748	0.739	0.722	0.706
17	0.907	0.889	0.872	0.857	0.843	0.830	0.817	0.806	0.794	0.784	0.774	0.765	0.756	0.739	0.723
18	0.914	0.898	0.882	0.868	0.855	0.842	0.830	0.819	0.808	0.798	0.789	0.780	0.771	0.754	0.739
19	0.921	0.906	0.891	0.878	0.865	0.853	0.842	0.831	0.821	0.811	0.802	0.793	0.785	0.769	0.754
20	0.928	0.913	0.899	0.886	0.874	0.863	0.852	0.842	0.832	0.823	0.814	0.806	0.798	0.782	0.768
25	0.950	0.939	0.929	0.919	0.910	0.901	0.892	0.884	0.876	0.869	0.862	0.855	0.848	0.835	0.823
30	0.963	0.955	0.947	0.940	0.932	0.925	0.919	0.912	0.906	0.900	0.894	0.888	0.882	0.872	0.861
35	0.972	0.966	0.959	0.954	0.948	0.942	0.937	0.931	0.926	0.921	0.916	0.911	0.907	0.898	0.889
40	0.978	0.973	0.968	0.963	0.958	0.954	0.949	0.945	0.941	0.936	0.932	0.928	0.924	0.917	0.909
45	0.982	0.978	0.974	0.970	0.966	0.962	0.959	0.955	0.951	0.948	0.944	0.941	0.938	0.931	0.925
50	0.985	0.982	0.979	0.975	0.972	0.969	0.966	0.963	0.959	0.956	0.954	0.951	0.948	0.942	0.937
60	0.990	0.987	0.985	0.982	0.980	0.978	0.975	0.973	0.971	0.968	0.966	0.964	0.962	0.958	0.954
70	0.992	0.990	0.989	0.987	0.985	0.983	0.981	0.980	0.978	0.976	0.974	0.973	0.971	0.968	0.965
80	0.994	0.993	0.991	0.990	0.988	0.987	0.986	0.984	0.983	0.981	0.980	0.979	0.977	0.975	0.972
90	0.995	0.994	0.993	0.992	0.991	0.990	0.988	0.987	0.986	0.985	0.984	0.983	0.982	0.980	0.978
100	0.996	0.995	0.994	0.993	0.992	0.991	0.991	0.990	0.989	0.988	0.987	0.986	0.985	0.983	0.982

are required to achieve a site-wide false positive rate of 5 % than for a single verification resample; hence, the preferred methods are pass one verification resample or pass one of two verification resamples. Also note that nonparametric limits requiring passage of two verification resamples will result in need for a larger number of background samples than are typically available (see 6.3.3.1) **(1)**.

5.1.4 *False Positive and False Negative Rates:*

5.1.4.1 Conduct simulation study based on current mon-

itoring network, constituents, detection frequencies, and distributional form of each monitoring constituent (see Appendix B of Ref **(4)**). The specific objectives of the simulation study are to determine if the false positive and false negative rates of the current monitoring program as a whole are acceptable and to determine if changes in verification resampling plans or choice of nonparametric versus Poisson prediction limits or inter-well versus intra-well comparison strategies will improve the overall performance of the detection monitoring program.

5.1.4.2 Project frequency of which verification resamples will be required and false assessments for site as a whole for each monitoring event based on the results of the simulation study. In this way the owner/operator will be able to anticipate the required amount of future sampling.

5.1.4.3 As a general guideline, a site-wide false positive rate of 5 % and a false negative rate of approximately 5 % for differences on the order of three to four standard deviation units are recommended. Note that USEPA recommends simulating the most conservative case of a release that effects a single constituent in a single downgradient well. In practice, multiple constituents in multiple wells will be impacted, therefore, the actual false negative rates may be considerably smaller than estimates obtained by means of simulation.

5.1.5 *Use of DLs and QLs in Ground-Water Monitoring:*

5.1.5.1 The DLs indicate that the analyte is present in the sample with confidence.

5.1.5.2 The QLs indicate that the true quantitative value of the analyte is close to the measured value.

5.1.5.3 For analytes with estimated concentration exceeding the DL but not the QL, it can be concluded that the true concentration is greater than zero; however, uncertainty in the instrument response is by definition too large to make a reliable quantitative determination. Note that in a qualitative sense, values between the DL and QL are greater than values below the DL, and this rank ordering can be used in a nonparametric method.

5.1.5.4 If the laboratory-specific DL for a given compound is 3 µg/L, and the QL for the same compound is 6 µg/L, then a detection of that compound at 4 µg/L could actually represent a true concentration of anywhere between 0 and 6 µg/L. The true concentration may well be less than the DL (1,2,11).

5.1.5.5 Direct comparison of a single value to a maximum concentration level (MCL), or any other concentration limit, is not adequate to demonstrate noncompliance unless the concentration is larger than the QL.

5.1.5.6 Verification resampling applies to this case as well.

6. Report

6.1 This section provides a description of the specific statistical methods referred to in this guide. Note that specific recommendations for any given facility require an interdisciplinary site-specific study that encompasses knowledge of the facility, it's hydrogeology, geochemistry, and study of the false positive and false negative error rates that will result. Performing a correct statistical analysis, such as nonparametric prediction limits, in the wrong situation (for example, when there are too few background measurements) can lead to erroneous conclusions.

6.2 *Upgradient Versus Downgradient Comparisons:*

6.2.1 *Case One—Compounds Quantified in All Background Samples:*

6.2.1.1 Test normality of distribution using the multiple group version of the Shapiro-Wilk test applied to n background measurements (12). The multiple group version of the Shapiro-Wilk test takes into consideration that background measurements are nested within different back-

ground monitoring wells, hence the original Shapiro-Wilk test does not directly apply.

NOTE—Background wells used for inter-well comparsons may in some cases include wells that are not hydraulically upgradient of the site.

6.2.1.2 Alternatively, residuals from the mean of each upgradient well can be pooled together and tested using the single group version of the Shapiro-Wilk test (13).

6.2.1.3 The need for a multiple group test to incorporate spatial variability among upgradient wells also raises the question of validity of upgradient versus downgradient comparisons. Where significant spatial variability exists, it may not be possible to obtain a representative upgradient background, and intra-well comparisons may be required. A one-way analysis of variance (ANOVA) applied to the upgradient well data provides a good way of testing for significant spatial variability.

6.2.1.4 If normality is not rejected, compute the 95 % prediction limit as follows:

$$\bar{x} + t_{[n-1,\alpha]} s \sqrt{1 + \frac{1}{n}}$$

where:

$$\bar{x} = \sum_{i=1}^{n} \frac{x_i}{n}$$

$$s = \sqrt{\sum_{i=1}^{n} \frac{(x_i - \bar{x})^2}{n - 1}}$$

α = false positive rate for each individual test,

$t_{[n-1,\alpha]}$ = one-sided $(1 - \alpha)$ 100 % point of Student's t distribution on $n - 1$ df, and

n = number of background measurements. Select α as the minimum of 0.01 or one of the following:

(*1*) Pass the first or one of one verification resample:

$$\alpha = (1 - 0.95^{1/k})^{1/2}$$

(*2*) Pass the first or one of two verification resamples:

$$\alpha = (1 - 0.95^{1/k})^{1/3}$$

(*3*) Pass the first or two of two verification resamples:

$$\alpha = \sqrt{1 - 0.95^{1/k}} \sqrt{1/2}$$

where:

k = number of comparisons (that is, monitoring wells times constituents (see 5.2.2 of Ref (4)).

6.2.1.5 Note that these formulas for computing the adjusted individual comparison α all ignore two sources of dependence: comparisons for a given constituent are all made against the same background and concentrations of the indicator constituents may be positively correlated over time. Solution of the first problem has been provided by Refs (1) and (14) and has provided detailed tabulation of factors that can be used in computing the exact prediction limits. In terms of the second problem, constituents that are highly correlated (based on pairwise correlations) could be eliminated, not from the statistical analysis, but from the total set of comparisons used to compute α, leading to more powerful and realistic prediction limits.

6.2.1.6 If normality is rejected, take natural logarithms of the n background measurements and recompute the multiple group Shapiro-Wilk test.

6.2.1.7 If the transformation results in a nonsignificant G statistic (that is, the values $log_e(x)$) are normally distributed compute the lognormal prediction limit as follows:

$$\exp\left(\overline{y} + t_{[n-1,\alpha]} s_y \sqrt{1 + \frac{1}{n}}\right)$$

where:

$$\overline{y} = \sum_{i=1}^{n} \frac{log_e(x_i)}{n}$$

and:

$$s_y = \sqrt{\sum_{i=1}^{n} \frac{(log_e(x_i) - \overline{y})^2}{n-1}}$$

6.2.1.8 If log transformation does not bring about normality (that is, the probability of G is less than 0.01), compute nonparametric prediction limits (Option—Compute normal prediction limit).

6.2.2 *Case Two—Compounds Quantified in at Least 50 % of All Background Samples:*

6.2.2.1 Apply the multiple group Shapiro-Wilk test to the n_1 quantified measurements only.

6.2.2.2 If the data are normally distributed compute the mean of the n background samples as follows:

$$\overline{x} = \left(1 - \frac{n_0}{n}\right)\overline{x}'$$

where:

$\overline{x}'$ = average of the n_1 detected values, and

n_0 = number of samples in which the compound is not detected. The standard deviation is:

$$s = \sqrt{\left(1 - \frac{n_0}{n}\right)s'^2 + \frac{n_0}{n}\left(1 - \frac{n_0 - 1}{n-1}\right)\overline{x}'^2}$$

where s' is the standard deviation of the n_1 detected measurements. The normal prediction limit can then be computed as previously described. This method is due to Aitchison (see 2.2.2 of Ref (**4**) and (**15**)). Note that this method imputes nondetects as zero concentrations.

6.2.2.3 A good alternative to Aitchison's method is Cohen's maximum likelihood estimator (**16**). Extensive tables and computational details are also provided in Gibbons, 1991. A useful approach to selecting between the two methods is described in 2.2.1 of Ref (**4**).

6.2.2.4 If the multiple group Shapiro-Wilk test reveals that the data are lognormally distributed, replace $\overline{x}'$ with $\overline{y}'$ and s' and s'_y in the equations for $\overline{x}$ and s. The lognormal prediction limit may then be computed as previously described.

NOTE—This adjustment only applies to positive random variables. The natural logarithm of concentration less than 1 are negative and therefore the adjustment does not apply. For this reason we add 1 to each value (for example, $log_e(x_i + 1) \geq 0$), compute the prediction limit on a log scale and then subtract one from the antilog of the prediction limit.

6.2.2.5 If the data are neither normally or lognormally distributed, compute a nonparametric prediction limit. (Option—compute normal prediction limit).

6.2.3 *Case Three—Compounds Quantified in Less Than 50 % of All Background Samples:*

6.2.3.1 In this application, the nonparametric prediction limit is the largest concentration found in n upgradient measurements (see 4.2.1 of Ref (**8**)).

6.2.3.2 Gibbons (**18,19**) has shown that the confidence associated with this decision rule, following one or more verification resamples, is a function of the multivariate extension of the hypergeometric distribution (see 5.2.3 of Ref (**8**)).

6.2.3.3 Complete tabulations of confidence levels for $n = 4, \ldots, 100$, $k = 1, \ldots, 100$ future comparisons (for example, monitoring wells), and a variety of verification resampling plans are presented in (**1**). For example with five monitoring wells and ten constituents (that is, 50 comparisons), 40 background measurements would be required to provide 95 % confidence (see 5.2.3 of Ref (**4**)). Table 1 displays confidence levels for a single verification resample.

6.2.3.4 As an option to the nonparametric prediction limits, compute Poisson prediction limits. Poisson prediction limits are useful for those cases in which there are too few background measurements to achieve an adequate site-wide false positive rate using the nonparametric approach. Gibbons (**19**) derived the original Poisson prediction limit. Cameron (**20**) found that use of a normal multiplier in place of Student's t-distribution resulted in a more powerful test, thus the Poisson prediction limit is:

$$Poisson\ PL = y/n + \frac{z^2}{2n} + z/n\ \sqrt{y(1 + n) + z^2/4}$$

where y is the sum of the detected measurements or the quantification limit for those samples in which the constituent was not detected, and z is the $(1 - \alpha)$ 100 upper percentage point of the normal distribution, where α is computed as in 6.2.1.4.

NOTE—If the Poisson prediction unit is less than the quantification limit, recompute the prediction limit substituting the quantification limit for the nondetects.

6.3 *Intra-Well Comparisons:*

6.3.1 One particularly good method for computing intra-well comparisons is the combined Shewart-CUSUM control chart (see 6.1 in Ref (**4**)). The method is sensitive to both gradual and rapid releases and is also useful as a method of detecting "trends" in data. Note that this method should be used on wells unaffected by the landfill. There are several approaches to implementing the method, and in the following, one useful way is described as well as discussion of some statistical properties.

6.3.2 *Assumptions:*

6.3.2.1 The combined Shewart-CUSUM control chart procedure assumes that the data are independent and normally distributed with a fixed mean μ and constant variance σ^2. The most important assumption is independence, and as a result, wells should be sampled no more frequently than quarterly. In some cases, where ground water moves relatively quickly, it may be possible to accelerate background sampling to eight samples in a single year; however, this should only be done to establish background and not for routine monitoring. The assumption of normality is somewhat less of a concern, and if problematic, natural log or square root transformation of the observed data should be adequate for most practical applications. For this method, nondetects can be replaced by the quantification limit without serious consequence. This procedure should only be applied to those constituents that are detected at least in 25 % of all samples, otherwise, σ^2 is not adequately defined.

6.3.2.2 When large intra-well background databases are available, (for example, three years or more of at least semiannual monitoring) obvious cyclic or trend patterns can be removed from both the baseline data and from the future data to be plotted on the chart. Similarly, when the background database consists of eight or more background measurements, use of Aitchison's (15) or Cohen's (16) methods for computing the background mean and standard deviation can be used in place of simple imputation of the quantification limit.

6.3.3 *Nondetects:*

6.3.3.1 For those well and constituent combinations in which the detection frequency is less than 25 %, the data should be displayed graphically until a sufficient number of measurements are available to provide 99 % confidence (that is, 1 % false positive rate) for an individual well and constituent using a nonparametric prediction limit, which in this context is the maximum detected value out of the n historical measurements. As previously discussed this amounts to 13 background samples for 1 resample, 8 background samples for pass 1 of 2 resamples and 18 background samples for pass 2 of 2 resamples. If nonparametric prediction limits are to be used for intra-well comparisons of rarely detected constituents, 2 verification resamples will often be required, and failure will only be indicated if both measurements exceed the limit (that is, the maximum of the first 8 samples).

6.3.3.2 Note that these background sample sizes provide 99 % confidence for a single future comparison and not all of the wells and constituents for which they will actually be applied. Adjustment for multiple comparisons will require even larger background sample sizes that may not be possible to obtain at most facilities. In light of this, the recommendations in 6.3.3.1 provide a minimum requirement.

6.3.3.3 For those cases in which the detection frequency is greater than 25 %, substitute the QL (or where there are multiple QLs, the median QL) for the nondetects. In this way, changes in quantification limits do not appear to be significant trends.

6.3.3.4 If nothing is detected in 8, 13, or 18 independent samples (depending on resampling strategy), use the quantification limit as the nonparametric prediction limit.

6.3.3.5 As in the previously described inter-well comparisons, optional use of Poisson prediction limits as an alternative to nonparametric prediction limits for rarely detected constituents (that is, less than 25 % detects) is recommended when the number of background measurements is small. Poisson prediction limits can be computed after eight background measurements regardless of detection frequency.

6.3.4 *Procedure:*

6.3.4.1 Require that at least eight historical independent samples are available to provide reliable estimates of the mean μ and standard deviation σ, of the constituent's concentration in each well.

6.3.4.2 Select the three Shewart-CUSUM parameters, h, (the value against which the cumulative sum will be com-

pared), c (a parameter related to the displacement that should be quickly detected), and SCL (the upper Shewart limit that is the number of standard deviation units for an immediate release). Lucas (21) and Starks (22) suggest that $c = 1$, $h = 5$, and $SCL = 4.5$ are most appropriate for ground-water monitoring applications. This sentiment is echoed by USEPA in their interim final guidance document (23).

6.3.4.3 Denote the new measurement at time-point t_i as x_i and compute the standardized value z_i:

$$z_i = \frac{x_i - \bar{x}}{s}$$

where $\bar{x}$ and s are the mean and standard deviation of at least eight historical measurements for that well and constituent (collected in a period of no less than one year).

6.3.4.4 At each time period, t_i, compute the cumulative sum S_i, as:

$$S_i = \max[0, (z_i - c) + S_{i-1}]$$

where: $\max[A, B]$ is the maximum of A and B, starting with $S_0 = 0$.

6.3.4.5 Plot the values of S_i (y-axis) versus t_i (x-axis) on a time chart. Declare an "out-of-control" situation on sampling period t_i if for the first time, $S_i \geq h$ or $z_i \geq SCL$. Any such designation, however, must be verified on the next round of sampling, before further investigation is indicated.

6.3.4.6 The reader should note that unlike prediction limits that provide a fixed confidence level (for example, 95 %) for a given number of future comparisons, control charts do not provide explicit confidence levels, and do not adjust for the number of future comparisons. The selection of $h = 5$, $SCL = 4.5$ and $c = 1$ is based on USEPA's own review of the literature and simulations 21,22, and 23). The USEPA indicates that these values "allow a displacement of two standard deviations to be detected quickly." Since 1.96 standard deviation units corresponds to 95 % confidence on a normal distribution, we can have approximately 95 % confidence for this test method as well. In practice, setting $h = SCL = 4.5$ results in a single limit with no compromise in leak detection capabilities.

6.3.4.7 In terms of plotting the results, it is more intuitive to plot values in their original metric (for example, microgram per litre) rather than in standard deviation units. In this case, $h = SCL = \bar{x} + 4.5s$, and the S_i are converted to the concentration metric by the transformation $S_i * s + \bar{x}$, noting that when normalized (that is, in standard deviation units) $\bar{x} = 0$ and $s = 1$ so that $h = SCL = 4.5$ and $S_i * 1 + 0 = S_i$. Note that when $n \geq 12$ recompute the mean and standard deviation and adjust the control limits $h = SCL = 4.0$ and $c = 0.75$.

6.3.5 *Outliers:*

6.3.5.1 From time to time, inconsistently large or small values (outliers) can be observed due to sampling, laboratory, transportation, transcription errors, or even by chance alone. Verification resampling will tremendously reduce the probability of concluding that an impact has occurred if such an anomalous value is obtained for any of these reasons. However, nothing has eliminated the chance that such errors might be included in the historical measurements for a particular well and constituent. If such erroneous values (either too high or too low) are included in the historical database, the result would be an artificial increase in the magnitude of the control limit, and a corresponding increase in the false negative rate of the statistical test (that is, conclude that there is no site impact when in fact there is).

6.3.5.2 To remove the possibility of this type of error, the historical data are screened for each well and constituent for the existence of outliers (see 6.2 in Ref (4)) using the well-known method described by Dixon (24). These outlying data points are indicated on the control charts (using a different symbol), but are excluded from the measurements that are used to compute the background mean and standard deviation. In the future, new measurements that turn out to be outliers, in that they exceed the control limit, will be dealt with by verification resampling in downgradient wells only.

6.3.5.3 This same outlier detection algorithm is applied to each upgradient well and constituent to screen outliers for inter-well comparisons as well.

6.3.6 *Existing Trends:*

6.3.6.1 If contamination is preexisting, trends will often be observed in the background database from which the mean and variance are computed. This will lead to upward biased estimates and grossly inflated control limits. To remove this possibility, first screen the background data for each well and constituent for trend using Sen's nonparametric estimate of trend (25). Confidence limits for this trend estimate are given by Gilbert (26). A significant trend is one in which the 99 % lower confidence bound is greater than zero. In this way, even preexisting trends in the background dataset will be detected.

6.3.6.2 When significant trends in background are found, their source must be identified prior to continuation of detection monitoring since they may be evidence of a prior site impact. If the source of the trend is found to be unrelated to the facility, then an alternative indicator constituent may be required for that well or all wells at the facility.

6.3.7 *Note on Verification Sampling:*

6.3.7.1 It should be noted that when a new monitoring value is an outlier, perhaps due to a transcription error, sampling error, or analytical error, the Shewart and CUSUM portions of the control chart are affected quite differently. The Shewart portion of the control chart compares each individual new measurement to the control limit, therefore, the next monitoring event measurement constitutes an independent verification of the original result. In contrast, however, the CUSUM procedure incorporates all historical values in the computation, therefore, the effect of the outlier will be present for both the initial and verification sample: hence the statistical test will be invalid.

6.3.7.2 For example, assume $\bar{x} = 50$ and $s = 10$. On Quarter 1 the new monitoring value is 50, so $z = (50 - 50)/10 = 0$ and $S_i = \max[0, (z - 1) + 0] = 0$. On Quarter 2, a sampling error occurs (that is, documented as an error after review of chain of custody) and the reported value is 200, yielding $z = (200 - 50)/10 = 15$ and $S_i = \max[0, (15 - 1) + 0] = 14$, that is considerably larger than 4.5; hence an initial exceedance is recorded. On the next round of sampling, the previous result is not confirmed, because the result is back to 50. Inspection of the CUSUM, however, yields $z = (50 - 50)/10 = 0$ and $S_i = \max[0, (0 - 1) + 14] = 13$, that would be taken as a confirmation of the exceedance, when in fact,

no such confirmation was observed. For this reason, the verification must replace the suspected result in order to have an unbiased confirmation.

6.3.8 *Updating the Control Chart*—As monitoring continues and the process is shown to be in control, the background mean and variance should be updated periodically to incorporate these new data. Every year or two, all new data that are in control should be pooled with the initial samples and $\bar{x}$ and s recomputed. These new values of $\bar{x}$ and s will then be used in constructing future control charts. This updating process should continue for the life of the facility or monitoring program, or both (see 6.1 in Ref (8)).

6.3.9 *An Alternative Based on Prediction Limits*—An alternative approach to intra-well comparisons involves computation of well-specific prediction limits. Prediction limits are somewhat more sensitive to immediate releases but less sensitive to gradual releases than the combined Shewart-CUSUM control charts. Prediction limits are also less robust to deviations from distributional assumptions (1).

7. Restriction of Background Samples

7.1 Certain states have interpreted the regulations as indicating that background be confined to the first four samples collected in a day or a semiannual monitoring event or a year. This conflicts with federal regulation and guidance. The first approach (that is, four samples in a day) violates the assumption of independence and confounds day to day temporal and seasonal variability with potential contamination. As an analogy, consider setting limits on yearly ambient temperatures in Chicago by taking four temperature readings on July 4th. On that day the temperature varied between 78 and 82°F (26 and 28°C) yielding a prediction interval from 70 to 90°F (21 to 32°C). In January, the temperature in Chicago can be −20°F (−28°C). Clearly, in this example restriction of background leads to nonrepresentative prediction of future measurements. In the second approach restricting establishment of background to the first four events taken in six months underestimates the component of seasonal variability and can lead to elevated false positive or false negative rates. The net result is that comparisons of background water quality in the summer may not be representative of downgradient ground-water quality in the winter (for example, disposal of road salts increasing specific conductivity in the winter). In the third approach in which background is restricted to the first four quarterly measurements, independence is typically not an issue and background versus point of compliance monitoring well comparisons are not confounded with season for that year, however, background from this year may not reflect temporal variability in future years (for example, a drought condition). In addition, as previously pointed out in the temperature illustration, restriction of background to only four samples dramatically increases the size of the statistical prediction limit thereby increasing the false negative rate of the test (that is, the prediction limit is over five standard deviation units above the background mean concentration). The reason for this is that the uncertainty in the true mean concentration covers the majority of the normal distribution. As such, virtually any mean and standard deviation could be obtained by chance alone. If by chance the values are low, false positive results will occur. If by chance the values are high, false negative results will occur. By increasing the background sample size, uncertainty in the sample-based mean and standard deviation decrease as does the size of the prediction limit, therefore both false positive and false negative rates are minimized.

7.2 In light of these considerations, it is always in the best interest to have the largest available background database consisting of independent and representative measurements. Two possible strategies used to obtain a larger background database are add background wells to the monitoring system (this also facilitates characterization of spatial variability) and update the background database at appropriate intervals (that is, either continuously for inter-well or every year or two for intra-well) with new measurements that are determined to belong to the same background population.

8. Keywords

8.1 control charts; detection monitoring; ground water; prediction limits; statistics; waste disposal facilities

REFERENCES

(1) Gibbons, R. D., *Statistical Methods for Ground-Water Monitoring*, John Wiley & Sons, 1994.
(2) Currie, *Analytical Chemistry, 40,* 1968, pp. 586–593.
(3) Koorse, *Environmental Law Reporter, 19,* 1989, pp. 10211–10222.
(4) USEPA, "Addendum to Interim Final Guidance Document," *Statistical Analysis of Ground-Water Monitoring Data at RCRA Facilities,* July, 1992.
(5) USEPA 40CFR Part 264: "Statistical Methods for Evaluating Ground-Water Monitoring from Hazardous Waste Facilities; Final Rule," *Federal Register,* 53, 196, 1988, pp. 39720–39731.
(6) USEPA 40CFR 258.53(h)(2).
(7) USEPA 40CFR 258.53(h)(1).
(8) USEPA 40CFR 258.53(h)(5).
(9) USEPA 40CFR 258.53(h)(3).
(10) USEPA 40CFR 258.55.
(11) Hubaux, A., and Vos, G., *Analytical Chemistry, 42,* 1970, pp. 849–855.
(12) Wilk, M. B., and Shapiro, S. S., *Technometrics, 10,* No. 4, 1968, pp. 825–839.
(13) Shapiro, S. S., and Wilk, M. B., *Biometrika, 52,* 1965, pp. 591–611.
(14) Davis, C. B. and McNichols, R. J., Technometrics, 29, 1987, pp. 359–370.
(15) Aitchison, J., *Journal of the American Statistical Association, 50,* 1955, pp. 901–908.
(16) Cohen, A. C., *Technometrics, 3,* 1961, pp. 535–541.
(17) Gibbons, R. D., *Ground Water, 28,* 1990, pp. 235–243.
(18) Gibbons, R. D., *Ground Water, 29,* 1991, pp. 729–736.
(19) Gibbons, R. D., *Ground Water, 25,* 1987, pp. 572–580.
(20) Cameron, K., (1995) EPA/530-R-93-003.
(21) Lucas, J. M., *Journal of Quality Technology, 14,* 1982, pp. 51–59.
(22) Starks, T. H., "Evaluation of Control Chart Methodologies for RCRA Waste Sites," *USEPA Technical Report CR814342-01-3,* 1988.
(23) *Statistical Analysis of Ground-Water Monitoring Data at RCRA Facilities,* 1989.
(24) *Biometrics, 9,* 1953, pp. 74–89.
(25) Sen, P. K., *Journal of the American Statistical Association, 63,* 1968, pp. 1379–1389.
(26) Gilbert, R. O., *Statistical Methods for Environmental Pollution Monitoring,* Van Nostrand Reinhold, New York, 1987.

PART 4.   WASTE/CONTAMINANT CHARACTERIZATION AND SAMPLING

4.1 GENERAL GUIDANCE

Standard Guide for
General Planning of Waste Sampling[1]

This standard is issued under the fixed designation D 4687; the number immediately following the designation indicates the year of original adoption or, in the case of revision, the year of last revision. A number in parentheses indicates the year of last reapproval. A superscript epsilon (ε) indicates an editorial change since the last revision or reapproval.

INTRODUCTION

The analysis and testing of solid waste requires collection of adequately sized, representative samples. Wastes are found in various locations and physical states. Therefore, each sampling routine must be tailored to fit the waste and situation. Wastes often occur as nonhomogeneous mixtures in stratified layers or as poorly mixed conglomerations. For example, wastes are commonly stored or disposed of in surface impoundments with stratified or layered sludges covered by ponded wastewater. In these situations, the collector may be faced with sampling the wastewater, the sludge, and some depth of soil beneath the sludges. Collecting representative samples in these situations requires a carefully assessed, well-planned, and well-executed sampling routine.

Currently, Subcommittee D34.01 is working on practices for sampling wastes from a variety of different sampling locations and situations. Also in progress is a practice for containerization, preservation, and holding times for waste samples. As these documents are approved by ASTM, reference to these standards will be made in this general guide on waste sampling. Further, Subcommittee D34.01 recommends this guide be used in conjunction with the new waste sampling practices when available in print by ASTM.

1. Scope

1.1 This guide provides information for formulating and planning the many aspects of waste sampling (see 1.2) which are common to most waste sampling situations.

1.2 The aspects of sampling which this guide addresses are as follows:

	Section
Sampling plans	4
Safety plans	5
Quality assurance considerations	6
General sampling considerations	7
Preservation and containerization	8
Cleaning equipment	9
Labeling and shipping procedures	10
Chain-of-custody procedure	11

1.3 This guide does not provide comprehensive sampling procedures for these aspects, nor does it serve as a guide to any specific application. It is the responsibility of the user to assure that the procedures used are proper and adequate.

1.4 *This standard does not purport to address all of the safety concerns, if any, associated with its use. It is the responsibility of the user of this standard to establish appropriate safety and health practices and determine the applicability of regulatory limitations prior to use.* For more specific precautionary statements see 3.2, 3.3, and Section 5.

2. Referenced Documents

2.1 *ASTM Standard:*
E 122 Practice for Choice of Sample Size to Estimate a Measure of Quality for a Lot or Process[2]
2.2 *Other Document:*
EPA-SW-846 Test Methods for Evaluating Solid Waste, Physical/Chemical Methods[3]

3. Significance and Use

3.1 The procedures covered in this guide are general and provide the user with information helpful for writing sampling plans, safety plans, labeling and shipping procedures, chain-of-custody procedures, general sampling procedures, general cleaning procedures, and general preservation procedures.

3.2 For purposes of this guide, it is assumed that the user has knowledge of the waste being sampled and the possible safety hazards.

3.3 This guide is not to be used when sampling sites or wastes where safety hazards are unknown. In such cases, the user must use other more appropriate procedures.

4. Sampling Plans

4.1 A sampling plan is a scheme or design to locate sampling points so that suitable representative samples descriptive of the waste body can be obtained. Development

[1] This guide is under the jurisdiction of ASTM Committee D-34 on Waste Management and is the direct responsibility of Subcommittee D34.01 on Sampling and Monitoring.
Current edition approved March 15, 1995. Published May 1995. Originally published as D 4687 – 87. Last previous edition D 4687 – 87.

[2] *Annual Book of ASTM Standards*, Vol 14.02.
[3] Available from Superintendent of Documents, U.S. Printing Office, Washington, DC 20402.

of sampling plans requires the following:

4.1.1 Review of background information about the waste and site.

4.1.2 Knowledge of the waste location and situation.

4.1.3 Decisions as to the types of samples needed.

4.1.4 Decisions as to the sampling design required.

4.2 Background data on the waste is extremely helpful in preassessment of the waste's composition, hazards, and extent. (See Notes 1 and 2.)

NOTE 1—If after researching the available background information the user cannot obtain from the material enough information about the waste to determine the probable composition and probable hazards, then the user should use other procedures. Such situations are beyond the scope of this guide.

NOTE 2—The background information is needed to determine necessary safety equipment, safety procedures, sampling equipment and sampling design, and procedures to be used.

4.2.1 Possible sources of information on the site and waste include the following:

4.2.1.1 File searches of state and local records including waste manifests, waste approvals, land permit applications.

4.2.1.2 File searches of generator records (if the generator can be identified) including chemical analyses, safety data sheets, design drawings, and manufacturing process information.

4.2.1.3 File searches of treatment, storage, disposal, and transport facilities. Records involved with handling the waste.

4.2.1.4 Researching published data concerning the site such as scientific journal articles, EPA publications, and newspaper stories. Newspapers are the most likely source but the information is seldom very technical.

4.2.1.5 Interviews of key people such as past and present employees of the site or generator, state and local officials, residents of the area, etc.

4.2.1.6 Aerial photographs provide a historical record of the site development. Many federal agencies conduct aerial surveys that are available to the public. Some of these agencies include the following:

(a) U. S. Department of Agriculture

(b) Soil Conservation Service (USDA-SCS).

(c) U. S. Geological Survey.

(d) U. S. Forest Service.

(e) National Air and Space Administration (NASA).

(f) National Oceanic and Atmospheric Administration (NOAA).

(g) National Weather Service.

(h) Corps of Engineers.

(i) Agricultural Stabilization and Conservation Service.

4.2.1.7 Published maps can also provide a historical record of the site development such as topographic, soil, and county maps.

4.3 Waste location and site conditions greatly influence a sampling plan. The most common waste locations may include lagoons, landfills, pipes, point discharges, piles, drums, bins, tanks, and trucks. The site conditions include the physical condition of the waste; that is, whether it is a solid (granular, consolidated, or cohesive), liquid (slurry or flowable sludge), or gas, and it describes under what conditions it was disposed; that is, does it exist as a multiphased waste in a lagoon, tank or drum; is it stratified solids in a

lagoon; is it a poorly mixed concoction of municipal garbage and hazardous sludges; or a landfill containing barrels of unknown waste.

4.3.1 Based on these considerations, the collector will have to decide what must be sampled. Each situation is different and requires the best judgement of the user in writing such a plan.

4.4 The types of samples that may be collected are most commonly either composite or single samples. The sample collector must decide considering the complexity of the waste location, the situation, and the financial resources, and what types of samples will best provide representative samples for reliable measurements.

4.4.1 A composite sample, sometimes referred to as a batch sample, is a well-mixed collection of subsamples of the same waste taken from different points. A composite sample is used most commonly in determining an average measure of a parameter. Generally, composite samples are taken when differences in the waste exist because of stratification, or because of the simultaneous deposition of different wastes such as in a landfill.

4.4.2 A single sample is a well-mixed sample taken from a single point. It is used to measure a particular parameter or parameter set at a given point or within a unique homogeneous layer or throughout the strata at one or several locations.

4.5 Sampling plans or schemes should be carefully thought out, well in advance of sampling. The most common sampling schemes involve the selection of sampling points using a judgement, a coordinate system, or a grid system.

4.5.1 *Judgement Samples*—This system is commonly used when, because of resource restraints, multiple samples cannot be collected. They are collected by deciding through visual observation or knowledge of the site where a representative sample may be collected. This type of design can be very effective if the collector is familiar and knowledgeable about the site, and if the goal of sampling is merely to establish whether a waste meets some set criteria.

4.5.2 *Coordinate Sampling System*—This system uses a one or two coordinate system and involves collecting samples at random points from the origin of the coordinates. Random numbers can be generated using random number tables available in most statistic texts. The origin of the coordinate system is normally placed at some corner of the site and marked off in steps, feet, yards, etc. for sampling landfills, waste piles, and lagoons. For storage areas containing barrels, the numbers of barrels from the origin are often used as intervals along the coordinate. For sampling from a flowing stream the origin may be taken as time-zero (start), and samples are collected at random time intervals over the period of interest.

4.5.3 *Grid System*—This system also involves taking samples at regular intervals, grid points, along an imaginary grid system laid out over the site. The number of sampling points will vary with the size of the grid. Such sampling schemes are used when a statistically sound sampling program is required. They should be used only when the waste body is known to be homogenous, or when the strata have been defined. If the waste is stratified, a separate grid system may be required for each stratum.

4.6 The proper number of samples required in a statistically sound sampling program can be estimated. This can be done using Eq 1 and by estimating the sample composition and variance either from a pilot sampling effort or knowledgeable judgement. The number of samples required, n, to achieve the desired precision in waste composition is estimated using fundamental statistical concepts, as follows (financial constraints not considered):[3]

$$n = (t^2_{0.80}S^2)/d^2 \qquad (1)$$

where:

n = appropriate number of samples to be collected;

$t^2_{0.80}$ = square of the tabulated value of student's t for a two-sided confidence interval and a coverage probability of 0.80 for the unknown mean, with the degrees of freedom defined for the S^2 used to estimate the population variance, σ^2;

S^2 = preliminary estimate of σ^2 obtained from previous samplings, a pilot sampling effort or other information such as the likely range of the population values;

d = deviation to be exceeded only in two cases out of ten in repeated sampling for the quantity $|\bar{X} - T|$, the difference in absolute value between the sample average and a threshold value such as a regulatory limit;

$\bar{X}$ = preliminary estimate of sample average; and

T = threshold value, often the regulatory limit.

4.6.1 The variables in Eq 1 are appropriate only for a given waste type. Therefore, the appropriate number of samples n, required to achieve the desired precision is also applicable only to that same waste type. If two or more waste types are present in the impoundment, either as strata or other segregated wastes, then a value for n should be calculated for each waste.

4.6.2 Although the use of Student's t distribution is based on an underlying normal distribution for the measurements, the robustness of the t statistic for many applications may be relied upon here. If ancillary information seems to indicate that normality may not be a good assumption, then a goodness of fit test should be performed to determine if the assumption of a normal distribution is reasonable. The Lilliefors goodness of fit test, as it applies to the pilot sampling presented here, is described in the Appendix. This test involves examining the data from a sampling and analysis program in order to test the hypothesis that the data are distributed normally. If the Lilliefors test shows the contention of normality is acceptable, it does *not* mean that the parent population is normal. But it does mean that the Student's t distribution does not appear to be an unreasonable approximation to the true unknown distribution. If the Lilliefors test shows that a normal distribution does not adequately fit the data, then further pilot sampling will be required to adequately determine the spatial distributions in the impoundment.

4.6.3 The following hypothetical example illustrates the use of Eq 1:

4.6.3.1 A preliminary study of barium levels in sludge collected from a lagoon generated values of 86, 90, 98, and 104 ppm for barium in four sludge samples. Based on these values and a knowledge of the processes producing the waste, the sludge is judged to be homogeneous (not stratified) within the lagoon. Therefore, preliminary estimates of $\bar{X}$ and s^2 are calculated as follows:

$$\bar{X} = \frac{\sum\limits_{i=1}^{n} X_i}{n} = \frac{86 + 90 + 98 + 104}{4} = 94.50, \text{ and}$$

$$s^2 = \frac{\sum\limits_{i=1}^{n} X_i^2 - \left(\sum\limits_{i=1}^{n} X_i\right)^2 \Big/ n}{n-1}$$

$$= \frac{35\ 916.00 - 35\ 721.00}{3} = 65.00$$

4.6.3.2 The deviation not to be exceeded for measured barium in the sludge samples, d, is chosen as 5.50 ppm, that is, the difference between the sample average, $\bar{X}$ or 94.5, and the threshold limit, T or 100.0 for barium (assuming 100.0 is the regulatory threshold for barium) is 5.50 ppm.

4.6.3.3 The value of $t_{0.80}$ is obtained from tabulated values of Student's t, as shown in Table 1. Although an assumption of a t distribution would seem to be restrictive, it can be shown that even non-normal populations possessing bell-shaped distributions can be closely approximated by a t distribution. From the preliminary study $n = 4$, and the degrees of freedom, $n - 1$, is 3. Therefore,

TABLE 1 Tabulated Values of Student's t for Evaluating Solid Wastes

Degrees of Freedom, $(n-1)$[A]	Tabulated t Value[B]
1	3.078
2	1.886
3	1.638
4	1.533
5	1.476
6	1.440
7	1.415
8	1.397
9	1.383
10	1.372
11	1.363
12	1.356
13	1.350
14	1.345
15	1.341
16	1.337
17	1.333
18	1.330
19	1.328
20	1.325
21	1.323
22	1.321
23	1.319
24	1.318
25	1.316
26	1.315
27	1.314
28	1.313
29	1.311
30	1.310
40	1.303
60	1.296
120	1.289
	1.282

[A] Degrees of freedom, df, are equal to the number of samples, n, collected from a solid waste less one.

[B] Tabulated t values are for a two-tailed confidence interval and a probability of 0.80 (the same values are applicable to a one-tailed confidence interval and a probability of 0.90).

$$t_{0.80} = 1.638$$

4.6.3.4 The appropriate number of sludge samples to be collected from the lagoon is,

$$n = t^2_{0.80}s^2 \Big/ d^2 = \frac{(1.638^2)(65.00)}{5.50^2} = 5.77,$$

or six. That number of samples (plus extra for protection against poor preliminary estimates of $\bar{X}$ and s^2) is collected from the lagoon.

5. Hazards

5.1 Proper safety precautions must always be observed when sampling wastes. Persons collecting samples must be aware that the waste can be a strong sensitizer and can be corrosive, flammable, explosive, toxic, and capable of releasing extremely poisonous gases. The background information obtained about the waste should be helpful in deciding the extent of safety precautions to be observed and in choosing protective equipment to be used. The information obtained should be checked for hazardous properties against such references as "Dangerous Properties of Industrial Materials" the "March Index," the "Condensed Chemical Dictionary," and the "Toxic and Hazardous Industrial Chemicals Safety Manual for Handling and Disposal with Toxicity and Hazardous Data."

NOTE 3—The following safety precautions are not comprehensive. Rather, they provide additional guidance on health and safety to complement professional judgment and experience.

5.2 Personnel should wear protective equipment when response activities involve known or suspected atmospheric contamination; when vapors, gases, or airborne particulates may be generated; or when direct contact with skin-affecting substances may occur. Respirators can protect lungs, gastrointestinal tract, and eyes against air toxicants. Chemical-resistant clothing can protect the skin from contact with skin-destructive and -absorbable chemicals. Good personal hygiene limits or prevents ingestion of material.

5.2.1 Equipment to protect the body against contact with known or anticipated chemical hazards has been divided into four categories according to the degree of protection afforded:

5.2.1.1 *Level A*—Should be worn when the highest level of respiratory, skin, and eye protection is needed.

5.2.1.2 *Level B*—Should be selected when the highest level of respiratory protection is needed, but a lesser level of skin protection. Level B protection is the minimum level recommended on initial site entries until the hazards have been further defined by on-site studies and appropriate personnel protection utilized.

5.2.1.3 *Level C*—Should be selected when the type(s) of airborne substance(s) is (are) known, the concentrations(s) is measured, and the criteria for using air-purifying respirators are met.

5.2.1.4 *Level D*—Should not be worn on any site with respiratory or skin hazards. It is primarily a work uniform providing minimal protection.

5.2.2 The level of Protection selected should be based primarily on the following:

5.2.2.1 Type(s) and measured concentration(s) of the chemical substance(s) in the ambient atmosphere and its toxicity and

5.2.2.2 Potential or measured exposure to substances in air, splashes of liquids, or other direct contact with material due to work being performed.

5.2.2.3 In situations where the type(s) of chemical(s), concentration(s), and possibilities of contact are not known, the appropriate Level of Protection must be selected based on professional experience and judgment until the hazards can be better characterized.

5.2.3 *Level A Protection—Personnel Protective Equipment:*

(*a*) Pressure-demand, self-contained breathing apparatus, approved by the Mine Safety and Health Administration (MSHA) and National Institute of Occupational Safety and Health (NIOSH),

(*b*) Fully encapsulating chemical-resistant suit,

(*c*) Coveralls,[4]

(*d*) Long cotton underwear,[4]

(*e*) Gloves (outer), chemical-resistant,

(*f*) Gloves (inner), chemical-resistant,

(*g*) Boots, chemical-resistant, steel toe and shank. (Depending on suit construction, worn over or under suit boot),

(*h*) Hard hat[4] (under suit),

(*i*) Disposable protective suit, gloves, and boots[4] (worn over fully encapsulating suit), and

(*j*) Two-way radio communications (intrinsically safe).

5.2.3.1 The fully encapsulating suit provides the highest degree of protection to skin, eyes, and respiratory system if the suit material is resistant to the chemical(s) of concern during the time the suit is worn or at the measured or anticipated concentrations, or both. While Level A provides maximum protection, the suit material may be rapidly permeated and penetrated by certain chemicals from extremely high air concentrations, splashes, or immersion of boots or gloves in concentrated liquids or sludges. These limitations should be recognized when specifying the type of chemical-resistant garment. Whenever possible, the suit material should be matched with the substance it is used to protect against.

5.2.3.2 Many toxic substances are difficult to detect or measure in the field. When such substances (especially those readily absorbed by or destructive to the skin) are known or suspected to be present and personnel contact is unavoidable, Level A protection should be worn until more accurate information can be obtained.

5.2.4 *Level B Protection—Personnel Protective Equipment:*

(*a*) Pressure-demand, self-contained breathing apparatus (MSHA/NIOSH approved),

(*b*) Chemical-resistant clothing (overalls and long-sleeved jacket; coveralls; hooded, one- or two-piece chemical-splash suit; disposable chemical-resistant coveralls),

(*c*) Coveralls,[4]

(*d*) Gloves (outer), chemical-resistant,

(*e*) Gloves (inner), chemical-resistant,

(*f*) Boots, chemical-resistant, steel toe and shank,

(*g*) Boots (outer), chemical-resistant (disposable, worn over permanent boots),[4]

---

[4] Equipment is optional.

(h) Hard hat (face shield),[4] and

(i) Two-way radio communications (intrinsically safe).

5.2.4.1 Level B equipment provides a high level of protection to the respiratory tract, but a somewhat lower level of protection to skin. The chemical-resistant clothing required in Level B is available in a wide variety of styles, materials, construction detail, permeability, etc. These factors all affect the degree of protection afforded. Therefore, a specialist should select the most effective chemical-resistant clothing (and fully encapsulating suit) based on the known or anticipated hazards or job function, or both.

5.2.4.2 For initial site entry and reconnaissance at an open site, approaching whenever possible from the upwind direction, Level B protection (with good quality, hooded, chemical-resistant clothing) should protect response personnel, providing the conditions described in selecting Level A are known or judged to be absent.

5.2.5 *Level C Protection—Personnel Protective Equipment:*

(a) Full-face, air purifying, canister-equipped respirator (MSHA/NIOSH approved),

(b) Chemical-resistant clothing (coveralls; hooded, two-piece chemical splash suit; chemical-resistant hood and apron; disposable chemical-resistant coveralls),

(c) Coveralls,[4]

(d) Gloves (outer), chemical-resistant,

(e) Gloves (inner), chemical-resistant,[4]

(f) Boots, chemical resistant, steel toe and shank,

(g) Boots (outer), chemical-resistant (disposable, worn over permanent boots),[4]

(h) Hard hat (face shield),[4]

(i) Escape mask[4], and

(j) Two-way radio communications (intrinsically safe).

5.2.5.1 Level C protection is distinguished from Level B by the equipment used to protect the respiratory system, assuming the same type of chemical-resistant clothing is used. The main selection criterion for Level C is that conditions permit wearing air-purifying devices.

5.2.5.2 Total unidentified vapor/gas concentrations of 5 ppm above background require Level B protection. Only a qualified individual should select Level C (air-purifying respirators) protection for continual use in an unidentified vapor/gas concentration of background to 5 ppm above background.

5.2.6 *Level D Protection—Personnel Protective Equipment:*

(a) Coveralls,

(b) Gloves,[4]

(c) Boots/shoes, leather or chemical-resistant, steel toe and shank,

(d) Boots, chemical-resistant (disposable worn over permanent boots),[4]

(e) Safety glasses or chemical splash goggles,[4]

(f) Hard hat (face shield),[4] and

(g) Escape mask.[4]

5.2.6.1 Level D protection is primarily a work uniform. It should be worn in areas where: (1) only boots can be contaminated, or (2) there are no inhalable toxic substances.

5.3 Personnel should not eat, drink, or smoke during or after sampling until after decontamination steps are taken.

Sampling personnel should be trained in safety aspects of hazardous waste sampling.

5.4 Testing air emission for determining the vapor/gas concentrations can be accomplished through the use of a portable organic vapor analyzer. The probe should be held 1 to 2 in. above the sampling point. Follow manufacturers operating instructions for proper calibration, use, and care.

6. Quality Assurance Considerations

6.1 Quality assurance for solid waste sampling should include adherence to the sampling plan and safety plan and in some cases, the use of quality control samples.

6.2 The sampling and safety plans should be well formulated before any actual sampling is attempted. The plans must be consistent with the objectives of the sampling. The sampling plan must include the selected points of sampling and the intended number, volumes, and types of samples to be taken. The safety plan should address the proper clothing and protective equipment, all known hazards associated with the sampling activities, and the measures to be taken to avoid these hazards.

6.3 Four types of quality control samples relate to the quality assurance of field sampling: (1) field blanks, (2) split samples, (3) field rinsates, and (4) field spikes. The selection of the types of quality control samples to be used should be made prior to the sampling event and included in the sampling plan. The nature of the sampling, the intended uses of the data, and the material being sampled all impact upon the selection of quality control samples to be used in an event.

6.3.1 *Field Blanks:*

6.3.1.1 Field blanks are samples prepared in the laboratory using reagent water or other blank matrix and sent with the sampling team. These samples are exposed to the sampling environment and returned with the samples to the laboratory for analysis. The purpose of the field blank is to verify that none of the analytes of interest measured in the field samples resulted from contamination of the samples during sampling.

6.3.1.2 The sampling plan should normally include a minimum of one field blank for each procedure for each sampling event. These samples can be submitted blind to the laboratory to challenge their analytical system or can be shipped with the instruction to hold them unless there is a reason to suspect sample contamination. The submission of blind field blanks would normally be reserved for those situations where the competency of the analytical laboratory was unproven (that is, where a new laboratory was being utilized).

6.3.2 *Split Samples:*

6.3.2.1 Split samples are used to challenge the analytical laboratory performance. Split samples are also used when two different parties are sampling the same site and verification of analytical results is necessary. A split sample is prepared by subsampling a homogenous sample into two or more portions and submitting each portion separately to the analytical laboratory.

6.3.2.2 For liquid matrixes, the material should be placed in a large, clean container and stirred or swirled to ensure thorough mixing of the medium prior to subsampling. For solid media, a sufficient quantity must be removed, mixed with clean utensils, and subsampled. Sufficient mixing of the sample should be accomplished to ensure that stratification of analytes is avoided.

NOTE 4: Caution—If volatile organics are a concern, homogenizing in open containers will likely result in losses of volatiles.

6.3.2.3 Split samples are treated as separate study samples, carried through the entire sample handling procedures, and submitted to the analytical laboratory without distinguishing identification. Split samples are an indication of the precision of the analytical procedures. For comments on sampling precision, see 4.6; the definition of acceptable levels of precision is the responsibility of the user.

6.3.2.4 Where feasible, each sampling event should include a minimum of one split sample for each type of media or location sampled. Where the data are intended for demonstration of data quality to an outside agency, replicates should be included at a greater frequency, up to 10 % of the total number of samples collected.

6.3.3 *Field Rinsates*—Field rinsates are samples collected in the field by filling a sample collection vessel, such as a well bailer with reagent water or other blank matrix, and then transferring this water to the proper sample bottles. It may be necessary in some instance to fill the collection vessel a number of times to ensure enough water is collected for analysis. The purpose of a field rinsate is to ensure that sampling equipment cleaned in the field is not cross contaminating samples through improper cleaning techniques. These types of samples should be taken at least once for each procedure for each sampling event when field cleaning is performed. If only one such sample is taken it should be collected just prior to the last sample.

6.3.4 *Field Spikes:*

6.3.4.1 Field spikes are samples collected in the field and spiked with compounds of interest or related compounds. These samples are used to check on the potential for loss of analyte on shipping and for recovery of analytes from a particular medium. The field spike is prepared by adding a known amount of the spiking material to a known amount of the matrix and mixing thoroughly. Where a liquid medium is to be collected, the spiking material may be added to the collection container at the laboratory and the sample medium added to the container. For a solid matrix, the material should be added in the field and thoroughly mixed through the matrix prior to closing and sealing the container.

6.3.4.2 Field spikes are normally not required. Instances when a field spike may be desired include where preservation techniques are in question and the integrity of the analytes at the laboratory is not known, when there is a question concerning matrix effects, and when the results from the analytical laboratory for a particular analyte or class of analytes is in question.

6.3.4.3 Field spikes should be submitted blind to the laboratory in the same manner as outlined for the split samples. These samples should be carried through all stages of the sampling and sample handling process as the actual study samples to ensure that they truly indicate the integrity of the samples collected.

7. General Sampling Consideration

7.1 Sampling equipment must be selected that is chemically compatible with the type of waste and type of analyses.

Generally, plastic sampling equipment is not suitable for waste containing or to be analyzed for organic parameters. Stainless steel, glass, and plastic are generally acceptable for most samples to be analyzed for inorganics. It is up to the user to ensure that the equipment will not contaminate or bias the analyses.

7.2 The sampling equipment must be capable of extracting a sample from the desired location, depth, or point and at the same time provide protection from cross-contamination during sampling. For instance, one very common problem is extracting a sludge sample from beneath a top layer of wastewater or sludge without contaminating the sample with the overlying wastewater or sludge. This situation, as well as many others, requires special equipment. The collector is therefore faced in many instances with having to fabricate the needed equipment.

7.3 Recommended sampling procedures are for collection of samples from the edge of ponds or lagoons or from piers or catwalks. Sampling from boats is not recommended and should be attempted only if the collector knows the waste poses no real health problem and every possible safety measure has been taken.

7.4 Tanks and drums containing unknown substances also pose potential health risks for the collector due to the possibility of fire, explosion, or the release of deadly gases upon opening. Therefore, it is recommended that in these situations only spark-proof, remote opening devices be used and only fully trained and experienced personnel attempt to do this.

7.5 Representative samples are intended to reflect the true makeup of the population. Composite sampling is one way to help achieve representativeness in a cost- and resource-efficient way. Frequently overlooked but important problems with composite sampling of waste materials include the following:

7.5.1 Loss of volatile components during the mixing process,

7.5.2 Reactivity of dissimilar materials combined into a single composite,

7.5.3 Collecting the correct number and size of aliquots to form the composite, and

7.5.4 Properly homogenizing and subsampling to reduce the amount of material sent to the laboratory.

7.6 The laboratory should provide guidance to the field sampling team to avoid losses due to volatilization or reactivity. Guidelines based on geostatistical principles for forming the composite from individual aliquots should be provided to sampling personnel. Sample homogenization and subsampling in the field should only be attempted by qualified personnel using appropriate equipment. If these are not available, the individual aliquots should be sent to the laboratory for processing. All composite samples or aliquots, or both, intended for forming a composite should be clearly marked so that the laboratory can mix and subsample appropriately.

7.6.1 If low chemical concentrations of a parameter to be tested are expected, take a large volume of sample.

7.6.2 If high chemical concentrations of a parameter to be tested are expected, smaller volumes will suffice.

7.7 When possible, it is recommended that sampling proceed from the least contaminated to the most contaminated areas to reduce problems of cross contamination.

8. Preservation and Containerization of Samples

8.1 Water sample preservation techniques cannot be used for waste samples but are sometimes confused with them.

NOTE 5—Subcommittee D34.01 on Sampling and Monitoring is currently balloting a draft practice on preservation of waste and when approved by the ASTM membership, it will be referenced here. Container specifications for waste samples are found in the same draft practice mentioned above and will be similarly referenced.

9. Cleaning Equipment

9.1 All sampling equipment must be cleaned before use and if possible, it is preferable to have them lab cleaned. Improper cleaning causes cross-contamination of samples. Use of disposable samplers is a very simple way of eliminating the cleaning problem. When samplers are lab cleaned, it is equally important to protect the samples from contamination by wrapping, packaging or containerizing in clean, non-contaminating material.

9.2 Wash equipment cleaned in the lab with a warm detergent solution, rinse several times with tap water, rinse with deionized water and air dry. If organic analyses are to be run, wash with a warm detergent solution (it may first be necessary to wipe with an absorbent cloth to eliminate residues), rinse several times with tap water, rinse with deionized water, rinse with an appropriate organic solvent (the solvent should be, if possible, the extracting solvent) and oven dry at 105°C for at least an hour.

9.3 Cleaning procedures in the field are the same except that tap water rinses, solvent rinses, and drying may not be practical. The user must be careful in using a solvent rinse without oven drying as it may permeate other samples if shipped in the same container or it may react with constituents in the waste. It should be kept in mind that field cleaning requires the capability to carry a large amount of water.

10. Packaging and Package Marking

10.1 An indelible label should be secured to the container identifying the sample. The label should contain or reference the following information:
10.1.1 Name and location of site,
10.1.2 Date and time of sampling,
10.1.3 Location of sampling,
10.1.4 Sample number,
10.1.5 Description and disposition of sample,
10.1.6 Name of sampling personnel,
10.1.7 If possible, full weight of sample and container upon shipping,
10.1.8 Type of preservative, and
10.1.9 Analytical requirements.

10.2 Pack the sample container securely in a shipping container. Generally, if the sample is to be analyzed for volatile organics, it should be packed in ice and cooled to 4°C (See Section 8). A minimum-maximum thermometer packed with the sample container is valuable where knowledge of temperature extremes is critical.

10.3 For shipping purposes, samples must be classified as environmental or hazardous material (waste) samples. In general, environmental samples are collected offsite (for example, from streams, ponds, or wells) and are not expected to be grossly contaminated with high levels of hazardous materials. Waste samples (for example, material from drums or bulk storage tanks; obviously contaminated lagoons, ponds, soils; and leachates from hazardous waste sites) are considered hazardous. Hazardous materials are subject to shipping regulation (for example, International Air Transportation Association (IATA) or U.S. Department of Transportation (USDOT)) and should be packaged and shipped by a person trained in those requirements. If the substance in the sample is known or can be identified, package, mark, and ship according to the specific instructions for that material. Further detail on the shipment of hazardous materials is beyond the scope of this guide, and professional assistance should be sought.

10.4 Make arrangements for handling, logging in, adequate storage, and analysis of the sample at its destination.

10.5 All information pertinent to a field survey or sampling activity must be recorded in a logbook. This should be bound, preferably with consecutively numbered waterproof pages. Where the same information is routinely collected, preprinted logbooks are suggested. Information should be recorded in indelible ink, all entry errors should be crossed out with a single line and initialed, and all entries should be dated and signed at least daily. Entries in the logbook should include information such as the following:
10.5.1 Identification of sampling plan (by reference),
10.5.2 Location of sampling,
10.5.3 Name and address of field contact,
10.5.4 Producer of waste and address,
10.5.5 Type of process producing waste (if known),
10.5.6 Type of waste (for example, sludge, wastewater),
10.5.7 Suspected waste composition, including concentrations,
10.5.8 Number and volume of samples taken,
10.5.9 Purpose of sampling (for example, surveillance, contract number),
10.5.10 Description of sampling point and sampling methods,
10.5.11 Date and time of collection,
10.5.12 Preservation used, if any (include ice),
10.5.13 Analytical parameter to be measured,
10.5.14 Unique sample identification number(s),
10.5.15 Sample destination, how transported, and name of transporter,
10.5.16 References, such as maps or photographs of the sampling site,
10.5.17 Field observations (for example, ambient temperature, wind conditions, or other site specific conditions),
10.5.18 Field measurements (for example, pH, explosivity) and results, and
10.5.19 Names of personnel on sampling team.

10.6 Record sufficient information so that anyone could reconstruct the sampling without reliance on the collector's memory. The contents of the logbook should be specified in the sampling plan, a standard practice or similar document. The logbook must be protected and kept in a safe place. Additional detail may be needed than listed above depending on the technical objectives of the sampling project.

11. Chain-of-Custody Procedure

11.1 A chain-of-custody procedure should be developed for samples that may be used in legal proceedings. It is recommended that legal counsel be consulted to assist in developing an individual chain-of-custody record. Chain-of-custody procedures are used to ensure sample integrity and to ensure the sample will provide legally and technically defensible data.

11.2 Samples should be collected in accordance with sampling procedures designated in the sampling plan if the sample is to maintain its legal integrity.

11.3 A legal seal should be attached to the immediate sample container in a manner that the sample cannot be opened without breaking the seal. The seal should have a unique number written across it or the signature of the sampler. Be sure to record the unique legal seal number in the log book.

11.4 The sample should be kept in view or within limited access or locked storage until custody is relinquished and formal documentation of the transfer is completed. The collector should initiate documents at the source of the sample and start the chain-of-custody procedure. Documentation should include the following:

11.4.1 Sample number and legal seal numbers,

11.4.2 Site name and location,

11.4.3 Date and time sampled,

11.4.4 Date sent to laboratory,

11.4.5 Name of sampler,

11.4.6 Information describing source of sample and sample,

11.4.7 Sampling method,

11.4.8 Preservation techniques,

11.4.9 Condition of legal seal upon delivery and name of receiving person,

11.4.10 Method of shipment,

11.4.11 Dates of all activities, and

11.4.12 Signature of persons delivering and receiving samples.

APPENDIX

(Nonmandatory Information)

X1. THE LILLIEFORS TEST FOR GOODNESS OF FIT

X1.1 The data consist of a random sample $X_1, X_2, \ldots, X_n$ of size, n, associated with some unknown distribution function, denoted by $F(x)$. Compute the sample mean:

$$\frac{1}{n} \sum_{i=1}^{n} X_i \qquad \bar{X} = \qquad \text{(X1.1)}$$

for use as an estimate of μ, and compute:

$$s = \sqrt{\frac{1}{n-1} \sum_{i=1}^{n} (X_i - \bar{X})^2} \qquad \text{(X1.2)}$$

as an estimate of σ. Then compute the "normalized" sample values Z_i, defined as follows:

$$Z_i = \frac{X_i - \bar{X}}{s} \quad i = 1, 2, \ldots, n \qquad \text{(X1.3)}$$

The test statistic is computed from the Z_i's instead of from the original random sample.

TABLE X1.1 Normal Distribution[A]

NOTE—This table was abridged from Tables 3 and 4 by Pearson, E. S., and Hartley, H. E., "Biometric Tables for Statisticians," Cambridge University Press, Cambridge, England, pp. 111–112, 1962.

w_p	P	w_p	P	w_p	P
−3.7190	0.0001	−0.4677	0.32	0.5244	0.70
−3.2905	0.0005	−0.4399	0.33	0.5534	0.71
−3.0902	0.001	−0.4125	0.34	0.5828	0.72
−2.5758	0.005	−0.3853	0.35	0.6128	0.73
−2.3263	0.01	−0.3585	0.36	0.6433	0.74
−2.1701	0.015	−0.3319	0.37	0.6745	0.75
−2.0537	0.02	−0.3055	0.38	0.7063	0.76
−1.9600	0.025	−0.2793	0.39	0.7388	0.77
−1.8808	0.03	−0.2533	0.40	0.7722	0.78
−1.7507	0.04	−0.2275	0.41	0.8064	0.79
−1.6449	0.05	−0.2019	0.42	0.8416	0.80
−1.5548	0.06	−0.1764	0.43	0.8779	0.81
−1.4758	0.07	−0.1510	0.44	0.9154	0.82
−1.4395	0.075	−0.1257	0.45	0.9542	0.83
−1.4051	0.08	−0.1004	0.46	0.9945	0.84
−1.3408	0.09	−0.0753	0.47	1.0364	0.85
−1.2816	0.10	−0.0502	0.48	1.0803	0.86
−1.2265	0.11	−0.0251	0.49	1.1264	0.87
−1.1750	0.12	0.0000	0.50	1.1750	0.88
−1.1264	0.13	0.0251	0.51	1.2265	0.89
−1.0803	0.14	0.0502	0.52	1.2816	0.90
−1.0364	0.15	0.0753	0.53	1.3408	0.91
−0.9945	0.16	0.1004	0.54	1.4051	0.92
−0.9542	0.17	0.1257	0.55	1.4395	0.925
−0.9154	0.18	0.1510	0.56	1.4758	0.93
−0.8779	0.19	0.1764	0.57	1.5548	0.94
−0.8416	0.20	0.2019	0.58	1.6449	0.95
−0.8064	0.21	0.2275	0.59	1.7507	0.96
−0.7722	0.22	0.2533	0.60	1.8808	0.97
−0.7388	0.23	0.2793	0.61	1.9600	0.975
−0.7063	0.24	0.3055	0.62	2.0537	0.98
−0.6745	0.25	0.3319	0.63	2.1701	0.985
−0.6433	0.26	0.3585	0.64	2.3263	0.99
−0.6128	0.27	0.3853	0.65	2.5758	0.995
−0.5828	0.28	0.4125	0.66	3.0902	0.999
−0.5534	0.29	0.4399	0.67	3.2905	0.9995
−0.5244	0.30	0.4677	0.68	3.7190	0.9999
−0.4959	0.31	0.4959	0.69		

[A] The entries in this table are quantiles w_p of the standard normal random variable W, selected so $P(W<w_p) = p$ and $P(W>w_p) = 1 - p$.

X1.2 *Assumptions*—The sample is a random sample.

X1.3 *Hypotheses:*

X1.3.1 H_0—The random sample has the normal distribution, with unspecified mean and variance.

X1.3.2 H_1—The distribution function of the X_i's is non-normal.

X1.4 *Test Statistic*—Ordinarily the test statistic is the usual two-sided Kolmogorov test statistic, defined as the maximum vertical distance between the empirical distribution function of the X_i's and the normal distribution function with mean $\bar{X}$ and standard deviation s, as given by Eqs X1.1 and X1.2. However, the following method of computing the test statistic is slightly easier, and is equivalent to the method indicated in X1.1.

X1.4.1 Draw a graph of the standard normal distribution function, and call it $F'(x)$. Table X1.1 may be of assistance. Also draw a graph of the empirical distribution function of the normalized sample, the Z_i's defined by Eq X1.3, using the same set of coordinates as was used above for $F'(x)$. Find the maximum vertical distance between the two graphs, $F'(x)$ and the empirical distribution function which we shall call $S(x)$. This distance is the test statistic. That is, the Lilliefors test statistic T_2 is defined by the following equation:

$$T_2 = \sup |F'(x) - S(x)| \qquad \text{(X1.4)}$$

X1.5 *Decision Rule*—Reject H_o at the approximate level of significance of α if T_2 exceeds the $l - \alpha$ quantile as given in Table X1.2.

X1.5.1 The same data used to calculate n, the number of samples required, in Section 5 will be used to illustrate the Lilliefors test.

X1.5.2 Barium levels in four sludge samples were 86, 90, 98, and 104 ppm. The sample values, X_i, are arranged from smallest to largest, and converted to Z_i by subtracting $\bar{X} = 94.50$, and dividing by $S = \sqrt{65.00} = 8.06$.

X_i	Z_i
86	−1.05
90	−0.56
98	0.43
104	1.18

X1.5.2.1 The null hypothesis of normality is tested with the Lilliefors test statistic

$$T_2 = \sup |F'(x) - S(x)|$$

where $F'(x)$ is the standard normal distribution function and $S(x)$ is the empirical distribution function of the Z_i's. Figure X1.1 shows the curves representing $F'(x)$ (generated from Table X1.1) and $S(x)$. The maximum vertical distance between $F'(x)$ and $S(x)$ is seen from Fig. X1.1 to occur at $x = −1.05$, where $S(x)$ equals 0 and $F'(x)$ equals 0.15, and so T_2 equals 0.15.

X1.5.2.2 The Lilliefors test calls for rejection of H_o at $\alpha = 0.05$ if T_2 exceeds its 0.95 quantile, which is given in Table X1.2 as follows:

$$W_{0.95} = \frac{0.381}{\sqrt{n}} = \frac{0.381}{\sqrt{4}} = 0.19$$

TABLE X1.2 Quantiles of the Lilliefors Test Statistic[A]

NOTE—This table was adapted from Table 1 of Lilliefors, H. W., "On the Kolmogorov-Smirnov Test for Normality with Mean and Variance Unknown," *Journal of American Statistical Association*, Vol 62, pp. 399–402, 1967.

	$p = 0.80$	0.85	0.90	0.95	0.99
Sample size $n = 4$	0.300	0.319	0.352	0.381	0.417
5	0.285	0.299	0.315	0.337	0.405
6	0.265	0.277	0.294	0.319	0.364
7	0.247	0.258	0.276	0.300	0.348
8	0.233	0.244	0.261	0.285	0.331
9	0.223	0.233	0.249	0.271	0.311
10	0.215	0.224	0.239	0.258	0.294
11	0.206	0.217	0.230	0.249	0.284
12	0.199	0.212	0.223	0.242	0.275
13	0.190	0.202	0.214	0.234	0.268
14	0.183	0.194	0.207	0.227	0.261
15	0.177	0.187	0.201	0.220	0.257
16	0.173	0.182	0.195	0.213	0.250
17	0.169	0.177	0.189	0.206	0.245
18	0.166	0.173	0.184	0.200	0.239
19	0.163	0.169	0.179	0.195	0.235
20	0.160	0.166	0.174	0.190	0.231
25	0.142	0.147	0.158	0.173	0.200
30	0.131	0.136	0.144	0.161	0.187
Over 30	$\dfrac{0.736}{\sqrt{n}}$	$\dfrac{0.768}{\sqrt{n}}$	$\dfrac{0.805}{\sqrt{n}}$	$\dfrac{0.886}{\sqrt{n}}$	$\dfrac{1.031}{\sqrt{n}}$

[A] The entries in this table are the approximate quantiles w_p of the Lilliefors test statistic T_2. Reject H_O at the level α if T_2 exceeds $w_{1-\alpha}$ for the particular sample size n.

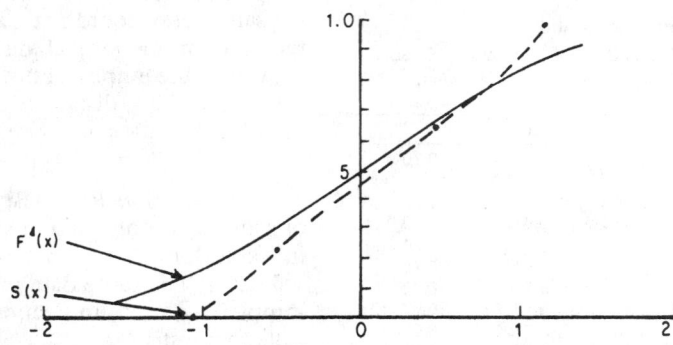

NOTE—Because T_2 equals 0.15, and is less than 0.19, the null hypothesis is accepted. This means that the normal distribution is a reasonable approximation of the true unknown distribution.

FIG. X1.1 Graphs of $F'(x)$ and $S(x)$ from Example

Standard Practice for
Generation of Environmental Data Related to Waste Management Activities: Quality Assurance and Quality Control Planning and Implementation[1]

This standard is issued under the fixed designation D 5283; the number immediately following the designation indicates the year of original adoption or, in the case of revision, the year of last revision. A number in parentheses indicates the year of last reapproval. A superscript epsilon (ε) indicates an editorial change since the last revision or reapproval.

1. Scope

1.1 Environmental data generation efforts are composed of four parts: (*1*) establishment of data quality objectives (DQOs); (*2*) design of field measurement and sampling strategies and specification of laboratory analyses and data acceptance criteria; (*3*) implementation of sampling and analysis strategies; and (*4*) data quality assessment. This practice addresses the planning and implementation of the sampling and analysis aspects of environmental data generation activities (Parts (*1*) and (*2*) above).

1.2 This practice defines the criteria that must be considered to assure the quality of the field and analytical aspects of environmental data generation activities. Environmental data include, but are not limited to, the results from analyses of samples of air, soil, water, biota, waste, or any combinations thereof.

1.3 DQOs should be adopted prior to application of this practice. Data generated in accordance with this practice are subject to a final assessment to determine whether the DQOs were met. For example, many screening activities do not require all of the mandatory quality assurance (QA) and quality control (QC) steps found in this practice to generate data adequate to meet the project DQOs. The extent to which all of the requirements must be met remains a matter of technical judgement as it relates to the established DQOs.

1.4 This practice presents extensive management requirements designed to ensure high-quality environmental data. The words "must," "shall," "may," and "should" have been selected carefully to reflect the importance placed on many of the statements made in this practice.

1.5 *This standard does not purport to address all of the safety problems, if any, associated with its use. It is the responsibility of the user of this standard to establish appropriate safety and health practices and determine the applicability of regulatory limitations prior to use.*

NOTE 1—A complete table of contents of this practice is given in Appendix X1.

2. Referenced Documents

2.1 *ASTM Standards:*
D 1129 Definitions of Terms Relating to Water[2]

E 1187 Terminology Relating to Laboratory Accreditation[3]

2.2 *U.S. Environmental Protection Agency Documents:*[4]
SW-846, *Test Methods for Evaluating Solid Waste*, Vol 1, Third Edition (NTIS No. PB88239223/LL), November 1986

QAMS-005/80 (NTIS No. PB83170514/LL), *Interim Guidelines and Specifications for Preparing Quality Assurance Project Plans*, Office of Monitoring Systems and Quality Assurance, December 29, 1980

EPA/QAMS, *Development of Data Quality Objectives, Description of Stages I and II*, July 16, 1986

QAMS 004/80 (NTIS No. PB83219667/LL), *Guidelines and Specifications for Preparing Quality Assurance Program Plans*, Office of Monitoring Systems and Quality Assurance, September 20, 1980

2.3 Other documents related to the subject matter of this practice are cited in Appendix X2. This list is not intended to be comprehensive.

3. Terminology

3.1 *Definitions*—The terms most applicable to this practice have been defined in Definitions D 1129 and E 1187.

3.2 *Descriptions of Terms Specific to This Standard:*

3.2.1 *background sample*—a sample taken from a location on or proximate to the site of interest and used to document baseline or historical information.

3.2.2 *collocated samples*—independent samples collected as close as possible to the same point in space and time and intended to be identical.

3.2.3 *data quality objectives* (DQOs)—statements on the level of uncertainty that a decision maker is willing to accept in the results derived from environmental data (see EPA/QAMS, July 16, 1986).

3.2.4 *environmental data generation activity*—tasks associated with the production of environmental data, including planning, sampling, and analysis.

3.2.5 *equipment rinsate (equipment blank)*—a sample of analyte-free media that has been used to rinse the sampling equipment. This blank is collected after the completion of decontamination and prior to sampling and is useful for documenting the adequate decontamination of sampling equipment.

3.2.6 *field blank*—a sample of analyte-free media similar

---

[1] This practice is under the jurisdiction of ASTM Committee D-34 on Waste Management and is the direct responsibility of Subcommittee D34.02 on Physical and Chemical Characterization.
Current edition approved Aug. 15, 1992. Published February 1993.
[2] *Annual Book of ASTM Standards*, Vol 11.01.

[3] *Annual Book of ASTM Standards*, Vol 14.02.
[4] Available from Superintendent of Documents, Government Printing Office, Washington, DC 20402.

to the sample matrix that is transferred from one vessel to another or exposed to the sampling environment at the sampling site. This blank is preserved and processed in the same manner as the associated samples and is used to document contamination in the sampling and analysis process.

3.2.7 *field duplicates*—collocated samples that are analyzed independently and are useful in documenting the precision of the sampling and analytical process.

3.2.8 *laboratory control sample*—a known matrix spiked with compound(s) representative of the target analytes and used to document laboratory performance.

3.2.9 *material blank*—a sample composed of construction materials such as those used in well installation, well development, pump and flow testing, and slurry wall construction. Examples of these materials are bentonite, sand, drilling fluids, and source and purge water. This blank documents the contamination resulting from use of the construction materials.

3.2.10 *matrix duplicate*—an intralaboratory split sample used to document the precision of a procedure in a given sample matrix.

3.2.11 *matrix spike*—an aliquot of sample spiked with a known concentration of target analyte(s) and used to document the bias of an analytical process in a given sample matrix. The spiking occurs prior to sample preparation and analysis.

3.2.12 *matrix spike duplicates*—intralaboratory split samples spiked with identical concentrations of target analyte(s) and used to document the precision and bias of a procedure in a given sample matrix. The spiking occurs prior to sample preparation and analysis.

3.2.13 *method blank*—an analyte-free media, to which all reagents are added in the same volumes or proportions used in sample processing. The method blank must be carried through the complete sample preparation and analytical procedure and is used to document contamination resulting from the analytical process.

3.2.14 *project*—single or multiple data collection activities that are related through the same planning sequence.

3.2.15 *project planning documents*—all documents related to the definition of the environmental data collection activities associated with a project.

3.2.16 *quality assurance program plan (QAPP)*—an orderly assemblage of management policies, objectives, principles, and general procedures by which an organization involved in environmental data generation activities outlines how it intends to produce data of known quality.

3.2.17 *quality assurance project plan (QAPjP)*—an orderly assemblage of detailed procedures designed to produce data of sufficient quality to meet the DQOs for a specific data collection activity.

3.2.18 *reference material*—a material containing known quantities of target analytes in either solution or a homogeneous matrix and used to document the bias of the analytical process.

3.2.19 *split samples*—aliquots of sample taken from the same container and analyzed independently. These are usually taken after mixing or compositing and are used to document intra- or interlaboratory precision.

3.2.20 *standard addition*—the practice of adding a known amount of an analyte to a sample immediately prior to analysis, typically used to evaluate matrix effects.

3.2.21 *standard operating procedures (SOPs)*—the established written procedures of a given organization. Special project plans may require procedures different from the established SOPs.

3.2.22 *surrogate*—an organic compound that is similar to the target analyte(s) in chemical composition and behavior in the analytical process, but is not normally found in environmental samples.

3.2.23 *trip blank*—a sample of analyte-free media taken from the laboratory (or appropriate point of origin) to the sampling site and returned to the laboratory unopened. A trip blank is used to document the contamination attributable to shipping and field handling procedures and is also useful in documenting the contamination of volatile organics samples.

4. Summary of Practice

4.1 This practice describes the criteria and activities for field and laboratory organizations involved in generating environmental data in terms of human and physical resources, QA and QC procedures, and documentation requirements depending on the DQOs.

5. Significance and Use

5.1 Environmental data are often required for making regulatory and programmatic decisions. These data must be of known quality commensurate with their intended use.

5.2 Data generation efforts involve the following: establishment of the DQOs; design of the project plan to meet the DQOs; implementation of the project plan; and assessment of the data to determine whether the DQOs have been met.

5.3 Certain minimal criteria must be met by the field and laboratory organizations generating environmental data. Additional activities may be required based on the DQOs of the data collection effort.

5.4 This practice defines the criteria for field and laboratory organizations generating environmental data and identifies some other activities that may be required based on the DQOs.

5.5 This practice emphasizes the importance of communication among those involved in establishing DQOs, planning and implementing the sampling and analysis aspects of environmental data generation activities, and assessing data quality.

5.6 Environmental field operations are discussed in Section 7, and environmental laboratory operations are discussed in Section 8.

6. Project Specification

6.1 Project activities should be defined prior to the start of any field or laboratory activities. At a minimum, project specifications should address the following topics:

6.2 *Data Quality Objectives*—DQOs for the data generation activity should be defined prior to the initiation of field and laboratory work. It is desirable that the field and laboratory organizations be aware of the DQOs so that the personnel conducting the work are able to make informed decisions during the course of the project.

6.3 *Project Plan*—The project should be designed to meet

the DQOs, and the project plan should define the following:

6.3.1 *Project Objectives*—Project objectives provide background information, state reasons for the data collection effort, identify any regulatory programs governing data collection, define specific objectives for each sampling location, and describe the intended uses for the data.

6.3.2 *Project Management*—A person(s) shall be designated as having responsibility and authority for the following: (*1*) developing project documents that implement the DQOs; (*2*) selecting field and laboratory organizations to conduct the work; (*3*) coordinating communication among the field and laboratory organizations and government agencies, as required; and (*4*) reviewing and assessing the final data.

6.3.3 *Sampling Requirements*—Sampling locations, equipment, and procedures and sample preservation and handling requirements shall be specified.

6.3.4 *Analytical Requirements*—The analytical procedures, analyte list, required detection limits, and required precision and bias values shall be specified. Regulatory requirements and DQOs shall be considered when developing the specifications.

NOTE 2—This does not imply that the specified analytical requirements can be met.

6.3.5 *Quality Assurance and Quality Control Requirements*—The QA and QC requirements shall address both field and laboratory activities. The means for controlling false positives and false negatives shall be specified. Standard practices for field and laboratory operations as described in Sections 7 and 8 of this practice shall be required.

6.3.5.1 *Field Quality Control*—The types and frequency of field QC samples to be collected, including field blanks, trip blanks, equipment rinsates, field duplicates, background samples, reference materials, material blanks, and split samples, shall be specified. Control parameters for field activities shall also be described (see 7.6.4).

6.3.5.2 *Laboratory Quality Control*—The types and frequency of use of laboratory QC samples, such as laboratory control samples, laboratory blanks, matrix spikes, matrix duplicates, and matrix spike duplicates, shall be specified. Any specific performance criteria shall be specified. Data validation criteria shall be defined.

6.4 *Project Documentation*—All documents required for planning, implementing, and evaluating the data collection effort shall be specified. These may include, although not limited to, a statement of work, technical and cost proposals, work plan, sampling and analysis plan, quality assurance project plan (QAPjP), health and safety plan, community relations plan, documents required by regulatory agencies, requirements for raw field and analytical records, technical reports assessing the environmental data, and records retention policy. Planning documents shall specify the required level of document control and identify the personnel having access. Document formats that may be required to ensure that all data needs are satisfied shall be specified. In addition, a project schedule that identifies critical milestones and completion dates should be available.

7. Standard Practices for Environmental Field Operations

7.1 *Purpose*—The field organization must conduct its operations in such a manner as to provide reliable information that meets the DQOs. To achieve this goal, certain minimum policies and procedures must be implemented in order to meet the DQOs.

7.2 *Organization*—The field organization shall be structured such that each member of the organization has a clear understanding of his or her duties and responsibilities and the relationship of those responsibilities to the total effort. The organizational structure, functional responsibilities, levels of authority, job descriptions, and lines of communication for activities shall be established and documented. One person may cover more than one organizational function.

7.2.1 *Management*—The management of the field organization is responsible for establishing organizational, operational, health and safety, and QA policies. Management shall ensure that the following requirements are met: (*1*) the appropriate methodologies are followed, as documented in the standard operating procedures (SOPs); (*2*) personnel clearly understand their duties and responsibilities; (*3*) each staff member has access to appropriate project documents; (*4*) any deviations from the project plan are communicated to project management; and (*5*) communication occurs between the field, laboratory, and project management, as specified in the project plan. Management shall foster an attitude within the organization that emphasizes the importance of quality and supports implementation of the quality assurance program plan (QAPP).

7.2.2 *Quality Assurance Function*—The organization shall appoint a person or persons to be responsible for monitoring field operations in order to ensure that the site facilities, equipment, personnel, procedures, practices, and documentation are in conformance with the organization's QAPP and any applicable QAPjP. The QA monitoring function should be entirely separate from, and independent of, personnel engaged in the work being monitored. The QA function shall be responsible for the QA review, as per 7.7.

7.2.3 *Personnel*—It is the responsibility of the organization to establish personnel qualifications and training requirements for all positions. Each member of the organization shall possess the education, training, technical knowledge and experience, or a combination thereof, to enable that individual to perform his or her assigned functions. Personnel qualifications shall be documented in terms of education, experience, and training. Training shall be provided for all staff members, as necessary, so that they can perform their functions properly.

7.2.4 *Subcontractors*—The use of subcontractors shall not jeopardize data quality. Therefore, subcontractors shall comply with the requirements of Sections 7 and 8, as appropriate to the specific task(s) they are performing.

7.3 *Field Logistics:*

7.3.1 *General*—Sampling site facilities shall be examined prior to the start of work in order to ensure that all required items are available. The actual sampling area shall be examined to ensure that trucks, drilling equipment, and personnel have access to the site. Security, health and safety, and protection of the environment shall be controlled at the site support areas and sampling site.

7.3.2 *Field Measurements*—Project planning documents shall both address the type of field measurements to be performed and plan for the appropriate area to perform the work. Planning documents shall address ventilation, protection from extreme weather and temperatures, access to stable power, and provisions for water and gases of required purity. Plans shall be made to identify and supply applicable safety equipment, as specified in the project health and safety plan.

7.3.3 *Sample Handling, Shipping, and Storage Area*—The determination of whether sample shipping is necessary shall be made during project planning. This need is established by evaluating the analyses required, holding times, and location of the site and laboratory. Shipping or transporting of the samples to a laboratory shall be completed in a timely manner, ensuring that the laboratory is allowed sufficient time to perform its analysis within any required holding times.

7.3.3.1 Samples shall be packaged, labeled, and documented in an area that minimizes sample contamination and provides for safe storage. The level of custody and whether sample storage is required shall be outlined in the planning documents.

7.3.4 *Chemical Storage*—Safe storage areas for solvents, reagents, standards, and reference materials shall be adequate to preserve their identity, concentration, purity, and stability prior to use.

7.3.5 *Decontamination*—Decontamination of sampling equipment may be performed at the location at which sampling occurs, prior to transfer to the sampling site, or in designated areas near the sampling site. Project documentation shall specify where this work will be performed and how it will be accomplished. If decontamination is to be conducted at the site, water and solvents of appropriate purity shall be available. The method of accomplishing decontamination and the materials, solvents, and water purity shall be specified in planning documents or standard operating procedures (SOPs).

7.3.6 *Waste Storage Area*—Waste materials may be generated during both the sampling process and on-site or *in situ* analysis. Planning documents and SOPs shall outline the method for storage and disposal of these waste materials. Adequate facilities shall be provided for the collection and storage of all wastes. These facilities shall be operated so as to minimize environmental contamination. Waste storage and disposal facilities shall comply with applicable federal, state, and local regulations.

7.3.7 *Data Storage Area*—Planning documents shall specify the location of long- and short-term storage for field records. The storage environment shall be maintained to ensure the integrity of the data. Access shall be limited to authorized personnel only.

7.4 *Equipment and Instrumentation:*

7.4.1 *Equipment and Instrumentation*—The equipment, instrumentation, and supplies required at the sampling site shall be appropriate to accomplish the activities planned. The equipment and instrumentation shall meet the requirements of pertinent specifications, methods, and SOPs. Before the field staff arrives at the site, a list of required items shall be prepared and checked to ensure availability at the site.

7.4.2 *Maintenance and Calibration of Equipment and Instrumentation*—An SOP or operation and maintenance manual shall set forth the methods, materials, and schedules to be used in the routine inspection, cleaning, maintenance, testing, and calibration of the equipment and instrumentation

used in performing geophysical, analytical, or *in situ* measurements. For common malfunctions, procedures or manuals may outline typical problems, methods of trouble-shooting, and possible corrective actions to be taken. Procedures shall designate a person(s) or organizations responsible for maintenance and calibration. Records of all inspections, maintenance, repairs, testing, and calibration shall be maintained.

7.5 *Standard Operating Procedures*—The organization shall have written SOPs for all procedures performed routinely that affect data quality. SOPs shall be available for the following areas and shall contain the information described:

7.5.1 *Sample Management*—These SOPs describe the numbering and labeling system, chain-of-custody procedures, and tracking of samples from collection to shipment or relinquishment to the laboratory. Sample management also includes the specification of holding times, volume of sample required by the laboratory, preservatives, and shipping requirements.

7.5.2 *Reagent and Standard Preparation*—These SOPs describe the procedures used to prepare standards and reagents. Information should be included concerning the specific grades of materials used in reagent and standard preparation, appropriate glassware and containers for preparation and storage, labeling and record keeping for stocks and dilutions, and safety precautions to be taken.

7.5.3 *Decontamination*—These SOPs describe the procedures used to clean field equipment before and during the sample collection process. The SOPs should include the cleaning materials used, the order of washing and rinsing with the cleaning materials, requirements for protecting or covering cleaned equipment, procedures for disposing of cleaning materials, and safety considerations.

7.5.4 *Sample Collection Procedures*—SOPs for sample collection procedures shall describe how the procedures are actually performed in the field and shall not be a simple reference to standard test methods, unless a procedure is performed exactly as described in the published test method. If possible, industry-recognized test methods from source documents published by the U.S. Environmental Protection Agency, ASTM, U.S. Department of the Interior, National Water Well Association, American Petroleum Institute, or other recognized organizations should be used. The SOP for sample collection procedures should include the following information:

7.5.4.1 Applicability of the procedure,

7.5.4.2 Equipment and reagents required,

7.5.4.3 Detailed description of the procedures to be followed in collecting the samples,

7.5.4.4 Common problems encountered,

7.5.4.5 Precautions to be taken, and

7.5.4.6 Health and safety considerations.

7.5.5 *Equipment Calibration and Maintenance*—These SOPs describe the procedures used to ensure that field equipment and instrumentation are in working order. The SOPs describe calibration and maintenance procedures and schedules, maintenance logs, service contracts or service arrangements for equipment, and spare parts available in-house. The calibration and maintenance of field equipment and instrumentation should generally be in accordance with manufacturers' specifications and shall be documented.

7.5.6 *Field Measurements*—These SOPs describe all

methods used in the field to determine a chemical or physical parameter. The SOPs shall address criteria from Section 8, as appropriate.

7.5.7 *Corrective Action*—These SOPs describe procedures used to identify and correct deficiencies in the sample collection process. These should include specific steps to take in correcting deficiencies, such as performing additional decontamination of equipment, resampling, or additional training of field personnel in methods procedures. The SOP shall specify that each corrective action must be documented with a description of the deficiency, the corrective action taken, and the person(s) responsible for implementing the corrective action.

7.5.8 *Data Reduction and Validation*—These SOPs describe procedures used to compute the results from field measurements and to review and validate these data. They should include all formulas used to calculate the results and procedures used to verify independently that the field measurement results are correct.

7.5.9 *Reporting*—These SOPs describe the process for reporting the results of field activities.

7.5.10 *Records Management*—These SOPs describe the procedures for generating, controlling, and archiving field records. The SOPs should describe the responsibilities for record generation and control and the policies for record retention, including type, time, security, and retrieval and disposal authorities. Records should include project-specific and field operations records.

7.5.10.1 Project-specific records relate to field work performed for a group of samples. Project records may include correspondence, chain-of-custody, field notes, all reports issued as a result of the work, project planning documents, and procedural SOPs used.

7.5.10.2 Field operations records document overall field operations. These records may include equipment performance and maintenance logs, personnel files, general field SOPs, and corrective action reports.

7.5.11 *Waste Disposal*—These SOPs describe policies and procedures for the disposal of waste materials resulting from field operations. The disposal of all wastes must conform to federal, state, and local regulations, including those associated with the Resource Conservation and Recovery Act, Superfund Act Reauthorization and Amendments, Department of Transportation, and Occupational Safety and Health Administration.

7.5.12 *Health and Safety*—These SOPs describe policies and procedures designed both to provide a safe and healthy working environment for field personnel and to comply with federal and state regulations.

7.6 *Field Quality Assurance and Quality Control Requirements:*

7.6.1 *Quality Assurance Program Plan*—The field organization shall have a written QAPP that describes the organization's QA policy. The plan shall specify the responsibilities of the field management and field staff and the QA function in the areas of QA and QC, and it shall also describe the QC procedures followed by the organization (see EPA QAMS-004/80 for an example).

7.6.2 *Quality Assurance Project Plan*—Some projects, particularly those that are large or complex, require a QAPjP. The QAPjP details the QA and QC goals and protocol for a specific data collection activity to ensure that the data generated by sampling and analysis activities are of quality commensurate with their intended use. QAPjP elements should include a discussion of the quality objectives of the project, identification of those involved in the data collection and their responsibilities and authorities, enumeration of the QC procedures to be followed, and reference to the specific SOPs that will be followed for all aspects of the project. Elements may be added or removed, as required by the project or the end-user of the data (see EPA QAMS-005/80 for an example).

7.6.3 *Control Samples*—Control samples are QC samples that are introduced into a process to monitor the performance of the system. Control samples, which may include blanks, duplicates, spikes, analytical standards, and reference materials, can be used in different phases of the overall process, beginning with sampling and continuing through transportation, storage, and analysis. The types of control samples used, and the frequency of usage, are dependent on the DQOs of the data collection effort and must be specified for each project.

7.6.4 *Procedures for Establishing Acceptance Criteria*—Procedures shall be in place for establishing acceptance criteria for field activities, as required in the project planning documents. Acceptance criteria may be qualitative or quantitative. Field events or data that fall outside of the established acceptance criteria may indicate a problem with the sampling process that must be investigated.

7.6.5 *Deviations*—Any activity not performed in accordance with the SOPs or project planning documents is considered a deviation from the plan. Deviations from the plan may or may not affect data quality. All deviations from the plan shall be documented as to the extent of, and reason for, the deviation.

7.6.6 *Corrective Action*—Errors, deficiencies, deviations, or field events or data that fall outside the established acceptance criteria require investigation. Corrective action may be necessary to resolve the problem and restore proper functioning to the system in some instances. Investigation of the problem and any subsequent corrective action taken shall be documented.

7.6.7 *Data Handling Procedures:*

7.6.7.1 *Data Reduction*—All field measurement data are reduced according to protocol described in the appropriate SOP. Computer programs used for data reduction shall be validated before use and verified on a regular basis. All information used in the calculations shall be recorded to enable reconstruction of the final result at a later date.

7.6.7.2 *Data Review*—All data are reviewed according to SOPs to ensure that the calculations are correct and to detect transcription errors. Spot checks are performed on computer calculations to verify program validity.

7.6.7.3 *Data Reporting*—Data are reported in accordance with the requirements of the end-user.

7.7 *Quality Assurance Review:*

7.7.1 *General*—The QA review consists of internal and external assessments to ensure that both QA and QC procedures are in use and field staff conform to these procedures. Planning documents shall specify the requirements for internal, external, and on-site assessment. These documents shall specify the frequency and documentation of these assessments.

7.7.1.1 *Internal Assessment*—Personnel responsible for performing field activities are responsible for continually monitoring individual compliance with the QA and QC programs and planning documents. A QA officer or an appropriate management designee shall review the field results and findings for compliance with the QA and QC programs and planning documents. The results of this internal assessment should be reported to management with requirements for a plan to correct the observed deficiencies.

7.7.1.2 *External Assessment*—The field staff may be reviewed by personnel external to the organization. The results of the external assessment should be submitted to management with requirements for a plan to correct the observed deficiencies.

7.7.1.3 *On-Site Evaluation*—On-site evaluations may be conducted as part of both internal and external assessments. On-site evaluations may include, but are not limited to, a complete review of the facilities, staff, training, instrumentation, SOPs, methods, field analysis, sample collection, QA and QC policies, and procedures related to the generation of environmental data. Records of each evaluation shall be maintained until superseded or according to policy. These records should include the date of the evaluation, area or site, areas reviewed, person performing the evaluation, findings and problems, actions recommended and taken to resolve the problems, and scheduled date for re-inspection. Any problems identified that are likely to affect data integrity shall be brought to the attention of management immediately.

7.7.2 *Evaluation of Field Records*—The review of field records shall be conducted by one or more persons knowledgeable in the field activities, evaluating the following subjects at a minimum:

7.7.2.1 *Completeness of Field Reports*—This review ensures that all requirements for field activities in the planning documents have been fulfilled, that complete records exist for each field activity, and that the procedures specified in the planning documents have been implemented. The emphasis on field documentation will help assure sample integrity and sufficient technical information to recreate each field event. The results of this completeness check shall be documented, and environmental data affected by incomplete records shall be identified.

7.7.2.2 *Identification of Invalid Samples*—This review involves interpretation and evaluation of the field records to detect problems affecting the representativeness of environmental samples. Examples of items that could indicate invalid samples include improper well development, improperly screened wells, instability of pH or conductivity, and collection of volatiles near combustion engines. The field records shall be evaluated against planning documents and SOPs. The reviewer shall document the sample validity and identify the environmental data associated with poor or incorrect field work.

7.7.2.3 *Correlation of Field Test Data*—The results of field measurements obtained by more than one method shall be compared. For example, surface geophysics may be surveyed using both ground penetrating radar and a resistivity survey.

7.7.2.4 *Identification of Anomalous Field Test Data*—Anomalous field test data should be identified. For example, a water temperature for one well that is five degrees higher than any other well temperature in the same aquifer should be noted. The impact of anomalous field measurement results on the associated environmental data shall be evaluated.

7.7.2.5 *Validation of Field Analysis*—All data from field analysis that are generated *in situ* or from a mobile laboratory shall be validated per 8.7.2. The results of the validation shall be reported. The report shall discuss whether the QC checks meet the acceptance criteria and whether corrective actions were taken for any analysis performed when acceptance criteria were not met.

7.7.3 *Quality Assurance Reports to Management*—The QA program shall provide for the periodic reporting of pertinent QA and QC information to management to allow assessment of the overall effectiveness of the QA program. There are three major types of QA reports to management:

7.7.3.1 *Report on Measurement Quality Indicators*—This report shall include the assessment of QC data (such as that generated per 7.6.3) gathered over the period, the frequency of repeating work due to unacceptable performance, and corrective action taken.

7.7.3.2 *Report on Quality Assurance Assessments*—This report shall be submitted immediately following any internal or external on-site evaluations or upon receipt of the results of any performance evaluation studies. The report shall include the results of the assessment and the plan for correcting identified deficiencies.

7.7.3.3 *Report on Key Quality Assurance Activities During the Period*—A report shall be delivered to management summarizing key QA activities during the period. The report shall stress measures that are being taken to improve data quality and shall include a summary of the significant quality problems observed and corrective actions taken. The report shall also include a summary of involvements in resolution of quality issues with clients or agencies, QA organizational changes, and notice of the distribution of any revised documents controlled by the QA function.

7.8 *Field Records*—Records provide direct evidence and support for the necessary technical interpretations, judgments, and discussions concerning project activities. These records, particularly those that are anticipated for use as evidentiary data, must directly support current or ongoing technical studies and activities and must provide the historical evidence necessary for later reviews and analyses. Records shall be legible, identifiable, and retrievable and protected from damage, deterioration, or loss. Field records generally consist of bound field notebooks with pre-numbered pages, sample collection forms, personnel qualification and training forms, sample location maps, equipment maintenance and calibration forms, chain-of-custody forms, sample analysis request forms, and field change request forms. All records shall be completed with black, waterproof ink. Procedures for reviewing, approving, and revising field records must be defined clearly, with the lines of authority included. At a minimum, all documentation errors shall be corrected by drawing a single line through the error and initialing by the responsible individual, along with the date of change. The correction is written adjacent to the error.

Deviations from field SOPs shall be documented.

7.8.1 *Personnel Training and Qualification Records*—It is the responsibility of the organization to establish personnel qualifications and training requirements. Each staff member shall possess the education, training, technical knowledge, and experience, or a combination thereof, to enable that individual to perform his or her assigned functions. Personnel qualifications shall be documented in terms of education, experience, and training. Training shall be provided for all staff members so that they can perform their functions properly.

7.8.2 *Standard Operating Procedures*—SOPs shall be available to those performing the task outlined. Any revisions to field SOPs shall be written and distributed to all affected individuals to ensure the implementation of changes. The areas covered by SOPs are given in 7.5.

7.8.3 *Quality Assurance Plans*—The QAPP and all applicable QAPjPs shall be on file.

7.8.4 *Equipment Maintenance*—Maintenance procedures shall be defined clearly and written for each measurement system and required support equipment. When maintenance is necessary, it shall be documented in either standard forms or in logbooks. A history of the maintenance record of each system serves as an indication of the adequacy of maintenance schedules and parts inventory.

7.8.5 *Calibration and Traceability of Standards and Reagents*—Calibration is a reproducible reference base to which all sample measurements can be correlated. A sound calibration program shall include provisions for documentation of the frequency, conditions, standards, and records reflecting the calibration history of a measurement system. The accuracy of calibration standards is an important point to consider because all data will be in reference to the standards used. A program for verifying and documenting the accuracy of all working standards against primary grade standards shall be followed routinely.

7.8.6 *Sample Collection and Tracking Records*—To ensure maximum utility of the sampling effort and resulting data, documentation of the sampling protocol, as performed in the field, is essential. At a minimum, sample collection records shall contain the persons conducting the activity, sample number, sample location, equipment used, climatic conditions, documentation of adherence to protocol, and unusual observations. The actual sample collection record is usually one of the following: a bound field notebook with prenumbered pages, a pre-printed form, or digitized information on a computer tape or disc.

7.8.6.1 Sample tracking records involving the possession of samples from the time at which they are obtained until they are relinquished shall be documented with the following minimum information: (*1*) project name; (*2*) signatures of the samplers; (*3*) sample number, date and time of collection, and grab or composite sample designation; (*4*) signatures of the individuals involved in sample transfer; and (*5*) the air bill or other shipping number, if applicable.

7.8.7 *Maps and Drawings*—Project planning documents and reports often contain maps. The maps are used to document the location of sample collection points and monitoring wells, and as a means of presenting environmental data. Information used to prepare maps and drawings is normally obtained through field surveys, property surveys,

surveys of monitoring wells, aerial photography, or photogrammetric mapping. The final, approved maps shall have a revision number and date and shall be subject to the same controls as other project records.

7.8.8 *Results from Control Samples*—Documentation for the collection of QC samples, such as field, trip, and equipment rinsate blanks, duplicate samples, spikes, and reference materials, shall be maintained.

7.8.9 *Correspondence*—Project correspondence can provide evidence supporting technical interpretations. Correspondence pertinent to the project shall be kept and placed in the project files.

7.8.10 *Deviations*—Field changes and deviations from the planning documents shall be reviewed and approved by either the authorized personnel who performed the original technical review or their designees. All deviations from the procedural and planning documents shall be recorded in the site log.

7.8.11 *Final Report*—The final report shall summarize the field activities, data, results of deviations from the planning documents, and interpretation of the data. The planning documents shall outline the items to be included in the report, which may include any special formats required, QC reporting requirements, conclusions, and recommendations.

7.9 *Documentation Storage:*

7.9.1 *Documentation Archive*—Procedures shall be established to ensure that documents required to recreate the sampling, analysis, and reporting of information are stored. These documents may include, but are not limited to, planning documents, SOPs, logbooks, field data records, sample tags and labels, chain-of-custody records, photographs, and any other information noted in 7.8.

7.9.2 *Storage Time*—The length of storage time for field records shall comply with regulatory requirements, organizational policy, or project requirements, whichever is/are more stringent.

7.9.3 *Filing System*—The control of records is essential in providing evidence of technical adequacy and quality for all project activities. These records shall be identified, retrievable, and organized to prevent loss.

7.9.4 *Personnel Authorized to Enter Archive*—Access to project files shall be controlled to restrict unauthorized personnel from having free and open access. An authorized access list shall be prepared for the project files and shall name the personnel who have unrestricted access to the files.

8. Standard Practices for Environmental Laboratory Operations

8.1 *Purpose*—Each laboratory must conduct its operations in such a way as to provide reliable information. To achieve this goal, certain minimum policies and procedures must be implemented.

8.2 *Organization*—The laboratory shall be structured such that each member of the organization has a clear understanding of his or her duties and responsibilities and the relationship of those responsibilities to the total effort. The organizational structure, functional responsibilities, levels of authority, job descriptions, and lines of communication for activities shall be established and documented. The laboratory shall also maintain a current list of accreditations from government agencies or private associations.

8.2.1 *Management*—The management of the laboratory is responsible for establishing organizational, operational, health and safety, and QA policies. These responsibilities include the following: oversight of personnel selection, development, and training; review, selection, and approval of analysis methods; and development, implementation, and maintenance of a QA program. Management shall foster an attitude within the organization that emphasizes the importance of quality and supports implementation of the QAPP.

8.2.2 *Quality Assurance Function*—The laboratory shall appoint a person or persons to be responsible for monitoring laboratory operations to ensure that the facilities, equipment, personnel, methods, practices, and documentation are in conformance with the laboratory QAPP and any applicable QAPjP(s). The QA monitoring function shall be entirely separate from, and independent of, personnel engaged in direct supervision or performance of the work being monitored. The QA function shall inspect records to ensure that analyses have been performed correctly and within the proper time frame; maintain copies of the QAPP and QAPjPs pertaining to all analyses; perform assessments of the laboratory to ensure adherence to the QAPP; periodically submit written status reports to management, noting any problems and the corrective actions taken; ensure that any deviations from the approved QAPP, QAPjP, or SOPs have been authorized and documented properly; and ensure that the results reported reflect the raw data accurately. The responsibilities of the QA function and procedures used in conducting those responsibilities shall be in writing and shall be maintained.

8.2.3 *Personnel*—It is the responsibility of the organization to establish personnel qualifications and training requirements for all positions. Each member of the organization shall possess the education, training, technical knowledge, and experience, or a combination thereof, which enables that individual to perform his or her assigned functions. Personnel qualifications shall be documented in terms of education, experience, and training. Training shall be provided for all staff members, as necessary, so that they can perform their functions properly.

8.2.4 *Subcontractors*—The use of subcontractors shall not jeopardize data quality. Subcontractors shall therefore comply with the requirements of Sections 7 and 8, as appropriate to the specific task(s) they are performing.

8.3 *Facilities:*

8.3.1 *General*—Each laboratory shall be of a size and construction suitable to facilitate proper conduct of the analyses. Adequate bench space or working area per analyst shall be provided. The space requirement per analyst depends on the equipment or apparatus being used, the number of samples the analyst is expected to handle at any one time, and the number of operations to be performed concurrently by a single analyst. The laboratory shall be well-ventilated, adequately lit, free of dust and drafts, protected from extreme temperatures, and offer access to a stable source of power. Laboratories shall be designed so that there is adequate separation of functions in order to ensure that no laboratory activity has an adverse effect on the analyses. The laboratory may require specialized facilities such as a perchloric acid hood or glovebox.

8.3.2 *Sample Handling, Receiving, and Storage Area*—As necessary to ensure safe and secure storage and prevent contamination or misidentification, there shall be adequate facilities for the receipt and storage of samples. The level of custody required and any special requirements for storage, such as refrigeration and lighting, shall be described in the planning documents.

8.3.3 *Chemical Storage*—Storage areas for reagents, solvents, standards, and reference materials shall be adequately safe to preserve their identity, concentration, purity, and stability.

8.3.4 *Laboratory Operations Area*—Separate spaces for laboratory operations and appropriate ancillary support shall be provided, as necessary, for the performance of routine and specialized procedures.

8.3.5 *Waste Storage Area*—Adequate facilities shall be provided for the collection and storage of all wastes, and these facilities shall be operated so as to minimize environmental contamination. Waste storage and disposal facilities shall comply with applicable federal, state, and local regulations.

8.3.6 *Data Storage Area*—Space shall be provided for the storage and retrieval of all documents, as specified in 8.8. The storage environment shall be maintained to assure the integrity of the materials stored. Access shall be limited to authorized personnel only.

8.4 *Equipment and Instrumentation:*

8.4.1 *Equipment and Instrumentation*—Equipment and instrumentation shall meet the requirements and specifications of the specific test methods and other SOPs. The laboratory shall maintain an equipment and instrument description list that includes the manufacturer, model number, year of purchase, accessories, and any modifications, updates, or upgrades that have been made.

8.4.2 *Maintenance and Calibration of Equipment and Instrumentation*—Equipment and instrumentation shall be adequately inspected, cleaned, maintained, tested, and calibrated, as required in the SOP or operations manual. SOPs or manuals shall specify the identification and repair of common maintenance problems. Procedures shall designate a person(s) or organizations responsible for maintenance. Records of all inspections, maintenance, repairs, testing, and calibration shall be maintained.

8.5 *Standard Operating Procedures*—The laboratory should have written SOPs for all laboratory functions that affect data quality. Procedures and methods shall be performed in the laboratory as described in the SOPs. Any modification of an SOP made during a data collection activity must be documented. SOPs shall be available for the following areas and shall, at a minimum, contain the information described:

8.5.1 *Sample Management*—These SOPs describe the receipt, handling, and storage of samples.

8.5.1.1 *Sample Receipt and Handling*—These SOPs describe precautions to be used in opening sample shipment containers, as well as procedures used to perform the following tasks: verify that chain-of-custody has been maintained, examine samples for damage, check for proper preservatives and temperature, assign the testing program, and log samples into the laboratory sample streams.

8.5.1.2 *Sample Scheduling*—These SOPs describe procedures and criteria used for scheduling work in the laboratory, including procedures used to ensure that holding time requirements are met.

8.5.1.3 *Sample Storage*—These SOPs describe the storage conditions for all samples and procedures used both to verify and document daily storage temperature, and to ensure that custody of the samples is maintained while in the laboratory.

8.5.2 *Reagent and Standard Preparation*—These SOPs detail procedures used to prepare standards and reagent mixtures. In addition, these SOPs shall specify requirements concerning the following: purity of materials used, including water; appropriate glassware and containers; record keeping and labeling, including dating; procedures used to verify concentration, purity, and stability; and safety precautions necessary to meet the requirements of the DQOs or test methods.

8.5.3 *General Laboratory Techniques*—These SOPs detail all of the essentials of laboratory operations not addressed in other SOPs. These techniques include, but are not limited to, glassware cleaning procedures, operation of analytical balances, pipetting techniques, and use of volumetric glassware.

8.5.4 *Analytical Methods*—SOPs for analytical methods for sample analysis shall be a description of how the analysis is actually performed in the laboratory and not a simple reference to standard test methods, unless the analysis is performed exactly as described in the published test method. If possible, industry-recognized test methods from source documents published by the U.S. Environmental Protection Agency, American Public Health Association, ASTM, the National Institute for Occupational Safety and Health, or other recognized organizations should be used. The SOP for analytical methods should include the following:

8.5.4.1 Sample preparation and analysis procedures, including the applicable holding time, extraction, digestion, or preparation steps, as appropriate to the method; procedures for determining the appropriate dilution to analyze; and any other information required to perform the analysis accurately and consistently.

8.5.4.2 Instrument standardization, including the concentration and frequency of analysis of calibration standards, linear range of the method, and calibration acceptance criteria.

8.5.4.3 Raw data recording requirements and documentation, including the sample identification number, analyst, data verification analyst, date of analysis and verification, and computational method(s).

8.5.4.4 Detection and reporting limits for all analytes in the method.

8.5.4.5 Reference to the applicable QC SOPs and any specific exceptions or additions.

8.5.5 *Equipment Calibration and Maintenance*—These SOPs describe procedures used to assure or verify that the laboratory equipment and instrumentation are in working order. The SOPs describe calibration and maintenance procedures and schedules, maintenance logs, service contracts or service arrangements for all equipment, and spare parts available in-house. Calibration and maintenance of the laboratory equipment and instrumentation shall be in accordance with manufacturers' specifications and shall be documented.

8.5.6 *Quality Control Data*—These SOPs detail the type,

purpose, and frequency of QC samples analyzed in the laboratory and establish acceptance criteria. They should include information on applicability of the QC sample to the analytical process, statistical treatment of the data, and responsibility of laboratory staff and management in generating and using the data.

8.5.7 *Corrective Action*—These SOPs describe procedures used to identify and correct deficiencies in the analytical process. These include specific steps to take in correcting deficiencies, such as the preparation of new standards and reagents, recalibration and restandardization of equipment, reanalysis of samples, or additional training of laboratory personnel in methods and procedures. The SOP shall specify that each corrective action must be documented with a description of the deficiency, corrective action taken, and person(s) responsible for implementing the corrective action.

8.5.8 *Data Reduction and Validation*—These SOPs describe the procedures used to review and validate the data. They should include procedures for computing and interpreting the results from QC samples and procedures used to verify independently that the analytical results are correct. In addition, routine procedures used to monitor precision and bias, including evaluations of reagent, field, and trip blanks, calibration standards, control samples, duplicate and matrix spike samples, and surrogate recovery should be detailed in an SOP.

8.5.9 *Reporting*—These SOPs describe the process for reporting the analytical results.

8.5.10 *Records Management*—These SOPs describe the procedures for generating, controlling, and archiving laboratory records. The SOPs should detail the responsibilities for record generation and control and the policies for record retention, including type, time, security, and retrieval and disposal authorities. Records shall include project-specific and laboratory operations records.

8.5.10.1 Project-specific records related to analyses performed for a group of samples shall be maintained. These records may include an index of documents, correspondence, chain-of-custody records, request for analysis, calibration records, raw and finished analytical and QC data, data reports, and project planning documents.

8.5.10.2 Laboratory operations records, which document the overall laboratory operation, shall be maintained. These records may include the following: laboratory notebooks, instrument performance and maintenance logs in bound notebooks with prenumbered pages; laboratory benchsheets; software documentation; control charts; reference material certification; personnel files; laboratory SOPs; and corrective action reports.

8.5.11 *Waste Disposal*—These SOPs describe policies and procedures for the disposal of chemicals, including standard and reagent solutions, process waste, and samples. The disposal of these materials shall conform to federal, state, and local regulations, including those associated with the Resource Conservation and Recovery Act, Superfund Act Reauthorization and Amendments, Department of Transportation, and Occupational Safety and Health Administration.

8.5.12 *Health and Safety*—These SOPs describe policies and procedures designed both to provide a safe and healthy working environment for laboratory staff and to comply with federal and state regulations.

8.6 *Laboratory Quality Assurance and Quality Control Procedures:*

8.6.1 *Quality Assurance Program Plan*—The laboratory shall have a written QAPP that describes the organization's QA policy. The plan shall specify the responsibilities of the laboratory staff and management and the QA function in the areas of QA and QC and describe the QC procedures followed by the laboratory, including the following: (*1*) use of control samples; (*2*) statistical and mathematical basis for assigning warning and rejection limits; (*3*) detection of shifts, trends, or biases; (*4*) how an out-of-control condition is detected; (*5*) how control is re-established; (*6*) how out-of-control events and corrective actions are documented; (*7*) how reporting limits are established, including their dependence on serial dilutions and sample size; (*8*) how instrument calibration and maintenance logs, corrective action reports, and routine QC data summaries are maintained; and (*9*) QA assessment procedures used (see EPA QAMS-004/80 for an example).

8.6.2 *Quality Assurance Project Plan*—Some projects, particularly those that are large or complex, require a QAPjP. The QAPjP details the QA and QC goals and protocol for a specific data collection activity in order to ensure that the data generated by sampling and analysis activities are of quality commensurate with their intended use. QAPjP elements should include a discussion of the quality objectives of the project, identification of those involved in data collection and their responsibilities and authorities, enumeration of QC procedures to be followed, and reference to the specific SOPs that will be followed for all aspects of the project. Elements may be added or removed, as required by the project or the end-user of the data (see EPA QAMS-005/80 for an example).

8.6.3 *Method Proficiency*—The laboratory shall have procedures for demonstrating proficiency with each analytical method used in the laboratory. These shall include procedures for demonstrating the precision and bias of the method as performed by the laboratory and procedures for determining the method detection limit (MDL). All terminology, procedures, and frequency of determinations associated with the laboratory's establishment of one MDL and the reporting limit shall be well-defined and well-documented. Documented precision, bias, and MDL information shall be maintained for all methods performed in the laboratory.

8.6.4 *Laboratory Control Procedures*—The laboratory shall have procedures for demonstrating that it is in control within laboratory established limits or project specified limits during each data collection activity.

8.6.4.1 *Laboratory Control Samples*—The laboratory shall analyze control samples for each analytical method. A laboratory control sample consists of a control matrix spiked with analytes representative of the target analytes. Laboratory control sample(s) shall be analyzed with each batch of samples processed in order to verify the precision and bias of the analytical process. The results of the laboratory control sample(s) are compared to the control limits established for both precision and bias to determine the usability of the data (see 8.6.6). Analytical data generated with laboratory control samples that fall within the prescribed limits are judged to be generated while the laboratory was in control. Data generated with laboratory control samples that fall outside of the established control limits are judged to be generated during an "out of control" situation. These data are considered suspect and must be repeated or reported with qualifiers.

8.6.4.2 *Method Blank*—To assess contamination levels in the laboratory, a method blank shall be run with each batch of samples processed, unless it is not appropriate for the method. The laboratory shall have guidelines in place for accepting or rejecting data based on the level of blank contamination.

8.6.5 *Determination of Matrix Effects*—The laboratory shall have procedures for documenting the effect of the matrix on method performance. The type of matrix-specific information required for a specific data collection activity is dependent on the DQOs of the activity.

8.6.5.1 *Matrix Spikes*—Procedures shall be in place for determining the bias of the method in the matrix unless not appropriate for the method. These procedures should include the analysis of matrix spikes, use of surrogates for organic methods, and method of standard additions for metal and inorganic methods. The frequency of use of these techniques shall be based on the DQOs of the data collection activity.

8.6.5.2 *Matrix Duplicates and Matrix Spike Duplicates*—Procedures shall be in place for determining the precision of the method in the matrix. These procedures should include the analysis of matrix duplicates or matrix spike duplicates, or both. The frequency of use of these techniques shall be based on the DQOs of the data collection activity.

8.6.5.3 *Sample-Specific Detection Limit*—Procedures shall be in place for determining the MDL in a specific sample or group of samples. The frequency of use of these procedures shall be based on the DQOs of the data collection activity.

8.6.6 *Procedures for Establishing Control Limits:*

8.6.6.1 Procedures shall be in place for establishing and updating the control limits for analysis control. Control limits shall be established in order to evaluate laboratory precision and bias based on the analysis of control samples (see 8.6.4.1). Control limits for bias are typically based on the historical mean recovery plus or minus three standard deviation units, and control limits for precision range from zero (no difference between duplicate control samples) to the historical mean relative percent difference plus three standard deviation units.

8.6.6.2 Procedures shall be in place for monitoring historical performance, and they should include graphical (control charts) or tabular presentations of the data.

8.6.7 *Deviations*—Any activity not performed in accordance with laboratory SOPs or project planning documents is considered a deviation from the plan. Deviations from the plan may or may not affect data quality. All deviations from the plan shall be documented as to the extent of, and reason for, the deviation.

8.6.8 *Corrective Action*—Errors, deficiencies, deviations, or laboratory events or data that fall outside of the established acceptance criteria require investigation. Corrective action may be necessary in some instances to resolve the problem and to restore proper functioning to the analytical system. Investigation of the problem and any subsequent corrective action taken shall be documented.

8.6.9 *Data Handling Procedures:*

8.6.9.1 *Data Reduction*—Data resulting from the analyses of samples are reduced according to protocol described in the laboratory SOPs. Computer programs used for data reduction shall be validated before use and verified on a regular basis. All information used in the calculations shall be recorded in order to enable reconstruction of the final result at a later date. This information may include the weight or volume of sample used, percent dry weight for solids, extract volume, dilution factor used, and blank- or background-correction protocol followed.

8.6.9.2 *Data Review*—All data are reviewed by a second analyst or supervisor according to laboratory SOPs in order to ensure that calculations are correct and to detect transcription errors. Spot checks are performed on computer calculations to verify program validity. Errors detected in the review process are referred back to the analyst(s) for corrective action.

8.6.9.3 *Data Reporting*—Data are reported in accordance with the requirements of the end-user. The data report may include the following:

(*1*) Laboratory name and address;

(*2*) Sample information (including unique sample identification, sample collection date and time, date of sample receipt, and date(s) of sample preparation and analysis);

(*3*) Analytical results reported with appropriate significant figures;

(*4*) Detection limits that reflect dilutions, interferences, or correction for equivalent dry weight;

(*5*) Method reference;

(*6*) Appropriate QC results as described in 8.6.4 and 8.6.5; and

(*7*) Data qualifiers with appropriate references and narrative on the quality of the results.

8.7 *Quality Assurance Review:*

8.7.1 *General*—The QA program shall provide for routine evaluations of the effectiveness of laboratory QA and QC procedures and conformance of the laboratory to these procedures. These evaluations can be performed by persons internal or external to the organization and can involve on-site evaluations or performance evaluation studies, or both.

8.7.1.1 *Internal Assessment*—The analyst is responsible for continually monitoring individual compliance with the QA and QC programs. In addition, on some frequent basis, the QA officer, or an appropriate management designee, must review the laboratory data and operations for compliance with the QA and QC programs. The results of the internal assessment should be reported to management with requirements for a plan to correct observed deficiencies.

8.7.1.2 *External Assessment*—The laboratory should periodically be subject to evaluation by an assessor who is independent of the laboratory. This may be a representative of a government agency or other independent organization. The results of the external assessment should be submitted to management with requirements for a plan to correct the observed deficiencies.

8.7.1.3 *On-Site Evaluation*—On-site evaluations should be conducted as part of both internal and external assessments. An on-site evaluation includes a complete evaluation of the facilities, staff, instrumentation, SOPs, analytical methods, sample management procedures, and QA and QC policies and procedures as they relate to the generation of data for specific analytical and regulatory applications. Records of each evaluation shall be maintained by the laboratory. These records should include the date of the evaluation, area(s) or analyses inspected, person performing the evaluation, findings and problems, actions recommended and taken to resolve existing problems, and scheduled date for re-inspection. Any problems identified during the course of an on-site evaluation that are likely to affect data integrity shall be brought to the attention of management immediately.

8.7.1.4 *Performance Evaluation Studies*—Performance evaluation (PE) studies are used to measure the performance of the laboratory on unknown samples. PE samples are typically submitted to the laboratory as blind samples by an independent, outside source. The results are compared to predetermined acceptance limits set by the submitting agency. PE samples can also be submitted to the laboratory by the laboratory QA officer or an appropriate management designee as part of an internal assessment of laboratory performance. Records of all PE studies shall be maintained by the laboratory. Problems identified as a result of participation in PE studies shall immediately be investigated and corrected.

8.7.2 *Evaluation of Laboratory Data*—The evaluation of laboratory data shall be conducted by one or more persons knowledgeable in laboratory activities. Such evaluations often occur in the QA section of the laboratory. Data evaluation should include the following subjects:

8.7.2.1 *Completeness of Laboratory Data*—This ensures that (*1*) all samples and analyses required by the project planning documents have been processed, (*2*) complete records exist for each analysis and the associated QC samples, and (*3*) the procedures specified in project planning documents and SOPs have been implemented. The results of the completeness check should be documented.

8.7.2.2 *Evaluation of Data with Respect to Detection Limits*—Analytical results should be compared to required detection limits. Any detection limits that exceed regulatory limits or action levels, as specified in the project planning documents, should be identified.

8.7.2.3 *Evaluation of Data with Respect to Control Limits*—The results of QC and calibration check samples should be compared to control criteria. Data not within the control limits require corrective action, and reviewers should check that both corrective action reports and the results of reanalysis are available. Samples associated with out-of-control QC data should be identified in a written record or the data review, and an assessment of the utility of such analytical results should be recorded.

8.7.2.4 *Review of Holding Time Data*—Sample holding times should be compared to those required by the project planning documents, and all deviations should be noted.

8.7.2.5 *Evaluation of Performance Evaluation Results*—PE study results can be helpful in evaluating the impact of out-of-control conditions. Recurring trends or problems evident in PE studies should be documented in the review, and their effect on environmental data should be evaluated.

8.7.2.6 *Correlation of Laboratory Data*—The results of data obtained from related laboratory tests, such as purgeable organic hallides (POX) and volatile organics, shall be documented and the significance of differences discussed in the review reports.

8.7.3 *Quality Assurance Reports to Management*—The QA program shall provide for the periodic reporting of pertinent QA and MQC information to management in order to allow management to assess the overall effectiveness of the QA program. Three examples of QA reports to management are as follows:

8.7.3.1 *Report on Measurement Quality Indicators*—This report should include the assessment of QC data gathered over the period, the frequency of analyses repeated due to unacceptable QC performance, and, if possible, the reason for the unacceptable performance and corrective action taken.

8.7.3.2 *Report on Quality Assurance Assessments*—This report should be submitted immediately following any internal or external on-site evaluation or upon receipt of the results of any PE studies. The report should include the results of the assessments and the plan for correcting the identified deficiencies.

8.7.3.3 *Report on Key Quality Assurance Activities During the Period*—A report summarizing key QA activities during the period should be delivered to management. The report should stress measures that are being taken to improve data quality and include a summary of the significant quality problems observed and corrective actions taken. The report should also include a summary of any changes in certification and accreditation status; involvements in the resolution of quality issues with clients or agencies; QA organizational changes; and notice of the distribution of revised documents controlled by the QA organization (that is, SOPs, QAPP).

8.8 *Laboratory Records:*

8.8.1 All information relevant to environmental QA and QC activities relating to laboratory facilities, equipment, personnel, methods, and practices shall be documented. Copies of the SOPs, and the equipment manuals provided by the manufacturer, shall be accessible in the workplace.

8.8.2 Handwritten test data shall be recorded legibly in ink in laboratory notebooks or on designated benchsheets. Each page shall be signed and dated by the person who performed the analysis and entered the data. Corrections shall be made by drawing a single line through the information to be changed and initialling and dating the change. The reason for the change shall be indicated.

8.8.3 Strip-chart recorder printouts shall be signed by the person who performed the instrumental analysis. If corrections need to be made in computerized data, a system parallel to the corrections for handwritten data shall be in place.

8.8.4 Records of sample management shall be available to permit the re-creation of an analytical event for review in the case of an audit or investigation of a dubious result.

8.8.5 *Personnel Training and Qualification Records*—It is the responsibility of the organization to establish personnel qualifications and training requirements. Each staff member shall possess the education, training, technical knowledge, and experience, or a combination thereof, to enable that individual to perform his or her assigned functions. Personnel qualifications shall be documented in terms of education, experience, and training. Training shall be provided for all staff members so that they can perform their functions properly.

8.8.6 *Standard Operating Procedures*—SOPs shall be written and available for all methods and practices.

8.8.7 *Quality Assurance Plans*—The QAPP and all applicable QAPjPs shall be on file.

8.8.8 *Equipment Maintenance*—Maintenance and performance logs shall be kept on all equipment and instrumentation and must include the following: (*1*) name of the equipment and manufacturer; (*2*) model and serial number of the equipment; (*3*) date the equipment was placed in service; (*4*) instructions for proper maintenance procedures; (*5*) dates of the maintenance and performance checks and the names of personnel performing them; and (*6*) if nonroutine repairs were performed as a result of a malfunction, nature and cause of the malfunction, and name of the person performing the repair.

8.8.9 *Proficiency*—Proficiency information on all parameters reported shall be maintained and shall include the following: (*1*) precision; (*2*) bias; (*3*) method detection limits; (*4*) spike recovery, where applicable; (*5*) surrogate recovery, where applicable; (*6*) checks on reagent purity, where applicable; and (*7*) checks on glassware cleanliness, where applicable.

8.8.10 *Traceability*—Traceability of standards, reagents, and weights shall be documented, with the traceability to national or international sources indicated.

8.8.11 *Calibration*—Calibration and standardization data records for each instrument and method shall include the following: (*1*) data of the last calibration; (*2*) calibration history; (*3*) frequency of calibration; (*4*) outside sources of calibration, if used, for example, the manufacturer; (*5*) date of preparation, expiration date, and name of the person performing the preparation of standards; and (*6*) written procedures for instrument calibration.

8.8.12 *Sample Management*—All required records pertaining to sample management shall be maintained and updated regularly. These include chain-of-custody forms, sample receipt forms, and sample disposition records.

8.8.13 *Original Data*—The raw data and calculated results for all samples shall be maintained in laboratory notebooks and logs, benchsheets, files, and other sample tracking or data entry forms. Instrumental output shall be stored in a computer file or a hardcopy report.

8.8.14 *Quality Control Data*—The raw data and calculated results for all QC and field samples and standards shall be maintained in the manner described in 8.8.13. QC samples include, but are not limited to, control samples, method blanks, matrix spikes, and matrix spike duplicates.

8.8.15 *Correspondence*—Project correspondence can provide evidence supporting technical interpretations. Correspondence pertinent to the project shall be kept and placed in the project files.

8.8.16 *Final Report*—A copy of the final report submitted to the sponsor of the work, or other designated individual, shall be retained.

8.9 *Document Storage:*

8.9.1 A written policy concerning the minimum length of time that documentation is to be retained shall be in place and understood by the appropriate personnel, including the sponsor of the work. Deviations from this policy shall be documented and readily available to those individuals responsible for the deletion or destruction of expired documentation.

8.9.2 Documentation shall be stored securely in a facility that adequately addresses or minimizes its deterioration for the length of time it is to be retained. A system allowing for the expedient retrieval of information shall exist.

8.9.3 Access to archived information shall be controlled to restrict unauthorized personnel from having free and open access. An authorized access list shall be prepared and shall name the personnel who have unrestricted access to the archived information.

8.9.4 All accesses to archived information shall be documented. This documentation shall include the name of the individual, date, reason for accessing the data, and all changes, deletions, or withdrawals that may have occurred.

8.9.5 If a facility conducting testing or an archive contract facility goes out of business before the time period specified in the policy (8.9.1), all documentation shall be transferred in whole to the archives of the sponsor of the work.

9. Data Quality Assessment

9.1 The assessment of environmental data occurs in two phases. Field records and analytical data are reviewed during the first phase to identify whether the data are accurate and defensible. The data are interpreted in the second phase with respect to meeting DQOs.

9.2 Technical reports of environmental data collection efforts should summarize the information contained in the field records and the results of the laboratory data review, as described in 7.7.2 and 8.7.2. This information should be used to identify clearly the data that are not representative of environmental conditions or that have been generated using poor field or laboratory practices.

9.3 The combined field and laboratory data are then subject to a final assessment to determine whether the DQOs have been met. Although the data quality assessment process is beyond the scope of this practice, a general description of the items that should be considered in the assessment process is given below.

9.4 Data quality assessment typically includes, but is not limited to the following:

9.4.1 Evaluation of field duplicate results;

9.4.2 Comparison of the results of all field blanks, trip blanks, and equipment rinsates, with the full data set to provide information concerning contaminants that may have been introduced during sampling or shipping;

9.4.3 Evaluation of matrix effects to assess performance of the analytical method with respect to the sample matrix and determine whether the data have been biased high or low due to matrix effects;

9.4.4 Integration of the field and laboratory data with geological, hydrogeological and meteorological data to provide information on the extent of contamination; and

9.4.5 Comparison of precision, bias, completeness, representativeness, and defensibility of the data generated with that required to meet the DQOs.

APPENDIXES

(Nonmandatory Information)

X1. TABLE OF CONTENTS

Section

Section

X2. REFERENCE LIST

X2.1 *ASTM Standards:*

D 1192 Specification for Equipment for Sampling Water and Steam[2]

D 1357 Practice for Planning the Sampling of the Ambient Atmosphere[5]

D 2777 Practice for Determination of Precision and Bias of Applicable Methods of Committee D-19 on Water[2]

D 3370 Practices for Sampling Water[2]

D 3614 Guide for Laboratories Engaged in Sampling and Analysis of Atmospheres and Emissions[5]

D 3670 Guide for Determination of Precision and Accuracy of Methods of Committee D-22[5]

D 3694 Practices for Preparation of Sample Containers and for Preservation of Organic Constituents[6]

D 3856 Guide for Evaluating Laboratories Engaged in Sampling and Analysis of Water and Waste Water[2]

D 4210 Practice for Interlaboratory Quality Control Procedures and a Discussion on Reporting Low-Level Data[2]

D 4411 Guide for Sampling Fluvial Sediment in Motion[6]

D 4447 Guide for the Disposal of Laboratory Chemicals and Samples[7]

D 4448 Guide for Sampling Groundwater Monitoring Wells[7]

D 4489 Practice for Sampling of Waterborne Oils[6]

D 4687 Guide for General Planning of Waste Sampling[7]

E 29 Practice for Using Significant Digits in Test Data to Determine Conformance with Specifications[3]

E 122 Practice for Choice of Sample Size to Estimate a Measure of Quality for a Lot or Process[3]

E 178 Practice for Dealing with Outlying Observations[3]

X2.2 *U.S. Environmental Protection Agency Documents:*[4]

A Compendium of Superfund Field Operations Methods, EPA/540/P-87/001 (NTIS No. PB88181557/LL), December 1987.

Blackman, Benson, Hardy, and Fisher, "Enforcement and Safety Procedures for Evaluation of Hazardous Waste Disposal Sites," *Management of Uncontrolled Hazardous Waste Sites,* National Enforcement Investigations Center, Denver, CO, 1980.

Characterization of Hazardous Waste Sites: A Methods Manual, Vol II, *Available Sampling Methods,* EPA-600/4-83-040 (NTIS No. PB84-12692/LL), Environmental Monitoring and Support Laboratory, Las Vegas, NV, 1983.

Characterization of Hazardous Waste Sites: A Methods Manual, Vol II, *Available Sampling Methods,* 2nd Edition, EPA 600/4-84-076 (NTIS No. PB85-168771/LL), December 1984.

Data Quality Objectives for Remedial Response Activities: 1987a-OSWER Directive 9355, 0-14 and 1987-OSWER Directive 9355, 0-7A; EPA 540/G-87/004 and 003; and NTIS-PB88-131388, PB88131370/LL.

DeVera, Emil R., Simmons, Bart P., Stephens, Robert D., and Storm, David L., *Samplers and Sampling Procedures for Hazardous Waste Streams,* EPA-600/2-80-018 (NTIS No. PB80-135353/LL), Municipal Environmental Research Laboratory, Cincinnati, OH, 1980.

Field Screening Methods Catalog: User's Guide, USEPA/540-2-88/005 (NTIS PB89-134159/LL), September 1988.

Guidance for Conducting Remedial Investigations and Feasibility Studies Under CERCLA, OSWER Directive 9335.3-01, Draft (NTIS No. PB89184626/LL), 1988.

Guide for Decontamination of Buildings, Structures and Equipment at Superfund Sites, EPA-600/2-85-028 (NTIS No. PB85-201234/LL), Hazardous Waste Engineering Research Laboratory, Cincinnati, OH, 1985.

Handbook for Analytical Quality Control in Water and Wastewater Laboratories, EPA-600/4-79-019 (NTIS No. PB297451/LL), Environmental Monitoring and Support Laboratory, Washington, DC, 1979.

Handbook for Sampling and Sample Preservation of Water and Wastewater, EPA-600/4-82-029 (NTIS No. PB83124503/LL), Environmental Monitoring and Support Laboratory, Cincinnati, OH, 1982.

Kerr, R. S., *Practical Guide for Groundwater Sampling,* EPA-600/2-85/104 (NTIS No. PB86-137304/LL), Environmental Research Laboratory, Ada, OK, 1985.

Manual of Ground-Water Quality Sampling Procedures, EPA-600/2-81-160 (NTIS No. PB82103045/LL), 1981.

Manual of Water Well Construction Practices, EPA-570/9-75-001 (NTIS No. PB267371/LL), National Technical Information Service, 1977.

Methods for Chemical Analysis of Water and Wastes, EPA-600/4-79-020 (NTIS No. PB84-128677/LL), Environmental Monitoring and Support Laboratory, Cincinnati, OH, 1983.

Methods for Chemical Analysis of Water and Wastes, EPA-600/4-79-020 (NTIS PB84128677/LL), March 1979.

NPDES Compliance Inspection Manual, 4211-545/11828, 1984.

NPDES Compliance Inspection Manual (NTIS No. PB88-221098/LL), Office of Water Enforcement and Permits (EN-338), Washington, DC, May 1988.

Occupational Safety and Health Guidance Manual for Hazardous Waste Site Activities, DHHS (NIOSH) Publication No. 85-115 (NTIS No. PB87-162855/LL), NIOSH, OSHA, USCG, EPA, October 1985.

Preparation of State Lead Remedial Investigation Quality Assurance Project Plans for Region V, Office of Monitoring Systems and Quality Assurance, Washington, DC, April 4, 1984.

RCRA Groundwater Monitoring Technical Enforcement Document, 0-172-437, 1986.

Scalf, Marion R., et al., *Manual of Ground Water Sampling Procedures,* U.S. EPA and National Well Water Association, Worthington, OH, 1981, pp. 92 and 93.

Soil Sampling Quality Assurance User's Guide, EPA-600/4-84-043 (NTIS No. PB84198621/LL), 1984.

Standard Operating Safety Guides, Office of Emergency and Remedial Response, Hazardous Response Support Division, November 1984.

[5] *Annual Book of ASTM Standards,* Vol 11.03.
[6] *Annual Book of ASTM Standards,* Vol 11.02.
[7] *Annual Book of ASTM Standards,* Vol 11.04.

X2.3 *Other Government Documents:*[4]

Barcelona, Gibb, and Miller, *A Guide to the Selection of Materials for Monitoring, Well Construction and Ground Water Sampling* (NTIS No. PB84-141779/LL), Champaign, IL, August 1983.

Chemistry Quality Assurance Handbook, Vol I, U.S. Department of Agriculture Food Safety and Inspection Service, July 2, 1987.

Environmental Survey Manual, DOE/EH-0053, Vols 1 to 5, U.S. Department of Energy, Washington, DC, January 1989.

National Handbook of Recommended Methods for Water-Data Acquisition, U.S. Department of Interior, U.S. Geological Survey, Reston, VA, 2 Vols, 1977.

Soil Survey Handbook, 0-336-718:QL3, U.S. Department of Agriculture, 1981.

Soil Taxonomy, 0-470-728, U.S. Department of Agriculture, 1975.

X2.4 *Books:*

Barth, D. S., and Mason, B. J., *Soil Sampling Quality Assurance and the Importance of an Exploratory Study,*

ACS Symposium Series 267: Environmental Sampling for Hazardous Waste, ACS Publications American Chemical Society, Washington, DC, 1984.

Compton, R. R., *Manual of Field Geology,* John Wiley and Sons, Inc., New York, NY, 1962.

Driscoll, Fletcher, G., *Groundwater and Wells,* 2nd Edition, Johnson Division, St. Paul, MN, 1987, pp. 885–891.

Keith, L. D., *Principles of Environmental Sampling,* American Chemical Society, Washington, DC, 1988.

Kratochvil, B., and Taylor, J. K., "Sampling for Chemical Analysis," *Analytical Chemistry,* American Chemical Society, Washington, DC, July 1981, pp. 924A–938A.

Quality Control in Remedial Site Investigation: Hazardous and Industrial Solid Waste Testing, 5th Vol, *ASTM STP 925,* C. L. Perket, Ed., ASTM, Philadelphia, PA, 1986.

RCRA Ground Water Monitoring Technical Enforcement Guidance Document, NWWA/EPA Series, National Water Well Association, Dublin, OH, September 1986.

Taylor, J. K., *Quality Assurance of Chemical Measurements,* Lewis Publishers, Inc., Chelsea, MI, 1987.

Standard Practice for
Generation of Environmental Data Related to Waste Management Activities: Development of Data Quality Objectives[1]

This standard is issued under the fixed designation D 5792; the number immediately following the designation indicates the year of original adoption or, in the case of revision, the year of last revision. A number in parentheses indicates the year of last reapproval. A superscript epsilon (ϵ) indicates an editorial change since the last revision or reapproval.

1. Scope

1.1 This practice covers the process of development of data quality objectives (DQOs) for the acquisition of environmental data. Optimization of sampling and analysis design is a part of the DQO process. This practice describes the DQO process in detail. The various strategies for design optimization are too numerous to include in this practice. Many other documents outline alternatives for optimizing sampling and analysis design. Therefore, only an overview of design optimization is included. Some design aspects are included in the practice's examples for illustration purposes.

1.2 DQO development is the first of three parts of data generation activities. The other two aspects are (1) implementation of the sampling and analysis strategies and (2) data quality assessment. This guide should be used in concert with Practice D 5283, which outlines the quality assurance (QA) processes specified during planning and used during implementation.

1.3 Environmental data related to waste management activities include, but are not limited to, the results from the sampling and analyses of air, soil, water, biota, or waste samples, or any combinations thereof.

1.4 The DQO process should be developed and initiated prior to the application of planning, implementation, and assessment of sampling and analysis activities.

1.5 This practice presents extensive requirements of management, designed to ensure high-quality environmental data. The words "must" and "shall" (requirements), "should" (recommendation), and "may" (optional), have been selected carefully to reflect the importance placed on many of the statements in this practice. The extent to which all requirements will be met remains a matter of technical judgement.

1.6 The values stated in SI units are to be regarded as the standard. The values given in parentheses are for information only.

1.7 *This standard does not purport to address all of the safety concerns, if any, associated with its use. It is the responsibility of the user of this standard to establish appropriate safety and health practices and determine the applicability of regulatory limitations prior to use.*

2. Referenced Documents

2.1 *ASTM Standards:*
C 970 Guide for Sampling Special Nuclear Materials in Multi-Container Lots[2]
C 1215 Guide for Preparing and Interpreting Precision and Bias Statements in Test Method Standards Used in the Nuclear Industry[2]
D 5283 Practice for Generation of Environmental Data Related to Waste Management Activities: Quality Assurance and Quality Control Planning and Implementation[3]

3. Terminology

3.1 *Definitions*—Where applicable, the originating reference is associated with the definition and follows the definition in boldface type.

3.1.1 *accuracy (see bias)*—(1) bias; (2) the closeness of a measured value to the true value; (3) the closeness of a measured value to an accepted reference or standard value.
D 5283

3.1.1.1 *Discussion*—For many investigators, *accuracy* is attained only if a procedure is both precise and unbiased (see *bias*). Because this blending of *precision* and *accuracy* can lead to confusion, ASTM requires a statement on *bias* instead of *accuracy*.
D 5283

3.1.2 *action level*—the numerical value that causes the decision maker to choose one of the alternative actions (for example, compliance or noncompliance). It may be a regulatory threshold standard, such as maximum contaminant level for drinking water, a risk-based concentration level, a technological limitation, or reference-based standard.
EPA QA/G-4 (1)[4]

3.1.3 *bias* (see *accuracy*)—the difference between the population mean of the test results and an accepted reference value.
C 1215, D 5283

3.1.3.1 *Discussion*—Bias represents a constant error as opposed to a *random error*. A method *bias* can be estimated by the difference (or relative difference) between a measured average and an accepted standard or reference value. The data from which the estimate is obtained should be statistically analyzed to establish *bias* in the presence of *random error*. A thorough *bias* investigation of a measurement

---

[1] This practice is under the jurisdiction of ASTM Committee D-34 on Waste Management and is the direct responsibility of Subcommittee D34.02 on Physical and Chemical Characterization.
Current edition approved Nov. 10, 1995. Published January 1996.

[2] *Annual Book of ASTM Standards*, Vol 12.01.
[3] *Annual Book of ASTM Standards*, Vol 11.04.
[4] The boldface numbers in parentheses refer to the list of references at the end of this practice.

procedure requires a statistically designed experiment to repeatedly measure, under essentially the same conditions, a set of standards or reference materials of known value that cover the range of application. *Bias* often varies with the range of application and should be reported accordingly.

C 1215, D 5283

3.1.4 *confidence interval*—an interval used to bound the value of a population parameter with a specified degree of confidence (this is an interval that has different values for different samples). **C 1215**

3.1.4.1 *Discussion*—When providing a *confidence interval*, analysts should give the number of observations on which the interval is based. The specified degree of confidence is usually 90, 95, or 99 %. The form of a *confidence interval* depends on underlying assumptions and intentions. *Confidence intervals* are usually taken to be symmetric, but that is not necessarily so, as in the case of *confidence intervals* for *variances*. **C 1215**

3.1.5 *confidence level*—the probability, usually expressed as a percent, that a *confidence interval* will contain the parameter of interest (see discussion of *confidence interval*).

3.1.6 *data quality objectives (DQOs)*—qualitative and quantitative statements derived from the DQO process describing the decision rules and the uncertainties of the decision(s) within the context of the problem(s).

3.1.6.1 *Discussion*—DQOs clarify the study objectives, define the most appropriate type of data to collect, determine the most appropriate conditions from which to collect the data, and establish acceptable levels of decision errors that will be used as the basis for establishing the quantity and quality of data needed to support the decision. The DQOs are used to develop a sampling and analysis design.

3.1.7 *data quality objectives process*—a quality management tool based on the scientific method and developed by the U.S. Environmental Protection Agency (EPA) to facilitate the planning of environmental data collection activities. The DQO process enables planners to focus their planning efforts by specifying the use of the data (the decision), decision criteria (action level), and decision maker's acceptable decision error rates. The products of the DQO process are the DQOs. **EPA QA/G-4**

3.1.7.1 *Discussion*—DQOs result from an iterative process between the decision makers and the technical team to develop qualitative and quantitative statements that describe the problem and the certainty and uncertainty that decision makers are willing to accept in the results derived from the environmental data. This acceptable level of uncertainty should then be used as the basis for the design specifications for project data collection and data assessment. All of the information from the first six steps of the DQO process are used in designing the study and assessing the data adequacy.

3.1.8 *decision error:*

3.1.8.1 *false negative error*—this occurs when environmental data mislead decision maker(s) into not taking action specified by a decision rule when action should be taken.

3.1.8.2 *false positive error*—this occurs when environmental data mislead decision maker(s) into taking action specified by a decision rule when action should not be taken.

3.1.9 *decision rule*—a set of directions in the form of a conditional statement that specify the following: (*1*) how the sample data will be compared to the action level, (*2*) which

decision will be made as a result of that comparison, and (*3*) what subsequent action will be taken based on the decisions.

3.1.10 *precision*—a generic concept used to describe the dispersion of a set of measured values. **D 5283**

3.1.10.1 *Discussion*—It is important that some quantitative measure be used to specify *precision*. A statement such as "the *precision* is 1.54 g" is useless. Measures frequently used to express *precision* are standard deviation, relative standard deviation, variance, repeatability, reproducibility, confidence interval, and range. In addition to specifying the measure and the *precision*, it is important that the number of repeated measurements upon which the estimated *precision* is based also be given. **D 5283**

3.1.11 *quality assurance (QA)*—an integrated system of management activities involving planning, quality control, quality assessment, reporting, and quality improvement to ensure that a process or service (for example, environmental data) meets defined standards of quality with a stated level of confidence. **EPA QA/G-4**

3.1.12 *quality control (QC)*—the overall system of technical activities whose purpose is to measure and control the quality of a product or service so that it meets the needs of users. The aim is to provide quality that is satisfactory, adequate, dependable, and economical. **EPA QA/G-4**

3.1.13 *random error*—(*1*) the chance variation encountered in all measurement work, characterized by the random occurrence of deviations from the mean value; (*2*) an error that affects each member of a set of data (measurements) in a different manner. **D 5283**

3.1.14 *risk*—the probability or expectation that an adverse effect will occur.

3.1.14.1 *Discussion*—Risk is frequently used to describe the adverse effect on health or on economics. Health-based *risk* is the probability of induced diseases in persons exposed to physical, chemical, biological, or radiological insults over time. This *risk* probability depends on the concentration or level of the insult, which is expressed by a mathematical model describing the dose and *risk* relationship. *Risk* is also associated with economics when decision makers have to select one action from a set of available actions. Each action has a corresponding cost. The *risk* or expected loss is the cost multiplied by the probability of the outcome of a particular action. Decision makers should adopt a strategy to select actions that minimize the expected loss.

3.1.15 *standard deviation*—the square root of the sum of the squares of the individual deviations from the sample average divided by one less than the number of results involved. **D 5283**

$$S = \sqrt{\frac{\sum_{j=1}^{n} (X_j - \overline{X})^2}{n-1}}$$

where:
S = sample standard deviation,
n = number of results obtained,
X_j = *j*th individual result, and
$\overline{X}$ = sample average.

3.1.15.1 *Discussion*—The use of the *standard deviation* to describe *precision* implies that the uncertainty is independent of the measurement value. The practice of associating the ±

symbol with *standard deviation* (or RSD) is not recommended. The ± symbol denotes an interval. The *standard deviation* is not an interval, and it should not be treated as such.   **D 5283**

4. Summary of Practice

4.1 This practice describes the process of developing and documenting the DQO process and the resulting DQOs. This practice also outlines the overall environmental study process as shown in Fig. 1. It must be emphasized that any specific study scheme must be conducted in conformity with applicable agency and company guidance and procedures.

4.2 For example, the investigation of a Superfund site would include feasibility studies and community relations plans, which are not a part of this practice.

5. Significance and Use

5.1 Environmental data are often required for making regulatory and programmatic decisions. Decision makers must determine whether the levels of assurance associated with the data are sufficient in quality for their intended use.

5.2 Data generation efforts involve three parts: development of DQOs and subsequent project plan(s) to meet the DQOs, implementation and oversight of the project plan(s), and assessment of the data quality to determine whether the DQOs were met.

5.3 To determine the level of assurance necessary to support the decision, an iterative process must be used by decision makers, data collectors, and users. This practice emphasizes the iterative nature of the process of DQO development. Objectives may need to be reevaluated and

modified as information related to the level of data quality is gained. This means that DQOs are the product of the DQO process and are subject to change as data are gathered and assessed.

5.4 This practice defines the process of developing DQOs. Each step of the planning process is described.

5.5 This practice emphasizes the importance of communication among those involved in developing DQOs, those planning and implementing the sampling and analysis aspects of environmental data generation activities, and those assessing data quality.

5.6 The impacts of a successful DQO process on the project are as follows: (*1*) a consensus on the nature of the problem and the desired decision shared by all the decision makers, (*2*) data quality consistent with its intended use, (*3*) a more resource-efficient sampling and analysis design, (*4*) a planned approach to data collection and evaluation, (*5*) quantitative criteria for knowing when to stop sampling, and (*6*) known measure of risk for making an incorrect decision.

6. Data Quality Objective Process

6.1 The DQO process is a logical sequence of seven steps that leads to decisions with a known level of uncertainty (Fig. 1). It is a planning tool used to determine the type, quantity, and adequacy of data needed to support a decision. It allows the users to collect proper, sufficient, and appropriate information for the intended decision. The output from each step of the process is stated in clear and simple terms and agreed upon by all affected parties. The seven steps are as follows:

(*1*) Stating the problem,
(*2*) Identifying possible decisions,
(*3*) Identifying inputs to decisions,
(*4*) Defining boundaries,
(*5*) Developing decision rules,
(*6*) Specifying limits on decision errors, and
(*7*) Optimizing data collection design.

All outputs from steps one through six are assembled into an integrated package that describes the project objectives (the problem and desired decision rules). These objectives summarize the outputs from the first five steps and end with a statement of a decision rule with specified levels of the decision errors (from the sixth step). In the last step of the process, various approaches to a sampling and analysis plan for the project are developed that allow the decision makers to select a plan that balances resource allocation considerations (personnel, time, and capital) with the project's technical objectives. Taken together, the outputs from these seven steps comprise the DQO process. The relationship of the DQO process to the overall project process is shown in Fig. 2. At any stage of the project or during the field implementation phase, it may be appropriate to reiterate the DQO process, beginning with the first step based on new information. See Refs (**2,3**) for examples of the DQO process.

6.2 *Step 1—Stating the Problem:*

6.2.1 *Purpose*—The purpose of this step is to state the problem clearly and concisely. The first indication that a problem (or issue) exists is often articulated poorly from a technical perspective. A single event or observation is usually cited to substantiate that a problem exists. The identity and

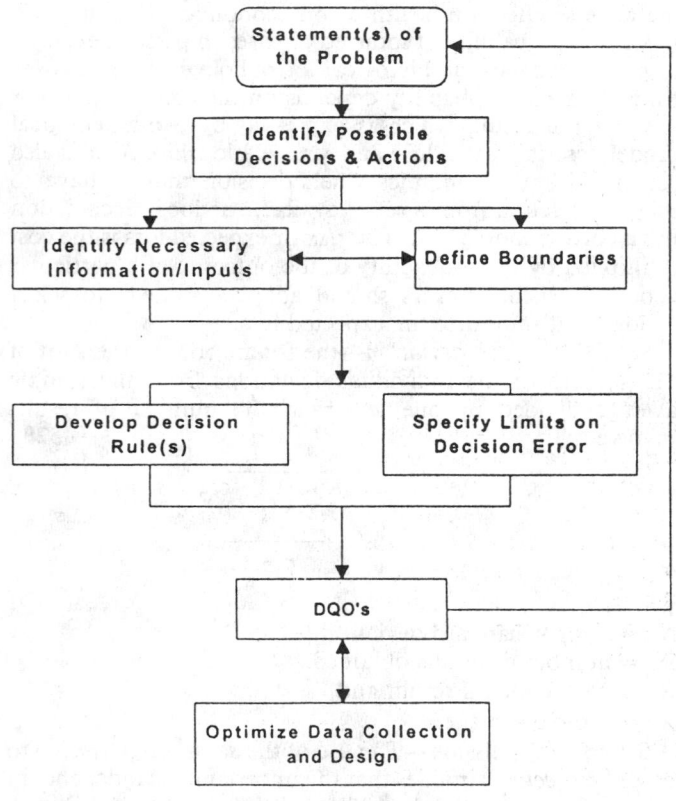

FIG. 1 DQO Process

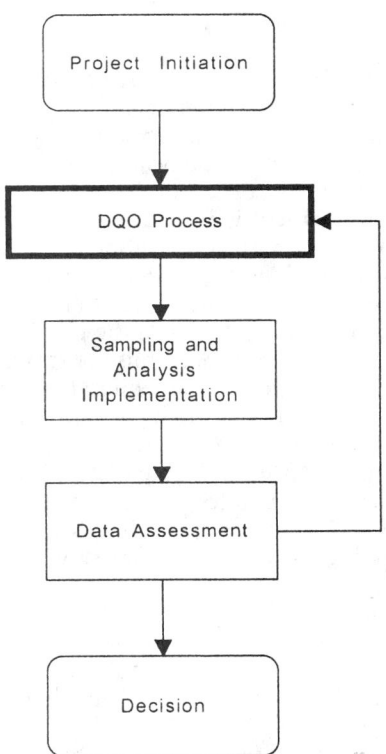

FIG. 2 DQOs Process and Overall Decision Process

roles of key decision makers and technical qualifications of the problem-solving team may not be provided with the first notice. Only after the appropriate information and problem-solving team are assembled can a clear statement of the problem be made.

6.2.2 *Activities:*

6.2.2.1 *Assembling of all Pertinent Information*—The necessary first action to describe a problem is to verify the conditions that indicate a problem exists. The pertinent information should be assembled during this phase of problem definition. A key source is any historical record of events at the site where the problem is believed to exist. This enables the decision makers to understand the context of the problem. A series of questions need to be developed concerning the problem.

(*1*) What happened (or could happen) that suggests a problem?

(*2*) When did it (could it) happen?

(*3*) How did it (could it) happen?

(*4*) Where did it (could it) happen?

(*5*) Why did it (could it) happen?

(*6*) How bad is (might be) the result or situation?

(*7*) How fast is (might be) the situation changing?

(*8*) What is (could be) the impact on human health and the environment?

(*9*) Who was (could be) involved?

(*10*) Who knows (should know) about the situation?

(*11*) Has anything been (might anything be) done to mitigate the problem?

(*12*) What contaminants are (could be) involved?

(*13*) How reliable is the information?

(*14*) What regulations could or should apply?

(*15*) Is there any information that suggests there is not a problem?

This list of potential information is not exhaustive, and there may be other data applicable to the definition of the problem.

6.2.2.2 *Identification of the DQO Team*—Even as information is being gathered, it is necessary to begin assembling a team of decision makers and technical support personnel to organize and evaluate the information. These individuals become the core of the DQO team and may be augmented by others as information and events dictate. The identities and roles of the DQO team members are usually determined by the decision makers who have either jurisdiction over the site and personnel or financial resources that will be used in resolving the problem. The DQO team is usually made up of the following key individuals:

(*1*) *Site Owners or Potentially Responsible Parties*—These individuals have authority to commit personnel and financial resources to resolve the problem and have a vital interest in the definition of the problem and possible decisions.

(*2*) *Representatives of Regulatory Agencies*—These individuals are usually responsible for enforcing the standards that have been exceeded, leading to classifying the observations or events as a problem. Additionally, they have an active role in characterizing the extent of the problem, approving any proposed remedial action, and concurring that the action mitigated the problem.

(*3*) *Project Manager*—This individual generally has the responsibility for overseeing resolution of the problem. This person may represent either the regulatory agency or the potentially responsible parties.

(*4*) *Technical Specialists*—These individuals have the expertise to assess the information and data to determine the nature and extent of the potential problem and may become key players in the design and implementation of proposed decisions.

It is important that these individuals be assembled early in the process and remain actively involved to foster good communications and to achieve consensus among the DQO team on important decision-related issues.

6.2.3 *Outputs:*

6.2.3.1 *Statement of Problem and Context*—Once the initial information and data have been collected, organized, and evaluated, the conclusions of the DQO team should be documented. If it is determined that no problem exists, the conclusion must be supported by a summary of the existing conditions and the standards or regulatory conditions that apply to the problem.

(*1*) If a problem is found to exist, the reasons must be stated clearly and concisely. Any standards or regulatory conditions that apply to the situation must be cited. If the initial investigation concludes that the existing conditions are the result of a series of problems, the DQO team should attempt to define as many discrete problems (or issues) as possible.

(*2*) The following are examples of problem statements:

(*a*) A former pesticide formulation facility is for sale, but it is unknown whether it meets local environmental standards for property transfer.

(*b*) An industrial site is known to be contaminated with

low levels of lead, but it is unknown whether levels are below risk-based standards.

(*c*) Most of a vacant lot is believed to be uncontaminated with PCBs (<2 ppm), but it is unknown whether abandoned, leaky transformers in the vacant lot make it necessary to remove any of the top layer of soil.

(*d*) The former industrial site has contaminated soil areas that may be contaminating ground water, and it is necessary to decide which type of monitoring program will satisfy local health requirements.

(*e*) The city would like to use local ground water on an athletic field near a Superfund site, but must know how this water will impact the health of the athletes and spectators.

(*3*) Complex problems should be broken down into manageable smaller problems that are linked together to form the final decision. As an example, the sale of a piece of property may involve solving the following problems:

(*a*) Is the site contaminated? If yes, then,

(*b*) Is off-site disposal required? If no, then

(*c*) Which of two allowable on-site treatment options should be used?

6.2.3.2 *Identification of Resources*—As the nature and magnitude of the problem is being documented, the decision makers should be conferring to determine the type and amount of resources that can be committed. Preliminary budget, personnel assignments, and schedule should be established. Preliminary milestones, timelines, and approvals should be documented and concurred upon by affected decision makers. The DQO team leader and technical specialists should be included in these discussions where possible. At a minimum, they should be kept informed of these issues so their impact can be anticipated in the definition of the problem.

(*1*) Figure 3 shows the primary components of the problem statement step. After this step is completed, the DQO team moves on to the next step, where the process to

resolve the problem continues.

(*2*) It is important to remember that the DQO process is an iterative one. New information is collected as projects proceed. The DQO team members associated with the problem-statement step should remain involved with the DQO process. If new data, unavailable to the DQO team during the development of the problem statement, demonstrates that the statement is incomplete or otherwise inadequate, the problem statement should be reconsidered.

6.3 *Step 2—Identifying Possible Decisions:*

6.3.1 *Purpose*—The purpose of this step is to identify the possible decision(s) that will address the problem once it has been stated clearly. Multiple decisions are required when the problem is complex. Information required to make decisions and to define the domain or boundaries of the decision will be determined in later steps (6.4 and 6.5, respectively). Each potential decision is tested to ensure that it is worth pursuing further in the process. A series of one or more decisions will result in actions that resolve the problem. The activities that lead to identifying the decision(s) are shown in Fig. 3 and discussed in 6.3.2.

6.3.2 *Activities:*

6.3.2.1 *Listing of Possible Decisions*—All possible decisions concerning the problem should be listed. Choices should not be eliminated at this time. Possible decision statements are presented in the form of a series of questions that, when answered, result in actions that will resolve the problem. Examples of questions related to problems given in 6.2.3 (Step 1) are as follows:

(*1*) Are possible contaminants on the site below regulatory thresholds?

(*2*) Must all of the surface soil be remediated to less than 5 ppm lead?

(*3*) Can only locations with PCB levels above 2 ppm be remediated?

(*4*) Will a ground water monitoring program at the site capable of detecting contaminants at the 5-ppm level satisfy regulatory requirements?

(*5*) Will a single monitoring point on or near the athletic field be sufficient?

6.3.3 *Output*—After all possible decisions that might be made have been documented, those determined to be most appropriate to resolve the problem should be prioritized by the DQO team in decreasing order of level of effort (available resources and technical challenge). Justification for the rankings should be provided. The recommended sequence in which the decisions are made should also be listed. In cases in which a complex decision statement has been broken down into a series of simpler decisions, the DQO team should identify whether the individual decisions should be addressed sequentially or in parallel. After the possible decisions have been identified, the DQO team focuses on gathering the information necessary to formulate the decision statements in Step 3 (6.4).

6.4 *Step 3—Identifying Inputs to Decisions:*

6.4.1 *Purpose*—The answers to each of the questions identified by the previous step in the DQO process must be resolved with data. Figure 4 shows the key activities that lead to development of the data requirements. This sequence of activities must be performed for each question. Note that the limits of the study (or boundary conditions) are determined

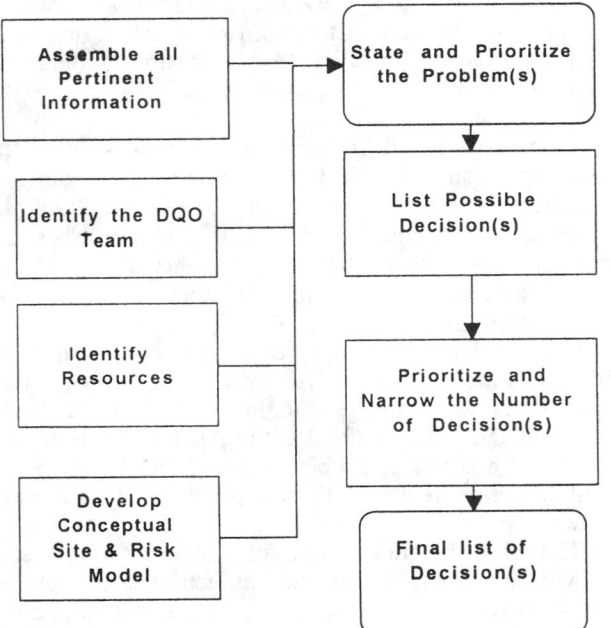

FIG. 3 Stating the Problem and Identifying the Decisions

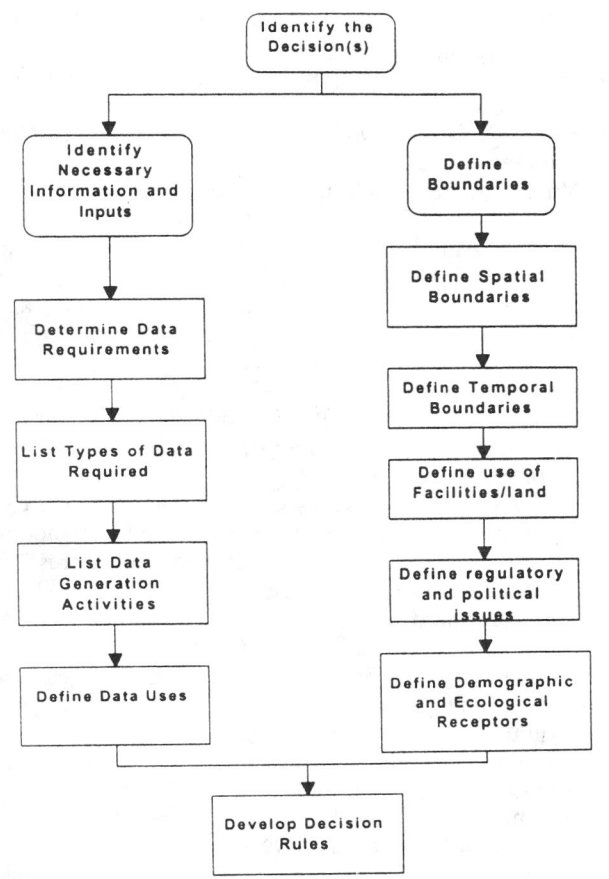

FIG. 4 Determination of Information Inputs and Study Boundaries

in a parallel step identified as "define boundaries" in Fig. 1. This is another type of data requirement and is discussed in 6.4.

6.4.2 *Activities:*

6.4.2.1 *Determination of Data Requirements*—At this stage of the process, it is important to carefully examine the complete set of data requirements needed to support each of the decisions. Each possible decision to be made should be considered independently of others to ensure that no omissions have occurred. After all possible questions concerning the decisions have been considered, group the data requirements together to determine overall data needs for the project. It may be possible to plan efficiencies in collecting and processing data to meet multiple needs and thereby lower overall project costs or reduce the time necessary to meet important milestones, or both.

(*1*) When considering whether specific information is needed for making a decision, test the data to ensure that it is appropriate for the decision statement. If no use of the data can be identified, it may be extraneous to the needs.

(*2*) The following list is indicative of some of the information needs that may be considered for each decision. It is not inclusive of all important data, but it provides examples common to many environmental problems.

(*a*) What regulatory limits may be associated with the problem or regulatory issue?

(*b*) Does contamination exceed regulatory limits?

(*c*) What tests must be performed for the type of waste in question?

(*d*) What are the hydrogeological considerations?

(*e*) What populations are at risk?

(*f*) What are the ecological considerations?

(*g*) What process knowledge is available?

(*h*) What historical/background data (past uses or spills) are available?

(*i*) What are the budget constraints?

(*j*) What is the time schedule?

(*k*) What potential health, political, and social factors must be considered?

(*l*) What is the potential for legal action?

(*m*) Who is the end-user of the data?

(*n*) What data validation criteria will be used?

(*o*) What, if any, limitations exist on the data collection process (detection limits, matrix interferences, or no known measurement technology)?

6.4.3 *Outputs:*

6.4.3.1 The DQO team must specify data needs for each problem/decision that has been identified in the first two steps.

6.4.3.2 List the types of data required. Some example data types include, but are not limited to, the following:

(*1*) Chemical,

(*2*) Physical (including site hydrogeology and meteorology),

(*3*) Biological,

(*4*) Toxicological,

(*5*) Historical,

(*6*) Economic (time, budget, and manpower),

(*7*) Demographic,

(*8*) Toxicity characteristics, and

(*9*) Fate and transport model output.

6.4.3.3 *Listing of Data Generation Activities*—Determine which data can be acquired from historical records and which new data must be obtained in the field or laboratory, or both. If the DQO team determines that no new data are necessary to make a decision, they should document their reasoning. If new information is necessary, activities that will be required to generate inputs (data) affecting the decision should be listed. Examples of these include, but are not limited to, the following:

(*1*) Assembly of historical data,

(*2*) Sampling and chemical analysis,

(*3*) Physical testing, and

(*4*) Modeling.

6.4.3.4 *Definition of Data Use(s)*—Each set of data will be used for some purpose. This purpose must be defined. For example, will action levels for contaminants be determined by a risk-based calculation, by reference dose, or by predefined threshold values established by regulators? If so, ensure that data requirements are consistent with the criteria against which they will be compared. Data collected at the parts per million level may not be useful if they are to be compared to criteria at the parts per billion level.

6.5 *Step 4—Defining Boundaries:*

6.5.1 *Purpose*—This step of the DQO process determines the boundaries to which the decisions will apply. Boundaries establish limits on the data collection activities identified in Step 3 (6.4). These boundaries include, but are not limited to, spatial boundaries (physical and geographical), temporal boundaries (time periods), demographic, regulatory, political, and budget. The activities for this step of the DQO process are shown in Fig. 4.

6.5.2 *Activities:*

6.5.2.1 *Definition of Spatial Boundaries*—Define the boundaries of the total area and smallest increment of concern. Examples of items affecting the boundary definition are as follows:

(*1*) Horizontal or lateral areas,

(*2*) Vertical boundaries (depth/height),

(*3*) Discrete locations (hot spots),

(*4*) Media/matrix (air, soil, water, biota, and waste),

(*5*) Number of containers of waste, and

(*6*) Volume.

6.5.2.2 *Definition of Temporal Boundaries (Time Period)*—This activity determines the time interval over which environmental data will be collected for use in the decision-making process. If current or future real-time data are used to represent or model previous conditions, the basis of these assumptions or models must be documented and agreed upon between the decision makers and the technical team. The same constraint is also placed on the extrapolation of historical or real-time data, or both, to future time periods.

(*1*) The duration of new data collection activities must be established. In addition, the following factors should be considered:

(*a*) Availability and reliability of existing historical data,

(*b*) Access to the site or impacted area,

(*c*) Exposure potential, and

(*d*) Budgetary constraints.

6.5.2.3 *Definition of the Demographic Receptors*—The DQO team must frequently define the receptor population that may be effected. All affected populations and the mode of their anticipated exposure should be identified. These populations include the following:

(*1*) *Known/Anticipated Population(s)*—Human (children, adults, age, gender, and so forth), plant/animal (wetlands, endangered species, and so forth), and global;

(*2*) Population activity pattern(s); and

(*3*) Exposure pathway for each population.

6.5.2.4 *Definition of Nontechnical Boundaries*—Decision makers also have to consider nontechnical boundaries that can impact the resolution of the problem seriously. These nontechnical boundaries include the following:

(*1*) Regulatory considerations, and

(*2*) Political or legal action(s).

6.5.3 *Outputs*—The results from each of the activities in this step must be documented. Care must be taken to identify which boundary conditions apply to each decision being made. It may be that similar information is needed for several decisions but different boundary conditions may apply. It is important that decision makers understand and concur on the boundaries; otherwise, the ability to make decisions may be compromised.

6.6 *Step 5—Developing Decision Rules:*

6.6.1 *Purpose:*

6.6.1.1 The purpose of this step is to integrate outputs from previous steps into a set of statements that describe the logical basis for choosing among alternative outcomes/results/actions. These statements are decision rules that define the following:

(*1*) How the sample data will be compared to the action level,

(*2*) Which decision(s) will be made as a result of that comparison, and

(*3*) What subsequent action(s) will be taken based on the decisions.

6.6.1.2 The formats for these rules are either "if (criterion) . . ., then (action)" statements or a decision tree, as shown in Fig. 5. The decision criteria should be stated as clearly and concisely as possible. The rule(s) must contain both a decision point (or action level) and an action. The decision rule is generated through a cooperative effort among the DQO team. If an acceptable decision rule cannot be formulated, the process returns to the appropriate previous step of the DQO process.

6.6.1.3 Decision rules usually contain the following elements: measurement of interest, sample statistic, action level, and a resultant action. "Measurement of interest" is the variable or attribute to be measured. It can be concentration of a contaminant, volume/mass of a waste, or physical property, such as flash point of a waste. "Sample statistic" is the quantity computed from the sample data. It can be average value, median, present/absent, or some other expression of quantity. If that data are not normally distributed, statistical methods based on other distributions or nonparametric methods can be used.

6.6.1.4 The "action level" is the limit against which the sample statistic will be compared. Depending on whether the action level is exceeded or not, the specified action will result. If the action level equals the regulatory threshold, the probability of a false positive error equals the probability of a false negative error. For unequal probabilities of the decision

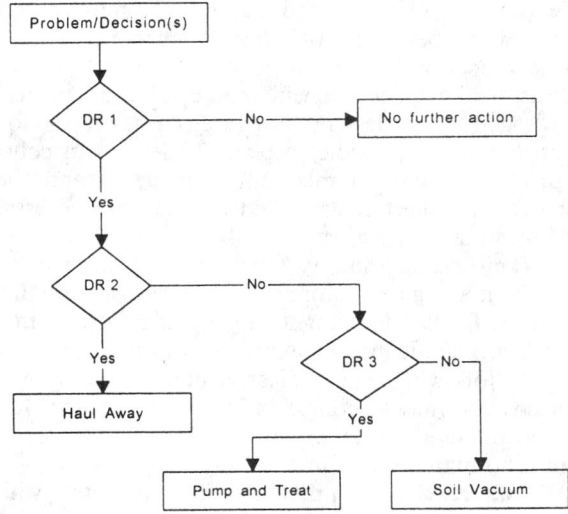

FIG. 5 Decision Tree for Three Sequential Decision Rules (DRs)

errors, the action level can be either less or greater than the regulatory threshold. The degree to which the action level is different from the regulatory threshold depends on the acceptable level of uncertainty for the decision errors that the decision makers are willing to accept. The action level is determined by the levels of false positive error, false negative error, measurement variability, and number of samples. Derivation of an action level for given level of false positive and false negative error is included as part of Appendix X1.

6.6.1.5 The decision rule is completed by stating the "resultant action" to be taken based on comparison of the sample statistic with the action level.

6.6.1.6 Two illustrations of general decision rule formats are as follows:

(1) "If the average concentration of a contaminant in waste is greater than the action level for that contaminant, then the waste will be classified as a 'hazardous' waste and will be disposed of according to the governing regulations."

(2) "If the average concentration of a contaminant in a waste is lower than the action level for that contaminant, then the waste is classified as 'nonhazardous' and there are no special limitations placed on the disposal options."

6.6.1.7 In this illustration, the measurement of interest is "concentration of a contaminant." The sample statistic is the "average concentration." The action level is some value to be specified. The resultant action is "disposal according to governing regulations." There may be separate decision rules for each medium, each domain (site), or other designated collections of data.

6.6.1.8 The action level may be an observation or occurrence in some cases. An example of this type of decision rule is as follows:

(1) If soil exhibits a visible dark spot as compared to the surrounding soil, use the portable organic monitor to screen for organics in the dark spot.

6.6.2 *Activities*—The activities that must be completed to establish a decision rule are shown in Fig. 6.

6.6.2.1 *Determination of Measurement of Interest*—A clear expression of the measurement (parameter) upon which the decision is based must be provided.

6.6.2.2 *Specification of Action Level*—The sample statistic

of the measurement or observation of interest that initiates the agreed-upon action must be specified. The determination of the action level for any decision is a combination of the total variability in the data acquisition process and the level of decision errors that decision makers will accept in the final decision. The role of decision makers and decision errors is discussed in 6.7 (Step 6), and the derivation of an action level is illustrated in Appendix X1.

6.6.2.3 *Specification of Sample Statistic (if Applicable)*—Prior to the statement of a decision rule, it is necessary to determine how the sample statistic will be calculated and expressed (units of measure). The statistical approach chosen can be the average, mean, median, high, low, range, present/absent, and so forth. The unit of measurement must correspond to those of the decision criteria, and the limit of detection (measurement) must be lower than the action level. A statistic may or may not be applicable in stating observations.

6.6.2.4 *Specification of Mode of Comparison*—After the sample statistic is derived from historical or real-time data and an action level has been identified, they must be compared. This comparison is usually stated as greater than . . ., less than . . ., equal to . . ., or present/absent. Depending on the results of the comparison, a specific action is indicated by the decision rule.

6.6.2.5 *Specification of Action*—When the result of the comparison of the sample statistic with the action level is known, an action must be specified. It should be sufficient to resolve the problem. In complex situations, the action may direct decision makers to another problem (addressed by an additional set of DQOs) that must also be resolved. This type of logical pathway is described frequently as a decision tree. These situations should have been identified in Step 2 (6.3). Figure 5 shows the decision tree derived from the application of a set of three sequential decision rules.

6.6.3 *Outputs*—An example showing the application of a decision rule is presented in Appendix X1. Some additional examples of decision rules that might apply to waste problems and possible actions discussed in 6.2 and 6.3, respectively, are given as follows:

6.6.3.1 If the historical record of site monitoring activities shows the absence of any regulated constituent above 1 ppm, then the site can be left as is.

NOTE—A value of 1 ppm selected for this example only.

6.6.3.2 If site characterization indicates that 20 % of the soil (top 30 cm) is contaminated above 5 ppm lead, then the entire soil layer (1 m) must be remediated.

6.6.3.3 If site characterization data show that 95 % of the total surface area (10 cm deep) of the site contains less than 2 ppm PCB, then only those areas exceeding that value need to be remediated.

6.6.3.4 If the levels of contaminants found in the monthly ground water monitoring program total less than 1000 ppm in each well, then no additional corrective action needs to be instituted.

6.6.3.5 If no contaminate above 1 ppm is observed in a ground water monitoring well located downgradient and within 100 m of the site boundary during monthly monitoring events, then additional monitoring wells will not be required.

6.7 *Step 6—Specifying Limits on Decision Errors:*

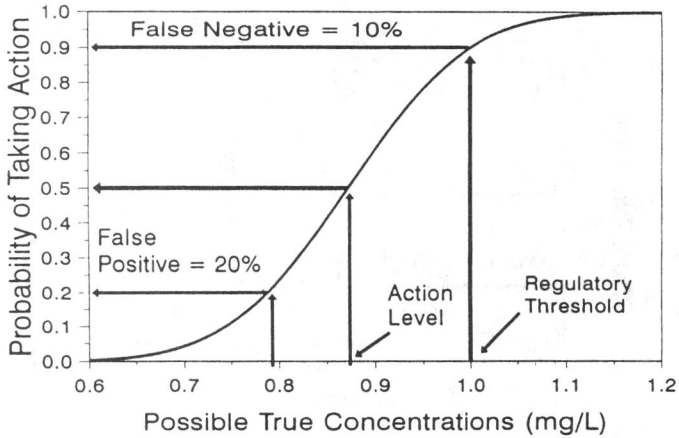

FIG. 6 Decision Rule Development

6.7.1 *Purpose*—An essential part of the DQO process is to establish the degree of uncertainty (decision errors) that decision makers are prepared to accept in making a decision concerning the problem (**Refs 4–6**). The purpose of this step is to define the acceptable decision errors based on a consideration of the consequences of making an incorrect decision. The perspective of the decision makers or baseline assumption must be stated clearly, that is, the site is considered contaminated or the site is not contaminated. There are two kinds of decision errors: false positive error and false negative error.

6.7.2 *Activities:*

6.7.2.1 *Specifications of Decision Errors*—It should be understood that, when a decision is made based on empirical data, there is no way to reduce either type of decision error to zero. Furthermore, there is usually a tradeoff between the two decision errors, meaning that a lower false negative error would lead to a higher false positive error, and vice-versa (for a given amount of data or number of samples). Decision makers should understand the consequences of decision errors and the tradeoffs between a false positive error and a false negative error. Error rates (false positive and false negative errors) must be specified relative to an agreed-upon concentration regulatory threshold or health-risk level.

6.7.2.2 *Consequences of an Incorrect Decision*—The random variability for empirical data is composed of sample variability and measurement variability. Sampling variability is composed of both environmental variability (for example, spatial, temporal, matrix, and so forth) and sample collection variability. Measurement variability is a function of extraction efficiencies, matrices effects, and analyte interferences. Taken together, sample variance and measurement variance components comprise the total variability in the data that contributes to errors in the decision under consideration. Decision makers must make an a priori judgement regarding how often they are willing to be wrong because of data variability. This uncertainty is the "acceptable error" in the decision. In the context of a decision designed to be protective of human health, they can be wrong by taking a prescribed action when none was necessary (false positive error), or they can fail to take action when it was necessary (false negative error).

6.7.2.3 *False Positive Error*—If the true concentration is lower than the regulatory threshold, but the decision makers conclude that the waste is hazardous because the sample average concentration is equal to or higher than the action level, then a false positive error has been made. The consequence of this error is that the nonhazardous waste will be remediated or disposed of according to stricter requirements than required by regulations. A false positive error is undesirable because it will incur unnecessary costs and result in inefficiency.

6.7.2.4 *False Negative Error*—If the true concentration is equal to or greater than the regulatory threshold, but the decision makers conclude that the waste is nonhazardous because the sample average concentration is below the action level, then a false negative error has been made. The consequence of this error is that the waste will be disposed of by a less stringent method. This error is undesirable because the allowed waste management method may allow consequences harmful to health or the environment.

6.7.2.5 The relationship between the probability of taking action on a decision rule and the possible true value of the measurement of interest is illustrated graphically by a decision performance curve in Fig. 7 based on the example described in Appendix X1. The decision performance curve depends on the decision makers' willingness to accept false positive and false negative errors, the total variability of the measurement process, the number of samples, and a regulatory threshold. The interval between the action level and the regulatory threshold represents the range of possible true measurement values over which decision makers are willing to take more than a 50 % chance of sending a nonhazardous waste to a regulated landfill to ensure a specified false negative error. The curve is derived from the following:

(*1*) Acceptable errors (either a false positive error or a false negative error) agreed upon between the decision makers,

(*2*) Total variability of the system,

(*3*) Number of samples analyzed, and

(*4*) Statistical distribution of sample data (normal, log-normal and so forth).

6.7.2.6 In some cases, the action level may equal a regulatory level, or risk level. In these cases, all of the decision makers must understand that the value of a false positive error and false negative error associated with making a decision are equal.

6.7.2.7 Specification of false positive error and false negative error is typically made on the basis of the relative importance of the consequence of an incorrect decision of either type. If the costs of environmental disposal or remediation are substantial and the potential environmental impact is relatively minor, then the emphasis may be on the control or reduction of false positive error (cost control). If the reverse is the case, then the emphasis may be on the control or reduction of the false negative error (control of environmental risk and liability). This important issue must be negotiated and resolved on a case-by-case basis for each problem identified in Step 1 by all decision makers.

6.7.2.8 This curve and several others that illustrate the relationships between these factors are discussed in the example in Appendix X1.

6.7.2.9 *Control of Decision Errors*—While decision errors cannot be eliminated, their errors can be reduced by (*1*)

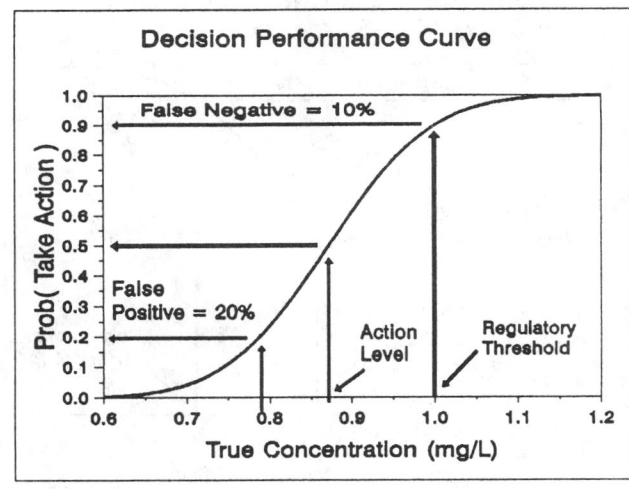

FIG. 7 Decision Performance Curve for Appendix X1 Example

reducing measurement errors (sampling or analytical variabilities, or both) or (2) increasing the number of samples taken. These issues relate to optimization of the study design and are covered in Step 7 (see 6.8).

6.7.3 *Output*—The rational and acceptable errors for both the false positive and false negative errors for each decision from Step 1 must be documented.

6.7.4 *DQO Summary:*

6.7.4.1 *Purpose:*

(1) The purpose of this step is to present the results of the DQO process clearly and concisely, in a form usable for optimizing data collection design (6.8; Step 7). This presentation of the DQOs and the complete documentation of the outputs and logic from which they were derived is essential for the initiation of data collection design.

(2) The DQOs are derived from the outputs of all of the preceding steps in the DQO process. Each output is important. However, the uncertainty on the decision and the decision rules incorporate the decision, boundaries, and inputs required to generate a sampling design. Indeed, the uncertainties on the decisions, together with the respective decision rules, are the primary results of the DQO process for a particular problem.

6.7.4.2 *Activities:*

(1) Activities include the establishment of a framework in which the decision rule(s) and associated limits on decision error are expressed as the DQO(s) supported by the documented logic and outputs of the previous steps of DQO process development. Within this decision framework, the DQOs can be improved and refined through an iterative process that includes use of and further evaluation of the following:

(a) Problem statement,

(b) Possible decisions,

(c) Inputs,

(d) Definition of spatial and temporal boundaries,

(e) Development of decision rule(s), and

(f) Acceptance of limits on decision error.

(2) Establishment of the DQOs by integration of concise decision rule(s) with their associated limits on decision error and the documentation of the DQO process is critical in facilitating understanding of the risk of making the wrong decision by the decision makers.

6.7.4.3 *Outputs:*

(1) Primary outputs consist of clear and concise presentation of the DQO process and complete documentation of the logic involved in development of the decision rules and associated limits on decision errors.

(2) As a useful tool, the DQO process can be integrated graphically into a typical decision tree or logic flow diagram that clearly indicates actions to be taken as the result of implementation of the decision rule(s) (see Fig. X1.1). These diagrams and associated descriptive text are effective formats for use during the optimization of data collection design and are important elements in project work plans.

(3) For example, the following are DQO summaries from Appendix X1: To make the following decision for the "cadmium incineration waste problem" with a false positive error not to exceed 20 % and a false negative error not to exceed 10 %. If the mean cadmium concentration in the toxicity characteristic leaching procedure (TCLP) extract is

equal to or >1 mg/L, then dispose of the fly ash load in a suitable landfill. If the mean cadmium concentration in the TCLP extract is <1 mg/L, then dispose of the fly ash load in a sanitary landfill.

6.7.4.4 *Application of Data Quality Objectives:*

(1) The DQOs are applied on a day-to-day basis by incorporating the decision errors into the action level. This makes the decision rule easier to use. To apply DQOs, statisticians apply statistical methods such as those used in the example in Appendix X1 to calculate an action level that takes into account the acceptable decision uncertainty.

(2) The applied DQOs from Appendix X1 are as follows:

(a) If the average concentration of cadmium is ≥ 0.87 mg/L, then dispose of the waste fly ash in a hazardous waste landfill; and

(b) If the average concentration of cadmium is <0.87 mg/L, then dispose of the waste fly ash in a sanitary landfill.

6.7.4.5 *Decision Tree Format*—In decision tree format, the DQOs are presented along with the actions and tasks that are required in the data collection design step (see Fig. 5).

6.8 *Step 7—Optimizing Data Collection Design:*

6.8.1 Prior to beginning this step of the process, the output from the first six steps must be assembled and provided to DQO team members who will undertake to optimize the actual sampling design for data collection. Care should be taken to separate the factual material from the DQO team's assumptions or estimates, or both, of factors important to development of the output from each step. The data collection effort must gather sufficient data to confirm (if possible/feasible) the accuracy of these assumptions.

6.8.2 *Purpose:*

6.8.2.1 The objective of this step is to generate the most resource-effective sampling design that will provide adequate data for decisions to be made. In this step, sampling designs are developed based on the outputs of the first six steps of the process, assumptions made during those steps, and applicable statistical techniques.

6.8.2.2 An understanding of the sources of variability and levels of uncertainty is essential in developing the sampling design alternatives. The focus of the DQO process is the balancing of the limits of decision errors against the resources available to complete the project. Many of the sampling design alternatives will address different strategies for balancing the different types of decision errors with the resources available (time, money, and personnel) to resolve the problem.

6.8.2.3 Once sampling designs are developed, the sampling design alternatives and required resources for each should be presented to the decision makers. These alternatives allow for an understanding of the benefits and resource commitments to each sampling design. If a resource-effective sampling design to provide adequate data for the decision rule cannot be found among the sampling design alternatives, it may be necessary to alter the decision or revise the inputs into the DQO process. This decision is the responsibility of the decision makers and requires that all DQO team members be involved. New members may be added if, in the opinion of the decision makers, their expertise is needed to develop acceptable DQOs.

6.8.3 *Activities*—The activities involved in the development of an optimal sampling design and chemical analyses are shown in Fig. 8.

6.8.3.1 *Summary of Information*—The data collectors should summarize any previous data and the outputs from the previous six steps of the DQO process. This allows data collectors to remain focused on the decision makers' needs in design optimization.

6.8.3.2 *Development of Sampling Design Alternatives*— Alternative sampling designs must be based on DQOs, which were developed with an understanding of measurement variability and the resources available for resolving the problem. Design alternatives must address the degree of representation of any one sample within the problem boundaries. This is accomplished by selecting from among the sampling designs those that best describe the system. These include, but are not limited to, random, sequential random, systematic, and stratified sampling designs.

(*1*) Probabilities of selecting an appropriate sample are related to the type of sampling design. An equal probability of selecting a sample implies a random sample design. Selecting unequal probabilities for sample selection implies a stratified sample design. The more heterogeneous the sampling units, the more likely unequal probabilities will be assigned to the sample. Furthermore, the more heterogeneous the waste site, the more useful historical or process information is in assessing the sampling design alternatives. The participation of a qualified statistician is critical in this process.

(*2*) Variability may also be introduced during sample

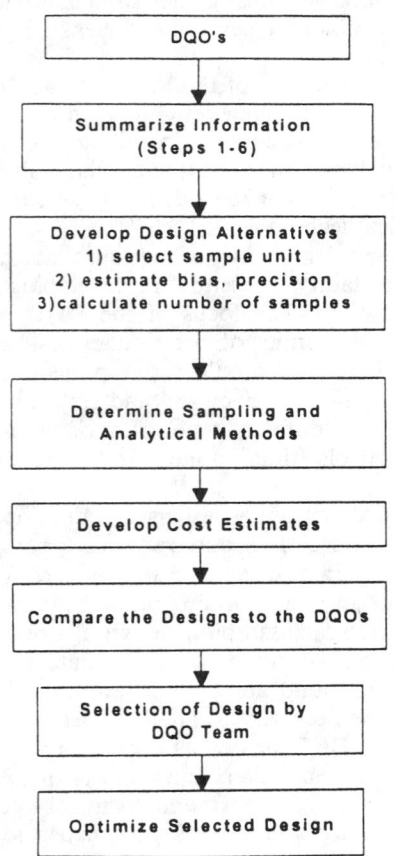

FIG. 8 **Optimization of Sample Design**

handling and preparation procedures that may be necessary between field sampling and analytical methods. Consideration of the important factors impacting sample variability should occur during the design process.

6.8.3.3 *Determination of Analytical Chemistry Methods*— The alternative analytical chemistry methods as documented during the DQO process must be considered. Factors that affect selecting alternative methods include, but are not limited to, the following:

(*1*) Detection limits versus action levels;

(*2*) Matrix effects on detection limits, bias, and variability; and

(*3*) Sample amount available (volume or weight).

6.8.3.4 *For Each Sampling Design Alternative, Selection of the Sample Unit that Satisfies the DQOs*—Sampling units include drums, tanks, an area within a grid, a boring location on a grid, a depth interval in a boring, or any other appropriate defined physical unit from which material can be obtained. Different sampling units may and often will be appropriate for different materials or locations. The sampling unit may depend on logistical and resource issues, such as whether the material will be disposed by drum or truck or the amount of material that can be excavated.

6.8.3.5 *For Each Sampling Design Alternative, Calculation of the Optimal Number of Samples that Satisfies the DQOs*—Typically, samples are collected from each sample unit for chemical analyses. Using the mathematical expressions for sampling design optimization, solve for the optimal number of samples that meet the uncertainty limits on the decision errors specified in the DQOs. Selection of the number of samples is an iterative process. Initial selection of the number of samples may be based on different project criteria (for example, budget, precision limits, and so forth). These initial calculations should be examined to determine whether they are adequate for the specified decision errors. In addition, preliminary sample designs may be required for better estimates of mean concentrations and measurement variability for optimal planning of larger sample designs.

6.8.3.6 *For Each Sampling Design Alternative, Development of Cost Estimates*—The estimates should relate the total cost of sampling and chemical analyses for alternative sampling designs. These cost functions may take into account such items as the cost of remediation or waste disposal by sample unit. This enables the decision makers to assess whether sampling and chemical analyses are more cost effective than proceeding with cleanup or disposal with minimal data collection.

6.8.4 *Outputs*—The list of sampling design alternatives is submitted to the decision makers for selection. After selection of the final sampling design, document the operational details and theoretical assumptions of the selected sampling design in a final sampling and chemical analyses plan. The documentation should include the sampling plan, sampling and analytical chemistry procedures, data assessment procedures, quality control requirements, and overall project quality assurance requirements.

7. Documentation of the Data Quality Objective Process

7.1 The following statements and information document the outputs of the specific DQO process used to develop the DQOs. The DQOs are meaningless if they are not connected with the specific problem and other qualifying information used to develop them.

7.2 DQO process documentation summaries can vary from problem to problem, but most will include information such as the following:

7.2.1 Facility name, location, and process;

7.2.2 List of decision makers, affiliations, and responsibilities for this project.

7.2.3 Statement of the problem.

7.2.4 Summary of logic for the decisions chosen for consideration. For each problem there must be at least one decision.

7.2.5 Information and inputs such as those given in 6.4.2. There should be appropriate inputs to allow generation of

the data to make a decision. It may be useful to establish separate decisions for each matrix (that is, soil, sediment, and water).

7.2.6 Defined boundaries, which should be addressed for each decision. It may be useful to segregate the boundaries by matrix.

7.2.7 Decision rules, which should incorporate appropriate boundaries. The rules may be stated by matrix.

7.2.8 Limits on decision error. The rationale or assumptions upon which decision error estimates are based should be documented.

8. Keywords

8.1 data quality objectives; DQOs; project planning; waste analysis; waste testing

APPENDIX

(Nonmandatory Information)

X1. DQO CASE STUDY—CADMIUM-CONTAMINATED FLY ASH WASTE

X1.1 *Background:*

X1.1.1 A municipal waste incineration facility located in the Midwest routinely removes "fly ash" from its flue gas scrubber system and disposes it in a sanitary landfill. It was determined previously that the ash was nonhazardous under hazardous waste regulations. However, the incinerator has recently begun treating a new waste stream. As a result, a local environmental public interest group asked that the ash be retested and evaluated for hazardous waste compliance before it is disposed. The group is primarily concerned that the ash may contain hazardous levels of cadmium due to the new waste sources. The facility manager has agreed to test the ash and decided to use the DQOs process to help guide decision making throughout the project. Although not constrained by cost, the facility is interested in minimizing expenditures.

X1.1.2 The 40 CFR Part 261 RCRA toxicity characteristic criteria (7) for determining whether a solid waste is hazardous requires collection of a "representative portion" of the waste and performance of TCLP. During this process, the solid fly ash will be "extracted" or mixed in an acid solution for 18 h. The extraction liquid will then be subjected to tests for specific metals.

X1.1.3 Since the impact of this new waste stream is not known, a preliminary study was conducted to determine the variability of the concentration of the contaminants. Random samples were collected from the first 20 truckloads. Since process knowledge of the waste stream indicated that cadmium was the only toxicity characteristic (TC) constituent in the waste, these samples were analyzed individually for cadmium using TCLP. The results were expressed as the average concentration along with the standard deviation.

X1.2 *Data Quality Objective Development*—The following is an example of the outputs from each step in the DQO process.

X1.2.1 *Statement of the Problem:*

X1.2.1.1 *Identification of the DQO Team*—The plant manager assembled a DQO team consisting of himself and a

representative of the current disposal facility staff. The two of them subsequently assembled the additional DQO team members.

(*1*) The decision makers on the DQO team included the incinerator owner and incineration plant manager, and a representative of the environmental public interest group, in which a representative of the community in which the ash is currently being disposed. The technical staff included a statistician, toxicologist, and chemist with sampling experience.

X1.2.1.2 *Statement of the Problem*—The problem is to determine whether any loads of fly ash are hazardous with cadmium under RCRA regulations using TCLP testing. If a load is hazardous, it must be disposed of in a RCRA landfill.

X1.2.2 *Identification of Possible Decisions:*

X1.2.2.1 *Decision*—Determine whether the concentration of cadmium in TCLP leachate from waste fly ash exceeds the regulatory RCRA standards.

X1.2.2.2 *Statement of the Actions that Could Result from the Decision:*

(*1*) If the average concentration of cadmium is greater than or equal to the action level, dispose of the waste fly ash in a RCRA landfill.

(*2*) If the average concentration of cadmium is less than the action level, dispose of the waste fly ash in a sanitary landfill.

X1.2.3 *Identification of Inputs to Decisions*—The DQO team identified the following inputs or information needed for the decision rules:

X1.2.3.1 *Preliminary Study Information*—Since the concern is with a new waste stream, the DQO team ordered a pilot study of the fly ash to determine the variability in the concentration of cadmium between loads of fly ash leaving the facility. They have determined that each load is fairly homogeneous. However, there is a high variability between loads due to the nature of the waste-stream. Most of the fly ash produced is not a RCRA hazardous waste and may be disposed of in a sanitary landfill. Because of this, the company has decided that testing each individual waste load before it leaves the facility would be the most economical. In that way, they could send loads of ash that exceeded the regulated cadmium concentrations to the higher-cost RCRA landfills and continue to send the others to the sanitary landfill.

(1) The study showed that the standard deviation of the cadmium concentration within a load was $S_w = 0.4$ mg/L, and the standard deviation of the cadmium concentration between loads was $S_b = 1.4$ mg/L. Sample and quality control data indicate that a normal distribution can be assumed.

X1.2.3.2 *Identification of Contaminants of Concern, Matrix, and Regulatory Limits*—The DQO team identified the following factors critical to the problem:

(1) *Contaminants of Concern*—Cadmium soluble in the TCLP extract.

(2) *Sample Matrix*—Fly ash.

(3) *Regulatory Threshold*—1 mg/L.

X1.2.3.3 *Specific Project Budget and Time Constraints*—The incinerator plant manager has requested that all stages of the operation be performed in a manner that minimizes the cost of sampling, chemical analysis, and waste disposal. However, no formal cost constraints have been implemented.

(1) The environmental public interest group has threatened to file a lawsuit for violation of environmental regulations if testing does not proceed within a "reasonable time-frame."

(2) The waste does not pose a threat to humans or the environment while contained in the trucks. Additionally, since the fly ash is not subject to change, disintegration, or alteration, the chemical properties of the waste do not warrant any temporal constraints. However, in order to expedite decision making, the DQO team has placed deadlines on sampling and reporting. The fly ash waste will be tested within 48 h of being loaded onto waste hauling trailers. The analytical results from each sampling round should be completed and reported within five working days of sampling.

X1.2.3.4 *Identification of the Testing Methods*—In this case, 40 CFR Part 261, Appendix II specified the TCLP Method SW 846, Method 1311 (8). The leachate must be analyzed by an appropriate method. Potential methods of characterizing the leachate for cadmium include, but are not limited to, SW 846, Methods 6010, 6020, 7130, or 7131.

X1.2.4 *Inputs to Be Determined:*

X1.2.4.1 *Method Validation and Quality Control (QC)*—The analytical method accuracy and precision and method detection limits in leachate from the fly ash matrix must be determined. The QC samples must be specified.

X1.2.4.2 *Identification of Sampling Procedure or Devices*—The following must be determined:

(1) Number of samples,

(2) Sampling methods for composite or grab samples of ash, and

(3) The QC requirements for sampling.

X1.2.5 *Definition of the Boundaries*—Define a detailed description of the spatial and temporal boundaries of the decision, characteristics that define the environmental media and objects or people of interest, and any practical considerations for the study.

X1.2.5.1 *Specification of the Characteristics that Define the Sample Matrix*—The fly ash should not be mixed with any other constituents except the water used for dust control.

X1.2.5.2 *Identification of Spatial Boundaries*—The variability between loads was greater than within a load; therefore, a decision will be made on each load. The waste fly ash will be tested after it has been deposited in the trailer used by the waste hauler. Separate decisions regarding the toxicity of the fly ash will be made for each load of ash leaving the incinerator facility. Each load of ash should fill the waste trailer at least 70 %. In cases in which the trailer is filled less than 70 %, the trailer must wait on-site until more ash is produced and can fill the trailer to the appropriate capacity.

X1.2.5.3 *Identification of Temporal Boundaries (Including the Time Frame Over Which the Study Should Be Conducted)*—The waste does not pose a threat to humans or the environment while contained in the trucks. However, in order to expedite decision making, the DQO team has placed deadlines for reaching a decision. The fly ash waste will be tested and a disposal decision made within 48 h of being loaded onto waste hauling trucks.

X1.2.6 *Development of Decision Rules*—The arithmetic mean of sample results will be compared to the action level.

X1.2.6.1 *Decision Rule:*

(1) If the average concentration of cadmium in a truck load is equal to or greater than the action level, then dispose of the waste fly ash in a RCRA landfill; or

(2) If the average concentration of cadmium in a truck load is less than the action level, then dispose of the waste fly ash in a sanitary landfill.

Note that the DQO team will decide that the action level is less than the regulatory level in order to meet a 10 % false negative error for concentrations at the regulatory level of 1 mg/L.

X1.2.7 *Specification of Limits on Decision Errors:*

X1.2.7.1 The decision makers specify acceptable decision errors based on the consequences of making an incorrect decision. Both types of decision errors have negative consequences.

(1) *False Positive Error* (declaring the load hazardous when it is not)—If the true cadmium concentration is below 1 mg/L, but the average measured cadmium concentration is above the action level, the nonhazardous fly ash waste will be sent to a RCRA landfill. The consequence of a false positive error is that the company will have to pay additional cost to dispose of the waste with a cadmium concentration between the action level and regulatory threshold at a RCRA facility as opposed to a less expensive method of disposal in a sanitary landfill.

(2) *False Negative Error* (declaring the load nonhazardous when it is hazardous)—If the true cadmium concentration is equal to or greater than 1 mg/L, but the average measured cadmium concentration is below the action level, the hazardous fly ash waste will be sent to a sanitary landfill. The consequence of a false negative error is that the fly ash waste may be disposed of in a manner that will be harmful to human health or the environment. Legal consequences and subsequent remedial costs are also possible consequences.

X1.2.7.2 The purpose of this stage of the process is to

specify the probabilities of making incorrect decisions that are acceptable to decision makers. The DQO team must agree on which type of decision error is of greater concern, either a false positive error or false negative error.

X1.2.7.3 For this example, the DQO team is more concerned about a false negative error because of the increased liability due to sending potentially hazardous waste to a sanitary landfill. The DQO team set a value for the false negative error of 10 % when the true concentration is 1 mg/L. The false negative error is a greater concern because of the perceived increased liability due to sending potentially hazardous waste to a sanitary landfill. This level is determined based on the comfort of the decision makers accepting the risk associated with calling a hazardous waste nonhazardous.

X1.2.7.4 *Data Quality Objective Summary*—Application of the DQOs on a day-to-day basis depends on (*1*) selecting the number of samples and (*2*) quantifying the action level for the decision rule. The decision performance curves are used to visually compare the desired decision errors versus the possible true cadmium concentrations for different numbers of samples.

(*1*) The uncertainty for the DQOs can be quantified by calculating the action level based on a false negative error of 10 % when the true cadmium concentration of a TCLP extract for a fly ash load has a value of the regulatory threshold (1 mg/L).

(*2*) To begin the early phases of design optimization, the DQO team determined how the environmental data should be summarized and used in the decision. The DQO team identified that the mean concentration of cadmium from each load would be compared to the action level. The background data indicated that a normal distribution can be used to calculate the action level. A normal distribution is an appropriate probability model for the preliminary data. A false negative error less than 50 % implies that an action level will be lower than the regulatory threshold.

(*3*) How the statisticians on the DQO team calculated the action level for the project is shown as follows. The action level is dependent on variables such as regulatory threshold, standard deviation, false negative error, and number of samples. Changing one variable will affect the value of the action level. Another iteration through the last DQO process steps must be made if any of these changes are made.

X1.2.7.5 *Concentration Range and Action Level*—The DQO team examined the concentration data from the first 20 analyses and determined that a reasonable concentration range to examine was between 0.6 and 1.3 μg/L. The DQO team agreed that the action level should be based on a 10 % false negative error at the regulatory threshold. This implies that the action level will be less than the regulatory threshold. Paragraph X1.2.8 describes the calculations for several action levels corresponding to different numbers of samples in the decision performance curve, using the standard deviation, the limits of error, and the desired false negative error. The decision performance curve will be calculated to determine the action level and review the performance of the decision rule. To calculate the decision performance curve, decision makers use the following steps:

(*1*) *Step 1—Number of Samples:*

(*a*) Selecting the number of samples is always difficult because imperfect knowledge is available concerning the variability of the measurement process for the selected sample matrix. All calculations for the number of samples are approximations. Different methods can be used to determine the number of samples. For the cadmium example, an initial selection of the number of samples is determined by an estimation method that specifies the precision limits on determining the concentration in the TCLP extract. Another sample size method would be based on the decision performance curve that examines the effect of a different number of samples on the decision errors. This decision method for number of samples is investigated in X1.2.8. Another method would be to calculate the number of samples for specified values of the measurement standard deviation, action level, and false positive error and false negative error. This procedure is illustrated in Guides C 970 and C 1215.

(*b*) For the initial fly ash waste loads, chemists on the DQO team would like to verify that their instrument is calibrated for the proper concentration range. They want to estimate the true cadmium concentration in the TCLP extract with an uncertainty of ±0.2 mg/L. In addition, the decision makers are willing to allocate resources to learn that the true cadmium concentration is in this interval with a confidence of 95 %. The number of samples for these precision limits can be approximated by a normal probability distribution. Another approximation to the number of samples could use an iterative method for a Student's *t*-distribution rather than the normal distribution. This more general assumption usually adds only two or three samples beyond the normal distribution used herein.

(*c*) The number of samples (*n*) is calculated by the following equations (**9, 10**), with $L = 0.2$ mg/L, $\sigma = S_w = 0.4$ mg/L, and $\alpha = 0.05$ (or $Z_{\alpha/2} = 1.960$ for a 95 % confidence level):

$$n = \left(\frac{Z_{\alpha/2}\ \sigma}{L}\right)^2$$

$$n = \left(\frac{1.960 \times 0.4}{0.2}\right)^2 \approx 16 \qquad \text{(X1.1)}$$

where:

n = number of samples,

L = limit of error on the average (for example, 0.2 mg/L),

$1 - \alpha$ = probability level for the confidence interval for $\alpha = 0.05$, and then $1 - \alpha = 0.95$ confidence interval,

σ = standard deviation of the measurement process (for example, 0.4 mg/L), and

$Z_{\alpha/2}$ = $\alpha/2$ percentile point of normal probability distribution (for example, $Z_{\alpha/2} = Z_{0.025}$). Common normal percentile values are given in Table X1.1.

(*2*) *Step 2—Action Level*—The action level value for the decision rule is determined by controlling the false negative error established in the DQO process. The quantification of the action level used a value of 0.10 (or 10 %) for the probability of the false negative error and 16 samples to determine the average cadmium concentration from the TCLP extracts. The probability calculations are based on an approximating normal probability distribution for the cadmium concentration measurements. This approximating normal probability assumes a mean = RT = 1.0 mg/L and a

TABLE X1.1 Common Normal Percentile Points

$Z_{0.20}$	$Z_{0.10}$	$Z_{0.05}$	$Z_{0.025}$	$Z_{0.01}$	$Z_{0.005}$
0.842	1.282	1.645	1.960	2.326	2.576

standard deviation = S_w = 0.4 mg/L. The 10 % percentile point for the standardized normal probability distribution is $Z_{0.10}$ = 1.282 (see Table X1.1). The probability (Pr) for the false negative error evaluated at RT is as follows:

Pr (false negative error)

= Pr (average < AL when the true concentration = RT) = 0.10

or

$$Pr(FN) = Pr \left[\frac{average - RT}{S_w/\sqrt{n}} < \frac{AL - RT}{S_w/\sqrt{n}} \right] = 0.10,$$

$$\frac{AL - RT}{S_w/\sqrt{n}} = -Z_{0.10},$$

$$AL = RT - Z_{0.10} \frac{S_w}{\sqrt{n}}.$$

AL = 1.0 mg/L − (1.282)(0.4 mg/L)/4 = 1.0 mg/L − 0.13 mg/L,

$$AL = 0.87 \text{ mg/L}. \quad (X1.2)$$

where:
AL = action level,
RT = regulatory threshold,
S_w = standard deviation of the measurement process estimated from a sufficient number of samples, and
$Z_{0.10}$ = tabulated 10 % percentile point from a standard normal distribution (see Table X1.1).

Therefore, the decision rule is as follows:

(a) If (average concentration of cadmium) ≥ 0.87 mg/L, the fly ash load is considered to be a RCRA waste and will be disposed of in a RCRA landfill; or

(b) If (average concentration of cadmium) < 0.87 mg/L, the fly ash load is not considered to be a RCRA waste and will be disposed of in a sanitary landfill.

X1.2.7.6 *Decision Tree Format*—Figure X1.1 shows the decision tree format for the DQOs, along with the action level and tasks that are required in the data-collection design step.

(1) *Step 3—True Concentration Corresponding to the False Positive Error:*

(a) Calculate the true concentration (θ mg/L < RT) that corresponds to a probability for the false positive error of

20 % using an action level of AL = 0.87 mg/L. This calculation again uses the approximating normal probability distribution for the cadmium concentration measurements. For the specified false positive error, the approximating normal probability assumes a mean = θ mg/L (to be determined), a standard deviation = S_w = 0.4 mg/L, and the number of samples = 16. The 20 % percentile point for the standardized normal probability distribution is $Z_{0.20}$ = 0.842 (see Table X1.1).

Pr (false positive error) = Pr {average ≥ AL
when the true concentration = θ < RT} = 0.20

or

$$Pr(FP) = Pr \left[\frac{average - \theta}{S_w/\sqrt{n}} \geq \frac{AL - \theta}{S_w/\sqrt{n}} \right] = 0.20,$$

$$\frac{AL - \theta}{S_w/\sqrt{n}} = +Z_{0.20},$$

$$\theta = AL - Z_{0.20} \frac{S_w}{\sqrt{n}}.$$

θ = 0.87 mg/L − (0.842)(0.4 mg/L)/4 = 0.87 mg/L − 0.08 mg/L,

$$\theta = 0.79 \text{ mg/L}. \quad (X1.3)$$

where:
AL = action level,
RT = regulatory threshold, and
$Z_{0.20}$ = tabulated 20 % percentile point from a standard normal distribution (see Table X1.1).

(b) The decision performance curve would have a probability of taking an action (that is, sending fly ash waste to a RCRA landfill) of 0.20 at a true cadmium concentration of θ = 0.79 mg/L. The possible true cadmium concentration values in the interval (0.79 and 1.0 mg/L) represent values that cause the decision rule to send fly ash waste to a RCRA landfill even though the true concentration is below the regulatory threshold. This interval can be reduced by increasing the number of samples, changing the false negative error, or changing the false positive error.

(2) *Step 4—Drawing the Decision Performance Curve:*

(a) Draw the decision performance curve by using the standardized normal probability distribution. The standardized normal probability distribution is defined as a normal probability distribution with mean = 0 and standard deviation = 1.0. There are many tables and computer programs that can be used to calculate probabilities for a standardized normal random variable, Z. A normal random variable, X, with mean = μ and standard deviation = σ can be transformed to a standardized normal random variable by $Z = (X - \mu)/\sigma$.

Prob (action) = Pr (average ≥ AL when the true concentration = θ)

$$Prob (action) = 1.0 - Prob \left(Z \leq \frac{AL - \theta}{S_w/\sqrt{n}} \right),$$

$$Prob (action) = 1.0 - Prob \left(Z \leq \frac{0.87 - \theta}{0.1} \right). \quad (X1.4)$$

(b) Figure X1.2 is a plot of the decision performance curve generated by calculating a Prob (action) value using the standard normal probability distribution for each possible true concentration value θ. The decision performance curve can frequently be drawn freehand if three pairs of

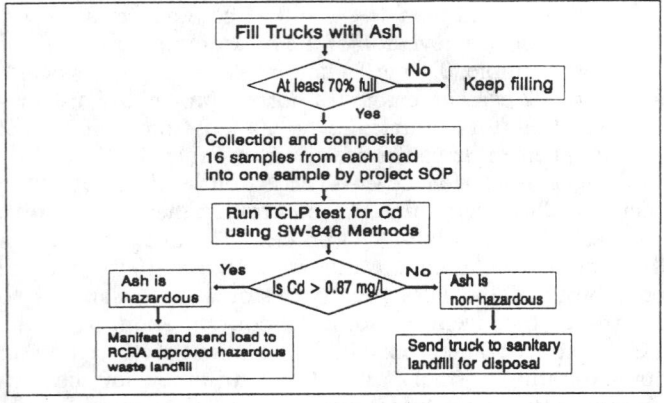

FIG. X1.1 Decision Tree for the Cadmium Example

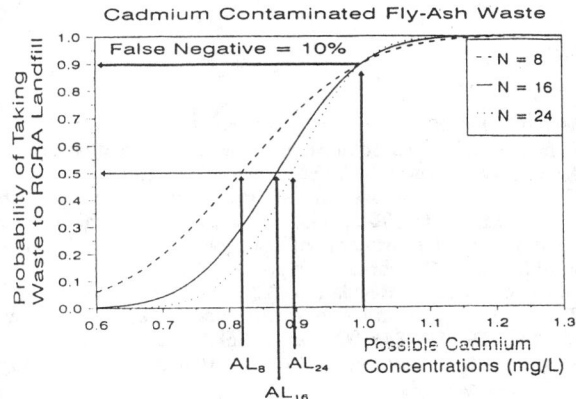

FIG. X1.2 Decision Performance Curves for Cadmium Example

(concentration and probability) values are determined: ((RT, 1 − Pr (false negative error)), (AL, 0.50), and (θ, Pr (false positive error)).

X1.2.8 *Optimizing Data Collection and Design*—The decision makers will select the lowest-cost sampling design that is expected to achieve the DQOs. The series of designs for sampling the fly ash waste will be generated by the statisticians on the DQO team. The choice of sampling plan will be decided by consensus.

X1.2.8.1 *Decision Performance Curve*—The decision performance curve in Fig. X1.2 plots the probability of taking action (disposing of the waste in a RCRA landfill) versus different possible values for the true concentration in the TCLP extract. The DQO process specified a probability of 0.10 for the false negative error when the true concentration is at the RT. This specified false negative error implies that the decision performance curve will have a probability of taking action equal to 0.90 when the true concentration is equal to RT. If the true concentration value is equal to the value of the action level (0.87 mg/L), there is a probability of taking action of 0.50. The DQO team can also determine the true concentration for a specified false positive error from the decision performance curve.

(*1*) Figure X1.2 shows three decision performance curves for three different numbers of samples (8, 16, and 24). All three decision performance curves meet the specified probability for the false negative error of 0.10 at a true concentra-

tion equal to RT. The purpose of these curves is to assess the effects of taking more or fewer samples on the action level and the false positive error. This analysis can be used to update applying the decision rule. For example, the decision makers concluded that eight additional samples (that is, 24) does not improve the AL value and false positive error sufficiently to justify the increase in cost.

X1.2.8.2 *Implementation*—Cadmium concentration values from the TCLP extracts will be collected over a long time period because this waste stream is a continuous process. The decision makers will establish a QC program to monitor the cadmium concentration values for process changes. After every 30 fly ash loads, the process variability will be re-estimated and new values for the number of samples and action level will be considered. This strategy becomes part of the decision process.

X1.2.8.3 *Documentation of the Data Quality Objective Process*—The following statements and information document the outputs of the specific DQO process used to develop the above-stated DQOs. These objectives are meaningless if they are not connected with the specific problem and other qualifying information used in the DQO development.

(*1*) The DQO team required that the documentation be a concise summary of the following information:

(*a*) Facility name, location, and process;

(*b*) List of DQO team members, affiliations, and responsibilities for this project;

(*c*) Statement of the problem;

(*d*) Logic for the solutions chosen for consideration;

(*e*) Information and inputs required by the DQO team to make the decision, including sample matrix, preliminary study results, sampling methods required, and use of each input in reaching a decision.

(*f*) Defined boundaries;

(*g*) Decision logic in rule or decision tree format; and

(*h*) Assumptions made regarding the decision error and any information used to generate preliminary action levels and the number of samples.

(*2*) All meetings held by the DQO team should be documented. The meeting minutes should include the attendees, information used to generate each step of the process, and rationale used to make final agreements on the decision logic, boundaries, inputs, and decision errors.

REFERENCES

(1) *Guidance for Planning for Data Collection in Support of Environmental Decision Making Using the Data Quality Objectives Process*, EPA QA/G-4, U.S. Environmental Protection Agency, Washington, DC, 1993.

(2) Blacker, S., and Maney, J., "The DQO Process: System DQO Planning Process and the Analytical Laboratory," *Environmental Testing and Analysis*, July/August 1993.

(3) Neptune, D., Blacker, S., Fairless, B., and Ryti, R., "Application of Total Quality Principles to Environmental Data Operations," *National Energy Division Conference*, American Society of Quality Control, 1990.

(4) *A Rationale for the Assessment of Errors in the Sampling of Soils*, EPA-600/4-90-013, Environmental Monitoring Systems Laboratory, U.S. Environmental Protection Agency, Washington, DC, May 1990.

(5) *Methods for Evaluating the Attainment of Cleanup Standards Volume 1: Soils and Solid Media*, U.S. Environmental Protection Agency, Washington, DC, February 1989.

(6) *Characterizing Heterogeneous Wastes: Methods and Recommendations*, EPA 600/R-92/033, U.S. Environmental Protection Agency and Department of Energy Office of Technology, Washington, DC, February 1992.

(7) 40 Code of Federal Regulations (CFR), Part 261, 1995.

(8) *Test Methods for Evaluating Solid Waste*, 3rd ed., with Updates I, II, IIA, IIB, SW-846, Office of Solid Waste and Emergency Response, U.S. Environmental Protection Agency, Washington, DC, January 1992.

(9) Cochran, W. G., *Sampling Techniques*, 3rd ed., John Wiley and Sons, Inc., New York, NY, 1977.

(10) Desu, M. M., and Raghavarao, D., *Sample Size Methodology*, Academic Press, San Diego, CA, 1990.

Standard Guide for
Representative Sampling for Management of Waste and Contaminated Media[1]

This standard is issued under the fixed designation D 6044; the number immediately following the designation indicates the year of original adoption or, in the case of revision, the year of last revision. A number in parentheses indicates the year of last reapproval. A superscript epsilon (ϵ) indicates an editorial change since the last revision or reapproval.

1. Scope

1.1 This guide covers the definition of representativeness in environmental sampling, identifies sources that can affect representativeness (especially bias), and describes the attributes that a representative sample or a representative set of samples should possess. For convenience, the term "representative sample" is used in this guide to denote both a representative sample and a representative set of samples, unless otherwise qualified in the text.

1.2 This guide outlines a process by which a representative sample may be obtained from a population. The purpose of the representative sample is to provide information about a statistical parameter(s) (such as mean) of the population regarding some characteristic(s) (such as concentration) of its constituent(s) (such as lead). This process includes the following stages: (1) minimization of sampling bias and optimization of precision while taking the physical samples, (2) minimization of measurement bias and optimization of precision when analyzing the physical samples to obtain data, and (3) minimization of statistical bias when making inference from the sample data to the population. While both bias and precision are covered in this guide, major emphasis is given to bias reduction.

1.3 This guide describes the attributes of a representative sample and presents a general methodology for obtaining representative samples. It does not, however, provide specific or comprehensive sampling procedures. It is the user's responsibility to ensure that proper and adequate procedures are used.

1.4 The assessment of the representativeness of a sample is not covered in this guide since it is not possible to ever know the true value of the population.

1.5 Since the purpose of each sampling event is unique, this guide does not attempt to give a step by step account of how to develop a sampling design that results in the collection of representative samples.

1.6 Appendix X1 contains two case studies, which discuss the factors for obtaining representative samples.

1.7 *This standard does not purport to address all of the safety concerns, if any, associated with its use. It is the responsibility of the user of this standard to establish appropriate safety and health practices and determine the applicability of regulatory limitations prior to use.*

2. Referenced Documents

2.1 *ASTM Standards:*
D 3370 Practices for Sampling Water from Closed Conduits[2]
D 4448 Guide for Sampling Groundwater Monitoring Wells[3]
D 4547 Practice for Sampling Waste and Soils for Volatile Organics[3]
D 4700 Guide for Soil Sampling from the Vadose Zone[4]
D 4823 Guide for Core-Sampling Submerged, Unconsolidated Sediments[5]
D 5088 Practice for Decontamination of Field Equipment Used at Nonradioactive Waste Sites[6]
D 5792 Practice for Generation of Environmental Data Related to Waste Management Activities: Development of Data Quality Objectives[3]
D 5956 Guide for Sampling Strategies for Heterogeneous Wastes[3]
D 6051 Guide for Composite Sampling and Field Subsampling for Environmental Waste Management Activities[3]

3. Terminology

3.1 *analytical unit, n*—the actual amount of the sample material analyzed in the laboratory.

3.2 *bias, n*—a systematic positive or negative deviation of the sample or estimated value from the true population value.

3.2.1 *Discussion*—This guide discusses three sources of bias—sampling bias, measurement bias, and statistical bias.

There is a sampling bias when the value inherent in the physical samples is systematically different from what is inherent in the population.

There is a measurement bias when the measurement process produces a sample value systematically different from that inherent in the sample itself, although the physical sample is itself unbiased. Measurement bias can also include any systematic difference between the original sample and the sample analyzed, when the analyzed sample may have been altered due to improper procedures such as improper sample preservation or preparation, or both.

There is a statistical bias when, in the absence of sampling bias and measurement bias, the statistical procedure produces a biased estimate of the population value.

Sampling bias is considered the most important factor affecting inference from the samples to the population.

[1] This guide is under the jurisdiction of ASTM Committee D-34 on Waste Management and is the direct responsibility of Subcommittee D34.01 on Sampling and Monitoring.
Current edition approved Nov. 10, 1996. Published January 1997.

[2] *Annual Book of ASTM Standards*, Vol 11.01.
[3] *Annual Book of ASTM Standards*, Vol 11.04.
[4] *Annual Book of ASTM Standards*, Vol 04.08.
[5] *Annual Book of ASTM Standards*, Vol 11.02.
[6] *Annual Book of ASTM Standards*, Vol 04.09.

3.3 *biased sampling, n*—the taking of a sample(s) with prior knowledge that the sampling result will be biased relative to the true value of the population.

3.3.1 *Discussion*—This is the taking of a sample(s) based on available information or knowledge, especially in terms of visible signs or knowledge of contamination. This kind of sampling is used to detect the presence of localized contamination or to identify the source of a contamination. The sampling results are not intended for generalization to the entire population. This is one form of authoritative sampling (see *judgment sampling.*)

3.4 *characteristic, n*—a property of items in a sample or population that can be measured, counted, or otherwise observed, such as viscosity, flash point, or concentration.

3.5 *composite sample, n*—a combination of two or more samples.

3.6 *constituent, n*—an element, component, or ingredient of the population.

3.6.1 *Discussion*—If a population contains several contaminants (such as acetone, lead, and chromium), these contaminants are called the constituents of the population.

3.7 *Data Quality Objectives, DQOs, n*—qualitative and quantitative statements derived from a DQO process describing the decision rules and the uncertainties of the decision(s) within the context of the problem(s) (see Practice D 5792).

3.8 *Data Quality Objective Process*—a quality management tool based on the Scientific Method and developed by the U.S. Environmental Protection Agency to facilitate the planning of environmental data collection activities. The DQO process enables planners to focus their planning efforts by specifying the use of data (the decision), the decision criteria (action level), and the decision maker's acceptable decision error rates. The products of the DQO process are the DQOs (see Practice D 5792).

3.9 *error, n*—the random or systematic deviation of the observed sample value from its true value (see *bias* and *sampling error*).

3.10 *heterogeneity, n*—the condition or degree of the population under which all items of the population are not identical with respect to the characteristic(s) of interest.

3.10.1 *Discussion*—Although the ultimate interest is in the statistical parameter such as the mean concentration of a constituent of the population, heterogeneity relates to the presence of differences in the characteristics (for example, concentration) of the units in the population. It is due to the presence of fundamental heterogeneity (or fundamental error)[7] in the population that sampling variance arises. Degree of sampling variance defines the degree of precision in estimating the population parameter using the sample data. The smaller the sampling variance is, the more precise the estimate is. See also *sampling error.*

3.11 *homogeneity, n*—the condition of the population under which all items of the population are identical with respect to the characteristic(s) of interest.

3.12 *judgment sampling, n*—taking of a sample(s) based

on judgment that it will more or less represent the average condition of the population.

3.12.1 *Discussion*—The sampling location(s) is selected because it is judged to be representative of the average condition of the population. It can be effective when the population is relatively homogeneous or when the professional judgment is good. It may or may not introduce bias. It is a useful sampling approach when precision is not a concern. This is one form of authoritative sampling (see *biased sampling.*)

3.13 *population, n*—the totality of items or units of materials under consideration.

3.14 *representative sample, n*—a sample collected in such a manner that it reflects one or more characteristics of interest (as defined by the project objectives) of a population from which it is collected.

3.14.1 *Discussion*—A representative sample can be a single sample, a collection of samples, or one or more composite samples. A single sample can be representative only when the population is highly homogeneous.

3.15 *representative sampling, n*—the process of obtaining a representative sample or a representative set of samples.

3.16 *representative set of samples, n*—a set of samples that collectively reflect one or more characteristics of interest of a population from which they were collected. See *representative sample.*

3.17 *sample, n*—a portion of material that is taken for testing or for record purposes.

3.17.1 *Discussion*—Sample is a term with numerous meanings. The scientist collecting physical samples (for example, from a landfill, drum, or monitoring well) or analyzing samples considers a sample to be that unit of the population that was collected and placed in a container. A statistician considers a sample to be a subset of the population, and this subset may consist of one or more physical samples. To minimize confusion, the term *sample*, as used in this guide, is a reference to either a physical sample held in a sample container, or that portion of the population that is subjected to in situ measurements, or a set of physical samples. See *representative sample.*

3.17.1.1 The term *sample size* also means different things to the scientist and the statistician. To avoid confusion, terms such as sample mass/sample volume and number of samples are used instead of sample size.

3.18 *sampling error*—the systematic and random deviations of the sample value from that of the population. The systematic error is the *sampling bias*. The random error is the *sampling variance*.

3.18.1 *Discussion*—Before the physical samples are taken, potential sampling variance comes from the inherent population heterogeneity (sometimes called the "fundamental error," see *heterogeneity*). In the physical sampling stage, additional contributors to sampling variance include random errors in collecting the samples. After the samples are collected, another contributor is the random error in the measurement process. In each of these stages, systematic errors can occur as well, but they are the sources of bias, not sampling variance.

3.18.1.1 Sampling variance is often used to refer to the total variance from the various sources.

3.19 *stratum, n*—a subgroup of the population separated

---

[7] Pitard, F. F., "*Pierre Gy's Sampling Theory and Sampling Practice: Heterogeneity, Sampling Correctness and Statistical Process Control*," 2nd ed., CRC Press Publishers, 1993.

in space or time, or both, from the remainder of the population, being internally similar with respect to a target characteristic of interest, and different from adjacent strata of the population.

3.19.1 *Discussion*—A landfill may display spatially separated strata, such as old cells containing different wastes than new cells. A waste pipe may discharge into temporally separated strata of different constituents or concentrations, or both, if night-shift production varies from the day shift. In this guide, strata refer mostly to the stratification in the concentrations of the same constituent(s).

3.20 *subsample, n*—a portion of the original sample that is taken for testing or for record purposes.

4. Significance and Use

4.1 Representative samples are defined in the context of the study objectives.

4.2 This guide defines the meaning of a representative sample, as well as the attributes the sample(s) needs to have in order to provide a valid inference from the sample data to the population.

4.3 This guide also provides a process to identify the sources of error (both systematic and random) so that an effort can be made to control or minimize these errors. These sources include sampling error, measurement error, and statistical bias.

4.4 When the objective is limited to the taking of a representative (physical) sample or a representative set of (physical) samples, only potential sampling errors need to be considered. When the objective is to make an inference from the sample data to the population, additional measurement error and statistical bias need to be considered.

4.5 This guide does not apply to the cases where the taking of a nonrepresentative sample(s) is prescribed by the study objective. In that case, sampling approaches such as judgment sampling or biased sampling can be taken. These approaches are not within the scope of this guide.

4.6 Following this guide does not guarantee that representative samples will be obtained. But failure to follow this guide will likely result in obtaining sample data that are either biased or imprecise, or both. Following this guide should increase the level of confidence in making the inference from the sample data to the population.

4.7 This guide can be used in conjunction with the DQO process (see Practice D 5792).

4.8 This guide is intended for those who manage, design, and implement sampling and analytical plans for waste management and contaminated media.

5. Representative Samples

5.1 Samples are taken to infer about some statistical parameter(s) of the population regarding some characteristic(s) of its constituent(s) of interest. This is discussed in the following sections.

5.2 *Samples*—When a representative sample consists of a single physical sample, it is a sample that by itself reflects the characteristics of interest of the population. On the other hand, when a representative sample consists of a set of physical samples, the samples collectively reflect some characteristics of the population, though the samples individually may not be representative. In most cases, more than one physical sample is necessary to characterize the population,

because the population in environmental sampling is usually heterogeneous.

5.3 *Constituents and Characteristics*—A population can possess many constituents, each with many characteristics. Usually it is only a subset of these constituents and characteristics that are of interest in the context of the stated problem. Therefore, samples need to be representative of the population only in terms of these constituent(s) and characteristic(s) of interest. A sampling plan needs to be designed accordingly.

5.4 *Parameters*—Similarly, samples need to be representative of the population only in the parameter(s) of interest. If the interest is only in estimating a parameter such as the population mean, then composite samples, when taken correctly, will not be biased and therefore constitute a representative sample (regarding bias) for that parameter. On the other hand, if the interest happens to be the estimation of the population variance (of individual sampling units), another parameter, then the variance of the composite samples is a biased estimate of the population variance and therefore is not representative. (It is to be noted that composite samples are often used to increase the precision in estimating the population mean and not to estimate the population variance of individual sampling units.)

5.5 *Population*—Since the samples are intended to be representative of a population, a population must be well defined, especially in its spatial or temporal boundaries, or both, according to the study objective.

5.6 *Representativeness*—The word "reflects" in this guide is used to mean a certain degree of low bias and high precision when comparing the sample value(s) to the population value(s). This is a broad definition of sample representativeness used in this guide. A narrower definition of representativeness is often used to mean simply the absence of bias.

5.6.1 *Bias*—Bias is sometimes mistakenly taken to be "a difference between the observed value of a physical sample and the true population value." The correct definition of bias is "a *systematic* (or consistent) difference between an observed (sample) value and the true population value." The word "systematic" here implies "on the average" over a set of physical samples, and not a single physical sample. Recall that sampling error consists of the random and systematic deviations of a sample (or estimated) value from that of the population. Although random deviations may occur on occasions due to imprecision in the sampling or measurement processes, or both, they balance out on the average and lead to no systematic difference between the sample (or estimated) value and the population value. The random deviation corresponds to the observation of "a random difference between a single physical sample value and the true population value," which can be randomly positive or negative, and is not a bias. On the other hand, a persistent positive or negative difference is a systematic error and is a bias.

5.6.1.1 In order to assess bias, the true population value must be known. Since the true population value is rarely known, bias cannot be quantitatively assessed. However, this guide provides an approach to identifying the potential sources of bias and general considerations for controlling or minimizing these potential biases.

5.6.2 *Precision*—Precision has to do with the level of confidence in estimating the population value using the sample data. If the population is totally homogeneous and the measurement process is flawless, a single sample will provide a completely precise estimate of the population value. When the population is heterogeneous or the measurement process is not totally precise, or both, a larger number of samples will provide a more precise estimate than a smaller number of samples.

5.6.2.1 In the case of bias, the goal in environmental sampling is its absence. In the case of precision, the goal in sampling will depend on factors such as:

(*1*) The precision level needed to achieve the desired levels of decision errors, both false positive and false negative errors,

(*2*) If the true value is known or suspected to be well below the regulatory limit, high precision in the samples may not be needed, and

(*3*) The study budget.

5.6.2.2 Note that the second item applies similarly to bias as well.

5.6.2.3 Since bias, especially during sampling, can be very large when proper procedures are not followed, it is considered to be the first necessary condition for sample representativeness. On the other hand, precision can be more or less controlled, for example, by increasing the number of samples taken or by decreasing the sampling or measurement variabilities, or both.

5.6.2.4 The optimal number of samples to take to achieve a desired level of precision is typically an issue in optimization of a sampling plan. Therefore, the precision issue will be covered only briefly in this guide.

6. A Systematic Approach to Representative Sampling

6.1 A systematic approach is one that first defines the desired end result and then designs a process by which such a result can be obtained. In representative sampling, the desired end result is a sample or a set of samples that achieves desired levels of low bias and high precision.

6.2 A representative sampling process is described in Fig. 1. The key components in the process are described in this section.

6.3 *Study Objective*—A sampling plan is designed according to a defined problem or a stated study objective. The samples are then collected according to the sampling plan. Generally, the study objective dictates that representative samples be taken for the purpose of inference about the population. In that case, these samples will need to be collected according to this guide in order for the inference to be valid. Occasionally, the objective is merely to detect the presence of a contaminant or to obtain a "worst case" sample. In that case, an authoritative sampling approach (biased sampling or judgment sampling) may be taken and this guide does not apply.

6.4 *Population*—A population consists of the totality of items or units of materials under consideration (Compilation of ASTM Standard Definitions, 1990). Its boundaries (spatial or temporal, or both) are defined according to the problem statement. This population is usually called the *target population*. In order to solve the stated problem, samples must be taken from the target population.

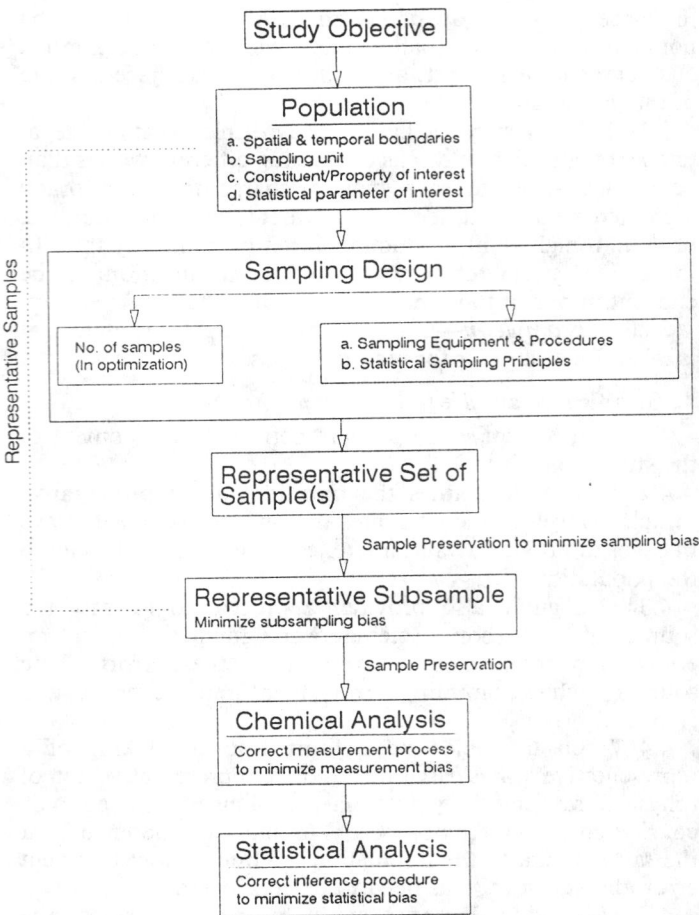

FIG. 1 A Systematic Approach to Representative Sampling

6.4.1 *Sampled Population*—Sometimes some parts of the target population may not be amenable to sampling due to factors such as accessibility. The boundaries of the target population actually sampled due to factors such as incomplete accessibility define the sampled population.

6.4.1.1 Although the samples taken from the sampled population may be representative of the sampled population, they may not be representative of the target population. In this case, potential exists that the samples taken from the sampled population may systematically deviate from the true value of the target population, thereby introducing bias when making inference from the samples to the target population.

6.4.1.2 When the boundaries of the target and sampled populations are not identical, some possible solutions are:

(*1*) The parties to the decision-making may agree that the sampled population is a sufficient approximation to the target population. A sampling plan can then be designed to take representative samples from the "sampled population,"

(*2*) Qualifications on the sampling results are made based on the differences between the two populations. Some professional judgment may have to be exercised here, and

(*3*) Redefine the problem by considering what problem is solvable based on the observed differences between the two populations.

6.4.1.3 Occasionally, the sampled population is chosen on purpose to be different from the target population. For example, an investigator may be interested in the lead content in the sludge of a surface impoundment (the target population). He may decide to take samples from the sludge near the inlet (sampled population). Thus, the impoundment is the target population, while the inlet area is the sampled population. If the interest is in the target population, then this is an example of a biased sampling approach. On the other hand, the involved parties may decide to redefine the target population to include only the inlet area. Then the target population and the sampled population are identical. Again, the definition of a population depends on the problem statement.

6.4.1.4 In yet other circumstances, an investigator may take only a sample from the population. The following cases are possible:

(1) This one physical sample can be a sample from a biased sampling approach, for the purpose of detecting the presence of a contaminant or identifying the source of contamination. Therefore, it is not a representative sample due to its bias,

(2) This one physical sample can be a sample from judgment sampling, for the purpose of estimating the average condition of the population. Bias may or may not exist depending to some degree on the expertise of the sampler,

(3) This sample can be viewed as a population itself if the investigator is interested in the sample alone and a result from this sample is not to be used to infer to areas outside the sample. In this case, no bias exists, and

(4) If this sample is the composite of a few samples taken from the population, bias is likely to be minimal if the original samples are carefully taken.

6.4.2 *Decision Unit*—Often a population may be divided into several exposure units, cleanup units, or strata. If the environmental management decision is to be made for the entire population as a whole, representative samples can be obtained by designs such as a stratified random sampling design. Here the entire population is the decision unit. On the other hand, if the decision is to be made on each unit or stratum, then each unit or stratum is the decision unit. In this case, representative sample(s) need to be taken from each unit or stratum as if the unit or stratum is the population.

6.4.2.1 If the units or strata are relatively small in size or too numerous to take many samples per unit or stratum, composite sample(s) can be taken from each unit or stratum to increase precision without introducing bias. Alternatively, if precision is not a concern and there is sufficient professional expertise to avoid bias, a judgment sample(s) can be taken from each unit or stratum.

6.4.3 *Heterogeneity*—Heterogeneity is discussed in greater detail in Guide D 5956.

6.4.3.1 The degree and extent of population heterogeneity affect potential bias and precision in the samples. Population heterogeneity can be viewed at least in three different ways:

(1) When the population is heterogeneous in a random manner in only the distribution of the concentration, but not in the physical materials such as particle sizes, designs such as a simple random sampling design will generally produce samples with minimal bias. Its precision will then depend on the number of samples taken,

(2) When the population is randomly heterogeneous in

concentrations due to large differences in the materials such as particle size, a simple random sampling design may still be effective if the sample volume/weight and sampling equipment are chosen to accommodate the largest particles and thereby prevent introduction of bias, and

(3) If the population is systematically heterogeneous, such as the presence of stratification in concentrations, then a simple random sampling design may not be biased, but will be less precise than an alternative design such as stratified random sampling.

6.4.3.2 Heterogeneity in the population affects the sampling variance. Sampling variance is a function of factors such as the population heterogeneity and the sample volume or weight. It is clear that the more heterogeneous the population is, the larger the inherent sampling variance is. It is also clear that samples of smaller volume or weight will have a higher sampling variance than those with greater volume or weight. However, the reduction in sampling variance due to increased volume or weight may eventually reach a limit. Determination of the optimal sample volume or weight is beyond the scope of this guide.[7]

6.4.3.3 The proper procedure is to first determine the right sample volume or weight, then to determine the number of samples needed for the chosen sample volume or weight.

6.4.3.4 Since stratification as a phenomenon of population heterogeneity is fairly common, it is discussed in greater details as follows.

6.4.4 *Stratification*—There are generally three types of stratification affecting sample representativeness. One is a stratification in the distribution of the contaminant concentration distribution alone. The second is a stratification in sampling materials or matrices alone. The third is a combination of both types. Stratification of any type is not a big problem regarding sample representativeness if each stratum is a decision unit. In that case, the units in a stratum are by definition relatively similar, apart from the random variations in concentrations. A simple random sampling design can be used to obtain representative samples (unbiased) for each stratum. The question of sample representativeness becomes more complicated when a decision is to be made over all the strata in the population.

6.4.4.1 *A Single Representative Sample in A Stratified Population*—When the objective is to obtain a single (physical) representative sample of all the strata, the sample must be a composite of individual samples from the strata (for example, at least one individual sample per stratum). Here the volumes or weights of the individual samples should be proportional to the relative stratum sizes. The composite sample so obtained would be unbiased. However, since there is only one composite sample, precision of the composite sample cannot be estimated. If there are existing data on the precision of the individual samples in the strata, then the precision of the composite sample can be inferred from the precision of the individual samples by theoretical or empirical relationship. See Guide D 6051.

6.4.4.2 *A Representative Set of Samples*—When the population is stratified, a set of samples obtained by statistical designs such as stratified random sampling, where the number of samples to be taken from the strata are proportional to the relative sizes of the strata, is unbiased and more precise than a set of samples taken without considering the stratification.

6.4.5 *Parameter(s) of Interest*—This refers to the statistical parameter such as mean or variance of the population. It is often used with a characteristic such as concentration of a constituent(s) of the population. An example is the mean (parameter) concentration (characteristic) of lead (constituent). Another example is a population of mixture of silt-size calcium carbonate particles and large cobble-size particles of calcium carbonate. The interest here could be in the mean (parameter) particle size or chemical composition (characteristic) of calcium carbonate (constituent), depending on the study objective.

6.5 *Develop A Sampling Design*—The objectives of a sampling design are to minimize bias and achieve a desired level of precision. Precision and bias are an issue at various stages of the process of inferring from the samples to the population. The first stage is the act of obtaining the physical samples. The second stage is the act of analyzing the physical samples and translating them into data. The third stage is the use of statistical method to infer from the sample data to the population. At the first stage, the main concerns are sampling precision and bias. At the second stage, the concerns are measurement of precision and bias. At the third stage, the concern is statistical bias.

6.5.1 At the first stage of obtaining physical samples, the issues of precision and bias are sometimes grouped together as sampling design issues.

6.5.2 Bias at this stage is often called the sampling bias. Sampling bias is the systematic difference between the value inherent in the physical samples and the true population value. The word "inherent" is used because at this point the physical samples have not been translated into data.

6.5.3 The phrase "systematic difference" implies a persistent difference in long-term average or expectation, not the occasional random difference. Representative samples, apart from the issue of precision, are obtained when this long-term expected difference is zero or nearly so.

6.5.4 Since the true population value is typically not known, sampling bias cannot be assessed. However, efforts to minimize sampling bias can be attempted in at least two areas:

6.5.4.1 *Proper Statistical Sampling Design*—Statistical sampling design has to do with where and how samples are to be taken, where equal probability of selecting any of the units or items in the population is often a primary requirement. If the probability of selection is not equal, it is highly likely that bias will have been introduced into the physical samples so obtained. Depending on the layout of the population, designs such as simple random sampling or stratified random sampling can be used.

6.5.4.2 *Proper Sampling Procedures and Sampling Equipment*—This includes proper procedures for compositing, subsampling, sample preparation and preservation, and proper use of the chosen sampling equipment. This is a major source affecting precision and bias, especially bias.

6.5.5 In the case of precision, it can be controlled by things such as the number of samples taken, the use of composite samples, or more precise sampling techniques. Often, the number of samples to take is considered the key design issue. Some considerations regarding precision are:

6.5.5.1 If a population is relatively small compared to the sample mass/volume and the distribution of the characteristic of interest is random, it may be appropriate to collect a smaller number of samples by a random or systematic sampling approach, and

6.5.5.2 If a population is relatively large compared to sample mass/volume and the characteristic of interest is not randomly distributed (for example, stratified), a greater number of samples and a stratified sampling approach may be needed.

6.5.6 *Compositing*—Compositing is the combination of two or more individual physical samples into a single sample. It is often used to reduce the analytical costs, while maintaining or increasing precision relative to the individual samples (see Guide D 6051). Bias may or may not be introduced in compositing, depending on the study objective and the physical means of compositing. For example:

6.5.6.1 If the study calls for the estimation of the population variance (or standard deviation) of individual samples, then composite samples will surely underestimate the population variance, and

6.5.6.2 If the physical means of compositing changes the characteristics of the samples, then bias may have been introduced (unless such changes are part of the study design).

6.6 *Subsampling*—Sampling bias can be introduced in subsampling unless the same proper sampling protocol is followed as in taking samples from the original population.

6.6.1 *Discussion*—After the physical samples have been obtained and before they are measured, bias can be prevented by following proper sample preservation and preparation procedures. It is not important whether these procedures are viewed as part of the sampling process or as part of the measurement process. It is only important in following the proper procedures to prevent bias.

6.7 *Measurement of Precision and Bias:*

6.7.1 The measurement process, like the sampling process, also consists of a random error and a systematic error. The random errors define the degree of measurement precision, and the systematic error defines the degree of measurement bias.

6.7.2 Like sampling precision, measurement precision is controlled by things such as the number of replicate analyses performed per sample and refinements of the analytical method.

6.7.3 Measurement bias is a systematic difference between the sample value produced by the measurement process and the true population value, assuming that the physical samples are unbiased before the analysis. The bias can come from contamination, loss or alteration of the sample materials, systematic errors in the measurement device, or from systematic human errors.

6.7.4 Often the measurement bias can be reasonably estimated in a laboratory testing setting when the true value is known. Laboratory samples spiked with known quantities of a chemical or certified reference standard can often be used to assess potential measurement bias. Minimization or adjustment for such estimable bias in the measurement process is essential in order to obtain data that are unbiased. When estimation of bias is not possible, care in measurement protocol and training is probably the only recourse.

6.7.4.1 *Discussion*—It is important to note that, when

inferring from the sample data to the population, all the sources of imprecision, including sampling, subsampling, and measurement, need to be combined. The process of accumulating these sources of variation is sometimes called the "propagation of errors." The determination of the optimal numbers of samples, subsamples, and replicates are an issue of optimization and is not covered in this guide.

6.8 *Statistical Bias*—Statistical bias can result from an inappropriate sampling design or inappropriate estimation procedures, or both:

6.8.1 *Selection Bias from Sampling Design*—In the course of taking the sample, if the population units do not have the same probability of being selected, bias can be introduced. This bias can be prevented or minimized when a statistical sampling design is carefully selected, based on the study objective and the layout of the population. Some possible designs are the simple random sampling design and the stratified random sampling design.

6.8.2 *Estimation of Bias from Estimation Procedures*— This bias occurs when the expected value of the statistical estimator does not equal the true value.

6.8.2.1 Estimation bias can occur when the wrong statistical distribution of the data is used. For example, if the normal distribution assumption is used when the true data distribution is lognormal, the interval estimate of the mean concentration will be an biased estimate against the true interval. Thus, the expected value of the estimator will not be equal to the true value. To avoid this potential bias, it is wise to check the data distribution.

6.8.2.2 Estimation bias can also occur when a wrong statistical estimator is used. For example, if the sum of squares of deviations from the sample mean divided by the number of samples (that is, $\Sigma_{i=1,n} (x_i - \bar{x})^2/n$) is used to estimate the population variance, then this estimator is biased (its mathematical expected value is not equal to the population variance). If its denominator is modified to be ($n -1$), then it is an unbiased estimator. For an unbiased statistical estimator, the reader is advised to check with a statistician.

7. Attributes of Representative Samples

7.1 The attributes of a representative (physical) sample or a representative set of (physical) samples can be described in the chronological order in which samples are taken. Note that these attributes apply only to how representative the physical samples are of the population. This corresponds to the upper half of Fig. 1.

7.2 *Design Considerations:*

7.2.1 A well-defined target population. The target population includes all the population units as determined from the stated problem.

7.2.2 The sampled population equals the target population in their spatial or temporal boundaries, or both. The sampled population consists of the population units directly available for measurement.[8]

7.2.2.1 When all the population units in the target population are accessible and directly available for measurement,

then the sampled population is identical to the target population in its spatial or temporal boundaries, or both.

7.2.2.2 When not all the population units are directly available for measurement, then the inference from the sample is made to the sampled population, not the target population.

7.2.3 Size (weight or volume) of the sampling unit is well defined.

7.2.3.1 The population can be divided into various sizes (weight or volume) of population units. The size of the sampling unit is the size of the population unit most appropriate for the sampling purposes.

7.2.3.2 The appropriate size of the sample is determined by degree of heterogeneity of the materials to be sampled, such as particle size or shape.

7.3 *Sampling and Measurement Considerations:*

7.3.1 Correct sampling procedures are followed to minimize sampling bias.

7.3.1.1 Absence or minimization of bias is a key attribute of representative samples. Sampling bias can be minimized by following correct sampling procedures. Correct sampling procedures have two components.

(*1*) A sampling procedure that maximizes the potential of population units having equal probability of selection as sampled, and

(*2*) Correct sampling procedures. This includes the selection of appropriate equipment and proper use of that equipment.

7.3.2 Sample integrity is maintained during sampling and before chemical analysis.

7.3.3 If subsampling is performed, correct sampling procedures are followed to minimize sampling bias.

7.3.4 Sample preparation errors such as contamination and loss or alteration of constituents are prevented or minimized.

7.3.5 The samples, in the end, collectively reflect the target population within the context of the problem.

7.3.6 These attributes can be summarized into three broad categories:

7.3.6.1 A well-defined population,

7.3.6.2 Correct sampling procedures, and

7.3.6.3 Samples collected in the context of the stated problem.

8. Practical Considerations

8.1 *Sampling Equipment*—The choice of appropriate sampling equipment can be crucial to the task of collecting a representative sample or a representative set of samples. Depending on the goals of the sampling activity, the sampling device used should minimize bias by having certain characteristics and capabilities, such as:

8.1.1 The ability to access and extract from every location in the target population,

8.1.2 The ability to collect a sample of proper shape,

8.1.3 The ability to collect a sufficient mass or volume of sample such that the distribution of particle sizes in the population are represented, and

8.1.4 The ability to collect a sample without the addition or loss of contaminants of interest.

8.2 *Equipment Design*—The improper design of sampling equipment may result in the collection of samples that

---

[8] Gilbert, Richard O., *Statistical Methods for Environmental Pollution Monitoring*, Van Nostrand Reinholt Co., New York, NY 1987.

are not representative of the population.

8.2.1 An example of equipment design influencing sampling results is samplers which exclude certain sized particles from a soil matrix or waste pile sample. The shape of some scoops may influence the distribution of particle sizes collected from a sample. Dredges used to collect river or estuarine sediments may also exclude certain sized particles, particularly the fines fraction which may contain a significant percentage of some contaminants such as polynuclear aromatic hydrocarbons (PAHs). Specific considerations in equipment design are outlined as follows.

8.2.1.1 *Sample Volume Capabilities*—Most sampling devices will provide adequate sample volume. However, the sampling equipment volumes should be compared to the volume necessary for all required analyses and the additional amount necessary for quality control (QC), split and repeat samples. Taking more than one aliquot to obtain an adequate sample volume can impact the representativeness of a sample.

8.2.1.2 *Compatibility*—It is important that sampling equipment, other equipment that may come in contact with samples (such as gloves, mixing pans, knives, spatulas, spoons, etc.) and sample containers be constructed of materials that are compatible with the matrices and analytes of interest. Incompatibility may result in the contamination of the sample and the degradation of the sampling equipment.

8.2.1.3 *Decontamination (see Practice D 5088) and Reuse*—Inadequate decontamination of sampling equipment can result in contamination of the sample and affects its representativeness. Due to design, some equipment is very difficult to adequately decontaminate. In some instances, it may even be desirable to either dispose of sampling equipment after use or to dedicate the equipment to a sampling point.

8.3 *Sampling Procedure*—Inappropriate use of sampling equipment is one of the largest sources of sampling bias. While it is beyond the scope of this guide to discuss it in depth, examples of how bias can be introduced during the sampling procedure are discussed in the following paragraphs. This guide does not provide comprehensive sampling procedures. It is the responsibility of the user to ensure that proper and adequate procedures are used.

8.3.1 *Ground Water*—For a more comprehensive discussion of sampling ground water refer to Guide D 4448.

8.3.1.1 Ground-water samples are usually collected through an in-place well, either temporarily or permanently installed. The following is a list of concerns that should be considered when collecting a ground-water sample.

(1) The well should be purged before collecting samples in order to clear the well of stagnant water which is not representative of aquifer conditions. Purging and sampling rates can cause chemical or physical changes in the water.

(2) Purging can be performed in such a way that the entire column of water is not removed. The best method for avoiding this situation is by lowering a pump or bailer into the top of the column of water.

(3) Bailing may stir up sediment in the well if conducted too vigorously. Increased turbidity can result in a higher metal content in the sample than in a non-turbid sample.

(4) Samples for volatile organic analysis should be collected in a fashion that minimizes agitation of the sample.

(5) Wells with in-place plumbing must also be purged. Samples should be collected immediately following purging. In order to collect a sample representative of ground water, samples should be collected before the water travels through any hoses or in-line treatment devices.

8.3.2 *Surface Water and Sediment*—For a more comprehensive discussion of sampling surface water and sediment, refer to Practice D 3370 and Guide D 4823. General and specific sampling concerns for collection of surface water and sediment samples are as follows:

8.3.2.1 *General Considerations:*

(1) Although bridges and piers may provide access for water and sediment sampling, these structures can also alter the nature of water flow and thus influence sediment deposition or scouring. Depending on the construction materials, these structures can contaminate samples collected in the immediate vicinity.

(2) Wading for water samples should be done with caution since bottom deposits are easily disturbed resulting in increased sediment in surface water samples and a removal of fines from the sediment sample.

8.3.2.2 *Rivers, Streams, and Creeks:*

(1) A good location to collect a vertically mixed surface water sample is immediately downstream of a riffle area. This location is also a likely area for deposition of sediment since the greatest deposition occurs where stream velocity slows down.

(2) Horizontal (cross-channel) mixing occurs in constrictions in the channel. However, this is a poor sediment sample collection area because of scouring.

(3) Surface water samples will be affected by point sources, such as tributaries and industrial and municipal effluents.

(4) Locations immediately upstream or downstream from the confluence of two streams or rivers may not immediately mix, and at times, due to possible back flow, can upset the normal flow patterns.

(5) Unless a stream is extremely turbulent, it is nearly impossible to measure the effect of a waste discharge or tributary immediately downstream of the source. Inflow frequently "hugs" the stream bank with very little cross-channel mixing for some distance. Samples from quarter points across a stream may miss the wastes altogether and reflect only the quality of water upstream from the waste source. Samples collected within the portion of the cross section containing the wastes would indicate excessive effects of the wastes with respect to the river as a whole.

(6) When sampling tributaries, care should be exercised to avoid collecting water from the main stream that may flow into the mouth of the tributary on either the surface or bottom.

8.3.2.3 *Lakes, Ponds, and Impoundments:*

(1) Stratification of surface water is of greater concern in standing water. For example: A turbidity difference may occur vertically where a highly turbid river enters a lake, and each layer of the stratified water column may need to be considered. In addition, stratification may be caused by water temperature difference; cooler, heavier river water is beneath the warmer lake water.

(2) Dredges used to collect sediment samples can displace and miss lighter materials if allowed to drop freely.

(3) Core samplers used to sample vertical columns of sediment are useful when there is a need to know the history of sediment deposition. Coring devices also minimize the disturbance of fines at the sediment-water interface. However, coring devices can only sample a relatively small surface area. Depending on the core diameter, larger particles may be excluded and a single aliquot may not be sufficient for analytical needs.

8.3.3 *Soils*—For more detailed information, refer to Practice D 4547 and Guide D 4700. General areas of concern for sampling soils are as follows:

8.3.3.1 Soil samples for purgeable organic analyses should be collected with a minimum disturbance of the sample.

8.3.3.2 Samples for VOA analysis should not be mixed.

8.3.3.3 Two potential problems are associated with compositing soil samples. Low concentrations of contaminants present in individual aliquots may be diluted to the extent that the total composite concentration is below the minimum quantification limit. In addition, depending on the soil type, it can be very difficult to produce a homogeneous mixture.

8.3.4 *Waste*—Wastes referred to in this section include any liquid, solid, or sludge from pits, ponds, lagoons, waste piles, landfills, and open or closed containers such as drums, tank trucks, and storage tanks.

8.3.4.1 Any of these units may have multiple phases (floating solids, different density liquid phases, and sludge) and one or all of them may need to be sampled.

8.3.4.2 If sampling from access valves or ports on an open or closed container, care should be taken to be sure that the desired layer is sampled. For example, bottom sampling ports would allow only the heavier contents to be sampled while surface or top sampling would allow only sampling of the lighter layers.

8.4 *Subsampling (Field):*

8.4.1 Different analyses require different types of bottles and preservation. For multiple analyses of the same waste stream, this may require subsampling in the field. Subsampling in the laboratory may require many of the same procedures; however, laboratory subsampling is beyond the scope of this guide.

8.4.1.1 Samples for organic analyses should always be taken from the first material collected. This minimizes loss of volatile organics during handling of the material.

8.4.1.2 If necessary, place the appropriate volume of material in a tray or other suitable container to composite. The volume is dependent on the needed analyses, and should be specified by the analytical laboratory.

8.4.1.3 Transfer the material into the required containers for analyses. If subsampling takes place, then the analytical sample is the final portion of the material subsampled from the original sampling unit and analyzed in the laboratory.

8.4.2 In subsampling, the original sampling unit can be considered as the population and the correct sampling procedures must be followed to ensure a representative subsample.

9. Keywords

9.1 bias; contaminated media; precision; representative; sample; waste; waste management

APPENDIX

(Nonmandatory Information)

X1. TWO CASE STUDIES OF REPRESENTATIVE SAMPLING

X1.1 Case Study One—Waste Pile Investigation

X1.1.1 *Background*—An industrial facility has managed recovery furnace slag and baghouse dust in a waste pile located on the site. No active management was occurring with the waste pile. No buried containers or extremely heterogenous material (debris) was suspected of being present in the waste pile based on facility records and interviews of personnel.

X1.1.1.1 Lead and cadmium were the constituents of concern based on process knowledge, and the possibility for the waste being hazardous by means of the Toxicity Characteristic (TC) Rule was the regulatory consideration. No preliminary information on the variability of lead and cadmium within the piles was available. The potential for off-site migration of contaminants by means of a drainage ditch that leads to a stream adjacent to the facility was an immediate concern.

X1.1.2 *Phase 1: Objective*—The primary objective of the initial investigation was to determine if the slag and baghouse dust in the waste piles were characteristic for lead via the Toxicity Characteristic Rule. A secondary objective was to provide preliminary information on potential migration and transport of contaminants from the waste piles off site.

X1.1.2.1 The sampling design for this initial investigation utilized a judgmental sampling strategy to provide a preliminary estimate of the lead and cadmium concentrations in the waste pile, the variability of contaminant concentrations in the pile, and the potential for leaching using the TCLP. Four areal composite samples were collected from the surface (0 to 6 in.) at the four quadrants of the waste pile. Borings were completed at the center of each area that was sampled on the surface. Each four-foot interval was analyzed to assess vertical variability.

X1.1.2.2 The following environmental samples were also collected using a judgmental approach:

(1) Several soil samples in the vicinity of the waste pile,

(2) Sediment upstream and downstream in a stream that borders the facility,

(3) Sediment in a ditch which contained run-off from the pile, and

(4) Two background soil samples.

X1.1.2.3 *Results*—Zinc, copper, cadmium, and lead were all elevated (compared to background) in the samples collected from the waste piles. Since lead and cadmium are TC Rule constituents, the TCLP was completed, and the lead results exceeded the regulatory level of 5 mg/L. Cadmium was just under the regulatory level of 1.0 mg/L. Lead and cadmium concentrations in the soil near the waste piles were 2 to 3 times above background, and the drainage ditch and downstream sediment sample also had elevated lead and cadmium levels.

X1.1.2.4 *Conclusion*—The waste piles contain slag and baghouse dust that is hazardous for lead. The waste pile requires further characterization to determine the variability in the pile. The presence of lead and cadmium in soils and the stream sediment downstream of the facility was confirmed and should be further investigated to determine the extent of contaminant transport.

X1.1.3 *Phase 2: Objective*—The sampling design utilized a systematic grid approach. This design will delineate horizontal and vertical variability in lead and cadmium concentrations. The Phase 1 investigation also provided a good estimate of the anticipated variability in the waste pile.

X1.1.3.1 The number of samples required to adequately characterize the waste pile was calculated based on the anticipated variability, the regulatory level of concern, and the specified confidence interval. The grid sizes were then adjusted to accommodate the projection on the required number of samples. Composite samples were collected within each grid cell based on one center point and eight points on the compass (45 deg intervals) equidistant from the center point.

X1.1.3.2 Twenty percent of the grids were designated for vertical characterization (at the grid center) at four-foot intervals, as well as surface (0 to 6 in.) sample collection. Additionally, ten percent of the grids were randomly designated for duplicate sampling (using a different aliquot pattern within the cell) to check the preliminary estimate on the variability.

X1.1.3.3 Additional environmental sampling was conducted that included a systematic sampling design for the stream adjacent to the facility with sediment samples collected at 100-ft intervals. A systematic approach was also used for the drainage ditch (50-ft intervals), with judgmental samples being collected at any location where visible staining was observed.

X1.1.3.4 *Results*—The results supported the initial investigation with lead consistently exceeding the TC Rule regulatory level; cadium was consistently below the regulatory level. Vertical differences in the lead and cadmium concentrations were not significant. Lead and cadmium were detected at elevated concentrations (relative to background)

in the adjacent stream at a point downstream of the confluence with the drainage ditch.

X1.1.3.5 *Conclusion*—The waste pile was characteristic for lead and subject to Subtitle C of RCRA. There was no significant variability with depth, although several gradients were noticed across the grid (horizontally) based on lead concentration (scan) results.

X1.2 Case Study Two—Drum Sampling

X1.2.1 *Background*—An industry has two areas where drums of waste have been stored. One area is a warehouse adjacent to an off-line plating process that contains less than 25 drums (55 gal). The drums have manufacturers' labels indicating they contain an acid solution, and all of the drums are similar in appearance. A second area is a covered shed that has an estimated 100 drums from a variety of processes, several of which are no longer in use at the facility. Information on the content of these drums is not available.

X1.2.2 *Objective*—The objective of the initial investigation was to survey both of the storage areas for safety purposes, assess and record information on the drums, and open drums that were candidates for screening. All drums that were opened were surveyed using an organic vapor analyzer (PID, FID), pH paper, halogen detector, cyanide detector, and radiation meter.

X1.2.2.1 A judgmental sampling design was utilized in the warehouse where the anticipated variability was low. Based on the site screening (pH measurement), six samples were collected for pH analysis from the warehouse.

X1.2.2.2 The drums in the shed were screened in a similar fashion. A variety of results were obtained which included elevated pH, high organic vapor readings, and so forth. A simple random sampling design was used which called for the collection of 15 samples, with five from each major group of drums based on the screening (five corrosives, five potential ignitables with no halogens, and five with elevated halogen readings).

X1.2.2.3 *Results*—The warehouse samples were all corrosive with pH values from 1 to 2 S.U. The shed samples resulted in the collection of five corrosive wastes, three that were both ignitable and characteristic for non-halogenated TC Rule constituents, and two that were ignitable and characteristic for halogenated constituents. In summary, of the 15 drums sampled, 10 contained hazardous waste.

X1.2.2.4 *Conclusions*—All of the drums in the warehouse are subject to Subtitle C of RCRA. The drums in the shed require further assessment due to the fact that several of those sampled did not contain hazardous waste.

ASTM D 6044

ASTM Designation: D 6051 – 96

Standard Guide for
Composite Sampling and Field Subsampling for
Environmental Waste Management Activities[1]

This standard is issued under the fixed designation D 6051; the number immediately following the designation indicates the year of original adoption or, in the case of revision, the year of last revision. A number in parentheses indicates the year of last reapproval. A superscript epsilon (ϵ) indicates an editorial change since the last revision or reapproval.

1. Scope

1.1 Compositing and subsampling are key links in the chain of sampling and analytical events that must be performed in compliance with project objectives and instructions to ensure that the resulting data are representative. This guide discusses the advantages and appropriate use of composite sampling, field procedures and techniques to mix the composite sample and procedures to collect an unbiased and precise subsample(s) from a larger sample. It discusses the advantages and limitations of using composite samples in designing sampling plans for characterization of wastes (mainly solid) and potentially contaminated media. This guide assumes that an appropriate sampling device is selected to collect an unbiased sample.

1.2 The guide does not address: where samples should be collected (depends on the objectives) (see Guide D 6044), selection of sampling equipment, bias introduced by selection of inappropriate sampling equipment, sample collection procedures or collection of a representative specimen from a sample, or statistical interpretation of resultant data and devices designed to dynamically sample process waste streams. It also does not provide sufficient information to statistically design an optimized sampling plan, or determine the number of samples to collect or calculate the optimum number of samples to composite to achieve specified data quality objectives (see Practice D 5792). Standard procedures for planning waste sampling activities are addressed in Guide D 4687.

1.3 The sample mixing and subsampling procedures described in this guide are considered inappropriate for samples to be analyzed for volatile organic compounds. Volatile organics are typically lost through volatilization during sample collection, handling, shipping and laboratory sample preparation unless specialized procedures are used. The enhanced mixing described in this guide is expected to cause significant losses of volatile constituents. Specialized procedures should be used for compositing samples for determination of volatiles such as combining directly into methanol (see Practice D 4547).

1.4 *This standard does not purport to address all of the safety concerns, if any, associated with its use. It is the responsibility of the user of this standard to establish appropriate safety and health practices and determine the applicability of regulatory limitations prior to use.*

2. Referenced Documents

2.1 *ASTM Standards:*
C 702 Practice for Reducing Samples of Aggregate to Testing Size[2]
D 1129 Terminology Relating to Water[3]
D 4439 Terminology for Geosynthetics[4]
D 4547 Practice for Sampling Waste and Soils for Volatile Organics[5]
D 4687 Guide for General Planning of Waste Sampling[5]
D 5088 Practice for Decontamination of Field Equipment Used at Nonradioactive Waste Sites[4]
D 5792 Practice for Generation of Environmental Data Related to Waste Management Activities: Development of Data Quality Objectives[5]
D 6044 Guide for Representative Sampling for Management of Wastes and Contaminated Media[5]
E 856 Definitions of Terms and Abbreviations Relating to Physical and Chemical Characteristics of Refuse-Derived Fuel[5]

3. Terminology

3.1 *Definitions:*

3.1.1 *composite sample, n*—a combination of two or more samples.   **D 1129**

3.1.2 *sample, n*—a portion of material taken from a larger quantity for the purpose of estimating properties or composition of the larger quantity.   **E 856**

3.1.3 *specimen, n*—a specific portion of a material or laboratory sample upon which a test is performed or which is taken for that purpose.   **D 4439**

3.1.4 *subsample, n*—a portion of a sample taken for the purpose of estimating properties or composition of the whole sample.

3.1.4.1 *Discussion*—a subsample, by definition, is also a sample.

4. Summary of Guide

4.1 This guide describes how the collection of composite samples, as opposed to individual samples, may be used to: more precisely estimate the mean concentration of a waste analyte in contaminated media, reduce costs, efficiently determine the absence or possible presence of a hot spot (a highly contaminated local area), and, when coupled with retesting schemes, efficiently locate hot spots. Specific proce-

[1] This guide is under the jurisdiction of ASTM Committee D-34 on Waste Management and is the direct responsibility of Subcommittee D34.01 on Sampling and Monitoring.
Current edition approved Dec. 10, 1996. Published February 1996.

[2] *Annual Book of ASTM Standards*, Vol 04.02.
[3] *Annual Book of ASTM Standards*, Vol 11.01.
[4] *Annual Book of ASTM Standards*, Vol 04.09.
[5] *Annual Book of ASTM Standards*, Vol 11.04.

dures for mixing a sample(s) and collecting subsamples for transport to a laboratory are provided.

5. Significance and Use

5.1 This guide provides guidance to persons managing or responsible for designing sampling and analytical plans for determining whether sample compositing may assist in more efficiently meeting study objectives. Samples must be composited properly, or useful information on contamination distribution and sample variance may be lost.

5.2 The procedures described for mixing samples and obtaining a representative subsample are broadly applicable to waste sampling where it is desired to transport a reduced amount of material to the laboratory. The mixing and subsampling sections provide guidance to persons preparing sampling and analytical plans and field personnel.

5.3 While this guide generally focuses on solid materials, the attributes and limitations of composite sampling apply equally to static liquid samples.

6. Attributes of Composite Sampling for Waste Characterization

6.1 In general, the individual samples to be composited should be of the same mass, however, proportional sampling may be appropriate in some cases depending upon the objective. For example, if the objective is to determine the average drum concentration of a contaminant, compositing equals volumes of waste from each drum would be appropriate. If the objective is to determine average contaminant concentration of the waste contained in a group of drums, the volume of each sample to be composited should be proportional to the amount of waste in each drum. Another example of proportional sampling is estimating the contaminant concentration of soil overlying an impermeable zone. Soil cores should be collected from the surface to the impermeable layer, regardless of core length.

6.2 The principal advantages of sample compositing include: reduction in the variance of an estimated average concentration (1),[6] increasing the efficiency of locating/identifying hot spots (2), and reduction of sampling and analytical costs (3). These main advantages are discussed in the following paragraphs. However, a principle assumption needed to justify compositing is that analytical costs are high relative to sampling costs. In general, appropriate use of sample compositing can:

6.2.1 reduce inter-sample variance, that is, improve the precision of the mean estimation while reducing the probability of making an incorrect decision,

6.2.2 reduce costs for estimating a total or mean value, especially where analytical costs greatly exceed sampling costs (also may be effective when analytical capacity is a limitation),

6.2.3 efficiently determine the absence or possible presence of hot spots or hot containers and, when combined with retesting schemes, identify hot spots, as long as the probability of hitting a hot spot is low,

6.2.4 be especially useful for situations, where the nature of contaminant distribution tends to be contiguous and non-random and the majority of analyses are "non-detects" for the contaminant(s) of interest, and

6.2.5 provide a degree of anonymity where population, rather than individual statistics are needed.

6.3 *Improvement in Sampling Precision*—Samples are always taken to make inferences to a larger volume of material, and a set of composite samples from a heterogeneous population provides a more precise estimate of the mean than a comparable number of discrete samples. This occurs because compositing is a "physical process of averaging." Averages of samples have greater precision than the individual samples. Likewise, a set of composite samples is always more precise than an equal number of individual samples. Decisions based on a set of composite samples will, for practical purposes, always provide greater statistical confidence than for a comparable set of individual samples.

6.3.1 If an estimated precision of a mean is desired, then more than one composite sample is needed; a standard deviation cannot be calculated from one composite sample. However, the precision of a single composite sample may be estimated when there are data to show the relationship between the precision of the individual samples that comprise the composite sample and that of the composite sample. The precision (standard deviation) of the composite sample is approximately the precision of the individual samples divided by the square root of the number of individual samples in the composite.

6.4 *Example 1*—An example of how a single composite sample can be used for decision-making purposes is given here. Assume a regulatory limit of 1 mg/kg and a standard deviation of 0.5 mg/kg for the individual samples. If the concentration of a site is estimated to be around 0.6 mg/kg, how many individual samples should be composited to have relatively high confidence that the true concentration does not exceed the regulatory limit when only one composite sample is used? Assuming the composite is well mixed, then the precision of a composite is a function of the number of samples as follows:

Number of Individual Samples in Composite	Precision (standard deviation ÷ $\sqrt{n}$) of One Composite Sample
2	0.35
3	0.29
4	0.25
5	0.22
6	0.20

Thus, if six samples are included in a composite, the composite concentration of 0.6 mg/kg is two standard deviations below the regulatory limit. Therefore, if the composite concentration is actually observed to be in the neighborhood of 0.6 mg/kg, we can be reasonably confident (approximately 95 %) that the concentration of the site is below the regulatory limit, using only one composite sample.

6.5 *Example 2*—Another example is when the standard deviation of the individual samples in the previous example is relatively small, say 0.1 mg/kg. Then the standard deviation of a composite of 6 individual samples is 0.04 mg/kg (0.1 mg/kg divided by the square root of 6 = 0.04 mg/kg), a very small number relative to the regulatory limit of 1 mg/kg. In this case, simple comparison of the composite concentration to the regulatory limit is often quite adequate for decision-making purposes.

---

[6] The boldface numbers in parentheses refer to a list of references at the end of this guide.

6.5.1 The effectiveness of compositing depends on the relative magnitude of sampling and analytical error. When sampling uncertainty is high relative to analytical error (as is usually assumed to be the case) compositing is very effective in improving precision. If analytical errors are high relative to field errors, sample compositing is much less effective.

6.5.2 Because compositing is a physical averaging process, composite samples tend to be more normally distributed than the individual samples. The normalizing effect is frequently an advantage since calculation of means, standard deviations and confidence intervals generally assume the data are normally distributed. Although environmental residue data are commonly non-normally distributed, compositing often leads to approximate normality and avoids the need to transform the data.

6.5.3 The spatial design of the compositing scheme can be important. Depending upon the locations from which the individual samples are collected and composited, composites can be used to determine spatial variability or improve the precision of the parameter being estimated. Figures 1 and 2 represent a site divided into four cells. Composite all samples with the same number together. The sampling approach in Fig. 1 is similar to sample random sampling, except they are now composite samples. Each composite sample in this case is a representative sample of the entire site, eliminates cell-to-cell variability, and leads to increased precision in estimating the mean concentration of the site. If there is a need to estimate the cell-to-cell variability, then the approach in Fig. 2 is suitable. In addition, if the precision of estimating the mean concentration of the cell is needed, multiple composite samples should be collected from that cell.

6.6 *Effect on Cost Reduction*—Because the composite samples yield a more precise mean estimate than the same number of individual samples, there is the potential for substantial cost saving. Given the higher precision associated with composite samples, the number of composite samples required to achieve a specified precision is smaller than that required for individual samples. This cost saving opportunity is especially pronounced when the cost of sample analysis is high relative to the cost of sampling, compositing, and analyzing.

6.7 *Hot Container/Hot Spot Identification and Retesting Schemes*—Samples can be combined to determine whether an individual sample exceeds a specified limit as long as the action limit is relatively high compared with the actual detection limit and the average sample concentration. Depending on the difficulty and probability of having to resample, it may be desirable to retain a split of the discrete samples for possible analysis depending on the analytical results from the composite sample.

1	2	4	3
4	3	2	1
4	2	1	4
3	1	2	3

FIG. 1 Example of Composing Across a Site

1	1	2	2
1	1	2	2
3	3	4	4
3	3	4	4

FIG. 2 Example of Within Cell Compositing

6.8 *Example 3*—One hundred drums are to be examined to determine whether the concentration of PCBs exceeds 50 mg/kg. Assume the detection limit is 5 mg/kg and most drums have non-detectable levels. Compositing samples from ten drums for analysis would permit determining that none of the drums in the composite exceed 50 mg/kg as long as the concentration of the composite is <5 mg/kg. If the detected concentration is >5 mg/kg, one or more drums may exceed 50 mg/kg and additional analyses of the individual drums are required to identify any hot drum(s). The maximum number of samples that can theoretically be composited and still detect a hot sample is the limit of concern divided by the actual detection limit (for example, 50 mg/kg ÷ 5 mg/kg = 10).

6.9 *Example 4*—Assume background levels of dioxin are non detectable, and the analytical detection limit is 1 μg/kg and the action level is 50 μg/kg. The site is systematically gridded (the most efficient sampling design for detecting randomly distributed hot spots) using an appropriate design, and cores to a depth of 10 cm are collected. Composite samples are collected since analytical costs for dioxin are high. In theory, groups of up to 50 samples could be composited and if the resultant concentration were <1 μg/kg, all samples represented in the composite should be below 50 μg/kg. If the contaminant concentration is >1 μg/kg, one or more spots may exist that exceed 50 μg/kg in the area covered by the composite sample although the precise location and areal extent would not be known without further sampling and analyses. Compositing fewer samples would probably be more practical, however.

6.9.1 The relative efficiency of compositing individual samples to detect a hot spot depends on the probability of a "hot" discrete sample being used to form a composite sample. According to Garner et al. (1), if the probability can be estimated as low, say 1 %, the optimum number of samples to composite is about ten, which would result in a cost saving of about 80 % (assuming there is no detection limit problem). When the probability of collecting a sample from a hot spot rises to 10 %, the optimal number of samples to composite is 4, which results in a 40 % cost savings. By the time the probability of sampling a hot spot rises to 40 %, there is no cost benefit to compositing. Other resampling and testing schemes are possible and may lead to somewhat different cost saving potentials.

7. Limitations of Composite Sampling

7.1 The principal limitations of sample compositing involve the loss of the discrete information contained in a single sample and the potential for dilution of the contaminants in a sample with uncontaminated material; however, in that case, the dilution factor can be used to estimate the maximum number of samples that can be composited. The following situations may not lend themselves to cost-effective sample compositing:

7.1.1 When the integrity of individual sample values change because of compositing, for example, chemical interaction occurs between constituents in the samples being combined or volatiles are lost during mixing,

7.1.2 Where the composite sample cannot be properly mixed and subsampled or the whole composite sample cannot be analyzed,

7.1.3 When the goal is to detect hotspots and a large proportion of the samples are expected to test positive for an attribute, compositing and retesting schemes may not be cost effective,

7.1.4 When analytical costs are low relative to sampling costs (for example, in situ field portable X-ray fluorescence takes only 30 s with no sample preparation so analytical costs/sample are very low), and

7.1.5 When regulations specify that a grab sample must be collected (usually a composite sample covering a limited area is still preferred from a technical standpoint).

8. Sample Mixing Procedures

8.1 Prior to sample mixing, project-specific instructions should be followed regarding sample collection, which may include removal of extraneous sample materials such as twigs, grass, rocks, etc. If samples are sieved or large materials are removed, it may be necessary to record the mass of materials removed for later estimation of contaminant concentration in the original sample. According to particulate sampling theory (4,5) the following sample masses are adequate to represent the corresponding maximum size particles in the sample with a relative standard deviation of 15 %.

Sample Mass, g	Maximum Particle Size, cm
5	0.170
50	0.37
100	0.46
500	0.79
1000	1.0
5000	1.7

8.1.1 Frequently it is necessary to mix an individual or composite sample and obtain a representative subsample(s) for transport to the analytical laboratory. This occurs when multiple containers of the identical material are desired (for example, separate sample jars for metals, semivolatile organics, etc. are desired) or when the original sample (or composite sample) size is greater than accepted by the laboratory. Even when the original sample volume is acceptable, it may be desirable to thoroughly mix the sample prior to transport to an analytical laboratory. However, some samples that have been well mixed in the field may segregate during shipment to the laboratory.

8.1.2 A laboratory typically collects a 0.5 to 30 g specimen (100 g for some extraction tests) from the sample for analysis. Specimens are frequently collected from the surface material in the container or after minimal mixing. Such procedures are inadequate to obtain a small representative specimen from a 100 to 300 g sample. Special mixing and subsampling procedures are necessary to obtain a representative subsample unless the sample is already homogenous.

Field mixing should be considered essential unless it is known that the sample in the container is homogeneous or it is known that the laboratory will homogenize the sample and collect a representative specimen. To help ensure that an unbiased and precise specimen is collected, the analytical laboratory should be provided instructions (preferably with the sample shipment) on homogenizing and obtaining a specimen for analysis. Few laboratories follow good sample homogenizing and specimen collection practices. To meet both sampling and analytical objectives, field and analytical personnel, and the end-user of the data must be aware of the laboratories standard practices for handling, mixing, and obtaining a specimen or specify such practices with the sample shipment.

8.1.3 To avoid subsampling it may be possible to collect a small sample (or composite samples) directly into the sample container that is delivered to the laboratory (**Caution:** small sample sizes may result in bias by excluding large particles). While no field mixing and subsampling is needed as long as the laboratory homogenizes the sample, it may be advisable to mix such samples anyway (see 8.1.2).

8.1.4 Soil, sediment, sludge and waste samples collected for purgeable/volatile organic compounds' analyses should *not* be mixed and subsampled using procedures described in this guide but other specialized procedures such as combining samples directly into methanol (see Practice D 4547) may be appropriate.

8.1.5 A significant problem with analyzing very small samples is that the smaller the volume of sample actually extracted or analyzed, the less representative that sample may be unless thoroughly mixed/homogenized and subsampled. Therefore, sample compositing without thorough mixing can nullify the potential benefits of compositing.

8.1.6 Methods that may be applicable to field mixing, depending on the matrix, include hand mixing in a pan, sieving, particle size reduction, kneading, etc. For highly heterogeneous waste such as municipal refuse, field comminution (grinding) may be needed. Some of these methods may be inappropriate if trace levels of contamination are a primary concern. The use of disposable equipment for mixing should be considered to minimize field decontamination problems. Field personnel should use care to ensure that samples do not become contaminated during the sampling, mixing and subsampling process.

8.1.7 Once a sample has been collected, it may have to be split into separate containers for different analyses. A true split of soil, sediment, or sludge samples may be difficult to accomplish under field conditions.

8.1.8 The following are some common methods for mixing soils, sludges, etc. While it is not always possible to determine that a sample is adequately mixed, following standard procedures and observing sample texture, color, and particle distribution are practical methods. While some materials cannot be homogenized, following the subsampling procedures in Section 9 will help ensure that a representative subsample is collected. Under certain conditions, some of the procedures that follow are applicable when trace level contaminants are of concern.

8.1.8.1 *Pan Mixing/Quartering*—One common method of mixing is referred to as quartering. Place the material in a glass or stainless steel sample pan and divide into quarters.

Mix each quarter separately, then mix all quarters into the center of the pan. Repeat this procedure several times until the sample is adequately mixed (usually a minimum of three repetitions). If round bowls are used for sample mixing, adequate mixing is achieved by stirring the material in a circular fashion and occasionally turning the material over.

8.1.8.2 *Mixing Square*—Combine samples through a non-contaminating screen into an appropriate clean mixing container. Mix in the container and pour onto a 1 metre square of non-contaminating material such as plastic for metals analyses or polytetrafluoroethylene for organics. Roll the sample backward and forward on the sheet while alternately lifting and releasing opposite side corners of the sheet. This is appropriate for flowable granular materials (6). If polytetrafluoroethylene sheeting is used, this procedure could be acceptable for trace level contaminants.

8.1.8.3 *Kneading*—Place the sample in a non-contaminating bag and knead as in bread making to mix the sample. This may be appropriate for viscous or clay-like materials. If a non-contaminating bag is used, this approach would be acceptable for trace level contaminants.

8.1.8.4 *Sieving and Mixing*—If a laboratory requires a small specimen (1 to 30 g) or if less than a specific particle size is required, disruption of aggregated particles or sieving, or both, followed by mixing may be needed. Sieving allows only those particles below a desired size to pass through the sieve into a mixing pan for subsequent mixing and subsampling into containers. Sieving works best with relatively dry granular materials. Sieving and the exclusion of large particles can result in very biased results and should only be conducted when designed into a sampling plan.

8.1.8.5 *Particle Size Reduction*—When particle size reduction is appropriate and trace contaminants are of concern, non-contaminating materials compatible with objectives should be used (for example, glass, ceramic, stainless steel). Other materials may be acceptable if trace levels of contaminants are not a concern. The reduction method can be as simple as using a hammer to break apart large pieces into smaller pieces that are either acceptable to the labora-

tory or that can pass through a sieve. This method of reduction creates a great deal of fine material which may or may not be included in the sample container, and could introduce bias. More complex reducers, such as ball mills, ceramic plate grinders, etc., are available, but usually require relatively dry samples and thorough decontamination to avoid cross contamination. Such a process may be more appropriately conducted in a laboratory.

8.1.9 With thorough decontamination (see Practice D 5088) of the particle size reducer, sieve and the mixing pan, these procedures could be acceptable for trace level contaminants.

8.1.10 *Other Mixing Equipment*—Riffle splitters, coning and quartering, etc., involve equipment and materials that are difficult to decontaminate, and awkward to use on a routine basis for waste management sampling. Since these procedures are not routinely used, the devices are not considered in this guide. However, procedures for coning and quartering, and the use of riffle splitters are described in Practice C 702 and could be modified for subsampling contaminated media.

9. Field Subsampling Procedures

9.1 If mixing procedures could ensure a truly homogenous sample, subsampling would be simple. Mixing of various particle sizes may, however, cause the particles to segregate according to size, and improper subsampling could introduce bias. Since homogeneity is frequently not achieved, appropriate subsampling procedures should be used by field personnel to provide representative subsamples. The procedures that follow are appropriate for collecting a representative sample from a larger sample. As noted previously, riffle splitters and coning and quartering procedures can also be used for subsampling as well as mixing (see Practice C 702).

9.1.1 *Rectangular Scoop*—As the final step of mixing, the material is arranged in a pile along the long axis of the rectangular pan. A flat bottomed scoop with vertical sides is moved across the entire width of the short axis of the pile to collect a swath of sample (Fig. 3). Multiple evenly-spaced

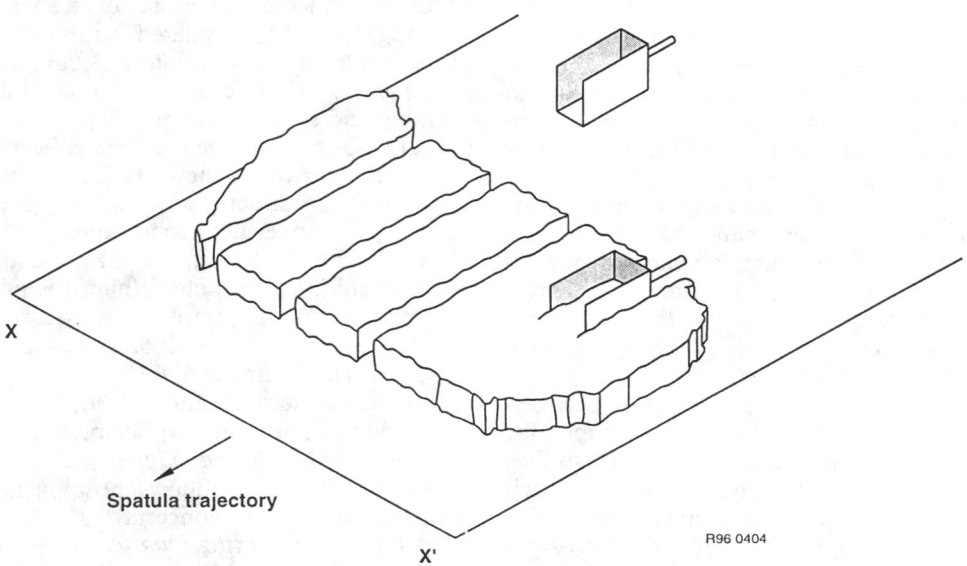

X

X'

Spatula trajectory

R96 0404

FIG. 3 Rectangular Scoop as Used to Collect Swaths for Subsample

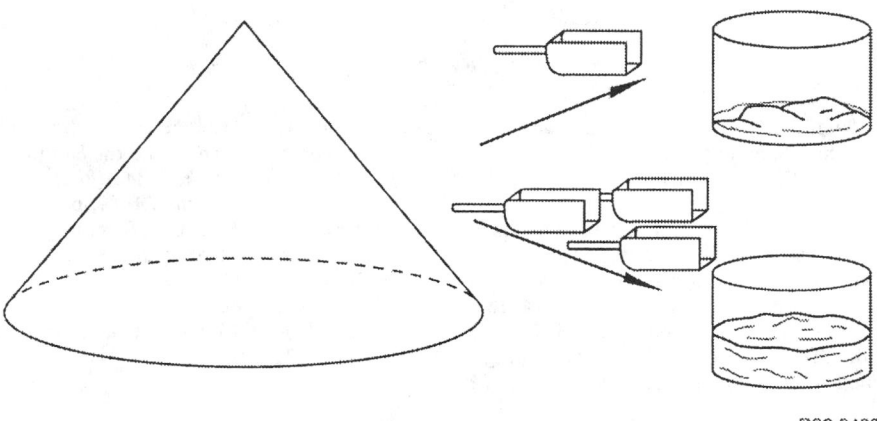

FIG. 4 Alternate Scoop Subsampling Technique

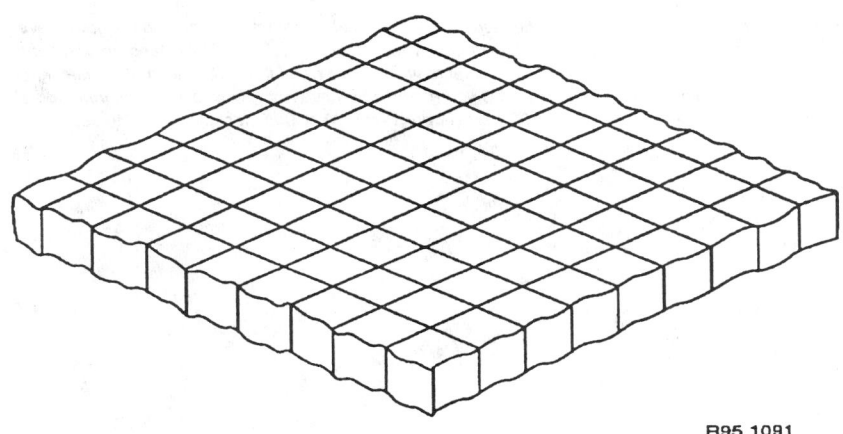

FIG. 5 Slab Cake Subsampling Technique

swaths are collected until the subsample container is full. Multiple containers are filled by rearranging the remaining material and collecting swaths as just described.

9.1.2 *Alternate Scoop*—The volume of material required for filling sample containers is compared to the volume of the mixed sample. Scoops of mixed material are placed in the sample container(s) or are discarded, that is, three scoops are discarded for every scoop saved when collecting a 25 % subsample (Fig. 4). Care should be taken that each scoop of material is of the same size and is collected in a consistent manner to minimize bias **(5)**.

9.1.3 *Slab-cake*—The cohesive or clay-like materials as discussed in 8.1.8.3 on kneading. The sample can be flattened, cut into cubes (Fig. 5) and the cubes randomly or systematically combined into subsample(s) **(5)**. The subsample should be re-kneaded before shipment to the laboratory unless it can be ensured that the laboratory will homogenize the subsample before collecting a specimen.

10. Keywords

10.1 composite; compositing; hot spot; particle size reduction; sample; sampling; subsample; subsampling

REFERENCES

(1) Garner et al., "Composite Sampling for Environmental Monitoring in Principals of Environmental Sampling," in *Principles of Environmental Sampling*, American Chemical Society (ACS), Keith, L., ed., 1988, pp. 363–374.

(2) Mack, G. A., and Robinson, P. E., "Use of Composited Samples to Increase the Precision and Probability of Detection of Toxic Chemicals," in *Environmental Applications of Chemometrics*, ACS Symposium Series 292, ACS, Washington, DC, 1985, pp. 174–183.

(3) Rajagopal, R., and Williams, L. R., "Economics of Sample Compositing as a Screening Tool in Ground Water Monitoring," *Ground Water Monitoring Review*, 1989, pp. 186–192.

(4) Ramsey, C. A., Ketterer, M. E., and Lowery, J. H., "Application of Gy's Sampling Theory to the Sampling of Solid Waste Materials," in *Proceedings of the EPA Fifth Annual Waste Testing and Quality Assurance Symposium*, 1989, p. II-494.

(5) Pitard, F. F., "Pierre Gy's Sampling Theory and Practice," *Sampling Correctness and Sampling Practice*, Volume II, CRC Press, Boca Raton, FL, 1989, p. 247.

(6) U.S. Environmental Protection Agency (EPA), *Description and Sampling of Contaminated Soils, A Field Pocket Guide*, EPA/625/12-91/002, 1991.

Standard Guide for
Sampling Strategies for Heterogeneous Wastes[1]

This standard is issued under the fixed designation D 5956; the number immediately following the designation indicates the year of original adoption or, in the case of revision, the year of last revision. A number in parentheses indicates the year of last reapproval. A superscript epsilon (ε) indicates an editorial change since the last revision or reapproval.

1. Scope

1.1 This guide is a practical, nonmathematical discussion for heterogeneous waste sampling strategies. This guide is consistent with the particulate material sampling theory, as well as inferential statistics, and may serve as an introduction to the statistical treatment of sampling issues.

1.2 This guide does not provide comprehensive sampling procedures, nor does it serve as a guide to any specification. It is the responsibility of the user to ensure appropriate procedures are used.

1.3 *This standard does not purport to address all of the safety concerns, if any, associated with its use. It is the responsibility of the user of this standard to establish appropriate safety and health practices and determine the applicability of regulatory limitations prior to use.*

2. Terminology

2.1 *Definitions of Terms Specific to This Standard:*

2.1.1 *attribute, n*—a quality of samples or a population.

2.1.1.1 *Discussion*—Homogeneity, heterogeneity, and practical homogeneity are population attributes. Representativeness and intersample variance are sample attributes.

2.1.2 *characteristic, n*—a property of items, a sample or population that can be measured, counted, or otherwise observed.

2.1.2.1 *Discussion*—A characteristic of interest may be the cadmium concentration or ignitability of a population.

2.1.3 *component, n*—an easily identified item such as a large crystal, an agglomerate, rod, container, block, glove, piece of wood, or concrete.

2.1.4 *composite sample, n*—a combination of two or more samples.

2.1.4.1 *Discussion*—When compositing samples to detect hot spots or whenever there may be a reason to determine which of the component samples that constitute the composite are the source of the detected contaminant, it can be helpful to composite only portions of the component samples. The remainders of the component samples then can be archived for future reference and analysis. This approach is particularly helpful when sampling is expensive, hazardous, or difficult.

2.1.5 *correlation, n*—the mutual relation of two or more things.

2.1.6 *database, n*—a comprehensive collection of related data organized for quick access.

2.1.6.1 *Discussion*—Database as used in this guide refers to a collection of data generated by the collection and analysis of more than one physical sample.

2.1.7 *data quality objectives (DQO), n*—DQOs are qualitative and quantitative statements derived from the DQO process describing the decision rules and the uncertainties of the decision(s) within the context of the problem(s).

2.1.8 *data quality objective process, n*—a quality management tool based on the scientific method and developed by the U.S. Environmental Protection Agency to facilitate the planning of environmental data collection activities.

2.1.8.1 *Discussion*—The DQO process enables planners to focus their planning efforts by specifying the use of the data (the decision), the decision criteria (action level) and the decision maker's acceptable decision error rates. The products of the DQO process are the DQOs.

2.1.9 *heterogeneity, n*—the condition of the population under which items of the population are not identical with respect to the characteristic of interest.

2.1.10 *homogeneity, n*—the condition of the population under which all items of the population are identical with respect to the characteristic of interest.

2.1.10.1 *Discussion*—Homogeneity is a word that has more than one meaning. In statistics, a population may be considered homogeneous when it has one distribution (for example, if the concentration of lead varies between the different items that constitute a population and the varying concentrations can be described by a single distribution and mean value, then the population would be considered homogeneous). A population containing different strata would not have a single distribution throughout, and in statistics, may be considered to be heterogeneous. The terms *homogeneity* and *heterogeneity* as used in this guide, however, reflect the understanding more common to chemists, geologists, and engineers. The terms are used as described in the previous definitions and refer to the similarity or dissimilarity of items that constitute the population. According to this guide, a population that has dissimilar items would be considered heterogeneous regardless of the type of distribution.

2.1.11 *item, n*—a distinct part of a population (for example, microscopic particles, macroscopic particles, and 20-ft long steel beams).

2.1.11.1 *Discussion*—The term *component* defines a subset of items. Components are those items that are easily identified as being different from the remainder of items that constitute the population. The identification of components may facilitate the stratification and sampling of a highly stratified population when the presence of the characteristic of interest is correlated with a specific component.

2.1.12 *population, n*—the totality of items or units under consideration.

[1] This guide is under the jurisdiction of ASTM Committee D-34 on Waste Management and is the direct responsibility of Subcommittee D34.01 on Sampling and Monitoring.

Current edition approved Oct. 10, 1996. Published December 1996.

2.1.13 *practical homogeneity, n*—the condition of the population under which all items of the population are not identical. For the characteristic of interest, however, the differences between individual physical samples are not measurable or significant relative to project objectives.

2.1.13.1 *Discussion*—For practical purposes, the population is homogeneous.

2.1.14 *random, n*—lack of order or patterns in a population whose items have an equal probability of occurring.

2.1.14.1 *Discussion*—The word *random* is used in two different contexts in this guide. In relation to sampling, random means that all items of a population have an equal probability of being sampled. In relation to the distribution of a population characteristic, random means that the characteristic has an equal probability of occurring in any and all items of the population.

2.1.15 *representative sample, n*—a sample collected in such a manner that it reflects one or more characteristics of interest (as defined by the project objectives) of a population from which it was collected.

2.1.15.1 *Discussion*—A representative sample can be (*1*) a single sample, (*2*) a set of samples, or (*3*) one or more composite samples.

2.1.16 *sample, n*—a portion of material that is taken for testing or for record purposes.

2.1.16.1 *Discussion*—Sample is a term with numerous meanings. The scientist collecting physical samples (for example, from a landfill, drum, or waste pipe) or analyzing samples, considers a sample to be that unit of the population collected and placed in a container. In statistics, a sample is considered to be a subset of the population, and this subset may consist of one or more physical samples. To minimize confusion the term *physical sample* is a reference to the sample held in a sample container or that portion of the population that is subjected to in situ measurements. One or more physical samples, *discrete samples*, or aliquots are combined to form a *composite sample*. The term *sample size* has more than one meaning and may mean different things to the scientist and the statistician. To avoid confusion, terms such as sample mass or sample volume and number of samples are used instead of sample size.

2.1.17 *sample variance, n*—a measure of the dispersion of a set of results. Variance is the sum of the squares of the individual deviations from the sample mean divided by one less than the number of results involved. It may be expressed as $s^2 = \Sigma (x_i - \bar{x})^2/(n - 1)$.

2.1.18 *sampling, n*—obtaining a portion of the material concerned.

2.1.19 *stratum, n*—a subgroup of a population separated in space or time, or both, from the remainder of the population, being internally consistent with respect to a target constituent or property of interest, and different from adjacent portions of the population.

2.1.19.1 *Discussion*—A landfill may display spatially separated strata since old cells may contain different wastes than new cells. A waste pipe may discharge temporally separated strata if night-shift production varies from the day shift. Also, a waste may have a contaminant of interest associated with a particular component in the population, such as lead exclusively associated with a certain particle size.

2.1.19.2 *Discussion*—Highly stratified populations consist of such a large number of strata that it is not practical or effective to employ conventional sampling approaches, nor would the mean concentration of a highly stratified population be a useful predictor (that is, the level of uncertainty is too great) for an individual subset that may be subjected to evaluation, handling, storage, treatment, or disposal. *Highly stratified* is a relative term used to identify certain types of nonrandom heterogeneous populations. Classifying a population according to its level of stratification is relative to the persons planning and performing the sampling, their experience, available equipment, budgets, and sampling objectives. Under one set of circumstances a population could be considered highly stratified, while under a different context the same population may be considered stratified.

2.1.19.3 *Discussion*—The terms *stratum* and *strata* are used in two different contexts in this guide. In relation to the population of interest, *stratum* refers to the actual subgroup of the population (for example, a single truck load of lead-acid batteries dumped in the northeast corner of a landfill cell). In relation to sampling, *stratum* or *strata* refers to the subgroups or divisions of the population as assigned by the sampling team. When assigning sampling strata, the sampling team should maximize the correlation between the boundaries of the assigned sampling strata and the actual strata that exist within the population. To minimize confusion in this guide, those strata assigned by the sampling team will be referred to as *sampling strata*.

3. Significance and Use

3.1 This guide is suitable for sampling heterogeneous wastes.

3.2 The focus of this guidance is on wastes; however, the approach described in this guide may be applicable to non-waste populations, as well.

3.3 Sections 4 through 9 describe a guide for the sampling of heterogeneous waste according to project objectives. Appendix X1 describes an application of the guide to heterogeneous wastes. The user is strongly advised to read Annex A1 prior to reading and employing Sections 4 through 9 of this guide.

3.4 Annex A1 contains an introductory discussion of heterogeneity, stratification, and the relationship of samples and populations.

3.5 This guide is intended for those who manage, design, or implement sampling and analytical plans for the characterization of heterogeneous wastes.

4. Sampling Difficulties

4.1 There are numerous difficulties that can complicate efforts to sample a population. These difficulties can be classified into four general categories:

4.1.1 Population access problems making it difficult to sample all or portions of the population;

4.1.2 Sample collection difficulties due to physical properties of the population (for example, unwieldy large items or high viscosity);

4.1.3 Planning difficulties caused by insufficient knowledge regarding population size, heterogeneity of the contaminant of interest, or item size, or a combination thereof; and,

4.1.4 Budget problems that prevent implementation of a workable, but too costly, sampling design.

4.2 The difficulties included in the first three categories are a function of the physical properties of the population being sampled. The last sampling difficulty category is a function of budget restraints that dictate a less-costly sampling approach that often results in a reduced number of samples and a reduced certainty in the estimates of population characteristics. Budget restraints can make it difficult to balance costs with the levels of confidence needed in decision making. These difficulties may be resolved by changing the objectives or sampling/analytical plans since population attributes or physical properties of the population can seldom be altered. Documents on DQOs discuss a process for balancing budgets with needed levels of confidence.

4.3 Population access and sample collection difficulties often are obvious, and therefore, more likely either to be addressed or the resulting limitations well-documented. A field notebook is likely to describe difficulties in collecting large items or the fact that the center of a waste pile could not be accessed.

4.4 Population size, heterogeneity, and item size have a substantial impact on sampling. The cost and difficulty of accurately sampling a population usually is correlated with the knowledge of these population attributes and characteristics. The least understood population attribute is heterogeneity of the characteristic of interest. If heterogeneity is not known through process knowledge, then some level of preliminary sampling or field analysis is often required prior to sampling design.

4.5 Sampling of any population may be difficult. However, with all other variables being the same, nonrandom heterogeneous populations are usually more difficult to sample. The increased difficulty in sampling nonrandom heterogeneous populations is due to the existence of unidentified or numerous strata, or both. If the existence of strata are not considered when sampling a nonrandom heterogeneous population, the resulting data will average the measured characteristics of the individual strata over the entire population. If the different strata are relatively similar in composition, then the mean characteristic of the population may be a good predictor for portions of the population and will often allow the project-specific objectives to be achieved. As the difference in composition between different strata increases, average population characteristics become less useful in predicting composition or properties of individual portions of the population. In this latter case, when possible, it is advantageous to sample the individual strata separately, and if an overall average of a population characteristic is needed, it can be calculated mathematically using the weighted averages of the sampling stratum means (7).

5. Stratification

5.1 Strata can be thought of as different portions of a population, which may be separated in time or space with each portion having internally similar concentrations or properties, which are different from adjacent portions of the population (that is, concentrations/properties are correlated with space, time, component, or source). Figure 1 is a graphical depiction of different types of strata.

5.1.1 A landfill may display spatially separated strata since old cells may contain different wastes than new cells (stratification over space);

5.1.2 A waste pipe may discharge temporally separated strata if night-shift production varies from the day shift (stratification over time);

5.1.3 Lead-acid batteries will constitute a strata separate from commingled soil if lead is the characteristic of interest (stratification by component); and,

5.1.4 Drums from an inorganic process may constitute a different strata from those co-disposed drums generated by an organic process (a subtype of stratification by component

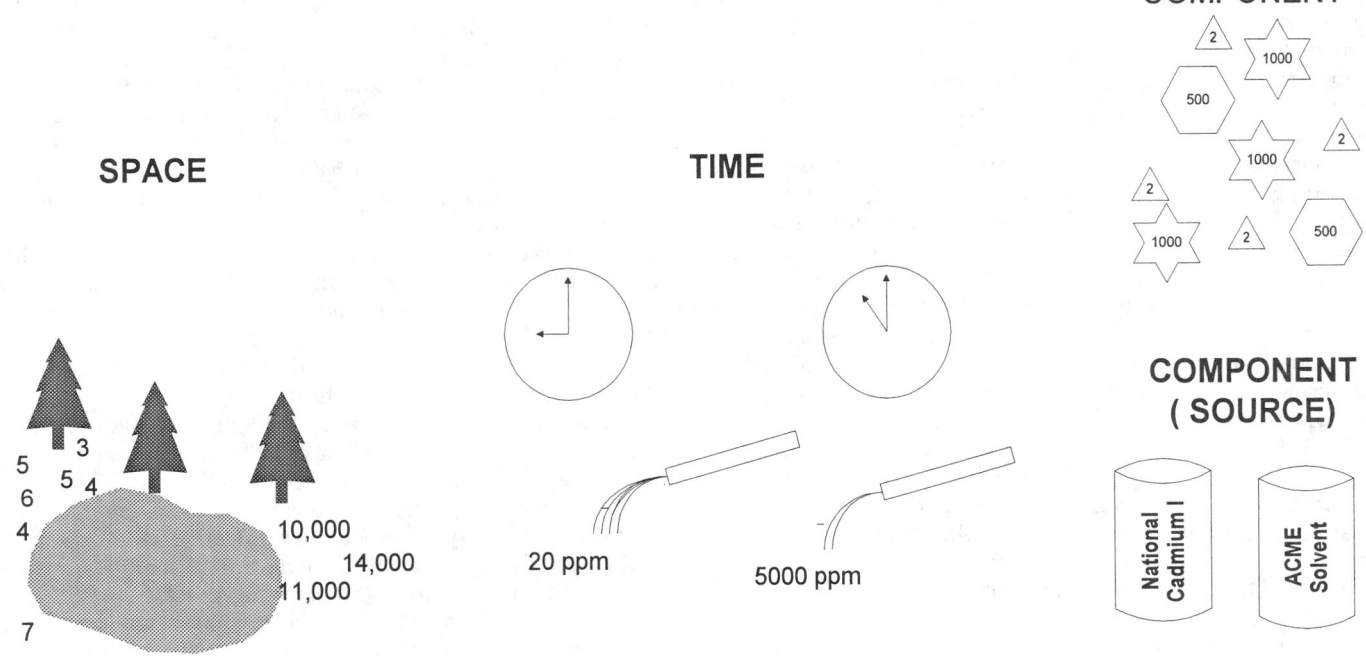

FIG. 1 Types of Stratified Heterogeneous Wastes

referred to as stratification by source).

5.2 Different strata often are generated by different processes or a significant variant of the same process. The different origins of the strata usually result in a different concentration distribution and mean concentration.

5.3 Highly stratified populations, a type of nonrandom heterogeneous populations, have so many strata that they become difficult to sample and characterize. Classifying a population according to its level of stratification is a relative issue pertaining to the persons planning and performing the sampling, their experience, available equipment, and budgets. Highly stratified populations are such that it is not practical or effective to employ conventional sampling approaches to generate a representative database, nor would the mean concentration of a highly stratified population be a useful predictor (that is, the level of uncertainty is too great) for an individual subset that may be subjected to evaluation, handling, storage, treatment, or disposal.

NOTE 1—An example of a highly stratified population is a landfill, a candidate for remediation, that is contaminated with the pure and very viscous Aroclor 1260 and with solutions containing varying concentrations of Aroclor 1260. (Aroclor 1260 is viscous and can exist as globules of the pure Aroclor.) The detected concentration of Aroclors in analytical subsamples would reflect a highly stratified population if some samples contained globules of pure 1260, while other samples contained soils that came in contact with solvents containing varying concentrations of 1260. Highly nonrandom heterogeneous populations have numerous strata, each of which contain different distributions of contaminants or item sizes, or both, such that an average value for the population would not be useful in predicting the composition or properties of individual portions of the waste (that is, statistically speaking, the variance and standard error of the mean will be large).

A second and more visually obvious example of a highly stratified population would be a landfill that is filled with unconfined sludge, building debris, laboratory packs, automobile parts, and contained liquids with the constituent of interest having different concentrations in each strata.

5.4 Certain populations do not display any obvious temporal or spatial stratification, yet the distribution of the target characteristic is excessively erratic. For these populations it may be helpful to consider stratification of the population by component. Stratification by component is applied to populations that contain easily identifiable items, such as large crystals or agglomerates, rods, blocks, gloves, pieces of wood, or concrete. Separating a population into sampling strata according to components is useful when a specific kind of component is distributed within the population and when a characteristic of interest is correlated with the component. Stratification by source (for example, organic process waste drums versus inorganic process waste drums) is a type of component stratification. Stratification by component is an important mechanism for understanding the properties of component-heterogeneous populations and for designing appropriate sampling and analytical efforts.

5.4.1 Component strata are not necessarily separated in time or space but are usually intermixed and the properties or composition of the individual components are the basis of stratification. For example, automobile batteries that are mixed in an unrelated waste would be a component that could constitute an individual strata if lead was a target characteristic. If one were to sequester the batteries, they would have a consistent distribution that was different from the rest of the waste.

5.4.2 There is usually no purpose in stratifying by component if different components have similar concentrations of the target characteristic or if the components are small enough such that the different components are represented in the chosen sample size. Even when components have similar composition, however, stratification and use of separate sampling strategies by component may be useful when the different components are so physically different that they cannot all be sampled with the same technique.

5.4.3 A primary objective for employing a stratified sampling strategy is to improve the precision of population parameters such as population means by dividing the population into homogeneous strata. The precision of the population parameters will increase as the sampling strata boundaries, chosen by the sampling team, more closely overlay the actual physical strata that exist within the population.

6. Sampling of Highly Stratified Heterogeneous Wastes

6.1 Sections 6 through 9 focus on the sampling of highly stratified wastes, a type of heterogeneous waste. It is strongly advised that Annex A1 be read and studied prior to the use of this guide. Annex A1 discusses heterogeneity and the relationship between samples and populations.

6.2 Nonrandom heterogeneous wastes contain two or more strata. Stratification of a waste does not always complicate the sampling process; at times, could simplify sampling. Highly stratified populations, however, contain such a large number of strata that they become difficult to sample and characterize. Use of the word *highly* and the classification of wastes according to their level of stratification is a relative issue pertaining to the persons planning and performing the sampling, their experience, available equipment, budgets, and objectives. Highly stratified wastes are such that it is not practical or effective to employ conventional sampling approaches, nor would the mean concentration of a highly stratified waste be a useful predictor (that is, the level of uncertainty is too great) for an individual subset that may be subjected to evaluation, handling, storage, treatment, or disposal.

6.3 A structured approach to sampling planning, such as the DQO process, is a useful approach for the sampling of all wastes regardless of their level of heterogeneity. The first step in characterizing any heterogeneous waste is to gather all available information, such as the need for waste sampling; objectives of waste sampling; pertinent regulations, consent orders, and liabilities; sampling, shipping, laboratory, health, and safety issues; generation, handling, treatment, and storage of the waste; existing analytical data and exacting details on how it was generated; and treatment and disposal alternatives. This information will be used in the planning of the sampling and analytical effort.

6.4 If enough information is available, the planning process may uncover the existence of stratification that may prevent achievement of objectives. If information is lacking, a preliminary sampling/analytical effort may identify and evaluate variability. It is not cost-effective to characterize highly stratified waste by conventional methods, which becomes apparent during the planning process.

6.5 Sections 7 through 9 consider approaches that lessen the impact of stratification and allow for more cost-effective sampling. Some of these approaches require changes in objectives, waste handling or disposal methods, and some require compromises, but all approaches require the above types of information.

6.6 Heterogeneity is a necessary condition for the existence of strata. Wastes can be heterogeneous in particle size or in composition, or both, allowing for the existence of the following:

6.6.1 Strata of different-sized items of similar composition,

6.6.2 Strata of similar-sized items of different composition, and,

6.6.3 Strata of different-sized items and different composition.

7. Strata of Different-Sized Items With Similar Composition

7.1 Wastes having stratification due only to different-sized items will by definition have the same composition or property (that is, for compositional characteristics there is no significant intersample variance and no correlation with space, time, or component) throughout its different strata. The different-sized items may be separated in space or in time. Unless one is attempting to measure particle size for which there is significant intersample variance, this type of population is the simplest of the highly stratified waste types to characterize. All items in these types of wastes usually are generated by the same process (for example, the discussion of silver nitrate powder and crystals in Annex A1), which is the reason for similar composition across all item sizes. These types of wastes, which are compositionally homogeneous and only heterogeneous in item size, are not commonly encountered.

7.2 The complexity of dealing with these types of wastes is in proving that the waste has similar composition across the varying item sizes. This determination can be made by using process knowledge or by sampling the different-sized items to determine if there are significant compositional differences. If the determination is made using knowledge of the waste, it is advisable to perform limited sampling to confirm the determination. The characterization process is greatly simplified once a determination has been made that the waste has similar composition or properties across the various item sizes. The sampling and subsequent analysis can be performed on items that are readily amenable to the sampling and analytical process, and the resulting data can be used to characterize the waste in its entirety.

7.3 It is important to periodically verify the assumption that the different-sized items are composed of materials having the same concentration levels and distributions of the contaminant of interest. This verification is especially important when there are any changes to the waste generation, storage, treatment, or disposal processes. Similarity of composition between items has to be verified for each characteristic of interest. The effect of different-sized items also must be considered when measuring properties, such as the leachability of waste components.

8. Strata of Similar-Sized Items and Different Composition

8.1 Stratification due only to composition or property (that is, there is a correlation of composition or property with time, space, or component) by definition necessitates that item sizes will be consistent across different strata. The strata may be separable in space, time, or by component or source. Identifying and sampling the individual strata may simplify the characterization process. An example of this waste type is a long-term accumulation of wastewater sludge produced by the processing of materials having different composition, through the same waste-generation process (that is, batch-processing that results in waste having uniform item size but different composition from batch to batch).

8.2 Wastes having uniform item size and different composition or properties can be sampled using the same strategy as described for waste containing strata having different composition and different item size (see Section 9).

9. Strata of Different-Sized Items and Different Composition

9.1 Wastes having excessive stratification due to both composition/property and item size (that is, particle size and composition or property, or both, are correlated with time, space, or component) are usually the most difficult wastes to characterize. The difficulty in sampling highly stratified waste can result from:

9.1.1 Various item sizes and waste consistency that makes sampling difficult and conventional sampling approaches cost prohibitive;

9.1.2 Extraordinary concentration gradients between different components or innumerable strata that lead to such excessive variance in the data, that project objectives cannot be achieved; and,

9.1.3 Wastes that exhibit the properties in 9.1.1 and 9.1.2.

9.2 Figure 2 summarizes an approach to characterizing these types of highly stratified wastes. If a waste is highly stratified, conventional methods of sampling will not allow objectives to be achieved cost-effectively. To sample cost-effectively a highly stratified waste, one must use a nonconventional approach, such as modification of the sampling, sample preparation, or analytical phase of the process. If after modifying the sampling and analysis, the objectives still cannot be achieved in a cost-effective manner, then the original plan of waste handling, treatment, or disposal has to be examined and changed so the waste can be characterized according to new and achievable objectives.

9.3 *Design of the Sampling Approach:*

9.3.1 The first efforts to resolve the difficulty in characterizing a highly stratified waste are focused usually on sampling. A strategy for designing a sampling plan for such highly stratified waste may include the following five steps:

9.3.1.1 Use a planning process such as the DQO process to identify the target characteristics, the population boundaries, the statistic of interest, confidence levels, and other critical issues.

9.3.1.2 Determine whether characteristics of interest are correlated with item size, space, time, components, or sources.

9.3.1.3 Determine if any waste components or strata can be eliminated from consideration during sampling because they do not contribute significantly to the target characteristic.

9.3.1.4 Determine if small items in a stratum represent the stratum, as well as large more difficult to sample items. If yes, sample the smaller items, and only track the volume/

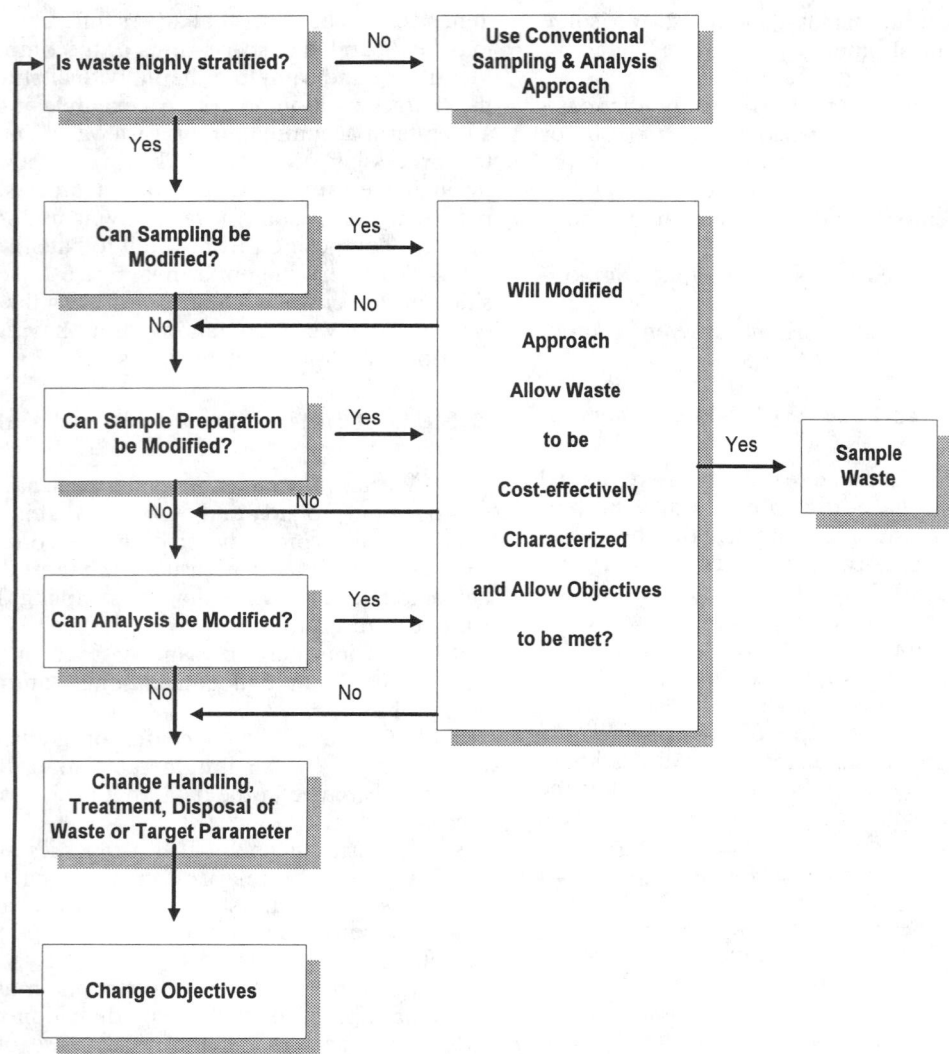

FIG. 2 Approach for the Characterization of Heterogeneous Wastes

mass contribution of the larger items.

9.3.1.5 Determine if the target characteristic is innate or surface adsorbed. Is the target characteristic surface adsorbed, which would allow the material to be sampled representatively by wipe sampling? Can large items be wiped and smaller items extracted, leached, or digested? Can waste be stratified according to impervious and nonimpervious waste and sampled and analyzed accordingly?

9.3.1.6 It is essential that all assumptions (that is, any correlations) be verified at least by knowledge of the waste, and preferably confirmed by sampling and analysis.

9.3.2 All steps taken to optimize sampling should be well-documented.

9.3.3 Appendix X1 contains a case study that applies the above process for optimizing sampling to highly stratified waste. If optimization of sampling design is not sufficient by itself to allow the project objectives to be met cost-effectively, changes to sample preparation or analysis should be considered.

9.4 *Modification of the Sample Preparation Method:*

9.4.1 Information gleaned from the analysis of samples is used to make inferences regarding population attributes. The perception of population homogeneity, as indicated by no significant intersample variance, or the perception of population heterogeneity (that is, as indicated by significant intersample variance) is analytical sample-mass dependent. Usually, the larger the sample mass/volume subjected to analysis the more representative the analytical sample. To improve representativeness of analytical samples and to accommodate large-sized items, conventional sample preparatory methods can be altered. All modifications of methods should be well-documented.

9.4.2 In the laboratory, the term *sample preparation* is commonly meant to include two separate steps: the sub-sampling of a field sample to generate an analytical sample, and the preparation of the analytical sample for subsequent analysis.

9.4.3 Regarding subsampling, the previously discussed logic for field sampling (see 9.3) is applicable also for the generation of analytical subsamples. Knowledge of concentration distributions within the waste can be used to simplify subsampling by considering the following:

9.4.3.1 Using process knowledge or the results of testing to eliminate any waste components or strata that do not contribute significantly to the concentration of the target compound;

9.4.3.2 Using process knowledge or the results of testing to discriminate against large items, and only select small items when small items represent the waste, as well as the large items; and,

9.4.3.3 Using process knowledge or the results of testing to restrict sampling to surface wipes of larger items and the extraction or digestion of fines if surface contamination is the source of the target characteristic.

9.4.4 If the approaches in 9.4.3.1 to 9.4.3.3 are not applicable to a field sample, the field sample will have to be subjected to particle size reduction (PSR) prior to subsampling or the sample preparation method will have to be modified to accommodate the entire field sample.

NOTE 2—Prior to modifying a sample preparatory method, it is advisable to consult the end user of the data to see if modifications could have any adverse affects. For example, PSR could dramatically alter leaching data.

9.4.5 The PSR is useful for handling field samples, which have items too large for proper representation in an analytical subsample. The intent of PSR is to decrease the maximum item size of the field sample so that the field sample then can be split or subsampled, or both, to generate a representative subsample. The difficulties in applying PSR to waste samples are the following:

9.4.5.1 Not all materials are easily amenable to PSR (for example, stainless steel artifacts);

9.4.5.2 Adequate PSR capabilities and capacities do not exist in all laboratories;

9.4.5.3 The PSR can change the properties of material (for example, leachability);

9.4.5.4 The PSR can be a source of cross-contamination;

9.4.5.5 The PSR often is not applicable to volatile and labile compounds; and,

9.4.5.6 Large mass/volumes may have to be shipped, handled, and disposed.

9.4.6 Modification of sample preparative methods can include the extraction, digestion, or leaching of much larger sample masses than specified. The advantage of this approach is that the characteristic of interest from a larger and more representative sample mass is dissolved into a relative homogeneous extract or digestate that is more suitable for subsampling. This approach is particularly important for volatile organic compounds that may suffer from substantial losses if subjected to PSR. For volatile organic compound analysis, larger portions of the wastes can be subjected to methanol extraction or possibly the entire field sample can be subjected to heated headspace analysis as one sample or as a series of large aliquots, or possibly the entire field sample can be preserved in the field with an equal volume of methanol or methanol/water solution.

9.5 *Modification of Analytical Method:*

9.5.1 The analytical phase of a sampling and analytical program allows another opportunity to simplify the characterization of a highly stratified waste. Examples of different classes of analytical methods are:

9.5.1.1 Screening methods,

9.5.1.2 Portable methods,

9.5.1.3 Field laboratories methods,

9.5.1.4 Nonintrusive methods,

9.5.1.5 Nondestructive methods,

9.5.1.6 Innovative methods, and

9.5.1.7 Fixed laboratory methods.

9.5.2 Screening, portable, and field laboratory methods have the distinct advantage that they allow for the cost-effective analysis of more samples. These methods generate more data, making it easier to detect correlations between concentration levels and waste strata or components. Also, some screening methods may analyze a larger sample volume than what is traditionally analyzed in a fixed laboratory.

9.5.3 Nonintrusive methods (for example, geophysical methods) can be useful when there are health and safety issues regarding exposure to the waste. These methods also may be used to evaluate large-volume wastes qualitatively or semiquantitatively.

9.5.4 Nondestructive methods are useful in that the integrity of the samples is maintained for additional analyses or evidence, or both.

9.5.5 Innovative methods may provide more cost-effective or timely results or improve sensitivity or accommodate larger and more representative sample sizes.

9.5.6 Fixed laboratory methods usually have the advantage of regulatory approval, established quality assurance/quality control requirements and often greater sensitivity than that achievable by screening, portable, or field laboratory methods.

9.6 *Modification of the Waste Handling, Treatment, Disposal Plan:*

9.6.1 If modifications to sampling, sample preparation, and analysis are not appropriate for a given waste, or are appropriate but still do not allow the objectives to be met cost-effectively, then the reasoning behind the original program must be reconsidered. It may be possible to achieve the program objectives by means of an alternative approach. For example, a change in waste treatment, handling, or disposal technologies may require analysis for different characteristics or may allow for simplified sampling. Alternatively, the waste population could be defined differently by employing smaller remediation or exposure units that would be sampled separately as opposed to characterizing the entire population. The need behind the waste characterization objectives has to be examined and an approach for simplifying the characterization process devised. This process is addressed in the optimization step of the planning process.

9.6.2 For example, consider a hypothetical waste that must be evaluated prior to waste disposal to determine if it is hazardous. An initial attempt to characterize the waste failed to meet the objective, indicated that the waste was highly stratified, and proved that portions of the waste are hazardous. After reviewing this preliminary information and the costs to attempt a defensible characterization of the waste, it could be decided that it is resource and cost-effective to consider all the waste hazardous and treat it as a hazardous waste by incineration. Under this scenario, the sampling and analytical requirements change, requiring simplified testing for general characteristics prior to incineration, and more comprehensive analysis of the less heterogeneous and more easily sampled incinerator ash to determine if it is within compliance.

9.7 *Changing Objectives*—If the project objectives are not met and none of the strategies can be changed or modified, the objectives need to be reconsidered. After changing the objectives, the sampling and analysis plans also should be adjusted. These iterations will continue until the project objectives can be met.

10. Keywords

10.1 analysis; heterogeneity; homogeneity; nonrandom; populations; random; sample preparation; samples; sampling; strata; stratified; stratum

ANNEX

(Mandatory Information)

A1. DISCUSSION OF HETEROGENEITY AND STRATIFICATION OF WASTES AND RELATIONSHIP OF SAMPLES AND POPULATIONS

A1.1 *Introduction*—This annex contains a practical non-mathematical discussion of issues pertinent to heterogeneous waste sampling. The discussion deals with heterogeneity, stratification, and the relationship of samples and populations in sampling design. It is consistent with sampling theory and statistics and may serve as an introduction to the statistical treatment of sampling issues (see Refs **1–10**).[2] The content of this annex is applicable to the sampling of wastes regardless of their degree of heterogeneity.

A1.2 *Population Attributes:*

A1.2.1 A population is the total collection of items to be studied. Theoretically, the classification of a population as being homogeneous or heterogeneous is straightforward. If all of the items in the population are identical, then the population is homogeneous. If one or more of the items are dissimilar, the population is heterogeneous. Theoretical homogeneity, the equivalent to nonheterogeneity, is a unique state of absolute uniformity for all items in the population

while heterogeneity is a variable attribute that can range from a population, which is almost homogeneous (that is, homogeneous for applied purposes) to a population that displays dissimilarity between all items of the population.

A1.2.2 According to the theoretical definition for homogeneity, virtually all real-world populations would be heterogeneous. From a practical perspective, however, as the level of heterogeneity approaches the state of homogeneity, populations can be considered homogeneous for applied purposes. References to the homogeneity of a population are usually made in light of this applied meaning, that is, for practical purposes, the population is homogeneous (practical homogeneity).

A1.2.3 The attributes of homogeneity and heterogeneity are relative. Heterogeneity and homogeneity are a function of the specified chemical constituent, property, particle size, visual appearance, sampling objectives, and the sample mass/volume. The same population can be homogeneous with regards to one constituent or property, and at the same time be heterogeneous with regards to another constituent or property.

---

[2] The boldface numbers in parentheses refer to the list of references at the end of this standard.

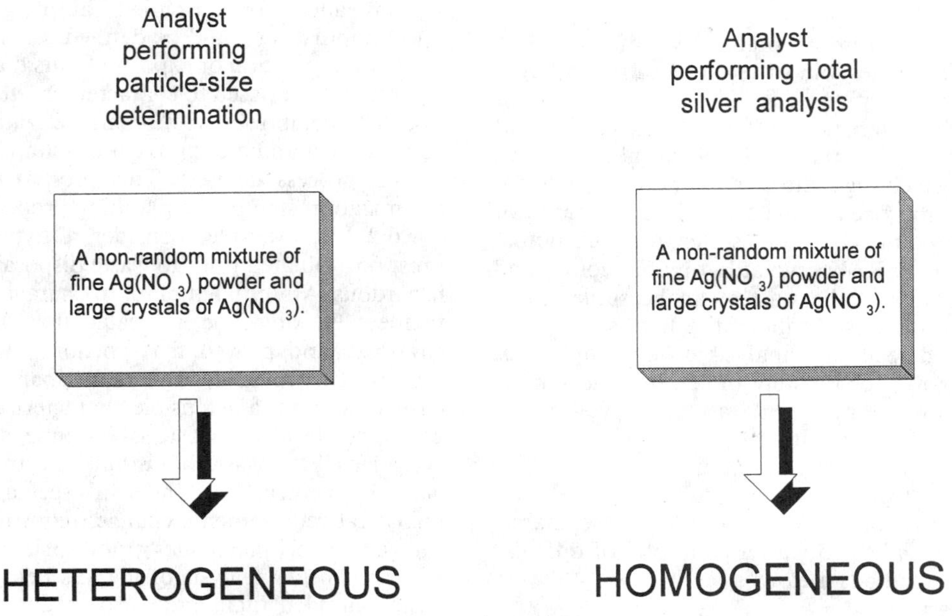

Analyst
performing
particle-size
determination

Analyst
performing Total
silver analysis

A non-random mixture of fine Ag(NO$_3$) powder and large crystals of Ag(NO$_3$).

A non-random mixture of fine Ag(NO$_3$) powder and large crystals of Ag(NO$_3$).

HETEROGENEOUS

HOMOGENEOUS

FIG. A1.1 Heterogeneity Relative to Objectives

A1.2.3.1 Consider a nonrandom mixture of silver nitrate, some of which is a powder and the remainder is in the form of large crystals (see Fig. A1.1). The population is heterogeneous when considering particle size or homogeneous when silver content is of interest.

A1.2.3.2 Following comprehensive emission spectroscopic and titrimetric analyses of uranium metal, a chemist may find the population to be homogeneous while the nuclear chemist analyzing for U^{235} and U^{238} would find the same population to be isotopically heterogeneous (see Fig. A1.2).

A1.2.3.3 Decisions regarding heterogeneity also can be a function of the analytical method used to process samples. If one method (AAS-graphite furnace atomic absorption spectroscopy) is more sensitive and has method detection limits (MDL) that are lower than the other (X-ray fluorescence field screening), what was originally thought to be a homogeneous waste may be found to be a heterogeneous waste (see Fig. A1.3).

A1.2.4 Two population attributes are the causative factors for heterogeneity. The primary attribute is referred to as compositional heterogeneity, and the secondary attribute is distributional heterogeneity.

A1.2.4.1 Compositional heterogeneity occurs when the concentration of the targeted constituent or targeted property varies from item to item. This compositional or property difference between items is a requisite for a heterogeneous population, that is, dissimilar items must be present for heterogeneity to exist.

A1.2.4.2 Distributional heterogeneity results from differences in the spatial distribution of dissimilar items resulting in microscopic or macroscopic concentration gradients or property gradients, or both. Compositional heterogeneity is a necessary condition for the existence of distributional heterogeneity. Distributional heterogeneity is a population attribute, and if a population is defined differently (that is, change the population boundaries), the distributional heterogeneity for the expanded or smaller population may differ.

A1.2.5 Compositional and distributional heterogeneity are the underlying causes for the more commonly understood types of random heterogeneity and nonrandom heterogeneity. Random and nonrandom are the terms that will be used in the remainder of this guide to describe the different types of heterogeneity. The introduction of compositional and distributional heterogeneity is to assist those who may want to investigate further the particulate material sampling theory.

A1.2.5.1 Random heterogeneity is that type of heterogeneity that occurs when dissimilar items are randomly distributed throughout the population.

A1.2.5.2 Nonrandom heterogeneity is that type of heterogeneity that occurs when dissimilar items in the population are nonrandomly distributed. In a nonrandom heterogeneous population, similar items or similar concentrations are grouped into strata. This type of population, also is referred to as a *stratified population*. The terms *stratified population* and *nonrandom heterogeneous populations* are interchangeable. Strata are separated from other strata by time or space or correlated with different components or waste sources. This guidance focuses on sampling strategies for a particular type of stratified waste referred to as *highly stratified*.

A1.3 *Physical Sample Attributes:*

A1.3.1 To characterize a population, it must be subjected to evaluation. The population can be characterized with great certainty if all population elements are evaluated for the characteristic of interest. Populations, however, are usually so large that the entire population cannot be subjected to evaluation. Practically and economically it makes more sense to collect a number of samples and compile the analytical results in a database that is used to make inferences regarding the population.

A1.3.2 Due to the different meanings assigned to the term *sample* and to minimize confusion the term *physical sample* is used throughout this discussion. Physical sample is a reference to the sample held in a sample container or that portion of the population that is subjected to in situ measurements. The term *sample size* also can have different meanings. Although use of multi-word terms can appear wordy, to avoid confusion, specific terms such as sample

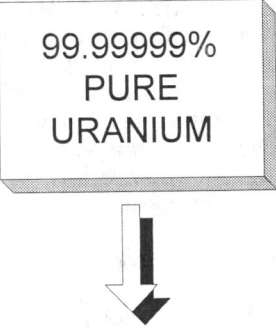

RADIOCHEMIST   **CHEMIST**

99.99999% PURE URANIUM   99.99999% PURE URANIUM

HETEROGENEOUS   **HOMOGENEOUS**

FIG. A1.2 Heterogeneity Relative to Perspective

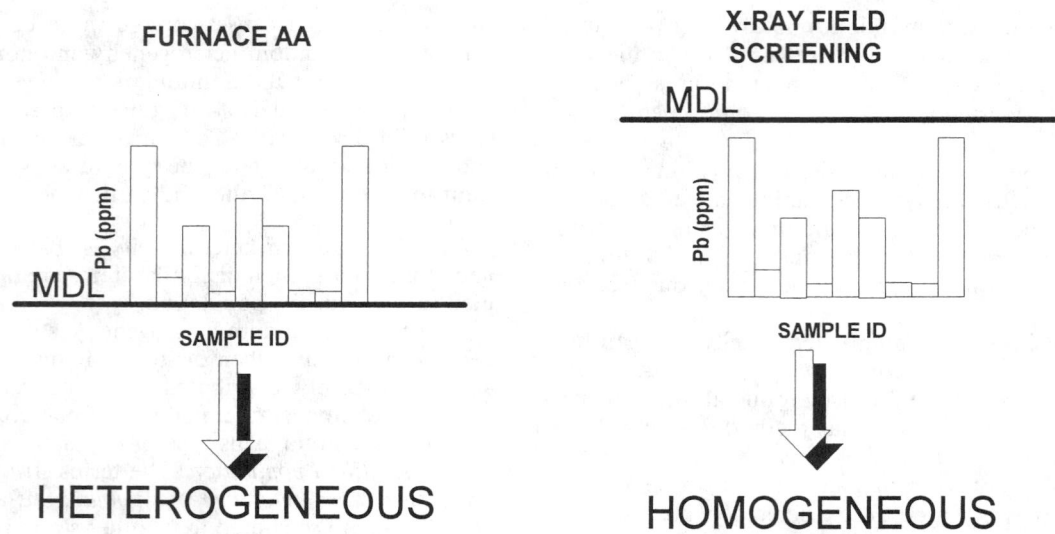

FIG. A1.3 Heterogeneity Relative to Method Detection Limits (MDLs)

mass, sample volume, and number of samples are used.

A1.3.3 The accuracy of inferences made to populations are dependent on how well the physical samples represent the population characteristic of interest. The term *representativeness* usually is associated with mean concentrations. Physical samples, however, also are used to measure other statistical parameters of the population, such as variance, trends, and proportions.

A1.3.4 Sampling of a theoretically homogeneous population always results in physical samples that represent the characteristics of the population, assuming that the sampling process itself does not introduce contamination or allows for selective loss of waste constituents. The lack of variance in a homogeneous population ensures all physical samples collected from the population are identical and representative of the population.

A1.3.5 The meaning of the term *representative sample* is susceptible to misinterpretation since it connotes a single sample. The difficulty in collecting a single physical sample that represents a population increases with increasing heterogeneity. When trying to represent a heterogeneous population, it is more appropriate to collect a number of physical samples. If the physical samples are collected according to a properly designed plan, the population is better represented by the characteristics associated with the entire set of physical samples. Such a set of physical samples would be referred to as a representative set of physical samples.

A1.3.6 To properly represent a characteristic of a heterogeneous population, more than one physical sample usually is required. Samples collected from a heterogeneous population will display intersample variance. Intersample variance measured between different physical samples results from the following:

A1.3.6.1 Differences in the composition of items between sampling locations;

A1.3.6.2 Differences in how these items are distributed throughout the population; and,

A1.3.6.3 Sampling and analytical errors that ideally will be minimal so that the true intersample variance can be measured.

A1.3.7 The intersample variance may be used to make inferences about the homogeneity or heterogeneity of the population. The accuracy of these inferences will be a function of the sampling design and the quality of the sampling efforts used to collect the samples and of the analytical efforts used to generate associated data.

A1.4 *Populations and Samples:*

A1.4.1 Homogeneity and heterogeneity are population attributes estimated by the evaluation of physical samples. Representativeness of a population characteristic and intersample variance are sample attributes. Physical samples are used to measure the homogeneity and heterogeneity of a population.

NOTE 3—If the entire physical sample is analyzed, the heterogeneity of the physical sample is not relevant. Physical samples only are assigned attributes of heterogeneity or homogeneity when they are being subsampled since at this time the physical sample is the population whose characteristics must be represented in the subsample.

A1.4.2 Physical samples are collected from the population, evaluated, and the resulting information is employed to make inferences regarding the entire population. The value of physical samples is related directly to how accurately they represent the population characteristics of interest. The value of the inferences about a population are only as good as the associated samples. To properly represent a population characteristic, sampling location, sample mass, sample collection methods, the number of physical samples and compositing of physical samples are controlled.

A1.4.3 In a nonrandom heterogeneous population, the concentration of a target constituent (for example, arsenic) or the degree to which a property (for example, ignitability) is expressed is correlated with time, space, component, or waste source. Conversely, the constituent or property displays no correlation with time, space, component, or waste source in a random heterogeneous population.

A1.4.4 Samples collected from nonrandom heterogeneous populations, therefore, display a correlation of constituent concentrations or properties with time, space, components, or waste source and less intersample variance when samples are collected from the same stratum. Samples collected from random heterogeneous populations display a significant amount of intersample variance but no correlation of concentration or property with space, time, component, or waste source. In summary:

Homogeneous	no significant intersample variance
Random heterogeneous	significant intersample variance
Nonrandom heterogeneous	significant intersample variance and correlation of concentration/property with time, space component, or waste source

A1.4.5 Table A1.1 summarizes the attributes of physical samples and populations, as well as the inferences that can be made from variance and concentration information. Figure A1.4 illustrates the process of using variance and concentration information to classify the type of heterogeneity.

A1.4.6 The relationship of physical samples to a population is explored in the following example. This example is designed to show the role physical samples play in the evaluation of population characteristics. In particular, this example emphasizes the impact of sample mass, particle size, and sample collection on sample representativeness and the resulting inferences for population characteristics.

A1.4.7 The population consists of a 2-L waste container that has 1-g nuggets of cadmium randomly distributed throughout an otherwise homogeneous and cadmium-free matrix. The cadmium-free matrix has a substantially smaller particle size than that of the cadmium nuggets. The cadmium nuggets constitute 37 % of the waste on a weight basis. The waste is composed of dissimilar particles resulting, in compositional differences and allowing distributional differences within the population. It is assumed that after collection, the entire physical sample is analyzed for cadmium.

Characteristics of interest: cadmium concentrations
Statistical parameters of interest: mean and standard deviation

A1.4.7.1 The following information pertains to Sampling Design No. 1 (Fig. A1.5).

Physical samples mass: 0.1 g
Sampling locations chosen randomly
Number of samples: 10
Sample collection device: a small spatula
Cadmium data: all 10 samples had concentrations less than 0.2 mg/kg
Average <0.2 mg/kg ± 0 mg/kg (<0.2 mg/kg = Method Detection Limit, MDL)

The lack of variance between physical samples (that is, no significant intersample variance) indicates falsely that the

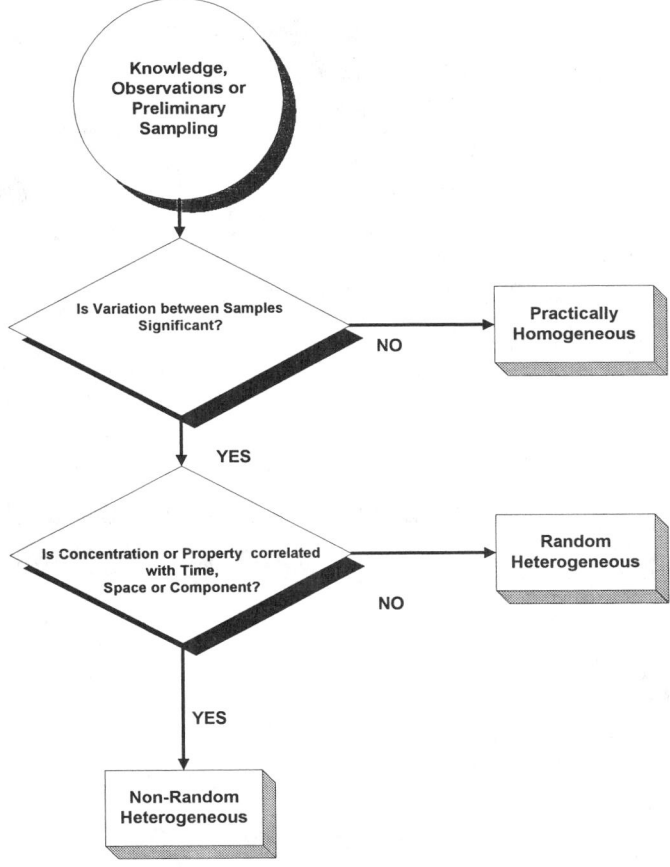

FIG. A1.4   Process for Classifying Type of Heterogeneity By Measurement or Process Knowledge

population is homogeneous with regards to cadmium. This is an incorrect evaluation since the physical samples are not representative of the population as a result of the sample collection method, which discriminated against the larger cadmium particles.

A1.4.7.2 The following information pertains to Sampling Design No. 2 (Fig. A1.6).

Physical samples mass: 1 g
Sampling locations chosen randomly

TABLE A1.1   Population and Sample Attributes

Population Attribute	Sample Description	Sample Attribute	Inference
Homogeneous (theoretical homogeneity)	All samples contain only identical items.	No significant intersample variance. No correlation of concentration or properties with time, space, component, or waste source.	Samples are representative of a homogeneous population.
Practical homogeneity	All samples contain dissimilar items, but each sample contains similar proportions.	No significant intersample variance. No correlation of concentration or properties with space, time, component, or waste source.	Samples are representative of a homogeneous population.
Random heterogeneous	All samples contain dissimilar items, each sample has different proportions, but these proportions are not correlated with time, space, or components.	Significant intersample variance. No correlation of concentration or properties with space, time, component, or waste source.	Samples are representative of a random heterogeneous population.
Nonrandom heterogeneous (stratified)	All samples contain dissimilar items, each sample has different proportions and these proportions are correlated with time, space, components, or source.	Significant intersample variance. Correlation of concentration or properties with space, time, component, or waste source.	Samples are representative of a nonrandom heterogeneous population.

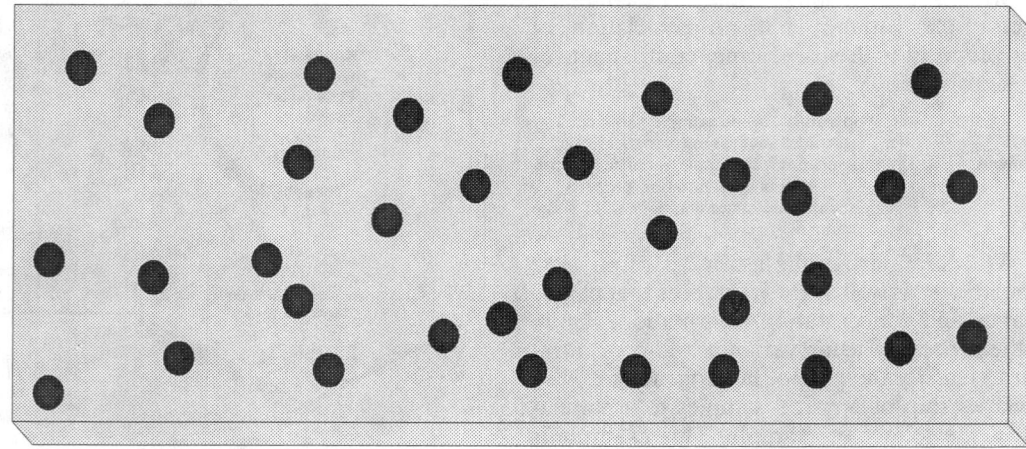

Population - 2 liter waste container (37% cadmium) with 1 gram cadmium nuggets in a cadmium-free matrix

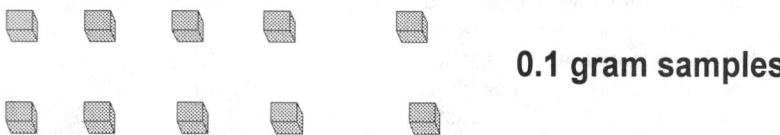

0.1 gram samples

Average = <0.2 mg/Kg +/- 0 (<0.2 mg/Kg = Method Detection Limit)

FIG. A1.5   Sampling Design No. 1

Number of samples: 10
Sample collection device: a spatula approximately 10 × larger than the one used in the previous example
Cadmium data: 100 %, <0.2 mg/kg, <0.2 mg/kg, <0.2 mg/kg, <0.2 mg/kg, 100 %, <0.2 mg/kg, <0.2 mg/kg, <0.2 mg/kg, <0.2 mg/kg (<0.2 mg/kg = Method Detection Limit, MDL)
Average: 20 ± 42 %

The variance between physical samples (that is, existence of significant intersample variance) indicates that the population is heterogeneous with regards to cadmium. Although more representative of the population characteristic than Sampling Design No. 1, this design also suffers from an sample collection error since the large cadmium particles were not collected unless they were aligned perfectly with the sampling device. Since only two samples had detected cadmium concentrations, there is a 25 % chance that these two samples could have occurred in the top half or the bottom half of the waste container (that is, there is a significant correlation of concentration to space). If this had occurred, the incorrect assumption may have been made that the population was nonrandomly heterogeneous (stratified) with a stratum of pure cadmium in half the container with the other half of the container consisting of cadmium-free material. These samples do not properly represent the population characteristic.

A1.4.7.3 The following information pertains to Sampling Design No. 3 (Fig. A1.7).

Physical samples mass: 30 g
Sampling locations chosen randomly
Number of samples: 10

Sample collection device: tube with a diameter that can easily accommodate a number of cadmium particles and can take a core from top to bottom
Cadmium data: 35 %, 50 %, 39 %, 24 %, 32 %, 47 %, 43 %, 27 %, 29 %, 44 %
Average: 37 ± 8.9 %

The sample mass required by this sampling design allowed for proper extraction and evaluation of the resulting physical samples yielding a database that was representative of the population (that is, there is significant intersample variance and that concentration is not correlated with space). The waste would be considered randomly heterogeneous with regards to cadmium.

A1.4.8 The previous three designs for sampling the same population showed how the perception of population heterogeneity can be affected by sampling design. The usefulness of using physical samples to make inferences regarding population heterogeneity varies according to the ability of the sampling device and the resulting sample mass to accommodate all the different-sized items of a population and the ability of all collected physical samples, as a set, to accommodate representative amounts of all constituents of the population.

A1.4.9 These previous sampling designs assumed that the entire physical sample was subjected to analysis. Practical experience indicates that most physical samples will be subjected to subsampling prior to analysis. Figure A1.8 graphically depicts the common relationship between populations, physical samples, subsamples, and data. If subsampling is employed, then subsamples are the windows through which the population is viewed, and the subsamples

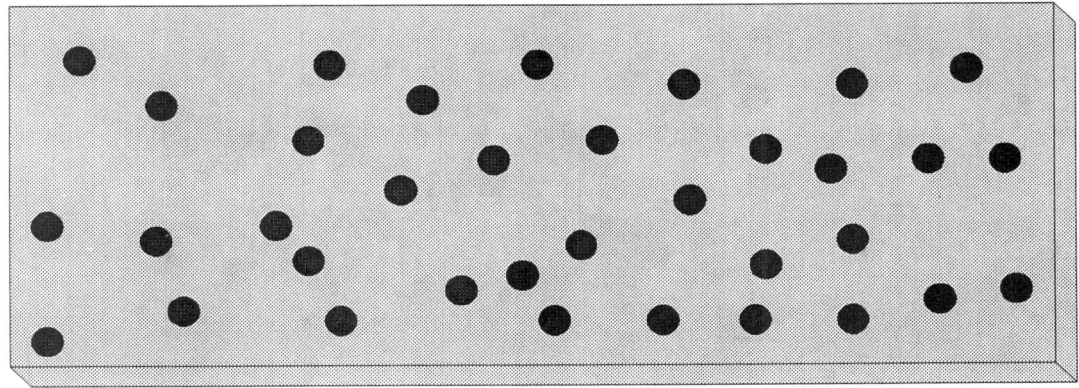

Population - 2 liter waste container (37% cadmium) with 1 gram cadmium nuggets in a cadmium-free matrix

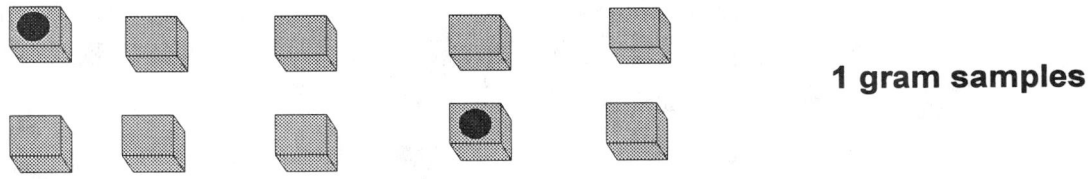

1 gram samples

Average = 20% mg/Kg +/- 42%
Possible Stratification
FIG. A1.6 Sampling Design No. 2

will be used to make inferences including those regarding the homogeneity, random heterogeneity, or nonrandom heterogeneity of the population. Subsampling, when required, becomes an additional critical step that must be implemented properly to ensure the accuracy of inferences.

A1.5 *Population Attributes and Sampling Design:*

A1.5.1 The relationship between physical samples and populations clearly implies that knowledge of population attributes and use of this knowledge should decrease the bias and increase the precision of sampling. Table A1.2 and Fig. A1.9 respectively tabularize and depict the relationship between critical sampling design decisions and population attributes, planning information and specifications gleaned from planning processes such as the DQO process, and, analytical requirements.

A1.5.2 In addition to budget constraints, the following information, to the extent that it is known, should be considered during sampling design:

A1.5.2.1 *Population Attributes:*

(1) Heterogeneity—Heterogeneity of the population in terms of the characteristic of interest; homogeneous, randomly heterogeneous, or nonrandomly heterogeneous (stratified).

(2) Item Size—The size of items present in the population including items that may or may not contain the characteristic of interest.

(3) Population Accessibility—The ability or inability to access all portions of the population for purposes of sampling.

A1.5.2.2 *DQOs:*

(1) Statistic—The mean, mode, variance, proportion, or other measure of a population which is of interest.

(2) Level of Confidence—The specified level of confidence that decisions will be correct. In other words, the maximum decision error rate that is acceptable to the decision maker.

(3) Boundaries—The temporal and spatial boundaries of the population that is to be studied.

A1.5.2.3 *Analytical Requirements:*

(1) Analytical Sample Volume/Mass—The sample volume/mass needed to prepare and analyze physical samples.

(2) Analyte/Media Integrity—The handling, containerization, preservation, and shipping procedures required to maintain the physical, compositional, and legal integrity of the physical samples.

A1.5.3 *Sample Mass or Volume*—The appropriate sample mass or volume will be determined by considering the size of the largest items contained within the population, the heterogeneity of the population, and the optimum sample mass/volume for preparation and analysis. Knowledge of item sizes contained within a population and their content of the characteristic of interest are needed to choose the correct sample mass/volume. Bias can result if certain item sizes are discriminated against during sampling. The correct sample mass/volume will accommodate all item sizes or be chosen such that the impact of any discrimination is accounted for and understood. The variance of data caused by local heterogeneity of the population may be controlled by using a properly sized sampling device and by taking greater sample volumes or masses.

A1.5.4 *Sampling Locations*—Sampling locations are a function of the population boundaries, the accessibility of all portions of the population, and the type of heterogeneity.

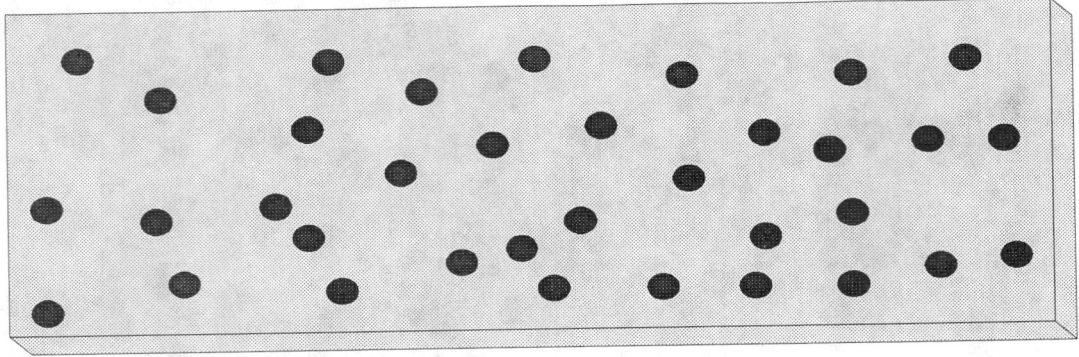

Population - 2 liter waste container (37% cadmium) with 1 gram cadmium nuggets in a cadmium-free matrix

30 gram samples

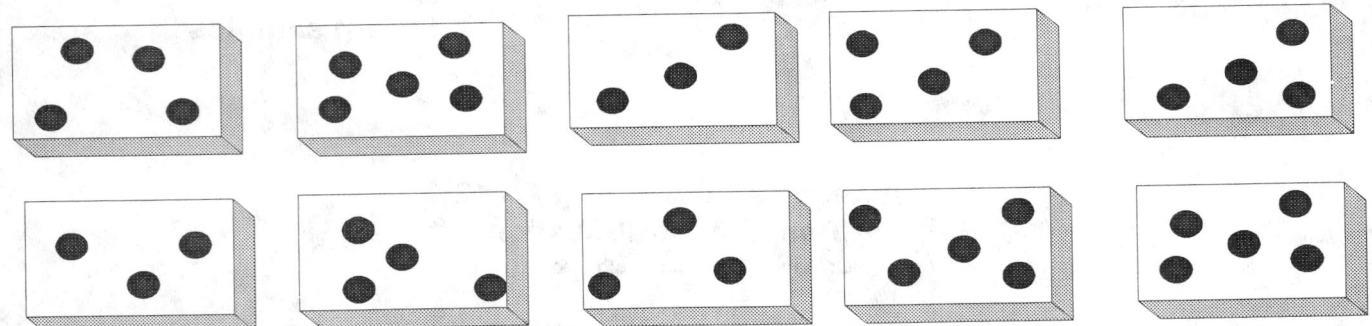

Average = 37% +/- 8.9%

FIG. A1.7 Sampling Design No. 3

Other than background and other reference samples, sampling usually is restricted to those accessible areas within the population boundaries.

NOTE 4—Since sampling of inaccessible portions of the population is not possible, any extrapolation of sampling/analytical data to these areas must be well-documented. Extrapolation to unsampled areas is a judgment call and not a statistically valid inference. The type of heterogeneity may impact the sample locations since the existence or potential existence of strata may alter the sampling strategy for choosing sampling locations, for example, simple random versus stratified random.

A1.5.5 *Number of Samples*—The number of samples collected is determined after considering the population heterogeneity and information and specifications generated during the initial stages of the planning process, that is, the statistic of interest, the levels of uncertainty in decision-making and the population boundaries. If a population is not substantially larger than the physical sample and the distribution of the characteristic of interest is randomly heterogeneous, it may be appropriate to collect a fewer number of samples by a random or systematic sampling procedure. If a population is relatively large as compared to the physical sample and the characteristic of interest is nonrandomly distributed (stratified), a greater number of samples and a stratified sampling approach may be needed to achieve similar levels of precision and bias and the specified confi-

dence level.

A1.5.6 *Composite Versus Discrete*—A composite sample is made by combining one or more discrete samples (physical samples) into one sample. Compositing has the potential advantage of yielding a more accurate estimate of average concentrations or average properties. Compositing, however, has the potential disadvantage of losing pertinent variance information and the possibility of diluting hot spots such that the average results fall below thresholds or limits of detection. The chosen statistical approach for data evaluation, the acceptable level of uncertainty, the type of heterogeneity, and budgets are considered when deciding between the use of composite and discrete samples.

A1.5.7 *Sampling Devices*—The choice of sampling devices is made after determining the analytical-sample requirements, the size of the largest items that must be accommodated by the sampling device, the accessibility of sampling locations, sample integrity, and reactivity of sampling-device materials.

A1.5.8 Comprehensive knowledge of population attributes is infrequent and the degree of knowledge varies from population to population. However, the more thorough the planning process and the better the understanding of a population's attributes, the more likely that samples and associated data will be representative of the characteristic of interest.

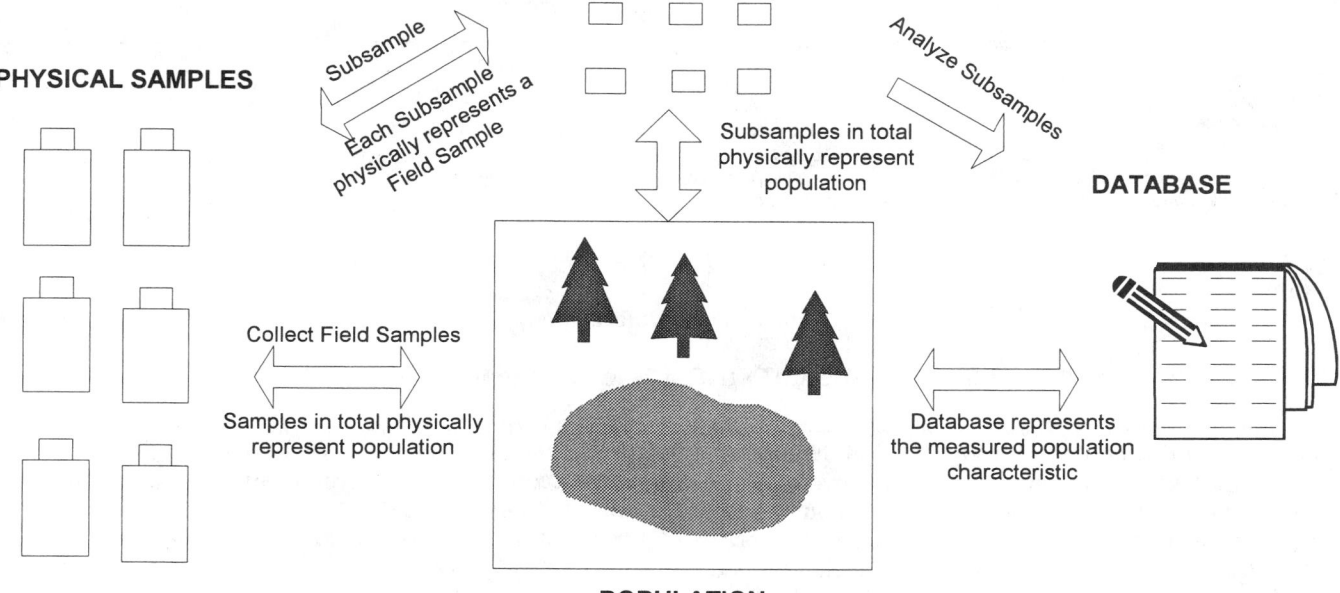

FIG. A1.8 Process of Using Physical Samples to Measure a Characteristic of the Population

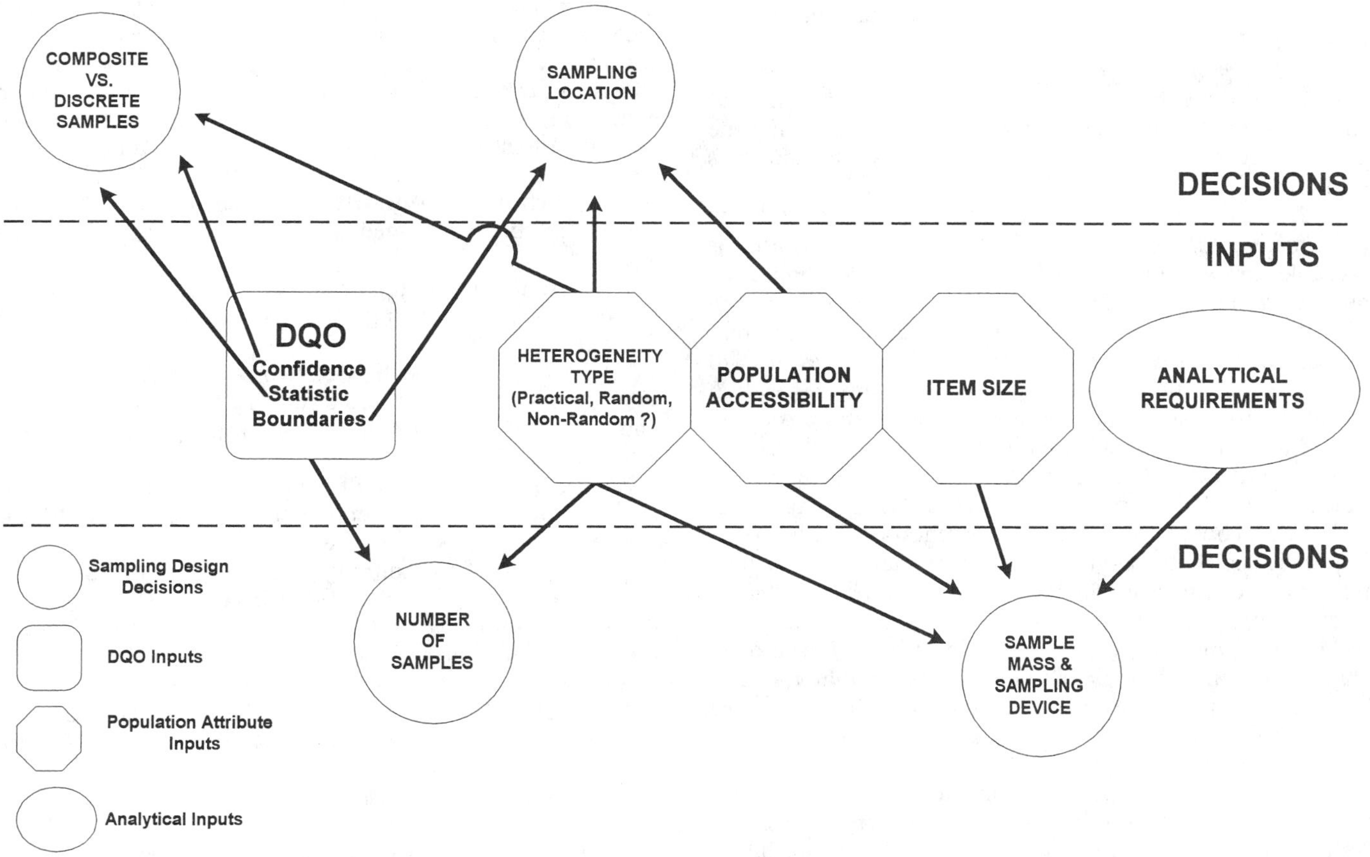

NOTE—Regulatory requirements addressed during the DQO process may impact sampling design.

FIG. A1.9 Role of Population Attributes, DQOs, and Analytical Requirements in Optimizing Sampling Designs

TABLE A1.2 Role of Population Attributes, DQOs, and Analytical Requirements in Optimizing Sampling Designs[A]

| | Inputs into Decision-Making Process | | | | | | |
Sampling Design Decisions	DQO Confidence Level	DQO Statistic	DQO Boundaries	Heterogeneity Type	Population Accessibility	Item Size	Analytical Requirements
Number of samples	X[B]	X	X	X			
Sampling location			X	X	X		
Sample mass/volume				X		X	X
Sampling device					X	X	X
Composite versus discrete	X	X		X			

[A] Regulatory requirements addressed during the DQO process may impact sampling design.
[B] X indicates that attribute or requirement will impact the design decision.

APPENDIX

(Nonmandatory Information)

X1. CASE STUDY FOR DESIGN OF A SAMPLING APPROACH FOR A HIGHLY STRATIFIED WASTE

X1.1 The following is a hypothetical scenario of how sample design can be optimized for highly stratified waste.

X1.1.1 A storage area contains 4000 drums of waste generated over a 15-year period. The drum contents are highly stratified and contain a myriad of wastes from process waste; destruction and construction debris such as wood, concrete; laboratory wastes including broken glassware, paper, or empty bottles. The initial stages of the planning process identified beryllium and solvents as the target characteristics and the mean and variance as the statistics of interest. In addition, groundwater modeling indicated that the storage area is the source of a plume contaminated with solvents and beryllium.

X1.2 *Sampling Design:*

X1.2.1 *What are the target characteristics?* Solvents and beryllium are the target characteristics identified during the planning process. The average and variance are the statistics of interest.

X1.2.2 *Are the target characteristics correlated with an identifiable strata or source?* The source of beryllium is traceable to one process, whose waste easily should be identifiable even when drum markings are not legible. The solvents are likewise traceable to a machine shop that disposed of its waste in easily identified drums. Testing will have to be performed to determine if there is any correlation with item size, space, time, or components in the waste.

X1.2.3 *Can any waste components or strata be eliminated from consideration during sampling?* Historical information indicated that 400 drums of construction debris were generated during construction of a new warehouse. The information indicates that the virgin nature of the materials may make these drums candidates for less intensive sampling or no sampling. Likewise, the source of beryllium contamination is a beryllium sludge that exists in drums by itself or in drums commingled with shredded packing material and laboratory wastes that were generated during physical testing of the beryllium product. Since the materials commingled with the beryllium waste are known not to be a source of contamination, the commingled material can be discriminated against during sampling, and only the beryllium sludge sampled and the volume contribution of the commingled material noted.

X1.2.4 *Are contamination levels correlated with item size?* Some of the older beryllium sludge has dried and formed a cementaceous aggregate of different item sizes. Since the sludge is known to be homogeneous within a batch, by process knowledge and preliminary sampling data, sampling can be restricted to the more easily sampled, smaller item sizes.

X1.2.5 *Is contamination innate or surface adsorbed?* The waste from the machine shop consists of varied material from fine metallic filings to large chunks of metal and out-of-specification metal product. Since the only contamination in the machine shop is solvents and cutting oils and the waste matrix is impervious, the contamination is surface adsorbed in nature. Sampling of these wastes, therefore, will consist of the sampling of fines that will be subjected to extraction, wipe sampling of the large metallic objects, and notation of the volume contributions of the different item sizes. It is essential that all assumptions (that is, any correlations) be verified at least by knowledge of the waste and preferably confirmed by exploratory sampling and analyses.

X1.3 In the preceding hypothetical case, the proposed strategy for characterizing the 4000 drums resulted in the following:

X1.3.1 The identification of two large strata that constitute the majority of the waste (that is, the beryllium sludge and the solvent and cutting oil contaminated machine shop waste),

X1.3.2 The elimination of the need to sample 10 % of the drums (that is, the construction debris) if preliminary testing verifies waste disposal information,

X1.3.3 Simplified sampling of the beryllium commingled waste by restricting sampling to the beryllium sludge and not the other commingled materials if preliminary testing verifies waste disposal information,

X1.3.4 Simplified sampling of the cementaceous beryllium sludge by limiting sampling to the more easily sampled small items if preliminary testing verifies waste disposal information.

X1.4 Simplifying the sampling of the machine wastes, since the source of contamination is surface adsorbed and not innate to the waste materials if preliminary testing, verifies waste disposal information.

REFERENCES

(1) "Characterizing Heterogeneous Wastes," EPA 600/R-92/033, February 1992.

(2) "Environmental Monitoring Issues," EPA/600/R-93/033, March 1993.

(3) "Guidance for the Data Quality Objectives Process," EPA QA/G-4, September 1994.

(4) "Preparation of Soil Sampling Protocols: Sampling Techniques and Strategies," EPA 600/R-92/128, July 1992.

(5) "Test Methods for Evaluating Solid Waste Physical/Chemical Methods," EPA SW-846, 3rd Edition, March 1995.

(6) Pitard, F. F., *Pierre Gy's Sampling Theory and Sampling Practice*, CRC Press, Ann Harbor, MI, 1993.

(7) Gilbert, Richard O., *Statistical Methods for Environmental Pollution Monitoring*, Van Nostrand Reinholt Co., New York, NY, 1987.

(8) Keith, L. H., *Environmental Sampling and Analysis*, Lewis Publisher, Inc., Chelsea, MI, 1991.

(9) Keith, L. H., *Principles of Environmental Sampling*, ACS, Washington, DC, 1991.

(10) Taylor, J. K., *Statistical Techniques for Data Analysis*, Lewis Publishers, Inc, Chelsea, MI, 1990.

Standard Practice for
Decontamination of Field Equipment Used at Nonradioactive Waste Sites[1]

This standard is issued under the fixed designation D 5088; the number immediately following the designation indicates the year of original adoption or, in the case of revision, the year of last revision. A number in parentheses indicates the year of last reapproval. A superscript epsilon (ε) indicates an editorial change since the last revision or reapproval.

1. Scope

1.1 This practice covers the decontamination of field equipment used in the sampling of soils, soil gas, sludges, surface water, and ground water at waste sites which are to undergo both physical and chemical analyses.

1.2 This practice is applicable only at sites where chemical (organic and inorganic) wastes are a concern and is not intended for use at radioactive or mixed (chemical and radioactive) waste sites.

1.3 Procedures are included for the decontamination of equipment which comes into contact with the sample matrix (sample contacting equipment) and for ancillary equipment that has not contacted the portion of sample to be analyzed (non-sample contacting equipment).

1.4 This practice is based on recognized methods by which equipment may be decontaminated. When collecting environmental matrix samples, one should become familiar with the site specific conditions. Based on these conditions and the purpose of the sampling effort, the most suitable method of decontamination can be selected to maximize the integrity of analytical and physical testing results.

1.5 This practice is applicable to most conventional sampling equipment constructed of metallic and synthetic materials. The manufacturer of a specific sampling apparatus should be contacted if there is concern regarding the reactivity of a decontamination rinsing agent with the equipment.

1.6 *This standard does not purport to address the safety problems associated with its use. It is the responsibility of the user of this standard to establish appropriate safety and health practices and determine the applicability of regulatory limitations prior to use.*

2. Referenced Document

2.1 *ASTM Standard:*
D 653 Terminology Relating to Soil, Rock, and Contained Fluids[2]

3. Terminology

3.1 *Definitions:*

3.1.1 *contaminant*—an undesirable substance not normally present or an unusually high concentration of a naturally occurring substance in water or soil.

3.1.2 *control rinse water*—water used for equipment washing and rinsing having a known chemistry.

3.1.3 *decontamination*—the process of removing 'or reducing to a known level undesirable physical or chemical constituents, or both, from a sampling apparatus to maximize the representativeness of physical or chemical analyses proposed for a given sample.

3.1.4 *non-sample contacting equipment*—related equipment associated with the sampling effort, but that does not directly contact the sample (for example, augers, drilling rods, excavations machinery).

3.1.5 *quality assurance/quality control (QA/QC)*—the efforts completed to evaluate the accuracy and precision of a sampling or testing procedure, or both.

3.1.6 *sample contacting equipment*—equipment that comes in direct contact with the sample or portion of sample that will undergo chemical analyses or physical testing (for example, ground water well bailer, split-spoon sampler, soil gas sampling probe).

3.1.7 For definitions of other terms used in this practice, see Terminology D 653.

4. Summary of Practice

4.1 Two different procedures are presented for the decontamination of sample-contacting and non-sample contacting equipment. The procedures have been developed based on a review of current state and federal guidelines, as well as a summary of commonly employed procedures. In general, sample contacting equipment should be washed with a detergent solution followed by a series of control water, desorbing agents and deionized water rinses. Nonsample contacting equipment should be washed with a detergent solution and rinsed with control water. Although such techniques may be difficult to perform in the field, they may be necessary to most accurately evaluate low concentrations of the chemical constituent(s) of interest.

4.2 Prior to initiating a field program that will involve equipment decontamination, a site specific equipment decontamination protocol should be prepared for distribution to the individuals involved with the particular sampling program. Information to be presented in the protocol should include:

4.2.1 Site location and description,

4.2.2 Statement of the sampling program objective and desired precision and accuracy, that is, is sampling effort for gross qualitative evaluation or for trace concentration, parameter specific evaluations,

4.2.3 Summary of available information regarding soil types, hydrogeology and anticipated chemistry of the materials to be sampled,

[1] This practice is under the jurisdiction of ASTM Committee D-18 on Soil and Rock and is the direct responsibility of Subcommittee D18.14 on Geotechnics of Waste Management.
Current edition approved June 29, 1990. Published September 1990.
[2] *Annual Book of ASTM Standards*, Vol 04.08.

4.2.4 Listing of equipment to be used for sampling and materials needed for decontamination,

4.2.5 Detailed step by step procedure for equipment decontamination for each piece or type of equipment to be utilized and procedures for rinse fluids containment and disposal as appropriate,

4.2.6 Summary of QA/QC procedures and QA/QC samples to be collected to document decontamination completeness including specific type of chemical analyses and their associated detection limit, and

4.2.7 Outline of equipment decontamination verification report.

5. Significance and Use

5.1 An appropriately developed, executed and documented equipment decontamination procedure is an integral and essential part of waste site investigations. The benefits of its use include:

5.1.1 Minimizing the spread of contaminants within a study area and from site to site,

5.1.2 Reducing the potential for worker exposure by means of contact with contaminated sampling equipment, and

5.1.3 Improved data quality and reliability.

5.2 This practice is not a substitute for a well-documented Quality Assurance/Quality Control (QA/QC) program. Because the ultimate test of a decontamination procedure is its ability to minimize erroneous data, a reasonable QA/QC program must be implemented.

5.3 This practice may not be applicable to all waste sites. When a sampling effort is completed to determine only the general range of chemical concentrations of interest less rigorous decontamination procedures can be adequate. Investigators should have the flexibility to modify the decontamination procedures with due consideration for the sampling objective or if QA/QC documentation supports alternative decontamination methods.

5.4 At sites where the reactivity of sampling equipment to decontamination washes creates concern for the generation of undesirable chemical by-products, the use of dedicated sampling equipment should be considered.

5.5 This practice, where applicable, should be used before, between, and after the completion of sampling events.

6. Reagents

6.1 *Detergent*, non-phosphate detergent solution.[3]

6.2 *Acid rinse (inorganic desorbing agent)*, 10 % nitric or hydrochloric acid solution-made from reagent grade nitric or hydrochloric acid and deionized water (1 % is to be applied to low-carbon steel equipment).

6.3 *Solvent rinse (organic desorbing agent)*, isopropanol, acetone, or methanol; pesticide grade.

6.4 *Control rinse water*, preferably from a water system of known chemical composition.

6.5 *Deionized water*, organic-free reagent grade.

7. Procedure for Sample Contacting Equipment

7.1 At a minimum, sample contacting equipment should be washed with a detergent solution and rinsed with control water.

7.2 For programs requiring more rigorous decontamination to meet the sampling or QA/QC objectives, the following procedures are indicated:

7.2.1 Wash with detergent solution, using a brush made of inert material to remove any particles or surface film.

7.2.1.1 For equipment that, because of internal mechanism or tubing cannot be adequately cleaned with a brush, the decontamination solutions should be circulated through the equipment.

7.2.2 Rinse thoroughly with control water.

7.2.3 Rinse with an inorganic desorbing agent (may be deleted if samples will not undergo inorganic chemical analysis).

7.2.4 Rinse with control water.

7.2.5 Rinse with organic desorbing agent (may be deleted if samples will not undergo organic chemical analyses).

7.2.6 Rinse with deionized water.

7.2.7 Allow equipment to air dry prior to next use.

7.2.8 Wrap equipment for transport with inert material (aluminum foil or plastic wrap) to direct contact with potentially contaminated material.

7.3 *Nonsample Contact Equipment:*

7.3.1 Clean the equipment with portable power washer or steam cleaning machine. Alternatively, hand wash with brush using detergent solution.

7.3.2 Rinse with control water.

7.3.3 The more rigorous decontamination procedures may be employed if necessary to meet sampling or QA/QC objectives.

7.4 Depending on site conditions, it may be appropriate to contain spent decontamination rinse fluids. If this is the case the appropriate vessel[4] for fluid containment should be used depending on the ultimate disposition of the material.

7.5 Depending on site conditions, it may be desirable to perform all equipment decontamination at a centralized location as opposed to the location where the equipment was used. If this is the case, care must be taken to transport the equipment to the decontamination area such that the spread of contaminants is minimized.

8. Quality Assurance/Quality Control

8.1 It is important to document the effectiveness of the decontamination procedure. To that end the projects QA/QC program should include provisions for the collection of samples to evaluate the completeness of a specific decontamination procedure. This could include:

8.1.1 Collection of rinse or wipe samples before the initial equipment decontamination prior to its use for sampling to establish a base line level of contaminants residing on or in the equipment,

8.1.2 Collection of final rinse or wipe samples after equipment decontamination following its use, and

8.1.3 The frequency of sampling to demonstrate the completeness of equipment decontamination is dependent upon objectives of the project as they relate to QA/QC. At a

---

[3] Alquinox or Liquinox or similar solution has been found suitable for this purpose.

[4] A drum approved by the Department of Transportation or similar container has been found suitable for this purpose.

minimum it is recommended after every ten decontamination washings.

9. Report

9.1 The activities completed for each equipment decontamination should be documented in writing. Included in this report should be the following information:

9.1.1 Site location, date, time, and weather,

9.1.2 Sample location where equipment was employed,

9.1.3 Location where decontamination was performed,

9.1.4 Individuals performing the decontamination,

9.1.5 Decontamination procedures,

9.1.6 Source of materials (solutions) used for decontamination,

9.1.7 Handling of rinse fluids and accumulates solids, if any, and

9.1.8 QA/QC sampling performed and analytical results of QA/QC samples whether completed in the field or laboratory subsequent to sampling event.

10. Keywords

10.1 contaminant; decontamination; sampling; waste

Standard Practice for
Decontamination of Field Equipment Used at Low Level Radioactive Waste Sites[1]

This standard is issued under the fixed designation D 5608; the number immediately following the designation indicates the year of original adoption or, in the case of revision, the year of last revision. A number in parentheses indicates the year of last reapproval. A superscript epsilon (ε) indicates an editorial change since the last revision or reapproval.

1. Scope

1.1 This practice covers the decontamination of field equipment used in the sampling of soils, soil gas, sludges, surface water, and ground water at waste sites known or suspected of containing low level radioactive wastes.

1.2 This practice is applicable at sites where low level radioactive wastes are known or suspected to exist. This practice may also be applicable for the decontamination of equipment used in known or suspected transuranic, or mixed wastes when used by itself or in conjunction with Practice D 5088.

1.3 Procedures are contained in this practice for the decontamination of equipment that comes into contact with the sample matrix (sample contacting equipment), and for ancillary equipment that has not contacted the sample, but may have become contaminated during use (non-contacting equipment).

1.4 This practice is applicable to most conventional sampling equipment constructed of metallic and hard, smooth synthetic materials. Materials with rough or porous surfaces, or having a high sorption rate should not be used in radioactive waste sampling due to the difficulties with decontamination.

1.5 In those cases where sampling will be periodically performed, such as sampling of wells, consideration should be given to the use of dedicated sampling equipment if legitimate concerns exist for the production of undesirable or unmanageable waste byproducts, or both, during the decontamination of tools and equipment.

1.6 This practice does not address regulatory requirements for personnel protection or decontamination, or for the handling, labeling, shipping or storing of wastes, or samples. Specific radiological release requirements and limits must be determined by users in accordance with local, state and federal regulations.

1.7 For additional information see DOE Publication DOE/EH-0256T, DOE Order 5480.5, DOE Order 5480.11, and 10CFR, Part 834.

1.8 The values stated in SI units are to be regarded as the standard.

1.9 *This standard does not purport to address all of the safety concerns, if any, associated with its use. It is the responsibility of the user of this standard to establish appropriate safety and health practices and determine the applica-*bility *of regulatory limitations prior to use.* Specific precautionary statements are given in Section 6.

2. Referenced Documents

2.1 *ASTM Standards:*
D 5088 Practice for the Decontamination of Field Equipment Used at Nonradioactive Waste Sites[2]
E 1168 Guide for Radiological Protection Training for Nuclear Facility Workers[3]
2.2 *United States Department of Energy Standards:*
DOE Publication DOE/EH-0256T Radiological Control Manual[4]
DOE Order 5480.5 Radiation Protection of the Public and the Environment[4]
DOE Order 5480.11 Radiological Protection for Occupational Workers[4]
2.3 *United States Code of Federal Regulations:*
10CFR, Part 834, "Radiological Protection for Occupational Workers"[4]

3. Terminology

3.1 *Descriptions of Terms Specific to This Standard:*
3.1.1 *as low as reasonable achievable (ALARA)*—an approach to radiological control to manage exposures to the work force and to the general public at levels as low as is reasonable, taking into account social, technical, economic, practical and public policy. ALARA has the objective of maintaining doses at a level far below applicable controlling limits.

3.1.2 *barrier*—a physical separation, such as a fence, wall, or temporary enclosure to prevent uncontrolled access and release from an area.

3.1.3 *contamination*—either fixed or removable radioactive materials in or on an item.

3.1.4 *contamination reduction corridor*—a defined pathway through a hazardous waste site where decontamination occurs.

3.1.5 *decontamination*—the process of removing or reducing to a known level undesirable physical, chemical, or radiological constituents from equipment. Decontamination of sample contacting equipment maximizes the representativeness of the physical, chemical, or radioactive analyses proposed for a given sample.

3.1.6 *fixed contamination*—radioactive material that

---

[1] This practice is under the jurisdiction of ASTM Committee D-18 on Soil and Rock and is the direct responsibility of Subcommittee D18.14 on Geotechnics of Waste Management.
Current edition approved Sept., 15, 1994. Published October 1994.

[2] *Annual Book of ASTM Standards*, Vol 04.09.
[3] *Annual Book of ASTM Standards*, Vol 12.02.
[4] Available from Superintendent of Documents, U.S. Government Printing Office, Washington, DC 20402.

cannot be readily removed from surfaces by nondestructive means, such as casual contact, wiping, brushing, or washing.

3.1.7 *inorganic desorbing agents*—acid rinse solutions, typically of 10 % nitric or hydrochloric acid solutions made from reagent grade nitric or hydrochloric acid and deionized water (1 % should be applied to low-carbon steel equipment).

3.1.8 *mixed wastes*—wastes containing both radioactivity (as defined by the Atomic Energy Act) and quantities of Resource Conservation and Recovery Act (RCRA) listed wastes.

3.1.9 *non-contacting equipment*—equipment used in and around the sampling that may become contaminated, but that does not contact the sample at anytime. Examples would include drilling rigs, hand tools, drill rods, excavation equipment, or barrier materials.

3.1.10 *organic desorbing agents*—solvent rinse solutions of isopropanol, acetone, hexane, or methanol; pesticide grade.

3.1.11 *QC water (control rinse water)*—water having a known chemistry, free (below detection levels) of organic or radiological constituents. Deionized water of reagent grade is normally sufficient.

3.1.12 *radioactive waste*—waste containing radioactive elements or activation regulated under the Atomic Energy Act, and is of negligible economic value, considering the cost of recovery. Waste is classified into three levels, all of which are harmful. The classifications are:

3.1.12.1 *low level waste*—wastes usually containing small amounts of radioactivity in a large amount of material. Typically the radioactivity dissipates in a relatively short period of time, anywhere between 500 and 600 years, although some low level wastes may remain radioactive for longer periods. Examples of Low Level Wastes are Uranium mining and mill tailings, soils, equipment, sludges, or liquids contaminated with or mixed with radioactive materials. Naturally Occurring Radioactive Materials (NORM) also fall into this classification. Typical examples of NORM low level wastes include uranium and thorium bearing sludges from water purification plants, high grade uranium ores, and petroleum pipeline sludges.

3.1.12.2 *mid level (transuranic) wastes*—wastes containing contamination with radioactive man-made elements having atomic weights greater than uranium (hence the name trans (or beyond) uranic). Examples of mid level wastes include liquids, sludges, resins, or soils and equipment contaminated or mixed with plutonium or other man-made alpha emitting radionuclides with half-lives of greater than 20 years and concentrations greater than 100 nCi/g at the time of assay.

3.1.12.3 *high level wastes*—wastes of highly concentrated radionuclides with long half-lives. Examples of high level wastes include spent nuclear fuels, nuclear fuel reprocessing wastes, syrups, and resins.

3.1.13 *removable contamination*—radioactive material that can be removed from surfaces by nondestructive means, such as brushing, wiping, or washing.

3.1.14 *rinse water*—water having a known chemistry. Deionized or distilled water may be used when small quantities are required. When large quantities are required, potable water of a chemistry known to be free (below

detection levels) of radioactive or chemical constituents can be used.

3.1.15 *sample contacting equipment*—equipment and tools that physically come in contact with a sample and that could allow cross-contamination from one sample to another. Examples include drive cylinders, bailers, sample handling, equipment, pumps, and sampling tubes.

3.1.16 *survey*—a radiation measurement with instrumentation to evaluate and assess the presence of radioactive materials or other sources of radiation under a specific set of conditions, (also known as frisking).

3.1.17 *unrestricted release limit*—the maximum contamination that an item may exhibit to be released for uncontrolled use by the public. Release limits differ, based on the type of radioactive materials and the amount and type of emissions (gamma, alpha, beta).

3.1.18 *wipe test*—a radiation detection test performed to determine the amount of removable radioactive material per 100 cm² surface area by wiping with a dry filter or soft absorbent paper with moderate pressure and then assessing the amount of radioactivity with an instrument of appropriate efficiency. A radiological survey and a wipe test is generally required for release of any equipment from a radiological area to an uncontrolled area or for unrestricted use, (also known as swipe test).

4. Summary of Practice

4.1 This practice provides guidance and details for the development of a site and sampling event specific decontamination plan for use in the decontamination of field equipment used during sampling or other activities in areas known, or suspected of containing low-level radioactive wastes. Four techniques or test methods are provided, with the selection and use based on the type of contamination and the difficulty of removal.

4.2 Approaches and procedures are provided for decontamination of two classifications of equipment, sample-contacting and non-contacting.

4.3 This practice includes the principles of ALARA and waste minimization as well as the protection of sample data quality.

5. Significance and Use

5.1 The primary objectives of work at low-level radioactive waste sites are the protection of personnel, prevention of the spread of contamination, minimization of additional wastes, protection of sample data quality, and the unconditional release of equipment used.

5.2 Preventing the contamination of equipment used at low-level radioactive waste sites and the decontamination of contaminated equipment are key aspects of achieving these goals.

5.3 This practice provides guidance in the planning of work to prevent contamination and when necessary, for the decontamination of equipment that has become contaminated. The benefits include:

5.3.1 Minimizing the spread of contamination within a site and preventing the spread outside of the work area.

5.3.2 Reducing the potential exposure of workers during the work and the subsequent decontamination of equipment.

5.3.3 Minimizing the amounts of additional wastes generated during the work, including liquid, or mixed wastes, including separation of the waste types, such as protective clothing, cleaning equipment, cleaning solutions, and protective wraps and drapes.

5.3.4 Improving the quality of sample data and reliability.

5.4 This practice may not be applicable to all low-level radioactive waste sites, such as sites containing low-level radioactive wastes mixed with chemical or reactive wastes. Field personnel, with assistance from trained radiological control professionals, should have the flexibility to modify the decontamination procedures with due consideration for the sampling objectives, or if past experience supports alternative procedures for contamination protection or decontamination.

5.5 This practice does not address the monitoring, protection, or decontamination of personnel working with low-level radioactive wastes.

5.6 This practice does not address regulatory requirements that may control or restrict work, the need for permits or regulatory approvals, or the accumulation or handling of generated wastes.

6. Hazards

6.1 Equipment decontamination activities involving radioactive constituents provide numerous opportunities for personnel contamination and radiation exposure, the uncontrolled spread of contamination, and the unnecessary generation of additional radioactive or mixed wastes.

6.2 Personnel involved in the decontamination of field equipment used in a known or suspected radiologically contaminated site must be trained and qualified in the work being performed and in emergency procedures.

6.3 Any work performed in a known or suspected radiologically contaminated site should be under the continuous control of a trained radiation control technician.

6.4 Strict controls around the work area must be maintained at all times to prevent the access or egress of personnel, equipment, or samples to prevent unnecessary exposure, uncontrolled releases of contaminated equipment or personnel, and unnecessary contamination of equipment. The controls will include barriers, such as fences, temporary building, or other enclosures to prevent access or egress without proper monitoring and decontamination.

6.5 Personnel working in a radiologically contaminated area have the potential for receiving radiation exposure as well as internal and external contamination. Personnel shall be trained to the Site Specific Health and Safety Plan which specifies the required training, personnel protection, and dosimetry equipment required.

6.6 Some decontamination solutions may be hazardous to humans, or may be incompatible with personnel protective clothing normally worn. For example, organic solvents or acids may permeate or degrade protective clothing or equipment. Protective clothing worn during decontamination should be selected for wet work involving the specific chemicals and solutions to be used.

6.7 Chemicals and solutions used during decontamination may be hazardous. Personnel involved should be properly trained and provided with Material Safety Data Sheets (MSDSs), and the appropriate emergency equipment.

6.8 Some equipment will degrade or produce deleterious reactions when in contact with decontamination solutions. Equipment and decontamination solution compatibility and resistance should be considered when selecting equipment and the decontamination method.

6.9 Decontamination methods may be incompatible with hazardous substances being removed and cause reactions that produce heat, toxic fumes, or explosions. The potential for incompatible material reactions should be evaluated as a part of the decontamination process selection.

7. General Procedures

7.1 Adequate planning is required prior to any activity in an area known or suspected to contain low-level radioactive or other wastes and contamination. The development of an equipment decontamination plan should be a part of the activity planning. All personnel involved in the work should be familiar with the plan and trained in the specific decontamination procedures. Work and decontamination planning should include the following:

7.1.1 The site location, conditions, known areas of surface and subsurface contamination.

7.1.2 The type, activity level, potential locations of mixed, chemical, or reactive contamination or wastes,

7.1.3 The location of other physical hazards, such as underground utilities, overhead powerlines, and existing waste storage locations.

7.1.4 Emergency responses plans, including emergency decontamination of personnel or equipment, site evacuation and accountability, and response to fire, explosion, or other situations that may occur.

7.1.5 Equipment required to prevent contamination, decontaminate equipment, contain spills, or store contaminated equipment.

7.1.6 Adequacy of monitoring and safety personnel and equipment for the anticipated work, during both normal or emergency conditions,

7.1.7 Assignment of responsibilities, including defining responsibilities for safety, quality, and the work processes being planned.

7.1.8 Establishing the work control area barriers, signage, and controls for personnel, tools, and equipment entering the work area, and the contamination release limits that will be required for equipment, samples, tools, personnel, and wastes to be released from the control area. Personnel and equipment access and release log requirements should be considered, along with requirements for various types of work control permits,

7.1.9 Establishing the decontamination location(s) for the various tools, equipment, samples and personnel equipment, including contamination reduction corridors and decontamination pads for large equipment such as backhoes, drilling rigs, or trucks and vehicles. The benefits of decontaminating equipment near the point of use should be weighed against the risks of transporting equipment to a central decontamination location and potentially spreading contamination.

7.1.10 Establish a waste disposal plan for how the anticipated wastes will be stored, both temporarily and long-term, the anticipated means of disposal, storage, labeling, or manifesting, and the organizations and individuals who will be responsible. Evaluation of the need for, and benefits of the work or samples should be balanced against the costs and difficulty of handling the wastes that may be generated prior to performing any work.

7.1.11 Providing personnel and equipment resources for environmental monitoring requirements. These may include air monitoring or controls for airborne or windblown contamination, surface runoff controls, or other specific weather work restrictions, such as restricting or stopping work during or before expected windy or wet conditions.

7.1.12 Responsibilities and sequencing for the decontamination and removal of equipment and personnel, removal of barriers, storage of both solid and liquid wastes, and the return of the site to a pre-work condition.

7.1.13 Identifying the records that will be required and assigning responsibilities for the completion, review, protection, and retention of records that will be generated during the work, including decontamination. Typical records include (but are not limited to): records of the survey equipment, survey equipment calibration and operational checks, process and effluent monitoring, environmental monitoring, tool and waste monitoring equipment, types and amounts of waste generated, sample identification and analyses that can be used to characterize the wastes.

7.2 Waste minimization should be an integral part of the planning and work processes. The following waste minimization considerations should be factored into the work planning process:

7.2.1 Preventing the spread of contamination by sequencing work from the least contaminated to the most contaminated areas.

7.2.2 Maintaining a high level of site housekeeping and cleanliness.

7.2.3 Selecting materials and equipment that are easily cleaned and decontaminated for use within a contaminated area. Generally, these are hard, nonporous materials, or materials with protective coatings or paints. Prohibit the use of soft, or porous materials. The use of greases, solvents, or other chemicals should be restricted or prohibited whenever possible in radiologically contaminated areas due to their proclivity to become contaminated or to created mixed wastes. Some types of plastic will attract radiological contamination due to static electricity and their use should be restricted.

7.2.4 Pre-clean all equipment prior to entry into a radiologically contaminated area. Transporting soils, greases, or other materials into the controlled areas will only increase the amount of decontamination required and generate additional wastes. Clean equipment should be wrapped until use, particularly if it has potential to come in contact with samples. If unused, decontamination prior to release will be eliminated. A best management practice is to radiologically verify equipment and tools are free of contamination prior to entry into a controlled area, particularly if the past use of the equipment is not known.

7.2.5 If an activity will be repeated frequently, such as sampling, consider the use of dedicated equipment that will remain in place and not require decontamination or become waste. In other cases, evaluate the availability of equipment which is already contaminated and can be used without contacting samples, such as drilling rigs, auger flights, or hand tools rather than bring clean equipment into the controlled areas.

7.2.6 Perform pre-work preparatory activities outside of the contaminated areas, such as taping labels to sample containers, wrapping, draping, or sealing equipment.

7.2.7 Minimize the use of liquids that have the potential to become contaminated. When the use of liquids is required, consider the possibilities for re-use, such as using final rinse waters later for initial washes. The use of evaporation ponds, filters, or other treatments may be used to reduce the amount of liquid wastes. Liquids from the various stages of decontamination, or from differing sections of a contamination reduction corridor should be kept separate for later reuse or treatment.

7.2.8 Survey any wastes prior to disposal or storage to prevent mixing clean and contaminated equipment or wastes. Drapes, wraps, tape, and other materials may not be contaminated and need not be disposed of as radiological wastes. In cases where only a portion is contaminated, the contaminated portion can be cut away or separated for disposal as radiologically contaminated wastes, with the remainder disposed of as uncontaminated waste.

7.2.9 When liquids or other materials are to be used for decontamination, verify that they are radiologically clean prior to use. Avoid using materials that contain significant amounts of naturally occurring radioactive matter and that may not be released from the work area.

7.2.10 Work areas around samples should be draped, or covered to prevent transport or spread of contamination that may affect the sample data quality.

7.2.11 Use rubber-tired equipment whenever possible. Avoid the use of tracked equipment, that has the tendency to spread contamination and are difficult to fully decontaminate.

8. Procedure for Sample Contacting Equipment

8.1 Decontaminate sample contacting equipment immediately after use and prior to use on the next sampling. If liquids are sampled, complete the decontamination before the liquids have dried on the surfaces, particularly on the internal of tubing, pumps, and other difficult to clean equipment. Prior to the initial use of sample contacting equipment, take baseline surveys, wipe tests, and collection of rinsate samples to establish an initial radiological and chemical baseline for the sampling equipment.

8.2 Determine effectiveness of equipment decontamination between uses within the work control area by performing radiation surveys to verify that contamination above background has been removed. Equipment that has been decontaminated must be both surveyed and wipe tested to verify that release limits have been met. Equipment that is found by the wipe test to be above release limits may be decontaminated again until releasable, or be handled and transported as radiologically contaminated equipment. If equipment is of high value, it may be possible to further decontaminate to release limits at a later date using chemicals, ultrasonic cleaning, electro-polishing, or other techniques outside the scope of this practice.

8.3 Four test methods of sample contacting equipment decontamination are presented. These test methods may be used individually or in combination to achieve a higher level of decontamination. The selection of test methods or combination of test methods will depend on the type of contaminants, the level of decontamination required, and the purpose of the decontamination. Remove loose or visible contamination by wiping or brushing. Pressurized water, or preferably steam cleaning may be appropriate to perform initial decontamination of equipment. The four test methods are:

8.4 *Method A: Decontamination Using QC Water*—This method is considered the minimum decontamination effort used to clean equipment. Its use is limited to those cases where the sampling equipment has contacted only relatively minimal contamination and there has been little likelihood of contact with organic substances.

8.4.1 *Apparatus:*

8.4.1.1 *QC Rinse Water*, adequate supplies.

8.4.1.2 *Wash bottles*, or pressure sprayer to dispense QC rinse water.

8.4.1.3 *Lint Free tissues or wipes.*

8.4.1.4 *Brushes and scrapers*, made of inert materials, which are free of contamination.

8.4.2 *Procedure:*

8.4.2.1 Remove any solid material from the equipment by scraping or brushing with implements made of inert, nonabsorbent materials.

8.4.2.2 Thoroughly rinse the piece of equipment with QC rinse water using a pressure sprayer or pressure from a wash bottle.

8.4.2.3 For equipment such as tubing and pumps that cannot be easily dismantled for cleaning with a pressure sprayer or wash bottle, circulate QC rinse water through the equipment.

8.4.2.4 Survey equipment for detectable radiation above the initial baseline. Collect QC water rinsate, or wipe samples for verification of decontamination effectiveness.

8.4.2.5 Allow to air dry or dry the equipment with lint-free wipes or tissues. Minimize the use of lint-free wipes or tissues to reduce waste production.

8.4.2.6 Store the equipment that will minimize possible recontamination by surface or atmospheric contaminants. This may be accomplished by bagging, wrapping, or covering the equipment.

8.4.3 This method represents the minimum amount of decontamination that should be performed. Cross-contamination of samples is possible if organic or other substances were not removed physically, or by the rinse water.

8.5 *Method B: Decontamination Using Detergent and Rinse Water*—This decontamination method is used when material being sampled is not easily removed, is not removed using Method A, or tends to absorb onto the equipment. This method employs a mild detergent wash that can chemically remove contaminants.

8.5.1 *Apparatus:*

8.5.1.1 *Control Water and QC Rinse Water*, adequate supplies.

8.5.1.2 *Pressure Sprayer or Wash Bottle.*

8.5.1.3 *Lint-Free Wipes or Tissues.*

8.5.1.4 *Brushes and scrapers*, made of inert materials that are free of contamination.

8.5.1.5 *Detergent*, phosphate-free, biodegradable, and soluble in hot or cold water.[5]

8.5.2 *Procedure:*

8.5.2.1 Remove any solid material from the equipment by scraping or brushing with implements made of inert, nonabsorbent materials.

8.5.2.2 Wash and scrub the equipment thoroughly with the detergent using a brush.

8.5.2.3 Rinse the equipment thoroughly using rinse water.

8.5.2.4 For equipment such as tubing and pumps that cannot be easily dismantled for cleaning with a pressure sprayer, wash bottle or submersion, circulate the detergent solution through the equipment, followed by a QC rinse water rinse.

8.5.2.5 Survey equipment for detectable radiation above the initial baseline. Collect QC rinse water rinsate, or wipe samples for verification of decontamination effectiveness.

8.5.2.6 Allow to air dry or dry the equipment with lint-free wipes or tissues. Minimize the use of lint-free wipes or tissues to reduce waste production.

8.5.2.7 Store the equipment that will minimize possible recontamination by surface or atmospheric contaminants. This may be accomplished by bagging, wrapping, or covering the equipment.

8.5.3 This method has less potential for cross-contamination of samples than Method A, but may not be adequate to decontaminate equipment that is grossly contaminated. In these cases, Methods C and D should be added as necessary to achieve complete decontamination.

8.6 *Method C: Decontamination Using an Organic Desorbing Agent*—This method should be used in cases when the possibility of organic contamination in addition to radiological contamination has occurred or is suspected, or when Method A or B is not sufficient to successfully decontaminate the equipment. The choice of organic desorbing agent will depend on the kind of organic contaminant present and the analytical requirements of the samples being collected. Generally, a pesticide grade of methanol is recommended (methanol does not interfere with gas chromatography/mass spectroscopy (GC/MS) analysis); however, stronger desorbing agents like acetone or hexane may be required for complete decontamination.

8.6.1 *Apparatus:*

8.6.1.1 *Rinse Water and QC Rinse Water*, adequate supply.

8.6.1.2 *Pressure Sprayer or Wash Bottles.*

8.6.1.3 *Wipes or Lint-Free Tissues.*

8.6.1.4 *Brushes and Scrapers*, made of inert materials that are free of contamination.

8.6.1.5 *Organic Desorbing Agent*, such as methanol, acetone, isopropanol, or hexane; pesticide grade.

8.6.2 *Procedure:*

8.6.2.1 Remove any solid material from the equipment by scraping or brushing with implements made of inert, nonabsorbent materials.

---

[5] Isoclean, Alquinox, or Liquinox, have been found suitable for this purpose.

8.6.2.2 Wash with rinse water using a pressure sprayer or wash bottle.

8.6.2.3 Wash with the organic desorbing agent using a pressure sprayer or wash bottle. Precautions should be taken to protect skin from contact with organic agents.

8.6.2.4 Rinse with QC rinse water.

8.6.2.5 For equipment such as tubing and pumps that cannot be easily dismantled for cleaning with a pressure sprayer, wash bottle or submersion, circulate the decontamination solutions and rinses in the order listed through the equipment, followed by a QC water rinsing.

8.6.2.6 Survey equipment for detectable radiation above the initial baseline. Collect QC rinse water rinsate, or wipe samples for verification of decontamination effectiveness.

8.6.2.7 Dry the equipment with wipes or lint-free tissues, or allow to air dry.

8.6.2.8 Store the equipment that will minimize possible recontamination by surface or atmospheric contaminants. This may be accomplished by bagging, wrapping, or covering the equipment.

8.6.3 This method has less potential for cross contamination of samples than Method A or B, but has a potential risk of contaminating the samples with organic desorbing agents.

8.7 *Method D: Decontamination Using an Inorganic Desorbing Agent*—This method should be used to decontaminate when organic substances and radiological contamination have been absorbed onto sampling equipment and Methods A and B are not sufficient to remove the substances. This decontamination method has the least potential for cross-contamination of a sample with inorganic elements from the site, but has the potential for contaminating the sample with inorganic desorbing agents. Additionally, desorbing agents, such as the acids used in this test method may dissolve or leach elements or compounds from the next sample, and care must be taken to remove all acidic residues from the sampling equipment.

8.7.1 *Apparatus:*

8.7.1.1 *Rinse Water and QC Rinse Water*, adequate supplies.

8.7.1.2 *Pressure Sprayer or Wash Bottles.*

8.7.1.3 *Wipes or Lint-Free Tissues.*

8.7.1.4 *Brushes and Scrapers*, made of inert materials that are free of contamination.

8.7.1.5 *Adequate Supplies of Inorganic Desorbing Agents*—An inorganic desorbing agent may be a 10 % nitric or 10 % hydrochloric acid solution made from reagent grade stock and deionized water. For decontamination of low-carbon steels, a 1 % nitric solution should be used. Other acids or combinations of acids prepared in a similar fashion may be appropriate.

8.7.2 *Procedure:*

8.7.2.1 Remove any solid material from the equipment by scraping or brushing with inert, nonabsorbent tools.

8.7.2.2 Wash the equipment thoroughly with QC rinse water using a pressure sprayer, wash bottles, or immersion.

8.7.2.3 Wash the equipment thoroughly with an inorganic desorbing agent using a pressure sprayer or wash bottle. Take precautions to protect skin from contact with the inorganic desorbing agents and from inhalation of fumes or mists.

8.7.2.4 Thoroughly rinse the equipment using rinse water, followed by QC rinse water.

8.7.2.5 For equipment such as tubing and pumps that cannot be easily dismantled for cleaning with a pressure sprayer, wash bottle or submersion, circulate the decontamination solutions and rinses in the order listed through the equipment, followed by a QC rinse water rinsing.

8.7.2.6 Survey equipment for detectable radiation above the initial baseline. Collect QC rinse water rinsate, or wipe samples for verification of decontamination effectiveness.

8.7.2.7 Dry the equipment with wipes or lint-free tissues, or allow to air dry.

8.7.2.8 Store the equipment that will minimize possible recontamination by surface or atmospheric contaminants. This may be accomplished by bagging, wrapping, or covering the equipment.

9. Procedure for Non-Sample Contacting Equipment

9.1 Non-sample contacting equipment should be decontaminated whenever the equipment is to be moved from the controlled work area. Partial decontamination may be appropriate if the equipment is to be moved from one controlled area to another, provided there is no loose contamination that could fall from the equipment, or at the completion of a day's work.

9.2 Remove any visible material adhering to the surface by wiping, brushing, or brooming. In some cases, this may be facilitated by allowing the equipment to air dry. Use soft brushes and avoid the use of hard wire brushes on soft materials that may embed the contamination. Survey the surface using the instrument(s) appropriate for the type of contamination (alpha, beta, or gamma). If no areas are above release limit, perform a wipe test for verification of the survey.

9.3 If areas are found by survey to be above release limits, observe those areas for any visible material adhering to the surface, discoloration, or corrosion and further wipe or brush the areas. Survey the areas again. If no areas are above release limits, perform wipe tests for verification of the survey.

9.3.1 If small areas are still found to be contaminated above release limits, moisten a lint-free paper wipe or towel with rinse water and wipe the area. Alternately, a small spray bottle can be used to wet the surface, and the area wiped dry. Survey to determine if equipment meets release criteria. Perform wipe tests to verify the survey.

9.3.2 For larger equipment, the equipment can either be dipped in a tub, bucket, or tank of water and wiped or air dried. Larger pieces of equipment can also be cleaned using high pressure/low volume water cleaners, or preferably steam cleaners using clean water, and brushing and wiping. Repeat cleaning and surveys until survey indicates that equipment meets release criteria. Perform a wipe test to verify the survey.

9.4 For equipment such as tubing and pumps that cannot be easily dismantled for cleaning, clean the exterior as described above and circulate rinse water through the equipment. Survey all exposed surfaces and verify by wipe test that release limits have been met. If equipment fails survey, repeat the cleaning using hot water or steam, and survey until equipment meets release limits. Dismantling may be required to allow for survey and wipe tests for verification prior to unrestricted release.

9.5 Heavy equipment, such as bulldozers, trucks, backhoes, drill rigs, drilling tools are difficult to decontaminate. Accordingly, equipment of this type should be wrapped, draped, or otherwise protected with disposable covers to prevent contamination. Prior to use, a baseline survey of the equipment should be conducted to determine if the equipment is contaminated from prior uses, and if so, to what extent. However, contamination will be unavoidable in many cases and decontamination will be required, either before, or after use, or both. Decontamination of heavy equipment is usually accomplished by physical cleaning, by high pressure water, or preferably steam cleaning with water or detergent solutions, or both, on a large decontamination pad or contamination reduction corridor. The decontamination method that will be used is a function of the degree and nature of the contaminate, and the degree of decontamination that must be achieved. As a general rule, wet contamination will be kept wet and dry contamination will be kept dry. Wetting some compounds may cause chemical reactions that will react with the equipment and produce undesirable substances that are more difficult to remove and later handle as wastes. The following general steps may be used for decontaminating heavy equipment:

9.5.1 Physically remove any bulk material adhering to the contaminated item by use of a wire brush, stiff bristle brush, or scraper. Dry cleaning can be further accomplished using vacuum cleaners equipped with High Efficiency Particulate Air (HEPA) filters, or sand blasting equipment designed to collect all sand and blasting debris. Other tools, such as vacuuming needle guns have also been used successfully. All parts of the equipment, including undercarriage, chassis, and cab must be decontaminated. Air filters should be surveyed and if contaminated, removed and disposed of. A thorough visual inspection of the equipment, supplemented by surveys should be used to determine if the decontamination is successful or whether additional repeat decontamination is required.

9.5.2 For wet cleaning, use water or a nonphosphate detergent solution (under pressure if necessary) to assist in the final removal of bulk materials.

9.6 Survey the equipment to verify that decontamination has been successful. If the survey indicates that equipment is clean, perform verification surveys and wipe tests. If the equipment survey still indicates the presence of contamination, continue decontamination using the following steps.

9.7 Steam clean the equipment, taking care to clean all surfaces.

9.8 Rinse with water.

9.9 Survey equipment to verify that decontamination has been successful. If the survey indicates that the equipment is still contaminated, repeat the cleaning process. For persistent or grossly contaminated areas, the use of a pesticide grade methanol rinse may be applied using wash bottles or pressure sprayers.

9.10 For unrestricted release, perform survey and wipe tests to verify that the equipment has been decontaminated to release limits.

9.11 Decontamination of heavy equipment will typically generate large volumes of contaminated liquids. Decontamination should take place on pads constructed for the purpose, with curbs, berms, and slopes to collect, control and contain the liquids. Cleaning and rinse solutions may be kept separate for later treatment. In some cases, the cleaning or rinse solutions may be recycled and reused. It is important to control the volume of decontamination solutions when expensive measures must be used for their disposal.

10. Quality Assurance/Quality Control

10.1 This practice is not a substitute for a well documented and implemented QA/QC program. The test of a decontamination procedure is the ability to evaluate and minimize erroneous sample data due to cross-contamination or nonrepresentative contamination.

10.2 *Sample Contacting Equipment*—The effectiveness of the decontamination process must be determined and documented to assess and protect the quality of the data derived from the samples. A project Quality Assurance/Quality Control (QA/QC) program should be in place prior to any sampling activity. The QA/QC program should include provisions for the collection of samples to evaluate the completeness and effectiveness of contamination prevention and decontamination. These provisions may include:

10.2.1 Requirements for pre-activity verification that radiological or chemical contamination does not exist, or if present, at what type and levels of contaminate residues residing in or on the sample contacting equipment.

10.2.2 Requirements for the collection of QC rinse water rinsate samples before initial sampling and periodically thereafter to determine the effectiveness of the decontamination.

10.2.3 The frequency of the sampling to demonstrate the completeness of sample contacting equipment decontamination is dependent on the data quality objectives of the project. As a minimum, it is recommended that a rinsate sample be collected and analyzed after every ten decontamination washings.

10.2.4 The use of trip blanks should be considered to determine whether contamination is being introduced into the samples at any point during the sampling process.

10.2.5 Field audits or surveillances of the sampling activities, including decontamination should be a part of a comprehensive QA/QC program. Field observation will ensure that decontamination practices are performed consistently and in accordance with procedures at all times, including when rinsate samples are taken, and that the rinsate samples are representative of the sampling process.

10.2.6 QA/QC in radiological work includes the verification that adequate and correct surveys, wipe tests, area controls, and personnel monitoring are performed to protect the personnel involved and to prevent environmental releases.

10.2.7 Requirements for documentation should be included in the sampling and analysis plan, health and safety plan, and the QA plans. Reviews of all completed documentation should be performed as a part of the sampling activity, and as a part of the data quality assessment that verifies the quality of the data.

10.3 *Non-Sample Contacting Equipment*—Non-sample contacting equipment contamination should have no effect on the quality of data derived from the samples, however, the QA/QC program should recognize the importance of controlling the spread of contamination, the minimization of wastes, and the consequences of unknowingly releasing radiologically contaminated equipment to the public. The QA/QC program should include planning for decontamination activities, the verification and documentation during field activities that contamination has been controlled and that decontamination efforts are effective and complete.

11. Keywords

11.1 contaminate; contamination; control rinse water; control water; decontaminate; decontamination; equipment; radioactive; radiological; wastes; sampling

Standard Guide for
Sampling Chain-of-Custody Procedures[1]

This standard is issued under the fixed designation D 4840; the number immediately following the designation indicates the year of original adoption or, in the case of revision, the year of last revision. A number in parentheses indicates the year of last reapproval. A superscript epsilon (ε) indicates an editorial change since the last revision or reapproval.

1. Scope

1.1 This guide contains a comprehensive discussion of potential requirements for a sample chain-of-custody program and describes the procedures involved in sample chain-of-custody. The purpose of these procedures is to provide accountability for and documentation of sample integrity from the time samples are collected until sample disposal.

1.2 These procedures are intended to document sample possession during each stage of a sample's life cycle, that is, during collection, shipment, storage, and the process of analysis.

1.3 Sample chain-of-custody is just one aspect of the larger issue of data defensibility (see 3.2.2 and Appendix X1).

1.4 A sufficient chain-of-custody process, that is, one that provides sufficient evidence of sample integrity in a legal or regulatory setting, is situationally dependent. The procedures presented in this guide are generally considered sufficient to assure legal defensibility of sample integrity. In a given situation, less stringent measures may be adequate. It is the responsibility of the users of this guide to determine their exact needs. Legal counsel may be needed to make this determination.

1.5 Because there is no definitive program that guarantees legal defensibility of data integrity in any given situation, this guide provides a description and discussion of a comprehensive list of possible elements of a chain-of-custody program, all of which have been employed in actual programs but are given as options for the development of a specific chain-of-custody program. In addition, within particular chain-of-custody elements, this guide proscribes certain activities to assure that if these options are chosen, they will be implemented properly.

1.6 *This standard does not purport to address all of the safety concerns, if any, associated with its use. It is the responsibility of the user of this standard to establish appropriate safety and health practices and determine the applicability of regulatory limitations prior to use.*

2. Referenced Documents

2.1 *ASTM Standards:*
D 1129 Terminology Relating to Water[2]

D 3325 Practice for Preservation of Waterborne Oil Samples[3]
D 3370 Practices for Sampling Water from Closed Conduits[2]
D 3694 Practices for Preparation of Sample Containers and for Preservation of Organic Constituents[3]
D 3856 Guide for Good Laboratory Practices in Laboratories Engaged in Sampling and Analysis of Water[2]
D 4210 Practice for Intralaboratory Quality Control Procedures and a Discussion on Reporting Low Level Data[2]
D 4841 Practice for Estimation of Holding Time for Water Samples Containing Organic and Inorganic Constituents[2]

2.2 *U.S. EPA Standard:*
U.S. EPA Good Automated Laboratory Practices[4]

3. Terminology

3.1 *Definitions*—For definitions of terms used in this guide, refer to Terminology D 1129.

3.2 *Descriptions of Terms Specific to This Standard:*

3.2.1 *custody*—physical possession or control. A sample is under custody if it is in possession or under control so as to prevent tampering or alteration of its characteristics.

3.2.2 *data defensibility*—a process that provides sufficient assurance, both legal and technical, that assertions made about a sample and its measurable characteristics can be supported to an acceptable level of certainty. See Appendix X1 for a discussion of the elements of a data defensibility process.

3.2.3 *sample*—a portion of an environmental or source matrix that is collected and used to determine the characteristics of that matrix.

3.2.4 *sample chain-of-custody*—a process whereby a sample is maintained under physical possession or control during its entire life cycle, that is, from collection to disposal.

3.2.5 *sample chain-of-custody record*—documentation providing evidence that physical possession or control was maintained during sample chain-of-custody.

4. Summary of Guide

4.1 This guide addresses chain-of-custody procedures as they relate to field practices, shipping methods, and laboratory handling of samples.

5. Significance and Use

5.1 Chain-of-custody procedures are a necessary element in a program to assure one's ability to support data and

---

[1] This practice is under the jurisdiction of ASTM Committee D-19 on Water and is the direct responsibility of Subcommittee D19.02 on General Specifications, Technical Resources, and Statistical Methods.

Current edition approved Dec. 10, 1995. Published February 1996. Originally published as D 4840 – 88. Last previous edition D 4840 – 88 (1993)[ε1].

[2] *Annual Book of ASTM Standards*, Vol 11.01.

[3] *Annual Book of ASTM Standards*, Vol 11.02.

[4] Available from Superintendent of Documents, Government Printing Office, Washington, DC.

conclusions adequately in a legal or regulatory situation, but custody documentation alone is not sufficient. A complete data defensibility scheme should be followed.

5.2 In applying the sample chain-of-custody procedures in this guide, it is assumed that all of the other elements of data defensibility have been applied, if applicable.

6. Procedure

6.1 *Facility Chain-of-Custody Standard Operating Procedure*—Each organization should have a chain-of-custody procedure document. This document should spell out in detail the specific procedures utilized at this facility to achieve sample chain-of-custody. It should contain copies of all the forms used in the chain-of-custody process and detailed instructions for their use. It should be kept current and revisions tracked. This guide may serve as a template for the chain-of-custody procedure document.

6.2 *Sample Collection Phase:*

6.2.1 *Custody Assignment*—A single field sampling person should be assigned responsibility for custody of samples. An alternate custodian should also be assigned to cover the prime custodian's absence. As few people as possible should handle samples. The assigned field sampler should be personally responsible for the care and custody of the samples collected until they are properly transferred. While samples are in their custody, field personnel should be able to testify that no one was able to tamper with the samples without their knowledge.

6.2.2 *Documentation/Field Custody Forms:*

6.2.2.1 Standard forms should be designed and available for recording custody information related to field sample handling. The forms may be designed to handle one sample or multiple samples. A single sample form may allow room for laboratory chain-of-custody.

6.2.2.2 In any sampling effort, there is field information related to sample collection and field measurements that are recorded. This information is not specifically part of chain-of-custody, but part of the larger aspect of data defensibility. This information may be recorded on chain-of-custody forms or other forms specific for the purpose. Record keeping may be simplified if separate forms are used.

6.2.2.3 It may be useful to print field forms on polyethylene or other plastic coated paper to keep them from being affected by water or chemicals. An indelible ink, paint, or crayon should be used to enter information on the forms.

6.2.2.4 Spaces for the following information should be on the form:

(*a*) Sample identifying name.

(*b*) Sampling location ID, sampling point ID, date, and sampling time interval.

(*c*) Signatures of sampling personnel and signatures of all personnel handling and receiving the samples.

(*d*) Project identification code (if applicable).

(*e*) Preservation (to alert lab personnel): amount and type.

(*f*) Number of containers (where field sub-sampling occurs). Indicate number of replicates if there are multiple containers of the same sample.

(*g*) Field notes.

(*h*) Analyses desired (may be required in some situations).

(*i*) Sample type: grab, composite, etc.

Example forms are shown in Appendix X2.

6.2.2.5 Freight bills, post office receipts, and bills of lading

should be retained as part of the permanent custody documentation.

6.2.3 *Sample Labeling:*

6.2.3.1 Sample labels may be in the form of adhesive labels or tags, or both. Tags have the advantage of being removable to become part of the record keeping process, although their inadvertent loss or inappropriate removal may leave the sample without documentation. Labels should be made of waterproof paper and indelible ink should be used to make entries. Alternatively, sample information may be written directly on the sample container, as long as the writing can be done indelibly. Containers should be free from other labels and other writing to prevent any confusion. If both tags and labels are used, care should be taken to ensure that the information on both is identical.

6.2.3.2 Labels or tags should be filled out just before or immediately after sample collection. Labels should contain spaces for the following information:

(*a*) Project identification code (if applicable).

(*b*) Sample identifying name (exactly as it appears on the chain-of-custody record).

(*c*) Sampling location ID, sampling point ID, and sampling time interval.

(*d*) Safety considerations (if applicable).

(*e*) Analysis schedule or schedule code (if applicable).

(*f*) Company or agency name.

An example label is shown in Appendix X2.

6.2.4 *Sample Sealing:*

6.2.4.1 Sample custody seals of waterproof adhesive paper may be used to detect unauthorized tampering with samples prior to receipt by the lab. When seals are used, they shall be applied so that it is necessary to break them in order to open the sample container.

6.2.4.2 Electrical (vinyl) tape may be used to prevent bottle closures from loosening in transit. Tape should be applied before any custody seals are applied.

NOTE—Electrical tape should not be used to seal vials used for volatile organic analyses due to the potential for sample contamination.

6.2.5 *Field Transfer of Custody and Shipment:*

6.2.5.1 Package samples properly for shipment and transport them to the laboratory for analysis. Special care should be taken when packaging in glass. It is important that all laws and regulations related to the transport of materials have been adequately addressed before shipping samples.

6.2.5.2 When employing a common carrier, the use of padlocks or custody seals on shipping containers should be considered. If padlocks are employed, the keys shall be shipped separately from the samples. Alternatively, padlocks may be sent unfastened to the field and the keys can be retained by the laboratory sample custodian (see 6.3.2.1). A separate custody record should accompany each shipment. Enter the method of shipment, courier name(s), and other pertinent information in the "remarks" section on the custody record.

6.2.5.3 If sent by mail, register the package with return receipt requested.

6.2.5.4 When transferring the possession of samples, the individuals relinquishing and the individuals receiving the samples should sign, date, and note the time on the custody record. Document any opening and closing of the sample containers on the custody record. Provisions should be made for receipt of samples at nonstandard hours, such as nights and weekends by nonlaboratory personnel. Shipping documents, with noted time of receipt and receipt by whom, should be made part of the custody record.

6.3 *Laboratory Handling and Analysis Phase:*

6.3.1 *Documentation—Laboratory Custody Forms:*

6.3.1.1 The sample chain-of-custody record in the laboratory is traditionally maintained on paper forms. Based on the data defensibility needs of the organization, it may be possible to maintain the laboratory record in an electronic format. Various computer systems, such as a laboratory information management systems (LIMS) or other electronic data management systems, may meet the data integrity needs. It is the responsibility of each organization to assure that an electronic record system meets these needs. Users of such systems are encouraged to assure compliance of their electronic data system with the U.S. EPA Good Automated Laboratory Practices. All references to laboratory custody record forms in this guide should be understood to refer to either paper or electronic documents.

6.3.1.2 Design a form for the recording of chain-of-custody information related to sample possession in the laboratory. If samples are to be split and distributed to multiple analysts, multiple forms will be needed to accompany the sample splits. Transfer sample identification information to the forms accompanying the splits exactly as it appears on the primary receipt laboratory chain-of-custody form. If an LIMS label is used for the sample splits, a duplicate should be placed on the chain-of-custody form that accompanies them. Example forms are shown in Appendix X2.

6.3.2 *Laboratory Sample Receipt and Handling:*

6.3.2.1 In the laboratory, assign a sample custodian(s) to receive the samples. It is preferable to assign one person the primary responsibility to receive samples as the sample custodian for the laboratory. A second person should serve only as an alternate.

6.3.2.2 Upon receipt of a sample, the custodian should inspect the condition of the sample and the custody sample seal, if used. If sample seals are used, record condition on chain-of-custody record. Reconcile the information on the sample label against that on the chain-of-custody record. The temperature of the samples should be recorded on the chain-of-custody record. If samples are not delivered in a cooler, indicate on record. If pH adjustment to preserve the sample was done in the field, the pH of the samples should be checked and recorded on the chain-of-custody record.

6.3.2.3 If a sample container is leaking, note it on the custody record. The custodian, along with the supervisor responsible for the analytical work, should decide whether the leaky sample is valid. If seals are used, the custodian should examine whether the sample seal is intact or broken, since a broken seal may mean sample tampering and may make analytical results inadmissible as evidence in court. Any discrepancies between the information on the sample label and seal and the information on the chain-of-custody record should be resolved before the sample is assigned for analysis. This effort might require communication with the sample collector. Record the results of any such investigation.

6.3.2.4 After processing the sample, (splitting, logging, preserving) record all sample splits on the laboratory chain-of-custody form. When the sample is logged, the sample identifying information should be transcribed exactly as it appears on the field chain-of-custody form. If custody transfer to analytical staff will not occur immediately or if sample processing is delayed, the samples should be transferred to the custody lockup (see 6.3.3). Record all transfers to and from a lockup on the chain-of-custody form. The custody form should remain with the sample.

6.3.3 *Laboratory Security:*

6.3.3.1 In some situations, legally defensible custody in the laboratory has been achieved without regulating possession within the laboratory but rather by assuring controlled and restricted access to the laboratory facility through keying, guarding access points, and other measures. Sufficiency of security measures for legal defensibility can only be assessed on a case by case basis and should involve legal counsel.

6.3.3.2 Within the laboratory, a secure, locked location (a refrigerator or freezer), if appropriate, should be available. Multiple locations may be necessary to provide access to analysts after they receive their portions of the sample.

6.3.3.3 Limit the number of keys to locked locations and maintain control over them. Limiting keys to laboratory supervisors or providing multiple lockups assigned to specific analysts are appropriate options. Limiting access to samples provides greater security against accidental mishandling of samples.

6.3.3.4 As an alternative to secure lockups, tamperproof seals may also be used in the laboratory. Note any application of seals and their removal on the chain-of-custody forms.

6.3.4 *Analyst Sample Receipt and Handling:*

6.3.4.1 When analytical staff take possession of their samples or sample aliquots, they should acknowledge receipt on the primary laboratory chain-of-custody form.

6.3.4.2 When an analyst takes possession of a sample split, he or she should also receive the accompanying chain-of-custody form. At that time, the analyst should inspect the condition of the sample and the sample seal, if used, and reconcile the information on the sample label against that on the chain-of-custody form.

6.3.4.3 While a sample is in their custody, analysts should be able to testify that no one tampered with the sample without their knowledge. If the sample, a portion of the sample, or processed sample such as a digestate will be held for an extended period of time, the analyst should store it in a security lockup and record all such transfers on the chain-of-custody form.

6.3.4.4 At such time as there is no further need for the sample, it should be disposed of properly and the disposal recorded. If the sample or processed sample is to be retained, it may be transferred to appropriate personnel. This transfer should be recorded on both the analyst custody form and the primary laboratory custody form. The primary custody form then accompanies the sample until its disposal.

6.3.5 *Interlaboratory Transfer:*

6.3.5.1 On some occasions, another laboratory will be performing analytical work that is not directly a part of the project plan, that is, data from this laboratory is not planned to be part of the data defensibility scheme. An example might be when a facility discharge is being monitored and the facility laboratory wishes a split of the sample. Under these circumstances, the chain-of-custody record remains with the owner. Prepare a receipt (an example receipt is shown in Appendix X2) for these samples and mark to indicate with whom the samples are being split. The person relinquishing the samples to the other laboratory should request the signature of a representative of the appropriate party acknowledging receipt of the samples. If a representative is unavailable or refuses to sign, note this in the "received by" section. Complete this form and give a copy to the owner, operator, or agent in charge. The original is retained by the project supervisor. When appropriate, as in the case where the representative is unavailable, the custody record should contain a statement that the sample splits were delivered to the designated location at a designated time.

6.3.5.2 On some occasions, the sample may have to be split with another laboratory in order to obtain all of the necessary analytical information required in the study plan. In this case, identical chain-of-custody procedures should be employed at the alternate laboratory. Transfer of custody of the split should be handled in like fashion to that used to an intralaboratory transfer (see 6.3.4).

7. Keywords

7.1 chain of custody; custody; data defensibility; validation

APPENDIXES

(Nonmandatory Information)

X1. DISCUSSION OF THE ELEMENTS OF DATA DEFENSIBILITY

X1.1 Data defensibility can be thought of as "proof" that a sample represents the material from which it was taken; that the sample integrity was maintained; that the measurements made on the sample produced valid results; and, that the documentation of the "proof" (custody records, data sheets, etc.) is a factual record. Data defensibility involves the following:

X1.1.1 The use of proper procedures (for sample collection, preservation, analysis, etc.),

X1.1.2 Protection of samples from inappropriate alteration (from tampering, loss, mishandling, etc.), that is, chain-of-custody,

X1.1.3 The use of proper record collection, record handling, and record security procedures, and

X1.1.4 Accurate documentation of all sample related information.

X1.2 There are six principal elements of data defensibility besides chain-of-custody. For a discussion of many of these elements, see Data Validation in Guide D 3856.

X1.2.1 *Project Setup and Preparation*—The production of data on environmental and source samples for the purpose of drawing valid conclusions requires good experimental design. Aspects of the project from sample collection to data interpretation shall be designed from a valid model.

X1.2.2 *Measurement Methods*—Measurements, both field determinations and lab analyses, shall be made using validated techniques with known levels of uncertainty. Use of methods such as those produced by ASTM Committee D-19 can provide assurance that the procedures used will produce useful information.

X1.2.3 *Sample Collection Methods*—Sample results can only be as good as the sample analyzed. It is vital that the sample analyzed be representative of the designated variables in the environmental matrix of concern. It should not be inferred that the experimental design is appropriate or representative for any other environmental variables than those designated in the experimental design. Containers shall be made of appropriate materials and properly cleaned. See Practices D 3370, specific test methods, and other practices related to sampling procedures for more information.

X1.2.4 *Sample Processing and Handling Methods*—During the course of a sample's life cycle, a variety of sample processing techniques shall be employed, such as sample splitting and preservation. Valid procedures shall be employed to maintain sample integrity. See Practices D 4841, D 3694, D 3325, and specific test methods for more information.

X1.2.5 *Data Recording, Archiving, and Retrieval Methods*—Information collected and observations made shall be correctly, legibly, and safely recorded. After a project is completed and information recorded, it is important that this record be safe from tampering and can be reliably retrieved.

X1.2.6 *Quality Control and Quality Assurance Procedures*—During stages of information generation, processes shall be maintained in a state of statistical control so that data uncertainties can be quantified. In addition, there shall be an "external" audit procedure to assure that the quality control procedures are effective. See Guide D 3856, Practice D 4210, and specific test methods for more information.

X2. EXAMPLE FORMS

X2.1 See Figs. X2.1 through X2.5.

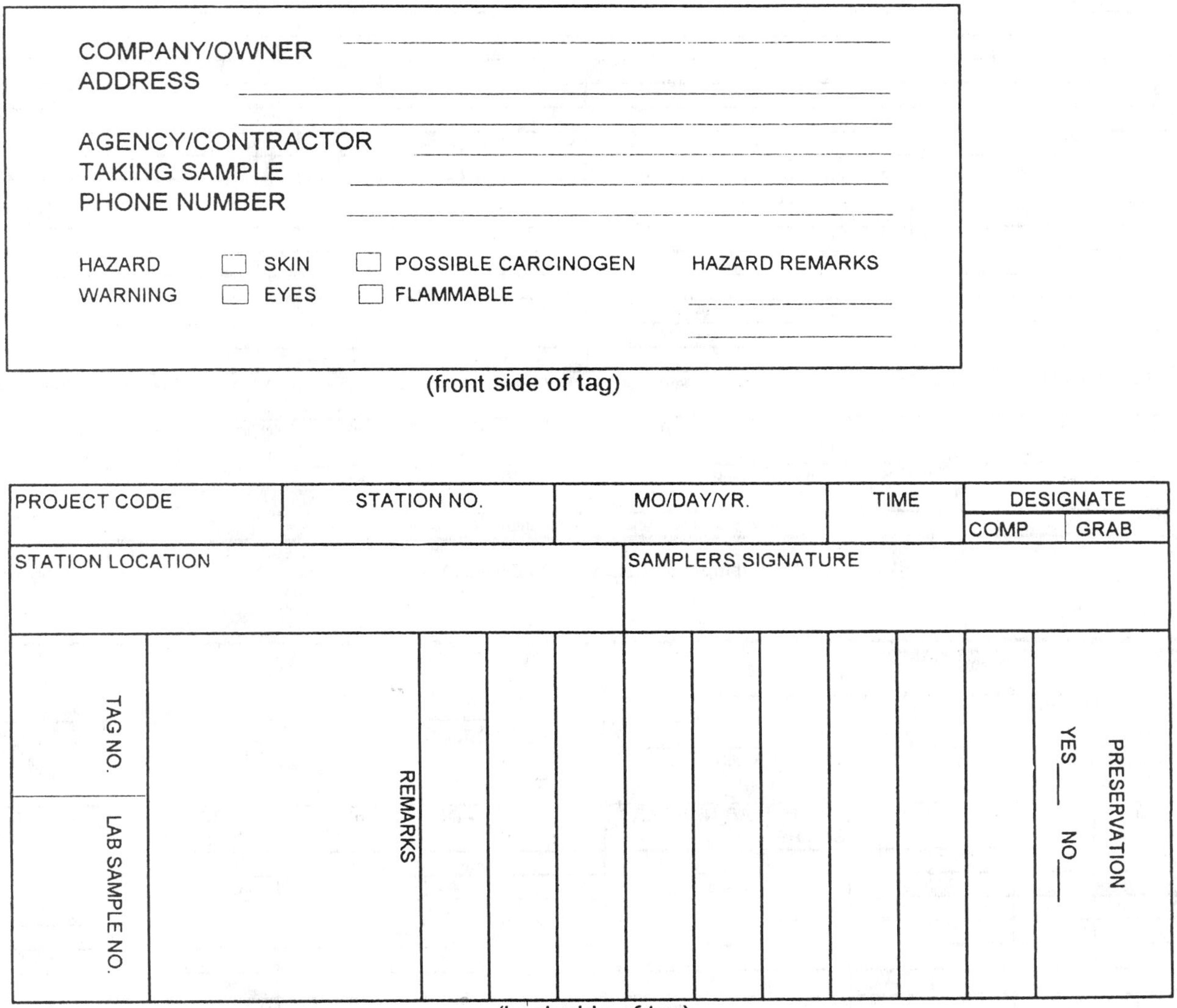

```
┌─────────────────────────────────────────────────────────────┐
│                                                               │
│   COMPANY/OWNER    ─────────────────────────────────          │
│   ADDRESS          ─────────────────────────────────          │
│                    ─────────────────────────────────          │
│                                                               │
│   AGENCY/CONTRACTOR   ───────────────────────────────         │
│   TAKING SAMPLE       ───────────────────────────────         │
│   PHONE NUMBER        ───────────────────────────────         │
│                                                               │
│                                                               │
│   HAZARD      ☐ SKIN    ☐ POSSIBLE CARCINOGEN   HAZARD REMARKS │
│   WARNING     ☐ EYES    ☐ FLAMMABLE             ──────────     │
│                                                 ──────────     │
│                                                               │
└─────────────────────────────────────────────────────────────┘
```

(front side of tag)

PROJECT CODE	STATION NO.	MO/DAY/YR.	TIME	DESIGNATE	
				COMP	GRAB
STATION LOCATION		SAMPLERS SIGNATURE			

TAG NO.	REMARKS										PRESERVATION YES ___ NO ___
LAB SAMPLE NO.											

(back side of tag)

FIG. X2.1 Example of Sample Identification Tag

PROJ. NO.		PROJECT NAME					NO. OF CON- TAINERS									REMARKS
SAMPLERS: (Signature)																
STA. NO.	DATE	TIME	COMP	GRAB	STATION LOCATION											

Relinquished by: (Signature)	Date/Time	Received by: (Signature)	Relinquished by: (Signature)	Date/Time	Received by: (Signature)
Relinquished by: (Signature)	Date/Time	Received by: (Signature)	Relinquished by: (Signature)	Date/Time	Received by: (Signature)
Relinquished by: (Signature)	Date/Time	Received for Laboratory by:	Date/Time	Remarks	

Distribution Original Accompanies Shipment Copy to Coordinator Field files

FIG. X2.2   Example of Field Sample Chain of Custody Record

PROJ. NO.	PROJECT NAME		Name of Facility		
SAMPLERS: (Signature)					
Split Samples Offered ( ) Accepted ( )Declined			Facility Location		
STA NO	DATE	TIME	COMP	GRAB	SPLIT SAMPS.	TAB NUMBERS	STATION DESCRIPTION	NO OF CONTAINERS	REMARKS

STA NO	DATE	TIME	COMP	GRAB	SPLIT SAMPS.	TAB NUMBERS	STATION DESCRIPTION	NO OF CONTAINERS	REMARKS

Transferred by: (Signature)		Received by: (Signature)		Telephone
Date	Time	Title	Date	Time

FIG. X2.3   Example of Receipt for Samples

Side 1

FIELD SAMPLE COLLECTION CHAIN-OF-CUSTODY FORM

A. Field Record

Field Sample Identification:_____ No. Containers:_____Project No:_____Sampling Location:_____
Collected By:_____ Witness:_____ Date(s):_____Time(s):_____
Preservation:_____ Notes:_____

B. Transfer Record

Relinquished By	Received By	Date	Time	Sample Condition/Observations

LABORATORY SAMPLE RECEIPT CHAIN-OF-CUSTODY FORM

A. Logging: Lab Sample I.D.:_____ By:_____ Date:_____ Time:_____

B. Transfer Record: (within sample receipt area)

Action	Delivered To	By	Date	Time	Cooler Access By

Potential Actions:
1. Sample/subsamples to locked cooler
2. Sample/subsample from locked cooler
3. Person to person custody transfer of samples/subsamples.
4. Disposal of sample/subsample.

C. Subsampling: By_____ Date:_____ Time:_____

D. Transfer to Analytical Lab

Aliquot	Analyses	Preservation	Container		Delivered By	Received By	Date	Time
0**								
1								
2								
3								
4								
5								
6								

**For use when field sample is not subsampled.

NOTE: ALL NAMES ON THIS FORM MUST BE ENTERED AS ORIGINAL SIGNATURES
NOTE: DELIVER A PHOTOCOPY OF THIS FORM (2 SIDED) WITH EACH ALIQUOT.

FIG. X2.4 Example Field, Laboratory Receipt, and Laboratory Sample Chain-of-Custody Record (Two-Sided)

Side 2

LABORATORY ANALYTICAL AREA CHAIN-OF-CUSTODY FORM

NOTE: ALL NAMES ON THIS FORM MUST BE ENTERED AS ORIGINAL SIGNATURES

Place computer label here

A. Condition At Receipt:______________________________

B. Handling/Transfer Record

Action	Delivered To	Delivered By	Date	Time	Cooler Access By

Additional Notes: ______________________________

Potential Actions:
1. Sample/subsamples to locked cooler
2. Sample/subsamples from locked cooler (indicate purpose)
3. Sample alterations (specify)
4. Subsampling of aliquot.
5. Re-analysis of aliquot.
6. Person-to-person custody transfer of samples/subsamples
7. Disposal of sample/subsamples

FIG. X2.5 Side Two of Custody Record

The American Society for Testing and Materials takes no position respecting the validity of any patent rights asserted in connection with any item mentioned in this standard. Users of this standard are expressly advised that determination of the validity of any such patent rights, and the risk of infringement of such rights, are entirely their own responsibility.

This standard is subject to revision at any time by the responsible technical committee and must be reviewed every five years and if not revised, either reapproved or withdrawn. Your comments are invited either for revision of this standard or for additional standards and should be addressed to ASTM Headquarters. Your comments will receive careful consideration at a meeting of the responsible technical committee, which you may attend. If you feel that your comments have not received a fair hearing you should make your views known to the ASTM Committee on Standards, 100 Barr Harbor Drive, West Conshohocken, PA 19428.

4.2 SPECIFIC SAMPLING PROCEDURES

Standard Guide for
Sampling Waste Piles[1]

This standard is issued under the fixed designation D 6009; the number immediately following the designation indicates the year of original adoption or, in the case of revision, the year of last revision. A number in parentheses indicates the year of last reapproval. A superscript epsilon (ε) indicates an editorial change since the last revision or reapproval.

1. Scope

1.1 This guide provides guidance for obtaining representative samples from waste piles. Guidance is provided for site evaluation, sampling design, selection of equipment, and data interpretation.

1.2 Waste piles include areas used primarily for waste storage or disposal, including above-grade dry land disposal units. This guide can be applied to sampling municipal waste piles.

1.3 This guide addresses how the choice of sampling design and sampling methods depends on specific features of the pile.

1.4 *This standard does not purport to address all of the safety concerns, if any, associated with its use. It is the responsibility of the user of this standard to establish appropriate safety and health practices and determine the applicability of regulatory limitations prior to use.*

2. Referenced Documents

2.1 *ASTM Standards:*

D 1452 Practice for Soil Investigation and Sampling by Auger Borings[2]

D 1586 Test Method for Penetration Test and Split-Barrel Sampling of Soils[2]

D 1587 Practice for Thin-Walled Tube Geotechnical Sampling of Soils[2]

D 4547 Practice for Sampling Waste and Soils for Volatile Organics[3]

D 4687 Guide for General Planning of Waste Sampling[3]

D 4700 Guide for Soil Sampling from the Vadose Zone[2]

D 4823 Guide for Core-Sampling Submerged, Unconsolidated Sediments[4]

D 5088 Practice for Decontamination of Field Equipment Used at Nonradioactive Sites[5]

D 5314 Guide for Soil Gas Monitoring in the Vadose Zone[5]

D 5451 Practice for Sampling Using a Trier Sampler[3]

D 5518 Guide for Acquisition of File Aerial Photography and Imagery for Establishing Historic Site-Use and Surficial Conditions[5]

D 5730 Guide to Site Characterization for Environmental Purposes with Emphasis on Soil, Rock, the Vadose Zone and Ground Water[5]

3. Terminology

3.1 *Definitions of Terms Specific to This Standard:*

3.1.1 *hot spots*—strata that contain high concentrations of the characteristic of interest and are relatively small in size when compared with the total size of the materials being sampled.

3.1.2 *representative sample*—a sample collected such that it reflects one or more characteristics of interest (as defined by the project objectives) of the population from which it was collected.

3.1.2.1 *Discussion*—A representative sample can be a single sample, a set of samples, or one or more composite samples.

3.1.3 *waste pile*—unconfined storage of solid materials in an area of distinct boundaries, above grade and usually uncovered. This includes the following:

3.1.3.1 *chemical manufacturing waste pile*—a pile consisting primarily of discarded chemical products (whether marketable or not), by-products, radioactive wastes, or used or unused feedstocks.

3.1.3.2 *scrap metal or junk pile*—a pile consisting primarily of scrap metal or discarded durable goods such as appliances, automobiles, auto parts, or batteries.

3.1.3.3 *trash pile*—a pile of waste materials from municipal sources, consisting primarily of paper, garbage, or discarded nondurable goods that contain or have contained hazardous substances. It does not include waste destined for recyclers.

4. Significance and Use

4.1 This guide is intended to provide guidance for sampling waste piles. It can be used to obtain samples for waste characterization related to use, treatment, or disposal; to monitor an active pile; to prepare for closure of the waste pile; or to investigate the contents of an abandoned pile.

4.2 Techniques used to sample include both in-place evaluations of the pile and physically removing a sample. In-place evaluations include techniques such as remote sensing, on-site gas analysis, and permeability.

4.3 Sampling strategy for waste piles is dependent on the following:

4.3.1 Project objectives including acceptable levels of error when making decisions;

4.3.2 Physical characteristics of the pile, such as its size and configuration, access to all parts of it, and the stability of the pile;

4.3.3 Process that generated the waste and the waste characteristics, such as hazardous chemical or physical properties, whether the waste consists of sludges, dry powders or granules, and the heterogeneity of the wastes;

4.3.4 History of the pile, including dates of generation,

---

[1] This guide is under the jurisdiction of ASTM Committee D-34 on Waste Management and is the direct responsibility of Subcommittee D34.01 on Sampling and Monitoring.

Current edition approved Oct. 10, 1996. Published December 1996.

[2] *Annual Book of ASTM Standards*, Vol 04.08.

[3] *Annual Book of ASTM Standards*, Vol 11.04.

[4] *Annual Book of ASTM Standards*, Vol 11.02.

[5] *Annual Book of ASTM Standards*, Vol 04.09.

methods of handling and transport, and current management methods;

4.3.5 Regulatory considerations, such as regulatory classification and characterization data;

4.3.6 Limits and bias of sampling methods, including bias that may be introduced by waste heterogeneity, sampling design, and sampling equipment.

4.4 It is recommended that this guide be used in conjunction with Guide D 4687, which addresses sampling design, quality assurance, general sampling considerations, preservation and containerization, cleaning equipment, packaging, and chain of custody.

4.5 A case history of the investigation of a waste pile is included in Appendix X1.

5. Site Evaluation

5.1 Site evaluations are performed to assist in designing the most appropriate sampling strategy. An evaluation may consist of on-site surveys and inspections, as well as a review of historical data. Nonintrusive geophysical and remote sensing methods are particularly useful at this stage of the investigation (see Guide D 5518). Table 1 summarizes the effects that various factors associated with the waste pile, such as the history of how the pile was generated, have upon the strategy and design of the sampling plan. The strategic and design considerations are discussed as well.

5.2 *Generation History*—The waste pile may have been created over an extended time period. A remote sensing method that is very useful in establishing historical management practices for waste piles is aerial imagery. Aerial photographs are widely available and may be used to determine the history of a waste pile, sources of waste, and the presence and distribution of different strata. Satellite imagery could be used for larger waste piles.

5.2.1 The date of generation could be important with respect to the types of processes that generated the waste, the characteristics of the waste, the distribution of the constituents, and regulatory concerns.

5.2.2 The type of process that generated the waste will determine the types of constituents that may be present in the waste pile. Chemical variability will influence the number of samples that are required to characterize the waste pile unless a directed (biased) sampling approach is acceptable.

5.2.3 The delivery method of the material to the waste pile could influence the concentrations of the constituents, affect the overall shape of the pile, or create physical dissimilarity within the waste pile through sorting by particle size or density.

5.2.4 If the pile is under current management and use, the variability in constituent types and concentrations may be affected. Current management activities also may influence the regulatory status of the waste pile.

5.2.5 Regulatory considerations will typically focus on waste identification questions, in other words is the material a solid waste that should be regulated and managed as a hazardous waste (1).[6] This may involve a limited, directed sampling approach, particularly if a regulatory agency is conducting the investigation. A more comprehensive sampling design may be required to determine if the waste classifies as hazardous. Remediation efforts and questions regarding permits may focus on characterizing the entire pile, possibly as the removal of material is occurring. It should be noted that concentrations of contaminants near regulatory levels may increase the number of samples required to meet the objectives of the investigation. These regulatory levels could be those established to determine if a waste is hazardous, or "cleanup" levels set for a removal or remediation.

5.3 *Physical Characteristics of Pile*—Several physical characteristics of the waste pile must be considered during the site evaluation. Variability in size, shape, and stability of the pile affects access to it to obtain samples as well as safety considerations. Physical variability will influence the number of samples that are required to characterize the waste pile unless a directed (biased) sampling approach is considered to be acceptable. Techniques that might be used include resistivity and seismic refraction (for determining the depth of very large piles).

5.3.1 The size of the waste pile will influence the sampling strategy in that increasing size is often accompanied by increased variability in the physical characteristics of the waste pile. The number of samples, however, that are needed to characterize a waste pile adequately will typically be a function of the study objectives as well as the inherent variability of the pile.

5.3.2 The shape of the waste pile can influence the sampling strategy by limiting access to certain locations within the pile, and if it is topologically complex it is difficult to lay out a sampling grid. Also, a waste pile may extend vertically both above and below grade, making decisions regarding the depth of sample collection difficult.

5.3.3 The stability of the waste pile also can limit access to both the face and the interior of the pile. The use of certain types of heavier sampling equipment also could be limited by the ability of the pile to bear the weight of the equipment.

5.4 *Waste Characteristics:*

5.4.1 The constituents could include inorganics, volatile organic compounds (VOCs), and semivolatile organic compounds (including pesticides and polychlorinated biphenyls (PCBs)) (see Practice D 4547). Speciality analyses may be warranted, such as leaching tests or analyses for dioxin/

TABLE 1 Strategy Factors

Waste Pile Factors	Strategic Considerations	Design Considerations
Generation history	Date of generation	Analysis required
	Types of processes	Location of samples
	Characteristics by process	
	Delivery method	
	Current management	
	Regulatory considerations	
Physical characteristics of pile:	Physical variability of pile	Number of samples
– size	Access	Location of samples
– shape	Safety	Equipment selection
– stability		
Waste characteristics	Constituents present	Number of samples
	Constituent distribution	Analysis required
	Heterogeneity	Location of samples
	– physical variability	Representative samples
	– chemical variability	Equipment selection

---

[6] The boldface numbers in parentheses refer to the list of references at the end of this standard.

furans or explosive compounds. Soil gas sampling is a minimally intrusive technique that may detect the presence and distribution of volatile organic compounds in soils and in porous, unconsolidated materials. Appropriate applications for soil gas monitoring are identified in Guide D 5314.

5.4.2 The distribution of constituents in the waste pile could be influenced by changes in the manufacturing process which resulted in changes in the composition of the waste; the length of time the material has remained in the pile (particularly for VOCs); the mode of delivery of the waste materials to the pile; and management practices, such as mixing together wastes from more than one process.

5.4.3 Physical and chemical variabilities would include variability in the chemical characteristics of the material within the pile, as well as variability in particle size, density, hardness, whether brittle or flexible, moisture content, consolidated, or unconsolidated. The variability may be random or found as strata of materials having different properties or containing different types or concentrations of constituents.

5.4.3.1 Geophysical survey methods may be used on piles to estimate physical homogeneity, which may or may not be related to chemical homogeneity, and to detect buried objects, both of which may need to be considered during the development of the sampling design and the safety plan for the investigation. The most suitable technique for detecting nonmetallic objects is electromagnetics. Ground-penetrating radar, a more sophisticated and complex technique, also may be considered. Electromagnetic techniques are suited particularly to large piles that contain leachate plumes (for example, mine tailings) or for the detection of large discontinuities in a pile (for example, different types of wastes or the transition from a disposal area to background soils). For metallic objects, metal detectors and magnetometers are useful and relatively easy to use in the field.

5.5 *Potential Investigation Errors:*

5.5.1 Equipment selection can bias sampling results even if the equipment is used properly. Bias can result from the incompatibility of the materials that the sampling equipment is made of with the materials being sampled. For example, the equipment could alter the characteristics of the sample. Some equipment will bias against the collection of certain particles sizes, and some equipment cannot penetrate the waste pile adequately.

5.5.2 Equipment, use, and operation can introduce error (bias) into the characterization of a waste pile. Sampling errors typically are caused when certain particle sizes are excluded, when a segment of the waste pile is not sampled, or when a location outside the pile is inadvertently sampled.

5.5.3 When stratification, layering, or solid phasing occurs it may be necessary to obtain and analyze samples of each of the distinct phases separately to minimize sampling bias. Care should be taken when sampling stratified layers to minimize cross contamination. Proper decontamination procedures should be used for all sampling equipment (see Practice D 5088).

5.5.4 Statistical bias includes situations where the data are not normally distributed or when the sampling strategy does not allow the potential for every portion of the pile to be sampled.

6. Sampling Strategy

6.1 Developing a strategy for sampling a waste pile requires a thorough examination of the site evaluation factors listed in Section 5. The location and frequency of sampling (number of samples) should be outlined clearly in the sampling plan, as well as provisions for the use of special sampling equipment, access of heavy equipment to all areas of the pile, if necessary, and so forth.

6.1.1 *Representative Sampling*—The collection of a representative set of samples from a waste pile typically will be complicated by the presence of a number of the site evaluation factors (2,3).

6.1.2 *Heterogeneous Wastes*—Waste piles may be homogeneous, for applied purposes, or may be quite heterogeneous in particle size and contaminant distribution. If the particle sizes of the material in the waste pile and the distribution of contaminants are known, or can be estimated, then less sampling may be necessary to define the properties of interest in the waste pile. An estimate of the variability in contaminant distribution may be based on process knowledge or determined by preliminary sampling (4). The more heterogeneous the waste pile is, the greater the planning and sampling requirements.

6.1.3 *Strata and Hot Spots*—A waste pile also could contain strata that have less internal variation in physical properties or concentrations of chemical constituents than the remainder of the waste pile (2,5). For example, strata may be present in a waste pile due to changes in the process that generated the waste, or if different processes at a facility contribute waste to different parts of the waste pile. A stratified sampling strategy would consider this situation by conducting independent sampling of each stratum, which could reduce the number of samples required. These strata could be in specific areas of the waste pile (4). Also, hot spots may be present in the waste pile that are unique in composition (2,5).

6.2 *Specific Sampling Strategies:*

6.2.1 Although the most appropriate method for evaluating material in waste piles is to sample at or immediately following the point of generation (for example, conveyor belt), most sampling problems involve existing or in-place waste piles. Therefore, the following discussion will focus on in-place waste piles. Sampling strategies available for waste piles include directed or judgmental sampling, simple random sampling, stratified random sampling, systematic grid sampling, and systematic sampling over time (2,6). General concerns about the collection of a representative sample, the existence of potential heterogeneity in the waste pile, the presence of strata within the waste pile, and the existence of distinct hot spots within the waste pile may also influence the selection of an appropriate sampling strategy and development of the sampling plan (5). The following paragraphs provide an introduction to determining the appropriate number of samples to collect and the sampling strategies available for determining sample locations.

6.2.2 *Determining the Frequency or Number of Samples*—The frequency of sampling or the number of samples to collect typically will be based on several factors including the study objectives, properties of wastes in the pile, degree of confidence required, access to sampling points, and budgetary constraints. Practical guidance for determining the number of samples is included in Guide D 4687 and Refs (2, 3).

6.2.3 *Directed Sampling*—Directed sampling (Fig. 1) is based on the judgment of the investigator and will not result necessarily in a sample that reflects the characteristics of the entire waste pile. Directed sampling also is called judgmental sampling, authoritative sampling, or nonprobability sampling. The experience of the investigator often is the basis for sample collection, and, depending on the study objectives, bias should be recognized as a potential problem. For preliminary screening investigations of a waste pile and for certain regulatory investigations, however, directed sampling may be appropriate. A directed sampling strategy could call for the collection of a composite sample from the surface area or the collection of discrete grabs at the surface of the pile (see Fig. 1). Directed sampling would typically focus on worst case conditions in a waste pile, for example, the most visually contaminated area or most recently generated waste.

6.2.4 *Simple Random Sampling*—Simple random sampling (Fig. 2) ensures that each element in the waste pile has an equal chance of being included in the sample (2). This may be the method of choice when, for purposes of the investigation, the waste pile is randomly heterogeneous (5). If the waste pile contains trends or patterns of contamination, a stratified random sampling or systematic grid sampling strategy would be more appropriate (2) (see 6.2.5 and 6.2.6).

6.2.4.1 A simple random approach could use a grid with random grids selected for sample collection (see Fig. 2). Note that the grid size could be selected based on the number of samples that are required (some guidance suggests having at least ten times the number of grids as samples required). Once the grid is overlaid and the sampling locations are selected, the decision must be made to collect either a discrete grab sample (surface), a composite of surface samples taken from predesignated locations within the grid cell (based on compass points), a vertical composite to a specified depth, or discrete grab samples at specified depths. If discrete grab samples are desired at specified depths, they typically would be collected at the same location as the bore hole is advanced into the pile. Figure 2 illustrates the collection of vertical composites at each of the randomly selected locations.

6.2.5 *Stratified Random Sampling*—Stratified random sampling (see Fig. 3) may be useful when distinct strata or homogeneous subgroups are identified within the waste pile (2). The strata may be located in different areas of the pile or may be comprised of different layers (see Fig. 3). This approach is useful when the individual strata may be considered internally homogeneous or at least have less

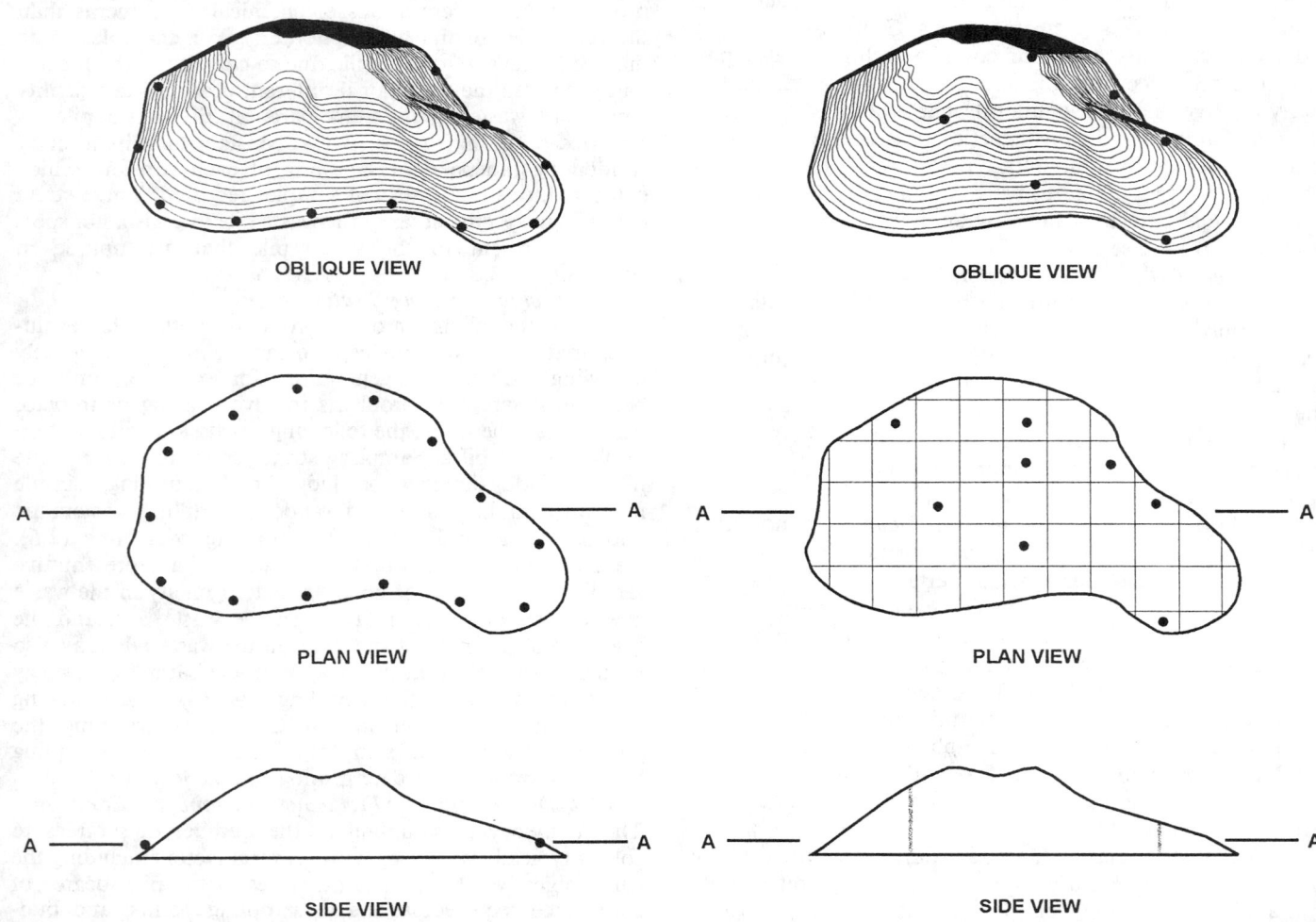

OBLIQUE VIEW

OBLIQUE VIEW

PLAN VIEW

PLAN VIEW

SIDE VIEW

SIDE VIEW

FIG. 1 Waste Pile Sampling Strategy—Directed Sampling

FIG. 2 Waste Pile Sampling Strategy—Simple Random Sampling

internal variation in what would otherwise be considered a heterogeneous waste pile (2). Information on the waste pile usually is required to establish the location of individual strata unless process knowledge or changes in the composition of the material is obvious, such as with discoloration or with the type of waste. The grid may be utilized for sampling several horizontal layers if the strata are oriented horizontally (4). A simple random sampling approach then is used within each stratum. The use of a stratified random sampling strategy may result in the collection of fewer samples. Figure 3 illustrates a scenario where the number of samples collected in each stratum varies (plan view), and discrete grabs are collected in each boring at predesignated depths (side view).

6.2.6 *Systemic Grid Sampling*—Systematic grid sampling (see Fig. 4) involves the collection of samples at fixed intervals and is useful when the contamination is assumed to be distributed randomly (2). This method also is commonly used with waste piles when estimating trends or patterns of contamination or when the objective is to locate hot spots. This approach may not be acceptable if the entire waste pile is not accessible or if the sampling grid locations become phased with variations in the distribution of contaminants within the waste pile (6). It also may be useful for identifying

the presence of strata within the pile. The grid and starting points should be laid out randomly over the waste pile, yet the method allows for rather easy location of exact sample locations by means of the grid (see Fig. 4). The same considerations discussed in 6.2.4 concerning the depth of each sample (surface, vertical composite, discrete grabs at depth) also should be considered. Figure 4 illustrates the collection of vertical composites at each grid, which could be difficult and costly. Also note that the grid size typically would be adjusted according to the number of samples that are required.

6.2.7 *Systematic Sampling Over Time*—Systematic sampling over time at the point of generation is useful if the material is being sampled from a conveyor belt or being delivered by means of truck or pipeline to the waste pile. The sampling interval can be determined on a time basis, for example, every hour from a conveyor belt or pipeline discharge, or from every third truck load. The time between intervals is influenced by the factors addressed in 6.2.2.

6.2.8 *Alternative Approach*—In many cases, an objective of waste pile characterization is to determine the impact of the pile on the environment. At times this may be accomplished more easily by sampling the routes by which contaminants are dispersed from the pile than through direct sampling of the pile, especially for piles that are difficult to

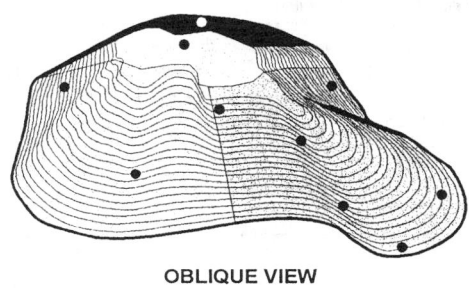

OBLIQUE VIEW

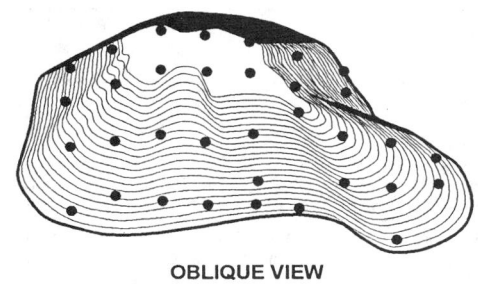

OBLIQUE VIEW

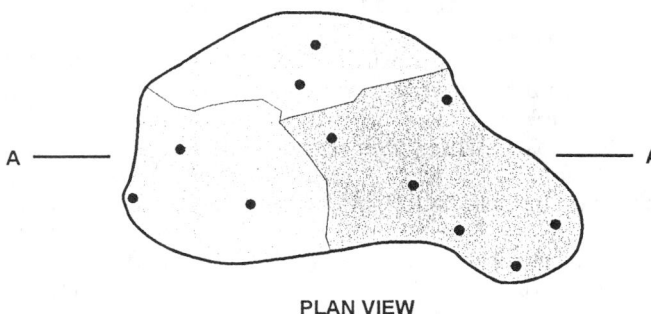

PLAN VIEW

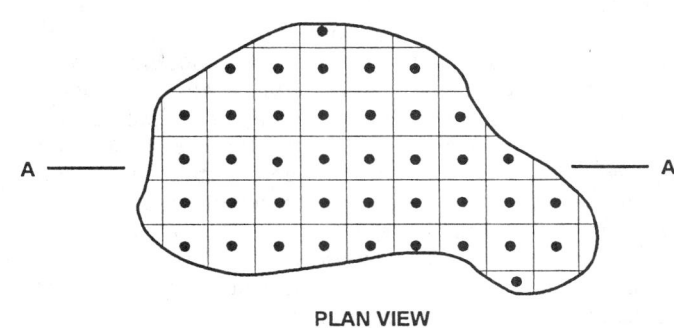

PLAN VIEW

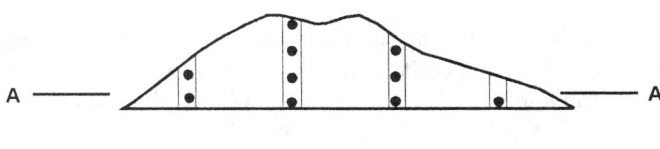

SIDE VIEW

FIG. 3 Waste Pile Sampling Strategy—Stratified Random Sampling

SIDE VIEW

FIG. 4 Waste Pile Sampling Strategy—Systematic Grid Sampling

TABLE 2 Sampling Devices Suitable for Waste Piles[A]

Location and Waste Type	Sampling Devices	ASTM Standard	Limitations
Subsurface Powdered, granular, or soil-like solids; sludges	split-barrel push coring device	D 1586 D 1587 D 4700 D 4823	Limited application for sampling moist and sticky solids, or particles with diameter 0.6 cm (0.25 in.) or more. Depth limitation of about 1 m.
	trier	D 5451	May not retain core sample of very dry granular materials. Not applicable to sampling solid wastes with particle diameter >½ the diameter of the sampling tube.
	auger	D 1452 D 4700	Does not collect undisturbed sample.
	thin-walled tube sampler	D 4823 D 4700	Collects relatively undisturbed core. Difficult to use on gravelly or rocky soils.
	drill rigs		Used for geoenvironmental exploration. To minimize sample contamination, avoid those using a water-based drilling fluid.
	soil gas samplers	D 5314	Used for volatile organic compounds.
Surface Powdered, granular, or soil-like solids; sludges	trowel or scoop	D 4700	Not applicable to sampling deeper than 8 cm (3 in.). Difficult to obtain reproducible mass of sample. May exclude certain particle sizes, especially large aggregates.
Slag	hammer/chisel Impact device		Changes particle size.

[A] This table is not all inclusive; other equipment may be used.

TABLE 3 Excavation and Removal Equipment for Waste Piles

Excavation and Removal Equipment	General Excavation	Ability to Excavate Hard and Compacted Material	Soil Hauling	Mixing of Solids, Soil	Spreading Cover	Site Maneuverability
Wheel or crawler Mounted backhoe	A[A]	A	B[B]/O[C]	A	A	A/B
Wheel or crawler Mounted front-end loader	A	A	A/B	A	A	A/B
Skid steer loader	A	B	B	A	B	A
Bulldozer	A	A	O	O	A	B

[A] A = Good choice. Equipment is fully capable of performing function listed.
[B] B = Secondary choice. Equipment is marginally capable of performing function listed.
[C] O = Not applicable or poor choice.

characterize. For example, ground water up-and-down gradient from the pile could be sampled to check for ground water contamination. The vadose zone below the pile also might be sampled to detect leachate (and potential ground water contamination) through soil sampling, vacuum lysimeters, or soil gas. Surface water and sediment in drainage channels down gradient from the pile also might be sampled. Surface soils, air samples, and contaminants deposited on vegetation can be used as indicators of atmospheric transport of contaminants from the pile, including both particulate and volatile materials. Such approaches will seldom replace pile sampling completely, but they may reduce the number of pile samples needed to make remedial action decisions (see Guide D 5730), also Refs (7,8,9).

7. Selection of Sampling Equipment

7.1 Wastes in piles are often complex, multiphase mixtures of solids and semisolids. The wastes can range from powders to granules to large, heterogeneous solid fragments and can cover many acres in area. No single type of sampler can be used to collect representative samples of all types of waste from piles. Large, thick piles may require drill rigs to obtain samples from depth. The sampling of gases from within the pile requires other types of equipment. Table 2 lists typical waste types and the corresponding recommended samplers to use.

7.2 Sampling at depth from inside the pile may require heavy equipment designed for excavation or removal of soil or rock. Table 3 lists such equipment and its applications for sampling waste piles (10).

7.3 Sampling equipment should be constructed of materials that are compatible with the waste to be sampled. Compatibility refers to the physical durability, lack of chemical reactivity with the waste, and lack of potential for contamination of the waste with analytes of concern. Typical materials of construction include stainless steel, plastic, and glass.

8. Data Use

8.1 The decisions that will be made based upon the data must be identified early in the planning process since these affect the approach to the problem and how the data will be evaluated. Decisions affecting waste classification, closure, and post-closure issues, are examples of the uses of the data. Methods to determine the volume of contaminated material in a pile or pile strata may be needed. Standard mathematical formulas for calculating the volume of a cone, cylinder, various prisms, and so forth, may be used.

8.1 *Statistical Considerations:*

8.1.1 Data quality assessment (DQA) methods are used to evaluate the data for any anomalies and to evaluate the assumptions for statistical evaluation. The statistician makes use of both subjective judgment (graphical analysis for identification of trends and anomalies) and statistical models and inference (for example, outlier detection, autocorrelation estimation) in the investigation of data for validity of the assumptions needed to make a statistical test. Classical statistical models assume that the samples collected from the population of interest are independent and have an identical probability distribution (that is, normal distribution with constant mean and variance). Random sampling is a method to ensure independence. The probability distributional assumptions are part of DQA that will determine if the classical statistical model is appropriate for the collected data. For directed sampling, the sampling is subjective and the sample results are typically judged on a qualitative basis.

8.1.2 Simple random sampling will provide an unbiased estimate of the average waste concentration, that is, an estimate of the mean. This unbiased estimate is independent of the geometry of the pile and of the distribution of the concentration of the contaminants, but it may not have the smallest variance. Other sampling designs, such as systematic grid sampling or stratified random sampling, may provide an average that has a smaller variance. If the waste pile has uneven topography, the calculation of the mean concentration of the pile should be a volume-weighted average, using core volume as the weighting factor to reduce the variance of the estimated mean.

8.1.2.1 For simple random sampling and systematic grid sampling designs, histogram and normal probability plots of the sample data can be used to judge if the data conform to normal distribution. If not, there are several alternatives. First, the classical statistical model may still be considered robust for the decision-making process. Second, a transformation of the data may approximate a normal distribution of the data. For example, logarithmic transformation will normalize data that are lognormal originally. If the data are lognormal, the question of whether to use the arithmetic mean or the geometric mean for decision-making purposes must be decided. Third, an alternative statistical model based on nonparametric methods, but which uses weaker assumptions, may be proposed to analyze the decision-making process. It may be advisable to consult a statistician.

8.1.2.2 For the stratified random sampling design, the test of normality is not straightforward. Generally, it requires a mathematical model to take out the strata effects first, then test for normality using the residuals. A statistician should be consulted.

8.1.2.3 In any of these cases, alternative consequences of the level of uncertainty can be calculated prior to collecting the data. These alternatives can be used by decision-makers to select the best strategy to minimize the environmental risks.

9. Keywords

9.1 piles; sampling; waste

APPENDIX

(Nonmandatory Information)

X1. WASTE PILE—A CASE HISTORY

X1.1 Background—The waste pile was generated by a facility that produces brass alloys from scrap metal. The by-product from this operation was slag, which was generated in the recovery furnace. The slag was ground subsequently in a ball mill prior to being reintroduced into the recovery furnace. A large amount of the ground slag was disposed of in a waste pile which covered about one acre. No active management was occurring with the waste pile. No buried containers or extremely heterogenous material (unground slag) was suspected of being present in the waste pile based on facility records and interviews of personnel.

X1.1.1 Lead and cadmium were the constituents of concern based on process knowledge, and the possibility for the waste being hazardous was the regulatory consideration. The potential for off-site migration of contaminants was also an immediate concern, and this was considered in the development of the Phase 1 study design. Figure X1.1 shows a site map of the facility and the slag pile. Figure X1.2 shows a computer enhancement of the slag pile, and Fig. X1.3 shows a topographic view of the pile.

X1.2 Phase 1:

X1.2.1 *Objective*—The primary objective of the initial investigation was to determine if the slag in the waste pile classified as hazardous based on the concentration of lead and cadmium in a leach test. A secondary objective was to provide preliminary information on the potential migration and transport of contaminants from the waste pile off-site. The sampling plan for this initial investigation utilized a directed sampling strategy to provide a preliminary estimate of the lead concentration in the waste, the variability of contaminant concentrations in the pile, and the potential for leaching using the applicable leaching procedure mandated in regulations. Four composite samples were collected from the surface (0 to 15 cm or 0 to 6 in.) of the waste pile at locations within the four quadrants. The following environmental samples were also collected:

X1.2.1.1 Several soil samples in the vicinity of the waste pile,

X1.2.1.2 Sediment upstream and downstream in a stream which borders the facility,

X1.2.1.3 Sediment in a ditch which contained runoff from the pile, and

X1.2.1.4 Two background soil samples.

X1.2.2 Figure X1.4 shows the Phase 1 sampling locations within the slag pile, and Fig. X1.5 shows the same sampling locations on the topographic map of the pile.

X1.2.3 *Results*—Zinc, copper, cadmium, and lead were

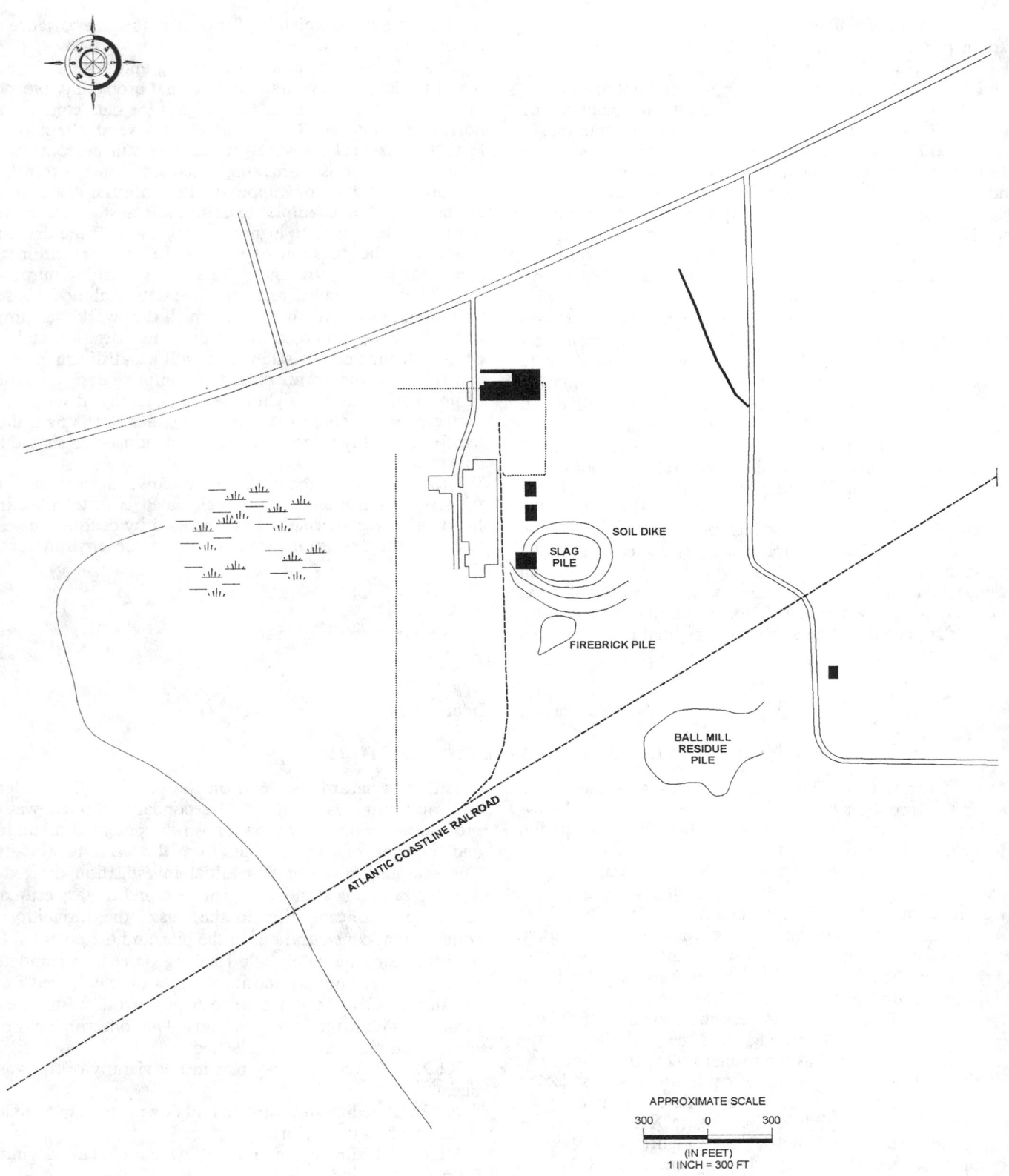

FIG. X1.1 Site Map

all elevated (compared to background) in the samples collected from the waste pile, and the concentrations did not appear to vary significantly between the samples. Since lead and cadmium are regulated constituents, a leach test was completed, and the lead results exceeded the regulatory level of 5 mg/L. Cadmium was just under the regulatory level of

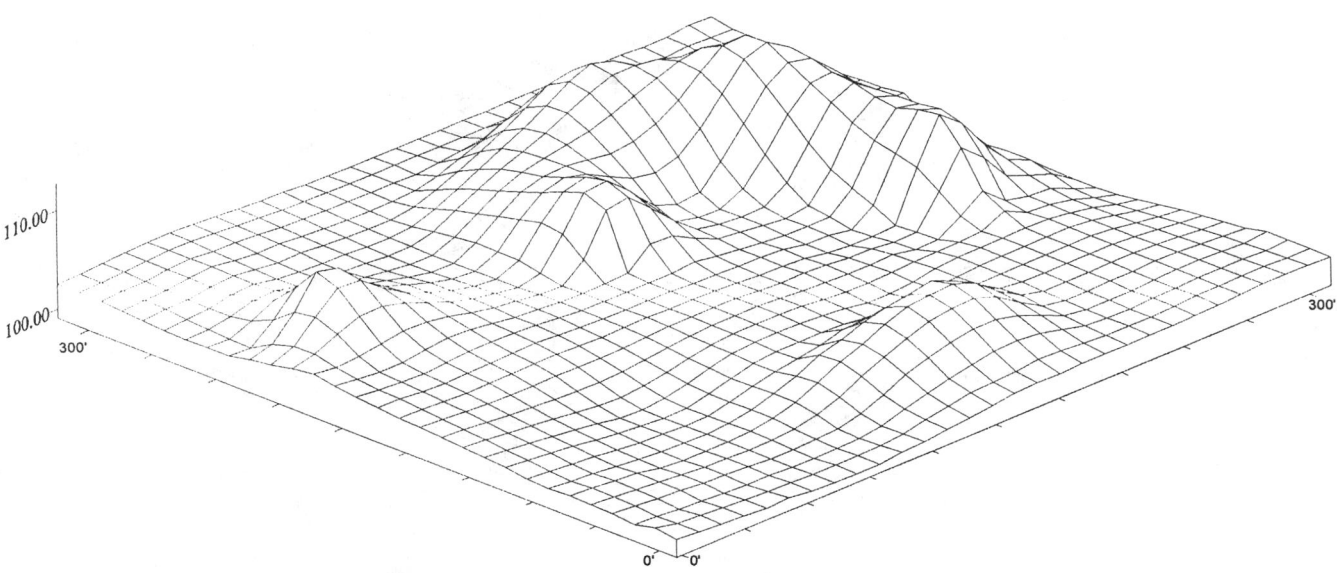

FIG. X1.2 Computer Enhancement of the Slag Pile (Front View) Scale 1:1:2

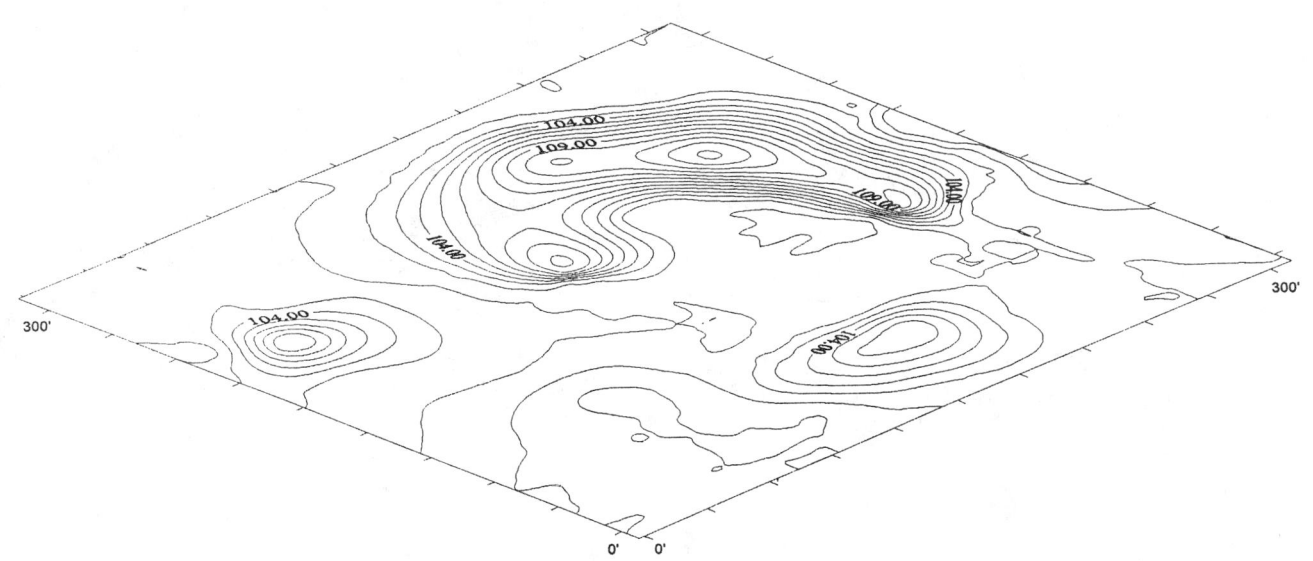

FIG. X1.3 Topographic View of the Slag Pile

1.0 mg/L. Lead and cadmium concentrations in the soil were 2 to 3 times above background, and the drainage ditch and downstream sediment sample also had elevated lead and cadmium levels.

X1.2.4 *Conclusion*—The waste pile contained slag that is hazardous for lead. The waste pile required further characterization to determine the variability in the pile. The presence of lead and cadmium in soils and the stream sediment downstream of the facility was confirmed and should be investigated further to determine the extent of contaminant transport.

X1.3 Phase 2:

X1.3.1 *Objective*—The objective is to characterize the waste pile further using a systematic grid sampling design. This design will delineate horizontal and vertical variability in lead and cadmium concentrations. The Phase 1 investigation also provided a good estimate of the anticipated variability in the waste pile. The number of samples required to characterize the waste pile adequately was calculated based on the average concentration, the anticipated variability, the regulatory level of concern, and the specified confidence interval. The grid size then was adjusted to accommodate the projection on the required number of samples. Composite samples were collected within each grid cell based on one center point and eight points on the compass (45° intervals) equidistant from the center point. Ten percent of the grids were designated for vertical as well as surface (0 to 15 cm or 0 to 6 in.) sample collection. Additionally, 10 % of the grids were designated randomly for duplicate sampling (using a different aliquot pattern) to check the preliminary estimate on the variability. Additional

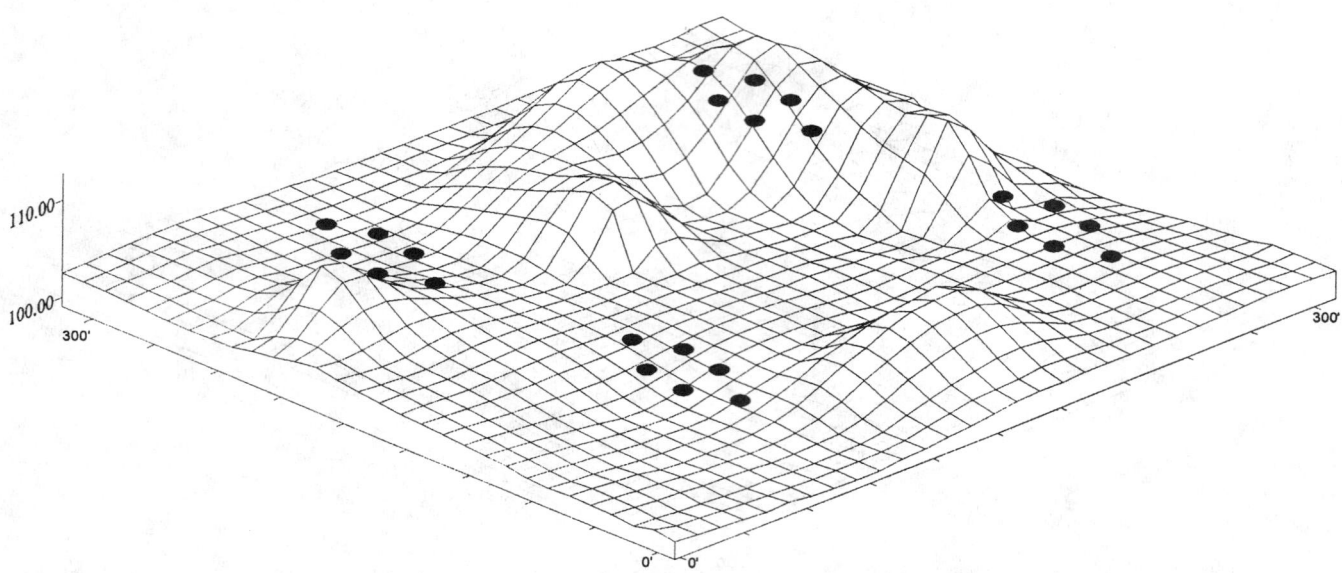

FIG. X1.4   Front View of the Slag Pile Showing Sampling Locations Scale 1:1:2

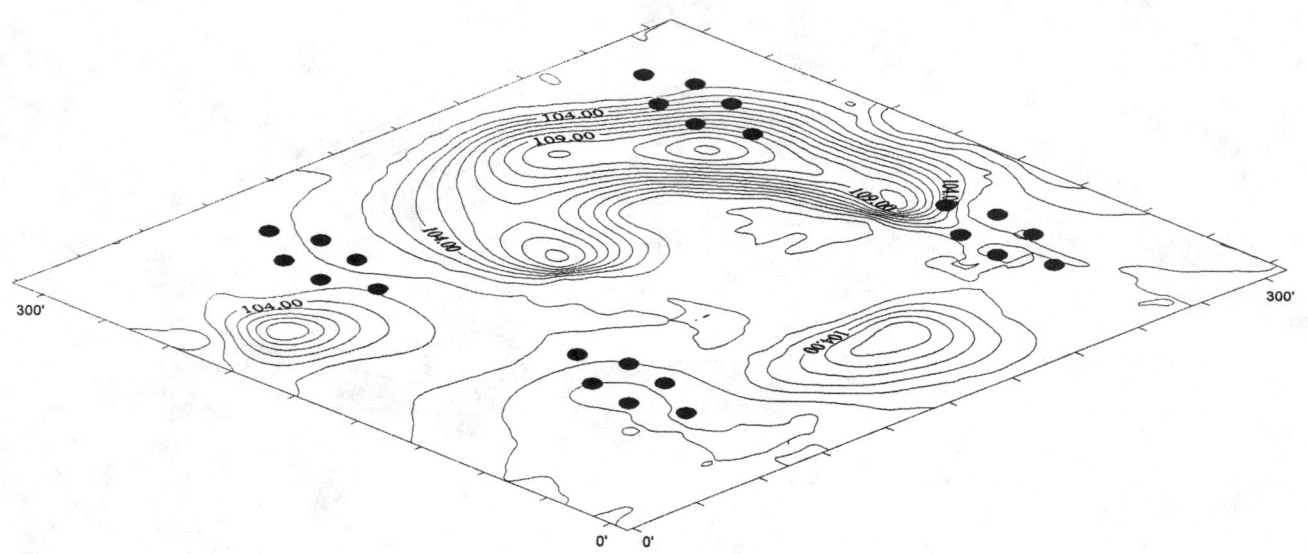

FIG. X1.5   Topographic View of the Slag Pile Showing Sampling Locations

environmental sampling was conducted but will not be covered in this discussion.

X1.3.2 *Results*—The results supported the initial Phase 1 investigation with lead consistently exceeding the regulatory level. Cadmium consistently was below the regulatory level.

X1.3.3 *Conclusion*—The waste pile was characteristic for lead and classified as hazardous according to the applicable regulations. There was no significant variability with depth, although several gradients were noticed across the grid based on lead concentration (scan) results.

X1.4 Phase 3:

X1.4.1 *Objective*—The objective is to determine the

volume of the waste pile in order to estimate both the disposal cost and the total amount of the civil penalty to be charged to the owner of the pile. The waste pile was surveyed using standard surveying techniques.

X1.4.2 *Results*—The results were used to calculate the volume using geometric principles. Also, a computer program was utilized which constructs contours based on the surveying information. The computer program was used as a check of the manual method, which produced a result that was 10 % higher in volume than the computer program.

X1.4.3 *Conclusion*—For penalty calculation purposes, the smaller estimate was utilized; however, the actual treatment and disposal costs could reflect the larger estimate.

REFERENCES

(1) U.S. Environmental Protection Agency (EPA). 1986. *Test Methods for Evaluating Solid Waste*; SW-846, 3rd Edition. EPA/530/SW-846 (NTIS PB88-239223); First update, 3rd edition. EPA/530/SW-846.3-1 (NTIS PB89-148076). Current edition and updates available on a subscription basis from U.S. Government Printing Office, Stock #955-001-00000-1.

(2) Gilbert, R. O., "Statistical Methods for Environmental Pollution Monitoring," Van Nostrand Reinhold Co., 1987.

(3) Ford, P. J., and Turina, P. J., *Characterization of Hazardous Waste Sites—A Methods Manual, Vol 1: Site Investigations*, EPA 600/4-84/075, (NTIS PB85-215960), 1985.

(4) U.S. Environmental Protection Agency (EPA), *Compendium of ERT Waste Sampling Procedures*, Section 5.0 Waste Pile Sampling, SOP No. 2017, EPA 540/P-91/008, OSWER Directive 9360.4-07, January 1991.

(5) Pitard, F., *Pierre Gy's Sampling Theory and Sampling Practice*, Vol 1: Heterogeneity and Sampling, Chemical and Rubber Company (CRC) Press, 1989.

(6) U.S. Environmental Protection Agency (EPA), *Characterizing Heterogeneous Wastes: Methods and Recommendations*, EPA 600/R-92/033, (NTIS PB92-216894) February 1992. [Also published as hardback Smoley Edition: Rupp and Joens (1993).]

(7) Keith, L., *Principles of Environmental Sampling*, Ed. ACS, 1988.

(8) U.S. Environmental Protection Agency (EPA). *Guidance Document on the Statistical Analysis of Ground Water Monitoring Data at RCRA Facilities*, Office of Solid Waste, 1993.

(9) McCoy and Associates, Inc., "Soil Sampling and Analysis—Practices and Pitfalls," Hazardous Waste Consultant, Vol 10, No. 6, Lakewood, CO, 1992.

(10) PEI Associates, 1991, *Survey of Materials-Handling Technologies Used at Hazardous Waste Sites*, EPA 540/2-91/010, June 1991. (NTIS PB91-186924), 225 pp.

Standard Practice for
Sampling Unconsolidated Waste From Trucks[1]

This standard is issued under the fixed designation D 5658; the number immediately following the designation indicates the year of original adoption or, in the case of revision, the year of last revision. A number in parentheses indicates the year of last reapproval. A superscript epsilon (ϵ) indicates an editorial change since the last revision or reapproval.

1. Scope

1.1 This practice covers several methods for collecting waste samples from trucks. These methods are adapted specifically for sampling unconsolidated solid wastes in bulk loads using several types of sampling equipment.

1.2 *This standard does not purport to address all of the safety concerns, if any, associated with its use. It is the responsibility of the user of this standard to establish appropriate safety and health practices and determine the applicability of regulatory limitations prior to use.* See Section 6 for specific precautionary statements.

2. Referenced Documents

2.1 *ASTM Standards:*
D 4687 Guide for General Planning of Waste Sampling[2]
D 4700 Guide for Soil Sampling from the Vadose Zone[3]
D 5088 Practice for Decontamination of Field Equipment Used at Non-radioactive Waste Sites[4]
D 5283 Practice for Generation of Environmental Data Related to Waste Management Activities: Quality Assurance and Quality Control Planning and Implementation[2]
D 5633 Practice for Sampling with a Scoop[2]

3. Summary of Practice

3.1 The truck and its contents are inspected and appropriate sampling equipment is selected. A clean sampling device is then used to scoop, core, or auger into the waste material. The sample or samples are collected and transferred to a sample container. The sampling device is then cleaned and decontaminated or disposed of.

4. Significance and Use

4.1 This practice is intended for use in the waste management industries to collect samples of unconsolidated waste from trucks. The sampling procedures described are general and should be used in conjunction with a site-specific work plan.

4.2 The purpose of collecting waste samples directly from a truck (rather than the waste source) is to verify (usually with screening analyses) that the waste contained in the truck is the same or similar material from a waste source that has been previously characterized and approved for treatment or disposal, or both.

5. Terminology

5.1 *Descriptions of Terms Specific to This Standard:*

5.1.1 *authoritative sampling*—a sample selected without regard to randomization.

5.1.2 *paperwork*—all required documentation, which may include manifests, waste profiles, sample labels, site forms, etc.

5.1.3 *screening analysis*—a preliminary qualitative or semi-quantitative test that is designed to give the user rapid and specific information about a waste that will aid in determining waste identification, process compatibility, and safety in handling.

5.1.4 *unconsolidated*—for solid material, the characteristic of being uncemented or uncompacted, or both, and easily separated into smaller particles.

5.1.5 *waste profile*—specific information about the waste including its properties and composition, chemical constituents, waste codes, transportation information, etc.

5.1.6 *work plan*—a plan specific to a particular site, for conducting activities specified in the plan.

6. Safety Precautions

6.1 Safety precautions must always be observed when sampling waste. The work plan must include a Worker Health and Safety section, because there are potential hazards associated with working around trucks as well as their potentially hazardous contents.

6.2 Truck sampling should be conducted from a properly designed platform to allow the sampler to safely access the truck bed with a minimum of difficulty.

7. Sampling Design

7.1 Truck sampling can be conducted for many different purposes. It is important that the purpose be integrated into the sample design. If the purpose of sampling is to characterize the waste, the sample should be collected from the waste source during the loading or unloading of the truck. This allows access to all portions of the material in the truck. If the purpose is to determine if the material in the truck conforms to a waste profile (that is, waste material that has previously been characterized), then a less rigorous sampling approach can be used. Because of the difficulties of sampling the material in the truck in situ, (authoritative) grab samples are usually collected from the top portion of the material and subjected to screening type analysis. This method will quickly demonstrate that the sampled material (top portion) does or does not match the waste profile.

7.2 A work plan should be prepared describing the

---

[1] This practice is under the jurisdiction of ASTM Committee D-34 on Waste Management and is the direct responsibility of Subcommittee D34.01 on Sampling and Monitoring.
Current edition approved Jan. 15, 1995. Published March 1995.
[2] *Annual Book of ASTM Standards*, Vol 11.04.
[3] *Annual Book of ASTM Standards*, Vol 04.08.
[4] *Annual Book of ASTM Standards*, Vol 04.09.

sampling locations, number of samples, depth of sampling and type of sampling equipment (see Practice D 5283 and Guide D 4687).

NOTE—Because of limited access to the truck bed for sampling, the samples collected are usually near-surface samples. There is a possibility that the material in the middle or on the bottom of the bed is different.

8. Pre-Sampling

8.1 *Basic Pre-Sampling Practices:*

8.1.1 Review all paperwork.

8.1.2 Access the truck by way of the sampling platform so that the waste can be visually inspected to confirm agreement with the paperwork and identify any obvious discrepancies (such as free liquids, etc.).

8.2 *Sampling Equipment:*

8.2.1 *Selection:*

8.2.1.1 Select the sampling equipment and sample containers appropriate for the waste in the truck, in accordance with the work plan or site-specific procedure. See Guide D 4687 for information on sample container selection.

8.2.1.2 The sampling equipment, sample preparation equipment, sample containers, etc., must be clean, dry, and inert to the material being sampled. Before use, all equipment including sample containers shall be inspected to ensure they are clear of obvious dirt and contamination and are in good working condition. Visible contamination shall be removed, and the equipment shall be decontaminated with the appropriate rinse materials. Prior to use, all cleaned equipment should be protected from contamination.

8.2.2 *Materials of Construction:*

8.2.2.1 Sampling devices are usually made of stainless steel, brass, or aluminum.

8.2.2.2 Sample containers should be made of plastic, glass, or other nonreactive materials (see Guide D 4687).

8.3 *Generic Equipment List*—The following is a general identification of equipment required for sampling unconsolidated waste from trucks.

8.3.1 *Scoop*, with extension handle.

8.3.2 *Trier.*

8.3.3 *Auger.*

8.3.4 *Concentric tube thief*, single slot, split tube, Missouri trier.

8.3.5 *Thin-walled tube.*

8.3.6 *Barrel auger.*

8.3.7 *Sample collection sheet.*

8.3.8 *Sample containers*, with lids and liners.

8.3.9 *Chain of custody forms.*

8.3.10 *Paperwork and site forms.*

8.3.11 *Sample labels.*

8.3.12 *Cloths or wipes.*

9. Sampling

9.1 *Basic Sampling Practices:*

9.1.1 Access the truck by way of the sampling platform and collect the required number of samples using techniques in accordance with 9.2.

9.1.2 Place the collected material in a sample container.

9.1.3 Close the sample container.

9.1.4 Wipe the outside of the sample container. Dispose of the wipe cloth properly.

9.1.5 Note on site forms all relevant conditions and physical characteristics associated with the collection of the sample.

9.1.6 Fill out all required paperwork for each sample, as required by the work plan.

9.1.7 Complete and attach the label to the side of the sample container after the sample has been collected.

9.2 *Sampling with a Concentric Tube Thief:*

9.2.1 *General Description*—This device consists of two tubes, one fitting snugly inside the other (see Fig. 1). The bottom end of the outer tube is fitted with a point. Oblong holes are cut through both tubes. The holes are opened or closed by rotating the inner tube. Concentric tube samplers are commercially available up to 6 ft (1.8 m) long and several inches (centimeters) in diameter.

9.2.2 Concentric tube samples have a limited application for sampling trucks. Materials that are not free-flowing such as those that are hard packed, moist, or finely powdered will not enter this type of sampler under normal field conditions. Sampling of materials containing granules or particles exceeding one third of the slot width should not be attempted because bridging may occur.

9.2.3 Insert the tube into the material and push with uniform force to the bottom of the truck or until refusal. Rotate the concentric tubes to the open position, thereby allowing the sample to flow into the inner tube. Wiggle the sampler several times and rotate the tubes to the closed position. Withdraw the sampler. Place the sampling device immediately over a sample collection sheet and release the sample by rotating the concentric tubes to the open position. A sample can normally be removed from the thief with a spatula or similar instrument (reamer) and placed in the sample container.

9.3 *Sampling with a Thin-Walled Tube Sampler:*

9.3.1 *General Description*—Tube samplers may vary in length, diameter, and material of construction (see Fig. 2). The material to be sampled must be of a physical consistency (cohesive solid material) to be cored and retrieved with the tube. Materials with particles larger than one third of the inner diameter of the tube should not be sampled with that particular device. The length of the tube will depend on the desired sampling depth (see Guide D 4700). The tube is attached to a length of solid or tubular rod. The upper end of this rod is threaded to accept a handle or extension rods. This sampler can be used to collect samples of unconsolidated clay-like materials.

9.3.2 The tube sampler is pushed into the material to be sampled by applying downward force on the unit's handle. Once the sampler has reached the bottom of the sampling interval, it is twisted to break the continuity at the tip. The

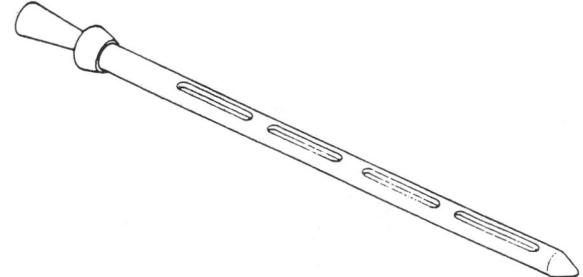

FIG. 1 Concentric Tube Thief

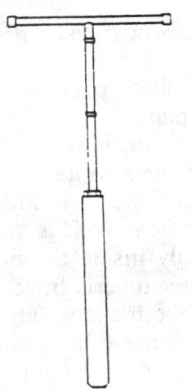

FIG. 2　Thin-walled Tube

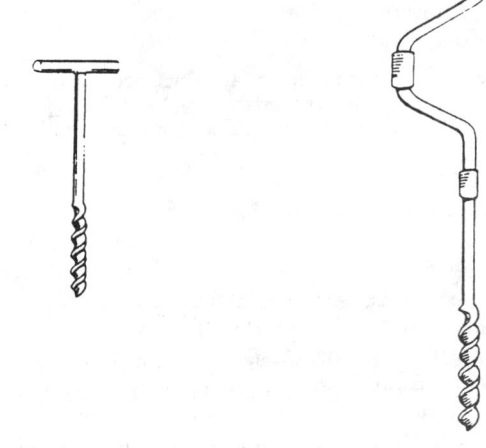

FIG. 4　Auger

sampler is pulled from the material and the sample is extruded into the sample container. Samples are extruded by forcing a rod through the tube.

9.4 *Sampling with a Trier Sampler:*

9.4.1 *General Description*—The trier is a metal or plastic tube from which one third to one half of the wall of the tube has been removed to form a slot along its entire length (see Fig. 3). This device can be up to 4 ft long (1.2 m) and should have a sharp, angled point at its lower end. The material to be sampled must have a physical consistency like a soil or similar fine-grained cohesive material. Sampling procedures can be found in Practice D 5451.

9.4.2 The trier is pushed vertically into the material and rotated one or two times to cut a core. The core is pulled out of the hole and removed from the trier with a spatula or similar instrument and placed in the sample container.

9.5 *Sampling with an Auger:*

9.5.1 *General Description*—The screw or ship auger is essentially a small diameter (for example, 1.5 in.; 3.8 cm) wood auger from which the cutting side flanges and tip have been removed. The auger is welded onto a length of solid or tubular rod. The upper end of this rod is threaded to accept a handle or extension rods (see Fig. 4 and Guide D 4700).

9.5.2 An auger can be used for collecting a disturbed sample of unconsolidated material from the truck. The auger is rotated manually or with a power source into the waste material. The operator may have to apply downward force to embed the auger; afterwards, the auger screws itself into the material. The auger is advanced to its full length, then pulled and removed. Material from the deepest interval is retained on the auger flights. The sample is collected from this ex-

tracted portion. Augers can be used to sample hard or compacted solid wastes or soil. Augers, like triers, are equipped with crossbars, facilitating the penetration of the waste.

9.6 *Sampling with a Barrel Auger:*

9.6.1 *General Discussion*—Dimensions and construction of a barrel auger will vary. A barrel auger typically consists of a stainless steel or carbide steel auger tip (orchard bit), a stainless steel cylinder, a bailed cap, an extension, and a cross handle (see Fig. 5 and Guide D 4700). A thin-walled internal sleeve may be used to contain the sample.

9.6.2 Barrel augers can be used for collecting discreet samples of disturbed material from various depths. The auger is rotated to advance the barrel into the truck load. The operator may have to apply downward pressure to keep the auger advancing. When the barrel is filled, the unit is withdrawn from the waste material and the sample is collected from the barrel.

9.7 *Sampling with a Scoop:*

9.7.1 *General Description*—Scoops must be of a size and shape suitable for the quantity and size of the particles to be sampled (see Fig. 6). The scoop is used for collecting equal portions at random spots at or near the surface of the waste. An extension to the scoop is often employed to assist the sampler in safely collecting the sample. For waste containing fragments or chunks, a scoop may be the only method

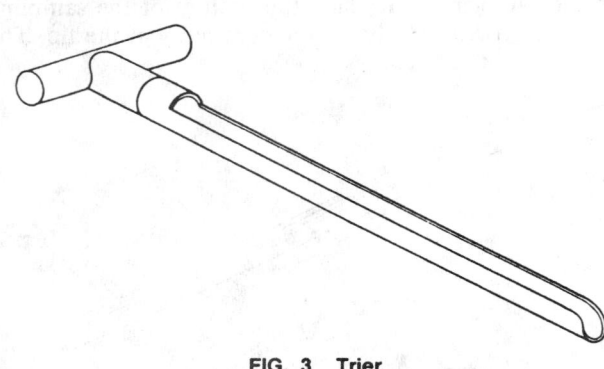

FIG. 3　Trier

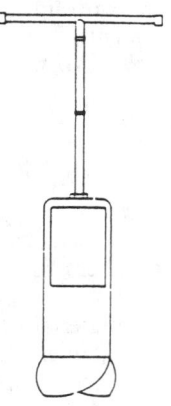

FIG. 5　Barrel Auger

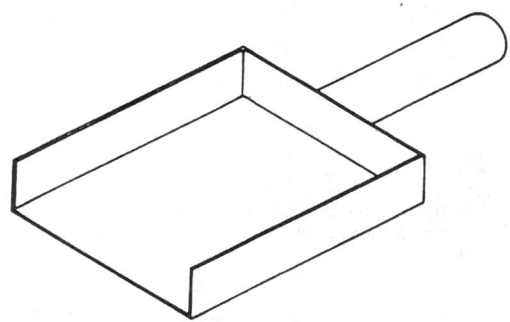

FIG. 6 Example of a Scoop

capable of retaining the material in a sampling device. A scoop may be used in conjunction with the sampling devices previously described, or as the primary sampling device. Sampling procedures can be found in Practice D 5633.

9.7.2 Attach the scoop to an extension of appropriate length and collect the sample.

10. Post Sampling

10.1 Transport the sample with appropriate chain of custody to the laboratory.

10.2 Remove all sampling equipment from the truck sampling area.

10.3 Transfer all direct contacting reusable equipment to a pre-designated decontamination area. Decontaminate the equipment according to the protocol established in the work plan (see Practice D 5088). Decontaminated sampling equipment should be protected from contamination. This may include, but not be limited to storage in aluminum foil, plastic bags, polytetrafluoroethylene film, or other means of protection that will not impact sample quality or the intended analyses.

10.4 Properly dispose of all used (disposable) contacting equipment.

11. Data Quality Objectives

11.1 The objectives for sampling and testing unconsolidated waste from trucks should be specified in the work plan (see Practice D 5283 and Guide D 4687).

12. Keywords

12.1 auger; barrel auger; concentric tube thief; sampling; scoop; thin-walled tube sampler; trier; truck sampling; waste

Standard Practice for
Sampling with a Scoop[1]

This standard is issued under the fixed designation D 5633; the number immediately following the designation indicates the year of original adoption or, in the case of revision, the year of last revision. A number in parentheses indicates the year of last reapproval. A superscript epsilon (ε) indicates an editorial change since the last revision or reapproval.

1. Scope

1.1 This procedure covers the method and equipment used to collect surface and near-surface samples of soils and physically similar materials using a scoop.

1.2 *This standard does not purport to address all of the safety concerns, if any, associated with its use. It is the responsibility of the user of this standard to establish appropriate safety and health practices and determine the applicability of regulatory limitations prior to use.*

2. Referenced Documents

2.1 *ASTM Standards:*
D 4687 Guide for General Planning of Waste Sampling[2]
D 5088 Practice for Decontamination of Field Equipment Used at Nonradioactive Waste Sites[3]
D 5283 Practice for Generation of Environmental Data Related to Waste Management Activities: Quality Assurance and Quality Control Planning and Implementation[2]

2.2 *Other Documents:*
Pierre Gy's Sampling Theory and Sampling Practice, Francis F. Pitard[4]

3. Summary of Practice

3.1 The top layers of material are removed down to the required sample depth using a shovel or other suitable equipment. A clean scoop is then used to collect the actual sample, which is placed in a sample container.

4. Significance and Use

4.1 This practice is intended for use in collecting samples of contaminated soils and similar materials.

4.2 Scoops are used primarily for collecting samples near the surface. Subsurface samples can be obtained by first removing higher layers using a shovel or other suitable equipment and collecting the sample with the scoop.

4.3 Because of their simplicity, scoops are useful in taking samples of waste materials where decontamination or disposal is a problem with other types of sampling equipment. Scoops are also suitable for use in rapid screening programs, pilot studies, and other semi-quantitative investigations.

4.4 Samples should be collected in accordance with an appropriate work plan (see Practice D 5283 and Guide D 4687).

5. Sampling Equipment

5.1 A shovel or other suitable equipment can be used for the initial removal of overburden material. This equipment should be manufactured from material that is compatible with the soil or waste to be sampled. The scoop must be manufactured from material that is compatible with the soil or waste to be sampled and the required test or analysis to be performed. For most hazardous waste sampling, either a disposable plastic scoop or a reusable stainless steel or polytetrafluoroethylene-coated scoop is suitable.

5.2 The design of the scoop is important to minimize sampling error, that is, all the material intended as the sample can be collected and placed in the sample container and is not lost as the scoop is systematically lifted from the source to the sample container (see Pierre Gy's Sampling Theory and Sampling Practice and Fig. 1).

5.3 For measurement of sample depth, a ruler or tape measure can be used.

6. Sample Containers

6.1 Plastic, glass, or other nonreactive containers should be used. Refer to Guide D 4687 for information on sample containers.

7. Procedure

7.1 Record all relevant information and observations about the sample location.

7.2 Use a shovel or other suitable equipment to remove any overburden material down to the level specified in the work plan.

7.3 Measure to the depth at which the sample will be collected with a ruler or tape measure. Record this information in the field log book.

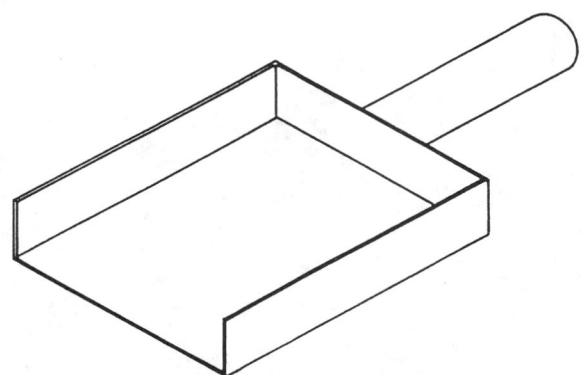

FIG. 1 Example of a Scoop

---

[1] This practice is under the jurisdiction of ASTM Committee D-34 on Waste Management and is the direct responsibility of Subcommittee D34.01 on Sampling and Monitoring.
Current edition approved Nov. 15, 1994. Published January 1995.
[2] *Annual Book of ASTM Standards*, Vol 11.04.
[3] *Annual Book of ASTM Standards*, Vol 04.09.
[4] Available from CRC Press, Inc., 2000 Corporate Blvd., NW, Boca Raton, FL 33431.

7.4 Remove the thin layer of material that was in contact with the overburden removal equipment and discard it using a clean scoop. The work plan will define if the scoop may or may not be reused to collect the actual sample.

7.5 Collect a suitable volume of sample with the scoop (the same scoop can be used to collect multiple scoopfuls to obtain sufficient volume to fill the container). Use a new (or decontaminated) scoop for each sample. Transfer the sample into the suitable container. Close the sample container and complete and attach the sample label.

7.6 Complete the field log book and chain-of-custody form.

7.7 Decontaminate the reusable equipment in accordance with the protocol specified in the work plan (see Practice D 5088).

8. Keywords

8.1 sampling; scoop; soil sampling; waste

Standard Practice for
Sampling Using a Trier Sampler[1]

This standard is issued under the fixed designation D 5451; the number immediately following the designation indicates the year of original adoption or, in the case of revision, the year of last revision. A number in parentheses indicates the year of last reapproval. A superscript epsilon (ε) indicates an editorial change since the last revision or reapproval.

1. Scope

1.1 This practice covers sampling using a trier. A trier resembles an elongated scoop as shown in Fig. 1. The trier is used to collect samples of granular or powdered materials that are moist or sticky and have a particle diameter less than one-half the diameter of the trier.

1.2 The trier can be used as a vertical coring device only when it is certain that a relatively complete and cylindrical sample can be extracted.

1.3 *This standard does not purport to address all of the safety problems, if any, associated with its use. It is the responsibility of the user of this standard to establish appropriate safety and health practices and determine the applicability of regulatory limitations prior to use.*

2. Referenced Documents

2.1 *ASTM Standards:*
D 4687 Guide for General Planning of Waste Sampling[2]
D 5088 Practice for the Decontamination of Field Equipment Used at Non-Radioactive Waste Sites[3]
D 5283 Practice for Generation of Environmental Data Related to Waste Management Activities: Quality Assurance and Quality Control Planning and Implementation[2]

3. Summary of Practice

3.1 As a coring device, the trier is pushed into the material to be sampled and is turned to cut the core. The core is then removed from the hole.

4. Significance and Use

4.1 This practice is applicable to sampling soils and similar fine-grained cohesive materials. This practice is to be used by personnel who are to acquire the samples.

4.2 This practice should be used in conjunction with Guide D 4687, which covers sampling plans, safety, quality assurance, preservation, decontamination, labeling, and chain-of-custody procedures; Practice D 5088, which covers the decontamination of field equipment used at waste sites; and Practice D 5283, which covers project specifications and practices for environmental field operations.

5. Sampling Equipment

5.1 The trier should be made from materials that are

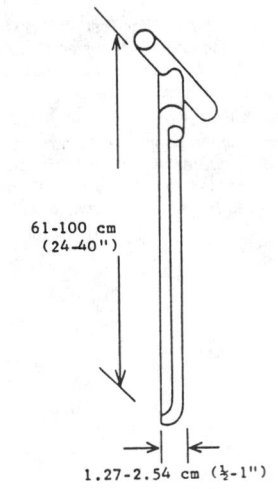

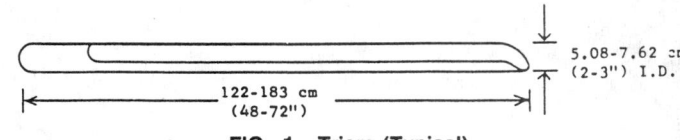

FIG. 1 Triers (Typical)

compatible with the substances being sampled and with the tests or analyses to be performed. Either stainless steel or polytetrafluoroethylene-coated metal will be suitable for most situations (see Fig. 1).

6. Sample Containers

6.1 Plastic, glass, or other nonreactive containers should be used. Refer to Guide D 4687 for further information on containers.

7. Procedure

7.1 Record appropriate information and observations on the sample location.

7.2 If sampling soils, remove surface vegetation or debris, or both, from the area of sample extraction.

7.3 For core sampling, proceed as follows:

7.3.1 Insert the trier approximately perpendicular to the surface of the material and rotate (one or two times) to cut a core.

7.3.2 Pull the core out of the hole slowly. Do not allow additional overburden material to become part of the sample. Inspect the core surface, and note the appearance of

[1] This practice is under the jurisdiction of ASTM Committee D-34 on Waste Management and is the direct responsibility of Subcommittee D34.01 on Sampling and Monitoring.
Current edition approved Aug. 15, 1993. Published October 1993.
[2] *Annual Book of ASTM Standards*, Vol 11.04.
[3] *Annual Book of ASTM Standards*, Vol 04.08.

any irregularities (for example, scratches, pock marks, and pebbles). Also inspect the core for breakage. If breakage has occurred and the core does not satisfy minimum length requirements, discard it and extract another from an immediately adjacent location.

7.4 Transfer the sample into a suitable container using a clean spatula. Seal the container and affix a label.

7.5 Complete the field logbook and chain-of-custody forms.

7.6 Decontaminate the used equipment in accordance with Practice D 5088.

8. Keywords

8.1 sampling; soil sampling; trier; waste

Standard Practices for
Sampling Wastes from Pipes and Other Point Discharges[1]

This standard is issued under the fixed designation D 5013; the number immediately following the designation indicates the year of original adoption or, in the case of revision, the year of last revision. A number in parentheses indicates the year of last reapproval. A superscript epsilon (ϵ) indicates an editorial change since the last revision or reapproval.

1. Scope

1.1 Those practices provide guidance for obtaining samples of waste at discharge points from pipes, sluiceways, conduits, and conveyor belts. The following are included:

	Sections
Practice A—Liquid or Slurry Discharges	7 through 9
Practice B—Solid or Semisolid Discharges	10 through 12

1.2 These practices are intended for situations in which there are no other applicable ASTM sampling methods (see Practices D 140 and D 75) for the specific industry.

1.3 These practices do not address flow and time-proportional samplers and other automatic sampling devices.

1.4 Samples are taken from a flowing waste stream or moving waste mass and, therefore, are descriptive only within a certain period. The length of the period for which a sample is descriptive will depend on the sampling frequency and compositing scheme.

1.5 It is recommended that these practices be used in conjunction with ASTM Guide D 4687.

1.6 *This standard does not purport to address all of the safety concerns, if any, associated with its use. It is the responsibility of the user of this standard to establish appropriate safety and health practices and determine the applicability of regulatory limitations prior to use.* See Section 5 for more information.

2. Referenced Documents

2.1 *ASTM Standards:*
D 75 Practice for Sampling Aggregates[2]
D 140 Practice for Sampling Bituminous Materials[2]
D 4687 Guide for General Planning of Waste Sampling[3]
E 882 Guide for Accountability and Quality Control in the Chemical Analysis Laboratory[4]
2.2 *Other Document:*
EPA-SW-846 Test Methods for Evaluating Solid Waste, Physical/Chemical Methods[5]

3. Summary of Practices

3.1 The variability of the waste stream is first determined based on (*1*) knowledge of the processes producing the stream, or (*2*) the results of a preliminary investigation of the waste stream's variability. A sampling design is then developed that considers the waste stream's variability, the time frame the sample is to represent, and the precision and accuracy required for waste analysis or testing. The actual sampling procedure consists of obtaining several grab samples from the moving stream or mass for analysis or testing.

4. Significance and Use

4.1 The procedure outlined in these practices are guides for obtaining descriptive samples of solid, semisolid and liquid waste from flowing streams, and incorporate many of the same procedures and equipment covered in the Referenced Documents. These practices by themselves will not necessarily result in the collection of samples representative of the total waste mass. The degree to which samples describe a waste mass must be estimated by application of appropriate statistical methods and measures of quality assurance. It is recommended that those practices be used in conjunction with Guide D 4687.

5. Hazards

5.1 In all sampling practices, safety should be the first consideration. Personnel involved in the sampling should be fully aware of, and take precautions against, the presence of toxic or corrosive gases, the potential for contact with toxic or corrosive liquids or solids, and the dangers of moving belts, conveyors, or other mechanical equipment. Guidance on waste sampling safety can be found in Guide D 4687.

6. Sampling Design

6.1 The frequency of sampling and the number of composites required to obtain a sample of the waste will depend on the following:
6.1.1 Time variability of the waste composition,
6.1.2 Time span which the sample is to represent, and
6.1.3 Precision of waste analysis that is required, for example, if a hazardous constituent is present in the waste at levels near the regulatory limit or another limit of concern, then better precision will be required than if the levels are well below or well above the limits of concern.
6.2 The processes that produce the waste will largely dictate the variability in the composition of the waste. If the processes are known to be constant and reliable, then fewer samples should be required than from a highly variable process.
6.3 To obtain a descriptive sample of the waste, the concentration levels and approximate variation in the waste composition should first be estimated. In some cases, a rough estimate can be made based on knowledge of the processes that produce the waste. In other cases, results from previous

[1] These practices are under the jurisdiction of ASTM Committee D-34 on Waste Management and are the direct responsibility of Subcommittee D34.01 on Sampling and Monitoring.
Current edition approved Nov. 24, 1989. Published January 1990.
[2] *Annual Book of ASTM Standards*, Vol 04.03.
[3] *Annual Book of ASTM Standards*, Vol 11.04.
[4] *Annual Book of ASTM Standards*, Vol 03.05.
[5] Available from Superintendent of Documents, U.S. Government Printing Office, Washington, DC 20402.

sampling efforts can be used to estimate waste composition and variability. A preliminary pilot sampling effort may be necessary to establish the waste composition prior to designing the primary sampling program. Procedures for estimating sample variability and for establishing a sampling design are provided in Guide D 4687.

6.4 The sampling design should include quality assurance procedures. At the least, this should include the following:

6.4.1 Sample handling quality control by carrying a blank sample through all of the sampling and analytical steps, and

6.4.2 User should be aware of the laboratories' internal quality control procedures. More rigorous quality control/quality assurance procedures may be required depending on the particular goals of the sampling program. For further information on quality control/quality assurance, see Guide E 882 and EPA SW-846.

6.5 A sampling plan should be prepared prior to sampling. The plan should describe such things as (a) safety procedures; (b) sampling design, including number and location of samples; (c) quality assurance procedures; (d) apparatus; (e) sampling procedures; and (f) sampling labeling. The details of the sampling procedure should consider all aspects of the specific discharge, including pipe diameter, velocity, rate of discharge, solids content of the discharge, requirements for grab or composite samples, and ultimate use of the analytical data.

PRACTICE A—LIQUID OR SLURRY DISCHARGES

7. Apparatus

7.1 *Dipper Sampler*—For slurry and liquid discharges, a dipper type sampler should be employed. One example of this type of sampler is depicted in Fig. 1. The dipper can be varied in size depending on the flow rate from the pipe or sluiceway. This procedure should not be used for high stream flow velocities or rates (>100 gal/min) because problems will arise in physically holding the dipper in the stream. Stream dimensions, size and shape of pipe, should also be considered in addition to flow rate. The sample should be taken across the full opening of the stream in as short a time as possible. This will minimize the effect of changes in composition of the waste stream due to density differentiation, laminar flow, and the like.

8. Procedure

8.1 Clean the beaker and container for compositing sample by methods appropriate for the analysis to be performed. Cleaning the equipment is especially important to prevent cross contamination between different waste types. In some cases, it may be necessary or simpler to dedicate equipment to a specific waste type.

8.2 Assemble the sampler by clamping the beaker to the pole.

8.3 Make sure that the sampler matches the dimensions of the discharge stream if at all possible. If not, pass the dipper in one sweeping motion through the discharge stream at a rate such that the dipper is filled in one pass. Make enough passes to cover the entire cross sectional area of the discharge stream.

8.4 If the entire discharge width cannot be covered in one pass additional passes will be needed. Begin each additional pass at the ending point of the previous pass. It may be necessary to make a few trial passes or practice runs before actually sampling the discharge. If compositing of samples is required by the sampling plan it may be preferable and sometimes necessary to have compositing performed in the laboratory. Include specific procedures for compositing in the sampling plan. If samples have not been appropriately composited in the field, clearly indicate this information on the sample label, in the field log book, and on the analytical request form.

8.5 In cases of high flow (>100 gal/min), a probe may be inserted upstream of the discharge to obtain a sample. Design the probe to collect a representative cross section of the flowing stream. See Fig. 2 for an example sampling probe.

8.6 Use preservation techniques appropriate for the analyses or testing to be conducted.

9. Packaging and Package Marking

9.1 An indelible label should be secured to the container identifying the sample. The label should contain or reference the following information:

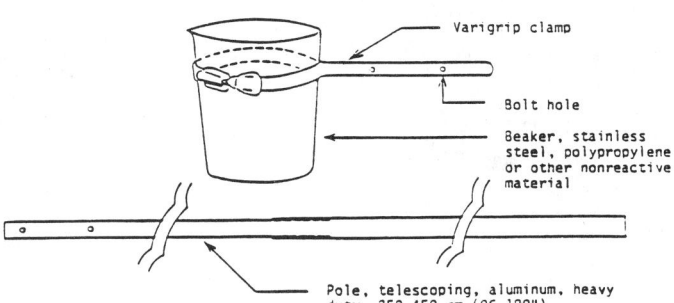

FIG. 1   Dipper Sampler (Source: Test Methods for Evaluating Solid Wastes, U.S. EPA, S2–846, July 1982)

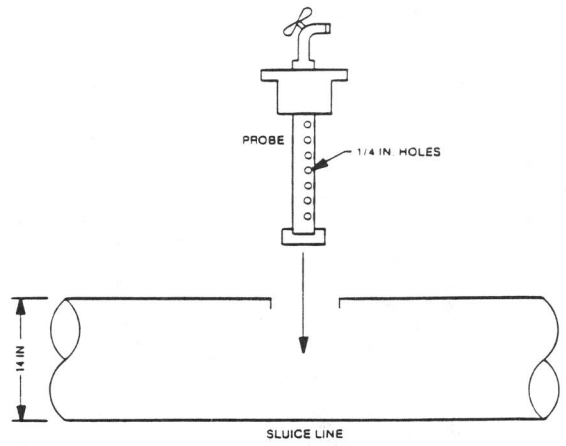

FIG. 2   In–Pipe Sampling Probe

9.1.1 Name and location of site,

9.1.2 Date and time of sampling,

9.1.3 Location of sampling,

9.1.4 Sample number,

9.1.5 Description and disposition of sample,

9.1.6 Name of sampling personnel,

9.1.7 Type of preservative, and

9.1.8 Sampling conditions, and analytical requirements.

9.2 Pack the sample container securely in a shipping container. If required for preservation of analytes, the sample container should be packed in ice and cooled to 4°C. A minimum-maximum thermometer should be packed with the samples.

9.3 Follow DOT (Department of Transportation) shipping regulations.

9.4 Make arrangements for handling, logging in, adequate storage and analysis of the sample at its destination. If warranted, follow chain-of-custody protocol.

PRACTICE B—SOLID OR SEMISOLID DISCHARGES

10. Apparatus

10.1 *Scoop or Shovel Sampler*—For solid or semisolid discharges from belts, a scoop or shovel should be used, made of material compatible with the waste stream and inert to the analytes. Where routine sampling is to be performed, a scoop may be designed to match the width and contour of the belt. In this way, a single time increment sample may be taken in one scoop.

11. Procedure

11.1 Clean the scoop or shovel as appropriate for the desired chemical analysis to prevent cross contamination. In some cases, it may be necessary to dedicate equipment to specific waste types or waste strata.

11.2 Sample at a convenient point along the belt.

11.3 Sample the waste with the scoop or shovel making sure to sample across the entire width of the belt. Be sure all fines and any liquid are also included representatively in the scooped sample.

11.4 Repeat sampling depending on the uniformity of the waste. Composite the grab samples and transfer a well-mixed portion to a chemically compatible container and seal. It may be preferable or necessary to have compositing performed in the laboratory. Include specific procedures for compositing in the sampling plan.

11.5 If compositing has not been appropriately accomplished prior to shipment, clearly indicate on the sample label, in the field log book, and on the analytical request form.

12. Sampling Labeling and Shipping

12.1 Refer to Section 9.

Standard Practice for
Sampling Waste and Soils for Volatile Organics[1]

This standard is issued under the fixed designation D 4547; the number immediately following the designation indicates the year of original adoption or, in the case of revision, the year of last revision. A number in parentheses indicates the year of last reapproval. A superscript epsilon (ε) indicates an editorial change since the last revision or reapproval.

1. Scope

1.1 This practice describes field sampling of solid wastes for subsequent volatile organics analysis in the laboratory. This practice is also intended to apply to soils and sediments that may contain volatile waste constituents.

1.2 Both the collection of the sample and the method of containing the sample for shipment to the laboratory are considered.

1.3 This practice concerns only sampling methods to be used in the field; it does not cover laboratory preparation of containers or solutions or other laboratory techniques related to processing or analysis of the samples.

1.4 It is recommended that this standard be used in conjunction with Guide D 4687.

1.5 *This standard does not purport to address all of the safety problems, if any, associated with its use. It is the responsibility of the user of this standard to establish appropriate safety and health practices and determine the applicability of regulatory limitations prior to use.* For specific precautionary statements, see Section 6.

2. Referenced Documents

2.1 *ASTM Standards:*
D 3550 Practice for Ring-Lined Barrel Sampling of Soils[2]
D 4687 Guide for General Planning of Waste Sampling[3]

3. Terminology

3.1 *Description of Terms Specific to This Standard:*

3.1.1 *material*—for the purposes of this practice, material covers any soil or wastes that are solid.

3.1.2 *sample*—the portion of the waste or soil material that is initially collected using the techniques described in this practice; portions of the waste or soil, in generic terms.

3.1.3 *subsample*—the portion of the waste or soil that is collected by subdividing or trimming of the initial sample.

3.1.4 *waste*—for the purposes of this practice, waste covers any discarded material that is solid in form.

4. Summary of Practice

4.1 Samples of soils or wastes can be obtained with minimal loss of volatile organic constituents. Materials may be either sampled from ground surface or test pits or obtained by using down-hole coring devices. These samples may be shipped in metal rings (that is, hollow metal cylinders) directly to the laboratory, or they may be subsampled by trimming or by using a small coring cylinder. With the coring method, the coring cylinder is driven into the waste or soil surface to remove the solid material without exposure to the air. The subsample is then extruded from the cylinder directly into a sample container. This method does not apply to cemented material or material with fragments coarse enough to interfere with proper coring techniques; such samples are trimmed before handling.

4.2 Subsamples obtained in the field are contained so as to prevent loss of volatiles using one of the two following methods: (1) the subsample is stored in a glass bottle with methanol; or (2) the subsample is stored in a vial designed to minimize loss of volatiles (for example, a specially adapted VOA vial). Advantages and disadvantages to both methods are discussed in Sections 7 and 8.

5. Significance and Use

5.1 The objective of this practice is to provide procedures for obtaining samples which will result in analytical data representative of the concentrations and compositions of volatile compounds actually present in the waste or soil. The procedure also allows for correlation of the analytical data with other properties of the waste or soil materials. Several factors are identified that influence the objective of this procedure.

5.1.1 *Loss of Volatiles:*

5.1.1.1 Loss of volatile organics during sample collection, handling, and shipping affects the concentrations detected by the laboratory. Comparison of the field testing of volatiles (using a gas chromatograph) with subsequent laboratory testing of the solids or ground water from the same zone suggests that losses can be significant, but are not necessarily due to one particular part of the sampling and analysis process. The principal mechanisms of loss are volatilization of the compounds and biodegradation. Susceptibility of the various compounds to these losses varies. Both the actual concentrations and the relative amounts of the compounds detected can be affected. In some cases, the loss of a compound or the formation of other compounds not actually present in the waste can occur. Compound gain and loss will result in analyses that are both unrepresentative of field conditions and subject to ambiguities in interpretation.

5.1.2 *Selection of Samples for Analysis*—The choice of representative samples is of particular concern in waste materials and soils in which heterogeneities are significant. Interpretation of the analytical data is generally improved if the individual(s) most familiar with the site can describe and select the sample(s) to be analyzed in the laboratory.

5.1.3 *Analytical Considerations*—The method of sample handling and containment is dependent on the method to be

[1] This practice is under the jurisdiction of ASTM Committee D-34 on Waste Disposal and is the direct responsibility of Subcommittee D34.01 on Sampling and Monitoring.
Current edition approved Aug. 15, 1991. Published October 1991.
[2] *Annual Book of ASTM Standards*, Vol 04.08.
[3] *Annual Book of ASTM Standards*, Vol 11.04.

used in the laboratory for analysis of the volatile compounds. The laboratory methods are addressed here only insofar as they affect the collection method and influence the objective stated above.

5.1.4 Other factors affecting the interpretation of data are as follows: sample size, sample matrix, whether the subsample analyzed is representative of the entire sample, potential losses during handling of the sample using the laboratory procedure, and detection limit.

5.2 This practice should be used in conjunction with Practice D 3550 and Guide D 4687.

6. Safety

6.1 Proper safety precautions must always be observed when sampling solid waste or contaminated soil. For general guidelines to safety precautions, refer to Guide D 4687 and Practice D 3350. These standards, however, should only complement the judgement of an experienced professional.

7. Sampling

7.1 *Introduction:*

7.1.1 This section is intended to define general sampling guidelines to be applied to a variety of possible materials and conditions. Many of the specific materials and conditions, however, are not addressed in this document. Specific sampling methods are presented for granular materials (for example, contaminated soils and non-cemented solid wastes) and materials that are cemented or of sufficient cohesion to make driven samplers impractical. The procedures are intended to allow flexibility in the following:

7.1.1.1 The means of collection (for example, from test pits, surface sampling, and sampling during drilling),

7.1.1.2 The selection of a method suited for the individual requirements of a project or the conditions encountered at a particular site, and

7.1.1.3 The design and dimensions of the actual sampling equipment.

7.2 *General Methods:*

7.2.1 The sampling procedure should be completed in a minimum amount of time, with the least possible handling of the sample before it is sealed in a container.

7.2.2 Rough trimming of the sample in the field should be considered if cross-contamination of the surface of the sample from other waste or soil strata is likely to occur during collection. Significant contamination of the sample surfaces can lead to redistribution of the volatiles throughout the sample during shipping and storage, which will result in misleading analytical data. The reduction of surface contamination errors by trimming should be balanced with the potential losses from volatilization during trimming operations.

7.2.3 If possible, the sample should be inspected visually and its characteristics logged. Adjacent samples that appear to have similar physical properties should be retained for testing to determine or verify the relevant properties of the solid materials (for example, grain size distribution). Ideally, the sample itself or another sample of the same material should be available for inspection and notation of the following: general appearance and color, presence of oils or other visible signs of contamination, grain-size distribution, volatile organics, and so forth. In the case of samples that are

collected directly into metal rings for shipment to the laboratory such inspection may not be possible (see 7.3.3), and alternative procedures (examination of the exposed ends of the core or adjacent samples) should be used. Selection of representative samples for volatiles analysis is aided greatly by information obtained from field testing of other samples from the same stratum.

7.3 *Granular or Uncemented Materials:*

7.3.1 Granular materials may be collected from the ground surface, the walls of test pits, blocks of wastes or soils, or by using split barrel or other sampling devices during the drilling of soil borings.

7.3.2 In the case of samples collected from test pits, ground surface, or larger blocks of soils or waste, the surface of the sample should be trimmed to remove contaminants from other waste or soil strata or to remove surface layers that may have already lost volatiles. This removal of surface layers can be accomplished by scraping the surface using a clean spatula or knife.

7.3.3 *Collection of Samples in Metal Rings:*

7.3.3.1 Samples from split barrels or similar devices may be collected in precleaned metal rings inserted in the sampling barrel, such as those described in Practice D 3550. The exposed ends of the solid or waste in the ring or in adjacent samples are used to log the sampled materials. The ends of the ring containing the soil or waste are then covered quickly and sealed in the field using an inert material (for example, TFE-fluorocarbon sheets or aluminum foil with tightly wrapped sealing tape or screw-on metal caps). Extrusion of a subsample for analysis is accomplished by the analytical laboratory and does not allow for logging or field testing of interior layers of the sample by on-site personnel.

7.3.3.2 This method should not be used in the case of poor sample recovery (that is, when the metal rings are not completely filled with the material to be sampled), due to the potential losses of compounds by volatilization into the headspace of the rings.

7.3.4 *Subsampling in the Field Using Metal Coring Cylinders:*

7.3.4.1 Samples taken from test pits, ground surface, or larger blocks of soils or wastes, or samples that have been removed from down-hole sampling devices, are subsampled using a small metal coring cylinder. Removal of the subsample to be sent to the laboratory is accomplished using a cleaned metal cylinder, open at both ends, which is driven into the solid material. This method allows the field sampling personnel to inspect the surrounding solid material, log its properties, and perform field tests on the excess sample. However, this method may be impractical for certain types of solid materials that are difficult to core or extrude.

7.3.4.2 The metal coring cylinder used for subsampling should be sharpened by grinding to allow greater ease in driving the sampler. The optimum diameter of the cylinder depends on the following: size of the sampling jar, dimensions of the original sample, particle size of the solid materials (for example, gravel size particles would require larger samplers), and volume of subsample required. It is anticipated that a number of cylinders of different diameters should be available to the field personnel for selection of the optimum size.

7.3.4.3 In general, the outside diameter of the coring

cylinder should be smaller than the inside diameter of the mouth of the sampling container to avoid contamination of the outside threads of the bottle, which may result in a bad seal.

7.3.4.4 The coring cylinder containing the subsample of material for analysis can be removed by excavation (surface or test pit) or by cleaning away the excess sample with a spatula or a clean, disposable towel. The solid materials around the cylinder are used (1) to log the properties of the sample, (2) to aid in determining whether the subsample is representative of the horizon to be sampled, or (3) for additional tests (for example, grain size analysis, field testing of total volatiles, field gas chromatography). If the subsample is not extruded from the cylinder immediately, it should be sealed temporarily, by covering the ends with aluminum foil and TFE-fluorocarbon tape, and stored on ice or similarly cooled.

NOTE 1—Aluminum foil may be unsuitable in very alkaline environments.

7.3.4.5 The subsample is extruded using a cleaned rod to push the subsample out of the cylinder. The subsample is extruded directly into the sampling container. If the subsample is to be placed in a 40-mL vial, ideally, the length of subsample collected in the cylinder should be greater than the height of the vial, so that the vial can be filled in one operation. Extrusion should be performed rapidly and soon after sampling to reduce volatilization and redistribution of volatiles, which may result from contaminated subsample ends.

NOTE 2—Extrusion of cooled subsamples under controlled conditions is preferred if it can be performed on-site or within a short period of time (that is, four to six h) after sample collection.

7.4 *Cemented Solid Wastes:*

7.4.1 *Subsampling Cemented Material by Trimming*— The solid wastes or contaminated soils may be so hard that the coring cylinder cannot be driven into the wastes. Subsamples of such wastes and soils may be collected by trimming the larger sample with a cleaned tool to a size that can be placed in the sampling jar. Although some loss of volatiles can be expected, the losses should be less than in loose, granular solids due to the lower surface area exposed to the atmosphere. Collection, trimming, and containment of the subsample should be accomplished in the least amount of time practical.

8. Handling

8.1 *General:*

8.1.1 In the case of materials not subsampled in the field (that is, those collected in metal rings inserted in down-hole sampling devices), the sample is retained in the metal ring and sealed as described in 6.3.3. Both the sample and the metal ring are then shipped to the laboratory, where the subsample is extruded for analysis.

8.1.2 In the case of materials subsampled in the field, glass containers with inert caps should be used for storage and shipment. To retard volatilization and biodegradation, subsamples should be placed on ice or similarly cooled as soon as practical. Two alternative methods of containing subsamples for shipment to the laboratory are outlined.

These methods have different applications, and advantages and disadvantages.

8.2 *Method 1—Methanol Container:*

8.2.1 This method can be used for a wide range of cases, but it is particularly useful for situations in which (1) larger samples are desired to obtain a composite representation of the volatiles concentration and composition, (2) high detection limits can be tolerated, or (3) biodegradation is a concern.

8.2.2 Sample containers consist of wide mouth, 8-oz. glass jars with TFE-fluorocarbon-lined lids. An aliquot of 100 mL of an appropriate analytical grade of methanol is added to the organic-free jar by the laboratory that supplies the jar, by the sample collector, or by a third party. The solid waste or soil is added to the jar containing the methanol to a specified level in the jar. This level is defined by the party responsible for preparing the sampling jars and is equivalent to the level that would correspond to the addition of approximately 100 g of the soil or waste (at an assumed specific gravity). The actual mass of material added to the jar is determined later by comparison with a tare weight, by the analytical laboratory. The jar containing the methanol should not be left open unnecessarily.

8.2.3 The volatile compounds are more soluble in the methanol than in the soil water, which results in transfer of the volatiles from the solid to the methanol for analysis. Addition of methanol to the subsample in the field allows a longer contact time with the subsample, which improves the extraction efficiency. In addition, extraction of volatiles is performed with a larger subsample than used in some methods (100 g versus 5 g with a heated He purge), which results in a more representative determination of the concentrations of the volatiles.

8.2.4 This method permits splitting of the sample into several jars, or compositing by placing several aliquots of the solid waste (from the coring cylinder described above) in one sample jar. The methanol reduces volatilization during repeated opening and closing of the jar for each subsample and serves as a medium for extracting volatiles from each subsample added to the jar. The physical mixing used in other types of analyses, with its potential for volatilization and incomplete mixing, is avoided with this method. This method also allows for multiple laboratory analyses of the same sample.

8.2.5 Since the partial pressure of the volatiles over the methanol is very low, losses by volatilization are reduced. The methanol also inhibits microbial activity, reducing losses from biodegradation.

8.2.6 The primary disadvantages of this method are (1) the need for laboratory cooperation in preparing tared sample jars, (2) possible shipping restrictions (if the methanol volume is sufficient to qualify the samples as flammable materials), and (3) the reduction of sensitivity of the gas chromatograph/Hall detector (if such detectors are used by the method).

8.3 *Method 2—Dry Container:*

8.3.1 This method involves placement of a subsample of the solid in a tared 40-mL glass container (or a size compatible with the analytical instrumentation) for shipment to the laboratory. This container is modified by the addition of a cap that allows direct connection of the container with the purge and trap device in the laboratory, so that removal of the subsample is not required for analysis. Subsample weights are determined in the laboratory.

8.3.2 The samples are extruded into the clean, organic-free jar from the coring cylinder, ideally in one operation, to minimize opening and closing of the jar and the potential loss from volatilization. This method is preferable for cases in which the methanol method is not desired (due to shipping limitations, detection limit requirements, or laboratory restrictions) or if only small samples are available for analysis (for example, if only a thin horizon of contaminated material exists or a limited zone of contaminated material is the target for analysis).

8.3.3 The primary disadvantages of this method are (1) the requirement for specialized containers, (2) the inability to perform additional analyses of the same sample, and (3) the small size of the sample, which can reduce the representativeness of the sample.

9. Packaging and Package Marking

9.1 An indelible label identifying the sample should be secured to the container. The label should contain or reference the following information:

9.1.1 Name and location of site,

9.1.2 Date and time of sampling,

9.1.3 Location of sampling,

9.1.4 Sample number,

9.1.5 Description and disposition of sample,

9.1.6 Name of sampling personnel,

9.1.7 Type of preservative, and

9.1.8 Sampling conditions and analytical requirements.

9.2 Pack the sample container securely in a shipping container. The sample container should be packed in ice and cooled to 4°C. A min/max thermometer should be packed with the samples.

9.3 Follow DOT (Department of Transportation)[4] shipping regulations.

9.4 Make arrangements for handling, logging in, adequate storage and analysis of the sample or subsample at its destination. If warranted, follow chain-of-custody protocol.

---

[4] Available from the Superintendent of Documents, U.S. Government Printing Office, Washington, DC 20402.

Standard Practices for
Preparation of Sample Containers and for Preservation of Organic Constituents[1]

This standard is issued under the fixed designation D 3694; the number immediately following the designation indicates the year of original adoption or, in the case of revision, the year of last revision. A number in parentheses indicates the year of last reapproval. A superscript epsilon (ε) indicates an editorial change since the last revision or reapproval.

1. Scope

1.1 These practices cover the various means of (*1*) preparing sample containers used for collection of waters to be analyzed for organic constituents and (*2*) preservation of such samples from the time of sample collection until the time of analysis.

1.2 The sample preservation practice is dependent upon the specific analysis to be conducted. See Section 9 for preservation practices listed with the corresponding applicable general and specific constituent test method. The preservation method for waterborne oils is given in Practice D 3325. Use of the information given herein will make it possible to choose the minimum number of sample preservation practices necessary to ensure the integrity of a sample designated for multiple analysis. For further considerations of sample preservation, see the *Manual on Water*.[2]

1.3 *This standard does not purport to address all of the safety concerns, if any, associated with its use. It is the responsibility of the user of this standard to establish appropriate safety and health practices and determine the applicability of regulatory limitations prior to use.* For specific hazard statements, see 6.7, 6.24, and 8.1.3.

2. Referenced Documents

2.1 *ASTM Standards:*
D 1129 Terminology Relating to Water[3]
D 1193 Specification for Reagent Water[3]
D 1252 Test Methods for Chemical Oxygen Demand (Dichromate Oxygen Demand) of Water[4]
D 1783 Test Methods for Phenolic Compounds in Water[4]
D 2036 Test Methods for Cyanides in Water[4]
D 2330 Test Method for Methylene Blue Active Substances[4]
D 2579 Test Methods for Total and Organic Carbon in Water[4]
D 2580 Test Method for Phenols in Water by Gas-Liquid Chromatography[4]
D 2908 Practice for Measuring Volatile Organic Matter in Water by Aqueous-Injection Gas Chromatography[4]
D 3113 Test Methods for Sodium Salts of EDTA in Water[4]

D 3325 Practice for Preservation of Waterborne Oil Samples[4]
D 3371 Test Method for Nitriles in Aqueous Solution by Gas-Liquid Chromatography[4]
D 3534 Test Method for Polychlorinated Biphenyls (PCBs) in Water[4]
D 3590 Test Methods for Total Kjeldahl Nitrogen in Water[4]
D 3695 Test Method for Volatile Alcohols in Water by Direct Aqueous-Injection Gas Chromatography[4]
D 3871 Test Method for Purgeable Organic Compounds in Water Using Headspace Sampling[4]
D 3921 Test Method for Oil and Grease and Petroleum Hydrocarbons in Water[4]
D 3973 Test Method for Low-Molecular Weight Halogenated Hydrocarbons in Water[4]
D 4129 Test Method for Total and Organic Carbon in Water by High-Temperature Oxidation and Coulometric Detection[4]
D 4165 Test Method for Cyanogen Chloride in Water[4]
D 4193 Test Method for Thiocyanate in Water[4]
D 4281 Test Method for Oil and Grease (Fluorocarbon Extractable Substances) by Gravimetric Determination[4]
D 4282 Test Method for Determination of Free Cyanide in Water and Wastewater by Microdiffusion[4]
D 4374 Test Methods for Cyanide in Water—Automated Methods for Total Cyanide, Dissociable Cyanide, and Thiocyanate[4]
D 4515 Practice for Estimation of Holding Time for Water Samples Containing Organic Constituents[4]
D 4657 Test Method for Polynuclear Aromatic Hydrocarbons in Water[4]
D 4744 Test Method for Organic Halides in Water by Carbon Adsorption Microcoulometric Detection[4]
D 4763 Practice for Identification of Chemicals in Water by Fluorescence Spectroscopy[4]
D 4779 Test Method for Total, Organic, and Inorganic Carbon in High Purity Water by Ultraviolet (UV) or Persulfate Oxidation, or Both, and Infrared Detection[4]
D 4839 Test Method for Total Carbon and Organic Carbon in Water by Ultraviolet, or Persulfate Oxidation, or Both, and Infrared Detection[4]
D 4841 Practice for Estimation of Holding Time for Water Samples Containing Organic and Inorganic Constituents[3]
D 4983 Test Method for Cyclohexylamine, Morpholine, and Diethylaminoethanol in Water and Condensed Steam by Direct Aqueous Injection Gas Chromatography[4]

---

[1] These practices are under the jurisdiction of ASTM Committee D-19 on Water and are the direct responsibilities of Subcommittee D19.06 on Methods for Analysis for Organic Substances in Water.
Current edition approved Dec. 10, 1996. Published March 1997. Originally published as D 3694 – 78. Last previous edition D 3694 – 95.
[2] *Manual on Water*, ASTM STP 442, ASTM, 1969.
[3] *Annual Book of ASTM Standards*, Vol 11.01.
[4] *Annual Book of ASTM Standards*, Vol 11.02.

D 5175 Test Method for Organohalide Pesticides and Polychlorinated Biphenyls in Water by Microextraction and Gas Chromatography[4]

D 5176 Test Method for Total Chemically Bound Nitrogen in Water by Pyrolysis and Chemiluminescence Detection[4]

D 5315 Test Method for N-Methyl-Carbamoyloximes and N-Methylcarbamates in Water by Direct Aqueous Injection HPLC with Post-Column Derivation[4]

D 5316 Test Method for 1,2-Dibromoethane and 1,2-Dibromo-3-Chloropropane in Water by Microextraction and Gas Chromatography[4]

D 5317 Test Method for the Determination of Chlorinated Organic Acid Compounds in Water by Gas Chromatography with an Electron Capture Detector[4]

D 5412 Test Method for Quantification of Complex Polycyclic Aromatic Hydrocarbon Mixtures or Petroleum Oils in Water[4]

D 5475 Test Method for Nitrogen and Phosphorus Containing Pesticides in Water by Gas Chromatography with a Nitrogen-Phosphorus Detector[4]

D 5790 Test Method for Measurement of Purgeable Organic Compounds in Water by Capillary Column Gas Chromatography/Mass Spectrometry[4]

D 5812 Test Method for Determination of Organochlorine Pesticides in Water by Capillary Column Gas Chromatography[4]

3. Terminology

3.1 *Definitions*—For definitions of terms used in this practice, refer to Terminology D 1129.

4. Significance and Use

4.1 There are four basic steps necessary to obtain meaningful analytical data: preparation of the sample container, sampling, sample preservation, and analysis. In fact these four basic steps comprise the analytical method and for this reason no step should be overlooked. Although the significance of preservation is dependent upon the time between sampling and the analysis, unless the analysis is accomplished within 2 h after sampling, preservation is preferred and usually required.

5. Apparatus

5.1 *Forced Draft Oven*, capable of operating at 275 to 325°C.

5.2 *Sample Bottle*, borosilicate or flint glass.

NOTE 1—High density polyethylene (HDPE) bottles and caps have been demonstrated to be of sufficient quality to be compatible for all tests except pesticides, herbicides, polychlorinated biphenyls, and volatile organics. However, this bottle cannot be recycled.

5.3 *Sample Bottle Cap*, TFE-fluorocarbon or aluminum foil-lined.

NOTE 2—Even these liners have some disadvantages. TFE is known to collect some organic constituents, for example, PCBs. Aluminum foil will react with samples that are strongly acid or alkaline. Clean TFE liners as described in 7.1. Replace aluminum foil with new foil after each use.

5.4 *Sample Vial*, glass.

5.5 *Septa*, PTFE-faced with screw cap lid and matching aluminum foil disks.

6. Reagents and Materials

6.1 *Purity of Reagents*—Reagent grade chemicals shall be used in all tests. Unless otherwise indicated, it is intended that all reagents shall conform to the specifications of the Committee on Analytical Reagents of the American Chemical Society.[5] Other grades may be used, provided it is first ascertained that the reagent is of sufficiently high purity to permit its use without lessening the accuracy of the determination.

6.2 *Purity of Water*—Unless otherwise indicated, reference to water shall be understood to mean reagent water conforming to Specification D 1193, Type II and demonstrated to be free of specific interference for the test being performed.

6.3 *Acetic Acid Buffer Solution* (pH 4)—Dissolve 6.0 g of sodium acetate in 75 mL of water. Add 30 mL of glacial acetic acid, with stirring.

6.4 *Acetone*.

6.5 *Acid Buffer Solution* (pH 3.75)—Dissolve 125 g of potassium chloride and 70 g of sodium acetate trihydrate in 500 mL of water. Add 300 mL of glacial acetic acid and dilute to 1 L.

6.6 *Ascorbic Acid*.

6.7 *Chromic Acid Cleaning Solution*—To a 2-L beaker, add 35 mL of saturated sodium dichromate solution followed by 1 L of sulfuric acid (sp gr 1.84) with stirring. **Warning**—Use rubber gloves, safety goggles, and protective clothing when preparing and handling this corrosive cleaning agent that is a powerful oxidant. Store the reagent in a glass bottle with a glass stopper.

6.8 *Detergent*, formulated for cleaning laboratory glassware.

6.9 *Hydrochloric Acid*—Concentrated HCl (sp gr 1.19).

6.10 *Hydrochloric Acid* (1+2)—To 200 mL of water, carefully add 100 mL of hydrochloric acid (see 6.9). Store in a glass-stoppered reagent bottle.

6.11 *Ice*, crushed wet.

6.12 *Lead Acetate Test Paper*.

6.13 *Lead Acetate Solution*—Dissolve 50 g of lead acetate in water and dilute to 1 L.

6.14 *Lead Carbonate*, powdered.

6.15 *Lime, Hydrated*, powdered.

6.16 *Mercuric Chloride*.

6.17 *Monochloroacetic Acid Buffer* (pH 3)—Prepare by mixing 156 mL of chloroacetic acid solution (236.2 g/L) and 100 mL of potassium acetate solution (245.4 g/L).

6.18 *Nitric Acid*—Concentrated HNO_3 (sp gr 1.42).

6.19 *Phosphate Buffer*—Dissolve 138 g of sodium dihydrogen phosphate in water and dilute to 1 L. Refrigerate this solution.

6.20 *Phosphate Solution*—Dissolve 33.8 g of potassium dihydrogen phosphate in 250 mL of water.

6.21 *Phosphoric Acid*—Concentrated H_3PO_4 (sp gr 1.83).

[5] *Reagent Chemicals, American Chemical Society Specifications*, American Chemical Society, Washington, DC. For suggestions on the testing of reagents not listed by the American Chemical Society, see *Analar Standards for Laboratory Chemicals*, BDH Ltd., Poole, Dorset, U.K., and the *United States Pharmacopeia and National Formulary*, U.S. Pharmacopeial Convention, Inc. (USPC), Rockville, MD.

TABLE 1 Recommended Preservation Practice for General Organic Constituent Test Methods

NOTE—The container preparation procedures described in Sections 7 and 8 should yield bottles of sufficient quality to be compatible with the test methods listed in this table. However, a sample bottle blank should be obtained to establish the fact.

Test Method(s)	Recommended Practice
D 1252, Oxygen demand, chemical	9.1.1.1 sulfuric acid or 9.1.1.3 sodium bisulfate or 9.1.3 refrigeration
D 2579, Organic carbon, total, by combustion-infrared or reduction-FID	9.1.3 refrigeration
D 3921, Oil and grease, petroleum hydrocarbons	9.1.1.1 sulfuric acid (1+1) or 9.1.1.3 sodium bisulfate
D 4129, Total and organic carbon, oxidation coulometric	9.1.3 refrigeration and 9.1.4 hermetically sealing, zero headspace
D 4281, Oil and grease, gravimetric	9.1.1.1 sulfuric acid or 9.1.1.2 hydrochloric acid
D 4744, Organic halides, by carbon absorption-microcoulometry	9.1.1.8 nitric acid and 9.1.2 sodium sulfite and 9.1.3 refrigeration
D 4763, Identification by fluorescence	9.1.3 refrigeration
D 4779, Carbon, total, organic and inorganic in high-purity water	9.1.1.8 nitric acid and 9.1.3 refrigeration
D 4839, Carbon, total and organic in water	9.1.3 refrigeration and either 9.1.1.1[A] sulfuric acid or 9.1.1.5 phosphoric acid or 9.1.1.8 nitric acid
D 5176 Nitrogen, total chemically-bound, pyrolysis-chemiluminescence	9.1.1.1 sulfuric acid or 9.1.1.2 hydrochloric acid, and 9.1.3 refrigeration
PS 48, Oil and grease (solvent extractable substances) by gravimetric determination	9.1.1.1 sulfuric acid (1+1) or 9.1.1.2 hydrochloric acid (1+1)

[A] Acidification can be used only when organic carbon alone is being determined. If total carbon is of interest, the sample must not be acidified; refrigeration is the only appropriate preservation technique.

6.22 *Phosphoric Acid Solution* (1+1)—Dilute 1 vol of phosphoric acid (sp gr 1.83).

6.23 *pH Paper*, narrow range for pH < 2, pH > 12, and pH 5 to 7.

6.24 *Potassium Iodide–Starch Test Paper.*

6.25 *Sodium Bisulfate.*

6.26 *Sodium Bisulfite Solution*—Dissolve 2 g of sodium bisulfite in 1 L of water and adjust to pH 2 by the slow addition of H_2SO_4 (1+1). **Warning**—Prepare and use this reagent in a well ventilated hood to avoid exposure to SO_2 fumes.

6.27 *Sodium Sulfite Solution* (0.1 M)—Transfer approximately 10.3 g of sodium sulfite to a 1-L volumetric flask. Dilute to volume with water.

6.28 *Sodium Thiosulfate.*

6.29 *Sodium Hydroxide Pellets.*

6.30 *Mercuric Chloride* (10 mg/mL)—Dissolve 100 mg of $HgCl_2$ in reagent water and dilute to 10 mL.

6.31 *Sulfuric Acid* (1+1)—Slowly and carefully add 1 vol of sulfuric acid (see 6.27) to 1 vol of water, stirring and cooling the solution during addition.

7. Preparation of HDPE Sample Bottles

7.1 Wash the bottles with two 100-mL portions of HCl (1+2) and rinse with three 100-mL portions of water. These volumes of wash and rinse portions are recommended for 1-L sample bottles; therefore, use proportionate volumes for washing and rinsing sample bottles of a different volume.

8. Preparation of Glass Sample Bottles and Vials

8.1 *Solvent-Detergent/Chromic Acid Preparation of Glass Sample Bottles:*

8.1.1 Rinse the container with 100 mL of dilute detergent or acetone. For some residues, a few alternative detergent and acetone rinses may be more satisfactory. Then rinse at least three times with tap water followed by a reagent water rinse to remove the residual detergent or acetone, or both.

8.1.2 Rinse the container with 100 mL of chromic acid solution, returning the chromic acid to its original container after use. Then rinse with at least three 100-mL portions of tap water followed by a reagent water rinse.

8.1.3 Rinse the container with 100 mL of $NaHSO_3$ solution to remove residual hexavalent chromium. **Warning**—Carry out this step in a hood to prevent exposure to SO_2 fumes.

8.1.4 Rinse the container with water until sulfurous acid and its vapors have been removed. Test rinsings for acid with a pH meter or an appropriate narrow range pH paper. Rinsings should have a pH approximately the same as the water used for rinsing.

8.1.5 When the last trace of $NaHSO_3$ has been removed, wash with three additional 100-mL portions of water. Allow to drain. This procedure is for 1-L sample containers, therefore, use proportionate volumes for washing and rinsing sample containers of a different volume.

8.1.6 Heat for a minimum of 4 h (mouth up) in a forced draft oven at 275 to 325°C. Upon cooling, fit the bottles with caps and the vials with septa.

NOTE 3—For some tests, heating may not be required. Refer to the individual method to determine the necessity for this treatment.

8.2 *Machine Washing Glass Sample Bottles and Vials:*

NOTE 4—Machine washing of narrow mouth sample bottles may not yield acceptable results.

8.2.1 Rinse the container with 100 mL of chromic acid solution, returning the chromic acid to its original container after use. Then rinse with at least three 100-mL portions of tap water.

8.2.2 Machine wash in accordance with the machine manufacturer's instructions using a detergent and 90°C water.

8.2.3 Remove the bottles from the machine and rinse them with two 100-mL portions of HCl (1+2), followed with three 100-mL portions of water.

8.2.4 Heat for a minimum of 4 h (mouth up) in a forced draft oven at 275 to 325°C. Upon cooling, fit the bottles with caps and the vials with septa (see Note 3).

9. Sample Preservation

9.1 Depending upon the type of analysis required, use any one or a combination of the following methods of sample preservation (see Tables 1, 2, and 3 and Annexes A1 and A2).

9.1.1 Adjust the pH. An adjustment to neutral pH is usually prescribed when chemical reactions, such as hydrolysis, are to be avoided. Adjustment to an extreme pH, for example, <2, is usually prescribed to inhibit biological activity for biodegradable organic chemicals.

NOTE 5—To confirm the adjustment of the pH of samples to the proper value, place a drop of sample on an appropriate pH test paper or measure with a pH meter.

9.1.1.1 *Sulfuric Acid*—To the sample bottle partially filled with sample, slowly add 2 mL of H_2SO_4 (sp gr 1.84) and mix

TABLE 2 Recommended Preservation Practice for Specific Organic Constituent Test Methods

Test Method(s)	Recommended Practice
D 1783 Phenolic compounds by 4-AAP	9.1.3 refrigeration and 9.1.1.1 sulfuric acid or 9.1.1.3 sodium bisulfate or 9.1.1.2 hydrochloric acid or 9.1.1.5 phosphoric acid
D 2036[A] Cyanide	9.1.1.4 sodium hydroxide and in presence of chlorine 9.1.2 chlorine removal
D 2330 Alkyl benzene sulfonate	9.1.1.1 sulfuric acid or 9.1.1.3 sodium bisulfate
D 2580 Phenols by gas liquid chromatography	9.1.3 refrigeration
D 2908 Volatile organic matter in water by aqueous injection gas chromatography (DAIGC)	9.1.3 refrigeration and 9.1.1.1 sulfuric acid or 9.1.1.3 sodium bisulfate
D 3086 Pesticides, organochlorine	9.1.3 refrigeration
D 3113 Ethylenediaminetetraacetate	9.1.3 refrigeration
D 3371 Nitriles by DAIGC	9.1.3 refrigeration and 9.1.1.1 sulfuric acid or 9.1.1.3 sodium bisulfate
D 3534 Polychlorinated biphenyls	9.1.3 refrigeration
D 3590 Total nitrogen, Kjeldahl	9.1.3 refrigeration and 9.1.1.1 sulfuric acid
D 3695 Volatile alcohols by DAIGC	9.1.3 refrigeration and 9.1.1.1 sulfuric acid or 9.1.1.3 sodium bisulfate
D 3871 Purgeable organic compounds	9.1.2 chlorine removal, 9.1.3 refrigeration, 9.1.4 hermetically sealing
D 3973 Low molecular weight hydrocarbons	9.1.3 refrigeration, and in presence of chlorine 9.1.2 chlorine removal and 9.1.4 hermetically sealing, zero headspace
D 4165 Cyanogen chloride	9.1.1.6 phosphate buffer to ph 8.0 to 8.5
D 4193 Thiocyanate	9.1.1.2 acid or 9.1.1.4 base
D 4282 Free cyanide	9.1.3 refrigeration, 9.1.1.4 sodium hydroxide, and 9.1.5 in dark
D 4374 Total dissociable cyanide, automated	9.1.3 refrigeration and 9.1.1.4 sodium hydroxide, in dark
D 4657 Polynuclear aromatic hydrocarbons	9.1.3 refrigeration, 9.1.1.2 acid, or 9.1.1.4 sodium hydroxide (to pH 6.0 to 8.0), 9.1.2 chlorine removal
D 4983 Cyclohexamine, morpholine, diethanolamine by DAI	9.1.1.9 phosphate/phosphoric acid and 9.1.3 refrigeration.
D 5175 Organohalide pesticides and PCBs	9.1.2 chlorine removal and 9.1.3 refrigeration
D 5315 Carbamate pesticides	9.1.2 chlorine removal, 9.1.3 refrigeration
D 5316 EDB and DBCP	9.1.1.2 chlorine removal, 9.1.2 chlorine removal, 9.1.3 refrigeration, hermetic seal
D 5317 Chlorinated Acids	9.1.2 chlorine removal, 9.1.3 refrigeration, 9.1.6 mercuric chloride
D 5412 PAH mixtures	9.1.3 refrigeration (5C)
D 5475 Nitrogen/phosphorus pesticides	9.1.2 chlorine removal, 9.1.3 refrigeration, 9.1.6 mercuric chloride
D 5790 Purgeable organic compounds by GC/MS	9.1.1.2 hydrochloric acid, 9.1.2 chlorine removal, 9.1.3 refrigeration, 9.1.4 hermetic seal
D 5812 Organochlorine pesticides by GC	9.1.2 chlorine removal, 9.1.3 refrigeration, 9.1.6 mercuric chloride

[A] See Annex A1 for alternative treatment if the sample is suspected to contain sulfide or a high concentration of carbonate.

thoroughly. Confirm that the pH is less than 2. If the pH is greater than 2, add additional acid until the pH is less than 2. This procedure is based on a 1-L sample bottle; therefore, use proportionate volumes for sample bottles with a different volume.

9.1.1.2 *Hydrochloric Acid*—To a sample bottle partially filled with sample, add 6 mL of HCl (sp gr 1.19) while swirling the bottle. After the acid addition, confirm that the pH is less than 2. If the pH is greater than 2, add additional acid to lower the pH to less than 2. This procedure is for a 1-L sample bottle; therefore, use proportionate volumes for sample bottles with a different volume.

9.1.1.3 *Sodium Bisulfate*—To a sample bottle partially filled with sample, add approximately 9 g of $NaHSO_4$. Mix to dissolve and confirm that the pH is less than 2. If the pH is greater than 2, add additional $NaHSO_4$ until the pH is less than 2. This procedure is based on a 1-L sample bottle, therefore, use proportionate amounts for sample bottles with a different volume.

9.1.1.4 *Sodium Hydroxide*—Adjust the sample pH to above 12 using NaOH (pellets). Store the sample away from light.

9.1.1.5 *Phosphoric Acid*—To a sample bottle partially filled with sample, slowly add 2 mL of phosphoric acid (sp gr 1.83) and mix thoroughly. Confirm that the pH is less than 2. If the pH is greater than 2, add additional acid until the pH is less than 2. This procedure is based on a 1-L sample bottle; therefore, use proportionate volumes for sample bottles with a different volume.

9.1.1.6 *Phosphate Buffer*—Reduce pH to 8.0 to 8.5 range with careful additions of phosphate buffer.

9.1.1.7 *Acid Buffer*—Add 4 mL per 100 mL of sample.

9.1.1.8 *Nitric Acid*—To a sample bottle partially filled with sample, slowly add 2 mL of nitric acid (sp gr 1.42) and mix thoroughly. Confirm that the pH is less than 2. If the pH is greater than 2, add additional acid until the pH is less than 2. This procedure is based on a 1-L sample bottle; therefore, use proportionate volumes for sample bottles with a different volume.

9.1.1.9 *Phosphate/Phosphoric Acid*—Add approximately 1 mL phosphate solution followed by a few drops of 1+1 phosphoric acid solution to 115 mL of water sample to bring the pH to approximately 3.

9.1.1.10 *Monochloroacetic Acid Buffer*—Add 1.8 mL of monochloroacetic acid buffer solution (pH 3) to a 60-mL sample bottle prior to filling.

9.1.1.11 *Biocide*—Add mercuric chloride to the sample bottle in amounts to produce a concentration of 10 mg/L.

9.1.2 *Chlorine Removal*—Chlorine is added to water supplies and discharges as a disinfectant and oxidant for organic compounds. If the chlorine is not eliminated at the time of sampling, chlorination of organics present in the sample may occur; that is, trihalomethanes and chlorophenols will form, causing a positive interference for these analytes. Test a drop of the sample with potassium iodide-starch paper; a blue color indicates the need for the following treatment: add ascorbic acid, dissolving a few crystals at a time, until a drop of sample produces no color on the indicator paper. Then add an additional 0.05 g of ascorbic acid[6] per litre of sample. Alternatively, sodium sulfite solution (0.1 M) is used; typically, 0.2 mL/100 mL of sample is sufficient. Sodium thiosulfate (3 mg/40 mL) is used to remove chlorine from pesticide/PCB samples. Sodium thiosulfate (3 mg/40 mL) is used to remove chlorine from pesticide/PCB samples.

9.1.3 *Refrigeration at 4°C*— Samples are cooled to reduce biological activity on the organic chemicals. Cool the sample to 4°C immediately after sampling using a wet ice water bath. During storage or shipment, or both, maintain the sample at 4°C. Prior to the analysis, raise the sample temperature to room temperature using a water bath with a temperature no

---

[6] Do not use ascorbic acid when organic carbon is to be determined.

TABLE 3 Maximum Holding Times Allowed by ASTM Test Methods and EPA Regulations

Test Method(s)	Holding Times	
	In Standard	In 40 CFR 136[A]
D 1252 Oxygen demand, chemical	24 h if not acidified	28 days
D 1783 Phenolic compounds by 4-AAP	28 days	28 days
D 2036 Cyanide	None stated	NA
D 2330 Alkyl benzene sulfonate	7 days	NA[B]
D 2579 Organic carbon, total by combustion-infrared or reduction-FID	None stated	28 days
D 2580 Phenols by gas liquid chromatography	Keep to minimum	28 days
D 2908 Volatile organic matter in water by aqueous injection gas chromatography (DAIGC)	Keep to minimum	NA
D 3113 Ethylenediaminetetracetate	15 min recommended	NA
D 3371 Nitriles by gas liquid chromatography by DAIGC	Keep to minimum	NA
D 3534 Polychlorinated biphenyls	None stated	7 days for extraction, 40 days after extraction
D 3590 Total nitrogen, Kjeldahl	28 days	28 days
D 3695 Volatile alcohols by DAIGC	None stated	NA
D 3871 Purgeable organic compounds	None stated	14 days
D 3921 Oil and grease, petroleum hydrocarbons	None stated	28 days
D 3973 Low molecular weight hydrocarbons	15 days	14 days
D 4129 Total and organic carbon, oxidation coulometry	None stated	28 days
D 4165 Cyanogen chloride	Immediate analysis recommended	NA
D 4193 Thiocyanate	None stated	NA
D 4281 Oil and grease, gravimetric	Up to 2 months	28 days
D 4282 Free cyanide	Immediate analysis recommended	14 days
D 4374 Total dissociable cyanide, automated	Immediate analysis recommended	14 days
D 4657 Polynuclear aromatic hydrocarbons	7 days for extraction, 30 days for analysis	7 days for extraction, 40 days after extraction
D 4744 Organic halides, by carbon absorption-microcoulometry	7 days	NA
D 4763 Identification by fluorescence	None stated	NA
D 4779 Carbon, total, organic and inorganic, in high purity water	None stated	28 days
D 4839 Carbon, total and organic, in water	None stated	28 days
D 4983 Cyclohexamine, morpholine diethanolamine by DAI	None stated	NA
D 5175 Organohalide pesticides and PCBs	7 days	7 days
D 5176 Nitrogen chemically bound	None stated	NA
D 5315 N-methyl-carbamoyloximes and N-methylcarbamates by HPLC	28 days	NA
D 5316 1,2-dibromoethane and 1,2-dibromo-3-chloropropane by GC	28 days	NA
D 5317 Chlorinated organic acid compounds by GC	14 days for extraction, 28 days for analysis	7 days for extraction, 40 days after extraction
D 5475 Nitrogen and phosphorous containing pesticides by GC	14 days for extraction, 14 days for analysis	NA
D 5790 Purgeable organic compounds by GC/MS	14 days	14 days (refer to Part 136, Table II for exceptions)
D 5812 Organochlorine pesticides by GC	7 days for extraction, 14 days for analysis	7 days for extraction, 40 days after extraction

[A] Title 40, Code of Federal Regulations, Part 136 (40 CFR 136), by the U.S. Environmental Protection Agency.
[B] NA = Not applicable.

more than 5°C above room temperature. If the sample temperature is not adjusted, then an appropriate temperature-volume correction must be made.

9.1.4 *Hermetically Sealing (Purgeable Organics)*—Add required preservatives to a sample vial. Fill the vial to overflowing so that a convex meniscus forms at the top. Place a septum, PTFE side down, carefully on the opening of the vial, displacing the excess water. Seal the vial with the screw cap and invert to verify the seal by demonstrating the absence of air bubbles (zero headspace).

9.1.5 *Minimize Photodecomposition*—Collect samples in dark bottles and store in the dark.

9.1.6 *Mercuric Chloride*—Mercuric chloride (1 mL of a 10 mg/mL mercuric chloride solution) should be added to a 1-L sample bottle prior to sample collection if biological degradation of the target analytes may occur. Mercuric chloride is a highly toxic chemical and must be handled with caution. Samples containing mercuric chloride must be disposed of properly.

10. Sample Holding Times

10.1 Table 3 lists maximum holding times prescribed in ASTM standards for measuring organic compounds in water. The applicable holding times cited in the U.S. EPA "Guidelines Establishing Test Procedures for the Analysis of Pollutants Under the Clean Water Act"[7] are also included in Table 3.

10.2 Samples that exceed holding times should be discarded rather than analyzed.

10.3 Holding times for organic constituents are highly dependent upon the chemical and biological composition of the sample. Sample holding times for a specific matrix may be determined by using the procedure described in Practice D 4515 or D 4841. These practices are particularly useful when sampling, transporting, or scheduling complications make it desirable to have holding times beyond those prescribed in the test method.

11. Keywords

11.1 organic constituents; sample containers; sample preservation

---

[7] "Guidelines Establishing Test Procedures for the Analysis of Pollutants Under the Clean Water Act," Title 40, Code of Federal Regulations, Part 136.3(e), written by the U.S. Environmental Protection Agency, available from the Superintendent of Documents, U.S. Government Printing Office, Washington, DC 20401.

ANNEXES

(Mandatory Information)

A1. ALTERNATIVE TREATMENT STEPS

A1.1 The following is a list of additional or alternative treatment steps that are required in the presence of specific interfering constituents:

Test Method(s)	Interfering Constituent	Recommended Treatment for Preservation
D 2036 Cyanide	sulfide	Sulfide in the sample can convert CN^- to SCN^-, especially at high pH. Before stabilizing the sample by raising the pH for the preservation of the cyanide content, test for sulfide by placing a drop of sample on lead acetate test paper previously moistened with acetic acid buffer solution (pH 4). Darkening of the paper indicates the presence of sulfide. (The simultaneous presence of both sulfide and oxidizing agents is not anticipated. Oxidizing agents should be also removed before sample preservation.)
		Sulfide is removed by adding lead acetate solution (50 g/L) a drop at a time; retest on the test paper and continue until a negative paper test has been read, add 1 drop in excess and immediately filter out the black lead sulfide precipitate. If sulfide content is high, add instead powdered lead carbonate to avoid significant reduction of the pH.
		After sulfide removal, continue with 9.1.1.5.
	high carbonate content	When sampling effluents such as coal gasification wastes, atmospheric emission scrub waters, and other high carbonate content wastes, use hydrated lime in the powder form to stabilize the sample instead of NaOH. Add slowly with stirring to raise the pH to 12 to 12.5. Decant the sample into the sample bottle after the precipitate has been settled. (High carbonate content affects the distillation procedure by causing excessive gasing when the acid is added. The CO_2 released also may significantly reduce the NaOH content in the adsorbent.)

A2. PRESERVATION OF COMPOSITE SAMPLES

A2.1 When composite samples are collected, the appropriate preservation reagents must be added to the compositing vessel prior to collection. If the preservation requirements call for refrigeration, the sample must be refrigerated during collection. The collection time for a single composite sample shall not exceed 24 h. If longer sampling periods are necessary, a series of composite samples shall be collected.

Standard Practice for
Sampling Surface Soil for Radionuclides[1]

This standard is issued under the fixed designation C 998; the number immediately following the designation indicates the year of original adoption or, in the case of revision, the year of last revision. A number in parentheses indicates the year of last reapproval. A superscript epsilon (ε) indicates an editorial change since the last revision or reapproval.

[ε1] NOTE—Keywords were added editorially in November 1995.

1. Scope

1.1 This practice covers the sampling of surface soil for the purpose of obtaining a sample representative of a particular area for subsequent chemical analysis of selected radionuclides.

1.2 *This standard does not purport to address all of the safety concerns, if any, associated with its use. It is the responsibility of the user of this standard to establish appropriate safety and health practices and determine the applicability of regulatory limitations prior to use.*

2. Referenced Documents

2.1 *ASTM Standards:*
D 420 Guide to Site Characterization for Engineering, Design, and Construction Purposes[2]
D 1129 Terminology Relating to Water[3]
E 380 Practice for Use of the International System of Units (SI) (the Modernized Metric System)[4]

3. Terminology

3.1 *Definition:*
3.1.1 *sampling*—obtaining a representative portion of the material concerned (see Terminology D 1129).

4. Summary of Practice

4.1 Guidance is provided for the collection of soil samples to a depth of 50 mm. Ten core samples are collected in a specified pattern and composited to obtain sufficient sample so as to be representative of the area.

5. Significance and Use

5.1 Soil provides a source material for the determination of selected radionuclides and serves as an integrator of the deposition of airborne materials. Soil sampling should not be used as the primary measurement system to demonstrate compliance with applicable radionuclides in air standards. This should be done by air sampling or by measuring emission rates. Soil sampling does serve as a secondary system, and in many cases, is the only available avenue if insufficient air sampling occurred at the time of an incident. For many insoluble radionuclides, the primary exposure pathway to the general population is by inhalation. The resuspension of transuranic elements has received considerable attention (**1, 2**)[5] and their measurement in soil is one means of establishing compliance with the U.S. Environmental Protection Agency (EPA) guidelines on exposure to transuranic elements. Soil sampling can provide useful information for other purposes, such as plant uptake studies, total inventory of various radionuclides in soil due to atmospheric nuclear tests, and the accumulation of radionuclides as a function of time. A soil sampling and analysis program as part of a pre-operational environmental monitoring program serves to establish baseline concentrations. Consideration was given to these criteria in preparing this practice.

5.2 Soil collected by this practice and subsequent analysis is used to monitor radionuclide deposition of emissions from nuclear facilities. The critical factors necessary to provide this information are sampling location, time of sampling, frequency of sampling, sample size, and maintenance of the integrity of the sample prior to analysis. Since the soil is considered to be a heterogeneous medium, multipoint sampling is necessary. The samples must represent the conditions existing in the area for which data are desired.

6. Apparatus

6.1 *Sampling Instrument*[6]—In order to standardize the sample collection, it is suggested that the coring tool be that instrument used by golf courses to place the hole in the putting green. This instrument is commercially available at reasonable cost, has approximately a 0.105-m diameter barrel, and can take samples down to 0.3 m. An illustration of the sampling instrument and its use is provided in Fig. 1.

6.2 *Sample Container,* such as metal cans with lids, plastic bags, etc.

6.3 *Meter Stick.*

6.4 *Small Scoop.*

7. Sampling

7.1 *Introduction*—The sampling depth for this practice is the top 50 mm of soil. Experience has shown this depth is best for this purpose (**3**) and provides samples for the analysis of deposited radionuclides following a recent airborne release. The difference in concentration from previously collected samples at the same locations would be a measure of the contamination. If the purpose of the sampling is to measure the total amount of a radionuclide deposited onto the soil, that is, from fallout of previous atmospheric nuclear tests, then sampling must be conducted to an 0.3-m depth. It

---

[1] This practice is under the jurisdiction of ASTM Committee C-26 on Nuclear Fuel Cycle and is the direct responsibility of Subcommittee C26.05 on Test Methods.
Current edition approved June 29, 1990. Published August 1990. Originally published as C 998 – 83. Last previous edition C 998 – 83.
[2] *Annual Book of ASTM Standards*, Vol 04.08.
[3] *Annual Book of ASTM Standards*, Vol 11.01.
[4] *Annual Book of ASTM Standards*, Vol 14.02.

[5] The boldface numbers in parentheses refer to the list of references at the end of this practice.
[6] Model 28200 Scalloped Style of the Standard Manufacturing Company of Cedar Falls, IA, or its equivalent, has been found satisfactory for this purpose.

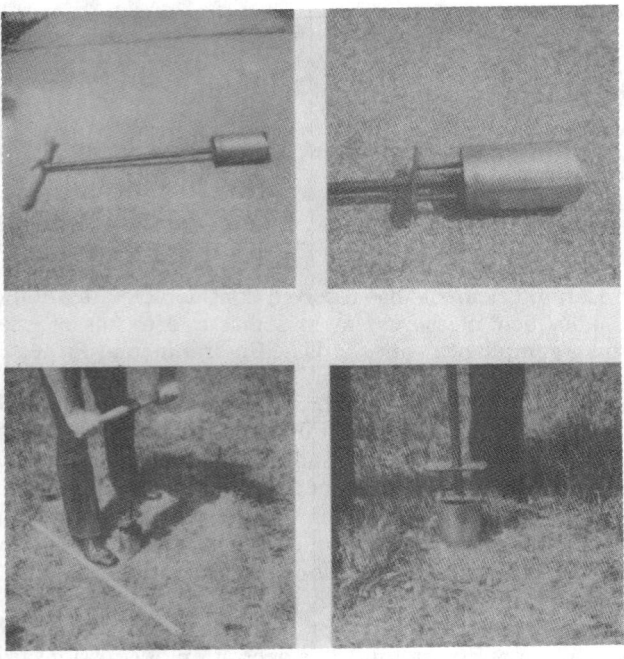

FIG. 1   Soil Sampling Instrument and Use

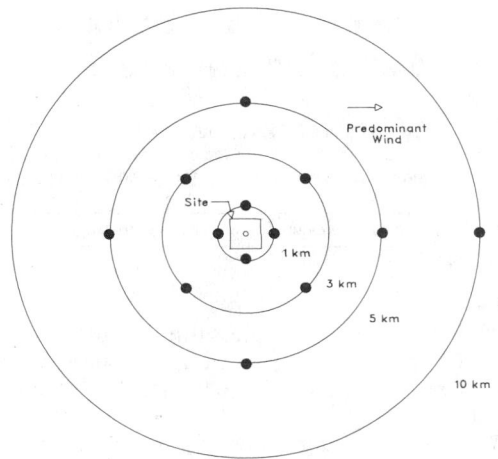

FIG. 2   Soil Sampling Pattern

is recommended by the EPA (2) that soil sampling for plutonium be the top 10 mm of soil. Although this may be a desirable depth for resuspension studies in certain parts of the country that have powdery, dry, loose, sandy soils, in most areas, the vegetative cover and root mat make this an unworkable sampling depth. Because the data may be used in various ways, it is important to accurately record the sample location, the depth of the sample, and the sample weight. In order to obtain sufficient sample to be representative of the area, due to the inherent heterogeneity of soil, it is recommended that a total sampling area of greater than 0.05 m² be collected as described in Section 8.

7.2 *Site Selection:*

7.2.1 As an idealized guideline, each site should be selected on the basis that the soil appears, or was known to have been, undisturbed for a number of years. Open, level, grassy areas that are mowed at reasonable intervals, such as public parks, are suitable choices. The site should have moderate to good permeability and there should be little or no runoff during heavy rains. The site should not be near enough to buildings, trees, or other obstructions that it is sheltered or shielded. High earthworm activity or aeration of the root zone may result in uneven mixing of the surface soil and, therefore, this type of site should be avoided. Care should be taken not to select a site that is fertilized or watered with sources that may add radioactive materials to the soil, that is, some fertilizers have high uranium concentrations. It is important to be able to accurately describe the location at which the sample was collected if it becomes necessary to return and resample the location.

7.2.2 The number of sites sampled is determined by the purpose of the sampling and the information required from the particular analysis. If the sampling is part of a preoperational survey around a facility, one acceptable distribution is that proposed in HASL-300 (4) and depicted in Fig. 2. This

distribution of 13 sampling sites extending up to 10 km in the downwind direction from the facility should be adequate to provide the background concentration of the nuclides of interest. Sampling for other purposes may require other distribution of sites, while sampling to define the distribution of a nuclide from a specific incident would require extensive knowledge of meteorological and climatological factors. It is important that the purpose of the sampling dictate the sample distribution.

8. Procedure

8.1 *Sampling Procedure:*

8.1.1 Select the sampling location based on Section 7.

8.1.2 Measure out two 1-m² areas, about 3 m apart.

8.1.3 Remove all vegetation to a height of 10 to 20 mm above the soil and save if desired.

8.1.4 Collect soil from the center and each corner of the two 1-m² areas.

8.1.5 Insert the sampling tool to a depth of 50 mm below the soil surface and remove the soil plug.

8.1.6 Place the soil plug and residual vegetation and roots in an appropriate container.

8.1.7 Repeat the procedure until the ten cores are collected. Composite the ten cores as one sample.

8.1.8 Label the container with such information as location, time, date, collector, depth of core, and area sampled.

8.1.9 Clean the sampling tools in water and detergent and dry before collecting the next sample.

8.2 *Sampling Rationale*—The intent of the sampling procedure is to define the operational steps necessary to collect a representative sample from a desired location. The selection of the sampling tool should be dictated by local soil conditions as it is not the intent of this practice to identify one instrument to the exclusion of all others. However, two common procedures, or variations thereof, are most frequently used. These two procedures are the core procedure and the ring procedure. Because of the large variation in soil types, the core method described in HASL-300 (4) is recommended where applicable, and a ring method used by the Nevada Applied Ecology Group (NAEG) is offered as an option (5) for dry, sandy soils. The concepts and techniques in this practice are applicable to most situations requiring sampling surface soil for radionuclides.

8.3 *Core Procedure*—The collection of ten cores will sample about 0.086 m² of soil surface. Composite the ten cores to produce a single sample of about 4 to 5 kg. Most soils contain sufficient moisture to be cohesive and the plug can be removed intact. For some types of dry, loose soils, wetting the ground by sprinkling prior to sampling may allow the plug to be removed. Place the plugs in a container, seal, and carefully label. Clean the sampling tools in water and detergent and dry before proceeding to the next sample collection site.

8.4 *Ring Procedure*—For the dry, loose, sandy soil for which the core method is not applicable, press a ring, 100 mm in diameter and 50 mm deep, into the soil. Remove the soil inside the ring with a small scoop to a depth of 50 mm and place into a container. Repeat this until a total of ten cores are collected, using the procedure outlined in Section 7 for sample location selection. Clean the sampling tools in water and detergent and dry before proceeding to the next sample collection site.

9. Discussion

9.1 Either method works well for fine-grained soils, but difficulties occur with rocky soils. For samples in which plutonium is the element of interest, the rocks may be considered voids in the sample and usually are discarded during sample preparation. If this is the case, larger numbers of cores, and therefore larger areas, should be sampled to ensure that the sample is representative of the site.

9.2 The sampling techniques described in this practice will provide sufficient information to allow the calculation of results in terms of deposition per unit area or concentration. If the sampling is part of a routine monitoring program, it may be necessary to repeat the sampling at each location and compare results to determine the effect of facility operation.

10. Keywords

10.1 environmental; radionuclides; sampling; soil

REFERENCES

(1) "Proposed Guidance on Dose Limits for Persons Exposed to Transuranium Elements in the General Environment," Environmental Protection Agency 520/4-77-016, October 1977.

(2) "Persons Exposed to Transuranium Elements in the Environment," Federal Register, Vol 42, No. 230, Nov. 30, 1977.

(3) "Measurements of Radionuclides in the Environment: Sampling and Analysis of Plutonium in Soil," Atomic Energy Commission Regulatory Guide 4.5, May 1974.

(4) Harley, J. H., ed, "EML Procedures Manual," D.O.E. Report HASL-300, August 1979.

(5) Fowler, F. B., Gilbert, R. O., and Essington, E. H., "Sampling of Soils for Radioactivity: Philosophy, Experience, and Results," Atmospheric-Surface Exchange of Particulate and Gaseous Pollutants, ERDA Symposium Series 38, 1974, pp 706–727.

Standard Practices for the Measurement of Radioactivity[1]

This standard is issued under the fixed designation D 3648; the number immediately following the designation indicates the year of original adoption or, in the case of revision, the year of last revision. A number in parentheses indicates the year of last reapproval. A superscript epsilon (ε) indicates an editorial change since the last revision or reapproval.

1. Scope

1.1 These practices cover a review of the accepted counting practices currently used in radiochemical analyses. The practices are divided into four sections:

	Section
General Information	5 to 10
Alpha Counting	11 to 21
Beta Counting	22 to 32
Gamma Counting	33 to 40

1.2 The general information sections contain information applicable to all types of radioactive measurements, while each of the other sections is specific for a particular type of radiation.

1.3 *This standard does not purport to address all of the safety concerns, if any, associated with its use. It is the responsibility of the user of this standard to establish appropriate safety and health practices and determine the applicability of regulatory limitations prior to use.*

2. Referenced Documents

2.1 *ASTM Standards:*
D 1066 Practice for Sampling Steam[2]
D 1129 Terminology Relating to Water[2]
D 1943 Test Method for Alpha Particle Radioactivity of Water[3]
D 2459 Test Method for Gamma Spectrometry of Water[4]
D 3084 Practice for Alpha Spectrometry of Water[3]
D 3085 Practice for Measurement of Low-Level Activity in Water[3]
D 3370 Practices for Sampling Water[2]
D 3649 Test Method for High-Resolution Gamma-Ray Spectrometry of Water[3]
E 380 Practice for Use of the International System of Units (SI) (the Modernized Metric System)[5]

3. Terminology

3.1 *Definitions:*
3.1.1 For definitions of terms used in these practices, refer to Terminology D 1129. For an explanation of the metric system, including units, symbols, and conversion factors, see Practice E 380.

4. Summary of Practices

4.1 The practices are a compilation of the various counting techniques employed in the measurement of radioactivity. The important variables that affect the accuracy or precision of counting data are presented. Because a wide variety of instruments and techniques are available for radiochemical laboratories, the types of instruments and techniques to be selected will be determined by the information desired. In a simple tracer application using a single radioactive isotope having favorable properties of high purity, energy, and ample activity, a simple detector will probably be sufficient and techniques may offer no problems other than those related to reproducibility. The other extreme would be a laboratory requiring quantitative identification of a variety of radionuclides, preparation of standards, or studies of the characteristic radiation from radionuclides. For the latter, a variety of specialized instruments are required. Most radiochemical laboratories require a level of information between these two extremes.

4.2 A basic requirement for accurate measurements is the use of accurate standards for instrument calibration. With the present availability of good standards, only the highly diverse radiochemistry laboratories require instrumentation suitable for producing their own radioactive standards. However, it is advisable to compare each new standard received against the previous standard.

4.3 Thus, the typical laboratory may be equipped with proportional or Geiger-Mueller counters for beta counting, sodium iodide or germanium detectors, or both, in conjunction with multichannel analyzers for gamma spectrometry, and scintillation counters suitable for alpha- or beta-emitting radionuclides.

5. Significance and Use

5.1 This practice was developed for the purpose of summarizing the various generic radiometric techniques, equipment, and practices that are used for the measurement of radioactivity.

GENERAL INFORMATION

6. Experimental Design

6.1 In order to properly design valid experimental procedures, careful consideration must be given to the following;
6.1.1 radionuclide to be determined,
6.1.2 relative activity levels of interferences,
6.1.3 type and energy of the radiation,
6.1.4 original sample matrix, and
6.1.5 required accuracy.
6.2 Having considered 6.1.1 through 6.1.5, it is now possible to make the following decisions:

---

[1] These practices are under the jurisdiction of ASTM Committee D-19 on Water and are the direct responsibility of D19.04 on Methods of Radiochemical Analysis.
Current edition approved April 15, 1995. Published June 1995. Originally published as D 3648 – 78. Last previous edition D 3648 – 78 (1987)[ε1].
[2] *Annual Book of ASTM Standards*, Vol 11.01.
[3] *Annual Book of ASTM Standards*, Vol 11.02.
[4] Discontinued—see 1987 *Annual Book of ASTM Standards*, Vol 11.02.
[5] *Annual Book of ASTM Standards*, Vol 14.02 (Excerpts in Vol 11.02).

6.2.1 chemical or physical form that the sample must be in for radioassay,

6.2.2 chemical purification steps,

6.2.3 type of detector required,

6.2.4 energy spectrometry, if required,

6.2.5 length of time the sample must be counted in order to obtain statistically valid data,

6.2.6 isotopic composition, if it must be determined, and

6.2.7 size of sample required.

6.3 For example, gamma-ray measurements can usually be performed with little or no sample preparation, whereas both alpha and beta counting will always require chemical processing. If low levels of radiation are to be determined, very large samples and complex counting equipment may be necessary.

6.3.1 More detailed discussions of the problems and interferences are included in the sections for each particular type of radiation to be measured.

7. Apparatus

7.1 *Location Requirements:*

7.1.1 The apparatus required for the measurement of radioactivity consists, in general, of the detector and associated electronic equipment. The latter usually includes a stable power supply, preamplifiers, a device to store or display the electrical pulses generated by the detector, or both, and one or more devices to record information.

7.1.2 Some detectors and high-gain amplifiers are temperature sensitive; therefore, changes in pulse amplitude can occur as room temperature varies. For this reason, it is necessary to provide temperature-controlled air conditioning in the counting room.

7.1.3 Instrumentation should never be located in a chemical laboratory where corrosive vapors will cause rapid deterioration and failure.

7.2 *Instrument Electrical Power Supply*—Detector and electronic responses are a function of the applied voltage; therefore, it is essential that only a very stable, low-noise electrical supply be used or that suitable stabilization be included in the system.

7.3 *Shielding:*

7.3.1 The purpose of shielding is to reduce the background count rate of a measurement system. Shielding reduces background by absorbing some of the components of cosmic radiation and some of the radiations emitted from material in the surroundings. Ideally, the material used for shielding should itself be free of any radioactive material that might contribute to the background. In practice, this is difficult to achieve as most construction materials contain at least some naturally radioactive species (such as potassium-40, members of the uranium and thorium series, etc.). The thickness of the shielding material should be such that it will absorb most of the soft components of cosmic radiation. This will reduce cosmic-ray background by approximately 25 %. Shielding of beta- or gamma-ray detectors with anticoincidence systems can further reduce the cosmic-ray or Compton scattering background for very low-level counting.

7.3.2 Detectors have a certain background counting rate from naturally occurring radionuclides and cosmic radiation from the surroundings; and from the radioactivity in the detector itself. The background counting rate will depend on

the amounts of these types of radiation and on the sensitivity of the detector to the radiations.

7.3.3 In alpha counting, low backgrounds are readily achieved since the short range of alpha particles in most materials makes effective shielding easy. Furthermore, alpha detectors are quite insensitive to the electromagnetic components of cosmic and other environmental radiation.

7.4 *Care of Instruments:*

7.4.1 The requirements for and advantages of operating all counting equipment under conditions as constant and reproducible as possible have been pointed out earlier in this section. The same philosophy suggests the desirability of leaving all counting equipment constantly powered. This implies leaving the line voltage on the electrical components at all times. The advantage to be gained by this practice is the elimination of the start-up surge voltage, which causes rapid aging, and the instability that occurs during the time the instrument is coming up to normal temperature.

7.4.2 A regularly scheduled and implemented program of maintenance is helpful in obtaining satisfactory results. The maintenance program should include not only checking the necessary operating conditions and characteristics of the components, but also regular cleaning of the equipment.

7.5 *Sample and Detector Holders*—In order to quantify counting data, it is necessary that all samples be presented to the detector in the same "geometry." This means that the samples and standards should be prepared for counting in the same way so that the distance between the source and the detector remains as constant as possible. In practice, this usually means that the detector and the sample are in a fixed position. Another configuration often used is to have the detector in a fixed position within the shield, and beneath it a shelf-like arrangement for the reproducible positioning of the sample at several distances from the detector.

7.6 *Special Instrumentation*—This section covers some radiation detection instruments and auxiliary equipment that may be required for special application in the measurement of radioactivity in water.

7.6.1 *4-π Counter:*

7.6.1.1 The 4-π counter is a detector designed for the measurement of the absolute disintegration rate of a radioactive source by counting the source under conditions that approach a geometry of 4-π steradians. Its most prevalent use is for the absolute measurement of beta emitters (1, 2).[6] For this purpose, a gas-flow proportional counter similar to that in Fig. 1 is common. It consists of two hemispherical or cylindrical chambers whose walls form the cathode, and a looped wire anode in each chamber. The source is mounted on a thin supporting film between the two halves, and the counts recorded in each half are summed. An argon (90 %)-methane (10 %) gas mixture can be used; however, pure methane gives flatter and longer plateaus and is preferred for the most accurate work. The disadvantage is that considerably higher voltages, about 3000 V, rather than the 2000 V suitable for argon-methane, are necessary. As with all gas-filled proportional counters, very pure gas is necessary for very high detector efficiency. The absence of electronegative

---

[6] The boldface numbers in parentheses refer to the list of references appended to these practices.

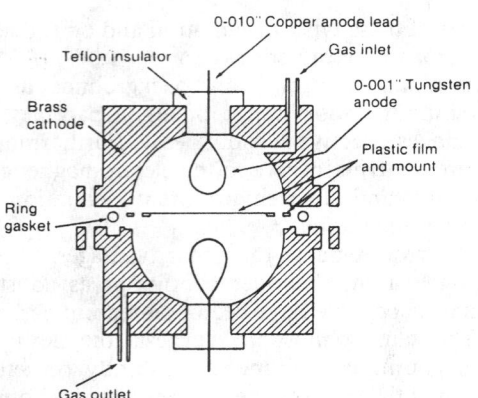

Teflon insulator — 0-010" Copper anode lead — Gas inlet — 0-001" Tungsten anode — Brass cathode — Plastic film and mount — Ring gasket — Gas outlet

FIG. 1 The 4π-Counting Chamber

gases that attach electrons is particularly important since the negative pulse due to electrons is counted in this detector. Commercial chemically pure (cp) gases are ordinarily satisfactory, but they should be dried for best results. A high-voltage power supply for the detector, an amplifier, discriminator, and a scaler complete the system.

7.6.1.2 To convert counting rate to disintegration rate, the principal corrections required are for self-absorption in the source and for absorption in the support film. The support film should be as thin as practicable to minimize absorption of beta particles emitted in the downward direction. Polyester film with a thickness of about 0.9 mg/cm² is readily available and easily handled. However, it is too thick for accurate work with the lower energy beta emitters. For this purpose, thin films (≈5 to 10 μg/cm²) are prepared by spreading a solution of a polymer in an organic solvent on water. VYNS (1), Formvar (2), and Tygon (3) plastics have been used for this purpose.

7.6.1.3 The films must be made electrically conducting (since they are a part of the chamber cathode) by covering them with a thin layer (2 to 5 μg/cm²) of gold or palladium by vacuum evaporation. The absorption loss of beta particles in the film must be known. Published values can be used, if necessary, but for accurate work an absorption curve using very thin absorbers should be taken (1). The "sandwich" method, in which the film absorption is calculated from the decrease in counting rate that occurs when the source surface is covered with a film of the same thickness as the backing film, is suitable for the higher beta energies.

7.6.1.4 The source itself must be very thin and deposited uniformly on the support to obtain negligible self-absorption. Various techniques have been used for spreading the source; for example, the evaporation of ⁶³Ni-dimethylglyoxime onto the support film (1), the addition of a TFE-fluorocarbon suspension (3), collodial silica, or insulin to the film as spreading agents, and hydrolysis (2). Self-absorption in the source or mount can be measured by 4-π beta-gamma coincidence counting (4, 5). The 4-π beta counter is placed next to a sodium iodide scintillation crystal, or a portion of the chamber wall is replaced by a sodium iodide crystal, and the absolute disintegration rate is evaluated by coincidence counting (6, 7). By adding a suitable beta-gamma tracer, the method has been used for pure beta as well as beta-gamma emitters (8). Accurate standardization of pure low-energy

beta emitters (for example, ⁶³Ni) is difficult, and the original literature should be consulted by those inexperienced with this technique.

7.6.1.5 Photon (gamma and strong X-ray) scintillation counters with geometries approaching 4-π steradians can be constructed from NaI(Tl) crystals in either of two ways. A well crystal (that is, a cylindrical crystal with a small axial hole covered with a second crystal) will provide nearly 4-π geometry for small sources, as will two solid crystals placed very close together with a small source between them. The counts from both crystals are summed as in the gas-flow counter. The deviation for 4-π geometry can be calculated from the physical dimensions. For absolute gamma-ray counting, the efficiency of the crystal for the gamma energy being measured and the absorption in the crystal cover must be taken into account. Additional information on scintillation counting is given in 7.6.4. The liquid scintillation counter is also essentially a 4-π counter for beta particles, since nearly all the radiations are emitted into and interact with the detecting medium.

7.6.2 *Low-Geometry Counters*—This type of instrument is particularly useful for the absolute counting of alpha particles. The alpha emitter, in the form of a very thin solid source, is placed at a distance from the detector such that only a small fraction (<1 %) of the alpha particles are emitted in a direction to enter the counter. This solid angle is obtained from the physical measurements of the instrument. The space between the source and the detector is evacuated to eliminate the loss of alpha particles by absorption in air. The detector can be any counter that is 100 % efficient for all alpha particles that enter the sensitive volume—a gas-flow proportional counter with a window that is thin (approximately 1 mg/cm²) compared to the range of the alpha particles or the semiconductor alpha detector with a 1-mg/cm² covering. The advantages of this instrument for absolute alpha counting are that: (1) the effect of absorption of alpha particles in the source itself is kept to a minimum since only particles that travel the minimum distance in the source enter the detector (particles that have longer paths in the source are emitted at the wrong angle, and (2) back-scattered alpha particles (those that are emitted into the source backing and are reflected back up through the source) lose sufficient energy so that they cannot enter the detector. One such instrument is described in Reference (9).

7.6.3 *Internal Gas Counters:*

7.6.3.1 The internal gas counter is so named because the radioactive material, in the gaseous state, is placed inside a counting chamber and thus becomes part of the counting gas itself. It is useful for high-efficiency counting of weak beta- and X-ray emitters. The radiations do not have to penetrate a counter window or solid source before entering the sensitive volume of a detector. The counter may be an ionization chamber, or it may be operated in the Geiger or proportional mode. Most present-day instruments are of the latter type, and they generally take the form of a metal or metal-coated glass cylinder as a cathode with a thin anode wire running coaxially through it and insulated from the cylinder ends. A wire through the wall makes electrical contact to the cathode. The counter has a tube opening through which it may be connected to a gas-handling system for filling. The purity of the gas is important for efficient and reproducible counting, particularly in the proportional mode.

7.6.3.2 In a modification of the internal gas counter, scintillation counting has been used in place of gas-ionization counting. The inner walls of the chamber are coated with a scintillation material and the radioactive gas introduced. An optical window is made a part of the chamber, and the counting is done by placing this window on a multiplier phototube to detect the scintillations. This system is particularly useful for counting radon gas with zinc sulfide as the scintillator. Additional details on internal gas counting may be found in Reference (10).

7.6.4 *Spectrometers and Energy-Dependent Detectors:*

7.6.4.1 The availability of energy-dependent detectors (detectors whose output signal is proportional to the energy of the radiation detected) that are easy to operate and maintain and have good resolution makes it possible to measure not only the total activity of a radioactive sample but the energy spectrum of the nuclear radiations emitted. Nuclear spectrometry is most useful for alpha particles, electromagnetic radiation (gamma and X rays), and conversion electrons, since these radiations are emitted with discrete energies. Beta spectra have more limited use since beta particles are emitted from a nucleus with a continuous energy distribution up to a characteristic maximum (E-max), making a spectrum containing several different beta emitters difficult to resolve into its components. The advantages of spectrometric over total activity measurements of radioactive sources are increased selectivity, sensitivity, and accuracy because nuclide identification is more certain, interference from other radioactive nuclides in the sample is diminished or eliminated, and counter backgrounds are reduced since only a small portion of the total energy region is used for each radiation. The detectors for alpha spectra are gridded ion-chambers and silicon semiconductor detectors. These are described in Practice D 3084. Gridded ion-chambers are no longer available commercially and must be constructed by the user. A variety of semiconductors can be purchased, and these detectors have essentially replaced ion-chambers for alpha spectrometry, although the chambers have the advantages of high efficiency (nearly 50 %) for large-area sources.

7.6.4.2 The principal detectors used for gamma-ray spectrometry are thallium-activated sodium iodide scintillation crystals, NaI(Tl), and lithium-drifted germanium semiconductors, Ge(Li). For X rays and very low energy gamma rays, lithium-drifted silicon semiconductor Si(Li), intrinsic Ge, and gas-filled thin (approximately 1 mg/cm^2) window proportional counters are used. Sodium iodide is hygroscopic, so the crystal must be hermetically sealed, and the entire crystal-phototube package must be light-tight. The complete spectrometer also requires a high-voltage power supply for the phototube (usually operated at 800 to 1000 V), a preamplifier, linear amplifier, pulse-height analyzer, and output recorder. The crystal is packaged in aluminum or stainless steel. The portion of the cover through which gamma rays enter is normally thinner than the rest of the package in order to reduce low-energy photon attenuation. Sodium iodide crystals are available in a large range of sizes and shapes, from 25 by 25-mm cylinders to hemispheres and cylinders at least 305 mm in diameter. Information on the types of crystal packages and mountings that can be used is available from the manufacturers.

7.6.4.3 Germanium and silicon detectors are junction-type semiconductor devices in which a large sensitive region has been produced by drifting metallic lithium into germanium or silicon under the influence of an electric field at an elevated temperature (100 to 400°C). The crystal functions as a "solid ion chamber" when a high voltage is applied. Because of the high mobility of the small lithium atom, the Ge(Li) detector must be kept cold to prevent the diffusion of lithium out of the crystal. In addition, in order to obtain high resolution, the detector must be operated at low temperatures to reduce thermal noise. At room temperature, sufficient free electrons will be present in the crystal to obscure the measurement of gamma and X rays (but not of alpha particles). Consequently, the detectors are operated and kept at liquid nitrogen temperatures by a cryostat consisting of a metallic cold-finger immersed in a Dewar flask containing liquid nitrogen. If the Ge(Li) detector is allowed to warm to room temperature for a short time, its resolution will deteriorate and in an hour or so it will lose sufficient lithium so that it cannot function as a detector without redrifting. The detector is kept hermetically sealed in a vacuum to prevent impurities from condensing on the surface and lowering its resistance and to reduce heat transfer from the room to the crystal. Aluminum is the usual covering, and a molecular sieve pump is incorporated into the system to maintain the vacuum. The electronic components required to obtain spectra are similar to those for sodium iodide crystals, except that because smaller pulses must be measured, high-quality electronics are needed. The complete system includes a high-voltage bias supply for the detector (up to 5000 V for large depletion volumes), a preamplifier, amplifier (usually charge-sensitive), biased amplifier (if needed), pulse height analyzer, and recording device.

7.6.4.4 A gamma ray entering either a NaI(Tl) crystal or a semiconductor detector may lose all or part of its energy in the detector. In the former case, through multiple Compton interactions or the photoelectric effect, a full energy peak is obtained. Otherwise, only part of the energy will be observed and a Compton continuum spectrum is seen. An alternative process for high-energy gamma rays (>1.02 MeV) is pair production, in which an electron-position pair is produced, and gamma-ray peaks are observed at 0.511-MeV intervals below the full energy peak. The two most important operating characteristics of gamma detectors are efficiency and resolution. The "peak-to-Compton" or "peak-to-valley" ratio is frequently given in the literature and is related to both efficiency and resolution. These parameters should be specified by the manufacturer and the conditions under which they were measured should be given.

7.6.4.5 The resolution of sodium iodide crystals is usually specified in terms of the width of the full-energy gamma-ray peak at half its maximum—the "full width at half maximum" (FWHM). This is shown graphically in the gamma-ray spectrum in Fig. 2. The resolution improves with increasing energy and the standard for comparison is usually the 0.662-MeV gamma ray emitted in the decay of ^{137}Cs. Good sodium iodide crystals have resolutions in the range of 6.5 to 7 % for ^{137}Cs. Detection efficiency for the same geometry and window thickness is a function of several parameters and much published information on efficiencies for various energies, detector sizes, source-to-detector distances, and other variables is available (11). The efficiency for gamma-ray detection may be expressed in various ways. Of primary interest in spectrometry is the full peak efficiency—the fraction of incident gamma rays that give a full-energy peak for a particular source-detector configuration. For a 102-mm thick NaI(Tl) crystal, with the source on the surface (zero distance), this fraction is approximately 0.24 for the 0.66-MeV gamma ray of ^{137}Cs and approximately 0.14 for the 1.33-MeV gamma ray of ^{60}Co. The "peak-to-valley" or "peak-to-Compton" ratio is the ratio of counts at the maximum height of the full-energy peak to the counts at the minimum of the Compton continuum (Fig. 2). A high ratio indicates narrow peaks, that is, good resolution, for that particular efficiency. The Compton spectrum does not give useful information in gamma-ray spectrometry and can be considered as "noise." The ratio varies with energy and is frequently given for the 1.33-MeV peak of ^{60}Co. It increases as the crystal size increases, and, after passing through a minimum, increases as the source-to-detector distance increases, since a larger fraction of the gamma rays pass through the full depth of the crystal. A peak-to-valley ratio of 12:1 for a crystal is very good. This ratio can be increased by anti-coincidence shielding to cancel Compton events as described in 7.6.5. The efficiency of silicon for gamma rays is considerably less than sodium iodide because of its lower atomic number (the efficiency for photoelectric absorption of gamma rays is proportional to Z^5) and lower density (the density of NaI is 3.7 g/cm^3 and of silicon 2.4 g/cm^3).

7.6.4.6 For a 1-MeV gamma ray, the total absorption coefficient is about 2 mm^{-1} for sodium iodide, 1.5 mm^{-1} for silicon, and 3 mm^{-1} for germanium. However, germanium detectors are not yet available in sizes approaching that of sodium iodide. The efficiency of a Ge(Li) detector is generally expressed by comparison with that of a 76 by 76-mm cylindrical NaI(Tl) detector. Comparison is made between the full-energy peak efficiencies for the 1.33-MeV gamma ray of ^{60}Co when the source is 250 mm from the detector. A

germanium detector with a volume of 35 cm^3 has an efficiency approximately 5 % that of a 76 by 76-mm NaI(Tl) crystal. Larger Ge(Li) detectors are available with relative efficiencies of 25 to 30 %.

7.6.4.7 There are limitations in the efficiency of the light production and collection processes in the sodium-iodide-photomultiplier system that make its resolution inferior to that of semiconductor detectors. One important factor is that about 500 eV are required to produce an electron at the photocathode in a sodium iodide detector system, while the average energy to produce the analogous electron-hole pair in silicon is only 3.5 eV and in germanium 2.8 eV. The resolution of semiconductor detectors does not change greatly with energy. Presently available germanium detectors have resolutions of 1.5 to 2.8 keV at 1.33 MeV and are from 3 to 30 % efficient compared to a 76 by 76-mm NaI(Tl) detector. This greater resolution makes this detector the one of choice for gamma-ray spectrometry and cancels to some extent the higher efficiency available from sodium iodide. Since the pulses from a single photopeak are spread over a much smaller energy range in germanium than in sodium iodide, the background under the peak is much less. This means that for small sources of moderately energetic gamma rays, germanium is more sensitive than sodium iodide. This is indicated in Table 1, where the efficiencies and backgrounds of 76 by 76-mm sodium iodide crystal and a 35-cm^3 (5.5 % efficiency) germanium detector are compared.

7.6.4.8 Spectra of beta particles and conversion electrons can be obtained with sodium iodide and semiconductor detectors sufficiently thick (a few centimetres) to absorb the particles completely. One disadvantage of sodium iodide and cooled lithium-drifted semiconductors is their relatively thick entrance windows. Other semiconductor detectors, particularly the silicon surface barrier type, have thin entrance windows and can be used for beta particles at room temperature or better at 0°C. The 5 to 10-keV resolution for 600-keV electrons is better than the 12 to 30-keV resolution for 5-MeV alpha particles.

7.6.4.9 Good spectra of low-energy beta particles, conversion electrons, and X rays can be obtained with a gas-flow proportional counter provided that a linear preamplifier is used. The resolution is intermediate between NaI and Ge. To reduce backscattering, the chamber should be made of low Z material. A counter constructed of a cylinder of graphite-impregnated plastic, poly(methyl methacrylate) ends, and a thin coaxial center wire gives good spectra for such radiations (11). A hole is cut into the outer wall and covered with aluminized polyester film to provide a thin entrance window. Argon (90 %)-methane (10 %) is a suitable counting gas.

7.6.4.10 Organic scintillators, such as anthracene and polystyrene polymerized with scintillating compounds, are also useful for beta spectrometry. They are packaged with a phototube in a manner similar to sodium iodide crystals. Liquid scintillation mixtures also give beta spectra, and the output of a commercial liquid scintillation counter can be fed into a multichannel pulse-height analyzer to obtain a spectrum (2). A spectrum of ^{210}Pb ^{210}Bi ^{210}Po in Fig. 3 shows the resolution obtainable by liquid scintillation counting of aqueous samples in a dioxane-based solution. The ^{210}Bi curve is from a beta particle, and the ^{210}Po peak is from an alpha particle. Organic scintillators are preferable to sodium

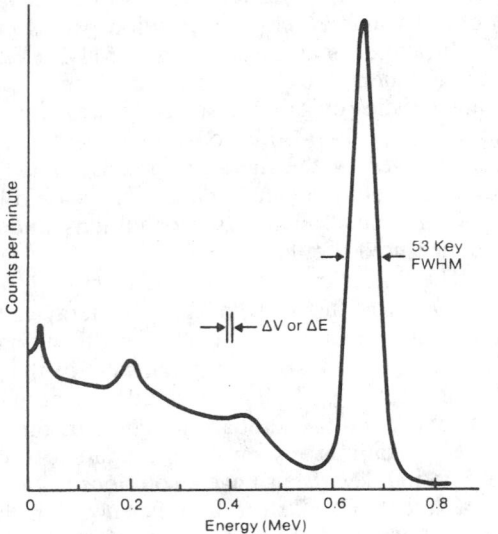

FIG. 2   Pulse Height or Energy Spectrum of Cesium-137

TABLE 1 Comparative Performance of NaI(Tl) and Ge(Li) Gamma-Ray Detectors

Note—

NaI =	76 by 76-mm cylindrical detector.	Detection limit =	the number of photons emitted from the source whose net count equals twice the counting error, or
Ge(Li) =	35-cm³ active volume, 5.5 % efficiency.		
A =	small source placed on detector.		
B =	57 by 57 by 57-mm thick source place on detector.		$$N + N_B + 2(N - N_B)^{1/2}$$
Counting efficiency =	percent of photons emitted from the source that give a full-energy peak.		where N is the total number of counts recorded when the source is measured and N_B is the total number of counts recorded when the background is measured.
Shielding =	152 mm of iron, 3.2 mm of lead.		
Counting time =	one 30 000-s count for both source and background.		

Photon Energy Detector	Background (cps) Under Peak	Counting Efficiency, %		Detection Limit, photons/s	
		A	B	A	B
0.14 MeV					
NaI	24	26	18	2.4	3.5
Ge	0.7	12	4	0.92	2.8
0.66 MeV					
NaI	20	14	9	4.1	6.3
Ge	0.11	1.3	0.68	3.6	6.8
1.33 MeV					
NaI	8	5.8	3.8	6.2	9.5
Ge	0.055	0.75	0.38	4.5	8.9

iodide for beta spectrometry because less backscattering occurs.

7.6.4.11 The output pulses of any energy-dependent detector, after linear amplification, must be sorted out according to energy to obtain the spectrum of incident radiation. The high resolution available in detectors requires analyzers with hundreds of channels to realize their full resolving power. The amplified pulse is digitized by an analog-to-digital converter (ADC), and the resulting number for a particular pulse is recorded in a pulse counter whose location is determined by digital circuitry. This makes it possible to use a digital computer to count and store in its memory the number of pulses in each channel. This conversion and storage is relatively slow, and the analyzer is blocked from processing a second pulse until the previous processing is completed. The time required to process a pulse increases with channel number. The instruments now available are sufficiently fast for almost all water measurement purposes. Some loss of pulse information is acceptable, as the analyzers measure and record "live time" fairly accurately. Thus, the counting time recorded by the analyzer will be the actual time it was in a condition to receive detector pulses, and not the elapsed time. To maintain good accuracy, the activity of the sample should be adjusted to give live times of 90 % or more. A computer may be permanently combined with the ADC to operate only in the pulse-height analyzer mode ("hard wired"), or a separate and larger computer is "soft wired" to the ADC and can also be programmed for

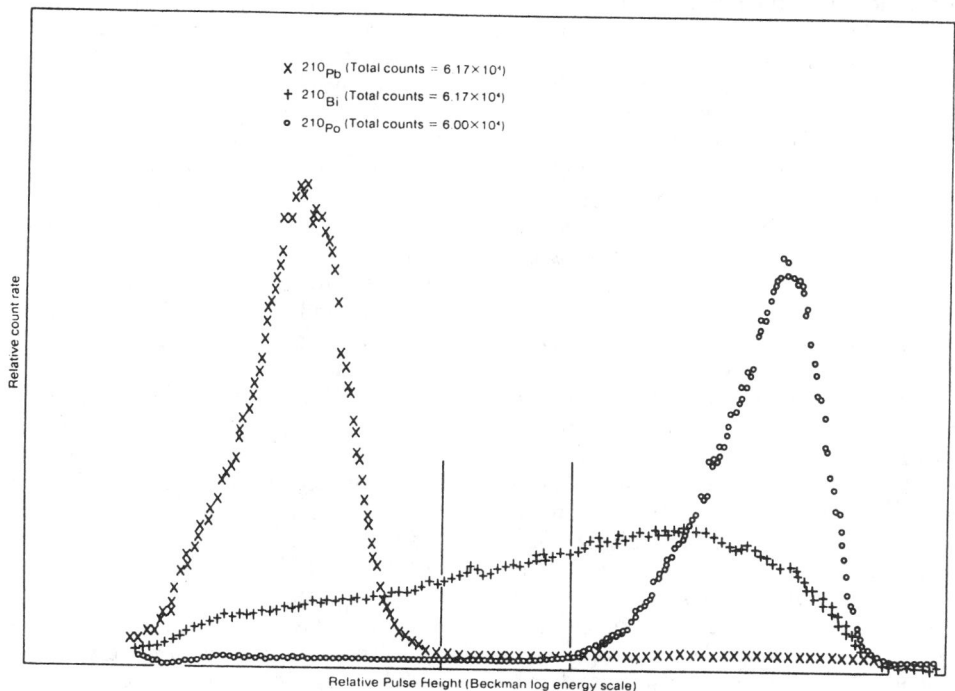

X ^{210}Pb (Total counts = 6.17×10⁴)

+ ^{210}Bi (Total counts = 6.17×10⁴)

o ^{210}Po (Total counts = 6.00×10⁴)

Relative count rate

Relative Pulse Height (Beckman log energy scale)

FIG. 3 Spectrum of ^{210}Pb ^{210}Bi ^{210}Po

operations other than pulse height analysis, such as data reduction and spectrum resolution. In either case, this type of analyzer makes possible automatic digital readout on printers, paper, and magnetic tape, automatic spectrum plotting, cathode-ray tube curve and digital presentation, and internal data reduction.

7.6.4.12 All multichannel pulse-height analyzers currently available are transistorized, and are fairly reliable instruments and relatively easy to operate. Their maintenance and repair is, however, a specialized skill similar to other computer repair. In comparing analyzers, some of the important specifications to consider are the number of channels, count capacity, stability, live-time accuracy, linearity, type of pulse input acceptable, and ADC speed. The minimum number of channels useful for NaI gamma-spectrometry is 128; Ge(Li) detectors should be used with at least a 1000-channel analyzer and alpha and beta spectra can profitably use 100 to 400 channels, depending on the energy range to be covered. Analyzers with 4096 channels are fairly common, and larger analyzers are available for special purposes.

7.6.4.13 Semiconductor detectors require low-noise, charge-sensitive amplifiers. Because of their excellent resolution, semiconductor detectors are often used with a biased amplifier following the main amplifier to isolate a portion of the spectrum for analysis. This makes it possible to use smaller analyzers than would otherwise be necessary.

7.6.5 *Anti-Coincidence Counters:*

7.6.5.1 Substantial background reduction can be achieved in beta and gamma counters by surrounding or covering the sample detector with another detector also sensitive to beta or gamma radiation, and connecting them electronically so that any pulse appearing in both detectors is cancelled and not recorded as a count. This is usually referred to as anti-coincidence shielding, and is recommended for obtaining very low backgrounds. This type of counter was used for many years in directional studies of cosmic rays, and was first applied to reducing the background of beta counters by Libby in his study of natural ^{14}C. The thick metal shielding (lead, iron, or mercury) ordinarily used to reduce cosmic-ray and gamma-ray background must also be present, and is placed outside the anti-coincidence shielding. Gas-filled beta detectors are generally shielded by gas-filled detectors, and such anti-coincidence shielding is effective primarily against the particulate component of cosmic rays. The anti-coincidence shielding for beta counters may consist of a number of long Geiger tubes ("cosmic-ray counters") surrounding the sample detector or a large (approximately 152 mm square) gas-flow detector, with several anode wires so the entire area of the counter is sensitive, placed just above the sample detector. For counting solid beta sources, the sample detector has a diameter of 25 to 51 mm. Surrounding these counters on all six sides there is frequently a layer of high-purity copper to absorb gamma rays emitted from the outermost shielding, and 102 to 152 mm of lead or iron on all six sides. This is the form usually taken by the commercially available anti-coincidence shielded beta counters. Plastic or inorganic scintillators could also be used as the anti-coincidence shielding.

7.6.5.2 Anti-coincidence shielding of gamma-ray detectors operates in a similar way, and is particularly useful in reducing the Compton continuum background of gamma rays (12). Gamma rays that undergo Compton scattering and produce a pulse in both the detector and the anti-coincidence shield are cancelled electronically. Ideally, only those gamma rays that are completely absorbed in the sample detector itself produce a count that is recorded with the total energy of the gamma ray (full-energy peak). There are second-order effects that prevent complete elimination of Compton scattering, but the improvement is substantial. The anti-coincidence shield can be a large sodium iodide or plastic scintillator suitably attached to phototubes. They usually have a large annular hole into which the sample detector, a smaller sodium iodide detector, or germanium iodide detector is placed (13, 14).

7.6.6 *Coincidence Counters:*

7.6.6.1 In coincidence counting, two or more radiation detectors are used together to measure the same sample, and only those nuclear events or counts that occur simultaneously in all detectors are recorded. The coincidence counting technique finds considerable application in studying radioactive decay schemes; but in the measurement of radioactivity, the principal uses are for the standardization of radioactive sources and for counter background reduction.

7.6.6.2 Coincidence counting is a very powerful method for absolute disintegration rate measurement (6, 15). Both alpha and beta emitters can be standardized if their decay schemes are such that β-γ, γ-γ, β-β, α-β, or α-X-ray coincidence occur in their decay. Gamma-gamma coincidence counting with two sodium iodide crystals, and the source placed between them, is an excellent method of reducing the background from Compton scattered events. Its use is limited, of course, to counting nuclides that emit two photons in cascade (which are essentially simultaneous), either directly as in ^{60}Co, by annihilation of positrons as in ^{65}Zn, or by immediate emission of a gamma ray following electron capture decay. If the crystals are operated with single-channel pulse-height analyzers to limit the events recorded from each crystal to one of the full-energy peaks of the photons being emitted, then essentially only those photons will be counted. Non-coincident pulses of any energy in either one of the crystals will be cancelled, including cosmic-ray photons in the background and degraded or Compton scattered photons from higher energy gamma rays in the sample. Thus, the method reduces interference from other gamma emitters in the sample. If, instead of single-channel analyzers, two multichannel analyzers are used to record the complete spectrum from each crystal, singly and in coincidence, then the complete coincident gamma-ray spectrum can be obtained with one measurement. The efficiency for coincidence counting is low since it is the product of the individual efficiencies in each crystal, but the sensitivity is generally improved because of the large background reduction (16). This technique is often referred to as two-parameter or multidimensional gamma-ray spectrometry.

7.6.6.3 Additional background improvement is obtained if the two crystals are surrounded by a large annular sodium iodide or plastic scintillation crystal connected in anti-coincidence with the two inner crystals. In this case a gamma ray that gives a pulse, but is not completely absorbed in one of the two inner crystals, and also gives a pulse in the surrounding crystal, is cancelled electronically (13, 16). This provides additional reduction in the Compton scattering background. Lithium-drifted germanium detectors may be used in place of the inner sodium iodide crystals for improved resolution and sensitivities (14).

7.7 All of the equipment described in Section 7 is available commercially.

8. Sampling

8.1 Collect the sample in accordance with Practice D 1066 or Section 14.3 of Practices D 3370.

8.2 Sample an appropriate volume depending on the expected concentration of radioactivity in the water. For precise measurements without long counting times, it is advisable to count an aliquot that contains at least 40 dps of radioactivity.

8.3 Chemical treatment of samples to prevent biological or algal growth is not recommended and should be avoided unless essential. When necessary, select the reagents used to avoid chemical interaction with the radioactive species in the sample. Analyze samples promptly.

8.4 Chemical treatment of samples to retain radioactive species in solution may be used but carefully select the specific treatment. The use of oxidizing acids such as HNO_3 is not recommended when iodide is present since it may be oxidized to iodine and lost or be absorbed into the plastic containers if they are used. In some cases, extreme chemical treatment may be used to keep a particular chemical species in solution; examples are strongly alkaline conditions to hold molybdenum and ruthenium in solution, or acid conditions with fluoride ion to keep zirconium in solution. The addition of an acid such as hydrochloric is generally desirable to reduce hydrolysis and the loss of activity on container walls. Frequently, samples will contain insoluble material. In such cases, treat the sample by one of the following methods:

8.4.1 Filter the insoluble material and analyze both the filtrate and insoluble matter on the filter. During filtration, some material may be sorbed onto the filter and assumed to be insoluble when in fact it is soluble.

8.4.2 Centrifuge the sample and analyze both phases. Wash the insoluble phase with distilled water to remove all soluble material without dissolving the insoluble fraction.

8.4.3 In either of the above separations when the total activity is required, the insoluble matter may be dissolved and recombined with the soluble fraction. When radioactivity is left on the walls of the sample container, desorb it and add it to the sample.

8.5 Composite samples may be made by mixing aliquots of successive samples collected by an automatic sampler. Analysis of such composite samples yields average results only and loses information on short-term effects.

9. Instrument Operation and Control

9.1 The following procedures ensure that counting equipment is functioning properly and remains in calibration.

9.2 *Establishing Counter Characteristics:*

9.2.1 The first step in instrument control is to establish the operating characteristics of the system. Carefully measure the efficiency for counting the nuclide of interest under the conditions to be employed. Select counting conditions, that is, optimize gain, discriminator setting, and voltage for the radionuclide of interest. Set the operating voltage so that any

change in counting rate is minimized for a given voltage fluctuation. Adjust the discriminators to exclude noise and unwanted interferences from the nuclides being counted. Make adjustment to optimize the signal-to-noise ratio. When the counting conditions have been selected, monitor known interferences to determine such things as the effect of counting betas in an alpha counter or alpha pulses in a beta proportional counter, etc.

9.2.2 Make daily performance checks and maintain a log for each instrument. This log should include the count for a standard and a background. When the counting rate differs statistically from the expected performance, perform additional counting to determine if the counter is malfunctioning. High background can indicate either an instrumental problem or counter contamination.

9.2.3 Certified standards are available from several suppliers. Most solution standards have the pH controlled and carrier added to ensure that hydrolysis or sorption, or both, do not change the concentration of the solution. When dilutions are made it is important to maintain the stability of the solution by diluting with a proper matrix. Store the standard in a container that minimizes evaporation by loss either through the walls or out of the stopper.

9.2.4 If a planchet is prepared as a standard, place it in a suitable container for storage, which will prevent the surface containing the activity from being contacted. A recommended practice is the preparation of two standards, using one and storing the second for periodic checks to see that the working standard has not been altered.

9.3 *Counter Control and Tolerance Charts:*

9.3.1 Evaluate the daily standard counts made on any counter on a statistical or tolerance basis. The best way to do this is to maintain a control chart on each counter. A control chart is a graph showing the number of counts recorded in a fixed counting period against the day of the year. Select a radioactive source having a suitable emission rate to give several thousands of counts in a relatively short counting time. Each measurement should be at least 10 000 counts in a given measurement time period. Determine the initial entry, $\bar{N}$, from the average of at least ten measurements over a period of days. For a statistical control chart, enter the error bands of $\pm 2S_x$ and $\pm 3S_x$ ($S_x = \sqrt{N}$) and draw lines on the chart that allow for decay of the standard over the year. For a tolerance chart, select the tolerance band for $\bar{N}$, based on needs, to which the counter will be held, for example, from ± 1 % or ± 3 % of the $\bar{N}$, etc. The tolerance band is equivalent to the $\pm 3S_x$ control band of a control chart. Draw control chart and tolerance band lines on the chart that allow for decay of the radioactive standard over the year. An example of a control chart is shown in Fig. 4.

9.3.2 For a statistical control chart, enter the result of the standard count in the control chart and take the following action:

9.3.2.1 If the result is inside the $\pm 2S_x$ band, consider the counter to be in control.

9.3.2.2 If the result lies outside the $\pm 2S_x$ band, but inside the $\pm 3S_x$ band, consider the counter to be in control but flag this result.

9.3.2.3 If the result lies outside the $\pm 3S_x$ band, consider the counter out of control. Corrective action is needed if repeated counts remain outside the $\pm 3S_x$ band.

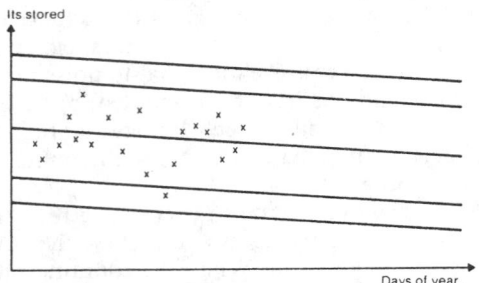

N̄ should be about 10⁵ counts whenever possible

FIG. 4   Typical Counter Control Chart

9.3.3 For a tolerance chart, enter the result of the standard count in the tolerance chart and take the following action:

9.3.3.1 If the result lies outside the tolerance band, consider the counter out of control. Corrective action is needed if repeated counts remain outside the tolerance band.

9.3.4 For alpha and gamma spectrometry, it is also important to monitor for system resolution (see Fig. 5).

9.3.5 In addition to the control charts made on each counter, keep all pertinent information about the system in a log book or permanent file.

Counter Logs

α and β Counters	*γ Spectrometers*
standard counts	standard counts
background counts	standard resolution
system changes	background counts
control charts	control charts

10. Counting Statistics

10.1 Each nuclear disintegration **(6, 17, 18)** is a completely random and independent process. Established methods of statistical analysis are available to describe the random phe- nomenon of nuclear disintegration. The total number of par- ticles counted in a time period can be shown to deviate from the average in accordance with the expression, $N \pm \sqrt{N}$ where N is the number of counts in the counting period t. Similar behavior is found in the background count rate for any counter and the recorded background will deviate from the average in accordance with the expression $N_B \pm \sqrt{N_B}$ where N_B is the number of counts recorded in the counting period t_B. Then find the net counting rate as follows:

$$N_x = \text{Net count rate} = \left(\frac{N}{t} - \frac{N_B}{t_B}\right) \qquad (1)$$

The standard deviation (uncertainty at the 68.3 % confidence level) of the Net count rate, N_x, is defined as follows:

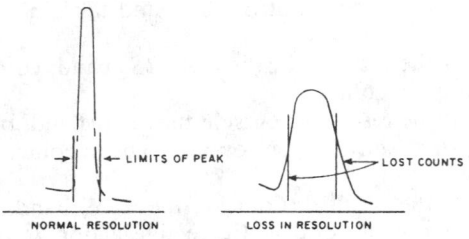

FIG. 5   System Resolution in Alpha and Gamma Spectrometry

$$S_x = \sqrt{(N/t^2) + (N_B/t_B{}^2)} \qquad (2)$$

or for the net count rate with its standard deviation:

$$\text{Net count rate} = \left(\frac{N}{t} - \frac{N_B}{t_B}\right) \pm \sqrt{\frac{N}{t^2} + \frac{N_B}{t_B{}^2}} \qquad (3)$$

In practice, many counts will be made that deviate from the average by more or less than the standard deviation, since S_x by definition would give those limits $X - S_x$ and $X + S_x$, which would include approximately 68 % of all observed values. In terms of the population of observations between limits, one can express the chance of probability of a result being between the limits $X - KS_x$ and $X + KS_x$ as "confidence levels," where K is simply a multiplier for S_x. In evaluating the effect of K for total counts (per measurement period) greater than ~60, one can tabulate K and the "confidence level."

Confidence Level, %	K
68.3	1.0
90	1.65
95	1.96
99	2.58
99.7	3.0

10.2 The 95 % confidence limit is frequently used and has been generally accepted since errors outside this band are considered statistically significant. In practice, K is often rounded to 2.0 and the term $2S_x$, or 2σ, error is often used. In some cases the $3S_x$, or 3σ, error is used and gives the 99.7 % confidence limits.

10.3 *Precision*—There is a measure of the reproducibility of a measurement. There are a number of items that affect the precision of radioactivity measurements, the more important of these are as follows:

10.3.1 *Position*—For point sources the observed radioactivity is proportional to the reciprocal of the square of the distance of the source from the detector. Measure all sources being directly compared at the same source-to-detector distance.

10.3.2 *Radiation Scattering*—Changes in the sample support and backing material can seriously affect the precision of radiation measurements, particularly that of beta radiation. Standards and samples should have the same backing material. The use of a sample support rack constructed of a material with low atomic number will reduce the effect of side-scattering. Scattering from backing material can be minimized by the use of a thin, low atomic number material.

10.3.3 *Background*—Measure this radioactivity with no sample near the detector and subtract from all measurements of gross sample activity. This requirement becomes more critical as the net sample activity becomes small with respect to the background. Perform routine periodic measurements of the background to check for possible detector contamination or malfunction.

10.3.4 *Absorption*—Alpha and beta radiations are partly absorbed by the sample and by all substances separating the sample from the detector. This effect is usually small for gamma radiation but beta and alpha radiation are seriously affected. If sources of the same atomic number and mass are compared on the same counter in the same geometry, absorption will be constant under these conditions but should not be ignored. Since the sample mass may vary significantly, construct calibration curves to correct for changes in self-absorption.

10.3.5 *Quenching*—This is any process that reduces the photon output in a scintillation system; this in turn reduces the measured count rate. Quenching can be caused by such things as sample color and chemicals in the sample. The need to correct for this phenomenon can be avoided if samples of the same color and same chemical composition are compared. If this is not possible, most instrument manufacturers and texts describe methods for construction of calibration curves to correct for this phenomenon (19, 20, 21).

10.4 *Overall Uncertainty in a Determination*—Report measurement results with the estimated overall measurement uncertainties shown. There are two approaches to this: (*1*) rigid propagation of uncertainties, which is not sound in practice since the individual components are not well characterized, and (*2*) a combination of known uncertainties. In combining uncertainties, use the following relationship:

$$S_x = \sqrt{S_1^2 + S_2^2} \qquad (4)$$

where:
S = overall uncertainty of the measurement (RSD),
S_1 = random counting uncertainty (RSD), and
S_2 = other random uncertainties (RSD).

The work can be divided into two main classes, as follows:

10.4.1 *Gamma Spectrometry*—Assuming that the integrity of the sample is known, the sampling and treatment effects are at a minimum and then only two significant terms are present. These are accuracy of calibration and precision of counting:

$$S_{\gamma spec} = \sqrt{S_c^2 + S_a^2} \qquad (5)$$

where:
$S\gamma_{spec}$ = overall error for the measurement (Relative Standard Deviation or RSD),
S_c = calibration uncertainty (RSD), and
S_a = counting uncertainty (RSD).

10.4.2 *Separative Work*—The sample treatment introduces uncertainty into the measurement. One estimate of great merit is as follows:

$$S_x = \sqrt{S_M^2 + S_a^2 + S_c^2} \qquad (6)$$

where:
S_x = overall uncertainty for the measurement (RSD),
S_M = intrinsic precision of the method (RSD),
S_a = counting uncertainty (RSD), and
S_c = calibration uncertainty (RSD).

The intrinsic precision may be determined by doing a single-operator precision (SOP) test at three or four activity levels.

SOP Test Data

Level, dps	S_x(%RSD)	S_a(%RSD)	S_M(%RSD)	
0.2	8	S_{a1}	S_{M1}	Average
2	4	S_{a2}	S_{M2}	$\bar{S}_M$
200	2	S_{a3}	S_{M3}	

The use of methods that have been round-robin tested will provide a method where S_x is known.

10.5 *Minimum Detectable Activity:*

10.5.1 The minimum detectable activity (MDA) is a statistical measure of the sensitivity of a counting determination. In the analysis of environmental samples or discharge samples, the sensitivity obviously becomes an important and often critical item. To determine if a measured sample count rate is different than the instrument background, the measured sample count rate is evaluated against the decision (or critical) level as defined by Currie (38). The decision level is defined as the "quantity of analyte at or above which a decision is made that a positive quantity of analyte is present." The industry standard has set the probability of erroneously reporting a detectable nuclide in an appropriate blank or sample at 0.05. Under this conversion, the decision level is mathematically defined as:

$$\text{Decision Level Count Rate} = DLCR = 1.645 \times S_o \qquad (7)$$

where S_o is defined as $s_B \times \sqrt{2}$ and S_B is the standard deviation of the background count rate. The above $DLCR$ equation assumes paired observations, for example, the sample and appropriate blank (background) are counted for the same length of time. The *a priori* minimum detectable activity (38) is defined as "The amount of a radionuclide, which, if present in a sample, would be detected with a 0.05 probability of non-detection while accepting a 0.05 probability of false detection (erroneously detecting that radionuclide in an appropriate blank sample)." The *a priori* minimum detectable activity, MDA, is mathematically defined as:

$$MDA = \frac{2.71}{T \times K} + 4.65 \frac{S_B}{K} \qquad (8)$$

where S_B is defined above, T is the sample counting time, and K includes the efficiency of detection of the counter as defined in the subsequent sections, the unit activity conversion factor, and the appropriate branching ratio of the emission under consideration. The MDA is expressed in units of Bq. Again this mathematical definition assumes paired observations. The *a priori* MDA is a "before the fact" calculation, or, more specifically, assumed to be a nominal MDA without prior knowledge of the activity level or nuclide interferences within the sample. For the *a priori* MDA to be truly representative of the radiation measurement system's detection capability for a given matrix, the use of the appropriate blank with interferences is required. However, since prior knowledge of the sample constituents and interferences is not usually available, the *a priori* MDA is typically a nominal value defining the system's capability for a given appropriate blank.

10.5.2 Where complex NaI(Tl) spectra are being analyzed by a weighted least squares unfolding analysis and the components that are being sought are not completely separable by instrument resolution, consider the *a priori* MDA detection limit as 3.29 times the standard deviation (68 % confidence level) of the nuclide's activity when a blank matrix is evaluated. The blank matrix should be evaluated using the library of nuclides routinely applied for such matrix unfolding applications.

10.5.3 In computing S_x, consider the sample and background counting rates, that is:

$$S_x = \sqrt{(N/t^2) + (N_B/t_B^2)} \text{ cps} \qquad (9)$$

If the sample and background counting times are equal ($t = t_B$), use a slightly different form:

$$S_x = \sqrt{N + N_B} \text{ counts} \qquad (10)$$

11. Calculation and Symbols

11.1 To calculate the amount of activity of a substance, X, use the following general method:

$$C = 1/VY \left[(N/t) - (N_B/t_B) \right] \qquad (11)$$

where:

N = number of counts accumulated,
t = sample counting period, s,
N_B = number of background counts accumulated,
t_B = background counting period, s,
C = net counts per second per millilitre, cps/mL,
V = volume of sample, mL, and
Y = fractional recovery of species (unity in methods where complete recovery is assumed).

11.2 Calculate the disintegration rate (D) as follows:

$$D = (C - R)/E \qquad (12)$$

where:

D = disintegrations per second per millilitre, dps/mL or Bq,
E = fractional efficiency of counter, cps/dps, and
R = reagent blank correction (cps) measured on an actual blank sample.

11.3 In gamma spectrometry use the following:

$$E = GP \qquad (13)$$

where:

G = fractional abundance of the gamma ray concerned, and
P = photopeak detection efficiency (fraction).

In beta counting use the following:

$$E = f_d f_a f_s f_m \qquad (14)$$

where:

f_d = detector efficiency factor for a source of given beta energy,
f_a = absorber factor for the total absorber thickness (detector window + air space + absorber + cover) for a source of given beta energy,
f_s = self absorption and backscatter factor for a given precipitate thickness and given beta energy, and
f_m = number of charged particles per disintegration.

The concentration of substance A may be calculated as follows:

$$A = \frac{D}{3.7 \times 10^4} \qquad (15)$$

where A = concentration in microcuries per millilitre (µCi/mL). For decay correction, use the expression:

$$D = D° \exp(-0.693 \, t/T) \qquad (16)$$

where:

$D°$ = disintegration rate at time zero, that is, reference time,
t = elapsed time between measurement and reference time, and
T = half-life of the nuclide.

Parent-daughter relationships are commonly shown by use of subscripts 1, 2, etc., for count rates, disintegration rates, or values of t and T.

ALPHA COUNTING

12. Scope

12.1 This practice covers the measurement of the alpha particle radioactivity of water. It is applicable to alpha emitters having energies above 3 MeV at activity levels above 0.02 dps per sample.

13. Summary of Method

13.1 Alpha particles are characterized by intense loss of energy in passing through matter. This intense loss of energy is used in differentiating alpha radioactivity from other types through the dense ionization or intense scintillation it produces. This high rate of loss of energy in passing through matter, however, also makes sample preparation conditions for alpha counting more stringent than is necessary for other types of radiation.

14. Alpha Detectors

14.1 Alpha radioactivity is normally measured by one of several types of detectors in combination with suitable electronic components. The detector devices most used are ionization chambers, proportional counters, silicon semiconductor detectors, and scintillation counters. The associated electronic components in all cases would include high-voltage power supplies, preamplifiers, amplifiers, scalers, and recording devices.

15. Detection Technique

15.1 In all of these systems, the initial event is converted to an electrical pulse that is amplified to a voltage sufficient to operate the scaler mechanism where provision is made for recording each pulse. The number of pulses per unit of time is directly related to the disintegration rate of the test sample. The efficiency of the system can be determined by counting standards prepared in the same manner as the samples. An arbitrary efficiency factor can be defined in terms of a different radionuclide, such as natural uranium, polonium-210, plutonium-239, or americium-241.

16. Measurement Variables

16.1 The measured alpha-counting rate from a sample will depend on a number of variables. The most important of these variables are geometry, source diameter, self-absorption, absorption in air and detector window, coincidence losses, and backscatter. These are discussed in detail in the literature (2, 3) and in many cases can be measured or corrected for by holding conditions constant during the counting of samples and standards. These effects may be described by the following relationship:

$$\text{dps} = \text{cps} \, (G_p)(f_{bs})(f_{aw})(f_d)(f_{ssa})(f_c) \qquad (17)$$

where:

cps = recorded counts per second, corrected for background,
dps = alpha disintegrations per second,
G_p = point source geometry, which is the solid angle subtended by the sensitive area of the detector. The effect of this variable is eliminated by maintaining a constant geometry for both standard and sample counts.

f_{bs} = backscatter factor, or ratio of cps with sample backing to cps without backing. This is not important in alpha counting since backscatter is small. In samples mounted on copper or stainless-steel planchets for counting 2π geometry, backscatter may be taken as 2 % (backscatter factor equals 1.02) without any serious error. In samples mounted on platinum, backscatter may be taken as 4 %.

f_{aw} = factor to correct for losses due to absorption in air or window of external counters.

f_d = factor to correct for dispersion of the source from a point configuration. The effect of this variable is eliminated by preparing the standard in the same configuration as the samples.

f_{ssa} = factor to correct for the absorption and scattering of alpha particles in the sample and its mount. This is covered further in Section 19.

f_c = factor to correct for losses due to the resolving time of the detector and its associated electronics.

f_c = $(1 - nt)$ where n is the observed counting rate, cpm, and t is the resolving time in minutes.

16.2 Alpha counters have low backgrounds and high efficiencies. However, some counters are easily contaminated internally and care should be taken to avoid contamination. Silicon detectors operated in vacuum may become contaminated due to recoil from sources. Recoil contamination can be eliminated by maintaining an air absorber of 12 µg/cm² between the source and the detector (4).

17. Interference

17.1 Some alpha counters are sensitive to beta radiation with a degree of efficiency depending on the detector (2, 3). In these cases, electronic discrimination is often used to eliminate the smaller pulses due to beta particles.

18. Apparatus

18.1 *Ionization Chambers*—Alpha particles entering the sensitive region of an ionization chamber produce dense ionization of the counting gas. The electrons are collected at the anode, thus developing a voltage pulse. Gridded ion chambers are used for alpha spectrometry but have been replaced by solid-state detectors. The major advantage of the ion chamber is its ability to measure large area samples with essentially a 2π geometry. The chambers are operated at an over-pressure with gas mixtures such as 10 % methane-90 % argon or 2 % ethylene-98 % argon. The peak resolution of such a detector is about 50-keV FWHM (see Practice D 3084).

18.2 *Proportional Counters:*

18.2.1 Alpha particles entering the sensitive region of a proportional counter produce ionization of the counting gas. In this case, the electrons are accelerated towards the anode, producing secondary ionization and developing a large voltage pulse by gas amplification. Proportional counters are usually operated in the "limited proportional" range, where the total ionization is proportional to the primary ionization produced by the alpha particle.

18.2.2 Proportional detectors are generally constructed of stainless steel or aluminum (see Fig. 6). No additional shielding is required for alpha proportional counting. The counter should be capable of accepting mounts up to 51 mm in diameter. Proportional detectors are available in two types, either with or without a window between the sample and the counting chamber. The manufacturer's specifications for either type should include performance estimates of background count rate, length and slope of the voltage plateau, and efficiency of counting a specified electrodeposited standard source, along with the type of gas used in the tests. For a window flow counter, the window thickness, in milligrams per square centimetre, should also be specified. With windowless low counter the sample and sample mount should be made of an electrical conductor in order to avoid erratic behavior due to static charge buildup.

18.2.3 Alpha emitters are counted with proportional instruments in 2π, or 50 %, configuration. Two-π geometry is obtained by placing the sample on a flat planchet inside the detector. Half the alpha particles are emitted downward into the planchet, of which approximately 2 % are backscattered into the upward direction. The other half are emitted upward into the gas volume.

18.2.4 Typical parameters for the alpha windowless flow counter are: background count rate—10 counts/h; length of voltage plateau—300 V; and slope of voltage plateau—1 %/100 V for an electrodeposited source. For a window flow counter, typical values are: window thickness—1 mg/cm²; background count rate—10 counts/h; length of voltage plateau—300 V; slope of voltage plateau—1 %/100 V for an electrodeposited source; and efficiency—35 to 40 % for an electrodeposited source. Gases commonly used in both types of alpha proportional counters are 10 % methane-90 % argon, pure methane, or pure argon.

18.3 *Scintillation Counters*

18.3.1 In a scintillation counter, the alpha particle transfers energy to a scintillator, such as zinc sulfide (silver activated). The transfer of energy to the scintillator results in the production of light at a wavelength characteristic to the scintillator, and with an intensity proportional to the energy transmitted from the alpha particle. The scintillator medium is placed in close proximity to the cathode of a multiplier phototube; light photons from the scintillator strike the photocathode and electrons are emitted. The photoelectrons are amplified by the multiplier phototube and a voltage pulse is produced at the anode.

18.3.2 The counter size is limited by the multiplier phototube size, a diameter of 51 mm being the most common. Two types of systems may be employed. In the first, the phosphor is optically coupled to the multiplier phototube and either covered with a thin (<1 mg/cm²) opaque window or enclosed in a light-proof sample changer. With the sample placed as close as possible to the scintillator, efficiencies approaching 40 % may be obtained. The second system employs a bare multiplier phototube housed in a light-proof assembly. The sample is mounted in contact with a disposable zinc sulfide disk and placed on the phototube for counting. This system gives efficiencies approaching 50 %, a slightly lower background, and less chance of counter contamination.

18.3.3 A major advantage of alpha scintillation counting is that the sample need not be conducting. For a 51-mm multiplier phototube with the phosphor coupled to the tube, typical values obtained are: background count rate—0.006 counts/s; and efficiency for an electrodeposited standard

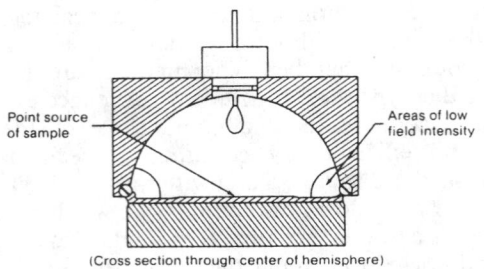

Point source of sample

Areas of low field intensity

(Cross section through center of hemisphere)

FIG. 6 Typical Chamber Geometry for Proportional Detector

source—35 to 40 %. With a disposable phosphor mounted on the sample, typical values are: background count rate—0.003 counts/s; and efficiency for an electrodeposited standard source—45 to 50 %. For both systems, voltage plateau length is 150 V with a slope of 5 %/100 V.

18.3.4 Liquid scintillation counting of alpha emitters with a commercially available instrument overcomes many of the problems inherent in other techniques (6, 9, 10, 23). Typical background counting rates range from 0.1 to 0.2 counts/s. Sample preparation involves mixing the sample aliquot with a suitable liquid scintillator solution or gel phosphor before counting. Planchet preparation is eliminated, volatile components are retained, and the completely enclosed sample cannot contaminate the counting chamber. The sample is uniformly distributed in the scintillator so there is no self-absorption, resulting in a counting efficiency of almost 100 %. Because of the high alpha energies, considerable chemical quenching effects can be tolerated before counting efficiency is reduced. Coincidence losses are small in liquid scintillation counting at count rates up to 2×10^4 cps. For samples that contain both alpha and high-energy beta emitters, difficulties do arise in distinguishing between the two. The problem is due primarily to the broad continuum of beta energy distribution up to the maximum energy and the poor resolution of liquid scintillation spectrometers. This problem is aggravated because the light yield per million electronvolt of alpha particles in most liquid scintillators is approximately tenfold lower than a beta particle of equivalent energy, putting the pulses from alphas and high-energy betas in the same region. Correction for beta activity may be made by certain mathematical or graphical techniques. It is preferable to separate the alpha emitter from the bulk of the beta activity by chemistry.

18.4 *Semiconductor Detectors*

18.4.1 The semiconductor detectors used for alpha counting are essentially solid-state ionization chambers. The ionization of the gas in an ionization chamber by alpha particles produces electron-ion pairs, while in a semiconductor detector electron-hole pairs are produced. The liberated charge is collected by an electric field and amplified by a charge-sensitive amplifier. In general, silicon surface barrier detectors are used for alpha counting. These detectors are *n*-type base material upon which gold is evaporated to make a contact. The semiconductor material must have a high resistivity since the background is a function of the leakage current. This leakage current is present in an electric field since the starting material is a semiconductor, not an insulator. The reversed bias that is applied reduces the leakage current and a "depletion layer" of free-charge

carriers is created. This layer is very thin and the leakage current is extremely low; therefore, the interactions of photons with the detector will have negligible effect. Since the detector shows a linear response with energy, any interactions of beta particles with the detector can be eliminated by electronic discrimination. The semiconductor is of special interest in alpha counting where spectrometric measurements may be made since the average energy required to produce an electron-hole pair in silicon is 3.5 ± 0.1 eV compared to the 25 to 30 eV needed to produce an ion pair in a gridded ionization chamber. Consequently, silicon detectors provide much improved resolution and also normally have lower background count rates.

18.4.2 The detector size is generally less than 25 mm in diameter since the resolution decreases and cost increases with detector size. For best results, the sample should be electrodeposited to make a lower mass source. The detector is operated in a vacuum chamber. Typical backgrounds range from 8×10^{-5} to 2×10^{-4} counts/s. Additional specifications for detectors may be found in Practice D 3084.

19. Absorption

19.1 The range of alpha particles is a few centimetres in air and a few thousandths of a centimetre in solids. Even fairly short path lengths in air and passage through thin windows will result in the absorption of some fraction of the alpha radiation.

20. Calibration

20.1 Calibrate alpha-counting equipment for specific nuclide measurement using a standard of similar alpha energy measured under exactly the same conditions as the sample that is to be counted.

20.2 When the gross alpha activity of a sample is to be measured (Test Method D 1943), calibrate the counting equipment using a standard. The standard should contain the same weight and distribution of solids as the sample and be mounted in an identical manner. If the samples contain variable amounts of solids or carrier, prepare a calibration curve relating the weight of solids present to counting efficiency. Express the efficiency factor (E) as a fraction of the disintegration rate of the standard.

20.3 Calibrated standards of plutonium-239, americium-241, polonium-210, thorium-228, radium-226, uranium-233, uranium-235, and uranium (natural) are readily available either from the National Institute of Standards and Technology or commercial organizations. Calibrated standards of other alpha-emitting radionuclides may be procured from the above suppliers upon special request.

21. Source Preparation

21.1 Appropriately mount the sample whose alpha activity is to be determined for the final measurement. This can be accomplished in a number of ways, some of which are dependent upon the counting method used. In all cases, however, mount the sample in a uniform and reproducible manner.

21.2 The most straightforward and commonly used method of preparing a sample for alpha counting is by evaporation onto a suitable planchet. Evaporated sources are usually counted in end-window or windowless proportional counters or in zinc sulfide scintillation counters. Producing a thin, uniform, and reproducible source by this technique is difficult and considerable care is necessary. Carry out evaporations slowly to avoid spattering and usually perform under infrared lamps.

21.3 Mounting of solid-free samples resulting from radiochemical separations of specific nuclides is best accomplished by evaporating the sample in the center of a planchet. Define an area in the center of the planchet and keep the sample within this area during evaporation. This can usually be accomplished by carefully evaporating the sample in small portions. Samples that contain solids present the problem that the material does not evaporate uniformly but tends to deposit in crystals and aggregates, producing uneven deposits consisting of rings, ridges, or localized amounts of solids. This will be the case when water samples are prepared for alpha counting without chemical separation of specific nuclides. These samples are best mounted by allowing the sample to evaporate on the entire surface of the planchet. This will produce a more uniform source and minimize the self-absorption effects. Surfactants can be used to spread the sample during evaporation. Dissolution of the solids on the planchet with water and re-evaporation may be necessary to obtain a uniformly distributed source.

21.4 Samples that contain carriers can be mounted using a precipitation technique. Keep the amounts of carrier used small (<10 mg) so as to minimize self-absorption. Precipitate the carrier by suitable chemical procedures and mount the precipitate for counting by either filtration through a filter or by evaporation of a slurry of the precipitate onto a planchet. The final surface density of such a deposit should be less than 1 mg/cm^2. Count these samples in an end-window proportional counter or in a zinc-scintillation counter because they are nonconductors.

21.5 Filtration of a precipitate onto a filter produces a more uniform and reproducible source than the evaporation technique. When evaporating a slurry of a precipitate onto a planchet, take care to obtain a uniform and reproducible source. However, this is usually more easily obtained with a slurry than when evaporating a solution.

21.6 Electrodeposition is the most effective and widely used technique for preparation of a thin, uniform, and reproducible source for alpha counting. Electroplated sources can be counted in any alpha-counting equipment except, of course, a liquid scintillation counter. The chemical separation of specific nuclides from the sample is required prior to electrodeposition. Electrodeposit the alpha-emitting element onto a flat metal disk. Platinum, stainless steel, nickel, and tantalum have been used for this purpose. Information on electrodeposition of alpha emitters has been published (24–26). The electrodeposition technique produces the most suitable sources for alpha spectrometry. One disadvantage of this technique is that for some samples quantitative deposition is not always obtained and a recovery correction factor must be applied. Electrodeposition followed by alpha spectrometry is particularly useful for alpha analysis in which yield measurements are determined by the addition of a known amount of an isotope of the same element to be measured (see Practice D 3084).

21.7 The use of liquid scintillation counting avoids some of the difficulties encountered in preparing samples for solid source counting, but has the disadvantage of blindly counting all alphas, as does any gross method. Self-absorption problems are eliminated and counting efficiencies are 90 to 100 %. Prepare samples for liquid scintillation counting by dissolving the sample in a scintillation mixture (27, 28) or extracting the alpha-emitting element into a scintillation mixture containing the extractant (29).

22. Calculation

22.1 This method is useful for comparing activities of a group of samples as in a tracer experiment. The following equation may be used to calculate the results:

$$\text{Activity, } C = \frac{1}{VY}\left(\frac{N}{t} - \frac{N_B}{t_B}\right) \quad (18)$$

where:
N = number of counts accumulated,
t = sample counting period, s,
N_B = number of background counts accumulated,
t_B = background counting period, s,
C = net counts per second per millilitre, cps/mL,
V = volume of sample, mL, and
Y = recovery of species (unity in methods where complete recovery is assumed).

22.2 The disintegration rate or concentration in microcuries may be calculated as follows:

$$\text{Disintegration rate, } D = C/E \quad (19)$$

where:
D = disintegrations per second per millilitre, dps/mL or Bq, and
E = efficiency of counter (fraction), cps/dps.

$$\text{Concentration, } A = D/(3.7 \times 10^4) \quad (20)$$

where A = concentration in microcuries per millilitre (µCi/ml).

22.3 Results may also be reported in terms of equivalent natural uranium activity, employing an efficiency determined by use of a natural uranium reference standard.

$$\text{Alpha activity equivalent to natural uranium,}$$
$$\text{Bq/mL} = C/E_u \quad (21)$$

where E_u = efficiency of counter for natural uranium (fraction).

22.4 To properly evaluate the result, calculate the uncertainty associated with the concentration in accordance with the recommended procedure in Section 10.

BETA COUNTING

23. Scope

23.1 This practice covers the measurement of the beta particle radioactivity of water. It also covers the general techniques used to prepare and measure the activity resulting from radiochemical separation of specific nuclides or groups of nuclides in water samples. It is applicable to beta emitters with activity levels above 0.4 dps per sample for most counting systems. For samples of lower activity, see Practice D 3085 for specific techniques. The method is not applicable to samples containing radionuclides that are volatile under conditions of the analysis. General information on radioactivity and measurement of radiation may be found in the literature (7, 23).

24. Summary of Method

24.1 Beta radioactivity may be measured by one of several types of instruments that provide a detector and a combined amplifier, power supply, and scaler. The most widely used detectors are proportional or Geiger-Mueller counters, but scintillation systems offer certain advantages.

24.2 Among the gas ionization-type detectors, the proportional-type counter is preferable because of the shorter resolving time and greater stability of the instrument. For preparing solid sources from water samples for beta activity measurement, the test sample is reduced to the minimum weight of solid material having measurable beta activity by precipitation, ion exchange, or evaporation techniques. For measuring solid sources resulting from individual radiochemical separation procedures, the precipitate is appropriately mounted for counting.

24.3 Beta particles entering the sensitive region of the detector produce ionization or scintillation photons that are converted into an electrical pulse suitable for counting. The number of pulses per unit time is directly related to the disintegration rate of the sample by an overall efficiency factor. This factor combines the effects of sample-to-detector geometry, sample self-shielding, backscatter, absorption in air and in the detector window (if any), and detector efficiency. Because most of these individual components in the overall beta-particle detection efficiency factor vary with beta energy, the situation can become complex when a mixture of beta emitters is present in the sample. The overall detection efficiency factor may be empirically determined with prepared standards of composition identical to those of the sample specimen, or an arbitrary efficiency factor can be defined in terms of a single standard such as cesium-137 or other nuclide. Gross counts can provide only a very limited amount of information and therefore should be used only for screening purposes or to indicate trends.

24.4 Liquid scintillation counting avoids many sources of error associated with counting solid beta sources: such as (1) self-absorption, (2) backscattering, (3) loss of activity during evaporation due to volatilization or spattering, and (4) variable detection efficiency over a wide beta-energy range. In addition to the greatly improved accuracy offered by liquid scintillation counting, sample preparation time and counting times are significantly shorter. Sample preparation involves only adding a sample aliquot to the scintillator or gel phosphor. Because every radioactive atom is essentially surrounded by detector molecules, the probability of detection is quite high even for low-energy beta particles. Radionuclides having maximum beta energies of 200 keV or more are detected with essentially 100 % efficiency. Liquid scintillation can, at times, be disadvantageous due to chemiluminescence, phosphorescence, quenching, or the typically higher backgrounds.

24.5 Organic scintillators, such as p-terphenyl plus a wave shifter in a plastic monomer, are polymerized to form sheet material of any desired thickness. The plastic phosphor counting system (8) has its widest use as a beta particle detector for separated, solid samples rather than for beta spectrometry applications.

24.6 The plastic beta scintillator phosphor is mounted directly on the sample and is discarded after counting. The phosphor-sample sandwich is placed in direct contact with the photomultiplier tube yielding essentially a 2π configuration. Since the output pulse of the detector system is energy dependent, the counting efficiency for a given phosphor thickness of 0.25 mm yields the highest counting efficiency with the lowest background.

24.7 Solid samples (precipitates from radiochemical separations) containing 3 to 5 mg/cm² of stable carrier are measured in such a system. For yttrium-90 a solid sample of this type would have a counting efficiency of 45 to 50 %.

24.8 A plastic scintillator/phosphor system with a 25-mm photomultiplier tube shielded with 12.7 mm of lead has background in the order of 4×10^{-2} cps. For very low backgrounds, about 4×10^{-3} cps, the photomultiplier tube and sample assembly are fitted into a well-type hollow anode Geiger tube operated in anti-coincidence. The entire assembly is then placed in a heavy shield.

24.9 The system has many advantages but reduction of background is probably most important. The reduction occurs since the scintillation does not see the surrounding mechanical components of the counter. The additional advantage of keeping the counter itself free from contamination by enclosing the phosphor-sample sandwich is also important.

24.10 A note of caution is advisable at this point. Any beta particle detection system, whether internal gas counters or scintillation counters, will detect alpha particles. It is not possible to electronically discriminate against all the alpha pulses.

24.11 If a sample is suspected of containing alpha activity, a separate alpha measurement must be made to determine the alpha contribution to the beta measurement.

25. Solid Source Counting

25.1 The observed count rate for a solid source is the result of the interaction of many variables. The most important of these are the effects of geometry, backscatter, radiation, source diameter, self-scatter and self-absorption, absorption in air and the detector window (for external counters), and coincidence counting losses. These effects have been discussed (17, 23) and in many cases can be reduced or corrected for by counting test standards and samples under identical conditions. For absolute measurements of a single, specific nuclide, apply appropriate correction factors. These effects may be described by the following relation:

$$\text{cps} = \text{dps}\,(G_p)(f_{bs})(f_{aw})(f_d)(f_{ssa})(f_c) \tag{22}$$

where:

cps = recorded counts per second, corrected for background,

dps = disintegrations per second (Bq) yielding beta particles,

G_p = source-to-detector geometry, which is the solid angle subtended by the sensitive area of the detector. This is corrected for most easily by maintaining constant counting configuration and geometry for standards and sample measurements.

f_{bs} = backscatter factor, or ratio of cps with backing to cps without backing. This is the phenomenon in which particles originally emitted away from the detector are scattered back toward it by reflection off the source-backing material. Its value ranges from 1 to almost 2, depending on the thickness and atomic number of the backing material. The backscatter error may be minimized by using identical backing material for counting standard and samples.

f_{aw} = factor to correct for losses due to absorption in the air and window of external counters. It is equal to the ratio of the actual counting rate to that which would be obtained if there were no absorption by the air and window between the source and the sensitive volume of the detector. It is related to the absorption coefficient and density of the absorber by the approximate equation: f_{aw} = exp $(-\mu x)$, where μ = absorption coefficient, in square centimetres per milligram, and x = absorber density in milligrams per square centimetre. In practice, it can be accounted for, together with the geometric factor, by maintaining the geometry of the counting configuration constant for standards and samples.

f_d = factor to correct a disk-source counting rate to the counting rate of the same activity as a point source on the same axis of the system. This can be corrected for by preparing and counting the standard in the same configuration as the samples.

f_{ssa} = factor to correct for the absorption and scatter of beta particles within the material accompanying the radioactive element. This is further discussed in 28.1.

f_c = factor to correct the counting rate for instrument-resolving time losses and defined by the simplified equation f_c = $(1 - nt)$, where n = observed counts per second and t = instrument resolving time in seconds. Generally, the sample size or source-to-detector distance is selected to preclude counting losses. Such losses are negligible for count rates of less than 13 000 counts per second if proportional detectors are used. Information on the effect of instrument-resolving time on the sample count rate, as well as methods for determining the resolving time of the counting systems, may be found in the literature (23).

26. Liquid Scintillation Counting

26.1 The observed count rate for a liquid scintillation sample is directly related to the beta (plus conversion electron) and positron emission rate in most cases. The important exceptions are: (1) beta emitters whose maximum energy is below 200 keV, and (2) counting systems wherein quenching decreases the expected photon yield, thereby decreasing the overall detection efficiency significantly below 100 %. Low-energy beta emitters, such as tritium or carbon-14, can be measured accurately only when the appropriate detection efficiency factor has been determined with a known amount of the same radionuclide counted under identical conditions. Quenching losses are greatest at low beta energies. Quenching may be evaluated by comparison to known-quench standards of the same radionuclide, using the channel ratio technique, or with other techniques as

described in the manufacturer's instructions. For absolute measurements of low-energy beta emitters, apply appropriate correction factors as follows:

$$Bq = \frac{cps}{(fe)(fq)} \qquad (23)$$

where:

Bq = disintegrations per second yielding beta particles,

cps = recorded counts per second, corrected for background,

fe = detection efficiency factor (see Note); observed cps/dps ratio for known standard of some radionuclide counted under identical conditions with no quenching, and

fq = quench correction factor; observed ratio of cps quenched to cps unquenched for known standard of same radionuclide with equivalent quenching.

NOTE—The detection efficiency may be assumed to be unity for beta emitters having maximum energies of 200 keV or greater.

26.2 In tracer studies or tests requiring only measurements in which data are expressed relative to a defined standard, the individual correction factors cancel whenever sample composition, sample weight, and counting configuration and geometry remain constant during the standardization and tests.

26.3 The limit of sensitivity for both Geiger-Mueller and proportional counters is a function of the background counting rate. Massive shielding or anti-coincidence detectors and circuitry, or both, are generally used to reduce the background counting rate to increase the sensitivity (23). For a more complete discussion of this, see Practice D 3085.

27. Interferences

27.1 *Solid Sources:*

27.1.1 Material interposed between the test sample and the detector, as well as increasing mass in the radioactive sample itself, produces significant losses in sample counting rates. Since the absorption of beta particles in the sample solids increases with increasing mass and varies inversely with the maximum beta energy, sample residue must remain constant between related test samples and should duplicate the residue of the evaporated standard.

27.1.2 Most beta-particle counters are somewhat sensitive to alpha, gamma, and X-ray radiations, with the degree of efficiency dependent upon the type of detector (23). The effect of the interfering radiations on the beta counting rate is more easily evaluated with external-type counters where appropriate absorbers can be placed between the sample and the detector to evaluate the effects of interfering radiation.

27.2 *Liquid Scintillation*—Liquid scintillation samples are subject to interference by substances that quench or enhance the scintillation process. This includes any chemiluminescence that changes the photon yield. Substances such as oxidants, organohalides, ketones, and aldehydes are to be particularly avoided. A second type of quenching is color quenching, in which a colored solution impairs the light collection efficiency for the photons produced.

28. Apparatus

28.1 *Beta Particle Counter*—The end-window Geiger-Mueller tube and the internal or external proportional gas-flow chambers are the two most prevalent types of detectors. Other types of detectors include scintillators and solid-state detectors. The material used in the construction of the detector and its surroundings should contain a minimal level of detectable radioactivity. If the detector is of the window-type, the window thickness may be used in calculating beta-ray attenuation; however, direct calibration of the entire counting system with standards is recommended. The manufacturer should provide all settings and data required for reliable and accurate operation of the instrument. Detectors requiring external positioning of the test sample shall include a support of low-density material (aluminum or plastic), which ensures a reproducible geometry between the sample and the detector. Because different sample-to-detector geometries are convenient for differing sample activity levels, the sample support may provide several fixed positions ranging from 5 to 100 mm from the detector.

28.2 *Liquid Scintillation*—Liquid scintillation counting systems use an organic phosphor as the primary detector. This organic phosphor is combined with the sample in an appropriate solvent that achieves a uniform dispersion. A second organic phosphor often is included in the liquid scintillation "cocktail" as a wavelength shifter. The wavelength shifter efficiently absorbs the photons of the primary phosphor and re-emits them at a longer wavelength more compatible with the multiplier phototube. Liquid scintillation counting systems use either a single multiplier phototube or two multiplier phototubes in coincidence. The coincidence counting arrangement is less likely to accept a spurious noise pulse that occurs in a single phototube, and thus provides lower background. The requirement that both multiplier phototubes respond to each has a slight effect on the overall detection efficiency of betas with E-max >200 keV; however, system response to beta E-max <200 keV will be significant. The need to minimize detectable radioactivity in the detector and its surroundings is likewise important in liquid scintillation counting. To achieve this, scintillation-grade organic phosphors and solvents are prepared from low carbon-14 materials such as petroleum. The counting vials are of low potassium glass or plastic to minimize counts due to potassium-40. Liquid scintillation provides a fixed geometry from a given size counting vial and liquid volume.

28.3 *Detector Shield*—The detector assembly should be surrounded by an external radiation shield of massive metal equivalent to approximately 5100 mm of lead and lined with 3.2-mm thick aluminum. The material of construction should be of minimal detectable radioactivity. The shield should have a door or port for inserting or removing specimens. Detectors having other than completely opaque windows are light-sensitive. The design of the shield and its openings should eliminate direct light paths to the detector window; beveling of the door and the opening generally is satisfactory. Liquid scintillation counting systems must provide an interlock that protects the photocathode of the multiplier phototube from light when the sample counting chamber is opened.

28.4 *Associated Electronic Equipment*—The high-voltage power supply amplifier, scaler, and mechanical register normally are contained in a single chassis. The power supply and amplifier sections are matched with the type of detector to produce satisfactory operating characteristics and to provide sufficient range in adjustments to maintain stable conditions. The scaler should have a capacity for storing and visually displaying at least 9×10^5 counts. The instrument should have an adjustable input sensitivity matched to that of the detector, and variable high-voltage power supply. (An adjustable power supply and meter are unnecessary for liquid scintillation systems.) Counting chambers of Geiger-Mueller and proportional counters contain a suitable counting gas and an electrode. Counting rates that exceed 200 cps must be corrected for dead-time loss when using a Geiger-Mueller tube. As the applied voltage to the electrode is increased, the counting chamber exhibits responses that are characteristic of a particular voltage region. At low voltages of the order of 100 V, there is no multiplication of the ionization caused by a charged particle. At voltages approaching 1000 V, there is appreciable amplification of any ionization within the counting chamber; however, the size of the output pulse is proportional to the amount of initial ionization. When operated in this voltage region, the device is known as a proportional counter. Usually, there is a region at least 100 V wide, known as a plateau, wherein the count rate of a standard is relatively unaffected. The operating voltage for proportional counters is selected to approximate the middle of this plateau in order to maintain stable responses during small voltage shifts. The plateau region is determined by counting a given source at voltage settings that differ by 25 or 50 V. The number of counts at each setting is recorded, and the resultant counts versus voltage are plotted as shown in Fig. 7. Voltage plateau curves are to be remeasured periodically to ensure continued instrument stability, or whenever an instrument malfunction is indicated. If the voltage is increased beyond the proportional region into the 1500 to 2000-V region, the pulse size increases and the dependence on the initial ionization intensity disappears. This is the beginning of the Geiger-counting region, where a single ion pair produces the same large pulse as an intense initial ionization.

28.5 *Alpha Interference*—Alpha particle interference can be substantial and must be considered with any type of beta counter. One technique involves placing a thin absorber between the solid source and the detector. The absorber diameter should exceed that of the detector window. The absorber should be placed against the window to minimize beta particle scatter. Any absorber that stops alpha particles will also attenuate low-energy beta particles somewhat. For example, an aluminum absorber of 7 mg/cm² will absorb 48 % of beta particles of 0.35-MeV maximum energy. The alpha-particle absorber is not recommended for use with internal beta-particle detectors, especially when either the composition or activity ratios of the radioelements or radioactivity level might vary significantly between samples. Chemical separation of the alpha- and beta-particle emitters produces a higher degree of accuracy for internal detector measurements. Published information on beta-particle absorption (23) should be used as a guide for use of an absorber. In liquid scintillation spectra, the alpha component appears as a peak on the beta continuum and thus provides a basis for resolving the two (30).

28.6 *Self-Absorption and Backscatter:*
28.6.1 Radioisotopes emit radiation uniformly in all directions. If the radioisotope is intimately mixed with the

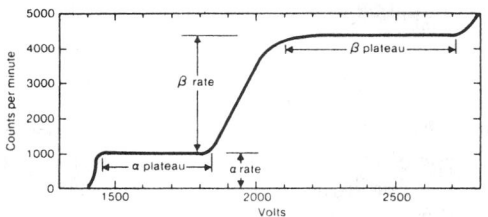

FIG. 7 Counting Rate as a Function of Applied Voltage for a Proportional Counter Exposed to a Source Emitting Both Alpha and Beta Particles

detector, as in the case with liquid scintillation counting, essentially all beta particles interact with the detector. With solid source counting, however, the source and detector usually are related by a solid angle of 2π or less. Even so, the number of beta particles reaching the detector can be significantly affected by other factors.

28.6.2 When a radioactive source decays by beta emission, the beta particles (electrons) are emitted with a distribution of energies ranging from zero to definite maximum value. The lower energies are absorbed by the sample itself and this self-absorption is particularly large for certain nuclides, such as carbon-14 and sulfur-35, in which all the beta particles are of low energy. It is therefore desirable to make solid samples as thin as possible.

28.6.3 Self-absorption corrections can be made, if necessary. A series of counts is made on a number of weighed samples of homogeneous material, with the material spread uniformly over the sample mount. The data are corrected for background and graphed on a semilog plot, with the logarithm of the counting rate per milligram of sample plotted against the number of milligrams. The resulting curve can be extrapolated to zero sample weight to give the counting rate per milligram corrected for self-absorption (3).

28.6.4 Samples consisting of the final product from radiochemical separations generally are of the same approximate weight from determination to determination because they usually contain the same amount of carrier, and the self-absorption is essentially constant. Corrections can be made by running a standard amount of activity of the nuclide through the procedure, with sample activities expressed in terms of this reference standard.

28.6.5 Beta particles are easily deflected in passing through matter. Some beta particles emitted in a direction away from the detector can be deflected by the source-backing material and scattered toward the detector. This phenomenon leads to an increased number of beta particles reaching the detector over that expressed from geometric considerations alone. Backscatter increases rapidly with increasing thickness and atomic number of the backing material. Backscatter effects may be held constant by maintaining sample size and mounts constant for relative measurements. If backscatter corrections must be made, these corrections can be determined by counting a sample first on thin film mount, and then recounting when backed by the desired backing material. Backscatter does have the effect of increasing the number of particles that reach the detector, thereby increasing the counting efficiency.

29. Calibration and Standardization

29.1 *Calibration and Standardization for General Measurements—Solid Sources*—Place a known amount of cesium-137 or other activity (approximately 20 to 200 Bq, with the smaller amount more appropriate for Geiger-Mueller detectors) in a volume of water containing salts equivalent to those of the test samples and prepare for counting as directed. Throughout the experiment, the evaporation, mounting, counting, and the density of the plate solid of this reference must be identical with those of the test samples. Count for the length of time necessary to produce the desired statistical reliability. Then express combined efficiency factor as a percentage of the disintegration rate of the reference standard.

29.1.1 **Caution:** This factor is inaccurate for beta particles whose energies differ appreciably from those of cesium-137; nevertheless, count rates for solid beta sources of undefined energy are often reported relative to the cesium-137 standard. The fact that the 662-keV gamma ray of barium-137 is significantly converted accounts for the fact that about 1.10 electrons leave the source for every disintegration. For this reason, calibrate cesium-137 standards in terms of combined beta plus electron emission rate.

29.2 *Liquid Scintillation*—Add a known amount of cesium-137 (approximately 370 Bq) to the liquid scintillation "cocktail." The volume and composition of the cesium solution and the "cocktail" should simulate those normally used. For samples without serious quenching problems the counting efficiency should approach 100 %. (Note in 29.1 that cesium-137 standards must be calibrated in terms of beta plus conversion electron emission rate.)

29.3 *Calibration and Standardization for Tracer Experiments*—Add a known quantity of activity of a reference solution of the tracer (approximately 200 Bq) to a nonradioactive standard test sample and process.

30. Source Preparation

30.1 *Solid Sources:*

30.1.1 Appropriately mount the sample whose activity is being determined for the final measurement. The exact form of the mount is somewhat dependent upon the particular instrument for counting. In general, most counters will accept either flat plate or dish mounts.

30.1.2 Water samples can be evaporated on either flat disks or dishes, with the dishes being preferred for high solids samples. The dishes should have a flat or concentrically ringed bottom whose diameter is no greater than that of the detector window, preferably having 3-mm high side walls with the angle between the dish bottom and side equal to or greater than 120 deg to reduce side-wall scattering. Sample dishes with verticle side walls may be used but the exact positioning of these dishes relative to the detector is very important. This factor becomes critical for dishes having the same diameter as the detector. Dishes having side walls more than 3 mm in height are not recommended. Dishes should be of a material that will not corrode under the plating conditions and should be of uniform surface density, preferably great enough to reach backscatter saturation (23).

30.1.3 Samples resulting from radiochemical separations are customarily in the form of a precipitate on a filter. These can be mounted on plates, dishes, or other suitable backing.

Fix the filter paper to the backing with a double-backed adhesive strip or a liquid adhesive, and cover to prevent sample loss and contamination of the counting equipment. Such covering can be accomplished with a cover of transparent plastic tape or film, or by spraying the precipitate with a collodion. Do not use nonconducting coverings in internal flow counters.

30.2 *Liquid Sources:*

30.2.1 Liquid samples of sufficiently high activity level may be added directly to an appropriate amount of liquid scintillator solution. For aqueous samples the liquid scintillator solution must contain a solvent, such as dioxane, which is miscible with water, or a surfactant to maintain a single phase with immiscible solvents. Low-activity liquid samples may be concentrated by evaporation before addition to the liquid scintillator solution.

30.2.2 Liquid samples are sometimes concentrated or separated, or both, by extraction into a suitable organic solvent. This solvent often is suitable for direct addition to the liquid scintillator solution. Caution is advised to avoid solvents that cause substantial quenching.

31. Radioactive Decay and Decay Curves

31.1 Decay curves are useful for identifying radionuclides by their characteristic half-lives and in determining the presence and number of interfering radioactive substances. When two or more beta-emitting isotopes are present in a source, an analysis of the decay curve is required to resolve them since the semilog plot of the disintegration rate versus time is a curve whose slope changes as the isotopic composition changes. After sufficient time the longest-lived activity predominates. The half-life of this long-lived component can be determined from extrapolation of the straight line portion of the curve to t_0, the original count rate for that radionuclide, as shown in Fig. 8. In this graph, the extrapolated line of the 4.6-day half-life component subtracted from the original curve yields a curve that represents all other components; this residual curve may be processed again in the same way to reduce (in principle) any complex decay curve into its component parts. In actual practice, uncertainties in the observed data place a practical limit to a three-component system except in unusual cases, and often a two-component system may not be satisfactorily resolved because of similar decay constants. Since the residual curve in Fig. 8 (1-day half-life) does not deviate from the straight line, and the original curve is concave except in the final portion, a two-component system of independently-decaying activities is indicated. In accumulating decay data, care must be taken to avoid disturbing the source and to locate the sample in the same geometrical position for each measurement. If the instrument is calibrated for the specific characteristics of the source and for the energies of the radiations measured, the count rate shown at the ordinate intercept of the extrapolated straight lines at t_0 will be equal to the disintegration rate at t_0. However, this technique requires specific beta-energy knowledge for each component of the sample, as well as counting-instrument standardization at these same energies under identical conditions. In practice, gamma spectrometry is preferred (see Test Method D 2459) for resolving mixtures of all radionuclides except pure beta emitters. An additional complication in decay-curve interpretation is the growth of radioactive daughters. See Ref (17, 23) for a discussion of this.

32. Beta Particle Maximum Energy

32.1 When a radioactive source decays by beta emission, the beta particles (electrons) are emitted with a distribution of energies ranging from zero to a definite maximum value. The maximum energy of the electrons is characteristic of a particular nuclide and is the "beta energy" shown in nuclear data tables. Determining the approximate maximum energy of the beta particles from a radioactive source thus aids in radionuclide identification.

32.2 The combined effects of a continuous beta spectrum and scattering produces an approximately exponential absorption law for a beta-particle source of a given maximum energy. In interposing varying thicknesses of aluminum between the source and the detector and plotting the count rate versus the thickness of aluminum absorber in milligrams per square centimetre on semilog paper, an empirical absorption curve is produced. The plotted data can be compared to a standard absorption curve determined for a particular nuclide to check for other radionuclides. If the sample and absorber are as close as possible to the detector and only one radionuclide is present, the absorption curve is nearly a straight line until it tails into the constant background count rate of the instrument, which in most cases includes the gamma rays from the source. Visual inspection of the point at which the beta activity is not detectable above the total background gives an approximate thickness value, although this is usually low. Better results can be obtained by subtracting the background from the total absorption curve, as shown in Fig. 9, to obtain the 220-mg/cm^2 component. The point at which the straight-line extrapolation intersects the count-rate error of the background is the absorber thickness value. The maximum beta energy can then be obtained from the range energy relation curve (Fig. 10). If the lower energy region of the curve deviates from the straight line extrapolation to zero absorber thickness, a second component is indicated. The values on the extrapolated line subtracted from the second curve produced the 40-mg/cm^2 component. The sum of the count rate for both components at zero absorber thickness is the true count rate.

32.3 The best absorption curve is obtained with the absorber against the detector and the source positioned for minimum clearance of the thickest absorber to be used. The source must not be moved after the first count rate, which is obtained with no absorber between the source and the detector. The activity of the source should be close to the maximum allowed by the resolving time of the instrument. Absorption coefficients given in the literature and half-thickness data utilize the initial portions of the absorption curve and are not as accurate as the total absorption determination. More accurate methods for determination of the beta particles' range, from absorption curves, are the comparison methods of Feather (23) and Harley, et al (31).

32.4 Many liquid scintillation counting systems provide an output suitable for accumulation and storage of beta spectra of known maximum energy. To obtain the most accurate beta maximum energy measurement, the pulse height-energy relationship of the system must be calibrated (32). With the aid of this relationship, the pulse height distribution of the sample may be converted to a Kurie plot (33) and extrapolated to obtain maximum beta energy. Excellent agreement with beta-energy spectrometry values has been obtained with this technique (34). A Kurie plot is shown in Fig. 11.

33. Calculation

33.1 Results may be expressed in observed counts per minute per millilitre for comparing relative activities of a group of samples as in tracer experiments.

$$C = \frac{1}{V}\left[\frac{N}{t} - \frac{N_B}{t_B}\right] \qquad (24)$$

where:

N = number of counts accumulated,
t = sample counting period, s,
N_B = number of background counts accumulated,
t_B = background counting period, s,
V = volume of initial sample, mL, and
C = net counts per second per millilitre (cps/mL).

33.2 Results may also be expressed in terms of equivalent cesium-137 activity, using the empirical efficiency determined for the reference standard:

$$D_{Cs} = C/E_{Cs} \qquad (25)$$

where:

C = net counts per second per millilitre (cps/mL), and

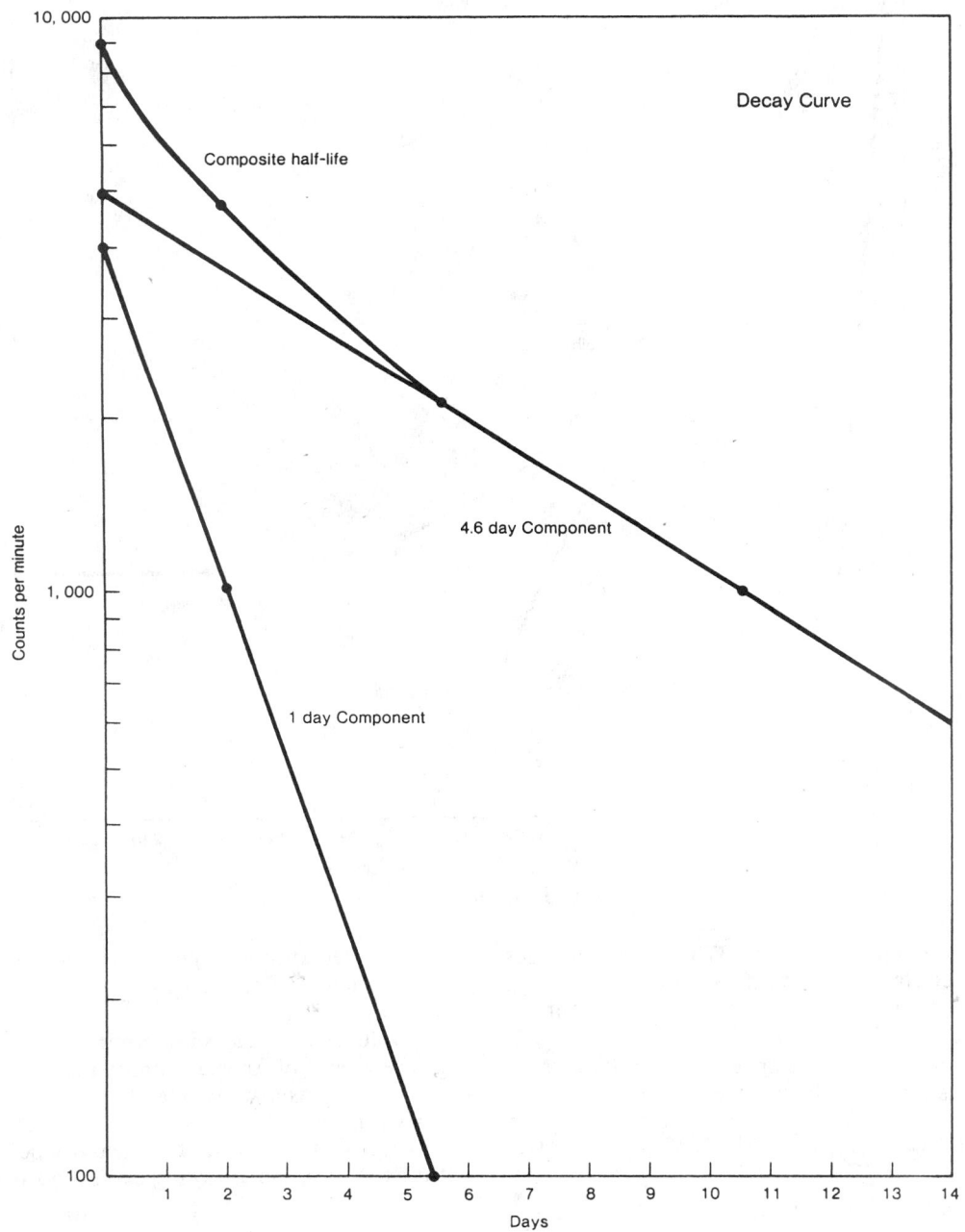

FIG. 8 Decay Curve

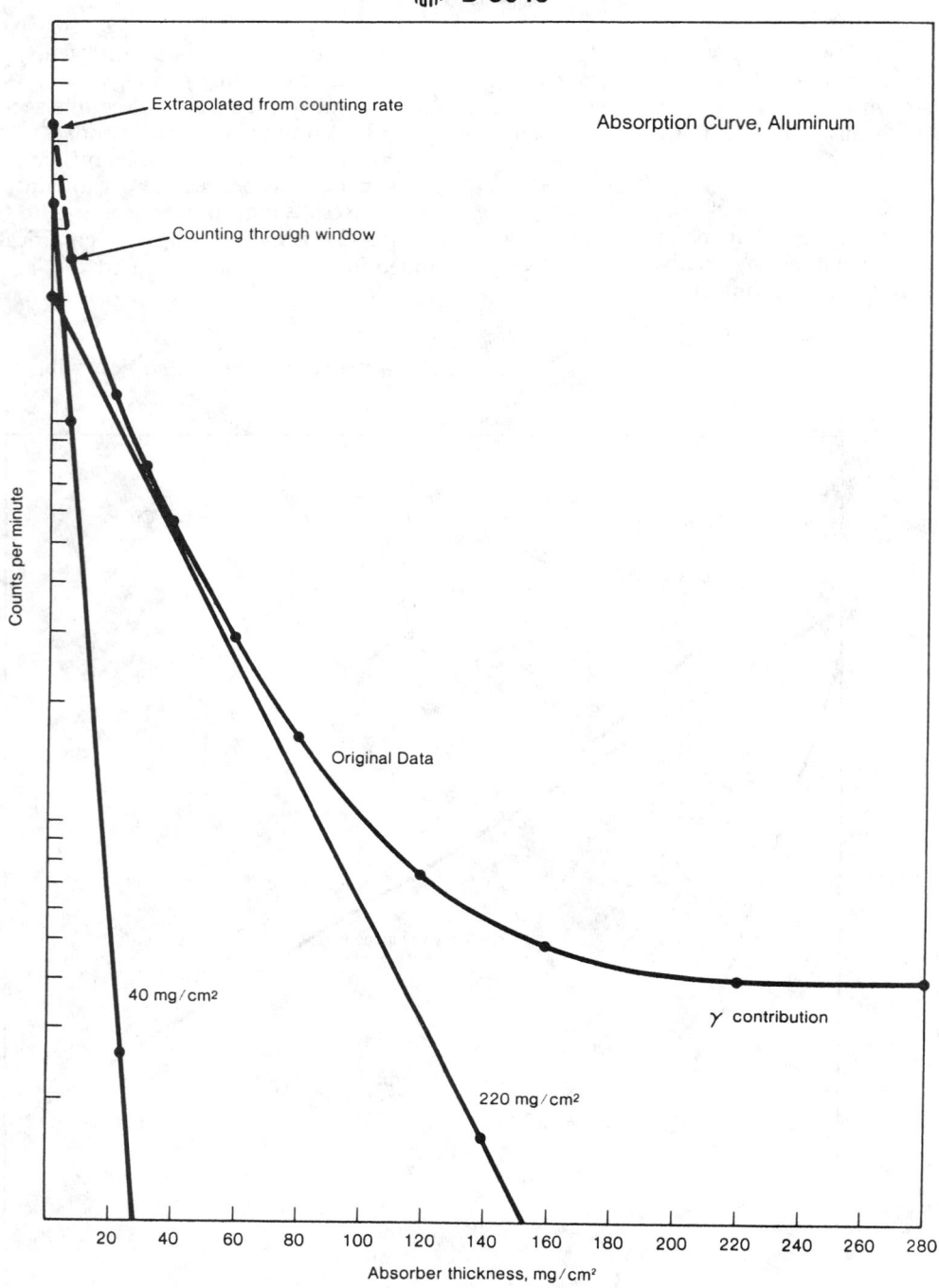

FIG. 9 Absorption Curve

E_{Cs} = efficiency for counting cesium-137 beta particles plus conversion electrons (count rate ÷ emission rate).

33.3 If it is known that only one nuclide is present, its disintegration rate may be determined by use of the efficiency factor measured with a reference standard of the same nuclide. The results may be calculated as follows:

$$D = \frac{1}{EVY}\left[\frac{N}{t} - B\right] \qquad (26)$$

where:
N = number of counts accumulated,
B = background in counts per second (cps),

D = disintegration rate per second per millilitre (Bq/mL),
E = efficiency of counter for the specific nuclide (counts ÷ disintegrations),
V = volume of test specimen, mL,
Y = recovery of species (unity in cases where complete recovery is assumed), and
t = count period, s.

33.4 If it is desired that the activity be expressed in units of microcuries, the following expression is used:

$$A = D/(3.7 \times 10^4) \qquad (27)$$

where:

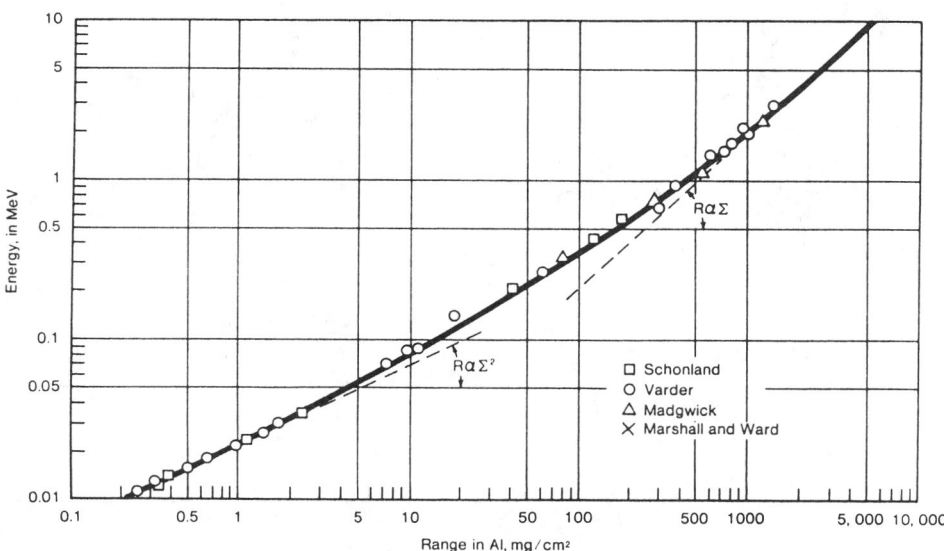

NOTE—Experimental values by several observers on monoenergetic electrons are shown. For monoenergetic electrons, the range coordinate refers to the extrapolated range. For continuous beta-ray spectra, the energy coordinate refers to the end-point energy E . . . , and the range coordinate becomes the maximum range. The smooth curve represents the empirical relationship, developed by Katz and Penfeld (36).

FIG. 10 Empirical Range-Energy Relationship for Electrons Absorbed in Aluminum

A = concentration, μCi/mL,
D = disintegration rate per second per millilitre (Bq/mL), and
3.7×10^4 = conversion factor, disintegration per second per microcurie.

GAMMA COUNTING

34. Scope

34.1 This practice covers the measurement of gross gamma radioactivity of water. Since gamma radiation is a penetrating form of radiation, it can be used for samples of any form and geometry as long as standards of the same form are available and are counted at the same geometry to calibrate the detector. Because of this penetrating nature, small variations in sample density or sample thickness are usually not significant and gamma counting is the preferred method in general radiochemical work. When a standard cannot be obtained in the matrix being counted, a correction for the different absorption in the matrices must be made.

34.2 Since different nuclides emit distinct and constant spectra of gamma radiation, the use of an energy-discriminating system provides identification and measurement of all the components present in a mixture of radioactivity. This technique is covered in Test Method D 2459. Gamma counting and gamma spectrometry are applicable to levels of about 0.4 dps or above. General information on gamma-ray detectors and gamma counting is covered in the literature (13, 23).

35. Summary of Practice

35.1 Gamma counting is generally carried out using solid detectors since a gas-filled detector will not provide adequate stopping power for energetic gammas. In solids such as NaI(Tl) or CsI, the gammas interact by excitation of atoms and energy is transferred to orbital electrons and then released as light photons when the orbits are refilled. The

scintillations are easily detected and amplified into useable electrical pulses by a multiplier phototube. The NaI(Tl) detector is the recommended detector for gross gamma counting.

35.1.1 In semiconductor detectors such as Si(Li) and Ge(Li), the gamma photons produce electron-hole pairs and the electrons are collected by an applied electrical field. A charge-sensitive preamplifier is used to detect the charge transferred and produce a useable electrical pulse. The semiconductor detectors are widely used in gamma spectrometry (see Test Method D 2459) and are not reviewed in this practice.

35.2 The output pulses from the multiplier phototube or preamplifier are directly proportional to the amount of energy deposited, which could either be total and included in the photopeak, or fractional and included in the continuum or escape peaks, in the detector by the incident photon. The pulses may be counted using a scaler or analyzed by pulse height to produce a gamma-ray spectrum.

35.3 Gamma photons interact with the detector by three distinct processes. The photoelectric effect results in complete absorption of the photon energy and produces the full energy or photopeak shown. The Compton effect results in a particle absorption of the photo energy and a scattered photon of lower energy results. The scattered photon carries energy away and the Compton continuum shown in Fig. 12 results. The third interaction is pair production, which occurs at energies above 1.02 MeV and results in the disappearance of the photon when an electron-positron pair is produced. The electron and positron give up their kinetic energy to the detector and the resulting electron joins the electron population of the detector; the positron, however, is annihilated in combining with an electron and produces two gamma photons of 0.511 MeV each. One or both of the 0.511-MeV photons may escape from the detector without interacting and the "single escape" and "double escape"

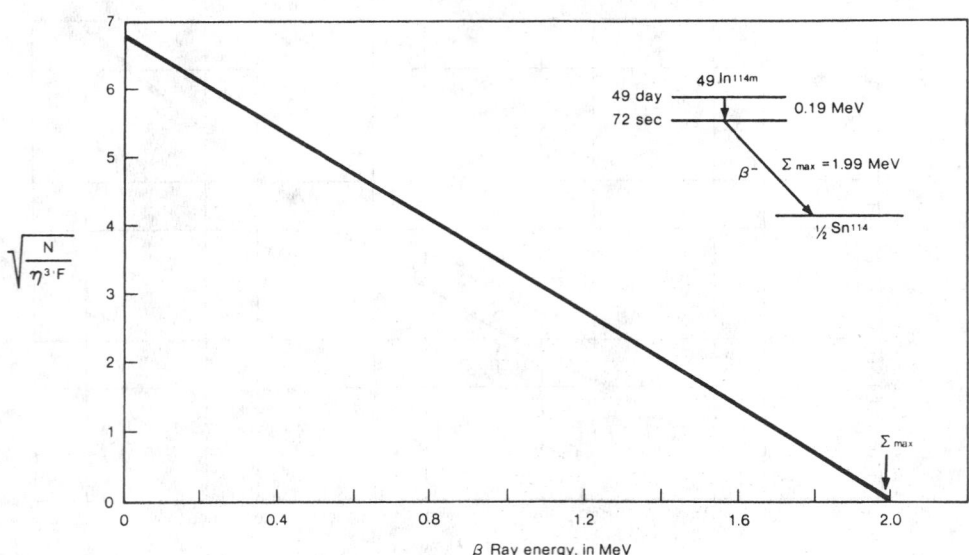

NOTE—Note especially that N means N(n), the number of beta rays in a momentum interval, Δ, of constant size. The horizontal coordinate is the kinetic energy, E, which corresponds to the midpoint of the momentum interval n + Δ_n. When spectral data give a straight line, such as this one, the N(n) is in agreement with the Fermi momentum distribution. The intercept of this straight line, on the energy axis, gives the disintegration energy E max (E_s), if the rest mass of the neutrino is zero (37).

FIG. 11 Kurie Plot of the Beta Spectrum in ¹¹⁴In

peaks shown in Fig. 13 result. The Comptons, from a higher energy photon, always present an interference problem in the counting of gamma photons and appropriate corrections must be made for this effect. Pair production can also be considered as an interference since the escape peaks may have an energy equal to the lower energy gamma of interest.

35.4 The change of the absorption coefficient with gamma energy results in a wide variation of detection efficiency. The detection efficiency falls rapidly as gamma energy increases for a fixed size of detector. Two other important effects are seen as a result of the variation of the absorption coefficient; firstly, low-energy photons may be absorbed in massive samples, such as large bottles of water, and erroneous results may be obtained. A similar absorption effect is seen in Ge(Li) systems where the can around the detector acts as an absorber for very low-energy gammas and the efficiency passes through a maximum usually around 100 KeV. The second result is that for low-energy gammas a thin detector may be as efficient as a much thicker one since the low-energy gammas are easily stopped in the thin detector.

35.5 Because of this variation in efficiency and the possible interferences from other activities, gross gamma counting is only reliable when used to compare standards and samples of the same nuclide. The use of gross gamma monitoring systems should be avoided when possible and, in all cases, proper allowance must be made for the lack of accuracy.

36. Interferences

36.1 The natural background present at all locations is detected very efficiently by gamma detectors and presents a significant interference that must be reduced by the use of shielding. Low-radioactivity-level lead or steel should be used for a 76 by 76-mm NaI detector. Approximately 102 mm of lead or 150 mm of steel shielding produces an acceptable background for most work. Details of shield designs are

given by Heath (35). Lead shields are a source of X rays when high-activity samples are counted. This X ray emission can be reduced in amount and energy by using a graded liner, such as 1.6 mm of cadmium and 0.4 mm of copper, the copper being nearer the detector. The cadmium strongly absorbs the lead X rays and in turn emits its own X rays at a much lower energy. These are strongly absorbed by the copper. Any residual copper X rays are usually below the energy level of interest.

36.2 The "Compton" and "Pair Production" effects discussed in 35.3 can be very significant interferences and must be corrected for.

36.3 At high count rates "random sum peaking" may occur. Two absorptions may occur within the resolving time of the detector and are summed and seen as one pulse. For a detector of resolving time, t, and a count rate of A counts per unit time, the time window available for summing is $2At$ (since the count summed could occur as early as t before or as late as t after the other count) and the probability of another count at any time is simply A. Therefore, the sum count rate will be $2A^2t$ in unit time. Random summing is strongly dependent on the count rate A and, if summing occurs, it can be reduced by increasing the sample to detector distance.

36.3.1 Well counters that have very high efficiencies are prone to summing since, for a given source strength, the count rate is higher than for a detector of lower efficiency. For moderate and high-source strengths, the trade-off is a poor one and the well counter is best suited for low-level work where its high efficiency is an important advantage.

36.4 Cascade summing may occur when nuclides that decay by a gamma cascade are counted. Cobalt-60 is an example; 1.17 MeV and 1.33 MeV from the same decay may enter the detector and be absorbed, giving a 2.50-MeV sum peak. Cascade summing may be reduced and eventually eliminated by increasing the source-to-detector distance.

36.5 The resolution of a gamma detector is the effective limit to its utility even when complex data-reduction methods are used. A typical 76 by 76-mm NaI(Tl) detector will give full widths at half maximum (FWHM) of approximately 60 keV at 662-keV gamma energy and approximately 90 keV at 1.33-MeV gamma energy.

37. Apparatus

37.1 *Sodium Iodide Detector Assembly*—A cylindrical 76 by 76-mm NaI detector is activated with about 0.1 % thallium iodide, with or without an inner sample well, optically coupled to a multiplier phototube, and hermetically sealed in a light-tight container. The NaI(Tl) crystal should contain less than 5 ppm of potassium and be free of other radioactive materials. In order to establish freedom from radioactive materials, the manufacturer shall supply a gamma spectrum of the background of the detector between 0.08 and 3.0 MeV. The resolution of the detector for the cesium-137 gamma at 0.662 MeV should be less than 60 keV FWHM or less than 9 %.

37.2 *Shielding* for the detector shall be constructed as needed from low-radioactivity-level lead or steel. A thickness equivalent to 102 mm of lead usually provides adequate shielding. If the shield is made of lead a graded liner should be used (35.1) unless the shield-to-detector distance is more than 250 mm.

37.3 *High-Voltage Power Supply*—500 to 2000 V dc regulated to 0.1 % with a ripple of not more than 0.01 %.

37.4 *Preamplifier*—Linear amplifier system to amplify the

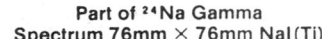

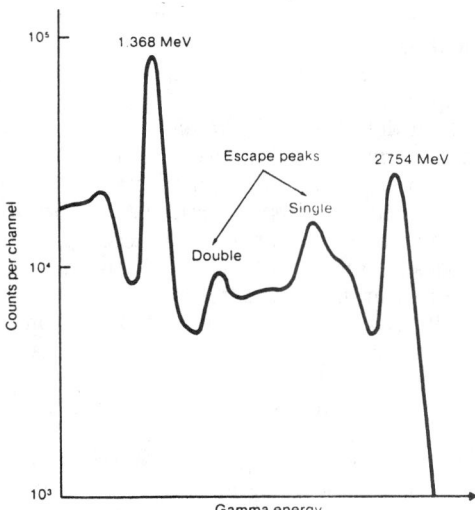

FIG. 13 Single and Double Escape Peaks

output from the multiplier phototube to a maximum output of 10 V.

37.5 *Analyzer with Scaler and Timer*—A single-channel discrimination system will accept all or any part of the output from the amplifier and pass it to the scaler. Any pulses lying outside the preset limits are rejected. The lower limit is usually referred to as the "threshold" and the difference between the two limits is the "window."

37.6 *Sample Mounts and Containers* may consist of any reproducible geometry container that is commercially available. Other considerations are cost, ease of use, disposal, and effective containment of radioactivity for the protection of the workplace and personnel from contamination.

37.7 *Geometry Control System*—A system of shelves or supports for the various sample containers that allows the user to place a sample in any of several preset and reproducible geometries.

37.8 *Beta Absorber*—A beta absorber of 3 to 6 mm of aluminum, beryllium, or poly(methyl methacrylate) should completely cover the upper face of the detector to prevent betas from reaching the detector.

38. Energy Efficiency Relationship

38.1 Because of the rapid falloff in gamma absorption as gamma energy rises, the detection efficiency shows a similar effect, and Fig. 14 shows the efficiency versus gamma energy plot for a 76 by 76-mm NaI(Tl) detector. The portion of the curve at low energy shows that as the absorption coefficient increases geometry becomes the limiting factor. The maximum efficiency seen is well below 50 % due to the presence of the beta absorber and the containment of the detector. The 76 by 76-mm NaI(Tl) detector is the most widely used size and a large amount of data are available in the open literature on the use of this size and results obtained. Heath (35) has written a comprehensive review and supplied many gamma-ray spectra in both graphical and digital form.

38.2 Other sizes of detectors may be used. However, the following should be noted:

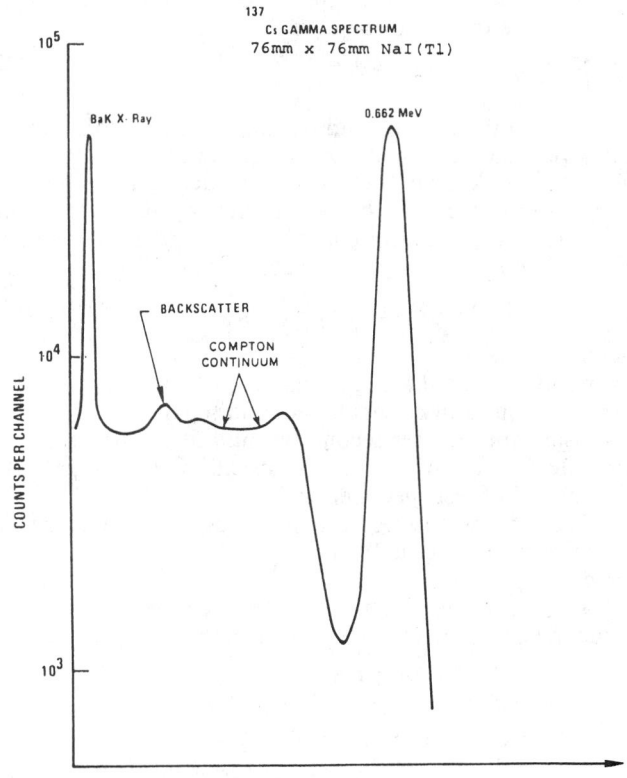

FIG. 12 Compton Continuum

38.2.1 Smaller detectors, such as 38 by 38 mm, will give efficiencies that are low and fall off more rapidly as gamma energy increases. Small or thin detectors are useful for the measurement of low-energy gammas since they are less responsive to high-energy gammas and the interference from Compton effects is reduced.

38.2.2 Larger detectors will give higher efficiencies and less fall off as gamma energy increases. Larger detectors are useful for situations where the highest attainable efficiency is desired and for the assembly of complete absorption detectors. The increase in efficiency is accompanied by an increased background count rate and an increase in the probability of summing in the detector.

38.2.3 Well detectors will give very high efficiencies, up to about 80 % for low- and moderate-energy gammas. The well detector is useful for low levels of activity and the background of a well detector is essentially the same as that of a plain cylindrical detector of the same overall dimensions. Summing becomes a definite problem at high activities since both random and cascade summing result from the high efficiencies and the high geometry of the well detector.

39. Calibration and Standardization

39.1 Calibrate the system for energy or gain by using at least two sources of different gamma energies. Cesium-137 at 662 keV and cobalt-60 at 1.17 MeV and 1.33 MeV are generally adequate. Use these sources to establish two parameters; (1) the required MeV or keV per volt of amplifier output, and (2) the condition that 0 MeV is equal to 0-V output. Carry out subsequent efficiency calibrations with the same gain settings.

39.2 Obtain efficiency calibrations by counting known amounts of single nuclides in the geometry and matrix that will be used for the unknown samples. Standards are available from the National Institute of Standards and Technology and other sources. Take extreme care in the preparation of subsamples of standards and, whenever possible, use gravimetric dilution and division techniques. Obtain the threshold and window levels used in volts using the energy calibration data obtained previously.

39.3 In comparing a standard and samples of the same nuclide, accurate results are possible. When an arbitrary standard is used to estimate the activity of a mixture of nuclides, the data are not accurate since the real detection efficiency varies with energy. Because of this gross gamma count, data obtained on mixtures of nuclides are only useful as an indication of approximate activity. Such data can be used for sample screening to select aliquot sizes for gamma spectrometry, comparison of process or waste samples with each other, and as a general screening for liquid wastes.

39.4 Report data obtained using an arbitrary standard for a mixture of nuclides as being based on a "calibration using nuclide-X" or as being "relative to nuclide-X." Data obtained by using this technique *should not be reported as dpm or Bq units without qualification.*

40. Source Preparation

40.1 Prepare samples in an identical manner to the standards and count in the same geometry under identical conditions. For gamma counting, samples may be in liquid, solid, or gaseous form contained in suitable containers.

40.2 The evaporation of liquid samples to dryness before counting is not necessary. However, samples that have been evaporated to dryness for gross beta counting can also be gamma counted. Evaporation may also be used as a sample concentration technique, especially with pure water such as evaporator condensates, where a large sample may be evaporated to small volume and counted with a higher efficiency.

40.3 Gaseous samples can be collected in any type of gas container but take care to ensure that all the gas containers of a particular type have a constant wall thickness. Wall thickness variation can be a significant problem in counting radioactive xenons that have gammas of 80 and 250-keV energy.

41. Calculation

41.1 Results may be expressed in observed counts per minute per millilitre for comparing relative activities of a group of samples as in tracer experiments:

$$C = \frac{1}{V}\left[\frac{N}{t} - \frac{N_B}{t_B}\right] \qquad (28)$$

where:
N = number of counts accumulated,
t = sample counting period, s,
N_B = number of background counts accumulated,
t_B = background counting period, s,
V = volume of initial sample, mL, and
C = net counts per second per millilitre (cps/mL).

41.2 Results also may be expressed in terms of equivalent cesium-137 activity, using the empirical efficiency determined for the reference standard:

$$D_{Cs} = C/E_{Cs} \qquad (29)$$

where:
C = net counts per second per millilitre (cps/mL), and
ECs = plus conversion electrons (cps/Bq).

41.3 If it is known that only one nuclide is present, its disintegration rate may be determined by use of the efficiency factor measured with a reference standard of the same nuclide. The results may be calculated as follows:

$$D = \frac{1}{EVY}\left[\frac{N}{t} - B\right] \qquad (30)$$

where:
N = number of counts accumulated,
B = background in counts per second (cps),
D = disintegrations per second per millilitre (Bq/mL),
E = efficiency of counter for the specific nuclide (cps/Bq),
V = volume of test specimen, mL,
Y = recovery of species (unity in cases where complete recovery is assumed), and
t = count period, s.

41.4 If it is desired that the activity be expressed in units of microcuries, use the following expression:

$$A = D/(3.7 \times 10^4) \qquad (31)$$

where:
A = concentration, μCi/ml,
D = disintegration rate per second per millilitre (Bq/mL), and
3.7×10^4 = conversion factor, disintegrations per second per microcurie.

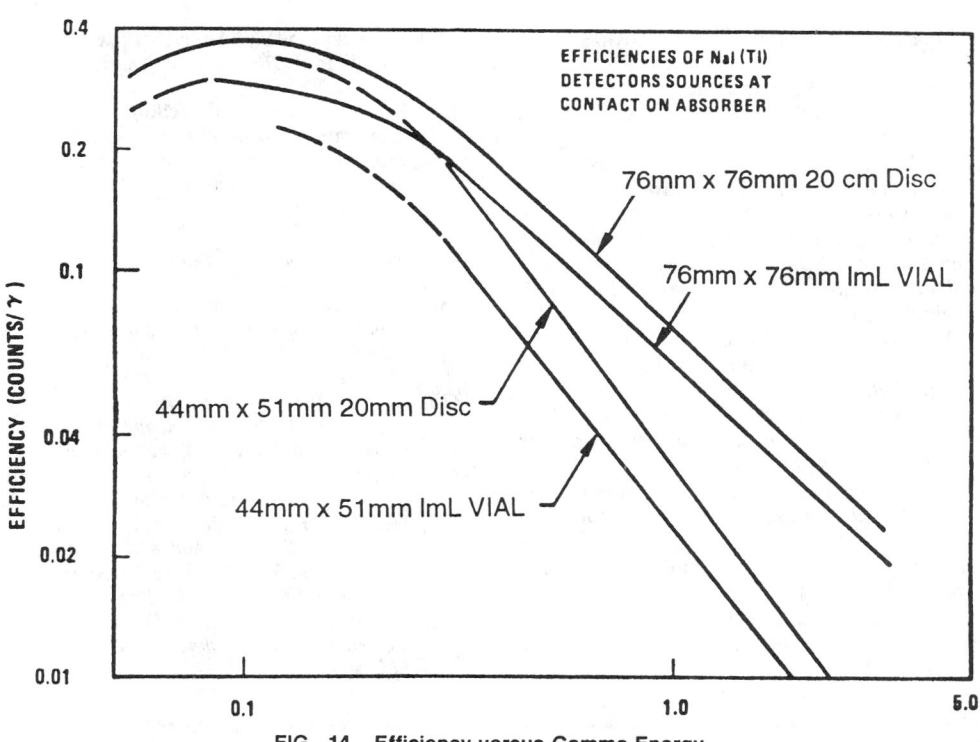

FIG. 14   Efficiency versus Gamma Energy

42. Keywords

42.1 alpha counting; beta counting; gamma-ray counting; radioactivity detection; radioactivity measurement; water

REFERENCES

(1) Pate, B. D., and Yaffe, L., "Disintegration Rate Determination by 4π Counting," *Canadian Journal of Chemistry*, Vol 33, 1955, pp. 15, 610, 929, and 1656.

(2) Blanchard, R., Kahn, B., and Birkoff, R. D., "The Preparation of Thin, Uniform Radioactive Sources by Surface Absorption and Electrodeposition," *Health Physics*, Vol 2, 1960, p. 246.

(3) Hallden, N. A., and Fisenne, I. M., "Minimizing Self-Absorption in 4π-8 counting, "*International Journal of Applied Radiation and Isotopes*, Vol 14, 1963, p. 529.

(4) Merritt, J. S., Taylor, J. G. V., and Campion, P. J., "Self-Absorption in Sources Prepared for 4π-β Counting," *Canadian Journal of Chemistry*, Vol 37, 1956, p. 1109.

(5) Gunnick, R., Colby, L. J., and Cobble, J. W., *Analytical Chemistry*, Vol 31, 1959, p. 796.

(6) Friedlander, G., Kennedy, J. W., and Miller, J. M., *Nuclear and Radiochemistry*, John Wiley and Sons, New York, 1964, 2nd ed.

(7) Price, W. J., *Nuclear Radiation Detection*, McGraw-Hill, New York, N.Y., 1964, 2nd ed.

(8) Campion, P. J., Taylor, J. G. V., and Merritt, J. S., "The Efficiency Tracing Technique for Eliminating Self-Absorption Errors in 4π Beta Counting," *International Journal of Applied Radiation and Isotopes*, Vol 8, 1960, p. 8.

(9) Curtis, M. L., Heyd, J. W., Olt, R. G., and Eichelberger, J. F., *Nucleonics*, Vol 13, May, 1955, p. 38.

(10) Watt, D. E., and Ramsden, O., *High Sensitivity Counting Techniques*, The MacMillan Company, New York, N.Y., 1964.

(11) Crouthamel, C. E., Adams, F., and Dams, R., *Applied Gamma-Ray Spectrometry*, Pergamon Press, New York, N.Y., 2nd ed., 1970.

(12) Nielson, J. M., *Gamma-Ray Spectrometry in Physical Methods of Chemistry*, A. Weissberger and B. W., Rossiter, eds. Vol. 1, Part III D, Chap. X, John Wiley and Sons, Inc. New York, N.Y., 1972.

(13) Perkins, R. W., *Nuclear Instruments and Methods*, 1965, p. 33. Wogman, N. A. and Perkins, R. W., *ibid*, 1968.

(14) Cooper, J. A., Ranticelit, L. A., Perkins, R. W., Haller, W. A., and Jackson, A. L., "An Anti-Coincidence Shielded Ge(Li) Gamma-Ray Spectrometer and Its Application to Neutron Activation Analyses", *Report BNWL-SA-2009*, Pacific Northwest Laboratory, Richland, Wash., 1968.

(15) *Metrology of Radionuclides, Proceedings of a Symposium*, Oct. 14–16, 1959, International Atomic Energy Agency, Vienna.

(16) Nielsen, J. M., and Kornberg, H. A., "Multidimensional Gamma-Ray Spectrometry and Its Use in Biology," *Radioisotope Sample Measurement Techniques in Medicine and Biology, Proceedings of a Symposium*, May 24–28, 1965, International Atomic Energy Agency, Vienna, p. 3.

(17) Lapp, R. E., and Andrews, H. L., *Nuclear Radiation Physics*, Prentice Hall, New York, 1948. Englewood Cliffs, N.J., 2nd ed., 1954.

(18) Jarrett, Allen A., "Statistical Methods Used in the Measurement of Radioactivity with Some Useful Graphs and Nomographs," *AECU 262*.

(19) Bell, C. G., and Hayes, F. N., *Liquid Scintillation Counting*, Pergamon Press, New York, 1958.

(20) Birks, J. B., *The Theory and Practice of Scintillation Counting*, Pergamon Press, New York, 1964.

(21) Bransome, E., Jr., ed. *The Current Status of Liquid Scintillation Counting*, Greene and Stralton, New York, 1970.

(22) *Environmental Radioactivity Surveillance Guide,* ORP/SID 72-2 USEPA 1972.

(23) Friedlander, G., Kennedy, J. W., and Miler, J., *Nuclear and Radiochemistry,* John Wiley and Sons, Inc., New York, NY, 2nd ed., 1964. Overman, R. T., and Clark, H. M., *Radioisotope Techniques,* McGraw-Hill Book Co., Inc., New York, NY, 1960. Price, W. J., *Nuclear Radiation Detection,* McGraw-Hill Book Co., Inc., New York, NY, 1964. Flynn, K. F., Glendenin, L. E., and Prodi, V. *Absolute Counting of Low Energy Beta Emitters Using Liquid Scintillation Counting Techniques in Organic Scintillators and Liquid Scintillation Counting,* D. L. Horrocks and Chim-Tzu Peng, eds., Academic Press, New York, 1971, p. 687.

(24) Sill, C. W., and Olson, D. G., "Sources and Prevention of Recoil Contamination of Solid-State Alpha Detector," *Analytical Chemistry,* Vol 42, 1970, p. 1956.

(25) Puphal, K. W., and Olson, D. R., "Electrodeposition of Alpha Emitting Nuclides from a Mixed Oxalate-Chloride Electrolyte:, *Analytical Chemistry,* Vol 44, 1972.

(26) Mitchell, R. F., "Electrodeposition of Actinide Elements at Tracer Concentrations," *Analytical Chemistry,* Vol 32, 1960, pp. 326–328.

(27) Bell, C. G., and Hayes, F. N., *Liquid Scintillation Counting,* Pergamon Press, New York, NY, 1958.

(28) Moghissi, A. A., "Low Level Counting by Liquid Scintillation," *Organic Scintillators and Liquid Scintillation Counting,* Academic Press, New York, NY, 1971.

(29) Horrocks, D. L., "Alpha Particle Energy Resolution in a Liquid Scintillator," *Review of Scientific Instruments,* Vol 55, 1964, p. 334.

(30) Bogen, D. C., and Welford, G. A., "Application of Liquid Scintillation Spectrometry for Total Beta and Alpha Assay," *Proceedings of the International Symposium of Rapid Methods for Measuring Radioactivity in the Environment, LAEA-SM-14/3,* 1971, p. 383.

(31) Harley, J. H., Hallden, N. A., and Fisenne, I. M., "Beta Scintillation Counting with Thin Plastic Phosphors," *Nucleonics,* NUCLA, Vol 20, 1962, p. 59.

(32) Flynn, K. F., Glendenin, E., Steinberg, E. P., and Wright, P. M., "Pulse Height-Energy Relations for Electrons and α-Particles in a Liquid Scintillator," *Nuclear Instruments and Methods,* Vol 27, 1964, p. 13.

(33) "Tables for the Analyses of β Spectra, *National Bureau of Standards Applied Mathematics Series Reports No. 13,* U.S. Government Printing Office, Washington, DC, 1952.

(34) Flynn, K. F., and Glendenin, L. E., "Half-Life and β-Spectrum of Rubidium-87," Physics Review, Vol 116, 1959, p. 744.

(35) Heath, R. L., "Scintillation Spectrometry Gamma-Ray Spectrum Catalog" *IDO 16880,* and *ANCR-1000.*

(36) Katz and Penfeld, "Reviews," *Modern Physics,* Vol 24, 1952, p. 28.

(37) Lawson and Cork, *Physics Review,* Vol 57, 1940, p. 982.

(38) Currie, L. A., "Limits for Qualitative Detection and Quantitative Determination," *Analytical Chemistry,* Vol 40, No. 3, 1968, pp. 586–593.

Standard Guide for
Sampling of Drums and Similar Containers by Field Personnel[1]

This standard is issued under the fixed designation D 6063; the number immediately following the designation indicates the year of original adoption or, in the case of revision, the year of last revision. A number in parentheses indicates the year of last reapproval. A superscript epsilon (ε) indicates an editorial change since the last revision or reapproval.

1. Scope

1.1 This guide covers information, including flow charts, for field personnel to follow in order to collect samples from drums and similar containers.

1.2 The purpose of this guide is to help field personnel in planning and obtaining samples from drums and similar containers, using equipment and techniques that will ensure that the objectives of the sampling activity will be met. It can also be used as a training tool.

1.3 *This standard does not purport to address all of the safety concerns, if any, associated with its use. It is the responsibility of the user of this standard to establish appropriate safety and health practices and determine the applicability of regulatory limitations prior to use.*

2. Referenced Documents

2.1 *ASTM Standards:*
C 783 Practice for Core Sampling of Graphite Electrodes[2]
D 1452 Practice for Soil Investigation and Sampling by Auger Borings[3]
D 1586 Test Method for Penetration Test and Split-Barrel Sampling of Soils[3]
D 1587 Practice for Thin-Walled Tube Geotechnical Sampling of Soils[3]
D 2113 Practice for Diamond Core Drilling for Site Investigation[3]
D 4448 Guide for Sampling Groundwater Monitoring Wells[4]
D 4687 Guide for General Planning of Waste Sampling[4]
D 4700 Guide for Soil Sampling from the Vadose Zone[3]
D 4823 Guide for Core-Sampling Submerged, Unconsolidated Sediments[5]
D 4840 Guide for Sample Chain of Custody Procedures[6]
D 5088 Practice for Decontamination of Field Equipment Used at Nonradioactive Waste Sites[7]
D 5283 Practice for Generation of Environmental Data Related to Waste Management Activities: Quality Assurance and Quality Control Planning and Implementation[4]

D 5358 Practice for Sampling with a Dipper or Pond Sampler[4]
D 5451 Practice for Sampling Using a Trier Sampler[4]
D 5495 Practice for Sampling with a Composite Liquid Waste (COLIWASA) Sampler[4]

3. Terminology

3.1 *Definitions:*
3.1.1 *bung, n*—usually a 2-in. (5-cm) or ¾-in. (1.3-cm) diameter threaded plug specifically designed to close a bung hole.
3.1.2 *bung hole, n*—an opening in a barrel or drum through which it can be filled, emptied or vented.
3.1.3 *consolidated solid, n*—*as used in this guide*, a compact solid not easily compressed or broken into smaller portions.
3.1.4 *drum, n*—*when used in the flow charts in this guide*, the word implies any drum, barrel or non-bulk container of 5 to 110 gal (19 to 400 L) capacity.
3.1.5 *representative sample, n*—a sample collected such that it reflects one or more characteristics of interest of the lot or population from which it was collected.
3.1.6 *sample, n*—one or more items or portions collected from a lot or population.
3.1.7 *sampler, n*—the device used to obtain a sample.
3.1.8 *sludge, n*—*as used in this guide*, any mixture of solids that settles out of solution; sludges contain liquids that are not apparent as free liquids.
3.1.9 *unconsolidated solid, n*—*as used in this guide*, uncemented or uncompacted material that is easily separated into smaller portions.
3.1.10 *work plans, n*—plans that are specific to sampling at a particular site; examples are Health and Safety Plans and Sampling and Analysis Plans.

4. Summary of Guide

4.1 This guide uses a decision-tree format to lead persons intending to sample waste materials from drums and similar containers through a series of questions. The answers to the questions result in recommended actions, including the selection of appropriate sampling equipment. Brief instructions on the use of the equipment are included.

4.2 This guide addresses commonly used sampling equipment and devices; it is not intended to cover all that might be purchased or custom made.

5. Significance and Use

5.1 This guide is intended to assist field personnel in obtaining samples from drums and similar containers for laboratory analysis. The costs associated with sampling and

[1] This guide is under the jurisdiction of ASTM Committee D-34 on Waste Management and is the direct responsibility of Subcommittee D34.01 on Sampling and Monitoring.
Current edition approved Dec. 10, 1996. Published February 1997.
[2] *Annual Book of ASTM Standards,* Vol 15.01.
[3] *Annual Book of ASTM Standards,* Vol 04.08.
[4] *Annual Book of ASTM Standards,* Vol 11.04.
[5] *Annual Book of ASTM Standards,* Vol 11.02.
[6] *Annual Book of ASTM Standards,* Vol 11.01.
[7] *Annual Book of ASTM Standards,* Vol 04.09.

analysis make it essential that samples be taken correctly before submitting them for chemical analysis or physical testing, or both. Incorrect sampling can invalidate resulting data.

5.2 This guide may be used by personnel who have no formal workplan. It draws their attention to issues that must be addressed before, during, and after taking a sample. It provides guidance in choosing the sampling technique and equipment suitable for specific situations. It can serve as a training tool for those who are unfamiliar with sampling. It is recommended that this guide be used as a supplement to a written workplan.

5.3 Some sections of this guide contain flow charts (see Figs. 1 through 5) that must be worked through, starting from the top of each page. By answering the questions in the diamond-shaped boxes, and following the appropriate arrows, the person planning to sample will be guided towards the most suitable procedures and equipment. The numbers at the bottom of some boxes refer to corresponding paragraphs in the text, which provide information to help the person sampling answer the questions.

5.4 Figures 6 through 15 are examples of types of equipment. Similar devices that do the same job in the same way

are not intended to be excluded.

6. Objectives of Sampling

6.1 The purpose of sampling is to collect a representative sample of all or part of the contents of the drum or similar container, to determine the physical and chemical characteristics of those contents (see Fig. 1). This information may then be used to:

6.1.1 Select suitable methods of treatment and disposal of the contents,

6.1.2 Provide evidence for use in a court of law,

6.1.3 Comply with regulations, such as those for the transportation of hazardous materials,

6.1.4 Confirm that the drums contain what is written on the label, manifest or other type of documentation, and

6.1.5 Find out if any drums in a lot contain different materials from the majority.

6.2 In most cases there is a written plan that describes the work to be done (Guide D 4687). In other cases, there is no written plan and the instructions are only verbal.

6.3 If the objectives of sampling are unclear or unknown to the field personnel, they should question their supervisor or project manager about the objectives. Well-informed field

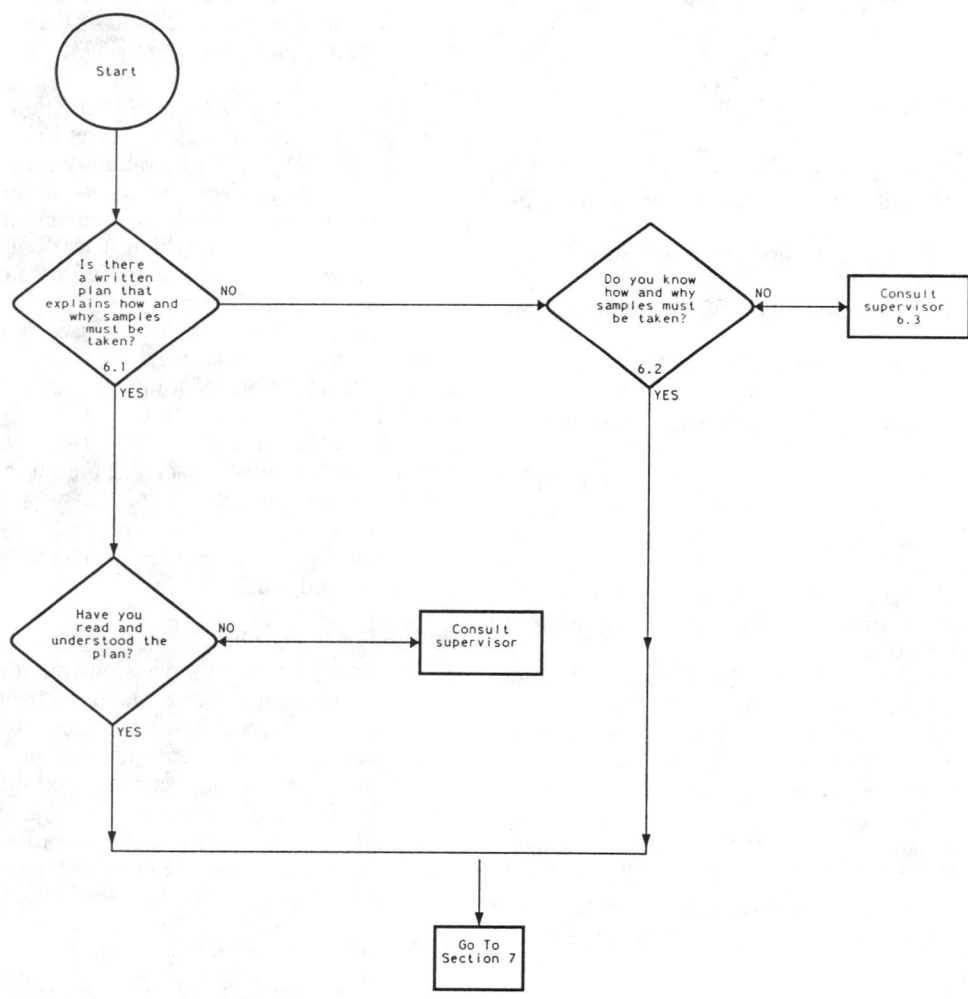

NOTE—This flow chart should be used with Section 6 in the text.

FIG. 1 Objectives of Sampling

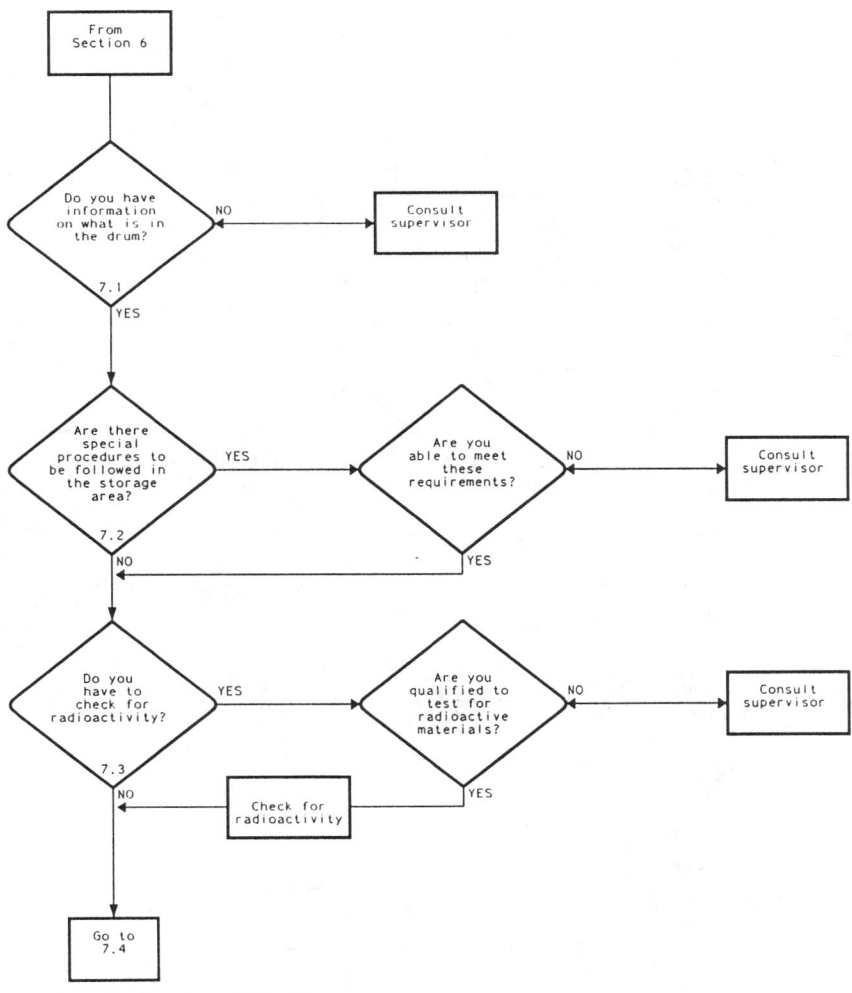

NOTE—This flow chart should be used with Sections 7.1 to 7.3 in the text.

FIG. 2 Pre-Sampling Inspection

personnel are then alert to unforeseen circumstances or events that might invalidate the samples.

7. Pre-Sampling Inspection

7.1 Information about the contents of the drums may be available from (see Fig. 2):

7.1.1 Previous analysis of drum contents from the same source,

7.1.2 The supplier/source of the material in the drums,

7.1.3 Manifest (shipping) documents,

7.1.4 Labels and other markings on the drums, or

7.1.5 Knowledge of the waste generating process.

7.2 Personnel doing the pre-sampling and sampling must be aware of any special procedures that are to be followed at a given site. Workplans include a worker health and safety section because there are potential hazards associated with opening drums as well as with potentially hazardous contents.[8,9] Examples of special procedures are change of

clothing, use of safety equipment of various kinds, evacuation procedures, fire and explosion procedures and vehicle cleaning procedures such as water washing before leaving the site or storage area, and many others that would be site or storage specific.

7.3 If you are certain that the drum does not contain radioactive material and the workplan does not require you to check for radioactivity, proceed to 7.4.

7.3.1 Many facilities are not licensed to handle radioactive materials and are legally obliged to prove that they do not knowingly accept them. Some facilities are licensed to handle radioactive materials; they need to have a measure of how radioactive the material is for the safety of their workers.

7.3.2 Hand-held monitors that check for radioactivity should always be used if you suspect that radioactive material might be present or if the workplan requires it. It is important that the monitor has been calibrated correctly, according to the manufacturer's instructions. Monitoring

---

[8] *Drum Handling Practices at Hazardous Waste Sites*, EPA/600/2-86/013, January 1986.

[9] *Field Sampling Procedures Manual*, Third Edition, New Jersey Department of Environmental Protection, Division of Hazardous Site Mitigation, February 1988.

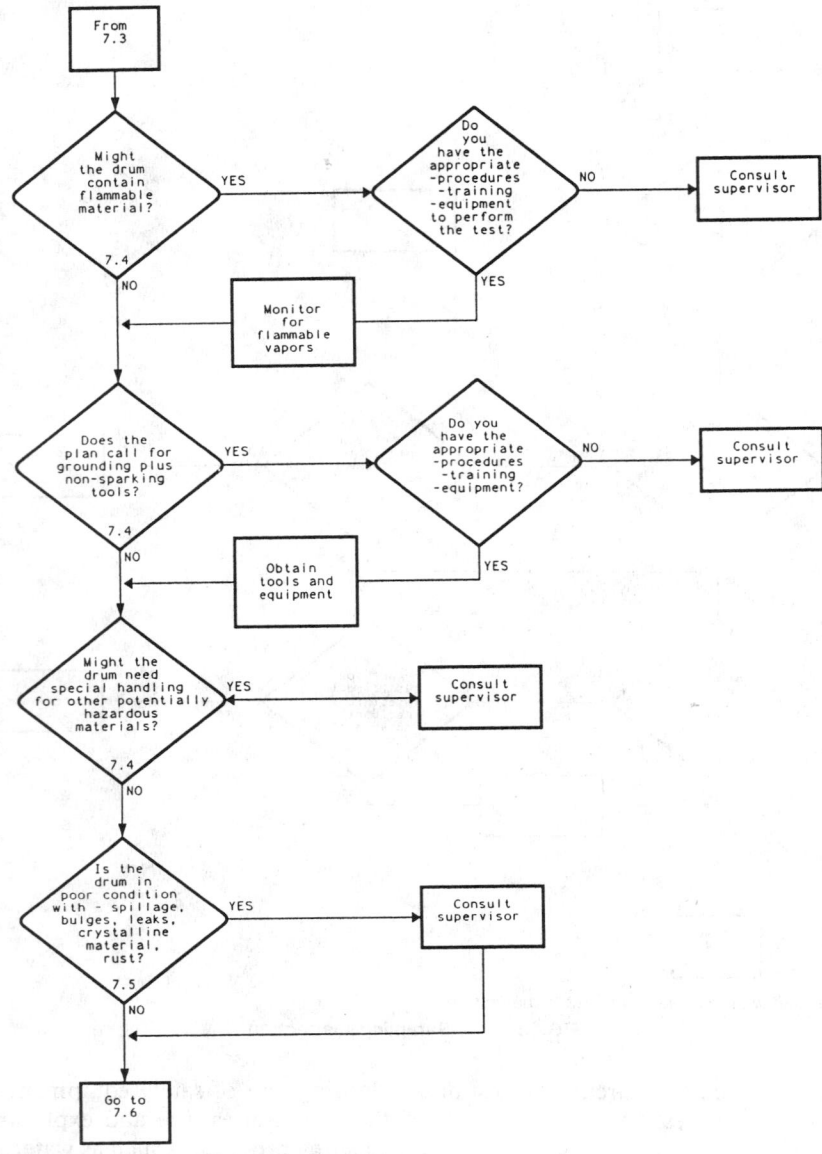

NOTE—This flow chart should be used with Sections 7.3 to 7.5 in the text.

FIG. 2 Pre-Sampling Insection (Continued)

should be done only by those with the appropriate written procedures, training and equipment.

7.3.3 It is prudent to monitor a storage area before entering it. If radioactive material is found to be present when it should not be, leave the area immediately, post warning signs to alert other workers, and consult your supervisor.

7.4 Drums may contain flammable materials, strong oxidizers or reducing materials, light-sensitive materials, corrosive acids or bases, and materials sensitive to moisture. All of these drums require special handling, including segregation.

7.4.1 Many solvents, like benzene, evaporate into air space in and around the drum where the vapour may be easily ignited.

7.4.2 If you are sampling a potentially flammable or unknown material, non-sparking tools should be used and the drums should be grounded.

7.4.3 If the drums are stored in a closed room or confined space, the air in the area should be tested by a hand-held monitor to check for flammable vapors. It is important that the monitor has been calibrated according to the manufacturer's instructions. The monitoring should be done by those with the appropriate training and written procedures.

NOTE 1: **Warning**—Flammable materials should be sampled in a well-ventilated area. There are other safety considerations that must be considered regarding confined spaces. It may be necessary to check for explosivity or oxygen levels.

7.4.4 Labels on drums of waste materials may not be accurate. Unless the drums come from a reliable source, for example, the generator of the material and the process that created the waste are known to you, it is prudent to assume that the labels may not match the contents.

NOTE 2: **Precaution**—Attempting to open a drum that is in poor condition can expose a worker to the possibility of a serious, even fatal,

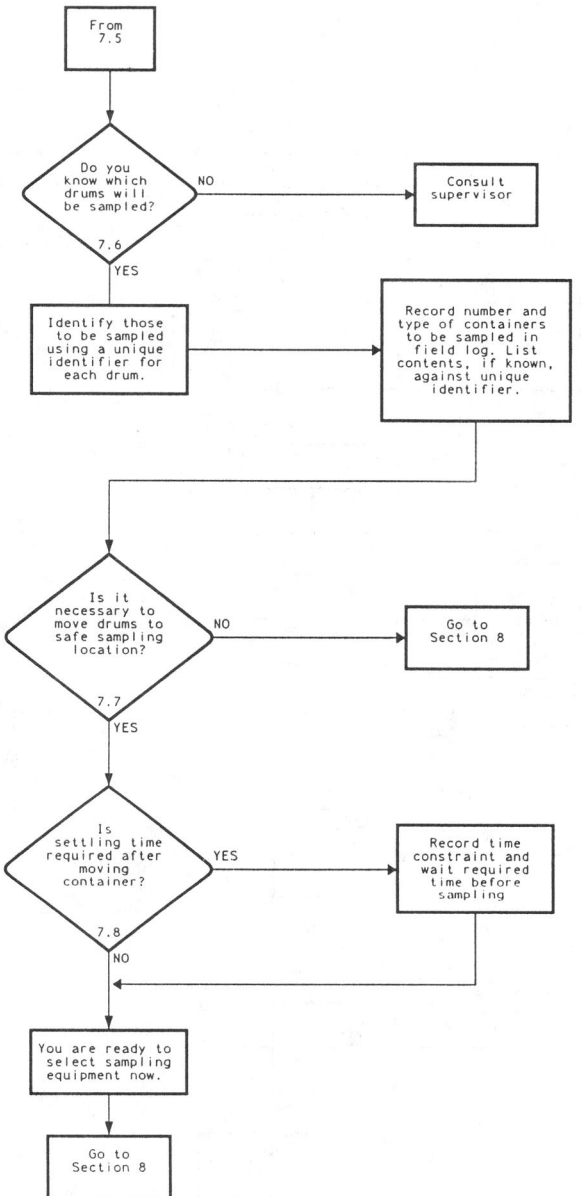

NOTE—This flow chart should be used with Sections 7.6 to 7.8 in the text.

FIG. 2   Pre-Sampling Inspection (Continued)

accident. Special precautions should be taken when the bungs are rusted or corroded since the drum top may give way, exposing the worker to vapor or liquid. Overpacking before sampling should be considered for drums in poor condition.

7.5 It is not always necessary to sample every drum in a lot. The workplans provide direction as to how many, and possibly which, drums should be sampled. Each drum that will be sampled must be identified in a unique way in case a second sample has to be taken later. Colored labels, crayons, paint sticks or pens, or stencilled paint can be used to identify drums. Any new identification system should not cover the existing labels or identifiers.

7.6 Sometimes drums have to be moved to another location for sampling; this is known as "drum staging". This is required if:

7.6.1 Sampling the drum in its present location poses a high risk to surrounding property and individuals,

7.6.2 The drum cannot be accessed for sampling in its current location, or

7.6.3 Exposure to climatic conditions alter the sample, for example, formation of ice; or create a health and safety risk, for example, the sun heating a drum containing solvents.

7.7 The physical condition of drums must be evaluated before attempting to open or move them (see Note 2). Drum carriers, which lock on the drum lip, should not be used to move the drum if the condition is poor.

7.8 Materials in layers, such as oil with water, can become mixed together when moved. If you want to sample each layer separately, the material may need time to settle before opening and sampling the drum.

8. Selection of Suitable Sampling Procedure

8.1 The physical state(s) of the material(s) being sampled is an important criterion when sampling (see Fig. 3).

8.2 A drum containing one liquid, such as water, or a mixture of liquids, such as a stable emulsion like hand cream, that does not separate into two layers regardless of time, is said to contain one phase. A drum containing two liquids, such as oil and water, which form two distinct layers when they are not stirred is said to contain two phases.

8.3 When it is necessary to know the amounts of solid and liquid layers in a drum, a calibrated measuring device or the sampling equipment (for example, a COLIWASA) can be inserted into the opened drum (see Section 12) and the liquid level measured.

8.4 Although sludges behave like sticky solids and are not usually pumped, they can contain quite a high proportion of liquids, such as oil or water, which is not visible as free liquid.

8.5 An unconsolidated solid is a material like sand or a powder. A consolidated solid consists of material, like sandstone or concrete.

8.5.1 A drum containing mixed materials, such as disposable personal protective equipment and laboratory supplies, is treated as one with unconsolidated solids.

8.6 If the waste material is likely to attack the sampling equipment, such as an acid corroding a metal thief:

8.6.1 The equipment may partially dissolve, adding constituents, such as metals to the sample. Faulty conclusions may be drawn about the composition of the sample, leading to costly and unnecessary remedial actions, and

8.6.2 The equipment will have to be replaced frequently, adding costs to the project.

8.7 When selecting equipment, it is important to be aware of the limitations of the tools. The design of some equipment can result in part of the material not being sampled. For example, if the size of the opening(s) that allow the sample to enter a trier is smaller than some of the particles in the drum, the sample will not be representative.

8.7.1 Volatile organic constituents are likely to be lost if the sampling equipment causes a buildup of heat or agitation of the sample, as will exposure to air for more than a very short time or storage in a sample container with a headspace above the sample.

8.7.2 See Table 1 for more information on the limitations of sampling equipment.

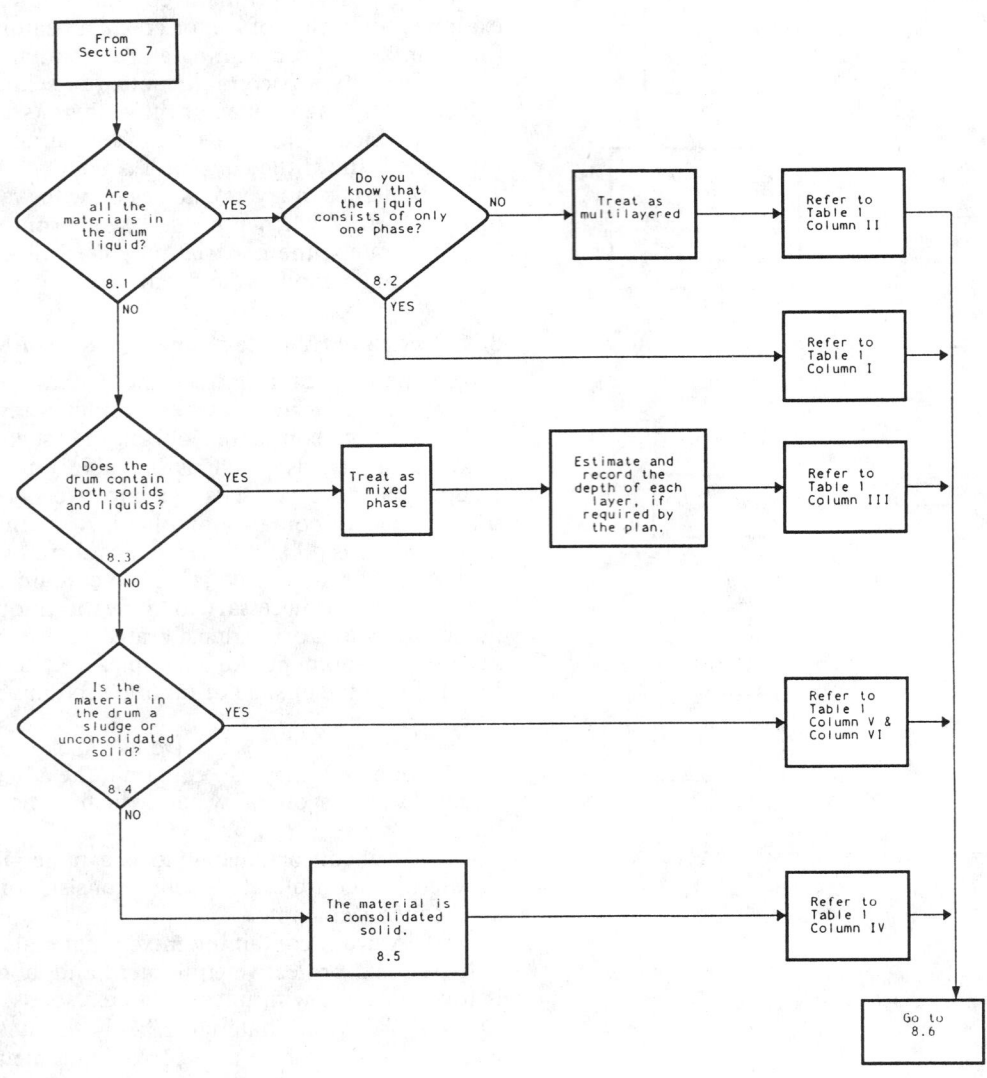

NOTE—This flow chart should be used with Sections 8.1 to 8.5 in the text.

FIG. 3 Selection of Suitable Sampling Procedure

8.8 When the quantity of material that is removed from a drum by the equipment will not provide enough for the laboratory to perform all the required tests, a number of subsamples must be taken and combined. In doing this it is critical not to disturb layers; this could result in an unsuitable sample.

8.9 Pumps may require electricity to operate. If the sampling location is outdoors, they may need to be protected from the weather. If flammable vapors were observed in 7.4, consult your supervisor about sources of ignition, such as pumps, electrical connections and switches.

8.10 Consideration should be given to having a separate clean sampling device for each drum since this eliminates cross-contamination and may be more efficient than

cleaning one sampling device after each drum. It is generally easier to clean the sampling devices and other equipment in the laboratory or another suitable place where solvent disposal and drying equipment are readily available.

8.11 It is worthwhile to make a comparison of the costs of cleaning or decontaminating equipment, or both, including disposal of the cleaning agents, versus using disposable equipment when selecting and preparing equipment. The initial purchase price of the equipment may also be a factor in the selection. There may be more than one suitable device (see Table 2).

8.12 Tables 1 and 2 list equipment that is commonly used for sampling liquid and solid wastes, and their limitations. They contain more than one type of equipment for some

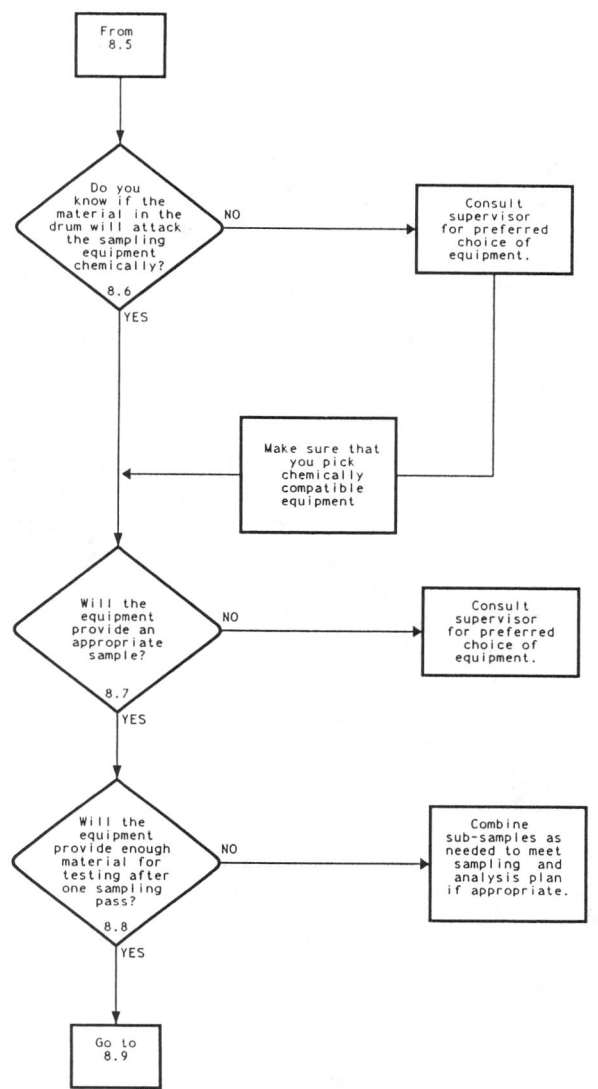

NOTE—This flow chart should be used with Sections 8.6 to 8.8 in the text.

FIG. 3 Selection of Suitable Sampling Procedure (Continued)

types of waste materials but are not intended to include all possible equipment or devices.

8.12.1 Other factors that will guide you in making the final choice follow in Section 13.

9. Preparation of Sampling Equipment

9.1 Damaged sampling equipment can affect the sample, for example, dull cutting edges or chipped glass parts. It may also be a safety hazard for the worker (see Fig. 4).

9.2 Cleaning and decontamination procedures should be identified in the plan (see Practice D 5088).

9.3 Check the workplan for preservatives, refrigeration, holding times, type and size of sample container, packaging, and shipping requirements. If the samples are going to be analyzed for volatile organic compounds, such as gasoline, special bottles must be used.

9.4 If the samples are going to be analyzed for trace amounts of volatile organic compounds, such as solvents, plan to fill and seal the special sample bottles required for this analysis first, before filling the other sample containers.

9.5 Personal safety equipment will vary depending on the hazards associated with the task. It must comply with the health and safety requirements of the organization that employs you.

9.5.1 Basic safety equipment includes:

9.5.1.1 Safety glasses,

9.5.1.2 Synthetic rubber gloves,

9.5.1.3 Safety shoes or boots, and

9.5.1.4 Protective clothing.

9.5.2 More hazardous situations may require equipment such as:

9.5.2.1 Hard hat,

9.5.2.2 Respirators,

9.5.2.3 Face shields,

9.5.2.4 Chemically-resistant suits,

9.5.2.5 Self-contained breathing apparatus, and

9.5.2.6 Two way radio communication.

9.5.3 If appropriate, segregate the drums and surrounding area from casual intrusion by using barricades or caution tape.

9.5.4 When sampling inside buildings, ventilation may be desirable.

9.6 Records that associate a sample with a drum are usually required. A field technician's log and chain of custody forms are commonly used (see Section 10).

10. Report

10.1 *Chain of Custody Forms:*

10.1.1 The purpose of chain of custody forms is to show that the samples analyzed are the same ones that were collected. They are required for regulatory purposes. They serve as legal documentation that sample integrity was maintained. When complete, they should show that there were no lapses in accountability. It is not always necessary to use chain of custody procedures but some form of sample tracking is necessary.

10.1.2 A chain of custody form may originate with the laboratory at the time the sample containers are prepared or in the field after the sample has been taken. Each time the samples change hands, the chain of custody form must be signed. The originator of the chain of custody keeps one copy to confirm that the samples were passed on and to whom; they may have been given directly to the laboratory or to a courier. The completed form, with all remaining copies intact, must accompany the samples. Copies of the custody forms will be returned to the parties involved, to confirm that the samples have been received at the laboratory.

10.1.3 Sometimes security seals are placed over the caps of empty, clean sample containers and signed off by the laboratory. The person sampling must break these seals in order to fill the containers. Security seals may also be attached after the sample container has been filled. The date, time of sampling and name of the person sampling are then written on the seals. The purpose of the seal is to indicate possible tampering with the sample.[10] (See Guide D 4840.)

10.2 *Field Log:*

---

[10] *Guide to the Collection and Submission of Samples for Laboratory Analysis,* 6th Edition, Ontario Ministry of the Environment and Energy, April 1989.

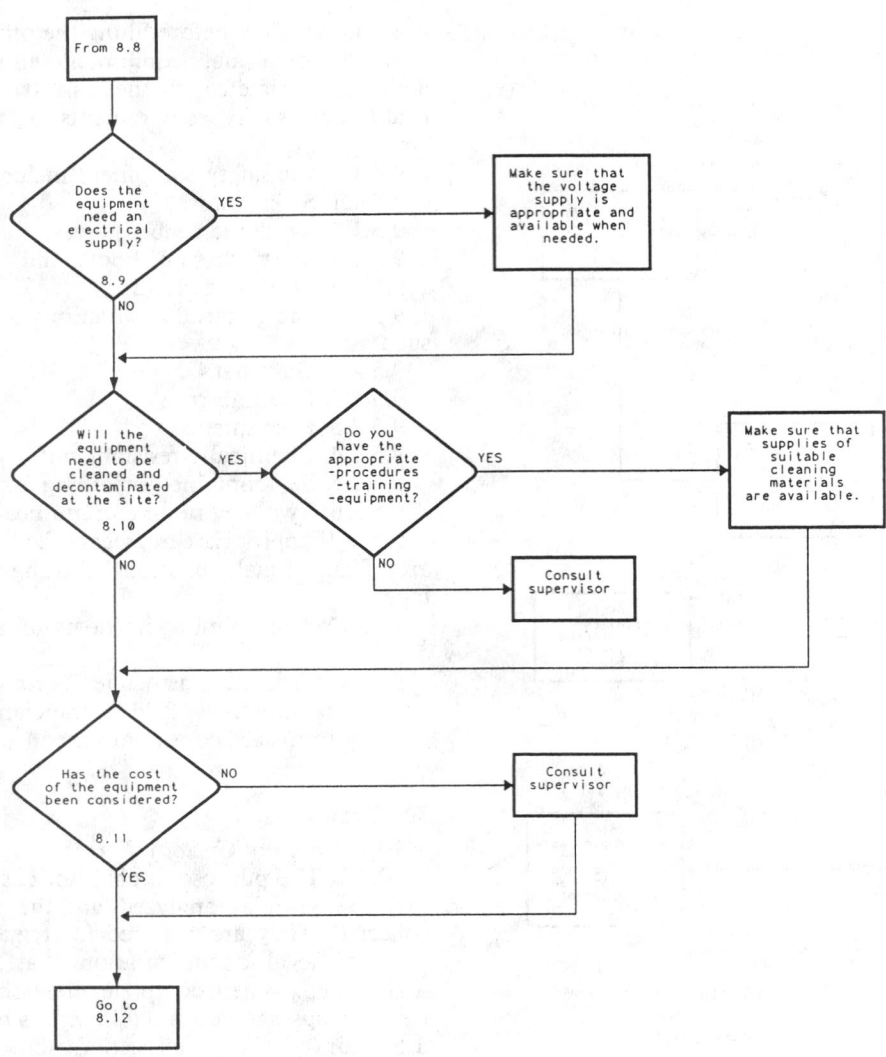

NOTE—This flow chart should be used with Sections 8.9 to 8.11 in the text.

FIG. 3 Selection of Suitable Sampling Procedure (Continued)

10.2.1 Ideally, a field log is maintained in a bound book with printed page numbers. Information should be recorded in indelible ink. Errors should be crossed out with a single line and initialled. Items normally documented include:

10.2.1.1 Type of waste, for example, sludge, wastewater,

10.2.1.2 Suspected waste composition, including concentrations,

10.2.1.3 Number and volume of sample taken,

10.2.1.4 Description of sampling point and sampling method,

10.2.1.5 Date, time and location of each sample collected,

10.2.1.6 Preservatives used, if any (including ice),

10.2.1.7 Analytical parameters to be measured,

10.2.1.8 Unique sample identification number,

10.2.1.9 Types and number of QC samples,

10.2.1.10 Equipment used to obtain samples,

10.2.1.11 Field observations; unusual events that may have an impact on the sample(s),

10.2.1.12 Field measurements and results (for example, pH),

10.2.1.13 Signature of person who took the sample, and

10.2.1.14 Names of personnel on sampling team.

10.2.2 Additional information may be specified in the workplan. (See Practice D 5283.)

10.3 *Labels:*

10.3.1 Information written on labels may include:

10.3.1.1 Sample ID number,

10.3.1.2 Name of person sampling,

10.3.1.3 Initials or signature of person sampling,

10.3.1.4 Date and time of sampling,

10.3.1.5 Sample location,

10.3.1.6 Sampling information (for example, grab or composite),

10.3.1.7 Preservative/preservation required,

10.3.1.8 Special instructions, and

10.3.1.9 Analysis request.

10.3.2 Labels may be put on containers before or after sampling. Consult the workplan.

11. Mixing of Liquids, With or Without Solids, Before Sampling

11.1 Thorough mixing of the material prior to sampling is

TABLE 1 Sampling Equipment Limitations

Sampling Equipment	Variables	Limitations
Auger (see 13.1)	Cross-contamination	Will disturb matrix during sampling.
	General use	May not reach all layers of the drum if the contents are multi-layered, and so may bias results.
	Stability of drum	Equipment may penetrate the bottom of the drum.
COLIWASA (see 13.3)	Angle of use	The angle of descent must be perpendicular to the surface of the material, or the resulting sample may be biased.
	Bottom layer	This sampler cannot be used to sample the material in the bottom of the drum. The actual depth of unsampled material varies depending on the COLIWASA used.
	Speed of use	If the speed of descent is too fast, the materials inside the tube will not be at the same level as those outside the tube, causing incorrect proportions in the sample.
	Speed of use	If the speed of descent is too fast, the layers of multi-layered materials will be disturbed.
Dipper (see 13.5)	General use	This sampler is used to grab materials from the surface of a drum, so the resulting sample may be biased.
Drum Thief (see 13.6)	Material of sampler	If sampler is made of glass, chips or cracks can cause an imperfect seal.
	Angle of use	The angle of descent must be perpendicular to the surface of the material, or the resulting sample may be biased.
	Bottom layer	May lose the bottom layers of material during sampling. The depth of the unsampled material varies with density, surface tension and viscosity of the material being sampled, for example, chlorosolvents, such as chloroform, are difficult to sample whereas water is easy.
	Consistency	With viscous materials, more material may end up on the outside of the tube than inside it.
	Speed of use	If the speed of descent is too fast, the liquid in the tube will not be at the same level as the liquid outside the tube, causing incorrect proportions in the sample.
Concentric Tube Thief (see 13.4)	General use	Can be used only for dry, powdery or granular materials.
	Particle size	Excludes certain particle sizes, including those that are over one third the slot width of the sampler.
Impact Devices Chipper Hammer Chisel	General use	Can be used only with consolidated solids.
	General use	These samplers are used to break off chunks of a consolidated solid from the surface only and so may bias resulting sample.
Push Coring Devices Split Barrel (see 13.2)	Media	Materials to be sampled must be moist enough to remain in the device.
	Stability of drum	Equipment may penetrate the bottom of the drum.
Push Coring Devices Thin-walled	Media	Tube will be damaged if pushed past the point of refusal.
	Consistency	Retrieving an intact core is difficult if the material is not sticky. Cannot be used in coarse, rocky materials.
	Stability of drum	Care required when used at bottom of drum to prevent penetration, particularly when auger tip is used.
Pump, Peristaltic (see 13.7)	Particle size	It is critical to realise that this sampler pumps suspendable solids, not heavy sludge. An example would be rainwater with suspended solids.
Rotating Core Device Concrete Corer (see 13.2)	General use	Used to sample consolidated solids only. May not get all the layers of a multi-layered drum, so may bias the sample.
	Power source	This device must be attached to a power drill.
Scissors and Tongs	General use	Used to cut pieces of unconsolidated solids within a drum, for example, to cut dark spots off gloves or rags to look for "worst case" areas. May not provide sample representative of all "parts" of the drum.
Scoop, Spoon and Trowel (see 13.8)	Particle size	May exclude certain particle sizes, especially large aggregates.
	General use	Used to grab material from the top of a drum and so may bias results.
Syringe Sampler (see 13.9)	Speed of use	The sampler must be lowered slowly into the drum to minimize mixing.
	Angle of use	The angle of descent should be perpendicular to the surface of the material to be sampled.
	Media	Material to be sampled must be viscous enough to remain in the device when the coring tip is used. The valve tip cannot be used with viscous materials.
Trier (see 13.10)	Consistency	Material will not be held in place during removal if it is not sticky.
	Cross-contamination	There may be some contamination of the material within the bounds of the trier by the material outside the bounds.

not always possible or necessary. When deciding whether or not to mix, the following points should be considered.

11.1.1 Does the plan require that the materials in the drum be mixed before sampling? Or does the plan prohibit mixing?

11.1.2 If the material is hazardous and mixing it will create a safety problem, such as allowing acidic fumes to escape, do not mix.

11.1.3 The results of the laboratory tests that the sample will undergo may be influenced by mixing. Examples are (1) if the material contains volatile compounds that could be lost during the mixing process, do not mix, and (2) if a thin layer of solids on the bottom of the drum contain the metals of concern, mixing is necessary.

11.1.4 If the sampling plan has been designed such that several smaller samples will be taken from various locations and combined into one sample, or if numerous discrete samples will be taken, there is less need to mix the material

TABLE 2 Equipment Selection

Equipment or Device	ASTM Standard	One Liquid Layer	Two or More Layers of Liquids	Both Liquid and Solid Layers	Consolidated Solids[A]	Unconsolidated Solids[A]	Sludge[A]	See Paragraph
Column Number		I	II	III	IV	V	VI	
Auger	D 4700	B	—	—	X[C]	X	—	13.1
	D 1452							
Chipper, Hammer and Chisel		—	—	—	X	N[D]	—	—
COLIWASA	D 5495	X	X	N	—	—	N	13.3
Concentric Tube Thief		—	—	—	—	X	—	13.4
Dipper	D 5358	X	—	N	—	—	N	13.5
Drum Thief		X	X	X	—	—	X	13.6
Peristaltic Pump	D 4448	X	X	N	—	—	N	13.7
Push Coring Devices:	D 1586							
	D 1587							
Split Barrel	D 4700	—	—	—	N	X	N	13.2
Thin-walled Tube	D 4823	—	—	—	N	X	—	
Butterfly Valve		—	—	N	—	N	X	
Rotating Coring Device	C 783	—	—	—	X	—	—	13.2
	D 2113							
Scissors and Tongs		—	—	—	—	X	—	—
Scoop, Spoon and Trowel		N	—	—	—	X	N	13.8
Syringe Sampler		N	N	X	—	—	X	13.9
Trier	D 5451	—	—	—	—	N	N	13.10

[A] The suitability of the sampling device depends on the consistency of the material to be sampled.
[B] Equipment is probably unsuitable.
[C] Equipment may usually be used with this type of waste.
[D] Not equipment of choice but may be used at discretion of supervisor.

because the sampling plan has addressed variations in the material.

11.1.5 Sampling equipment designed to take a profile of the waste (for example, COLIWASA) does not require the material to be mixed first.

11.1.6 If the drum was moved or agitated before sampling, samples may be taken only when the contents of the drum are either completely settled or are thoroughly mixed.

11.2 If the materials are going to be mixed, do so in such a way that no contamination is introduced to the drum and sample container and no material is lost. It should be performed in a safe manner (for example, don't roll a 200-L drum back and forth with the hope of mixing the contents—it rarely works and the potential for injury is high).

11.3 A recirculating pump can be used to mix a liquid material in, for example, a 200-L drum.

12. Opening the Drum

12.1 The bung, ring or other fastening device that secures the lid should be removed slowly, allowing any pressure or vacuum to equalize (see Fig. 5).

12.1.1 Pails with "snap-on" lids may be difficult to open. Be careful to avoid splashing the contents when opening them.

12.1.2 If the top of the drum is dished inward (dimpled), solid or liquid materials, or both, may have accumulated there. The lid may "pop" when equalizing pressure, spraying the person sampling with any material that has accumulated on the lid. To avoid this, material on top of the drum must be removed before opening the drum.

12.1.3 If the top of the drum is bulging, opening it too quickly may cause the top to fly upward, possibly injuring the person sampling.

12.1.4 If there is evidence of a chemical reaction or sudden pressure buildup (for example, fumes escaping, heat build-up, rocking drum), the person sampling should leave the area immediately and decide if remote drum opening equipment should be used.

12.1.5 Drums should be grounded before being opened, if grounding is required.

12.2 Drums should be opened, sampled and closed individually to minimize possible volatilization of organic compounds and also exposure of the person sampling to the material(s).

12.3 When opening a drum with a bung, if the lid is dimpled or bulging, loosen the smaller bung first. This will make it easier to control the release of pressure. Also, cover a manual bung wrench with a cloth to control potential liquid spray. Then remove the large bung.

12.3.1 When opening a drum with a removable lid, slowly loosen the bolt with a manual wrench or an air impact wrench. If the lid is dimpled or bulging, allow the pressure to equalize before completely loosening the bolt. Remove the bolt, the ring and the lid.

12.3.2 When opening a pail with a removable lid (side-lever lock ring), slowly release the lever. If the lid is dimpled or bulging, allow the pressure to equalize before loosening the ring sufficiently for lid removal.

13. How to Use the Equipment

13.1 *Auger:*

13.1.1 *Description*—An auger samples hard or packed solid materials or soil. It consists of sharpened spiral blades attached to a hard metal central shaft (see Fig. 6).

13.1.2 *Steps to Extract a Sample:*

13.1.2.1 Collect the sample by rotating the handle of the auger in a clockwise direction while applying slight downwards pressure.

13.1.2.2 Continue turning until the desired depth has been reached. Pull the auger straight up out of the material.

13.1.2.3 Remove the material that has been withdrawn in the screw thread of the auger, and place it in the container.

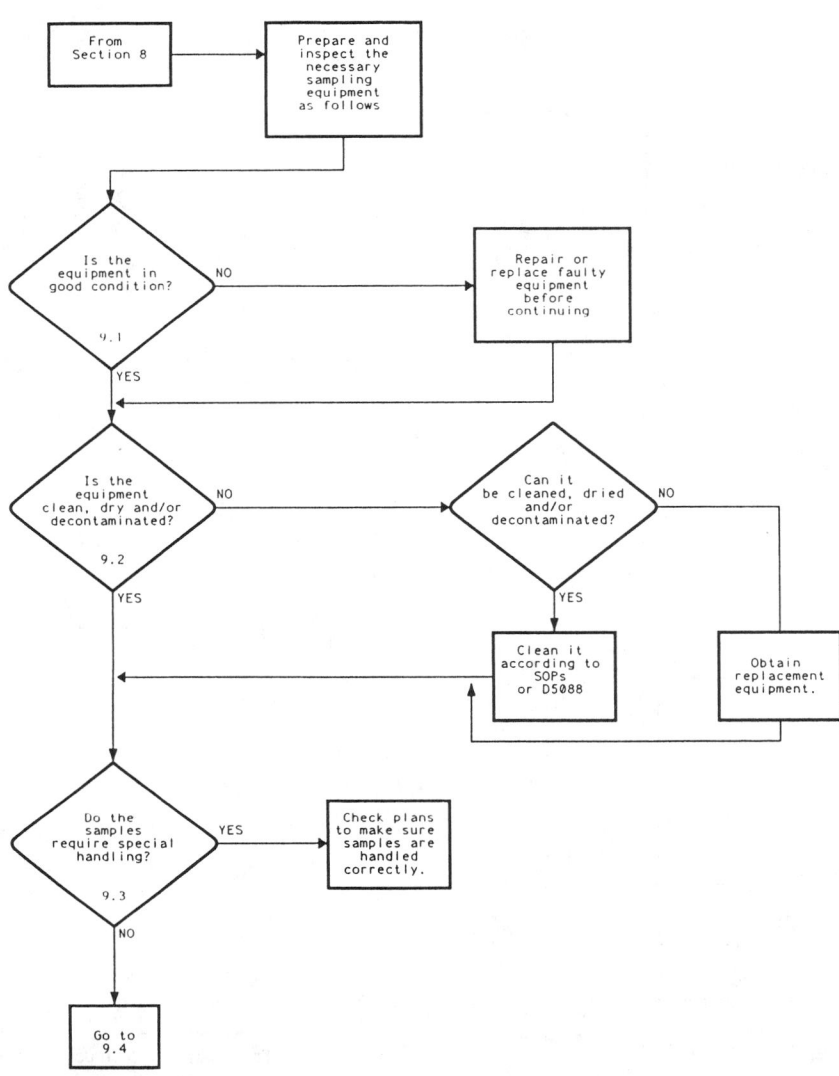

NOTE—This flow chart should be used with Sections 9.1 to 9.3 in the text.

FIG. 4 Preparation of Sampling Equipment

This is usually done with the auger placed on a clean plastic sheet.

13.1.2.4 Transfer the material to the appropriate container for analysis.

13.2 *Coring Devices:*

13.2.1 *Description*—A coring device samples hard or packed solid waste or soil (see Fig. 7). They are available with coring tips on all types and auger tips on some. The coring tip cuts a sample a little smaller than the inside diameter of the sampler body. The sampler is connected to a metal extension that allows the sampler to be driven into the material either with downward hand pressure or with a slide hammer. When a sampler with an auger tip is used, the sampler is rotated into the material being sampled, using the extension and attached cross handle. The coring sampler with butterfly valve is designed for use in sludges and liquid-saturated materials. The valve closes and prevents loss of sludges and semi-solids when raised from the sampling medium.

13.2.1.1 A stainless steel or plastic liner may be inserted into the hollow tube prior to use.

NOTE 3—Plastic liners may be inappropriate if the sample will be analyzed for organics.

13.2.2 *Steps to Extract a Sample:*

13.2.2.1 Place the coring device on the surface of the material to be sampled.

13.2.2.2 Twist the device slightly, and apply downward pressure to begin penetration. This steadies the device for the next step.

13.2.2.3 Use the slide hammer when manual pressure is no longer effective to drive the coring device into the material, until the desired depth is reached. Always mark the extension with a reference point equivalent to when the tip is near the bottom of the container. Take care not to overdrive the sampler to prevent damage to the bottom of the drum or container.

13.2.2.4 Twist the device again to break the core off from the bottom.

13.2.2.5 Pull the coring device straight up out of the material.

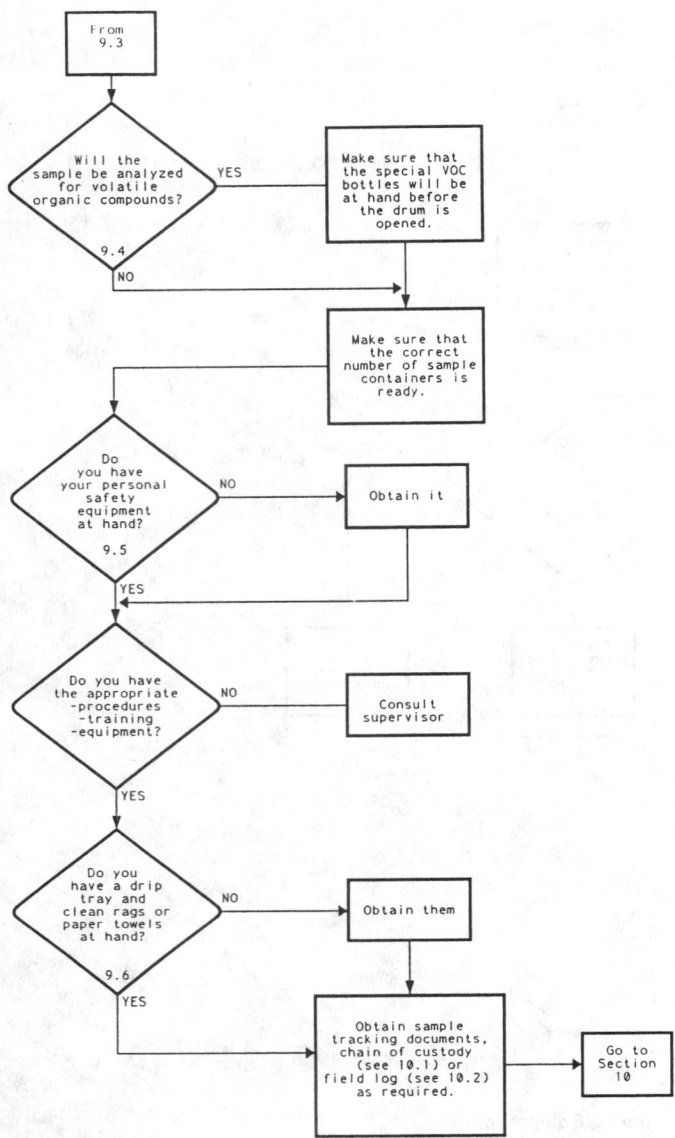

NOTE—This flow chart should be used with Sections 9.4 to 10.2 in the text.

FIG. 4 Preparation of Sampling Equipment (Continued)

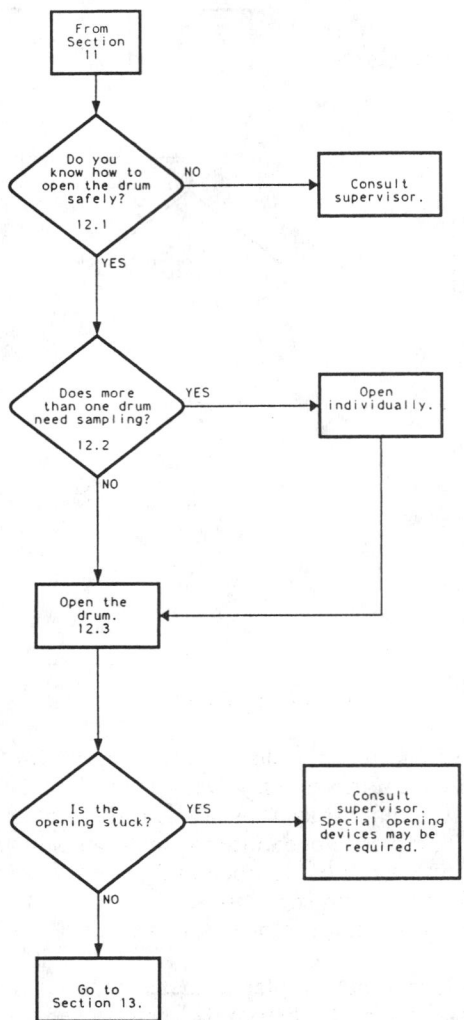

NOTE—This flow chart should be used with Sections 12.1 to 12.3 in the text.

FIG. 5 Opening the Drum

13.2.2.6 Open the device. Either cap off the liner, if present, or remove the material and place it into the appropriate sample container for analysis. This is usually done with the coring device placed on a clean plastic sheet.

13.3 *Composite Liquid Waste Sampler (COLIWASA):*

13.3.1 *Description*—COLIWASAs are available commercially with different types of stoppers and mechanisms, but they all operate using the same principle (see Fig. 8). They can be constructed of materials such as polyvinyl chloride (PVC), glass, metal or polytetrafluoroethylene (PTFE). They usually consist of two sections. The outer section is a sleeve that may be tapered at the end. The inner section is a rod with some type of stopper on the end. When the inner section is fitted inside the outer section, a seal is formed, and the unit is locked. There are variations in the mechanisms of opening and closing COLIWASAs. They can be used to obtain samples from specific depths. Sometimes it is neces-

sary to enter the drum a number of times with this sampler to obtain enough sample to meet the needs of the analytical laboratory.

13.3.2 *Steps to Extract a Sample:*

13.3.2.1 Place the COLIWASA in the open position.

13.3.2.2 Lower the COLIWASA slowly into the liquid, keeping it vertical at all times, making sure that the level of the liquid inside and outside the sampler tube remain about the same.

13.3.2.3 When the unit touches the bottom of the container, or has reached the desired depth, close the COLIWASA.

13.3.2.4 Remove the sampler from the liquid with one hand while wiping the outer tube with a disposable cloth or rag with the other hand.

13.3.2.5 Open the COLIWASA over the sample container.

13.4 *Concentric Tube Thief:*

13.4.1 *Description*—A concentric tube thief (grain sampler) (see Fig. 9) is used to sample free-flowing, dry granules or powdery materials whose particle diameter is less than

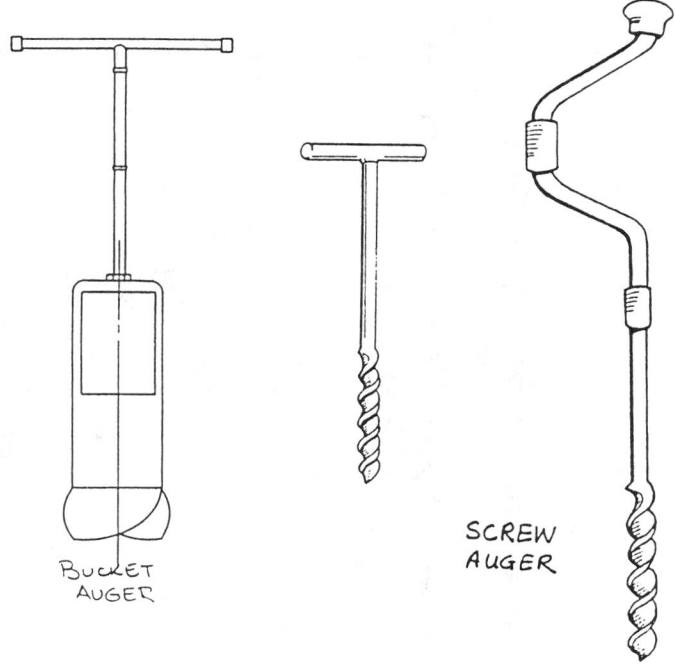

BUCKET AUGER

SCREW AUGER

FIG. 6 Augers

one-third the width of the slots in the sampler. It consists of two slotted concentric tubes, usually made of stainless steel or brass. The outer tube has a conical pointed tip that permits the sampler to penetrate the material being sampled. The inner tube is rotated to open and close the sampler.

13.4.2 *Steps to Extract a Sample:*

13.4.2.1 Ensure the sampler is in the closed position to start.

13.4.2.2 Insert the sampler into the material so that the largest cross-section of material will be sampled.

13.4.2.3 Open the unit by rotating the inner tube.

13.4.2.4 Jiggle the sample to encourage the material to enter the sampler.

13.4.2.5 Close the unit by rotating the inner tube and remove the sampler from the material.

13.4.2.6 Lay the sampler on a clean surface with the slots facing upwards. This is usually done on a clean plastic sheet. Note: some types of grain samplers allow the two tubes to separate from each other, allowing for easier removal of the sample.

13.4.2.7 Remove the sample and place in appropriate container for analysis.

13.5 *Dipper:*

13.5.1 *Description*—A dipper samples single phased liquids (see Fig. 10). It consists of a glass, metal or plastic beaker clamped to the end of a two- or three-piece telescoping aluminum or fiberglass pole that serves as a handle. Samples are taken at, or just below, the surface.

13.5.2 *Steps to Extract a Sample:*

13.5.2.1 Submerge the dipper into the material slowly, to cause minimum surface disturbance.

13.5.2.2 Allow the beaker to fill and slowly bring it to the surface.

13.5.2.3 Slowly pour the contents into the sample container.

13.6 *Drum Thief:*

13.6.1 *Description*—A drum thief is used to sample liquids (see Fig. 11). It consists of an open-ended tube, usually made of glass or stainless steel. Narrow bore tubes can be used for liquids with low surface tension, such as chlorinated solvents, and wider bore tubes can be used to sample sludges.

13.6.2 *Steps to Extract a Sample:*

13.6.2.1 Slowly lower the thief into the liquid, keeping it vertical at all times, until it hits touches the bottom of the container.

13.6.2.2 Place your thumb, or a stopper, over the top to create a vacuum. This will hold the sample in the tube while it is removed from the container. Use caution as this form of vacuum does not always hold the sample in place. If the material releases prematurely from the bottom of the thief, this equipment is unsuitable.

13.6.2.3 Place the thief over the appropriate container for analysis and release the vacuum by removing your thumb or the stopper.

13.7 *Peristaltic Pumps:*

13.7.1 *Description*—Peristaltic pumps (see Fig. 12), in which a cam rotates against flexible tubing, thereby moving the material inside the tube through it, are used for sampling liquids and slurries (see Guide D 4448). Hosing should be compatible with the material being sampled. General-use plastic tubing works well for low hazard situations where the sample is water-based, contains few or no organic constituents, and will not be analyzed for organics. High hazard materials and situations where organic compounds are present, and which may be analyzed, should be sampled using fluorocarbon resin tubing.

13.7.2 *Steps to Extract a Sample:*

13.7.2.1 Place the end of the inlet tubing within the container at the depth from which the sample will be extracted.

13.7.2.2 Place the end of the outlet tubing within the container and turn on the pump.

13.7.2.3 Once the sample is adequately mixed, turn off the pump and place the end of the outlet tubing into the appropriate container for analysis.

13.7.2.4 Turn on the pump and collect the sample into the container.

13.8 *Scoop, Spoon, Trowel:*

13.8.1 *Description*—Scoops, spoons and trowels (see Fig. 13) are used to sample a wide variety of materials, including sludge, soil, powder or hard solid waste.

13.8.2 *Steps to Extract a Sample:*

13.8.2.1 Collect sample from the appropriate depth by digging and rotating the sampler. Trowels may not be suitable for collecting solids from depths, or if digging is limited by the sides of the container.

13.8.2.2 Transfer the material to the appropriate container for analysis.

13.9 *Syringe Sampler:*

13.9.1 *Description*—A syringe sampler (see Fig. 14) is used to sample thick liquids, sludges and tar-like substances. It can also draw samples when only a small amount remains at the bottom of a drum or tank. Syringe samplers are available commercially. They usually include a manually operated piston assembly consisting of a T-handle, lock nut, control rod (polytetrafluoroethylene-covered aluminum rod

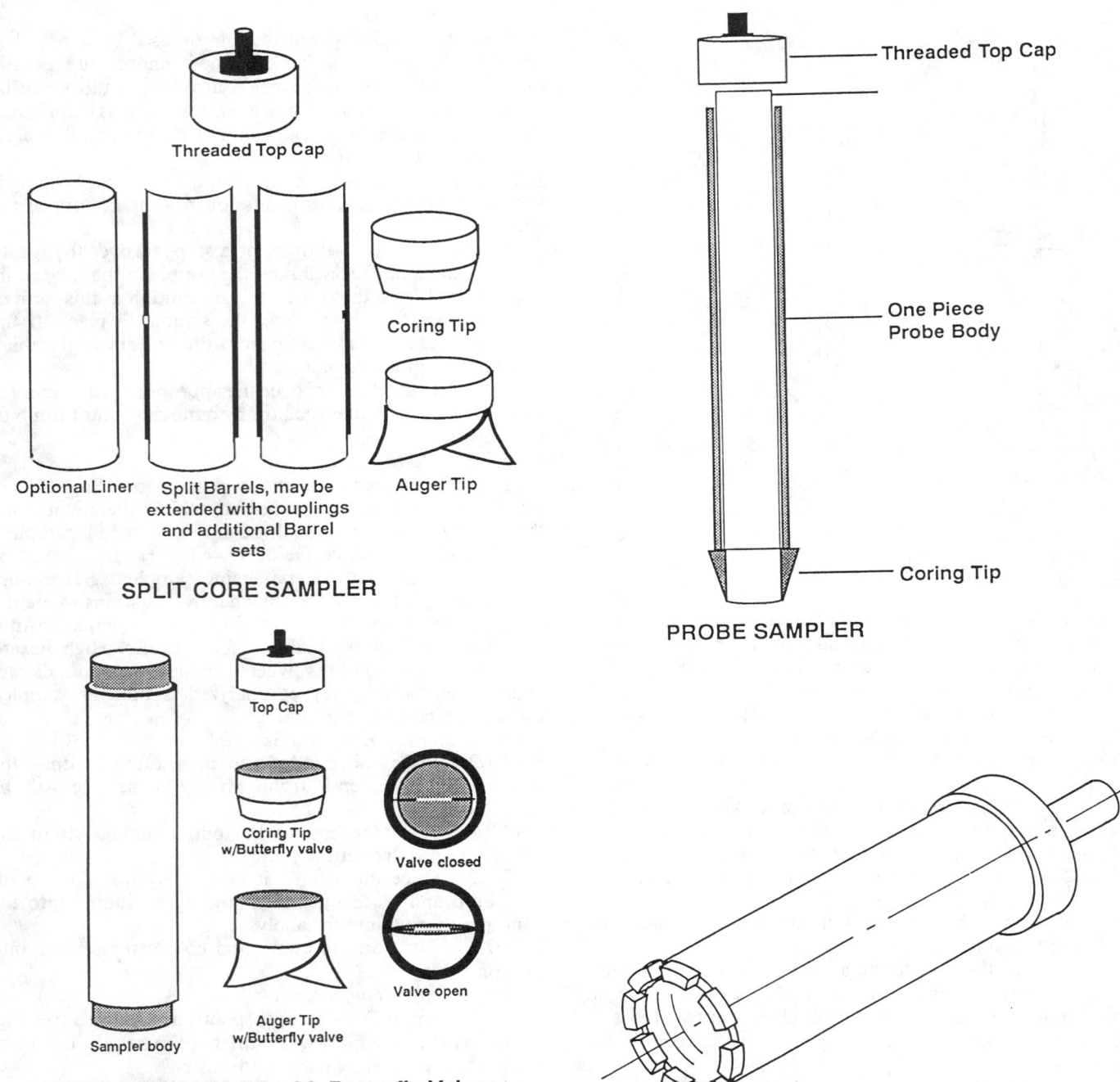

FIG. 7 Coring Devices

facilitates operation of the piston), piston assembly, sampling tube and two tips for the lower end, one with a closeable valve and one with a coring tip.

13.9.2 *Steps to Extract a Sample:*

13.9.2.1 Make sure that the top and bottom fittings are secured to the sampling tube and the valve tip, if used, is open. The piston assembly should be at the lower end of the sampler.

13.9.2.2 Slowly lower the syringe sampler into the drum or tank until it contacts the surface of the material to be sampled.

13.9.2.3 To collect a point sample, hold the sampler body, loosen the lock nut and gradually raise the T-handle to draw the sample into the sampler.

13.9.2.4 To collect a core sample with the coring tip, loosen the lock nut, hold the T-handle and slowly push the sampler body down into the material.

13.9.2.5 When the desired sample is obtained, hand tighten the lock nut to secure the piston rod.

13.9.2.6 Remove the sampler from the drum with one hand while wiping the sampler body with a disposable cloth or rag with the other hand.

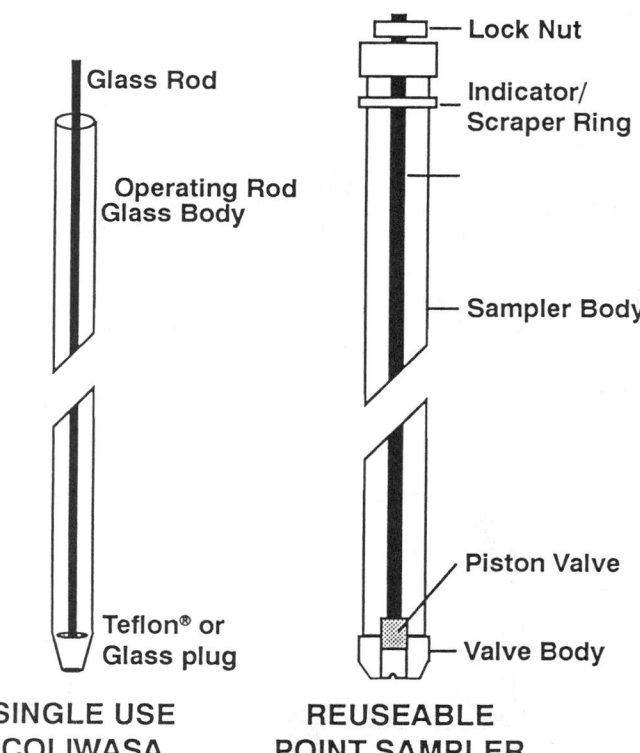

Glass Rod

Operating Rod
Glass Body

Teflon® or
Glass plug

**SINGLE USE
COLIWASA**

Lock Nut

Indicator/
Scraper Ring

Sampler Body

Piston Valve

Valve Body

**REUSEABLE
POINT SAMPLER**

FIG. 8   COLIWASA

FIG. 10   Dipper

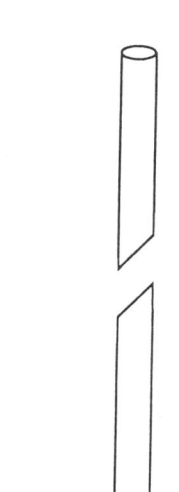

FIG. 11   Drum Thief

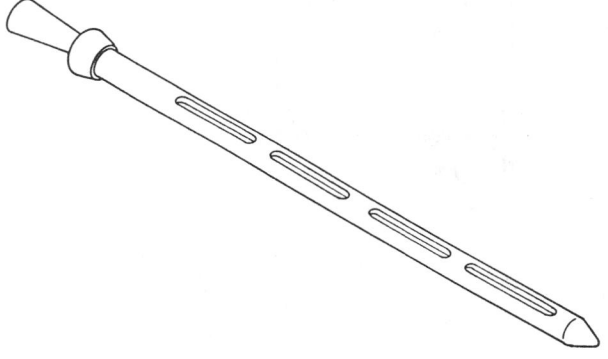

FIG. 9   Concentric Tube Thief

13.9.2.7 Hold the syringe sampler over the sample container; open the valve tip, if fitted and closed.

13.9.2.8 Loosen the lock nut and force the sample out of the sampler by depressing the piston rod.

13.10 *Trier:*

13.10.1 *Description*—A trier is used to sample moist or sticky solids with a particle diameter less than one-half the diameter of the trier. It consists of a handle and a tube cut in half lengthwise, with a sharpened tip that allows the sampler to cut into sticky materials and loosen solids.

13.10.2 *Steps to Extract a Sample:*

13.10.2.1 Hold the trier either horizontally or with the handle end tilted slightly downwards.

13.10.2.2 Insert the trier into the material to be sampled, at this angle.

13.10.2.3 Cut a core of the material by rotating the trier once or twice.

13.10.2.4 Stop the rotation with the open face pointing upwards.

13.10.2.5 Slowly remove the trier and empty the contents into the appropriate container for analysis.

14. Keywords

14.1 drum; sampling; waste

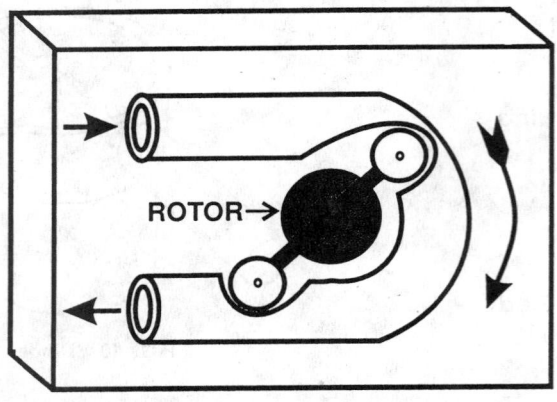

PERISTALTIC PUMP

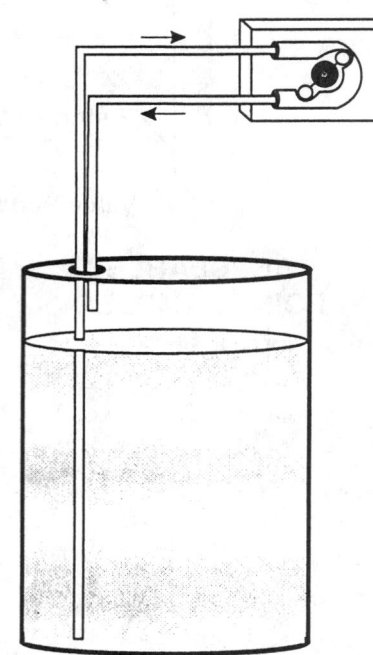

FIG. 12 Peristaltic Pumps

Stainless Steel Scoops
sizes #2,3,6,8,12 denote ozs.

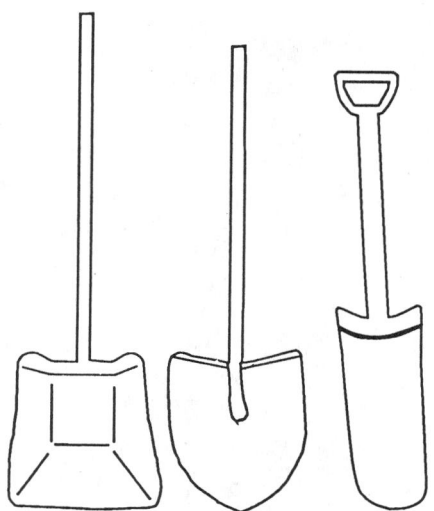

Stainless Steel Shovels

FIG. 13   Scoop, Spoon, Trowel

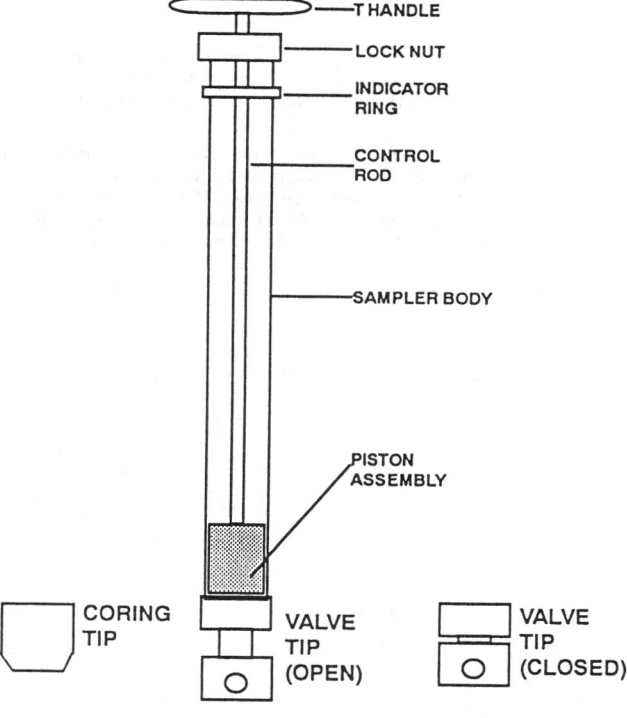

T HANDLE

LOCK NUT

INDICATOR RING

CONTROL ROD

SAMPLER BODY

PISTON ASSEMBLY

CORING TIP

VALVE TIP (OPEN)

VALVE TIP (CLOSED)

FIG. 14   Syringe Sampler

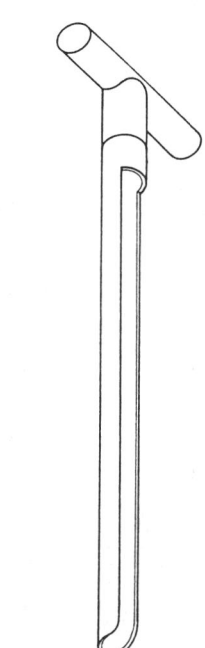

FIG. 15   Trier

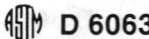 **D 6063**

Standard Practice for
Sampling Unconsolidated Solids in Drums or Similar Containers[1]

This standard is issued under the fixed designation D 5680; the number immediately following the designation indicates the year of original adoption or, in the case of revision, the year of last revision. A number in parentheses indicates the year of last reapproval. A superscript epsilon (ε) indicates an editorial change since the last revision or reapproval.

1. Scope

1.1 This practice covers typical equipment and methods for collecting samples of unconsolidated solids in drums or similar containers. These methods are adapted specifically for sampling drums having a volume of 110 U.S. gal (416 L) or less. These methods are applicable to hazardous material, product, or waste. Specific sample collection and handling requirements should be described in the site-specific work plan.

1.2 The values stated in inch-pound units are to be regarded as the standard. The values given in parentheses are for information only.

1.3 *This standard does not purport to address all of the safety concerns, if any, associated with its use. It is the responsibility of the user of this standard to establish appropriate safety and health practices and determine the applicability of regulatory limitations prior to use.*

2. Referenced Documents

2.1 *ASTM Standards:*
C 702 Practice for Reducing Samples of Aggregate to Testing Size[2]
D 4547 Practice for Sampling Waste and Soils for Volatile Organics[3]
D 4687 Guide for General Planning of Waste Sampling[3]
D 4700 Guide for Soil Sampling from the Vadose Zone[4]
D 5088 Practice for the Decontamination of Field Equipment Used at Non-Radioactive Waste Sites[4]
D 5283 Practice for Generation of Environmental Data Related to Waste Management Activities: Quality Assurance and Quality Control Planning[3]
D 5451 Practice for Sampling with a Trier Sampler[3]
E 300 Practice for Sampling Industrial Chemicals[5]
D 5633 Practice for Sampling With a Scoop[3]
2.2 *NSC Document:*
Accident Prevention Manual for Industrial Operations, 1992[6]
2.3 *Government Documents:*[7]

Drum Handling Practices at Hazardous Waste Sites, EPA/600/2-86/013, January 1986
Soil Sampling and Analysis for Volatile Compounds, EPA/540/4-91/001, February 1991
Occupational Safety and Health Guidance Manual for Hazardous Waste Site Activities, National Institute for Occupational Safety and Health (NIOSH), Occupational Safety and Health Administration (OSHA), U.S. Coast Guard (USCG), and U.S Environmental Protection Agency (EPA), October 1985

3. Terminology

3.1 *Definitions:*

3.1.1 *bonding*—touching the sample equipment to the drum to form an electrically conductive path to minimize potential electrical differences between the sampling equipment and the drum, reducing the buildup of static electricity.

3.1.2 *bung*—usually a 2-in. (5.1-cm) or ¾-in. (1.3-cm) diameter threaded plug designed specifically to close a bung hole.

3.1.3 *bung hole*—an opening in a barrel or drum through which it can be filled, emptied, or vented.

3.1.4 *deheading*—removal of the lid of a closed-head drum; usually accomplished with a drum deheader.

3.1.5 *drum*—implies any drum, barrel, or non-bulk container of 5 to 110 U.S gal (19 to 416 L) capacity.

3.1.6 *pail*—a small container, usually with a capacity of 5 U.S gal. Pails typically have bungs or spouts, or the entire lid can be removed.

3.1.7 *paperwork*—all required site documentation, which may include the manifests, waste profiles, material safety data sheets (MSDS), site forms, sample labels, custody seals, and chain of custody forms.

3.1.8 *unconsolidated*—for solid material, the characteristic of being uncemented or uncompacted, or both, and separated easily into smaller particles.

3.1.9 *work plan*—a plan specific to a particular site; for conducting activities specified in the plan.

4. Summary of Practice

4.1 The drum and its contents are inspected, and appropriate sampling equipment is selected. A clean sampling device is then used to auger, scoop, or core into the unconsolidated solid material to be sampled. The sample is collected and placed in a sample container. The sampling device is then either disposed of or cleaned and decontaminated.

5. Significance and Use

5.1 This practice is intended for use in collecting samples

[1] This practice is under the jurisdiction of ASTM Committee D-34 on Waste Management and is the direct responsibility of Subcommittee D34.01 on Sampling and Monitoring.
Current edition approved June 15, 1995. Published August 1995. Originally published as D 5680 – 95. Last previous edition D 5680 – 95.
[2] *Annual Book of ASTM Standards*, Vol 04.02.
[3] *Annual Book of ASTM Standards*, Vol 11.04.
[4] *Annual Book of ASTM Standards*, Vol 04.08.
[5] *Annual Book of ASTM Standards*, Vol 15.05.
[6] Available from National Safety Council, P.O. Box 558, Itasca, IL 60143-0558.
[7] Available from the Superintendent of Documents, U. S. Government Printing Office, Washington, DC 20402.

of unconsolidated solid materials from drums or similar containers, including those that are unstable, ruptured, or compromised otherwise. Special handling procedures (for example, remote drum opening, overpressurized drum opening, drum deheading, etc.) are described in *Drum Handling Practices at Hazardous Waste Sites.*

6. Interferences

6.1 The condition of the materials to be sampled and the condition and accessibility of the drums will have a significant impact on the selection of sampling equipment.

7. Pre-Sampling

7.1 *General Principles and Precautions:*

7.1.1 Samples should be collected in accordance with an appropriate work plan (Practice D 5283 and Guide D 4687). This plan must include a worker health and safety section because there are potential hazards associated with opening drums as well as potentially hazardous contents. See *Occupational Safety and Health Guidance Manual for Hazardous Waste Site Activities* for information on health and safety at hazardous waste sites.

7.1.2 Correct sampling procedures must be applied to the conditions as they are encountered. It is impossible to specify rigid rules describing the exact manner of sample collection because of unknowns associated with each solid sampling situation. It is essential that the samples be collected by a trained and experienced sampler because of the various conditions under which drummed solids must be sampled.

7.1.3 To be able to make probability or confidence statements concerning the properties of a sampled lot, the sampling procedure must allow for some element of randomness in selection because of the possible variations in the material. The sampler should always be on the alert for possible biases arising from the use of a particular sampling device or from unexpected segregation within the material.

7.1.4 All auger, trier, thief, and scoop methods may fail a prime sampling requirement: that of random selection of sample fractions. Scoops are limited to use at or near the top surface. Augers, triers, and thiefs are normally inserted in a present pattern. Particles on the bottom or along the sides of the drum may consequently never have an opportunity to be included in a sample. Sample particles should be selected by techniques that will minimize variation in measured characteristics between the available fractions and the resulting sample (Practice C 702).

7.1.5 The sampling equipment, sample preparation equipment, sample containers, etc. must be clean, dry, and inert to the material being sampled. All equipment, including sample containers, shall be inspected before use to ensure that they are clear of obvious dirt and contamination and are in good working condition. Visible contamination shall be removed, and the equipment shall be decontaminated with the appropriate rinse materials. Decontaminated sampling equipment should be protected from contamination. This may include, but not be limited to, storage in aluminum foil, plastic bags, polytetrafluoroethylene (PTFE) film, or other means of protection that will not impact the sample quality of intended analysis.

7.2 *Basic Pre-Sampling Practices:*

7.2.1 Review all paperwork.

7.2.2 Select the sampling equipment and sample containers appropriate for the material in the drum, as detailed in the work plan.

7.2.3 Enter the work zone.

7.2.4 Inspect all drums to be sampled visually. Note any abnormal conditions, including rust marks, stains, bulges, or other signs of pressurization or leaks that may require special handling. The work plan should define clearly the limiting condition under which special handling procedures shall be initiated. See *Drum Handling Practices at Hazardous Waste Sites* for information on opening overpressurized drums and the use of remotely operated drum opening equipment.

7.2.5 Stage the drums to be sampled in a designated work area if they cannot be sampled in their current location. See *Drum Handling Practices at Hazardous Waste Sites* for further information on staging turns.

7.2.5.1 Move the drums to upright, stable positions if necessary. Sufficient space shall be left between drums to prevent movement hazards.

7.2.5.2 Number or identify uniquely all drums to be sampled.

7.2.6 Perform a detailed inspection of individual drums.

7.2.6.1 Record all relevant information from drum labels, markings, data sheets, etc. in the field log book or on forms specified in the work plan.

7.2.6.2 Make sure there are no discrepancies with existing paperwork.

7.2.7 Slowly loosen the ring that secures the lid, or loosen the bung allowing any pressure or vacuum to equalize.

7.2.7.1 *Precautionary Notes:*

(*1*) If the drum or pail appears to be under positive or negative pressure (that is, a slight bulge or dimple in the lid), control the release of pressure until it has equalized. For example, if the drum or pail is equipped with bungs, loosen the smaller bung first since doing so will make it easier to control the release of pressure.

(*2*) If the top of the drum is dished inward (dimpled), it may "pop" when equalizing pressure, spraying the sampler with any material that is sitting on top of the drum.

(*3*) If there is evidence of a chemical reaction or sudden pressure buildup, the sampler should leave the area immediately and evaluate whether remote drum opening equipment should be used.

(*4*) For flammable or explosive materials, the drum and sampling equipment should be grounded if the generation of static electricity while opening or sampling the drum is a possibility. The drum and sampling equipment should be grounded to a ground stake or to an existing ground (building ground, grounded water pipes, etc.). New sampling equipment may have some residual static electrical charge due to the materials in which they are packed and shipped. The work plan should specify whether grounding is necessary. See *Accident Prevention Manual for Industrial Operations* for information on grounding and bonding.

7.2.7.2 Drums should be opened, sampled, and closed individually to minimize the risk of exposure.

7.2.7.3 *Drums (or Pails) with Bungs*—Loosen the large bung slowly. Use non-sparking tools.

7.2.7.4 *Drums with Removable Lids*—Loosen the ring slowly with a manual wrench or air impact wrench. Use non-sparking tools.

7.2.7.5 *Pails with Removable Lids (Side-Lever Lock Ring)*—Release the lever slowly.

7.2.7.6 *Pails with Removable Lids (Snap-On)*—Pry the lid loose slowly with a pail lid opener.

7.2.8 Manual or remote puncturing or deheading will be required if the drum (or pail) has a stuck bung or the lid cannot be removed. See *Drum Handling Practices at Hazardous Waste Sites* for further information on manual or remote drum opening.

7.2.9 Any discrepancy discovered (such as evidence of free liquid) upon opening the drum should be recorded in the field log book.

7.3 *Sampling Equipment—Selection:*

7.3.1 Table 1 summarizes selection criteria for equipment by the material to be sampled.

7.3.2 *Sampling Equipment, Materials of Construction*—Sampling devices will usually be made of stainless steel, brass, or aluminum. Devices using permanent coatings or liners (such as PTFE) may be subject to abrasion, leading to contamination of the sample.

7.3.3 *Generic Equipment List*—A general list of equipment used for sampling unconsolidated solids follows:

7.3.3.1 Scoop.
7.3.3.2 Trier.
7.3.3.3 Auger.
7.3.3.4 Concentric tube thief (single slot, multi-slot, grain probe, and missouri trier).
7.3.3.5 Thin-walled tube.
7.3.3.6 Scissors.
7.3.3.7 Tongs.
7.3.3.8 Hammer and chisel.
7.3.3.9 Cloths or wipes, or both.
7.3.3.10 Spatula.
7.3.3.11 Sample containers, lids, and liners.
7.3.3.12 Chain of custody forms.
7.3.3.13 Field log books.
7.3.3.14 Sample labels.
7.3.3.15 Sample cooler.
7.3.3.16 Ice or gel ice.
7.3.3.17 Grounding cables with alligator clips and emery cloth.
7.3.3.18 Portable monitoring equipment (combustible gas indicator, organic vapor detectors, radiation survey meter, etc.).

7.3.4 Equipment needed to open drums should be non-sparking (brass or beryllium copper) and include, but not be limited to, the following:

7.3.4.1 Bung wrenches (one straight and one bent),
7.3.4.2 Flatblade screwdriver,
7.3.4.3 Breaker bar (1/2 in. (1.3 cm)),
7.3.4.4 Ratchet (1/2 in. (1.3 cm)),
7.3.4.5 Speed handle (1/2 in. (1.3 cm)),
7.3.4.6 Adjustable wrenches (10 and 12 in. (25 and 30 cm)),
7.3.4.7 Air impact wrench and sockets, and
7.3.4.8 Pail lid opener.

8. Sample Collection

8.1 *Basic Sampling Practices:*

8.1.1 Bond the sampling equipment to the drum, if specified in the work plan.

8.1.2 Note the physical characteristics, including any discrepancies (such as free liquid).

8.1.3 Collect the required number of samples from the drum.

8.1.3.1 See Practice D 4547 and *Soil Sampling and Analysis for Volatile Compounds* for the collection of samples for volatile analysis.

8.1.4 Place the collected material in a sample container.

8.1.5 Close the sample container.

8.1.6 Wipe the outside of the sample container. Dispose of the wipe cloth properly.

8.1.7 Record in the field log book all relevant conditions and physical characteristics associated with the collection of each sample.

8.1.8 Fill out all required paperwork for each sample, as required by the work plan.

8.1.9 Complete and attach the label to the side of the sample container before or after sampling, as directed by the work plan. The sample label should include the following:

(*1*) Sample ID number,
(*2*) Name of sampler,
(*3*) Sampler's initials or signature,
(*4*) Date and time of sampling, and
(*5*) Sample location.

8.1.9.1 The sample label can also include the following:

(*1*) Sampling information (for example, grab, composite, etc.),
(*2*) Preservative and preservation required,
(*3*) Special instructions, and
(*4*) Analysis request.

8.2 *Sampling Using a Scoop:*

8.2.1 *General Description*—A plastic or metal scoop is used for collecting approximately equal fractions at random intervals at or near the surface of the material (see Fig. 1). This sampling tool should be of a size and shape suitable for the quantity and size of the particles to be sampled. Scoop sampling provides best results if the material is uniform.

8.2.2 *Operation and Use*—A thin layer of material is removed with the scoop and discarded. A suitable volume of material is collected with the scoop and transferred to a sample container.

8.3 *Sampling with a Thin-Walled Tube Sampler:*

8.3.1 This sampler can be used to collect samples of unconsolidated material that is usually moist or cohesive and may be powdery or granular.

8.3.2 *General Description*—Tube samplers may vary in diameter, length, and material of construction (see Fig. 2).

TABLE 1 Selection Criteria for Equipment

Equipment	ASTM Standard	Cohesive Solid	Sheet, Cloth, or Chunk Material	Dry Flowable Solids	Moist Flowable Solids
Scoop	D 5633	X[A]	X	X	X
Auger	D 4700	X	...[B]	...	...
Trier	D 5451	X	...	...	X
Thin-walled tube	D 4700	X	...	X[D]	X[D]
Hammer and chisel		N[C]	N	...	...
Scissors and tongs		...	X	...	...
Concentric tube thief	E 300	...	...	X	...

[A] X = equipment may be used with this type of waste.
[B] Equipment is probably unsuitable.
[C] N = not equipment of choice, but may be used.
[D] Sampling equipment with retaining device.

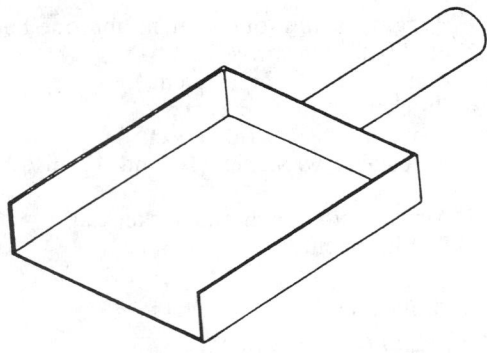

FIG. 1   Scoop

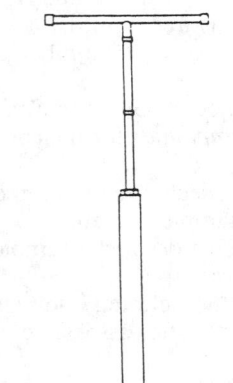

FIG. 2   Thin-Walled Tube Sampler

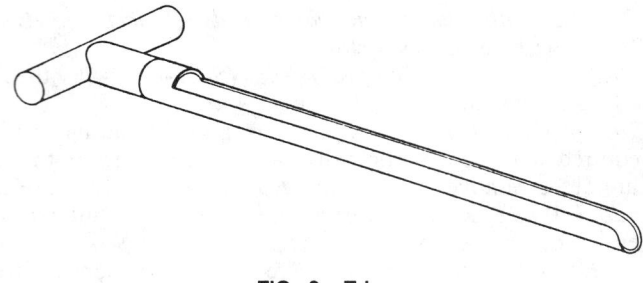

FIG. 3   Trier

The material to be sampled must be of a physical consistency (cohesive solid material) to be cored and retrieved with the tube. Materials with particles larger than one-third of the inner diameter of the tube should not be sampled with this particular device. The length of the tube will depend on the desired sampling depth (Guide D 4700). The tube is attached to a length of a solid or tubular rod. The upper end of this rod is threaded to accept a handle or extension rods. This sampler can be used to collect samples of unconsolidated clay-like materials.

8.3.3 *Operation and Use*—The sampler is pushed into the material to be sampled by applying downward force on the unit's handle. Once the sampler has reached the bottom of the sampling interval, it is twisted to break the continuity at the tip. The sampler is pulled from the material, and the sample material is extruded into the sample container. Samples are extruded by forcing a rod through the tube.

8.4 *Sampling with a Trier Sampler:*

8.4.1 *General Description*—The trier is a metal or plastic tube from which one-third to one-half of the wall of the tube has been removed to form a slot along its entire length (see Fig. 3). This device can be up to 4 ft (1.2 m) long and should have a sharp, angled point at its lower end. The trier can be used as a coring device in drums of solids such as soils and similar fine-grained cohesive materials.

8.4.2 *Operation and Use*—The trier is pushed vertically into the material and rotated one or two times to cut a core. The core is pulled out of the hole and transferred to a sample container (Practice D 5451).

8.5 *Sampling with an Auger:*

8.5.1 *General Description*—The screw or ship auger is essentially a small-diameter (for example, 1.5-in. (3.8-cm)) wood auger from which the cutting side flanges and tip have been removed. The auger is welded onto a length of solid or tubular rod. The upper end of this rod is threaded to accept a handle or extension rods. An auger can be used for collecting a disturbed sample of unconsolidated material in drums (see Fig. 4 and Guide D 4700).

8.5.2 *Operation and Use*—The auger is rotated manually or with a power source into the material to be sampled. The operator may have to apply downward force to embed the auger; the auger screws itself into the material afterwards. The auger is advanced to its full length and then pulled and removed. Material from the deepest interval is retained on the auger flights. Sample material can be collected from the flights using a spatula.

8.6 *Sampling with a Hammer and Chisel, Scissors, or Tongs:*

8.6.1 *General Description*—A hammer is used to impact a hardened steel chisel to break the unconsolidated material into chips, flakes, and chunks suitable for collection with a scoop. Scissors are used in combination with tongs to collect samples of material that are clothlike, elastic, paperlike, etc.

8.6.2 *Operation and Use*—These tools are used as necessary to collect sample material from the drum. A hammer and chisel have been found useful in sampling drums for which particle size reduction is necessary. This method is not recommended for samples requiring volatile organics analysis.

8.7 *Sampling with a Concentric Tube Thief:*

8.7.1 *General Description*—This device consists of two tubes, one fitting snugly inside the other (see Fig. 5). The bottom end of the outer tube is fitted with a point. Oblong holes are cut through both tubes. The holes are opened or closed by rotating the inner tube. They are constructed as either single compartment (single-slot and multi-slot tube thiefs) or multiple compartment (Missouri trier or Grain Probe) sampling devices. Concentric tube thiefs are commercially available up to 6 ft (1.8 m) long and several in. (cm) in diameter. Concentric tube thiefs have a limited application for sampling drums. Material that is not free-flowing, such as those that are hard packed, moist, or finely powdered, will not enter this sampler under normal field conditions. Sampling of materials containing granules or particles exceeding one-third of the slot width should not be attempted because bridging may occur. These devices cannot sample the bottom of a drum because of their pointed ends.

8.7.2 *Operation and Use*—Insert the tube into the mate-

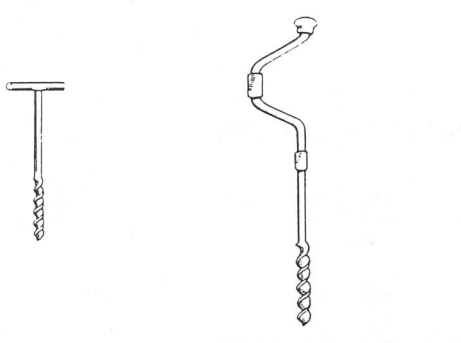

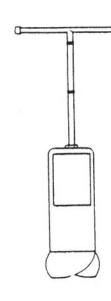

FIG. 4 Augers (Typical)

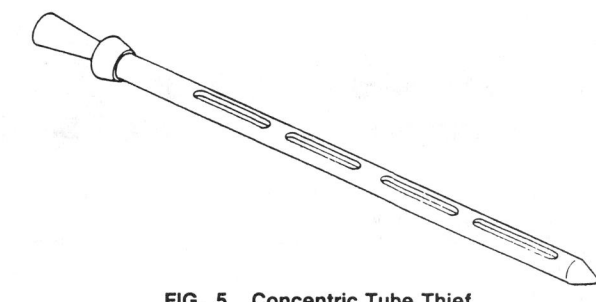

FIG. 5 Concentric Tube Thief

rial, and push with uniform force to the bottom of the drum or until refusal. Rotate the concentric tubes to the open position, thereby allowing the sample to flow into the inner tube. Wiggle the sampler several times, and rotate the tubes to the closed position. Withdraw the sampler. Place the sampling device immediately over a sample collection sheet, and release the sample by rotating the concentric tubes to the open position. A sample can normally be removed from the thief with a spatula or similar instrument (reamer) and placed in the sample container.

9. Post-Sampling

9.1 Remove all sampling equipment from the work zone.

9.2 Transfer all reusable equipment that was in contact with the waste to a pre-designated decontamination area. Decontaminate the equipment according to the protocol established in the work plan (Practice D 5088). Decontaminated sampling equipment should be protected from con-

tamination. This may include, but not be limited to, storage in aluminum foil, plastic bags, PTFE film, or other means of protection that will not impact the sample quality or intended analysis.

9.3 Dispose of all used (disposable) contacting equipment.

10. Data Quality Objectives

10.1 The objectives for sampling and testing of unconsolidated solid material should be specified in the work plan.

11. Quality Control

11.1 Quality control (QC) (for example, equipment blanks, trip blanks, and duplicates) must be collected as required by the work plan. These QC samples shall be evaluated to provide a determination of the sampling quality and reliability of the resulting analytical data.

12. Keywords

12.1 Auger; concentric tube thief; scoop; thin-wall tube; trier; unconsolidated solids; waste

Standard Practice for
Sampling Consolidated Solids in Drums or Similar Containers[1]

This standard is issued under the fixed designation D 5679; the number immediately following the designation indicates the year of original adoption or, in the case of revision, the year of last revision. A number in parentheses indicates the year of last reapproval. A superscript epsilon (ε) indicates an editorial change since the last revision or reapproval.

1. Scope

1.1 This practice covers typical equipment and methods for collecting samples of consolidated solids in drums or similar containers. These methods are adapted specifically for sampling drums having a volume of 110 U.S. gal (416 L) or less. These methods are applicable to hazardous material, product, or waste. Specific sample collection and handling requirements should be described in the site-specific work plan.

1.2 The values stated in inch-pound units are to be regarded as the standard. The values given in parentheses are for information only.

1.3 *This standard does not purport to address all of the safety concerns, if any, associated with its use. It is the responsibility of the user of this standard to establish appropriate safety and health practices and determine the applicability of regulatory limitations prior to use.*

2. Referenced Documents

2.1 *ASTM Standards:*
C 702 Practice for Reducing Samples of Aggregate to Testing Size[2]
C 783 Practice for Core Sampling of Graphite Electrodes[3]
D 4547 Practice for Sampling Waste and Soils for Volatile Organics[4]
D 4687 Guide for General Planning of Waste Sampling[4]
D 4700 Guide for Soil Sampling from the Vadose Zone[5]
D 5088 Practice for the Decontamination of Field Equipment Used at Non-Radioactive Waste Sites[5]
D 5283 Practice for Generation of Environmental Data Related to Waste Management Activities: Quality Assurance and Quality Control Planning[4]
2.2 *NSC Document:*
Accident Prevention Manual for Industrial Operations, 1985[6]
2.3 *Government Documents:*[7]
Drum Handling Practices at Hazardous Waste Sites, EPA/600/2-86/013, January 1986

Occupational Safety and Health Guidance Manual for Hazardous Waste Site Activities, National Institute for Occupational Safety and Health (NIOSH), Occupational Safety and Health Administration (OSHA), U.S. Coast Guard (USCG), and U.S. Environmental Protection Agency (EPA), October 1985

3. Terminology

3.1 *Definitions:*
3.1.1 *bonding*—touching the sampling equipment to the drum to form an electrically conductive path to minimize potential electrical differences between the sampling equipment and the drum, reducing the buildup of static electricity.
3.1.2 *bung*—usually a 2-in. (5.1-cm) or ¾-in. (1.3-cm) diameter threaded plug designed specifically to close a bung hole.
3.1.3 *bung hole*—an opening in a barrel or drum through which it can be filled, emptied, or vented.
3.1.4 *consolidated*—the characteristic of being cemented or compacted, or both, and not separated easily into smaller particles.
3.1.5 *deheading*—removal of the lid of a closed-head drum; usually accomplished with a drum deheader.
3.1.6 *drum*—implies any drum, barrel, or non-bulk container of 5 to 110 U.S. gal (19 to 416 L) capacity.
3.1.7 *pail*—a small container, usually with a capacity of 5 U.S. gal (19 L). Pails typically have bungs or spouts, or the entire lid can be removed.
3.1.8 *paperwork*—all required site documentation, which may include the manifests, waste profiles, material safety data sheets (MSDS), site forms, sample labels, custody seals, and chain of custody forms.
3.1.9 *work plan*—a plan, specific to a particular site, for conducting activities specified in the plan.

4. Summary of Practice

4.1 The drum and its contents are inspected, and appropriate sampling equipment is selected. A clean device is then used to auger, chisel, chip, or core into the consolidated solid material to be sampled. The sample is collected and placed in a sample container. The sampling device is then cleaned and decontaminated or disposed of.

5. Significance and use

5.1 This practice is intended for use in collecting samples of consolidated or compacted materials from drums or similar containers, including those that are unstable, ruptured, or compromised otherwise. Special handling procedures (for example, remote drum opening, overpressurized drum opening, drum deheading, etc.) are described in *Drum Handling Practices at Hazardous Waste Sites.*

[1] This practice is under the jurisdiction of ASTM Committee D-34 on Waste Management and is the direct responsibility of Subcommittee D34.01 on Sampling and Monitoring.
Current edition approved June 15, 1995. Published August 1995. Originally published as D 5679 – 95. Last previous edition D 5679 – 95.
[2] *Annual Book of ASTM Standards,* Vol 04.02.
[3] *Annual Book of ASTM Standards,* Vol 15.01.
[4] *Annual Book of ASTM Standards,* Vol 11.04.
[5] *Annual Book of ASTM Standards,* Vol 04.08.
[6] Available from National Safety Council, P.O. Box 558, Itasca, IL 60143-0558.
[7] Available from Superintendent of Documents, U.S. Government Printing Office, Washington, DC 20402.

6. Interferences

6.1 The condition of the materials to be sampled and the condition and accessibility of the drums will have a significant impact on the selection of sampling equipment.

7. Pre-Sampling

7.1 *General Principles and Precautions:*

7.1.1 Samples should be collected in accordance with an appropriate work plan (Practice D 5283 and Guide D 4687). This plan must include a worker health and safety section because there are potential hazards associated with opening drums as well as potentially hazardous contents. See *Occupational Safety and Health Guidance Manual for Hazardous Waste Site Activities* for information on health and safety at hazardous waste sites.

7.1.2 Correct sampling procedures must be applied to the conditions as they are encountered. It is impossible to specify rigid rules describing the exact manner of sample collection because of unknowns associated with each solid sampling situation. It is essential that the samples be collected by a trained and experienced sampler because the various conditions under which drummed solids must be sampled.

7.1.3 To be able to make probability or confidence statements concerning the properties of a sampled lot, the sampling procedure must allow for some element of randomness in selection because of the possible variations in the material. The sampler should always be on the alert for possible biases arising from the use of a particular sampling device or from unexpected segregation within the material.

7.1.4 All augering, chipping, or flaking sampling methods may fail a prime sampling requirement: that of random selection of sample fractions. Particles on the bottom or along the sides of the drum may consequently never have an opportunity to be included in a sample. Sample particles should be selected by techniques that will minimize variation in measured characteristics between the available fractions and the resulting sample (Practice C 702).

7.1.5 The sampling equipment, sample preparation equipment, sample containers, etc. must be clean, dry, and inert to the material being sampled. All equipment, including sample containers, must be inspected before use to ensure that they are clear of obvious dirt and contamination and in good working condition. Visible contamination must be removed, and the equipment must be decontaminated with the appropriate rinse materials. Decontaminated sampling equipment should be protected from contamination. This may include, but not be limited to, storage in aluminum foil, plastic bags, polytetrafluoroethylene (PTFE) film, or other means of protection that will not impact the sample quality or intended analysis.

7.2 *Basic Pre-Sampling Practices:*

7.2.1 Review all paperwork.

7.2.2 Select the sampling equipment and sample containers appropriate for the material in the drum, as detailed in the work plan.

7.2.3 Enter the work zone.

7.2.4 Inspect all drums to be sampled visually. Note any abnormal conditions, including rust marks, stains, bulges, or other signs of pressurization or leaks that may require special handling. The work plan should define clearly the limiting conditions under which special handling procedures shall be initiated. See *Drum Handling Practices at Hazardous Waste Sites* for information on opening overpressurized drums and the use of remotely operated drum opening equipment.

7.2.5 Stage the drums to be sampled in a designated work area if they cannot be sampled in their current location. See *Drum Handling Practices at Hazardous Waste Sites* for further information on staging drums.

7.2.5.1 Move the drums to upright stable positions if necessary. Sufficient space shall be left between drums to prevent movement hazards.

7.2.5.2 Number or identify uniquely all drums to be sampled.

7.2.6 Perform a detailed inspection of individual drums.

7.2.6.1 Record all relevant information from the drum labels, markings, data sheets, etc. in the field log book or on forms specified in the work plan.

7.2.6.2 Make sure there are no discrepancies with existing paperwork.

7.2.7 Slowly loosen the ring that secures the lid or loosen the bung, allowing any pressure or vacuum to equalize.

7.2.7.1 *Precautionary Notes:*

(1) If the drum or pail appears to be under positive or negative pressure (that is, a slight lid bulge or dimple), control the release of pressure until it has equalized. For example, if the drum or pail is equipped with bungs, loosen the smaller bung first since doing so will make it easier to control the release of pressure.

(2) If the tope of the drum is dished inward (dimpled), it may "pop" when equalizing pressure, spraying the sampler with any material that is sitting on top of the drum.

(3) If there is evidence of a chemical reaction or sudden pressure buildup, the sampler should leave the area immediately and evaluate whether remote drum opening equipment should be used.

(4) For flammable or explosive materials, the drum and sampling equipment should be grounded if the generation of static electricity while opening or sampling the drum is a possibility. The drum and sampling equipment should be grounded to a ground stake or to an existing ground (building ground, grounded water pipes, etc.). New sampling equipment may have some residual static electrical charge due to the materials in which they are packed and shipped. The work plan should specify whether grounding is necessary. See *Accident Prevention Manual for Industrial Operations* for information on grounding and bonding.

7.2.7.2 Drums should be opened, sampled, and closed individually to minimize the risk of exposure.

7.2.7.3 *Drums (or Pails) with Bungs*—Loosen the large bung slowly. Use non-sparking tools.

7.2.7.4 *Drums with Removable Lids*—Loosen the ring slowly with a manual wrench or air impact wrench. Use non-sparking tools.

7.2.7.5 *Pails with Removable Lids (Side-Lever Lock Ring)*—Release the lever slowly.

7.2.7.6 *Pails with Removable Lids (Snap-On)*—Pry the lid loose slowly with a pail lid opener.

7.2.8 Manual or remote puncturing or deheading will be required if the drum (or pail) has a stuck bung or the lid cannot be removed. See *Drum Handling Practices at Hazardous Waste Sites* for further information on manual or remote drum opening.

7.2.9 Any discrepancy discovered (such as evidence of free liquid) upon opening the drum should be recorded in the field log book.

7.3 *Sampling Equipment—Selection:*

7.3.1 Table 1 summarizes selection criteria for equipment by the material to be sampled.

7.3.2 *Sampling Equipment, Materials of Construction—* Sampling devices will usually be made of stainless steel, brass, aluminum, or plastic. Devices using permanent coatings or liners (such as PTFE) may be subject to abrasion, leading to contamination of the sample.

7.3.3 *Generic Equipment List—*A general list of equipment used for sampling consolidated solids follows:

7.3.3.1 Scoop.

7.3.3.2 Rotating corer.

7.3.3.3 Thin-wall tube sampler.

7.3.3.4 Chipper.

7.3.3.5 Hammer and chisel.

7.3.3.6 Auger.

7.3.3.7 Pry bars.

7.3.3.8 Wipes or cloths, or both.

7.3.3.9 Spatula.

7.3.3.10 Sample containers, lids, and liners.

7.3.3.11 Sample labels.

7.3.3.12 Chain of custody forms.

7.3.3.13 Field log books.

7.3.3.14 Sample cooler.

7.3.3.15 Ice or gel ice.

7.3.3.16 Grounding cables with alligator clips and emery cloth.

7.3.3.17 Portable monitoring equipment (combustible gas indicator, organic vapor detector, radiation survey meter, etc.).

7.3.4 Equipment needed to open drums should be non-sparking (brass or beryllium copper) and include, but not be limited to, the following:

7.3.4.1 Bung wrenches (one straight and one bent),

7.3.4.2 Flathead screwdriver,

7.3.4.3 Breaker bar (½ in. (1.3 cm)),

7.3.4.4 Ratchet (½ in. (1.3 cm)),

7.3.4.5 Speed handle (½ in. (1.3 cm)),

7.3.4.6 Adjustable wrenches (10 and 12 in. (25 and 30 cm)).

7.3.4.7 Air impact wrench and sockets, and

7.3.4.8 Pail lid opener.

8. Sample Collection

8.1 *Basic Sampling Practices:*

8.1.1 Bond the sampling equipment to the drum, if specified in the work plan.

8.1.2 Note the physical characteristics, including any discrepancies (such as free liquid).

8.1.3 Collect the required number of samples from the drum.

8.1.3.1 See Practice D 4547 for the collection of samples for volatile analysis.

8.1.4 Place the collected material in a sample container.

8.1.5 Close the sample container.

8.1.6 Wipe the outside of the sample container. Dispose of the wipe cloth properly.

8.1.7 Record in the field log book all relevant conditions associated with the collection of each sample.

8.1.8 Fill out all required paperwork for each sample, as required by the work plan.

8.1.9 Complete and attach the label to the side of the sample container before or after sampling, as directed by the work plan. The sample label should include the following:

(*1*) Sample ID number,

(*2*) Name of sampler,

(*3*) Sampler's initials or signature,

(*4*) Date and time of sampling, and

(*5*) Sampling location.

8.1.9.1 The sample label can also include the following:

(*1*) Sampling information (for example, grab, composite, etc.),

(*2*) Preservative and preservation required,

(*3*) Special instructions, and

(*4*) Analysis request.

8.2 *Sampling Using a Rotating Corer—*The rotating corer can be as simple as a cylinder attached to an electric drill with the crown modified for cutting (see Fig. 1 and Practice C 783) or as complex as a double metal tube fitted onto a diamond-impregnated coring bit, mounted on a portable stand. The double metal tube corer mounted on a portable stand has the capability of collecting a full-depth core of the drum contents. This procedure describes the single metal tube corer attached to an electric drill.

8.2.1 *General Description—*The rotating corer is usually 1 to 1½ ft (0.3 to 0.5 m) long, with available diameters of approximately 2 to 6 in. (5.1 to 15.2 cm). The corer is driven by suitable equipment, such as a portable electric drill.

8.2.2 *Operation and Use—*Place the coring apparatus over the area to be cored. Turn the coring mechanism on, and place the coring bit in continuous contact with the solid material by supplying uniform and continuous pressure. Continue the operation until the solid material is bored to the specified depth and the resulting core is forced into the

TABLE 1   Selection Criteria for Equipment

Equipment	ASTM Standard	Visually Homogeneous	Heterogeneous
Auger	D 4700	X[A]	X
Chipper, hammer, and chisel	...	X	X
Rotating corer	C 783	X	X
Thin-walled tube	D 4700	N[B]	...[C]

[A] X = equipment usually may be used with this type of waste.

[B] N = not equipment of choice but may be used (for example, a clay-like material).

[C] Equipment is probably not suitable.

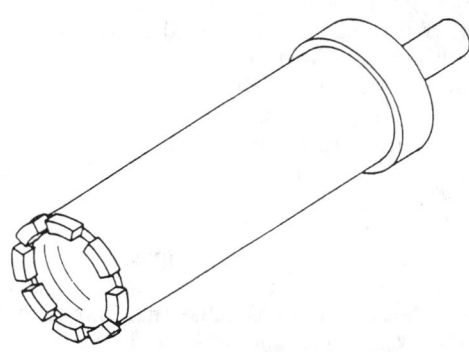

FIG. 1   Rotating Corer

tube. Withdraw the corer. Record the actual length of the core. The tube can be capped and transferred to the analytical laboratory. This is the desirable method for samples requiring volatile organics analysis. The sample core can also be extruded into a form-fitting sample container.

8.3 *Sampling with a Thin-Wall Tube Sampler:*

8.3.1 *General Description*—Thin-wall tube samplers may vary in diameter, length, and material of construction (see Fig. 2). The material to be sampled must be of a physical consistency (cohesive solid material) to be cored and retrieved with the tube. Materials with particles larger than one-third of the inner diameter of the tube should not be sampled with that particular device. The length of the tube will depend on the desired sampling depth (Guide D 4700). The tube is attached onto a length of solid or tubular rod. The upper end of this rod is threaded to accept a handle or extension rods. This sampler can be used to collect samples of consolidated clay-like materials.

8.3.2 *Operation and Use*—The sampler is pushed into the material to be sampled by applying downward force on the unit's handle. Once the sampler has reached the bottom of the sampling interval, it is twisted to break the continuity at the tip. The sampler is pulled from the material, and the sample material is extruded into the sample container. Samples are extruded by forcing a rod through the tube.

8.4 *Sampling with a Chipper:*

8.4.1 *General Description*—A hardened steel bit or knife is fitted to a pneumatic hammer (see Fig. 3). The hammer forces the bit or knife to break the consolidated material into chips, flakes, and chunks suitable for collection with a scoop. This method is not recommended for samples requiring volatile organics analysis.

8.4.2 *Operation and Use*—Insert the chipper into the drum. Place the chipping knife directly on the material to be sampled. Activate the pneumatic hammer. Break the consolidated material into manageably sized chips, flakes, and chunks. Remove the chipper. Collect a sample from the drum with a scoop, and transfer to the sample container.

NOTE 1—See 7.1.4 and the site work plan for information on selecting chipped particles for the sample.

8.5 *Sampling with a Hammer and Chisel:*

8.5.1 *General Description*—A hammer is used to impact a hardened steel chisel to break the consolidated material into chips, flakes, and chunks suitable for collection with a scoop.

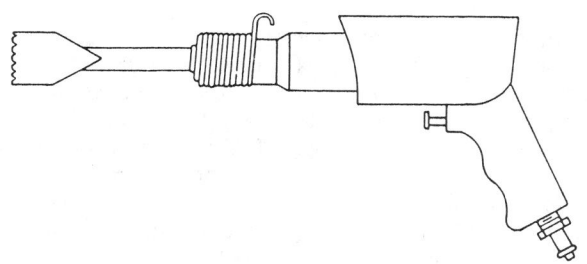

FIG. 3 Pneumatic Chipper

8.5.2 *Operation and Use*—These tools are used as necessary to collect sample material from the drum. A hammer and chisel have been found useful in sampling drums for which surface sampling is required or particle size reduction is necessary. This method is not recommended for samples requiring volatile organics analysis.

NOTE 2—See 7.1.4 and the site work plan for information on selecting chipped particles for the sample.

8.6 *Sampling with an Auger:*

8.6.1 *General Description*—The screw or ship auger is essentially a small-diameter (for example, 1½-in. (3.8-cm)) wood auger from which the cutting side flanges and tip have been removed. The auger is welded onto a length of solid or tubular rod. The upper end of this rod is threaded to accept a handle or extension rods. An auger can be used for collecting a disturbed sample of consolidated material in drums (See Fig. 4 and Guide D 4700).

8.6.2 *Operation and Use*—The auger is rotated manually or with a power source into the material to be sampled. The operator may have to apply downward force to embed the auger; the auger screws itself into the material afterwards. The auger is advanced to its full length and then pulled up and removed. Material from the deepest interval is retained on the auger flights. Sample material can be collected from the flights using a spatula.

9. Post-Sampling

9.1 Remove all sampling equipment from the work zone.

9.2 Transfer all reusable equipment that was in contact with the waste to a pre-designated decontamination area.

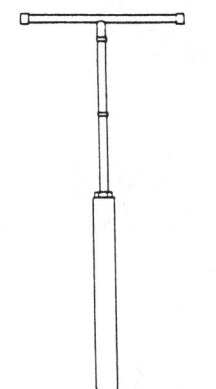

FIG. 2 Thin-Walled Tube

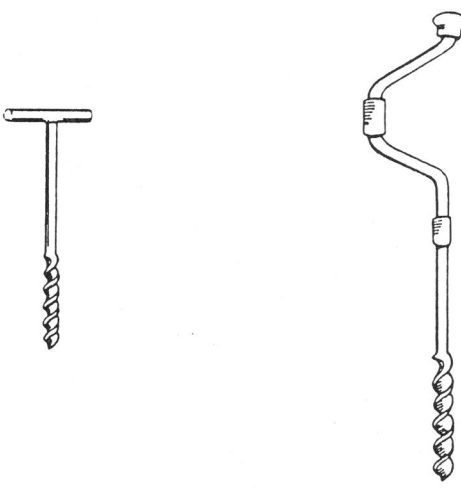

FIG. 4 Augers (Typical)

Decontaminate the equipment according to the protocol established in the work plan (Practice D 5088). Decontaminated sampling equipment should be protected from contamination. This may include, but not be limited to, storage in aluminum foil, plastic bags, PTFE film, or other means of protection that will not impact the sample quality or intended analysis.

9.3 Dispose of all used (disposable) contacting equipment.

10. Data Quality Objectives

10.1 The objectives for sampling and testing of consoli-

dated solid material should be specified in the work plan.

11. Quality Control

11.1 Quality Control (QC) samples (for example, equipment blanks, trip blanks, and duplicates) must be collected as required by the work plan. These QC samples must be evaluated to provide a determination of the sampling quality and reliability of the resulting analytical data.

12. Keywords

12.1 auger; chipper; consolidated solids; hazardous waste; rotating corer sampling; thin-wall tube

Standard Practice for
Sampling Single or Multilayered Liquids, With or Without Solids, in Drums or Similar Containers[1]

This standard is issued under the fixed designation D 5743; the number immediately following the designation indicates the year of original adoption or, in the case of revision, the year of last revision. A number in parentheses indicates the year of last reapproval. A superscript epsilon (ε) indicates an editorial change since the last revision or reapproval.

1. Scope

1.1 This practice covers typical equipment and methods for collecting samples of single or multilayered liquids, with or without solids, in drums or similar containers. These methods are adapted specifically for sampling drums having a volume of 110 gal (416 L) or less. These methods are applicable to hazardous material, product, or waste. Specific sample collection and handling requirements should be described in the site-specific work plan.

1.2 The values stated in inch-pound units are to be regarded as the standard. The values given in parentheses are for information only.

1.3 *This standard does not purport to address all of the safety concerns, if any, associated with its use. It is the responsibility of the user of this standard to establish appropriate safety and health practices and determine the applicability of regulatory limitations prior to use.* Specific precautionary statements are given in 7.2.7.1 and Notes 1 and 2.

2. Referenced Documents

2.1 *ASTM Standards:*
D 4687 Guide for General Planning of Waste Sampling[2]
D 5088 Practice for the Decontamination of Field Equipment Used at Non-Radioactive Waste Sites[3]
D 5283 Practice for Generation of Environmental Data Related to Waste Management Activities: Quality Assurance and Quality Control Planning and Implementation[2]
D 5495 Practice for Sampling with a Composite Liquid Waste Sampler (COLIWASA)[2]
2.2 *Other Documents:*
Drum Handling Practices at Hazardous Waste Sites, EPA/600/2-86/013, PB 165362, January 1986[4]
Accident Prevention Manual for Industrial Operations, 1992[5]
Occupational Safety and Health Guidance Manual for Hazardous Waste Site Activities, No. 85-115, October 1985[4]

3. Terminology

3.1 *Definitions:*
3.1.1 *bonding*—touching the sampling equipment to the drum to form an electrically conductive path to minimize potential electrical differences between the sampling equipment and drum, reducing the buildup of static electricity.
3.1.2 *bung*—usually a 2-in. (5.1-cm) or ¾-in. (1.3-cm) diameter threaded plug designed specifically to close a bung hole.
3.1.3 *bung hole*—an opening in a barrel or drum through which it can be filled, emptied, or vented.
3.1.4 *deheading*—removal of the lid of a closed-head drum; it is usually accomplished with a drum deheader.
3.1.5 *drum*—implicitly any drum, barrel, or non-bulk container of 5 to 110-gal (19 to 416-L) capacity.
3.1.6 *pail*—a small container, usually of 5-gal (19-L) capacity. Pails typically have bungs or spouts, or the entire lid can be removed.
3.1.7 *paperwork*—all required site documentation, which may include the manifests, waste profiles, material safety data sheets (MSDS), site forms, sample labels, seals, and chain of custody forms.
3.1.8 *sludge*—any mixture of solids that settles out of solution. Sludges contain liquids that are not apparent as free liquids.
3.1.9 *work plan*—a plan specific to a particular site; it is for conducting activities specified in the plan.

4. Summary of Practice

4.1 The drum and its contents are inspected, and appropriate sampling equipment is selected. A clean sampling device is lowered slowly into the liquid to be sampled. After the material has entered the device, it is removed from the drum. The contents of the device are discharged into a sample container. The sampling device is then either disposed of or cleaned and decontaminated.

5. Significance and Use

5.1 This practice is intended for use in collecting samples of single and multilayered liquids, with or without solids, from drums or similar containers, including those that are unstable, ruptured, or otherwise compromised. Special handling procedures (for example, remote drum opening, overpressurized drum opening, drum deheading, etc.) are described in *Drum Handling Practices at Hazardous Waste Sites.*

6. Interferences

6.1 The condition of the materials to be sampled, and the

---

[1] This practice is under the jurisdiction of ASTM Committee D-34 on Waste Management and is the direct responsibility of Subcommittee D34.01 on Sampling and Monitoring.
Current edition approved Aug. 15, 1995. Published October 1995.
[2] *Annual Book of ASTM Standards*, Vol 11.04.
[3] *Annual Book of ASTM Standards*, Vol 04.09.
[4] Available from National Technical Information Service, 5285 Port Royal Road, Springfield, VA 22161.
[5] Available from National Safety Council, P.O. Box 558, Hasca, IL 60143-0558.

condition and accessibility of the drums, will have a significant impact on the selection of sampling equipment.

7. Pre-Sampling

7.1 *General Principles and Precautions:*

7.1.1 Samples should be collected in accordance with an appropriate work plan (Practice D 5283 and Guide D 4687). This plan must include a worker health and safety section because there are potential hazards associated with opening drums as well as potentially hazardous contents. See the *Occupational Safety and Health Guidance Manual for Hazardous Waste Site Activities* for information on health and safety at hazardous waste sites.

7.1.2 Correct sampling procedures must be applied to conditions as they are encountered. It is impossible to specify rigid rules describing the precise manner of sample collection because of unknowns associated with each liquid sampling situation. It is essential that the samples be collected by a trained and experienced sampler because of the various conditions under which drummed liquids must be sampled.

7.1.3 To be able to make probability or confidence statements concerning the properties of a sampled lot, the sampling procedure must allow for some element of randomness in selection because of possible variations in the material. The sampler should always be on the alert for possible biases arising from the use of a particular sampling device or from unexpected segregation within the material.

7.1.4 The sampling equipment, sample preparation equipment, sample containers, etc. must be clean, dry, and inert to the material being sampled. All equipment, including sample containers, must be inspected before use to ensure that they are clear of obvious dirt and contamination and are in good working condition. Visible contamination must be removed, and the equipment must be decontaminated with the appropriate rinse materials. Decontaminated sampling equipment should be protected from contamination. This may include, but not be limited to, storage in aluminum foil, plastic bags, polytetrafluoroethylene (PTFE) film, or other means of protection that will not impact the sample quality or intended analysis.

7.2 *Basic Pre-Sampling Practices:*

7.2.1 Review all paperwork.

7.2.2 Select the sampling equipment and sample containers appropriate for the material in the drum, as detailed in the work plan.

7.2.3 Enter the work zone.

7.2.4 Inspect all drums to be sampled visually. Note any abnormal conditions (for example, rust marks, stains, bulges, or other signs of pressurization or leaks) that may require special handling. The work plan should clearly define the limiting conditions under which special handling procedures shall be initiated. See *Drum Handling Practices at Hazardous Waste Sites* for information on opening overpressurized drums and the use of remotely operated drum opening equipment.

7.2.5 Stage the drums to be sampled in a designated work area if they cannot be sampled in their current location. See *Drum Handling Practices at Hazardous Waste Sites* for further information on staging drums.

7.2.5.1 Move the drums to upright stable positions if

necessary. Sufficient space shall be left between drums to prevent movement hazards.

7.2.5.2 Allow adequate time for the drum contents to stabilize if movement of a drum is required. The settling time is dependent on the type of material expected.

7.2.5.3 Number or identify uniquely all drums to be sampled.

7.2.6 Perform a detailed inspection of individual drums.

7.2.6.1 Record all relevant information from drum labels, markings, data sheets, and so forth, in the field log book or on forms specified in the work plan.

7.2.6.2 Verify that there are no discrepancies with existing paperwork.

7.2.6.3 Any discovered inconsistency from the paperwork (such as evidence of crystals on the drum exterior) should be noted in the field log book.

7.2.7 Slowly remove the bung or loosen the ring that secures the lid, allowing any pressure or vacuum to equalize.

7.2.7.1 *Precautionary Notes:*

(1) If the drum or pail appears to be under positive or negative pressure (that is, a slight bulge or dimple in the lid), control the release of pressure until it has equalized. For example, if the drum or pail is equipped with bungs, loosen the smaller bung first since doing so will make it easier to control the release of pressure.

(2) Pails equipped with snap-on lids may be difficult to open. Care must be exercised when opening to minimize the potential of splashing of the contents.

(3) If the top of the drum is dished inward (dimpled), it may "pop" when equalizing pressure, spraying the sampler with any material that is sitting on top of the drum.

(4) If there is evidence of a chemical reaction or sudden pressure buildup, the sampler should leave the area immediately and evaluate whether remote drum opening equipment should be used.

(5) For flammable or explosive materials, the drum and sampling equipment should be grounded if the generation of static electricity while opening or sampling the drum is a possibility. The drum and sampling equipment should be grounded to a ground stake or to an existing ground (building ground, grounded water pipes, etc.). New glass, plastic thiefs, or composite liquid waste samplers (COLIWASAs) may have some residual static electrical charge due to the materials in which they are packed and shipped. The work plan should specify whether grounding is required. See the *Accident Prevention Manual for Industrial Operations* for information on grounding and bonding.

7.2.7.2 Drums should be opened, sampled, and closed individually to minimize the risk of volatilization and exposure.

7.2.7.3 *Drums (or Pails) with Bungs*—When using a manual bung wrench, cover it with a wipe or cloth to control potential liquid spray. Use non-sparking tools.

7.2.7.4 *Drums with Removable Lids*—Loosen the ring slowly with a manual wrench or air impact wrench. Use non-sparking tools.

7.2.7.5 *Pails with Removable Lids (Side-Lever Lock Ring)*—Release the lever slowly.

7.2.7.6 *Pails with Removable Lids (Snap-On)*—Pry the lid loose slowly with a pail lid opener.

7.2.8 Manual or remote puncturing or deheading will be

required if the drum has a stuck bung or the lid cannot be removed. See *Drum Handling Practices at Hazardous Waste Sites* for further information on manual or remote drum opening.

7.2.9 If required, insert a measuring rod (graduated in litres or gallons) into the drum to measure the liquid volume and determine the presence of solids at the bottom and estimate their percentage. (If minimal disturbance of the contents is required, the measuring rod can be inserted in the vent bung hole when working with a bung-top-drum.) The rod can be graduated in litres or gallons for a specific size drum, or it can be graduated in linear units (inches, centimetres, and so forth), with the liquid depth converted to volume using an appropriate volume conversion. The measuring rod should be nonreactive to the waste being contacted.

NOTE 1: **Caution**—Before inserting the measuring rod into the drum, touch the rim gently with the rod (bonding) opposite from the bung to equalize any static charge that the drum may exhibit. The work plan should specify whether bonding is required.

7.2.9.1 For many liquids, the sampling equipment can serve as a substitute measuring device. This can be accomplished by measuring the length of the liquid column as it is being held over the drum and applying an appropriate volume conversion (for example, 1 in. (2.54 cm) equals 1.7 gal (6.43 L) in a 55-gal (208-L) drum).

NOTE 2: **Caution**—The sampling equipment or measuring rod should be at or near the temperature of the drummed liquid to minimize any reaction caused by temperature differences.

7.3 *Sampling Equipment, Selection*— Table 1 summarizes selection criteria for equipment by the material to be sampled.

7.4 *Sampling Equipment, Materials of Construction*— Each of the sampling devices listed should be constructed from materials that are inert to any materials that may be encountered at a specific site. These devices are usually made of glass, stainless steel, aluminum, brass, or plastic. Devices with permanent coatings or liners of an inert nonreactive material, such as PTFE, may be substituted, if approved by the work plan.

7.5 *Generic Equipment List:*
7.5.1 A list of equipment generally required for sampling liquids follows:
7.5.1.1 Sample containers, lids, and liners;
7.5.1.2 Sample labels;
7.5.1.3 COLIWASAs, drum thiefs, sludge samplers, or equivalent devices;
7.5.1.4 Measuring rods;
7.5.1.5 Chain of custody forms;

TABLE 1 Selection Criteria for Equipment

Equipment	ASTM Standard	One Liquid Layer	Two or More Liquid Layers	Liquid and Solid (Sludge) Layers
Drum thief		X[A]	X	X
COLIWASA	D 5495	X	X	N[B]
Syringe-type sampler		N	N	X
Coring-type sludge sampler		—[C]	—	X

[A] X = equipment may usually be used with this type of waste.
[B] N = not the equipment of choice, but it may be used.
[C] — = equipment is probably unsuitable.

7.5.1.6 Field log books;
7.5.1.7 Sample cooler;
7.5.1.8 Wipes or cloths, or both;
7.5.1.9 Ice or gel ice;
7.5.1.10 Grounding cables with alligator clips and emery cloth; and
7.5.1.11 Portable monitoring equipment (combustible gas indicator, organic vapor detectors, radiation survey meter, etc.).
7.5.2 Equipment needed to open drums should be non-sparking (brass or beryllium copper) and include, but not be limited to, the following:
7.5.2.1 Bung wrenches (one straight and one bent),
7.5.2.2 Flathead screwdriver,
7.5.2.3 Breaker bar (½ in. (13 mm)),
7.5.2.4 Ratchet (½ in. (13 mm)),
7.5.2.5 Speed handle (½ in. (13 mm)),
7.5.2.6 Adjustable wrenches (10 and 12 in. (25 and 30 cm)),
7.5.2.7 Air impact wrench and sockets, and
7.5.2.8 Pail lid opener.

8. Sample Collection

8.1 *Basic Sampling Practice:*
8.1.1 Bond the sampling equipment to the drum, if specified in the work plan.
8.1.2 Collect a sample from the drum. Whenever possible, do not sample where the measuring rod has been inserted; however, bung-type drums might not permit avoidance of the disturbed region.
8.1.3 Note the physical characteristics, including any discrepancies (such as solidified contents or crystalline material).
8.1.4 Place the collected material in a sample container.
8.1.5 Close the sample container.
8.1.6 Wipe the outside of the sample container. Dispose of the wipe cloth properly.
8.1.7 Record in the field log book all of the relevant conditions and physical characteristics associated with the sample.
8.1.8 Fill out all of the required paperwork for each sample, as required by the work plan.
8.1.9 Complete and attach a label to the side of the sample container before or after sampling, as directed by the work plan. The sample label should include the following:
(*1*) Sample ID number,
(*2*) Name of sampler,
(*3*) Sampler's initials or signature,
(*4*) Date and time of sampling, and
(*5*) Sample location.
The sample label can also include the following:
(*1*) Sampling information (for example, grab or composite),
(*2*) Preservative or preservation required,
(*3*) Special instructions, and
(*4*) Analysis request.
8.2 *Sampling with a Drum Thief:*
8.2.1 *General Description*—A tube of small diameter, which yields a vertical representation of the contents of a drum when lowered and sealed (see Fig. 1).

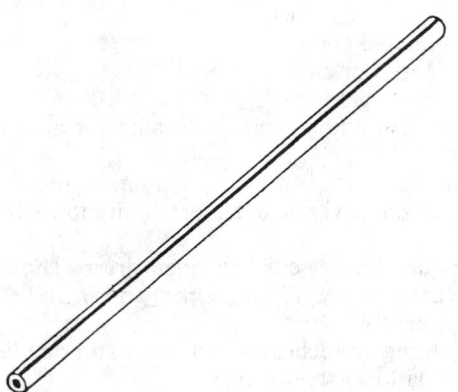

FIG. 1 Drum Thief

NOTE 3—When sampling liquids of high specific gravity, it may be difficult to retain the entire sample in the drum thief. A smaller-diameter drum thief may overcome this problem. The use of a COLIWASA or similar device may be necessary if the problem persists.

8.2.2 *Operation and Use*—Slowly insert the tube vertically until it reaches either the bottom of the drum or the liquid layer to be sampled. The sampling device should be lowered at a rate that permits the liquid level inside and outside the tube to be approximately the same.

NOTE 4—Multiple sample increments are usually necessary to provide enough sample volume for analysis and quality control (QC). Drum contents will become increasingly disturbed with each successive insertion of the drum thief.

8.2.2.1 Cover the top of the tube with the thumb or a rubber stopper to form a seal. Use gloves or a stopper, as described in the work plan.

8.2.2.2 Withdraw the tube carefully.

8.2.2.3 Use a clean cloth or paper towel to wipe the tube as it is being extracted from the liquid, to prevent unnecessary dripping.

8.2.2.4 Note the proportions of any layers or solids.

8.2.2.5 Place the bottom end of the tube into the sample container, and release the contents slowly.

8.3 *Sampling with a COLIWASA:*

8.3.1 *General Description*—A glass, plastic, or metal tube with an end closure that can be opened while the tube is immersed in the waste to be sampled (see Practice D 5495). The COLIWASA will yield a vertical representation of a drum's contents when immersed in the open position into a drum (see Fig. 2).

NOTE 5—Multiple sample increments are usually necessary to provide enough sample volume for analysis and QC. Drum contents will become increasingly disturbed with each successive insertion of the COLIWASA.

8.4 *Sampling with a Syringe-Type Sampler:*

8.4.1 *General Description*—A tube with a manually operated piston that can be used as a syringe for high-viscosity liquids or as a coring device for sludge (see Fig. 3).

8.4.2 *Operation and Use*—(1) For high-viscosity liquids, the tube is lowered to the sampling point and the piston is pulled out to collect the sample. (2) For sludge, the tube is lowered to the surface of the sludge. The sampler body is pushed into the sludge while allowing the piston to move up within the sampler body.

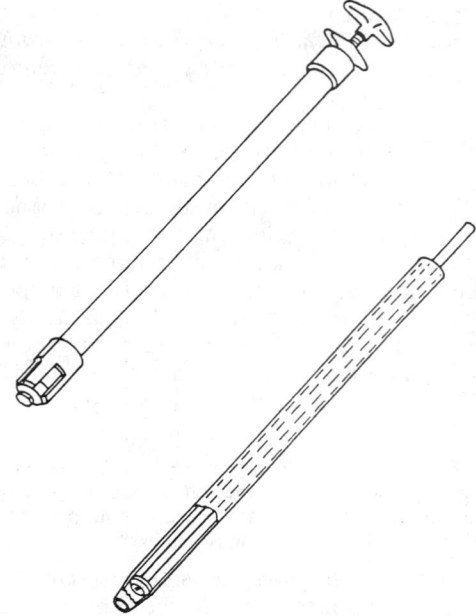

FIG. 2 COLIWASAs (Typical)

8.4.2.1 Assemble with the piston at the lower end of the sampler body. Attach the bottom valve (for high-viscosity liquids) or the coring tip (for sludge).

(1) For high-viscosity liquids, lower to the sampling point and withdraw the piston to collect the sample. Close the bottom valve by pushing against the side or bottom of container with the sampler body.

(2) For sludge, lower to the surface of the material to be sampled. Push the sampler body into the material while allowing the piston to move up within the sampler body.

8.4.2.2 Use a clean cloth or paper towel to wipe the sampler body as it is being extracted from the liquid or sludge, to prevent unnecessary dripping.

8.4.2.3 Transfer the sample into the sample container by opening the bottom valve, if fitted, and pushing the piston down.

8.5 *Sampling with a Coring-Type Sampler:*

8.5.1 *General Description*—A coring-type sampler consists of a cylinder, a coring tip (or auger tip) with a retaining device, a top cap, and an extension with a cross handle (see Fig. 4). A thin-walled internal sleeve may be used to contain the sample.

8.5.2 *Operation and Use*—The coring-type sampler is pushed (pushed and rotated with an auger tip) into the sludge to collect the sample and removed. The retaining device allows the sludge to enter the cylinder when pushing the sampler. The retaining device closes to hold the sludge in the cylinder while removing the sampler.

8.5.3 Remove the top cap and transfer the sample from the cylinder into a sample container. If equipped with an internal sleeve, remove the top cap and place an end cap on the internal sleeve. Invert, remove the internal sleeve from the cylinder, and place an end cap on the open end of the sleeve.

9. Post-Sampling

9.1 Remove all sampling equipment from the work zone.

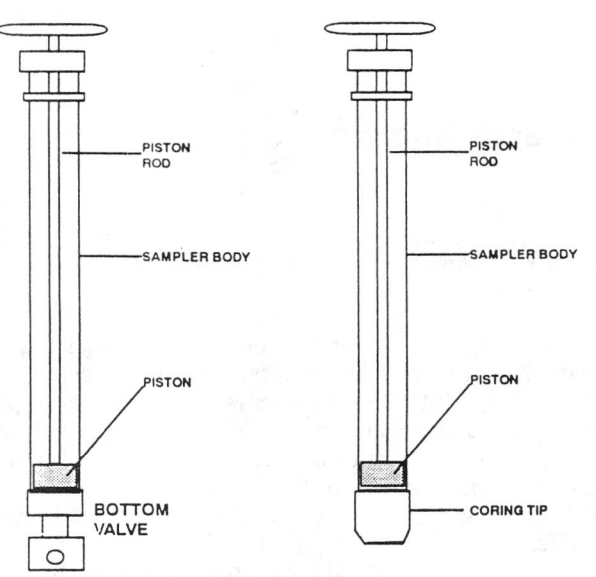

FIG. 3 Syringe-Type Sampler (Typical)

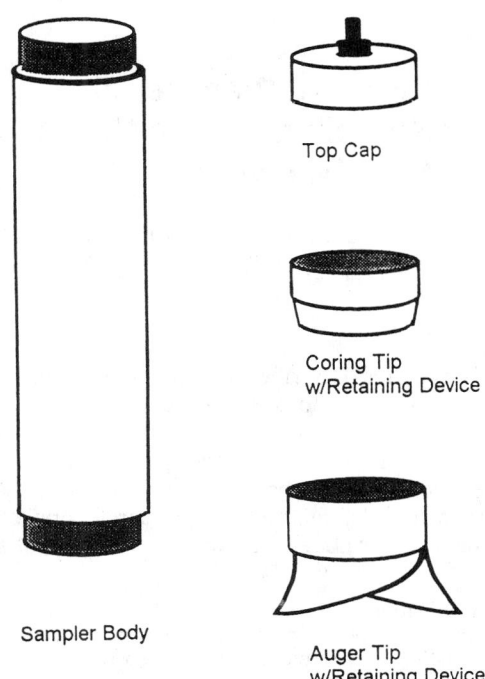

FIG. 4 Corer-Type Sampler (Typical)

9.2 Transfer all reusable sampling equipment that was in contact with the waste to a predesignated decontamination area. Decontaminate the equipment according to the protocol established in the work plan (Practice D 5088). Decontaminated sampling equipment should be protected from contamination. This may include, but not be limited to, storage in aluminum foil, plastic bags, PTFE film, or other means of protection that will not impact the sample quality or intended analysis.

9.3 Dispose properly of all used (disposable) contacting equipment.

10. Data Quality Objectives

10.1 The objectives for sampling and testing of liquids and sludges should be specified in the work plan.

11. Quality Control

11.1 Quality Control (QC) samples (for example, equipment blanks, trip blanks, and duplicates) must be collected as required by the work plan. These QC samples must be evaluated to provide a determination of the quality of the sampling and reliability of the resulting analytical data.

12. Keywords

12.1 COLIWASA; drum; drum thief; liquid; pail; sampling; sludge sampler; waste

Standard Practice for
Sampling With a Composite Liquid Waste Sampler (COLIWASA)[1]

This standard is issued under the fixed designation D 5495; the number immediately following the designation indicates the year of original adoption or, in the case of revision, the year of last revision. A number in parentheses indicates the year of last reapproval. A superscript epsilon (ε) indicates an editorial change since the last revision or reapproval.

1. Scope

1.1 This practice describes the procedure for sampling liquids with the composite liquid waste sampler, or "COLIWASA." The COLIWASA is an appropriate device for obtaining a representative sample from stratified or unstratified liquids. Its most common use is for sampling containerized liquids, such as tanks, barrels, and drums. It may also be used for pools and other open bodies of stagnant liquid.

NOTE—A limitation of the COLIWASA is that the stopper mechanism may not allow collection of approximately the bottom inch of material, depending on construction of the stopper.

1.2 The COLIWASA should not be used to sample flowing or moving liquids.

1.3 *This standard does not purport to address all of the safety problems, if any, associated with its use. It is the responsibility of the user of this standard to establish appropriate safety and health practices and determine the applicability of regulatory limitations prior to use.*

2. Referenced Documents

2.1 *ASTM Standards:*
D 4687 Guide for General Planning of Waste Sampling[2]
D 5088 Practice for Decontamination of Field Equipment Used at Nonradioactive Waste Sites[3]
D 5283 Practice for Generation of Environmental Data Related to Waste Management Activities: Quality Assurance and Quality Control Planning and Implementation[2]

3. Summary of Practice

3.1 A clean device is slowly lowered into the liquid to be sampled. After it has filled, the bottom of the sampling tube is closed and the device is retrieved. The contents are subsequently discharged into a sample container.

4. Significance and Use

4.1 This practice is applicable to sampling liquid wastes and other stratified liquids. The COLIWASA is used to obtain a vertical column of liquid representing an accurate cross-section of the sampled material. To obtain a representative sample of stratified liquids, the COLIWASA should be open at both ends so that material flows through it as it is slowly lowered to the desired sampling depth. The COLIWASA must not be lowered with the stopper in place. Opening the stopper after the tube is submerged will cause material to flow in from the bottom layer only, resulting in gross over-representation of that layer.

4.2 This practice is to be used by personnel acquiring samples.

4.3 This practice should be used in conjunction with Guide D 4687 which covers sampling plans, safety, QA, preservation, decontamination, labeling and chain-of-custody procedures; Practice D 5088 which covers decontamination of field equipment used at waste sites; and Practice E 5283 which covers project specifications and practices for environmental field operations.

5. Sampling Equipment

5.1 COLIWASA's are available commercially with different types of stoppers and locking mechanisms, but they all operate using the same principle. They can also be constructed from materials such as polyvinylchloride (PVC), glass, metal, or polytetrafluoroethylene (PTFE). A traditional model of the COLIWASA is shown in Fig. 1 [de Vera et al.][4]; however, the design can be modified or adapted, or both, to meet the needs of the sampler. COLIWASA's must be selected that are constructed of materials compatible with the waste being sampled and with the analyses or tests to be performed. Due to the unknown nature of most containerized liquid wastes, COLIWASA's made of glass or polytetrafluoroethylene are best for general use.

6. Sample Containers

6.1 Plastic, glass or other nonreactive containers should be used. Refer to Guide D 4687 for further information on containers.

7. Procedure

7.1 Make certain the COLIWASA is clean and functioning properly. It is essential that the stopper at the bottom of the sampling tube closes securely.

7.2 Open the COLIWASA by placing the stopper mechanism in the open position.

7.3 Lower the COLIWASA into the liquid slowly so that

---

[1] This practice is under the jurisdiction of ASTM Committee D-34 on Waste Management and is the direct responsibility of Subcommittee D34.01 on Sampling and Monitoring.
Current edition approved Jan. 15, 1994. Published March 1994.
[2] *Annual Book of ASTM Standards*, Vol 11.04.
[3] *Annual Book of ASTM Standards*, Vol 04.08.

[4] de Vera, E. R., Simmons, B. P., Stephens, R. C., and Storm, D. L., "Samplers and Sampling Procedures for Hazardous Waste Streams," EPA-600/2-80-018, January 1980.
[5] Ford, P. J., Turina, P. J., and Seeley, D. E., "Characterization of Hazardous Waste Sites—A Methods Manual: Volume II," Available Sampling Methods, Second Edition, EPA-600/4-84-076, December 1984.

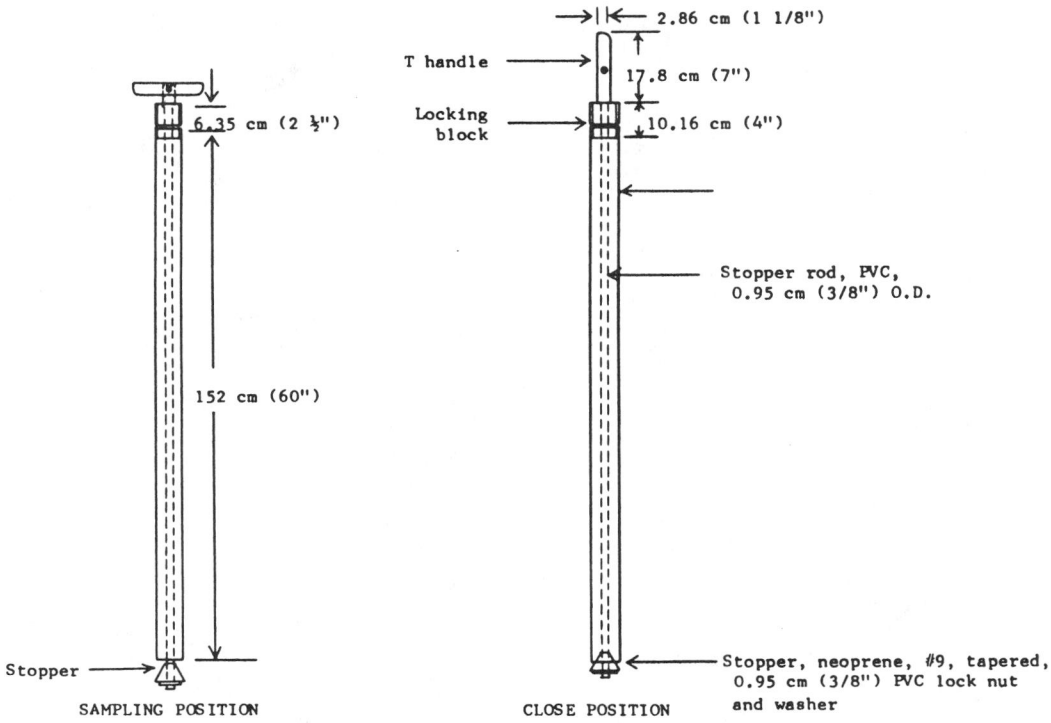

Pipe, PVC, translucent,
4.13cm (1 5/8") I.D.,
4.76cm (1 7/8") O.D.

2.86 cm (1 1/8")

T handle

17.8 cm (7")

Locking block

10.16 cm (4")

6.35 cm (2 ½")

Stopper rod, PVC,
0.95 cm (3/8") O.D.

152 cm (60")

Stopper

Stopper, neoprene, #9, tapered,
0.95 cm (3/8") PVC lock nut
and washer

SAMPLING POSITION   CLOSE POSITION

FIG. 1 Composite Liquid Waste Sampler (COLIWASA) (Typical)

the levels of the liquid inside and outside the sampler tube remain about the same. If the level of the liquid in the sample tube is lower than that outside the sampler, the sampling rate is too fast and a nonrepresentative sample will result.

7.4 Use the stopper mechanism to close the COLIWASA when it reaches the desired depth in the liquid.

7.5 Withdraw the sampler from the liquid with one hand while wiping the sampler tube with a disposable cloth or rag with the other hand.

7.6 Carefully discharge the sample into a suitable container by slowly opening the stopper mechanism while the lower end of the COLIWASA is positioned in the sample container.

7.7 Seal the sample container; attach the label and seal; record in the field logbook; and complete the chain-of-custody record.

7.8 Decontaminate the used equipment in accordance with Practice D 5088.

8. Keywords

8.1 COLIWASA; drum sampling; liquid sampling; sampling; waste

PART 5. ATMOSPHERIC CHARACTERIZATION AND SAMPLING

5.1 FIELD MEASUREMENTS

Standard Practice for
Determining the Operational Comparability of Meteorological Measurements[1]

This standard is issued under the fixed designation D 4430; the number immediately following the designation indicates the year of original adoption or, in the case of revision, the year of last revision. A number in parentheses indicates the year of last reapproval. A superscript epsilon (ϵ) indicates an editorial change since the last revision or reapproval.

1. Scope

1.1 Sensor systems used for making meteorological measurements may be tested for laboratory accuracy in environmental chambers or wind tunnels, but natural exposure cannot be fully simulated. Atmospheric quantities are continuously variable in time and space; therefore, repeated measurements of the same quantities as required by Practice E 177 to determine precision are not possible. This practice provides standard procedures for exposure, data sampling, and processing to be used with two measuring systems in determining their operational comparability (**1, 2**).[2]

1.2 The procedures provided produce measurement samples that can be used for statistical analysis. Comparability is defined in terms of specified statistical parameters. Other statistical parameters may be computed by methods described in other ASTM standards or statistics handbooks (**3**).

1.3 Where the two measuring systems are identical, that is, same make, model, and manufacturer, the operational comparability is called functional precision.

1.4 Meteorological determinations frequently require simultaneous measurements to establish the spatial distribution of atmospheric quantities or periodically repeated measurement to determine the time distribution, or both. In some cases, a number of identical systems may be used, but in others a mixture of instrument systems may be employed. The procedures described herein are used to determine the variability of like or unlike systems for making the same measurement.

1.5 *This standard does not purport to address the safety concerns, if any, associated with its use. It is the responsibility of the user of this standard to establish appropriate safety and health practices and determine the applicability of regulatory limitations prior to use. (See 8.1 for more specific safety precautionary information.)*

2. Referenced Documents

2.1 *ASTM Standards:*
D 1356 Terminology Relating to Sampling and Analysis of Atmospheres[3]

E 177 Practice for Use of the Terms Precision and Bias in ASTM Test Methods[4]

3. Terminology

3.1 For additional definitions of terms, refer to Terminology D 1356.

3.2 *Definitions of Terms Specific to This Standard.*

3.2.1 *difference (D)*—the difference between the derived mean (d) of a set of samples and the true mean (μ) of the population:

$$D = d - \mu \tag{1}$$

3.2.2 *systematic difference (d)*—the mean of the differences in the measurement by the two systems:

$$d = \frac{1}{N} \sum_{i=1}^{N} (X_{ai} - X_{bi}) \tag{2}$$

3.2.3 *operational comparability (C)*—the root mean square (rms) of the difference between simultaneous readings from two systems measuring the same quantity in the same environment:

$$C = \pm \sqrt{\frac{1}{N} \sum_{i=1}^{N} (X_{ai} - X_{bi})^2} \tag{3}$$

where:
X_{ai} = ith measurement made by one system,
X_{bi} = ith simultaneous measurement made by another system, and
N = number of samples used.

3.2.3.1 *functional precision*—the operational comparability of identical systems.

3.2.4 *estimated standard deviation of the difference (s)*—a measure of the dispersion of a series of differences around their mean.

$$s = \pm \sqrt{C^2 - d^2} \tag{4}$$

3.2.5 *skewness (M)*—the symmetry of the distribution (the third moment about the mean).

$$M = \frac{\sum_{i=1}^{N} ((X_{ai} - X_{bi}) - d)^3}{N^3} \tag{5}$$

$M = 0$ for normal distribution.

3.2.6 *kurtosis (K)*—the peakedness of the distribution (the fourth moment about the mean), $K = 3$ for normal distribution.

[1] This practice is under the jurisdiction of ASTM Committee D-22 on Sampling and Analysis of Atmospheres and is the direct responsibility of Subcommittee D22.11 on Meteorology.

Current edition approved Oct. 10, 1996. Published December 1996. Originally published as D 4430 – 84. Last previous edition D 4430 – 84 (1990).

[2] The boldface numbers in parentheses refer to the list of references at the end of this practice.

[3] *Annual Book of ASTM Standards*, Vol 11.03.

[4] *Annual Book of ASTM Standards*, Vol 14.02.

3.2.7 *response time (T)*—the time required for the change in output of a measuring system to reach 63 % of a step function change in the variable being measured.

$$K = \frac{\sum\limits_{i=1}^{N} ((X_{ai} - X_{bi}) - d)^4}{N^4} \qquad (6)$$

3.2.8 *identical systems*—systems of the same make and model produced by the same manufacturer.

3.2.9 *resolution (r)*—the smallest change in an atmospheric variable that is reported as a change in the measurement.

4. Summary of Practice

4.1 The systems to be compared must make measurements within a cylindrical volume of the ambient atmosphere not greater than 10 m in horizontal diameter. The vertical extent of the volume must be the lesser of 1 m or one-tenth H, where H is the height above the earth's surface of the base of the volume. The sample volume must be selected to ensure homogeneous distribution of the variable being measured.

4.2 For some measurements (for example, visibility) the horizontal distance or the height (for example, cloud height) may be the variable of interest. In the first case, one of the two dimensions of horizontal distance is minimized and may not exceed 10 m while all other criteria remain the same. In the second case, all criteria for position and sampling described in 4.1 remain unchanged and the measured height is treated as if it were an atmospheric variable. The physical dimension of some measuring systems may exceed the spatial limits of 4.1 (for example, a rotating beam ceilometer with a 200-m baseline). In those cases the systems must be installed so that the measurements are obtained from within the volume specified in 4.1.

4.3 Samples are taken in pairs and the time interval between the pairs of samples must be no less than four times the response time (4T) of the measuring systems (**4**).

4.4 The time between members of a pair of measurements must be as small as possible, but must not exceed one tenth the response time.

4.5 The root mean square (rms) of the measurement differences is calculated to provide operational comparability or functional precision of the systems.

4.6 Measurement differences may change with the magnitude of the measurement (for example, the absolute value of the difference in the measurement of wind speed by two systems may be greater or smaller at high-wind speeds than at low-wind speeds). To test the data for such dependence, the range of measurements shall be divided into no less than three class intervals and each class shall have a sufficient number of samples to represent the class. The change in rms difference between classes indicates the dependence of the measurement difference on the magnitude of the measurement.

5. Significance and Use

5.1 This practice provides data needed for selection of instrument systems to measure meteorological quantities and to provide an estimate of the precision of measurements made by such systems.

5.2 This practice is based on the assumption that the repeated measurement of a meteorological quantity by a sensor system will vary randomly about the true value plus an unknowable systematic difference. Given infinite resolution, these measurements will have a Gaussian distribution about the systematic difference as defined by the Central Limit Theorem. If it is known or demonstrated that this assumption is invalid for a particular quantity, conclusions based on the characteristics of a normal distribution must be avoided.

6. Interferences

6.1 Exposure of the systems shall be such as to avoid interference from sources, structures, or other conditions that may produce a gradient in the measurement across the sample volume.

6.2 A mutual interference by systems may produce a systematic difference (d) or bias that would not occur if one system were used by itself. That bias is not a part of the comparability and must be reported separately.

6.3 A systematic difference greater than one increment of resolution must be investigated by interchanging the position of the sensors with an equal number of samples taken in each position. If the bias changes sign, it is due to the exposure and must be reported separately.

7. Apparatus

7.1 The apparatus used is the combination of sensor systems for which the operational comparability or functional precision is to be determined plus the data-processing equipment required to extract the data and calculate the statistical parameters.

8. Precautions

8.1 Safety precautions accompanying the sensor systems must be followed.

8.2 *Technical Precautions:*

8.2.1 Measurement-system mutual electrical interference must be minimized.

8.2.2 Use of this practice is based on a statistical analysis of the distribution of differences used to calculate operational comparability. Mean, standard deviation, skewness, and kurtosis of the distribution are reported to facilitate such analysis.

9. Sampling

9.1 Samples are collected in pairs from two sensors sampling the free ambient atmosphere.

9.2 Samples are collected from a cylindrical volume of the free atmosphere as defined in 4.1.

9.3 The distance between sensors should be the smallest distance that avoids sensor interaction but must meet 9.2.

9.4 The time between pairs of samples (X_{ai}, X_{bi}, and $X_{ai} + 1$, $X_{bi} + 1$) must be equal to or greater than four times the response time (4T) of the sensor system. The nature of atmospheric data is such that time intervals between pairs of samples as long as an hour or more may be desirable.

9.5 The time between members of a pair of samples (X_{ai} and X_{bi}) must not exceed one tenth of the response time ($T/10$).

9.6 The comparability determined is limited to the range

of atmospheric conditions encountered. The number of samples cannot be too large. The minimum number of samples that must be exceeded is found by using the criteria for 99.7 % or greater confidence that the difference (D) between the derived mean of a set of samples (d) and the true mean (μ) of the population of all samples is less than or equal to three times the standard deviation ($3s$) about the mean, divided by the square root of the number of samples in the set. To calculate D the estimated standard deviation (s) is used to provide:

$$D \leqq \frac{3s}{\sqrt{N}} \qquad (7)$$

9.6.1 The sampling is not complete until D is less than or equal to one increment of resolution (r) of the system being tested. Stated another way, the number of samples needed N_n must be:

$$N_n \geqq \left(\frac{3s}{r}\right)^2 \qquad (8)$$

10. Preparation

10.1 The systems to be compared must be prepared for operation individually according to manufacturer's instructions.

10.2 Deliberate readjustment to obtain identical simultaneous readings shall be avoided.

11. Procedure

11.1 Install two or more meteorological measuring systems so that they are measuring the free ambient atmosphere from a cylindrical volume as defined in 4.1.

11.2 Record a measurement from each system separated by no more than $T/10$-s time interval.

11.3 Repeat 11.2 at a time interval at least four times the response time ($4T$) of the particular systems being tested. If systems with different response times are being compared, the longest shall be used to determine the minimum allowable time between pairs of samples. The period between the readings may be much larger than four times the response time ($4T$) for practical and operational reasons. It is advisable to choose both the time period between readings and the total period over which the determination is made long enough to include a wide sample of naturally occurring meterological phenomena at the site.

11.4 Continue sampling until at least N_n samples have been obtained where:

$$N_n \geqq \left(\frac{3s}{r}\right)^2 \qquad (9)$$

11.5 Divide the range of measurement into no less than three class intervals. Continue sampling until the number of samples in each interval (Ni) is:

$$N_i \geqq \left(\frac{2s}{r}\right)^2 \qquad (10)$$

11.6 Test the data for dependence between the difference measured and the magnitude of the measurement.

11.7 Calculate the skewness (M) (see 3.1) and the kurtosis (K) (see 3.1) of the frequency distribution of the differences.

12. Reports

12.1 Report C, the two-system operational comparability.

12.2 Report d, the systematic difference in the measurement by the two systems.

12.3 Report N, the number of samples used to calculate C and d.

12.4 Report t, the time interval between pairs of samples.

12.5 Report the range of measurements across which sampling was made.

12.6 Report on the dependence between the sample difference measured and the magnitude of the measurement.

12.7 Report any evidence of system interaction that would affect the systematic difference d.

12.8 Report M, the skewness of the frequency distribution of the differences.

12.9 Report K, the peakedness of the frequency distribution of the differences.

12.10 Report date and time of most recent calibration.

12.11 Report r, the resolution of the measurements.

12.12 Report date and time of beginning of data-gathering period.

12.13 Report date and time of end of data-gathering period.

13. Precision and Bias

13.1 Sample sizes have been chosen to assure a 99.7 % confidence level for C and d within the resolution of the measurements.

14. Keywords

14.1 atmosphere; functional precision; measurement comparisons; meteorological measurements

REFERENCES

(1) Hoehne, W. E., "Standardizing Functional Tests," *IEEE Transactions on Geoscience Electronics*, Vol GE-11, No. 2, April 1973.

(2) Stone, R. J., "National Weather Service Automated Observational Networks and the Test and Evaluation Division Functional Testing Program," *Fourth Symposium on Meteorological Observations and Instrumentation*, Denver, Colorado, April 10–14, 1978.

(3) Natrella, Mary Gibbon, "Experimental Statistics," *National Bureau of Standards Handbook*, Vol 91, August 1, 1963.

(4) Haykin, Simon S., *Communication Systems*, John Wiley & Sons, New York, NY, 1978.

Standard Test Methods for Measuring Surface Atmospheric Pressure[1]

This standard is issued under the fixed designation D 3631; the number immediately following the designation indicates the year of original adoption or, in the case of revision, the year of last revision. A number in parentheses indicates the year of last reapproval. A superscript epsilon (ϵ) indicates an editorial change since the last revision or reapproval.

1. Scope

1.1 These methods cover the measurement of atmospheric pressure with two types of barometers: the Fortin-type mercurial barometer and the aneroid barometer.

1.2 In the absence of abnormal perturbations, atmospheric pressure measured by these methods at a point is valid everywhere within a horizontal distance of 100 m and a vertical distance of 0.5 m of the point.

1.3 Atmospheric pressure decreases with increasing height and varies with horizontal distance by 1 Pa/100 m or less except in the event of catastrophic phenomena (for example, tornadoes). Therefore, extension of a known barometric pressure to another site beyond the spatial limits stated in 1.2 can be accomplished by correction for height difference if the following criteria are met:

1.3.1 The new site is within 2000 m laterally and 500 m vertically.

1.3.2 The change of pressure during the previous 10 min has been less than 20 Pa.

The pressure, P_2 at Site 2 is a function of the known pressure P_1 at Site 1, the algebraic difference in height above sea level, $h_1 - h_2$, and the average absolute temperature in the space between. The functional relationship between P_1 and P_2 is shown in 10.2. The difference between P_1 and P_2 for each 1 m of difference between h_1 and h_2 is given in Table 1 and 10.4 for selected values of P_1 and average temperature.

1.4 Atmospheric pressure varies with time. These methods provide instantaneous values only.

1.5 The values stated in SI units are to be regarded as the standard.

1.6 *This standard does not purport to address all of the safety concerns, if any, associated with its use. It is the responsibility of the user of this standard to establish appropriate safety and health practices and determine the applicability of regulatory limitations prior to use.* Specific safety precautionary statements are given in Section 7.

2. Referenced Documents

2.1 *ASTM Standards:*
D 1356 Terminology Relating to Sampling and Analysis of Atmospheres[2]
D 3249 Practice for General Ambient Air Analyzer Procedures[2]

E 380 Practice for Use of the International System of Units (SI) (the Modernized Metric System)[3]

3. Terminology

3.1 Pressure for meteorological use has been expressed in a number of unit systems including inches of mercury, millimetres of mercury, millibars, and others less popular. These methods will use only the International System of Units (SI), as described in Practice E 380.

3.1.1 Much of the apparatus in use and being sold reads in other than SI units, so for the convenience of the user the following conversion factors and error equivalents are given.

3.1.1.1 The standard for pressure (force per unit area) is the pascal (Pa).

3.1.1.2 One standard atmosphere at standard gravity (9.80665 m/s^2) is a pressure equivalent to:
29.9213 in. Hg at 273.15 K
760.000 mm Hg at 273.15 K
1013.25 millibars
14.6959 lbf/in.^{2}
101325 Pa or 101.325 kPa

3.1.1.3 1 Pa is equivalent to:
0.000295300 in. Hg at 273.15 K
0.00750062 mm Hg at 273.15 K
0.01000000 millibars
0.000145037 lbf/in.^{2}
0.000009869 standard atmospheres

3.2 *standard gravity*—as adopted by the International Committee on Weights and Measures, an acceleration of 9.80665 m/s^2 (see 10.1.3).

3.3 The definitions of all other terms used in these methods can be found in Terminology D 1356 and Practice D 3249.

4. Summary of Methods

4.1 The instantaneous atmospheric pressure is measured with two types of barometers.

4.2 Method A utilizes a Fortin mercurial barometer. The mercury barometer has the advantage of being fundamental in concept and direct in response. The disadvantages of the mercury barometer are the more laborious reading procedure than the aneroid barometer, and the need for temperature correction.

4.3 Method B utilizes an aneroid barometer. The aneroid barometer has the advantages of simplicity of reading, absence of mercury, no need for temperature compensation by the observer, and easy detection of trend of change. The main

---

[1] These test methods are under the jurisdiction of ASTM Committee D-22 on Sampling and Analysis of Atmospheres and are the direct responsibility of Subcommittee D22.11 on Meteorological Measurements.

Current edition approved Nov. 10, 1995. Published January 1996. Originally pub- lished as D 3631 – 77. Last previous edition D 3631 – 84 (1990).

[2] *Annual Book of ASTM Standards*, Vol 11.03.

[3] *Annual Book of ASTM Standards*, Vol 14.02.

TABLE 1 Selected Values

Average Temperature, $\frac{T_1 + T_2}{2}$	Pressure P_1, Pa				
	110 000	100 000	90 000	80 000	70 000
	Correction to P_1, Pa/m, positive if $h_1 > h$, negative if $h_1 < h$				
230	16	15	13	12	10
240	16	14	13	11	10
250	15	14	12	11	10
260	14	13	12	11	9
270	14	13	11	10	9
280	13	12	11	10	9
290	13	12	11	9	8
300	13	11	10	9	8
310	12	11	10	9	8

disadvantages of the aneroid barometer are that it is not fundamental in concept as the mercury barometer, and it requires calibration periodically against a mercury barometer.

5. Significance and Use

5.1 Atmospheric pressure is one of the basic variables used by meteorologists to describe the state of the atmosphere.

5.2 The measurement of atmospheric pressure is needed when differences from "standard" pressure conditions must be accounted for in some scientific and engineering applications involving pressure dependent variables.

5.3 These methods provide a means of measuring atmospheric pressure with the accuracy and precision comparable to the accuracy and precision of measurements made by governmental meteorological agencies.

6. Apparatus

6.1 *Fortin Barometer*, which is a mercurial barometer consisting of a glass tube containing mercury with an adjustable cistern and an index pointer projecting downward from the roof of the cistern. The mercury level may be raised or lowered by turning an adjustment screw beneath the cistern.

6.1.1 To provide acceptable measurements, the specifications of 6.1.2 through 6.1.11 must be met.

6.1.2 Maximum error at 100 000 Pa ± 30 Pa.

6.1.2.1 Maximum error at any other pressure for a barometer whose range: (*a*) does not extend below 80 000 Pa ± 50 Pa (*b*) extends below 80 000 Pa ± 80 Pa.

6.1.2.2 For a marine application the error at a point must not exceed ± 50 Pa.

6.1.3 Difference between errors over an interval of 10 000 Pa or less ± 30 Pa.

6.1.4 Accuracy must not deteriorate by more than ±50 Pa over a period of a year.

6.1.5 It must be transportable without loss of accuracy.

6.1.6 A mercurial barometer must be able to operate at ambient temperatures ranging from 253 to 333 K (−20 to 60°C) and must not be exposed to temperatures below 253 K (−38°C). It must be able to operate over ambient relative humidities ranging from 0 to 100 %.

6.1.7 A thermometer with a resolution of 0.11 K and a precision and accuracy of 0.05 K must be attached to the barrel of the barometer.

6.1.8 The actual temperature for which the scale of a mercury barometer is designed to give true readings (at standard gravity) must be engraved on the barometer.

6.1.9 If the evacuated volume above the mercury column

can be pumped, the head vacuum must be measured with a gage such as a McLeod gage or a thermocouple gage and reduced to 10 Pa or less.

6.1.10 The meniscus of a mercurial barometer must not be flat.

6.1.11 The axis of the tube must be vertical (that is, aligned with the local gravity vector).

6.2 *Precision aneroid barometer*, consisting of an evacuated elastic capsule coupled through mechanical, electrical, or optical linkage to an indicator.

6.2.1 To provide acceptable measurements, an aneroid barometer must meet the specifications of 6.2.2 through 6.2.7.

6.2.2 Resolution of 50 Pa or less.

6.2.3 Precision of ±50 Pa.

6.2.4 Accuracy of ±50 Pa root mean square error with a maximum observed error not to exceed 150 Pa throughout the calibration against a basic standard.

6.2.5 Temperature compensation must be included to prevent a change in reading of more than 50 Pa for a change of temperature of 30 K.

6.2.6 The accuracy must not deteriorate by more than ±100 Pa over a period of a year.

6.2.7 The hysteresis must be sufficiently small to ensure that the difference in reading before a 5000-Pa pressure change and after return to the original value does not exceed 50 Pa.

6.3 *Static Pressure Head*—Atmospheric pressure-measuring instruments may be installed inside an enclosed space. The pressure in the space must, however, be directly coupled to the pressure of the free atmosphere and not artificially affected by heating, ventilating, or air-conditioning equipment, or by the dynamic effects of wind passage.

6.3.1 The *Manual of Barometry* (1)[4] describes these effects. For barometers with a static port they can be overcome with a static pressure vent, such as that described by Gill (2), mounted outside and beyond the influence of the building. It is practical to consider an external static vent installation if and only if the pressure in the building differs by more than 30 Pa from true pressure. The pressure difference due to a ventilating or air conditioning system, or both can be determined from pressure readings taken with a precision aneroid barometer inside and outside the building on calm days when the ventilating and air conditioning system is in operation. The existence of pressure errors due to the dynamic effects of wind on the building can often be diagnosed by careful observation of a fast response barometer in the building during periods of gusty winds.

6.3.2 The significant pressure field near a building in wind can extend to a height of 2.5 times the height of the building and to a horizontal distance up to 10 times the height of the building to the leeward. It may be impractical to locate a static vent beyond this field but the following considerations must be made:

6.3.2.1 The static vent must *not* be located on a side of the building;

6.3.2.2 The distance from the building must be as large as practical;

---

[4] Boldface numbers in parentheses refer to references at the end of these methods.

6.3.2.3 The length of the tube connecting the vent to the barometer must be minimized;

6.3.2.4 To avoid blockages, a vertical run of connecting tube is preferable to a horizontal run; and

6.3.2.5 The connecting tube system must include moisture traps and drainage slopes on horizontal runs.

6.3.3 The tubing used to connect the vent to the barometer has a minimum allowable internal diameter that is a function of the ambient static pressure, the volume of the air chambers associated with the instrument making the pressure measurement, the length of the tube between the static head and the barometer, the viscosity of the air in the tubing and connected equipment. The time lag constant must not exceed 1 s so that for pressure and temperature of the zero pressure altitude in the standard atmosphere, the inside diameter d of the tubing connecting the static pressure head with the barometer must be such that

$$d > (7.21 \times 10m^{-9} \, LV)^{1/4} \qquad (1)$$

where:

L = length of the tube, m,
V = volume of the air capacity of the pressure responsive instrument and any connected air chambers within the system together with one half the volume of the tubing, m³, and
d = inside diameter of the tubing, m.

When this calculation is made the minimum allowable inside diameter will frequently be 5 mm or less. It is often more convenient to use tubing larger than this size, and use of such larger tubing enhances the value of the static head and makes it applicable to a wider range of temperatures and pressures.

7. Safety Precautions

7.1 Mercury vapor is poisonous even in small quantities and prolonged exposure can produce serious physical impairment. Installation and use of a mercury barometer should include adequate ventilation and avoidance of skin contact (1, 3).

7.2 Do not store or operate a mercurial barometer at temperatures below 235 K (−38°C), the freezing point of mercury.

7.3 A broken tube, cistern, or bag will release mercury. Carefully collect, place, and seal all of this mercury in a strongly made nonmetallic container.

8. Calibration and Standardization

8.1 A barometer is calibrated by comparing it with a secondary standard traceable to one of the primary standards at locations listed in Table 2.

8.2 For the United States this standard is maintained by the National Institute of Standards and Technology, Gaithersburg, MD 20899.

8.3 Except in the case of catastrophic phenomena (for example, tornadoes) the horizontal pressure gradient at the earth's surface is less than 1 Pa/100 m so that the pressure at two instruments within 100 m of each other horizontally will not differ by an amount large enough to measure with instruments suggested for this method. Instruments separated by a vertical distance of less than 0.5 m may be compared without correcting for height difference.

8.3.1 Calibration of one or more barometers that do not

TABLE 2 Regional Standard Barometers

Region	Location	Category
I	Pretoria, South Africa	A_r
II	Calcutta, India	B_r
III	Rio de Janeiro, Brazil	A_r
	Buenos Aires, Argentina	B_r
	Maracay, Venezuela	B_r
IV	Washington, DC, (Gaithersburg, Md.), USA	A_r
V	Melbourne, Australia	A_r
VI	London, United Kingdom	A_r
	Leningrad, U.S.S.R.	A_r
	Paris, France	A_r
	Hamburg, Federal Republic of Germany	A_r

A_r—A barometer that has been selected by regional agreement as a reference standard for barometers of that region and is capable of independent determination of pressure to an accuracy of ±5 Pa.

B_r—A working standard barometer with known errors established by comparison with a primary or secondary standard. Such barometers are used in a region where the National meteorological services of the region agree to use them as the standard barometer for the region in the event that a barometer of category A_r is unavailable.

Taken from Annex 3, of *Guide to Meteorological Instruments and Observing Practice, World Meteorological Organization.*

produce mutual interference with the standard or each other can be accomplished by simple comparison with traveling or fixed standards by methods described in Refs (1), (4), and (5). If the instruments used can cause mutual interference (for example, electronic instruments) use isolation barriers that freely transmit atmospheric pressure.

8.4 Calibration is done by making a number of comparisons between the instrument being calibrated and the standard under a broad range of pressures.

8.5 Calibration records include pressure readings from the barometers; temperature readings from the attached thermometers; wind speed and gustiness (observed in accordance with methods described in Refs (4) or (5)); corrections for gravity, temperature, and instrumental error; the elevation above mean sea level of the zero point of the barometers; the latitude; the longitude; the name of the place; and the dates and times of observations.

8.6 Aneroid barometers are equipped with a means of setting the mechanism during calibration and comparison.

8.7 Protect all barometers from violent mechanical shock and explosive changes in pressure. A barometer subjected to either of these must be recalibrated.

8.8 Maintain the vertical and horizontal temperature gradients across the instruments at less than 0.1 K/m. Locate the instrument so as to avoid direct sunlight, drafts, and vibration.

9. Procedures

9.1 For synoptic meteorological observations determine the latitude and longitude of the station to the nearest second of arc and the height above mean sea level to the nearest 0.03 m. A method for such determination is described in the *Manual of Barometry* (1).

NOTE 1—This information is not needed for nonsynoptic purposes when pressure is being measured by Method B or by Method A when the local acceleration of gravity is known.

9.2 *Method A, Fortin Mercurial Barometer:*

NOTE 2—The method for measuring atmospheric pressure from a mercurial barometer is described in detail in 3.1.3 through 3.1.6 of the

World Meteorological Guide to Meteorological Instruments and Practices (4).

9.2.1 Read the temperature T from the thermometer attached to the barrel to the nearest 0.1 K.

9.2.2 Lower the mercury level in the cistern until it clears the index pointer. Raise the level slowly until a barely discernible dimple appears on the surface of the mercury.

9.2.3 Tap the barrel near the top of the mercury column.

9.2.4 Set the vernier so that the base just cuts off light at the highest point of the meniscus (the curved upper surface of the mercury column) and carefully avoid parallax error.

9.2.5 Read the height of the mercury Column B from the barometer in the manner appropriate to the vernier scale used to the equivalent of the nearest 10 Pa. Apply appropriate corrections as described in Section 10.

9.3 *Method B, Aneroid Barometer:*

9.3.1 Always read an aneroid barometer when it is in the same position (vertical or horizontal) as when calibrated.

9.3.2 Immediately before an aneroid barometer with mechanical linkage is read tap its case lightly to overcome bearing drag.

9.3.3 Read the aneroid barometer to the nearest equivalent of 10 Pa.

10. Calculations

10.1 For Method A using a Fortin-type barometer with brass scales, determine the temperature correction by means of the following equation or an appropriate table (6):

$$C_t = (0.04452345 - 0.000163T)B \tag{2}$$

where:

T = temperature, K,
C_t = correction at temperature T, and
B = observed barometer reading at temperature T, Pa.

10.1.1 Correct the reading by applying the temperature correction and instrumental correction as follows:

$$B_1 = B + C_t + C_i \tag{3}$$

where:

B_1 = barometer reading reduced to standard temperature and corrected for instrumental errors but not reduced to standard gravity and
C_i = instrumental error determined by calibration.

10.1.2 Correct for gravity as follows:

$$B_n = B_1 \frac{g_{\phi,H}}{g_n} \tag{4}$$

where:

B_n = barometric pressure at standard gravity (g_n) and standard temperature, 288.15 K (15°C), and corrected for instrumental errors,
$g_{\phi,H}$ = local acceleration of gravity in m/s² at the station latitude ϕ and station elevation H above sea level, and

g_n = standard acceleration of gravity, which is 9.80665 m/s².

10.1.3 The local acceleration of gravity may be calculated by the method described in Section 3.8 of the *Guide to Meteorological Instruments and Observing Practices* (4), Table 168 of the *Smithsonian Meteorological Tables* (6), determined by direct measurement with a gravimeter or obtained from government or academic institutions. If the value is reported by the Potsdam system the value $g_{\phi,H}$ is obtained by subtracting 0.00013 m/s².

$g_{\phi,H}$ = local acceleration of gravity in m/s² at the station latitude ϕ and station elevation H above sea level to be used for meteorological purposes.

$(g_{\phi,H})_P$ = measured gravity by the Potsdam system.

10.2 If the atmospheric pressure P_1, height h_1, and atmospheric temperature T_1 at some Site 1 and the height h_2 at a Site 2 are known then the atmospheric pressure P_2 at Site 2 can be calculated from the following equation:

$$P_2 = P_1 \exp \frac{0.068332(h_1 - h_2)}{(T_1 + T_2)} \tag{5}$$

where:

P_1 = pressure at Site 1, Pa,
P_2 = pressure at Site 2, Pa,
h_1 = height above mean sea level of Site 1, m,
h_2 = height above mean sea level of Site 2, m,
T_1 = atmospheric temperature at site 1, K, and
T_2 = atmospheric temperature at Site 2, K.

10.3 Table 1 provides a solution for selected values of $\dfrac{T_1 + T_2}{2}$ and P_1. For lateral distances less than 200 m and vertical distances less than 500 m, P_2 may be obtained from P_1 by adding the correction shown in Table 1 for each 1 m of height difference between h_1 and h_2.

11. Precision and Bias

11.1 The agreement between a single corrected reading using the Fortin-type mercurial barometer and reference measurements using primary and secondary standards has been found to be within 20 Pa (4). The precision of repeated measurements made with a single instrument is ±10 Pa (4).

11.2 The agreement between single readings of aneroid barometers and reference measurements using primary and secondary standards has been found to be within ±50 Pa (4). The precision of repeated measurements made with a single instrument is ±50 Pa (4).

12. Keywords

12.1 aneroid barometer; atmospheric pressure; barometer; barometry; Fortin-type mercurial barometer; mercurial barometer; pressure

REFERENCES

(1) *Manual of Barometry*, Vol 1, First Edition 1963, U.S. Department of Commerce, Weather Bureau, U.S. Department of Air Force, Air Weather Service, U.S. Department of Navy, Naval Weather Service, Washington, DC.

(2) Gill, Gerald C., *Development and Testing of a No-Moving-Parts Static Pressure Inlet for Use on Ocean Buoys*, University of Michigan, Ann Arbor, MI 1976.

(3) Occupational Safety and Health Standards Subpact Z—Toxic and Hazardous Substances, Section 1910.1000 Air Contaminants, Table Z-2. 29 Code of Federal Regulations.

(4) "Guide to Meteorological Instruments and Observing Practices," *World Meteorological Organization*, WMOBA, No. 8, TP3, Fourth Edition 1971, Secretariat of WMO, Geneva, Switzerland.

(5) *Federal Meteorological Handbook No. 1, Surface Observations*, U.S. Department of Commerce, U.S. Department of Defense, U.S. Department of Transportation, effective July 1, 1975, available from Superintendent of Documents, U.S. Government Printing Office, Washington, DC 20402.

(6) List, R. J. (compiler), *Smithsonian Meteorological Tables*, Sixth Revised Edition, 1949, Fourth Reprint issued 1968, Smithsonian Institution Publications, SIPMA, Washington, DC.

Designation: D 4230 – 83 (Reapproved 1996)$^{\epsilon 1}$

Standard Test Method of
Measuring Humidity with Cooled-Surface Condensation (Dew-Point) Hygrometer[1]

This standard is issued under the fixed designation D 4230; the number immediately following the designation indicates the year of original adoption or, in the case of revision, the year of last revision. A number in parentheses indicates the year of last reapproval. A superscript epsilon (ϵ) indicates an editorial change since the last revision or reapproval.

$^{\epsilon 1}$ NOTE—Section 14 was added editorially in April 1996.

1. Scope

1.1 This test method covers the determination of the thermodynamic dew- or frost-point temperature of ambient air by the condensation of water vapor on a cooled surface. For brevity this is referred to in this method as the condensation temperature.

1.2 This test method is applicable for the range of condensation temperatures from 60°C to −70°C.

1.3 This test method includes a general description of the instrumentation and operational procedures, including site selection, to be used for obtaining the measurements and a description of the procedures to be used for calculating the results.

1.4 This test method is applicable for the continuous measurement of ambient humidity in the natural atmosphere on a stationary platform.

1.5 *This standard does not purport to address all of the safety concerns, if any, associated with its use. It is the responsibility of the user of this standard to establish appropriate safety and health practices and determine the applicability of regulatory limitations prior to use.* For specific precautionary statements, see Section 8.

2. Referenced Documents

2.1 *ASTM Standards:*
D 1356 Terminology Relating to Sampling and Analysis of Atmospheres[2]
D 3631 Test Methods for Measuring Surface Atmospheric Pressure[2]
D 4023 Terminology Relating to Humidity Measurements[2]

3. Terminology

3.1 *Definitions:*

3.1.1 For definitions of terms used in this test method, refer to Terminology D 4023.

3.1.2 For definitions of other terms in this method, refer to Terminology D 1356.

3.2 *Definitions of Terms Specific to This Standard:*

3.2.1 *nonhygroscopic material*—material that neither absorbs nor retains water vapor.

3.2.2 *mirror (front surface)*—a polished surface, usually a metallic surface, on which condensates are deposited.

3.3 *Symbols:*

e = vapor pressure of water vapor in moist air.
e_i = saturation pressure of water vapor in equilibrium with the plane surface of ice.
e_w = saturation pressure of water vapor in equilibrium with the plane surface of water.
f = enhancement factor.
p_a = ambient pressure.
p_c = mirror chamber pressure.
r = mixing ratio.
t = ambient air temperature.
T_d = thermodynamic dew- or frost-point temperature.
U_i = relative humidity with respect to ice.
U_w = relative humidity with respect to water.
x_v = mole fraction of water vapor.
x_{vi} = saturation mole fraction of water vapor with respect to ice.
x_{vw} = saturation mole fraction of water vapor with respect to water.

4. Summary of Test Methods

4.1 The ambient humidity is measured with a dew- and frost-point hygrometer.

4.2 The mirror or some other surface on which the condensate is deposited is provided with the means for cooling and heating, detection of condensate, and the measurement of the temperature of the mirror surface.

5. Significance and Use

5.1 Humidity data is important for the understanding and interpretation of a number of phenomena. Atmospheric water vapor affects precipitation; the formation of dew and fog; the prediction of frosts damaging to agriculture; the potential danger of forest fires; and the propagation of electromagnetic energy. It affects evaporation from rivers, lakes, reservoirs, oceans, and snow and ice surfaces. It affects the transpiration of moisture from soils, growing crops, and forest.

6. Interferences

6.1 This method is not applicable if other constituents in the atmosphere condense before water vapor.

7. Apparatus

7.1 *Dew-point hygrometers*, specifically designed for meteorological observations are available commercially. A sche-

[1] This test method is under the jurisdiction of ASTM Committee D-22 on Sampling and Analysis of Atmospheres and is the direct responsibility of Subcommittee D22.11 on Meteorology.
Current edition approved Feb. 25, 1983. Published October 1983.
[2] *Annual Book of ASTM Standards*, Vol 11.03.

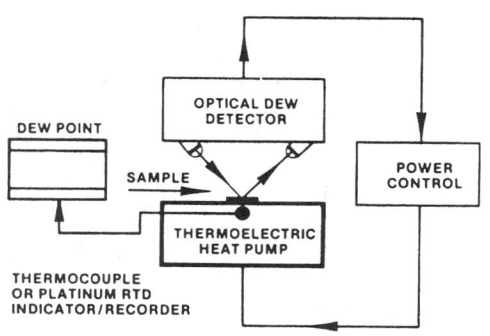

FIG. 1 Schematic of a Thermoelectric Cooled Condensation Hygrometer

matic arrangement of a typical optical dew-point hygrometer is shown in Fig. 1.

7.1.1 The sample air flows through a small chamber.

7.1.2 Within the chamber is a mirror or surface on which the condensate can be deposited.

7.1.3 A beam of light from an incandescent lamp, light emitting diode or other suitable light source shines on the mirror.

7.1.4 Dew or frost is detected with an electro-optic device.

7.1.5 The mirror is cooled by a Peltier thermoelectric element. Peltier cooling is a convenient method for unattended and automatic instruments.

7.1.6 Preferred methods of sensing mirror temperature are resistance thermometers, thermistors, and thermocouples.

7.1.6.1 The temperature sensors shall be attached to or embedded in the mirror to measure the temperature of the surface of the mirror.

7.1.7 Suitable control circuitry shall be provided to maintain a constant quantity of condensate on the mirror.

7.1.8 Suitable provisions shall be provided to compensate for the contamination of the surface of the mirror.

7.2 *Auxiliary Equipment:*

7.2.1 Provision shall be provided for assuring air flow past the dew-point mirror.

7.2.2 Readout instrumentation is available with the dew-point hygrometer.

8. Precautions

8.1 *Safety Precautions:*

8.1.1 The hygrometer shall be packaged in a suitable enclosure for application in industrial or outdoor environment.

8.1.2 Electrical connectors and cables shall be suitable for the outdoor environment.

8.1.3 Appropriate voltage surge protection circuitry must be incorporated.

8.2 *Technical Precautions:*

8.2.1 The accuracy of a cooled-surface condensation hygrometer is degraded by the presence of water-soluble materials. A mirror-cleaning schedule, consistent with the contamination rate, is necessary to maintain the initial calibration accuracy. The user must determine the required maintenance schedule for the specific site, by comparison of calibrations made before and after cleaning.

8.2.2 Caution in performing this method should be taken if the indicated mirror temperature is between 0°C and

−30°C. Below freezing, the initial formation of the condensate on the surface of a mirror may be either dew or frost. In the case of nonfiltered atmospheric air, the supercooled water usually does not persist long on a mirror surface and quickly changes to frost. The only positive method for determining the state of the condensate is by visual observation of the mirror surface.

8.2.2.1 The following illustrates the magnitude of the error involved when dew or frost is not differentiated: The saturation vapor pressure of supercooled water at −30°C corresponds to saturation vapor pressure of ice at −27.2°C; dew point of −20°C corresponds to frost point of −18.0°C; −10°C dew point corresponds to frost point of −8.9°C. (The frost point temperature is approximately 90 % of the dew-point temperature in degrees Celsius.)

8.2.3 A positive method for identifying the state of the condensate is to visually observe the condensate on the mirror with the aid of a microscope or other optical magnifier.

8.2.4 A finite length of time is required for the condensate to deposit on the mirror and for the hygrometer to reach equilibrium with the ambient humidity. The response of the hygrometer depends on the humidity of the ambient air, and on such factors as the ventilation rate of the ambient air past the mirror, the sensitivity of the condensate detector, and the maximum cooling rate of the hygrometer. The worst case occurs during the initial dew-point reading after clearing the mirror of all condensates. The time it takes the hygrometer to reach equilibrium after clearing the mirror will vary from instrument to instrument. As an illustration of the magnitude of this time, the following are approximate times required by a hygrometer to reach equilibrium after clearing the mirror.

8.2.4.1 For dew points warmer than 0°C: 5 min after clear.

8.2.4.2 For dew points 0°C to −20°C: 5 to 20 min after clear.

8.2.4.3 For dew points −20°C to −40°C: 20 min to 1 h after clear.

8.2.4.4 For dew points −40°C to −60°C: 1 h to 2 h after clear.

8.2.4.5 For dew points −60°C to −70°C: 2 h to 6 h after clear.

8.2.5 The pressure differential between the mirror chamber and the ambient shall not be greater than 0.5 % that is, not more than 500 Pa.

8.2.6 The thermometer must measure the temperature of the mirror surface and not be influenced by the ambient air temperature.

8.2.7 All materials, which come into contact with the sample air before it reaches the dew-point mirror, shall be nonhygroscopic. Metal, glass, polytetrafluoroethylene, or stabilized polypropylene are examples of suitable materials. Polyvinyl chloride tubing must be avoided.

9. Sampling

9.1 Automatic dew-point hygrometers provide an output which may be recorded continuously. Modern data loggers sample temperature-sensor output periodically, convert the analog sensor signal to a digital form, and store the data. The proper sampling interval depends on the data application (see 13.2).

9.2 Locate a blower or pump, which can be used to move the air sample through the mirror chamber, downstream of the dew-point mirror.

9.3 Select the site or location so that the measurement data represents the water vapor content of the ambient atmosphere.

9.3.1 Select the location so that it is normally not influenced by a local water vapor source. (Of course, if the purpose is to measure the effects of a local source that is, water cooling ponds, etc., then it is necessary to locate the site downwind from the source.)

9.3.2 Place the automatic dew-point hygrometer away from any paved surfaces that may be wet, the immediate influence of trees and buildings, and as far as practicable, not too close to steep slopes, ridges, cliffs, or hollows. Avoid dusty areas.

9.3.3 Mount the instrument over a surface which is representative of the area.

9.4 The successful application of this method requires that all the materials which come in contact with the sample air upstream of the dew-point mirror be nonhygroscopic.

9.5 The materials which come in contact with the sample air upstream of the dew-point mirror might be wetted by rain, dew, or frost; for example, dew forming on a surface in the early morning. Design the sampling system to minimize these deleterious effects.

10. Calibration

10.1 Provide the calibration data for the thermometer, used for measuring the condensation temperature with the hygrometer. Consult the manufacturer's operating manual for calibrating the thermometer readout instrumentation.

10.2 The cooled-surface condensation (dew-point) method is considered to be an absolute or fundamental method for measuring humidity. This method requires an accurate measurement of the temperature of the surface of the dew-point mirror. It is not uncommon for the dew-point temperature to be more than 35 K colder than the ambient air temperature. To measure this temperature accurately, without being influenced by the warmer ambient and the colder heat-sink temperature, requires careful placement of the dew-point thermometer.

10.3 Therefore, in addition to the temperature calibration of the thermometer, (see 10.1), a humidity calibration must also be performed to verify the proper operation of the hygrometer (see Annex A1). The following are additional examples of factors that can affect the accuracy of the measurement: extraneous thermal emfs, heat leakage through the thermometer leads, self-heating of the thermometer, poor thermal contact, temperature gradient across the mirror, etc.

11. Procedure

11.1 *Selection of Sampling Site*—Select sampling site as indicated in 3.3 and also in 1.3.2 of the World Meteorological Organization, *Guide to Meteorological Instrument and Observing Practices* (1).[3]

11.2 Consult the manufacturer's operating manual for start-up procedures.

11.3 Perform necessary calibration as indicated in Section 10. The dew-point thermometer will not undergo large shifts (.05°C) in calibration unless it is subjected to physical shock. If the thermometer read-out instrumentation is subjected to

---

[3] The boldface numbers in parentheses refer to the references at the end of this method.

varying ambient temperatures, the read-out instrumentation checks must be over the expected range of ambient temperatures. The frequency with which which these checks are required will be determined by the stability of the readout instrumentation.

11.4 Check and verify that all necessary variables are measured and recorded to compute the humidity in the desired unit(s) see also 8.2.1.

NOTE 1—In general, it is recommended that ambient temperature and pressure (the pressure in the mirror chamber should not differ from the ambient pressure by more than 0.5 %) and the dew-point temperature be measured and recorded. The ambient pressure is to be measured according to Test Methods D 3631. This will enable other users of the data to calculate in the different units of humidity.

12. Calculations

12.1 In the meteorological range of pressure and temperature, the saturation vapor pressure of the pure water phase and of the moist air will be assumed to be equal. This assumption will introduce an error of approximately 0.5 % of reading or less. See Appendix X1.

12.2 Calculate the ambient relative humidity with respect to water as follows.

$$(U_w)_{t,p_a} = \frac{p_a}{p_c} \frac{e(t_d)}{e_w(t)} 100 \%$$

where:

$(U_w)_{t,p_a}$ = relative humidity with respect to water, %, at temperature t and pressure p_a,

p_a = average barometric pressure, Pascal (Pa) during the sampling period,

p_c = average absolute pressure, Pa, in the dew-point mirror chamber during the sampling period,

$e(t_d)$ = saturation vapor pressure, Pa, at condensation temperature t_d, °C, where t_d is the average value during the sampling period, see Note 2, and

$e_w(t)$ = saturation vapor pressure, Pa, over water at ambient temperature t, °C, where t is the average value during the sampling period. See Appendix X3.

NOTE 2—If the condensate on the mirror is water (dew), use the saturation vapor pressure over water, Appendix X3, corresponding to the condensation temperature t_d. If the condensate is ice (frost), use the saturation vapor pressure over ice, Appendix X4, corresponding to the condensation temperature t_d.

12.3 Calculate the relative humidity with respect to ice as follows:

$$(U_i)_{t,p_a} = \frac{p_a}{p_c} \frac{e(t_d)}{e_i(t)} 100 \%$$

where:

$(U_i)_{t,p_a}$ = relative humidity with respect to ice, %, at temperature t and pressure p_a,

$e_i(t)$ = saturation vapor pressure, Pa, over ice at ambient temperature t, °C, where t is the average value during the sampling period, see Appendix X4, and

$p_a, p_c, e(t_d)$ = see 12.2

12.4 Calculate the mixing ratio as follows:

$$r = 0.622 \, e(t_d)/[p_c - e(t_d)]$$

where:

r = mixing ratio and

$e(t_d), p_c$ = see 12.2.

12.5 Calculate parts per million by mass as follows:

$$ppm_m = r \times 10^6$$

12.6 Calculate parts per million by volume as follows:

$$ppm_v = e(t_d)/[p_c - e(t_d)] \times 10^6$$

13. Precision and Bias

13.1 The estimated precision for this method is valid only for *constant* ambient humidity and the precision varies with the condensation temperature as shown in Fig. 2. The precision is based on single-laboratory and multioperator-device test.

13.2 The estimated bias of the dew-point hygrometer, as shown in Fig. 2, is valid only for constant ambient humidity. This is true especially for low-condensation temperatures. The bias varies from ±0.4°C for condensation temperatures above freezing to ±2.0°C at condensation temperature −70°C. All uncertainties are at the 95 % confidence level.

13.3 If the standard deviation is equal to or less than the values listed in 13.3.1, the ambient humidity is assumed to be sufficiently constant so that the bias curve given in Fig. 2 is valid.

13.3.1 The following criteria shall be used to determine whether the ambient humidity is constant:

Condensation Temperature (°C)	Duration of Sampling Time (min)	Number of Readings Taken Over Equally Spaced Time Intervals	Calculated Standard Deviation (°C)
−70	120	10 to 25	±0.5
−60	90	10 to 25	±0.4
−50	60	10 to 25	±0.3
−40	30	10 to 25	±0.2
−25	20	10 to 25	±0.15
−10	15	10 to 25	±0.1
60	15	10 to 25	±0.1

13.4 The following is an example on use of the table in 13.3.1:

13.4.1 If the indicated condensation temperature is −50°C, take 10 to 25 readings taken at equally spaced time interval for a period of approximately 60 min, compute the average value (in this case −50°C) and the standard deviation. If this calculated standard deviation is equal to or less than ±0.3°C, the ambient humidity is assumed to be constant and the bias of the reading at −50°C is ±1.2°C (Fig. 2). See also 8.2.2.

14. Keywords

14.1 dew-point; humidity; hygrometer; saturation; temperature, dew-point; vapor pressure

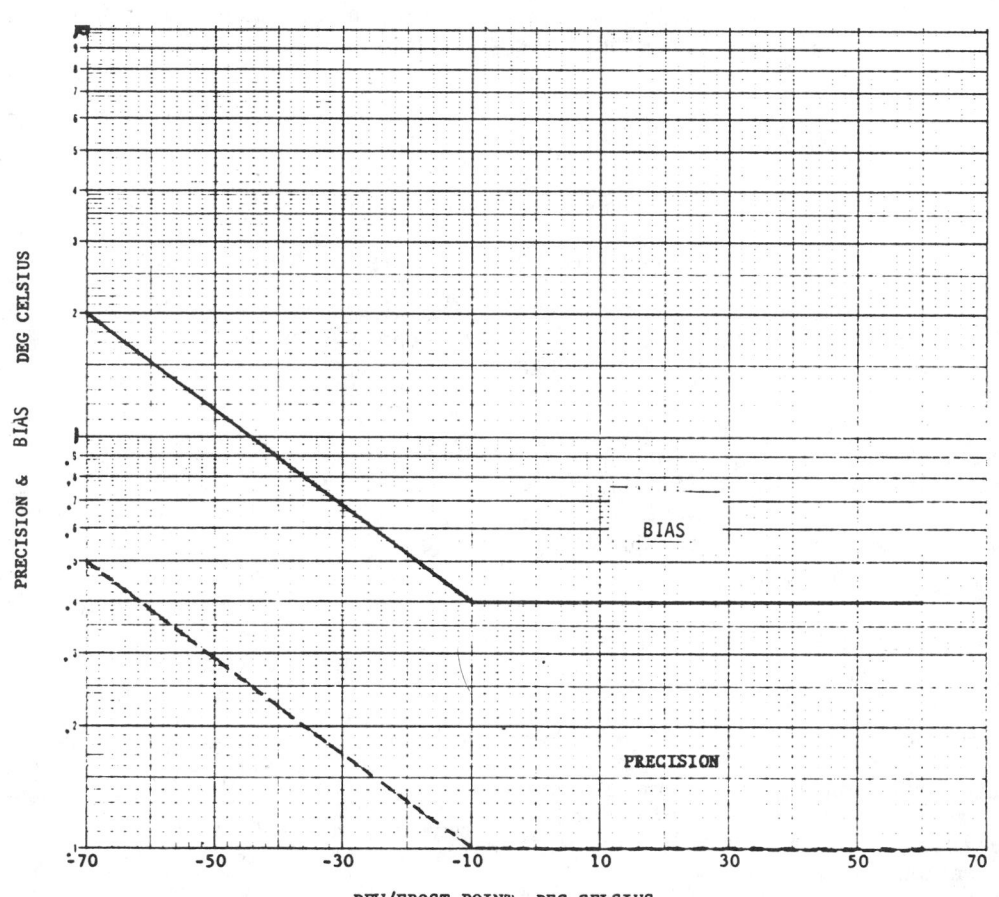

FIG. 2 Precision and Bias Versus Dew/Frost Point

ANNEXES

(Mandatory Information)

A1. LABORATORY CALIBRATION OF DEW-POINT HYGROMETER

A1.1 An accurate method for the calibration of the hygrometer is to test the instrument with a humidity generator that produces air of known humidity (2).

A1.2 An alternative method is by direct comparison with a secondary or working standard hygrometer when both instruments are subjected to the same, preferably constant, humidity.

NOTE A1.1—Secondary standard hygrometers are characterized by long term repeatability and predictable behavior when verified to be performing properly. Working standard hygrometers are characterized by satisfactory (which meet the users requirements) precision and stability when calibrated against a humidity generator or intercompared with a secondary standard.

A2. FIELD CALIBRATION OF DEW-POINT HYGROMETER

A2.1 Install a secondary or working standard hygrometer adjacent to the working hygrometer and run an intercomparison test. It is very important that the standard and the working hygrometers are sampling the same air mass. The exhaust air from both hygrometers must not be mixed with the intake air sample.

A2.2 If there is a fixed bias between the readings of the two hygrometers, interchange the positions of the two instruments to determine if the bias is due to sampling problems.

A2.3 The intercomparison test should continue until constant humidity indications, as defined in 13.2, are seen on both instruments. If the sample condensation temperature is steady, the duration of the test will be shorter than when there are large fluctuations in humidity.

A2.4 The tests should be performed during periods when the relative humidity is low and also during periods when the relative humidity is high. (A one-point verification check is sufficient if the working hygrometer had been calibrated prior to the installation in the field.)

APPENDIXES

(Nonmandatory Information)

X1. SATURATION VAPOR PRESSURE OF MOIST AIR[4]

X1.1 Moist air does not behave precisely like an ideal gas (3). The product of the mole fraction of the water vapor and the pressure p_a is not generally equal to the pressure that would be exerted by the water vapor if it existed alone in the same volume as the mixture had at the same temperature. The effective saturation pressure of water vapor in equilibrium with a plane surface of liquid or solid water in the presence of an admixed gas differs from the saturation vapor pressure of the pure phase and is expressed as follows:

$$f = x_w p_a/e_s = (1 - x_a)p_a/e_s \qquad (X1.1)$$

where:

f = enhancement factor, see Appendix X2,
x_w, x_a = mole fractions of the water vapor and of the air, respectively, in the saturated mixture,
p_a = total pressure above the surface of the condensed phase (water or ice),
e_s = pure phase saturation vapor pressure of water substance at the temperature of saturation, and
$x_w p_a$ = effective water vapor pressure at the given pressure and temperature of saturation.

X1.2 The percentage relative humidity with respect to water of a moist air at absolute pressure p_a and temperature t is defined as follows:

$$U_x = 100(x_v/x_{vw})_{p_a,t} \% \quad and \qquad (X1.2)$$

using Eq X1.1, Eq X1.2 can be expressed in the following form:

$$U_x = 100 \frac{p_a f(p_c, t_d)e(t_d)}{p_c f_w(p_a, t)e_w(t)} \% \qquad (X1.3)$$

where:

$p_a, p_c, e(t_d)$ and $e_w(t)$ = see 12.2,
$f(p_c, t_d)$ = the enhancement factor at pressure p_c and condensation temperature t_d. If the condensate on the mirror is water use f_w (water) and if the condensate is ice use f_i (ice)—see X2 for values of f_w and f_i, and
$f_w(p_a, t)$ = the enhancement factor for water at pressure p_a and air temperature t. See X2 for values of f_w.

X1.3 Similarly, the percentage relative humidity with respect to ice is as follows:

$$U_i = 100 \frac{p_a f(p_c, t_d)e(t_d)}{p_c f_i(p_a, t)e_i(t)} \qquad (X1.4)$$

where:

$p_a, p_c, e(t_d)$ = see 12.2,
$e_i(t)$ = see 12.2,

---

[4] The fundamental concepts and definitions of moist air properties are described in detail by Harrison (3).

$f(p_c,t_d)$ = see X1.2, and
$f_i(p_a,t)$ = the enhancement factor for ice at pressure p_a and temperature t. See X4 for values of f_i.

X1.4 The mixing ratio for moist air is as follows:

$$r = 0.622 f(p_c,t_d)e(t_d)/[p_c - f(p_c,t_d)e(t_d)]$$

where:

p_c and $e(t_d)$ = see 12.2, and
$f(p_c,t_d)$ = see X1.2.

X1.5 The parts per million by volume is expressed as follows:

$$ppm_v = f(p_c,t_d)e(t_d)/[p_c - f(p_c,t_d)e(t_d)] \times 10^6$$

X2. ENHANCEMENT FACTORS FOR WATER VAPOR IN AIR[5]

X2.1 Equations which explicitly express the enhancement factors for water vapor are given in *Functional Equations for the Enhancement Factors for CO₂-Free Moist Air* (4). These equations are approximation to the formulation of Hyland (5) and provide the means of obtaining enhancement factors with very modest computational facilities. See Table X2.1.

NOTE X2.1—Hyland does not assign uncertainties below −50°C and for water below 0°C. Despite the questionable validity of extrapolating his formulation in these areas, namely the region of supercooled water and the region of ice below −50°C, there is no formulation for f in these regions that is more valid at this time.

[5] The derivation of the enhancement factors is described in detail in *A Correlation for the Second Interaction Virial Coefficients and Enhancement Factors for Moist Air* (5).

TABLE X2.1   Enhancement Factors

°C	f_w (water)				f_i (ice)			
	Total Pressure, MPa							
	0.025	0.050	0.100	0.150	0.025	0.050	0.100	0.150
70	1.0000	1.0029	1.0060	1.0077	...	...	...	...
60	1.0011	1.0036	1.0057	1.0071	...	...	...	...
50	1.0021	1.0036	1.0052	1.0065	...	...	...	...
40	1.0022	1.0032	1.0047	1.0060	...	...	...	...
30	1.0020	1.0028	1.0043	1.0056	...	...	...	...
20	1.0017	1.0025	1.0040	1.0054	...	...	...	...
10	1.0015	1.0023	1.0039	1.0055	...	...	...	...
0	1.0013	1.0022	1.0038	1.0056	1.0013	1.0022	1.0040	1.0057
−10	1.0011	1.0020	1.0039	1.0058	1.0012	1.0022	1.0041	1.0060
−20	1.0010	1.0020	1.0041	1.0062	1.0012	1.0022	1.0044	1.0065
−30	1.0010	1.0021	1.0044	1.0067	1.0012	1.0024	1.0047	1.0071
−40	1.0010	1.0022	1.0048	1.0073	1.0013	1.0026	1.0052	1.0078
−50	1.0010	1.0024	1.0052	1.0080	1.0014	1.0029	1.0058	1.0086
−60	...	...	...	...	1.0016	1.0032	1.0064	1.0097
−70	...	...	...	...	1.0018	1.0036	1.0072	1.0109

X3. SATURATION VAPOR PRESSURE OVER WATER

X3.1 The saturation vapor pressure of the pure phase over plane surface of pure water for temperatures 0 to 100°C was obtained from Wexler's 1976 formulation (6). See Table X3.1. Other suitable saturation vapor pressure tables are given in the Smithsonian Meteorological Tables (7), International Meteorological Tables (8) and ASHRAE Handbook and Product Directory (10).

X3.2 The tabulated vapor pressure for temperatures below 0°C (supercooled water) were obtained by the extrapolation of Wexler's 1976 formulation (6) down to −50°C. See Table X3.2. Despite the questionable validity of extrapolating the formulation in these areas, there is no formulation for saturated vapor pressures for supercooled water that is more valid at this time.

TABLE X3.1   Saturation vapor pressure over water (IPTS—68)

Temp °C	.0	0.1	0.2	0.3	0.4	0.5	0.6	0.7	0.8	0.9
	Pascal									
0	611.213[A]	615.667	620.150	624.662	629.203	633.774	638.373	643.003	647.662	652.350
1	657.069	661.819	666.598	671.408	676.249	681.121	686.024	690.958	695.923	700.920
2	705.949	711.010	716.103	721.228	726.386	731.576	736.799	742.055	747.344	752.667
3	758.023	763.412	768.836	774.294	779.786	785.312	790.873	796.469	802.100	807.766
4	813.467	819.204	824.977	830.786	836.631	842.512	848.429	854.384	860.375	866.403
5	872.469	878.572	884.713	890.892	897.109	903.364	909.658	915.991	922.362	928.773
6	935.223	941.712	948.241	954.810	961.419	968.069	974.759	981.490	988.262	995.075
7	1 001.93	1 008.83	1 015.76	1 022.74	1 029.77	1 036.83	1 043.94	1 051.09	1 058.29	1 065.52
8	1 072.80	1 080.13	1 087.50	1 094.91	1 102.37	1 109.87	1 117.42	1 125.01	1 132.65	1 140.33
9	1 148.06	1 155.84	1 163.66	1 171.53	1 179.45	1 187.41	1 195.42	1 203.48	1 211.58	1 219.74

[A] Metastable state.

TABLE X3.1 *Continued*

Temp °C	.0	0.1	0.2	0.3	0.4	0.5	0.6	0.7	0.8	0.9
						Pascal				
10	1 227.94	1 236.19	1 244.49	1 252.84	1 261.24	1 269.68	1 278.18	1 286.73	1 295.33	1 303.97
11	1 312.67	1 321.42	1 330.22	1 339.08	1 347.98	1 356.94	1 365.95	1 375.01	1 384.12	1 393.29
12	1 402.51	1 411.79	1 421.11	1 430.50	1 439.93	1 449.43	1 458.97	1 468.58	1 478.23	1 487.95
13	1 497.72	1 507.54	1 517.43	1 527.36	1 537.36	1 547.42	1 557.53	1 567.70	1 577.93	1 588.21
14	1 598.56	1 608.96	1 619.43	1 629.95	1 640.54	1 651.18	1 661.89	1 672.65	1 683.48	1 694.37
15	1 705.32	1 716.33	1 727.41	1 738.54	1 749.75	1 761.01	1 772.34	1 783.73	1 795.18	1 806.70
16	1 818.29	1 829.94	1 841.66	1 853.44	1 865.29	1 877.20	1 889.18	1 901.23	1 913.34	1 925.53
17	1 937.78	1 950.10	1 962.48	1 974.94	1 987.47	2 000.06	2 012.73	2 025.46	2 038.27	2 051.14
18	2 064.09	2 077.11	2 090.20	2 103.37	2 116.61	2 129.92	2 143.30	2 156.75	2 170.29	2 183.89
19	2 197.57	2 211.32	2 225.15	2 239.06	2 253.04	2 267.10	2 281.23	2 295.44	2 309.73	2 324.10
20	2 338.54	2 353.07	2 367.67	2 382.35	2 397.11	2 411.95	2 426.88	2 441.88	2 456.94	2 472.13
21	2 487.37	2 502.70	2 518.11	2 533.61	2 549.18	2 564.85	2 580.59	2 596.42	2 612.33	2 628.33
22	2 644.42	2 660.59	2 676.85	2 693.19	2 709.62	2 726.14	2 742.75	2 759.45	2 776.23	2 793.10
23	2 810.06	2 827.12	2 844.26	2 861.49	2 878.82	2 896.23	2 913.74	2 931.34	2 949.04	2 966.82
24	2 984.70	3 002.68	3 020.74	3 038.91	3 057.17	3 075.52	3 093.97	3 112.52	3 131.16	3 149.90
25	3 168.74	3 187.68	3 206.71	3 225.85	3 245.08	3 264.41	3 283.85	3 303.38	3 323.02	3 342.76
26	3 362.60	3 382.54	3 402.59	3 422.73	3 442.99	3 463.34	3 483.81	3 504.37	3 525.05	3 545.83
27	3 566.71	3 587.71	3 608.81	3 630.02	3 651.33	3 672.76	3 694.29	3 715.94	3 737.69	3 759.56
28	3 781.54	3 803.63	3 825.83	3 848.14	3 870.57	3 893.11	3 915.77	3 938.54	3 961.42	3 984.42
29	4 007.54	4 030.77	4 054.12	4 077.59	4 101.18	4 124.88	4 148.71	4 172.65	4 196.71	4 220.90
30	4 245.20	4 269.63	4 294.18	4 318.85	4 343.64	4 368.56	4 393.60	4 418.77	4 444.06	4 469.48
31	4 495.02	4 520.69	4 546.49	4 572.42	4 598.47	4 624.65	4 650.96	4 677.41	4 703.98	4 730.68
32	4 757.52	4 784.48	4 811.58	4 838.81	4 866.18	4 893.68	4 921.32	4 949.09	4 976.99	5 005.04
33	5 033.22	5 061.53	5 089.99	5 118.58	5 147.32	5 176.19	5 205.20	5 234.36	5 263.65	5 293.09
34	5 322.67	5 352.39	5 382.26	5 412.27	5 442.43	5 472.73	5 503.18	5 533.78	5 564.52	5 595.41
35	5 626.45	5 657.64	5 688.97	5 720.46	5 752.10	5 783.89	5 815.83	5 847.93	5 880.17	5 912.58
36	5 945.13	5 977.84	6 010.71	6 043.73	6 076.91	6 110.25	6 143.75	6 177.40	6 211.22	6 245.19
37	6 279.33	6 313.62	6 348.08	6 382.70	6 417.48	6 452.43	6 487.54	6 522.82	6 558.26	6 593.87
38	6 629.65	6 665.59	6 701.71	6 737.99	6 774.44	6 811.06	6 847.85	6 884.82	6 921.95	6 959.26
39	6 996.75	7 034.40	7 072.24	7 110.24	7 148.43	7 186.79	7 225.33	7 264.04	7 302.94	7 342.02
40	7 381.27	7 420.71	7 460.33	7 500.13	7 540.12	7 580.28	7 620.64	7 661.18	7 701.90	7 742.81
41	7 783.91	7 825.20	7 866.67	7 908.34	7 950.19	7 992.24	8 034.47	8 076.90	8 119.53	8 162.34
42	8 205.36	8 248.56	8 291.96	8 335.56	8 379.36	8 423.36	8 467.55	8 511.94	8 556.54	8 601.33
43	8 646.33	8 691.53	8 736.93	8 782.54	8 828.35	8 874.37	8 920.59	8 967.02	9 013.66	9 060.51
44	9 107.57	9 154.84	9 202.32	9 250.01	9 297.91	9 346.03	9 394.36	9 442.91	9 491.67	9 540.65
45	9 589.84	9 639.25	9 688.89	9 738.74	9 788.81	9 839.11	9 889.62	9 940.36	9 991.32	10 042.51
46	10 093.92	10 145.56	10 197.43	10 249.52	10 301.84	10 354.39	10 407.18	10 460.19	10 513.43	10 566.91
47	10 620.62	10 674.57	10 728.75	10 783.16	10 837.82	10 892.71	10 947.84	11 003.21	11 058.82	11 114.67
48	11 170.76	11 227.10	11 283.68	11 340.50	11 397.57	11 454.88	11 512.45	11 570.26	11 628.32	11 686.63
49	11 745.19	11 804.00	11 863.07	11 922.38	11 981.96	12 041.78	12 101.87	12 162.21	12 222.81	12 283.66
50	12 344.78	12 406.16	12 467.79	12 529.70	12 591.86	12 654.29	12 716.98	12 779.94	12 843.17	12 906.66
51	12 970.42	13 034.46	13 098.76	13 163.33	13 228.18	13 293.30	13 358.70	13 424.37	13 490.32	13 556.54
52	13 623.04	13 689.82	13 756.88	13 824.23	13 891.85	13 959.76	14 027.95	14 096.43	14 165.19	14 234.24
53	14 303.57	14 373.20	14 443.11	14 513.32	14 583.82	14 654.61	14 725.69	14 797.07	14 868.74	14 940.72
54	15 012.98	15 085.55	15 158.42	15 231.59	15 305.06	15 378.83	15 452.90	15 527.28	15 601.97	15 676.96

A Metastable state.

TABLE X3.1 *Continued*

Temp °C	.0	0.1	0.2	0.3	0.4	0.5	0.6	0.7	0.8	0.9
					Pascal					
55	15 752.26	15 827.87	15 903.79	15 980.02	16 056.57	16 133.42	16 210.59	16 288.07	16 365.87	16 443.99
56	16 522.43	16 601.18	16 680.26	16 759.65	16 839.37	16 919.41	16 999.78	17 080.47	17 161.49	17 242.84
57	17 324.51	17 406.52	17 488.86	17 571.52	17 654.53	17 737.86	17 821.53	17 905.54	17 989.88	18 074.57
58	18 159.59	18 244.95	18 330.66	18 416.71	18 503.10	18 589.84	18 676.92	18 764.35	18 852.13	18 940.26
59	19 028.74	19 117.58	19 206.76	19 296.30	19 386.20	19 476.45	19 567.06	19 658.03	19 749.35	19 841.04
60	19 933.09	20 025.51	20 118.29	20 211.43	20 304.95	20 398.82	20 493.07	20 587.69	20 682.68	20 778.05
61	20 873.78	20 969.90	21 066.39	21 163.25	21 260.50	21 358.12	21 456.13	21 554.51	21 653.28	21 752.44
62	21 851.98	21 951.91	22 052.23	22 152.93	22 254.03	22 355.52	22 457.40	22 559.68	22 662.35	22 765.42
63	22 868.89	22 972.75	23 077.02	23 181.69	23 286.76	23 392.23	23 498.12	23 604.40	23 711.10	23 818.20
64	23 925.72	24 033.65	24 141.99	24 250.74	24 359.91	24 469.50	24 579.51	24 689.93	24 800.78	24 912.04
65	25 023.74	25 135.85	25 248.39	25 361.36	25 474.76	25 588.58	25 702.84	25 817.53	25 932.66	26 048.22
66	26 164.21	26 280.64	26 397.52	26 514.83	26 632.58	26 750.78	26 869.42	26 988.51	27 108.04	27 228.02
67	27 348.46	27 469.34	27 590.68	27 712.46	27 834.71	27 957.41	28 080.57	28 204.19	28 328.26	28 452.80
68	28 577.81	28 703.28	28 829.21	28 955.61	29 082.48	29 209.82	29 337.64	29 465.92	29 594.68	29 723.92
69	29 853.63	29 983.82	30 114.49	30 245.65	30 377.28	30 509.40	30 642.01	30 775.10	30 908.68	31 042.75
70	31 177.32	31 312.37	31 447.92	31 583.97	31 720.51	31 857.55	31 995.09	32 133.14	32 271.68	32 410.73
71	32 550.29	32 690.35	32 830.93	32 972.01	33 113.61	33 255.71	33 398.34	33 541.48	33 685.13	33 829.31
72	33 974.01	34 119.23	34 264.97	34 411.24	34 558.03	34 705.36	34 853.21	35 001.59	35 150.51	35 299.96
73	35 449.95	35 600.47	35 751.54	35 903.14	36 055.29	36 207.98	36 361.21	36 514.99	36 669.32	36 824.20
74	36 979.63	37 135.61	37 292.15	37 449.24	37 606.89	37 765.10	37 923.87	38 083.21	38 243.10	38 403.56
75	38 564.59	38 726.19	38 888.36	39 051.10	39 214.41	39 378.30	39 542.76	39 707.80	39 873.42	40 039.63
76	40 206.41	40 373.78	40 541.74	40 710.28	40 879.42	41 049.14	41 219.46	41 390.37	41 561.88	41 733.99
77	41 906.69	42 080.00	42 253.91	42 428.42	42 603.54	42 779.27	42 955.61	43 132.55	43 310.11	43 488.29
78	43 667.08	43 846.48	44 026.51	44 207.16	44 388.43	44 570.33	44 752.85	44 936.00	45 119.77	45 304.18
79	45 489.23	45 674.91	45 861.22	46 048.17	46 235.76	46 424.00	46 612.87	46 802.39	46 992.56	47 183.38
80	47 374.85	47 566.97	47 759.74	47 953.17	48 147.25	48 342.00	48 537.40	48 733.47	48 930.20	49 127.60
81	49 325.67	49 524.40	49 723.81	49 923.89	50 124.64	50 326.08	50 528.19	50 730.98	50 934.45	51 138.61
82	51 343.45	51 548.98	51 755.20	51 962.11	52 169.72	52 378.01	52 587.01	52 796.70	53 007.10	53 218.20
83	53 430.00	53 642.50	53 855.72	54 069.64	54 284.28	54 499.63	54 715.69	54 932.47	55 149.97	55 368.19
84	55 587.13	55 806.80	56 027.20	56 248.32	56 470.17	56 692.76	56 916.08	57 140.13	57 364.92	57 590.45
85	57 816.73	58 043.74	58 271.51	58 500.02	58 729.27	58 959.28	59 190.05	59 421.57	59 653.84	59 886.87
86	60 120.67	60 355.23	60 590.55	60 826.64	61 063.50	61 301.12	61 539.52	61 778.70	62 018.65	62 259.38
87	62 500.89	62 743.18	62 986.26	63 230.12	63 474.78	63 720.22	63 966.45	64 213.48	64 461.31	64 700.93
88	64 959.35	65 209.58	65 460.61	65 712.45	65 965.09	66 218.55	66 472.82	66 727.90	66 983.80	67 240.52
89	67 498.06	67 756.42	68 015.60	68 275.62	68 536.46	68 798.13	69 060.64	69 323.98	69 588.15	69 853.17
90	70 119.03	70 385.73	70 653.28	70 921.67	71 190.91	71 461.01	71 731.96	72 003.76	72 276.42	72 549.95
91	72 824.33	73 099.58	73 375.70	73 652.68	73 930.54	74 209.27	74 488.87	74 769.35	75 050.71	75 332.95
92	75 616.07	75 900.08	76 184.98	76 470.77	76 757.44	77 045.02	77 333.49	77 622.86	77 913.13	78 204.30
93	78 496.38	78 789.36	79 083.26	79 378.06	79 673.78	79 970.42	80 267.97	80 566.45	80 865.85	81 166.17
94	81 467.42	81 769.60	82 072.71	82 376.75	82 681.73	82 987.65	83 294.51	83 602.31	83 911.06	84 220.75
95	84 531.40	84 842.99	85 155.54	85 469.05	85 783.51	86 098.94	86 415.33	86 732.68	87 051.00	87 370.29
96	87 690.56	88 011.80	88 334.01	88 657.20	88 981.38	89 306.54	89 632.68	89 959.82	90 287.94	90 617.06
97	90 947.17	91 278.28	91 610.39	91 943.50	92 277.62	92 612.74	92 948.87	93 286.02	93 624.18	93 963.35
98	94 303.54	94 644.76	94 986.99	95 330.26	95 674.55	96 019.87	96 366.23	96 713.62	97 062.05	97 411.51
99	97 762.02	98 113.58	98 466.18	98 819.83	99 174.54	99 530.30	99 887.11	100 244.99	100 603.93	100 963.93
100	101 324.99									

A Metastable state

TABLE X3.2 Saturation Vapor Pressure Over Supercooled Water (IPTS—68) Extrapolated Values See X3.2

Temperature °C	.0	0.1	0.2	0.3	0.4	0.5	0.6	0.7	0.8	0.9
	Pascal									
−50	6.445									
−49	7.217	7.136	7.056	6.977	6.899	6.821	6.745	6.669	6.593	6.519
−48	8.037	7.983	7.895	7.807	7.720	7.634	7.549	7.465	7.382	7.299
−47	9.020	8.921	8.823	8.726	8.630	8.534	8.440	8.347	8.255	8.163
−46	10.07	9.958	9.850	9.742	9.636	9.531	9.427	9.323	9.221	9.120
−45	11.22	11.10	10.98	10.87	10.75	10.63	10.52	10.40	10.29	10.18
−44	12.50	12.37	12.24	12.11	11.98	11.85	11.72	11.60	11.47	11.35
−43	13.91	13.76	13.62	13.47	13.33	13.19	13.05	12.91	12.77	12.64
−42	15.46	15.30	15.14	14.98	14.82	14.67	14.51	14.36	14.21	14.06
−41	17.17	16.99	16.82	16.64	16.47	16.30	16.13	15.96	15.79	15.63
−40	19.05	18.85	18.66	18.47	18.28	18.09	17.90	17.72	17.53	17.35
−39	21.11	20.89	20.68	20.47	20.26	20.05	19.85	19.65	19.44	19.24
−38	23.37	23.14	22.90	22.67	22.44	22.21	21.99	21.77	21.54	21.33
−37	25.85	25.59	25.34	25.08	24.83	24.58	24.34	24.09	23.85	23.61
−36	28.57	28.29	28.01	27.73	27.45	27.18	26.91	26.64	26.38	26.11
−35	31.54	31.23	30.93	30.62	30.32	30.02	29.73	29.43	29.14	28.86
−34	34.80	34.46	34.12	33.79	33.46	33.13	32.81	32.49	32.17	31.86
−33	38.35	37.98	37.61	37.25	36.89	36.53	36.18	35.83	35.48	35.14
−32	42.22	41.82	41.42	41.03	40.63	40.24	39.86	39.48	39.10	38.72
−31	46.45	46.01	45.58	45.15	44.72	44.29	43.87	43.45	43.04	42.63
−30	51.06	50.58	50.11	49.64	49.17	48.71	48.25	47.79	47.34	46.90
−29	56.08	55.56	55.04	54.53	54.02	53.52	53.02	52.52	52.03	51.54
−28	61.53	60.97	60.40	59.85	59.30	58.75	58.20	57.67	57.13	56.60
−27	67.47	66.85	66.24	65.63	65.03	64.44	63.85	63.26	62.68	62.10
−26	73.91	73.24	72.58	71.92	71.27	70.62	69.98	69.34	68.71	68.09
−25	80.90	80.18	79.46	78.74	78.04	77.33	76.64	75.95	75.26	74.58
−24	88.48	87.70	86.92	86.14	85.38	84.62	83.86	83.11	82.37	81.63
−23	96.70	95.85	95.00	94.17	93.34	92.51	91.69	90.88	90.08	89.28
−22	105.60	104.68	103.76	102.85	101.95	101.06	100.18	99.30	98.42	97.56
−21	115.22	114.23	113.24	112.26	111.28	110.32	109.36	108.41	107.46	106.53
−20	125.63	124.55	123.48	122.42	121.37	120.33	119.29	118.26	117.24	116.23
−19	136.88	135.71	134.56	133.41	132.27	131.14	130.02	128.91	127.81	126.72
−18	149.01	147.76	146.51	145.28	144.05	142.83	141.62	140.42	139.23	138.05
−17	162.11	160.76	159.41	158.08	156.75	155.44	154.13	152.84	151.56	150.28
−16	176.23	174.77	173.32	171.88	170.45	169.04	167.63	166.24	164.85	163.48
−15	191.44	189.87	188.31	186.76	185.22	183.69	182.18	180.67	179.18	177.70
−14	207.81	206.12	204.44	202.77	201.12	199.48	197.84	196.22	194.62	193.02
−13	225.43	232.61	221.80	220.01	218.23	216.46	214.71	212.96	211.24	209.52
−12	244.37	242.41	240.47	238.54	236.63	234.73	232.84	230.97	229.11	227.26
−11	264.72	262.62	260.53	258.46	256.40	254.36	252.34	250.32	248.32	246.34
−10	286.57	284.32	282.08	279.85	277.64	275.45	273.28	271.11	268.97	266.84
−9	310.02	307.60	305.19	302.81	300.44	298.09	295.75	293.43	291.13	288.84
−8	335.16	332.56	329.99	327.43	324.89	322.37	319.86	317.38	314.90	312.45
−7	362.10	359.32	356.56	353.82	351.10	348.40	345.71	343.04	340.40	337.77
−6	390.95	387.98	385.02	382.09	379.18	376.28	373.40	370.55	367.71	364.90
−5	421.84	418.66	415.49	412.35	409.23	406.14	403.06	400.00	396.96	393.95
−4	454.88	451.47	448.09	444.73	441.40	438.08	434.79	431.52	428.27	425.04
−3	490.19	486.56	482.94	479.35	475.78	472.24	468.72	465.23	461.75	458.30
−2	527.93	524.04	520.18	516.34	512.54	508.75	504.99	501.25	497.54	493.86
−1	568.22	564.07	559.95	555.85	551.78	547.74	543.73	539.74	535.78	531.84
−0	611.21	606.79	602.39	598.02	593.68	589.37	585.08	580.82	576.60	572.39

X4. SATURATION VAPOR PRESSURE OVER ICE

X4.1 The saturation vapor pressure of the pure phase over plane surface of pure ice for temperatures 0 to −100°C was obtained from Wexler's 1976 formulation (10). See Table X4.1. Other suitable saturation vapor pressure tables are given in the Smithsonian Meteorological Tables (7), International Meteorological Tables (8) and ASHRAE Handbook and Product Directory (9).

TABLE X4.1 Saturation vapor pressure over ice

Temperature °C	0.0	0.1	0.2	0.3	0.4	0.5	0.6	0.7	0.8	0.9
	Pascal									
−0	611.153	606.140	601.164	596.225	591.323	586.458	581.630	576.837	572.081	567.360
−1	562.675	558.025	553.411	548.830	544.285	539.774	535.297	530.853	526.444	522.067
−2	517.724	513.414	509.136	504.891	500.679	496.498	492.349	488.232	484.146	480.091
−3	476.068	472.075	468.112	464.180	460.278	456.406	452.564	448.751	444.968	441.213
−4	437.488	433.791	430.123	426.483	422.871	419.287	415.731	412.202	408.700	405.226
−5	401.779	398.358	394.964	391.597	388.256	384.940	381.651	378.387	375.149	371.936
−6	368.748	365.585	362.446	359.333	356.244	353.179	350.138	347.121	344.128	341.158
−7	338.212	335.289	332.389	329.512	326.658	323.826	321.017	318.230	315.465	312.722
−8	310.001	307.302	304.624	301.967	299.332	296.717	294.124	291.551	288.998	286.467
−9	283.955	281.464	278.992	276.540	274.108	271.696	269.303	266.929	264.575	262.239
−10	259.922	257.624	255.345	253.084	250.841	248.617	246.410	244.222	242.051	239.898
−11	237.762	235.644	233.543	231.459	229.393	227.343	225.310	223.293	221.293	219.309
−12	217.342	215.391	213.456	211.537	209.633	207.745	205.873	204.017	202.175	200.349
−13	198.538	196.742	194.961	193.194	191.442	189.705	187.982	186.274	184.579	182.899
−14	181.233	179.581	177.942	176.318	174.706	173.109	171.524	169.953	168.396	166.851
−15	165.319	163.800	162.294	160.801	159.320	157.852	156.396	154.952	153.521	152.101
−16	150.694	149.299	147.915	146.544	145.184	143.835	142.498	141.173	139.858	138.555
−17	137.263	135.982	134.713	133.453	132.205	130.968	129.741	128.524	127.318	126.123
−18	124.938	123.763	122.598	121.443	120.298	119.163	118.038	116.923	115.817	114.721
−19	113.634	112.557	111.489	110.431	109.381	108.341	107.310	106.288	105.275	104.271
−20	103.276	102.289	101.311	100.341	99.380 9	98.428 4	97.484 3	96.548 5	95.621 0	94.701 6
−21	93.790 4	92.887 2	91.992 0	91.104 7	90.225 3	89.353 7	88.489 8	87.633 6	86.785 0	85.943 9
−22	85.110 4	84.284 2	83.465 5	82.654 0	81.849 8	81.052 8	80.262 9	79.480 1	78.704 3	77.935 5
−23	77.173 5	76.418 4	75.670 1	74.928 6	74.193 7	73.465 5	72.743 8	72.028 6	71.319 9	70.617 6
−24	69.921 7	69.232 1	68.548 7	67.871 6	67.200 5	66.535 6	65.876 8	65.223 9	64.577 0	63.936 0
−25	63.300 8	62.671 5	62.047 9	61.430 0	60.817 8	60.211 2	59.610 1	59.014 6	58.424 5	57.839 9
−26	57.260 7	56.686 8	56.118 2	55.554 8	54.996 6	54.443 6	53.895 8	53.353 0	52.815 2	52.282 4
−27	51.754 6	51.231 7	50.713 6	50.200 3	49.691 9	49.188 2	48.689 2	48.194 8	47.705 1	47.219 9
−28	46.739 3	46.263 2	45.791 6	45.324 4	44.861 6	44.403 1	43.948 9	43.499 1	43.053 4	42.612 0
−29	42.174 8	41.741 7	41.312 6	40.887 7	40.466 7	40.049 8	39.636 8	39.227 8	38.822 6	38.421 3
−30	38.023 8	37.630 1	37.240 2	36.854 0	36.471 4	36.092 6	35.717 3	35.345 7	34.977 6	34.613 1
−31	34.252 1	33.894 5	33.540 4	33.189 7	32.842 3	32.498 3	32.157 7	31.820 3	31.486 2	31.155 4
−32	30.827 7	30.503 2	30.181 9	29.863 7	29.548 6	29.236 5	28.927 5	28.621 5	28.318 5	28.018 5
−33	27.721 4	27.427 2	27.135 8	26.847 4	26.561 7	26.278 9	25.998 8	25.721 5	25.446 9	25.175 1
−34	24.905 9	24.639 4	24.375 5	24.114 2	23.855 5	23.599 3	23.345 7	23.094 7	22.846 1	22.599 9
−35	22.356 3	22.115 0	21.876 2	21.639 7	21.405 6	21.173 9	20.944 4	20.717 3	20.492 4	20.269 8
−36	20.049 4	19.831 2	19.615 2	19.401 4	19.189 8	18.980 3	18.772 9	18.567 5	18.364 3	18.163 1
−37	17.964 0	17.766 9	17.571 7	17.378 6	17.187 4	16.998 2	16.810 8	16.625 4	16.441 9	16.260 3
−38	16.080 5	15.902 5	15.726 4	15.552 1	15.379 5	15.208 8	15.039 7	14.872 5	14.706 9	14.543 0
−39	14.380 9	14.220 4	14.061 5	13.904 3	13.748 8	13.594 8	13.442 4	13.291 6	13.142 4	12.994 7
−40	12.848 6	12.704 0	12.560 9	12.419 2	12.279 1	12.140 4	12.003 2	11.867 4	11.733 0	11.600 0
−41	11.468 5	11.338 3	11.209 5	11.082 0	10.955 9	10.831 1	10.707 6	10.585 4	10.464 5	10.344 9
−42	10.226 6	10.109 5	9.993 66	9.879 03	9.765 63	9.653 43	9.542 43	9.432 60	9.323 95	9.216 46
−43	9.110 11	9.004 90	8.900 82	8.797 85	8.695 98	8.595 21	8.495 52	8.396 90	8.299 34	8.202 83
−44	8.107 36	8.012 92	7.919 50	7.827 08	7.735 67	7.645 25	7.555 80	7.467 33	7.379 81	7.293 25
−45	7.207 63	7.122 94	7.039 17	6.956 31	6.874 36	6.793 30	6.713 13	6.633 84	6.555 42	6.477 85
−46	6.401 14	6.325 26	6.250 22	7.176 01	6.102 62	6.030 03	5.958 24	5.887 25	5.817 04	5.747 61
−47	5.678 94	5.611 04	5.543 89	5.477 49	5.411 82	5.346 88	5.282 67	5.219 17	5.156 38	5.094 29
−48	5.032 90	4.972 19	4.912 16	4.852 80	4.794 11	4.736 08	4.678 70	4.621 96	4.565 87	4.510 40
−49	4.455 56	4.401 34	4.347 73	4.294 73	4.242 33	4.190 52	4.139 30	4.088 66	4.038 60	3.989 10
−50	3.940 17	3.891 79	3.843 97	3.796 69	3.749 96	3.703 75	3.658 08	3.612 93	3.568 29	3.524 17

TABLE X4.1 *Continued*

Temperature °C	0.0	0.1	0.2	0.3	0.4	0.5	0.6	0.7	0.8	0.9
	Pascal									
−51	3.480 56	3.437 44	3.394 83	3.352 70	3.311 06	3.269 90	3.229 21	3.189 00	3.149 25	3.109 96
−52	3.071 13	3.032 75	2.994 81	2.957 31	2.920 25	3.883 62	2.847 42	2.811 65	2.776 28	2.741 34
−53	2.706 80	2.672 66	2.638 93	2.605 59	2.572 65	2.540 09	2.507 91	2.476 11	2.444 69	2.413 64
−54	2.382 96	2.352 63	2.322 67	2.293 06	2.263 81	2.234 90	2.206 33	2.178 10	2.150 21	2.122 65
−55	2.095 42	2.068 52	2.041 93	2.015 67	1.989 72	1.964 08	1.938 74	1.913 71	1.888 98	1.864 55
−56	1.840 42	1.816 57	1.793 01	1.769 74	1.746 74	1.724 03	1.701 59	1.679 42	1.657 52	1.635 89
−57	1.614 52	1.593 40	1.572 55	1.551 95	1.531 60	1.511 50	1.491 65	1.472 04	1.452 66	1.433 53
−58	1.414 63	1.395 96	1.377 52	1.359 31	1.341 33	1.323 56	1.306 02	1.288 69	1.271 57	1.254 67
−59	1.237 97	1.221 49	1.205 20	1.189 12	1.173 24	1.157 56	1.142 07	1.126 78	1.111 67	1.096 76
−60	1.082 03	1.067 49	1.053 12	1.038 94	1.024 94	1.011 11	0.997 462	0.983 980	0.970 668	0.957 524
	Millipascal									
−61	944.545	931.731	919.079	906.587	894.253	882.076	870.053	858.183	846.465	834.895
−62	823.473	812.196	801.064	790.074	779.225	768.514	757.941	747.504	737.201	727.030
−63	716.990	707.079	697.297	687.640	678.109	668.700	659.414	650.248	641.200	632.270
−64	623.457	614.758	606.172	597.698	589.335	581.081	572.935	564.895	556.961	549.131
−65	541.403	533.778	526.252	518.826	511.497	504.265	497.128	490.086	483.137	476.280
−66	469.514	462.838	456.250	449.750	443.337	437.009	430.765	424.605	418.527	412.530
−67	406.613	400.776	395.017	389.335	383.730	378.200	372.745	367.363	362.054	356.817
−68	351.650	346.553	341.525	336.566	331.674	326.848	322.088	317.393	312.761	308.193
−69	303.688	299.244	294.860	290.537	286.273	282.068	277.920	273.829	269.795	265.816
−70	261.892	258.023	254.206	250.443	246.732	243.072	239.463	235.904	232.394	228.934
−71	225.521	222.157	218.389	215.567	212.342	209.161	206.025	202.933	199.885	196.879
−72	193.916	190.994	188.114	185.274	182.475	179.715	176.994	174.311	171.667	169.060
−73	166.491	163.958	161.461	158.999	156.573	154.182	151.824	149.501	147.210	144.953
−74	142.728	140.535	138.373	136.243	134.143	132.074	130.035	128.025	126.044	124.092
−75	122.168	120.273	118.404	116.563	114.749	112.961	111.200	109.464	107.753	106.068
−76	104.407	102.771	101.159	99.570 5	98.005 3	96.463 1	94.943 7	93.446 8	91.972 0	90.519 0
−77	89.087 5	87.677 2	86.287 9	84.919 2	83.570 9	82.242 7	80.934 2	79.645 3	78.375 7	77.125 0
−78	75.893 0	74.679 5	73.484 2	72.306 9	71.147 2	70.005 0	68.880 0	67.772 0	66.680 7	65.605 9
−79	64.547 3	63.504 7	62.478 0	61.466 8	60.471 0	59.490 4	58.524 6	57.573 6	56.637 1	55.714 9
−80	54.806 7	53.912 5	53.032 0	52.164 9	51.311 2	50.470 6	49.642 9	48.828 0	48.025 6	47.235 6
−81	46.457 8	45.692 1	44.938 1	44.195 6	43.465 2	42.745 8	42.037 0	41.340 5	40.654 1	39.978 5
−82	39.313 5	38.658 8	38.014 4	37.380 0	36.755 6	36.141 0	35.536 1	34.940 7	34.354 6	33.777 8
−83	33.210 1	32.651 4	32.101 4	31.560 2	31.027 6	30.503 4	29.987 5	29.479 9	28.980 3	28.488 6
−84	28.004 9	27.528 8	27.060 3	26.599 4	26.145 8	25.699 5	25.260 3	24.828 2	24.403 1	23.984 8
−85	23.573 2	23.168 3	22.769 9	22.378 0	21.992 4	21.613 1	21.239 9	20.872 8	20.511 6	20.156 3
−86	19.806 8	19.463 0	19.124 9	18.792 2	18.465 0	18.143 2	17.826 6	17.515 2	17.209 0	16.907 7
−87	16.611 5	16.320 1	16.033 6	15.751 7	15.474 6	15.202 0	14.933 9	14.670 3	14.411 1	14.156 2
−88	13.905 5	13.659 0	13.416 6	13.178 3	12.944 0	12.713 5	12.487 0	12.264 2	12.045 2	11.829 9
−89	11.618 2	11.410 0	11.205 4	11.004 2	10.806 5	10.612 0	10.420 9	10.233 0	10.048 3	9.866 80
−90	9.688 33	9.512 90	9.340 47	9.170 98	9.004 39	8.840 64	8.679 71	8.521 53	8.366 07	8.213 29
−91	8.063 13	7.915 56	7.770 53	7.628 01	7.487 95	7.350 31	7.215 06	7.082 16	6.951 56	6.823 23
−92	6.697 14	6.573 24	6.451 50	6.331 89	6.214 37	6.098 90	5.985 46	5.874 01	5.764 51	5.656 94
−93	5.551 26	5.447 45	5.345 46	5.245 28	5.146 86	5.050 19	4.955 23	4.861 95	4.770 33	4.680 34
−94	4.591 95	4.505 13	4.419 86	4.336 12	4.253 87	4.173 10	4.093 77	4.015 86	3.939 35	3.864 22
−95	3.790 44	3.717 99	3.646 85	3.576 99	3.508 39	3.441 03	3.374 90	3.309 97	3.246 21	3.183 61
−96	3.122 16	3.061 82	3.002 58	2.944 43	2.887 34	2.831 29	2.776 27	2.722 26	2.669 24	2.617 20
−97	2.566 12	2.515 97	2.466 76	2.418 45	2.371 03	2.324 50	2.278 82	2.234 00	2.190 00	2.146 83
−98	2.104 45	2.062 87	2.022 07	1.982 02	1.942 73	1.904 17	1.866 34	1.829 21	1.792 79	1.757 04
−99	1.721 98	1.687 57	1.653 81	1.620 69	1.588 20	1.556 32	1.525 05	1.494 37	1.464 28	1.434 76
−100	1.405 80									

REFERENCES

(1) *Guide to Meteorological Instruments and Observing Practices*, World Meteorological Organization, WMO No. 8, TP3, Fourth Edition, 1971, Secretariat of WMO, Geneva, Switzerland.

(2) Hasegawa, S., and Little, J. W., "The NBS Two-Pressure Humidity Generator, Mark 2" *Journal of Research National Bureau of Standards* (U.S.), 81A, No. 1, Jan. to Feb. 1977, pp. 81–88.

(3) Harrison, L. P., "Fundamental Concepts and Definitions Relating to Humidity," *Humidity and Moisture*, Vol. 3, Arnold Wexler (ed.), Reinhold Publishing Corp., NY, 1964.

(4) Greenspan, L., "Functional Equations for the Enhancement Factors for CO_2—Free Moist Air," *Journal of Research National Bureau of Standards* (U.S.), 80A, Jan. to Feb. 1976, pp. 41–44.

(5) Hyland, R. W., "A Correlation for the Second Interaction Virial Coefficients and Enhancement Factors for Moisture Air," *Journal of Research National Bureau of Standards* (U.S.), 79A, July to Aug. 1975, 551–560.

(6) Wexler, A., "Vapor Pressure Formulation for Water in Range 0 to 100°C. A Revision," *Journal of Research National Bureau of Standards* (U.S.), 80A, Sept. to Dec. 1976, pp. 775–785.

(7) *Smithsonian Meteorological Tables*, Smithsonian Press, Washington, D.C., 1968.

(8) *International Meteorological Tables*, World Meteorological Organization, WMO No. 188, TP94, 1966, Secretariat of WMO, Geneva, Switzerland.

(9) *American Society of Heating, Refrigeration and Air Conditioning Engineers Handbook and Product Directory*, 1977 Fundamentals, American Society for Heating and Refrigeration, Atlanta, GA 30329.

(10) Wexler, A., "Vapor Pressure Formulation for Ice," *Journal of Research National Bureau of Standards* (U.S.), 81A, Jan. to Feb. 1977, pp. 5–20.

Standard Practices for
Measuring Surface Wind and Temperature by Acoustic Means[1]

This standard is issued under the fixed designation D 5527; the number immediately following the designation indicates the year of original adoption or, in the case of revision, the year of last revision. A number in parentheses indicates the year of last reapproval. A superscript epsilon (ε) indicates an editorial change since the last revision or reapproval.

1. Scope

1.1 This practice covers procedures for measuring one-, two-, or three-dimensional vector wind components and sonic temperature by means of commercially available sonic anemometer/thermometers that employ the inverse time measurement technique. This practice applies to the measurement of wind velocity components over horizontal terrain using instruments mounted on stationary towers. This practice also applies to speed of sound measurements that are converted to sonic temperatures but does not apply to the measurement of temperature by the use of ancillary temperature devices.

1.2 The values stated in SI units are to be regarded as the standard.

1.3 *This standard does not purport to address all of the safety concerns, if any, associated with its use. It is the responsibility of the user of this standard to establish appropriate safety and health practices and determine the applicability of regulatory limitations prior to use.*

2. Referenced Documents

2.1 *ASTM Standards:*
D 3631 Test Methods for Measuring Surface Atmospheric Pressure[2]
D 4230 Test Method of Measuring Humidity with Cooled-Surface Condensation (Dew Point) Hygrometer[2]
E 337 Test Method for Measuring Humidity with a Psychrometer (the Measurement of Wet- and Dry-Bulb Temperatures)[2]
E 380 Practice for Use of the International System of Units (SI) (The Modernized Metric System)[3]

3. Terminology

3.1 *Definitions:*

3.1.1 *acceptance angle (±α, deg)*—the angular distance, centered on the array axis of symmetry, over which the following conditions are met: (*a*) wind components are unambiguously defined, and (*b*) flow across the transducers is unobstructed or remains within the angular range for which transducer shadow corrections are defined.

3.1.2 *acoustic pathlength (d, (m))*—the physical distance between transducer transmitter-receiver pairs.

3.1.3 *sampling period(s)*—the record length or time interval over which data collection occurs.

3.1.4 *sampling rate (Hz)*—the rate at which data collection occurs, usually presented in samples per second or Hertz.

3.1.5 *sonic anemometer/thermometer*—an instrument consisting of a transducer array containing paired sets of acoustic transmitters and receivers, a system clock, and microprocessor circuitry to measure intervals of time between transmission and reception of sound pulses.

DISCUSSION—The fundamental measurement unit is transit time. With transit time and a known acoustic pathlength, velocity or speed of sound, or both, can be calculated. Instrument output is a series of quasi-instantaneous velocity component readings along each axis or speed of sound, or both. The speed of sound and velocity components may be used to compute sonic temperature (T_s), to describe the mean wind field, or to compute fluxes, variances, and turbulence intensities.

3.1.6 *sonic temperature (T_s), (K))*—an equivalent temperature that accounts for the effects of temperature and moisture on acoustic wavefront propagation through the atmosphere.

DISCUSSION—Sonic temperature is related to the velocity of sound c, absolute temperature T, vapor pressure of water e, and absolute pressure P by (1).[4]

$$c^2 = 403T (1 + 0.32e/P) = 403T_s \qquad (1)$$

(Guidance concerning measurement of P and e are contained in Test Methods D 3631, D 4230, and E 337.)

3.1.7 *transducer shadow correction*—the ratio of the *true* along-axis velocity, as measured in a wind tunnel or by another accepted method, to the instrument along-axis wind measurement.

DISCUSSION—This ratio is used to compensate for effects of along-axis flow shadowing by the transducers and their supporting structure.

3.1.8 *transit time (t, (s))*—the time required for an acoustic wavefront to travel from the transducer of origin to the receiving transducer.

3.2 *Symbols:*

B	(dimensionless)	squared sums of sines and cosines of wind direction angle used to calculate wind direction standard deviation
c	(m/s)	speed of sound
d	(m)	acoustic pathlength
e	(Pa)	vapor pressure of water
f	(dimensionless)	compressibility factor
P	(Pa)	ambient pressure
t	(s)	transit time
T	(K)	absolute temperature, K
T_s	(K)	sonic temperature, K

---

[1] This practice is under the jurisdiction of ASTM Committee D-22 on Sampling and Analysis of Atmospheres and is the direct responsibility of Subcommittee D22.11 on Meteorology.
Current edition approved March 15, 1994. Published May 1994.
[2] *Annual Book of ASTM Standards*, Vol 11.03.
[3] *Annual Book of ASTM Standards*, Vol 14.02.

[4] The boldface numbers in parentheses refer to the list of references at the end of this practice.

γ	(dimensionless)	specific heat ratio (c_p/c_v)
M	(g/mol)	molar mass of air
n	(dimensionless)	sample size
R^*	(J/mol·K)	the universal gas constant
u	(m/s)	velocity component along the determined mean wind direction
u_s	(m/s)	velocity component along the array u axis
v	(m/s)	velocity component crosswind to the determined mean wind direction
v_s	(m/s)	velocity component along the array v axis
w	(m/s)	vertical velocity
WS	(m/s)	wind speed computed from measured velocity components
θ	(deg)	determined mean wind direction with respect to true north
θr	(deg)	wind direction measured in degrees clockwise from the sonic anemometer $+v_s$ axis to the along-wind u axis
α	(deg)	acceptance angle
ϕ	(deg)	orientation of the sonic anemometer axis with respect to the true north
σ_θ	(deg)	standard deviation of wind azimuth angle

3.3 *Units*—Units of measurement used should be in accordance with Practice E 380.[5]

4. Summary of Practice

4.1 A calibrated sonic anemometer/thermometer is installed, leveled, and oriented into the expected wind direction to ensure that the measured along-axis velocity components fall within the instrument's acceptance angle.

4.2 The wind components measured over a user-defined sampling period are averaged and subjected to a software rotation into the mean wind. This rotation maximizes the mean along-axis wind component and reduces the mean cross-component v to zero.

4.3 Mean horizontal wind speed and direction are computed from the rotated wind components.

4.4 For the sonic thermometer, the speed of sound solution is obtained and converted to a sonic temperature.

4.5 Variances, covariances, and turbulence intensities are computed.

5. Significance and Use

5.1 Sonic anemometer/thermometers are used to measure turbulent components of the atmosphere except for confined areas and very close to the ground. This practice applies to the use of these instruments for field measurement of the wind, sonic temperature, and atmospheric turbulence components. The quasi-instantaneous velocity component measurements are averaged over user-selected sampling times to define mean along-axis wind components, mean wind speed and direction, and the variances or covariances, or both, of individual components or component combinations. Covariances are used for eddy correlation studies and for computation of boundary layer heat and momentum fluxes. The sonic anemometer/thermometer provides the data required to characterize the state of the turbulent atmospheric boundary layer.

5.2 The sonic anemometer/thermometer array shall have a sufficiently high structural rigidity and a sufficiently low coefficient of thermal expansion to maintain an internal alignment to within ±0.1°. System electronics must remain stable over its operating temperature range; the time counter oscillator instability must not exceed 0.01 % of frequency.

Consult with the manufacturer for an internal alignment verification procedure.

5.3 The calculations and transformations provided in this practice apply to orthogonal arrays. References are also provided for common types of non-orthogonal arrays.

6. Interferences

6.1 Mount the sonic anemometer probe for an acceptance angle into the mean wind. Wind velocity components from angles outside the acceptance angle may be subject to uncompensated flow blockage effects from the transducers and supporting structure, or may not be unambiguously defined. Obtain acceptance angle information from the manufacturer.

6.2 Mount the sonic array at a distance that exceeds the acoustic pathlength by a factor of at least 2π from any reflecting surface.

6.3 To obtain representative samples of the mean wind, the sonic array must be exposed at a representative site. Sonic anemometer/thermometers are typically mounted over level, open terrain at a height of 10 m above the ground. Consider surface roughness and obstacles that might cause flow blockage or biases in the site selection process.

6.4 Carefully measure and verify array tilt angle and alignment. The vertical component of the wind is usually much smaller than the horizontal components. Therefore, the vertical wind component is highly susceptible to cross-component contamination from tilt angles not aligned to the chosen coordinate system. A typical coordinate system may include establishing a level with reference to either the earth or to local terrain slope. Momentum flux computations are particularly susceptible to off-axis contamination (2). Calculations and transformations (Section 9) for sonic anemometer data are based on the assumption that the mean vertical velocity (w) is not significantly different from zero. Arrays mounted above a sloping surface may require tilt angle adjustments. Also, avoid mounting the array close (within 2 m) to the ground surface where velocity gradients are large and w may be nonzero.

6.5 The transducers are tiny microphones and are, therefore, sensitive to extraneous noise sources, especially ultrasonic sources at the anemometer's operating frequency. Mount the transducer array in an environment free of extraneous noise sources.

6.6 Sonic anemometer/thermometer transducer arrays contribute a certain degree of blockage to flow. Consequently, the manufacturer should include transducer shadow corrections as part of the instrument's data processing algorithms or define, an acceptance angle beyond which valid measurements cannot be made, or both.

6.7 Ensure that the instrument is operated within its velocity calibration range and at temperatures where thermal sensitivity effects are not observed.

6.8 This practice does not address applications where moisture is likely to accumulate on the transducers. Moisture accumulation may interrupt transmission of the acoustic signal, or possibly damage unsealed transducers. Consult the manufacturer concerning operation in adverse environments.

---

[5] Excerpts from Practice E 380 are included in Vol 11.03.

7. Sampling

7.1 The basic sampling rate of a sonic anemometer is on the order of several hundred hertz. Transit times are averaged within the instrument's software to produce basic measurements at a rate of 10 to 20 Hz, which may be user-selectable. This sampling is done to improve instrument measurement precision and to suppress high frequency noise and aliasing effects. The 10 to 20-Hz sample output in a serial digital data stream or through a digital to analog converter is the basic unit of measurement for a sonic anemometer.

7.2 Select a sampling period of sufficient duration to obtain statistically stable measurements of the phenomena of interest. Sampling periods of at least 10 min duration usually generate sufficient data to describe the turbulent state of the atmosphere during steady wind conditions. Sampling periods in excess of 1 h may contain undesired trends in wind direction.

8. Procedure

8.1 Perform system calibration in a zero wind chamber (refer to the manufacturer's instructions).

8.2 Mount the instrument array on a solid, vibration-free platform free of interferences.

8.3 Select an orientation into the mean flow within the instrument's acceptance angle. Record the orientation angle with a resolution of 1°. Use a leveling device to position the probe to within ±0.1° of the vertical axis of the chosen coordinate system.

NOTE 1: Caution—Wind measurements using a sonic anemometer should only be made within the acceptance angle.

8.4 Install cabling to the recording device, and keep cabling isolated from other electronics noise sources or power cables to minimize induction or crosstalk.

8.5 As a system check, collect data for several sequential sampling periods (of at least 10-min duration over a period of at least 1 h) during representative operating conditions. Examine data samples for extraneous spikes, noise, alignment faults, or other malfunctions. Construct summary statistics for each sampling period to include means, variances, and covariances; examine these statistics for reasonableness. Compute 1-h spectra and examine for spikes or aliasing affecting the −5/3 spectral slope in the inertial subrange.

NOTE 2—Calculations and transformations presented in this practice are based on the assumption of a zero mean vertical velocity component. Deviation of the mean vertical velocity component from zero should not exceed the desired measurement precision. Alignment or data, reduction software modifications not addressed in this practice may be needed for locations where w is nonzero.

8.6 Recalibrate and check instrument alignment at least once a week, whenever the instrument is subjected to a significant change in weather conditions, or when transducers or electronics components are changed or adjusted.

8.7 Check for bias, especially in w, using a data set collected over an extended time period. The array support structure, topography, and changes in ambient temperature may produce biases in vertical velocity w. Procedures described in (3) are recommended for bias compensation.

NOTE 3: Caution—Uncompensated flow distortion due to the acoustic array and supporting structure is likely for vertical wind angles in excess of ±15°.

9. Calculations and Transformations

9.1 Each sonic anemometer provides wind component measurements with respect to a coordinate system defined by its array axis alignment. Each array design requires specific calculations and transformations to convert along-axis measurements to the desired wind component data. The calculations and transformations are applicable to orthogonal arrays. References (4), (5), and (6) provide information on common non-orthogonal arrays. Obtain specific calculations and transformation equations from the manufacturer.

9.2 Figure 1 illustrates a coordinate system applicable to orthogonal array sonic anemometers. The usual wind component sign convention is as follows:

9.2.1 An along-axis wind component entering the array from the front will have a positive sign ($+u_{si}$).

9.2.2 A cross-axis wind component entering the array from the left will have a positive sign ($+v_{si}$).

9.2.3 A vertical wind component entering the array from the bottom will have a positive sign ($+w_{si}$).

9.2.4 The subscript s refers to a wind component measured with respect to the sonic array axes, and the subscript i refers to the ith individual measurement. Array orientation (ϕ) is measured clockwise from true north, as illustrated in Fig. 1.

9.3 Sonic anemometers employing the inverse time ($1/t$) measurement technique obtain velocity by subtracting the inverse transit times of acoustic pulses traveling in opposite directions along an acoustic path. A quasi-instantaneous along-axis velocity component is (Ref (5)) calculated as follows:

$$u_{si} = \frac{d}{2}\left[\frac{1}{t_1} - \frac{1}{t_2}\right] \qquad (2)$$

where d is the acoustic pathlength and t_1 and t_2 are the along-axis acoustic pulse transit times. Similar equations provide cross-axis and vertical-axis velocity components.

9.4 The data of interest for sonic anemometer wind measurement will often be the mean wind speed and direction, or the individual components that are used to calculate variances and covariances, or both. A coordinate rotation is required to obtain these data from the measured u_{si} and v_{si}. A three-dimensional coordinate notation would also include w_{si}.

9.5 *Mean Wind Speed (WS)*—Mean wind speeds of interest may be the vector wind speed required for trajectory calculations, or the scalar wind speed required for dispersion modeling. The horizontal vector mean wind speed is defined as the square root of the sum of the squares of mean along-axis and cross-axis horizontal velocity components. That is, for a user-defined time interval,

$$\overline{WS}\text{ (vector)} = [(\bar{u}_s)^2 + (\bar{v}_s)^2]^{0.5} \qquad (3)$$

where $\bar{u}_s$ and $\bar{v}_s$ are the mean along- and cross-axis wind components defined by:

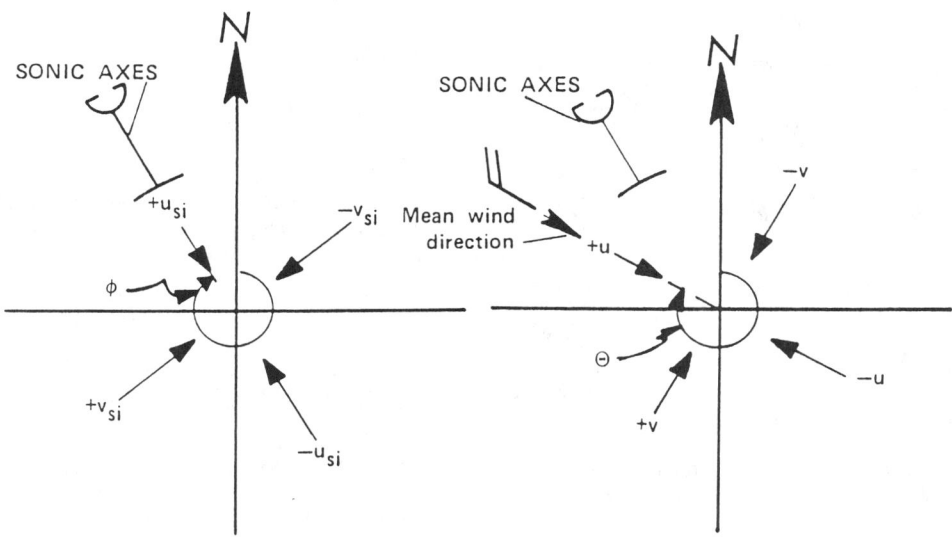

NOTE—This sonic anemometer array coordinate system is oriented with respect to true north.

FIG. 1   Sonic Anemometer Array Coordinate System

$$\bar{u}_s = \frac{1}{n}\left(\sum_{i=1}^{n} u_{si}\right) \qquad (4)$$

$$\bar{v}_s = \frac{1}{n}\left(\sum_{i=1}^{n} v_{si}\right) \qquad (5)$$

Sample size is represented by n. The scalar mean horizontal wind speed is the square root of the sum of the squares of the individual horizontal velocity components divided by sample size.

$$\overline{WS} \text{ (scalar)} = \frac{1}{n}\left(\sum_{i=1}^{n} [u^2_{si} + v^2_{si}]^{0.5}\right) \qquad (6)$$

9.6 *Mean Wind Direction*—A FORTRAN two-argument arc tangent function ATAN2D is used to define a rotated mean wind direction θ_r measured in degrees clockwise from the $+v_s$ array axis to the along wind (u) axis as

$$\bar{\theta}_r = \text{ATAN2D}\,(\bar{u}_s/\bar{v}_s) \qquad (7)$$

The mean wind direction $\bar{\theta}$, defined with respect to true north, is obtained by adding θ_r to the sonic anemometer axis orientation (ϕ) minus 90°.

$$\bar{\theta} = \bar{\theta}_r + -90° \qquad (8)$$

9.7 If wind azimuth angles are normally distributed, the standard deviation of the wind azimuth angle (σ_θ) can be calculated in a computationally efficient manner using the unit vector method (7).

$$\sigma_\theta = \arcsin\,[(1 - B^2)^{0.5}] \qquad (9)$$

where B^2 is obtained from sines and cosines of individual wind angles.

$$B^2 = \left(\frac{1}{n}\sum_{i=1}^{n} sin\,\theta_{si}\right)^2 + \left(\frac{1}{n}\sum_{i=1}^{n} cos\,\theta_{si}\right)^2 \qquad (10)$$

To achieve a representative sample size while minimizing the influences of long-term wind-direction trends on σ_θ, at least 10-min averaged σ_θ calculations are recommended (8).

9.8 The mean along-wind and cross-wind components are defined in terms of $\bar{\theta}_r$ as:

$$\bar{u} = \bar{u}_s \sin \bar{\theta}_r + \bar{v}_s \cos \bar{\theta}_r \qquad (11)$$

$$\bar{v} = \bar{u}_s \cos \bar{\theta}_r + \bar{v}_s \sin \bar{\theta}_r = 0 \qquad (12)$$

9.9 Sonic anemometer/thermometers employing the inverse time measurement technique obtain a speed of sound solution (usually on the vertical axis of an orthogonal array) using the sum of the inverse transit times of acoustic pulses traveling in opposite directions along the acoustic path. A solution for speed of sound obtained from the vertical axis is

$$c = \left[\frac{d^2}{4}\left(\frac{1}{t_1} + \frac{1}{t_2}\right)^2 + u^2 + v^2\right]^{0.5} \qquad (13)$$

A sonic temperature (T_s) solution is obtained from the speed of sound equation

$$T_s = \frac{Mc^2}{\gamma fR^*} \qquad (14)$$

where M is the molar mass of the air, γ is the specific heat ratio, f is the compressibility factor, and R^* is the universal gas constant. M, γ, and f are slowly varying functions of temperature and humidity.

9.10 Variances and covariances for orthogonal arrays can be computed using θ_r, T_s, and the unrotated u_s and v_s. Commonly used variances (covariances) are given by the mean of the squares (mean of the products) minus the square of the individual means (product of the means), as defined in 9.1.10.1 to 9.10.6. Note that products of means containing $\bar{v}$ are zero.

9.10.1 *Along-wind Velocity Variance:*

$$\overline{u'u'} = (\overline{uu}) - (\overline{u})(\overline{u}) = (\overline{u_su_s})\sin^2\bar{\theta}_r$$
$$+ 2(\overline{u_sv_s})\sin \bar{\theta}_r \cos \bar{\theta}_r + \overline{v_sv_s}\cos^2\bar{\theta}_r - (\overline{u_s})(\overline{u_s})\sin^2\bar{\theta}_r \qquad (15)$$
$$- 2(\overline{u_s})(\overline{v_s})\sin \bar{\theta}_r \cos \bar{\theta}_r - (\overline{v_s})(\overline{v_s})\cos^2\bar{\theta}_r$$

9.10.2 *Cross-wind Velocity Variance:*

$$\overline{v'v'} = (\overline{vv}) = (\overline{v_sv_s})\sin^2\bar{\theta}_r - 2(\overline{u_sv_s})\sin \bar{\theta}_r \cos \bar{\theta}_r + (\overline{u_su_s})\cos^2\bar{\theta}_r \qquad (16)$$

9.10.3 *Vertical Velocity Variance:*

$$\overline{w'w'} = (\overline{ww}) - (\overline{w})(\overline{w}) \tag{17}$$

9.10.4 Covariance of Along-wind and Vertical Velocities (Stress):

$$\overline{u'w'} = (\overline{uw}) - (\overline{u})(\overline{w}) = (\overline{u_sw})\sin\overline{\theta}_r + (\overline{v_sw})\cos\overline{\theta}_r$$
$$- (\overline{u_s})(\overline{w})\sin\overline{\theta}_r - (\overline{v_s})(\overline{w})\cos\overline{\theta}_r \tag{18}$$

9.10.5 Covariance of Sonic Temperature and Vertical Velocity:

$$\overline{w'T_s} = (\overline{wT'_s}) - (\overline{w})(\overline{T_s}) \tag{19}$$

9.10.6 Covariance of Along-wind and Cross-wind Velocities:

$$\overline{u'v'} = (\overline{uv}) = (\overline{u_sv_s} - \overline{v_sv_s})\sin\overline{\theta}_r\cos\overline{\theta}_r + \overline{u_sv_s}\cos^2\overline{\theta}_r \tag{20}$$

10. Keywords

10.1 acceptance angle; scalar wind; sonic anemometer; sonic temperature; sonic thermometer; speed of sound; vector wind; velocity variance

REFERENCES

(1) Kaimal, J. C., and Gaynor, J. E., "Another Look at Sonic Thermometry," *Boundary Layer Meteorology*, Vol 56, 1991, pp. 401–410.

(2) Kaimal, J. C., and Haugen, D. A., "Some Errors in the Measurement of Reynolds Stress," *Journal of Applied Meteorology*, Vol 8, 1969, pp. 460–462.

(3) Skibin, D., Kaimal, J. C., and Gaynor, J. E., "Measurement Errors in Vertical Wind Velocity at the Boulder Atmospheric Observatory," *Journal of Atmospheric and Oceanic Technology*, Vol 2, 1985, pp. 598–604.

(4) Coppin, P. A., and Taylor, K. J., "A Three Component Sonic Anemometer/Thermometer System for General Micrometeorological Research," *Boundary Layer Meteorology*, Vol 27, 1983, pp. 27–42.

(5) Hanafusa, T., Fujitani, T., Kobori, Y., and Mitsuta, Y., "A New Type of Sonic Anemometer-Thermometer for Field Operation," *Papers in Meteorology Geophysics*, Vol 33, 1982, pp. 1–19.

(6) Zhang, S. F., Wyngaard, J. C., Businger, J. A., and Oncley, S. P., "Response Characteristics of the U.W. Sonic Anemometer," *Journal of Atmospheric and Oceanic Technology*," Vol 3, 1986, pp. 315–323.

(7) Haugen, D. A., "A Simplified Method for Automatic Computation of Turbulent Wind Direction Statistics," *Journal of Applied Meteorology*, Vol 2, 1963, pp. 306–308.

(8) EPA, "On-Site Meteorological Program Guidance for Regulatory Modeling Applications," EPA-450/4-87-013, 1987, Office of Air Quality Planning and Standards. Research Triangle Park, NC 27711.

Standard Practice for
Characterizing Surface Wind Using a Wind Vane and Rotating Anemometer[1]

This standard is issued under the fixed designation D 5741; the number immediately following the designation indicates the year of original adoption or, in the case of revision, the year of last revision. A number in parentheses indicates the year of last reapproval. A superscript epsilon (ε) indicates an editorial change since the last revision or reapproval.

1. Scope

1.1 This practice covers a method for characterizing surface wind speed, wind direction, peak one-minute speeds, peak three-second and peak one-minute speeds, and standard deviations of fluctuation about the means of speed and direction.

1.2 This practice may be used with other kinds of sensors if the response characteristics of the sensors, including their signal conditioners, are equivalent or faster and the measurement uncertainty of the system is equivalent or better than those specified below.

1.3 The characterization prescribed in this practice will provide information on wind acceptable for a wide variety of applications.

NOTE 1—This practice builds on a consensus reached by the attendees at a workshop sponsored by the Office of the Federal Coordinator for Meteorological Services and Supporting Research in Rockville, MD on Oct. 29–30, 1992.

1.4 *This standard does not purport to address all of the safety concerns, if any, associated with its use. It is the responsibility of the user of this standard to establish appropriate safety and health practices and determine the applicability of regulatory limitations prior to use.*

2. Referenced Documents

2.1 *ASTM Standards:*
D 1356 Terminology Relating to Sampling and Analysis of Atmospheres[2]
D 5096 Test Method for Determining the Performance of a Cup Anemometer or Propeller Anemometer[2]
D 5366 Test Method for Determining the Dynamic Performance of a Wind Vane[2]

3. Terminology

3.1 *Definitions of Terms Specific to This Standard:*

3.1.1 *aerodynamic roughness length* (z_0, m)—a characteristic length representing the height above the surface where extrapolation of wind speed measurements, below the limit of profile validity, would predict the wind speed would become zero (1).[3] It can be estimated for direction sectors from a landscape description.

3.1.2 *damped natural wavelength* (λ_d, m)—a characteristic of a wind vane empirically related to the delay distance and the damping ratio. See Test Method D 5366 for test methods to determine the delay distance and equations to estimate the damped natural wavelength.

3.1.3 *damping ratio* (η, dimensionless)—the ratio of the actual damping, related to the inertial-driven overshoot of wind vanes to direction changes, to the critical damping, the fastest response where no overshoot occurs. See Test Method D 5366 for test methods and equations to determine the damping ratio of a wind vane.

3.1.4 *distance constant* (L, m)—the distance the air flows past a rotating anemometer during the time it takes the cup wheel or propeller to reach $(1 - 1/e)$ or 63 % of the equilibrium speed after a step change in wind speed. See Test Method D 5096.

3.1.5 *maximum operating speed* (u_m, m/s)—*as related to anemometer*, the highest speed as which the sensor will survive the force of the wind and perform within the accuracy specification.

3.1.6 *maximum operating speed* (u_m, m/s)—*as related to wind vane*, the highest speed at which the sensor will survive the force of the wind and perform within the accuracy specification.

3.1.7 *standard deviation of wind direction* (σ_θ, degrees)—the unbiased estimate of the standard deviation of wind direction samples about the mean horizontal wind direction. The circular scale of wind direction with a discontinuity at north may bias the calculation when the direction oscillates about north. Estimates of the standard deviation such as suggested by (2) are acceptable.

3.1.8 *standard deviation of wind speed* (σ_u, m/s)—the estimate of the standard deviation of wind speed samples about the mean wind speed.

3.1.9 *starting threshold* (u_0, m/s)—*as related to anemometer*, the lowest speed at which the sensor begins to turn and continues to turn and produces a measurable signal when mounted in its normal position (see Test Method D 5096).

3.1.10 *starting threshold* (u_0, m/s)—*as related to system*, the indicated wind speed when the anemometer is at rest.

3.1.11 *starting threshold* (u_0, m/s)—*as related to wind vane*, the lowest speed at which the vane can be observed or measured moving from a 10° offset position in a wind tunnel (see Test Method D 5366).

3.1.12 *wind direction* (θ, degrees)—the direction, referenced to true north, from which air flows past the sensor location if the sensor or other obstructions were absent. The wind direction distribution is characterized over each 10-min

---

[1] This practice is under the jurisdiction of ASTM Committee D-22 on Sampling and Analysis of Atmospheres and is the direct responsibility of Subcommittee D22.11 on Meteorology.
Current edition approved May 10, 1996. Published July 1996.
[2] *Annual Book of ASTM Standards*, Vol 11.03.
[3] The boldface numbers in parentheses refers to the list of references at the end of this standard.

period with a scalar (non-speed weighted) mean, standard deviation, and the direction of the peak 1-min average speed. The circular direction range, with its discontinuity at north, requires special attention in the averaging process. A unit vector method is an acceptable solution to this problem.

3.1.12.1 *Discussion*—Wind vane direction systems provide outputs when the wind speed is below the starting threshold for the vane. For this practice, report the calculated values (see 4.3 or 4.4) when more than 25 % of the possible samples are above the wind vane threshold and the standard deviation of the acceptable samples, σ_θ, is 30° or less, otherwise report light and variable code, 000.

3.1.13 *wind speed* (u, m/s)—the speed with which air flows past the sensor location if the sensor or other obstructions were absent. The wind speed distribution is characterized over each 10-min period with a scalar mean, standard deviation, peak 3-s average, and peak 1-min average.

3.2 For definitions of additional terms used in this practice, refer to Terminology D 1356.

4. Summary of Practice

4.1 *Siting of the Wind Sensors:*

4.1.1 The wind sensor location will be identified by an unambiguous label which will include either the longitude and latitude with a resolution of 1 s of arc (about 30 m or less) or a station number which will lead to that information in the station description file. When redundant sensors or microscale network stations (for example, airport runway sensors) are available, they will have individual labels which unambiguously identify the data they produce.

4.1.2 The anemometer and wind vane shall be located at a 10-m height above level or gently sloping terrain with an open fetch of at least 150 m in all directions, with the largest fetch possible in the prevailing wind direction. Compromise is frequently recognized and acceptable for some sites. Obstacles in the vicinity should be at least ten times their own height distant from the wind sensors.

4.1.3 The wind sensors shall preferably be located on top of a solitary mast. If side mounting is necessary, the boom length should be at least three times the mast width. In the undesirable case that locally no open terrain is available and the measurement is to be made above some building, then the wind sensor height above the roof top should be at least 1.5 times the lesser of the maximum building height and the maximum horizontal dimension of the major roof surface.

In this case, the station description file shall indicate the height above ground level (AGL) of the highest part of the building, the height of the wind sensors above ground, AGL, and the height of the wind sensors above roof level. Site characteristics shall be documented in sectors no greater than 45 nor smaller than 30 in width around the wind sensors. The near terrain may be characterized with photographs, taken at wind sensor height if possible, aimed radially outward at labeled central angles, with respect to true north. Average roughness of the nearest 3 km of each sector shall be characterized according to the roughness class as tabulated above (3). The z_0 numbers in Table 1 are typical and not precise statements.

4.1.4 Important terrain features at distances larger than 3 km (hills, cities, lakes, and so forth, within 20 km) shall be identified by sector and distance. Additional information, such as aerial photographs, maps, and so forth, pertinent to the site, is recommended to be added to the basic site documentation.

NOTE 2—Cameras using 35-mm film in the landscape orientation will have the following theoretical focal length to field angle relationships:

> 50 mm yields 40°
> 40 mm yields 48°
> 28 mm yields 66°

Prints or transparencies may not utilize the total theoretical width of the image. It is desirable to label known angles in the photograph. For example, a 45° sector photograph could have a central label of 360 with marker flags located at 337.5° and 022.5° true.

4.2 *Characteristics of the Wind Systems*—There are two categories of sensor design within this practice. *Sensitive* describes sensors commonly applied for all but extreme wind conditions. *Ruggedized* describes sensors intended to function during extreme wind conditions. The application of this practice requires the starting threshold (u_0) of both the wind vane and the anemometer to meet the same operating range category.

4.2.1 *Operating Range:*

Category	Starting Threshold, u_0	Maximum Speed, u_m
Sensitive	0.5 m/s	50 m/s
Ruggedized	1.0 m/s	90 m/s

4.2.2 *Dynamic Response Characteristics*—Dynamic response characteristics of the measurement system may include both the sensor response and a measurement circuit contribution. The specified values are for the entire measure-

TABLE 1 Characterizations Extracted from Wieringa, J. (3)

No.	z_0, m	Landscape Description
1:	0.0002 Sea	Open sea or lake (irrespective of the wave size), tidal flat, snow-covered flat plain, featureless desert, tarmac and concrete, with a free fetch of several kilometres.
2:	0.005 Smooth	Featureless land surface without any noticeable obstacles and with negligible vegetation; for example, beaches, pack ice without large ridges, morass, and snow-covered or fallow open country.
3:	0.03 Open	Level country with low vegetation (for example, grass) and isolated obstacles with separations of at least 50 obstacle heights; for example, grazing land without windbreaks, heather, moor and tundra, runway area of airports.
4:	0.10 Roughly open	Cultivated area with regular cover of low crops, or moderately open country with occasional obstacles (for example, low hedges, single rows of trees, isolated farms) at relative horizontal distances of at least 20 obstacle heights.
5:	0.25 Rough	Recently developed young landscape with high crops or crops of varying heights, and scattered obstacles (for example, dense shelter-belts, vineyards) at relative distances of about 15 obstacle heights.
6:	0.5 Very rough	Old cultivated landscape with many rather large obstacle groups (large farms, clumps of forest) separated by open spaces of about 10 obstacle heights. Also low-large vegetation with small interspaces, such as bushland, orchards, young densely planted forest.
7:	1.0 Closed	Landscape totally and quite regularly covered with similar-size large obstacles, with open spaces comparable to the obstacle heights; for example, mature regular forests, homogeneous cities, or villages.
8:	>2 Chaotic	Centers of large towns with mixture of low-rise and high-rise buildings. Also irregular large forests with many clearings.

ment system, including sensors and signal conditioners (4). It is expected that the characteristics of the sensors, which can be independently determined by the referenced Test Methods D 5096 and D 5366, will not be measurably altered by the circuitry.

Anemometer	Distance constant, L	<5 m
Wind vane	Damping ratio, η	>0.3
Wind vane	Damped natural wavelength, λ_d	<10 m

4.2.3 Measurement Uncertainty:

Wind speed	Between 0.5 (or 1) and 10 m/s	±0.5 m/s
Wind speed	>10 m/s	5 % of reading
Wind direction	Degrees of arc to true north	±5° (see Note 5)

NOTE 3—The relative accuracy of the position of the vane with respect to the sensor base should be less than ±3° for averaged samples. The bias of the sensor base alignment to true north should be less than ±2°.

4.2.4 Measurement Resolution:

	Average	Standard Deviation
Wind speed	0.1 m/s	0.1 m/s
Wind direction	1°	0.1°

4.2.5 *Sampling*—Periods of time, specified as the averaging intervals, are fixed clock periods and not running or overlapping intervals, except for the three-second gust. Outputs must be continuously and uniformly sampled during the reporting period. Incomplete data must be identified.

| Wind speed | 1 to 3 s (see Note 4) |
| Wind direction | 1 to 3 s (see Note 5) |

NOTE 4—A true 3-s average wind speed results from counting the output pulses of the anemometer transducer for 3 s. If a pulse-generating transducer is not used, a suitable sampling rate and averaging method is required to produce a true 3-s average.

NOTE 5—A sample of the wind direction may be used ONLY when the sample of wind speed is at or above the wind direction starting threshold.

4.3 *Standard Data Output for Archives*—Time labels should use the ending time of the interval. If a different labeling method is consistently used, it must be defined. The data outputs are listed as follows:

4.3.1 Ten-minute scalar averaged wind speed.

4.3.2 Ten-minute unit vector or scalar averaged wind direction.

4.3.3 Fastest 3-s gust during the 10-min period.

4.3.4 Time of the fastest 3-s gust during the 10-min period.

4.3.5 Fastest 1-min scalar averaged wind speed during the 10-min period (fastest minute).

4.3.6 Average wind direction for the fastest 1-min wind speed.

4.3.7 Standard deviation of the wind speed samples (1 to 3 s) about the 10-min mean speed (σ_u).

4.3.8 Standard deviation of the wind direction samples (1 to 3 s) about the 10-min mean direction (σ_θ).

4.4 *Optional Condensed Data Output for Archives*—Some networks will not be able to save eight 10-min data sets (48 values plus time and identification) each hour. For those cases, an abbreviated or condensed alternative is provided. When the condensed output is employed the following outputs are required.

4.4.1 Sixty-minute scalar averaged wind speed.

4.4.2 Sixty-minute unit vector or scalar averaged wind direction.

4.4.3 Fastest 3-s gust during the 60-min period.

4.4.4 Wind direction for the fastest 3-s gust.

4.4.5 Fastest 1-min scalar averaged wind speed during the 60-min period.

4.4.6 Average wind direction for the fastest 1-min wind speed.

4.4.7 Ending time of the fastest 1-min wind speed.

4.4.8 Root-mean-square of six 10-min standard deviations of the wind speed samples about their 10-min mean speeds.

4.4.9 Root-mean-square of six 10-min standard deviations of the wind direction samples about their 10-min mean directions.

4.5 *Nonstandard Data Outputs for Archives*—When some, but not all, of the required outputs are reported from a station which meets all of the measurement and sensor performance specifications, they may be reported as conforming to the standard with missing data. Stations which report all the standard outputs but do not meet the measurement specifications may not claim to meet this practice.

5. Significance and Use

5.1 This practice will characterize the distribution of wind with a maximum of utility and a minimum of archive space. Applications of wind data to the fields of air quality, wind engineering, wind energy, agriculture, oceanography, forecasting, aviation, climatology, severe storms, turbulence and diffusion, military, and electrical utilities are satisfied with this practice. When this practice is employed, archive data will be of value to any of these fields of application. The consensus reached for this practice includes representatives of instrument manufacturers which provides a practical acceptance of these theoretical principles used to characterize the wind.

6. Sampling Techniques

6.1 The longest sampling interval used in this practice is 3 s. It is possible to satisfy the requirement for a 3-s average wind speed and a 3-s sample wind direction by using a strategy which takes data into the system processor each 3 s. This generates 200 values for calculating the standard deviations for each 10-min period, when all samples are above the starting threshold speed. A better characterization of the peak 3-s speed comes from faster sampling. A 1-s sampling period is preferred, when possible, to find the peak 3-s speed from a running average rather than the clock dependent average necessary with 3-s sampling. The 1-s sampling generates 600 values for calculating the standard deviations for each 10-min period.

7. System Operational Considerations and Requirements

7.1 The mounting design and protective measures taken should protect the measurement system from hostile environments such as high winds, icing, lightning, salt, or dust particles. The following considerations will optimize the value of these data taken during destructive storms.

7.2 *Survivability*—The support hardware must be designed to survive the maximum speed range of the sensor. To ensure this performance, the support structure with a instruments installed should withstand the forces of wind speeds 25 % higher than the measurement maximum. For maximum data recovery, the power system must have backup resources to record all wind data when primary power sources fail.

7.3 *Special Data Recovery*—Provisions can be made to save all the highest time resolution data during periods of destructive storms. This special recording should begin when either the 1-min average speed exceeds 20 m/s or when the 3-s average speed exceeds 25 m/s. The special recording should end 1 h after the last trigger event is observed. This process should be automatic and the data survival should be independent of commercial power.

8. Data Quality

8.1 *Quality Assurance:*

8.1.1 All calibrations or audits should use standard methods, such as those found in ASTM standards or described in (5). All calibrations should be documented in site logs and should specify the calibration authority, such as NIST, to which calibration instruments can be traced or referenced, when necessary. Of special importance is the starting threshold for both wind speed and wind direction sensors which will predictably degrade with bearing wear and contamination.

8.1.2 Calibrations and audits verify performance at one point in time. The data should also be routinely inspected to validate the performance of the measurement system between calibrations or audits. At a minimum, range tests and rate-of-change tests should be automatically performed on machine-processible data. Discrepancies found, flagged, and responded to with corrective action should be documented and noted in the site log.

8.2 *Data Availability:*

8.2.1 Data quality is judged by the ability to learn all the necessary details about where and how the data were collected. A station file must be maintained and made available to data users. The operators of the measurement systems are responsible for gathering the necessary information, maintaining a station log on site, and transmitting the information in a standard format to a data archive such as National Climatic Data Center (NCDC). Then, the data user may acquire copies of the data and the support documentation from the same source.

8.2.2 The support documentation must include the following:

8.2.2.1 Station name and identification number,

8.2.2.2 Station location in longitude and latitude or equivalent,

8.2.2.3 Sensor type (sensitive or ruggedized),

8.2.2.4 Date of first continuous operation,

8.2.2.5 Siting information including,

(*1*) Sensor heights, AGL,

(*2*) Building top height, AGL, if appropriate,

(*3*) Surface roughness analysis by sector with analysis date,

(*4*) Site photographs with date (five-year repeat cycle),

(*5*) Tower size and distance of sensors from centerline, if appropriate, and

(*6*) Size and bearing of nearby obstructions to flow.

8.2.2.6 Measurement system description, including model and serial numbers,

8.2.2.7 Date and results of calibrations and audits,

8.2.2.8 Date and description of repairs and upgrades,

8.2.2.9 Data flowchart with sample rates and averaging methods,

8.2.2.10 Statement of exceptions to standard requirements, if any, and

8.2.2.11 Software documentation of all generated statistics.

9. Keywords

9.1 anemometer; fastest minute; peak gust; Sigma Theta; Sigma U; wind direction; wind speed; wind vane

REFERENCES

(**1**) Wieringa, J., "Representative Roughness Parameters for Homogeneous Terrain," *Boundary-Layer Meteorology*, Vol 63, 1993, pp. 323–363.

(**2**) Yamartino, R. J., "A Comparison of Several 'Single-Pass' Estimates of the Standard Deviation of Wind Direction," *Journal of Climate Applied Meteorology*, Vol 23, 1984, pp. 1362–1366.

(**3**) Wieringa, J., "Updating the Davenport Roughness Classification," *Journal of Wind Engineering Industrial Aerodynamics*, Vol 41, 1992, pp. 357–368.

(**4**) Snow, J. T., Lund, D. E., Conner, M. D., Harley, S. B., and Pedigo, C. B., "The Dynamic Response of a Wind Measuring System," *Journal of Atmospheric and Oceanic Technology*, Vol 6, 1989, pp. 140–146.

(**5**) "Quality Assurance Handbook for Air Pollution Measurement Systems," *Meteorological Measurements*, Vol IV, EPA/600/4-90/003, U.S. Environmental Protection Agency, Office of Research and Development, AREAL, Research Triangle Park, NC 27711, 1989, p. 207.

5.2 GENERAL SAMPLING

ASTM Designation: D 5111 – 95

Standard Guide for
Choosing Locations and Sampling Methods to Monitor Atmospheric Deposition at Non-Urban Locations[1]

This standard is issued under the fixed designation D 5111; the number immediately following the designation indicates the year of original adoption or, in the case of revision, the year of last revision. A number in parentheses indicates the year of last reapproval. A superscript epsilon (ε) indicates an editorial change since the last revision or reapproval.

1. Scope

1.1 This guide assists individuals or agencies in identifying suitable locations and choosing appropriate sampling strategies for monitoring atmospheric deposition at non-urban locations. It does not purport to discuss all aspects of designing atmospheric deposition monitoring networks.

1.2 The guide is suitable for use in obtaining estimates of the dominant inorganic constituents and trace metals found in acidic deposition. It addresses both wet and dry deposition and includes cloud water, fog and snow.

1.3 The guide is best used to determine estimates of atmospheric deposition in non-urban areas although many of the sampling methods presented can be applied to urban environments.

2. Referenced Documents

2.1 *ASTM Standards:*
D 1356 Terminology Relating to Atmospheric Sampling and Analysis of Atmospheres[2]
D 1357 Practice for Planning the Sampling of the Ambient Atmospheres[2]
D 1605 Practices for Sampling Atmospheres for Analysis of Gases and Vapors[3]
D 3249 Practice for General Ambient Air Analyzer Procedures[2]
D 4841 Practice for Estimation of Holding Time for Water Samples Containing Organic and Inorganic Constituents[4]
D 5012 Guide for Preparation of Materials Used for the Collection and Preservation of Atmospheric Wet Deposition[2]

3. Terminology

3.1 *Definitions:*

3.1.1 *cloud water*—an aggregate of condensed water vapor or ice crystals that are suspended in the free atmosphere. Cloud water droplet sizes are typically less than those of precipitation, measuring between 1 and 100 μm in diameter.

3.1.2 *denuder*—a sampling device designed to collect atmospheric gases that employs the diffusion of the gas to a surface where it can be collected as the principle of its operation. In practice, most denuder designs have difficulty distinguishing atmospheric aerosols from atmospheric gases.

3.1.3 *dew*—water vapor that has condensed onto a surface near the ground because of radiational cooling of that surface to a temperature that is below the dew point of the air surrounding the surface.

3.1.4 *dry deposition*—all forms of deposition derived from the net vertical transfer of chemical species to a surface that are not the result of precipitation. Dry deposition includes both turbulent diffusion and gravitational settling. Dew and frost are anomalous forms of dry deposition which rely upon a near surface, condensation process as their principle means of effecting the net vertical transfer.

3.1.5 *fog*—a visible aggregate of condensed water vapor or ice crystals suspended in the atmosphere near the earth's surface. Fog differs from cloud water only in that it resides very close to the earth's surface.

3.1.6 *frost*—ice crystals resulting from the direct sublimation of water vapor onto a surface that is below freezing. Frost is due to radiational cooling and only occurs when the temperature of the air in contact with the surface falls below the freezing point of water.

3.1.7 *snow*—a solid form of wet deposition composed of white or translucent ice crystals chiefly in complex hexagonal form and often agglomerated into snowflakes.

3.2 *Description of Terms Specific to This Standard:*

3.2.1 *collocated sampling*—the use of more than one sampling device within a monitoring site.

3.2.2 *event sampling*—a special form of intermittent sampling (Terminology D 1356) where the duration of a sampling period is defined as a single, discrete occurrence of precipitation, dew, fog or frost.

3.2.3 *fetch*—a vector within the local area which describes the direction and area of, or within, an air mass that will be sampled by a sampling device.

3.2.4 *filter-pack*—a sampling device comprised of one or more filters in series where each filter is designed to sample an atmospheric chemical species or remove interferences to a subsequent filter. Filters may be of different design; material; or be coated or impregnated to obtain the specificity of chemical species required.

3.2.5 *inferential sampling*—an indirect sampling method that utilizes a mathematical model to quantify an unmeasurable or difficult to measure property of atmospheric deposition.

3.2.6 *local area*—an area of a few square kilometers which describes an area of common vegetation, land-surface form and land use surrounding the monitoring site and defines the local characteristics surrounding the sampling device, see Fig. 1.

[1] This guide is under the jurisdiction of ASTM Committee D-22 on Sampling and Analysis of Atmospheres and is the responsibility of D 22.06 on Atmospheric Deposition.

Current edition approved April 15, 1995. Published June 1995. Originally published as D 5111 – 90. Last previous edition D 5111 – 90.

[2] *Annual Book of ASTM Standards*, Vol 11.03.

[3] *Discontinued—See 1991 Annual Book of ASTM Standards*, Vol 11.03.

[4] *Annual Book of ASTM Standards*, Vol 11.01.

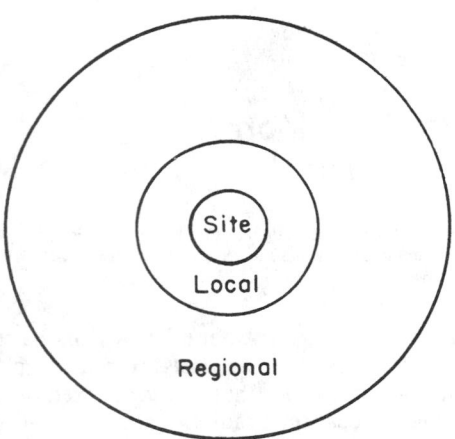

FIG. 1 Diagram of Siting Guidelines

3.2.7 *monitoring site*—a radius of a few decameters which immediately surrounds the sampling device, see Fig. 1.

3.2.8 *regional area*—an area between the local area and a threshold that defines where any single local area characteristic can not be distinguished from regional characteristics, see Fig. 1.

3.2.9 *sequential sampling*—withdrawal of a portion of the atmosphere over a period of time with continuous analysis or with separation of the desired material continuously and in a linear form. Such a sample may be obtained with a considerable concentration of the contaminant but it still indicates fluctuations in that property which occur during the period of sampling (Terminology D 1356; see *sample, running*).

3.2.10 *surrogate surface sampling*—a sampling technique that utilizes an artificial surface to estimate dry deposition. Ideally, the artificial surface chosen will approximate the real surface's roughness and wetness properties. In practice this is impossible. Therefore, comparisons of the surrogate surface to the real surface must always be done as a part of the technique.

3.2.11 *wet deposition*—the deposition of water from the atmosphere in the form of hail, mist, rain, sleet and snow. Deposits of dew, fog and frost are excluded (Terminology D 1356; see *precipitation, meteorological*).

3.3 The definitions of other terms used in this guide can be found in Terminology D 1356.

4. Significance and Use

4.1 The guide consolidates into one document, siting criteria and sampling strategies used routinely in various North American atmospheric deposition monitoring programs.

4.2 The guide leads the user through the steps of site selection, sampling frequency and sampling equipment selection, and presents quality assurance techniques and other considerations necessary to obtain a representative deposition sample for subsequent chemical analysis.

4.3 The guide extends Practice D 1357 to include specific guidelines for sampling atmospheric deposition including acidic deposition.

5. Summary of Guide

5.1 The guide assists the user in establishing siting guide-

lines and in choosing sampling frequencies and sampling devices for atmospheric deposition monitoring. Special considerations for the monitoring of specific types of atmospheric deposition are discussed.

5.2 A worksheet is provided to assist the user in documenting the final siting criteria and sampling strategy chosen—see Appendix X1.

5.3 The guide references site selection and sampling documents of some of the currently operating deposition monitoring networks in North America (Appendix X2).

6. Sampling Locations

6.1 *General Requirements:*

6.1.1 General requirements for choosing atmospheric deposition sampling locations follow Practice D 1357. This guide should be used in conjunction with that document.

6.1.2 A standardized site description questionnaire should be developed and completed during the site selection process. The questionnaire will describe the chosen location in detail. Examples of these questionnaires can be found in Refs (1-3).[5]

6.1.3 Figure 1 illustrates the concentric organization of location guidelines used in this document. Monitoring site requirements are common to all types of monitoring stations, while regional area requirements invoke a combination of monitoring site, local area and regional area guidelines. Which guidelines within each area category are chosen and whether all area categories are used will depend upon the purpose of the monitoring effort.

6.1.4 Some specific atmospheric deposition sample types require that additional criteria be met. These are identified towards the end of each sampling location section with an appropriate key word; DRY for dry deposition; FOG for fog; etc. Guidelines that contain no key word are common to all types of deposition monitoring within their monitoring site, local area, or regional area grouping.

6.1.5 The user of this guide should use all of the guidelines listed for the deposition type being monitored and all of the guidelines that are not deposition type specific. Exceptions to the use of all of the guidelines should be noted on the worksheet in X1 of the guide and be accompanied with a brief exclusion statement.

6.2 *Regional Area Guidelines:*

6.2.1 Regional area guidelines should be based upon a consensus interpretation of the concept of regional representativeness by the monitoring project management. Regions may be identified based upon physiography, meteorology, demography or some other more specific goal of the monitoring project. Ground-based concepts of representativeness, such as the ecological classifications of Bailey and others (4, 5) or areas sensitive to acid deposition, are often more easily defined than meteorological concepts which tend to be highly variable both spatially and temporally. For this reason definitions of regional representativeness based heavily upon meteorological phenomena are best developed *a posteriori* using mathematical and statistical models (6).

6.2.2 When developing regional area guidelines, distance criteria should reflect the thresholds where any characteris-

---

[5] Boldface numbers in parentheses refer to references at the end of this guide.

tics of a local area become indistinguishable from those of other local areas and are instead typical of the area that will be declared a region.

6.2.3 Population centers of greater than 10 000 should be at least 10 km from the sampling device. This distance should be increased dramatically if the sampling device is located downwind of the center in the prevailing wind direction.

6.2.4 All industrial and natural sources of emissions greater than 10 000 tons per annum of each analyte of interest should be at least 10 km from the sampling device. This distance should be increased dramatically if the sampling device is located downwind of the source in the prevailing wind direction.

6.2.5 Complex terrain should be avoided unless its influence is necessary to meet the specific goal of the monitoring effort.

6.3 *Local Area Guidelines:*

6.3.1 The local area surrounding a monitoring site should describe a small geographic area where land use, topography and meteorology are common and representative of the regional area. No single emission source should dominate the air quality at the site except as it typifies the common emission characteristics of the regional area. Ideal sites will be located in areas where land use practices are not expected to change over the course of the monitoring effort.

6.3.2 Emission source amounts, their frequency and intensity, and meteorological diversity will dominate the actual influence of each guideline on samples collected in any monitoring program. Because of this, local area guidelines are typically the portion of a site selection plan that is not met. A relaxation of the guidelines can be tolerated when the impact of non-compliance on program objectives can be quantified.

6.3.3 Monitoring sites should be located away from population centers. A recommended distance is 1 km per 1000 persons.

6.3.4 Intensive agricultural and waste treatment activities should be more than 500 meters from the sampling device. Dairy operations, crop cultivation, especially in areas where chemical applications are used and solid waste and wastewater treatment facilities are of particular concern.

6.3.5 Transportation related sources of emissions should be no closer than 100 meters from the sampling device. Parking lots, unpaved roadways and high volume vehicular, railroad and airplane traffic are of particular concern. One hundred meters is a minimum acceptable distance cited by some of the existing atmospheric monitoring networks (See X2.3., X2.4, X2.5). The distance should be increased in proportion to increases in traffic volume and diversity. One kilometer is considered adequate under most conditions.

6.3.6 The open or surface storage of agricultural or industrial products should be kept at least 100 m from the sampling device. Examples of these products would include salt and sand piles, fuels and chemicals.

6.3.7 *Dry*—For methods employing the estimation or use of atmospheric fluxes (see 9.2.3. and 9.2.9.), the surface micro-meteorology and surface composition should be as uniform as possible within 500 m of the sampling device.

NOTE 1—The success of tower based eddy correlation techniques and many other dry deposition techniques utilizing deposition velocity estimates, are dependent upon the uniformity of the upwind surface roughness and wetness. If the upwind micro-meteorology and surface characteristics enhance the turbulent mixing of the parameter of interest as it approaches the point sampling then the distance requirements for fetch can sometimes be relaxed. If on the other hand deposition rate estimates are expected to be small, the fetch distance requirements may need to be increased ((7) see 9.2.3.).

6.3.8 *Dry*—For methods employing the estimation of atmospheric fluxes, see 9.2.3, the sampling device should be located at least 5 km from prominent discontinuities in terrain such as large bodies of water, isolated hills or valleys and cliffs.

6.4 *Monitoring Site Guidelines:*

6.4.1 Monitoring sites should be located on naturally vegetated or grassed, open, level areas. Ground cover should be homogeneous and the area should slope no more than 15 %.

6.4.2 The distance from the sampling device to any object greater than the height of the sampling device should be at least twice the height of the object (2:1). This will ensure that no object or structure will project onto the sampling device with an angle greater than 30° from the horizontal.

6.4.3 With the exception of wind shields, objects with sufficient mass to deflect the wind or otherwise change the aerodynamic properties of the sampling device should be located no closer than 2 m from the sampling device.

NOTE 2—Wind shields are considered to be an integral part of the sampling device in this guide.

6.4.4. Residential structures should be outside of a 30° cone of the prevailing wind direction.

6.4.5 Sampling devices should be oriented towards the annual averaged prevailing wind. In the absence of site specific wind direction information projects should standardize the orientation of the device to one direction.

6.4.6 Seasonal vegetation should be maintained at a level that is at least 1 meter below the orifice of the sampling device to a distance that defines one-half of the monitoring site.

6.4.7 Grazing animals and the cultivation of agricultural crops should not be permitted within the monitoring site.

6.4.8 All activities not directly related to sampling should be discouraged within the monitoring site.

6.4.9 *Snow*—The sampling device should be located in a setting that is sheltered from the wind. Locating the monitoring site within a forest clearing or installing a wind shield around the sampling device improves snow capture (8).

NOTE 3—Wind speeds in excess of 1 m/sec significantly reduce the efficiency of snow sampling devices (8). Light, dry snows are the most difficult to sample. Reducing or eliminating the wind around the sampling device by either shielding the device or locating the device below the vegetation canopy improves snow capture and eliminates re-entrainment of already collected samples.

6.4.10 *Dry*—For methods utilizing towers in the estimation of atmospheric fluxes (section 9.2.3.), the tower heights should be standardized and be at least 5 meters above the surface of interest (for example, forest canopy and agricultural crops). For measurements over bare ground this distance may need to be doubled.

6.4.11 *Dry*—Methods utilizing micro-meteorological measurements in the estimation of atmospheric deposition require stricter slope requirements of 5 % and stricter projection requirements of 5:1, see 6.4.1 and 6.4.2.

7. General Sampling Requirements

7.1 Once the goals of the monitoring effort have been established and site locations have been identified, sampling frequency and sampling equipment decisions can be made. Location decisions should be made in advance of sampling decisions since, in addition to cost, the latter are almost always limited by site availability and accessibility.

7.2 The choice of a sampling method for atmospheric deposition monitoring oftentimes will be a compromise brought about by the availability of a suitable site, the ability of a particular sampling device to selectively measure the deposition type and chemical species of interest, and the differences in cost of implementing some of the available techniques. The selection of sampling intervals and sampling devices may be an iterative exercise especially if a wide variety of chemical species are of interest.

7.3 Users of this guide should recognize that all of the sampling techniques mentioned in this guide are not directly comparable and may not be interchangeable. Comparability, especially in the area of dry deposition, has only recently begun. For projects requiring a wide variety of deposition estimates or short sampling intervals, this often means selecting multiple methods.

7.4 Projects requiring comparability or additivity of estimates derived from more than one method should establish the level of uncertainty in using this approach.

8. Sampling Frequency

8.1 *Continuous Sampling:*

8.1.1 Continuous Sampling in the context of atmospheric deposition monitoring is a combination of both continuous and instantaneous sampling (Terminology D 1356). It is frequently used for the estimation of ambient air concentrations in sampling techniques which compute dry deposition rates. Continuous measurements are typically the most expensive form of measurements to obtain since they most always require sophisticated instrumentation and a high level of expertise to minimize and troubleshoot periods of nonsampling.

8.1.2 Continuous sampling should only be considered when instantaneously sampled deposition data are necessary, as in dose response types of effects studies, when averaged information is necessary to statistically reduce error estimates, or, as in the calculation of dry deposition rates, instantaneous sampling results must be paired with instantaneous meteorological measurements.

8.1.3 General recommendations for continuous ambient air analyzers are given in Practice D 3249.

8.2 *Cumulative Sampling:*

8.2.1 Cumulative samples represent a temporal composite or integration of the parameter being monitored. The length of time a sample accumulates in the sampling device can be adjusted to match the temporal resolution required in the monitoring program. Intervals of days through months are typical for wet deposition and hours through weeks are typical for dry deposition. Cumulative sampling is the most widely used technique in both wet and dry deposition.

8.2.2 When using cumulative sampling, attention must be paid to the possibility of sample degradation that can occur during the accumulating time period. Short accumulation times are recommended, especially when samples are not preserved. Both loss and transformation of chemical species have been observed in cumulative samples (9, 10).

8.2.3 Cumulative sampling can be used to reduce the number of samples collected and analyzed along with their associated costs, and to increase the sensitivity of a method by averaging over time. Filter packs, denuders and impingers all use the principle of cumulative sampling.

8.3 *Event Sampling:*

8.3.1 Event sampling is a special form of intermittent sampling used to collect liquid deposition from discrete occurrences of precipitation, dew, frost, and fog.

8.3.2 Event sampling is used for studying atmospheric processes and for determining noncumulative effects of atmospheric deposition on agricultural and natural ecosystems. Event sampling is especially useful when monitoring objectives are associated with episodic phenomena such as storm types, direction or intensity, or when the tracking of a parameter through time and space is required.

8.3.3 Event sampling is less susceptible to the sample integrity problems associated with cumulative sampling (9, 10). This is especially noticeable when events are of short duration (for example, less than days).

8.3.4 Event sampling is not an effective monitoring frequency when the predominant sample collected contains too small an amount of analyte mass for analysis or consistently produces analyte concentrations below the method detection limit. The cost of standby time (time waiting for events to occur) should also be considered when selecting event sampling frequencies.

8.4 *Sequential Sampling:*

8.4.1 Sequential sampling is used to characterize within event variability. Sequential sampling strategies typically break events into consecutive, equal-volume or equal-time subsamples of the event. Like event sampling, sequential sampling is limited to liquid deposition types and is used to study atmospheric processes.

8.4.2 Sequential sampling should only be used when project goals emphasize within event variability as more or equally as important as between event variability. It is seldom considered for long-term monitoring.

8.4.3 All of the cautions of event sampling—see 8.3.3 and 8.3.4—also apply to sequential sampling.

9. Sampling Devices and Techniques

9.1 *Wet Deposition:*

9.1.1 *General Characteristics*—Wet deposition sampling devices typically consist of a precipitation detector or sensor and a mechanically operated lid which covers a sample container or inlet. The sensor detects the presence of water and activates the mechanical lid which exposes the sample container or inlet to precipitation. At the cessation of precipitation the lid returns to a position which protects the sample container from dry deposition. Any sampling system that has the ability to capture wet-only precipitation and protect the captured sample from dry deposition can be used.

9.1.2 A wet deposition collector is designed to capture a representative sample of precipitation for subsequent chemical analysis and prevent this captured precipitation from mixing with other forms of deposition. Because the emphasis of the design is towards representative chemistry and not necessarily on the quantification of precipitation amount,

D 5111

the collector should not be relied upon for estimates of precipitation volume (see 9.1.4).

9.1.3 *Precipitation Sensors*—Most precipitation sensors work on a resistance principle, interrupting or establishing an electrical current when their surface becomes wet. Heating devices in the sensor attempt to evaporate the accumulated water from the sensor surface, returning the sensor to a dry status. The resistance setting, surface area and heating rate of the sensor determines the sensitivity of the sensor to the wetness created during a precipitation, fog, dew or frost event. A sensor's selectiveness towards these deposition types can be altered by altering the resistance, heating rate, temperature and surface area of the sensor, and by controlling the response time between the electrical detection of wetness and the activation of the collector's mechanical system. Sensor resistance, heating rates and temperature, surface area and activation delay capability should be chosen to emphasize the goals of the monitoring project.

NOTE 4—It should be recognized, that all wet deposition collectors capture small amounts of dew, frost and fog and that the sensor plays a critical role in determining the sensitivity of the collector to these forms of deposition. The sensor design should restrict the collection of dew, frost and fog and improve the likelihood of collecting a wet-deposition-only sample.

9.1.4 *Collector Sampling Efficiency*—Because a wet deposition collector relies on the movement of a mechanical lid to expose its sampling container or activate its sampling system, it does not always open or close in perfect sequence with the initiation or cessation of precipitation. Differences in the collector's aerodynamic design and its' sensor's characteristics will largely determine the suitability of the collector for certain geographic and climatic conditions. Snow accompanied by wind is especially difficult to sample. The performance of the collector under the geographic and seasonal climatic conditions in which it will operate should be established so that sampling biases towards specific deposition types or conditions are minimized and so that the sampling efficiency of the device is characterized. Collector sampling efficiency is best determined by comparing the volume in the wet deposition collector to the volume of a collocated national meteorological service or World Meteorological Organization (WMO) rain gage.

9.1.5 *Sample Integrity Considerations*—The construction and workings of the mechanical lid which covers the wet deposition sample is a critical component of the wet deposition collector. The lid should be designed in such a manner as to seal the deposition sample during periods of dry deposition in order to minimize the chance for sample degradation due to evaporation, diffusion, thermal decomposition and wind-borne contamination. No device will perform ideally, but careful attention to the workings and materials of construction of the lid will improve the representativeness of the wet deposition sample collected.

9.2 *Dry Deposition:*

9.2.1 *General Principles*—Sampling techniques used for the measurement of dry deposition are as varied as the substances they are designed to quantify. Each requires a different level and area of expertise and has only limited flexibility for sampling a variety of chemical species and deposition types. Most techniques represent innovative approaches to resolving the critical needs of dry deposition sampling in

specific locations (forests vs grasslands) or for specific receptors (such as a plant species or blocks of marble).

9.2.2 Many of the dry deposition estimation techniques involve the measurement of one or more chemical or meteorological parameter and the application of some type of mathematical model to quantify dry deposition. The model may be a simple difference equation or a complex sequence of equations designed to simulate natural science theories under different conditions of meteorology, and other site specific physiographic features.

NOTE 5—The use of a model to estimate dry deposition requires that meteorological and chemical measurement techniques be available for the time and spatial scales used by the model and for the complexity of the terrain. It is further required that the combination of model and measurements be specific to the surface that is receiving the deposition. Models and measurements for most chemical species under various terrain and meteorological conditions are still rather limited. For a more complete compilation of current applications the reader is referred to *Proceedings of the NAPAP Workshop on Dry Deposition*, Harpers Ferry, March 25–27, 1986 edited by Hicks, Wesely, Lindberg and Bromberg (7).

9.2.3 *Inferential Methods*—These methods require micrometeorological measurements, air concentration measurements, a suitable mathematical model (11, 12) and an explicit assumption of the true deposition velocity (V_d) of the chemical species being monitored. The methods result in deposition rate estimates which can be used to establish deposition loading. The methods are often limited by the lack of chemical analysis techniques or the costs of utilizing continuous gas analyzers. The current models tend to be very specific to terrain, surface composition and surface condition. When choosing these techniques users should be certain that 1) there is an available model for the chemical species and time scale of interest 2) there is a chemical measurement technique for the chemical state and sampling frequency required 3) sampling sites meet the surface composition and condition requirements of the model and 4) the assumptions implicit to the chosen model are appropriate for the proposed project.

9.2.4 For studies requiring chemical measurements spanning days or weeks, continuous meteorological measurements are typically integrated as appropriate to the model. Impingers (Practices D 1605), denuders (13) and filter packs (13) are the chemical sampling methods of choice. Changes in micro-meteorological conditions and in surface characteristics (wetness, stage of growth, etc) during the sampling period however, often results in abnormal or unrealistic integrations of meteorological or surface parameters causing the models to run outside of their designed limits.

9.2.5 When the cost and availability of continuous gas analyzers are feasible, usually for short duration projects at a single site, models can be chosen to make maximum use of meteorological and surface driven parameters. Dry deposition estimates utilizing inferential techniques are best suited to studies where the monitoring site has flat, homogeneous terrain, and where the meteorology is relatively uniform throughout a day.

9.2.6 *Methods Based Upon Surface Analysis*—Surface analysis methods estimate the deposition loading to surfaces during non-precipitation periods either by direct analysis of the surface that is receiving the deposition [foliar extraction, surrogate surfaces (7, 14)] or by using techniques that compute dry deposition by difference [throughfall/stemflow

1297

estimation, runoff/catchment mass balancing, snow sampling (7, 14)]. The methods are usually confined to specific ecological regions (forests, watersheds) and seasons (with foliage, with snow) where representative surfaces are available and conventional surface sampling techniques can be applied.

NOTE 6—Siting requirements given in 6.4.1, 6.4.3 and 6.4.5 are meaningless and therefore unnecessary for surface analysis methods using natural, living surfaces (for example, foliar extraction, throughfall, etc) as the deposition collector. Requirement 6.4.2 should also be examined for suitability.

9.2.7 Surrogate surface techniques are effective in quantifying chemical species where gravitational settling of large particles dominate the dry deposition process. Therefore, users of surrogate surfaces should evaluate the representativeness of the surrogate under various climatic conditions which might alter the surrogate's aerodynamics or other surface conditions. When used in lieu of vegetation, surrogates should also be evaluated as to how they perform in contrast to the vegetation's normal physiological processes (that is, leaf movement, changes in stomatal status, etc).

9.2.8 Users of techniques which derive dry deposition estimates by difference, need to ensure that integration times, usually storms through seasons, are long enough to produce calculated differences that are in excess of the sum of the errors associated with each measured variable used in the difference calculation. Techniques using differences are most effective when the chemical composition of all other components of the equation except dry deposition can be easily characterized and are well understood.

9.2.9 *Atmospheric Flux Methods*—Current methods [eddy correlation, eddy accumulation, vertical gradient, aerometric mass balance, etc (7, 14)] like the inferential methods, also require micro-meteorological measurements and air concentration measurements. These methods however, compute deposition velocities (V_d) as a part of their model rather than assume them.

9.2.10. Because the deposition velocities are being computed within the technique, instrumentation requirements tend to be more rigorous. The lack of inexpensive accurate and fast response sensors both for meteorological and chemical parameters, and the increased need for more detailed and timely surface characterization information (roughness and wetness) tend to limit the techniques to specific research situations.

9.3 *Fog/Cloud Water Collectors:*

9.3.1 Sampling equipment designed to collect fog and cloud water work on the principles of interception and/or impaction (15). Small diameter (5-410 μm) filaments are used to passively or dynamically intercept cloud or fog water and direct it towards a storage container. Mechanical rotation or moderate volume air movement (nominally 1.5 m³/min utilizing positive or negative pressure) is often used to increase the collection efficiency of the device. Sampling intervals may range from hours through events.

9.3.2 Like wet deposition collectors, fog and cloud water collectors are designed to collect a representative sample for subsequent chemical analysis. They may not be a reliable way of quantifying fog or cloud water volumes.

9.4 *Bulk Sampling:*

9.4.1 Bulk sampling is the combined collection of both wet and dry deposition into a single container. Bulk sampling describes a wide variety of collection designs from simple open containers to the sub-surface sampling of snow and ice. Each attempts to collect total deposition over some unit of time. Typically collection is passive employing collectors constructed from a bucket or funnel. These containers collect all forms of deposition that come in contact with their exposed surfaces.

9.4.2 Bulk sampling has the advantage of simplicity, being a passive device that requires no power. In many instances however, it has been shown to be susceptible to chemical contamination from other than atmospheric sources (bird droppings, twigs, etc) and to evaporation losses. In some instances, bulk collectors may not preserve the speciation of collected deposition, especially if the sampling device allows free interaction between both the wet and dry components of the deposition process and the atmosphere.

9.4.3 With the possible exceptions of snow sampling and in cases where dry deposition is accomplished almost exclusively by gravitational settling of coarse particles, bulk deposition collectors are not recommended. If used, it is recommended that sampling be initiated just prior to a precipitation event and concluded within 12 hours after the cessation of precipitation.

9.4.4 Bulk sampling has been used effectively to estimate the deposition of snow (16). Differences between samples taken from recent snow events versus those of previous snow events (snow boards) have also been used to apportion wet and dry components of deposition. Snow and ice cores have also been shown to be effective methods of obtaining bulk samples when sampling times are considerate of the changes in chemical composition that take place during freeze/thaw cycles (17).

10. Quality Control/Quality Assurance

10.1 *Sample Containers:*

10.1.1 Sampling containers, both those used for sample collection and sample transportation should be chosen to minimize sample contamination, chemical species transformation, microbial activity and liquid evaporation. When appropriate rigid containers can not be found for the sampling device chosen, bag liners may be utilized. Bags have the advantages of being disposable and have a more uniform surface chemical composition than can be obtained by recycling glass or plastic ware. Bag manufacturing lots however, need to be monitored to ensure chemical uniformity over the life of the project. All containers used in the routine sampling of wet deposition should be prepared in accordance with Guide D 5012.

10.1.2 Many of the sample containers used in dry deposition techniques are an integral part of or comprise the sampling device. For this reason dry deposition sample containers and filters should be prepared according to the specific demands of the technique.

10.1.3 Filter media used in many dry deposition techniques should be standardized and bought in large lots if possible. Analysis of each lot should be conducted to document any changes in the amount or frequency of false positive analyte loading.

10.1.4 Contamination and analysis errors may be further minimized by choosing a single container for both collection and transport of the deposition sample. Large containers however may be expensive to transport.

10.1.5 Color and texture as well as composition (glass, plastic, etc) should be considered when choosing sample containers and filter media. The choices should be standardized throughout the life of the project. Dark colors promote the absorption of heat when exposed to sunlight. This might be beneficial in some situations to minimize freezing or encourage thawing. Surface roughness and porosity, especially in dry deposition techniques, can alter the yield of sampled material or complicate chemical extraction techniques.

10.1.6 Bag liners and other types of single-use containers, including coated filters, impingers and denuders used in dry deposition techniques should be monitored for changes in their chemical contribution to the analyte measurements.

10.2 *Sample Preservation:*

10.2.1 Sample preservation techniques may be incorporated into sampling device designs (that is, refrigeration) or implemented as a part of the standard sampling protocols. Multiple preservation techniques may need to be applied to a collected sample. The use of these techniques may necessitate the aliquoting of the sample at or near the time of collection and prior to sample transport. Aliquoting may also be necessary to meet the holding time requirements of some analytes (Practice D 4841).

10.2.2 Common techniques employed include filtration, refrigeration, acidification and the addition of biocides. The different techniques are discussed in Guide D 5012.

10.3 *Site Audits:*

10.3.1 Site description questionnaires, see 6.1.2, should periodically be reviewed to ensure their continued compliance with established site criteria. Reviews should include a re-evaluation of all emission sources and any local changes in land use or condition.

10.3.2 Routine collector maintenance, along with the use of standard reference materials and calibration checks, should be incorporated into the project's quality assurance program. A wet deposition collector's sensors, lid and drive mechanism should be checked at the end of each sampling interval to ensure proper operation. Wet deposition and bulk collection sample volumes should constantly be compared to national meteorological service or WMO approved rain gages to maintain a record of the collector's sampling efficiency.

10.3.3 Many dry deposition methods require that air handling equipment and other meteorological equipment calibration be maintained. Error analysis of the measurement system should be monitored to ensure that results are significantly above noise levels. This is especially true for methods deriving estimates by difference.

10.4 *System Performance Audits:*

10.4.1 A review of the frequency and magnitude of false-positive measurements of deposition should be conducted periodically to quantify, in parts and as a whole, the contribution of analyte that is associated with violations of sample integrity (collector lid seals, sample container contamination, etc), analytical bias (due to the collector or chemical analysis methodology) and data transformation and summarization (signal integration biases, detection limit values becoming real values, the fate of low volume or analyte events). Both field and laboratory contributions to false-positive measurement should be examined.

10.4.2 Collocated sampling equipment within the monitoring project as well as between monitoring programs should be implemented to allow system wide estimates of measurement precision and bias. The co-location protocol should duplicate the entire monitoring protocol through data management. Duplicate sampling is also useful for overall error estimation.

10.4.3 Sampling completeness statistics should be computed for annual and seasonal comparisons of representativeness (18). The completeness figures provide the percent of deposition collected as compared to a referenced amount, define the percent of samples that are ultimately valid after using the chosen protocols and establish the percentage of samples containing various levels or degrees of severity of caveats in the declared data set.

11. Keywords

11.1 atmospheric deposition; dry deposition monitoring; sampling methods; siting guidelines; wet deposition

APPENDIXES

(Nonmandatory Information)

X1. WORKSHEET FOR CHOOSING SAMPLING LOCATIONS AND SAMPLING METHODS FOR MONITORING ATMOSPHERIC DEPOSITION AT NON-URBAN LOCATIONS

Project Title: Project Objectives:		

Type and Method of Sampling: Sampling Frequency: Comments	Circle One on Each Line rain snow dry fog cloud water dew/frost wet-only snow-only bulk fog/cloud water inferential surface analysis flux continuous cumulative event sequential

Monitoring Site and Area Diameters	Monitoring Site Local Area Regional Area

ASTM Siting Modifications: Section	Changes	Stricter ?

Equipment Needed: Analyte	Type of Sampling Device	Operator Time/WK

Totals	Equipment Costs	Operator Time

X2. REFERENCES FOR LOCATION CRITERIA AND SAMPLING METHODS USED IN EXISTING ACIDIC DEPOSITION MONITORING NETWORKS

X2.1 Acid Precipitation In Ontario Study-Daily Network (APIOS-D)

Chan, W. H., Orr, D. B., and Vet, R. J. *Acid Precipitation In Ontario Study-An Overview: The Event Wet/Dry Deposition Network.* API 002/82/ISBN 0-7743-7304-0. Ontario Ministry of the Environment, Toronto. Summer 1982.

X2.2 Canadian Air and Precipitation Monitoring Network (CAPMoN)

Canadian Air and Precipitation Monitoring Network. *Technical Manual: Canadian Air and Precipitation Monitoring Network*, TM 09-01-01. Atmospheric Environment Service, Environment Canada, Downsview, Ontario. January 1984.

X2.3 Electric Power Research Institute's Utility Acid

Precipitation Study (EPRI/UAPSP)

Topol, L. E. *Utility Acid Precipitation Study Program: Network Description and Measurements for 1981 thru 1987*, UAPSP 117, Utility Acid Precipitation Study Program. Washington, DC, October 1989.

X2.4 National Atmospheric Deposition Program/National Trends Network (NADP/NTN)

Bigelow, David S., *Instruction Manual: NADP/NTN Site Selection and Installation*, National Atmospheric Deposition Program, Natural Resource Ecology Laboratory, Colorado State University, Ft. Collins, CO 80523, July 1984.

Bigelow, D. S. and Dossett, S. R. *NADP Instruction Manual: Site Operations.* National Atmospheric Deposition

Program, Natural Resource Ecology Laboratory, Colorado State University, Ft. Collins, CO. April 1988.

X2.5 National Dry Deposition Network (NDDN)

Porter, L. F. *Guidelines for the Design, Installation,* *Operation and Quality Assurance for Dry Deposition Monitoring.* EPA 600/3-88-047. U. S. Environmental Protection Agency. October 1988.

REFERENCES

(1) Bigelow, David S., *Instruction Manual: NADP/NTN Site Selection and Installation*, National Atmospheric Deposition Program, Natural Resource Ecology Laboratory, Colorado State University, Ft. Collins, CO 80523, July 1984, p. 40.

(2) Topol, L. E., Lev-On, M., Flanagan, J., Schwall, R. J. and Jackson, A. E., *Quality Assurance Manual For Precipitation Measurement Systems*, Environmental Monitoring Systems Laboratory, Office Of Research and Development, U.S. Environmental Protection Agency, Research Triangle Park, NC, January, 1985, p. 172.

(3) Vet, R. J., Stevens, D. L. M., and Butler, E. D., *A Summary of the Canadian Air and Precipitation Monitoring Network (CAPMoN) Siting and Installation Program*, Concord Scientific Corporation, Toronto, Ontario, March 31, 1983, p. 142.

(4) Bailey, R. G., *Description Of The Ecoregions Of The United States.* USDA, Forest Service, Intermountain Region, Ogdan, UT 1978.

(5) Bailey, R. G. and Cushwa, C. T., *Description Of The Ecoregions Of North America After The Classification of J. M. Crowley.* (map). USDI, Fish and Wildlife Service, Eastern Energy and Land Use Team, Kearneyville, WV 1981.

(6) Nappo, C. J. et al., *The Workshop On The Representativeness Of Meteorological Observations*, June 1981, Boulder, CO. Bull American Meteorological Society, Vol. 63, No. 7, 1982, pp. 761–764.

(7) Goodison, B. E., Ferguson, H. L., and McKay, G. A., *Handbook of Snow-Principles, Processes, Management and Use.* Gray, D. M. and Male, D. H. (ed). Pergamon Press, Willowdale, Ontario. 1981, pp. 191–274.

(8) Sisterson, D. L., Wurfel, B. E., and Lesht, B. M., *Chemical Differences Between Event And Weekly Precipitation Samples In Northeastern Illinois.* Atmospheric Environment, Vol. 19, 1985, pp. 1453–1469.

(9) de Pena, R. G., Walker, K. C., Lebowitz, L. and Micka, J. G., *Wet Deposition Monitoring-Effect Of Sampling Period.* Atmospheric Environment Vol. 19, 1985, pp. 151–156.

(10) Hicks, B. B., Wesely, M. L., Lindberg, S. E., and Bromberg, S. M., (eds). *Proceedings of the NAPAP Workshop On Dry Deposition*, March 25–27, 1986, Harpers Ferry, WV, 1986. Available through: Librarian, NOAA/ATDD, PO Box 2456, Oak Ridge, TN 37831.

(11) Davidson, C. I. and Yee-Lin W., 'Dry Deposition Of Particles And Vapors' in *Acid Precipitation Volume 2, Sources, Emissions, and Mitigation.* D. C. Adriano (ed). Advances in Environmental Sciences Series, Springer-Verlag, NY 1988.

(12) Stevens, R. K., "Review Of Methods To Measure Chemical Species That Contribute To Acid Dry Deposition," in *Proceedings: Methods For Acidic Deposition Measurement*, EPRI EA-4663, EPA 600/9-86/014, USGS Contract 54000-5622, Electric Power Research Institute, Palo Alto, CA, 1986.

(13) Hicks, B. B., Balidocchi, D. D., Housker, R. P., Hutchison, B. A., Matt, D. R., McMillen, R. T., and Satterfield, L. C., "On The Use of Monitored Air Concentrations to Infer Dry Deposition," NOAA Technical Memorandum, ERL-ARL, 1985, 65 pp.

(14) Wesley, M. L. and Lesht, B. M., "Comparison of RADAM Dry Deposition Algorithms With a Site Specific Method for Inferring Dry Deposition," *Water, Soil and Air Pollution*, 1989.

(15) Hering, S. V. et al., "Field Intercomparison of Five Types of Fogwater Collectors," *Environmental Science and Technology*, Vol 21, no. 7, 1987, pp 654–663.

(16) Laird, L. B., Taylor, H. E., and Kennedy, V. C., "Snow Chemistry of the Cascade-Sierra Nevada Mountains," *Environmental Science and Technology*, Vol 20, No. 3, 1986, pp 275–290.

(17) Sanberg, D. K., McQuaker, N. R., Vet, R. J., and Weibe, H. A., "SnowPack: Recommended Methods For Sampling And Site Selection," Federal/Provincial Research and Monitoring Coordinating Committee (RMCC)—Quality Assurance SubGroup, Environment Canada, September, 1986.

(18) Unified Deposition Data Base Committee, "A Unified Wet Deposition Data Base for Eastern North America: Data Screening, Calculation Procedures, and Results for Sulphates and Nitrates (1980)," Report to the Canadian Federal-Provincial Research and Monitoring Coordinating Committee, 1986, 120 pp.

Standard Practice for
Planning the Sampling of the Ambient Atmosphere[1]

This standard is issued under the fixed designation D 1357; the number immediately following the designation indicates the year of original adoption or, in the case of revision, the year of last revision. A number in parentheses indicates the year of last reapproval. A superscript epsilon (ϵ) indicates an editorial change since the last revision or reapproval.

1. Scope

1.1 The purpose of this practice is to present the broad concepts of sampling the ambient air for the concentrations of contaminants. Detailed procedures are not discussed. General principles in planning a sampling program are given including guidelines for the selection of sites and the location of the air sampling inlet.

1.2 Investigations of atmospheric contaminants involve the study of a heterogeneous mass under uncontrolled conditions. Interpretation of the data derived from the air sampling program must often be based on the statistical theory of probability. Extreme care must be observed to obtain measurements over a sufficient length of time to obtain results that may be considered representative.

1.3 The variables that may affect the contaminant concentrations are the atmospheric stability (temperature-height profile), turbulence, wind speed and direction, solar radiation, precipitation, topography, emission rates, chemical reaction rates for their formation and decomposition, and the physical and chemical properties of the contaminant. To obtain concentrations of gaseous contaminants in terms of weight per unit volume, the ambient temperature and atmospheric pressure at the location sampled must be known.

1.4 *This standard does not purport to address all of the safety concerns, if any, associated with its use. It is the responsibility of the user of this standard to establish appropriate safety and health practices and determine the applicability of regulatory limitations prior to use.*

2. Referenced Documents

2.1 *ASTM Standards:*
D 1356 Terminology Relating to Sampling and Analysis of Atmospheres[2]
D 3249 Practice for General Ambient Air Analyzer Procedures[2]
D 3614 Guide for Laboratories Engaged in Sampling and Analysis of Atmospheres and Emissions[2]

2.2 A list of references are appended to this practice which provide greater details including background information, air quality modeling techniques, and special purposes air sampling programs (**1**).[3]

3. Terminology

3.1 *Definitions*—For definitions of terms used in this practice, refer to Terminology D 1356.

4. Summary of Practice

4.1 This practice describes the general guidelines in planning for sampling the ambient air for the concentrations of contaminants.

5. Significance and Use

5.1 Since the analysis of the atmosphere is influenced by phenomena in which all factors except the method of sampling and analytical procedure are beyond the control of the investigator, statistical consideration must be given to determine the adequacy of the number of samples obtained, the length of time that the sampling program is carried out, and the number of sites sampled. The purpose of the sampling and the characteristics of the contaminant to be measured will have an influence in determining this adequacy. Regular, or if possible, continuous measurements of the contaminant with simultaneous pertinent meteorological observations should be obtained during all seasons of the year. Statistical techniques may then be applied to determine the influence of the meteorological variables on the concentrations measured (**2**).

5.2 Statistical methods may be used for the interpretation of all of the data available (**2**). Trends of patterns and relationships between variables of statistical significance may be detected. Much of the validity of the results will depend, however, on the comprehensiveness of the analysis and the location and contaminant measured. For example, if 24-h samples of suspended particulate matter are obtained only periodically (for example, every 6 or 8 days throughout the year), the geometric mean of the measured concentrations is representative of the median value assuming the data are log normally distributed. The geometric mean level may be used to compare the air quality at different locations at which such regular but intermittent observations of suspended particulate matter are made.

6. Basic Principles

6.1 The choice of sampling techniques and measurement methodology, the characteristics of the sites, the number of sampling stations, and the amount of data collected all depend on the objectives of the monitoring program. These objectives may be one or more of the following:

6.1.1 Air quality assessment including determining maximum concentration,

6.1.2 Health and vegetation effects studies,

6.1.3 Trend analysis,

6.1.4 Evaluation of pollution abatement programs,

[1] This practice is under the jurisdiction of ASTM Committee D-22 on Sampling and Analysis of Atmospheres and is the direct responsibility of Subcommittee D22.03 on Ambient Atmospheres and Source Emissions.
Current edition approved Jan. 15, 1995. Published March 1995. Originally published as D 1357 – 55 T. Last previous edition D 1357 – 94.
[2] *Annual Book of ASTM Standards*, Vol 11.03.
[3] The boldface numbers in parentheses refer to the list of references at the end of this practice.

6.1.5 Establishment of air quality criteria and standards by relating to effects,

6.1.6 Enforcement of control regulations,

6.1.7 Development of air pollution control strategies,

6.1.8 Activation of alert or emergency procedures,

6.1.9 Land use, transportation, and energy systems planning,

6.1.10 Background evaluations, and

6.1.11 Atmospheric chemistry studies.

6.2 In order to cover all the variable meteorological conditions that may greatly affect the air quality in an area, air monitoring for lengthy periods of time may be necessary to meet most of the above objectives.

6.3 The topography, demography, and micrometeorology of the area as well as the contaminant measured, must be considered in determining the number of monitoring stations required in the area. Photographs and a map of the locations of the sampling stations is desirable in describing the sampling station.

6.4 Unless the purpose of the sampling programs is site specific, the sites monitored should, in general, be selected so as to avoid undue influence by any local source that may cause local elevated concentrations that are not representative of the region to be characterized by the data.

6.5 Monitoring sites for determining the impact on air quality by individual sources should be selected, if possible, so as to isolate the effect of the source being considered. When there are many sources of the contaminant in the area, the sites sampled should be strategically located so that with wind direction data obtained simultaneously near the sites, the monitoring results will provide evidence of the contributions of the individual sources. Multiple samplers or monitors operating simultaneously upwind and downwind from the source are often very valuable and efficient.

7. Meteorological Factors

7.1 The meteorological parameters that are most important in an atmospheric sampling program are:

7.1.1 Wind direction and speed, the degree of persistence in direction, and gustiness;

7.1.2 Temperature and its changes with height above ground; the mixing height, that is, the height above ground that the pollutants will diffuse to during the afternoon; and

7.1.3 Solar radiation and hours of sunshine, humidity, precipitation, and barometric pressure. These parameters are important in assessing the pollution potential of an area and should be considered in the planning of a monitoring program and in the interpretation of the data. Pertinent meteorological and climatological information may be obtained from the local weather department. In many localities, however, the micrometeorology may be unique and meteorological investigations to provide data specific to the area may be needed.

7.2 The influences of each of the meteorological parameters important to air quality are discussed in detail. The methods of carrying out the related meteorological investigations are also discussed (3, 4, 5).

8. Topographical Factors

8.1 Topography can influence the contaminant concentrations in the atmosphere. For example, a valley will cause persistence in wind directions and intensify low-level noc-turnal inversions that will limit the dispersion of pollutants emitted into it. Mountains or plateaus may act as barriers affecting the flow of air as well as the contaminant concentrations in their vicinity. Consideration should be given to the influence of these features as well as that of large lakes, the sea, and oceans (2, 3).

9. Sampling Procedure and Siting Concepts

9.1 The choice of procedure for the air sampling is dependent on the contaminant to be measured. See Practice D 3249 for recommendations for general ambient air analyzer procedures. ASTM recommended methods have been published for most of the common contaminants that are sampled. Automatic instruments providing a continuous record of the concentrations of the contaminant should be utilized whenever possible to save manpower and increase efficiency. Very often factors such as temperature, humidity, and vibrations, as well as the power line voltage can influence the output of the air monitoring instrument and these should be controlled.

9.2 The monitors must be supplied with sample air that represents the ambient air under investigation. Careful consideration should be given to the sample conveying system. A duct system is often utilized for this purpose. There should be as few abrupt enlargements and elbows as possible, as these may affect the uniformity and hence the concentration of the contaminants measured. The material for the duct should be such that there will be little or no interaction between it and the contaminants in the air sampled. Employ temperature control to limit the condensation forming in the sampling lines. Take the samples from straight sections of the duct with the inlet lead to the monitoring instrument as short as possible.

9.3 The following guidelines are recommended for sampling locations unless site specific measurements are desired. The height of the inlet to the sampling duct should be normally from 2.5 to 5 m above ground whenever possible. The height of the inlet above the sampling station structure or vegetation adjacent to the station should be greater than 1 m. Sampling should preferably be through a vertical inlet with an inverted cone over the opening. For a horizontal inlet, there should be a minimum of 2 m from the face of the structure. For access to representative ambient air in the area sampled, the elevation angle from the inlet to the top of nearby buildings should be less than 30°. To be representative of the area in which a large segment of the population is exposed to contaminants emitted by automobiles, the inlet should be at a distance greater than 15 m from the nearest high-volume traffic artery. Photochemical oxidants or ozone samplers should be located at distances greater than 50 m from high traffic locations. Particulate matter samplers should be sited at locations that are greater than 200 m from unpaved streets or roads (6, 7, 8).

10. Apparatus

10.1 Details of the apparatus or instruments employed in sampling the air or carrying out associated meteorological investigations are discussed in other ASTM methods and recommendations.

11. Plan of Sampling Procedure

11.1 The procedure for sampling should be undertaken in the following steps:

11.1.1 A general exploratory survey of the area including the topography, an inventory of sources for the contaminants, the height of their emissions, traffic, and land use data.

11.1.2 A preliminary meteorology analysis to identify wind direction frequencies, wind velocity, and temperature-height profiles.

11.1.3 Exploratory short-term temporary sampling requires a number of temporary sampling stations, to determine the need and the best sites for extensive long term monitoring. Using air quality models, and the input of emission inventory and meteorological information for the area obtained in 11.1.1 and 11.1.2, an estimate for the levels of air quality over the area may be calculated. The model results will provide guidance in determining the locations for monitors that will measure the maximum levels and the number of monitors required to characterize the air quality in the area of concern (9, 10).

12. Quality Assurance

12.1 Quality assurance programs include all of the activities necessary to provide measurement data at a requisite precision and accuracy. An air quality assurance program should be developed and implemented for every air monitoring program to ensure compatibility of data both externally and between stations in the monitoring network.

12.2 Guidelines for quality assurance programs are given in Practice D 3614. The quality assurance program should include all the following elements to the extent to which they are applicable:

12.2.1 Sampling and analytical procedures should be specified, using standard methods such as ASTM methods where applicable and appropriate to the objectives to be achieved.

12.2.2 Calibration procedures should be specified in the standard methods. For a continuing program, frequency of calibration re-checks should be specified.

12.2.3 Data collection and recording procedures should be specified, to identify responsibility for record keeping, analysis of recorder chart records, conversion to computer format, and method and frequency of reporting. Specifications should also be included for safeguarding equipment, samples, and records to maintain the chain of evidence if data are required for legal purposes.

12.2.4 Sample shipping and storage procedures should be documented, especially if the methods used impose any limits on time delay prior to analysis, refrigeration, or other storage conditions, or any other precautions in sample handling.

12.2.5 Methods of computation after analysis or automatic data recording should be specified, including data validation procedures used to minimize errors in computation or record keeping.

12.2.6 Independent audits of the entire measurement program by an outside agency are helpful, in which parallel sampling and analysis is conducted independently to verify the precision and accuracy of the final results. Simultaneous analysis of ambient atmospheres can be used, as well as independent measurements using cylinder gases or other calibration standards.

12.2.7 Interlaboratory tests (also called "round-robin" tests) are useful as a means of confirming the precision and accuracy of the methods used, and also make it possible for each organization to check its own performance against other laboratories or organizations engaged in the same type of activity.

12.3 For a continuing monitoring program, a manual is needed to formalize and document the quality assurance practices so that consistent operation of the entire monitoring program is assured over an extended period of time regardless of changes in personnel. This manual should document the methods used and the various sampling, calibration and other procedures mentioned above.

12.4 All monitoring equipment and apparatus should be carefully calibrated using highly reliable calibration standards. Such standards should be traceable to the National Institute of Standards and Technology, as far as possible, to provide a means for intercalibration for the various stations in a monitoring network. The calibration procedures should be described in the quality assurance manual.

13. Keywords

13.1 ambient atmospheres; sampling

REFERENCES

(1) Calvert, S., and Englund, H. M., *Handbook of Air Pollution Technology*, New York, John Wiley & Sons, 1984.

(2) Kornreich, L. D., Proceedings of the Symposium on Statistical Aspects of Air Quality Data. EPA-650/4-74-038, October 1974.

(3) Munn, R. E., *Descriptive Micrometeorology*, Academic Press, New York, NY, 1966.

(4) Slade, D. H., "Meteorology and Atomic Energy," TID-24180 U.S. Atomic Energy Commission, National Bureau of Standards; U.S. Dept. of Commerce, Springfield, VA, 1968.

(5) Morris, A. L., and Barras, R. C., *Air Quality Meteorology and Ozone*, ASTM Special Technical Publication, *STP 653*, Philadelphia, 1978.

(6) OAQPS No. 1.2-012. "Guidance for Air Quality Monitoring Network Design and Instrument Siting," U.S.-E.P.A., September 1975.

(7) EPA-450/3-75-077, "Selecting Sites for Carbon Monoxide Monitoring," September 1975.

(8) EPA-450/3-77-013, "Optimum Site Exposure Criteria for SO_2 Monitoring," April 1977.

(9) Seinfeld, J. H., "Optimal Locations of Pollutant Monitoring Stations in an Airshed," *Atmospheric Environment*, Pergamon Press, Vol 6, 1972, pp. 847–858.

(10) Benarie, M. M. *Urban Air Pollution Modeling*, MIT Press, Cambridge, MA 1980.

Standard Practice for
General Ambient Air Analyzer Procedures[1]

This standard is issued under the fixed designation D 3249; the number immediately following the designation indicates the year of original adoption or, in the case of revision, the year of last revision. A number in parentheses indicates the year of last reapproval. A superscript epsilon (ϵ) indicates an editorial change since the last revision or reapproval.

This standard has been approved for use by agencies of the Department of Defense. Consult the DoD Index of Specifications and Standards for the specific year of issue which has been adopted by the Department of Defense.

1. Scope

1.1 This practice is a general guide for ambient air analyzers used in determining air quality.

1.2 The actual method, or analyzer chosen, depends on the ultimate aim of the user: whether it is for regulatory compliance, process monitoring, or to alert the user of adverse trends. If the method or analyzer is to be used for federal or local compliance, it is recommended that the method published or referenced in the regulations be used in conjunction with this and other ASTM methods.

1.3 *This standard does not purport to address all of the safety concerns, if any, associated with its use. It is the responsibility of the user of this standard to establish appropriate safety and health practices and determine the applicability of regulatory limitations prior to use. For specific hazard statements, see Section 6.*

2. Referenced Documents

2.1 *ASTM Standards:*
D 1356 Terminology Relating to Sampling and Analysis of Atmospheres[2]
D 1357 Practice for Planning the Sampling of the Ambient Atmosphere[2]
D 3609 Practice for Calibration Techniques Using Permeation Tubes[2]
D 3670 Guide for Determination of Precision and Bias of Methods of Committee D-22[2]
E 177 Practice for Use of the Terms Precision and Bias in ASTM Test Methods[3]
E 200 Practice for Preparation, Standardization, and Storage of Standard and Reagent Solutions for Chemical Analysis[4]

3. Terminology

3.1 *Definitions:*

3.1.1 For definitions of terms used in this practice other than those following, refer to Terminology D 1356.

3.1.2 *analyzer*—the instrumental equipment necessary to perform automatic analysis of ambient air through the use of physical and chemical properties and giving either cyclic or continuous output signal.

3.1.2.1 *analyzer system*—all sampling, analyzing, and readout instrumentation required to perform ambient air quality analysis automatically.

3.1.2.2 *sample system*—equipment necessary to provide the analyzer with a continuous representative sample.

3.1.2.3 *readout instrumentation*—output meters, recorder, or data acquisition system for monitoring analytical results.

3.1.3 *full scale*—the maximum measuring limit for a given range of an analyzer.

3.1.4 *interference*—an undesired output caused by a substance or substances other than the one being measured. The effect of interfering substance(s), on the measurement of interest, shall be expressed as: ($\pm$) percentage change of measurement compared with the molar amount of the interferent. If the interference is nonlinear, an algebraic expression should be developed (or curve plotted) to show this varying effect.

3.1.5 *lag time*—the time interval from a step change in the input concentration at the analyzer inlet to the first corresponding change in analyzer signal readout.

3.1.6 *linearity*—the maximum deviation between an actual analyzer reading and the reading predicted by a straight line drawn between upper and lower calibration points. This deviation is expressed as a percentage of full scale.

3.1.7 *minimum detection limit*—the smallest input concentration that can be determined as the concentration approaches zero.

3.1.8 *noise*—random deviations from a mean output not caused by sample concentration changes.

3.1.9 *operating humidity range of analyzer*—the range of ambient relative humidity of air surrounding the analyzer, over which the analyzer will meet all performance specifications.

3.1.9.1 *operating humidity range of sample*—the range of ambient relative humidity of air which passes through the analyzer's sensing system, over which the monitor will meet all performance specifications.

3.1.10 *operational period*—the period of time over which the analyzer can be expected to operate unattended within specifications.

3.1.11 *operating temperature range of analyzer*—the range of ambient temperatures of air surrounding the analyzer, over which the monitor will meet all performance specifications.

3.1.11.1 *operating temperature range of sample*—the range of ambient temperatures of air, which passes through

[1] This practice is under the jurisdiction of ASTM Committee D-22 on Sampling and Analysis of Atmospheres and is the direct responsibility of Subcommittee D 22.03 on Ambient Analyzers.
Current edition approved Jan. 15, 1995. Published March 1995. Originally published as D 3249 – 73 T. Last previous edition D 3249 – 90.
[2] *Annual Book of ASTM Standards*, Vol 11.03.
[3] *Annual Book of ASTM Standards*, Vol 14.02.
[4] *Annual Book of ASTM Standards*, Vol 15.05.

the analyzer's sensing system, over which the analyzer will meet all performance specifications.

3.1.12 *output*—a signal that is related to the measurement, and intended for connection to a readout or data acquisition device. Usually this is an electrical signal expressed as millivolts or milliamperes full scale at a given impedance.

3.1.13 *precision*—see Practice D 3670.

3.1.13.1 *repeatability*—a measure of the precision of the analyzer to repeat its results on independent introductions of the same sample at different time intervals. This is that difference between two such single instrument results, obtained during a stated time interval, that would be exceeded in the long run in only one case in twenty when the analyzer is operating normally.

3.1.13.2 *reproducibility*—a measure of the precision of different analyzers to repeat results on the same sample.

3.1.14 *range*—the concentration region between the minimum and maximum measurable limits.

3.1.15 *response time*—the time interval from a step change in the input concentration at the analyzer inlet to an output reading of 90 % of the ultimate reading.

3.1.16 *rise time*—response time minus lag time.

3.1.17 *span drift*—the change in analyzer output over a stated time period, usually 24 h of unadjusted continuous operation, when the input concentration is at a constant, stated upscale value. Span drift is usually expressed as a percentage change of full scale over a 24-h operational period.

3.1.18 *zero drift*—the change in analyzer output over a stated time period of unadjusted continuous operation when the input concentration is zero; usually expressed as a percentage change of full scale over a 24-h operational period.

4. Summary of Practice

4.1 A procedure for ambient air analyzer practices has been outlined. It presents definitions and terms, sampling information, calibration techniques, methods for validating results, and general comments related to ambient air analyzer methods of analysis. This is intended to be a common reference method which can be applied to all automatic analyzers in this category.

5. Significance and Use

5.1 The significance of this practice is adequately covered in Section 1.

6. Hazards

6.1 Each analyzer installation should be given a thorough safety engineering study.[5]

6.2 Electrically the analyzer system as well as the indi-

TABLE 1 Typical Analyzer Calibration Using Spot Samples

Sample No.	Analyzer Results	ASTM Test Method Average	Difference
1	9.5	9.3	−0.2
2	9.5	9.4	−0.1
3	9.5	9.2	−0.3
4	9.7	9.5	−0.2
5	9.6	9.5	−0.1
6	9.5	9.3	−0.2
7	9.6	9.4	−0.2
8	9.4	9.2	−0.2
9	9.5	9.3	−0.2

Calibration offset = −0.2 (average difference)

vidual components shall meet all code requirements for the particular area classification.

6.2.1 All analyzers using 120-V, a-c, 60-Hz, 3-wire systems should observe proper polarity and should not use mechanical adapters for 2-wire outlets.

6.2.2 The neutral side of the power supply at the analyzer should be checked to see that it is at ground potential.

6.2.3 The analyzer's ground connection should be checked to earth ground for proper continuity.

6.2.4 Any analyzer containing electrically heated sections should have a temperature-limit device.

6.2.5 The analyzer, and any related electrical equipment (the system), should have a power cut-off switch, and a fuse or breaker, on the "hot" side of the line(s) of each device.

6.3 Full consideration must be given to safe disposal of the analyzer's spent samples and reagents.

6.4 Pressure relief valves, if applicable, shall be provided to protect both the analyzer and analyzer system.

6.5 Precautions should be taken when using cylinders containing gases or liquids under pressure. Helpful guidance may be obtained from Ref (1), (2), (3), (4), and (5).[6]

6.5.1 Gas cylinders must be fastened to a rigid structure and not exposed to direct sun light or heat.

6.5.2 Special safety precautions should be taken when using or storing combustible or toxic gases to ensure that the system is safe and free from leaks.

7. Installation of Analyzer System

7.1 Assure that information required for installation and operation of the analyzer system is supplied by the manufacturer.

7.2 Study operational data and design parameters furnished by the supplier before installation.

7.3 Review all sample requirements with the equipment supplier. The supplier must completely understand the application and work closely with the user and installer. It is absolutely necessary to define carefully all conditions of intended operation, components in the atmosphere to be analyzed, and expected variations in sample composition.

7.4 Choose materials of construction in contact with the ambient air sample to be analyzed to prevent reaction of materials with the sample, sorption of components from the sample, and entrance of contaminants through infusion or diffusion (6, 7, 8, 9).

[5] The user, equipment supplier, and installer should be familiar with requirements of the National Electrical Code, any local applicable electrical code, U.L. Safety Codes, and the Occupational Safety & Health Standards (Federal Register, Vol 36, No. 105, Part II, May 29, 1971). Helpful guidance may also be obtained from API RP500, "Classification of Areas for Electrical Installations in Petroleum Refineries"; ISA RP12.1, "Electrical Instruments in Hazardous Atmospheres"; ISA RP12.2, "Intrinsically Safe and Nonincendive Electrical Instruments"; ISA RP12.4, "Instrument Purging for Reduction of Hazardous Area Classification"; and AP RP550, "Installation of Refinery Instruments and Control Systems, Part II."

[6] The boldface numbers in parentheses may be found in the Reference section at the end of this method.

7.4.1 Choose materials of construction and components of the analyzer system to withstand the environment in which it is installed.

7.4.2 Avoid the use of pipe-thread compounds in favor of polytetrafluorethylene tape.

7.5 Select the sampling point so as to provide a representative and measurable sample as close as possible to the sample system and analyzer (see Practice D 1357).

7.5.1 Provide a convenient access to the entire analyzer system.

7.5.2 Provide a necessary connection for introducing standard samples or withdrawing laboratory check samples immediately upstream of the analyzer sampling system.

7.6 Sample lines should be as short as practical.

7.6.1 Install the analyzer's exhaust so that no liquid or gas pressure buildup will occur. Provide proper venting, as far as possible from the sampling point.

7.7 After the installation has been completed, allow the analyzer to stabilize before testing performance specifications.

8. Calibration

8.1 One of the most important steps in analyzer operation is proper calibration of the instrument. Various calibration techniques may be used depending on the sample's physical or chemical property requiring measurement. Frequency of calibration depends largely on the application, degree of accuracy, and reliability expected. Perform calibration using spot samples (ambient) or a standard reference sample and utilize the analyzer adjustments as recommended by the manufacturer. Consult the supplier to determine the calibration procedure necessary for the particular analysis involved as preliminary instrument adjustments using zero and upscale standards may be necessary. Charts and calibration curves are essential and should be routinely verified.

8.1.1 In all cases, standard used for calibration purposes must be as representative as possible of the atmosphere to be analyzed, but cannot always contain all potential interfering substances.

8.2 *Spot Sample Calibration Method*—A sample is removed from the sampling line close to the analyzer inlet during a period when the sample flowing through the line is of uniform composition and the analyzer readout has reached an equilibrium value.

8.2.1 When this condition is reached, withdraw a sample from the inlet stream for analysis using the appropriate ASTM test method for the component of interest.

8.2.2 For most applications, a minimum of nine samples are required, and these shall be withdrawn each cycle for intermittent analyzers or for continuous analyzers after a stable response is achieved.

8.2.3 After each spot sample has been removed, record it as to time, sample number, date and corresponding analyzer readout. This equivalent readout is used in establishing a single calibration point.

8.2.4 Each spot sample must be analyzed in duplicate using the corresponding ASTM test method and the two results averaged. The standard deviation for the spot sample is calculated as the difference (larger value minus the smaller value) divided by $\sqrt{2}$. If this standard deviation exceeds the test method repeatability limit, r, (see Practice E 177) then that test average must be discarded. (This assumes that a repeatability limit has been determined for the test method and the laboratory conducting the test. This rejection criterion will discard 5 % of the spot sample results even if the test method is operating properly.)

8.2.5 Determine the amount of calibration offset by averaging the deviations, as shown in Table 1, and correct the analyzer readout accordingly. It may be necessary to review the manufacturer's recommended procedure for making calibration offset adjustments.

8.3 *Standard Sample Calibration Method*—Use a standard reference sample in accordance with the ASTM test method chosen, or by generating a known sample concentration, using NIST calibrated permeation tubes (see Practice D 3609).

8.3.1 A standard sample benchmark analysis is made by averaging the results of at least nine determinations using the corresponding ASTM test method. This average value is acceptable for benchmark analysis only if the corresponding standard deviation is lower in magnitude than the test method's repeatability limit, r, (see Practice E 177).

8.3.2 Check all operating parameters of the system in accordance with the instrument specifications and data for specific analysis. Allow sufficient time for the analyzer to reach equilibrium as indicated by a stable output.

8.3.3 Introduce the standard reference sample into the analyzer using the recommended instrument operational procedure. Activate the readout equipment.

8.3.4 After sufficient standard has been allowed to flow through the analyzer, adjust the readout to conform with the benchmark value. This establishes a single calibration point.

8.3.5 Continue introducing standard sample and record analysis after a stable response is achieved, or for each cycle if an intermittent analyzer is used, until repeatable data are recorded.

8.3.6 Discard any standard when any change in composition is detected (see Practice E 200).

9. Procedure

9.1 Begin sampling of the atmosphere.

9.2 Check all operating parameters in accordance with the application engineering data and method for specific analysis.

9.3 Observe the sample analysis as indicated by the readout equipment after the analyzer has been thoroughly purged with the sample.

9.4 If it is desirable to validate the analyzer spot sample results, refer to the procedure given in Appendix X1.

9.5 After the analyzer is placed in service there is a continuing need to observe periodically that the original calibration remains valid. Achieve this by applying either the spot sample or standard sample technique used in calibration. The results logged over a period of time will indicate whether or not the analyzer remains within acceptable limits of calibration. Frequency with which these checks are required will be determined by the stability of the analyzer. If the record indicates frequent recalibration to be necessary, make a thorough investigation of the analyzer system to determine the cause of instability.

9.6 Successful operation of the analyzer system depends to a large extent on the amount of maintenance provided. Type of analyzer, complexity of the system, and condition of the sample stream usually determine the maintenance requirements.

10. Calculation

10.1 Each individual analyzer system, and ASTM test method chosen, determines the necessary calculations on the output signal. Most analyses are recorded as direct readouts based on instrument calibration. However, in some cases, the measurement sensitivity range is involved and scale factors are necessary to determine the final results. This is usually a simple multiplication step.

11. Report

11.1 Reports should include information on the analyzer system, calibration or validation used, and analysis of the sample over the time period involved. A report form is described in Appendix X2.

12. Precision

12.1 Preferably, each analyzer system method should include its own precision section based on cooperative test program results. This section would then incorporate the expected limit of deviation of test results from a determined value and be reported as repeatability and reproducibility.

13. Keywords

13.1 ambient air analyzers; ambient air quality

REFERENCES

(1) *Safe Handling of Compressed Gases*, Pamphlet P-1, Compressed Gas Association, Inc., New York, NY.

(2) *Compressed Gases, Safe Practices*, Pamphlet No. 95, National Safety Council, Chicago, IL.

(3) "Handbook of Laboratory Safety," CRC Press, Boca Raton, FL, 1971.

(4) Sax, N. Irving, *Dangerous Properties of Industrial Materials*, 3rd Edition, 1968, Reinhold Book Corp., New York, NY.

(5) Matheson Gas Data Book—Sixth Edition, Matheson Gas Products, East Rutherford, NJ, 1980.

(6) Lebovits, Alexander, "Permeability of Polymers to Gases, Vapors and Liquids," *Modern Plastics*, March 1966, pp. 139–210.

(7) Hendrickson, E. R., "Air Sampling and Quantity Measurement," in Air Pollution (A. C. Stern, Ed.), Vol II, Academic Press, New York, NY, 1968, p. 23.

(8) Wilson, K. W., and Buchberg, H., *Industrial Engineering Chemical 50*, 1958, p. 1705.

(9) Baker, R. A., and Doerr, R. C., *Intl. Journal Air Pollution 2*, 1959, p. 142.

(10) Institute of Petroleum, Code of Practice—The Calibration, Verification and the Reporting of Analyzer Performance.

APPENDIX

X1. ANALYZER VALIDATION

X1.1 *Spot Sample Method:*

X1.1.1 Establish the validity of the spot sample data by comparing the analyzer results with the ASTM laboratory method results using the paired Student's "t" test.

$$t_c = \frac{(\overline{X}_i - \overline{X}_r)\sqrt{n}}{s}$$

where:

X_i = instrument values for the ith sample,
X_r = individual benchmark values,
$\overline{X}_i$ = average instrument value,
$\overline{X}_r$ = average benchmark value,

n = number of spot samples corresponding to instrument and benchmark results, and

$$s = \sqrt{\frac{\Sigma(X_i - X_r)^2 - [\Sigma(X_i - X_r)^2]/n}{n - 1}}$$

X1.1.2 Compare t_c with the values of "t" given in Table X1 for the number of degrees of freedom (df), $n - 1$, used in the calculation. If t_c is equal to or less than the tabulated value for "t," the instrument value can be considered valid. If t_c is greater than the tabulated value of "t," the instrument value differs from the benchmark value and the instrument results cannot be considered valid.

TABLE X1 Student's "t" Test for 95 % Confidence Level

Degrees of Freedom	"t"	Degrees of Freedom	"t"
1	12.706	17	2.110
2	4.303	18	2.101
3	3.182	19	2.093
4	2.776	20	2.086
5	2.571	21	2.080
6	2.447	22	2.074
7	2.365	23	2.069
8	2.306	24	2.064
9	2.262	25	2.060
10	2.228	30	2.042
11	2.201	40	2.021
12	2.179	50	2.008
13	2.160	60	2.000
14	2.145	120	1.980
15	2.131	200	1.960
16	2.120		

X2. REPORT FORM

Name of organization: _____

Date of analysis: _____

Time of analysis: _____

Barometric pressure: _____

Sample temperature: _____

Sample humidity: _____

Sample flow rate: _____

ASTM test method duplicated: _____

Analyzer type, model, serial No.: _____

Readout type, model, serial No.: _____

Analyzer range used: _____

Analyzer reading, uncorrected: _____

Analyzer corrections required, type and amount: _____

Analyzer reading, corrected: _____

Calibration validation used: _____

Date calibrated and validated: _____

5.3 SPECIFIC SAMPLING PROCEDURES

Standard Practice for
Sampling Atmospheres to Collect Organic Compound Vapors (Activated Charcoal Tube Adsorption Method)[1]

This standard is issued under the fixed designation D 3686; the number immediately following the designation indicates the year of original adoption or, in the case of revision, the year of last revision. A number in parentheses indicates the year of last reapproval. A superscript epsilon (ε) indicates an editorial change since the last revision or reapproval.

1. Scope

1.1 This practice covers a method for the sampling of atmospheres for determining the presence of certain organic vapors by means of adsorption on activated charcoal using a charcoal tube and a small portable sampling pump worn by a worker. A list of some of the organic chemical vapors that can be sampled by this practice is provided in Annex A1. This list is presented as a guide and should not be considered as absolute or complete.

1.2 This practice does not cover any method of sampling that requires special impregnation of activated charcoal or other adsorption media.

1.3 *This standard does not purport to address all of the safety concerns, if any, associated with its use. It is the responsibility of the user of this standard to establish appropriate safety and health practices and determine the applicability of regulatory limitations prior to use. A specific safety precaution is given in 9.4.*

2. Referenced Documents

2.1 *ASTM Standards:*
D 1356 Terminology Relating to Atmospheric Sampling and Analysis of Atmospheres[2]
D 3687 Practice for Analysis of Organic Compound Vapors Collected by the Activated Charcoal Tube Adsorption Method[2]
2.2 *NIOSH Standard:*
CDC-99-74-45 Documentation of NIOSH Validation Tests[3]
HSM-99-71-31 Personnel Sampler Pump for Charcoal Tubes; Final Report[3]
2.3 *OSHA Standard:*
CFR 1910 General Industrial OSHA Safety and Health Standard[4]

3. Terminology

3.1 For definitions of terms used in this method, refer to Terminology D 1356.

3.2 Activated charcoal refers to properly conditioned coconut-shell charcoal.

4. Summary of Practice

4.1 Air samples are collected for organic vapor analysis by aspirating air at a known rate through sampling tubes containing activated charcoal, which adsorbs the vapors.

4.2 Instructions are given to enable the laboratory personnel to assemble charcoal tubes suitable for sampling purposes.

4.3 Instructions are given for calibration of the low flow-rate sampling pumps required in this practice.

4.4 Information on the correct use of sampling devices is presented.

4.5 Practice D 3687 describes a practice for the analysis of these samples.

5. Significance and Use

5.1 Promulgations by the Federal Occupational Safety and Health Administration (OSHA) in 29 CFR 1910.1000 designate that certain organic compounds must not be present in workplace atmospheres at concentrations above specific values.

5.2 This practice, when used in conjunction with Practice D 3687, will provide the needed accuracy and precision in the determination of airborne time-weighted average concentrations of many of the organic chemicals given in 29 CFR 1910.1000, CDC-99-74-45 and HSM-99-71-31.

5.3 A partial list of chemicals for which this method is applicable is given in Annex A1, along with their OSHA permissible exposure limits.

6. Interferences

6.1 Water mist and vapor can interfere with the collection of organic compound vapors. Humidity greater than 60 % can reduce the adsorptive capacity of activated charcoal to 50 % for some chemicals (**1**).[5] Presence of condensed water droplets in the sample tube will indicate a suspect sample.

7. Apparatus

7.1 *Charcoal Tube:*

7.1.1 A sampling tube consists of a length of glass tubing containing two sections of activated charcoal which are held in place by nonadsorbant material and sealed at each end.

7.1.1.1 Sampling tubes are commercially available. The tubes range in size from 100/50 to 800/400 mg, which means the tubes are divided into two sections with the front section

[1] This practice is under the jurisdiction of ASTM Committee D-22 on Sampling and Analysis of Atmospheres, and is the direct responsibility of Subcommittee D22.04 on Analysis of Workplace Atmospheres.
Current edition approved Jan. 15, 1995. Published March 1995. Originally published as D 3686 – 78. Last previous edition D 3686 – 89.
[2] *Annual Book of ASTM Standards*, Vol 11.03.
[3] Available from the U.S. Department of Commerce, National Technical Information Service, Port Royal Road, Springfield, VA 22161.
[4] Available from Superintendent of Documents, U.S. Government Printing Office, Washington, DC 20402.

[5] The boldface numbers in parentheses refer to the list of references at the end of this standard.

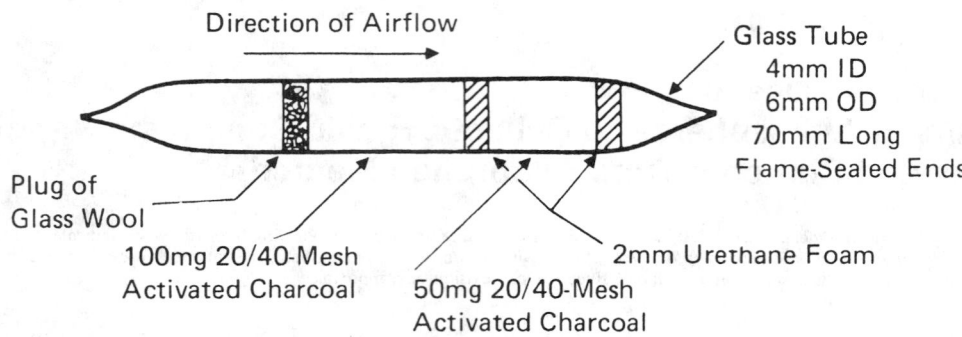

FIG. 1 Activated Charcoal Adsorption Sampling Tube

containing 100 to 800 mg of activated charcoal and the back section containing 50 to 400 mg of activated charcoal. The 100/50-mg tube ((2, 3, 4) and Fig. 1) which is the one most frequently used, consists of a glass tube, 70-mm long, 6-mm outside diameter, 4-mm inside diameter, and contains two sections of 20/40 mesh-activated charcoal but separated by a 2-mm section of urethane foam. The front section of 100 mg is retained by a plug of glass wool, and the back section of 50 mg is retained by either a second 2-mm portion of urethane foam or a plug of glass wool. Both ends of the tube are flame-sealed.

NOTE 1—Urethane foam is known to adsorb certain pesticides (5), for which this practice is contraindicated.

7.1.1.2 When it is desirable to sample highly volatile compounds for extended periods, or at a high volume flow rate, a larger device capable of efficient collection can be used, provided the proportions of the tube and its charcoal contents are scaled similarly to the base dimensions, to provide nominally the same linear flow rate and contact time with the charcoal bed.

7.1.2 The back portion of the sampler tube, which may contain between 25 and 100 % of the mass of activated charcoal present in the front section, adsorbs vapors that penetrate the front section and serves as a warning that breakthrough may have occurred. (Annex A1 gives recommended maximum tube loading information for many chemicals.)

7.1.2.1 Should analysis of the back portion show it to contain more than 10 % of the amount found in the front section, the possibility exists that solvent vapor penetrated both sections of charcoal, and the sample must be considered suspect. These percentages apply to 100/50-mg tubes. For other size tubes having disproportionate amounts of charcoal in the front and back sections, the percentages used to indicate potential breakthrough must be adjusted to take into account different ratios of charcoal. If results from the analysis of suspect samples are used to calculate vapor concentrations, the results must be reported as equal to or greater than the calculated concentrations. In such cases, the test must be repeated for confirmation of vapor concentration.

NOTE 2—Reportings from suspect samples would have significance when health standards are clearly exceeded and the amount by which they are exceeded is academic. (See 9.5.)

7.1.3 The adsorptive capacity and desorption efficiency of different batches of activated charcoal may vary. Commercial tubes, if used, should be purchased from the same batch

and in sufficient number to provide sampling capacity for a definite period of time. *Care must be taken to have enough tubes from the same batch for a given study.*

7.1.3.1 The desorption efficiency and contamination level of a batch of tubes should be determined, following the procedure outlined in Practice D 3687 for activated charcoal. A random selection of at least five charcoal tubes from a specified lot should be taken for these checks.

7.1.4 Pressure drop across the sampling tube should be less than 25 mm Hg (3.3 kPa) at a flow rate of 1000 mL/min and less than 4.6 mm Hg (0.61 kPa) at a flow rate of 200 mL/min.

7.1.5 Charcoal sampling tubes prepared in accordance with this practice and with sealed glass ends may be stored indefinitely.

7.2 *Sampling Pumps:*

7.2.1 Any pump whose flow rate can be accurately determined and be set at the desired sampling rate is suitable. Primarily though, this practice is intended for use with small personal sampling pumps.

7.2.2 Pumps having stable low flow rates (10 to 200 mL/min) are preferable for long period sampling (up to 8 h) or when the concentration of organic vapors is expected to be high. Reduced sample volumes will prevent exceeding the adsorptive capacity of the charcoal tubes. (Suggested flow rates and sampling times are given in Annex A1 for anticipated concentration ranges. Sample volumes are also discussed in 9.5.)

7.2.3 Pumps are available that will provide stable flow rates between ±5 %. Pumps should be calibrated before and after sampling. If possible, flow rates should be checked during the course of the sampling procedure.

7.2.4 All sampling pumps must be carefully calibrated with the charcoal tube device in the proper sampling position. (See Annex A2 for calibration procedure.)

8. Reagents

8.1 *Activated Coconut-Shell Charcoal*—Prior to being used to make sampling devices the charcoal should be heated in an inert gas to 600°C and held there for 1 h. Commercially available coconut charcoal (20/40 mesh) has been found to have adequate adsorption capacity. Other charcoals can be used for special applications.

9. Sampling with Activated Charcoal Samplers

9.1 *Calibration of the Sampling System*—Calibrate the sampling system, including pump, flow regulator, tubing to be used, and a representative charcoal tube (or an equivalent induced resistance) with a primary or calibrated secondary flow-rate standard to ±5%.

9.1.1 A primary standard practice is given for the calibration of low flow-rate pumps in Annex A2 and Fig. A2.1.

9.2 Break open both ends of the charcoal tube to be used for sampling, ensuring that each opening is at least one half the inside diameter of the tube.

9.3 Insert the charcoal tube into the sampling line, placing the back-up section nearest to the pump. At no time should there be any tubing ahead of the sampling tubes.

9.4 For a breathing zone sample, fasten the sampling pump to the worker, and attach the sampling tube as close to the worker's breathing zone as possible. Position the tube in a vertical position to avoid channeling of air through the adsorber sections.

NOTE 3: **Warning**—Assure that the presence of the sampling equipment is not a safety hazard to the worker.

9.4.1 Turn on the pump and adjust the flow rate to the recommended sampling rate.

9.4.2 Record the flow rate and starting time or, depending on the make of pump used, the register reading.

9.5 *Sampling Volumes*—The minimum sample volume will be governed by the detection limit of the analytical method, and the maximum sample volume will be determined by either the adsorptive capacity of the charcoal or limitations of the pump battery.

9.5.1 One method of calculating required sample volumes is to determine first the concentration range, over which it is important to report an exact number, for example from 0.2 to 2 times the permissible exposure concentration, and then calculate the sample volumes as follows:

Minimum sample volume, m^3

$$= \frac{\text{minimum detection limit, mg}}{0.2 \times \text{permissible exposure limit,mg/m}^3}$$

Maximum sample volume, m^3

$$= \frac{\text{tube capacity for vapors, mg}}{2 \times \text{permissible exposure limit, mg/m}^3}$$

9.5.2 Select a sampling rate that, in the sampling time desired, will result in a sample volume between the minimum and maximum calculated in 9.5.1.

9.5.2.1 Generally a long sampling time at a low flow rate is preferable to short-term high-volume sampling. This is consistent with the fact that most health standards are based on 8-h/day time-weighted averages of exposure concentrations.

9.5.2.2 A sample flow rate of less than 10 mL/min, however, should not be used. Calculations based upon diffusion coefficients for several representative compounds indicate that sampling at less than 10 mL/min may not give accurate results.[6]

9.5.2.3 Approximate sample volumes and sample times are given in Annex A1.

9.5.3 When spot checks are being made of an environment, a sample volume of 10 L is adequate for determining vapor concentrations in accordance with exposure guidelines.

9.6 At the end of the sampling period recheck the flow rate, turn off the pump, and record all pertinent information: time, register reading, and if pertinent, temperature, barometric pressure, and relative humidity.

9.6.1 Seal the charcoal tube with the plastic caps provided.

9.6.2 Label the tube with the appropriate information to identify it.

9.7 At least one charcoal sampling tube should be presented for analysis as a field blank with every 10 or 15 samples, or for each specific inspection or field study.

9.7.1 Break the sealed ends off the tube and cap it with the plastic caps. Do not draw air through the tube, but in all other ways treat it as an air sample.

9.7.2 The purpose of the field blank is to assure that if the sampling tubes adsorb vapors extraneous to the sampling atmosphere, the presence of the contaminant will be detected.

9.7.3 Results from the field blanks shall not be used to correct sample results. If a field blank shows contamination, the samples taken during the test must be assumed to be contaminated.

9.8 *Calculation of Sample Volume:*

9.8.1 For sample pumps with flow-rate meters:

$$\text{Sample volume, mL} = f \times t \left(\sqrt{\frac{P_1}{P_2} \times \frac{T_2}{T_1}} \right)$$

where:

f = flow rate sampled, mL/min,
t = sample time, min,
P_1 = pressure during calibration of sampling pump (mm Hg or kPa)
P_2 = pressure of air sampled (mm Hg or kPa)
T_1 = temperature during calibration of sampling pump (K), and
T_2 = temperature of air sampled (K).

9.8.2 For sample pumps with counters:

Sample volume, mL

$$V = \frac{(R_2 - R_1) \times V}{I} \times \frac{P_1}{760} \times \frac{298}{T_1 + 273}$$

where:

R_2 = final counter reading,
R_1 = beginning counter reading,
V = volume, (1) mL-count (1)
P_1 = barometric pressure, mm Hg,
T_1 = temperature, °C, and
V = total sample volume, mL.

10. Handling and Shipping of Samples Collected on Charcoal Sampling Tubes

10.1 There is a paucity of information on the possible fate of the many different chemical species that can be collected in activated charcoal and the variety of conditions to which these samples may be exposed. Good practice suggests the following:[7]

---

[6] Heitbrink, W. A., "Diffusion Effects Under Low Flow Conditions," American Industrial Hygiene Association Journal, Vol 44, No. 6, 1983, pp. 453–462.

[7] Two studies that present information pertinent to this section are:
Saalwaechter, A. T., et al, "Performance Testing of the NIOSH Charcoal Tube Technique for the Determination of Air Concentrations of Organic Vapors," *American Industrial Hygiene Association Journal,* Vol 38, No. 9, September 1977, pp. 476–486.
Hill, R. H., Jr., et al, "Gas Chromatographic Determination of Vinyl Chloride in Air Samples Collected on Charcoal," *Analytical Chemistry,* Vol 48, No. 9, August 1976, pp. 1395–1398.

10.1.1 Samples should be capped securely and identified clearly.

10.1.2 Samples collected in charcoal tubes should not be kept in warm places or exposed to direct sunlight.

10.1.3 Samples of highly vaporous or low-boiling materials, such as vinyl chloride, should be stored and transported in dry ice.

10.1.4 At present there are no published test data on the effect of conditions in aircraft cargo holds on capped samples. The preferred procedure is to carry the samples on board.

10.1.5 Samples should be shipped as soon as possible, stored under refrigeration until they are analyzed, and analyzed if possible within five working days.

10.1.6 Migration or equilibration of the sampled material within the sampling tube during prolonged or adverse storage or handling could be interpreted as break-through. This can be prevented by separating the front and back sections immediately after sampling, by having each section in a separate tube and capping them separately.

10.1.7 In some situations, circumstances and facilities may permit making up calibration standards at the facility where the study is being made and submitting these standards as quality control checks. (See Practice D 3687 for recommended procedure for making up standards.)

10.1.8 Bulk solvent samples should never be shipped or stored with the collected air samples.

11. Keywords

11.1 activated charcoal tube; air monitoring; charcoal tube; organic vapors; sampling and analysis; workplace atmospheres

ANNEXES

(Mandatory Information)

A1. INFORMATION OF SOME ORGANIC COMPOUND VAPORS THAT CAN BE COLLECTED ON COCONUT-SHELL CHARCOAL
(100/50 mg tubes)

Substance PEL ppm-mg/m³ [A]	Recommended Sampling Rate, mL/min to Detect Approximately 15 to 200 % of PEL in Time Given [B]			Recommended Maximum Tube Loading, mg [D]	Approximate Desorption Efficiency % [E]	Eluent	GC Column [F]	CV_T [G]
	2h	4h	8h					
Acetone, 1000–2400	10	[c]	[c]	9	86 ± 10	CS_2	3	0.082
Acetonitrile, 40–70	50	25	25	2.7				0.072
Allyl alcohol, 2–4.8	200	100	50	<0.4	89 ± 5	CS_2 + 5 % 2-propanol	2	0.11
n-Amyl acetate, 100–525	50	25	10	15	86 ± 5	CS_2	4	0.051
sec-Amyl acetate, 125–650	50	25	10	15.5	91 ± 10	CS_2	4	0.071
Isoamyl alcohol, 100–360	50	25	10	10		CS_2 + 5 % 2-propanol	2	0.077
Benzene, 10–31.3	100	100	50		96	CS_2	1	0.060
Benzyl chloride, 1–5	—	200	200	<0.4	90 ± 5	CS_2	2	0.096
Butadiene, 1000–2200	10	[c]	[c]	4		CS_2	1	0.058
2-Butoxy ethanol, 50–240	100	50	25		99 ± 5	methylene chloride + 5 % methanol	2	0.060
n-Butyl acetate, 150–710	50	25	10	15	95	CS_2	4	0.069
sec-Butyl acetate, 200–950	50	25	10	15	91 ± 5	CS_2	4	0.054
tert-Butyl acetate, 200–950	50	25	10	12.5	94 ± 5	CS_2	4	0.091
Butyl alcohol, 100–300	100	50	25	10.5	88 ± 5	CS_2 + 1 % 2-propanol	2	0.065
sec-Butyl alcohol, 150–450	50	25	10	6	93 ± 5	CS_2 + 1 % 2-propanol	2	0.066
tert-Butyl alcohol, 100–300	50	25	10	5	90 ± 5	CS_2 + 1 % 2-propanol	2	0.075
Butyl glycidyl ether, 50–270	100	50	25	11.5	86 ± 10	CS_2		0.074
p-tert-Butyl toluene, 10–60	100	50	25	2.5	100+	CS_2	2	0.067
Camphor, 0.32–2	200	100	50	13.4	98 ± 5	CS_2 + 1 % methanol	2	0.074
Carbon disulfide, 20–60	200	100	50		95	benzene	8	0.059
Carbon tetrachloride, 10–65	200	100	50	7.5	97 ± 5	CS_2	1	0.092
Chlorobenzene, 75–350	50	25	10	15.5	90 ± 5	CS_2	2	0.056
Chlorobromomethane 200–1050	25	10	[c]	9.3	94 ± 5	CS_2	2	0.061
Chloroform, 50–240	100	50	25	11	96 ± 5	CS_2	1	0.057
Cumene, 50–245	50	25	10	11	100+	CS_2	2	0.059
Cyclohexane, 300–1050	25	10	[c]	6.3	100+	CS_2	3	0.066
Cyclohexanol, 50–200	100	50	25	10	99 ± 5	CS_2 + 5 % 2-propanol	2	0.080
Cyclohexanone, 50–200	100	50	25	13	78 ± 5	CS_2	2	0.062
Cyclohexene, 300–1015	25	10	[c]		100+	CS_2	3	0.073
Diacetone alcohol, 50–240	100	50	25	12	77 ± 10	CS_2 + 5 % 2-propanol	2	0.101
o-Dichlorobenzene 50–300	50	25	10	15	85 ± 5	CS_2	6	0.067
1,1-Dichloroethane, 100–405	50	25	10	7.5	100+	CS_2	2	0.057

A1 *Continued*

Substance PEL ppm-mg/m³ A	Recommended Sampling Rate, mL/min to Detect Approximately 15 to 200 % of PEL in Time Given B			Recommended Maximum Tube Loading, mg D	Approximate Desorption Efficiency % E	Eluent	GC Column F	CV_T G
	2h	4h	8h					
1,2-Dichloroethylene, 200–790	25	10	c	5.1	100+	CS₂	2	0.052
p-Dioxane, 100–360	100	50	25	13	91 ± 5	CS₂	1	0.054
Dipropylene glycol methyl ether, 100–600	25	10	c		75 ± 15	CS₂	2	0.064
2-Ethoxyethyl acetate, 100–500	50	25	10	19	74 ± 10	CS₂	4	0.062
Ethyl acetate, 400–1400	25	10	c	12.5	89 ± 5	CS₂	4	0.058
Ethyl acrylate, 25–100	200	100	50	<5	95 ± 5	CS₂	4	0.054
Ethyl alcohol, 1000–1885	c	c	c	2.6	77 ± 10	CS₂ + 1 % 2-butanol	2	0.065
Ethyl benzene, 100–435	200	100	50	16	100+	CS₂	2	0.041
Ethyl bromide, 200–890	100	50	25	7.1	83 ± 5	isopropanol	2	0.054
Ethyl butyl ketone, 50–230	50	25	10	<5.5	93 ± 5	CS₂ + 1 % methanol	2	0.086
Ethyl ether 400–1210	10	c	c	7.5	98 ± 5	ethyl acetate	3	0.053
Ethyl formate, 100–300	50	25	10	4.8	80 ± 10	CS₂	1	0.074
Ethylene dibromide, 20–155	100	50	25	<10.7	93 ± 5	CS₂	2	0.077
Ethylene dichloride, 50–202.5	100	50	25	12	95 ± 5	CS₂	6	0.079
Glycidol, 50–150	100	50	25	22.5	90 ± 5	tetrahydrofuran	2	0.080
Heptane, 500–2000	10	c	c	12.5	96 ± 5	CS₂	6	0.056
Hexane, 500–1800	10	c	c	11	94 ± 5	CS₂	1	0.062
Isoamyl acetate, 100–525	50	25	10	16.5	90 ± 5	CS₂	4	0.056
Isoamyl alcohol, 100–360	50	25	10	10	99 ± 5	CS₂ + 5 % 2-propanol	2	0.065
Isobutyl acetate, 150–700	50	25	10	14	92 ± 5	CS₂	4	0.065
Isobutyl alcohol, 100–305	50	25	10	10.5	84 ± 10	CS₂ + 1 % 2-propanol	2	0.073
Isopropyl acetate, 250–950	25	10	c	13	85 ± 5	CS₂	4	0.067
Isopropyl alcohol 400–985	25	10	c	5.6	94 ± 5	CS₂ + 1 % 2-butanol	2	0.064
Isopropyl glycidyl ether, 50–240	100	50	25	10.5	80 ± 10	CS₂	2	0.067
Mesityl oxide, 25–100	100	50	25	4.8	79 ± 5	CS₂ + 1 % methanol	2	0.071
Methyl acetate, 200–610	25	10	c	7	88 ± 5	CS₂	1	0.055
Methyl acrylate, 10–35	200	100	50	<1.5	80 ± 10	CS₂	4	0.066
Methylal, 1000–3110	10	c	c	11.5	78 ± 10	hexane	3	0.06
Methyl amyl ketone, 100–465	50	25	10	7.5	80 ± 10	CS₂ + 1 % methanol	2	0.061
Methyl butyl ketone, 100–410	50	25	10	2.0	79 ± 10	CS₂	2	0.053
Methyl cellosolve, 25–80	100	50	25	10	97 ± 5	methylene chloride + 5 % methanol	2	0.068
Methyl cellosolve acetate, 25–120	100	50	25	5	76 ± 10	CS₂	4	0.068
Methyl chloroform, 350–1900	25	10	c	18	98+	CS₂	6	0.054
Methyl cyclohexane, 500–2000	10	c	c		95 ± 5	CS₂	1	0.052
Methyl ethyl ketone, 200–590	50	25	10	9.5	89 ± 10	CS₂	2	0.072
Methyl isobutyl carbinol, 25–105	200	100	50	5.7	99 ± 5	CS₂ + 5 % 2-propanol	2	0.080
a-Methyl styrene, 100–480	100	50	25	21	91 ± 5	CS₂	2	0.054
Methylene chloride, 500–1740	10	c	c	9.3	95 ± 5	CS₂	1	0.073
Naphtha (coal tar), 100–400	100	50	25	14.8	88 ± 5	CS₂	7	0.051
n-octane, 500–2350	10	c	c	15	93 ± 5	CS₂	1	0.060
Pentane, 1000–2950	10	c	c	9	96 ± 5	CS₂	1	0.055
2-Pentanone, 200–700	25	10	c		88 ± 5	CS₂	2	0.063
Perchloroethylene, 100–680	50	25	10	29	95 ± 5	CS₂	6	0.052
Petroleum distillates, 500–2000	10	c	c	12.3	96 ± 5	OS₂	6	0.052
Phenyl ether vapor, 1–7		200	200	0.6	90 ± 5	CS₂	2	0.070
Phenyl glycidyl ether, 10–59	100	50	25	12.5	97 ± 5	CS₂	2	0.057
n-Propyl acetate, 200–840	50	25	10	14.5	93 ± 5	CS₂	4	0.056
n-Propyl alcohol, 200–490	50	25	10	9	87 ± 5	CS₂ + 1 % 2-propanol	2	0.075
Propylene dichloride, 75–350	50	25	10	5	97 ± 5	CS₂	2	0.056
Propylene oxide, 100–240	25	10	c	2	90 ± 5	CS₂	3	0.085
Pyridine, 5–15	200	100	50	<7.3	70 ± 10	CS₂	2	0.059
Stoddard solvent, 500–2950	10	c	c	13	96 ± 5	CS₂	7	0.052
Styrene (monomer), 100–425	100	50	25	18	87 ± 5	CS₂	2	0.057
1,1,1,2-Tetrachloro-2,2-difluoro-ethane, 500–4170	10	c	c	19.5	100+	CS₂	2	0.069
1,1,1,2-Tetrachloro-1,2-difluoro-ethane, 500–4170	10	c	c	26	96 ± 5	CS₂	2	0.054
Tetrahydrofuran, 200–590	25	10	c	7.5	92 ± 5	CS₂	3	0.055
1,1,2-Trichloroethane, 10–55	100	50	25	5	96 ± 5	CS₂	6	0.057
Trichloroethylene, 100–535	100	50	25	21	96 ± 5	CS₂	6	0.082
1,1,2-Trichloro-1,2,2-trifluoroethane, 1000–7660	10	c	c	20	100+	CS₂	5	0.07
Turpentine, 100–560	50	25	10	13	96 ± 5	CS₂	7	0.055
Vinyl toluene, 100–480	100	50	25	17	85 ± 10	CS₂	2	0.058

A1 *Continued*

[A] *Substances*—The list does not contain all compounds for which the method is applicable. It lists only those for which reliable data could be obtained. PEL-Federal Permissible Exposure Limits, as given in the *Federal Register*, June 1974, and updated May 1976. These values, which may be either ceiling limits or 8-h/day average exposure limits, depending on the compound, are presented to give guidance in selecting sampling rates and times. These values are subject to change by the Federal Occupational Safety and Health Administration.

[B] *Recommended Sampling Rate*—The suggested sampling rates for the different sampling periods are sufficient to provide a tube loading of at least 0.01 mg when concentrations are 15 % of the PEL., but will not exceed the recommended tube loading when atmosphere are 200 % of the PEL. These figures are based on the 100-mg coconut-shell charcoal tubes described in this practice.

[C] Sample rates of less than 10 mL/min are not recommended. Shorter sampling periods are required.

[D] *Recommended Maximum Tube Loading*—These values are conservative, to allow for high humidity or the presence of other substances which reduce the normal tube capacity.

[E] *Approximate Desorption Efficiency*—These figures are given only as guides for carrying out system calibrations. Actual desorption efficiencies should always be determined at the time of analysis, and any significant deviation should be regarded as a possible indication of a systematic error in the analytical technique. The figure given for desorption efficiency is an average figure. The desorption efficiency for a compound will vary with the amount; in most cases, the desorption efficiency will be lower for reduced tube loadings.

[F] *Gas Chromatographic Columns*—key:

1—20-ft × 1/8 in: ss packed with 10 % FFAP on Chromosorb W AW
2—10-ft × 1/8 in: ss packed with 10 % FFAP on Chromosorb W AW
3—4-ft × 1/4 in: ss packed with 60/80 Porapak Q
4—10-ft × 1/8 in: ss packed with 5 % FFAP on Supelcoport
5—6-ft × 1/4 in: ss packed with 60/80 Porapak Q
6—10-ft × 1/8 in: ss packed with 10 % OV-101 on Supelcoport
7—6-ft × 1/8 in: ss packed with 1.5 % OV-101 on Chromosorb W AW
8—6-ft × 1/4 in: Glass column packed with 5 % OV-17 on Supelcoport

[G] CV_T—Coefficient of variation (that is, relative standard deviation) of the total (net) error in the method (including variability of the pump).

$$CV_T = (CV_{A+\overline{DE}}^2 + CV_S^2 + CV_P^2)^{1/2}$$

where:

CV_{A+DE} = coefficient of variation of a single future assay including error in the desorption efficiency factor $\overline{DE}$,
CV_S = coefficient of variation due to sampling errors (not including variable of the pump) along with variability in true desorption efficiency from tube-to-tube, and
CV_P = coefficient of variation due to pump ($CV_P = 0.05$ assumed).

Acknowledgements: The information in this table comes from NIOSH Standards Completion Program.[8] We gratefully acknowledge NIOSH's contribution to this table, by making available previously unpublished CV_T data, and we acknowledge having used summaries of SCP data prepared by MDA Scientific, Inc., Park Ridge, IL, SKC Corp., Eighty-Four, PA, and Supelco, Inc., Bellefonte, PA.

---

[8] Taylor, D. G., Kupel, R. E., and Bryant, J. M., "Documentation of NIOSH Validation Tests," DHEW (NIOSH), Pub. No. 77-185. Available from National Technical Information Service, Springfield, VA 22161 (PB274-248).

A2. METHOD FOR CALIBRATION OF SMALL VOLUME AIR PUMPS

A2.1 Using a buret that approximately represents a 1-min sampling volume, assemble the apparatus as shown in Fig. A2.1 using any good soap bubble solution as a source of the film. Make sure all connections are tight.

A2.1.1 It is advisable to check the volume of burets used for calibrating sampling pumps by weighing the volume of water contained in the buret and calculating the true volume.

A2.1.2 Make sure the batteries of the pump are charged.

A2.2 Prime the surface of the cylinder with bubble solution by drawing repeated films up the tube until a single film travels to the desired mark.

A2.3 With a stop watch, time the travel of a single film from an initial zero mark to a selected volume mark. Note the time and repeat this procedure at least three times.

A2.4 Calculate the sampling rate of the pump, correcting the air volume to 25°C and 760 mm Hg (101.3 kPa), using the ambient barometric pressure.

A2.5 Replace the charcoal tube sampler with another one selected at random, and repeat the calibration sequence.

A2.5.1 Sampling tubes should consistently meet the pressure drop criterion given in 7.1.4.

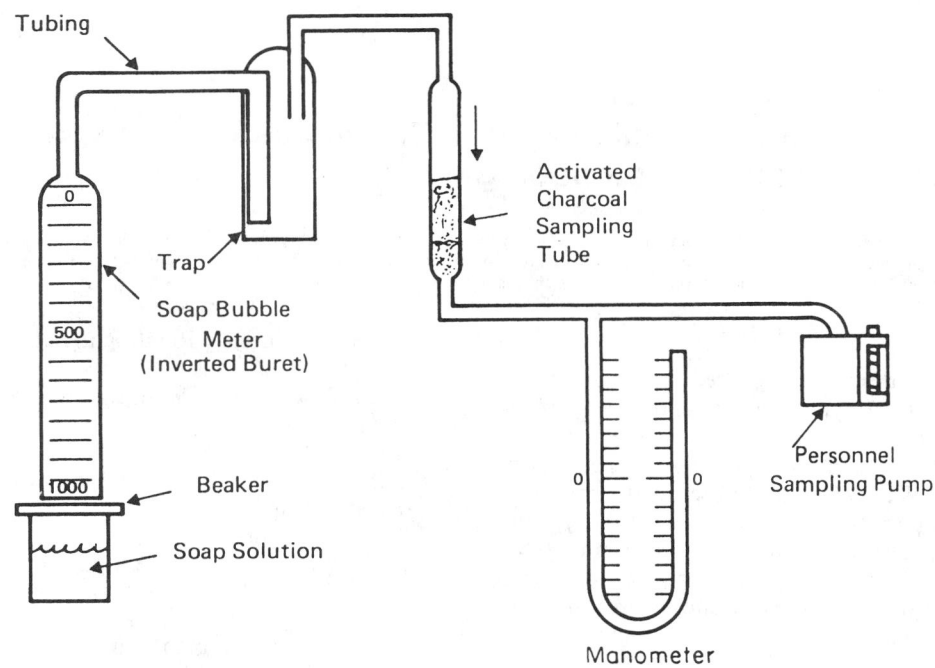

FIG. A2.1 Calibration Setup for Personnel Sampling Pump with Activated Charcoal Sampling Tube

REFERENCES

(1) "Second NIOSH Solid Sorbents Roundtable," Ed. E. V. Ballou, 1976, NIOSH Publication No. 76–193.

(2) White, L. D., Taylor, D. G., Mauer, P. A., and Kupel, R. E., "A Convenient Optimized Method for the Analysis of Selected Solvent Vapors in the Industrial Atmosphere," *American Industrial Hygiene Association Journal,* Vol 31, 1970, pp. 225–232.

(3) Otterson, E. J., and Guy, C. U., "A Method of Atmospheric Solvent Vapor Sampling on Activated Charcoal in Connection with Gas Chromatography," *Transactions of the 26th Annual Meeting of the American Conference of Governmental Industrial Hygienists,* Philadelphia, 1964, pp. 37–46.

(4) Reid, F. H., and Halpin, W. R., "Determination of Halogenated and Aromatic Hydrocarbons in Air by Charcoal Tube and Gas Chromatography," *American Industrial Hygiene Association Journal,* July-August 1968, pp. 390–396.

(5) Turner, B. C., and Glotfelty, D. E., "Field Sampling of Pesticide Vapors with Polyurethane Foam," *Analytical Chemistry,* Vol 49, 1977, pp. 7–10.

Standard Test Method for
Determination of Volatile Organic Chemicals in Atmospheres (Canister Sampling Methodology)[1]

This standard is issued under the fixed designation D 5466; the number immediately following the designation indicates the year of original adoption or, in the case of revision, the year of last revision. A number in parentheses indicates the year of last reapproval. A superscript epsilon (ε) indicates an editorial change since the last revision or reapproval.

1. Scope

1.1 This test method describes a procedure for sampling and analysis of volatile organic compounds (VOCs) in ambient, indoor, or workplace atmospheres. The test method is based on the collection of air samples in passivated stainless steel canisters. The VOCs are subsequently removed from the canisters, separated by gas chromatography, and measured by a mass spectrometric detector. This test method describes procedures for sampling into canisters to final pressures both above and below atmospheric pressure (respectively referred to as pressurized and subatmospheric pressure sampling).[2]

1.2 This test method is applicable to specific VOCs that have been tested and determined to be stable when stored in pressurized and subatmospheric pressure canisters. Numerous compounds, many of which are chlorinated VOCs, have been successfully tested for storage stability in pressurized canisters (1, 2).[3] While not as extensive, documentation is currently also available demonstrating stability of VOCs in subatmospheric pressure canisters.

1.3 The organic compounds that have been successfully collected in pressurized canisters by this test method are listed in Table 1. These compounds have been successfully measured at the parts per billion by volume (ppbv) level. This test method is applicable to concentrations of VOC from the detection limit to 300 ppb by volume. Above this concentration samples require dilution with dry ultra high purity nitrogen or air.

1.4 *This standard does not purport to address all of the safety concerns, if any, associated with its use. It is the responsibility of the user of this standard to establish appropriate safety and health practices and determine the applicability of regulatory limitations prior to use.* Safety practices should be part of the user's SOP manual.

2. Referenced Documents

2.1 *ASTM Standards:*

D 1356 Terminology Relating to Sampling and Analysis of Atmospheres[4]
D 1357 Practice for Planning and Sampling of the Ambient Atmosphere[4]
E 260 Practice for Packed Column Gas Chromatography[5]
E 355 Practice for Gas Chromatography Terms and Relationships[5]

2.2 *Other Documents:*

U.S. Environmental Protection Agency Technical Assistance Document for Sampling and Analysis of Toxic Organic Compounds in Ambient Air (3)
Laboratory and Ambient Air Studies (4–20)

3. Terminology

3.1 *Definitions*—For definitions of terms used in this test method, refer to Terminology D 1356. Other pertinent abbreviations and symbols are defined within this practice at point of use.

3.2 *Descriptions of Terms Specific to This Standard:*

3.2.1 *absolute canister pressure*—$Pg + Pa$, where Pg = gage pressure in the canister. (KPa, psi) and Pa = barometric pressure (see 5.2).

3.2.2 *absolute pressure*—pressure measured with reference to absolute zero pressure (as opposed to atmospheric pressure), usually expressed as kPa, mm Hg, or psia.

3.2.3 *certification*—the process of demonstrating with humid zero air and humid calibration gases that the sampling systems components and the canister will not change the concentrations of sampled and stored atmospheres.

3.2.4 *cryogen*—a refrigerant used to obtain very low temperatures in the cryogenic trap of the analytical system. A typical cryogen is liquid argon (bp −185.7°C) or liquid nitrogen (bp −195°C).

3.2.5 *dynamic calibration*—calibration of an analytical system using calibration gas standard concentrations generated by diluting known concentration compressed gas standards with purified, humidified inert gas.

3.2.5.1 *Discussion*—Such standards are in a form identical or very similar to the samples to be analyzed. Calibration standards are introduced into the inlet of the sampling or analytical system in the same manner as authentic field samples.

3.2.6 *gage pressure*—pressure measured above ambient atmospheric pressure (as opposed to absolute pressure). Zero gage pressure is equal to ambient atmospheric (barometric) pressure.

---

[1] This test method is under the jurisdiction of ASTM Committee D-22 on Sampling and Analysis of Atmospheres and is the direct responsibility of Subcommittee D22.05 on Indoor Air.

Current edition approved Dec. 10, 1995. Published February 1996. Originally published as D 5466 – 93. Last previous edition D 5466 – 93.

[2] This test method is based on EPA Compendium Method TO-14, "The Determination of Volatile Organic Compounds (VOCs) in Ambient Air Using SUMMA Passivated Canister Sampling and Gas Chromatographic Analysis," May 1988.

[3] The **boldface** numbers in parentheses refer to the list of references at the end of the standard.

[4] *Annual Book of ASTM Standards*, Vol 11.03.
[5] *Annual Book of ASTM Standards*, Vol 14.02.

TABLE 1 Volatile Organic Compounds Known to Have Been Analyzed by the Canister Method

Compound (Synonym)	Formula	Molecular Weight	Boiling Point (°C)	Melting Point (°C)	Cas Number
Freon 12 (Dichlorodifluoromethane)	Cl_2CF_2	120.91	−29.8	−158.0	
Methyl chloride (Chloromethane)	CH_3Cl	50.49	−24.2	−97.1	74-87-3
Freon 114 (1,2-Dichloro-1,1,2,2-tetrafluoroethane)	$ClCF_2CClF_2$	170.93	4.1	−94.0	
Vinyl chloride (Chloroethylene)	$CH_2{=}CHCl$	62.50	−13.4	−1538.0	75-01-4
Methyl bromide (Bromomethane)	CH_3Br	94.94	3.6	−93.6	74-83-9
Ethyl chloride (Chloroethane)	CH_3CH_2Cl	64.52	12.3	−136.4	75-00-3
Freon 11 (Trichlorofluoromethane)	CCl_3F	137.38	23.7	−111.0	
Vinylidene chloride (1,1-Dichloroethene)	$C_2H_2Cl_2$	96.95	31.7	−122.5	75-35-4
Dichloromethane (Methylene chloride)	CH_2Cl_2	84.94	39.8	−95.1	75-09-2
Freon 113 (1,1,2-Trichloro-1,2,2-trifluoroethane)	CF_2ClCCl_2F	187.38	47.7	−36.4	
1,1-Dichloroethane (Ethylidene chloride)	CH_3CHCl_2	98.96	57.3	−97.0	74-34-3
cis-1,2-Dichloroethylene	$CHCl{=}CHCl$	96.94	60.3	−80.5	
Chloroform (Trichloromethane)	$CHCl_3$	119.38	61.7	−63.5	67-66-3
1,2-Dichloroethane (Ethylene dichloride)	$ClCH_2CH_2Cl$	98.96	83.5	−35.3	107-06-2
Methyl chloroform (1,1,1,-Trichloroethane)	CH_3CCl_3	133.41	74.1	−30.4	71-55-6
Benzene (Cyclohexatriene)	C_6H_6	78.12	80.1	5.5	71-43-2
Carbon tetrachloride (Tetrachloromethane)	CCl_4	153.82	76.5	−23.0	56-23-5
1,2-Dichloropropane (Propylene dichloride)	$CH_3CHClCH_2Cl$	112.99	96.4	−100.4	78-87-5
Trichloroethylene (Trichloroethene)	$ClCH{=}CCl_2$	131.29	87	−73.0	79-01-6
cis-1,3-Dichloropropene (cis-1,3-dichloropropylene)	$CH_3CCl{=}CHCl$	110.97	76		
trans-1,3-Dichloropropene (cis-1,3-Dichloropropylene)	$ClCH_2CH{=}CHCl$	110.97	112.0		
1,1,2-Trichloroethane (Vinyl trichloride)	$CH_2ClCHCl_2$	133.41	113.8	−36.5	79-00-5
Toluene (Methyl benzene)	$C_6H_5CH_3$	92.15	110.6	−95.0	108-88-3
1,2-Dibromoethane (Ethylene dibromide)	$BrCH_2CH_2Br$	187.88	131.3	9.8	106-93-4
Tetrachloroethylene (Perchloroethylene)	$Cl_2C{=}CCl_2$	165.83	121.1	−19.0	127-18-4
Chlorobenzene (Phenyl chloride)	C_6H_5Cl	112.56	132.0	−45.6	108-90-7
Ethylbenzene	$C_6H_5C_2H_5$	106.17	136.2	−95.0	100-41-4
m-Xylene (1,3-Dimethylbenzene)	$1,3\text{-}(CH_3)_2C_6H_4$	106.17	139.1	−47.9	
p-Xylene (1,4-Dimethylxylene)	$1,4\text{-}(CH_3)_2C_6H_4$	106.17	138.3	13.3	
Styrene (Vinyl benzene)	$C_6H_5CH{=}CH_2$	104.16	145.2	−30.6	100-42-5
1,1,2,2-Tetrachloroethane	$CHCl_2CHCl_2$	167.85	146.2	−36.0	79-34-5
o-Xylene (1,2-Dimethylbenzene)	$1,2\text{-}(CH_3)_2C_6H_4$	106.17	144.4	−25.2	
1,3,5-Trimethylbenzene (Mesitylene)	$1,3,5\text{-}(CH_3)_3C_6H_6$	120.20	164.7	−44.7	108-67-8
1,2,4-Trimethylbenzene (Pseudocumene)	$1,2,4\text{-}(CH_3)_3C_6H_6$	120.20	169.3	−43.8	95-63-6
m-Dichlorobenzene (1,3-Dichlorobenzene)	$1,3\text{-}Cl_2C_6H_4$	147.01	173.0	−24.7	541-73-1
Benzyl chloride (α-Chlorotoluene)	$C_6H_5CH_2Cl$	126.59	179.3	−39.0	100-44-7
o-Dichlorobenzene (1,2-Dichlorobenzene)	$1,2\text{-}Cl_2C_6H_4$	147.01	180.5	−17.0	95-50-1
p-Dichlorobenzene (1,4-Dichlorobenzene)	$1,4\text{-}Cl_2C_6H_4$	147.01	174.0	53.1	106-46-7
1,2,4-Trichlorobenzene Hexachlorobutadiene (1,1,2,3,4,4-Hexachloro-1,3-butadiene)	$1,2,4\text{-}Cl_3C_6H_3$	181.45	213.5	17.0	120-82-1

3.2.7 *megabore column*—chromatographic column having an internal diameter (I.D.) greater than 0.50 mm.

3.2.7.1 *Discussion*—The Megabore column is a trademark of the J & W Scientific Co. For purposes of this test method, Megabore refers to chromatographic columns with 0.53 mm I.D.

3.2.8 *MS-SCAN*—the GC is coupled to a Mass Spectrometer (MS) programmed to scan all ions over a preset range repeatedly during the GC run.

3.2.8.1 *Discussion*—As used in the current context, this procedure serves as a qualitative identification and characterization of the sample.

3.2.9 *MS-SIM*—the GC is coupled to a MS programmed to acquire data for only specified ions and to disregard all others. This is performed using SIM coupled to retention time discriminators. The GC-SIM analysis provides quantitative results for selected constituents of the sample gas as programmed by the user.

3.2.10 *pressurized sampling*—collection of an air sample in a canister with a (final) canister pressure above atmospheric pressure, using a sample pump.

3.2.11 *qualitative accuracy*—the ability of an analytical system to correctly identify compounds.

3.2.12 *quantitative accuracy*—the ability of an analytical system to correctly measure the concentration of an identified compound.

3.2.13 *static calibration*—calibration of an analytical system using standards in a form different than the samples to be analyzed.

3.2.13.1 *Discussion*—An example of a static calibration would be injecting a small volume of a high concentration standard directly onto a GC column, bypassing the sample extraction and preconcentration portion of the analytical system.

3.2.14 *subatmospheric sampling*—collection of an air sample in an evacuated canister to a (final) canister pressure below atmospheric pressure, with or without the assistance of a sampling pump.

3.2.14.1 *Discussion*—The canister is filled as the internal canister pressure increases to ambient or near ambient pressure. An auxiliary vacuum pump may be used as part of the sampling system to flush the inlet tubing prior to or during sample collection.

4. Summary of Test Method

4.1 Both subatmospheric pressure and pressurized sampling modes use an evacuated canister. A sampling line less than 2 % of the volume of the canister or a pump-ventilated sample line are used during sample collection. Pressurized

TABLE 2 Ion/Abundance and Expected Retention Time for Selected VOCs Analyzed by GC-MS-SIM

Compound	Ion/Abundance (amu/% base peak)	Expected Retention Time (min)
Freon 12 (Dichlorodifluoromethane)	85/100 87/31	5.01
Methyl chloride (Chloromethane)	50/100 52/34	5.69
Freon 114 (1,2-Dichloro-1,1,2,2-tetrafluoroethane)	85/100 135/56 87/33	6.55
Vinyl chloride (Chloroethene)	62/100 27/125 64/32	6.71
Methyl bromide (Bromomethane)	94/100 96/85	7.83
Ethyl chloride (Chloroethane)	64/100 29/140 27/140	8.43
Freon 11 (Trichlorofluoromethane)	101/100 103/67	9.97
Vinylidene chloride (1,1-Dichloroethylene)	61/100 96/55 63/31	10.93
Dichloromethane (Methylene chloride)	49/100 84/65 86/45	11.21
Freon 113 (1,1,2-Trichloro-1,2,2-trifluoroethane)	151/100 101/140 103/90	11.60
1,1-Dichloroethane (Ethylidene dichloride)	63/100 27/64 65/33	12.50
cis-1,2-Dichloroethylene	61/100 96/60 98/44	13.40
Chloroform (Trichloromethane)	83/100 85/65 47/35	13.75
1,2-Dichloroethane (Ethylene dichloride)	62/100 27/70 64/31	14.39
Methyl chloroform (1,1,1-Trichloroethane)	97/100 99/64 61/61	14.62
Benzene (Cyclohexatriene)	78/100 77/25 50/35	15.04
Carbon tetrachloride (Tetrachloromethane)	117/100 119/97	15.18
1,2-Dichloropropane (Propylene dichloride)	63/100 41/90 62/70	15.83
Trichloroethylene (Trichloroethane)	130/100 132/92 95/87	16.10
cis-1,3-Dichloropropene	75/100 39/70 77/30	16.96
trans-1,3-Dichloropropene (1,3-Dichloro-1-propene)	75/100 39/70 77/30	17.49
1,1,2-Trichloroethane (Vinyl trichloride)	97/100 83/90 61/82	17.61
Toluene (Methyl benzene)	91/100 92/57	17.86
1,2-Dibromoethane (Ethylene dibromide)	107/100 109/96 27/115	18.48
Tetrachloroethylene (Perchloroethylene)	166/100 164/74 131/60	19.01
Chlorobenzene (Benzene chloride)	112/100 77/62 114/32	19.73

TABLE 2 Continued

Compound	Ion/Abundance (amu/% base peak)	Expected Retention Time (min)
Ethylbenzene	91/100 106/28	20.20
m,p-Xylene (1,3/1,4-dimethylbenzene)	91/100 106/40	20.41
Styrene (Vinyl benzene)	104/100 78/60 103/49	20.81
1,1,2,2-Tetrachloroethane (Tetrachloroethane)	83/100 85/64	20.92
o-Xylene (1,2-Dimethylbenzene)	91/100 106/40	20.92
4-Ethyltoluene	105/100 120/29	22.53
1,3,5-Trimethylbenzene (Mesitylene)	105/100 120/42	22.65
1,2,4-Trimethylbenzene (Pseudocumene)	105/100 120/42	23.18
m-Dichlorobenzene (1,3-Dichlorobenzene)	146/100 148/65 111/40	23.31
Benzyl chloride (α-Chlorotoluene)	91/100 126/26	23.32
p-Dichlorobenzene (1,4-Dichlorobenzene)	146/100 148/65 111/40	23.41
o-Dichlorobenzene (1,2-Dichlorobenzene)	146/100 148/65 111/40	23.88
1,2,4-Trichlorobenzene	180/100 182/98 184/30	26.71
Hexachlorobutadiene (1,1,2,3,4,4-Hexachloro-1,3-butadiene)	225/100 227/66 223/60	27.68

sampling requires an additional pump to provide positive pressure to the sample canister. A sample of air is drawn through a sampling train comprising components that regulate the rate and duration of sampling into a precleaned and pre-evacuated SUMMA® passivated canister.

4.2 After the air sample is collected, the canister isolation valve is closed, the canister is removed from the sampler, an identification tag is attached to the canister, and the canister is transported to a laboratory for analysis.

4.3 Upon receipt at the laboratory, the data on the canister tag are recorded and the canister is attached to a pressure gage which will allow accurate measurement of the final canister pressure. During analysis, water vapor may be reduced in the gas stream by a Nafion dryer (if applicable), and the VOCs are then concentrated by collection in a cryogenically-cooled trap. The cryogen is then removed and the temperature of the trap is raised. The VOCs originally collected in the trap are revolatilized, separated on a GC column, then detected by a mass spectrometer. Compound identification and quantitation are performed with this test method.

4.4 A mass spectrometric detector (MS coupled to a GC) is the principal analytical tool used for qualitative and quantitative analysis because it allows positive compound identification. MS detectors include, but are not limited to, magnetic sector mass analyzers, guadrupole mass filters, combined magnetic sector-electrostatic sector mass analyzers, time-of-flight mass analyzers and ion trap mass spectrometers.

4.4.1 *Comparison of GC/MS–Full Scan and GC/MS-SIM:*

4.4.1.1 *GC/MS–Full Scan:*

(*1*) Positive nontarget compound identification possible,

(*2*) Less sensitivity than GC/MS-SIM,

(*3*) Greater sample volume may be required compared to SIM,

(*4*) Resolve co-eluting interfering ions is possible,

(*5*) Positive compound identification,

(*6*) Quantitative determination of compounds on calibration list, and

(*7*) Qualitative and semiquantitative determination of compounds not contained on calibration list.

4.4.1.2 *GC/MS-SIM:*

(*1*) Can't identify non-target compounds,

(*2*) Less operator interpretation, and

(*3*) Higher sensitivity than GC/MS–full scan.

4.4.2 The GC/MS–full scan option uses a capillary column GC coupled to a MS operated in a scanning mode and supported by spectral library search routines. This option offers the nearest approximation to unambiguous identification and covers a wide range of compounds as defined by the completeness of the spectral library. GC/MS-SIM mode is limited to a set of target compounds which are user defined and is more sensitive than GC/MS-SCAN by virtue of the longer dwell times at the restricted number of m/z values. As the number of ions monitored simultaneously in a GC/MS-SIM analysis increases, the sensitivity of this technique approaches GC/MS-SCAN. The practical limit for GC/MS-SIM is reached at about 4 to 5 ions monitored simultaneously.

5. Significance and Use

5.1 VOCs are emitted into the ambient, indoor, and workplace atmosphere from a variety of sources. In addition to the emissions from the use of various products, appliances, and building materials, fugitive or direct emissions from ambient sources such as manufacturing processes further complicate air composition. Many of these VOC compounds are acute or chronic toxins. Therefore, their determination in air is necessary to assess human health impacts.

5.2 The use of canisters is particularly well suited for the collection and analysis of very volatile, stable compounds in atmosphere (for example, vinyl chloride). This test method collects and analyzes whole gas samples and is not subject to high volatility limitations.

5.3 VOCs can be successfully collected in passivated stainless steel canisters. Collection of atmospheric samples in canisters provides for: (*1*) convenient integration of air samples over a specific time period (for example, 8 to 24 h), (*2*) remote sampling and central laboratory analysis, (*3*) ease of storing and shipping samples, (*4*) unattended sample collection, (*5*) analysis of samples from multiple sites with one analytical system, (*6*) dilution or additional sample concentration to keep the sample size introduced into the analytical instrument within the calibration range, (*7*) collection of sufficient sample volume to allow assessment of measurement precision or analysis, or both, of samples by

several analytical systems, and (*8*) can be performed in remote access areas using a vacuum regulator flow controller if electricity is not available.

5.4 Interior surfaces of the canisters are treated by the SUMMA® passivation process,[6] in which a pure chrome-nickel oxide is formed on the surface.

5.5 This test method can be applied to sampling and analysis of compounds that can be quantitatively recovered from the canisters. The typical range of VOC applicable to this test method are ones having saturated vapor pressures at 25°C greater than 15 Pa (10^{-1} mm Hg).

5.6 Recovery and stability studies must be conducted on any compound not listed in Table 1 before expanding the use of this test method to additional compounds.

6. Interferences and Limitations

6.1 For those applications where a membrane dryer (for example, Nafion) is used, interferences can occur in sample analysis if moisture accumulates in the dryer (see 10.1.1.2). An automated cleanup procedure that periodically heats the dryer to about 100°C while purging with zero air eliminates any moisture buildup. This procedure does not degrade sample integrity.

NOTE 1—Removing moisture from samples is not necessary with GC/MS systems that are differentially pumped and which do not employ membrane drying apparatus.

6.2 Contamination may occur in the sampling system if canisters are not properly cleaned before use. Additionally, all other sampling equipment (for example, pump and flow controllers) must be thoroughly cleaned to ensure that the filling apparatus will not contaminate samples. Instructions for cleaning the canisters and certifying the field sampling system are described in 11.1 and 11.2, respectively. In addition, sufficient system and field blank samples shall be analyzed to detect contamination as soon as it occurs.

6.3 If the GC/MS analytical system employs a Nafion permeable membrane dryer or equivalent to remove water vapor selectively from the sample stream, polar organic compounds will permeate this membrane concurrently with the moisture. Consequently, the analyst must calibrate his or her system with the specific organic constituents under examination. For quantitative analysis of polar compounds analytical systems may not employ Nafion permeable membrane dryers.

7. Apparatus

7.1 Subatmospheric pressure and pressurized canister sampling systems are commercially available and have been certified for VOC testing in air (**20**). Several configurations of standard hardware can be used successfully as canister sampling units.

7.1.1 *Subatmospheric Pressure* (see Fig. 1).

---

[6] SUMMA® process is a registered trademark of Molectrics Inc., 4000 East 89th St., Cleveland, OH 44105.

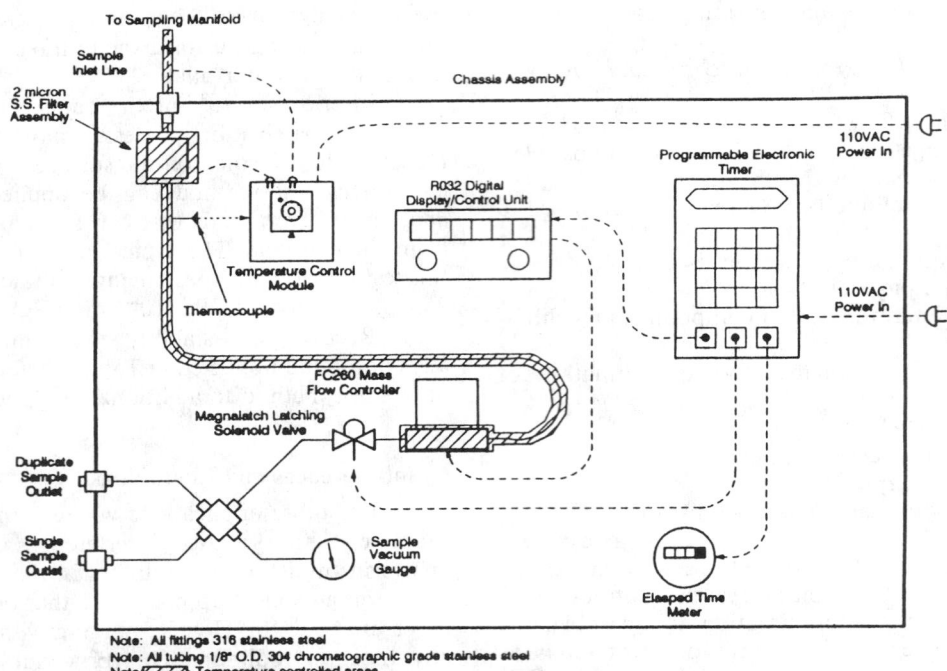

FIG. 1 Sampler Configuration For Subatmospheric Pressure Canister Sampling

7.1.1.1 *Inlet Line*, Stainless steel tubing to connect the sampler to the sample inlet.

7.1.1.2 *Canister*, Leak-free stainless steel pressure vessels of desired volume (for example, 6 L), with valve and passivated interior surfaces.[7]

7.1.1.3 *Vacuum/Pressure Gage*,[8] Capable of measuring vacuum (−100 to 0 kPa or 0 to 30 in Hg) and pressure (0 to 200 kPa or 0 to 30 psig) in the sampling system. Gages shall be tested clean and leak tight.

7.1.1.4 *Mass Flow Meter and Controller*,[9] Capable of maintaining a constant flow rate (±10 %) over a sampling period of up to 24 h and under conditions of changing temperature (20 to 40°C) and humidity.

7.1.1.5 *Filter*, 7-μm sintered stainless-steel in-line filter.[10]

7.1.1.6 *Electronic Timer*,[11] For unattended sample collection.

7.1.1.7 *Solenoid Valve*, Electrically operated, bi-stable solenoid valve with fluoroelastimer[12] seat and o-rings, or low temperature solenoid valve.

7.1.1.8 *Tubing and Fittings*, Chromatographic grade stainless steel tubing and fittings for interconnections. All such materials in contact with sample, analyte, and support gases prior to analysis shall be chromatographic grade stainless steel.

7.1.1.9 *Heater*, thermostatically controlled to maintain temperature inside insulate sampler enclosure above am-

bient temperature if needed.[13]

7.1.1.10 *Fan*, For cooling sampling system, if needed.[14]

7.1.1.11 *Theromstat*[15]—Automatically regulates fan operation, if needed.

7.1.1.12 *Maximum-Minimum Thermometer*, Records highest and lowest temperatures during sampling period.[16]

7.1.1.13 *Shut-Off Valve*[17]—Stainless steel—leak free, for vacuum/pressure gage.

7.1.1.14 *Auxiliary Vacuum Pump*—optional, continuously draws air to be sampled through the inlet manifold at 10 L/min or higher flow rate. Sample is extracted from the manifold at a lower rate, and excess air is exhausted. The use of higher inlet flow rates dilutes any contamination present in the inlet and reduces the possibility of sample contamination as a result of contact with active adsorption sites on inlet walls. Pump is not necessary if the intake manifold represents less than 5 % of the final sample.

7.1.1.15 *Elapsed Time Meter*[18]—To measure duration of sampling.

7.1.1.16 *Optional Fixed Orifice, Capillary, Adjustable Micrometering Valve, or Vacuum Regulator*, May be used in lieu of the electronic flow controller for grab samples or short duration time-integrated samples. Such systems require manual activation and deactivation. In this standard, application of a pumpless simple orifice sampler is appropriate only in situations where samples consume 60 % of the total capacity of the canister used for collection. Typically this limits the sample duration to a maximum of 8 h per 6 L

[7] Scientific Instrumentation Specialists, Inc., P.O. Box 8941, Moscow, ID 83843, or Andersen Samplers, Inc., 4215-C Wendell Dr., Atlanta, GA 30336.

[8] Matheson, P.O. Box 136, Morrow, GA.

[9] Tylan Corp., 19220 S. Normandie Ave., Torrance, CA 90502.

[10] Nupro Co., 4800 E. 345th St., Willoughby, OH 44094.

[11] Paragon Elect. Co., 606 Parkway Blvd., P.O. Box 28, Twin Rivers, WI 54201.

[12] VITON® fluoroelastimer, trademark of E. I. du Pont de Nemours, has been found satisfactory. An equivalent can be used.

[13] Watlow Co., Pfafftown, NC.

[14] EG&G Rotron, Woodstock, NY, Model SUZAI.

[15] Elmwood Sensors, Inc., Pawtucket, RI.

[16] Thomas Scientific, Brooklyn Thermometer Co., Inc.

[17] Nupro Co., 4800 E. 345th St., Willoughby, OH 44094.

[18] Conrac, Cramer Div., Old Saybrook, CT.

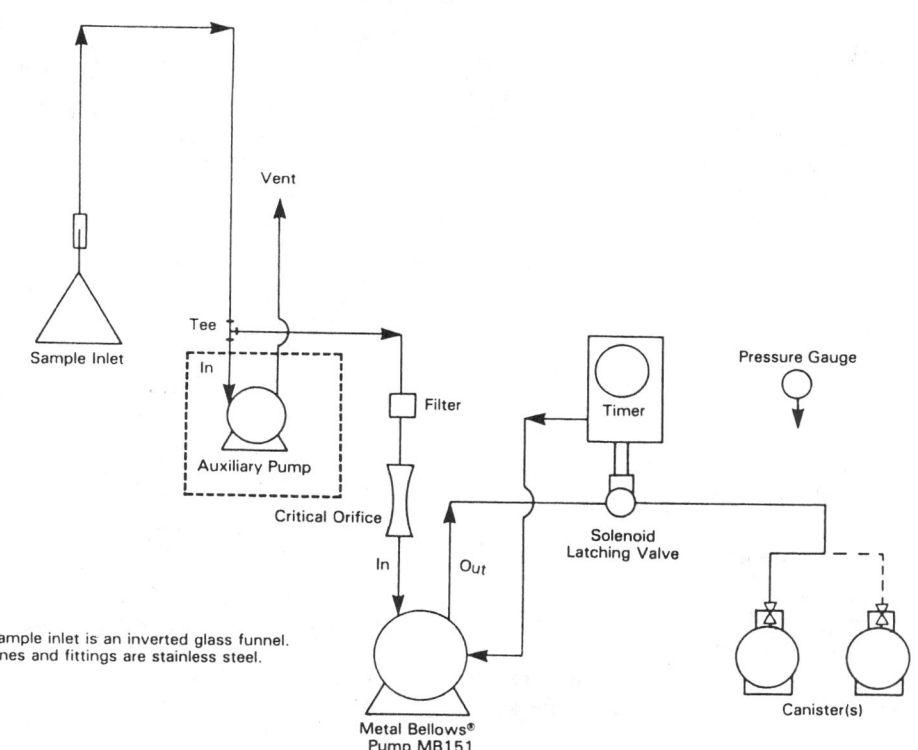

FIG. 2 Alternative Sampler Configuration for Pressurized Canister Sampling

canister or 20 h per 15 L canister.

7.1.2 *Pressurized*—See Figs. 1 and 2.

7.1.2.1 *Sample Pump*, Stainless steel pump head, metal bellows type[19] capable of 200 kPa output pressure. Pump must be free of leaks, clean, and uncontaminated by oil or organic compounds.

NOTE 2—Alternative sampling systems have been developed. The Rasmussen sampler (**21**) is illustrated in Fig. 2. This flow system uses, in order, a pump, a mechanical flow regulator, and a mechanical compensating flow restrictive device. In this configuration the pump is purged with a large sample flow, thereby eliminating the need for an auxiliary vacuum pump to flush the sample inlet. The Radian sampler (**20**) is illustrated in Fig. 3. This system draws air directly through a mass flow controller into an evacuated canister. Interferences using either configuration have been minimal.

7.1.2.2 *Other Supporting Materials*—All other components of the pressurized sampling system (Figs. 1, 2, and 3) are similar to components discussed in 7.1.1.1 through 7.1.1.16.

7.2 *Sample Analysis Equipment:*

7.2.1 *GC/MS-Analytical System (Full Scan and SIM).*

7.2.1.1 The GC/MS-SCAN analytical system must be capable of acquiring and processing data in the MS–full scan mode. The GC/MS-SIM analytical system must be capable of acquiring and processing data in the MS-SIM mode.

7.2.1.2 *Gas Chromatograph*, Capable of sub-ambient temperature programming for the oven, with other standard features such as gas flow regulators, automatic control of valves and integrator, etc. Flame ionization detector optional.

7.2.1.3 *Chromatographic Detector*, Mass spectrometric detector equipped with computer and appropriate software. The GC/MS is set in the SCAN mode, where the MS screens the sample for identification and quantitation of VOC species.

7.2.1.4 *Cryogenic Trap with Temperature Control Assembly*—Refer to 10.1.1.3 for complete description of trap and temperature control assembly. Traps may be built into the gas chromatograph by the manufacturer or added to existing units.[20]

7.2.1.5 *Electronic Mass Flow Controllers (3)*, to maintain constant flow for carrier gas and sample gas and to provide analog output to monitor flow anomalies.[21]

7.2.1.6 *Vaccum Pump*, General purpose laboratory pump, capable of evacuating a known volume reservoir (which will be used for sample transfer) or for drawing the desired sample volume through the cryogenic trap.

7.2.1.7 *Chromatographic Grade Stainless Steel Tubing and Stainless Steel Plumbing Fittings*—Refer to 7.1.1.8 for description.

7.2.1.8 *Chromatographic Column*, To provide compound separation such as shown in Table 3.

NOTE 3—Other columns (for example, DB-624) can be used as long as the system meets user needs. The wider Megabore column (that is, 0.53 mm I.D.) is less susceptible to plugging as a result of trapped water, thus eliminating the need for a Nafion dryer in the analytical system. The Megabore column has sample capacity approaching that of a packed column, while retaining much of the peak resolution traits of narrower columns (that is, 0.32 mm I.D.).

---

[19] Metal Bellows Corp., 1075 Providence Highway, Sharon, MA 02067.

[20] Nutec Corporation, 2142 Geer St., Durham, NC 27704.
[21] Tylan, 0–100 or 0–20 mL/min.

TABLE 3 General GC and MS Operating Conditions

Chromatography	
Column	Hewlett-Packard OV-1 crosslinked methyl silicone (50 m by 0.31-mm I.D., 17 μm film thickness), or equivalent
Carrier Gas	Helium (2.0 cm³/min at 250°C)
Injection Volume	Constant (1–3 μL)
Injection Mode	Splitless
Temperature Program	
Initial Column Temperature	−50°C
Initial Hold Time	2 min
Program	8°C/min to 150°C
Final Hold Time	15 min
Mass Spectrometer	
Mass Range	18 to 250 amu
Scan Time	1 s/scan
EI Condition	70 eV
Mass Scan	Follow manufacturer's instruction for selecting mass selective detector (MS) and selected ion monitoring (SIM) mode
Detector Mode	Multiple ion detection
FID System (Optional)	
Hydrogen Flow	30 cm³/min
Carrier Flow	30 cm³/min
Burner Air	400 cm³/min

7.2.1.9 *Stainless Steel Vacuum/Pressure Gage (optional)*, capable of measuring vacuum (−100 to 0 kPa) and pressure (0–200 kPa) in the sampling system. Gages shall be tested clean and leak tight.

7.2.1.10 *Cylinder Pressure Stainless Steel Regulators*, standard, two-stage cylinder regulators with pressure gages for helium, zero air, nitrogen, and hydrogen gas cylinders.

7.2.1.11 *Gas Purifiers (4)*, Molecular sieve or carbon used to remove organic impurities and moisture from gas streams.[22]

7.2.1.12 *Low Dead-Volume Tee or Press Fit Splitter (optional)*, used to split the exit flow from the GC column.[23]

7.2.1.13 *Nafion Dryer (optional)*, consisting of Nafion tubing coaxially mounted within larger tubing.[24] Refer to 10.1.1.2 for description.

7.2.1.14 *Six-Port Gas Chromatographic Valve.*

7.3 *Canister Cleaning System (see Fig. 3):*

7.3.1 *Vacuum Pump*, capable of evacuating sample canister(s) to an absolute pressure of less than 0.064 kPa (0.5 mm Hg).

7.3.2 *Manifold*, made of stainless steel with connections for simultaneously cleaning several canisters.

7.3.3 *Shut-Off Valve(s)*, On-off toggle valves.

7.3.4 *Stainless Steel Vacuum Gage*, capable of measuring vacuum in the manifold to an absolute pressure of 0.00064 kPa (0.05 mm Hg) or less.

7.3.5 *Cryogenic Trap (2 required)*, made of stainless steel U-shaped open tubular trap cooled with liquid nitrogen, for air purification purposes to prevent contamination from back diffusion of oil from vacuum pump and to provide clean, zero air to sample canister(s).

7.3.6 *Stainless Steel Pressure Gages (2)*, 0 to 350 kPa (0 to 50 psig) to monitor zero air pressure.

7.3.7 *Stainless Steel Flow Control Valve*, to regulate flow of zero air into canister(s).

FIG. 3 Canister Cleanup Apparatus

7.3.8 *Humidifier Consisting of Pressurizable Water Bubbler*, (typically a SUMMA® passivated canister equipped with dip tube and dual valves). Humidifier contains high performance liquid chromatography (HPLC) grade deionized water.

7.3.9 *Isothermal Oven (optional)*, for heating canisters.[25]

NOTE 4—Oven temperature must not exceed 80°C during cleaning to avoid degradation of the passivated canister surface on repeated cleaning.

7.4 *Calibration System and Manifold (see Fig. 4):*

7.4.1 *Calibration Manifold*, chromatographic grade stainless steel or glass manifold (125 mm I.D. by 660 mm), with sampling ports and internal mixing for flow disturbance to ensure proper mixing.

7.4.2 *Humidifier*, 500-mL impinger flask containing HPLC grade deionized water.

7.4.3 *Electronic Mass Flow Controllers*, one 0 to 5 L/min and one 0 to 50 mL/min.[26]

7.4.4 *TFE–Fluorocarbon Filter(s)*, 47-mm TFE–Fluorocarbon filter for particulate control.

8. Reagents and Materials

8.1 Gas cylinders of helium, hydrogen, nitrogen, and zero air ultrahigh purity grade.

8.2 Gas calibration standards—cylinder(s) containing approximately 10 ppmv of each of the following compounds of interest:

vinyl chloride	1,2-dibromoethane
vinylidene chloride	tetrachloroethylene
1,1,2-trichloro-1,2,2-trifluoroethane	chlorobenzene
p-dichlorobenzene	benzyl chloride
chloroform	hexachloro-1,3-butadiene
1,2-dichloroethane	methyl chloroform
benzene carbon	tetrachloride
toluene	trichloroethylene
dichlorodifluoromethane	cis-1,3-dichloropropene
methyl chloride	trans-1,3-dichloropropene
ethylbenzene	1,2-dichloro-1,1,2,2-tetrafluoroethane
1,2,4-trichlorobenzene	o-dichlorobenzene
methyl bromide	o-xylene

---

[22] Hewlett-Packard, Rt. 41, Avondale, PA 19311.
[23] Alltech Associates, 2051 Waukegan Rd., Deerfield, IL 60015.
[24] Perma Pure Products, 8 Executive Drive, Toms River, NJ 08753.

[25] Fisher Scientific, Pittsburgh, PA.
[26] Tylan Corporation, 23301-TS Wilmington Ave., Carson, CA 90745.

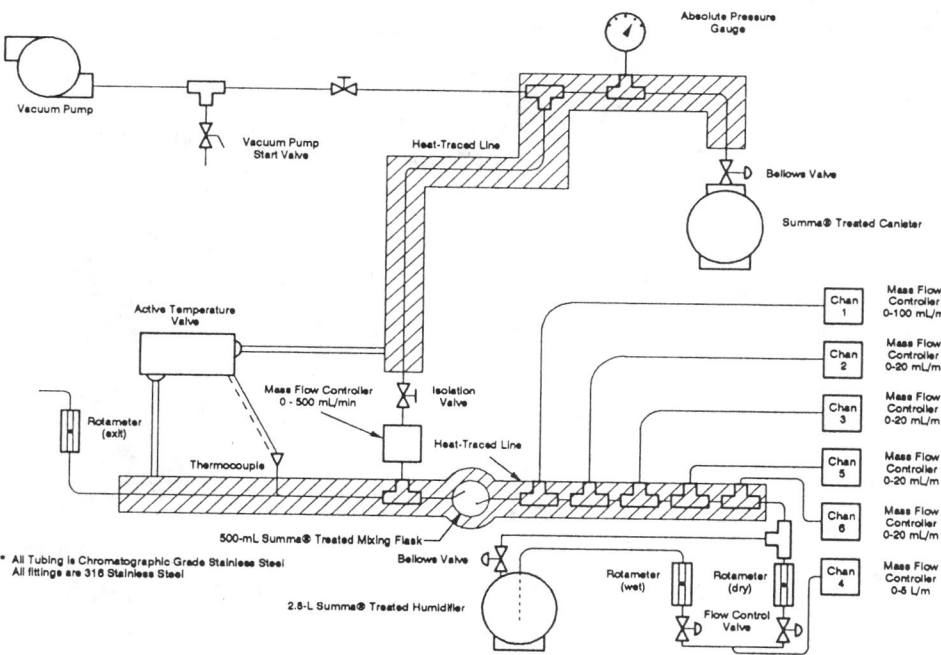

FIG. 4 Schematic of Calibration System and Manifold for (a) Analytical System Calibration, (b) Testing Canister Sampling System and (c) Preparing Canister Transfer Standards

ethyl chloride	m-xylene
fluorotrichloromethane	p-xylene
dichloromethane	styrene
1,1-dichloroethane	1,1,2,2-tetrachloroethane
cis-1,2-dichloroethylene	1,3,5-trimethylbenzene
1,2-dichloropropane	1,2,4-trimethylbenzene
1,1,2-trichloroethane	m-dichlorobenzene

8.2.1 The cylinder(s) shall be traceable to a National Institute of Standards and Technology (NIST) Standard Reference Material (SRM) or to a NIST/EPA approved Certified Reference Material (CRM). The components may be purchased in one cylinder or may be separated into different cylinders. Refer to manufacturer's specification for guidance on purchasing and mixing VOCs in gas cylinders. Those compounds purchased should match one's own target list.

8.3 *Liquid Nitrogen (bp −195.8°C)*, used only for clean air traps and GC oven coolant, and sample concentration traps requiring active control to maintain −185.7°C.

8.4 *Liquid Argon (bp −185.7°C)*, for sample traps that are not actively controlled to −185.7°C.

8.5 *Gas Purifiers*—Molecular sieve or carbon, connected in-line between hydrogen, nitrogen, and zero air gas cylinders and system inlet line, to remove moisture and organic impurities from gas streams.

8.6 *Deionized Water*—High performance liquid chromatography (HPLC) grade,[27] ultrahigh purity (for humidifier).

8.7 *4-Bromofluorobenzene*—Used for tuning GC/MS.

8.8 *Methanol*—For cleaning sampling system components, reagent grade.

9. Sampling System

9.1 *System Description:*

9.1.1 *Subatmospheric Pressure Sampling*—See Figs. 1 or 2.

9.1.1.1 In preparation for subatmospheric sample collection in a canister, the canister is evacuated to 64 Pa (5.0 mm Hg) or less. When opened to the atmosphere containing the VOCs to be sampled, the differential pressure causes the sample to flow into the canister. This technique may be used to collect grab samples (duration of 10 to 30 s) or time-integrated samples (duration of 12 to 24 h) taken through a flow-restrictive inlet (for example, mass flow controller, vacuum regulator, or critical orifice).

9.1.1.2 With a critical orifice flow restrictor, there will be a decrease in the flow rate if the pressure approaches atmospheric. However, with a mass flow controller the subatmospheric sampling system can be increased since the restrictor size can be adjusted. For example, an electronic flow controller with a flow rate range of 0 to 50 cc/min can maintain a constant (less than 5 % change) flow rate of 5 cc/min from full vaccum to within 7 kPa (1.0 psi) below ambient pressure.

9.1.2 *Pressurized Sampling*—See Fig. 1.

9.1.2.1 Pressurized sampling is used when longer-term integrated samples or higher volume samples are required. The sample is collected in a canister using a pump and flow control arrangement to achieve a typical 100–200 kPa (15–30 psig) final canister pressure. For example, a 6-L evacuated canister can be filled at 7.1 mL/min for 24 h to achieve a final pressure of about 150 kPa (15 psig).

NOTE 5—Collection of pressurized samples in humid environments may result in condensation of water in sampling canisters. This water may decrease the recovery of polar compounds from the canister.

9.1.2.2 In pressurized canister sampling, a metal bellows type pump draws in air from the sampling manifold to fill and pressurize the sample canister.

---

[27] Fisher Scientific, Pittsburgh, PA.

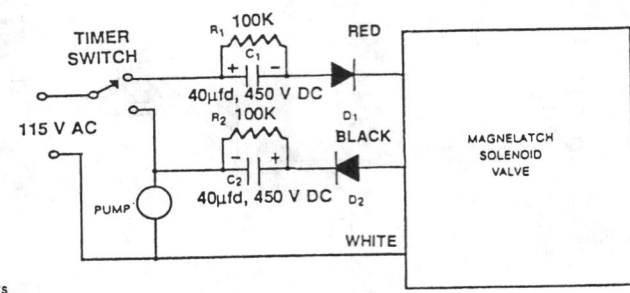

COMPONENTS
Capacitor C₁ and C₂ - 40 uf, 450 VDC (Sprague Atom® TVA 1712 or equivalent)
Resister R₁ and R₂ - 0.5 watt, 5% tolerance
Diode D₁ and D₂ - 1000 PRV, 2.5 A (RCA, SK 3081 or equivalent)

SIMPLE CIRCUIT FOR OPERATING MAGNELATCH VALVE

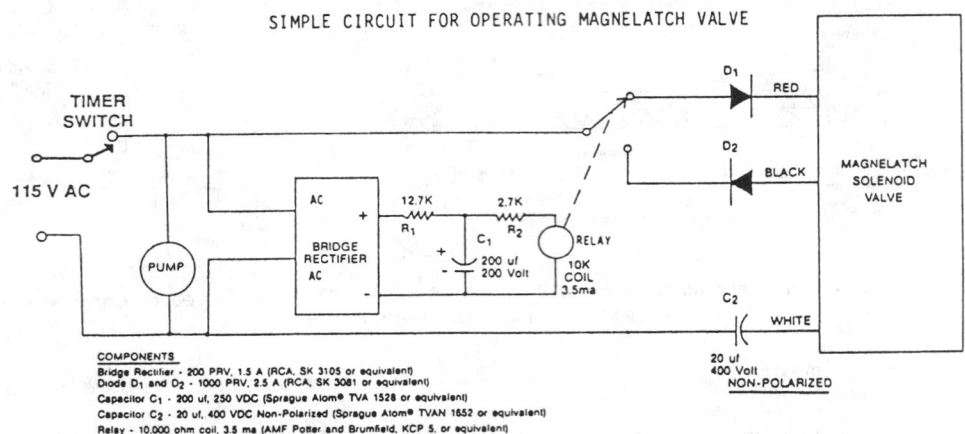

COMPONENTS
Bridge Rectifier - 200 PRV, 1.5 A (RCA, SK 3105 or equivalent)
Diode D₁ and D₂ - 1000 PRV, 2.5 A (RCA, SK 3081 or equivalent)
Capacitor C₁ - 200 uf, 250 VDC (Sprague Atom® TVA 1528 or equivalent)
Capacitor C₂ - 20 uf, 400 VDC Non-Polarized (Sprague Atom® TVAN 1852 or equivalent)
Relay - 10,000 ohm coil, 3.5 ma (AMF Potter and Brumfield, KCP 5, or equivalent)
Resister R₁ and R₂ - 0.5 watt, 5% tolerance

FIG. 5 Electrical Pulse Circuits for Driving Skinner Magnelatch Solenoid Valve with a Mechanical Timer

9.1.3 *All Samplers:*

9.1.3.1 A flow control device is chosen to maintain a constant flow into the canister over the desired sample period. This flow rate is determined so the canister is filled (to about 88.1 kPa for subatmospheric pressure sampling or to about one atmosphere above ambient pressure for pressurized sampling) over the desired sample period. The flow rate can be calculated by:

$$F = (P \times V)/(t \times 60) \qquad (1)$$

where:
F = flow rate, mL/min,
P = final canister pressure, atmospheres absolute. P is approximately equal to [(kPa gage)/100] + 1,
V = volume of the canister, mL, and
t = sample period, h.

9.1.3.2 For example, if a 6-L canister is to be filled to 200 kPa (2 atmospheres) absolute pressure in 24 h, the flow rate can be calculated by:

$$F = (2 \times 6000)/(24 \times 60) = 8.3 \text{ mL/min} \qquad (2)$$

9.1.3.3 For automatic operation, the timer is wired to start and stop the pump at appropriate times for the desired sample period. The timer must also control the solenoid valve, to open the valve when starting the pump and close the valve when stopping the pump.

9.1.3.4 The use of a latching solenoid, or low temperature valve, avoids any substantial temperature rise occurring with a conventional, normally energized solenoid during the entire sample period. The temperature rise in the valve could cause outgassing of organic compounds from the Viton valve seat material which must be avoided to reduce background. The Skinner Magnelatch solenoid requires an electronic timer that can be programmed for short (5 to 60 s) ON periods. A simple electrical pulse circuit for operating the Skinner Magnelatch solenoid valve with a conventional mechanical timer is illustrated in Fig. 5.

9.1.3.5 The connecting lines between the sample inlet and the canister shall be as short as possible to minimize their volume. The flow rate into the canister shall remain relatively constant over the entire sampling period (see 9.1.1.2).

9.1.3.6 As an option, a second electronic timer (see 7.1.1.6) may be used to start the auxiliary pump several hours prior to the sampling period to flush and condition the inlet line.

9.1.3.7 Prior to use, each sampling system must pass a humid zero air certification (see 12.2.2). All plumbing shall be checked carefully for leaks. The canisters must also pass a humid zero air certification before use (see 12.1).

9.2 *Sampling Procedure:*

9.2.1 The sample canister shall be cleaned and tested according to the procedure in 12.1.

9.2.2 A sample collection system is assembled as shown in Fig. 1 (and Fig. 2) and must meet certification requirements as outlined in 11.2.3.

NOTE 6—The sampling system shall be contained in an appropriate enclosure when ambient samples are collected.

9.2.3 Prior to locating the sampling system, the user may want to perform "screening analyses" by taking quick grab samples over a short period of time. The information gathered from the screening samples is used to determine the potential concentration range for analysis and identify potential interferents with the GC/MS analysis. Screening samples should be analyzed using the procedure in this standard. Sampling is performed using a simple sampler described in 7.1.1.16.

9.2.4 Immediately prior to any sample collection record the ambient temperature, humidity, and pressure where the sampler is located.

NOTE 7—The following discussion is related to Fig. 1.

9.2.5 To verify correct sample flow, a "practice" (evacuated) canister is used in the sampling system. Attach a certified mass flow meter to the inlet line of the manifold, just in front of the filter. Open the canister. Start the sampler and compare the reading of the certified mass flow meter to the sampler mass flow controller. The values shall agree within ±10 %. If not, the sampler mass flow meter shall be recalibrated or the sampler must be repaired if a leak is found in the system.

NOTE 8—For a subatmospheric sampler, the flow meter and practice canister are needed. For the pump-driven system, the practice canister is not needed if the flow can be measured as supplied to the canister.

NOTE 9—Mass flow meter readings may drift. Check the zero reading carefully and add or subtract the zero reading when reading or adjusting the sampler flow rate, to compensate for any zero drift. Adjust the desired canister flow rate to the proper value after a 2 min warm up period, using the sampler flow control unit controller (for example, 3.5 mL/min for 24 h, 7.0 mL/min for 12 h). Measure and record the actual final flow.

9.2.6 Turn the sampler off and reset the elapsed time meter to 000.0.

NOTE 10—Any time the sampler is turned off, wait at least 30 s to turn the sampler back on.

9.2.7 Disconnect the "practice" canister and certified mass flow meter. Attach a clean certified (see 12.1) canister to the system.

9.2.8 Open the canister valve and vacuum/pressure gage valve.

9.2.9 Record the pressure/vacuum in the canister as indicated by the sampler vacuum/pressure gage.

9.2.10 Close the vacuum/pressure gage valve and reset the maximum/minimum thermometer to current temperature. Record time of day and elapsed time meter readings.

9.2.11 Set the electronic timer to begin and stop the sampling period at the appropriate times. Sampling commences and stops by the programmed electronic timer.

9.2.12 After the desired sampling period, record the maximum, minimum, current interior temperature and current ambient or indoor temperature. Record the current reading from the flow controller, the ambient or indoor humidity and pressure.

9.2.13 At the end of the sampling period, briefly open and close the vacuum/pressure gage valve on the sampler and record pressure/vacuum in the canister.

NOTE 11—For a subatmospheric sampling system, if the canister is at atmospheric pressure when the final pressure check is performed, the sampling period may be suspect. This information shall be noted on the sampling field data sheet. Time of day and elapsed time meter readings are also recorded.

9.2.14 Close the canister valve. Disconnect the sampling line from the canister and remove the canister from the sampling system. For a subatmospheric system, connect a certified mass flow meter to the inlet manifold in front of the in-line filter and attach a "practice" canister to the valve of the sampling system. Record the final flow rate.

NOTE 12—Attaching a mass flow meter and recording the flow rate is not necessary if the initial and final canister vacuum are recorded.

NOTE 13—For a pressurized system, the final flow may be measured directly before the sampler is turned off.

9.2.15 Attach an identification tag to the canister. Record canister serial number, sample number, location, and date on the tag.

10. Analytical System

10.1 *System Description:*

10.1.1 *GC/MS System (Full Scan and SIM):*

10.1.1.1 The analytical system is comprised of a GC equipped, with a mass-spectrometric detector set to operate in the full scan mode or SIM mode (see Fig. 4). The GC/MS is set up for automatic, repetitive analysis. The GC system is comprised of a GC equipped with an OV-1 capillary column (0.32 mm by 50 m), or equivalent. The system also includes a computer and appropriate software for data acquisition, data reduction, and data reporting. In operation, an air sample (usually 250 to 800 mL) is recovered from the canister and routed to the analytical system. The sample air may be passed through a Nafion dryer; however, many polar compounds are not identified using this drying procedure. Sample is routed through a chromatographic valve, then into a cryogenic trap. Concentration of compounds based upon a previously installed calibration table is reported by an automated data reduction program. In full scan mode the GC/MS acquires mass spectral data by continuously scanning a range of masses typically between 35 and 250 amu. The SIM system is programmed to acquire data for only the target compounds and to disregard all others. The sensitivity is 1 ppb by volume for a 500 mL air sample.

10.1.1.2 SIM analysis is based on a combination of retention times and relative abundances of selected ions (see Table 2). These qualifiers are stored on the hard disk of the GC/MS computer and are compared to sample data for identification of each chromatographic peak. The retention time qualifier is determined to be ±0.10 min of the library retention time of the compound. The acceptance level for relative abundance is determined to be ±15 % of the expected abundance, except for vinyl chloride and methylene chloride, which is determined to be ±25 %. Three ions are measured for most of the forty compounds. When compound identification is made by the computer, any peak that fails any of the qualifying tests is flagged (for example, with an asterisk). All the data shall be manually examined by the analyst to determine the reason for the flag and whether the compound can be reported as found. While this adds some subjective judgment to the analysis, computer-generated identification problems must be clarified by an experienced operator. Manual inspection of the quantitative results must also be performed to verify concentrations outside the expected range. To realize the maximum sensitivity of SIM, retention time windows shall be chosen for each compound or group of compounds so that the number of ions monitored during a scan is kept to three or four.

TABLE 4 Response Factors (ppbv/area count) and Expected Retention Time for GC-MS-SIM Analytical Configuration

Compounds	Response Factor (ppbv/area count)	Expected Retention Time (minutes)
Freon 12	0.6705	5.01
Methyl chloride	4.093	5.64
Freon 114	0.4928	6.55
Vinyl chloride	2.343	6.71
Methyl bromide	2.647	7.83
Ethyl chloride	2.954	8.43
Freon 11	0.5145	9.87
Vinylidene chloride	1.037	10.93
Dichloromethane	2.255	11.21
Trichlorotrifluoroethane	0.9031	11.50
1,1-Dichloroethane	1.273	12.50
cis-1,2-Dichloroethylene	1.363	13.40
Chloroform	0.7911	13.75
1,2-Dichloroethane	1.017	14.39
Methyl chloroform	0.7078	14.62
Benzene	1.236	15.04
Carbon tetrachloride	0.5880	15.18
1,2-Dichloropropane	2.400	15.83
Trichloroethylene	1.383	16.10
cis-1,3-Dichloropropene	1.877	16.96
trans-1,3-Dichloropropene	1.338	17.49
1,1,2-Trichloroethane	1.891	17.61
Toluene	0.9406	17.86
1,2-Dibromoethane (EDB)	0.8662	18.48
Tetrachloroethylene	0.7357	19.01
Chlorobenzene	0.8558	19.73
Ethylbenzene	0.6243	20.20
m,p-Xylene	0.7367	20.41
Styrene	1.888	20.80
1,1,2,2-Tetrachloroethane	1.035	20.92
o-Xylene	0.7498	20.92
4-Ethyltoluene	0.6181	22.53
1,3,5-Trimethylbenzene	0.7088	22.65
1,2,4-Trimethylbenzene	0.7536	23.18
m-Dichlorobenzene	0.9643	23.31
Benzyl chloride	1.420	23.32
p-Dichlorobenzene	0.8912	23.41
o-Dichlorobenzene	1.004	23.88
1,2,4-Trichlorobenzene	2.150	26.71
Hexachlorobutadiene	0.4117	27.68

TABLE 5 4-Bromofluorobenzene Key Ions and Ion Abundance Criteria

Mass	Ion Abundance Criteria
50	15 to 40 % of mass 95
75	30 to 60 % of mass 95
95	Base Peak, 100 % Relative Abundance
96	5 to 9 % of mass 95
173	<2 % of mass 174
174	>50 % of mass 95
175	5 to 9 % of mass 174
176	>95 % but <101 % of mass 174
177	5 to 9 % of mass 176

10.1.1.3 A Nafion permeable membrane dryer may be used to remove water vapor selectively from the sample stream. The permeable membrane consists of Nafion tubing (a copolymer of tetrafluoroethylene and fluorosulfonyl monomer) that is coaxially mounted within larger tubing. The sample stream is passed through the interior of the Nafion tubing, allowing water (and other light, polar compounds) to permeate through the walls into a dry air purge stream flowing through the annular space between the Nafion and outer tubing. To prevent excessive moisture build-up and any memory effects in the dryer, a cleanup procedure involving periodic heating of the dryer (100°C for 20 min) while purging with dry zero air (500 mL/min) shall be implemented as part of the user's standard operating procedure (SOP) manual. The clean-up procedure is repeated during each analysis (see Section 14, Ref (7)). Removal of water with a Nafion type dryer shall not be performed for compounds other than those on the list in Table 1 unless recovery studies are performed to validate analysis of these compounds. Polar compounds are particularly susceptible to loss through the Nafion membrane interface.

NOTE 14—Recent studies have indicated no substantial loss of targeted VOCs utilizing the above water clean-up procedure (7). This cleanup procedure is particularly useful when employing cryogenic preconcentration of VOCs with subsequent GC analysis because excess accumulated water can cause trap and column blockage and also adversely affect detector precision. This is a particular problem using GC/MSD systems. In addition, the improvement in water removal from the sampling stream will allow analyses of much larger volumes of sample air in the event that greater system sensitivity is required for targeted compounds.

NOTE 15—While a differentially pumped GC/MS analytical system does not need a Nafion dryer for drying the sample gas stream, such a dryer may be used with GC/Mass Selective Detector (GC/MSD) type GC/MS systems because GC/MSD units are far more sensitive to excessive moisture than the GC/MS analytical systems. Moisture can adversely affect detector precision.

10.1.1.4 The cryogenic sample trap is heated to at least 120°C and no more than 200°C in approximately 60 s and the analyte is injected onto the DB-5 capillary column (0.53 mm by 60 m). Rapid heating of the trap provides efficient transfer of the sample components onto the gas chromatographic column. Upon sample injection onto the column, the MS computer is signaled by the GC computer to begin detection of compounds which elute from the column. The gas stream from the GC is scanned within a preselected range of atomic mass units (amu). For detection of compounds in Table 1, the range shall be 18 to 250 amu, resulting in a 1.5 Hz repetition rate. Six scans per eluting chromatographic peak are provided at this rate. Automated computer peak selection, or manual selection of each target compound is performed according to the instrument manufacturer's specifications. A library search is then performed and up to ten of the best matches for each peak are listed. A qualitative characterization of the sample is provided by this procedure.

10.1.1.5 Packed metal tubing is used for reduced temperature trapping of VOCs. The cooling unit is comprised of a 32 mm outside diameter (O.D.) nickel tubing loop packed with 60–80 mesh borosilicate glass beads.[28]

NOTE 16—The nickel tubing loop can be placed in an aluminum or brass block containing a tube heater (500 to 1000 watt) or wound onto a cylindrically formed tube heater (250 watt). A cartridge heater (25 watt) is required for the cylindrically wound trap. This low watt heater is sandwiched between pieces of metal plate at the trap inlet and outlet to provide additional heat to eliminate cold spots in the transfer tubing. Rapid heating (−178 to +120°C in 55 s) is accomplished by direct thermal contact between the heater and the trap tubing. Cooling of aluminum or brass mounted traps is achieved by immersion in liquid cryogen. Cooling of cylindrically wound traps is achieved by vaporization around or submersion of the trap in the cryogen. In the shell, efficient cooling (+120 to −178°C in 225 s) is facilitated by confining the vaporized cryogen to the small open volume surrounding the trap assembly. The trap assembly and chromatographic valve are mounted

_____

[28] Nutec Corporation, 2142 Geer St., Durham, NC 27704.

on a baseplate fitted into the injection and auxiliary zones of the GC on an insulated pad directly above the column oven when used with the Hewlett-Packard 5880 GC.

10.1.1.6 As an option, the analyst may wish to split the gas stream exiting the column with a low dead-volume tee, passing one-third of the sample gas (1.0 mL/min) to the mass selective detector and the remaining two-thirds (2.0 mL/min) through a flame ionization detector, as illustrated as an option in Fig. 4. The use of the specific detector (MS-SCAN) coupled with the nonspecific detector (FID) enables enhancement of data acquired from a single analysis. In particular, the FID provides the user with the following:

(*1*) Semi-real time picture of the progress of the analytical scheme,

(*2*) Confirmation by the concurrent MS analysis of other labs that can provide only FID results, and

(*3*) Ability to compare GC-FID with other analytical laboratories with only GC-FID capability.

10.2 *GC/MS-SCAN-SIM System Performance Criteria:*

10.2.1 *GC/MS System Operation:*

10.2.1.1 Prior to analysis, assemble and check the GC/MS system according to manufacturer's instructions.

10.2.1.2 Table 3 outlines general operating conditions for the GC/MS-SCAN-SIM system with optional FID.

10.2.1.3 Challenge the GC/MS system with humid zero air (see 11.2.2). The humid zero air challenged must contain less than 0.2 ppb by volume of targeted VOCs prior to sample analysis.

10.2.2 *Daily GC/MS Tuning:*

10.2.2.1 At the beginning of each day or prior to a calibration, tune the GC/MS system to verify that acceptable performance criteria are achieved.

10.2.2.2 For tuning the GC/MS, introduce gas from a cylinder containing 4-bromofluorobenzene by way of a sample loop valve injection system. Obtain a background corrected mass spectrum of 4-bromofluorobenzene and check that all key ion criteria are met. BFB calibration requirements are listed in Table 4. If the criteria are not achieved, the analyst shall retune the mass spectrometer and repeat the test until all criteria are achieved.

NOTE 17—Some systems allow auto-tuning to facilitate this process. The key ions and ion abundance criteria that must be met are illustrated in Table 5. Analysis shall not begin until all those criteria are met.

10.2.2.3 The performance criteria shall be achieved before any samples, blanks or standards are analyzed. If any key ion abundance observed for the daily 4-bromofluorobenzene mass tuning check differs by more than 10 % absolute abundance from that observed during the previous daily tuning, retune the instrument or reanalyze the sample or calibration gases, or both, until the above condition is met.

10.2.2.4 The GC/MS tuning standard may also be used to assess GC column performance (chromatographic check) and as an internal standard.

10.2.3 *GC/MS Calibration:*

10.2.3.1 *Initial Calibration*—Initially, a multipoint dynamic calibration (three to five levels plus humid zero air) is performed on the GC/MS system, before sample analysis, with the assistance of a calibration system (see Fig. 4). The calibration system uses National Institute of Standards and Technology (NIST) traceable standards or NIST/EPA CRMs in pressurized cylinders [containing a mixture of the targeted

VOCs at nominal concentrations of 10 ppm by volume in nitrogen (8.2)] as working standards to be diluted with humid zero air. The contents of the working standard cylinder(s) are metered (2 mL/min) into the heated mixing chamber where they are mixed with a 2 L/min humidified zero air gas stream to achieve a nominal 10 ppb by volume per compound calibration mixture (see Fig. 4). This nominal 10 ppb by volume standard mixture is allowed to flow and equilibrate for a minimum of 24 h. After the equilibration period, the gas standard mixture is sampled and analyzed by the real-time GC/MS system (see Fig. 4(a) and 7.2.1). The results of the analyses are averaged, flow audits are performed on the mass flow meters and the calculated concentration compared to generated values. After the GC/MS is

TABLE 6 Precision Results for Canister VOC Method (Example)

Compound	Standard Deviation	% CV	Instrument Detection Limit
1,3-butadiene	0.15	12.5	0.20
vinyl chloride	0.11	12.3	0.38
propylene	0.18	16.8	0.95
chloromethane	0.13	12.4	0.48
chloroethane	0.12	7.8	0.56
bromomethane	0.07	18.5	0.22
methylene chloride	0.44	49.7	0.23
trans-1,2-dichloroethane	0.22	16.4	0.66
1,1-dichloroethane	0.08	6.3	0.26
chloroprene	0.08	8.2	0.26
bromochloromethane	0.06	4.3	0.23
chloroform	0.26	6.1	0.81
1,1,1-trichloroethane	0.20	15.9	0.72
carbon tetrachloride	0.03	9.2	0.09
benzene	0.04	9.0	0.12
1,2-dichloroethane	0.08	5.2	0.21
trichloroethene	0.04	14.0	0.15
1,2-dichloropropane	0.07	6.1	0.16
bromodichloromethane	0.14	9.7	0.46
trans-1,3-dichloropropene	0.07	5.7	0.23
toluene	0.17	24.5	0.52
n-octane	0.32	22.7	1.01
cis-1,3-dichloropropene	0.05	8.4	0.14
1,1,2-trichloroethane	0.31	12.0	0.96
tetrachloroethene	0.08	19.0	0.27
dibromochloromethane	0.04	26.3	0.11
chlorobenzene	0.07	7.5	0.22
ethylbenzene	0.23	7.9	0.73
m-/p-xylene	0.41	11.1	1.03
styrene	0.15	37.9	0.46
o-xylene	0.23	16.2	0.71
bromoform	0.03	6.5	0.10
1,1,2,2-tetrachloroethane	0.09	6.7	0.22
m-dichlorobenzene	0.09	8.7	0.27
p-dichlorobenzene	0.04	12.9	0.11
o-dichlorobenzene	0.14	8.9	0.38
1,1-dichloroethene	0.05	4.2	0.17
1,2-dichloroethene	0.08	6.7	0.24
cis-1,2-dichloroethene	0.06	5.0	0.19
Halocarbon 11 (trichlorofluoromethane)	0.06	4.8	0.18
Halocarbon 113 (1,1,2-trichloro-1,2,2-trifluoroethane)	0.06	5.3	0.20
Halocarbon 114 (1,2-dichloro-1,1,2,2-tetrafluoroethane)	0.09	7.5	0.27
Halocarbon 12 (dichlorodifluoromethane)	0.08	6.9	0.26
acetonitrile	0.20	16.7	0.63
acrylonitrile	0.09	7.2	0.27
benzyl chloride	0.06	4.8	0.18
4-ethyltoluene	0.13	10.9	0.41
1,2,4-trichlorobenzene	0.32	26.8	1.01
1,2,4-trimethylbenzene	0.17	13.8	0.52
1,3,5-trimethylbenzene	0.13	10.6	0.40
hexachloro-1,3-butadiene	0.32	16.8	1.01

TABLE 7 NIST Traceable GC/MS Audit Results

Compound	Audit #1164			Audit #1252			Audit #1366			Audit #1496		
	Ref. (ppb by volume)	Rep. (ppb by volume)	% Bias	Ref. (ppb by volume)	Rep. (ppb by volume)	% Bias	Ref. (ppb by volume)	Rep. (ppb by volume)	% Bias	Ref. (ppb by volume)	Rep. (ppb by volume)	% Bias
Vinyl chloride	3.6	2.8	−22.0	4.9	4.5	−8.16	3.6	3.2	−11.11	2.4	2.1	−12.20
Bromomethane	3.6	3.5	−2.8	4.8	4.2	−12.50	3.5	3.1	−11.43	2.4	1.1	−54.17[A]
Methylene chloride	7.2	7.9	4.2	9.8	9.1	−7.14	7.2	6.7	−6.94	4.9	5.6	14.29
trans-1,2-Dichloroethylene	6.9	7.0	1.4	9.3	7.6	−18.28	7.6	5.4	−20.59	4.9	5.1	8.51
1,1-Dichloroethane	3.8	3.3	−13.0	5.1	3.9	−23.53	3.9	2.9	−27.03	2.6	2.6	0.00
Chloroform	3.5	3.8	8.6	4.8	5.4	12.50	3.5	4.3	22.86	2.4	2.9	20.83
1,1,1-Trichloroethane	3.6	4.1	14.0	4.8	4.9	2.08	3.6	4.0	11.11	2.4	2.7	12.50
Carbon tetrachloride	3.3	3.4	3.0	4.5	3.6	−20.00	3.3	3.7	12.12	2.3	2.6	13.04
Benzene	7.3	7.0	−4.1	9.9	10.5	6.06	7.3	8.2	12.33	4.9	4.5	−8.16
Trichloroethylene	3.6	3.6	0.0	10.0	10.1	1.00	3.6	6.0	66.67	2.4	2.1	−12.50
1,2-Dichloropropane	7.4	7.2	−2.7	4.9	5.9	20.41	7.3	8.8	20.55	5.0	4.1	−18.00
Toluene	3.8	4.1	7.9	5.1	4.6	−9.80	3.8	3.6	−5.26	2.6	2.5	−3.85
Tetrachloroethylene	3.8	5.3	39.0[A]	5.2	5.6	7.69	3.8	4.6	21.05	2.6	2.8	7.69
Chlorobenzene	7.6	9.5	25.0	10.3	6.6	−35.92[A]	7.5	5.5	−26.67	5.1	4.4	−13.72
Styrene	3.7	2.0	−46.0[A]	5.0	4.6	−8.00	3.7	0.5	−86.49	2.5	2.4	−4.00
o-Xylene	8.8	8.0	−9.1	12.0	8.9	−25.83	8.8	6.9	−21.59	6.0	6.1	1.67
Ethylbenzene	7.8	6.8	−13.0	10.5	7.2	−31.43	7.7	5.7	−25.97	5.3	5.1	−3.77

[A] Greater than the ±30 % data quality objectives.

calibrated at three to five concentration levels, a second humid zero air sample is passed through the system and analyzed. The second humid zero air test is used to verify that the GC/MS system is certified clean (less than 0.2 ppb by volume of target compounds).

NOTE 18—Alternative approaches for generation of calibration standards are acceptable as long as the calibration range (0–100 ppb by volume) and humidity are accurately maintained.

10.2.3.2 As an alternative, a multipoint humid static calibration (three to five levels plus zero humid air) can be performed on the GC/MS system. During the humid static calibration analyses, three (3) SUMMA® passivated canisters are filled each at a different concentration between 1 to 20 ppb by volume from the calibration manifold using a pump and mass flow control arrangement [see Fig. 4(c)]. The canisters are then delivered to the GC/MS to serve as calibration standards. The canisters are analyzed by the MS in the SIM mode, each analyzed twice. The expected retention time and ion abundance (see Table 2 and Table 5) are used to verify proper operation of the GC/MS system. A calibration response factor is determined for each analyte, as illustrated in Table 5.

10.2.3.3 *Routine Calibration*—The GC/MS system is calibrated daily (and before sample analysis) with a one point calibration. The GC/MS system is calibrated either with the dynamic calibration procedure [see Fig. 4(a)] or with a 6 L SUMMA® passivated canister filled with humid calibration standards from the calibration manifold (see 10.2.3.2). After the single point calibration, the GC/MS analytical system is challenged with a humidified zero gas stream to ensure the analytical system returns to specification (less than 0.2 ppb by volume of selective organics).

10.3 *Analytical Procedures:*

10.3.1 *Canister Receipt:*

10.3.1.1 The overall condition of each sample canister is observed. Each canister must be received with an attached sample identification tag.

10.3.1.2 Each canister is recorded in the dedicated laboratory logbook. Also noted on the identification tag are date received and initials of recipient.

NOTE 19—A log containing the usage and history of each canister shall be kept. This historical record will assist in ensuring that canisters used for source sampling are not mixed with canisters used for indoor or ambient air. Samples used for high level standards or to acquire high level VOC samples should be flagged to receive individual blanking quality control checks after sample analysis.

10.3.1.3 The pressure of the canister is checked by attaching a pressure gage to the canister inlet. The canister valve is opened briefly and the pressure (kPa, psig) is recorded.

NOTE 20—If pressure is <83 kPa (<12 psig), the user may wish to pressurize the canisters, as an option, with zero grade nitrogen up to 137 kPa (20 psig) to ensure that enough sample is available for analysis. However, pressurizing the canister can introduce additional error, increase the minimum detection limit (MDL), and is time consuming. The user must consider these limitations as part of his program objectives before pressurizing. Final cylinder pressure is recorded.

10.3.1.4 If the canister pressure is increased, a dilution factor (DF) is calculated and recorded on the sampling data sheet:

$$DF = Y_a/X_a \qquad (3)$$

where:

TABLE 8 Average Performance on Audits

	Average Deviation	Standard	n
Vinyl chloride	−3.9	14.4	8
Bromomethane	5.5	19.7	8
Methylene chloride	5.9	9.9	8
trans-1,2-Dichloroethylene	−4.2	10.9	8
1,1-Dichloroethane	−7.9	12.5	8
Chloroform	15.9	5.3	5
1,1,1-Trichloroethane	8.9	5.6	8
Carbon tetrachloride	6.0	12.6	8
Benzene	5.6	12.5	8
Trichloroethylene	9.1	24.2	8
1,2-Dichloropropane	7.0	15.0	8
Toluene	1.6	12.3	8
Tetrachloroethylene	18.3	15.1	7
Chlorobenzene	4.2	22.4	6
Styrene	25.7	30.9	7
o-Xylene	−13.7	10.8	4
Ethylbenzene	−13.7	9.7	8

X_a = canister pressure absolute before dilution, kPa, psia and
Y_a = canister pressure absolute after dilution, kPa, psia.
After sample analysis, detected VOC concentrations are multiplied by the dilution factor to determine concentration in the sampled air.

10.3.2 *GC/MS-SCAN and SIM Analysis:*

10.3.2.1 When the MS is placed in the full scan mode (SCAN) all ions are scanned between the preset windows form monitoring. The characteristic mass spectrum of any compound or group of compounds reaching the MS detector are recorded and can be interpreted for both qualitative identification and quantitative determination. In the SIM mode of operation, the MS monitors only preselected ions, rather than scanning all masses continuously between two mass limits. As a result, increased sensitivity and improved quantitative analysis can be achieved at the expense of identifying and quantifying unknown compounds.

10.3.2.2 The analytical system shall be properly assembled, humid zero air certified (see 12.3), operated (see Table 3), and calibrated for accurate VOC determination.

10.3.2.3 The mass flow controllers are checked and adjusted to provide correct flow rates for the system.

10.3.2.4 The sample canister is connected to the inlet of the GC/MS-SCAN or GC/MS-SIM analytical system. For pressurized samples, a mass flow controller is placed on the canister, the canister valve is opened and the canister flow is vented past a tee inlet to the analytical system at a flow of 75 mL/min so that the inlet system up to the six-port sample injection valve is flushed with sample gas (typically 40 mL). The cryogenic trap is connected and verified to be operating properly while cooled with cryogen through the system.

NOTE 21—Flow rate is not as important as acquiring sufficient sample volume.

10.3.2.5 Sub-ambient pressure samples are connected directly to the inlet. The sample system is evacuated prior to collecting the sample on the concentration loop.

10.3.2.6 The GC oven and cryogenic trap (inject position) are cooled to their set points of −50°C and −178°C, respectively.

10.3.2.7 As soon as the cryogenic trap reaches its lower set point of −178°C, the six-port chromatographic valve is turned to its fill position to initiate sample collection.

10.3.2.8 A ten minute collection period of canister sample is utilized.

NOTE 22—More or less canister sample is used for analysis depending on the sensitivity of the mass detection unit and the concentration of the target analytes in the sample.

10.3.2.9 After the sample is preconcentrated in the cryogenic trap, the GC sampling valve is cycled to the inject position and the cryogenic trap is heated. The trapped analytes are thermally desorbed onto the head of the OV-1 or equivalent capillary column (0.31 mm I.D. by 50 m length). The GC oven is programmed to start at −50°C and after 2 min to heat to 150°C at a rate of 8°C per minute.

10.3.2.10 Upon sample injection onto the column, the MS is signaled by the computer to start data acquisition. In the SCAN mode, the eluting carrier gas passing through the mass spectrometer source is scanned from 35 to 250 amu, resulting in a 1.5 Hz repetition rate. This corresponds to about 6 scans per eluting chromatographic peak.

10.3.2.11 The individual analyses are handled in three phases: data acquisition, data reduction, and data reporting.

10.3.2.12 Primary identification is based upon retention time and relative abundance of eluting ions as compared to the spectral library stored on the hard disk of the GC/MS data computer. In the SIM, the data acquisition software is set to monitor specific compound fragments at specific times in the analytical run. Data reduction is coordinated by the postprocessing program that is automatically accessed after data acquisition is completed at the end of the GC run. Resulting ion profiles are extracted, peaks are identified and integrated, and an integration report is generated by the computer software. A reconstructed ion chromatogram for hard copy reference is prepared by the program and various parameters of interest such as time, date, and integration constants are printed. At the completion of the program, the data reporting software is accessed. The appropriate calibration table is retrieved by the data reporting program from the computer's hard disk storage and the proper retention time and response factor parameters are applied to the macro program's integration file. With reference to certain pre-set acceptance criteria, peaks are automatically identified and quantified and a final summary report is prepared.

10.3.2.13 Approximately 64 min are required for each sample analysis, 15 min for system initialization, 14 min for sample collection, 30 min for analysis, and 5 min for post run equilibration, during which a report is printed.

10.3.2.14 The helium and sample mass flow controllers are checked and adjusted to provide correct flow rates for the system. Helium is used to purge residual air from the trap at the end of the sampling phase and to carry the revolatilized VOCs from the trap onto the GC column.

10.3.2.15 The concentration (ppb by volume) is calculated using the previously established response factors (see 10.2.3.2), as illustrated in Table 5.

NOTE 23—If the canister is diluted before analysis, an appropriate multiplier is applied to correct for the volume dilution of the canister (10.3.1.4).

11. Cleaning and Certification Program

11.1 *Canister Cleaning and Certification:*

11.1.1 All canisters must be clean and free of any contaminants before sample collection.

11.1.2 All canisters are leak tested by pressurizing them to approximately 200 kPa (30 psig) with zero air.

NOTE 24—The canister cleaning system in Fig. 3 can be used for this task. The initial pressure is measured, the canister valve is closed, and the final pressure is checked after 24 h. If leak tight, the pressure shall not vary more than ±13.8 kPa (±2 psig) over the 24-h period.

11.1.3 A canister cleaning system may be assembled as illustrated in Fig. 3. Cryogen is added to both the vacuum pump and zero air supply traps. The canister(s) are connected to the manifold. The vent shut-off valve and the canister valve(s) are opened to release any remaining pressure in the canister(s). The vacuum pump is started and the vent shut-off valve is then closed and the vacuum shut-off valve is opened. The canister(s) are evacuated to less than 0.064 kPa (5.0 mm Hg) for at least one hour.

NOTE 25—On a daily basis, or more often if necessary, the cryogenic traps shall be purged with zero air to remove any trapped water from previous canister cleaning cycles.

11.1.4 The vacuum and vacuum/pressure gage shut-off valves are closed and the zero air shut-off valve is opened to pressurize the canister(s) with humid zero air to approximately 200 kPa (30 psig). If a zero gas generator system is used, the flow rate may need to be limited to maintain the zero air quality.

11.1.5 The zero shut-off valve is closed and the canister(s) is allowed to vent down to atmospheric pressure through the vent shut-off valve. The vent shut-off valve is closed. Steps 11.1.3 through 11.1.5 are repeated two additional times for a total of three evacuation/pressurization cycles for each set of canisters.

11.1.6 At the end of the evacuation/pressurization cycle, the canister is pressurized to 200 kPa (30 psig) with humid zero air. Analyze the canister with the GC/MS or GC-FID-ECD analytical system. Any canister that has not tested clean (compared to direct analysis of humidified zero air of less than 0.2 ppb by volume of targeted VOCs) shall not be used. As a "blank" check of the canister(s) and cleanup procedure, analyze the final humid zero air fill of 100 % of the canisters until the cleanup system and canisters are proven reliable (less than 0.2 ppb by volume of target VOCs). This blank check may be reduced to one canister per batch after the blank criterion has been met on one entire batch.

11.1.7 Reattach the canister to the manifold and reevacuate to less than 0.064 kPa (5.0 mm Hg). The canister valve is closed. Remove the canister from the cleaning system and cap the canister connection with a stainless steel fitting. The canister is now ready for collection of an air sample. Attach an identification tag to the neck of each canister for field notes and chain-of-custody purposes. Retain the canister in this condition until used.

11.1.8 As an option to the humid zero air cleaning procedures, heat the canisters in an isothermal oven to no greater than 100°C using the apparatus described in 11.1.3.

NOTE 26—For sampling heavier, more complex VOC mixtures, the canisters shall be heated to 250°C during 11.1.3 through 11.1.7. Canister valves shall not be heated during this cleaning process. Once heated, the canisters are evacuated to 0.064 kPa (5.0 mm Hg). At the end of the heated/evacuated cycle, pressurize the canisters with humid zero air and analyze by the blanking procedures in this standard. Any canister that has not tested clean (less than 0.2 ppb by volume of targeted compounds) shall not be used. Once tested clean, reevacuate the canisters to 0.064 kPa (5 mm Hg) or less and retain in the evacuated state until used. Repeated heating of canisters may degrade the SUMMA® treated surface of the canister and premature degradation of samples may ensue. Periodic (yearly) check of calibration standard stability shall be performed in canisters that are repeatedly heat treated.

11.2 *Sampling System Cleaning and Certification:*

11.2.1 *Cleaning Sampling System Components:*

11.2.1.1 Sample components are disassembled and cleaned before the sampler is assembled. Nonmetallic parts are rinsed with HPLC grade deionized water and dried in a vacuum oven at 50°C. Typically, stainless steel parts and fittings are cleaned by placing them in a beaker of hexane in an ultrasonic bath for 15 min. This procedure is repeated with methanol as the solvent.

11.2.1.2 The parts are then rinsed with HPLC grade deionized water and dried in a vacuum oven at 100°C for 12 to 24 h.

11.2.1.3 Once the sampler is assembled, the entire system is purged with humid zero air for 24 h.

11.2.2 *Humid Zero Air Certification:*

11.2.2.1 The system is "certified" if no significant additions or deletions (less than 0.2 ppb by volume of targeted compounds) have occurred when challenged with the ultra high purity humidified air test stream. The cleanliness of the sampling system is determined by testing the sampler with humid zero air with an evacuated canister, as follows.

11.2.2.2 The calibration system and manifold are assembled as illustrated in Fig. 4. The sampler (with an evacuated sampling canister) is connected to the manifold and the zero air cylinder activated to generate a humid gas stream (2 L/min) to the calibration manifold [see Fig. 4(b)].

11.2.2.3 The humid zero gas stream passes through the calibration manifold, through the sampling system and collected in a clean evacuated sampling canister (see 10.2.2.1). After the sample (a minimum of 250 mL) is preconcentrated on the trap, the trap is heated and the VOCs are thermally desorbed onto the head of the capillary column. Since the column is at −50°C, the VOCs are cryofocussed on the column. Then, the oven temperature (programmed) increases and the VOCs begin to elute and are detected by a GC/MS (see 10.2). The analytical system must not detect greater than 0.2 ppb by volume of targeted VOCs in order for the sampling system to pass the humid zero air certification test. If the sampler passes the humid zero air test, it is then tested with humid calibration gas standards containing selected VOCs at concentration levels expected in field sampling (for example, 5.0 to 20 ppb by volume) as outlined in 11.2.3.

NOTE 27—As an alternative to save the use of GC/MS time, a GC-FID-ECD may be used for canister analysis of certification standards. Such a system must include a sample concentration interface identical to the one used for GC/MS analysis (10.3). It must also be calibrated with standards prepared in the same way as those used for GC/MS analysis. Retention time is verified with known standard compounds and used to identify certification standards or analyzed zero test samples.

11.2.3 *Sampler System Certification with Humid Calibration Gas Standards:*

11.2.3.1 Assemble the dynamic calibration system and manifold as illustrated in Fig. 4.

NOTE 28—The certification manifold will often become contaminated with certification compounds in the process of certifying samplers. Separate manifolds shall be used for zero certification and humid calibration gas challenge to avoid erroneous results caused by carryover of compounds in the test manifold. Alternatively, one manifold may be used; however, the manifold must be certified as clean prior to the start of a zero certification test.

NOTE 29—Manifold components and flow regulators must be heated during humid calibration gas standards certification to ensure complete vaporization of challenge gas components.

11.2.3.2 Verify that the calibration system is clean (less than 0.2 ppb by volume of targeted compounds) by sampling a humidified gas stream, without gas calibration standards, with a previously certified clean canister (see 12.1).

11.2.3.3 The assembled dynamic calibration system is certified clean if less than 0.2 ppb by volume of targeted compounds are found.

11.2.3.4 For generating the humidified calibration standards, the calibration gas cylinder(s) (see 8.2) containing nominal concentrations of 10 ppm by volume in nitrogen of selected VOCs are attached to the calibration system, as outlined in 10.2.3.1. The gas cylinders are opened and the gas mixtures are passed through 0 to 50 mL/min certified mass flow controllers to generate ppb levels of calibration standards.

11.2.3.5 After the appropriate equilibration period, attach the sampling system (containing a certified evacuated canister) to the manifold, as illustrated in Fig. 4(a).

11.2.3.6 Sample the dynamic calibration gas stream with the sampling system according to 9.2.1.

NOTE 30—To conserve generated calibration gas, bypass the canister sampling system manifold and attach the sampling system to the calibration gas stream at the inlet of the in-line filter of the sampling system so the flow will be less than 500 mL/min.

11.2.3.7 Concurrent with the sampling system operation, real time monitoring of the calibration gas stream is accomplished by collection of a canister sample connected directly to a mass flow controller and the certification test gas feed line. Analysis of this check sample must be performed to confirm the concentration of standard gas delivered to the samplers being certified.

11.2.3.8 At the end of the sampling period (normally same time period used for anticipated sampling), the sampling system canister is analyzed and compared to the reference canister results to determine if the concentration of the targeted VOCs was increased or decreased by the sampling system.

11.2.3.9 A recovery of between 85 % and 115 % is expected for the average of all targeted VOCs. Individual compounds must fall within the range of 80 % and 120 % for acceptable certification.

12. Performance Criteria and Quality Assurance

12.1 *Standard Operating Procedures (SOPs)*:

12.1.1 SOPs must be generated in each laboratory describing and documenting the following activities: (*1*) assembly, calibration, leak check, and operation of specific sampling systems and equipment used, (*2*) preparation, storage, shipment, and handling of samples, (*3*) assembly, leak-check, calibration, and operation of the analytical system, addressing the specific equipment used, (*4*) canister storage and cleaning, and (*5*) all aspects of data recording and processing, including lists of computer hardware and software used.

12.1.2 Specific stepwise instructions should be provided in the SOPs and should be readily available to and understood by the laboratory personnel conducting the work.

12.2 *Method Relative Accuracy and Linearity*:

12.2.1 Accuracy can be determined by injecting VOC standards (see 8.2) from an audit cylinder into a sampler. The contents are then analyzed for the components contained in the audit canister. Percent relative accuracy is calculated:

$$\% \text{ Relative Accuracy} = (X - Y)/X \times 100 \qquad (4)$$

where:

Y = concentration of the targeted compound recovered from sampler, and

X = concentration of VOC-targeted compound in the NIST-SRM or EPA-CRM audit cylinders.

12.2.2 If the relative accuracy does not fall between 80 and 120 %, the sampler should be removed from use, cleaned, and recertified according to initial certification

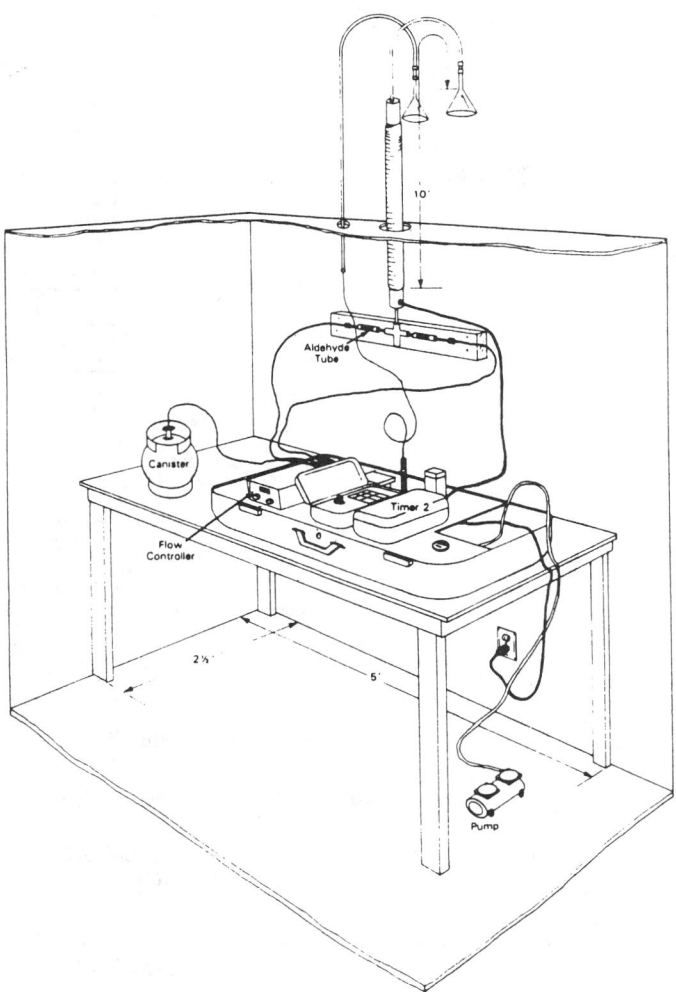

FIG. 6 Perspective View of UATMP Sampler

procedures outlined in 11.2.2 and 11.2.3.

12.3 *Method Modification*:

12.3.1 *Sampling*:

12.3.1.1 *Urban Air Toxics Sampler*—The sampling system described in this test method (Fig. 1) may be modified like the sampler in EPA's FY-90 Urban Air Toxics Pollutant Program (see Fig. 6).

12.3.1.2 *Analysis*:

(*1*) Heat inlet tubing from the calibration manifold to 50°C (same temperature as the calibration manifold) to prevent condensation on the internal walls of the system.

(*2*) The analytical strategy for this test method involves positive identification and quantitation by GC/MS-SCAN or -SIM mode. This is a highly specific and sensitive detection technique. Because a specific detector system (GC/MS-SCAN or -SIM) is more complicated and expensive than the use of non-specific detectors (GC-FID-ECD-PID), the analyst may perform a screening analysis and preliminary quantitation of VOC species in the sample, including any polar compounds, by utilizing the GC-multidetector (GC-FID-ECD-PID) analytical system prior to GC/MS analysis. This multidetector system can be used for approximate quantitation. The GC-FID-ECD-PID provides a "snapshot"

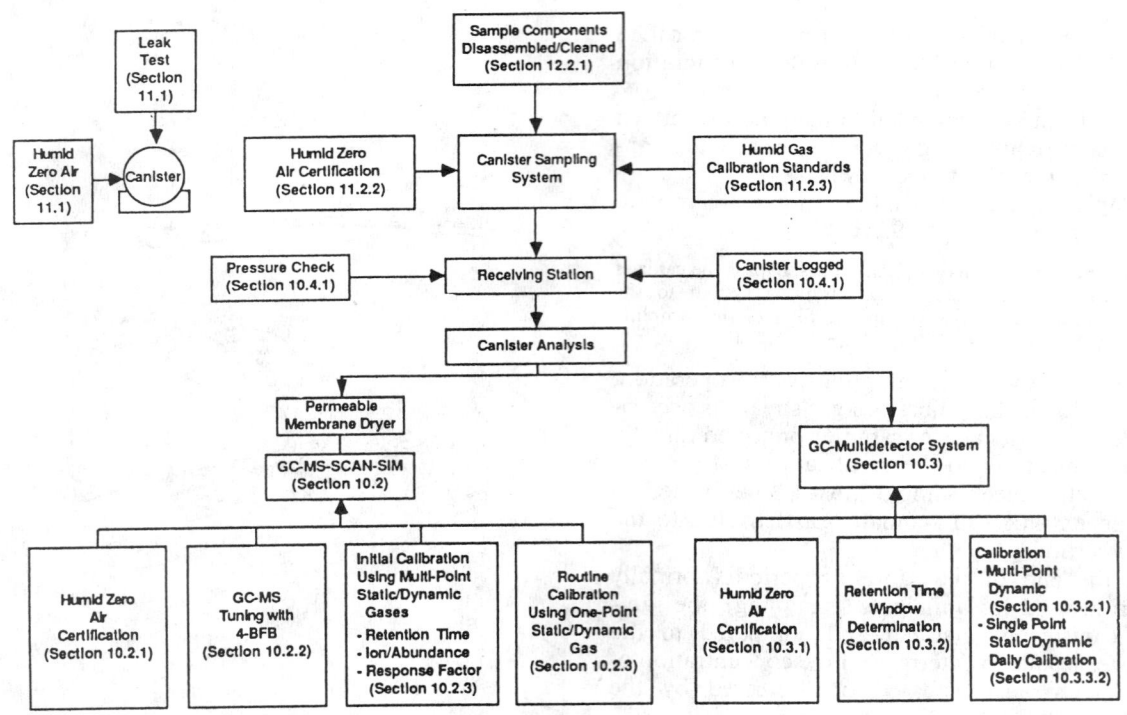

FIG. 7 System Quality Assurance/Quality Control (QA/QC) Activities Associated with Various Analytical Systems

of the constituents in the sample, allowing the analyst to determine:

(*a*) Whether the constituents are within the calibration range of the anticipated GC/MS-SCAN-SIM analysis or does the sample require further dilution, and

(*b*) Are there unexpected peaks which need further identification through GC/MS-SCAN or will GC/MS-SIM be adequate.

12.4 *Quality Assurance (See Fig. 7):*

12.4.1 *Sampling System:*

12.4.1.1 Paragraph 9.2 requires pre- and post-sampling measurements with a certified mass flow controller for flow verification of sampling system.

12.4.1.2 Paragraph 11.1 requires all canisters to be pressure tested to 200 kPa ± 14 kPa (30 psig ± 2 psig) over a period of 24 h.

12.4.1.3 Paragraph 11.1 requires that all canisters be certified clean (containing less than 0.2 ppb by volume of targeted VOCs) through a humid zero air certification program.

12.4.1.4 Paragraph 11.2.2 requires all sampling systems to be certified initially clean (containing less than 0.2 ppb by volume of targeted VOCs) through a humid zero air certification program.

12.4.1.5 Paragraph 11.2.3 requires all sampling systems to pass an initial humidified calibration gas certification [at VOC concentration levels expected in the field (for example, 0.5 to 20 ppb by volume)] with a percent recovery of greater than 90.

12.4.2 *GC/MS-SCAN-SIM System Performance Criteria:*

12.4.2.1 Paragraph 10.2.1 requires the GC/MS analytical system to be certified clean (less than 0.2 ppb by volume of targeted VOCs) prior to sample analysis, through a humid zero air certification.

12.4.2.2 Paragraph 10.2.2 requires the tuning of the GC/MS with 4-bromofluorobenzene (4-BFB) and that it meet the key ions and ion abundance criteria (10 %) outlined in Table 5.

12.4.2.3 Paragraph 10.2.3 requires both an initial multipoint humid static calibration (three levels plus humid zero air) and a daily calibration (one point) of the GC/MS analytical system.

12.4.2.4 Paragraph 10.2.3.3 requires that a calibration check sample in the mid range of the calibration curve is analyzed once each day or once every eight samples to ensure the calibration of the GC/MS is still valid and under control.

13. Precision and Bias

13.1 The precision of replicate gas sample analysis will vary depending on the volatile organic compound being determined. Typical precision reported as %CV should be ±30 %, for determinations made on the same sample over an 8-h period. If the GC/MS analysis does not meet or exceed these criteria, the instrument should be retuned and recalibrated. Precision measured for seven replicate analyses analyzed over a 10-h period are reported in Table 6.

13.2 Tests performed to measure the bias of this procedure have been conducted with a performance audit sample prepared by U.S. EPA referencable to a primary standard gas mixture prepared by NIST. Humidified gas performance standards were analyzed in a GC/MS system configured in accordance with this standard, following the procedure without the use of a Nafion membrane dryer. Four performance samples ranging in concentration from 2 to 10 ppb by volume were analyzed one sample every other month over a period of 10 months. The test results shown in Table 7 were obtained using this system.

14. Keywords

14.1 ambient atmospheres; analysis; atmospheres; canister sampling; gas chromatography–mass spectrometry; indoor atmospheres; sampling; volatile organic compounds; workplace atmospheres

REFERENCES

(1) Oliver, K. D., Pleil, J. D., and McClenny, W. A., "Sample Integrity of Trace Level Volatile Organic Compounds in Ambient Air Stored in SUMMA® Polished Canisters," *Atmospheric Environment*, 20:1403, 1986.

(2) Holdren, M. W., and Smith, D. L., "Stability of Volatile Organic Compounds While Stored in SUMMA® Polished Stainless Steel Canisters," Final Report, EPA Contract No. 68-02-4127, Research Triangle Park, NC, Battelle Columbus Laboratories, January, 1986.

(3) McClenny, W. A., et al., "Canister-Based Method for Monitoring Toxic VOCs in Ambient Air," JA&WA, Vol 41, No. 10, October 1991, pp. 1308–1318.

(4) Coutant, R. W., and McClenny, W. A., "Competitive Adsorption Effects and the Stability of VOC and PVOC in Canisters," Proceedings of the 1991 U.S. EPA/AW&MA International Symposium on Measurement of Toxic and Related Air Pollutants, May 1991, A&WMA Publication VIP-21, EPA 600/9-91/-18.

(5) Crist, H. L., "Assessing the Performance of Ambient Air Samplers for Volatile Organic Compounds," in *Monitoring Methods for Toxics in the Atmosphere, ASTM STP 1052*, W. L. Zielinski, Jr., and W. D. Dorko, Eds., ASTM, Philadelphia, PA, 1990, pp. 46–52.

(6) Riggin, R. M., *Technical Assistance Document for Sampling and Analysis of Toxic Organic Compounds in Ambient Air*, EPA-600/4-83-027, U.S. Environmental Protection Agency, Research Triangle Park, NC, 1983.

(7) Winberry, W. T., and Tilley, N. V., *Supplement to EPA-600/4-84-041: Compendium of Methods for the Determination of Toxic Organic Compounds in Ambient Air*, EPA-600/4-87-006, U.S. Environmental Protection Agency, Research Triangle Park, NC, 1986.

(8) McClenny, W. A., Pleil, J. D., Holdren, J. W., and Smith, R. N., "Automated Cryogenic Preconcentration and Gas Chromatographic Determination of Volatile Organic Compounds," *Analytical Chemistry*, 56:2947, 1984.

(9) Pleil, J. D., and Oliver, K. D., "Evaluation of Various Configurations of Nafion Dryers: Water Removal from Air Samples Prior to Gas Chromatographic Analysis," EPA Contract No. 68-02-4035, Research Triangle Park, NC, Northrop Services, Inc.—Environmental Sciences, 1985.

(10) Oliver, K. D., and Pleil, J. D., "Automated Cryogenic Sampling and Gas Chromatographic Analysis of Ambient Vapor-Phase Organic Compounds: Procedures and Comparison Tests," EPA Contract No. 68-02-4035, Research Triangle Park, NC, Northrop Services, Inc.—Environmental Sciences, 1985.

(11) McClenny, W. A., and Pleil, J. D., "Automated Calibration and Analysis of VOCs with a Capillary Column Gas Chromatograph Equipped for Reduced Temperature Trapping," *Proceedings of the 1984 Air Pollution Control Association Annual Meeting*, San Francisco, CA, June 24–29, 1984.

(12) McClenny, W. A., Pleil, J. D., Lumpkin, T. A., and Oliver, K. D., "Update on Canister-Based Samplers for VOCs," *Proceedings of the 1987 EPA/APCA Symposium on Measurement of Toxic and Related Air Pollutants*, May, 1987, APCA Publication VIP-8, EPA 600/9-87-010.

(13) Pleil, J. D., "Automated Cryogenic Sampling and Gas Chromatographic Analysis of Ambient Vapor-Phase Organic Compounds: System Design," EPA Contract No. 68-02-2566, Research Triangle Park, NC, Northrop Services, Inc.—Environmental Sciences, 1982.

(14) Oliver, K. D., and Pleil, J. D., "Analysis of Canister Samples Collected During the CARB Study in August 1986," EPA Contract No. 68-02-4035, Research Triangle Park, NC, Northrop Services, Inc.—Environmental Sciences, 1987.

(15) Pleil, J. D., and Oliver, K. D., "Measurement of Concentration Variability of Volatile Organic Compounds in Indoor Air: Automated Operation of a Sequential Syringe Sampler and Subsequent GC/MS Analysis," EPA Contract No. 68-02-4444, Research Triangle Park, NC, Northrop Services, Inc.—Environmental Sciences, 1987.

(16) Walling, J. F., "The Utility of Distributed Air Volume Sets When Sampling Ambient Air Using Solid Adsorbents," *Atmospheric Environ.*, 18:855–859, 1984.

(17) Walling, J. F., Bumgarner, J. E., Driscoll, J. D., Morris, C. M., Riley, A. E., and Wright, L. H., "Apparent Reaction Products Desorbed From Tenax Used to Sample Ambient Air," *Atmospheric Environment*, 20:51–57, 1986.

(18) *Portable Instruments User's Manual for Monitoring VOC Sources*, EPA340/1-88-015, U.S. Environmental Protection Agency, Office of Air Quality Planning and Standards, Washington, DC, June, 1986.

(19) McElroy, F. F., Thompson, V. L., and Richter, H. G., *A Cryogenic Preconcentration—Direct FID (PDFID) Method for Measurement of NMOC in the Ambient Air*, EPA-600/4-85-063, U.S. Environmental Protection Agency, Research Triangle Park, NC, August 1985.

(20) Dayton, D-P., Brymer, D. A., and Jongleux, R. F., "Canister Based Sampling Systems—A Performance Evaluation," *Proceedings of the 1990 EPA/APCA Symposium on Measurement of Toxic and Related Air Pollutants*, May, 1990 APCA Publication VIP-17, EPA 600/9-90/026.

(21) Rasmussen, R. A., and Lovelock, J. E., "Atmospheric Measurements Using Canister Technology," *J. Geophysical Research*, 83:8369–8378, 1983.

(22) Rasmussen, R. A., and Khalil, M. A. K., "Atmospheric Halocarbons: Measurements and Analysis of Selected Trace Gases," *Proc. NATO ASI on Atmospheric Ozone*, BO:209–231.

Standard Practice for
Measuring the Concentration of Toxic Gases or Vapors Using Detector Tubes[1]

This standard is issued under the fixed designation D 4490; the number immediately following the designation indicates the year of original adoption or, in the case of revision, the year of last revision. A number in parentheses indicates the year of last reapproval. A superscript epsilon (ε) indicates an editorial change since the last revision or reapproval.

1. Scope

1.1 This practice covers the detection and measurement of concentrations of toxic gases or vapors using detector tubes (1, 2).[2] A list of some of the gases and vapors that can be detected by this practice, their 1994–95 TLV values recommended by the ACGIH, and their measurement ranges are provided in Annex A1. This list is given as a guide and should be considered neither absolute nor complete.

1.2 *This standard does not purport to address all of the safety concerns, if any, associated with its use. It is the responsibility of the user of this standard to establish appropriate safety and health practices and determine the applicability of regulatory limitations prior to use.*

2. Referenced Documents

2.1 *ASTM Standard:*
D 1356 Terminology Relating to Sampling and Analysis of Atmospheres[3]
2.2 *Other Document:*
29 CFR 1910 Federal Occupational Safety and Health Standard Title 29[4]

3. Terminology

3.1 For definitions of terms used in this method, refer to Terminology D 1356.

4. Summary of Practice (3)

4.1 Detector tubes may be used for either short-term sampling (grab sampling; 1 to 10 min typically) or long term sampling (actively or passively; 1 to 8 h) of atmospheres containing toxic gases or vapors.

4.1.1 *Short-Term Sampling (Grab Sampling)* (4–18)—A given volume of air is pulled through the tube by a mechanical pump. If the substance for which the detector tube was designed is present, the indicator chemical in the tube will change color (stain). The concentration of the gas or vapor may be estimated by either (*a*) the length-of-stain compared to a calibration chart, or (*b*) the intensity of the color change compared to a set of standards.

4.1.2 *Long-Term Active Sampling (Long-Term Tubes)* (19–22)—A sample is pulled through the detector tube at a slow, constant flow rate by an electrical pump. The time-weighted average concentration of the gas or vapor is determined by correlating the time of sampling either with (*a*) the length-of-stain read directly from the calibration curve imprinted on the tube or (*b*) the intensity of the color change compared to a set of standards.

4.1.3 *Long-Term Passive Sampling (Diffusion or Dosimeter Tubes)* (25)—The contaminant molecules move into the tube according to Fick's First Law of Diffusion. The driving force is the concentration differential between the ambient air and the inside of the tube. The time-weighted average concentration of the gas or vapor is determined by dividing the indication on the tube by the number of hours sampled (1 to 10 h according to the manufacturers' instructions).

4.2 Instructions are given for the calibration of the sampling pumps required in this practice.

4.3 Information on the correct use of the detector tubes is presented.

5. Significance and Use

5.1 The Federal Occupational Safety and Health Administration, in 29 CFR 1910, designates that certain gases and vapors must not be present in workplace atmospheres at concentrations above specific values.

5.2 This practice will provide a means for the determination of airborne concentrations of certain gases and vapors given in 29 CFR 1910.

5.3 A partial list of chemicals for which this practice is applicable is presented in Annex A1.

5.4 This practice also provides for the sampling of gaseous atmospheres to be used for process control or other purposes (2, 23–25).

6. Interferences (26, 27)

6.1 Some common interferences for the various tubes are listed in the instruction sheets provided by the manufacturers.

7. Apparatus (28–31)

7.1 *Detector Tube*—A detector tube consists of a glass tube containing an inert granular material that has been impregnated with a chemical system which reacts with the gas or vapor of interest. As a result of this reaction, the impregnated chemical changes color. The granular material is held in place within the glass tube by porous plugs of a suitable inert material. The ends of the glass tube are flame-sealed to protect the contents during storage.

7.2 *Pump (32):*

---

[1] This practice is under the jurisdiction of ASTM Committee D-22 on Sampling and Analysis of Atmospheres and is the direct responsibility of Subcommittee D22.04 on Workplace Atmospheres.
Current edition approved Oct. 10, 1996. Published December 1996. Originally published as D 4490 – 85. Last previous edition D 4490 – 90.
[2] The boldface numbers in parentheses refer to the list of references at the end of this practice.
[3] *Annual Book of ASTM Standards*, Vol 11.03.
[4] *Code of Federal Regulations*, Part 1910.1000 Subpart 2 and Part 1926.55 Subpart D.

7.2.1 *Short-Term Sampling*—A mechanical, hand-operated, aspirating pump is used to draw the sample through the detector tube during the short-term sampling. Two types of pumps are commercially available: piston-operated and bellows-operated. The pumps have a capacity of 100 mL for a full pump stroke. By varying the number of pump strokes, the sample volume is controlled. Sampling pumps should be maintained and calibration checked periodically according to the manufacturer's instructions. The pumps shall be accurate to ±5 % of the volume stated.

7.2.2 *Long-Term Sampling*—Small electrical pumps having stable low flow rates (2 to 50 mL/min), are required for long-term sampling (2 to 8 h). Flow rates to be used with each detector tube are given by the manufacturers. As with the mechanical pumps, the electrical pumps must be maintained and calibrated regularly. Maintenance and calibration are performed using the instructions supplied by the manufacturer of the pump. The pump flow rate, and, therefore, the sampled volume, shall be accurate to ±5 % of the stated flow rate. With this system either area or personal monitoring can be accomplished.

7.3 *Accessories*—Several accessories are provided with detector tubes for special applications:

7.3.1 *Reactor Tubes*—These are tubes that are used in conjunction with detector tubes. Some gases and vapors, because of their low reactivity, are not easily detected by detector tubes alone. The reactor tubes consist of very powerful chemical reactants, which break down the unreactive compound into other more readily detectable substances, which standard detector tubes can detect. Thus, the reactor tube is placed upstream of the detector tube and the combination must be used for certain compounds as a detector tube system.

7.3.2 *Dryer Tubes*—Water vapor interferes with the detection of certain substances; therefore, dryer tubes are used upstream of the detector tube in these cases to remove the water vapor.

7.3.3 *Pyrolyzer*—A pyrolyzer is a hot wire instrument operated by batteries. Instructions for its use and maintenance are given in the manufacturers' instruction manuals. The purpose of the pyrolyzer, as with reactor tubes, is to break down difficult-to-detect compounds into other compounds more easily detected. The breakdown in this case is caused by heat. The pyrolzyer is particularly useful for organic nitrogen compounds, one of the products of breakdown being nitrogen dioxide, which is easily monitored.

7.3.4 *Remote Sampling Line*—When the sampling point is remote from the pump location, a length of nonreactive tubing can be attached to the pump with the detector tube attached to the other end of the tubing. This is useful for sampling in inaccessible or dangerous places.

7.3.5 *Cooling Unit*—The cooling unit consists of a length of metal tubing through which the sampled gas is pulled. Because of the high thermal conductivity of the metal tubing, the hot sampling gas is cooled sufficiently so that it will not destroy the indicator in the detector tube. The cooling unit must be placed upstream from the detector tube. Cooling units are particularly useful when sampling flue gases.

8. Reagents

8.1 The reagents used are specific for each tube, and, to detect a specific gas, may vary from manufacturer to manufacturer. The instruction sheets supplied by the manufacturers give the principal chemical reaction(s) that occur(s) in the tube, thus showing the reagent that is used to react with the gas or vapor to produce the color change.

9. Sampling with Detector Tubes

9.1. *General*—Detector tubes made by one manufacturer must not be used with pumps made by a different manufacturer (33). Each lot of detector tubes is calibrated at the manufacturer's plant, using their equipment. The pumps of other manufacturers have different flow characteristics that cause different lengths-of-stain resulting in erroneous readings.

9.2 *Procedure* (34)—The detector tube program should be conducted under the supervision of a trained professional such as a chemist or an industrial hygienist. Carefully follow the instruction sheet of the manufacturer for the proper use of each detector tube. In general, the instruction sheet will include the following information:

9.2.1 Storage conditions.

9.2.2 Shelf life.

9.2.3 Chemical reaction and color change.

9.2.4 Test procedure.

9.2.5 Significant interferences.

9.2.6 Temperature and humidity correction factors, if required.

9.2.7 Correction for atmospheric pressure.

9.2.8 Measurement range.

10. Accuracy of Detector Tubes

10.1 The Safety Equipment Institute (SEI) has a certification program for certain detector tubes used in short-term sampling. This program is similar to the NIOSH program for evaluating and certifying detector tube performance (35, 36). Under this program the tubes are required to meet an accuracy (95 % confidence level) of ±25 % between one and five times the SEI test concentration and ±35 % at one half the test concentration. The SEI test concentration is chosen as the Threshold Limit Value as defined by the American Conference of Governmental Industrial Hygienists for the test gas or vapor (37). The calculation of tube accuracy is based on a set of statistical procedures (38) and provides an estimate of accuracy under actual use conditions. The SEI Certified Equipment List should be consulted for the listing of approved units.[5]

10.2 In general, the accuracy of any detector tube depends on the construction and chemistry of the tube along with the actual composition of the test atmosphere and the conditions under which the tube is read. For gases and vapors not covered by the SEI program, detector tubes may or may not meet the accuracy requirements of the previous paragraph (39, 40). There is also some variation in accuracy between manufacturers tubes designed to detect a specific compound. Therefore the user should verify the accuracy with the tube

---

[5] Available from the Safety Equipment Institute, 1901 N. Moore St., Arlington, VA 22209.

manufacturer or run his own tests to determine accuracy (41–43). It must be emphasized that a correct estimate of accuracy can only be done by qualified operators and with careful attention to the generation and verification of test gas or vapor concentrations (44).

10.3 Because the accuracy of a detector tube towards a specific compound depends on the cross-sensitivity of the tube to other gases or vapors present in the test atmosphere, the manufacturer should be consulted for information on cross-sensitivity effects for the specific chemistry employed in their tube. Quite frequently, several different indicating chemistries for a specific compound are available. Proper choice of indicating chemistry can minimize the effect of a co-contaminant in the test atmosphere.

11. Keywords

11.1 air monitoring; detector tubes; dosimeter sampling; grab sampling; sampling and analysis; toxic gases and vapor; workplace atmospheres

ANNEX

(Mandatory Information)

A1. SOME COMPOUNDS THAT CAN BE MEASURED BY DETECTOR TUBES

A1.1 The measurement ranges shown in Table A1.1 are not for a single tube. They are for the lowest and highest concentrations listed in manufacturer's brochures. Values are given in ppm(v) unless otherwise indicated.

TABLE A1.1 Non-Exclusive List of Compounds Measurable by Detector Tubes

Substance	1994–1995 TLV ACGIH, ppm(v)	Measurement Range, ppm(v)	Substance	1994–1995 TLV ACGIH, ppm(v)	Measurement Range, ppm(v)
Acetaldehyde	25[A]	4–10 000	N,N-Dimethylaniline	5	0.5–10
Acetic acid	10	0.13–100	Dimethyl ether	. . .	100–12 000
Acetic anhydride	5	Qual.	Dimethyl formamide	10	0.8–90
Acetone	750	50–20 000	1,1-Dimethyl hydrazine	0.5[E]	0.25–3
Acetonitrile	40	. . .	N-Dimethyl nitrosamine	E	. . .
Acetylene	B,C	50–40 000	Dimethyl sulfate	0.1[E]	0.005–0.6
Acrolein	0.1	0.1–18 000	Dimethyl sulfide	. . .	1–100
Acrylonitrile	2[E]	0.13–35 000	Dioxane	25	13–25 000
Allyl chloride	1	Qual.	Epichlorohydrin	2	5–50
Ammonia	25	0.25–300 000	Ethanol amine	3	0.5–6
Amyl acetate (all isomers)	100–125	10–3000	2-Ethoxyethanol	5	Qual.
Amyl alcohols	. . .	100–2000	Ethyl acetate	400	20–15 000
Amyl mercaptan	. . .	. . .	Ethyl acrylate	5[E]	2–300
Aniline	2	0.5–60	Ethyl alcohol	1000	25–75 000
Antimony hydride, as Sb	0.5 mg/m³	0.05–3	Ethyl amine	5	1–60
Arsine	0.05	0.04–160	Ethyl benzene	100	5–1800
Benzene	10[E]	0.13–500	Ethyl bromide	5[E]	15–400
Benzyl bromide	. . .	. . .	Ethyl chloride	1000	50–8000
Benzyl chloride	1	0.2–8	Ethyl chloroformate	. . .	. . .
Bromine	0.1	0.2–30	Ethyl ether	400	20–10 000
Bromobenzene	. . .	30–720	Ethylene	B,C	0.1–5000
Bromoethane	. . .	15–400	Ethylenediamine	10	0.5–27
Bromoform	0.5	7–200	Ethyleneimine	0.5	. . .
1,3-Butadiene	2[E]	1–1200	Ethylene oxide	1[E]	0.4–40 000
Butane	800	25–8000	Ethyl glycol acetate	. . .	50–700
2-Butoxyethanol	25	Qual.	Ethyl mercaptan	0.5	0.5–1000
Butyl acetate (all isomers)	150–200	10–10 000	N-Ethyl morpholine	5	. . .
Butyl alcohol (all isomers)	50–100[A]	4–5100	Formaldehyde	0.3[A,E]	0.04–6400
Butyl amine (all isomers)	5[A]	2–36	Formic acid	5	1–2500
Butyl mercaptan	0.5	0.5–15	Furfuryl alcohol	10	Qual.
Carbon dioxide	5000	100–1 000 000	Heptane	400	20–2600
Carbon disulfide	10	0.65–4000	Hexane (all isomers)	50–500	20–12 000
Carbon monoxide	25	2–400 000	Hydrazine	0.1[E]	0.05–10
Carbon tetrachloride	5[F]	0.2–70	Hydrochloric acid	5[A]	0.1–5000
Carbonyl sulfide	. . .	0.5–60	Hydrocyanic acid	4.7[A]	0.2–30 000
Chlorine	0.5	0.05–1000	Hydrogen	B,C	500–20 000
Chlorine dioxide	0.1	0.05–20	Hydrogen fluoride as F	3[A]	0.25–100
Chlorobenzene	10	5–610	Hydrogen selenide as Se	0.05	1–600
Chlorobromomethane	200	5–180	Hydrogen sulfide	10	0.2–400 000
1 Chloro-1,1-difluoroethane	. . .	Qual.	Isoamyl alcohol	100	Qual.
Chlorodifluoromethane	1000	200–2800	Isobutyl acetate	150	10–1000
Chloroform	10[E]	2–500	Isobutyl alcohol	50	5–2900
1-Chloro-1-nitropropane	2	. . .	Isooctane	. . .	10–1500
Chloropicrin	0.1	0.1–15	Isopropyl acetate	250	20–12 000
β-Chloroprene	10	0.5–90	Isopropyl alcohol	400	20–50 000
Chromic acid, as Cr	0.05 mg/m[D]	0.1–0.5 mg/m³	Isopropylamine	5	2–30
Cumene	50	5–400	Mercury (inorganic)	0.025 mg/m³[G]	0.05–13.2 mg/m³
Cyanogen	10	. . .	Methyl acetate	200	100–30 000
Cyanides, as CN	5 mg/m³	2–15 mg/m³	Methyl acrylate	10	5–200
Cyanogen chloride	0.3[A]	0.25–5	Methylacrylonitrile	1	0.2–32
Cyclohexane	300	100–6000	Methyl alcohol	200	20–60 000
Cyclohexanol	50	5–500	Methylamine	5	1–60
Cyclohexylamine	10	2–38	Methyl bromide	5	0.5–600
Decaborane	0.05	. . .	Methyl chloride	50	. . .
Demeton	0.01	Qual.	Methyl chloroformate	. . .	. . .
Demetonmethyl	0.5 mg/m³	Qual.	Methyl chloroform	350	5–1500
Diborane	0.1	0.02–5	Methyl cyclohexanol (all isomers)	50	5–100
1,2-Dibromoethane	E	5–700	Methylene chloride	50[E]	25–3000
o-Dichlorobenzene	25	2.5–300	Methyl ethyl ketone (MEK)	200	20–50 000
p-Dichlorobenzene	10[F]	2.5–300	Methyl hydrazine	0.2[A,E]	0.2–10
Dichlorodifluoromethane (R-12)	1000	. . .	Methyl iodide	2[E]	5–40
1,1-Dichloroethane	100	8–300	Methyl isobutyl carbinol	25	. . .
1,2-Dichloroethane	10	5–720	Methyl isobutyl ketone (MIBK)	50	5–15 600
1,2-Dichloroethylene	200	5–500	Methyl mercaptan	0.5	0.25–1650
Dichloroethyl ether	5	. . .	Methyl methacrylate	100	5–700
1,1-Dichloro-1-nitroethane	2	. . .	Methyl styrene	50	10–500
Dichlorotetra fluoroethane (R-114)	1000	200–2800	Morpholine	20	Qual.
Dichlorovos	0.1	0.05	Nickel carbonyl	0.05	0.1–800
Diethylamine	5[G]	1–60	Nitric acid	2	0.1–700
Diethyl benzene	. . .	. . .	Nitric oxide	25	0.5–5000
Diisobutyl ketone	25	. . .	Nitroethane	100	. . .
Diisopropylamine	5	Qual.	Nitrogen dioxide	3	0.5–1000
Dimethyl acetamide	10	1.5–400	Nitroglycerine	0.05	Qual.
Dimethylamine	5	1–60	Nitroglycol	. . .	0.25

TABLE A1.1 Non-Exclusive List of Compounds Measurable by Detector Tubes *Continued*

Substance	1994–1995 TLV ACGIH, ppm(v)	Measurement Range, ppm(v)	Substance	1994–1995 TLV ACGIH, ppm(v)	Measurement Range, ppm(v)
Nitromethane	20	...	Sulfuric acid	1 mg/m³	1–5 mg/m³
1-Nitropropane	25	...	1,1,2,2-Tetrabromoethane	...	25–200
2-Nitropropane	10[E]	...	1,1,2,2-Tetrachloroethane	1	50–1000
Octane	300	10–5000	Tetrachloropropene	...	...
Oil (mist and vapour)	5 mg/m³	1–10 mg/m³	Tetrahydrofuran	200	20–50 000
Oxygen	...	1.5–24 vol. %	Tetrahydrothiophene	...	1–16
Ozone	0.1[A]	0.025–1000	Toluene	50	1–1800
Pentane	600	50–3900	2,4-Toluene diisocyanate	0.005	0.02–0.2
Perchloroethylene	25[F]	0.1–10 000	2,6-Toluene diisocyanate	...	0.02–0.2
Phenol	5	0.4–63	1,1,2-Trichloroethane	10	10–170
Phosgene	0.1	0.02–75	Trichloroethylene	50[H]	0.125–10 000
Phosphine	0.3	0.01–3000	Trichlorofluoromethane (R 11)	1000[A]	100–1400
Propane	B,C	100–20 000	1,2,3-Trichloropropane	10	10–1200
Propyl acetate (both isomers)	200	20–14 000	1,1,2-Trichloro-1,2,2-tri-fluoroethane (R 113)	1000	200–2800
Propylene	B,C	20–31 400	Triethylamine	1	1–60
Propylene dichloride	75	5–440	Trifluorobromomethane	1000	Qual.
Propylene imine	2[E]	0.25–3	Trimethylamine	5	1–30
Propylene oxide	20	500–50 000	Vinyl chloride	5[D]	0.025–10 000
Propyl mercaptan	...	0.5–10	Vinylidene chloride	5	0.4–600
n-Propyl nitrate	25	...	Vinyl pyridine	...	...
Pyridine	5	0.2–35	Water vapour	...	0.05–40 mg/L
Stibine	0.1	0.05–3	Xylene (all isomers)	100	5–2500
Styrene, monomer	50	2–1000			
Sulfur dioxide	2	0.1–80 000			

[A] Denotes ceiling limit.
[B] Simple asphyxiant.
[C] Explosive.
[D] Confirmed human carcinogen.
[E] Suspected human carcinogen.
[F] Animal carcinogen.
[G] Not classifiable as a human carcinogen.
[H] Not suspected as a human carcinogen.

REFERENCES

(1) *Air Sampling Instruments by the American Conference of Governmental Hygienists*, 4th ed., 1972.

(2) American Industrial Hygiene Association: *Direct Reading Colorimetric Indicator Tubes*, 1st ed., 1976.

(3) Collings, A. J., "Performance Standard for Detector Tube Units Used to Monitor Gases and Vapors in Working Areas," *Pure and Applied Chemistry*, Vol 54, pp. 1763–1767, 1982.

(4) Saltzman, B. E., *Direct Reading Colorimetric Indicators, Air Sampling Instruments for Evaluation of Atmospheric Contaminants*, fourth ed., American Conference of Governmental Industrial Hygienists, 1972.

(5) Ketcham, N. H., "Practical Experience with Routine Use of Field Indicators," *American Industrial Hygiene Association Journal*, Vol 23, p. 127, 1962.

(6) Linch, A. L. and H. Pfaff, "Carbon Monoxide—Evaluation of Exposure Potential by Personnel Monitor Surveys," *American Industrial Hygiene Association Journal*, Vol 32, p. 745, 1971.

(7) Kitagawa, T: "The Rapid Measurement of Toxic Gases and Vapors," Transactions of the 13th International Congress on Occupational Health, New York, NY, 1960.

(8) Ringold, A., Goldsmith, J. R., Helwig, H. L., Finn, R., and F. Scheute, "Estimating Recent Carbon Monoxide Exposures, A Rapid Method, *Archives of Environmental Health*" Vol 5, p. 38, 1963.

(9) Leichnitz, K., "Detector Tube Measuring Techniques," Ecomed, 1983.

(10) Beatty, R. L., "Methods for Detecting and Determining Carbon Monoxide," *Bureau of Mines Bulletin 557*, 1955.

(11) Ingram, W. T., "Personal Air Pollution Monitoring Devices," *American Industrial Hygiene Association Journal*, Vol 25, p. 298, 1964.

(12) Linch, A. L., *Evaluation of Ambient Air Quality by Personnel Monitoring*, CRC Press Inc., 1974.

(13) Shepherd, M., "Rapid Determination of Small Amounts of Carbon Monoxide," *Analytical Chemistry* Vol 19, pp. 77–81, 1947.

(14) Shepherd, M., Schuhmann, S., and M. V. Kilday, "Determination of Carbon Monoxide in Air Pollution Studies," *Analytical Chemistry* Vol 27, pp. 380–383, 1955.

(15) Shepherd, G. M., "Colorimetric Gas Detection," U.S. Patent No. 2 487 077, 1949.

(16) McConnaughey, P. W., "*Article for the Determination of Carbon Monoxide*," U.S. Patent No. 3 507 623, April 21, 1970.

(17) Littlefield, J. B., Yant, W. P., and L. B. Berger, "A Detector for Quantitative Estimation of Low Concentrations of Hydrogen Sulfide," Department of the Interior, U.S. Bureau of Mines Report, Vol 3276, 1935.

(18) Underhill, Dwight W., "New Developments in Dosimetry," Department of Industrial Environmental Health Science, University of Pittsburgh, Pittsburgh, PA, 1982.

(19) Jentzch, D., and D. A. Frazer, "A Laboratory Evaluation of Long Term Detector Tubes," *American Industrial Hygiene Association Journal*, Vol 42, pp. 810–823, 1981.

(20) Liechnitz, K., "Detector Tubes and Prolonged Air Sampling," *National Safety News*, April 1977.

(21) Liechnitz, K., "Detector Tubes for Long-Term Measurements," Annals of Occupational Hygiene, Vol 19, pp. 159–161, 1976.

(22) Portable Pump, Model C-210 Instruction Manual, Mine Safety Appliances Company; Revision 2 1983.

(23) Coldwell, B. B., and H. W. Smith, "Alcohol Levels in Body Fluids After Ingestion of Distilled Spirits," *Canadian Journal of Biochemistry*, Vol 37, p. 43, 1959.

(24) Turner, R. F. et al., "Evaluating Chemical Tests for Intoxication,"

(25) Hill, R. H., and D. A. Fraser, "Passive Dosimetry Using Detector Tubes," *American Industrial Hygiene Association Journal*, Vol 41, pp. 721–729, 1980.

(26) Ayer, H. E., and Saltzman, B. E., "Notes of Interferences by Oxides of Nitrogen with Estimations of Carbon Monoxide in Air by the NBS Indicating Tubes," *American Industrial Hygiene Association Journal*, Vol 20, pp. 337–338, 1959.

(27) McCammon, Charles S. Jr., et al., "The Effect of Extreme Humidity and Temperature on Gas Detector Tube Performance," *American Industrial Hygiene Association Journal*, Vol 43, pp. 18–25, 1982.

(28) Gas and Vapor Detection Products, National Draeger, Inc., Pittsburgh, PA 1984.

(29) Gas Detector Tubes, Sensidyne/Gastec, Sensidyou, Inc., Layo, FL.

(30) Gas Detector Tubes (T-102-5), Matheson Kitagawa, Matheson Safety Products, East Rutherford, NJ, 1982.

(31) Detector Tubes, Reagents and Accessories for Samplair® Pump (Data Sheet 08-01-02), MSA, Pittsburgh, PA, 1984.

(32) Samplair Pump, Model A (Data Sheet 08-02-02), MSA, Pittsburgh, PA, 1981.

(33) Colen, Frederick H., "A Study of the Interchangeability of Gas Detector Tubes and Pumps," Report No. TR-71, National Institute for Occupational Safety and Health, Morgantown, WV, June 15, 1972.

(34) ISO/TC 146/SC 2N55, "Determination of the Mass Concentration of Carbon Monoxide by Direct Indicating Detector Tubes," April 1981.

(35) "Certification of Gas Detector Tube Units," Federal Register, Vol 38, No. 88, p. 11458 May 8, 1973, or Code of Federal Regulations, Title 42, Part 84.

(36) Roper, P., "The NIOSH Detector Tube Certification Program," *American Industrial Hygiene Association Journal*, Vol 35, p. 438, 1974.

(37) *Threshold Limit Values for Chemical Substances in the Work Environment*, adopted by the A.C.G.I.H. with Intended Changes for 1989–1990, American Conference of Governmental Industrial Hygienists, Cincinnati, OH.

(38) Leidel, N. A., and K. A. Busch, "Statistical Methods for the Determination of Noncompliance with Occupational Health Standards," *National Institute for Occupational Safety and Health Technical Report*, 1975.

(39) Leesch, J. G., "Accuracy of Different Sampling Pumps and Detector Tube Combinations to Determine Phosphine Concentrations," *Journal of Economic Entomology*, Vol 75, pp. 899–905, 1982.

(40) McKee, Elmer S., and Paul W. McConnaughey, "Evaluation of Eight Frequently Used Detector Tubes," presentation at the *American Industrial Hygiene Conference*, Detroit, MI, 1984.

(41) Stead, F. M., and G. J. Taylor, "Calibration of Field Equipment from Air Vapor Mixtures in a Five Gallon Bottle," *Journal of Industrial Hygiene Toxicology*, Vol 29, p. 408, 1974.

(42) Setterlind, A. N., "Preparation of Known Concentrations of Gases and Vapors in Air," *American Industrial Hygiene Association Quarterly*, Vol 14, p. 113, 1953.

(43) Nelson, G. O., *Controlled Test Atmospheres, Principles and Techniques*, Ann Arbor Science Publishers, 1971.

(44) Brief, R. S., "Problems and Pitfalls in the Application and Use of Portable Direct-Reading Air Sampling Instruments, Proceedings of the National Safety Congress," Industrial Subject Sessions, p. 24, 1972.

Standard Practice for
Measuring the Concentration of Toxic Gases or Vapors Using Length-of-Stain Dosimeters[1]

This standard is issued under the fixed designation D 4599; the number immediately following the designation indicates the year of original adoption or, in the case of revision, the year of last revision. A number in parentheses indicates the year of last reapproval. A superscript epsilon (ε) indicates an editorial change since the last revision or reapproval.

1. Scope

1.1 This practice describes the detection and measurement of time weighted average (TWA) concentrations of toxic gases or vapors using length-of-stain colorimetric dosimeter tubes. A list of some of the gases and vapors that can be detected by this practice is provided in Annex A1. This list is given as a guide and should be considered neither absolute nor complete.

1.2 Length-of-stain colorimetric dosimeters work by diffusional sampling. The results are immediately available by visual observation; thus no auxiliary sampling, test nor analysis equipment are needed. The dosimeters, therefore, are extremely simple to use and very cost effective.

1.3 *This standard does not purport to address the safety problems associated with its use. It is the responsibility of the user of this standard to establish appropriate safety and health practices and determine the applicability of regulatory limitations prior to use.*

2. Referenced Documents

2.1 *ASTM Standards:*
D 1356 Terminology Relating to Sampling and Analysis of Atmospheres[2]
2.2 *Other Document:*
Federal Occupational Safety and Health Standard—Title 29 1910.1000 Subpart Z[3]

3. Terminology

3.1 For definitions of terms used in this practice, refer to Terminology D 1356.

4. Summary of Practice

4.1 Length-of-stain colorimetric dosimeters consist of a sealed glass tube containing a detector inside the tube (1–3).[4] The detector is either a paper strip or a length of granulated material impregnated with a reactive chemical that is sensitive to the particular gas for which the dosimeter is designed. To use the tube, one end is opened. The gas, if present, diffuses into the tube and reacts with the indicator paper or

gel, causing the latter to change color. Each lot of dosimeters is individually calibrated so that by measuring the length of stain and the time of exposure, the TWA concentration to which the dosimeter has been exposed can be determined directly and immediately.

4.2 Information on the correct use of length of stain dosimeter tubes is presented.

5. Significance and Use

5.1 The Federal Occupational Safety and Health Administration in 29 CFR 1910.1000 Subpart Z designates that certain gases and vapors present in work place atmospheres must be controlled so that their concentrations do not exceed specified limits.

5.2 This practice will provide a means for the determination of airborne concentrations of certain gases and vapors listed in 29 CFR 1910.1000.

5.3 A partial list of chemicals for which this practice is applicable is presented in Annex A1 with current Threshold Limit Values (TLV) (4) and typical measurement ranges for the selected chemicals as obtained from various manufacturer's specifications.

5.4 This practice may be used for either personal or area monitoring.

6. Interferences

6.1 The instructions may provide correction factors to be applied when certain interferences are present. Some common interfering gases or vapors for each dosimeter are listed in the instruction sheets for the dosimeter provided by the manufacturers.

7. Apparatus

7.1 *Dosimeter Tube:*

7.1.1 *General Description*—A length of stain dosimeter tube consists of a glass tube containing an inert granular material or strip of paper that has been impregnated with a chemical system that reacts with the gas or vapor of interest. As a result of this reaction, the impregnated chemical changes color. The granular material or paper strip is held in place within the glass tube by porous plugs of a suitable inert material. To protect the contents during storage, the ends of the glass tube are flame sealed or covered with plastic caps and the unit sealed in an impervious plastic pouch. In some cases, a scale is printed on the glass tube to make it easy to read the length of stain of reacted chemical; in other instances, a scale is provided external to the tube for reading the length of stain.

7.1.2 *Stability on Storage*—Stability on storage may vary

---

[1] This practice is under the jurisdiction of ASTM Committee D-22 on Sampling and Analysis of Atmospheres and is the direct responsibility of Subcommittee D22.04 on Workplace Atmospheres.
Current edition approved Nov. 30, 1990. Published February 1991. Originally published as D 4599 – 86. Last previous edition D 4599 – 86.
[2] *Annual Book of ASTM Standards*, Vol 11.03.
[3] Code of Federal Regulations, available from U.S. Government Printing Office, Washington, DC 20402.
[4] The boldface numbers in parentheses refer to the list of references appended to this practice.

depending on manufacturer and type of dosimeter, but most dosimeter tubes can be stored for at least 30 months with no deleterious effects.

7.2 *Tube Holders*—During use, the dosimeter tube is held in a lightweight, plastic holder. The tube holder protects the dosimeter during use and also helps to minimize effects of air currents on performance. The holder has a clip that allows it to be fastened to a collar or pocket during personal sampling or to some appropriate object during area sampling.

8. Reagents

8.1 The reagents used to impregnate the paper or granular material in the dosimeters are specific for each tube, and, to detect a specific gas or vapor, may vary from manufacturer to manufacturer. The instruction sheets supplied by the manufacturers usually give the principal chemical reaction(s) that occur(s) in the tube.

9. Diffusional Sampling Theory (1–3, 5, 6)

9.1 Fick's First Law of Diffusion states that the mass of material that diffuses is directly proportional to the diffusion coefficient of the material, the diffusional cross sectional area, the concentration gradient and the time, and inversely proportioned to the length of the diffusion path. Mathematically, this can be expressed as follows:

$$M = \frac{DA}{L} t(C - C_o)(10^{-6})$$

where:
M = mass of material adsorbed (mg),
D = diffusional coefficient (cm^2/min),
A = diffusion cross-sectional area (cm^2),
L = length of the diffusion path (cm),
C = concentration in the atmosphere (mg/m^3),
C_o = concentration at the surface of the adsorbent or reactant (mg/m^3),
t = exposure time (min), and
10^{-6} = conversion factor for cm^3 to m^3.

9.1.1 For most adsorbent and reactive devices, such as the impregnated material in length-of-stain dosimeters, C_o is assumed to be 0. Further, the mass transferred is proportional to the length of stain in these devices, or $M = bS$ where S is the length of stain and b is the proportionality constant. Also, the length of the diffusion path is equal to the distance from the opening of the dosimeter tube to the indicating material plus the length of stain of the indicating material, that is,

$$L = a + S$$

where:
a = the distance from the tube opening to the indicating material. Thus,

$$bS = [DA/(a + S)]Ct$$

or transforming and combining constants:

$$aS + S^2 = kCt$$

where:
$k = DA/b$.
k may be determined experimentally by plotting $aS + S^2$ versus Ct where t is expressed in h.

9.1.2 Once k is known, it may be used to determine the concentration under any set of stain length and time

combinations by using the formula in 9.1.1. The k value must be determined for each lot of tubes due to slight variations in materials and tube construction from lot to lot.

9.2 The theoretical discussion in 9.1 applies very well to dosimeters that contain paper type indicators. With this type of indicator, there is very little adsorption; and reaction between the indicator and the gas or vapor is complete because there is so little indicator and the diffusion rate is slow. With dosimeters that contain indicator impregnated on silica gel, alumina, or other adsorbent material, the behavior of the system is more complex, because of the possibility of adsorption and incomplete reaction as the diffusion front passes. The latter could be caused by a slow diffusion process of the gas or vapor into the pores of the gel to react with indicator sorbed in the pores. In spite of these complications, the behavior of the dosimeters can still be expressed by the equation, $aS + S^2 = KCt$, but in these cases, both a and K must be determined experimentally as empirical constants. In some cases, $a = o$ (6).

9.3 *Measurement Range* (7–10)—The measurement range of the various length-of-stain dosimeters is shown in Annex A1.

9.4 *Air Velocity*—The sampling rate of the dosimeter tubes is very slow (of the order of 0.1 cm^3/min); thus the "starving" effect in static air is not significant for these devices, so that air velocity is not critical. However, a stream of high velocity air should not be permitted to flow directly into the open end of the tube (parallel to the axis of the tube). The tube holder provides additional protection from turbulence within the dosimeter.

10. Sampling with Length-Of-Stain Dosimeter Tubes

10.1 *General*—Since these dosimeters work by diffusion, the procedure for using them is very simple. All that is necessary is to open one end of the dosimeter properly, place the opened tube into its holder, and fasten the holder to an object at a point where the sampling is to be done. Follow the instruction sheet of the manufacturer for the proper use of each dosimeter tube. The sampling starting time and ending time must be recorded so that the sampling time is known. This is needed to estimate the average concentration (TWA) over the sampling time.

10.2 *Estimating the Concentration from the Stain Length-Time Relationship:*

10.2.1 Figure 1 shows a series of calibration graphs for ammonia. Suppose a test is run for 6 h and the stain length after this exposure time is 35 mm. On the graph of Fig. 1, the intersection of the projections from 6 h and 35 mm is located-as shown by the dot. From the location of this point, the average concentration can be estimated. The point is 4/5 of the distance between the lines 12.5 and 25 ppm, thus it is 4/5 × 12.5 (the interval between 12.5 and 25.0) or 10 ppm above the 12.5 ppm line. The concentration would be 12.5 ± 10 = 22.5 ppm. Alternatively, the concentration could be calculated from the following equation if a and k are known:

$$C = (aS + S^2)/kt$$

In this case,
a = 15, and
k = 13.2.
Therefore,

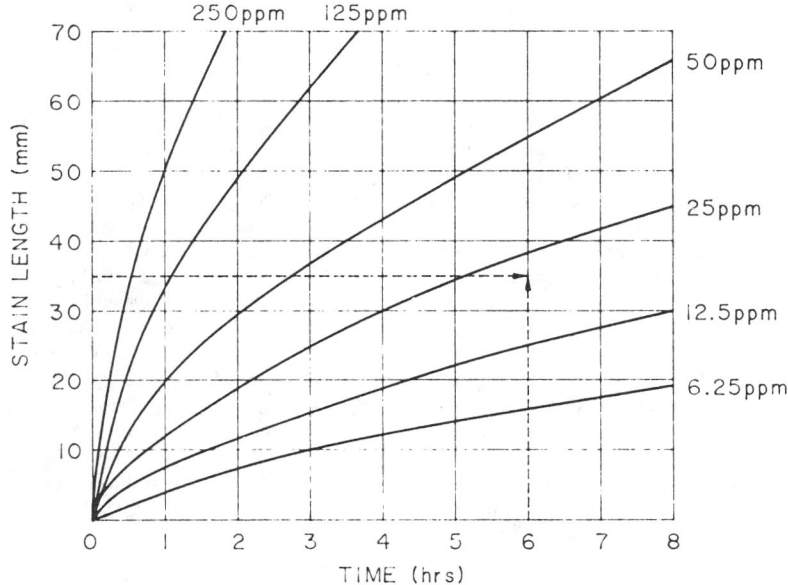

FIG. 1   Calibration of Ammonia Dosimeter, Lot 3, K = 13.2

$$C = [15(35) + (35)^2]/13.2(6) = (525 + 1225)/79.2 = 22.1 \text{ ppm}$$

10.2.2 There is approximately a 2 % error between these two calculations. This is due to the assumption that the concentration between the lines of the graph in Fig. 1 vary linearly with stain length, which is not quite correct. With some tubes, a ppm-h scale is printed on the tube. With these tubes, the ppm-h reading from the tube, at the end of stain, is divided by the sampling time in hours to get the TWA exposure.

11. Accuracy of Dosimeter Tubes (2, 3, 6, 11)

11.1 The accuracy of dosimeter tubes is generally within ±25 %; however, some tube types may vary from this, and specific tube accuracy may vary from lot to lot and manufacturer to manufacturer. Therefore, if users want to know the accuracy of a certain tube, they should check with the manufacturer for an accuracy statement or run their own tests to determine accuracy under their particular conditions of use.

ANNEX

(Mandatory Information)

A1.   SOME GASES AND VAPORS THAT CAN BE MEASURED BY LENGTH-OF-STAIN DOSIMETER TUBES

TABLE A1.1   Threshold Limit Values (4)

Substance	TWA Values[A] ppm	mg/m³	Measurement Range (ppm)
Amines	see Table A1.2		
Ammonia	25	17	1–125
Carbon Dioxide	5000	9000	300–40 000
Carbon Monoxide	50	57	5–500
Chlorine	0.5	1.5	0.1–20
Hydrogen Chloride	C5[B]	C7.5[B]	1–50
Hydrogen Cyanide	C10[B]	C11[B]	1–120
Hydrogen Sulfide	10	14	0.5–60
Nitrogen Dioxide	3	5.6	1–100
Sulfur Dioxide	2	5.2	0.5–50

[A] Values for ppm and mg/m³ are Time-Weighted Average (TWA).
[B] C = Ceiling limit, the concentration that should not be exceeded under any condition.

TABLE A1.2   Threshold Limit Values for Amines

Amines	TWA Values[A] ppm	mg/m³	Measurement Range (ppm)
n-Butyl	C5[B]	C15[B]	2–25
Cyclohexyl	10	41	5–50
Diethyl	10	30	5–50
Dimethyl	10	18	5–50
Dipropyl	—	—	5–50
Ethyl	10	18	5–50
n-Ethyl Morpholine	5	24	2–25
Isopropyl	5	12	2–25
Methyl	10	13	5–50
Triethyl	10	41	5–50
Trimethyl	10	24	5–50

[A] Values for ppm and mg/m³ are Time-Weighted Average (TWA).
[B] C = Ceiling limit, the concentration that should not be exceeded under any condition.

REFERENCES

(1) McConnaughey, Paul W., McKee, Elmer S., and Pritts, Irvin M., "Passive Colorimetric Dosimeter Tubes for Ammonia, Carbon Dioxide, Carbon Monoxide, Hydrogen Sulfide, Nitrogen Dioxide and Sulfur Dioxide," *American Industrial Hygiene Association Journal*, Vol 46, 1985, pp. 357–362.

(2) McKee, Elmer S., and McConnaughey, Paul W., "A Passive, Direct Reading, Length-of-Stain Dosimeter for Ammonia," *American Industrial Hygiene Association Journal*, Vol 46, 1985, pp. 407–410.

(3) McKee, Elmer S., and McConnaughey, Paul W., "Laboratory Validation of a Passive Length-of-Stain Dosimeter for Hydrogen Sulfide," presentation at the American Industrial Hygiene Conference, Detroit, MI, 1984, submitted to the *American Industrial Hygiene Association Journal*, 1985.

(4) "Threshold Limit Values for Chemical Substances in the Work Environment with Intended Change for 1989–1990," American Conference of Governmental Industrial Hygienists, Cincinnati, OH 1989.

(5) Palmes, E. D., and Lindenboom, R. H., "Ohm's Law, Fick's Law and Diffusion Samplers for Gases," Analytical Chemistry, Vol 51, 1974, pp. 2400–2401.

(6) Roberson, R. W., Matsunobu, K., Hoshino, F., and Komatsu, T., "Performance Testing of Sensidyne/Gastec Dosi Tubes for CO, H_2S, SO_2 and HCN," presented at the American Industrial Hygiene Conference (1985).

(7) "Gas and Vapor Detection Products," Data Sheet 12-84-35M, National Draeger, Inc., Pittsburgh, PA, 1984.

(8) "Vapor Gard® Inorganic Vapor Dosimeter Tubes," Data Sheet 08-00-36, MSA Pittsburgh, PA, 1983.

(9) "Dosimeter Tubes," 0185, Sensidyne, Largo, FL, 1985.

(10) "Passive Colorimetric Dosimeters," Form No. 8503, Willson Safety Products, Reading, PA, 1985.

(11) Willson Safety Products, Validation Report Passive Inorganic Colorimetric Dosimeter, 1985.

PART 6.   BIOLOGICAL SAMPLING

Standard Practice for
Aseptic Sampling of Biological Materials[1]

This standard is issued under the fixed designation E 1287; the number immediately following the designation indicates the year of original adoption or, in the case of revision, the year of last revision. A number in parentheses indicates the year of last reapproval. A superscript epsilon (ε) indicates an editorial change since the last revision or reapproval.

1. Scope

1.1 This practice presents the principles, state-of-the-art concepts and generally accepted methods for aseptic sampling of materials involved with or produced by biotechnical processes where contamination of either the sample or the source of the sample cannot be accepted. These processes could involve living organisms such as virus, bacteria, yeasts, and mammalian cells or biologically active constituents, such as enzymes and biochemicals that must exist in a noncontaminated state.

1.2 This practice also applies to the products from these bioprocesses that can be for human consumption, sterile or parental drug applications, which also require aseptic sampling to meet regulatory, current good manufacturing practices, or other quality control requirements.

1.3 **Warning**—Since some biotechnical processes could produce flammable products, this sampling practice should be applied only after taking into account all of the factors that are pertinent to an assessment of the fire hazard of a particular end use.

1.4 *This standard does not purport to address all of the safety concerns, if any, associated with its use. It is the responsibility of the user of this standard to establish appropriate safety and health practices and determine the applicability of regulatory limitations prior to use.*

2. Referenced Documents

2.1 *ASTM Standards:*
D 1356 Terminology Relating to Atmospheric Sampling and Analysis[2]
D 4177 Practice for Automatic Sampling of Petroleum and Petroleum Products[3]
E 884 Practice for Sampling Airborne Microorganisms at Municipal Solid-Waste Processing Facilities[4]

3. Terminology

3.1 The definitions covered in Terminology D 1356, Practice D 4177, and Practice E 884 are applicable to this practice.

3.2 *Definitions:*

3.2.1 *aseptic sampling*—sampling process in which no extraneous microorganisms or substances are introduced into the sample or its original bulk material as a result of the sampling system and activity.

3.2.2 *current good manufacturing practices (CGMP)*—current regulations published by the United States Food and Drug Administration (FDA) regarding manufacturing, processing, packaging and storing of drug and biological products.

3.2.3 *dead leg*—any inactive, trapped or stagnant zone of a biological fluid that is to be sampled aseptically where this liquid zone would not be representative of the bulk fluid that is to be sampled. This "dead leg" zone could deviate from the bulk system in oxygen content, nutrients levels, material composition, temperature, bacterial contamination, and other process variables that would prevent any sample drawn through this system from representing the bulk fluid quality to be tested.

DISCUSSION—This definition may be more restrictive than the FDA definition which is any unused pipe greater in length than six of its internal diameters. Since valve designs and presence of other devices in the sampling system must be considered in this aseptic sampling procedure, the entire sampling system from bulk fluid to sample container should be validated by using proper biological challenges to show that the intended sterility and sample quality objective can be met and reproduced within the prescribed limits of the specific process.

3.2.4 *pathogenic*—disease causing.

3.2.5 *sterile*—free of any living organism.

3.2.6 *validation*—the quality assurance evaluation of an item of equipment or overall process wherein the equipment or process, or both, is challenged to perform under the "worst case" conditions of process variables and applicable micro-organism contamination to meet preestablished acceptance criteria.

4. Summary of Practice

4.1 A general description of aseptic sampling-system design guidelines is included either to remove a representative sample of the bulk fluid for external testing, or to directly measure the fluid properties in-situ. Validation of sampling equipment and methods is also described. Suggested sample system designs are presented for consideration and application as appropriate to specific processes. Where possible, the advantages and disadvantages of different sample removal designs are presented. Fabrication and maintenance considerations are also discussed. The Appendix includes general guidelines for sterilization of aseptic sampling devices.

5. Significance and Use

5.1 This practice should be used for removing samples from biological processes in the laboratory or commercial manufacturing facilities where the sample system removes

[1] This practice is under the jurisdiction of ASTM Committee E-48 on Biotechnology and is the direct responsibility of Subcommittee E48.03 on Unit Processes and Their Control.
Current edition approved Feb. 24, 1989. Published April 1989.
[2] *Annual Book of ASTM Standards*, Vol 11.03.
[3] *Annual Book of ASTM Standards*, Vol 05.02.
[4] *Annual Book of ASTM Standards*, Vol 11.04.

the sample from the process for use in external testing.

5.2 This practice also addresses the sampling procedures required for in-situ measurements wherein the sample is not removed from the process but must represent the process material being tested. Generally, the in-situ measurement device is either sterilized separately from the process equipment and then inserted into the sterilized equipment, or the in-situ device is permanently mounted in the equipment and then sterilized together with the equipment.

5.3 Levels of contamination are not specified in this practice since each biological system and bioprocess can differ as to the amount and types of micro-organism, bacteria, virus, and other contaminants that can be allowed in the sample and process materials for acceptable operations under CGMP or similar requirements. With the properly designed micro-organism challenges to the sterile system, then the sample system can be tested and validated.

5.4 Since biological process samples can vary widely in sterility requirements, sample size, material composition, (liquid, vapor, slurry, etc.), stability, and other characteristics, the practices described herein are general and are to be applied as appropriate to each specific situation. These practices are limited to aseptic sampling conditions and are not intended to apply to containment of highly toxic or hazardous materials that require additional precautions to avoid exposure of the sample contents to the environment, or workers, where health and other safety considerations could require more stringent practices. Sample applications can include the following:

5.4.1 *External Testing of Removed Samples*: Liquid chromatography, spectroscopy (ultraviolet, infrared, and fluorescence), fiber optics, mass spectroscopy,

5.4.2 *Direct Measurement of In-Situ Sample*: Temperature, pressure, pH, etc.

6. Procedure for Aseptically Removing Samples from a Biological System

6.1 *General Criteria for Designing Sampling Systems:*

6.1.1 Sample removal devices should access the bulk material at the desired location:

6.1.1.1 If the sample is to be representative of the average quality of the bulk material, then sufficient agitation of the bulk material is required to ensure uniformity within the equipment. With a homogeneous, single-phase bulk material this uniformity can be achieved and validated using sufficient agitation. Caution must be exercised to evaluate the agitation in all locations of the bulk material since localized areas of insufficient agitation could affect accurate sampling.

6.1.1.2 With a heterogenous or multiple phase material that is encountered frequently with fermenter slurries of liquids and solids (such as living organisms), then a representative sampling method must be designed considering a uniform slurry from which a sample is withdrawn.

6.1.2 General criteria to be considered when removing the sample from the bulk material could include:

6.1.2.1 Obtain a fresh sample and avoid a "dead leg" of older material that could compromise the quality of the desired sample.

6.1.2.2 Consider the kinetics of the reactions that may require the sample to be treated with freezing, neutralization, filtration, or other appropriate processes, immediately after

collection, to maintain the sample quality at the actual time of sampling. This consideration would be specific to each biological process.

6.1.2.3 Consider the change of process conditions, (temperature, pressure, mixing efficiency, component concentration, etc.) from the bulk material to the sampling device and container. These changes should not affect the sample quality. Again, the sample should represent the bulk material source in all desired respects. If pH is the desired analytical measurement, then separation of some solids from the slurry in the sample container may be acceptable, where it would not be acceptable for a solids content determination. A rapid pressure drop or temperature change could adversely affect the living organisms if a live cell count is desired.

6.1.2.4 Consider the removal of solids by filtration or other means, to permit collecting a single phase sample for analysis where solids are not tolerated or could affect the quality of sample as it is prepared for analytical testing. If the solids are yeast or nutrients that could continue to react and change the time-value of the sample, then the effect of these components must be negated to have a reliable sample.

6.1.2.5 Validate these procedures and facilities using appropriate system challenges to document the variations in the sample results. This validation is performed after determining the specific sampling criteria for the process, the sampling method documented, the sampling apparatus constructed, the sample handling techniques documented, and the personnel trained in these procedures. Movement of the sampling device to other locations in the bulk material container, such as closer to mixing devices, etc., could be used to verify the optimum sampling device method and location for the specific process.

6.1.2.6 Consider particle size when designing the sample removing device, the sample container, and the method for removing the sample from the container, for the analytical testing. Plugging, separation of solids, and agglomeration, are some of the potential problems to be considered in the sampling system.

6.1.2.7 Reusing the sample device during the same batch requires the consideration of:

(*a*) Does carry over or residual material from the previous sample adversely affect the quality of the next sample? If flushing the sample device prior to collecting the next sample is adequate, then multiple samples can be taken through the sampler. For pH determination this procedure may be accurate, but may not be acceptable for live cell counts, pyrogen determinations, etc., where previous sample residue may be a detriment.

(*b*) Can the sampling device be sterilized between sample collections to purge previous sample residue? Some sampling devices are designed with methods to steam sterilize portions of the sample flow channel. If complete sterilization is required then careful attention is needed to ensure that all sample device components are sterilized that are in contact with the sample. Normally, a ball valve, rising stem valve, or similar close tolerance device, is in contact with the bulk material being sampled. Portions of this valve are very difficult to sterilize without the sampler being autoclaved or sterilized when the bulk container is being sterilized. Possibly a combined sterilization and thorough flush prior to collecting the next sample can achieve the sampling objectives.

If not, then a multiple sampling device system (one for each sample), may be required to effectively and economically ensure that an aseptic sampling device is used. When the sampling device is qualified and the aseptic system validated, then the sampling protocol would be used in routine operations.

6.1.2.8 Consider available quantity of sample required for purging and retained sample. Small reactors could be adversely affected by the quantity of material removed thereby upsetting the reaction kinetics and equilibrium. The procedure must consider the quantity of material purged around the sample container and the quantity diverted into the container.

6.1.2.9 Consider disposal requirements of purged material and exposure of this material to the environment and other living systems. Provide proper containment devices and disposal procedures that are consistent with regulatory rules such as CGMP and Good Laboratory Practice (GLP), operating procedures, and other regulations or guidelines.

6.1.2.10 Develop proper sample handling procedures from the sampling device through the analytical testing to disposal or retained sample storage. This includes proper sample identification, stable storage, repetitive uniform sample retrieval from the container for analytical tests, sample security, proper disposal if sample is no longer needed, and proper sterilization of the sample container if it is to be reused.

6.1.2.11 Design sample containers to withstand both the bulk process and the sample handling design conditions of temperature, pressure, composition, corrosivity, toxicity, flammability, other hazardous properties, slurry removal, and cleaning and sterilization procedures. If these containers are to be reused then develop and qualify a container testing procedure and schedule.

6.1.2.12 Consider disposable, nonreusable containers for these samples as appropriate to achieve the sampling objectives and design considerations listed in 6.1.

6.2 *Typical Sample Device Design Considerations for Externally Removed Samples:*

NOTE—The following general descriptions of frequently used aseptic sampling devices are presented with the caution that actual applications to a specific process must be tailored appropriately and validated as required to ensure that accurate samples are obtained and that the sampling objectives are attained that meet or exceed regulatory or other requirements.

6.2.1 *Flush-Mounted Sample Valves*—The following considerations apply to selecting the proper valve to remove samples aseptically.

6.2.1.1 Sample valve should minimize holdup of sample within the valve especially if subsequent samples are to be taken under aseptic conditions. Diaphraghm and ball valves generally have less holdup than gate or globe valves. Diaphragm valves also are normally sterilized easier than other standard design valves. Specially designed rising or lowering stem valves can reduce holdup further. Close tolerances between moving parts should minimize trapped sample materials.

6.2.1.2 Sample valves should have minimum lengths of piping connection on suction and discharge ports to avoid trapping sample material that cannot be removed and sterilized between samples.

6.2.1.3 If a sample valve is to be used for consecutive aseptic samples then sterilize the valve between samples. These and other criteria are described in 6.1.

6.2.1.4 Make sure that bulk fluids composition at valve inlet is representative of entire material to be sampled.

6.2.2 *Recycle Loop for Aseptic Sample*—Primary application is for homogeneous, single phase fluids.

6.2.2.1 Make sure that source of material entering sample tubing is representative of bulk of material being sampled.

6.2.2.2 Isokinetic sampling considerations should be followed where appropriate. Make sure that velocity of material entering and within the tubing or piping is adequate to maintain uniform composition throughout entire recycle loop especially at the location of the sample removal. If material is a slurry then the fluid design velocity, piping layout and fittings design, should maintain a uniform sample quality throughout the recycle loop. Avoid sample removal near wall of recycle pipe where velocity is lowest or near abrupt fluid direction changes that could distort the solids distribution in the slurry. Generally, turbulent flows will enhance uniform distribution within the material being sampled.

6.2.2.3 Specifically design sterilization procedures for the recycle loop and its sample removal system to ensure that full sterilization is achieved. Document sampling procedures for the aseptic sampling system validation.

6.2.3 *Draw Tube or Siphon Tube*—These devices are extremely difficult to use and to ensure that a fresh representative sample is obtained on a uniform basis. Primary application of this device applies to homogeneous, single phase fluids. If slurries are sampled then its composition could vary in the tube that is normally stagnant and not representative of the bulk material being sampled. Cellular or viscous materials could stick to the wall of the tubes and cause sampling errors. Excessive purging and proper velocity control may be needed on a repetitive basis to obtain a representative sample.

6.2.3.1 *Vertical Sample, Tube Design*, where fluid is pressured or pumped upward in the tubing should be designed to ensure that a fresh, not contaminated, representative sample is withdrawn using minimal purging. Velocity considerations are important to avoid settling of solids and flashing of volatile liquids if the pressure drop in the piping system causes vaporization and two phases of products. Avoid sample removal from side wall of tube as described in 6.2.2.2. Inert, sterile gas could be used for blowback of the sample tube to minimize stale sample accumulations. Other general sampling pipe systems, automatic sampling guidelines, and other information for consideration are presented in Practice D 4177, where appropriate.

6.2.3.2 *Vertical Sample Tube Design*, where fluid is pressured or pumped downward for removal should avoid designs where solids could accumulate and fill the tube while in the passive mode. This stratification and solids accumulation problem can be minimized by using a horizontal section at the tube inlet and sterile gas blowback as appropriate. Other considerations in 6.2.3 apply to this sampling method.

6.2.3.3 *Sample Tube Extensions*, to valves described in 6.2.1 also should follow the same guidelines as described in 6.2.3. These extensions permit withdrawing representative samples near sources of fluid agitation near mixers and away from the bulk container wall.

6.2.4 *Insertable Sterile Sample Probe or Draw Tube*— Several devices are included in this category such as (1) sterile syringe and membrane barrier, and (2) insertable sterile probe into system to withdraw sample.

6.2.4.1 *Sterile syringe,* can be used to penetrate a membrane at the sample port, extend the sample needle into the bulk of the material, remove the sample, and use the syringe as a sample container where applicable. With small diameter syringe needles, this sampling technique is usually limited to homogeneous, single phase system.

6.2.4.2 Inserting a sterile sample probe aseptically into the bulk container usually requires special sterilization techniques for the membrane or probe port entry device. Special sterilization fluids or steam would be needed to ensure that the sampling system is sterile.

6.2.5 *Containment of Sample Collection Emission*— When any sample is collected aseptically, design considerations should include containment of vapor purged when the sample container is filled, collection and disposal of any purged liquids, excess samples, and byproducts from resterilization. Glove boxes and other containment systems can be designed into the process and validated.

6.2.6 *Sample System Fabrication and Maintenance Considerations:*

6.2.6.1 Use all welded construction where permitted.

Threaded connections are more difficult to sterilize.

6.2.6.2 Minimize "dead legs" in piping system.

6.2.6.3 Select valves that minimize dead space for bacteria growth.

6.2.6.4 Avoid rough surfaces where microorganisms can grow uncontrolled. Grind welds smooth. Use top quality welding procedures.

6.2.6.5 Select proper materials of construction.

6.3 *Typical Sample Device Design Considerations for In-Situ Measurement Devices:*

6.3.1 Design in-situ measurement devices that are permanently attached to be sterilized with the overall system.

6.3.1.1 Design consideration for in-situ measurement devices such as pressure gages, thermocouples, pH probes should meet the same design conditions as for the bulk container sterilization.

6.3.1.2 Locate measurement device at the desired sample position. Ease of removal for maintenance is also an important consideration.

6.3.2 Design in-situ measurement devices that can be removed for sterilization with proper containment precautions for the specific material being sampled, while maintaining aseptic conditions during reinstallation. Each system would be designed to the specific process conditions. Proper sealing devices are needed to avoid leakage from the bulk material vessel or systems to the environment.

APPENDIX

Nonmandatory Information

X1. STERILIZATION GUIDELINES

X1.1 Know the organisms, and infectious strains that need to be removed by sterilization. Adjust the sterilization method and materials to effectively remove these undesirable components.

X1.2 Normally, saturated steam is used for sterilization since bacterial spores are more effectively destroyed. Superheated steam requires cooling to reach saturation quality for maximum effectiveness. Wet steam has a lower heat content and adds more water to the system that must be removed.

X1.3 Remove air from the sample system to avoid steam dilution and air pockets that prevent full sterilization. The saturated steam is the sterilizing medium. Air removal is required to avoid reducing the steam partial pressure which would be equivalent to superheating the steam that reduces its sterilization effectiveness.

X1.4 If parts of the sample system can be detached from the vessel, pipe or equipment item, then they can be sterilized separately in an autoclave or similar device.

X1.5 Sample device sterilization temperatures using saturated steam must be maintained accurately. External appendages of the sample system can increase heat transfer and cool faster than the main vessel that could prevent effective sterilization of the sampling system.

 **E 1287**

Standard Classification for
Sampling Phytoplankton in Surface Waters[1]

This standard is issued under the fixed designation D 4149; the number immediately following the designation indicates the year of original adoption or, in the case of revision, the year of last revision. A number in parentheses indicates the year of last reapproval. A superscript epsilon (ϵ) indicates an editorial change since the last revision or reapproval.

1. Scope

1.1 This classification covers both qualitative and quantitative techniques that are used commonly for the collection of phytoplankton. The particular techniques that are used during an investigation are dependent upon the study objectives. Of additional importance in the selection of a technique is the uneven distribution of organisms both temporally and spacially. This classification describes qualitative and quantitative ways of collecting phytoplankton from inland surface waters. Specifically, qualitative samplers include conical tow nets and pumps; quantitative samplers include the Clarke-Bumpus plankton sampler, Juday plankton trap, water sampling bottles, and depth-integrating samplers.

2. Referenced Document

2.1 *ASTM Standard:*
D 1129 Terminology Relating to Water[2]

3. Terminology

3.1 *Definitions*:

3.1.1 For definitions of terms used in this method refer to Terminology D 1129.

3.2 *Descriptions of Terms Specific to This Method*:

3.2.1 *phytoplankton*—is the community of suspended or floating, mostly microscopic plants that drift passively with water currents. Frequently, phytoplankton are differentiated on the basis of size. The generally accepted size ranges, as commonly used are (1):[3]

Macroplankton	>500 µm
Microplankton (net plankton)	10 to 500 µm
Nannoplankton	10 to 50 µm
Ultraplankton	<10 µm

4. Significance and Use

4.1 Because of the direct association of phytoplankton with the water and the water masses that move in response to wind-or-gravity-generated currents, the species composition and abundance of phytoplankton are related to water quality. Moreover, the phytoplankton directly affect water quality, notably dissolved oxygen, pH, concentrations of certain solutes, and optical properties. At times the abundance or presence of particular species of algae result in nuisance conditions (2).

4.2 Organisms of the phytoplankton communities are collected and studied for many reasons, and the techniques used will vary with the study objectives. In the design of a sampling program and in the selection of techniques, the investigator must take into consideration the uniqueness of each study area and the natural characteristics of phytoplankton communities.

4.3 The principal factors to consider when collecting phytoplankton are the uneven distribution, composition, and abundance of phytoplankton in space and time. Phytoplankton blooms can occur quickly and can be of short duration. Succession of taxa can occur in a matter of 1 to 2 weeks. Furthermore, phytoplankton abundance and composition can change abruptly in the horizontal plane. There also can be remarkable numerical and qualitative differences between depths. The heterogeneous abundance and composition can occur not only over small areas but also over large areas. The uneven distribution makes it difficult to collect a representative sample from a given area and makes replication of samples and, especially, an adequate vertical and horizontal sampling program essential (3).

5. Basis of Classification

5.1 Qualitative samplers include the conical tow nets and pumps. Quantitative samplers include the Clarke-Bumpus plankton sampler, Juday plankton trap, water-sampling bottles, and depth-integrating samplers.

5.2 *Conical Tow Nets*—Most qualitative samplers are cone-shaped nets constructed of silk bolting cloth or a synthetic material such as nylon. Nets should not be used for quantitative studies because they do not retain all the phytoplankton taxa; for example, nannoplankton and ultraplankton generally will pass through a net. Even so, nets are valuable collecting tools and excellent for many types of studies.

5.3 *Pumps*—Pumping systems of various kinds have been used to collect qualitative or semiquantitative samples of phytoplankton. Several papers summarizing these techniques have been published in the literature (4, 5, 6). Although a variety of pump apparatus have been used, the basic design consists of a pump, generally with a volume register, a base, and a concentrating net, such as a simple tow net sampler or Wisconsin net sampler. Water is pumped from a discrete depth and through the net. The sample is removed from the net.

5.4 *Clarke-Bumpus Plankton Sampler*—The sampler utilizes a net for the concentration of organisms and, as such, may be considered to be a semiquantitative sampler. It is quantitative in that the actual volume of water entering the

[1] This classification is under the jurisdiction of ASTM Committee E-47 on Biological Effects and Environmental Fate and is the direct responsibility of Subcommittee E47.08 on Biological Field Testing.

Current edition approved June 25, 1982. Published October 1982.

[2] *Annual Book of ASTM Standards*, Vol 11.01.

[3] The boldface numbers in parentheses refer to the references at the end of this classification.

sampler is measured by a calibrated flow meter.

5.5 *Juday Plankton Trap*—Like the Clarke-Bumpus plankton sampler, the Juday plankton trap utilizes a net for the concentration of organisms. The trap collects a discrete sized sample from a predetermined depth.

5.6 *Water-Sampling Bottles*—The closing water bottles, which are actuated by a messenger, are perhaps the most satisfactory and simple quantitative sampling device.

5.7 *Depth-Integrating Samplers*—Depth-integrating samplers are used to obtain a representative, quantitative sample of phytoplankton in the cross section of a stream. The sampler and sampling procedure compensates for the disparity of phytoplankton density in the cross section.

REFERENCES

(1) Wetzel, R. G., *Limnology*, W. B. Saunders Co., Philadelphia, PA, 1975, p. 743.

(2) Greeson, P. E., et al., "Methods for Collection and Analysis of Aquatic Biological and Microbiological Samples." *U.S. Geological Survey Technology of Water-Resources Investigations*, Book 5, Chapter A4, 1977, p. 332.

(3) National Academy of Sciences, "Recommended Procedures for Measuring the Productivity of Plankton Standing Stock and Related Oceanic Properties," National Academy Sciences, Washington, 1969, p. 59.

(4) Aron, W., "The Use of a Large Capacity Portable Pump for Plankton Sampling, with Notes on Plankton Patchiness," *Journal of Marine Research*, Vol 16, 1958, pp. 158–174.

(5) Gibbons, S. G., and Fraser, J. H., "The Centrifugal Pump and Suction Base as a Method of Collecting Plankton Samples," *Journal Construction Permanent International Explorer Merchants*, Vol 12, 1937, pp. 155–170.

(6) Weber, C. I., ed., "Biological Field and Laboratory Methods for Measuring the Quality of Surface Waters and Effluents," U.S. Environmental Protection Agency. EPA-670/4-73-001, 1973.

Classification for Fish Sampling[1]

This standard is issued under the fixed designation D 4211; the number immediately following the designation indicates the year of original adoption or, in the case of revision, the year of last revision. A number in parentheses indicates the year of last reapproval. A superscript epsilon (ϵ) indicates an editorial change since the last revision or reapproval.

1. Scope

1.1 This classification covers rotenone and antimycin which are used to collect or eradicate fish; numerous chemicals have been used but presently only rotenone and antimycin are EPA approved for this use.

2. Referenced Document

2.1 *ASTM Standard:*
D 4131 Practice for Sampling Fish with Rotenone[2]

3. Basis of Classification

3.1 *EPA-Approved Fish Toxicants.*

3.1.1 *Rotenone*—This fish toxicant is also known as derris or cube and is derived from roots of several plants of the family *Leguminosae*. Its action mode as a powerful respiratory inhibitor in fish starts with entrance into the blood stream via the gills and then by translocation to vital organs. Formulations are in the following general forms: a liquid containing 5 % rotenone, liquid with 2.5 % rotenone and synergists, and wettable powder. Persistence in the environment is seldom more than two weeks although it may remain longer in very cold, soft water.

3.1.2 *Antimycin*—This fish-toxicant is an antibiotic produced by Streptomyces and is known as Fintrol 5®, Fintrol 15®, and Fintrol-Concentrate®. The Fintrol 5® and 15® are coated sand-like grains for use in waters 5 to 15 ft (1.52 to 4.57 m) deep. The Fintrol-Concentrate® is a liquid for use in streams or shallow waters. Antimycin is effective in small concentrations (5 ppb) and enters fish through the gills to irreversibly block cellular respiration. It is nonpersistent in the environment.

4. Significance and Use

4.1 The significance of using chemical fish toxicants is that more complete population analyses or total eradication, or both, can be accomplished. Target species can be selectively eradicated by varying concentrations. This provides a very effective tool in fisheries investigations and management programs. Water conditions (that is, pH, temperature, alkalinity, etc.) and morphology can be limiting factors.

4.2 *Rotenone*—Rotenone used as a fish toxicant is highly versatile and can be used effectively to collect fish samples; to eradicate fish; and to selectively remove certain fish species.

4.2.1 Its effectiveness is reduced in cold <20°C, and dosage required increases with alkalinities. It may also eliminate food web organisms. Fish may be repulsed from treated areas.

4.3 *Antimycin*—Antimycin is versatile in the selective removal of scalefish or even more selectively against certain centrarchids (sunfish) and minnows.

4.3.1 Its effectiveness is reduced in water with pH above 8.5.

[1] This classification is under the jurisdiction of ASTM Committee E-47 on Biological Effects and Environmental Fate and is the direct responsibility of Subcommittee E47.08 on Biological Field Testing.

Current edition approved Nov. 26, 1982. Published March 1983.

[2] *Annual Book of ASTM Standards*, Vol 11.04.

Standard Guide for Selecting
Grab Sampling Devices for Collecting Benthic
Macroinvertebrates[1]

This standard is issued under the fixed designation D 4387; the number immediately following the designation indicates the year of original adoption or, in the case of revision, the year of last revision. A number in parentheses indicates the year of last reapproval. A superscript epsilon (ε) indicates an editorial change since the last revision or reapproval.

1. Scope

1.1 This guide covers selecting grab sampling devices for collecting benthic macroinvertebrates. (See Table 1)

1.2 The grab sampler when used correctly is a quantitative collecting device. It is designed to penetrate and grab or scoop a variety of substrates or sediment types from which macroinvertebrates are collected in freshwater, estuarine, and marine habitats.

2. Referenced Documents

2.1 *ASTM Standards:*
D 1129 Terminology Relating to Water[2]
D 4342 Practice for Collecting Benthic Macroinvertebrates with Ponar Grab Sampler[3]
D 4343 Practice for Collecting Benthic Macroinvertebrates with Ekman Grab Sampler[3]
D 4344 Practice for Collecting Benthic Macroinvertebrates with Smith-McIntyre Grab Sampler[3]
D 4345 Practice for Collecting Benthic Macroinvertebrates with Van Veen Grab Sampler[3]
D 4346 Practice for Collecting Benthic Macroinvertebrates with Okean 50 Grab Sampler[3]
D 4347 Practice for Collecting Benthic Macroinvertebrates with Shipek (Scoop) Grab Sampler[3]
D 4348 Practice for Collecting Benthic Macroinvertebrates with Holme (Scoop) Grab Sampler[3]
D 4401 Practice for Collecting Benthic Macroinvertebrates with Petersen Grab Sampler[3]
D 4407 Practice for Collecting Benthic Macroinvertebrates with Orange Peel Grab Sampler[3]

3. Terminology

3.1 *Definitions*—For definitions of terms used in this guide, refer to Terminology D 1129.

3.2 *Descriptions of Terms Specific to This Standard:*

3.2.1 *benthos*—the community of organisms living in or on the bottom or other substrate in an aquatic environment.

3.2.2 *grab*—any device designed to "bite" or "scoop" into the bottom sediment of a lake, stream, estuary, ocean, and similar habitats to sample the benthos. Grabs are samplers with jaws that are forced shut by weights, lever arms, springs, or cables. Scoops are grab samplers that scoop sediment with a rotating container.

3.2.3 *habitat*—the place where an organism lives, that is, mud, rock, shoreline, etc.

3.2.4 *macroinvertebrates*—benthic or substrate dwelling organisms visible to the unaided eye and retained on a U.S. Standard No. 30 (0.595-mm mesh openings) sieve. The standard sieve opening for marine benthic fauna is 1.0 mm, U.S. Standard No. 18 sieve. Examples of macroinvertebrates are aquatic insects, macrocrustaceans, mollusks, annelids, roundworms, flatworms, and echinoderms.

4. Significance and Use

4.1 Qualitative and quantitative samples of macroinvertebrates inhabiting sediments or substrates are usually taken by a grab sampler. Grab samplers, if used correctly, are devices that sample a unit area or volume of the habitat. They are used to obtain a quantitative estimate of the number of individuals and number of taxa of aquatic macroinvertebrates. In view of the advantages and limitations regarding the penetration of the sediment by many grab samplers and their closing mechanisms, it is not possible to recommend any single instrument as suitable for general use. However, the Petersen grab is considered the least effective bottom grab sampler and, therefore, has limited application. The type and size of the grab sampler or device selected for use will depend on such factors as the size of boat, hoisting gear available, the type of substrate or sediment to be sampled, depth of water, current velocity, and whether sampling is in sheltered areas or in open waters of large rivers, reservoirs, lakes, and oceans. A great variety of instruments have been described and choice of a grab sampler will depend largely on what is available, what is suitable for the sampling area, and what can be obtained without difficulty.

5. Descriptions of Grab Samplers

5.1 *Ponar Grab Sampler* (see Fig. 1) is designed to obtain quantitative samples of macroinvertebrates from sediments in lakes, rivers, estuaries, oceans, and similar habitats. This device is most useful for collecting benthic macroinvertebrates from coarse and hard substrates, such as coarse sand, gravel, and similar substrates, rather than soft sediments, such as mud, fine sand, or sludge. The sampler can be used in swift currents and deeper waters. The sampler is available in a range of sizes from 23 cm to 15 cm. For operating procedures, see Practice D 4342.

5.2 *Ekman Grab Sampler* (see Fig. 2) is designed to obtain quantitative samples of macroinvertebrates from soft sediments in lakes, estuaries, oceans, and similar habitats where

---

[1] This guide is under the jurisdiction of Committee E-47 on Biological Effects and Environmental Fate and is the direct responsibility of Subcommittee E47.08 on Biological Field Testing.

Current edition approved June 29, 1984. Published November 1984.

[2] *Annual Book of ASTM Standards*, Vol 11.01.

[3] *Annual Book of ASTM Standards*, Vol 11.05.

TABLE 1 Standard Classification of Grab Sampling Devices for Collecting Benthic Macroinvertebrates

Grab Sampling Device	Habitat Sampled	Substrate Type Sampled	Effectiveness of Sampling Device; Taxa Sampled	Advantages	Limitations	Preference or Recommendation	Selected Literature
Ponar Grab Standard	Freshwater lakes, rivers, and estuaries, reservoirs	Hard sediments, except hard clay; somewhat less efficient in softer sediments	Sample area 523 cm²; efficient and versatile; not entirely adequate for deep burrowing organisms in soft sediments; quantitative and qualitative sampling obtained; sediment inhabiting macroinvertebrates	Better penetration than other grabs; side plates and screens prevent washout and shock wave that accompany other grabs	Requires boat, winch, and cable; jaws can be blocked and part of sample lost	Better for quantitative sampling than Petersen grab	Brinkhurst (1)[A] (2) Elliot and Drake (3) Elliott and Tullet (4) Flannagan (5) Howmiller (6) Hudson (7) Lewis, Mason, Weber (8) Powers and Robertson (9) Weber (10)
Petite	Freshwater lakes, rivers, and estuaries, reservoirs	Hard sediments, except hard clay; somewhat less efficient in softer sediments	Sample area 232 cm²; efficient and versatile; not entirely adequate for deep burrowing organisms in soft sediments; sediment inhabiting macroinvertebrates	Better penetration than other grabs; side plates and screens prevent washout and shock wave that accompany other grabs; can be operated by hand	Jaws can be blocked and part of sample lost; insufficient in swiftly moving water ½ to 1 m/s velocity		
Ekman Grab Standard	Freshwater lakes, reservoirs, where there is little current; usually small bodies of water	Soft sediments only	Sample area 232 cm²; efficient in soft sediments; extra weights can be used for deeper penetration; quantitative and qualitative obtainable; sediment inhabiting macroinvertebrates	Can be operated by hand; can be operated in shallow, sand or mud bottom streams; comes in a range of sizes	Jaws can fail to penetrate; only partial cylinder cut from substrate, small surface area coverage jaws can be blocked and part of sample lost; inefficient in deep water or moderate to strong currents		Beatties (11) Burton and Flannagan (12) Ekman (13) (14) Elliott and Drake (4) Elliott and Tullett (4) Flannagan (5) Howmiller (6) Hudson (7) Lanz, (15) Lewis, Mason, Weber (8) Lind (16) Milbrink and Wiederholm (17) Rowe and Clifford (18)
Standard Tall	Same as above	Same as above	Sample area 232 cm² Same as above	Same as above	Same as above		Paterson and Fernando (19) Schwoerbel (20)
Large	Same as above	Same as above	Sample area 523 cm² Same as above	Same as above	Same as above		Rawson (21) Welch (22) Weber (10)
Extra Large	Same as above	Same as above	Sample area 929 cm²	Same as above	Same as above		
Petersen Grab	Freshwater lakes, reservoirs; adaptable to rivers, estuaries, and oceans	Sand, gravel, mud, clay	Sample penetration limited sample area from 0.06 to 0.099 m²; sediment inhabiting macroinvertebrates	Gives reasonable quantitative samples when used carefully; comes in a range of sizes	Fairly heavy; need boat and power winch; jaws maybe blocked by sand, etc.; inadequate for deep burrowing organisms; questionable value for strictly quantitative samples; hard to use in adverse weather conditions	Least preferred grab sampler	Barnes (23) Birkett (24) Brinkhurst, (25) Davis (26) Edmondson and Winberg (25) Elliott and Tullett (4) Holme and McIntyre (27) Hudson (7) Howmiller (6) Lewis, Mason, Weber (8) Lind (17) Petersen (28) Thorson (29) Welch (23) Weber, 1973 (10) Petersen and Boysen Jensen (30)

[A] The boldface numbers in parentheses refer to the list of references at the end of this guide.

Table 1 continued

TABLE 1 *Continued*

Grab Sampling Device	Habitat Sampled	Substrate Type Sampled	Effectiveness of Sampling Device; Taxa Sampled	Advantages	Limitations	Preference or Recommendation	Selected Literature
Smith-McIntyre Grab	Marine and estuaries; adaptable to large rivers, lakes	Sand, gravel, mud, clay, and similar substrates	Sample area limited to 0.1 m² with approximately 4 cm deep in hard sand; reasonably quantitative; sediment inhabiting macroinvertebrates	Reasonable quantitative samples; the trigger plates provide added leverage essential to its penetration of substrate	Heavy; need boat and power winch; spring-loaded jaws, hazardous; jaws can be blocked; inadequate for deep burrowing organisms	Widely acceptable sampling device for use in marine and estuary habitats	Carey and Heyamoto (31) Carey and Paul (32) Elliott and Tullett (4) Holme (33) (34) Hopkins (35) Hunter and Simpson (36) McIntyre (37) Smith and McIntyre (38) Tyler and Shackley (39) Wigley (40) Word (41)
Van Veen Grab	Marine and estuaries, adaptable to freshwater areas	Sand, gravel, mud, clay, and similar substrates	Sample area 0.1 m² and 0.2 m²; reasonable penetration; to depth of approximately 5–7 cm; sediment inhabiting macroinvertebrates	Jaws close tighter than Petersen grab; samples most sediment types; comes in a range of sizes	Need large boat, power winch and cable line; blockage of jaws may cause sample loss; not useful for deep burrowing organisms	Limited application	Barnes (23) Beukema (42) Birkett (24) Elliott and Drake (3) Elliott and Tullet (4) Holme (33) (34) Lassig (43) Longhurst (44) McIntyre (37) (45) Nichols and Ellison (46) Schwoerbel (20) Ursin (47) Wigley (40), Word (48) (49)
Orange-Peel Grab	Marine waters, deep lakes	Sandy substrates, cobble, rubble stone	Sample area 0.025 m²; penetration depth about 18 cm; qualitative sampler, not a satisfactory quantitative sampler; should not be used in critical quantitative work that is to be compared with results from other sampling areas; sediment inhabiting macro-invertebrates	Comes in a range of sizes	Need large boat, powered which and cable line; blocking of jaws may cause sample loss	Limited application; reconnaisance sampling only	Word (41) Briba and Reys (50) Elliott and Tullett (4) Hartman (51) Hopkins (35) Merna (52) Packard (53) Reish (54) Thorson (29) Word (48)
Okean 50 Grab	Marine, estuarine, also large rivers	Sand, gravel, mud, clay, similar substrates	Sample area 0.25 m²; must be lowered slowly for quantitative work; moderately deep penetration in hard sand; better for quantitative sampling than Petersen grab; sediment inhabiting macroinvertebrates	Moderately deep penetration in hard sand; gauze covered window at top of each bucket to allow water to escape while grab is closing; offer some resistance to swift currents; lowering of grab desirable for deep sea sampling; may also have hinged doors instead of screened windows; rapid rates of lowering are possible; comes in a range of sizes	Heavy; requires large boat, powered winch and cable line; jaws may be blocked and sample lost; not entirely adequate for deep burrowing organisms; must be lowered slowly for quantitative sampling		Elliott and Tullett (4) Holme (33) (34) Holme and McIntyre (27) Lisitsin and Udintsen (55) Zhadin (56)

Table 1 continued

TABLE 1 *Concluded*

Grab Sampling Device	Habitat Sampled	Substrate Type Sampled	Effectiveness of Sampling Device; Taxa Sampled	Advantages	Limitations	Preference or Recommendation	Selected Literature
Shipek Grab	Estuarine areas, also large freshwater lakes	Sand, gravel, mud, clay, and similar substrates	Sample area 20 cm², approximately 10 cm deep at center; sediment inhabiting macroinvertebrates	Scoop type sampler	Heavy; requires boat, powered winch and line; must be lowered on a near vertical line; inadequate for deep burrowing organisms; sampled area may be rather small for quantitative work	Limited application	Barnes (27) Elliott and Tullett (4) Flannagan (5) Holme (38) (39) Holme and McIntyre (31)
Holme Grab	Marine, estuarine areas, deep lakes	Sand, gravel, mud, clay, and similar substrates	Sample area 0.05 m², approximately 15 cm. in hard sand, etc., sediment inhabiting macroinvertebrates	Scoop type sampler; comes with a single scoop or double scoop	Heavy; requires boat, powered winch and line; springs of scoop may be difficult to reset; inadequate for deep burrowing organisms	Limited application	Barnes (27) Elliott and Tullett (4) Holme (52) (53) (39) Holme and McIntyre (31) Thorson (50)

there is little current. This device is most useful for collecting macroinvertebrates from soft sediments, such as very fine sand, mud, and sludge. The sampler is available in sizes of 15 cm, 23 cm, and 30 cm. For operating procedures, see Practice D 4343.

5.3 *Petersen Grab Sampler* is designed to obtain quantitative samples of macroinvertebrates from sediments in lakes, reservoirs, and similar habitats and is adaptable to rivers, estuaries, and oceans. This device (see Fig. 3) is useful for sampling sand, gravel, marl, and clay in swift currents and deep waters. This sampler is available in a range of sizes that will sample an area from 0.06 to 0.099 m². A concensus of aquatic biologists consider the use of this device the least preferable grab sampler and would use it only in limited applications. For operating procedures, see Practice D 4401.

5.4 *Smith-McIntyre Grab Sampler* (see Fig. 4) is designed to obtain quantitative samples of macroinvertebrates from sediments in rough weather in hard sand bottoms in lakes, streams, estuaries, and oceans. This device is useful for sampling macroinvertebrates from sand, gravel, mud, clay, and similar substrates. This device samples a surface area of 0.1 m². For operating procedures, see Practice D 4344.

5.5 *Van Veen Grab Sampler* (see Fig. 5) is designed to give quantitative samples of macroinvertebrates from sediments in estuaries, oceans, and similar habitats, and is adaptable to freshwater areas. This device is useful for sampling sand, gravel, mud, clay and similar substrates. This sampler is available in two sizes, 0.1 m² and 0.2 m². For operating procedures, see Practice D 4345.

5.6 *Orange-Peel Grab Sampler* (see Fig. 6) is designed to obtain quantitative samples of macroinvertebrates from sediments in marine waters and deep lakes. This device is useful for sampling sandy and similar substrates. The sampler is available in a range of sizes but the 1600 cm³ is generally used although larger sizes are available. For operating procedures, see Practice D 4407.

5.7 *Okean 50 Grab Sampler* (See Holme, 1971 for illustration (34)) is designed to obtain quantitative samples of sediment and macroinvertebrates primarily in marine, estuarine, and large rivers. This device is useful for collecting macroinvertebrates from sand, gravel, mud, clay, and similar substrates. The sampler is available in various sizes, generally a sampling area of 0.25 m². For operating procedures, see Practice D 4346.

5.8 *Shipek (Scoop) Grab Sampler* (see Fig. 7) is designed to obtain quantitative samples of macroinvertebrates from sediments in marine waters and large inland bodies of water. This device is useful for sampling macroinvertebrates from sand, gravel, mud, clay, and similar substrates. It is designed to take a sediment sample with a surface area of 20 cm² to approximately 10 cm deep at the center. For operating procedures, see Practice D 4347.

5.9 *Holme (Scoop) Grab Sampler* (see Fig. 8) is designed to obtain quantitative samples of sediment and macroinvertebrates primarily in marine and estuarine waters and large deep freshwater lakes. This device is useful for sampling macroinvertebrates from sand, gravel, mud, clay, and similar substrates. This sampler is designed to take a sediment sample with a surface area of 0.05 m² and approximately 15 cm deep at the center. The device comes with a single scoop or double scoops. For operating procedures, see Practice D 4348.

5.10 All samplers take discrete grabs or scoops of a defined area.

(a)

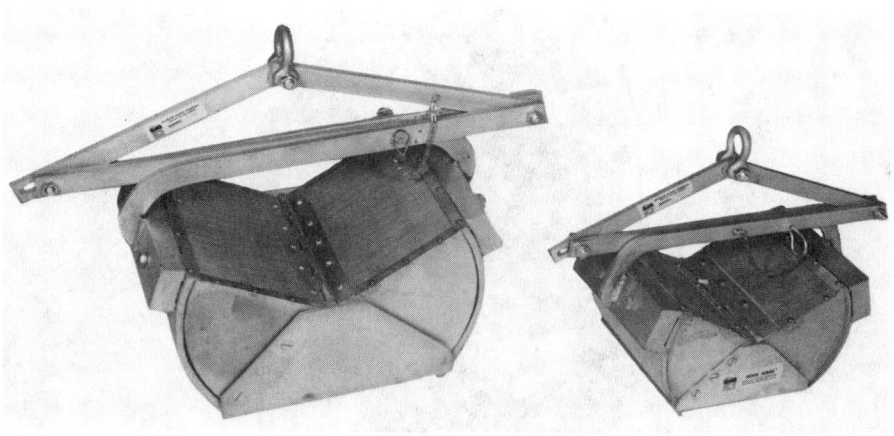

(b)   (c)

FIG. 1   Ponar Grabs. (a) Screen-Top Sediment Grab, Standard Design (Photograph courtesy of Kahl Scientific Instrument Corp.); (b) Screen-Top Wildco Ponar Grab, Standard Design; (c) Wildco Petite Ponar Grab (Photograph courtesy of Wildlife Supply Co.)

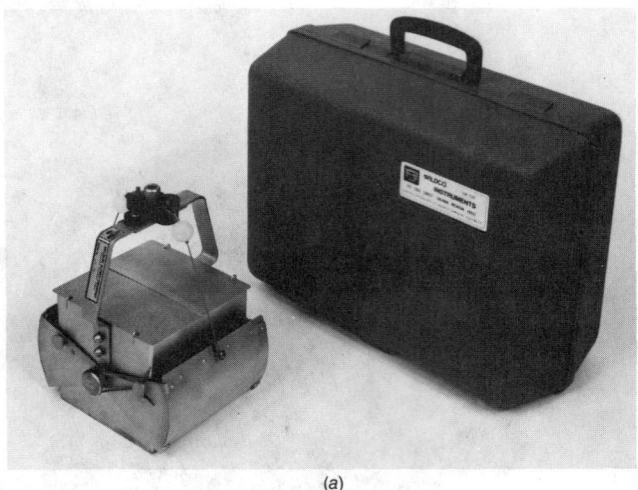

(a)

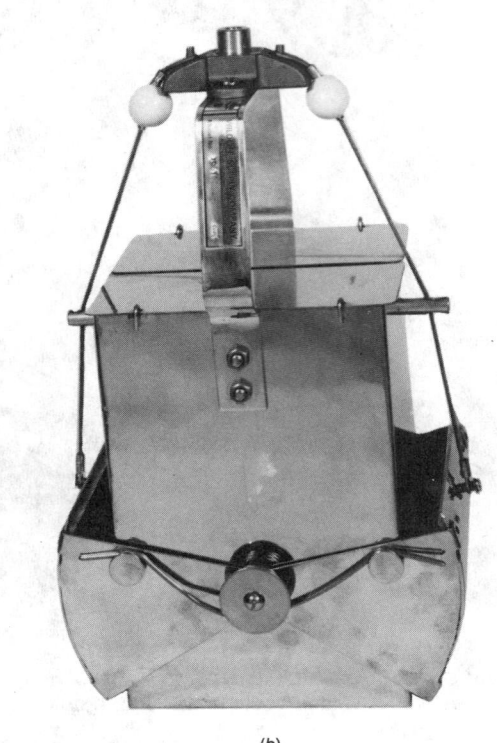

(b)

FIG. 2 Ekman Grabs. (*a*) Wildco Ekman Grab, Standard Design with Case; (*b*) Wildco Ekman Grab, tall design, (Photographs courtesy of Wildlife Supply Co.; (*c*) Ekman Box Sediment Grab (Birge-Ekman Design), (Photograph courtesy of Kahl Scientific Instrument Corp.) FIG. 2   continued

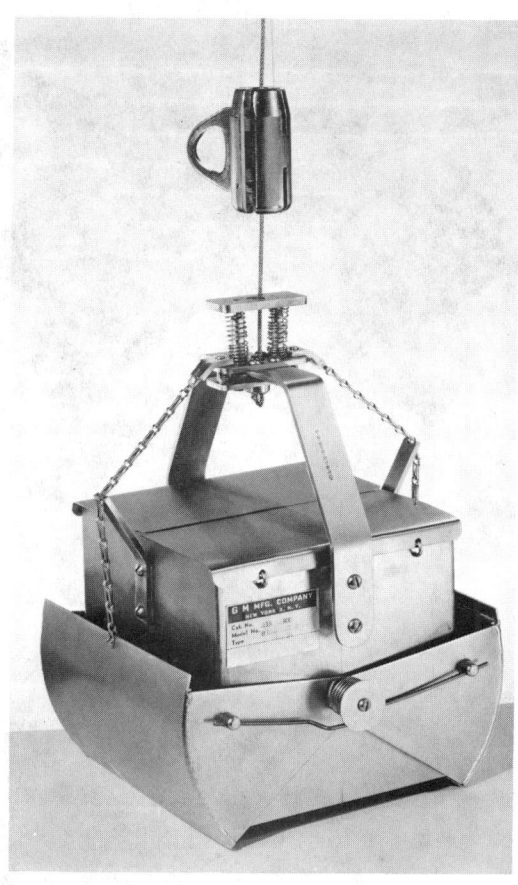

(c)

FIG. 2 (Concluded)

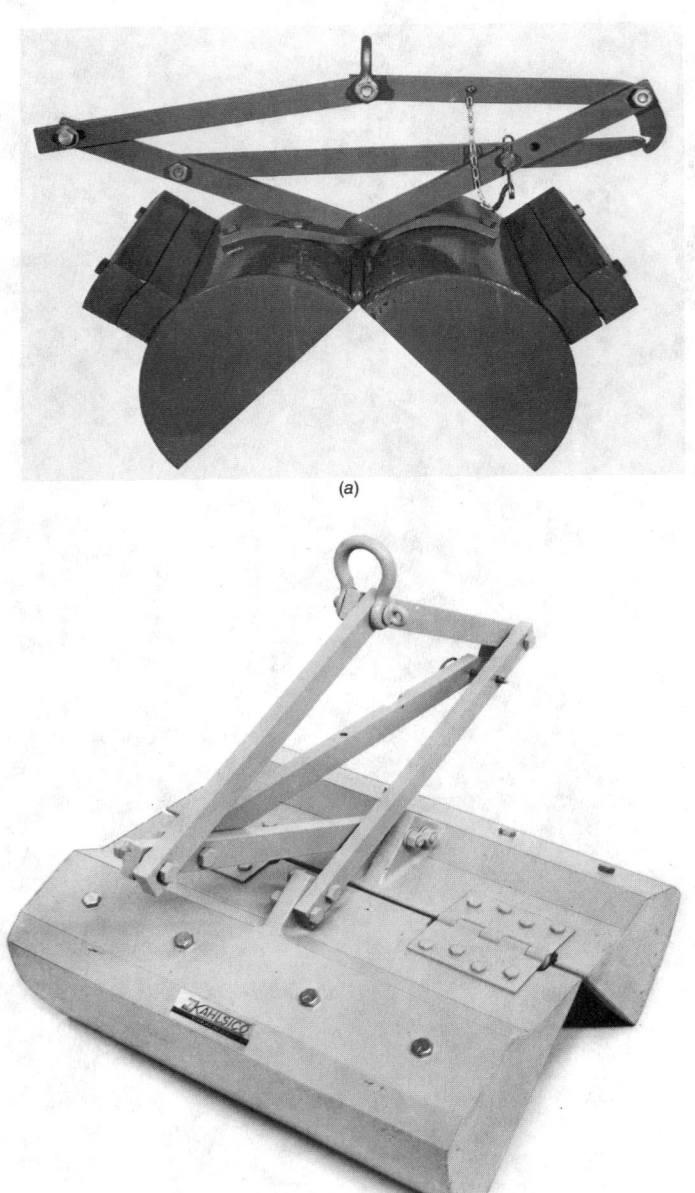

(a)

(b)

FIG. 3 Petersen Grabs (a) Wildco Petersen Grab (Photograph courtesy of Wildlife Supply Co.); (b) Kahl Petersen Grab (Photograph courtesy of Kahl Scientific Instrument Corp.)

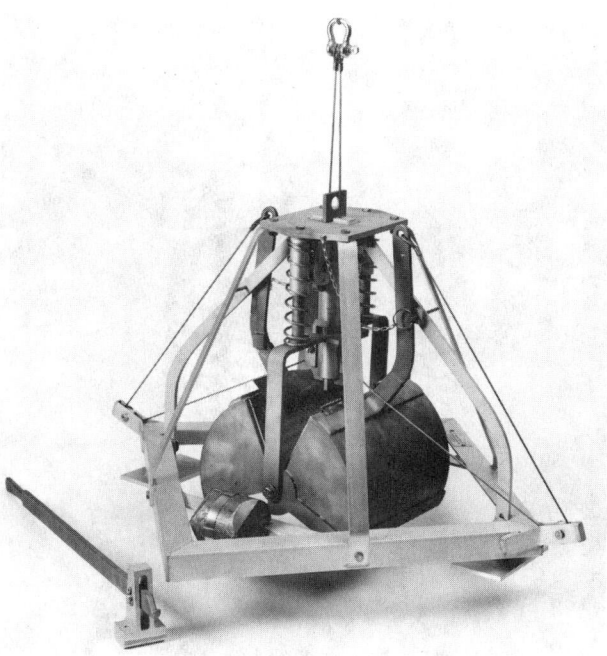

FIG. 4 Smith-McIntyre Grab (Photograph courtesy of Kahl Scientific Instrument Corp.)

FIG. 5 Van Veen Grab (Photograph courtesy of Kahl Scientific Instrument Corp.)

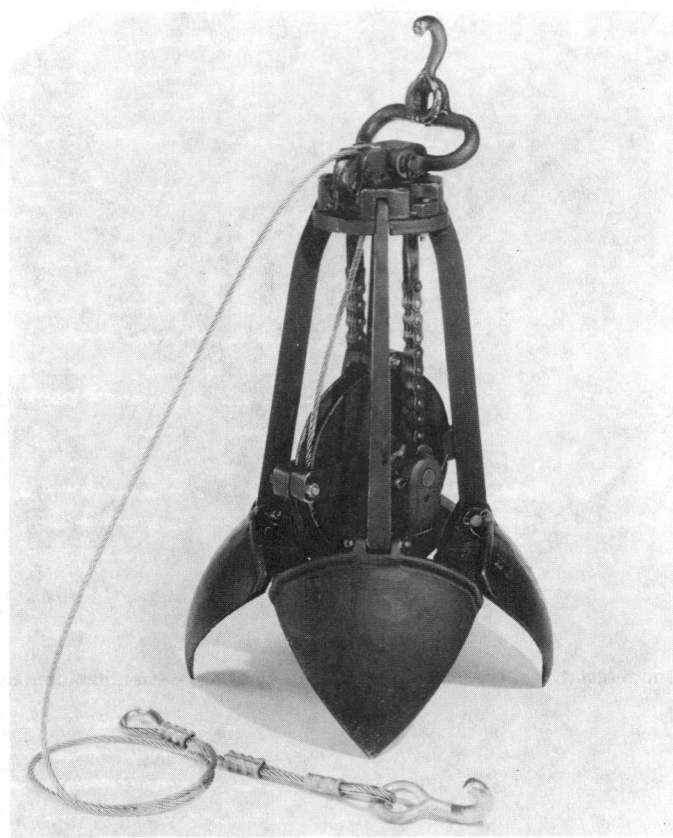

FIG. 6   Orange-Peel Grab (Photograph courtesy of Kahl Scientific Instrument Corp.)

FIG. 7   Shipek (Scoop) Grab (Photograph courtesy of Hydro Products.)

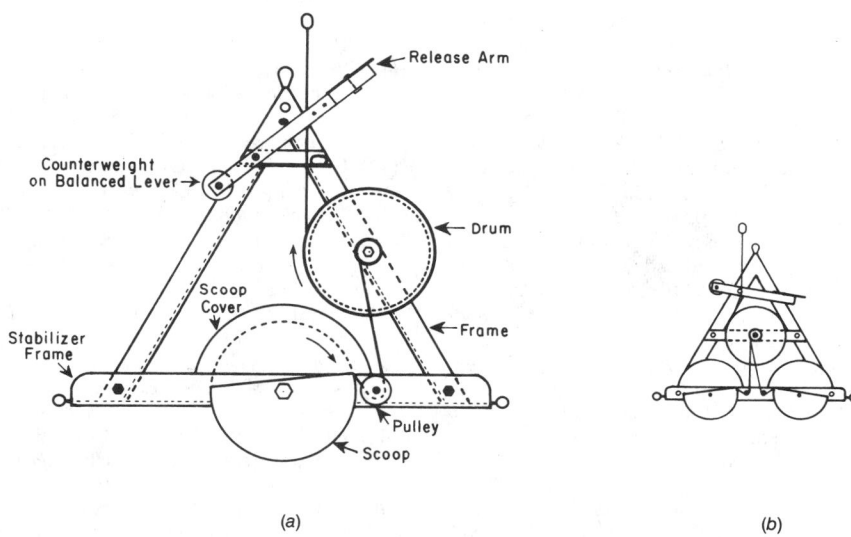

FIG. 8   Holme Grabs. (*a*) Single Holme; (*b*) Double Holme (See Holme and McIntyre (1971), pages 103–105)

REFERENCES

(1) Brinkhurst, R. O., "Sampling the Benthos" *Great Lakes Institute, Progress Report Number 32*, University of Toronto, Toronto, Canada, 1967.

(2) Brinkhurst, R. O., *The Benthos of Lakes*, St. Martin's Press, New York, 1974, p. 190.

(3) Elliott, J. M. and Drake, C. M., "A Comparative Study of Seven Grabs Used for Sampling Benthic Macroinvertebrates in Rivers," *Freshwater Biological Association*, Vol 11, 1981, pp. 99–120.

(4) Elliott, J. M. and Tullett, P. A., "A Bibliography of Samplers for Benthic Invertebrates," *Freshwater Biological Association*, Occasional Publication, No. 4, 1978, p. 61.

(5) Flannagan, J. F., "Efficiencies of Various Grabs and Corers in Sampling Freshwater Benthos," *Journal of Fisheries of Research Board of Canada*, Can., Vol 27, 1970, pp. 1631–1700.

(6) Howmiller, R. P., "A Comparison of the Effectiveness of Ekman and Ponar Grabs," *Transactions of the American Fisheries Society*, Vol 100, 1971, pp. 560–564.

(7) Hudson, P. L., "Quantitative Sampling with Three Benthic Dredges," *Transactions of the American Fisheries Society*, Vol 99, 1970, pp. 603–607.

(8) Lewis, P. A., Mason, W. T., Jr., and Weber, C. I., "Evaluation of Three Bottom Grab Samplers for Collecting River Benthos," *Ohio Journal of Science*, Vol 82, 1982, pp. 103–113.

(9) Powers, C. F. and Robertson, A., "Design and Evaluation of an All Purpose Benthos Sampler," *Special Report No. 30*, Great Lakes Research Div., University of Michigan, Ann Arbor, 1967, pp. 126–131.

(10) Weber, C. I. (ed.), "Biological Field and Laboratory Methods for Measuring the Quality of Surface Waters and Effluents," *Environmental Monitoring Series*, U.S. Environmental Protection Agency, EPA-670/4-73-001.

(11) Beattie, D. M., "A Modification of the Ekman-Birge Bottom Sampler for Heavy Duty," *Freshwater Biological Association*, Vol 9, 1979, pp. 181–182.

(12) Burton, W. and Flannagan, J. F., "An Improved-Ekman Type Grab," *Journal of Fisheries Research Board of Canada*, Vol 30, pp. 287–290.

(13) Ekman, S., "Neue Apparate Zur Quatitativen Und Quantitativen Erforschung der Bodenfauna der Seen," *Internationale Revue Der Gesamten Hydrobiologie und Hydrographie*, Vol 3, 1911, pp. 553–561.

(14) Ekman, S., "Ueber Die Festigkeit Der Marinen Sedimente Als Faktor Der Tieverbreitung," *Ein Beitrag Zur Ass. Analyse. Zoologiska Bidrag fran Uppsala Bidr. Fran. Uppsala*, Vol 25, 1947, pp. 1–20.

(15) Lanz, F., "Untersuchung Uber Die Vertikalverteilung Der Bodenfauna in Tiafen-sediment von Seen." Ein nauer Bodengreifer mit Zerteilingsvorrichtung, Verhardlungen der Internationalen Verein Limnologie Vol 5, 1931, pp. 323–261.

(16) Lind, O. T., *Handbook of Common Methods in Limnology*, The C.V. Mosby Co., St. Louis, Missouri, 1974, p. 154.

(17) Milbrink, G. and Wiederholm, T., "Sampling Efficiency of Four Types of Mud Bottom Samplers, *Oikos 24*, 1973, pp. 479–482.

(18) Rowe, G. T. and Clifford, C. H., "Modification of the Birge Ekman Box Corer for Use with SCUBA or Deep Submergence Research Vessels," *Limnology and Oceanography*, Vol 18, 1973, pp. 172–175.

(19) Paterson, C. G. and Fernando, C. H., "A Comparison of a Simple Corer and An Ekman Grab for Shallow-Water Benthos," *Journal of Fisheries Research Board of Canada*, Vol 23(3), 1971, pp. 365–368.

(20) Schwoerbel, J., "Methods of Hydrobiology," *Freshwater Biological Association*, Pergamon Press, Inc., New York, 1970, p. 356.

(21) Rawson, D. S., "An Automatic-Closing Ekman Dredge and Other Equipment for Use in Extremely Deep Water," *Special Publications of Limnology Society of America*, Vol 18, 1947, pp. 1–8.

(22) Welch, P. S., "*Limnological Methods*," McGraw Hill Co., New York, NY, 1948, p. 382.

(23) Barnes, H., "*Oceanography and Marine Biology, A Book of Techniques*," The Macmillan Co., New York, NY, 1959, p. 218.

(24) Birkett, L., "A Basic for Comparing Grabs," *Explor. Mer.*, Vol 23, 1958, pp. 202–207.

(25) Edmondson, W. T. and G. G. Winberg (eds.), "A Manual on Methods for the Assessment of Secondary Productivity in Fresh Water," *International Biological Program, (IBP) Handbook No. 17*, Blackwell Sci. Publ. Oxford and Edinburgh, 1971, p. 358.

(26) Davis, F. M., "Quantitative Studies on the Fauna of the Sea Bottom. No. 2., Results of the Investigation in the Southern North Sea 1921–1924.," *Fishery. Investigations, London Services II*, Vol

8(4), 1925, pp. 1–50. Edinburgh, 1971, p. 358.

(27) Holme, N. A. and A. D. McIntyre (eds.), "Methods for the Study of Marine Benthos." *International Biological Program, (IBP) IBP Handbook No. 16*, Blackwell Sci. Publ. Oxford and Edinburgh, 1971, p. 334.

(28) Petersen, C. G. J., "The Sea Bottom and Its Production of Fish Food." *Report of the Danmarks Biological Station 25*, 1918, p. 62.

(29) Thorson, G., "Sampling the Benthos." "Treatise on Marine Ecology and Paleoecology,"J. W. Hedgpeth ed., *Geological Society of America Memoirs 67*, Vol 1, 1957, pp. 61–73.

(30) Petersen, C. G. J. and P. Boysen Jensen, "Valuation of the Sea I Animal Life of the Sea Bottom, Its Food and Quantity," *Report of the Danmarks Biological Station 20*, 1911, p. 81.

(31) Carey, A. G., Jr. and Heyamoto, H. *Techniques and Equipment for Sampling Benthic Organisms*, Prutar, A. T. and Alverson, D. L. eds., The Columbia River Estuary and Adjacent Ocean Water: Bioenvironmental Studies, 1972, pp. 378–408.

(32) Carey, A. G. and Paul, R. R., "A Modification of the Smith-McIntyre Grab for Simultaneous Collection of Sediment and Bottom Water," *Limnological Oceanography*, Vol 13, 1968, pp. 545–549.

(33) Holme, N. A., "Methods of Sampling the Benthos," Advances in Marine Biology, F. S. Russel ed., London Academic, Vol 2, 1964, pp. 171–260.

(34) Holme, N. A., "Macrofauna Sampling," Holme, N. A. and McIntyre, A. D. eds., "Methods for the Study of Marine Benthos," *IBP Handbook No. 16*, 1971, pp. 80–130. Blackwell Scientific Publications Oxford and Edinburgh.

(35) Hopkins, T. L., "A Survey of Marine Bottom Samplers," M. Sears ed., Progress in Oceanography, Vol 2, 1964, pp. 213–256.

(36) Hunter, B. and Simpson, A. E., "A Benthic Grab Designed for Easy Operation and Durability," *Journal of Marine Biological Association of the United Kingdom*, Vol 56, 1976, pp. 951–957.

(37) McIntyre, A. D., "Efficiency of Marine Bottom Samplers," *In:* "Methods for the Study of Marine Benthos," Holme, N. A. and McIntyre, A. D. eds., *IBP Handbook* No. Vol 16, 1971, pp. 140–146.

(38) Smith, W. and McIntyre, A. D., "A Spring-Loaded Bottom Sampler," *Journal of Marine Biological Association of the United Kingdom*, Vol 33(1), 1954, pp. 257–264.

(39) Tyler, P. and Shackley, S. E., "Comparative Efficiency of the Day and Smith-McIntyre Grabs," *Estuarine and Coastal Marine Science*, Vol 6, 1978, pp. 439–445.

(40) Wigley, R. L., "Comparative Efficiency of Van Veen and Smith-McIntyre Grab Samplers as Revealed by Motion Pictures," *Ecology*, Vol 48, 1967, pp. 168–169.

(41) Word, J. Q., Kauwling, T. J., and Mearns, A. J., "A Comparative Field Study of Benthic Sampling Devices Used in Southern

California Benthic Survey," *Southern California Coastal Water Research Project*, 1500 E. Imperial Highway, El Segundo, CA, 1976.

(42) Beukema, J. J., "The Efficiency of the Van Veen Grab Compared with the Reineck Box Sampler," *Journal Du Conseil. Conseil Permanent International Pour L'Exploration De La Mer* (Copenhagen), Vol 35(3), 1974, pp. 319–327.

(43) Lassig, J., "An Improvement to the Van Veen Bottom Grab," *Journal of Du Conseil Conseil Permanent International Pour L'Exploration de La Mer*, Vol 29, 1965, pp. 352–353.

(44) Longhurst, A. R., "The Sampling Problem in Benthic Ecology," *Proceedings of New Zealand Ecological Society*, Vol 6, 1959, pp. 8–12.

(45) McIntyre, A. D., "The Use of Trawl, Grab and Camera in Estimating Marine Benthos, *Journal of Marine Biological Association of the United Kingdom*, U.K., Vol, 35, 1956, pp. 419–429.

(46) Nichols, M. M. and Ellison, R. L., "Light-Weight Bottom Sampler for Shallow Water, *Chesapeake Science*, Vol 7, 1966, pp. 215–217.

(47) Ursin, E., "Efficiency of Marine Bottom Samplers of the Van Veen and Petersen Types," *Meddelelser fra Danmarks Fiskeri og Havundersogelser*, N.S., Vol 17, 1954, pp. 1–8.

(48) Word, J. Q., "An Evaluation of Benthic Invertebrate Sampling Devices for Investigating Breeding Habits of Fish," Simenstad, Fish Food Habit Studies, C. A., and Lipovsky, S. J. eds., *University of Washington*, No. WSG-W077-2, 1976.

(49) Word, J. Q., "Biological Comparison of Grab Sampling Devices," *Southern California Coastal Water Res. Project (No. 77)*, 1977, pp. 189–194.

(50) Briba, C. and Reys, J. P., "Modifications D'Une Benne 'Orange Peel' Pour Des Prelevements Quantitatifs Du Benthos De Substrate Meubales," *Recueil des Travaux Institut d'Ecologie et de Biogeographie Academie Serbe des Sciences*, Endoume 41, 1966, pp. 117–121.

(51) Hartman, O., "Quantitative Survey of the Benthos of San Pedro Basin, Southern Calif. Part 1," Preliminary Results Allan *Hancock Pacific Expedition*, Vol 19, 1955, p. 185.

(52) Merna, J. W., "Quantitative Sampling with the Orange Peel Dredge," *Limnology Oceanography*, Vol 7, 1962, pp. 432–33.

(53) Packard, E. L., "A Quantitative Analysis of the Molluscan Fauna of San Francisco Bay," *University of California Publications in Zoology*, Vol 18, 1918, pp. 299–336.

(54) Reish, D. J., "Modification of the Hayward Orange Peel Bucket for Bottom Sampling," *Ecology*, Vol 40, 1959, pp. 502–503.

(55) Lisitsyn, A. P. and Udintsev, G. B., "New Model Dredges," Trudy Vsesoyuznogo Gidrologicheskogo Obshchestua Vol 6, 1955, pp. 217–222.

(56) Zhadin, V. I., *Methods of Hydrobiological Investigation*, Moskva. Vysshaya Shkola, 1960, p. 191.

Standard Guide for
Selecting Stream-Net Sampling Devices for Collecting Benthic Macroinvertebrates[1]

This standard is issued under the fixed designation D 4556; the number immediately following the designation indicates the year of original adoption or, in the case of revision, the year of last revision. A number in parentheses indicates the year of last reapproval. A superscript epsilon (ε) indicates an editorial change since the last revision or reapproval.

ᵉ¹ NOTE—Section 7 was added editorially in March 1995.

1. Scope

1.1 The Surber, portable invertebrate box, Hess, Hess stream bottom, and stream-bed fauna samples are qualitative and most are reasonably quantitative collecting devices. They are operated by hand. They provide for outlining a definite unit-area, for collecting the macroinvertebrates within the area, and are equipped with a net to retain organisms. They are designed to be placed by hand onto or in some cases into mud, sand, gravel, or rubble substrate types, in shallow streams, or shallow areas of rivers.

1.2 The drift net sampler is a qualitative and quantitative collecting device used to capture drifting organisms in flowing waters.

1.3 *This standard does not purport to address all of the safety concerns, if any, associated with its use. It is the responsibility of the user of this standard to establish appropriate safety and health practices and determine the applicability of regulatory limitations prior to use.*

2. Referenced Documents

2.1 *ASTM Standards:*
D 1129 Terminology Relating to Water[2]
D 4557 Practice for Collecting Benthic Macroinvertebrates with Surber and Related Type Samplers[3]
D 4558 Practice for Collecting Benthic Macroinvertebrates with Drift Nets[3]

3. Terminology

3.1 *Definitions*—For definitions of terms used in this guide, refer to Terminology D 1129.

3.2 *Descriptions of Terms Specific to This Standard:*

3.2.1 *benthos*—the community of organisms living in or on the bottom or other substrate in an aquatic environment.

3.2.2 *habitat*—the place where an organism lives, for example, mud, rocks, shoreline, twigs, riffle, pool, etc.

3.2.3 *macroinvertebrates*—benthic or substrate dwelling organisms visible to the unaided eye and retained on a U.S. Standard No. 30 (0.595-mm mesh openings) sieve. The standard sieve opening for marine benthic fauna is 1.0 mm, U.S. Standard No. 18 sieve. Examples of macroinvertebrates are aquatic insects, macrocrustaceans, mollusks, annelids, roundworms, and echinoderms.

3.2.4 *microhabitat*—a smaller and more restricted habitat, for example, certain position on a rock, certain particle size sediment, etc.

3.2.5 *stream-net sampler*—a lotic collecting device, fitted with a net of various sizes, that collect organisms into it by flowing water passing through the sampler.

4. Significance and Use

4.1 The significance of using stream-net samplers is to collect macrobenthos inhabiting a wide range of habitat types from shallow flowing streams or shallow areas in rivers. The stream-net devices (Surber, portable invertebrate box, Hess, Hess stream bottom, and stream-bed fauna samplers) are unit area samplers used for collecting benthic organisms in certain types of substrates. These devices are hand operated and permit collections of qualitative or reasonably quantitative samples of benthic macroinvertebrates from flowing shallow waters. They are used to obtain quantitative estimates of the standing crop, for example, biomass, number of individuals and number of taxa of benthic macroinvertebrates per unit area of stream bottom. Drift nets are another type of qualitative and quantitative sieving device that are useful for collecting benthic macroinvertebrates that either actively or passively enter the water column from all types of substrates in flowing waters. These devices are used to determine the drift of benthic organisms from a variety of substrate types at one time.

5. Description of Stream-Net Samplers

5.1 The Surber sampler (see Fig. 1) is designed to obtain a qualitative or quantitative sample of macroinvertebrates from a unit area. The device is used in shallow flowing streams and shallow areas of rivers with mud, sand, gravel, or rubble substrates. Modification of its basic design has resulted in other sampling devices, such as the portable invertebrate box sampler (see Fig. 2). The latter closed-box-type sampler is preferred, if available. A variety of mesh sizes is available and mesh size should be selected based on the objectives of the study; the finer the mesh, the more organisms (instars) will be collected. These devices sample an area of 0.1 m². For operating procedures, see Practice D 4557.

5.2 The Hess (cylindrical) sampler (see Fig. 3) is designed to obtain a qualitative or quantitative sample of macroinvertebrates from a unit area. The device is used in shallow flowing streams and shallow areas of rivers with mud, sand, gravel, or rubble substrates. Modification of its basic design has resulted in other sampling devices, such as the Hess stream bottom sampler (see Fig. 4) and stream-bed

[1] This guide is under the jurisdiction of ASTM Committee E-47 on Biological Effects and Environmental Fate and is the direct responsibility of Subcommittee E47.08 on Biological Field Testing.
Current edition approved Dec. 27, 1985. Published May 1986.
[2] *Annual Book of ASTM Standards*, Vol 11.01.
[3] *Annual Book of ASTM Standards*, Vol 11.04.

TABLE 1 Standard Classification of Stream-Net Samplers for Collecting Benthic Macroinvertebrates

Stream-Net Samplers	Habitat Sampled	Substrate Type Sampled	Effectiveness of Sampling Device; Taxa Sampled	Advantages	Limitations	Preference or Recommendation	Selected Literature
Surber sampler	Shallow, flowing waters, depth recommended	Mud, sand, gravel, or rubble substrates	Depends on experience and ability of user; area sampled 0.1 m², performance depends on current and substrate; size of macroinvertebrates collected depends on mesh size; variety of mesh sizes may be used.	Easily transported or constructed; samples a unit area; partial screen enclosure	Does not produce quantitative samples consistently; clogging with sand or algae; difficult to set in some substrate types, that is, large rubble; cannot be used efficiently in still or deep water.	Can be modified to fit difficult situations.	Elliot and Tullett (1) [A] Ellis and Rutter (2) Lane (3) Merritt, Cummins, and Resh (4) Needham and Usinger (5) Pollard and Kinney (6) Rutter and Ettinger (7) Resh (8) Rutter and Poe (9) Surber (10) (11) Welch (12) Kroger (13)
Portable invertebrate box sampler	Same as above	Same as above	Same as above	Same as above; completely enclosed; prevents escape of organisms; stable platform; can be used in weed beds.	Same as above	Same as above	Resh, et al (14)
Hess sampler	Same as above	Same as above	Same as above	Same as above; completely enclosed; prevents escape of organisms; can be used in weed beds.	Same as above	Same as above	Canton and Chadwick (15) Elliott and Tullett (1) Hess (16) Merritt, Cummins, and Resh (4) Pollard and Kinney (6) Resh (8) Usinger (17) Welch (12) Resh, et al (14)
Hess stream bottom sampler	Shallow, flowing waters, depth recommended	Mud, sand, gravel, or rubble substrates	Depends on experience and ability of user; area sampled 0.1 m²; performance depends on current and substrate; size of macroinvertebrates collected depends on mesh size; variety of mesh sizes may be used.	Easily transported, or constructed; samples a unit area completely enclosed; prevents escape of organisms; can be used in weed beds.	Does not produce quantitative samples consistently; clogging with sand or algae; difficult to set in some substrate types that is, large rubble; cannot be used efficiently in still or deep water.	Can be modified to fit difficult situations	
Stream-bed fauna sampler	Same as above	Same as above	Same as above	Same as above	Same as above	Same as above	
Drift nets	Flowing rivers and stream	Drifting benthic macroinvertebrates from all substrate types.	Effective in collecting all taxa which drift in the water column; performance depends on current velocity and sampling period; size of macroinvertebrates collected depends on mesh size used.	Low sampling error; less time, money, effort; collects macroinvertebrates from all substrates; usually collects more taxa.	Unknown where organisms come from; terrestrial species may make up a large part of sample in summer and periods of wind and rain.	Limited application	Allen (18) Allan and Russek (19) Bailey (20) Berner (21) Chaston (22) Clifford (23) (24) Cushing (25) (26) Dimond (27) Edington (28) Elliott (29) (30) (31) (32) (33) (34) Ferrington (35) Hales and Gaufin (36) Hildebrand (37) Holt and Waters (38) Hynes (39)

[A] The boldface numbers in parentheses refer to the list of references at the end of this guide.

Table 1 continued

TABLE 1 *Concluded*

Stream-Net Samplers	Habitat Sampled	Substrate Type Sampled	Effectiveness of Sampling Device; Taxa Sampled	Advantages	Limitations	Preference or Recommendation	Selected Literature
Drift nets	Flowing rivers and stream	Drifting benthic macroinvertebrates from all substrate types.	Effective in collecting all taxa which drift in the water column; performance depends on current velocity and sampling period; size of macroinvertebrates collected depends on mesh size used.	Low sampling error; less time, money, effort; collects macroinvertebrates from all substrates; usually collects more taxa.	Unknown where organisms come from; terrestrial species may make up a large part of sample in summer and periods of wind and rain.	Limited application	Keefer and Maughan (40) Larimore (41)Larkin and McKone (42) Lehmkuhl and Anderson (43) McLay (44) Merritt, Cummins, and Resh (4) Minshall and Winger (45) Modde and Schulmbach (46) Muller (47) (48) Mullican, Sansing, and Sharber (49) Mundie (50) (51) Pearson and Franklin (52) Pearson and Kramer (53) (54) Pearson, Kramer, and Franklin (55) Pfitzer (56) Radford and Hartland-Rowe (57) Reisen and Prins (58) Resh (8) Resh, et. al (14) Spence and Hynes (59) Tanaka (60) Tranter and Smith (61) Waters (62) (63) (64) (65) (66) (67) (68) (69) (70) (71) Weber (72) Wilson and Bright (73) Winner, Boesel, and Farrell (74) Wojtalik and Waters (75)

fauna sampler (see Fig. 5). A variety of mesh sizes is available, and mesh size should be selected based on the objectives of the study; the finer the mesh, the more organisms (instars) will be collected. The area sampled by these devices is dependent on their diameter and is comparable to 5.1. These devices sample an area of 0.1 m². For operating procedures, see Practice D 4557.

5.3 The drift net sampler (see Fig. 6) is designed to obtain qualitative and quantitative samples of macroinvertebrates which drift in flowing streams and rivers with a velocity of not less than 0.05 m/s. Drift nets vary in size, but the type recommended for use in water pollution surveys or other ecological assessments has an upstream opening of 15 by 30 cm, and the collection bag is 1.3 m long. A variety of mesh sizes is available, and mesh size should be selected based on the objectives of the study; the finer the mesh, the more organisms (instars) will be collected. For operating proce-

dures, see Practice D 4558.

6. Basis of the Guide

6.1 The Surber, portable invertebrate box, Hess, Hess stream bottom, and stream-bed fauna samplers are designed for use in riffles of shallow flowing streams or shallow areas of rivers. They sample a definite area that can be worked by hand for the collection of benthic macroinvertebrates. See Practice D 4557.

6.2 Drift net samplers are designed to collect emigrating or dislodged benthic macroinvertebrates inhabiting all substrate types in flowing streams and rivers and is used to determine drift density and drift rate. See Practice D 4558.

6.3 For the description and operating procedures of the stream-net samplers, see Practices D 4557 and D 4558 and the following:

6.3.1 Table 1, summary of stream-net samplers, including:

6.3.1.1 Surber sampler and portable invertebrate box sampler,

6.3.1.2 Hess (cylindrical) sampler, Hess stream bottom, and stream-bed fauna sampler, and

6.3.1.3 Drift nets.

6.3.2 Figures of stream-net samplers, including:

6.3.2.1 Figures 1(*a*) and 1(*b*), Surber samplers,

FIG. 1(*a*) Surber Sampler (Illustration courtesy of Kahl Scientific Instrument Corp., P.O. Box 1166, El Cajon, CA 92022-1166)

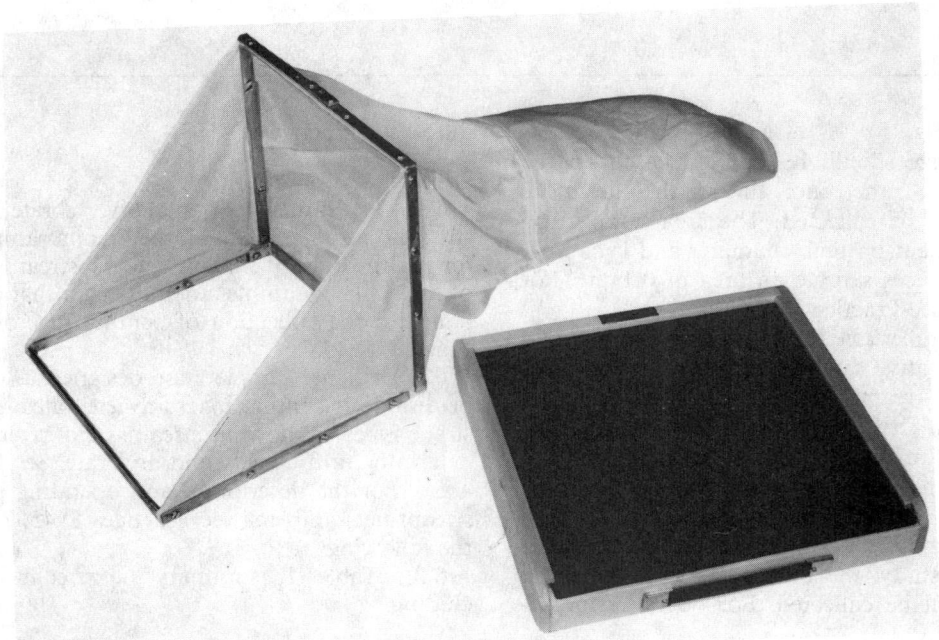

FIG. 1(*b*) Surber Sampler (Photograph courtesy of Wildlife Supply Co., 301 Cass St., Saginaw, MI 48602)

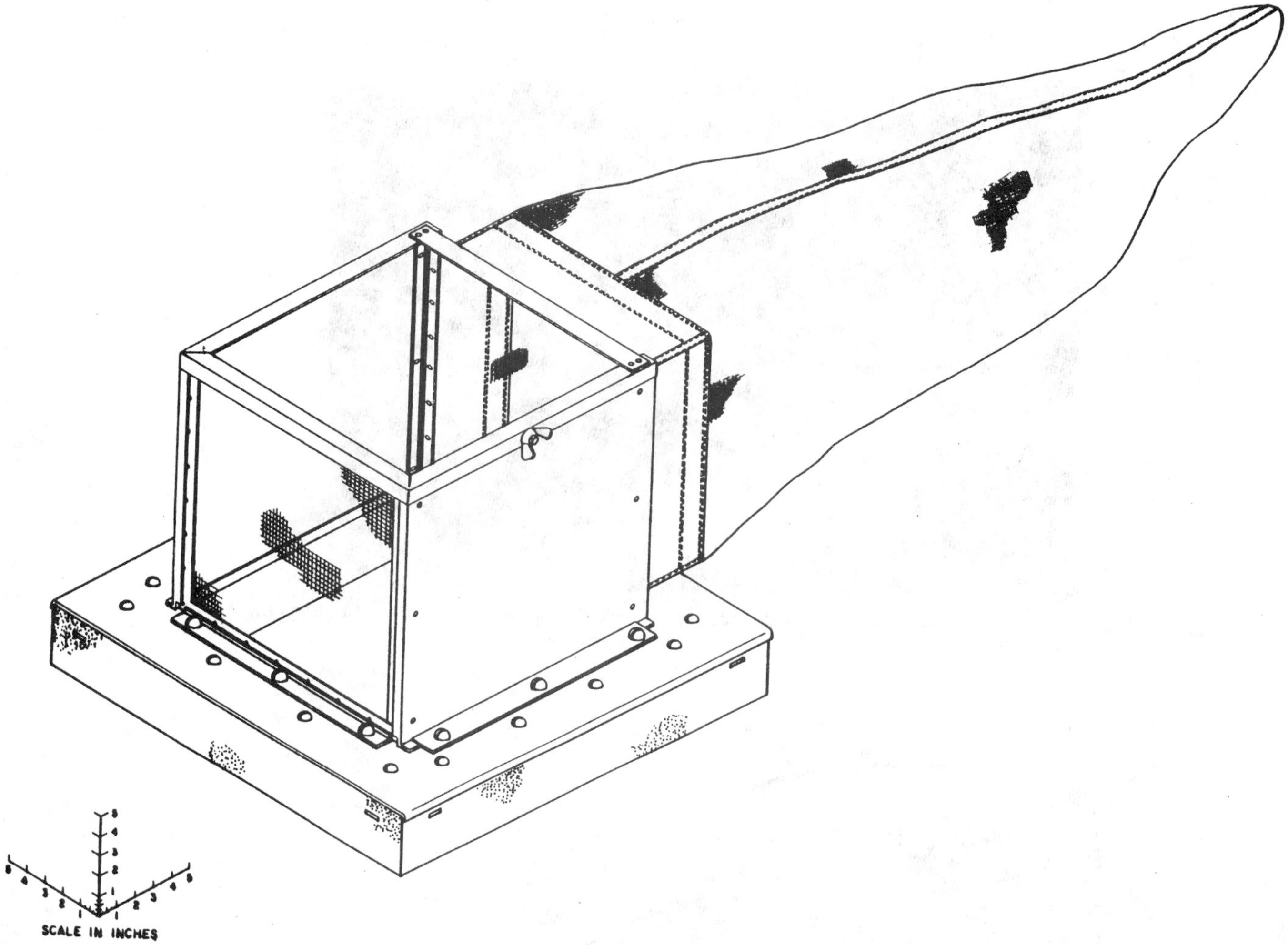

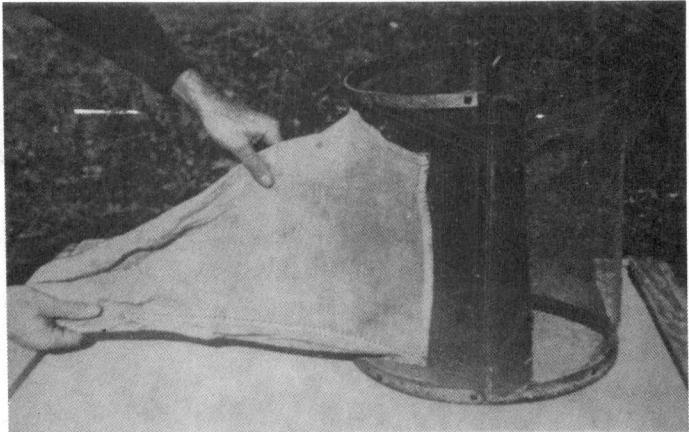

FIG. 2 Portable Invertebrate Box Sampler (Illustration courtesy of Ellis-Rutter Associates, P.O. Box 401, Punta Gorda, FL 33950)

6.3.2.2 Figure 2, portable invertebrate box sampler,
6.3.2.3 Figure 3, Hess (cylindrical) sampler,
6.3.2.4 Figure 4, Hess stream bottom sampler,
6.3.2.5 Figure 5, stream-bed fauna sampler, and
6.3.2.6 Figures 6(*a*) and 6(*b*), drift nets.

7. Keywords

7.1 benthic macroinvertebrates; hess; rivers; shallow streams; stream-bed fauna; stream-net sampling devices; stream bottom; surber

FIG. 3 Hess Sampler (Photograph courtesy of Billy G. Isom)

FIG. 4 **Hess Stream Bottom Sampler** (Photograph courtesy of Wildlife Supply Co., 301 Cass St., Saginaw, MI 48602)

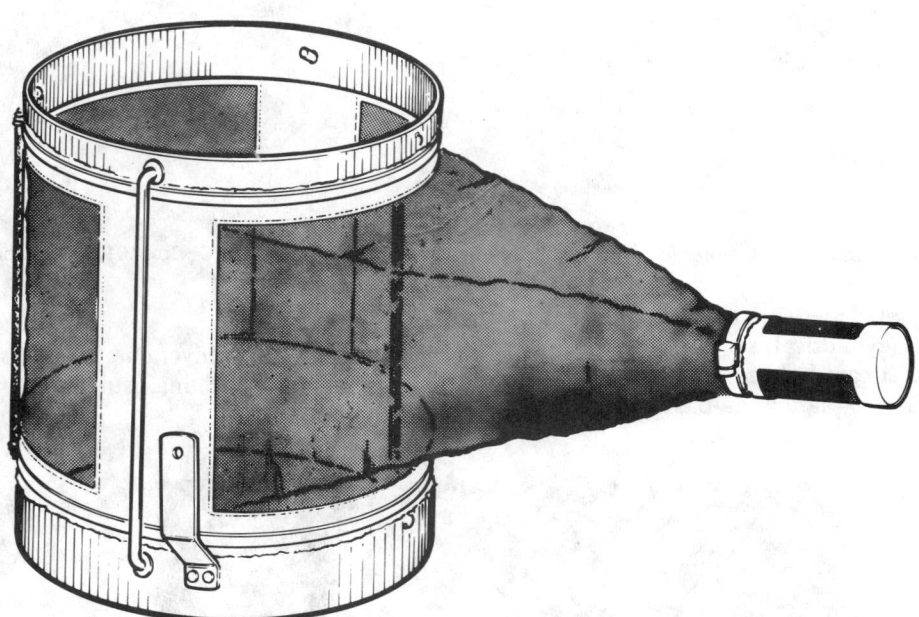

FIG. 5 **Stream-Bed Fauna Sampler** (Photograph courtesy of Kahl Scientific Instrument Corp., P.O. Box 1166, El Cajon, CA 92022-1166)

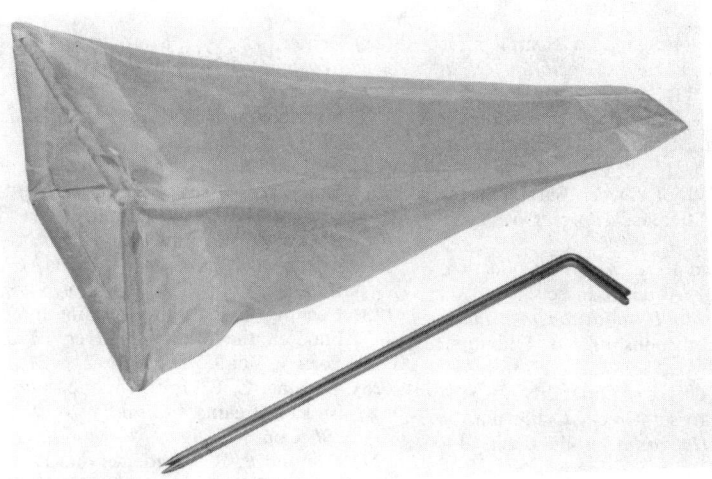

FIG. 6(a) **Drift Net** (Photograph courtesy of Wildlife Supply Co., 301 Cass St., Saginaw, MI 48602)

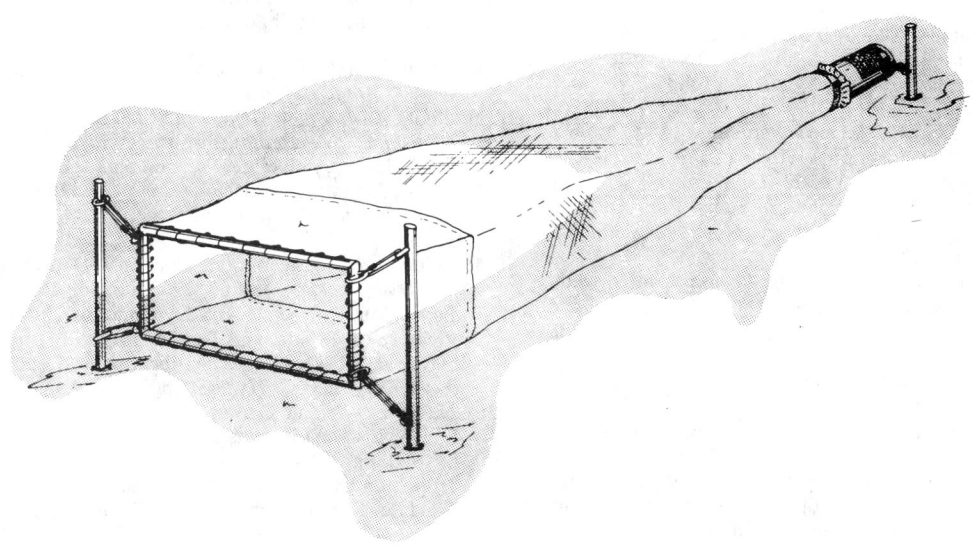

FIG. 6(b) **Drift Net** (Photograph courtesy of Kahl Instrument Corp., P.O. Box 1166, El Cajon, CA 92022-1166)

REFERENCES

(1) Elliott, J. M., and Tullett, P. A., "A Bibliography of Samplers for Benthic Invertebrates," *Freshwater Biological Association Occasional Publication*, No. 4, 1978, pp. 9–11.

(2) Ellis, R. H., and Rutter, R. P., *Portable Invertebrate Box Sampler*, Ellis-Rutter Association, Punta Gorda, FL, advertisement circular, 1973, 4 pp.

(3) Lane, E. D., "An Improved Method of Surber Sampling for Bottom and Drift Fauna in Small Streams," *Progressive Fish-Culturist*, Vol 36, pp. 20–22.

(4) Merritt, T. W., Cummins, K. W., and Resh, V. H., "Collecting, Sampling, and Rearing Methods for Aquatic Insects," R. W. Merritt and K. W. Cummins (eds.), *An Introduction to Aquatic Insects of North America*, Kendall-Hunt Publishing Co., Dubuque, IA, 1984, pp. 11–26.

(5) Needham, P. R., and Usinger, R. L., "Variability in the Macrofauna of a Single Riffle in Prosser Creek, California, as Indicated by the Surber Sampler," *Helgardia*, Vol 24, No. 14, 1956.

(6) Pollard, J. E., and Kinney, W. L., *Assessment of Macroinvertebrate Monitoring Techniques in an Energy Development Area: A Test of the Efficiency of the Macrobenthic Sampling Methods in the White River*, Environmental Monitoring and Support Laboratory, U.S. Environmental Protection Agency, Las Vegas, NV, 1979.

(7) Rutter, R. P., and Ettinger, W. S., "Method for Sampling Invertebrate Drift from a Small Boat," *Progressive Fish-Culturist*, Vol 39, No. 1, pp. 49–52.

(8) Resh, V. H., "Sampling Variability and Life History Features: Basic Considerations in the Design of Aquatic Insect Studies," *Journal of Fisheries Research Board of Canada*, Vol 36, 1979, pp. 290–311.

(9) Rutter, R. P., and Poe, T. P., "Macroinvertebrate Drift in Two Adjoining Southeastern Pennsylvania Streams," *Proceedings of Pennsylvania Academy of Science*, Vol 52, 1978, pp. 24–30.

(10) Surber, E. W., "Rainbow Trout and Bottom Fauna Production in One Mile of Stream," *Transactions of the American Fisheries Society*, Vol 66, 1937, pp. 193–202.

(11) Surber, E. W., "Procedure in Taking Stream Bottom Samples with the Stream Square Foot Bottom Sampler," *Proceedings, A Conference Southeastern Association Game Fisheries Commission*, Vol 23, 1970, pp. 587–591.

(12) Welch, P. S., *Limnological Methods*, McGraw-Hill Co., New York, NY, 1948.

(13) Kroger, R. L., "Underestimation of Standing Crop by the Surber Sampler," *Limnology Oceanography*, Vol 17, 1972, pp. 475–478.

(14) Resh, V. H., et al., "Quantitative Methods for Evaluating the Effects of Geothermal Energy Development on Stream Benthic Communities at the Geysers, California," California Water Resources Center, University of California, Contribution No. 190, ISSN 0575-4941, Davis, CA, 1984, 57 pp.

(15) Canton, S. P., and Chadwick, C. W., "A New Modified Hess Sampler," *Progressive Fish-Culturist*, Vol 46, No. 1, 1984, pp. 57–59.

(16) Hess, A. D., *New Limnological Sampling Equipment*, Limnological Society of America Special Publication No. 6, Ann Arbor, MI, 1941, 5 pp.

(17) Usinger, R. L. (ed.), *Aquatic Insects of California*, University of California Press, Los Angeles, CA, 1963.

(18) Allen, J. D., "The Size Composition of Invertebrate Drift in a Rocky Mountain Stream," *Oikos*, Vol 43, 1984, pp. 68–76.

(19) Allan, D., and Russek, E., "The Quantification of Stream Drift," *Canadian Journal of Fisheries and Aquatic Sciences*, Vol 42, 1985, pp. 210–215.

(20) Bailey, R. G., "Observations on the Nature and Importance of Organic Drift in a Devon River," *Hydrobiologia*, Vol 27, 1964, pp. 353–367.

(21) Berner, L. M., "Limnology of the Lower Mississippi River," *Ecology*, Vol 32, No. 1, 1951, pp. 1–12.

(22) Chaston, I., "The Light Threshold Controlling the Periodicity of Invertebrate Drift," *Journal of Animal Ecology*, Vol 38, No. 1, 1969, pp. 171–180.

(23) Clifford, H. F., "Drift of Invertebrates in an Intermittent Stream Draining Marshy Terrain of West-Central Alberta," *Canadian Journal of Zoology*, Vol 50, 1972a, pp. 985–991.

(24) Clifford, H. F., "A Year's Study of the Drifting Organisms in a Brownwater Stream of Alberta, Canada," *Canadian Journal of Zoology*, Vol 50, 1972b, pp. 975–983.

(25) Cushing, C. E., "Filter-Feeding Insect Distribution and Planktonic Food in the Montreal River," *Transactions American Fisheries Society*, Vol 92, 1963, pp. 216–219.

(26) Cushing, C. E., Jr., "An Apparatus for Sampling Drifting Organisms in Streams," *Journal of Wildlife Management*, Vol 28, No. 3, 1964, pp. 592–594.

(27) Dimond, J. B., "Evidence that Drift of Stream Benthos is Density Related," *Ecology*, Vol 48, No. 5, 1967, pp. 855–857.

(28) Edington, J. M., "The Effect of Water Flow on Populations of Net-Spinning Trichoptera," *Mitteilingen Internationale Vereines Limnologie*, Vol 13, 1965, pp. 40–48.

(29) Elliott, J. M., "Daily Fluctuations of Drift Invertebrates in a Dartmoor Stream," *Nature*, (London), Vol 205, 1965, pp. 1127–1129.

(30) Elliott, J. M., "Diel Periodicity in Invertebrate Drift and the Effect of Different Sampling Periods," *Oikos*, Vol 20, No. 2, 1969, pp. 524–528.

(31) Elliott, J. M., "Methods of Sampling Invertebrate Drift in Running Water," *Annales De Limnology*, Vol 6, No. 2, 1970, pp. 133–159.

(32) Elliott, J. M., *Some Methods for the Statistical Analysis of Samples of Benthic Invertebrates*, Freshwater Biological Association, U.K. Ferry House, Ambleside, Westmoreland, England, 1971, 144 pp.

(33) Elliott, J. M., and Minshall, C. W., "The Invertebrate Drift in the River Dudden, English Lake District," *Oikos*, Vol 19, 1968, pp. 39–52.

(34) Elliott, J. M., "Invertebrate Drift in a Dartmoor Stream," *Archiv fuer Hydrobiologie*, Vol 63, 1967, pp. 202–237.

(35) Ferrington, L. C., Jr., "Drift Dynamics of Chironomidae Larvae: I. Preliminary Results and Discussion of Importance of Mesh Size and Level of Taxonomic Identification in Resolving Chironomidae diel Drift Patterns," *Hydrobiologia*, Vol 114, 1984, pp. 215–227.

(36) Hales, D. C., and Gaufin, A. R., Comparison of Two Types of Stream Insect Drift Nets, *Limnology Oceanography*, Vol 14, No. 3, 1969, pp. 459–461.

(37) Hildebrand, S. G., "The Relation of Drift to Benthos Density and Food Level in an Artificial Stream," *Limnology Oceanography*, Vol 19, No. 6, 1974, pp. 951–957.

(38) Holt, C. S., and Waters, T. F., "Effect of Light Intensity on the Drift of Stream Invertebrates," *Ecology*, Vol 48, No. 2, 1967, pp. 225–234.

(39) Hynes, H. B. N., *The Ecology of Running Waters*, University of Toronto Press, Suffolk, Great Britain, 1970, 555 pp.

(40) Keefer, L., and Maughan, O. E., "Effects of Headwater Impoundment and Channelization on Intertebrate Drift," *Hydrobiologia*, Vol 127, 1985, pp. 161–169.

(41) Larimore, R. W., "Stream Drift as an Identification of Water Quality," *Transactions American Fisheries Society*, Vol 103, No. 3, 1974, pp. 507–517.

(42) Larkin, P. A., and McKone, D. W., "An Evaluation by Field Experiments of the McLay Model of Stream Drift," *Canadian Journal of Fisheries and Aquatic Science*, Vol 42, 1985, pp. 909–918.

(43) Lehmkuhl, D. M., and Anderson, N. H., "Microdistribution and Density as Factors Affecting the Downstream Drift of Mayflies," *Ecology*, Vol 53, No. 4, 1972, pp. 661–667.

(44) McLay, C., "A Theory Concerning the Distance Travelled by Animals Entering the Drift of a Stream," *Journal of Fisheries Research Board of Canada*, Vol 27, No. 2, 1970, pp. 359–370.

(45) Minshall, G. W., and Winger, P. V., "The Effect of Reduction in Stream Flow on Invertebrate Drift," *Ecology*, Vol 49, No. 3, 1968, pp. 580–582.

(46) Modde, T. C., and Schmulbach, J. C., "Seasonal Changes in the Drift and Benthic Macroinvertebrates in the Unchannelized Missouri River in South Dakota," *Proceedings of South Dakota Academy of Science*, Vol 52, 1973, pp. 118–139.

(47) Muller, K., "An Automatic Stream Drift Sampler," *Limnology Oceanography*, Vol 10, 1965, pp. 483–485.

(48) Muller, K., "Stream Drift as a Chronobiological Phenomenon in Running Water Ecosystems," *Annual Review of Ecology and Systematics*, Vol 5, 1974, pp. 309–323.

(49) Mullican, H. N., Sansing, H. T., and Sharber, J. F., Jr., *A Biological Evaluation of the Caney Fork River in the Tail Waters of Center Hill Reservoir*, Tennessee Stream Pollution Control Board, Tennessee Department of Public Health, Nashville, TN, 1967, 19 pp.

(50) Mundie, J. H., "A Sampler for Catching Emerging Insects and Drifting Materials in Streams," *Limnology Oceanography*, Vol 9, No. 3., 1964, pp. 456–459.

(51) Mundie, J. H., "The Diurnal Activity of the Larger Invertebrates at the Surface of Lac La Ronge," *Canadian Journal of Zoology*, Saskatchewan, Canada, Vol 37, 1959, pp. 945–956.

(52) Pearson, W. D., and Franklin, D. R., "Some Factors Affecting Drift Rates of *Baetis* and *Simulidae* in a Large River," *Ecology*, Vol 49, No. 1, 1968, pp. 75–81.

(53) Pearson, W. D., and Kramer, R. H., "A Drift Sampler Driven by a Waterwheel," *Limnology Oceanography*, Vol 14, No. 3, 1969, pp. 462–465.

(54) Pearson, W. D., and Kramer, R. H., "Drift and Production of Two Aquatic Insects in a Mountain Stream," *Ecological Monographs*, Vol 42, No. 3, 1972, pp. 365–385.

(55) Pearson, W. D., Kramer, R. H., and Franklin, R. D., "Macroinvertebrates in the Green River Below Flaming Gorge Dam, 1964–65 and 1967," *Proceedings*, Utah Academy of Sciences, Arts, and Letters, Vol 45, 1968, pp. 148–167.

(56) Pfitzer, D. W., "Investigations of Waters Below Storage Reservoirs in Tennessee," *Transactions*, 19th North American Wildlife Conference, 1954, pp. 271–282.

(57) Radford, D. S., and Hartland-Rowe, R., "A Preliminary Investigation of Bottom Fauna and Invertebrate Drift in an Unregulated and Regulated Stream in Alberta," *Journal of Applied Ecology*, Vol 8, 1971, pp. 883–903.

(58) Reisen, W. K., and Prins, R., "Some Ecological Relationships of the Invertebrate Drift in Praters Creek, Pickens County, SC," *Ecology*, Vol 53, No. 5, 1972, pp. 876–884.

(59) Spence, J. A., and Hynes, H. B. N., "Differences in Benthos Upstream and Downstream of an Impoundment," *Journal of Fisheries Research Board of Canada*, Vol 28, No. 1, 1971, pp. 35–43.

(60) Tanaka, H., "On the Daily Change of the Drifting of Benthic Animals in Stream, Especially on the Types of Daily Change Observed in Taxonomic Groups of Insects," *Bulletin of Freshwater Fisheries Research Laboratory*, Tokyo, Vol 9, 1960, pp. 13–24.

(61) Tranter, D. J., and Smith, P. E., "Filtration Performance," *In: Zooplankton Sampling, Monographs on Oceanographic Methodology*, UNESCO, Geneva, 1968, pp. 27–35.

(62) Waters, T. F., "Standing Crop and Drift of Stream Bottom Organisms," *Ecology*, Vol 42, No. 3, 1961, pp. 532–537.

(63) Waters T. F., "Diurnal Periodicity in the Drift of Stream Invertebrates," *Ecology*, Vol 43, No. 2, 1962, pp. 316–320.

(64) Waters, T. F., "Recolonization of Denuded Stream Bottom Areas by Drift," *Transactions of the American Fisheries Society*, Vol 93, No. 3, 1964, pp. 311–315.

(65) Waters, T. F., "Interpretation of Invertebrate Drift in Streams," *Ecology*, Vol 46, No. 3, 1965, pp. 327–334.

(66) Waters, T. F., "Production Rate, Population Density, and Drift of a Stream Invertebrate," *Ecology*, Vol 47, No. 4, 1966, pp. 595–604.

(67) Waters, T. F., "Diurnal Periodicity in the Drift of a Day-Active Stream Invertebrate," *Ecology*, Vol 49, No. 1, 1968, pp. 152–153.

(68) Waters, T. F., "Invertebrate Drift-Ecology and Significance to Stream Fisheries," *In: Symposium Salmon and Trout in Streams*, T. C. Northcote, ed., H. R. MacMillan Lectures in Fisheries, University of British Columbia, Vancouver, 1969, pp. 121–134.

(69) Waters, T. F., "Subsampler for Dividing Large Samples of Stream Invertebrate Drift," *Limnology Oceanography*, Vol 14, No. 5, 1969b, pp. 813–815.

(70) Waters, T. F., "The Drift of Stream Insects," *Annual Review of Entymology*, Vol 17, 1972, pp. 253–272.

(71) Waters, T. F., and Knapp, R. J., "An Improved Stream Bottom Fauna Sampler," *Transactions of the American Fisheries Society*, Vol 90, 1961, pp. 225–226.

(72) Weber, C. I. (ed.), "Biological Field and Laboratory Methods for Measuring the Quality of Surface Waters and Effluents," *Environmental Monitoring Series*, U.S. Environmental Protection Agency, *EPA-670/4-73-001*, 1973.

(73) Wilson, R. S., and Bright, P. L., "The Use of Chironomid Pupal Exuvis for Characterizing Streams," *Freshwater Biology*, Vol 3, 1973, pp. 283–302.

(74) Winner, R. W., Boesel, M. W., and Farrell, M. P., "Insect Community Structure as an Index of Heavy-Metal Pollution in Lotic Ecosystems," *Canadian Journal of Fisheries and Aquatic Science*, Vol 37, 1980, pp. 647–655.

(75) Wojtalik, T. A., and Waters, T. F., "Some Effects of Heated Water on the Drift of Two Species of Stream Invertebrates," *Transactions of the American Fisheries Society*, Vol 99, No. 4, 1970, pp. 782–788.

APPENDIX

Major ASTM Guides and Practices

A.1 Waste Characterization and Sampling

A.1.1 General Guidance

D 4687-95	Guide for General Planning of Waste Sampling (Vol. 11.04)
D 5956-96	Guide for Sampling Strategies for Heterogeneous Wastes (Vol. 11.04).
D 6009-96	Guide for Sampling Waste Piles (Vol. 11.04).
D 6044-96	Guide for Representative Sampling and Management of Waste and Contaminated Media (vol. 11.04).
D 6051-96	Guide for Composite Sampling and Field Subsampling For Environmental Waste Management Activities (Vol. 11.04).
D 6063-96	Guide for Sampling of Drums and Similar Containers By Field Personnel (Vol. 11.04).

A.1.2 Specific Sampling Procedures

D 4489-95	Practices for Sampling of Waterborne Oils (Vol. 11.02).
D 4547-91	Practice for Sampling Waste and Soils for Volatile Organics (Vol. 11.04).
D 4823-95	Guide for Core-Sampling Submerged, Unconsolidated Sediments (Vol. 11.02).
D 5013-89 (1993)	Practices for Sampling Wastes from Pipes and Other Point Discharges (Vol. 11.04).
D 5358-93$^{\epsilon 1}$	Practice for Sampling with a Dipper or Pond Sampler (Vol. 11.04).
D 5451-93	Practice for Sampling Using a Trier Sampler (Vol. 11.04).
D 5495-94	Practice for Sampling with a Composite Liquid Waste Sampler (COLIWASA) (Vol. 11.04).
D 5633-94	Practice for Sampling with a Scoop (Vol. 11.04).
D 5658-95	Practice for Sampling Unconsolidated Waste from Trucks (Vol. 11.04).
D 5679-95a	Practice for Sampling Consolidated Solids in Drums or Similar Containers (Vol. 11.04).
D 5680-95a	Practice for Sampling Unconsolidated Solids in Drums or Similar Containers (Vol. 11.04).
D 5743-95	Practice for Sampling Single or Multilayered Liquids, With or Without Solids, in Drums or Similar Containers (Vol. 11.04).

A.2 Environmental Site Characterization

A.2.1 General Guidance

D 5730-96	Guide to Site Characterization for Environmental Purposes With Emphasis on Soil, Rock, the Vadose Zone and Ground Water (Vol. 4.09).
D 5995-96	Guide for Environmental Site Characterization in Cold Regions (Vol. 4.09).
D 420-93	Guide for Site Characterization for Engineering, Design, and Construction Purposes (Vol. 4.08).
E 1689-95	Guide for Developing Conceptual Site Models for Contaminated Sites (Vol. 11.05)
PS 3-95	Guide for Accelerated Site Characterization for Confirmed or Suspected Petroleum Releases (Vol. 11.04).
PS 85-96	Guide for Expedited Site Characterization of Hazardous Waste Contaminated Sites (Vol. 4.09).

A.2.2 Aerial Photography and Imagery

D 5518-94	Guide for Acquisition of File Aerial Photography and Imagery for Establishing Historic Site-Use and Surficial Conditions (Vol. 4.09)

A.2.3 Data Elements

D 5911-96	Practice for a Minimum Set of Data Elements to Describe a Soil Sampling Site (Vol. 4.09).
D 5474-93	Guide for Selection of Data Elements for Ground-Water Investigations (Vol. 4.09).
D 5254-92	Practice for the Minimum Set of Data Elements to Identify a Ground Water Site (Vol. 4.09).
D 5408-93	Guide for the Set of Data Elements to Describe a Ground-Water Site, Part 1-Additional Identification Descriptors (Vol. 4.09).
D 5409-93	Guide for the Set of Data Elements to Describe a Ground-Water Site, Part 2-Physical Descriptors (Vol. 4.09).
D 5410-93	Guide for the Set of Data Elements to Describe a Ground-Water Site, Part 3-Usage Descriptors (Vol. 4.09).

A.2.4 Geologic and Hydrogeologic Characterization

D 5979-96	Guide for Conceptualization and Characterization of Ground-Water Flow Systems (Vol. 4.09).
D 6030-96	Guide to Selection of Methods for Assessing Ground Water or Aquifer Sensitivity and Vulnerability (Vol. 4.09).
D 5717-95	Guide for Design of Ground-Water Monitoring Systems in Karat and Fractured-Rock Aquifers (Vol. 4.09).
D 5980-96	Guide for Selection and Documentation of Existing Wells for Use in Environmental Site Characterization and Monitoring (Vol. 4.09).
D 6067-96	Guide for Using the Electronic Cone Pentrometer for Environmental Site Characterization (Vol. 4.09).
D 5434-93	Guide for Field Logging of Subsurface Explorations of Soil and Rock (Vol. 4.09).
D 4043-96	Guide for Selection of Aquifer-Test Field and Analytical Procedures in Determination of Hydraulic Properties by Well Techniques (Vol. 4.08).
D 5126-90	Guide for Comparison of Field Methods for Determining Hydraulic Conductivity in the Vadose Zone (Vol. 4.09).
D 6000-96	Guide for the Presentation of Water-Level Information From Ground Water Sites (Vol. 4.09).

A.2.5 Geophysical Methods

D 5753-95	Guide for Planning and Conducting Borehole Geophysical Logging (Vol. 4.09).
D 5777-95	Guide for Using the Seismic Refraction Method for Subsurface Investigation (Vol. 4.09).

A.2.6 Drilling Methods

D 5781-95	Guide for Use of Dual-Wall Reverse-Circulation Drilling for Geoenvironmental Exploration and the Installation of Subsurface Water-Quality Monitoring Devices (Vol. 4.09).
D 5782-95	Guide for Use of Direct Air-Rotary Drilling for Geoenvironmental Exploration and the Installation of Subsurface Water-Quality Monitoring Devices (Vol. 4.09).
D 5783-95	Guide for Use of Direct Rotary Drilling with Water-Based Drilling Fluid for Geoenvironmental Exploration and the Installation of Subsurface Water-Quality Monitoring Devices (Vol. 4.09).
D 5784-95	Guide for Use of Hollow-Stem Augers for Geoenvironmental Exploration and the Installation of Subsurface Water-Quality Monitoring Devices (Vol. 4.09).
D 5872-95	Guide for Use of Casing Advancement Drilling Methods for Geoenvironmental Exploration and the Installation of Subsurface Water-Quality Monitoring Devices (Vol. 4.09).
D 5875-95	Guide for Use of Cable-Tool Drilling and Sampling Methods for Geoenvironmental Exploration and Installation of Subsurface Water-Quality Monitoring Devices (Vol. 4.09).
D 5876-95	Guide for Use of Direct Rotary Wireline Casing Advancement Drilling Methods for Geoenvironmental Exploration and Installation of Subsurface Water-Quality Monitoring Devices (Vol. 4.09).

A.3 Ground Water Monitoring Wells (see also drilling methods in A.2.6 above)

D 5092-90 (1995)	Practice for Design and Installation of Ground Water Monitoring Wells in Aquifers (Vol. 4.09).
D 5787-95	Practice for Monitoring Well Protection (Vol. 4.09).
D 5521-94	Guide for Development of Ground-Water Monitoring Wells in Granular Aquifers (Vol. 4.09).
D 4750-87 (1993)[e1]	Test Method for Determining Subsurface Liquid Levels in a Borehole or Monitoring Well (Observation Well) (Vol. 4.09).
D 5978-96	Guide for Maintenance and Rehabilitation of Ground-Water Monitoring Wells (Vol. 4.09).
D 5299-92	Guide for the Decommissioning of Ground Water Wells. Vadose Zone Monitoring Devices, Boreholes and Other Devices for Environmental Activities (Vol. 4.09).

A.4 Ground Water Sampling

D 5903-96	Guide for Planning and Preparing for a Ground-Water Sampling Event (Vol. 4.09).
D 4448-85a (1992)	Guide for Sampling Groundwater Monitoring Wells (Vol. 11.04).
D 6001-96	Guide for Direct Push Water Sampling for Geoenvironmental Investigations (Vol. 4.09).

A.5 Vadose Zone Monitoring

D 4700-91	Guide for Soil Sampling from the Vadose Zone (Vol. 4.08).
D 3404-91	Guide for Measuring Metric Potential in the Vadose Zone Using Tensiometers (Vol. 4.08).
D 4696-92	Guide for Pore-Liquid Sampling from the Vadose Zone (Vol. 4.08).
D 5314-92	Guide for Soil Gas Monitoring in the Vadose Zone (Vol. 4.09).
D 5299-92	Guide for the Decommissioning of Ground Water Wells, Vadose Zone Monitoring Devices, Boreholes and Other Devices for Environmental Activities (Vol. 4.09).

A.6 Sample Handling

D 5088-90	Practice for Decontamination of Field Equipment Used at Nonradioactive Waste Sites (Vol. 4.09).
D 4547-91	Practice for Sampling Waste and Soils for Volatile Organics (Vol. 11.04)

D 4840-95	Guide for Sampling Chain of Custody Procedure (Vol. 11.01).
D 4841-88 (1993)[ε1]	Practice for Estimation of Holding Time for Water Samples Containing Organic and Inorganic Constituents (Vol. 11.01).
D 4220-95	Practice for Preserving and Transporting Soil Samples (Vol. 4.08).
D 5079-90	Practice for Preserving and Transporting Rock Core Samples (Vol. 4.09).

A.7 QA/QC

D 5283-92	Practice for Generation of Environmental Data Related to Waste Management Activities: QA/QC Planning and Implementation (Vol. 11.04).
D 5612-94	Guide for Quality Planning and Field Implementation of a Water Quality Measurement Program (Vol. 11.01).
D 5792-95	Practice for Generation of Environmental Data Related to Waste Management Activities: Development of Data Quality Objectives (Vol. 11.04).
D 5851-95	Guide for Planning and Implementing a Water Monitoring Program (Vol. 11.02).

A.8 Data Analysis and Contingency Planning

A.8.1 Statistical Analysis of Ground Water Quality Data

| PS 64-96 | Guide for Developing Appropriate Statistical Approaches for Ground-Water Detection Monitoring Programs (Vol. 4.09). |

A.8.2 Graphic Analysis of Ground Water Quality Data

D 5738-95	Guide for Displaying the Results of Chemical Analyses of Ground Water for Major Ions and Trace Elements—Diagrams for Single Analyses (Vol. 4.09).
D 5754-95	Guide for Displaying the Results of Chemical Analyses of Ground Water for Major Ions and Trace Elements—Trilinear and Other Multi-Coordinate Diagrams (Vol. 4.09).
D 5877-95	Guide for Displaying the Results of Chemical Analyses of Ground Water for Major Ions and Trace Elements—Diagrams Based on Data Analytical Calculations (Vol. 4.09).
D 6036-96	Guide for Displaying the Results of Chemical Analyses of Ground Water for Major Ions and Trace Elements—Use of Maps (Vol. 4.09).

A.8.3 Geostatistical Analysis of Environmental Data

D 5549-94	Guide for Reporting Geostatistical Site Investigations (Vol. 4.09).
D 5922-96	Guide for Analysis of Spatial Variation in Geostatistical Site Investigations (Vol. 4.09).
D 5923-96	Guide for the Selection of Kriging Methods in Geostatistical Site Investigations (Vol. 4.09).
D 5924-96	Guide for the Selection of Simulation Approaches in Geostatistical Site Investigations (Vol. 4.09).

A.8.4 Contingency Planning

| D 5745-95 | Guide for Developing and Implementing Short-Term Measures or Early Actions for Site Remediation (Vol. 11.04). |

Index

ASTM Standards Related to
Environmental Site Characterization

This index covers the standards and related material appearing in this volume(s) or section only. The boldface references represent the ASTM designations. For multiple volume or section indexes, the volume in which the standard appears is included in parentheses. A triangle (Δ) preceding the ASTM designation denotes that Adjunct Material for the standard is available separately; the Adjunct No. is given in the standard. A Combined Index, covering the standards appearing in all volumes of the *1997 Annual Book of ASTM Standards*, is issued as Volume 00.01.

Alphabetization in the index is letter-for-letter, with no consideration to punctuation or word division. Initial prepositions of (indented) subentries are ignored for alphabetization.

In the preparation of indexes, every attempt has been made to index standards on three levels: (1) by main subject, using general and specific search terms; (2) by tests or other significant sections of ASTM standards; and (3) by cross-references to locate main subject entry terms. (*See also* references are abbreviated as *Sa* and appear under main entry terms.) The following examples illustrate ASTM's method of indexing.

INDEX TERMS FOR SPECIFICATIONS

Steel pipe—carbon steel
   Seamless and Welded Steel Pipe for Low-Temperature Service,
      Specification for, **A 333/333M (01.01)**

Steel tube—ferritic stainless steel
   Seamless and Welded Ferritic/Austenitic Stainless Steel
      Tubing for General Service, Specification for,
      A 789/789M (01.01)

INDEX TERMS FOR TESTS

Gold—electrodeposited coating
   Installing Corrugated Aluminum Structural Plate Pipe for
      Culverts and Sewers, Practice for, **B 789 (02.02)**

Effective elastic parameter (E_eff)
   Determining the Effective Elastic Parameter for X-Ray
      Diffraction Measurements of Residual Stress,
      Test Method for, **E 1426 (03.01)**

CROSS-REFERENCES

Water—drinking
      See   **Drinking water**
PDB (pressure design basis)
      See   **Pressure design basis (PDB)**
      Sa   **Pressure testing**

Index of ASTM Standards

A

Above-grade dry land disposal units
Sampling Waste Piles, Guide for, **D 6009**

Accelerated site characterization
Accelerated Site Characterization for Confirmed or Suspected Petroleum Releases, Provisional Guide for, **PS 3**

Acceptance angle
Measuring Surface Wind and Temperature by Acoustic Means, Practices for, **D 5527**

Accuracy
See **Precision**

Acid-dissociable cyanides
See **Cyanides**

Acid waste
See **Waste materials/processing**

Acoustical tests
Measuring Surface Wind and Temperature by Acoustic Means, Practices for, **D 5527**

Acoustic logging
Planning and Conducting Borehole Geophysical Logging, Guide for, **D 5753**

Activated charcoal adsorption method
Sampling Atmospheres to Collect Organic Compound Vapors (Activated Charcoal Tube Adsorption Method), Practice for, **D 3686**

Activated sludge
See **Sludge**

Administrators guide
Planning and Implementing a Water Monitoring Program, Guide for, **D 5851**

Admixtures—soil
See **Soil stabilization**

Adsorption
Sampling Atmospheres to Collect Organic Compound Vapors (Activated Charcoal Tube Adsorption Method), Practice for, **D 3686**

Aggregate—soil-aggregate
Field Determination of Water (Moisture) Content of Soil by the Calcium Carbide Gas Pressure Tester Method, Test Method for, **D 4944**
Water Content of Soil and Rock in Place by Nuclear Methods (Shallow Depth), Test Method for, **D 3017**

Agricultural drainage
See **Land drainage piping systems**

Air analysis
See **Atmospheric analysis**

Air analysis—calibration of equipment
See **Calibration—atmospheric analysis instrumentation**

Air analysis—sampling
See **Sampling—atmospheric analysis**

Air entry permeameter
Comparison of Field Methods for Determining Hydraulic Conductivity in the Vadose Zone, Guide for, **D 5126**

Air-rotary drilling
Use of Direct Air-Rotary Drilling for Geoenvironmental Exploration and the Installation of Subsurface Water-Quality Monitoring Devices, Guide for, **D 5782**

Air sampling/analysis
Sa **Sampling—atmospheric analysis**
Generation of Environmental Data Related to Waste Management Activities: Development of Data Quality Objectives, Practice for, **D 5792**

Alkylbenzene sulfonates
Preparation of Sample Containers and for Preservation of Organic Constituents, Practices for, **D 3694**

Alpha particle radioactivity—water
Measurement of Radioactivity, Practices for, **D 3648**

Ambient atmospheric analysis
General Ambient Air Analyzer Procedures, Practice for, **D 3249**
Measuring Humidity With Cooled-Surface Condensation (Dew-Point) Hygrometer, Test Method of, **D 4230**
Planning the Sampling of the Ambient Atmosphere, Practice for, **D 1357**

Amines content
Preparation of Sample Containers and for Preservation of Organic Constituents, Practices for, **D 3694**

Anemometer
Characterizing Surface Wind Using a Wind Vane and Rotating Anemometer, Practice for, **D 5741**

Anemometers
Measuring Surface Wind and Temperature by Acoustic Means, Practices for, **D 5527**

Aneroid barometers
Measuring Surface Atmospheric Pressure, Test Methods for, **D 3631**

Antimycin
Fish Sampling, Classification for, **D 4211**

Aqueous slurries
See **Slurries**

Aquifers
Design and Installation of Ground Water Monitoring Wells in Aquifers, Practice for, **D 5092**
Design of Ground-Water Monitoring Systems in Karst and Fractured-Rock Aquifers, Guide for, **D 5717**
Determining Subsurface Liquid Levels in a Borehole or Monitoring Well (Observation Well), Test Method for, **D 4750**
Development of Ground-Water Monitoring Wells in Granular Aquifers, Guide for, **D 5521**
Methods for Measuring Well Discharge, Guide for, **D 5737**
Sampling Groundwater Monitoring Wells, Guide for, **D 4448**
Selection of Aquifer-Test Method in Determining of Hydraulic Properties by Well Techniques, Guide for, **D 4043**
Selection of Methods for Assessing Ground Water or Acquifer Sensitivity and Vulnerability, Guide to, **D 6030**

Archeological investigations
Using the Seismic Refraction Method for Subsurface Investigation, Guide for, **D 5777**

Architectural acoustical materials
See **Acoustical tests**

Area
Measurement of Morphologic Characteristics of Surface Water Bodies, Guide for, **D 4581**

Area integration sampling
Sampling Fluvial Sediment in Motion, Guide for, **D 4411**

Artesian well
See **Aquifers**

Aseptic sampling
Aseptic Sampling of Biological Materials, Practice for, **E 1287**

Assessment
Environmental Site Assessments: Phase I Environmental Site Assessment Process, Practice for, **E 1527**
Environmental Site Assessments: Transaction Screen Process, Practice for, **E 1528**

Atmospheric analysis
Choosing Locations and Sampling Methods to Monitor Atmospheric Deposition at Non-Urban Locations, Guide for, **D 5111**
Measuring the Concentration of Toxic Gases or Vapors Using Detector Tubes, Practice for, **D 4490**
Measuring the Concentration of Toxic Gases or Vapors Using Length-of-Stain Dosimeters, Practice for, **D 4599**

Atmospheric analysis—calibration of instruments
See **Calibration—atmospheric analysis instrumentation**

Atmospheric analysis—meteorological measurements
Determining the Operational Comparability of Meteorological Measurements, Practice for, **D 4430**

B

C

Index of ASTM Standards

Index of ASTM Standards

D

F

N

Quality assurance (QA)—water
Measurement Program, Guide for, **D 5612**

R

Radiation exposure
Decontamination of Field Equipment Used at Low Level Radioactive Waste Sites, Practice for, **D 5608**

Radiation exposure—soil
Water Content of Soil and Rock in Place by Nuclear Methods (Shallow Depth), Test Method for, **D 3017**

Radioactive waste materials
Decontamination of Field Equipment Used at Low Level Radioactive Waste Sites, Practice for, **D 5608**

Radioactive water analysis
Sampling Water-Formed Deposits, Practice for, **D 887**

Radiochemical analysis
Measurement of Radioactivity, Practices for, **D 3648**
Sampling Surface Soil for Radionuclides, Practice for, **C 998**

Radiological examination
Decontamination of Field Equipment Used at Low Level Radioactive Waste Sites, Practice for, **D 5608**

Radionuclides
Sampling Surface Soil for Radionuclides, Practice for, **C 998**

Raster
Content of Digital Geospatial Metadata, Specification for, **D 5714**

Rate of emission
Sampling Surface Soil for Radionuclides, Practice for, **C 998**

Rate of pressure rise
See **Pressure testing**

Rating curve
Developing a Stage-Discharge Relation for Open Channel Flow, Practice for, **D 5541**

Raw refuse/sewerage
See **Resource recovery**

RBCA process
Risk-Based Corrective Action Applied at Petroleum Release Sites, Guide for, **E 1739**

Recharge studies
Measuring Matric Potential in the Vadose Zone Using Tensiometers, Guide for, **D 3404**

Reconnaissance surveys
Site Characterization for Engineering, Design and Construction Purposes, Guide to, **D 420**
Site Characterization for Environmental Purposes With Emphasis on Soil, Rock, the Vadose Zone and Ground Water, Guide for, **D 5730**

Reference sediments
Collection, Storage, Characterization, and Manipulation of Sediments for Toxicological Testing, Guide for, **E 1391**

Refraction
Using the Seismic Refraction Method for Subsurface Investigation, Guide for, **D 5777**

Regional representativeness
Choosing Locations and Sampling Methods to Monitor Atmospheric Deposition at Non-Urban Locations, Guide for, **D 5111**

Regulatory compliance audits
Environmental Regulatory Compliance Audits, Practice for, **PS 11**

Rehabilitation
Maintenance and Rehabilitation of Ground-Water Monitoring Wells, Guide for, **D 5978**

Relative explosion pressure
See **Pressure testing**

Replication
Estimation of Holding Time for Water Samples Containing Organic and Inorganic Constituents, Practice for, **D 4841**

Representative sample
Composite Sampling and Field Subsampling for Environmental Waste Management Activities, Guide for, **D 6051**
Development of Ground-Water Monitoring Wells in Granular Aquifers, Guide for, **D 5521**
Sampling Waste Piles, Guide for, **D 6009**

Reservoir sedimentation data
Measurement of Morphologic Characteristics of Surface Water Bodies, Guide for, **D 4581**

Residual radioactivity
See **Radiation exposure**

Resistivity logging
Planning and Conducting Borehole Geophysical Logging, Guide for, **D 5753**

Resonance neutron flux
See **Fast neutron flux/fluence**

Resource recovery
Determining Subsurface Liquid Levels in a Borehole or Monitoring Well (Observation Well), Test Method for, **D 4750**

Respiring bacteria
See **Bacteria/bacterial control**

Ring-lines barrel sampling
Ring-Lined Barrel Sampling of Soils, Practice for, **D 3550**

Risk analysis—environmental
Developing Conceptual Site Models for Contaminated Sites, Guide for, **E 1689**
Risk-Based Corrective Action Applied at Petroleum Release Sites, Guide for, **E 1739**

Rivers
Sa **Fresh water**
Planning and Implementing a Water Monitoring Program, Guide for, **D 5851**

Rock
Design and Installation of Ground Water Monitoring Wells in Aquifers, Practice for, **D 5092**
Diamond Core Drilling for Site Investigation, Practice for, **D 2113**
Preserving and Transporting Rock Core Samples, Practices for, **D 5079**
Site Characterization for Engineering, Design and Construction Purposes, Guide to, **D 420**
Site Characterization for Environmental Purposes With Emphasis on Soil, Rock, the Vadose Zone and Ground Water, Guide for, **D 5730**
Water Content of Soil and Rock in Place by Nuclear Methods (Shallow Depth), Test Method for, **D 3017**

Rock-mass rating (RMR) system
Using Rock-Mass Classification Systems for Engineering Purposes, Guide for, **D 5878**

Rock material field classification procedure (RMFCP)
Using Rock-Mass Classification Systems for Engineering Purposes, Guide for, **D 5878**

Rock quality designation (RQD)
Using Rock-Mass Classification Systems for Engineering Purposes, Guide for, **D 5878**

Rock structure rating (RSR) system
Using Rock-Mass Classification Systems for Engineering Purposes, Guide for, **D 5878**

Rocky shores
See **Oil spill control systems**

Rotary drilling
Sa **Borehole drilling**
Diamond Core Drilling for Site Investigation, Practice for, **D 2113**
Soil Investigation and Sampling by Auger Borings, Practice for, **D 1452**

Rotating anemometer
Characterizing Surface Wind Using a Wind Vane and Rotating Anemometer, Practice for, **D 5741**

Y

100 Barr Harbor Drive
West Conshohocken, PA 19428-2959
Phone: (610) 832-9693
FAX: (610) 832-9555

European Office
27-29 Knowl Piece, Wilbury Way
Hitchin, Herts SG4 OSX, England
Phone: 1462-437933
FAX: 1462-433678

1997 INDIVIDUAL MEMBE

SEE FEE AND BENEFITS ON REVERS

APPLICATION IS MADE FOR MEMBERSHIP IN ASTM:

INDIVIDUAL MEMBERSHIP	AN INDIVIDUAL MEMBER SHALL PAY AN ANNUAL FEE AND MAY PARTICIPATE ON ASTM TE(COMMITTEES REPRESENTING HIS OR HER EMPLOYER UNLESS UNASSOCIATED WITH ANY O

PLEASE PRINT ALL INFORMATION CLEARLY. DO NOT EXCEED THE CAPACITY OF EACH LINE.

LAST NAME	FIRST NAME / INITIAL
AFFILIATION	
FACILITY	
STREET	
P.O. BOX	
CITY	STATE / ZIP
COUNTRY	
PHONE	( ) — EXTENSION
FAX	( ) —
E-MAIL	
JOB TITLE	

NOTE: IF YOUR AFFILIATION IS A SUBSIDIARY, PLEASE IDENTIFY THE PARENT ORGANIZATION.

COMMITTEE APPLICATION REQUEST

THIS APPLICATION IS FOR SOCIETY MEMBERSHIP ONLY. A TECHNICAL COMMITTEE REQUIRES A SEPARATE AP
PLEASE SEND APPLICATIONS FOR THE FOLLOWING COMMITTEE(S):

☐ ENCLOSED IS PAYMENT.
ALL CHECKS MADE PAYABLE TO ASTM
IN U.S. FUNDS ON U.S. BANKS.

☐ EFT: CORESTATES PHILADELPHIA NATIC
P.O. BOX 7618, PHILADELPHIA, PA
ACCT# 0109-9766 ROUTE # 031000

OPTIONAL METHODS OF PAYMENT

AMOUNT $_____

☐ AMERICAN EXPRESS   ☐ MASTER CARD   ☐ VIS

ACCOUNT NO. (ALL DIGITS)   EXPIRATION DATE   SIGNATURE

MAIL TO: ASTM   ATTN: MEMBER SERVICES

Ann̄tive Fee $65.00*†

Pa₂STM Technical Committees   Yes

C₍ook of ASTM Standards (Vols. 1-72) $4,700.00 (Prepaid)†

C₅ at Member Price   No Limit

₍on to **ASTM STANDARDIZATION NEWS**   1 Free

ll also receive:

Symposium Registration Fees

blications at Member Prices

Information about Standards & Activities

ear is January 1 to December 31. Fees are payable in advance and are not prorated.
will become effective upon payment of fees.

inistrative fee payments made to ASTM are not tax deductible as charitable contributions for
eral income tax purposes.

anadian Members – Please add 7% GST – Registration Number R129162244.

NOTE: Membership benefits are subject to change.